CASSIDY AND ALLANSON'S MANAGEMENT OF GENETIC SYNDROMES

CASSIDY AND ALLANSON'S MANAGEMENT OF GENETIC SYNDROMES

Fourth Edition

Edited by

JOHN C. CAREY
University of Utah
Salt Lake City, UT, USA

AGATINO BATTAGLIA
IRCCS Stella Maris Foundation
Pisa, Italy

DAVID VISKOCHIL
University of Utah
Salt Lake City, UT, USA

SUZANNE B. CASSIDY
University of California,
San Francisco, CA, USA

WILEY Blackwell

This edition first published 2021
© 2021 John Wiley & Sons, Inc.

Edition History
John Wiley & Sons Ltd: 1e, 2001; 2e, 2004; 3e, 2010

All rights reserved. No part of this publication may be reproduced, stored in a retrieval system, or transmitted, in any form or by any means, electronic, mechanical, photocopying, recording or otherwise, except as permitted by law. Advice on how to obtain permission to reuse material from this title is available at http://www.wiley.com/go/permissions.

The right of John C. Carey, Agatino Battaglia, David Viskochil and Suzanne B. Cassidy to be identified as the authors of this work has been asserted in accordance with law.

Registered Offices
John Wiley & Sons, Inc., 111 River Street, Hoboken, NJ 07030, USA
John Wiley & Sons Ltd, The Atrium, Southern Gate, Chichester, West Sussex, PO19 8SQ, UK

Editorial Office
The Atrium, Southern Gate, Chichester, West Sussex, PO19 8SQ, UK

For details of our global editorial offices, customer services, and more information about Wiley products visit us at www.wiley.com.

Wiley also publishes its books in a variety of electronic formats and by print-on-demand. Some content that appears in standard print versions of this book may not be available in other formats.

Limit of Liability/Disclaimer of Warranty
While the publisher and authors have used their best efforts in preparing this work, they make no representations or warranties with respect to the accuracy or completeness of the contents of this work and specifically disclaim all warranties, including without limitation any implied warranties of merchantability or fitness for a particular purpose. No warranty may be created or extended by sales representatives, written sales materials or promotional statements for this work. The fact that an organization, website, or product is referred to in this work as a citation and/or potential source of further information does not mean that the publisher and authors endorse the information or services the organization, website, or product may provide or recommendations it may make. This work is sold with the understanding that the publisher is not engaged in rendering professional services. The advice and strategies contained herein may not be suitable for your situation. You should consult with a specialist where appropriate. Further, readers should be aware that websites listed in this work may have changed or disappeared between when this work was written and when it is read. Neither the publisher nor authors shall be liable for any loss of profit or any other commercial damages, including but not limited to special, incidental, consequential, or other damages.

Library of Congress Cataloging-in-Publication data applied for

Hardback (9781119432678)

Cover Design: Wiley
Cover Images: Blue DNA Helix © nechaev-kon/Getty Images, Stethoscope © LevKing/iStock.com

Set in 10/12pt Times by SPi Global, Pondicherry, India
Printed and bound in Singapore by Markono Print Media Pte Ltd
10 9 8 7 6 5 4 3 2 1

We dedicate this book
To our families
Leslie, Patrick, and Andrew Carey
Emi and Chiara Battaglia
Barbara, Richard, Joseph, and Kayla Viskochil
Joshua Cassidy and Christopher, Adam and Alexandra Visher

For all they taught us, for their tolerance, and for their love and encouragement

AND

To our patients
Who have inspired us with their resilience and their living with the unique challenges of rare conditions

CONTENTS

FOREWORD TO THE FOURTH EDITION		xi
FOREWORD TO THE THIRD EDITION		xiii
FOREWORD TO THE SECOND EDITION		xv
FOREWORD TO THE FIRST EDITION		xvii
PREFACE		xix
LIST OF CONTRIBUTORS		xxi
INTRODUCTION		xxvii
1	**Aarskog Syndrome** *Roger E. Stevenson*	1
2	**Achondroplasia** *Richard M. Pauli and Lorenzo Botto*	9
3	**Alagille Syndrome** *Henry C. Lin and Ian D. Krantz*	31
4	**Albinism: Ocular and Oculocutaneous Albinism and Hermansky–Pudlak Syndrome** *C. Gail Summers and David R. Adams*	45
5	**Angelman Syndrome** *Charles A. Williams and Jennifer M. Mueller-Mathews*	61
6	**Arthrogryposis** *Judith G. Hall*	75
7	**ATR-X: α Thalassemia/Mental Retardation-X-Linked** *Richard J. Gibbons*	93
8	**Bardet–Biedl Syndrome** *Anne M. Slavotinek*	107

9 Beckwith–Wiedemann Syndrome and Hemihyperplasia — 125
Cheryl Shuman and Rosanna Weksberg

10 Cardio-Facio-Cutaneous Syndrome — 147
Maria Inês Kavamura and Giovanni Neri

11 CHARGE Syndrome — 157
Donna M. Martin, Christine A. Oley, and Conny M. van Ravenswaaij-Arts

12 Coffin–Lowry Syndrome — 171
R. Curtis Rogers

13 Coffin–Siris Syndrome — 185
Tomoki Kosho and Noriko Miyake

14 Cohen Syndrome — 195
Kate E. Chandler

15 Cornelia de Lange Syndrome — 207
Antonie D. Kline and Matthew Deardorff

16 Costello Syndrome — 225
Bronwyn Kerr, Karen W. Gripp, and Emma M.M. Burkitt Wright

17 Craniosynostosis Syndromes — 241
Elizabeth J. Bhoj and Elaine H. Zackai

18 Deletion 1p36 Syndrome — 253
Agatino Battaglia

19 Deletion 4p: Wolf–Hirschhorn Syndrome — 265
Agatino Battaglia

20 Deletion 5p Syndrome — 281
Antonie D. Kline, Joanne M. Nguyen, and Dennis J. Campbell

21 Deletion 22q11.2 (Velo-Cardio-Facial Syndrome/DiGeorge Syndrome) — 291
Donna M. McDonald-McGinn, Stephanie Jeong, Michael-John McGinn II, Elaine H. Zackai, and Marta Unolt

22 Deletion 22q13 Syndrome: Phelan–McDermid Syndrome — 317
Katy Phelan, R. Curtis Rogers, and Luigi Boccuto

23 Denys–Drash Syndrome, Frasier Syndrome, and WAGR Syndrome (*WT1*-related Disorders) — 335
Joyce T. Turner and Jeffrey S. Dome

24 Down Syndrome — 355
Aditi Korlimarla, Sarah J. Hart, Gail A. Spiridigliozzi, and Priya S. Kishnani

25 Ehlers–Danlos Syndromes — 389
Brad T. Tinkle

26 Fetal Alcohol Spectrum Disorders — 405
H. Eugene Hoyme and Prachi E. Shah

27 Fetal Anticonvulsant Syndrome — 425
Elizabeth A. Conover, Omar Abdul-Rahman, and H. Eugene Hoyme

28 Fragile X Syndrome and Premutation-Associated Disorders — 443
Randi J. Hagerman

29	**Gorlin Syndrome: Nevoid Basal Cell Carcinoma Syndrome** *Peter A. Farndon and D. Gareth Evans*	**459**
30	**Hereditary Hemorrhagic Telangiectasia** *Jonathan N. Berg and Anette D. Kjeldsen*	**475**
31	**Holoprosencephaly** *Paul Kruszka, Andrea L. Gropman, and Maximilian Muenke*	**487**
32	**Incontinentia Pigmenti** *Dian Donnai and Elizabeth A. Jones*	**505**
33	**Inverted Duplicated Chromosome 15 Syndrome (Isodicentric 15)** *Agatino Battaglia*	**515**
34	**Kabuki Syndrome** *Sarah Dugan*	**529**
35	**47,XXY (Klinefelter Syndrome) and Related X and Y Chromosomal Conditions** *Carole Samango-Sprouse, John M. Graham Jr, Debra R. Counts, and Jeannie Visootsak*	**539**
36	**Loeys–Dietz Syndrome** *Aline Verstraeten, Harry C. Dietz, and Bart L. Loeys*	**563**
37	**Marfan Syndrome** *Uta Francke*	**577**
38	**Mowat–Wilson Syndrome** *David Mowat and Meredith Wilson*	**597**
39	**Myotonic Dystrophy Type 1** *Isis B.T. Joosten, Kees Okkersen, Baziel G.M. van Engelen, and Catharina G. Faber*	**611**
40	**Neurofibromatosis Type 1** *David Viskochil*	**629**
41	**Noonan Syndrome** *Judith E. Allanson and Amy E. Roberts*	**651**
42	**Oculo-Auriculo-Vertebral Spectrum** *Koenraad Devriendt, Luc De Smet, and Ingele Casteels*	**671**
43	**Osteogenesis Imperfecta** *An N. Dang Do and Joan C. Marini*	**683**
44	**Pallister–Hall Syndrome and Greig Cephalopolysyndactyly Syndrome** *Leslie G. Biesecker*	**707**
45	**Pallister–Killian Syndrome** *Emanuela Salzano, Sarah E. Raible, and Ian D. Krantz*	**717**
46	**Prader–Willi Syndrome** *Shawn E. McCandless and Suzanne B. Cassidy*	**735**
47	**Proteus Syndrome** *Leslie G. Biesecker*	**763**
48	***PTEN* Hamartoma Tumor Syndrome** *Joanne Ngeow and Charis Eng*	**775**

49	**Rett Syndrome** *Eric E. Smeets*	791
50	**Robin Sequence** *Howard M. Saal*	807
51	**Rubinstein–Taybi Syndrome** *Leonie A. Menke and Raoul C. M. Hennekam*	823
52	**Silver–Russell Syndrome** *Emma L. Wakeling*	837
53	**Smith–Lemli–Opitz Syndrome** *Alicia Latham and Christopher Cunniff*	851
54	**Smith–Magenis Syndrome** *Ann C.M. Smith and Andrea L. Gropman*	863
55	**Sotos Syndrome** *Trevor R.P. Cole and Alison C. Foster*	895
56	**Stickler Syndrome** *Mary B. Sheppard and Clair A. Francomano*	915
57	**Treacher Collins Syndrome and Related Disorders** *Marilyn C. Jones*	927
58	**Trisomy 18 and Trisomy 13 Syndromes** *John C. Carey*	937
59	**Tuberous Sclerosis Complex** *Laura S. Farach, Kit Sing Au, and Hope Northrup*	957
60	**Turner Syndrome** *Angela E. Lin and Melissa L. Crenshaw*	977
61	**VATER/VACTERL Association** *Benjamin D. Solomon and Bryan D. Hall*	995
62	**Von Hippel–Lindau Syndrome** *Samantha E. Greenberg, Luke D. Maese, and Benjamin L. Maughan*	1005
63	**Williams Syndrome** *Colleen A. Morris and Carolyn B. Mervis*	1021

INDEX   1039

FOREWORD TO THE FOURTH EDITION

Almost a generation has gone by since *Management of Genetic Syndromes* was conceived and the first edition published. That volume provided up-to-date, expert-authored, practical information on clinical features, natural history, medical concerns, and management of 30 conditions, each with an incidence between 1/600 and 1/60,000. Thus, most were considered rare conditions, and unlikely to be on the radar of primary care practitioners or non-genetics specialists unless/until that physician had an affected individual in their practice. Most, if not all, chapter authors had worked closely for many years with a family support group, accumulating a wealth of knowledge on the myriad associated medical consequences and natural history. While diagnostic testing was available for about two-thirds, the remaining conditions could only be diagnosed by recognition of a pattern that might include congenital anomalies, differences in stature, appearance and/or development, and specific health problems.

Over the course of the last 20 years, enormous changes have occurred in the technologies available for testing and interpretation of results, the way we approach pattern recognition as it applies to differences in appearance, the options for management and, occasionally, treatment of associated health concerns, and, perhaps most importantly, the way the medical profession and society view rare diseases. In short, nothing is the same!

Technology has moved away from analysis of chromosomes and individual genes to testing of the genome using microarrays and whole exome and whole genome sequencing. When a diagnosis is suspected, focused testing remains the gold standard. However, ordering a test that evaluates multiple genes that cause conditions with overlapping features, or screening the patient's entire genome if a testable diagnosis is not suspected, is now not only possible but increasingly cost effective. Clinical diagnosis relies on painstaking assessment and recognition of specific physical differences, the phenotype. It is just as important to describe that phenotype in an accurate and consistent manner. The development of a human phenotype ontology with more than 10,000 terms, each of which describes a phenotypic variation seen in human disease, is of great importance for the classification and comparison of affected individuals. Recognition of differences in facial appearance, some quite subtle, has moved from an "art" to a "science". Initially grounded in the experience and observation skills of the clinician, pattern recognition was enhanced by simple linear measurements and then revolutionized by 3D photography, computer modeling, machine learning, and artificial intelligence. Syndrome recognition aided by the use of images taken by a point-and-shoot camera or even a mobile phone now has an established place in clinical practice.

Over the last 20 years, there has been increasing recognition that rare diseases, in aggregate, are common, and estimated to affect 200 million people worldwide. A substantive proportion of the approximately 7000 known rare diseases are due to the altered function of single genes. For an individual with a rare genetic disease, timely provision of a molecularly confirmed diagnosis is critical for many reasons. It shortens the diagnostic odyssey, improves disease management by identifying specific health risks and preventing unnecessary or harmful diagnostic interventions and treatments, and it enables recurrence risk counseling and reproductive choice. This is the cornerstone of precision medicine. Understanding how genes function and interact and concomitant insight into biological mechanisms will not only enhance diagnosis and medical care for individuals with rare diseases, but will transform our understanding of health and disease. Documentation of pathogenic variations in the same gene in at least two unrelated individuals with the same rare disease is necessary to confirm the identity of a novel disease gene. As we study diseases of increasing rarity, a single academic centre or program is unlikely to achieve this standard. Thus diagnosis (and treatment) of individually rare disorders is best addressed by broad international collaboration. To this end, the International Rare Diseases Research Consortium (IRDiRC) was established in 2011, just a year after publication of the third edition of *Management of Genetic Syndromes*. It now brings together nearly 50 organizations in 18 countries invested in rare disease research. In addition to enabling the formation of scientific groups focused on diagnostics and therapies, close association with private-sector pharmaceutical and biotechnology companies has increased funding, data discovery and data sharing. International databases for genotype-driven and phenotype-driven matching of individuals with unsolved rare diseases have been developed, with an overarching platform that connects the data silos. Critically, partnership with national and continental patient advocacy organizations from Africa, Asia, Australia, Europe, and North America ensures the gene discovery research remains grounded in patient/family experience and relevance.

This fourth edition of *Management of Genetic Syndromes* includes five new conditions, bringing the total to 63 and more than doubling the original cohort of disorders. The thoughtful and consistent layout is unchanged, as it has proved to be a popular, effective, easy to use format. Since individuals with syndromes will be found in all medical practices and will benefit most when their provider is comfortable with the issues that need to be addressed to assure the best medical and quality of life outcomes, the information herein is no longer the purview of the genetics specialist, but is relevant to many other specialists and primary care providers. We know that barriers to the integration of genomic services into these practices include lack of time, knowledge, and confidence in the skills required. Easy access to current, reliable, practical, evidence-based information facilitates the effective transition of genomic medicine into primary care and non-genetics specialty practice. Increasingly, families play a role in the management of what is commonly a life-long chronic condition and, in many cases, they become more knowledgeable than their doctor. Although *Management of Genetic Syndromes* is not written for the public, its clear and practical presentation style, with management sections that have a system by system, checklist-like format, allows families and providers to follow together the issues to be addressed, in the right way and at the right time. Links to consumer health information are also provided for each condition. Hopefully, as *Management of Genetic Syndromes* evolves to include an alternative electronic format, individual chapters will be available to families and patient advocacy groups.

Management of Genetic Syndromes remains a model for how to organize information in a meaningful format and context for physicians and families. The hype about gene discovery and effective treatment of genetic disorders persists, albeit moving closer to reality in a few circumstances. In the meantime, *Management of Genetic Syndromes* delivers the goods on what is really known about management and treatment of these 63 disorders. I have had the good fortune and honour to be involved in the previous three editions as both an editor and an author. The new editorial team has delivered another great edition of this must-have book.

JUDITH E. ALLANSON

Department of Paediatrics, University of Ottawa and Department of Genetics, Children's Hospital of Eastern Ontario

FOREWORD TO THE THIRD EDITION

Cassidy and Allanson have done it again: produced a new edition of the one must-have book on management of genetic disorders for health care providers of all specialties. To incorporate advances in medical genetics into their practices, clinicians need an expert-authored resource that provides up-to-date information on available diagnostic approaches and practical day-to-day, age-oriented management. *Management of Genetic Syndromes* does not require that clinicians become genetics experts or fluent in genetics lingo. It is written with the knowledge that persons with inherited disorders are found in all medical practices and, similar to people with other medical conditions, these individuals will benefit most when their health care providers are comfortable with the issues that need to be addressed to assure the best medical and quality-of-life outcomes. This book presents to clinicians in primary care and specialty practice the information necessary to allow the clinician to decide for their patients with rare inherited disorders which care is within the scope of his or her practice and which specific needs should be referred out to other specialists.

Management of Genetic Syndromes is a boon to busy primary care practitioners who, I am told, have 90 seconds in which to answer a question brought up during a patient visit. If clinicians do not have a reliable, easy-to-use resource, those questions will go unanswered. The logical division of chapters by disorder and the thoughtful and consistent layout of each chapter into sections on diagnosis first (how can you provide disorder-specific care if you can't be sure that you have the correct diagnosis?) followed by detailed management issues by organ system for all ages allows the busy clinician to hone in on an authoritative answer in a predictable "place." Eliminating the guess work about specific care issues is tremendously valuable to busy clinicians who want to assure the best care for their patients, but cannot take the time to second guess the exact needs for an individual with a one-of-a-kind disorder in their practice. Similar to all quality information resources, *Management of Genetic Syndromes* provides citations to more detailed documentation of diagnostic and management recommendations for those clinicians with the time or inclination to learn more.

In these days of hype on pending cutting-edge treatment for genetic disorders and "personalized" medicine, clinicians need a filter that can separate what is really known about treatment and what is hypothesis-driven wishful thinking for which no prescription can be written. *Management of Genetic Syndromes* provides this filter, thus assuring clinicians and families that clinicians have at their fingertips information that will be most useful.

Although the promise of the Human Genome Project to provide gene-based therapy for inherited disorders is still a long way from reality, other aspects of the discoveries of the molecular basis of inherited disorders have benefited those with and at risk for inherited disorders. One example is surveillance of those at risk for a potential complication of an inherited disorder, which enables early diagnosis and, hence treatment to improve outcome. For example, in families with an inherited cancer predisposition, such as a hereditary colon cancer syndrome, at-risk relatives benefit from knowing who has inherited the family-specific mutation and who has not, so that those at greatest risk are screened using disease-specific protocols starting at the appropriate age and those who are not at increased risk are advised to follow population-based screening protocols. *Management of Genetic Syndromes* emphasizes the practical approach to the risk-defining use of molecular genetic testing with outcome-oriented surveillance. The reader does not need to be familiar with the jargon or principles of molecular genetics to understand how to use this approach for the benefit of patients in his or her practice.

Those with genetic disorders and their families often appreciate transparency in the care that they receive and they want access to the same information as their health care providers. The workman-like, practical approach to management in this book provides a "checklist-like" view that enables clinician and patient to follow together the issues to be addressed and their timelines. The chapters in *Management of Genetic Syndromes* are excellent "handouts" at clinic visits. In my academic clinical practice of medical genetics, my colleagues and I have on hand a ready supply of copies of the chapters of *Management of Genetic Syndromes*, which we read before the clinic visit and then provide to families at the time of their clinic appointment and to the referring clinicians with the clinic note. We know that, although the primary audience for this book is not affected individuals and their families, and, therefore, it was not written at the appropriate level for this audience, the clear, no-nonsense presentation style makes the content accessible to those families seeking to partner with their physician in their care.

Increasingly, families play a key role in the management of their inherited disorder, which most commonly is a chronic lifelong condition that may affect other family members of all generations. Consumer-oriented health information sources have grown exponentially with the discovery of the molecular genetic basis of inherited disorders, the growing use of the Internet, and the development of hundreds of disease-oriented patient advocate groups. Consumer health information resources, which often provide the most practical day-to-day information available for patients and their families, are a valuable adjunct to clinic visits. The essential role of consumer health information is acknowledged by *Management of Genetic Syndromes* by providing information on these resources in an easy-to-find location at the end of each chapter.

Management of Genetic Syndromes is an unparalleled medical genetics information resource for students, be they medical students, residents in primary care fields or specialty fields, or participants in continuing medical education. It is the one book I tell them to buy.

When I see the *Management of Genetic Syndromes* in a clinician's office, I respect that clinician for taking the initiative to anticipate the needs of his or her patients with rare inherited disorders and know that the clinician, his or her patient, and the patient's family will be grateful for the practical approach of this trusted colleague on the bookshelf.

ROBERTA A. PAGON, MD

University of Washington and Seattle Children's Hospital

FOREWORD TO THE SECOND EDITION

It was not very many years ago that the coupling of the terms "management" and "genetic syndromes" would have been regarded as an oxymoron. With the exception of the inborn errors of metabolism, the notion of managing genetic disorders would have been considered quite foreign and of managing genetic *syndromes*, by which we mean conditions in which several organ systems and/or parts of the body are affected, even more so. The principal role of the medical geneticist was to diagnose these conditions as best as he or she could. Management, such as it was, was essentially symptomatic and was usually left to primary care physicians and medical specialists with little direct knowledge of the syndromes themselves. The literature on genetic syndromes reflected this situation. It was, for the most part, descriptive, and the emphasis was on diagnosis. Although many admirable reference books on diagnosis were written, most notable of which was (and still is), *Smith's Recognizable Patterns of Human Malformations*, it was frequently difficult to find definitive information about how to manage these conditions once the diagnoses had been made.

However, much has changed recently with regard to genetic syndromes, with perhaps the most important change being societal, not medical or scientific. It is now generally accepted that persons with genetic syndromes, whether associated with mental retardation or not, *should*, if possible, be treated. This was not always so, and a graphic example of how thinking has altered is provided by Down syndrome, certainly one of the quintessential genetic syndromes. Within my professional lifetime, there has been a shift from exclusion from society, generally by institutionalization, to rearing at home, educational inclusion, and participation in all aspects of daily life. Similarly, a policy of nonintervention, often with certain death, when major heart or gastrointestinal abnormalities were present has been replaced by aggressive surgical correction. Guidelines for the prevention of known complications have been developed, and their implementation is now commonplace. As a result, these changes have led, even without any specific therapy for Down syndrome, to an increase in lifespan, better cognitive development, and an overall improvement in the quality of life, both physically and socially.

In addition to the attitudinal shift, there have been many medical and scientific advances that have altered our approach to genetic syndromes. The mutations that cause many of the monogenic or contiguous gene syndromes are now known, and more are being discovered almost daily. The functions of the genes that these mutations affect are gradually being elucidated. For the aneuploidies, the mapping of the human genome is providing information about how many and which genes are at dosage imbalance. All of this has changed genetic syndromes from being curiosities that could not be understood to disorders that can be rationally approached in terms of cause and potential therapy, another and quite major change in attitude. This information has also led to the development of molecularly based tests that are greatly improving disease diagnosis and are permitting discrimination among conditions that had hitherto been confused with one another. In the future, this genetic information promises to lead to therapies that are tailored to individual diseases. In addition, medical diagnostic procedures and therapeutic approaches have become much more powerful. These include, for example, the various forms of imaging, surgical techniques such as for complex congenital heart defects or ambiguous genitalia, and highly specific and potent pharmacological agents. And, finally, more is continually being learned about the long-term consequences of genetic syndromes—about their natural histories—which is essential if comprehensive approaches to management are to be developed.

So, if societal attitudes have changed and genetic and medical information and capabilities are rapidly expanding, who should be undertaking the management of persons with genetic syndromes? Who should be reading this book? There is no simple answer to this question because in a sense each syndrome must be dealt with on its own merits. Given the multitude of systems that these syndromes may affect and the different combinations of abnormalities that may occur in one compared with another, the approach to management needs to be quite flexible. Nevertheless, someone must be responsible for the overall coordination of care. Who this will be will depend on local circumstances, but the important thing is that it be someone who is knowledgeable and willing to act in the interests of the affected individual.

In most instances, persons with genetic syndromes are usually managed by a mix of genetic professionals, primary care physicians, and medical and other specialists. By "genetic professional" I mean medical geneticists, genetic counselors and genetic nurses, and laboratory geneticists who have special knowledge about and experience in dealing with a large number of genetic syndromes that are individually quite uncommon or rare. For the most part, genetic professionals have traditionally been engaged in the diagnosis and counseling of these conditions. Unlike the situation with inherited metabolic disorders, in which geneticists do participate directly in therapy, their involvement in the therapeutic aspects of the management of genetic syndromes has generally involved referrals to appropriate specialists for specific forms of medical or surgical therapy. Primary care physicians, in addition to providing day-to-day care of individuals with genetic syndromes, often act as intermediaries in the referral process. And, beyond this list of medical personnel, a variety of other professionals and social and educational organizations, both governmental and voluntary, also provide many services to affected individuals and their families.

In some instances, the medical specialists, genetic professionals, and allied health professionals work together in multidisciplinary clinics devoted to individual disorders (e. g., Marfan or Down syndrome) or groups of related disorders (craniofacial anomalies or skeletal dysplasias) or perhaps even to birth defects more generally. These clinics provide a coordinated approach to management that is usually more efficient from the point of view of providers and of affected individuals and their families than is possible when many independent providers are involved in the care of the patient and may be a model for the provision of services in the future.

Regardless of how the services are organized and of who is actually coordinating management, many providers with different degrees of knowledge about any particular condition are likely to be involved. It is, therefore, essential that each understand what he or she is dealing with and what will be required to properly care for the affected individual and his or her family, and it is here that this volume, *Management of Genetic Syndromes,* uniquely fills a void that has long existed in the literature on genetic syndromes. Gathered together within a reasonably compact volume are authoritative descriptions written for a diverse readership of the management of over 50 of the most common conditions that fall within the rubric of genetic syndromes (including two that are primarily teratogenic, but are usually grouped with the others). The concept of what is entailed in management is broadly interpreted. Therefore, each chapter begins with considerations of etiology, pathogenesis, genetics, and diagnosis (including diagnostic criteria, testing, and differential diagnosis), all of which are necessary if the patient and his or her condition are to be fully understood. These are then followed by detailed discussions of what might be considered to be at the heart of management—the evaluation of each of the relevant systems and the treatment of the abnormalities that are likely to be present. The chapter concludes with selected references and a listing of available support groups and other resources. The evaluation and treatment sections are greatly enhanced by the use of an outline form of presentation, with bullets to highlight individual points.

When it appeared in 2001, the first edition of this book was eagerly seized upon by the medical genetics community. The need was there, and there was nothing else like it. From my own personal experience and observation in a genetics service that handles a large number of persons with genetic syndromes, I can testify that the book rapidly proved to be of great value to all of the clinic personnel—geneticists and counselors, physicians and nonphysicians, students, residents, and fellows. The rapid appearance of this second edition indicates that my own experience has been more generally shared, and the near doubling of the number of conditions covered will make the book even more valuable than before. Given the rapid progress that is being made in genetics and medicine and in the ability to diagnose and treat genetic syndromes, it is likely that frequent revisions will be required.

CHARLES J. EPSTEIN

Department of Pediatrics
University of California,
San Francisco

FOREWORD TO THE FIRST EDITION

This is a book whose time has come. Genetic disorders and syndromes are usually thought of as being rare, and yet for affected individuals, their families, and their primary and specialty care physicians, it is essential to have reliable information about the natural history and management of the specific disorders.

The thirty conditions described in this book may seem rare (with incidences between 1 in 600 and 1 in 60,000). However, when you put together all the individuals cases or a particular condition in North America, in Europe, and in the world, a very large number of affected individual will benefit from the information in this book. In the past it has been difficult to bring together information of this type about specific disorders, and that is why this book fills a very important niche. It becomes a model for how to organize information that is needed for the families and primary care providers to manage the many, many other genetic disorders, congenital anomalies, and syndromes that are known to occur. The book is written in understandable language appropriate for families and for primary care and specialty physicians. It is major contribution.

Over the last two decades, remarkable progress has been made with regard to developing diagnostic tests and unraveling the human genome. Within the next few years all of the human genes will have been defined. The next major goal in genetics will be to understand how genes interact and function, both in the course of development and over a lifetime. In addition to the remarkable progress in basic and clinical genetics, there has been increasing communication and access to information. Through the Internet, the public has access to research reports and data that were usually not readily available in the past. However, it is essential to put that information into a meaningful form and context. That is exactly what this book does. The communication explosion has allowed the networking of researchers and families. The development of parent/lay support groups has led to a cooperation between researchers and families that has helped to define the natural history and the variation that can be seen in a specific disorder.

What every family and physician wants is to provide the best care possible for the affected individual. Nobody wants to miss the opportunity for that individual to reach his or her full potential, to benefit from a useful therapy, or to avoid a complication. Parents need an understanding of what will happen over time so that they can plan. They don't want to waste money and effort going from expert to expert or doing test after test. They need a realistic approach to what they should expect both in childhood and adulthood. They also usually want to know whether there is some risk of recurrence of the condition in their other children, in other family members, and in the affected individual's offspring. They want to know whether prenatal diagnosis is available, and they want to know the spectrum of variation that can occur. The beauty of this new book is that it provides that kind of information for each specific disorder in a logical and understandable form. Most families and physicians will focus in on the chapter relevant to a specific individual. However, they can't help but glance at other chapters and see the remarkable spectrum of complications that are not present in the disorder of interest to them. They are likely to benefit from this broader perspective.

Most pediatricians will have heard of all thirty disorders; however, some primary care and specialty physicians may not have heard of a specific disorder until they have the affected individual in their practice. The book should help to alert health care professionals to consider these conditions and should lead to appropriate testing to make a correct diagnosis, reducing the time it takes to make a specific diagnosis. Two-thirds of the conditions in this book have a specific diagnostic test, but the other one-third require "pattern recognition" and an alert, trained health care professional to consider the diagnosis.

It can be expected that additional advances will be made over the next few decades leading to better understanding and better management. So this book is already dated! There is still a lot to be learned! In fact, every family and every affected individual will contribute to that increased knowledge by giving feedback to the authors. Disorder-specific parent/lay support groups will continue to play in important role in improving our understanding. The authors of each chapter have worked together with the support groups and are very aware that it is the process of working together with these groups and the members' willingness to provide information that has led to present-day understanding. We are all very grateful to each of the parents and affected individuals who have taken part in studies that have advanced our knowledge.

To write a book about management, it is necessary to know the natural history of the disorder. The authors of each of these chapters have a wealth of experience and knowledge that has been collected over the least couple of decades. Understanding the natural history not only tells us what to expect at various ages but also how to recognize various complications. It is important to understand the natural history of the condition to determine whether various therapies actually improve the outcome. It is important to understand the natural history to recognize subgroups representing the variability and heterogeneity within the disorder. It is important to understand the natural history to learn the mechanisms that lead to the disorder, e.g., what sort of gene is likely to be involved? Where is the mutation in the gene? How does that mutation relate to severity of complications? How big is the deletion? Does that size relate to severity of complications? How does this gene act against the background of other genes or pathways? Is it possible to recognize a cellular mechanism leading to this disorder? Are there parent-of-origin effects on the expression of the gene or the mutation rate? Are there hot spots that have markedly increased mutation rates? Does the place on the chromosome where the gene lies put it at increased risk for mutation? These are only a few of the questions we hope to answer over the next few decades.

No one is more motivated than the family or the affected individual to learn about these disorders. It is important for them to be as knowledgeable as possible. The families of an affected person usually know more about the condition than most of the physicians they visit. It is important for families to continue to ask questions and to gain as much knowledge as possible to ensure the best outcome for the affected individual. It is important for families and affected individuals to keep their own records about the affected individual, such as a notebook of their visits to health care facilities, copies of the reports, and the results of the tests that have been done. It is also important to keep a photographic record of changes over time.

Once a family or an affected individual becomes involved in collecting information about the disorder, they often develop quite creative ideas that challenge the standard way of thinking about the disorder. Part of the advantage of participating in a support group is that those ideas then can be shared with the medical advisors and researchers and may lead to new knowledge.

Much of our understanding of these disorders is based on the manifestations in childhood, on feeding, on growth and development, and on social skills. However, information on adults is also beginning to accumulate and has been included in this book. In some conditions there is a stable situation, in others there is improvement with aging, and in still others deterioration can be expected. For many of the conditions described in this book, behavioral patterns have been recognized.

How should a family and their primary care physician use the experts? It would be impossible for the authors of these chapters to see every individual with the condition, but it is usually helpful for a family and the affected individual to see a clinical geneticist, to visit a developmental center, or to use the multidisciplinary team that is available in their area. Over the years, specialty clinics to deal with specific conditions have been developed. At some time it is probably appropriate to visit such a clinic at least once to review the affected individual's progress and to consider any special complications or responses. On the other hand, it is very important to have a knowledgeable primary care physician who cares for day-to-day medical needs and is aware of the unique complications of the condition.

The parent/lay support groups form an international network keeping up with new information on the specific disorders, and new information is sure to come. Some new information will come through organized studies of natural history; other data will come through clinical trails of new therapies; and further information will come from basic work on cellular mechanisms and biochemical pathways. For many of these disorders animal models will be developed, such as mice with the specific disorder, so that various therapies can be considered before trails in human beings. We live in a very exciting age and can anticipate major advances over the next few decades for each of the disorders described in this book. The international network of families, affected individuals, and researchers should and will communicate about new ideas, innovative approaches, and better understanding about these conditions.

We have begun to enter an era of evidence-based medicine. Only by having natural history information is it possible to understand the benefits of new interventions and therapies. We will hope that this book is outdated very rapidly because of such new developments, but in the meanwhile this book on management of common genetic syndromes is extremely welcome to families and health care providers alike.

JUDITH G. HALL

Professor and Head, Department of Pediatrics
University of British Columbia and
British Columbia Children's Hospital
Professor, Medical Genetics
James and Annabelle McCreary Professor University of
British Columbia

PREFACE

This book is designed to assist primary care physicians, medical specialists, other care providers, and families in assuring optimal care for individuals who have multiple problems that are components of genetic syndromes. It represents the combined experience and knowledge of many experts in medical genetics and related fields, each of whom has spent years participating in the diagnosis and clinical management of a specific genetic syndrome. Most of the chapter authors have conducted major clinical research on "their" respective disorders.

The syndromes selected for inclusion in this book are those that are sufficiently common as to be regularly encountered in clinics specializing in genetics, development, neurology, psychiatry, cancer, or craniofacial disorders. Many of these disorders will not have been seen in the practice of most primary care physicians or non-genetics specialists. When they are encountered, the physician typically has little knowledge of how to confirm the diagnosis, identify the associated problems and clinical manifestations, and optimally care for the affected individual. This lack of knowledge is due only partly to infrequent exposure to the disorder. For many of these conditions, few publications devoted to management have been published, and a search for this knowledge is extremely time-consuming, often provides incomplete information, and is frequently futile. This book was designed to provide that knowledge, based on the medical literature and the cumulative experience of an expert or experts on each condition. As a result, a proportion of the information found in this source will be personal experience or observation. For only a few of these disorders is there an established "standard of care" based on controlled trials or outcome studies, though some disorders have recommendations for management based on consensus. Where available, reference to evidence-based studies and other published sources has been included; where unavailable, reference to the "personal experience" or "personal observation" of the author(s) has been noted, to reflect non-peer-reviewed information.

Deciding on which disorders to include is no mean task, and there are some disorders for which there is little accumulated experience in management. In addition to more than 60 genetic (or probably genetic) conditions, we have included two teratogenic disorders, fetal alcohol syndrome and fetal anticonvulsant syndrome, because of their frequency and because genetic factors influence susceptibility.

The editors hope that this continues to be a useful text to primary care physicians, nurse practitioners, medical geneticists, genetic counselors, and many other medical specialists, educators, and providers of care for the individuals and families affected with these genetic syndromes. Like those with more frequent medical conditions, these individuals deserve the best possible medical, educational and psychological care.

We appreciate the guidance and assistance of our contacts at John Wiley & Sons. Most importantly, we thank the contributing authors and the many individuals who gave permission to have their photographs published in this book and who participated in the clinical research that provided the information for its content.

JOHN C. CAREY
AGATINO BATTAGLIA
DAVID VISKOCHIL
SUZANNE B. CASSIDY

LIST OF CONTRIBUTORS

Omar Abdul-Rahman Division of Genetic Medicine, Munroe Meyer Institute, University of Nebraska Medical Center, Omaha, Nebraska, USA

David R. Adams Deputy Director for Clinical Genomics, Office of the Clinical Director, National Human Genome Research Institute and Undiagnosed Diseases Program, National Institutes of Health, Bethesda, Maryland, USA

Judith E. Allanson Department of Genetics, University of Ottawa, Children's Hospital of Eastern Ontario, Ottawa, Ontario, Canada

Kit Sing Au Division of Medical Genetics, Department of Pediatrics, McGovern Medical School, University of Texas Health Science Center at Houston, Houston, Texas, USA

Agatino Battaglia IRCCS Stella Maris Foundation, Pisa, Italy; Division of Medical Genetics, Department of Pediatrics University of Utah School of Medicine, Salt Lake City, Utah, USA; and Division of Medical Genetics, Department of Pediatrics, Sanford School of Medicine, University of South Dakota, Sioux Falls, South Dakota, USA

Jonathan N. Berg Department of Clinical Genetics, Ninewells Hospital and Medical School, Dundee, UK

Elizabeth J. Bhoj The Children's Hospital of Philadelphia and The Perelman School of Medicine at the University of Pennsylvania School of Medicine, Philadelphia, Pennsylvania, USA

Leslie G. Biesecker Medical Genomics and Metabolic Genetics Branch, National Human Genome Research Institute, Bethesda, Maryland, USA

Luigi Boccuto Greenwood Genetic Center – Greenville, Greenville, South Carolina, USA; Clemson University, Clemson, Pickens, South Carolina, USA

Lorenzo D. Botto Division of Medical Genetics, Department of Pediatrics, University of Utah, Salt Lake City, Utah, USA

Dennis J. Campbell Retired College of Education, University of South Alabama, Chair Professional Advisory Board 5P Minus Society, Auburn, Alabama, USA

John C. Carey Division of Medical Genetics, Department of Pediatrics, University of Utah Health Salt Lake City, Utah, USA

Suzanne B. Cassidy Division of Medical Genetics, Department of Pediatrics, University of California, San Francisco, San Francisco, California, USA

Ingele Casteels Department of Ophthalmology, University of Leuven, Leuven, Belgium

Kate E. Chandler Manchester Centre for Genomic Medicine, St Mary's Hospital, Manchester University NHS Foundation Trust, Manchester, UK and Division of Evolution and Genomic Sciences, School of Biological Sciences, Faculty of Biology, Medicines and Health, University of Manchester, Manchester, UK

Trevor R.P. Cole Clinical Genetics Unit, Birmingham Women's Hospital, Edgbaston, Birmingham, UK

Elizabeth A. Conover Division of Genetic Medicine, Munroe Meyer Institute, University of Nebraska Medical Center, Omaha, Nebraska, USA

Debra R. Counts Pediatric Endocrinology, Sinai Hospital, Baltimore, Maryland, USA

Melissa L. Crenshaw Division of Genetics, Johns Hopkins All Children's Hospital, Johns Hopkins University School of Medicine, St. Petersburg, Florida, USA

Christopher Cunniff Division of Medical Genetics, Weill Cornell Medical College, New York, USA

An N. Dang Do Office of the Clinical Director, *Eunice Kennedy Shriver* National Institute of Child Health and Human Development, National Institutes of Health, Bethesda, Maryland, USA

Matthew Deardorff Division of Human Genetics, The Children's Hospital of Philadelphia, and Department of Pediatrics, the Perelman School of Medicine at the University of Pennsylvania, Philadelphia, Pennsylvania, USA

Koenraad Devriendt Center for Human Genetics, University of Leuven, Leuven, Belgium

Luc De Smet Department of Orthopaedic Surgery, University of Leuven, Leuven, Belgium

Harry C. Dietz Howard Hughes Medical Institute, Baltimore, MD, USA; and Institute of Genetic Medicine, Johns Hopkins University School of Medicine, Baltimore, Maryland, USA

Jeffrey S. Dome Children's National Hospital and the George Washington University School of Medicine and Health Sciences, NW, Washington, DC, USA

Dian Donnai Manchester Centre for Genomic Medicine, Manchester University NHS Foundation Trust and University of Manchester, Manchester Academic Health Sciences Centre, Manchester, UK

Sarah Dugan Clinical Genomics and Predictive Medicine, Providence Medical Group, Spokane, Washington, USA

Charis Eng Genomic Medicine Institute, Cleveland Clinic, Ohio, USA and Department of Genetics and Genome Sciences, Case Western Reserve University, Cleveland, Ohio, USA

D. Gareth Evans Manchester Centre for Genomic Medicine, St Mary's Hospital, Manchester Academic Health Centre (MAHSC), Division of Evolution and Genomic Sciences, University of Manchester, Manchester, UK

Catharina G. Faber Department of Neurology, Maastricht University Medical Center, Maastricht, The Netherlands

Laura S. Farach Division of Medical Genetics, Department of Pediatrics, McGovern Medical School, University of Texas Health Science Center at Houston, Houston, Texas, USA

Peter A. Farndon Clinical Genetics Unit, Birmingham Women's Hospital, Edgbaston, Birmingham, UK

Alison C. Foster Clinical Genetics Unit, Birmingham Women's Hospital, Edgbaston, Birmingham, UK

Uta Francke Departments of Genetics and Pediatrics, Stanford University Medical Center, Stanford, California, USA

Clair A. Francomano Department of Medical and Molecular Genetics, Indiana University School of Medicine, Indianapolis, Indiana, USA

Richard J. Gibbons MRC Molecular Haematology Unit, Weatherall Institute of Molecular Medicine, University of Oxford, John Radcliffe Hospital, Oxford, UK

John M. Graham Jr Department of Pediatrics, Cedars Sinai Medical Center, Harbor-UCLA Medical Center, Department of Pediatrics, David Geffen School of Medicine at UCLA, Los Angeles, California, USA

Samantha E. Greenberg Huntsman Cancer Institute, Salt Lake City, Utah, USA

Karen W. Gripp Division of Medical Genetics, Department of Pediatrics, S. Kimmel Medical College, Thomas Jefferson University, Philadelphia, Pennsylvania, and A.I. duPont Hospital for Children, Wilmington, Delaware, USA

Andrea L. Gropman Department of Neurology, Division of Neurogenetics and Neurodevelopmental Disabilities, Children's National Health System, Washington, DC, USA and Medical Genetics Branch, National Human Genome Research Institute, National Institutes of Health, Bethesda, Maryland, USA

Randi J. Hagerman Endowed Chair in Fragile X Research, Distinguished Professor of Pediatrics, UC, Davis Health, Sacramento, California, USA

Bryan D. Hall Division of Clinical/Biochemical Genetics and Dysmorphology, Department of Pediatrics, University of Kentucky and Kentucky Clinic, Lexington, Kentucky, USA; and Greenwood Genetic Center, Greenwood, South Carolina, USA

Judith G. Hall Departments of Pediatrics and Medical Genetics, British Columbia's Children's Hospital, Vancouver, British Columbia, Canada

Sarah J. Hart Department of Pediatrics, Duke University Medical Center, Durham, North Carolina, USA

Raoul C.M. Hennekam Department of Pediatrics, Academic Medical Center, University of Amsterdam, Amsterdam, The Netherlands

H. Eugene Hoyme Departments of Pediatrics and Medicine, The University of Arizona College of Medicine, Tucson, Arizona, USA; Sanford Children's Genomic Medicine Consortium and Sanford Imagenetics, Sanford Health, Sioux Falls, South Dakota, USA

Stephanie Jeong The Children's Hospital of Philadelphia and The Perelman School of Medicine at the University of Pennsylvania, Philadelphia, Pennsylvania, USA

Elizabeth A. Jones Manchester Centre for Genomic Medicine, Manchester University NHS Foundation Trust and University of Manchester, Manchester Academic Health Sciences Centre, Manchester, UK

Marilyn C. Jones Department of Pediatrics, University of California, San Diego, and Cleft Palate and Craniofacial Treatment Programs, Rady Children's Hospital, San Diego, California, USA

Isis B.T. Joosten Department of Neurology, Maastricht University Medical Center, Maastricht, The Netherlands

Maria Inês Kavamura Medical Genetics Center, UNIFESP, São Paulo, Brazil

Bronwyn Kerr Genomic Medicine, Manchester University Hospitals NHS Foundation Trust, Manchester, UK

Priya S. Kishnani Department of Pediatrics, Duke University Medical Center, Durham, North Carolina, USA

Anette D. Kjeldsen Department of Otorhinolaryngology Odense University Hospital, Denmark

Antonie D. Kline Clinical Genetics, Harvey Institute for Human Genetics, Greater Baltimore Medical Center, Baltimore, Maryland, USA

Aditi Korlimarla Department of Pediatrics, Duke University Medical Center, Durham, North Carolina, USA

Tomoki Kosho Center for Medical Genetics, Shinshu University Hospital, Matsumoto, Japan; and Department of Medical Genetics, Shinshu University School of Medicine, Matsumoto, Japan

Ian D. Krantz The Perelman School of Medicine at the University of Pennsylvania and Division of Human Genetics and Genomics, The Children's Hospital of Philadelphia, Pennsylvania, USA

Paul Kruszka Medical Genetics Branch, National Human Genome Research Institute, National Institutes of Health, Bethesda, Maryland, USA

Alicia Latham Division of Medical Genetics, Weill Cornell Medical College, and Division of Clinical Genetics, Memorial Sloan Kettering Cancer Center, New York, USA

Angela E. Lin Medical Genetics Unit, MassGeneral Hospital for Children, Harvard University School of Medicine, Boston, Massachusetts, USA

Henry C. Lin Division of Gastroenterology, Doernbecher Children's Hospital and Oregon Health and Science University, Portland, Oregon, USA

Bart L. Loeys Center of Medical Genetics, Faculty of Medicine and Health Sciences, University of Antwerp and Antwerp University Hospital, Antwerp, Belgium; and Department of Human Genetics, Radboud University Medical Centre, Nijmegen, The Netherlands

Luke D. Maese Hematology/Oncology, Department of Pediatrics, Huntsman Cancer Institute, University of Utah, Salt Lake City, Utah, USA

Joan C. Marini Section on Heritable Disorders of Bone and Extracellular Matrix, *Eunice Kennedy Shriver* National Institute of Child Health and Human Development, National Institutes of Health, Bethesda, Maryland, USA

Donna M. Martin The University of Michigan Medical School, United States

Benjamin L. Maughan Huntsman Cancer Institute, Salt Lake City, Utah, USA

Shawn E. McCandless Section of Genetics and Metabolism, Department of Pediatrics, University of Colorado Anschutz Medical Campus and Children's Hospital Colorado, Aurora, Colorado, USA

Donna M. McDonald-McGinn The Children's Hospital of Philadelphia and The Perelman School of Medicine at the University of Pennsylvania, Philadelphia, Pennsylvania, USA

Michael-John McGinn II The Children's Hospital of Philadelphia and The Perelman School of Medicine at the University of Pennsylvania, Philadelphia, Pennsylvania, USA

Leonie A. Menke Department of Pediatrics, Academic Medical Center, University of Amsterdam, Amsterdam, The Netherlands

Carolyn B. Mervis Department of Psychological and Brain Sciences, University of Louisville, Louisville, Kentucky, USA

Noriko Miyake Yokohama City University Graduate School of Medicine, Yokohama, Japan

Colleen A. Morris Division of Genetics, Department of Pediatrics, University of Nevada, Reno School of Medicine; UNLV Ackerman Center for Autism and Neurodevelopment Solutions, Las Vegas, Nevada, USA

David Mowat Centre for Clinical Genetics, Sydney Children's Hospital, Randwick, School of Women's and Child Health, University of New South Wales, Sydney, Australia

Maximilian Muenke Medical Genetics Branch, National Human Genome Research Institute, National Institutes of Health, Bethesda, Maryland, USA

Jennifer M. Mueller-Mathews Division of Genetics and Metabolism, Department of Pediatrics, University of North Carolina, Chapel Hill, North Carolina, USA

Giovanni Neri Fondazione Policlinico Universitario A. Gemelli IRCCS, Istituto di Medicina Genomica, Università Cattolica del Sacro Cuore, Rome, Italy and Self Research Institute, Greenwood Genetic Center, Greenwood (SC), USA

Joanne Ngeow Lee Kong Chian School of Medicine, Nanyang Technological University Singapore and Cancer Genetics Service, National Cancer Centre Singapore

Joanne M. Nguyen McGovern Medical School, The University of Texas Health Science Center at Houston, Houston, Texas, USA

Hope Northrup Division of Medical Genetics, Department of Pediatrics, McGovern Medical School, University of Texas Health Science Center at Houston, Houston, Texas, USA

Kees Okkersen Department of Neurology, Radboud University Medical Center, Nijmegen, The Netherlands

Christine A. Oley West Midlands Regional Genetics Service, Birmingham Women's Hospital, Edgbaston, Birmingham, UK

Richard M. Pauli Midwest Regional Bone Dysplasia Clinic, University of Wisconsin-Madison, Madison, Wisconsin, USA

Katy Phelan Florida Cancer Specialists and Research Institute, Fort Myers, Florida, USA

Sarah E. Raible Division of Human Genetics and Genomics, The Children's Hospital of Philadelphia, Pennsylvania, USA

Amy E. Roberts Department of Cardiology and Division of Genetics, Department of Medicine, Boston Children's Hospital, Boston, Massachusetts, USA

R. Curtis Rogers Greenwood Genetic Center – Greenville, Greenville, South Carolina, USA; Clemson University, Clemson, Pickens, South Carolina, USA

Howard M. Saal Division of Human Genetics, Cincinnati Children's Hospital Medical Center, Department of Pediatrics, University of Cincinnati College of Medicine, Cincinnati, Ohio, USA

Emanuela Salzano Division of Human Genetics and Genomics, The Children's Hospital of Philadelphia, Pennsylvania, USA

Carole Samango-Sprouse Department of Pediatrics, George Washington University, Washington, DC, USA; and Department of Human and Molecular Genetics, Florida International University, Miami, Florida, USA

Prachi E. Shah Division of Developmental and Behavioral Pediatrics and the Center for Human Growth and Development, Department of Pediatrics, the University of Michigan Medical School, Ann Arbor, Michigan, USA

Mary B. Sheppard Department of Family and Community Medicine, Department of Surgery, Department of Physiology, Saha Cardiovascular Research Center, University of Kentucky, Lexington, Kentucky, USA

Cheryl Shuman Department of Genetic Counselling, The Hospital for Sick Children and Department of Molecular Genetics, University of Toronto, Toronto, Ontario, Canada

Anne M. Slavotinek Division of Medical Genetics, Department of Pediatrics, University of California, San Francisco, San Francisco, California, USA

Eric E. Smeets Rett Expertise Center – Governor Kremers Center, Maastricht University Medical Center, Maastricht, The Netherlands; and Department of Pediatrics, Maastricht University Medical Center, Maastricht, The Netherlands

Ann C.M. Smith Office of the Clinical Director, National Human Genome Research Institute, National Institutes of Health, Bethesda, Maryland, USA

Benjamin D. Solomon National Human Genome Research Institute, Bethesda, Maryland, USA

Gail A. Spiridigliozzi Department of Psychiatry and Behavioral Sciences, Duke University Medical Center, Durham, North Carolina, USA

Roger E. Stevenson Curry Chair in Genetic Therapeutics, Greenwood Genetic Center, Greenwood, South Carolina, USA

C. Gail Summers Departments of Ophthalmology & Visual Neurosciences and Pediatrics, University of Minnesota, Minneapolis, Minnesota, USA

Brad T. Tinkle Peyton Manning Children's Hospital, Indianapolis, Indiana, USA

Joyce T. Turner Children's National Hospital and the George Washington University School of Medicine and Health Sciences, NW, Washington, DC, USA

Marta Unolt The Children's Hospital of Philadelphia and The Perelman School of Medicine at the University of Pennsylvania, Philadelphia, Pennsylvania, USA; and Ospedale Bambino Gesù and La Sapienza University, Rome, Italy

Baziel G.M. van Engelen Department of Neurology, Radboud University Medical Center, Nijmegen, The Netherlands

Conny M. van Ravenswaaij-Arts University of Groningen, The Netherlands

Aline Verstraeten Center of Medical Genetics, Faculty of Medicine and Health Sciences, University of Antwerp and Antwerp University Hospital, Antwerp, Belgium

David Viskochil Division of Medical Genetics, Department of Pediatrics, University of Utah, Salt Lake City, Utah, USA

Jeannie Visootsak Ovid Therapeutics, New York, USA

Emma L. Wakeling Consultant in Clinical Genetics, North East Thames Regional Genetics Service, Great Ormond Street Hospital for Children NHS Foundation Trust, London, UK

Rosanna Weksberg Division of Clinical and Metabolic Genetics, The Hospital for Sick Children and Department of Pediatrics/Institute of Medical Science, University of Toronto, Toronto, Ontario, Canada

Charles A. Williams Division of Genetics and Metabolism, Department of Pediatrics, University of Florida School of Medicine, Gainesville, Florida, USA

Meredith Wilson Department of Clinical Genetics, Children's Hospital at Westmead, Westmead, Discipline of Genomic Medicine, University of Sydney, Sydney, Australia

Emma M.M. Burkitt Wright Genomic Medicine, Manchester University Hospitals NHS Foundation Trust, Manchester, UK

Elaine H. Zackai The Children's Hospital of Philadelphia and The Perelman School of Medicine at the University of Pennsylvania School of Medicine, Philadelphia, Pennsylvania, USA

INTRODUCTION

JOHN C. CAREY
Department of Pediatrics, Division of Medical Genetics, University of Utah Health, Salt Lake City, Utah, USA

AGATINO BATTAGLIA
The Stella Maris Clinical Research Institute for Child and Adolescent Neurology and Psychiatry, Calambrone, Pisa, Italy, and Division of Medical Genetics, Department of Pediatrics University of Utah School of Medicine, Salt Lake City, Utah, USA; and Division of Medical Genetics, Department of Pediatrics, Sanford School of Medicine, University of South Dakota, Sioux Falls, South Dakota, USA

DAVID VISKOCHIL
Department of Pediatrics, Division of Medical Genetics, University of Utah Health, Salt Lake City, Utah, USA

SUZANNE B. CASSIDY
Department of Pediatrics, Division of Medical Genetics, University of California, San Francisco, San Francisco, California and Division of Genetics and Genomic Medicine, University of California, Irvine, Irvine, California, USA

This book primarily focuses on genetic disorders involving syndromes of congenital malformations and neurodevelopmental disabilities. Because of the coverage in other published resources of biochemical conditions, the book does not include the disorders due to established inborn errors of metabolism. Rather, we cover the management of selected conditions labeled by Epstein and colleagues as the "inborn errors of development" (Erickson and Wynshaw-Boris 2016).

The need for a comprehensive and current overview of the practice guidelines for genetic syndromes commonly seen in primary care settings, specialty clinics, and the general medical genetics clinic has never been more evident than in the second decade of the 21st century. Because of the ongoing advances in the molecular underpinnings of classical syndromes and the recently described conditions, current diagnostic testing and understanding of pathogenesis have established a new era in phenotype analysis, knowledge of natural history, management options, and in many cases pharmacotherapeutic treatments.

MANAGEMENT OF SYNDROMES AND THE "CENTRAL DOGMA" OF MEDICAL GENETICS

The management of genetic syndromes comprises two principal components: health supervision guidelines and treatment of the manifestations of the disorder. Comprehensive phenotype analysis, including the understanding of the natural history of the condition, logically informs the development of clinical guidelines and treatment modalities. This progression from phenotype to natural history to management can be considered the "central dogma of medical genetics" (following the theme of the central dogma of biology, DNA→RNA→protein) (see Figure 1).

The modern analysis of phenotype has reached a new level of depth in the genomics era (Carey 2017). Moreover, the study of the natural history of genetic syndromes has been a time-honored endeavor dating back to the 1960s but brought to the center of attention of the genetics field by Hall's seminal paper (1988). Hall defined natural history as "an account of all of the consequences of that disorder" over

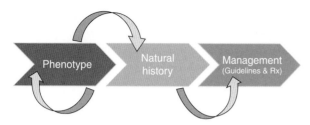

FIGURE 1 Modelled after the central dogma of molecular biology, the "dogma" of medical genetics underlies the process of establishment of the phenotype of a particular condition, including the characterization of its natural history that is so vital to the creation of management guidelines and treatment.

time. Comprehensive and objective review of the clinical manifestations (consequences) of the syndromes covered in this book will be one of the major components of each chapter as outlined in the next section. However, it is important to emphasize that the study of the natural history of these syndromes is daunting: all of the conditions covered herein would be considered "rare diseases". Investigations of the natural history of a rare condition are laden with ascertainment and publication bias, making it difficult to establish the actual frequency of manifestations of the condition. It is challenging to determine the true outcome without multi-center studies and national or international registries because of the small numbers of individuals with the disorder seen clinically by any one investigator. Ensuring accurate diagnosis by all investigators can also complicate collaborative studies for some disorders.

Development of clinical guidelines for health supervision is also difficult because of the lack of randomized clinical trials to determine the efficacy of an intervention or diagnostic modality. While consensus guidelines for a few selected conditions have been published by various academies and societies (e.g., the American Academy of Pediatrics, the American College of Medical Genetics and Genomics), most guidelines for care (as in this book) are based on narrative reviews performed by "experts" combined with the few existing published studies aimed at providing an unbiased approach to management for the condition.

THE ORGANIZATION OF THIS BOOK

Each chapter of this book is dedicated to the diagnosis and management of a specific syndrome (or closely related syndromes) that is encountered with regularity in specialty programs and often in primary care practice. The authors of each chapter are acknowledged "experts" who have considerable personal experience in the management of the disorder. Each chapter thus contains unpublished information based on that experience and on the author's personal approach to management in addition to a comprehensive, unbiased, and current review of published information. Whenever available, evidence-based treatments are included. Each chapter format is similar, providing general information on incidence and inheritance, pathogenesis and etiology, diagnostic criteria and testing, and differential diagnosis. The myriad manifestations of each syndrome are presented system by system, with emphasis on the manifestations, evaluation, management, and prognosis. The first two "systems" in every chapter are "growth and feeding" and "development and behavior." After these, the systems relevant to the specific disorder are discussed, usually in order of importance for that disorder. Every attempt has been made to include whatever is known about the disorder in adulthood. Each chapter concludes with a listing of family support organizations and selected resources available to families and professionals in print and electronic formats. Photographs of physical findings important for diagnosis or management are provided, and in some chapters figures of other aspects, including genetic basis and pathogenesis, are provided. Selected references stressing management issues and citations of the best review articles have been included.

This introductory chapter is designed to inform the reader about genetics-related terms used in this book, inheritance patterns, general methods for genetic testing, measurement methods, and the role of the medical geneticist and genetic counselor in the care of genetic disorders. It also provides some important references to additional resources of information about genetic disorders, differential diagnoses, genetic testing, and support organizations.

While we have sought to place the chapters in alphabetical order by name, for ease of locating, some chapters pose challenges in that regard. In particular, this is true of the disorders that are caused by a chromosomal abnormality and also have an associated name, most of which are deletion syndromes. In this edition, we have clustered the chromosomal syndromes under "deletion" (e.g., Deletion 4p for Wolf–Hirschhorn syndrome, Deletion 22q11.2 for Velo-Cardio-Facial/DiGeorge syndrome). The disorders with more than one causative genetic mechanism remain under the commonly used name (e.g., Smith–Magenis syndrome and Prader–Willi syndrome). While we realize that this organization is not perfect, we hope that this will facilitate finding the reader's chapter of interest.

CATEGORIZATION OF DISORDERS

The descriptive language for patterns of anomalies is somewhat unique to the field of dysmorphology and deserves a brief review. There are a number of different types of patterns of malformation (see Jones et al. 2013). The term **syndrome** is used specifically to describe a broad alteration of morphogenesis in which the simultaneous presence of more than one malformation or functional defect is known or thought to be the result of a single etiology (Hennekam et al. 2011). Its use implies that the group of malformations and/or physical or cognitive differences has been observed repeatedly in a fairly consistent, recognizable, and unique pattern. The initial definition of any syndrome occurs after the publication of several similar case reports. It becomes refined over time as newly described individuals suggest the inclusion of additional abnormalities and the exclusion of others. Thus, a syndrome comes to be delineated by the coexistence of a small but variable number of "hallmark" abnormalities, whereas several other manifestations may be observed at lower frequencies. Even after a particular syndrome is well established and even defined genetically, the inherent variability or rarity can make clinical diagnosis difficult.

In a specific individual, one or more of the hallmark features of a disorder may be absent and yet the person is affected. This has become evident as genetic testing has advanced and demonstrated the broadness of the clinical spectrum for many disorders. It is important to stress that many syndromes are not associated with intellectual disability. Generally, no one manifestation or anomaly is pathognomonic of a syndrome, and even experienced dysmorphologists may disagree about diagnosis. Often, the individual clinician will have had little direct experience of the syndrome. In this environment, the addition of objective methods of evaluation may be useful. Available techniques include direct measurement (anthropometry), measurements from standard photographs (photogrammetry), facial recognition technologies, and radiologic assessment (cephalometry). Each method has advantages and disadvantages, and each has its proponents (for further details, see Allanson 1997; Carey 2017).

The term **sequence** is used to designate a series of secondary anomalies resulting from a cascade of events initiated by a single malformation, dysplasia, deformation, or disruption (Hennekam et al. 2011; Spranger et al. 1982). A well-known example is the Robin sequence, in which the initiating event is micrognathia. The small mandible then precipitates glossoptosis (posterior and upward displacement of the tongue in the pharynx) with resultant incomplete fusion of the palatal shelves. The initiating event may be a malformation of the mandible or a deformation caused by in utero constraint and thus inhibiting normal growth of the mandible. The individual components of a sequence may well involve quite disparate parts of the body. For example, lower limb joint contractures and bilateral equinovarus deformity are commonly observed in a child with a meningomyelocele because of the altered movement of the lower extremities during intrauterine development.

An **association** is a non-random occurrence in two or more individuals of multiple anomalies not known to represent a sequence or syndrome (Spranger et al. 1982). These anomalies are found together more often than expected by chance alone, demonstrating a statistical relationship but not necessarily a known causal one. For example, the VATER (also known as VACTERL) association (see Chapter 61); the best known example of this pattern, represents the simultaneous occurrence of two or more malformations that include vertebral anomalies, anal atresia/stenosis, heart defects, tracheoesophageal fistula, radial ray defects, and renal and limb abnormalities. An association has limited prognostic significance, and the degree of variability may pose diagnostic problems for the clinician. Most affected children will not have all the anomalies described, which makes establishment of minimal diagnostic criteria difficult. Recognition of an association is useful in that it can guide the clinician, after discovery of two or more component malformations, toward a directed search for the additional anomalies. Associations are generally sporadic within a family and have a low empirical recurrence risk. It is most important to remember that associations are diagnoses of exclusion. Any child with multiple anomalies affecting several systems, with or without growth and/or intellectual disability, should first be assessed to exclude a specific syndrome diagnosis and, lacking such a diagnosis, should have cytogenomic SNP microarray analysis (see explanation of genetic tests below).

The term, **spectrum,** is used to refer to a few selected patterns, including OAV spectrum (see Chapter 42 and Jones et al. 2013). The designation has never been formally defined by any consensus or working group but indicates a recognizable pattern of multiple anomalies that displays a particularly wide spectrum of manifestations of unknown etiology and pathogenesis that is not discrete enough as an entity to represent a syndrome. Like an association, a spectrum comprises a heterogeneous group of conditions that may be sorted out in the future.

MEASUREMENTS

Selected measurements, with comparison to normal standards, may be helpful in confirming the subjective impression of an abnormality. Common craniofacial dimensions, which provide details about facial shape and size, include head circumference, inner and outer canthal distances, and ear length, position, and rotation. Evaluation of stature should include height (length), upper and lower body segment (measured from the pubis), arm span, hand length, palm length, and foot length. Normal standards for these and a wide variety of other standardized measurements can be found in the *Handbook of Physical Measurements* (Gripp et al. 2013), *Growth References: Third Trimester to Adulthood* (Saul et al. 1998), *Smith's Recognizable Patterns of Human Malformation* (Jones et al. 2013), and Gorlin's *Syndromes of the Head and Neck* (Hennekam et al. 2010); however, ethnic background, for which norms may vary, should be taken into consideration. Increasingly, standard growth curves are being developed for particular syndromes.

Many syndrome-specific standards have been compiled and are referenced in the chapters of this book.

The best approach to documenting dysmorphic features is to photograph them. Photographs also facilitate consultations with colleagues and consultants by providing objective evidence of the individual's physical findings. They can be compared with examples of other syndromes in photographic databases such as POSSUM and the London Dysmorphology Database (see below). The prudent clinician will often adopt an attitude of "watchful waiting" if the diagnosis is not apparent at the first assessment (Aase 1990). As children's facial and body features evolve with time, they may appear more typical of a syndrome, and serial photographs provide documentation of these changes. There is great value to reassessment of the individual with multiple anomalies whose diagnosis is unclear, because there is significant diagnostic yield (Battaglia et al. 1999). The "art" of dysmorphology is eloquently discussed by Aase (1990).

COMMON GENETIC TERMINOLOGY

With the recent rapid advances in human genetics has come a proliferation of terms whose meaning may be unclear to some practitioners. Therefore, a summary of the common terms relating to genes and chromosomes and the major inheritance patterns is in order.

Genes are the individual pieces of coding information that we inherit from our parents, the blueprint, as it were, for an organism. It is estimated that ~20,000 genes are required to develop and "operate" a human being, with alternative mRNA transcript processing leading to multiple isoforms providing even more regulatory control points for this primary **exome**. Individual genes occur in pairs, one inherited from each parent. The balance of the expression of these genes is extremely delicate, with significant abnormality resulting when this balance is disturbed for some genes. Variant forms of the same gene are known as **alleles**, and variation may have no apparent phenotypic effect or even minor variation could have major consequences, depending on the specific gene and many other factors. When a variant has no phenotypic effect, it is often called a **polymorphism** and if a variant exerts minor phenotype effect it may be called a **hypomorphic allele**.

Some syndromes are caused by a permanent structural or sequence change (**variant** or **mutation**) involving a single gene. When variants cause human disease, the mutation is referred to as a **pathogenic variant**. When intragenic variants lead to deficiency of its encoded protein, this is referred to as **haploinsufficiency**. This happens when an intragenic mutation results in decreased gene expression or a failure to produce the gene product, which can be a so-called **null mutation**. Other mutations cause adverse effects by either interfering with normal protein function or causing a new adverse effect, and such variants are called **dominant negative mutations**, which often result when a structurally abnormal protein is formed. Most intragenic mutations result in alteration of the sequence and/or length of the bases composing the gene code. Such alterations that result in the substitution of one amino acid for another is a **missense mutation**. A DNA sequence change that alters the code for an amino acid leading to premature translation termination is a **nonsense mutation** that can lead to either nonsense-mediated decay of the mRNA transcript or a shortened peptide. An unusual form of mutation that is present in a number of neurogenetic disorders, such as fragile X syndrome, myotonic dystrophy, Huntington disease, and the spinocerebellar ataxias, among others, is expansion of **trinucleotide repeats**. Some genes contain within them a string of three bases repeated a number of times. For example, CGG is repeated up to 50 times in the normal *FMR1* gene (CGGCGGCGG...). Under certain circumstances, this number expands, resulting in an increase in the number of such repeated triplets of bases. Thus, in individuals who are affected with fragile X syndrome, an X-linked cause of intellectual disability, there may be hundreds of such repeated triplets. This triplet repeat expansion interferes with the normal function of the gene, causing abnormality (in this case, intellectual disability). In fragile X syndrome, the gene is inactivated if the expansion exceeds a certain number of repeats. Myotonic dystrophy type 1 is another example of a trinucleotide repeat expansion disorder.

In recent years, some new types of changes in the genetic apparatus have been recognized to cause human disorders. An **epigenetic mutation** is a biochemical change of the DNA that does not alter the base sequence but rather modifies the gene's expression. This generally includes methylation of bases or changes in chromatin structure that alters access to genomic DNA by transcription factors resulting in altered gene transcription that changes normal protein production. These changes can be heritable. Epigenetic modification of some DNA is normal, but perturbations or changes in dosage of that modification have been shown to result in disorders such as Prader–Willi syndrome, Angelman syndrome, Beckwith–Wiedemann syndrome, and Silver–Russell syndrome. Such changes are described in more detail in those individual chapters.

The nomenclature for genes and gene products (proteins) can be confusing, despite the best efforts toward a logical approach. The names of genes are often put in italics, and these may represent an abbreviation of the name of the disorder, the name of the protein, or a function of the protein or the gene. For example, the gene causing neurofibromatosis type 1 is called *NF1*, and the protein is named neurofibromin, whereas the gene for Angelman syndrome, *UBE3A*, is named for its protein product, ubiquitin-protein ligase E3A, which is one of a family of enzymes that are part of the intracellular protein degradation process. The gene responsible for fragile X syndrome is called *FMR1* (*f*ragile X-linked *m*ental *r*etardation 1), and the protein is called FMRP (*f*ragile X-linked *m*ental *r*etardation *p*rotein). Information on the genes is included in the chapters for those who are interested, but aside from genetic testing purposes, it is not critical to know the nomenclature to understand and treat the disorder.

The Genome Nomenclature Committee operates under the auspices of Human Genome Organization (HUGO) to approve gene names and symbols, and almost 33,000 symbols have been approved. This function is funded as a nonprofit organization by the US National Human Genome Research Institute (NHGRI) and the Wellcome Trust (UK) (www.genenames.org). The Locus Reference Genomic project is an effort to implement universal reference sequences for reporting of gene variants based on the conventions developed by the Human Genome Variation Society (MacArthur et al. 2014). Once a DNA sequence variant is established as benign, likely benign, variant of uncertain significance, likely pathogenic, or pathogenic, various databases can be screened in an attempt to establish a genotype–phenotype correlation, which could be helpful in stratifying individuals with their respective genetic condition for ongoing management. Databases harboring annotated sequence variants are not uniform and vary by genetic syndromes. Periodic surveillance for new mutation-specific information with respect to clinical care is an ongoing vigil that should be integrated in follow-up evaluations.

Human genes are "packaged" into 46 **chromosomes**, of which normally 23 chromosomes are transmitted to the offspring in the egg from the mother and 23 in the sperm from the father. One pair of chromosomes, the **sex chromosomes**, differs between males and females. Females have two copies of the X chromosome, whereas males have one copy, the second sex chromosome being the Y chromosome with a largely different set of genes. The remaining 22 pairs, the **autosomes**, do not differ between males and females. The autosomes are numbered in a standard way from largest to smallest. The location of a specific gene on a chromosome is called the **locus** (the plural is **loci**). Some of the syndromes described in this book are caused by the presence of an entire extra chromosome (e.g., Down syndrome, Klinefelter syndrome), called **trisomy**, or duplication of a segment of a chromosome (e.g., some cases of Beckwith–Wiedemann syndrome) called **microduplication**. Others occur because of loss of an entire chromosome (e.g., Turner syndrome), **monosomy**, or loss of part of a chromosome (e.g., WAGR, velo-cardio-facial syndrome), called **microdeletion**. Microdeletions or microduplications of chromosomal segments lead to altered gene expression due to the resulting genomic imbalance. Deletions are more compromising than duplications, and lead to haploinsufficiency of the genes within the deleted region.

The terms that clinical geneticists use to describe a body part may be unfamiliar to some readers. They have gradually evolved in a haphazard and uncoordinated manner and have been critically reviewed in the series of articles referred to as the Elements of Morphology (Allanson et al. 2009; Biesecker et al. 2009; Carey et al. 2009; Hall et al. 2009; Hennekam et al. 2009; Hunter et al. 2009). While we have strived to use lay language wherever possible, there may be descriptive terms in these chapters that require definition. In the series of articles cited above, the reader will find preferred terms for each feature of the head and face, and hands and feet, with a definition and description of how to observe and measure (where possible) the feature. Each term is accompanied by at least one photograph.

PATTERNS OF INHERITANCE

An alteration in a gene can be dominant or recessive. A **dominant** gene mutation only needs to be present in one allele of the gene pair to have a clinically evident impact. Any individual with an autosomal dominant gene mutation will have a one in two chance to pass it on to his or her child, male or female, with each pregnancy. An example is achondroplasia. In achondroplasia, the affected child frequently has two average-stature parents, indicating that the mutation occurred in the egg or sperm that was involved in the conception. This is referred to as a **new mutation** or a **de novo mutation**. Rarely, an apparently normal couple will have more than one child with the same apparently new mutation in an autosomal dominant gene. This suggests that the mutation is present in some of the cells of the germ line (gonads) but not in most other cells of the body of one parent. This is known as **germ line** (or **gonadal**) **mosaicism**. When a parent has a gonadal cell line with a dominant mutation, the recurrence risk is significantly greater than the background population risk for a second child with a new mutation but less than the 50% risk expected if the parent had the mutation in all cells of the body and manifested the condition. Several different dominant disorders have been documented to recur in more than one child of an unaffected parent because of germ line mosaicism. Alternately, the autosomal dominant mutation may be carried in a proportion of a parent's somatic cells as well as the germ line. In this situation, the manifestations of the condition may differ, being milder, segmental, or focal. This **somatic mosaicism** may manifest as a streaky alteration in skin pigmentation. Somatic and germ line mosaicism, at the level of the gene or chromosome, occur after conception.

An autosomal **recessive** gene mutation, when present in a single copy in an individual, will be hidden. Such a person is known as a "**carrier**" and will be unaffected. If, by chance, a person inherits an abnormal allele for an autosomal recessive disorder from each carrier parent, there is no normal gene partner and the two altered alleles will cause symptoms and signs, for example, cystic fibrosis. When each parent carries a recessive mutation for the same disorder, the chance that they both will pass on the mutation to their child, who is then affected, is 25%.

Recessive genes on the X chromosome have different consequences in males and females. A mutated recessive allele on the X chromosome will tend to have little impact in a female, because there is a second, normal copy of the gene on the second X chromosome of the pair to compensate for the mutant allele. In contrast, in the male, a mutation of a **recessive X-linked gene** will have an impact because the genes on the Y chromosome are different from those on the X, and no second gene copy exists. That male must pass the mutated X-linked gene to all his daughters but to none of his sons, because he passes his Y chromosome to his sons. Some disorders are **X-linked dominant**, and females will also be affected. However, males are generally more severely affected in such disorders.

In certain areas of the genetic code, genes behave differently if they have been inherited from the father (**paternally inherited**) rather than from the mother (**maternally inherited**). Only one copy of that gene may be active, whereas the other is inactivated, usually by a process of DNA methylation of the silenced gene. This is a normal manner in which dosage of some genes is controlled. These genes, whose action differs depending on the parent of origin, are said to be **imprinted**. Disorders may occur if the active gene is altered or missing, but not if the silenced allele is altered or missing. More can be learned about this phenomenon in the chapters on the imprinted disorders Angelman syndrome (Chapter 5), Beckwith–Wiedemann syndrome (Chapter 9), Prader–Willi syndrome (Chapter 46), and Silver–Russell syndrome (Chapter 52). A more detailed account of patterns of inheritance, imprinting, and mosaicism can be found in any standard text of human or medical genetics, such as those listed under **additional resources** below.

GENETIC TESTING

Several terms used in this book in describing genetic tests are likely to be unfamiliar to some readers. For some disorders, the appropriate test is a **chromosome analysis** (or **karyotype**, which is an ordered display of an individual's chromosomes). Chromosomes are analyzed by special staining techniques that result in visibility of dark and light bands, which are designated in a standardized way from the centromere, or major constriction. The short arm of the chromosome is called "p", the long arm is called "q", and bands are numbered up from the centromere on the p arm and down from the centromere on the q arm. Each band is further subdivided according to areas within the bands or between them. Thus, the deletion found in velo-cardio-facial syndrome is in the first band of the q arm of chromosome 22 and is designated del22(q11.2). A standard chromosome analysis has at least 450 bands, which is quite adequate for numerical chromosome anomalies like Down syndrome. For some disorders, however, the anomaly cannot be seen reliably on standard chromosome analysis and requires special handling while being processed called **high-resolution banding**. An alternative term, **prometaphase banding**, is used because the cell growth during culturing is adjusted to maximize the number of cells in prometaphase, where the chromosomes are much less condensed and thus longer, rather than in metaphase, where cell growth is stopped in standard chromosome studies. High-resolution banding often has 550 to 800 bands and allows much more detailed analysis.

Another technique combines chromosome analysis with the use of fluorescence-tagged molecular markers (called probes) that are applied after the chromosome preparation is produced. This method is called **fluorescence *in situ* hybridization**, or **FISH**, and relies on the phenomenon of hybridization (intertwining) of complementary pieces of DNA. Thus, to test whether there is a microdeletion that is not visible using chromosome analysis alone, a fluorescence-tagged DNA probe complementary to the deleted material is applied to the chromosome preparation. If the chromosome material is present in the normal amount, a fluorescent signal will be visible at that site under the fluorescence microscope; if the normal chromosome material is absent (deleted), there will be no fluorescent signal. FISH is a powerful tool not only for diagnosing relatively common microdeletion or microduplication disorders but also for identifying the origin of extra chromosome material that cannot be identified by inspection alone and for sorting out the origin of the components of a **translocation** (structural rearrangement of chromosomal material).

Smaller deletions and duplications are being more frequently identified by the technique called **array comparative genomic hybridization** (commonly abbreviated to **array CGH**, or just **CGH**). With the application of single nucleotide polymorphisms (**SNPs**) in recent years, the technique is now referred to as cytogenomic microarray (**CMA**). This diagnostic tool merges traditional chromosome analysis with molecular diagnostics. Array CGH and CMA allow for sensitive and specific detection of copy number variations of submicroscopic chromosomal regions, even intragenic imbalances, throughout the entire human genome.

Other types of genetic testing rely exclusively on molecular diagnostic methodologies. **Polymerase chain reaction (PCR)** is a powerful technique for amplifying small quantities of a segment of DNA so that it can be analyzed for genomic balance and sequence differences in specific alleles. PCR is used for testing for many genetic disorders that are due to a recurring mutation (such as achondroplasia) or a finite number of common mutations. It can also be used to identify the presence of alterations in the normal methylation pattern in imprinted disorders. **Southern blot** techniques are more time consuming; they involve breaking DNA into small pieces using restriction enzymes and then separating them out using gel electrophoresis and analyzing whether there is a deviation in the distance that a segment of the DNA travels through a gel matrix, indicating that its size is different from usual. Both PCR and Southern blotting usually involve the use of **DNA markers**, or **probes**. These are small segments of DNA complementary to an area of interest. One special type of probe takes advantage of the fact that DNA normally contains many runs of repeated base pairs, such as CACACACACA..., which are located throughout the genome and generally have no phenotypic consequences. These are called **microsatellites**. The number of repeats at these loci are inherited as alleles like a genetic trait. Unlike single nucleotide polymorphisms, there are multiple variations in the number of repeated doublets or triplets, which can be scored by length to designate a signature of bi-allelic repeats at specific loci in the genome. These so-called **microsatellite markers** form the basis for paternity testing and are also used for diagnostic testing of neighboring genes or the genes within which they occur, although they are not the mutation of the relevant gene that causes disease.

The nomenclature for **markers** is a bit more uniform than that for genes. Markers are indicated by the letter D (standing for DNA), followed by the number of the chromosome they are on, followed by the letter S (standing for single copy) and the number representing the numerical order in which they were identified. Thus, D15S10 was the 10th marker to be identified on chromosome 15.

The delineation of DNA marker locations served as the framework for the Human Genome Project, which ultimately transformed approaches to gene identification and molecular confirmation of genetic syndrome diagnosis. Knowledge of unique DNA sequence at mapped loci enabled rapid and sensitive molecular testing to confirm suspected diagnoses. A quantum leap in genetic testing was the development of **NGS (next generation sequencing)** whereby DNA from patients could be screened for sequence variants in days to weeks rather than months to years (Rehm et al. 2013). Not only did this technique increase sensitivity for detection of pathogenic mutations in single genes, it enabled labs to develop panels to screen many genes for a given phenotype. Genetic syndromes with overlapping manifestations could be easily screened simultaneously, for example individuals with a mandibulofacial dysostosis could be screened for multiple genes that, when mutated, cause Treacher Collins syndrome (Chapter 57) or other conditions within the differential. The convenience of NGS led to the development of wider and wider testing panels, ultimately making **whole exome sequence (WES)** analysis a reality in clinical care.

The availability of WES at lower cost has led to the emergence of a "genotype first" approach when a genetic syndrome is not diagnosed clinically. Most of the genetic syndromes in this book have distinctive enough phenotypes that even individuals with attenuated manifestations can be recognized with experience. However, individuals with milder manifestations may not be so easily recognized, and a broader molecular screening approach leads to the diagnosis of individuals on the milder end of the syndrome spectrum. This application of "genotype first" has led to a recognition of phenotype expansion for some of the disorders in this book, which may lead to a better understanding of specific management issues on a genotype-phenotype basis.

The methodologies for genetic testing have been optimized for many of the syndromes presented here, and the technical and complex nature of this testing is beyond the scope of this book. The interested reader is referred to the list of glossaries at the end of this chapter. The most accessible, detailed, and current of these glossaries is to be found online at the GeneTests web site (*https://www.ncbi.nlm.nih.gov/gtr/*).

ROLE OF THE MEDICAL GENETICIST AND GENETIC COUNSELOR

Many syndromes are relatively rare, and any individual physician may have limited personal experience. Medical geneticists, on the other hand, frequently have considerable experience of many affected individuals and have ready access to additional information through the genetics literature, specialized databases, and interactive meetings. In many instances, genetic testing results in a laboratory determination of **VUS (variant of uncertain significance)** (Richards et al. 2015). Ascertaining additional phenotype descriptions, both medical history and physical manifestations, may enable medical geneticists to develop a strong case for the laboratory director to convert a VUS to a likely pathogenic or even pathogenic variant. In addition, medical geneticists and genetic counselors review the details of pathogenic variants to tease out genotype–phenotype correlations that might lead to different approaches in care, for example surveillance for malignancy or evolving organ insufficiency.

The myriad manifestations of each of the syndromes included in this book often require the care of many diverse specialties. The geneticist can assist in diagnosis, testing, and counseling of affected individuals and their family members as a consultant to the nongenetics physician and can orchestrate coordination of care to focus on the whole child or adult. Anticipatory guidance is particularly helpful in providing families a sense of what lies ahead and how they can be proactive in care. The role of the geneticist extends beyond the individual to involve the care and well-being of the entire family. The primary care physician is encouraged to consult medical geneticists to assist in the management of individuals with multiple anomaly syndromes, especially those with behavioral phenotypes. Finally, integration of management guidelines with family needs and expectations is an important component of multispecialty care.

An important facet of the care of individuals with syndromes and their families is genetic counseling. This is the provision of non-directive information about the diagnosis and its implications not only for the individual (prognosis) but also for the family (reproductive risks and options). It includes knowledge of the inheritance pattern, likelihood of recurrence in a future pregnancy, and prenatal diagnostic options. Referral to relevant community resources, such as patient support groups, brochures and websites, and financial, social, and educational services, can also be made during this process. It is recommended that families with a confirmed diagnosis be referred to information provided by these groups even prior to consultation with a genetic counselor or medical geneticist, to avert having them do an internet search. Internet searches often result in biased or mistaken information. Assisting the individual and/or family to understand the condition and its impact, provide optimal care, and adapt to the existence of a chronic and complex disorder are all part of the process of genetic counseling. Adjustment to a new diagnosis may put considerable strain on a family, and providing emotional support for the family is paramount. Genetic counseling is usually provided either by medical geneticists or by genetic counselors, who are Masters-prepared professionals knowledgeable about genetic disorders and their inheritance. Genetic counselors are well trained to determine genetic risks and are adept in assisting with emotional and psychological adjustments necessitated for optimal outcome.

REFERENCES, ADDITIONAL RESOURCES, AND WEB SITES

Additional information concerning the included disorders, as well as explanations of inheritance information and diagnostic testing, may be found in **standard texts** on genetics and genetic disorders. A few particularly useful **texts and references, including those cited above** are listed below.

Aase JM (1990) *Diagnostic Dysmorphology*, 1st ed. New York: Kluwer Academic/Plenum Publishers.

Allanson JE (1997) Objective techniques for craniofacial assessment: What are the choices? *Am J Med Genet* 70:1–5.

Allanson JE, Cunniff C, Hoyme HE, McGaughran J, Muenke M, Neri G (2009) Elements of Morphology: Standard terminology for the head and face. *Am J Med Genet* 149A:6–28.

Battaglia A, Bianchini E, Carey JC (1999) Diagnostic yield of the comprehensive assessment of developmental delay/mental retardation in an institute of child neuropsychiatry. *Am J Med Genet* 82:60–66.

Biesecker LG, Aase JM, Clericuzio C, Gurrieri F, Temple K, Toriello H (2009) Elements of morphology: Standard terminology for the hands and feet. *Am J Med Genet* 149A:93–127.

Carey JC (2017) Phenotype analysis of congenital and neurodevelopmental disorders in the next generation sequencing era. *Am J Med Genet, Part C* 175C:320–328.

Carey JC, Cohen MM Jr, Curry C, Devriendt K, Holmes L, Verloes A (2009) Elements of Morphology: Standard terminology for the lips, mouth, and oral region. *Am J Med Genet* 149A:77–92.

Erickson RP Wynshaw-Boris A (2016) *Epstein's Inborn Errors of Development*, 3rd ed. New York: Oxford University Press.

Gripp KW, Slavotinek AM, Hall JG, Allanson JE (2013) *Handbook of Physical Measurements*, 3rd ed., Oxford: Oxford University Press.

Hall JG (1988) The value of the study of natural history in genetic disorders and congenital anomaly syndromes. *J Med Genet* 25:434–444.

Hall BD, Graham JM Jr, Cassidy SB, Opitz JM (2009) Elements of Morphology: Standard terminology for the periorbital region. *Am J Med Genet* 149A:29–39.

Hennekam RCM, Biesecker LG, Allanson JE, Hall JG, Opitz JM, Temple IK, Carey JC (2011) Elements of Morphology: general terms for congenital anomalies. *Am J Med Genet Part A* 161A:2726–2733.

Hennekam RCM, Cormier-Daire V, Hall J, Méhes K, Patton M, Stevenson R (2009) Elements of Morphology: Standard terminology for the nose and philtrum. *Am J Med Genet* 149A:61–76.

Hennekam RCM, Krantz ID, Allanson JE (2010) *Gorlin's Syndromes of the Head and Neck*, 5th ed. New York: Oxford University Press.

Hunter A, Frias J, Gillessen-Kaesbach G, Hughes H, Jones K, Wilson L (2009) Elements of morphology: Standard terminology for the ear. *Am J Med Genet* 149A:40–60.

Jones KL, Jones MC, Del Campo M (2013) *Smith's Recognizable Patterns of Human Malformation*, 7th ed. Philadelphia: Saunders/Elsevier.

King RA, Rotter J, Motulsky AH (2002) *The Genetic Basis of Common Disease*, New York: Oxford University Press.

MacArthur JA, Morales J, Tully RE, Astashyn A, Gil L, Bruford EA, Larsson P, Flicek P, Dalgleish R, Maglott DR, Cunningham F (2013) Locus Reference Genomic: reference sequences for the reporting of clinically relevant sequence variants. *Nucleic Acids Research* 42: D873–D878.

Nussbaum RL, McInnes RR, Willard HF (2016) *Genetics in Medicine*, 8th ed. Philadelphia: Saunders/Elevier.

Pyeritz RE, Korf BR, Grody WW (2018) *Emery and Rimoin's Principles and Practice of Medical Genetics*, 7th ed. Academic Press/Elsevier.

Rehm HL, Bale SJ, Bayrak-Toydemir P, Berg JS, Brown KK, Deignan JL, Friez MJ, Funke BH, Hegde MR, Lyon E, Working Group of the American College of Medical Genetics and Genomics Laboratory Quality Assurance Committee (2013) ACMG clinical laboratory standards for next-generation sequencing. *Genet Med* 15:733–747.

Richards S, Aziz N, Bale S, Bick D, Das S, Gastier-Foster J, Grody WW, Hegde M, Lyon E, Spector E, Voelkerding K, Rehm HL, ACMG Laboratory Quality Assurance Committee (2015) Standards and guidelines for the interpretation of sequence variants: a joint consensus recommendation of the American College of Medical Genetics and Genomics and the Association for Molecular Pathology. *Genet Med* 17:405–424.

Saul RA, Seaver LH, Sweet KM, Geer JS, Phelan MC, Mills CM (1998) *Growth References: Third Trimester to Adulthood*, 2nd ed. Greenwood: Greenwood Genetic Center.

Scriver CR, Beaudet AL, Valle D, Sly WS (2001). *The Metabolic and Molecular Bases of Inherited Disease*, 8th ed. New York: McGraw-Hill.

Spranger J, Benirschke K, Hall JG, Lenz W, Lowry RB, Opitz JM, Pinsky L, Schwarzacher HG, Smith DW (1982) Errors of morphogenesis: Concepts and terms. Recommendations of an International Working Group. *J Pediatr* 100:160–165.

In addition, important **online resources** on genetic disorders are readily available, including:

- Online Mendelian Inheritance in Man (OMIM) (*https://www.omim.org*) is a comprehensive catalogue of inherited disorders.
- GeneReviews (*https://www.ncbi.nlm.nih.gov/books/NBK1116/*) provides current online information on diagnosis, testing, and management of genetic disorders.
- Face2Gene (*https://www.face2gene.com/*) is a smartphone accessible resource for facial diagnosis in syndromes.

For those with a deeper interest, there are **electronic databases** that aid in diagnosis and provide photographs and references concerning not only common but also rare genetic disorders. These must be purchased, and include:

- London Dysmorphology Database (*https://www.face2gene.com/lmd-library-london-medical-database-dysmorphology/*) operated out of Face2Gene
- POSSUM (Pictures of Standard Syndromes and Undiagnosed Malformations) (*http://www.possum.net.au*).

A **resource of laboratories** doing specialized diagnostic testing, both clinically and for research, for genetic disorders and syndromes is:

- GeneTests (*https://www.ncbi.nlm.nih.gov/gtr/*) provides information and links on the laboratories around the world providing testing of genetic disorders.

Further information on individual syndromes for practitioners or families can be obtained from other **online resources**, including:

- National Organization for Rare Diseases (NORD) (*https://rarediseases.org*)
- March of Dimes/Birth Defects Foundation (*http://www.marchofdimes.org*)
- The Alliance of Genetic Support Groups (*http://www.geneticalliance.org*)
- Orphanet (*https://www.orpha.net/consor/cgi-bin/index.php*)
- The UNIQUE Support Group (https://www.rarechromo.org/).

1

AARSKOG SYNDROME

ROGER E. STEVENSON
Curry Chair in Genetic Therapeutics, Greenwood Genetic Center, Greenwood, South Carolina, USA

INTRODUCTION

Aarskog syndrome is one of the most clinically distinctive phenotypes among the hereditary syndromes. Manifestations in the facial morphology, skeleton, and genitalia comprise a clinically useful diagnostic triad, present from birth. The condition was first described in a Norwegian family by the pediatrician Dagfinn Aarskog in 1970 and in an American family by the geneticist Charles Scott in 1971. Aarskog syndrome is an X-linked condition with fully expressed manifestations in males and subtle findings in many carrier females. The causative gene, *FGD1*, located at Xp11.22 was identified in 1994 (Pasteris et al. 1994).

Although the condition has been given the more descriptive names of facial-genital-digital syndrome and faciogenital dysplasia, the designations Aarskog syndrome and Aarskog–Scott syndrome have retained greatest favor.

Incidence

Over 250 affected individuals with Aarskog syndrome have been reported, providing a rich descriptive literature and precluding the reporting of most currently identified cases (Aarskog, 1970; Scott 1971; Furukawa et al. 1972; Porteous and Goudie 1991; Fryns 1992; Teebi et al. 1993; Fernandez et al. 1994; Stevenson et al. 1994; Orrico et al. 2004, 2005, 2007, 2015; Shalev et al. 2006; Bottani et al. 2007; Pérez-Coria et al. 2015; Griffin et al. 2016). Many multigenerational pedigrees have been identified because of the X-linked inheritance pattern, the presence of distinctive external manifestations, and the absence of lethal manifestations. Aarskog syndrome has been reported worldwide and from most ethnic and racial groups. There does not appear to be an increased rate in any subpopulation. Among those clinically diagnosed, only about 20% are found to have *FGD1* mutations, suggesting overdiagnosis or genetic heterogeneity among these cases.

The wide recognition and large number of ascertained cases notwithstanding, a reliable prevalence for Aarskog syndrome is not known. An estimate of 1/25,000 live births has been made (Orrico et al. 2015). Subtle manifestations permit many, perhaps most, cases to go undiagnosed.

Lifespan is said to be normal, but this too has not been documented by systematic study. Survival into the eighth decade is found within the reported pedigrees.

Diagnostic Criteria

The diagnosis of Aarskog syndrome is based on clinical findings. In most cases, the pattern of craniofacial, skeletal and genital manifestations is sufficient to make the diagnosis (Table 1.1). Identification of an X-linked pattern of inheritance and the presence of subtle findings in carrier females assist in familial cases. A responsible gene has been identified and testing for sequencing variants and deletions/duplications is currently available in many diagnostic laboratories.

Manifestations of Aarskog syndrome are present from birth. Head size is usually normal, but may appear large in relation to the face and the body. The facial appearance is distinctive and

---

Cassidy and Allanson's Management of Genetic Syndromes, Fourth Edition.
Edited by John C. Carey, Agatino Battaglia, David Viskochil, and Suzanne B. Cassidy.
© 2021 John Wiley & Sons, Inc. Published 2021 by John Wiley & Sons, Inc.

TABLE 1.1 Clinical manifestations in Aarskog syndrome

Finding	<25%	25–50%	50–75%	>75%
Craniofacial				
Broad forehead			+	
Widow's peak			+	
Hypertelorism				+
Ptosis			+	
Downward eye slant			+	
Large cornea			+	
Other ocular abnormality	+			
Maxillary hypoplasia			+	
Small cupped ears				+
Fleshy earlobes			+	
Short nose				+
Anteverted nares				+
Long philtrum				+
Wide mouth				+
Thin upper lip				+
Dental abnormalities				+
Crease below lower lip				+
Small mandible				
Skeletal				
Brachydactyly				+
Syndactyly			+	
Short finger V			+	
Clinodactyly V			+	
Short bulbous toes			+	
Single palmar crease			+	
Pectus excavatum			+	
Metatarsus adductus			+	
Joint laxity			+	
Cervical spine abnormalities		+		
Genital				
Shawl scrotum				+
Undescended testes			+	
Inguinal hernia			+	
Umbilical prominence/hernia		+		
Other				
Short stature				+
Long trunk			+	
Cardiac defect	+			

Source: Adapted from Furukawa et al. (1972), Berman et al. (1975), Grier et al. (1983), Porteous and Goudie (1991), Tsukahara and Fernandez (1994), Fernandez et al. (1994), and Gorski et al. (2000).

in most individuals is diagnostic (Figure 1.1). Changes are present in the upper, middle, and lower portion of the face. Increased width of the forehead, widow's peak, ocular hypertelorism, down-slanted palpebral fissures, and blepharoptosis are the major features of the upper face. A short nose with anteverted nares and small simply formed ears that may protrude are the major features of the midface. Ears often have a broad insertion. The midface may be hypoplastic, but this is rarely of sufficient degree to dominate the appearance of the face. The mouth is wide and the lips thin with a V-shaped configuration at the middle of the upper lip's vermilion border and a subtle upturn to the corners of the mouth. The chin is small. A transverse crease is often present below the lower lip. Individuals have been reported with generous anterior fontanel and facial clefting, which may represent coincidental findings (Völter et al. 2014; Pariltay et al. 2016).

Musculoskeletal findings should be anticipated in most cases (Fryns 1992; Stevenson et al. 1994; Gorski et al. 2000; Al-Semari et al. 2013; Griffin et al. 2016). The hands and fingers are short, often with a single palmar crease and mild cutaneous webbing between the digits. Distinctive posturing of the lateral four fingers with flexion at the metacarpophalangeal joints and the distal interphalangeal joints and hyperextension at the proximal interphalangeal joint can be elicited in the childhood and teen years (Figure 1.2). The feet and toes are short and midfoot varus deformation is commonly present. Overall stature is reduced, falling below the third centile in childhood and returning to the lower normal centiles after puberty. Pectus excavatum occurs in many cases. The joints are typically hyperextensible.

Radiographs may show asynchronous bone age, small middle phalanges of the fifth fingers, short long bones with wide metaphyses, small ilia and ossification anomalies of the cervical spine.

Altered appearance of the genitalia may also be helpful in diagnosis. One or both testes may fail to descend into the scrotal sac. The scrotum tends to surround the penis giving a so-called "shawl scrotum" appearance (Figure 1.3).

Early developmental milestones are usually normal and ultimate cognitive abilities encompass an extended range. Hyperactivity and attention deficit may interfere with academic performance. More severe mental illness has been rarely reported (Trevizol et al. 2015).

Since most literature reports describe affected individuals in childhood, the clinician must be aware of the changing phenotype with age (Fryns, 1992; Stevenson et al. 1994) (Figure 1.1). Three changes tend to obscure the diagnosis in adults. The face elongates, and with this elongation prominence of the forehead and hypertelorism become less apparent. With puberty, growth improves, with adult height in most cases being above the third centile. Pubic hair obscures the presence of shawl scrotum, one of the key clinical findings during childhood.

Typically, female carriers show subtle manifestations. They tend to be shorter than non-carrier sisters and generally have mild craniofacial changes including hypertelorism and fullness of the tip of the nose. They may also exhibit brachydactyly and the typical posturing of the digits with hyperextension of the proximal interphalangeal joints and flexion of the distal joints.

INTRODUCTION 3

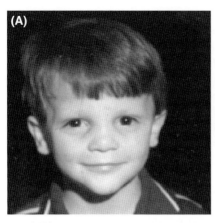

FIGURE 1.1 Aarskog syndrome. Facial findings in three individuals. (A) Four-year-old with broad forehead, hypertelorism, down-slanting palpebral fissures, cupped ears, and wide mouth. (B) Seventeen-year-old with prominent forehead, ptosis, broad nasal root, and down-slanting palpebral fissures. (C) Sixty-year-old with elongation of face, ptosis, and cupped ears, but less apparent widening of the midface and forehead.

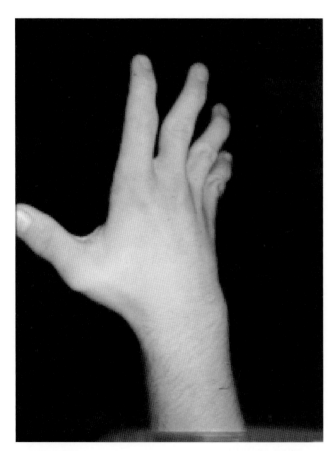

FIGURE 1.2 Characteristic posturing of extended fingers.

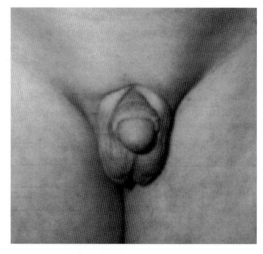

FIGURE 1.3 Shawl scrotum.

Etiology, Pathogenesis, and Genetics

From the initial description, Aarskog syndrome has been recognized as an X-linked disorder. Linkage analysis placed the gene in the pericentromeric region of the X chromosome and a causative gene, *FGD1*, was isolated with positional cloning techniques utilizing an X;8 translocation in which a female carrier had manifestations of Aarskog syndrome (Bawle et al. 1984; Pasteris et al. 1994; Stevenson et al. 1994). The gene, located at Xp11.22, encodes a 961 amino acid protein from 19 exons that span 100kb of the X chromosome.

FGD1, a guanine nucleotide exchange factor, exerts its influence, at least in part, by activating the Rho GTPase, Cdc42 (Gorski et al. 2000). The family of Rho GTPases are expressed widely in embryonic tissues, contributing to the morphology of these tissues through organization of the actin cytoskeleton and a number of other cellular components and processes (Gao et al. 2011; Hou et al. 2003; Genot et al. 2012; Pedigo et al. 2016). Studies in mice indicate that *fdg1*, the homologous mouse gene, is expressed exclusively in skeleton, specifically during periods of incipient and active ossification of both endochondral and intramembranous bones (Gorski et al. 2000). Postnatally, expression is also found in perichondrium, joint capsules and cartilage. These expression studies give little insight into the role of the gene in the development of non-skeletal tissues, e.g, brain, eye, and genital system.

Over 100 different gene alterations, the majority of which lead to truncated proteins, have been identified. The probands include representatives from the original families reported by Aarskog (16bp insertion at nucleotide 519, exon 3) and by Scott (C to T transition at nucleotide 577, exon 3). Mutations of all classes have been found, have been distributed throughout the gene, both within and outside of the major functional domains of the gene, and no genotype–phenotype correlation has emerged. The only whole gene duplications also encompassed adjacent genes (Grams et al. 2016; El-Hattab et al. 2011).

Diagnostic Testing

Demonstration of a mutation in *FGD1* confirms the clinical diagnosis of Aarskog syndrome. Many diagnostic laboratories offer *FGD1* gene sequencing (as part of X-linked intellectual disability panels, whole exome or gene-specific sequencing) as well as deletion/duplication analysis. About 20% of individuals with the clinical diagnosis will have *FGD1* mutations (Schwartz et al. 2000; Orrico et al. 2004, 2015). Mutations have been found more frequently in familial cases and in instances with subtle expression in obligate female carriers.

Differential Diagnosis

None of the manifestations of Aarskog syndrome can be considered pathognomonic, but the manifestations in composite are unlikely to lead to confusion with many other disorders. Although there are other X-linked syndromes with hypertelorism (Atkin–Flaitz, Simpson–Golabi–Behmel, Alpha–Thalassemia Intellectual Disability (*ATRX*, see Chapter 7), Opitz FG, and telecanthus-hypospadias syndromes), other somatic features are useful in distinguishing these syndromes (Stevenson et al. 2012). There are several autosomal syndromes that may be considered in the differential diagnosis.

Noonan syndrome (see Chapter 40) poses the greatest problem in differential diagnosis. Both Aarskog and Noonan syndromes are readily called to mind by clinicians, have short stature, craniofacial and skeletal changes of similar nature, and may have mild cognitive disability. In most individuals, however, they may be readily separable. Persons with Noonan syndrome commonly have webbed neck and cardiac defects. In Aarskog syndrome, the hands and feet are more distinctive, and the shawl scrotum is a helpful finding. Noonan syndrome affects males and females equally. In Aarskog syndrome, females may have subtle craniofacial findings and may be shorter than non-affected sisters. The Noonan-associated disorders – Leopard, Noonan-neurofibromatosis, Costello (see Chapter16) and cardio-facio-cutaneous (see Chapter10) syndromes – are unlikely to be confused with Aarskog syndrome. Noonan syndrome is linked to 12q24, and mutations in *PTPN11* have been demonstrated in about half of those affected. There is considerable genetic heterogeneity with sequencing variants found in *KRAS, NRAS, SOS1, SOS2, RAF1, BRAF, RIT1, LZTR1, SHOC2,* and *LBL* in Noonan and Noonan-like syndromes.

In Teebi hypertelorism syndrome, the facial manifestations – prominent forehead, hypertelorism, short nose with long philtrum and, in some cases, widow's peak and ptosis – are similar to those in Aarskog syndrome (Tsai et al. 2002). Male-to-male transmission, normal stature in affected males and females, the presence of cardiac malformations, and absence of shawl scrotum are distinguishing features. Mutations in *SPECC1L* have been reported in the Teebi hypertelorism syndrome (Bhoj et al. 2015).

Robinow syndrome has similar facial findings, particularly in infancy and early childhood, but can be distinguished by the inheritance pattern (both autosomal dominant and autosomal recessive forms occur), mesomelic or acromesomelic limb shortening and penis hypoplasia (Patton and Afzal, 2002). Mutations in the *ROR2* gene have been found in autosomal recessive Robinow syndrome and mutations in *WNT5A, DVL1* and *DVB3* have been found in autosomal dominant Robinow syndrome.

MANIFESTATIONS AND MANAGEMENT

Growth and Feeding

Although growth may be quite variable in Aarskog syndrome, it is exceptional for stature to be above average. Most affected males, in fact, grow slowly during infancy and childhood and reach only the lower centiles of general population height in adolescence and adulthood. The tendency to short stature notwithstanding, obesity is not a problem of note.

Birth measurements are usually within the normal range. Head growth continues at a normal rate throughout childhood, although the head may appear disproportionately large in relation to the face and body. Length, however, falls below the third centile within the first few years of life and remains there until puberty. Except for delay in its onset, sexual maturation is normal. The accompanying growth acceleration leads to adult heights in the lower centiles, usually between 160 and170 cm.

Hormonal function is generally normal. Bone age lags several years behind height age, giving an expanded period for catch up growth. A number of males have received treatment with growth hormone (Petryk et al. 1999; Darendeliler et al. 2003). Although an increase in the rate of growth could be documented during therapy, it is not assured that adult height would be significantly above what it would have reached without therapy.

Evaluation

- Growth should be monitored by taking measurements of growth parameters at all routine assessments.
- In cases where statural growth appears more severely impaired, i.e. progressive fall below -3 SD, bone age determination, thyroid function, and growth hormone measurements may be indicated.

Management

- No therapy for growth delay and pubertal delay in usually required. Stature ultimately reaches the low/normal centiles in adolescence and adulthood and growth may continue throughout the second decade.

Development and Behavior

Detailed observations of childhood developmental milestones and neurobehavioral manifestations in Aarskog syndrome have not been reported but a gestalt of neurodevelopmental function may be gained from the case report literature. Early childhood motor and speech development usually proceed in normal fashion, and in so doing predict that intellectual function will be normal as well. In a minority, early developmental milestones lag behind age-peers and it is this minority that will likely show impaired cognitive function at maturity.

Overall there appears to be a shift into the lower half of the intellectual curve among males with Aarskog syndrome. Fryns (1992) reports that about 10% of affected males have moderate intellectual disability (IQ < 50) and about 20% have mild intellectual disability (IQ 50–70). In contrast, Logie and Porteous (1998) found intellectual disability to be exceptional, their males exhibiting a wide range of intellectual function with IQs between 62 and 128. The balance of reports in which objective IQ testing is available is in general agreement with 20–30% of males having IQ levels below 70. The clinician should give an attentive eye to developmental progress as an early predictor of school performance. Intellectual disability of moderate–severe degree should prompt a search for coexisting reasons for the impairment.

During childhood, attention deficient and hyperactivity may pose equal schooling challenges. Attention deficit commonly occurs among those with normal cognitive function, but more so among those with subnormal intellectual function. Spontaneous resolution at the time of adolescence is the rule.

Evaluation

- Assessment of developmental progress should be an integral part of routine evaluation of children with Aarskog syndrome.
- Testing of cognitive function and behavior using standardized tests are indicated in individuals who show signs of developmental lag, learning disability, attention deficit disorder or autism spectrum disorder.
- A careful search for a coexisting disorder should be considered in those children with moderate–severe developmental delay.

Management

- Infant stimulation and other early interventions may benefit young children with developmental delay.
- During the school year, special education may be necessary for those with learning difficulties.
- Behavioral therapy and/or a trial of pharmacologic agents such as methylphenidate are indicated for attention deficit and hyperactivity, although as yet there have been no reports of controlled trials of these approaches. Behavior tends to normalize after puberty.

Ophthalmologic

In additional to hypertelorism, downward slanting of the palpebral fissures and ptosis, prominence of the corneas, strabismus, astigmatism, hyperopia, limitation of upward gaze, and tortuosity of the retinal blood vessels have been reported (Kirkham et al. 1975; Brodsky et al. 1990).

Evaluation

- Vision and strabismus screening should be conducted at all routine clinical evaluations.
- Ophthalmologic assessment is recommended in all children prior to school entry and in all who appear to have vision impairment or other ocular abnormality on screening assessment.

Management

- Glasses are commonly required for correction of hyperopia
- Occlusion therapy or surgical correction of strabismus should be initiated as soon as detected, as in the general population.
- Significant ptosis may be surgically corrected for both vision and cosmetic reasons.

Dental

Dental abnormalities have included delayed eruption, broad central incisors, hypodontia, crowding, and excessive caries (Halse et al. 1979; Reddy et al. 1999; Ahmed et al. 2016).

Evaluation

- Routine dental examination should begin in the preschool years and careful attention should be given to tooth alignment.

- Referral for orthodontic evaluation should be made by the early school years.

Management

- Instruction in dental care with flossing, brushing, and use of fluoride containing toothpaste should be given early to prevent cavities and gum disease.
- Standard orthodontic treatment may be required for misaligned teeth.

Musculoskeletal

Skeletal findings including brachydactyly, pectus excavatum, and midfoot varus are common. Skeletal findings should be anticipated in most cases (Fryns, 1992; Stevenson et al. 1994; Gorski et al. 2000; Griffin et al. 2016). Overall stature is reduced with height during childhood often falling at or below the third centile. The trunk is long in relation to the limbs. The hands and fingers are short and there may be some mild cutaneous webbing between the fingers. The fingers are often held in a distinctive position with flexion at the joint between the hand and the fingers, over extension at the proximal joint of the finger and flexion at the distal joint (Figure 1.2). This hand posturing becomes more obvious when there is an attempt to extend and spread the fingers. Often there is only a single palmar crease. The toes are also short and tend to have bulbous tips, and non-fixed midfoot varus deformation occurs commonly.

All of the joints may be unusually hyperextensible. Malformations and excessive movement of the cervical spine may lead to impingement on the spinal cord. Pectus excavatum occurs in many cases.

Mild short stature is common and is often accompanied by small hands and mild webbing between the fingers. Radiographic findings may include asynchronic delay of bone age, shortening of the long bones with widening of the metaphyses, hypoplasia of the middle phalanges of the fifth fingers, small ilia with anteverted femoral heads, and developmental abnormalities of the cervical spine (Lizcano-Gil et al. 1994; Petryk et al. 1999; Gorski et al. 2000).

Evaluation

- Clinical evaluation should include standard assessment of all components of the skeleton with particular attention to cervical vertebrae and metatarsus adduction.

Management

- Progressive casting will usually correct metatarsus adduction. Recalcitrant cases may require surgical correction.
- Intervention for joint laxity is usually not indicated unless it involves the cervical spine (see Neurologic, below).

Neurologic

Neurological complications related to malformation and instability of the cervical spine are the most serious and perhaps the least appreciated finding in Aarskog syndrome. Hypoplasia of the first cervical vertebra and the odontoid, fusion of vertebral bodies, prolapse of intervertebral discs, and ligamentous laxity may lead to cord compression in childhood or adult life (Scott, 1971; Stevenson et al. 1994; Gorski et al. 2000). Delay of motor milestones, gait instability, paresthesia, hyperreflexia, and pain may indicate the possibility of cord compression.

Evaluation

- Evaluation for signs of cervical cord impingement should be an integral part of each examination. Appropriate care should be taken to exclude causes for neurological signs and symptoms.
- Radiographs of the cervical spine should be taken to detect anomalies that may predispose to cervical instability and cord impingement.
- Magnetic resonance imaging of the cervical spine is helpful in locating the site and degree of cord compression.

Management

- Referral for consideration of surgical stabilization of the cervical spine is appropriate in the presence of signs of cord compression due to cervical instability. The surgical procedure is the same as in any other circumstance.
- Standard traction therapy or surgery benefit individuals with herniation of the intervertebral discs.

Genitourinary

Genital abnormalities include shawl scrotum and undescended testes. Hernias occur commonly in the inguinal region and less commonly at the umbilicus. Pubertal development may be delayed but full sexual maturation and fertility should be anticipated.

Puberty begins spontaneously, although delayed for several years in many cases, and sexual maturation is otherwise normal.

Evaluation

- Genital examination with particular attention to undescended testes and inguinal hernias should be included in all routine clinical evaluations.
- Stages of sexual maturation should be documented beginning in the teen years.

Management

- A standard approach to management of cryptorchidism is appropriate. This includes a trial of human chorionic gonadotropin injection or surgical correction prior to school age.

- Surgical evaluation and repair of inguinal hernias follows standard practice in the general population.
- Reassurance regarding pubertal development and fertility may be given. Rarely is pubertal delay sufficient to warrant hormone induction.

RESOURCES

Support Groups

Alliance of Genetic Support Groups

4301 Connecticut Ave, NW, Suite 404,
Washington, DC 20008, USA
Phone: (202) 966-5557.
Fax: (202) 966-8553.
Website: http://www.geneticalliance.org

National Organization for Rare Disorders (NORD)

PO Box 8923, New Fairfield
CT 06812-8923, USA
Phone: (203) 746-6518 or (800) 999-6673
Fax: (203) 746-6481
Website: http://www.rarediseases.org

REFERENCES

Aarskog D (1970) A familial syndrome of short stature associated with facial dysplasia and genital anomalies. *J Pediatr* 77(5): 856–861.

Ahmed A, Mufeed A, Ramachamparambathu AK, Hasoon U (2016) Identifying Aarskog Syndrome. *J Clin Diagn Res* 10(12): ZD09–ZD11.

Al-Semari A, Wakil SM, Al-Muhaizea MA, Dababo M, Al-Amr R, Alkuraya F, Meyer BF (2013) Novel FGD1 mutation underlying Aarskog-Scott syndrome with myopathy and distal arthropathy. *Clin Dysmorphol* 22(1): 13–17.

Bawle E, Tyrkus M, Lipman S, Bozimowski D (1984) Aarskog syndrome: full male and female expression associated with an X-autosome translocation. *Am J Med Genet* 17(3): 595–602.

Berman PA, Desjardins C, Fraser FC (1975) Inheritance of the Aarskog syndrome. Birth Defects Orig Artic Ser X(7): 151–159.

Bhoj EJ, Li D, Harr MH, Tian L, Wang T, Zhao Y, Qiu H, Kim C, Hoffman JD, Hakonarson H, Zackai EH (2015) Expanding the SPECC1L mutation phenotypic spectrum to include Teebi hypertelorism syndrome. *Am J Med Genet Part A* 167A(11): 2497–2502.

Bottani A, Orrico A, Galli L, Karam O, Haenggeli CA, Ferey S, Conrad B (2007) Unilateral focal polymicrogyria in a patient with classical Aarskog-Scott syndrome due to a novel missense mutation in an evolutionary conserved RhoGEF domain of the faciogenital dysplasia gene FGD1. *Am J Med Genet Part A* 143A(19): 2334–2338.

Brodsky MC, Keppen LD, Rice CD, Ranells JD (1990) Ocular and systemic findings in the Aarskog (facial-digital-genital) syndrome. *Am J Ophthalmol* 109(4): 450–456.

Darendeliler F, Larsson P, Neyzi O, Price AD, Hagenas L, Sipila I, Lindgren AC, Otten B, Bakker B, Board KI (2003) Growth hormone treatment in Aarskog syndrome: analysis of the KIGS (Pharmacia International Growth Database) data. *J Pediatr Endocrinol Metab* 16(8): 1137–1142.

El-Hattab AW, Bournat J, Eng PA, Wu JB, Walker BA, Stankiewicz P, Cheung SW, Brown CW (2011) Microduplication of Xp11.23p11.3 with effects on cognition, behavior, and craniofacial development. *Clin Genet* 79(6): 531–538.

Fernandez I, Tsukahara M, Mito H, Yoshii H, Uchida M, Matsuo K, Kajii T (1994) Congenital heart defects in Aarskog syndrome. *Am J Med Genet* 50(4): 318–322.

Fryns JP (1992) Aarskog syndrome: the changing phenotype with age. *Am J Med Genet* 43(1–2): 420–427.

Furukawa CT, Hall BD, Smith DW (1972) The Aarskog syndrome. *J Pediatr* 81(6): 1117–1122.

Gao L, Gorski JL, Chen CS (2011) The Cdc42 guanine nucleotide exchange factor FGD1 regulates osteogenesis in human mesenchymal stem cells. *Am J Pathol* 178(3): 969–974.

Genot E, Daubon T, Sorrentino V, Buccione R (2012) FGD1 as a central regulator of extracellular matrix remodelling--lessons from faciogenital dysplasia. *J Cell Sci* 125(Pt 14): 3265–3270.

Gorski JL, Estrada L, Hu C, Liu Z (2000) Skeletal-specific expression of Fgd1 during bone formation and skeletal defects in faciogenital dysplasia (FGDY; Aarskog syndrome). *Developmental Dynamics : An Official Publication of the American Association of Anatomists* 218(4): 573–586.

Grams SE, Argiropoulos B, Lines M, Chakraborty P, McGowan-Jordan J, Geraghty MT, Tsang M, Eswara M, Tezcan K, Adams KL, Linck L, Himes P, Kostiner D, Zand DJ, Stalker H, Driscoll DJ, Huang T, Rosenfeld JA, Li X, Chen E (2016) Genotype-phenotype characterization in 13 individuals with chromosome Xp11.22 duplications. *Am J Med Genet Part A* 170A(4): 967–977.

Griffin LB, Farley FA, Antonellis A, Keegan CE (2016) A novel FGD1 mutation in a family with Aarskog-Scott syndrome and predominant features of congenital joint contractures. *Cold Spring Harbor Molecular Case Studies* 2(4): a000943.

Halse A, Bjorvatn K, Aarskog D (1979) Dental findings in patients with Aarskog syndrome. *Scand J Dent Res* 87(4): 253–259.

Hou P, Estrada L, Kinley AW, Parsons JT, Vojtek AB, Gorski JL (2003) Fgd1, the Cdc42 GEF responsible for Faciogenital Dysplasia, directly interacts with cortactin and mAbp1 to modulate cell shape. *Human Mol Genet* 12(16): 1981–1993.

Kirkham TH, Milot J, Berman P (1975) Ophthalmic manifestations of Aarskog (facial-digital-genital) syndrome. *Am J Ophthalmol* 79(3): 441–445.

Lizcano-Gil LA, Garcia-Cruz D, Cantu JM, Fryns JP (1994) The facio-digito-genital syndrome (Aarskog syndrome): a further delineation of the distinct radiological findings. *Genetic Counseling (Geneva, Switzerland)* 5(4): 387–392.

Logie LJ, Porteous ME (1998) Intelligence and development in Aarskog syndrome. *Archiv Dis Child* 79(4): 359–360.

Orrico A, Galli L, Buoni S, Hayek G, Luchetti A, Lorenzini S, Zappella M, Pomponi MG, Sorrentino V (2005) Attention-deficit/hyperactivity disorder (ADHD) and variable clinical expression of Aarskog-Scott syndrome due to a novel FGD1 gene mutation (R408Q). *Am J Med Genet Part A* 135(1): 99–102.

Orrico A, Galli L, Cavaliere ML, Garavelli L, Fryns JP, Crushell E, Rinaldi MM, Medeira A, Sorrentino V (2004) Phenotypic and molecular characterisation of the Aarskog-Scott syndrome: a survey of the clinical variability in light of FGD1 mutation analysis in 46 patients. *Eur J Hum Genet* 12(1): 16–23.

Orrico A, Galli L, Clayton-Smith J, Fryns JP (2015) Clinical utility gene card for: Aarskog-Scott Syndrome (faciogenital dysplasia) - update 2015. *Eur J Hum Genet* 23(4): 558.

Orrico A, Galli L, Obregon MG, de Castro Perez MF, Falciani M, Sorrentino V(2007)Unusually severe expression of craniofacial features in Aarskog-Scott syndrome due to a novel truncating mutation of the FGD1 gene. *Am J Med Genet Part A* 143A(1): 58–63.

Pariltay E, Hazan F, Ataman E, Demir K, Etlik O, Ozbek E, Ozkan B (2016) A novel splice site mutation of FGD1 gene in an Aarskog-Scott syndrome patient with a large anterior fontanel. *J Pediatr Endocrinol Metab* 29(9): 1111–1114.

Pasteris NG, Cadle A, Logie LJ, Porteous ME, Schwartz CE, Stevenson RE, Glover TW, Wilroy RS, Gorski J L (1994) Isolation and characterization of the faciogenital dysplasia (Aarskog-Scott syndrome) gene: a putative Rho/Rac guanine nucleotide exchange factor. *Cell* 79(4): 669–678.

Patton MA, Afzal AR (2002) Robinow syndrome. *J Med Genet* 39(5): 305–310.

Pedigo NG, Van Delden D, Walters L, Farrell CL (2016) Minireview: Role of genetic changes of faciogenital dysplasia protein 1 in human disease. *Physiol Genom* 48(7): 446–454.

Pérez-Coria M, Lugo-Trampe JJ, Zamudio-Osuna M, Rodriguez-Sanchez IP, Lugo-Trampe A, de la Fuente-Cortez B, Campos-Acevedo LD, Martinez-de-Villarreal LE (2015) Identification of novel mutations in Mexican patients with Aarskog-Scott syndrome. *Mol Genet Genom Med* 3(3): 197–202.

Petryk A, Richton S, Sy JP, Blethen SL (1999) The effect of growth hormone treatment on stature in Aarskog syndrome. *J Pediatr Endocrinol Metab* 12(2): 161–165.

Porteous ME, Goudie DR (1991) Aarskog syndrome. *J Med Genet* 28(1): 44–47.

Reddy P, Kharbanda OP, Kabra M, Duggal R (1999) Dental and craniofacial features of Aarskog syndrome: report of a case and review of literature. *J Clin Pediatr Dent* 23(2): 155–159.

Schwartz CE, Gillessen-Kaesbach G, May M, Cappa M, Gorski J, Steindl K Neri, G. (2000) Two novel mutations confirm FGD1 is responsible for the Aarskog syndrome. *Eur J Hum Genet* 8(11): 869–874.

Scott CI (1971) Unusual facies, joint hypermobility, genital anomaly and short stature: a new dysmorphic syndrome. *Birth Defects Orig Art Ser* VII(6): 240–246.

Shalev SA, Chervinski E, Weiner E, Mazor G, Friez MJ, Schwartz CE (2006) Clinical variation of Aarskog syndrome in a large family with 2189delA in the FGD1 gene. *Am J Med Genet Part A* 140(2): 162–165.

Stevenson RE, May M, Arena JF, Millar EA, Scott CI, Jr. Schroer, RJ. Simensen, RJ Lubs, HA Schwartz, CE (1994) Aarskog-Scott syndrome: confirmation of linkage to the pericentromeric region of the X chromosome. *Am J Med Genet* 52(3): 339–345.

Stevenson RE, Schwartz CE, Rogers RC (2012) *Atlas of X-Linked Intellectual Disability Syndromes*, Second Edition. New York, NY: Oxford University Press.

Teebi AS, Rucquoi JK, Meyn MS (1993) Aarskog syndrome: report of a family with review and discussion of nosology. *Am J Med Genet* 46(5): 501–509.

Trevizol AP, Sato IA, Dias DR, de Barros Calfat EL, de Carvalho Tasso B, Alberto RL, Cordeiro Q, Shiozawa P (2015) Aarskog-Scott syndrome presenting with psychosis: A case study. *Schizophrenia Res* 165(1): 108–109.

Tsai AC, Robertson JR, Teebi AS (2002) Teebi hypertelorism syndrome: report of a family with previously unrecognized findings. *Am J Med Genet* 113(3): 302–306.

Tsukahara M, Fernandez GI (1994) Umbilical findings in Aarskog syndrome. *Clin Genet* 45(5): 260–265.

Völter C, Martinez R, Hagen R, Kress W (2014) Aarskog-Scott syndrome: a novel mutation in the FGD1 gene associated with severe craniofacial dysplasia. *Eur J Pediatr* 173(10): 1373–1376.

2

ACHONDROPLASIA

RICHARD M. PAULI
Midwest Regional Bone Dysplasia Clinic University of Wisconsin-Madison, Madison, Wisconsin, USA

LORENZO D. BOTTO
Division of Medical Genetics, Department of Pediatrics, University of Utah, Salt Lake City, Utah, USA

INTRODUCTION

Incidence

The external physical features of achondroplasia have been recognized for millennia and are well represented in artwork from diverse cultures in all parts of the world (Enderle et al. 1994). It is the most common, and still most readily recognizable, of the skeletal dysplasias (also known as bone dysplasias, chondrodysplasias, and osteochondrodystrophies), with best estimates of birth prevalence around 1 in 25,000–30,000 (Waller et al. 2008). Although achondroplasia is an autosomal dominant single-gene disorder, most cases are sporadic.

Most individuals with achondroplasia can be expected to have a normal life expectancy. Nevertheless, they are at some increased risk for premature death (Hecht et al. 1987; Wynn et al. 2007; Simmons et al. 2014), including sudden and unexpected deaths in infancy. Most early deaths (in the first year of life) likely arise secondary to acute foraminal compression of the upper cervical cord or lower brain stem (see below) (Pauli et al. 1984). In addition, mean survival is about 10 years less than that of the general population (Hecht et al. 1987; Wynn et al. 2007), with much of the difference related to cardiovascular mortality in adulthood (Wynn et al. 2007). In addition, there may be an increased risk related to pedestrian and motor vehicular trauma (Hashmi et al. 2018).

Diagnostic Criteria

Well-defined clinical and radiologic features allow for virtual certainty of diagnosis in all infants with achondroplasia. External physical characteristics include disproportionately short limbs, particularly the proximal or rhizomelic (upper) segment of the arms; short fingers often held in a typical "trident" configuration with fingers deviating distally; moderately enlarged head; depressed nasal bridge; and modestly constricted chest (Figure 2.1). In all infants in whom the diagnosis is suspected, radiographic assessment is mandatory. Features that are most helpful in distinguishing achondroplasia from other short-limb disorders include small skull base and foramen magnum; narrowing rather than widening of the interpediculate distance in the lumbar spine (although not present in infancy) and short vertebral bodies; square iliac wings, flat acetabulae, narrowing of the sacrosciatic notch and a characteristic radiolucency of the proximal femora (Figure 2.2); short, thick long bones; flared metaphyses; and short proximal and middle phalanges (Langer et al. 1967). Although both the clinical and the radiologic features evolve with age, virtually all instances of diagnostic uncertainty will arise in the neonate.

Prenatal diagnosis of achondroplasia using ultrasonographic criteria can be exceedingly difficult (Patel and Filly 1995; Hatzaki et al. 2011) particularly because bone foreshortening is often not evident until about 20–24 weeks of gestation (Patel and Filly 1995). Other alternatives, including

---

Cassidy and Allanson's Management of Genetic Syndromes, Fourth Edition.
Edited by John C. Carey, Agatino Battaglia, David Viskochil, and Suzanne B. Cassidy.
© 2021 John Wiley & Sons, Inc. Published 2021 by John Wiley & Sons, Inc.

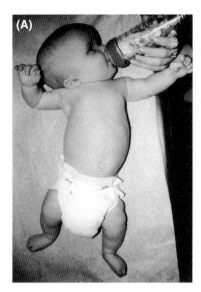  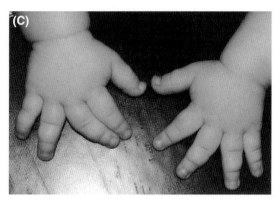

FIGURE 2.1 Primary clinical features of infants and children with achondroplasia. (A) General appearance in a child of about two months of age. (B) Facial features, which include a high forehead and depressed nasal bridge. (C) Hands, which show not only shortening of the fingers but also the typical trident configuration (with increased distance particularly between the third and fourth fingers).

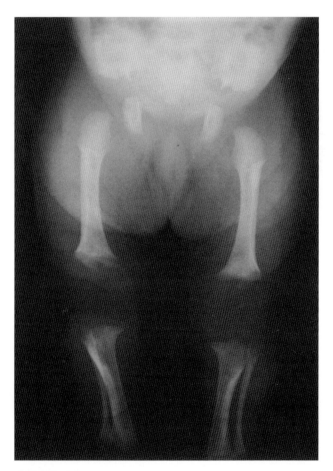

FIGURE 2.2 Anteroposterior radiograph of an infant with achondroplasia showing, in particular, aberrant shape of the ilia and the characteristic radiolucency of the proximal femora.

molecular diagnosis earlier in gestation, before ultrasonographic manifestations are evident, are not generally applicable, because of the predominance of sporadic occurrences.

Etiology, Pathogenesis, and Genetics

Most, and perhaps all, of the clinical characteristics and medical complications of achondroplasia are explicable on the basis of abnormalities of growth of cartilaginous bone or disproportionate growth of cartilaginous bone when compared with other contiguous tissues.

The molecular origin of this defect in cartilaginous bone development has been elucidated (Laederich and Horton 2010). All instances of achondroplasia arise from a mutation in one copy of the fibroblast growth factor receptor 3 (*FGFR3*) gene (Shiang et al. 1994; Bellus et al. 1995) and, more remarkably, virtually always from the same nucleotide substitution at the same site in the *FGFR3* gene, specifically the c.1138G>A, rarely the c.138G>C variant, causing a glycine to arginine substitution at amino acid position 380 (p.Gly380Arg) (Bellus et al. 1995). *FGFR3* is one of four receptors for a large set of growth factors. When *FGFR3* is mutated, as in achondroplasia, its normal inhibitory function is constitutively activated (i.e. turned on whether or not a fibroblast growth factor has bound to it), resulting in increased inhibition of growth of cartilage cells (Laederich and Horton 2010).

Because achondroplasia is an autosomal dominant disorder, offspring of affected individuals will have a 50% chance to be affected. However, three additional considerations are important in understanding the transmission of achondroplasia. First, although dominant, most instances of achondroplasia arise from new, spontaneous mutations (Orioli et al. 1995); this means that most individuals with achondroplasia are born to parents of average stature. The frequency with which these new mutations arise is correlated with advanced paternal age (Orioli et al. 1995), and, in fact, virtually all of the achondroplasia new mutations arise in the father's germinal cells (Wilkin et al. 1998). Second, there are rare, unexpected instances of recurrence of achondroplasia in siblings born to unaffected parents (Mettler and Fraser 2000). Most likely these arise because of germinal mosaicism in the father secondary to favored selection of spermatogonia carrying the *FGFR3* mutation (Goriely and Wilkie 2012). Nonetheless, recurrence risk in this situation is very low – certainly less than 1% and possibly much lower. Finally, individuals of small stature often marry one another. Offspring whose parents both have achondroplasia not only are at a 50% risk to have achondroplasia but have an additional 25% risk to receive two copies of the abnormal allele, resulting in homozygous achondroplasia, a usually lethal condition (Pauli 1983). If the reproductive partner of an individual with achondroplasia has a different skeletal dysplasia, expectations in offspring can be exceedingly complex (Flynn and Pauli 2003).

Diagnostic Testing

Because virtually all instances of achondroplasia arise from substitutions at the same base pair (Bellus et al. 1995), molecular testing is straightforward and is available commercially. However, the vast majority of affected individuals can be unequivocally diagnosed on the basis of clinical and radiologic features (personal observation). Consultation with a clinician and/or radiologist with expertise in diagnosing skeletal dysplasias would seem to be a more reasonable approach, with *FGFR3* molecular testing reserved for those rare instances in which diagnosis is in doubt (personal observation).

See **Pregnancy** section for a brief discussion of utilization of molecular testing for pathogenic variants of *FGFR3* for prenatal diagnosis.

Differential Diagnosis

In the broadest sense, achondroplasia may be considered in any individual with disproportionately short stature characterized by greater shortening of the limbs and a relatively normal size trunk. In practice, only four such diagnoses should cause any confusion: on the severe end of the spectrum, thanatophoric dysplasia, homozygous achondroplasia, SADDAN syndrome, and on the milder end of the spectrum, hypochondroplasia.

Thanatophoric dysplasia is nearly always a lethal disorder, usually in the perinatal period. It results in profound shortening of the limbs, marked macrocephaly, and marked chest constriction as well as characteristic radiologic features (Maroteaux et al. 1967). Infants with this disorder most often die from respiratory insufficiency as a result of either a constricted chest or central apnea related to profound stenosis of the foramen magnum (personal observation). Homozygous achondroplasia closely resembles thanatophoric dysplasia clinically, but has radiologic characteristics that are distinct both from thanatophoric dysplasia and from heterozygous achondroplasia (Pauli 1983). Of course, if an infant born to parents both of whom have achondroplasia has very severe bony changes, the differentiation should not be difficult. The SADDAN syndrome is characterized by bony changes nearly as severe as those in thanatophoric dysplasia, plus developmental retardation and acanthosis nigricans (Bellus et al. 1999). However, in infancy, before the onset of acanthosis nigricans and before developmental abnormalities can be identified, distinguishing SADDAN syndrome from achondroplasia and thanatophoric dysplasia may be difficult without molecular evaluation.

On the opposite end of the severity spectrum, achondroplasia may need to be differentiated from hypochondroplasia. Individuals with hypochondroplasia (Hall and Spranger 1979) have features resembling achondroplasia, but in most cases the manifestations are uniformly milder. There is a virtual continuum of both clinical and radiologic features ranging from typical achondroplasia to severe hypochondroplasia to mild hypochondroplasia to normal. It is now recognized that the similarities of these disorders are understandable at a molecular level. Each typically arises from different mutations of the *FGFR3* gene resulting in different severity of constitutive activation of the specific FGF receptor (Laederich and Horton 2010).

Assessment solely on clinical and radiologic grounds is virtually always sufficient to differentiate achondroplasia from all but hypochondroplasia. Only in this case might molecular testing be needed, looking for the common achondroplasia *FGFR3* mutation. If the mutation is present, diagnosis of achondroplasia is confirmed. If it is absent, then the potentially more difficult task of identifying a molecular basis for a presumptive diagnosis of hypochondroplasia could be pursued. With the advent of new sequencing technology, such a task is becoming simpler, quicker and cheaper.

Solely based upon its commonly used name, one might suspect that pseudoachondroplasia would share many features with achondroplasia. However, they share virtually nothing in common except for small stature and rhizomelic shortening of the limbs.

MANIFESTATIONS AND MANAGEMENT

Unlike many of the disorders discussed in this book, guidelines for care of children with achondroplasia have been previously generated (Trotter et al. 2005) and other clinical reviews with varying emphases are available (Ireland et al. 2012b, Pauli 2012, Wright and Irving 2012).

The recommendations made here differ only modestly from the published guidelines, based on materials that have been published since those guidelines were generated and on the authors' personal observations.

Members of the Medical Advisory Board of Little People of America have special expertise regarding diagnosis, assessment and management of individuals with achondroplasia (see **Resources**). Optimal care should include involvement of such an individual, or others with similar expertise, as a periodic consultant working in concert with the family physician, pediatrician, or internist who assumes major responsibility for the general care of an individual with achondroplasia.

Growth and Feeding

Moderate to marked short stature is universal in achondroplasia. Ultimate adult heights vary from about 120 to 145 cm (47 to 57 in.) with a mean of 130 cm (51 in.) in males and from 115 to 137 cm (45 to 54 in.) with a mean of 125 cm (49 in.) in females. Standard growth charts for achondroplasia have been generated (Horton et al. 1978, Hoover-Fong et al. 2017) and modified for easier use (Greenwood Genetics Center 1988) (Figure 2.3A).

In general, individuals with achondroplasia are tall enough to function effectively in most environments with modest adaptive modifications (e.g., appropriately placed stools, seating modification). Reaching may on occasion be problematic, particularly for those with very short arms and concomitant limitation of elbow extension (see below).

Most of those providing care for individuals with achondroplasia concur that obesity is prevalent (Hecht et al. 1988). Excess weight gain appears to begin in early childhood (Hecht et al. 1988). Indirect methods suggest that between 13 and 43% of adults with achondroplasia are obese (3–8 times the general population rate of obesity at the time of that study) (Hecht et al. 1988), although no rigorous studies of body fat content have been published. Excess weight may contribute to risks related to neurologic and orthopedic complications (Hoover-Fong et al. 2007) and may, in part, account for the modest excess mortality demonstrated in adults (Hecht et al. 1987).

Evaluation

- Length or height should be measured at each childhood contact with a health care provider. These measures should be plotted on achondroplasia-specific growth charts. Only in this way can growth be used as a nonspecific measure of well-being in children with achondroplasia (as it is used in average-statured children).
- Standard weight curves (Hoover-Fong et al. 2007) (see Figure 2.3B) and weight-by-height curves (Hunter et al. 1996a) (see Figure 2.3C) have been generated. Note that these do not reflect optimal weight and do not exclude individuals who themselves may be obese. Nevertheless, they are of great value in assessing whether an individual with achondroplasia exceeds the norms for this population. These curves are also of value in identifying failure-to-thrive that may accompany respiratory compromise in infants and young children with achondroplasia.
- Body Mass Index (BMI) standards specific for achondroplasia are available (Hoover-Fong et al. 2008) (see Figure 2.3D). Inappropriate use of standards for the general population will define virtually every individual with achondroplasia as being obese.
- Nutritional assessment and referral for management are indicated in those becoming obese.

Management

- Currently there is no effective treatment that will reverse the decremental growth found in achondroplasia.
- Exceptional interventions, such as growth hormone therapy or extended limb lengthening, may be elected by some families. A series of studies have assessed the effects of growth hormone therapy on children with achondroplasia (Miccoli et al. 2016). Most have shown at least transient increases in growth velocity, with diminishing effect over time. Final adult heights in those having undergone long-term growth hormone treatment were increased around 3 cm (less than 1 ½ in.) (Harada et al. 2017).
- Even more controversial is the use of extended limb lengthening (Schiedel and Rodl 2012). Various techniques have been employed to achieve increases in height of 30 cm (12 in.) or more through osteotomy (cutting through the bone to be lengthened) and distraction (stretching of the bone as the fractured region heals). This is a demanding undertaking and does entail some risk. The Medical Advisory Board of the Little People of America has developed a position statement regarding extended limb lengthening (available at http://www.lpaonline.org/ell-position-statement . Experience suggests that, at least in North America, most parents embrace the philosophy of modifying the environment to accommodate the child (see below) rather than attempting to modify the child (personal observation).
- Early parental counseling is appropriate, encouraging high-volume, low-calorie food snacks, not using food as reward, and involvement in age-appropriate and safe physical activities. Caloric need and energy expenditure are less in individuals with achondroplasia (Takken et al. 2007). As a rough rule of thumb, calorie need is typically about two thirds of that of an individual of average stature (personal observation). Typical interventions to prevent or treat obesity are usually effective. Bariatric surgery has been successfully carried out (Carneiro et al. 2007; personal observation).

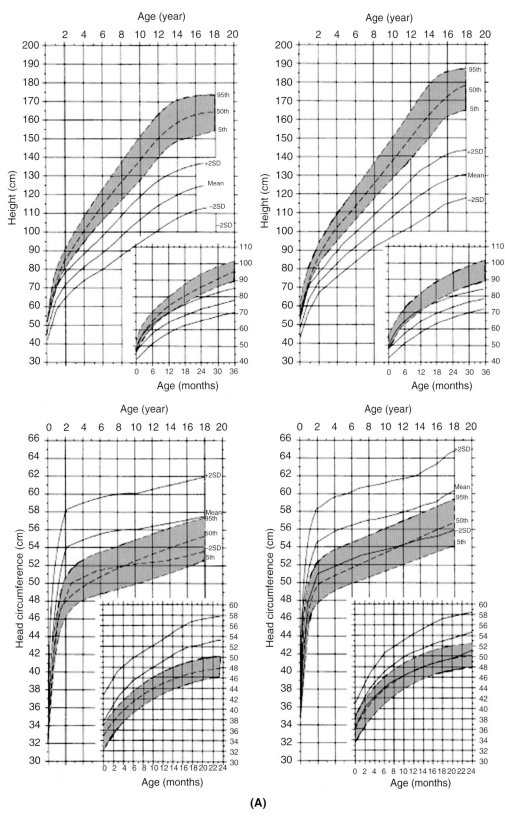

FIGURE 2.3 (A) Standard linear growth grids for females (*upper left*) and males (*upper right*) and standard head circumference grids for females (*lower left*) and males (*lower right*). (Reproduced with permission from Greenwood Genetics Center, 1988.) (B) Standard weight-by-age curves for males 0–36 months (*upper left*), males 2–16 years (*upper right*), females 0–36 months (*lower left*) and females 2–16 years (*lower right*). (Reproduced with permission from Hoover-Fong et al. 2007.) (C) Standard weight-by-height curves for males (*upper left and right*) and females (*lower left and right*). (Reproduced with permission from Hunter et al., 1996.) (D) BMI by age for males and females (Reproduced with permission from Hoover-Fong et al., 2008.)

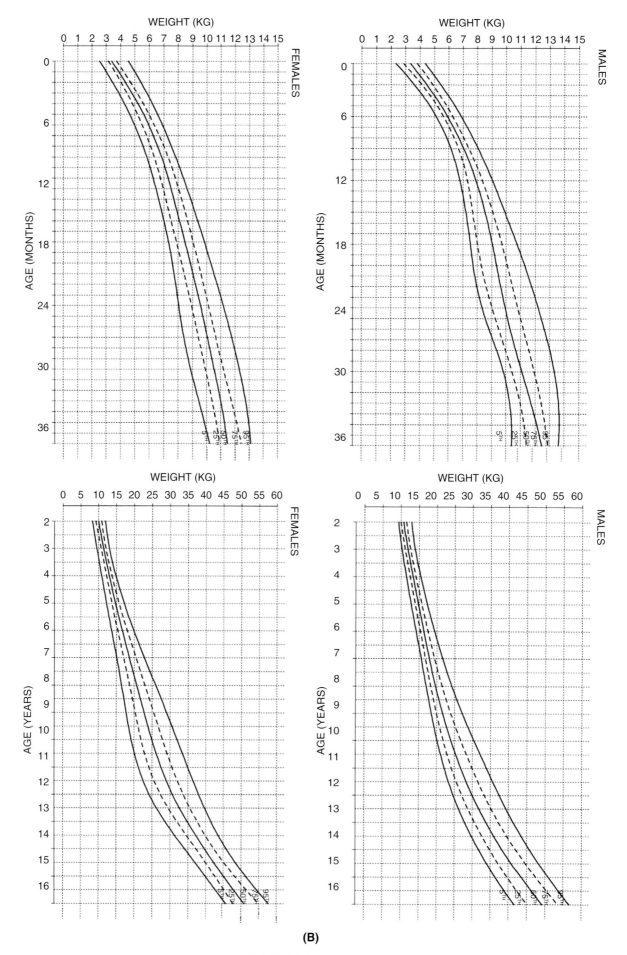

FIGURE 2.3 (*Continued*)

MANIFESTATIONS AND MANAGEMENT 15

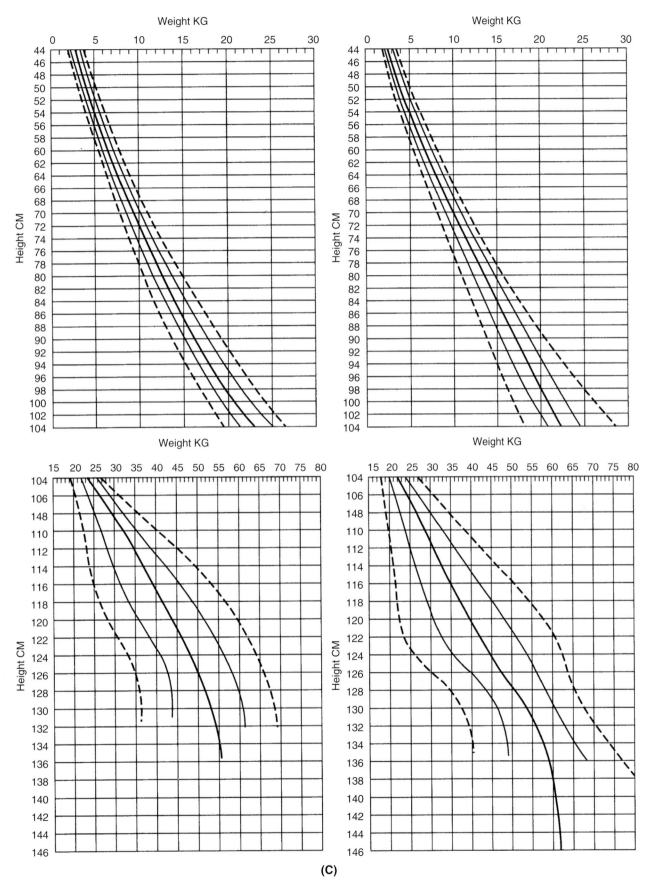

(C)

FIGURE 2.3 (*Continued*)

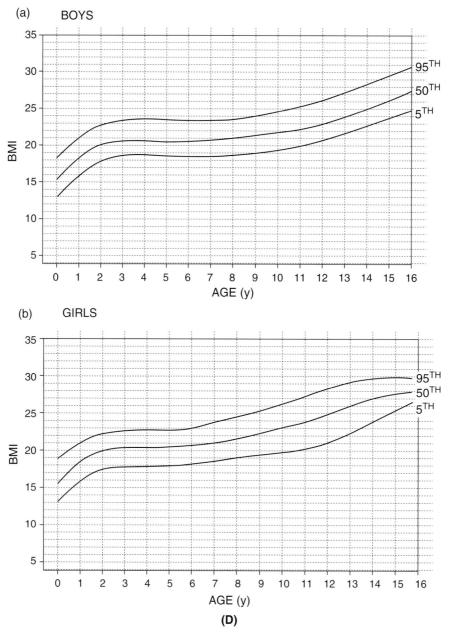

FIGURE 2.3 (*Continued*)

- Various adaptive devices may be needed because of biophysical differences such as short stature, limited reach, and hypotonia. These include appropriate use of stools, adaptations of school furniture, modifications for toileting, use of a reacher for wiping after toileting, and so on. Adults, too, can benefit from modest adaptations of their work and home environment (Crandall et al. 1994). A large body of information is available concerning practical issues of adaptations, much of it through Little People of America (see Resources).
- Adaptation also will be needed for adolescents learning to drive. Most will require pedal extenders. All who sit less than about 12 in. (about 30 cm) from the steering wheel will need medical justification to disable the airbag because of demonstrated risk of airbag deployment in small statured individuals (Roberts et al. 1999).

Development and Behavior

Overall, individuals with achondroplasia have normal cognitive development and cognitive function (Hecht et al. 1991). However, when compared with appropriate controls they do have specific differences in development. First, there is clear evidence that children with achondroplasia uniformly have both delayed and unusual motor development (Fowler et al. 1997, Ireland et al. 2012a), principally on the basis of biophysical differences related to stature, disproportion, macrocephaly, joint hypermobility, and hypotonia (Fowler et al. 1997). Gross motor delays vary, but, for example, on average independent walking is attained at around 18 months of age (Fowler et al. 1997, Ireland et al. 2012a). Second, a disproportionate number of children have language-related problems (Brinkmann et al. 1993, Ireland et al. 2012a); this is most likely caused, at least in part, by an extremely high frequency of middle ear disease and either persistent or fluctuating hearing loss (Brinkmann et al. 1993; personal observation). Most common are expressive language delays, which with appropriate speech and language interventions normalize by school age. Finally, a small minority of children with achondroplasia will be more seriously delayed, demonstrate significant learning disabilities, and may have autism spectrum disorders and/or a cognitive disability. The frequency of such problems has not yet been well documented but is almost certainly no greater than 10% (personal observation).

Evaluation

- In infancy, care providers must recognize the anticipated differences in development, particularly motoric development (Hecht et al. 1991, Fowler et al. 1997, Ireland et al. 2012a) and provide reassurance to parents when appropriate.

- Standards have been generated that can be used to compare a child with others with the same diagnosis, the most helpful of which is shown in Figure 2.4. Every child with achondroplasia who diverges significantly from these published standards should be referred for further assessment.
- Marked, persistent hypotonia and gross motor delays, in particular, may be indicative of problems at the craniocervical junction requiring acute multidisciplinary assessment (Pauli et al. 1995).
- Assessment for adaptive needs, particularly at early school age, is crucial.

Management

- Standard interventions are used if delayed development is noted.

Neurologic

Hydrocephalus Nearly all children with achondroplasia are macrocephalic (Horton et al. 1978). Neuroimaging will most often demonstrate both ventriculomegaly and increased extra-axial (subarachnoid) fluid volume, but considerable controversy exists over whether these features represent clinically significant hydrocephalus that needs treatment (Steinbok et al. 1989) or a benign process requiring no intervention (Pierre-Kahn et al. 1980). Preponderant opinion favors the latter.

Skill	25th-90th %ile for children with achondroplasia	Average age for average statured
Sit without support	9-20.5	5.5
Pull to stand	12-20	7.5
Stand alone	16-29	11.5
Walk	14-27	12
Reach	6-15	3.5
Pass object	8-14	6
Bang two objects	9-14	8.5
Scribble		13.5

FIGURE 2.4 Developmental milestones in infants and young children with achondroplasia. (Reproduced with permission from Ireland et al. 2012b)

Nevertheless, using conservative criteria, about 5% of children with achondroplasia will develop symptomatic increased intracranial pressure that requires intervention (King et al. 2009). It seems reasonable to differentiate benign ventriculomegaly and clinically significant hydrocephalus even if, as seems the case, they both arise secondary to increased intracranial venous pressure because of jugular foraminal stenosis (Pierre-Kahn et al. 1980; Steinbok et al. 1989).

Because of increased extra-axial fluid volume and increased vascular flow, infants with achondroplasia also appear to have increased risk of rupture of bridging vessels and consequent development of subdural hematomata (personal observation). Although the risk is small, most probably in the range of 1–2% (personal observation), it needs to be recognized that this sequela does not necessarily imply non-accidental trauma in these children.

Evaluation

- At the time of diagnosis, every child with achondroplasia should have assessment of ventricular size and volume of extra-axial fluid by computerized tomography or magnetic resonance imaging.
- All children should have serial head circumference measurements plotted on achondroplasia-specific grids (Horton et al. 1978) (Figure 2.3) every 1–2 months in the first year of life and at each well-child visit thereafter until around 5 years of age.
- The physician should be aware of the signs and symptoms that suggest increased intracranial pressure – accelerating head growth, bulging and tense fontanel, lethargy or irritability, unexplained vomiting, apparent headache, etc. Because of the etiology of the hydrocephalus in achondroplasia (Pierre-Kahn et al. 1980; Steinbok et al. 1989), sudden increase in the prominence of superficial veins over the scalp or eye region (reflecting increased ancillary blood flow bypassing the pressure gradient at the jugular foramina) should be sought. Likewise, parents should be taught the major signs and symptoms of hydrocephalus and be encouraged to have their child evaluated emergently should such concerns arise.
- Should acceleration of head growth arise or should signs and symptoms suggesting increased intracranial pressure develop, then the child should have magnetic resonance imaging of the brain, which can be compared with imaging obtained at the time of first diagnosis, and ophthalmologic examination for assessment of presence of papilledema.

Management

- If features of increased intracranial pressure develop, but subsequently no acute symptoms persist, and if imaging shows preservation of the subarachnoid space and lack of evidence for periventricular parenchymal edema, the physician, in consultation with a neurologist or neurosurgeon, may elect a period of watchful waiting. This compensated hydrocephalus (benign ventriculomegaly) is probably of no consequence.
- In instances where symptomatic hydrocephalus has unequivocally developed, standard ventriculoperitoneal shunting is appropriate. With appropriate intervention in those few requiring intervention, prognosis for hydrocephalus in general should be favorable. There will be the usual risks and complications related to the placement of a ventriculoperitoneal shunt.
- A few children have been treated by third ventriculostomy (Swift et al. 2012). That this is at least sometimes effective suggests that flow restriction at the level of the foramen magnum may be a second way in which hydrocephalus may develop (Mukherjee et al. 2014). If proven generally effective, this procedure could be a practical alternative particularly in low resource settings where prompt access to facilities able to manage acute shunt malfunction is limited.
- Decompression of a jugular foramen has only rarely been attempted (Lundar et al. 1990).

Cervicomedullary Junction Constriction Every infant with achondroplasia has a small foramen magnum as a result of abnormality of growth of the endochondral cranial base (Yang et al. 1977). The foramen magnum is not only small but often misshapen, resembling a "keyhole" rather than being round or ovoid. The asynchrony of growth of the foramen magnum compared with its contents (spinal cord and blood vessels) causes risks for either apnea-associated death or high cervical myelopathy and paralysis.

Pauli et al. (1984) and others (e.g., Reid et al. 1987) have demonstrated that the diminution of foraminal size may be of considerable consequence to the health and survival of infants with achondroplasia. Without appropriate assessment and intervention some infants will die unexpectedly in the first 1–2 years of life (Pauli et al. 1984; Hecht et al. 1987). These apneic deaths likely arise because of vascular compression at the craniocervical junction, resulting in damage to the respiratory control centers in the lower medulla (Pauli et al. 1984; Reid et al. 1987). Such deaths may mimic sudden infant death syndrome but are more likely to be daytime deaths and deaths associated with uncontrolled head movement (Pauli et al. 1984). Without evaluation or management, risk of death is certainly increased over the general population (Pauli et al. 1984; Simmons et al. 2014; Hashmi et al. 2018) and may approach 10% (Hecht et al. 1987), although experience suggests that a more realistic figure may be in the 2–3% range (personal observation).

Compression of the cervicomedullary cord can also result in high cervical myelopathy, most often presenting in young children with disproportionate and long-persistent hypotonia, weakness, hyperreflexia, asymmetric reflexes, and ankle clonus (Pauli et al. 1995) and far less frequently in older children and adults. This, too, appears to arise through hypoxic injury or traumatic compression (Hecht et al. 1984). No accurate estimate of the frequency of these complications is available.

Less frequently, children with achondroplasia and cervicomedullary junction compression may present with chronic, otherwise unexplained respiratory signs and symptoms without demonstrable neurological concomitants (Reid et al. 1987).

Because complications arise from disproportionate growth of bone compared with the neural tissue it surrounds, efforts at evaluation and management are aimed at minimizing neural damage and, when necessary, decompressing the constraining bone. With appropriate assessment and selective intervention, prognosis is excellent, not only for survival but also for survival without sequelae (Pauli et al. 1995). Quality of life assessment has demonstrated the benefit of selective decompressive surgery (Ho et al. 2004).

Evaluation

A prospective study provides support for comprehensive evaluation of every infant with achondroplasia regarding the risk for cervicomedullary junction constriction to identify those who may be at higher than average risk (Pauli et al. 1995). Evaluation should be completed at the time of initial diagnosis. Although differences in opinion and in practice persist, a 'standard' evaluation (based on the objective evidence available in the literature) should include the following:

- Complete a neurological history.
- Perform a careful neurological examination, preferably by a physician with experience in achondroplasia or child neurology, including judging whether severity of hypotonia is outside the expected range for infants with achondroplasia.
- Complete neuroimaging. This can either be computerized tomography with thin cuts and bone windows through the foramen magnum [necessary to obtain measures of the size of the foramen that can be compared with diagnosis-specific standards (Hecht et al., 1989) and usually obtainable in infants without sedation] or magnetic resonance imaging [avoiding radiation exposure, with better visualization of neural structures including assessment for parenchymal changes, but no direct measure of foramen magnum size (Brühl et al. 2001), absence of standards with which to compare, and usually requiring sedation or anesthesia]. In some instances, multipositional magnetic resonance imaging will demonstrate serious abnormality that otherwise would be undetected by routine imaging (Danielpour et al. 2007). Note that calvarial ultrasound cannot assess foraminal structures and should be abandoned as a screening tool in assessing infants with achondroplasia. Advances in imaging are arising at a rapid pace; the clinician should consult with a pediatric neuroradiologist regarding the most appropriate methods of assessment.
- Do overnight multichannel polysomnographic evaluation (sleep study) looking specifically for evidence of central apnea in a sleep center accustomed to assessing infants.
- If initial findings of the comprehensive assessment discussed above are reassuring, no further investigations are needed. However, the infant should continue to have careful periodic monitoring of neurological and respiratory history and clinical and neurological reexaminations about every 6 months.
- In infants with worrisome features [the most highly predictive of which are persisting hypotonia, increased reflexes or clonus in the legs, foramen magnum measurements below −1 SD for achondroplasia, and central hypopneas (Pauli et al. 1995)], magnetic resonance imaging should be completed (with or without flow studies).
- While no other protocol has been prospectively assessed, various alternatives that have been promoted include: (a) magnetic resonance imaging as an initial evaluative step; (b) use of three dimensional computerized tomography; (c) stepwise protocols, such as only initially completing clinical assessment and polysomnography and on those bases deciding whether to undertake any kind of neuroimaging.

Management

- Parents should be counseled regarding the importance of careful neck support, using a solid-back stroller and an infant head and neck pillow, and avoidance of umbrella strollers, automatic swings, doorway jumpers and so on, which may precipitate uncontrolled head movement around a constricted foramen magnum. Automatic swings in which an infant is in a sitting or near-sitting position are particularly risky; at least six sudden apneic deaths in infants with achondroplasia have been observed associated with their use (Pauli et al. 1984; personal observation). Acute life threatening events also have arisen multiple times associated with placement in car seats (Collins and Choi 2007; personal observation) and good head and neck support while in a car seat is critical.
- If unequivocal evidence for cord compression is present (as defined above), then surgical decompression should be done immediately (Bagley et al. 2006). With such criteria, 10–12% of all children with achondroplasia will undergo suboccipital and cervical decompressive surgery (King et al. 2009; Menezes 2009; Ireland et al. 2012b; personal observation).

- All individuals of all ages should be considered to have relative cervical spinal stenosis and, on this basis, to have increased risk related to severe head and neck trauma. Certain physical activities should be strongly discouraged, including full-contact American football, full-contact ice hockey, rugby, downhill skiing, trampoline, dive rolls, vaulting or other gymnastics in which full body-weight impact on the head or neck is likely, hanging upside down from knees or feet, diving from diving boards, and heading in soccer (personal observations).

Respiratory

Infants with achondroplasia have smaller than average thoraces (Stokes et al. 1983; Hunter et al. 1996b) as well as clinical evidence of increased compliance of the rib cage (personal observation). It has been suggested that such features may result in decreased effective lung volumes, decreased respiratory reserve, and increased probability of chronic hypoxemia (Stokes et al. 1983; Reid et al. 1987; Mogayzel et al. 1998). Although this suggestion in part has been rebutted (Tasker et al. 1998), there remain a few young children with achondroplasia who show chronic hypoxemia often accompanied by failure-to-thrive (Stokes et al. 1983; personal observation). Living at high altitude predisposes to this complication (personal observation). The frequency of such hypoxemia in isolation is unknown. Appropriate evaluation and treatment should prevent sequelae of chronic hypoxemia, including effects on cognition. Without treatment, secondary difficulties related to cor pulmonale might be life threatening,

Snoring, when isolated, is a virtually uniform feature in individuals with achondroplasia of all ages and should not be assumed to reflect clinically significant obstruction of the upper airway. Similarly, virtually all infants with achondroplasia perspire excessively, including in sleep, and this in isolation, too, should not be taken as an indicator of respiratory issues. Although it is generally conceded that obstructive apnea is common in achondroplasia, prevalence estimates in the literature vary widely – from about 20 to 100% (Tenconi et al. 2017). Most estimates have suffered from small sample size and/or ascertainment and referral bias. Published sequential series suggest that real prevalence of clinically significant obstructive apnea is probably around 30–40% (Sisk et al. 1999; Collins and Choi 2007; Afsharpaiman et al. 2011).

Several factors contribute to the exceedingly high frequency of obstructive apnea. There is hypoplasia of the cranial base and midface resulting in the diminution of airway size (Stokes et al. 1983). Then, with physiologic hypertrophy of the lymphatic ring, obstructive apnea may result, thereby explaining why its onset is so frequently between 2 and 10 years of age. In addition, muscular obstruction presumed to be secondary to abnormal innervation may contribute to some instances of serious obstructive apnea in young children with achondroplasia (Tasker et al. 1998); this, in turn, may arise from stenosis of the jugular foramina or hypoglossal canals caused by constriction of the calvarial base (Tasker et al. 1998). Airway malacia also is common in children with achondroplasia, demonstrable in about 5% of children with this diagnosis (Dessoffy et al. 2014). Gastroesophageal reflux is also sometimes of considerable importance (Stokes et al. 1983; Tasker et al. 1998). Any of these causes can be compounded by coexisting obesity.

In most children who are adequately assessed and treated, obstructive apnea will resolve without long-term sequelae. In some, and in many adults, long-term use of continuous positive airway pressure at night will be essential both to maintain appropriate oxygen saturations and to prevent re-emergence of fragmented sleep.

Evaluation

- Marked tachypnea or failure to thrive in infancy should alert the clinician to the possibility of significant restrictive pulmonary disease.
- Night-time oximetry should be assessed as part of polysomnographic evaluation as discussed above. Many healthy infants with achondroplasia of less than a year of age will display frequent transient dips of oxygen saturation into the 85–90% range (personal observation).
- Persistent hypoxemia or desaturations below 85% require further assessment, including pulmonary history and pulmonary evaluation.
- Evaluation of oxygen saturations by spot oximetry during waking hours (e.g., active alert, feeding, crying) is also appropriate.
- Chest circumference measurements compared with achondroplasia-specific standards (Hunter et al. 1996b) may be of some utility. It is likely that those with the smallest chests are at greatest risk for restrictive pulmonary complications and should be monitored more closely for this.
- Parents of affected children should be taught to monitor for signs and symptoms of obstruction during sleep and should be questioned at each medical visit regarding changes in these signs and symptoms. In adults, the sleep partner or other individual should observe breathing characteristics in sleep for significant features at least yearly. Clinical characteristics in sleep that should be sought include neck hyperextension, loud snoring, glottal stops, observed apneic pauses, deep, compensatory sighs, self-arousals, enuresis, and night-time emesis (Sisk et al. 1999; personal observation). In addition, daytime symptoms including excessive irritability, hypersomnolence or awake respiratory distress may be of relevance (Sisk et al. 1999). New problems with learning and/or behavior may arise secondary to disrupted sleep, too.

- Physical assessment, in particular the severity of tonsillar (and, by implication, adenoidal) hypertrophy, should be carried out.
- If there is suspicion that clinically significant obstructive apnea is occurring, then overnight polysomnographic studies must be completed.
- When significant obstructive apnea is demonstrated, assessment for right ventricular hypertrophy and pulmonary hypertension should be carried out.
- In the presence of obstructive apnea, otolaryngologic assessment of the nasopharynx and oropharynx should be completed to help determine whether surgical intervention is likely to be of benefit.
- Simultaneous assessment of neurological status is critical to rule out those less frequent instances when upper airway obstruction is caused by central nervous system dysfunction (Reid et al. 1987; Tasker et al. 1998).

Management

- In those with small chests, significant hypoxemia, and no other identified cause of respiratory problems, transient oxygen supplementation can be effective in blunting the hypoxemic episodes and in allowing resolution of failure-to-thrive (Stokes et al. 1983; personal observation).
- Intervention for obstructive apnea, if determined to be present, is a graded series of options that should be pursued in a stepwise fashion depending on response. In children and some adults, initial management should be tonsillectomy and adenoidectomy (Sisk et al. 1999; Tenconi et al. 2017). In those who are obese, weight loss efforts should begin as well.
- Should follow-up polysomnography demonstrate persistent clinically significant obstruction then positive airway pressure [continuous positive airway pressure (CPAP) or bilevel positive airway pressure (BiPAP)] can be used effectively in both adults and children (Afsharpaiman et al. 2011; Tenconi et al. 2017).
- Additional surgical intervention may occasionally be considered should positive airway pressure be ineffective or not tolerated, including uvulectomy and modified uvulopharyngopalatoplasty (Sisk et al. 1999) although their roles are unclear; in rare instances, midface distraction surgery has been undertaken (Elwood et al. 2003).
- Finally, only in a small minority will temporary tracheostomy be needed – about 2% of all individuals with this diagnosis (personal observation).
- Interventions in adults with achondroplasia are similar, although it will be less frequent that adenotonsillar hypertrophy is central and more frequent that obesity is a complicating factor.

Ears and Hearing

Newborn hearing screening is usually normal in babies with achondroplasia. Although middle ear dysfunction is generally accepted as a frequent complication in achondroplasia (Tunkel et al. 2012), no well-designed, prospective study has been completed. More than 50% of children with achondroplasia have middle ear dysfunction sufficiently severe to require myringotomy and ventilation tube placement (Berkowitz et al. 1991), and around 40% of individuals of all ages have significant hearing impairment (Tunkel et al. 2012). Before surgery most children who have myringotomy and tube placement have had either fluctuating or persisting hearing loss of a severity that could interfere with normal language acquisition (personal observation). If not aggressively sought and appropriately treated, hearing loss may be a major contributing factor to language and speech delays.

Outcome depends on aggressiveness of assessment and appropriateness of intervention.

Evaluation

- Audiometric and tympanometric assessment should be completed first at approximately 8–12 months of age and then every 6–12 months throughout preschool years (Trotter et al. 2005) and should be completed less frequently in older children and adults as well.
- A high level of suspicion should be maintained for middle ear problems throughout childhood.

Management

- Medical management of middle ear infections has, by and large, been ineffective (personal observation). Therefore, aggressive use of myringotomy and tube placement is encouraged. Experience suggests that, in those requiring tube placement once, sufficient autonomous eustachian tube function to allow for normalization of middle ear function usually does not occur until around 7–8 years of age (personal observation). Therefore, in those children, the maintenance and replacement of ventilating tubes until that age is appropriate. Those placing pressure-equalizing tubes need to be aware of the increased likelihood of encountering jugular bulb dehiscence in children with achondroplasia (Pauli and Modaff 1999).
- During periods of documented or suspected hearing loss, standard interventions are appropriate (e.g., preferential seating at school, en face communication).
- Some individuals may show substantial speech and language delay. In those instances, referral for speech and language therapy is indicated.
- Some individuals will have persistent, sufficiently severe hearing loss either in childhood or as adults that amplification is necessary (Tunkel et al. 2012).

Musculoskeletal

Kyphosis Transient non-congenital kyphotic deformity at the thoracolumbar junction of the spine is present in 90–95% of young infants with achondroplasia (Kopits 1988b; Pauli et al. 1997). In most, it spontaneously resolves after the assumption of orthograde posture (Margalit et al. 2016). However, about 10% of adults with achondroplasia have a fixed, angular kyphosis that can result in serious neurological sequelae because of tethering of the spinal cord (Kopits 1988b, Pauli et al. 1997). Beighton and Bathfield (1981) first suggested that positioning early in life could be determinative in whether the flexible and transient kyphosis of infancy becomes fixed. Other factors contributing to the development of a kyphosis include trunk hypotonia, ligamentous laxity, and macrocephaly, all of which result in an infant with achondroplasia slumping forward if placed in a sitting position (Pauli et al. 1997). If long periods are spent in such a position, remodeling in response to anomalous forces results in anterior wedging of vertebrae and fixed kyphosis (Pauli et al. 1997).

In a sequential series involving 71 infants with achondroplasia, prohibition of unsupported sitting was demonstrated to be effective in decreasing the probability of a fixed kyphosis developing (Pauli et al. 1997). Furthermore, using a generated algorithm (see Figure 2.5), the frequency with which a fixed kyphosis of medical significance arises could be reduced to zero (Pauli et al. 1997). With appropriate care, this is by and large a preventable problem. The use of the protocol summarized below should prevent virtually all clinically significant instances of thoracolumbar kyphosis; failures are rare (Margalit et al. 2016) and usually have been because of failed compliance (personal observation).

Evaluation

- Clinical evaluation of the infant's spine should occur about every 6 months through the first 3 years of life, with particular emphasis on the severity of persisting kyphosis when the child is placed prone.
- If the kyphosis is moderate or marked, then radiographic assessment – sitting lateral and cross-table prone (or cross-table supine over a bolster) lateral X-rays of the thoracolumbar spine – should be obtained.

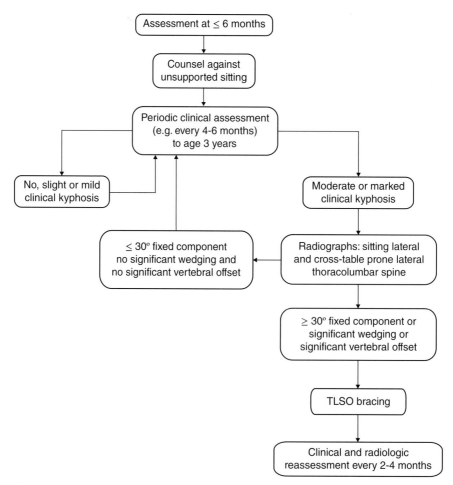

FIGURE 2.5 Algorithm for the assessment and prevention of fixed angular kyphosis (originally published in Pauli et al., 1997a).

- In adolescents or adults who have not been treated in the manner recommended here, clinical assessment should include examination of the spine. If a kyphosis is easily palpated, possible consequences should be sought through careful history (e.g., focal back pain, weakness, exercise-induced pain, dysesthesia of the legs, and bowel or bladder problems) and a lateral standing thoracolumbar spine radiograph should be obtained.

Management

- Parents of infants should be counseled against allowing unsupported sitting for at least the first 12 months of life, to encourage prone positioning work, and to avoid devices that cause disadvantageous positioning such as soft infant carriers, umbrella-style strollers, and canvas seats.
- In infants and young children, if the prone radiograph shows that there is an irreversible curve of >30°, the family should be referred for the consideration of bracing using a thoracolumbosacral orthosis (TLSO) (Kopits 1988b; Pauli et al. 1997) (Figure 2.5).
- In those for whom such anticipatory medical care has not been provided, or in whom it has failed, surgical intervention for kyphosis may be needed (Kopits 1988b; Lonstein 1988; Ain and Browne 2004).
- Additional information and instruction for parents, orthotists, and physicians are available online (https://lpamrs.memberclicks.net/medical-info; https://lpamrs.memberclicks.net/assets/kyphosis%20education%20for%20parents.doc).

Lumbosacral Spinal Stenosis Stenosis of the entire spinal canal is uniformly present in individuals with achondroplasia (Jeong et al. 2006). Although problems related to cervical spinal stenosis are of greater concern in infants and young children, lumbosacral spinal stenosis is more commonly problematic in adolescents and adults. Exercise-induced intermittent spinal claudication (pain, dysesthesias, and, less frequently, motor changes only precipitated by activity and rapidly resolved with rest) is present in most adults with achondroplasia if symptoms are carefully sought (personal observation). However, unless pain is severe or activity markedly compromised, surgical treatment for claudication is elective.

In contrast, spinal stenosis that causes either nerve root or cord compression can lead to serious and irreversible sequelae (Pyeritz et al. 1987). Often, signs and symptoms develop in the third or fourth decade of life (Kahanovitz et al. 1982; Pyeritz et al. 1987), and may affect nearly half of young and middle-aged adults (Kahanovitz et al. 1982). Only rarely does lumbosacral spinal stenosis cause serious problems in children with achondroplasia (Schkrohowsky et al. 2007). Onset of symptoms is likely related to the development of additional factors that can exacerbate the intrinsic spinal stenosis, including intervertebral disk disease and degenerative arthritis (Kahanovitz et al. 1982; Pyeritz et al. 1987). In addition, those with significant thoracolumbar kyphosis (above) and/or severe hyperlordosis of the lumbar spine are likely to be at greater risk to develop symptomatic stenosis (Kahanovitz et al. 1982; Pyeritz et al. 1987).

Evaluation

- Every adolescent and adult should have periodic evaluation for signs and symptoms of lumbosacral spinal stenosis. Symptoms to be sought include numbness, dysesthesias, radicular pain, leg weakness, clumsiness, changes in gait, or problems with bladder or bowel continence (Pyeritz et al. 1987). Examination should include complete motor (assessing for weakness or strength asymmetry, abnormal deep tendon reflexes) and sensory evaluation of the legs.
- If abnormalities are discovered, and particularly if there is change in neurological findings over time, referral for neurological and neurosurgical assessment should be made. At that time, neuroimaging (computerized tomography, magnetic resonance imaging, and/or magnetic resonance myelography) should help in the assessment of the anatomic severity, level, and associated factors of the stenosis.

Management

- Those with severe and/or progressive spinal stenosis require urgent neurosurgical intervention through extended and wide posterior laminectomy (Pyeritz et al. 1987; Lonstein 1988; Carlisle et al. 2011).
- More complex surgery with both anterior and posterior approaches may be needed in those who also have severe, angulated thoracolumbar kyphosis (Lonstein 1988).
- Decompression laminectomy usually results in some improvement of symptoms and function (Pyeritz et al. 1987), although perioperative complications are frequent (Ain et al. 2008). Long-term outcomes are less uniformly positive, with only 50% of affected individuals showing long-term benefit (Pyeritz et al. 1987). Additional surgery is often necessary (Pyeritz et al. 1987).

Knee Instability Nearly all young children with achondroplasia have unstable knees with both genu recurvatum and moderate mediolateral instability. The recurvatum deformity usually results in hyperextension of between 20° and 70° (Kopits 1988a; personal observation). It appears to arise primarily from abnormalities of growth of the tibial plateau (Kopits 1988a, Brooks et al. 2016). It is usually most severe in the second year of life (Kopits 1988a) and rarely requires intervention but does contribute to the motor delays seen in young children compared to their peers (Fowler et al. 1997). On occasion, however, there may be frank tibiofemoral subluxation [about 1–2% of young children (personal observation)]. Likewise, although mediolateral instability may result in local discomfort associated with orthograde activity, probably related more to the need for voluntary muscle stabilization than to the instability per se (personal observation), only rarely is this severe enough in itself to require any substantial intervention. Virtually in all individuals, these features remit with increasing age and usually are well compensated or resolved by maturity (personal observation).

Evaluation

- Historical and physical determination of the severity of knee hyperextensibility and whether symptoms appear to be arising from it should be sought in young children with achondroplasia.
- History and physical examination should assess for disadvantageous transitions from sit to stand over a hyperextended knee. This frequently is seen in young children with marked knee instability and may worsen its severity.
- A history of activity-precipitated pain over the lateral or posterior knee should be sought.
- When such knee pain is associated with moderate or severe varus deformity, it should precipitate additional referral for orthopedic surgical assessment.

Management

- Parents of those with asymptomatic mild or moderate knee instability should be reassured that no intervention is needed.
- In those with disadvantageous transitions, use of low stools, rather than having the child sit on the floor while playing, will entrain rising over bent rather than hyperextended knees.
- In those in whom knee instability causes recurrent pain (most often seen after a day of physical activity), rest, warmth, massage, and non-steroidal anti-inflammatory medications can be used.
- Rarely, transient bracing for severe knee instability may allow for more normal gross motor development (personal observation). Note that bracing is not indicated in the treatment of varus deformity and, indeed, may exacerbate that problem (Kopits 1988a).

Varus Deformity A majority of children (personal observation) and about 93% of adults (Kopits 1988a) have bowleg deformity. About 70% of children evaluated in one survey have clinically relevant varus (personal observation). It is often asymmetric (Inan et al. 2006). About a quarter of children will need surgery for bowleg deformity (Kopits 1988a). Although referred to as tibia varus, the deformity is usually more complex than a simple lateral bow (Inan et al. 2006), involving tibia varus, tibia recurvatum, and internal tibial torsion often accompanied by genu recurvatum and lateral instability of the knees and, occasionally, by knee subluxability (Stanley et al. 2002; personal observation). With appropriate management, no long-term sequelae should be anticipated. If untreated, worsening pain, increasing disability, and secondary joint damage arise (Kopits 1980), although individuals with achondroplasia may be at less risk for malalignment-precipitated osteoarthritis than is the general population (Tang et al. 2016).

Evaluation

- Examination should include assessment of the child while standing. Serial measurements of unloaded and loaded distances between the knees, mid-tibiae, and medial malleoli are helpful in assessing whether deformity is stable, progressing, or accelerating (personal observation).
- Evaluation of whether the three weight-bearing joints remain "in plumb" (Kopits 1980) (Figure 2.6) is helpful in deciding if pediatric orthopedic assessment is warranted. If a child is out of plumb when standing, then further evaluation is needed.
- Assessment of severity of knee instability and evaluation of gait for lateral knee thrust (sudden outward displacement of the knee with weight bearing) (Kopits 1980; personal observation) should be accomplished. Presence of a thrust, too, warrants pediatric orthopedic assessment.

Management

- For those with varus deformity of sufficient severity to be out of plumb and symptomatic, or in whom a marked knee thrust has developed, surgery is indicated (Kopits 1980; personal observation). Surgery in those who are out of plumb but essentially asymptomatic is elective. In most of those in need of surgery, standard intervention is valgus producing and derotational osteotomies (Kopits 1980).
- Although most often valgus producing and derotational osteotomies are completed in childhood or adolescence, similar surgical intervention can be carried out in adults as well.
- Guided growth using eight-plates (Stevens 2007) is being used more frequently in children with achondroplasia (McClure et al. 2017). This entails applying tension band plates and screws to the proximal tibiae and, sometimes, the distal femora, which induces

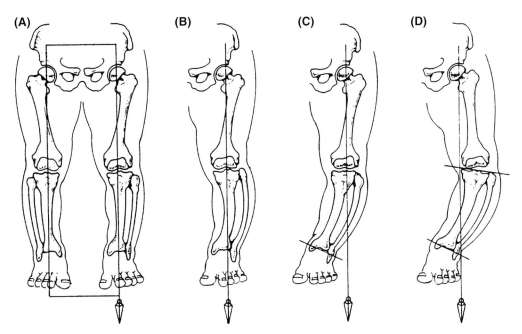

FIGURE 2.6 Diagrammatic representation of increasing severity (B–D; A is normal) of varus deformity. The situation illustrated in D and sometimes that in C will require surgical intervention. (Reprinted with permission from Johns Hopkins University Press).

gradual correction of the varus. In some hands, impressive results have been accomplished (P. Stevens, personal communication, 2015), including, unexpectedly, correction of internal torsion as well as the lateral bow. Whether this should supersede more invasive surgeries and in what circumstances are currently unsettled questions.

Other Musculoskeletal Complications A discoid lateral meniscus seems to be more common in individuals with achondroplasia, and may be another source of leg pain in this population (Akyol et al. 2015; Hoernschemeyer et al. 2016).

With assumption of standing and walking, most children develop a hyperlordosis (swayback) of moderate or severe degree. This is usually asymptomatic. It may exacerbate risk for lumbosacral spinal stenosis in adulthood (Kahanovitz et al. 1982; Pyeritz et al. 1987). Local pain may develop at the apex of the lordosis in both children and adults (personal observation). In a few children the horizontal sacrum and consequently superficial coccyx may result in chronic coccydynia.

Hypermobility of the shoulders is a virtually constant feature but is only infrequently symptomatic (personal observation). Even in those with inferior subluxability, pain is uncommon (Kopits 1988a).

The elbows are the one exception to the generalized joint hypermobility seen in individuals with achondroplasia. Although individuals often display marked overpronation and oversupination (personal observation), limitation of extension often develops in early childhood (Kopits 1988a), frequently of about 20–60°. Less often, there is posterolateral radial head dislocation (Kopits 1988a) causing more severe limitation of extension as well as of pronation and supination. When present, this may further limit functionally effective reach (e.g., for toileting).

Wrists are usually hypermobile in childhood. Some individuals have remarkable dorsoventral subluxability (personal observation). The intrinsic joints of the hands are also usually hypermobile and may limit fine motor endurance, particularly in young children (personal observation).

Evaluation

- An assessment of the degree of hyperlordosis should be made, and a history of coccyx pain should be elicited.
- Evaluation for joint hypermobility should be done, and a history of subluxation or pain should be elicited.

Management

- Physical therapy for lower abdominal muscle strengthening and teaching of "tucking under" of the buttocks (pelvic rotation) may be warranted in those children with the most marked hyperlordosis (personal observation). In children with very severe hyperlordosis, this may be of some benefit in decreasing the probability of symptomatic lumbosacral spinal stenosis (personal observation).
- When chronic coccydynia is present, it can be treated by padding sewn into the underwear (personal observation).

- Surgical intervention for elbow limitation generally is not indicated. Although humeral lengthening is occasionally considered, it is rarely warranted; instead adaptive devices (such as bottom wipers) should be used as needed.
- Discomfort and fatigue of the wrists when doing fine motor tasks, because of hypermobility, may be relieved by using a simple stabilizing brace, if this proves particularly problematic (personal observation).

Dental

Structural abnormalities of the midface and jaw are exceedingly common in individuals with achondroplasia, including, in particular, midface hypoplasia and relative mandibular overgrowth. In addition, the palate is often narrow and anteriorly V-shaped (personal observation). The result is a high frequency of malocclusion and crowding of the teeth, including palisading of the upper incisors, anterior open bite, crossbite, and prognathism (personal observation). Outcome depends upon timely assessment and intervention.

Evaluation

- In addition to routine pediatric dental care, children with apparent bite abnormalities should be referred for orthodontic assessment as early as 5–6 years of age.

Management

- Options can include palatal expansion or other early management, traditional orthodontic manipulations, and, in instances of exceedingly severe midface hypoplasia, consideration of a Le Fort I facial advancement procedure.

Anesthetic Risks

Related to many of the problems already discussed, both children and adults with achondroplasia are likely to face one or more operative procedures. Anesthesia presents certain special risks in most dwarfing disorders (Berkowitz et al. 1990). Achondroplasia is no exception. Primary areas of concern include cervical spinal stenosis and consequent risks related to extremes of positioning while sedated or anesthetized; obstructive apnea with or without cor pulmonale and concomitant risks of post-sedation obstruction or post-extubation pulmonary edema; reduced airway size and possible restrictive lung disease in young children (Berkowitz et al. 1990). With appropriate care, the complication rate should approach that of the general population.

Evaluation

- Anesthesiologists should be made aware of the potential risks and of the availability of the excellent review by Berkowitz et al. (1990).

Management

- Many pediatric surgeons elect (appropriately) to admit young children with achondroplasia for an overnight stay even for surgery that usually is done as a same-day procedure because of the risk of airway-related complications.

Pregnancy

Extraordinarily little information is available on the risks in pregnancy of women with achondroplasia (Allanson and Hall 1986). On the basis of a questionnaire administered to a self-selected convenience sample of 87 women with achondroplasia (Allanson and Hall 1986), the following information appears to be secure. First, women with achondroplasia can continue pregnancies to term, presumably because of relatively normal trunk size. Second, complications during pregnancy are relatively infrequent but may include risk for worsening of neurological symptoms related to increasing hyperlordosis and maternal respiratory failure. Predicting who may develop respiratory compromise has not been possible. In two women of similar size and similar baseline pulmonary status, one developed respiratory failure in the early third trimester, whereas the other successfully carried twins to term (personal observation). Except for the small possibility of maternal respiratory failure requiring early delivery, successful pregnancies should be anticipated. Because of cephalopelvic disproportion, all women with achondroplasia need to have delivery by Caesarian section.

Prenatal testing by *FGFR3* molecular analysis may be elected, principally in two circumstances. First, when both parents have achondroplasia it can be used to distinguish homozygous achondroplasia from other possible outcomes (Shiang et al. 1994; Gooding et al. 2002). Second, when a sporadic short-limbed dwarfing condition is discovered by ultrasound, the presence or absence of achondroplasia as the cause of limb shortening can be determined in this manner. This will now often be undertaken as part of a skeletal dysplasia molecular panel. With developments in cell-free DNA analysis using maternal serum, non-invasive prenatal molecular testing will likely be another emerging diagnostic option in pregnancies at risk (Chitty et al. 2015).

Evaluation

- Pre-pregnancy counseling related to options for prenatal diagnosis should be provided to couples where one member or both members has/have achondroplasia (Gooding et al. 2002).
- Women with achondroplasia should be considered at high risk particularly related to the possibility of respiratory compromise later in pregnancy. Baseline pulmonary function studies may provide a basis for

monitoring respiratory status as the pregnancy progresses (Allanson and Hall 1986).
- Careful follow-up for this and other maternal complications is essential.

Management

- Women should be counseled to anticipate a scheduled Cesarean delivery without a trial of labor.
- There is current controversy and no consensus about appropriate anesthetic management – general, spinal, or epidural – for Cesarean delivery in these women (Dubiel et al. 2014).

Pathway-driven Therapies

Although no treatment directed toward correcting the molecular pathology that causes achondroplasia is yet available, a number of approaches are being explored (Ornitz and Legeai-Mallet 2017). For one of these, in which a C-type natriuretic peptide analog is used, human trials have begun (www.biomarin.com/products/clinical-trials/bmn-111-for-achondroplasia/), and others are being developed. The overall goal is, or should be, prevention of the many potential clinical issues associated with achondroplasia as discussed in this chapter, and better quality of life through the lifespan.

RESOURCES

Little People of America (LPA)

Support group for individuals of marked small stature and for their families; many secondary resources are available through the national LPA and its local district and chapter personnel. Similar organizations now exist in many other countries (accessible through the LPA Web site).

250 El Camino Real, Suite 208
Tustin, CA 92780, USA
Phone: 1-888-LPA-2001
Email: info@lpaonline.org
Website: http://www.lpaonline.org

LPA Medical Resource Center

https://lpamrs.memberclicks.net/index.php

LPA Medical Advisory Board

William G. Mackenzie, M.D., Chairr
Alfred I. DuPont Hospital for Children
PO Box 269
1600 Rockland Road
Wilmington, DE 19899, USA
Website: http://www.lpaonline.org/index.php?option=com_content&view=article&id=106

Books and Pamphlets

Riggs S (2001) *Never Sell Yourself Short*. Morton Grove, IL: Albert Whitman. This is a photo essay about a 14-year-old boy with achondroplasia that is appropriate for families and for late primary and middle school age children.

Lovell P, Carrow D (2001) Stand Tall Molly Lou Melon. New York, NY: G.P. Putnam's Sons. For preschool and early school age children. It is not explicitly about dwarfism, but does explore dealing with difference.

Roberts J, Robinson C (2014) The Smallest Girl in the Smallest Grade. New York, NY: G.P. Putnam's Sons. Appropriate for preschool and kindergarten, this book addresses in rhyme themes of exclusion and teasing.

Arnold J, Klein B (2016, 2017) Life is Short (No Pun Intended) and Think Big: Overcoming Obstacles with Optimism. New York, NY: Howard Books. Two books from the 'Little Couple' of The Learning Channel.

Campbell J, Dorren N (2006) *It's a Whole New View. A Beginner's Guide for New Parents of a Child with Dwarfism*. LPA Inc. Available without charge from Little People of America and may be downloaded from an LPA Web site PDF at http://www.lpaonline.org/mc/page.do?sitePageId=44398&ordId=lpa .

Adelson BM (2005) Dwarfism: Medical and Psychosocial Aspects of Profound Short Stature, Baltimore: Johns Hopkins University Press, 368 pages.

Scott C, Mayeux N, Crandall R, Weiss J (1994) *Dwarfism. The Family and Professional Guide*. Irvine, CA: Short Stature Foundation, 195 pages.

Ablon J (1988) Living with Differences. *Families with Dwarf Children*. New York: Praeger (Greenwood), 194 pages.

To Celebrate: Understanding Developmental Differences in Young Children with Achondroplasia (1997). Madison, WI: Midwest Regional Bone Dysplasia Clinic. Available from the Midwest Regional Bone Dysplasia Clinic; Clinical Genetics Center; University of Wisconsin-Madison; 1500 Highland Ave., Madison, WI 53705-2280.r

Little People, Big Schools: Preparing the School for Your Young Child with Short Stature (1997) Madison, WI: Midwest Regional Bone Dysplasia Clinic. Available from the Midwest Regional Bone Dysplasia Clinic; Clinical Genetics Center; University of Wisconsin-Madison; 1500 Highland Ave., Madison, WI 53705-2280.

REFERENCES

Afsharpaiman S, Sillence DO, Sheikhvatan M, Ault JE, Waters K (2011) Respiratory events and obstructive sleep apnea in children with achondroplasia: investigation and treatment outcomes. *Sleep Breath* 15:755–761.

Ain MC, Browne JA (2004) Spinal arthrodesis with instrumentation for thoracolumbar kyphosis in pediatric achondroplasia. *Spine (Phila Pa 1976)* 29:2075–2080.

Ain MC, Chang TL, Schkrohowsky JG, Carlisle ES, Hodor M, Rigamonti D (2008) Rates of perioperative complications associated with laminectomies in patients with achondroplasia. *J Bone Joint Surg Am* 90:295–298.

Akyol Y, Averill LW, Atanda A, Kecskemethy HH, Bober MB, Mackenzie WG (2015) Magnetic resonance evaluation of the knee in children and adolescents with achondroplasia. *Pediatr Radiol* 45:888–895.

Allanson, JE, Hall JG (1986) Obstetric and gynecologic problems in women with chondrodystrophies. *Obstet Gynecol* 67:74–78.

Bagley CA, Pindrik JA, Bookland MJ, Camara-Quintana JQ, Carson BS (2006) Cervicomedullary decompression for foramen magnum stenosis in achondroplasia. *J Neurosurg* 104:166–172.

Beighton P, Bathfield CA (1981) Gibbal achondroplasia. *J Bone Joint Surg Br* 63-B:328–329.

Bellus GA, Hefferon TW, Ortiz de Luna RI, Hecht JT, Horton WA, Machado M, Kaitila I, McIntosh I, Francomano CA (1995) Achondroplasia is defined by recurrent G380R mutations of FGFR3. *Am J Hum Genet* 56:368–373.

Bellus GA, Bamshad MJ, Przylepa KA, Dorst J, Lee RR, Hurko O, Jabs EW, Curry CJ, Wilcox WR, Lachman RS, Rimoin DL, Francomano CA (1999) Severe achondroplasia with developmental delay and acanthosis nigricans (SADDAN): phenotypic analysis of a new skeletal dysplasia caused by a Lys650Met mutation in fibroblast growth factor receptor 3. *Am J Med Genet* 85:53–65.

Berkowitz ID, Raja SN, Bender KS, Kopits SE (1990) Dwarfs: pathophysiology and anesthetic implications. *Anesthesiology* 73:739–759.

Berkowitz RG, Grundfast KM, Scott C, Saal H, Stern H, Rosenbaum K (1991) Middle ear disease in childhood achondroplasia. *Ear Nose Throat J* 70:305–308.

Brinkmann G, Schlitt H, Zorowka P, Spranger J (1993) Cognitive skills in achondroplasia. *Am J Med Genet* 47:800–804.

Brooks JT, Bernholt DL, Tran KV, Ain MC (2016) The Tibial Slope in Patients With Achondroplasia: Its Characterization and Possible Role in Genu Recurvatum Development. *J Pediatr Orthop* 36:349–354.

Brühl K, Stoeter P, Wietek B, Schwarz M, Humpl T, Schumacher R, Spranger J (2001) Cerebral spinal fluid flow, venous drainage and spinal cord compression in achondroplastic children: impact of magnetic resonance findings for decompressive surgery at the cranio-cervical junction. *Eur J Pediatr* 160:10–20.

Carlisle ES, Ting BL, Abdullah MA, Skolasky RL, Schkrohowsky JG, Yost MT, Rigamonti D, Ain MC (2011) Laminectomy in patients with achondroplasia: the impact of time to surgery on long-term function. *Spine (Phila Pa 1976)* 36:886–892.

Carneiro JR, da Silveira VG, Vasconcelos AC, de Souza LL, Xerez D, da Cruz GG, Quaresma JC, Macedo RG, de Oliveira JE (2007) Bariatric surgery in a morbidly obese achondroplasic patient – use of the 6-minute walk test to assess mobility and quality of life. *Obes Surg* 17:255–257.

Chitty LS, Mason S, Barrett AN, McKay F, Lench N, Daley R, Jenkins LA (2015) Non-invasive prenatal diagnosis of achondroplasia and thanatophoric dysplasia: next-generation sequencing allows for a safer, more accurate, and comprehensive approach. *Prenat Diagn* 35:656–662.

Collins WO, Choi SS (2007) Otolaryngologic manifestations of achondroplasia. *Arch Otolaryngol Head Neck Surg* 133:237–244.

Crandall R, Crosson T, Scott CI, Mayeux N, Weiss J (1994) Dwarfism. The Family and Professional Guide. Irvine, CA, Short Stature Foundation.

Danielpour M, Wilcox WR, Alanay Y, Pressman BD, Rimoin DL (2007) Dynamic cervicomedullary cord compression and alterations in cerebrospinal fluid dynamics in children with achondroplasia. Report of four cases. *J Neurosurg* 107:504–507.

Dessoffy KE, Modaff P, Pauli RM (2014) Airway malacia in children with achondroplasia. *Am J Med Genet A* 164A:407–414.

Dubiel L, Scott GA, Agaram R, McGrady E, Duncan A, Litchfield KN (2014) Achondroplasia: anaesthetic challenges for caesarean section. *Int J Obstet Anesth* 23:274–278.

Elwood ET, Burstein FD, Graham L, Williams JK, Paschal M (2003) Midface distraction to alleviate upper airway obstruction in achondroplastic dwarfs. *Cleft Palate Craniofac J* 40:100–103.

Enderle A, Meyerhofer D, Unverfehrt G (1994) Small People-Great Art. Restricted Growth from an Artistic and Medical Viewpoint. Hamm, Germany, Artcolor.

Flynn MA, Pauli RM (2003) Double heterozygosity in bone growth disorders: four new observations and review. *Am J Med Genet A* 121A:193–208.

Fowler ES, Glinski LP, Reiser CA, Horton VK, Pauli RM (1997) Biophysical bases for delayed and aberrant motor development in young children with achondroplasia. *J Dev Behav Pediatr* 18:143–150.

Gooding HC, Boehm K, Thompson RE, Hadley D, Francomano CA, Biesecker BB (2002) Issues surrounding prenatal genetic testing for achondroplasia. *Prenat Diagn* 22:933–940.

Goriely A, Wilkie AO (2012) Paternal age effect mutations and selfish spermatogonial selection: causes and consequences for human disease. *Am J Hum Genet* 90:175–200.

Greenwood Genetics Center (1988) Growth References from Conception to Adulthood. Clinton, SC, Jacobs.

Hall BD, Spranger J (1979) Hypochondroplasia: clinical and radiological aspects in 39 cases. *Radiology* 133:95–100.

Harada D, Namba N, Hanioka Y, Ueyama K, Sakamoto N, Nakano Y, Izui M, Nagamatsu Y, Kashiwagi H, Yamamuro M, Ishiura Y, Ogitani A and Seino Y (2017) Final adult height in long-term growth hormone-treated achondroplasia patients. *Eur J Pediatr* 176:873–879.

Hashmi SS, Gamble C, Hoover-Fong J, Alade AY, Pauli RM, Modaff P, Carney M, Brown C, Bober MB, Hecht JT (2018) Multi-center study of mortality in achondroplasia. *Am J Med Genet* 176:2359–2364.

Hatzaki A, Sifakis S, Apostolopoulou D, Bouzarelou D, Konstantinidou A, Kappou D, Sideris A, Tzortzis E, Athanassiadis A, Florentin L, Theodoropoulos P, Makatsoris C, Karadimas C, Velissariou V (2011) FGFR3 related skeletal dysplasias diagnosed prenatally by ultrasonography and molecular analysis: presentation of 17 cases. *Am J Med Genet A* 155A:2426–2435.

Hecht JT, Butler IJ, Scott, Jr CI (1984) Long-term neurological sequelae in achondroplasia. *Eur J Pediatr* 143:58–60.

Hecht JT, Francomano CA, Horton WA, Annegers JF (1987) Mortality in achondroplasia. *Am J Hum Genet* 41:454–464.

Hecht JT, Hood OJ, Schwartz RJ, Hennessey JC, Bernhardt BA, Horton WA (1988) Obesity in achondroplasia. *Am J Med Genet* 31:597–602.

Hecht JT, Thompson NM, Weir T, Patchell L, Horton WA (1991) Cognitive and motor skills in achondroplastic infants: neurologic and respiratory correlates. *Am J Med Genet* 41:208–211.

Ho NC, Guarnieri M, Brant LJ, Park SS, Sun B, North M, Francomano CA, Carson BS (2004) Living with achondroplasia: quality of life evaluation following cervico-medullary decompression. *Am J Med Genet A* 131:163–167.

Hoernschemeyer DG, Atanda, Jr A, Dean-Davis E, Gupta SK (2016) Discoid meniscus associated with achondroplasia. *Orthopedics* 39:e498–503.

Hoover-Fong JE, McGready J, Schulze KJ, Barnes H, Scott CI (2007) Weight for age charts for children with achondroplasia. *Am J Med Genet A* 143A:2227–2235.

Hoover-Fong JE, Schulze KJ, McGready J, Barnes H, Scott CI (2008) Age-appropriate body mass index in children with achondroplasia: interpretation in relation to indexes of height. *Am J Clin Nutr* 88:364–371.

Hoover-Fong JE, McGready J, Schulze KJ, Alade AY, Scott CI (2017) A height-for-age growth reference for children with achondroplasia: Expanded applications and comparison with original reference data. *Am J Med Genet A* 173:1226–1230.

Horton WA, Rotter JI, Rimoin DL, Scott CI, Hall JG (1978) Standard growth curves for achondroplasia. *J Pediatr* 93:435–438.

Hunter AG, Hecht JT, Scott, Jr CI (1996a) Standard weight for height curves in achondroplasia. *Am J Med Genet* 62:255–261.

Hunter AG, Reid CS, Pauli RM, Scott, Jr CI (1996b) Standard curves of chest circumference in achondroplasia and the relationship of chest circumference to respiratory problems. *Am J Med Genet* 62:91–97.

Inan M, Thacker M, Church C, Miller F, Mackenzie WG, Conklin D (2006) Dynamic lower extremity alignment in children with achondroplasia. *J Pediatr Orthop* 26:526–529.

Ireland PJ, Donaghey S, McGill J, Zankl A, Ware RS, Pacey V, Ault J, Savarirayan R, Sillence D, Thompson E, Townshend S, Johnston LM (2012a) Development in children with achondroplasia: a prospective clinical cohort study. *Dev Med Child Neurol* 54:532–537.

Ireland PJ, Johnson S, Donaghey S, Johnston L, Ware RS, Zankl A, Pacey V, Ault J, Savarirayan R, Sillence D, Thompson E, Townshend S, McGill J (2012b) Medical management of children with achondroplasia: evaluation of an Australasian cohort aged 0-5 years. *J Paediatr Child Health* 48:443–449.

Jeong S T, Song HR, Keny SM, Telang SS, Suh SW, Hong SJ (2006) MRI study of the lumbar spine in achondroplasia. A morphometric analysis for the evaluation of stenosis of the canal. *J Bone Joint Surg Br* 88:1192–1196.

Kahanovitz N, Rimoin DL, Sillence DO (1982) The clinical spectrum of lumbar spine disease in achondroplasia. *Spine (Phila Pa 1976)* 7:137–140.

King JA, Vachhrajani S, Drake JM, Rutka JT (2009) Neurosurgical implications of achondroplasia. *J Neurosurg Pediatr* 4:297–306.

Kopits SE (1980) Genetics clinics of The Johns Hopkins Hospital. Surgical intervention in achondroplasia. Correction of bowleg deformity in achondroplasia. *Johns Hopkins Med J* 146:206–209.

Kopits SE (1988a) Orthopedic aspects of achondroplasia in children. *Basic Life Sci* 48:189–197.

Kopits SE (1988b) Thoracolumbar kyphosis and lumbosacral hyperlordosis in achondroplastic children. *Basic Life Sci* 48:241–255.

Laederich MB, Horton WA (2010) Achondroplasia: pathogenesis and implications for future treatment. *Curr Opin Pediatr* 22:516–523.

Langer LO, Jr, Baumann PA, Gorlin RJ (1967) Achondroplasia. *Am J Roentgenol Radium Ther Nucl Med* 100:12–26.

Lonstein JE (1988) Treatment of kyphosis and lumbar stenosis in achondroplasia. *Basic Life Sci* 48:283–292.

Lundar T, Bakke SJ, Nornes H (1990) Hydrocephalus in an achondroplastic child treated by venous decompression at the jugular foramen. Case report. *J Neurosurg* 73:138–140

Margalit A, McKean G, Lawing C, Galey S, Ain MC (2016) Walking out of the curve: thoracolumbar kyphosis in achondroplasia. *J Pediatr Orthop* 2016 Epub ahead of print.

Maroteaux P, Lamy M, Robert JM (1967) [Thanatophoric dwarfism]. *Presse Med* 75:2519–2524.

McClure PK, Kilinc E, Birch JG (2017) Growth modulation in achondroplasia. *J Pediatr Orthop* 37:e384–e387.

Menezes AH (2009) Editorial. Achondroplasia. *J Neurosurg Pediatr* 4:295–296; discussion 296.

Mettler Gand FC Fraser (2000) Recurrence risk for sibs of children with "sporadic" achondroplasia. *Am J Med Genet* 90:250–251.

Miccoli M, Bertelloni S, Massart F (2016) Height outcome of recombinant human growth hormone treatment in achondroplasia children: A meta-analysis. *Horm Res Paediatr* 86:27–34.

Mogayzel PJ, Jr, Carroll JL, Loughlin GM, Hurko O, Francomano CA, Marcus CL (1998) Sleep-disordered breathing in children with achondroplasia. *J Pediatr* 132:667–671.

Mukherjee D, Pressman BD, Krakow D, Rimoin DL, Danielpour M (2014) Dynamic cervicomedullary cord compression and alterations in cerebrospinal fluid dynamics in children with achondroplasia: review of an 11-year surgical case series. *J Neurosurg Pediatr* 14:238–244.

Orioli IM, Castilla EE, Scarano G, Mastroiacovo P (1995) Effect of paternal age in achondroplasia, thanatophoric dysplasia, and osteogenesis imperfecta. *Am J Med Genet* 59:209–217.

Ornitz DM, Legeai-Mallet L (2017) Achondroplasia: Development, pathogenesis, and therapy. *Dev Dyn* 246:291–309.

Patel MD, Filly RA (1995) Homozygous achondroplasia: US distinction between homozygous, heterozygous, and unaffected fetuses in the second trimester. *Radiology* 196:541–545.

Pauli RM, Conroy MM, Langer Jr LO, McLone DG, Naidich T, Franciosi R, Ratner IM, Copps SC (1983) Homozygous achondroplasia with survival beyond infancy. *Am J Med Genet* 16:459–473.

Pauli RM, Scott CI, Wassman, Jr ER, Gilbert EF, Leavitt LA, Ver Hoeve J, Hall JG, Partington MW, Jones KL, Sommer A, Feldman W, Langer LO, Rimoin DL, Hecht JT, Lebovitz R (1984) Apnea and sudden unexpected death in infants with achondroplasia. *J Pediatr* 104:342–348.

Pauli RM, Horton VK, Glinski LP, Reiser CA (1995) Prospective assessment of risks for cervicomedullary-junction compression in infants with achondroplasia. *Am J Hum Genet* 56:732–744.

Pauli RM, Breed A, Horton VK, Glinski LP, Reiser CA (1997) Prevention of fixed, angular kyphosis in achondroplasia. *J Pediatr Orthop* 17:726–733.

Pauli RM, Modaff P (1999) Jugular bulb dehiscence in achondroplasia. *Int J Pediatr Otorhinolaryngol* 48:169–174.

Pauli RM (2012) Achondroplasia. *GeneReviews [Internet]*. MP. Adam, HH Ardinger, RA Pagon and ED Wallace. Seattle, WA, University of Washington, Seattle https://www.ncbi.nlm.nih.gov/books/NBK1152/.

Pierre-Kahn A, Hirsch JF, Renier D, Metzger J, Maroteaux P (1980) Hydrocephalus and achondroplasia. A study of 25 observations. *Child's Brain* 7:205–219.

Pyeritz RE, Sack, Jr GH, Udvarhelyi GB (1987) Thoracolumbosacral laminectomy in achondroplasia: long-term results in 22 patients. *Am J Med Genet* 28:433–444.

Reid CS, Pyeritz RE, Kopits SE, Maria BL, Wang H, McPherson RW, Hurko O, Phillips, 3rd JA, Rosenbaum AE (1987) Cervicomedullary compression in young patients with achondroplasia: value of comprehensive neurologic and respiratory evaluation. *J Pediatr* 110:522–530.

Roberts D, Pexa C, Clarkowski B, Morey M, Murphy M (1999) Fatal laryngeal injury in an achondroplastic dwarf secondary to airbag deployment. *Pediatr Emerg Care* 15:260–261.

Schiedel F, Rodl R (2012) Lower limb lengthening in patients with disproportionate short stature with achondroplasia: a systematic review of the last 20 years. *Disabil Rehabil* 34:982–987.

Schkrohowsky JG, Hoernschemeyer DG, Carson BS, Ain MC (2007) Early presentation of spinal stenosis in achondroplasia. *J Pediatr Orthop* 27:119–122.

Shiang R, Thompson LM, Zhu YZ, Church DM, Fielder TJ, Bocian M, Winokur ST, Wasmuth JJ (1994) Mutations in the transmembrane domain of FGFR3 cause the most common genetic form of dwarfism, achondroplasia. *Cell* 78:335–342.

Simmons K, Hashmi SS, Scheuerle A, Canfield M, Hecht JT (2014) Mortality in babies with achondroplasia: revisited. *Birth Defects Res A Clin Mol Teratol* 100:247–249.

Sisk EA, Heatley DG, Borowski BJ, Leverson GE, Pauli RM (1999) Obstructive sleep apnea in children with achondroplasia: surgical and anesthetic considerations. *Otolaryngol Head Neck Surg* 120:248–254.

Stanley G, McLoughlin S, Beals RK (2002) Observations on the cause of bowlegs in achondroplasia. *J Pediatr Orthop* 22:112–116.

Steinbok P, Hall J, Flodmark O (1989) Hydrocephalus in achondroplasia: the possible role of intracranial venous hypertension. *JNeurosurg* 71:42–48.

Stevens PM (2007) Guided growth for angular correction: a preliminary series using a tension band plate. *J Pediatr Orthop* 27:253–259.

Stokes DC, Phillips JA, Leonard CO, Dorst JP, Kopits SE, Trojak JE, Brown DL (1983) Respiratory complications of achondroplasia. *J Pediatr* 102:534–541.

Swift D, Nagy L, Robertson B (2012) Endoscopic third ventriculostomy in hydrocephalus associated with achondroplasia. *J Neurosurg Pediatr* 9:73–81.

Takken T, van Bergen MW, Sakkers RJ, Helders PJ, Engelbert RH (2007) Cardiopulmonary exercise capacity, muscle strength, and physical activity in children and adolescents with achondroplasia. *J Pediatr* 150:26–30.

Tang J, Su N, Zhou S, Xie Y, Huang J, Wen X, Wang Z, Wang Q, Xu W, Du X, Chen H, Chen L (2016) Fibroblast growth factor receptor 3 inhibits osteoarthritis progression in the knee joints of adult mice. *Arthritis Rheumatol* 68:2432–2443.

Tasker RC, Dundas I, Laverty A, Fletcher M, Lane R, Stocks J (1998) Distinct patterns of respiratory difficulty in young children with achondroplasia: a clinical, sleep, and lung function study. *Arch Dis Child* 79:99–108.

Tenconi R, Khirani S, Amaddeo A, Michot C, Baujat G, Couloigner V, De Sanctis L, James S, Zerah M, Cormier-Daire V, Fauroux B (2017) Sleep-disordered breathing and its management in children with achondroplasia. *Am J Med Genet A* 173:868–878.

Trotter TL, Hall JG, American Academy of Pediatrics Committee on Genetics (2005) Health supervision for children with achondroplasia. *Pediatrics* 116:771–783.

Tunkel D, Alade Y, Kerbavaz R, Smith B, Rose-Hardison D, Hoover-Fong J (2012) Hearing loss in skeletal dysplasia patients. *Am J Med Genet A* 158A:1551–1555.

Waller DK, Correa A, Vo TM, Wang Y, Hobbs C, Langlois PH, Pearson K, Romitti PA, Shaw GM, Hecht JT (2008) The population-based prevalence of achondroplasia and thanatophoric dysplasia in selected regions of the US. *Am J Med Genet A* 146A:2385–2389.

Wilkin DJ, Szabo JK, Cameron R, Henderson S, Bellus GA, Mack ML, Kaitila I, Loughlin J, Munnich A, Sykes B, Bonaventure J, Francomano CA (1998) Mutations in fibroblast growth-factor receptor 3 in sporadic cases of achondroplasia occur exclusively on the paternally derived chromosome. *Am J Hum Genet* 63:711–716.

Wright MJ, Irving MD (2012) Clinical management of achondroplasia. *Arch Dis Child* 97:129–134.

Wynn J, King TM, Gambello MJ, Waller DK, Hecht JT (2007) Mortality in achondroplasia study: a 42-year follow-up. *Am J Med Genet A* 143A:2502–2511.

Yang SS, Corbett DP, Brough AJ, Heidelberger KP, Bernstein J (1977) Upper cervical myelopathy in achondroplasia. *Am J Clin Pathol* 68:68–72.

3

ALAGILLE SYNDROME

HENRY C. LIN

Division of Gastroenterology, Doernbecher Children's Hospital and Oregon Health and Science University, Portland, Oregon, USA

IAN D. KRANTZ

The Perelman School of Medicine at the University of Pennsylvania and Division of Human Genetics and Genomics, The Children's Hospital of Philadelphia, Pennsylvania, USA

INTRODUCTION

Alagille syndrome (ALGS), also known as syndromic bile duct paucity or arteriohepatic dysplasia, is a multisystem autosomal dominant developmental disorder with highly variable expression. This syndrome was first reported by Alagille in 1969 with emphasis on the hepatic manifestations (Alagille D 1969; Alagille et al. 1987). Subsequently Watson and Miller, focusing more on the cardiac findings, described the same entity (Watson and Miller 1973). In 1975, Alagille formally described the syndrome that carries his name and specific diagnostic criteria were established (Alagille et al. 1975). Alagille syndrome is the most common form of familial cholestatic liver disease. In addition to the liver manifestations, Alagille syndrome is characterized by abnormalities of the heart, eye, and skeleton, and a characteristic facial appearance. The kidney, vascular, and central nervous system are also affected in a smaller percentage of affected individuals (Emerick et al. 1999; Emerick et al. 2005). Two disease causing genes have been identified: *JAGGED1* (*JAG1*) and *NOTCH2*, with the majority of mutations in *JAG1* and 2% of mutations in *NOTCH2*. (Li et al. 1997; Oda, Elkahloun et al. 1997; McDaniell et al. 2006; Saleh et al. 2016).

Incidence

The prevalence of ALGS was initially reported as 1 in 70,000 live births, although this is most likely an underestimate as individuals were ascertained based solely on the finding of neonatal liver disease and individuals without apparent liver disease were not scored as affected (Danks et al. 1977). In addition, molecular testing has identified many individuals with a disease-causing mutation that do not have clinically significant liver disease, making the condition likely to be much more prevalent (Kamath et al. 2003). Currently, the estimated frequency of ALGS is 1 in 30,000 (Spinner et al. 2019; Saleh et al. 2016).

ALGS is the second most common cause of intrahepatic cholestasis in infancy. The mortality of Alagille syndrome varies according to the involvement of each particular organ system in an individual. Chronic cholestatic liver disease is a main cause of morbidity. Overall mortality is between 10–20%, with liver disease, cardiac disease, and vascular accidents being the leading causes (Kamath et al. 2004; Spinner et al. 2019). The 20-year predicted life expectancy has been estimated at 75% (Emerick et al. 1999), 80% for people without liver transplantation, and 60% for people who required liver transplantation. However, the presence of

---

Cassidy and Allanson's Management of Genetic Syndromes, Fourth Edition.
Edited by John C. Carey, Agatino Battaglia, David Viskochil, and Suzanne B. Cassidy.
© 2021 John Wiley & Sons, Inc. Published 2021 by John Wiley & Sons, Inc.

an intracardiac lesion severely limits survival. Cardiac disease accounts for nearly all the early mortality in ALGS. Individuals with intracardiac disease have approximately 40% survival to 6 years of life compared to 95% survival in individuals with ALGS and no intracardiac lesions (Emerick et al. 1999). Furthermore, cardiovascular disease contributes significantly to the morbidity of the disorder, and has been implicated in the increased post-transplantation mortality seen in some series.

Diagnostic Criteria

With the availability of molecular testing for mutations in *JAG1* and *NOTCH2*, the broad phenotypic variability of ALGS can be fully appreciated. Traditionally, the clinical diagnosis of Alagille syndrome has been based on the criteria established by Alagille (Alagille et al. 1975; Alagille et al. 1987) and includes the histological finding of paucity of the interlobular bile ducts on liver biopsy, in association with a minimum of three of five major clinical features: chronic cholestasis, cardiac disease, skeletal anomalies, ocular abnormalities and characteristic facial features. Bile duct paucity on liver biopsy has been considered to be the most important and constant feature of ALGS. It is important to note that paucity may not be apparent in infancy in many individuals ultimately shown to have ALGS. Overall, bile duct paucity is present in about 89% of affected individuals (Emerick et al. 1999). While the characteristic findings used to establish a clinical diagnosis are seen in the majority of individuals with ALGS, several other organs and structures have also been noted to be involved to a lesser degree: kidney, pancreas, vascular system, and the extremities (Figure 3.1).

An even wider spectrum of clinical involvement (including intestinal abnormalities, orofacial clefts, hearing loss, and mental retardation) may be seen in individuals with ALGS individuals with deletions of chromosome 20p12 (Krantz et al. 1997). The wide spectrum of clinical variability in ALGS is most notable in studies of affected family members (Spinner et al. 2019). Kamath et al. (2003) evaluated 53 *JAG1* variant-confirmed relatives of affected individuals, and almost half of these relatives (25/53) did not meet clinical diagnostic criteria despite having a pathogenic *JAG1* mutation (Kamath et al. 2003).

Hepatic disease dominates the presentation of most individuals with ALGS and it is usually evident in the neonatal period as conjugated hyperbilirubinemia. The diagnosis of ALGS in neonates with cholestasis requires a careful physical examination and a thorough biochemical and radiologic evaluation. Laboratory findings are indicative of a defect in biliary excretion and most commonly include elevations of serum bile acids, conjugated bilirubin, alkaline phosphatase, cholesterol, and gamma-glutamyl transpeptidase (GGT), Serum aminotransferases are not typically markedly elevated. Hypercholesterolemia and triglyceridemia may be profound in the presence of severe cholestasis. If the diagnosis of ALGS is clear, based on the presentation of high-GGT cholestasis in association with other syndromic features, a liver biopsy is not mandatory to confirm the diagnosis. The most important differential diagnosis that must be excluded is biliary atresia. Excretion of nuclear tracer into the duodenum via hepatobiliary scintigraphy can be helpful to eliminate biliary atresia from consideration; however, non-excretion is common in ALGS as well. If there is any diagnostic dilemma, a liver biopsy becomes necessary.

The liver biopsy in ALGS classically shows intrahepatic bile duct paucity (Figure 3.1), although the diagnostic histopathological lesion of paucity is progressive and may not be evident in the newborn period. In one large series, bile duct paucity was only evident in 60% of infants younger than 6 months, but present in 95% of older individuals (Emerick et al. 1999).

Cardiac involvement is almost universal in ALGS (Emerick et al. 1999; McElhinney et al. 2002). Right-sided lesions predominate and the most common structural abnormality is pulmonary artery stenosis. The presence of a murmur in a child with cholestasis should prompt a cardiologic evaluation. The possibility of peripheral pulmonary stenosis should be specifically evaluated as it may be missed on routine echocardiography.

Facial gestalt is an important diagnostic criterion for ALGS. It occurs almost universally in mutation-positive probands and relatives (Kamath et al. 2003). The constellation of facial features seen in ALGS includes a prominent forehead, deep set eyes with moderate hypertelorism, pointed chin, and concave nasal ridge (previously called saddle nose) or straight nose with a bulbous tip. The combination of these features gives the face a triangular appearance (Figure 3.1). The facies are dynamic and evolve with age, with some of the features being more subtle in the newborn period because of subcutaneous facial adipose tissue. The features are most characteristic in the toddler to preadolescent period, but begin to change around adolescence to a very typical, but perhaps less recognized, appearance in adulthood (Figure 3.1) (Kamath et al. 2002). The chin in the adult becomes prognathic, with less prominence of the forehead, almost inverting the emphasis of the features seen in childhood. It has been suggested that there is interobserver variability in identification of these features and that they are not specific to ALGS but rather are possibly due to cholestasis (cholestasis facies) (Sokol et al. 1983). However, a study of the ability of dysmorphologists to differentiate the facies of individuals with ALGS from other forms of congenital cholestasis indicates that the facial features in ALGS are readily distinguishable from those in other forms of cholestasis (Kamath et al. 2002).

Ophthalmologic involvement most commonly manifests as posterior embryotoxon, a prominent centrally positioned Schwalbe's ring (Figure 3.1) that is visualized by slit-lamp examination. Posterior embryotoxon has been reported in up to 89% of individuals and is therefore important

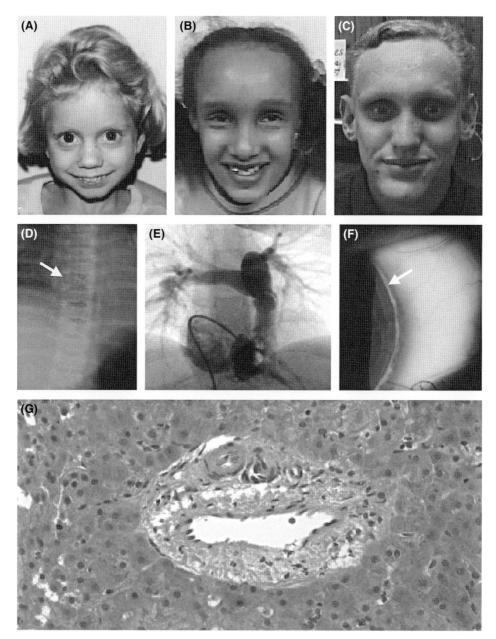

FIGURE 3.1 Clinical findings associated with Alagille syndrome. (A)–(C) Typical facial features are shown in a child (A), pre-teen (B) and adult (C). Note broad forehead with prominent supraorbital ridges, deep-set eyes, straight nose with broad tip and prominent pointed chin that becomes more prognathic with age. (D) Butterfly thoracic vertebra indicated by arrow. (E) Cardiac catheterization demonstrating branch pulmonic stenosis. (F) Slit lamp exam of the eye illustrating the finding of posterior embryotoxon (indicated by arrow). (G) Liver biopsy from a 2-month-old male with Alagille syndrome reveals several arteries without accompanying bile ducts in this otherwise unremarkable portal tract (hematoxylin-eosin, x200).

diagnostically although it is of no consequence for visual acuity. Posterior embryotoxon also occurs in the general population with a frequency of 8–15% (Waring et al. 1975), therefore its value as a diagnostic tool is limited.

The most common skeletal abnormality in ALGS is butterfly vertebrae (Rosenfield et al. 1980) (Figure 3.1). Butterfly vertebrae are usually asymptomatic radiologic findings and can be detected on plain radiographic films of the thoracic vertebrae. The frequency of butterfly vertebrae in reported cases of ALGS ranges from 22 to 87% (Alagille et al. 1987; Emerick et al. 1999).

Diagnosis of the syndrome can be hindered by the highly variable expressivity of the clinical manifestations (Krantz et al. 1999). Large series of affected individuals reported by different groups have demonstrated differing frequencies of the manifestations of ALGS (Hoffenberg et al. 1995; Emerick et al. 1999; Lykavieris et al. 2001). This variability has been observed both within and between families. Diagnosis can also be complicated by the presence in the general population of several of the diagnostic features, such as posterior embryotoxon or heart murmurs, which are present in 6% of all newborns.

Familial studies and molecular testing, have identified individuals with very subtle or isolated findings of ALGS (Krantz et al. 1999; Kamath et al. 2003). Molecular identification has allowed modification of the diagnostic criteria. For the proband in a family, it is reasonable to apply the original Alagille criteria (Alagille et al. 1975; Alagille et al. 1987). In infants less than 6 months of age, when paucity of the intrahepatic bile ducts is commonly absent, three or four clinical features are adequate to make the diagnosis. In families with one definite proband, other family members with even one manifestation are likely to be carrying a mutation in *JAG1*.

Etiology, Pathogenesis and Genetics

The familial nature of ALGS was recognized from the first descriptions of the disorder. Watson and Miller (1973) studied five affected families and discussed the possible dominant inheritance and variable expressivity of this disorder. Alagille et al. (1975) reported that 3 of their 15 patients had siblings with neonatal cholestasis. Throughout all of the reports, transmission was consistent with an autosomal dominant pattern of inheritance, but the penetrance of the disorder was reduced (likely due to decreased recognition of the sub-clinically affected individuals in a family) and expressivity quite variable (Shulman et al. 1984; Dhorne-Pollet et al. 1994; Spinner et al. 1994).

The finding of a cytogenetically visible deletion or translocation of chromosome 20 in multiple individuals with ALGS led to the assignment of Alagille syndrome to 20p12 (Byrne et al. 1986; Anad et al. 1990; Spinner et al. 1994). While the percentage of individuals with a chromosomal deletion or rearrangement was found to be quite low (less than 7%) (Krantz et al. 1997), these individuals were instrumental in establishing the precise localization of the genomic region containing the disease gene. In 1997, two groups were able to demonstrate that *JAG1* was physically located within the commonly deleted region on the short arm of chromosome 20, and that point mutations in *JAG1* were found in people with ALGS (Li et al. 1997; Oda et al. 1997).

JAG1 is a cell surface protein that functions in the Notch signaling pathway. This pathway was first described in Drosophila melanogaster and the name Notch derives from the characteristic notched wing found in flies carrying only one functioning copy of the gene. *JAG1* serves as a ligand for the four Notch receptors (Notch 1, 2, 3 and 4) and activation of the Notch pathway is involved in cell fate determination.

The *JAG1* gene is located within band 20p12. The *JAG1* cDNA (GenBank accession # 4557678) is 6 Kb with a coding region of 3657 nucleotides (Oda et al. 1997). At the genomic level, *JAG1* occupies 36,000 base pairs of DNA sequence and is comprised of 26 exons.

Using single strand conformation polymorphism analysis, previous studies have demonstrated *JAG1* mutations in about 70% of individuals who meet clinical criteria (Krantz et al. 1998; Crosnier et al. 1999; Spinner et al. 2001). However, with the more straightforward (and less costly) approach of direct sequencing, *JAG1* mutations can be identified in close to 95% of people with ALGS (Warthen et al. 2006). To date, more than 500 *JAG1* mutations have been identified in affected individuals. The mutations are distributed across the entire coding region of the *JAG1* gene with no hotspots. Sixty-five percent of mutations are intragenic and 72% of these are protein truncating (frameshift and nonsense). Nine percent of affected individuals have splicing mutations and 9% have missense mutations (Krantz et al. 1998; Crosnier et al. 1999).

If a pathogenic mutation is not identified, or if individuals do not meet clinical diagnostic guidelines, deletion and duplication analysis can be performed (Spinner et al. 2019). Across all studies, total gene deletions have been identified in 3–7% of affected individuals. In some cases, mosaicism has been identified in a parent with no clinically evident signs of ALGS (Giannakudis et al. 2001; Laufer-Cahana et al. 2002).

JAG1 is highly expressed in many of the affected organ systems in Alagille syndrome (Crosnier et al. 2000), but the penetrance and expressivity of ALGS are not uniform across patients or organ systems (Shulman et al. 1984). For the inherited *JAG1* mutations, the variable expressivity in ALGS is evident as over 40% of *JAG1* mutations were found in relatively healthy parents who underwent genetic analysis following the diagnosis of a family member (Krantz et al. 1997; Crosnier et al. 2000). In most cases, no genotype-phenotype correlation between mutation type or origin and clinical manifestations has been reported. One exception to this appears to be a *JAG1* missense mutation (G274D) that segregates with isolated cardiac disease in a large family (Eldadah et al. 2001). Expression and functional studies of this mutant have demonstrated that some of the G274D protein molecules are normally processed and transported to the cell surface where they function appropriately, while some of them are incorrectly processed and transported. These results suggest that while haploinsufficiency for *JAG1* is associated with the well-characterized phenotype of ALGS, the leaky G274D mutant, which allows more *JAG1* protein to reach the cell surface, is associated with a cardiac-specific phenotype (Lu et al. 2003). Therefore, cardiac development appears to be more sensitive to *JAG1* dosage than liver development. This is the first *JAG1* mutation identified with a phenotypic correlation.

Other potential phenotypes with a genetic correlation can be traced to 20p12.2 microdeletions in individuals with ALGS who have developmental delay or hearing loss. In one study, 10 out of 21 individuals with deletions of the short arm of chromosome 20 had developmental delay and hearing loss (Kamath et al. 2009).

The lack of consistent phenotypes both within and between families with the same *JAG1* mutations suggests that there are modifiers of the ALGS phenotype. These modifiers could be either genetic or environmental. Other support for the presence of potential genetic modifiers comes from work in the mouse. A mouse knockout of the *JAG1* gene has been reported (Xue et al. 1999). Homozygotes for the *JAG1* null allele die early from vascular defects and *JAG1* mutant heterozygous mice exhibit only eye defects. Mice that are doubly heterozygous for *Jag1* and *Notch2* mutations are an excellent model for ALGS (McCright et al. 2002). These mice are jaundiced and have bile duct, heart, eye and kidney abnormalities that are similar to those seen in individuals with ALGS. This work led to the screening of a *JAG1* mutation-negative cohort of people with ALGS for alterations in *NOTCH2*, encoding a known receptor of *JAG1* peptide.

To date, 12 pathogenic variants in *NOTCH2* have been identified in individuals with clinical features of ALGS. These mutations include splice site, frameshift, nonsense, and missense mutations. Individuals with *NOTCH2* mutations appear to have a prominent renal phenotype and there also appears to be a trend towards less cardiac involvement and decreased penetrance of vertebral abnormalities and facial features as compared to individual with *JAG1* mutations (Kamath et al. 2012; Kamath et al. 2013). However, the cohort of individuals with AGLS due to *NOTCH2* mutations is too small to draw conclusions from these observations (McDaniell et al. 2006).

Genetic Counseling

After the identification of an affected individual, parental evaluation should be carried out. If parents do not carry the mutation or deletion identified in their child with ALGS, their risk of recurrence is very low, as germline mosaicism remains a possibility but has not been documented in ALGS to date. Even so, some individuals without clinical manifestations of ALGS have been identified as mosaic for point mutations and deletions (Giannakudis et al. 2001; Laufer-Cahana et al. 2002). Targeted mutation analysis of family members of the proband, regardless of clinical assessment, is recommended. Even within families segregating a single mutation, the expressivity of the disorder has been found to range from mild to severe (Shulman et al. 1984; Li et al. 1997; Kamath et al. 2002).

Prenatal diagnosis can be carried out in at-risk individuals if a *JAG1* or *NOTCH2* mutation is known in an affected member of the family. It is important to remember that there is no way to predict the severity of the clinical features if a mutation is found prenatally.

Diagnostic Testing

The diagnosis of ALGS is largely based on the clinical criteria described above. Molecular testing for deletions of or mutations within *JAG1* is a useful adjunct for diagnosis in atypical cases or in individuals with subtle manifestations. FISH (fluorescence in situ hybridization) detects deletions in 5–7% of individuals with ALGS and direct sequencing identifies *JAG1* mutations in about 95% of clinically diagnosed individuals (Warthen et al. 2006). Clinical molecular analysis of *JAG1* and *NOTCH2* is available on a commercial and research basis. Once a *JAG1* or *NOTCH2* mutation is identified in a proband, it is straightforward and inexpensive to test parents and other relatives for the same mutation. Mutations are inherited in 30–50% of affected individuals and occur de novo in the rest. While low, the risk of recurrence to subsequent siblings of an individual with an apparent de novo pathogenic mutation variant is greater than in the general population due to the possibility of germline mosaicism. Prenatal testing is possible if the pathogenic mutation in an affected family member is known but cannot predict the severity of clinical manifestations (Spinner et al. 2019).

Differential Diagnosis

Many other forms of cholestatic liver disease (biliary atresia, alpha 1 antitrypsin deficiency, cystic fibrosis) may present with conjugated hyperbilirubinemia in the neonatal period. ALGS is most easily confused with extrahepatic disorders such as biliary atresia or metabolic disorders that present with elevated gamma-glutamyl transferase. Commonly, it can be misdiagnosed as biliary atresia due to the overlap of biochemical, scintigraphic and cholangiographic features. However, in ALGS, the pattern of histologic involvement of bile ducts is different and therefore liver biopsy is often a useful component of the evaluation in those who either do not have access to molecular testing or in whom the testing fails to identify a pathogenic mutation in either *JAG1* or *NOTCH2*. In biliary atresia, bile duct proliferation is the typical histologic lesion, and paucity is extremely rare at diagnosis. By contrast, in Alagille syndrome, proliferation is rare, and paucity is nearly always present by 6 months of age. However, as the evaluation of neonatal cholestasis commonly occurs in the first two months of life as the diagnosis of biliary atresia is time-sensitive, the histologic findings may be non-diagnostic, and operative or endoscopic cholangiography may be required.

ALGS must also be distinguished from other syndromes in which right-sided cardiac defects and vertebral anomalies co-exist, such as deletion 22q11.2 (see Chapter 21) and VATER association (see Chapter 61). Posterior embryotoxon has also been reported in 69% of individuals with deletion 22q11.2 (McDonald-McGinn et al. 1999) (see Chapter 21).

Mutations in *NOTCH2* have also been found to cause Hadju–Cheney syndrome (acroosteolysis with osteoporosis and changes in the skull and mandible and the phenotypically overlapping serpentine fibula–polycystic kidney syndrome). While Hadju–Cheney syndrome is quite distinct clinically from ALGS, clinicians should be aware of the molecular connection between these two diagnoses. Interestingly all mutations causative of Hadju–Cheney syndrome have been localized to exon 34 of the *NOTCH2* gene (Simpson et al. 2011; Isidor et al. 2011).

MANIFESTATIONS AND MANAGEMENT

Growth and Feeding

Severe growth retardation is seen in 50–87% of affected individuals (Alagille et al. 1987; Hoffenberg et al. 1995; Emerick et al. 1999). It is particularly evident in the first four years of life and especially in individuals with severe liver disease phenotype. Resting energy expenditure is not increased in prepubertal children with ALGS and poor growth seems more likely to be due to poor caloric intake or underlying liver disease (Rovner et al. 2006). Malnutrition due to malabsorption is a major factor in this failure to thrive, and cachexia can be severe in ALGS. There appear to be limitations in linear growth even when protein-calorie malnutrition is not evident. This may be due, in part, to long bone and spine abnormalities associated with ALGS. Endocrine abnormalities are not common in ALGS. Individuals with growth failure appear to be insensitive to exogenous growth hormone (Bucuvalas et al. 1993). Many affected adults appear to have short stature, although a systematic study of adult height has not been completed.

Poor growth also impacts quality of life for individuals with ALGS. A study by the Childhood Liver Disease Research Network reported that children with ALGS had lower health-related quality of life (HRQOL) scores compared to healthy children. In particular, that HRQOL is associated with growth failure (Kamath et al. 2015).

Evaluation

- Height and weight should be measured at each health care visit during childhood and plotted on standard growth charts. Dietary records and nutritional assessments are helpful to ascertain caloric intake.
- Fat-soluble vitamin levels should be checked, especially in individuals with more severe liver disease phenotype.
- Renal tubular acidosis and pancreatic insufficiency are other treatable causes of growth failure and should be investigated with urinalysis and fecal fat excretion estimation, respectively.

Management

- Malnutrition and growth failure should be treated with aggressive nutritional therapy.
- With severe liver disease phenotype, there will be significant malabsorption of long-chain fats, therefore formulas supplemented with medium-chain triglycerides have some nutritional advantage.
- Supplemental feeding via a nasogastric or gastrostomy tube may be indicated if adequate nutrition cannot be taken orally.
- A percutaneous gastrostomy tube is contraindicated in a child with significant splenomegaly.
- Fat soluble vitamin supplementation is warranted in most individuals with significant liver disease.

Development and Behavior

In the earlier reports of this syndrome, Alagille noted significant though mild mental retardation (IQ 60–80) in 9 of 30 individuals studied (Alagille et al. 1975). A more recent study (Emerick et al. 1999) demonstrated mild delays in gross motor skills in 16% and mild mental retardation in only 2%. This decreased incidence of developmental and cognitive involvement in the later studies is most likely secondary to earlier disease recognition and more aggressive medical, surgical, and nutritional management and intervention.

ALGS does not appear to have a behavioral phenotype.

Evaluation

- Assessment of developmental skills should be done at routine well child visits.
- A child demonstrating delay in development should be thoroughly assessed in a standard way, optimally by a child development specialist.

Management

- Standard interventions should be initiated if developmental problems are ascertained.

Gastrointestinal

The majority of symptomatic individuals present in the first year of life, and most typically with jaundice, which resolves later in childhood in about half of these infants. Conjugated hyperbilirubinemia is typically seen in the neonatal period. The magnitude of the hyperbilirubinemia is minor compared to the degree of cholestasis. Cholestasis is manifest by pruritus, which is among the most severe in any chronic liver disease. It is rarely present before three to five months of age but is seen in nearly all children by the third year of life, even in those who are anicteric. Hepatomegaly is recognized in 93–100% of individuals with ALGS and is common in infancy. Splenomegaly is unusual early in the course of the disease but is eventually found in up to 70% of affected individuals. Synthetic liver failure is extremely uncommon in the first year of life.

Severe cholestasis results in the formation of xanthomas, characteristically on the extensor surfaces of the fingers, the palmar creases, nape of the neck, popliteal fossa, buttocks, and around inguinal region. The lesions persist throughout childhood but may gradually disappear after ten years of age. Formation of xanthomas correlates with a serum cholesterol greater than 500 mg/dl. Hypercholesterolemia and hypertriglyceridemia may be profound, reaching levels exceeding 1,000 mg/ml and 2,000 mg/ml respectively.

The most common laboratory abnormalities are elevations of serum bile acids, conjugated bilirubin, alkaline phosphatase, and gamma-glutamyl transpeptidase, which suggest a defect in biliary excretion in excess of the abnormalities in hepatic metabolism or synthesis. There are elevations of the serum aminotransferases, although hepatic synthetic function is usually well preserved. Nevertheless, progression to cirrhosis and hepatic failure is recognized in approximately 20% of individuals with ALGS.

Liver biopsy classically shows intrahepatic bile duct paucity, although the diagnostic histopathological lesion of duct paucity is progressive and may not be evident in the newborn period. Bile duct paucity is present in 80-100% of liver biopsies from individuals with ALGS, but is evident in only 60% of infants younger than six months (Alagille et al. 1987; Hoffenberg et al. 1995; Emerick et al. 1999; Quiros-Tejeira et al. 1999). Depending on when a biopsy is performed, there may be a broad range of histologic findings including portal fibrosis and, rarely, bile duct proliferation (Novotny et al. 1981). Identifying bile duct paucity on biopsy is therefore dependent on the age of the individual at the time of biopsy, the site from which the biopsy is taken, and the expertise of the pathologist.

Based on the available literature, progression to cirrhosis and liver failure occurs in a significant portion of affected individuals. Management has minimal impact on the natural history of liver disease, and 21 to 47% of individuals with ALGS-associated liver disease ultimately require a liver transplant (Hoffenberg et al. 1995; Emerick et al. 1999; Quiros-Tejeira et al. 1999). There is currently no way to predict which individuals with ALGS and neonatal liver disease will progress to end-stage liver disease and require transplantation. Indications for transplantation include synthetic liver dysfunction, intractable portal hypertension, bone fractures, severe pruritus, xanthomata, and growth failure (Piccoli and Spinner 2001). Overall, the estimated survival with native liver in ALGS has been reported to be 51% at 10 years and 38% at 20 years (Lykavieris et al. 2001). Individuals with ALGS do well post-transplantation with one-year patient survival rates of 87.5% as compared to 96% for individuals with biliary atresia. While transplant outcomes are fairly comparable to other chronic liver diseases, the slightly lower survival rate in ALGS can be attributed to vascular complications such as hepatic artery thrombosis and portal vein thrombosis (Kamath et al. 2012). Living related donors should be considered with caution as on two occasions apparently unaffected living-related transplant donors were found intraoperatively to be unsuitable due to bile duct paucity (Gurkan et al. 1999). Thus, related donors should undergo targeted genetic testing to assess for gene carrier status (Kamath et al. 2010; Turnpenny and Ellard 2012).

Pancreatic insufficiency can also be seen in ALGS (Chong et al. 1989; Emerick et al. 1999). Jejunal and ileal atresia and stenosis, malrotation and microcolon have occasionally been identified in affected individuals (Piccoli and Spinner 2001), particularly those with large deletions of 20p12.

Evaluation

- The initial evaluation of a jaundiced infant suspected to have ALGS should include laboratory testing.
- Liver biopsy may be required (see above).
- Further evaluation of children with an established diagnosis is dependent on the severity of the hepatic involvement.
- Individuals with the mildest and most stable disease should have annual laboratory testing including bilirubin, hepatic transaminases, gamma-glutamyl transpeptidase, clotting studies, serum bile acids, and fat soluble vitamin levels.
- In the setting of stable liver disease and an established diagnosis of ALGS, there is no indication to repeat a liver biopsy, even if paucity was not evident on the initial biopsy.
- Signs of pancreatic insufficiency or gastrointestinal obstruction should be evaluated in a standard manner.

Management

- Cholestasis is commonly profound. Bile flow may be stimulated with the choleretic ursodeoxycholic acid, but in many individuals the pruritus continues unabated.
- Care should be taken to keep the skin hydrated with emollients, and fingernails should be trimmed.
- Therapy with antihistamines may provide some relief from pruritis, but many individuals require additional therapy with agents such as rifampin or naltrexone.
- Biliary diversion has been successful in a number of individuals and should be considered before transplantation for intractable prutitus and xanthomas (Emerick and Whitington 2002; Mattei et al. 2006).

- Fat soluble vitamin deficiency is present to a variable degree in most individuals. Multivitamin preparations may not provide the correct ratio of fat soluble vitamins, and thus vitamins are best administered as individual supplements.
- Administration of vitamin A is not generally recommended as toxicity is largely hepatic.
- Individuals with splenomegaly should be fitted for a spleen guard, which should be worn for physical activities. In general, extreme contact sports should be avoided by individuals with significant splenomegaly.
- Liver transplantation is indicated for synthetic liver dysfunction, intractable portal hypertension, bone fractures, severe pruritus, severe xanthomata, and growth failure (Piccoli and Spinner 2001). Transplantation reportedly becomes necessary in 21%-50% of individuals with hepatic manifestations in infancy (Hoffenberg et al. 1995; Emerick et al. 1999) with post-transplant survival ranging from 79%-100% (Cardona et al. 1995; Hoffenberg et al. 1995; Emerick et al. 1999). Individuals with ALGS are good candidates for transplantation, although morbidity and mortality post-transplant is influenced by the degree of cardiopulmonary involvement.
- Pharmacologic and dietary treatment modalities are summarized in Table 3.1.
- Other gastrointestinal abnormalities should be treated in a standard fashion.

Cardiovascular

Studies have reported congenital heart disease to be present in 81%-100% of individuals with ALGS (Alagille et al. 1987; Deprettere et al. 1987; Emerick et al. 1999; McElhinney et al. 2002). The pulmonary vasculature (pulmonary valve, artery and/or its branches) is most commonly involved, with branch pulmonary artery stenosis being the most prevalent, found in 76% (Figure 3.1) (McElhinney et al. 2002). Intracardiac lesions are seen in 24% (Emerick et al. 1999) with tetralogy of Fallot being the most common complex cardiac malformation (12%) (McElhinney et al. 2002). The frequency of more clinically severe forms of tetralogy of Fallot is greater in ALGS, as 40% of those with ALGS and tetralogy of Fallot have pulmonary atresia (Silberbach et al. 1994; McElhinney et al. 2002). Other cardiac defects seen in association with ALGS include: ventricular septal defects, atrial septal defects, aortic stenosis, coarctation of the aorta, truncus arteriosus, and anomalous pulmonary venous return (Silberbach et al. 1994; McElhinney et al. 2002).

While the majority of cardiovascular malformations are hemodynamically insignificant, the more severe malformations have accounted for the majority of early mortality in some series of individuals with ALGS (Deprettere et al. 1987; Emerick et al. 1999; McElhinney et al. 2002). In the series reported by Emerick et al. (1999), the mortality rate was 33% for tetralogy of Fallot and 75% for tetralogy of Fallot with pulmonary atresia, which is significantly higher than mortality rates for non-syndromic tetralogy of Fallot (77%-89% for tetralogy of Fallot alone and 58% for tetralogy of Fallot with pulmonary atresia) (Vobecky et al. 1993). This higher rate may be biased due to the co-occurrence of significant hepatic disease in many of the individuals with ALGS reported in these series. However, many of those with *JAG1* mutations may have isolated congenital heart defects without clinically relevant hepatic involvement (Krantz et al. 1999). The prevalence of *JAG1* mutations amongst a large cohort of individuals with non-syndromic tetralogy of Fallot has been reported as 4% (Smith et al. 1998). Le Caignec et al. (2002) have also described a cohort with familial deafness, congenital heart defects, and posterior embryotoxon who share a mutation in the first epidermal growth factor-like domain of *JAG1* (Le Caignec et al. 2002). These findings prompted Bauer et al. (2010) to hypothesize that *JAG1* may be a candidate gene for additional cardiac findings and that the wider utility of whole genome analysis may identify modifying factors to development of isolated cardiac disease in *JAG1* mutation positive individuals (Bauer et al. 2010).

Cardiac disease accounts for nearly all the early mortality in ALGS. Individuals with intracardiac disease have approximately 40% survival to 6 years of life compared to 95% survival in individuals with ALGS and no intracardiac lesions (Emerick et al. 1999). Furthermore, cardiovascular disease contributes significantly to the morbidity of the disorder, and has been implicated in the increased post-transplantation mortality seen in some series.

Vascular anomalies in ALGS have been noted since the earliest descriptions, and pulmonary artery involvement is one of the most common manifestations. Other structural vascular anomalies seen in ALGS are renal artery stenosis, coarctation of the aorta, aortic aneurysms, and cerebral vasculature problems, which include aneurysms of the basilar, carotid, and middle cerebral arteries (Hoffenberg et al. 1995; Kamath et al. 2004; Emerick et al. 2005). Moyamoya disease, a progressive intracranial arterial occlusive disease, has also been described (Rachmel et al. 1989; Emerick et al. 1999; Woolfenden et al. 1999; Connor et al. 2002; Kamath et al. 2004). MRI angiography can detect these vascular lesions and one cohort of individuals with ALGS has been studied prospectively using magnetic resonance angiography, with 23% of asymptomatic patients having vascular anomalies. Thus, the current recommendation is for all individuals with ALGS to have a baseline screening MRI angiography.

Intracranial bleeding is increasingly recognized as a significant cause of morbidity and mortality in ALGS, with occurrences as high as 16% (Hoffenberg et al. 1995;

TABLE 3.1 Overview of therapeutic modalities in Alagille syndrome

Symptom	Pharmacologic therapy	Dietary and other therapies
Fat malabsorption	Medium chain triglycerides (added to diet)	Optimize carbohydrate and protein intake
Fat soluble vitamin deficiency (vitamin D, E, and A levels should be routinely monitored, and PT/PTT should be monitored as an indicator of vitamin K deficiency; inability to correct coagulopathy in some patients may indicate severe synthetic liver dysfunction)	Vitamin K (oral/intramuscular) Vitamin D (oral/intramuscular) (absorption of vitamin D may be enhanced by administration of d-α-tocopheryl polyethylene glycol-1000 succinate (TPGS)) Vitamin E (oral) (TPGS-soluble preparation)	
Pruritus	Ursodeoxycholic acid (choleretic; paradoxically appears to exacerbate pruritus in some individuals) Antihistamines Rifampin (exact mechanism unknown) Cholestyramine (binds bile salts and prevents reabsorption; some individuals develop severe acidosis on this therapy) Naltrexone (opioid antagonist) Phenobarbital (?efficacy)	Hydrate skin with emollients Trim fingernails Ultraviolet therapy (?efficacy) Biliary diversion
Decreased bone density/Osteoporosis	Calcium supplements	Annually monitor bone density with DEXA scans

Emerick et al. 1999). Fatality rates for these events are 30%-50%. Intracranial Bleeds have been reported to be epidural, subdural, subarachnoid and intraparenchymal. The majority are spontaneous and not associated with a clear predisposing event, although some have been temporally related to minor head trauma or coagulopathy.

A retrospective review of a large cohort of individuals with ALGS demonstrated almost 10% with a vascular lesion or event (Kamath et al. 2004). In addition to the type of vessel anomalies described above, this study also identified abnormalities of the renal vessels and aorta, amongst others. These have also been documented in many case reports (Rachmel, 1989 #1149; Woolfenden et al. 1999; Connor et al. 2002). These reports demonstrated that the vasculopathy of ALGS is not confined to the central nervous system but is a more widespread phenomenon.

It is possible that disruptions of the Notch signaling pathway may interfere with vasculogenesis and/or the maintenance of vascular integrity as mutations in a gene coding for another member of the Notch signaling pathway, the Notch 3 receptor, result in CADASIL (cerebral autosomal-dominant arteriopathy with subcortical infarcts and leukoencephalopathy) syndrome. CADASIL is an adult-onset disorder characterized by strokes and dementia that results from an angiopathy involving primarily the small cerebral arteries (Joutel et al. 1996). Furthermore, the *Jag1* knockout mouse is lethal in the early embryonic period due to vascular anomalies of the developing yolk sac (Xue et al. 1999).

Evaluation

- At the time of diagnosis every affected individual should have an echocardiogram and evaluation by a cardiologist.
- Further invasive evaluation, such as cardiac catheterization, is dictated by the echocardiographic findings.
- There is a lack of prospective data as yet to determine the optimal management plan for vascular anomalies in ALGS. The current recommendation is for all asymptomatic individuals to have a screening head and neck MRI/MRA as a baseline when the individual is old enough to tolerate the MRI without the need for sedation.
- The presence of neurologic signs or symptoms or a traumatic head injury should prompt careful neurological evaluation and appropriate imaging.

Management

- The management of cardiac manifestations is dictated by the severity of the anomaly and is standard for the relevant cardiac lesion.
- Individuals with mild peripheral pulmonary stenosis, the most common cardiac finding in ALGS, may be followed with echocardiography on an annual basis only.
- Non-surgical invasive techniques have been used successfully for ALGS, including valvuloplasty, balloon dilatation and stent implantation.

- Heart-lung transplantation has been successfully performed in combination with liver transplantation in a child with ALGS.
- Specific treatment of a vascular abnormality is targeted to any structural anomaly detected and follows standard practice.

Genitourinary

Renal anomalies have been reported in 23 to 74% of individuals with ALGS. These include structural abnormalities such as solitary kidney, ectopic kidney, bifid pelvis and duplicated ureters, small kidneys, and unilateral and bilateral multicystic and dysplastic kidneys. Additionally, functional abnormalities such as renal tubular acidosis in infancy, neonatal renal insufficiency, fatal juvenile nephronophthisis, lipidosis of the glomeruli, tubulointerstitial nephropathy, and adult-onset renal insufficiency and failure have been reported (LaBrecque et al. 1982; Hyams et al. 1983; Alagille et al. 1987; Habib et al. 1987; Tolia et al. 1987; Emerick et al. 1999; Kamath et al. 2013). Renal vascular disease (arterial stenosis) that may result in systemic hypertension has also been noted (Berard et al. 1998; Quiros-Tejeira et al. 1999).

Evaluation

- Biochemical evaluation of renal function should be accomplished with routine laboratory testing at least annually.
- A baseline renal ultrasound is recommended to detect any structural anomalies.
- Hypertension should prompt renal investigation as a potential cause.
- Renal tubular acidosis should be sought in any child with ALGS and growth failure.
- Renal function should be reassessed during the evaluation for hepatic transplantation.
- In individuals with pronounced renal involvement in whom a *JAG1* mutation is not identified, *NOTCH2* mutational analysis should be considered.

Management

- Treatment of renal disease in ALGS is targeted to the specific anomaly or disease, and follows standard practice.

Ophthalmologic

Larger studies have reported the prevalence of ophthalmologic findings in Alagille syndrome to be 56%-88%. Ocular abnormalities in ALGS include structural anomalies of the cornea, iris, retina, and optic disc, as well as being secondary to vitamin deficiencies. The majority of these involve defects of the anterior chamber (posterior embryotoxon, Axenfeld anomaly, Rieger anomaly), and retinal pigmentary changes (Alagille et al. 1987; Deprettere et al. 1987; Emerick et al. 1999; Hingorani et al. 1999). Overall, visual prognosis is good. The most common ophthalmologic finding is posterior embryotoxon, which is seen in up to 90% of patients with ALGS (Hingorani et al. 1999). Posterior embryotoxon (a prominent centrally positioned Schwalbe's ring) (Figure 3.1) is best visualized by slit-lamp examination (Alagille et al. 1987; Deprettere et al. 1987; Emerick et al. 1999; Hingorani et al. 1999). Posterior embryotoxon also has an 8-15% prevalence in of the general population (Kamath 2007). This finding is generally not of clinical significance, although it is important diagnostically.

In one series, ocular ultrasound examination was performed on 20 children with ALGS, and optic disc drusen was seen in 90%, suggesting that ocular ultrasound might aid in clinical diagnosis (Nischal et al. 1997).

Evaluation

- An ophthalmologic evaluation forms part of the diagnostic process in ALGS and should be accomplished at diagnosis in all affected individuals.
- Though posterior embryotoxon generally has no impact on vision, anterior chamber defects are rarely associated with glaucoma and therefore ophthalmologic follow-up with ocular pressure monitoring every one or two years is recommended.
- Ocular ultrasound may aid in the diagnosis of optic disc drusen.

Management

- Posterior embryotoxon requires no specific treatment.
- Any other treatment is targeted to the specific abnormality and follows standard practice.

Musculoskeletal

Butterfly vertebrae, which result from clefting abnormalities of the vertebral body (Figure 3.1), are the most common skeletal abnormality reported in ALGS, with a frequency of 22–87% (Rosenfield et al. 1980; Alagille et al. 1987; Emerick et al. 1999). Butterfly vertebrae are usually asymptomatic radiologic findings. The incidence of butterfly vertebrae in the general population is unknown but presumed to be very low. Butterfly vertebrae are frequently an incidental finding that carries no structural significance and are generally asymptomatic. Although relatively uncommon, butterfly vertebrae are also seen in a wide range of disorders including Crouzon syndrome (see Ch. 17), Deletion 22q11 syndrome (Ch. 21), and VATER syndrome (vertebral defects, anal atresia, tracheoesophageal fistula, radial and radial defects) (Ch.61)

(McDonald-McGinn et al. 1999; Kamath 2007). Other reported skeletal anomalies in ALGS include narrowing of interpeduncular spaces in the lumbar spine (50%), pointed anterior process of C1, spina bifida occulta, fusion of adjacent vertebrae, hemivertebrae, bony connections between ribs, and short fingers (Watson and Miller 1973; Rosenfield et al. 1980; Alagille et al. 1987; Deprettere et al. 1987).

Severe metabolic bone disease with osteoporosis and pathologic fractures is common in ALGS. Recurrent fractures, particularly of the femur, have been cited as a major indication for hepatic transplantation. A number of factors may contribute to osteopenia and fractures, including severe chronic malnutrition, vitamin D and vitamin K deficiency, chronic hepatic and renal disease, magnesium deficiency, and pancreatic insufficiency (Heubi et al. 1997; Piccoli and Spinner 2001).

Evaluation

- Radiologic evaluation of the thoracic vertebrae is important as a diagnostic tool to help make the diagnosis of ALGS, but has no clinical significance and is not necessary in the absence of symptoms.
- Monitoring of bone density with dual beam X-ray absorptiometry scans should be performed once every year or two starting at 5–7 years.

Management

- Butterfly vertebrae require no specific treatment.
- The presence of poor bone density or a history of fractures necessitates aggressive supplementation of calories, vitamin D and vitamin K.
- Recurrent pathological fractures have also been considered an indication for liver transplantation.

Miscellaneous

Other repeatedly reported findings in ALGS have included delayed puberty, high-pitched voice (Alagille et al. 1987), hearing loss (LaBrecque et al. 1982; Hingorani et al. 1999), supernumerary digital flexion creases (Kamath et al. 2002), and craniosynostosis (Kamath et al. 2002). Tracheal and bronchial stenosis, otitis media, chronic sinusitis, macrocephaly, hypothyroidism, and insulin-dependent diabetes are also found (Piccoli and Spinner 2001). Several of these expanded manifestations have been reported in individuals with ALGS who have a deletion of chromosome 20p12 that encompasses the *JAG1* gene and other genes in the region. A number of sporadic reports of hepatic lesions, most commonly hepatocellular carcinoma, have been described (Ennaifer et al. 2016). While this is concerning, more formal evaluations about the risk of developing neoplastic liver disease is needed before population screening guidelines can be recommended.

RESOURCES

Support Groups

The Alagille Syndrome Alliance (regular publication: Liverlink)

Website: http://www.alagille.org/

The Alagille Syndrome Diagnostic Center at The Children's Hospital of Philadelphia

Website: http://www.chop.edu

REFERENCES

Alagille D, A Estrada, M Hadchouel, M Gautier, M Odievre, JP Dommergues (1987) Syndromic paucity of interlobular bile ducts (Alagille syndrome or arteriohepatic dysplasia): review of 80 cases. *J Pediatr* 110(2):195–200.

Alagille D HE, Thomassin N (1969) L'atresie des voies biliaires extrahepatiques permeables chez l'enfant. *J Par Pediatr* 301–318.

Alagille D, M Odievre, M Gautier, JP Dommergues (1975) Hepatic ductular hypoplasia associated with characteristic facies, vertebral malformations, retarded physical, mental, and sexual development, and cardiac murmur. *J Pediatr* 86(1):63–71.

Anad F, J Burn, D Matthews, I Cross, BC Davison, R Mueller, M Sands, DM Lillington, E Eastham (1990) Alagille syndrome and deletion of 20p. *J Med Genet* 27(12):729–737.

Bauer RC, AO Laney, R Smith, J Gerfen, JJ Morrissette, S Woyciechowski, J Garbarini, K M Loomes, ID Krantz, Z Urban, BD Gelb, E Goldmuntz, NB Spinner (2010) Jagged1 (JAG1) mutations in patients with tetralogy of Fallot or pulmonic stenosis. *Hum Mutat* 31(5):594–601.

Bucuvalas JC, JA Horn, L Carlsson, WF Balistreri, SD Chernausek (1993) Growth hormone insensitivity associated with elevated circulating growth hormone-binding protein in children with Alagille syndrome and short stature. *J Clin Endocrinol Metab* 76(6):1477–1482.

Byrne JL, MJ Harrod, JM Friedman, PN Howard-Peebles (1986) del(20p) with manifestations of arteriohepatic dysplasia. *Am J Med Genet* 24(4):673–678.

Cardona J, D Houssin, F Gauthier, D Devictor, J Losay, M Hadchouel, O Bernard (1995) Liver transplantation in children with Alagille syndrome--a study of twelve cases. *Transplantation* 60(4):339–342.

Chong SK, J Lindridge, C Moniz, AP Mowat (1989) Exocrine pancreatic insufficiency in syndromic paucity of interlobular bile ducts. *J Pediatr Gastroenterol Nutr* 9(4):445–449.

Connor SE, D Hewes, C Ball, JM Jarosz (2002) Alagille syndrome associated with angiographic moyamoya. *Childs Nerv Syst* 18(3-4):186–190.

Crosnier C, C Driancourt, N Raynaud, S Dhorne-Pollet, N Pollet, O Bernard, M Hadchouel, M Meunier-Rotival (1999) Mutations in JAGGED1 gene are predominantly sporadic in Alagille syndrome. *Gastroenterology* 116(5):1141–1148.

Crosnier C, P Lykavieris, M Meunier-Rotival, M Hadchouel (2000) Alagille syndrome. The widening spectrum of arteriohepatic dysplasia. *Clin Liver Dis* 4(4):765–778.

Danks DM, PE Campbell, I Jack, J Rogers, AL Smith (1977) Studies of the aetiology of neonatal hepatitis and biliary atresia. *Arch Dis Child* 52(5):360–367.

Deprettere A, B Portmann, AP Mowat (1987) Syndromic paucity of the intrahepatic bile ducts:diagnostic difficulty; severe morbidity throughout early childhood. *J Pediatr Gastroenterol Nutr* 6(6):865–871.

Dhorne-Pollet S, JF Deleuze, M Hadchouel, C Bonaiti-Pellie (1994) Segregation analysis of Alagille syndrome. *J Med Genet* 31(6):453–457.

Eldadah ZA, A Hamosh, NJ Biery, RA Montgomery, M Duke, R Elkins, HC Dietz (2001) Familial Tetralogy of Fallot caused by mutation in the jagged1 gene. *Hum Mol Genet* 10(2):163–169.

Emerick KM, I D Krantz, BM Kamath, C Darling, DM Burrowes, NB Spinner, PF Whitington, DA Piccoli (2005) Intracranial vascular abnormalities in patients with Alagille syndrome. *J Pediatr Gastroenterol Nutr* 41(1):99–107.

Emerick KM, EB Rand, E Goldmuntz, ID Krantz, NB Spinner, DA Piccoli (1999) Features of Alagille syndrome in 92 patients: frequency and relation to prognosis. *Hepatology* 29(3):822–829.

Emerick KM., PF Whitington (2002) Partial external biliary diversion for intractable pruritus and xanthomas in Alagille syndrome. *Hepatology* 35(6):1501–1506.

Ennaifer R, L Ben Farhat, M Cheikh, H Romdhane, I Marzouk, N Belhadj (2016) Focal liver hyperplasia ina apatient with Alagille syndrome: Doagnpstic difficulties. A case report. *Int J of Surg Case Reports* 25:55–61.

Giannakudis J, A Ropke, A Kujat, M Krajewska-Walasek, H Hughes, JP Fryns, A Bankier, D Amor, M Schlicker, I Hansmann (2001) Parental mosaicism of JAG1 mutations in families with Alagille syndrome. *Eur J Hum Genet* 9(3):209–216.

Habib R, JP Dommergues, MC Gubler, M Hadchouel, M Gautier, M Odievre, D Alagille (1987) Glomerular mesangiolipidosis in Alagille syndrome (arteriohepatic dysplasia) *Pediatr Nephrol* 1(3):455–464.

Heubi J E, JV Higgins, EA Argao, RI Sierra, BL Specker (1997) The role of magnesium in the pathogenesis of bone disease in childhood cholestatic liver disease: a preliminary report. *J Pediatr Gastroenterol Nutr* 25(3):301–306.

Hingorani M, KK Nischal, A Davies, C Bentley, A Vivian, AJ Baker, G Mieli-Vergani, AC Bird, WA Aclimandos (1999) Ocular abnormalities in Alagille syndrome. *Ophthalmology* 106(2):330–337.

Hoffenberg EJ, M R Narkewicz, JM Sondheimer, DJ Smith, A Silverman, RJ Sokol (1995) Outcome of syndromic paucity of interlobular bile ducts (Alagille syndrome) with onset of cholestasis in infancy. *J Pediatr* 127(2):220–224.

Hyams JS, M M Berman, BH Davis (1983) Tubulointerstitial nephropathy associated with arteriohepatic dysplasia. *Gastroenterology* 85(2):430–434.

Isidor B, P Lindenbaum, O Pichon, S Bezieau, C Dina, S Jacquemont, D Martin-Coignard, C Thauvin-Robinet, M Le Merrer, J-L Mandel, A David, L Faivre, V Cormier-Daire, R Redon, C Le Caignec (2011) Truncating mutations in the last exon of NOTCH2 cause a rare skeletal disorder with osteoporosis. *Nat Genet* 43:306–308.

Joutel A, C Corpechot, A Ducros, K Vahedi, H Chabriat, P Mouton, S Alamowitch, V Domenga, M Cecillion, E Marechal, J Maciazek, C Vayssiere, C Cruaud, EA Cabanis, MM Ruchoux, J Weissenbach, JF Bach, MG Bousser, E Tournier-Lasserve (1996) Notch3 mutations in CADASIL, a hereditary adult-onset condition causing stroke and dementia. *Nature* 383(6602):707–710.

Kamath BM, L Bason, DA Piccoli, ID Krantz, NB Spinner (2003) Consequences of JAG1 mutations. *J Med Genet* 40(12):891–895.

Kamath BM, RC Bauer, KM Loomes, G Chao, J Gerfen, A Hutchinson, W Hardikar, G Hirschfield, P Jara, ID Krantz, P Lapunzina, L Leonard, S Ling, VL Ng, PL Hoang, DA Piccoli, NB Spinner (2012) NOTCH2 mutations in Alagille syndrome. *J Med Genet* 49(2):138–144.

Kamath BM, Z Chen, R Romero, EM Fredericks, EM Alonso, R Arnon, J Heubi, PM Hertel, SJ Karpen, KM Loomes, KF Murray, P Rosenthal, KB Schwarz, G Subbarao, JH Teckman, YP Turmelle, KS Wang, AH Sherker, RJ Sokol, JC Magee, N Childhood Liver Disease Research (2015) Quality of Life and Its Determinants in a Multicenter Cohort of Children with Alagille Syndrome. *J Pediatr* 167(2):390–396 e393.

Kamath BM, KM Loomes, RJ Oakey, KE Emerick, T Conversano, NB Spinner, DA Piccoli, ID Krantz (2002) Facial features in Alagille syndrome: specific or cholestasis facies? *Am J Med Genet* 112(2):163–170.

Kamath BM, K M Loomes, RJ Oakey, ID Krantz (2002) Supernumerary digital flexion creases: an additional clinical manifestation of Alagille syndrome. *Am J Med Genet* 112(2):171–175.

Kamath BM, KM Loomes, DA Piccoli (2010) Medical management of Alagille syndrome. *J Pediatr Gastroenterol Nutr* 50(6):580–586.

Kamath BM, SN, Piccoli DA (2007) Alagille syndrome. In: *Liver Disease in Children* Suchy FJ, Balistreri WF (eds). Cambridge, Cambridge University Press. pp. 326–345.

Kamath BM, N B Spinner, KM Emerick, AE Chudley, C Booth, DA Piccoli, ID Krantz (2004) Vascular anomalies in Alagille syndrome: a significant cause of morbidity and mortality. *Circulation* 109(11):1354–1358.

Kamath BM, NB Spinner, ND Rosenblum (2013) Renal involvement and the role of Notch signalling in Alagille syndrome. *Nat Rev Nephrol* 9(7):409–418.

Kamath BM, BD Thiel, X Gai, LK Conlin, PS Munoz, J Glessner, D Clark, DM Warthen, TH Shaikh, E Mihci, DA Piccoli, SF Grant, H Hakonarson, ID Krantz, NB Spinner (2009) SNP array mapping of chromosome 20p deletions: genotypes, phenotypes, and copy number variation. *Hum Mutat* 30(3):371–378.

Kamath BM, W Yin, H Miller, R Anand, EB Rand, E Alonso, J Bucuvalas, T Studies of Pediatric Liver (2012) Outcomes of liver transplantation for patients with Alagille syndrome: the studies of pediatric liver transplantation experience. *Liver Transpl* 18(8):940–948.

Krantz ID, RP Colliton, A Genin, EB Rand, L Li, DA Piccoli, NB Spinner (1998) Spectrum and frequency of jagged1 (JAG1) mutations in Alagille syndrome patients and their families. *Am J Hum Genet* 62(6):1361–1369.

Krantz ID, DA Piccoli, NB Spinner (1997) Alagille syndrome. *J Med Genet* 34(2):152-157.

Krantz ID, DA Piccoli, NB Spinner (1999) Clinical and molecular genetics of Alagille syndrome. *Curr Opin Pediatr* 11(6):558–564.

Krantz ID, EB Rand, A Genin, P Hunt, M Jones, AA Louis, JM Graham, Jr, S Bhatt, DA Piccoli, NB Spinner (1997) Deletions of 20p12 in Alagille syndrome: frequency and molecular characterization. *Am J Med Genet* 70(1):80–86.

Krantz ID, R Smith, RP Colliton, H Tinkel, EH Zackai, DA Piccoli, E Goldmuntz, NB Spinner (1999) Jagged1 mutations in patients ascertained with isolated congenital heart defects. *Am J Med Genet* 84(1):56–60.

LaBrecque DR, FA Mitros, RJ Nathan, KG Romanchuk, GF Judisch, GH El-Khoury (1982) Four generations of arteriohepatic dysplasia. *Hepatology* 2(4):467–474.

Laufer-Cahana, A, ID Krantz, LD Bason, FM Lu, DA Piccoli, NB Spinner (2002) Alagille syndrome inherited from a phenotypically normal mother with a mosaic 20p microdeletion. *Am J Med Genet* 112(2):190–193.

Le Caignec C, M Lefevre, JJ Schott, A Chaventre, M Gayet, C Calais, JP Moisan (2002) Familial deafness, congenital heart defects, and posterior embryotoxon caused by cysteine substitution in the first epidermal-growth-factor-like domain of jagged 1. *Am J Hum Genet* 71(1):180–186.

Li L, ID Krantz, Y Deng, A Genin, AB Banta, CC Collins, M Qi, BJ Trask, WL Kuo, J Cochran, T Costa, ME Pierpont, EB Rand, DA Piccoli, L Hood, NB Spinner (1997) Alagille syndrome is caused by mutations in human Jagged1, which encodes a ligand for Notch1. *Nat Genet* 16(3):243–251.

Lu F, JJ Morrissette, NB Spinner (2003) Conditional JAG1 mutation shows the developing heart is more sensitive than developing liver to JAG1 dosage. *Am J Hum Genet* 72(4):1065–1070.

Lykavieris P, M Hadchouel, C Chardot, O Bernard (2001) Outcome of liver disease in children with Alagille syndrome: a study of 163 patients. *Gut* 49(3):431–435.

Mattei P, D von Allmen, D Piccoli, E Rand (2006) Relief of intractable pruritus in Alagille syndrome by partial external biliary diversion. *J Pediatr Surg* 41(1):104–107; discussion 104–107.

McCright B, J Lozier, T Gridley (2002) A mouse model of Alagille syndrome: Notch2 as a genetic modifier of Jag1 haploinsufficiency. *Development* 129(4):1075–1082.

McDaniell R, DM Warthen, PA Sanchez-Lara, A Pai, ID Krantz, DA Piccoli, NB Spinner (2006) NOTCH2 mutations cause Alagille syndrome, a heterogeneous disorder of the notch signaling pathway. *Am J Hum Genet* 79(1):169–173.

McDonald-McGinn DM, R Kirschner, E Goldmuntz, K Sullivan, P Eicher, M Gerdes, E Moss, C Solot, P Wang, I Jacobs, S Handler, C Knightly, K Heher, M Wilson, JE Ming, K Grace, DDriscoll, P Pasquariello, P Randall, D Larossa, BS Emanuel, EH Zackai (1999) The Philadelphia story: the 22q11.2 deletion: report on 250 patients. *Genet Couns* 10(1):11–24.

McElhinney DB, ID Krantz, L Bason, DA Piccoli, KM Emerick, NB Spinner, E Goldmuntz (2002) Analysis of cardiovascular phenotype and genotype-phenotype correlation in individuals with a JAG1 mutation and/or Alagille syndrome. *Circulation* 106(20):2567–2574.

Morrissette JD, RP Colliton, NB Spinner (2001) Defective intracellular transport and processing of JAG1 missense mutations in Alagille syndrome. *Hum Mol Genet* 10(4):405–413.

Nischal KK, M Hingorani, CR Bentley, AJ Vivian, AC Bird, AJ Baker, AP Mowat, G Mieli-Vergani, WA Aclimandos (1997) Ocular ultrasound in Alagille syndrome: a new sign. *Ophthalmology* 104(1):79–85.

Novotny NM, RK Zetterman, DL Antonson, JA Vanderhoof (1981) Variation in liver histology in Alagille's syndrome. *Am J Gastroenterol* 75(6):449–450.

Oda T, AG Elkahloun, BL Pike, K Okajima, ID Krantz, A Genin, DA Piccoli, PS Meltzer, NB Spinner, FS Collins, SC Chandrasekharappa (1997) Mutations in the human Jagged1 gene are responsible for Alagille syndrome. *Nat Genet* 16(3):235–242.

Piccoli DA, NB Spinner (2001) Alagille syndrome and the Jagged1 gene. *Semin Liver Dis* 21(4):525–534.

Quiros-Tejeira RE, ME Ament, MB Heyman, MG Martin, P Rosenthal, TR Hall, SV McDiarmid, JH Vargas (1999) Variable morbidity in alagille syndrome: a review of 43 cases. *J Pediatr Gastroenterol Nutr* 29(4):431–437.

Rachmel A, A Zeharia, M Neuman-Levin, R Weitz, R Shamir, G Dinari (1989) Alagille syndrome associated with moyamoya disease. *Am J Med Genet* 33(1):89–91.

Rosenfield, NS, M J Kelley, PS Jensen, E Cotlier, AT Rosenfield, CA Riely (1980) Arteriohepatic dysplasia: radiologic features of a new syndrome. *AJR Am J Roentgenol* 135(6):1217–1223.

Rovner AJ, VA Stallings, DA Piccoli, AE Mulberg, BS Zemel (2006) Resting energy expenditure is not increased in prepubertal children with Alagille syndrome. *J Pediatr* 148(5):680–682.

Saleh M, BM Kamath, D Chitayat (2016) Alagille syndrome: clinical perspectives. *Appl Clin Genet* 9:75–82.

Shulman SA, JS Hyams, R Gunta, RM Greenstein, SB Cassidy (1984) Arteriohepatic dysplasia (Alagille syndrome): extreme variability among affected family members. *Am J Med Genet* 19(2):325–332.

Silberbach M, D Lashley, MD Reller, WF Kinn, Jr, A Terry, CO Sunderland (1994) Arteriohepatic dysplasia and cardiovascular malformations. *Am Heart J* 127(3):695–699.

Simpson MA, M.D Irving, E Asilmaz, M.J Gray, D Dafou, F.V Elmslie, S Mansour, S.E Holder, CE Brain, B. Burton, K. Kim, RM Pauli, S Aftimos, H Stewart, CA Kim, M Holder-Espinasse, SP Robertson, WM Drake, RC Trembath (2011) Mutations in NOTCH2 cause Hajdu-Cheney syndrome, a disorder of severe and progressive bone loss. *Nat Genet* 43:303–305.

Sokol RJ, JE Heubi, WF Balistreri (1983) Intrahepatic cholestasis facies: is it specific for Alagille syndrome? *J Pediatr* 103(2):205–208.

Spinner NB, RP Colliton, C Crosnier, ID Krantz, M Hadchouel, M Meunier-Rotival (2001) Jagged1 mutations in alagille syndrome. *Hum Mutat* 17(1):18–33.

Spinner NB, Gilbert MA, Loomes KM, et al. (2013) Alagille Syndrome. *GeneReviews*. Available from: https://www.ncbi.nlm.nih.gov/books/NBK1273/

Spinner NB, EB Rand, P Fortina, A Genin, R Taub, A Semeraro, DA Piccoli (1994) Cytologically balanced t(2;20) in a two-generation family with alagille syndrome: cytogenetic and molecular studies. *Am J Hum Genet* 55(2):238–243.

Tolia V, RS Dubois, FB Watts, Jr., E Perrin (1987) Renal abnormalities in paucity of interlobular bile ducts. *J Pediatr Gastroenterol Nutr* 6(6):971–976.

Tsai EA, MA Gilbert, CM Grochowski, LA Underkoffler, H Meng, X Zhang, MM Wang, H Shitaye, KD Hankenson, D Piccoli, H Lin, BM Kamath, M Devoto, NB Spinner, KM Loomes (2016) THBS2 is a candidate modifier of liver disease severity in Alagille syndrome. *Cell Mol Gastroenterol Hepatol* 2(5):663–675 e662.

Turnpenny PD, S Ellard (2012) Alagille syndrome: pathogenesis, diagnosis and management. *Eur J Hum Genet* 20(3):251–257.

Vobecky SJ, WG Williams, GA Trusler, JG Coles, IM Rebeyka, J Smallhorn, P Burrows, R Gow, RM Freedom (1993) Survival analysis of infants under age 18 months presenting with tetralogy of Fallot. *Ann Thorac Surg* 56(4):944–949; discussion 949–950.

Waring GO, 3rd, MM Rodrigues, PR Laibson (1975) Anterior chamber cleavage syndrome. A stepladder classification. *Surv Ophthalmol* 20(1):3–27.

Warthen DM, EC Moore, BM Kamath, JJ Morrissette, PA Sanchez-Lara, DA Piccoli, ID Krantz, NB Spinner (2006) Jagged1 (JAG1) mutations in Alagille syndrome: increasing the mutation detection rate. *Hum Mutat* 27(5):436–443.

Watson GH., V Miller (1973) Arteriohepatic dysplasia: familial pulmonary arterial stenosis with neonatal liver disease. *Arch Dis Child* 48(6):459–466.

Woolfenden AR, GW Albers, GK Steinberg, JS Hahn, DC Johnston, K. Farrell (1999) Moyamoya syndrome in children with Alagille syndrome: additional evidence of a vasculopathy. *Pediatrics* 103(2):505–508.

Xue Y, X Gao, CE Lindsell, CR Norton, B Chang, C Hicks, M Gendron-Maguire, EB Rand, G Weinmaster, T Gridley (1999) Embryonic lethality and vascular defects in mice lacking the Notch ligand Jagged1. *Hum Mol Genet* 8(5):723–730.

4

ALBINISM: OCULAR AND OCULOCUTANEOUS ALBINISM AND HERMANSKY–PUDLAK SYNDROME

C. Gail Summers
Departments of Ophthalmology & Visual Neurosciences and Pediatrics, University of Minnesota, Minneapolis, Minnesota, USA

David R. Adams
Deputy Director for Clinical Genomics, Office of the Clinical Director, National Human Genome Research Institute and Undiagnosed Diseases Program, National Institutes of Health, Bethesda, Maryland, USA

INTRODUCTION

The term albinism refers to a group of congenital genetic conditions resulting from an inability of the pigment cell (melanocyte) to synthesize normal amounts of melanin pigment. It is caused by mutations in 18 or more different genes (Giebel et al. 1990; Huizing et al. 2017). Reduced melanin synthesis in the melanocytes of the skin, hair, and eyes produces oculocutaneous albinism (OCA), whereas a reduction primarily involving the iris and retinal pigment epithelium of the eyes produces ocular albinism (OA1, sometimes published as XLOA). Hermansky–Pudlak syndrome (HPS) includes the triad of OCA, a mild bleeding diathesis, and a subtype-dependent ceroid storage disease affecting primarily the lungs and the gastrointestinal system (Gahl et al. 1998; Huizing et al. 2008).

Incidence

OCA is the most common inherited disorder of generalized hypopigmentation, with an estimated frequency of 1:20,000 in most populations (King et al. 2001). OCA has been described in all ethnic groups and in all animal species, making it one of the most widely distributed genetic disorders in the animal kingdom. Estimates of the frequency of OCA in different populations are not precise, and reliable data are not readily available. The two common types of OCA are type 1 (OCA1) (Figure 4.1) and type 2 (OCA2). OCA1 is the most common type recognized in most studies (Tomita et al. 2000; Hutton and Spritz, 2008a, 2008b). OCA2 is found in all populations. OCA2 has an estimated frequency of 1/1,000–1/15,000 in sub-Saharan African populations (Hong et al. 2006; Lund et al. 2007) and a lower frequency in the African-American population. It has been suggested that OCA2 is the most common type of albinism in the world in absolute number affected, primarily because of its high frequency in equatorial Africa (Puri et al. 1997; Kerr et al. 2000), but this is a speculation and accurate epidemiologic data are not available. The frequency in non-African populations is unknown. Several smaller isolated populations have also been reported to have a high frequency of OCA2, and this is thought to represent a founder effect (Woolf 1965; Yi et al. 2003).

OCA3 has been identified in the South African population and is infrequent in other populations (Boissy et al. 1996; Manga et al. 1997; Hutton and Spritz, 2008a, 2008b). OCA4 is the second most common type of OCA in the Japanese population after OCA1 and is infrequent in other populations (Newton et al. 2001; Inagaki et al. 2004, 2005; Rundshagen et al. 2004; Sengupta et al. 2007; Hutton and Spritz, 2008b). OCA5 and OCA6 are defined by

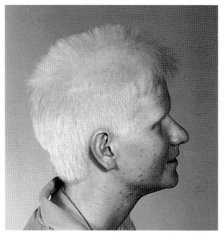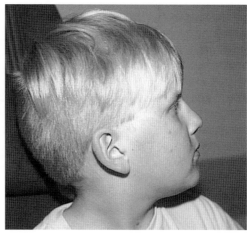

FIGURE 4.1 Typical pigmentary features of oculocutaneous albinism type 1 (OCA1).

molecular findings that have not been conclusively substantiated (Kausar, et al. 2013; Wei et al. 2013). OCA7 was identified in a Faroese cohort and has been detected only rarely in other populations (Gronskov et al. 2013). Hermansky–Pudlak syndrome (HPS) is a rare condition in most populations. At least 10 different types of HPS have been identified based on the gene involved. In the Puerto Rican population, however, two types of HPS are found with increased frequency, including HPS1 with an estimated frequency of approximately 1/1800, and HPS3, which is somewhat less frequent (Huizing and Gahl 2002; Huizing et al. 2008; Huizing et al. 2017). Both appear to result from founder mutations in this population. Chediak–Higashi syndrome is rare, is associated with hypopigmentation, and will not be discussed further (Introne et al. 1999). Other rare causes of hypopigmentation have been described (Tey 2010).

OA1 (ocular albinism) is less common than OCA1 or OCA2, with an estimated frequency of 1/50,000–60,000 (Rosenberg and Schwartz 1998; Roma et al. 2007). It is X-linked. No other types of ocular albinism have been identified. The term autosomal recessive ocular albinism has been used to refer to individuals who have ocular but minimal cutaneous features of OCA (O'Donnell et al. 1978; Fukai et al. 1995; Hutton and Spritz, 2008a). Some have been found to have a pigmenting type of OCA with genetic testing (Summers et al. 1996; Hutton and Spritz 2008a; Hutton and Spritz 2008b; Gronskov et al. 2009). However, several types of OCA have minimal cutaneous hypopigmentation, indicating that autosomal recessive OA refers to the phenotype and is not a precise term for a type of albinism.

Changes in mortality in albinism are related to skin cancer risk. In most of the US and Northern European areas, adequate skin protection from clothing, hats and sunscreens greatly reduces the frequency of skin cancer. In equatorial parts of the world (Africa and Asia), mortality in individuals with albinism is increased because of inadequate skin protection in an environment with a high ultraviolet exposure.

Diagnostic Criteria

Cutaneous and ocular hypopigmentation are not sufficient to define albinism because there are conditions with cutaneous or ocular hypopigmentation that are not part of the albinism spectrum (e.g., piebaldism, vitiligo, gyrate atrophy, and choroideremia). The precise definition of albinism includes ocular and cutaneous hypopigmentation associated with characteristic and specific changes in the development and function of the eyes and the optic nerves; the ocular changes are necessary to make this diagnosis.

All types of albinism have similar changes in the eye, and these are the features that link all of the types of OCA and OA1 under a single broad definition. The ophthalmic findings in albinism are characteristic of the group as a whole, but the severity can be variable. The cardinal diagnostic features of albinism are nystagmus, reduced visual acuity, iris transillumination, foveal hypoplasia, and poor stereovision associated with abnormal optic nerve connections between the retina and the optic cortex (Summers, et al. 1996). Strabismus and abnormal head postures are frequently associated with albinism.

The types of OCA and OA1 are usually defined by the gene involved, and not by the amount of hair, skin, or eye pigment present. The cutaneous hypopigmentation associated with each type of OCA and OA1 varies by genotype and ethnicity. However, in cases where mutations allow for any pigmentation, there can be substantial variation among individuals with the same genotype, including family members. As a result, clinical accuracy in diagnosis can be difficult because of overlap in degrees of hypopigmentation between types. Terms such as "tyrosinase-positive" and "tyrosinase-negative" and "partial" and "complete" are no longer used in the clinical diagnosis of albinism.

Males with OA1 have normal cutaneous pigmentation that may be slightly lighter than other family members and are recognized because of the ocular symptoms or because of the family history. Persons with OCA are recognized by the changes in their skin, hair, and eye pigmentation, by the ocular symptoms, or both, depending on the degree of cutaneous hypopigmentation. For example, OCA in a Caucasian child may be identified at birth because of the marked reduction or absence of hair, skin, and iris pigmentation, in the first few months of life because of the development of nystagmus, or in early childhood because of reduced acuity or the presence of head posturing to damp nystagmus and/or improve binocular alignment. OCA is more obvious at an early age in a family from an ethnic group with darker constitutional cutaneous pigmentation such as Asian or African families.

HPS refers to a group of disorders that present with OCA, a bleeding diathesis related to platelet dysfunction, and various systemic manifestations such as pulmonary fibrosis or granulomatous enterocolitis (Grucela et al. 2006; Pierson et al. 2006; Huizing et al. 2008). They often present with nystagmus, reduced vision, and easy bruising and/or epistaxis.

Etiology, Pathogenesis, and Genetics

OCA1–7 result from mutations in genes that encode proteins involved in the formation of melanin in the melanocyte. The melanocyte is present in its normal location (dermal-epidermal junction, base of hair follicle, posterior iris epithelium, retinal pigment epithelium and ciliary body, and striae vascularis and other areas of the ear), but the amount of melanin that forms within the pigment granule known as the melanosome is reduced. The different types of HPS are associated with impaired melanosome biogenesis and transfer (Huizing et al. 2000).

In the melanosome, melanin forms in a pathway that starts with the conversion of tyrosine to dopaquinone by tyrosinase, the most important enzyme in the pathway (Kushimoto et al. 2003; Schweikardt et al. 2007). At this point, the pathway divides and ultimately forms brown/black eumelanin, after the action of two additional enzymes, yellow-red pheomelanin, or a mixture of the two types of melanin (Barsh, 1996; Healy et al. 2001).

Genes related to albinism are involved in the enzymatic steps in the pathway, in forming the melanosome and maintaining its internal environment suitable for melanin synthesis, or in the transport of tyrosinase and other proteins from the endoplasmic reticulum to the developing melanosome through the process of vesicular transport (Kushimoto et al. 2003; Watabe et al. 2003; Huizing et al. 2008). The types of albinism, including HPS, are listed in Table 4.1. Many of the albinism-associated genes have been found to be important in normal pigmentation (Sturm 2006; Han et al. 2008).

All types of OCA, including all types of HPS, are autosomal recessive in inheritance. OA1 is X-linked recessive in inheritance, as mentioned earlier.

Reduction of melanin in skin in OCA and HPS produces sensitivity to the ultraviolet radiation of the sun and reduction or loss of the ability to tan. Acute sun exposure produces a sunburn, and chronic exposure produces thickened skin (pachydermia), premature aging of the skin, and either basal cell or squamous cell carcinoma (Okoro 1975; King et al. 1980; Hong et al. 2006; Adegbidi et al. 2007). Skin cancer is common in individuals with OCA in Africa but not common in most other parts of the world where sun exposure is less and protective clothing and sunscreen are available. The reduction in melanin in the hair in OCA and HPS produces white to light brown hair.

The major effect on quality of life caused by the reduced ability to form melanin is in the eyes. The iris stromal melanocytes and pigmented epithelium on the posterior surface of the iris form little or no melanin. The iris is blue or light brown/tan and transilluminates with globe illumination with a penlight or with slit lamp biomicroscopy. The retinal pigment epithelium forms little or no melanin, often resulting in visualization of the choroidal vessels beneath the retina when it is viewed with an ophthalmoscope. The fovea is typically hypoplastic, but an annular reflex is associated with relatively better vision, as is granular melanin pigment in the macula (Harvey et al. 2006, 2008; dem Hagen 2007; Lee et al. 2001; McCafferty et al. 2015). Finally, the decussation of the optic nerves at the chiasm is disrupted, with extra nerve fibers crossing from the temporal retina to the contralateral side of the brain, resulting in poor stereovision and functional monocularity frequently associated with alternating strabismus (Summers 1996; Lee et al. 2001; dem Hagen 2007). This is felt to be the etiology of a positive angle kappa (visual axis located nasal to the center of the pupil) in albinism, noted with monocular fixation on a penlight (Merrill et al. 2004; Brodsky and Frey 2004). A positive angle kappa can diminish the appearance of an esotropia and increase the appearance of an exotropia. It is not known how the loss of melanin in the developing eye leads to these developmental changes, and this is a current area of intense investigation (Reis et al. 2007; Rymer et al. 2007; McAllister et al. 2010; Wilk et al. 2014). Nystagmus, foveal hypoplasia, and misrouting of the optic nerve fibers contribute to the typical reduction in vision despite refractive correction.

Diagnostic Testing

Diagnostic testing for all types of albinism includes a careful pigment history, a physical examination, and a complete ophthalmologic examination. OCA is usually obvious from the cutaneous hypopigmentation, but this is not always true, particularly in lightly pigmented families of Northern European origin. The cutaneous pigment status of a child

TABLE 4.1 Types of albinism

Type	Name	Protein/*gene* involved	Function
OCA1	Oculocutaneous albinism 1	Tyrosinase/*TYR*	Enzyme for melanin biosynthesiss
OCA2	Oculocutaneous albinism 2	OCA2/*OCA2*	Melanosomal pH/solute regulation
OCA3	Oculocutaneous albinism 3	Tyrosinase-related protein 1/*TYRP1*	Enzyme for melanin biosynthesis
OCA4	Oculocutaneous albinism 4	Solute carrier family 45, member 2/*SLC45A2*, formerly *MATP*	Melanosomal pH/solute regulation
OCA5	Oculocutaneous albinism 5	Mapped region on chromosome 4*	Unknown
OCA6	Oculocutaneous albinism 6	Solute carrier family 24, member 5/*SLC24A5*	Hypothesized to have a role in calcium transport
OCA7	Oculocutaneous albinism 7	Leucine-rich melanocyte differentiation-associated protein /*LRMDA*, formerly *C10orf11*	Unknown
HPS1	Hermansky–Pudlak syndrome 1	HPS1/*HPS1*	Lysosomal-related organelle biogenesis
HPS2	Hermansky–Pudlak syndrome 2	Adapter-related protein complex 3, Beta-1 subunit/*AP3B1*	Lysosomal-related organelle biogenesis
HPS3	Hermansky–Pudlak syndrome 3	HPS3/*HPS3*	Unknown
HPS4	Hermansky–Pudlak syndrome 4	HPS4/*HPS4*	Lysosomal-related organelle biogenesis
HPS5	Hermansky–Pudlak syndrome 5	HPS5/*HPS5*	Lysosomal-related organelle biogenesis
HPS6	Hermansky–Pudlak syndrome 6	HPS6/*HPS6*	Lysosomal-related organelle biogenesis
HPS7	Hermansky–Pudlak syndrome 7	Dystrobrevin-binding protein 1/*DTNBP1*	Lysosomal-related organelle biogenesis
HPS8	Hermansky–Pudlak syndrome 8	Biogenesis of lysosome-related organelles complex 1, subunit 3/*BLOC1S3*	Lysosomal-related organelle biogenesis
HPS9	Hermansky–Pudlak syndrome 9	Biogenesis of lysosome-related organelles complex 1, subunit 6/*BLOC1S6*	Lysosomal-related organelle biogenesis
HPS10	Hermansky–Pudlak syndrome 10	Adaptor-related protein complex 3, delta-1 subunit/*AP3D1*	Lysosomal-related organelle biogenesis
CHS	Chediak–Higashi syndrome	*CHS1*	Melanosome formation
OA1	Ocular albinism 1	G protein-coupled receptor 143/*GPR143*, formerly *OA1*	Melanosome environment

* Molecular testing not currently available

who develops nystagmus at birth or within the first few months of life needs to be evaluated carefully to determine if the child is lighter than other members of the family. Such differences may be equivocal in lightly pigmented families. The presence of cutaneous hypopigmentation and nystagmus should lead directly to an ophthalmologic examination. Mothers of boys with cutaneous pigment and the ocular features of albinism should be examined for pigmentary mosaicism, seen in the obligate carrier of OA1; the mosaicism, called Lyonization in X-linked disorders, represents the expression of both the normal and mutated X chromosomes in the retina. Further diagnostic testing with next generation sequencing may be performed to confirm the type of albinism (Summers et al. 1996; Gronskov et al. 2009).

An eye examination by an experienced eye care professional establishes the diagnosis in individuals with nystagmus. An occasional child presents with nystagmus, moderate cutaneous hypopigmentation, and a blunted fovea, and in this case, a visual evoked potential study demonstrating the misrouting of the optic nerves at the chiasm can be performed to support the diagnosis of albinism. Molecular testing may also be performed. Brain imaging is not indicated in the routine evaluation of a child with albinism. The electroretinogram and routine color vision testing are normal in albinism, in contrast to retinal dystrophies that can also show an immature fovea. Hearing is normal in all types of albinism (Table 4.1). Although the melanocyte is found in the inner ear, loss of melanin in the melanocyte does not appear to alter hearing. This is in contrast to conditions in which deafness is associated with a lack of melanocytes in the ear (e.g., Waardenburg syndrome). Albinism and deafness have been reported to occur together, but the number of cases is limited and the diagnosis of albinism rather than hypopigmentation is not always compelling or there appears to be coexpression of two separate traits in the family (Ziprkowski et al. 1962; Lezirovitz et al. 2006).

Quantitative studies of hair or skin melanin pigment are not indicated in the evaluation of an individual with albinism. Hairbulb studies of melanin formation (hairbulb incubation test) or tyrosinase activity (hairbulb tyrosinase assay) are no longer performed because of their imprecision.

A history of bleeding or pulmonary fibrosis in an individual with nystagmus, foveal hypoplasia, reduced visual acuity, and strabismus suggests Hermansky–Pudlak syndrome.

Platelet function studies in persons with HPS indicate a storage pool deficiency with abnormal secondary aggregation, and whole mount electron microscopy shows an absence of dense bodies (this is the best single test for HPS) (Gerritsen et al. 1979; Witkop et al. 1987). The recent identification of at least ten genes associated with HPS has raised the question of routine platelet testing for all individuals with albinism, but this is not indicated at the present time (Huizing and Gahl 2002; Huizing et al. 2008). HPS is rare except in Puerto Rico, and testing for HPS is not generally available except in academic centers. The vast majority of persons with albinism do not have HPS; however, any child or adult with OCA and an unusual bleeding history (e.g., recurrent epistaxis, hematuria, or excessive bruising) should be evaluated for HPS, and the diagnosis should be considered in any adult with albinism and pulmonary fibrosis or inflammatory bowel disease (Brantly et al. 2000; Pierson et al. 2006; Grucela et al. 2006). Initial testing involves electron microscopy of platelets, and if an absence of dense bodies is noted, molecular diagnosis is performed to identify the type of HPS as associated systemic manifestations vary, and some individuals require additional evaluation for gastrointestinal and pulmonary manifestations of the disorder (Huizing et al. 2017).

The differentiation between the various types of OCA and HPS can be suggested by the history of cutaneous pigmentation, but molecular studies are necessary to accurately define the type except in certain circumstances (King et al. 2003a; Hutton and Spritz 2008, 2008b). Those with lifelong absence of melanin pigment have OCA1A, and males with relatively normal cutaneous pigment but the ocular features of albinism, combined with a history of X-linked inheritance and/or the finding of pigmentary mosaicism in the retina of the mother, have OA1. The molecular diagnosis of OCA1 involves testing of the tyrosinase gene, *TYR* (chromosomal locus 11q14-q21), which is clinically available.

The gene for OCA2 (*OCA2*), previously called the *P* gene, is the only gene known to be associated with OCA2. OCA2 appears to function as a chloride channel. Preliminary data suggest that it has a role in modulating melanosomal pH. Sequence analysis and testing for the 2.7 kb deletion found in individuals of African heritage are available on a clinical basis. No functional assays are available. An understanding of the phenotypic range of OCA2 has come from the recent molecular studies of *OCA2* on chromosome 15q. Associated presentations include the classic "tyrosinase-positive" OCA, brown OCA, and the hypopigmentation found in Prader–Willi syndrome (see Chapter 46) and Angelman syndrome (see Chapter 5).

OCA3 is caused by mutations in the *TYRP1* gene (tyrosinase-related protein 1, at chromosome 9q23) (Boissy et al. 1996; Sarangarajan and Boissy 2001). OCA3 was originally described in the African and African-American individuals and has been described infrequently in individuals from other areas (Boissy et al. 1996; Manga et al. 1997; Forshew et al. 2005; Rooryck et al. 2006; Hutton and Spritz 2008a). Linkage studies mapped "rufous" or "red oculocutaneous albinism" to the *TYRP1* locus in the South African population, and a common deletion mutation (1104delA) of the *TYRP1* gene was found (Manga et al. 1997).

OCA4 is caused by mutations in the *SLC45A2* gene (formerly known as *MATP* gene, encoding membrane-associated transporter protein on chromosome 5p) (Newton et al. 2001). Molecular testing for OCA types 3 and 4 is available.

OCA5 does not have an associated gene. It was reported based on a single consanguineous family (Kauser et al. 2013).

OCA6 is a hypothesized albinism type associated with the gene *SLC24A5* (Wei et al. 2013). The SLC24A5 gene product has been convincingly shown to be associated with pigment variation in several model organisms. As a result, it has frequently been put forward as a candidate gene in individuals with albinism. Sequencing is clinically available, usually as part of comprehensive albinism-gene-sequencing panels.

OCA7 was described in a consanguineous family and five additional unrelated individuals from the Faroe Islands (Gronskov et al. 2013). Due to the small number of identified individuals, the phenotypic spectrum is difficult to document with certainty. Sequencing is clinically available, usually as part of comprehensive albinism-gene-sequencing panels.

There are at least ten types of HPS, and each type is associated with mutations in a gene encoding a protein involved in lysosomal-related organelle biogenesis (Huizing et al. 2008). These proteins form complexes known as biogenesis of lysosomal-related organelle complex 1, 2, and 3 (BLOC 1, 2, and 3), and the clinical presentation varies with the BLOC that is dysfunctional. Molecular genetic testing of all HPS genes is clinically available, usually as a part of comprehensive albinism-gene-sequencing panels. For HPS1, the *HPS1* gene (chromosomal locus 10q23.1-q23.3, BLOC 3) has common mutations in individuals of Puerto Rican ancestry. Homozygosity for a 3.9 kb deletion has been identified in these individuals (Anikster et al. 2001). Molecular genetic testing for a 1303 + 1G > A splice site mutation in *HPS3* (BLOC 2) is also available on a clinical basis. Homozygosity for this mutation has been identified in individuals of Ashkenazi Jewish ancestry only (Huizing et al. 2001).

Testing for changes in the *GPR143* (formerly *OA1*) gene, located at Xp22.3, is also available clinically. These genes are summarized in Table 4.1. Males with nystagmus but without the ocular findings of albinism may also be tested for changes in the *FRMD7* gene to establish the correct diagnosis (Tarpey et al. 2006).

Prenatal diagnosis of OCA, OA1, or HPS by molecular genetic testing is possible in at-risk pregnancies when the gene and the disease-causing mutation has been identified in an affected family member. This can be done using DNA extracted from cells obtained from chorionic villus sampling or amniocentesis. Pre-implantation genetics is also available.

Differential Diagnosis

An individual who is developmentally and physically normal except for cutaneous hypopigmentation, nystagmus, foveal hypoplasia, reduced visual acuity, and strabismus has albinism, and there are no other diagnostic possibilities.

Accurate clinical diagnosis of OCA or OA1 versus "normal" in a young lightly pigmented male in a family with no history of albinism can be difficult if the family constitutional pigmentation is light and includes "tow-headed" children who do not have albinism. Examination of a heterozygous mother may show a variegated retinal pigment pattern, called pigmentary mosaicism (present in approximately 80% of heterozygous females), and a skin biopsy of a person with OA1 will show macromelanosomes (Gillespie and Covelli 1963; Charles et al. 1992, 1993); however, macromelanosomes may be difficult to identify on routine skin biopsy specimens and their absence can only be confirmed by a dermatopathologist.

Nystagmus also occurs because of sensory disorders such as optic nerve hypoplasia or optic atrophy, or motor abnormalities. Careful ophthalmic examination of normal ocular pigment eliminates the diagnosis of albinism. Foveal hypoplasia seen in retinal dystrophies is not associated with deficient ocular pigment, and diagnosis can be confirmed by an abnormal electroretinogram. Those with foveal hypoplasia type 1 (*PAX6* gene at 11p13) or type 2 (*SLC38A8* gene at 16q.23.3) may also have nystagmus and foveal hypoplasia. Nystagmus and foveal hypoplasia occur in aniridia, but the presence of only an iris remnant establishes the diagnosis. Very rarely, the obligate carrier of OA1 shows reduced vision and nystagmus. Other ocular causes of photosensitivity include corneal opacities, glaucoma, and retinal dystrophies.

MANIFESTATIONS AND MANAGEMENT

Growth and Feeding

There are no growth or feeding abnormalities associated with albinism.

Development and Behavior

Development generally follow normal patterns in all of the types of albinism listed in Table 4.1 (Kutzbach et al. 2007, 2008). Some gross- and fine-motor milestones may be mildly and temporarily delayed due to visual impairment, e.g. navigation of stairs. Visual development may be delayed. Rare individuals with albinism associated with more severe and/or persistent developmental abnormalities have been reported but are thought to represent a chance combination of albinism with other conditions (Lezirovitz et al. 2006; Reich et al. 2008). Attention deficit/hyperactivity disorder (ADHD) is more frequent in individuals with albinism than in the general population, unrelated to the extent of visual impairment (Kutzbach et al. 2007). With an increased size of font, either with enlargement of print or electronic increase in font size, reading skills are acquired, though reading fluency may be reduced (Merrill et al. 2011; MacDonald et al. 2012). Reading may be more difficult because of eye fatigue if print size is small. Children and adults with albinism, particularly those with OCA1A who have severe reduction in pigmentation, can develop a sense of isolation because of their appearance, and counseling may be indicated to overcome this.

Despite a reduction in visual acuity, individuals with albinism usually show normal educational achievement and career choice is only rarely limited by a requirement for normal vision.

Evaluation

- Reading acuity with different sizes of print should be assessed in school-aged children with albinism.
- Evaluation for attention deficit/hyperactivity disorder should take place if relevant symptoms are present by history.
- The physician should seek evidence of a sense of isolation in children with OCA.

Management

- For developmental and school-related adaptations related to visual problems, see below, **Ophthalmologic**.
- Treatment for attention deficit/hyperactivity disorder is standard.
- Psychological counseling may be beneficial for those with a sense of isolation.

An individualized education plan (IEP) can help to manage educational expectations (see School Kit at http://www.albinism.org).

Dermatologic

The range of phenotypes in OCA1 is continuous from total absence (OCA1A) to varying amounts of cutaneous pigmentation (OCA1B) that can appear nearly normal. Classic OCA1A with life-long white hair, white skin, and blue eyes is widely recognized in all populations and described throughout recorded human history because of the obvious absence of melanin (Garrod 1908a; Pearson et al. 1911; Taylor 1978). Although this is the most obvious type of oculocutaneous OCA, recent studies have demonstrated that there is a wide phenotypic range in OCA associated with tyrosinase gene mutations (King et al. 2003a; Hutton and Spritz 2008a, 2008b). This has become apparent from molecular studies demonstrating that several types of OCA previously described as separate entities (e.g., yellow albinism, minimal pigment albinism, and platinum albinism) are, in fact, related to mutations of the tyrosinase gene and represent components of a phenotypically broad spectrum. The major determinant in the OCA1 phenotype is the amount or type of residual activity of the enzyme produced by the mutant allele(s), but the constitutional pigment background of the affected individual also plays a role (King et al. 2003b).

Individuals with albinism who are diagnosed at birth are often recognized due the presence of marked hypopigmentation (King et al. 2003a). The hair color in any newborn child with scant scalp hair can be difficult to determine because very lightly pigmented (light yellow or blond) hair is often called "white" if most members of a family have dark hair. This even includes the term "tow-headed," an old English term that refers to flaxen-colored or very light blond, almost white, hair present in many northern European individuals at birth.

Individuals with OCA1A are born with true white hair and skin and blue eyes, and there is no change as they mature. The phenotype is the same in all ethnic groups and at all ages. No pigmented lesions develop in the skin, although amelanotic nevi can be present. Hypomorphic mutations of the tyrosinase gene that result in some residual enzyme activity produce OCA1B. The amount of residual enzyme function varies but is sufficient to produce variable amounts of iris, hair, and skin pigment (Giebel et al. 1991; Fukai et al. 1995; Spritz et al. 1997; Matsunaga et al. 1999). Hair color usually turns from white or nearly white at birth to varying shades of golden blond indicating that the major melanin being synthesized is pheomelanin, but some individuals eventually develop brown hair. Skin color remains white in many with OCA1, and a definite tan with sun exposure can develop in those with OCA1B. The ability to tan correlates with the amount of hair pigment that develops and with the pigment pattern in the family.

OCA2 is the second common type of OCA. The phenotypic feature of OCA2 that usually separates this from OCA1 is the common presence of pigmented hair at birth. Individuals with OCA2 often develop localized skin pigment (nevi, freckles, and lentigines), often in sun-exposed regions of the skin, and many accumulate pigment in their hair and eyes during their lifetime. OCA2 in Caucasian individuals usually presents with blond hair, white skin, and blue or lightly pigmented irides. The hair can be very lightly pigmented at birth, having a light yellow/blond or nearly white color, or be more pigmented with a definite blond or golden blond color, and the OCA2 phenotype can be indistinguishable from OCA1B. Red hair has been described with OCA2 (King et al. 2003b). OCA2 is reported in other populations (Suzuki and Tomita, 2008).

OCA2 in African-American and African individuals usually presents with yellow hair at birth that remains yellow through life, although the color may darken. Pigmented nevi, lentigines, and freckles develop, but a tan does not develop with exposure to the sun (Kromberg et al. 1989; Bothwell, 1997). African-American and African individuals may also have brown skin and hair, in a pattern that is similar to a type of OCA originally described in the African population as brown OCA (King et al. 1985; King and Rich 1986; Manga et al. 2001). With the brown OCA phenotype in African and African-American individuals, the hair and skin color are light brown and the irides are gray to tan at birth.

The phenotype of OCA3 in South African individuals includes red or reddish brown skin, ginger or reddish hair, and hazel or brown irides (Kromberg et al. 1990; Manga et al. 1997). The phenotype of German and Pakistani individuals with OCA3 is similar to classic OCA2 (Forshew et al. 2005; Rooryck et al. 2006).

OCA4 was originally described in a single Turkish individual having a phenotype similar to a lightly pigmented individual with OCA2 (Newton et al. 2001). There were no defining characteristics, and pigmented hair and white skin were present. Subsequent studies have shown a diverse phenotype in Japanese individuals, where this type of oculocutaneous albinism is common (Inagaki et al. 2004, 2005, 2006; Suzuki and Tomita 2008) and infrequently in individuals from other areas (Rundshagen et al. 2004; Sengupta et al. 2007; Hutton and Spritz 2008b).

The hallmark of the cutaneous pigment phenotype in HPS is variability. Affected individuals can have marked cutaneous hypopigmentation similar to that of OCA1A, whereas others have white skin and yellow or blond hair similar to OCA1B or OCA2, and still others have only moderate cutaneous hypopigmentation suggesting OA rather than OCA. Variation is seen within as well as between families, and the within-family variation is more marked than in other types of albinism. Affected individuals in Puerto Rico have hair color that varies from white to yellow to brown (Witkop et al. 1989; Gahl et al. 1998; Shotelersuk and Gahl 1998; Huizing et al. 2008). Skin is white and definitely lighter than normally pigmented individuals in this population. Freckles can be present in the sun-exposed regions (face, neck, arms, and hands), and can coalesce into large areas that look like normal dark skin pigment, but tanning does not occur. To the contrary, the presence of OCA may not be obvious in an individual of Puerto Rican ancestry with brown hair, skin pigment in exposed areas, and brown irides unless the cutaneous pigmentation is compared with unaffected family members (who are generally more darkly pigmented) and the ocular features of albinism are recognized (Gahl et al. 1998; Shotelersuk et al. 1998; Huizing et al. 2008). Affected individuals have been identified in other populations, and the phenotype shows the same degree of variation in pigmentation as is found in Puerto Rico. Hair color varies from white to brown, and this correlates with the ethnic group. The skin is light and iris color varies from blue to pigmented.

Long-term (i.e. over many years) exposure to the sun of lightly pigmented skin can result in coarse, rough, thickened skin, solar keratoses (premalignant lesions), and skin cancer. Both basal cell carcinoma and squamous cell carcinoma can develop (Hong et al. 2006; Adegbidi et al. 2007). Skin cancer is unusual in individuals with OCA in North America because of the availability of sunscreens, the social acceptability of wearing clothes that cover most of the exposed skin, and the fact that individuals with albinism often do not spend a great deal of time outside in the sun. Skin cancer is common in other parts of the world, such as sub-Saharan Africa, where sun exposure is more constant and protective mechanisms are not traditionally used or are not easily available (Introne eet al. 2014). Although skin melanocytes are present in individuals with HPS, melanoma is rare.

Evaluation

- Careful annual skin examination for evidence of cancerous or precancerous skin lesions is essential.

Management

- Educational counseling of affected individuals concerning avoidance of sun exposure is extremely important. For all type of oculocutaneous albinism, the cutaneous hypopigmentation requires skin protection from ultraviolet radiation. Sun exposure as short as 5–10 minutes can be significant in very sensitive individuals, and exposure of 30 minutes or more is usually significant in less sensitive individuals.
- Physical methods, including long-sleeve shirts, long pants, and hats with a wide brim, are excellent for avoidance of sun exposure, but these are often under used because of fashion or age.
- Sunscreens are effective in protecting the skin and should be used whenever possible. The sun protection factor (SPF) rating of a sunscreen should be greater than 30 for good protection and the sunscreen labeling should include the words "broad spectrum" (a US Food and Drug Administration term indicating protection from both UVA and UVB light). For those with very sensitive skin, total blocks with sun protection factor values of 50+ are appropriate.
- Latitude is important in ultraviolet exposure, and an individual in New Jersey can tolerate more sun than one in Florida. Sand reflects ultraviolet rays, and it is possible to become burned when sitting in the shade on a beach. The greatest intensity of ultraviolet light occurs at the summer solstice and between the hours of 10 am and 2 pm standard time, and protection or avoidance of the sun in these periods can greatly reduce ultraviolet exposure.
- Treatment of skin cancer does not differ from that in the general population.

Ophthalmologic

The ocular features of albinism are always present and help identify an individual as having albinism, even when the cutaneous pigment appears normal for ethnic background. For all types of albinism, conjugate pendular nystagmus develops within a few weeks after birth and may have a large amplitude that diminishes over time. As persons with albinism mature, they may learn to use a head posture to damp the nystagmus and improve vision. It is rare for a person with albinism to not have nystagmus. Alternating strabismus and absent stereoacuity are found in most individuals with albinism. Strabismus is rarely associated with the development of amblyopia. A small group of individuals with albinism and more macular melanin pigment and rudimentary foveal development have some degree of stereovision (Lee et al. 2001).

Irides vary in color, but are often blue, and iris transillumination can be detected and graded with slit-lamp biomicroscopy (Summers et al. 1988). Examination of the fundus invariably shows foveal hypoplasia, with only an occasional rudimentary annular reflex being present. Many individuals are photosensitive, related in part to reduced ocular pigment. Careful inspection with indirect ophthalmoscopy may show finely granular gray-black pigment in the macula, but most often no melanin pigment is found throughout the fundus. When present, vision is relatively better. Optical coherence tomography, which determines foveal thickness, may be a more sensitive approach to evaluation of foveal hypoplasia in albinism, in comparison with macular transparency (Seo et al. 2007; Harvey et al. 2008, Mohammad et al. 2011; Sepulveda-Vazquez et al. 2014). Individuals with albinism typically have reduced vision, often from almost normal to legal blindness with best-corrected visual acuity less than 20/200. Visual acuity has rarely been reported to be better than 20/40 and can vary between affected family members (Castronuovo et al. 1991; Summers, 1996; Lee et al. 2001). Reduced visual acuity significantly impacts quality of life in albinism (Kutzbach et al. 2009c). In young children, grating acuity determined using the Teller acuity cards (gray cards that have a square box of high contrast vertical stripes on one end, presented as cards with progressively smaller stripes until the child no longer preferentially looks toward them as they can no longer be visually resolved) is reduced at ages one, two, and three (Whang et al. 2002) and does not correlate well with eventual letter acuity (Louwagie et al. 2006). High refractive errors are common and require early correction.

In OCA1A, the irides are translucent and appear pink or red in ambient light early in life and usually become a darker blue or a gray-blue color with time, but no melanin accumulates in the eye. In OCA1B, the irides can develop light tan/golden pigment, particularly in the inner third of the iris, and melanin pigment in the posterior iris epithelium is present with transillumination.

In OCA2–4 and HPS1–10, the irides may be transparent or pigmented at birth or early in life. All of the ocular features of albinism are not always present in OCA3. The ocular features of OCA4 are similar to other pigmenting types of OCA (Inagaki et al. 2004, 2005; Suzuki and Tomita 2008).

The pigmentation phenotype of OCA1 and OCA2 is usually similar among siblings (Preising et al. 2007). There may be small differences in degrees of hair pigmentation, sun sensitivity, and ocular features such as nystagmus, but visual acuity does not usually vary markedly between affected siblings (Heinmiller et al. 2016).

The eye changes of Hermansky–Pudlak syndrome are identical to those found in other types of albinism. The ocular pigment phenotype in HPS is extremely variable.

The clinical features of OA1 include a male with normal cutaneous pigmentation and the ocular features of albinism (Charles et al. 1993).

Evaluation

- It is recommended that all individuals with albinism have an ophthalmologic examination at diagnosis and be followed to maintain appropriate refractive correction. Once in school, annual examinations, including assessment of refractive error, are needed.

Management

- Correction of refractive errors can improve visual acuity, and glasses and/or contact lenses are required in most cases to improve vision and binocular alignment (Anderson et al. 2004).
- As the child matures, low vision aids, including bifocals, magnifiers, and telescopes, may be beneficial (Collins and Silver 1990). Aids such as handheld magnifying devices or bioptic lenses are helpful adjuncts in the care of visually impaired individuals with albinism. Enlargement of font size on electronic media is often preferred.
- In school, children should receive preferential seating at the front of the classroom, and they frequently require copies of board work or overhead projections at their desk. When smart boards are used, content can be synced with an electronic device on the child's desk.
- Affected individuals can hold materials closely to see them without hurting the eyes.
- A vision consultant or teacher for the visually impaired can evaluate the needs of an affected child in the classroom. Vision therapy and eye exercises have not been shown to be beneficial to educational performance or visual acuity.
- Because altered head posture can be used to damp nystagmus and improve vision, it should not be discouraged. Occasionally, altered head posture is severe enough to warrant consideration of eye muscle surgery to change head posture and alter the characteristics of the nystagmus.
- In some, a modest improvement in visual acuity has been documented following extraocular muscle surgery for nystagmus (Helveston et al. 1991; Egbert et al. 1995). Strabismus surgery is usually not required but can be considered to restore alignment if the strabismus is marked.
- Because of their photosensitivity, many individuals with albinism prefer dark sunglasses, photochromic lenses, or light-shielded spectacles, in addition to a cap or hat with a large brim or visor to shield them from sunlight. However, some prefer to go without dark glasses because they report that they reduce vision.
- Opaque contact lenses have been used to reduce light exposure but are usually unsuccessful, probably because of excessive light transmission through the sclera, choroid, and retina. Studies of the use of yellow filters for photosensitivity and brightness enhancement have not been shown to be helpful, although occasionally an individual may prefer them. Prosthetic iris implants have been suggested to manage iris translucency, iris glare, and photophobia; controlled studies of this approach are not available, but they likely fail to completely reduce symptoms, similar to colored contact lenses (Karatza et al. 2007).
- Some affected people will be able to obtain a driver's license, with vision requirements varying by location-specific regulations. Bioptics are permitted in some locations. Restrictions to licenses often include a reduction in maximum speed and use of corrective lenses, but limitation to daylight driving is not necessary.

Hematologic

The bleeding diathesis in HPS results from a deficiency of storage granules in the platelets (i.e. storage pool-deficient platelets). Storage granules or dense bodies are absent, and this is associated with a deficiency of serotonin, adenine nucleotides, and calcium in the platelet (White et al. 1973; Rao et al. 1974; Gerritsen et al. 1977, 1979). As a result, platelets in HPS do not show irreversible secondary aggregation when stimulated with agents that normally produce this response. This deficiency produces mild hemorrhagic episodes in many affected individuals, including easy bruisability, epistaxis, hemoptysis, gingival bleeding with brushing or dental extraction, prolonged bleeding after circumcision, and postpartum bleeding (Gahl et al. 1998). Serious bleeding can occur and can be life-threatening. Typically, cuts bleed longer than usual but heal normally. Bruising generally first occurs at the time of ambulation. Epistaxis occurs in childhood and diminishes after adolescence. Menstrual cycles may be heavy and irregular.

Individuals with colitis may bleed excessively per rectum. In rare cases, childbirth, trauma or surgery can precipitate exsanguination.

Evaluation

- The sine qua non for diagnosis of HPS is absence of dense bodies on whole-mount electron microscopy of platelets (Witkop et al. 1987). Platelet dense bodies, which contain ADP, ATP, serotonin, calcium, and phosphate, release their contents upon stimulation to attract other platelets. This process constitutes the secondary aggregation response, which cannot occur in the absence of the dense bodies. There are normally 4–8 dense bodies per platelet, but none in individuals with HPS.
- Coagulation studies reveal that the secondary aggregation response of platelets is impaired and the bleeding time is generally prolonged.
- Coagulation factor activity and platelet counts are normal.
- OA1 and OCA not associated with HPS have normal function of platelet dense bodies and therefore do not require coagulation studies.

Management

- Platelets in an individual with HPS do not self-activate normally, but can be successfully activated by factors released from transfused platelets. Therefore, platelet transfusions tend to be highly effective if warrented by severe bleeding.
- Management includes prevention of bleeding and prompt treatment of bleeding.
- Prevention relies on avoidance of all aspirin-containing products and conducting life in a manner designed to minimize the chance of a bleeding episode.
- Humidifiers may reduce the frequency of nosebleeds.
- Birth control pills can limit the duration of menstrual periods.
- Treatment of minor cuts includes placing thrombin-soaked gelfoam over an open wound that fails to clot spontaneously.
- For more invasive trauma, such as wisdom tooth extraction, 1-desamino-8-D-arginine vasopressin (DDAVP, 0.2 mcg kg^{-1} in 50 mL of normal saline) can be given as a 30 minute intravenous infusion just before the procedure.
- For extensive surgeries or protracted bleeding, platelet or red blood cell transfusions may be required, with periodic check of coagulation studies.
- Individuals with HPS should consider wearing a medical alert bracelet that explicitly describes the functional platelet defect, because the standard tests for bleeding dysfunction (platelet count, prothrombin time, and partial thromboplastin time) are normal in HPS.

Pulmonic

The third component of the HPS triad (after OCA and bleeding diathesis) is the production of ceroid, a yellow waxy material found in urine of affected individuals, and present in many tissues throughout the body when analyzed at autopsy (Ohbayashi et al. 1995; Sakuma et al. 1995). The accumulation of ceroid in the lungs and gastrointestinal tract is associated with the clinical manifestations involving these tissues, but the extent to which this occurs varies by type of HPS.

Interstitial pulmonary fibrosis (Garay et al. 1979; Pierson et al. 2006) causes progressive restrictive lung disease with a highly variable course that typically produces symptoms in the early thirties and progresses to death within a decade (Gahl et al. 1998). Studies have shown that the lamellar bodies in the type 2 pneumocytes, containing surfactant, have a similar origin to platelet dense bodies, and this may be related to the pulmonary fibrosis seen in HPS (Nakatani et al. 2000; Lyerla et al. 2003).

Evaluation

- Pulmonary fibrosis occurs more frequently in some types of HPS (1, 2, and 4) than in others. Therefore, risk counseling benefits from a molecular determination of the HPS type.
- Careful history and physical examination relative to the pulmonary system should be part of every routine or sick visit to the care provider.
- Evaluation does not differ from that of suspected or diagnosed pulmonary fibrosis in the general population.

Management

- Before the development of pulmonary fibrosis, attention should be paid to maximizing pulmonary function. This entails avoiding cigarette smoke, prompt treatment of pulmonary infections, immunization with influenza and pneumococcal vaccines, and engaging in routine moderate exercise.
- No successful therapy for or prophylaxis against the pulmonary fibrosis of HPS exists at this time, with the exception of lung transplantation (Lederer et al. 2005); the risk of bleeding can be significant with major surgery. Newer drugs to halt or reverse pulmonary fibrosis are being evaluated (Vicary et al. 2016). A study of pirfenidone showed no effect compared to controls (O'Brien et al. 2011).
- Steroids are often tried, but with no apparent beneficial effect.
- When the pulmonary disease becomes severe, oxygen therapy can be palliative.

Gastrointestinal

The gastrointestinal changes of HPS are among the most severe clinical manifestations. The development of granulomatous colitis, presenting with abdominal pain and bloody diarrhea in a child or an adult, has been described in many individuals with HPS (Schinella et al. 1980; Sherman et al. 1989; Mahadeo et al. 1991; Grucela et al. 2006) and may be related to the ceroid present in the tissues. A bleeding granulomatous colitis resembling Crohn disease presents, on average, at 15 years of age, with wide variability (Schinella et al. 1980; Gahl et al. 1998). The colitis is severe in 15% of individuals. The etiology of colitis is unknown, and immunologic studies do not show an abnormality.

Evaluation

- History of gastrointestinal symptoms or rectal bleeding should be sought at each medical visit.
- Evaluation of these symptoms is not different from that in the general population when colitis is suspected.

Management

- Treatment of the colitis does not differ from that in the general population.
- The colitis may respond to steroids and other anti-inflammatory agents.
- When severe, colectomy is occasionally required.

Miscellaneous

Cardiomyopathy and renal failure have also been reported to result from ceroid production in individuals with HPS (Witkop et al. 1989), although renal and cardiac function are usually normal (Ohbayashi et al. 1995).

Evaluation

- If cardiac or renal symptoms are noted, referral to cardiology or nephrology is recommended.

Management

- Treatment of cardiomyopathy and renal failure are not different from that in the general population.

RESOURCES

Support Groups

The National Organization of Albinism and Hypopigmentation (NOAH)

PO Box 959
East Hampstead, New Hampshire 03826-0959, USA
Phone: (800) 473-2310
Fax: (603) 887-2310
Email: noah@albinism.org
Website: www.albinism.org

The Hermansky–Pudlak Syndrome Network

Website: www.hpsnetwork.org

The PanAmerican Society for Pigment Cell Research

Website: http://www.paspcr.org/

The World Albinism Alliance

Website: www.worldalbinism.org

REFERENCES

Adegbidi H, Yedomon H, Atadokpede F, Balley-Pognon MC, do Ango-Padonou F (2007) Skin cancers at the National University Hospital of Cotonou from 1985 to 2004. *Int J Dermatol* 46 (Suppl 1):26–29.

Anderson J, Lavoie J, Merrill K, King RA, Summers CG (2004) Efficacy of spectacles in persons with albinism. *J AAPOS* 8:515–520.

Anikster Y, Huizing M, White J, Shevchenko YO, Fitzpatrick DL, Touchman JW, Compton JG, Bale SJ, Swank RT, Gahl WA, Toro JR (2001) Mutation of a new gene causes a unique form of Hermansky–Pudlak syndrome in a genetic isolate of central Puerto Rico. *Nat Genet* 28:376–380.

Barsh GS (1996) The genetics of pigmentation: From fancy genes to complex traits. *Trends Genet* 12:299–305.

Boissy RE, Zhao H, Oetting WS, Austin LM, Wildenberg SC, Boissy YL, Zhao Y, Strum RA, Hearing VJ, King RA, Nordlund JJ (1996) Mutation in and lack of expression of tyrosinase-related protein-1 (TRP-1) in melanocytes from an individual with brown oculcutaneous albinism: A new subtype of albinism classified as "OCA3". *Am J Hum Genet* 48:1145–1156.

Bothwell JE (1997) Pigmented skin lesions in tyrosinase-positive oculocutaneous albinos: A study in black South Africans. *Int J Dermatol* 36:831–836.

Brantly M, Avila NA, Shotelersuk V, Lucero C, Huizing M, Gahl WA (2000) Pulmonary function and high-resolution CT findings in patients with an inherited form of pulmonary fibrosis, Hermansky–Pudlak syndrome, due to mutations in HPS-1. *Chest* 117:129–136.

Brodsky MC, Fray KJ (2004) Positive angle kappa: a sign of albinism in patients with congenital nystagmus. *Am J Ophthalmol* 137:625–629.

Castronuovo S, Simon JW, Kandel GL, Morier A, Wolf B, Witkop CJ, Jenkins PL (1991) Variable expression of albinism within a single kindred. *Am J Ophthalmol* 111:419–426.

Charles SJ, Moore AT, Grant JW, Yates JRW (1992) Genetic counseling in X-linked ocular albinism. Clinical features of the carrier state. *Eye* 6:75–79.

Charles SJ, Green JS, Grant JW, Yates JRW, Moore AT (1993) Clinical features of affected males with Xlinked ocular albinism. *Br J Ophthalmol* 77:222–227.

Collins B, Silver J (1990) Recent experiences in the management of visual impairment in albinism. *Ophthalmic Paediatr Genet* 11:225–228.

dem Hagen EA (2007) Pigmentation predicts the shift in the line of decussation in humans with albinism. *Eur J Neurosci* 25:503–511.

Egbert JE, Anderson JH, Summers CG (1995) Increased duration of low retinal slip velocities following retroequatorial placement of horizontal recti. *J Pediatr Ophthalmol Strabismus* 32:359–363.

Forshew T, Khaliq S, Tee L, Smith U, Johnson CA, Mehdi SQ, Maher ER (2005) Identification of novel TYR and TYRP1 mutations in oculocutaneous albinism. *Clin Genet* 68:182–184.

Fukai K, Holmes SA, Lucchese NJ, Siu VM, Weleber RG, Schnur RE, Spritz RA (1995) Autosomal recessive ocular albinism associated with a functionally significant tyrosinase gene polymorphism. *Nat Genet* 9:92–95.

Gahl WA, Brantly M, Kaiser-Kupfer MI, Iwata F, Hazelwood S, Shotelersuk V, Duffy LF, Kuehl EM, Troendle J, Bernardini I (1998) Genetic defects and clinical characteristics of patients with a form of oculocutaneous albinism (Hermansky–Pudlak syndrome). *N Engl J Med* 338:1258–1264.

Gahl WA, Brantly M, Troendle J, Avila NA, Padua A, Montalvo C, Cardona H, Calis KA, Gochuico B (2002) Effect of pirfenidone on the pulmonary fibrosis of Hermansky–Pudlak syndrome. *Mol Genet Metab* 76:234–242.

Garay SM, Gardella JE, Fazzini EP, Goldring RM (1979) Hermansky–Pudlak syndrome. Pulmonary manifestations of a ceroid storage disorder. *Am J Med* 66:737–747.

Garrod AE (1908) Croonian lectures on inborn errors of metabolism. Lecture 1. *Lancet* 2:1–7.

Gerritsen SM, Akkerman JWN, Nijmeijer B, Sixma JJ, Witkop CJ, White J (1977) The Hermansky–Pudlak syndrome: Evidence for a lower 5-hydroxytryptamine content in platelets of heterozygotes. *Scand J Haematol* 18:249–256.

Gerritsen SM, Akkerman JWN, Staal G, Roelofsen B, Koster JF, Sixma JJ (1979) Biochemical studies in Hermansky–Pudlak syndrome. *Scand J Haematol* 23:161–168.

Giebel LB, Strunk KM, King RA, Hanifin JM, Spritz RA (1990) A frequent tyrosinase gene mutation in classic, tyrosinase-negative (type IA) oculocutaneous albinism. *Proc Natl Acad Sci USA* 87:3255–3258.

Giebel LB, Tripathi RK, Strunk KM, Hanifin JM, Jackson CE, King RA, Spritz RA (1991) Tyrosinase gene mutations associated with type IB ("yellow") oculocutaneous albinism. *Am J Hum Genet* 48:1159–1167.

Gillespie FD, Covelli B (1963) Carriers of ocular albinism with and without ocular changes. *Arch Ophthalmol* 70:209–213.

Gronskov K, Ek J, Sand A, Scheller R, Bygum A, Brixen K, Brondum-Nielsen K, Rosenberg T. (2009) Birth prevalence and mutation spectrum in Danish patients with autosomal recessive albinism. *Invest Ophthalmol Vis Sci* 50:1058–1064.

Gronskov, K. Dooley, C. M. Ostergaard, E. Kelsh, R. N. Hansen, L. Levesque, M. P. Vilhelmsen, K. Mollgard, K. Stemple, D. L. Rosenberg, T. (2013) Mutations in C10orf11, a melanocyte-differentiation gene, cause autosomal-recessive albinism. *Am. J. Hum. Genet.* 92: 415–421.

Grucela AL, Patel P, Goldstein E, Palmon R, Sachar DB, Steinhagen RM (2006) Granulomatous enterocolitis associated with Hermansky–Pudlak syndrome. *Am J Gastroenterol* 101:2090–2095.

Han J, Kraft P, Nan H, Guo Q, Chen C, Qureshi A, Hankinson SE, Hu FB, Duffy DL, Zhao ZZ, Martin NG, Montgomery GW, Hayward NK, Thomas G, Hoover RN, Chanock S, Hunter DJ (2008) A genome-wide association study identifies novel alleles associated with hair color and skin pigmentation. *PLoS Genet* 4:e1000074.

Harvey PS, King RA, Summers CG, (2006) Spectrum of foveal development in albinism detected with optical coherence tomography. *J AAPOS* 10:237–242.

Harvey PS, King RA, Summers CS (2008) Foveal depression and albinism. *Ophthalmology* 115:756–757.

Healy E, Jordan SA, Budd PS, Suffolk R, Rees JL, Jackson IJ (2001) Functional variation of MC1R alleles from red-haired individuals. *Hum Mol Genet* 10:2397–2402.

Heinmiller LJ, Holleschau A, Summers CG (2016) Concordance of visual and structural features between siblings with albinism. *J AAPOS.* 20:34–36.

Helveston EM, Ellis FD, Plager DA (1991) Large recession of the horizontal recti for treatment of nystagmus. *Ophthalmology* 98:1302–1305.

Hong ES, Zeeb H, Repacholi MH (2006) Albinism in Africa as a public health issue. *BMC Public Health* 6:212.

Huizing M, Gahl WA (2002) Disorders of vesicles of lysosomal lineage: The Hermansky–Pudlak syndromes. *Curr Mol Med* 2:451–467.

Huizing M, Anikster Y, Gahl WA (2000) Hermansky–Pudlak syndrome and related disorders of organelle formation. *Traffic* 1:823–835.

Huizing M, Anikster Y, Fitzpatrick DL, Jeong AB, D'Souza M, Rausche M, Toro JR, Kaiser-Kupfer MI, White JG, Gahl WA (2001) Hermansky–Pudlak syndrome type 3 in Ashkenazi Jews and other non-Puerto Rican patients with hypopigmentation and platelet storage-pool deficiency. *Am J Hum Genet* 69:1022–1032.

Huizing M, Helip-Wooley A, Westbroek W, Gunay-Aygun M, Gahl WA (2008) Disorders of lysosome-related organelle biogenesis: Clinical and molecular genetics. *Annu Rev Genomics Hum Genet* 9:359–386.

Huizing M, Malicdan MCV, Gochuico BR, Gahl WA. (2017) Hermansky–Pudlak Syndrome. In: Adam MP, Ardinger HH, Pagon RA, Wallace SE, Bean LJH, Mefford HC, Stephens K, Amemiya A, Ledbetter N, editors. GeneReviews® [Internet]. Seattle (WA): University of Washington, Seattle; 1993-2017.2000 Jul 24 [updated 2017 Oct 26].

Hutton SM, Spritz RA (2008a) A comprehensive genetic study of autosomal recessive ocular albinism in Caucasian patients. *Invest Ophthalmol Vis Sci* 49:868–872.

Hutton SM, Spritz RA (2008b) Comprehensive analysis of oculocutaneous albinism among non-Hispanic Caucasians shows that OCA1 is the most prevalent OCA type. *J Invest Dermatol* 128:2442–2450.

Inagaki K, Suzuki T, Shimizu H, Ishii N, Umezawa Y, Tada J, Kikuchi N, Takata M, Takamori K, Kishibe M, Tanaka M, Miyamura Y, Ito S, Tomita Y (2004) Oculocutaneous albinism type 4 is one of the most common types of albinism in Japan. *Am J Hum Genet* 74:466–471.

Inagaki K, Suzuki T, Ito S, Suzuki N, Fukai K, Horiuchi T, Tanaka T, Manabe E, Tomita Y (2005) OCA4: Evidence for a founder effect for the p.D157N mutation of the MATP gene in Japanese and Korean. *Pigment Cell Res* 18:385–388.

Introne W, Boissy RE. Gahl WA (1999) Clinical, molecular, and cell biological aspects of Chediak-Higashi syndrome. *Mol Genet Metab* 68:283–303.

Karatza EC, Burk SE, Snyder ME, Osher RH (2007) Outcomes of prosthetic iris implantation in patients with albinism. *J Cataract Refract Surg* 33:1763–1769.

Kausar T Bhatti, MA Ali M Shaikh, RS Ahmed ZM (2013) OCA5, a novel locus for non-syndromic oculocutaneous albinism, maps to chromosome 4q24. (Letter) *Clin. Genet.* 84: 91–93.

Kerr R, Stevens G, Manga P, Salm S, John P, Haw T, Ramsay M (2000) Identification of P gene mutations in individuals with oculocutaneous albinism in sub-Saharan Africa. *Hum Mutat* 15:166–172.

King RA, Rich SS (1986) Segregation analysis of brown oculocutaneous albinism. *Clin Genet* 29:496–501.

King RA, Creel DJ, Cervenka J, Okoro AN, Witkop CJ (1980) Albinism in Nigeria with delineation of new recessive oculocutaneous type. *Clin Genet* 17:259–270.

King RA, Lewis RA, Townsend D, Zelickson A, Olds DP, Brumbaugh JA (1985) Brown oculocutaneous albinism. Clinical, ophthalmological, and biochemical characterization. *Ophthalmology* 92:1496–1505.

King RA, Hearing VJ, Creel DJ, Oetting WS (2001) Albinism. In: *The Metabolic & Molecular Bases of Inherited Disease*, Scriver CR et al. 8th ed. New York: McGraw-Hill.

King RA, Pietsch J, Fryer JP, Savage S, Brott MJ, Russell-Eggitt I, Summers CG, Oetting WS (2003a) Tyrosinase gene mutations in oculocutaneous albinism 1 (OCA1): Definition of the phenotype. *Hum Genet* 113:502–513.

King RA, Pietsch J, Fryer JP, Savage S, Brott MJ, Russell-Eggitt I, Summers CG, Oetting WS (2003b) MC1R mutations modify the classic phenotype of oculocutaneous albinism type 2 (OCA2). *Am J Hum Genet* 73:638–645.

Kiprono SK, Chaula BM, Beltraminelli H (2014) Histological review of skin cancers in African Albinos: a 10-year perspective. *BMC Cancer* 14:157 doi: 10.1186/1471-2407-14-157.

Kromberg JGR, Castle D, Zwane EM, Jenkins T (1989) Albinism and skin cancer in Southern Africa. *Clin Genet* 36:43–52.

Kromberg JGR, Castle DJ, Zwane EM, Bothwell J, Kidson S, Bartel P, Phillips JI, Jenkins T (1990) Red or rufous albinism in Southern Africa. *Ophthalmic Pediatr Genet* 11:229–235.

Kushimoto T, Valencia JC, Costin GE, Toyofuku K, Watabe H, Yasumoto K, Rouzaud F, Vieira WD, Hearing VJ (2003) The Seiji memorial lecture: The melanosome: An ideal model to study cellular differentiation. *Pigment Cell Res* 16:237–244.

Kutzbach B, Summers CG, Holleschau AM, King RA, MacDonald JT (2007) The prevalence of attention-deficit/hyperactivity disorder among persons with albinism. *J Child Neurol* 22:1342–1347.

Kutzbach BR, Summers CG, Holleschau AM, MacDonald JT (2008) Neurodevelopment in children with albinism. *Ophthalmology* 115:1805–1808.

Kutzbach BR, Merrill K, Hogue K, Downes S, Holleschau AM, MacDonald T, Summers CG (2009) Evaluation of vision-specific quality of life in albinism. *J AAPOS* 13:191–195.

Lederer DJ, Kawut SM, Sonett JR, Vakiani E, Seward SL, Jr, White JG, Wilt JS, Marboe CC, Gahl WA, Arcasov SM (2005) Successful bilateral lung transplantation for pulmonary fibrosis associated with the Hermansky–Pudlak syndrome. *J Heart Lung Transplant* 24:1697–1699

Lee KA, King RA, Summers CG, (2001) Stereopsis in patients with albinism: Clinical correlates. *J AAPOS* 5:98–104.

Lezirovitz K, Nicastro FS, Pardono E, Abreu-Silva RS, Batissoco AC, Neustein I, Spinelli M, Mingroni-Netto RC, (2006) Is autosomal recessive deafness associated with oculocutaneous albinism a "coincidence syndrome"? *J Hum Genet* 51:716–720.

Louwagie CR, Jensen AA, Christoff A, Holleschau AM, King RA, Summers CG (2006) Correlation of grating acuity with letter recognition acuity in children with albinism. *J AAPOS* 10:168–172.

Lund PM, Maluleke TG, Gaigher I, Gaigher MJ (2007) Oculocutaneous albinism in a rural community of South Africa: A population genetic study. *Ann Hum Biol* 34:493–497.

Lyerla TA, Rusiniak ME, Borchers M, Jahreis G, Tan J, Ohtake P, Novak EK, Swank RT (2003) Aberrant lung structure, composition, and function in a murine model of Hermansky–Pudlak syndrome. *Am J Physiol Lung Cell Mol Physiol* 285: L643–L653.

MacDonald JT, Kutzbach BR, Holleschau AM, Wyckoff S, Summers CG (2012) Reading skills in children and adults with albinism—the role of visual impairment. *J Pediatr Ophthalmol Strabismus* 49:184–188.

Mahadeo R, Markowitz J, Fisher S, Daum F (1991) Hermansky–Pudlak syndrome with granulomatous colitis in children. *J Pediatr* 118:904–906.

Manga P, Kromberg JG, Box NF, Sturm RA, Jenkins T, Ramsay M (1997) Rufous oculocutaneous albinism in southern African blacks is caused by mutations in the TYRP1 gene. *Am J Hum Genet* 61:1095–1101.

Manga P, Kromberg JGR, Turner A, Jenkins T, Ramsay M (2001) In southern Africa, brown oculocutaneous albinism (BOCA) maps to the OCA2 locus on chromosome 15q: P-gene mutations identified. *Am J Hum Genet* 68:782–787.

Matsunaga J, Dakeishi-Hara M, Tanita M, Nindl M, Nagata Y, Nakamura E, Miyamura Y, Kikuchi K, Furue M, Tomita Y (1999) A splicing mutation of the tyrosinase gene causes yellow oculocutaneous albinism in a Japanese patient with a pigmented phenotype. *Dermatology* 199:124–129.

McAllister JT, Dubis AM, Tait DM, Ostler S, Rha J, Stepien KE, Summers CG, Carroll J (2010) Arrested development: High-resolution imaging of foveal morphology in albinism. *Vis Research* 50:810–817.

McCafferty BK, Summers CG, King RA, Wilk MA, McAllister JT, Stepien KE, Dubis AM, Brilliant MH, Anderson JL, Carroll J (2015) Clinical insights into foveal morphology in albinism. *J Pediatr Ophthalmol Strabismus.* 52:167–172.

Merrill, KS, Lavoie JD, King RA, Summers CG (2004) Positive angle kappa in albinism. *J AAPOS* 8:237–239.

Merrill K, Hogue K, Downes S, Holleschau AM, Kutzbach BR, MacDonald JT, Summers CG (2011) Reading acuity in albinism: Evaluation with MNREAD charts. *J AAPOS* 15:29–32.

Mohammad S, Gottlob I, Kumar A, Thomas M, Degg C, Sheth V, Proudlock FA (2011) The functional significance of foveal abnormalities in albinism measured using spectral-domain optical coherence tomography. *Ophthalmology* 118:1645–1652.

Nakatani Y, Nakamura N, Sano J, Inayama Y, Kawano N, Yamanaka S, Miyagi Y, Nagashima Y, Ohbayashi C, Mizushima M, Manabe T, Kuroda M, Yokoi T, Matsubara O (2000) Interstitial pneumonia in Hermansky–Pudlak syndrome: Significance of florid foamy swelling/degeneration (giant lamellar body degeneration) of type-2 pneumocytes. *Virchows Arch* 437:304–313.

Newton JM, Cohen-Barak O, Hagiwara N, Gardner JM, Davisson MT, King RA, Brilliant MH (2001) Mutations in the human orthologue of the mouse underwhite gene (uw) underlie a new form of oculocutaneous albinism, OCA4. *Am J Hum Genet* 69:981–988.

O'Brien K, Troendle J, Gochuico BR, Markello TC, Salas J, Cardona H, Yao J, Bernardini I, Hess R, Gahl WA (2011) Pirfenidone for the treatment of Hermansky–Pudlak syndrome pulmonary fibrosis. *Mol Genet Metab* 103:128–134.

O'Donnell FE Jr, King RA, Green WR, Witkop CJ Jr (1978) Autosomal recessively inherited ocular albinism: A new form of ocular albinism affecting females as severely as males. *Arch Ophthalmol* 96:1621–1625.

Ohbayashi C, Kanomata N, Imai Y, Ito H, Shimasaki H (1995) Hermansky–Pudlak syndrome: A case report with analysis of auto-fluorescent ceroid-like pigments. *Gerontology* 41 (Suppl 2):297–303.

Okoro AN (1975) Albinism in Nigeria: A clinical and social study. *Br J Dermatol* 92:485–492.

Pearson K, Nettleship E, Usher CH (1911) *A Monograph on Albinism in Man: Drapers' Company Research Memoirs, Biometric Series VI.* London Department of Applied Mathematics, Dulau and Co. Limited.

Pierson DM, Ionescu D, Qing G, Yonan AM, Parkinson K, Colby TC, Leslie K (2006) Pulmonary fibrosis in Hermansky–Pudlak syndrome. A case report and review. *Respiration* 73:382–395.

Preising MN, Forster H, Tan H, Lorenz B, de Jong PT, Plomp AS (2007) Mutation analysis in a family with oculocutaneous albinism manifesting in the same generation of three branches. *Mol Vis* 13:1851–1855.

Puri N, Durham-Pierre D, Aquaron R, Lund PM, King RA, Brilliant MH (1997) Type 2 oculocutaneous albinism (OCA2) in Zimbabwe and Cameroon: Distribution of the 2.7-kb deletion allele of the P gene. *Hum Genet* 100:651–656.

Rao GHR, White JG, Jachimowicz AA, Witkop CJ (1974) Nucleotide profiles of normal and abnormal platelets by high-pressure liquid chromatography. *J Lab Clin Med* 84:839–850.

Reich S, Keitzer R, Schmidt RE, Jacobs R, Varnholt V, Buck D, Herold R, Renz H (2008) Oculocutaneous albinism accompanied by minor morphologic stigmata and reduced number and function of NK cells. A new variant of NK cell defect? *Eur J Pediatr* 167:1175–1182.

Reis RA, Ventura AL, Kubrusly RC, de Mello MC, de Mello FG (2007) Dopaminergic signaling in the developing retina. *Brain Res Rev* 54:181–188.

Roma C, Ferrante P, Guardiola O, Ballabio A, Zollo M (2007) New mutations identified in the ocular albinism type 1 gene. *Gene* 402:20–27.

Rooryck C, Roudaut C, Robine E, Müsebeck J, Arveiler B (2006) Oculocutaneous albinism with TYRP1 gene mutations in a Caucasian patient. *Pigment Cell Res* 19:239–242.

Rosenberg T, Schwartz M (1998) X-linked ocular albinism: Prevalence and mutations—A national study. *Eur J Hum Genet* 6:570–577.

Rundshagen U, Zühlke C, Opitz S, Schwinger E, Käsmann-Kellner B (2004) Mutations in the MATP gene in five German patients affected by oculocutaneous albinism type 4. *Hum Mutat* 23:106–110.

Rymer J, Choh V, Bharadwaj S, Padmanabhan V, Modilevsky L, Jovanovich E, Yeh B, Zhang Z, Guan H, Payne W, Wildsoet CF (2007) The albino chick as a model for studying ocular developmental anomalies, including refractive errors, associated with albinism. *Exp Eye Res* 85:431–442.

Sakuma T, Monma N, Satodate R, Satoh T, Takeda R, Kuriya S-I (1995) Ceroid pigment deposition in circulating blood monocytes and T lymphocytes in Hermansky–Pudlak syndrome: An ultrastructural study. *Pathol Int* 45:866–870.

Sarangarajan R, Boissy RE (2001) Tyrp1 and oculocutaneous albinism type 3. *Pigment Cell Res* 14:437–444.

Schinella RA, Greco MA, Cobert BL, Denmark LW, Cox RP (1980) Hermansky–Pudlak syndrome with granulomatous colitis. *Ann Int Med* 92:20–23.

Schweikardt T, Olivares C, Solano F, Jaenicke E, García-Borron JC, Decker H (2007) A three-dimensional model of mammalian tyrosinase active site accounting for loss of function mutations. *Pigment Cell Res* 20:394–401.

Sengupta M, Chaki M, Arti N, Ray K (2007) SLC45A2 variations in Indian oculocutaneous albinism patients. *Mol Vis* 13:1406–1411.

Seo JH, Yu YS, Kim JH, Choung HK, Heo JW, Kim SJ (2007) Correlation of visual acuity with foveal hypoplasia grading by optical coherence tomography in albinism. *Ophthalmology* 114:1547–1551.

Sepulveda-Vazquez HE, Villanueva-Mendoza C, Zenteno JC, Villegas-Ruiz V, Pelcastre-Luna E, Garcia-Aguirre G (2014) Macular optical coherence tomography findings and GPR143 mutations in patients with ocular albinism. *Int Ophthalmol* 343:1075-1081.Sherman A, Genuth L, Hazzi CG, Balthazar EJ, Schinela RA (1989) Perirectal abscess in the Hermansky–Pudlak syndrome. *Am J Gastroenterol* 84:552–556.

Shotelersuk V, Gahl WA (1998) Hermansky–Pudlak syndrome: Models for intracellular vesicle formation. *Mol Genet Metab* 65:85–96.

Shotelersuk V, Hazelwood S, Larson D, Iwata F, Kaiser-Kupfer MI, Kuehl E, Bernardini I, Gahl WA (1998) Three new mutations in a gene causing Hermansky–Pudlak syndrome: Clinical correlations. *Mol Genet Metab* 64:99–107.

Spritz RA, Ho L, Furumura M, Hearing VJ (1997) Mutational analysis of copper binding by human tyrosinase. *J Invest Dermatol* 109:207–212.

Sturm RA (2006) A golden age of human pigmentation genetics. *Trends Genet* 22:464–468.

Summers CG, et al. (1988) Hermansky–Pudlak syndrome: Ophthalmic findings. *Ophthalmology* 95:545–554.

Summers CG (1996) Vision in albinism. *Trans Am Ophthalmol Soc* 94:1095–1155.

Summers CG, Oetting WS, King RA (1996) Diagnosis of oculocutaneous albinism with molecular analysis. *Am J Ophthalmol* 121:724–726.

Suzuki T, Tomita Y (2008) Recent advances in genetic analyses of oculocutaneous albinism types 2 and 4. *J Dermatol Sci* 51:1–9.

Tarpey P, Thomas S, Sarvananthan N, Mallya U, Lisgo S, Talbot CJ, Roberts, EO Awan M, Surendran M, McLean RJ, Reinecke RD, Langmann A, and 30 others. (2006) Mutations in FRMD7, a newly identified member of the FERM family, cause X-linked idiopathic congenital nystagmus. *Nature Genet*. 38: 1242–1244, 2006. Note: Erratum: (2011) *Nature Genet*. 43: 720 only.

Taylor WOG, (1978) Visual disabilities of oculocutaneous albinism and their alleviation. *Trans Ophthalmol Soc UK* 98:423–445.

Tey HL. (2010) A practical classification of childhood hypopigmentation disorders. *Acta Derm Venereol* 90:6–11.

Tomita Y, Miyamura Y, Kono M, Nakamura R, Matsunaga J (2000) Molecular bases of congenital hypopigmentary disorders in humans and oculocutaneous albinism 1 in Japan. *Pigment Cell Res* 13 (Suppl 8):130–134.

Vicary GW, Vergne Y, Santiago-Comier A, Young LR, Roman J (2016) Pulmonary fibrosis in Hermansky–Pudlak syndrome. *Ann Am Thorac Soc* 13:1839–1846.

Watabe H, Valencia JC, Yasumoto KI, Kushimoto T, Ando H, Vieira WD, Mizoguchi M, Appella E, Hearing VJ (2003) Regulation of tyrosinase processing and trafficking by organellar pH and by proteasome activity. *J Biol Chem* 279:7971–7981.

Wei A-H, Zang D-J, Zhang Z, Liu X-Z, He X. Yang L, Wang Y, Zhou Z-Y, Zhang M-R, Dai L-L, Yang X-M, Li W (2013) Exome sequencing identifies SLC24A5 as a candidate gene for nonsyndromic oculocutaneous albinism. *J. Invest. Derm* 133: 1834–1840.

Whang SJ, King RA, Summers CG (2002) Grating acuity in albinism in the first three years of life. *J AAPOS* 6:393–396.

White JG, Witkop CJ, Gerritsen SM (1973) The Hermansky–Pudlak syndrome: Ultrastructure of bone marrow macrophages. *Am J Pathol* 70:329–344.

Wilk MA, McAllister JT, Cooper RF, Dubis AM[4], Patitucci TN, Summerfelt P, Anderson JL, Stepien KE, Costakos DM, Connor TB Jr, Wirostko WJ, Chiang PW, Dubra A, Curcio CA, Brilliant MH, Summers CG, Carroll J (2014) Relationship between foveal cone specialization and pit morphology in albinism. *Invest Ophthalmol Vis Sci* 55:4186–4198.

Witkop CJ, Krumwiede M, Sedano H, White JG (1987) The reliability of absent platelet dense bodies as a diagnostic criterion for Hermansky–Pudlak syndrome. *Am J Hematol* 26:305–300.

Witkop CJ, Quevedo WC, Fitzpatrick TB, King RA, (1989) Albinism. In: *The Metabolic Basis of Inherited Diseases*, Scriver CR et al. 6th ed. New York: McGraw-Hill.

Woolf CM (1965) Albinism among Indians in Arizona and New Mexico. *Am J Hum Genet* 17:23–35.

Yi ZH, Garrison N, Cohen-Barak O, Karafet TM, King RA, Erickson RP, Hammer MF, Brilliant MH (2003) A 122.5-kilobase deletion of the P gene underlies the high prevalence of oculocutaneous albinism type 2 in the Navajo population. *Am J Hum Genet* 72:62–72.

Ziprkowski L, Krarowski A, Adam A, Costeff H, Sade J (1962) Partial albinism and deaf mutism. *Arch Dermatol* 86:530–539.

5

ANGELMAN SYNDROME

CHARLES A. WILLIAMS
Division of Genetics and Metabolism, Department of Pediatrics, University of Florida School of Medicine, Gainesville, Florida, USA

JENNIFER M. MUELLER-MATHEWS
Division of Genetics and Metabolism, Department of Pediatrics, University of North Carolina, Chapel Hill, North Carolina, USA

INTRODUCTION

Angelman syndrome (AS) is a genetic neurobehavioral condition characterized by severe to profound developmental delay, ataxic gait, absence of or severe truncation of speech, seizures, and spontaneous bouts of laughter or apparent happy facial grimacing (Angelman 1965). Since the syndrome's initial description, many affected individuals have been identified and the clinical manifestations are relatively well known (Bird 2014; Clayton-Smith and Laan 2003; Dagli et al. 2012).

Angelman syndrome has been reported throughout the world among divergent racial groups. Although the exact incidence of AS is unknown, an estimate of between 1 in 12,000 to 1 in 24,000 seems reasonable (Mertz et al. 2013).

Lifespan does not appear to be dramatically shortened. A 74-year-old individual with AS has been reported (Philippart and Minassian 2005), and many individuals in their fourth or fifth decades of life are known.

Diagnostic Criteria

Angelman syndrome is usually not recognized at birth or in early infancy because the developmental problems are nonspecific during this time. Parents may first suspect the diagnosis after learning about Angelman syndrome through social media web sites or by encountering another family having a child with the condition. The most common age of diagnosis is between two and five years of age, when the characteristic behaviors and features become most evident, particularly as walking begins. A summary of the developmental and clinical features has been published for the purpose of establishing clinical criteria to assist in making the diagnosis (Williams et al. 2006). These features are detailed in Tables 5.1 and 5.2. All of the features do not need to be present for the clinical diagnosis to be made.

Etiology, Pathogenesis, and Genetics

Angelman syndrome results from disruption of the function of *UBE3A*, encoding the E6-associated protein (E6AP) ubiquitin ligase protein (Kishino et al.1997). *UBE3A* maps to chromosome region 15q11.2-q13, and demonstrates imprinting only in neurons, by showing predominant maternal allele expression. However, the total neuron levels of E6AP may be normal (Hillman et al. 2017). There is biallelic expression in glia and in other somatic tissues.

There are four categories of genetic abnormalities that lead to disruption of *UBE3A* and cause Angelman syndrome:

- **Deletions of 15q11.2-q13** on the maternally derived chromosome (65–75% of cases). Those with this deletion have an approximate 5–7 Mb deletion. Chromosome microarray analysis can distinguish

Cassidy and Allanson's Management of Genetic Syndromes, Fourth Edition.
Edited by John C. Carey, Agatino Battaglia, David Viskochil, and Suzanne B. Cassidy.
© 2021 John Wiley & Sons, Inc. Published 2021 by John Wiley & Sons, Inc.

TABLE 5.1 Developmental history and laboratory findings in Angelman syndrome

Normal prenatal and birth history with normal head circumference; absence of major birth defects; developmental delay evident by 6–12 months of age

Delayed but forward progression of development (no loss of skills)

Normal metabolic, hematologic, and chemical laboratory profiles

Structurally normal brain (MRI or CT may have mild cortical atrophy or dysmyelination)

TABLE 5.2 Clinical features of Angelman syndrome

Consistent (100%)

Developmental delay, functionally severe

Speech impairment, none or minimal use of words; receptive and non-verbal communication skills higher than verbal ones

Movement or balance disorder, usually ataxia of gait and/or tremulous movement of limbs

Behavioral uniqueness: any combination of frequent laughter/smiling; apparent happy demeanor; easily excitable personality, often with hand flapping movements; hypermotoric behavior; short attention span

Frequent (>80%)

Delayed, disproportionate growth in head circumference, usually resulting in microcephaly (absolute or relative) by age 2

Seizures, onset usually <3 years of age

Abnormal electroencephalogram, characteristic pattern with largeamplitude slow-spike waves

Associated (20–80%)

Flat occiput; occipital groove

Tongue thrusting; suck/swallowing disorders

Feeding problems and/or muscle hypotonia in infancy

Prognathia

Wide mouth, wide-spaced teeth

Frequent drooling, protruding tongue

Excessive chewing/mouthing behaviors

Strabismus

Hypopigmented skin, light hair, and eye color compared with family in deletion-positive cases

Hyperactive lower extremities, deep tendon reflexes

Uplifted, flexed arm position especially during ambulation

Wide-based gait with out-going (pronated or valgus-positioned) ankles

Increased sensitivity to heat

Abnormal sleep-wake cycles and diminished need for sleep

Attraction to/fascination with water; and crinkly items such as certain papers and plastics

Abnormal food related behaviors

Obesity (in the older child; more common in nondeletion classes)

Scoliosis

Constipation

between the larger class I deletion (that includes four non-imprinted genes) and the smaller class II deletion. These two deletions comprise 90% of deletion cases (see Figure 5.1). The common deletion breakpoints causing Angelman syndrome can be the same breakpoints that cause Prader–Willi syndrome, when the deletion occurs on the paternally derived chromosome 15. In addition, there are rare families with unique chromosome 15 translocations or with smaller deletions within 15q11.2-q13 that disrupt *UBE3A* and cause Angelman syndrome. Stretches of repeated copies of non-functional genes and other genetic elements map to both the proximal and distal breakpoints, predisposing to unequal recombination and result in deletions, as well as duplications (duplications are not associated with an Angelman syndrome phenotype) (Sahoo et al. 2007). Figure 5.2 illustrates the appearance of individuals with genetically proven Angelman syndrome.

- **Paternal uniparental disomy** of chromosome 15 (3–7% of cases). Individuals with Angelman syndrome due to paternal uniparental disomy (UPD) may have a somewhat milder phenotype (i.e. lower incidence of seizures) than that observed in Angelman syndrome caused by other types of genetic mechanisms (Bird 2014; Lossie et al. 2001; Tan et al. 2011).

- **Imprinting defects** (3% of cases). This subset of individuals with Angelman syndrome have defects in the mechanism(s) involved in the imprinting process that is normally operative during gametogenesis and early embryogenesis. Defects in the imprinting center (IC), located within the promoter region of *SNRPN* (that maps centromeric to *UBE3A* within 15q11.2-q13) can perturb the normal DNA methylation of *SNRPN*. This can then, as a long range effect via antisense transcription (Meng et al. 2012), lead to impairment of *UBE3A* function. Microdeletions in the IC have been identified in some Angelman syndrome individuals, but the great majority of those who have imprinting defects, do not have any detectable DNA sequence anomaly in the IC (Buiting et al. 2003; Dagli et al. 2012).

- ***UBE3A* mutations** (5–11% of cases). Sequence analysis of individuals with Angelman syndrome who have intragenic mutations, reveals that the majority of mutations are protein-truncating (Lossie et al. 2001; Malzac et al. 1998), although many are missense. It is possible that individuals with mutations causing less functional impairment in *UBE3A* may show some, but not all, the clinical features associated with Angelman syndrome. A few individuals with Angelman syndrome have been found to have complete or relatively large partial deletions of *UBE3A*.

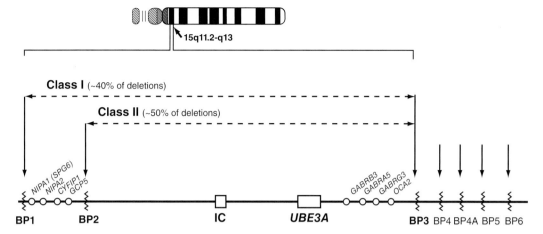

FIGURE 5.1 The chromosome deletion region, 15q11.2-q13, is illustrated with breakpoints (BP) indicated by arrows for the common class I and class II deletions. The Angelman gene, *UBE3A*, and the imprinting center (IC) is indicated by open boxes. Genes located within the class I deletion and between breakpoints BP1 and BP2 are listed. Other genes distal to *UBE3A* are listed and include *OCA2*. Less frequent breakpoints (BP4-BP6) are indicated. Typically, these less frequent breakpoints all begin at BP1 and extend to one of these distal regions. See text for details.

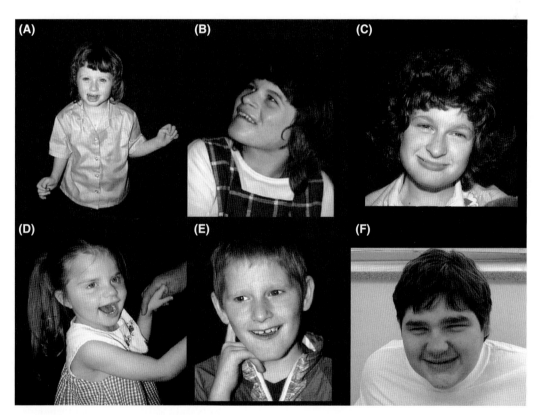

FIGURE 5.2 Individuals with Angelman syndrome: age, in years, is in parenthesis: A(7), B(16), C(29), D(5), E(7) and F(17); A-E have the common 15q11.2-q13 class I or II deletions; F has an imprinting defect.

Biopathogenesis

Most research focuses on the ubiquitin ligase function of *UBE3A*, although there is also a steroid receptor recognition functional site that is located within the gene (El Hokayem and Nawaz 2014). Isoforms of E6AP indicate both nuclear and cytosolic locations, probably indicating both a nuclear gene regulatory and a localized synaptic, post-translational role. Multiple protein targets have been identified that involve E6AP direct ubiquitination and there appears be many (perhaps hundreds) of additional proteins that are associated with and/or somehow affected by E6AP action. Actions of the currently known targeted proteins do not provide a very clear insight into the complex neurological phenotype of Angelman syndrome. Mouse models of Angelman syndrome recapitulate the core features of the human disorder including learning deficits, ataxia, seizures and the neuron-imprinting phenomenon. The mouse model has facilitated research on unsilencing of the paternal *UBE3A* allele, as well as research on vector-delivered gene insertion into neurons. In addition, human induced pluripotent stem cells (IPSCs) can be differentiated into neurons, enabling *UBE3A* and E6AP studies in these imprinted human cells. Current research into these areas of biopathogenesis is exciting, although a review of this is beyond the scope of this chapter (Fink et al. 2017; Galligan et al. 2015; Tan and Bird 2016).

Genotype–Phenotype Correlation

Individuals with 15q11.2-q13 deletions have a more pronounced phenotype, especially in terms of seizure severity and movement disorder. The mildest Angelman syndrome phenotype occurs in those with the non-deletion imprinting defect who show evidence of mosaicism on peripheral blood DNA methylation studies. Although there are somewhat subtle clinical differences among those with *UBE3A*, UPD, and imprinting defects, the core features of Angelman syndrome are present in all of the four genetic subtypes.

When Angelman syndrome is caused by large deletion, skin and eye hypopigmentation usually result. This occurs because a gene involved in melanogenesis, *OCA2*, is also deleted (see Figure 5.1) and E6AP (which is presumably in half dosage in all non-neuronal tissues) appears to act as an transcription factor in the regulation of melanogenesis (Toyofuku et al. 2002). The *OCA2* gene appears to be important in intracellular transport of tyrosinase and possibly other related proteins (Toyofuku et al. 2002). In some people with Angelman syndrome, this hypopigmentation can be so severe that a form of albinism is suspected. However, not all of those with deletions of *OCA2* are obviously hypopigmented; they may only have relatively lighter skin color than either parent.

Diagnostic Testing

Diagnostic testing can detect most causes of Angelman syndrome. No combination of testing can detect all causes.

- **DNA methylation**: Unaffected individuals have a methylated and an unmethylated *SNRPN* allele in both the Southern blot analysis (Glenn et al. 1996) and methylation-specific PCR assay. Individuals with AS caused by the large deletion of 15q11.2-q13, uniparental disomy, or an imprinting defect have only an unmethylated (i.e. "paternal") contribution (i.e. an abnormal parent-specific DNA methylation imprint). Although DNA methylation testing is the most sensitive single test for diagnosis of AS, it will not detect AS caused by *UBE3A* mutations. Also, methylation analysis cannot identify the molecular class (15q11.2-q13 deletion, uniparental disomy, or IC defect) to which the person with AS belongs, although newer test methods are beginning to address this matter. It will, however, diagnose approximately 80% of individuals with classic AS. A negative test does not exclude the diagnosis.

- **Chromosome microarray analysis**: This test method is currently commonly performed in the field of diagnostic clinical genetics, and it is now the best method to identify deletions at 15q11.2-q13. It has generally replaced the need to assay for the deletion using florescence in situ hybridization (FISH) probes. About 70% of those with AS will have a large deletion. Chromosome microarray analysis can also identify the size of the deletion to determine class I, class II, or other deletion types. Routine G-banded chromosome analysis will typically not detect the 15q deletion and will thus yield a false-negative result. Individuals suspected to have Angelman syndrome, but who have negative DNA methylation test and *UBE3A* mutations analysis, should still undergo a chromosome microarray study. This test might detect other non-Angelman syndrome causing deletions or duplications that mimic the clinical features of Angelman syndrome.

- **DNA polymorphisms**: Analysis of microsatellite DNA in the parents and affected individual can determine the parental origin of 15q11.2-q13 and the rest of chromosome 15. It can detect deletions and uniparental disomy, but it will not detect individuals with *UBE3A* mutations or those with non-deletion IC defects. Individuals with IC defects will have biparental inheritance of 15q11.2-q13 but abnormal DNA methylation.

- **Imprinting center analysis**: About 10–20% of imprinting defects are due to microdeletions (6–200 kb) that occur within the Angelman syndrome IC. The nature of the other 80–90% of imprinting defects is thought to be an epigenetic mutation (one affecting imprinting but not the DNA sequence), occurring during maternal oogenesis or in early embryogenesis (Buiting et al. 2003). Characterization of the imprinting defect as either an IC deletion or epigenetic defect is available in only a few clinical laboratories.

- **ature sequence analysis**: When the DNA methylation test is normal, *UBE3A* sequence analysis should be considered for individuals with clinical features of Angelman syndrome. A few individuals with Angelman syndrome have complete or partial deletions of *UBE3A*, or have intragenic deletions. Some chromosome microarray platforms may be able to detect some of these deletions.

The choice and sequence of laboratory testing will vary based on the experience of the physician, clinical presentation, availability of laboratory resources and other factors. Diagnostic testing approaches have been published (Ramsden et al. 2010). In general, the first line of testing for a suspected case is DNA methylation analysis. When an Angelman syndrome-specific methylation pattern is found, further evaluation such as a chromosome microarray or fluorescence in situ hybridization and/or DNA polymorphism analysis is required to identify the molecular class of Angelman syndrome (deletion, uniparental disomy, or IC defect). When a deletion is verified, no further diagnostic evaluation typically occurs. Additional study may be needed to provide genetic counseling (such as further study of the mother's chromosome 15 to rule out a cryptic rearrangement). Finally, there remains a group of individuals with the apparent clinical phenotype of AS that have negative genetic studies, including *UBE3A* mutation analysis. Accordingly, the clinician must be cautious about excluding the diagnosis of AS when genetic testing is negative but the clinical picture is compelling.

Genetic Counseling

Most cases of Angelman syndrome result from typical large de novo deletions of 15q11.2-q13 and are expected to have a low (<1%) risk of recurrence. AS caused by uniparental disomy that has occurred in the absence of a parental translocation, is likewise expected to have a <1% risk of recurrence. A mother carrying a structurally abnormal but functionally balanced chromosome complement can be at increased risk for having a child with a 15q11.2-q13 deletion or uniparental disomy (e.g., via generation of maternal gamete that was nullisomic for chromosome 15, with subsequent post-zygotic "correction" to paternal disomy). These situations will have case-specific recurrence risks. To clarify these recurrence risks, mothers of a child with a deletion should be offered the option of having a G-banded chromosome study with 15q11.2 FISH probes to determine if they carry a chromosomal rearrangement.

In instances where there is no identifiable large deletion or uniparental disomy, the risk for recurrence may be as high as 50% due to either a maternally inherited IC defect or a *UBE3A* mutation. Individuals with AS who have none of the above abnormalities comprise a significant proportion of cases (perhaps 5–10%) and some of these families may be at a 50% recurrence risk. Clinical misdiagnoses will also be represented in this group. The causal heterogeneity of Angelman syndrome, as well as the detection of maternal germ cell mosaicism (Malzac et al. 1998), can make it difficult to provide recurrence risk estimation. Formal genetic counseling is advised for those families seeking such estimates (Stalker and Williams 1998).

Differential Diagnosis

Infants with Angelman syndrome commonly present with non-specific psychomotor delay and/or seizures and the differential diagnosis is often broad and non-specific (Tan et al. 2014), encompassing such entities as cerebral palsy, pervasive developmental disorder and static encephalopathy. The presence of hypotonia and seizures may raise the possibility of an inborn error of metabolism or a defect in oxidative phosphorylation such as a mitochondrial encephalomyopathy. Subsequent testing for these abnormalities normally includes urine organic acids, serum amino acids, plasma acylcarnitine profiles, and mitochondrial enzyme and DNA mutation screens. Some infants with Angelman syndrome may be suspected of having myopathy, although the typical presence of brisk deep tendon reflexes suggests that the lower motor neuron and muscle cell units are normal. Subsequent muscle biopsy with routine histology and electron microscopy studies and electromyogram, are normal or show mildly abnormal, nonspecific findings.

Seizures and severe speech impairment in infants with AS can resemble those seen in the Rett syndrome (see Chapter 49), but children with AS do not lose purposeful use of their hands. The distinction between these two syndromes is usually resolved by age 3–4 years, when children with AS are progressing developmentally but those with Rett syndrome are clearly at a developmental plateau or have apparent regression. Non-classical variants of Rett syndrome do occur and can mimic AS, even in older children (Watson et al. 2001).

It is unusual for infants with AS to have a dysmorphic facial appearance or to have any congenital anomalies, so chromosomal syndromes are usually not suspected. However, terminal deletions involving 22q13 (see Chapter 22) and other chromosome abnormalities have been described (Williams et al. 2001) in individuals with findings of AS.

Chromosome microarray study occasionally detects other deletions or duplications whose clinical features can mimic those of Angelman syndrome. Infants with AS who do have some degree of apparent facial dysmorphology, usually are only manifesting unique parental traits, accentuated by the child's microcephaly and behavioral abnormalities. Some infants may have a happy affect with paroxysms of laughter. Rarely, entities such as Williams syndrome (see Chapter 63) may be initially considered but are quickly ruled out by a complete history and physical examination. Occasionally, an infant with AS will be misdiagnosed as having Prader–Willi syndrome (see Chapter 46) when the diagnosis is actually Angelman syndrome, because of the 15q11.2-q13 deletion involving the maternally and not the paternally derived chromosome. This is possible because both disorders display infantile hypotonia. DNA methylation testing will distinguish between the two.

Older children with non-specific cerebral palsy are often referred for evaluation for Angelman syndrome because they exhibit gait ataxia, happy affect, and abnormal speech. However, most occurrences of cerebral palsy do not manifest the extent of tremulousness, jerkiness, and the ballismic-like limb movements seen in AS. Some degree of expressive speech is usually present in those with cerebral palsy, whereas speech remains extremely disrupted in AS (only minimal sounds), even in the face of relatively good attention and socialization.

Mowat–Wilson syndrome (see Chapter 38) may present with symptoms suggestive of AS (Zweier et al. 2005). Affected individuals may have a prominent mandible and microcephaly. A happy affect, speech deficits and constipation are features observed in both syndromes. Mutations in the gene *ZEB2* are causative. Mowat–Wilson syndrome typically results from a de novo dominant mutation. Recently, an X-linked disorder has been identified that can mimic the happy affect, seizures, and behavioral mannerisms of AS (Gilfillan et al. 2008). In males with Angelman syndrome-like characteristics and family histories consistent with X-linked inheritance, mutation testing for the *SLC9A6* gene should be considered.

MANIFESTATIONS AND MANAGEMENT

Growth and Feeding

The prenatal and birth periods are usually uncomplicated. Newborns appear to be physically well formed with normal birth weight and head circumference. In the first 6–12 months, weight gain may be slow because of feeding difficulties and low muscle tone, but the growth rate in length is usually preserved. Average height is lower than the mean for children in the general population, but most children with Angelman syndrome will have a height that falls within the normal range. Familial factors will influence growth and ultimate height, so that a child with AS born from taller parents is likely to be taller than the average height of all children with AS.

Children with AS often have low or near normal subcutaneous fat but by late childhood, some increased weight gain can occur and mild to moderate obesity can be seen (Clayton-Smith 1993). It appears to occur more often in those with IC defects, paternal UPD or *UBE3A* mutations (Tan et al. 2011). Those with uniparental disomy and imprinting defects demonstrate greater height and head circumference growth rates compared to those with deletions or *UBE3A* mutations (Brennan et al. 2015; Lossie et al. 2001).

Feeding problems are frequent but not generally severe. They usually manifest in infants as difficulty in sucking or swallowing (Zori et al. 1992). Tongue movements may be uncoordinated, with thrusting and generalized oral-motor incoordination. There may be trouble initiating sucking and sustaining breast-feeding. The feeding difficulties may first present to the physician as a problem of poor weight gain or as failure to thrive. Gastroesophageal reflux is relatively common but rarely associated with severe failure to thrive, emesis or aspiration pneumonia. Occasionally, gastrostomy tube placement and fundoplication are required for treatment.

During adolescence, puberty may be delayed by 1–3 years, but sexual maturation occurs normally with development of normal secondary sexual characteristics.

Evaluation

- In infants and children, the cause of feeding problems (gastroesophageal reflux, oromotor incoordination) should be evaluated, beginning with an occupational therapy assessment.
- Height, weight and head circumference should be measured and plotted at each visit in childhood.
- Weight control should be monitored periodically in adults.

Management

- No growth-related treatments are needed.
- Feeding difficulties should prompt feeding therapy, often carried out by occupational therapists.
- Occasionally, surgery is needed to treat reflux. The surgical technique is the same as that used in the general population.
- Obesity should be treated with individualized diet and exercise, but obesity in this syndrome is often very difficult to treat. Severe degrees of obesity are rare.

Development and Behavior

Cognitive Development In general, cognitive abilities are severely to profoundly impaired in Angelman syndrome. However, because of their receptive ability to understand some language and their outgoing, prosocial personality, children with AS often distinguish themselves from those with other causes of severe developmental impairment. Developmental testing is compromised in individuals with Angelman syndrome due to attention deficits, language and speech impairments and hypermotoric behaviors. It is possible that the actual cognitive abilities in Angelman syndrome are higher than indicated from developmental testing. Nevertheless, the developmental delay is still consistently in the functionally severe range. Formal psychometric testing seems to indicate a ceiling for developmental achievement at around the 24–30 month range (Peters et al. 2004).

Young adults with AS are usually socially adept and respond to many basic or socially crucial cues and interactions. Because of their interest in people, they establish rewarding friendships and communicate a broad repertoire of feelings and sentiments. They participate in group events, household chores, and the activities of daily living. They enjoy many recreational activities such as watching TV, going on outings and field trips and participating in water/pool-based games.

There is a wide range in developmental outcome, so that not all individuals with Angelman syndrome attain the above-noted skills. A few will be more impaired in terms of their intellectual deficiency and social interactions. This seems especially the case in those with autistic-like symptoms or with difficult-to-control seizures. With optimal environmental stimulation and consistent behavioral intervention, children with Angelman syndrome show improved developmental progress. Young adults with Angelman syndrome continue to learn and do not have deterioration in their mental abilities.

Speech and Language The speech disorder in Angelman syndrome has a somewhat typical evolution. Babies and young infants cry less often and have decreased cooing and babbling. A single apparent word, such as "mama", may develop around 10–18 months, but it is used infrequently and indiscriminately without symbolic meaning. By 2–3 years of age, it is clear that speech is delayed but it may not be evident how little the child with AS is verbally communicating. Crying and other vocal outbursts may also be reduced. By 3 years of age, higher-functioning children with AS are initiating some type of non-verbal language. Some point to body parts and indicate some of their needs by use of simple gestures, but they are much better at following and understanding commands. Others, especially those with severe seizures or extreme hyperactivity, cannot be attentive enough to achieve the first stages of communication, such as establishing sustained eye contact. The non-verbal language skills in those with Angelman syndrome vary greatly, with the most advanced individuals being able to learn some sign language and use such aids as picture-based communication boards. Some affected children seem to have enough comprehension to be able to speak, but in even the highest functioning, conversational speech does not develop (Summers et al. 1995). It has been reported that a few individuals speak one to three words (Clayton-Smith 1993). In a survey of 47 individuals, it was reported that 39% spoke up to four words. Whether these words were used meaningfully, was not noted (Buntinx et al. 1995). The exception to this profile is observed in some individuals who have mosaicism for an IC defect and have considerable speech (Fairbrother et al. 2015; Le Fevre et al. 2017).

Behavior The first evidence of the distinctive behavior of Angelman syndrome may be the onset of early or persistent social smiling at the age of 1–3 months. Giggling, chortling, and constant smiling soon develop and appear to represent normal reflexive laughter, but cooing and babbling are delayed or reduced. Later, several types of facial or behavioral expressions characterize the infant's personality. A few have pronounced laughing that is truly paroxysmal or contagious – "bursts of laughter" occurred in 70% in one study (Buntinx et al. 1995). More often, happy grimacing and a happy disposition are the predominant behaviors. In rare cases, the apparent happy disposition is fleeting, as irritability and motor hyperactivity are the prevailing personality traits – crying, shrieking, screaming, or short guttural sounds may then be the predominant behaviors. At times, apparent aggressive behaviors occur such as pinching, grabbing, biting, slapping, and hitting. These behaviors often represent attention seeking, but if they are not addressed early on, they can become persistent. Consistent social disapproval and a structured behavioral modification program will eventually be successful in managing this problem.

The laughter in AS can be a socially appropriate, emotive response to a humorous situation. More often it appears to be mostly a nonspecific behavioral event. Most reactions to stimuli, physical or mental, are accompanied by laughter or laughter-like facial grimacing. People with AS do, however, experience the entire spectrum of emotionality.

Children with AS are notorious for putting everything in their mouths. In early infancy, hand sucking (and sometimes foot sucking) is frequent. Later, most exploratory play is by oral manipulation and chewing. The tongue appears to be of normal shape and size, but in 30–50%, persistent tongue protrusion is a distinctive feature. For the child with AS who has protruding tongue behavior, the problem remains throughout childhood and can persist into adulthood. Drooling is usually associated and its treatment is problematic, often requiring use of bibs.

The behavioral profile that defines the syndrome, generally continues without much change into the adult years. Parents and caregivers do report dealing with challenging behaviors that include sensory stimulation and attention concerns, apparent stubbornness associated with avoidance of undesired situations, and issues related to preference and want frustrations (Larson et al. 2015). These issues appear to become more problematic in the teenage and adult years. At times, apparent anxiety seems to best characterize the genesis or underpinning of some of these behaviors.

Hyperactivity Many young children with Angelman syndrome have some component of hypermotoric behavior, with males and females appearing equally affected. Infants and toddlers may have seemingly ceaseless activity, constantly keeping their hands or toys in their mouth, moving from object to object. Attention span can be so short that social interaction is prevented because the child cannot attend to facial and other social cues. Development progress will be difficult unless more attentiveness occurs. In extreme cases, the constant movement can cause accidental bruises and abrasions. Grabbing, pinching, and biting in older children have also been noted and may be heightened by the hypermotoric activity. Hypermotoric activity in Angelman syndrome usually lasts throughout childhood but gradually improves. It is most severe between 3 and 10 years of age.

Sleep Abnormalities Decreased need for sleep and abnormal sleep/wake cycles are often seen in those with Angelman syndrome (Allen et al. 2013; Bird, 2014). A commonly observed pattern is one of initially being able to get to sleep but then having early morning awakening with several hours of awake activities, then a returning to sleep until a later morning awakening. At times, sleep wake cycles can be reversed, creating a problem of daytime somnolence. Families often construct safe, monitored bedroom areas to accommodate the interim awake hours. Sleep difficulties seem to improve in older children and young adults, although some continue to have sleep disturbance throughout their life. Having said this, there also is a significant group of those with the syndrome who have relatively good sleep and in those situations parents do not report family stress because of sleep abnormalities.

Evaluation

- Formal evaluation to detect hyperactivity or attention deficit disorder is not necessary. Parental report and observation in the examination room or in the home or school environment is sufficient for evaluation.
- Occurrences of marked exacerbation of restlessness and hypermotoric behavior should prompt a careful medical evaluation to detect occult illness such as otitis media, urinary tract infection, or dental abscess.
- Clinical psychological evaluation may be helpful, especially in the evaluation of a family's adjustment to a sleep or nighttime behavioral problem. Often, sleep/wake cycles are reversed and family dynamics have accordingly adjusted to this, sometimes without the family's awareness.

Management

- The severe developmental delay in AS mandates that a full range of early training and enrichment programs be made available.
- Unstable or non-ambulatory children may benefit from physical therapy. Special adaptive chairs or positioners may be required at various times, especially for hypotonic or extremely ataxic children.
- Occupational therapy may help improve fine motor and oral-motor control.
- Speech and communication therapy is essential and should focus on non-verbal methods of communication. Augmentative communication aids, such as picture cards or communication boards, should be used at the earliest appropriate time.
- Extremely active and hypermotoric children with AS will require special provisions in the classroom. Teacher's aides or assistants may be needed to integrate the child into the classroom. Children with attention deficits and hyperactivity need room to express themselves and to "grapple" with their hypermotoric activities. The classroom setting should be structured, in its physical design and its curricular program, so that the active child with AS can fit in or adjust to the school environment. Individualization and flexibility are important factors.
- Consistent behavior modification in the school and at home can enable the child with AS to be toilet trained (schedule trained) and to perform most self-help skills related to eating, dressing, and performing general activities in the home (Summers et al. 1992). Use of medications such as scopolamine patches (only for age >12 years, 1.5 mg/patch, remove after 72 hours) or glycopyrrolate (0.04–0.1 mg/kg/dose every 4–8 hours in children and 1–2 mg/dose, twice or three times daily in adults) to dry secretions may be beneficial for drooling, but often these medications do not provide an adequate long-term effect (personal experience).
- Tongue protrusion is not so extensive that tongue reduction surgery is performed, although that is an option in severe cases. Likewise, surgical reimplantation of salivary gland ducts could be an option in cases of severe drooling, but this has yet to be reported in AS.
- Most affected people do not receive drug therapy for hyperactivity, and the hypermotoric behavior is tolerated, for better or worse, by the parents and the school (personal experience). Some parents report that changes in diet can affect activity levels, but no consistent dietary associations seem apparent. Some may benefit from use of stimulant medications such as methylphenidate, dexedrine, and combinations of amphetamine and dextroamphetamine (such as Adderal®), whereas others are given antihistamines such as diphenhydramine and hydroxyzine. Clonidine and risperidone have also been used.
- Treatment of anxiety may be helpful if that problem is felt to be an underlying issue for challenging behaviors. Clobazam, buspirone, lorazepam, and clonazepam are agents often used in AS for treatment of anxiety.
- A structured environment, consistent behavioral modification, and often one-on-one personal interaction may be required in the school or home to deal with the hyperactivity and provide viable developmental training.
- Many families construct safe but confining bedrooms to accommodate disruptive nighttime wakefulness.
- Many infants and children with AS do not receive sleep medications. Those who do, may not require long-term

use. Most commonly used medications to treat sleep disorders in the syndrome include melatonin, clonidine, and trazodone.
- A behavioral psychology consultation may prove helpful in teaching the family how to implement consistent sleep and daytime schedules (Summers et al. 1992).

Neurologic

Microcephaly The prevalence of absolute microcephaly in deletion-positive individuals varies from 88% (Zori et al. 1992) to 34%, and may be as low as 15–25% when non-deletion cases are also included (Lossie et al. 2001). A significant proportion of affected individuals, however, will not develop microcephaly. Many individuals with Angelman syndrome have a head circumference less than the 25th percentile by age 3 years, often accompanied by a flattened occiput. Brain MRI typically shows delayed myelination in combination with microcephaly, but there are no gyral anomalies in brain structure or in gross pathology (Harting et al. 2009; Wilson et al. 2011). A diffusion tensor study suggested nonspecific defects in language pathways (Wilson et al. 2011). Severe microcephaly is very unusual; most have normal cranial growth rates in later childhood but maintain low percentiles or mild absolute microcephaly on the growth points.

Seizures The seizure prevalence in Angelman syndrome has been reported to be as high as 90%, but this is probably an overestimation. Fewer than 25% develop seizures before 12 months of age; most have onset before 3 years, but initial occurrence in older children or in teenagers is not exceptional (Zori et al. 1992). The most common type of seizures observed are atypical absences and myoclonic seizures (Thibert et al. 2009; Thibert et al. 2013). Less frequently, generalized extensor tonic seizures or flexor spasms occur (Minassian et al. 1998). The seizures can be of any type and occur in all genetic classes of AS. Most seizures are effectively controlled with anticonvulsants but intractable epilepsy can occur. At times, seizures may be difficult to recognize or distinguish from the child's usual tremulousness, hyperkinetic limb movements, or attention deficits.

Electroencephalograms (EEGs) do vary at different times, and apparently normal tracings are not infrequent, especially in young infants. The typical EEG is often more abnormal than expected in terms of the clinical picture and the EEG may suggest seizures when in fact there appear to be none clinically. Typically, there is a symmetrical, high-voltage slow (Delta) wave activity persisting for most of the record (unrelated to drowsiness) and very-large-amplitude slow activity at 2–3 cycles per second, occurring in runs and more prominent anteriorly. In addition, spikes or sharp waves, mixed with large-amplitude 3–4 cycles per second components, are seen posteriorly and are usually provoked by passive eye closure (Fiumara et al. 2010; Galvan-Manso et al. 2005; Minassian et al. 1998). Recently, power spectral analysis of delta waves in Angelman syndrome reveals quantitative findings considered to be a biomarker for Angelman EEG phenotype (Sidorov et al. 2017).

At times, tremulousness in AS may be of such severity to suggest a seizure disorder. However, these tremors may represent a type of cortical myoclonus without synchronizing EEG changes (Goto et al. 2015). In teenagers and young adults, these tremors can become episodic and more severe in frequency to such an extent that they suggest a new onset seizure disorder or other reason for neurological deterioration. The tremors can interfere with activities of daily living. The long-term course of these myoclonic-like tremors appears to be chronic, with a waxing and waning occurrence.

Gait and Movement Disorders In early childhood, the mildly impaired child can have almost normal walking. There may be only mild toe-walking or an apparent prancing gait. This may be accompanied by a tendency to lean or lurch forward. During running, the tendency to lean forward is accentuated and the arms are held uplifted. For these children, balance and coordination does not appear to be a major problem. More severely affected children can be very stiff and robot-like, or extremely shaky and jerky when walking. Voluntary movements are often irregular, varying from slight jerkiness to uncoordinated coarse movements that prevent walking, feeding, and reaching for objects. Although they can crawl fairly effectively, affected individuals may become rigid or appear anxious when placed in the standing position. The legs are kept wide based and the feet are flat and turned outward. This, accompanied by uplifted arms, flexed elbows, and downward turned hands, produces the characteristic gait and posture of Angelman syndrome. Most of those with Angelman syndrome are so ataxic and jerky that walking is often delayed until 3 or 4 years of age, when they are better able to compensate motorically for the jerkiness. About 10% may fail to achieve walking (Clayton-Smith, 1993). Although many children with AS require a lengthy period of training and assistance, the prognosis for independent ambulation is good.

Evaluation

- Microcephaly in Angelman syndrome seems to be a reflection of generalized decrease in brain growth, so magnetic resonance imaging and/or computed tomography imaging does not appear warranted unless other factors (e.g., focal neurological findings or unexplained psychomotor regression) are present.
- An EEG is recommended during the early phase of diagnosis and treatment.
- Orthopedic, physical, and occupational therapy evaluations may be needed.

Management

- There is no agreement as to the optimal seizure medication, but there are patterns of use that are more frequent. The most commonly used ones are clobazam, lamotrigine, and levetiracetam (Larson et al. 2015; Thibert et al. 2009). Almost all people with AS have breakthrough seizures and seizure exacerbations. However, most are controlled with a single anticonvulsant medication and the prognosis for adequate seizure control is good. Perhaps 10–15% will have poorly controlled seizures that require more than two anticonvulsant medications and often require periodic hospitalizations. Many adults have infrequent seizures and can be tapered off medications, but after a long period of abatement, seizures may recur. Some children with uncontrollable seizures have been placed on a ketogenic diet, and this has proven helpful in some but not all cases of intractable seizures. Recently, improved seizure control in AS has been achieved using a low glycemic diet (Grocott et al. 2017). Carbamezapine, although not contraindicated, is infrequently used compared to other common anticonvulsants. Vigabatrin and tiagabine (anticonvulsants that increase brain GABA levels) are contraindicated in Angelman syndrome. For unknown reasons, carbamazepine, vigabatrine and tiagabine can cause development of other seizure types or non-convulsive status epilepticus. This paradoxical seizure development is not limited to patients with AS (Pel et al. 2008).
- Children with AS are at risk for medication overtreatment because their movement abnormalities or attention deficits can be mistaken for seizures and because electroencephalographic abnormalities can persist even when seizures are controlled.
- For some children with AS who have a cerebral palsy-like presentation, there may be significant heel cord tightening. Tendon lengthening surgery can be helpful.
- Walkers are helpful for many until independent walking begins.
- Ongoing physical and occupational therapies are often needed throughout childhood.

Ophthalmologic

Surveys of individuals with Angelman syndrome demonstrate a 30–60% incidence of strabismus. Strabismus appears to be more common in a number of genetic disorders that cause ocular hypopigmentation, because pigment in the retina is crucial to normal development of the optic nerve pathways. Long-term retinal function, anterior chamber function, and other measures of ocular health appear to be normal. Keratoconus has been noted secondary to persistent behavior of eye rubbing or gouging.

Evaluation

- Referral to an ophthalmologist is indicated at diagnosis and then every one to two years for all affected individuals, regardless of the genetic type of AS.

Management

- Management of strabismus, correction of any visual deficit and, where appropriate, patching and surgical adjustment of the extraocular muscles in AS is similar to that in other children. The hypermotoric activities of some children with AS will make wearing patches or glasses difficult. Surgical outcome for strabismus repair is probably equivalent to that in other conditions involving developmental delay; most cases are successful.

Musculoskeletal

Physical therapy is usually helpful in improving ambulation. Some affected individuals have marked out-toeing and pes planus, thus have a valgus gait, often with subluxing ankles. Scoliosis can develop in adolescence. It is especially a problem in those who are non-ambulatory.

Evaluation

- Foot and ankle positioning should be checked as the child starts to mobilize.
- The spine should be checked for scoliosis during early adolescence.

Management

- Sometimes ankle bracing or surgical intervention may be needed to properly align the legs.
- Scoliosis should be treated with early bracing to prevent progression. Surgical correction or stabilization may be necessary for severe cases.

RESOURCES

In this day and age of social media, Angelman syndrome has a new platform for advocacy, research, and education. Like never before, families, health care providers, researchers, and many other supporters are connected and collaborating through vast social networks on an international scale. A quick search on the internet reveals support groups throughout the world and presence on Facebook, Twitter, Snapchat, Pinterest, LinkedIn, and YouTube. A few of the international support organizations include:

Argentina

Casa Angelman
Website: http://www.casaangelman.org/

Australia

Angelman Syndrome Association Australia
Website: http://www.angelmansyndrome.org/

Canada

Canadian Angelman Syndrome Society
Website: http://www.angelmancanada.org/

Angelman Syndrome Foundation of Quebec
Website: http://angelman.ca/

China

Hong Kong Angelman Syndrome Foundation
Website: http://www.hkasf.org/tc/

France

Association Francophone du Syndrome d'Angelman (AFSA)
Website: http://www.angelman-afsa.org/

Syndrome Angelman France
Website: http://www.syndromeangelman-france.org/?lang=en

Germany

Angelman e.V.
Website: http://www.angelman.de/

Italy

Associazione Angelman
Website: http://www.associazioneangelman.it/

Netherlands

Angelman Syndrome Nederland
Website: https://www.angelmansyndroom.nl/

United Kingdom

Angelman Syndrome Support, Education and Research Trust (ASSERT)
Website: http://angelmanuk.org/

United States

Angelman syndrome Foundation
Website: https://www.angelman.org/

Foundation for Angelman Syndrome Therapeutics
Website: http://cureangelman.org/

For a list of additional international support organizations, you can visit the International AS Resources page on the Angelman Syndrome Foundation website at: https://www.angelman.org/resources-education/international-as-resources/.

ACKNOWLEDGMENTS

This document was developed in part with grant funding from the Raymond C. Philips Research and Education Contract, Children's Medical Services, Department of Health, State of Florida.

REFERENCES

Allen KD, Kuhn BR, DeHaai KA, Wallace DP (2013) Evaluation of a behavioral treatment package to reduce sleep problems in children with Angelman Syndrome. *Res Dev Disabil* 34(1):676–686.

Angelman H (1965) 'Puppet' Children. A report of three cases. *Dev Med Child Neurol* 7:681–688.

Bird LM (2014) Angelman syndrome: review of clinical and molecular aspects. *Appl Clin Genet* 7:93–104.

Brennan ML, Adam MP, Seaver LH, et al. (2015) Increased body mass in infancy and early toddlerhood in Angelman syndrome patients with uniparental disomy and imprinting center defects. *Am J Med Genet A* 167A(1):142–146.

Buiting K, Gross S, Lich C, Gillessen-Kaesbach G, el-Maarri O, Horsthemke B (2003) Epimutations in Prader-Willi and Angelman syndromes: a molecular study of 136 patients with an imprinting defect. *Am J Hum Genet* 72(3):571–577.

Buntinx IM, Hennekam RC, Brouwer OF, Stroink H, Beuten J, Mangelschots K, Fryns JP (1995) Clinical profile of Angelman syndrome at different ages. *Am J Med Genet* 56(2):176–183.

Clayton-Smith J (1993) Clinical research on Angelman syndrome in the United Kingdom: observations on 82 affected individuals. *Am J Med Genet* 46(1):12–15.

Clayton-Smith J, Laan L (2003) Angelman syndrome: a review of the clinical and genetic aspects. *J Med Genet* 40(2):87–95.

Dagli A, Buiting. K, Williams CA (2012) Molecular and Clinical Aspects of Angelman Syndrome. *Mol Syndromol* 2(3-5):100–112.

El Hokayem J, Nawaz Z (2014) E6AP in the brain: one protein, dual function, multiple diseases. *Mol Neurobiol* 49(2):827–839.

Fairbrother LC, Cytrynbaum C, Boutis P, Buiting K, Weksberg R, Williams C (2015) Mild Angelman syndrome phenotype due to a mosaic methylation imprinting defect. *Am J Med Genet A* 167(7):1565–1569.

Fink JJ, Robinson TM, Germain ND, et al. (2017) Disrupted neuronal maturation in Angelman syndrome-derived induced pluripotent stem cells. *Nat Commun* 8:15038.

Fiumara A, Pittala A, Cocuzza M, Sorge G (2010) Epilepsy in patients with Angelman syndrome. *Ital J Pediatr* 36(1):31.

Galligan JT, Martinez-Noel G, Arndt V, Hayes S, Chittenden TW, Harper JW, Howley PM (2015) Proteomic Analysis and Identification of Cellular Interactors of the Giant Ubiquitin Ligase HERC2. *J Proteome Res* 14(2):953–966.

Galvan-Manso M, Campistol J, Conill J, Sanmarti FX (2005) Analysis of the characteristics of epilepsy in 37 patients with the molecular diagnosis of Angelman syndrome. *Epilep Disord* 7(1) 19–25.

Gilfillan GD, Selmer KK, Roxrud I, et al. (2008) SLC9A6 mutations cause X-linked mental retardation, microcephaly, epilepsy, and ataxia, a phenotype mimicking Angelman syndrome. *Am J Hum Genet* 2(4):1003–1010.

Glenn CC, Saitoh S, Jong MT, Filbrandt MM, Surti U Driscoll DJ, Nicholls RD (1996) Gene structure, DNA methylation, and imprinted expression of the human SNRPN gene. *Am J Hum Genet* 58(2):335–346.

Goto M, Saito Y, Honda R, et al. (2015) Episodic tremors representing cortical myoclonus are characteristic in Angelman syndrome due to UBE3A mutations. *Brain Dev* 37(2):216–222.

Grocott OR, Herrington KS, Pfeifer HH, Thiele EA, Thibert RL (2017) Low glycemic index treatment for seizure control in Angelman syndrome: A case series from the Center for Dietary Therapy of Epilepsy at the Massachusetts General Hospital. *Epilep Behav* 68:45–50.

Harting I, Seitz A, Rating D, Sartor K, Zschocke J, Janssen B, … Wolf NI (2009) Abnormal myelination in Angelman syndrome. *Eur J Paediatr Neurol* 13(3):271–276.

Hillman PR, Christian SGB, Doan R, Cohen ND, Konganti K, Douglas K, … Dindot SV (2017) Genomic imprinting does not reduce the dosage of UBE3A in neurons. *Epigenet Chromatin* 10:27.

Kishino T, Lalande M, Wagstaff J (1997) UBE3A/E6-AP mutations cause Angelman syndrome. *Nat Genet* 15(1):70–73.

Larson AM, Shinnick JE, Shaaya EA, Thiele EA, Thibert RL (2015) Angelman syndrome in adulthood. *Am J Med Genet A* 167(2):331–344.

Le Fevre A, Beygo J, Silveira C, Kamien B, Clayton-Smith J, Colley A, … Dudding-Byth T (2017) Atypical Angelman syndrome due to a mosaic imprinting defect: Case reports and review of the literature. *Am J Med Genet A* 173(3):753–757.

Lossie AC, Whitney MM, Amidon D, Dong HJ, Chen P, Theriaque D, … Driscoll DJ (2001) Distinct phenotypes distinguish the molecular classes of Angelman syndrome. *J Med Genet* 38(12) 834–845.

Malzac P, Webber H, Moncla A, Graham J. M, Kukolich M, Williams C, Pagon RA, … Wagstaff J (1998) Mutation analysis of UBE3A in Angelman syndrome patients. *Am J Hum Genet* 62(6):1353–1360.

Meng L, Person RE, Beaudet A (2012) Ube3a-ATS is an atypical RNA polymerase II transcript that represses the paternal expression of Ube3a. *Hum Mol Genet* 21(13):3001–3012.

Mertz LG, Christensen R, Vogel I, Hertz JM, Nielsen KB, Gronskov K, Ostergaard JR (2013) Angelman syndrome in Denmark. birth incidence, genetic findings, and age at diagnosis. *Am J Med Genet A* 161A(9):2197–2203.

Minassian BA, DeLorey TM, Olsen RW, et al. (1998) Angelman syndrome: correlations between epilepsy phenotypes and genotypes. *Ann Neurol* 43(4):485–493.

Pelc K, Boyd SG, Cheron G, Dan B (2008) Epilepsy in Angelman syndrome. *Seizure: Eur J Epilepsy* 17(3):211–217.

Peters SU, Goddard-Finegold J. Beaudet AL, Madduri N, Turcich M, Bacino CA (2004) Cognitive and adaptive behavior profiles of children with Angelman syndrome. *Am J Med Genet A* 128(2):110–113.

Philippart M, Minassian BA (2005) Angelman syndrome from infancy to old age (abstract) *Am J Hum Genet* 79(Suppl):605.

Ramsden SC, Clayton-Smith J, Birch R, Buiting K (2010) Practice guidelines for the molecular analysis of Prader-Willi and Angelman syndromes. *BMC Med Genet* 11:70.

Sahoo T, Bacino CA, German JR, et al. (2007) Identification of novel deletions of 15q11q13 in Angelman syndrome by array-CGH: molecular characterization and genotype-phenotype correlations. *Eur J Hum Genet* 15(9):943–949.

Sidorov MS, Deck GM, Dolatshahi M, Thibert RL, Bird LM, Chu CJ, Philpot BD (2017) Delta rhythmicity is a reliable EEG biomarker in Angelman syndrome: a parallel mouse and human analysis. *J Neurodev Disord* 9:17.

Stalker HJ, Williams CA (1998) Genetic counseling in Angelman syndrome: the challenges of multiple causes. *Am J Med Genet* 77(1):54–59.

Summers JA, Allison DB, Lynch PS, Sandler L (1995) Behaviour problems in Angelman syndrome. *J Intel Disabil Res* 39(Pt 2):97–106.

Summers JA, Lynch PS, Harris JC, Burke JC, Allison DB, Sandler L (1992) A combined behavioral/pharmacological treatment of sleep-wake schedule disorder in Angelman syndrome. *J Dev Behav Pediatr* 13(4):284–287.

Tan WH, Bacino CA, Skinner SA, Anselm I, Barbieri-Welge R, Bauer-Carlin A, … Bird, LM (2011) Angelman syndrome: Mutations influence features in early childhood. *Am J Med Genet A* 155A(1):81–90.

Tan WH, Bird LM (2016) Angelman syndrome: Current and emerging therapies in 2016. *Am J Med Genet C* 172(4) 384–401.

Tan WH, Bird L.M, Thibert RL, Williams CA (2014) If not Angelman, what is it? A review of Angelman-like syndromes. *Am J Med Genet A* 164A(4):975–992.

Thibert RL, Conant KD, Braun EK, Bruno P Said RR, Nespeca MP, Thiele EA (2009) Epilepsy in Angelman syndrome: a questionnaire-based assessment of the natural history and current treatment options. *Epilepsia* 50(11):2369–2376.

Thibert RL, Larson AM, Hsieh DT, Raby AR, Thiele EA (2013) Neurologic manifestations of Angelman syndrome. *Pediatr Neurol* 48(4):271–279.

Toyofuku K, Valencia JC, Kushimoto T, Costin GE, Virador VM, Vieira WD, … Hearing, V, J (2002) The etiology of oculocutaneous albinism (OCA) type II: the pink protein modulates the processing and transport of tyrosinase. *Pigment Cell Res* 15(3):217–224.

Watson P, Black G, Ramsden S, Barrow M, Super M, Kerr B, Clayton-Smith J (2001) Angelman syndrome phenotype associated with mutations in MECP2, a gene encoding a methyl CpG binding protein. *J Med Genet* 38(4):224–228.

Williams CA, Beaudet AL, Clayton-Smith J, Knoll JH, Kyllerman M, Laan LA, … Wagstaff J (2006) Angelman syndrome 2005: updated consensus for diagnostic criteria. *Am J Med Genet A* 40(5):413–418.

Williams CA, Lossie A, Driscoll D (2001) Angelman syndrome: mimicking conditions and phenotypes. *Am J Med Genet* 101(1):59–64.

Wilson BJ, Sundaram SK, Huq AH, et al.(2011) Abnormal language pathway in children with Angelman syndrome. *Pediatr Neurol* 44(5):350–356.

Zori RT, Hendrickson J, Woolven S, Whidden EM, Gray B, Williams CA (1992) Angelman syndrome: clinical profile. *J Child Neurol* 7(3):270–280.

Zweier C, Thiel CT, Dufke A, Crow YJ, Meinecke P, Suri M, … Rauch A (2005) Clinical and mutational spectrum of Mowat-Wilson syndrome. *Eur J Med Genet* 48(2):97–111.

6

ARTHROGRYPOSIS

JUDITH G. HALL

Departments of Pediatrics and Medical Genetics, British Columbia's Children's Hospital, Vancouver, British Columbia, Canada

INTRODUCTION

Arthrogryposis is the term used to describe multiple contractures of the joints in more than one area of the body that are present at birth. Historically, the term *arthrogryposis multiplex congenita* has been used to describe multiple congenital contractures: (arthro = joint, gry = curved, multiplex = multiple, congenita = present at birth). It implies a type of contracture that is non-progressive and relatively rigid (Hall and Vincent 2014; Hall 2013). Many of the affected joints respond to physical therapy and orthopedic procedures; however, without continued intervention, very often the affected joint returns to the position present at birth (Staheli et al. 1998; van Bosse et al. 2017). Over the years, it has become apparent that there are many different conditions that include multiple congenital contractures (Hall 1997; Hall 2013). Thus, the term arthrogryposis is used as a sign or as a general category of disorders. For the purpose of this chapter, the term will be used to encompass many different conditions with varied causes that include multiple congenital contractures in at least two different parts of the body.

The challenge of arthrogryposis for the clinician is to try to make a specific diagnosis since the natural history, recurrence risk, and optimum therapies can then be determined. Since this book is oriented towards management, an effort will be made to provide common themes for the care of individuals born with different types of multiple contractures (Hall 2013, 2014b, 2014c).

Incidence

Approximately 1% of children are born with some type of contracture (Hall, 2013): clubfoot occurs in 1 in 300 births, camptodactyly (flexed fingers) in 1 in 200 births, and dislocated hips in 1 in 200 births. However, several population-based studies have determined that the incidence of multiple congenital contractures in different body areas is between 1 in 3000 and 1 in 6000 live births (Lowry et al. 2010; Hall and Vincent 2014). The most common type of arthrogryposis is Amyoplasia which occurs in 1/10,000 live births (Hall et al. 2014) (Figure 6.1). Over 400 specific disorders have associated arthrogryposis (Hall 2013; Hall and Kiefer 2016). Over 350 genes have been identified for these conditions.

Diagnostic Criteria

By definition, arthrogryposis involves multiple congenital contractures in different body areas.

Most varieties of arthrogryposis involve all four limbs and multiple joints. Jaw and spine may also be involved. There may be significant variability within a specific condition such that some affected children have primarily multiple congenital contractures and others with the same disorder have only hypotonia or clubfoot For example, in the autosomal recessive condition cerebro-oculo-facial syndrome (COFS), one child in a family may be born with contractures and the next is born hypotonic without contractures (Graham et al. 2001). Infants who are

---

Cassidy and Allanson's Management of Genetic Syndromes, Fourth Edition.
Edited by John C. Carey, Agatino Battaglia, David Viskochil, and Suzanne B. Cassidy.
© 2021 John Wiley & Sons, Inc. Published 2021 by John Wiley & Sons, Inc.

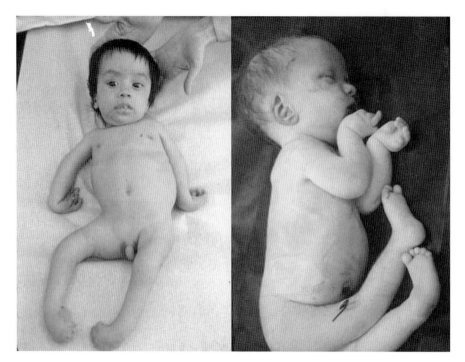

FIGURE 6.1 Note the varying positions of the limbs seen in infants with amyoplasia. Fixed, extended elbows with internal rotation of the shoulders is suggestive of amyoplasia. Hyperextension of joints is a poor prognostic sign since the joint is likely not to have formed in a normal way.

severely hypotonic *in utero* may not move enough to avoid developing contractures, particularly in the large joints. The specific type of arthrogryposis is diagnosed by a combination of pregnancy history, family history, the presence of other anomalies, the specific position of contractures at birth, and various laboratory and genetic tests (Reed et al. 1985; Clarren and Hall 1983; Darin 2000; Hall and Vincent 2014; Hall and Kiefer 2016). A discussion of the various historical data, signs and symptoms important for determining the specific condition associated with arthrogryposis is presented below.

Decreased intrauterine movement, which causes arthrogryposis, can lead to a series of secondary abnormalities that may be present to a greater or lesser extent. The full-blown picture is called the fetal akinesia deformation sequence and involves many body systems. Since it was first described by Pena and Shokeir, it is also referred to as the "Pena–Shokeir phenotype" (Hall 1986; Witters et al. 2002; Hall 2009; Hall 2010a; Nayak et al. 2014). The features include:

1. Contractures of joints
2. Polyhydramnios, apparently secondary to a lack of fetal swallowing, which then may lead to lack of maturity of the intestinal tract because of immobility, secondary short length of the gut, and feeding problems after birth
3. Pulmonary hypoplasia, secondary to lack of fetal breathing *in utero* with failure of maturation of alveoli and surfactin
4. A short umbilical cord because it is not stretched by fetal movement, which can lead to avulsion from the placenta and problems during the birth process
5. Intrauterine growth restriction, since apparently normal in utero growth of long bones requires movement, and there is decreased muscle mass
6. Osteoporosis and failure of modeling of bones (particularly the joint surfaces) predisposing them to fractures at birth
7. Craniofacial abnormalities including micrognathia, cleft palate (hard and/or soft palate), underdevelopment of the maxilla, prominence of the bridge of the nose, and a depressed tip of the nose.

Arthrogryposis may be recognized prenatally for many reasons. Often, the mother will have noted decreased fetal movement (Fahy and Hall 1990), or routine prenatal ultrasound may demonstrate clubfoot, nuchal edema or thin ribs. Then, real time ultrasound may demonstrate fixed contractions in several body areas (Hellmund et al. 2016). However, 75% of cases of four-limb amyoplasia are missed prior to 20 weeks, since movement is not routinely evaluated prenatally (Filges and Hall 2014). Making the prenatal diagnosis of a specific type of arthrogryposis is almost impossible unless there is a positive family history (Hall 1997; Witters et al. 2001; Hall 2014b, 2014c; Hall and Kiefer 2016), allowing "DNA diagnosis".

Etiology, Pathogenesis, and Genetics

Many types of arthrogryposis have a genetic basis (Hall 1985, 1998, 2010a, 2010b, 2013, 2014b). Some occur as an isolated event in families and are of unknown etiology (e.g., amyoplasia), while a third group has been recognized to have environmental etiologies (Hall and Reed 1982a; Hall 1988). Some types occur most frequently in certain ethnic groups (Bayram et al. 2016). Most types do not have a gender preference, although several different forms of X-linked arthrogryposes have been described only in males (Hall et al. 1982c; Hunter et al. 2013).

It appears that any in utero process that leads to decreased fetal movement can secondarily lead to fetal contractures. Swinyard (1963) described this as a "collagen" response to the lack of joint movement, and indeed thickening of the joint capsule and surrounding tissue is usually seen and must be considered when undertaking therapy. The earlier in fetal development decreased movement occurs, the more severe and immobilizing will be the contractures (Hall 1985b, 2010).

The fetus, just like the child and the adult, develops stiffness in a joint if it is not used regularly and moved through a full range of motion. In humans, movement of the limbs begins between eight and nine weeks gestation and normally continues throughout the pregnancy. There are many different reasons that a fetus does not move or stops moving normally; these comprise the various mechanisms that underlie the joint contractures. The list includes the following:

1. Neurological abnormalities, including dysfunction or structural anomalies of the central nervous system, the peripheral nerves, and the neuromuscular endplate (Gaitanis et al. 2010; Engel and Sine 2005; Michalk et al. 2008; Hoffman et al. 2006).
2. Muscular abnormalities, including both structural and functional abnormalities of muscle. Myopathic etiologies may be more common than the previous estimation of 5%, particularly for the types of arthrogryposis involving the distal part of the limbs (Hall et al. 1982; Sung et al. 2003, McMillin et al. 2014).
3. Insufficient space to allow fetal in utero movement because of uterine abnormality such as bicornuate uterus or a uterine fibroid, in the presence of additional fetus(es), such as in multiple births, or when there is decreased amniotic fluid such as with amniotic fluid leakage (Hall 2014b).
4. Connective tissue disorders such as a chondrodysplasia, an abnormally formed joint, misplaced tendons, abnormalities of cartilage, or abnormally constrictive skin.
5. Certain maternal illnesses, including myotonic dystrophy, myasthenia gravis, various infections and metabolic imbalances (Engel and Sine 2005; Xavier-Neto et al. 2017).
6. Maternal medication or drug use during pregnancy, including misoprostol, curare, muscle relaxants, and cocaine (Hall and Reed 1982a).
7. Maternal injuries, such as a motor vehicle accident or attempted termination of pregnancy in the first trimester (Hall 1996, 2012a);
8. Compromise of the vascular flow to the placenta or fetus (e.g., amyoplasia, distal arthrogryposis 2E) (Hall et al. 1982; Reid et al. 1986).

There are over 400 disorders with arthrogryposis, many of which have specific syndrome designations and known etiologies. The genes for over 350 of the inherited forms have been discovered (Hall 2013; Hall and Kiefer, 2016). Amyoplasia, which is the most common type of arthrogryposis, appears to be a completely sporadic condition without a genetic basis, but may have additional anomalies associated with vascular compromise (Reid et al. 1986; Hall et al. 2014, 1982b; Bamshad et al. 1996). Chromosomal mosaicism and submicroscopic deletions have often been seen in individuals with multiple congenital contractures who have developmental delay and in whom no other etiology has been found.

One clinical approach to making a specific diagnosis is by separating disorders into three groups: (Hall 2013):

1. Those in which primarily only the limbs are involved
2. Those in which the limbs plus other body systems are involved
3. Those with severe central nervous system dysfunction or that are associated with death in utero and at a young age.

An extensive review using this approach is available (Hall 2013). If a specific *condition associated with* arthrogryposis is diagnosed, the recurrence risk will be straightforward. However, if a specific diagnosis is not made, the empiric recurrence risk is approximately 3% in general and slightly higher (7%) for those with central nervous system involvement (Hall 2014).

Diagnostic Testing

Until recently, laboratory and/or functional diagnostic testing was not particularly useful in arthrogryposis. Muscle biopsy from an involved area as compared to an uninvolved area (using appropriate staining and electron microscopic studies), nerve conduction studies, creatine phosphokinase, chromosome analysis, and molecular studies may each have value. Table 6.1 lists the recommended evaluations for an individual with arthrogryposis. Microarray, gene panels, exome sequencing, and whole genome sequencing are appropriate for consideration. Many of the known disorders now have specific mutational analysis available to confirm the diagnosis.

TABLE 6.1 Approach to the individual with congenital contractures

Clinical Evaluation

History
- *Pregnancy*

Maternal illness, chronic or acute (e.g., diabetes, myasthenia gravis, myotonic dystrophy)
Maternal infection (e.g., rubella, rubeola, coxsackie, enterovirus, Akabane, Zika)
Fever (>39 °C, determine timing in gestation)
Nausea (e.g., viral encephalitis, position of baby in utero)
Drugs (e.g., curare, robaxin, alcohol, phenytoin, addictive drugs, misoprostol)
Fetal movement (e.g., polyhydramnios, fetal kicking in one place, "rolling", decreased)
Oligohydramnios, chronic leakage of amniotic fluid
Polyhydramnios, hydrops
Trauma during pregnancy (e.g., blow to the abdomen, attempted termination, car accident)
Other complications during pregnancy (e.g., bleeding, abnormal lie, threatened abortion, suspected lost twin)
Prenatal diagnostic procedures (e.g., early amniocentesis, CVS, ultrasound studies)

- *Delivery*

Presentation (e.g., abnormal lie, breech, transverse)
Length of gestation
Traumatic delivery (e.g., CNS abnormality, spasticity, fracture)
Intrauterine mass (e.g., twin, fibroid)
Abnormal uterine structure or shape (e.g., bicornuate, septum)
Abnormal placenta, membranes, or cord length or position
Time of year, geographic location

- *Family history*

Increased incidence of congenital contractures
Marked variability within family
Change of contractures with time – degenerate versus improve
Hyperextensibility or hypotonia present in family member
Myotonic dystrophy, myasthenia gravis in parents (particularly mother)
Consanguinity
Advanced parental age
Increased stillbirths or miscarriages
If more than one affected, if there is increasing severity consider maternal antibodies to fetal neurotransmitters

Examination
- *Newborn*

Standard measurements (weight, length, head circumference)
Description and photography of contractures
Which limbs and joints
Proximal versus distal
Flexion versus extension
Amount of limitation (fixed versus passive versus active movement)
Characteristic position at rest
Severity (firm versus some give)
Complete fusion or ankylosis versus soft tissue contracture
Presence or absence of flexion creases

TABLE 6.1 (Continued)

- *Other anomalies*

Genitalia (e.g., cryptorchidism, labial hypoplasia or aplasia, microphallus, hypospadias)
Limbs (e.g., pterygium, shortening, webs, cord wrapping, absent patella, dislocated radial heads, dimples)
Jaw anomaly (e.g., micrognathia, trismus)
Facies (e.g., asymmetry, flat bridge of nose, stork mark, ptosis, movement)
Scoliosis or kyphosis (fixed or flexible)
Dimples (over specific joints or bones)
Skin (e.g., hemangiomas, defects, hirsuitism, dimples, creases, amniotic bands)
Dermatoglyphics (e.g., absent, distorted, crease abnormalities)
Hernias (e.g., inguinal and umbilical), abdominal wall defect
Other features of fetal akinesia sequence:
Intrauterine growth restriction
Pulmonary hypoplasia
Craniofacial anomalies (micrognathia, hypertelorism, cleft palate, depressed tip of nose, high bridge of nose)
Functional short gut with feeding problem
Short umbilical cord

- *Malformations*

Skull (e.g., craniosynostosis, asymmetry, microcephaly)
Central nervous system (e.g., structural malformation, seizures, mental retardation)
Eyes (e.g., small, corneal opacities, malformed, ptosis, strabismus)
Palate (e.g., high, cleft, submucous cleft)
Limb (e.g., amputations, synostosis)
Tracheal or laryngeal clefts and/or stenosis
Heart (e.g., congenital structural anomalies versus cardiomyopathy)
Lungs (e.g., hypoplasia versus weak intercostal muscles and/or hypoplastic diaphragm)
Changes in vasculature (e.g., hemangiomas, cutis mamorata, blue cold distal limbs)
Genitourinary (e.g., structural anomalies of kidneys, ureters, bladder)
Other visceral anomalies

- *Neurologic abnormalities*

Vigorous versus lethargic
Tone (hypotonic, hypertonic)
Deep tendon reflexes (present versus absent, slow versus fast)
Sensation (intact or not)

- *Muscle*

Mass (normal versus decreased)
Texture (soft versus firm)
Fibrous fatty bands
Changes with time
Abnormalities of smooth, striated, cardiac, eye, craniofacial, diaphragm, intercostal muscles.

- *Connective tissue*

Skin (e.g., soft, doughy, thick, extensible)
Subcutaneous (e.g., decreased fat, increased fat, calcifications)

TABLE 6.1 (*Continued*)

Hernias (inguinal, umbilical, diaphragmatic or eccentric)
Joints (e.g., thickness of capsule, symphalangism)
Tendon attachment and length

Natural History
- *Changes with time*

Developmental landmarks (motor versus social and language)
Intellectual abilities
Growth of affected limbs
Progression of contractures
Course (lethal versus central nervous system damage versus stable versus improvement)
Asymmetry (decreases or progresses)
Trunk versus limb changes
Socialization
Feeding problems

- *Response to therapy*

Spontaneous improvement
Response to physical therapy
Response to casting
Which surgery at which time
Development of motor strength proportionate to limb size
Abnormal reaction to drugs

Laboratory Evaluations
- Radiographs if:
  Bony anomalies (e.g., gracile, fusions, extra or missing patella, carpals and tarsals)
  Disproportionate
  Scoliosis
  Ankylosis or synostosis
  Dislocation (e.g., hips, radial head, patella)
- Computed tomography or magnetic resonance imaging of brain or muscle if neurological abnormalities suspected or to document muscle mass obscured by contractures
- Ultrasound evaluation of CNS (if young)
- Ultrasound of heart or abdominal organs
- Ultrasound of muscles to establish potential muscle quantity
- Chromosome studies if:
  Multiple system involvement
  Central nervous system abnormality (eye, microcephaly, mental retardation, lethargic, degenerative)
  Streaky or segmental involvement
  Consider fibroblast studies if mental retardation present and lymphocytes are chromosomally normal
- Consider exome or whole genome sequencing
- Documentation of range of motion and position with photographs
- Video of movement (including facial) demonstrating range of movement and strength. Repeat at regular intervals
- Immunologic studies – IgM in newborn
- Muscle biopsy in normal and affected areas at time of surgery to distinguish myopathic from neuropathic from endplate disorders (electron microscopy, histopathology, mitochondria, storage, etc.). Rule out ragged red fibers
- Electromyogram in normal and affected area
- Nerve conduction in normal and affected area

TABLE 6.1 (*Continued*)

- Serum creatine phosphokinase if:
  Generalized weakness
  Doughy or decreased muscle mass
  Progressively worse
- Consider spinal muscular atrophy (SMA) if hypotonia, may need to consider chromosomal deletion on 5q involving the SMA gene

Autopsy

Visceral anomalies, other deformations or disruptions
Central nervous system – brain neuropathology
Spinal cord (number and size of anterior horn cells, presence or absence of tracts at various levels)
Peripheral nerve, ganglion, myelin
Eye (neuropathology)
Muscle tissue from different muscle groups and types – electron microscopy, special strains, fiber size, rule out ragged red fibers
Diaphragm for thickness or hernia
Are fibrous bands replacing muscle?
Are tendons in normal position?
Are tendon attachments normal?
Is there cartilaginous or bony fusion?

Gene Mutation Studies

Studies if suspect specific disorder where gene has been identified as in:
- Distal arthrogryposes
- Multiple pterygium syndromes
- Lethal congenital contracture syndromes
- Myasthenia syndromes

Microarray in intellectual disability or CNS structural anomaly
Exome sequencing and/or whole genome sequencing

Differential Diagnosis

The differential diagnosis of arthrogryposis is extensive. Table 6.2 provides an overview according to clinical findings. The reader is also directed to other reviews (Hall 2013, 2014c, 2015; Hall and Kiefer 2016).

MANIFESTATIONS AND MANAGEMENT

Growth and Feeding

Children with multiple congenital contractures are smaller than expected for age and family, generally at the third centile for their familial height (Staheli et al. 1998). Growth curves have been developed for various types of arthrogryposis (Hall 1998, 2010a, 2010b). Some types of arthrogryposis, apparently those involving neurologic etiology, commonly have obvious shortening of the limbs as adults. For instance, children with severe amyoplasia involving only the upper limbs usually have shortening and decreased muscle mass of the arms. Similarly, in individuals with amyoplasia involving

TABLE 6.2 Specific organ or area of involvement in various disorders with arthrogryposis

ARACHNODACTYLY

Blepharophimosis arachnodactyly syndrome – dislocated radial heads
Congenital contractural arachnodactyly
Congenital Marfan syndrome
Marden–Walker syndrome
Van Benthem syndrome
Van den Ende-Gupta syndrome.

ARMS ONLY

Amyoplasia
Shalev arthrogryposis
Baraitser camptodactyly

BONE ABNORMALITIES

Angulation of long bone – apex of angulation has dimples
Bruck syndrome – easily fractured
Camptomelic dysplasia
Caudal deficiency and asplenia
Coalitions – many types
Conradi–Hunermann syndrome (chondrodysplasia punctata)
Diastrophic dysplasia
Dyggve–Melchior–Clausen syndrome
Dyssegmental dysplasia
Freeman–Sheldon dysplasia
Geleophysic dysplasia
Humeroradial synostosis
Kniest syndrome
Liebenberg syndrome
Lenz–Majewski dysplasia
Lower limb and pelvic dysplasia (AR)
Megalocornea and skeletal dysplasia
Mesomelic dysplasia
Metaphyseal dysplasia
Metatropic dysplasia
Osteogenesis imperfecta
Osteolysis – Teebi syndrome
Parastremmatic dysplasia
Patella aplasia – hypoplasia syndrome
Pseudodiastropic dysplasia
Radio-ulnar synostosis
Rhizomelic chondrodysplasia punctata
Saul–Wilson dysplasia
Spondyloepiphyseal dysplasia congenita
Symphalangism – many types
Van den Ende-Gupta syndrome

BOWEL ABNORMALITY

Amyoplasia – gastroschisis
FG syndrome – imperforate anus
Mitochondrial disorders – atresias
Shalev arthrogryposis – umbilical hernia
VATER association

TABLE 6.2 *(Continued)*

CARDIAC ABNORMALITY

Camptodactyly, pericarditis, synovitis
Camptodactyly, Tel-Hashomer –congenital heart disease
Cardiac deficiency and asplenia
Cardiomelic syndrome
Congenital Marfan syndrome – aortic incompetency
Contractural arachnodactyly
Ectodermal dysplasia and cardiomyopathy
Geleophysic dysplasia
Holt–Oram syndrome
Larsen syndrome
Mitochondrial defects
Myhre muscle hypertrophy
Nemaline myopathy
Neuropathic, Israeli-Arab
PHAVER syndrome
Pfeiffer
    cardio-cranial syndrome (type V acro-cephalo-syndactyly)

CHEST DEFORMITY

Adducted thumbs syndrome
Camptodactyly, Guadalajara – pectus excavatum
Congenital Marfan syndrome
Contractural arachnodactyly – pectus excavatum
Fryns syndrome – absent diaphragm
Hoepffner syndrome- pectus excavatum
King Denborough syndrome
Multiple pterygium, Escobar type – restrictive lung disease
Neuromuscular disease of larynx – absent arytenoid cartilage.
Pena–Shokeir phenotype
Rutledge syndrome – diaphragmatic hernia
Schwartz–Jampel syndrome – pectus excavatum
Spondylothoracic dysostosis
Van Bentham syndrome
Van Biervielt syndrome

CLEFT PALATE

Aase–Smith syndrome
Adducted thumbs syndrome
Bartsocas–Papas syndrome
Bixler microcephaly
Camptomelic dysplasia
Diastrophic dysplasia
Distal arthrogryposis 2C
Dyssegmental dysplasia
Ectodermal dysplasia and cleft lip/palate
Focal femoral dysplasia
Fryns syndrome
Gordon syndrome
King Denborough syndrome
Kniest syndrome
Larsen syndrome
Multiple pterygium, Escobar type
Oculo-auriculo-vertebral spectrum

TABLE 6.2 (Continued)

Oral-cranial-digital dysplasia
Oto-palatal-digital syndrome
Popliteal pterygium syndrome, Gorlin Type
Rudiger syndrome
Sonoda syndrome

VSR syndrome

CENTRAL NERVOUS SYSTEM STRUCTURAL ANOMALY

Aase–Smith syndrome
Adducted thumbs syndrome – dysmyelination
Basal ganglion syndrome – choreoathetosis
Bixler microcephaly
Blepharophimosis, mental retardation, Dandy–Walker malformation – cerebellar anomalies
Caudal deficiency and asplena – agenesis corpus callosum
Clasped thumb syndrome
Coffin Siris syndrome
Dandy–Walker syndrome
FG – agenesis corpus callosum
Fowler syndrome – proliferative vasculopathy, hydranencephaly, hydrocephaly
Fukuyama muscular dystrophy – hetereotopias, cerebral atrophy, lissencephaly, polymicrogyria
Genito patellar syndrome – absent corpus callosum
Johnson hyperkeratosis and decreased posterior columns
Lissencephaly with fetal akinesia
Martsolf syndrome - Arnold–Chiari malformation
Miller–Dieker syndrome – lissencephaly
Multiple pterygium, lethal type – cerebellar hypoplasia
Neu-Laxova syndrome– lissencephaly, agenesis corpus callosum
PEHO – small cerebellum, spongy vacuolation of cortex; loss of Purkinje cells
Tuberous sclerosis
van den Ende-Gupta syndrome
Walker–Warburg – lissencephaly, dysmyelination
X-linked anterior horn cell loss

CRANIOSYNOSTOSIS

Adducted thumbs syndrome
Antley–Bixler syndrome
Campomelic dysplasia
Pfeiffer-cardio-cranial syndrome
Symphalangism/brachydactyly
Trigonencephaly syndrome
VSR syndrome

DEAFNESS

Deafness and camptodactyly
Diastrophic dysplasia
Distal arthrogryposis and deafness
Kniest syndrome
Multiple synostosis syndrome
Myhre muscle hypertrophy
Oculo-auriculo-vertebral spectrum
Ohno blepharophimosis syndrome

TABLE 6.2 (Continued)

Oto-palato-digital syndrome
Waardenburg syndrome

DENTAL

Distal arthrogryposis with absent teeth buds
Ectodermal dysplasia – cardiomyopathy, hypoplastic teeth
Ectodermal dysplasia cleft palate – dental anomaly
Lenz–Majewski syndrome – abnormal enamel, microdontia
Oculo-dento-digital syndrome – hypoplastic enamel microdontia
Ohno blepharophimosis - hypoplasia
Oto-palatal-digital syndrome- adontia

DISLOCATED/ABNORMAL RADIAL HEAD

Antley–Bixler syndrome
Blepharophimosis arachnodactyly syndrome
Humero-radial synostosis
Ives microcephaly
Liebenberg syndrome
Nievergelt–Pearlman syndrome
Ophthalmo-cranial digital syndrome
Radio-humeral synostosis
Roberts syndrome
Rutledge syndrome
Symphalangia/brachydactyly syndrome

EARS ABNORMAL

Blepharophimosis syndrome – Dandy–Walker malformation, auricular pits
Bowen–Conradi syndrome – large ears
Contractual arachnodactyly – over folded top
Diastrophic dysplasia – calcified cartilage
Oculo-auriculo-vertebral spectrum
Ohno-blepharophimosis
Oto-onycho-peroneal syndrome – unfolded ears
PHAVER syndrome

EYE ANOMALY

Blepharophimosis arachnodactyly syndrome
Blepharophimosis, MR, Dandy–Walker
Bowen–Conradi syndrome – cloudy cornea
Camptodactyly, Guadalajara type– microcornea microphthalmia
Camptodactyly, Kilic type – myopia, ptosis, medial fibrosis
COFS – cataracts, microphthalmia, retinal pigmentation
Conradi–Humermann syndrome – cataracts
Distal arthrogryposis 2B – ophthalmoplegia
Duane's retraction syndrome with contractures – inability to abduct the eye
Ectodermal involvement and caudal appendage – blepharophimosis
Freeman–Sheldon syndrome – ptosis
Fryns syndrome – cloudy cornea
Fukuyama muscular dystrophy – optic atrophy
Kniest syndrome – myopia
Mietens syndrome – corneal opacity, strabismus, nystagmus

(continued)

TABLE 6.2 (Continued)

Marden–Walker syndrome – blepharophimosis, ptosis
Marfan syndrome – dislocated lens
Martsoff syndrome – cataracts
Megalocornea and skeletal anomalies
Möbius syndrome – lack of eye movement
Neu-Laxova syndrome – open eyes, hypertelorism, exophthalmoses
Oculo-auriculo-vertebral spectrum
Oculo-dental-digital syndrome – sunken eyes
Ohno blepharophimosis
Ophthalmo-mandibular-melic dysplasia – corneal opacities
Ophthalmoplegia, retinitis pigmentosa, mental retardation – ophthalmoplegia, retinitis pigmentosa
PEHO – optic atrophy
Restrictive dermopathy – open eyes
Schwartz–Jampel syndrome– myopia
Spastic paraplegia, Goldblatt type
Spondylo-epiphyseal dysplasia – myopia
van den Ende-Gupta syndrome
Walker–Warburg syndrome – retinal aplasia, microphthalmia, cataract
Weill–Marchesani syndrome – dislocated lenses
Winchester syndrome – corneal opacity

FACIAL MOVEMENT DECREASED

Camptodactyly, London
Distal arthrogryposis 2B – ophthalmoplegia, decreased facial movement
Distal arthrogryposis with facial involvement (Sheldon–Hall)
Freeman–Sheldon syndrome – pursed mouth
Marden–Walker syndrome
Maternal myasthenia gravis
Möbius syndrome
Multiple pterygium, Escobar type
Myotonic dystrophy
Schwartz–Jampel syndrome

GENITAL ANOMALY

Campomelic dysplasia
Ectodermal dysplasia and clefts – hypospadias
Multiple pterygium, Escobar type
Popliteal pterygium, Gorlin type

LUNG HYPOPLASIA

Eagle–Barrett syndrome
Finnish anterior horn cell
Fetal akinesia deformation sequence
Fowler syndrome
Lethal congenital contracture syndrome I–III

JAW ANOMALY

Cardiomelic syndrome
Distal arthrogryposis 2E – trismus
Blepharophimosis arachnodactyly syndrome

TABLE 6.2 (Continued)

Distal arthrogryposis with facial involvement (Sheldon–Hall)
Freeman–Sheldon syndrome
Ophthalmo-mandibulo-melic syndrome – fusion of temporomandibular joint
Schwartz–Jampel syndrome
Trismus pseudocamptodactyly syndrome

LEGS ONLY AFFECTED

Amyoplasia – legs only
Autosomal dominant – Fleury
Autosomal recessive – Sarralde, Ray
Kuskokwim syndrome
Meningomyelocele, spina bifida
X-linked arthrogryposis – Zori

LIMB DEFICIENCY

Amyoplasia – distal loss
Attempted termination of pregnancy
Focal femoral dysplasia – proximal loss
Hanhart syndrome
Holt–Oram syndrome – radial ray loss
Ives microcephaly
Multiple synostosis syndrome – distal loss
Oculo-dental syndrome – hypoplasia
Oto-onycho-peroneal syndrome – fibular aplasia
Poland anomaly – distal loss
Popliteal pterygium syndrome, Gorlin type
Roberts syndrome
Symphalangism/brachydactyly syndrome – distal loss

LIVER

Gaucher disease infantile type
Nezeloff syndrome/arthrogryposis-renal dysfunction-cholestasis syndrome
Phosphofructokinase deficiency
Zellweger syndrome

MALIGNANT HYPERTHERMIA

King Denborough syndrome
Multiple pterygium and malignant hyperthermia
Myopathic types of arthrogryposis

MUSCLES

Central core myopathy
Congenital fiber type disproportion
Continuous muscle discharge and tibulation
Distal arthrogryposis 2B – firm muscles, ragged red fibers
Fukuyama congenital muscle dystrophy
Myhre muscle hypertrophy
Nemeline myopathy
Neuromuscular disease of larynx
Schwartz–Jampel syndrome – myotonia
Stiff man/stiff baby syndrome

TABLE 6.2 (Continued)	TABLE 6.2 (Continued)
NOSE Bixler microcephaly – large nose Bowan–Conradi syndrome – prominent nose Ectodermal involvement with caudal appendage – bulbous nose Freeman–Sheldon syndrome– notched alae nasi Hoepffner syndrome – pinched nose **NEUROLOGICAL** Carbohydrate deficient, glycoprotein syndrome – ataxia, hypotonia Neurosensory sensory defect with arthrogryposis PEHO – hypsarrhythmia, hypotonia, hyperreflexion Spastic paraplegia syndrome **PAIN** Myopathies Noonan type arthrogryposis – night cramps **PATELLA – APLASIA OR HYPOPLASIA** Camptomelic dysplasia Coffin–Siris syndrome Ear, patella, short stature syndrome Genitopatellar syndrome Guadalajara camptodactyly Nail patella syndrome Patella aplasia, hypoplasia syndrome **RENAL** ARC Eagle–Barrett syndrome Freeman–Sheldon syndrome Genital patellar syndrome – hyphonephosis Nail-patella syndrome Potters syndrome – agenesis Rudiger syndrome – hydronephrosis, urethral stenosis Schinzel–Giedion syndrome – hydronephrosis Sonoda – hydronephrosis Tuberous sclerosis Zellweger syndrome **SEIZURES** Basal ganglion disease Bixler microcephaly FG syndrome King Denborough syndrome Miller–Dieker syndrome PEHO – hypsarrhythmia Phosphofructokinase deficiency Saul–Wilson skeletal dysplasia Tuberous sclerosis X-linked arthrogryposis – anterior horn cell loss	**SHIN DIMPLES** Angulation of long bones Autosomal recessive arthrogryposis – inherited type Camptomelic dysplasia X-linked arthrogryposis type 3 **SKIN** Absence of dermal ridges syndrome – milia Bartsocas–Papas syndrome – abnormal hypoplastic skin around the mouth and anus Camptodactyly, London – icthyosis Conradi–Hunermann syndrome – alopecia, follicularis Ectodermal dysplasia and cardiomyopathy – woolly, sparse hair, dry skin Ectodermal dysplasia and contractures – hyperkeratosis, pili torti, dry skin Ectodermal dysplasia and clefting – sparse hair Ectodermal involvement and caudal appendage – slow growing hair Ehlers–Danlos VIB – hyperextensibility, bruisable Johnson hyperkeratosis and decreased posterior columns Lenz–Majewski syndrome – thin with prominent vessels Leprechaunism – loose, dry, hirsute Neu-Laxova syndrome – constrictive, ichthyotic skin Oto-onycho-peroneal syndrome – absent or dysplastic nails Popliteal pterygium, Gorlin type – absent nails and lip webs Proteus syndrome – lipomas, harmatomas, hemangiomas Puretic–Murray syndrome – fibromatosis Restrictive dermopathy Schinzel–Giedion syndrome – hypertrichosis, abundant skin Tuberous sclerosis – adenoma, depigmentation, shagreen patches Waardenburg syndrome – white patches Winchester syndrome – malar flush, thick facial skin **SYNDACTYLY** Absence of dermal ridges syndrome Amyoplasia Bartsocas–Papas syndrome Neu-Laxova syndrome **THUMBS, PROXIMAL OR ABNORMAL** Adducted thumbs syndrome Clasped thumb and mental retardation Clasped thumb syndrome Deafness and camptodactyly Dystrophic dysplasia – "hitch-hiker" thumb Fryns syndrome Hand-foot-uterus syndrome – small Holt–Oram syndrome – small Martsoff syndrome MASA syndrome Ophthalmo-cranial-digital/Juberg–Haywood syndrome X-linked arthrogryposis type 2

(continued)

TABLE 6.2 (*Continued*)

TONGUE

Hanhart syndrome – small
Möbius syndrome – hypoplastic

TRISMUS

Carney variant
Distal arthrogryposis 2E
Distal arthrogryposis – Sheldon–Hall
Freeman–Sheldon syndrome
Trismus pseudocamptodactyly

VERTEBRAL ANOMALY

Distal arthrogryposis 2D
Dyssegmental dysplasia
Ehlers-Danlos VIB – scoliosis
Larsen syndrome
Mitochondria defects
Multiple pterygium, Escobar type – segmentation defects
PHAVER syndrome
Sacral agenesis
Spondytothoracic dysostosis
VATER association
X-linked arthrogryposis, Zori

WEBBING AND PTERYGIA

Amyoplasia
Antecubital pterygium
Bruck syndrome
Distal arthrogryposis with facial involvement (Sheldon–Hall)
King Denborough syndrome
Kuskokwim syndrome
Multiple ptergyium, Escobar type – progressive
Popliteal pterygium syndrome – Gorlin type

---

only their lower limbs, the lower segment is obviously short. Because of lack of movement, the bones in arthrogryposis often have less calcification and can be recognized, even in utero, to be osteoporotic. Children with arthrogryposis may have other causes of growth restriction such as growth hormone deficiency, but that is rare.

Many children with arthrogryposis have problems with feeding. This appears to be related to lack of musculature or lack of maturation of intestinal muscle coordination. Many affected children have aversion to solids and need help to develop regular swallowing. Often, improvement in gastrointestinal function is observed over the first six months. Because of immobility, many older individuals have constipation and require bowel training.

For some types of arthrogryposis, obesity can be a problem. The children are thought by their physicians and family to be underweight because their limbs look so thin. The families may then over feed the children with the best of intentions. In fact, most individuals with arthrogryposis have decreased muscle mass and would be expected to be below the usual weight for their length/height (approximately 5–15% below expected weight for height or age since the muscle of the limbs is normally responsible for that much body weight). It is important during childhood to monitor the amount of subcutaneous fat and reduce caloric intake if there is excess subcutaneous fat (Hall 1998) in order to avoid the physical difficulties and morbidities associated with obesity

Evaluation

- Regular height and weight measurements and photographs should be recorded at least every six months in order to document growth and joint positioning during the first few years. After that, at least every two years. These should be compared to the norms for the condition where possible (Hall 1998) and not to standard growth charts, since it can be expected that children with all forms of arthrogryposis will be short and underweight when compared to age standards for the general population.
- Awareness that feeding problems are frequently present in children with arthrogryposis should lead to early standard evaluation.

Management

- Most children with arthrogryposis will not require treatment for their growth. However, care must be given not to over feed young children in order to avoid obesity.
- Some children require tube feeding early and occasionally, gastrostomy is needed in the first six months.
- Therapy will need to be individually tailored to the child's diagnosis.

Development and Behavior

The development and behavior of individuals with arthrogryposis is very much dependent on the type of arthrogryposis they have. For instance, children with severe central nervous system dysfunction are often unable to interact with and respond to their caregivers in a normal way; whereas children with amyoplasia tend to be assertive, interactive, and very engaged in solving their own problems. Determining whether there is a structural rather than functional basis for those children with central nervous system dysfunction can be helpful in understanding which modalities of therapy to use and the expected long-term outcome (see Neurologic, below). In other children, such as with amyoplasia, the ventricles may seem slightly dilated and the spinal cord may appear slightly small. This is not necessarily cause for concern and may reflect fewer neurons but excellent intellectual capacity can be preserved (Hall et al. 2014). Obviously, if limbs cannot be moved because of lack of functional muscle or nerve, there will be delay in motor milestones, but this is not necessarily reflective of future cognitive ability. It should be assumed that, until appropriate evaluations can be done at an older age, children with arthrogryposis will be able to achieve independence and therefore they should be given every opportunity for normal development (Dillon et al. 2009; Nouraei et al. 2017). Nevertheless, approximately one-third will have severe developmental delay and intellectual impairment and may die at an early age or require custodial care.

Behavioral difficulties may be specific to the underlying cause of arthrogryposis and so generalizations cannot be made.

Evaluation

- Developmental assessment should be accomplished at regular well child visits using standard screening methods. Individuals with arthrogryposis can be expected to have motor delays related to their contractures, but social development may well be normal.
- Some children with central nervous system dysfunction develop spasticity or hypotonia on top of their multiple congenital contractures.
- Evaluation of joint range of motion and position of the limbs at rest should be documented on a regular basis in order to track the effects of therapy.
- A muscle biopsy should sample both normal and affected muscle to distinguish myopathic from neuropathic etiologies. Appropriate stains for muscle, nerve, myelin, and end plate should be used, and electron microscopy studies are needed for distinguishing various myopathies versus disuse atrophy of the muscle.
- Magnetic resonance imaging of muscle mass helps to determine whether the muscle mass that is present can be mobilized through physical therapy.
- Computed tomography or magnetic resonance imaging of the central nervous system, including spinal cord, should be performed in all children with arthrogryposis at least once after myelinization is complete in order to evaluate central nervous system structures. After age 2 years is appropriate.

Management

- Spasticity and hypotonia are treated in a standard manner.
- When residual muscle is present, extensive physiotherapy may be beneficial.

Intervention and education for children with developmental delays are standard or, in some cases, specific to the underlying disorder.

Musculoskeletal

This is the area usually requiring most attention for individuals with arthrogryposis. In general, joints that are fixed or contracted should be mobilized as much as possible during the first four months of life, when they are most responsive. This relates to preserving muscle tissue so that it does not atrophy from non-use. Also, it ensures the joint surfaces are molded by normal use and movement, with a mildly curved surface rather than the flat joint surface of non-use. It also appears that contractures are most responsive to stretching in the first four months of life.

Several specific types of arthrogryposis have limitation of the jaw movement. The size of the oral opening is particularly important for dental care, feeding and anesthesia. Aspiration, in particular, can be life threatening when there is limitation of jaw opening. Parents need to be aware of this possibility. Temporomandibular join (TMJ) surgery is not recommended, although jaw stretching may gradually increase the aperture. A few individuals with arthrogryposis on the basis of myopathies may have a malignant hyperthermia reaction to anesthesia.

The natural history of the various types of arthrogryposis will dictate response to therapies. For instance, in diastrophic dysplasia, too vigorous movement of joints of the fingers leads to calcification of the cartilage and ultimately fusion, while in most other types of congenital contractures stretching is essential to mobilizing the joint. Determining the etiology of the decreased movement is also important in terms of the expectation for improvement and response to therapy. For instance, if the onset of limitation of movement is in the last trimester, most infants will have relatively good response to physical therapy, while if the decreased fetal movement has been present since the beginning of the second trimester, the contractures are likely to be quite severe and resistant to therapy.

In many forms of arthrogryposis, the bones are gracile and osteoporotic. Thus, iatrogenic fractures are not uncommon. Almost 10% of affected infants have fractures during the birth process or during the first week of life.

The back is often involved in arthrogryposis and stiffness is frequent. In forms of arthrogryposis with hypotonia or severe weakness, kyphosis and scoliosis are frequent. These can lead to life threatening pulmonary compromise.

The hips are frequently dislocated in various forms of arthrogryposis and will need appropriate treatment.

Arthritis in joints which have had congenital contractures is frequent, starting in the 20s. This appears to be a degenerative osteoarthritis.

Evaluation

- All children with arthrogryposis will need a complete evaluation when first diagnosed by an orthopedist familiar with arthrogryposis, and then followed by an

- orthopedist. Most often the orthopedist will do a skeletal survey and follow affected areas with repeated X-rays
- An assessment of jaw opening is important at diagnosis.
- Careful documentation, with photographs, of the position and range of joints is needed at diagnosis. Evaluation of joint range of motion and position of the limbs at rest should be documented on a regular basis in order to track the effects of physical therapy.
- A search should be made for webbing (pterygia), dimples, or excessively tight skin.
- Examination of flexion creases may provide a clue to the timing and onset of decreased joint movement.
- Early and regular monitoring of the spine, particularly at puberty, is important.
- A muscle biopsy at surgery should sample both normal and affected muscle to distinguish myopathic from neuropathic etiologies. Appropriate stains for muscle, nerve and end plate should be used and electron microscopy studies are needed to distinguish various myopathies from disuse atrophy of the muscle and endplate disorders, since therapies would be different.
- Magnetic resonance or ultrasound imaging of muscle mass helps to determine whether there is muscle mass present that can be mobilized through physical therapy.
- Most affected children will need physical therapy, and so rehabilitation evaluation will be needed.

Management

- The philosophy behind treatment is to maximize independent function. This involves alignment and achieving joint positions of function (Sells et al. 1996; Szoke et al. 1996; Staheli, 1998; Bevan et al. 2007; Miller and Sawatzky 2017; Steen 2017; Komolkin et al. 2017).
- Therapy will need to be individually tailored to the child's diagnosis and limitations (Villard et al. 2017; Binkiewicz-Glinska et al. 2016).
- Standard physical and occupational therapies, as well as orthopedic intervention, are recommended depending on diagnosis.
- The therapies used are different at different times during the individual's life and may be different for individual types of arthrogryposis. When residual muscle is present, extensive physiotherapy may be beneficial.
- When arthrogryposis is prenatally diagnosed there is an opportunity for therapy aimed at increasing fetal movement. No controlled studies are available.
- Maternal physical activity, deep breathing, and caffeine all increase fetal movement.
- A program of in utero "treatment" in moderation may be undertaken starting in the second trimester including (1) maternal exercise with two or three vigorous 15 minute walks each day, (2) 10 deep breaths four times each day and (3) 3–4 cups of coffee, tea, or caffeinated soft drinks each day (only after the first trimester).
- Early delivery (perhaps at 36 weeks if lungs are mature) may also have benefit since physical therapy can begin sooner, and the contractures are likely to be less severe.
- Osteopenia/osteoporosis is often present at birth and responds to weight bearing spontaneously. It does not require special therapy, but care should be given to avoid iatrogenic fractures.
- During the first year, the goal of treatment is to increase the range of motion at the affected joints whenever possible and to maintain the acquired range of motion through splinting in a way that positions the infant appropriately and also allows and encourages active movement. This requires helping parents to feel comfortable and be knowledgeable about the condition and the therapies. Often, alternative casting and active stretching works best.
- Stretching of joints with contractures should be undertaken as early as possible. Physical therapy is extremely important in mobilizing joints, maintaining muscle tissue that is present, and maintaining positions that have been achieved through surgery. Some children will respond to the stretching of joints through physical therapy alone. However, the stretching may be quite difficult and painful.
- Physical therapy consultation from professionals familiar with children, and contractures in children, is required to teach parents how to do the stretching properly.
- Surgery is frequently needed to place a limb into a position of function; however, the Ponseti method (of sequential casting) has worked in milder cases (Oishi et al. 2017).
- Staheli (1998) has observed that the primary deformity tends to recur and active range of motion and regular splinting (particularly night splinting) are important to maintain the range of motion achieved by surgery. Range of motion using gentle but full pressure must be done several times a day and the majority of this is done by the parents or caregivers. Night splinting may be useful it if is tolerated up until puberty.
- Short periods of casting provide a longer period of stretch and maintains gains from passive movement; however, lengthy casting may lead to muscle atrophy.
- Through the use of casts, splints and other prosthetic devices, the position of the function can usually be maintained (Salminger et al. 2016).
- Night splints are particularly important for maintaining the range of motion.

- Experience suggests that surgical procedures at the temporomandibular joint only make jaw limitation worse and lead to scarring and further jaw contracture.
- Simple stretching of the jaw during the first 4-6 months of life can increase the amount of jaw opening that then can be maintained by splinting with a soft rubber insert.
- Standard therapy must be undertaken for kyphosis and scoliosis since they may compromise pulmonary function.
- Dislocated hips should be treated in a standard fashion, at first with stretching and triple diapers, and, later if necessary, by surgery (Szoke et al. 1996).
- Therapy must be individualized to the affected individual and the cause or type of arthrogryposis, and will include all involved areas. Ideally, there will be as few hospitalizations as possible by combining surgical procedures.
- Alerting the anesthesiologist to the increased risk of malignant hyperthermia in some types of arthrogryposis, particularly myopathies, will allow standard prophylaxis for and treatment of malignant hyperthermia if it occurs (Froster-Iskenius et al, 1988; Ma and Yu 2017; Gleich et al. 2017). Some individuals with limited jaw opening will be difficult to intubate.
- Many affected children will need special consideration as they enter school. Consultation with the school prior to entry and on a regular basis is appropriate (Staheli et al. 1998).

Neurologic

Some of the conditions associated with arthrogryposis are associated with abnormalities of the central nervous system. These will often have additional neurologic manifestations, such as developmental delay (see Development and Behavior), hypotonia, spasticity, and/or seizures.

Evaluation

- A complete neurologic evaluation is indicated at diagnosis to establish baseline neurologic functioning.
- A complete annual neurologic evaluation is important during childhood since certain types of problems, such as ophthalmoplegia, can develop later or be progressive. These progressive features can be extremely important for the designation of a specific diagnosis and for recognition of the natural history of particular types of arthrogryposis.

Management

- Spasticity and hypotonia are treated in a standard manner.
- Other neurological problems, such as seizures, are also treated in a standard manner unless dictated otherwise by their underlying condition.

- The rare form of Escobar syndrome due to mutations in the embryonic neuroreceptor may respond to drug therapy activating the adult receptor (Hoffman et al. 2006).

Ophthalmologic

The eyes are involved in some forms of arthrogryposis with ophthalmoplegia, ptosis, cataracts or retinal changes reported, although in the majority, the eyes are normal. The presence of esotropia is relatively common and may be suggestive of a central nervous system abnormality. Cataracts are evident in some metabolic forms of multiple congenital contractures. Retinal or macular structural abnormalities may reflect a failure of central nervous system development or maturation. Scars may signify congenital infection.

Evaluation

- Careful ophthalmologic evaluation at diagnosis and every two years during childhood seems appropriate since subtle changes may impair function disproportionately.

Management

- If abnormalities are found, standard therapy should be undertaken.

Respiratory

Some individuals with arthrogryposis have significantly compromised pulmonary function from birth. This seems to relate to lack of in utero breathing and failure to "stretch" and mature the lungs. Tracheomalacia is also seen. Improvement is often seen in the first four months of life. The respiratory problems may be related to lack of normal movement of the intercostal or diaphragmatic muscles. Children with severe pulmonary hypoplasia may not be able to be weaned off the respirator and in circumstance where the decision is made to continue intervention, treatment may require permanent tracheostomy and ventilation. A few types of arthrogryposis develop compromised pulmonary function secondary to scoliosis. Lack of physical activity and the development of scoliosis may lead to additional respiratory compromise.

Evaluation

- Since respiratory compromise is common in arthrogryposis, early evaluation targeted at determining the specific problem is appropriate.
- Pulmonary function studies and ultrasound assessment of diaphragmatic movement should be undertaken at diagnosis if respiratory compromise is present.

Management

- Supportive therapy may be all that is needed until the lungs can grow and mature.
- There is a small subgroup of individuals with several types of arthrogryposis who have severe respiratory compromise. They may require long-term or even permanent respiratory support. Life expectancy may be compromised for these patients.

Ears and Hearing

Many children with arthrogryposis develop chronic otitis media. In some types of arthrogryposis this is associated with a high palate, an overt or submucous cleft palate. In other types of arthrogryposis, otitis media, and chronic fluid in the middle ear may be due to being in a horizontal position for extended periods of time. In these individuals, there is theoretically no structural reason for the increase in otitis, but it certainly needs to be treated and monitored in the first few years of life to avoid permanent hearing loss. Some forms of arthrogryposis have associated deafness due to structural anomalies, e.g., ossicular fusion, and/or inner ear dysplasias. Early recognition may help to identify the specific type of arthrogryposis.

Evaluation

- Newborn screening and then regular hearing tests every six months in early childhood and specific testing after upper respiratory tract infection during the first two years is appropriate. Thereafter, suspicion of the possibility of serous otitis should lead to evaluation.

Management

- Therapies for otitis media and chronic fluid in the middle ear are standard.
- Educational planning should take any hearing deficit into consideration.

RESOURCES

Parent Group Information

Australia

The Australian Arthrogryposis Group (TAAG) Inc
http://www.taag.org.au/
Jacqueline Brand (Secretary)
Phone: 02 4938 8060
Email: berriga@telstra.com

Canada

Shriner's Hospital for Children
c/o Noémi Dahan-Oliel
1529 Cedar Avenue
Montreal, QC H3G 1A6, Canada
Phone: 514-842-4464 ext. 2278
Email: ndahan@shrinenet.org

Denmark

Landsforeningen for armdefekte/bendefekte og AMC
Website: http://www.arm-bendefekte-amc.dk/?id=2

France

Alliance Arthrogypose
Email: contact2008@arthrogrypose.fr
Website: http://www.arthrogrypose.fr/pub/index.php

Germany

Interessengemeinschaft Arthrogryposis (IGA) e.V
Email: info@arthrogryposis.de
Website: http://arthrogryposis.de/

Ireland

Arthrogryposis Association of IrelandThe Secretary
Cor-na-GarkKilcormac
County Offaly, Ireland
Phone: 057-9135152
E-mail: enquiries@arthrogryposis.ie
Website: http://www.arthrogryposis.ie

New Zealand

The Arthrogryposis Group New Zealand
Jean Parsons
Secretary TAG-NZ
Glenone Lodge, 308 Mill Rd
R.D. Alfriston, Auckland, New ZealandPhone: 09 266 9999
E-mail: arthrogryposis.info@nzord.org.nz
Website: http://www.nzord.org.nz/support_groups/a/arthrogryposis_group_of_new_zealand_tag-nz

Norway

Landsforeningen for Arthrogryposis Multiplex Congenita
Berit Evensen
Phone: 35 51 01 30
E-mail: leder@amc-info.com
Website: http://www.arthrogryposis-alliance.eu/

Spain

Asociación para las Deficiencias que Afectan al Crecimiento y al Desarrollo Manuel Villalobos 41 41009 Sevilla, SPAIN
Phone: 08-7657846
Website: https://enfermedades-raras.org/index.php?option=com_content&view=article&id=652&Itemid=116

Sweden

AMC-föreningen i Sverige
AMC Society in Sweden
Nilstorp Way 51
18,147 Indore, Sweden
Phone: 08-7657846
E-mail: amcforeningen@gmail.com
Website: https://amcforeningen.se/

United Kingdom

The Arthrogryposis Group (UK)
Sharon Baker
E-mail: SMBakerTAGUK@aol.com
Website: http://www.tagonline.org.uk/

United States

AMC Support, Inc.
PO Box 6291
Spartanburg, SC 29304, USA
E-mail: bod@amcsupport.org
Website: http://www.amcsupport.org

REFERENCES

Bamshad M, Jorde LB, Carey JC (1996) A revised and extended classification of the distal arthrogryposes. *Am J Med Genet* 65(4)L277–281.

Bayram Y, Karaca E, Coban Akdemir Z, Yilmaz EO, Tayfun GA, Aydin H, Torun D, Bozdogan ST, Gezdirici A, Isikay S, Atik MM, Gmabin T, Harel T, El-Hattab AW, Charng WL, Pehlivan D, Jhangiani SN, Muzny DM, Karaman A, Celik T, yuregir OO, Yildirim T, Bayhan IA, Boerwinkle E, Gibbs RA, Elcioglu N, Tuysuz B, Lupski JR (2016) Molecular etiology of arthrogryposis in multiple families of mostly Turkish origin. *J Clin Investig* 126(2):762–768.

Bevan WP, Hall JG, Bamshad M, Staheli LT, Jaffe KM, Song K (2007) Arthrogryposis multiplex congenita (amyoplasia): an orthopaedic perspective. *J Pediatr Orthop* 27:594–600.

Binkiewicz-Glinska A, Wierzba J, Szurowska E, Ruckeman-Dziurdzinska K, Bakula S, Sokolow M, Renska A (2016) Arthrogryposis multiplex congenita – multidisciplinary care – including own experience. *Medycyna Wieku Rozwojowego* 20(3):191–196.

Clarren SK, Hall JG (1983) Neuropathologic findings in the spinal cords of 10 infants with arthrogryposis. *J Neurol Sci* 58:89–102.

Darin N (2000) *Neuromuscular Disorders in Childhood. Epidemiology and Characterization of a New Myopathy*. Institute for the Health of Women and Children, Department of Pediatrics, The Queen Silvia Children's Hospital, Goteborg University, Goteborg, Sweden.

Dillon ER, Bjornson KF, Jaffe KM, Hall JG, Song K (2009) Ambulatory activity in youth with arthrogryposis: a cohort study. *J Pediatr Orthop* 29:214–217.

Engel AG, Sine SM (2005) Current understanding of congenital myasthenic syndromes. *Curr Opin Pharmacol* 5:308–321.

Fahy MJ, Hall JG (1990. A retrospective study of pregnancy complications among 828 cases of arthrogryposis. *Genetic Counseling* 1:3–11.

Filges I, Hall JG, Friedman JM (2014) Exome sequencing identifies mutations in KIF14 as a novel cause of an autosomal recessive lethal cilopathy phenotype. *Clin Genet* 86:220–228.

Froster-Iskenius, Waterson JR, Hall JG (1988) A recessive form of congenital contractures and torticollis associated with malignant hyperthermia. *J Med Genet* 25:104–112.

Gaitanis JN, McMillan HJ, Wu A, Darras BT (2010) Electrophysiologic evidence for anterior horn cell disease in Amyoplasia. *Pediatr Neurol* 43(2):142–147.

Gleich SJ, Tien M, Schoreder DR, Hanson AC, Flick R, Nemergut ME (2017) Anesthetic outcomes of children with arthrogryposis syndromes: no evidence of hyperthermia. *Anesthesia and Analgesia* 3(124):908–914.

Graham JM Jr, Ayane-Yeboa K, Raams A, Appeldoorn E, Kleijer WJ, Garritsen VH, Busch D, Edersheim TG, Jaspers NGJ (2001) Cerebro-oculo-facio-skeletal syndrome with a nucleotide excision-repair defect and a mutated *XPD* gene, with prenatal diagnosis in a triplet pregnancy. *Am J Hum Genet* 69: 291–300.

Hall JG, Reed SD and Greene G (1982) The distal arthrogryposes: Delineation of new entities - Review and nosologic discussion. *Am J Med Genet* 11:185–239.

Hall JG (1985a) Genetic aspects of arthrogryposis. *Clinical Orthopedics and Related Research* 184:44-53.

Hall JG (1985b) In utero movement and use of limbs are necessary for normal growth: A study of individuals with arthrogryposis. In: *Endocrine Genetics and Genetics of Growth*, Papadatos J and Bartsocas CS, eds. New York: Alan R Liss Inc, pp. 155–162.

Hall JG (1986) Invited Editorial Comment: Analysis of Pena Shokeir phenotype. *Am J Med Genet* 25:99–117.

Hall JG (1996) Arthrogryposis associated with unsuccessful attempts at termination of pregnancy. *Am J Med Genet* 63:293–300.

Hall JG (1997) Arthrogryposis multiplex congenita etiology, genetics, classification, diagnostic approach, and general aspects. *J Ped Orthop B* 6:159–166.

Hall JG (1998) Overview of Arthrogryposis. In: *In: Arthrogryposis: A Text Atlas*, Staheli LT, Hall JG, Jaffe KM, Paholke DO eds, Cambridge: Cambridge University Press, pp. 1–25.

Hall JG (2009) Pena Shokeir phenotype (fetal akinesia deformation sequence) revisited. *Birth Defects Res A* 85:677–694.

Hall JG (2010a) Arthrogryposis. In: *Management of Genetic Syndromes*, 3rd ed. Cassidy SB and Allanson JE eds, John Wiley & Sons, Hoboken, NJ, Chapter 7, pp. 81–96.

Hall JG (2010b) Syndromes with congenital contractures. In: *Gorlin's Syndromes of the Head and Neck*, 5th ed. Hennekam RCM, Krantz ID and Allanson JE eds, Oxford University Press, NY.

Hall JG (2010c) The importance of muscle movement for normal craniofacial development. *J Craniofac Mov* 21:1336–1338.

Hall JG (2012a) Arthrogryposis (multiple congenital contractures) associated with failed termination of pregnancy. *Am J Med Genet* 158A:2214–2220.

Hall JG (2012b) *Revisiting limb pterygium syndromes. Science Sake. Festschrift Professor al Chirstos SP Bartsokas*. Honorary volume for Professor Christos S. Bartsocas. Athens, Greece: ZHTA Medical Publications, pp. 373–385.

Hall JG (2012c) Uterine structural anomalies and arthrogryposis-death of an urban legend. *Am J Med Genet* 161A:82–88.

Hall JG (2013) Arthrogryposes (Multiple Congenital Contractures). In: *Emery and Rimoin's Principle and Practice of Medical Genetics*. 6th ed. Rimoin DL, Pyeritz RE Korf BR eds. Churchill Livingstone: New York, Chapter 161, pp.1–161.

Hall JG (2014a) Amyoplasia involving only the upper limbs or only involving the lower limbs with review of the relevant differential diagnoses. *Am J Med Genet A* 164:859–873.

Hall JG (2014b) Arthrogryposis (multiple congenital contractures): diagnostic approach to etiology, classification, genetics, and general principles. *Eur J Med Genet* 57:464–472.

Hall, JG (2014c) Congenital contractures: Emphasizing multiple congenital contractures (MCC) – Arthrogryposis. In: Hudgins and Toriello's Signs and Symptoms of Genetic Disease. Hudgens L and Enns GM eds. *Oxford University Press: New York, Chapter* 27, pp. 420–439.

Hall JG (2014d) Oligohydramnios sequence revisited in relationship to arthrogryposis, with distinctive skin changes. *Am J Med Genet* 164A:2775–2792.

Hall JG (2015) Arthrogryposis. In: Human Malformations and Associated Anomalies, 3rd ed. Stevenson R and Hall JG eds. *Oxford University Press, NY, Chapter* 1, pp. 104–112.

Hall JG, Aldinger KA, Tanaka KI (2014) Amyoplasia revisited. *Am J Med Genet* 164:700–730.

Hall JG, Kiefer J (2016) Arthrogryposis as a syndrome: gene ontology analysis. *Mol Syndromol* 7:101–109.

Hall JG, Reed SD (1982) Teratogens associated with congenital contractures in humans and in animals. *Teratology* 25:173–191.

Hall JG, Reed SD, Greene G (1982b) The distal arthrogryposes: Delineation of new entities – review and nosologic discussion. *Am J Med Genet* 11:185–239.

Hall JG, Reed SD, Scott CI, Rogers JG, Jones KL, Camarano, A (1982c) Three distinct types of X-linked arthrogryposis seen in 6 families. *Clin Genet* 21:81–97.

Hall JG, Vincent A (2014) Arthrogryposis. In: *Neuromuscular Diseases of Infancy, Childhood, Adolescence - a Clinicians' Approach*, 2nd ed. Jones H, De Vivo DC, Darris BT, eds. Elsevier Inc.: London, *UK, Chapter* 7, pp. 123–141.

Hellmund A, Berg C, Geipel A, Muller A, Gembruch U (2016) Prenatal diagnosis of fetal akinesia deformation sequence (FADS): a study of 79 consecutive cases. *Arch Gynec Obstr* 294(4):697–707.

Ho CA, Karol LA (2008) The utility of knee releases in arthrogryposis. *J Pediatr Orthop* 28(3):307–313.

Hoffmann K, Muller JS, Stricker S, Megarbane A, Rajab A, Lindner TH, Cohen M,

Chouery E, Adaimy L, Ghanem I, Delague V, Boltshauser E, Talim B, Horvath R,

Robinson PN, Lochmüller H, Hübner C, Mundlos S (2006) Escobar syndrome is a prenatal myasthenia caused by disruption of the acetylcholine receptor fetal gamma subunit. *Am J Hum Genet* 79: 303–312.

Hunter JM, Kiefer J, Balak CD, Jooma S, Ahearn ME, Hall JG, Baumbach-Reardon L (2015) Review of X-linked syndromes with arthrogryposis or early contractures-aid to diagnosis and pathway identification. *Am J Med Genet* 167A:931–973.

Komolkin I, Ulrich EV, Agranovich OE, van Bosse HJP (2017) Treatment of scoliosis associated with arthrogryposis multiplex congenita. *J Pediatr Orthop* 37(5):S24–S28.

Lowry RB, Sibbald B, Bedard T, Hall JG (2010) Prevalence of multiple congenital contractures including arthrogryposis multiplex congenita in Alberta, Canada and a strategy for classification and coding. *Birth Defects Res A Clin Mol Teratol* 88:1057–1061.

Ma L, Yu X (2017) Arthrogryposis multiplex congenita: classification, diagnosis, perioperative case, and anesthesia. *Fronteras en Medicina* 11(1):48–52.

McMillin MJ, Beck AE, Chong JX, Shively KM, Buckingham KJ, Gildersleeve HI, Aracena MI, Aylsworth AS, Bitoun P, Carey JC, Clericuzio CL, Crow YJ, Curry CJ, Devriendt K, Everman DB, Fryer A, Gibson K, Giovannucci Uzielli ML, Graham JM Jr, Hall JG, Hecht JT, Heidenreich RA, Hurst JA, Irani S, Krapels IP, Leroy JG, Mowat D, Plant GT, Robertson SP, Schorry EK, Scott RH, Seaver LH, Sherr E, Splitt M, Stewart H, Stumpel C, Temel SG, Weaver DD, Whiteford M, Williams MS, Tabor HK, Smith JD, Shendure J, Nickerson DA; University of Washington Center for Mendelian Genomics, Bamshad MJ (2014) Mutations in PIEZO2 Cause Gordon Syndrome, Marden-Walker Syndrome, and Distal Arthrogryposis Type 5. *Am J Hum Genet* 95:734–744.

Michalk A, Stricker S, Becker Jm Rupps R, Pantzar T, Miertus J, botta G, Naretto VG, m Janetzki C, Yaqoob N, Ott C-E, Seelow D, Wieczorek D, Fiebig B, Wirth B, Hoopmann M, Walther, M, Körber F, Blakenburg M, Mundlos S, Heller R, Hoffmann K (2008) Acetylcholine receptor pathway mutations explain various fetal akinesia deformation sequence disorders. *Am J Hum Genet* 82:464–476.

Miller R, Sawatzky B (2017) Outcomes at 2-year minimum follow up of shoulder, elbow and wrist surgery in individuals with arthrogryposis multiplex congenita. *J Clin Exp Orthop* 3(29):1–8.

Nayak SS, Kadavigere R, Mathew M, Kumar P, Hall JG, Girisha KM (2014) Fetal akinesia deformation sequence: expanding the phenotypic spectrum. *Am J Hum Genet* 164A(10):2643–2648.

Nouraei H, Sawatzky B, MacGillivray M, Hall JG (2017) Long-term functional and mobility outcomes for individuals with arthrogryposis multiple congenita. *Am J Med Genet A* 173(5):1270–1278.

Oishi SN, Agranovich O, Pajardi GE, Novelli C, Baindurashvili AG, Trofimova SI, Abdel-Ghani H, Kochenova E, Prosperpio G, Jester A, Yilmaz G, Senaran H, Kose O, Butler L (2017) Treatment of the upper extremity contracture/deformities. *J Pediatr Orthop* 37(5):S9–S15.

Reed SD, Hall JG, Riccardi VM, Aylsworth A, Timmons, C (1985) Chromosomal abnormalities associated with congenital contractures (arthrogryposis). *Clin Genet* 27:353–372.

Reid COMV, Hall JG, Anderson C, Bocian M, Carey J, Costa T, Curry C, Greenberg F, Horton W, Jones M, Lafer C, Larson E,

Lubinsky M, McGillivray B, Pembry M, Popkin J, Seller M, Siebert V, Verhagen A (1986) Association of amyoplasia with gastroschisis, bowel atresia, and defects of the muscular layer of the trunk. *Am J Med Genet* 24:701–710.

Salminger S, Roche AD, Sturma A, Hruby LA, Aszmann OC (2016) Improving arm function by prosthetic limb replacement in a patient with severe arthrogryposis multiplex congenita. *J Rehab Med* 48(8):725–728.

Sells JM, Jaffe KM, Hall JG (1996) Amyoplasia, the most common type of arthrogryposis: the potential for good outcome. *Pediatr* 97:225–231.

Staheli LT, Hall JG, Jaffe KM, Paholke DO (1998) *Arthrogryposis: A Text Atlas*. Cambridge University Press; Cambridge, UK.

Staheli LT (1998) Orthopedic management principles. In: *Arthrogryposis: A Text Atlas*, Staheli LT, Hall JG, Jaffe KM, Paholke DO, eds. Cambridge: Cambridge University Press, pp. 27–43.

Steen U, Christensen E, Samargian A (2017) Adults living with Amyoplasia: function, psychosocial aspects, and the benefit of AMC support groups. *J Pediatr Orthop* 37(5):S31–S32.

Sung SS, Brassington AE, Grannatt K, Rutherford A, Whitby FG, Krakowiak PA, Jorde LB, Carey JC, Bamshad M (2003) Mutations in genes encoding fast-twitch contractile proteins cause distal arthrogryposis syndromes. *Am J Hum Genet* 72:681–690.

Swinyard CA (1963) Multiple congenital contractures (arthrogryposis): Nature of the syndrome and hereditary considerations. *Proceedings of "The Second International Congress of Human Genetics"* 3:1397–1398.

Szoke G, Staheli LT, Jaffe K, Hall JG (1996) Medial-approach open reduction of hip dislocation in amyoplasia-type arthrogryposis. *J Pediatr Orthop* 16:127–130.

van Bosse HJP, Ponten E, Wada A, Agranovich OE, Kowalczyk B, Lebel E, Senaran H, Derevianko DV, Vavilov MA, Petrova EV, Barsukov DB, Batkin SF, Eylon S, Kenis VM, Stepanova YV, Bulaev DS, Yilmaz G, Kose O, Trofimova SI, Durgut F (2017) Treatment of the lower extremity contracture/deformities. *J Pediatr Orthop* 37(5):S16–S23.

Villard L, Nordmark-Andersson E, Crowley B, Straub V, Bertoli M (2017) Multidisciplinary clinics. *J Pediatr Orthop* 37(5):S29–S30.

Witters I, Moerman P, Fryns JP (2002) Fetal akinesia deformation sequence: A study of 30 consecutive in utero diagnoses. *Am J Med Genet* 113:23–38.

Witters I, Moerman PH, Van Assche FA, Fryns JP (2001) Cystic hygroma colli as the first echographic sign of the fetal akinesia sequence. *Genet Counseling* 12:91–94.

Xavier-Neto J, Carvalho M, Pascoalino BD, Cardoso AC, Costa AM, Pereira AH, Santos LN, Saito A, Marques RE, Smetana JH, Consonni SR, Bandeira C, Costa VV, Bajgelman MC, de Oliveira PS, Cordeiro MT, Gonzales Gil LH, Pauletti BA, Granato DC, Paes Leme AF, Freitas-Junior L, Holanda de Freitas CB, Teixeira MM, Bevilacqua E, Franchini K (2017) Hydrocephalus and arthrogryposis in an immunocompetent mouse model of ZIKA teratogeny: a developmental study. *PLoS Neglected Tropical Diseases* 11(2):e0005363.

7

ATR-X: α THALASSEMIA/MENTAL RETARDATION-X-LINKED

RICHARD J. GIBBONS

MRC Molecular Haematology Unit, Weatherall Institute of Molecular Medicine, University of Oxford, John Radcliffe Hospital, Oxford, UK

INTRODUCTION

The rare association of α thalassemia and intellectual disability was recognized by Weatherall and colleagues (1981). It was known that α thalassemia arises when there is a defect in the synthesis of the α globin chains of adult hemoglobin (HbA, $\alpha_2\beta_2$). When they described three children with severe learning difficulties, α thalassemia and a variety of developmental abnormalities, their interest was stimulated by the unusual nature of the α thalassemia. The children were of North European origin, where α thalassemia is uncommon, and although one would have expected to find clear signs of this inherited anemia in their parents, it appeared to have arisen de novo in the affected offspring. It was thought that the combination of α thalassemia, intellectual disability and the associated developmental abnormalities represented a new syndrome and that a common genetic defect might be responsible for the diverse clinical manifestations. This conjecture has been confirmed, and what has emerged is the identification of two quite distinct syndromes: ATR-16, a contiguous gene syndrome in which the α globin genes are deleted along with a variable amount of DNA on chromosome 16p13.3, and ATR-X, which results from mutation of a gene on the X chromosome encoding a putative chromatin remodeling factor.

ATR-X syndrome has a complex phenotype that, in addition to severe intellectual disability and α thalassemia, includes characteristic facial dysmorphism, genital abnormalities and a variety of other features. This indicates the importance of this gene product in the efficient transcription of an extensive repertoire of genes.

Incidence

It is likely that this syndrome is substantially under-recognized. Over 200 affected families have been identified worldwide and individuals have been reported in most racial groups. A Japanese study made an estimate of the prevalence of 1/30,000–1/40,000 (Wada et al. 2013), which is consistent with the observation that variants in ATRX were one of the most common causes of intellectual disability (Grozeva et al. 2015). There are no long-term longitudinal data on this relatively newly described syndrome, but a number of affected individuals are fit and well in their 30s and 40s.

Early childhood appears to be a vulnerable time, with 20/33 (61%) of the known deaths occurring under the age of five years. Four deaths have occurred due to aspiration of vomitus. This may be related to the fact that gastroesophageal reflux and vomiting are often more severe in the early years.

Diagnostic Criteria

The cardinal features of ATR-X syndrome are severe intellectual disability, a characteristic facial appearance, genital abnormalities, and α thalassemia (see Table 7.1). Definitive diagnostic criteria have not been established. Distinctive facial traits are most readily recognized in early childhood

---

Cassidy and Allanson's Management of Genetic Syndromes, Fourth Edition.
Edited by John C. Carey, Agatino Battaglia, David Viskochil, and Suzanne B. Cassidy.
© 2021 John Wiley & Sons, Inc. Published 2021 by John Wiley & Sons, Inc.

and the gestalt is probably secondary to facial hypotonia (Figure 7.1). The frontal hair is often upswept, there is telecanthus, epicanthal folds, depressed nasal bridge and midface hypoplasia, and a small triangular upturned nose with the alae nasi extending below the columella and septum. The upper lip is tented and the lower lip full and everted giving the mouth a "carp-like" appearance. The frontal incisors are frequently widely spaced, the tongue protrudes and there is prodigious drooling. The ears may be simple, slightly low set and posteriorly rotated. Genital anomalies vary in severity from cryptorchidism alone through male pseudohermaphroditism. Short stature is common as are minor skeletal malformations including clinodactyly, brachydactyly, tapered digits, joint contractures, chest wall deformity, kyphosis, scoliosis, carus and valgus deformities of the foot, and pes planus. Less common are cardiac defects, abnormalities of the renal/urinary system and asplenia.

Etiology, Pathogenesis, and Genetics

Generally (see below) only males are affected with ATR-X syndrome, and case reports of families with multiple affected individuals indicated that this is an X-linked condition. Linkage analysis in 16 families mapped the disease interval to Xq13.1-q21.1. Using a candidate gene approach, mutations were identified in a previously found gene known as *XH2/XNP* (Gibbons et al. 1995b). The HUGO nomenclature committee proposed that the gene be renamed *ATRX*.

Gene Characteristics and Function: It is now known that the *ATRX* gene spans about 300 kilobases of genomic DNA and contains 35 exons (Gibbons et al. 2008). It encodes at least two alternatively spliced approximately 10.5 kb mRNA transcripts which differ at their 5′ ends and are predicted to give rise to slightly different proteins of 265 and 280 kDa, respectively. A further transcript of approximately 7 kb represents an isoform that retains intron 11 and truncates at this point. This truncated protein isoform, ATRXt, is conserved between mouse and man (Garrick et al. 2004).

Within the N-terminal region lies a complex cysteine-rich segment (ADD, Figure 7.2). The ADD domain comprises a plant homeodomain-like zinc finger and an additional putative zinc-binding C_2C_2 motif just upstream (Gibbons et al. 1997). The functional significance of the ADD segment is demonstrated by the high degree of conservation between human and mouse (97 of 98 amino acids) and the fact that it represents a major site of mutations in patients with ATR-X syndrome, containing approximately 50% of all mutations (Figure 7.2 and see below) despite only accounting for 4% of the coding sequence.

The central portion of the molecule contains motifs that identify the ATRX protein as a novel member of the SNF2 subgroup of a superfamily of proteins. This group of proteins is characterized by the presence of seven highly conserved co-linear helicase motifs. Other members of the SNF2 subfamily are involved in a wide variety of cellular functions, including the regulation of transcription, control of the cell cycle, DNA repair, and mitotic chromosome segregation. Recent studies indicate that ATRX functions with the histone chaperone DAXX to insert the histone variant H3.3 into repetitive regions of the genome (Goldberg et al. 2010). G-rich repeats targeted by ATRX targets may take up DNA secondary structures such as G-quadruplex forms and ATRX may play a role in helping resolve them or prevent their formation by restoring nucleosomal organization (Law et al. 2010).

Functional Consequences of Mutations: Mutations in ATRX have been associated with a wide range of biological effects: altered patterns of DNA methylation and histone modifications, aberrant chromosome congression in mitosis and segregation in meiosis, telomere dysfunction and changes in gene expression. Somatic mutations are frequently found in a subgroup of cancers that maintain their telomeres by the telomerase-independent alternative lengthening of telomere pathway; ATRX appears to function as a tumor suppressor. Four reports of osteosarcoma from three ATR-X families with germline mutations in ATRX have recently been published (Ji et al. 2017; Smolle et al. 2017) (and personal observation). One possible unifying hypothesis, for which there is growing support, is that ATRX facilitates the replication of repetitive regions of DNA and in its absence DNA secondary structures stall DNA replication leading to diverse biological epi-phenomena (reviewed in (Clynes et al. 2013)).

Allelic Conditions: Since its identification in 1995, the *ATRX* gene has emerged as the disease gene for numerous forms of syndromal X-linked intellectual disability: X-linked α thalassemia/mental retardation or ATR-X (Gibbons

TABLE 7.1 Clinical findings in 168 individuals with ATR-X syndrome

Clinical finding	Total[a]	%
Profound intellectual disability	160/167[b]	96
Characteristic face	138/147	94
Skeletal abnormalities	128/142	90
HbH inclusions	130/147	88
Neonatal hypotonia	88/105	84
Genital abnormalities	119/150	79
Microcephaly	103/134	77
Gut dysmotility	89/117	76
Short stature	73/112	65
Seizures	53/154	34
Cardiac defects	32/149	21
Renal/urinary abnormalities	23/151	15

[a] Total represents the number of affected individuals on whom appropriate information is available and includes persons who do not have α thalassemia but in whom *ATRX* mutations have been identified.
[b] One person is too young (<1 year) to assess degree of mental retardation. HbH = hemoglobin H

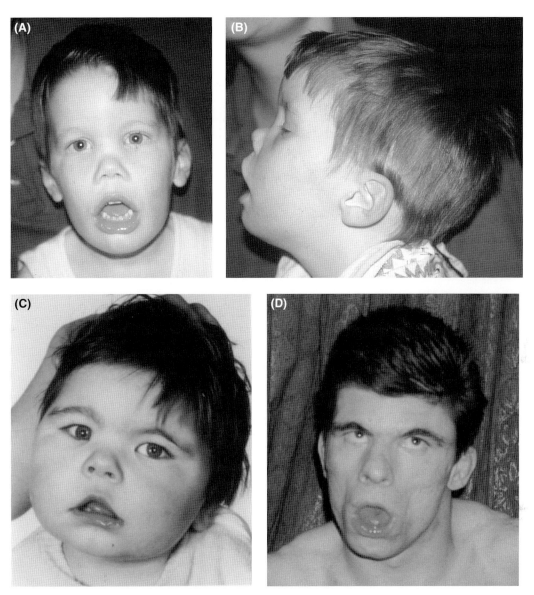

FIGURE 7.1 Four-year-old male with ATR-X syndrome showing (A) the typical facial appearance, (B) midface retrusion, (C) individual aged 1.5 years, and (D) same individual aged 23 years. (Taken from Gibbons, et al. 1995a by permission of John Wiley & Sons.)

et al. 1995b), Carpenter–Waziri syndrome (Abidi et al. 1999), Holmes–Gang syndrome (Stevenson et al. 2000a), Smith–Fineman–Myers syndrome (Villard et al. 2000), Chudley–Lowry syndrome (Abidi et al. 2005), and X-linked intellectual disability-arch fingerprints-hypotonia syndrome (Basehore et al. 2015). It is also the disease gene for X-linked mental retardation with spastic paraplegia (Lossi et al. 1999). It has been estimated that approximately 10% of named X-linked intellectual disability syndromes are candidates for allelism based on phenotypic findings. These include Brooks, Miles–Carpenter, Proud, Vasquez, Young–Hughes, and X-linked mental retardation-psoriasis syndromes (Stevenson et al. 2000b). Future studies to establish the full range of disease-causing mutations will facilitate genetic counseling and elucidate functionally important aspects of the protein.

Genotype/phenotype Correlations: To date 113 different mutations have been documented in 182 separate families (Figure 7.2) (Gibbons et al 2008) and most of these are missense mutations. These missense mutations are clustered in two regions, the ADD (zinc-finger) domain and the helicase domain. One particular mutation, 736C>T (R246C), is found in approximately one quarter of all affected individuals. Analysis of the mutations and their resulting phenotypes allow important conclusions to be drawn.

It has become clear that there is a phenotypic spectrum and the severity of some aspects of the phenotype,

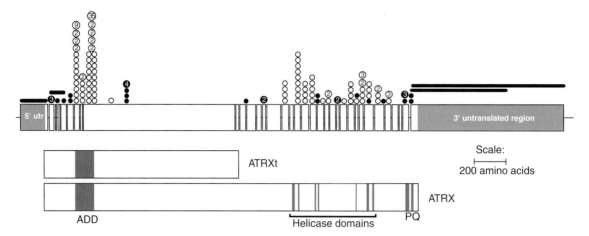

FIGURE 7.2 Schematic diagram of the complete *ATRX* cDNA. The boxes represent the 35 exons (excluding the alternatively spliced exon 7), thin horizontal lines represent the introns (not to scale). The largest open reading frame (ORF) is shown as an open box. The principal domains, the zinc finger motif (ADD) and the highly conserved helicase motif, are indicated, as are the P box (P) and a glutamine rich region (Q). In the upper part of the figure, 113 different *ATRX* mutations are illustrated. The positions of the mutations are shown by circles: filled circles represent mutations (nonsense or leading to a frameshift) that would cause protein truncation; unfilled circles represent missense mutations; deletions are indicated by horizontal dashed lines. Recurrent mutations are illustrated by larger circles and the number of independent families is indicated.

intellectual disability, gross motor disability, genital abnormality, and α thalassemia, is quantifiable to some degree (Gibbons et al. 2008). One particular mutation, R37X, is often associated with mild or moderate intellectual disability (Basehore et al 2015; Guerrini et al. 2000). Mutations in the ADD domain correlate with more severe psychomotor impairment than mutations in the helicase region (Badens et al. 2006a; Gibbons et al 2008). Mutations in the C-terminus are associated with more severe urogenital abnormalities and this region may play a specific role in urogenital development (Figure 7.3). Variability in the severity of the α-thalassemia in ATR-X syndrome is, however, less related to the nature of the ATRX mutation and more dependent on the length of a tandem repeat adjacent to the α globin locus to which ATRX binds (Law et al. 2010).

Structural and functional studies have shown that most ATRX mutations affect protein stability and therefore lead to reduced steady-state levels of ATRX protein in the cells (Argentaro et al. 2007; Mitson et al. 2011). A curious finding is that none of the truncating mutations that have been studied lead to complete loss of full-length ATRX but there is "rescue" by alternative splicing that leads to skipping of the stop mutation to some degree (Gibbons et al. 2008). It is possible that complete absence of ATRX, a true null, may be lethal as is the case in the mouse ATRX knockout model.

Diagnostic Testing

For the clinician and the family of a child with intellectual disability, the important consideration is accurate and speedy diagnosis. Research over the last 20 years has provided an array of useful investigations for the identification of ATR-X syndrome and its clinical variants. A search for hemoglobin H (HbH) inclusions is an easy and rapid first line test that will confirm the diagnosis in many cases as they are found in 85% of ATR-X cases. The most sensitive test for HbH inclusions uses light microscopy to detect red cells containing HbH inclusions (Figure 7.4) after incubation of venous blood with 1% brilliant cresyl blue for 4–24 hours at room temperature. HbH is unstable and cells with inclusions may be more difficult to find in blood samples drawn more than two days before testing. When the family history and phenotype are strongly suggestive of ATR-X, a careful search for inclusions should be made in all the affected individuals as their presence may vary between different affected individuals within a family. It is important to note that, in most, there is insufficient HbH to be detected by electrophoresis. A normal full blood count does not exclude the diagnosis as the red cell parameters may be remarkably normal.

The observation of methylation abnormalities in ATR-X potentially adds a new, valuable and complementary functional screening test.

If such functional testing confirms the clinical suspicion then the next step is to determine the underlying mutation in the *ATRX* gene. If the functional tests are negative and the index of suspicion is high, it is still worth proceeding to mutation analysis. However, with the introduction of high throughput DNA sequencing of whole genomes, exomes or gene panels for intellectual disability many cases are being identified where the diagnosis has not been considered but a variant is found in *ATRX*. The challenge here is to determine whether the variant is pathological. In addition to standard

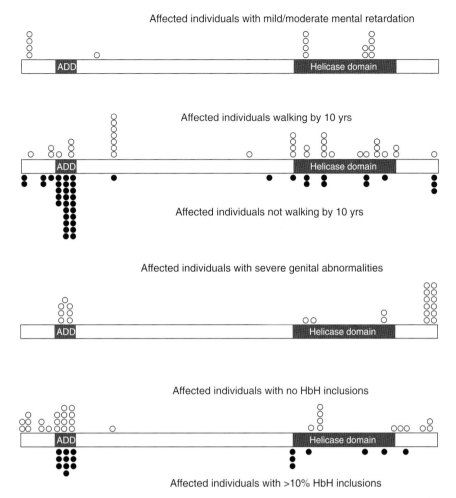

FIGURE 7.3 The positions of mutations are shown for affected individuals with a particular phenotype. (Taken from Gibbons, et al. 2008 by permission of John Wiley & Sons.)

practice in assessing the pathological potential of sequence variants (Richards et al. 2015) it is worth noting that missense mutations are more likely to be pathological if located in the ADD or ATPase domains, but functional testing is extremely valuable as is the clinical picture.

Large intragenic duplications of *ATRX* detected by array comparative genomic hybridization (aCGH) have been described in three families with individuals with ATR-X syndrome (Cohn et al. 2009; Thienpont et al. 2007). Presumably the duplications lead to disruption of the *ATRX* gene, but no mapping was performed in these people to determine if this was the case. PCR amplification and sequencing of genomic DNA may not detect duplications and may miss mutations that lead to abnormal splicing of ATRX mRNA; in persons where functional testing confirms the diagnosis but no mutation is identified, then it is worth considering checking gene dosage by aCGH or multiplex ligation-dependent probe amplification (MLPA) or abnormal splicing in cDNA by reverse transcription PCR.

Identification and Counseling of Female Carriers: ATR-X syndrome is a recessive X-linked condition. With the exception of two heterozygous females with intellectual disability (Badens et al. 2006b; Wada et al. 2005), female carriers are phenotypically and intellectually normal and therefore additional tests are required to determine a female's genotype. Skewed X-inactivation has been utilized as a marker for carriers of the ATRX mutation. However, 5–10% of normal females have skewed X-inactivation and occasional female carriers have a balanced pattern (Gibbons et al. 1992), therefore this method should be used with caution. Twenty-five percent of obligate female carriers exhibit rare cells with HbH inclusions, but a negative result does not exclude the carrier state. Since the identification of the gene causing ATR-X, mutation detection has become the mainstay of carrier identification. In families in which the causative mutation has not been identified, linked markers may be used to identify whether descendants of an obligate carrier have inherited the disease-associated haplotype.

For the female who has been identified as a carrier, there is a 50% risk of passing on the disease allele with each pregnancy but, since only males are clinically affected, the risk of

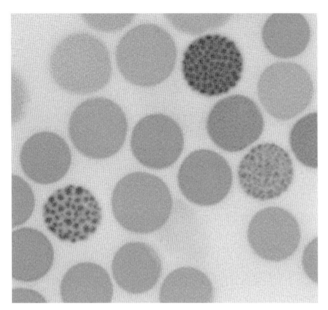

FIGURE 7.4 Photomicrograph of the peripheral blood of an individual with ATR-X syndrome showing three cells containing HbH inclusions.

having an affected child is 25% for each pregnancy. Prenatal or preimplantation genetic diagnosis for such at risk females is feasible.

The principal issue when counseling is determining the risk of recurrence for families with a sporadic case of ATR-X. One small study showed that 17/20 (85%) mothers of sporadic cases were carriers (Bachoo and Gibbons 1999).

Germline mosaicism has been reported in ATR-X syndrome. Thus, despite a negative mutation test, a mother of an affected boy may still be at risk of further affected offspring (Bachoo and Gibbons 1999). It is thus advisable to offer all mothers of affected children prenatal diagnosis even when they are mutation negative.

Differential Diagnosis

The diagnosis of ATR-X syndrome is relatively straightforward in males who present with typical clinical features and HbH inclusions. However, where the hematology has not been checked or HbH inclusions are absent, diagnostic difficulty may arise. Coffin–Lowry syndrome (see Chapter 12) may be confused with ATR-X syndrome, particularly in early childhood. Distinguishing features are the down-slanting palpebral fissures, broad nose, pudgy tapering digits, absence of genital abnormalities, and the frequent presence of carrier manifestations in Coffin–Lowry syndrome. There is also phenotypic overlap with Angelman syndrome (see Chapter 5), Smith–Lemli–Opitz (see Chapter 53) and Pitt–Hopkins syndromes (Takano et al. 2011). There are readily available diagnostic tests for these disorders.

MANIFESTATIONS AND MANAGEMENT

There have been no systematic studies of the treatment of the various problems that arise in ATR-X syndrome and consequently the information below is anecdotal.

Growth and Feeding

Infantile hypotonia is very common and is associated with considerable difficulty with sucking. Gavage (nasogastric tube) feeding may be required for a number of weeks to assure adequate nutrition. Persistent feeding problems affect some children and these may be associated with gastroesophageal reflux (see below). Where nutrition is severely compromised, a feeding gastrostomy may be required.

Short stature is seen in two thirds of affected individuals. Longitudinal data are available in only a few of them. In some individuals growth retardation is apparent throughout life, whereas in others it has become manifest at a later stage, e.g., at the time of the pubertal growth spurt.

Episodes of food and drink refusal may occur and are discussed under Gastrointestinal (below).

Evaluation

- Adequacy of feeding ability should be assessed in diagnosed infants.
- Growth parameters should be closely monitored and plotted on growth charts during clinical assessments.

Management

- Nutrition consultation for assurance of adequate caloric intake may be needed. Where adequate nutrition is not possible with standard feeding or nasogastric tube feeding, then a gastrostomy should be considered.
- Episodes of food and drink refusal may require hospital admission and intravenous fluids. The possibility of a gastrointestinal problem should be considered (see below).

Development and Behavior

In early childhood, all milestones are delayed. In a cohort of 83 affected children for whom information was available, 45% had learned to walk by the age of 9; all those who did eventually learn to walk did so by the age of 15 years (Figure 7.5, unpublished data). Most have no speech, although there are several individuals with a handful of words or signs. They frequently have only situational understanding, and are dependent on caregivers for almost all activities of daily living. Only partial bowel and bladder control may be attained. Nevertheless, some children have relatively good performance as measured by other

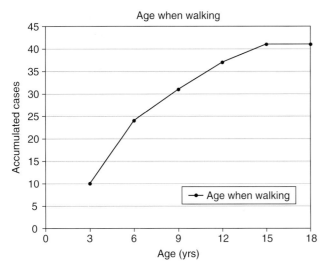

FIGURE 7.5 A measure of gross motor function is whether the children are able to walk and the age at which this is achieved. The appropriate information was available for 41 affected individuals who learnt to walk. The graph shows the age at which this was achieved. Approximately 75% of these were able to walk by the age of 9 years.

parameters, and this may vary from child to child in a sibship. A child with no discernable words may lead a parent by the hand to desired food, turn on the television or even operate the DVD player. New skills continue to be acquired through childhood and adulthood.

More recent reports, however, point to a wider spectrum of intellectual handicap than previously thought. A mutation in the *ATRX* gene was identified in the family originally described by Carpenter and colleagues (Carpenter et al. 1999). All affected males have moderate intellectual disability and exhibit expressive language delay though no psychometric evaluation is available. Guerrini and colleagues have reported a mutation (c.109C>T; p.R37C) in an Italian family with four affected male cousins, one of whom has profound intellectual disability, whereas the others have IQs of 41, 56, and 58 (Guerrini et al 2000). A similar, variable but generally milder phenotype was identified in five unrelated kindreds with this same mutation (Basehore et al 2015). The basis for this marked variation is unknown. Generally, affected individuals continue to acquire new skills, though a brief period of neurological deterioration has been reported in three people.

No systematic study of behavior has been carried out in ATR-X syndrome. Consequently, most reports of behavioral characteristics are anecdotal. Nevertheless, a thumbnail sketch of the mannerisms of this condition is slowly emerging (Kurosawa et al. 1996; Wada et al. 1998) and this may be helpful diagnostically. Affected individuals are usually described by their parents as content and of a happy disposition. They exhibit a wide range of emotions that are usually appropriate to their circumstances. There have been reports, however, of unprovoked emotional outbursts with sustained laughing or crying. There may be emotional fluctuation with sudden switches between almost manic-like excitement or agitation to withdrawal and depression. In several instances the episodes of crying have been thought to be associated with pain, possibly of a gastrointestinal origin (see below).

Whereas many of affected individuals are affectionate to their caregivers and appreciate physical contact, some exhibit autistic-like behavior: they appear to be in a world of their own, show little interest or even recognition of those around them, and avoid eye contact. The latter behavior may be associated with unusual and persistent postures.

Affected persons may be restless, exhibiting choreoathetotic-like movements. Frequently, they put their hands into their mouths and may induce vomiting. Sometimes, they engage in self-injurious behavior, biting or hitting themselves. They may hit, push or squeeze their necks with their hands to the point of cyanosis, a state they may also achieve through breath-holding. Repetitive stereotypic movements may be manifest and these may vary from pill-rolling or hand flapping to spinning around on one spot while gazing into a light. These characteristic behaviors are reminiscent of Angelman syndrome (see Chapter 5) and may lead to diagnostic confusion.

Individuals with ATR-X syndrome are for the most part sociable, interacting well with others, particularly close members of the family or primary caregivers. Nevertheless, aggressive behavior towards others, for example siblings, may occur, including hair pulling, pinching and scratching. This often occurs when attention is paid to others. Changes in routine may lead to severe anxiety.

Affected males may have episodes when they become hyperactive, highly agitated and vocal. These episodes, which can be very prolonged with screaming lasting days, may be accompanied by head banging or other self-injurious behavior. The cause is unclear.

Evaluation

- Evaluation of developmental skills from infancy ensures that the appropriate intervention services are introduced as early as possible.

Management

- Infant stimulation, early intervention and special education are important to optimize abilities.
- Individualized education and therapies are essential especially in facilitating communication.
- Affected individuals may show improvement in socialization with one-to-one therapy.
- Anxiolytic or anti-psychotic medication may be effective in treating severe behavior problems where other measures have failed.

Gastrointestinal

Recurrent vomiting, regurgitation or gastro-esophageal reflux, particularly in early childhood, are common findings. In two recent studies a barium meal revealed that the affected children had episodic gastric pseudovolvulus (Martucciello et al. 2006; Watanabe et al. 2014). In this condition the stomach does not have the normal system of peritoneal ligaments and has a propensity to twist around itself leading to gastric outlet obstruction and secondary gastro-esophageal reflux. An apparent reluctance to swallow has been reported by several parents and probably reflects the uncoordinated swallowing that was observed radiologically in two well-studied cases (personal experience). Impaired gastro-esophageal motility in ATR-X was reported by Watanabe et al. (2014). Aspiration is commonly implicated as a cause of death in early childhood.

Evidence suggests that affected individuals are susceptible to peptic ulceration. Esophagitis, esophageal stricture and peptic ulcer have been observed endoscopically in single individuals. In five individuals, an upper gastrointestinal bleed was observed, one of which required transfusion (haemoglobin, 5 g dL^{-1}) (Gibbons et al. 1995a; Martucciello et al 2006). Pain resulting from peptic ulceration or esophagitis are possible explanations for the episodes of persistent crying and food refusal reported by a number of parents.

Drooling is very common in ATR-X, particularly in young children. Many mothers will describe their sons soaking several bibs during the course of the day. The open mouth associated with facial hypotonia no doubt is an important factor, as is their reluctance to swallow even with a mouth full of saliva.

Constipation occurs often, and in some individuals is a major management problem. Martucciello *et al.* (2006) demonstrated ultra-short Hirshsprung disease and colonic hypoganglionosis in two affected children. The authors reviewed 128 cases of ATR-X and found that hospital admissions for recurrent ileus were reported in two, and reduced intestinal mobility was observed radiologically in four. This may be a consequence of a widespread abnormality in the enteric nervous system, leading to abnormal gut motility. Two of the individuals required partial resection of the ileum after developing ischemia of the small bowel, which in one was attributed to a volvulus. Volvulus was reported in an additional person. One child required a right hemicolectomy following an episode of necrotizing enterocolitis at 13 days of age (reviewed in [Martucciello et al 2006]). The importance of recognizing volvulus in ATR-X was stressed in a subsequent report (Horesh et al. 2015).

Evaluation

- If the individual refuses to eat or appears to be in pain, such as having episodes of prolonged crying, a possible gastric or gastrointestinal cause should be thoroughly investigated, given the frequency and severity of the reported problems.
- Evaluation of recurrent vomiting should be standard. There is a strong case for 24 hour pH monitoring and barium study of the upper gastrointestinal tract in children with this condition.
- If constipation is resistant to conservative management then rectal biopsy should be considered to exclude ultra-short segment Hirshsprung disease and colonic hypoganglionosis. Histochemical evaluation for acetylcholinesterase (AChE) of the rectal biopsies is required.

Management

- Initial treatment for gastro-esophageal reflux should be standard: appropriate treatment of helicobacter pylori infection or with H2-receptor antagonists or proton pump inhibitors should be instigated.
- In severe cases of gastro-esophageal reflux, surgical treatment by fundoplication may be required. If gastric pseudo-volvulus is occurring, then anterior gastropexy should be considered in addition.
- Numerous methods have been tried to control drooling. In other conditions, anticholinergics are commonly used to reduce production of saliva. However, they often cause reduced gastrointestinal motility (which may already be abnormal in males with ATR-X) and may exacerbate constipation so should be used with caution in ATR-X syndrome. Botulinum toxin type A (Botox) injection of the salivary glands might be tried, but reports are scant and the treatment needs repeating. The surgical options of redirecting the submandibular ducts or removing the glands themselves may be considered.
- Initial treatment for constipation should be standard. Adequate hydration is important in preventing constipation, as is the use of bulking agents in the diet and the regular use of osmotic laxatives such as lactulose.
- Where ultra-short segment Hirshsprung disease is identified, the therapy of choice is sphinctero-myectomy if dilation of the internal sphincter proves ineffective.

Neurologic

Although the head circumference is usually normal at birth, post-natal microcephaly usually develops. Macrocephaly has not been reported.

Computed tomography may show mild cerebral atrophy. However, a recent study using magnetic resonance imaging revealed white matter changes in all persons (Wada et al. 2013). In a retrospective study 17/27 individuals with ATR-X had non-specific and non-progressive brain atrophy; 11/27 localized (especially around the trigones) and 1/27 widespread white matter abnormalities; 4/27 delayed myelination; 1/27 severe and rapidly progressive cortical brain atrophy. Partial or complete agenesis of the corpus callosum was also reported. There have been three autopsy reports. The brain was small in each; in two the morphology was normal, in one the temporal gyri on the right were indistinct and there was hypoplasia of the cerebral white matter (personal experience).

In one report, electroencephalogram changes were consistent with encephalitis (Donnai et al. 1991). The family originally reported by Holmes and Gang (1984) was subsequently shown to have an *ATRX* mutation (Stevenson et al 2000a). All three affected males from this family died in childhood and the death of one was attributed to encephalitis.

As affected individuals age there is often a tendency toward spasticity. One report described a family with an *ATRX* mutation where affected members had spastic paraplegia from birth (Lossi et al. 1999).

Seizures occur in approximately one third of people with ATR-X and most frequently are clonic/tonic or myoclonic in nature. In the main, seizures respond well to standard therapy. Some affected individuals exhibit jerking movements which, though appearing to be seizures, are not associated with epileptiform activity on EEG (personal experience).

Evaluation

- Thorough neurologic evaluation should be part of all routine care.
- Regular evaluations of the need for physical therapy are needed.
- Electroencephalogram may need to be carried out with video recording to correlate seizure activity and abnormal movements.

Management

- Ongoing physical therapy can ameliorate spasticity.
- Approach to seizure control is standard.

Hematology

Although initially the presence of α thalassemia was one of the defining elements of the syndrome, it is clear that there is considerable variation in the hematological manifestations associated with *ATRX* mutations. A number of families have been identified in which some or all of the affected members have no signs of α thalassemia (Villard et al. 1996a; Villard et al. 1996b). Among 118 individuals in whom the diagnosis has been confirmed by the identification of an *ATRX* mutation, 15% did not have detectable HbH inclusions (Gibbons et al 2008). Nevertheless, the test for α thalassemia is simple and, when positive, quickly establishes the diagnosis. The most sensitive test uses light microscopy to detect red cells containing HbH inclusions (Figure 7.4). It is important to note that, in most individuals with ATR-X, there is insufficient HbH to be detected by electrophoresis. The hematology is often surprisingly normal considering the presence of α thalassemia. Neither the hemoglobin concentration nor mean cell hemoglobin are as severely affected as in the classical forms of α thalassemia that are associated with *cis*-acting mutations in the α globin complex, and this probably reflects the different pathophysiology of the conditions.

Evaluation

- Obtain a full blood count including red cell indices to see if there is hypochromic microcytic anemia.
- Request staining for HbH inclusions in red cells after incubation in brilliant cresyl blue solution. A positive result indicates α thalassemia consistent with ATR-X syndrome.
- If α thalassemia is present, check parental blood counts. In ATR-X, the parental full blood counts are normal. If either parent has a hypochromic microcytic anemia, consider the possibility that α thalassemia is being inherited as an independent trait.

Management

- In ATR-X syndrome the anemia is mild and does not require treatment. Iron is not indicated unless iron stores are shown to be low.

Genitourinary

Genital abnormalities are seen in 80% of affected individuals. These may be very mild, such as undescended testes or deficient prepuce, but the spectrum of abnormality extends through hypospadias and micropenis to ambiguous or female external genitalia. The most severely affected children, who are clinically defined as male pseudohermaphrodites, are usually raised as females. In such cases, no Müllerian structures are present and dysgenetic testes or streak gonads have been found intra-abdominally (Ion et al. 1996; Wilkie et al. 1990). Of particular interest is the finding that these abnormalities breed true within families (McPherson et al. 1995). Puberty is frequently delayed, and in a few teens appears to be arrested. Curiously, premature adrenarche has been noted in two children.

Structural abnormalities of the kidneys and ureters are well described and may predispose to urinary tract infections. Renal abnormalities (hydronephrosis, renal hypoplasia or agenesis, polycystic kidney, vesico-ureteric reflux) may present with recurrent urinary tract infections. Death from renal failure has occurred in two individuals and was thought to have been a consequence of previous urinary infections (personal experience).

Evaluation

- The possibility of cryptorchidism should be assessed in all affected children.
- The urinary system should be imaged with ultrasound at diagnosis.
- Urine should be cultured when there is symptomatology such as fever or pain on urination.

Management

- Orchidopexy should be carried out as required at the standard age.
- Intra-abdominal testes, which are usually dysgenetic, should be removed because of the long-term risk of malignancy.
- Urinary tract infections are treated in a standard manner.
- Prophylactic antibiotics are indicated if urinary tract anomalies are present to prevent long-term damage to the kidneys.

Musculoskeletal

A wide range of relatively mild skeletal abnormalities have been noted, some of which are probably secondary to hypotonia and immobility (Gibbons et al. 1995a). Fixed flexion deformities, particularly of the fingers, are common. Other abnormalities of the fingers and toes that have been observed are: clinodactyly, brachydactyly, tapering of the fingers, drum stick phalanges, cutaneous syndactyly, overlapping of the digits, and a single child with a bifid thumb. Foot deformities occur in 29% and include pes planus, talipes equinovarus, and talipes calcaneo-valgus.

Almost a third of affected persons have kyphosis and/or scoliosis, and chest wall deformity has been seen in 10 people. Sacral dimples were present in three individuals, radiological spina bifida in two and other abnormalities of the vertebrae in five. Only a few individuals have had thorough radiological investigation. In those who have, the most common findings were delayed bone age and coxa valga.

Evaluation

- A careful musculoskeletal exam should be done at diagnosis and well child visits throughout childhood. This should include a clinical examination for scoliosis/kyphosis.

Management

- Treatment of musculoskeletal anomalies, when appropriate, is standard.

Ears and Hearing

Sensorineural deafness has previously been considered a feature that distinguishes ATR-X syndrome from the allelic condition Juberg–Marsidi syndrome (Saugier-Veber et al. 1995). However, of the 13 individuals with a documented sensorineural hearing deficit, seven have ATR-X syndrome, i.e. presence of α thalassemia.

Evaluation

- Standard distraction tests and, if suspect, auditory evoked responses should be done.

Management

- Hearing loss should be managed as for anyone.

Cardiovascular

A wide range of cardiac abnormalities have been noted: septal defects (10 cases); patent ductus arteriosus (6); pulmonary stenosis (3); aortic stenosis (2); tetralogy of Fallot (2), and single cases of transposition of the great arteries, dextrocardia with situs solitus and aortic regurgitation.

Evaluation

- Careful auscultation for a cardiac murmur is appropriate
- Echocardiography should be considered at the time of diagnosis.

Management

- Management of cardiac defects is the same as for any affected individual.

Dermatologic

A combination of drooling and constant placement of hands into the mouth can lead to maceration of the skin and the development of ulceration or fissures despite the regular application of emollients.

Evaluation

- When maceration occurs, a dermatological consultation is advised.

Management

- Repetitive behavior is particularly difficult to alter and may necessitate the temporary use of arm splints in younger children to speed the recovery of the skin.
- Emollients should be used to soothe and hydrate the skin.

Ophthalmologic

Refractive abnormalities, in particular myopia, are common and in some individuals there may be high myopia (more than −10 dioptres). Strabismus may be present. Pale discs or optic atrophy are frequently observed. Rarely the individual may be blind.

Evaluation

- A formal ophthalmologic evaluation is appropriate at diagnosis and regularly thereafter.

Management

- Ocular problems should be treated as in the general population

Asplenia

Absence of the spleen has been described in two unrelated ATR-X families and in one individual who presented with recurrent pneumococcal infections (Leahy et al. 2005).

Evaluation

- A blood film may reveal Howell–Jolly bodies. Abdominal ultrasound will determine if the spleen is missing

Management

- Antibody prophylaxis and vaccination to prevent pneumococcal and meningococcal infections are recommended if the spleen is missing.

CANCER PREDISPOSITION

Three individuals with osteosarcoma from two ATR-X families with germline mutations in ATR-X have recently been reported (Ji et al 2017; Smolle et al 2017). One of them developed two independent primary tumors. However, in a series of 260 people with ATR-X syndrome only two with cancer have been reported: one with a bowel cancer and one with osteosarcoma (personal observation).

Evaluation

- There is no indication for tumor surveillance. However painful lesions of the bones should be investigated to exclude osteosarcoma.

Management

- Malignancy should be treated as in the general population.

RESOURCES

Support group

Facebook page: https://www.facebook.com/groups/163849465337/

English language version of Dutch ATR-X Syndrome Foundation: http://atrxsyndroom.nl/?lang=en

Parents discussion forum: http://groups.yahoo.com/group/atr-x/

Information online

http://www.imm.ox.ac.uk/mhu/atrx.html
www.genetests.org

REFERENCES

Abidi F, Schwartz CE, Carpenter NJ, Villard L, Fontes M, Curtis M (1999) Carpenter-Waziri syndrome results from a mutation in XNP. *Am J Med Genet* 85(3):249–251.

Abidi FE, Cardoso C, Lossi AM, Lowry RB, Depetris D, Mattei MG, Lubs HA, Stevenson RE, Fontes M, Chudley AE, Schwartz CE (2005) Mutation in the 5' alternatively spliced region of the XNP/ATR-X gene causes Chudley-Lowry syndrome. *Eur J Hum Genet* 13(2):176–183.

Argentaro A, Yang JC, Chapman L, Kowalczyk MS, Gibbons RJ, Higgs DR, Neuhaus D, Rhodes D (2007) Structural consequences of disease-causing mutations in the ATRX-DNMT3-DNMT3L (ADD) domain of the chromatin-associated protein ATRX. *Proc Natl Acad Sci USA* 104(29):11939–11944.

Bachoo S, Gibbons RJ (1999) Germline and gonosomal mosaicism in the ATR-X syndrome. *Eur J Hum Genet* 7(8):933–936.

Badens C, Lacoste C, Philip N, Martini N, Courrier S, Giuliano F, Verloes A, Munnich A, Leheup B, Burglen L, Odent S, Van Esch H, Levy N (2006a) Mutations in PHD-like domain of the ATRX gene correlate with severe psychomotor impairment and severe urogenital abnormalities in patients with ATRX syndrome. *Clin Genet* 70(1):57–62.

Badens C, Martini N, Courrier S, DesPortes V, Touraine R, Levy N, Edery P (2006b) ATRX syndrome in a girl with a heterozygous mutation in the ATRX Zn finger domain and a totally skewed X-inactivation pattern. *Am J Med Genet A* 140(20):2212–2215.

Basehore MJ, Michaelson-Cohen R, Levy-Lahad E, Sismani C, Bird LM, Friez MJ, Walsh T, Abidi F, Holloway L, Skinner C, McGee S, Alexandrou A, Syrrou M, Patsalis PC, Raymond G, Wang T, Schwartz CE, King MC, Stevenson RE (2015) Alpha-thalassemia intellectual disability: variable phenotypic expression among males with a recurrent nonsense mutation - c.109C>T (p.R37X). *Clin Genet* 87(5):461–466.

Carpenter NJ, Qu Y, Curtis M, Patil SR (1999) X-linked mental retardation syndrome with characteristic "coarse" facial appearance, brachydactyly, and short stature maps to proximal Xq. *Am J Med Genet* 85(3):230–235.

Clynes D, Higgs DR, Gibbons RJ (2013) The chromatin remodeller ATRX: a repeat offender in human disease. *Trends Biochem Sci* 38(9):461–466.

Cohn DM, Pagon RA, Hudgins L, Schwartz CE, Stevenson RE, Friez MJ (2009) Partial ATRX gene duplication causes ATR-X syndrome. *Am J Med Genet A* 149A(10):2317–2320.

Donnai D, Clayton-Smith J, Gibbons RJ, Higgs DR (1991) The non-deletion a thalassaemia/mental retardation syndrome. Further support for X linkage. *J Med Genet* 28:742–745.

Garrick D, Samara V, McDowell TL, Smith AJ, Dobbie L, Higgs DR, Gibbons RJ (2004) A conserved truncated isoform of the ATR-X syndrome protein lacking the SWI/SNF-homology domain. *Gene* 326:23–34.

Gibbons RJ, Bachoo S, Picketts DJ, Aftimos S, Asenbauer B, Bergoffen J, Berry SA, Dahl N, Fryer A, Keppler K, Kurosawa K, Levin ML, Masuno M, Neri G, Pierpont ME, Slaney SS, Higgs DR (1997) Mutations in a transcriptional regulator (hATRX) establish the functional significance of a PHD-like domain. *Nat Genet* 17:146–148.

Gibbons RJ, Brueton L, Buckle VJ, Burn J, Clayton-Smith J, Davison BCC, Gardner RJM, Homfray T, Kearney L, Kingston HM, Newbury-Ecob R, Porteous MEP, Wilkie AOM, Higgs DR (1995a) The clinical and hematological features of the X-linked a thalassemia/mental retardation syndrome (ATR-X). *Am J Med Genet* 55:288–299.

Gibbons RJ, Picketts DJ, Villard L, Higgs DR (1995b) Mutations in a putative global transcriptional regulator cause X-linked mental retardation with a-thalassemia (ATR-X syndrome). *Cell* 80:837–845.

Gibbons RJ, Suthers GK, Wilkie AOM, Buckle VJ, Higgs DR (1992) X-linked a thalassemia/mental retardation (ATR-X) syndrome: Localisation to Xq12-21.31 by X-inactivation and linkage analysis. *Am J Hum Genet* 51:1136–1149.

Gibbons RJ, Wada T, Fisher C, Malik N, Mitson M, Steensma D, Goudie D, Fryer A, Krantz I, Traeger-Synodinos J (2008) Mutations in the chromatin associated protein ATRX. *Hum Mutat* 29(6):796–802.

Goldberg AD, Banaszynski LA, Noh KM, Lewis PW, Elsaesser SJ, Stadler S, Dewell S, Law M, Guo X, Li X, Wen D, Chapgier A, DeKelver RC, Miller JC, Lee YL, Boydston EA, Holmes MC, Gregory PD, Greally JM, Rafii S, Yang C, Scambler PJ, Garrick D, Gibbons RJ, Higgs DR, Cristea IM, Urnov FD, Zheng D, Allis CD (2010) Distinct factors control histone variant H3.3 localization at specific genomic regions. *Cell* 140(5):678–691.

Grozeva D, Carss K, Spasic-Boskovic O, Tejada MI, Gecz J, Shaw M, Corbett M, Haan E, Thompson E, Friend K, Hussain Z, Hackett A, Field M, Renieri A, Stevenson R, Schwartz C, Floyd JA, Bentham J, Cosgrove C, Keavney B, Bhattacharya S, Italian XlMRP, Consortium UK, Consortium G, Hurles M, Raymond FL (2015) Targeted Next-Generation Sequencing Analysis of 1,000 Individuals with Intellectual Disability. *Hum Mutat* 36(12):1197–1204.

Guerrini R, Shanahan JL, Carrozzo R, Bonanni P, Higgs DR, Gibbons RJ (2000) A nonsense mutation of the ATRX gene causing mild mental retardation and epilepsy. *Ann Neurol* 47(1):117–121.

Horesh N, Pery R, Amiel I, Shwaartz C, Speter C, Guranda L, Gutman M, Hoffman A (2015) Volvulus and bowel obstruction in ATR-X syndrome-clinical report and review of literature. *Am J Med Genet A* 167A(11):2777–2779.

Ion A, Telvi L, Chaussain JL, Galacteros F, Valayer J, Fellous M, McElreavey K (1996) A novel mutation in the putative DNA Helicase XH2 is responsible for male-to-female sex reversal associated with an atypical form of the ATR-X syndrome. *Am J Hum Genet* 58:1185–1191.

Ji J, Quindipan C, Parham D, Shen L, Ruble D, Bootwalla M, Maglinte DT, Gai X, Saitta SC, Biegel JA, Mascarenhas L (2017) Inherited germline ATRX mutation in two brothers with ATR-X syndrome and osteosarcoma. *Am J Med Genet A* 173(5):1390–1395.

Kurosawa K, Akatsuka A, Ochiai Y, Ikeda J, Maekawa K (1996) Self-induced vomiting in X-linked a-thalassemia/mental retardation syndrome. *Am J Med Genet* 63:505–506.

Law MJ, Lower KM, Voon HP, Hughes JR, Garrick D, Viprakasit V, Mitson M, De Gobbi M, Marra M, Morris A, Abbott A, Wilder SP, Taylor S, Santos GM, Cross J, Ayyub H, Jones S, Ragoussis J, Rhodes D, Dunham I, Higgs DR, Gibbons RJ (2010) ATR-X Syndrome Protein Targets Tandem Repeats and Influences Allele-Specific Expression in a Size-Dependent Manner. *Cell* 143(3):367–378.

Leahy RT, Philip RK, Gibbons RJ, Fisher C, Suri M, Reardon W (2005) Asplenia in ATR-X syndrome: a second report. *Am J Med Genet A* 139(1):37–39.

Lossi AM, Millan JM, Villard L, Orellana C, Cardoso C, Prieto F, Fontes M, Martinez F (1999) Mutation of the XNP/ATR-X gene in a family with severe mental retardation, spastic paraplegia and skewed pattern of X inactivation: demonstration that the mutation is involved in the inactivation bias. *Am J Hum Genet* 65(2):558–562.

Martucciello G, Lombardi L, Savasta S, Gibbons RJ (2006) The gastrointestinal phenotype of ATR-X syndrome. *Am J Med Genet* 140(11):1172–1176.

McPherson E, Clemens M, Gibbons RJ, Higgs DR (1995) X-linked alpha thalassemia/mental retardation (ATR-X) syndrome. A new kindred with severe genital anomalies and mild hematologic expression. *Am J Med Genet* 55:302–306.

Mitson M, Kelley LA, Sternberg MJ, Higgs DR, Gibbons RJ (2011) Functional significance of mutations in the Snf2 domain of ATRX. *Hum Mol Genet* 20(13):2603–2610.

Richards S, Aziz N, Bale S, Bick D, Das S, Gastier-Foster J, Grody WW, Hegde M, Lyon E, Spector E, Voelkerding K, Rehm HL, Committee ALQA (2015) Standards and guidelines for the interpretation of sequence variants: a joint consensus recommendation of the American College of Medical Genetics and Genomics and the Association for Molecular *Pathol Genet Med* 17(5):405–424.

Saugier-Veber P, Munnich A, Lyonnet S, Toutain A, Moraine C, Piussan C, Mathieu M, Gibbons RJ (1995) Letter to the Editor: Lumping Juberg-Marsidi syndrome and X-linked a-thalassemia/mental retardation syndrome? *Am J Med Genet* 55:300–301.

Smolle MA, Heitzer E, Geigl JB, Al Kaissi A, Liegl-Atzwanger B, Seidel MG, Holzer LA, Leithner A (2017) A novel mutation in

ATRX associated with intellectual disability, syndromic features, and osteosarcoma. *Pediatr Blood Cancer* 64:e26522.

Stevenson RE, Abidi F, Schwartz CE, Lubs HA, Holmes LB (2000a) Holmes-Gang syndrome is allelic with XLMR-hypotonic face syndrome. *Am J Med Genet* 94(5):383–385.

Stevenson RE, Schwartz CE, Schroer RJ (2000b) X-linked Mental Retardation. New York; Oxford: Oxford University Press.

Takano K, Tan WH, Irons MB, Jones JR, Schwartz CE (2011) Pitt-Hopkins syndrome should be in the differential diagnosis for males presenting with an ATR-X phenotype. *Clin Genet* 80(6):600–601.

Thienpont B, de Ravel T, Van Esch H, Van Schoubroeck D, Moerman P, Vermeesch JR, Fryns JP, Froyen G, Lacoste C, Badens C, Devriendt K (2007) Partial duplications of the ATRX gene cause the ATR-X syndrome. *Eur J Hum Genet* 15(10):1094–1097.

Villard L, Fontes M, Ades LC, Gecz J (2000) Identification of a mutation in the XNP/ATR-X gene in a family reported as Smith-Fineman-Myers syndrome. *Am J Med Genet* 91(1):83–85.

Villard L, Lacombe D, Fontés M (1996a) A point mutation in the XNP gene, associated with an ATR-X phenotype without a-thalassemia. *Europ J Hum Genet* 4:316–320.

Villard L, Toutain A, Lossi A-M, Gecz J, Houdayer C, Moraine C, Fontès M (1996b) Splicing mutation in the ATR-X gene can lead to a dysmorphic mental retardation phenotype without a-thalassemia. *Am J Hum Genet* 58:499–505.

Wada T, Ban H, Matsufuji M, Okamoto N, Enomoto K, Kurosawa K, Aida N (2013) Neuroradiologic features in X-linked alpha-Thalassemia/mental retardation syndrome. *Am J Neuroradiol*.

Wada T, Nakamura M, Matsushita Y, Yamada M, Yamashita S, Iwamoto H, Masuno M, Imaizumi K, Kuroki Y (1998) Three Japanese children with X-linked alpha-thalassemia/mental retardation syndrome (ATR-X). *No To Hattatsu* 30(4):283–289.

Wada T, Sugie H, Fukushima Y, Saitoh S (2005) Non-skewed X-inactivation may cause mental retardation in a female carrier of X-linked alpha-thalassemia/mental retardation syndrome (ATR-X): X-inactivation study of nine female carriers of ATR-X. *Am J Med Genet A* 138(1):18–20.

Watanabe T, Arai K, Takahashi M, Ohno M, Sato K, Fuchimoto Y, Wada T, Ida S, Kawahara H, Kanamori Y (2014) Esophagogastric motility and nutritional management in a child with ATR-X syndrome. *Pediatr Int* 56(4):e48–51.

Weatherall DJ, Higgs DR, Bunch C, Old JM, Hunt DM, Pressley L, Clegg JB, Bethlenfalvay NC, Sjolin S, Koler RD, Magenis E, Francis JL, Bebbington D (1981) Hemoglobin H disease and mental retardation. A new syndrome or a remarkable coincidence? *New Engl J Med* 305:607–612.

Wilkie AOM, Zeitlin HC, Lindenbaum RH, Buckle VJ, Fischel-Ghodsian N, Chui DHK, Gardner-Medwin D, MacGillivray MH, Weatherall DJ, Higgs DR (1990) Clinical features and molecular analysis of the a thalassemia/mental retardation syndromes. II. Cases without detectable abnormality of the a globin complex. *Am J Hum Genet* 46:1127–1140.

8

BARDET–BIEDL SYNDROME

ANNE M. SLAVOTINEK

Division of Medical Genetics, Department of Pediatrics, University of California, San Francisco, San Francisco, California, USA

INTRODUCTION

Laurence–Moon syndrome was first described by two English ophthalmologists in a family of four siblings with pigmentary retinopathy, cognitive disability, hypogenitalism, and spastic paraparesis (Laurence and Moon 1866). Bardet (1920) and Biedl (1922) separately reported individuals with retinopathy, polydactyly, obesity, developmental disability, and hypogenitalism, describing Bardet–Biedl syndrome (BBS). The co-occurrence of retinopathy, intellectual disability, and hypogenitalism led to the initial assumption that these reports concerned the same condition (Solis-Cohen and Weiss 1924). However, as mutations in the *PNPLA6* gene have now been described in individuals with Laurence–Moon syndrome and Oliver–McFarlane syndrome (Hufnagel et al. 2015), these entities are considered distinct from BBS and Laurence–Moon syndrome is not further considered in this chapter.

Several excellent clinical reviews of BBS have been published in the past, including those of Green et al. (1989), O'Dea et al. (1996), and Beales et al. (1999). In recent years, publications have focused more on the molecular genetic etiology of BBS and the role of the causative genes in the pathogenesis of clinical features that are part of this complex disease, such as obesity. Nevertheless, recent comprehensive clinical reviews include those by Forsythe and Beales (2015), Khan et al. (2016), and Suspitsin and Imyanitov (2016).

Incidence

The incidence of BBS ranges from 1 in 140,000 to 1 in 160,000 live births in the North American and European populations (Beales et al. 1999). However, BBS has a higher incidence in selected populations, likely due to founder effects. In Newfoundland, the frequency of BBS is 1 in 17,500 (Green et al. 1989), but more than eight separate mutations have been identified without evidence of a common mutation (Katsanis et al. 2001b). In the mixed Arab population of Kuwait, BBS has an estimated minimum prevalence of 1 in 36,000 and 1 in 13,500 in two studies of those with Bedouin ethnicity (Farag and Teebi 1988, 1989). There were 13 individuals with BBS registered among 48,000 inhabitants of the Faroe Islands, leading to a disease frequency estimate of 1 in 3700 (Hjortshøj et al. 2009). There has been no evidence of race-specific variation in the clinical features (Katsanis et al. 2001b).

Sex ratios in favor of both males (Beales et al. 1999) and females (Green et al. 1989) have been described, but there is no known sex predilection. The incidence of consanguinity was 8% in a population consisting of affected Caucasian (95%) and Indian individuals (5%) (Beales et al. 1999). However, frequencies of consanguinity as high as 87% have been noted in populations with increased rates of inbreeding (Farag and Teebi 1988; Green et al. 1989). Twins have been described.

There are few reports on life expectancy in BBS. The majority of individuals in the literature were evaluated at less than 40 years of age (Beales et al. 1999), and the relative

---

Cassidy and Allanson's Management of Genetic Syndromes, Fourth Edition.
Edited by John C. Carey, Agatino Battaglia, David Viskochil, and Suzanne B. Cassidy.
© 2021 John Wiley & Sons, Inc. Published 2021 by John Wiley & Sons, Inc.

paucity of reported older individuals with BBS may reflect ascertainment bias or increased mortality (Beales et al. 1999). Although the diagnosis has been established in the sixth decade of life, a new diagnosis of retinopathy is rare by late adulthood. The age at the time of death in BBS in one study ranged from 1 to 63 years, and 25% were deceased by 44 years of age (O'Dea et al. 1996), although normal life expectancy has also been reported. Causes of death in BBS have included polycystic kidney disease, chronic renal failure, and renal carcinoma with metastatic disease (O'Dea et al. 1996; Riise 1996; Tobin and Beales 2007). In recent years, the structural renal and urinary tract anomalies and renal dysfunction that are reported to affect 53–82% of people with BBS have been increasingly recognized as causes of morbidity and mortality (Forsythe et al. 2016). Myocardial infarction, congestive heart failure, diabetic coma, pulmonary embolism, and the complications of morbid obesity can also lead to mortality in patients with BBS (Escallon et al. 1989).

Diagnostic Criteria

Diagnostic criteria for BBS initially included at least four of the five features of rod-cone dystrophy, polydactyly, obesity, "mental retardation", and hypogenitalism (Schachat and Maumenee 1982). Retinal dystrophy was considered invariable. Following the reviews of Harnett et al. (1988) and Green et al. (1989), renal malformations were included in the diagnostic criteria, whereas hypogonadism in females and intellectual disability were deemed less important. The diagnostic criteria were rewritten to include the five major features of Schachat and Maumenee (1982), renal malformations as a sixth major feature, and additional minor features in an effort to encompass the phenotypic pleiotropy and variability (Beales et al. 1999) (Table 8.1).

Many of the major clinical features of BBS are age-dependent, and the diagnosis is rarely established at birth without a family history. At birth, polydactyly, renal anomalies and/or cystic kidneys, hepatic fibrosis, genital abnormalities, and heart malformations can be present (Dippell and Varlam, 1998; Karmous-Benailly et al. 2005). Obesity may be noted as early as the neonatal period (Bauman and Hogan 1973). Developmental delays can present in early childhood and visual impairment is generally detected in childhood or teenage years (O'Dea et al. 1996). In early adulthood, hypertension, diabetes mellitus, and renal failure may bring the individual to medical attention. The average age at diagnosis in a study of 109 individuals was 9 years (Beales et al. 1999). Phenotypic overlap with other ciliopathies can also confound the diagnosis (Suspitsin and Imyanitov 2016). It may be particularly difficult to diagnose BBS in an obese child with developmental delay without polydactyly until the visual impairment is manifest (Ross and Beales 2007).

TABLE 8.1 Diagnostic criteria for Bardet–Biedl syndrome

Primary features
Four of the following primary features are required to be present
    Rod-cone dystrophy
    Polydactyly
    Obesity
    Learning difficulties
    Hypogenitalism in males
    Renal anomalies
Or
Three primary features plus two secondary features are required
Secondary Features
    Speech disorder or speech delay
    Strabismus or cataracts or astigmatism
    Brachydactyly or syndactyly
    Developmental delay
    Polyuria and polydipsia (nephrogenic diabetes insipidus)
    Ataxia and/or poor coordination
    Mild spasticity (especially lower limbs)
    Diabetes mellitus
    Dental crowding or hypodontia or small roots or a high palate
    Left ventricular hypertrophy or congenital heart disease
    Hepatic fibrosis

Source: Beales et al. (1999).

A summary of the physical features in BBS is provided in Table 8.2. A characteristic pattern of craniofacial dysmorphism has been suggested, including deep-set eyes, hypertelorism with downslanting palpebral fissures, a depressed nasal bridge with anteverted nares, prominent nasolabial folds, a long philtrum, and a thin upper lip vermilion (Beales et al. 1999) (Figure 8.1). Macrocephaly has been found in almost half of the affected individuals (Bauman and Hogan 1973). Adults with BBS may have a prominent forehead and males may have premature balding (Beales et al. 1999). A high palate was found in 40 of 45 individuals, and orofacial clefting has been noted (Beales et al. 1999).

Intrafamilial variation in the expression of BBS has been considered to be as substantial as interfamilial variation and has been observed for all of the major features of the disease, short stature, and dental anomalies (Green et al. 1989; Elbedour et al. 1994; Riise et al. 1997).

Etiology, Pathogenesis, and Genetics

BBS is an autosomal recessive or triallelic disease with substantial genetic heterogeneity. Currently there are 21 known genes (Table 8.3) in which biallelic deleterious sequence variants can cause BBS, together accounting for about 80% of affected individuals, but still implying that additional BBS genes may be identified (Stoetzel et al. 2007; Suspitsin and Imyanitov 2016). *BBS1* and *BBS10* are the most commonly mutated genes in Europeans, with *BBS1* accounting for 20–25% of mutations and *BBS10* accounting for 15–20%

TABLE 8.2 Summary of phenotypic features in Bardet–Biedl syndrome

Phenotypic feature	Green et al. (1989)	Beales et al. (1999)
Ocular manifestations		
Rod-cone dystrophy	28/28 (100%)	102/109 (93%)
Blindness	27/28 (96%)	—
Myopia	12/16 (75%)	—
Nystagmus	14/27 (52%)	—
Glaucoma	5/23 (22%)	—
Posterior subcapsular cataracts	12/27 (44%)	—
Mature cataracts	8/27 (30%)	—
Digital manifestations		
Postaxial polydactyly (PAP)	18/31 (58%)	75/109 (69%)
PAP of hands and feet	7/18 (39%)	23/109 (21%)
PAP of hands only	2/18 (11%)	9/109 (8%)
PAP of feet only	9/18 (50%)	23/109 (21%)
Brachydactyly	13/26 (50%)[a]	51/109 (46%)
Syndactyly	—	9/109 (8%)
Macrodactyly	4/24 (17%)	—
Obesity and growth		
Height		
Males	7/10 < 50th	—
Females	9/15 < 50th	—
Weight		
Males	12/28 > 95th	—
Females	(Both sexes)	—
Overweight	—	78/109 (72%)[b]
Obese	—	56/109 (52%)
Renal manifestations		
Calyceal and/or parenchymal cysts	13/21 (62%)	6/57 (10%)
Calyceal clubbing/anomalies	20/21 (95%)	6/57 (10%)
Fetal lobulation	20/21 (95%)	7/57 (12%)
Scarring	5/21 (24%)	7/57 (12%)
Cortical loss	6/21 (29%)	—
Unilateral agenesis	—	2/57 (4%)
Dysplastic kidneys	—	3/57 (5%)
Renal calculi	—	1/57 (2%)
Vesicoureteric reflux	3/20 (15%)	5/57 (9%)
End-stage renal failure	3/32 (9%)	6/109 (5%)
Development and central nervous system features		
Developmental delay	—	55/109 (50%)
Learning difficulties	9/22 (41%)[c]	68/109 (62%)[d]
Behavioral problems	100%[e]	59/109 (54%)
Speech deficit	—	36/109 (33%)
Ataxia, poor coordination	—	43/109 (40%)
Hypogenitalism (males)		
Hypogonadism	7/8 (88%)	60/62 (89%)
Cryptorchidism	—	8/62 (13%)
Delayed puberty	—	19/62 (31%)
Other		
Cardiac disease	—	8/109 (7%)
Dental anomalies	—	29/109 (27%)
Hearing loss[f]	—	26/109 (24%)
Diabetes mellitus (type II)	9/20 (45%)	7/109 (6%)

[a] Anthropomorphic measurements.
[b] Overweight defined as BMI > 25 kg m^{-2}; obese defined as BMI > 30 kg m^{-2}.
[c] Verbal IQ < 70.
[d] Usually mild to moderate (Beales et al. 1999).
[e] Inappropriate mannerisms and shallow affect.
[f] Includes conductive and mixed hearing loss.

of mutations in this population (Stoetzel et al. 2007; Sheffield et al. 2008; Forsythe and Beales 2015; Suspitsin and Imyanitov 2016). Several genes are responsible for 5–10% of families (*BBS2*, *BBS6*, *BBS9*, *BBS12*) (Stoetzel et al. 2007; Forsythe and Beales 2015; Suspitsin and Imyanitov 2016). The remaining genes are much less frequently involved and mutations in some genes, such as *BBS11*, are exceptionally rare (Chiang et al. 2006; Forsythe and Beales 2015; Suspitsin and Imyanitov 2016).

Recurrent deleterious sequence variants are relatively rare in BBS, but one common missense mutation, p.(Met390Arg) in *BBS1* has been found in the homozygous state in 16 of 60 unrelated probands from North America (Mykytyn et al. 2002). This mutation accounts for up to 80% of mutations in the *BBS1* gene and is present in 18–32% individuals with BBS (Ross and Beales 2007). A mutation with an insertion of a single nucleotide that is predicted to cause a frameshift (c.271_273insT, predicting p.Cys91Leu*fs**4) and premature protein truncation in *BBS10*

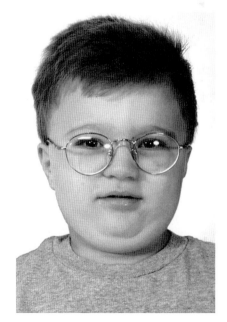

FIGURE 8.1 Facial view of a child with Bardet–Biedl syndrome at 2 1/2 years of age showing round facial appearance.

TABLE 8.3 Genetic Heterogeneity in Bardet–Biedl syndrome

Gene	Locus	Exons	Mutational spectrum	Protein/subcellular localization	Frequency[a]	References
BBS1	11q13	17	MS; NX; FS; SPL	593 Aa; BBSome	23–39%	Mykytyn et al. (2002)
BBS2	16q21	17	MS; NX; FS	721 Aa; BBSome	8–9%	Nishimura et al. (2001)
BBS3, ARL6	3p12	9	MS; NX	186 Aa; Ras superfamily of small GTP-binding proteins; BBSome assembly	0.4–2%	Chiang et al. (2004)
BBS4	15q23	16	MS; NX; FS: SPL; Indel	519 Aa; BBSome	1-2%	Mykytyn et al. (2001), Katsanis et al. (2002)
BBS5	2q31	12	SPL; NX; Indel	342 Aa; BBSome	0.4–3%	Li et al. (2004)
BBS6	20p12	6	MS; NX; FS	570 Aa; chaperonin complex	4–6%	Katsanis et al. (2000), Slavotinek et al. (2000), Stone et al. (2000)
BBS7	4q27	19	FS; MS; Indel	672 Aa; BBSome	1.5–2%	Badano et al. (2003)
BBS8, TTC8	14q32.1	16	Indel; SPL	531 Aa; BBSome	1-1.2%	Ansley et al. (2003)
BBS9, B1, PTHB1	7p14	25	MS; NX; SPL; FS	879–916 Aa; BBSome	6%	Nishimura et al. (2005)
BBS10	12q	2	MS; FS; NX	723 Aa; Chaperonin complex	20%	Stoetzel et al. (2006)
BBS11, TRIM32	9q31-34.1	2	MS	650 Aa; E3 ubiquitin ligase activity	0.1-0.4%	Chiang et al. (2006)
BBS12	4q27	2	FS; NX; MS; Indel	710 Aa; Chaperonin complex	5%	Stoetzel et al. (2007)
BBS13, MKS1	17q23	18	Indel	559 Aa; role in basal body, ciliogenesis	4.5%	Leitch et al. 2008
BBS14, CEP290	12q21.3	54	MS; Indel	2481 Aa; centrosomal protein; ciliary entry of BBSome	1%	Leitch et al. 2008
BBS15, C2ORF86, WDPCP	2p15	12	MS; SPL	3326 Aa; WD repeat-containing and planar cell polarity effector protein fritz homologue	1%	Kim et al. 2010
BBS16, SDCCAG8	1q43-q44	18	MS; SPL; NX; Indel	713 Aa; basal body; centrosomal associated protein	1%	Otto et al. 2010
BBS17, LZTFL1	3p21.31	10	MS; Indel	299 Aa; negative regulator of the BBSome	Unknown	Marion et al. 2012
BBS18, BBIP1	10q25.2	4	NX	92 Aa; BBSome	Unknown	Scheidecker et al. 2014
BBS19, IFT27	22q12	9	MS	186 Aa; intraflagellar transport	Unknown	Aldahmesh et al. 2014
BBS20, IFT172	2p23.3	44	SPL; Indel	1749 Aa; intraflagellar transport	Unknown	Bujakowska et al. 2015
BBS21, NPHP1	2q13	20	Indel; NX; MS	732 Aa; basal body anchoring, primary cilium assembly	Unknown	Lindstrand et al. 2014

Abbreviations: Aa, amino acids; MS, missense variant; FS, frameshift variant; Indel, insertion deletion; NX, nonsense variant; SPL, splice-site variant; WD = tryptophan-aspartic acid; BBSome = Bardet–Biedl syndrome 'ome'. [a] *Resources:* Beales et al. (2001), Katsanis (2004), Stoetzel et al. (2006), Chiang et al. (2006), and Stoetzel et al. (2007), Forsythe and Beales (2015), Khan et al. (2016), Suspitsin and Imyanitov (2016).

also accounts for 46% of mutant *BBS10* alleles and can be identified in approximately 10% of all individuals with BBS (Stoetzel et al. 2006).

Many of the BBS causative genes affect only a few families, complicating attempts at a phenotype-genotype analysis (Khan et al. 2016). Several reports have noted a milder phenotype with less renal anomalies, obesity, hypogenitalism, and developmental differences for those affected with *BBS1* variants (Hjortshøj et al. 2010). People with *BBS1* variants also have less severe retinal disease (Daniels et al. 2012). Polydactyly has not been observed to date in those with *BBS16* or *BBS18* variants (Khan et al. 2016). Other observations have revealed that individuals with *BBS10* variants had a significantly higher age and sex standardized body mass index (BMI-Z), greater visceral adiposity, increased insulin resistance, and a higher frequency of urogenital anomalies compared to patients with *BBS1* variants (Feuillan et al. 2011; Castro-Sanchez et al. 2015). Those with variants affecting the chaperonin genes, *BBS6*, *BBS10*, and *BBS12*, have also had more severe renal disease (Imhoff et al. 2011). Finally, a higher frequency of cognitive impairment was noted in people with *BBS12* variants compared to people with *BBS1* variants (Castro-Sanchez et al. 2015).

With numerous causative genes, it has long been hypothesized that the BBS genes and proteins would function in a common pathway, and it is now known that the majority of the BBS genes are involved in motile and immotile sensory cilia formation or function (Tobin and Beales 2007). *BBS8* was the first BBS gene to be linked to ciliary function (Ansley et al. 2003). The BBS8-predicted protein has eight tetratricopeptide motifs with significant similarity to a prokaryotic domain, *pilF*, and is localized to the centrosome in human cells (Ansley et al. 2003). Immunoprecipitation experiments show that the BBS8 protein is able to bind PCM1, a protein that interacts with the BBS4 protein and that is localized to ciliary basal bodies. An orthologous *BBS8* transcript in *Caenorhabditis elegans* is also expressed in ciliated cells, suggesting that the BBS8 protein and other BBS proteins may have a role in ciliary biogenesis and/or function (Ansley et al. 2003). Other BBS proteins participate in intraflagellar transport, functioning as adaptors for protein loading at the cytoplasmic end of the cilia and in retrograde transport of proteins from the cilium to the cytoplasm.

It is now known that many of the BBS proteins are members of a complex called the BBSome, including *BBS1*, *BBS2*, *BBS4*, *BBS5*, *BBS7*, *BBS8*, *BBS9*, and *BBS18*, whereas *BBS3*, *BBS6*, *BBS10*, *BBS11*, and *BBS12* are not part of the BBSome (Nachury et al. 2007; Khan et al. 2016; Suspitsin and Imyanitov 2016). The BBSome localizes to the cilium membrane and is required for ciliogenesis by interacting with the Rab8GTP/GDP exchange factor (Nachury et al. 2007; Khan et al. 2016). The BBSome also functions in the localization of select transmembrane proteins to, and removal from, the cilium (Vaisse et al. 2016). Other BBS proteins are critical for BBSome function, including *BBS3* (*ARL6*) that recruits the BBSome to the cell membrane and *BBS17* (*LZTFL1*) that regulates the entry of the BBSome into cilia (Khan et al. 2016). *BBS14* (*CEP290*) has been hypothesized to regulate the integrity of the BBSome (Barbelanne et al. 2015). Other BBS genes function in the intraflagellar transport complex that controls bidirectional motility along the axoneme, or central strand of a cilium, that is critical for ciliary function (Nachury et al. 2007; Wei et al. 2012; Khan et al. 2016).

In contrast, the *BBS6*, *BBS10*, and *BBS12* genes are all encode proteins that function as chaperonins (Stoetzel et al. 2007). Chaperonins are involved in folding and stabilizing other proteins, leading to the hypothesis that the age-dependent features of BBS could manifest because of a progressive decline in levels of correctly folded proteins as a result of chaperonin malfunction. Interestingly, the α-transducin protein is a substrate for type II chaperonin proteins and loss of function of the α-transducin gene in homozygous knockout mice results in photoreceptor degeneration (Calvert et al. 2000). The von Hippel–Lindau protein (see Chapter 62) also requires the involvement of chaperonins to form an active complex, providing a possible link between renal disease and BBS (Beales et al. 2000).

It has been proposed that the inheritance in as many as 10% of BBS families is triallelic, following the detection of three deleterious alleles from two different loci in affected individuals (Katsanis et al. 2001a, 2002; Beales et al. 2003; Katsanis 2004; Ross and Beales 2007). This hypothesis stemmed from attempts to explain the high frequency of heterozygous mutations in the *BBS6* and *BBS2* genes in affected individuals (Katsanis et al. 2001a; Slavotinek et al. 2002). An integration of haplotype analysis and mutation screening for the *BBS2* and *BBS6* genes in 163 individuals showed that three mutant alleles were present in probands from four pedigrees, whereas individuals who had biallelic, deleterious variants in the *BBS2* gene were unaffected (Katsanis et al. 2001a). Similar analyses for *BBS4* suggested that this gene also participates in triallelic inheritance with *BBS1* and *BBS2* genes (Katsanis et al. 2002). Further evidence for oligogenic inheritance in BBS was later obtained after screening the *BBS1* gene for mutations, with at least four families demonstrating two mutations in *BBS1* and one mutation in another BBS gene (Beales et al. 2003). However, it is now recognized that triallelic inheritance is relatively uncommon and the majority of affected individuals with deleterious variants in one of the causative genes for BBS have autosomal recessive inheritance (Suspitsin and Imyanitov 2016).

Phenotypic Manifestations in First-degree Relatives of Individuals with Bardet–Biedl Syndrome Clinical features of BBS have been noted in the first-degree relatives of affected individuals (Croft and Swift 1990; Croft et al. 1995), but their significance remains controversial (O'Dea et al. 1996). Obesity, hypertension, diabetes, and renal disease were observed in obligate carriers in a five-generation family pedigree (Croft and Swift, 1990) and deafness, cataracts and night blindness have also been described (Urben and Baugh 1999). The presence of phenotypic manifestations in first-degree relatives may be particularly significant for renal disease. Congenital renal malformations were detected in 5 out of 123 (4.1%) relatives of probands together with a significant increase in the incidence of renal agenesis (Beales et al. 2000). In view of the risk of hypertension and renal failure, it has been suggested that first-degree relatives of probands should be screened with renal sonography for occult renal tract malformations (Beales et al. 1999, 2000). Regarding vision, 32 obligate carriers of BBS variants had no visual impairment (Héon et al. 2005). However, three out of six carriers in a different study demonstrated significant functional abnormalities on multifocal electroretinogram, despite a normal fundus (Kim et al. 2007). Screening of first-degree relatives for visual abnormalities remains discretionary.

Genetic Counseling BBS is inherited as an autosomal recessive condition in most affected individuals. The chance of recurrence for a further affected sibling being born to parents who have an affected child is 25% with this mode of inheritance and both males and females can be affected. For individuals with BBS, recurrence in offspring is rare, unless marriage occurs within the family or within a small community. Clinical genetic testing for mutations should be offered if the parents opt for prenatal diagnosis.

Diagnostic Testing

BBS can be diagnosed using clinical criteria (see Diagnostic Criteria; Table 8.1). A testing strategy comprising sequential testing of the two most commonly mutated genes, *BBS1* and *BBS10*, followed by screening of the rarer genes, was once recommended (Sheffield et al. 2008) but exome sequencing or specific BBS-gene panels are clinically available. A clinical diagnosis can suffice to start management recommendations if genetic testing is not available.

Differential Diagnosis

BBS shows phenotypic overlap with several syndromes (Table 8.4) and diagnostic confusion can result from the delayed appearance of age-dependent manifestations in BBS together with the clinical variability. There is overlap with Joubert syndrome and other ciliopathies, although the characteristic cerebellar and brain stem malformations associated with Joubert syndrome, including vermis hypoplasia and the distinctive molar tooth sign on cranial magnetic resonance imaging, are absent in BBS (Forsythe and Beales 2015). Alström syndrome comprises retinal dystrophy, obesity, diabetes mellitus, cardiomyopathy, and neurosensory deafness but does not include polydactyly, learning disability or hypogonadism (Russell-Eggitt et al. 1998). Alström syndrome is caused by mutations in the *ALMS1* gene and thus is molecularly distinct from BBS (Collin et al. 2002; Hearn et al. 2002). Similarity to Pallister–Hall syndrome (Chapter 44) was noted in an individual with BBS who had a hypothalamic hamartoma and to Meckel–Gruber syndrome in families with cystic renal dysplasia and polydactyly (Gershoni-Baruch et al. 1992; Karmous-Benailly et al. 2005).

In females, overlap with McKusick–Kaufman syndrome comprising hydrometrocolpos or female hypogenitalism, polydactyly, and congenital heart disease is important because of the difference in visual and intellectual prognosis (Schaap et al. 1998). Individuals with McKusick–Kaufman syndrome do not develop many of the age-dependent manifestations of BBS, and the diagnosis of McKusick–Kaufman syndrome cannot be reliably made until a female with hydrometrocolpos and polydactyly is at least 5 years of age (David et al. 1999; Slavotinek and Biesecker 2000).

MANIFESTATIONS AND MANAGEMENT

Growth and Feeding

Obesity is common in BBS and is present in 72–96% of affected individuals, although the reported frequency can be influenced by the definition of obesity, and ascertainment bias has been postulated (Beales et al. 1999; Sheffield et al. 2008). Increased weight can be noticeable from birth and 38% of infants with BBS have a birth weight greater than the 90th centile (Bauman and Hogan 1973). The distribution of adipose tissue is truncal and rhizomelic, affecting the proximal portions of limbs, and although the obesity is variable, the degree can reach morbid proportions (Green et al. 1989; Sheffield et al. 2008). Females with BBS have had significantly greater body mass indexes than their affected brothers (O'Dea et al. 1996). Hyperphagia is not present to the extent seen in Prader–Willi syndrome (see Chapter 46), although it has been reported as a significant problem in childhood (Beales et al. 1999).

The mechanism for obesity in BBS has been studied in homozygous null mice for the *Bbs2*, *Bbs4*, and *Bbs6* genes and was associated with hyperleptinemia and leptin resistance in all three murine models (Rahmouni et al. 2008). As many of the leptin-responsive neurons in mice are ciliated, aberrant ciliary function caused by mutations in the BBS genes could lead to defects in neuronal cilia that perturb the signaling associated with the leptin receptor (Rahmouni et al. 2008). In addition, decreased locomotor activity has been noted in BBS null mice, similar to a lower level of physical activity that has been noted in people with BBS compared to healthy controls (Grace et al. 2003). This suggests that obesity in BBS results from both decreased energy expenditure and increased energy intake due to leptin resistance.

Evaluation

- Growth charts should be used to plot height, weight, and head circumference at diagnosis and at regular visits throughout childhood.
- Body mass index (BMI) should also be calculated for diagnosis and monitoring of obesity. In adults, weight and BMI should also be assessed at medical visits.
- Dietary evaluation should be performed if obesity is present and a dietary logbook should be maintained.
- Complications of obesity should be anticipated through annual measurements of blood pressure, blood glucose and hemoglobinA1C levels, serum cholesterol, and lipid levels. Guidelines for the age to start this testing have not been definitively established, but many physicians start monitoring from the age at diagnosis.

Management

- Obesity in people with BBS has been responsive to caloric restriction (O'Dea et al. 1996). Appetite suppressants and surgical interventions have not been assessed in formal trials but can be considered if there are no contraindications.
- Nutritional education with a calorie-restricted diet and the involvement of a nutritionist may be helpful.

TABLE 8.4 Differential diagnosis of Bardet–Biedl syndrome.

Manifestation	Bardet–Biedl syndrome	Laurence–Moon syndrome	Alström syndrome	Meckel–Gruber syndrome	McKusick–Kaufman syndrome
Rod-cone dystrophy	+	+	+	–	–
Coloboma	–	–	–	–	–
Polydactyly	+	–	+/–	+	+
Obesity	+	–	+	–	–
Learning disabilities	+	+	–	–	–
Hypogenitalism	+	+	+/–	+	+
Renal anomalies	+	+	–	+	+/–
Diabetes mellitus	+/–	–	+	–	–
Spasticity	+/–	+	–	–	–
Encephalocele	–	–	–	–	–
Cardiac disease	+/–	–	+	–	+
Deafness	+/–	–	+	–	–

Abbreviations: +, common or cardinal feature; +/–, less common feature; –, absent.

- Regular exercise programs are important for weight control.
- Behavioral and family therapy may be required to assist families in managing the obesity.
- Complications of obesity, such as diabetes mellitus and hyperlipidemia, should be treated as in the general population.

Development and Behavior

Many of the first studies on intellectual development in BBS did not compensate for the visual impairment when learning potential was assessed (Green et al. 1989). It has been widely recognized that developmental delays or intellectual disability are not invariable and there is familial and individual variation in intellectual capabilities (Green et al. 1989). Many adults are capable of attaining independent living skills (Forsythe and Beales 2015). In one study, mild to moderate learning disabilities were found in 62% of affected individuals and 50% attended special school (Beales et al. 1999). There may be selective cognitive deficits. Performance abilities have measured higher than verbal skills (Green et al. 1989). Other studies have found impaired fine motor function and olfaction (Brinckman et al. 2013). Speech dysfunction is common and has been characterized by scanning speech, articulation disorders with errors in substitution, hypernasal speech or a breathy voice quality (Beales et al. 1999; Urben and Baugh 1999). Severe deficits and autism are relatively rare (Ross and Beales 2007). Many affected individuals are able to remain in mainstream education with support for visual impairment (Sheffield et al. 2008).

Behavioral changes have included emotional outbursts and hyperactivity, frustration, inflexibility, and a preference for routines, obsessive-compulsive symptoms, and alterations in affect (Green et al. 1989; Beales et al. 1999).

Psychiatric problems in individuals with BBS are common and have included anxiety, mood disorders, depression, psychosomatic manifestations, and bipolar disorder (Ross and Beales 2007). The criteria for a psychiatric diagnosis were met by 14 of 46 (30%) individuals with BBS in one longitudinal study with an average patient age of 44 years (Moore et al. 2005). Schizophrenia was noted in 2% (Bauman and Hogan 1973; Beales et al. 1999).

Evaluation

- A full developmental and behavioral assessment by a specialist should be performed at the time of initial evaluation or diagnosis. The type of assessment should be tailored to age and ability and may range from a clinical evaluation of milestones to formal assessment of intellectual capabilities by a clinical psychologist. Visual status should be considered in selecting assessment measures.
- Speech assessment by a qualified individual should be accomplished after age 2 years.
- Further evaluation and monitoring by a developmental pediatrician or specialist at a child development clinic may be needed. Similarly, the opinion of a behavioral psychologist or speech therapist may be sought.
- Issues related to transitioning to adulthood are critical in those with significant cognitive disability. Again, considerations related to visual impairment must be addressed.

Management

- Enrollment in early intervention programs and provision of appropriate educational services and a statement of educational needs should be commenced after an assessment that indicates developmental delay.

Adaptation of school programming to learning disabilities, visual impairment, and behavior problems is necessary.
- Involvement of a speech therapist may be needed.
- Severe behavior problems, when present, should be referred to a behavior specialist for management, which is standard.
- Psychiatric problems are treated in a standard manner.

Ophthalmologic

Visual impairment affects more than 90% of people with BBS (Khan et al. 2016). The ophthalmologic findings in BBS have been considered to be the most important physical features in establishing the diagnosis (Schachat and Maumenee 1982). The ocular signs are primarily retinal degeneration with a rod-cone dystrophy and pigmentary retinal changes associated with myopia and early involvement of the macula causing loss of central and peripheral vision (Green et al. 1989; Riise et al. 1996a, 1996b), although rod-cone dystrophy has also been described. Onset of ocular symptoms is usually between 4 and 9 years of age and consists of night blindness, visual impairment in bright light, light aversion, loss of peripheral vision, or a combination of these symptoms (Riise et al. 1996b; Beales et al. 1999). Maculopathy is present in all affected individuals by the second decade of life (Fulton et al. 1993), and 73% of individuals over 30 years of age are unable to count fingers (Green et al. 1989). The progression to blindness can be rapid and is strongly correlated with age, although individual variation can occur (Fulton et al. 1993). Blindness can develop from 5 to 43 years, with the mean age being 15–18 years (O'Dea et al. 1996; Beales et al. 1999). However, the course of visual impairment can be unpredictable, and one man was reported to have stable vision in his fourth decade (Osusky et al. 1991). A study of the eye phenotype in 10 individuals with the p.(Met390Arg) variant in *BBS1* showed a wide spectrum of retinal disease, ranging from a subtle maculopathy to rod-cone dysfunction with a negative electroretinogram waveform, to loss of central or peripheral vision in the most severely affected individuals (Azari et al. 2006). Severity varied in families and was unrelated to age, suggesting that factors other than the disease-causing mutations may be contributory (Azari et al. 2006).

The fundoscopic appearance in BBS includes the changes of typical and atypical retinal pigmentation with dense, "bone spicule-like" pigmentation, surface retinal wrinkling, rounded or sparse clumps of pigmentation, pale optic disks, optic atrophy, and attenuated vessels (Figures 8.2A and B) (Green et al. 1989). Cone-rod dystrophy can also be present without pigmentary changes, especially early in life ("sine pigmenti") (Héon et al. 2005; Deffert et al. 2007; Sheffield et al. 2008). The appearance of the fundus does not predict vision (Schachat and Maumenee 1982; Green et al. 1989). Electroretinograms show an early loss of signals from both rods, which are affected first, and cones, with reduced amplitudes (Riise et al. 1996b), but abnormalities in electroretinograms may be delayed until after 14 months of age (Ross and Beales, 2007).

Nystagmus due to visual loss or neurological disease has been stated both as common (Riise et al. 1996a) and rare (Sheffield et al. 2008). Mean age of onset of nystagmus is 14 years, and abnormal eye movements are also correlated with increasing age, although nystagmus was an early presenting feature in one child (Riise et al. 1996a). Other ocular complications have included significant refractive errors (Sheffield et al. 2008), strabismus, cataracts in particular involving the posterior subcapsular area (Schachat and Maumenee 1982; Green et al. 1989; Sheffield et al. 2008), impairment of color vision, glaucoma (Green et al. 1989), macular edema and degeneration (Beales et al. 1999), microphthalmia (Bauman and Hogan 1973) and Duane retraction syndrome, ptosis, and keratoconus.

Evaluation

- Immediate referral is recommended to a retinal specialist or an ophthalmologist for formal testing with assessment of visual acuity, visual fields, refractive errors, and fundoscopy with dilatation at initial evaluation to determine baseline ophthalmologic status. Fundoscopic photographs can be filed for later reference. Electroretinogram and visual evoked responses and optical coherence tomography should also be performed.
- Regular ophthalmologic review under a retinal specialist or ophthalmologist is recommended.

Management

- The need for visual aids should be assessed, and they should be obtained and used as appropriate.
- Referral to support services for the blind is indicated.
- Treatment should be initiated for other factors that may impede vision, such as cataract removal and carbonic anhydrase inhibition for macular edema.
- Educational programming should be adapted for the expectation of progressive visual impairment.
- Anticipation of the needs of a blind adult should be incorporated into educational programming.
- Experimental approaches with retinal cell transplantation and direct cortical stimulation with retinal implantation are being investigated.

Musculoskeletal

The digital manifestations in BBS were reviewed by Rudling et al. (1996). Postaxial polydactyly is present in 69% of affected Europeans, and it is commonly osseous (Green et al. 1989; Beales et al. 1999). Mesoaxial or insertional polydactyly has rarely been observed. However, complex digital anomalies may be more common than isolated polydactyly, and polysyndactyly has been found in 93–98% of affected individuals (Green et al. 1989; Beales et al. 1999), with partial syndactyly most commonly involving the second and third toes (Ross and Beales 2007). Brachydactyly involving the metacarpals, metatarsals, and phalanges has been noted on radiographic evaluation in more than 90% of individuals (Green et al. 1989; Rudling et al. 1996). Other features are broad hands and feet, proximal insertion of the thumbs, fifth finger clinodactyly, a prominent sandal gap, and flat distal joint surfaces of the metacarpal and metatarsal bones on radiographs (Rudling et al. 1996; Beales et al. 1999).

Skull defects have been reported several times (Biedl et al. 1922; Lee et al. 1986; Wei et al. 1998). Kyphoscoliosis (Farag and Teebi, 1988), hip subluxation and joint laxity (Beales et al. 1999), tibia vara and tibia valga (Farag and Teebi 1988; Ross and Beales 2007), epiphyseal dysgenesis, and preaxial polydactyly are rare. Both advanced and delayed bone age have been reported (Bauman and Hogan 1973).

Evaluation

- The hands and feet should be examined carefully for extra digits or scars from digit removal.
- The configuration of the digits should be inspected for syndactyly and brachydactyly, and appropriate measurements should be recorded.
- Plain radiographs can be considered when polydactyly is present and may be useful in assessing syndactyly, brachydactyly and digit configuration prior to surgery. Podiatry assessment may be appropriate.
- Examination of the spine and limbs for kyphoscoliosis or joint subluxation should be performed at routine visits, with radiological studies and orthopedic referral as appropriate.

Management

- Additional digits should be removed by a surgeon with appropriate expertise after consideration of limb function and appearance. Removal before age 2 years is common.
- Cutaneous syndactyly is generally not treated unless it impairs function.

Genitourinary

Renal Tract Malformations Structural renal malformations have included persistent fetal lobulation, calyceal clubbing or blunting, cysts or diverticulae, cortical atrophy, cortical scarring, and renal agenesis (Harnett et al. 1988; Elbedour et al. 1994; Beales et al. 1999). Malrotation and malposition of the kidneys (Elbedour et al. 1994), renal ectopia and horseshoe kidneys are rare (Beales et al. 1999). Mesangial proliferation and sclerosis, glomerular basement membrane disease, glomerulosclerosis, and tubulo-interstitial fibrosis have been reported. Cystic tubular disease is also frequent (Khan et al. 2016). Prenatal diagnosis has been accomplished by the visualization of polysyndactyly and large, echogenic kidneys resembling infantile cystic kidney disease on ultrasound scan at 16 weeks of gestation or later (Gershoni-Baruch et al. 1992; Dar et al. 2001).

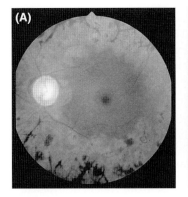
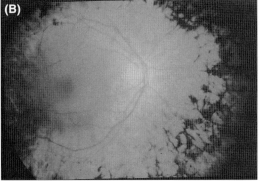

FIGURE 8.2 (A and B) Photographs of pigmentary retinopathy as seen in Bardet–Biedl syndrome, showing pallor of the optic disks, attenuation of the retinal vessels, and "bone spicule" pigmentation. (Photographs courtesy of Dr. Jacques Duncan, Department of Ophthalmology, University of California, San Francisco.)

Renal impairment may be evident as early as 2 years of age (O'Dea et al. 1996). Polydipsia and polyuria caused by reduced concentrating ability due to reduced responsiveness to vasopressin with nephrogenic diabetes may be the presenting renal abnormality (Herman and Siegel 1999). Rare features include microaneurysms and vascular occlusions, nephrolithiasis (Beales et al. 1999), vesicoureteric reflux, renal tubular acidosis (Harnett et al. 1988), and cystinuria. There may be an increased tendency to renal tract infections.

Hypertension that can be of renal origin is present in 50–66% of affected individuals and is usually manifest by the fourth decade of life (Harnett et al. 1988; O'Dea et al. 1996). Chronic renal failure is reported in 15–55% of individuals with BBS (Harnett et al. 1988; Collins et al. 1994) and has been treated by dialysis. Renal transplant has been successfully performed, in up to 10% of people with BBS in one cohort (Tobin and Beales 2007). Steroid sparing antirejection regimes are recommended (Sharifian et al. 2007). However, transplant was complicated in two cases by the development of morbid obesity attributed to steroid immunosuppressants (Collins et al. 1994) and in one person who subsequently developed a primary central nervous system lymphoma (Ersoy et al. 2005).

Hypogenitalism Hypogenitalism has been more commonly observed in males with BBS (Green et al. 1989; Katsanis et al. 2001b). Reported anomalies have included cryptorchidism, hypospadias, and a small penis (Beales et al. 1999). Pubertal development can also be delayed (Beales et al. 1999). Normal serum testosterone levels and low testosterone levels with high basal FSH levels have been described (Green et al. 1989). Testicular biopsy has shown fibrosis and degeneration of the tubules (Pagon et al. 1982).

In females, structural malformations have involved both the upper and the lower genital tract, including hydrometrocolpos and hematocolpos, vaginal and/or uterine atresia, duplex uterus, transverse vaginal septum, persistent urogenital sinus, ectopic urethra, vesicovaginal fistula or absent urethral opening, and hypoplasia of the ovaries and fallopian tubes (Stoler et al. 1995; Mehrotra et al. 1997; David et al. 1999; Slavotinek and Biesecker 2000). Irregular menstruation is frequent (Green et al. 1989). Primary ovarian failure (O'Dea et al. 1996) is rare. Hormonal investigations have shown low estrogen levels in women of reproductive age and high prolactin and luteinizing hormone levels (Green et al. 1989). However, to our knowledge there are no published data on hormone replacement in BBS. Affected females and two affected males have had children, and secondary sexual characteristics are usually normal in both sexes (Green et al. 1989; Ross and Beales 2007).

Renal Cell Carcinoma and Bardet–Biedl Syndrome Disseminated clear cell renal carcinoma was diagnosed in one 30-year-old male reported to have BBS because of retinitis pigmentosa and polydactyly (Zaldivar et al. 2008). Clear cell renal carcinoma was also diagnosed in 3 out of 180 (1.6%) parents of individuals with BBS (Beales et al. 1999, 2000). Although one of the parents had a mutation in the von Hippel–Lindau disease gene (see Chapter 62) (likely to be causative of the renal carcinoma, and thus carrier status for BBS was probably incidental), the finding of renal cancer in two relatives indicated an 11-fold increased risk over the population incidence for adults less than 55 years of age (Beales et al. 2000). However, the apparently increased risk of renal cancer was not verified in a Danish study examining the incidence of cancer. No renal cancers were observed in 116 people with BBS or their 428 blood relatives (Hjortshøj et al. 2007). The authors theorized that the prior reported increased risk for renal cell cancers was related to specific BBS genotypes or mutations or the occurrence of end-stage renal cystic disease (Hjortshøj et al. 2007). Screening guidelines for renal carcinoma in affected individuals and relatives have not been established.

Evaluation

- Measurements of blood pressure, serum electrolytes, renal function with urea and creatinine, urinalysis for glucose, protein and osmolarity, and renal ultrasonography should be obtained at the time of evaluation or diagnosis.
- Timely investigation of symptoms that may indicate renal disease, such as hematuria or dysuria, is important. An abdominal magnetic resonance imaging scan may be required to detect calyceal abnormalities. Nephrogenic diabetes insipidus may be overlooked and information concerning fluid intake and output should be sought together with tests of renal concentrating activity, if appropriate.
- Routine referral to a nephrologist is recommended.
- If investigations do not indicate renal disease at initial evaluation, individuals should be monitored with twice yearly urinalysis and annual measurement of blood pressure and renal function (Beales et al. 1999).
- The external and internal genitalia in females and the external genitalia in males should be carefully examined.
- Pelvic ultrasound is indicated to image the internal female genitalia and for cryptorchidism in males.
- Referral to an endocrinologist may be appropriate. Sex hormone levels are not routinely measured.

Management

- Renal toxins should be avoided.
- Antibiotic therapy for urinary tract infections and prophylactic antibiotic therapy for vesicoureteric reflux is indicated.
- Therapy for hypertension should be initiated as in the general population.

- A nephrologist should assess and manage chronic renal failure. This should include dialysis treatment and assessment for renal transplantation. Management of renal failure does not differ from that in the general population.
- Surgical intervention may be required for structural renal malformations.
- Surgical repair of structural abnormalities such as hydrometrocolpos is indicated, using the same techniques as in the general population.
- Contraceptive advice should be offered to all females with BBS, and sex education is appropriate for both sexes.

Cardiovascular

The cardiac abnormalities in BBS were specifically addressed in one study that evaluated 22 Bedouin individuals with echocardiography (Elbedour et al. 1994). Structural malformations were present in seven (32%) and comprised a bicuspid aortic valve, atrial septal defect, pulmonic valve stenosis, and tricuspid incompetence (Elbedour et al. 1994). A wide variety of other structural abnormalities have been found, including tetralogy of Fallot, ventricular septal defect, single ventricle with transposition of the great vessels, and hypoplasia of the aorta (Bauman and Hogan, 1973). Cardiomyopathy and hypertrophy of the interventricular septum may be relatively frequent (Elbedour et al. 1994; Beales et al. 1999). Dextrocardia and situs inversus (either complete or involving the abdominal viscera) have been reported but are relatively uncommon (Lorda-Sanchez et al. 2000; Deffert et al. 2007).

Evaluation

- Cardiac examination should be performed at initial evaluation or diagnosis, optimally including referral to a cardiologist for auscultation, electrocardiogram, and echocardiogram.
- Screening guidelines regarding the later development of cardiomyopathy have not been established.

Management

- Appropriate and standard supportive management and surgical intervention for structural heart disease and cardiomyopathy should take place.
- Prophylactic antibiotics may be indicated for surgical and dental procedures in individuals with structural cardiac malformations.

Endocrine

The incidence of type II diabetes mellitus in individuals with BBS has been reported to be as high as 45% (Green et al. 1989; O'Dea et al. 1996), and a relationship between the development of insulin resistance and obesity in BBS has been postulated (Escallon et al. 1989). Diabetes developed by 50 years of age in half of the individuals in one study (O'Dea et al. 1996). Insulin was required for treatment in some, although insulin dependence in BBS is rare (Escallon et al. 1989; Iannello et al. 2002).

Please see the Genitourinary section for a discussion of hypogonadism.

Other endocrine abnormalities are rare, although hypothyroidism and an elevated thyroid-stimulating hormone (TSH) have been recorded (Özer et al. 1995; Beales et al. 1999; Slavotinek et al. 2002).

Evaluation

- Fasting blood glucose, hemoglobinA1C, and lipid levels should be measured at initial evaluation or diagnosis, starting with the development of obesity in early childhood, and checked annually.
- A glucose tolerance test should be considered if blood sugars are elevated.
- Referral to an endocrinologist should be considered for management of insulin resistance or diabetes, if present.
- The complications of diabetic disease should be evaluated, as in the general population.
- Measurement of thyroid function is indicated when symptoms and signs are suggestive of thyroid disease.
- Testing of pituitary gland function may be warranted in the presence of peripheral hormone abnormalities.

Management

- Diabetes mellitus and hyperlipidemia should be treated as in the general population.
- Thyroid hormone abnormalities should be treated in a standard manner.

Ears and Hearing

An increased incidence of deafness has been noted in individuals with BBS and their relatives. Hearing loss was reported in 24% of persons with BBS and was most commonly conductive (Beales et al. 1999). Otitis media causing conductive deafness is common in affected children (Ross and Beales 2007). Bifid epiglottis and choanal stenosis are rare.

Evaluation

- Audiologic testing and an otolaryngology assessment should be performed at diagnosis and can be repeated at the discretion of the audiologist, depending on age and family history.

Management

- Prompt treatment for ear infections is required.
- Consultation with an otolaryngologist for surgical intervention may be required for recurrent otitis media or anomalies such as choanal stenosis.

Dental

Hypodontia, small teeth, and short roots of the permanent teeth are significantly more common in individuals with BBS than controls. The second premolars in the mandible and the maxilla are the most frequently affected teeth. Malocclusion, dental crowding, enamel hypodontia, oligodontia, and gingival overgrowth have been reported (Beales et al. 1999).

Evaluation

- A dental assessment for hygiene, dental crowding, and hypodontia should be performed at the first evaluation or at diagnosis.
- Routine dental follow-up should be stressed.

Management

- Attention to dental hygiene is important to keep dentition as healthy as possible.
- Appropriate and standard intervention is indicated for malocclusion, extractions or dental crowding.

Gastrointestinal

Ductal plate malformations with cystic dilatation of intra- and extra-hepatic bile ducts (Tsuchiya et al. 1977) and hepatic fibrosis (Pagon et al. 1982; Nakamura et al. 1990) have been observed in BBS. Other liver findings include perilobular and periportal fibrosis with shortened bile ducts, increased length of the bile ducts and biliary cirrhosis (Khan et al. 2016). Hepatic disease may be congenital (Tsuchiya et al. 1977) but is usually diagnosed in childhood or adolescence. Cholelithiasis and cholecystitis are rare (Beales et al. 1999). Intestinal malformations are uncommon, although Hirschsprung disease (Islek et al. 1996; Lorda-Sanchez et al. 2000) and anal stenosis (Beales et al. 1999) have been reported. There is one reported case with splenic lobulations consistent with a situs defect.

Evaluation

- Symptoms related to the gastrointestinal system should trigger prompt evaluation for structural and/or functional abnormalities.

Management

- Abnormalities should be addressed as in the general population.

Neurologic

Ataxia and impaired coordination have been noted in up to 86% of individuals with BBS (Ross and Beales 2007). Cerebral imaging has been performed infrequently in BBS. Hypoplasia of the cerebellar vermis has been described in affected individuals (Baskin et al. 2002). There have been reports of a mega cisterna magna (Baskin et al. 2002), craniopharyngioma, and pituitary adenomas (Riise et al. 2002). Enlargement or atrophy of the gyri and agenesis or hypoplasia of the corpus callosum have been noted (Bauman and Hogan 1973). Structural brain malformations should thus not preclude the diagnosis of BBS. Seizures are uncommon (Bauman and Hogan 1973).

Anosmia has been reported in BBS (Kulaga et al. 2004) and reduced olfaction was noted in two members of a five-generation family who were homozygous for a deletion in the *BBS4* gene (Iannoccone et al. 2005). The olfactory receptor neuron is ciliated and emanates immotile cilia, consistent with the demonstration of this phenotypic feature in people with a ciliopathy (Kulaga et al. 2004). Reduced olfaction has variable expressivity and delayed onset, and the influence of chronic sinusitis and recurrent otitis media on the ability to smell is unknown (Kulaga et al. 2004).

Evaluation

- If seizures are suspected, electroencephalogram should be performed.
- A thorough neurologic examination should be performed with attention to gait and coordination at the time of initial evaluation or diagnosis.
- Testing for reduced olfaction can be done if lack of olfaction is reported.

Management

- Treatment of seizures is standard.
- Physical and occupational therapies can be offered for ataxia and coordination defects.

RESOURCES

Internet

OMIM database (Online Mendelian Inheritance in Man)
National Center for Biotechnology Information and Johns Hopkins University
Website: *https://www.ncbi.nlm.nih.gov/omim/*

BARDET–BIEDL SYNDROME
OMIM entry 209900
Website: *http://omim.org/entry/209900*

GeneReview Entry
Website: *https://www.ncbi.nlm.nih.gov/books/NBK1363/*

PubMed

Website: *https://www.ncbi.nlm.nih.gov/pubmed/*

Laboratories involved in Bardet–Biedl syndrome testing and research

GTR: Genetic Testing Registry
Website: *https://www.ncbi.nlm.nih.gov/gtr/*

Support Groups

Bardet–Biedl Family Association

People served: Families with Bardet–Biedl syndrome
Services provided: Support and information
Website: https://www.bardetbiedl.org/

Foundation Fighting Blindness (United States)

People served: Families dealing with visual impairment
Services provided: Support and information
11435 Cronhill Drive. Owings Mill, MD 21117-2220, USA
Website: *www.blindness.org*

REFERENCES

Aldahmesh MA, Li Y, Alhashem A, Anazi S, Alkuraya H (2014) IFT27, encoding a small GTPase component of IFT particles, is mutated in a consanguineous family with Bardet-Biedl syndrome. *Hum Mol Genet* 23:3307–3315.

Ansley SJ, Badano JL, Blacque OE, Hill J, Hoskins BE, Leitch CC, Kim JC, Ross AJ, Eichers ER, Teslovich TM, Mah AK, Johnsen RC, Cavender JC, Lewis RA, Leroux MR, Beales PL, Katsanis N (2003) Basal body dysfunction is a likely cause of pleiotropic Bardet-Biedl syndrome. *Nature* 425: 628–633.

Azari AA, Aleman TS, Cideciyan AV, Schwartz SB, Windsor EA, Sumaroka A, Cheung AY, Steinberg JD, Roman AJ, Stone EM, Sheffield VC, Jacobson SG (2006) Retinal disease expression in Bardet-Biedl syndrome-1 (BBS1) is a spectrum from maculopathy to retina-wide degeneration. *Invest Ophthalmol Vis Sci* 47: 5004–5010.

Badano JL, Ansley SJ, Leitch CC, Lewis RA, Lupski JR, Katsanis N (2003) Identification of a novel Bardet-Biedl syndrome protein, BBS7, that shares structural features with BBS1 and BBS2. *Am J Hum Genet* 72:650–658.

Barbelanne M, Hossain D, Chan DP, Peränen J, Tsang WY (2015) Nephrocystin proteins NPHP5 and Cep290 regulate BBSome integrity, ciliary trafficking and cargo delivery. *Hum Mol Genet* 24: 2185–2200.

Bardet G (1920) Sur un syndrome d'obesite infantile avec polydactylie et retinite pigmentaire (contribution a l'etude des formes cliniques de l'obesite hypophysaire). PhD thesis, Paris.

Baskin E, Kayiran SM, Oto S, Alehan F, Agildere AM, Saatci U (2002) Cerebellar vermis hypoplasia in a patient with Bardet-Biedl syndrome. *J Child Neurol* 17:385–387.

Bauman ML, Hogan GR (1973) Laurence-Moon-Biedl syndrome. Report of two unrelated children less than 3 years of age. *Am J Dis Child* 126:119–126.

Beales PL, Warner AM, Hitman GA, Thakker R, Flinter FA (1997) Bardet-Biedl syndrome: A molecular and phenotypic study of 18 families. *J Med Genet* 34:92–98.

Beales PL, Elcioglu N, Woolf AS, Parker D, Flinter FA (1999) New criteria for improved diagnosis of Bardet-Biedl syndrome: Results of a population survey. *J Med Genet* 36:437–446.

Beales PL, Reid HA, Griffiths MH, Maher ER, Flinter FA, Woolf AS (2000) Renal cancer and malformations in relatives of patients with Bardet-Biedl syndrome. *Nephrol Dial Transplant* 15:1977–1985.

Beales PL, Katsanis N, Lewis RA, Ansley SJ, Elcioglu N, Raza J, Woods MO, Green JS, Parfrey PS, Davidson WS, Lupski JR (2001) Genetic and mutational analyses of a large multiethnic Bardet-Biedl cohort reveal a minor involvement of BBS6 and delineate the critical intervals of other loci. *Am J Hum Genet* 68:606–616.

Beales PL, Badano JL, Ross AJ, Ansley SJ, Hoskins BE, Kirsten B, Mein CA, Froguel P, Scambler PJ, Lewis RA, Lupski JR, Katsanis N (2003) Genetic interaction of BBS1 mutations with alleles at other BBS loci can result in non-Mendelian Bardet-Biedl syndrome. *Am J Hum Genet* 72:1187–1199.

Biedl A (1922) Ein geshwisterpaar mit adiposo-genitaler dystrophie. *Deutsch Med Wochenschr* 48:1630.

Brinckman DD, Keppler-Noreuil KM, Blumhorst C, Biesecker LG, Sapp JC, Johnston JJ, Wiggs EA (2013) Cognitive, sensory, and psychosocial characteristics in patients with Bardet-Biedl syndrome. *Am J Med Genet A* 161A:2964–2971.

Bujakowska KM, Zhang Q, Siemiatkowska AM, Liu Q, Place E, Falk MJ, Consugar M, Lancelot ME, Antonio A, Lonjou C, Carpentier W, Mohand-Saïd S, den Hollander AI, Cremers FP, Leroy BP, Gai X, Sahel JA, van den Born LI, Collin RW, Zeitz C, Audo I, Pierce EA (2015) Mutations in IFT172 cause isolated retinal degeneration and Bardet-Biedl syndrome. *Hum Mol Genet* 24:230–242.

Calvert PD, Krasnoperova NV, Lyubarsky AL, Isayama T, Nicolo M, Kosaras B, Wong G, Gannon KS, Margolskee RF, Sidman RL, Pugh EN Jr, Makino CL, Lem J (2000) Photo-transduction in transgenic mice after targeted deletion of the rod transducin alpha-subunit. *Proc Natl Acad Sci USA* 97:13913–13918.

Castro-Sánchez S, Álvarez-Satta M, Cortón M, Guillén E, Ayuso C, Valverde D (2015) Exploring genotype-phenotype relationships in Bardet-Biedl syndrome families. *J Med Genet* 52:503–513.

Chiang AP, Nishimura D, Searby C, Elbedour K, Carmi R, Ferguson AL, Secrist J, Braun T, Casavant T, Stone EM, Sheffield VC (2004) Comparative genomic analysis identifies an ADP-ribosylation factor-like gene as the cause of Bardet-Biedl syndrome (BBS3). *Am J Hum Genet* 75:475–484.

Chiang AP, Beck JS, Yen H-J, Tayeh MK, Scheetz TE, Swiderski RE, Nishimura DY, Braun TA, Kim K-YA, Huang J, Elbedour K, Carmi R, Slusarski DC, Casavant TL, Stone EM, Sheffield VC (2006) Homozygosity mapping with SNP arrays identifies TRIM32, an E3 ubiquitin ligase, as a Bardet-Biedl syndrome gene (BBS11). *Proc Natl Acad Sci USA* 103:6287–6292.

Collin GB, Marshall JD, Ikeda A, So WV, Russell-Eggitt I, Maffei P, Beck S, Boerkoel CF, Sicolo N, Martin M, Nishina PM, Naggert JK (2002) Mutations in ALMS1 cause obesity, type 2 diabetes and neurosensory degeneration in Alstrom syndrome. *Nat Genet* 31:74–78.

Collins CM, Mendoza SA, Griswold WR, Tanney D, Lieberman E, Reznik VM (1994) Pediatric renal transplantation in Laurence-Moon-Biedl syndrome. *Pediatr Nephrol* 8:221–222.

Croft JB, Swift M (1990) Obesity, hypertension, and renal disease in relatives of Bardet-Biedl syndrome sibs. *Am J Med Genet* 36:37–42.

Croft JB, Morrell D, Chase CL, Swift M (1995) Obesity in heterozygous carriers of the gene for the Bardet-Biedl syndrome. *Am J Med Genet* 55:12–15.

Daniels AB, Sandberg MA, Chen J, Weigel-DiFranco C, Fielding Hejtmancic J, Berson EL (2012) Genotype phenotype correlations in Bardet-Biedl syndrome. *Arch Ophthalmol* 130:901–907.

Dar P, Sachs GS, Carter SM, Ferreira JC, Nitowsky HM, Gross SJ (2001) Prenatal diagnosis of Bardet-Biedl syndrome by targeted second-trimester sonography. *Ultrasound Obstet Gynecol* 17:354–356.

David A, Bitoun P, Lacombe D, Lambert JC, Nivelon A, Vigneron J, Verloes A (1999) Hydrometrocolpos and polydactyly: A common neonatal presentation of Bardet-Biedl and McKusick-Kaufman syndromes. *J Med Genet* 36:599–603.

Deffert C, Niel F, Mochel F, Barrey C, Romana C, Souied E, Stoetzel C, Goossens M, Dollfus H, Verloes A, Girodon E, Gerard-Blanluet M (2007) Recurrent insertional polydactyly and situs inversus in a Bardet-Biedl syndrome family. *Am J Med Genet A* 143:208–213.

Dippell J, Varlam DE (1998) Early sonographic aspects of kidney morphology in Bardet-Biedl syndrome. *Pediatr Nephrol* 12:559–563.

Elbedour K, Zucker N, Zalzstein E, Barki Y, Carmi R (1994) Cardiac abnormalities in the Bardet-Biedl syndrome: Echocardiographic studies of 22 patients. *Am J Med Genet* 52:164–169.

Ersoy A, Kahvecioglu S, Bekar A, Aker S, Akdag I, Dilek K (2005) Primary central nervous system lymphoma in a renal transplant recipient with Bardet-Biedl syndrome. *Transplant Proc* 37:4323–4325.

Escallon F, Traboulsi EI, Infante R (1989) A family with the Bardet-Biedl syndrome and diabetes mellitus. *Arch Ophthalmol* 107:855–857.

Farag TI, Teebi AS (1988) Bardet-Biedl and Laurence-Moon syndromes in a mixed Arab population. *Clin Genet* 33:78–82.

Farag TI, Teebi AS (1989) High incidence of Bardet-Biedl syndrome among the Bedouin. *Clin Genet* 36:463–464.

Feuillan PP, Ng D, Han JC, Sapp JC, Wetsch K, Spaulding E, Zheng YC, Caruso RC, Brooks BP, Johnston JJ, Yanovski JA, Biesecker LG (2011) Patients with Bardet-Biedl syndrome have hyperleptinemia suggestive of leptin resistance. *J Clin Endocrinol Metab* 96:E528–E535.

Forsythe E, Beales PL. *Bardet-Biedl syndrome*. In: Adam MP, Ardinger HH, Pagdon RA, Wallace SE, Bean LJH, Mefford HC, Stephens K, Amemiya A, Ledbetter N, editors. GeneReviews® [online]. Seattle (WA): University of Washington, Seattle; 1993-2017. 2003 Jul 14 [updated 2015 Apr 23].

Fulton AB, Hansen RM, Glynn RJ (1993) Natural course of visual functions in the Bardet-Biedl syndrome. *Arch Ophthalmol* 111:1500–1506.

Gershoni-Baruch R, Nachlieli T, Leibo R, Degani S, Weissman I (1992) Cystic kidney dysplasia and polydactyly in 3 sibs with Bardet-Biedl syndrome. *Am J Med Genet* 44:269–273.

Grace C, Beales P, Summerbell C, Jebb SA, Wright A, Parker D, Kopelman P (2003) Energy metabolism in Bardet-Biedl syndrome. *Int J Obes Relat Metab Disord* 27:1319–1324.

Green JS, Parfrey PS, Harnett JD, Farid NR, Cramer BC, Johnson G, Heath O, McManamon PJ, O'Leary E, Pryse-Phillips W (1989) The cardinal manifestations of Bardet-Biedl syndrome, a form of Laurence-Moon-Biedl syndrome. *N Engl J Med* 321:1002–1009.

Harnett JD, Green JS, Cramer BC, Johnson G, Chafe L, McManamon P, Farid NR, Pryse-Phillips W, Parfrey PS (1988) The spectrum of renal disease in Laurence-Moon-Biedl syndrome. *N Engl J Med* 319:615–618.

Hearn T, Renforth GL, Spalluto C, Hanley NA, Piper K, Brickwood S, White C, Connolly V, Taylor JF, Russell-Eggitt I, Bonneau D, Walker M, Wilson DI (2002) Mutation of ALMS1, a large gene with a tandem repeat encoding 47 amino acids, causes Alstrom syndrome. *Nat Genet* 31:79–83.

Héon E, Westall C, Carmi R, Elbedour K, Panton C, Mackeen L, Stone EM, Sheffield VC (2005) Ocular phenotypes of three genetic variants of Bardet-Biedl syndrome. *Am J Med Genet* 132A:283–287.

Herman TE, Siegel MJ (1999) Special imaging casebook. *Neonatal Bardet-Biedl syndrome with renal anomalies and hydrocolpos with vesicovaginal fistula. J Perinatol* 19:74–76.

Hjortshøj TD, Grønskov K, Rosenberg T, Brøndum-Nielsen K, Olsen JH (2007) Risk for cancer in patients with Bardet-Biedl syndrome and their relatives. *Am J Med Genet* 143A:1699–1702.

Hjortshøj TD, Grønskov K, Brøndum-Nielsen K, Rosenberg T (2009) A novel founder BBS1 mutation explains a unique high prevalence of BardetBiedl syndrome in the Faroe Islands. *Br J Ophthalmol* 93:409–413.

Hjortshøj TD, Grønskov K, Philp AR, Nishimura DY, Sheffield VC, Rosenberg T, Brøndum-Nielsen K (2010) Bardet-Biedl syndrome in Denmark – report of 13 novel sequence variations in six genes. *Hum Mutat* 31:429–436.

Hufnagel RB, Arno G, Hein ND, Hersheson J, Prasad M, Anderson Y, Krueger LA, Gregory LC, Stoetzel C, Jaworek TJ, Hull S, Li A, Plagnol V, Willen CM, Morgan TM, Prows CA, Hegde RS, Riazuddin S, Grabowski GA, Richardson RJ, Dieterich K, Huang T, Revesz T, Martinez-Barbera JP, Sisk RA, Jefferies C, Houlden H, Dattani MT, Fink JK, Dollfus H, Moore AT, Ahmed ZM (2015) Neuropathy target esterase impairments cause Oliver-McFarlabe and Laurence-Moon syndromes. *J Med Genet* 52:85–94.

Iannaccone A, Mykytyn K, Persico AM, Seaerby CC, Baldi A, Jablonski MM, Sheffield VC (2005) Clinical evidence of decreased olfaction in Bardet-Biedl syndrome caused by a deletion in the BBS4 gene. *Am J Med Genet* 132A:343–346.

Iannello S, Bosco P, Cavaleri A, Camuto M, Milazzo P, Belfiore F (2002) A review of the literature of Bardet-Biedl disease and report of three cases associated with metabolic syndrome and diagnosed after the age of fifty. *Obes Rev* 3:123–135.

Imhoff O, Marion V, Stoetzel C, Durand M, Holder M, Sigaudy S, Sarda P, Hamel CP, Brandt C, Dollfus H, Moulin B (2011) Bardet-Biedl syndrome: a study of the renal and cardiovascular phenotypes in a French cohort. *Clin J Am Soc Nephrol* 6:22–29.

Islek I, Kucukoduk S, Erkan D, Bernay F, Kalayci AG, Gork S, Kandemir B, Gurses N (1996) Bardet-Biedl syndrome: Delayed diagnosis in a child with Hirschsprung disease. *Clin Dysmorphol* 5:271–273.

Karmous-Benailly H, Martinovic J, Gubler MC, Sirot Y, Clech L, Ozilou C, Auge J, Brahimi N, Etchevers H, Detrait E, Esculpavit C, Audollent S, Goudefroye G, Gonzales M, Tantau J, Loget P, Joubert M, Gaillard D, Jeanne-Pasquier C, Delezoide AL, Peter MO, Plessis G, Simon-Bouy B, Dollfus H, Le Merrer M, Munnich A, Encha-Razavi F, Vekemans M, Attié-Bitach T (2005) Antenatal presentation of Bardet-Biedl syndrome may mimic Meckel syndrome. *Am J Hum Genet* 76:493–504.

Katsanis N (2004) The oligogenic properties of Bardet-Biedl syndrome. *Hum Mol Genet* 13(Spec No 1):R65–R67.

Katsanis N, Beales PL, Woods MO, Lewis RA, Green JS, Parfrey PS, Ansley SJ, Davidson WS, Lupski JR (2000) Mutations in MKKS cause obesity, retinal dystrophy and renal malformations associated with Bardet-Biedl syndrome. *Nat Genet* 26:67–70.

Katsanis N, Ansley SJ, Badano JL, Eichers ER, Lewis RA, Hoskins BE, Scambler PJ, Davidson WS, Beales PL, Lupski JR (2001a) Triallelic inheritance in Bardet-Biedl syndrome, a Mendelian recessive disorder. *Science* 293:2256–2259.

Katsanis N, Lupski JR, Beales PL (2001b) Exploring the molecular basis of Bardet-Biedl syndrome. *Hum Mol Genet* 10:2293–2299.

Katsanis N, Eichers ER, Ansley SJ, Lewis RA, Kayserili H, Hoskins BE, Scambler PJ, Beales PL, Lupski JR (2002) BBS4 is a minor contributor to Bardet-Biedl syndrome and may also participate in triallelic inheritance. *Am J Hum Genet* 71:22–29.

Khan SA, Muhammad N, Khan MA, Kamal A, Rehman ZU, Khan S (2016) Genetics of human Bardet-Biedl syndrome, an update. *Clin Genet* 90:3–15

Kim LS, Fishman GA, Seiple WH, Szlyk JP, Stone EM (2007) Retinal dysfunction in carriers of Bardet-Biedl syndrome. *Ophthalmic Genet* 28:163–168.

Kim SK, Shindo A, Park TJ, Oh EC, Ghosh S, Gray RS, Lewis RA, Johnson CA, Attie-Bittach T, Katsanis N, Wallingford JB (2010) Planar cell polarity acts through septins to control collective cell movement and ciliogenesis. *Science* 329:1337–1340.

Kulaga HM, Leitch CC, Eichers ER, Badano JL, Lesemann A, Hoskins BE, Lupski JR, Beales PL, Reed RR, Katsanis N (2004) Loss of BBS proteins causes anosmia in humans and defects in olfactory cilia structure and function in the mouse. *Nat Genet* 36:994–998.

Laurence J, Moon R (1866) Four cases of retinitis pigmentosa, occurring in the same family, and accompanied by general imperfections of development. *Ophthalmol Rev* 2:32–41.

Lee CS, Galle PC, McDonough PG (1986) The Laurence-Moon-Bardet-Biedl syndrome. Case report and endocrinologic evaluation. *J Reprod Med* 31:353–356.

Li JB, Gerdes JM, Haycraft CJ, Fan Y, Teslovich TM, May-Simera H, Li H, Blacque OE, Li L, Leitch CC, Lewis RA, Green JS, Parfrey PS, Leroux MR, Davidson WS, Beales PL, Guay-Woodford LM, Yoder BK, Stormo GD, Katsanis N, Dutcher SK (2004) Comparative genomics identifies a flagellar and basal body proteome that includes the BBS5 human disease gene. *Cell* 117:541–552.

Lindstrand A, Davis EE, Carvalho CM, Pehlivan D, Willer JR, Tsai IC, Ramanathan S, Zuppan C, Sabo A, Muzny D, Gibbs R, Liu P, Lewis RA, Banin E, Lupski JR, Clark R, Katsanis N (2014). Recurrent CNVs and SNVs at the NPHP1 locus contribute pathogenic alleles to Bardet-Biedl syndrome. *Am J Hum Genet* 94:745-754.

Lorda-Sanchez I, Ayuso C, Ibanez A (2000) Situs inversus and Hirschsprung disease: Two uncommon manifestations in Bardet-Biedl Syndrome. *Am J Med Genet* 90:80–81.

Marion V, Stutzmann F, Gérard M, De Melo C, Schaefer E, Claussmann A, Hellé S, Delague V, Souied E, Barrey C, Verloes A, Stoetzel C, Dollfus H (2012) Exome sequencing identifies mutations in LZTFL1, a BBSome and smoothened trafficking regulator, in a family with Bardet-Biedl syndrome with situs inversus and insertional polydactyly. *J Med Genet* 49:317–321.

Mehrotra N, Taub S, Covert RF (1997) Hydrometrocolpos as a neonatal manifestation of the Bardet-Biedl syndrome. *Am J Med Genet* 69:220.

Moore SJ, Green JS, Fan Y, Bhogal AK, Dicks E, Fernandez BA, Stefanelli M, Murphy C, Cramer BC, Dean JC, Beales PL, Katsanis N, Bassett AS, Davidson WS, Parfrey PS (2005) Clinical and genetic epidemiology of Bardet-Biedl syndrome in Newfoundland: A 22-year prospective, population-based, cohort study. *Am J Med Genet A* 132:352–360.

Mykytyn K, Braun T, Carmi R, Haider NB, Searby CC, Shastri M, Beck G, Wright AF, Iannaccone A, Elbedour K, Riise R, Baldi A, Raas-Rothschild A, Gorman SW, Duhl DM, Jacobson SG, Casavant T, Stone EM, Sheffield VC (2001) Identification of the gene that, when mutated, causes the human obesity syndrome BBS4. *Nat Genet* 28:188–191.

Mykytyn K, Nishimura DY, Searby CC, Shastri M, Yen HJ, Beck JS, Braun T, Streb LM, Cornier AS, Cox GF, Fulton AB, Carmi R, Luleci G, Chandrasekharappa SC, Collins FS, Jacobson SG, Heckenlively JR, Weleber RG, Stone EM, Sheffield VC (2002) Identification of the gene (BBS1) most commonly involved in Bardet-Biedl syndrome, a complex human obesity syndrome. *Nat Genet* 3:435–438.

Nachury MV, Loktev AV, Zhang Q, Westlake CJ, Peränen J, Merdes A, Slusarski DC, Scheller RH, Bazan JF, Sheffield VC, Jackson PK (2007) A core complex of BBS proteins cooperates with the GTPase Rab8 to promote ciliary membrane biogenesis. *Cell* 129:1201–1213.

Nakamura F, Sasaki H, Kajihara H, Yamanoue M (1990) Laurence-Moon-Biedl syndrome accompanied by congenital hepatic fibrosis. *J Gastroenterol Hepatol* 5:206–210.

Nishimura DY, Searby CC, Carmi R, Elbedour K, Van Maldergem L, Fulton AB, Lam BL, Powell BR, Swiderski RE, Bugge KE, Haider NB, Kwitek-Black AE, Ying L, Duhl DM, Gorman SW, Héon E, Iannaccone A, Bonneau D, Biesecker LG, Jacobson SG, Stone EM, Sheffield VC (2001) Positional cloning of a novel gene on chromosome 16q causing Bardet-Biedl syndrome (BBS2). *Hum Mol Genet* 10:865–874.

Nishimura DY, Swiderski RE, Searby CC, Berg EM, Ferguson AL, Hennekam R, Merin S, Weleber RG, Biesecker LG, Stone EM,

Sheffield VC (2005) Comparative genomics and gene expression analysis identifies BBS9, a new Bardet-Biedl syndrome gene. *Am J Hum Genet* 77:1021–1033.

O'Dea D, Parfrey PS, Harnett JD, Hefferton D, Cramer BC, Green J (1996) The importance of renal impairment in the natural history of Bardet-Biedl syndrome. *Am J Kidney Dis* 27:776–783.

Osusky R, Alsaadi AH, Farpour H (1991) [Case report of Laurence-Moon-Bardet-Biedl syndrome]. *Klin Monatsbl Augenheilkd* 198:445–446.

Otto EA, Hurd TW, Airik R, Chaki M, Zhou W, Stoetzel C, Patil SB, Levy S, Ghosh AK, Murga-Zamalloa CA, van Reeuwijk J, Letteboer SJ, Sang L, Giles RH, Liu Q, Coene KL, Estrada-Cuzcano A, Collin RW, McLaughlin HM, Held S, Kasanuki JM, Ramaswami G, Conte J, Lopez I, Washburn J, Macdonald J, Hu J, Yamashita Y, Maher ER, Guay-Woodford LM, Neumann HP, Obermüller N, Koenekoop RK, Bergmann C, Bei X, Lewis RA, Katsanis N, Lopes V, Williams DS, Lyons RH, Dang CV, Brito DA, Dias MB, Zhang X, Cavalcoli JD, Nürnberg G, Nürnberg P, Pierce EA, Jackson PK, Antignac C, Saunier S, Roepman R, Dollfus H, Khanna H, Hildebrandt F (2010) Candidate exome capture identifies mutation of SDCCAG8 as the cause of a retinal-renal ciliopathy. *Nat Genet* 42:840–850.

Özer G, Yuksel B, Suleymanova D, Alhan E, Demircan N, Onenli N (1995) Clinical features of Bardet-Biedl syndrome. *Acta Paediatr Jpn* 37:233–236.

Pagon RA, Haas JE, Bunt AH, Rodaway KA (1982) Hepatic involvement in the Bardet-Biedl syndrome. *Am J Med Genet* 13:373–381.

Rahmouni K, Fath MA, Seo S, Thedens DR, Berry CJ, Weiss R, Nishimura DY, Sheffield VC (2008) Leptin resistance contributes to obesity and hypertension in mouse models of Bardet-Biedl syndrome. *J Clin Invest* 18:1458–1467.

Ross AJ., Beales PL. (2007) Bardet-Biedl syndrome. Gene Reviews. Clinics Website: *http://www.geneclinics.org*; updated 2007.

Riise R (1996) The cause of death in Laurence-Moon-Bardet-Biedl syndrome. *Acta Ophthalmol Scand Suppl* 219:45–47.

Riise R, Andreasson S, Wright AF, Tornqvist K (1996a) Ocular findings in the Laurence-Moon-Bardet-Biedl syndrome. *Acta Ophthalmol Scand* 74:612–617.

Riise R, Andreasson S, Tornqvist K (1996b) Full-field electroretinograms in individuals with the Laurence-Moon-Bardet-Biedl syndrome. *Acta Ophthalmol Scand* 74:618–620.

Riise R, Andreasson S, Borgastrom MK, Wright AF, Tommerup N, Rosenberg T, Tornqvist K (1997) Intrafamilial variation of the phenotype in Bardet-Biedl syndrome. *Br J Ophthalmol* 81:378–385.

Riise R, Tornqvist K, Wright AF, Mykytyn K, Sheffield VC (2002) The phenotype in Norwegian patients with Bardet-Biedl syndrome with mutations in the BBS4 Gene. *Arch Ophthalmol* 120:1364–1367.

Rudling O, Riise R, Tornqvist K, Jonsson K (1996) Skeletal abnormalities of hands and feet in Laurence-Moon-Bardet-Biedl (LMBB) syndrome: A radiographic study. *Skeletal Radiol* 25:655–660.

Russell-Eggitt IM, Clayton PT, Coffey R, Kriss A, Taylor DS, Taylor JF (1998) Alstrom syndrome. Report of 22 cases and literature review. *Ophthalmology* 105:1274–1280.

Schaap C, ten Tusscher MP, Schrander JJ, Kuijten RH, Schrander-Stumpel CT (1998) Phenotypic overlap between McKusick-Kaufman and Bardet-Biedl syndromes: Are they related? *Eur J Pediatr* 157:170–171.

Schachat AP, Maumenee IH (1982) Bardet-Biedl syndrome and related disorders. *Arch Ophthalmol* 100:285–288.

Scheidecker S, Etard C, Pierce NW, Geoffroy V, Schaefer E, Muller J, Chennen K, Flori E, Pelletier V, Poch O, Marion V, Stoetzel C, Strähle U, Nachury MV, Dollfus H (2014) Exome sequencing of BardetBiedl syndrome patient identifies a null mutation in the BBSome subunit BBIP1 (BBS18). *J Med Genet* 51:132–136.

Sharifian M, Dadkhah-Chimeh M, Einollahi B, Nafar M, Simforoush N, Basiri A, Otukesh H (2007) Renal transplantation in patients with Bardet-Biedl syndrome. *Arch Iran Med* 10:339–342.

Sheffield VC, Zhang Q, Heon E, Stoen EM, Carma R (2008) The Bardet-Biedl syndromes. In: *Inborn Errors of Development*, 2nd ed. Epstein CJ, Erickson RP, Wynshaw-Boris A. eds, Oxford University Press, pp. 1371–1378.

Slavotinek AM, Biesecker LG (2000) Phenotypic overlap of McKusick-Kaufman syndrome with Bardet-Biedl syndrome: A literature review. *Am J Med Genet* 95:208–215.

Slavotinek AM, Stone EM, Mykytyn K, Heckenlively JR, Green JS, Héon E, Musarella MA, Parfrey PS, Sheffield VC, Biesecker LG (2000) Mutations in MKKS cause Bardet-Biedl syndrome. *Nat Genet* 26:15–16.

Slavotinek AM, Searby C, Al-Gazali L, Hennekam RC, Schrander-Stumpel C, Orcana-Losa M, Pardo-Reoyo S, Cantani A, Kumar D, Capellini Q, Neri G, Zackai E, Biesecker LG (2002) Mutation analysis of the MKKS gene in McKusick-Kaufman syndrome and selected Bardet-Biedl syndrome patients. *Hum Genet* 110:561–567.

Solis-Cohen S, Weiss E (1924) Dystrophia adiposagenitalis, with atypical retinitis pigmentosa and mental deficiency, possible of cerebral origin: A report of four cases in one family. *Trans Assoc Am Phys* 39:356–358.

Stoetzel C, Laurier V, Davis EE, Muller J, Rix S, Badano JL, Leitch CC, Salem N, Chouery E, Corbani S, Jalk N, Vicaire S, Sarda P, Hamel C, Lacombe D, Holder M, Odent S, Holder S, Brooks AS, Elcioglu NH, Silva ED, Rossillion B, Sigaudy S, de Ravel TJ, Lewis RA, Leheup B, VerloesA, Amati-Bonneau P, Mégarbané A, Poch O, Bonneau D, Beales PL, Mandel JL, Katsanis N, Dollfus H (2006) BBS10 encodes a vertebrate-specific chaperonin-like protein and is a major BBS locus. *Nat Genet* 38:521–524.

Stoetzel C, Muller J, Laurier V, Davis EE, Zaghloul NA, Vicaire S, Jacquelin C, Plewniak F, Leitch CC, Sarda P, Hamel C, de Ravel TJ, Lewis RA, Friederich E, Thibault C, Danse JM, Verloes A, Bonneau D, Katsanis N, Poch O, Mandel JL, Dollfus H (2007) Identification of a novel BBS gene (BBS12) highlights the major role of a vertebrate-specific branch of chaperonin-related proteins in Bardet-Biedl syndrome. *Am J Hum Genet* 80: 1–11.

Stoler JM, Herrin JT, Holmes LB (1995) Genital abnormalities in females with Bardet-Biedl syndrome. *Am J Med Genet* 55:276–278.

Stone DL, Slavotinek A, Bouffard GG, Banerjee-Basu S, Baxevanis AD, Barr M, Biesecker LG (2000) Mutation of a gene encoding a putative chaperonin causes McKusick-Kaufman syndrome. *Nat Genet* 25:79–82.

Suspitsin EN, Imyanitov EN (2016) Bardet-Biedl syndrome. *Mol Syndromol* 7:62–71.

Tayeh MK, Yen HJ, Beck JS, Searby CC, Westfall TA, Griesbach H, Sheffield VC, Slusarski DC (2008) Genetic interaction between Bardet-Biedl syndrome genes and implications for limb patterning. *Hum Mol Genet* 17:1956–1967.

Tobin JL, Beales PL (2007) Bardet-Biedl syndrome: beyond the cilium. *Pediatr Nephrol* 22:926–923.

Tsuchiya R, Nishimura R, Ito T (1977) Congenital cystic dilation of the bile duct associated with Laurence-Moon-Biedl-Bardet syndrome. *Arch Surg* 112:82–84.

Urben SL, Baugh RF (1999) Otolaryngologic features of Laurence-Moon-Bardet-Biedl syndrome. *Otolaryngol Head Neck Surg* 120:571–574.

Vaisse C, Reiter JF, Berbari NF (2017) Cilia and Obesity. *Cold Spring Harb Perspect Biol* 9(7).

Wei LJ, Pang X, Duan C, Pang X (1998) Bardet-Biedl syndrome: A review of Chinese literature and a report of two cases. *Ophthalmic Genet* 19:107–109.

Wei Q, Zhang Y, Li Y, Zhang Q, Ling K, Hu J (2012) The BBSome controls IFT assembly and turn around in cilia. *Nat Cell Biol* 14:950–957.

Zaldivar RA, Neale MD, Evans WE, Pulido JS (2008) Asymptomatic renal cell carcinoma as a finding of Bardet-Biedl syndrome. *Ophthalmic Genet* 29:33–35.

9

BECKWITH–WIEDEMANN SYNDROME AND HEMIHYPERPLASIA

CHERYL SHUMAN

Department of Genetic Counselling, The Hospital for Sick Children and Department of Molecular Genetics, University of Toronto, Toronto, Ontario, Canada

ROSANNA WEKSBERG

Division of Clinical and Metabolic Genetics, The Hospital for Sick Children and Department of Pediatrics/Institute of Medical Science, University of Toronto, Toronto, Ontario, Canada

INTRODUCTION

Beckwith (1998a) collated a comprehensive history of overgrowth and related syndromes. In his review, he includes a case report from 1861 of an individual with findings suggestive of Beckwith–Wiedemann syndrome (BWS), and a ceramic figure from West Mexico dating back to 200 BCE to 200 CE with macroglossia and a possible umbilical defect. Although there were numerous early reports of individuals with manifestations of BWS, a syndromic designation was defined when Beckwith (1963) reported three unrelated children with omphalocele, hyperplasia of the kidneys and pancreas, and fetal adrenal cytomegaly. The following year, Wiedemann (1964) published a report of siblings with omphalocele, macroglossia, and macrosomia. The triad of omphalocele (exomphalos), macroglossia, and gigantism considered pathognomonic of this newly described syndrome, generated the early designation EMG syndrome. The syndrome is now commonly referred to as Beckwith–Wiedemann syndrome (BWS).

In 1822, Meckel first documented hemihypertrophy in the medical literature; the first clinical case report by Wagner appeared in 1839 (Ringrose et al. 1965). Hemihypertrophy, referring to increased cell size, was widely used until recently to describe "unilateral overgrowth of the body, including the structures of the head, trunk and limbs" (Viljoen et al. 1984). "Hemihyperplasia" replaced the term hemihypertrophy, when it was recognized that the condition usually involved an abnormality of cell proliferation restricted to one or more regions of the body leading to asymmetric overgrowth involving bone and/or soft tissue (Cohen 1989; Clericuzio and Martin 2009). Recently, the term isolated lateralized overgrowth has been proposed to describe this finding (Kalish et al. 2017a). Isolated hemihyperplasia is a diagnosis of exclusion because hemihyperplasia/body asymmetry can be a feature of numerous genetic conditions (Hoyme et al. 1998; Clericuzio and Martin 2009) as outlined in the Differential Diagnosis section below.

Incidence

The population incidence of BWS is estimated to be 1/10,000–1/13,700, with equal incidence in males and females (Pettenati et al. 1986; Mussa et al. 2013). This is likely an underestimate, as individuals with milder phenotypes may not be identified. For hemihyperplasia, the incidence is estimated to be 1/86,000 (Parker and Skalko 1969), with some authors reporting a higher frequency in females

---

Cassidy and Allanson's Management of Genetic Syndromes, Fourth Edition.
Edited by John C. Carey, Agatino Battaglia, David Viskochil, and Suzanne B. Cassidy.
© 2021 John Wiley & Sons, Inc. Published 2021 by John Wiley & Sons, Inc.

(Hoyme et al. 1998). Some individuals who present with isolated hemihyperplasia may, in fact, have Beckwith–Wiedemann syndrome with reduced expressivity. Evidence for this comes from several findings occurring in both Beckwith–Wiedemann syndrome and isolated hemihyperplasia. These include (1) increased birth weight (mean 3.8 kg); (2) specific renal anomalies (e.g. medullary sponge kidney); and (3) a well-documented increase in risk for embryonal tumors, especially Wilms tumor and hepatoblastoma (Hoyme et al. 1998; Niemitz et al. 2005; Shuman et al. 2006).

Diagnostic Criteria

The variable clinical expressivity in BWS has made it challenging to develop consensus diagnostic criteria. Generally, the presence of at least three of the major findings or two major and one minor finding, as detailed below, is required for a clinical diagnosis. With fewer manifestations, such as macroglossia with umbilical hernia, the differential diagnosis should include BWS, and consideration should be given to molecular testing. The option of tumor surveillance should be considered even when molecular testing in blood is negative, because a proportion of individuals with BWS and somatic mosaicism (see below) may only carry the tumor risk-associated genetic alteration in target organs.

Major findings associated with BWS include macrosomia (prenatal and/or postnatal gigantism), hemihyperplasia, macroglossia (typically present at birth but also reported to develop postnatally) (Chitayat et al. 1990a), abdominal wall defect (omphalocele, umbilical hernia), embryonal tumors (e.g. Wilms tumor and hepatoblastoma), cytomegaly of the fetal adrenal cortex, ear anomalies (anterior linear lobe creases, posterior helical pits), visceromegaly, renal abnormalities, cleft palate, and positive family history (Pettenati et al. 1986; Weng et al. 1995a). Because the phenotype may be variable even within a family, pedigree review should survey parental birth weights, history of abdominal wall defect, increased tongue size or tongue surgery, and other features of BWS. In adults, the most helpful physical manifestations include prominence of the jaw, enlarged tongue, ear creases and pits, and evidence of repaired omphalocele. Abdominal ultrasound may be helpful in evaluating abnormalities of kidneys and other abdominal organs in other family members suspected to have BWS. Adult heights are usually normal, and other manifestations may be subtle or even surgically altered; hence, early childhood photographs are useful adjuncts to family assessment and estimation of recurrence risk.

Additional diagnostic manifestations include pregnancy-related findings (polyhydramnios, pre-eclampsia, placentomegaly, placental mesenchymal dysplasia) (Wilson et al. 2008), prematurity, neonatal hypoglycemia, cardiomegaly and occasional structural cardiac anomalies, nevus simplex or other vascular malformation, advanced bone age, diastasis recti, and characteristic facies with mid-facial hypoplasia (Figure 9.1A). This characteristic facial appearance tends to regress over time, especially if macroglossia and the attendant prognathism are mild or treated (Figure 9.1B).

Most individuals with BWS have a good prognosis for long-term physical health and cognitive development, but in some, there are serious and life-threatening medical issues.

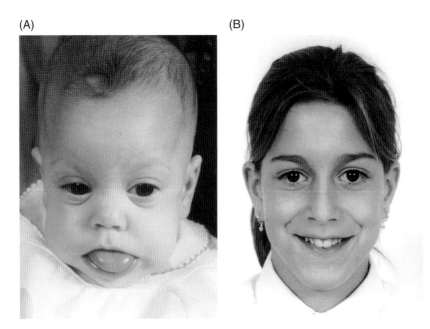

FIGURE 9.1 Girl with Beckwith–Wiedemann syndrome. (A) At age 6 months demonstrating nevus flammeus, prominent eyes, malar hypoplasia, and macroglossia. (B) At age 10 years. This photo is post-partial glossectomy and demonstrates only a few residual facial features (e.g. prominent chin).

Within this more severe group, perinatal complications involving prematurity, persistent hypoglycemia, cardiomyopathy, tumors, and/or severe macroglossia may lead to death. The frequency of early demise is likely lower than the previously quoted figure of 20%, given current approaches to medical management; however, there remains an increased rate of death in children with BWS over that in the general population (Pettenati et al. 1986; Weng et al. 1995a; Smith et al. 2007).

When making a diagnosis of isolated hemihyperplasia, one must also consider the possibility of hemihypoplasia, in which the smaller body part is not the 'normal side' but rather hypoplastic. Molecular testing may be useful in defining hemihyperplasia versus hemihypoplasia (see below). Hemihyperplasia can involve a single organ or region of the body or several regions. When several regions are involved, these may be on one side of the body (ipsilateral) or opposite sides (contralateral). The degree of asymmetry is variable and may be mild in appearance. When asymmetry is limited to one limb, a measurable difference of greater than 1 cm in length and/or a significant measurable difference in girth can be used to support a diagnosis of hemihyperplasia (Figure 9.2). Because hemihyperplasia can be subtle, and because some degree of asymmetry exists in the normal population, there is a "gray zone" making it difficult to clinically define the significance of asymmetry in some individuals. Once the finding of asymmetric overgrowth is established, other clinical features may point to a diagnosis of BWS or to other related diagnoses (see Differential Diagnosis).

Etiology, Pathogenesis, and Genetics

BWS is currently understood to be a complex, multigenic disorder caused by a number of different genetic (DNA sequence) and epigenetic (DNA methylation, histone modification) alterations that result in transcriptional dysregulation of growth regulatory genes on chromosome 11p15 (Figures 9.3 and 9.4) (Li et al. 1998). The genetic/epigenetic heterogeneity of BWS is challenging; however, defining the etiology of BWS is important for medical management and can be facilitated by categorizing these individuals according to family history, karyotype, and molecular data. This is further elaborated in the section below on approaches to molecular testing.

The chromosome 11p15 region associated with BWS, spanning 1000 kb, contains several imprinted genes implicated in the condition (Figure 9.3). Most mammalian autosomal genes are expressed from both the maternally and paternally inherited copies of a chromosome pair. Genomic imprinting is an epigenetic phenomenon whereby the two alleles of a gene are differentially modified such that only one parental allele, parent-specific for a given gene, is normally expressed. Genomic imprinting is regulated by epigenetic mechanisms (extrinsic to changes in primary DNA sequence), including DNA methylation, histone modifications, and non-coding RNAs. Imprinted genes, clustered in distinct regions on chromosomes, are associated with imprinting centers (IC) that control resetting of closely linked imprinted genes during transmission through the germline. During gametogenesis, imprinting marks from the previous generation are erased and imprinting is reset according to the sex of the transmitting parent. Imprinting centers, also termed differentially methylated regions (DMRs), demonstrate differential methylation of the parental alleles and regulate the expression of imprinted genes *in cis* (on the same chromosome) over large distances.

Many types of parent-of-origin-specific and dosage-sensitive molecular alterations are observed in BWS (see Table 9.1). These include paternal uniparental disomy, preferential maternal transmission of BWS in autosomal dominant pedigrees, and parent-of-origin effects in chromosome abnormalities associated with BWS. These data are consistent with the findings of alterations in imprinted genes on 11p15 in BWS. Therefore, to understand the pathophysiology of BWS, one must take account of the relative dosage, as well as the parent-of-origin of imprinted genomic regions. These two factors can be used to assess the number of transcriptionally active or transcriptionally silent alleles of each BWS-associated growth regulatory gene.

The regulation of imprinted genes on chromosome 11p15 is shown in Fig. 9.3. Chromosome 11p15 houses two imprinted domains, each having an imprinting center (IC) and a non-coding RNA. Our current understanding of the role of some of these imprinted genes in BWS is outlined below using currently recommended nomenclature (Monk et al. 2016).

In the telomeric imprinted domain (domain 1), the imprinting center, IC1 (*H19/IGF2*:IG-DMR), regulates transcription of two genes, *H19* (a non-coding RNA) and insulin-like growth

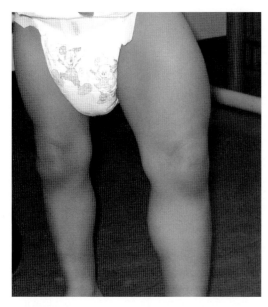

FIGURE 9.2 Hemihyperplasia involving legs.

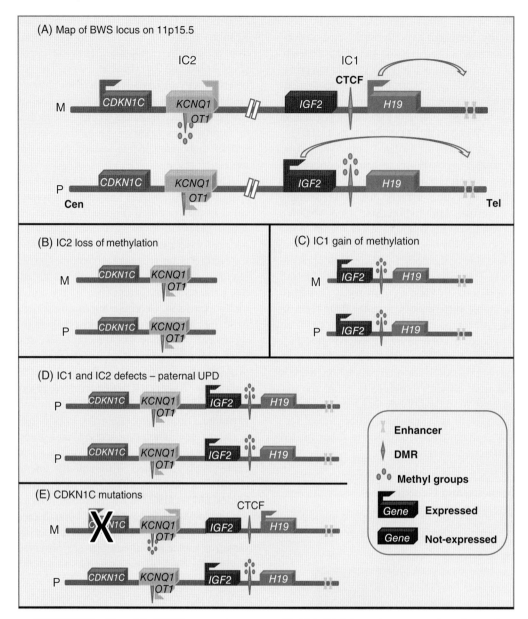

FIGURE 9.3 Map of the BWS locus on chromosome 11p15.5. (A) Schematic representation of normal parent of origin-specific imprinted allelic expression. Note, only the region that is altered is shown in (B) and (C). (D) BWS caused by paternal uniparental disomy (UPD). (E) BWS caused by pathogenic sequence variants in the *CDKN1C* gene. IC – imprinting center (IC1-H19/IGF2:IG-DMR; IC2-KCNQ1OT1:TSS-DMR). Cen – centromere; Tel – telomere; P – paternal; M – maternal; DMR – differentially methylated region; OT1, KCNQ1OT1, refers to the KCNQ1 antisense transcript. The image is not drawn to scale. *(From Choufani et al [2010]; republished with permission from John Wiley and Sons).*

factor-2 (*IGF2*). In the centromeric domain (domain 2), the imprinting center, IC2 (*KCNQ1OT1*:TSS-DMR), maps to the promoter region of the non-coding RNA *KCNQ1OT1*. IC2 regulates the mono-allelic expression of *KCNQ1OT1* as well as that of several other imprinted genes including *CDKN1C*. Other regulatory elements such as enhancers are also involved in the transcriptional regulation of imprinted genes including non-coding RNAs (Heide et al. 2017).

Molecular alterations associated with BWS can occur in either domain 1 and/or domain 2 on chromosome 11p15.5. These alterations involve DNA methylation alone (epigenetic) or DNA methylation alterations occurring in conjunction with genomic alterations.

Molecular Alterations Involving Domain 1

Gain of Maternal Methylation at IC1 Normally, IC1 is methylated on the paternal chromosome and unmethylated on the maternal chromosome. In 5% of individuals with BWS, gain of maternal methylation at IC1 is associated with loss of *H19* expression and bi-allelic *IGF2* expression (Table 9.1). *H19* is a maternally-expressed gene encoding a biologically

INTRODUCTION 129

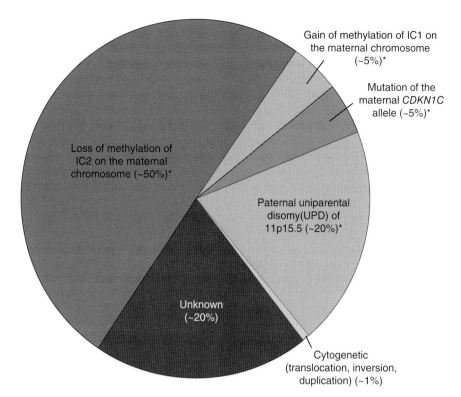

FIGURE 9.4 Frequency of genetic alterations detected in individuals with Beckwith–Wiedemann syndrome
*These molecular subgroups defined by DNA methylation abnormalities can have an underlying genomic alteration. Such genomic aberrations are most common for gain of methylation of IC1 and less common for loss of methylation at IC2. Genomic aberrations, limited to the BWS critical region on chromosome 11p15.5, can be detected by MS-MLPA (methylation-sensitive multiplex ligation-dependent probe amplification), SNP (single nucleotide polymorphism) array or targeted sequencing. From Shuman C, Beckwith JB, Weksberg R (2016) Beckwith-Wiedemann Syndrome © 1993–2017 University of Washington, https://www.ncbi.nlm.nih.gov/books/NBK1394/ Republished with permission from GeneReviews.

active noncoding messenger ribonucleic acid (mRNA) that may function as a tumor suppressor (Raveh et al. 2015). Gain of methylation at IC1 can be associated with genomic (DNA) alterations (microdeletions/microduplications or single nucleotide variants in *OCT4/SOX* binding sites) in some individuals with BWS. These may be either inherited or de novo and are subsequently heritable in future generations.

Bi-allelic Expression of IGF2   Insulin-like growth factor-2 (*IGF2*) is a paternally expressed embryonic growth factor. Disruption of *IGF2* imprinting (leading to bi-allelic expression) is observed in some individuals with BWS (Weksberg et al. 1993) as well as in multiple embryonal tumors, including Wilms tumor (Ogawa et al. 1993; Rainier et al. 1993; Scott et al. 2008a). Loss of *IGF2* imprinting may occur in concert with loss of *H19* expression and IC1 gain of methylation. However, *IGF2* bi-allelic expression may also occur independently of IC1 and *H19* dysregulation.

Molecular Alterations Involving Domain 2

Mutation of CDKN1C   The *CDKN1C* gene encodes the $p57^{KIP2}$ protein, a member of the cyclin-dependent kinase inhibitor gene family, which negatively regulates cell proliferation. It is both a tumor-suppressor gene and a potential negative regulator of fetal growth. Mutations in this gene are reported in 5–10% of individuals with BWS as the first case in the family. These mutations, in association with clinical manifestations of omphalocele (Lam et al. 1999) and cleft palate (Hatada et al. 1997; Li et al. 2001), are found in approximately 40% of individuals with BWS who have a positive family history (O'Keefe et al. 1997; Lam et al. 1999).

Loss of Maternal Methylation at IC2   Normally, IC2 on the maternally derived chromosome is methylated, whereas IC2 on the paternally derived chromosome is unmethylated. *KCNQ1OT1* is an imprinted transcript that is antisense to *KCNQ1*. The promoter of *KCNQ1OT1* is differentially methylated and represents the IC2 imprinting control element on human chromosome 11p15.5 (Lee et al. 1999; Smilinich et al. 1999). Loss of maternal methylation at IC2 is seen in 50% of individuals with BWS (Table 9.1) (Lee et al. 1999; Smilinich et al. 1999). In some cases, potentially heritable microduplications or microdeletions are associated with alterations of DNA methylation at the BWS locus (Baskin et al. 2014; Heide et al. 2017).

TABLE 9.1 Beckwith–Wiedemann syndrome: genetic and epigenetic molecular groups

Molecular group	Imprinted domain	Frequency	Heritability*	Recurrence risk
IC2 (*KCNQ1OT1:TSS-DMR*) loss of maternal methylation:	2	50%		
• Without genomic alteration			Sporadic	Low
• With genomic alteration			Heritable (de novo or parentally transmitted)	As high as 50%
Paternal uniparental disomy	1, 2	20%	Sporadic	Very low
CDKN1C mutation	2	• 5% in sporadic cases		Low but gonadal mosaicism theoretically possible
		• 40% in inherited cases	Almost exclusively maternal transmission	50% if maternally transmitted, unknown if paternally transmitted[a]
IC1 (*H19/IGF2:IG-DMR*) gain of maternal methylation:	1	5%		
• Without genomic alteration			Sporadic	Low
• With genomic alteration			Heritable (de novo or parentally transmitted)	50% if parent carries alteration[b]
Chromosome 11p15 duplication – visible cytogenetically	1, 2	<1%	Heritable (de novo or paternally transmitted)	Increased if father carries 11p15 translocation/rearrangement[c]
11p15 chromosome translocation/inversion	2	<1%	Heritable (de novo or maternally transmitted)	As high as 50% if translocation/inversion is maternally transmitted
Positive family history – no molecular alteration identified	ND	ND	Heritable	May be as high as 50% and parent of origin specific

Note: Overall, 85% of individuals with Beckwith–Wiedemann syndrome are the first in the family to be affected (sporadic), whereas 15% are associated with vertical transmission (i.e. have an affected parent). The molecular groups in this table are not mutually exclusive.

* Heritability in this table refers to recurrence risk for sibs. Individuals with de novo genomic alterations will have a significant risk for transmitting these alterations to the next generation.

Abbreviations: ND – not determined.

[a] Rare cases of paternal transmission are reported.
[b] Risk can also apply to Silver–Russell syndrome (see Chapter 52) depending on the sex of transmitting parent (see Vals et al. 2015)
[c] Specific figures are not known.

Maternally Transmitted Translocations/Inversions of Chromosome 11p15 There are rare (1%) de novo and maternally transmitted translocations/inversions of 11p15.5 associated with BWS. Translocation/inversions almost always disrupt the *KCNQ1OT1* gene (Smilinich et al. 1999), but most do not demonstrate DNA copy number changes, and the majority are not associated with DNA methylation alterations. Individuals with 11p15 translocations or inversions exhibit typical features of BWS.

Molecular Alterations Involving Domain 1 and Domain 2 In some individuals with BWS, the molecular alterations span both domain 1 and domain 2, altering the expression of multiple genes. Thus, both maternally expressed growth-suppressor genes and paternally expressed growth-promoter genes are affected. This is seen in individuals with BWS who have chromosome 11p15 chromosomal duplications or chromosome 11p15 uniparental disomy.

Paternally Derived Duplications of Chromosome 11p15
Paternally derived duplications of chromosome 11p15.5 associated with BWS are rare (1%). In those with cytogenetically visible 11p15 duplications, the duplications span both domain 1 and domain 2. Such individuals have atypical clinical features and a significant risk of developmental delay (Waziri et al. 1983; Slavotinek et al. 1997).

Paternal Uniparental Disomy of Chromosome 11 About 20% of individuals with BWS demonstrate paternal uniparental disomy as they have two paternally derived copies of chromosome 11p15 and no maternal contribution for that region. The 11p15 chromosome region is always involved, but the extent of the uniparental disomy varies considerably (Slatter et al. 1994).

All reported cases of BWS with uniparental disomy exhibit somatic mosaicism. This implies that uniparental disomy arises post-zygotically via somatic recombination, and may be found only in some tissues, for example, in fibroblasts or renal tissue but not lymphocytes. Because most somatic tissues are not available for testing, the quoted frequency of uniparental disomy in BWS should be considered an underestimate.

For paternal uniparental disomy of 11p15, the BWS-associated clinical phenotype and tumor predisposition are likely caused by the combination of increased expression of paternally expressed growth-promoter genes (e.g., *IGF2*) and loss of maternally expressed growth-suppressor genes (e.g., *H19*). Figure 9.3 illustrates this model, depicting the gene dosages of paternally expressed growth-promoter genes and maternally expressed growth-suppressor genes.

Unknown Molecular Etiology   In approximately 15–20% of individuals with BWS, the molecular etiology is not identified. The underlying mutation may be in regions of genes already implicated in the pathophysiology of BWS but not currently screened by molecular analysis, or there may be somatic mosaicism. Mutations in other independent genes should be considered.

Molecular Alterations Associated with Isolated Hemihyperplasia

In apparently isolated hemihyperplasia, somatic mosaicism is the sentinel finding making the search for an underlying molecular defect difficult. Reported molecular etiologies in cases of isolated hemihyperplasia, with or without embryonal tumors, include somatic mosaicism for paternal uniparental disomy for 11p15, chromosomal abnormality, and methylation alterations at either IC1 or IC2 (Clericuzio and Martin 2009). A proportion of individuals with hemihyperplasia likely represent "mild" BWS based on the risk of embryonal tumor development, the known spectrum of tumors, and some variably associated clinical features (high birth weight and renal findings) (Hoyme et al. 1998).

Unique Clinical and Molecular Findings in Beckwith–Wiedemann Syndrome

Discordant Female Monozygotic Twins and Loss of Methylation at IC2   An enigmatic clinical group consists of monozygous ("identical") twin pairs with BWS. More than 35 such twin pairs have been reported, most commonly female and discordant for BWS (Weksberg et al. 2002). Furthermore, the incidence of female monozygotic twinning in BWS syndrome is dramatically increased compared with that in the general population. However, a small number of cases of male monozygotic twins are discordant for BWS, and both male and female monozygotic twin pairs concordant for BWS have been reported (Leonard et al. 1996; Weksberg et al. 2002; Smith et al. 2006). Male monozygotic twins have a variety of molecular defects as seen in singleton cases of BWS. In contrast, in five female monozygotic twin pairs discordant for BWS, skin fibroblasts for each affected twin had an imprinting defect at *KCNQ1OT1* on 11p15, whereas the unaffected twin maintained normal imprinting (Weksberg et al. 2002). We proposed that in female monozygotic twins discordant for BWS, loss of imprinting at *KCNQ1OT1* is etiologically tied to the twinning process, both events occurring at a critical stage of preimplantation development (Weksberg et al. 2002).

Offspring of Parents with Sub-fertility Treated with Assisted Reproductive Technology

Data generated from both retrospective and prospective studies suggest an increased risk for imprinting disorders, most notably BWS, in offspring born to parents with sub-fertility who conceived the pregnancy using assisted reproductive technologies. The most compelling aspect of the evidence is that the vast majority of such children demonstrate a specific chromosome 11p15 molecular alteration, loss of maternal methylation at IC2 (Maher et al. 2003). Of interest, the increased rate of children with Angelman syndrome (see Chapter 5) born following sub-fertility/assisted reproductive technology also demonstrates a specific and parallel molecular alteration – loss of maternal methylation at the imprinting center on chromosome 15q11-13 (Ludwig et al. 2005). This single mechanism – loss of maternal methylation at two independent loci – supports the hypothesis that sub-fertility/assisted reproductive technology is associated with an increased rate of imprinting disorders (Cortessis et al. 2018)

Beckwith–Wiedemann Syndrome can be Associated with Methylation Defects at Non-chromosome 11 Imprinting Centers

Investigations of BWS with loss of maternal methylation at IC2 identified a subset of individuals who also demonstrate methylation aberrations at other imprinted loci across the genome (Bliek et al. 2008; Lim et al. 2008; Rossignol et al. 2006). This is often referred to as multi-locus imprinting disorder (MLID) (Eggermann et al. 2014). This has generated interest in defining genes on other chromosomes that act *in trans* (i.e. at genomic regions other than chromosome 11) to regulate groups of imprinted genes across the genome. One such gene has been proposed for an individual with BWS who demonstrated loss of methylation at *KvDMR* as well as other imprinted loci. In this case, the investigators demonstrated a recessive mutation in *NALP2* in the mother as a putative cause for the imprinting defects at multiple loci (Meyer et al. 2009).

MLID has also been associated with imprinting disorders other than BWS. For example, transient neonatal diabetes mellitus is associated with loss of maternal methylation at the imprinted gene *PLAGL1* on chromosome 6q24. The MLID in transient neonatal diabetes mellitus can be caused by a mutation in a zinc-finger protein gene (*ZFP57*) (Mackay et al. 2008). However, MLID in BWS has not been associated with mutations in this gene.

Genome-wide epigenetic studies in BWS will be crucial to understanding the interaction across different chromosomal regions in the development of the BWS phenotype. Such studies might identify currently unknown molecular (genetic or epigenetic) abnormalities and define new molecular defects in the 20% of affected individuals who currently elude molecular characterization. Such findings could also shed light on new molecular pathways to explain the phenotypic variability seen in BWS.

Recurrence Risks for Beckwith–Wiedemann Syndrome and Isolated Hemihyperplasia

Elucidation of the molecular etiology is critical for defining recurrence risks for families (Table 9.1). The risk of recurrence for BWS is believed to be low in the absence of a genomic alteration or a positive family history. The chance to have a child with BWS for those families presenting with a positive family history is likely increased, while the chance for those with identified genomic abnormalities may be increased or decreased depending on the molecular finding (see Diagnostic Testing). Consideration should also be given to the risk for Silver–Russell syndrome (see Chapter 52) based on the opposite imprinting defects on chromosome 11p15.5 in BWS versus Silver–Russell syndrome locus; that is, a genomic alteration causing BWS can confer a risk of Silver–Russell syndrome in offspring, depending on the sex of the transmitting parent (Vals et al. 2015).

The recurrence risk for isolated hemihyperplasia is low but would be dependent on the underlying etiology. Assessment should be undertaken for potential associated genetic syndromes in individuals presenting with apparently isolated hemihyperplasia prior to providing genetic counseling regarding recurrence risks.

Diagnostic Testing

Diagnostic testing is useful for confirming the diagnosis of BWS and determining recurrence risks. BWS can be categorized into distinct genetic and epigenetic groups (Table 9.1). In about 85% of individuals, BWS presents as the first case in the family. In 10–15% of karyotypically normal individuals with BWS, parent-of-origin-specific autosomal dominant inheritance is evident. Heritable cytogenetically visible chromosome abnormalities of 11p15 occur in approximately 1–2% of individuals with BWS. A systematic approach to the multiple diagnostic testing options for BWS is important (Eggermann et al. 2016), and all individuals should have a high-resolution banding karyotype to detect rare de novo and maternally transmitted translocations/inversions (1%). Translocations/inversions almost always disrupt *KCNQ1OT1* (Smilinich et al. 1999), and are not usually detectable by the molecular tests outlined below that assess copy number changes or DNA methylation changes on chromosome 11p15. A widely utilized testing modality for BWS is based on methylation-specific multiplex ligation-dependent probe amplification (MS-MLPA) to evaluate both DNA methylation and DNA dosage for multiple probes across the chromosome 11p15 region (Scott et al. 2008b). MS-MLPA technology is currently the most sensitive clinical testing method for detecting the majority of epigenetic and genetic abnormalities associated with BWS. It detects microdeletions, microduplications, and changes in DNA methylation in the 11p15.5 region, including those associated with uniparental disomy (UPD). However, it does not detect all genomic alterations associated with BWS (e.g. *OCT4/SOX* single nucleotide variants). In some cases, methylation abnormalities at IC1 and/or IC2 are difficult to interpret. In particular, given the recent reports regarding the limitations of MS-MLPA to detect somatic mosaicism for 11p15 paternal uniparental disomy, parallel testing using SNP arrays should be undertaken concurrently to assist in the interpretation of the IC1 and IC2 methylation data (Keren et al. 2013; Russo et al. 2016; Brzezinski et al. 2017). If MS-MLPA testing is negative, a screen for *CDKN1C* mutations should be undertaken, even in the absence of specific clinical indications (e.g. positive family history or cleft palate). For individuals with BWS and a cleft palate, a screen for *CDKN1C* mutations can be prioritized. *CDKN1C* sequence analysis is available in some clinical diagnostic laboratories, and mutations are identified in cases of BWS that are sporadic (5%) and in autosomal dominant pedigrees (40%) (Li et al. 2001).

If a genomic abnormality (e.g. CDKN1C pathogenic variant, microdeletion, etc.) is found in a child with BWS, parents should be offered testing. If the genomic alteration is not detected in either parent, one should still consider prenatal testing for recurrence in view of the theoretical possibility of gonadal mosaicism, although there are no published reports of recurrence for parents who do not have a constitutional mutation.

Because all cases of uniparental disomy associated with BWS reported to date involve somatic mosaicism, failure to detect uniparental disomy in one tissue (usually leukocytes) is not conclusive. One should consider obtaining another tissue, e.g. buccal swab, or skin, especially in the event of surgery. Uniparental disomy for 11p15 may be found in many tissues of individuals with BWS or may be limited to "normal" kidney tissue surrounding a Wilms tumor in a phenotypically normal child. The presence of mosaicism for 11p15 uniparental disomy would confer a low recurrence risk, as this results from a postzygotic event.

Methylation alterations associated with BWS, such as IC1 gain of methylation and loss of methylation at IC2, can be detected by several technologies other than MS-MLPA that are currently available in clinical laboratories. These include Southern blotting and PCR-based assays. In the case of negative test results, it is important for the clinician to explore with the diagnostic laboratory whether there are options for further molecular testing. Such tests are less robust than MS-MLPA because they target fewer regions of 11p15 for assessment of dosage and methylation. Furthermore, they often do not address the possible presence of genomic alterations coexisting with the methylation alteration, for example, IC1 deletion and IC1 gain of methylation. Such a dual finding confers a high recurrence risk, whereas IC1 gain of methylation in the absence of a genomic alteration is associated with a low recurrence risk.

Loss of imprinting at *IGF2* is seen in multiple molecular subgroups but is not currently used as a primary molecular diagnostic test. *IGF2* imprinting status and expression studies remain research tools and should not be considered part of the routine diagnostic work-up for BWS. *IGF2* allelic expression studies can be undertaken only on tissues expressing *IGF2*, which include skin fibroblast samples, but not leukocytes, and only for individuals informative (i.e. heterozygous) for transcribed *IGF2* polymorphisms to enable interpretation of expression patterns from the two alleles.

Constitutional chromosome 11p15 alterations involving IC1 and/or IC2 have been reported not only in BWS, but also in a spectrum of related clinical disorders, such as isolated hemihyperplasia (Clericuzio and Martin 2009) and isolated Wilms tumor (Scott et al. (2008a).

Differential Diagnosis

There are a number of endocrine disorders and overgrowth syndromes that should be considered in the differential diagnosis of children presenting with macrosomia or other features of BWS. Of immediate concern are the possibilities of maternal diabetes mellitus during pregnancy and congenital hypothyroidism, which should be considered and investigated. Other manifestations uncommon in BWS may suggest other diagnoses. Several syndromes with phenotypes overlapping that of BWS are discussed below. Some individuals demonstrating overgrowth do not fit into any of these defined syndromes; clearly, there are many other unspecified overgrowth syndromes and nonsyndromic causes of overgrowth already defined and new ones rapidly being identified using next generation sequencing.

Simpson–Golabi–Behmel syndrome shares the following manifestations with BWS: macrosomia, visceromegaly, macroglossia, and renal cysts. Manifestations of Simpson–Golabi–Behmel syndrome not seen in BWS include coarse features, cleft lip, high frequency of cardiac defects (Lin et al. 1999), supernumerary nipples, polydactyly, and other skeletal anomalies. Simpson–Golabi–Behmel syndrome, like BWS, has an increased risk of neonatal mortality, and an increased risk for developing embryonal tumors, including Wilms tumor and hepatoblastoma. Simpson–Golabi–Behmel syndrome is caused by pathogenic variants in an X-linked gene, *GPC3*, encoding an extracellular proteoglycan (glypican-3) that functions in growth control regulation during development.

Perlman syndrome is defined by macrosomia, increased risk of neonatal mortality, intellectual disability, nephroblastomatosis, and a high incidence of bilateral Wilms tumor usually occurring in the first year of life. There is a characteristic facial appearance: round face, upsweep of anterior scalp hair, depressed nasal bridge, and micrognathia. The molecular basis of Perlman syndrome involves homozygous deletions or compound heterozygous pathogenic variants in *DIS3L2* on chromosome 2q37 (Morris et al. 2013).

Costello syndrome (see Chapter 16) overlaps clinically with BWS only in the neonatal period, with neonates presenting with "overgrowth" because of edema and cardiac defects. Over time, they can be distinguished easily from BWS by their distinctive facial coarsening and failure-to-thrive (Johnson et al. 1998; Van Eeghen et al. 1999). Missense variants in *HRAS* are detected in approximately 80–90% of individuals with Costello syndrome.

Sotos syndrome (see Chapter 55) is characterized by overgrowth involving height and head circumference as well as a typical facial appearance and developmental disability. The majority of individuals with Sotos syndrome (80–90%) have a detectable variant or deletion in *NSD1*.

Weaver syndrome presents with overgrowth involving height and head circumference and variable intellectual disability. Individuals with Weaver syndrome have pathogenic variants in *EZH2, EED or SUZ12*.

Mosaic genome-wide paternal uniparental isodisomy presents with features overlapping those in BWS (e.g. large for gestational age, macroglossia, hypoglycemia, increased tumor risk, etc.) as well as features noted in other imprinting disorders including developmental delay (Inbar-Feigenberg et al. 2013; Kalish et al. 2013).

Hemihyperplasia may be a feature of a number of syndromes other than BWS, and careful evaluation for findings suggestive of other syndromes should be undertaken prior to assigning a diagnosis of isolated hemihyperplasia. Such syndromes include: neurofibromatosis type 1 (see Chapter 40), Klippel–Trenauney–Weber syndrome, Proteus syndrome (see Chapter 47), CLOVES, MCAP, McCune–Albright syndrome, epidermal nevus syndrome, triploid/diploid mixoploidy, Maffucci syndrome, and osteochrondomatosis or Ollier disease. Hemihyperplasia of the face, either isolated or as part of BWS, should be carefully distinguished from plagiocephaly (asymmetric cranial shape).

MANIFESTATIONS AND MANAGEMENT

Most children with BWS have the same developmental trajectory as children in the general population and many of the distinctive facial features become less apparent in early childhood. Some children experience challenges associated with some of the BWS-associated findings, for example the presence of significant hemihyperplasia, hyperinsulinemia, macroglossia, medullary sponge kidney disease, etc. While the majority of children with BWS do not develop tumors, ongoing surveillance remains a critical component of health management. Long-term studies of health outcomes in adults with BWS are not available; however, anecdotally, these adults do not appear to experience significant new health issues.

Growth and Feeding

A significant percentage of individuals with BWS have birth weights and lengths at about the 97th centile for gestational age (Brioude et al. 2013). Onset of rapid growth can occur from the prenatal period to as late as 1 year of age (personal experience). Overgrowth is not an absolute requisite for clinical diagnosis and absence of overgrowth can be associated with concomitant methylation defects at other imprinting centers (Bliek et al. 2008a). Head circumference varies and often does not parallel the increased height percentiles. Macrocephaly out of keeping with parental head circumference warrants consideration of alternative overgrowth diagnoses, as listed above.

In BWS, the increased rate of somatic growth typically continues through the first few years of life. Growth generally parallels the normal growth curve (Pettenati et al. 1986), and bone age is usually not advanced. In some studies, growth rate decreases somewhat beyond mid-childhood (Weng et al. 1995a). Adult heights range from the 50th centile to the 97th centile (Pettenati et al. 1986; Weng et al. 1995a) and are likely influenced by familial heights.

Hemihyperplasia occurs in about 25% of individuals with BWS. Hemihyperplasia may not be evident at birth and may become more marked in the first few years of life (Elliott and Maher, 1994).

Macroglossia, typically involving length and bulk, is a common finding in BWS as it likely generates an ascertainment bias. Macroglossia can occasionally lead to serious difficulties with feeding, obstructive sleep apnea, and/or speech articulation (see also Craniofacial below)

Evaluation

- Height, weight, and head circumference should be measured annually. Familial heights, especially parental, should be obtained and considered in determining whether macrosomia is truly present.
- If hemihyperplasia is observed, measurements of affected regions, both length and girth, should be taken regularly. If the leg length discrepancy is greater than 1 cm, referral to orthopedics is indicated.
- It is important to follow children with hemihyperplasia over several years to evaluate relative growth velocity of the two limbs so that surgical treatment, if necessary, can be carried out at the optimal time.
- If hemihyperplasia involves the limbs or trunk, clinical examination of the spine is recommended for evaluation of scoliosis. Referral to an orthopedist is indicated for monitoring and management of scoliosis.
- If significant feeding difficulties are noted, evaluation by a feeding specialist is recommended.
- Significant macroglossia and/or facial hemihyperplasia may require referral to a craniofacial surgeon.

Management

- Feeding difficulties encountered because of macroglossia may be ameliorated by use of a longer nipple such as those used for babies with cleft palate.
- Surgery to reduce tongue size may be considered by the craniofacial surgical team/appropriate experts (see Craniofacial section).
- Rarely, nasogastric tube feedings are indicated for a period of time.
- There is no specific treatment required for macrosomia.
- For children with leg length discrepancy, annual orthopedic follow-up throughout childhood is important to determine whether shoe lifts are appropriate and if surgical intervention is indicated prior to puberty.
- Epiphysiodesis of the hyperplastic leg may be considered when that leg attains the final length predicted for the normal leg (typically undertaken just before puberty).

Development and Behavior

Contrary to early reports, development is usually normal in individuals with BWS unless there are serious complications associated with prematurity, episodes of severe hypoxia, a period of uncontrolled or undetected hypoglycemia, or chromosomal duplication involving 11p15. As well, there are rare reports of abnormalities of the posterior fossa; these are associated with IC2 molecular alterations or pathogenic variants in *CDKN1C* (Gardiner et al. 2012; Brioude et al. 2015).

Some parents have commented that their children with BWS are behaviorally different from their other children; however, this was not confirmed in a small experimental sample (personal observation). A UK study indicated an increased incidence of emotional and behavioral difficulties, including autism spectrum disorder, in children with BWS (Kent et al. 2008), but this has not been replicated. Follow up studies using standardized testing for autism spectrum disorder as well as molecular testing to firmly establish a diagnosis, are indicated to elucidate this association.

For those with isolated hemihyperplasia, which is etiologically heterogeneous, an increased incidence of developmental delay (15–20%) is reported (Viljoen et al. 1984; Ringrose et al. 1965). There may be some ascertainment bias in this figure, as hemihyperplasia is more likely to be identified if there are other associated findings, such as developmental delay. The association of hemihyperplasia with developmental delay may be due to a different underlying genetic syndrome or somatic mosaicism for a chromosome abnormality.

Evaluation

- Developmental screening should be part of every routine visit for children with either BWS or hemihyperplasia.
- Individuals with suspected BWS and developmental delay should be carefully evaluated for brain malformations of the posterior fossa detectable on magnetic resonance imaging (MRI), for other syndromes, or for chromosome abnormalities (for the latter group, this may include chromosome studies of the skin).
- Those with developmental delay should have a careful, complete developmental assessment.

Management

- Any individual with developmental delay associated with BWS and/or hemihyperplasia should be offered standard interventions, such as infant stimulation programs, occupational and physical therapy, and individualized education programs.

Cardiovascular

The reported incidence of structural cardiac malformations ranges from 9 to 34%, with about half involving cardiomegaly (Pettenati et al. 1986; Elliott and Maher, 1994; Mussa et al. 2016). Other reported malformations include hypoplastic left heart, mild pulmonary stenosis, and persistent foramen ovale (Elliott et al. 1994). Cardiomegaly of early infancy in BWS usually resolves spontaneously. Cardiomyopathy has been rarely reported and can be severe and lethal despite current interventions (Smith et al. 2007). The prognosis for other structural cardiac anomalies depends on the specific defect identified and current treatment options.

Evaluation

- In BWS, there should be a high index of suspicion for cardiac problems; standard cardiac evaluation should precede any surgical or dental procedure.
- If a cardiac abnormality is suspected on clinical evaluation, a comprehensive and systematic cardiac evaluation is recommended, including electrocardiogram and echocardiogram.
- If a conduction defect is found, the alternate diagnoses of Costello syndrome (Chapter 16) or Simpson–Golabi–Behmel syndrome should be considered.

Management

- Treatment of cardiac abnormalities is the same as for the general population.

Endocrine

Hypoglycemia is reported to occur in approximately 30–50% of babies with BWS (Pettenati et al. 1986; Engstrom et al. 1988; Mussa et al. 2016). This often resolves in the neonatal period; however, some cases persist well beyond this time period and require pharmacologic management and possibly surgical treatment. The underlying cause of the hypoglycemia appears to be hyperinsulinemia and islet cell hyperplasia. This may be related to 11p15 uniparental disomy and/or other mechanisms involved in dysregulation of genes on 11p15. Mosaicism for 11p15 uniparental disomy is one of the recognized pathogenic mechanisms causing focal nesidioblastosis of the pancreas (de Lonlay et al. 1997).

No data are available concerning long-term outcome of hypoglycemia associated with BWS. In general, children without BWS who have had significant hypoglycemia have significantly smaller head circumferences and more neurological deficits, including lower IQ scores at 5–7 years of age, than their unaffected counterparts. Neonates with seizures secondary to hypoglycemia tend to have the worst overall neurological prognosis (Halamek and Stevenson, 1998). Newborns with asymptomatic hypoglycemia, although not completely without risk for sequelae, have the best prognosis. None of these studies adequately assesses the risks faced by neonates who do not experience seizure activity (Halamek and Stevenson 1998).

Other abnormal laboratory findings noted in BWS include hypocalcemia (4.6%) and hyperlipidemia (2.3%) (Engstrom et al. 1998). Hypothyroidism has been reported in several individuals with BWS (Martinez y Martinez et al. 1985).

Evaluation

- Any neonate suspected of having BWS should be screened for hypoglycemia for the first few days of life. Serial blood glucose measurements to detect asymptomatic hypoglycemia as well as frequent examination for clinical signs of symptomatic hypoglycemia are recommended. The level at which neonatal hypoglycemia becomes clinically important, warranting intervention, is poorly defined (Boluyt et al. 2006; Burns et al. 2008). Hypoglycemia may persist in a small percentage of babies, requiring longer-term monitoring.

- If early discharge is planned, health care professionals should advise parents about the typical clinical manifestations and treatment of hypoglycemia and should ensure that rapid access to medical care is available.
- If hypoglycemia persists beyond the first few days, evaluation by a pediatric endocrinologist is suggested because hypoglycemia, in a few cases, can be refractory to standard first-line treatment.
- A high index of suspicion should be maintained for hypocalcemia and hypothyroidism. Evaluation is standard.

Management

- Either intravenous or oral glucose should be administered while awaiting laboratory confirmation of hypoglycemia if clinical manifestations of hypoglycemia are present. The route of administration should be guided by the level of serum glucose and the severity of the clinical manifestations.
- Any individual treated for hypoglycemia should be carefully followed with serial glucose determinations at 15–30 minute intervals until the level rises to 60–100 mg dL^{-1}.
- A number of drugs, such as diazoxide or somatostatin, can be used for refractory hypoglycemia (Halamek and Stevenson, 1998) and surgical intervention for focal nesidioblastosis of the pancreas may be required.
- Early intervention for hypocalcemia or hypothyroidism is indicated. Treatment for such issues in children with BWS is standard.

Craniofacial

Distinctive facial features of BWS include macroglossia, anterior ear lobe creases, posterior helical pits, facial nevus simplex, prominent eyes with infraorbital creases, and broad lower jaw leading to square-face shape. Macroglossia can occasionally lead to serious difficulties with respect to feeding and respiratory complications. Macroglossia may also lead to difficulties with speech articulation and local trauma once the teeth come in (Tomlinson et al. 2007). Later childhood problems may include malocclusion, as mandibular, or lower jaw, growth accelerates in response to tongue size. A prognathic jaw may develop even after tongue-reduction surgery, meriting consideration of further surgical intervention (personal observation). In addition, hemihyperplasia may affect one side of the face and/or tongue, leading to an asymmetric appearance.

Evaluation

- Concerns about speech difficulties as a result of macroglossia should preferably be assessed by a speech pathologist familiar with the macroglossia associated with BWS and its natural history.
- Referral to respiratory medicine and/or otolaryngology along with a sleep study should be considered for children with significant macroglossia, especially if there is any concern about obstructive sleep apnea.
- Issues related to facial appearance are optimally assessed by a multidisciplinary team including plastic surgeons, speech pathologists, and orthodontists. It is recommended that health care providers who have experience with the natural history of BWS be selected so that growth patterns and potential long-term impact of macroglossia and/or facial hemihyperplasia are anticipated and medical, dental, and surgical interventions can be optimally planned.

Management

- Children with only mild-to-moderate macroglossia tend to be able to better accommodate their tongues as their facial bone structures grow, so that they can keep their tongues fully in their mouths. They can be followed longitudinally by an experienced craniofacial team.
- Partial tongue resection is generally considered for children anticipated to encounter significant orthodontic and/or cosmetic concerns or, rarely, severe airway obstruction (Weng et al. 1995b). In such cases, early consultation with an experienced surgeon should be organized. Many procedures have been described for tongue reduction. The most common ones involve excision of either the central or the anterior portions of the tongue (Tomlinson et al. 2007; Heggie et al. 2013).
- Orthodontic treatment as well as plastic surgery may be considered when there is significant prognathism.

Gastrointestinal

Abdominal wall defects are a common manifestation in children with BWS. Such defects include omphalocele, umbilical hernia, and diastasis recti. Other less common findings are inguinal hernia, prune-belly sequence, and gastrointestinal malformations including atresia, stenosis, and malrotation.

The outcome for omphalocele surgery for children with BWS depends on the size of the defect and on whether or not the liver is involved, which is similar to outcome for children without BWS. In addition, the prognosis may be affected by the presence of associated medical and surgical complications. Early detection and surgical intervention of gastrointestinal malformations are important for a good outcome.

Visceromegaly is a common finding and may involve any or all of the following: liver, spleen, pancreas, kidneys, and adrenals. Adrenocortical cytomegaly is a cardinal pathologic feature of BWS. Renal findings are further discussed in a separate section below. Functional problems associated with enlarged liver and spleen are generally not reported. However, hyperplastic changes in the pancreas are associated with persistent hypoglycemia (see **Endocrine**).

Diaphragmatic eventrations, which are uncommon, may be detected during ultrasound screening for embryonal tumors.

Evaluation

- Evaluation of abdominal wall defects and visceromegaly, including standard physical examination and abdominal ultrasound, should be undertaken at the time of diagnosis.
- There should be a high index of suspicion for gastrointestinal malformations, which may require imaging for diagnosis of stenosis, atresia, or malrotation and referral to a gastroenterologist or general surgeon.
- Ultrasound findings of an abnormal relationship between the superior mesenteric artery and superior mesenteric vein may indicate a possible malrotation. They should be investigated with an upper GI series.

Management

- In general, abdominal wall defects and gastrointestinal malformations are amenable to early surgical treatment with good outcomes. Standard surgical intervention is indicated.
- Surgical risks may result from hypoglycemia or intubation difficulties related to macroglossia or to any associated cardiac abnormalities.

Genitourinary

Individuals with BWS frequently have renal anomalies, including nephromegaly, which may be unilateral or bilateral. Reports also note the following findings: renal medullary dysplasia, duplicated collecting system, nephrocalcinosis, medullary sponge kidney, cystic changes, hydronephrosis, nephrolithiasis, and renal diverticulae (Choyke et al. 1998; Borer et al. 1999; Goldman et al. 2002).

Prognosis for renal function is generally good. However, it is important to identify individuals at high risk for loss of renal function from congenital malformation, nephrocalcinosis, medullary sponge kidney, or medullary dysplasia, especially in children who develop Wilms tumor. Active surveillance and appropriate treatment are required to maximize long-term renal function.

Enlargement of the bladder, uterus, phallus, clitoris, ovaries, and testes have been reported. The natural history of other genitourinary organomegalies is not well documented but has not been reported as problematic. Given the natural history of other types of somatic overgrowth in BWS, surveillance without active intervention should be the first course of action.

Evaluation

- Complete abdominal ultrasound is recommended at the time of diagnosis which includes an assessment of the integrity of the renal tract.
- If congenital malformations are identified, further imaging studies may be indicated and referral to a pediatric nephrologist and/or urologist is suggested.
- Evaluation of urinary calcium excretion (calcium-to-creatinine ratio) is indicated, especially if ultrasound findings suggest possible nephrocalcinosis (Goldman et al. 2003). When abnormalities of urinary calcium-to-creatinine ratio are found, referral to a pediatric nephrologist is recommended.
- Reported symptoms of polyuria and polydipsia should prompt evaluation for medullary sponge kidney disease as renal medullary dysplasia may lead to reduction in renal concentrating ability.
- Prospective studies are underway to evaluate whether individuals with BWS over 8 years of age should continue to have annual clinical assessments with renal ultrasound. The primary reason for study is to define the risk of developing renal changes such as nephrocalcinosis or medullary sponge kidney at a later age. The risk for such renal complications is believed to be low.

Management

- Surgical intervention may be indicated for congenital malformations of the renal tract, as for the general population.
- Treatment of nephrocalcinosis consists first of increased fluid intake with possible addition of diuretics.

Neoplasia

Children with BWS and/or hemihyperplasia are predisposed to certain malignancies. The overall risk for tumor development in individuals with BWS is estimated to be 7.5% with a range from 4 to 28% (Tan and Amor 2006; Maas et al. 2016; Mussa et al. 2016; Brzezinski et al. 2017). Most of the increased tumor risk in both BWS and isolated hemihyperplasia occurs in the first 5–8 years of life. Although the oldest reported individual with BWS who developed a Wilms tumor was 10 years 2 months, 96% of all Wilms tumors in a series of 121 individuals with BWS presented by 8 years (Beckwith, 1998b).

The tumors reported in BWS are primarily embryonal, such as Wilms tumor, hepatoblastoma, rhabdomyosarcoma, adrenocortical carcinoma, and neuroblastoma. Also seen are a wide variety of other tumors, both malignant and benign (Sotelo-Avila et al. 1980; Wiedemann, 1983; Maas et al. 2016). Wilms tumor is the most common embryonal tumor detected in children with BWS. (Brioude et al. 2018). Several clinical factors appear to be associated with Wilms tumor development in BWS. These include the presence of hemihyperplasia (Wiedemann 1983), nephromegaly (DeBaun et al. 1998), and nephroblastomatosis, which refers to the diffuse or multifocal presence of nephrogenic rests (Beckwith et al. 1990). Beckwith and colleagues proposed the term *nephrogenic rest* for a focus of abnormally persistent nephrogenic cells that can be induced to form a Wilms tumor. Specifically, hypertrophic nephrogenic rests rather than dormant, microscopic rests, confer a risk for neoplasia.

Data indicate that specific molecular subgroups within BWS carry different tumor risks and susceptibilities for specific tumor profiles. Individuals with 11p15 uniparental disomy and IC1 gain-of-methylation carry the highest tumor risk overall, and they most commonly develop Wilms tumors, whereas those with loss-of-methylation at IC2 have a lower tumor risk and are primarily, though not exclusively, susceptible to non-Wilms tumors (Bliek et al. 2001; Weksberg et al. 2001; Maas et al. 2016; Mussa et al. 2016; Brzezinski et al. 2017). Some clinicians and researchers advocate that tumor surveillance strategies should incorporate these data such that children with BWS and loss-of-methylation at IC2 not be offered tumor screening (Maas et al. 2016; Mussa et al. 2016). However, at this time, we, and others, do not recommend revising existing clinical management recommendations for the following reasons. First, children with loss-of-methylation defects at IC2 have a low but significantly increased risk over the general population to develop Wilms tumor. As noted below, these children also have an increased risk for other embryonal tumors such as hepatoblastoma. Second, early detection of Wilms tumor optimizes treatment approaches including nephron sparing surgery, which is of particular importance given the potential for individuals with BWS to experience long term renal problems (Romao et al. 2012). In one study, the outcome in children with BWS was not improved through regular abdominal ultrasound screening (Craft et al. 1995). However, another report found that a regular screening protocol reduced the percentage of stage III and IV Wilms tumors at diagnosis (Choyke et al. 1999), which would positively impact long term outcomes. Third, even the most robust molecular testing currently available may not detect paternal uniparental disomy in the presence of low-level somatic mosaicism (Brzezinski et al. 2017; Kalish et al. 2017), and these individuals are at increased risk for Wilms tumor and other neoplasms. Fourth, while some concern exists regarding the potential for 'scanxiety' (Malkin et al. 2017), a recent study indicates that parents of children with BWS felt that surveillance decreased their overall worry and was not burdensome for their families (Duffy et al. 2018).

Ultimately these issues should be contextualized for each family in terms of mounting a surveillance program that respects the needs of the child and family.

No specific data are available for long-term survival of children with BWS who have tumors. In general, it is appropriate to counsel those with BWS-associated Wilms tumor, hepatoblastoma, or other tumors, that the prognosis is not known to be different from that for children without BWS. Although one study reported a better prognosis for 10 children with BWS and Wilms tumor than for Wilms tumor alone (Vaughan et al. 1995), in another small cohort there was no difference in prognosis for the two groups (personal experience). Prognosis is generally very good (>80%) for long-term survival. The best prognostic indicators are smaller tumor size, absence of anaplasia, and absence of metastatic spread.

Hepatoblastoma is the second most commonly detected tumor type detected in approximately 2% of children with BWS (Mussa and Ferrero 2015; Brioude et al. 2018). Notably, hepatoblastoma is significantly represented in children with either IC2 loss of methylation or uniparental disomy (Maas et al. 2016). Tumor stage and localization are predictors for prognosis and survival; i.e. those that can be totally resected due to lower stage and localized disease have excellent survival rates (Duffy et al. 2017). Alpha-fetoprotein assay is a sensitive indicator for hepatoblastoma, with elevated levels often predating ultrasound visualization (Clericuzio et al. 2003). Mussa et al. (2019) recently evaluated chronological age for hepatoblastoma development in children with BWS as compared with those without BWS and found that the mean age for tumor development was lower in 76 children with BWS. They found that 100% of cases were diagnosed before 30 months of age. Therefore, they recommend hepatoblastoma screening, including both alpha-fetoprotein assay and abdominal ultrasound be undertaken to 30 months of age to support early detection. This age is younger than the currently recommended 4 years and further prospective data may support the revision of this guideline.

In the future, accurate estimates of tumor risks correlated to molecular subgroups may be achieved with prospective ascertainment of large numbers of children with BWS who have undergone optimized molecular testing using multiple platforms and the potential testing of multiple tissues. These data may support evidence-based molecular stratification of surveillance approaches for BWS; however, such data are not yet validated for clinical implementation.

For children with isolated hemihyperplasia, the risk for tumor development is reported to be approximately 5.9% (Hoyme et al. 1998; Clericuzio and Martin 2009). The anatomic site of the tumor does not always correlate with the laterality of the hemihyperplasia (Hoyme et al. 1998). For hemihyperplasia, the reported range of tumor types overlaps significantly with that of BWS (Hoyme et al. 1998). This again suggests that some individuals with hemihyperplasia may represent a *forme fruste* of BWS and that the majority of tumors arise from genetic events that overlap the underlying mechanisms responsible for BWS, including predisposition to neoplasia. However, some tumors associated with hemihyperplasia, such as leiomyosarcoma, are more likely to be related to disorders other than BWS.

Evaluation

- Children with either suspected or diagnosed BWS or isolated hemihyperplasia should be followed on a quarterly (every 3 months) basis with abdominal ultrasound until the age of 8 years (Beckwith, 1998b). It is likely that these recommendations will be refined in the near future; e.g. reduce the term of full abdominal ultrasound to approximately 3 years of age followed by targeted renal ultrasound to age 8 years. Common ultrasound findings include hepatomegaly, nephromegaly, and splenomegaly. As ultrasound technology continues to improve, findings such as "bulky" pancreas and/or mesenteric nodes may be detected more commonly. Generally, such findings are not associated with neoplasm, but they should be carefully followed for several intervals of ultrasound screening.
- Masses detected in the liver or kidney must be distinguished from lesions such as hemangiomata. Depending on the type of lesion, appropriate imaging studies such as CT or MRI can be used for better definition.
- A baseline MRI study should be considered for individuals entering a tumor surveillance program (Clericuzio et al. 1992; Beckwith, 1998b). As imaging modalities continue to improve this may become more or less relevant in detecting early neoplastic changes.
- Consultation with an oncologist and/or relevant subspecialist may be useful to evaluate specific imaging findings.
- If hypertrophic nephrogenic rests are detected or suspected on imaging, careful follow-up should be undertaken. In the future, better MRI technology should facilitate visualization of macroscopic (hypertrophic) nephrogenic rests (Gylys-Morin et al. 1993), possibly identifying individuals with BWS who are most likely to develop Wilms tumor.
- In some centers, parents are advised to perform abdominal palpation for tumor surveillance. Some concern has been raised that this might place undue pressure on the parent–child relationship and lead to feelings of guilt in the event that a mass was not detected via palpation. However, some parents may feel empowered by becoming more actively involved in their child's medical management.
- Screening for neuroblastoma with urinary homovanillic acid (HVA) and vanilmandilic acid (Chitayat et al. 1990b), as well as chest X-ray, has been suggested, but such screening is generally not incorporated into baseline tumor screening protocols because of the relatively low risk for this tumor (Tan and Amor 2006). However, in an unwell child, or if enlarged mesenteric lymph nodes are detected on abdominal ultrasound, such tests are warranted.
- α-Fetoprotein (AFP) should be measured in children with BWS or isolated hemihyperplasia every three months to 4 years of age (Tan and Amor, 2006; Kalish et al. 2017). As noted above, recent data for children with BWS may support reducing the duration of AFP screening to approximately 3 years (Mussa et al. 2019). Serum AFP levels tend to be somewhat higher in children with BWS in the first year of life (Everman et al. 2000), but the most important indicator for management is whether the serum AFP is falling or rising. When serum AFP is elevated, a high index of suspicion must be maintained. Follow-up at monthly intervals should be scheduled with repeat serum AFP (to determine whether it is falling), liver function studies, and repeat imaging, including chest X-ray. In any case of a rising serum AFP, an exhaustive search for an underlying tumor, including germ cell tumors, is indicated. Consultation with an oncologist may also be useful.
- Clinically unaffected monozygotic co-twins of affected individuals should also be followed with tumor surveillance because of the possibility of somatic mosaicism or seeding of BWS-positive cells due to vascular anastomoses in utero.

Management

- For all tumors detected, treatment follows standard oncology protocols. As part of this treatment, surveillance for other tumors should be continued, e.g. a child being treated for hepatoblastoma at age 2 years should continue surveillance for Wilms tumor to age 8 years. Children with BWS appear to respond well to chemotherapeutic agents, and the nephrogenic rests in the kidneys of such children have shown a marked reduction in size in response to chemotherapy (Regalado et al. 1997). Treatment should be aimed at preserving as much functional renal tissue as possible.

Pregnancy and Perinatal Period

When a fetus has BWS, there is a high incidence (about 50%) of premature birth (Weng et al. 1995a), polyhydramnios (about 50%), and fetal macrosomia (about 90%) (Elliott et al. 1994). Other notable features are enlarged placenta with many averaging almost twice the normal weight for gestational age (Weng et al. 1995a), placental mesenchymal dysplasia (Wilson et al. 2008), and long umbilical cord.

Prognosis depends on the severity of the presenting perinatal problem. There remains an increased perinatal mortality rate associated with BWS (Pettenati et al. 1986; Weng et al. 1995a; Smith et al. 2007).

Evaluation

- Fetal evaluation in pregnancies with suspected fetal BWS (e.g. detection of fetal omphalocele) should include serial ultrasounds and biophysical profiles. Molecular testing should be considered, as it may confirm a suspected diagnosis (Wilkins-Haug et al. 2009).
- Given the increased risk of fetal macrosomia and maternal preeclampsia/eclampsia, women suspected of carrying a fetus with BWS should be closely monitored. Perinatal management should be undertaken in a high-risk unit.
- Some families may wish to undertake prenatal testing for pregnancy and delivery management, and others may consider whether or not to continue with the pregnancy. If molecular testing is pursued for pregnancies undertaken subsequent to a child with BWS or for those pregnancies identified to be at increased risk based on fetal findings, consideration should be given to the testing approach. For methylation analysis, DNA extracted from amniotic fluid is currently more reliable than DNA extracted from chorionic villi, though false negative results have been reported (Eggermann et al. 2015).
- If a genomic lesion has been identified in a previously affected child or an affected parent (e.g., *CDKN1C* mutation or a microdeletion), the possibility of prenatal diagnosis by chorionic villus sampling (CVS) or amniocentesis may be considered in a subsequent pregnancy.
- If a non-invasive screening approach is preferred, this could include maternal serum α-fetoprotein at 16 weeks gestation and monitoring with high-resolution ultrasound at 19–20 weeks and 32 weeks gestation to identify an abdominal wall defect, to assess growth parameters (macrosomia is usually not detectable until late in the second trimester), and to detect organomegaly, renal anomalies, cleft palate, cardiac abnormality, and macroglossia. In addition, there has been one report of an early ultrasound, between 10 and 14 weeks gestation, showing increased nuchal translucency and omphalocele in a fetus later found to have BWS (Souka et al. 1998).
- The same surveillance recommendations noted above are made for pregnancies where one parent has BWS.
- Even in the absence of obvious clinical findings on prenatal investigation, the newborn in a family identified to be at increased risk based on family history or prenatal findings should be considered at-risk and monitored for neonatal hypoglycemia as outlined above.

Management

- When BWS is suspected in pregnancy, delivery planning should anticipate possible perinatal complications, such as polyhydramnios, pre-eclampsia, prematurity, macrocephaly, large birth weight, macroglossia, and hypoglycemia. Less frequent complications include hypocalcemia or polycythemia (19.5%). Management of each of the complications is not different from that in the general population.
- Management may need to include the delivery of an infant with an omphalocele. Thus, it is preferable to plan the delivery in a medical center equipped to handle such issues.

RESOURCES

Beckwith–Wiedemann Children's Foundation International
Website: https://www.beckwithwiedemann.org

Beckwith–Wiedemann Support Group
Website: https://www.bwssupport.org

ACKNOWLEDGEMENT

We would like to formally acknowledge Dr. Bruce Beckwith's seminal contributions in defining both the phenotype and the pathophysiology of BWS as well as his longstanding commitment to individuals with BWS and their families. Dr. Beckwith contributed enormously to the evidence based strategies underpinning management guidelines and his indelible imprint continues to benefit patients, clinicians and scientists.

REFERENCES

Baskin B, Choufani S, Chen YA, Shuman C, Parkinson N, Lemyre E, Micheil Innes A, Stavropoulos DJ, Ray PN, Weksberg R (2014a) High frequency of copy number variations (CNVs) in the chromosome 11p15 region in patients with Beckwith-Wiedemann syndrome. *Hum Genet* 133:321–30.

Baskin B, Choufani S, Chen YA, Shuman C, Parkinson N, Lemyre E, Micheil Innes A, Stavropoulos DJ, Ray PN, Weksberg R. (2014b) High frequency of copy number variations (CNVs) in the chromosome 11p15 region in patients with Beckwith-Wiedemann syndrome. *Hum Genet* 133:321–30.

Beckwith JB (1963) Extreme cytomegaly of the adrenal fetal cortex, omphalocele, hyperplasia of kidneys and pancreas, and Leydig-cell hyperplasia: Another syndrome? *Abstract, Western Society for Pediatric Research, Los Angeles, November* 11, 1963.

Beckwith JB, Kiviat NB, Bonadio JF (1990) Nephrogenic rests, nephroblastomatosis, and the pathogenesis of Wilms' tumor. *Pediatr Pathol* 10:1–36.

Beckwith JB (1998a) Vignettes from the history of overgrowth and related syndromes. *Am J Med Genet* 79:238–248.

Beckwith JB (1998b) Nephrogenic rests and the pathogenesis of Wilms tumor: Developmental and clinical considerations. *Am J Med Genet* 79:268–273.

Bliek J, Maas SM, Ruijter JM, Hennekam RCM, Alders M, Westerveld A, Mannens MMAM (2001) Increased tumour risks for BWS patients correlates with aberrant *H19* and not *KCNQ1OT1* hypomethylation in familial cases of BWS. *Hum Mol Genet* 10:467–476.

Bliek J, Verde G, Callaway J, Maas SM, De Crescenzo A, Sparago A, Cerrato F, Russo S, Ferraiuolo S, Rinaldi MM, Fischetto R, Lalatta F, Giordano L, Ferrari P, Cubellis MV, Larizza L, Temple IK, Mannens MM, Mackay DJ, Riccio A (2008) Hypomethylation at multiple maternally methylated imprinted regions including PLAGL1 and GNAS loci in Beckwith-Wiedemann syndrome. *Eur J Hum Genet* 17:611–619.

Borer JG, Kaefer M, Barnewolt CE, Elias ER, Hobbs N (1999) Renal findings on radiological followup of patients with Beckwith-Wiedemann syndrome. *J Urol* 161:235–239.

Boluyt N, van Kempen A, Offringa M (2006) Neurodevelopment after neonatal hypoglycemia: A systematic review and design of an optimal future study. *Pediatrics* 117:2231–2243.

Brioude F, Lacoste A, Netchine I, Vazquez MP, Auber F, Audry G, Gauthier-Villars M, Brugieres L, Gicquel C, Le Bouc Y, Rossignol S (2013) Beckwith-Wiedemann syndrome: growth pattern and tumor risk according to molecular mechanism, and guidelines for tumor surveillance. *Horm Res Paediatr* 80:457–65.

Brioude F, Netchine I, Praz F, Le Jule M, Calmel C, Lacombe D, Edery P, Catala M, Odent S, Isidor B, Lyonnet S, Sigaudy S, Leheup B, Audebert-Bellanger S, Burglen L, Giuliano F, Alessandri JL, Cormier-Daire V, Laffargue F, Blesson S, Coupier I, Lespinasse J, Blanchet P, Boute O, Baumann C, Polak M, Doray B, Verloes A, Viot G, Le Bouc Y, Rossignol S (2015) Mutations of the Imprinted CDKN1C Gene as a Cause of the Overgrowth Beckwith-Wiedemann Syndrome: Clinical Spectrum and Functional Characterization. *Hum Mutat* 36:894–902

Brioude F, Kalish JM, Mussa A, Foster AC, Bliek J, Ferrero GB, Boonen SE, Cole T, Baker R, Bertoletti M, Cocchi G, Coze C, De Pellegrin M, Hussain K, Ibrahim A, Kilby MD, Krajewska-Walasek M, Kratz CP, Ladusans EJ, Lapunzina P, Le Bouc Y, Maas SM, Macdonald F, Õunap K, Peruzzi L, Rossignol S, Russo S, Shipster C, Skórka A, Tatton-Brown K, Tenorio J, Tortora C, Grønskov K, Netchine I, Hennekam RC, Prawitt D, Tümer Z, Eggermann T, Mackay DJG, Riccio A, Maher ER. (2018) Expert consensus document: Clinical and molecular diagnosis, screening and management of Beckwith-Wiedemann syndrome: an international consensus statement. *Nat Rev Endocrinol* 14:229–249.

Brzezinski J, Shuman C, Choufani S, Ray P, Stavropoulos DJ, Basran R, Steele L, Parkinson N, Grant R, Thorner P, Lorenzo A, Weksberg R (2017) Wilms tumour in Beckwith-Wiedemann Syndrome and loss of methylation at imprinting centre 2: revisiting tumour surveillance guidelines. *Eur J Hum Genet* 25:1031–1039.

Burns CM, Rutherford MA, Boardman JP, Cowan FM (2008) Patterns of cerebral injury and neurodevelopmental outcomes after symptomatic neonatal hypoglycemia. *Pediatrics* 122:65–74.

Chitayat D, Rothchild A, Ling E, Friedman JM, Couch RM, Yong SL, Baldwin VJ, Hall JG (1990a) Apparent postnatal onset of some manifestations of the Wiedemann-Beckwith syndrome. *Am J Med Genet* 36:434–439.

Chitayat D, Friedman JM, Dimmick JE (1990b) Neuroblastoma in a child with Wiedemann-Beckwith syndrome. *Am J Med Genet* 35:433–436.

Choufani S, Shuman C, Weksberg R. 2010. Beckwith–Wiedemann syndrome. *Am J Med Genet Part C* 153C:343–354.

Choyke PL, Siegel MJ, Oz O, Sotelo-Avila C, DeBaun MR (1998) Nonmalignant renal disease in pediatric patients with Beckwith-Wiedemann syndrome. *AJR Am J Roentgenol* 171:733–737.

Choyke PL, Siegel MJ, Craft AW, Green DM, DeBaun MR (1999) Screening for Wilms tumor in children with Beckwith-Wiedemann syndrome or idiopathic hemihypertrophy. *Med Pediatr Oncol* 32:196–200.

Clericuzio CL, D'Angio GJ, Duncan M, Green DM, Knudson AG Jr (1992) Summary and Recommendations of the Workshop held at the First International Conference on Molecular and Clinical Genetics of Childhood Renal Tumors, Albuquerque, New Mexico May 14–16.

Clericuzio CL, Chen E, McNeil DE, O'Connor T, Zackai EH, Medne L, Tomlinson G, DeBaun M. (2003) Serum alpha-fetoprotein screening for hepatoblastoma in children with Beckwith-Wiedemann syndrome or isolated hemihyperplasia. *J Pediatr* 143:270–272

Clericuzio CL and Martin RA (2009) Diagnostic criteria and tumor screening for individuals with isolated hemihyperplasia. *Genet Med* 11:220–2

Cohen MM Jr, (1989) In: Advances in Human Genetics Harris H, Hirschhorn K, ed. New York: Plenum, Vol. 18, pp. 181–303; Addendum, pp. 373–376.

Cohen AS, Tuysuz B, Shen Y, Bhalla SK, Jones SJ, Gibson WT. (2015) A novel mutation in EED associated with overgrowth. *J Hum Genet* 60:339–42.

Cortessis VK, Azadian M, Buxbaum J, Sanogo F, Song AY, Sriprasert I, Wei PC, Yu J, Chung K, Siegmund KD, (2018) Comprehensive meta-analysis reveals association between multiple imprinting disorders and conception by assisted reproductive technology. *J Assist Reprod Genet* 35: 943–952

Craft AW, Parker L, Stiller C, Cole M (1995) Screening for Wilms' tumour in patients with aniridia, Beckwith syndrome, or hemihypertrophy. *Med Pediatr Oncol* 24:231–234.

de Lonlay P, Fournet JC, Rahier J, Gross-Morand MS, Poggi-Travert F, Foussier V, Bonnefont JP, Brusset MC, Brunelle F, Robert JJ, Nihoul-Fekete C, Saudubray JM, Junien C (1997) Somatic deletion of the imprinted 11p15 region in sporadic persistent hyperinsulinemic hypoglycemia of infancy is specific of focal adenomatous hyperplasia and endorses partial pancreatectomy. *J Clin Invest* 100:802–807.

Duffy KA, Deardorff MA, Kalish JM (2016) The Utility of Alpha Feto-protein Screening in Beckwith-Wiedemann Syndrome (2017) *Am J Med Genet A* 173:581–58

Duffy KA, Grand KL, Zelley K, Kalish JM (2018) Tumour Screening in Beckwith-Wiedemann Syndrome: Parental Perspectives. *J Genet Couns* 27:844–853

Eggermann K, Bliek J, Brioude F, Algar E, Buiting K, Russo S, Tümer Z, Monk D, Moore G, Antoniadi T, Macdonald F, Netchine I, Lombardi P, Soellner L, Begemann M, Prawitt D, Maher ER, Mannens M, Riccio A, Weksberg R, Lapunzina P, Grønskov K, Mackay DJ, Eggermann T (2016) EMQN best practice guidelines for the molecular genetic testing and reporting of chromosome 11p15 imprinting disorders: Silver-Russell and Beckwith-Wiedemann syndrome. *Eur J Hum Genet* 24:1377–87.

Eggermann T, Soellner L, Buiting K, Kotzot D. (2015) Mosaicism and uniparental disomy in prenatal diagnosis. *Trends Mol Med* 21:77–87.

Eggermann T, Heilsberg AK, Bens S, Siebert R, Beygo J, Buiting K, Begemann M, Soellner L (2014) Additional molecular findings in 11p15-associated imprinting disorders: an urgent need for multi-locus testing. *J Mol Med (Berl)* 92:769–77.

Elliott M, Maher ER (1994) Beckwith-Wiedemann syndrome. *J Med Genet* 31:560–564.

Elliott M, Bayly R, Cole T, Temple IK, Maher ER (1994) Clinical features and natural history of Beckwith-Wiedemann syndrome: Presentation of 74 new cases. *Clin Genet* 46:168–174.

Engstrom W, Lindham S, Schofield P (1998) Wiedemann-Beckwith syndrome. *Eur J Pediatr* 147:450–457.

Everman DB, Shuman C, Dzolgonovski B, O'Riordan MA, Weksberg R, Robin NH (2000) Serum alpha-fetoprotein levels in Beckwith-Wiedemann Syndrome. *J Pediatr* 137:123–127.

Gardiner K1, Chitayat D, Choufani S, Shuman C, Blaser S, Terespolsky D, Farrell S, Reiss R, Wodak S, Pu S, Ray PN, Baskin B, Weksberg R (2012) Brain abnormalities in patients with Beckwith-Wiedemann syndrome. *Am J Med Genet* 158A:1388–94.

Gibson WT, Hood RL, Zhan SH, Bulman DE, Fejes AP, Moore R, Mungall AJ, Eydoux P, Babul-Hirji R, An J, Marra MA; FORGE Canada Consortium, Chitayat D, Boycott KM, Weaver DD, Jones SJ (2012) Mutations in EZH2 cause Weaver syndrome. *Am J Hum Genet* 90:110–118.

Goldman M, Smith A, Shuman C, Caluseriu O, Wei C, Steele L, Ray P, Sadowski P, Squire J, Weksberg R, Rosenblum ND (2002) Renal abnormalities in Beckwith-Wiedemann syndrome are associated with 11p15.5 uniparental disomy. *J Am Soc Nephrol* 8:2077–2084.

Goldman M, Shuman C, Weksberg R, Rosenblum N (2003) Hypercalciuria in BWS. *J Peds* 142:205–207.

Gylys-Morin V, Hoffer FA, Kozakewich H, Shamberger RC (1993) Wilms tumor and nephroblastomatosis: Imaging characteristics at gadolinium-enhanced MR imaging. *Radiology* 188:517–521.

Halamek LP, Stevenson DK (1998) Neonatal hypoglycemia, part II: Pathophysiology and therapy. *Clin Pediatr (Phila)* 37:11–16.

Hatada I, Nabetani A, Morisaki H, Xin Z, Ohishi S, Tonoki H, Niikawa N, Inoue M, Komoto Y, Okada A, Steichen E, Ohashi H, Fukushima Y, Nakayama M, Mukai T (1997) New p57KIP2 mutations in Beckwith-Wiedemann syndrome. *Hum Genet* 100:681–683.

Heggie AA1, Vujcich NJ, Portnof JE, Morgan AT (2013) Tongue reduction for macroglossia in Beckwith Wiedemann syndrome: review and application of new technique. *Int J Oral Maxillofac Surg* 42:185–91.

Heide S, Chantot-Bastaraud S, Keren B, Harbison MD, Azzi S, Rossignol S, Michot C, Lackmy-Port Lys M, Demeer B, Heinrichs C, Newfield RS, Sarda P, Van Maldergem L, Trifard V, Giabicani E, Siffroi JP, Le Bouc Y, Netchine I, Brioude (2017) Chromosomal rearrangements in the 11p15 imprinted region: 17 new 11p15.5 duplications with associated phenotypes and putative functional consequences. *J Med Genet* 104919: 1–9

Hoyme H, Seaver LH, Jones KL, Procopio F, Crooks W, Feingold M (1998) Isolated hemihyperplasia (hemihypertrophy): Report of a prospective multicenter study of the incidence of neoplasia and review. *Am J Med Genet* 79:274–278.

Inbar-Feigenberg M, Choufani S, Cytrynbaum C, Chen YA, Steele L, Shuman C, Ray PN, Weksberg R (2013) Mosaicism for genome-wide paternal uniparental disomy with features of multiple imprinting disorders: diagnostic and management issues. *Am J Med Genet* 161A:13–20.

Imagawa E, Albuquerque EVA, Isidor B, Mitsuhashi S, Mizuguchi T, Miyatake S, Takata A, Miyake N, Boguszewski MCS, Boguszewski CL, Lerario AM, Funari MA, Jorge AAL, Matsumoto N (2018) Novel SUZ12 mutations in Weaver-like syndrome. *Clin Genet* 94:461–466.

Johnson JP, Golabi M, Norton ME, Rosenblatt RM, Feldman GM, Yang SP, Hall BD, Fries MH (1998) Costello syndrome: Phenotype, natural history, differential diagnosis, and possible cause. *J Pediatr* 133:441–448.

Kalish JM, Conlin LK, Bhatti TR, Dubbs HA, Harris MC, Izumi K, Mostoufi-Moab S, Mulchandani S, Saitta S, States LJ, Swarr DT, Wilkens AB, Zackai EH, Zelley K, Bartolomei MS, Nichols KE, Palladino AA, Spinner NB, Deardorff MA (2013) Clinical features of three girls with mosaic genome-wide paternal uniparental isodisomy. *Am J Med Genet* 161A:1929–39.

Kalish JM, Biesecker LG, Brioude F, Deardorff MA, Di Cesare-Merlone A, Druley T, Ferrero GB, Lapunzina P, Larizza L, Maas S, Macchiaiolo M, Maher ER, Maitz S, Martinez-Agosto JA13, Mussa A, Robinson P, Russo S, Selicorni A, Hennekam RC (2017a) Nomenclature and definition in asymmetric regional body overgrowth. *Am J Med Genet* 173A:1735–1738.

Kalish JM, Doros L, Helman LJ, Hennekam RC, Kuiper RP, Maas SM, Maher ER, Nichols KE, Plon SE, Porter CC, Rednam S, Schultz KAP, States LJ, Tomlinson GE, Zelley K, Druley TE (2017) Surveillance Recommendations for Children with Overgrowth Syndromes and Predisposition to Wilms Tumors and Hepatoblastoma. *Clin Cancer Res* 23:e115–e122.

Lam WW, Hatada I, Ohishi S, Mukai T, Joyce JA, Cole TR, Donnai D, Reik W, Schofield PN, Maher ER (1999) Analysis of germline *CDKN1C (p57KIP2)* mutations in familial and sporadic Beckwith-Wiedemann syndrome (BWS) provides a novel genotype-phenotype correlation. *J Med Genet* 36:518–523.

Lee MP, DeBaun MR, Mitsuya K, Galonek HL, Brandenburg S, Oshimura M, Feinberg AP (1999) Loss of imprinting of a paternally expressed transcript, with antisense orientation to *KVLQT1*, occurs frequently in Beckwith-Wiedemann syndrome and is independent of insulin-like growth factor II imprinting. *Proc Natl Acad Sci USA* 96:5203–5208.

Leonard NJ, Bernier FP, Rudd N, Machin GA, Bamforth F, Bamforth S, Grundy P, Johnson C (1996) Two pairs of male monozygotic twins discordant for Wiedemann-Beckwith syndrome. *Am J Med Genet* 61:253–257.

Li M, Squire JA, Weksberg R (1998) Molecular genetics of Wiedemann-Beckwith syndrome. *Am J Med Genet* 79:253–259.

Li M, Squire J, Shuman C, Atkin J, Pauli R, Amith A, Chitayat D, Weksberg R (2001) Imprinting status of 11p15 genes in Beckwith-Wiedemann syndrome patients with *CDKN1C* mutations. *Genomics* 74:370–376.

Lim D, Bowdin SC, Tee L, Kirby GA, Blair E, Fryer A, Lam W, Oley C, Cole T, Brueton LA, Reik W, Macdonald F, Maher ER (2008) Clinical and molecular genetic features of Beckwith-Wiedemann syndrome associated with assisted reproductive technologies. *Hum Reprod* 24:741–747.

Lin AE, Neri G, Hughes-Benzie R, Weksberg R (1999) Cardiac anomalies in the Simpson-Golabi-Behmel syndrome. *Am J Med Genet* 83:378–381.

Ludwig M, Katalinic A, Gross S, Sutcliffe A, Varon R, Horsthemke B (2005) Increased prevalence of imprinting defects in patients with Angelman syndrome subfertile couples. *J Med Genet* 42:289–291.

Kent L, Bowdin S, Kirby GA, Cooper WN, Maher ER (2008) Beckwith Wiedemann syndrome: A behavioral phenotype-genotype study. *Am J Med Genet* 147B:1295–1297.

Keren B, Chantot-Bastaraud S, Brioude F, Mach C, Fonteneau E, Azzi S, Depienne C, Brice A, Netchine I, Le Bouc Y, Siffroi JP, Rossignol S (2013) SNP arrays in Beckwith-Wiedemann syndrome: an improved diagnostic strategy. *Eur J Med Genet* 56:546–50.

Maas SM, Vansenne F, Kadouch DJM, Ibrahim A, Bliek J, Hopman S, Mannens MM, Merks JHM, Maher ER, Hennekam RC (2016) Phenotype, cancer risk, and surveillance in Beckwith–Wiedemann syndrome depending on molecular genetic subgroups. *Am J Med Genet* 9999A:1–13.

Mackay DJ, Callaway JL, Marks SM, White HE, Acerini CL, Boonen SE, Dayanikli P, Firth HV, Goodship JA, Haemers AP, Hahnemann JM, Kordonouri O, Masoud AF, Oestergaard E, Storr J, Ellard S, Hattersley AT, Robinson DO, Temple IK (2008) Hypomethylation of multiple imprinted loci in individuals with transient neonatal diabetes is associated with mutations in ZFP57. *Nat Genet* 40:949–951.

Maher ER, Brueton LA, Bowdin SC, Luharia A, Cooper W, Cole TR, Macdonald F, Sampson JR, Barrett C, Reik W, Hawkins MM (2003) Beck-Wiedemann syndrome and assisted reproduction technology (ART). *J Med Genet* 40:62–64.

Malkin D, Nichols KE, Schiffman JD, Plon SE, Brodeur GM (2017) The Future of Surveillance in the Context of Cancer Predisposition: Through the Murky Looking Glass. *Clin Cancer Res* 23:e133–e137.

Martinez y Martinez R, Ocampo-Campos R, Perez-Arroyo R, Corona-Rivera E, Cantu JM (1985) The Wiedemann-Beckwith syndrome in four sibs including one with associated congenital hypothyroidism. *Eur J Pediatr* 143:233–235.

Meyer E, Lim D, Pasha S, Tee LJ, Rahman F, Yates JRW, Woods CG, Reik W, Maher ER (2009) Germline mutation in NLRP2 (NALP2) in a familial imprinting disorder (Beckwith-Wiedemann syndrome). *PLOS Genet* 5:1–5.

Monk D, Morales J, den Dunnen JT, Russo S, Court F, Prawitt D, Eggermann T, Beygo J, Buiting K, Tümer Z (2016) Recommendations for a nomenclature system for reporting methylation aberrations in imprinted domains. *Epigenetics* 13:117–121.

Morris MR, Astuti D, Maher ER (2013) Perlman syndrome: overgrowth, Wilms tumor predisposition and DIS3L2. *Am J Med Genet* 163C:106–13

Mussa A, Russo S, De Crescenzo A, Chiesa N, Molinatto C, Selicorni A, Richiardi L, Larizza L, Silengo MC, Riccio A, Ferrero GB (2013) Prevalence of Beckwith-Wiedemann syndrome in North West of Italy. *Am J Med Genet* 161A:2481–6

Mussa A, Ferrero GB (2015) Screening Hepatoblastoma in Beckwith-Wiedemann Syndrome: A Complex Issue. *J Pediatr Hematolo Oncol* 37:627

Mussa A, Russo S, De Crescenzo A, Freschi A, Calzari L, Maitz S, Macchiaiolo M, Molinatto C, Baldassarre G, Mariani M, Tarani L, Bedeschi MF, Milani D, Melis D, Bartuli A, Cubellis MV, Selicorni A, Cirillo Silengo M, Larizza L, Riccio A, Ferrero GB (2016) (Epi)genotype-phenotype correlations in Beckwith-Wiedemann syndrome. *Eur J Hum Genet* 24:183–90.

Mussa A, Duffy KA, Carli D, Ferrero GB, Kalish JM. (2019), Defining an optimal time window to screen for hepatoblastoma in children with Beckwith-Wiedemann syndrome. *Pediatr Blood Cancer* 66:1–4

Niemitz EL1, Feinberg AP, Brandenburg SA, Grundy PE, DeBaun MR (2005) Children with idiopathic hemihypertrophy and Beckwith-Wiedemann syndrome have different constitutional epigenotypes associated with Wilms tumor. *Am J Hum Genet* 77:887–91.

Ogawa O, Eccles MR, Szeto J, McNoe LA, Yun K, Maw MA, Smith PJ, Reeve AE (1993) Relaxation of insulin-like growth factor II gene imprinting implicated in Wilms' tumour. *Nature* 362:749–751.

O'Keefe D, Dao D, Zhao L, Sanderson R, Warburton D, Weiss L, Anyane-Yeboa K, Tycko B (1997) Coding mutations in *p57KIP2* are present in some cases of Beckwith-Wiedemann syndrome but are rare or absent in Wilms tumors. *Am J Hum Genet* 61:295–303.

Parker DA, Skalko RJ (1969) Congenital asymmetry: Report of 10 cases with associated developmental abnormalities. *Pediatrics* 44:584–589.

Pettenati MJ, Haines JL, Higgins RR, Wappner RS, Palmer CG, Weaver DD (1986) Wiedemann-Beckwith syndrome: Presentation of clinical and cytogenetic data on 22 new cases and review of the literature. *Hum Genet* 74:143–154.

Raveh E, Matouk IJ, Gilon M, Hochberg A (2015) The H19 long non-coding RNA in cancer initiation, progression and metastasis – a proposed unifying theory. *Mol Cancer* 14: 184–198

Rainier S, Johnson LA, Dobry CJ, Ping AJ, Grundy PE, Feinberg AP (1993) Relaxation of imprinted genes in human cancer. *Nature* 362:747–749.

Regalado JJ, Rodriguez MM, Toledano S (1997) Bilaterally multicentric synchronous Wilms' tumor: Successful conservative treatment despite persistence of nephrogenic rests. *Med Pediatr Oncol* 28:420–423.

Ringrose RE, Jabbour JT, Keele DK, (1965) *Hemihypertrophy*. *Pediatrics* 36:434–448.

Romão RL1, Pippi Salle JL, Shuman C, Weksberg R, Figueroa V, Weber B, Bägli DJ, Farhat WA, Grant R, Gerstle JT, Lorenzo AJ (2012) Nephron sparing surgery for unilateral Wilms tumor in children with predisposing syndromes: single center experience over 10 years. *J Urol* 188:1493–8.

Rossignol S, Steunou V, Chalas C, Kerjean A, Rigolet M, Viegas-Pequignot E, Jouannet P, Le Bouc Y, Gicquel C (2006) The epigenetic imprinting defect of patients with Beckwith-Wiedemann syndrome born after assisted reproductive technology is not restricted to the 11p15 region. *J Med Genet* 43:902–907.

Russo S, Calzari L, Mussa A, Mainini E, Cassina M, Di Candia S, Clementi M, Guzzetti S, Tabano S, Miozzo M, Sirchia S, Finelli P, Prontera P, Maitz S, Sorge G, Calcagno A, Maghnie M, Divizia MT, Melis D, Manfredini E, Ferrero GB, Pecile V, Larizza L (2016) A multi-method approach to the molecular diagnosis of overt and borderline 11p15.5 defects underlying Silver-Russell and Beckwith-Wiedemann syndromes. *Clin Epigenetics* 8:23.

Scott RH, Douglas J, Baskcomb L, Huxter N, Barker K, Hanks S, Craft A, Gerrard M, Kohler JA, Levitt GA, Picton S, Pizer B, Ronghe MD, Williams D; Factors Associated with Childhood Tumours (FACT) Collaboration, Cook JA, Pujol P, Maher ER, Birch JM, Stiller CA, Pritchard-Jones K, Rahman N (2008a) Constitutional 11p15 abnormalities, including heritable imprinting center mutations, cause nonsyndromic Wilms tumor. *Nat Genet* 40:1329–1334.

Scott RH, Douglas J, Baskcomb L, Nygren AO, Birch JM, Cole TR, Cormier-Daire V, Eastwood DM, Garcia-Minaur S, Lapunzina P, Tatton-Brow K, Bliek J, Maher ER, Rahman N (2008b) Methylation-specific multiplex ligation-dependent probe amplification (MS-MLPA) robustly detects and distinguished 11p15 abnormalities associated with overgrowth and growth retardation. *J Med Genet* 45:106–113.

Shuman C, Smith AC, Steele L, Ray PN, Clericuzio C, Zackai E, Parisi MA, Meadows AT, Kelly T, Tichauer D, Squire JA, Sadowski P, Weksberg R (2006) Constitutional UPD for chromosome 11p15 in individuals with isolated hemihyperplasia is associated with high tumor risk and occurs following assisted reproductive technologies. *Am J Med Genet* 140A:1497–1503.

Slatter RE, Elliott M, Welham K, Carrera M, Schofield PN, Barton DE, Maher ER (1994) Mosaic uniparental disomy in Beckwith-Wiedemann syndrome. *J Med Genet* 31:749–753.

Slavotinek A, Gaunt L, Donnai D (1997) Paternally inherited duplications of 11p15.5 and Beckwith-Wiedemann syndrome. *J Med Genet* 34:819–826.

Smilinich NJ, Day CD, Fitzpatrick GV, Caldwell GM, Lossie AC, Cooper PR, Smallwood AC, Joyce JA, Schofield PN, Reik W, Nicholls RD, Weksberg R, Driscoll DJ, Maher ER, Shows TB, Higgins MJ (1999) A maternally methylated CpG island in *KvLQT1* is associated with an antisense paternal transcript and loss of imprinting in Beckwith-Wiedemann syndrome. *Proc Natl Acad Sci USA* 96:8064–8069.

Smith AC, Rubin T, Shuman C, Estabrooks L, Aylsworth AS, McDonald MT, Steele L, Ray PN, Weksberg R (2006) New chromosome 11p15 epigenotypes identified in male monozygotic twins with Beckwith-Wiedemann syndrome. *Cytogenet Genome Res* 113:313–317.

Smith AC, Shuman C, Chitayat D, Steele L, Ray PN, Bourgeois J, Weksberg R (2007) Severe presentation of Beckwith-Wiedemann syndrome associated with high levels of constitutional paternal uniparental disomy for chromosome 11p15. *Am J Med Genet* 143A:3010–3015.

Sotelo-Avila C, Gonzalez-Crussi F, Fowler JW (1980) Complete and incomplete forms of Beckwith-Wiedemann syndrome: Their oncogenic potential. *J Pediatr* 96:47–50.

Souka AP, Snijders RJ, Novakov A, Soares W, Nicolaides KH (1998) Defects and syndromes in chromosomally normal fetuses with increased nuchal translucency thickness at 10-14 weeks of gestation. *Ultrasound Obstet Gynecol* 11:391–400.

Tan TY, Amor DJ (2006). Tumour surveillance in Beckwith-Wiedemann syndrome and hemihyperplasia: a critical review of the evidence and suggested guidelines for local practice. *J Paed Child Health.* 42(9):486–490.

Tomlinson JK, Morse SA, Bernard SP, Greensmith AL, Meara JG (2007) Long-term outcomes of surgical tongue reduction in Beckwith-Wiedemann syndrome. *Plast Reconstr Surg* 119:992–1002.

Vals MA, Kahre T, Mee P, Muru K, Kallas E, Žilina O, Tillmann V, Õunap K (2015) Familial 1.3-Mb 11p15.5p15.4 Duplication in Three Generations Causing Silver-Russell and Beckwith-Wiedemann Beckwith–Wiedemann Syndromes. *Mol Syndromol* 6:147–51.

Van Eeghen AM, Van Gelderen I, Rhennekam RC (1999) Costello syndrome: Report and review. *Am J Med Genet* 82:187–193.

Vaughan WG, Sanders DW, Grosfeld JL, Plumley DA, Rescoria FJ, Scherer LR 3rd, West KW, Breitfeld PP (1995) Favorable outcome in children with Beckwith-Wiedemann syndrome and intraabdominal malignant tumors. *J Pediatr Surg* 30:1042–1044.

Viljoen D, Pearn J, Beighton P (1984) Manifestations and natural history of idiopathic hemihypertrophy: A review of eleven cases. *Clin Genet* 26:81–86.

Waziri M, Patil SR, Hanson JW, Bartley JA (1983) Abnormality of chromosome 11 in patients with features of Beckwith-Wiedemann syndrome. *J Pediatr* 102:873–876.

Weksberg R, Shen DR, Fei YL, Song QL, Squire J (1993) Disruption of insulin-like growth factor 2 imprinting in Beckwith-Wiedemann syndrome. *Nat Genet* 5:143–150.

Weksberg R, Nishikawa J, Caluseriu O, Fei YL, Shuman C, Wei C, Steele L, Cameron J, Smith A, Ambus I, Li M, Ray PN, Sadowski P, Squire J (2001) Tumor development in the Beckwith-Wiedemann syndrome is associated with a variety of constitutional molecular 11p15 alterations including imprinting defects of *KCNQ1OT1*. *Hum Mol Genet* 10:2989–3000.

Weksberg R, Shuman C, Caluseriu O, Amith AC, Fei YL, Nishikawa J, Stockley TL, Best L, Chitayat D, Olney A, Ives E, Schneider A, Bestor TH, Li M, Sadowski P, Squire J (2002) Discordant KCNQ1TO1 imprinting in sets of monozygotic twins discordant for Beckwith-Wiedemann syndrome. *Hum Mol Genet* 11:1317–1325.

Weng EY, Moeschler JB, Graham JM Jr (1995a) Longitudinal observations on 15 children with Wiedemann-Beckwith syndrome. *Am J Med Genet* 56:366–373.

Weng EY, Mortier GR, Graham JM Jr (1995b) Beckwith-Wiedemann syndrome. *An update and review for the primary pediatrician. Clin Pediatr (Phila)* 34:317–326.

Wiedemann H-R (1964) Complexe malformatif familial avec hernie ombilicale et macroglossie, un "syndrome nouveau". *J Genet Hum* 13:223–232.

Wiedemann H-R (1983) Tumours and hemihypertrophy associated with Wiedemann-Beckwith syndrome. *Eur J Pediatr* 141:129.

Wilkins-Haug L, Porter A, Hawley P, Benson CB (2009) Isolated fetal omphalocele, Beckwith-Wiedemann syndrome, and assisted reproductive technologies. *Birth Defects Res A Clin Mol Teratol* 85:58–62.

Wilson M, Peters G, Bennetts B, McGillivray G, Wu ZH, Poon C, Algar E (2008) The clinical phenotype of mosaicism for genome-wide paternal uniparental disomy: Two new reports. *Am J Med Genet* 146A:137–148.

10

CARDIO-FACIO-CUTANEOUS SYNDROME

MARIA INÊS KAVAMURA
Medical Genetics Center, UNIFESP, São Paulo, Brazil

GIOVANNI NERI
Fondazione Policlinico Universitario A. Gemelli IRCCS, Istituto di Medicina Genomica, Università Cattolica del Sacro Cuore, Rome, Italy and Self Research Institute, Greenwood Genetic Center, Greenwood (SC), USA

INTRODUCTION

Incidence

There are no epidemiologic studies providing an estimate of the incidence or prevalence of cardio-facio-cutaneous syndrome (CFC syndrome). Since 1986, when the syndrome was first described (Baraitser and Patton 1986; Reynolds et al. 1986), more than 400 unpublished cases are known to CFC International, Inc., a family support group operating worldwide (*http://www.cfcsyndrome.org*). This number probably accounts for a fraction of cases, because it excludes most mildly affected individuals. Almost all affected individuals to date are sporadic, most likely the result of new dominant mutations, as supported by the observation of an association with older paternal age (Roberts et al. 2006).

Life expectancy is shortened on average, with early death caused by severe cardiac involvement in some. Comprehensive reviews of CFC syndrome are published (Roberts et al. 2006; Narumi et al. 2007; Armour and Allanson 2008).

Diagnostic Criteria

Clinical diagnosis can be established in CFC syndrome based on the history, cardinal manifestations, and evolution of phenotype, which are well delineated and present in the great majority of patients. A checklist of manifestations that comprises the CFC Index (Kavamura et al. 2002) can be used to phenotype the individual being assessed within the cardio-facio-cutaneous syndrome population. A comprehensive guideline has been developed for clinical diagnosis and management of CFC syndrome, described by specialty, for common versus rare manifestations that may occur (Pierpont et al. 2014).

Pregnancy can be uneventful, but is often marked by polyhydramnios. Hypotonia and serious feeding difficulties, resulting in failure-to-thrive, are hallmarks of the early postnatal period. Cardinal manifestations include: moderate-to-severe psychomotor disability, hypotonia, short stature with relative macrocephaly, distinctive facial appearance (high forehead, bitemporal narrowing, downslanting and wide-spaced palpebral fissures, ptosis, short nose with depressed bridge, low-set ears with creases on the lobes, sparse curly hair, sparse or absent eyebrows and lashes) (Figure 10.1), ectodermal findings (especially follicular keratosis, hyperkeratosis, palmoplantar keratosis, hyperelastic skin, cutaneous vascular malformations, pigmented nevi), and congenital heart defects (pulmonic stenosis, atrial septal defects, and cardiomyopathy being the most frequent). Neurologic abnormalities are frequent with seizures being present in 50% of cases. Orthopedic conditions such as scoliosis, pectus deformities and dysfunctional gait are also commonly seen in CFC syndrome. Craniosynostosis has been observed as part of the CFC phenotype (Shimojima et al. 2016; Ueda et al. 2017).

Cassidy and Allanson's Management of Genetic Syndromes, Fourth Edition.
Edited by John C. Carey, Agatino Battaglia, David Viskochil, and Suzanne B. Cassidy.
© 2021 John Wiley & Sons, Inc. Published 2021 by John Wiley & Sons, Inc.

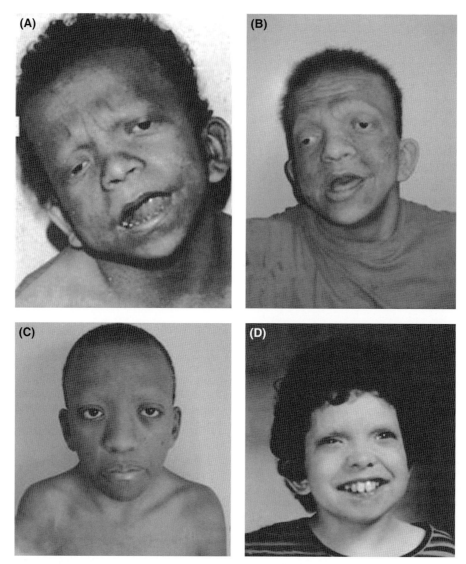

FIGURE 10.1 Three individuals with cardio-facio-cutaneous syndrome. (A) and (B) Male with a *BRAF* mutation, at ages 2 and 30 years; (C) male with a *MEK2* mutation at age 6 years; and (D) female with a *MEK1* mutation at age 12 years.

Etiology, Pathogenesis, and Genetics

Since the original description in 1986, multiple genes causing CFC syndrome have been identified. In 2006, Rodriguez-Viciana et al. studied a cohort of 23 individuals with CFC syndrome and found 11 different *BRAF* mutations in 18, *MEK1* mutations in two, and a *MEK2* mutation in one affected individual. All were de novo missense mutations with a gain-of-function effect. Simultaneously, Niihori et al. (2006) studied a non-overlapping group of 43 people with CFC syndrome and found eight different de novo missense mutations of *BRAF* in 16 of them. Pathogenic *KRAS* mutations have been reported in individuals with a clinical diagnosis of CFC syndrome (Matsubara 2006; Niihori et al. 2006; Schubbert et al. 2006; Nystrom et al. 2008) and have also been reported with the Noonan syndrome phenotype (Schubbert et al. 2006).

KRAS, BRAF, MEK1, and *MEK2* are part of the RAS/MAPK signaling pathway, which is implicated in growth factor-mediated cell proliferation, differentiation, and apoptosis, and plays a crucial role in embryonic development. *RAS* genes encode GTP-binding proteins that function as on-off switches to activate or inhibit downstream molecules. Dysregulation of the RAS/MAPK pathway is the common underlying mechanism of the phenotypically related Noonan syndrome (*PTPN11, SOS1, KRAS, RAF1, SHOC2, RIT1, LZTR1*, and *SOS2*), Costello syndrome (*HRAS*) and CFC syndrome (*BRAF, MEK1, MEK2, KRAS*) (Niihori et al. 2006).

Diagnostic Testing

Multigene panel testing is the preferred method to start the molecular investigation for CFC syndrome. All genes associated with Rasopathies can be tested at once, therefore allowing the establishment of early molecular diagnosis, even prenatal (Pierpont et al. 2014; Mucciolo et al. 2016). However, this technique is not available worldwide and in the absence of such, mutation analysis of *BRAF*, *MEK1*, *MEK2* and *KRAS* is recommended to confirm the diagnosis of CFC syndrome.

BRAF is the most frequently mutated gene (68.5%). Therefore, a screening strategy for individuals with clinical signs of CFC syndrome should start with analysis of *BRAF* (exons 6, 11, 12, 13, 14, 15, and 16), followed by *MEK1* (exons 2 and 3) and *MEK2* (exons 2, 3, and 7) (Schulz et al. 2008), and finally *KRAS*. If all results are negative and the features are typical of CFC syndrome, the entire coding sequence of the genes should be analyzed. Approximately 10–30% of individuals with clinical manifestations of CFC syndrome do not have mutations in any of the above genes (Narumi et al. 2007; Schulz et al. 2008), suggesting that at least one additional gene is yet to be discovered.

Differential Diagnosis

Noonan syndrome (Chapter 41) and Costello syndrome (Chapter 16), especially the former, are the main differential diagnoses, and there is marked phenotypic overlap among the three conditions (see Table 10.1). Individuals with Noonan syndrome have normal intelligence or mild psychomotor involvement, less common and less severe failure-to-thrive, a history of bleeding disorder, and autosomal dominant inheritance in some families. About 50% have mutations in *PTPN11* (Tartaglia et al. 2001), 10% in *SOS1* (Zenker et al. 2007), 8% in *RAF1* (Razzaque et al. 2007), and about 1% in *KRAS* (Carta et al. 2006; Schubbert et al. 2006). Individuals with *SOS1* mutations commonly have ectodermal manifestations very similar to those in CFC syndrome, including keratosis pilaris, sparse eyebrows, and curly hair. The prevalence of keratosis pilaris and curly hair is significantly higher in those with *SOS1* mutations than in those with *PTPN11* mutations (58% versus 6% and 78% versus 34%, respectively). Differences between Costello syndrome and CFC syndrome include papillomata, which can be perinasal, perioral and perianal in location, a higher likelihood of deep palmar and plantar creases, a significant risk of chaotic atrial arrhythmia or multifocal atrial tachycardia, and an increased frequency of malignant tumors such as rhabdomyosarcoma, transitional cell carcinoma of the bladder and neuroblastoma (Gripp et al. 2002), which have only rarely been described in CFC syndrome. *HRAS* mutations are found in more than 90% of individuals with Costello syndrome (Rauen 2006).

MANIFESTATIONS AND MANAGEMENT

Pregnancy

Pregnancies are often marked by polyhydramnios (77%) of a severity that may require reduction by amniocentesis. Prematurity is also common, occurring in 49% of pregnancies (Armour and Allanson, 2008). Prenatal ultrasound may ascertain the fetus with macrosomia, ventriculomegaly, increased nuchal thickness or frank hydrops (Armour and Allanson, 2008).

TABLE 10.1 Differential diagnosis of CFC and other conditions. Adapted from Pierpont et al. 2014

Syndrome	Common features with CFC	Differences with CFC
NS	Hypertelorism, downslanting palpebral fissures, ptosis, short stature, relative macrocephaly, PVS, HCM, atrial septal defect, and some with cognitive delay.	Facial features less coarse, lower incidence of severe feeding problems, fewer cutaneous features such as follicular hyperkeratosis, sparse eyebrows, ulerythema ophryogenes, and less incidence of marked cognitive delay.
CS	Coarse facial features, curly hair, broad nasal bridge, downslanting palpebral fissures, epicanthal folds, PVS, HCM, pectus deformity, short stature, and intellectual disability.	Papillomata of the face or perianal regions, multifocal atrial tachycardia, ulnar deviation of wrist and fingers, and loose skin.
NS with multiple lentigines (formerly LEOPARD syndrome)	Short stature, hypertelorism, PVS, HCM, and some with cognitive delay.	Multiple skin lentigines, frequent sensorineural deafness, cardiac conduction abnormalities, and less incidence of marked cognitive delay.
NS with loose anagen hair	Triangular facies, hypertelorism, high forehead, sparse thin hair, short stature, eczema, dry skin, and macrocephaly.	After infancy facial features less coarse, mitral valve dysplasia, and less incidence of marked cognitive delay. Hair anomaly (easily pluckable, sparse, thin, slow growth).

Growth and Feeding

Measurements at birth tend to be normal. Feeding problems start soon after birth, with poor suck, aspiration, gastroesophageal reflux, oral aversion, hyperemesis, and gastrointestinal dysmotility, leading to serious and long-lasting failure-to-thrive. These problems are significant enough to require nasogastric tube feeding, or gastrostomy tube placement. Assisted feeding may be required into middle childhood or beyond. Weight lies below the normal curve for age.

Short stature is found in 78% of individuals (Kavamura et al. 2002) and either relative or absolute macrocephaly is a consistent manifestation (Armour and Allanson 2008). Both delayed and precocious puberty have been reported, without clear etiologies.

Evaluation

- Clinical nutritional assessment is recommended at the time of diagnosis.
- Discomfort during feeding should lead to an evaluation for gastroesophageal reflux.
- Feeding difficulties should be investigated in a standard fashion.

Management

- Thickened feeds, upright positioning with feeding, oral medication or Nissen fundoplication may be beneficial in the management of reflux.
- Feeding difficulties and associated failure-to-thrive should be managed as in the general population. Referral to a dysphagia clinic is an option.
- Assisted feeding may be necessary to avoid malnutrition and maximize growth. It may be required into late childhood or rarely, into adolescence.

Development and Behavior

Psychomotor delay is usually manifest early in life. Hypotonia is near universal and contributes to delayed motor skill acquisition. Average age for standing is 27.9 months and for walking, 36.8 months (Armour and Allanson 2008). Speech delays are invariable and first words are rarely present before age 2 years. Less than 50% of affected individuals are able to converse in full sentences (Armour and Allanson 2008). In a cohort of 39 affected children studied by Yoon et al. (2007), all had delayed gross motor, fine motor, and language milestones. Seven never achieved independent ambulation or language acquisition and were totally dependent for all activities of daily living. Intellectual impairment, usually moderate-to-severe, is present in almost all individuals with CFC syndrome (Yoon et al. 2007). Most children require special education, although some are integrated into a normal classroom.

Behavior is variable. Many children display tactile defensiveness, short attention span, irritability, stubbornness, and obsessive behavior. A minority is aggressive to others (Armour and Allanson 2008). Nonetheless, parents often comment on warm and loving personalities and enjoyment of social interactions, particularly music and dance.

Higher risk of psychological problems has been reported, with attention deficits, social difficulties and unusual behaviors, such as obsessive thoughts and repetitive acts among other behavior findings (Pierpont and Wolford 2016).

Autism spectrum disorder (ASD) has also been identified in children with CFC syndrome. Using the Collaborative Program for Excellence in Autism criteria, eight out of nine (88.9%) of the patients studied met the criteria for ASD, with equal sex distribution (Garg et al. 2017).

Evaluation

- Complete developmental assessment should be carried out at diagnosis and periodically throughout childhood. The nonverbal Raven Progressive Matrices Test is particularly useful when there is speech delay.
- Neuropsychological assessment to test memory, visual-spatial working memory, constructional, and planning skills is important to detect strengths and weaknesses and inform educational strategies.

Management

- Special individualized educational programs are generally recommended.
- Occupational therapy, physical therapy, and speech therapy should be offered early.
- Social inclusion should be sought as much as possible.
- Stimulation should focus on adaptative behavior skills, to create independence in daily activities.

Gastroenterology

Severe constipation, despite normal intestinal ganglion cells, requiring daily medication and occasional disimpaction, has been reported (Herman and McAlister 2005). This is a frequent and long-lived issue that was observed in 71% of those with molecular confirmation (Armour and Allanson 2008). Additional structural anomalies of the gastrointestinal tract have been reported, such as malrotation, antral foveolar hyperplasia, and anal stenosis (McDaniel and Fujimoto 1997).

Evaluation

- Physical examination should be directed to identifying causes of constipation: the abdomen should be examined for masses, distension, absent bowel sounds and anal examination is indicated.

- Imaging studies should be performed if constipation is persistent and fails to respond to conservative treatment or when dysmotility or malformation is suspected.

Treatment

- Nutritional education is important to avoid constipation.
- Surgical correction may be indicated for gastrointestinal anomalies such as anal stenosis. These are standard interventions.

Cardiovascular

One or more cardiac abnormalities are found in 75% of individuals with CFC syndrome (Armour and Allanson 2008). Pulmonary valve stenosis is the most frequent finding, present in 45% of cases. Atrial septal defect is found in almost 30%, whereas ventricular septal defects are found in 22%. Myocardial disease, most often hypertrophic cardiomyopathy, is present in 40% (Roberts et al. 2006).

Little is known about the natural history of the hypertrophic cardiomyopathy. This condition can be rapidly progressive in infancy or remain stable for many years. It may develop late in childhood, resolve, remain stable or progress. Symptomatic hypertrophic cardiomyopathy in infancy often has a poor prognosis and can be a cause of early death. A variety of other structural defects has been reported, including left-sided defects, but none is individually common. Arrhythmias are quite uncommon.

Grade 3 pulmonary hypertension was observed in a 21-year-old male with CFC syndrome who died suddenly. A murmur had been heard in early childhood and attributed to mild pulmonary valve stenosis. Hypertrophic cardiomyopathy was not diagnosed until shortly before his death. Autopsy confirmed hypertrophic cardiomyopathy and lung findings that were compatible with pulmonary hypertension (Roberts et al. 2006).

Evaluation

- Clinical evaluation with electrocardiogram and echocardiogram should be performed at the time of diagnosis. Periodic reassessment will be guided by the findings.
- Echocardiographic follow-up is indicated every 1–2 years, regardless of previous results and clinical manifestations, as hypertrophic cardiomyopathy can develop at any age and may not be accompanied by obvious clinical manifestations.

Management

- Surgical intervention for outflow valvular obstruction is recommended once right ventricular pressure exceeds 80 mm Hg.
- Balloon valvuloplasty is recommended as the first palliation for valve stenosis; however, open pulmonary valvotomy may be needed.
- β-Blockers or calcium channel blockers have been most frequently used in the treatment of obstructive cardiomyopathy (Ishizawa et al. 1996).
- Where response to drug therapy is poor, standard surgical myomectomy or heart transplantation may be indicated (Sharland et al. 1992).
- When a cardiac structural defect is present, bacterial prophylaxis may be required for procedures that can promote bacteremia, such as dental treatment or surgery.

Neurologic

Neurologic involvement is extensive, and affects structure and/or function of the cortex, brainstem, and ventricular system. Brain magnetic resonance imaging of 32 individuals with CFC syndrome showed ventriculomegaly in 14, but only 3 required ventriculo-peritoneal shunts. Prominent Virchow–Robin spaces, abnormalities of myelination, type I Chiari malformation, arachnoid cyst, and subependymal gray matter heterotopia were also noted (Yoon et al. 2007). Generalized hypotonia is universal and observed from birth.

Seizures were reported in 15 of 39 individuals with mutation-positive CFC syndrome (Yoon et al. 2007). Mean age of onset was 3 years. The most common types were tonic-clonic and absence seizures, followed by complex partial and infantile spasms. The most common findings on electroencephalogram were generalized spike activity, followed by hypsarrhythmia and epileptiform discharges.

Evaluation

- Neurologic examination should be carried out at every clinical assessment.
- MRI and electroencephalogram should be carried out at diagnosis or when new signs or symptoms of neurologic dysfunction are noted.

Management

- Anticonvulsant therapy does not differ from the general population.
- Treatment of hydrocephalus, when required, is standard.

Dermatologic

Cutaneous or adnexal abnormalities are present in 100% of individuals with CFC syndrome. Pigmented nevi (76%), follicular hyperkeratosis of arms, legs and face (73%), and hyperkeratosis (61%) are the most frequent cutaneous findings. Sparse, slow-growing, curly hair is one of the hallmarks of CFC syndrome, present in 85% of affected individuals (Roberts et al. 2006). Other frequent cutaneous findings are cutaneous vascular malformations (47%), eczema (44%), ichthyosis (30%), and café-au-lait spots (27%) (Armour and Allanson 2008). We have observed palmoplantar hyperkeratosis and limb lymphedema in the teenage/adult years (Kavamura and Neri, personal observation). Histologic findings are non-specific, with ichthyosis and hyperkeratosis of sweat glands and hair follicles.

Some patients show persistent and progressive lymphatic dysplasia, presenting a characteristic pattern of neck webbing, bilateral lower limb lymphedema, genital swelling with chylous reflux, intestinal lymphangiectasia and chylotorax. Lymphocintigraphy shows reflux and rerouting of lymphatic drainage and venous duplex scans demonstrates incompetent veins (Joyce et al. 2016).

Twenty-year follow-up of some affected individuals has shown that, with age, skin dryness and follicular hyperkeratosis tend to improve. However, palmoplantar hyperkeratosis and limb lymphedema may become more severe. The latter, and its complications, were the cause of death of an affected teenager. Fungal and bacterial infection of the toenails can spread systemically causing septic shock (personal observation). Good results in lymphedema treatment are seen with compression garments. No malignant transformation of pigmented nevi to melanoma has been reported so far.

Evaluation

- Full dermatologic examination should be performed at the time of diagnosis and at each periodic health examination, at least annually.

Management

- All pigmented nevi should be examined annually and excised if indicated using the same guidelines as in the general population.
- Treatment of skin and nail infection is standard, and has particular importance when lymphedema is present.
- Topical emollients for treatment of dry skin (preparations containing urea 10–15% and/or petrolatum) and keratolytic preparations for hyperkeratosis and follicular keratosis (preparations with salicylic acid, α-hidroxy acids or retinoids) have been used with good results.

Ophthalmologic

Ocular manifestations are common in CFC syndrome and have been well documented by Young (2003), who personally examined 25 affected individuals (reviewed in Roberts et al. 2006). There is a considerable range of visual function among those affected. Loss of vision to the level of only light perception is described but rare, whereas decreased visual acuity is quite common. The most common refractive error, myopia, is present in 20% of the affected individuals. Strabismus is present in about one third, exotropia being more common than esotropia. Nystagmus is present in about one third. The only structural eye finding that has been noticed in more than one individual is optic nerve hypoplasia or dysplasia, which is reported in 45% (Armour and Allanson 2008). One individual had optic nerve atrophy, possibly because of increased intracranial pressure. There are isolated cases of cataracts, vertical strabismus, dissociated vertical deviation and inferior oblique muscle overaction, nasolacrimal duct obstruction, and keratoconus. Most individuals have difficulty with depth perception and binocular function. Although this may be the result of visual processing challenges, an overlying strabismus or nystagmus can also contribute to fixation and tracking abnormalities (Roberts et al. 2006).

Evaluation

- Ophthalmologic evaluation should be performed at the time of diagnosis, and periodically thereafter, depending on the findings.

Management

- Treatment of strabismus and refractive errors are standard.
- Treatment of high intracranial pressure to avoid brain and optic nerve atrophy is standard.

Genitourinary

Cryptorchidism and kidney/bladder abnormalities are relatively common, present in 38 and 20% of cases, respectively. Renal abnormalities include hydronephrosis (prenatal or postnatal in onset), duplex collecting system, enlarged kidneys, and vesicoureteric reflux (Armour and Allanson 2008).

Evaluation

- Abdominal ultrasound is recommended at the time of diagnosis to exclude malformations.
- Recurrent urinary tract infection should prompt standard evaluation for vesico-ureteric reflux.
- Genital evaluation looking for cryptorchidism is important at diagnosis.

Management

- Treatment of cryptorchidism is the same as in the general population.
- Infections of the urinary tract malformations should be treated with antibiotics in a standard manner.

Musculoskeletal

Bone age is significantly delayed and osteopenia is occasionally observed (Herman and McAlister 2005). Joint hyperextensibility and pectus excavatum/carinatum were the most common musculoskeletal features, present in 63%, in the review by Armour and Allanson (2008), followed by scoliosis (33%), kyphosis (23%), and joint contractures (19%). Other reported features in that study included wide anterior fontanelle, pes planus, crowded or overlapping toes, clinodactyly, broad hands and feet, broad thumb/first toe, and syndactyly.

Giant cell lesions are benign tumors that most frequently affect the jaws, leading to the clinical picture of cherubism, but they can also occur in other bones or soft tissues. Multiple giant cell lesions, repeatedly reported in Noonan syndrome, are also described in three individuals with CFC syndrome and *BRAF* mutations (Neumann et al. 2009).

Craniosynostosis has been reported in CFC patients with *BRAF*, *MEK2* and *KRAS* mutations. Further studies are needed to elucidate the pathogenic mechanism responsible for craniosynostosis mediated by the Ras/MAPK signaling pathway (Shimojima et al. 2016; Ueda et al. 2017).

Evaluation

- Evaluation of the spine should be part of each health assessment.
- If scoliosis or kyphosis is present, orthopedic evaluation and follow-up is recommended.
- Bone densitometry should be performed if osteopenia is suspected.
- If mandibular enlargement is noted, radiologic evaluation looking for giant cell lesions is warranted.

Management

- Physiotherapy is recommended as early as possible to reduce the likelihood or extent of scoliosis and kyphosis and treat contractures.
- Osteopenia treatment is standard.
- Giant cell lesions must be followed and receive standard treatment (cryotherapy or surgery) if they do not regress spontaneously or if they cause pain or other symptoms.

Neoplasia

There has been no report of a significant association between CFC syndrome and neoplasia; however, three individuals with molecularly confirmed CFC syndrome have developed malignancies. Acute lymphoblastic leukemia was diagnosed in two affected individuals. The first child was diagnosed at 5 years of age (van den Berghe and Hennekam, 1999). Years later, a *BRAF* p.G469E mutation was identified, confirming CFC syndrome (Niihori et al. 2006). The second child was diagnosed at 1 year and 9 months of age when he presented with hepatosplenomegaly and right testicular involvement, confirmed by biopsy. He was treated and responded well, being healthy at the age of 9 years and 3 months. He had a *BRAF* p.E501G mutation (Makita et al. 2007). Both *BRAF* mutations have been demonstrated in other individuals with CFC syndrome without malignancy. Hepatoblastoma was reported in a 3-year-old boy with CFC syndrome and a *MEK1* p.Y130C mutation who had undergone a cardiac transplant at age 8 months for hypertrophic cardiomyopathy. The patient died shortly after an intracardiac mass was diagnosed as metastatic hepatoblastoma. In this case, it is unclear whether the posttransplant immune-suppressive therapy played a role in tumor development (Al-Rahawan et al. 2007).

BRAF is a proto-oncogene and somatic mutations of *BRAF* have been identified in 7% of cancers, particularly melanoma, colon, and thyroid cancer. The mutation p.V600E is frequently identified in these cancers (Makita et al. 2007).

A study that compared the frequency of child cancer among the Noonan, Costello and CFC syndrome population with the German childhood cancer registry reached the conclusion that there is an intermediate cancer risk associated with Rasopathies. The risk is higher than in the general population but not as high as in familial cancers. No cancer cases were detected in CFC syndrome patients in this study (Kratz et al. 2015).

Evaluation

- The low risk of malignancy does not warrant a particular surveillance protocol.

Management

- Leukemia and solid tumors, if detected, are treated in a standard manner.

Hematologic

Bleeding disorders and easy bruising have not been a significant problem in CFC syndrome. The authors have experience of one newborn with thrombocytopenia and another with frequent nosebleeds, which improved with cauterization.

Evaluation

- Routine clotting studies are not recommended in asymptomatic patients.

Management

- Bleeding diatheses are treated in a standard manner.

RESOURCES

Brochures and Newsletters

CFC International

This volunteer, non-profit family support group hosts the site: *http://www.cfcsyndrome.org* and contains: cardio-facio-cutaneous syndrome brochure, parent guide, newsletter, photogallery, research information, conference news. Email: *info@cfcsyndrome.org*

Medical Information

http://www.ncbi.nlm.nih.gov/entrez/dispomim.cgi?id=115150

http://www.orpha.net/data/patho/GB/uk-CardioFacioCutan.pdf

REFERENCES

Al-Rahawan MM, Chute DJ, Sol-Chuch K, Gripp KW, Stabley DL, Mc Daniel NL, Wilson WG, Waldron PE (2007) Hepatoblastoma and heart transplantation in a patient with cardio-facio-cutaneous syndrome. *Am J Med Genet* 143:1481–1488.

Armour CM, Allanson JE (2008) Further delineation of cardio-facio-cutaneous syndrome: clinical features of 38 individuals with proven mutations. *J Med Genet* 45:249–254.

Baraitser M, Patton MA (1986) A Noonan-like short stature syndrome with sparse hair. *J Med Genet* 24:9–13.

Bertola DR, Pereira AC, Brasil AS, Albano LM, Kim CA, Krieger JE (2007) Further evidence of genetic heterogeneity in Costello syndrome: involvement of the *KRAS* gene. *J Hum Genet* 52:521–526.

Carta C, Pantaleoni F, Bocchinfuso G, Stella L, Vasta I, Sarkozy A, Digilio C, Palleschi A, Pizzuti A, Grammatico P, Zampino G, Dallapiccola B, Gelb BD, Tartaglia M (2006) Germline missense mutations affecting *KRAS* isoform B are associated with a severe Noonan syndrome phenotype. *Am J Hum Genet* 79:129–135.

Garg S, Brooks A, Burns A, Burkitt-Wright E, Kerr B, Huson S, Emsley R, Green J (2017) Autism spectrum disorder and other neurobehavioural comorbidities in rare disorders of the Ras/MAPK pathway. *Dev Med Child Neurol* 59:544–549.

Gripp KW, Scott CI Jr, Nicholson L, McDonald-McGinn DM, Ozeran JD, Jones MC, Lin AE, Zackai EH (2002) Five additional Costello patients with rhabdomyosarcoma: proposal for a tumor screening protocol. *Am J Med Genet* 108:80–87.

Herman TE, McAlister WH (2005) Gastrointestinal and renal abnormalities in cardio-faciocutaneous syndrome. *Pediatr Radiol* 35:202–205.

Ishizawa A, Oho S, Dodo H, Katori T, Homma SI (1996) Cardiovascular abnormalities in Noonan syndrome: The clinical findings and treatment. *Acta Paediatr Jpn* 38:84–90.

Joyce S, Goirdon K, Brice G, Ostergaard P, Nagaraja R, Short J, Moore S, Ortimer P, Mansour S (2016) The lymphatic phenotype in Noonan and cardiofaciocutaneous syndrome. *Eur J Hum Genet* 25:690-696.

Kavamura MI, Peres CA, Alchorne MM, Brunoni D (2002) CFC index for the diagnosis of cardiofaciocutaneous syndrome. *Am J Med Genet* 112:12–16.

Kratz CP, Franke L, Peters H, Kohlschmidt N, Kazmierczak B, Finckh U, Bier A, Elchhorn B, Blank C, Kraus C, Kohlhase J, Pauli S, Wildhardt G, Kutsche K, Auber B, Christmann A, Bachmann N, Mitter D, Cremer FW, Mayer K, Daumer-Haas C, Nevinny-Stickel-Hinzpeter C, Oeffner F, Schluter G, Gencik M, Uberlacker B, Lissewski C, Schanze I, Greene MH, Spix C, Zenker M (2015) Cancer spectrum and frequency among children with Noonan, Costello, and cradio-facio-cutaneous syndromes. *Br J Cancer* 112:1392-1397.

Makita Y, Narumi Y, Yoshida M, Niihori T, Kure S, Fujieda K, Matsubara Y, Aoki Y (2007) Leukemia in cardio-facio-cutaneous (CFC) syndrome: a patient with a germline mutation in *BRAF* proto-oncogene. *J Pediatr Hematol Oncol* 29:287–290.

Matsubara Y (2006) Germline *KRAS* and *BRAF* mutations in cardio-facio-cutaneous syndrome. *Nat Genet* 38:294–296.

McDaniel CH, Fujimoto A (1997) Intestinal malrotation in a child with cardio-facio-cutaneous syndrome. *Am J Med Genet* 70:284–286.

Mucciolo M, Dello Russo C, D'Emidio L, Mesoraca A, Giorlandino C (2016) Next generation sequencing approach in a prenatal case of cardio-facio-cutaneous syndrome. *Int J Mol Sci* 17:952.

Narumi Y, Aoki Y, Niihori T, Neri G, Cave H, Verloes A, Nava C, Kavamura MI, Okamoto N, Kurosawa K, Hennekam RC, Wilson LC, Gillessen-Kaesbach G, Wieczorek D, Lapunzina P, Ohashi H, Makita Y, Kondo I, Tsuchiya S, Ito E, Sameshima K, Kato K, Kure S, Matsubara Y (2007) Molecular and clinical characterization of cardio-facio-cutaneous (CFC) syndrome: Overlapping clinical manifestations with Costello syndrome. *Am J Med Genet* 143:799–807.

Neumann TE, Allanson J, Kavamura I, Kerr B, Neri G, Noonan J, Cordeddu V, Gibson K, Tzschach A, Krüger G, Hoeltzenbein M, Goecke TO, Kehl HG, Albrecht B, Luczak K, Sasiadek MM, Musante L, Peters H, Tartaglia M, Zenker M, Kalscheuer V (2009) Multiple giant cell lesions in patients with Noonan syndrome and cardio-facio-cutaneous syndrome. *Eur J Hum Genet* 17:420–425.

Niihori T, Aoki Y, Narumi Y, Neri G, Cave H, Verloes A, Okamoto N, Hannekam RC, Gillessen-Kaesbach G, Wieczorek D, Kavamura MI, Kurosawa K, Ohashi H, Wilson L, Heron D, Bonneau D, Corona G, Kaname T, Naritomi K, Baumann C, Matsumoto N, Kato K, Kure S, Matsubara Y (2006) Germline *KRAS* and *BRAF* mutations in cardio-facio-cutaneous syndrome. *Nat Genet* 38:294–296.

Nystrom AM, Ekvall S, Berglund E, Björkqvist M, Braathen G, Duchen K, Enell H, Holmberg E, Holmlund U, Olsson-Engman M, Annerén G, Bondeson ML (2008) Noonan and cardio-facio-cutaneous syndromes: Two clinically and genetically overlapping disorders. *J Med Genet* 45:500–506.

Pierpont EI, Wolford M (2016) Behavioral functioning in cvardio-faciocutaneous syndrome: risk factors and impact on parenting experience. *Am J Med Genet* 170:1974–1988.

Pierpont MEM, Magoulas PL, Adi S, Kavamura MI, Neri G, Noonan J, Pierpont EI, Reinker K, Roberts AE, Shankar S, Sullivan J, Wolford M, Conger B, Santa Cruz M, Rauen KA (2014) Cardio-facio-cutaneous syndrome: clinical features, diagnosis, and management guidelines. *Pedriatics* 134:1149–1162.

Rauen K (2006) Distinguishing Costello versus cardio-facio-cutaneous syndrome: BRAF mutations in patients with a Costello phenotype. *Am J Med Genet A* 140:1681–1683.

Razzaque MA, Nishizawa T, Komoike Y, Yagi H, Furutani M, Amo R, Kamisago M, Momma K, Katayama H, Nakagawa M, Fujihara Y, Matsushima M, Mizuno K, Tokuyama M, Hirota H, Muneuchi J, Higashinakagawa T, Matsuoka R (2007) Germline gain-of-function mutations in *RAF1* cause Noonan syndrome. *Nat Genet* 39:1013–1017.

Reynolds JF, Neri G, Hermann JP, Blumberg B, Coldwell JG, Miles PV, Opitz JM (1986) New multiple congenital anomalies/mental retardation syndrome with cardio-facio-cutaneous involvement. *Am J Med Genet* 23:413–427.

Roberts A, Allanson J, Jadico SK, Kavamura MI, Noonan J, Opitz JM, Young T, Neri G (2006) The Cardio-Facio-Cutaneous (CFC) syndrome: A review. *J Med Genet* 43:833–842.

Rodriguez-Viciana P, Tetsu O, Tidyman WE, Estep AL, Conger BA, Cruz MS, McCormick F, Rauen K (2006) Germline mutations in genes within the MAPK pathway cause cardio-facio-cutaneous syndrome. *Science* 311:1287–1290.

Schubbert S, Zenker M, Rowe SL, Boll S, Klein C, Bollag G, van der Burgt I, Musante L, Kalscheuer V, Wehner LE, Nguyen H, West B, Zhang KY, Sistermans E, Rauch A, Niemeyer CM, Shannon K, Kratz CP (2006) Germline *KRAS* mutations cause Noonan syndrome. *Nat Genet* 38:331–336.

Schubbert S, Bollag G, Shannon K (2007) Deregulated Ras signaling in developmental disorders: New tricks for an old dog. *Curr Opin Genet Dev* 17:15–22.

Schulz AL, Albrecht B, Arici C, van der Burgt I, Buske A, Gillessen-Kaesbach G, Hellen R, Horn D, Hubner CA, Korenke GC, Konig R, Kress W, Kruger G, Meinecke P, Mucke J, Plecko B, Rossier E, Schinzel A, Schulze A, Seemanova E, Seidel H, Spranger S, Tuysuz B, Uhrig S, Wieczorek D, Kutsche K, Zenker M (2008) Mutation and phenotypic spectrum in patients with cardio-facio-cutaneous and Costello syndrome. *Clin Genet* 73:62–70.

Sharland M, Burch M, McKenna WM, Patton MA (1992) A clinical study of Noonan syndrome. *Arch Dis Child* 67:178–183.

Shimojima K, Ondo Y, Matsufuji M, Sano N, Tsuru H, Oyoshi T, Higa N, Tokimura H, Arita K, Yamamoto T (2016) Concurrent occurrence of an inherited 16p13.11 microduplication and a de novo 19p13.3 microdeletion involving MAP2K2 in a patient with developmental delay, distinctive facial features, and lambdoid synostosis. *Eur J Med Genet* 59:559–563.

Tartaglia M, Mehler EL, Goldberg R, Zampino G, Brunner HG, Kremer H, van der Burgt I, Crosby AH, Ion A, Jeffrey S, Kalidas K, Patton MA, Kucherlapati RS, Gelb BD (2001) Mutations in *PTPN11*, encoding the protein tyrosine phosphatase SHP-2, cause Noonan syndrome. *Nat Genet* 29:465–468.

Ueda K, Yaoita M, Niihori T, Aoki Y, Okamoto N (2017) Craniosynostosis in patients with RASopathies: accumulating clinical evidence for expanding the phenotype. *Am J Med Genet* 173:2346-2352.

van den Berghe H, Hennekam RC (1999) Acute lymphoblastic leukaemia in a patient with cardiofaciocutaneous syndrome. *J Med Genet* 36:799–800.

Yoon G, Rosenberg J, Blaser S, Rauen KA (2007) Neurological complications of cardio-facio-cutaneous syndrome. *Dev Med Child Neurol* 49:894–899.

Young TL (2003) *Cardio-facio-cutaneous syndrome conference ophthalmic findings summary, Rockville, Maryland,* June 2003. Available at: www.cfcsyndrome.org/conferencesummary.

Zenker M, Lehmann K, Schulz AL, Barth H, Hansmann D, Koenig R, Korinthenberg R, Kreiss-Nachtsheim M, Meinecke P, Morlot S, Mundlos S, Quante AS, Raskin S, Schnabel D, Wehner LE, Kratz CP, Horn D, Kutsche K (2007) Expansion of the genotypic and phenotypic spectrum in patients with *KRAS* germline mutations. *J Med Genet* 44:131–5.

11

CHARGE SYNDROME

Donna M. Martin
The University of Michigan Medical School, United States

Christine A. Oley
West Midlands Regional Genetics Service, Birmingham Women's Hospital, Edgbaston, Birmingham, UK

Conny M. van Ravenswaaij-Arts
University of Groningen, The Netherlands

INTRODUCTION

The various abnormalities that comprise the CHARGE syndrome were first described by Hall in 1979, although there had been several reports in the 1950s and 1960s of children with choanal atresia and congenital heart disease, coloboma and congenital heart disease, and coloboma and choanal atresia (Hall 1979). Pagon et al. (1981) reported a further 21 cases and coined the acronym "CHARGE" to describe this pattern of anomalies (Table 11.1) (Pagon et al. 1981). Good reviews and large case series are available (Bergman et al. 2011; Blake et al. 1998; Davenport et al. 1986; Janssen et al. 2012; Jongmans et al. 2006; Lalani et al. 2006; Legendre et al. 2017; Oley et al. 1988; Pagon et al. 1981; Sanlaville and Verloes 2007; Tellier et al. 1998; Zentner et al. 2010).

Incidence and Mortality

Many hundreds of individuals have now been reported, and CHARGE syndrome is one of the most common multiple anomaly conditions encountered by clinicians. Prevalence is thought to be at least 1/10,000 births (Blake et al. 1998; Sanlaville and Verloes 2007). Issekutz et al. in 2005 estimated the prevalence to be 1/8500 births in Canadian provinces (Issekutz et al. 2005). Janssen et al. estimated a birth prevalence of 1/12,000–1/17,000 in the Netherlands (Janssen et al. 2012).

Males and females are equally likely to be affected, and in molecularly confirmed cohorts there is a female preponderance (Lalani et al. 2006). Although the most consistent features are those represented by the letters in the acronym CHARGE (**c**oloboma, congenital **h**eart defects, choanal **a**tresia, **r**etardation of growth, developmental delay, **g**enital abnormalities, **e**ar abnormalities and deafness), there are additional abnormalities that occur frequently. These include semicircular canal abnormalities, facial palsy, renal abnormalities, orofacial clefts, and tracheoesophageal fistula. Often, the presence of these additional anomalies can be useful as confirmatory evidence of the diagnosis. However, the total clinical spectrum and the criteria for the diagnosis of CHARGE syndrome continue to be refined.

Neonates with CHARGE syndrome frequently have multiple life-threatening medical problems. If a complex congenital heart defect is present, the mortality rate can be significant (30–40%), even though death may not be directly related to the underlying congenital heart disease (Wyse et al. 1993). Outlook for survival is poor if more than one of the following three features is present: cyanotic cardiac lesion, bilateral posterior choanal atresia, or tracheoesophageal fistula (Blake et al. 1990). Early mortality and morbidity seem largely to be related to underlying pharyngeal and laryngeal incoordination that result in aspiration of secretions. In a study of 47 affected individuals, Tellier et al. in 1998 reported a mortality

---

Cassidy and Allanson's Management of Genetic Syndromes, Fourth Edition.
Edited by John C. Carey, Agatino Battaglia, David Viskochil, and Suzanne B. Cassidy.
© 2021 John Wiley & Sons, Inc. Published 2021 by John Wiley & Sons, Inc.

TABLE 11.1 Common features observed in CHARGE syndrome. Prevalence (centiles rounded) of features based on three cohorts with molecularly confirmed CHARGE syndrome (Bergman et al. 2011; Lalani et al. 2006; Zentner et al. 2010)

Feature	%
External ear anomaly	95
Semicircular canal anomaly	95
Coloboma	80
Choanal atresia	50
Cleft lip and/or palate	40
Cranial nerve dysfunction (VII, VIII, others)	95
Feeding difficulties	80
Facial palsy	60
Anosmia/hyposmia	80
Genital hypoplasia	70
Congenital heart defect	80
Tracheoesophageal anomaly	25
Developmental delay	100
Intellectual disability	80
Growth retardation	55

rate of 49%, mostly before 6 months of age. In this study, the poor life-expectancy correlated with male gender, central nervous system malformation, bilateral choanal atresia, and tracheoesophageal fistula (Tellier et al. 1998). Issekutz et al. in 2005 reported a high mortality in infants with atrioventricular septal defects and in infants with ventriculomegaly and brainstem and/or cerebellar anomalies (Issekutz et al. 2005). They also showed that feeding difficulties were a major cause of morbidity at all ages. Bergman et al. compared clinical features of a cohort with CHARGE syndrome who died after the neonatal period, but before the age of 10 years, with a cohort of survivors beyond the age of 10; the deceased children were more likely to have a congenital heart defect and gastroesophageal reflux disease (GERD) (Bergman et al. 2010a). Thus, health care providers should be aware of the mortality risk of swallowing difficulties and GERD in CHARGE syndrome. In this respect, it is important to mention that congenital arch vessel anomalies are common in CHARGE syndrome and may result in an increased risk for choking and aspiration if the esophagus is compressed (Corsten-Janssen et al. 2016). However, given the natural tendency to ascertain more severely affected patients and the advances in diagnostics since the 1990s, infants who survive the newborn period may have a better prognosis than previously reported.

Diagnostic Criteria

Clinical diagnosis of the CHARGE syndrome is made using criteria that have been refined several times. Following the first reports of what was then known as the CHARGE association, it was proposed that at least four of the major features included in the acronym must be present to make a confident diagnosis (Oley et al. 1988) and that these should include either coloboma or choanal atresia (Pagon et al. 1981). In 1998, Blake et al. proposed that the diagnosis be based on the presence of major and minor criteria (Blake et al. 1998).

Until recently, the most commonly used clinical criteria are the Blake criteria and the Verloes criteria (Lalani et al. 1993; Verloes 2005). In 2016, Hale et al. updated these by including pathogenic *CHD7* variants as a major criterion to be able to include those individuals with a milder presentation of the syndrome (Hale et al. 2016). All three sets of clinical criteria make a distinction between major and minor characteristics: Tables 11.2 and 11.3 respectively. The number of characteristics that must be present to make a diagnosis slightly differs between the three sets of criteria, but in general the following applies:

- If a definitely pathogenic *CHD7* variant is present, one other major characteristic is sufficient.
- In the absence of *CHD7* testing or a negative test result, a clinical diagnosis can be established if at least three major, or two major and three minor characteristics are present*.

*Verloes also defined atypical CHARGE syndrome: two major, or one major and three minor characteristics.

Several reviews document the frequency of the major features (Bergman et al. 2011; Blake et al. 1998; Harvey et al. 1991; Janssen et al. 2012; Oley et al. 1988; Pagon et al. 1981; Tellier et al. 1998; Zentner et al. 2010). The figures given in Table 11.1 are based on three large cohorts with *CHD7*-related CHARGE syndrome. Colobomas involving iris, retina, or optic disk can be bilateral or unilateral. They are found in 75–90% of affected individuals. Vision loss varies according to the size and location of the coloboma. Congenital heart defects occur in 75–90%. A wide range of different heart defects is seen with a relative over-representation of conotruncal defects and atrioventricular septal defects (Corsten-Janssen et al. 2013a). Choanal atresia or stenosis is found in 35–60% of individuals with CHARGE syndrome and is frequently bilateral. Growth retardation is documented in 35–70%, while developmental delays occur in 75 to 100% of individuals. Genital abnormalities have mainly been reported in males, with a frequency of 60–70%. Ear anomalies and/or deafness occur in 90–100%. The external ears are quite distinctive, described as short and wide, low-set, protruding, lop or cup-shaped, and simple with absence of the ear lobe and triangular concha

TABLE 11.2 Major diagnostic characteristics of CHARGE syndrome

Ocular coloboma
Choanal atresia or stenosis or orofacial clefting
Cranial nerve dysfunction (I anosmia, VII facial palsy, VIII deafness, IX/X swallowing problems)
Abnormal external, middle or inner ears, including hypoplastic semicircular canals
Pathogenic *CHD7* variant

TABLE 11.3 Minor characteristics of CHARGE syndrome

Developmental delay (delayed milestones)
Heart or esophagus abnormalities
Dysphagia/feeding difficulties
Structural brain abnormalities
Hypothalamo-hypophyseal dysfunction (gonadotropin or growth hormone deficiency) and genital anomalies
Renal anomalies
Skeletal/limb anomalies
Facial features (square face)

(Figures 11.1 and 11.2). The antihelix is unusual and distinctive (Fig. 11.1). Earlier studies suggested that semicircular canal hypoplasia occurs in more than 90% of individuals with CHARGE syndrome (Amiel et al. 2001; Lin et al. 1990; Sanlaville and Verloes 2007). A more recent study showed that among temporal bone CT scans from individuals with CHARGE syndrome, only two ears (one patient) among 84 (42 patients) exhibited normal anatomy (Vesseur et al. 2016c).

Other relatively common findings include facial asymmetry, unilateral facial palsy (Figures 11.1B and 11.2C), cleft lip and/or palate, swallowing dysfunction, gastroesophageal reflux, esophageal atresia and/or tracheoesophageal fistula, and renal anomalies (including vesicoureteric reflux, hydronephrosis, rotated kidneys, and small kidneys) (Davenport et al. 1986; Oley et al. 1988). Limb anomalies have been reported in about 30% (Sanlaville and Verloes 2007) and include clinodactyly, polydactyly, hypoplastic nails, and foot deformities. In 2007, Van der Laar et al. reported three individuals with CHARGE syndrome and severe limb anomalies including monodactyly, tibial aplasia, and bifid femora (Van de Laar et al. 2007).

Etiology, Pathogenesis, and Genetics

Before recognition that the *CHD7* gene was implicated in the causation of CHARGE syndrome, the pattern was considered by most to be a heterogeneous group of anomalies that occur together more often than would be expected by chance alone; this formed the basis for applying the prior designation of "association" (Davenport et al. 1986; Lubinsky 1994; Verloes 2005). There was some evidence that the various anomalies seen may result from abnormal neural crest cell growth, migration, differentiation, or survival (Siebert et al. 1985), which was later confirmed by studies on the function of CHD7 (Bajpai et al. 2010). Most early reported patients with CHARGE were sporadic, and a number of factors supported the view that a genetic abnormality was involved including (1) concordance of phenotype in monozygotic twins and discordance in dizygotic twins, (2) absence of definite identified environmental factors, (3) a significantly higher paternal age at conception than in the general population, (4) the existence of chromosomal anomalies in some cases, and finally, (5) rare familial forms (Hughes et al. 2014; Jongmans et al. 2008; Metlay et al. 1987; Oley et al. 1988; Tellier et al. 1998).

In 2004, mutations in the chromodomain helicase DNA-binding protein gene, *CHD7* at 8q12.1, were reported to be a major cause of CHARGE syndrome (Vissers et al. 2004). Mutations in *CHD7* have been found in up to 65–90% of individuals with CHARGE syndrome (Bergman et al. 2011; Janssen et al. 2012; Jongmans et al. 2009; Lalani et al. 2006; Legendre et al. 2017; Vissers et al. 2004; Zentner et al. 2010). The majority of individuals with the combination of coloboma, choanal atresia, and hypoplastic semicircular canals have mutations in *CHD7* (Lalani et al. 2006). The published mutations in *CHD7* are scattered throughout the gene. There is no clear genotype/phenotype relationship, although missense mutations in general result in a milder phenotype (Bergman et al. 2012b). Mutations have been shown to be dominantly inherited in some families, with wide clinical variability.

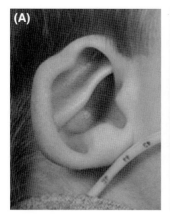

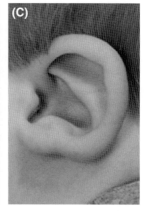

FIGURE 11.1 Fifteen-month old boy with CHARGE syndrome due to a truncating *CHD7* variant (frameshift). He has a square-shaped face with right-sided facial palsy (B), coloboma of the left iris and dysmorphic ears (A), (C) with notched helix, prominent crus of antihelix [especially in (A)], absence of triangular fossa, triangular-shaped conchae, prominent antitragus and absent ear lobes.

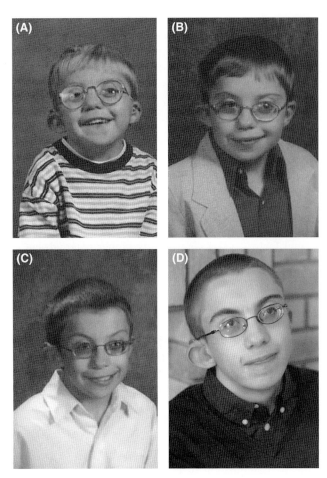

FIGURE 11.2 Evolving facial features in CHARGE syndrome. A boy with CHARGE photographed at age 5 (A), 10 (B), 13 (C) and 19 (D) years. Facial features include anteriorly protruding, low-set ears and asymmetric orbicularis oris muscle tone.

Somatic and germ line mosaicism have been observed (Jongmans et al. 2008).

CHD7 belongs to a large family of evolutionarily conserved genes encoding proteins thought to play a role in chromatin organization. It is likely to have an important role in regulating early embryonic development and cell-cycle control (Sanlaville and Verloes 2007). Mutations in *CHD7* have also been found in individuals with Kallmann syndrome who were negative for mutations in *KAL1*, *FGFR1*, *PROK2*, and *PROKR2* genes (Balasubramanian and Crowley 2017; Bergman et al. 2012a; Jongmans et al. 2009; Xu et al. 2017), and first reported by Kim et al 2008 (Kim et al. 2008). The main features of Kallmann syndrome are hypogonadotropic hypogonadism and anosmia or hyposmia, which are also frequently found in CHARGE syndrome. In individuals with Kallmann syndrome, pathogenic *CHD7* missense mutations are more likely than *CHD7* truncating variants (Marcos et al. 2014). It is speculated that *CHD7* may influence either expression or the actions of the genes causing Kallmann syndrome (Jongmans et al. 2009).

Diagnostic Testing

The diagnosis of CHARGE syndrome is primarily established using clinical criteria. Molecular testing for *CHD7* is useful to confirm the diagnosis and to provide information for recurrence risk and prenatal diagnosis. When the diagnosis is suspected but *CHD7* analysis is normal, it is prudent to rule out a chromosome anomaly by microarray analysis Guidelines for when to perform *CHD7* analysis have been developed (Bergman et al. 2011; van Ravenswaaij-Arts et al. 2015); however, with the decrease in both costs and turn-around time an increasingly number of individuals with CHARGE syndrome will likely be diagnosed by either whole exome or genome sequencing, especially in neonates with multiple congenital anomalies.

Differential Diagnosis

Although CHARGE syndrome is considered to be etiologically heterogeneous, in over 90% of typical CHARGE patients a pathogenic variant of the *CHD7* gene can be found. Detection rates are lower, typically around 60%, in cohorts of patients who are clinically suspected of CHARGE syndrome (reviewed in (Janssen et al. 2012)). Both chromosomal abnormalities and single-gene defects are implicated in the diferential diagnosis. Although some of the features seen in CHARGE syndrome are also seen in thalidomide embryopathy and diabetic embryopathy, there have been no reports of these being implicated as the cause of CHARGE syndrome.

Several chromosomal abnormalities have been reported in individuals with the constellation of anomalies seen in CHARGE syndrome, including patients presenting as CHARGE syndrome who had a microdeletion of 22q11.2 (Chapter 21) (Clementi et al. 1991; Devriendt et al. 1998). In a subsequent study, Corsten-Janssen et al. identified five individuals with a clinical diagnosis of CHARGE syndrome who had a 22q11.2 deletion. Conversely, the authors also found truncating *CHD7* variants in five patients who presented with a 2q11.2 deletion phenotype (Corsten-Janssen et al. 2013b). Other chromosomal imbalances that may mimic CHARGE syndrome have been reviewed (Janssen et al. 2012); therefore, a cytogenomic microarray should be performed in any child with CHARGE syndrome who tests negative for a *CHD7* pathogenic variant.

There are a number of monogenic disorders that show clinical resemblance to CHARGE syndrome. The most important ones are Kabuki syndrome (Chapter 34, *KMT2D*), syndromic micropthalmia (*BMP4, SOX2, OTX2*), Pallister–Hall syndrome (Chapter 45, *GLI3*), mandibulofacial dysostosis (*EFTUD2*), Mowat–Wilson syndrome (Chapter 38,

ZEB2), Kallmann syndrome type 2 (*FGFR1*) and *USP9X*-related disorder (Reijnders et al. 2016). Oculo-auriculo-vertebral spectrum (OAV) (Chapter 42) shares some features with CHARGE association (Van Meter and Weaver 1996); however, OAV spectrum is frequently associated with eyelid coloboma, which is rarely seen in CHARGE syndrome.

MANIFESTATIONS AND MANAGEMENT

For optimal management, individuals with CHARGE syndrome should ideally be referred to a center that can undertake all areas of specialist treatment and management. This helps in long-term follow-up, it benefits the parents and caregivers, and, most importantly, it offers an integrated, multidisciplinary approach. In this way, surgical management can be combined with clinical investigations that also require a general anesthetic. This requires active coordination and cooperation between the appropriate specialties. In 2017, a comprehensive checklist for the follow-up of individuals with CHARGE syndrome was published (Trider et al. 2017).

Growth and Feeding

Although the majority of individuals with CHARGE syndrome have appropriate birth weight and length, over 70% experience postnatal growth retardation, with weight and length falling below the third centile. Feeding difficulties are a major feature in infants with CHARGE syndrome and can persist into childhood and even into adulthood (Blake et al. 2005). Feeding can be associated with weak sucking and chewing, coughing, choking, nasal regurgitation, aspiration, and/or gastroesophageal reflux. Aspiration and swallowing dysfunction are common and are mainly the result of cranial nerve IX/X abnormalities (Dobbelsteyn et al. 2008). Detailed overviews of gastrointestinal problems that occur in CHARGE syndrome provide guidelines in care (Blake and Hudson 2017; Macdonald et al. 2017).

Early feeding problems are likely to improve and catch-up growth to the low normal range can occur. However, some adults with CHARGE syndrome avoid certain foods that are difficult to swallow. Also, problematic feeding behaviors like pocketing of food in cheeks is often seen. Therefore, individualized evaluation of feeding behavior should be a part of the standard otolaryngologic and feeding team care (Hudson et al. 2015).

Children who maintain their weight in the early months are often the ones who have nutritional intervention (usually in the form of gastrostomy tube feeding). In the others, catch-up growth by 3 years of age and normal height velocity are reported to be achieved (Harvey et al. 1991). Many children with severe cardiac problems show poor early growth. In most individuals, ultimate height is at or below the third centile. Growth hormone deficiency has been reported in a few affected individuals (Pinto et al. 2005; Tellier et al. 1998). Gonadotropin deficiency, however, is common and highly associated with anosmia/hyposmia (Bergman et al. 2010b). There is no correlation between the degree of growth retardation and the malformations present in individuals with CHARGE syndrome.

Evaluation

- Growth parameters should be measured and plotted on growth curves at routine well-child visits to assess their adequacy.
- Because gastroesophageal reflux and swallowing difficulties occur commonly, early assessment of feeding should be instituted by a multidisciplinary team, including gastroenterologists, speech and language pathologists, occupational therapists, and dietitians.
- Esophageal pH studies for reflux, barium swallow, and endoscopy may need to be considered. Be aware of the possible presence of a vascular ring that may compress the esophagus (Corsten-Janssen et al. 2016).
- If growth remains poor in early childhood, then growth hormone secretion should be assessed in a standard way.

Management

- Feeding problems in infants with CHARGE syndrome continue to be one of the most difficult management issues. The majority of children will need nasogastric tube feeding at some point (Dobbelsteyn et al. 2008). Because of prolonged feeding problems, many will require gastrostomy tube feeding.
- Aggressive medical management of gastroesophageal reflux is usually necessary. Some children will require a fundoplication procedure for persistent reflux.
- Consider referral to a clinic with expertise in dysphagia.
- In children with ongoing growth deficiency but normal growth hormone levels, growth hormone may be beneficial.

Development and Behavior

It is difficult to determine the frequency and degree of developmental delay in individuals with CHARGE syndrome from published reports because of differences in ascertainment and frequent lack of formal psychometric testing. Raqbi et al. showed that the intellectual performance of individuals with CHARGE syndrome ranged from major learning disability with no speech to almost normal (Raqbi et al. 2003). Only 25% of the studied group had a poor intellectual outcome. In a report by Jongmans et al., 8 of the 32 individuals with a *CHD7* mutation had an IQ above 70 (Jongmans et al. 2006). When assessed, learning difficulties have been reported in 60–100%. However, some of the reports suggesting that most children have learning problems did not involve long-term follow-up. Early motor delay is seen in the majority of affected individuals with severe visual problems and vestibular abnormalities. Deafness and the effects of chronic illness contribute greatly to the motor delay. However, early gross motor delay does not necessarily indicate a worse cognitive prognosis.

To achieve as normal a level of functioning as possible it is essential that all of the individual's deficits be identified and treated as early as possible, particularly the hearing, visual, motor, and language deficits. Potentially beneficial interventions for these deficits may be overlooked by surgical teams intent on achieving survival after a major cardiac, choanal, or tracheoesophageal repair.

The diagnosis of intellectual disability should be made with caution in anyone with impaired hearing and vision, as these individuals are inherently understimulated. Deafness, if not recognized early, may present with delay in language; poor vision may cause delay in fine motor skills. Children with both defective hearing and vision may present at an earlier age with delay in gross motor skills. Problems with balance occur in almost all children due to hypoplastic semicircular canals. Some individuals have autism spectrum disorder. Intellectual disability may be present but should only be diagnosed when the extent of the sensory deficit is known and has been taken into account during developmental testing and when the child is in an adequate educational program. Severe intellectual disability/learning disability appears to be uncommon. Improvement in development is often achieved once parents realize that intellectual disability is not inevitable.

Because it is now clear that intellectual disability is less common than previously thought, "retardation" should not be used as one of the cardinal features of the CHARGE syndrome. Instead, consideration for CHARGE should be given to children with growth or developmental delays as a cardinal fact (Blake et al. 1998). It may appear that a child is going to have significant problems because early intellectual assessments look particularly at gross and fine motor development as well as speech development. Thus, there can be a discrepancy between the results of these early assessments and later cognitive abilities once motor performance and language skills have improved.

Microcephaly, brain malformation, and extensive bilateral colobomas resulting in low vision seem to be the only findings predictive of poor intellectual outcome (Raqbi et al. 2003). The presence of central nervous system malformation is strongly associated with choanal atresia. There is a predominance of forebrain anomalies, particularly arrhinencephaly (Hoch et al. 2017). Anomalies of the olfactory tracts and bulbs are common (Chalouhi et al. 2005).

Children and adults with CHARGE syndrome frequently exhibit moderate-to-severe behavioral difficulties. They are often diagnosed with obsessive-compulsive disorder, attention deficit disorder, and autism

Developmental outcome in many affected individuals will be influenced greatly by the presence of structural brain abnormalities, the presence and severity of visual and hearing deficits, early management of sensory deficits, adequate feeding and nutrition, adverse perinatal factors, coexisting medical and surgical problems, and implementation of early intervention strategies and family supports. Primary care providers need to be particularly alert to the problems of visual and hearing deficits, which may be difficult to detect in infancy or may develop during later childhood.

Evaluation

- Thorough assessment of vision and hearing as well as early referral for appropriate developmental and educational services within the first few months of life are essential.
- If there are multiple sensory deficits, then evaluation should be carried out by specially trained therapists.

Management

- Early evaluation and therapy for developmental delays are vital and should be accomplished by a team of specialists including developmental pediatricians, speech and language therapists, occupational therapists, and physiotherapists, as well as therapists dealing with deaf–blind children.
- Hearing aids/cochlear implants and intensive speech/language therapy should be initiated for children with hearing loss as soon as it is identified.
- If visual correction is needed, it should be accomplished expeditiously. If this is not possible, then an intensive program for the visually handicapped should be instituted.
- If brain imaging is performed, this should be as complete as possible to avoid multiple procedures. Compehensive guidelines are available (de Geus et al. 2017).

Eyes and Vision

The major ocular feature of the CHARGE syndrome is coloboma, which occurs in 80% of affected individuals. These are frequently bilateral and can involve iris, retina, or optic disk. They rarely involve the iris alone. Eyelid notches (often incorrectly called colobomas) occur infrequently (Davenport et al. 1986). Vision loss varies according to the size and location of the coloboma, with a range from no impairment to severe visual loss with lack of light perception (Russell-Eggitt et al. 1990). However, over 50% of children with the CHARGE syndrome have some visual impairment (Kaplan 1989).

Other ocular abnormalities are also common. These include microphthalmia, which has been described in about 30% (Kaplan 1989); nystagmus, which may be related to macular or optic nerve involvement or may be central in origin and can be horizontal, vertical, or rotatory (Russell-Eggitt et al. 1990); and strabismus. Refractive errors are also common. More rare abnormalities include atresia of lacrimal canaliculi, blocked nasolacrimal duct, and congenital glaucoma. True ptosis is uncommon, and more often pseudoptosis is associated with microphthalmia or orbital asymmetry (Russell-Eggitt et al. 1990). Delayed visual maturation has been reported occasionally, particularly in infants who are severely ill from various causes (Davenport et al. 1986).

Evaluation

- It is important for the clinician to be aware of the ophthalmologic features of CHARGE syndrome, because some, such as chorioretinal coloboma, may be undetected until complicated by retinal detachment. Chorioretinal coloboma often occurs without an iris defect and may not be easily detected without a formal ophthalmologic evaluation.
- Early examination by an ophthalmologist is important to determine whether chorioretinal colobomas or other eye anomalies are present. If ocular malformations are detected, then appropriate ophthalmologic follow-up should be arranged.
- Assessment of visual function is important to maximize potential for education.
- Even if initial ophthalmologic assessment is normal, visual acuity should be measured periodically in infancy and early childhood because of the risk of amblyopia as a result of anisometropia and strabismus.
- Prompt ophthalmological evaluation should be pursued after any head trauma due to increased risk of retinal detachment.

Management

- Corrective lenses or low-vision aids may improve visual performance.
- If strabismus is present, then appropriate standard treatment by occlusion or surgery may be necessary.
- Surgery should be performed in a standard manner for retinal detachment or cataracts.

Heart and Aortic Arch Vessels

Between 70 and 80% of individuals with CHARGE syndrome have congenital heart defects. The severity and spectrum of congenital heart defects varies, with no specific defect predominating, although there is a preponderance of atrioventricular septal defects and conotruncal defects. Patent ductus arteriosus is a common finding, either alone or associated with more severe lesions (Corsten-Janssen et al. 2013a).

Conotruncal anomalies (e.g., tetralogy of Fallot, double-outlet right ventricle, and truncus arteriosus) are found in approximately 30% of those with congenital heart disease (Corsten-Janssen et al. 2013a). Aortic arch anomalies (e.g., vascular ring, aberrant subclavian artery, and interrupted aortic arch) are reported in 36% in one study (Lin et al. 1987), and 14% in another (Corsten-Janssen et al., 2016). It is important to diagnose these anomalies since arch vessel anomalies can cause serious morbidity due to compression of the esophagus (Corsten-Janssen et al. 2016). About 75% of congenital heart defects require surgical management, with the remaining 25% being comparatively mild: small atrial septal defect, ventricular septal defect, pulmonary stenosis, patent ductus arteriosus, or a combination of these (Wyse et al. 1993).

Evaluation

- If a neonate presents with cyanosis, it could be because of congenital heart disease, choanal atresia, or a combination. If the child remains cyanotic despite an oral airway being established, the cause is likely to be congenital heart disease. The child should then be stabilized and transported to a tertiary referral center.
- All individuals with CHARGE syndrome should undergo a full cardiac examination at diagnosis, including visualisation of the aortic arch vessels, even if the heart appears to be normal.
- Careful imaging is essential before surgery, as complex and multiple cardiovascular abnormalities can be present that may not be easily detectable.
- Although it may be possible to plan cardiac surgery on the basis of echocardiographic findings alone, cardiac catheterization may be necessary to identify anomalous subclavian vessels and/or to rule out peripheral pulmonary stenosis.

Management

- Cardiac surgery is no different in individuals with CHARGE syndrome than in those with the same cardiac lesion in the general population. However, there is a greater risk of anesthesia complications (Trider et al. 2017).
- Caution must be exercised if prostaglandin administration is indicated to maintain ductal patency before surgery, as there may be high mortality in neonates with CHARGE syndrome who receive prosta-

glandins (Blake et al. 1998). These should only be administered once a neonate is adequately ventilated, has good venous access, and cardiac anatomy has been well defined.
- Individuals with CHARGE syndrome may require an early shunt procedure before a later definitive repair.
- Outcome is optimized when collaboration between specialist surgical teams allows necessary procedures to be performed using the minimum of anesthetics.

Airway and Breathing

Choanal atresia is caused either by failure to develop patency between the nasal cavity and nasopharynx, which normally occurs between the 25th and 45th day of embryonic life, or by an abnormality in the migration of cephalic neural crest after neural tube closure (Kaplan 1989). It can involve both bone and soft tissue and result in either partial or total obstruction of the posterior nasal airway. Computed tomography scan evidence suggests that choanal atresia in CHARGE syndrome may lead to a narrower posterior choanal region with a more contracted nasopharynx than in individuals who have isolated choanal atresia. Between 35 and 60% of individuals with CHARGE syndrome have choanal atresia or stenosis. It is usually bilateral and can be bony or membranous. A female predominance has been reported (Hengerer and Strome 1982; Tellier et al. 1998). Choanal atresia is rare in the presence of cleft palate (Sanlaville and Verloes 2007). The typical presentation of bilateral choanal atresia is respiratory distress in the neonatal period, usually very soon after birth. There may be a history of polyhydramnios in the pregnancy. Because newborns are obligate nose breathers, unilateral choanal atresia may also present as respiratory distress, but this often occurs later and is less severe. The usual presentation of unilateral choanal atresia is noisy breathing, persistent nasal discharge, or feeding difficulty, which may not be detected until after the early neonatal period.

Apart from choanal atresia, other upper airway abnormalities are common and include laryngomalacia, subglottic stenosis, and laryngeal clefts or webs (Stack and Wyse 1991). Tracheoesophageal anomalies are seen in 25%. Laryngomalacia may result in such significant airway problems that tracheostomy is needed. Underlying pharyngeal and laryngeal incoordination may result in aspiration of secretions.

If micrognathia is present, intubation may be difficult because of the combination of small jaw and anterior larynx. Intubation may continue to be a problem with increasing age. With subglottic stenosis, either congenital or acquired, the airway will usually increase in size with age (Stack and Wyse 1991). It is generally advised to combine surgical procedures because of their increased risk of post-operative complications and intubation problems (Bergman et al. 2010a).

Evaluation

- Typically, the nares are patent, and direct visualization of the posterior nasal space may not show evidence of deeper obstruction. In the delivery room, inability to pass a standard nasogastric or suction catheter beyond the nares should alert the physician to the possibility of choanal atresia.
- The diagnosis of atresia may be inferred using simple clinical tests of airway patency such as observing misting on a metal spatula, auscultation of the nares, or using a wisp of cotton wool.
- High-resolution axial computed tomography scanning with contrast is the investigation of choice to diagnose choanal atresia or stenosis (Morgan and Bailey 1990). There have been reports of misleading computed tomography scans caused by accumulation of mucus in the nasal cavities, so careful preparation using vasoconstrictor drops and standard nasal suction is essential if high-quality images are to be obtained.

Management

- Initial management is establishment of the airway using an oral airway or, failing that, by endotracheal intubation.
- A tracheostomy is only indicated if there is an associated anomaly preventing oral intubation.
- Three surgical approaches have been advocated in the treatment of choanal atresia, transpalatal, transnasal, and transseptal. There is no 'gold standard'; however, many surgeons prefer the transnasal endoscopic approach (Riepl et al. 2014). Although the transpalatal route allows good access, it is associated with longer operating time, greater blood loss, longer convalescence, and stunted palatal growth in some individuals (Morgan and Bailey 1990). Some studies found no difference in the success rate compared with the transnasal route (Black et al. 1983), but used the transpalatal approach only in older individuals who had a failed transnasal repair. The advantages of the transnasal repair are preservation of the developing hard palate, shorter operating time, and minimal blood loss. After transnasal correction using a diamond burr, polyethylene tube stents are normally left in situ for 6 weeks, followed by dilation where necessary. The overall success rate has been up to 94% (Morgan et al. 1993). Transnasal approach with a rigid endoscope and electrically powered instrumentation, rather than a manual drill, without prolonged stenting after surgery, gives an 80% success rate (Van Den Abbeele et al. 2002). The transnasal approach is now considered by most to be the technique of choice. The reoperation rate is much

higher in CHARGE syndrome (Stack and Wyse 1991) than in isolated choanal atresia.

Ears and Hearing

Distinctive ear anomalies and/or deafness are reported in 90–100% of affected individuals. The external ear anomalies in CHARGE syndrome are sufficiently distinctive that a tentative diagnosis can be made based on those features alone, and a search for the other associated anomalies instituted. Although the ear abnormality is frequently bilateral, asymmetry is common. The most typical ear findings are low-set ears (64%), asymmetry (34%), small or absent lobes (32%), posterior angulation (25%), and "square" shape (Tellier et al. 1998). The patterns of more severe ear anomalies appear distinct, with Kaplan describing most ears in CHARGE syndrome as "cupped, trumpet, lop, or pixie-like" (Figures 11.1 and 11.2) (Kaplan 1989).

When present, hearing loss may be variable, although the most common finding is moderate-to-severe progressive mixed loss. The conductive component can be attributed to ossicular anomalies and/or middle ear effusion (Vesseur et al. 2016c). Ossicular defects may affect hearing, primarily in the low frequencies, but deficits in the mid- to high-frequency range have also been reported. Conductive loss secondary to middle ear effusion may persist beyond childhood. The sensorineural component can be mild to severe and tends to be greatest in the high frequencies. The resultant audiogram yields a "wedge-shaped" pattern, with a low-frequency conductive loss and concomitant high-frequency sensorineural loss. The bone conduction threshold curve may slope downward from low to high frequencies, whereas the air conduction threshold curve is flat (Brown and Israel 1991). The conductive and sensorineural losses are usually progressive, although the rate of progression tends to be slow and it may arrest (Davenport et al. 1986). The presence of facial nerve palsy is reliable in predicting the presence of sensorineural hearing loss (Edwards et al. 2002). A study of hearing loss and cognitive abilities showed that 38% (18/47) individuals with CHARGE syndrome have moderate hearing loss, and that hearing loss and cognitive delays have a significant impact on language acquisition (Vesseur et al. 2016b). Cochlear implantation should be considered in all individuals with CHARGE syndrome and severe sensorineural hearing loss, after careful work-up including CT and MRI scans, audiometry, and assessment by a specialized multidisciplinary team (Choo et al. 2017; Song et al. 2011; Vesseur et al. 2016a).

Vestibular abnormalities are found in the majority of affected individuals. Characteristic abnormalities demonstrated by temporal bone computed tomography (CT) or magnetic resonance imaging (MRI) scans include hypoplastic incus, decreased number of turns to the cochlea (Mondini defect), and, particularly, absent semicircular canals (Amiel et al. 2001; Choo et al. 2017; Collins and Buchman 2002; Vesseur et al. 2016c). The auditory nerves can be seriously hypoplastic, which may affect the mode of treatment; however, the remaining auditory function is often difficult to assess. If a child does not respond on cochlear implantation (CI), an auditory brainstem implant (ABI) maybe an alternative (Vesseur et al. 2018, PMID 29494474).

Evaluation

- All individuals with CHARGE syndrome should have an audiologic evaluation at diagnosis.
- Initial assessment consists of examination of the tympanic membranes to exclude a middle ear effusion.
- Brain stem-evoked audiometry may subsequently be used to help assess the level of hearing.
- CT or MRI of temporal bone is used to exclude ossicular or inner ear abnormalities.
- Assessment of hearing is often difficult in children who have multisensory deprivation, and this must be undertaken by a team of specialists.
- Regular audiologic assessment is recommended, at least yearly, because the hearing loss may be progressive and there may be ongoing problems with middle ear effusions.

Management

- Chronic serous otitis media is common in CHARGE syndrome and may require early and repeated myringotomies with insertion of ventilation tubes. These tubes may make an enormous difference to quality of life, and tube insertion needs to be undertaken early, because the first 2 years of life are vitally important in the development of speech.
- Involvement of a teacher of the deaf at an early age will allow counseling for parents and offer better provision of educational needs.
- Most individuals with CHARGE syndrome and hearing loss are candidates for hearing aids. However, there are several factors that reduce the success rate. Malformations and softness of the pinnae, as well as small ear canals, make fitting of hearing aids difficult. Many individuals have sufficiently large cochlear losses that bone conduction aids will not provide adequate amplification as an alternative. The children are often supplied with hearing aids too late, and this can result in poor compliance, especially if they have acquired the habit of disregarding sounds.
- Cochlear implants are useful in many individuals (MacArdle et al. 2002). There may be significant technical challenges in children with CHARGE syn-

drome because of variations in temporal bone anatomy, but early implantation is associated with improved outcomes (Bauer et al. 2002; Vesseur et al. 2016a; Vesseur et al. 2016b).
- If cochlear nerve aplasia is seen on MRI and the child does not attain speech perception, an auditory brain implant should be considered (Vesseur et al. 2018, PMID 29494474)

Genitourinary Tract

The presence of micropenis and undescended testes makes the early diagnosis of CHARGE syndrome easier in boys. Hypospadias is uncommon, and penile agenesis and microtestis are rare. Genital anomalies in females are uncommon, but hypoplastic labia and clitoris, and atresia of the uterus, cervix, and vagina have been reported (Blake et al. 2005).

Pubertal development in both sexes may be delayed or absent and may be correlated with pituitary or hypothalamic deficiency. Some individuals with recognized CHARGE syndrome have entered puberty, and for those, fertility is usually normal. Parent-to-child transmission has been described in this group. Individuals with hypogonadotropic hypogonadism can be treated by gonadotropins to try to restore fertility, but there are no data available of success rates.

Renal anomalies are common, occurring in about 25% (Blake et al. 1998). These include hydronephrosis, renal hypoplasia, duplex kidneys, vesicoureteral reflux, and single kidney. There is a significant association between the presence of a facial palsy and renal anomalies (Blake et al. 1998).

Adrenal function is seldomly abnormal (Wong et al. 2016).

Evaluation

- If micropenis and cryptorchidism are present, assessment of hypothalamic-pituitary function is indicated. Referral to pediatric endocrinology could be considered. Basal follicle-stimulating hormone (FSH), luteinizing hormone (LH), and testosterone levels should be measured, preferably around 6 weeks after birth (mini-puberty). At a later age, a 3–5 day intramuscular stimulation test with human chorionic gonadotropin (hCG) to evaluate testicular function for testosterone production, as well as an intravenous stimulation test with luteinizing hormone-releasing hormone (LHRH) to evaluate the integrity of the hypothalamic-pituitary axis should be considered. If cryptorchidism is present, assessment of testicular position by pelvic/abdominal ultrasound is recommended.
- Evaluation of growth hormone is indicated only based on growth curves.
- Moderate delay in the onset of puberty is common in CHARGE syndrome, even when pituitary function appears normal. This "normal" delay should be taken into consideration when contemplating investigation. If the initial physical changes of puberty are not present by age 13 years in girls or age 14 years in boys, evaluation of the hypothalamic-pituitary-gonadal axis should be considered. This would include bone age and gonadotropin levels, both before and after gonadotropin-releasing hormone stimulation.
- Anosmia and hypogonadotropic hypogonadism are highly correlated in CHARGE syndrome. If anosmia is present then the child is most likely not going to enter puberty spontaneously and hormone replacement therapy should be considered (Bergman et al. 2010b).
- Urinary tract screening, including at least renal ultrasound and, if indicated, a voiding cystourethrogram, should be arranged in infancy or at diagnosis (Ragan et al. 1999).

Management

- In a boy with micropenis, one or two short courses of testosterone therapy in infancy and childhood should augment penile size into normal range for age (Bin-Abbas et al. 1999). Adequate penile growth with this form of therapy may avoid some of the potential psychological issues associated with small penis. Replacement therapy at age of puberty may result in a near-normal adult-size penis.
- In cryptorchidism surgery between 6 and 12 months of age is the preferred mode of treatment (Ritzen 2008). The long-term outlook for fertility remains largely unknown.
- Individuals of both sexes with CHARGE syndrome may require hormone therapy to achieve puberty. This could involve intramuscular testosterone or subcutaneous gonadotropins in males. In females, low-dosage estrogen therapy initially followed by cyclic estrogen–progesterone therapy is preferred.

Immunology

Recurrent infections have been reported in patients with CHARGE syndrome and are mostly related to anatomic anomalies (e.g. middle ear infections) and neurological dysfunction (i.e. aspiration due to poor swallowing). In addition, abnormalities of the immune system include mild T-cell dysfunction to severe combined immune deficiency (reviewed in (Wong et al. 2015b)). Given the clinical overlap with the 22q11 deletion syndrome (Chapter 21), it is not surprising that T-cell dysfunction occurs (Chopra et al. 2009; Wong et al. 2015a), althought this is not confirmed by others (Hsu et al. 2016; Mehr et al. 2017).

Evaluation

- Specialized immunological evaluation (B- and T-cell numbers and vaccination responses) if patient has recurrent infections.

Management

- Consider booster vaccines if vaccination response is insufficient (Wong et al. 2015a).

ACKNOWLEDGMENTS

We thank all the individuals and families who generously participate in research on CHARGE syndrome and those who consented for photography. DMM is supported by NIH R01 DC009410, R01 DC014456, and by the Donita B. Sullivan, MD Research Professorship in Pediatrics and Communicable Diseases. D.M.M. is Chair and CMAV is a member of the Scientific Advisory Board for the International CHARGE Syndrome Foundation.

RESOURCES

USA Support Group: CHARGE Syndrome Foundation, Inc.

141 Middle Neck Road
Sands Point, New York 11050, USA
Phone: (516) 684 4720
Fax: (516) 883 9060
Email: *info@chargesyndrome.org*
Website: *http://www.chargesyndrome.org*

UK Support Group: CHARGE Family Support Group

Carol Thomas Burnside
50 Commercial Street
Slaithwaite
Huddersfield, HD7 5JZ, UK
Phone: (01484) 844202
Email: *cajthomas@btinternet.com*
Website: *http://www.chargesyndrome.org.uk*

Australasia Support Group: CHARGE Syndrome Association of Australasia

Phone: Australia: +61 (0) 7 3357 1191
Phone: New Zealand: +64 (0) 3 379 6305
Email: *admin@chargesyndrome.org.au*
Website: *https://www.chargesyndrome.org.au/*

German CHARGE Syndrome Support Group

Website: *http://www.charge-syndrom.de/*

French CHARGE Syndrome Support Group

Website: *http://www.associationcharge.fr/*

REFERENCES

Amiel J, Attiee-Bitach T, Marianowski R, Cormier-Daire V, Abadie V, Bonnet D, Gonzales M, Chemouny S, Brunelle F, Munnich A, Manach Y, Lyonnet S (2001) Temporal bone anomaly proposed as a major criteria for diagnosis of CHARGE syndrome. *Am J Med Genet* 99(2):124–127.

Bajpai R, Chen DA, Rada-Iglesias A, Zhang J, Xiong Y, Helms J, Chang CP, Zhao Y, Swigut T, Wysocka J (2010. CHD7 cooperates with PBAF to control multipotent neural crest formation. *Nature* 463(7283):958–962.

Balasubramanian R, Crowley WF, Jr (2017) Reproductive endocrine phenotypes relating to CHD7 mutations in humans. *Am J Med Genet C Semin Med Genet* 175(4):507–515.

Bauer PW, Wippold FJ, 2nd, Goldin J, Lusk RP (2002) Cochlear implantation in children with CHARGE association. *Arch Otolaryngol Head Neck Surg* 128(9):1013–1017.

Bergman JE, Blake KD, Bakker MK, du Marchie Sarvaas GJ, Free RH, van Ravenswaaij-Arts CM (2010a) Death in CHARGE syndrome after the neonatal period. *Clin Genet* 77(3):232–240.

Bergman JE, Bocca G, Hoefsloot LH, Meiners LC, van Ravenswaaij-Arts CM (2010b) Anosmia Predicts Hypogonadotropic Hypogonadism in CHARGE Syndrome. *J Pediatr.* 158:474–479.

Bergman JE, de Ronde W, Jongmans MC, Wolffenbuttel BH, Drop SL, Hermus A, Bocca G, Hoefsloot LH, van Ravenswaaij-Arts CM (2012a) The results of CHD7 analysis in clinically well-characterized patients with Kallmann syndrome. *J Clin Endocrinol Metab* 97(5):E858–862.

Bergman JEH, Janssen N, Hoefsloot LH, Jongmans MCJ, Hofstra RMW, van Ravenswaaij-Arts CMA (2011) CHD7 mutations and CHARGE syndrome: the clinical implications of an expanding phenotype. *J Med Genet* 48(5):334–342.

Bergman JEH, Janssen N, van der Sloot AM, de Walle HEK, Schoots J, Rendtorff ND, Tranebjaerg L, Hoefsloot LH, van Ravenswaaij-Arts CMA, Hofstra RMW (2012b) A novel classification system to predict the pathogenic effects of CHD7 missense variants in CHARGE syndrome. *Hum Mutat* 33(8):1251–1260.

Bin-Abbas B, Conte FA, Grumbach MM, Kaplan SL (1999) Congenital hypogonadotropic hypogonadism and micropenis: effect of testosterone treatment on adult penile size why sex reversal is not indicated. *J Pediatr* 134(5):579–583.

Black RJ, Pracy R, Evans JN (1983) Congenital posterior choanal atresia. *Clin Otolaryngol Allied Sci* 8(4):251–255.

Blake KD, Davenport SL, Hall BD, Hefner MA, Pagon RA, Williams MS, Lin AE, Graham JM, Jr (1998) CHARGE association: an update and review for the primary pediatrician. *Clin Pediatr* 37(3):159–173.

Blake KD, Hudson AS (2017) Gastrointestinal and feeding difficulties in CHARGE syndrome: A review from head-to-toe. *Am J Med Genet C Semin Med Genet.* 175:496–506.

Blake KD, Russell-Eggitt IM, Morgan DW, Ratcliffe JM, Wyse RK (1990) Who's in CHARGE? *Multidisciplinary management of patients with CHARGE association.* I65(2):217–223.

Blake KD, Salem-Hartshorne N, Daoud MA, Gradstein J (2005) Adolescent and adult issues in CHARGE syndrome. *Clin Pediatr* 44(2):151–159.

Brown DP, Israel SM (1991) Audiologic findings in a set of fraternal twins with CHARGE association. *J Am Acad Audiol* 2(3):183–188.

Chalouhi C, Faulcon P, Le Bihan C, Hertz-Pannier L, Bonfils P, Abadie V (2005) Olfactory evaluation in children: application to the CHARGE syndrome. *Pediatrics* 116(1):e81–88.

Choo DI, Tawfik KO, Martin DM, Raphael Y (2017) Inner ear manifestations in CHARGE: Abnormalities, treatments, animal models, and progress toward treatments in auditory and vestibular structures. *Am J Med Genet C Semin Med Genet.* 175:439–449.

Chopra C, Baretto R, Duddridge M, Browning MJ (2009) T-cell immunodeficiency in CHARGE syndrome. *Acta Paediatr* 98(2):408–410.

Clementi M, Tenconi R, Turolla L, Silvan C, Bortotto L, Artifoni L (1991) Apparent CHARGE association and chromosome anomaly: chance or contiguous gene syndrome. *Am J Med Genet* 41(2):246–250.

Collins WO, Buchman CA (2002) Bilateral semicircular canal aplasia: a characteristic of the CHARGE association. *Otology & neurotology: official publication of the American Otological Society, American Neurotology Society (and) European Academy of Otology and Neurotology* 23(2):233–234.

Corsten-Janssen N, Kerstjens-Frederikse WS, du Marchie Sarvaas GJ, Baardman ME, Bakker MK, Bergman JE, Hove HD, Heimdal KR, Rustad CF, Hennekam RC, Hofstra RM, Hoefsloot LH, Van Ravenswaaij-Arts CM, Kapusta L (2013a) The cardiac phenotype in patients with a CHD7 mutation. *Circ Cardiovasc Genet* 6(3):248–254.

Corsten-Janssen N, Saitta SC, Hoefsloot LH, McDonald-McGinn DM, Driscoll DA, Derks R, Dickinson KA, Kerstjens-Frederikse WS, Emanuel BS, Zackai EH, van Ravenswaaij-Arts CM (2013b) More Clinical Overlap between 22q11.2 Deletion Syndrome and CHARGE Syndrome than Often Anticipated. *Mol Syndromol* 4(5):235–245.

Corsten-Janssen N, van Ravenswaaij-Arts CMA, Kapusta L (2016) Congenital arch vessel anomalies in CHARGE syndrome: A frequent feature with risk for co-morbidity. *Int J Cardiol Heart Vasc* 12:21–25.

Davenport SL, Hefner MA, Mitchell JA (1986) The spectrum of clinical features in CHARGE syndrome. *Clin Genet* 29(4):298–310.

de Geus CM, Free RH, Verbist BM, Sival DA, Blake KD, Meiners LC, van Ravenswaaij-Arts CMA (2017) Guidelines in CHARGE syndrome and the missing link: Cranial imaging. *Am J Med Genet C Semin Med Genet* 175(4):450–464.

Devriendt K, Swillen A, Fryns JP (1998) Deletion in chromosome region 22q11 in a child with CHARGE association. *Clin Genet* 53(5):408–410.

Dobbelsteyn C, Peacocke SD, Blake K, Crist W, Rashid M (2008) Feeding difficulties in children with CHARGE syndrome: prevalence, risk factors, and prognosis. *Dysphagia* 23(2):127–135.

Edwards BM, Kileny PR, Van Riper LA (2002) CHARGE syndrome: a window of opportunity for audiologic intervention. *Pediatrics* 110(1 Pt 1):119–126.

Hale CL, Niederriter AN, Green GE, Martin DM (2016) Atypical phenotypes associated with pathogenic CHD7 variants and a proposal for broadening CHARGE syndrome clinical diagnostic criteria. *Am J Med Genet A* 170A(2):344–354.

Hall BD (1979) Choanal atresia and associated multiple anomalies. *J Pediatr* 95(3):395–398.

Hartshorne TS, Grialou TL, Parker KR (2005) Autistic-like behavior in CHARGE syndrome. *Am J Med Genet A* 133(3):257–261.

Hartshorne TS, Stratton KK, Brown D, Madhavan-Brown S, Schmittel MC (2017) Behavior in CHARGE syndrome. *Am J Med Genet C Semin Med Genet* 175(4):431–438.

Harvey AS, Leaper PM, Bankier A (1991) CHARGE association: clinical manifestations and developmental outcome. *Am J Med Genet* 39(1):48–55.

Hengerer AS, Strome M (1982) Choanal atresia: a new embryologic theory and its influence on surgical management. *Laryngoscope* 92(8 Pt 1):913–921.

Hoch MJ, Patel SH, Jethanamest D, Win W, Fatterpekar GM, Roland JT, Jr., Hagiwara M (2017) Head and Neck MRI Findings in CHARGE Syndrome. *Am J Neuroradiol.* 38:2357–2363.

Hsu P, Ma A, Barnes EH, Wilson M, Hoefsloot LH, Rinne T, Munns C, Williams G, Wong M, Mehr S (2016) The Immune Phenotype of Patients with CHARGE Syndrome. *J Allergy Clin Immunol Pract* 4(1):96–103 e102.

Hudson A, Colp M, Blake K (2015) Pocketing of food in cheeks during eating in an adolescent with CHARGE syndrome. *J Paediatr Child Health* 51(11):1143–1144.

Hughes SS, Welsh HI, Safina NP, Bejaoui K, Ardinger HH (2014) Family history and clefting as major criteria for CHARGE syndrome. *Am J Med Genet A* 164A(1):48–53.

Issekutz KA, Graham JM, Jr., Prasad C, Smith IM, Blake KD (2005) An epidemiological analysis of CHARGE syndrome: preliminary results from a Canadian study. *Am J Med Genet A* 133(3):309–317.

Janssen N, Bergman JE, Swertz MA, Tranebjaerg L, Lodahl M, Schoots J, Hofstra RM, van Ravenswaaij-Arts CM, Hoefsloot LH (2012) Mutation update on the CHD7 gene involved in CHARGE syndrome. *Hum Mutat* 33(8):1149–1160.

Jongmans MC, Hoefsloot LH, van der Donk KP, Admiraal RJ, Magee A, van de Laar I, Hendriks Y, Verheij JB, Walpole I, Brunner HG, van Ravenswaaij CM (2008) Familial CHARGE syndrome and the CHD7 gene: a recurrent missense mutation, intrafamilial recurrence and variability. *Am J Med Genet A* 146A(1):43–50.

Jongmans MCJ, Admiraal RJ, van der Donk KP, Vissers LELM, Baas AF, Kapusta L, van Hagen JM, Donnai D, de Ravel TJ, Veltman JA, van Kessel AG, De Vries BBA, Brunner HG, Hoefsloot LH, van Ravenswaaij CMA (2006) CHARGE syndrome: the phenotypic spectrum of mutations in the CHD7 gene. *J Med Genet* 43(4):306–314.

Jongmans MCJ, van Ravenswaaij-Arts CMA, Pitteloud N, Ogata T, Sato N, Claahsen-van der Grinten HL, van der Donk K, Seminara S, Bergman JEH, Brunner HG, Crowley WF,

Hoefsloot LH (2009) CHD7 mutations in patients initially diagnosed with Kallmann syndrome - the clinical overlap with CHARGE syndrome. *Clin Genet* 75(1):65–71.

Kaplan LC (1989) The CHARGE association: choanal atresia and multiple congenital anomalies. *Otolaryngol Clin North Am* 22(3):661–672.

Kim HG, Kurth I, Lan F, Meliciani I, Wenzel W, Eom SH, Kang GB, Rosenberger G, Tekin M, Ozata M, Bick DP, Sherins RJ, Walker SL, Shi Y, Gusella JF, Layman LC (2008) Mutations in CHD7, encoding a chromatin-remodeling protein, cause idiopathic hypogonadotropic hypogonadism and Kallmann syndrome. *Am J Hum Genet* 83(4):511–519.

Lalani SR, Hefner MA, Belmont JW, Davenport SLH (1993) CHARGE Syndrome. *GeneReviews(R)*. https://www.ncbi.nlm.nih.gov/books/NBK1117/

Lalani SR, Safiullah AM, Fernbach SD, Harutyunyan KG, Thaller C, Peterson LE, McPherson JD, Gibbs RA, White LD, Hefner M, Davenport SL, Graham JM, Bacino CA, Glass NL, Towbin JA, Craigen WJ, Neish SR, Lin AE, Belmont JW (2006) Spectrum of CHD7 Mutations in 110 Individuals with CHARGE Syndrome and Genotype-Phenotype Correlation. *Am J Hum Genet* 78(2):303–314.

Legendre M, Abadie V, Attie-Bitach T, Philip N, Busa T, Bonneau D, Colin E, Dollfus H, Lacombe D, Toutain A, Blesson S, Julia S, Martin-Coignard D, Genevieve D, Leheup B, Odent S, Jouk PS, Mercier S, Faivre L, Vincent-Delorme C, Francannet C, Naudion S, Mathieu-Dramard M, Delrue MA, Goldenberg A, Heron D, Parent P, Touraine R, Layet V, Sanlaville D, Quelin C, Moutton S, Fradin M, Jacquette A, Sigaudy S, Pinson L, Sarda P, Guerrot AM, Rossi M, Masurel-Paulet A, El Chehadeh S, Piguel X, Rodriguez-Ballesteros M, Ragot S, Lyonnet S, Bilan F, Gilbert-Dussardier B (2017) Phenotype and genotype analysis of a French cohort of 119 patients with CHARGE syndrome. *Am J Med Genet C Semin Med Genet* 175(4):417–430.

Lin AE, Chin AJ, Devine W, Park SC, Zackai E (1987) The pattern of cardiovascular malformation in the CHARGE association. *Am J Dis Child* 141(9):1010–1013.

Lin AE, Siebert JR, Graham JM, Jr (1990) Central nervous system malformations in the CHARGE association. *Am J Med Genet* 37(3):304–310.

Lubinsky MS (1994) Properties of associations: identity, nature, and clinical criteria, with a commentary on why CHARGE and Goldenhar are not associations (editorial; comment). *Am J Med Genet* 49(1):21–25.

MacArdle BM, Bailey C, Phelps PD, Bradley J, Brown T, Wheeler A (2002) Cochlear implants in children with craniofacial syndromes: assessment and outcomes. *Int J Audiol* 41(6):347–356.

Macdonald M, Hudson A, Bladon A, Ratcliffe E, Blake K (2017) Experiences in feeding and gastrointestinal dysfunction in children with CHARGE syndrome. *Am J Med Genet A* 173(11):2947–2953.

Marcos S, Sarfati J, Leroy C, Fouveaut C, Parent P, Metz C, Wolczynski S, Gerard M, Bieth E, Kurtz F, Verier-Mine O, Perrin L, Archambeaud F, Cabrol S, Rodien P, Hove H, Prescott T, Lacombe D, Christin-Maitre S, Touraine P, Hieronimus S, Dewailly D, Young J, Pugeat M, Hardelin JP, Dode C (2014) The prevalence of CHD7 missense versus truncating mutations is higher in patients with Kallmann syndrome than in typical CHARGE patients. *J Clin Endocrinol Metab* 99(10):E2138–2143.

Mehr S, Hsu P, Campbell D (2017) Immunodeficiency in CHARGE syndrome. *Am J Med Genet C Semin Med Genet* 175(4):516–523.

Metlay LA, Smythe PS, Miller ME (1987) Familial CHARGE syndrome: clinical report with autopsy findings. *Am J Med Genet* 26(3):577–581.

Morgan D, Bailey M, Phelps P, Bellman S, Grace A, Wyse R (1993) Ear-nose-throat abnormalities in the CHARGE association. *Arch Otolaryngol Head Neck Surg* 119(1):49–54.

Morgan DW, Bailey CM (1990) Current management of choanal atresia. *Int J Pediatr Otorhinolaryngol* 19(1):1–13.

Oley CA, Baraitser M, Grant DB (1988) A reappraisal of the CHARGE association. *J Med Genet* 25(3):147–156.

Pagon RA, Graham JM, Jr., Zonana J, Yong SL (1981) Coloboma, congenital heart disease, and choanal atresia with multiple anomalies: CHARGE association. *J Pediatr* 99(2):223–227.

Pinto G, Abadie V, Mesnage R, Blustajn J, Cabrol S, Amiel J, Hertz-Pannier L, Bertrand AM, Lyonnet S, Rappaport R, Netchine I (2005) CHARGE syndrome includes hypogonadotropic hypogonadism and abnormal olfactory bulb development. *J Clin Endocrinol Metab* 90(10):5621–5626.

Ragan DC, Casale AJ, Rink RC, Cain MP, Weaver DD (1999) Genitourinary anomalies in the CHARGE association. *J Urol* 161(2):622–625.

Raqbi F, Le Bihan C, Morisseau-Durand MP, Dureau P, Lyonnet S, Abadie V (2003) Early prognostic factors for intellectual outcome in CHARGE syndrome. *Dev Med Child Neurol* 45(7):483–488.

Reijnders MR, Zachariadis V, Latour B, Jolly L, Mancini GM, Pfundt R, Wu KM, van Ravenswaaij-Arts CM, Veenstra-Knol HE, Anderlid BM, Wood SA, Cheung SW, Barnicoat A, Probst F, Magoulas P, Brooks AS, Malmgren H, Harila-Saari A, Marcelis CM, Vreeburg M, Hobson E, Sutton VR, Stark Z, Vogt J, Cooper N, Lim JY, Price S, Lai AH, Domingo D, Reversade B, Study DDD, Gecz J, Gilissen C, Brunner HG, Kini U, Roepman R, Nordgren A, Kleefstra T (2016) De Novo Loss-of-Function Mutations in USP9X Cause a Female-Specific Recognizable Syndrome with Developmental Delay and Congenital Malformations. *Am J Hum Genet* 98(2):373–381.

Riepl R, Scheithauer M, Hoffmann TK, Rotter N (2014) Transnasal endoscopic treatment of bilateral choanal atresia in newborns using balloon dilatation: own results and review of literature. *Int J Pediatr Otorhinolaryngol* 78(3):459–464.

Ritzen EM (2008) Undescended testes: a consensus on management. *Eur J Endocrinol* 159 Suppl 1:S87–90.

Russell-Eggitt IM, Blake KD, Taylor DS, Wyse RK (1990) The eye in the CHARGE association. *Br J Ophthalmol* 74(7):421–426.

Sanlaville D, Verloes A (2007) CHARGE syndrome: an update. *Eur J Hum Genet* 15(4):389–399.

Siebert JR, Graham JM, Jr., MacDonald C (1985) Pathologic features of the CHARGE association: support for involvement of the neural crest. *Teratology* 31(3):331–336.

Song MH, Cho HJ, Lee HK, Kwon TJ, Lee WS, Oh S, Bok J, Choi JY, Kim UK (2011) CHD7 mutational analysis and clinical considerations for auditory rehabilitation in deaf patients with CHARGE syndrome. *PLoS One* 6(9):e24511.

Stack CG, Wyse RK (1991) Incidence and management of airway problems in the CHARGE Association. *Anaesthesia* 46(7):582–585.

Tellier AL, Cormier-Daire V, Abadie V, Amiel J, Sigaudy S, Bonnet D, de Lonlay-Debeney P, Morrisseau-Durand MP, Hubert P, Michel JL, Jan D, Dollfus H, Baumann C, Labrune P, Lacombe D, Philip N, LeMerrer M, Briard ML, Munnich A, Lyonnet S (1998) CHARGE syndrome: report of 47 cases and review. *Am J Med Genet* 76(5):402–409.

Trider CL, Arra-Robar A, van Ravenswaaij-Arts C, Blake K (2017) Developing a CHARGE syndrome checklist: Health supervision across the lifespan (from head to toe). *Am J Med Genet A* 173(3):684–691.

Van de Laar I, Dooijes D, Hoefsloot L, Simon M, Hoogeboom J, Devriendt K (2007) Limb anomalies in patients with CHARGE syndrome: an expansion of the phenotype. *Am J Med Genet A* 143A(22):2712–2715.

Van Den Abbeele T, Francois M, Narcy P (2002) Transnasal endoscopic treatment of choanal atresia without prolonged stenting. *Arch Otolaryngol Head Neck Surg* 128(8):936–940.

Van Meter TD, Weaver DD (1996) Oculo-auriculo-vertebral spectrum and the CHARGE association: clinical evidence for a common pathogenetic mechanism. *Clin Dysmorphol* 5(3):187–196.

van Ravenswaaij-Arts CM, Blake K, Hoefsloot L, Verloes A (2015) Clinical utility gene card for: CHARGE syndrome - update 2015. *Eur J Hum Genet* 23(11).

Verloes A (2005) Updated diagnostic criteria for CHARGE syndrome: a proposal. *Am J Med Genet A* 133(3):306–308.

Vesseur A, Free R, Langereis M, Snels C, Snik A, Ravenswaaij-Arts C, Mylanus E (2016a) Suggestions for a Guideline for Cochlear Implantation in CHARGE Syndrome. *Otology & Neurotology: official publication of the American Otological Society, American Neurotology Society (and) European Academy of Otology and Neurotology* 37(9):1275–1283.

Vesseur A, Langereis M, Free R, Snik A, van Ravenswaaij-Arts C, Mylanus E (2016b) Influence of hearing loss and cognitive abilities on language development in CHARGE Syndrome. *Am J Med Genet A* 170(8):2022–2030.

Vesseur AC, Verbist BM, Westerlaan HE, Kloostra FJJ, Admiraal RJC, van Ravenswaaij-Arts CMA, Free RH, Mylanus EAM (2016c) CT findings of the temporal bone in CHARGE syndrome: aspects of importance in cochlear implant surgery. *Eur Arch Otorhinolaryngol* 273(12):4225–4240.

Vissers LELM, van Ravenswaaij CMA, Admiraal R, Hurst JA, de Vries BBA, Janssen IM, van der Vliet WA, Huys EHLPG, de Jong PJ, Hamel BCJ, Schoenmakers EFPM, Brunner HG, Veltman JA, van Kessel AG (2004) Mutations in a new member of the chromodomain gene family cause CHARGE syndrome. *Nat Genet* 36(9):955–957.

Wong MT, Lambeck AJ, van der Burg M, la Bastide-van Gemert S, Hogendorf LA, van Ravenswaaij-Arts CM, Scholvinck EH (2015a) Immune Dysfunction in Children with CHARGE Syndrome: A Cross-Sectional Study. *PLoS One* 10(11): e0142350.

Wong MT, Scholvinck EH, Lambeck AJ, van Ravenswaaij-Arts CM (2015b) CHARGE syndrome: a review of the immunological aspects. *Eur J Hum Genet* 23(11):1451–1459.

Wong MT, van Ravenswaaij-Arts CM, Munns CF, Hsu P, Mehr S, Bocca G (2016) Central Adrenal Insufficiency Is Not a Common Feature in CHARGE Syndrome: A Cross-Sectional Study in 2 Cohorts. *J Pediatr* 176:150–155.

Wyse RK, al-Mahdawi S, Burn J, Blake K (1993) Congenital heart disease in CHARGE association. *Pediatr Cardiol* 14(2):75–81.

Xu C, Cassatella D, van der Sloot AM, Quinton R, Hauschild M, De Geyter C, Fluck C, Feller K, Bartholdi D, Nemeth A, Halperin I, Pekic Djurdjevic S, Maeder P, Papadakis G, Dwyer AA, Marino L, Favre L, Pignatelli D, Niederlander NJ, Acierno J, Jr., Pitteloud N (2017) Evaluating CHARGE syndrome in congenital hypogonadotropic hypogonadism patients harboring CHD7 variants. Genet med

Zentner GE, Layman WS, Martin DM, Scacheri PC (2010) Molecular and phenotypic aspects of CHD7 mutation in CHARGE syndrome. *Am J Med Genet A* 152A(3):674–686.

12

COFFIN–LOWRY SYNDROME

R. Curtis Rogers

Greenwood Genetic Center – Greenville, Greenville, South Carolina, USA; Clemson University, Clemson, Pickens, South Carolina, USA

INTRODUCTION

Incidence

Coffin–Lowry syndrome (CLS) is a well-established X-linked dominantly inherited intellectual disability disorder that has a distinctive facial appearance and associated clinical signs that may be helpful in its diagnosis. It was initially described by Coffin et al. in 1966. Lowry et al. (Lowry et al. 1971) reported a three-generation family with similar features and intellectual disability. Males were most commonly affected and it was surmised to be a dominant disorder, most likely X-linked since there was no male-to-male transmission. Temtamy et al. (1975) recognized that the case reports were consistent with a single disorder and suggested the eponym Coffin–Lowry syndrome. Well over 100 affected individuals have been described in the literature since the original report (Hunter 2002). Delaunoy et al. (2001) have reported screening 250 unrelated people with "clinical features suggestive of Coffin–Lowry syndrome" for mutations in the responsible gene, *RPS6KA3* (ribosomal S6 kinase), and a further 65 affected individuals have been studied by Abidi and Schwartz (unpublished data, cited in Hunter et al. 2002).

There have been no systematic studies to determine the incidence of Coffin–Lowry syndrome in individuals with intellectual disability. The estimated minimum birth prevalence for CLS is 1 in 40,000–50,000 live births (Marques Pereira et al. 2010), and there does not appear to be any specific ethnic predisposition.

Likewise, there are no data regarding the optimized life expectancy in CLS. A review of 111 affected individuals found that 12 of 89 affected males and 1 of 22 affected females were deceased at the time they were reported (Hunter 2002). The mean age at death for males was 20.5 years with a range of 13–34 years, and the woman died at the age of 48 years. Several manuscripts also mention affected sibs who had died, but they were not described in detail. The cause of death was not always clear, but cardiac, respiratory, skeletal (kyphoscoliosis), neurological factors, and neoplasia each played roles. Subsequently, Coffin (2003) reported the deaths, at 18 years of age, of two of the original cases; one from acute aspiration of food, and the other from pneumonia and pulmonary abscess. Both had evidence of chronic pulmonary and cardiac disease. A further unreported individual with chronic mixed obstructive and central sleep apnea died of postoperative respiratory complications following jaw advancement surgery.

Diagnostic Criteria

Coffin–Lowry syndrome is characterized by intellectual disability, variable microcephaly, typical facial appearance, hyperextensible hands that have fleshy, tapering digits, progressive kyphoscoliosis, cortical hyperostosis, and other minor radiological changes. The diagnosis requires a clinical suspicion and can be confirmed, but not ruled out (see below), by mutation analysis. No formal diagnostic criteria have been developed. Older individuals typically have severe to profound intellectual disability, but the level of intellectual disability may be more variable in younger individuals. Beyond infancy the diagnosis in the affected male is usually

Cassidy and Allanson's Management of Genetic Syndromes, Fourth Edition.
Edited by John C. Carey, Agatino Battaglia, David Viskochil, and Suzanne B. Cassidy.
© 2021 John Wiley & Sons, Inc. Published 2021 by John Wiley & Sons, Inc.

suspected based on the characteristic facial features. However, with the advent of DNA-based diagnosis it has become clear that the diagnosis is more challenging in some individuals (Field et al. 2006). Microcephaly is a common but not universal finding. The forehead and supraorbital ridges are prominent. Hypertelorism and downslanting palpebral fissures are often marked, but either or both may be absent. The nose has a low bridge and blunt tip and the nares are small because of thick alae nasi and nasal septum. The mouth tends to be held open, often showing small widely spaced teeth. The lips are full and patulous with eversion of the lower lip (Figure 12.1). Although the ears may appear large, they generally measure within normal limits. The face appears mildly coarsened in childhood, becoming coarser with age and often taking on a somewhat pugilistic appearance.

Some of the most characteristic signs of CLS are seen in the hands. The fingers taper distally and they are extremely hyperextensible (Figure 12.2). The hands, including the fingers, have a soft, fleshy fullness. A horizontal palmar crease frequently is found across the hypothenar area. Molecularly diagnosed males without the characteristic hand changes have been reported. Pectus carinatum and/or excavatum and progressive kyphoscoliosis are common, albeit non-specific, findings in CLS.

Radiological signs are nonspecific but may provide additional support when making a diagnosis. These include cranial hyperostosis with large frontal sinuses, kyphoscoliosis, narrow disk spaces, degenerative Scheuermann-type spinal changes with anterior beaked vertebrae that are most prominent at the thoracolumbar junction, narrow iliac wings, and metacarpal pseudo-epiphyses with poorly modeled middle and tufted distal phalanges (Hunter et al. 1982; Gilgenkrantz et al. 1988).

The presence of a normal allele in females who are heterozygous for a pathogenic variant in the causative gene

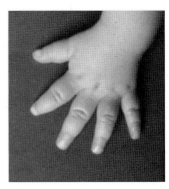

FIGURE 12.2 The characteristic hand, as seen in a young child, showing tapering and fleshy digits, which are hyperextensible.

mitigates the clinical manifestations. However, there are severely affected females and it is assumed that inactivation of one X-chromosome (lyonization) results in the observed clinical variation, although as yet there are no supporting data from lymphocyte-based skewed X-inactivation studies. Quintela et al. reported a classical phenotype in a female with deletion of the *RPS6KA3* gene. Physical examination of women in the matrilineal line of males with CLS may show mild signs typical of the syndrome (Figures 12.3 and 12.4). Careful assessment of intellect may show varying degrees of impairment that generally fall between that of affected males and non-gene-carriers in the family (Simensen et al. 2002). Genetic testing can play an important role, especially if a known mutation is present in the family.

The appearance of many syndromes becomes more developed and distinctive with increasing age. This may be particularly true for CLS as several authors commented about the difficulty of making the diagnosis at a young age (Wilson and Kelly 1981; Vles et al. 1984; Touraine et al. 2002). The younger child may show a prominent forehead, hypertelorism, the characteristic nasal findings, and relatively full lips (Figure 12.5). The fingers do not show the typical adult characteristics but may taper somewhat and tend to be hyperextensible. Fullness of the forearm, due to an excess of subcutaneous fat, may be a useful sign in the infant with CLS (Hersh et al. 1984).

Etiology, Pathogenesis, and Genetics

Coffin–Lowry syndrome was long thought to be an X-linked condition because of its variable and usually milder expression in females, and affected males in multiplex families were related through either normal or mildly affected women. This was confirmed by early linkage studies that assigned CLS to the region of Xp21-pter (Partington et al. 1988). Further studies narrowed the region to Xp22.2-22.1, and Trivier et al. (1996) used a candidate gene approach to identify pathogenic mutations in *RPS6KA3* [*RSK2, p90(rsk)*]. Delaunoy et al. (2001) used the single-strand conformation

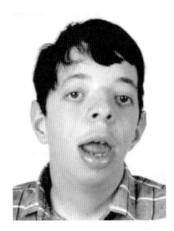

FIGURE 12.1 Young adult male with Coffin–Lowry syndrome illustrating prominent forehead, downslanting palpebrae, hypertelorism, anteverted nares, moderately thickened nasal septum, some malar flatness, patulous lips, everted lower lip, small teeth, and poorly formed helices.

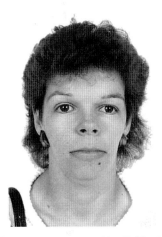

FIGURE 12.3 Adult woman with Coffin–Lowry syndrome showing a prominent forehead and supraorbital ridges, wide-spaced eyes, mildly downslanting palpebrae, a short nose, full and everted lower lip.

FIGURE 12.4 Sporadic affected girl who, in addition to features seen in Figure 12.3, illustrates well the typical wide nasal septum and patulous lips.

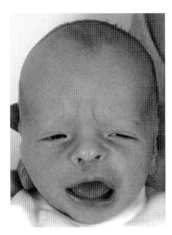

FIGURE 12.5 Young male infant with Coffin–Lowry syndrome showing a prominent forehead, a short nose with a low nasal bridge, and a large mouth with a full and everted lower lip.

polymorphism (SSCP) technique for an initial study of 250 unrelated individuals who had been diagnosed clinically, and they found 71 different mutations among 86 (34%). A subset of 26 of them, in whom no mutation had been found, was later studied using Western blot analysis and an in vitro kinase assay (Zeniou et al. 2002). Seven novel mutations were found. No mutations were found when the promoter region of *RPS6KA3* was studied. Currently, sequencing is estimated to detect a mutation in about 50% of clinically diagnosed individuals (Delaunoy et al. 2006). This relative lack of success in finding disease-causing mutations has led to the suggestion that there are other genes that may result in similar CLS phenotypes (Zeniou et al. 2002). Clinical details and photographs are not available for the individuals who were reported in these studies and thus it is difficult to assess how well they fit the CLS phenotype. Undetected intragenic deletions and duplications may be contributing to the etiology. Additionally, a deep intronic variant was reported in a severe form of CLS, and thus, RNA analysis may be indicated in individuals with the strong clinical suspicion of CLS (Schneider et al. 2013). Marques Pereira et al. (2007) have reported the first instance of a tandem duplication within the *RPS6KA3* gene in an individual with CLS, and, noting the high frequency of Alu sequences within the gene, they suggest these may be relatively common events.

Molecular assessments of genes thought to interact in the *RPS6KA3* pathway in people who have tested negative for *RPS6KA3* have identified some coding changes (C.E. Schwartz, personal communication), but the significance of these mutations is unproven, and the clinical details and photographs are not yet available for the individuals who were studied. Thus, it is difficult to determine how well they fit the classic Coffin–Lowry syndrome. Clinical data on a series of *RPS6KA3* mutation-positive and mutation-negative individuals suggest that the presence of certain clinical signs, such as the fleshy and tapering fingers, hypertelorism and downslanting palpebral fissures, may help distinguish the *RPS6KA3*-positive group (F. Abibi and C.E. Schwartz, personal communication). At this time there are no published linkage data from a well-described CLS family that are compatible with a second Coffin–Lowry syndrome locus.

All males and a significant proportion of females with CLS do not reproduce, therefore, as expected, a high proportion (70–80%) represent sporadic cases, and these may include affected females (Delaunoy et al. 2001).

Of the 71 mutations initially reported, 38% were missense, 20% nonsense, 18% splicing errors, and 21% small insertions or deletions. Updated data show over 140 different pathogenic variants with only slight changes in the original observation of mutation categories; 30% missense, 15% nonsense, 20% splicing, and 30% short deletions or insertions (Marques Pereira et al. 2010). Information recently collated in the Human Gene Mutation Database (HGMD) shows 180 different variants in *RPS6KA3*, with more than

160 occurring in individuals clinically diagnosed with CLS. The variant categories were similar to those previously reported. Pathogenic variants in *RPS6KA3* were also found in 17 individuals described as having intellectual disability and/or autism and one with multiple congenital anomalies. About 60–65% reported mutations result in premature termination of translation (protein truncation). The mutations range across the gene, with the exception of exons 1 and 2, and, in general, do not show any positional genotype/phenotype correlations. However, there may be some trend for missense mutations to be less deleterious. Harun et al. (2001), in a small study of seven individuals, found a correlation between IQ and the degree of loss of *RPS6KA3*-mediated *CREB*-peptide phosphorylation in lymphocytes (see below). A missense mutation in *RPS6KA3* has also been found in the family reported with non-syndromic mental retardation 19 (MRX19) (Merienne et al. 1999) and in families with milder signs of CLS (Manouvrier-Hanu et al. 1999; Field et al. 2006). The MRX19 mutation resulted in an 80% reduction in *RPS6KA3* activity, rather than the usual 100% loss seen in classical CLS.

The *RPS6KA3*-encoded gene product, RPSKA3, is a growth-factor-regulated serine-threonine kinase. Its open reading frame has 22 exons that result in a protein with 740 amino acids (Delaunoy et al. 2001). Humans have four RPS6K proteins that share 80–85% homology and are closely related in structure and function. Two non-identical catalytic domains are connected by a linker, and both are required for maximal activity (Trivier et al. 1996; Yntema et al. 1999). These kinases act at the downstream effectors of the ras-mitogen-activated protein kinase (Ras/MAPK) pathway, through direct interaction with the extracellular signal-kinase (ERK) in response to a variety of stimuli including: growth factors, insulin, oncogenic transformations, and neurotransmitters. Loss of RSK2 activity in drosophila impairs synaptic function and axonal transport in motor neurons (Beck et al. 2015). Animal models of CLS and RSK2-associated neurologic dysfunction were reviewed by Fischer and Raabe (2018). Recently, Park et al. (2016) demonstrated in human cells that depletion of RSK2 is associated with a reduction in mitotic spindle stabilization and irregular distribution of mitotic spindles. Through this pathway and the epidermal growth factor (EGF)-stimulated phosphorylation of histone H3, RPS6KA3 has a role in stimulating the cell cycle between G0 and G1. RPS6KA3 also appears to act in the protein kinase C and adenyl-cyclase pathways (Harun et al. 2001). Cells from individuals with CLS have been shown to have defective EGF-stimulated phosphorylation of S6 (Trivier et al. 1996), H3 (Sassone-Corsi et al. 1999), and cAMP-responsive element-binding protein (CREB) (Harun et al. 2001). As yet, the direct relationship between mutations in *RPS6KA3* and the cognitive impairment and physical characteristics of CLS is not known. However, the role of CREB in assuring neuronal survival and in converting short- to long-term memory makes this pathway a plausible explanation for the intellectual deficit(s) in CLS. Zeniou et al. (2002) have shown that the highest expression of *Rsk2* in adult mouse brain is in areas with high synaptic activity in the neocortex, hippocampus, and Purkinje cells; all areas that are key to cognitive functioning.

Males with CLS, unless mildly affected, are not expected to reproduce because of their degree of intellectual impairment. Any child they fathered would be a carrier, if female, and unaffected, if male. Normal or more mildly affected female carriers are at 50% risk to transmit the non-functional gene to their offspring, resulting in all males and a significant proportion of females who receive the pathogenic mutant allele being clinically affected. Therefore, once the diagnosis of CLS has been made in a family member, it becomes critically important to determine if other females in the sibship and/or matrilineal line are themselves at risk to have affected children. Careful physical examination looking for mild craniofacial, hand, and radiological signs of CLS and judicious use of genetic testing should help to best inform at-risk family members. When a pathogenic mutation has been identified in either an affected or carrier individual, preimplantation genetic diagnosis (PGD) or prenatal diagnosis by mutation analysis is possible.

Diagnostic Testing

DNA testing of *RPS6KA3* has only identified pathogenic variants in about 50% of clinically diagnosed cases, and variants are widely dispersed as unique variants along the gene. As linkage studies have not shown evidence of genetic heterogeneity, it is likely that the low detection rate is in large part the result of testing individuals who do not have CLS. As yet, there are no large studies of mutation-negative patients using approaches such as MLPA (multiplex ligation-dependent probe amplification) or comparative genomic hybridization (CGH)-array to determine the frequency of duplications (Marques Pereira et al. 2007) or deletions. Bertini et al. (2015) reported a 625 kb microduplication at Xp22.12 that disrupted expression of *RPS6KA3* in a child with mild intellectual disability. The in vitro functional ribosomal S6 kinase assay (Merienne et al. 1998; Delaunoy et al. 2001; Zeniou et al. 2002) is a potentially useful initial functional diagnostic test, but it requires a fibroblast or lymphoblastoid cell line, and it cannot be used reliably in females because random X-inactivation gives a broad range of results. Micheli et al. (2007) have refined a diagnostic approach using CREB-peptide substrate and successfully screened a male and two clinically affected females. The sensitivity for mildly or clinically unaffected women needs to be evaluated further, but it could be used as an initial screen in symptomatic patients. Unfortunately, methylation signature studies do not provide a characteristic pattern that can aid in diagnosis (Sadikovic, personal communication).

Selection for genetic testing should be based upon clinical circumstances. A positive test will confirm, but a negative test cannot rule out the diagnosis because of the current low rate of variant detection in published series. It is likely that a large proportion of those individuals who have a negative in vitro kinase assay, and negative DNA mutation screening do not have CLS, although, as discussed, some negative studies may still result from current technical limitations in the standard approaches to testing. Once a mutation is known in a family member, targeted genetic testing can be extremely helpful in establishing if other female family members carry the mutation and are at risk to have affected children. However, failure to detect a known variant in the mother of an affected individual does not eliminate the risk for future pregnancies because germ line mosaicism has been reported (Jacquot et al. 1998; Horn et al. 2001). Prenatal diagnosis should still be offered to these women.

DIFFERENTIAL DIAGNOSIS

The characteristic appearance of Coffin–Lowry syndrome becomes more apparent with age, and the differential diagnosis is a challenge either in a young child or in a female who has less marked clinical signs. Children with α-thalassemia-mental retardation (ATR-X) (see Chapter 7) have a flat face, hypertelorism, a low nasal bridge, and a large mouth with everted lower lip and small teeth. However, they lack the characteristic nasal shape and hand changes seen in CLS. Instead, they have a tented or upside-down V-shaped upper lip, and a high incidence of genital abnormalities. Females are unlikely to show signs of ATR-X. The finding of a non-iron deficient microcytic anemia and hemoglobin H bodies is also diagnostic of ATR-X, although the facial appearance and intellectual disability can be seen without the α thalassemia.

Vealle et al. (1979) reported an individual as having Börjeson–Forssman–Lehmann syndrome (Börjeson et al. 1962), but it is clear from the photograph that he had CLS. Those with Börjeson–Forssman–Lehmann syndrome have a coarse face, small and anteverted nose, and the forehead and supraorbital ridges tend to be prominent. The ears measure large. The hands may resemble those seen in CLS; however, the overall facial gestalt, obesity, gynecomastia, and hypogenitalism are distinguishing features.

Young children with Williams syndrome have a prominent forehead, low nasal bridge with anteverted nares, a wide mouth with full lips and small teeth (see Chapter 63). However, there are many distinguishing signs including bitemporal narrowness, a bulbous nasal tip without alar fullness, and stellate irides that identify individuals with Williams syndrome, and they lack the typical hand changes seen in CLS. Developmental progress is likely to be significantly more advanced in Williams syndrome than in most individuals with CLS.

Sotos syndrome (see Chapter 55) and Fragile-X syndrome (see Chapter 28) have been considered to be in the differential diagnosis of CLS (Plomp et al. 1995; Touraine et al. 2002), but the overlap in clinical manifestations is minimal. Onset of overgrowth and macrocephaly in Sotos syndrome is in contrast to the usual growth failure of CLS, and the face shape with the high forehead, receding hairline, lengthening face and jaw are quite distinctive. The hands in those with Sotos syndrome often have a soft fullness, but they tend to be large and to not have the tapering of digits and degree of hyperextensibility associated with CLS. Individuals with Fragile-X syndrome may have prominent ears that appear large but their facial appearance is generally subtle and unlike that seen in CLS. Their hands are unlike those seen in CLS, and simple DNA testing for the expanded trinucleotide repeat is readily available.

McCandless et al. (2000) reported a family in which the members who carried a del10(q25.1q25.3) had findings suggestive of CLS. Thus, cytogenetic studies are warranted in atypical or doubtful cases before considering the more expensive genetic testing for CLS.

MANIFESTATIONS AND MANAGEMENT

A brief summary of the major management recommendations detailed below is provided in Table 12.1.

Growth and Feeding

Significant abnormalities of prenatal growth are not typical of CLS and most affected newborns have a normal length, weight, and head circumference. Postnatal growth failure can be expected within the first few months, and most males with CLS drop below the third centile in length. Radiographs (Touraine et al. 2002) and measurements of body proportions (Hunter 2002) suggest that short stature involves the limbs disproportionately. There are no specific feeding difficulties and a weight appropriate to height is usually maintained. Some older males, perhaps on the basis of relative inactivity, have obesity, which should be avoided in terms of general health and because it may compromise the management of some of the long-term complications in CLS, such as progressive kyphoscoliosis. About half of the males will develop a head circumference less than the third centile (Hunter et al. 1982). The majority of heterozygous females show normal growth, although those with more severe manifestations may have growth failure and microcephaly.

Evaluation

- There are no specific growth charts for CLS. Length/height, weight, and head circumference should be plotted regularly on standard pediatric growth curves. Any significant subsequent deviation from the individual's established growth pattern is cause for concern.
- Attention should be paid to weight relative to height and this is most readily done using a weight-for-height or body mass index graph.

Management

- There is no treatment for the relative growth failure associated with CLS.
- Feeding difficulties, significant changes in established growth pattern, or emerging obesity should be treated in a standard manner.

TABLE 12.1 Suggested approach to management for Coffin–Lowry syndrome.

System	At diagnosis	From infancy to age 6 years	Age 7 to adulthood	Adult
Growth and feeding	Graph length, weight, and head circumference (HC)	Graph length, height, weight, and HC every 6 months; expect fall to <3rd% by 18 months	Graph length, weight, and HC every year; evaluate if continued growth deceleration	Assess weight for height; assure balanced diet
Development and behavior	Complete assessment; intervention plan	Q6 months reassess; goals of communication, gait, self-help	Reassess yearly; goals of communication, gait, self-help; plan for long-term care; evaluate significant behavioral changes	Reassess every 5 years; maximize independence/participation; work to maintain and improve skills; evaluate significant behavioral changes
Neurological	History and neurological examination	Q6 months history and neurological examination; gait, bowel, bladder, reflexes, changed tone; ask about seizures and drop-attacks	Yearly history and neurological examination; gait, bowel, bladder, reflexes, changed tone; ask about drop-attacks and seizures	Yearly history and neurological examination; gait, bowel, bladder, reflexes, changed tone; ask about drop-attacks and seizures
Ophthalmologic	Detailed examination including refraction and fundoscopy	If prior test normal, repeat biannually	If prior test normal, repeat every 3–4 years; look for blepharitis, eye rubbing	If prior test normal repeat, every 5 years; look for blepharitis, eye rubbing
Ears and hearing	Developmental age appropriate assessment	Annual otological examination and hearing assessment	Biannual otological examination and hearing assessment	5-yearly otological examination and hearing assessment
Cardiovascular	Physical examination and ECG; consider baseline echocardiogram	If prior normal, repeat physical examination annually	If prior normal, repeat physical examination annually; ECG and baseline echocardiogram by age 10 years	If prior normal, repeat physical examination annually; ECG and echocardiogram every 5–10 years
Respiratory	History and physical	Annual history (airway obstruction/apnea) and physical	Annual history (airway obstruction/apnea) and physical	Annual history (airway obstruction/apnea) and physical
Musculoskeletal	Physical examination; attention to chest and spine	Annual physical examination; attention to chest and spine; refer to treat kyphoscoliosis	Annual physical examination; attention to chest and spine; refer to treat kyphoscoliosis	Annual physical examination; attention to chest and spine; refer to treat kyphoscoliosis
Dental	Age appropriate examination	Teach dental care including gum hygiene	Reinforce dental care; plan possible interventions for anomalies	Assure dental care including gum hygiene; intervene as needed
Other	Full clinical/genetic evaluation of family; family psychosocial assessment; intervention as needed	Assure genetics is understood; assure family has adequate support; assure other affected members have care	Assure family members understand genetic risk and, when appropriate, have been offered testing; inquire about behavioral changes in other family members	Assure family members understand genetic risk and, when appropriate, have been offered testing; inquire about behavioral changes in other family members

Development and Behavior

CLS is first and foremost an intellectual disability syndrome. Many clinical reports do not provide detailed information about the developmental profile of the affected individuals and simply describe affected individuals as severely or profoundly delayed. Touraine et al. (2002) report a mean age of walking of 3 years. A review of individuals reported in the literature found that, although a majority of males were profoundly or severely disabled, many affected individuals were moderately delayed and a few were considered mildly delayed (Hunter, 2002). When affected persons were subdivided by age, the proportion of individuals diagnosed as moderately or mildly impaired decreased with increasing age. Although the number of cases were fewer, this trend was also noted in the families reported by Simensen et al. (2002) in which extensive developmental testing was performed. There are a number of possible explanations for these findings. A significant number of affected persons have been institutionalized and this proportion is higher among older individuals. Although one could hypothesize that these individuals might have been presented with fewer developmental opportunities, no such evidence is available, and a comparison of the institutionalized and non-institutionalized individuals did not identify obvious differences. It remains possible that, as originally questioned, there is a true deterioration of intellect in CLS (Coffin et al. 1966; Procopis and Turner 1972; Temtamy et al. 1975). However, there is no evidence of continued deterioration in those adults who have been restudied at a later date (Partington et al. 1988; Hunter 2002). The most likely explanation for this apparent deterioration with age is that tests in younger children simply overestimate their ultimate developmental potential. Although details of specific assessments and long-term follow-up were not provided, Touraine et al. (2002) have shown that with appropriate intervention services the intellectual disability is only moderate in most affected individuals and "most patients in our experience can acquire a substantial capability of oral communication". Although outcomes may not always be this encouraging, these comments certainly emphasize that every effort must be made to provide opportunities for individuals with CLS to reach their maximum potential. In general, females will be developmentally normal or show an intermediate degree of impairment.

There have been no systematic studies of behavior in CLS. Parents and caregivers generally describe even severely disabled individuals as happy, friendly and largely cooperative (Hunter 2002; Touraine et al. 2002). Behavior may deteriorate in situational stress environments such as new surroundings, unfamiliar staff, or crowding (Hunter 2002).

Although the numbers are small and psychiatric illness is common in the general population, there is growing evidence of an increased risk of psychosis in persons with CLS, especially in heterozygous females (reviewed in Hunter 2002), even in women who have little physical or intellectual evidence of CLS. A rate of 8.8% was found among a total of 22 affected females, 38 carrier mothers and 8 affected sisters (Hunter 2002). One of two women studied by Micheli et al. (2007) was described as having a "psychosis" and one of two affected sisters reported by Wang et al. (2006) carried a diagnosis of schizophrenia.

Evaluation

- Upon diagnosis, arrangements should be made for a detailed developmental assessment and this should be repeated at appropriate intervals to monitor progress and update planned intervention.
- Pediatric/neurological evaluation should assure that there are no cardiac, musculoskeletal or focal neurological impediments to development.
- Hearing loss can interfere with the acquisition of communication skills. Developmental-age-appropriate audiological evaluations should be carried out yearly in infants and young children and should be continued, with reduced frequency, in older children and adults.
- There should be a thorough evaluation of the ability of the family to cope and provide the appropriate psychological and social environment for the child with CLS. Support may be required, especially in those 20–30% of families where the mutant allele has been transmitted through a mother with some intellectual disability. Such families may become marginalized in society and have very limited financial and personal resources.
- Care should be taken to assure that female relatives who may be heterozygotes do not have mild intellectual disability that has gone unrecognized and/or untreated.
- There needs to be awareness among caregivers of the possible increased rate of psychiatric illness in gene carriers so that significant changes in behavior are noted and evaluated further.
- If significant behavioral changes occur, situational, physical, and psychiatric causes need to be considered.

Management

- Early intervention with a developmental team is advised. This should include individuals capable of assessing the spectrum of abilities in a child who may have significant communication problems, speech and language therapists capable of presenting the gamut of approaches to communication, occupational and physical therapists, and pediatric expertise to cope with any health concerns that could affect development.
- Progress should be evaluated on a regular basis commensurate with the age of the individual, and the plan adapted as required.

- Intervention is required throughout the school years and, because little is known about the best approach to learning in CLS, educators should not shy from experimenting with different approaches and publicizing their outcomes.
- Every effort should be made to support the family in what is likely to be the significant challenge of raising a child with severe intellectual impairment and potential physical problems. This should include respite care. The need for social support may be especially extensive in those families where additional family members (e.g., mothers or sisters) also have intellectual disability.
- Treatment of behavioral problems, where present, is standard.
- Psychiatric illness should be treated in a standard manner.

Neurologic

Major malformations of the brain are not a common feature of CLS, suggesting that the impact of *RPS6KA3* is confined largely to submicroscopic neuronal interrelationships and/or functioning. Early neuroimaging studies have usually been normal or have shown non-specific increased ventricular and/or subarachnoid and Virchow–Robin spaces. The latter are considered a sign of brain aging and have shown an association with age and cognitive function. Hydrocephalus has been diagnosed in a few cases, but in some early instances may represent over-interpretation of a large ventricle in the presence of a prominent forehead (Hunter et al. 1982). Callosal anomalies, including agenesis and thinning, and periventricular white matter abnormalities have been reported (Özden et al. 1994; Soekarman and Fryns 1994; Kondoh et al. 1998; Wang et al. 2006; Tos et al. 2015; Miyata et al. 2018). The three sibs reported by Wang et al. (2006) had vermis hypoplasia. Minor changes in gyral patterning, simplified gyri, an area of polymicrogyria, and abnormal cortical lamination were reported at autopsy by Coffin (2003). To date, there is a single report of a 6-year-old boy with CLS who had multiple focal, frontal and parasagittal hypodensities diagnosed by a magnetic resonance imaging (MRI) scan (Kondoh et al. 1998). Although MRI data remain limited, O'Riordan et al. (2006) mention deep white matter high signal areas on T2 and FLAIR images, some of which suggested dilated perivascular spaces, and Wang et al. (2006) noted cerebrospinal fluid signals in the deep cortex. In the first quantitative MRI study reported, Kesler et al. (2007) showed reduced overall brain volume, affecting gray and white matter, in three male and three female affected members of two families. Maximum volumetric changes were seen in the temporal lobes, cerebellum, and hippocampus. Of interest, the hippocampus was significantly small in one family, but large in another family studied. The authors acknowledge that this was a small and preliminary study but raised the possibility of correlation between the magnitude of the changes and developmental abilities.

The neurological examination may be compromised because of intellectual impairment and difficulty with communication. A wide range of neurological signs have been reported including loss of muscle mass and strength, notably in the legs, both increased and decreased tendon reflexes, foot drop, decreased amplitude and latency on nerve conduction studies, and neurogenic changes on muscle biopsy (reviewed in Hunter 2002). More important, a significant number of affected individuals have been reported with neurological complications that may be progressive and even life-threatening. Progressive paraplegia (Machin et al. 1987; Hunter 2002) and loss of the ability to walk (Procopis and Turner 1972; Partington et al. 1988; Hunter 2002) have been noted. Ishida et al. (1992) attributed progressive spasticity in three males to calcification and thickening of the ligamenta flava, resulting in spinal stenosis. Another individual has also been noted to have a thick ligamentum flavum, but in that case the neurological complications were ascribed to a congenital cervical canal stenosis (Hunter 2002). A case of foramen magnum compression has been reported (Upadia et al. 2017). One case each of stroke and sleep apnea treated by tracheostomy has also been reported (Hunter 2002), and an affected individual with mixed central and obstructive sleep apnea was described. Seizures have been reported in CLS and Touraine et al. (2002) state that seizures affected one-third of affected individuals. Recurrent non-convulsive status epilepticus was described in a 45-year-old male with episodes of decreased alertness (Gschwind et al. 2015).

In 1990, Padley et al. (1990) described a type of exaggerated startle response that may be unique to CLS, and additional cases have been described (Crow et al. 1998; Fryns and Smeets 1998; Caraballo et al. 2000; Hunter 2002; Nakamura et al. 2005; O'Riordan et al. 2006; Havaligi et al. 2007; Arslan et al. 2014; Touraine et al. 2002; Ishida et al. 1992; Plomp et al. 1995). Thus, the phenomenon may be quite common with estimates ranging from 10 to 20% of cases. The episodes affect both sexes and usually begin between 4 and 17 years of age with a mean age of onset of 8.6 years (Nakamura et al. 2005). Unexpected tactile or auditory stimuli or excitement cause a fall, often backwards, that is associated with a brief loss of muscle tone in the lower limbs but no loss of consciousness. The electroencephalogram remains normal, but the electromyogram in the lower limbs shows electrical silence for about 60–80 ms after the stimulus (Crow et al. 1998; Caraballo et al. 2000; Nakamura et al. 2005). These findings are distinct from cataplexy, hyperekplexia, and negative myoclonus and are generally referred to as stimulus-induced drop attacks (SIDAs). An unusual evolution of neurophysiological and clinical signs was observed in one of two males with SIDAs, as reported by Nelson and Hahn (2003). At the age of 6 years, the stimulus-induced episodes were accompanied by typical loss of lower extremity tone, but by the age of 11 years the episodes had changed to brief myoclonic jerks and tonic spasm, with increased tonic electromyogram activity, more typical of hyperekplexia. Fryns and Smeets (1998) noted that in two brothers the increase in frequency of these "drop-attacks" paralleled the progression of their kyphoscoliosis, and that the attacks stopped in one brother after the surgical correction of his spinal defect. However, a direct link between these attacks and kyphoscoliosis is not likely, as at least one male with drop-attacks did not have kyphoscoliosis (Hunter 2002). Obstructive sleep apnea has been associated with SIDAs and a 12 year old improved after correction of his airway obstruction with tracheostomy (Imataka et al. 2016). It is of note that, thus far, all *RPS6KA3* mutations in affected individuals with SIDA result in early termination of translation of the N-terminal kinase domain, and no missense mutations have been reported (Nakamura et al. 2005).

In some individuals, these attacks are relatively infrequent; however, in others they have been so frequent that the person has been placed in a wheelchair to prevent injury. Response to treatment has been variable, and, in some cases, poor. Standard anticonvulsive therapy has generally been ineffective (Caraballo et al. 2000) and a wide range of medications have been tried; including tiagabine, lamotrigine, phenobarbital, felbamate, tricyclics, serotonin uptake inhibitors, carbamazepine, valproic acid, and clomipramine. Benzodiazepine compounds may be of benefit in controlling the frequency of these episodes, although, over time, the dose may need to be increased (Touraine et al. 2002; Nakamura et al. 2005; Arslan et al. 2014). Havaligi et al. (2007) reported good response to sodium oxybate after failed response to a number of medications.

Evaluation

- Severe intellectual disability with concomitant problems in communication may hinder the neurological evaluation. There should therefore be a high index of suspicion for neurological complications.
- Regular periodic neurological evaluation should be carried out with special attention to changes in gait, bladder and/or bowel control, and focal neurological signs such as changes in deep tendon reflexes and tone.
- History suggestive of disturbances in sleep and sleep apnea should be sought.
- Electrophysiological studies such as somatosensory evoked responses, electromyography and nerve conduction studies, and magnetic resonance imaging examination of the spine may help to resolve the cause of apparent neurological changes.
- Careful observation of the circumstances and actual events, combined with electroencephalogram and electromyogram studies, should allow distinction between seizures and drop-attacks. The possibility that other systems, such as cardiovascular, might cause syncope should also be considered.

Management

- Surgical intervention may be required in cases where significant spinal stenosis is found. To date, results have not always been encouraging, but, in part, this may reflect delays in diagnosis that result in irreversible changes.
- Progression to severe scoliosis and obesity may contribute to neurological symptoms and also complicate therapy, and every effort should be made to prevent their occurrence.
- Seizures should be treated with standard therapies.
- Drop-attacks may respond to benzodiazepines and/or other medications. In addition, efforts should be made to modify the environment so as to reduce stimuli that may trigger these events. In extreme and unresponsive cases protection with a helmet or even use of a wheelchair may be required.

Ophthalmologic

Significant ocular disease in CLS is uncommon, and in a clinical cohort include two cases with cataract, two with retinal changes, one with bilateral optic atrophy, and one with unexplained exophthalmos (reviewed in Hunter 2002). It is probable that thorough evaluation of vision is rarely carried out. In a follow-up of four individuals first reported in 1982, three were found to have an idiopathic chronic irritation and mild inflammation of the eyelids (Hunter 2002). There might be potential for chronic rubbing to cause intraocular damage.

Evaluation

- Affected individuals should undergo a complete eye examination, including refraction, at the time of diagnosis. The interval between subsequent evaluations will depend upon initial findings but there should be periodic review.

Management

- Standard ophthalmologic interventions should be applied as needed. Even individuals with significant intellectual disability may learn to wear glasses and benefit from correction of refractive errors.
- At present, the significance and frequency of eyelid irritation is unclear and there have not been published attempts to alleviate the problem. Artificial tears and/or local steroids might be tried. There should be intervention if eye rubbing has potential to cause intraocular damage.

Ears and Hearing

Both conductive and sensorineural hearing loss have been reported in CLS (Hunter et al. 1982; Collacott et al. 1987; Hartsfield et al. 1993; Higashi and Matsuki 1994; Sivagamasundari et al. 1994; Rosanowski et al. 1998; Sculerati, 2000; Hunter 2002). It is not possible to estimate the prevalence of hearing loss in CLS from these reports because audiological evaluation has not always been performed, and there may be a reporting bias (Hartsfield et al. 1993; Higashi and Matsuki 1994; Rosanowski et al. 1998). From a review of the literature, Hunter (2002) found 14 males and one female with mixed or sensorineural hearing loss and added two individuals with chronic conductive loss. Touraine et al. (2002) estimated that significant hearing impairment is found in one third of affected individuals, but they did not provide data. Hearing loss was reported in 30% by Pereira et al. (2010). The hearing loss appears causally heterogeneous. One case had a malformation of the labyrinth (Higashi and Matsuki 1994). Cases may cluster in families and be of later onset after documented normal hearing has been reported (Rosanowski et al. 1998).

Evaluation

- Developmental age appropriate hearing tests should be performed at the time of diagnosis. Lack of cooperation in testing may require the use of electrophysiological testing (brainstem auditory evoked response) beyond the usual age.
- Periodic monitoring for middle ear infections and effusions with tympanometry and/or pneumo-otoscopy is important.
- The occurrence of later onset hearing loss requires periodic re-evaluation of hearing. There are no evidence-based studies on which to form recommendations but annual review in young children could become less frequent with age unless concern is raised in prior studies.

Management

- Standard treatment should be applied in the case of middle ear disease.
- Hearing aides can be successfully introduced as required.
- FM amplification and/or preferential seating in school may be beneficial.

Cardiovascular

Cardiovascular anomalies occur in a significant proportion of individuals with CLS, and that they can play a causal or contributing role in the excess of premature death (reviewed in Hunter, 2002). It is highly probable that many persons reported in the literature have not had a careful cardiac evaluation or echocardiogram. Thus, the 13.5% of males and 4.5% of females reported with cardiovascular anomaly should be considered a minimum estimate. Hemodynamically important mitral valve incompetence and/or prolapse seems to be the most common anomaly, and, in two cases, autopsy showed associated short chordae (Machin et al. 1987; Massin et al. 1999). Other abnormalities include idiopathic cardiomyopathy, myocarditis, tricuspid valve anomalies, dilated aorta with abnormal aortic valve, pulmonary artery dilatation, and two cases of congestive heart failure of unspecified cause.

Evaluation

- There are no systematic longitudinal studies on which to base recommendations.
- A competent cardiovascular assessment should be performed at the time of diagnosis and should include an electrocardiogram. A baseline echocardiogram may be obtained, especially if a pectus or kyphoscoliosis interferes with the cardiac assessment. At a minimum, it should be carried out by the age of 10 years. In the absence of concern about a cardiovascular anomaly, studies should be repeated every 5–10 years, as some of the abnormalities that have been reported would likely have later onset.

Management

- Abnormalities should be treated as per the general population, as appropriate to the specific findings.

Respiratory

There is no evidence that primary lung disease is a significant component of CLS. Pulmonary function may be compromised by severe progressive scoliosis, pectus deformities, chronic heart disease or aspiration associated with seizures. Machin et al. (1987) found panacinar emphysema in two adult male sibs who came to autopsy, but the significance of these findings and whether they are actually related to CLS is not clear. There is one report of an enlarged trachea (Hunter, 2002) and one reported death from pneumonia (Partington et al. 1988). Severe restrictive lung disease has been reported in a male with CLS (Venter et al. 2019).

Evaluation

- Affected individuals should be followed and screened for heart disease and kyphoscoliosis that may lead to respiratory compromise.
- A history suggestive of obstructive or central apnea should be sought.

Management

- Problems that may lead to respiratory compromise should be evaluated and treated in a standard appropriate and timely fashion.

Musculoskeletal

Motor development is delayed in CLS and a stiff, flat-footed, somewhat broad-based gait often develops. There are a number of minor radiographic variants such as short, poorly modeled middle and tufted distal phalanges, narrow iliac wings, cervical ribs, and cranial hyperostosis with large frontal sinuses that may be useful in diagnosis but are of no clinical importance. Pectus carinatum and/or excavatum are common but are not usually clinically symptomatic. Muscular development and tone may be poor, but it is important to be sure this is not progressive and due to neurological complications.

Kyphoscoliosis is very common in CLS and has the potential to cause significant problems with mobility and general care, pain, and cardiorespiratory compromise that may cause premature death. A review of reported cases found that 47% of males and 31.8% of affected females had kyphoscoliosis (Hunter 2002). The kyphoscoliosis was uncommon in young children and it increased in prevalence and severity with age. However, even some older individuals did not develop this complication. The development of kyphoscoliosis in CLS may relate to the underlying associated dysplastic vertebral anomalies that are commonly seen and joint/ligamentous laxity. Higher rates were reported in a small series by Herrera-Soto et al. (2007), but this was from a population selected by referral to orthopedics. Welborn et al. (2018) reviewed the literature which included 6 affected males and described delayed skeletal maturity and severe progressive thoracic lordosis, thoracolumbar kyphosis and scoliosis; and suggested this natural history supports a lower threshold for surgical intervention.

Evaluation

- Careful evaluation for kyphosis and scoliosis should be part of an annual physical assessment.
- Early referral for orthopedic evaluation and management should be made if vertebral changes are found.

Management

- Early referral to a developmental physiotherapist may be considered in the hope of helping motor development and particularly to try to develop a more normal gait.
- In the rare instance where pectus excavatum is so severe as to raise a health concern, a cardiorespiratory assessment should be made and consideration given to surgical intervention.
- Every effort should be made to arrest the progression of kyphoscoliosis. Surgery should not be delayed to the point of cardiorespiratory compromise because deaths have occurred in males with CLS undergoing anesthesia and surgery in these circumstances. Standard treatment of kyphoscoliosis is recommended.

Dental

The dental anomalies that have been reported in CLS are varied and are among the most common findings seen in this condition. They comprise malocclusion including open bite, malpositioning, hypodontia, small, abnormally shaped and widely spaced teeth, as well as accelerated or delayed primary eruption, and premature loss of teeth (reviewed in Hunter 2002). In some cases, premature loss of teeth may be caused by hypoplastic cementum and sparse peridontal membrane fibers (Day et al. 2000). *RPS6KA3* is necessary for proper cementum formation and *Rsk2*-deficient mice display cementum hypoplasia and hypomineralization (Koehne et al. 2016). In an individual report by Igari et al. (2006), premature loss was preceded by tooth "elongation" and root exposure followed by rapid resorption of the roots.

Evaluation

- Referral for pediatric dental care, at least by the age at which primary dentition is expected to be complete (3–4 years), is important to minimize subsequent problems and for long-term planning.
- Adequate fluoridation should be assured.

Management

- Early instruction of caregivers in dental hygiene, which may actually be facilitated by the small widely spaced teeth, is important. Some males may be taught to participate in their dental care.
- Particular attention should be paid to factors, such as gum disease, that may exacerbate premature loss of teeth.
- Orthodontic planning should take into account the possible susceptibility to premature loss of teeth.

Gastrointestinal

A number of gastrointestinal anomalies have been reported in single individuals with CLS and it is not clear whether these are rare complications or simply coincidental concurrence (reviewed in Hunter 2002). The findings have included anterior placement of the anus, pyloric stenosis, rectal prolapse, multiple jejunal diverticuli, and sigmoid colon diverticuli with reduced ganglion cells. The last two anomalies were found at autopsy (Machin et al. 1987). Evaluation should be guided by the symptoms and signs and treatment based on the outcome of investigations.

Neoplasia

CLS is not considered to be a syndrome with a propensity to malignancy. Among 111 males and 38 obligate heterozygotes reported, one male and one female died of Hodgkin's disease, one mother with clinical signs of CLS had a Wilms tumor, and a monozygous twin of an affected male died of a posterior fossa tumor (Hunter, 2002). Given the pathways in which RPS6KA3 acts, it is possible that pathogenic mutations might increase the susceptibility to malignancy. Current data do not warrant specific surveillance.

ACKNOWLEDGEMENT

Alasdair G.W. Hunter authored this chapter on Coffin–Lowry syndrome for the first three editions of Management of Genetic Syndromes, and he graciously provided his template for my revisions of his previous chapter on Coffin–Lowry syndrome in the third edition.

RESOURCES

Support Group

Coffin–Lowry Syndrome Foundation
Attn: Mary C. Hoffman
675 Kalmia Pl NW
Issaquah, WA 98027, USA
Email: CoffinLowry@gmail.com
In the UK:
CONTACT
209 City Road, London EC1V 1JN, UK
Phone: 020 7608 8700

Websites

Coffin–Lowry Syndrome Foundation
Website: *http://www.clsf.info/*
GeneReviews
Coffin–Lowry Syndrome
Website: *http://www.geneclinics.org*

NINDH Coffin–Lowry Information Page

Website: *www.ninds.nih.gov/health_and_medical/disorders/coffin_lowry.htm*

Several detailed clinical reviews of this syndrome (Hunter et al. 1982; Young 1988; Hunter, 2002; Touraine et al. 2002; Marques Pereira et al. 2010; Rogers and Abidi 2018) have been published.

REFERENCES

Arslan EA, Ceylander S, Turanli G (2014) Stimulus-induced myoclonus treated effectively with clonazepam in genetically confirmed Coffin-Lowry syndrome. *Epilep Behav Case Rep* 2:196–198.

Beck K, Ehmann N, Andlauer TFM, Ljaschenko D, Streker K, Fischer M, Kittel RJ, Raabe T (2015) Loss of the Coffin-Lowry syndrome-associated gene *RSK2* alters ERK activity, synaptic function, and axonal transport in Drosophila motorneurons. *Dis Mod Mechan* 8:1389–1400.

Bertini V, Cambi F, Bruno R, Toschi B, Forli F, Berrettini S, Simi P, Valetto A (2015) 625 kb microduplication at Xp22.12 including RPS6KA3 in a child with mild intellectual disability *J Hum Genet* 60:777–780.

Börjeson M, Forssman H, Lehmann O (1962) An X-linked, recessively inherited syndrome characterized by grave mental deficiency, epilepsy and endocrine disorder. *Acta Med Scand* 171:13–21.

Caraballo R, Tesi Rocha A, Medina C, Fejerman N (2000) Drop episodes in Coffin-Lowry syndrome: An unusual type of startle response. *Epileptic Disord* 2:173–176.

Coffin GS (2003) Postmortem findings in the Coffin-Lowry syndrome. *Genet Med* 5:187–193.

Coffin GS, Siris E, Eldridge C, Wegienka LC (1966) Mental retardation with osteocartilaginous anomalies. *Am J Dis Child* 112:205–213.

Collacott RA, Warrington JS, Young ID (1987) Coffin-Lowry syndrome and schizophrenia: A family report. *J Ment Defic Res* 31:199–207.

Crow YJ, Zuberi SM, McWilliam R, Tolmie JL, Hollman A, Pohl K, Stephenson JBP (1998) Cataplexy and muscle ultrasound abnormalities in Coffin-Lowry syndrome. *J Med Genet* 35: 94–98.

Day P, Cole B, Welbury R (2000) Coffin-Lowry syndrome and premature tooth loss: A case report. *ASDC J Dent Child* 67:148–150.

Delaunoy J-P, Abidi F, Zeniou M, Jacquot S, Merienne K, Pannetier S, Schmitt M, Schwartz CE, Hanauer A (2001) Mutations in the X-linked *RSK2* gene (RPS6KA3) in patients with Coffin-Lowry syndrome. *Hum Mutat* 17:103–116.

Delaunoy J-P, Dubos A, Marques Pereira P, Hanauer A (2006) Identification of novel mutations in the *RSK2* gene (RPS6KA3) in patients with Coffin-Lowry syndrome. *Clin Genet* 70:161–166.

Field M, Tarpey P, Boyle J, Edkins S, Goodship J, Luo Y, Moon J, Teague J, Stratton MR, Futreal PA, Wooster R, Raymond FL, Turner G (2006) Mutations in the RSK2(RPS6KA3) gene cause Coffin-Lowry syndrome and nonsyndromic X-linked mental retardation. *Clin Genet* 70:509–515.

Fischer M, Raabe T (2018) Animal Models for Coffin-Lowry syndrome: RSK2 and nervous system dysfunction. *Front Behav Neurosci* 12:1–7.

Fryns JP, Smeets E (1998) Cataplexy in Coffin-Lowry syndrome. *J Med Genet* 35:702.

Getz J, Lower K (2002) The BFL syndrome gene: Finding the mutation in eight families. In: *New Genes on the X-chromosome Affecting Mental Development*. Newcastle, NSW, Australia: John Hunter Hospital.

Gilgenkrantz S, Mujica P, Gruet P, Tridon P, Schweitzer F, Nivelon-Chevallier A, Nivelon JL, Couillault G, David A, Verloes A, Lambotte C, Piussan CH, Mathieu M (1988) Coffin-Lowry syndrome: A multicenter study. *Clin Genet* 34:230–245.

Gschwind M, Foletti G, Baumer A, Bottani A, Novy J (2015) Recurrent nonconvulsive status epilepticus in a patient with Coffin-Lowry syndrome. *Mol Syndromol* 6:91–95.

Hartsfield JK Jr, Hall BD, Grix AW, Kousseff BG, Salzar JF, Haufe SMW (1993) Pleiotropy in Coffin-Lowry syndrome: Sensorineural hearing deficit and premature tooth loss as early manifestations. *Am J Med Genet* 45:552–557.

Harun KH, Alemi L, Johnston MV (2001) Cognitive impairment in Coffin-Lowry syndrome correlates with reduced *RSK2* activation. *Neurology* 56:207–214.

Havaligi N, Matadeen-Ali C, Khurana DS, Marks H, Kothare SV (2007) Treatment of drop attacks in Coffin-Lowry syndrome with the use of sodium oxybate. *Pediatr Neurol* 37:373–374.

Herrera-Soto JA, Santiago-Cornier A, Segal LS, Ramirez N, Tamai J (2007) The musculoskeletal manifestations of the Coffin-Lowry syndrome. *J Pediatr Orthop* 27:85–89.

Hersh JH, Weisskopf B, DeCoster C (1984) Forearm fullness in Coffin-Lowry syndrome: A misleading yet possible early diagnostic clue. *Am J Med Genet* 18:195–199.

Higashi K, Matsuki C (1994) Coffin-Lowry syndrome with sensorineural deafness and labyrinthine anomaly. *J Laryngol Otol* 108:147–148.

Horn D, Delaunoy JP, Kunze J (2001) Prenatal diagnosis in Coffin-Lowry syndrome demonstrates germinal mosaicism confirmed by mutation analysis. *Prenat Diagn* 21:881–884.

Hunter AGW (2002) Coffin-Lowry syndrome: A 20 year follow-up and review of long term outcomes. *Am J Med Genet* 111:289–294.

Hunter AGW, Partington MW, Evans JA (1982) The Coffin-Lowry syndrome. *Experience from four centres. Clin Genet* 21:321–335.

Hunter AGW, Schwartz C, Abidi F (2002) Coffin Lowry syndrome. In: *GeneReviews: Genetic Diseases Online Reviews at GeneTests-GeneClinics* [database online]. Copyright University of Washington, Seattle. Available at http://www.geneclinics. org.

Igari K, Hozumi Y, Monma Y, Mayanagi H (2006) A case of Coffin-Lowry syndrome with premature exfoliation of primary teeth. *Int J Paediatr Dent* 16:213–217.

Imataka G, Nakajima I, Goto K, Konno W, Hirabayashi H, Arisaka O (2016) Drop episodes improved ater tracheostomy: a case of Coffin-Lowry syndrome associated with obstructive sleep apnea syndrome. *Eur Rev Med Pharmacol Sci* 20:498–501.

Ishida Y, Oki T, Ono Y, Nogami H (1992) Coffin-Lowry syndrome with calcium pyrophosphate crystal deposition in the ligamenta flava. *Clin Orthop Rel Res* 275:144–151.

Jacquot S, Merienne K, Trivier E, Zeniou M, Pannetier S, Hanauer A (1998) Coffin-Lowry syndrome: Current status. *Am J Med Genet* 85:214–215.

Kesler SR, Simensen RJ, Voeller K, Abidi F, Stevenson RE, Schwartz CE, Reiss AL (2007) Altered neurodevelopment associated with mutations of RSK2: A morphometric MRI study of Coffin-Lowry syndrome. *Neurogenetics* 8:143–147.

Koehne T, Jeschke A, Petermann F, Seitz S, Neven M, Peters S, Luther J, Schweizer M, Schinke T, Kahl-Nieke B, Amling M, DavidJP (2016) Rsk2, the kinase mutated in Coffin-Lowry syndrome, controls cementum formation. *J Dent Res* 95:752-760.

Kondoh T, Matsumoto T, Ochi M, Sukegawa K, Tsuji Y (1998) New radiological finding by magnetic resonance imaging of the brain in Coffin-Lowry syndrome. *J Hum Genet* 43:59–61.

Kurotaki N, Imaizumi N, Harada N, Masuno M, Kondoh M, Nagai T, Ohashi H, Naritomi K, Tsukahara M, Makita Y, Sugimoto T, Sonoda T, Hasegawa T, Chinen Y, Tomita H, Kinoshita A, Mizuguchi T, Yoshiura K, Ohta T, Kishino T, Fukushima Y, Niikawa N, Matsumoto N (2002) Haploinsufficiency of NSD1 causes Sotos syndrome. *Nat Genet* 30:365–366.

Lowry B, Miller JR, Fraser FC (1971) A new dominant gene mental retardation syndrome. *Am J Dis Child* 121:496–500.

Machin GA, Walther GL, Fraser VM (1987) Autopsy findings in two adult siblings with Coffin-Lowry syndrome. *Am J Med Genet Suppl* 3:303–309.

Manouvrier-Hanu S, Amiel J, Jacquot S, Merienne K, Moerman A, Coeslier A, Labarriere F, Vallée L, Croquette MF, Hanauer A (1999) Unreported *RSK2* missense mutation in two male sibs with an unusually mild form of Coffin-Lowry syndrome. *J Med Genet* 36:775–778.

Marques Pereira P, Heron D, Hanauer A (2007) The first large duplication of the RSK2 gene identified in a Coffin-Lowry syndrome patients. *Hum Genet* 122:541–543.

Marques Pereira P, Schneider A, Pannetier S, Heron D, Hanauer A (2010) Coffin-Lowry syndrome. *Eur J Hum Genet* 18:627-633.

Massin MM, Radermecker MA, Verloes A, Jacquot S, Grenade TH (1999) Cardiac involvement in Coffin-Lowry syndrome. *Acta Paediatr* 88:468–470.

McCandless SE, Schwartz S, Morrison S, Garlapati K, Robin NH (2000) Adult with an interstitial deletion of chromosome 10 [del(10)(q25.1q25.3)]: Overlap with Coffin-Lowry syndrome. *Am J Med Genet* 95:93–98.

Merienne K, Jacquot S, Trivier E, Pannetier S, Rossi A, Scott C, Schinzel A, Castellan C, Kress W, Hanauer A (1998) Rapid immunoblot and kinase assay tests for a syndromal form of X linked form of mental retardation: Coffin-Lowry syndrome. *J Med Genet* 35:890–894.

Merienne K, Jacquot S, Pannetier S, Zeniou M, Bankier A, Gecz J, Mandel J-L, Mulley J, Sassone-Corsi P, Hanauer A (1999) A missense mutation in RPS6KA3 (*RSK2*) responsible for non-specific mental retardation. *Nat Genet* 22:13–14.

Micheli V, Sestini S, Parri V, Fichera M, Romano C, Ariani F, Longo I, Mari F, Bruttini M, Renieri A, Meloni I (2007) *RSK2* enzymatic assay as a second level diagnostic tool in Coffin-Lowry syndrome. *Clin Chim Acta* 384:35–40.

Miyata Y, Saida K, Kumada S, Miyake N, Mashimo H, Nishida Y, Shirai I, Kurihara E, Nakata, Y, Matsumoto N (2018) Periventricular small cystic lesions in a patient with Coffin-Lowry syndrome who exhibited a novel mutation in the RPS6KA3 gene. *Brain Dev* 40:566–569.

Nakamura M, Yamagata T, Mori M, Momoi MY (2005) *RSK2* gene mutations in Coffin-Lowry syndrome with drop episodes. *Brain Dev* 27:114–117.

Nelson GB, Hahn JS (2003) Stimulus-induced drop episodes in Coffin-Lowry syndrome. *Pediatrics* 111:e197–e202.

O'Riordan S, Patton M, Schon F (2006) Treatment of drop attacks in Coffin-Lowry syndrome. *J Neurol* 253:109–110.

Özden A, Dirik E, Emel A, Sevinc N (1994) Callosal dysgenesis in a patient with Coffin-Lowry syndrome. *Indian J Pediatr* 61:101–103.

Padley S, Hodgson SV, Sherwood T (1990) The radiology of Coffin-Lowry syndrome. *Br J Radiol* 63:72–75.

Park Y, Nam H, Do M, Lee J (2016) The p90 ribosomal kinase 2 specifically affects mitotic progression by regulating the basal level, distribution and stability of mitotic spindles. *Exp Mol Med* 48:e250.

Partington MW, Mulley JC, Sutherland GR, Thode A, Turner G (1988) A family with the Coffin-Lowry syndrome revisited: Localization of Coffin-Lowry syndrome to Xp21-pter. *Am J Med Genet* 30:509–521.

Plomp AS, De Die-Smulders CEM, Meinecke P, Ypma-Verhulst JM, Lissone DA, Fryns JP (1995) The Coffin-Lowry syndrome at different ages and symptoms in female carriers. *Genet Counsel* 6:259–268.

Procopis PG, Turner B (1972) Mental retardation, abnormal fingers, and skeletal anomalies: Coffin's syndrome. *Am J Dis Child* 124:258–261.

Quintela I, Barros-Angueira F, Perez-Gay L, Castro-Gago M, Carracedo A, Eiris-Punaj J (2015) Deletion of the RPS6KA3 gene in a female with a classical phenotype of Coffin-Lowry syndrome including stimulus-induced drop attacks. *Rev Neurol* 61:94–96.

Rogers RC, Abidi FE (2018) Coffin Lowry syndrome. In: *GeneReviews: Genetic Diseases Online Reviews at GeneTests-GeneClinics [database online].* Copyright University of Washington, Seattle. Available at http://www.geneclinics.org.

Rosanowski F, Hoppe U, Pröschel U, Eysholdt U (1998) Late onset sensorineural hearing loss in Coffin-Lowry syndrome. *ORL* 60:224–226.

Sassone-Corsi P, Mizzen CA, Cheung P, Crosjo C, Monaco L, Jacquot S, Hanauer A, Allis CD (1999) Requirement of Rsk-2 for epidermal growth factor-activated phosphorylation of histone H3. *Science* 285:886–891.

Schneider A, Maas S, Hennekam R, Hanauer A (2013) Identification of the first deep intronic mutation in the RPS6KA3 gene in a patient with a severe form of Coffin-Lowry syndrome. *Eur J Med Genet* 56:150–152.

Sculerati N (2000) Analysis of a cohort of children with sensory hearing loss using the SCALE systematic nomenclature. *Laryngoscope* 110:787–798.

Simensen RJ, Abidi F, Collins JS, Schwartz CE, Stevenson RE (2002) Cognitive function in Coffin-Lowry syndrome. *Clin Genet* 61:299–304.

Sivagamasundari U, Fernando H, Jardine P, Rao JM, Lunt P, Jayewardene SLW (1994) The association between Coffin-Lowry syndrome and psychosis: A family study. *J Intellect Disabil Res* 38:469–473.

Soekarman D, Fryns JP (1994) Corpus callosum agenesis in Coffin-Lowry syndrome. *Genet Couns* 5:77–80.

Temtamy SA, Miller JD, Dorst JP, Hussels-Maumenee I, Salinas C, Lacassie Y, Kenyon KR (1975) The Coffin-Lowry syndrome: A simply inherited trait comprising mental retardation, facio-digital anomalies and skeletal involvement. *Birth Defects OAS* IX:133–152.

Tos T, Alp MY, Aksoy A, Ceylander S, Hanauer A (2015) A familial case of Coffin-Lowry syndrome caused by RPS6KA3 c.898C>T mutation associated with multiple abnormal brain imaging findings. *Genet Couns* 26:47-51.

Touraine R-L, Zeniou M, Hanauer A (2002) A syndromic form of X-linked mental retardation: The Coffin-Lowry syndrome. *Eur J Pediatr* 161:179–187.

Trivier E, De Cesare D, Jacquot S, Pannetier S, Zackai E, Young I, Mandel J-L, Sassone-Corsi P, Hanauer A (1996) Mutations in the kinase Rsk-2 associated with Coffin-Lowry syndrome. *Nature* 384:567–570.

Upadia J, Oakes J, Hamm A, Hurst ACE, Robin NH (2017) Foramen magnum compression in Coffin-Lowry syndrome: A case report. *Am J Med Genet* 173A:1087–1089.

Vealle RM, Brett EM, Rivinus TM, Stephens R (1979) The Börjeson-Forssman-Lehmann syndrome: A new case. *J Ment Def Res* 23:231–242.

Venter F, Evans A, Fontes C, Stewart C (2019) Sever restrictive lung disease in one of the oldest documented males with Coffin-Lowry syndrome. *J Investig Med High Impact Case Rep* 7:2324709618820660.

Vles JSH, Haspeslagh M, Raes MM, Fryns JP, Casaere P, Eggermont E (1984) Early clinical signs in Coffin-Lowry syndrome. *Clin Genet* 26:448–452.

Wang Y, Martinez JE, Wilson GL, He XY, Tuck-Muller CM, Maertens P, Wertelecki W, Chen TJ (2006) A novel *RSK2* (RPS6KA3) gene mutation associated with abnormal brain MRI findings in a family with Coffin-Lowry syndrome. *Am J Med Genet* A140:1274–1279.

Welborn M, Farrell S, Knott P, Mayekar E, Mardjetko S (2018) The natural history of spinal deformity in patients with Coffin-Lowry. *J Child Ortho* 12:70–75.

Wilson WG, Kelly TE (1981) Brief clinical report: Early recognition of the Coffin-Lowry syndrome. *Am J Med Genet* 8:215–220.

Yntema HG, van den Helm B, Kissing J, van Duijnhoven G, Poppelaars F, Chelly J, Moraine C, Fryns JP, Hamel BC, Heilbronne H, Pander HJ, Brunner HG, Ropers HH, Cremers FP, van Bokhoven H (1999) A novel ribosomal S6-kinase (RSK4; PRS6KA6) is commonly deleted in patients with complex X-linked mental retardation. *Genomics* 62:332–343.

Young ID (1988) The Coffin-Lowry syndrome. *J Med Genet* 25:344–348.

Zeniou M, Pannetier S, Fryns J-P, Hanauer A (2002) Unusual splice-site mutations in the *RSK2* gene and suggestion of genetic heterogeneity in Coffin-Lowry syndrome. *Am J Hum Genet* 70:1421–1433.

13

COFFIN–SIRIS SYNDROME

Tomoki Kosho
Center for Medical Genetics, Shinshu University Hospital, Matsumoto, Japan; and Department of Medical Genetics, Shinshu University School of Medicine, Matsumoto, Japan

Noriko Miyake
Yokohama City University Graduate School of Medicine, Yokohama, Japan

INTRODUCTION

Coffin–Siris syndrome (CSS) was originally reported by Drs. Grange S. Coffin and Evelyn Siris from Sonoma State Hospital in San Francisco (Coffin and Siris 1970). Three girls were described to have growth impairment and severe developmental delay, and lack the nails and terminal phalanges of the fifth fingers. After a report of another case by Bartsocas and Tsiantos (1970), Weiswasser et al. (1973) described two other patients, and proposed that these patients had a specific clinical entity designated as "Coffin–Siris syndrome" (Weiswasser et al. 1973). Carey and Hall (1978) reported five additional patients and established the clinical entity (Carey and Hall 1978).

Incidence

To date, more than 100 patients with molecularly confirmed CSS have been reported (Schrier Vergano et al. 2016), although the precise incidence has not been calculated. Indeed, this figure is likely an underestimate because some patients may be undiagnosed or not come to medical attention (Schrier Vergano et al. 2016). Furthermore, *ARID1B*, the most common causative gene in individuals called CSS, is also one of the most common genes observed in a cohort of individuals with moderate to severe intellectual disabilities (Grozeva et al. 2015), some of whom show subtle features of CSS. Very recently, a large-scale international study involving many collaborators reported that clinical features in 79 patients documented to have *ARID1B*-related CSS and in 64 patients with *ARID1B*-related intellectual disabilities demonstrate a clinical spectrum, and patients in both groups should be managed similarly (van der Sluijs et al. 2018).

Diagnostic Criteria

Consensus diagnostic criteria for CSS have not been established. Based on a literature review and description of additional 18 patients, Fleck et al. (2001) proposed that the minimal criteria included some degree of developmental delay, a coarse facial appearance, hypertrichosis, hypoplastic, or absent fifth fingernails or toenails, and hypoplastic or absent fifth distal phalanges (Fleck et al. 2001). Other findings that support the diagnosis included feeding difficulties, frequent infections, delayed dentition, and heart defects. A recent review by Schrier Vergano et al. (2016) summarized suggestive findings for CSS: fifth-digit nail/phalanx hypoplasia/aplasia (other digits including toes may also be affected); developmental or cognitive delay of variable degree; facial features (a wide mouth with thick, everted upper lips, broad nasal bridge with broad nasal tip, thick eyebrows, and long eyelashes); hypotonia of central origin; hypertrichosis (back, arms, or face); and sparse scalp hair (Schrier Vergano et al. 2016).

Cassidy and Allanson's Management of Genetic Syndromes, Fourth Edition.
Edited by John C. Carey, Agatino Battaglia, David Viskochil, and Suzanne B. Cassidy.
© 2021 John Wiley & Sons, Inc. Published 2021 by John Wiley & Sons, Inc.

Etiology, Pathogenesis and Genetics

At least seven genes responsible for CSS have been identified: *ARID1A*, *ARID1B*, *SMARCA4*, *SMARCB1*, *SMARCE1*, *SOX11*, and *DPF2* (Santen et al. 2012; Tsurusaki et al. 2012; Tsurusaki et al. 2014a; Vaslileiou et al. 2018). These genes encode components of the BRG1- and BRM-associated factor (BAF) complex (*ARID1A*, *ARID1B*, *SMARCA4*, *SMARCB1*, *SMARCE1*, and *DPF2*), or act downstream of the BAF complex (*SOX11*) (Santen et al. 2012; Tsurusaki et al. 2012; Tsurusaki et al. 2014a). Based on published literature, we estimate that 55–70% of patients with Coffin–Siris syndrome can be explained by any one of the known seven responsible genes (Santen et al. 2013; Wieczorek et al. 2013; Tsurusaki et al. 2014a). Among these genes, pathogenic mutations are most frequently identified in *ARID1B*. Further, these genes are known to be heterozygously mutated (i.e. show autosomal dominant inheritance). Since the fitness of this syndrome is estimated to be low due to the medically serious phenotype, the majority of patients are sporadic and due to de novo mutations. The syndrome is thought to show complete penetrance as the only known genes for Coffin–Siris syndrome show autosomal dominant inheritance. Based on the genetic evidence, loss-of-function mechanisms of *ARID1A*, *ARID1B*, and *SOX11* mutations, and gain-of-function or dominant-negative effects of *SMARCB1*, *SMARCA4*, and *SMARCE2* mutations form the basis of the syndrome. A dominant-negative effect of *DPF2* was suggested, but it remains unclear (Vasileiou et al. 2018). In addition, frameshift mutations in *ARID2* and 9p duplication involving *SMARCA2* have been reported in two patients, each who showed a Coffin–Siris syndrome-like phenotype (Miyake et al. 2016; Bramswig et al. 2017). However, more genetic and clinical data are needed to conclude that these conditions can be classified as CSS. Some affected siblings have been reported; thus autosomal recessive inheritance in Coffin–Siris syndrome may also account for rare cases (Carey and Hall 1978; Haspeslagh et al. 1984). However, no genes that are responsible for Coffin–Siris syndrome inherited in a recessive fashion have been identified so far.

Diagnostic Testing

Diagnosis of Coffin–Siris syndrome is established based on identification of a heterozygous pathogenic variant in one of the related genes (Schrier Vergano et al. 2016). Serial single-gene testing attaches greater importance to *ARID1B*, while next-generation sequencing-based multi-gene panel testing (including related genes) is reasonable for this genetically heterogeneous syndrome. Microdeletion involving any of the responsible genes can be detected by copy number analysis including chromosomal microarray testing.

Differential Diagnosis

The following diseases are considered as differential diagnoses of CSS.

1. Nicolaides–Baraitser syndrome

    This is an autosomal dominant disorder caused by heterozygous mutations in *SMARCA2* (Van Houdt et al. 2012), which is also a component of the BAF complex. *SMARCA2* mutations identified in this syndrome are of the non-truncating type, and a dominant-negative or gain-of-function effect has been suggested to cause this syndrome (Sousa and Hennekam 2014). Clinical features of this syndrome are similar to CSS. Sparse scalp hair is frequently observed (97%). Prominent interphalangeal joints are characteristic (84.7%), which are not usually observed in CSS individuals. Coarse face, sparse scalp hair, and joint anomalies become prominent with age. Small fifth fingers or toes are observed in a few patients. Neither absent nails of the fifth fingers/toes nor small or absent distal phalanges are observed (Sousa and Hennekam 2014).

2. Borjeson–Forssman–Lehmann syndrome

    This is an X-linked recessive disease caused by *PHF6* (Lower et al. 2002), which is characterized by short stature, moderate obesity, hypogonadism, delayed puberty, and severe intellectual disability. Some female patients have been reported, but the majority of females with heterozygous *PHF6* mutations show no phenotypic abnormalities due to marked X-inactivation of the X chromosome with *PHF6* mutation (Crawford et al. 2006).

3. Brachymorphism-onychodysplasia-dysphalangism (BOD) syndrome

    This is thought to be an autosomal dominant disorder, although the responsible gene has not yet been identified. Clinical features are similar to CSS, and the only difference is thought to be severity of intellectual disability: mild in BOD syndrome, and moderate to severe in CSS. Thus, these two conditions are considered to be allelic (Brautbar et al. 2009). Molecular analysis of presumed persons with BOD syndrome may clarify this issue in the future.

4. Deafness, onychodystrophy, osteodystrophy, mental retardation, and seizures (DOORS) syndrome

    This is an autosomal recessive syndrome caused by *TBC1D24* mutations (Campeau et al. 2014). Intellectual disability or developmental delay, seizures, short distal phalanges, and small or absent nails are common. Unlike CSS, no common facial features have been reported, while half of the individuals with biallelic *TBC1D24* mutations show a broad nasal bridge. Increased urinary 2-oxoglutaric acid excretion is often observed (Campeau and Hennekam 2014).

5. Cornelia de Lange syndrome (see Chapter 15).

   Typical features of this syndrome include prenatal onset growth deficiency, developmental delay, characteristic facial features (bushy and arched eyebrows, synophrys, long curly eyelashes, upturned nose, long philtrum, thin upper lip vermilion, and downturned angles of mouth), microbrachycephaly, and micromelia. Clinodactyly of the fifth fingers is observed in 74% of cases (Jones 2006). At least five responsible genes are known: *NIPBL*, *SMC3*, and *RAD21*, showing autosomal dominant inheritance, and *SMC1A* and *HDAC8* with X-linked dominant inheritance (Tonkin et al. 2004; Deardorff et al. 2007; Deardorff et al. 2012; Harakalova et al. 2012).

6. Mabry syndrome

   Mabry syndrome is also known as hyperphosphatasia with mental retardation-1, which is caused by biallelic variants of *PIGV* (Krawitz et al. 2010). This gene encodes the PIGV protein, which plays an important role in glycosylphosphatidylinsitol biosynthesis. Severe intellectual disability, seizures, and brachytelephalangy (hypoplastic nails in some patients) are observed in Mabry syndrome and CSS, but hyperphosphatesia is usually observed in Mabry syndrome and not in CSS.

MANIFESTATIONS AND MANAGEMENT

Kosho and Okamoto (2014) provided a detailed and comprehensive review of the clinical features of patients with CSS due to mutations in *SMARCB1*, *SMARCA4*, *SMARCE1*, and *ARID1A* (Kosho and Okamoto 2014). Santen *et al.* (2014) provided a similar review for patients with *ARID1B* mutations (Santen and Clayton-Smith 2014). Vergano and Deardorff (2014) and Schrier Vergano *et al.* (2016) reviewed the management of patients with CSS (Vergano and Deardorff 2014; Schrier Vergano 2016). Including additional reports of patients with *SMARCB1*, *SMARCA4*, *SMARCE1*, *ARID1A*, or *ARID1B* mutations (Gossai et al. 2015; Zarate et al. 2016), as well as those with *SOX11* mutations (Tsurusaki et al. 2014a; Hempel et al. 2016; Okamoto et al. 2017), the clinical features classified according to causative gene are summarized in Table 13.1. A large cohort reported by Santen and Clayton-Smith (2014) included patients with intellectual disability syndrome, not originally suspected as CSS.

Growth and Feeding

Growth is mildly impaired prenatally, especially length, and more severely and frequently impaired postnatally, including weight and occipitofrontal circumference (OFC). Sucking and/or feeding difficulties are observed in most patients, yet less frequently in patients with *ARID1B* mutations (Kosho et al. 2013; Kosho and Okamoto 2014; Santen and Clayton-Smith 2014). Most patients require tube feeding, but can occasionally be weaned off (Kosho and Okamoto 2014).

Evaluation

- Height, weight, and OFC should be monitored at each visit: in infancy, every 1–3 months, while in childhood, every 6 months, and in school years, annually (Vergano and Deardorff 2014).
- If sucking and/or feeding difficulties as well as significant growth impairment are present, extensive evaluation is warranted for the possibility of nutritional, metabolic, endocrinological, immunological, cardiac, respiratory, gastrointestinal, or neurological abnormalities (Vergano and Deardorff 2014).

Management

- If sucking and/or feeding difficulties as well as significant growth impairment are present, feeding therapy, nutritional supplementation, and tube feeding (including gastrostomy placement) should be considered (Vergano and Deardorff 2014; Schrier Vergano 2016). Some centers have dysphagia/feeding clinics that could be a possible referral.

Development and Behavior

Developmental delay and/or intellectual disabilities are severe in 40%, moderate in a half, and mild in 10% of individuals. People with *SMARCB1* or *ARID1A* mutations are likely to show a severe delay, while those with *SMARCE1* or *ARID1B* mutations are likely to have a moderate delay (Kosho et al. 2013; Kosho and Okamoto 2014; Santen and Clayton-Smith 2014; Tsurusaki et al. 2014b; Gossai et al. 2015; Hempel et al. 2016; Zarate et al. 2016; Okamoto et al. 2017). In a cohort of individuals with *SMARCB1*, *SMARCA4*, *SMARCE1*, or *ARID1A* mutations, overall median age of independent sitting was 18 months and independent walking was 36 months (Kosho and Okamoto 2014).

All show a variable degree of speech impairment. Absence of speech is observed in the majority of persons with *SMARCB1* or *ARID1A* mutations, but only in approximately 10% of those with *ARID1B* or *SOX11* mutations (Kosho et al. 2013; Kosho and Okamoto 2014; Santen and Clayton-Smith 2014; Tsurusaki et al. 2014b; Gossai et al. 2015; Hempel et al. 2016; Zarate et al. 2016; Okamoto et al. 2017). Less than half of individuals show behavioral abnormalities including hyperactivity, self-injurious behavior, short attention span, and obsession. Some have a diagnosis of autism spectrum disorder (Kosho et al. 2013; Kosho and Okamoto 2014; Santen and Clayton-Smith 2014; Tsurusaki et al. 2014b; Gossai et al. 2015; Hempel et al. 2016; Zarate et al. 2016; Okamoto et al. 2017).

TABLE 13.1 Comparison of clinical features of patients caused by mutations in each gene.

Gene	SMARCB1	SMARCA4	SMARCE1	ARID1A	ARID1B	SOX11	Total
Patient number	14	12	6	8	≈100#	13	≈150
Growth and feeding							
Prenatal growth							
Birth weight <−2 SD or 3P	27% (3/11)	27% (3/11)	50% (3/6)	17% (1/6)		11% (1/9)	29% (10/34)
Birth length <−2SD or 3P	50% (3/6)	50% (3/6)	50% (2/4)	0% (0/2)		33% (1/3)	44% (8/18)
Birth OFC <−2 SD or 3P	0% (0/6)	17% (1/6)	67% (2/3)	100% (1/1)		33% (1/3)	25% (4/16)
Postnatal growth at the last observation							
Weight <−2 SD or 3P	78% (7/9)	33% (4/12)	60% (3/5)	50% (3/6)		40% (4/10)	53% (17/32)
Height <−2 SD or 3P	100% (11/11)	67% (8/12)	50% (3/6)	25% (2/8)		45% (5/11)	65% (24/37)
OFC <−2 SD or 3P	91% (10/11)	91% (10/11)	50% (2/4)	29% (2/7)		69% (9/13)	73% (24/33)
Sucking	100% (1111)	92% (11/12)	83% (5/6)	100% (7/7)	65% (37/57)	89% (8/9)	77% (79/102)
Craniofacial features							
Sparse sculp hair	91% (10/11)	42% (5/12)	83% (5/6)	60% (3/5)	57% (33/58)	75% (3/4)	61% (59/96)
Hypertrichosis	73% (8/11)	100% (12/12)	100% (3/3)	100% (7/7)	95% (56/59)	100% (3/3)	94% (89/95)
Thick eyebrows	100% (12/12)	75% (9/12)	67% (4/6)	75% (6/8)	92% (54/59)	0% (0/2)	86% (85/99)
Long eyelashes	100% (12/12)	83% (10/12)	67% (2/3)	100% (6/6)	85% (51/60)	50% (1/2)	86% (82/95)
Ptosis	55% (6/11)	75% (9/12)	33% (1/3)	43% (3/7)	22% (13/59)	50% (1/2)	35% (33/94)
Wide nasal bridge	55% (6/11)	0% (0/11)	0% (0/3)	80% (4/5)	23% (11/47)		27% (21/77)
Flat nasal bridge	18% (2/11)	45% (5/11)	33% (2/6)	0% (0/5)	47% (22/47)	50% (1/2)	39% (32/82)
Long philtrum	40% (4/10)	27% (3/11)	50% (1/2)	14% (1/7)	51% (28/55)	0% (0/4)	42% (37/89)
Thin upper lip vermilion	73% (8/11)	27% (3/11)	67% (2/3)	50% (3/6)	51% (29/57)	0% (0/2)	50% (45/90)
Thick lower lip vermilion	75% (9/12)	83% (10/12)	100% (6/6)	86% (6/7)	81% (46/57)	83% (5/6)	82% (82/100)
Cleft palate	20% (2/10)	33% (4/12)	40% (2/5)	33% (2/6)	7% (4/58)	0% (0/2)	15% (14/93)
Skeletal-limb features							
Hypoplastic 5th fingers or toes	73% (8/11)	100% (12/12)	100% (5/5)	86% (6/7)	73% (43/59)$	100% (2/2)	79% (76/96)
Hypoplastic 5th fingernails or toenails	100% (12/12)	100% (12/12)	100% (4/4)	88% (7/8)	81% (39/48)	100% (7/7)	89% (81/91)
Hypoplastic other fingernails and toenails	73% (8/11)	50% (5/10)	83% (5/6)	75% (6/8)		100% (3/3)	71% (27/38)
Prominent interphalangeal joints	44% (4/9)	27% (3/11)	50% (1/2)	20% (1/5)	32% (16/50)		32% (25/77)
Prominent distal phalanges	75% (6/8)	50% (5/10)	33% (1/3)	20% (1/5)	37% (19/52)		41% (32/78)
Scoliosis	78% (7/9)	10% (1/10)	33% (1/3)	29% (2/7)	29% (16/56)	80% (4/5)	34% (31/90)
Internal complications							
Cardiovascular	45% (5/11)	42% (5/12)	80% (4/5)	38% (3/8)	~22% (~12/54)	33% (1/3)	~32% (~30/93)
Gastrointestinal	70% (7/10)	67% (8/12)	50% (1/2)	57% (4/7)	~19% (~8/43)	67% (4/6)	40% (~32/80)
Genitourinary	45% (5/11)	25% (3/12)	25% (1/4)	29% (2/7)	38% (9/24)	100% (3/3)	38% (23/61)
Hernia	88% (7/8)	55% (6/11)	0% (0/2)	25% (1/4)	11% (3/28)	0% (0/2)	31% (17/55)
Hearing and vision							
Hearing impairment	75% (6/8)	33% (4/12)	40% (2/5)	33% (2/6)	14% (8/56)	0% (0/2)	25% (22/89)
Visual impairment	56% (5/9)	45% (5/11)	67% (2/3)	75% (3/4)	34% (19/56)	83% (5/6)	44% (39/89)
Immunology							
Frequent infection	89% (8/9)	67% (8/12)	50% (3/6)	60% (3/5)	49% (28/57)	0% (0/2)	55% (50/91)
Neurology							
Hypotonia	75% (9/12)	73% (8/11)	50% (3/6)	88% (7/8)	75% (42/56)	50% (1/2)	74% (70/95)
Seizures	80% (8/10)	17% (2/12)	33% (2/6)	29% (2/7)	23% (14/60)	0% (0/2)	29% (28/97)
Structural CNS abnormalities	100% (10/10)	86% (6/7)	33% (3/3)	88% (7/8)	27% (13/48)	33% (1/3)	51% (40/79)
Development and intelligence							
Developmental delay and ID							
Severe	75% (9/12)	55% (6/11)	33% (2/6)	71% (5/7)	25% (11/44)	33% (1/3)	41% (34/83)
Moderate	17% (2/12)	36% (4/11)	67% (4/6)	0% (0/7)	64% (28/44)	33% (1/3)	47% (39/83)
Mild	8% (1/12)	9% (1/11)	0% (0/6)	29% (2/7)	11% (5/44)	33% (1/3)	12% (10/83)
Speech impairment							
No words	80% (8/10)	36% (4/11)	60% (3/5)	83% (5/6)	13% (6/45)	8% (1/12)	30% (27/89)
Behavior							
Behavioral abnormalities	50% (4/8)	88% (7/8)	67% (2/3)	60% (3/5)	~33% (~16/48)	100% (1/1)	~45% (~33/73)

#, including ID syndrome; $, small 5th finger and/or absent distal phalanx; CNS, central nervous system; ID, intellectual disability; mild, including low normal to mild; moderate, including mild to moderate; P, percentile; SD, standard deviation; severe, including moderate to severe.

Evaluation

- Developmental assessment is recommended at every visit, with formal tests for developmental or intellectual status planned at specific ages, e.g., 3 years old and/or 6 years old before elementary school.
- Behavioral assessment is recommended at every visit, with formal tests for autism spectrum disorder planned if suspected symptoms are present.

Management

- Early intervention is recommended, including physical therapy, occupational therapy, and speech therapy (Vergano and Deardorff 2014).

Neurologic

Most people show hypotonia. Other neurological features include seizures and structural abnormalities of the central nervous system, typically abnormal corpus callosum (hypoplasia or agenesis), followed by cerebellar hypoplasia and Dandy–Walker malformation (Kosho and Okamoto 2014; Santen and Clayton-Smith 2014; Hempel et al. 2016).

Evaluation

- Following initial diagnosis, brain imaging is recommended (Vergano and Deardorff 2014).
- If patients have seizures, electroencephalograms and brain imaging are recommended.

Management

- Seizures are treated according to the pattern, severity, and electroencephalographic abnormalities.

Craniofacial

Characteristic craniofacial features in most people include hypertrichosis, thick eyebrows, long eyelashes, and thick lower lip vermilion (Figure 13.1). Those with the recurrent mutation "p.Lys364del" of *SMARCB1* show a characteristic facial coarseness: in early childhood, a round face with thick and arched eyebrows, short nose with bulbous tip and anteverted nostrils, long philtrum, small mouth, and microretrognathia; while in later years, a broad nasal bridge without anteverted nostrils, broad philtrum, large tongue, and protruding jaw. Individuals with *SMARCA4* mutations have a less coarse face with characteristic patterns of the philtrum (short or long/ broad) and upper lip vermilion (everted or thin), as well as a pointed chin, especially in older ages. Those with *ARID1A* mutations show some coarseness with a short nose and characteristic patterns of the philtrum (short or long/broad) and lower lip vermilion (thick, everted) (Kosho et al. 2013; Kosho and Okamoto 2014). Individuals with *ARID1B* mutations typically show a progressive coarseness including prominent eyelashes, thick eyebrows, large mouth, and thick lower lip vermilion (Kosho et al. 2013; Santen and Clayton-Smith 2014). Those with *SOX11* mutations share wide mouths and thick lips (Tsurusaki et al. 2014a; Hempel et al. 2016). Overall, 15% of patients have palatal abnormalities requiring medical intervention, including cleft palate, submucous cleft palate, or bifid uvula.

Evaluation

- Following initial diagnosis, palatal abnormalities should be investigated.

Management

- Routine management is offered if a cleft palate is found, including plastic surgery, feeding and/or speech therapy, and orthodontic intervention.

Skeletal-limb

Most individuals have hypoplastic fifth fingers or toes and hypoplastic fifth fingernails or toenails. One third of them have scoliosis, which is more frequently observed in those with *SMARCB1* or *SOX11* mutations (Kosho et al. 2013; Kosho and Okamoto 2014; Santen and Clayton-Smith 2014; Hempel et al. 2016).

Evaluation

- Whole spine radiographs are recommended yearly (Vergano and Deardorff 2014). Referral to orthopedists is considered if scoliosis is found.

Management

- Orthopedic intervention for progressive scoliosis, including braces and corrective surgery, is managed depending on intellectual and behavioral abnormalities.

Cardiovascular

One third of individuals have cardiovascular complications including ventricular septal defects, atrial septal defects, pulmonary stenosis, dextrocardia, coarctation of aorta, and aortic stenosis (Kosho et al. 2013; Kosho and Okamoto 2014; Santen and Clayton-Smith 2014; Okamoto et al. 2017).

Evaluation

- Following initial diagnosis, cardiovascular screening is recommended, including chest radiograph,

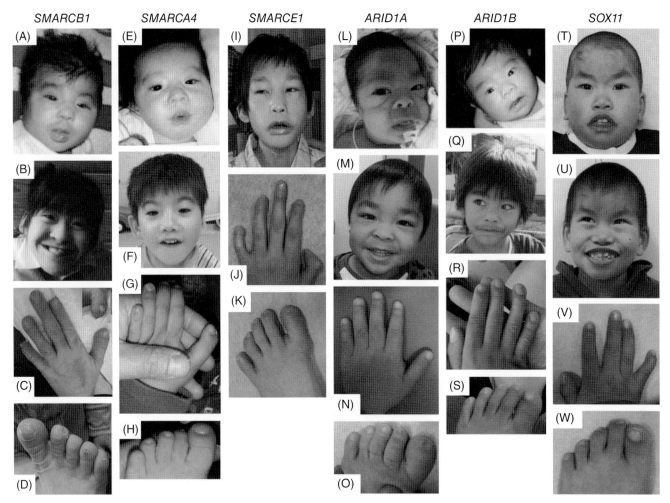

FIGURE 13.1 Clinical photographs of individuals with Coffin–Siris syndrome caused by various genes. A girl with a "p.Lys364del" mutation in *SMARCB1* at age 2 months (A), 18 years (B), and 21 years (C), (D). A boy with a *SMARCA4* mutation at age 4 months (E) and 5 years and 1 month (F). Another boy with a *SMARCA4* mutation at age 3 years and 7 months (G), (H). A girl with a *SMARCE1* mutation at age 14 years (I), (J), (K). A boy with an *ARID1A* mutation in the neonatal period (L) and at age 3 years and 4 months (M), (N), (O). A boy with an *ARID1B* mutation in the neonatal period (P), 6 years and 8 months (Q), and 6 years and 9 months (R), (S). A boy with a *SOX11* mutation at age 5 years (T), (U), (V), (W). [Figures other than (J)–(O) and (T)–(W), originally published in *Am J Med Genet* Part A 161A: 1221–37.]

electrocardiogram, and echocardiography (Vergano and Deardorff 2014; Schrier Vergano et al. 2016).

Management

- Medical intervention for cardiovascular complications are managed as usual with consideration of the individual's general physical conditions.

Respiratory

Several individuals with *SMARCA4* or *ARID1A* mutations have laryngomalacia or bronchomalacia (Kosho et al. 2013; Kosho and Okamoto 2014).

Evaluation

- Laryngomalacia and/or bronchomalacia should be considered if the individuals have respiratory or feeding impairments. Referral to pulmonology and/or otolaryngology for assessment by laryngoscopy and/or bronchoscopy should be considered.

Management

- Laryngomalacia and/or bronchomalacia should be treated according to the symptoms, including positive pressure ventilation and/or tracheostomy. Referral to otolaryngology or a pulmonary specialist can occur as needed.

Gastrointestinal

In total, 40% of individuals have gastrointestinal complications including gastroesophageal reflux, pyloric stenosis, and constipation (Kosho et al. 2013; Kosho and Okamoto 2014; Hempel et al. 2016).

Evaluation

- Investigation for gastroesophageal reflux and pyloric stenosis is recommended if individuals have suggestive symptoms such as vomiting, desaturation, or aspiration with/without lower respiratory tract infection.

Management

- Gastroesophageal reflux is treated according to the symptoms, including thickened food in mild cases, oral medications, and duodenal tube placement or Nissen fundoplication in severe cases. Referral to gastroenterology could occur as indicated.
- Pyloric stenosis is usually treated by surgical correction.
- Constipation is treated with glycerin enema and/or oral laxatives.

Genitourinary

One third of individuals have genitourinary complications including cryptorchidism in males (Kosho et al. 2013; Kosho and Okamoto 2014; Santen and Clayton-Smith 2014; Tsurusaki et al. 2014a; Okamoto et al. 2017). Inguinal hernia is occasionally found (Kosho et al. 2013; Kosho and Okamoto 2014; Santen and Clayton-Smith 2014).

Evaluation

- Following initial diagnosis, renal ultrasonography is recommended to evaluate structural kidney or genitourinary anomalies (Vergano and Deardorff 2014; Schrier Vergano 2016). Physical examination for cryptorchidism is also recommended in males.

Management

- Cryptorchidism is treated with consideration of the individual's general physical conditions relevant to other cardiovascular or respiratory complications.

Immunological

Approximately half of individuals suffer from frequent infections, including those of the respiratory tract and urinary tract. Most individuals with *SMARCB1* mutations exhibit frequent infections, including one who was diagnosed with immunodeficiency and treated by regular administration of intravenous antibiotics (Kosho et al. 2013; Kosho and Okamoto 2014; Santen and Clayton-Smith 2014; Tsurusaki et al. 2014b; Gossai et al. 2015; Zarate et al. 2016).

Evaluation

- In the presence of frequent infections, comprehensive immunological assessments including neutrophil count, immunoglobulin, and complement. Assessment of aspiration and/or vesicoureteral reflux are recommended according to the symptoms (Schrier Vergano 2016).

Management

- Routine immunization is recommended in all individuals.
- If someone is found to have a specific type of immunodeficiency, aspiration, or vesicoureteral reflux, treatment is planned in the routine manner.

Hearing and Vision

A quarter of individuals show a variable degree of hearing impairment, which is frequent in those with *SMARCB1* mutations and less frequent with *ARID1B* mutations. More than 40% have various types of visual impairment (e.g., myopia, strabismus, optic disc coloboma), which is frequent in those with *SOX11* or *ARID1A* mutations and less frequent with *ARID1B* mutations (Kosho et al. 2013; Kosho and Okamoto 2014; Santen and Clayton-Smith 2014; Tsurusaki et al. 2014b; Gossai et al. 2015; Zarate et al. 2016).

Evaluation

- Following initial diagnosis, comprehensive ophthalmological and otological investigation is recommended, followed every 1–3 years by screening of hearing and vision (Vergano and Deardorff 2014; Schrier Vergano 2016).

Management

- Hearing aid and corrective lenses for refractive errors and strabismus are considered as needed.

Malignancy

Although there has been increasing evidence to suggest that BAF complex genes play a significant role in tumor suppression (Biegel et al. 2014), only two individuals have been reported to date with neoplasms. A male child with an

ARID1A mutation had a hepatoblastoma because of abdominal distension at 1 year and 10 months. The tumor had arisen from the right lobe of the liver, and progressed with increased levels of liver transaminases, causing paralytic ileus and death at 2 years and 3 months (Kosho et al. 2013). The other report represented an adult male individual, who was found to have schwannomatosis after presenting with acute spinal cord compression caused by spinal schwannomas. In addition to a constitutional missense mutation in *SMARCB1*, loss of 22q (including *SMARCB1* and *NF2*) as well as a mutation in the remaining *NF2* allele were detected in tumor tissue as the second to fourth hit. Bevacizumab followed by lapatinib with surgical resection relieved the symptoms (Gossai et al. 2015).

Evaluation

- Presence or absence of tumor predisposition in individuals with mutations in BAF complex genes, and the necessity for tumor surveillance in this condition, remains to be clarified (Kosho and Okamoto 2014).
- Aggressive radiological and biochemical investigations should be performed if the person shows symptoms suggestive of tumor occurrence.

Management

- Tumors are treated with consideration of the individual's general physical conditions relevant to other cardiovascular or respiratory complications.

RESOURCES

Genetic and Rare Diseases Information Center (GARD)
PO Box 8126, Gaithersburg MD 20898-8126, USA
Phone: 888-205-2311; 888-205-3223 (TTY); 301-251-4925
Fax: 301-251-4911
Email: GARDinfo@nih.gov
Website: https://rarediseases.info.nih.gov/

American Association on Intellectual and Developmental Disabilities (AAIDD)
501 3rd Street Northwest, Suite 200, Washington DC 20001, USA
Phone: 202-387-1968
Fax: 202-387-2193
Email: sis@aaidd.org
Website: www.aaidd.org

Coffin-Siris Syndrome Foundation
Websites: https://www.coffinsiris.org/
https://www.facebook.com/Coffin-Siris-Syndrome-389657281170288/

REFERENCES

Bartsocas CS, Tsiantos AK (1970) Mental retardation with absent fifth fingernail and terminal phalanx. *Am J Dis Child* 120:493–494.

Biegel JA, Busse TM, Weissman BE (2014) SWI/SNF chromatin remodeling complexes and cancer. *Am J Med Genet C* 166C:350–366.

Bramswig NC, Caluseriu O, Ludecke HJ, Bolduc FV, Noel NC, Wieland T, Surowy HM, Christen HJ, Engels H, Strom TM, Wieczorek D (2017) Heterozygosity for ARID2 loss-of-function mutations in individuals with a Coffin-Siris syndrome-like phenotype. *Hum Genet* 136:297–305.

Brautbar A, Ragsdale J, Shinawi M (2009) Is this the Coffin-Siris syndrome or the BOD syndrome? *Am J Med Genet A* 149A:559–562.

Campeau PM, Hennekam RC (2014) DOORS syndrome: phenotype, genotype and comparison with Coffin-Siris syndrome. *Am J Med Genet C* 166C:327–332.

Campeau PM, Kasperaviciute D, Lu JT, Burrage LC, Kim C, Hori M, Powell BR, Stewart F, Felix TM, van den Ende J, Wisniewska M, Kayserili H, Rump P, Nampoothiri S, Aftimos S, Mey A, Nair LD, Begleiter ML, De Bie I, Meenakshi G, Murray ML, Repetto GM, Golabi M, Blair E, Male A, Giuliano F, Kariminejad A, Newman WG, Bhaskar SS, Dickerson JE, Kerr B, Banka S, Giltay JC, Wieczorek D, Tostevin A, Wiszniewska J, Cheung SW, Hennekam RC, Gibbs RA, Lee BH, Sisodiya SM (2014) The genetic basis of DOORS syndrome: an exome-sequencing study. *Lancet Neurol* 13:44–58.

Carey JC, Hall BD (1978) The Coffin-Siris syndrome: five new cases including two siblings. *Am J Dis Child* 132:667–671.

Coffin GS, Siris E (1970) Mental retardation with absent fifth fingernail and terminal phalanx. *Am J Dis Child* 119:433–439.

Crawford J, Lower KM, Hennekam RC, Van Esch H, Megarbane A, Lynch SA, Turner G, Gecz J (2006) Mutation screening in Borjeson-Forssman-Lehmann syndrome: identification of a novel de novo PHF6 mutation in a female patient. *J Med Genet* 43:238–243.

Deardorff MA, Kaur M, Yaeger D, Rampuria A, Korolev S, Pie J, Gil-Rodriguez C, Arnedo M, Loeys B, Kline AD, Wilson M, Lillquist K, Siu V, Ramos FJ, Musio A, Jackson LS, Dorsett D, Krantz ID (2007) Mutations in cohesin complex members SMC3 and SMC1A cause a mild variant of cornelia de Lange syndrome with predominant mental retardation. *Am J Hum Genet* 80:485–494.

Deardorff MA, Wilde JJ, Albrecht M, Dickinson E, Tennstedt S, Braunholz D, Monnich M, Yan Y, Xu W, Gil-Rodriguez MC, Clark D, Hakonarson H, Halbach S, Michelis LD, Rampuria A, Rossier E, Spranger S, Van Maldergem L, Lynch SA, Gillessen-Kaesbach G, Ludecke HJ, Ramsay RG, McKay MJ, Krantz ID, Xu H, Horsfield JA, Kaiser FJ (2012) RAD21 mutations cause a human cohesinopathy. *Am J Hum Genet* 90:1014–1027.

Fleck BJ, Pandya A, Vanner L, Kerkering K, Bodurtha J (2001) Coffin-Siris syndrome: review and presentation of new cases from a questionnaire study. *Am J Med Genet* 99:1–7.

Gossai N, Biegel JA, Messiaen L, Berry SA, Moertel CL (2015) Report of a patient with a constitutional missense mutation in

SMARCB1, Coffin-Siris phenotype, and schwannomatosis. *Am J Med Genet A* 167A:3186–3191.

Grozeva D, Carss K, Spasic-Boskovic O, Tejada MI, Gecz J, Shaw M, Corbett M, Haan E, Thompson E, Friend K, Hussain Z, Hackett A, Field M, Renieri A, Stevenson R, Schwartz C, Floyd JA, Bentham J, Cosgrove C, Keavney B, Bhattacharya S, Hurles M, Raymond FL (2015) Targeted Next-Generation Sequencing Analysis of 1,000 Individuals with Intellectual Disability. *Hum Mutat* 36:1197–1204.

Harakalova M, van den Boogaard MJ, Sinke R, van Lieshout S, van Tuil MC, Duran K, Renkens I, Terhal PA, de Kovel C, Nijman IJ, van Haelst M, Knoers NV, van Haaften G, Kloosterman W, Hennekam RC, Cuppen E, Ploos van Amstel HK (2012) X-exome sequencing identifies a HDAC8 variant in a large pedigree with X-linked intellectual disability, truncal obesity, gynaecomastia, hypogonadism and unusual face. *J Med Genet* 49:539–543.

Haspeslagh M, Fryns JP, van den Berghe H (1984) The Coffin-Siris syndrome: report of a family and further delineation. *Clin Genet* 26:374–378.

Hempel A, Pagnamenta AT, Blyth M, Mansour S, McConnell V, Kou I, Ikegawa S, Tsurusaki Y, Matsumoto N, Lo-Castro A, Plessis G, Albrecht B, Battaglia A, Taylor JC, Howard MF, Keays D, Sohal AS, Kuhl SJ, Kini U, McNeill A (2016) Deletions and de novo mutations of SOX11 are associated with a neurodevelopmental disorder with features of Coffin-Siris syndrome. *J Med Genet* 53:152–162.

Jones KL (2006) *SMITH'S Recognizable Patterns of Human Malformation*. Philadelphia, PA: Elsevier Saunders.

Kosho T, Okamoto N (2014) Genotype-phenotype correlation of Coffin-Siris syndrome caused by mutations in SMARCB1, SMARCA4, SMARCE1, and ARID1A. *Am J Med Genet C* 166C:262–275.

Kosho T, Okamoto N, Ohashi H, Tsurusaki Y, Imai Y, Hibi-Ko Y, Kawame H, Homma T, Tanabe S, Kato M, Hiraki Y, Yamagata T, Yano S, Sakazume S, Ishii T, Nagai T, Ohta T, Niikawa N, Mizuno S, Kaname T, Naritomi K, Narumi Y, Wakui K, Fukushima Y, Miyatake S, Mizuguchi T, Saitsu H, Miyake N, Matsumoto N (2013) Clinical correlations of mutations affecting six components of the SWI/SNF complex: detailed description of 21 patients and a review of the literature. *Am J Med Genet A* 161A:1221–1237.

Krawitz PM, Schweiger MR, Rodelsperger C, et al. (2010) Identity-by-descent filtering of exome sequence data identifies PIGV mutations in hyperphosphatasia mental retardation syndrome. *Nat Genet* 42:827–829.

Lower KM, Turner G, Kerr BA, Mathews KD, Shaw MA, Gedeon AK, Schelley S, Hoyme HE, White SM, Delatycki MB, Lampe AK, Clayton-Smith J, Stewart H, van Ravenswaay CM, de Vries BB, Cox B, Grompe M, Ross S, Thomas P, Mulley JC, Gecz J (2002) Mutations in PHF6 are associated with Borjeson-Forssman-Lehmann syndrome. *Nat Genet* 32:661–665.

Miyake N, Abdel-Salam G, Yamagata T, Eid MM, Osaka H, Okamoto N, Mohamed AM, Ikeda T, Afifi HH, Piard J, van Maldergem L, Mizuguchi T, Miyatake S, Tsurusaki Y, Matsumoto N (2016) Clinical features of SMARCA2 duplication overlap with Coffin-Siris syndrome. *Am J Med Genet A* 170:2662–2670.

Okamoto N, Ehara E, Tsurusaki Y, Miyake N, Matsumoto N (2017) Coffin-Siris syndrome and cardiac anomaly with a novel SOX11 mutation. *Congenit Anom* 58:105–107.

Santen GW, Aten E, Sun Y, Almomani R, Gilissen C, Nielsen M, Kant SG, Snoeck IN, Peeters EA, Hilhorst-Hofstee Y, Wessels MW, den Hollander NS, Ruivenkamp CA, van Ommen GJ, Breuning MH, den Dunnen JT, van Haeringen A, Kriek M (2012) Mutations in SWI/SNF chromatin remodeling complex gene ARID1B cause Coffin-Siris syndrome. *Nat Genet* 44:379–380.

Santen GW, Aten E, Vulto-van Silfhout AT, Pottinger C, van Bon BW, van Minderhout IJ, Snowdowne R, van der Lans CA, Boogaard M, Linssen MM, Vijfhuizen L, van der Wielen MJ, Vollebregt MJ, Breuning MH, Kriek M, van Haeringen A, den Dunnen JT, Hoischen A, Clayton-Smith J, de Vries BB, Hennekam RC, van Belzen MJ (2013) Coffin-Siris syndrome and the BAF complex: genotype-phenotype study in 63 patients. *Hum Mutat* 34:1519–1528.

Santen GW, Clayton-Smith J (2014) The ARID1B phenotype: what we have learned so far. *Am J Med Genet C* 166C:276–289.

Schrier Vergano S, Wieczorek D, Wollnik B, Matsumoto N, Deardorff MA (2016) Coffin-Siris Syndrome. GeneReviews® [Online].

Sousa SB, Hennekam RC (2014) Phenotype and genotype in Nicolaides-Baraitser syndrome. *Am J Med Genet C* 166C:302–314.

Tonkin ET, Wang TJ, Lisgo S, Bamshad MJ, Strachan T (2004) NIPBL, encoding a homolog of fungal Scc2-type sister chromatid cohesion proteins and fly Nipped-B, is mutated in Cornelia de Lange syndrome. *Nat Genet* 36:636–641.

Tsurusaki Y, Koshimizu E, Ohashi H, Phadke S, Kou I, Shiina M, Suzuki T, Okamoto N, Imamura S, Yamashita M, Watanabe S, Yoshiura K, Kodera H, Miyatake S, Nakashima M, Saitsu H, Ogata K, Ikegawa S, Miyake N, Matsumoto N (2014a) De novo SOX11 mutations cause Coffin-Siris syndrome. *Nat Commun* 5:4011.

Tsurusaki Y, Okamoto N, Ohashi H, Kosho T, Imai Y, Hibi-Ko Y, Kaname T, Naritomi K, Kawame H, Wakui K, Fukushima Y, Homma T, Kato M, Hiraki Y, Yamagata T, Yano S, Mizuno S, Sakazume S, Ishii T, Nagai T, Shiina M, Ogata K, Ohta T, Niikawa N, Miyatake S, Okada I, Mizuguchi T, Doi H, Saitsu H, Miyake N, Matsumoto N (2012) Mutations affecting components of the SWI/SNF complex cause Coffin-Siris syndrome. *Nat Genet* 44:376–378.

Tsurusaki Y, Okamoto N, Ohashi H, Mizuno S, Matsumoto N, Makita Y, Fukuda M, Isidor B, Perrier J, Aggarwal S, Dalal AB, Al-Kindy A, Liebelt J, Mowat D, Nakashima M, Saitsu H, Miyake N (2014b) Coffin-Siris syndrome is a SWI/SNF complex disorder. *Clin Genet* 85:548–554.

van der Sluijs EPJ, Jansen S, Vergano SA, Adachi-Fukuda M, Alanay Y, AlKindy A, Baban A, Bayat A, Beck-Wödl S, Berry K, Bijlsma EK, Bok LA, Brouwer AFJ, van der Burgt I, Campeau PM, Canham N, Chrzanowska K, Chu YWY, Chung BHY, Dahan K, De Rademaeker M, Destree A, Dudding-Byth T, Earl R, Elcioglu N, Elias ER, Fagerberg C, Gardham A, Gener B, Gerkes EH, Grasshoff U, van Haeringen A, Heitink KR, Herkert JC, den Hollander NS, Horn D, Hunt D, Kant SG,

Kato M, Kayserili H, Kersseboom R, Kilic E, Krajewska-Walasek M, Lammers K, Laulund LW, Lederer D, Lees M, López-González V, Maas S, Mancini GMS, Marcelis C, Martinez F, Maystadt I, McGuire M, McKee S, Mehta S, Metcalfe K, Milunsky J, Mizuno S, Moeschler JB, Netzer C, Ockeloen CW, Oehl-Jaschkowitz B, Okamoto N, Olminkhof SNM, Orellana C, Pasquier L, Pottinger C, Riehmer V, Robertson SP, Roifman M, Rooryck C, Ropers FG, Rosello M, Ruivenkamp CAL, Sagiroglu MS, Sallevelt SCEH, Sanchis Calvo A, Simsek-Kiper PO, Soares G, Solaeche L, Mujgan Sonmez F, Splitt M, Steenbeek D, Stegmann APA, Stumpel CTRM, Tanabe S, Uctepe E, Utine GE, Veenstra-Knol HE, Venkateswaran S, Vilain C, Vincent-Delorme C, Vulto-van Silfhout AT, Wheeler P, Wilson GN, Wilson LC, Wollnik B, Kosho T, Wieczorek D, Eichler E, Pfundt R, de Vries BBA, Clayton-Smith J, Santen GWE. The ARID1B spectrum in 143 patients: from nonsyndromic intellectual disability to Coffin-Siris syndrome. *Genet Med* 21:1295–1307.

Van Houdt JK, Nowakowska BA, Sousa SB, van Schaik BD, Seuntjens E, Avonce N, Sifrim A, Abdul-Rahman OA, van den Boogaard MJ, Bottani A, Castori M, Cormier-Daire V, Deardorff MA, Filges I, Fryer A, Fryns JP, Gana S, Garavelli L, Gillessen-Kaesbach G, Hall BD, Horn D, Huylebroeck D, Klapecki J, Krajewska-Walasek M, Kuechler A, Lines MA, Maas S, Macdermot KD, McKee S, Magee A, de Man SA, Moreau Y, Morice-Picard F, Obersztyn E, Pilch J, Rosser E, Shannon N, Stolte-Dijkstra I, Van Dijck P, Vilain C, Vogels A, Wakeling E, Wieczorek D, Wilson L, Zuffardi O, van Kampen AH, Devriendt K, Hennekam R, Vermeesch JR (2012) Heterozygous missense mutations in SMARCA2 cause Nicolaides-Baraitser syndrome. *Nat Genet* 44:445–449, S441.

Vergano SS, Deardorff MA (2014) Clinical features, diagnostic criteria, and management of Coffin-Siris syndrome. *Am J Med Genet C* 166C:252–256.

Weiswasser WH, Hall BD, Delavan GW, Smith DW (1973) Coffin-Siris syndrome. Two new cases. *Am J Dis Child* 125:838–840.

Wieczorek D, Bogershausen N, Beleggia F, Steiner-Haldenstatt S, Pohl E, Li Y, Milz E, Martin M, Thiele H, Altmuller J, Alanay Y, Kayserili H, Klein-Hitpass L, Bohringer S, Wollstein A, Albrecht B, Boduroglu K, Caliebe A, Chrzanowska K, Cogulu O, Cristofoli F, Czeschik JC, Devriendt K, Dotti MT, Elcioglu N, Gener B, Goecke TO, Krajewska-Walasek M, Guillen-Navarro E, Hayek J, Houge G, Kilic E, Simsek-Kiper PO, Lopez-Gonzalez V, Kuechler A, Lyonnet S, Mari F, Marozza A, Mathieu Dramard M, Mikat B, Morin G, Morice-Picard F, Ozkinay F, Rauch A, Renieri A, Tinschert S, Utine GE, Vilain C, Vivarelli R, Zweier C, Nurnberg P, Rahmann S, Vermeesch J, Ludecke HJ, Zeschnigk M, Wollnik B (2013) A comprehensive molecular study on Coffin-Siris and Nicolaides-Baraitser syndromes identifies a broad molecular and clinical spectrum converging on altered chromatin remodeling. *Hum Mol Genet* 22:5121–5135.

Zarate YA, Bhoj E, Kaylor J, Li D, Tsurusaki Y, Miyake N, Matsumoto N, Phadke S, Escobar L, Irani A, Hakonarson H, Schrier Vergano SA (2016) SMARCE1, a rare cause of Coffin-Siris Syndrome: Clinical description of three additional cases. *Am J Med Genet A* 170:1967–1973.

14

COHEN SYNDROME

KATE E. CHANDLER

Manchester Centre for Genomic Medicine, St Mary's Hospital, Manchester University NHS Foundation Trust, Manchester, UK and Division of Evolution and Genomic Sciences, School of Biological Sciences, Faculty of Biology, Medicines and Health, University of Manchester, Manchester, UK

INTRODUCTION

Incidence

Cohen syndrome is an autosomal recessive inherited disorder presenting with multi-system features in association with intellectual disability. More than 200 people have now been reported in the literature, and overall the incidence is estimated at about 1 in 100,000. In a large cohort of individuals referred with unexplained developmental delay and intellectual disability it was the fifth most common diagnosis after Fragile X syndrome, occurring at a frequency of 0.7% (Rauch et al. 2006). It is described from many different ethnic backgrounds but occurs at a higher prevalence in some geographical regions because of a founder effect of the causative genetic mutation. Areas of higher incidence include Finland (Norio 1994), the Ohio Amish (where a prevalence of up to 1 in 500 has been reported, Falk et al. 2004), and the Irish traveler community (Murphy et al. 2007). In areas where individuals are frequently homozygous for a common founder mutation, the phenotype shows greater homogeneity (Douzgou and Petersen 2011). The earlier suggestion that a Jewish subtype of Cohen syndrome exists (Sack and Friedman 1986) was disputed because of the highly variable phenotype reported in these individuals and the inconsistent presence of ophthalmic and hematological features (Chandler and Clayton-Smith 2002).

Lifespan: There is no evidence to suggest that life span in Cohen syndrome is significantly shortened and several affected individuals are known to have lived into their sixth decade, although long term follow up is not well documented. Kivitie-Kallio et al. (1999a) reported the presence of left ventricular failure in women over the age of 40 years, but this has not been a consistent finding in other cohorts studied. Impaired glucose tolerance and type 2 diabetes is reported in a number of people with Cohen syndrome and seems to be linked to insulin response dysregulation and abnormal pattern of fat distribution, specifically truncal obesity with a large waist circumference and slim extremities, (Limoge et al. 2015). Low high-density lipoprotein (HDL) values with normal low-density lipoprotein (LDL) levels, triglyceride, and cholesterol have also been described suggesting a possible increased risk of heart disease and metabolic syndrome. Consequently, recommendations for ongoing management of Cohen syndrome include annual screening for diabetes and monitoring of blood pressure and lipid levels, in particular HDL (Limoge et al. 2015).

Diagnostic Criteria

Cohen syndrome was first described in 1973 (Cohen 1973) in three individuals (two of whom were siblings) with intellectual disability, facial dysmorphism, microcephaly, and truncal obesity, hypotonia and joint hyperextensibility. The first published diagnostic criteria for Cohen syndrome were based on observations in a Finnish cohort (Kivitie-Kallio and Norio 2001) and comprised a set of five essential features:

Cassidy and Allanson's Management of Genetic Syndromes, Fourth Edition.
Edited by John C. Carey, Agatino Battaglia, David Viskochil, and Suzanne B. Cassidy.
© 2021 John Wiley & Sons, Inc. Published 2021 by John Wiley & Sons, Inc.

1. Non-progressive psychomotor delay, motor clumsiness, and microcephaly
2. Typical facial features including wave-shaped eyelids, short philtrum, thick hair, and low hairline (Figures 14.1–14.3)
3. Childhood hypotonia and hyperextensibility of the joints
4. Retinochoroidal dystrophy and myopia by 5 years of age
5. Periods of isolated neutropenia.

Additional findings supportive of the diagnosis included the following:

- Long/thick eyelashes, thick eyebrows, prominent root of nose, prominent upper incisors, and high-arched palate (Figs. 14.1 and 14.2)
- Relatively enlarged corpus callosum on brain MRI
- Low-voltage EEG in those over 14 years of age
- Typical metacarpophalangeal pattern profile
- Slender and short fingers (Figure 14.4)
- A wide gap between first and second toes
- Early lens opacities
- An almost total absence of maladaptive behavior (cheerful disposition).

The clinical and genetic homogeneity of the Finnish cohort raised questions as to the applicability of these diagnostic criteria to other Cohen syndrome populations. A subsequent UK study of greater genetic heterogeneity refined the criteria (Chandler et al. 2003a). The diagnosis of Cohen syndrome was deemed appropriate if the individual has moderate

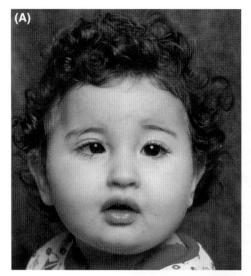
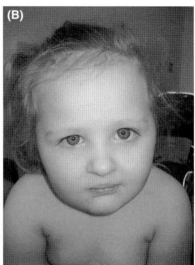
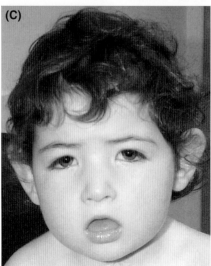

FIGURE 14.1 (A)–(C) Making the diagnosis of Cohen syndrome in the young child can be difficult as the facial appearance is less distinctive than later in life. They have a "doll-like" appearance with a thick head of hair, striking "wave-shaped" eyes with thick eyelashes, and a small mouth with full vermilion, but the philtrum is not necessarily short.

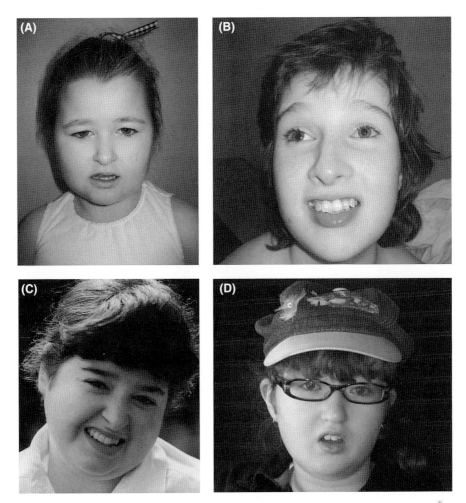

FIGURE 14.2 (A)–(D) By mid-childhood, the facial gestalt of Cohen syndrome is more evident, with a short upturned philtrum, snarling grimace on smiling, and a curved and convex nasal bridge. Almost invariably, glasses are worn as a result of significant myopia.

to severe learning difficulties and at least two of the following major criteria:

1. Typical facial gestalt (Figures 14.1, 14.2, 14.5), characterized by:
   - Thick hair, eyebrows, and eyelashes
   - Wave-shaped downward-slanting palpebral fissures
   - Convex nasal profile
   - Short upturned philtrum with grimacing expression on smiling.
2. Progressive myopia and pigmentary retinopathy.
3. Neutropenia (defined as less than 2×10^{-9} mm^{-3}).

Additional supportive diagnostic criteria were proposed, namely:

1. Microcephaly
2. Early-onset myopia
3. Truncal obesity, with slender extremities by age 8 years (Figure 14.3)

4. Joint hyperextensibility in the hands, feet, knees, and ankles
5. An overfriendly, cheerful disposition
6. A high-pitched voice.

Following identification of the *VPS13B* gene these criteria were modified and individuals were considered as having Cohen syndrome when six of the following eight criteria were present (Kolehmainen et al. 2004):

1. Developmental delay
2. Microcephaly
3. Typical facial gestalt
4. Obesity and slender extremities
5. Overly sociable behavior
6. Joint hypermobility
7. High myopia and/or retinal dystrophy (by age 5 years)
8. Intermittent neutropenia.

The strongest predictors of a diagnosis of Cohen syndrome with a confirmed mutation in *VPS13B* have since

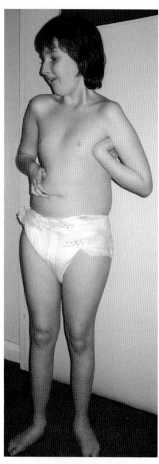

FIGURE 14.3 A 14-year-old girl with Cohen syndrome demonstrating the truncal obesity and slender limbs that are typical by the mid-childhood years.

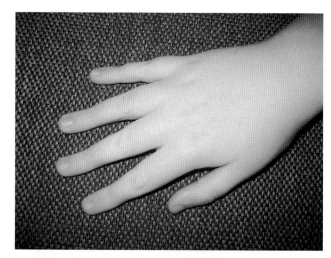

FIGURE 14.4 The hands of individuals with Cohen syndrome are narrow with slender fingers that taper from the proximal interphalangeal joint.

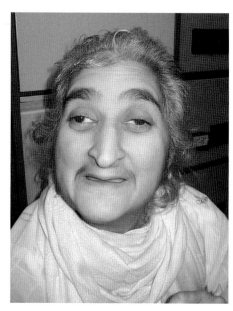

FIGURE 14.5 In middle age, individuals with Cohen syndrome can have a prematurely aged appearance with early graying of the hair.

been shown to be neutropenia and retinal dystrophy (Chehadeh et al. 2010). The facial gestalt of Cohen syndrome is less distinct in pre-school children and the typical body habitus has not yet developed, making the diagnosis in young children difficult (El Chehadeh-Djebbar et al. 2013) The inclusion of neutrophil count and electroretinography is therefore recommended when investigating a young child in whom a diagnosis of Cohen syndrome is being considered.

Etiology, Pathogenesis, and Genetics

Cohen syndrome is an autosomal recessive disorder caused by mutations within the *VPS13B* gene on chromosome 8q22-q23 (Kolehmainen et al. 2003). It contains 66 exons and undergoes alternative splicing with four different termination codons. The longest transcript, from 62 exons, encodes a putative transmembrane protein comprising 4022 amino acids and has a complex domain structure. The exact function of VPS13B is still unknown but its strong homology to *Saccharomyces cerevisiae* Vps13p protein suggests a role in intracellular vesicle-mediated sorting and transport of proteins (Velayos-Baeza et al. 2004). VPS13B is localized to the Golgi apparatus and has an essential role in Golgi integrity and function (Seifert et al. 2011). One of the key functions of the Golgi complex is post-translational modification by glycosylation of newly synthesized proteins. People with Cohen syndrome have been shown to have an unusual pattern of glycosylation of serum proteins (with significant accumulation of agalactosylated and asialylated fucosylated structures), suggesting Cohen syndrome may be due to a defect of glycosylation (Duplomb et al. 2014). More than 100 mutations have been reported, predominantly nonsense or frameshift mutations leading to truncated protein and functional null allele (Kolehmainen et al. 2003, Hennies et al. 2004, Seifert et al. 2006). Missense mutations have been reported, often in association with a milder phenotype (Seifert et al. 2006). An adult female with two *VPS13B* splicing mutations presented with neutropenia, retinopathy, microcephaly, and mild facial dysmorphism but no intellectual disability or truncal obesity (Gueneau et al. 2013), suggestive of a dose effect of residual normal VPS13B production leading to a less severe phenotype. There does not appear to be a mutational hotspot in *VPS13B* (Seifert et al. 2009) and many mutations are family specific. There are common founder mutations in certain populations, such as the Finnish, Amish, Irish communities, and Greek Islands (Kolehmainen et al. 2003; Falk et al. 2004; Murphy et al. 2007; Bugiani et al. 2008). Most individuals with Cohen syndrome demonstrate homozygosity or compound heterozygosity for *VPS13B* mutations, although in some a second mutation remains undetected. Intragenic copy number variations (CNV) are common, identified in up to a third of cases, and so should be screened for by multiplex ligation-dependent probe amplification (MLPA) or high-resolution array-comparative genomic hybridization (CGH) (Balikova et al. 2009; Parri et al. 2010). Large-scale deletions are three times more common than duplications (El Chehadeh-Djebbar et al. 2011). Overall, *VPS13B* mutations and deletions are detected in 88% of affected individuals (Balikova et al. 2009).

There is no strong genotype-phenotype correlation, and variability in phenotype occurs even within families, where affected individuals have the same mutations (Horn et al. 2000; Chandler et al. 2003b).

Cohen syndrome follows an autosomal recessive inheritance pattern. Carrier testing and prenatal diagnosis are possible where pathogenic mutations have been identified in the proband. Parents of an affected child are heterozygote mutation carriers and are asymptomatic. They have a 25% risk of having an affected child with each subsequent pregnancy.

Diagnostic Testing

All individuals in whom a diagnosis of Cohen syndrome is being considered should have the following evaluations:

(i) Detailed assessment, preferably by a clinical dysmorphologist, to delineate the presence of the typical dysmorphic features, body habitus, growth parameters, and behavioral pattern.

(ii) Differential white cell count to look for neutropenia. As this is an intermittent feature of the condition, it may not be present at the time of testing. Repeated blood count evaluation is not indicated but review of past results may reveal a previously low neutrophil count.

(iii) Full ophthalmic assessment by an experienced pediatric ophthalmologist, including electrodiagnostic testing (electroretinogram) to look for evidence of a retinal dystrophy. In infants, this should be repeated annually if the diagnosis is suspected but the examination is normal because electroretinograms may be technically difficult to carry out in young children.

Definitive diagnosis of Cohen syndrome can be made by identification of biallelic pathogenic mutations and/or copy number variants in *VPS13B* on molecular genetic testing.

Differential Diagnosis

The differential diagnosis of Cohen syndrome includes the group of conditions in which global developmental delay and learning disability are associated with impaired vision and/or obesity. These include:

- Bardet–Biedl syndrome (see Chapter 8) can be distinguished from Cohen syndrome by a different facial appearance, the presence of polydactyly and often brachydactyly, a slightly later onset of retinal dystrophy (age 7–8 years), and the absence of neutropenia. The intellectual impairment in Bardet–Biedl syndrome is generally milder. Obesity occurs at an earlier age and is usually generalized rather than truncal. Renal disease is a significant cause of morbidity and mortality in that disorder.
- Prader–Willi syndrome (see Chapter 46), although associated with learning disability and hypotonia, does not have retinopathy or neutropenia as a feature. Obesity is of earlier onset, is associated with hyperphagia and progresses to be a major cause of morbidity and mortality in adults.
- Alström syndrome is associated with obesity, pigmentary retinopathy, and diabetes, but learning disability is not usually a feature while deafness is a characteristic feature.
- 1p36 subtelomeric deletion (see Chapter 18) can be associated in some individuals with truncal obesity and learning disability of a similar degree to that seen in Cohen syndrome. The prominent eyebrows seen in deletion 1p36 also lead to similarities in the facial features. Additional malformations such as cleft lip and palate and congenital heart disease are often present, and pigmentary retinopathy is not a feature of 1p36 deletion.
- In maternal uniparental disomy of chromosome 14, childhood-onset obesity, learning disability and joint laxity are often present, but the learning disability is mild and pigmentary retinopathy is not a feature.
- Mirrhosseini–Holmes–Walton (1972) reported two brothers in association with intellectual disability,

retinal dystrophy, microcephaly, and joint hyperextensibility. Rogers and Stevenson (1984) described siblings with mental retardation, short stature, premature graying of the hair, and joint laxity. Both conditions are now believed to be allelic to Cohen syndrome.

MANIFESTATIONS AND MANAGEMENT

Growth and Feeding

Birth weights on average are low for gestation (third centile). Failure-to-thrive may occur in the neonatal period because of significant feeding difficulties, particularly poor sucking reflex (Kivitie-Kallio and Norio, 2001). In the majority of babies this improves over the first months of life, although rarely infants have been known to require percutaneous gastrostomy until age 18 months.

A typical growth pattern in Cohen syndrome includes exaggerated weight gain, especially in the truncal region, from the age of 8 years. No specific endocrine cause has been found for the weight gain observed. Although some children have an excessive appetite, hyperphagia is not usually present and they do not show a tendency for over-consumption of sweet, fatty or salty foods (Limoge et al. 2015). Truncal obesity with increased waist circumference but slender arms and legs is the characteristic body habitus in Cohen syndrome, and generalized obesity with a high body mass index is uncommon (Kivitie-Kallio et al. 1999a, Limoge et al. 2015). Most individuals with Cohen syndrome are of short stature, and height less than 0.4 centile is not unusual, especially after puberty (Chandler et al. 2003a).

Pubertal development is often delayed, but it is unclear as to whether this is related to Cohen syndrome per se, as delayed puberty may be seen more frequently in individuals with learning disability. Abnormalities of the pituitary-gonadal axis have not been confirmed and when investigated anterior pituitary, thyroid, and adrenocortical functions have been normal (Kivitie-Kallio et al. 1999a).

Evaluation

- Height, weight, and head circumference should be measured and plotted on standard growth charts at initial assessment and at subsequent routine clinical visits. Close monitoring of growth parameters is particularly important during infancy to assess for failure-to-thrive.
- If failure-to-thrive is present, standard assessments for a cause should take place before assuming that it is related to Cohen syndrome.

Management

- Nasogastric tube feeding is often needed for feeding difficulties in the neonatal period.
- Protracted feeding problems lasting over a year are rare but may require insertion of a percutaneous gastrostomy feeding tube.
- Caloric intake should be adjusted in a standard manner for infants with failure-to-thrive.
- Advice from a dietician can be valuable from the mid-childhood years to prevent worsening of truncal obesity.
- An exercise regime should be encouraged as part of weight management.

Development and Behavior

All children with Cohen syndrome have global developmental delay evident in the first year of life. Studies have identified the median age of sitting unsupported as 12 months, and of walking independently as 2.5 years (Kivitie-Kallio and Norio 2001; Chandler et al. 2003a). Speech delay is often marked but varies considerably between individuals. In the UK study, the median age for first spoken words was 2.5 years and for speaking in short sentences 5 years (Chandler et al. 2003a). Although most affected individuals learn to speak, their competence is limited, with comprehension of speech better than expression (Thomaidis et al. 1999; Kivitie-Kallio and Norio 2001).

Learning difficulties requiring special educational provision are a consistent feature of Cohen syndrome. Although the severity can vary even within the same family, the level of intellectual disability is mostly described as severe (Horn et al. 2000; Chandler et al. 2003b). When reported, intelligence quotients range between 30 and 70 (Norio et al. 1984; Thomaidis et al. 1999). Profound cognitive impairment is reported in up to 20% (Kivitie-Kallio 1999b).

People with Cohen syndrome are described as having a distinctive behavioral phenotype, with a sociable and cheerful disposition (Cohen et al. 1973; Partington and Anderson 1994). Formal psychometric assessment of individuals with Cohen syndrome has shown that they score highly in positive domains such as self-direction, responsibility, and socialization, but stereotypic behavior and unusual mannerisms are common (Kivitie-Kallio et al. 1999b). However, hyperactivity, attention deficit, pervasive developmental disorder and autistic spectrum behavior have all been reported in Cohen syndrome (Thomaidis et al. 1999; Horn et al. 2000; Chandler et al. 2003b; Douzgou and Petersen 2011; Rafiq et al. 2015). Indeed, several studies report *VPS13B* mutations identified through exome sequencing in cohorts of individuals with autism (ASD) or intellectual disability with autistic features, and prevalence is particularly high in consanguineous or multiplex families with ASD where the detection rate was 1.2% (Yu et al. 2013; Ionitao-Laza et al. 2014).

Evaluation

- All children with Cohen syndrome require ongoing assessment of their level of developmental delay and

learning difficulties, preferably by a multidisciplinary team, which optimally includes a community pediatrician, speech and language therapist, physiotherapist, occupational therapist, and educational psychologist.
- Neuropsychological assessments can be helpful in clarifying the level of learning difficulties and the presence of specific behavioral traits.
- Adults with Cohen syndrome will require individual assessment for determination of appropriate residential placement and the need and possibility of sheltered employment.

Management

- Early intervention after assessment by a multidisciplinary team is needed. The input of physiotherapists and occupational therapists is particularly important for addressing the gross and fine motor delay in Cohen syndrome.
- Speech therapy is required as language skills are often significantly impaired.
- All affected children have special educational needs as a result of their learning difficulties. Individual educational programs focusing on the development of practical daily living and independence skills are recommended.
- Consideration should be given to the visual deficits in planning developmental and educational programs.
- Assurance of appropriateness of living and employment situation is needed for adults.
- There is no reported experience with medication for the behavioral problems of Cohen syndrome. Hence, there are no specific recommendations for treatment of these problems.

Ophthalmologic

Ophthalmic abnormalities are a cardinal feature of Cohen syndrome. Affected individuals characteristically present with early-onset myopia and progressive retinopathy (Norio et al. 1984; Warburg et al. 1990; Chandler et al. 2002a). A wide range of ophthalmic abnormalities has been reported including strabismus, astigmatism, ptosis, nystagmus, myopia, oval pupil, and iris atrophy (Resnick et al. 1986; Warburg et al. 1990; Thomaidis et al. 1999; Taban et al. 2007). Congenital abnormalities of the eye are rare but include microphthalmia and microcornea, and choroidal and retinal colobomata (Cohen et al. 1973; Chandler et al. 2002a; Taban et al. 2007).

Myopia in Cohen syndrome is attributed to high corneal and lenticular refractive power in the presence of a normal axial length (Summanen et al. 2002). It usually starts in the preschool years and progresses over an average of 15 years to high-grade myopia, with a median refractive error of −11 diopters. Individuals registered in the US National Cohen Syndrome Database had their first ophthalmologic visit leading to their first pair of glasses at an average age of 4.5 years (Taban et al. 2007).

Retinal pigmentary changes commence around the macula to give a "bull's eye maculopathy" in children as young as 3 years (Chandler et al. 2002a). Chorioretinal dystrophy with characteristic fundus changes and isoelectric electroretinogram (ERG) is evident in most affected individuals over the age of 5 years (Kivitie-Kallio et al. 2000). By 10 years of age, most individuals with Cohen syndrome have widespread pigmentary retinopathy with symptoms of nyctalopia (night blindness) and progressive visual field loss.

Detailed ophthalmologic keratometry and biometry measurements indicate a shallower anterior chamber and a thicker lens that worsens with increasing age in individuals with Cohen syndrome (Summanen et al. 2002). Low corneal thickness has also been reported when detailed pachymetry is performed (Douzgou et al. 2011), indicative of an anterior segment dysgenesis.

The severity and rate of progression of visual handicap varies between individuals, even within the same families (Chandler et al. 2002a). Severe ophthalmic complications requiring enucleation was described in two siblings who presented with total retinal detachment and optic atrophy (Hennies et al. 2004). Early onset, high-grade cataracts with associated vitreal abnormalities and high risk of blindness (20%) was reported in a Greek isolate with Cohen syndrome (Douzgou and Petersen, 2011). Keratoconus and corneal ectasia resulting in severe visual deterioration was reported in teenage as well as older people with Cohen syndrome (Khan et al. 2006; Douzgou et al. 2011). Other important ophthalmic complications in Cohen syndrome include atrophic irides and lens subluxation (Chandler et al. 2002a). By adult life, and particularly into the fourth decade, visual handicap is usually marked, and individuals are registered as blind or severely visually impaired (Kivitie-Kallio et al. 2000; Seifert et al. 2006).

Evaluation

- Referral to a pediatric ophthalmologist for detailed examination, including electrophysiological testing, and keratometry and pachymetry measurements, should be arranged in all children with a diagnosis of Cohen syndrome.
- A general or specialist ophthalmologist should be involved in regular ophthalmologic evaluation of affected adults.

Management

- Early detection and assessment of a child's visual problems play an important role in preparing the family

regarding their child's visual prognosis. It provides invaluable information for both the parents and educational team around which they can make appropriate decisions and plan for the child's special needs.
- Regular monitoring of changes in vision is needed to identify correctable complications, such as high myopia, glaucoma, cataracts, and lens subluxation, as well as to advise about deterioration.
- When appropriate, the individual should be registered as visually impaired and information given on visual aids and entitlement to financial benefits.
- Ophthalmologic intervention for the eye complications of Cohen syndrome does not differ from that for others with similar problems.

Neurologic

Neonatal hypotonia is a common finding in Cohen syndrome and contributes both to the early feeding difficulties and delay in motor milestone (Kivitie-Kallio and Norio; 2001).

Microcephaly is not always present at birth but usually becomes apparent during the first 3 years of life (Partington and Anderson 1994).

Intracranial abnormalities are not a common feature in Cohen syndrome despite evidence from the mouse models that the *VPS13B* gene is widely expressed in the cerebellum and cerebral cortex (Mochida et al. 2004). An increased diameter of the body of the corpus callosum was identified in Finnish individuals with Cohen syndrome (Kivitie-Kallio et al. 1998) but has not been substantiated in other cohorts. Five non-Finnish individuals with Cohen syndrome were reported to have abnormalities of the brain on magnetic resonance imaging scan (MRI), including a relatively large corpus callosum as well as mild cerebellar vermis hypoplasia and pontocerebellar atrophy, although no images were shown in the publications (Mochida et al. 2004; Katzaki et al. 2007). Cerebellar hypoplasia has been reported in two individuals with Cohen syndrome from a consanguineous Pakistani family with a homozygous *VPS13B* mutation and in a Finnish individual with the common ancestral mutation (Waite et al. 2010).

Seizures are rarely described in those with Cohen syndrome (Cohen et al. 1973; Thomaidis et al. 1999). A single individual with focal right polymicrogyria associated with left hemi-generalized clonic seizures and left hemiparesis has been reported (Coppola et al. 2003). EEG showed a continuous spike and wave electrical status epilepticus over the right hemisphere during slow-wave sleep in this individual. It is not known whether this finding was coincidental or associated with the diagnosis of Cohen syndrome.

Evaluation

- Assessment of hypotonia and joint laxity should be conducted in all affected individuals.

- Neuroimaging with MRI or computed tomography scans is not necessary as a routine investigation.
- EEG should only be performed if there are concerns about epileptic seizures.

Management

- The presence of hypotonia should trigger referral for support from ongoing physiotherapy.
- Seizures are rare in Cohen syndrome, but, if suspected, referral to a pediatric neurologist for appropriate assessment and management is indicated.

Endocrine and Immunologic

Diabetes mellitus (DM) is reported in a number of individuals with Cohen syndrome although its actual incidence is not documented. The pattern of truncal obesity may predispose affected individuals to an increased risk of impaired glucose tolerance and type II diabetes, especially as adults (Limoge et al 2015).

Other autoimmune disorders have been rarely reported, namely autoimmune hemolytic anemia, celiac disease and Hashimoto's thyroiditis (personal experience).

Evaluation

- Annual screening for diabetes in a standard manner should be carried out from puberty.
- If there are symptoms suggestive of thyroid or gastrointestinal dysfunction or anemia, the possibility of an autoimmune disorder should be considered and investigated in a standard manner.

Management

- If there is evidence of diabetes, thyroiditis, malabsorption, or anemia, treatment should take place in a standard manner, often involving a physician with the appropriate specialist knowledge, such as an endocrinologist, immunologist, or hematologist.

Hematologic

Neutropenia is a characteristic finding in Cohen syndrome (Kivitie-Kallio et al. 1997; Chandler et al. 2003a). However, its presence as an obligate feature has been disputed, particularly in non-Europeans (Hennies et al. 2004). Neutropenia may be detected from birth but can be intermittent (Fryns et al. 1996). It can be associated with a normal total white cell count, potentially leaving it unnoticed if a differential count is not requested. Neutropenia is an isolated finding with no other blood cell lineages affected. It does not progress and there is no evidence for associated malignant development. Bone marrow analysis shows normal cellularity (Kivitie-Kallio et al. 1997; Olivieri et al. 1998).

Individuals with Cohen syndrome frequently suffer from minor skin and oral infections. Recurrent infections of the upper and lower respiratory tract have been described in people with Cohen syndrome from the Irish travelling community (Murphy et al. 2007). However, despite a significant neutropenia, severe life-threatening infections are uncommon (Olivieri et al. 1998; Hennies et al. 2004) and patients rarely require hospitalization.

Evaluation

- Neutropenia can be a key diagnostic marker, especially in younger individuals, and a differential white cell count should be requested. However, repeated measurements and other hematological investigations are not warranted.

Management

- Treatment interventions to improve the neutrophil count are rarely required, as individuals with Cohen syndrome are usually well, despite their neutropenia.
- In a few individuals who have had recurrent and severe infections, repeated infusions of granulocyte colony-stimulating factor (G-CSF) in combination with prophylactic antibiotics have been used to good effect (personal experience). Referral to a hematologist for management supervision is recommended for these individuals.

Cardiovascular

Cardiac abnormalities have been reported in a few individuals with Cohen syndrome (Norio et al. 1984; Partington and Anderson 1994). However, detailed cardiac assessment in Finnish individuals with Cohen syndrome confirmed normal cardiac anatomy, although there was evidence of decreasing left ventricular function with advancing age, especially in females over 40 years of age (Kivitie-Kallio et al. 1999a). Low high-density lipoprotein (HDL) values with normal low-density lipoprotein levels (LDL), triglyceride, and total cholesterol have been described suggesting a possible increased risk of heart disease and metabolic syndrome in Cohen syndrome (Limoge et al 2015).

Evaluation

- Cardiac assessment including echocardiogram to assess left ventricular function is recommended in women with Cohen syndrome over the age of 40 years or in any affected individual with symptoms suggestive of cardiac failure. Periodic re-assessment every 5 years is recommended for this group.
- Routine echocardiogram is not indicated.
- Annual monitoring of blood pressure and lipid levels, in particular HDL, is recommended in adults.

Management

- Treatment for individuals with Cohen syndrome found to have cardiac disease or metabolic syndrome is as for all people.

Respiratory

Stridor resulting from laryngomalacia is common in the neonatal period, but usually resolves spontaneously (Chandler et al. 2003a). Rarely, it may persist and be complicated by severe respiratory difficulties requiring tracheostomy. In such cases, etiologies include laryngeal stenosis and vocal cord paresis (Chandler et al. 2002b). A high-pitched cry and voice are often reported although its precise etiology is unknown.

Evaluation

- Individuals with stridor should be referred for assessment by a pediatric otolaryngologist.
- If severe or prolonged, examination under anesthesia should be performed to investigate for laryngeal anomalies such as stenosis.

Management

- Conservative management only is often sufficient for individuals with Cohen syndrome who have stridor related to laryngomalacia, as it usually resolves with time.
- In those in whom stridor is so severe that it compromises respiratory function, tracheostomy insertion may be necessary.

Musculoskeletal

Joint laxity is very common in Cohen syndrome, particularly of the fingers, ankles, knees, and elbow joints (Chandler et al. 2003a). Joint dislocation may occur, particularly affecting the knees and may require surgical intervention (personal experience). Kyphoscoliosis and pectus carinatum are frequent findings in older individuals and those from the Greek Isolates (Kivitie-Kallio et al. 1999a; Douzgou and Petersen 2011).

Evaluation

- Assessment of the degree of joint hyperextensibility should be carried out in all individuals with Cohen syndrome using the Beighton score (Beighton et al. 1997; see also Chapter 25).

- Those with joint instability or dislocation should be referred for specialist assessment to a rheumatologist and orthopedic surgeon.

Management

- Physiotherapy and exercises, such as swimming, to promote joint stability are advisable.
- Weight control should be encouraged in all individuals with Cohen syndrome, but especially those with joint instability.
- In rare individuals, orthopedic surgical management for joint dislocations may be needed. Their indication and procedures are standard.

Dental

An altered periodontal bacterial flora has been found in individuals with Cohen syndrome resulting in increased susceptibility to periodontitis and premature tooth loss (Alaluusua et al. 1997).

Evaluation

- Regular dental assessment is recommended.

Management

- Good oral hygiene should be encouraged to prevent gingivitis and premature tooth loss. This should include regular brushing, antiseptic mouthwashes, and treatment with antibiotics as required.
- Standard orthodontic treatment may be required if there is dental malalignment.

RESOURCES

The following Web sites provide reviews and information on Cohen syndrome:

GeneReviews

Free website for disease information
Website: https://www.ncbi.nlm.nih.gov/books/NBK1482/

Orphanet

A portal for rare diseases
Website: *http://www.orpha.net*
https://www.orpha.net/data/patho/GB/uk-cohen.pdf

National Organization for Rare Disorders, USA

Website: *https://rarediseases.org/rare-diseases/cohen-syndrome/*

USA Cohen syndrome organization

Email: info@cohen_syndrome.org
Website: *www.cohensyndrome.org*
Facebook: *https://www.facebook.com/Cohen-Syndrome-Association-106234886074462/*

Contact

UK based support group for families with disabled children
https://contact.org.uk/advice-and-support/health-medical-information/conditions/c/cohen-syndrome/

Association Maladie Rare Syndrome de Cohen Internationale

BP 152
ROOY
24101 Bergerac
France
Website: *www.orpha.net*

REFERENCES

Alaluusua S, Kivitie-Kallio S, Wolf J, Haavio M-L, Asikainen, Pirinen S (1997) Periodontal findings in Cohen syndrome with chronic neutropenia. *J Periodontol* 68:473–478.

Balikova I, Lehesjoki AE, de Ravel TJ, Thienpont B, Chandler KE, Clayton-Smith J, Traskelin AL, Fryns JP, Vermeesch JR (2009) Deletions in the *VPS13B* (*COH1*) gene as a cause of Cohen syndrome. *Hum Mutat* 30:E845–E854.

Beighton P, de Paepe A, Steinman B, Tsipouras P, Wenstrup RJ (1998) Ehlers-Danlos syndromes: Revised nosology, Villefranche, 1997. *Am J Med Genet* 77:31–37.

Bugiani M, Gyftodimou Y, Tsimpouka P, Lamantea E, Katzaki E, d'Adamo P, Nakou S, Georgoudi N, Grigoriadou M, Tsina E, Kabolis N, Milani D, Pandelia E, Kokotas H, Gasparini P, Giannoulia-Karantana A, Renieri A, Zeviani M, Petersen MB (2008) Cohen syndrome resulting from a novel large intragenic COH1 deletion segregating in an isolated Greek island population. *Am J Med Genet A* 146A(17):2221–6.

Chandler KE, Clayton-Smith J (2002) Does a Jewish type of Cohen syndrome truly exist? *Am J Med Genet* 111:453–454.

Chandler KE, Biswas S, Lloyd IC, Parry N, Clayton-Smith J, Black GCM (2002a) The ophthalmic findings in Cohen Syndrome. *Br J Ophthalmol* 86:1395–1398.

Chandler KE, McKee S, Fryer A (2002b) Cohen syndrome is associated with significant laryngeal anomalies. *J Med Genet* 39:S33.

Chandler KE, Kidd A, Al-Gazali L, Black GCM, Clayton-Smith J (2003a) Diagnostic criteria, clinical characteristics and natural history of Cohen Syndrome. *J Med Genet* 40:233–241.

Chandler KE, Moffett M, Clayton-Smith J, Baker GA (2003b) Neuropsychological assessment of a group of UK patients with Cohen syndrome. *Neuropediatrics* 34:7–13.

Cohen MM Jr, Hall BD, Smith DW, Graham CB, Lampert KJ (1973) A new syndrome with hypotonia, obesity, mental deficiency and facial, oral, ocular and limb anomalies. *J Pediatr* 83:280–284.

Coppola G, Federico RR, Epifanio G, Tagliente F, Bravaccio C (2003) Focal polymicrogyria, continuous spike-and-wave discharges during slow-wave sleep and Cohen syndrome: A case report. *Brain Dev* 25:446–449.

Douzgou S, Petersen MB. (2011) Clinical variability of genetic isolates of Cohen syndrome. *Clin Genet* 79(6):501-6

Douzgou S, Samples JR, Georgoudi N, Petersen MB. (2011) Ophthalmic findings in the Greek isolate of Cohen syndrome. *Am J Med Genet A* 155A(3):534-9

Duplomb L, Duvet S, Picot D, Jego G, El Chehadeh-Djebbar S, Marle N, Gigot N, Aral B, Carmignac V, Thevenon J, Lopez E, Rivière JB, Klein A, Philippe C, Droin N, Blair E, Girodon F, Donadieu J, Bellanné-Chantelot C, Delva L, Michalski JC, Solary E, Faivre L, Foulquier F, Thauvin-Robinet C. (2014) Cohen syndrome is associated with major glycosylation defects. *Hum Mol Genet* 23(9):2391–9.

El Chehadeh-Djebbar S, Faivre L, Moncla A, Aral B, Missirian C, Popovici C, Rump P, Van Essen A, Frances AM, Gigot N, Cusin V, Masurel-Paulet A, Gueneau L, Payet M, Ragon C, Marle N, Mosca-Boidron AL, Huet F, Balikova I, Teyssier JR, Mugneret F, Thauvin-Robinet C, Callier P. (2011) Djebbar The power of high-resolution non-targeted array-CGH in identifying intragenic rearrangements responsible for Cohen syndrome. *J Med Genet* 48(11):e1.

El Chehadeh-Djebbar S, Blair E, Holder-Espinasse M, Moncla A, Frances AM, Rio M, Debray FG, Rump P, Masurel-Paulet A, Gigot N, Callier P, Duplomb L, Aral B, Huet F, Thauvin-Robinet C, Faivre L.(2013) Changing facial phenotype in Cohen syndrome: towards clues for an earlier diagnosis. *Eur J Hum Genet* 21(7):736–42.

Falk MJ, Feiler HS, Neilson DE, Maxwell K, Lee JV, Segall SK, Robin NH, Wilhelmsen KC, Traskelin AL, Kolehmainen J, Lehesjoki AE, Wiznitzer M, Warman ML (2004) Cohen syndrome in the Ohio Amish. *Am J Med Genet A* 128:23–28.

Fryns JP, Legius E, Devriendt K, Meire F, Standaert L, Baten E, van den Berghe H (1996) Cohen syndrome: The clinical symptoms and stigmata at a young age. *Clin Genet* 49:237–241.

Gueneau L, Duplomb L, Sarda P, Hamel C, Aral B, Chehadeh SE, Gigot N, St-Onge J, Callier P, Thevenon J, Huet F, Carmignac V, Droin N, Faivre L, Thauvin-Robinet C. (2014) Congenital neutropenia with retinopathy, a new phenotype without intellectual deficiency or obesity secondary to VPS13B mutations. *Am J Med Genet A* 164A(2):522–7

Hennies HC, Rauch A, Seifert W, Schumi C, Moser E, Al-Taji E, Tariverdian G, Chrzanowska KH, Krajewska-Walasek M, Rajab A, Giugliani R, Neumann TE, Eckl KM, Karbasiyan M, Reis A, Horn D (2004) Allelic heterogeneity in the *COH1* gene explains clinical variability in Cohen syndrome. *Am J Hum Genet* 75:138–145.

Horn D, Krebsova A, Kunze J, Reis A (2000) Homozygosity mapping in a family with microcephaly, mental retardation, and short stature to a Cohen syndrome region on 8q21.3-8q22.1: Redefining a clinical entity. *Am J Med Genet* 92:285–292.

Ionitao-Laza I, Capanu M, De Rubeis S, McCallum K, Buxbaum JD (2014) Identification of rare causal variants in sequence-based studies: methods and applications to VPS13B, a gene involved in Cohen syndrome and autism. *PLoS Genet* 10(12):e1004729.

Khan A, Chandler K, Pimenides D, Black GC, Manson FD (2006) Corneal ectasia associated with Cohen syndrome: A role for COH1 in corneal development and maintenance? *Br J Ophthalmol* 90:390–391.

Kivitie-Kallio S, Norio R (2001) Cohen syndrome: Essential features, natural history, and heterogeneity. *Am J Med Genet* 102:125–135.

Kivitie-Kallio S, Rajantie J, Juvonen E, Norio R (1997) Granulocytopenia in Cohen syndrome. *Br J Haemat* 98:308–311.

Kivitie-Kallio S, Autti T, Salonen O, Norio R (1998) MRI of the brain in the Cohen syndrome: A relatively large corpus callosum in patients with mental retardation and microcephaly. *Neuropediatrics* 29:298–301.

Kivitie-Kallio S, Eronen M, Lipsanen-Nyman M, Marttinen E, Norio R (1999a) Cohen syndrome: Evaluation of its cardiac, endocrine and radiological features. *Clin Genet* 56:41–50.

Kivitie-Kallio S, Larsen A, Kajasto K, Norio R (1999b) Neurological and psychological findings in patients with Cohen syndrome: A study of 18 patients aged 11 months to 57 years. *Neuropediatrics* 30:181–189.

Kivitie-Kallio S, Summanen P, Raitta C, Norio R (2000) Ophthalmologic findings in Cohen syndrome—A long term follow up. *Ophthalmology* 107:1737–1745.

Kolehmainen J, Black GC, Saarinen A, Chandler K, Clayton-Smith J, Traskelin AL, Perveen R, Kivitie-Kallio S, Norio R, Warburg M, Fryns JP, de la Chapelle A, Lehesjoki AE (2003) Cohen syndrome is caused by mutations in a novel gene, *COH1*, encoding a transmembrane protein with a presumed role in vesicle-mediated sorting and intracellular protein transport. *Am J Hum Genet* 72:1359–1369.

Kolehmainen J, Wilkinson R, Lehesjoki AE, Chandler K, Kivitie-Kallio S, Clayton-Smith J, Träskelin AL, Waris L, SaarinenA, Khan J, Gross-Tsur V, Traboulsi EI, Warburg M, Fryns JP, Norio R, Black GC, Manson FD. (2004) Delineation of Cohen syndrome following a large-scale genotype-phenotype screen. *Am J Hum Genet* 75(1):122–7.

Limoge F, Faivre L, Gautier T, Petit JM, Gautier E, Masson D, Jego G, El Chehadeh-Djebbar S, Marle N, Carmignac V, Deckert V, Brindisi MC, Edery P, Ghoumid J, Blair E, Lagrost L, Thauvin-Robinet C, Duplomb L. (2015) Insulin response dysregulation explains abnormal fat storage and increased risk of diabetes mellitus type 2 in Cohen Syndrome. *Hum Mol Genet* 24(23):6603–13.

Mirrhosseini SA, Holmes LB, Walton DS. (1972) Syndrome of pigmentary retinal degeneration, cataract, microcephaly, and severe mental retardation *J Med Genet* 9(2):193–6.

Mochida GH, Rajab A, Eyaid W, Lu A, Al-Nouri D, Kosaki K, Noruzinia M, Sarda P, Ishihara J, Bodell A, Apse K, Walsh CA (2004) Broader geographical spectrum of Cohen syndrome due to *COH1* mutations. *J Med Genet* 41:e87–e90.

Murphy AM, Flanagan O, Dunne K, Lynch SA (2007) High prevalence of Cohen syndrome amongst Irish Travellers. *Clin Dysmorphol* 16:257–259.

Norio R (1994) Cohen syndrome is neither uncommon nor new. *Am J Med Genet* 53:202–203.

Norio R, Raitta C, Lindahl E (1984) Further delineation of the Cohen syndrome; report on chorioretinal dystrophy, leukopenia and consanguinity. *Clin Genet* 25:1–14.

Olivieri O, Lombardi S, Russo C, Corrocher R (1998) Increased neutrophil adhesive capability in Cohen syndrome, an autosomal recessive disorder associated with granulocytopenia. *Haematologica* 83:778–782.

Parri V, Katzaki E, Uliana V, Scionti F, Tita R, Artuso R, Longo I, Boschloo R, Vijzelaar R, Selicorni A, Brancati F, Dallapiccola

B, Zelante L, Hamel CP, Sarda P, Lalani SR, Grasso R, Buoni S, Hayek J, Servais L, de Vries BB, Georgoudi N, Nakou S, Petersen MB, Mari F, Renieri A, Ariani F (2010) High frequency of COH1 intragenic deletions and duplications detected by MLPA in patients with Cohen syndrome. *Eur J Hum Genet* 18(10):1133–40.

Partington M, Anderson D (1994) Mild growth retardation and developmental delay, microcephaly, and a distinctive facial appearance. *Am J Med Genet* 49:247–250.

Rafiq MA, Leblond CS, Saqib MA, Vincent AK, Ambalavanan A, Khan FS, Ayaz M, Shaheen N, Spiegelman D, Ali G, Amin-ud-Din M, Laurent S, Mahmood H, Christian M, Ali N, Fennell A, Nanjiani Z, Egger G, Caron C, Waqas A, Ayub M, Rasheed S, Forgeot d'Arc B, Johnson A, So J, Brohi MQ, Mottron L, Ansar M, Vincent JB, Xiong L (2015) Novel VPS13B Mutations in three large Pakistani Cohen syndrome families suggests a Baloch variant with autistic-like features. *BMC Med Genet* 16:41.

Rauch A, Hoyer J, Guth S, Zweier C, Kraus C, Becker C, Zenker M, Hüffmeier U, Thiel C, Rüschendorf F, Nürnberg P, Reis A, Trautmann U (2006) Diagnostic yield of various genetic approaches in patients with unexplained developmental delay or mental retardation. *Am J Med Genet A* 140(19):2063–74.

Resnick K, Zuckerman J, Cotlier E (1986) Cohen syndrome with bull's eye macular lesion. *Ophthalmic Paediatr Genet* 7:1–8.

Rogers RC, Stevenson RE (1984) Mental retardation, premature graying, and ligamentous laxity in four siblings. *Proc Greenwood Genet Cent* 3:11–14.

Sack J, Friedman E (1986) The Cohen syndrome in Israel. *Israel J Med Sci* 22:766–770.

Seifert W, Holder-Espinasse M, Spranger S, Hoeltzenbein M, Rossier E, Dollfus H, Lacombe D, Verloes A, Chrzanowska KH, Maegawa GH, Chitayat D, Kotzot D, Huhle D, Meinecke P, Albrecht B, Mathijssen I, Leheup B, Raile K, Hennies HC, Horn D (2006) Mutational spectrum of *COH1* and clinical heterogeneity in Cohen syndrome. *J Med Genet* 43:e22.

Seifert W, Holder-Espinasse M, Kühnisch J, Kahrizi K, Tzschach A, Garshasbi M, Najmabadi H, Walter Kuss A, Kress W, Laureys G, Loeys B, Brilstra E, Mancini GM, Dollfus H, Dahan K, Apse K, Hennies HC, Horn D (2009) Expanded mutational spectrum in Cohen syndrome, tissue expression, and transcript variants of COH1. *Hum Mutat* 30(2):E404–20.

Seifert W, Kühnisch J, Maritzen T, Horn D, Haucke V, Hennies HC. (2011) Cohen syndrome-associated protein, COH1, is a novel, giant Golgi matrix protein required for Golgi integrity. *J Biol Chem* 286(43):37665–75.

Summanen P, Kivitie-Kallio S, Norio R, Raitta C, Kivela T (2002) Mechanism of myopia in Cohen syndrome mapped to chromosome 8q22. *Invest Ophthalmol Vis Sci* 43:1686–1693.

Taban M, Memoracion-Peralta DS, Wang H, Al-Gazali LI, Traboulsi EI. (2007) Cohen syndrome: report of nine cases and review of the literature, with emphasis on ophthalmic features. *J AAPOS* 11(5):431–7.

Thomaidis L, Fryssira H, Katsarou E, Metaxatou C (1999) Cohen syndrome: Two new cases in siblings. *Eur J Pediatr* 158:838–841.

Velayos-Baeza A, Vettori A, Copley RR, Dobson-Stone C, Monaco AP (2004) Analysis of the human VPS13 gene family. *Genomics* 84(3):536-49.

Waite A, Somer M, O'Driscoll M, Millen K, Manson FD, Chandler KE (2010) Cerebellar hypoplasia and Cohen syndrome: a confirmed association. *Am J Med Genet A* 152A(9):2390–3.

Warburg M, Pedersen SA, Horlyk H (1990) The Cohen syndrome. Retinal lesions and granulocytopenia. *Ophthalmic Paediatr Genet* 11:7–13.

Yu TW, Chahrour MH, Coulter ME, Jiralerspong S, Okamura-Ikeda K, Ataman B, Schmitz-Abe K, Harmin DA, Adli M, Malik AN, D'Gama AM, Lim ET, Sanders SJ, Mochida GH, Partlow JN, Sunu CM, Felie JM, Rodriguez J, Nasir RH, Ware J, Joseph RM, Hill RS, Kwan BY, Al-Saffar M, Mukaddes NM, Hashmi A, Balkhy S, Gascon GG, Hisama FM, LeClair E, Poduri A, Oner O, Al-Saad S, Al-Awadi SA, Bastaki L, Ben-Omran T, Teebi AS, Al-Gazali L, Eapen V, Stevens CR, Rappaport L, Gabriel SB, Markianos K, State MW, Greenberg ME, Taniguchi H, Braverman NE, Morrow EM, Walsh CA (2013) Using whole-exome sequencing to identify inherited causes of autism. *Neuron* 77(2):259–73.

15

CORNELIA DE LANGE SYNDROME

ANTONIE D. KLINE
Clinical Genetics, Harvey Institute for Human Genetics, Greater Baltimore Medical Center, Baltimore, Maryland, USA

MATTHEW DEARDORFF
Division of Human Genetics, The Children's Hospital of Philadelphia, and Department of Pediatrics, the Perelman School of Medicine at the University of Pennsylvania, Philadelphia, Pennsylvania, USA

INTRODUCTION

Cornelia de Lange syndrome (CdLS) is a rare developmental malformation syndrome characterized by learning issues, small stature, limb abnormalities, and distinctive craniofacial features. Diagnosis is made clinically and relies on medical history and physical examination. The disorder is genetically heterogeneous, with seven different causative genes identified to date: *NIPBL, SMC1A, SMC3, RAD21, HDAC8, ANKRD11* and *BRD4*. Mutations in one of these genes (most commonly *NIPBL*) can be detected in about 75% of individuals with a definite clinical diagnosis (Mannini et al. 2014). Because of the range of overlapping features, the condition has been referred to as CdLS spectrum disorder (Deardorff et al. 2016). Within CdLS there is a wide range of severity. At the most severe end are markedly growth-retarded infants with congenital diaphragmatic hernia and/or severe limb deficiency defects and global cognitive impairment with little or no speech. Mildly affected individuals may have no major malformations, near normal intelligence, and are able to live independently and reproduce.

The eponymous name of the diagnosis recognizes the Dutch pediatrician Cornelia de Lange, who described two children with a distinct pattern of malformations (de Lange 1933). The syndrome is also called Brachmann–de Lange syndrome in recognition of a German doctor who reported an earlier individual (Brachmann 1916). An even earlier article has come to light, described article by Vrolik, a Dutch anatomist working in Amsterdam in the late nineteenth century (Oostra et al. 1994). There are now many hundreds of reported individuals. The terms Amsterdam dwarfism and Typus Degenerativus Amstelodamensis were used historically but have been discarded.

Incidence

The published estimates of incidence for Cornelia de Lange syndrome vary from 1 in 10,000 to 1 in 100,000 live births (Barisic et al. 2008; Opitz 1985). A complete ascertainment of all known individuals with CdLS in Denmark over a five year period demonstrated an incidence of 1 in 50,000 (Beck and Fenger, 1985); from the European Surveillance of Congenital Anomalies database, prevalence of the more severe phenotype was found to be 1 in 81,000 births (Barisic et al. 2008). An unpublished study of the entire phenotypic spectrum in the north of England found an incidence of CdLS of 1 in 37,000 (Dr. M. Ireland, personal communication). It is suspected that there are many mildly affected individuals who remain undiagnosed and not included in these figures.

Diagnostic Criteria

Diagnosis is made on a clinical basis, although a pathogenic mutation in one of the seven known associated genes

---

Cassidy and Allanson's Management of Genetic Syndromes, Fourth Edition.
Edited by John C. Carey, Agatino Battaglia, David Viskochil, and Suzanne B. Cassidy.
© 2021 John Wiley & Sons, Inc. Published 2021 by John Wiley & Sons, Inc.

provides useful confirmation of the clinical diagnosis. The facial gestalt is the most widely accepted criterion for the diagnosis of CdLS (Figure 15.1)

The facial features are typically striking in that they dominate the facial features associated with the genetic and ethnic background of the individual. Affected individuals have microbrachycephaly with a short neck and a low anterior and posterior hairline. The eyebrows are thin, well-defined and arched with synophrys; bushy eyebrows are not typically characteristic. The eyes are usually normally set but occasionally can be downslanting. The eyelashes are often long, lush, and curled. Ptosis of the upper eyelids is common and results in a compensatory backward tilt of the head in many cases. The ears may be normally set or posteriorly angulated and low set, and may have a thickened helix and appear large in relation to the face. At birth the nasal bridge is usually broad and depressed and the nares are anteverted. The philtrum is usually long, smooth, and prominent. The vermilion of the lips is usually thin with some having a small central V-shaped indentation in the upper vermilion border. The mouth is small and crescent-shaped with the corners turned down. The mandible is usually smaller than expected for the size of the face and some individuals can have severe micrognathia. The palate is often high and, in a significant minority of individuals, cleft or incompletely fused. Teeth may be widely spaced or absent. The supraorbital ridges and zygomatic arches are poorly developed. With increasing age, the nasal bridge becomes more prominent, and the nares less anteverted; the face is elongated and the chin becomes squarer (Figure 15.1D). Figures 15.1C–D and Figures 15.2A–E show individuals with a pathogenic variant in six of the seven associated cohesin genes.

Growth, development, and behavior are major criteria for diagnosis. Typically, all growth parameters are below the

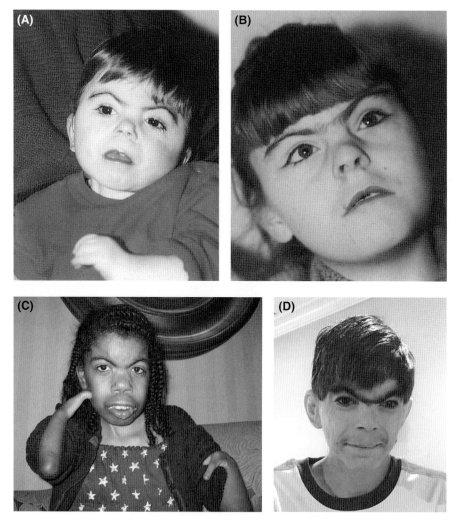

FIGURE 15.1 – Facial features of children with more classical Cornelia de Lange syndrome, typical of those from mutations in *NIPBL*. (A) and (B) Note neat, arched eyebrows with a "pencilled" quality, long, thick eyelashes, depressed nasal bridge, broad nasal root, anteverted nares, long prominent philtrum and small chin. (C) 17-year-old female with frameshift mutation in the *NIPBL* gene. Note similar facial features, bilateral single digits with shortened forearms. (D) 48-year-old male with frameshift mutation in *NIPBL* with typical adult features.

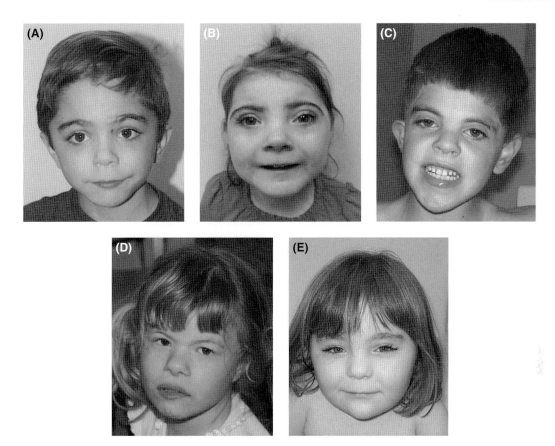

FIGURE 15.2 Similar facial features but milder in those with mutations in other cohesin genes: (A) 3-year-old male with *SMC1A* mutation. (B) Almost 3-year-old female with *SMC3* mutation. (C) 8-year-old male with *RAD21* mutation. (D) 7-year-old female with *HDAC8* mutation. (E) 3-year-old female with *ANKRD11* mutation.

fifth centile for age, with the head circumference being below the second centile for age, although there are many individuals whose parameters fall within the normal range. Global developmental delays or intellectual disability are common; however, there are individuals who have normal intelligence with specific learning disabilities. Behavioral issues are highly characteristic and can range from severe self-injurious behavior, aggression, and autistic-like features to extreme shyness and obsessive-compulsive tendencies.

The characteristic pattern of malformation and/or dysmorphic features of the limbs is also very helpful in the diagnosis. Hands, when intact, are nearly always small by measurement. The hands have relatively small and proximally placed thumbs, single palmar creases, and brachyclinodactyly of the fifth fingers. On radiological examination short first and fifth metacarpals with relatively long third and fourth metacarpals are useful diagnostic findings, as are hypoplasia and subluxation or dislocation of the radial head causing limited extension of elbows (Filippi 1989). A discriminative score based on the metacarpophalangeal profile has been devised (Halal and Preus 1979). Phocomelia, ulnar hypoplasia/aplasia, and oligodactyly are less commonly seen but very characteristic when present. Feet are small, and partial syndactyly of toes 2 and 3 are common. Pectus excavatum, club feet and hip dislocation are other common musculoskeletal features. Bunion, scoliosis, and/or hip dysplasia may develop (Kline et al. 2007a).

The neurosensory system and other major body systems are helpful minor diagnostic criteria. Eye findings, hearing loss, seizures, generalized hypertrichosis, and cutis marmorata are included. The presence of gastrointestinal malformation, diaphragmatic hernia, cleft palate, and genitourinary anomalies, as well as the highly prevalent gastroesophageal reflux are included in the other systems. Without positive molecular testing, facial features plus two major system criteria, or one major and two or more minor body system criteria, are sufficient to make the diagnosis. Molecular testing tends to be reserved for diagnostic dilemmas or if relevant for prenatal diagnostic or other familial purposes.

Diagnostic criteria have been published (Kline et al. 2007b; Selicorni et al. 2007), and the most recent consensus statement (Kline et al. 2018) describes a scoring system for diagnosis Several other authors have considered minimal diagnostic criteria for CdLS. Preus and Rex (1983) sought to define the syndrome more accurately using numerical taxonomy. They evaluated 207 different clinical characteristics in 48 individuals who had been referred with a possible diagnosis of CdLS using cluster analysis to identify 30 discriminative characteristics. Ireland and colleagues (1993) considered four craniofacial features to be most specific to the syndrome: the characteristic eyebrows, long, smooth philtrum, thin lips, and crescent-shaped mouth. Those features that were felt to be non-specific were hypertrichosis, synophrys, and bushy eyebrows. Jackson et al. (1993) stated that the facial features and limb anomalies, particularly small hands and feet, were most useful for accurate diagnosis, although no diagnostic criteria suffice for certainty of diagnosis.

The most recent consensus diagnostic criteria (Table 15.1) include six cardinal features giving 2 points each and seven suggestive features giving 1 point each. The cardinal features are: synophrys (and/or thick eyebrows), nose findings (short nose, upturned nasal tip, and/or concave nasal ridge), long and/or smooth philtrum, mouth findings (thin upper lip and/or down-turned corners), oligodactyly and/or adactyly, and congenital diaphragmatic hernia. The suggestive features are: global developmental delay and/or intellectual disability, prenatal growth retardation, postnatal growth retardation, microcephaly, small hands and/or feet, brachydactyly fifth finger and hirsutism. Diagnosis is confirmed and considered "classic" with a point score of 11 or higher (at least three of the cardinal features), and "non-classic" with a point score of 9 or 10 (with at least two of the cardinal features). Molecular testing is indicated for a point score of 4 to 8 (with at least one of the cardinal features). For a point score of below 4, molecular testing is not indicated and CdLS is unlikely (Kline et al. 2018)

Etiology, Pathogenesis, and Genetics

Pathogenic mutations have been identified in seven genes in CdLS: *NIPBL* (5p13.2) (Krantz et al. 2004; Tonkin et al. 2004), *SMC1A* (Xp11.22) (Musio et al. 2006; Borck et al. 2007; Deardorff et al. 2007), *SMC3* (10q25) (Deardorff et al. 2007), *RAD21* (8q24) (Deardorff et al. 2012a), *HDAC8* (Xq13.1) (Deardorff et al. 2012b), *ANKRD11* (16q24.3) (Ansari et al. 2014), and *BRD4* (19p13.12) (Olley et al. 2018). The proteins encoded by these genes strongly implicate dysfunction of the cohesin complex as the molecular basis of CdLS. The cohesin complex was identified and characterized through genetic and biochemical analysis of mutations affecting chromosome segregation in yeast. In *Saccharomyces cerevisiae* the cohesin complex exists as a heterotetrameric ring structure consisting of Smc1 (the human ortholog: *SMC1A*), Smc3 (*SMC3*), Scc1 (*RAD21*), and Scc3 (*STAG1/2/3*). NIPBL is the human ortholog of yeast Scc2, a key protein responsible for loading cohesin onto chromosomes. HDAC8, a deacetylase, plays an important role in deacetylating SMC3 to enable recycling of the cohesin ring. Thus, both *NIPBL* and *HDAC8* are regulatory proteins and *SMC1A*, *SMC3*, and *RAD21* are structural genes. In higher eukaryotes, cohesin has multiple functions in addition to chromosome cohesion, including regulation of gene expression, chromatin remodeling, and DNA repair (Strachan 2005; Misulovin et al. 2008).

Over 60% of individuals with CdLS encompassing both sporadic and familial cases have mutations in *NIPBL* that are detectable in blood. The gene *NIPBL*, or Nipped-B-like as the homologue for the Drosophila Nipped-B gene, is located on the short arm of chromosome 5 and produces a protein that facilitates enhancer–promotor communication, regulates developmental pathways and is homologous to proteins involved in sister chromatid cohesion, chromosome condensation and DNA repair. Most *NIPBL* mutations that cause a more severe clinical picture are de novo heterozygous loss-of-function mutations caused by frameshift, truncating, or splice site changes in the open reading frame. There is evidence that a reduction in the NIPBL protein leads to multiple gene expression changes, particularly genes involved in the heart and the gut, and this widespread change in expression in all body tissues most likely directly leads to the developmental delays and intellectual disability seen in CdLS (Kawauchi et al. 2016). Genotype–phenotype studies have shown that both milder and more severe individuals may have mutations in *NIPBL*, and those with missense mutations tend to be less involved than those with frameshift,

TABLE 15.1 Consensus diagnostic criteria for Cornelia de Lange syndrome (CdLS).

Features	Clinical findings	Points assigned
Cardinal features	Synophrys and/or thick eyebrows	2
	Short nose and/or concave nasal ridge and/or upturned nasal tip	2
	Long and/or smooth philtrum	2
	Thin upper lip vermilion and/or downturned corners of mouth	2
	Hand oligodactyly and/or adactyly	2
	Congenital diaphragmatic hernia	2
Suggestive features	Global developmental delay and/or intellectual disability	1
	Prenatal growth retardation (below 2 standard deviations)	1
	Postnatal growth retardation (below 2 standard deviations)	1
	Microcephaly (prenatal or postnatal)	1
	Small hands and/or small feet	1
	Brachydactyly fifth finger(s)	1
	Hirsutism	1

Legend: Classic CdLS > 11 points with at least three cardinal features, non-classic CdLS 9–10 points with at least two cardinal features. For 4–8 points, with at least one cardinal feature, molecular testing indicated. For <4 points, likely not CdLS.
Adapted from: Kline AD et al. Nature Reviews Genetics 19:649–66, 2018.

truncating or splice site mutations (Gillis et al. 2004; Bhuiyan et al. 2006). Mutations in *NIPBL* tend to cause more "classical" findings (Figures 15.1A-D). Although the majority of *NIPBL* mutations are de novo, familial cases have been described, and in these families the condition is inherited in an autosomal dominant fashion. Parental germline mosaicism for *NIPBL* mutations has been documented (Slavin et al. 2012), and likely explains a portion of the 1.5% empiric sibling recurrence risk that was calculated for counseling parents of affected children who had no evidence of CdLS themselves (Jackson et al. 1993). In addition, as high as 23% of individuals with "classical" CdLS features have mosaic *NIBPL* mutations detectable in buccal swabs, skin fibroblast or urine sediment specimens (Huisman et al. 2013; Ansari et al. 2014).

The genes *SMC1A* and *SMC3*, or the structural maintenance of chromosomes 1A and 3, both produce proteins which are part of the cohesin complex that help organize chromosomes in mitosis, DNA repair and transcriptional regulation. Approximately 5% of individuals with CdLS have mutations in *SMC1A*, located on the short arm of the X chromosome. They tend to have more normal somatic growth, normal to small hands, fairly atypical facial features (Figure 15.2A), with often a significant degree of cognitive impairment (Musio et al. 2006; Borck et al. 2007; Deardorff et al. 2007). The gene *SMC3* is located on the long arm of chromosome 10, and its mutations are typically de novo autosomal dominant. Approximately 1–2% of individuals with CdLS have mutations in *SMC3*, initially reported in a single individual with atypical features and a mild phenotype (Deardorff et al. 2007), and more recently as a series with generally milder involvement including slightly small stature, minimal facial characteristics (Figure 15.2B) and minimal organ system involvement (Gil-Rodriguez et al. 2015).

Only 1% of individuals with CdLS have a mutation in *RAD21* (Deardorff et al. 2012a), presenting with similar facial features (Figure 15.2C) and bushy eyebrows, mild skeletal findings, minimal organ system involvement, and fairly mild intellectual disability. This gene is located on the long arm of chromosome 8 and dominant mutations produce functional defects, which lead to decreased separation of sister chromatids and defective DNA repair. Four percent of individuals with CdLS have been identified to have mutations in *HDAC8*, located on the long arm of the X chromosome. This gene product helps recycle cohesin for the next cell cycle (Deardorff et al. 2012b). Mutations are expressed as X-linked dominant but milder in females, most of whom have skewed X-inactivation, and some minimally affected carrier mothers have been noted. Clinically, individuals with mutations in *HDAC8* present with milder facial features (Figure 15.2D), delayed closure of sutures of the skull with resultant large anterior fontanelles in infancy, few structural defects, less severe growth restriction, and varied intellectual disability (Deardorff et al. 2016).

The two newest genes said to cause CdLS are also present in a low percentage of affected individuals (Kline et al. 2018). Mutations in the *ANKRD11* gene, located on the long arm of chromosome 16, typically associated with KBG syndrome, have presented clinically as CdLS (Ansari et al. 2014; Parenti et al. 2016), with feeding difficulties, developmental delays particularly speech, gastroesophageal reflux, behavioral issues including self-injury and autistic features, small stature, similar facial features when younger, hirsutism, small hands with brachydactyly, and proximally placed thumbs (Figure 15.2E). Differentiating features include normocephaly, large teeth, hypertonia, and normal hearing in KBG syndrome. Both *ANKRD11* and the cohesin complex are involved in transcriptional regulation and likely overlap in function (Parenti et al. 2016; Kline et al. 2018). Missense and frameshift mutations in the *BRD4* gene, located on the short arm of chromosome 19, present clinically with milder facial features of CdLS, as well as slightly smaller stature and mildly delayed development (Olley et al. 2018). *BRD4* associates with the cohesin ring and interacts with *NIPBL*.

It is clear that other loci for CdLS remain to be found because about 25% of affected individuals in the CdLS spectrum still do not have a detectable mutation. A number of genes interacting with the cohesin complex can produce phenocopies (see Differential Diagnosis below). There is some clinical evidence for accelerated aging in CdLS (Kline et al. 2007a), which may be related to issues of DNA repair. There are, too, searches for potential treatments, both at the small molecule level (Deardorff et al. 2016) and in animal models (Kawauchi et al. 2016).

Diagnostic Testing

Molecular testing is available in North America, Europe and elsewhere individually and by panels, and, if positive, helps to confirm the diagnosis. Panel testing is available for the first five genes described, and larger panels for intellectual disability likely do not include *ANKRD11* and *BRD4*, although they will be detected on exome sequencing. Clinical diagnosis rests on recognition of the characteristic pattern of delayed growth and development, associated abnormalities of limb development and, in particular, the craniofacial features. However, individuals with atypical features are not uncommon and an objective test can be helpful if positive. This is particularly true with regard to prenatal diagnosis, which currently relies on detailed ultrasonography looking for structural abnormalities such as a cystic hygroma, limb reduction defect, diaphragmatic hernia, heart abnormality, or the characteristic facial profile. The facial profile on prenatal ultrasound consists of severe micrognathia with a long, bulging philtrum and small nose (Urban and Hartung 2001).

In 1983, Westergaard and colleagues reported the total absence of pregnancy-associated plasma protein-A (PAPP-A) in serial serum samples from a pregnant woman who later gave birth to a child with CdLS (Westergaard et al. 1983). Aitken et al. (1999) confirmed a reduction in PAPP-A and produced a table of likelihood ratios based on retrospective analysis of 19 second-trimester maternal serum samples. PAPP-A testing is no longer offered in the second trimester, although levels from the first trimester may prove to be a valuable additional marker for "high-risk" pregnancies. Clark et al. (2012) reviewed 53 pregnancies with prenatal details available and noted reduced PAPP-A and an increased nuchal translucency. Detailed ultrasonography remains the most appropriate basis for the prediction of CdLS during fetal life (Kliewer et al. 1993; Clark et al. 2012) if molecular diagnosis is unavailable.

Differential Diagnosis

The historically important differential diagnosis is a relatively consistent syndrome associated with partial duplication of the long arm of chromosome 3. This is characterized by intellectual disability and failure-to-thrive, although usually with normal birth weight and length. Facially there is a resemblance to CdLS in that hair extends over the forehead, the eyelashes may be prominent, the nasal bridge is depressed, and the nares are anteverted. The philtrum is usually long and prominent, although, unlike CdLS, it continues to show a central philtral groove. Micrognathia is also common, but it is unusual for the lips to be thin or the mouth crescent-shaped. The facial features that differentiate duplication 3q syndrome from CdLS are a sloping forehead, bushy eyebrows, hypertelorism, upslanting palpebral fissures, epicanthal folds, a broader nose, maxillary prognathism, and relatively normal lips. Malformations common in duplication 3q syndrome include central nervous system, eye, cardiac, and renal abnormalities, cleft palate, and genital hypoplasia. Craniosynostosis, camptodactyly, and talipes are the most commonly associated skeletal abnormalities. Chromosomal analysis of the affected child will confirm the diagnosis, as will array comparative genomic hybridization.

The most severe phenotype of CdLS shares features with Fryns syndrome, an autosomal recessive disorder characterized by a coarse face, diaphragmatic hernia (85%), cleft palate (30%), and distal limb hypoplasia (75%). Hypertrichosis, narrow palpebral fissures, flat nasal bridge, upturned nose, micrognathia, cardiac, renal, and genital abnormalities are common in both conditions. Fryns syndrome is distinguished by a short upper lip and macrostomia. In addition, polyhydramnios and premature labor are common in Fryns syndrome, and birth weight, length, and head circumference are usually normal for gestational age.

Fetal alcohol spectrum disorder (see Chapter 26) is characterized by intrauterine growth retardation, failure to thrive, and mild to severe developmental abnormalities. Craniofacial features overlap with those seen in CdLS, especially microcephaly, short palpebral fissures, a short, upturned nose, a smooth underdeveloped philtrum, and a thin upper lip. Cardiac defects similar to those seen in CdLS are also well recognized. The hands and feet are not small, although small distal phalanges are common. Speech is affected much more in CdLS than in fetal alcohol spectrum disorder, and individuals with the latter are only likely to be confused with milder CdLS.

Individuals with Bohring–Opitz syndrome can resemble those with CdLS. Findings include severe developmental delay, feeding difficulties and failure to thrive, microcephaly, trigonocephaly, abnormal palate, cardiac defects, gastrointestinal involvement, hypertrichosis, and central nervous system anomalies (Russell et al. 2015). Facial features are similar to that in CdLS, especially in the infant period, but proptosis and hypertelorism are common. Extremities tend to be normal although with external rotation of the shoulders and flexion of the wrists and fingers. One of the genes associated with Bohring–Opitz syndrome, *ASXL1*, also is involved in chromatin remodeling (Russell et al. 2015).

Several additional genes can also present with mutations that overlap the phenotype of CdLS in some individuals, and at other times a different condition. CHOPS syndrome is a recently described condition, characterized by *c*ognitive impairment and *c*oarse facies, *h*eart defects, *o*besity, *p*ulmonary involvement, *s*hort stature, and *s*keletal dysplasia. There are again many similar features between CHOPS syndrome and CdLS, but in the former, individuals tend to be overweight with very short stature, coarser facial features, and to have significant pulmonary involvement. CHOPS is caused by heterozygous pathogenic gain-of-function variants in the gene *AFF4*, a core component of the super elongation complex that mobilizes RNA polymerase 2 when it has been paused (Izumi et al. 2015). Both *AFF4* and *NIPBL* are involved in transcriptional regulation, and cohesin has been found to interact both with the super elongation complex and RNA polymerase 2 (Izumi et al. 2015).

Similarly, mutations in the gene *KMT2A* have been demonstrated to cause Wiedemann–Steiner syndrome, a condition with short stature, pattern of hirsutism most prominent on the elbows (also called "hairy elbow syndrome"), distinctive facial features including bushy eyebrows, brachydactyly and intellectual disability. A report by Yuan et al. (2015) describes patients clinically suspected to have Wiedemann–Steiner syndrome who were found to have mutations in *SMC1A* or *SMC3*, and patients clinically diagnosed as having CdLS, who were found to have mutations in *KMT2A*. In addition, other patients thought to resemble CdLS have had mutations in a recessive gene, *TAF6*, with similar overlapping features: small stature, microcephaly, hirsutism, cryptorchidism and skeletal changes, along with intellectual disability and seizures (Yuan et al. 2015).

Finally, Rubinstein–Taybi (see Chapter 51), Floating Harbor, Coffin–Siris (see Chapter 13), and Nicolaides–Baraitser syndromes, though distinctive, also have overlapping findings, including hirsutism, digit anomalies and short stature; all are due to variants in genes involved in chromatin-mediated transcription regulation and/or chromatin remodeling. The facies in Floating Harbor syndrome has more of a tubular nose with beaked nasal tip, wider mouth and shorter philtrum. Facial features in Coffin–Siris syndrome are coarser with broader mouths and fuller lips, and there may be hypoplastic nails. Sparse scalp hair is seen in Nicolaides–Baraitser syndrome, along with shallow orbits, prominent lower lip, and wide mouth.

Family history, details of the pregnancy, and birth weight aid in making the diagnosis of CdLS. Clinical evaluation will include assessment of facial and extremity anomalies by a physician experienced in dysmorphology to allow discrimination of the above entities from CdLS. SNP microarray may be done initially to rule out rearrangements; gene testing may follow.

MANIFESTATIONS AND MANAGEMENT

Growth and Feeding

Affected individuals have prenatal onset growth failure manifested by a significantly low birth weight (mean approximately 2300 g), short stature, and poor growth velocity throughout childhood. Typically, all growth parameters are below the fifth centile for age, with head circumference below the second centile for age, although there are individuals whose parameters fall within the normal range. Syndrome-specific growth charts have been developed (http://www.cdlsusa.org/?s=growth+chart) (Figure 15.3).

Feeding difficulties are present from birth, and it is not uncommon for affected infants to require nasogastric tube feeding with subsequent insertion of a gastrostomy tube. The role of gastrostomy feeding in improving nutritional intake and growth in CdLS has not been formally studied. Individuals with CdLS seem to grow at their own rate regardless of additional caloric intake. Gastroesophageal reflux is common (see below), and contributes to feeding problems and failure-to-thrive. Lack of coordination of oral musculature, small mouth and jaw, poor bowel motility, and difficulty in chewing may also play a part (Kline et al. 1993a). Final adult heights have been accurately collated and these remain very significantly below normal with a mean adult height for males and females of 155.8 cm and 131.1 cm, respectively (Kline et al. 1993a). Obesity has been noted in 20% of an aging population (Kline et al. 2007a) with no specific etiology identified.

Evaluation

- Head circumference, length, and weight should be measured at birth and every 6–12 months thereafter throughout childhood.

- Growth should be charted on CdLS-specific growth charts (http://www.cdlsusa.org/?s=growth+chart). If growth velocity is less than expected, investigations for gastrointestinal anomalies (see below), thyroid dysfunction, and growth hormone secretory anomalies should be considered.

Management

- Adequate caloric intake should be ensured through nasogastric feeding if the child is unable to bottle-feed.
- When severe gastroesophageal reflux and feeding difficulties are present, it may be necessary to place a gastrostomy tube.
- Growth hormone therapy is not usually indicated. Kousseff et al. (1993) found evidence of growth hormone deficiency in 4 of the 12 affected individuals and end-organ resistance to growth hormone in another individual. However, the authors concluded that the therapeutic use of human growth hormone may not be warranted in CdLS unless there was evidence of persistent hypoglycemia.
- Adequate exercise and healthy diet are recommended.

Development and Behavior

There is a broad spectrum of developmental abilities. Reported IQs in CdLS have ranged from 30 to 102 (Kline et al. 1993b), with the majority of individuals classified as having mild to moderate global intellectual disability with IQ scores 55–70 (Kline et al. 2007b). Motor delays are less severe than speech and language, and fine motor skills are a strength. Even in individuals with absent forearms and single digits, parents report remarkable dexterity. Most children reject prostheses when offered. The degree of intellectual disability roughly correlates with the gene involved, and individuals with mutations in *NIPBL*, *SMC1A* and *SMC3* tend to have more cognitive impairment (Parenti et al. 2016). Early intervention therapy is critical for maximizing developmental potential (Kline et al. 1993b).

Problems with communication and expressive language are of great concern to caregivers and health care professionals. There are many reports in the literature of speech being absent in individuals with CdLS. In a study of 116 affected individuals, the indicators of poor prognosis for the development of speech included: birth weight less than 2.27 kg, moderate to severe hearing impairment, upper limb malformations, poor social interactions, and severe motor delay (Goodban, 1993). Of those over the age of 4 years, 53% could construct sentences of two or more words, whereas 33% had up to two words. Four percent of individuals were thought to have language skills in the normal or low–normal range. Overall, the study found that expressive language was inferior to comprehension and that individuals with highly developed

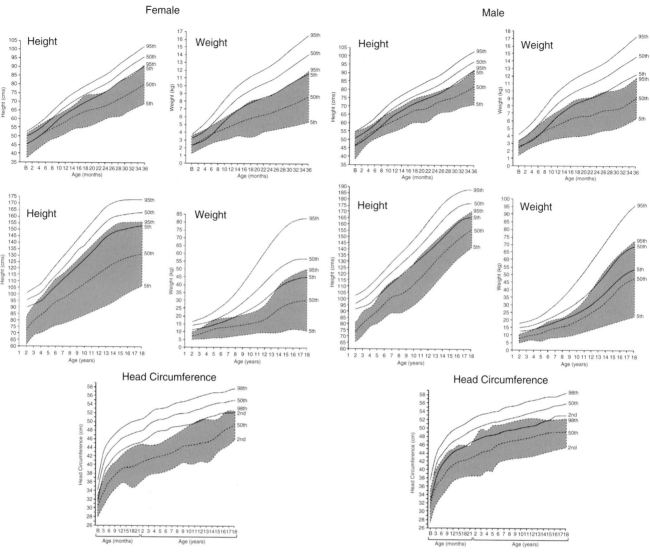

FIGURE 15.3 Male and female growth charts for height, weight, and head circumference. The solid lines represent the 95th, 50th, and 5th centile for the general population at different ages, and the shaded areas represent the same centiles for children with Cornelia de Lange syndrome.

vocabularies often had poor syntactic skills. Individuals were usually quiet and rarely talked even if they had well-developed vocabularies. This suggests that speech and language skills might be underestimated in CdLS. Poor eye contact, and limited attempts at initiation of speech have been noted. Although no systematic study of the use of sign language in CdLS is available, many parents report marked improvement in communication and behavior using this approach.

A low-pitched cry is characteristic of the syndrome in infancy, as is a deeper voice in older children. Characteristics of selective mutism have been reported in up to 40%, higher than in fragile X syndrome or autism spectrum disorder, as well as other syndromes (Moss et al. 2016).

Behavioral disturbances are common throughout life in CdLS. A wide variety of abnormal behaviors was reported in one study of 49 individuals with CdLS, including hyperactivity (40%), self-injury (44%), daily aggression (49%), and sleep disturbance (55%) (Berney et al. 1999). In a study of 56 individuals with CdLS based on maternal questionnaires, the variability of behavioral characteristics including hyperactivity, attention deficit disorder, anxiety, compulsive disorders, self-injurious behavior, and autistic-like features was found to correlate with clinical and functional aspects of the individuals, including age, cognitive level, and clinical findings (Basile et al. 2007). Self-injurious and self-restraint behaviors have been noted to have a significant association with compulsive behavior (Hyman et al. 2002). Self-injury has also been associated with particular environmental events (Hyman et al. 2002; Arron et al. 2006). Behavioral problems tend to worsen during adolescence (Kline and

Audette 2002) and specific psychiatric diagnoses, including self-injury, anxiety, attention-deficit disorder, autistic features, depression, and obsessive-compulsive behavior, can become apparent (Kline et al. 2007a). In addition, both socialization and features of communication decline with increasing age (Srivastava et al. 2014).

Gene mutation may play a role in some behavioral manifestations. A recent study of 34 individuals showed that individuals with mutations in *NIPBL* had lower mood, decreased interest and pleasure, and increased insistence on sameness with increasing age, as well as significantly lower self-help skills and fewer verbal skills, compared to those who were negative for *NIPBL* mutations (including *SMC1A*, *HDAC8* or no identified mutation) (Moss et al. 2017). Other behaviors showed no significant difference related to the presence of detectable *NIPBL* mutation, including repetitive behaviors, mood, challenging behaviors, activity, impulsivity, and autistic characteristics. Thus, there may be similarities in behavioral phenotype between those with and without the *NIPBL* mutation, when controlling for differences in self-help skills.

Behavior is correlated closely with the presence of autistic spectrum disorder and the degree of intellectual disability. The autistic features include lack of social relatedness and impassivity, rejection of physical contact, rigidity, and inflexibility to change. In infancy, low frequency of eye contact was thought to be a risk factor for poor developmental outcome based on video observations (Sarimski 2007). Previous studies have assessed behavior in CdLS in comparison to multiple other causes of intellectual disability (Oliver et al. 2008). The association between autistic spectrum disorder and CdLS was not solely accounted for by the degree of intellectual disability, and the profile of impairments in the syndrome may not be typical compared to idiopathic autistic spectrum disorder. Additionally, repetitive behaviors appear to contribute less to the profile of autistic spectrum disorder in individuals with CdLS than impairments in communication and social interaction (Oliver et al. 2012). Stereotypies, hyperactivity, and lethargy are some of the maladaptive traits noted in another study (Srivastava et al. 2014). A further study showed that 65% of individuals with CdLS met the cut-off score for autism using standard measures compared to individuals diagnosed with idiopathic autistic spectrum disorder and matched for receptive language and adaptive behavior (Moss et al. 2012). Those with CdLS showed significantly less repetitive behaviors, more eye contact, more gestures, less stereotyped speech and higher levels of anxiety than those with idiopathic autism spectrum disorder.

Individuals with CdLS exhibit socially motivated attention-soliciting behaviors, important in the consideration of behavioral management (Arron et al. 2006). The prognosis for the management of behavioral problems is improved if an underlying medical problem, particularly gastroesophageal reflux, is recognized and treated. Difficulties in communication often exacerbate the situation (see above) and the early use of sign language can be helpful. In the absence of such underlying problems, however, behavior remains a difficult management issue for families and care providers.

Evaluation

- Early referral should be made to an intervention program or child development center for the identification of appropriate special intervention and education needs.
- A speech and language therapist should evaluate communication skills.
- Hearing should be assessed regularly (see below).
- Any deterioration in behavior should be investigated promptly, and a medical cause such as gastroesophageal reflux should be considered.
- Consider psychiatric evaluation if symptoms of depression, obsessive-compulsive behaviors or anxiety become evident.

Management

- Early intervention and special education should be programmed as indicated by developmental and educational evaluations. These should include appropriate physical, occupational, and speech therapy.
- Vocational programs should be instituted after school age.
- Treatment of middle ear disease and hearing loss is standard (see below).
- Any medical problem such as gastroesophageal reflux or underlying behavioral deterioration should be treated in a standard manner.
- Long-standing behavioral problems are difficult to change but may be improved with behavioral modification. Guidance in techniques of behavior modification may be sought from a behavioral psychologist or developmental pediatrician.
- Despite the absence of formal studies prospectively evaluating behavior and medication use, anecdotally many individual regimens have been useful. For example, stimulants (e.g., methylphenidate) have been useful for attention deficit disorder with hyperactivity, anti-anxiety medications (e.g., buspirone) helpful for anxiety and other behavioral problems, and selective serotonin reuptake inhibitors (e.g., sertraline) have been helpful for mood issues.

Ears and Hearing

Some degree of hearing impairment is present in over 90% of individuals with CdLS (Sataloff et al. 1990). This has been shown to be sensorineural in origin and usually bilateral (Sataloff et al. 1990), although conductive hearing loss due to middle ear effusion alone or in combination with sensorineural deficit has also been noted. Early diagnosis and use of hearing aids can augment development of language. A more recent study (Janek et al. 2016) based on chart review of 78 individuals and parental survey of 35 individuals showed that up to 46% reported improvement in hearing over time, and 6% of those with severe to profound hearing loss initially on audiograms demonstrated subsequent normal hearing, with 14% demonstrating improvement. This is attributed to resolution of a conductive component, and possible presence of auditory neuropathy. Otitis media and chronic middle ear disease with subsequent conductive deafness are common and can be difficult to diagnose because of the narrow external auditory canals found in many individuals with CdLS. Middle ear disease is less of a problem later in childhood and in adulthood, although chronic sinusitis is noted in 39% of older individuals, and nasal polyps may be seen (Kline et al. 2007a).

Evaluation

- Auditory evoked brainstem responses should be assessed in infancy or at diagnosis in all individuals with CdLS and followed up if abnormal.
- Hearing should be checked every 6 months until age 6 years by routine audiology to detect conductive loss related to chronic middle ear disease.

Management

- Antibiotic treatment should be used in a standard manner for otitis media.
- Consider referral to an otolaryngologist or audiologist if hearing loss is detected.
- Hearing aids should be fitted early.
- Insertion of ventilatory tubes should be accomplished to treat chronic middle ear disease.
- Sinusitis and nasal polyps should be treated in a standard manner to minimize complications.

Craniofacial

Cleft palate occurs in 20% of individuals with CdLS. Cleft lip does not occur, or is unrelated. Submucous cleft palate also occurs and may be under-reported; in a study of 49 patients, submucous cleft palate was found in 14%, most undetected before evaluation in that study (Kline et al. 2007a). The prognosis for those with cleft palate is usually good, although affected individuals are less likely to develop speech than those with a normal palate.

Micrognathia and microstomia are well reported, and can lead to difficulty with intubation for anesthesia (Moretto et al. 2016). Dental crowding is very commonly associated with CdLS, as are small or absent teeth. There is often poor oral hygiene and dental caries may be problematic (Kline et al. 2007a). Bruxism is common. With aging, periodontal disease may occur.

Evaluation

- Physical examination of the palate should occur at diagnosis.
- If symptoms or signs suggest problems related to submucous cleft palate, referral for a specialist assessment is indicated.
- Routine dental assessment and cleaning should occur every 4–6 months beginning at age 2 years.
- Periodic dental examination under anesthesia may be required every 1–3 years.

Management

- Repair of the cleft palate is ideally accomplished in a unit or craniofacial team experienced in this area. The methods are no different from those in the general population.
- Speech therapy is indicated postoperatively to maximize the development of speech.
- Treatment of dental problems under anesthesia is often required. The methods are the same as those in the general population, but caution should be used with intubation.
- Anesthesia should be undertaken by a pediatric specialist expert in intubation. In a review of anesthesia and sedation used for procedures, Moretto et al. (2016) discussed complications; there have been adverse allergic reactions to Midazolam, which is no longer recommended for use in CdLS.

Musculoskeletal

A wide spectrum of limb abnormalities is seen in CdLS. The most commonly observed pattern includes small hands with short digits, proximally placed thumbs, single palmar creases, and clinodactyly of the fifth fingers (Filippi et al. 1989). A characteristic pattern is observed on the metacarpophalangeal index (see Diagnostic Criteria). Dermatoglyphic abnormalities are common, with an increased frequency of radial loops and arches and a decrease in the number of whorls. The total finger ridge count is low, and interdigital triradii are also common (Filippi 1989).

Severe limb deficiencies with missing hands and forearms occur in 16% of individuals and absent digits in 9% (Kline et al. 2007a). Even with severe limb deficiencies, many individuals have remarkably good fine motor skills.

Flexion contractures of the elbows are common and are usually secondary to deformation of the proximal metaphysis of the radius, which is often subluxed. Syndactyly, polydactyly, and duplication of the nail bed have been noted in CdLS but are rare malformations (Beck and Fenger 1985). The vast majority of severe deficiency defects involve only the upper limbs, predominantly the ulnar aspect. Few major lower limb defects have been reported although hip abnormalities are common. Scoliosis, leg length discrepancy, and Legg–Calve–Perthes disease may be acquired (Roposch et al. 2004; Kline et al. 2007a). Decreased bone density has been seen with aging (Kline et al. 2007a).

Evaluation

- Measurement of the metacarpophalangeal index and radiographs of the elbow may aid diagnosis.
- An experienced physical therapist should evaluate the range of movement at all joints.
- Bone densitometry (DEXA scan) should be considered if there is concern about osteopenia or osteoporosis.
- Refer to orthopedics if there is concern about Legg–Calve–Perthes disease.

Management

- There are no published reports of benefit from surgical approaches to limb anomalies, and even those with severe intellectual disability often have remarkably good fine motor control despite their anomalies.
- Early use of physical therapy is recommended to prevent joint contractures and improve motor skills.
- Management for Legg–Calve–Perthes disease has included rest and non-steroidal anti-inflammatory drugs (Roposch et al. 2004).

Gastrointestinal

Gastroesophageal reflux disease is by far the most common gastrointestinal problem in CdLS and is present in varying degrees in nearly all individuals. Untreated, it can lead to significant morbidity and mortality. The acquisition of esophageal stenosis and other sequelae of gastroesophageal reflux disease are common. Barrett esophagus has been reported in 10% of an aging population (Kline et al. 2007a), and the development of adenocarcinoma of the esophagus has been documented (DuVall and Walden 1996). Gastroesophageal reflux can present with failure-to-thrive, reduced growth velocity, anemia, recurrent pneumonia, apnea, or behavioral disturbance including hyperactivity (Luzzani et al. 2003). Gastroesophageal reflux may worsen during puberty (Kline and Audette 2002), may present at any age, and symptoms may wax and wane. A high index of suspicion is essential. Stories of the dramatic improvement in behavior and development are common in successfully treated individuals.

Malformations of the gastrointestinal tract presenting shortly after birth include pyloric stenosis in approximately 1% of affected individuals and, very rarely, an annular pancreas causing duodenal obstruction. In up to 10%, abnormalities of mesenteric insertion and failure of normal rotation of the bowel during embryonic development can lead to malrotation and have resulted in life-threatening volvulus with ischemia, and infarction of the bowel (Masumoto et al. 2001). Signs including severe abdominal pain, bilious vomiting, and hard abdomen on palpation should be treated emergently. Respiratory causes, including those related to reflux, and gastrointestinal obstruction and volvulus were the most common causes of death in a recent comprehensive series (Schrier et al. 2011). Several cases of diaphragmatic hernia have been seen, with high mortality rates. Rumination and chronic constipation are reported with aging (Kline et al. 2007a).

Evaluation

- A high index of suspicion for gastroesophageal reflux should be maintained when anemia, deterioration in behavior, or a decrease in growth velocity occurs. Symptoms occurring when the child lies down at night and the appearance of dystonic posturing are good clues.
- Early referral to a specialist in pediatric gastrointestinal disease should be considered.
- Barium studies are of little help in making the diagnosis of gastroesophageal reflux, unless present but some affected individuals will tolerate 24 hour monitoring with a pH probe.
- All affected individuals should undergo barium swallow followed to the duodenum to ascertain the presence of malrotation.
- Endoscopic examination under anesthesia is the most practical way to diagnose gastroesophageal reflux and, if possible, other evaluations such as hearing, eye, and dental examinations can be done simultaneously while under anesthesia.
- If invasive investigation is not practical then an on/off therapeutic trial of proton pump inhibitors and prokinetic medication can be used as an "investigation" to evaluate its potential effectiveness.
- Emergency evaluation for volvulus, in a standard manner, should be carried out if symptoms seem suspicious.

Management

- Initial treatment for gastroesophageal reflux may consist of advice on small, thickened feeds and keeping the individual as upright as possible after feedings, but this

is usually inadequate in Cornelia de Lange syndrome.
- Pharmacotherapy with antacids and/or H_2-receptor blockers can be tried. Proton pump inhibitors used alone or in combination can be highly effective.
- Fundoplication, occasionally with the insertion of a gastrostomy tube, may be necessary. Surgical issues are the same as in the general population.

Cardiovascular

Congenital heart defects are present in 14–20% of individuals with CdLS (Beck and Fenger 1985; Jackson et al. 1993). The most common abnormality is ventricular septal defect, followed by atrial septal defect, and stenosis of the pulmonary valve, either alone or in combination with a ventricular septal defect. A wide range of other cardiac defects have also been reported, including tetralogy of Fallot, coarctation of the aorta, hypoplastic left heart, single ventricle, and atrioventricular canal defect. Mild pulmonary branch stenosis is seen in 2% of more mildly affected individuals, but rarely requires treatment (Ireland 1996). The most severe cardiac defects are often seen in infants with diaphragmatic hernia and limb deficiency defects.

It should be noted that perioral blueness is very common in CdLS particularly in infants (Jackson et al. 1993) and may not, in itself, require further investigation. The prognosis for those with CdLS and congenital heart defects is related to the severity of the malformation and the degree to which surgical intervention can correct or palliate the hemodynamic status.

Evaluation

- All individuals with CdLS should have a cardiac assessment shortly after birth or at diagnosis, including echocardiography.

Management

- Most individuals with congenital heart disease no longer need to be given antibiotic prophylaxis before and during any procedure that induces a transient bacteremia, such as dental work, to prevent infective endocarditis, according to standard cardiologic recommendations.
- Referral for management should be made to a pediatric cardiology unit for all those with a cardiac defect.
- The indications for and methods of cardiac surgery are the same as in the general population.

Ophthalmologic

Ptosis, recurrent blepharitis, and myopia are common in CdLS (Levin et al. 1990; Wygnanski-Jaffe et al. 2005). Ptosis surgery may assist in motor development. The severity of ptosis has been correlated with specific mutations in *NIPBL* (Nallasamy et al. 2006). Nasolacrimal duct obstruction occurs and may require surgical correction. Nystagmus, mild microcornea, and/or strabismus can occur. Cataract and glaucoma have been reported rarely and may represent chance occurrences or be the result of self-injury. Children with severe myopia may be at higher risk for retinal detachment. Optic nerve pallor is uncommon, although a visually insignificant pigmented ring around the optic nerve is almost universal (Levin et al. 1990; Wygnanski-Jaffe et al. 2005).

Evaluation

- Referral for ophthalmologic assessment should be made in the first 6 months of life for all affected individuals.
- Regular ophthalmologic evaluations should occur every 6–12 months throughout life.

Management

- Correction of refractory errors should begin early because glasses are poorly tolerated if not fitted early.
- Surgery to correct ptosis is indicated if there is obstruction to the visual axis.
- Prompt treatment of conjunctivitis in a standard manner should occur.
- Baby shampoo eyelash scrubs may be helpful in relieving recurrent blepharitis and red eye discharge.
- Some children may require nasolacrimal duct probing for obstruction if conservative therapy fails.

Genitourinary

Structural anomalies of the kidney and genitourinary tract have been detected in 41% of individuals, with impaired renal function in up to one third of these (Selicorni et al. 2005). The most common abnormality of the renal tract observed in CdLS is pelvic dilatation, but hydronephrosis has also been seen. Vesicoureteric reflux has been noted in up to 12% of children (Jackson et al. 1993). Cryptorchidism is an extremely common finding in males with CdLS, seen in 83% in one study, and may occur in association with hypospadias, micropenis, and a hypoplastic scrotum (Kline et al. 2007a).

During early childhood the umbilicus and nipples are hypoplastic. Puberty may be slightly delayed in both sexes and can be incomplete. Secondary sexual development is often normal, although primary amenorrhea is seen in up to 25% of females (Kline and Audette 2002; Kline et al. 2007a). Most females have normal gynecologic examinations and pap smears. Individuals with milder forms of the syndrome should be presumed to be fertile (Russell et al. 2001).

Evaluation

- Ultrasound evaluation of the kidneys should be accomplished at diagnosis.
- Voiding cystourethrogram should be performed if any urinary tract infection develops or there is a functional question.
- Pubertal development should be followed to identify the time at which reproduction is possible.

Management

- Orchidopexy should be undertaken before 2 years of age, ideally between 6 and 18 months, with repair of any concomitant hypospadias being organized at the same time.
- If recurrent urinary tract infections result from kidney malformations, they should be treated with antibiotics and/or surgery as in the general population. However, the structural abnormalities of the kidneys that have been reported rarely require surgical intervention.
- The indications for and method of control of menses and prevention of pregnancy should be the same as in the general population, as should gynecologic care.
- Pap smears should be performed in females over the age of 21 years as recommended for routine screening purposes.
- Sex education should be provided for all adolescents and adults within the educational, clinical, and home environment.

Neurologic

Hypertonicity and hyperreflexia are common. Many caregivers report a high threshold for pain in affected individuals (Kline et al. 2001). Unusual hand posturing at eye level is very common. Seizures occur in 23% and tend to be easily managed medically (Kline et al. 2007b); they may be under-recognized. Sensory neuropathy has been reported, and there is a high threshold for pain (Kline et al. 2001). Sleep disturbance, particularly insomnia, and interrupted sleep pattern has been reported in up to 55% of individuals, with suspicion for circadian rhythm disorder (Berney et al. 1999).

Radiological brain findings may include: enlarged ventricles, including enlargement of basal cisterns; thinning or atrophy of white matter, particularly frontal lobes, with relative sparing of parietal lobes; brainstem and temporal lobe hypoplasia; and cerebellar vermal hypoplasia or agenesis (Kline et al. 2007b; Roshan Lal et al. 2016). In a comparison of MRI findings and behavior as reported on the aberrant behavior checklist, more abnormal structural changes were seen in individuals with milder behavioral involvement and higher lethargy scores, more consistent with autistic or depressive traits, and less abnormal brain changes in those with behavior exhibiting hyperactivity (Roshan Lal et al. 2016). Neuropathological examination has been reported in a few cases (Filippi et al. 1989; Vuilleumier et al. 2002). Commonly reported features include microcephaly with hypoplasia and poor myelination of the pyramidal tracts, as well as no anomalies. Several reports suggest a neuronal migration defect may be present with mild abnormalities of the gyri.

Evaluation

- Elucidation of the type and frequency of seizures is accomplished by clinical history and electroencephalography, as in the general population.
- Complaints of persistent pain or discomfort should be taken seriously in those with a high pain threshold.
- Brain imaging should be considered following an initial seizure.

Management

- Anticonvulsant medication should be used for seizures as in the general population.
- No studies have been undertaken to compare the use of different anticonvulsants.

Hematologic and Immunologic

Anemia is not uncommon in CdLS and may be an indicator of chronic gastroesophageal reflux. A recent study (Lambert et al. 2011) found an increased incidence of thrombocytopenia in 18%, and immune thrombocytopenia (ITP) in 8% of 85 individuals with clinically diagnosed CdLS and platelet assessments, indicating a potential underlying risk for bleeding.

Common infections include otitis media and sinusitis, with pneumonia occurring in the presence of gastroesophageal reflux and aspiration. In a study of 45 individuals with CdLS, Jyonouchi et al. (2013) reported that recurrent infections are common, and one third of the patients evaluated immunologically had antibody deficiency, many with decreased T-cells. Immune dysfunction was noted in the more severely involved patients.

Evaluation

- A full blood count is warranted if there is clinical evidence of anemia, bleeding or recurrent/intractable infections.
- Platelet count at diagnosis and if any history of unusual bleeding or bruising.
- Unexplained anemia should prompt full investigation of the gastrointestinal tract and bone marrow. In

patients with recurrent infections, consider obtaining total immunoglobulin levels (IgG, IgM, IgA), antibody titers to common vaccine antigens (tetanus, diphtheria, pneumococcus), and a complete blood count with differential. Consider referral to a clinical immunologist for further evaluation and management if counts are low.

Management

- Prompt treatment of gastroesophageal reflux is recommended to avoid or stop bleeding.
- Blood or platelet transfusion may be required in symptomatic pancytopenia or thrombocytopenia.
- Standard medical practice for idiopathic thrombocytopenia should be carried out.
- Standard medical practice for immunoglobulin deficiency should be carried out.

Dermatologic

Cutis marmorata and generalized hypertrichosis are very common features of CdLS (Jackson et al. 1993). Medical professionals should be aware that plucking or waxing of both synophrys and body hair is now commonly used by caregivers and affected individuals. Premature gray hair is frequently seen (Kline et al. 2007a). Cutis verticis gyrata has been reported in two older male patients.

Evaluation

- No routine evaluation is required.

Management

- Cosmetic therapy for hypertrichosis may be useful for some affected individuals.

ACKNOWLEDGMENTS

We are grateful to the families and individuals with CdLS whom we have encountered, and for their patience and participation. We are also grateful to the Cornelia de Lange Syndrome Foundation Executive Directors, staff and Board of Directors for their tireless and critical help. We would like to thank Dr. Laird Jackson for his expertise and encouragement related to CdLS. We are also grateful to Dr. Maggie Ireland, who wrote the first edition of this chapter, and Dr. David FitzPatrick, who helped collaborate on the second and third editions of this chapter. Dr. Deardorff's work was supported by National Institutes of Health grants K08HD055488 (NICHD) and the Doris Duke Charitable Foundation.

RESOURCES

Cornelia de Lange Syndrome Foundation
30 Tower Lane, #400
Avon, CT 06001, USA
Phone: 1 800 753 2357
Fax: 860 676 8337
Website: www.cdlsusa.org

Cornelia de Lange Syndrome Foundation of UK and Ireland
CdLS Foundation USA
PO Box 8368
Ripley, Derbyshire
DE5 4DA, UK
Phone: 01375 376439
Fax: 0207 536 8998
Email: info@cdls.org.uk
Website: www.cdls.org.uk

CdLS-World

CdLS-World is an international "hub" for worldwide organizations and communities united by Cornelia de Lange Syndrome. Country-specific contact information is available on the CdLS website: www.cdlsworld.org

Brochures

Facing the Challenges: A Guide to CdLS (2001, revised 2016): Booklet published by the Cornelia de Lange Syndrome Foundation USA and available online

CdLS UK CareCard: Summary of treatable complications written for health professionals, available from the Cornelia de Lange Syndrome Foundation, UK

Facts About Cornelia de Lange Syndrome: Fact sheet that can be obtained from the Cornelia de Lange Syndrome Foundation, UK or US

Cornelia de Lange Syndrome: A booklet produced by the Cornelia de Lange syndrome Foundation UK summarizing the main features of the syndrome

Cornelia de Lange Syndrome Foundation Family Album: Photographs and details of over 100 individuals with Cornelia de Lange syndrome whose parents are members of the Cornelia de Lange Syndrome Foundation USA

Online Resources

Cornelia de Lange Syndrome Foundation
Website: www.cdlsusa.org
Website: www.cdls.org.uk

National Organization for Rare Disorders (NORD)
Website: http://www.nord-rdb.com/~*orphan*

REFERENCES

Aitken DA, Ireland M, Berry E, Crossley JA, Macri JN, Burn J, Connor JM (1999) Second-trimester pregnancy associated plasma protein-A levels are reduced in Cornelia de Lange syndrome pregnancies. *Prenat Diagn* 19:706–710.

Ansari M, Poke G, Ferry Q, Williamson K, Aldridge R, Meynert AM, Bengani H, Chan CY, Kayserili H, Avci S, Hennekam RC, Lampe AK, Redeker E, Homfray T, Ross A, Falkenberg Smeland M, S Mansour Parker MJ, Cook JA, Splitt M, Fisher RB, Fryer A, Magee AC, Wilkie A, Barnicoat A, Brady AF, Cooper NS, Mercer C, Deshpande C, Bennett CP, Pilz DT, Ruddy D, Cilliers D, Johnson DS, Josifova D, Rosser E, Thompson EM, Wakeling E, Kinning E, Stewart F, Flinter F, Girisha KM, Cox H, Firth HV, Kingston H, Wee JS, Hurst JA, Clayton-Smith J, Tolmie J, Vogt J, Tatton-Brown K, Chandler K, Prescott K, Wilson L, Behnam M, McEntagart M, Davidson R, Lynch SA, Sisodiya S, Mehta SG, McKee SA, Mohammed S, Holden S, Park SM, Holder SE, Harrison V, McConnell V, Lam WK, Green AJ, Donnai D, Bitner-Glindzicz M, Donnelly DE, Nellaker C, Taylor MS and FitzPatrick DR (2014) Genetic heterogeneity in Cornelia de Lange syndrome (CdLS) and CdLS-like phenotypes with observed and predicted levels of mosaicism. *J Med Genet* 51:659-668.

Arron K, Oliver C, Hall S, Sloneem J, Forman D, McClintock K (2006) Effects of social context on social interaction and self-injurious behavior in Cornelia de Lange syndrome. *Am J Ment Retard* 111:184–192.

Barisic I, Tokic V, Loane M, Bianchi F, Calzolari E, Garne E, Wellesley D, Dolk H (2008) Descriptive epidemiology of Cornelia de Lange syndrome in Europe. *Am J Med Genet A* 146:51–59.

Basile E, Villa L, Selicorni A, Molteni M (2007) The behavioural phenotype of Cornelia de Lange Syndrome: A study of 56 individuals. *J Intellect Disabil Res*, 51:671–681.

Beck B, Fenger K (1985) Mortality, pathological findings and causes of death in the de Lange syndrome. *Acta Paediatr Scand* 74:765–769.

Berney TP, Ireland M, Burn J (1999) Behavioural phenotype of Cornelia de Lange syndrome. *Arch Dis Child* 81:333–336.

Bhuiyan ZA, Klein M, Hammond P, van Haeringen A, Mannens MM, Van Berckelaer-Onnes I, Hennekam RC (2006) Genotype-phenotype correlations of 39 patients with Cornelia de Lange syndrome: the Dutch experience. *J Med Genet* 40:568–75

Borck G, Zarhrate M, Bonnefont JP, Munnich A, Cormier-Daire V, Colleaux L (2007) Incidence and clinical features of X-linked Cornelia de Lange syndrome due to SMC1L1 mutations. *Hum Mutat* 28:205–206.

Brachmann W (1916) Ein Fall von symmetrischer Monodaktylie durch Ulnadefekt, mit symmetrischer Flughautbildung in den Ellenbeugen, sowie anderen Abnormitaten (Zwerghaftigkeit, Halsrippen, Behaarung). *Jahr Kinderheilkunde* 84:225–235.

Clark DM, Sherer I, Deardorff MA, Byrne JL, Loomes KM, Nowaczyk MJ, Jackson LG, Krantz ID. Identification of a prenatal profile of Cornelia de Lange syndrome (CdLS): a review of 53 CdLS pregnancies (2012) *Am J Med Genet A* 158A:1848–56.

Deardorff MA, Kaur M, Yaeger D, Rampuria A, Korolev S, Pie J, Gil-Rodriguez C, Arnedo M, Loeys B, Kline AD, Wilson M, Lillquist K, Siu V, Ramos FJ, Musio A, Jackson LS, Dorsett D, Krantz ID (2007) Mutations in cohesin complex members SMC3 and SMC1A cause a mild variant of cornelia de Lange syndrome with predominant mental retardation. *Am J Hum Genet* 80:485–494.

Deardorff MA, Wilde JJ, Albrecht M, Dickinson E, Tennstedt S, Braunholz D, Mönnich M, Yan Y, Xu W, Gil-Rodríguez MC, Clark D, Hakonarson H, Halbach S, Michelis LD, Rampuria A, Rossier E, Spranger S, Van Maldergem L, Lynch SA, Gillessen-Kaesbach G, Lüdecke HJ, Ramsay RG, McKay MJ, Krantz ID, Xu H, Horsfield JA, Kaiser FJ (2012a) RAD21 mutations cause a human cohesinopathy. *Am J Hum Genet* 90:1014–27.

Deardorff MA, Bando M, Nakato R, Watrin E, Itoh T, Minamino M, Saitoh K, Komata M, Katou Y, Clark D, Cole KE, De Baere E, Decroos C, Di Donato N, Ernst S, Francey LJ, Gyftodimou Y, Hirashima K, Hullings M, Ishikawa Y, Jaulin C, Kaur M, Kiyono T, Lombardi PM, Magnaghi-Jaulin L, Mortier GR, Nozaki N, Petersen MB, Seimiya H, Siu VM, Suzuki Y, Takagaki K, Wilde JJ, Willems PJ, Prigent C, Gillessen-Kaesbach G, Christianson DW, Kaiser FJ, Jackson LG, Hirota T, Krantz ID, Shirahige K (2012b) HDAC8 mutations in Cornelia de Lange syndrome affect the cohesin acetylation cycle. *Nature* 489:313–7.

Deardorff MA, Porter NJ, Christianson DW (2016) Structural aspects of HDAC8 mechanism and dysfunction in Cornelia de Lange syndrome spectrum disorders. *Protein Sci*, 25, 1965–1976.

de Lange C (1933) Sur Un Type Nouveau de Degeneration (Typus Amstelodamensis). *Arch Med Enfants* 36:713–719.

DuVall GA, Walden DT (1996) Adenocarcinoma of the esophagus complicating Cornelia de Lange syndrome. *J Clin Gastroenterol* 22:131–133.

Filippi G (1989) The de Lange syndrome. Report of 15 cases. *Clin Genet* 35:343–363.

Gillis LA, McCallum J, Kaur M, DeScipio C, Yaeger D, Mariani A, Kline AD, Li HH, Devoto M, Jackson LG, Krantz ID (2004) NIPBL mutational analysis in 120 individuals with Cornelia de Lange syndrome and evaluation of genotype-phenotype correlations. *Am J Hum Genet* 75:610–623.

Gil-Rodriguez MC, Deardorff MA, Ansari T, Tan CA, Parenti I, Baquero-Montoya C, Ousager LB, Puisac B, Hernández-Marcos M, Teresa-Rodrigo ME, Marcos-Alcalde I, Wesselink JJ, Lusa-Bernal S, Bijlsma EK, Braunholz D, Bueno-Martinez I, Clark D, Cooper NS, Curry CJ, Fisher R, Fryer A, Ganesh J, Gervasini C, Gillessen-Kaesbach G, Guo Y, Hakonarson H, Hopkin RJ, Kaur M, Keating BJ, Kibaek M, Kinning E, Kleefstra T, Kline AD, Kuchinskaya E, Larizza L, Li YR, Liu X, Mariani M, Picker JD, Pié Á, Pozojevic J, Queralt E, Richer J, Roeder E, Sinha A, Scott RH, So J, Wusik KA, Wilson L, Zhang J, Gómez-Puertas P, Casale CH, Ström L, Selicorni A, Ramos FJ, Jackson LG, Krantz ID, Das S, Hennekam RC, Kaiser FJ,

FitzPatrick DR, Pié J (2015) De novo heterozygous mutations in SMC3 cause a range of Cornelia de Lange syndrome-overlapping phenotypes. *Hum Mutat* 36:454–62.

Goodban MT (1993) Survey of speech and language skills with prognostic indicators in 116 patients with Cornelia de Lange syndrome. *Am J Med Genet* 47:1059–1063.

Halal F, Preus M (1979) The hand profile on de Lange syndrome: Diagnostic criteria. *Am J Med Genet* 3:317–323.

Huisman SA, Redeker EJ, Maas SM, Mannens MM, Hennekam RC (2013) High rate of mosaicism in individuals with Cornelia de Lange syndrome. *J Med Genet* 50:339–344.

Hyman P, Oliver C, Hall S (2002) Self-injurious behavior, self-restraint, and compulsive behaviors in Cornelia de Lange syndrome. *Am J Ment Retard* 107:146–154.

Ireland M (1996) Cornelia de Lange syndrome: Clinical features, common complications and long-term prognosis. *Curr Pediatr* 6:69–73.

Ireland M, Donnai D, and Burn J (1993) Brachmann-de Lange syndrome. Delineation of the clinical phenotype. *Am J Med Genet* 47:959–964.

Izumi K, Nakato R, Zhang Z, Edmondson AC, Noon S, Dulik MC, Rajagopalan R, Venditti CP, Gripp K, Samanich J, Zackai EH, Deardorff MA, Clark D, Allen JL, Dorsett D, Misulovin Z, Komata M, Bando M, Kaur M, Katou Y, Shirahige K, Krantz ID (2015) Germline gain-of-function mutations in AFF4 cause a developmental syndrome functionally linking the super elongation complex and cohesin. *Nat Genet* 47:338–44.

Jackson L, Kline AD, Barr MA, and Koch S (1993) de Lange syndrome: a clinical review of 310 individuals. *Am J Med Genet* 47:940–946.

Janek KC, Smith DF, Kline AD, Benke JR, Chen M-L, Kimball A, Ishman SL (2016) Improvement in hearing loss over time in Cornelia de Lange syndrome. *Int J Pediatr Otorhinolaryngol* 87:203–7

Jyonouchi S, Orange J, Sullivan KE, Krantz I, Deardorff M (2013) Immunologic features of Cornelia de Lange syndrome. *Pediatrics* 132:e484–9.

Kawauchi S, Santos R, Muto A, Lopez-Burks ME, *Schilling TF*, Lander AD, Calof AL (2016)

Using mouse and zebrafish models to understand the etiology of developmental defects in Cornelia de Lange syndrome. *Am J Med Genet C* 172:138–45.

Kliewer MA, Kahler SG, Hertzberg BS, Bowie JD (1993) Fetal biometry in the Brachmann-de Lange syndrome. *Am J Med Genet* 47:1035–1041.

Kline AD, Audette L (2002) Puberty and adolescence in CdLS: A survey of 67 patients. *Am J Hum Genet* 71:672.

Kline AD, Barr M, Jackson LG (1993a) Growth manifestations in the Brachmann-de Lange syndrome. *Am J Med Genet* 47:1042–1049.

Kline AD, Stanley C, Belevich J, Brodsky K, Barr M, Jackson LG (1993b) Developmental data on individuals with the Brachmann-de Lange syndrome. *Am J Med Genet* 47:1053–1058.

Kline AD, Krantz ID, Goldstein A, Koo B, Jackson LG (2001) Cornelia de Lange syndrome: Evidence for a sensorineuropathy. *Am J Hum Genet* 69:280.

Kline AD, Grados M, Sponseller P, Levy HP, Blagowidow N, Schoedel C, Rampolla J, Clemens K, Krantz I, Kimball A, Pichard C, Tuchman D (2007a) Natural history of aging in Cornelia de Lange syndrome. *Am J Med Genet C* 145:248–260.

Kline AD, Krantz ID, Sommer A, Kliewer M, Jackson LG, FitzPatrick DR, Levin AV, Selicorni A (2007b) Cornelia de Lange syndrome: Clinical review, diagnostic and scoring systems, and anticipatory guidance. *Am J Med Genet A* 143:1287–1296.

Kline AD, Moss JF, Selicorni A, Bisgaard A-M, Deardorff MA, Gillett PM, Ishman SL, Kerr LM, Levin AV, Mulder PA, Ramos FJ, Wierzba J, Ajmone PF, Axtell D, Blagowidow N, Cereda A, Costantino A, Cormier-Daire V, FitzPatrick D, Grados M, Groves L, Guthrie W, Huisman S, Kaiser FJ, Koekkoek G, Levis M, Mariani M, McCleery JP, Menke LA, Metrena A, O'Connor J, Oliver C, Pie J, Piening S, Potter C, Quaglio A, Redeker E, Richman D, Rigamonti C, Shi A, Tumer Z, Van Balkom IDC, Hennekam R (2018) Diagnosis and Management in Cornelia de Lange Syndrome: First International Consensus Statement. *Nature Rev Genetics* 19:649–666.

Kousseff BG, Thomson-Meares J, Newkirk P, Root AW (1993) Physical growth in Brachmann-de Lange syndrome. *Am J Med Genet* 47:1050–1052.

Krantz ID, McCallum J, DeScipio C, Kaur M, Gillis LA, Yaeger D, Jukofsky L, Wasserman N, Bottani A, Morris CA, Nowaczyk MJ, Toriello H, Bamshad MJ, Carey JC, Rappaport E, Kawauchi S, Lander AD, Calof AL, Li HH, Devoto M, Jackson LG (2004) Cornelia de Lange syndrome is caused by mutations in NIPBL, the human homolog of Drosophila melanogaster Nipped-B. *Nat Genet* 36:631–635.

Levin AV, Seidman DJ, Nelson LB, Jackson LG (1990) Ophthalmologic findings in the Cornelia de Lange syndrome. *J Pediatr Ophthalmol Strabismus* 27:94–102.

Luzzani S, Macchini F, Valade A, Milani D, Selicorni A (2003) Gastroesophageal reflux and Cornelia de Lange syndrome: Typical and atypical symptoms. *Am J Med Genet A* 119:283–287.

Mannini L, Cucco F, Quarantotti V, Krantz ID, Musio A (2014) Mutation spectrum and genotype-phenotype correlation in Cornelia de Lange syndrome. *Hum Mutat* 34:1589–96.

Masumoto K, Izaki T, Arima T (2001) Cornelia de Lange syndrome associated with cecal volvulus: Report of a case. *Acta Paediatr* 90:701–703.

Misulovin Z, Schwartz YB, Li XY, Kahn TG, Gause M, MacArthur S, Fay JC, Eisen MB, Pirrotta V, Biggin MD, Dorsett D (2008) Association of cohesin and Nipped-B with transcriptionally active regions of the Drosophila melanogaster genome. *Chromosoma* 117:89–102.

Moretto A, Scaravilli V, Ciceri V, Bosatra M, Giannatelli F, Ateniese B, Mariani M, Cereda A, Sosio S, Zanella A, Pesenti A, Selicorni A (2016) Sedation and general anesthesia for patients with Cornelia De Lange syndrome: A case series. *Am J Med Genet C* 172:222–8.

Moss J, Howlin P, Magiati I, Oliver C (2012) Characteristics of autistic spectrum disorder in Cornelia de Lange syndrome. *J Child Psychol Psychiatr* 53:883–91.

Moss J, Nelson L, Powis L, Waite J, Richards C, Oliver C (2016) A Comparative Study of Sociability in Angelman, Cornelia de Lange, Fragile X, Down and Rubinstein Taybi Syndromes and

Autism Spectrum Disorders. *Am J Intellect Dev Disabil* 121:465–486.

Moss J, Penhallow J, Ansari M, Barton S, Bourne D, FitzPatrick D, Goodship J, Hammond P, Roberts C, Welham A, Oliver C (2017) Genotype–phenotype correlations in Cornelia de Lange syndrome: behavioral characteristics and changes with age. *Am J Med Genet A* 173:1566–74.

Musio A, Selicorni A, Focarelli ML, Gervasini C, Milani D, Russo S, Vezzoni P, Larizza L (2006) X-linked Cornelia de Lange syndrome owing to SMC1L1 mutations. *Nat Genet* 38:528– 530.

Nallasamy S, Kherani F, Yaeger D, McCallum J, Kaur M, Devoto M, Jackson LG, Krantz ID, Young TL (2006) Ophthalmologic findings in Cornelia de Lange syndrome: A genotype-phenotype correlation study. *Arch Ophthalmol* 124:552–557.

Olley G, Ansari M, Bengani H, Grimes GR, Rhodes J, von Kriegsheim A, Blatnik A, Stewart FJ, Wakeling E, Carroll N, Ross A, Park SM, Bickmore WA, Pradeepa MM, FitzPatrick DR, Deciphering Developmental Disorders Study (2018) *Nat Genet* 50:329–332.

Oostra RJ, Baljet B, Hennekam RC (1994) Brachmann-de Lange syndrome "avant la lettre". *Am J Med Genet* 52:267–268.

Opitz JM (1985) The Brachmann-de Lange syndrome. *Am J Med Genet*, 22, 89–102.

Parenti I, Gervasini C, Pozojevic J, Graul-Neumann L, Azzollini J, Braunholz D, Watrin E, Wendt KS, Cereda A, Cittaro D, Gillessen-Kaesbach G, Lazarevic D, Mariani M, Russo S, Werner R, Krawitz P, Larizza L, Selicorni A, Kaiser FJ (2016) Broadening of cohesinopathies: exome sequencing identifies mutations in ANKRD11 in two patients with Cornelia de Lange-overlapping phenotype. *Clin Genet* 89:74–81.

Preus M, Rex AP (1983) Definition and diagnosis of the Brachmann-De Lange syndrome. *Am J Med Genet* 16:301–312.

Roposch A, Bhaskar AR, Lee F, Adedapo S, Mousny M, Alman BA (2004) Orthopaedic manifestations of Brachmann-de Lange syndrome: a report of 34 patients. *J Pediatr Orthop B* 13:118–22.

Roshan Lal TR, Roshan Lal TR, Kliewer MA, Lopes T, Rebsamen SL, O'Connor J, Grados MA, Kimball A, Clemens J, Kline AD (2016) Cornelia de Lange syndrome: correlation of brain MRI findings with behavioral assessment. *Am J Med Genet C* 172:190–7.

Russell B, Johnston JJ, Biesecker LG, Kramer N, Pickart A, Rhead W, Tan WH, Brownstein CA, Kate Clarkson L, Dobson A, Rosenberg AZ, Vergano SA, Helm BM, Harrison RE, Graham JM Jr (2015) Clinical management of patients with ASXL1 mutations and Bohring-Opitz syndrome, emphasizing the need for Wilms tumor surveillance. *Am J Med Genet A* 167A:2122–31.

Russell KL, Ming JE, Patel K, Jukofsky L, Magnusson M, Krantz ID (2001) Dominant paternal transmission of Cornelia de Lange syndrome: A new case and review of 25 previously reported familial recurrences. *Am J Med Genet* 104:267–276.

Sarimski K (2007) Infant attentional behaviours as prognostic indicators in Cornelia-de-Lange syndrome. *J Intellect Disabil Res* 51:697–701.

Sataloff RT, Spiegel JR, Hawkshaw M, Epstein JM, Jackson L (1990) Cornelia de Lange syndrome. Otolaryngologic manifestations. *Arch Otolaryngol Head Neck Surg* 116:1044–1046.

Schrier SA, Sherer I, Deardorff MA, Clark D, Audette L, Gillis L, Kline AD, Ernst L, Loomes K, Krantz ID, Jackson LG (2011) Causes of Death and Autopsy Findings in a Large Study Cohort of Individuals with Cornelia de Lange Syndrome and Review of the Literature. *Am J Med Genet A* 155:3007–3024.

Selicorni A, Sforzini C, Milani D, Cagnoli G, Fossali E, Bianchetti MG (2005) Anomalies of the kidney and urinary tract are common in de Lange syndrome. *Am J Med Genet A* 132:395–397.

Selicorni A, Russo S, Gervasini C, Castronovo P, Milani D, Cavalleri F, Bentivegna A, Masciadri M, Domi A, Divizia MT, Sforzini C, Tarantino E, Memo L, Scarano G, Larizza L (2007) Clinical score of 62 Italian patients with Cornelia de Lange syndrome and correlations with the presence and type of NIPBL mutation. *Clin Genet* 72: 98–108.

Slavin T, Labeznik N, Clark DM, Vengoechea J, Cohen L, Kaur M, Konczal L, Crowe CA, Corteville JE, Nowaczyk MJ, Byrne JL, Jackson LG, Krantz ID (2012) Clinical Report: Germline Mosaicism in Cornelia de Lange Syndrome. *Am J Med Genet A* 158A:1481–5.

Srivastava S, Landy-Schmitt C, Clark B, Kline AD, Specht M, Grados MJ (2014) Autism traits in children and adolescents with Cornelia de Lange syndrome. *Am J Med Genet A* 164A:1400–10.

Strachan T (2005) Cornelia de Lange syndrome and the link between chromosomal function, DNA repair and developmental gene regulation. *Curr Opin Genet Dev* 15:258–264.

Tonkin ET, Wang TJ, Lisgo S, Bamshad MJ, Strachan T (2004) NIPBL, encoding a homolog of fungal Scc2-type sister chromatid cohesion proteins and fly Nipped-B, is mutated in Cornelia de Lange syndrome. *Nat Genet* 36:636–641.

Urban M, Hartung J (2001) Ultrasonographic and clinical appearance of a 22-week-old fetus with Brachmann-de Lange syndrome. *Am J Med Genet* 102:73–75.

Vuilleumier N, Kovari E, Michon A, Hof PR, Mentenopoulos G, Giannakopoulos P, Bouras C (2002) Neuropathological analysis of an adult case of the Cornelia de Lange syndrome. *Acta Neuropathol Vuilleumier* 104:327–332.

Westergaard JG, Chemnitz J, Teisner B, Poulsen HK, Ipsen L, Beck B, Grudzinskas JG (1983) Pregnancy-associated plasma protein A: A possible marker in the classification and prenatal diagnosis of Cornelia de Lange syndrome. *Prenat Diagn* 3:225–232.

Wygnanski-Jaffe T, Shin J, Perruzza E, Abdolell M, Jackson LG, Levin AV (2005) Ophthalmologic findings in the Cornelia de Lange Syndrome. *J AAPOS* 9:407–415.

Yuan B, Pehlivan D, Karaca E, Patel N, Gambin T, Gonzaga-Jauregui C, Sutton VR, Yesil G, Bozdogan ST, Tos T, Koparir A, Koparir E, Beck CR, Gu S, Aslan H, Yuregir OO, Al Rubeaan K, Alneqeb D, Alshammari MJ, Bayram Y, Atik MM, Aydin H, Geckinli BB, Seven M, Ulucan H, Fenercioglu E, Ozen M, Jhangiani S, Muzny DM, Boerwinkle E, Tuysuz B, Alkuraya FS, Gibbs RA, Lupski JR (2015) Global transcriptional disturbances underlie Cornelia de Lange syndrome and related phenotypes. *J Clin Invest* 125:636–651.

16

COSTELLO SYNDROME

BRONWYN KERR
Genomic Medicine, Manchester University Hospitals NHS Foundation Trust, Manchester, UK

KAREN W. GRIPP
Division of Medical Genetics, Department of Pediatrics, S. Kimmel Medical College, Thomas Jefferson University, Philadelphia, Pennsylvania, and A.I. duPont Hospital for Children, Wilmington, Delaware, USA

EMMA M.M. BURKITT WRIGHT
Genomic Medicine, Manchester University Hospitals NHS Foundation Trust, Manchester, UK

INTRODUCTION

Costello syndrome (CS) is a rare multiple anomaly syndrome first described by Jack Costello, a New Zealand pediatrician (Costello 1971, 1977, 1996). The distinctive phenotype varies with age and may have features reminiscent of an overgrowth syndrome, connective tissue disorder, storage disorder, or premature aging disease. The strongest clinical overlap is with other disorders of the Ras/MAPK pathway, particularly cardio-facio-cutaneous (CFC) syndrome (see Chapter 10). The reader is directed to several reviews by Gripp and Lin (2006), Rauen (2006), Quezada and Gripp (2007), and Kerr (2009).

Incidence

Although a formal incidence has not been calculated, based on the experience of a diagnostic laboratory, the birth incidence of diagnosed cases in the UK over a ten year period was 1/374,000 (Gionnalatou 2014). Since a diagnostic test became available (see below), confirmed molecular diagnoses have been reported in the literature in over 150 individuals (Sol-Church and Gripp 2009). There is no gender preference. Confirmed cases have been reported from most continents and many different ethnic groups (Costello 1971, 1977; Sigaudy et al. 2000; Gregersen and Viljoen 2004; Aoki et al. 2005; White et al. 2005).

Formal lifetime table analyses have not been done, but survival to the fifth decade has been documented (White et al. 2005). Early mortality may result from severe hypertrophic cardiomyopathy (Digilio et al. 2007; Lo et al. 2008), respiratory disease or medical complications due to profound muscle weakness, and hypotonia (Burkitt Wright et al. 2012; van der Burgt et al. 2007).

Diagnostic Criteria

The original report of Costello syndrome described a condition with a distinctive face, moderate intellectual disability, poor postnatal growth with relative macrocephaly, loose skin of the hands and feet, and nasal papillomas. These remain the core criteria. Additional features included polyhydramnios, increased birth weight, severe feeding difficulty, and cardiovascular abnormalities. Musculoskeletal abnormalities include the distinctive, extreme hyperextensibility of the small joints of the hand and flexion and ulnar deviation at the wrist. Although these hand findings are characteristic, gene testing has confirmed that similar hand

---

Cassidy and Allanson's Management of Genetic Syndromes, Fourth Edition.
Edited by John C. Carey, Agatino Battaglia, David Viskochil, and Suzanne B. Cassidy.
© 2021 John Wiley & Sons, Inc. Published 2021 by John Wiley & Sons, Inc.

abnormalities can be seen in CFC syndrome (see Chapter 10) (Gripp et al. 2007).

The clinical manifestations of CS, with the most distinctive features highlighted, are shown in Figure 16.1. Facial features, and their evolution over time, are illustrated in Figures 16.2 and 16.4. Figure 16.3 demonstrates the characteristic hand features and postures. Mutation analyses (see below) have confirmed that the phenotype and natural history in most cases are homogenous and distinctive (Van Steensel et al. 2006; Kerr et al. 2006; Zampino et al. 2007). However, both milder and more severe, apparently rarer, variants have been delineated since the advent of routine gene testing (Van der Burgt et al. 2007; Gripp et al. 2008; Lo et al. 2008, Lorenz et al. 2012, Weaver et al. 2014, Gripp et al. 2015).

The distinctive facial appearance may not be easily recognized in the newborn (Digilio et al. 2007; Lo et al. 2008), or in patients with less typical phenotypes. In infancy and beyond, there is a tall forehead with temporal narrowing. The lower jaw is square, often with a small pointed chin, accentuated by full cheeks. Hair quality is variable, ranging from sparse to normal to abundantly curly. Telecanthus and epicanthal folds are present. Fleshy ears may be cupped or rotated posteriorly, with earlobes that often appear upturned and prominent. Lips are full and frequently exhibit vertical creases, and the mouth is often wide. In adolescence, the round face elongates and the chin lengthens, often appearing pointed. The mouth remains wide and the lips can be thick. Although the facial appearance in infancy and adulthood is striking, the face in mid-childhood and adolescence may be less characteristic (Figures 16.2 and 16.4). The use of 3D photographic analysis has shown an overlap between the facial features of CS with CFC syndrome (Rauen et al. 2008). Digital facial analysis through the Face2Gene application (www.face2gene.com) may identify the facial gestalt of CS, as well as the overlap with other related conditions such as Noonan syndrome (see Chapter 41) and CFC syndrome (see Chapter 10).

A cardiac abnormality can be detected in 80% of the individuals with Costello syndrome (Lin et al. 2002; Gripp et al. 2007; Lin et al. 2011). Anomalies include structural cardiovascular malformations (congenital heart defects), cardiac hypertrophy, and rhythm disturbances, especially atrial tachycardia. Although no single defect is pathognomonic for CS, the combination of neonatal cardiac hypertrophy, chaotic atrial tachycardia, and mild pulmonic valve stenosis strongly suggests this diagnosis.

Skin abnormalities are common and often distinctive. They include soft skin, excess palmar skin resulting in deep palmar creases, hyperkeratosis of palms and soles, premature aging, wrinkling, hyperpigmentation, acanthosis nigricans, and alopecia. Papillomas usually appear during childhood in the perinasal region, less commonly in the perianal region, torso, and extremities. Joint abnormalities are also common, with the characteristic hand and wrist posture the most diagnostically useful.

Etiology, Pathogenesis, and Genetics

The discovery of heterozygous mutations in *HRAS* as causative of CS (Aoki 2005) confirmed the findings of segregation analysis, which had supported autosomal dominant inheritance (Lurie, 1994). Gonadal mosaicism has been reported, (Zampino et al. 1993; Gripp et al. 2011). Transmission from a mosaic father to an affected son with CS has also been documented (Sol-Church et al. 2009). Other cases of somatic mosaicism have been reported (Gripp et al. 2006a; Gripp et al. 2006b; Sol-Church et al. 2009; Bertola et al. 2017).

Analyses of cohorts of individuals with a clinical diagnosis of CS (Aoki et al. 2005; Estep et al. 2006; Gripp et al. 2006a; Gripp et al. 2006b; Kerr et al. 2006) reported de novo *HRAS* mutations in 80–90%, with the remainder undiagnosed cases often affected with a closely related rasopathy such as CFC syndrome or Noonan syndrome. Absence of a pathogenic *HRAS* mutation on molecular testing is not consistent with a diagnosis of CS; however, as outlined above, rare instances of somatic mosaicism have been reported. Almost all of the mutations arise in the paternal germ line (Sol-Church et al. 2006; Gionnalatou et al. 2014), consistent with point mutations.

The *Ras* genes are proto-oncogenes that are often somatically mutated in various human cancers. Isoforms encoded by *KRAS*, *HRAS*, and *NRAS* act as signal transduction molecules, cycling between GTP-bound active and GDP-bound inactive states. Somatic mutations in *HRAS* have been identified in bladder carcinoma, thyroid cancer, and melanoma, and the constitutional mutations identified by Aoki et al. (2005) affect two amino acids previously found to be mutated in these human tumors. Monoallelic *HRAS* expression, typically due to loss of the wildtype allele, occurs in rhabdomyosarcomas from affected individuals (Estep et al. 2006; Robbins et al. 2016).

Substitutions of glycine 12 or 13 account for 95% of CS-associated mutations, with the most common change, c.34G>A, p.(Gly12Ser), accounting for over 80%. The p.(Gly12Ala) change is the second commonest mutation. G12S and G12A are usually associated with the classical phenotype. It remains unclear if the risk of malignancy is higher with a G12A mutation. A number of other mutations have been observed, involving these or other amino acid residues, each in only a few affected individuals, which, in some cases, are associated with unusual clinical presentations (Kerr et al. 2006; van der Burgt et al. 2006; Gripp et al. 2008; Lo et al. 2008). Table 16.1 summarizes the reported mutations. For more detailed discussion, the reader is referred to Sol-Church and Gripp (2009).

Diagnostic Testing

Although some of the distinctive features of Costello syndrome had been reported in a small number of individuals who were subsequently found to have mutations in *BRAF*,

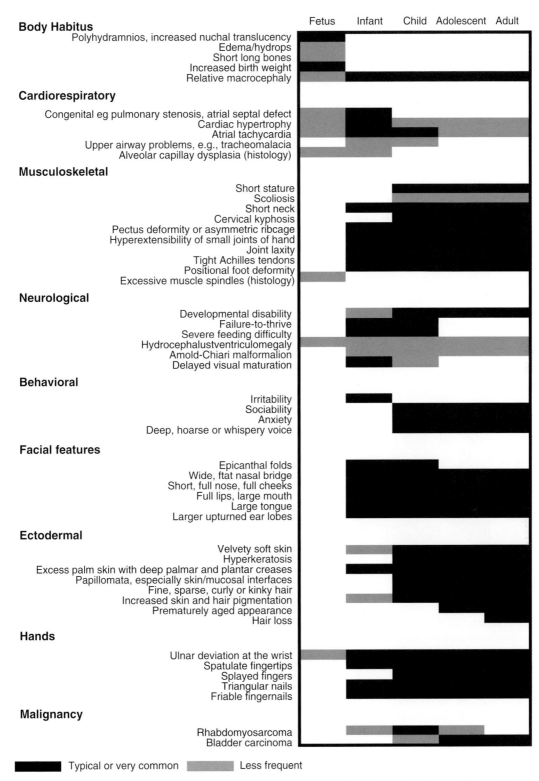

FIGURE 16.1 Characteristic manifestations of Costello syndrome. Manifestations within each organ system are shown, with the age at which they may become apparent. Dark gray denotes very common features, lighter gray the less commonly observed phenotypes.

KRAS, *MEK1*, or *MEK2* (Rauen 2006; Bertola et al. 2007; Gripp et al. 2007; Narumi et al. 2007; Nava et al. 2007), more rigorous subsequent phenotype analysis has confirmed that the clinical diagnosis of CS before mutation analysis was not secure in many of those individuals (Kerr et al. 2008). The significantly increased risk of malignancy is only

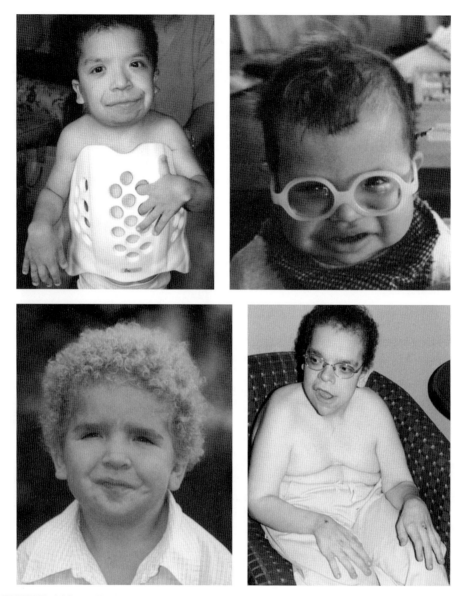

FIGURE 16.2 Individuals with Costello syndrome and typical facial appearance and other characteristic features. Top: Five-year-old boy wearing a "scoliosis jacket" (left) and 5-year-old boy wearing corrective lenses (right). Bottom: Six-year-old girl with fine curly hair (left) and 28-year-old man with striking finger laxity even in adulthood (right). (All photographs were graciously provided with the consent of the parents.)

seen in association with mutations in *HRAS*. The diagnosis of CS should be reserved for those with an *HRAS* mutation (Nava et al. 2007; Kerr et al. 2008; Schulz et al. 2008) because this uniquely predicts the increased malignancy risk, and the core phenotype associated with the most frequently observed mutations is relatively homogenous (Van Steesel et al. 2006; Zampino et al. 2007; Kerr et al. 2009).

Polyhydramnios occurs in most CS pregnancies. When detected by ultrasonography, this can lead to prenatal diagnosis, especially when accompanied by hydrops, nuchal thickening, short long bones, abnormal hand posture, ventriculomegaly, large size, and macrocephaly (Lin et al. 2009). The additional occurrence of fetal atrial tachycardia is especially significant, and may herald postnatal serious tachycardia. Prenatal diagnosis of CS with confirmative *HRAS* gene mutation testing has been reported (Kuniba et al. 2009; Hague et al. 2017).

Differential Diagnosis

The differential diagnosis of Costello syndrome is greatly influenced by the age of examination. In the newborn period, Beckwith-Wiedemann syndrome (see Chapter 9) and Simpson-Golabi-Behmel syndrome may be considered because of apparent overgrowth. However, the malformations

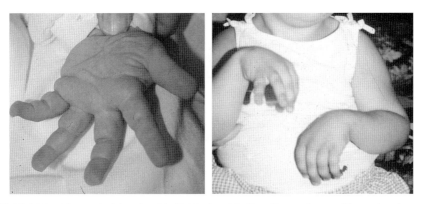

FIGURE 16.3 The typical hands of individuals with Costello syndrome, illustrating the excess palmar skin, and characteristic postures of hyperextension of the small joints of the hand, flexion, and ulnar deviation at the wrist.

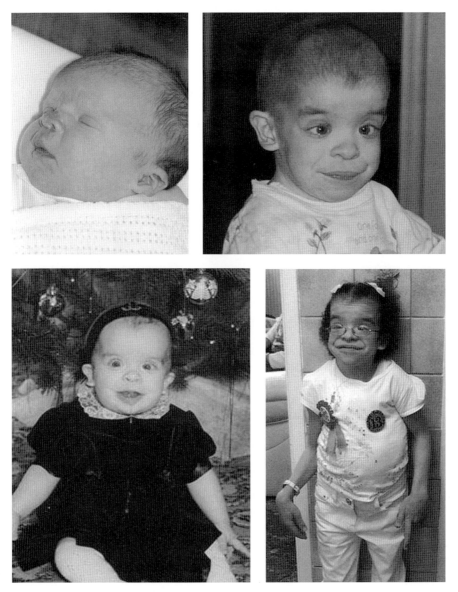

FIGURE 16.4 The changing appearance over time in a girl with Costello syndrome as a newborn, at age 18 months, 5 years, and 13 years.

TABLE 16.1 Germline *HRAS* mutations identified in Costello syndrome

Nucleotide change	Predicted protein effect	Associated phenotypes	Reference
c.34G>A	p.(Gly12Ser)	Classical CS; severe presentations reported	Aoki et al. 2005; Kerr et al. 2006
c.34G>T	p.(Gly12Cys)	Severe CS	Kerr et al. 2006; Lo et al. 2008
c.35G>A	p.(Gly12Asp)	Severe CS	Lo et al. 2008; Kuniba et al, 2009
c.35G>C	p.(Gly12Ala)	Classical CS	Aoki et al, 2005; Kerr et al. 2006
c.35G>T, c.35_36delinsTA, c.35_36delinsTT	p.(Gly12Val)	Severe CS	Van der Burgt et al. 2007; Burkitt Wright et al. 2012
c.35_36delinsAA	p.(Gly12Glu)	Severe CS	Kerr et al. 2006; Weaver et al. 2014
c.37G>A	p.(Gly13Cys)	Distinctive, milder CS phenotype	Sol-Church et al. 2006; Piccione et al. 2009; Gripp et al. 2011
c.38G>A	p.(Gly13Asp)	Rare	Aoki et al. 2005
c.64C>A	p.(Gln22Lys)	Rare, severe reported	van der Burgt et al. 2007
c.108_111dupAGA, c.110_111+1dupAGG	p.(Glu37dup)	Rare	Gremer et al. 2010
c.173C>T	p.(Thr58Ile)	Milder	Gripp et al. 2008
c.179G>A	p.(Gly60Asp)	Milder	Gripp et al. 2015
c.187G>A	p.(Glu63Lys)	Severe CS	van der Burgt et al. 2007
c.187_207dup	p.(Glu63_Asp69dup)	Rare	Lorenz et al. 2013
c.350A>G	p.(Lys117Arg)	Rare, severe reported	Kerr et al. 2006; Denayer et al. 2008
c.436G>A	p.(Ala146Thr)	Milder	Zampino et al. 2007
c.437C>T	p.(Ala146Val)	Milder	Gripp et al. 2008

seen in these syndromes and the absence of subsequent severe feeding difficulty and growth failure usually allow them to be differentiated from CS. A metabolic disorder may also be considered in individuals with hypoglycemia and multisystem disease (Lo et al. 2008).

The presentation of severe hypotonia may lead to consideration of a congenital myopathy (van der Burgt et al. 2007), or an inherited disorder of connective tissue, such as cutis laxa. These disorders can be readily excluded either by testing or by consideration of the natural history.

Clinical differentiation between Costello syndrome, CFC syndrome (Chapter 10) syndrome, and Noonan syndrome (Chapter 41) in the first years of life can be difficult, as facial features and facial and peripheral edema may be very similar. Feeding problems and failure-to-thrive are usually more severe in CS and CFC syndrome than in Noonan syndrome. When ulnar deviation at the wrist, marked small joint hyperextensibility in the hands, and striking excess palmar skin are present, the combination will usually differentiate CS from other conditions. A minor anomaly of the hand (Allanson et al. 2007; Ørstavik et al. 2007) is distal phalangeal creases, particularly obvious on the palmar surface of the thumb. Although first described in CS, it is now described in CFC syndrome and is a marker for all the disorders of the Ras/MAPK pathway.

A formal comparison of the clinical manifestations of Costello syndrome and CFC syndrome in a relatively small number of individuals achieved statistical significance for polyhydramnios, presence of more than one papilloma, and growth hormone deficiency, all being significantly more common in CS (Gripp et al. 2007).

Cardiovascular malformations (60%), and specifically pulmonic stenosis, which is seen in 70% of individuals with any cardiovascular anomaly, are more common in Noonan syndrome than in Costello syndrome (Gripp et al. 2007; Schulz et al. 2008). Chaotic atrial tachycardia is most often seen in young infants with Costello syndrome and improves over time (Lin et al. 2009). It is much less common in the related rasopathies (Lin et al. 2009; Levin et al. 2018). The frequency of hypertrophic cardiomyopathy is slightly higher in Costello syndrome (30%) compared with Noonan syndrome (20%) and CFC syndrome (10%).

MANIFESTATIONS AND MANAGEMENT

The management guidelines suggested here are based on "expert consensus" derived from the literature and supplemented by the shared experience of many families and professionals.

Growth and Feeding

Fetuses with CS often show increased nuchal thickness (Kerr et al. 1998; Gripp et al. 2002). Polyhydramnios is common and hydrops has been described. Decreased or increased fetal movement has been reported. Because major malformations are uncommon, prenatal ultrasonography rarely detects a structural anomaly.

Although some infants appear to have true macrosomia (elevated birth weight accompanied by a proportional increase in length and head circumference), many have elevated birth weight due to edema, which is rapidly lost after birth. Most will have birth weight and head circumference above the 50th centile. Virtually all infants with CS have severe feeding difficulty from birth. This is manifested as an inability to suck, and is often worsened by stridor resulting from tracheomalacia and excessive secretions. An increased frequency of pyloric stenosis has been observed (Gripp et al. 2008). The feeding problems are so severe that the children can appear marasmic in the first years of life. Most will require nasogastric or gastrostomy feeding. Because of gastroesophageal reflux and irritability, Nissen fundoplication is often performed, but will not decrease the need for supplemental feeding.

Children with CS generally become willing or able to take oral feeds, usually between 2 and 4 years. The first acceptable tastes are often strong flavors, such as ketchup or potato crisps. Paradoxically, older children or adults with CS can be described as being obsessed with food and eating. Speech onset frequently correlates with resolution of feeding difficulty.

Postnatal growth retardation, with preservation of head circumference, despite high normal birth weight and normal birth length, is a consistent feature of CS. It appears unrelated to whether or not adequate nutrition can be maintained in the first years of life. The reported adult height range is 122–154 cm for females (overall mean 139 cm, or 130 cm for those not treated with growth hormone) and 124–153 cm for males (mean 142 cm) (White et al. 2005). Delayed bone age is common, with final adult height reached at a mean age of 21 years (range 16–28 years) (White et al. 2005). CS-specific growth charts were developed based on individuals not exposed to growth hormone (Sammon et al. 2012).

Growth hormone deficiency appears common in CS, but the response to GH replacement is variable. Growth should be monitored on CS-specific growth charts, and investigations instituted if height velocity falters, as per standard protocols. Response to treatment should be evaluated as for other children undergoing treatment. In the absence of documented GH deficiency, there is no evidence that GH improves muscle mass in CS.

There have been concerns regarding the progression of cardiomyopathy or development of malignancy whilst on GH treatment. Bladder carcinoma has occurred in an individual with CS treated with growth hormone (Gripp et al. 2000) who developed multiple papillary transitional bladder carcinomas at age 16 years. A rhabdomyosarcoma occurred in another patient while on growth hormone treatment (Kerr et al. 2003).

Progression of cardiomyopathy after initiation of growth hormone treatment has occurred, although it is not known whether the relationship was causal or coincidental (Kerr et al. 2003). It is unknown whether the anabolic actions of growth hormone accelerate pre-existing cardiac hypertrophy or whether the dose is insufficient to promote myocyte hypertrophy (Lin et al. 2002). As a comparison, many studies of growth hormone in Noonan syndrome have only demonstrated one individual with worsening hypertrophic cardiomyopathy, although none of the study cohorts had pre-existing hypertrophy (Cotteril et al. 1996; Noordam et al. 2001; Padidela et al. 2008).

Recommendations for growth hormone treatment are to restrict its use to affected individuals with documented growth hormone deficiency, to perform regular tumor surveillance, and cardiac monitoring, and to titrate treatment to avoid a supraphysiological range (Stein et al. 2004).

Evaluation

- Careful monitoring of growth rate is recommended, with use of Costello syndrome-specific growth charts (Sammon et al. 2012).
- Early referral to a dietician experienced in pediatrics for appropriate nutritional evaluation.
- On-going assessment of the need for nasogastric or gastrostomy feeding is required for children with decreased growth velocity or difficulty feeding.
- The contribution of gastroesophageal reflux to feeding should be considered.
- Consideration should be given to evaluation of the growth hormone axis if growth rate decreases over time.

Management

- Early referral to a speech therapist or other appropriate specialist for help with feeding management.
- Assurance that caloric intake is adequate for growth.
- Treatment of gastroesophageal reflux, when present.
- Consider body positioning, a trial of hypoallergenic milks, and/or the use of prokinetic medication to treat feeding issues.
- Regular monitoring with echocardiography and adherence to recommended tumor surveillance protocols should be undertaken in individuals receiving growth hormone replacement therapy.

Development and Behavior

CS is associated with hypotonia and global developmental delay. Delayed visual maturation, along with the inability to feed and irritability in the first year, can lead to a misleading clinical impression of a poor outlook for intellectual functioning (Kawame et al. 2003). Serial assessment of a group of children, now all known to be *HRAS* mutation-positive, has shown a mean full score IQ of 57 (range 30–87), with most functioning in the mild-to-moderate range of intellectual disability (Axelrad et al. 2004, 2007). Recognition memory is a relative strength compared to verbal memory (Schwartz et al. 2017). Around 10% individuals with CS have scores in the low–average range.

Developmental milestones are all acquired late (Kawame et al. 2003), with a mean age of sitting of 23 months (10 months to 3 years), and mean age of walking alone of 4 years and 11 months (26 months to 9 years). Language development is frequently more severely delayed, particularly expressive language, but most will speak in full sentences by adulthood. Health permitting, many children will attend mainstream educational facilities, particularly until high school entry.

Although children and adults with CS are described as having a warm sociable personality, the early years of life are often characterized by extreme irritability with hypersensitivity to sounds and tactile stimuli, sleep disturbance, and excessive shyness with strangers (Kawame et al. 2003). This is independent of the presence of gastroesophageal reflux, and often improves between ages 2 and 4 years.

Assessment for evidence of autistic spectrum disorder may yield scores supporting the diagnosis in children under 5 years of age, but this will often reverse by school entry (Schwartz et al. 2017).

Anxiety is common in the school aged child and beyond (Axelrad et al. 2007, 2009).

Evaluation

- Early referral for developmental needs assessment.
- Early referral for speech therapy assessment (see feeding above).
- Formal hearing and visual assessment should be performed in the first year, as in any child with developmental delay, and regularly thereafter.
- Ongoing regular assessment of developmental status should be carried out with appropriate educational provision. This should include assessment of anxiety states and behavior.
- Adults should be screened for anxiety and depressive symptoms.

Management

- Referral to services for developmental disability for occupational therapy, physiotherapy, and early intervention.
- Refer to clinical psychology for management of anxiety, when present.
- Parental education and training, as required, to support augmented communication, management of difficult behavior, and anxiety.
- Consider the use of social stories, and educational strategies based on narrative. Augmented communication devices may be helpful, as may be the use of multiple-choice examinations.
- Individualized educational programming and planning for transition to adulthood is appropriate and does not differ because of the diagnosis of CS.
- Consider vocational training as part of transition discussion.

Cardiovascular

Congenital heart defects occur in 30% of individuals with CS, typically valvar pulmonic stenosis in 10%. Other valve anomalies, such as mitral valve prolapse or myxomatous or redundant valve leaflets may occur alone or as polyvalvar disease, and are reminiscent of the valve anomalies in mucopolysaccharidoses. Membranous ventricular septal defects or secundum-type atrial septal defects of mild-to-moderate severity are infrequent. The combination of atrial septal defect with pulmonic stenosis is rare, and helps distinguish CFC syndrome from CS (Gripp et al. 2007). Complex congenital heart defects, such as conotruncal defects, have not been reported, supporting the observation that CS is a dysplasia condition rather than a multiple malformation syndrome.

Cardiac hypertrophy is more common than reported in early series, occurring in about 60% of people with CS (Lin et al. 2011). The pattern of distribution may vary, but involves the left ventricle in one-half of the cases, a pattern consistent with hypertrophic cardiomyopathy. A smaller number of affected individuals have biventricular concentric hypertrophy, resembling what is seen in mucoplysaccharidoses. Severity ranges from a severe lethal form, to mild thickening of the left ventricle or interventricular septum. The natural history of hypertrophic cardiomyopathy in CS has not been completely defined. Onset in the first year, however, can be rapid and relatively asymptomatic.

Arrhythmia has been noted in 45% of individuals with CS, most often atrial tachycardia (94% of those with arrhythmia). Although the tachycardia may be reported broadly as supraventricular, paroxysmal or ectopic tachycardia, it is most distinctly chaotic or multifocal atrial tachycardia (also known as chaotic atrial rhythm). Because rhythm disturbance has occurred in children who did not have a congenital heart defect or hypertrophic cardiomyopathy, these are viewed as primary tachycardias (Lin et al. 2011; Levin et al. 2018). They differ from cardiomyopathy-associated dysrhythmias, which are typically ventricular. Fetal onset of tachycardia is documented (Lin et al. 2009). Of those who had a rhythm disturbance, approximately two-thirds have an additional abnormality, a congenital heart defect or hypertrophic cardiomyopathy. The cause of the tachycardia is unknown, though a conduction system dysplasia has been postulated.

A recently recognized category of cardiovascular disease is the vasculopathy of CS, which can affect both small- and large-bore arteries and manifests as systemic hypertension, coronary artery dysplasia, and aortic dilation, respectively (Lin et al. 2011)

The current information about congenital heart abnormalities in CS is fairly well established by several groups. However, future research is needed to delineate almost every other aspect of the natural history of these abnormalities, including age of onset, progression, and risk factors for morbidity, and differentiation from other syndromes in the Ras/MAPK pathway.

Evaluation

- Routine medical evaluation in individuals with CS at all ages should include clinical cardiac examination, including blood pressure measured in an age-appropriate way.
- At the time of diagnosis, referral for a baseline evaluation by a cardiologist is strongly recommended. The evaluation must include two-dimensional echocardiography with Doppler interrogation and electrocardiogram. The cardiologist should consider 24 hour ambulatory monitoring, based on age and clinical situation.
- Repeat evaluation every 3–6 months in the first two years of life.
- A reasonable schedule for routine cardiac follow-up would be every 2–3 years from age 2 until puberty, annual echo until age 20 and growth has ceased, and every 3–5 years thereafter.
- Before having any surgical procedure, careful attention should be given to cardiac status.

Management

- Pulmonic valve stenosis should be managed as in the general population, based on the gradient and clinical status. There is insufficient information to know whether balloon angioplasty can be performed with the same success as in an individual without CS.
- As in the general population, consider if antibiotic prophylaxis for subacute bacterial endocarditis is indicated.
- Hypertrophic cardiomyopathy may progress in severity, and require intensive medical and surgical intervention, including myomectomy. This is as in the general population.
- Chaotic (multifocal) atrial tachycardia and other arrhythmias may be a therapeutic challenge. Flecainide or amiodarone were reported to be more effective than propranolol or digoxin (Levin et al. 2018.) In addition to aggressive anti-arrhythmic medication, which is standard for the type of arrhythmia, catheter ablation or pacemaker may be required.

Respiratory

Excessive secretions are common in an infant or a young child with CS and may contribute to the feeding difficulty. Laryngomalacia, tracheomalacia, and bronchomalacia are common, and occasionally will be of sufficient severity to necessitate tracheostomy. Tracheomalacia and gastroesophageal reflux may contribute to sleep disturbance and irritability. Alveolar capillary dysplasia (Burkitt Wright et al. 2012), and pulmonary lymphangiectasia have also been reported.

Obstructive sleep apnea is common (Della Marca et al. 2006) in both childhood and adult life. Difficulty with intubation due to unusual upper airway anatomy has been reported (Dearlove and Harper 1997).

Evaluation

- Routine history taking should include specific questioning regarding respiratory pattern, sleep apnea, snoring, and excessive sleepiness.
- Airway evaluation and sleep studies should be undertaken in the presence of suggestive symptoms.

Management

- Airway management does not differ from that of the general population.
- General anesthesia should take place in centers with experience in management of difficult airways.

Musculoskeletal

Musculoskeletal anomalies are common in CS due to the ubiquitous hypotonia and joint laxity (Yassir et al. 2003; Detweiler et al. 2013; Reinker et al. 2011; Stevenson and Yang 2011). Short humeri and femurs have been noted prenatally, as has the characteristic wrist posture. Neonatal osteoporosis with enlargement of anterior ribs has been reported (Digilio et al. 2007). The digits and wrists are hyperextensible. Ulnar deviation at the wrist and of the fingers is characteristic, accompanied by wrist flexion and splaying of the fingers. The fingers are often short and broad, with nails that are deep-set and friable. In contrast, the elbows may have a flexion deformity, with decreased range of motion of the shoulders. Joint contractures may appear elsewhere. Radial head dislocation has been observed.

Analogous to hand findings, the feet may have an abnormal position. There can be congenital vertical talus, planovalgus feet, and overriding toes. Tight achilles tendons result in a tiptoed gait. There is a high rate of re-operation after surgery. The hips can be subluxed, either congenitally or developing in childhood and adolescence (Detweiler et al. 2013; Yassir et al. 2003). The gait is broad-based and shuffling.

Kyphoscoliosis is common and may occur early and be severe, necessitating surgical intervention (Yassir et al. 2003, Detweiler et al. 2013; Reinker et al. 2011; Stevenson and Yang 2011). Milder forms lead to a round-shouldered posture. Pectus excavatum has also been noted. The neck is short and the chest has an increased anteroposterior diameter.

Abnormal bone density with bone pain, vertebral compression fractures, and loss of height have been reported in adult life (White et al. 2005, Leoni et al. 2014; Stevenson et al. 2011; Detweiler et al. 2013) and contribute significantly to morbidity in older individuals. There has been one case report of abnormal bone density in early childhood, likely in association with growth hormone deficiency (Hou 2008). Long-term follow-up of the natural history of bone density and its management is important.

Evaluation

- All children with CS need careful serial physical evaluations of their large and small joints, rib cage, and spine. These should be performed by the primary care provider with referral to an orthopedic specialist, as indicated.
- Routine skeletal surveys for all affected individuals are not advised. Instead, imaging studies should be obtained based on specific findings or concerns.
- In adult life, consideration should be given to bone scanning to evaluate bone density if fractures or suggestive symptoms occur.

Management

- Splinting and physical therapy may be required to prevent joint contractures, such as fixed ulnar deviation of fingers and hands or elbow contractures.
- Several orthopedic problems may require surgical treatment, including vertical talus, hip subluxation, radial head dislocation, and kyphoscoliosis. Treatment is not different from that in the general population.
- Osteopenia should be managed as in the general population.

Neurologic

Macrocephaly, hypotonia, and global delays are to be expected in CS. In infancy, arching of the neck and back is common, and although not understood, it is usually not indicative of any particular pathology. Nystagmus associated with delayed visual maturation is usual, and often persists. A literature review of 38 individuals with a clinical diagnosis of CS (Delrue et al. 2003) found normal cerebral imaging in 25%. Crowding in the posterior fossa can progress to Chiari I malformation, which in turn may contribute to decreased spinal fluid flow with hydrocephalus and syrinx formation (White et al. 2005; Gripp et al. 2010). Tethered cord appears to be more common than in the typical population and may contribute to neurologic, urologic and musculoskeletal anomalies (Gripp et al. 2010).

Van Eeghen et al. (1999) noted that 75% of people with CS had electroencephalographic (EEG) abnormalities (usually non-specific) and 8% had seizures. In an *HRAS* mutation-positive series (Kerr et al. 2006), the seizure frequency was 11%.

A severe newborn presentation mimicking a congenital myopathy has been described in four affected babies, three of whom had less common *HRAS* mutations (van der Burgt et al. 2007). These babies had generalized hypotonia, variable contractures, absent movement, and areflexia. On muscle biopsy, excess muscle spindles were consistently reported in this series, leading to the description "congenital myopathy with excess of muscle spindles", but the prevalence of this abnormality in other individuals with CS is uncertain. A few additional cases with a muscular dystrophy phenotype were reported (Bolocan et al. 2014). Skeletal muscle pathology in CS was reviewed and compared to findings in the closely related CFC syndrome by Tidyman et al. (2011).

Evaluation

- Neurological examination should include serial head circumference measurements in infancy and childhood, with plotting on standard growth charts, as well as documentation of reflexes. The development of hyperreflexia should be investigated, as it may indicate development of spinal cord or posterior fossa pathology. Examination in early life should consider the possibility of craniosynostosis.
- Baseline cerebral imaging including the posterior fossa should be undertaken at diagnosis. A low threshold for repeat cranial imaging is advised in the presence of focal deficits, seizures, or rapidly growing head circumference.
- Spinal imaging should be considered in the presence of an alteration in gait, or abnormal neurological examination. This should include an MRI of the lower spine to evaluate for tethered cord, as well as single spine radiograph to count vertebral bodies.
- EEG testing should be obtained when there is a clinical suspicion of seizures.
- The possibility of hypoglycemia, secondary to an underlying endocrine or pancreatic abnormality, needs to be investigated in the presence of seizures (see below).

Management

- Individuals with cortical atrophy and ventriculomegaly will not need treatment.
- Those who have clinically significant hydrocephalus or symptomatic Chiari I malformation should be followed by a neurologist and/or neurosurgeon. Shunting or decompression may be indicated, as in the general population.
- Individuals with clinical seizures should be treated appropriately and in a standard fashion.
- Tethered cord likely benefits from surgical release.

Endocrine

People with CS have short stature and delayed bone age, and are often suspected to have growth hormone deficiency. This topic is reviewed in Growth and Feeding.

Abnormal glucose homeostasis has been mentioned in numerous reports, and is generally multifactorial in origin. Transient neonatal hypoglycemia is frequently seen, and responds to glucose infusions. Hypoglycemia has been associated with hyperinsulinism (Alexander et al. 2005) and, in one individual, with a nesidioblastosis-like lesion of the pancreas (Lo et al. 2008). A pancreatic nodule causing hyperinsulinemic hypoglycemia in an individual with the rare p.Gln22Lys *HRAS* mutation (Scheffield et al. 2015) resulted from loss of heterozygosity for 11p15.5 and duplication of the paternally derived mutant allele within the tumor (Gripp et al. 2016). Rarely, symptomatic hypoglycemia is identified in older affected individuals. Hypoglycemia has been reported with complete or partial growth hormone deficiency (Gripp et al. 2000; Gregerson and Viljoen 2005). Hypoglycemia may also occur after gastrostomy and fundoplication (Calabria 2011).

Pubertal development is often delayed in both males and females with CS (White et al. 2005). Menarche is frequently delayed (Martin and Jones 1991; Yetkin et al. 1998; White et al. 2005) and may be followed by irregular periods and secondary amenorrhea (Yetkin et al. 1998). Primary amenorrhea with normal breast development is not uncommon (Van Eeghen et al. 1999; White et al. 2005). There are no known pregnancies conceived by a woman with typical CS. Multiple intraductal breast papillomas and fibroadenosis are relatively common in adult women (White et al. 2005).

Precocious puberty appears relatively common, and should be treated as per standard protocols.

Evaluation

- Investigate hypoglycemia with blood levels below 50 mg dL^{-1} as in any child per current professional society guidelines.
- In the postnatal period, and at times of medical illness, a high level of suspicion for hypoglycemia should be maintained. Blood sugar levels below 70 mg dL^{-1} should be treated with a glucose infusion, as per guidelines above, and levels should be monitored.
- Monitor glucose carefully after gastrostomy and fundoplication.
- Throughout infancy and adolescence, hypoglycemia should be excluded as an underlying cause for seizures.
- CS is not known to be a multiple endocrinopathy. As clinically indicated, specific deficits, such as hypothyroidism, should be sought with appropriate evaluations.

Management

- Hypoglycemia should be treated with glucose infusion, as per standard guidelines.
- Treat underlying hyperinsulinism or growth hormone deficiency as per standard protocols.
- Delayed puberty should be investigated and managed in the standard fashion, but with special thought given to the possibility of gonadal tumors.

Ophthalmologic

Children with CS often have delayed visual maturation and nystagmus. Strabismus, keratoconus, and cataract can occur. Retinal dystrophy has been reported in two cases (Pierpont et al. 2017).

Evaluation

- Referral for an ophthalmology evaluation should be made in the first three months of life or at diagnosis, whichever is soonest.
- Annual ophthalmology assessment should continue throughout life.

Management

- Treatment of nystagmus, strabismus, and cataract do not differ from that in the general population.

Oncologic

A population-based study of childhood cancer has demonstrated that children with CS have a standardized incidence ratio for cancer of 42.4 (95% CI 5.1-153.2) (Kratz et al. 2015). The risk was predominantly for embryonal rhabdomyosarcoma, confirming earlier reports (Kerr et al. 1998; Gripp et al. 2002), but also neuroblastoma and bladder carcinoma. From a literature review of reported tumors (Gripp et al. 2002), most rhabdomyosarcomas showed embryonal histology with one alveolar and one pleomorphic subtype.

The age at presentation ranged from 6 months to 6 years. Of these tumors, eight originated from the abdomen, pelvis, or urogenital area. Neuroblastoma occurred in three individuals, ranging in age from 2 months to 4 years. Transitional cell carcinoma of the bladder, usually a tumor of old age, has been reported from adolescence.

A number of benign tumors have also been recorded in CS, including breast fibroadenomatosis and intraductal papillomas, choroid plexus papillomas, stomach polyp, parathyroid adenoma, and ganglioneuroblastoma (Zampino et al. 1993).

Gripp et al. (2002) suggested a tumor frequency of up to 17%, based on the 17 solid tumors reported in about 100 known cases and proposed a screening protocol. This was modified after reports of abnormal urinary catecholamines in the absence of neuroblastoma in individuals with Costello syndrome (Gripp et al. 2004; Bowron et al. 2005). While this screening protocol (see Evaluation) has not been proven to improve outcome, it is supported by other published expert opinion (Villani et al. 2017).

It remains unknown if people with CS have an increased, albeit lower relative risk of other tumors. Given the role of *HRAS* in the Ras pathway, consideration of CS as a tumor predisposition syndrome with appropriate evaluation of any individual presenting with unexplained or persistent symptoms is important.

Evaluation

- Clinical examination and ultrasound evaluation of the abdomen and pelvis every 3–4 months until age 8–10 years to identify rhabdomyosarcoma and abdominal neuroblastoma. Chest radiograph should also be considered (Villani et al. 2017).
- Urinalysis for hematuria annually starting at age 10 years to allow for early detection of bladder carcinoma.
- Any unexplained or prolonged symptoms should be evaluated promptly for underlying malignancy.

Management

- Treatment of identified tumors should follow the protocols used for malignancies. Treating oncologists should be made aware of the cardiac abnormalities, including the risk for atrial tachycardia and cardiac hypertrophy in individuals with CS.

Dermatologic

Skin abnormalities are common and often distinctive. The skin is soft, even velvety. Adolescents and adults have thinning and wrinkling of the skin, especially of the hands, contributing to the appearance of a much older individual. The creases of the palms and feet are deep, with excess palmar and plantar skin. There is often generalized hyperpigmentation, nevi, and vascular birthmarks as seen in CFC and Noonan syndromes. Acanthosis nigricans on the neck and other flexural regions can be striking. Adolescents may develop hyperkeratosis and callouses, which may invade the dorsal areas of hands and feet. These appear unrelated to friction or trauma. Skin over joints can be thickened by hyperkeratosis. Eczema is common and keratosis pilaris is frequent.

Individuals with CS typically manifest skin lesions called papillomas. What appear as tags or sessile nodules may be true papillomas, viral warts or keratoses. Papillomas may occur in the perianal, perioral, paranasal, and popliteal regions and can cause pruritus. These sites should be carefully treated to avoid trauma and bacterial contamination.

Scalp hair can be tightly curled, almost frizzy, resulting in a "cloud of curls." However, scalp hair can be unremarkable. Patchy alopecia has been noted. Nails may be friable or dystrophic with longitudinal striations.

Evaluation

- Skin should be regularly monitored by the individual, caregiver, and primary care doctor for tags, warts, dry skin, and acanthosis nigricans.
- Lesions that are large, intensely itchy, or dry or in unusual locations should be examined by a dermatologist.

Management

- Skin "tags" can be removed to avoid trauma or bacterial infection.
- Recurrent facial papillomas have been successfully managed with cryotherapy.
- If a wart is diagnosed, treatment may be necessary to minimize spread of the virus.
- Callouses, especially on soles, should be treated because they can be painful. Keratolytic preparations used locally can be very helpful, as can local debridement.
- Dystrophic nails should be tested for fungi and treated if necessary.

ACKNOWLEDGMENTS

We thank our many colleagues and the families and individuals with Costello syndrome in the Costello Syndrome support groups.

RESOURCES

Website: *http://www.costellokids.com*
Costello Syndrome Family Network (USA)
   Sandra Taylor (Executive Director)

1702 Tyndall Drive
Panama City, FL 32401
Email: *sandra@costellosyndromeusa.org*

International Costello Syndrome Support Group
Colin Stone
90 Parkfield Road North
New Moston
Manchester M40 3RQ, UK
Email: *c.stone@costellokids.com*

Association Française des Syndromes de Costello and CFC
48, rue de Chouiney
33170 Gradignan
France
Email: *afscostello@free.fr*

REFERENCES

Alexander S, Ramadan D, Alkhayyat H, Al-Sharkawi I, Backer KC, El-Sabban F, Hussain K (2005) Costello syndrome and hyperinsulinemic hypoglycaemia. *Am J Med Genet* 139:227–230.

Allanson J, Kavamura I, Neri G, Noonan J, Ross A, Kerr B (2007) Distal phalangeal creases: More evidence of this feature in disorders of the Ras signaling pathway. *Eur J Med Genet* 50:482–483.

Aoki Y, Niihori T, Kawame H, Kurosawa K, Ohashi H, Tanaka Y, Filocamo M, Kato K, Suzuki Y, Kure S, Matsubara Y (2005) Germline mutations in HRAS proto-oncogene cause Costello syndrome. *Nat Genet* 37:1038–1040.

Axelrad M, Glidden R, Nicholson L, Gripp KW (2004) Adaptive skills, cognitive and behavioral characteristics of Costello syndrome. *Am J Med Genet* 128A:396–400.

Axelrad ME, Nicholson L, Stabley DL, Sol-Church K, Gripp KW (2007) Longitudinal assessment of cognitive characteristics in Costello syndrome. *Am J Med Genet* 143A:3185–3193.

Axelrad ME, Schwartz DD, Fehlis JE, Hopkins E, Stabley DL, Sol-Church K, Gripp KW (2009) Longitudinal course of cognitive, adaptive, and behavioral characteristics in Costello syndrome. *Am J Med Genet A* 149A:2666–2672.

Bertola D, Buscarilli M, Stabley DL, Baker L, Doyle D, Bartholomew DW, sol-Church K, Gripp KW (2017) Phenotypic spectrum of Costello syndrome individuals harboring the rare HRAS mutation p.Gly13Asp. *Am J Med Genet* 173(5):1309-1318.

Bolocan A, Quijano-Roy S, Seferian AM, Baumann C, Allamand V, Richard P, Estournet B, Carlier R, Cave H, Gartioux C, Blin N, Le Moing AG, Gidaro T, Germain DP, Fardeau M, Voit T, Servais L, Romero NB (2014) Congenital muscular dystrophy phenotype with neuromuscular spindles excess in a 5-year-old girl caused by HRAS mutation. *Neuromuscular disorders* 24:993–998.

Burkitt Wright EM, Bradley L, Shorto J, McConnell VP, Gannon C, Firth HV, Park SM, D'Amore A, Munyard PF, Turnpenny PD, Charlton A, Wilson M, Kerr B (2012) Neonatal lethal Costello syndrome and unusual dinucleotide deletion/insertion mutations in HRAS predicting p.Gly12Val. *Am J Med Genet A.* 158 (5):1102–10.

Bowron A, Scott JG, Brewer C, Weir P (2005) Increased HVA detected on organic acid analysis in a patient with Costello syndrome. *J Inherit Metab Dis* 28:1155–1156.

Costello JM (1971) A new syndrome. *NZ Med J* 74:397A.

Costello JM (1977) A new syndrome: Mental subnormality and nasal papillomata. *Aust Paediatr J* 13:114–118.

Costello JM (1996) Costello syndrome: Update on the original cases and commentary. *Am J Med Genet* 62:199–201.

Cotteril AM, McKenna WJ, Brady AF, Sharland M, Elsawi M, Wamada M, Camacho Hubner C, Kelnar CJ, Dunger DB, Patton MA, Savage MO (1996) The short-term effects of growth hormone therapy on height velocity and cardiac ventricular wall thickness in children with Noonan's syndrome. *J Clin Endocrinol Metab* 81:2291–2297.

Dearlove A, Harper N (1997) Costello syndrome. *Paediatr Anaesth Lett* 7:476–478.

Della Marca G, Vasta I, Scarano E, Rigante M, DeFeo E, Mariotti P, Rubino M, Vollono C, Mennuni G, Tonali P, Zampino G (2006) Obstructive sleep apnea in Costello syndrome. *Am J Med Genet* 140A:257–262.

Delrue M-A, Chateil J-F, Arveiler B, Lacombe D (2003) Costello syndrome and neurological abnormalities. *Am J Med Genet* 123A:301–305.

Denayer E, Parret A, Chmara M, Schubbert S, Vogels A, Devriendt K, Frijns JP, Rybin V, de Ravel TJ, Shannon K, Cools J, Scheffzek K, Legius E (2008) Mutation analysis in Costello syndrome: functional and structural characterization of the HRAS p.Lys117Arg mutation. *Hum Mutat* 29:232–9.

Detweiler S, Thacker M, Hopkins E, Conway L, Gripp KW (2013) Orthopedic manifestations and implications for individuals with Costello Syndrome. *Am J Med Genet* 161A:1940–1949.

Digilio M, Sarkozy A, Capolino R, Testa M, Esposito G, de Zorzi A, Cutrera R, Marino B, Dallapiccola B (2007) Costello syndrome: Clinical diagnosis in the first year of life. *Eur J Pediatr* 167:621–628.

Estep AL, Tidyman WE, Teitell MA, Cotter PD, Rauen KA (2006) HRAS mutations in Costello syndrome: Detection of constitutional activating mutations in codons 12 and 13 and loss of wild type allele in malignancy. *Am J Med Genet* 140A:8–16.

Giannoulatou E, McVean G, Taylor IB, McGowan SJ, Maher GJ, Iqbal Z, Pfeifer SP, Turner I, Burkitt-Wright EM, Shorto J, Itani A, Turner K, Gregory L, Buck D, Rajpert-De Meyts E, Looijenga LH, Kerr B, Wilkie AO, Goriely A (2013) Contribution of intrinsic mutation rate and selfish selection to levels of de novo HRAS mutations in the paternal germline. *Proc Natl Acad Sci USA* 110(50);20152–7.

Gregersen N, Viljoen D (2004) Costello syndrome with growth hormone deficiency and hypoglycaemia: A new report and review of the endocrine associations. *Am J Med Genet* 129A:171–175.

Gremer L, De Luca A, Merbitz-Zahradnik T, Dallapiccola B, Morlot S, Tartaglia M, Kutsche K, Ahmadian MR, Rosenberger G (2010) Duplication of Glu37 in the switch I region of HRAS impairs effector/GAP binding and underlies Costello syndrome by promoting enhanced growth factor-dependent MAPK and AKT activation. *Hum Mol Gene.* 19:790–802

Gripp KW, Scott CI Jr, Nicholson L, Figueroa TE (2000) A second case of bladder carcinoma in a patient with Costello syndrome. *Am J Med Genet* 90:256–259.

Gripp KW, Scott CI Jr, Nicholson L, McDonald-McGinn DM, Ozeran JD, Jones MC, Lin AE, Zackai EH (2002) Five additional Costello syndrome patients with rhabdomyosarcoma: Proposal for a tumor screening protocol. *Am J Med Genet* 108:80–87.

Gripp KW, Kaware H, Viskochil DH, Nicholson L (2004) Elevated catecholamine metabolites in patients with Costello syndrome. *Am J Med Genet* 128A:48–51.

Gripp KW, Lin AE, Stabley DL, Nicholson L, Scott CI, Doyle D, Aoki Y, Matsubara Y, Zackhai EH, Lapunzina P, Gonzalez-Meneses A, Holbrook J, Agresta CA, Gonzalea I, Sol-Church K (2006a) *HRAS* mutation analysis in Costello syndrome: Genotype and phenotype correlation. *Am J Med Genet* 140A:1–7.

Gripp KW, Stabley DL, Nicholson L, Hoffman JD, Sol-Church K (2006b) Somatic mosaicism for an *HRAS* mutation causes Costello Syndrome. *Am J Med Genet* 140A:2163–2169.

Gripp KW, Lin AE (2006) Costello syndrome. GeneReviews. http://www.geneclinics.org. Updated Jan 12, 2012; accessed Feb 28, 2019.

Gripp K, Lin A, Rebolledo M, Wheeler P, Wilson W, Al-Rahawan M, Sol-Church K (2007) Further delineation of the phenotype resulting from *BRAF* or *MEK1* mutations helps differentiate cardio-facio-cutaneous syndrome from Costello syndrome. *Am J Med Genet* 143A:1472–1480.

Gripp KW, Innes MA, Axelrad ME, Gillan TL, Parboosingh JS, Davies C, Leonard NJ, Lapointe M, Doyle D, Catalano S, Nicholson L, Stabley DL, Sol-Church K (2008) Costello syndrome associated with novel germline *HRAS* mutations: An attenuated phenotype? *Am J Med Genet* 146A:683–690.

Gripp KW, Hopkins E, Doyle D, Dobyns WB (2010): High incidence of progressive postnatal cerebellar enlargement in Costello syndrome: Brain overgrowth associated with HRAS mutations as the likely cause of structural brain and spinal cord abnormalities. *Am J Med Genet* 152A:1161–1168.

Gripp KW, Hopkins E, Sol-Church K, Stabley DL, Axelrad ME, Doyle D, Dobyns WB, Hudson C, Johnson J, Tenconi R, Graham GE, Sousa AB, Heller R, Piccione M, Corsello G, Herman GE, Tartaglia M, Lin AE (2011): Phenotypic analysis of individuals with Costello syndrome due to HRAS p.G13C. *Am J Med Genet* 155A:706–716.

Gripp KW, Sol-Church K, Smpokou P, Graham GE, Stevenson DA, Hanson H, Viskochil DH, Baker LC, Russo B, Gardner N, Stabley DL, Kolbe V, Rosenberger G (2015) An attenuated phenotype of Costello syndrome in three unrelated individuals with a HRAS c.179G>A (p.Gly60Asp) mutation correlates with uncommon functional consequences. *Am J Med Genet* 167A (9):2085–97.

Gripp KW, Robbins KM, Sheffield BS, Lee AF, Patel MS, Yip S, Doyle D, Stabley D, Sol-Church K. (2016) Paternal uniparental disomy 11p15.5 in the pancreatic nodule of an infant with Costello syndrome: Shared mechanism for hyperinsulinemic hypoglycemia in neonates with Costello and Beckwith–Wiedemann syndrome and somatic loss of heterozygosity in Costello syndrome driving clonal expansion. *Am J Med Genet A* 170(3):559–64.

Hague J, Hackett G, Acerini C, Park SM (2017) Prenatal genetic diagnosis of Costello syndrome in a male fetus with recurrent HRAS mutation p.Gly12Ser. *Prenat Diagn* 37(4):409–411.

Hou JW (2008) Rapidly progressive scoliosis after successful treatment for osteopenia in Costello syndrome. *Am J Med Genet* 146A:393–396.

Kawame H, Matsui M, Kurosawa K, Matsuo M, Masuno M, Ohashi H, Fueki N, Aoyama K, Miyatsuka Y, Suzuki K, Akatsuka A, Ochiai Y, Fukushima Y (2003) Further delineation of the behavioral and neurologic features in Costello syndrome. *Am J Med Genet* 118A:8–14.

Kerr B, Eden OB, Dandamudi R, Shannon N, Quarrell O, Emmerson A, Ladusans E, Gerrard M, Donnai D (1998) Costello syndrome: Two cases with embryonal rhabdomyosarcoma. *J Med Genet* 335:136–139.

Kerr B, Einaudi MA, Clayton P, Gladman G, Eden T, Saunier P, Genevieve D, Philip N (2003) Should growth hormone be used in Costello syndrome? *J Med Genet* 40:e74.

Kerr B, Delrue M-A, Sigaudy S, Perveen R, Marche M, Burgelin I, Stef M, Tang B, Eden OB, O'Sullivan J, De Sandre-Giovannoli A, Reardon W, Brewer C, Bennett C, Quarell O, Fryer A, Donnai D, Stewart F, Raoul Hennekam Hélène Cavé Alain Verloes Philip N, Lacombe D, Levy N, Arveiler B, Black G (2006) Genotype-phenotype correlation in Costello syndrome; *HRAS* mutation analysis in 43 cases. *J Med Genet* 43:401–405.

Kerr B, Allanson J, Delrue M, Gripp K, Lacombe D, Lin A, Rauen K (2008) The diagnosis of Costello syndrome: Nomenclature in Ras/MAPK pathway disorders. *Am J Med Genet* 146A:1218–1220

Kerr B, (2009.) The clinical phenotype of Costello syndrome. In: *Noonan Syndrome and Related Disorders. Monographs in Human Genetics,* Zenker M, ed. Basel: Karger, Vol. 17, pp. 83–93.

Kratz CP, Franke L, Peters H, Kazmierczak B, Finckh U, Bier A, Eichhorn B, blank C, Kraus C, Kolhase J, Pauli S, Wildhardt G, Kutsche K, auber B, Christmann A, Bachmann N, Mitter D, Cremer FW, Mayer K, Daumer-Hass C, Nevinny-Stickel-Hinzpeter C, Oeffner F, Schülter G, Gencik M, Überlacker B Lissewski C, Schanze I, Greene MH, Spix C, Zenker M (2015) Cancer spectrum and frequency among children with Noonan, Costello and cardio-facio-cutaneous syndromes *Br J Cancer* 112:1392–1397.

Kuniba J, Pooh RK, Sasaki K, Shimokawa O, Harada N, Kondoh T, Egashira M, Moriuchi H, Yoshiura K, Niikawa N (2009) Prenatal diagnosis of Costello syndrome using 3D ultrasonography amniocentesis confirmation of the rare *HRAS* mutation G12D. *Am J Med Genet* 149A:785–787.

Leoni C, Stevenson DA, Martini L, De Sanctis R, Mascolo G, Pantaleoni F, De Santis S, La Torraca I, Persichilli S, Caradonna P, Tartaglia M, Zampino G (2014) Decreased bone mineral density in Costello syndrome. *Mol Genet Metab* 111(1);41–5.

Levin MD, Saitta SC, Gripp KW, Wenger TL, Ganesh J, Kalish JM, Epstein MR, Smith R, Czosek RJ, Ware SM, Goldenberg P, Myers A, Chatfield KC, Gillespie MJ, Zackai EH, Lin AE. 2018. Nonreentrant atrial tachycardia occurs independently of hypertrophic cardiomyopathy in RASopathy patients. *Am J Med Genet* 176:1711–1722.

Lin AE, Grossfeld PD, Hamilton R, Smoot L, Proud V, Wesberg R, Gripp K, Wheeler P, Picker J, Irons M, Zackai E, Scott CI, Nicholson L (2002) Further delineation of cardiac anomalies in Costello syndrome. *Am J Med Genet* 111:115–129.

Lin AE, O'Brien B, Demmer LA, Almeda KK, Blanco CL, Glasow PF, Berul CI, Hamilton R, Innes M, Lauzon JL, Sol-Church K, Gripp KW (2009) Prenatal features of Costello syndrome: Ultrasonographic findings and atrial tachycardia. *Prenat Diagn* 29(7):682–690.

Lin AE, Alexander ME, Colan SD, Kerr B, Rauen KA, Noonan J, Baffa J, Hopkins E, Sol-Church K, Limongelli G, Digilio MC, Marino B, Innes AM, Aoki Y, Silberbach M, Delrue MA, White SM, Hamilton RM, O'Connor W, Grossfeld PD, Smoot LB, Padera RF, Gripp KW (2011) Clinical, pathological, and molecular analyses of cardiovascular abnormalities in Costello syndrome: A Ras/ MAPK pathway syndrome. *Am J Med Genet A* 155:1–22.

Lo I, Brewer C, Shannon N, Shorto J, Tang B, Black G, Soo M, Ng D, Lam S, Kerr B (2008) Severe neonatal manifestations of Costello syndrome. *J Med Genet* 45:167–171.

Lorenz S, Petersen C, Kordaß U, Seidel H, Zenker M, Kutsche K (2012) Two cases with severe lethal course of Costello syndrome associated with HRAS p.G12C and p.G12D. *Eur J Med Genet* 55(11):615–9.

Lorenz S, Lissewski C, Simsek-Kiper PO, Alanay Y, Boduroglu K, Zenker M, Rosenberger G (2013) Functional analysis of a duplication (p.E63_D69dup) in the switch II region of HRAS: new aspects of the molecular pathogenesis underlying Costello syndrome. *Hum Mol Genet* 22:1643–53.

Lurie IW (1994) Genetics of the Costello syndrome. *Am J Med Genet* 52:358–359.

Martin RA, Jones KL (1991) Delineation of Costello syndrome. *Am J Med Genet* 41:346–349.

Narumi Y, Aoki Y, Niihori T, Neri G, Cave H, Verloes A, Nava C, Kavumara MI, Okamoto N, Kurasowa K, Hennekam RC, Wilson LC, Gillessen-Kaesbach G, Wieczorek D, Lapunzina P, Ohashi H, Makita Y, Kondo I, Tsuchiya S, Ito E, Sameshima K, Kato K, Kure S, Matsubara Y (2007) Molecular and clinical characterization of cardio-facio-cutaneous (CFC) syndrome: Overlapping clinical manifestations with Costello syndrome. *Am J Med Genet* 143A:799–807.

Nava C, Hanna N, Michot C, Pereira S, Pouvreau N, Niihori T, Aoki Y, Matsubara Y, Arveiler B, Lacombe D, Pasmant E, Parfait B, Baumann C, Heron D, Sigaudy S, Toutain A, Rio M, Goldenberg A, Leheup B, Verloes A, Cave H (2007) CFC and Noonan syndromes due to mutations in Ras/MAPK signaling pathway: Genotype/phenotype relationships and overlap with Costello syndrome. *J Med Genet* 44:763–771.

Noordam C, Draaisma JM, van den Nieuwenhof J, van der Burgt I, Otten BJ, Daniels O (2001) Effects of growth hormone treatment on left ventricular dimensions in children with Noonan's syndrome. *Horm Res* 56:110–113.

Ørstavik K, Tangeraas T, Molven A, Prescott TE (2007) Distal phalangeal creases—A distinctive dysmorphic feature in disorders of the Ras signalling pathway. *Eur J Med Genet* 50:155–158.

Padidela R, Camacho-Hübner C, Attie Km Savage MO (2008) Abnormal growth in Noonan syndrome: Genetic and endocrine features and optimal treatment. *Horm Res* 70:129–136.

Piccione M, Piro E, Pomponi MG, Matina F, Pietrobono R, Candela E, Gabriele B, Neri G, Corsello G (2009) A premature infant with Costello syndrome due to a rare G13C HRAS mutation. *Am J Med Genet A* 149A:487–9.

Pierpont ME, Richards M, Engel KW, Mendelsohn NJ, Summers CG (2017) Retinal dystrophy in two boys with Costello syndrome due to the HRAS p.Gly13Cys mutation. *Am J Med Genet* 173(5);1309–1318.

Quezada E, Gripp KW (2007) Costello syndrome and related disorders. *Curr Opin Pediatr* 19:636–644.

Rauen KA (2006) HRAS and the Costello syndrome. *Clin Genet* 71:101–108.

Rauen KA, Hefner E, Carrillo K, Taylor J, Messier L, Aoki Y, Gripp KW, Matsubara Y, Proud VK, Hammond P, Allanson JE, Delrue MA, Axelrad ME, Lin AE, Doyle DA, Kerr B, Carey JC, McCormick F, Silva AJ, Kieran KW, Hinek A, Nguyen TT, Schoyer (2008) Conference Report: Molecular aspects, clinical aspects, and possible treatment modalities for Costello syndrome: Syndrome Research Symposium 2007. *Am J Med Genet* 146A:125–127.

Reinker KA, Stevenson DA, Tsung A (2011) Orthopaedic conditions in Ras/MAPK related disorders. *J Pediatr Orthop* (5):599–605.

Robbins KM, Stabley DL, Holbrook J, Sahraoui R, Sadreameli A, Conard K, Baker L, Gripp KW, Sol-Church K. (2016) Paternal uniparental disomy with segmental loss of heterozygosity of chromosome 11 are hallmark characteristics of syndromic and sporadic embryonal rhabdomyosarcoma. *Am J Med Genet A* 170(3):559–64.

Sammon M, Doyle D, Hopkins E, Sol-Church K, Stabley D, McGready J, Schulze K, Alade Y, Hoover-Fong J, Gripp KW (2012): Normative growth charts for individuals with Costello syndrome. *Am J Med Genet A* 158:2692–2699.

Sheffield BS, Yip S, Ruchelli ED, Dunham CP, Sherwin E, Brooks PA, Sur A, Singh A, Human DG, Patel MS, Lee AF (2015) Fatal congenital hypertrophic cardiomyopathy and a pancreatic nodule morphologically identical to focal lesion of congenital hyperinsulinism in an infant with Costello syndrome: Case report and review of the literature. *Pediatr Dev Pathol* 18:237–244.

Schwartz DD, Katzenstein JM, Highley EJ, Stabley DL, Sol-Church K, Gripp KW, Axelrad ME (2017) Age related differences in prevalence of autistic spectrum disorder symptoms in children and adolescents with Costello syndrome. *Am J Med Genet* 173(5):1294–1300.

Schulz AL, Albrecht B, Arici C, van der Burgt I, Gillessen-Kaesbach G, Heller R, Horn D, Hübner CA, Korenke GC, König R, Kress W, Krüger G, Meinecke P, Mucke J, Plecko B, Rossier E, Schinzel A, Schulze A, Seemanova E, Seidel H, Spranger S, Tuyuz B, Wieczorek D, Kutsche K, Zenker M

(2008) Mutation and phenotypic spectrum in patients with cardio-facio-cutaneous and Costello syndrome. *Clin Genet* 73:62–70.

Sigaudy S, Vittu G, Vigneron J, Lacombe D, Moncia A, Flori E, Philip N (2000) Costello syndrome: Report of six patients including one with an embryonal rhabdomyosarcoma. *Eur J Pediatr* 159:139–142.

Sol-Church K, Gripp KW (2009) The molecular basis of Costello syndrome. In: *Noonan Syndrome and Related Disorders. Monographs in Human Genetics,* Zenker M, ed. Basel: Karger, Vol. 17, pp. 94–103.

Sol-Church K, Stabley DL, Nicholson L, Gonzalez IL, Gripp KW (2006) Paternal bias in parental origin of *HRAS* mutations in Costello syndrome. *Hum Mutat* 27:736–741.

Sol-Church K, Stabley DL, Demmer LA, Agbulos A, Lin AE, Smoot L, Nicholson L, Gripp KW (2009) Male to male transmission of Costello syndrome: G12S *HRAS* germline mutation inherited from a father with somatic mosaicism. *Am J Med Genet* 149A:315–321.

Stein RL, Legault L, Daneman D, Weksberg R, Hamilton J (2004) Growth hormone deficiency in Costello syndrome. *Am J Med Genet* 129A:166–170.

Stevenson DA, Yang F-C. 2011. The musculoskeletal phenotype of the RASopathies. *Am J Med Genet C* 157:90–103.

Tidyman WE, Lee HS, Rauen KA. 2011. Skeletal muscle pathology in Costello and cardio-facio-cutaneous syndromes: Developmental consequences of germline Ras/MAPK activation on myogenesis. *Am J Med Genet Part C* 157:104–114.

Van der Burgt I, Kupsky W, Stassou S, Nadroo A, Barroso C, Diem A, Kratz CP, Dvorsky R, Ahmadian MR, Zenker M (2007) Myopathy caused by HRAS germline mutations—Implications for disturbed myogenic differentiation in the presence of constitutive H-Ras activation. *J Med Genet* 44:459–462.

Van Eeghen AM, van Gelderen I, Hennekam RCM (1999) Costello syndrome: Report and review. *Am J Med Genet* 82:187–193.

Van Steensel MA, Vreeburg M, Van Ravenswaaiji-Arts CM, Biljsma E, Schrander-Stumpel CT, van Geel M (2006) Recurring HRAS mutation G12S in Dutch patients with Costello syndrome. *Exp Dermatol* 15:731–734.

Villani A, Greer M-L C, Kalish JM, nakagawara A, Nathanson KL, Pajtler KW, Pfister SM, Walsh MF, Wasserman JD, Zelley K, Kratz CP (2017) Recommendations for cancer surveillance in individuals with RASopathies and other rare genetic conditions with increased cancer risk. *Clin Cancer Res* 23(12):e83–e90.

Weaver KN, Wang D, Cnota J, Gardner N, Stabley D, Sol-Church K, Gripp KW, Witte D, Bove K, Hopkin R (2014) Early-lethal Costello syndrome due to rare HRAS tandem base substitution (c.35_36GC>AA; p.G12E) associated pulmonary vascular disease. *Pediatr Dev Pathol* 17:421–430.

White S, Graham JM, Kerr B, Gripp K, Weksberg R, Cytrynbaum C, Reeder JL, Stewart FJ, Edwards M, Wilson M, Bankier A (2005) The adult phenotype in Costello syndrome. *Am J Med Genet* 136A:128–305.

Yassir WK, Grottkau BE, Goldberg MJ (2003) Costello syndrome: Orthopedic manifestations and functional health. *J Pediatr Orthop* 23:94–98.

Yetkin I, Ayvaz G, Arslan M, Yilmaz M, Cakir N (1998) A case of Costello syndrome with endocrine features. *Ann Genet* 41:157–160.

Zampino G, Mastroiacovo P, Ricci R, Zollino M, Segni G, Martini-Neri ME, Neri G (1993) Costello syndrome: Further clinical delineation, natural history, genetic definition, and nosology. *Am J Med Genet* 47:176–183.

Zampino G, Pantaleoni F, Carta C, Cobellos G, Vasta I, Neri C, Pogna EA, De Feo E, Delogu A, Sarkozy A, Atzeri F, Selicorni A, Rauen KA, Cytrynbaum CS, Weksberg R, Dallapiccola B, Ballabio A, Gelb BD, Neri G, Tartaglia M (2007) Diversity, parental germline origin and phenotypic spectrum of *de novo HRAS* missense changes in Costello syndrome. *Hum Mutat* 28:265–272.

17

CRANIOSYNOSTOSIS SYNDROMES

ELIZABETH J. BHOJ AND ELAINE H. ZACKAI
The Children's Hospital of Philadelphia and The Perelman School of Medicine at the University of Pennsylvania School of Medicine, Philadelphia, Pennsylvania, USA

INTRODUCTION

When craniosynostosis is suspected, a thorough evaluation is necessary to determine which sutures are involved and if there is an underlying cause. The skull shape varies depending on which sutures are affected, with compensatory skull growth in those dimensions not restricted by the synostosis (Figure 17.1). Sagittal synostosis leads to an increase in the anterior-posterior diameter of the skull, called dolichocephaly or scaphocephaly. Unilateral coronal synostosis gives rise to an asymmetric skull shape, with retrusion of forehead on the affected side and protrusion of the unaffected side. This asymmetric skull is referred to as plagiocephaly. Bicoronal synostosis leads to a foreshortened anteroposterior diameter of the skull, termed brachycephaly, and an increased height, referred to as turricephaly. Premature fusion of the metopic suture causes wedging of the forehead with prominence of the metopic region, called trigonocephaly. An evaluation of the abnormal skull shape can be used to identify the fused suture or sutures, and radiographs or a computed tomography study of the skull can confirm the synostosis.

More than 200 craniosynostosis syndromes have been described, many of these are accompanied by limb abnormalities, suggesting common molecular pathways for craniofacial and limb development. Apert, Pfeiffer, Crouzon, Muenke, and Saethre–Chotzen are the most common of the craniosynostosis syndromes (see Table 17.1 for an overview). This chapter will focus on these entities.

Incidence

Craniosynostosis, the premature fusion of cranial sutures, is relatively common, with an estimated birth prevalence of 1 in 2000–2500. Most often the sagittal suture is affected, and the majority of these are of non-genetic origin. Coronal synostosis is the second most common form, and this may be an isolated finding (about two-thirds of cases) or the result of an underlying genetic cause (in about one-third of cases).

Diagnostic Criteria

There are no established diagnostic criteria for any of the craniosynostosis syndromes. Thus, diagnosis relies on identifying the pattern of abnormalities, and there is significant phenotypic overlap between syndromes.

Apert Syndrome Apert syndrome is considered the most severe of the craniosynostosis syndromes. In addition to the alterations of skull shape, a characteristic finding is symmetric and severe syndactyly affecting hands and feet (Fig 17.2). Osseous and cutaneous syndactyly may range from partial to complete fusion of the second, third, fourth, and to a lesser degree, fifth fingers. The thumb is typically not involved in the fusion, giving rise to the descriptive term "mitten hand". Feet typically show syndactyly affecting all toes, with a broad and medially deviated distal phalanx of the hallux less involved in the fusion than toes 2–5. Bicoronal synostosis causes a short anterior-posterior diameter of the skull with a tall and broad forehead. The synostosis is often

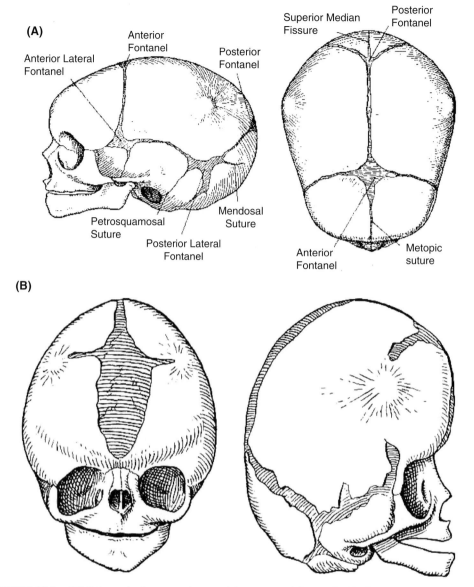

FIGURE 17.1 (A) Schematic demonstrating cranial sutures and fontanels. (B) Schematic demonstrating craniosynostosis with resultant brachycephaly. Reproduced, with permission, from Toriello (1993).

irregular, and large midline defects may be present, extending from the anterior to the posterior fontanel. Facial findings include a broad and flat forehead with supraorbital grooves, ocular hypertelorism and downslanting palpebral fissures, a depressed nasal root with a short, broad nose, and a long upper lip with thin vermillion. Cleft palate occurs in 43% (Slaney et al. 1996). Structural central nervous system abnormalities may include agenesis of the corpus callosum and ventriculomegaly. Mental deficiencies are present in the vast majority of affected individuals. For a more detailed review, see Cohen (2000, 2005).

Although the clinical diagnosis is based on the craniofacial findings in combination with the characteristic syndactyly, various organ systems can be affected. Congenital heart defects and genitourinary anomalies are each present in about 10%, and affected individuals should be evaluated appropriately for such anomalies. The upper airway may be compromised by choanal stenosis and reduced nasopharyngeal dimensions. Cohen and Kreiborg (1992) reported two cases of Apert syndrome and reviewed five from the literature with complete or partial cartilage sleeve abnormality of the trachea. Tracheal sleeve has become an increasingly recognized feature of craniofacial syndromes (Wenger et al. 2017).

Crouzon Syndrome Crouzon syndrome typically presents with bicoronal synostosis, although rarely pansynostosis develops in infancy or early childhood. Facial findings include hypertelorism, proptosis, midfacial hypoplasia, and a convex nasal ridge (Figure 17.3). Hands and feet do not

TABLE 17.1 Overview of the common craniosynostosis syndromes and the causal gene variants

Syndrome	Apert	Crouzon	Pfeiffer	Muenke	Saethre–Chotzen
Synostosis; skull findings	Bicoronal; irregular with ossification defects	Bicoronal; rarely pansynostosis or cloverleaf skull	Bicoronal; cloverleaf skull in type III	Uni- or bicoronal synostosis; rarely macrocephaly only	Bi- or unicoronal; rarely metopic
Facial findings	Hypertelorism, downslanting palpebral fissures; cleft palate	Proptosis, hypertelorism; beaked nose	Proptosis, hypertelorism, downslanting palpebral fissures	Mild facial findings, downslanting palpebral fissures	Ptosis, small ears with a prominent crus, facial asymmetry
Hand and foot abnormalities	Severe syndactyly of hands and feet	None	Broad and medially deviated thumbs and halluces; rarely symphalangism	Mild brachydactyly	Hallux valgus, partial duplication of the first halluces, mild syndactyly of hands and feet; brachydactyly
Other	Intellectual disability common; cardiac disease in 10%	None	Multiple malformations in types II and III	Possible learning disabilities	Intellectual disability common with microdeletion
Gene variants	*FGFR2* Ser252Trp or Pro253Arg	*FGFR2* variants, multiple	*FGFR1* Pro252Arg; *FGFR2* variants, multiple	*FGFR3* Pro250Arg	*TWIST* variants; rarely complete gene deletion

show abnormalities, although radiographic abnormalities, such as carpal fusion, may occasionally be present. This radiographic finding was considered to be diagnostic of Jackson–Weiss syndrome. Jackson–Weiss syndrome shares pathogenic variants in common with Crouzon and Pfeiffer syndromes, thus these separate clinical entities are allelic.

Pfeiffer Syndrome Pfeiffer syndrome consists of craniosynostosis with broad and medially deviated great toes and broad thumbs (Cohen 1993). Bicoronal synostosis causes the typical brachycephalic skull shape with a tall forehead, and proptosis and midfacial hypoplasia are often seen (Figure 17.4). Pfeiffer syndrome can be differentiated from Crouzon and Muenke syndromes based on the characteristic thumb and toe abnormalities.

Three clinical subtypes have been reported, with type I being by far the most common. Pfeiffer syndrome type 1 consists of craniosynostosis and the typical hand and foot abnormalities. Affected individuals do not typically show other major malformations or intellectual impairment. Pfeiffer syndrome type II presents at birth with cloverleaf skull, a severe skull abnormality that is the result of premature fusion of all cranial sutures. These individuals may have additional malformations and have a poor prognosis. Pfeiffer syndrome type III does not show cloverleaf skull, but craniosynostosis and its associated facial findings are severe, associated malformations are often present, and the prognosis is poor.

Tracheal cartilaginous sleeve is an airway malformation in which distinct tracheal rings cannot be identified. In place of normal cartilaginous arches a continuous segment of cartilage extends from below the subglottis to the carina or possibly to the mainstem bronchi. The resulting significant respiratory distress is likely to alter management, including more invasive interventions such as tracheostomy. In a recent series tracheal cartilaginous sleeves were found in 22% of patients with syndromic crainosynostosis and were found in 5/5 patients with the *FGFR2* W290C variant (Wenger 2017).

Muenke Syndrome The craniosynostosis syndrome associated with the Pro250Arg variant in fibroblast growth factor receptor 3 (FGFR3) was delineated based on the underlying variant (Muenke et al. 1997), and has been referred to as FGFR3-associated coronal synostosis syndrome, Muenke craniosynostosis, and Muenke syndrome (Cohen 2002). *FGFR3* is located at 4p16.3, and this gene is causally involved in the dwarfing conditions achondroplasia (see Chapter 2), hypochondroplasia, and thanatophoric dysplasia. Craniosynostosis in the FGFR3-associated synostosis syndrome is typically bicoronal but may be unicoronal (Gripp et al. 1998a), and macrocephaly without synostosis may occur occasionally (Figure 17.5). Brachydactyly and subtle radiographic changes, such as thimble-shaped middle phalanges, are associated. Mild to moderate, low to mid-frequency, mostly sensorineural, hearing loss has been seen in many individuals (Doherty et al. 2007). Mild learning disabilities have also been reported.

Saethre–Chotzen Syndrome Saethre–Chotzen syndrome was first described by Saethre (1931) and Chotzen (1932), and its variable clinical findings were reviewed by Pantke et al. (1975). In addition to coronal synostosis, other sutures

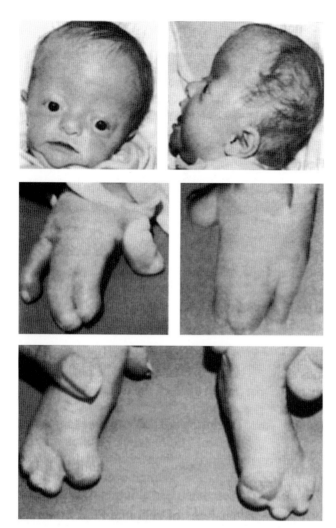

FIGURE 17.2 Baby with Apert syndrome demonstrating bicoronal synostosis with resultant brachycephaly, syndactyly of the hands and feet, and medial deviation of the thumbs and great toes as a result of a novel pathogenic variant of the IgII-IgIII linker region. Reproduced, with permission, from Oldridge et al. (1997).

may be involved and the skull shape can be asymmetric. Facial findings include ptosis, tear duct stenosis, small ears with a prominent crus and helical root, and facial asymmetry (Figure 17.6). Hands may show a single palmar flexion crease and mild cutaneous syndactyly and may be short. Broad or bifid hallucs in valgus position and syndactyly of toes may be present. Other findings, such as vertebral malformations or anal atresia, are much less common but consistent with a diagnosis of Saethre–Chotzen syndrome. Saethre–Chotzen syndrome is autosomal dominant with variable expressivity, and mildly affected family members can be identified by thorough examination. Most affected persons do not have neurodevelopmental abnormalities. Craniosynostosis is not an obligatory finding for the diagnosis; rather, affected family members may be identified by other typical findings and family history.

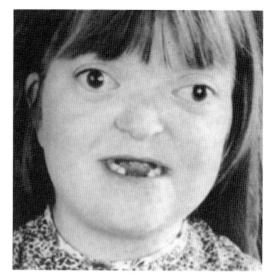

FIGURE 17.3 Proptosis, midfacial hypoplasia, and a beaked nose in a girl with Crouzon syndrome caused by an *FGFR2* pathogenic variant. Reproduced, with permission, from Reardon et al. (1994).

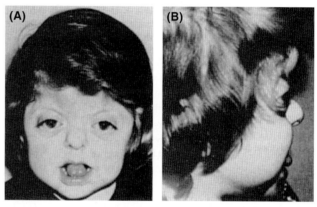

FIGURE 17.4 Girl with Pfeiffer syndrome whose features include bicoronal synostosis leading to severe ocular proptosis and midfacial hypoplasia caused by a splice site pathogenic variant in exon B of *FGFR2*. Reproduced, with permission, from Schell et al. (1995).

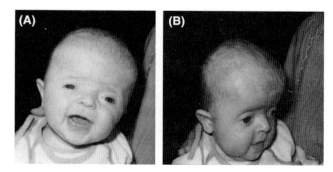

FIGURE 17.5 A baby with bicoronal synostosis caused by an *FGFR3* pathogenic variant. Reproduced, with permission, from Muenke et al. (1997).

INTRODUCTION

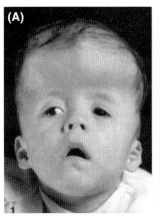 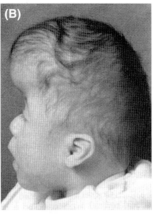

FIGURE 17.6 A baby with bicoronal synostosis, prominent forehead, hypertelorism, ptosis, downslanting palpebral fissures, and a prominent ear crus consistent with a clinical diagnosis of Saethre–Chotzen syndrome and confirmed by a *TWIST* deletion. Reproduced, with permission, from Reid et al. (1993).

Etiology, Pathogenesis, and Genetics

Apert Syndrome Apert syndrome is caused by pathogenic variants in the fibroblast growth factor receptor 2 (*FGFR2*) gene, located on chromosome 10 (10q26) (Wilkie et al.

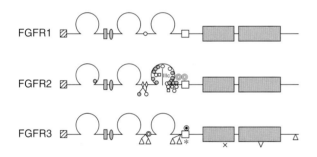

◊ Apert syndrome
○ Pfeiffer syndrome
● Crouzon syndrome
□ Jackson-Weiss syndrome
◉ Beare-Stevenson cutis gyrata syndrome
△ Thanatophoric dysplasia, Type I
▽ Thanatophoric dysplasia, Type II
∗ Achondroplasia
× Hypochondroplasia
⊛ Crouzonodermoskeletal syndrome
⊚ FGFR3 associated coronal synostosis syndrome

FIGURE 17.7 Fibroblast growth factor receptors 1, 2, 3. Most pathogenic variants for major craniosynostosis syndromes are in *FGFR2*. Note pathogenic variants for short limb skeletal dysplasia (achondroplasia, hypochondroplasia, thanatophoric dysplasia type I, and thanatophoric dysplasia type II) on *FGFR3*. Hatched square = signal peptide. Solid oblong = acid box. Solid oval = CAM homology domain. Open square = transmembrane domain. Long oblongs = kinase domains 1 and 2. Three loops from left to right are immuno-globulin-like domains (IgI, IgII, and IgIII). IIIc is the alternatively spliced form of second half of IgIII. For clarity, only a few of the many pathogenic variants for Crouzon syndrome and Pfeiffer syndrome are shown on *FGFR2*. Pathogenic variants for osteoglophonic dysplasia have been found on *FGFR1* IgIII (White et al. 2005). Reproduced, with permission, from Cohen (2000).

1995) (Figure 17.7). In more than 98% of affected individuals one of two point variants affecting adjacent amino acids is present, leading to a predicted amino acid change of either Ser252Trp or Pro253Arg (Bochukova et al. 2009). These pathogenic variants have been shown to cause reduced dissociation of the ligand/receptor complex, thus leading to increased signaling activity. Although Apert syndrome is an autosomal dominant condition with complete penetrance, most cases are the result of a de novo variant because individuals with Apert syndrome have significantly decreased reproductive fitness. The pathogenic variants are exclusively of paternal origin (Moloney et al. 1996), and there is an association of de novo variants causing Apert syndrome with advanced paternal age.

Crouzon Syndrome In about half of all cases of Crouzon syndrome, a pathogenic variant can be identified in the fibroblast growth factor receptor 2 (*FGFR2*) gene, located on chromosome 10 (10q26) (Passos-Bueno et al. 2008) (Figure 17.7). The vast majority of these are single nucleotide variants, and some are identical to variants identified in Pfeiffer syndrome (Kan et al. 2002). These pathogenic variants act through a gain-of-function mechanism and are thus inherited as an autosomal dominant condition. Because Crouzon syndrome does not affect intellectual abilities and reproductive fitness is not reduced, it is often inherited rather than the result of a de novo variant.

Pfeiffer Syndrome Pfeiffer syndrome is now known to be heterogeneous with pathogenic variants in *FGFR1* and *FGFR2* (Figure 17.7). Pfeiffer syndrome type I may be caused by one specific single nucleotide variant (Pro252Arg) in *FGFR1* (Muenke et al. 1994) or by one of numerous variants in *FGFR2* (Schell et al. 1995). The pathogenic variants in *FGFR2* can be identical to those identified in Crouzon syndrome and Pfeiffer syndrome types II and III. Pfeiffer syndrome type I can be inherited as an autosomal dominant trait, whereas Pfeiffer syndrome types II and III are typically caused by de novo autosomal dominant pathogenic variants (Passos-Bueno et al. 2008).

Muenke Syndrome By definition, individuals with Muenke syndrome carry the *FGFR3* Pro250Arg pathogenic variant (Figure 17.7). This *FGFR3* variant is homologous to the *FGFR1* Pro252Arg variant causing Pfeiffer syndrome and the *FGFR2* Pro253Arg variant seen in Apert syndrome. Like other *FGFR* pathogenic variants, the *FGFR3* Pro250Arg variant acts through a gain-of-function mechanism and can be inherited as an autosomal dominant condition. Because the phenotype can be mild, including unicoronal synostosis or macrocephaly only, careful evaluation of first-degree relatives is necessary to identify affected family members.

Saethre-Chotzen syndrome Pathogenic variants in the *TWIST* gene, a developmentally regulated transcription factor, have been identified in individuals with Saethre–Chotzen syndrome (El Ghouzzi et al. 1997; Howard et al. 1997). Complete deletions of the *TWIST* gene are found in some affected individuals (Johnson et al. 1998). Rarely, a translocation involving 7p21, the *TWIST* gene locus, causes Saethre–Chotzen syndrome, presumably because of a position effect (Reid et al. 1993). Haploinsufficiency for the functional gene product leads to the phenotypic findings.

Often Saethre–Chotzen syndrome is inherited as an autosomal dominant condition In contrast, individuals with a microdeletion of 7p21, encompassing the *TWIST* gene and presumably other genes, are more likely to have a de novo abnormality, and often show significant neurodevelopmental abnormalities (Johnson et al. 1998; Gripp et al. 2001).

Diagnostic Testing

Apert Syndrome Molecular testing (targeted Sanger sequencing) for the two common pathogenic variants, Ser252Trp and Pro253Arg in *FGFR2*, with reflex to sequencing and deletion/duplication analysis of *FGFR2* is recommended.

Crouzon/Pfeiffer Syndrome Molecular testing (targeted Sanger sequencing) for pathogenic variants in exons 8 and 10 in *FGFR2* is recommended with reflex to sequencing of exons 3, 5, 11, 14, 15, 16, and 17. If these are negative, then targeted sequencing for *FGFR3* pathogenic variant Pro250Arg can be considered. *FGFR1* sequencing can also be performed, especially in individuals with a mild presentation. A negative study does not preclude a clinical diagnosis of Crouzon/Pfeiffer syndrome.

Muenke Syndrome Targeted Sanger sequencing for the *FGFR3* Pro250Arg pathogenic variant is recommended.

Saethre–Chotzen Syndrome Molecular testing (targeted Sanger sequencing) for *TWIST* sequence changes and microdeletions is recommended. When these assays are normal, chromosome analysis may be considered because balanced translocation may exert a position effect on the *TWIST* gene.

In addition, there are many clinically available tests that sequence and perform duplication/deletion analysis of multiple genes involved in craniosynostosis syndromes (targeted gene panels). These may be performed using Sanger sequencing or next generation sequencing, usually based on the number of genes tested. If targeted testing is negative then clinical exome sequencing can be considered for one of the >200 Mendelian disorders that can present with syndromic craniosynostosis.

Differential Diagnosis

Isolated craniosynostosis, without an identifiable genetic cause, is the most common form of craniosynostosis and typically affects the sagittal suture. Unilateral coronal synostosis, causing anterior plagiocephaly is most often an isolated finding but can be seen in Muenke or Saethre–Chotzen syndromes. In apparently isolated craniosynostosis, a thorough dysmorphology examination and a complete family history may give clues to the genetic nature of the synostosis. Metopic synostosis, giving rise to trigonocephaly, can be seen in Saethre–Chotzen syndrome, but is heterogeneous in its cause. Metopic synostosis has been reported after intrauterine valproic acid exposure (Chapter 27), in numerous chromosome anomalies, and as an isolated finding. Comparative genomic hybridization array study may be considered in individuals with trigonocephaly.

Wilkie et al. (2006) discouraged genetic testing for sporadic nonsyndromic sagittal and metopic synostosis because they have found that the diagnostic yield is close to zero. In contrast, they recommend that all individuals with bicoronal and unicoronal synostosis have testing at least for the P250R pathogenic variant in *FGFR3*.

Craniosynostosis in combination with multiple congenital anomalies can be seen in less common craniosynostosis syndromes such as Antley–Bixler or Baller–Gerold syndromes. Craniofrontonasal syndrome, caused by pathogenic variants in the X-chromosome *EFNB1* gene (Wieland et al. 2004) may present with coronal synostosis in females. Thanatophoric dysplasia, a lethal short limb type of skeletal dysplasia caused by pathogenic variants in *FGFR3*, may present with cloverleaf skull (multiple suture synostosis). When additional malformations are noted, these can be used to identify rare craniosynostosis syndromes. It is also important to remember that craniosynostosis may be secondary to another pathologic process, such as primary microcephaly, rickets, hypercalcemia, hyperthyroidism, or various hematologic disorders. Lastly, chromosomal anomalies can present with craniosynostosis affecting multiple sutures and additional malformations.

MANIFESTATIONS AND MANAGEMENT

Growth and Feeding

Feeding difficulties may be related to cleft palate, which is seen in Apert syndrome. In infants, upper airway problems, including choanal stenosis or narrowing, may lead to feeding difficulties. Neurological abnormalities, for example those related to increased intracranial pressure, can also cause feeding difficulties and emesis.

In addition to the abnormal skull shape directly related to the craniosynostosis, macrocephaly may be seen in Muenke syndrome. Microcephaly caused by pansynostosis may occasionally be the presenting finding in Crouzon syndrome, when synostosis develops postnatally.

Linear growth may be slightly below average, most notably in Apert syndrome. Height velocity in Apert syndrome decelerates with ages, typically resulting in height at the low end of the normal range (Cohen and Kreiborg 1993). Relative short stature is occasionally seen in Saethre–Chotzen syndrome.

Evaluation

- Feeding difficulties in infants should prompt an evaluation for cleft palate and for choanal stenosis or atresia. A neurological cause, in particular related to raised intracranial pressure caused by the craniosynostosis, should be considered.
- A feeding team or speech pathology evaluation may be helpful in making treatment suggestions, regardless of the underlying cause.
- Head circumference should be measured regularly. A sudden increase may raise concern for hydrocephalus, and a CT study of the brain may be helpful.
- Height velocity should be assessed regularly throughout childhood and adolescence, and a significant change in the growth velocity may require further investigation.

Management

- If a specific underlying cause for feeding difficulties, such as cleft palate or choanal stenosis, can be identified, standard treatment should be directed accordingly.
- Surgery may be necessary to correct raised intracranial pressure or hydrocephalus. Standard procedures are used.

Development and Behavior

Although fine motor development may be hampered by the severe syndactyly in Apert syndrome, intellectual development is of greater concern. The expected range of abilities varies greatly for the different craniosynostosis syndromes. Many individuals with Apert syndrome are intellectually disabled, with the mean IQ reported in different studies as 62 (range 10–114) or 74 (range 52–89) (Lefebvre et al. 1986; Patton et al. 1988). In contrast, individuals with Crouzon syndrome and Pfeiffer syndrome type I typically have normal intellectual development. The more severe Pfeiffer syndromes, types II and III, usually show severe delays (Gripp et al. 1998b). In Muenke syndrome, mild learning disabilities may be present. Although most individuals with Saethre–Chotzen syndrome have normal intellectual abilities, some have learning difficulties. Individuals presenting with Saethre–Chotzen syndrome and severe developmental delay or intellectual disability are more likely to have a microdeletion encompassing the complete *TWIST* gene (Johnson et al. 1998; Gripp et al. 2001).

There are no unusual behaviors, other than those directly related to intellectual disability, that are characteristic of any of the craniosynostosis syndromes. It is important to consider self-esteem issues as a possible cause for behavior problems in older persons with craniofacial abnormalities.

Evaluation

- A proactive approach pursuing a full developmental evaluation should be used in individuals with Apert syndrome and Pfeiffer syndrome types II and III because severe developmental delays are likely to be present.
- In Muenke syndrome, an increased index of suspicion for learning disabilities should lead to a more thorough screening, followed by a full developmental evaluation if an area of concern is noted.
- As in all children with developmental problems, a hearing evaluation and ophthalmologic examination should be performed to rule out treatable problems causing the developmental delay.
- Awareness of the risk for self-esteem problems related to craniofacial abnormalities may lead to referral to a counselor or psychologist if problems are identified. Signs of such problems should be sought during physician visits.

Management

- Early intervention should begin as soon as possible for children with Apert syndrome and Pfeiffer types II and III.
- Hearing and vision problems should be addressed as needed and in a standard fashion.

Craniofacial

Abnormal skull shape is often the presenting sign of craniosynostosis. Physical examination may show ridging of the affected sutures, and the fontanels may be closed prematurely or may be enlarged. Based on the particular skull shape and the physical findings, one may suspect which sutures fused prematurely. Radiographs or a CT study (with three-dimensional reconstruction, if available) will then confirm the suspected fusion and may show associated skull defects. Once craniosynostosis is identified, referral to a craniofacial surgeon experienced in the field is indicated to arrange for surgical repair. Often the surgical procedure can be delayed for several months, and it may be planned in stages. The surgery is typically performed in collaboration with a neurosurgeon.

Individuals with craniosynostosis can have altered facial features including hypertelorism, proptosis, exophthalmous, maxillary hypoplasia, mandibular prognathism, or hypoplasia. Some people with Saethre–Chotzen have ptosis.

Evaluation

- Simple synostosis such as metopic or unilateral coronal can be assessed clinically. 3D CT is helpful for multi-suture craniosynostosis and useful for surgical planning in all cases.

- Most affected individuals benefit from evaluation by a craniofacial team. Craniofacial teams are composed of craniofacial surgeons and neurosurgeons, ophthalmologists, geneticists, orthodontists, speech therapists, audiologists, social workers, and other team members. Because people with craniosynostosis syndromes often require long-term management, and treatment is not finished until facial growth is completed in young adulthood, treatment planning and long-term follow-up are necessary and are handled most appropriately by an experienced team.
- Individuals with abnormalities in facial features resulting from craniosynostosis should be evaluated by plastic surgery and oculo-plastic surgery in a coordinated manner (Cunningham et al. 2007).

Management

- Treatment for the premature fusion of cranial sutures is suturectomy (Wilkie et al. 2010).
- Correction of the abnormal skull shape and the facial abnormalities involves complex procedures.
  - Often the last surgical procedure is delayed until facial growth is completed. As an example, a study of factors associated with the need for a reoperation in cases of apparently non-syndromic coronal synostosis demonstrated that the presence, or absence, of the Pro250Arg variant in *FGFR3* was the *only* significant prognostic factor. Pathogenic variant-positive cases had a re-operation rate of 20.7% compared with 4.3% for negative cases, a significant 4.9-fold difference (Wilkie et al. 2006).
  - Pathogenic variant-positive cases were reported to be more likely to require early intervention with a posterior release operation (at around 6 months) to prevent excessive frontal bulging (Thomas et al. 2005). A secondary major procedure is required for recurrent supraorbital retraction in at least 43% of individuals with *FGFR3* pathogenic variants. A secondary or tertiary extracranial forehead contouring should be anticipated in all individuals with craniosynostosis syndromes (Honnebier et al. 2008).
- Endoscopic strip craniectomies followed by cranial molding helmet is a minimally invasive procedure that is performed in young infants (Jimenez et al. 2002).
- Individuals with proptosis secondary to maxillary hypoplasia will need to undergo mid-face advancement surgery. Mid-face advancement surgery is usually not undertaken until the fifth year of life, at which point the facial skeleton is at a developmental maturity that will allow advancement using distraction osteogenesis.
- Affected individuals with significant degrees of orbital hypertelorism, such as those with midline facial clefts or craniofrontonasal dysplasia, will require medial translocation of their orbits by surgical osteotomies with resection of their excessive naso-fronto-ethmoid bones. This significantly morbid and surgically challenging procedure is not undertaken until the children are at least 5 years of age to ensure adequate surgical correction.
- Issues of malocclusion are not treated until children have reached skeletal maturity in their late adolescence and should be coordinated with the orthodontic team to allow for the establishment of a normal occlusion. As the occlusion is related to the position of the maxilla and mandible, a treatment plan should be formulated not only to achieve a normal dental occlusion but also to optimize the aesthetic outcome of the individual's appearance.

Otolaryngologic

Individuals with Apert syndrome and Pfeiffer syndrome types II and III are at an increased risk for tracheal cartilaginous sleeve, an airway malformation in which distinct tracheal rings cannot be identified. In place of normal cartilaginous arches, a continuous segment of cartilage extends from below the subglottis to the carina or possibly to the mainstem bronchi. This airway complication may be diagnosed by bronchoscopy.

Low- to mid-frequency hearing loss in Muenke syndrome may be of consequence in environments, such as the classroom, with ambient noise.

Evaluation

- Bronchoscopy with deep sedation or high-resolution CT of the neck and chest is recommended in all individuals with Pfeiffer and Apert syndromes.
- Regular audiometry evaluation is important in all individuals with craniosynostosis, particularly in Muenke syndrome.

Management

- Tracheostomy has been the mainstay of airway management for more severely affected patients (Hockstein et al. 2004).
- For individuals with Muenke syndrome who have hearing loss, special accommodation such as sound field amplification and preferential seating in the classroom can improve speech perception.

Ophthalmologic

The structural and functional integrity of the eyes can be affected in numerous ways in the craniosynostosis syndromes. Although ophthalmology is part of the craniofacial team approach, an ophthalmologic evaluation may be considered even before the individual with craniosynostosis has a full team appointment. It is important to identify papilledema, an indicator for increased intracranial pressure, as it may lead to optic atrophy. The abnormal skull shape typically gives rise to a shallow orbit, which in turn leads to proptosis of the globe. If severe proptosis prevents complete closure of the eyelids, the cornea will develop exposure keratitis and corneal ulcers, causing scarring and visual impairment. Less urgent problems include tear duct stenosis and ptosis, which are commonly seen in Saethre–Chotzen syndrome, and eye positional abnormalities, most often hypertelorism and downslanting palpebral fissures. The abnormal anatomy of the orbits and the orbital position often cause mechanical disturbances of the extraocular muscles, presenting with strabismus. Reconstructive craniofacial surgeries can change the anatomy of the orbits significantly, in turn affecting globe position and muscular function.

Evaluation

- An early ophthalmologic evaluation should be considered to identify urgent needs.
- Long-term regular follow-up, preferably by an experienced ophthalmologist working as a member of the craniofacial team, is warranted on an annual basis.

Management

- The immediate care of an infant with proptosis will include measures to ensure that there is no orbital exposure, which may cause corneal scarring.
- Tarsorrhaphies may be required to prevent exposure keratitis.
- Additional interventions should be planned with the craniofacial team, as craniofacial surgical procedures impact directly on ocular structure and function.
- Strabismus is treated in a standard manner.

Neurological

Pansynostosis or cloverleaf skull may cause increased intracranial pressure, requiring urgent neurosurgical intervention. Central nervous system malformations, including agenesis of the corpus callosum and ventriculomegaly, are seen with increased frequency in Apert syndrome and Pfeiffer syndrome types II and III.

Evaluation

- Once craniosynostosis is diagnosed, a thorough neurological evaluation and a neurosurgical consultation may be considered. Neurosurgical evaluation may be helpful if there is a concern about raised intracranial pressure.
- A brain imaging study to identify congenital structural anomalies is recommended in individuals with Apert syndrome and Pfeiffer syndrome types II and III.

Management

- Hydrocephaly should be treated in a standard manner. If increased intracranial pressure results from pansynostosis, it should be treated with suturectomies.

Musculoskeletal

Skeletal abnormalities are not uncommon, and include vertebral abnormalities in Saethre–Chotzen syndrome, radiohumeral synostosis in Pfeiffer syndrome, and shoulder girdle abnormalities with short humeri in Apert syndrome. Severe hand abnormalities, such as the syndactyly seen in Apert syndrome, can be present.

Evaluation

- A thorough physical examination for musculoskeletal abnormalities should be done when craniosynostosis is recognized.
- Individuals with significant hand malformations benefit from referral to a hand surgeon or orthopedist.

Management

- Surgical syndactyly release may be considered.
- Musculoskeletal malformations associated with craniosynostosis are treated in a standard manner.

Cardiovascular

Individuals with Apert syndrome and Pfeiffer syndrome types II and III are at increased risk for cardiac malformations. In about 10% of individuals with Apert syndrome structural cardiac anomalies, including ventricular septal defect and pulmonic stenosis, are present.

Evaluation

- A cardiology consultation or an echocardiogram should be considered to identify heart malformations in those with Apert syndrome and Pfeiffer syndrome types II and III.

Management

- Cardiac malformations should be treated in a standard manner.

Genitourinary

Individuals with Apert syndrome and Pfeiffer syndrome types II and III are at increased risk for genitourinary tract malformations. In about 10% of individuals with Apert syndrome structural anomalies, including hydronephrosis, polycystic kidneys, bicornuate uterus, vaginal atresia, and cryptorchidism, are present.

Evaluation

- Cryptorchidism should be sought on physical examination.
- An ultrasound of the urinary tract and pelvis will identify structural anomalies. It should be done in those with Apert syndrome and Pfeiffer II and III syndromes.

Management

- Genitourinary tract malformations associated with craniosynostosis syndromes should be treated in the standard manner.

RESOURCES

FACES: The National Craniofacial Association
Website: *http://www.faces-cranio.org/Disord/Cranio.htm*

National Association of Rare Diseases – Primary Craniosynostosis
Website: https://rarediseases.org/rare-diseases/primary-craniosynostosis

Craniosynostosis and Positional Plagiocephaly Support (CAPPS)
Website: https://www.cappskids.org

Orphanet: The Portal for Rare Diseases and Orphan Drugs (International)
Website: https://www.orpha.net/

REFERENCES

Bochukova EG, Roscioli T, Hedges DJ, Taylor IB, Johnson D, David DJ, Deininger PL, Wilkie AO (2009) Rare mutations of FGFR2 causing Apert syndrome: Identification of the first partial gene deletion, and an Alu element insertion from a new subfamily. *Hum Mutat* 30:204–11.

Chotzen F (1932) Eine eigenartige familiaere Entwicklungsstoerung (Akrocephalosyndactylie, Dysostosis craniofacialis und Hypertelorismus). *Monatschr Kinderheilkd* 55:97–122.

Cohen MM Jr (1993) Pfeiffer syndrome update, clinical subtypes and guidelines for differential diagnosis. *Am J Med Genet* 45:300–307.

Cohen MM Jr (2000) In: *Craniosynostosis: Diagnosis, Evaluation and Management*, 2nd ed., *Cohen MM Jr, MacLean RE eds*, New York: Oxford University Press.

Cohen MM Jr (2002) Fibroblast growth factor mutations. In: *Molecular Basis of Inborn Errors of Development*, Epstein CJ, Erickson RP, Wynshaw-Boris A, eds, New York: Oxford University Press.

Cohen MM Jr (2005) Editorial: Perspectives on craniosynstosis. *Am J Med Genet* 16A:313–326.

Cohen MM Jr Kreiborg S (1992) Upper and lower airway compromise in the Apert syndrome. *Am J Med Genet* 44:90–93.

Cohen MM Jr, Kreiborg S (1993) Growth pattern in Apert syndrome. *Am J Med Genet* 47:617–623.

Cunningham ML, Seto ML, Ratisoontorn C, Heike CL, Hing AV (2007) Syndromic craniosynostosis: From history to hydrogen bonds. *Orthod Craniofac Res* 10:67–81.

Doherty ES, Lacbawan F, Hadley DW, Brewer C, Zalewksi C, Kim HJ, Solomon B, Rosenbaum K, Domingo DL, Hart TC, Brooks BP, Immken L, Lowry RB, Kimonis V, Shanske AL, Sarquis Jehee F, Pasos Bueno MR, Knightly C, McDonald-McGinn DM, Zackai EH, Muenke M (2007) Muenke syndrome (*FGFR3*-related craniosynostosis): Expansion of the phenotype and review of the literature. *Am J Med Genet* 143A:3204–3215.

El Ghouzzi V, Le Merrer M, Perrin-Schmitt F, Lajeunie E, Benit P, Renier D, Bourgeois P, Bolcato-Bellemin A-L, Munnich A, Bonaventure J (1997) Mutations of the *TWIST* gene in the Saethre-Chotzen syndrome. *Nat Genet* 15:42–46.

Gripp KW, McDonald-McGinn DM, Gaudenz K, Whitaker LA, Bartlett SP, Glat PM, Cassileth LB, Mayro R, Zackai EH, Muenke M (1998a) Identification of a genetic cause for isolated unilateral coronal synostosis: A unique mutation in the fibroblast growth factor receptor 3. *Pediatrics* 132:714–716.

Gripp KW, Stolle CA, McDonald-McGinn DM, Markowitz RI, Barlett SP, Katowitz JA, Muenke M, Zackai EH (1998b) Phenotype of the fibroblast growth factor receptor 2 Ser351Cys mutation: Pfeiffer syndrome type III. *Am J Med Genet* 78:356–360.

Gripp KW, Kasparacova V, McDonald-McGinn DM, Bhatt S, Bartlett SP, Storm Al, Drumheller TC, Emanuel BS, Zackai EH, Stolle CA (2001) A diagnostic approach to identifying submicroscopic 7p21 deletions in Saethre-Chotzen syndrome: Fluorescence *in situ* hybridization and dosage-sensitive Southern blot analysis. *Genet Med* 3:102–108.

Hockstein NG, McDonald-McGinn DM, Zackai E, Bartlett S, Huff DS, Jacobs IN (2004) Tracheal anomalies in Pfeiffer syndrome. *Arch Otolaryngol Head Neck Surg* 130:1298–1302.

Honnebier MB, Cabiling DS, Hetlinger M, McDonald-McGinn DM, Zackai EH, Bartlett SP (2008) The natural history of patients treated for FGFR3-associated (Muenke-type) craniosynostosis. *Plast Reconstr Surg* 121:919–931.

Howard TD, Paznekas WA, Green ED, Chiang LC, Ma N, Ortiz de Luna RI, Garcia Delgado C, Gonzalez-Ramos M, Kline AD, Jabs EW (1997) Mutations in *TWIST*, a basic helix-loop-helix transcription factor, in Saethre-Chotzen syndrome. *Nat Genet* 15:36–41.

Jimenez DF, Barone CM, Cartwright CC, Baker L, (2002) Early management of craniosynostosis using endoscopic-assisted strip cranioectomies and cranial orthotic molding therapy. *Pediatrics* 110:97–104.

Johnson D, Horsley SW, Moloney DM, Oldridge M, Twigg SR, Walsh S, Barrow M, Njolstad PR, Kunz J, Ashworth GJ, Wall SA, Kearney L, Wilkie AOM (1998) A comprehensive screen for *TWIST* mutations in patients with craniosynostosis identifies a new microdeletion syndrome of chromosome band 7p21.1. *Am J Hum Genet* 63:1282–1293.

Kan SH, Elanko N, Johnson D, Cornejo-Roldan L, Cook J, Reich EW, Tomkins S, Verloes A, Twigg SR, Rannan-Eliya S, McDonald-McGinn DM, Zackai EH, Wall SA, Muenke M, Wilkie AOM (2002) Genomic screening of fibroblast growth-factor receptor 2 reveals a wide spectrum of mutations in patients with syndromic craniosynostosis. *Am J Hum Genet* 70:472–486.

Lefebvre A, DipSc FT, Arndt EM, Munro IR (1986) A psychiatric profile before and after reconstructive surgery in children with Apert's syndrome. *Br J Plast Surg* 39:510–513.

Moloney DM, Slaney SF, Oldridge M, Wall SA, Sahlin P, Stenman G, Wilkie AOM (1996) Exclusive paternal origin of new mutations in Apert syndrome. *Nat Genet* 13:48–53.

Muenke M, Schell U, Hehr A, Robin NH, Losken HW, Schinzel A, Pulleyn LJ, Rutland P, Reardon W, Malcolm S (1994) A common mutation in the fibroblast growth factor receptor 1 gene in Pfeiffer syndrome. *Nat Genet* 8:269–274.

Muenke M, Gripp KW, McDonald-McGinn DM, Gaudenz K, Whitaker LA, Bartlett SP, Markowitz RI, Robin NH, Nwokoro N, Mulvihill JJ, Losken HW, Mulliken JB, Guttmacher AE, Wilroy RS, Clarke LA, Hollway G, Ades LC, Haan EA, Mulley JC, Cohen MM Jr, Bellus GA, Francomano CA, Moloney DM, Wall SA, Wilkie AOM, Zackai EH (1997) A unique point mutation in the fibroblast growth factor receptor 3 gene (*FGFR3*) defines a new craniosynostosis syndrome. *Am J Hum Genet* 60:555–564.

Oldridge M, Lunt PW, Zackai EH, McDonald-McGinn DM, Muenke M, Moloney D, Twigg SRF, Heath JK, Howard TD, Hoganson G, Gagnon DM, Jabs EW, Wilkie AOM (1997) Genotype-phenotype correlation for nucleotide substitutions in the IgII-IgIII linker of FGFR2. *Hum Mol Genet* 6:137–143.

Pantke OA, Cohen MM Jr, Witkop CJ Jr, Feingold M, Schaumann B, Pantke HC, Gorlin RJ (1975) The Saethre-Chotzen syndrome. *Birth Defects Orig Art Ser* 11:190–225.

Passos-Bueno MR, Serti Eacute AE, Jehee FS, Fanganiello R, Yeh E. (2008) Genetics of craniosynostosis: genes, syndromes, mutations and genotype-phenotype correlations. *Front Oral Biol* 12:107–43.

Patton MA, Goodship J, Hayward R, Lansdown R (1988) Intellectual development in Apert's syndrome: A long term follow up of 29 patients. *J Med Genet* 25:164–167.

Reardon W, Winter RM, Rutland P, Pulleyn LJ, Jones BM, Malcolm S (1994) Mutations in the fibroblast growth factor receptor 2 gene cause Crouzon syndrome. *Nat Genet* 8:98–103.

Reid CS, McMorrow LE, McDonald-McGinn DM, Grace KJ, Ramos FJ, Zackai EH, Cohen MM Jr, Jabs EW (1993) Saethre-Chotzen syndrome with familial translocation at chromosome 7p22. *Am J Med Genet* 47:637–639.

Saethre H (1931) Ein Beitrag zum Turmschaedelproblem (Pathogenese, Erblichkeit und Symptomatologie). *Dtsch Z Nerven-heilkd* 117:533–555.

Schell U, Hehr A, Feldman GJ, Robin NH, Zackai EZ, de Die-Smulders C, Viskochil DH, Stewart JM, Wolff G, Ohashi H (1995) Mutations in *FGFR1* and *FGFR2* cause familial and sporadic Pfeiffer syndrome. *Hum Mol Genet* 4:323–328.

Slaney SF, Oldridge M, Hurst JA, Moriss-Kay GM, Hall CM, Poole MD, Wilkie AOM (1996) Differential effects of *FGFR2* mutations on a syndactyly and cleft palate in Apert syndrome. *Am J Hum Genet* 58:923–932.

Thomas GPI, Wilkie AOM, Richards PG, Wall SA (2005) FGFR3 P250R mutation increases the risk of reoperation in apparent 'nonsyndromic' coronal craniosynostosis. *J Craniofac Surg* 16:347–352.

Toriello H (1993) Cranium. In: Human Malformations and Related Anomalies. Vol. II. Stevenson RE, Hall JG, Goodman RM, eds., *New York: Oxford University Press*.

Wenger TL, Dahl J, Bhoj EJ, Rosen A, McDonald-McGinn D, Zackai E, Jacobs I, Heike CL, Hing A, Santani A, Inglis AF, Sie KC, Cunningham M, Perkins J (2017) Tracheal cartilaginous sleeves in children with syndromic craniosynostosis. *Genet Med* 19(1):62–68.

White KE, Cabral JM, Davis SI, Fishburn T, Evans WE, Ichikawa S, Fields J, Yu X, Shaw NJ, McLellan NJ, McKeown C, FitzPatrick D, Yu K, Ornitz DM, Econs MJ (2005) Mutations that cause osteoglophonic dysplasia define novel roles for FGFR1 in bone elongation. *Am J Hum Genet* 76:361–367.

Wieland I, Jakubiczka S, Muschke P, Cohen M, Thiele H, Gerlach KL, Adams RH, Wieacker P (2004) Mutations of the ephrin-B1 gene cause craniofrontonasal syndrome. *Am J Hum Genet* 74:1209–1215.

Wilkie AOM (1997) Craniosynostosis: Genes and mechanisms. *Hum Mol Genet* 6:1647–1656.

Wilkie AOM, Slaney SF, Oldridge M, Poole MD, Ashworth GJ, Hockley AD, Hayward RD, David DJ, Pulleyn LJ, Rutland P (1995) Apert syndrome results from localized mutations of FGFR2 and is allelic with Crouzon syndrome. *Nat Genet* 9:165–172.

Wilkie AOM, Bochukova EG, Hansen RMS, Taylor IB, Rannan-Eliya SV, Byren JC, Wall SA, Ramos L, Venacio M, Hurst JA, O'Rourke AW, Williams LJ, Seller A, Lester A (2006) Research Review: Clinical dividends from the molecular genetic diagnosis of craniosynostosis. *Am J Med Genet* 140A:2631–2639.

Wilkie AO, Byren JC, Hurst JA, Jayamohan J, Johnson D, Knight SJ, Lester T, Richards PG, Twigg SR, Wall SA. (2010) Prevalence and complications of single-gene and chromosomal disorders in craniosynostosis. *Pediatrics* 126:e391–400.

18

DELETION 1p36 SYNDROME

AGATINO BATTAGLIA

IRCCS Stella Maris Foundation, Pisa, Italy; Division of Medical Genetics, Department of Pediatrics University of Utah School of Medicine, Salt Lake City, Utah, USA; and Division of Medical Genetics, Department of Pediatrics, Sanford School of Medicine, University of South Dakota, Sioux Falls, South Dakota, USA

INTRODUCTION

Terminal deletion 1p36 was first reported in 1980 (Hain et al. 1980) as the result of malsegregation of a parental balanced translocation. Fifteen years later, two distinct phenotypes, differentiated by growth failure versus macrosomia, were proposed by Keppler-Noreuil et al. (1995) in their review of 13 individuals with small terminal deletions involving 1p36.22. However, most of these individuals had double segmental imbalances because of unbalanced translocations. Thus, it was difficult to distinguish the clinical characteristics due to deletion 1p36 from those caused by imbalance of the other chromosome. It was only in 1997 that this newly emerging clinical entity was brought to the attention of clinical geneticists through the description of 13 individuals with isolated deletion of 1p36 (Shapira et al. 1997). Several other case reports were published thereafter (Eugster et al. 1997; Riegel et al. 1999; Zenker et al. 2002; Heilstedt et al. 2003b). The comprehensive delineation of the clinical phenotype, the spectrum of epilepsy, and the natural history in deletion 1p36 syndrome were described in larger numbers of individuals in the last 12 years (Gajecka et al. 2007; Bahi-Buisson et al. 2008; Battaglia et al. 2008). Lack of familiarity with the clinical phenotype, along with the low yield from conventional cytogenetic studies, may have contributed, in the past, to a low ascertainment rate of affected individuals (Zenker et al. 2002; Battaglia et al. 2008).

Incidence

It is generally estimated that 1/5000 live births are diagnosed with deletion 1p36 syndrome (Heilstedt et al. 2003a), making it the most common terminal deletion observed in humans. Deletions occur with a female predilection of 2:1 (Bahi-Buisson et al. 2008; Battaglia et al. 2008; Shimada et al. 2015) and across all ethnicities (Gajecka et al. 2007). Deletions of 1p36 are estimated to be responsible for about 1% of all "idiopathic intellectual disability" (Giraudeau et al. 2001).

Diagnostic Criteria

There are no consensus diagnostic criteria. However, the diagnosis of 1p36 deletion syndrome is suggested by a distinct facial appearance, which includes straight eyebrows, deep-set eyes, midface retrusion, broad and depressed nasal bridge, long philtrum, pointed chin, and facial hypotonia (Figure 18.1). Additional features include microbrachycephaly, large and late-closing anterior fontanel, epicanthal folds, malformed ears, brachy/camptodactyly, and short feet (Shapira et al. 1997; Heilstedt et al. 2003b; Battaglia, 2005; Battaglia et al. 2008; Battaglia and Shaffer 2008). The distinct facial phenotype remains easy to recognize over time (Battaglia et al. 2008). Developmental delay/intellectual disability is present in all, and hypotonia in 95%. Seizures occur in 44–58% of affected individuals (Battaglia et al. 2008; Bahi-Buisson et al. 2008) (Table 18.1). Since the use

---

Cassidy and Allanson's Management of Genetic Syndromes, Fourth Edition.
Edited by John C. Carey, Agatino Battaglia, David Viskochil, and Suzanne B. Cassidy.
© 2021 John Wiley & Sons, Inc. Published 2021 by John Wiley & Sons, Inc.

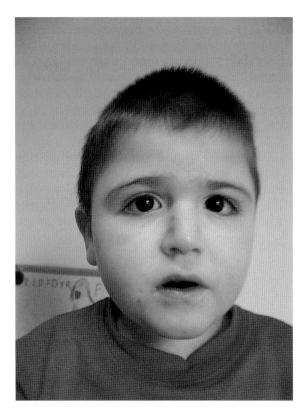

FIGURE 18.1 A 3-year-9-month-old boy, showing microbrachycephaly, straight eyebrows, deep-set eyes, strabismus, hypotonic face, midface retrusion, and pointed chin.

TABLE 18.1 Major clinical findings in 1p36 deletion syndrome.

Findings	Frequency
Distinct facial appearance	>75%
Intellectual disability	
Poor/absent speech	
Hypotonia	
Brachycamptodactyly	
Short feet	
Brain abnormalities	
Congenital heart defects	50–75%
Eye/vision problems	
Seizures	
Skeletal anomalies	25–50%
Sensorineural deafness	
Gastrointestinal anomalies	
Abnormalities of the external genitalia	
Behavior disorders	
Non-compaction cardiomyopathy	<25%
Renal anomalies	
Anal anomalies	
Hypothyroidism	

of array-CGH (comparative genomic hybridization) as the first-tier clinical diagnostic test for individuals with developmental delay/intellectual disability, a few children with developmental delay/intellectual disability harboring small interstitial deletions of 1p36 but lacking the distinct facial appearance have been observed (personal experience).

Etiology, Pathogenesis, and Genetics

The disorder is usually the result of partial loss of the most distal portion of the short arm of chromosome 1 (1p36). The commonly deleted region is the most distal 1p36.33-1p36.32 band, which is more gene rich than the remainder of 1p36. A possible threshold deletion-size (>11 Mb) effect for haplolethality has been postulated (Nicoulaz et al. 2011).

Four classes of rearrangements have been identified so far in deletion 1p36 syndrome. About 52% of individuals have a de novo terminal deletion; 29% an interstitial deletion; 12% show more complex chromosome rearrangements including more than one 1p36 deletion, or a 1p36 deletion with a 1p36 duplication, or a 1p36 deletion with triplications, insertions, and/or inversions of 1p36; and 7% show a derivative chromosome 1 with the 1p telomeric region replaced by another chromosome portion (Battaglia 2013).

The distal portion of the short arm of chromosome 1 is gene-rich; therefore, identification of specific genes involved in the phenotype is very difficult. Nonetheless, a number of genes have been identified as potential candidates for some of the characteristics. Gajecka et al. (2005) described two individuals with small terminal deletions associated with large duplications and triplications of 1p36. Both had craniosynostosis involving the metopic suture in one and the sagittal and coronal sutures in the other. As people with monosomy 1p36 have a large, late-closing anterior fontanel, whereas the ones with duplication or triplication have craniosynostosis, they proposed that a gene regulating cranial sutures might be located in 1p36. Comparison of rearrangements showed a 1.1 Mb region of overlap that, when triplicated, was associated with craniosynostosis and, when deleted, was associated with large, late-closing anterior fontanel. Within this region is the *MMP23B* gene, shown to have expression at the cranial sutures and suggested to play a role in regulating closure of the fontanels (Gajecka et al. 2005).

Mutations in the *Drosophila* genes encoding either a K$^+$ channel β-subunit or α-subunit genes cause epilepsy-like phenotypes (Yao and Wu 1999). In humans, mutation in the voltage-gated K$^+$ channel genes is the molecular basis of idiopathic epilepsies, such as the benign familial neonatal convulsions (Charlier et al. 1998; Singh et al. 1998). Consequently, the K$^+$ channel β-subunit gene, *KCNAB2*, located at 1p36, appears to be a good candidate gene for epilepsy in monosomy 1p36 syndrome. Eight individuals with del 1p36 and epilepsy or epileptiform EEG activity were shown to be deleted for the *KCNAB2* locus (Heilstedt et al. 2001), supporting a possible relation between loss of this gene and the development of seizures. Haploinsufficiency of this gene has been suggested to also contribute to developmental delay and intellectual disability (Perkowski and Murphy 2011). Another gene, *CHD5*, which encodes a neuron-specific protein regulating the expression of neuronal genes, has been suggested to be a modifier gene for intellectual disability (Potts et al. 2011).

A candidate gene for the neuropsychiatric and neurodevelopmental abnormalities in deletion 1p36 syndrome is the human γ-aminobutyric acid A receptor δ-subunit gene (*GABRD*) mapped to 1p36.33 (Windpassinger et al. 2002). It encodes a γ-aminobutyric acid (*GABA*) channel, the major inhibitory neurotransmitter in the mammalian brain. Within the most distal 2 Mb of chromosome 1p is the human *SKI* gene, a protooncogene that is deleted in all individuals with the syndrome. Experiments by Colmenares et al. (2002) on *Ski+* mice with a particular genetic background suggest that *SKI* is likely contributing to development of cleft lip/palate, depressed nasal bridge, eye and digit abnormalities, and muscle hypotonia in deletion 1p36 syndrome. Haploinsufficiency of the *PRDM16* gene, encoding a zinc finger transcription factor, has been shown to cause left ventricular noncompaction and dilated cardiomyopathy in 1p36 deletion syndrome (Arndt et al. 2013). Other candidate genes, such as *SKI, PRKCZ, RERE, UBE4B, MASP2,* and *CASZ1*, may contribute to the development of cardiomyopathy (Liu et al. 2014; Zaveri et al. 2014). Since several of these cardiac critical regions include more than one candidate gene, and large terminal and interstitial deletions often overlap more than one cardiac critical region, it is possible that a combination of multiple genes contribute synergistically to the cardiac phenotypes associated with many 1p36 deletions. Based on studies of mouse models, the haploinsufficiency of other genes located at 1p36 (*SPEN, PDPN, LUZP1, HSPG2*) has been shown to contribute to defects in cardiac, palatal, and brain development/function (Jordan et al. 2015). Haploinsufficiency of *PRKCZ* has also been suggested to impact proper brain development and function through encoding an atypical protein kinase C, which is necessary for mediating axon differentiation (Zhang et al. 2007).

Diagnostic Testing

The diagnosis of 1p36 deletion syndrome is made by detection of a deletion of the most distal band of the short arm of chromosome 1 (1p36). Standard G-banded cytogenetic analysis (at 400–550 band level), FISH, or array-CGH can all be used to detect deletions; however, the complexity of some deletions may only be detected by array-CGH. Once clinical observation suggests a possible/probable monosomy 1p36, the most suitable diagnostic testing to pursue is either FISH using at least two subtelomeric region-specific probes [Vysis 1p SUBTEL probe, Vysis p58 probe (Des Plaines, IL); D1Z2 Oncor probe or CEB 108/T7 (Illkirch, Graffenstaden, France)], or array-CGH. FISH can identify parental rearrangements and may detect terminal and interstitial deletions and derivative chromosomes. However, FISH cannot detect an interstitial deletion proximal to the probes used and cannot distinguish between a true terminal deletion and a more complex rearrangement, nor can it define the exact deletion size. Array-CGH detects DNA copy-number variations smaller than 5 Mb and has the potential to do all three (Shaffer 2005; Bejjani and Shaffer 2006; Shaffer et al. 2006). Array-CGH and FISH detect more than 95% of 1p36 deletions, whereas standard cytogenetics detects only 25% (Battaglia 2013).

Once the diagnosis is made in the proband, parental studies may be necessary. Prenatal diagnosis for at-risk pregnancies is available when recurrence risk is increased because of the presence of a parental rearrangement. Of note, germline mosaicism has also been reported (Gajecka et al. 2010; Di Donato et al. 2014).

Differential Diagnosis

The clinical phenotype and facial gestalt of individuals with 1p36 deletion is highly distinctive. However, some affected individuals may be misdiagnosed as a result of characteristics overlapping with other disorders, including Rett syndrome (see Chapter 49), Angelman syndrome (see Chapter 5), Prader–Willi syndrome (see Chapter 46), Smith–Magenis syndrome (see Chapter 54), Cohen syndrome (see Chapter 14), or Aicardi syndrome. Rett syndrome is characterized by postnatal deceleration of head growth, loss of purposeful hand skills between 1 and 4 years, hand stereotypies, evolving social withdrawal, communication dysfunction, loss of acquired speech, and cognitive and locomotion impairment. Angelman syndrome is characterized by microcephaly, severe developmental delay/intellectual disability, severe speech impairment, and gait ataxia. Prader–Willi syndrome is characterized by severe hypotonia and feeding difficulties in early infancy, cognitive impairment with delay in motor milestones and language development, and childhood onset of excessive eating and gradual development of morbid obesity if not controlled externally. Smith–Magenis syndrome is characterized by distinctive facial features, developmental delay, cognitive impairment, behavioral abnormalities, and a significant sleep disturbance. Cohen syndrome is an autosomal recessive inherited multi-system disorder characterized by intellectual disability, characteristic facial dysmorphism (including prominent maxillary central incisors), progressive retinal dystrophy, and intermittent neutropenia. Aicardi syndrome is an X-linked dominant disorder with lethality in males, characterized by agenesis of the corpus callosum, distinctive chorioretinal lacunae, and infantile spasms, associated with microcephaly, axial hypotonia, and appendicular hypertonia with spasticity, and developmental delay/intellectual disability. Recently, Fregeau et al. (2016) hypothesized that mutations in the *RERE* gene cause a genetic syndrome showing many of the phenotypes associated with proximal 1p36 deletion. Of note, *ADNP* mutation, causing the Helsmoortel–Van der Aa syndrome (Helsmoortel et al. 2014), can present as a deletion 1p36 phenocopy (personal observation).

MANIFESTATIONS AND MANAGEMENT

Early intervention and comprehensive lifelong management have a significant impact on the health and quality of life of affected individuals. Optimal management requires the involvement of a multidisciplinary team consisting of the professionals caring for each system, and the development of an individualized approach. Family members are critical components of the team, as they become the greatest experts concerning their own child's history, behavior, and needs. Many of the recommendations described herein are the result of several years of personal experience together with review of the international scientific literature.

Growth and Feeding

Growth deficiency of prenatal onset is almost always present in deletion 1p36 syndrome. Delayed bone age is reported in 22% (Battaglia et al. 2008).

Feeding difficulties, reported in just over two thirds (Gajecka et al. 2007), may be caused by hypotonia and/or oral facial clefts with related difficulty in sucking, poorly coordinated swallow with consequent aspiration, and/or gastroesophageal reflux and vomiting. Gastroesophageal reflux has been reported in over 50% of babies with deletion 1p36 (Gajecka et al. 2007; Unique 2013). Mild to severe oropharyngeal dysphagia has been observed on swallow studies in 72% of affected individuals (Heilstedt et al. 2003b). Gastrointestinal anomalies, including hypertrophic pyloric stenosis and intestinal malrotation with malposition of the cecum leading to a volvulus, are present in less than one third of the individuals (Gajecka et al. 2007; Battaglia et al. 2008), and can play a role in the feeding problems. However, feeding difficulties can be easily overcome in most cases, and postnatal growth usually follows the normal standards (personal experience). In spite of infantile feeding problems and poor weight gain, a few individuals developed truncal obesity and/or macrosomia, and hyperphagia in childhood similar to that seen in Prader–Willi syndrome (Wenger et al. 1988; Wargowski et al. 1991; Keppler-Noreuil et al. 1995; D'Angelo et al. 2010; Shimada et al. 2015).

Evaluation

- Head circumference, length, and weight should be measured during each health supervision visit and plotted on standard growth curves.
- The likelihood of feeding problems should be recognized and assessed, particularly in those with clefts. Referral to a dysphagia team for evaluation of swallowing difficulties may be useful.
- The presence of gastroesophageal reflux or gastrointestinal malformation as a potential factor contributing to feeding problems should be assessed in a standard manner.
- If a child with deletion 1p36 is not bottle feeding by 6 months of age or has persistent failure-to-thrive, consideration of gastrostomy tube placement is indicated to sustain weight gain, protect the airway, and manage gastroesophageal reflux.

Management

- Assurance of adequate caloric intake, if necessary, by increased caloric formulas and/or oral or nasogastric tube feeding, is indicated.
- Standard therapy/management for gastroesophageal reflux can be initiated when this is diagnosed. Nissen–Hill fundoplication has been successfully performed in a number of children (Unique 2013).
- Gastrostomy has been successfully performed in some babies (Unique 2013).
- Anesthetic risks for surgical procedures in children with structural heart defects require consultation by cardiology.

Development and Behavior

One of the three important challenges in 1p36 deletion syndrome is represented by development disability. Since the description of the first individual with monosomy 1p36 (Yunis et al. 1981), development has been regarded as being severely compromised, with significant impairment of expressive language. Only recent studies of larger numbers of individuals with this condition have allowed recognition of a more complete continuum of the phenotype, pointing out a much wider clinical spectrum than previously thought (Shapira et al. 1997; Gajecka et al. 2007; Battaglia et al. 2008; Brazil et al. 2014).

Recent reviews (Slavotinek et al. 1999; Wu et al. 1999; Battaglia, 2005; Brazil et al. 2014) indicate that there is greater acquisition of milestones than suggested by the older medical literature. Cognitive deficits are a hallmark of the syndrome (Shapira et al. 1997; Shaffer and Heilstedt 2001; Heilstedt et al. 2003b; Gajecka et al. 2007; Battaglia et al. 2008). Severe to profound intellectual disability is noted in about 88% of individuals, whereas 12% have a mild to moderate deficit (Battaglia et al. 2008). Expressive language, although absent in about 75% and limited to a few isolated words in 17%, could develop to the level of simple sentences in almost 8% of individuals studied by Battaglia et al. (2008). In a more recent study performed by Brazil et al. (2014) in adolescents and adults with 1p36 deletion through an electronic and anonymous survey, 38% were able to communicate in sentences. Communication skills, which are limited in the early years, tend to improve over time with development of signing ability (Battaglia , 2013).

Motor milestones are delayed. Head control is usually acquired between 5 and 15 months of age; just over two thirds can sit without support between 7 months and 3 years; and only 26% are able to walk independently by age 2–7 years. They tend to walk with a broad-based gait, which improves over time, and with poor swinging movements of the upper limbs. In the survey by Brazil et al. (2014), 80% were reportedly walking, some of them for at least short distances. In a recent study, the non-ambulatory individuals were shown to harbor 1p36 deletions larger than 6.2 Mb (Shimada et al. 2015).

Fine motor abilities are limited to scribble. Less than 50% of persons with deletion 1p36 are said to be toilet trained to various degrees. All individuals show a slow but constant improvement in all aspects of development over time (Battaglia et al. 2008; Brazil et al. 2014).

Stereotyped and self-abusive behavior with restricted interests, associated with a primordial exploration of objects and, less frequently, hyperphagia, are observed in about 30% to 50% of the individuals with del 1p36 (Battaglia et al. 2008; Brazil et al. 2014).

Battaglia (2013) discusses the challenges faced by parents of children with deletion 1p36 syndrome, directing primary care practitioners and pediatricians to approach long-term management in these children as they would in any child with disabilities.

Evaluation

- A thorough developmental evaluation by a specialist team should be performed at intervals appropriate for planning of early intervention, including physical, occupational, and speech therapies, and for educational and vocational planning.

Management

- Referral to early intervention programs is recommended.
- Enrollment in an individualized rehabilitation program that covers motor aspects (including oral motor and feeding therapy), cognition, communication, and socialization is highly recommended.
- Appropriate school placement, based on detailed developmental evaluation, is important.
- Planning for transition to adulthood, including vocational training and living situation, should begin in adolescence.

Neurologic

In addition to developmental disability, the other main challenge of infancy and childhood is the occurrence of seizures, which were reported to occur in 50–80% of infants and children with deletion 1p36 syndrome (Heilstedt et al. 2003b; Battaglia 2005; Gajecka et al. 2007). Recent studies of larger numbers of affected individuals document seizures in about 44–58% (Battaglia et al. 2008; Bahi-Buisson et al. 2008). Mean age at onset of seizures is 7 months (between first days of life and 2 years 8 months). First seizures are generalized in half the individuals, and are tonic, tonic-clonic, clonic/myoclonic, or partial in just over a third. The first seizures are infantile spasms in about 10–25%, associated with a typical or near-hypsarrhythmic electroencephalogram (EEG) (West syndrome). In 15% of these infants, the electroclinical picture evolves toward that of a Lennox–Gastaut-like syndrome. Isolated seizures, such as simple partial seizures or generalized tonic-clonic seizures, and epileptic apneas can be observed in a minority (Kanabar et al. 2012). In 30–60% of affected individuals, seizures stop between age 8 weeks and 25 years (Battaglia et al. 2008; Bahi-Buisson et al. 2008). Fewer than 15% of individuals have discontinued antiepileptic drugs (Battaglia et al. 2008; Bahi-Buisson et al. 2008).

A variety of EEG abnormalities have been observed in all individuals who have had serial studies, during both waking and sleep (Battaglia et al. 2008; Bahi-Buisson et al. 2008). These abnormal findings include poverty of the usual rhythmic activities, mainly over the posterior brain regions; asymmetry of slow activities; hypsarrhythmia or near-hypsarrhythmia; multifocal or generalized spikes, polyspikes, and spike/wave discharges; and slow background activity (Battaglia et al. 2008; Bahi-Buisson et al. 2008).

A variety of developmental abnormalities of the brain have been described in deletion 1p36 syndrome (Gajecka et al. 2007; Battaglia et al. 2008; Bahi-Buisson et al. 2008;

Dobyns et al. 2008). These include polymicrogyria; enlargement of the fronto-temporal opercula; leukoencephalopathy-like picture; cortical or generalized atrophy; prominent ventricles; anomalies or morphologic variants of the commissural structures; and, less frequently, Chiari 1 malformation. Periventricular nodular heterotopia has also been reported in a few individuals and lumbosacral hydromyelia in one (Neal et al. 2006; Saito et al. 2008; Descartes et al. 2011; Shiba et al. 2013). Cervicomedullary compression at the skull base has been detected in an infant with apneic spells (Shiba et al. 2013).

Generalized hypotonia, which on occasion may be marked, is present in almost all affected individuals. This central hypotonia appears to be of prenatal onset and persists throughout life. It causes decreased fetal movements with frequent abnormal fetal position often necessitating delivery by Cesarean section. It is almost invariably associated with poor suck with consequent failure-to-thrive during early years.

Evaluation

- A waking/sleeping video-electroencephalogram-polygraphic study is recommended in infancy and childhood to achieve the best characterization of seizures. This is of the utmost importance in infancy, when the parents or professionals report the occurrence of "spasms," because early diagnosis and treatment of "infantile spasms" associated with a typical or near-hypsarrhythmic EEG is crucial for prognosis.
- Regular standard follow-up of seizure status and pharmacotherapy is essential.
- Neurological examination, preferably by a child neurologist, is recommended for all infants, children, and adolescents.
- Brain MRI should be performed in all individuals with seizures because malformation of the cerebral cortex may modify prognosis and management.
- MRI should be performed in all infants with apneic spells to check for any cervicomedullary compression. Referral to a neurosurgeon should be considered thereafter.

Management

- In the author's experience, the "infantile spasms" can be well controlled by ACTH. Those treated with different antiepileptic drugs develop severe, refractory epilepsy (Bahi-Buisson et al. 2008).
- In most individuals, all other seizure types are well controlled by standard antiepileptic drugs, provided the first-choice drug is started as early as possible.
- Hypotonia should be treated as in the general population with physical therapy, which is essential to help increase strength and develop muscle tone.
- Neurosurgery for decompression of the involved structures, if indicated, follows standard practice.

Cardiovascular

Congenital heart malformations are present in about 75% of children with deletion 1p36 syndrome. In recent series, between 8 and 28% had an atrial septal defect; about 25% had ventricular septal defect; 20.5% had valvular anomalies, including bicommissural aortic valve, pulmonary valve stenosis/dysplasia/atresia, and mitral valve insufficiency; between 13 and 27% had patent ductus arteriosus; 24% had patent foramen ovale; 7.7% had tetralogy of Fallot; 5.1% had coarctation of the aorta; 2–3% had Ebstein anomaly; and 2% had infundibular stenosis of the right ventricle (Gajecka et al. 2007; Battaglia et al. 2008). Of note, between 27 and 31% had a history of cardiomyopathy in infancy and childhood (Gajecka et al. 2007; Battaglia et al. 2008). In a large series, cardiomyopathy was of the "non-compaction" type in 23% and the dilated type in 4% (Battaglia et al. 2008). Of note, later onset cardiomyopathy has been reported in three individuals at the age of 16, 21, and 25 years (Brazil et al. 2014; Lee et al. 2014). On the whole, the congenital cardiac anomalies are not complex and are amenable to repair.

Evaluation

- Cardiac evaluation is important at the time of diagnosis and should include auscultation, electrocardiogram, and echocardiography.
- Proactive heart function screening may be warranted thereafter.

Management

- In some instances, septal defects will close spontaneously.
- In all other cases, the defects can easily be surgically repaired, following standard practices.
- Decision concerning cardiac surgery should be made with active participation of parents or guardians, considering the other anomalies present in the child.
- "Non-compaction" cardiomyopathy responds well to standard pharmacotherapy (e.g., furosemide, captopril, and digoxin) (personal observation).

Respiratory

Several individuals with deletion 1p36 syndrome have frequent respiratory tract infections, usually overcome as they get older. Other abnormalities of the respiratory system are rare, although persistent buccopharyngeal membrane and small laryngeal structures, leading to respiratory distress, have been reported in one infant (Ferril et al. 2014).

Evaluation

- Evaluations do not differ from those in other children with similar symptoms.

Management

- Treatment follows standard practice.

Ophthalmologic

Strabismus is the most common ophthalmologic finding observed in monosomy 1p36 syndrome, with a frequency ranging from 35 to 67% (Gajecka et al. 2007; Battaglia et al. 2008). Refractive errors such as hypermetropia, myopia, or astigmatism occur in 23–40%, and nystagmus in 26.5% (Gajecka et al. 2007; Battaglia et al. 2008). Less frequent abnormalities, occurring in 3–6%, include unilateral cataract, retinal albinism, and unilateral optic nerve coloboma (Battaglia et al. 2008). Visual inattentiveness, defined as absence of attentive visual behavior with fixation and following movements, has been observed in 40–64% of children (Gajecka et al. 2007; Battaglia et al. 2008).

Evaluation

- Ophthalmology consultation is recommended at diagnosis even in the absence of overt anomalies.
- Follow-up evaluation will depend on the findings.

Management

- Ophthalmologic abnormalities are treated in a standard manner.
- Visual inattentiveness can be treated with an appropriate rehabilitation program (Bolognini et al. 2005).

Ears and Hearing

About 50% of individuals with deletion 1p36 syndrome show ear anomalies. The ears are poorly formed, simple, posteriorly rotated, or low set, and can be asymmetrical. Hearing loss occurs in 47–77% of individuals, and there is some evidence that it is progressive in some children (Gajecka et al. 2007; Battaglia et al. 2008; Unique, 2013). It is most often of the sensorineural type and bilateral, but it can also be conductive and mixed (Battaglia et al. 2008).

Evaluation

- Comprehensive audiologic and otologic evaluation of individuals should be performed at diagnosis and annually thereafter.
- Individuals with significant mental handicap can be tested with brainstem auditory-evoked responses and/or otoacoustic emissions. The advantage of brainstem auditory-evoked response testing over auditory field examination is the ability to rule out frequency-specific or unilateral sensorineural hearing loss.

Management

- A trial of hearing aids in older infants who have abnormal audiological evaluations is appropriate (personal experience). However, if the hearing aid does not appear to improve communication and is difficult for the child to tolerate, the decision to continue or discontinue should be individualized.
- Cochlear implants should be considered in those individuals severely affected by sensorineural hearing loss.
- Hearing loss should be considered in planning educational and vocational programming.

Musculoskeletal

A variety of musculoskeletal abnormalities occurs in deletion 1p36 syndrome, including medically significant malformations and minor anomalies of limbs and skeleton. Deformities of the feet include bilateral talipes valgus, although bilateral calcaneovalgus positioning seems to be rare. Congenital spinal stenosis and congenital fiber-type disproportion myopathy have been occasionally reported (Reish et al. 1995; Okamoto et al. 2002). Other skeletal anomalies include brachycephalic skull, which is associated with craniostenosis in a minority of subjects; rib anomalies, such as 11 ribs and bifid/fused/enlarged ribs; lower limb asymmetry; congenital hip dysplasia; valgus deformity of the femoral neck; thinning of the long bones; phalangeal hypoplasia of the hands; and cone-shaped epiphyses of the hands and feet (Battaglia et al. 2008). Polydactyly has also been reported occasionally (Keppler-Noreuil et al. 1995). Scoliosis is common in older children and is not necessarily related to structural defects of the vertebrae, although malformed vertebral bodies have also been reported (Battaglia et al. 2008). Scoliosis may progress between late childhood and early adolescence.

Evaluation

- Scoliosis should be evaluated clinically in older children at routine health supervision visits.

- Spine radiographs should be performed when clinical scoliosis is present.
- Referral to orthopedics should be considered on recognition of significant musculoskeletal abnormalities.

Management

- Casting for positional abnormalities of the feet follows standard practice.
- Because between 26 and 80% of individuals with deletion 1p36 syndrome walk independently, even if for just short distances, standard orthopedic surgery is recommended for foot deformities at an early age.
- Decisions about surgery for scoliosis may need to be made in the older child. If treatment is offered, it follows standard practice.

Genitourinary

A variety of structural defects of the genital and urinary tracts have been described in almost 22% of individuals with deletion 1p36 syndrome (Battaglia et al. 2008). These include unilateral renal pelvis with hydronephrosis of the upper pole, kidney ectopia with right kidney cyst, and unilateral pelvic ectasia.

Abnormalities of the genitalia, reported in almost 25% of individuals, include hypospadias, cryptorchidism, scrotal hypoplasia, and micropenis in males, and clitoral hyperplasia, small labia minora, labia majora hypertrophy, and uterine hypoplasia in females (Battaglia et al. 2008).

Of note, the bladder exstrophy-epispadias complex (BEEC) spectrum has been described in two children with deletion 1p36 (El-Hattab et al. 2010; Çöllü et al. 2016).

In a recent survey, recurrent kidney infections were reported in adolescents and adults with deletion 1p36 (Brazil et al. 2014).

Evaluation

- Renal ultrasound and serum creatinine are indicated in all individuals at diagnosis.
- Individuals with structural renal defects should be followed for infection and renal insufficiency by periodic serum creatinine, urinalysis, and urine culture.
- Voiding retrograde cystography and intravenous pyelography or scintigraphy may be necessary in the presence of abnormal structure or function.
- Urinalysis should also be done during infections where the site is unknown.

Management

- Management of urinary tract infection and genitourinary anomalies are the same as in the general population.

Gastrointestinal

Gastrointestinal anomalies, including hypertrophic pyloric stenosis and intestinal malrotation with malposition of the cecum leading to a volvulus are present in less than one third of the individuals (Gajecka et al. 2007; Battaglia et al. 2008). Annular pancreas has been reported in one Japanese child (Minami et al. 2005), and duodenal atresia in one female with a 1.75 Mb interstitial deletion (Rankin et al. 2009). Non-alcoholic steatohepatitis has been reported in two children, in one of whom del 1p36 was associated with dup Xp22 (Haimi et al. 2011; Nobili et al. 2018). Biliary atresia in a premature neonate with 1p36 deletion has been recently reported (Chawla et al. 2018); and type IV laryngotracheoesophageal cleft associated with type III esophageal atresia has been observed in another child with 1p36 deletion (Pelizzo et al. 2018).

Constipation has been reported in 40% to 65% of individuals (Gajecka et al. 2007; Unique 2013; Brazil et al. 2014).

Evaluation

- Significant feeding problems should trigger standard evaluations for possible gastrointestinal tract malformations, including imaging studies and/or referral to a gastroenterologist.
- History of constipation should be elicited during routine medical evaluations.

Management

- Treatment of gastrointestinal malformations is standard and will often necessitate surgery.
- Liver disease is treated in a standard manner.
- Management of constipation is the same as in any individual.

Endocrinologic

Between 15 and 20% of children with deletion 1p36 syndrome have been shown to have hypothyroidism (Gajecka et al. 2007; Battaglia et al. 2008). Growth hormone deficiency has been reported in one child (Shiba et al. 2013).

Evaluation

- Triiodothyronine, thyroxine, and thyrotropin levels should be evaluated at diagnosis, and periodically.
- Growth hormone studies should be performed, because growth hormone deficiency has recently been documented (Shiba et al. 2013).

Management

- Management of hypothyroidism and of growth hormone deficiency is the same as in any individual.

Dermatologic

Pemphigus vulgaris has been reported in one affected individual (Halpern et al. 2006), and cutis laxa in another (Zhang et al. 2018). Telangiectatic skin lesions and hyperpigmented macules have also been described (Keppler-Noreuil et al. 1995). Of note, dermatitis artefacta (a condition where skin lesions are self-inflicted) has been reported in an adult female (Winship & Braue 2014). Sacral/coccigeal dimples have been observed in 15% of the individuals (Battaglia et al. 2008)

Evaluation

- Evaluations do not differ from those in other children with similar symptoms.

Management

- Treatment should be decided on an individual basis.

Neoplasia

Three children with deletion 1p36 syndrome reported in the literature developed neuroblastoma by age 5–9 months (Laureys et al. 1990; Biegel et al. 1993; Anderson et al. 2001). Abdominal paraganglioma has recently been reported in a 24-year-old Japanese woman with 1p36 deletion syndrome (Murakoshi et al. 2017). However, no history of cancer has been reported in recent large series of individuals with deletion 1p36, including about 7% of adults, (Gajecka et al. 2007; Battaglia et al. 2008; Brazil et al. 2014; Shimada et al. 2015).

Adulthood

As of this writing, there is very little information on the natural history of deletion 1p36 or its manifestations in adulthood. From a recent electronic and anonymous survey (Brazil et al. 2014) including 19 adult individuals (sixteen of them aged 20 to 27 years), functional skills were better than previously reported. Forty-four percent were said to use complex speech abilities, and the majority to independently sit and walk. However, Unique's experience has been that individuals with the disorder "will normally need lifelong care and medical support and will at best achieve only limited independence" (Unique 2013).

RESOURCES

Chromosome Deletion Outreach, Inc.
PO Box 724
Boca Raton FL 33429-0724, USA
Phone: 888 CDO 6880 (888 236 6680); 561 395 4252 (family helpline)
Email: *info@chromodisorder.org*
Website: *www.chromodisorder.org*

1p36 Deletion
Support & Awareness
Website: *http://www.1p36dsa.org/*

Rare Chromosome Disorder Support Group,
The Stables, Station Rd West, Oxted, Surrey RH8 9EE, UK
Phone: +44(0)1883 723356
Email: info@rarechromo.org
Website: *www.rarechromo.org*

REFERENCES

Anderson J, Kempski H, Hill L, Rampling D, Gordon T, Michalski A (2001) Neuroblastoma in monozygotic twins—A case of probable twin-to-twin metastasis. *Br J Cancer* 85:493–496.

Arndt AK, Schafer S, Drenckhahn JD, Sabeh MK, Plovie ER, Caliebe, Klopocki E, Musso G, Werdich AA, Kalwa H, Heinig M, Padera RF, Wassilew K, Bluhm J, Harnack C, Martitz J, Barton PJ, Greutmann M, Berger F, Hubner N, Siebert R, Kramer H-H, Cook SA, MacRae CA, Klaassen S (2013) Fine mapping of the 1p36 deletion syndrome identifies mutation of PRDM16 as a cause of cardiomyopathy. *Am J Hum Genet* 93:67–77.

Bahi-Buisson N, Gutierrez-Delicado E, Soufflet C, Rio M, Cormier Daire V, Lacombe D, Heron D, Verloes A, Zuberi SM, Burglen L, Afenjar A, Moutard LM, Edery P, Dulac O, Nabbout R, Plouin P, Battaglia A (2008) Spectrum of epilepsy in terminal 1p36 deletion syndrome. *Epilepsia* 49:509–515.

Battaglia A (2005) Del 1p36 syndrome: A newly emerging clinical entity. *Brain Dev* 27:358–361.

Battaglia A. 1p36 Deletion Syndrome. 2008 Feb 1 [Updated 2013 Jun 6]. GeneReviews® [Online]. Available from: https://www.ncbi.nlm.nih.gov/books/NBK1191/

Battaglia A, Hoyme HE, Dallapiccola B, Zackai E, Hudgins L, McDonald-McGinn D, Bahi-Buisson N, Romano C, Williams CA, Braley LL, Zuberi SM, Carey JC (2008) Further delineation of deletion 1p36 syndrome in 60 patients: A recognizable phenotype and common cause of developmental delay and mental retardation. *Pediatrics* 121:404–410.

Bejjani BA, Shaffer LG (2006) Application of array-based comparative genomic hybridization to clinical diagnostics. *J Mol Diagn* 8:528–533.

Biegel JA, White PS, Marshall HN, Fujimori M, Zackai EH, Scher CD, Brodeur GM, Emanuel BS (1993) Constitutional 1p36 deletion in a child with neuroblastoma. *Am J Hum Genet* 52:176–182.

Bolognini N, Rasi F, Coccia M, Ladavas E (2005) Visual search improvement in hemianopic patients after audio-visual stimulation. *Brain* 128:2830–2842.

Brazil A, Stanford K, Smolarek T, Hopkin R. (2014). Delineating the phenotype of 1p36 deletion in adolescents and adults. *Am J Med Genet Part A* 164A: 2496–2503.

Charlier C, Sing NA, Ryan SG (1998) A pore mutation in a novel KQT-like potassium channel gene in an idiopathic epilepsy family. *Nat Genet* 18:53–55.

Chawla V, Anagnost MR, Eldemerdash AE, Reyes D, Scherr R, Ezeanolue K, Banfro F, Alhosh R. (2018). A Novel Case of Biliary Atresia in a Premature Neonate With 1p36 Deletion Syndrome. *J Investig Med High Impact Case Rep.* 6:2324709618790613.

Çöllü M, Yüksel Ş, Şirin BK, Abbasoğlu L, Alanay Y. (2016). Is 1p36 deletion associated with anterior body wall defects? *Am J Med Genet A* 170(7):1889–94.

Colmenares C, Heilstedt HA, Shaffer LG, Schwartz S, Berk M, Murray JC, Stavnezer E (2002) Loss of the SKI proto-oncogene in individuals affected with 1p36 deletion syndrome is predicted by strain-dependent defects in Ski$^{-/-}$ mice. *Nat Genet* 30:106–109.

D'Angelo CS, Kohl I, Varela MC, de Castro CIE, Kim CA, Bertola DR, Loureco CM, Koiffmann CP (2010) Extending the phenotype of monosomy 1p36 syndrome and mapping of a critical region for obesity and hyperphagia. *Am J Med Genet Part A* 152A:102–110.

Descartes M, Mikhail FM, Franklin JC, McGrath TM, Bebin M (2011) Monosomy 1p36.3 ans trisomy 19p13.3 in a child with periventricular nodular heterotopia. *Pediatr Neurol* 45:274–278.

Di Donato N, Klink B, Hahn G, Schrock E, Hackmann K (2014) Interstitial deletion 1p36.32 in two brothers with a distinct phenotype--overgrowth, macrocephaly and nearly normal intellectual function. *Eur J Med Genet* 57(9):494–497.

Dobyns WB, Mirzaa G, Christian SL, Petras K, Roseberry J, Clark GD ... Shaffer LG (2008) Consistent chromosome abnormalities identify novel polymicrogyria loci in 1p36, 2p16.1-p23.1, 4q21.21-q22.1, 6q26-q27, and 21q2. *Am J Med Genet A.* 146A:1637–1654.

El-Hattab AW, Skorupski JC, Hsieh MH, Breman AM, Patel A, Cheung SW, Craigen WJ (2010) OEIS complex associated with chromosome 1936 deletion: a case report and review. *Am J Med Genet Part A* 152A: 504–511.

Eugster EA, Berry SA, Hirsch B (1997) Mosaicism for deletion 1p36.33 in a patient with obesity and hyperphagia. *Am J Med Genet* 70:409–412.

Ferril GR, Barham HP, Prager JD (2014) Novel airway findings in a patient with 1p36 deletion syndrome. *Int J Pediatr Otorhinolaryngol.* 78(1):157–8.

Fregeau B, Kim BJ, Hernandez-Garcia A, Jordan VK, Cho MT, Schnur RE, Monaghan KG, Juusola J, Rosenfeld JA, Bhoj E, Zackai EH, Sacharow S, Barañano K, Bosch DGM, de Vries BBA, Lindstrom K, Schroeder A, James P, Kulch P, Lalani SR, van Haelst MM, van Gassen KLI, van Binsbergen E, Barkovich AJ, Scott DA, Sherr EH (2016) De novo mutations of RERE cause a genetic syndrome with features that overlap those associated with proximal 1p36 deletions. *Am J Hum Genet* 98:963–970.

Gajecka M, Yu W, Ballif BC, Glotzbach CD, Bailey KA, Shaw CA, Kashork CD, Heilstedt HA, Ansel DA, Theisen A, Rice R, Rice DPC, Shaffer LG (2005) Delineation of mechanisms and regions of dosage imbalance in complex rearrangements of 1p36 leads to a putative gene for regulation of cranial suture closure. *Eur J Hum Genet* 13:139–149.

Gajecka M, Mackay KL, Shaffer LG (2007) Monosomy 1p36 deletion syndrome. *Am J Med Genet C* 145C:346–356.

Gajecka M, Saitta SC, Gentles AJ, Campbell L, Ciprero K, Geiger E, Catherwood A, Rosenfeld JA, Shaikh T, Shaffer LG (2010) Recurrent interstitial 1p36 deletions: evidence for germline mosaicism and complex rearrangement breakpoints. *Am J Med Genet Part A* 152A:3074–3083.

Giraudeau F, Taine L, Biancalana V, Delobel B, Journel H, Missirian C, Lacombe D, Bonneau D, Parent P, Aubert D, Hauck Y, Croquette MF, Toutain N, Mattei MG, Loiseau HA, David A, Vergnaud G (2001) Use of a set of highly polymorphic minisatellite probes for the identification of cryptic 1p36.3 deletions in a large collection of patients with idiopathic mental retardation. *J Med Genet* 38:121–125.

Haimi M, Iancu TC, Shaffer LG, Lerner A (2011) Severe lysosomal storage disease of liver in del(1)(p36): a new presentation. *Eur J Med Genet* 54:209–213.

Hain D, Leversha M, Campbell N, Daniel A, Barr PA, Rogers JG (1980) The ascertainment and implications of an unbalanced translocation in the neonate. Familial 1:15 translocation. *Aust Paediatr J* 16:196–200.

Halpern AV, Bansal A, Heymann WR (2006) Pemphigus vulgaris in a patient with 1p36 deletion syndrome. *J Am Acad Dermatol* 55(Suppl 5):98–99.

Heilstedt HA, Burgess DL, Anderson AE, Chedrawi A, Tharp B, Lee O, Kashork CD, Starkey DE, Wu YQ, Noebels JL, Shaffer LG, Shapira SK (2001) Loss of the potassium channel beta-subunit gene, KCNAB2, is associated with epilepsy in patients with 1p36 deletion syndrome. *Epilepsia* 42:1103–1111.

Heilstedt HA, Ballif BC, Howard LA, Kashork CD, Shaffer LG (2003a) Population data suggest that deletions of 1p36 are a relatively common chromosome abnormality. *Clin Genet* 64:310–316.

Heilstedt HA, Ballif BC, Howard LA, Lewis RA, Stal S, Kashork CD, Bacino CA, Shapira SK, Shaffer LG (2003b) Physical map of 1p36, placement of breakpoints in monosomy 1p36, and clinical characterization of the syndrome. *Am J Hum Genet* 72:1200–1212.

Helsmoortel C, Vulto-van Silfhout AT, Coe BP, Vandeweyer G, Rooms L, van den Ende J.......Van der Aa N (2014) ASWI/SNF-related autism syndrome caused by de novo mutations in ADNP. *Nat Genet* 46:380–385.

Jordan VK, Zaveri HP, Scott DA (2015) 1p36 deletion syndrome: an update. *Appl Clin Genet*, 8, 189–200.

Kanabar G, Boyd S, Schugal A, Bhate S (2012) Multiple causes of apnea in 1p36 deletion syndrome include seizures. *Seizure* 21:402–406.

Keppler-Noreuil KM, Carroll AJ, Finley WH, Rutledge SL (1995) Chromosome 1p terminal deletion: report of new findings and confirmation of two characteristic phenotypes. *J Med Genet* 32:619–622.

Laureys G, Speleman F, Opdenakker G, Benoit Y, Leroy J (1990) Constitutional translocation t(1;17)(p36;q12-21) in a patient with neuroblastoma. *Genes Chromosomes Cancer* 2:252–254.

Lee J, Rinehart S, Polsani V (2014) Left Ventricular Noncompaction Cardiomyopathy: Adult Association with 1p36 Deletion Syndrome. *Methodist Debakey Cardiovasc J* 10(4): 258–259.

Liu Z, Li W, Ma X, et al. (2014) Essential role of the zinc finger transcription factor Casz1 for mammalian cardiac morphogenesis and development. *J Biolog Chem* 289:29801–29816.

Minami K, Boshi H, Minami T, Tamura A, Yanagawa T, Uemura S, Takifuji K, Kurosawa K, Tsukino R, Izumi G, Yoshikawa N (2005) 1p36 deletion syndrome with intestinal malrotation and annular pancreas. *Eur J Pediatr* 164:193–194.

Murakoshi M, Takasawa K, Nishioka M, Asakawa M, Kashimada K, Yoshimoto T, Yamamoto T, Takekoshi K, Ogawa Y, Shimohira M (2017) Abdominal paraganglioma in a young woman with 1p36 deletion syndrome. *Am J Med Genet A*. 173(2):495–500.

Neal J, Apse K, Sahin M, Walsh CA, Sheen VL (2006) Deletion of chromosome 1p36 is associated with periventricular nodular heterotopia. *Am J Med Genet* 140A:1692–1695.

Nicoulaz A, Rubi F, Lieder L, Wold R, Goeggel-Simonetti B, Steinlin M, Wiest R, Bonel HM, Shaller AGallati S, Conrad B (2011) Contiguous ~16 Mb 1p36 deletion: dominant features of classical distal 1p36 monosomy with haplo-lethality. *Am J Med Genet Part A* 155:1964–1968.

Nobili V, Mosca A, Francalanci P, Maria Cristina D, Dallapiccola B (2018) First case of nonalcoholic steatohepatitis in a child with del(1p36) and dup (Xp22): review of the literature. *Clin Dysmorphol* 27(2):42-45.

Okamoto N, Toribe Y, Nakajima T, Okinaga T, Kurosawa K, Nonaka I, Shimokawa O, Matsumoto N (2002) A girl with 1p36 deletion syndrome and congenital fiber type disproportion myopathy. *J Hum Genet* 47:556–559.

Pelizzo G, Puglisi A, Lapi M, Piccione M, Matina F, Busè M, Mura GB, Re G, Calcaterra V (2018) Type IV laryngotracheoesophageal cleft associated with type III esophageal atresia in 1p36 deletions containing the RERE gene: Is there a causal role for the genetic alteration? *Case Rep Pediatr*. Aug 29:4060527.

Perkowski JJ, Murphy GG (2011) Deletion of the mouse homolog of KCNAB2, a gene linked to monosomy 1p36, results in associative memory impairments and amygdala hyperexcitability. *J Neurosci* 31(1):46–54

Potts RC1, Zhang P, Wurster AL, Precht P, Mughal MR, Wood WH 3rd ... Pazin MJ (2011) CHD5, a brain-specific paralog of Mi2 chromatin remodeling enzymes, regulates expression of neuronal genes. *PLoS One* 6(9):e24515. Pazin MJ.

Rankin J, Allwood A, Canham N, Delmege C, Crolla J, Maloney V (2009) Distal monosomy 1p36: an atypical case with duodenal atresia and a small interstitial deletion. *Clin Dysmorphol* 18(4):222–4.

Reish O, Berry SA, Hirsch B (1995) Partial monosomy of chromosome 1p36.3: Characterization of the critical region and delineation of a syndrome. *Am J Med Genet* 59:467–475.

Riegel M, Castellan C, Balmer D, Brecevic L, Schinzel A (1999) Terminal deletion, del(1)(p36.3), detected through screening for terminal deletions in patients with unclassified malformation syndromes. *Am J Med Genet* 82:249–253.

Saito S, Kawamura R, Kosho T, Shimizu T, Aoyama K, Koike K, Wada T, Matsumoto N, Kato M, Wakui K, Fukushima Y (2008) Bilateral perisylvian polymicrogyria, periventricular nodular heterotopia, and left ventricular noncompaction in a girl with 10.5-11.1 Mb terminal deletion of 1p36. *Am J Med Genet A* 146A:2891–2897.

Shaffer LG (2005) American College of Medical Genetics guideline on the cytogenetic evaluation of the individual with developmental delay or mental retardation. *Genet Med* 7:650–654.

Shaffer LG, Lupski JR (2000) Molecular mechanisms for constitutional chromosomal rearrangements in humans. *Annu Rev Genet* 34:297–329.

Shaffer LG, Heilstedt HA (2001) Terminal deletion of 1p36. *Lancet* 358(Suppl):S9.

Shaffer LG, Kashork CD, Saleki R, Rorem E, Sundin K, Ballif BC, Bejjani BA (2006) Targeted genomic microarray analysis for identification of chromosome abnormalities in 1500 consecutive clinical cases. *J Pediatr* 149:98–102.

Shapira SK, McCaskill C, Northrup H, Spikes AS, Elder FF, Sutton VR, Korenberg JR, Greenberg F, Shaffer LG (1997) Chromosome 1p36 deletions: The clinical phenotype and molecular characterization of a common newly delineated syndrome. *Am J Hum Genet* 61:642–650.

Shiba N, Daza RA, Shaffer LG, Barkovich AJ, Dobyns WB, Hevner RF (2013) Neuropathology of brain and spinal malformations in a case of monosomy 1p36. *Acta Neuropathol Commun* 2013 Aug 2;1:45.

Shimada S, Shimojima K, Okamoto N, Sangu N, Hirasawa K, Matsuo M, Ikeuchi M, Shimakawa S, Shimizu K, Mizuno S, Kubota M, Adachi M, Saito Y, Tomiwa K, Haginoya K, Numabe H, Kako Y[17], Hayashi A, Sakamoto H, Hiraki Y, Minami K, Takemoto K, Watanabe K, Miura K, Chiyonobu T, Kumada T, Imai K, Maegaki Y, Nagata S, Kosaki K, Izumi T, Nagai T, Yamamoto T (2015) Microarray analysis of 50 patients reveals the critical chromosomal regions responsible for 1p36 deletion syndrome-related complications. *Brain Develop* 37:515–526.

Singh NA, Charlier C, Stauffer D (1998) A novel potassium channel gene, KCNQ2, is mutated in an inherited epilepsy of newborns. *Nat Genet* 18:25–29.

Slavotinek A, Shaffer LG, Shapira SK (1999) Monosomy 1p36. *J Med Genet* 36:657–663.

Unique (2013) 1p36 deletion syndrome. www.rarechromo.org

Wargowski D, Sekhon G, Laxova R, Thompson K, Kent C (1991) Terminal deletions of band 1p36: Emergence of two overlapping phenotypes. *Am J Hum Genet Suppl* 49:268.

Wenger SL, Steele MW, Becker DJ (1988) Clinical consequences of deletion 1p35. *J Med Genet* 25:263.

Windpassinger C, Kroisel PM, Wagner K, Petek E (2002) The human gamma-aminobutyric acid A receptor delta (GABRD) gene: Molecular characterization and tissue-specific expression. *Gene* 292:25–31.

Winship I, Braue A (2014) Dermatitis artefacta presenting as a recurrent skin eruption in a patient with 1p36 deletion syndrome. *Australas J Dermatol.* 55(1):90.

Wu YQ, Heilstedt HA, Bedell JA, May KM, Starkey DE, McPherson JD, Shapira SK, Shaffer LG (1999) Molecular refinement of the 1p36 deletion syndrome reveals size diversity and a preponderance of maternally derived deletions. *Hum Mol Genet* 8:313–321.

Yao WD, Wu CE (1999) Auxiliary hyperkinetic beta subunit of K+ channels. Regulation of firing properties and K+ currents in Drosophila neurons. *J Neurophysiol* 81:2472–2484.

Yunis E, Quintero L, Leibovici M (1981) Monosomy 1pter. *Hum Genet* 56:279–282.

Zaveri HP, Beck TF, Hernandez-Garcia A, Shelly KE, Montgomery T, van Haeringen A, Anderlid BM, Patel C, Goel H, Houge G, Morrow BE, Cheung SW, Lalani SR, Scott, DA (2014) Identification of critical regions and candidate genes for cardiovascular malformations and cardiomyopathy associated with deletions of chromosome 1p36. *PLOS One* 9(1):e85600

Zenker M, Rittinger O, Grosse KP, Speicher MR, Kraus J, Rauch A, Trautmann U (2002) Monosomy 1p36—A recently delineated, clinically recognizable syndrome. *Clin Dysmorphol* 11:43–48.

Zhang X, Zhu J, Yang GY, Wang QJ, Qian L, Chen YM, Chen F, Tao Y, Hu H-S, Wang T, Luo ZG (2007) Dishevelled promotes axon differentiation by regulating atypical protein kinase C. *Nat Cell Biol* 9:743–754.

Zhang Z, Wang J, Li N, Yao R, Chen J (2018) Cutis laxa in a patient with 1p36 deletion syndrome. *J Dermatol* 45(7):871–873.

19

DELETION 4p: WOLF–HIRSCHHORN SYNDROME

AGATINO BATTAGLIA

IRCCS Stella Maris Foundation, Pisa, Italy; Division of Medical Genetics, Department of Pediatrics, University of Utah School of Medicine, Salt Lake City, Utah, USA; and Division of Medical Genetics, Department of Pediatrics, Sanford School of Medicine, University of South Dakota, Sioux Falls, South Dakota, USA

INTRODUCTION

Wolf–Hirschhorn syndrome (WHS) is a multiple congenital anomaly/intellectual disability disorder that was first described by Cooper and Hirschhorn in 1961 in a child with defects of midline fusion in association with deletion of a B group chromosome. However, it was not brought to the attention of the genetics community until 1965, when Wolf et al. described a similar individual with 4p deletion together with Hirschhorn's original case (Hirschhorn et al. 1965). Over the following six years, with the description of additional affected individuals, the syndrome was introduced to the pediatric community (Guthrie et al. 1971). A more complete delineation of the phenotype was only possible during the last two decades because of the publication of larger series of individuals with WHS/4p deletion (Wilson et al. 1981; Preus et al. 1985; Battaglia et al. 1999a, 1999b, 2008).

Although there are different phenotypes associated with deletions of the short arm of chromosome 4, both distal and proximal to that causing WHS (Battaglia and Carey 2008), this chapter is devoted only to the classical disorder called Wolf–Hirschhorn syndrome.

Incidence

It is generally estimated that approximately 1 in 50,000 to 1 in 20,000 individuals is diagnosed with WHS, with a female predilection of 2:1 (Lurie et al. 1980; Maas et al. 2008). In an epidemiological study carried out in the UK, a minimum birth prevalence of 1 in 95,896 was found (Shannon et al. 2001). Based on numerous reports (Battaglia 1997; Battaglia and Carey 1998, 2000; Battaglia et al. 1999a, 1999b, 2009a, 2001; Wright et al. 1999) it is possible that the above frequency is an underestimation because many affected individuals are still unrecognized and misdiagnosed. Nowadays, this occurs because of the difficulty of recognizing the facial gestalt, particularly in older individuals (Battaglia et al. 2000) and in those with a "mild phenotype" (Rauch et al. 2001; Bayindir et al. 2013; Zollino et al. 2014).

Almost 30% of infants and children with WHS die within the first few years of life (two-thirds of them within the first year) as a result of birth anoxia, withdrawal of treatment after premature delivery, congenital anomalies (heart or kidney defects, pulmonary hypoplasia, or diaphragmatic hernia), or lower respiratory tract infections (Shannon et al. 2001). Sudden unexplained deaths have occurred in a minority between 1 and 15 years of age (Shannon et al. 2001; personal observation). Based on the author's experience, on literature data (Opitz 1995; Smith et al. 1995; Wheeler et al. 1995; Ogle et al. 1996; Schaefer et al. 1996; Battaglia et al. 2000; Lanters 2000; Battaglia et al. 2008, Ho et al. 2018) and on communication from colleagues (Hennekam and van Ravenswaaij, personal communications), individuals with 4p-syndrome survive well into adult life (aged 24–61 years). They have few additional health concerns.

Cassidy and Allanson's Management of Genetic Syndromes, Fourth Edition.
Edited by John C. Carey, Agatino Battaglia, David Viskochil, and Suzanne B. Cassidy.
© 2021 John Wiley & Sons, Inc. Published 2021 by John Wiley & Sons, Inc.

Diagnostic Criteria

The cardinal features of WHS, as delineated from the most recent literature, are prenatal and postnatal growth deficiency, the "Greek warrior helmet appearance of the nose" (that is, the broad prominent nasal bridge continuing down from the forehead), deficits in neurological function (especially hypotonia and seizures), and variable degrees of intellectual deficits (Battaglia and Carey 2000; Rauch et al. 2001; Zollino et al. 2003; Fisch et al. 2010; Battaglia et al. 2015). These and all other distinctive features have been well described in a number of reviews (Wilson et al. 1981; Battaglia et al. 2001, 2002).

The "Greek warrior helmet appearance of the nose" that is the most distinctive craniofacial feature is easily recognizable in most individuals with WHS from birth to childhood, becoming somewhat less evident at puberty and at adulthood (Battaglia et al. 2000) (Figure 19.1). The forehead is tall with a prominent glabella. The nose has a broad prominent root and bridge continuing down from the forehead, and a flattened, sometimes beaked tip, with occasional asymmetric nares. The eyebrows are usually highly arched and sparse in the medial half. The philtrum is short and sometimes smooth, and the mouth is usually distinctive, with a full upper lip that appears to be pulled up in the middle and curled downward at the corners. Microcephaly, hypertelorism, epicanthal folds, somewhat prominent eyes, and micrognathia complete the craniofacial picture in the vast majority of affected individuals (Wilson et al. 1981; Estabrooks et al. 1995; Battaglia et al. 1999a, 1999b, 2001, 2008, 2015; Battaglia and Carey, 2000). Craniofacial asymmetry is seen in 60% of affected individuals. In more than 80% the ears are poorly formed, simple, posteriorly angulated, or low-set, with an attached lobe and pits and tags. The cartilage of the outer ear can be extremely underdeveloped or even absent. Unilateral or bilateral clefts of the lip and/or palate and/or uvula are seen in just over one-third of individuals. Eyelid ptosis, usually bilateral, is observed in 50% (Battaglia and Carey 2000). Of note, the craniofacial features in WHS are milder in those individuals with small deletions (Hammond et al. 2012; Battaglia et al. 2015). A wide variety of other anomalies can also be seen, including abnormalities of the skin (70%), skeleton (60–100%), eyes (40%), teeth (50%), heart (50%), hearing (40%), genitourinary tract (25%), and the immune, gastrointestinal and respiratory systems.

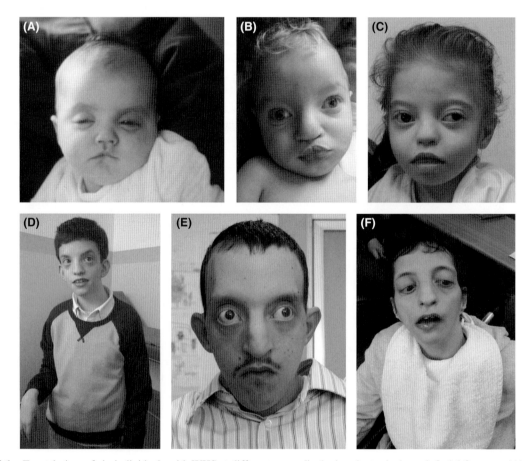

FIGURE 19.1 Frontal view of six individuals with WHS at different ages, displaying the typical craniofacial features: (A) age 4 months; (B) age 1 year 4 months; (C) age 5 years 3 months; (D) age 12 years 3 months; (E) age 28 years; (F) age 32 years 2 months.

A characteristic body habitus is noted, including hypotonic facies, sloping shoulders, marked reduction of subcutaneous fat mainly during infancy and childhood, and marked muscle hypotrophy involving the lower legs (Battaglia et al. 2000, 2008).

In addition to the above phenotype, a possible "mild Wolf–Hirschhorn syndrome phenotype" is emerging, based on literature reports (Rauch et al. 2001; Zollino et al. 2003, 2008; Hammond et al. 2012; Battaglia et al. 2015; Lozier et al. 2018) and personal observations.

Etiology, Pathogenesis, and Genetics

Wolf–Hirschhorn syndrome is usually the result of a deletion, or loss of material, of the distal portion of the short arm of chromosome 4. In the original cases and in individuals reported before 1970 (that is, before chromosome banding), the deletion encompassed about half of the short arm of chromosome 4. Lurie et al. (1980), based on about 100 cases, pointed out that the full phenotype could be appreciated when band 4p16 alone is deleted. Ten years later, Estabrooks et al. (1992) and Gandelman et al. (1992) proposed the notion of a critical region containing genes responsible for the WHS phenotype (WHSCR). Almost 50% of the deletions will not be detected by standard cytogenetic analysis (G-banding at less than the 500 band level) (Battaglia and Carey 2000; Battaglia et al. 2008). Therefore, chromosomal microarray (CMA) using oligonucleotide arrays or SNPs (single nucleotide polymorphisms) genotyping arrays is necessary to confirm the diagnosis in this situation. This observation accounts for delay in making the diagnosis in a number of individuals, and it has been partly responsible for many misdiagnoses encountered in others.

In their review of cytogenetically confirmed cases, Lurie et al. (1980) reported that about 75% of individuals with WHS will have a de novo isolated deletion of the short arm of chromosome 4; about 12% will have either a ring 4 chromosome, mosaicism, or a sporadic unbalanced translocation. The remaining 13% will have the 4p deletion as a result of a parental chromosome translocation. South et al. (2008a), using CMA to analyze individuals with WHS, found a much higher than expected frequency of unbalanced translocations (45%). Giglio et al. (2002) studied six individuals affected by WHS associated with der(4)t(4;8)(p16;p23), five of whom represented de novo cases and were of maternal origin. Interestingly, all five mothers were found to be heterozygous for both 4p and 8p submicroscopic inversions. Heterozygous inversions on 4p16 and 8p23 were also detected in 12.5% and 26% of control individuals, respectively. In about 85% of de novo deletions, the origin of the deleted chromosome is paternal (Tupler et al. 1992; Dallapiccola et al. 1993), whereas in almost two-thirds of the translocations, the mother carries the rearrangement (Bauer et al. 1985). The fact that de novo cases are paternal in origin and that unbalanced translocations are more often maternally inherited speaks against the possibility that imprinted genes may play a role in the expression of WHS. When an unbalanced translocation is identified, the parents are at increased risk of having another child with a chromosome abnormality (WHS or 4p trisomy with monosomy of the other involved chromosome, depending on the malsegregation of derivative chromosomes).

With the refinement of cytogenetic techniques, cryptic translocations were increasingly described in WHS (Bauer et al. 1985; Altherr et al. 1991; Reid et al. 1996; South et al. 2008a), most interestingly in familial cases. In addition, submicroscopic inversions have been observed in parents of individuals with WHS with apparently sporadic unbalanced translocations (Giglio et al. 2002). It is always recommended to perform state-of-the-art karyotypes on the parents of a child with WHS, in search of a possible translocation or other less common rearrangements. In an apparently de novo deletion (where parental chromosomes are normal), comparative genomic hybridization microarray (array-CGH) is advisable to exclude a cryptic unbalanced translocation (Altherr et al. 1991; Reid et al. 1996). However, it should be recognized that array-CGH analysis does not identify unbalanced translocations involving the acrocentric p-arms (South et al. 2008a).

The WHS clinical spectrum is complex and highly variable and there is now wide evidence that the core phenotype (the distinctive craniofacial features, pre/postnatal growth deficiency, intellectual disability, and seizures) is due to haploinsufficiency of several closely linked genes as opposed to a single gene (Maas et al. 2008; Zollino et al. 2008; Andersen et al. 2013; Ho et al. 2015). The proximal boundary of the WHSCR was defined by the analysis of two individuals showing the full core WHS phenotype and a 1.9 Mb terminal deletion of 4p16.3 including *WHSC1* (*NSD2*) and *LETM1* (Zollino et al. 2003; Rodriguez et al. 2005). The distal boundary of the WHSCR was established through the study of individuals with a full WHS phenotype and an interstitial deletion of 4p16 (Wright et al. 1997) and individuals with a terminal 4p16 deletion without the craniofacial characteristics of WHS (South et al. 2008b). However, following the identification of people with components of the core phenotype and more distal deletions, currently we think that WHS represents a true contiguous gene deletion syndrome with contribution of genes within a 1.5–1.6 Mb region in the ~0.4–1.9 Mb terminal 4p16.3 (van Buggenhout et al. 2004; South et al. 2007; Hammond et al. 2012).

To date the following genes that fall within this critical region have been characterized. *NSD2* (*WHSC1* Wolf–Hirschhorn syndrome candidate 1) (Stec et al. 1998) is a novel gene spanning 90 kb, two-thirds of which maps in the telomeric end of the critical region. Stec et al. (1998) detected expression of *WHSC1* in the telencephalon and rhombencephalon, spinal and trigeminal ganglia, the frontal face region, the jaw anlage, liver, adrenals, and the urogenital region of the mouse embryo. The protein domain identities and gene expression in early development suggest that it may play a role in normal development. Its deletion is likely to contribute to the WHS phenotype, mainly to the facial phenotype and poor growth, which constitute the core phenotype of WHS. Campos-Sanchez et al. (2017) have demonstrated a causal role for *NSD2* (*WHSC1)* in the immunodeficiency of WHS persons. *WHSC2* (Wolf–Hirschhorn syndrome candidate 2) (Wright et al. 1999), also known as *NELF-A,* spans 26.2 kb, and it is ubiquitously expressed. The identification of a mouse homolog, *Whsc2h*, and the location of this gene suggest that *WHSC2* encodes a protein that may play a role in the more global aspects of WHS. In particular, haploinsufficiency of *WHSC2* (*NELF-A*) and/or *SLBP* gene contributes to delayed cell-cycle progression and impaired DNA replication, likely responsible for microcephaly and pre-postnatal growth deficiency (Kerzendorfer et al. 2012) (Figure 19.2). *LETM1* was initially been proposed as a candidate gene for seizures. Its position immediately distal to the critical region means it is deleted in almost all affected individuals. It functions in ion exchange with potential roles in cell signaling and energy production (Dimmer et al.2008; Jiang et al. 2009; Kuum et al. 2012). However, a number of individuals have been described with 4p deletions including *LETM1* but without seizures, as well as having seizures with a 4p deletion excluding *LETM1* (Van Buggenhout et al. 2004; Faravelli et al. 2007; Maas et al. 2008; Misceo et al. 2012; Bayindir et al. 2013; Andersen et al. 2013). This would suggest that *LETM1* is not the only gene involved in the occurrence of seizures. Recent research has suggested that occurrence of seizures in WHS is probably enhanced synergistically by deletion of a combination of genes, such as *LETM1, CPLX1, CTBP1,* and *PIGG* (Bayindir et al. 2013; Shimizu et al. 2014; Zollino et al. 2014; Ho et al. 2016) (Figure 19.3). *CPLX1* belongs to a family of cytosolic proteins functioning in exocytosis of synaptic vescicles, and encodes complexin 1. Complexins are presynaptic regulatory proteins playing an important role in the modulation of neurotransmitters release (Cho et al. 2010). Mouse and rat models proved the transcriptional co-repressor *ctbp1* to be a relevant candidate gene for epilepsy. The hemizygosity of *CTBP1* in WHS should therefore be considered a potential contributor to the pathogenesis of seizures (Simon and Bergemann 2008). *PIGG* is involved in the biogenesis of GPI anchor proteins. Animal models showed that alterations in the biogenesis of GPI anchor proteins can alter expression of Na1.1 encoded by *SCN1A* (Nakano et al. 2010), responsible for an epileptic phenotype similar to that observed in WHS. *FGFRL1*, encoding a putative fibroblast growth factor (FGF) decoy receptor, has been implicated in the craniofacial development, axial and appendicular skeletal anomalies, and short stature of WHS (Engbers et al. 2009; Catela et al. 2009; Hammond et al. 2012). Besides *NSD2* (*WHSC1*) and *LETM1,* there are other oncogenes and putative oncogenes in the WHSCR, such as *GAK, TACC3, FGFR3*. Present research is focusing on the role of those genes in a possible increased cancer risk in WHS. Additional genes include *DFNA6,* and *MSX1. DFNA6,* one of the more than 25 genes responsible for non-syndromic hearing loss, has been mapped to 4p16.3 (Petit 1996), which might explain the sensorineural hearing deficit described in a number of individuals with WHS. *MSX1* gene has been found to be deleted

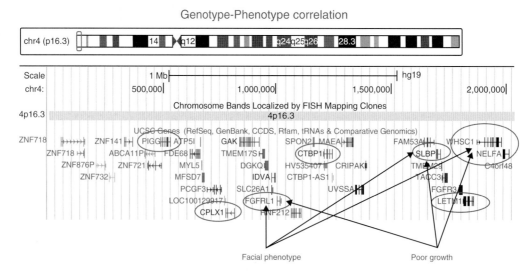

FIGURE 19.2 This diagram displays the candidate genes for craniofacial features and poor growth, on distal 4p. A.Battaglia et al. (2015). *Am J Med Genet Part C Semin Med Genet*; 169C: 216–223

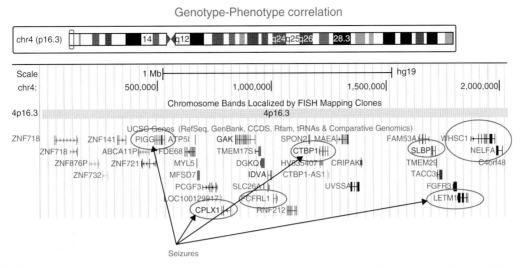

FIGURE 19.3 This diagram displays the candidate genes for seizures, on distal 4p. A.Battaglia et al. (2015) *Am J Med Genet Part C Semin Med Genet* 169C:216–223.

in people with WHS with oligodontia (Nieminen et al. 2003) (Figures 19.2 and 19.3).

Two other deletions of the short arm of chromosome 4 have been identified, responsible for two disorders that differ from WHS: the proximal interstitial deletion syndrome due to monosomy proximal to the WHSCR (Chitayat et al. 1995; Fryns 1995; Bailey et al. 2010), and a recently described entity due to monosomy of 4p16.3 distal to the WHSCR (South et al. 2008b).

A correlation between size of the deletion and severity of the clinical phenotype has been established. Wieczorek et al. (2000) and Zollino et al. (2000, 2008) have suggested a partial or a complete genotype–phenotype correlation. South et al. (2008a) have shown that clinical variation may be explained by cryptic unbalanced translocations, where trisomic material from the second chromosome can modify expected clinical manifestations. Shannon et al. (2001) reported a statistically significant relationship between deletion size and overall risk of death in de novo deletion cases; the reason for this could not be identified. More recent studies have shown a clear genotype–phenotype correlation in WHS (Hammond et al. 2012; Shimizu et al. 2014; Battaglia et al. 2015).

Diagnostic Testing

Definitive diagnosis of WHS is made by detecting deletion of the distal portion of the short arm of chromosome 4 involving band 4p16. Routine and high-resolution cytogenetic analyses detect approximately 50% of the deletions; FISH, using a probe within the critical region, detects more than 95% of deletions (Johnson et al. 1994; Zollino et al. 2003). Subtelomeric FISH screening (Knight et al. 2000) is probably the most sensitive and specific way to determine whether a cytogenetically visible deletion is due to an unbalanced translocation. Subtelomeric analysis can also identify the partner chromosome involved in a translocation. In the case of strong clinical suspicion with negative cytogenetic testing done more than five years previously, repeat testing using more up-to-date probes is recommended. Comparative genomic hybridization microarray (array-CGH) using oligonucleotide arrays or SNP genotyping arrays can detect the WHSCR deletion in more than 95% of probands, and has been shown to be a useful tool for detecting cryptic unbalanced translocations (South et al. 2008a), which occur frequently. The ability to size the deletion depends on the type of microarray used and the density of probes in the 4p16.3 region. Determining the size of the deletion in an individual would allow better prediction of their prognosis and more appropriate detailed guidelines for health supervision and anticipatory guidance for that individual.

Differential Diagnosis

The pattern of features in WHS is quite distinctive and should rarely be confused with other syndromes. The most frequent diagnoses overlapping with WHS include Seckel syndrome (Griffith et al. 2008), CHARGE syndrome (see Chapter 11), partial trisomy 4p syndrome (Zollino et al. 1999), Smith–Lemli–Opitz syndrome (see Chapter 53), Opitz–GBBB syndrome (De Falco et al. 2003; McDonald-McGinn et al. 2005), Malpuech syndrome (Malpuech et al. 1983), Lowry–MacLean syndrome (Lowry and MacLean

1977), Williams syndrome (see Chapter 63), Rett syndrome (see Chapter 49), Angelman syndrome (see Chapter 5), and Smith–Magenis syndrome (see Chapter 54).

Of note, Pitt–Rogers–Danks syndrome was described in four individuals (two of them sisters) with prenatal and postnatal growth retardation, microcephaly, a characteristic face, intellectual disability, and seizures. It was thought to be a possible new autosomal recessive syndrome (Pitt et al. 1984). In 1996 a distal 4p microdeletion, identical to that seen in WHS, was described in the siblings in the original report and in two other individuals (Clemens et al. 1996). It is likely that Pitt–Rogers–Danks syndrome and Wolf–Hirschhorn syndrome are the same condition (Donnai 1996; Battaglia and Carey 1998; Wright et al. 1998, 1999).

MANIFESTATIONS AND MANAGEMENT

Growth and Feeding

Growth deficiency of prenatal onset is the rule for WHS. In spite of adequate caloric and protein intake, all individuals continue to show short stature and slow weight gain (below third centile) later in life despite dietary enrichment (Estabrooks et al. 1995; Battaglia and Carey 1999; Battaglia et al. 2001, 2002; Rauch et al. 2001). Severely delayed bone age seems to be a hallmark of WHS, being present in all individuals studied (Battaglia and Carey 2000; Battaglia et al. 2008). Specific growth charts have been produced only for children 0–4 years of age (Antonius et al. 2008). Head circumference tends to be below the second centile in all affected individuals, except those with cryptic unbalanced translocations (South et al. 2008a).

Varying degrees of premature aging have been reported by some parents and were observed in a few affected adults at several US 4p-support group and Associazione Italiana Sindrome Wolf–Hirschhorn gatherings (personal observation).

Feeding difficulties constitute one of the two main challenges of infancy and childhood. The causes vary and include: oral facial clefts (in about 50%) and their related difficulty in sucking; poor suck because of central hypotonia; poorly coordinated swallow with consequent aspiration; and gastroesophageal reflux (Battaglia et al. 2001, 2002). Although transitory, gastroesophageal reflux is a consistent finding in infants with WHS and may explain irritability and recurrent respiratory tract infections. Aspiration during feeding or from reflux may be the cause of early death (Shannon et al. 2001). Gastrointestinal malformations can also play a role in feeding problems (see Gastroenterology).

Evaluation

- Head circumference, length, and weight should be measured during each health supervision visit and plotted on standard growth curves. Before 4 years of age, diagnosis-specific charts can be used (Antonius et al. 2008).
- The likelihood of feeding problems should be recognized and assessed, particularly in those with clefts.
- Referral to a dysphagia team for the evaluation of swallowing difficulties may be useful, as may a radiographic swallowing study.
- The presence of gastroesophageal reflux or a gastrointestinal malformation as a potential factor in feeding problems should be assessed in a standard manner.
- If a child with WHS is not bottle feeding by 6 months of age or has persistent failure-to-thrive, consideration of gastrostomy tube placement is indicated to sustain weight gain, protect the airway, and manage gastroesophageal reflux.
- Growth hormone studies should be performed, because partial growth hormone deficiency has recently been documented (unpublished personal observation).

Management

- Assurance of adequate caloric intake, if necessary, by increased caloric formulas and/or the use of the Haberman feeder, or oral or nasogastric tube feeding, is indicated.
- Standard therapy/management for gastroesophageal reflux can be initiated when this is diagnosed. Nissen–Hill fundoplication has been successfully performed in a number of children studied (Battaglia et al. 2001).
- Gastrostomy has been successfully performed in many children (Battaglia et al. 2001, 2008).
- Anesthetic risks for surgical procedures in children with structural heart defects require consultation by cardiology.

Development and Behavior

Until a few years ago, it was commonly stated in the literature that development is severely compromised in individuals with WHS, with no development of speech and minimal communication (Centerwall et al. 1975; Wilson et al. 1981; Stengel-Rutkowski et al. 1984; Estabrooks et al. 1995; O'Brien and Yule 1995). Only recent studies of large numbers of individuals with this condition have allowed recognition of a more complete continuum of the phenotype, pointing out a much wider clinical spectrum than previously thought (Battaglia and Carey 2000; Battaglia et al. 2001, 2002, 2008, 2015).

Battaglia and Carey (1999) discussed the challenges faced by parents of children with WHS, directing primary care practitioners and pediatricians to approach long-term management in these children as they would in any child with disabilities. More recent reviews (Battaglia and Carey 1999; Battaglia et al. 1999a, 1999b, 2001, 2002, 2008, 2015) indicate that there is greater acquisition of milestones than suggested by the older medical literature. Cognitive deficits are a hallmark of the syndrome (Centerwall et al. 1975; Wilson et al. 1981; Stengel-Rutkowski et al. 1984; Estabrooks et al. 1995; O'Brien and Yule 1995). Severe intellectual disabilities are noted in 65–70% of individuals, whereas 30–35% have a mild to moderate deficit. One child with low normal intelligence has been reported (Rauch et al. 2001). Academic performance is poor for cognitive ability, at least in those children with a deficit of lesser degree. Expressive language, although limited to guttural or disyllabic sounds in most individuals, can develop to the level of simple sentences in almost 6% (Battaglia and Carey 2000; Battaglia et al. 2008, 2015). Communication skills are present in most individuals with WHS and improve over time with signing ability. Motor milestones are delayed, but 45% of children walk by the age of 2–12 years, either independently (25%) or with support (20%). They tend to walk with a broad-based gait, which improves over time, and with poor swinging movements of the upper limbs. Sphincter control (mainly by day) is achieved in 10% of affected individuals, between age 4 and 14 years. Fine motor abilities are limited to scribble. Some degree of autonomy with eating, dressing and undressing, and simple household tasks is reached by 18% of children. All individuals show a slow but constant improvement in all aspects of development over time.

Almost all affected individuals interact and relate to their family members and to the people they know or meet (e.g., their peers, teachers, and professional caregivers). They show an intense interaction with adults in an attempt to draw their attention. Fisch et al. (2010) showed that among children with WHS, the profile of mean cognitive abilities and deficits is relatively flat and extends to all cognitive areas. Despite their limitations in cognitive skills and overall adaptive behavior, persons with WHS exhibit relative competence in socialization skills compared to their abilities in other adaptive behavior domains. Just over 50% of children with WHS exhibit inattentiveness and/or hyperactivity and unusual internalizing behavior (Fisch et al. 2010).

Different kinds of stereotypies have been observed in almost 50% of children and adolescents with WHS (Battaglia and Carey 2000; Battaglia et al. 2008, 2015), including hand washing or flapping, holding both hands in front of the face, patting oneself on the chest, rocking, head shaking, and stretching of the legs.

Sleep problems seem to be rather common in children with WHS during the early years, but are seldom worrisome (Battaglia and Carey 2000). Children tend to wake up between midnight and 4 a.m., start crying, and then generally stop at their carer's arrival in the bedroom. At times, cuddling is sufficient for them to settle down. On most occasions, when no medical factors (e.g., otitis media, gastroesophageal reflux, or eczema) are involved, such behavior tends to modify spontaneously, and all children are able to achieve a satisfactory overnight sleep. Where sleeping problems are reinforced by parental attention, modification of parental attention for that behavior has proven to be an effective intervention (Curfs et al. 1999).

Evaluation

- A thorough developmental evaluation by a specialist should be performed at intervals appropriate for planning of early interventions, including physical, occupational and speech therapies, and for educational and vocational planning.

Management

- Involvement in early intervention programs is recommended in the ongoing care of infants and children with WHS.
- Enrollment in an individualized rehabilitation program that covers motor aspects (including oral motor and feeding therapy), cognition, communication, and socialization, is highly recommended.
- Appropriate school placement, after full evaluation, is important.
- Planning for transition to adulthood, including vocational training and living situation, should begin in adolescence.

Neurologic

Besides feeding difficulties, the other main challenge of infancy and childhood is a seizure disorder, which occurs in 50–100% of infants and children with WHS (Guthrie et al. 1971; Centerwall et al. 1975; Stengel-Rutkowski et al. 1984; Battaglia and Carey 1999; Battaglia et al. 1999a, 1999b). The author's study of 87 individuals in Italy and the US documented seizures in about 95% of affected children. Age at onset varies between the neonatal period and 36 months of age with a peak incidence around 6–12 months of age. Seizures are either unilateral clonic or tonic, with or without secondary generalization, or generalized tonic-clonic from the onset. They are frequently triggered by fever, even low-grade fever, and can occur in clusters and last over 15 minutes. Other risk factors include respiratory or urinary tract infections, and, reportedly, tiredness and excitement in a few individuals (Battaglia et al. 2008). Unilateral or generalized clonic or tonic-clonic status epilepticus occurs in about 50–60% of children during the early years in spite of adequate pharmacotherapy (Battaglia et al. 2009; Ho et al. 2018). Over 30% of children develop atypical absences by age 1–6 years, often accompanied by a mild myoclonic component involving the eyelids and, less frequently, the eyeballs and the upper limbs. In 55% seizures stop by age 2–13 years (Battaglia et al. 2009). In a previous study, fewer than 20% of individuals had discontinued antiepileptic drugs (Battaglia et al. 2001). In a recent analysis of 22 Japanese individuals with WHS, Shimizu et al. (2014) observed a higher frequency of status epilepticus in those with larger deletions (>6 Mb).

Distinctive electroencephalographic (EEG) abnormalities have been observed in almost all individuals who have had serial studies, during waking and sleep (Battaglia et al. 1996, 2009, 2001). These abnormal findings include frequent, ill-defined, diffuse or generalized, high-amplitude sharp element spike/wave complexes at 2–3.5 Hz occurring in bursts lasting up to 25 s that are activated by slow wave sleep, and frequent high-amplitude spikes, polyspikes or wave complexes at 4–6 Hz, over the posterior third of the head, often triggered by eye closure. These diffuse or generalized abnormalities are often associated with atypical absences, whereas the ones localized posteriorly do not appear to be strictly related to seizures and can be seen for many years after seizures have stopped.

A variety of developmental abnormalities of the brain have been described in the few autopsy reports of WHS (Lazjuk et al. 1980; Gottfried et al. 1981). These include brain hypoplasia with narrow gyri, heterotopias, shortening of the H2 area of Ammon's horn, arrhinencephaly, microscopic dysplasias of the lateral geniculate nucleus and of the dentate and the inferior olivary nuclei, and dystopic dysplastic cerebellar gyri. Corpus callosum hypoplasia has been described as the most frequent abnormality in WHS, both at autopsy and on brain neuroimaging. Occasionally this has been associated with diffusely decreased volume of the white matter or marked hypoplasia or agenesis of the posterior lobes of the cerebellar hemispheres on brain computed tomography or brain magnetic resonance imaging (Battaglia et al. 2001, 2008). Microcephaly occurs in 90–96% of individuals (Battaglia et al. 2008; Zollino et al. 2008).

Generalized hypotonia, which on occasion may be marked, is present in all affected individuals. This central hypotonia appears to be of prenatal onset and persists throughout life. It causes decreased fetal movements with frequent abnormal fetal position, often necessitating delivery by Cesarean section. It is almost invariably associated with poor suck with consequent failure-to-thrive. Muscle hypotrophy, mainly of the lower limbs, is observed in most individuals with WHS and may be linked to the reduced movements in utero. Electromyography, nerve conduction velocities, and muscle biopsy are normal.

Evaluation

- A waking/sleeping video-EEG-polygraphic study is recommended in infancy and childhood, to achieve the best characterization of seizures. This is of utmost importance when the parents or professionals report the occurrence of "staring spells" or "eye-tics." On most occasions, these episodes are shown to be atypical absence spells.
- Regular follow-up of seizure status and pharmacotherapy is essential.
- Neurological examination is recommended for all infants, children, and adolescents.
- Neuroimaging should be performed in all individuals with seizures, because malformation of the cerebral cortex may modify prognosis and management.

Management

- Clonic or tonic-clonic seizures are usually effectively controlled by phenobarbital.
- Clonic or tonic-clonic status epilepticus can be effectively controlled by intravenous benzodiazepines such as diazepam or midazolam. On occasion, it may be necessary to add intravenous phenytoin.
- Absence or myoclonic status epilepticus can be effectively controlled by intravenous benzodiazepines.
- In the author's experience the atypical absence spells can be well controlled by valproic acid, alone or associated with ethosuccimide. On rare occasion a benzodiazepine such as clobazam or clonazepam may be needed. Carbamazepine and oxcarbazepine may severely worsen the electroclinical picture (Battaglia et al. 2009; Ho et al. 2018).
- Abnormalities of muscle tone should be treated as in the general population with physical therapy.

Cardiovascular

Congenital heart malformations are present in about 50–86% of individuals with WHS (Shimizu et al. 2014; Battaglia et al. 2015). In recent series, 25–60% had an atrial septal defect; 15–27% had pulmonary stenosis; 6% had ventricular septal defect; and patent ductus arteriosus was observed either isolated in 22% or associated with aortic insufficiency in 2% (Battaglia et al. 1999a, 1999b, 2008; Battaglia and Carey, 2000; Shimizu et al. 2014). In one series, one of 13 individuals had tetralogy of Fallot (Wilson et al. 1981). Another study found one of 32 deceased children with WHS had tetralogy of Fallot with a more complex anomaly involving a double outlet single ventricle with a single atrioventricular valve, large aorta, and small posterior pulmonary artery with pulmonary valve stenosis (Shannon et al. 2001). Congenital heart disease was the cause of death in five of the 32 children in that study (Shannon et al. 2001). On the whole, the congenital cardiac anomalies are usually not complex and are amenable to repair. More recently, Shimizu et al. (2014) observed a higher frequency of congenital heart defects in those individuals with larger deletions.

Evaluation

- Cardiac evaluation is important at the time of diagnosis and should include auscultation, electrocardiogram, and echocardiography.
- Decision concerning cardiac surgery should be made with active participation of parents or guardians, considering the other abnormalities present in the child.

Management

- In some instances, septal defects will close spontaneously.
- In all other cases, the defects can be readily surgically repaired, following standard practices.

Genitourinary

A variety of structural defects of the genital and urinary tracts have been described in almost 50% of individuals with WHS (Battaglia and Carey 2000; Battaglia et al. 2008; Shimizu et al. 2014). These include renal agenesis, either unilateral or bilateral, cystic dysplasia/hypoplasia, horseshoe kidney, and oligomeganephroma (Lazjuk et al. 1980; Tachdijan et al. 1992; Park and Chi 1993; Estabrooks et al. 1995; Shimizu et al. 2014). In a series of six affected individuals, renal structural imaging showed unilateral agenesis in 2/6, malrotation in 1/6, horseshoe kidney in 1/6, small rudimentary left kidney in 1/6, and hyperechoic kidneys with loss of corticomedullary differentiation on ultrasonography in 4/6. Abnormal renal function was observed in all. The only individual with normal imaging studies showed a decreased glomerular filtration rate, suggesting an underlying microscopic anomaly. High degree (III–IV) vesicoureteric reflux was diagnosed in 4/6 (Grisaru et al. 2000). Renal tubular acidosis has also occasionally been reported (Schaefer et al. 1996). In addition, bladder exstrophy and obstructive uropathy have been observed in 4–5% of individuals (Battaglia and Carey 2000). Deterioration of renal function in some individuals with WHS might lead to difficult management decisions (personal experience).

Abnormalities of the genitalia include hypospadias and cryptorchidism in almost 50% of males (Battaglia and Carey 2000; Battaglia et al. 2008), and clitoral aplasia or hyperplasia, streak gonads, absent uterus, and absent vagina in a number of females (Fryns, 1973; Lazjuk et al. 1980; Gonzales et al. 1981).

Evaluation

- Renal ultrasound and serum creatinine are indicated in all individuals at diagnosis.
- Individuals with vesicoureteric reflux and structural renal defects should be followed for infection and renal insufficiency by periodic serum blood urea nitrogen (BUN), creatinine, and cystatin C; urinalysis, creatinine clearance test, and urine culture.
- Voiding retrograde cystography and intravenous pyelography or scintigraphy may be necessary in the presence of abnormal renal structure or function.
- Urinalysis should also be done during infections where the site is unknown.
- Careful examination of genitalia for anomalies should occur at diagnosis. Referral of males with cryptorchidism to a urologist or surgeon should be made at diagnosis.

Management

- Antibiotic prophylaxis is indicated in the presence of vesicoureteric reflux.
- Management of urinary tract infection and renal obstruction are the same as in any individual.
- Management of genital anomalies is the same as in any individual.

Ears and Hearing

More than 80% of individuals with WHS show ear anomalies. The ears are poorly formed, simple, posteriorly angulated, or low set, and may have attached lobes. In some, the cartilage of the outer ear is extremely underdeveloped and may even be absent.

Hearing loss occurs in more than 40% of children with WHS (Battaglia and Carey 2000; Battaglia et al. 2008; Shimizu et al. 2014). Most often of the conductive type, and secondary to chronic serous otitis media, it is of a degree that can potentially affect language acquisition and educational achievement. Sensorineural hearing loss has also been reported in 10–16% (Estabrooks et al. 1994; Lesperance et al. 1998; Battaglia and Carey 2000; Battaglia et al. 2008). It is relevant that *DFNA6*, one of the more than 25 genes responsible for non-syndromic hereditary hearing impairment, has been mapped to the chromosome 4p16.3 region (Petit 1996).

Evaluation

- Comprehensive audiologic and otologic evaluation of children should be performed at diagnosis.
- Children with significant mental handicap can be tested with brainstem auditory evoked responses and/or otoacoustic emissions. The advantage of brainstem auditory evoked response testing over auditory field examination is its ability to rule out frequency-specific or unilateral sensorineural hearing loss.

Management

- A trial of hearing aids in older infants who have abnormal audiologic evaluations is appropriate

(personal experience). However, if the hearing aid does not appear to improve communication and is difficult for the child to tolerate, the decision to continue or discontinue should be individualized.
- Hearing loss should be considered in planning educational and vocational programming.

Ophthalmologic

Defects of the eye or optic nerve are present in 40–60% of people with WHS (Shimizu et al. 2014; Battaglia et al. 2015). The most typical are exodeviation (exotropia) and colobomas, which usually involve only the iris. Glaucoma is an infrequent occurrence in these individuals; but, when present, can be of severe degree, and constitutes a serious clinical challenge. Ptosis of the eyelids, usually bilateral, can be present in as many as 50% of affected individuals, and eyelid hypoplasia has been observed occasionally (Battaglia and Carey 2000). Nasolacrimal obstruction and cataract have also been observed (Shimizu et al. 2014; personal experience).

Evaluation

- Ophthalmology consultation is recommended at diagnosis.

Management

- Treatment of exodeviation, nasolacrimal obstruction, cataract, and glaucoma follows standard practice.
- Eyelid hypoplasia may require skin grafting.

Dental

More than 50% of individuals with WHS show altered tooth development (Battaglia and Carey 2000; Battaglia et al. 2008). This usually consists of delayed dental eruption, with persistence of deciduous teeth into adulthood in a minority (personal experience); taurodontism in the primary dentition, peg-shaped teeth, or agenesis of lower lateral incisors or upper canines (Breen 1998; Battaglia and Carey 2000; Battaglia et al. 2008).

Evaluation

- Careful orthodontic evaluation is recommended in children and adolescents at diagnosis.
- Routine dental evaluations, as in the general population, are appropriate.

Management

- Routine dental hygiene and care is indicated.
- Orthodontic treatment to prevent malocclusion follows standard practice.

Musculoskeletal

A variety of musculoskeletal abnormalities occur in WHS, including medically significant malformations and minor anomalies of limbs and skeleton. According to Lazjuk et al. (1980), orthopedic deformities occur in 82% of affected individuals when one takes into account infants who die during the perinatal period. Deformities of the feet varying from pes cavus (Coffin, 1968) to talipes equinovarus (Subrt et al. 1969) are most frequent. Other skeletal anomalies include craniostenosis with brachycephalic skull, platybasia, malformed occipital condyles, occipital stenosis, and several anomalies of the cervical vertebrae (butterfly vertebrae, agenesis of the anterior arch), underdevelopment of ossification centers in the cervical spine, anomalies of sternal and pubic bone ossification, clinodactyly, finger-like appearance of the thumb, thin fingers with bilateral overriding of the second finger onto the third, absence of the ossification nucleus for the ulnar styloid apophysis, proximal radioulnar synostosis, and malformed toes (double first phalanx of the first ray) (Magill et al. 1980; Battaglia and Carey 2000; personal experience). Unilateral split hand has also been reported occasionally (Bamshad et al. 1998; Shanske et al. 2010). Small iliac wings with coxa valga subluxans, hip dislocation, scoliosis, and kyphosis have also been observed (Battaglia and Carey 2000; Battaglia et al. 2008). Scoliosis is common in older children and is not necessarily related to structural defects of the vertebrae, although malformed vertebral bodies and accessory or fused ribs have also been reported (Stengel-Rutkowski et al. 1984; Sergi et al. 1998; personal experience). Scoliosis may progress between late childhood and early adolescence.

Evaluation

- The spine should be evaluated clinically in older children at routine health supervision visits.
- Spine radiographs should be performed when clinical scoliosis is present.
- Neuroimaging of the skull–brain and spine should be performed if unexpected neurological signs appear (such as hyperreflexia of the lower limbs), searching for the presence of possible anomalies of the base of the skull and cervical vertebrae with consequent kinking of the brainstem. Referral to a neurosurgeon should be considered thereafter.
- Referral to an orthopedist should be considered on recognition of significant musculoskeletal abnormalities.

Management

- The treatment of club foot follows standard practice.

- Because almost half the children with WHS walk by the age of 2–12 years, either independently or with assistance, standard orthopedic surgery is recommended for foot deformities at an early age.
- Decisions about surgery for scoliosis may need to be made in the older child. If treatment is offered, it follows standard practice.
- Neurosurgery for fixation of occipito-cervical anomalies with osteo-dural decompression of the posterior fossa follows standard practice.

Respiratory

The majority of individuals with WHS has frequent respiratory tract infections, mostly of the upper tract. These are sometimes related to aspiration and/or swallowing difficulties. Lower respiratory tract infections may occasionally be the cause of death in infants and children (Shannon et al. 2001). Other abnormalities of the respiratory system are quite rare but include bilobed or trilobed lungs and lung hypoplasia. Obstructive sleep apnea has been rarely reported (personal experience).

Evaluation

- Evaluations do not differ from those in other children with similar symptoms.

Management

- Treatment follows standard practice.

Gastroenterology

In addition to the functional problems of gastroesophageal reflux discussed in Growth and Feeding, malformations of the gastrointestinal tract can occur in WHS. Esophageal atresia, malrotation of the gut, diaphragmatic hernia with upward displacement of viscera or severe diaphragmatic defect with displacement of abdominal viscera into the chest have been reported occasionally in WHS (Carter et al. 1969; Eiben et al. 1988; Verloes et al. 1991; Tachdijan et al. 1992; Estabrooks et al. 1995; Sergi et al. 1998). Non-familial hypercholesterolemia (>220 mg dL^{-1}) has been reported in about 35% of individuals with WHS in whom serum cholesterol levels were examined (Shimizu et al. 2014; personal experience).

Evaluation

- Symptoms suggestive of gastrointestinal tract malfunction should prompt standard evaluations as appropriate.

Management

- Treatment of gastrointestinal malformations is standard and often requires surgical intervention.

Immunologic

B-cell defects (IgA or IgG2 subclass deficiency, isolated IgA deficiency, and impaired polysaccharide responsiveness) were found in 9 of 13 children studied by Hanley-Lopez et al. (1998). Campos-Sanchez et al. (2017) have demonstrated that hemizygous loss of *WHSC1* causes a progressive decline of lymphocyte numbers with age. The occurrence of common variable immunodeficiency might explain the recurrent respiratory tract infections.

Evaluation

- Careful evaluation of the immune and hematopoietic systems is recommended in all individuals with frequent and difficult to treat respiratory tract infections, at initial presentation and over the years.

Management

- Affected individuals may benefit from intravenous immunoglobulin infusions or prophylactic antibiotics.

Dermatologic

Midline scalp defects, although reported in both original patients (Cooper and Hirschhorn 1961; Wolf et al. 1965), seem to be an occasional finding in WHS (Centerwall et al. 1975; Wilson et al. 1981; Battaglia et al. 1999a, 1999b).

Skin changes such as capillary venous malformation, mainly of the forehead, and marble and/or dry skin are seen in about 70% of the affected individuals. Sacral dimples seem to be an almost constant feature (Battaglia and Carey 2000; Battaglia et al. 2008). Several people with WHS have been noted to have signs of premature aging (Battaglia et al. 2015).

Evaluation

- Evaluations do not differ from those in other children with similar symptoms.

Management

- Treatment should be decided on an individual basis.

Neoplasia

Following research on the natural history of WHS into adulthood, new concerns for long-term medical complications are emerging. Prunotto et al. (2013) reported on two individuals with WHS aged 20 and 11 years showing hepatic adenomas with evidence of progression. Battaglia et al. (2018) recently described four people with WHS in whom hepatic adenomas were diagnosed at the age of 15 years, 18 years, 22 years, and 32 years, respectively. Evolution of the lesions was quite different, being stable over time in two of them, needing medical intervention in one, and progressing to hepatocellular carcinoma in the remaining. A 2.5-year-old child with WHS with hepatoblastoma was reported by Bayan et al. (2017).

Given the rarity of WHS, the recent observation of the occurrence of hepatic neoplasms (mainly adenomas) in seven individuals, with progression in two of them, raises concern that hepatic neoplasms may be a component feature of WHS.

Inflammatory myofibroblastic bladder tumor has been observed in a further individual with WHS (Marte et al. 2013). A papillary carcinoma of the thyroid has been detected in a 30-year-old lady with WHS (personal experience).

Evaluation

- Monitoring for symptoms alone may be inadequate. Proactive screening with abdominal sonograms after age 10 years should be considered in adolescents with WHS.

Management

- Treatment should be decided on an individual basis.

Adulthood

Data on lifespan natural history are very limited. The preliminary findings of an on-going study of the natural history into adulthood of persons with WHS have shown that just over 50% require close to total care, constituting a significant burden for the family and society. Just over 40% are partly independent, requiring supervision on certain daily routines. However, a positive perspective is given by the overall good health enjoyed by almost 80% of the adults observed. Relevant clinical problems, other than those discussed above, have been detected in a minority, and include type 2 diabetes associated with hypercholesterolemia (~5%), Raynaud disease (~5%), esophagitis (~5%), severe thoracolumbar scoliosis (~20%), and cataracts before age 30 (~20%) (Battaglia et al. 2013).

RESOURCES

4p– Support Group

Amanda Lortz, Executive Director
1495 Forest Brooke Way, #262
Delaware, Ohio 43015
USA
Phone: 614-307-1092
Website: www.4p-supportgroup.org

Wolf Hirschhorn Syndrome Support Group UK

Phone: + 44 (0) 1634 264816
Email: whs@webk.co.uk
Website: https://www.wellaware.org.uk/organisation/wolf-hirschhorn-syndrome-support-group/

Wolf Hirschhorn Syndrome Trust

UK & Ireland
Website: http://whs4pminus.co.uk/

Chromosome Deletion Outreach, Inc.

PO Box 724
Boca Raton, FL 33429-0724
USA
Phone: 1-888-236-6880; 561-391-5098
Email: info@chromodisorder.org
Website: *https://chromodisorder.org/*

Associazione Italiana Sindrome Wolf-Hirschhorn (AISiWH)

c/o via Tiziano, 20
62010 Montecosaro (MC)
Italy
Phone: + 39 0733 864275
Email: *segreteria.aisiwh@gmail.com*
Website: *www.aisiwh.it*

Dutch Parent Support Group Wolf-Hirschhorn Syndrome

Federatie van Oudervereinigingen
Mieke van Leeuwen
PO Box 85276
3508 AG Utrecht
The Netherlands
Phone: + 31 (0) 30 2363767
Email: *m.vanleeuwen@fvo.nl*

Frambu- Senter for Sjeldne Funksjonshemninger

(Resource Centre for Rare Disorders)
Sandbakkveien 18, 1404 Siggerud- Oslo
Norway
Phone: + 47 64 85 60 00
Email: info@frambu.no
Website: www.frambu.no

REFERENCES

Altherr MR, Bengtsson U, Elder FF, Ledbetter DH, Wasmuth JJ, McDonald ME, Gusella JF, Greenberg F (1991) Molecular confirmation of Wolf-Hirschhorn syndrome with a subtle translocation of chromosome 4. *Am J Hum Genet* 49:1235–1242.

Andersen EF, Carey JC, Earl DL, Corzo D, Suttie M, Hammond P, South ST. (2013). Deletions involving genes WHSC1 and LETM1 may be necessary, but are not sufficient to cause Wolf–Hirschhorn Syndrome. *Eur J Hum Genet* 22:464–470.

Antonius T, Draaisma J, Levichenko E, Knoers N, Renier W, van Ravenswaaij C (2008) Growth charts for Wolf-Hirschhorn syndrome (0–4 years of age). *Eur J Pediatr* 167:807–810.

Bailey NG, South ST, Hummel M, Wenger SL (2010) Case report: cytogenetic and molecular analysis of proximal interstitial deletion of 4p, review of the literature and comparison with Wolf-Hirschhorn syndrome. *J Assoc Genet Technol* 36:5–10.

Bamshad M, O'Quinn JR, Carey JC (1998) Wolf-Hirschhorn syndrome and a split hand malformation. *Am J Med Genet* 75:351–354.

Battaglia A (1997) Sindrome di Wolf-Hirschhorn (4p-): una causa di ritardo mentale grave di difficile diagnosi. *Riv Ital Pediat (IJP)* 23:254–259.

Battaglia A, Carey JC (1998) Wolf-Hirschhorn syndrome and Pitt-Rogers-Danks syndrome. *Am J Med Genet* 75:541.

Battaglia A, Carey JC (1999) Health supervision and anticipatory guidance of individuals with Wolf-Hirschhorn syndrome. *Am J Med Genet* 89:111–115.

Battaglia A, Carey JC (2000) Update on the clinical features and natural history of Wolf-Hirschhorn syndrome (WHS): Experience with 48 cases. *Am J Hum Genet* 6:127.

Battaglia A, Carey JC (2008) Wolf-Hirschhorn syndrome and the 4p-related syndromes. *Am J Med Genet C Semin Med Genet* 148C:241–243.

Battaglia A, Carey JC, Tliompson JA, Filloux F (1996) EEG studies in the Wolf-Hirschhorn (4p-) syndrome. *EEG Clin Neurophysiol* 99:324.

Battaglia A, Carey JC, Cederholm P, Viskochil DH, Brothman AR, Galasso C (1999a) Natural history of Wolf-Hirschhorn syndrome: Experience with 15 cases. *Pediatrics* 103:830–836.

Battaglia A, Carey JC, Cederholm P, Viskochil DH, Brothman AR, Galasso C (1999b) Storia naturale della sindrome di Wolf-Hirschhorn: Esperienza con 15 casi. *Pediatrics* 11:236–242.

Battaglia A, Carey JC, Viskochil DH, Cederholm P, Opitz JM (2000) Wolf-Hirschhorn syndrome (WHS): A history in pictures. *Clin Dysmorph* 9:25–30.

Battaglia A, Carey JC, Wright TJ (2001) Wolf-Hirschhorn (4p-) syndrome. *Adv Pediatr* 48:75–113.

Battaglia A, Carey JC, South ST Wolf-Hirschhorn Syndrome (2002) Apr 29 [Updated 2015 Aug 20]. *GeneReviews®* [Online]. Available from: https://www.ncbi.nlm.nih.gov/books/NBK1183/

Battaglia A, Doccini V, Filippi T, Lortz A, Carey JC (2013) Wolf-Hirschhorn syndrome: natural history into adulthood. A preliminary study of twenty-four individuals. *Am Society Hum Genet, 63rd Annual Meeting, Boston, Massachusetts, Book of Abstracts* 418.

Battaglia A, Filippi T, South ST, Carey JC (2009) Spectrum of epilepsy and electroencephalogram patterns in Wolf-Hirschhorn syndrome: Experience with 87 patients. *Dev Med Child Neurol* 51:373–380.

Battaglia A, Filippi T, Carey JC (2008b) Update on the clinical features and natural history of Wolf-Hirschhorn (4p-) syndrome: Experience with 87 patients and recommendations for routine health supervision. *Am J Med Genet C* 148C:246–251.

Battaglia A, Carey JC, South ST (2015) Wolf-Hirschhorn syndrome: a review and update. *Am J Med Genet C* 169C:216–223.

Battaglia A, Calhoun A, Lortz A, Carey JC (2018) Risk of hepatic neoplasms in Wolf-Hirschhorn syndrome (4p-): four new cases and review of the literature. *Am J Med Genet A* 176:2389–2394.

Bauer K, Howard-Peebles PN, Keele D, Friedman JM (1985) Wolf-Hirschhorn syndrome owing to 1:3 segregation of a maternal 4;21 translocation. *Am J Med Genet* 21:351–356.

Bayan T, Aydin B, Yalcin B, Orhan D, Akyuz C (2017) Hepatoblastoma and Wolf-Hirschhorn syndrome: coincidence or a new feature of a rare disease? *Pediatr Int* doi:10.1111/ped.13345.

Bayindir B, Piazza E, Della Mina E, Limongelli I, Brustia F, Ciccone R, Veggiotti P, Zuffardi O, Dehghani MR (2013) Dravet phenotype in a subject with a der(4) t(4;8)(p16. 3;p23.3) without the involvement of the LETM1 gene. *Eur J Med Genet* 56:551–555.

Breen GH (1998) Taurodontism, an unreported dental finding in Wolf-Hirschhorn (4p-) syndrome. *J Dent Child* 65:344–345.

Campos-Sanchez E, Deleyto-Seldas N, Dominguez V, Carrillode-Santa-Pau E, Ura K, Rocha PP, Kim J, Aljoufi A, Esteve-Codina A, Dabad M, Gut M, Heyn H, Kaneda Y, Nimura K, Skok JA, Martinez-Frias ML, Cobaleda C (2017) Wolf-Hirschhorn Syndrome Candidate 1 is necessary for correct hematopoietic and B cell development. *Cell Reports* 19:1586–601.

Carter R, Baker E, Hayman D (1969) Congenital malformations associated with a ring 4 chromosome. *J Med Genet* 6:224–227.

Catela C, Bilbao-Cortes D, Slonimsky E, Kratsios P, Rosenthal N, te Welscher P (2009) Multiple congenital malformations of Wolf–Hirschhorn syndrome are recapitulated in Fgfrl1 null mice. *Dis Model Mech* 2:283–294.

Centerwall WR, Thompson WP, Allen IE, Fobes TC (1975) Translocation 4p-syndrome. *Am J Dis Child* 129:366–370.

Chitayat D, Ruvalcaba RHA, Babul R, Teshima IE, Posnik JC, Vekemans MJJ, Scarpelli H, Thuline H (1995) Syndrome of proximal interstitial deletion 4p15: Report of three cases and review of the literature. *Am J Med Genet* 55:147–154.

ChoRW, Song Y, Littleton JB (2010) Comparative analysis of Drosophila and mammalian complexins as fusion clamps and facilitators of neurotransmitter release. *Mol Cell Neurosci* 45:389–397.

Clemens M, Martsolf JT, Rogers JG, Mowery-Rushton P, Surti U, McPherson E (1996) Pitt-Rogers-Danks Syndrome: The result of a 4p microdeletion. *Am J Med Genet* 66:95–100.

Coffin GS (1968) A syndrome of retarded development with characteristic appearance. *Am J Dis Child* 115:698–702.

Cooper H, Hirschhorn K (1961) Apparent deletion of short arms of one chromosome (4 or 5) in a child with defects of midline fusion. *Mamm Chrom Nwsl* 4:14.

Curfs LMG, Didden R, Sikkema SPE, De Die-Smulders CE (1999) Management of sleeping problems in Wolf-Hirschhorn syndrome: A case study. *Genet Counsel* 10:345–350.

Dallapiccola B, Mandich P, Bellone E, Selicorni A, Mokin V, Ajmar F, Novelli G (1993) Parental origin of chromosome 4p deletion in Wolf-Hirschhorn syndrome. *Am J Med Genet* 47:921–924.

De Falco F, Cainarca S, Andolfi G, Ferrentino R, Berti C, Criado GR, Rittinger O, Dennis N, Odent S, Rastogi A, Liebelt J, Chitayat D, Winter R, Jawanda H, Ballabio A, Franco B, Meroni G (2003) X-linked Opitz syndrome. Novel mutations in the *MID1* gene and redefinition of the clinical spectrum. *Am J Med Genet* 120:222–228.

Dimmer KS, Navoni F, Casarin A, Trevisson E, Endele S, Winterpacht A, Salviati L, Scorrano L (2008) LETM1, deleted in Wolf–Hirschhorn syndrome is required for normal mitochondrial morphology and cellular viability. *Hum Mol Genet* 17:201–214.

Donnai D (1996) Editorial comment: Pitt-Rogers-Danks syndrome and Wolf-Hirschhorn syndrome. *Am J Med Genet* 66:101–103.

Eiben B, Leipoldt M, Schubbe I, Ulbrich R, Hansmann I (1988) Partial deletion of 4p in fetal cells not present in chorionic villi. *Clin Genet* 33:49–52.

Engbers H, van der Smagt JJ, van 't Slot R, Vermeesch JR, Hochstenbach R, Poot M (2009) Wolf–Hirschhorn syndrome facial dysmorphic features in a patient with a terminal 4p16. 3 deletion telomeric to the WHSCR and WHSCR 2 regions. *Eur J Hum Genet* 17:129–132.

Estabrooks LL, Lamb AN, Kirkman HN, Callanan RP, Rao KW (1992) A molecular deletion of distal chromosome 4p in two families with a satellited chromosome 4 lacking the Wolf-Hirschhorn syndrome phenotype. *Am J Hum Genet* 51:971–978.

Estabrooks LL, Lamb AN, Aylswortli AS, Callanan NP, Rao KW (1994) Molecular characterization of chromosome 4p deletions resulting in Wolf-Hirschhorn syndrome. *J Med Genet* 31:103–107.

Estabrooks LL, Breg RW, Hayden MR, Ledbetter DH, Myers RM, Wyandt HE, Yang-Feng TL, Hirschhorn K (1995) Summary of the 1993 ASHG ancillary meeting "Recent research on chromosome 4p syndromes and genes". *Am J Med Genet* 55:453–458.

Faravelli F, Murdolo M, Marangi G, Bricarelli FD, Di Rocco M, Zollino M. (2007) Mother toson amplification of a small subtelomeric deletion: A new mechanism of familialrecurrence in microdeletion syndromes. *Am J Med Genet Part A* 143A: 1169–1173.

Fisch GS, Grossfeld P, Falk R, Battaglia A, Youngbloum J, Simensen R (2010) Cognitive-behavioral features of Wolf-Hirschhorn syndrome and other subtelomeric microdeletions. *Am J Med Genet C* 154C:417-426.

Fryns JP (1973) The 4p-syndrome, with a report of two new cases. *Humangenetik* 19:99–109.

Fryns JP (1995) Syndrome of proximal interstitial deletion 4p15. *Am J Med Genet* 58:295–296.

Gandelman K-Y, Gibson L, Meyn MS, Yang-Feng TL (1992) Molecular definition of the smallest region of deletion overlap in the Wolf-Hirschhorn syndrome. *Am J Hum Genet* 51:571–578.

Giglio S, Calvari V, Gregato G, Gimelli G, Camanini S, Giorda R, Ragusa A, Guerneri S, Selicorni A, Stumm M, Tonnies H, Ventura M, Zollino M, Neri G, Barber J, Wieczorek D, Rocchi M, Zuffardi O (2002) Heterozygous submicroscopic inversions involving olfactory receptor-gene clusters mediate the recurrent t(4;8)(p16;p23) translocation. *Am J Hum Genet* 71:276–285.

Gonzales CH, Capelozzi VL, Vajntal A (1981) Brief clinical report: Pathologic findings in the Wolf-Hirschhorn (4p-) syndrome. *Am J Med Genet* 9:183–187.

Gottfried M, Lavine L, Roessmann U (1981) Neuropathological findings in Wolf-Hirschhorn (4p-) syndrome. *Acta Neuropathol (Berl)* 55:163–165.

Griffith E, Walker S, Martin CA, Vagnarelli P, Stiff T, Vernay B, Sanna NA, Saggar A, Hamel B, Earnshaw WC, Jeggo PA, Jackson AP, O'Driscoll M, (2008) Mutations in pericentrin cause Seckel syndrome with defective ATR-dependent DNA damage signaling. *Nat Genet* 40:232–236.

Grisaru S, Ramage IJ, Rosenblum ND (2000) Vesicoureteric reflux associated with renal dysplasia in the Wolf-Hirschhorn syndrome. *Pediatr Nephrol* 14:146–148.

Guthrie RD, Aase JM, Asper AC, Smith D (1971) The 4p-syndrome. *Am J Dis Child* 122:421–425.

Hammond P, Hannes F, Suttie M, Devriendt K,Vermeesch JR, Faravelli F, Forzano F, Parekh S, Williams S, McMullan D, South ST, Carey JC, Quarrell O (2012) Fine-grained facial phenotype-genotype analysis in Wolf–Hirschhorn syndrome. *Eur J Hum Genet* 20:33–40.

Hanley-Lopez J, Estabrooks LL, Steihm ER (1998) Antibody deficiency in Wolf-Hirschhorn syndrome. *J Pediatr* 133:141–143.

Hirschhorn K, Cooper H, Firschein IL (1965) Deletion of short arms of chromosome 4–5 in a child with defects of midline fusion. *Humangenetik* 1:479–482.

Ho K, Markham LM, Twede H, Lortz A, Olson LM, Sheng X, Weng C, Wassman ER, Newcomb TM, Carey JC, Battaglia A (2018) A survey of antiepileptic drug responses identifies drugs with efficacy for seizure control in Wolf-Hirschhorn syndrome. *Epilepsy Behav* 81:55-61.

Ho KS, South ST, Lortz A, Venkatasubramanian S, Hensel CH, Sedano MR, Vanzo RJ, Martin MM, Pfeiffer A, Calhoun A, Battaglia A, Carey JC (2015) Correlations of genotype to phenotypes in 48 individuals with Wolf-Hirschhorn syndrome. *American College of Medical Genetics and Genomics, Annual Clinical Genetics Meeting, Salt Lake City, UT, Book of Abstracts*.

Ho KS, South ST, Lortz A, Hensel CH, Sdano MR, Vanzo RJ, Martin MM, Peiffer A, Lambert CG, Calhoun A, Carey JC, Battaglia, A (2016) Chromosomal microarray testing identifies a 4p terminal region associated with seizures in Wolf-Hirschhorn syndrome. *J Med Genet* 53:256-263.

Jiang D, Zhao L, Clapham DE (2009) Genomewide RNAi screen identifies Letm1 as a mitochondrial Ca2þ/Hþ antiporter. *Science* 326:144–147

Johnson VP, Altherr MR, Blake JM, Keppen LD (1994) FISH detection of Wolf-Hirschhorn syndrome: Exclusion of D4F26 as critical site. *Am J Med Genet* 52:70–74.

Kerzendorfer C, Hannes F, Colnaghi R, Abramowicz I, Carpenter G, Vermeesch JR, O'Driscoll M (2012) Characterizing the functional consequences of haploinsufficiency of NELF-A (WHSC2) and SLBP identifies novel cellular phenotypes in Wolf-Hirschhorn syndrome. *Hum Mol Genet* 21:2181–93.

Knight SJ, Lese CM, Precht KS, Kuc J, Ning Y, Lucas S, Regan R, Brenan M, Nicod A, Lawrie NM, Cardy DL, Nguyen H, Hudson TJ, Riethman HC, Ledbetter DH, Flint J (2000) An optimized set of human telomere clones for studying telomere integrity and architecture. *Am J Hum Genet* 67:320–332.

Kuum M, Veksler V, Liiv J, Ventura-Clapier R, Kaasik A (2012) Endoplasmic reticulum potassium-hydrogen exchanger and small conductance calcium-activated potassium channel activities are essential for ER calcium uptake in neurons and cardiomyocytes. *J Cell Sci* 125:625–633.

Lanters LT (2000) *Nieuwsbrief Wolf-Hirschhorn* 6:1–2.

Lazjuk GI, Lurie IW, Ostrowskaja TI, Kirillova IA, Nedzved MK, Cherstvoy ED, Silyaeva NF (1980) The Wolf-Hirschhorn syndrome. II. Pathologic anatomy. *Clin Genet* 18:6–12.

Lesperance MM, Grundfast KM, Rosenbaum KN (1998) Otologic manifestations of Wolf-Hirschhorn syndrome. *Arch Otol Head Neck Surg* 124:193–196.

Lowry RB, MacLean JR (1977) Syndrome of mental retardation, cleft palate, eventration of diaphragm, congenital heart defect, glaucoma, growth failure and craniosynostosis. *Birth Defects* 13:203–228.

Lozier ER, Konovalov FA, Kanivets IV, Pyankov DV, Koshkin PA, Baleva LS, Sipyagina AE, Yakusheva EN, Kuchina AE, Korostelev SA (2018) De novo nonsense mutation in WHSC1 (NSD2) in patient with intellectual disability and dysmorphic features. *J Hum Genet* 63:919–922.

Lurie IW, Lazjuk CL, Ussova YI, Presman EB, Gurevich DB (1980) The Wolf-Hirschhorn syndrome. I. Genetics. *Clin Genet* 17:375–384.

Maas NMC, Van Buggenhout G, Hannes F, Thienpont B, Sanlaville D, Kok K …Vermeesch JR (2008) Genotype-phenotype correlation in 21 patients with Wolf-Hirschhorn syndrome using high resolution array comparative genome hybridisation (CGH). *J Med Genet* 45:71–80.

Magill HL, Shackelford GD, McAlister WH, Graviss ER, (1980) 4p- (Wolf-Hirschhorn) syndrome. *AJR Am J Roentgenol* 135:283–288.

Malpuech G, Demecocq F, Palcoux JB, Vanlieferighen P (1983) A previously undescribed autosomal recessive multiple congenital anomalies/mental retardation (MCA/MR) syndrome with growth failure, lip/palate cleft(s), and urogenital anomalies. *Am J Med Genet* 16:475–480.

Marte A, Indolfi P, Ficociello C, Russo D, Oreste M, Bottigliero G, Gualdiero G, Barone C, Vigliar E, Indolfi C, Casale F (2013) Inflammatory Myofibroblastic Bladder Tumor in a Patient with Wolf-Hirschhorn Syndrome. *Case Rep Urol* 2013:675059.

McDonald-McGinn DM, Emanuel BS, Zackai EH,22q11.2 deletion syndrome (2005) www.GeneTests.org.

Misceo D, Baroy T, Helle JR, Braaten O, Fannemel M, Frengen E (2012) 1.5 Mb deletion of chromosome 4p16.3 associated with postnatal growth delay, psychomotor impairment, epilepsy, impulsive behavior and asynchronous skeletal development. *Gene* 507:85–91.

Nakano Y, Fujita M, Ogino K (2010) Biogenesis of GPI-anchored proteins is essential for surface expression of sodium channels in zebrafish Rohon-Beard neurons to respond to mechanosensory stimulation. *Development* 137:1689–1698.

Nieminen, P., Kotilainen, J., Aalto, Y., Knuutila, S., Pirinen, S., Thesleff, I (2003) MSX1 gene is deleted in Wolf-Hirschhorn syndrome patients with oligodontia. *J Dent Res* 82: 1013–1017.

O'Brien G, Yule W (1995) *Behavioural Phenotypes*, Oxford University Press, pp. 205–206.

Ogle R, Sillence DO, Merrick A, Ell J, Lo B, Robson L, Smith A (1996) The Wolf-Hirschhorn syndrome in adulthood: Evaluation of a 24-year-old man with a rec (4) chromosome. *Am J Med Genet* 65:124–127.

Opitz JM (1995) Twenty-seven-year follow-up in the Wolf-Hirschhorn syndrome. *Am J Med Genet* 55:459–461.

Park SH, Chi JB (1993) Oligomeganephroma associated with 4p deletion type chromosomal anomaly. *Pediatr Pathol* 13:731–740.

Petit C (1996) Genes responsible for human hereditary deafness. *Nat Genet* 14:385–391.

Pitt DB, Rogers JG, Danks DM (1984) Mental retardation, unusual facies and intrauterine growth retardation—A new recessive syndrome? *Am J Med Genet* 19:307–313.

Preus M, Ayme S, Kaplan P, Vekemans M (1985) A taxonomic approach to the del (4p) phenotype. *Am J Med Genet* 21:337–345.

Prunotto G, Cianci P, Cereda A, Scatigno A, Fossati C, Maitz S, Biondi A, Selicorni A (2013) Two cases of hepatic adenomas in patients with Wolf-Hirchhorn syndrome: a new rare complication? *Am J Med Genet Part A* 161(7): 1759–1762.

Rauch A, Schellmoser S, Kraus C, Dorr HG, Trautmann U, Altherr MR, Pfeiifer RA, Reis A (2001) First known microdeletion within the Wolf-Hirschhorn-syndrome critical region refines genotype phenotype correlation. *Am J Med Genet* 99:338–342.

Reid E, Morrison N, Barron L, Boyd E, Cooke A, Fielding D, Tolmie JL (1996) Familial Wolf Hirschhorn syndrome resulting from a cryptic translocation: A clinical and molecular study. *J Med Genet* 33:197–202.

Ricke DO, Mundt MO, Buckingham JM, Deaven LL, Moyzis RK (1996) Whole genome sequence sampling and gene analysis and annotation of megabases of low redundancy human chromosome 16p13 Sample Sequencing Comparison Analysis (SCAN). *Microb Comp Genom* 1:264.

Rodríguez L, Zollino M, Climent S, Mansilla E, López-Grondona F, Martínez-Fernández ML, Murdolo M, Martínez-Frías ML. (2005) The new Wolf-Hirschhorn syndrome critical region (WHSCR-2): a description of a second case. *Am J Med Genet A*. 136:175–8

Schaefer BG, Kleimola CN, Stenson C, Daley SE, Farmer P, Holladay K (1996) *Wolf-Hirschhorn Syndrome (Deletion 4p): A Guidebook for families*. Omaha, NE: SOFT 18,13 and Related Disorders and Meyer Rehabilitation Institute, University of Nebraska Medical Center.

Sergi C, Schulze BRB, Hager HD, Beedgen B, Zilow E, Linderkamp O, Otto HF, Tariverdian G (1998) Wolf-Hirschhorn Syndrome: Case report and review of the chromosomal aberrations associated with diaphragmatic defects. *Pathologica* 90:285–293.

Shannon NL, Maltby EL, Rigby AS, Quarrell OWJ (2001) An epidemiological study of Wolf-Hirschhorn syndrome: Life expectancy and cause of mortality. *J Med Genet* 38:674–679.

Shanske AL, Yachelevich N, Ala-Kokko L, Leonard J, Levy B (2010) Wolf-Hirschhorn syndrome and ectrodactyly: new findings and a review of the literature. *Am J Med Genet A* 152A:203–208.

Shimizu K, Wakui K, Kosho T, Okamoto N, Mizuno S, Itomi K, Hattori S, Nishio K, Samura O, Kobayashi Y, Kako Y, Arai T, Tsutomu OI, Kawame H, Narumi Y, Ohashi H, Fukushima Y (2014) Microarray and FISHbased genotype–phenotype analysis of 22 Japanese patients with Wolf–Hirschhorn syndrome. *Am J Med Genet A* 164A:597–609.

Simon R, Bergemann AD (2008) Mouse models of Wolf-Hirschhorn syndrome. *Am J Med Genet C* 148C:275–280.

Smith SA, Walker AM, Monk AJ, Young ID (1995) Long-term survival in the Wolf Hirschhorn (4p-) syndrome. *J Int Disab Res* 39:83–86.

Somer M, Peippo M, Keinanen M (1995) Controversial findings in two patients with commercially available probe D4S96 for the Wolf-Hirschhorn syndrome. *Am J Hum Genet* 57:127.

South, S.T., Bleyl, S.B., Carey, J.C (2007) Two unique patients with novel microdeletions in 4p16.3 that exclude the WHS critical regions: Implications for critical region designation. *Am J Med Genet A* 143A:2137–2142.

South ST, Whitby H, Battaglia A, Carey JC, Brothman AR (2008a) Comprehensive analysis of Wolf-Hirschhorn syndrome using array CGH indicates a high prevalence of translocations. *Eur J Hum Genet* 16:45–52.

South ST, Hannes F, Fisch GS, Vermeesch JR, Zollino M (2008b) Pathogenic significance of deletions distal to the currently described Wolf-Hirschhorn syndrome critical regions on 4p16.3. *Am J Med Genet C* 148C:270–274.

Stec I, Wright TJ, van Ommen G-JBde Boer PA, van Haeringen A, Moorman AF, Altherr MR, den Dunnen JT (1998) *WHSC1*, a 90kb SET domain-containing gene, expressed in early development and homologous to a *Drosophila* dysmorphy gene maps in the Wolf-Hirschhorn syndrome critical region and is fused to *IgH* in t(4;14) multiple myeloma. *Hum Mol Genet* 7:1071–1082.

Stengel-Rutkowski S, Warkotsch A, Schimanek P, Stene J (1984) Familial Wolf's syndrome with a hidden 4p deletion by translocation of an 8p segment. Unbalanced inheritance from a maternal translocation (4;8)(pl5.3;p22). Case report, review and risk estimates. *Clin Genet* 25:500–521.

Subrt L, Blehova A, Sedlakova E (1969) Mewing cry in a child with the partial deletion of the short arm of chromosome no. 4. *Hum Genet* 8:242–248.

Tachdijan G, Fondacci C, Tapia S, Huten Y, Blot P, Nessmann C (1992) The Wolf-Hirschhorn syndrome in fetuses. *Clin Genet* 42:281–287.

Tupler R, Bortotto L, Buhler E, Aikan M, Malik NJ, Al Jadooa NB, Memo L, Maraschio P (1992) Paternal origin of de novo deleted chromosome 4 in Wolf-Hirschhorn syndrome. *J Med Genet* 29:53–55.

Van Buggenhout G, Melotte C, Dutta B, Froyen G, Van Hummelen P, Marynen P, Matthijs G, de Ravel T, Devriendt K, Fryns JP, Vermeesch JR (2004) Mild Wolf–Hirschhorn syndrome: Microarray CGH analysis of atypical 4p16.3 deletions enables refinement of the genotype-phenotype map. *J Med Genet* 41:691–698.

Verloes A, Schaaps JP, Herens C, Soyeur D, Hustin J, Dodinval P (1991) Prenatal diagnosis of cystic hygroma and chorioangioma in the Wolf-Hirschhorn syndrome. *Prenat Diagn* 11:129–132.

Wheeler PG, Weaver DD, Palmer CG (1995) Familial translocation resulting in Wolf-Hirschhorn syndrome in two related unbalanced individuals: Clinical evaluation of a 39-year-old man with Wolf Hirschhorn syndrome. *Am J Med Genet* 55:462–465.

Wieczorek D, Krause M, Majewski F, Albrecht B, Horn D, Riess O, Gillesen-Kaesbach G (2000) Effect of the size of the deletion and clinical manifestation in Wolf-Hirschhorn syndrome: Analysis of 13 patients with a de novo deletion. *Eur J Hum Genet* 8:519–526.

Wilson MG, Towner JW, Coffin GS, Ebbin AJ, Siris E, Brager P (1981) Genetic and clinical studies in 13 patients with the Wolf-Hirschhorn syndrome [del(4p)]. *Hum Genet* 59:297–307.

Wolf U, Reinwein H, Porsch R, Schroter R, Baitsch H (1965) Defizienz am den kurzen Armen eines chromosoms nr. 4. *Humangenetik* 1:397–413.

Wright TJ, Ricke DO, Denison K, Abmayr S, Cotter PD, Hirschhorn K, Keinanen M, McDonald McGinn D, Somer M, Spinner N, Yang-Feng T, Zachai E, Altherr MR (1997) A transcript map of the newly defined 165kb Wolf-Hirschhorn syndrome critical region. *Hum Mol Genet* 6:317–324.

Wright TJ, Clemens M, Quarrell O, Altherr MR (1998) Wolf-Hirschhorn and Pitt-Rogers-Danks syndromes caused by overlapping 4p deletions. *Am J Med Genet* 75:345–350.

Wright TJ, Altherr MR, Callen D, Hirschhorn K (1999) Reply to the letter to the editor by Partington and Turner "Wolf-Hirschhorn and Pitt-Rogers-Danks syndromes". *Am J Med Genet* 82:89–90.

Zollino M, Wright TJ, Di Stefano C, Tosolini A, Battaglia A, Altherr MR, Neri G (1999) "Tandem" duplication of 4p16.1p16.3 chromosome region associated with 4p16.3pter molecular deletion resulting in Wolf-Hirschhorn syndrome phenotype. *Am J Med Genet* 82:371–375.

Zollino M, Di Stefano C, Zampino G, Mastroiacovo P, Wright TJ, Sorge G, Selicorni A, Tenconi R, Zappala A, Battaglia A, Di Rocco M, Palka G, Pallotta R, Altherr MR, Neri G (2000) Genotype-phenotype correlations and clinical diagnostic criteria in Wolf-Hirschhorn syndrome. *Am J Med Genet* 94:254–261.

Zollino M, Lecce R, Fischetto R, Murdolo M, Faravelli F, Selicorni A, Buttè C, Memo L, Capovilla G, Neri G (2003) Mapping the Wolf-Hirschhorn syndrome phenotype outside the currently accepted WHS critical region and defining a new critical region, WHSCR-2. *Am J Hum Genet* 72:590–597.

Zollino M, Murdolo M, Marangi G, Pecile V, Galasso C, Mazzanti L, Neri G (2008) On the nosology and pathogenesis of Wolf-Hirschhorn syndrome: Genotype-phenotype correlation analysis of 80 patients and literature review. *Am J Med Genet C* 148:257–269.

Zollino M, Orteschi D, Ruiter M, Pfundt R, Steindl K, Cafiero C, Ricciardi S, Contaldo I, Chieffo D, Ranalli D, Acquafondata C, Murdolo M, Marangi G, Asaro A, Battaglia D (2014) Unusual 4p16.3 deletions suggest an additional chromosome region for the Wolf–Hirschhorn syndrome-associated seizures disorder. *Epilepsia* 55:849–857.

20

DELETION 5p SYNDROME

ANTONIE D. KLINE
Clinical Genetics, Harvey Institute for Human Genetics, Greater Baltimore Medical Center, Baltimore, Maryland, USA

JOANNE M. NGUYEN
McGovern Medical School, The University of Texas Health Science Center at Houston, Houston, Texas, USA

DENNIS J. CAMPBELL
Retired College of Education, University of South Alabama, Chair Professional Advisory Board 5P Minus Society, Auburn, Alabama, USA

INTRODUCTION

Deletion 5p syndrome was first described by Lejeune et al. in 1963. The characteristic cry, because of its unique and easily identifiable pitch, is often the factor contributing to identification. However, not all individuals with 5p deletions exhibit the cat-like cry, and there is a spectrum of involvement. Initially, children come to medical attention because of growth delay and the cry. Later, microcephaly, intellectual disability and behavioral issues such as aggression and self-injury comprise a less specific phenotype. Over 80% are diagnosed in the first year of life (Mainardi et al. 2006). Additionally, individuals who were not initially suspected of having 5p deletions are increasingly being identified and broadening the phenotypic spectrum (Zhu et al. 2016).

Incidence

The incidence has been reported to be as high as 1 in 15,000 live births with a prevalence of 1 in 50,000 (Higurashi et al. 1990; Niebuhr 1978), thus making it one of the more common chromosomal deletion syndromes. The variation in the rate can be attributed to the method of making this determination. For example, the rate of 1 in 15,000 comes from a study of 27,472 live births in a large Tokyo hospital from 1972 to 1985. During this period, two individuals were born with deletion 5p syndrome (Higurashi et al. 1990). Niebuhr (1978) surveyed facilities for individuals with intellectual disability in the Netherlands and tested all individuals who appeared to have the physical traits of deletion 5p syndrome. From this study, he concluded that the prevalence was about 1 in 50,000.

Diagnostic Criteria

The major diagnostic finding is a positive cytogenetic result involving the short arm of chromosome 5. There are no specific clinical diagnostic criteria. The unique cry is the most recognizable clinical feature. Other typical features include a slow growth rate, microcephaly, hypotonia, and mildly dysmorphic facies, including hypertelorism, downslanting palpebral fissures, epicanthal folds, low-set ears, broad nasal bridge, downturned corners of the mouth, and microretrognathia (Figures 20.1A and B). Short neck, preauricular tags, and dental malocclusion may be seen (Mainardi 2006; Wilkins et al. 1983; Niebuhr 1978). Extremity findings include single transverse palmar creases, clinodactyly, syndactyly, short metacarpals, and pes planus.

Cassidy and Allanson's Management of Genetic Syndromes, Fourth Edition.
Edited by John C. Carey, Agatino Battaglia, David Viskochil, and Suzanne B. Cassidy.
© 2021 John Wiley & Sons, Inc. Published 2021 by John Wiley & Sons, Inc.

FIGURE 20.1 Facial features in a female with deletion 5p syndrome (A) at 1 year 9 months of age; (B) at 25 years of age. Note the changing features with aging, including longer face, less prominent epicanthal folds and fuller lips.

Etiology, Pathogenesis, and Genetics

Deletions of 5p leading to the syndrome can be interstitial or terminal, occasionally involving the entire short arm. Terminal deletions make up 80–90% of cases, and interstitial deletions 3–5% (Mainardi et al. 2006). Most of the deletions are de novo, possibly arising from chromosome breakage during gamete formation in males (Mainardi et al. 2001). Ten to 15% arise from an unbalanced parental translocation (Mainardi et al. 2006). Less common parental rearrangements leading to deletion 5p are mosaicism (1.4%), inversions (0.5%) and ring chromosomes (0.5%) (Perfumo et al. 2000).

Although many individuals have similar breakpoints, there is no common recurring breakpoint (Mainardi et al. 2001). The size of the deletions ranges from 560 kb to 40 Mb (Elmakky et al. 2014; Gu et al. 2013; Simmons et al. 1995). The critical region for deletion 5p syndrome is said to be within 5p15.2, associated with the typical facial features and the cat-like cry, which extends into 5p15.3 (Wu et al. 2005; Zhang et al. 2005). The location on the chromosome related to cognitive function has been mapped to several regions, particularly 5p15.2, associated with the delta-catenin gene (Wu et al. 2005; Medina et al. 2000). Other specific genes have been described which, when in the hemizygous state, account for some of the findings in the deletion 5p syndrome (Nguyen et al. 2015). These include *SEMA5A*, associated with brain development, *TEB4*, associated with the cat-like cry, and *CTNND2*, associated with nervous system development; still others are conditionally hemizygous (Nguyen et al. 2015).

There have been several families documented with autosomal dominant transmission of a 5p deletion from parent to child (Zhang et al. 2016; Nguyen et al. 2014). Families who have multiple affected members show phenotypic heterogeneity among family members with the same 5p deletion (Cornish et al. 1999; Church et al. 1995; Fang et al. 2008; Walker et al. 1984). There are reports of children with many findings consistent with deletion 5p syndrome whose parents have the same deletion and yet do not appear to have any major symptoms, and whose siblings have variability in their growth and development (Church et al. 1995; Gersh et al. 1995). Members of families with maternally inherited deletions were observed to have milder growth and developmental delays than those with paternal deletions (Gersh et al. 1994). No imprinted genes have been identified; however, 80–83% of the 5p deletions in two large series were found to be paternal in origin (Church et al. 1995; Overhauser et al. 1990).

Genetic Counseling

Since the vast majority of 5p deletions occur de novo, the recurrence risk for those parents to have another affected child is less than 1%. If a parent is found to be affected and/or carry the deletion, then the recurrence risk is 50% for each future pregnancy. If a parent carries a balanced translocation, then the offspring have a 25% chance of inheriting the same rearrangement, a 50% chance of inheriting an unbalanced rearrangement, and a 25% chance of inheriting the unaffected normal chromosomes.

Diagnostic Testing

Younger individuals who have characteristic signs such as the typical dysmorphic features and monotonous cry can be identified clinically. Diagnosis can be confirmed with genetic testing of peripheral blood. Although most 5p deletions can be detected with a routine karyotype, advances in molecular techniques using a single nucleotide polymorphism (SNP) chromosomal microarray allow for better characterization of the chromosomal breakpoints. This offers the opportunity to identify which genes are missing and potentially contributing to the individual's phenotype, as well as reveal additional duplications or deletions not identified by former diagnostic methods.

Prenatal testing is available for parents carrying a known rearrangement involving the short arm of chromosome 5, either by chorionic villus sampling or amniocentesis. Prenatal screening methods, such as non-invasive prenatal screening, have the ability to detect deletions as well, although a recent study found no cases of deletion 5p (Petersen et al. 2016). Chromosomal microarray on a prenatal sample has also been able to detect a 5p microdeletion (Zhu et al. 2016; Nguyen et al. 2014).

Differential Diagnosis

When the classic constellation of dysmorphic features, hypotonia, and the typical cry is not evident or recognizable, then the differential diagnosis can be quite broad, especially with microcephaly and/or developmental delays. Other chromosomal disorders, including other deletion syndromes, can present similarly with mildly dysmorphic facial features, relatively normal growth, seizures, and intellectual disability. These could include particularly the 22q11 deletion syndrome (see Chapter 21), and Angelman syndrome (see Chapter 5). In fact, as individuals age, they are said to resemble others with Angelman syndrome, but with more self-injury (van Buggenhout et al. 2000). No other syndrome has the distinctive cry. Preauricular ear tags can be seen in other syndromes, particularly oculo-auriculo-vertebral spectrum (see Chapter 42).

MANIFESTATIONS AND MANAGEMENT

Growth and Feeding

Low birth weight and slow growth are frequently observed in deletion 5p syndrome. Although infants are smaller than normal at birth, they maintain a slow but consistent growth rate. The median head circumference and weight remain consistently near or below the fifth centile through adulthood. Prenatal and postnatal growth deficiency is common (52–70%) and continues life-long. Poor feeding occurs in 44% of infants (Mainardi et al. 2006; Niebuhr, 1978). Often infants have difficulty with sucking and swallowing. Older patients also have difficulty with feeding. Final adult height means are reported as 152.8 cm for females and 167.1 cm for males (Honjo et al. 2018).

Evaluation

- As part of routine developmental surveillance, growth parameters should be measured. The growth charts specific for cri du chat syndrome should be used to plot height and weight from 0–18 years of age, as well as head circumference from 0–15 years of age (Marinescu et al. 2000); see website https://fivepminus.org/wp-content/uploads/2016/10/growth-charts.pdf
- Slow growth may lead to suspicion of failure to thrive. If concerns regarding growth remain despite using diagnosis specific growth charts, appropriate standard investigations for growth delay should be considered.

Management

- Nutritional support and speech therapy should be offered early (Lefranc et al. 2016).
- Speech therapists can assist with swallowing studies and feeding therapy. For more severe feeding issues, oral, nasogastric, or gastrostomy tubes can be used and may be needed for several years.

Development and Behavior

Intellectual disability is common in deletion 5p syndrome. The severity of the cognitive delays varies, ranging from mild to profound, with most having moderate to severe intellectual disability. Verbal and performance IQ scores are comparable, and mean full-scale IQ score has been found to be 47.8 (Cornish et al. 1999). There appears to be a plateau and then a decline in intellectual functioning after age 10 years. It has been observed that individuals with larger deletions tend to have more significant cognitive delays (Mainardi et al. 2006). There are, however, conflicting reports regarding whether or not clinical severity correlates with the size of the deletion (Mainardi et al. 2001; Wilkins et al. 1983), and some individuals have apparently normal intellectual function (Church et al, 1995; Gersh et al. 1995).

Delayed motor skills and clumsiness have been reported. The median age for children to walk independently is 3 years old, with a range of 15 months to 7 years (Mainardi et al. 2006). The average ages reported for other motor developmental milestones include sitting up at 14 months, walking alone at 43 months, dressing at 78 months, and toilet-training at 90 months. Independent walking occurs in 72.2% (Honjo et al. 2018).

Speech and communication disorders are prevalent in deletion 5p syndrome. About 50% of individuals develop some speech and most communicate functionally using a variety of modes including gestures and sign language. Receptive language skills are far better than expressive (Cornish and Munir 1998). There are severe articulation problems, with laryngeal malformations and/or hypotonia likely contributing (Cornish et al. 1999). Despite their speech delays, children with deletion 5p are able to communicate their needs and socially interact with others, particularly with sign language (Cornish and Pigram 1996).

Little is available in the literature about adult patients with deletion 5p syndrome. A recent article based on parental survey (Honjo et al. 2018) includes descriptions of 23 adults up to age 40 years. Seventy percent are said to be dependent on caregivers for activities of daily living. Of these, 26% can brush their teeth independently, 22% can shower alone, 61% can eat independently and 57% have bowel or bladder control (Honjo et al. 2018). In another study, of 31 adults, 58% live at home with parents and the remainder live in group homes, rental homes and other unspecified locations (Nguyen et al. 2015).

Over 80–90% of children with deletion 5p syndrome have hyperactivity, with 70% displaying clinical features of attention deficit hyperactivity disorder (Cornish and Munir 1998; Dykens and Clarke 1997). Other common behaviors include poor concentration, frustration, impulsiveness, stubbornness, and temper tantrums. Aggressive behaviors (e.g. biting, pinching, hair pulling, and hitting), and self-injurious behaviors (e.g. head-banging, rubbing, self-biting, scratching, and skin-picking) are seen commonly with increasing age (Huisman et al. 2017; Collins and Cornish 2002; van Buggenhout et al. 2000). There may be characteristics of autism spectrum disorder such as hand flapping, rocking, repetitive movements, and obsessive attachments with or twirling of objects. Obsessive-compulsive disorder has also been reported. No gender differences for maladaptive behaviors have been noted (Nguyen et al. 2015). Behaviors can escalate due to pain, constipation, vision or hearing deficits, and sleep disorders (see Neurologic section).

Evaluation

- Children should be referred for full evaluation by a developmental specialist including assessment of splinter skills, or abilities that do not relate to other general tasks, as found in the autism spectrum.
- Individual evaluations with physical, occupational and speech therapists will be invaluable in identifying deficits on which to focus therapy.
- Evaluation by a behavioral specialist may also assist with assessing self-injurious behaviors.
- Hearing loss and vision impairment may influence ability to advance development and learn (see below), and should be addressed in school individualized education programs.

Management

- Early intervention, appropriate educational placement, as well as occupational, speech, and physical therapies facilitate progression of skills.
- Intervention for speech is recommended as early as possible and should include not only speech therapy and communication devices, but also sign language (Cornish et al. 1999).
- Psychological support for behavioral issues, and development of a behavior management plan, may be considered.
- Standard management for ADHD is used, specifically behavioral therapy and standard medications.

Craniofacial

Individuals with 5p deletions usually have mildly dysmorphic features, particularly when younger. Microcephaly and round facies are seen in infancy, with face elongating with age. Facial features include hypertelorism, epicanthal folds, downslanting palpebral fissures, broad nasal bridge, normal to low-set ears, downturned corners of the mouth, high palate, and microretrognathia (Mainardi 2006; Wilkins et al. 1980; Niebuhr 1978). Preauricular tags are common. Dysmorphic facial features become less striking with age (van Buggenhout et al. 2000) (see Figure 20.1B).

Individuals commonly have dental malocclusion, and may have an anterior open bite and enamel hypoplasia (Molina-García et al. 2016). There is no increased incidence of clefting.

Laryngeal malformations include a small and/or narrow laryngeal opening with an increased incidence of laryngomalacia, paralysis of the vocal cords, and an abnormal epiglottis (Guala et al. 2015). Limited cervical mobility is also noted. These provide risks for anesthesia.

Evaluation

- Ear tags may be evaluated by otolaryngology or plastic surgery.
- Children should start visiting a pediatric dentist when the first tooth appears or by their first birthday. Attention to hygiene should be made to avoid dental caries and periodontitis.
- Because of abnormalities of the craniofacial region (Corcuera-Flores et al. 2016) such as the mandible, larynx, epiglottis, and cervical spine/cranial base, individuals with deletion 5p syndrome should be evaluated prior to general anesthesia (Guala et al. 2015). There are otherwise no known anesthetic complications unique to this condition.

Management

- Preauricular tags can be removed by an otolaryngologist or plastic surgeon.
- Dental care should continue per standard guidelines such as recommended by the US American Academy of Pediatric Dentistry, www.aapd.org.

- Orthodontic evaluation and treatment, if tolerated, should be provided as would be for the general population.

Neurologic

Hypotonia is very common (72%) in individuals with deletion 5p syndrome (Mainardi et al. 2006; Niebuhr 1978). With time, hypotonia persists, although it may evolve into hypertonia during adolescence (Espirito Santo et al. 2016). Seizure disorders may occur in up to 15% of individuals (Mainardi et al. 2006). Microcephaly persists into adulthood. Up to 30% have brain findings such as hypoplasia or agenesis of the corpus callosum, periventricular leukomalacia, abnormalities of white matter myelination, cerebral and/or cerebellar atrophy, hydrocephalus, or pontine hypoplasia (Corrêa et al. 2017; Mainardi et al. 2006).

Up to half suffer from sleep disturbances, similar to other groups of individuals with intellectual disability. Complaints include difficulty falling asleep, difficulty staying asleep, abnormal breathing during sleep, and snoring (Esbensen and Schwichtenberg 2017). Sleep disturbances continue from childhood into adulthood and may worsen with age.

Evaluation

- Neurological examination should be performed for all individuals at each health supervision visit.
- Individuals who have or are suspected to have a seizure disorder should be evaluated with an electroencephalogram and neuroimaging.
- Sleep assessment may be considered for sleep disturbances, especially with snoring.

Management

- Hypotonia or hypertonia should be treated with standard practices including physical therapy. Because of hypotonia, precautions should be considered before muscle relaxants are given during general anesthesia for procedures or surgeries (Colover et al. 1972).
- Seizure disorders should be controlled appropriately in a standard manner with individualized pharmacotherapy regimens.
- Medications to promote improved sleep quality may be considered.

Cardiovascular

Cardiovascular anomalies occur in 18–36% of individuals with deletion 5p syndrome (Mainardi et al. 2006), higher in those with unbalanced translocations (Wilkins et al. 1983). The congenital heart defects include, most commonly: ventricular septal defect, patent ductus arteriosus, tetralogy of Fallot, and pulmonary atresia, with no increase in mortality following surgery (Hills et al. 2006). Congenital heart disease may increase risk for anesthesia.

Evaluation

- All individuals should have an echocardiogram at diagnosis, which should be followed if indicated.

Management

- Caution with anesthesia should be followed if the congenital heart disease is significant.
- Congenital heart disease should be managed appropriately per standard medical and/or surgical treatment.

Genitourinary

Renal (6–18%) and genitourinary anomalies (4–21%) may occur, with a higher percentage in those with deletion 5p syndrome due to translocations (Mainardi et al. 2006). Renal malformations include renal agenesis, renal hypoplasia, and hydronephrosis. Urinary tract infections are reported in about 3% of individuals (Mainardi et al. 2006).

Genital anomalies include cryptorchidism in males and, rarely, hypospadias, as well as hypoplastic genitalia (Mainardi et al. 2006). One third of adolescent and adult males were found to have small testes (Breg et al. 1970). Puberty occurs in children with deletion 5p syndrome at typical ages. Menstruation has been normal when reported (van Buggenhaut et al. 2000).

Evaluation

- Evaluation for genitourinary anomalies is recommended at diagnosis with a renal ultrasound.
- Individuals should be investigated for a urinary tract infection if the source of an infection or fever is unknown.

Management

- Surgical management of genitourinary anomalies should be as for the general population, with orchidopexy by 2 years of age.
- Treatment of urinary tract infections should follow standard medical care.
- Management of menses should be standard, following discussion with the family.

Ears and Hearing

Conductive hearing loss may occur in children due to chronic otitis media infections. Sensorineural hearing loss has also been reported in 8.4% (Nguyen et al. 2015). Chronic otitis media and/or serous otitis media lead to 15% of children having tympanostomy tubes placed. Hearing loss may present as inconsistency in responding to directions, difficulty with language and speech delay, as well as behavioral issues, such as frustration, aggression, and immature social interactions (Nguyen et al. 2015).

There is also a high incidence of hyperacusis, seen in 70–80% of children with deletion 5p syndrome (Nguyen et al. 2015). Heightened sensitivity to sounds can manifest behaviorally as severe agitation, stress, or being easily startled. Adults with hyperacusis have a greater reaction to sound ranges and pitches (Nguyen et al. 2015).

Evaluation

- Audiologic evaluation is recommended at diagnosis. Depending on the age and cooperativity of an individual, audiologic evaluation using brainstem auditory evoked responses or otoacoustic emissions may be required.
- Routine hearing exams should continue as for the general population and repeated if suspicion for hearing loss develops or maladaptive behavior changes occur.

Management

- Many children and adults can benefit from hearing aids, if tolerated.
- Appropriate treatments for ear infections should be provided as in the general population.
- Cognizance of environmental surroundings is important if maladaptive behaviors escalate, especially if the individual has hyperacusis.

Ophthalmology

Over 45% of affected individuals have ophthalmologic involvement (Nguyen et al. 2015). Vision abnormalities that are prevalent among individuals with deletion 5p syndrome include myopia (15%), strabismus (45–53%), cataracts (2%), optic nerve abnormalities (5–19%), and, less commonly, decreased tearing, optic atrophy, microspherophakia, and tortuous retinal blood vessels (Mainardi et al. 2006; Niebuhr 1978). Cataracts may develop and optic atrophy has been reported (van Buggenhout et al. 2000).

Evaluation

- Ophthalmologic evaluation is recommended at diagnosis.
- Routine vision exams should continue regularly, ideally every six months, and repeated if suspicion for decreased acuity develops or behavior changes.

Management

- Treatment of vision abnormalities should be provided as in the general population.

Musculoskeletal

Affected individuals typically have minor skeletal anomalies, including single transverse palmar creases, short metacarpals, clinodactyly, syndactyly, and/or pes planus deformity. Scoliosis occurs frequently (43%) in individuals with deletion 5p syndrome (Mainardi et al. 2006; Niebuhr 1978).

Evaluation

- Individuals should be screened for scoliosis with clinical examinations of the spine during each health supervision visit.
- If detected, spine radiographs should be obtained for further evaluation.
- Referral to orthopedic surgery for monitoring and treatment should be made if scoliosis develops.

Management

- Management options for scoliosis should be offered as for the general population.
- Physical therapy can be helpful for gait in individuals with pes planus deformities.

Gastroenterology

Gastrointestinal anomalies are reported in 4–21% of individuals with deletion 5p syndrome, and have included malrotation and Hirschsprung disease, which have led to intestinal obstruction (Ullah et al. 2017; Mainardi et al. 2006; Wilkins et al. 1983). Constipation is common and remains a significant problem into adulthood (Mainardi et al. 2006; Niebuhr 1978). Likely, hypotonia contributes to this.

Evaluation

- Gastrointestinal anomalies should be evaluated as in the general population.
- Constipation should be considered if an individual has maladaptive behavior changes but is unable to verbalize discomfort.
- Malrotation should be investigated if there is a presentation of intermittent bowel obstruction or volvulus.
- Hirschsprung disease should be considered if signs or symptoms suggest poor intestinal mobility or severe constipation occurs in infancy.

Management

- Constipation should be treated accordingly per standard medical protocol.
- Standard intervention for malrotation or Hirschsprung disease should be provided as in the general population.

Respiratory

Respiratory tract infections are common (52%) in individuals with deletion 5p syndrome (Mainardi et al. 2006; Niebuhr 1978). In addition to neonatal respiratory distress syndrome, there can be pneumonia and/or bronchitis. Hypotonia and swallowing difficulties may lead to aspiration-related respiratory tract infections.

Primary ciliary dyskinesia is associated with frequent infections such as respiratory distress in newborns, otitis media, sinusitis, pneumonia, and infertility. Primary ciliary dyskinesia is an autosomal recessive disorder caused by mutations in the *DNAH5* (dynein axonemal heavy chain 5) gene, which results in ciliary dysmotility. Because *DNAH5* is located on 5p, some individuals with 5p deletions who also have a *DNAH5* mutation have been affected with primary ciliary dyskinesia (Shapiro et al. 2014).

Because of possible airway abnormalities, individuals may develop sleep apnea. Snoring is the most common sleep disturbance in deletion 5p syndrome, but obstructive sleep apnea occurs as well (Maas et al. 2012).

Evaluation

- With a history of recurrent pneumonia, evaluation for aspiration should be made.
- Recurrent frequent sino-pulmonary infections should prompt consideration of evaluation for primary ciliary dyskinesia.
- Evaluation of sleep apnea should be made as for the general population, especially if snoring is present.

Management

- Infections should be treated accordingly per standard medical protocol.
- Sleep apnea management should be individualized depending on tolerance for the interventions.

Aging and Life Expectancy

Individuals may have premature graying of hair beginning as early as 15 years of age. With increasing age, facial features change, becoming longer with a broader mouth (Figure 20.1B). Microcephaly persists. Cataracts and optic atrophy may develop, as noted above. As previously mentioned, self-help skills occur in three-quarters of adults, and aggression and self-injury increase with aging.

Niebuhr (1978) completed early studies and reported that nearly 10% of the individuals with deletion 5p syndrome died young, with 75% in the first few months of life and 90% within the first year. Early literature often reported mistakenly that 90% of individuals died in the first year of life. In more recent studies, mortality rates were lower at 6.4%, with 36% of deaths occurring in the first month of life and 64% within the first year (Mainardi et al. 2006). Pneumonia, congenital heart defects, and respiratory distress syndrome are the most common causes of death. Those with unbalanced translocations that lead to a 5p deletion have a higher mortality rate (18.5%) than those with terminal deletions (4.8%) (Mainardi et al. 2006). If no major organ defects or other critical medical conditions exist, life expectancy appears to be normal with individuals known to survive at least into their 70s (Guala et al. 2017).

ACKNOWLEDGEMENTS

We acknowledge the generous contributions of the families and members of the Five p Minus Society, and the original work of Mary Esther Carlin, MD.

RESOURCES

5p Minus Society, an online family support group in the US for individuals with 5p deletions
Website: https://fivepminus.org/
Cri du Chat Syndrome in Genetics Home Reference
Website: https://ghr.nlm.nih.gov/condition/cri-du-chat-syndrome
Cri du Chat Syndrome in Genetic and Rare Diseases Information Center
Website: https://rarediseases.info.nih.gov/diseases/6213/cri-du-chat-syndrome

REFERENCES

Breg WR, Steele MW, Miller OJ, Warburton D, Allderdice PW (1970) The cri du chat syndrome in adolescents and adults: Clinical finding in 13 older patients with partial deletion of the short arm of chromosome no. 5 (5p-). *J Pediatr* 77:782–791.

Church DM, Bengtsson U, Nielsen KV, Wasmuth JJ, Niebuhr E (1995) Molecular definition of deletions of different segments of distal 5p that result in distinct phenotypic features. *Am J Hum Genet* 56:1162-72.

Collins MS, Cornish K (2002) A survey of the prevalence of stereotypy, self-injury and aggression in children and young adults with cri du chat syndrome. *J Intellect Disabil Res* 46:133–140.

Colover J, Lucas M, Comley JA, Roe AM (1972) Neurological abnormalities in the 'cri-du-chat' syndrome. *J Neurol Neurosurg Psychiatry* 35:711–719.

Corcuera-Flores J-R, Casttellanos-Cosano L, Torres-Lagares D, Serrera-Figallo MA, Rodriguez-caellero A, Machuca-Portillo G (2016) A systematic review of the oral and craniofacial manifestations of cri du chat syndrome. *Clin Anat* 29:555–60.

Cornish KM, Bramble D, Munir F, Pigram J (1999) Cognitive functioning in children with typical cri du chat (5p-) syndrome. *Develop Med Child Neurol* 41:263–6

Cornish KM, Munir F (1998) Receptive and expressive language skills in children with cri-du-chat syndrome. *J Commun Disord* 31:73–80.

Cornish KM, Pigram J (1996) Developmental and behavioural characteristics of cri du chat syndrome. *Arch Dis Child* 75:448–450.

Corrêa DG, Ventura N, Gasparetto EL (2017) Pontine hypoplasia in cri-du-chat syndrome: alterations in diffusion tensor imaging. *Child's Nerv Syst* 33:1241–2.

Dykens EM, Clarke DJ (1997) Correlates of maladaptive behavior in individuals with 5p- (cri du chat) syndrome. *Dev Med Child Neurol* 39:752–756.

Elmakky A, Carli D, Lugli L, Torelli P, Guidi B, Falcinelli C, Fini S, Ferrari F, Percesepe A (2014) A three-generation family with terminal microdeletion involving 5p15.33- 32 due to a whole-arm 5;15 chromosomal translocation with a steady phenotype of atypical cri du chat syndrome. *Eur J Med Genet* 57:145–150.

Esbensen AJ, Schwichtenberg AJ (2017) Sleep in Neurodevelopmental Disorders. *Int Rev Res Dev Disabil* 51:153–191.

Espirito Santo LD, Moreira LMA, Riegel M (2016) Cri-du-chat syndrome: clinical profile and chromosomal microarray analysis in six patients (2016) *Biomed Res Int* 2016:5467083

Fang JS, Lee KF, Huang CT, Syu CL, Yang KJ, Wang LH, Liao DL, Chen CH (2008) Cytogenetic and molecular characterization of a three-generation family with chromosome 5p terminal deletion. *Clin Genet* 73:585–90.

Gersh M, Goodart SA, Overhauser J (1994) Physical mapping of genetic markers on the short arm of chromosome 5. *Genomics* 24:577–9.

Gersh M, Goodart SA, Pasztor LM, Harris DJ, Weiss L, Overhauser J (1995) Evidence for a distinct region causing a cat-like cry in patients with 5p deletions. *Am J Hum Genet* 56:1404–10.

Guala A, Spunton M, Kalantari S, Kennerknecht I, Danesino C (2017) Neoplasia in cri du chat syndrome from Italian and German Databases. *Case Rep Genet* 2017:5181624.

Guala A, Spunton M, Mainardi PC, Emmig U, Acucella G, Danesino C (2015) Anesthesia in cri du chat syndrome: information on 51 Italian patients. *Am J Med Genet A* 167A:1168–70.

Higurashi M, Oda M, Iijima K, Iijima S, Takeshita T, Watanabe N, Yoneyama K (1990) Live- birth prevalence and follow-up of malformation syndromes in 27,472 newborns. *Brain Dev* 12:770–773.

Hills C, Moller JH, Finkelstein M, Lohr J, Schimmenti L (2006) Cri du chat syndrome and congenital heart disease: a review of previously reported cases and presentation of an additional 21 cases from the Pediatric Cardiac Care Consortium. *Pediatrics* 117:e924–7.

Honjo RS, Mello CB, Pimenta LSE, Nunes-Vaca EC, Benedetto LM, Khoury RBF, Befi-Lopes DM, Kim CA (2018) Cri du chat syndrome: characteristics of 73 Brazilian patients. *J Intellect Disabil Res* 62:467–73.

Huisman S, Mulder P, Kuijk J, Kersholt M, van Eeghen A, Leenders A, van Balkom I, Oliver C, Piening S, Hennekam R (2017) Self-injurious behavior. *Neurosci Biobehav Rev* 84:483–491.

Lefranc V, de Luca A, Hankard R (2016) Protein-energy malnutrition is frequent and precocious in children with cri du chat syndrome. *Am J Med Genet A* 170A:1358–1362.

Maas AP, Didden R, Korzilius H, Curfs LM (2012) Exploration of differences in types of sleep disturbance and severity of sleep problems between individuals with cri du chat syndrome, Down's syndrome, and Jacobsen syndrome: a case control study. *Res Dev Disabil* 33:1773–9.

Mainardi PC, Perfumo C, Cali A, Coucourde G, Pastore G, Cavani S, Zara F, Overhauser J, Pierluigi M, Bricarelli FD (2001) Clinical and molecular characterisation of 80 patients with 5p deletion: Genotype–phenotype correlation. *J Med Genet* 38:151–158.

Mainardi PC1, Pastore G, Castronovo C, Godi M, Guala A, Tamiazzo S, Provera S, Pierluigi M, Bricarelli FD (2006) The natural history of cri du chat syndrome. A report from the Italian Register. *Eur J Med Genet* 49:363–83.

Marinescu RC, Mainardi PC, Collins MR, Kouahou M, Coucourde G, Pastore G, Eaton-Evans J, Overhauser J (2000) Growth charts for cri-du-chat syndrome: an international collaborative study. *Am J Med Genet* 94:153–62.

Medina M, Marinescu RC, Overhauser J, Kosik KS (2000) Hemizygosity of delta-catenin (CTNND2) is associated with severe mental retardation in cri-du-chat syndrome. *Genomics* 63:157–64.

Molina-García A, Castellanos-Cosano L, Machuca-Portillo G, Posada-de la Paz M. 2016. Impact of rare diseases in oral health. *Med Oral Patol Oral Cir Bucal* 21:e587–94.

Niebuhr E. 1978. The cri du chat syndrome: Epidemiology, cytogenetics, and clinical features. *Hum Genet* 44:227–275.

Nguyen JM, Gamble C, Smith JL, Raia M, Johnson A, Czerwinski J (2014) Prenatal diagnosis of 5p deletion syndrome in a female fetus leading to identification of the same diagnosis in her mother. *Prenat Diagn* 34:1–4.

Nguyen JM, Qualmann KJ, Okashah R, Reilly A, Alexeyev MF, Campbell DJ (2015) 5p deletions: Current knowledge and future directions. *Am J Med Genet C* 169:224–38.

Overhauser J, Mcmahon J, Oberlender S, Carlin ME, Niebuhr E, Wasmuth JJ, Lee-Chen J (1990) Parental origin of chromosome 5 deletions in the cri-du-chat syndrome. *Am J Med Genet* 37:83–86.

Overhauser J, McMahon J, Oberlender S, Carlin ME, Niebuhr E, Wasmuth JJ, Lee-Chen J (1995) *Am J Hum Genet* 37:83–6.

Perfumo C, Cerruti Mainardi P, Cali A, Cou-courde G, Zara F, Cavani S, Overhauser J, Bricarelli FD, Pierluigi M (2000) The first three mosaic cri du chat syndrome patients with two rearranged cell lines. *J Med Genet* 37:967–972.

Petersen AK, Cheung SW, Smith JL, Bi W, Ward PA, Peacock S, Braxton A, van den Veyver IB, Breman AM (2017) Positive predictive value estimates for cell-free noninvasive prenatal screening from data of a large referral genetic diagnostic laboratory. *Am J Obstet Gynecol* 217(6):e1–691.

Shapiro AJ, Weck KE, Chao KC, Rosenfeld M, Nygren AO, Knowles MR, Leigh MW, Zariwala MA (2014) Cri du chat syndrome and primary ciliary dyskinesia: A common genetic cause on chromosome 5p. *J Pediatr* 165:858–861.

Ullah I, Mahajan L, and Magnuson D (2017) A newly recognized association of hirschsprung disease with cri-du-chat syndrome. *Am J Gastroenterol* 112:185–6.

van Buggenhout GJCM, Pijkels E, Holvoet M, Schaap C, BCG Hamel, Fryns JP (2000) Cri du chat syndrome: changing phenotype in older patients. *Am J Med Genet* 90:203–15.

Walker JL, Blank CE, Smith BA (1984) Interstitial deletion of the short arm of chromosome 5 in a mother and three children. *J Med Genet* 21:465–7.

Wilkins LE, Brown JA, Wolf B (1980) Psychomotor development in 65 home-reared children with cri-du-chat syndrome. *J Pediatr* 97:401–5.

Wilkins LE, Brown JA, Nance WE, Wolf B (1983) Clinical heterogeneity in 80 home-reared children with cri du chat syndrome. *J Pediatr* 102:528–33.

Wu Q, Niebuhr E, Yang H, Hansen L (2005) Determination of the "critical region" for cat-like cry of Cri-du-chat syndrome and analysis of candidate genes by quantitative PCR. *Eur J Hum Genet* 13:475–85.

Zhang B, Willing M, Grange DK, Shinawi M, Manwaring L, Vineyard M, Kulkarni S, Cottrell CE (2016) Multigenerational autosomal dominant inheritance of 5p chromosomal deletions. *Am J Med Genet A* 170A:583–593.

Zhu X, Li J, Ru T, Wang Y, Xu Y, Yang Y, Wu X, Cram DS, Hu Y (2016) Identification of copy number variations associated with congenital heart disease by chromosomal microarray analysis and next-generation sequencing. *Prenat Diagn* 36:321–7.

21

DELETION 22q11.2 (VELO-CARDIO-FACIAL SYNDROME/ DIGEORGE SYNDROME)

DONNA M. MCDONALD-MCGINN, STEPHANIE JEONG, MICHAEL-JOHN MCGINN, II, ELAINE H. ZACKAI
The Children's Hospital of Philadelphia and The Perelman School of Medicine at the University of Pennsylvania, Philadelphia, Pennsylvania, USA
MARTA UNOLT
The Children's Hospital of Philadelphia and The Perelman School of Medicine at the University of Pennsylvania, Philadelphia, Pennsylvania, USA; and Ospedale Bambino Gesù and La Sapienza University, Rome, Italy

INTRODUCTION

The 22q11.2 deletion has been identified in the majority of individuals with DiGeorge syndrome (de la Chapelle et al. 1981; Scambler et al. 1991; Driscoll et al. 1992), velocardiofacial syndrome (Driscoll et al. 1993), conotruncal anomaly face syndrome (Burn et al. 1993; Matsouka et al. 1994), and in some individuals with the autosomal dominant Opitz G/BBB syndrome (McDonald-McGinn et al. 1995; LaCassie and Arriaza, 1996) and Cayler Cardio-facial syndrome (Giannotti et al. 1994). Originally described as individual entities by a number of subspecialists who were concentrating on one particular area of interest, following the widespread use of fluorescence in situ hybridization (FISH) beginning in 1992, these syndromes are now collectively referred to by their chromosomal designation: 22q11.2 deletion.

Prevalence

The 22q11.2 deletion is the most common microdeletion syndrome, with an estimated prevalence of approximately 1/3000–1/6000 live births (Wilson et al. 1994; Devriendt et al. 1998; Goodship et al. 1998; Botto et al. 2003; Oskarsdóttir et al. 2004) and 1/992 fetuses without evidence of congenital heart disease or cleft palate (Grati et al. 2015). It is present in 1/68 child born with congenital heart disease (Wilson et al. 1994); it is the most common cause of syndromic palatal defects; and it is the second most common cause of developmental delay, accounting for about 2.4% of affected individuals (Rauch et al. 2006). It is so common, in fact, that affected first cousins have been identified with the 22q11.2 deletion by chance alone (Saitta et al. 2004). Furthermore, there have been a number of individuals found to have both the 22q11.2 deletion and concomitant diagnoses, including familial single gene disorders, such as Marfan syndrome; neurofibromatosis; craniosynostosis with a *FGFR3* mutation, and Ehlers–Danlos syndrome, as well as additional sporadic cytogenetic abnormalities including trisomy 8 mosaicism and trisomy 21. This high frequency supports the need to think broadly when evaluating individuals with features associated with deletion 22q11.2 and to consider deletion studies even in the presence of other underlying diagnoses (McDonald-McGinn et al. 2015).

In contrast to the early reports on individuals with DiGeorge syndrome, the mortality rate for people with the 22q11.2 deletion is low (4%), likely reflecting advances in cardiac care and infectious disease management. The median age of death is 4 months, most often secondary to complications of complex congenital heart disease (McDonald-McGinn et al. 2002; Repetto et al. 2014). Thus, as there is little effect on reproductive fitness and a 50% recurrence risk

---

Cassidy and Allanson's Management of Genetic Syndromes, Fourth Edition.
Edited by John C. Carey, Agatino Battaglia, David Viskochil, and Suzanne B. Cassidy.
© 2021 John Wiley & Sons, Inc. Published 2021 by John Wiley & Sons, Inc.

for affected individuals, a rise in the prevalence of the 22q11.2 deletion will likely be appreciated in the near future. Based on the limited data available from the literature, life expectancy for adults with 22q11.2 deletion seems to be shorter, with death occurring in the 40s. This may be related to the long-term complications of congenital cardiac anomalies or psychiatric illness (Bassett et al. 2009; Repetto et al. 2014) or to an unrelated factor not yet identified. However, further prospective studies are needed to define life expectancy and mortality risks in adults with 22q11.2 deletion syndrome.

Diagnostic Criteria

There are no established diagnostic criteria for 22q11.2 deletion syndrome. The most frequent significant clinical features in individuals with 22q11.2 deletion include the following: immunodeficiency, congenital heart disease, and palatal defects in approximately three-quarters, hypocalcemia in one-half; dysphagia in at least half; and renal anomalies in one-third, as well as developmental disabilities in over 90% (McDonald-McGinn et al. 1999) (Table 21.1). There is wide inter- and intra-familial variability, even between identical twins, which does not appear to be influenced by the parent-of-origin or whether it is familial or de novo (Yamagishi et al. 1998; McDonald-McGinn et al. 2001).

Overt cleft palate is found in 10% of affected individuals with the 22q11.2 deletion and cleft lip and palate is seen in 1–2% (McDonald-McGinn et al. 1999, 2002). These figures are significantly greater than the general population incidence of cleft palate (1/2500) and cleft lip and palate (1/800) (Fraser 1981). Structural renal abnormalities are seen in 31% of those with 22q11.2 deletion, thought to be the result of a renal developmental drive *CRKL1* located within the 22q11.2 deletion (Lopez-Rivera et al. 2017). Findings on ultrasound include renal agenesis (12%), hydronephrosis (5%), and multicystic/dysplastic kidneys (4%) (Wu et al. 2002; Kujat et al. 2006). Pre- and postaxial polydactyly of the hands has been observed in 4% of individuals with the deletion, whereas postaxial polydactyly of the feet has been seen in 1% (Ming et al. 1997). This is at least 10 times the general population incidence of polydactyly in both Caucasians and African-Americans. Additional limb defects include radial aplasia (Digilio et al. 1997), symbrachydactyly (Devriendt et al. 1997), absent/hypoplastic thumb (Cormier-Daire et al. 1995), tibial hemimelia, clubfoot and a terminal transverse defect of the upper extremity (personal experience). Common spine anomalies include hemivertebrae, butterfly vertebrae (Ming et al. 1997), C2–C3 fusion in 34% (Hamidi et al. 2014), and scoliosis (Homans et al. in press). Congenital diaphragmatic hernia is found in 1% of those with 22q11.2 deletion (McDonald-McGinn et al. 2002; Unolt et al. 2017), which is 20 times the incidence in the general population. Neural tube defects are occasionally reported in 22q11.2 deletion syndrome (Nickel and Magenis 1996), suggesting that they may occur more frequently than in the general population.

In adults, major clues to the diagnosis include congenital heart disease; palatal anomalies including velopharyngeal incompetence/hypernasal speech; a learning disability or intellectual deficit; psychiatric illness, most often schizophrenia; immune deficiency or a history of chronic infection; hypoparathyroidism; and characteristic minor facial dysmorphic features (Bassett et al. 2005; Vogels et al. 2014). Dysmorphia most often includes malar flatness, hooded eyelids, hypertelorism, upslanting palpebral fissures, auricular anomalies, a prominent nasal root with a fullness superior to the nasal tip and underdeveloped alae nasi, a nasal dimple or crease, and a narrow mouth or asymmetric crying facies (Figures 21.1–21.6) (McDonald-McGinn et al. 1997, 1999, 2001a). However, there appears to be a paucity of typical facial features in non-Caucasian individuals (Figure 21.7) and, therefore, the face may be less helpful in identifying affected African-American and Asian individuals (McDonald-McGinn et al. 2005; Kruszka et al. 2017). Less common features in adults may include a history of scoliosis, genitourinary anomalies including hypospadias (McDonald-McGinn et al. 1995), craniosynostosis (McDonald-McGinn et al. 2005a), laryngeal abnormalities (McDonald-McGinn et al. 1995), autoimmune disorders (idiopathic thrombocytopenic purpura, juvenile rheumatoid arthritis, psoriasis, vitiligo, Graves disease, autoimmune hemolytic anemia, autoimmune neutropenia) (Kawame et al. 2001; Lambert et al. 2017) and hearing loss, especially in conjunction with a learning disability or psychiatric illness (Ryan et al. 1997; McDonald-McGinn 2015).

Etiology, Pathogenesis, Genetics The majority of affected individuals (approximately 85%) have a standard (~3 Mb) LCR22A-LCR22D deletion encompassing ~50 functional

TABLE 21.1 Frequent clinical findings in deletion 22q11.2.

Feature	Frequency
Developmental disability	>90%
Congenital heart defect	76%
Palatal defects	76%
Immunodeficiency	77%
Hypocalcemia	55%
Renal anomalies	36%
Dysphagia	>35%
Scoliosis	25%
Schizophrenia	25%
Polyhydramnios	16%
Asymmetric crying facies	14%[a]
Polydactyly	4%
Congenital diaphragmatic hernia	1%

[a] Pasick et al. 2013.

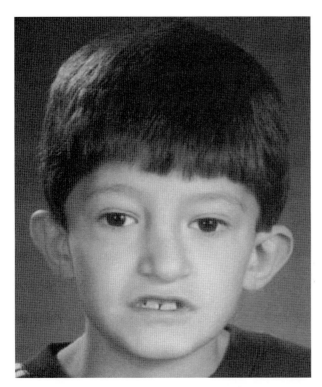

FIGURE 21.1 A 10-year-old Caucasian male with 22q11.2 deletion syndrome. Note malar flatness, protuberant ears, fullness superior to the nasal tip with somewhat hypoplastic alae nasi, short philtrum, a narrow mouth, and thin upper lip.

genes. The majority of remaining individuals have a smaller nested deletion (LCR22A-LCR22B, LCR22A-LCR22C, LCR22B-LCR22D, LCR22C-LCR22D) within the "DiGeorge critical region" (McDonald-McGinn 2015) (Figure 21.8). Standard and proximal deletions (LCR22A-LCR22B) (~7% of patients) include a number of well-studied genes including *UFD1L*, *COMT*, and *TBX1*. The latter is a member of the T-box family of genes, which has been shown to program similar features in haploinsufficient mouse models (Baldini 2005). Nested distal deletions (LCR22B-LCR22D and LCR22C-LCR22D) do not include *TBX1* (Yamagishi et al. 1998; McQuade et al. 1999; Shaikh et al. 2000; Garcia-Minaur et al. 2002; Rump et al. 2014; McDonald-McGinn 2015) but do include other important developmental genes. Of note, a few individuals with features of 22q11.2 deletion syndrome but without evidence of a copy number variant in this region have been reported with both gain and loss of function mutations in *TBX1* (Yagi et al. 2003; Zweier et al. 2007). Besides the association between congenital heart disease and *TBX1*, other examples of genotype–phenotype correlations have been reported including: haploinsufficiency of *CRKL* (v-crk avian sarcoma virus CT10 oncogene homologue-like) with cardiac anomalies and renal differences in individuals with nested distal deletions (Racedo et al. 2015; Lopez-Rivera et al. 2017), *SNAP29* (synaptosomal-associated protein 29 kDa) with cerebral dysgenesis and neuropathy (McDonald-McGinn et al. 2013), and *GP1BB* (platelet glycoprotein Ib β-polypeptide) in platelet dysfunction (Lambert et al. 2017). Nonetheless, variability in this condition is likely a result of a combination of factors including the multi-gene deletion, the sensitivity of individual genes within the 22q11.2 region to gene dosage variants in genes on the intact 22q11.2 allele, and additional 'modifying' variants outside the 22q11.2 region involving both protein-coding genes and regulatory mechanisms (McDonald-McGinn et al. 2001, 2015).

Most deletions (90%) are de novo (McDonald-McGinn et al. 2001a; McDonald-McGinn et al. 2015) and occur as a result of the inherent structure of chromosome 22q11.2.

FIGURE 21.2 An adult Caucasian female with 22q11.2 deletion syndrome. Note malar flatness, mildly upslanting palpebral fissures, thick helices with attached lobes, a prominent nasal root, fullness superior to the nasal tip with hypoplastic alae nasi and a nasal crease, and narrow mouth.

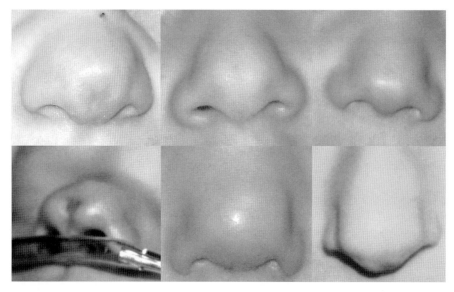

FIGURE 21.3 Nasal anomalies in six affected individuals demonstrating variable fullness superior to the nasal tip with hypoplastic alae nasi and a nasal dimple or crease, which at times is punctuated by a strawberry hemangioma.

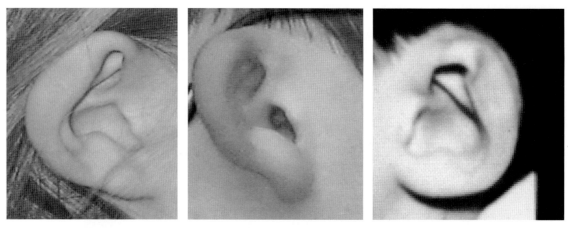

FIGURE 21.4 Variable auricular anomalies are seen in individuals with the 22q11.2 deletion, including thick overfolded helices, attached lobes, crumpled helices, protuberant ears, and small ears. Preauricular pits, tags, and microtia have also been reported (Digilio et al. 2009).

Specifically, there are segmental duplications (low copy repeats – LCR-A, LCR-B, LCR-C, LCR-D) that define the breakpoints, flank the deletion structurally, and make this region especially susceptible to rearrangements because of unequal meiotic crossovers and thus aberrant interchromosomal exchanges (Edelman et al. 1999; Shaikh et al. 2000; Saitta et al. 2004; McDonald-McGinn 2015) (Figures 21.8 and 21.9). One unaffected adult with somatic mosaicism has been identified (McDonald-McGinn et al. 2002) and there are several reports of germ line mosaicism where siblings of parents without the 22q11.2 deletion are both affected (Hatchwell et al. 1998; Kasprzak et al. 1998; Sandrin-Garcia et al. 2002) suggesting a small recurrence risk for all parents of a child with a de novo deletion. Lastly, like all contiguous gene deletion syndromes, anyone with the 22q11.2 deletion has a 50% recurrence risk.

Genetic Counseling Genetic counseling for an adult with 22q11.2 deletion syndrome may be challenging given the cognitive limitations in some affected individuals (McDonald-McGinn et al. 2001a). Furthermore, determining the risk to offspring for medically relevant findings such as congenital heart disease is complicated by the wide inter- and intra-familial variability, as well as ascertainment bias (McDonald-McGinn 2015). However, findings in 36 affected persons identified only following the diagnosis in their relative, including 23 parents of affected children and 13 children born before or after the diagnosis in the parent, revealed the following: 56% had no visceral anomalies (65%

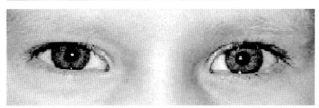

FIGURE 21.5 Variability in external eye findings. Note significant hooding of the eyelids with a decrease in visible eyelashes and epicanthal folds in the upper figure of a 5.5-year-old child, downslanting palpebral fissures in the middle child, and mild hooding with a mild upslant in the lower figure of a nearly 4-year-old child (Forbes et al. 2016).

of affected parents and 39% of children), congenital heart disease was noted in only 19%; overt cleft palate in 11%, hypocalcemic seizures in 6%, a laryngeal web in 3%, and schizophrenia in 3%. Data on educational outcome in 20 of these adults revealed that 70% were high school graduates. However, 30% required significant learning support in high school, and despite the high graduation rate, these adults had difficulty understanding the ramifications of the diagnosis, had a poor understanding of their recurrence risk, and had difficulty complying with treatment recommendations for themselves and their affected offspring. Despite these difficulties, many adults were found to be gainfully employed in areas such as culinary arts, agriculture, law enforcement/military, environmental services, supply chain, administrative support, early childhood education, and homemakers. In addition, a number of individuals had college degrees and even Master's level training.

Individuals with a "nested" deletion may have a milder cognitive phenotype, although the numbers may be too small to state this with certainty; however, these deletions are most definitely more frequently inherited (McDonald-McGinn et al. 2015; Rump et al. 2014).

Diagnostic Testing

Historically, the 22q11.2 deletion was identified using standard cytogenetic testing in a small number of individuals with DiGeorge syndrome (De la Chapelle et al. 1981). Subsequent laboratory advances utilizing FISH probes such as N25 or TUPLE within the commonly deleted "DiGeorge critical region" followed (Scambler et al. 1991; Driscoll et al. 1992; Desmaze et al. 1993) and allowed the identification of submicroscopic deletions. Currently, because FISH is limited to one target sequence within the DiGeorge critical region, newer techniques such as single nucleotide polymorphism (SNP) microarray and multiplex ligation-dependent probe amplification (MLPA) are more frequently utilized both to screen for abnormalities in people with multiple congenital anomalies and/or developmental delay and to precisely define the deletion breakpoints in individuals already identified with deletion 22q11.2 deletion syndrome (Kariyazono et al. 2001; Mantripragada et al. 2004).

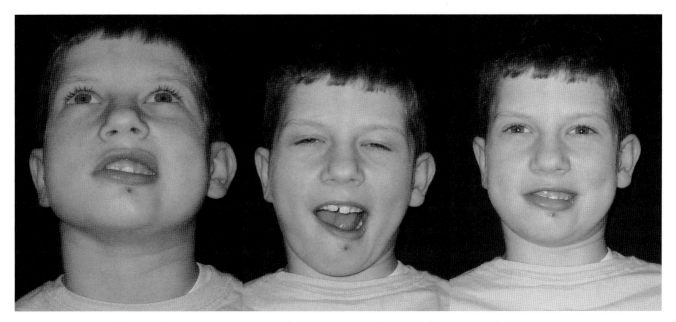

FIGURE 21.6 Asymmetric crying facies in an 8-year-10-month-old boy with deletion 22q11.2.

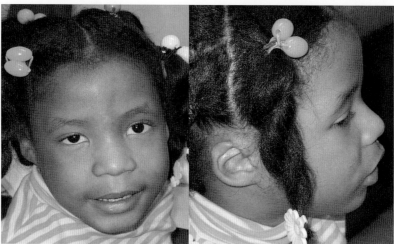

FIGURE 21.7 Top: African-American female with mild dysmorphic features including upslanting palpebral fissures, a somewhat broad nasal tip with the columella extending below the nares, and micrognathia. Bottom: African-American female with more pronounced dysmorphic features including hypertelorism, epicanthal folds, small ears with thick and overfolded helices and attached lobes, and a broad nasal root and tip.

Nested deletions, which exclude the N25 or TUPLE probes but lie within the DiGeorge critical region of 22q11.2, have been observed in several individuals (Garcia-Minaur et al. 2002; Rump et al. 2014), as have point mutations in *TBX1* (Yagi et al. 2003, Zweier et al. 2007). Thus, individuals with negative deletion studies and typical clinical features should be further investigated using a microarray, MLPA, and if those studies are negative, sequencing of *TBX1*.

Whole genome-wide microarray has the added benefit of not introducing bias by requiring pre-selection of a specific genomic region. This may be particularly useful in individuals with few or with atypical features. Indeed, dual diagnosis of other unrelated conditions has been reported (McDonald-McGinn et al. 2015).

There are a number of frequently identified findings in fetuses with the 22q11.2 deletion: congenital heart defects (especially those of conotruncal origin), cleft palate, renal anomalies, polyhydramnios, polydactyly, congenital diaphragmatic hernia, clubfoot, and neural tube defects (McDonald-McGinn et al. 2002, 2005; Unolt et al. 2017). Thus, deletion studies are recommended when these findings are seen on a level II (high-resolution) prenatal ultrasound. Polyhydramnios has been noted retrospectively in 16% of affected pregnancies (McDonald-McGinn et al. 2002). This is 16 times the general population incidence and may be attributable to the presence of fetal palatal anomalies, swallowing difficulties, or esophageal atresia (Digilio et al. 1997; Eicher et al. 2000). Non-invasive prenatal screening for fetal aneuploidy may lead to the diagnosis in the fetus as well as in a previously undiagnosed mother.

Recently, several technologies for DNA detection have been tested for sensitivity and specificity in the setting of newborn screening for 22q11.2 deletion (Pretto et al. 2015). Newborn screening could provide an unbiased incidence of

INTRODUCTION 297

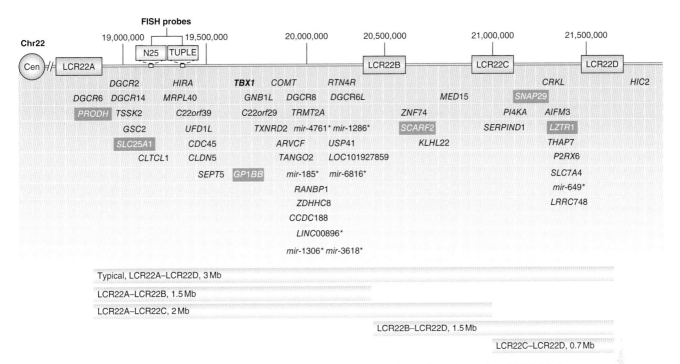

FIGURE 21.8 The majority of affected individuals (~85%) have a large (more than 3 Mb) deletion encompassing about 50 genes, whereas ~7% have a smaller (1.5–2 Mb) "nested proximal" deletion including TBX1. "Nested distal" deletions do not include TBX1 but do include other important developmental genes such as CRKL. From Nature Reviews Primers.

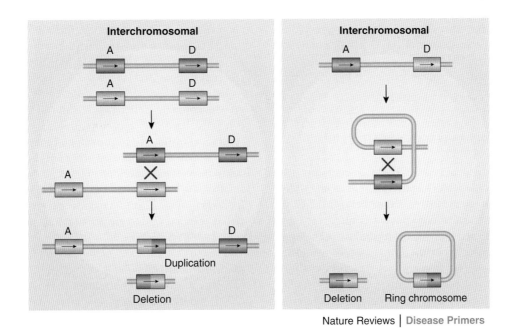

FIGURE 21.9 Non-homologous crossing over, because of blocks of duplicated sequences, results in a duplication on chromosome "A" and a deletion on chromosome "B".

22q11.2 deletion, give an opportunity for timely assessment and interventions, and obviate a protracted diagnostic odyssey. Interestingly, a recent study reported that newborn screening for severe combined immunodeficiency is now identifying infants with 22q11.2 deletion syndrome because of T-cell lymphopenia and pointed out how some clinical features may go unrecognized (Barry et al. 2017).

Differential Diagnosis

Structural anomalies found in individuals with a 22q11.2 deletion can also be observed as isolated findings in an otherwise typical individual. Syndromic diagnoses with overlapping features of the 22q11.2 deletion syndrome include the following: Smith–Lemli–Opitz syndrome (when polydactyly and cleft palate are present) (Chapter 53); Alagille syndrome (when butterfly vertebrae, congenital heart disease, and posterior embryotoxon are present) (Chapter 3); VATER/VACTERL association (when heart disease, vertebral, renal, and limb anomalies are present) (Chapter 61); oculo-auriculo-vertebral spectrum/Goldenhar syndrome (when ear anomalies, vertebral detects/butterfly vertebrae, heart disease, and renal anomalies are present) (Chapter 42); and Kabuki syndrome (when cleft palate, congenital heart disease, and butterfly vertebrae are present) (Chapter 34). Individuals with both unicoronal and bicoronal craniosynostosis in whom molecular causes for known craniosynostosis syndromes (Chapter 17) have been excluded, including mutations in *FGFR 1, 2, 3, TWIST*, and deletions in *TWIST* are also candidates for deletion studies (McDonald-McGinn et al. 2005a).

Individuals suspected of having a 22q11.2 deletion who have negative FISH studies may have a chromosome abnormality involving some other chromosomal region, including a cytogenetically visible deletion at 10p13-p14 or a mutation in *CHD7*, as seen with CHARGE syndrome associated with congenital heart disease, immunodeficiency, hypocalcemia, and hearing loss (see Chapter 11).

MANIFESTATIONS AND MANAGEMENT

Management of individuals with a 22q11.2 deletion is age- and symptom-specific (McDonald-McGinn et al. 1999; McDonald-McGinn et al. 2015). Nonetheless, there are some general recommendations (Bassett et al. 2011; Fung et al. 2015). All affected individuals, regardless of their age at presentation, benefit from the following evaluations: a cardiology evaluation, which often includes a chest X-ray, electrocardiogram, and echocardiogram; laboratory studies to assess the immune and endocrine systems including the presence of hypocalcemia/hypoparathyroidism and thyroid disease; a renal ultrasound; and parental 22q11.2 deletion studies to provide appropriate genetic counseling (McDonald-McGinn et al. 1999; McDonald-McGinn et al. 2015). Other evaluations are indicated in the appropriate sections below. Many individuals with a 22q11.2 deletion have complex medical needs and benefit from a multidisciplinary approach, much like the cleft palate team model, with one team leader designated to collate the multitude of specialist recommendations and provide the family with one unified message.

Growth and Feeding

Approximately 35–60% of children with the 22q11.2 deletion have significant dysphagia (Eicher et al. 2000; Mascarenhas et al. 2018), often requiring nasogastric tube feeding and/or gastrostomy tube placement. Feeding difficulties are independent of cardiac and palatal anomalies, and it is important to recognize that this aspect of the 22q11.2 deletion is often the most challenging for families. Further evaluation of such children frequently reveals a preponderance of nasopharyngeal reflux, prominence of the cricopharyngeal muscle, abnormal cricopharyngeal closure, and/or diverticulum. Thus, the underlying feeding problem in many children appears to be dysmotility in the pharyngoesophageal area, and esophagus (Figure 21.10) (Eicher et al. 2000). Abnormal swallowing with or without aspiration may be erroneously attributed to palatal or cardiac abnormalities, rather than to dysphagia related to dysmotility and abnormality of the oropharyngeal and cricoesophageal swallowing phase. Aspiration should be considered a possible cause for respiratory compromise or recurrent pulmonary infections and reactive airways disease in any affected child.

Constipation is a chronic feature in the majority of individuals with 22q11.2 deletion, and structural bowel anomalies such as intestinal malrotation, intestinal nonrotation, Hirschsprung disease, and feeding difficulties secondary to a vascular ring (McDonald-McGinn et al. 1999) are reported (Figure 21.11).

Individuals with 22q11.2 deletion occasionally have growth hormone deficiency leading to short stature. This generally becomes apparent after the first year of life. Additional symptoms of growth hormone deficiency, which may or may not be present in such individuals, include: delay in pubertal development with a high-pitched voice, small hands and feet, and underdeveloped male genitalia. Individuals with 22q11.2 deletion and growth hormone deficiency respond well to growth hormone therapy (Weinzimer et al. 1998).

There is an increase of hypothyroidism and hyperthyroidism among individuals with 22q11.2 deletion, and this may affect growth (Weinzeimer et al. 2001). Basset et al. (2005) reports that one in five affected adults have thyroid disease.

Irrespectively of feeding difficulties, growth hormone deficiency or other physical problems, growth in children with 22q11.2 deletion syndrome may be slower than in

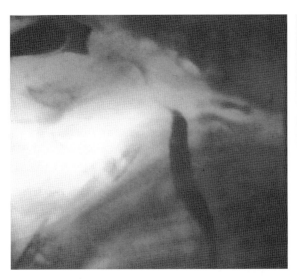
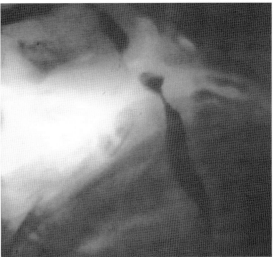

FIGURE 21.10 Barium swallow demonstrating esophageal dysmotility on the right compared with a normal scan on the left (image courtesy of Peggy Eicher, MD).

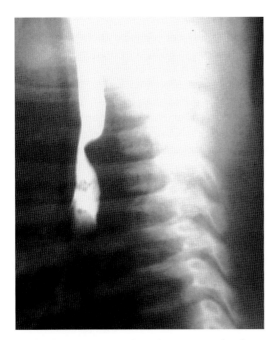

FIGURE 21.11 Barium swallow shows a vascular ring causing constriction of the esophagus.

general population. Catch up growth occurs usually at the school age, and, later, in adolescence and adulthood, the trend switches toward overweight and obesity (Habel et al. 2012; Digilio et al. 2001; Voll et al. 2017).

Evaluation

- At diagnosis a history of gastrointestinal and feeding abnormalities should be sought.
- Referral to a gastroenterologist or feeding specialist may be warranted.
- Structural or functional abnormalities may require imaging studies such as upper gastrointestinal contrast study with small bowel follow-through, pH probe, milk scan, or swallow study to precisely identify the underlying etiology.
- Feeding should be monitored on a regular basis during regular well-child visits.
- Stooling patterns should be ascertained at regular medical visits.
- Growth should be evaluated by obtaining length/height and weight at diagnosis and at yearly intervals.
- Children whose height/length is less than third centile by age 3 years should be evaluated for thyroid hormone abnormality.
- Evidence of growth failure should prompt evaluation for growth hormone deficiency in a standard manner, preferably by a pediatric endocrinologist.
- Overweight and obesity in adolescents and adults should prompt active monitoring for weight gain and metabolic side effects. Evaluation of thyroid function and of glucose tolerance is recommended.

Management

- Gastrointestinal symptoms are generally amenable to standard therapy, such as acid blockade, treatment of delayed gastric emptying and/or constipation, and facilitation of more mature oral motor skills through occupational therapy.
- Some children with gastroesophageal reflux will require fundoplication.
- Many children benefit from nasogastric or gastrojejunal tube feedings, primarily in the neonatal period but some individuals may need prolonged tube feedings.

- Growth hormone deficiency and thyroid hormone abnormalities should be treated in a standard manner.
- To prevent obesity, early and ongoing attempts to encourage healthy diet and exercise behaviors are recommended. Antipsychotic and other psychotropic medications choice should include balancing efficacy and side effects.

Development, Behavior, and Psychiatric

From a psychosocial, developmental, and cognitive perspective, it is well established that the "behavioral phenotype" in both children and adults with 22q11.2 deletion is quite broad (Gerdes et al. 1999; McDonald-McGinn et al. 2001; Solot et al. 2001; De Smedt et al. 2003; Swillen and McDonald-McGinn 2015). Young children with 22q11.2 deletion have delays in achieving motor milestones (mean age of walking is 18 months), delays in emergence of language (many are non-speakers at age 2–3 years), and autism/autistic spectrum disorders (in about 20%) (Fine et al. 2005). In a study of 28 toddlers assessed with standardized tests, mental development was average in 21%, mildly delayed in 32%, and significantly delayed in 46%, and motor development was average in 8%, mildly delayed in 13%, and significantly delayed in 79%. In a group of 12 preschoolers assessed using the WPPSI-R, the full-scale IQ was 78 ± 11, the mean performance IQ was 78 ± 14, and the mean verbal IQ was 82 ± 15. In a total language assessment, 16% were average, 44% were mildly delayed, and 40% were significantly delayed (Gerdes et al. 1999). In school-aged children, using the age-appropriate Weschler IQ test, the mean IQ is between 70 and 76 (Moss et al. 1999), with 18% attaining full-scale IQ scores in the average range, 20% in the low-average range, 32% in the borderline range, and 30% in the intellectual disability range (Moss et al. 1999). Structural brain abnormalities and/or hypoxic ischemic events are often associated with a more guarded development outcome.

A school-aged child with 22q11.2 deletion generally has an atypical neuropsychological profile across multiple domains, the most striking being a significantly higher verbal than performance IQ score. Moss et al. (1999) observed a greater than 10-point mean split between the verbal IQ and performance IQ in 66% of 80 school-aged children consistent with a non-verbal learning disability, which is quite rare in the general population. Because the full-scale IQ score alone does not accurately represent the abilities of many individuals with the 22q11.2 deletion, verbal and performance IQ scores often need to be considered separately. Affected individuals appear to exhibit relative strengths in the areas of rote verbal learning and memory, reading decoding, and spelling, whereas deficits are found in the areas of non-verbal processing, visual-spatial skills, complex verbal memory, attention, working memory, visual-spatial memory, and mathematics. This evidence of stronger verbal than visual memory skills and stronger reading than math skills also supports the presence of a non-verbal learning disorder which requires specific cognitive remediation, behavior management, and parental counseling (Moss et al. 1999; Swillen et al. 1999; Bearden et al. 2001). It is important to recognize that a substantial number of affected individuals also has language difficulties, including both receptive and expressive language (Gerdes et al. 1999).

Delayed speech and language development is one of the most consistent features in young people with 22q11.2 deletion and a source of great frustration for families (Gerdes et al. 1999; Swillen et al. 1999). Many children do not babble, and the average emergence of language is 2.5 years (Gerdes et al. 1999). In looking at total language development, Gerdes et al. (1999) reported on a sample of 40 children with 22q11.2 deletion aged 13–63 months, and found 40% to be significantly delayed, 44% mildly delayed, and 16% in the average range. Although both expressive and receptive language demonstrated a high degree of delay, expressive language was more impaired than receptive. In addition, such delays in expressive language remained significant even after controlling for cognitive levels. Nonetheless, children show remarkable improvement in speech and language skills by school age (Gerdes et al. 1999), perhaps as a result of intervention, and over time verbal skills paradoxically become strengths (Moss et al. 1995).

Cognitive and adaptive functioning may change over time, and repeated assessments are necessary, especially at key developmental stages. Indeed, several studies indicate that cognitive development varies with divergent trajectories and individuals with 22q11.2 deletion syndrome between 8 years and 24 years of age may show an average decline of 7 full scale IQ points, with verbal IQ often falling to meet the performance IQ (Duijff et al. 2012; Vorstman et al. 2015; Swillen and McDonald-McGinn 2015).

In addition to language difficulties, many young children also have significant articulation deficits and speech abnormalities such as a high-pitched voice, hoarseness, compensatory articulation errors, dysarthria, and velopharyngeal incompetence leading to hypernasal speech.

The psychiatric phenotype associated with 22q11.2 deletion includes disinhibition and impulsiveness on the one hand, and shyness and withdrawal on the other (Swillen et al. 1999). Anxiety, perseveration, and difficulty with social interactions are also common, along with autism and autistic spectrum disorders (Swillen et al. 1999; Fine et al. 2005; Vorstman et al. 2006; Schneider et al. 2014). Attention deficits have also been reported (Gerdes et al. 1999) and may or may not require behavioral or medical intervention. Additionally, thought problems such as muddled thinking, frequent or easy confusion, ruminations, and repetitive thoughts have been identified (Swillen et al. 1999). Of note, symptomatology may be aggravated by cognitive delays.

The incidence and type of frank psychiatric disease associated with 22q11.2 deletion, including schizophrenia, bipolar disorder, and depression, varies from study to study and has evolved over time (Popolos et al. 1996; Bassett et al. 2003; Baker et al. 2005; Schneider et al. 2014). Most recently a 20–30% prevalence of psychosis, most often schizophrenia, appears to be consistent across reports (Bassett et al. 2005; Vorstman et al. 2006), whereas "psychosis-like" symptoms have been identified in some adolescents and young adults (age 13–25 years) (Baker et al. 2005). No precursor symptoms have been identified to date. However, a recent longitudinal study by the International Consortium on Brain and Behavior in 22q11.2 Deletion Syndrome showed that an early cognitive decline is a robust indicator of the risk of developing a psychotic illness. According to this study in a total of 411 individuals with 22q11.2 deletion mean full-scale IQ at first cognitive assessment was lower in those who developed a psychotic disorder compared with those without a psychotic disorder. Moreover, while all children with 22q11.2 deletion showed a mild decline in IQ with increasing age, particularly in the domain of verbal IQ, in those who developed psychotic illness this decline was significantly steeper. This divergence of verbal IQ trajectories between those who subsequently developed a psychotic disorder and those who did not was distinguishable from age 11 years onward (Vorstman et al. 2015).

The behavioral phenotype in individuals with a 22q11.2 deletion must be examined further with a specific emphasis on the longitudinal course of symptoms (Vorstman et al. 2006) to provide more accurate information to affected individuals and their families. Conversely, our understanding of the biological etiology of psychosis and autism in the general population may well benefit greatly from the basic science research involving people with a 22q11.2 deletion.

Evaluation

At diagnosis
- An appreciation of the individual's cognitive level and adaptive functioning (social, school, or occupational and activities of daily living) is useful regardless of the age at diagnosis.

Infancy (0–1 year)
- Evaluation for delays in achievement of developmental milestones should be done using standard screening tools at each medical visit.
- Cranial imaging to evaluate for structural brain anomalies is indicated when delays are pronounced.
- A history of seizure activity should be sought.

Early Childhood (1–5 years)
- Developmental milestones should be assessed using standard screening tools at routine medical visits.
- A history of anxiety symptoms and social interaction difficulties should be obtained during medical visits.
- A formal psychological or psychiatric assessment may be indicated by evidence of significant difficulty with anxiety or social interaction or frank psychotic symptoms.

School Age
- Formal cognitive assessment is needed if not previously obtained and should include evaluation for specific strengths and weaknesses and for the presence of a learning disorder (e.g., visual-perception integration deficits, mathematics disorder).
- A psychiatric evaluation should be performed if symptoms of a psychiatric disorder are present (e.g., ADHD, autistic spectrum disorder, obsessive-compulsive disorder, anticipatory anxiety, or repetitive or preservative behaviors or concerns) to facilitate appropriate and timely treatment.
- Monitoring for rapid or excessive weight gain is recommended.

Adolescence
- Emergence of psychotic symptoms (delusions, hallucinations, or disorganized behaviors or speech) may occur in up to 25% of individuals with the 22q11.2 deletion. These symptoms should be actively elicited as they may not be readily self-disclosed.
- Social withdrawal, without psychotic symptoms, may also be present and should be elicited.

Adulthood
- A repeat cognitive assessment should be considered in adulthood, as the cognitive profile (strengths and weaknesses) may have changed from childhood.
- Assessment for mood, anxiety, and psychotic disorders is important.
- Continued monitoring for the development of psychiatric disorders is indicated.

Management

Infancy (0–1 year)
- Early developmental delays may not predict the eventual global cognitive outcome but early intervention services, to include physical, occupational, and speech therapy, are often useful. In particular, the introduction of sign language to prevent negative behaviors arising from frustrations surrounding communication difficulties is beneficial.
- All children with 22q11.2 deletion benefit from speech and language therapy beginning at the time of diagnosis or by 6 months of age whichever comes first.

Early Childhood (1–5 years)
- Sign language is often useful as an adjunct to verbal communication where needed to avoid frustration with associated behavioral difficulties (Solot et al. 2001).
- Although most children enter regular preschool, some may require extra assistance. Because of delayed development, a prolonged preschool or kindergarten may benefit some children.
- In some individuals, transition to a special school system, resource room support, or pairing with a "one-on-one" aide needs to be anticipated to ensure appropriate placement.

School Age
- Close contact between parents and teachers and the school are important for proper management.
- Special assistance in classes is often necessary.
- Children with a 22q11.2 deletion benefit from speech therapy for articulation defects including those that are compensatory due to velopharyngeal dysfunction (Solot et al. 2001).
- Appropriate physical activities, as in the general population, should be encouraged when cleared medically.

Adolescence
- Post-high school education and vocational placement need to be considered.
- Affected adolescents should receive appropriate sex education to address issues of intimacy, sexuality, sexual behavior, and risk of transmission to offspring, as they are vulnerable because of their cognitive deficits.
- Legal guardianship may need to be considered as well as potential government disability entitlements before transition from childhood to adulthood as a result of impairments in cognitive competence (abilities related to personal care, financial, medical treatment).
- Psychiatric disorders require treatment according to standard clinical practice. Medical conditions that may affect psychiatric presentations need to be considered, such as hypocalcemia, thyroid disorders, and seizures.

Adulthood
- Early treatment of psychiatric disorders is essential to minimize future psychiatric morbidity and social isolation or exclusion. Standard medical treatments for psychiatric disorders appear to be equally effective in people with 22q11.2 deletion (Bassett et al. 2003).
- Counseling regarding risk of transmission to offspring is strongly recommended.
- Social safety nets may be warranted, perhaps in the form of a committed family member, social worker, or community case worker.

Craniofacial

Palatal anomalies are an important feature of the 22q11.2 deletion as approximately 69% of individuals with the deletion have a confirmed palatal abnormality (McDonald-McGinn et al. 1999). The most common of these is velopharyngeal incompetence (VPI). It is notable, however, that the reported incidence of palatal anomalies varies widely, ranging from 9 to 98% (Ryan et al. 1997; McDonald-McGinn et al. 1999) depending on numerous factors such as the reporting technique, the diligence with which the diagnosis is sought, the age at which the individual is evaluated, and the inherent ascertainment bias of any single center (Kirschner 2005). The figures presented here are those derived from the authors' own center's experience, in which there is a multidisciplinary team approach to evaluation and management. They are provided as a general guideline for the reader.

Velopharyngeal incompetence may be the result of a structural problem such as a short palate, a functional problem such as hypotonia of the velopharyngeal musculature, or a combination of both. Submucosal clefts and/or a bifid uvula are also fairly prevalent (16 and 5%, respectively), whereas overt cleft palate and cleft lip/cleft lip and palate are less common, accounting for 11 and 2% of palatal anomalies observed, respectively (McDonald-McGinn et al. 1999). Conversely, 22q11.2 deletion does not appear to be associated with isolated non-syndromic cleft palate (Mingarelli et al. 1996), and therefore screening studies in older individuals with overt cleft palate are likely unwarranted.

Additional craniofacial anomalies that may result in an affected individual initially presenting to the cleft palate/craniofacial team include microtia and craniosynostosis.

Several well-described craniofacial features of the 22q11.2 deletion syndrome that generally do not require intervention may be clues to the diagnosis, including a variety of auricular abnormalities in addition to microtia/anotia, such as thick, overfolded, and often "crumpled" helices; protuberant ears with attached lobes; hooded eyelids; hypertelorism; malar flatness; fullness superior to the nasal tip with underdeveloped alae nasi, a nasal crease or groove with or without a small strawberry hemangioma; a small mouth; asymmetric crying facies; and micrognathia.

Obstructive sleep apnea is common in people with deletion 22q11.2 (~10%) due to structural anomalies including retrognathia. Its identification and treatment in this population may impact more positively than in otherwise healthy individuals because those with 22q11.2 deletion are more likely to have baseline abnormalities in cardiovascular and neuropsychological function, which may be exacerbated by the obstructive sleep apnea syndrome (Kennedy et al. 2014).

Evaluation

- All individuals with a 22q11.2 deletion should be evaluated by a cleft palate team, if available, or an otorhinolaryngologist (or plastic surgeon with experience in velopharyngeal dysfunction) if no team is available.
- Evaluation for overt cleft palate is indicated at diagnosis and evidence of velopharyngeal incompetence should be sought when speech emerges. Overt and classic submucosal clefts are readily diagnosable on physical examination/intraoral inspection, but detection of "occult submucosal cleft palate" may require nasendoscopic examination as will confirmation of velopharyngeal incompetence. Alternatively, videofluoroscopy can be used to document nasopharyngeal reflux.
- All children should have a comprehensive speech and language evaluation by a speech language pathologist with experience in cleft palate/craniofacial anomalies, preferably one who is an active member of a cleft palate/craniofacial team.

Management

- Surgical management of palatal anomalies is aimed at attaining normal speech outcome as early as possible. However, there is little consensus regarding the timing or type of procedure for submucosal cleft palate or velopharyngeal incompetence. The most common surgical procedures include a posterior pharyngeal flap and sphincter pharyngoplasty.
- Complications associated with repair of velopharyngeal incompetence include hemorrhage, dehiscence, oronasal fistula formation, pneumonia, and cardiac arrest (Kirschner 2005). Postoperative complications can include nasal airway obstruction and hyponasal speech which generally resolves once edema subsides; snoring which may cause some to become obligate mouth breathers; and, rarely, obstructive sleep apnea, especially in those with tonsillar hyperplasia.
- When the tonsils are enlarged, tonsillectomy is generally recommended before velopharyngeal incompetence surgery. Signs of postoperative obstructive sleep apnea include loud snoring, nocturnal arousals, and excessive daytime somnolence. Such symptoms should prompt further evaluation and may require a surgical revision of the velopharyngeal incompetence repair (Kirschner 2005).
- Polysomnographic monitoring for obstructive sleep apnea should be considered before and after velopharyngeal surgery (Kennedy et al. 2014).
- Velopharyngeal incompetence has been observed in individuals with 22q11.2 deletion following adenoidectomy, so the need for this procedure should be weighed against the chance of developing or worsening symptoms of velopharyngeal incompetence.
- Anomalies of the carotid arteries, specifically medial displacement of the internal carotid arteries at the level of the posterior pharynx, have been described in as many as 25% of people with 22q11.2 deletion, placing them at risk for injury at the time of pharyngoplasty. Preoperative imaging studies such as magnetic resonance angiography or CT may be useful in defining the vascular anatomy, thereby allowing for precise surgical planning to minimize the risk of intraoperative vascular injury (Kirschner 2005).
- Referral for speech/language therapy should be made by age 6 months. Therapy often includes the use of sign language.

Respiratory

Respiratory symptoms may result from a number of associated factors including congenital heart disease, asthma, laryngotracheoesophageal anomalies such as vascular ring and laryngeal web, congenital diaphragmatic hernia, tracheoesophageal fistula, and esophageal atresia (Sacca et al. 2017). Aspiration as a result of gastrointestinal dysmotility should be considered a possible cause for respiratory compromise or recurrent pulmonary infections and reactive airways disease in any affected child.

Evaluation

- Assessment for respiratory abnormalities is standard, and includes history, physical examination, and respiratory function studies.
- Laryngoscopy should be considered when stridor is present, as should echocardiogram and/or chest MRI.

Management

- Treatment of respiratory abnormalities will depend on the cause and is standard. (Please see appropriate sections dealing with those causes.)

Cardiovascular

Congenital heart defects are an important feature of individuals with 22q11.2 deletion as they are present in 75% (see Table 21.1) of affected individuals and are the major cause of mortality (over 90% of all deaths) in this condition (Ryan et al. 1997; McDonald-McGinn et al. 2001a; Repetto et al. 2014). The most frequent anomalies are conotruncal defects of the outflow tract of the heart, which are shown in Table 21.2. Importantly, conotruncal defects in people with 22q11.2 deletion are frequently associated with additional

TABLE 21.2 Most clinically significant cardiovascular abnormalities associated with chromosome 22q11.2 deletion syndrome

Congenital heart defect[a]	%
Tetralogy of Fallot	20–45%
Pulmonary atresia + ventricular septal defect	10–25%
Interrupted aortic arch	5–20 %
Truncus arteriosus	5–10%
Ventricular septal defects (conoventricular)	10–`50%
Isolated aortic arch anomalies	10%
Atrial septal defect	3.5%
Other[b]	10%

[a] Prevalence of conotruncal anomalies in individuals with 22q11.2 deletion syndrome based on a review of publications with large numbers of subjects (Ryan, et al. 1997; Botto, et al. 2003; Park, et al. 2007; Matsuoka, et al. 1998; McDonald-McGinn, et al. 1999; Marino, et al. 2001; Oskarsdottir, et al. 2004).
[b] Transposition of the great vessels, bicuspid aortic valve, pulmonary valve stenosis, isolated right pulmonary artery atresia, hypoplastic left heart syndrome, and A-V canal/heterotaxy.

cardiovascular anomalies as a distinctive recognizable pattern (e.g. discontinuity, diffuse hypoplasia or crossing of the pulmonary arteries, right or cranial aortic arch with or without aberrant left subclavian artery) (Marino, et al. 2001; Momma 2010). This is very important for correct differential diagnosis between syndromic and non-syndromic conotruncal defects, as well as for cardiac surgery management.

Besides congenital cardiovascular malformations, a subset of individuals with 22q11.2 deletion develops aortic root dilation. This may be isolated or associated with minor cardiovascular anomalies or with conotruncal defects. In several of them the dilation progresses (John et al. 2009).

Evaluation

- All individuals with a 22q11.2 deletion should be evaluated for structural cardiac abnormalities, including arch sidedness, using echocardiography and electrocardiography at diagnosis.
- Those with cardiac defects should be referred to and followed by a cardiologist appropriate to their age. In individulas with normal cardiac anatomy, electrocardiography and echocardiography should be repeated in adolescence, at initial assessment during transition from pediatric care, and during the biennial follow-up in adults with 22q11.2 deletion, especially when other major conditions (e.g. obesity, hypothyroidism) associated with increased cardiovascular defect risk are present (Bassett, et al. 2011; Fung, et al. 2015).

Management

- Follow-up and treatment are defect-specific and should be provided by a cardiologist.
- Surgical approaches for cardiac defects are standard.
- The majority of individuals with 22q11.2 deletion and a congenital heart defect requiring surgical interventions do quite well with a very low operative risk and an excellent long-term prognosis (Goldmuntz et al. 1998; Carotti et al. 2008). This depends on the preoperative anatomy and any additional confounding factors such as intrauterine growth retardation, immunodeficiency, laryngotracheoesophageal abnormalities, or structural brain abnormalities (e.g., polymicrogyria).
- Because of possible immune function deficiency, affected individuals are often treated with irradiated blood products to avoid the risk of fatal graft-versus-host disease, as well as with prophylactic antibiotics/anti-fungals depending on their preoperative immune status (Marino et al. 2005; Carotti et al. 2008).

Neuromuscular Although the majority of individuals with 22q11.2 deletion have a history of hypotonia in infancy and learning disabilities (Moss et al. 1999), specific neurologic manifestations are uncommon. Seizures are present in some individuals and are most often associated with hypocalcemia. However, in one study, 7% of 383 individuals with 22q11.2 deletion had unprovoked seizures (Kao et al. 2004). Several individuals have asymmetric crying facies although the etiology of this is unclear (Cayler 1969; Giannotti et al. 1994). Rarely, ataxia and atrophy of the cerebellum are observed (Lynch et al. 1995). Additional reported central nervous system abnormalities include multicystic white matter lesions of unknown significance, perisylvian dysplasia (Bingham et al. 1997), hypoplasia of the pituitary gland (Weinzimer et al. 1998), and polymicrogyria (Robin et al. 2006). Recent investigations utilizing functional MRI scans revealed significantly reduced posterior brain volumes relative to age- and sex-matched controls, with more significant white matter loss in the left occipital and left parietal regions compared with the frontal lobes (Bearden et al. 2004; Bish et al. 2004). Many of these changes in brain structure have been postulated to relate to the specific cognitive deficits exhibited in the area of working memory, executive function, visuospatial skill, language, and math performance. Overall, the pattern of CNS abnormalities is broad and overlaps with that seen in some cases of Opitz G/BBB syndrome (Guion-Almeida and Richieri-Costa, 1992; MacDonald et al. 1993).

Adults with 22q11.2 deletion are at risk of early-onset Parkinson disease, with typical symptom pattern, treatment response, and course of the disease. The common use of antipsychotics in persons with 22q11.2 deletion may delay diagnosis of Parkinson disease (Butcher et al. 2013).

Evaluation

- The presence of seizures or other neurological symptoms/signs should prompt referral to a neurologist.

- Individuals with 22q11.2 deletion should be monitored for the development of parkinsonian symptoms.

Management

- Seizures are treated in a standard manner.
- Specific follow-up and treatment of neurological abnormalities are standard.

Immunologic

Seventy-seven percent of affected individuals have an immunodeficiency regardless of their clinical presentation. Of these, 67% have impaired T-cell production, 23% have humoral defects, and 6% have IgA deficiency (Sullivan 2004). Nonetheless, the majority are only mildly immunocompromised and generally do not develop opportunistic or life-threatening infections. However, viral infections are often prolonged. Abnormal palatal anatomy and gastroesophageal reflux frequently lead to an increased susceptibility to upper airway bacterial infections and aspiration pneumonia, especially in those with congenital heart defects (Sullivan 2004). This combination of impaired T-cells and palatal abnormalities is associated with a high frequency of otitis media and sinusitis. Thus, adults and children over the age of 9 years often continue to have infections, including 25–33% with recurrent sinusitis or otitis media, and 4–7% with recurrent lower airway infections. Of note, recurrent infections seem to have no correlation with immunological status, suggesting that these infections have more to do with anatomic and functional airway problems than defects in the host defense. Despite these issues, very few school-aged children require active management for their immunodeficiency (Sullivan 2004).

In the general population, normal T-cell counts decline rapidly in the first year of life and decline slowly during the next few years. T-cell counts in many individuals with 22q11.2 deletion rise slightly in the first year and then decline more slowly than in children without the deletion (Sullivan 2004). Therefore, the T-cell counts in these individuals approach normal values as they age (Junker and Driscoll 1995). As a result, affected individuals with slight decreases in T-cell numbers have normal defenses against pathogens (Sullivan 2005). A recent study suggested that there is a relationship of the immunodeficiency to the deletion breakpoints, with the "nested distal" non-TBX1-inclusive deletions associated with better T-cell counts (Crowley et al. 2018).

Immunoglobulin levels are usually normal, although subtle immunoglobulin abnormalities may be noted. Hypogammaglobulinemia present in the first year of life usually resolves, and hypergammaglobulinemia may occur after age 5 years (Sullivan, 2005). Although the majority have normal antibody function and antibody avidity (Junker and Driscoll 1995), some have functional antibody defects (Eicher et al. 2000), whereas those with recurrent sinopulmonary infections have frequent immunoglobulin abnormalities, in particular impaired antibody responses to pneumococcal polysaccharide vaccine (Gennery et al. 2002).

Autoimmune disease is common; however, it does not correlate with severe T-cell dysfunction and it includes a range of pediatric diseases. Autoimmune cytopenias and juvenile rheumatoid arthritis appear to be the most common and may occur 20–100 times more frequently than in the general population (Verloes et al. 1998; Duke et al. 2000; Davies et al. 2001; Gennery et al. 2002). Juvenile rheumatoid arthritis is often polyarticular and may be difficult to manage (Davies et al. 2001). Autoimmune thyroid disease and other autoimmune abnormalities have also been described, and it is likely that the T-cell defect acts synergistically with other predisposing factors such as major histocompatibility complexes to cause autoimmune disease (Kawame et al. 2001). Selective IgA deficiency may occur in up to 10% and seems to be particularly common in those individuals with autoimmune problems including juvenile rheumatoid arthritis (Davies et al. 2001).

Evaluation

- In the newborn period, a complete blood count with differential and flow cytometry should be obtained.
- In infancy, CD3$^+$, CD4$^+$, CD8$^+$, CD19$^+$, and CD3$^-$/16$^+$/56$^+$ subsets and lymphocyte proliferation responses to phytohemagglutinin, pokeweed mitogen, and concanavalin A should be evaluated.
- Over age 1 year, in addition to the above, immunoglobulins and phytohemagglutinin (PHA) should be studied.
- Lymphocyte subset analysis, proliferative studies with recall antigen responses and measures of antibody production at 1 year of age, should be carried out in preschool children with recurrent infections or at diagnosis in the preschool period.

Management

- Live viral vaccines should be withheld in a child with 22q11.2 deletion and impaired T-cell function, an inability to produce functional antibodies, or markedly diminished peripheral blood T-cell counts.
- In the presence of antibody responses to killed vaccines, normal proliferative responses to mitogens and recall antigens, and a CD8 T-cell count of greater than 250 cells mm^{-3} at 1 year of age, live viral vaccines have been administered without sequelae.
- For those who have not received varicella vaccine and who are exposed to varicella, varicella-zoster immune globulin or acyclovir prophylaxis is indicated.

- When the hypogammaglobulinemia is severe or is accompanied by a defect in antibody function, intravenous immune globulin (IVIG) is warranted.
- For individuals with selective IgA deficiency and recurrent infection, a trial of prophylactic antibiotics can be given (Sullivan 2005).

Endocrine

Hypocalcemia/hypoparathyroidism is found in approximately 50% of individuals with the 22q11.2 deletion; it may present at birth or later in life during times of stress such as perioperatively, during puberty, in pregnancy, and with illness (McDonald-McGinn et al. 1999; Kapadia et al. 2008). Symptoms may include tremor, tetany, tachypnea, arrhythmia, irritability, dysphagia, and seizures. Calcium homeostasis generally normalizes with age, although recurrence of hypocalcemia in later childhood or adulthood has been reported.

Hypothyroidism and growth-hormone deficiency leading to short stature is discussed in the Growth and Feeding section.

Evaluation

- Periodic monitoring of calcium levels, including ionized calcium studies, should be performed in all individuals with deletion 22q11.2 at diagnosis, at least every five years thereafter, with stress such as at surgery and acute illness, and annually after the age of 12 years.
- Parathyroid hormone should be measured if hypocalcemia is identified.
- In infancy (0–12 months), a search for tremor, tetany, tachypnea, arrhythmia, irritability, dysphagia, and seizures is warranted at each health visit.
- Ongoing monitoring of thyroid health is recommended.

Management

- Treatment of hypocalcemia generally includes calcium replacement therapy as well as calcitriol/vitamin D while being vigilant to avoid nephrocalcinosis. Standard doses are appropriate.

Genitourinary Renal abnormalities have been identified in 31% of people with deletion 22q11.2, including single kidney/multicystic/dysplastic kidney (10%), hydronephrosis (5%), vesicoureteral reflux/irregular bladder (6%), and dysfunctional voiding (11%) (Wu et al. 2002. Additional findings have included small kidneys, renal calculi (often secondary to treatment for hypocalcemia), bladder wall thickening, horseshoe kidney, duplicated collecting system, renal tubular acidosis, and enuresis. These findings warrant nephrourological investigation in everyone with a 22q11.2 deletion.

Hypospadias (8%), undescended testes (6%) (McDonald-McGinn et al. 1995; Wu et al. 2002), and primary amenorrhea resulting from an absent uterus have also been observed (Garcia-Minaur et al. 2002; Sundaram et al. 2007).

Evaluation

- Renal ultrasound should be obtained at diagnosis.
- Some will require a voiding cystourethrogram to further define abnormalities.
- Examination of the genitalia for cryptorchidism or hypospadias at diagnosis is important for identifying abnormalities.

Management

- Some affected individuals will require prophylactic antibiotics and/or surgical correction because of structural anomalies of the urogenital system.

Ophthalmology Eye findings reported in 22q11.2 deletion include posterior embryotoxon (prominent, anteriorly displaced Schwalbe's line at the corneal limbus or edge) in 49%, tortuous retinal vessels in 34%, hooded eyelids in 20%, strabismus in 18%, ptosis in 4%, amblyopia in 4%, and tilted optic nerves in 1% (Forbes et al. 2007). In addition, sclerocornea, a non-progressive, non-inflammatory condition in which one or both corneas demonstrate some degree of opacification coupled with flattening of the normal corneal curvature, has been identified in several individuals (Binenbaum et al. 2008). As familial cases of sclerocornea have been reported previously, including both autosomal dominant and autosomal recessive pedigrees (Bloch 1965), the association with 22q11.2 deletion may provide some insight into the etiology of these eye findings. Specifically, the deletion may unmask an autosomal recessive condition, as it has previously with the hematological disorder, Bernard–Soulier syndrome (Budarf et al. 1995), or alternatively, the deletion may, in association with other genetic loci previously associated with sclerocornea, play multifactorial and interrelated roles in neural crest migration and differentiation (Binenbaum et al. 2008). The incidence of astigmatism, myopia, and hyperopia is comparable with that seen in the general population, whereas a small number of people with 22q11.2 deletion have cataracts and colobomas (Forbes et al. 2007).

Evaluation

- A comprehensive ophthalmologic evaluation is recommended in all children by age 3 years and again before entering school.
- Follow-up ophthalmologic examinations are dependent on the findings (Forbes et al. 2007).

Management

- Treatment of ocular abnormalities does not differ from that in the general population.

Ears and Hearing

Otolaryngologic abnormalities are an important feature of 22q11.2 deletion (Dyce et al. 2002). Many affected individuals have chronic otitis media (88%) and chronic sinusitis because of their underlying immune deficiency and palatal differences, resulting in conductive hearing loss (37%) (McDonald-McGinn et al. 1999). Sensorineural hearing loss and mixed hearing loss are also found, although the incidence is much lower than conductive loss (McDonald-McGinn et al. 1999). A Mondini malformation of the cochlea may also be present and should be considered in anyone with 22q11.2 deletion and unexplained meningitis.

Evaluation

- Monitoring of hearing, middle ear function, and the airway is critical during the first year of life.
- Audiometric assessment including tympanometry should be conducted at the time of diagnosis and every 6–12 months until school age by use of audiogram.
- Referral to an otolaryngologist should be made if hearing issues are present.

Management

- Treatment of hearing loss does not differ from the general population and may include the need for hearing aids.
- Treatment of chronic/recurrent otitis media and sinusitis is standard, except in the presence of immune deficiency (see Immunologic).
- Adenoidectomy can cause velopharyngeal incompetence; therefore, when adenoidectomy is medically necessary, partial adenoidectomy should be considered where possible to minimize this risk.

Musculoskeletal

Musculoskeletal anomalies have been observed in people with 22q11.2 deletion, including pre- and postaxial polydactyly of the upper extremity in 6%. Lower extremity abnormalities include postaxial polydactyly, clubfoot, overfolded toes, and syndactyly of toes 2 and 3 in 15% (Ming et al. 1997). Vertebral anomalies are seen in 19%, including butterfly vertebrae, hemivertebrae, and coronal clefts (Figure 21.12). Rib anomalies, most commonly supernumerary or absent ribs, are seen in 19%, and hypoplastic scapulae are present in 1.5% (Ming et al. 1997). Significant cervical spine abnormalities have been observed in approximately 50% of affected individuals studied, including C2–C3 fusion in 34%, fusion of the posterior elements only in 21%, and complete block vertebrae of C2–C3 in 13% (Hamidi et al. 2014). In addition, 56% of persons with cervical spine anomalies have been found to have instability on flexion and extension radiographs and 33% have increased motion at more than one vertebral level (Hamidi et al. 2014; Homans et al. 2017). A small subset (4 of 79) had increased C2–C3 segmental motion with anterior and posterior narrowing of the spinal canal, and two of the four underwent surgical stabilization. One of these required an emergency procedure following symptoms of spinal cord compression (Hamidi et al. 2014; Homans et al. 2017).

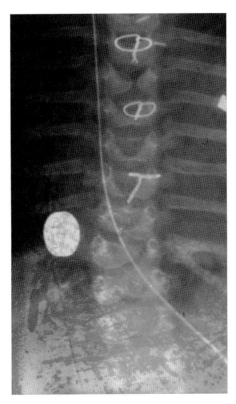

FIGURE 21.12 Butterfly vertebrae in a child with 22q11.2 deletion.

Evaluation

- Careful examination for evidence of skeletal anomalies is recommended at diagnosis.
- In infancy, chest X-ray is recommended to evaluate for thoracic and rib anomalies.
- Six-view cervical spine radiographs (AP, lateral, flexion, extension, open mouth, and skull base) are recommended in all affected individuals over the age of 4 years.
- The spine should be assessed clinically at routine medical visits through adolescence, with follow-up scoliosis X-ray series when examination indicates.

- Referral to an orthopedist is indicated for musculoskeletal findings potentially requiring intervention, such as scoliosis, cervical vertebral anomalies, clubfoot, and polydactyly.
- The cervical spine should be evaluated before surgery, when hyperextension of the neck may be required for intubation.

Management

- High-impact sports and tumbling activities should be avoided until the cervical spine has been cleared of abnormalities by radiographs.
- Treatment of musculoskeletal anomalies is standard.

RESOURCES

22q11.2 Society

Website: www.22qsociety.org

International 22q11.2 Foundation, Inc.

PO Box 532
Matawan, NJ 07747, USA
Phone: (877) 739-1849
Website: www.22q.org

USA and International

The "22q and You" Center
The Children's Hospital of Philadelphia
34th & Civic Center Blvd., Philadelphia, PA 19104, USA
Phone: (215) 590-2920
Email: genetics@email.chop.edu
Website: www.chop.edu

USA

Arizona

Phoenix, AZ – Phoenix Children's Hospital

Medical Director: Theresa Grebe, MD
1919 East Thomas Road, Phoenix, AZ 85016, USA
Phone: (602) 933-2252
Email: 22qcenter@phoenixchildrens.com
Website: http://www.phoenixchildrens.org/medical-specialties/genetics-and-metabolism/22q-clinic

California

Los Angeles, CA – Children's Hospital of Los Angeles

Director of Clinical Genetics: Pedro Sanchez, MD (pasanchez@chla.usc.edu)
8723 W Alden Drive SSB # 240 – Los Angeles, CA 90048, USA
Phone: (310) 614-3071
Website: http://www.chla.org/site/c.ipINKTOAJsG/b.4815711/k.9B11/Clinical_Genetics.htm#.VDbJcWea_zY

Sacramento, CA – 22q Healthy Minds Clinic at the UC Davis MIND Institute

Program Director: Tony Simon, PhD
2825 50th St Sacramento, CA 95817, USA
Phone: 916-703-0409
Email: hs-22q@ucdavis.edu
Website: http://www.ucdmc.ucdavis.edu/mindinstitute/research/cabil/healthymindsclinic.html

Colorado Aurora, CO – Chromosome 22q11.2 Deletion Syndrome Specialty Clinic at Children's Hospital of Colorado

Director: Naomi Meeks, MD
13123 East 16th Avenue, Aurora, CO 80045,USA
Phone: (303) 724-2370
Email: 22q@childrenscolorado.org
Website: https://www.childrenscolorado.org/conditions-and-advice/conditions-and-symptoms/conditions/chromosome-22q/

Georgia Atlanta, GA – Emory & Children's Healthcare of Atlanta, 22q Specialty Clinic

Medical Director: Dr. Lisa Kobrynski, Pediatric Immunologist (lkobryn@emory.edu)
Emory Children's Center, 2015 Uppergate Drive, Atlanta, GA 30322, USA
Phone: 404-785-2490
Website: https://www.choa.org/medical-services/genetics/22q

Illinois Chicago, IL, 22q11 Deletion Syndrome Program, Lurie Children's Hospital

Program Coordinators: Rachel Hickey, MS, CGC and Lauren Hitchens, APN
225 E. Chicago Avenue, Box 59, Chicago, IL 60611, USA
Phone: 312-227-6120
Website: https://www.luriechildrens.org/en/specialties-conditions/22q112-deletion-syndrome/

Oak Lawn, IL, 22q Diagnostic and Treatment Center at Advocate Children's Hospital

Co-Directors: Javeed Akhter, MD and Lydia Jazmines, MD
4440 West 95th St, Oak Lawn, IL 60453, USA
Phone: 708-684-2529
Website: https://www.advocatechildrenshospital.com/care-and-treatment/digeorge-diagnostic-and-treatment-center/

Massachusetts   Boston, MA – Massachusetts General Hospital New England Regional Center for 22q11
Clinical Director: Paula C. Goldenberg, MD, MSW, MSCE
Genetic Counselor: Punitha Kannan, MS, CGC
55 Fruit St, Boston, MA 02114
Phone: 617-726-1561
email: smassalski@partners.org
Website: http://www.massgeneral.org/children/services/treatmentprograms.aspx?id=1815

Missouri   Kansas City, MO – Children's Mercy
Mathew Feldt, DO and Susan Starling-Hughes, MS, CGC
2401 Gillham Road, Kansas City, MO 64108, USA
Phone: (816) 234-3290 (genetics), 816-960-8820 (endocrinology)
Email: superQexpress@cmh.edu
Website: https://www.childrensmercy.org/22q11.2-clinic/

New York   Bronx, NY - The 22q11.2 Center at CHAM in NYC
The Children's Hospital at Montefiore (CHAM) and the Albert Einstein College of Medicine in the Bronx, NY - Marion – Division of Genetics
Clinical Director: Robert Marion, MD
Email: marion@montefiore.org
Research Director: Bernice Morrow, PhD
Email: Bernice.morrow@einstien.yu.edu
Phone: call (718) 741-2323 to schedule an appointment or email Erica Kessler (coordinator) at Erica.Kessler@einstein.yu.edu

Syracuse, NY – 22q Center
Upstate University Hospital:
Medical Director: Robert Roger Lebel, MD, FACMG
Clinical Director: Eileen Marrinan, MS, MPH
Research Director: Wendy Kates, PhD
Syracuse University:
Clinical Director: Kimberly Lamparelli, PhD
Phone: 315-464-6580
Email: marrinae@upstate.edu
Website: http://www.upstate.edu/22q/

North Carolina   Durham, NC – 22q11 Deletion Syndrome Clinic at Duke Children's Hospital
Pediatric Medical Genetics Specialist: Vandana Shashi, MD
Genetic Counselor: Kelly Schoch, Ms, CGC (kelly.schoch@duke.edu)
300 Erwind Rd, Durham, NC 27705, USA
Phone: (919) 681-2524
Website:https://www.dukehealth.org/hospitals/duke-childrens-hospital/home#22q11_deletion

Ohio   Cincinnati, OH- 22q-VCFS Center at Cincinnati Children's Hospital Medical Center
Director of Clinical Genetics: Howard M. Saal, MD, FACMG
Co-Director: Robert Hopkin, MD
Genetic Counselor: Bettsy Leech, MS, CGC
3333 Burnet Ave MLC 4006, Cincinnati, OH 45229, USA
Phone: 513-803-1884
Websitehttp://www.cincinnatichildrens.org/service/t/22q-vcfs/default/
Email: 22Q-VCFS@cchmc.org

Columbus, OH – 22q Center at Nationwide Children's Hospital
Co-Director: Richard E. Kirschner, MD (Richard.Kirschner@NationwideChildrens.org)
Co-Director: Joan F. Atkin, MD, Pediatric Geneticist
Co-Director: Adriane Baylis, PhD, CCC-SLP
700 Children's Drive, Columbus, Ohio 43205, USA
Phone Number: 614-962-6373
Website:http://www.nationwidechildrens.org/22q-deletion-syndrome-center

Pennsylvania   Lewisburg, PA – 22q Developmental Clinic at Geisinger's Autism & Developmental Medicine Institute (ADMI)
Associate Director & Clinical Investigator: Brenda Finucane, MS
Contact: Barbara Haas Givler
Email: bahaasgivler@geisinger.edu
120 Hann Drive, 2nd Floor, Lewisburg, PA 17837, USA
Phone: 570-522-6287
Website: https://www.geisinger.org/patient-care/find-a-location/autism-development-medicine/22q-developmental-clinic

Philadelphia, PA – 22q and You Center at Children's Hospital of Philadelphia
Director: Donna McDonald-McGinn, MS, LCGC
Medical Director: Elaine H. Zackai, MD
Scientific Director: Beverly S. Emanuel, PhD
34th and Civic Center Blvd Philadelphia, PA 19104, USA
Phone: 215-590-2920
Email: genetics@email.chop.edu
Website: http://www.chop.edu/service/22q-and-you-center/home.html
Research Programs: Brain Behavior Consortium and Genetic Modifier Consortium

Washington

Seattle, WA – 22q Clinic at Seattle Children's Hospital
Director, Craniofacial Media Lab & Craniofacial Medicine: Carrie Heike, MD, MS (Carrie.Heike@SeattleChildrens.org)
4800 Sand Point Way NE Seattle, WA 98105, USA
Phone: (206) 987-2208
Website: https://www.seattlechildrens.org/clinics/craniofacial/services/22q-clinic/

Wisconsin **Milwaukee, WI – 22q11.2 Deletion Syndrome Specialty Clinic at Children's Hospital of Wisconsin**
Director: Steve Conley, MD (ENT)
Genetic Counselor: LuAnn Weik, MS, CGC
8915 W. Connell Ave, Milwaukee, WI 53226, USA
Phone: 414-266-3347
Website: http://www.chw.org/medical-care/genetics-and-genomics-program/programs-and-services/velocardiofacial-syndrome-vcfs-program/

International:

Australia **Ourimbah – The University of Newcastle**
Contacts: Linda Campbell, PhD (Linda.E.Campbell@newcastle.edu.au) and Dr Honey Heussler, MD (honey.heussler@gmail.com)
E1. 19 Science Offices
10 Chittaway Road, Ourimbah NSW 2258 Australia
Phone: +61 434 94490
Fax: +61 434 94404
Website: https://www.newcastle.edu.au/research-and-innovation/centre/configuration/ARCHIVED/disability-research-network/research/22q11.2-deletion-syndrome

Belgium **Leuven - Catholic University of Leuven**
Department of Human Genetics O & N Herestraat 49 - box 602
3000 Leuven, Belgium
Contact: Anne Swillen, PhD
Phone: +32 16 3 41483
Email: ann.swillen@med.kuleuven.be
Contact: Koen Devriendt, MD
Phone: +32 163 45903
Email: koen.devriendt@med.kuleuven.be
Website: https://gbiomed.kuleuven.be/english/research/50000622/laboratories/laboratory-for-behaviour-and-neurodevelopment

Canada **Toronto – Dalglish Family 22q Clinic**
Director: Anne Bassett, MD, FRCPC
Toronto General Hospital, 8 Norman Urquhart, Room 802
200 Elizabeth Street, Toronto, Ontario, M5G 2C4, Canada
Phone: +1 (416)-340-5145
Fax: (416)-340-5004
Email: 22q@uhn.ca
Website: http://www.22q.ca/

Chile **Santiago – Pontifical Catholic University of Chile**
Av Libertador Bernardo O Higgins 340, Santiago, Santiago Metropolitan Region, Chile
Phone: +56 2 2354 2000
Contact: Gabriela Repetto, MDEmail: grepetto@udd.cl

France: **Marseille - Timone Hospital**
Timone Hospital, 264 rue Saint-Pierre 13385 MARSEILLE CEDEX 5 FRANCE
Phone: +33 (0)4 91 38 66 36
Contact: Nicole Philip, MD
Email: Nicole.PHILIP@mail.ap-hm.fr
Website: http://fr.ap-hm.fr/service/departement-de-genetique-medicale-hopital-timone

Spain **Madrid - University Hospital La Paz**
University Hospital La Paz, Pº de la Castellana, 261 28046 – Madrid, Spain
Phone: +34 91 727 7000
Contact: Sixto Garcia-Minaur, MD
Email: sixto.garciamin@gmail.com
Website: http://www.madrid.org/cs/Satellite?language=es&pagename=HospitalLaPaz/Page/HPAZ_home

Mallorca – University Hospital Son Espases
Ctra. de Valldemossa 79, Palma City, 07010, Spain
Phone: + 34 871 205 000
Contacts: Damian Heine Suner, PhD; Jaume Morey Canyelles
Email: damian.heine@ssib.es
Website: –http://www.hospitalsonespases.es/index.php?option=com_k2&view=item&id=763:gen%C3%B3mica-de-la-salud&Itemid=804&lang=es

Ireland **Dublin – Beaumont Hospital**
Contact: Professor Kieran Murphy, MD - Psychiatrist
Beaumont Hospital, Beaumont Road, Dublin, Ireland
Phone: +353 1 809 3740
Fax: +353 1 837 6982
Email: KMurphy@rcsi.ie
Website: http://www.beaumont.ie

Israel **Ramat Aviv – Tel Aviv University**
Tel Aviv University POB 39040, Israel
Phone: +972 (0)3 530 2663
Contact: Doron Gothelf, MD - Psychiatrist
Email: gothelf@post.tau.ac.il

Italy **Rome – Pediatric Hospital Bambino Gesù**
Piazza S. Onofrio, 4-00165 Rome, Italy
Phone: +39-06-68592227

Contact: Maria Cristina Digilio, MD
Email: ten.gbpo@oiligid
Website: http://www.ospedalebambinogesu.it/home
Rome – University La Sapienza
Viale Regina Elena, 324 Rome, 00161 Italy
Contact: Bruno Marino, MD – Cardiologist
Email: bruno.marino@uniroma1.it

Japan   Tokyo – Ochanomizu University
2-1-1 Ohtsuka, bunkyo-ku Tokyo 112-8610 Japan
Phone: +81-3-5978-5587
Contact: Hiroshi Kawame, MD - Pediatric Clinical Geneticist
Email: kawame.hiroshi@ocha.ac.jp

Netherlands   **Maastricht – Maastricht University**
Minderbroedersberg 4-6, 6211 LK Maastricht, Netherlands
Phone: +31-43-3883584
Contact: Therese van Amelsvoort, MD - Psychiatrist
Email: t.vanamelsvoort@maastrichtuniversity.nl
Utrecht – University Medical Centre
UMC Utrecht Brain Center Rudolf Magnus Universiteitsweg 100 3584 CG Utrecht, Netherlands
Phone: +31 88 75 8180
Contact: Jacob Vorstman, MD, PhD – Child Psychiatrist
Email: j.a.s.vorstman@umcutrecht.nl
Website: http://research.umcutrechthersencentrum.nl

Poland   **Warsaw – Instytut Matki i Dziecka**
Instytut Matki i Dziecka, Department of Medical Genetics, Kasprzaka 17A, Warsaw, Poland
Phone: +48 22 3277140
Contact: Beata Nowakowska, PhD
Email: Beata.Nowakowska@med.kuleuven.be

Switzerland   **Geneva – University of Geneva**
University Medical Center Dept. of Neuroscience, room 7008 1 rue Michel-Servet, CH1211 Geneva 4, Switzerland
Phone: 0041 22 379 5381
Fax: +41 22 379 5402
Contact: Stephan Eliez, MD
Email: Stephan.eliez@unige.ch
Website: https://www.ige3.unige.ch/eliez.php

Sweden   **Gothenburg – Queen Silvia Children's Hospital**
Department of Pediatric Immunology, The Queen Silvia Children's Hospital
Sahlgrenska University Hospitaal Rondgatan 10 SE 41685 Goteborg Sweden
Phone: +46 31 343 5220
Contact: Solveig Oskarsdottir, MD
Email: janette.nystromy@vgregion.se

United Kingdom   **England**
London – Maudsley Hospital & Great Ormond Street Hospital
Bethlem Royal Hospital, Monks Orchard Road, Beckenham, London, BR3 3BX, UK
Phone: +44 2032 284653
Contact: Dr. Clodagh Murphy, Psychiatrist
Email: clodagh.m.murphy@kcl.ac.uk
Wales
Cardiff – Cardiff University
Cardiff University School of Medicine, Hadyn Ellis Building, Maindy Road, Cathays, Cardiff, CF24 4HQ, UK
Phone: +44 (2920 688320, ext. 88320
Contact: Professor Sir Michael Owen - Neurologist
Email: OwenMJ@cf.ac.uk
Contact: Marianne van den Bree, PhD - Psychiatrist
Email: VANDENBREEMB@cardiff.ac.uk
Website: https://www.cardiff.ac.uk/people/view/57797-owen-michael

REFERENCES

Baker K, Baldeweg T, Sivagnanasundaram S, Scambler P, Skuse D (2005) COMT Val108/158 Met modifies mismatch negativity and cognitive function in 22q11 deletion syndrome. *Biol Psychiatry* 58:23–31.

Baldini A (2005) Dissecting contiguous gene defects: TBX1. *Curr Opin Genet Dev* 15:279–284.

Barry JC, Crowley TB, Jyonouchi S, Heimall J, Zackai EH, Sullivan KE, McDonald-McGinn DM (2017) Identification of 22q11.2 Deletion Syndrome via Newborn Screening for Severe Combined Immunodeficiency. *J Clin Immunol* 37(5):476–485.

Bassett AS, Chow EWC, AbdeMalik P, Gheorghiu M, Husted J, Weksberg R (2003) The schizophrenia phenotype in 22q11 deletion syndrome. *Am J Psychiatry* 160:1580–1586.

Bassett AS, Chow EW, Husted J, Weksberg R, Caluseriu O, Webb GD, Gatzoulis MA (2005) Clinical features of 78 adults with 22q11 deletion syndrome. *Am J Med Genet* 138:307–313.

Bassett AS, Chow EW, Husted J, Hodgkinson KA, Oechslin E, Harris L, Silversides C. 2009. Premature death in adults with 22q11.2 deletion syndrome. *J Med Genet* 46:324–330.

Bassett AS, McDonald-McGinn DM, Devriendt K, Digilio MC, Goldenberg P, Habel A, Marino B, Oskarsdottir S, Philip N, Sullivan K, Swillen A, Vorstman J (2011) Practical guidelines for managing patients with 22q11.2 deletion syndrome. *J Pediatr* 159:332–339 e1.

Bearden CE, Woodin MF, Wang PP, Moss E, McDonald-McGinn D, Zackai E, Emanuel B, Cannon TD (2001) The neurocognitive phenotype of the 22q11.2 deletion syndrome: Selective deficit in visual-spatial memory. *J Clin Exp Neuropsychol* 23:447–464.

Bearden CE, van Erp TG, Monterosso JR, Simon TJ, Glahn DC, Saleh PA, Hill NM, McDonald-McGinn DM, Zackai E, Emanuel BS, Cannon TD (2004) Regional brain abnormalities in 22q11.2 deletion syndrome: Association with cognitive abilities and behavioral symptoms. *Neurocase* 10:198–206.

Binenbaum G, McDonald-McGinn DM, Zackai EH, Walker BM, Coleman K, Mach AM, Adam M, Manning M, Alcorn DM, Zabel C, Anderson DR, Forbes BJ (2008) Sclerocornea associated with the chromosome 22q11.2 deletion syndrome. *Am J Med Genet A* 146:904–909.

Bingham P, Zimmerman RA, McDonald-McGinn DM, Driscoll DA, Emanuel BS, Zackai EH (1997) Enlarged Sylvian fissures in infants with interstitial deletion of chromosome 22q11. *Am J Med Genet Neuropsychol Genet* 74:538–543.

Bish JP, Nguyen V, Ding L, Ferrante S, Simon TJ (2004) Thalamic reductions in children with chromosome 22q11.2 deletion syndrome. *Neuroreport* 15:1413–5141.

Bloch N (1965) The different types of sclerocornea, their hereditary modes and concomitant congenital malformations. *J Hum Genet* 14:133–172.

Botto LD, May K, Fernhoff PM, Correa A (2003) A population-based study of the 22q11.2 deletion: Phenotype, incidence, and contribution to major birth defects in the population. *Pediatrics* 112:101–107.

Budarf ML, Konkle BA, Ludlow LB, Michaud D, Li M, Yamashiro DJ, McDonald-McGinn D, Zackai EH, Driscoll DA (1995) Identification of a patient with Bernard-Soulier syndrome and a deletion in the DiGeorge/Velocardio-facial chromosomal region in 22q11. *Hum Mol Genet* 4:763–766.

Burn J, Takao A, Wilson D, Cross I, Momma K, Wadey R, Scambler P, Goodship J (1993) Conotruncal anomaly face syndrome is associated with a deletion within chromosome 22. *J Med Genet* 30:822–824.

Butcher NJ, Kiehl TR, Hazrati LN, Chow EW, Rogaeva E, Lang AE, Bassett AS (2013) Association between early-onset Parkinson disease and 22q11.2 deletion syndrome: identification of a novel genetic form of Parkinson disease and its clinical implications. *JAMA Neurol* 70(11):1359–66.

Carotti A, Digilio MC, Piacentini G, Saffirio C, Di Donato RM, Marino B. 2008. Cardiac defects and results of cardiac surgery in 22q11.2 deletion syndrome. *Dev Disabil Res Rev* 14:35–42.

Cayler GG (1969) Cardiofacial syndrome. Congenital heart disease and facial weakness, a hitherto unrecognized association. *Arch Dis Child* 44:69–75.

Cormier-Daire V, Iserin L, Theophile D, Sidi D, Vervel C, Padovani JP, Vekemans M, Munnich A, Lyonnet S (1995) Upper limb malformations in DiGeorge syndrome. *Am J Med Genet* 56:39–41.

Crowley B, Ruffner M, McDonald McGinn DM, Sullivan KE. Variable immune deficiency related to deletion size in chromosome 22q11.2 deletion syndrome (2018) *Am J Med Genet A* 176:2081–2086.

Davies K, Stiehm ER, Woo P, Murray KJ (2001) Juvenile idiopathic polyarticular arthritis and IgA deficiency in the 22q11 deletion syndrome. *J Rheumatol* 28:2326–2334.

De la Chapelle A, Herva R, Koivisto M, Aula P (1981) A deletion in chromosome 22 can cause DiGeorge syndrome. *Hum Genet* 57:253–256.

Desmaze C, Scambler P, Prieur M, Halford S, Sidi D, LeDeist F, Aurias A (1993) Routine diagnosis of DiGeorge by fluorescence *in situ* hybridization. *Hum Genet* 90:663–665.

De Smedt B, Swillen A, Ghesquiere P, Devriendt K, Fryns JP (2003) Pre-academic and early academic achievement in children with velocardiofacial syndrome (del22q11.2) of borderline or normal intelligence. *Genet Couns* 14:15–29.

Devriendt K, de Smet L, de Boeck K, Fryns JP (1997) DiGeorge syndrome and unilateral symbrachydactyly. *Genet Couns* 8:345–347.

Devriendt K, Fryns JP, Mortier G, van Thienen MN (1998) The annual incidence of DiGeorge/velocardiofacial syndrome. *J Med Genet* 35:789–790.

Digilio MC, Giannotti A, Marino B, Guadagni AM, Orzalesi M, Dallapiccola B (1997) Radial aplasia and chromosome 22q11 deletion. *J Med Genet* 34:942–944.

Digilio, M. C. et al (2001). Auxological evaluation in patients with DiGeorge/velocardiofacial syndrome (deletion 22q11.2 syndrome). *Genet Med* 3:30–33.

Digilio MC, McDonald-McGinn DM, Heike C, Catania C, Dallapiccola B, Marino B, Zackai EH (2009) Three patients with oculo-auriculo-vertebral spectrum and microdeletion 22q11.2. *Am J Med Genet A* 0(12):2860–4.

Driscoll DA, Budarf ML, Emanuel BS (1992) A genetic etiology for DiGeorge syndrome: Consistent deletions and microdeletions of 22q11. *Am J Hum Genet* 50:924–933.

Driscoll DA, Salvin J, Sellinger B, McDonald-McGinn D, Zackai EH, Emanuel BS (1993) Prevalence of 22q11 microdeletions in DGS and VCFS: Implications for genetic counseling and prenatal diagnosis. *J Med Genet* 30:813–817.

Duijff SN, Klaassen PW, de Veye HF, Beemer FA, Sinnema G, Vorstman JA (2012) Cognitive development in children with 22q11.2 deletion syndrome. *Br J Psychiatry* 200(6):462–8.

Duke SG, McGuirt WF Jr, Jewett T, Fasano MB (2000) Velocardiofacial syndrome: Incidence of immune cytopenias. *Arch Otolaryngol Head Neck Surg* 126:1141–1145.

Dyce O, McDonald-McGinn D, Kirschner RE, Zackai E, Young K, Jacobs IN (2002) Otolaryngologic manifestations of the 22q11.2 deletion syndrome. *Arch Otolaryngol Head Neck Surg* 128:1408–1412.

Edelman L, Pandita RK, Morrow BE (1999) Low-copy repeats mediate the common 3-Mb deletion in patients with velocardio-facial syndrome. *Am J Hum Genet* 64:1076–1086.

Eicher PS, McDonald-McGinn DM, Fox CA, Driscoll DA, Emanuel BS, Zackai EH (2000) Dysphagia in children with a 22q11.2 deletion: Unusual pattern found on modified barium swallow. *J Pediatr* 137:158–164.

Fine S, Weissman A, Gerdes M, Pinto-Martin J, Zackai E, McDonald-McGinn D, Emanuel B (2005) Autism spectrum disorders and symptoms in children with molecularly confirmed 22q11.2. *J Autism Dev Disabil* 35:461–470.

Forbes BJ, Binenbaum G, Edmond JC, DeLarato N, McDonald-McGinn DM, Zachai EH (2007) Ocular findings in the chromosome 22q11.2 deletion syndrome. *J AAPOS* 11:179–182.

Fraser FC, (1980) The genetics of common familial disorders– major genes or multifactorial? *Canad J Genet Cytol* 23:1–8.

Fung WL, Butcher NJ, Costain G, Andrade DM, Boot E, Chow EW, Chung B, Cytrynbaum C, Faghfoury H, Fishman L, García-Miñaúr S, George S, Lang AE, Repetto G, Shugar A, Silversides C, Swillen A, van Amelsvoort T, McDonald-McGinn DM, Bassett AS (2015) Practical guidelines for managing adults with 22q11.2 deletion syndrome. *Genet Med* 17:599–609.

Garcia-Minaur S, Fantes J, Murray RS, Porteous MEM, Strain L, Burns JE, Stephen J, Warner JP (2002) A novel atypical 22q11.2 distal deletion in father and son. *J Med Genet* 39:1–5.

Gennery AR, Barge D, O'Sullivan JJ, Flood TJ, Abinun M, Cant AJ (2002) Antibody deficiency and autoimmunity in 22q11.2 deletion syndrome. *Arch Dis Child* 86:422–425.

Gerdes M, Solot C, Wang PP, Jawad A, DaCosta AM, LaRossa D, Randall P, Goldmuntz B, Clark BJ III, Driscoll DA, Emanuel BS, McDonald-McGinn DM, Batshaw ML, Zackai EH (1999) Cognitive and behavioral profile of preschool children with chromosome 22q11.2 microdeletion. *Am J Med Genet* 85:127–133.

Giannotti A, Diglio MC, Marino B, Mingarelli R, Dallapiccola B (1994) Cayler cardiofacial syndrome and del 22q11: Part of the CATCH22 phenotype. *Am J Med Genet* 30:807–812.

Goldmuntz E, Clark BJ, Mitchell LE, Jawad AF, Cuneo BF, Reed L, McDonald-McGinn D, Chien P, Feuer J, Zackai EH, Emanuel BS, Driscoll DA (1998) Frequency of 22q11 deletions in patients with conotruncal defects. *J Am Coll Cardiol* 32:492–498.

Goodship J, Cross I, LiLing J, Wren C (1998) A population study of chromosome 22q11 deletions in infancy. *Arch Dis Child* 79:348–351.

Grati FR, Molina Gomes D, Ferreira JCPB, Dupont C, Alesi V, Gouas L, Horelli-Kuitunen N, Choy KW, García-Herrero S, de la Vega AG, Piotrowski K, Genesio R, Queipo G, Malvestiti B, Hervé B, Benzacken B, Novelli A, Vago P, Piippo K, Leung TY, Maggi F, Quibel T, Tabet AC, Simoni G, and Vialard F (2015) Prevalence of recurrent pathogenic microdeletions and microduplications in over 9500 pregnancies. *Prenat Diagn* 35:801–809.

Guion-Almeida ML, Richieri-Costa A (1992) CNS midline anomalies in the Opitz G/BBB syndrome: Report on 12 Brazilian patients. *Am J Med Genet* 43:918–928.

Habel A, McGinn MJ 2nd, Zackai EH, Unanue N, McDonald-McGinn DM. (2012) Syndrome-specific growth charts for 22q11.2 deletion syndrome in *Caucasian children Am J Med Genet A* 158A(11):2665–71.

Hamidi M, Nabi S, Husein M, Mohamed ME, Tay KY, McKillop S (2014) Cervical spine abnormalities in 22q11.2 deletion syndrome. *Cleft Palate Craniofac J* 51(2):230–3

Hatchwell E, Long F, Wilde J, Crolla J, Temple K (1998) Molecular confirmation of germ line mosaicism for a submicroscopic deletion of chromosome 22q11. *Am J Med Genet* 78:103–106.

Homans JF, Tromp IN, Colo D, Schlösser TPC, Kruyt MC, Deeney VFX, Crowley TB, McDonald-McGinn DM, Castelein RM (2017) Orthopaedic manifestations within the 22q11.2 Deletion syndrome: A systematic review. *Am J Med Genet A.* 176:2104–2120.

Homans JF, Baldew VGM, Brink RC, Kruyt MC, Houben ML, Deeney VFX, Crowley TB, Castelein RM, McDonald-McGinn DM (2018) Scoliosis within the 22q11.2 deletion syndrome: an observational study. *Archives of Disease in Childhood* 104:19–24.

John AS, McDonald-McGinn DM, Zackai EH, Goldmuntz E (2009) Aortic root dilation in patients with 22q11.2 deletion syndrome. *Am J Med Genet A* 149:939–942.

Junker AK, Driscoll DA (1995) Humoral immunity in DiGeorge syndrome. *J Pediatr* 127:231–237.

Kao A, Mariani J, McDonald-McGinn DM, Maisenbacher MK, Brooks-Kayal AR, Zackai EH, Lynch DR (2004) Increased prevalence of unprovoked seizures in patients with a 22q11.2 deletion. *Am J Med Genet A* 129:29–34.

Kapadia CR, Kim YE, McDonald-McGinn DM, Zackai EH, Katz LE (2008) Parathyroid hormone reserve in 22q11.2 deletion syndrome. *Genet Med* 10:224–228.

Kariyazono H, Ohno T, Ihara K, Igarashi H, Joh-o K, Ishikawa S, Hara T (2001) Rapid detection of the 22q11.2 deletion with quantitative real-time PCR. *Mol Cell Probes* 15:71–73.

Kasprzak L, Der Kaloustian VM, Elliott AM, Shevell M, Lejtenyi C, Eydoux P (1998) Deletion of 22q11 in two brothers with different phenotype. *Am J Med Genet* 75:288–291.

Kawame H, Adachi M, Tachibana K, Kurosawa K, Ito F, Gleason MM, Weinzimer S, Levitt-Katz L, Sullivan K, McDonald-McGinn DM (2001) Graves' disease in patients with 22q11.2 deletion. *J Pediatr* 139:892–895.

Kennedy WP, Mudd PA, Maguire MA, Souders MC, McDonald-McGinn DM, Marcus CL, Zackai EH, Solot CB, Mason TB, Jackson OA, Elden LM (2014) 22q11.2 Deletion syndrome and obstructive sleep apnea. *Int J Pediatr Otorhinolaryngol* 78(8):1360–4.

Kirschner RE, (2005) Palatal anomalies and velopharyngeal dysfunction associated with velo-cardio-facial syndrome. *Velo-Cardio-Facial Syndrome: A Model for Understanding Microdeletion Disorders*, Cambridge, UK: Cambridge University Press, pp.83–105.

Kruszka P, Addissie YA, McGinn DE, Porras AR, Biggs E, Share M, Crowley TB, Chung BH, Mok GT, Mak CC, Muthukumarasamy P, Thong MK, Sirisena ND, Dissanayake VH, Paththinige CS, Prabodha LB, Mishra R, Shotelersuk V, Ekure EN, Sokunbi OJ, Kalu N, Ferreira CR, Duncan JM, Patil SJ, Jones KL, Kaplan JD, Abdul-Rahman OA, Uwineza A, Mutesa L, Moresco A, Obregon MG, Richieri-Costa A, Gil-da-Silva-Lopes VL, Adeyemo AA, Summar M, Zackai EH, McDonald-McGinn DM, Linguraru MG, Muenke M (2017) 22q11.2 deletion syndrome in diverse populations. *Am J Med Genet A* 173(4):879–888.

Kujat A, Schulz MD, Strenge S, Froster UG (2006) Renal malformations in deletion 22q11.2 patients. *Am J Med Genet A* 140(14):1601–2.

LaCassie Y, Arriaza MI (1996) Letter to the Editor: Opitz GBBB syndrome and the 22q11.2 deletion syndrome. *Am J Med Genet* 62:318.

Lambert MP, Arulselvan A, Schott A, Markham SJ, Crowley TB, Zackai EH, McDonald-McGinn DM. (2018) The 22q11.2 deletion syndrome: Cancer predisposition, platelet abnormalities and cytopenias. *Am J Med Genet A* 176(10):2121–2127.

Lopez-Rivera E, Liu YP, Verbitsky M, et al. (2017) Genetic Drivers of Kidney Defects in the DiGeorge Syndrome. *N Engl J Med* 376:742–54.

Lynch DR, McDonald-McGinn D, Zackai EH, Emanuel BS, Driscoll DA, Whitaker LA, Fischbeck KA (1995) Cerebellar atrophy in a patient with velocardiofacial syndrome. *J Med Genet* 32:561–563.

Mantripragada KK, Tapia-Páez I, Blennow E, Nilsson P, Wedell A, Dumanski JP (2004) DNA copy-number analysis of the 22q11 deletion-syndrome region using array-CGH with genomic and PCR-based targets. *Int J Mol Med* 13:273–279.

Marino B, Digilio MC, Toscano A, Anaclerio S, Giannotti A, Feltri C, de Ioris MA, Angioni A, Dallapiccola B (2001) Anatomic patterns of conotruncal defects associated with deletion 22q11. *Genet Med* 3:45–48.

Marino B, Mileto F, Digilio MC, Carotti A, DiDonato R, (2005) Congenital cardiovascular disease and velocardiofacial syndrome. *Velo-Cardio-Facial Syndrome: A Model for Understanding Microdeletion Disorders*, Cambridge, UK: Cambridge University Press, pp. 47–82.

Matsouka R, Takao A, Kimura M, Imamura S-I, Kondo C, Joh-o K, Ikeda K, Nishibatake M, Ando M, Momma K (1994) Confirmation that the conotruncal anomaly face syndrome is associated with a deletion within 22q11.2. *Am J Med Genet* 53:285–289.

MacDonald MR, Schaefer GB, Olney AH, Tamayo M, Frías JL (1993) Brain magnetic resonance imaging findings in the Opitz G/BBB syndrome: Extension of the spectrum of midline brain anomalies. *Am J Med Genet* 46:706–711.

McDonald-McGinn DM, Driscoll DA, Bason L, Christensen K, Lynch D, Sullivan K, Canning D, Zavod W, Quinn N, Rome J, Paris Y, Weinberg P, Clark BJ, Emanuel BS, Zackai EH (1995) Autosomal dominant "Opitz" GBBB syndrome due to a 22q11.2 deletion. *Am J Med Genet* 59:103–113.

McDonald-McGinn DM, Kirschner R, Goldmuntz E, Sullivan K, Eicher P, Gerdes M, Moss E, Solot C, Wang P, Jacobs I, Handler S, Knightly C, Heher K, Wilson M, Ming JE, Grace K, Driscoll D, Pasquariello P, Randall P, LaRossa D, Emanuel BS, Zackai EH (1999) The Philadelphia Story: The 22q11.2 Deletion: Report on 250 Patients. *Genet Couns* 10:11–24.

McDonald-McGinn DM, Tonnesen MK, Laufer-Cahana A, Finucane B, Driscoll DA, Emanuel BS, Zackai EH (2001a) Phenotype of the 22q11.2 deletion in individuals identified through an affected relative: Cast a wide *FISH*ing net! *Genet Med* 3:23–29.

McDonald-McGinn DM, Driscoll DA, Tonnesen M, Sullivan K, Kirschner R, Goldmuntz E, Weinzimer S, Lynch D, Wang P, Moss E, Gerdes M, Solot C, Catania C, Saitta S, Emanuel BS, Zackai EH (2001) Parent of origin does not determine phenotype in the 22q11.2 deletion. *Am J Hum Genet* 69:285(A597).

McDonald-McGinn DM, Driscoll DA, Saitta S, Jawad A, Tonnesen M, Ming JE, Goldmuntz E, Canning D, Spinner N, Emanuel BS, Zackai EH (2002) Guidelines for prenatal detection of the 22q11.2 deletion. *Am J Hum Genet* 71:198(A173).

McDonald-McGinn DM, Minugh-Purvis N, Kirschner R, Jawad A, Tonnesen MK, Catanzaro JR, Driscoll D, LaRossa D, Emanuel B, Zackai EH (2005) The 22q11.2 deletion in African-American patients? An underdiagnosed population. *Am J Med Genet* 134:242–246.

McDonald-McGinn DM, Gripp KW, Kirschner RE, Maisenbacher MK, Hustead V, Schauer GM, Keppler-Noreuil KM, Ciprero KL, Pasquariello P Jr, LaRossa D, Bartlett SP, Whitaker LA, Zackai EH (2005a) Craniosynostosis: Another feature of the 22q11.2 deletion syndrome. *Am J Med Genet* 136:358–362.

McDonald-McGinn DM and Zackai EH (2008) Genetic counseling for the 22q11.2 deletion. *Develop Dis Res Rev* 14:69–74.

McDonald-McGinn DM, Sullivan KE, Marino B, Philip N, Swillen A, Vorstman JA, Zackai EH, Emanuel BS, Vermeesch JR, Morrow BE, Scambler PJ, Bassett AS (2015) 22q11.2 deletion syndrome. *Nat Rev Dis Primers* 1:15071.

McQuade L, Christodoulou J, Budarf M, Sachdev R, Wilson M, Emanuel B, Colley A (1999) Patient with a 22q11.2 deletion with no overlap of the minimal DiGeorge syndrome critical region (MDGCR). *Am J Med Genet* 86:27–33.

Ming JE, McDonald-McGinn DM, Megerian TE, Driscoll DA, Elics ER, Russell BM, Irons M, Emanuel BS, Markowitz RI, Zackai EH (1997) Skeletal anomalies in patients with deletions of 22q11. *Am J Med Genet* 72:210–215.

Mingarelli R, Digilio MC, Mari A (1996) The search for hemizygosity at 22q11 in patients with isolated cleft palate. *J Craniofac Genet Dev Biol* 16:118–121.

Momma K 2010. Cardiovascular anomalies associated with chromosome 22q11.2 deletion syndrome. *Am J Cardiol* 105:1617–1624.

Moss E, Wang PP, McDonald-McGinn DM, et al (1995) Characteristic cognitive profile in patients with a 22q11.2 deletion—Verbal IQ exceeds nonverbal IQ (abstract). Am J Hum Genet 57:SS91.

Moss EM, Batshaw ML, Solot CB, Gerdes M, McDonald-McGinn DM, Driscoll DA, Emanuel BS, Zackai EH, Wang PP (1999) Psychoeducational profile of the 22q11.2 microdeletion: A complex pattern. *J Pediatr* 134:193–198.

Nickel RE, Magenis RE (1996) Neural tube defects and deletions of 22q11. *Am J Med Genet* 66:25–27.

Oskarsdóttir S, Vujic M, Fasth A (2004) Incidence and prevalence of the 22q11 deletion syndrome: A population-based study in Western Sweden. *Arch Dis Child* 89:148–151.

Pasick C, McDonald-McGinn DM, Simbolon C, Low D, Zackai E, Jackson O (2013) Asymmetric crying facies in the 22q11.2 deletion syndrome implications for future screening. *Clin Pediatr* 52(12):1144–1148.

Popolos DF, Faedda GL, Veit S, Goldberg R, Morrow B, Kucherlapati R, Shprintzen RJ (1996) Bipolar spectrum disorders in patients diagnosed with velo-cardio-facial syndrome: Does a hemizygous deletion of chromosome 22q result in bipolar affective disorder? *Am J Psychiatry* 153:1541–1547.

Pretto D, Maar D, Yrigollen CM, Regan J, Tassone F (2015) Screening newborn blood spots for 22q11.2 deletion syndrome using multiplex droplet digital PCR. *Clin Chem* 61(1):182–90.

Racedo SE, McDonald-McGinn DM, Chung JH, et al. (2015) Mouse and human CRKL is dosage sensitive for cardiac outflow tract formation. *Am J Hum Genet* 96:235–44

Rauch A, Hoyer J, Guth S, Zweier C, Kraus C, Becker C, Zenker M, Hüffmeier U, Thiel C, Rüschendorf F, Nürnberg P, Reis A, Trautmann U (2006) Diagnostic yield of various genetic approaches in patients with unexplained developmental delay or mental retardation. *Am J Med Genet A* 140:2063–2074.

Repetto GM, Guzman ML, Delgado I, Loyola H, Palomares M, Lay-Son G, Vial C, Benavides F, Espinoza K, Alvarez P (2014) Case fatality rate and associated factors in patients with 22q11 microdeletion syndrome: a retrospective cohort study. *BMJ Open* 4(11):e005041.

Robin NH, Taylor CG, McDonald-McGinn DM, Zackai EH, Bingham P, Collins KJ, Earl D, Gill D, Granata T, Guerrini R, Katz N, Kimonis V, Lin JP, Lynch DR, Mohammed SN, Massey RF, McDonald M, Rogers RC, Splitt M, Stevens CA, Tischkowitz MD, Stoodley N, Leventeer RJ, Pilz DT, Dobyns WB (2006) Polymicrogyria and deletion 22q11.2 syndrome: Window to the etiology of a common cortical malformation. *Am J Med Genet* 140:2416–2425.

Ryan AK, Goodship JA, Wilson DI, Philip N, Levy A, Seidel H, Schuffenhauer S, Oechsler H, Belohradsky B, Prieur M, Aurias A, Raymond FL, Clayton-Smith J, Hatchwell E, McKeown C, Beemer FA, Dallapiccola B, Novelli G, Hurst JA, Ignatius J, Green AJ, Winter RM, Brueton L, Brondum-Nielsen K, Scambler (1997) Spectrum of clinical features associated with interstitial chromosome 22q11 deletions: A European collaborative study. *J Med Genet* 34:798–804.

Sacca R, Zur KB, Crowley TB, Zackai EH, Valverde KD, McDonald-McGinn DM (2017) Association of airway abnormalities with 22q11.2 deletion syndrome. *Int J Pediatr Otorhinolaryngol* 96:11–14.

Saitta SC, Harris SE, McDonald-McGinn DM, Emanuel BS, Tonnesen MK, Zackai EH, Seitz SC, Driscoll DA (2004) Independent *de novo* 22q11.2 deletions in first cousins with DiGeorge/velocardiofacial syndrome. *Am J Med Genet* 124:313–317.

Sandrin-Garcia P, Macedo C, Martelli LR (2002) Recurrent 22q11.2 deletion in a sibling suggestive of parental germline mosaicism in velocardiofacial syndrome. *Clin Genet* 61:380–383.

Scambler PJ, Carey AH, Wyse RK, Roach S, Dumanski JP, Nordenskjold M, Williamson R (1991) Microdeletions within 22q11 associated with sporadic and familial DiGeorge syndrome. *Genomics* 10:201–206.

Schneider M, Debbané M, Bassett AS, Chow EW, Fung WL, van den Bree M, Owen M, Murphy KC, Niarchou M, Kates WR, Antshel KM, Fremont W, McDonald-McGinn DM, Gur RE, Zackai EH, Vorstman J, Duijff SN, Klaassen PW, Swillen A, Gothelf D, Green T, Weizman A, Van Amelsvoort T, Evers L, Boot E, Shashi V, Hooper SR, Bearden CE, Jalbrzikowski M, Armando M, Vicari S, Murphy DG, Ousley O, Campbell LE, Simon TJ, Eliez S; International Consortium on Brain and Behavior in 22q11.2 Deletion Syndrome. (2014) Psychiatric disorders from childhood to adulthood in 22q11.2 deletion syndrome: results from the International Consortium on Brain and Behavior in 22q11.2 Deletion Syndrome. *Am J Psychiatry* 171(6):627–39.

Shaikh TH, Kurahashi H, Saitta SC, Hu P, Rose BA, Driscoll DA, McDonald-McGinn DM, Zackai EH, Budarf ML, Emanuel BS (2000) Chromosome 22-specific low copy repeats and the 22q11.2 deletion syndrome: Genomic organization and deletion endpoint analysis. *Hum Mol Genet* 9:489–501.

Solot CB, Gerdes M, Kirschner RE, McDonald-McGinn DM, Moss E, Woodin M, Aleman D, Zackai EH, Wang PP (2001) Communication issues in 22q11.2 deletion syndrome: Children at risk. *Genet Med* 3:67–71.

Sullivan KE (2004) The clinical, immunological, and molecular spectrum of chromosome 22q11.2 deletion syndrome and DiGeorge syndrome. *Curr Opin Allergy Clin Immunol* 4:505–512.

Sullivan KE (2005) Immunodeficiency in velo-cardio-facial syndrome. *Velo-Cardio-Facial Syndrome: A Model for Understanding Microdeletion Disorders*, Cambridge, UK: Cambridge University Press, pp. 123–134.

Sundaram UT, McDonald-McGinn DM, Huff D, Emanuel BS, Zackai EH, Driscoll DA, Bodurtha J (2007) Primary amenorrhea and absent uterus in the 22q11.2 deletion syndrome. *Am J Med Genet A* 143:2016–2018.

Swillen A, Devriendt K, Legius E (1999) The behavioral phenotype in velocardiofacial syndrome (VCFS): From infancy to adolescence. *Genet Couns* 10:79–88.

Swillen A, McDonald-McGinn D. (2015) Developmental trajectories in 22q11.2 deletion. *Am J Med Genet C* 169(2):172–81.

Tonnesen M, McDonald-McGinn DM, Valverde K, Zackai EH (2001) Affected parents with a 22q11.2 deletion: The need for basic and ongoing educational health, and supportive counseling. Am J Hum Genet Suppl 69:223(A241).

Unolt M, DiCairano L, Schlechtweg K, Barry J, Howell L, Kasperski S, Nance M, Adzick NS, Zackai EH, McDonald-McGinn DM (2017) Congenital diaphragmatic hernia in 22q11.2 deletion syndrome. *Am J Med Genet A* 173:135–142.

Verloes A, Curry C, Jamar M, Herens C, O'Lague P, Marks J, Sarda P, Blanchet P (1998) Juvenile rheumatoid arthritis and del (22q11) syndrome: A non-random association. *J Med Genet* 35:943–947.

Vogels A, Schevenels S, Cayenberghs R, Weyts E, Van Buggenhout G, Swillen A, Van Esch H, de Ravel T, Corveleyn P, Devriendt K (2014) Presenting symptoms in adults with the 22q11 deletion syndrome. *Eur J Med Genet* 57:157–162.

Voll SL, Boot E, Butcher NJ, Cooper S, Heung T, Chow EW, Silversides CK, Bassett AS (2017) Obesity in adults with 22q11.2 deletion syndrome. *Genet Med* 19(2):204–208

Vorstman JA, Morcus ME, Duijff SN, Klaassen PW, Heineman-de Boer JA, Beemer FA, Swaab H, Kahn RS, van Engeland H (2006) The 22q11.2 deletion in children: High rate of autistic disorders and early onset of psychotic symptoms. *J Am Acad Child Adolesc Psychiatry* 45:1104–1113.

Vorstman JA, Breetvelt EJ, Duijff SN, Eliez S, Schneider M, Jalbrzikowski M, Armando M, Vicari S, Shashi V, Hooper SR, Chow EW, Fung WL, Butcher NJ, Young DA, McDonald-McGinn DM, Vogels A, van Amelsvoort T, Gothelf D, Weinberger R, Weizman A, Klaassen PW, Koops S, Kates WR, Antshel KM, Simon TJ, Ousley OY, Swillen A, Gur RE, Bearden CE, Kahn RS, Bassett AS (2015) International Consortium on Brain and Behavior in 22q11.2 Deletion Syndrome. (2015) Cognitive decline preceding the onset of psychosis in patients with 22q11.2 deletion syndrome. *JAMA Psychiatry* 72(4):377–85.

Weinzimer SA, McDonald-McGinn DM, Driscoll DA, Emanuel BS, Zackai EH (1998) Growth hormone deficiency in patients with a 22q11.2 deletion: Expanding the phenotype. *Pediatrics* 101:929–932.

Weinzimer SA (2001) Endocrine aspects of the 22q11.2 deletion syndrome. *Genet Med* 3:19–22.

Wilson DI, Cross IE, Wren C (1994) Minimum prevalence of chromosome 22q11 deletions. *Am J Hum Genet Suppl* 55:A169.

Wu H-Y, Rusnack SL, Bellah RD, Plachter N, McDonald-McGinn DM, Zackai EH, Canning DA (2002) Genitourinary malformations in chromosome 22q11.2 deletion. *J Urol* 168:2564–2565.

Yagi H, Furutani Y, Hamad H, Sasaki T, Asakawa S, Minoshima S, Ichida F, Joo K, Kimura M, Imamura S (2003) Role of TBX1 in human 22q11.2 syndrome. *Lancet* 362:1366–1373.

Yamagishi H, Ishii C, Maeda J, Kojima Y, Matsuoka R, Kimura M, Takao A, Momma K, Matsuo N (1998) Phenotypic discordance in monozygotic twins with 22q11.2 deletion. *Am J Med Genet* 78:319–321.

Zweier C, Sticht H, Aydin-Jaylagul I, Campbell CE, Rauch A (2007) Human TBX1 missense mutations cause gain of function resulting in the same phenotype as 22q11.2 deletions. *Am J Hum Genet* 80:510–517.

22

DELETION 22q13 SYNDROME: PHELAN–MCDERMID SYNDROME

KATY PHELAN
Florida Cancer Specialists and Research Institute, Fort Myers, Florida, USA

R. CURTIS ROGERS AND LUIGI BOCCUTO
Greenwood Genetic Center – Greenville, Greenville, South Carolina, USA; Clemson University, Clemson, Pickens, South Carolina, USA

INTRODUCTION

Phelan–McDermid syndrome (deletion 22q13 syndrome, PMS) is characterized by neonatal hypotonia, normal growth, absent to severely delayed speech, global developmental delay, autism spectrum disorder or autistic-like behaviors, and minor dysmorphic features. The syndrome typically results from loss of the terminal portion of the long arm of chromosome 22, either by simple deletion or secondary to an unbalanced structural rearrangement. Disruption of 22q13 within the gene *SHANK3* and frameshift variants in *SHANK3* have also been shown to lead to features of PMS.

Incidence

Over 2100 individuals with deletion 22q13 syndrome are registered with the Phelan–McDermid Syndrome Foundation (Venice, Florida, 2017), yet this does not represent the total number of individuals diagnosed with the condition. The deletion remains significantly underdiagnosed, although with the increased utility of chromosomal microarray in the first tier of testing for individuals with intellectual impairment and/or autism spectrum disorders, the number of cases diagnosed with PMS has increased. Likewise, the availability of gene sequencing has led to an increase in the cases found to have pathogenic variants affecting *SHANK3*.

Kolevzon et al. (2014a) report that haploinsufficiency of *SHANK3* causes a monogenic form of autism spectrum disorder (ASD) in a minimum of 0.5% of individuals with ASD and causes moderate to profound intellectual disability in about 2% of affected children.

Diagnostic Criteria

The diagnosis of PMS is established by the unequivocal demonstration of deletion or disruption of 22q13, or a pathogenic variant of *SHANK3*, in a clinically significant proportion of cells. Chromosomal microarray (CMA) is currently the most effective method of detecting deletion of 22q13. Fluorescence in situ hybridization (FISH) can be used to confirm the presence of a deletion or unbalanced translocation and, depending on the size of the involved regions, chromosome studies can confirm the presence of an unbalanced translocation or ring chromosome. Due to the risk of neurofibromatosis 2 (NF2) in individuals with ring 22 (see Etiology), it is imperative that a microarray that detects an apparent simple deletion of 22q13 be followed by a chromosome analysis to look for a ring 22. Variants within *SHANK3* can be demonstrated by DNA sequencing.

Clinical diagnostic criteria for PMS have not been established. The prominent manifestations of this syndrome (neonatal hypotonia, absent or delayed speech, and

---

Cassidy and Allanson's Management of Genetic Syndromes, Fourth Edition.
Edited by John C. Carey, Agatino Battaglia, David Viskochil, and Suzanne B. Cassidy.
© 2021 John Wiley & Sons, Inc. Published 2021 by John Wiley & Sons, Inc.

developmental delay) are non-specific and may vary in severity as the individual ages (Phelan et al. 1992; Nesslinger et al. 1994). Deletion of 22q13 should be suspected in newborns with hypotonia of unexplained etiology. As the child gets older, muscle tone may improve and may not be a helpful clue to this diagnosis. Only 11% of affected children have short stature (less than third percentile) and 11% have microcephaly (Rollins et al. 2011). The absence of growth deficiency in the presence of ASD/ID may lead one to suspect this diagnosis.

Infants may babble at an appropriate age and possess a limited vocabulary until 3 or 4 years. At this age, many children seem to lose the ability to speak, although through aggressive therapy and communication training some may regain and increase their vocabulary. Nonetheless, speech will remain impaired throughout life. Reierson et al. (2017) observed an average age of 39 months for onset of language loss in PMS, notably later than the onset in ASD (about 2 years) and Rett syndrome (1–3 years). The majority of individuals have severe global developmental delay, although a small number of individuals with subtelomeric deletions are reported to have mild delays (Wong et al. 1995).

The craniofacial features of deletion 22q13 syndrome are relatively subtle (Figure 22.1A). The head tends to be long or dolichocephalic. Frequently observed facial features include thick eyebrows, puffy eyelids, and long thick eyelashes. There may be midface retrusion with a wide nasal bridge and puffy cheeks. The chin is often pointed and may become prominent with age. Ears are typically large and may be prominent or poorly formed. Less frequently observed features are epicanthal folds, high arched palate, long philtrum, and prominent alae nasi.

Facial appearance changes with age – the jaw becomes prominent and the bony brow becomes thicker (Figure 22.1B). Longitudinal studies are needed to monitor the progression in facial features.

A summary of the manifestations seen in deletion 22q13 syndrome and their frequency is provided in Table 22.1.

Etiology, Pathogenesis, and Genetics

PMS typically results from the loss of a segment from the distal long arm of chromosome 22. The size of the deletion is variable, ranging from fewer than 100 kilobases to over 9 Mb (Dhar et al. 2010; Luciani et al. 2003; Wilson et al. 2003). Males and females are equally likely to be affected. The parental origin of the deleted chromosome does not appear to exert an effect on the phenotype (Phelan et al. 2001). Most individuals with this syndrome (80%) have simple deletions of 22q13; there is a single break in the long arm resulting in loss of the genetic material distal to the break. As in other terminal deletion syndromes, there is an excess of paternally derived deletions of 22q13. Deletion of 22q13 occurs on the paternal chromosome 22 in 69–74% of cases (Luciani et al. 2003; Wilson et al. 2003). A few individuals have been reported to have interstitial deletions of 22 (Romain et al. 1990). Molecular studies demonstrate that these presumed interstitial deletions are often, in fact, terminal deletions. The risk of recurrence of a de novo deletion of 22q13 is not significantly increased over the general population risk. However, there have been reports of presumed parental gonadal mosaicism leading to the clinical features of PMS in two offspring of phenotypically normal parents (Moessner et al. 2007).

PMS also results from unbalanced translocations in which the distal long arm of 22 is lost and replaced by a segment from a second chromosome. About 80% of unbalanced translocations involving chromosome 22 are inherited, resulting from the malsegregation of a parental chromosome translocation. Multiplex families in which two or more children have inherited the derivative chromosome 22 from a translocation carrier parent have been reported (Bonaglia et al. 2011; Rodriguez et al. 2003; Su et al. 2011). Recombination of parental inversions and insertions involving a breakpoint in 22q13 has also led to affected offspring (Watt et al. 1985; Slavotinek et al. 1997; Babineau et al. 2006; Tagaya et al. 2008). Parent-to-child transmission has been reported in a case in which both mother and son had a direct insertion (22;7)(q13.3;q21.2q22.1) resulting in a submicroscopic deletion of 22q13.3. The son was also trisomic for a segment of 7q (Slavotinek et al. 1997).

Only about 20% of structural rearrangements leading to deletion 22q13 result from de novo chromosome rearrangements. One child with a de novo translocation between chromosomes 15 and 22 had features of both Prader–Willi syndrome and deletion 22q13 syndrome (Smith et al. 2000). The risk for future offspring with the same de novo structural aberration is negligible.

Ring chromosome 22 with the long arm breakpoint in q13 also produces this phenotype. First mentioned in 1968, over 100 cases of ring 22 have been described in the literature (Ishmael et al. 2003). Ring chromosomes typically result from two breaks within the same chromosome, one in the long arm and one in the short arm. The broken ends subsequently join to form a circle or ring-like structure. Ring chromosomes are unstable during cell division such that the ring chromosome frequently has a "mosaic" distribution. The ring chromosome may be lost in some cells and one or more copies may be seen in other cells. The proportion of cells containing the ring may vary from tissue to tissue. When the ring involves chromosome 22, the break in the short arm may result in loss of the short arms, stalks, and/or satellites of this chromosome. No phenotypic effect is expected to result from the loss of these regions in an acrocentric chromosome. If the long arm breakpoint occurs at band q13.3, a phenotype consistent with deletion 22q13 syndrome is produced. It is difficult to clearly establish breakpoints within ring chromosomes by cytogenetic methods. Thus, FISH, microarray-CGH, and/or molecular studies are required to define the exact breakpoint.

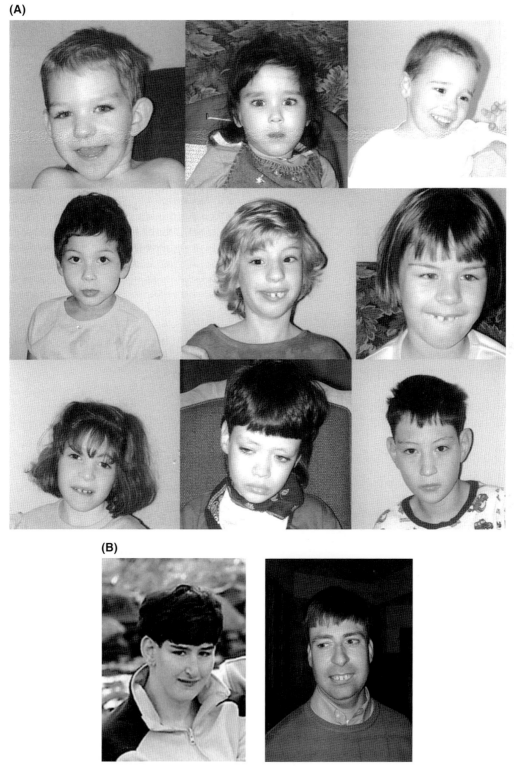

FIGURE 22.1 (A) Facial features of the deletion 22q13 syndrome. Top row (left to right): 2-year-old male, 2-year-10-month-old female, and 3-year-3-month-old male. Middle row: 5-year-5-month-old female, 6-year-10-month-old female, and 7-year-6-month-old female. Bottom row: 7-year-11-month-old female, 8-year-5-month-old male, and 10-year-1-month-old male. The first girl in the bottom row has der(22) t(19;22); the remaining individuals have simple deletions. (B) Facial features of a 32-year-old female (left) and 42-year-old male (right) with deletion of 22q13.

TABLE 22.1 Common manifestations of deletion 22q13 syndrome

>75%
Neonatal hypotonia
Global developmental delay
Absent or severely delayed speech
Normal growth
Autism spectrum disorder/autistic-like behavior
Decreased perception of pain
Mouthing/chewing behavior

>50%
Large, fleshy hands
Long eyelashes
Hyperextensibility
Poorly formed/large ears
Thick eyebrows
Full/puffy cheeks
Periorbital fullness
Deep-set eyes
Midface retrusion
Wide nasal bridge
Bulbous nose
Dysplastic toenails
Decreased perspiration/ overheating

>25%
Seizures
Dolichocephaly
Ptosis
Strabismus
Renal problems
Gastroesophageal reflux
Epicanthal folds
High-arched palate
Malocclusion/wide-spaced teeth
2/3 toe syndactyly
Lymphedema

Individuals with PMS resulting from ring 22 have an additional concern that does not affect other individuals with deletion of 22q13. Ring 22 has been associated with an increased risk of neurofibromatosis type 2 (NF2) (Tommerup et al. 1992). NF2 is characterized by vestibular schwannomas with accompanying tinnitus, hearing loss, and balance dysfunction; schwannomas of other cranial nerves; meningiomas; and other low grade brain malignancies (Tommerup et al. 1992; Tsilchorozidou et al. 2004). The gene for NF2 is a tumor suppressor mapped to 22q12.2; the loss of both alleles is necessary for tumorigenesis. Zirn et al. (2010) proposed a two-hit model for the development of meningioma in patients with ring 22. The first hit is the loss of the ring 22 during mitosis, as ring chromosomes are known to be unstable during cell division, making the individual hemizygous for the *NF2* allele. The second hit is a somatic variant of the remaining *NF2* allele. Other tumor suppressor genes, in addition to *NF2*, may reside on chromosome 22. Consequently, monosomy 22 may result in atypical cases of NF2 with multiple meningiomas (Arinami et al. 1986; Petrella et al. 1993) rather than the usual bilateral vestibular schwannomas. The major neurological manifestations of deletion 22q13 syndrome have historically been attributed to haploinsufficiency of the *SHANK3* gene (Wilson et al. 2003; Bonaglia et al. 2001). Bonaglia et al. (2001) reported a disruption of the *SHANK3* gene in a child with a de novo t(12;22)(q24.1;q13.3). The child had manifestations of the deletion 22q13 syndrome including hypotonia, developmental delay, speech delay, and minor dysmorphic features. *SHANK3* maps to 22q13.3, the region deleted in this syndrome, and codes for a scaffolding protein within the postsynaptic density of excitatory neurons. The postsynaptic density is a complex, multiprotein structure at the membrane of the postsynaptic neurons. It contains glutamate receptors, signaling proteins and scaffolding proteins, such as SHANK3. One of the functions of SHANK3 is to connect ion channels and receptors in the postsynaptic membrane to the cytoskeleton and to signal transduction pathways. The location and the function of the *SHANK3* gene make it a highly plausible candidate to explain the neurological manifestations of PMS.

Interstitial deletions of 22q13 suggest that genes other than *SHANK3* contribute to the cognitive and language delays in PMS. Two children with interstitial deletions of 22q13 proximal to, and not overlapping with, *SHANK3* had intellectual disability and severe language delay (Wilson et al. 2008). The children also presented with hypotonia, which, along with cognitive impairment and speech delay, is one of the most consistent findings in PMS. Interestingly, one of the children inherited the deletion from his mother, who had mild speech delay (Wilson et al. 2003). Nine additional interstitial deletions of 22q13, ranging from 2.7 to 6.9 Mb in size and proximal to *SHANK3*, were described in individuals with developmental and speech delay, hypotonia and feeding difficulties (Disciglio et al. 2014). These results suggest that haploinsufficiency for *SHANK3* is not the sole cause of the cognitive and speech delay associated with deletion of 22q13. Genes proximal to *SHANK3* also impact these neurological functions.

Genetic variants of *SHANK3* include point mutations, deletions, and/or translocations with breakpoints disrupting the gene sequence, and have been associated with a variety of phenotypes (Guilmatre et al. 2014). Such phenotypes include, other than PMS, isolated ASD (Durand et al. 2007; Moessner et al. 2007; Gauthier et al. 2009; Waga et al. 2011; Boccuto et al. 2013), and schizophrenia with moderate to severe intellectual disability and epilepsy in >25% of cases (Gauthier et al. 2010). SHANK3 levels were found to be decreased in brains of both patients and mouse models of Alzheimer disease (Gong et al. 2009; Pham et al. 2010). On the other hand, duplications of this gene have also been proved to be potentially deleterious, since they have observed in patients with high functioning ASD, attention deficit and hyperactivity disorder (ADHD), or schizophrenia (Guilmatre et al. 2014).

In this context of phenotypical variability, it is noteworthy that no correlation has been established so far with the SHANK3 domains carrying the variants and the associated clinical presentation. Therefore, the sole detection of a SHANK3 genetic variant, in the absence of a proper clinical and neurobehavioral evaluation of the individual, is not sufficient to establish a diagnosis of PMS.

Diagnostic Testing

The clinical presentation of the individual usually prompts a genetic work-up. A newborn with hypotonia or an older individual with developmental delay, ASD, and/or dysmorphic features, are candidates for cytogenomic SNP microarray (CMA). Deletion of 22q13 concomitant with a terminal duplication on a second chromosome might suggest an unbalanced translocation warranting follow-up with FISH or karyotyping. Parental FISH studies should be performed to rule out the presence of a balanced translocation or inversion of chromosome 22. Between 15 and 20% of cases of deletion 22q13 result from the malsegregation of a parental chromosome rearrangement. Detection of a parental rearrangement significantly increases the risk of recurrence in future pregnancies to between 10 and 25%. The birth of a previous affected child has established that chromosomally unbalanced, viable offspring are possible. The magnitude of the recurrence risk is influenced by the degree of imbalance attributable to the second chromosome involved in the rearrangement.

Following the detection of a simple deletion by CMA, chromosome analysis is required in all cases to determine if the deletion is due to a ring chromosome. Although this sounds contrary to conventional wisdom in terms of following an abnormal microarray with a screening karyotype, it is critical given the risk of NF2 to individuals with ring 22. It is essential that the referring physician or healthcare provider understand that the risk of NF2 is not dependent upon the NF2 locus being deleted. The NF2 locus is at 22q12.2 and is typically present in ring chromosomes. Because the ring is unstable it may be lost during mitosis leaving some cells monosomic for chromosome 22. If a pathogenic variant occurs in the single remaining copy of the NF2 allele in those monosomic cells, the individual may develop manifestations of NF2.

When there is high clinical suspicion of PMS but normal results by CMA, other techniques are available to pursue the diagnosis. MPLA (multiplex ligation-dependent probe amplification) can detect cryptic deletions of beyond the resolution of CMA, and DNA sequencing can be used to identify pathogenic variants of SHANK3, other genes associated with autism, or other potential causative genes localized to 22q13.3 (Phelan et al. 2016).

Deletion 22q13 has been diagnosed successfully by prenatal chromosome analysis of amniotic fluid and chorionic villi with confirmation by FISH. Prenatal diagnosis has detected both mosaic and non-mosaic deletions of 22q13 (Riegal et al. 2000; Phelan et al. 2001; Phelan 2001). Although very small deletions can be detected by CMA used for prenatal diagnosis, low-level mosaicism may be undetected. SNP-based non-invasive prenatal screening (NIPS) has been used to detect other microdeletion syndromes (Wapner et al. 2015), making this technology seem feasible to detect deletion of 22q13. An important caveat is that a positive NIPS should always be followed by confirmatory diagnostic testing and clinical evaluation.

Differential Diagnosis

PMS should be considered in all individuals with neonatal hypotonia of unknown etiology. Deletion of 22q13 and mutations or disruptions of SHANK3 should be excluded in individuals with ASD or autistic-like behaviors with intellectual impairment and absent or delayed speech. Sequencing studies of cohorts with ASD/ID have identified point mutations of SHANK3 in 71 unrelated individuals (De Rubeis et al. 2018).

With the increasing availability of next generation sequencing tools, a higher number of SHANK3 variants is expected to be detected, since this gene has been associated with several neurobehavioral disorders, such as ASD, schizophrenia, and ADHD (see Etiology, Pathogenesis, and Genetics). Unfortunately, due to lack of genotype–phenotype correlation, the sole detection of a SHANK3 variant is not sufficient to orient the diagnosis and should suggest a thorough clinical and neurobehavioral evaluation of the patient.

Individuals with cerebral palsy who have features consistent with PMS should also be tested. Cerebral palsy (CP) is a non-specific diagnosis that is applied to individuals with a variety of muscular and neurological phenotypes. When individuals have muscle weakness, incoordination, and intellectual impairment, the diagnosis of PMS should be considered.

Other diagnoses carried by individuals subsequently shown to have deletion 22q13 syndrome include atypical Angelman syndrome (see Chapter 5), Williams syndrome (see Chapter 63), trichorhinophalangeal syndrome, Smith–Magenis syndrome (see Chapter 54), velo-cardio-facial syndrome (see Chapter 21), and spastic paraplegia (Phelan et al. 2001). Hypoplastic toenails and long eyelashes are common features of PMS that may facilitate the diagnosis in toddlers with speech delay and/or an ASD diagnosis. Loss of speech at 3–4 years of age also distinguish this diagnosis from autism, where loss of speech typically occurs earlier in life (Reierson et al. 2017).

MANAGEMENT AND MANIFESTATIONS

The information in this section has been obtained from the published literature and, in significant part, through personal evaluations, interviews, and medical records of individuals participating in the Phelan–McDermid Syndrome Support Group and Foundation.

Growth and Feeding

Birth weight, length, and head circumference are typically appropriate for gestational age. The mean gestational age is 38.2 weeks. In contrast to the growth deficiency observed in more common microdeletion syndromes, most individuals with simple deletions of 22q13 have normal postnatal growth. In their study of 55 individuals ranging in age from 10 months to 40 years, Rollins et al. (2011) observed normal stature in 78% and tall stature (>95%) in 11% of individuals. With regard to head size, 68% of individuals were normocephalic, 21% were macrocephalic, and the remaining were microcephalic.

Feeding problems may occur during infancy related to neonatal hypotonia, but are not severe and usually consist of difficulty in sucking and/or swallowing. Gastroesophageal reflux is seen in just over 40% of individuals and cyclic vomiting is reported in about 25% (Phelan et al. 2001; Sarasua et al. 2014; Soorya et al. 2013). These conditions will be discussed further in the Gastrointestinal section.

Evaluation

- In infancy and childhood, height, weight and head circumference should be monitored at routine visits and plotted on age-appropriate growth charts.
- Feeding problems should be evaluated by an occupational therapist, speech pathologist, or other feeding specialist to determine if feeding therapy is warranted. Assessment for gastroesophageal reflux and cyclic vomiting should be considered in affected individuals with persistent feeding problems and failure to thrive.
- Neurosurgical consultation may be appropriate to review issues related to cyclic vomiting (see Neurologic).

Management

- For persistent vomiting, intravenous fluids may be required to prevent dehydration.
- Typically, gastroesophageal reflux is controlled by standard medications. No specific medication is recommended as various medications are successful in different individuals. In persistent cases, fundoplication with or without gastrostomy tube may be required.
- If arachnoid cyst or other factors are identified as causing cyclic vomiting, neurosurgical intervention should be considered.

Development and Behavior

While global developmental delay is characteristic of PMS, the severity of the delay tends to vary with deletion size (Sarasua et al. 2011; Zwanenburg et al. 2016). Individuals with smaller deletions are likely to be less severely affected than those with larger deletions, although the correlation between deletion size and the degree of impairment is not 100%, as individuals with the same size deletion may be vastly different in their degree of disability (Dhar et al. 2010). In addition, the delays are more pronounced in older individuals than in younger children, a trend referred to by Zwanenburg et al. (2016) as "growing into deficit". In their study of 33 Dutch children assessed with the Bayley-II-NL scale of infant development, Zwanenburg and colleagues (2016) reported global developmental delay with a maximum developmental age equivalent to 3–4.5 years.

Major motor milestones are impaired. Infants smile and roll over later than usual, with parents and caregivers attributing the delays to poor muscle tone. The mean age for rolling over is 8.2 months with a range of 3–24 months. Many children never crawl on all fours because of lack of coordination between arms and legs. Many resort to "commando" crawling, rolling, or sliding along the floor. For those who do crawl on all fours, the mean age for this accomplishment is 16 months, ranging from 7 to 36 months.

Significant delays in walking are generally of great concern to parents and caregivers. Lack of balance and decreased upper body strength related to poor muscle tone may contribute to delays in walking. Physical therapy and exercise may help strengthen the upper body muscles. Handholding, walkers, or other assistive devices are helpful in overcoming the balance problems. Most individuals ultimately achieve independent walking at a median age of 24 months and a range of 12 months to 8 years (Zwanenberg et al. 2016). Once walking is achieved, the gait is typically unsteady and may be characterized by a broad base with high stepping or toe walking (Kolevzon et al. 2014a).

Speech is absent or delayed in individuals with deletion 22q13 syndrome. Most infants babble and coo as expected. Beyond that, many individuals acquire no verbal expression. Others acquire a few words by 2 years of age and then seem to lose their speech by age 4 or 5 years. Still others will gain limited speech, but fail to achieve age-appropriate language skills. Sarasua et al. (2011) reported the effect of deletion size on speech with about 40% of those with a small deletion (<5.3 Mb) being able to speak in sentences; in contrast, the ability to speak in sentences was absent in those with larger deletions (>5.3 Mb). The ability to speak was also correlated with the degree of developmental delay, as reported by parents. Those who spoke in sentences were reported to have moderate delay while those without sentences had a severe to profound developmental delay score (Sarasua et al. 2014). Children who have not acquired verbal expression will use eye contact, gestures, waving, or sounds to make their needs or wants apparent. Sign language is not a viable option in many children because they lack the fine motor coordination required to make typical signs. Children with stronger gross motor and fine motor skills tend to have greater non-verbal communication skills because of their ability to point, gesture, and grasp. Many parents feel that the ability to communicate empowers the child and drives the child to achieve better expressive skills. The child experiences the power of non-verbal expression and is more attentive to acquiring verbal communication skills.

In most individuals with deletion 22q13, receptive language is more advanced that expressive language (Soorya et al. 2013). Good receptive language is demonstrated by the ability to understand simple commands, express appropriate emotions, and convey humor. The communication profile in PMS has been described as severe to profound expressive language deficiency with mild to moderate receptive language deficiency and mild to moderate oral motor skills (Webster et al. 2004).

Very few affected individuals have mastered toilet training, although parents of younger children are reporting limited success. Even though many children stay dry at night, they have difficulty expressing the need to toilet during the day. Philippe et al. (2008) reported that daytime bladder control was achieved in five out of eight children between the ages of 4 and 6 years and that night-time bladder control was achieved in two out of eight children between the ages of 4 and 8 years. It should be noted that six of the eight children in their study had deletions that were less than 2 Mb in size, four of which were reported to be 0.15 Mb. It is possible that children with smaller deletions may have more success with bladder control. Some individuals have had success with "scheduled" visits to the toilet, though it is a time-consuming process and its success is often dependent on how well the caregiver is trained rather than how well the affected individual is trained. Although they may urinate in the commode, they seem more reluctant to have a bowel movement there. Many affected individuals seem to lose interest in toilet training over time. They have periods of successful toileting interspersed with periods of incontinence. Diapers or pull-on underpants with moisture absorption are required throughout life. Because individuals with deletion 22q13 may be incontinent beyond the infant and toddler stage, the "changing tables" in most public restrooms are not long enough for them to lie down for a diaper change, or they may resist lying down. Caregivers have found success allowing the individual to stand in a bathroom stall or family changing room while their diaper or pants are changed.

Sleep disturbances occur in about 45% of children and adults with deletion 22q13, although sleep apnea is uncommon (Sarasua et al. 2014; Soorya et al. 2013; Philippe et al. 2008). There is difficulty getting to sleep at night and staying asleep, as is often the case in those with developmental disabilities and/or neurological dysfunction. Individuals with deletion 22q13 may require only 4 hours of sleep per night. Napping during the day only aggravates the problem of trying to keep the individual asleep throughout the night. Sleep deprivation may result in excess daytime sleepiness or morning headaches. People living with an individual with 22q13 may consequently also suffer from sleep deprivation.

Autism spectrum disorder (ASD) is present in up to 84% of individuals (Soorya et al. 2013). Others are described as having autistic-like behaviors. Toddlers and young children may have tactile defensiveness, but seem to outgrow this behavior. They may avoid eye contact with strangers, express anxiety in social situations, and demonstrate self-stimulatory behavior like flapping their hands or rocking. Other behavioral issues include persistent chewing and mouthing of non-food items (44–85%), teeth grinding (25%), and tongue thrusting (15%), all of which can contribute to dental problems that must be addressed (Phelan et al. 2016).

Hypthyroidism occurs in about 5% of individual and should be considered if there is a change in behavior, including lethargy, decreased cognition, loss of interest, and incoordination (Phelan et al. 2016). If these changes are observed testing for hypothyroidism, a known association, is warranted although a concern for regression, as discussed below must be considered if thyroid screening is normal. This is likely to be an underestimate of the incidence of hypothyroidism associated with PMS because so few of the individuals have had thyroid screens.

Aggressive behavior has been reported in about 25% of individuals. Aggression is manifested as hair pulling, biting, or hitting. The aggressive behavior typically occurs when the individual becomes frustrated and is often directed at the parent or caregiver. In other cases, a younger sibling or playmate may be the target of the aggression.

Although loss of speech in young children has been long recognized in PMS, motor and neurological regression was not initially recognized as a manifestation of the syndrome. Motor regression was first reported in a female with a 100 kb deletion who experienced decline in speech and reduction in daily living skills accompanied by loss of balance, ataxic gait, and urinary incontinence (Anderlid et al. 2002). At the time, it was not known if this was a sign of early dementia in this one individual or if neurologic and motor regression were a consequence of the deletion. A number of adults and adolescents undergoing cognitive decline, neurologic and motor deterioration, and developing late onset seizures were subsequently reported (Manning et al. 2004; Bonaglia et al. 2011; Willemsen et al. 2011; Denayer et al. 2012; Verhoeven et al. 2012; Soorya et al. 2013; Messias et al. 2013). In some cases, the decline occurred after an illness or a life changing experience, such as placement in a group home (Willemsen et al. 2011; Soorya et al. 2013). Atypical bipolar disorder, major depressive disorder with psychotic features, and schizophrenia with catatonia have also been reported in adolescents and adults (Denayer et al. 2012; Verhoeven et al. 2012; Messias et al. 2013). In their study of 32 individuals, Soorya et al. (2013) noted that parents reported skill regression as early as 15 months and as late as 17 years. Loss of skills included decline in language, hand movements, self-help skills, play skills, and social engagement. Loss of language was associated with onset of seizures in 55% of individuals. It should be noted that the individuals with decline in language did not meet the ADI-R criteria for language loss because their initial verbal skills were so poor.

The most comprehensive study to date of developmental regression in PMS was performed by Reierson et al. (2017). Using data from the ADI-R, these investigators assessed the frequency and characteristics of regression in 42 individuals ranging in age from 4 to 48 years. By parent report, regression affected 43% of individuals (18/42), with average age of onset around 6 years (76 months). Although an increased frequency of regression was observed in males compared to females, the difference was not statistically significant. The majority of individuals (61%) had not recovered their skills at the time of assessment; for those who had recovered skills, the time until recovery spanned from 1 month to 10 years. The ADI-R permits characterization of regression with the following results: loss of motor skills (56%) at average age of 8 years; loss of self-help skills (50%) at average age 4 years; loss of language (33%) at average age about 3 years; loss of social engagement and responsiveness (33%) at average age 5 years; loss of purposeful hand movements (28%) at average age 7 years; and loss of constructive or imaginative play (22%) at average age 7 years, In contrast to the early regression occurring around 2 years of age in ASD and between 1 and 3 years in Rett syndrome, the regression in PMS occurs later and has a greater effect on motor and self-help skills (Reierson et al. 2017). No association was found between the risk of regression and abnormal EEG or rate of seizures (Reierson et al. 2017).

Evaluation

- In infancy and early childhood, major milestones should be monitored. Neurological evaluation of fine motor coordination, gross motor skills, cranial nerves and deep tendon reflexes is recommended (Kolevzon et al. 2014a).
- Delays secondary to hypotonia should be evaluated and addressed in a standard manner, often including a physical therapy evaluation.
- Children should be assessed for autism or autistic-like features by a child development specialist, and/or a child psychiatrist. Standardized methods such as the ADI-R, ADOS-G and DSM-IV have been used (Soorya et al. 2013). The latter has recently been replaced by DSM-5 (2013).
- Referral to a speech pathologist to evaluate receptive and expressive language skills may be beneficial.
- Neuropsychological testing should be considered to determine if delay in speech is the result of delayed processing of auditory stimuli.
- To document the affected individual's pattern of sleeping and awakening, caregivers should record a sleep history.
- The individual should be referred for a sleep study if the sleep cycle is abnormal. Recognition of the specific sleep pattern disturbance may lead to a more specific treatment plan.
- A thyroid panel should be obtained on individuals who exhibit change in abilities, lethargy, regression, or other symptoms of hypothyroidism.

Management

- Early intervention programs designed to meet the child's developmental needs are beneficial to both parent and child.
- Communication through assistive technology provides the opportunity to communicate independently with the goal of verbal or nonverbal communication. Assistive technologies may include touch screen on a computer, voice-based system to allow the user to independently have a large vocabulary using icons, photographs of real objects or people, and manual manipulation of icons.
- Behavior modification programs with positive reinforcement may be successful in encouraging the affected individual to increase the frequency of desired behaviors. Parents have reported that applied behavior analysis (ABA) therapy has been helpful in modifying behavior and improving development.
- Non-specific medication for hyperactivity and self-stimulatory behavior may be indicated in some individuals with deletion 22q13. Therapy sessions may be more beneficial if provided more frequently and in shorter duration (Kolevzon et al. 2014a).
- Parents find that television, videotapes, DVDs, and music often have a soothing effect on the individual. However, each person differs and what soothes one may disturb the next.
- With early intervention and aggressive speech therapy, speech patterns improve.

- Body strengthening activities such as exercise and adaptive sports seem to increase the child's awareness of his or her surroundings and consequently improve the ability to speak.
- Speech therapy, occupational therapy, physical therapy, and introduction of assistive technologies should be available to improve communication.
- Sleep patterns may be improved by behavioral techniques implemented by the parents and other caretakers with the aim of promoting self-soothing skills that allow the child to fall and return to sleep independently. A dark, quiet, relatively cool environment with little visual and auditory stimuli will promote better sleep and establishing a consistent bedtime routine will provide a consistent sleep/wake schedule. Calming techniques may include soothing music or massage.
- A variety of medications have been used to improve sleep patterns in children with neurodevelopmental disorders (NDD). Results have been inconsistent but for the individual for whom melatonin is not beneficial, these medications may be helpful. A recent review of pediatric sleep disorders in children with NDD proves an excellent summary of the types of sleep disturbances and suggested management (Angriman et al. 2015).
- Melatonin (N-acetyl-5-methoxytryptamine) is a chronobiotic drug that helps in regulation of the sleep–wake cycle. It is typically used as a first line trial at an oral dose of 3–6 mg given an hour prior to bedtime. Lower doses may be effective in younger children. Melatonin can decrease sleep latency and increase total sleep time.
- Clonidine is a central and peripheral α-adrenergic agonist that acts on presynaptic and postsynaptic neurons and inhibits noradrenergic release and transmission. Clonidine is frequently prescribed for sleep disorders, but there are no well-controlled studies demonstrating its benefit. A dose of 0.1–0.3 mg, typically required to produce the postsynaptic effects on the central nervous system (CNS) α2-adrenergic receptors, may result in increased REM and total sleep. Clonidine may be given orally or transdermally with the transdermal route causing less adverse effects such as dizziness, fatigue, hypotension, and bradycardia. Side effects typically resolve over time.
- Gabapentin and benzodiazepines have also been used to promote more typical sleep patterns. Gabapentin easily crosses the blood–brain barrier, and increases brain synaptic gamma-amino butyric acid (GABA) in the CNS inhibitory neurons. It is typically well-tolerated at an oral dose of 5–15 mg kg^{-1} given at bedtime and many children with NDD have shown benefit. Benzodiazepines bind to the benzodiazepine subunit of the GABA chloride receptor complex, facilitating the action of the inhibitory neurotransmitter GABA. Due to potential side effects and dependence, benzodiazepines are infrequently used in children.
- If sleep apnea is diagnosed, routine protocols for sleep apnea should be implemented.
- If hypothyroidism is diagnosed, conventional treatment with synthetic T4 replacement should be initiated with periodic monitoring of T4 and TSH levels.
- Lithium has been used effectively in two patients with PMS and ASD due to *SHANK3* mutations (Serret et al. 2015; Egger et al. 2017). During adolescence both patients presented with catatonia, regression, and behavioral disorders after stressful events. Several attempts at treatment failed until lithium successfully reversed the regression, improved the behavioral issues, and recovered the patients to their pre-catatonia level of functioning (Serret et al. 2015). An adult with a *SHANK3* mutation was also effectively treated with lithium to stabilize his atypical bipolar mood disorder (Egger et al. 2017).
- Schmidt et al. (2008) reported the use of intranasal insulin therapy in six children with PMS. The investigators observed improved motor skills, cognition, and behavior in five of the six children. One child was discharged from the study after showing signs of regression. Anecdotal reports from parents indicate that behavior, sleep pattern, and motor development improved with intranasal insulin therapy. More systematic studies are needed to confirm these results and to assess the long-term effects of this treatment.
- The use of insulin-like growth factor-1 (IGF-1) to ameliorate the symptoms of PMS was conducted in a placebo-controlled, double-blind, crossover study involving nine children (Kolevzon et al. 2014b). While the children demonstrated significant improvement in social impairment and in restrictive behavior during the IGF-1 phase compared to the placebo phase, and suffered no serious adverse effects, this was a small trial and larger studies are needed to confirm the efficacy of this treatment (Kolevzon et al. 2014b).

Neurologic

Neonatal hypotonia is a hallmark of PMS, occurring in over 97% of individuals. Newborns appear "floppy" and may have trouble in feeding and swallowing. Infants may have a weak cry and poor head control. Decreased muscle tone and generalized muscle weakness lead to delays in motor skills. Poor tone often persists as joint hypermobility and poor reflexes. Children may display uncoordinated and ataxic movements. In individuals with ring 22, peripheral neuropathy is a more common presenting symptom in children than adults with NF2. If such symptoms are present, the individual should be carefully assessed by a neurologist (Evans et al. 2005). Vestibular schwannomas may have accompanying problems of balance and disorientation.

Seizures have been reported in up to 41% of individuals with deletion 22q13, including 22% with febrile seizures, 13% with non-febrile seizures, and 6% with both febrile and non-febrile seizures (Soorya et al. 2013). Many are febrile seizures, which do not appear to cause permanent damage and do not typically require medication. Grand mal seizures, focal seizures, and absence seizures have also been reported. The incidence of seizures in PMS has been reported to increase with age, with the occurrence of seizures at 11% for those 5 years of age or less; 26% between 5 and 9.9 years; 43% between 10 and 17.9 years; and 60% between 18 and 64 years of age (Sarasua et al. 2014). There has been one report of Lennox–Gestaut syndrome in a 16-year-old male with PMS (Lund et al. 2013) and several reports of Lennox–Gestaut or Lennox–Gestaut-like epilepsy among the membership of the PMS Foundation (Phelan et al. 2016). Brain imaging studies have found a variety of abnormalities with the overall incidence of defects ranging from just over 55% to 80% (Philippe et al. 2008; Dhar et al. 2010; Aldinger et al. 2013; Soorya et al. 2013). Findings include thinning, hypoplastic or atypical corpus callosum; delayed myelination or white matter atrophy; nonspecific white matter hyperintensities or gliosis; ventricular dilatation; enlargement of the sylvian fissure; periventricular nodular heterotopias; subarachnoid cysts; cerebellar vermis hypoplasia, and abnormal posterior fossa (Philippe et al. 2008; Dhar et al. 2010; Aldinger et al. 2013; Soorya et al. 2013). In their report of cerebellar and posterior fossa malformations in PMS, Aldinger et al. (2013) suggest that deletion of genes other than *SHANK3*, specifically *PLANXNB2* and *MAPK8IP2*, may contribute to cerebellar defects.

Arachnoid cysts, fluid-filled sacs that occur on the surface of the brain, occur in over 15% of individuals with deletion 22q13. This is significantly higher than the general population frequency of 1%. Most arachnoid cysts are present early in life, although they may not become symptomatic until adolescence. Small cysts may remain asymptomatic and are discovered incidentally when brain imaging is performed for another indication. Signs and symptoms of increased intracranial pressure may accompany larger cysts. These symptoms include incessant crying bouts, irritability, severe headaches, cyclic vomiting, and seizures. When the presence of an arachnoid cyst is suspected, brain MRI is warranted. Even if a child has had previous imaging that failed to show the presence of a cyst, the onset of symptoms of increased intracranial pressure should prompt a repeat of these studies. Sensory processing abnormalities are common with over 75% of individuals showing decreased reaction to painful stimuli, while about 50% are hypersensitive to tactile stimuli (Philippe et al. 2008; Sarasua et al. 2014; Soorya et al. 2013). Functional MRI (fMRI) studies were used to determine response to communicative versus non-communicative sounds in children with PMS compared to children with idiopathic ASD (iASD) (Ting Wang et al. 2016). Results for the PMS group indicated that communicative vocalizations evoked elective activity in the right superior temporal gyrus and in other regions of the brain associated with social cognition – the medial prefrontal cortex, insula, and inferior frontal gyrus. Non-communicative sounds did not evoke these responses in the PMS group. The iASD group showed selective activity in response to non-communicative vocalizations but did not show differential response to communicative vocalizations (Ting Wang et al. 2016). In addition to showing distinct differences between the PMS group and the iASD group, this study shows selective superior temporal gyrus activation in the right hemisphere only, rather than bilaterally as expected in normal individuals. Lack of bilateral response may contribute to the language and communication deficits of children with PMS (Ting Wang et al. 2016)

About half of the individuals with deletion 22q13 syndrome have sacral dimples compared with only about 2% of individuals in the general population. Sacral dimples that are higher than usual, appear to be deep or large, or are associated with other cutaneous markers, may be indicative of an underlying problem. Atypical sacral dimples should be further evaluated.

Evaluation

- The presence of neonatal hypotonia warrants evaluation by a neurologist.
- Head circumference should be measured and plotted on a growth chart at routine visits during childhood and in adults at diagnosis.
- Seizures should be evaluated by an electroencephalogram. Because brief EEG recording may fail to detect seizure activity in some individuals, Kolevzon et al. (2014a) recommend an overnight video EEG using the standard seizure detection programs. These investigators also observe that the overnight EEG may also aid in identifying the type of seizure, thereby guiding treatment decisions. Changes in behavior, loss of motor skills, and other regression should prompt repeating the EEG (Kolevzon et al. 2014a).
- Brain imaging such as MRI and CT scan are indicated for symptoms of increased intracranial pressure and/or suspicion of arachnoid cyst.
- MRI should be considered to assess brain growth and to detect possible CNS anomalies in individuals with microcephaly.
- For atypical sacral dimples without associated neurological problems, ultrasound is sufficient for characterization. When neurological problems such as seizures,

loss of movement, or lethargy are present, MRI is indicated.

- Children with ring 22 should have a head MRI beginning at age 10–12, earlier if they are symptomatic. The MRI should be repeated every two years up to age 20, then every three years for asymptomatic individuals. For patients with tumors, a comprehensive neurologic exam, including an MRI, should be performed at least annually until the tumor growth rate is established. Following the diagnosis of NF2, the extent of disease should be evaluated by head and spine MRI, hearing evaluation (including BAER), speech discrimination testing, ophthalmologic evaluation, and cutaneous examination (Evans 2011).

Management

- For hypotonia, the goal of increasing muscle tone should be pursued through physical therapy, occupational therapy, and other therapy programs.
- Recurrent seizures or epilepsy should be treated with standard anticonvulsant medication.
- In children with microcephaly, brain imaging studies should be obtained to determine if intervention is appropriate. Although there is no specific treatment for primary microcephaly, vision should be evaluated annually and appropriate physical and developmental therapy programs should be implemented to lessen the effects of intellectual and/or physical impairment.
- Ventriculomegaly and/or arachnoid cysts may require surgical placement of a shunt to relieve intracranial pressure.
- If a sacral dimple is infected, removal may be indicated.
- If a sacral dimple extends to the spinal cord, the need for surgical intervention should be assessed by a neurosurgeon and treated in a standard manner. It has been recommended that children with ring 22 be evaluated as recommended for individuals with NF2 and their offspring, including baseline and annual ocular, dermal, and neurologic examinations for children between 2 and 10 years of age with the addition of annual audiology screening and brain MRI every two years after age 10 (Lyons-Warren et al. 2017). Vestibular schwannoma may be approached with stereotactic radiosurgery using the gamma knife (Evans 2011).

Ears and Hearing

Children with deletion 22q13 are often described as having "apparent hearing loss." Their parents express concern that hearing loss is related to multiple ear infections in infancy and childhood. About 60% of children have recurrent ear infections and require tympanostomy tubes. However, rather than a true hearing deficiency, many individuals with deletion 22q13 have delayed processing time. The children hear a command but experience a lag before initiating action. Delayed response coupled with a difficulty in discriminating spoken commands from background noise may be misinterpreted as a hearing loss. Hearing evaluations are normal in about 80% of children.

For children with ring 22, hearing evaluations, including BAER testing, should begin at age 10–12 years. Tinnitus, hearing loss with speech discrimination, and balance dysfunction may signify the emergence of a vestibular schwannoma (Evans 2011).

Evaluation

- Hearing evaluation by an individual experienced in testing children with severe global delay is warranted in the child that has no expressive or receptive language skills.

Management

- Ear infections should be treated in a standard manner. Tympanostomy tubes may be required in the child with frequent ear infections.
- In children with a hearing deficit, appropriate management is warranted. Although the use of hearing aids has been successful in some children, others will not tolerate a hearing aid and continually remove it from their ear.
- For individuals with ring 22 and NF2, cochlear implants may be appropriate if hearing aids fail (Evans 2011).
- Because communication is already problematic in individual with PMS, all attempts should be made to preserve hearing/communication.

Ophthalmologic

Most individuals with deletion 22q13 syndrome appear to have normal vision, although hyperopia and myopia are reported. Ptosis is seen in almost 50% of individuals but is rarely, if ever, severe enough to require surgical correction. Strabismus occurs in about 35% of affected individuals. Glaucoma has been reported in one child, most likely an incidental finding unrelated to deletion of 22q13.

Cortical visual impairment has been reported in 6% of individuals, but is most likely underdiagnosed. Affected individuals exhibit extensive use of peripheral vision, difficulty in processing visual cluttered images, problems with depth perception, and a tendency to look away from an object before reaching for it. Visual function fluctuates, appearing to be better at times and worse at other times. Blindness and optic nerve hypoplasia have been associated with cortical visual impairment in deletion 22q13.

Evaluation

- Newborns with strabismus should have a careful eye examination. Older individuals should be referred to a specialist experienced in performing ocular examinations in developmentally delayed individuals. Ophthalmologic examinations should be performed by at least 3 years of age, and annually, thereafter.
- To evaluate cortical visual impairment, a number of therapists and physicians should be consulted. These include physical therapists, occupational therapists, orientation and mobility specialists, pediatric neurologists, pediatric ophthalmologists, and pediatric neurophysiologists, experienced with performing visual-evoked-potentials (VEPs) in children.

Management

- Patching the eye or other interventions may be effective in cases of strabismus.
- Specific visual problems should be managed in a standard manner.

Dental

Malocclusion and dental crowding are the most frequently encountered problems in deletion 22q13 syndrome. Poor muscle tone, constant chewing on non-food items, and tongue thrusting may contribute to malocclusion. Depending on the severity, malocclusion may interfere with swallowing and lead to excess drooling. Although most of the individuals with deletion 22q13 are non-verbal, malocclusion also tends to make verbalization more difficult.

Defective enamel formation, particularly on deciduous teeth, is common. Spotty or incomplete enamel formation may be attributed to recurrent antibiotic therapy, acid reflux, and extended use of bottle feedings. Accelerated tooth decay has also been reported.

Evaluation

- Early dental evaluation, including orthodontic consultation, should be performed.
- At least semiannual dental examinations should begin by age 3 years.

Management

- Parents and caregivers should be consistent about routine brushing of the individual's teeth.
- Oral motor therapy and speech therapy may aid in chewing and swallowing problems.
- Fluoride sealants should be applied to protect and strengthen the teeth.
- An orthodontist should be consulted at a standard time to determine if dental extraction or orthodontic therapy is warranted.
- Regular (at least semiannual) dental prophylaxis should occur.

Cardiovascular No significant cardiac abnormalities are consistently found in deletion 22q13 syndrome. Benign heart murmurs are noted and resolve spontaneously. Tricuspid valve regurgitation, atrial septal defect, patent ductus arteriosus, and total anomalous pulmonary venous return have been reported (Soorya et al. 2013; Jeffries et al. 2005), but the incidence of these defects is not well-defined. Adults with deletion 22q13 may be at risk for hypertension. Other age-related cardiovascular risks are unknown because of the paucity of known affected adults.

Evaluation

- The heart should be carefully auscultated at diagnosis. Heart murmurs or other cardiac abnormalities should be monitored or evaluated in a standard fashion with referral to a cardiologist, if appropriate.
- Individuals with PMS should be routinely monitored for hypertension.

Management

- If a cardiac abnormality is present, surgical intervention or medical treatment follows standard protocols.
- Hypertension should be treated by standard protocols

Musculoskeletal

Deletion 22q13 syndrome is characterized by normal body proportions and musculature. Just over 30% of individuals have syndactyly of the second and third toes. However, one out of five individuals with PMS and toe syndactyly has a parent who also demonstrates this trait. About 20% of individuals are reported to have fifth finger clinodactyly.

Over 60% of individuals with deletion 22q13 syndrome have relatively large and fleshy hands, a trait that is often evident in infants and toddlers (Sarasua et al. 2014). Feet may also seem large for age. The "puffiness" of the hands may be the result of edema. Lymphedema has been reported in about 25% (Nesslinger et al. 1994; Dhar et al. 2010 2010; Soorya et al. 2013; Sarasua et al. 2014). The lower extremities are more frequently affected than the upper extremities and the problem may worsen with age. The cause of lymphedema in deletion 22q13 syndrome is unclear, although both increased capillary permeability and delayed lymphatic drainage have been suggested as the potential culprit (McGaughran et al. 2010).

Scoliosis and, less frequently, lordosis have been reported in individuals with deletion 22q13. The degree of curvature is generally mild and does not require bracing or surgery.

Evaluation

- A physical therapist should be consulted to recommend exercises to improve muscle strength and joint mobility.
- The spine should be clinically evaluated annually.
- An orthopedist should be consulted in cases of scoliosis or lordosis. Observation and radiographs may be used to monitor changes in curvature.
- Parents are most likely to make the first observation of lymphedema at home and to report it to the physician. As a part of the annual physical examination, the physician should check for pitting edema and ask parents or caregiver if there has been any swelling of the extremities. If there is a history of intermittent swelling or if swelling is present at the time of examination, the length and girth of the limb should be measured and plotted at each visit.

Management

- No therapeutic intervention is generally required for syndactyly or clinodactyly.
- For lymphedema, massage and elevation of the affected limb may improve lymphatic flow and lessen swelling. The application of compressive stockings or bandages may be helpful to improve circulation in the affected limb. For severe lymphedema, use of a compression device with a pneumatic pump may be warranted. An example of such device is the Lympha Press™ system in which a special sleeve or boot composed of compression cells is placed on the affected limb. The boot covers the foot to the hip with overlapping cells that inflate and deflate in timed cycles, creating a peristaltic motion pushing fluid from the distal extremity to the body.
- Referral to a vascular surgeon may be warranted in cases of painful lymphedema.
- Recommendations of the orthopedist should be followed regarding the necessity for bracing or surgery in more severe cases of scoliosis or lordosis.

Gastrointestinal Chronic diarrhea and/or constipation are reported in almost 40% of individuals with PMS. Episodes of diarrhea with loose stools may alternate with periods of constipation. These symptoms seem to be unrelated to diet. It is possible that some medications taken for other reasons such as hyperactivity or sleeplessness may contribute to the diarrhea and constipation.

Gastroesophageal reflux is reported in 40–45% of the affected individuals (Sarasua et al. 2014; Soorya et al. 2013).

Reflux may resolve by 6–12 months of age. Smaller feeds, thickened formula, and careful positioning of the infant may alleviate the problem. In older individuals, reflux may be difficult to diagnose. Most people with deletion 22q13 have decreased perception of pain and thus do not indicate discomfort associated with reflux. Parents or caregivers may not be aware that the individual is experiencing reflux until repeated vomiting or hematemesis occurs.

Episodic vomiting has been reported in a small number of individuals. These episodes often require hospitalization for dehydration. The vomiting episodes tend to occur every few months and may be accompanied by headaches, lethargy, and/or dehydration. These findings should alert the clinician to the possibility of an arachnoid cyst (see Neurologic).

Evaluation

- A history of stooling patterns and evidence of reflux should be obtained at routine physician visits.
- A gastroenterologist should be consulted to evaluate constipation and/or diarrhea and gastroesophageal reflux that does not respond to management by the primary care physician.
- Episodic vomiting, if not caused by arachnoid cyst and accompanying increased intracranial pressure, should be evaluated by a gastroenterologist.

Management

- In cases of chronic constipation, increased fluid intake and stool softeners may be helpful.
- For chronic diarrhea, an increased fluid intake should be maintained to prevent dehydration.
- Modifications that may ameliorate the symptoms of gastroesophageal reflux include eating smaller meals, avoiding food that causes irritation, elevating the head of the bed, sleeping on the left side, and avoiding food intake within 2–3 hours of bedtime.
- If simple measures fail to ameliorate the symptoms of reflux, standard medical, and/or surgical approaches may be warranted.

Dermatologic

Two ectodermal derivatives, the nails and the sweat glands, are affected in deletion 22q13, suggesting the possibility of an ectodermal dysplasia. About 80% of individuals have dysplastic toenails and almost 60% have decreased sweating and a tendency to overheat. The toenails are thin, flaky, and tend to be ingrown. The fingernails are affected to a lesser degree. Although individuals do not perspire normally, the sweat glands have not been studied to determine if an abnormality is present. The hair and teeth do not appear to be affected.

Recurrent cellulitis has been reported in about 10% of individuals with deletion 22q13 syndrome. The inflammation typically occurs in the legs, which become hot, red, and tender. Accompanying symptoms are fever, chills or sweats, joint tenderness, and muscle aches. The cause of cellulitis is often unknown, as there is no apparent break or cut in the skin. Nonetheless, the cellulitis appears to be secondary to lymphedema. The accumulation of large volumes of fluid makes the affected limb more susceptible to infection. Parents commonly report limb swelling before the onset of cellulitis (personal observation).

Other skin problems, occasionally, reported in deletion 22q13 syndrome include café-au-lait spots, hypopigmented areas, eczema, ringworm, hirsutism, and skin tags.

Evaluation

- A diary of the occurrence of skin rashes and precipitating events should be kept to determine if there is an environmental trigger.
- Individuals who have recurrent rashes should be evaluated by a dermatologist.
- Evaluation of cellulitis may include a white blood cell count and blood cultures to exclude septicemia.

Management

- If perspiration is insufficient, the individual should be watched closely in warm weather to avoid dehydration and hyperthermia. Direct exposure to the sun should be avoided. Protective clothing and sunscreen should be used to prevent sunburn.
- If toenails become ingrown or infected, a dermatologist or podiatrist should be consulted regarding removal of the toenail.
- Oral antibiotics are indicated for treatment of cellulitis. The affected area should be elevated and heat or warm soaks applied. In severe cases, hospitalization for intravenous antibiotics may be required.

Respiratory

Over 40% of children report recurrent respiratory infections that may be related to aspiration and poor muscle tone. These include pneumonia, respiratory syncytial virus, bronchitis, and bronchiolitis. In a young child, respiratory problems may interfere with feeding and sleep. As the child ages, the respiratory infections subside.

Evaluation

- Chest radiograph to evaluate the cause of respiratory problems is warranted.

Management

- Treatment will be specific to the respiratory problem and may include supplemental oxygen, vaporizer, humidifier, inhaler, ear tubes, and pharmacotherapy.

Immunologic

No specific immunological deficiency has been associated with deletion 22q13, although frequent respiratory infections and recurrent ear infections occur. Recurrent infections seem to subside as the child becomes older. Commonly variable immune deficiency has rarely been observed.

Evaluation

- Referral to a pediatric pulmonologist and to an immunologist should be considered for recurrent infections.

Management

- Immune deficiencies should be treated by standard protocols.

Genitourinary Renal abnormalities have been reported in 38% of individuals with PMS and include cystic kidney, dysplastic kidneys, hydronephrosis, vesico-ureteric reflux, renal agenesis, extrarenal pelvis, bilateral horseshoe kidneys, and pyelectasis (Soorya et al. 2013; Jeffries et al. 2005). There is one report of Wilms tumor in a 22-month-old child. The child was found to have a multicystic kidney by prenatal ultrasound and the tumor affected the non-cystic kidney (Kirkpatrick et al. 2011). The PMS Foundation reports two additional children who had Wilms tumor. Over 10% of individuals with deletion 22q13 report recurrent urinary tract infections that are successfully treated with antibiotic therapy.

Genital abnormalities are not a constant finding in deletion 22q13. Hypospadias, undescended testes, and scrotal hydrocele have been reported in males. Females have had ovarian cysts, cliteromegaly, and precocious puberty. Other females have had hypoplastic labia and delayed puberty. Others go through puberty in a normal manner, including onset of menses.

Evaluation

- Individuals with recurrent urinary tract infections should be referred to a pediatric nephrologist and/or urologist for standard assessment.
- A baseline renal ultrasound is recommended for all individuals with deletion 22q13. The study should be performed as soon as possible after the diagnosis is

established. Additional studies may be suggested by the findings or clinical signs and symptoms.

- Precocious or delayed puberty should be investigated in a standard manner.

Management

- Urinary tract infections should be treated by standard antibiotic therapy.
- Malformations of the urinary system or genital system may require surgical intervention of a standard sort.
- Undescended testes should be brought down into the scrotum in a standard fashion.

RESOURCES

Support groups

Phelan-McDermid Syndrome Foundation
Website: *www.pmsf.org*
Chromosome 22 Central: support for all chromosome 22-related disorders
Website: */www.c22c.org/*
Unique: Rare Chromosome Disorder Support Group
Website: *www.rarechromo.org/*

Discussion groups for parents

www.facebook.com/Phelan-McDermid-Syndrome-Foundation-170674299632131/

Twitter: @Phelan_McDermid

Information online

Website: *www.genetests.org*
GeneReviews: 22q13.3 deletion syndrome
Website:*www.rarediseases.org/rare-diseases/phelan-mcdermid-syndrome*

REFERENCES

Aldinger KA, Kogen J, Kimonis V, Fernandez B, Horn D, Klopocki E, Chung B, Toutain A, Weksberg R, Millen KJ, Barkovich AJ, Dobyns WB (2013) Cerebellar and posterior fossa malformations in patients with autism-associated 22q13 terminal deletion. *Am J Med Genet A* 161A(1):131–6.

American Psychiatric Association (2013) *Diagnostic and Statistical Manual of Mental Disorders*. 5th ed. Arlington, VA: American Psychiatric Association.

Anderlid BM, Schoumans J, Annerén G, Tapia-Paez I, Dumanski J, Blennow E, Nordenskjöld M (2002) FISH-mapping of a 100-kb terminal 22q13 deletion. *Hum Genet* 110:439–43.

Angriman M, Caravale B, Novelli L, Ferri R, Bruni O (2015) Sleep in Children with Neurodevelopmental Disabilities. *Neuropediatrics* 46:199–210.

Arinami T, Hayashi N, Nagase H, Ogawa M, Nakamura Y (2006) Multifocal Meningiomas in a patient with a constitutional ring chromosome 22. *J Med Genet* 23:178–80.

Babineau T, Wilson HL, Dawson AJ, Chodirker BN, Der Kaloustian VM, Demaczuk S, McDermid HE (2006) Unusual dicentric chromosome 22 associated with a 22q13 deletion. *Am J Med Genet A* 140:2819–2823.

Boccuto L, Lauri M, Sarasua SM, Skinner CD, Buccella D, Dwivedi A, Orteschi D, Collins JS, Zollino M, Visconti P, Dupont B, Tiziano D, Schroer RJ, Neri G, Stevenson RE, Gurrieri F, Schwartz CE (2013) Prevalence of SHANK3 variants in patients with different subtypes of autism spectrum disorders. *Eur J Hum Genet* 21:310–316.

Bonaglia MC, Giorda R, Borgatti R, Felisari G, Gagliardi C, Selicorni A, Zuffardi O (2001) Disruption of the *ProSAP2* gene in a t (12;22)(q24.1;q13.3) associated with the 22q13.3 deletion syndrome. *Am J Hum Genet* 69:261–268.

Bonaglia MC, Giorda R, Beri S, De Agostini C, Novara F, Fichera M, Grillo L, Galesi O, Vetro A, Ciccone R, Bonati MT, Giglio S, Guerrini R, Osimani S, Marelli S, Zucca C, Grasso R, Borgatti R, Mani E, Motta C, Molteni M, Romano C, Donatella G, Reitano S, Baroncini A, Lapi E, Cecconi A, Guilia A, Patricelli MG, Pantaleoni C, D'Arrigo S, Riva D, Sciacca F, Bernadina BD, Zoccante L, Darra F, Termine C, Maserati E, Bigoni S, Priolo E, Bottani A, Gimelli S, Bena F, Brusco A, di Gregorio E, Bagnasco I, Guissani U, Nitsch L, Politi P, Martinez-Frias ML, Martinez-Fernandez ML, Martinez Guardia N, Bremer A, Anderlid B-M, Zuffardi O (2011) Molecular mechanisms generating and stabilizing terminal 22q13 deletions in 44 subjects with Phelan/McDermid syndrome. *PLoS Genet* 7(7):e1002173.

De Rubeis S, Siper PM, Durkin J, Weissman J, Muratet F, Halperin D, Del Pilar Trellis M, Yitzchak, F, Lozano R, Wang AT, Holder Jr JL, Betancur C, Buxbaum JD, Kolevzon A (2018) Delineation of the genetic and clinical spectrum of Phelan-McDermid syndrome caused by *SHANK3* point mutations. *Molec Autism* 9:31.

Denayer A, Van Esch H, de Rave; T. Frijns JP. Vam Biggenhout G, Vogels A, et al. (2012) Neuropsychopathology in 7 patients with the 22q13 deletion syndrome: presence of bipolar disorder and progressive loss of skills. *Mol Syndromol* 3:14–20.

Dhar SU, del Gaudio D, German JR, Peters SU, Ou Z, Bader PI, Berg JS, Blazo M, Brown CW, Graham BH, Grebe TA, Lalani S, Irons M, Sparagana S, Williams M, Phillips JA, Beaudet AL, Stankiewicz P, Patel A, Cheung Sw, Sahoo T (2010) 22q23 deletion syndrome: clinical and molecular analysis using array CGH. *Am J Med Genet* 152A(3):573–81.

Disciglio V, Rizzo CL, Mencarelli MA, Mucciola M, Marozza A, Di Marco C, Massarelli A, Canocchi V, Baldassarri M, Ndoni E, Frullanti E, Amabile S, Anderlid BM, Metcalfe K, Le Caignec C, David A, Fryer A, Boute O, Joris A, Greco D, Pecile V, Battini R, Novelli A, Fichera M, Romano C, Mari F, Renieri (2014) Interstitial 22q13 deletions not involving SHANK3 gene: a new contiguous gene syndrome. *Am J Med Genet A* 164(7):1666–76.

Durand CM, Betancur C, Boeckers TM, Bockmann J, Chaste P, Fauchereau F, Nygren G, Rastam M, Gillberg IC, Anckarsäter H,

Sponheim E, Goubran-Botros H, Delorme R, Chabane N, Mouren-Simeoni MC, de Mas P, Bieth E, Rogé B, Héron D, Burglen L, Gillberg C, Leboyer M, Bourgeron T (2007) Mutations in the gene encoding the synaptic scaffolding protein SHANK3 are associated with autism spectrum disorders. *Nat Genet* 39:25–7.

Egger JIM, Verhoueven WMA, Groenendijk-Reijenga R, Kant SG (2017) Phelan-McDermid syndrome due to a *SHANK3* mutation in an intellectually disabled adult male: successful treatment with lithium. *Brit Med J Case Report* 2017:bcr-2017-220778.

Evans DG (2011) Neurofibromatosis 2 in *GeneReviews®*. Available at https://www.ncbi.nlm.nih.gov/books/NBK1201/

Evans, DGR, Baser ME, O'Reilly B, Rowe J, Gleeson M, Saeed S, King A, Huson SM, Kerr R, Thomas N, Irving R, MacFarlane R, Ferner R, McLeod R, Moffat D, Ramsden R (2005) Management of the patient and family7 with neurofibromatosis 2: a consensus conference statement. *Brit J Neurosurg* 19(1):5–12.

Gauthier J, Spiegelman D, Piton A, Lafrenière RG, Laurent S, St-Onge J, Lapointe L, Hamdan FF, Cossette P, Mottron L, Fombonne E, Joober R, Marineau C, Drapeau P, Rouleau GA (2009) Novel de novo SHANK3 mutation in autistic patients. *Am J Med Genet B* 150B(3):421–4.

Gauthier J, Champagne N, Lafrenière RG, Xiong L, Spiegelman D, Brustein E, Lapointe M, Peng H, Côté M, Noreau A, Hamdan FF, Addington AM, Rapoport JL, Delisi LE, Krebs MO, Joober R, Fathalli F, Mouaffak F, Haghighi AP, Néri C, Dubé MP, Samuels ME, Marineau C, Stone EA, Awadalla P, Barker PA, Carbonetto S, Drapeau P, Rouleau GA; S2D Team (2010) De novo mutations in the gene encoding the synaptic scaffolding protein SHANK3 in patients ascertained for schizophrenia. *Proc Natl Acad Sci USA* 107(17):7863–8.

Gong Y, Lippa CF, Zhu J, Lin Q, Rosso AL (2009) Disruption of glutamate receptors at Shank-postsynaptic platform in Alzheimer's disease. *Brain Res* 1292:191–198.

Guilmatre A, Huguet G, Delorme R, Bourgeron T (2014) The emerging role of SHANK genes in neuropsychiatric disorders. *Dev Neurobiol* 74(2):113–22.

Ishmael HA, Cataldi C, Begleiter ML, Pasztor LM, Dasouki MJ, Butler MG (2003) Five new subjects with ring chromosome 22. *Clin Genet* 63:410–414.

Jeffries AR, Curran S, Elmslie F, Sharma A, Wenger S, Hummel M, Powell J (2005) Molecular and phenotypic characterization of ring chromosome 22. *Am J Med Genet A* 37:139–47.

Kirkpatrick BE, El-Khechen D (2011) A unique presentation of 22q13 deletion syndrome: multicystic kidneys, orfacial clefting, and Wilms tumor. *Clin Dysmorphol* 20(1):53–4.

Kolevzon A, Angarita B, Bush L, Ting Wang A, Yitzchak F, Yang A, Rapaport R, Saland J, Srivastava S, Farrell C, Edelmann LJ, Buxbaum JD (2014a) Phelan-McDermid syndrome: a review of the literature and practice parameters for medical assessment and monitoring. *J Neurodevel Dis* 6:39–51

Kolevzon A, Bush L, Ting Wang A, Halpern D, Frank Y, Grodberg D, Rapaport R, Tavassoli T, Chaplin W, Soorya L (2014b) A pilot controlled trial of insulin-like growth factor-1 in children with Phelan-McDermid syndrome. *Molec Autism* 5:54.

Luciani JJ, de Mas P, Depetris D, Mignon-Ravix C, Bottani A, Prieur M, Jonveaux P, Phillipe A, Bourrouillou G, de Martinville B, Delobel B, Vallee L, Croquette M-F, Mattei M-G (2003) Telomeric 22q13 deletions resulting from rings, simple deletions, and translocations: Cytogenetic, molecular, and clinical analyses of 32 new observations. *J Med Genet* 40:690–696.

Lund C, Brodtkorb E, Rosby O, ROdningen OK, Selmer KK (2013) Copy number variants in adult patients with Lenox-Gestaut syndrome features *Epilepsy Res* 105:110–7.

Lyons-Warren AM, Cheung W, Holder JL Jr (2017) Clinical reasoning: A common cause for Phelan-McDermid syndrome and neurofibromatosis type 2 : one ring to bind them (2017) *Neurology* 89:e205–2209.doi:10.1012

Manning MA, Cassidy SB, Clericuzio C, Cherry AM, Schwartz S, Hudgins L, Enns GM, Hoyme HE (2004) Terminal 22q deletion syndrome: a newly recognized cause of speech and language disability in the autism spectrum. *Pediatrics* 114:451–7.

McGaughran J, Hadwen T, Clark R (2010) Progressive edema leading to pleural effusions in a female with a ring chromosome 22 leading to a 22q13 deletion. *Clin Dysmorphol* 19(1):28–9.

Messias E, Kaley SN, McKelve KD (2013) Adult-onset psychosis and clinical genetics: a case of Phelan-McDermid syndrome *J Neuropschiatry Clin Neurosci* 25(4):E27.

Moessner R, Marshall CR, Sutcliffe JS, Skaug J, Pinto D, Vincent J, Swaigenbaum L, Fernandez B, Roberts W, Szatmari P, Scherer SW (2007) Contribution of SHANK3 mutations to autism spectrum disorders. *Am J Hum Genet* 81:1289–1297.

Nesslinger NJ, Gorski JL, Kurczynski TW, Shapira SK, Siegel-Bartelt J, Dumanski JP, Cullen RF, French BN, McDermid HE (1994) Clinical, cytogenetic, and molecular characterization of seven patients with deletions of chromosome 22q13.3. *Am J Hum Genet* 54:464–472.

Petrella R, Levine S, Wilmot PL, Ashar KD, Casamassima AC, Shapira LP (1993) Multiple meningiomas in a patient with a constitutional ring chromosome 22. *Am J Med Genet* 1993: 47:184–6.

Pham E, Crews L, Ubhi K, Hansen L, Adame A, Cartier A, Salmon D, Galasko D, Michael S, Savas JN, Yates JR, Glabe C, Masliah E (2010) Progressive accumulation of amyloid-beta oligomers in Alzheimer's disease and in amyloid precursor protein transgenic mice is accompanied by selective alterations in synaptic scaffold proteins. *FEBS J* 277:3051–3067.

Phelan K, Boccuto L, Sarasua S (2016) Phelan-McDermid syndrome: clinical aspects. In *Neuronal and Synaptic Dysfunction in Autism Spectrum Disorder and Intellectual Disability*, C Sala and C Verpelli (eds) London: Elsevier.

Phelan K, McDermid HE (2012) The 22q13.3 deletion syndrome (Phelan-McDermid syndrome). *Molec Syndromol* 2(3–5):186–201.

Phelan MC (2001) Prenatal diagnosis of mosaicism for deletion 22q13.3. *Prenat Diagn* 21:1100.

Phelan MC, Thomas GR, Saul RA, Rogers RC, Taylor HA, Wenger DA, McDermid HE (1992) Cytogenetic, biochemical, and molecular analyses of a 22q13 deletion. *Am J Med Genet* 43:872–876.

Phelan MC, Rogers RC, Saul RA, Stapleton GA, Sweet K, McDermid HE, Shaw SR, Claytor J, Willis J, Kelly DP (2001) Research review: 22q13 deletion syndrome. *Am J Med Genet* 101:91–99.

Philippe A, Boddaert N, Vaivre-Douret L, Robel L, Danon-Boileau, L, Malan V, et al. (2008) Neurobehavioural profile and brain imaging studies of the 22q13.3 deletion syndrome in childhood *Pediatrics* 122(2):e376–82.

Reierson G, Bernstein J, Froehlich-Santino W, Urban A, Purmann C, Berquist S, Jordan J, O'Hara R, Hallmayer J (2017) Characterizing regression in Phelan McDermid Syndrome (22q13 deletion syndrome). *J Psychiatr Res* 91:139–144.

Riegal M, Baumer A, Wisser J, Acherman J, Schinzel A (2000) Prenatal diagnosis of mosaicism for a del(22)(q13). *Prenat Diagn* 20:76–79.

Rodriguez L, Martinez Guardia N, Herens C, Jamar M, Verloes A, Lopez F, Santos Munos J, Martinez-Frias ML (2003) Subtle Trisomy 12q24.3 and subtle monosomy 22q13.3: three new cases and a review. *Am J med Genet A* 122A(2):119–224.

Rollins JD, Sarasua SM, Phelan K, DuPont BR, Rogers CR, Collins JS (2011) Growth in Phelan-McDermid syndrome. *Am J Med Genet A* 155:2324–6.

Romain DR, Goldsmith J, Cairney H, Columbano-Green LM, Smythe RH, Parfitt (1990) Partial monosomy for chromosome 22 in a patient with del(22)(pter- > q13.1::q13.33- > qter). *J Med Genet* 27:558–589.

Sarasua SM, Boccuto L, Sharp JL. Dwivedi A, Chen C-F, Rollins JD, Rogers CR, Phelan K, DuPont BR (2014a) Clinical and genomic evaluation of 201 patients with Phelan-McDermid syndrome. *Hum Genet* 133(7):847–59.

Sarasua SM, Dwivedi A, Boccuto L, Chen C-F, Sharp JL, Rollins JD, Collins JS, Rogers CR, Phelan K, DuPont BR (2014b) 22q13.2q13.32 genomic regions associated with severity of speech delay, and physical features in Phelan-McDermid syndrome. *Genet Med* 16(4):318–28.

Sarasua SM, Dwiviedi A, Boccuto L, Rollins JD, Chen C_F, Rogers RC, Phelan K, DuPont BR, Collins JS (2011) Association between deletion size and important phenotypes expands the genomic region of interest in Phelan-McDermid syndrome (22q13 deletion syndrome). *J Med Genet* 48(11):61–6.

Schmidt H, Werner K, Giese R, Hallschmid M, Enders A (2008) Intranasal insulin to improve developmental delay in children with 22q13 deletion syndrome: An exploratory clinical trial. *J Med Genet* Oct 23.

Serret S, Thummler S, Dor E, Vesperini S, Santos A, Askenazy F (2015) Lithium as a rescue therapy for regression and catatonia features in two SHANK3 patients with autism spectrum disorder: case reports. *BMC Psychiatr* 15:107.

Slavotinek A, Maher E, Gregory P, Rowlandson P, Huson SM (1997) The phenotypic effects of chromosome rearrangement involving bands 7q21.3 and 22q13.3. *J Med Genet* 34:857–861.

Smith A, Jauch A, St. Heaps L, Robson L, Kearney B (2000) Unbalanced translocation t(15;22) in "severe" Prader-Willi syndrome. *Ann Genet* 43:125–130.

Soorya L, Kolevzon A, Zweifach J, Lim T, Dobry Y, Schwartz L, Frank Y, Ting Wang A, Cai G, Parkhomenko E, Halpern D, Grodberg D, Angarita B, Willner JP, Yang A, Canitano R, Chaplin W, Betancur C, Buxbaum JD (2013) Prospective investigation of autism and genotype-phenotype correlations in 22q13 deletion syndrome and SHANK3 deficiency. *Mol Autism* 4:18.

Su P-H, Chen J-Y, Chen S-J (2011) Siblings with deletion 22q13.3 and trisomy 15q26 inherited from a maternally balanced translocation. *Pediatr Neonatol* 52:287–9.

Tagaya M, Mizuno S, Hayakawa M, Yokotsuka T, Shimizu S, Fujimaki H (2008) Recombination of a maternal pericentric inversion results in 22q13 deletion syndrome. *Clin Dysmorphol* 17:19–21.

Ting Wang A, Lim T, Jamison J, Bush Lauren, Soorya LV, Tavassoli T, Siper PM, Buxbaum JB, Kolevzon A (2016) Neural selectivity for communicative auditory signals in Phelan-McDermid syndrome. *J Neurodevel Disord* 8:5.

Tommerup N, Warburg M, Gieselmann, V, Hansen BR, Koch, J, Petersen BG (1992) Ring chromosome 22 and neurofibromatosis. *Clin Genet* 42:171–177.

Tsilchorozidou T, Menko FH, Lalloo F, Kidd A, De Silva R, Thomas H, Smith P, Malcolmson A, Dore J, Madan K, Brown A, Yovos JG, Tsaligopoulos M, Vogiatzis N, Baser ME, Wallace AJ, Evans DG (2004) Constitutional rearrangements of chromosome 22 as a cause of neurofibromatosis 2. *J Med Genet* 41:529–34.

Verhoeven WM, Egger JI, Willemsen MH, de Leijer GJ, Kleefstra T (2012) Phelan-McDermid syndrome in two adult brothers: atypical bipolar disorder as its psychopathological phenotype? *Neuropsychiatr Dis Treat* 8:175–9.

Wang X, McCoy PA, Rodriguiz RM, Pan Y, Je HS, Roberts AC, Kim CJ, Berrios J, Colvin JS, Bousquet-Moore D, Lorenzo I, Wu G, Weinberg RJ, Ehlers MD, Philpot BD, Beaudet AL, Wetsel WC, Jiang YH (2011) Synaptic dysfunction and abnormal behaviors in mice lacking major isoforms of Shank3. *Hum Mol Genet* 20:3093–3108.

Wapner RJ, Babiarz JE, Levy B, Stosic M, Zimmermann B, Sigurjonsson S, Wayham N, Ryan A, Banjevic M, Lacroute P, Hu J, Hall MP, Demko Z, Siddiqui A, Rabinowitz M, Gross SJ, Hill M, Benn P (2015) Expanding the scope of noninvasive prenatal testing: detection of fetal microdeletion syndromes. *Am J Obstet Gynecol* 212(3):332.e1–9.

Watt JL, Olson IA, Johnston AW, Ross HS, Couzin DA, Stephen GS (1985) A familial pericentric inversion of chromosome 22 with a recombinant subject illustrating a 'pure' partial monosomy syndrome. *J Med Genet* 22:283–287.

Webster KT, Raymond GV (2004) 22q13 deletion syndrome: a report of the language function in two cases. *J Med Speech – Lang Pathol* 12:42–6.

Willemsen MH. Rensen JHM, van Schronjenstein-Lantman de Valt HMJ, Hamel BCJ, Kleefstra T (2011) Adult phenotypes in Angelman- and Rett-like syndromes. *Mol Syndromol* 3:14–20.

Wilson HL, Wong ACC, Shaw SR, Tse W-Y, Stapleton GA, Phelan MC, Hu S, Marshall J, McDermid HE (2003) Molecular characterisation of the 22q13 deletion syndrome supports the role of haploinsufficiency of *SHANK3/PROSAP*2 in the major neurological symptoms. *J Med Genet* 40:575–584.

Wilson HL, Crolla JA, Walker D, Artifoni L, Dallapiccola B, Takano T, Vasudevan P, Huang S, Maloney V, Yobb T, Quarrell O, McDermid HE (2008) Interstitial 22q13 deletions: Genes other than SHANK3 have major effects on cognitive and language development. *Eur J Hum Genet* 16:1301–1310

Wong AC, Bell CJ, Dumanski JP, Budarf ML, McDermid HE (1995) Molecular characterization of a microdeletion at 22q13.3. *Am J Hum Genet* 57:A130.

Zirn B, Arning L, Bartels I, Shoukier M, Hoffjan G, Neubauer B, Hahn A (2010) Ring Chromosome 22 and neurofibromatosis type II: proof of a two-hit model for the loss of the *NF2* gene in the development of meningioma. *Clin Genet* 81:82–87

Zwanenburg RJ, Ruiter SAJ, van den Heuvel ER, Flapper BCT, Van Ravenswaaij-Arts MA (2016) Developmental Phenotype in Phelan-McDermid (22q13 deletion) syndrome: a systemic and prospective study in 34 children. *J Neurodevel Disord* 8:16.

23

DENYS–DRASH SYNDROME, FRASIER SYNDROME, AND WAGR SYNDROME (*WT1*-RELATED DISORDERS)

JOYCE T. TURNER AND JEFFREY S. DOME

Children's National Hospital and the George Washington University School of Medicine and Health Sciences, NW, Washington, DC, USA

INTRODUCTION

WT1-related disorders are a group of conditions associated with an aberrant or absent copy of the *WT1* gene. The most prevalent of the *WT1*-related disorders is WAGR syndrome (*W*ilms tumor, *a*niridia, *g*enital anomalies, and a *r*ange of developmental delays), a contiguous gene deletion syndrome involving *WT1* and neighboring genes. Denys–Drash syndrome (DDS) is classically defined as the clinical triad of incomplete male genital development, early onset nephropathy, and Wilms tumor. Frasier syndrome (FS) is defined as the triad of XY gonadal dysgenesis, childhood onset renal failure, and gonadoblastoma. The rarest of the *WT1*-related disorders, Meacham syndrome (MS), is characterized by genitourinary malformations in males, cardiac and pulmonary malformations, diaphragmatic hernia, and frequently, early lethality.

Incidence

The exact incidence of the *WT1*-related disorders is unknown, but they are all rare, with only dozens to several hundred cases collectively reported in the literature. Variation in onset of manifestations, phenotypic variability (even among family members), and incomplete penetrance makes it difficult to determine the true incidence of these conditions.

The syndrome with the greatest number of reported cases is WAGR syndrome. The co-occurrence of aniridia and Wilms tumor was first reported 50–60 years ago (Brusa and Torricelli 1953; Miller et al. 1964), and the most comprehensive reviews are by Turleau et al. (1984) and Fischbach et al. (2005). DDS was first described by Denys in 1967 and Drash in 1970 (Denys et al. 1967; Drash et al. 1970). The largest review to date includes 150 cases reported by Mueller in 1994. The clinical features of FS were first described in 1964, and the condition was later named by Moorthy et al. (Frasier et al. 1964; Moorthy et al. 1987). Meacham syndrome (MS) was first described in 1991. Only about one to two dozen individuals have been described with some harboring a *WT1* mutation (Meacham et al. 1991; Killeen et al. 2002; Suri et al. 2007).

Molecular genetic studies of populations of individuals with Wilms tumor or kidney disease provide further insight into the prevalence of *WT1*-related disorders. Segers et al. (2012) analyzed *WT1* variances in 109 individuals diagnosed with Wilms tumor. They reported that 11% (*n* = 12) harbored a *WT1* mutation, four (3.7%) had WAGR syndrome, one (0.92%) had DDS, four (3.7%) had isolated genitourinary malformations, and three (2.8%) had isolated bilateral Wilms tumor (Segers et al. 2012). This study did not include individuals with FS, as the majority do not present with Wilms tumor. In addition, a study reported by the National Wilms Tumor Study Group found that, among a cohort of individuals with Wilms tumor, 0.75% (*n* = 65/8533) had WAGR syndrome (Breslow et al. 2003). In another study among 115 individuals with steroid resistant nephrotic syndrome (SRNS), eight individuals (~7%) harbored a *WT1* pathogenic variant (Ruf et al. 2004). Furthermore, a study

---

Cassidy and Allanson's Management of Genetic Syndromes, Fourth Edition.
Edited by John C. Carey, Agatino Battaglia, David Viskochil, and Suzanne B. Cassidy.
© 2021 John Wiley & Sons, Inc. Published 2021 by John Wiley & Sons, Inc.

including 354 children with SRNS identified 21 children (6%) with a *WT1* pathogenic variant (Ahn et al. 2017).

Diagnostic Criteria

Each of the *WT1*-related disorders has typical features, yet there are no consensus diagnostic criteria. Phenotypic variation and incomplete penetrance make establishing a diagnosis challenging. Because of this, a thorough clinical evaluation and genetic testing are needed to (1) clarify the diagnosis, (2) identify at-risk individuals, and (3) provide close monitoring for those at risk for complications such as renal failure and neoplasia.

In WAGR syndrome, genital anomalies can be seen in both males and females. These vary in severity, yet are greater in males versus females. Aniridia may be the only presenting finding in some individuals, especially among genotypic females. Cognitive impairment varies based on the size of the deletion present. Affected individuals may also display autistic features (Xu et al. 2008; Yamamoto et al. 2013; Han et al. 2013). Additional findings observed in WAGR syndrome include nephropathy and renal failure (Breslow et al. 2005; Fischbach et al. 2005), obesity (Han et al. 2008; Rodríguez-López et al. 2013), and, less frequently, diaphragmatic hernia (Scott et al. 2005).

Infants identified as having aniridia (Figure 23.1) should be evaluated for WAGR syndrome versus isolated pathogenic variants in *PAX6* (Andrade et al. 2008). Aniridia is almost always present in WAGR syndrome although rare cases with absence of aniridia have been reported (Turleau et al. 1984; Fishbach et al. 2005). The clinical diagnosis of the syndrome can be made when aniridia and one other feature of the condition, such as genitourinary malformations or nephrotic syndrome, are present. External genital anomalies are common in males (Figure 23.2). Fischbach et al. reported that 60% of their cohort had cryptorchidism (Fischbach et al. 2005). Females with the WAGR syndrome (Figure 23.3) may go unrecognized if aniridia is the only feature present at birth. Males with unexplained genital anomalies should have an ophthalmologic evaluation.

Chromosomal microarray can determine if a contiguous gene deletion is present, the extent of the deletion, the identification of the genes within the deletion that may be of clinical significance, as well as chromosomal gender. The absence of a microdeletion would suggest that additional genetic testing may be beneficial to clarify the diagnosis. A diagnostic flow chart to guide genetic testing and identify those in greatest need for continued surveillance is shown in Figure 23.4.

While males with DDS and FS frequently have ambiguous genitalia or complete sex reversal, XX females with DDS or FS have normal female genitalia. Females with FS usually come to medical attention due to signs of renal failure, and those with DDS may present with Wilms tumor and/or renal failure (Swiatecka-Urban et al. 2001). Not everyone with DDS and FS present with all of the features of these conditions, as there is phenotypic variability (Yu et al. 2012; Yang et al. 2013c). For example, some individuals with DDS present with genitourinary anomalies plus Wilms tumor without renal failure (Köhler, 1999; Köhler et al. 2011), some have Wilms tumor and renal failure but no

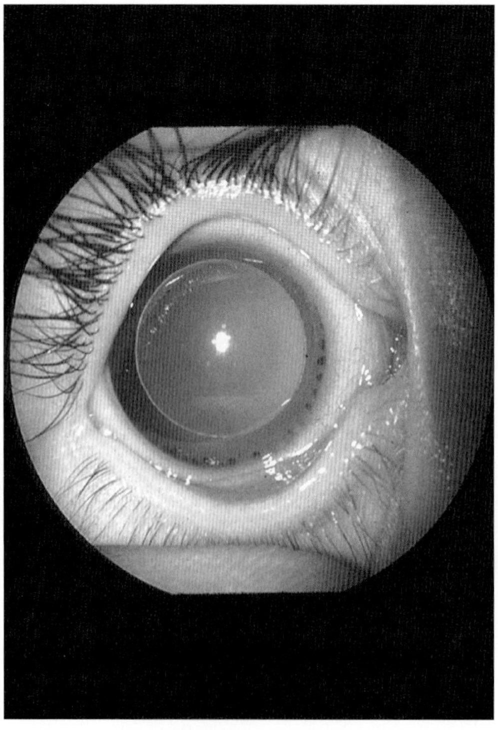

FIGURE 23.1 Aniridia.

FIGURE 23.2 Male ambiguous genitalia.

2012), and yet some have isolated SRNS (Ruf et al. 2004; Megremis et al. 2011; Yang et al. 2013a; Yang et al. 2013b; Guaragna et al. 2013; Lipska et al. 2013). On rare occurrence, individuals with DDS develop gonadoblastoma (Patel et al. 2013; Finken et al. 2015) and individuals with FS develop Wilms tumor (Barbosa et al. 2010). Incomplete penetrance has also been documented among this group of disorders. Family members of individuals with *WT1* mutations may harbor the familial gene change, but may not display any manifestations of the disorder whatsoever (Fenci et al. 2012; Zhu et al. 2013).

Meacham syndrome (MS) is typically associated with pseudohermaphroditism in males, cardiac and pulmonary malformations, diaphragmatic hernia, and affected individuals frequently die in infancy (Meacham et al. 1991; Killeen et al. 2002; Suri et al. 2007).

Absence of the typical triad of features seen in DDS (Wilms tumor, nephropathy, genitourinary abnormalities), FS (XY gonadal dysgenesis, renal failure, and gonadoblastoma), or all the medical findings that constitute the WAGR acronym may result in failure to recognize individuals with *WT1* aberrations (Little et al. 2004). Therefore, clinicians need to be attuned to recognize any part of the phenotypic spectrum of these disorders to provide optimal care to affected individuals. Individuals born with ambiguous genitalia and/or aniridia warrant a genetics consultation including a clinical evaluation,

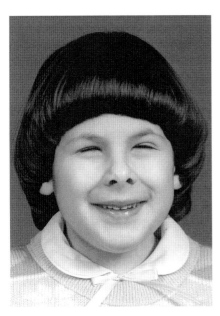

FIGURE 23.3 Facial phenotype of WAGR syndrome.

genitourinary anomalies (Ismaili et al. 2008). Infrequently, affected individuals display isolated disorders of sexual development (Huff, 1996; Royer-Pokora et al. 2004; Köhler et al. 2011; Segars et al. 2012), others develop Wilms tumor but lack syndromic features (Little et al. 2004; Segars et al.

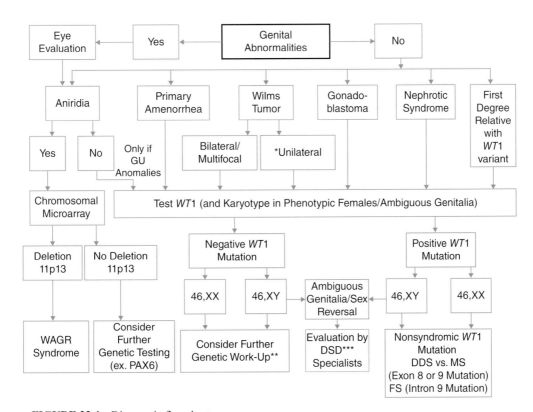

FIGURE 23.4 Diagnostic flowchart.
*Consider WT1 testing in context of family history and other medical findings
** No further work-up in first-degree relatives negative for known familial mutation
***DSD- Disorders of Sexual Differentiation

chromosomal analysis, and *WT1* molecular testing (Köhler et al. 1999; Ruf et al. 2004; Andrade et al. 2008; Capito et al. 2011; Finken et al. 2015). Identifying individuals with *WT1* aberrations is crucial to monitoring for neoplasms and nephropathy. The classic features seen in DDS, FS, MS, and WAGR syndrome are listed in Table 23.1.

The absence of a congenital anomaly makes it more challenging to identify those with *WT1* pathogenic variants prior to presenting with amenorrhea, Wilms tumor, gonadoblastoma, proteinuria, focal segmental glomerulosclerosis (FSGS), diffuse mesangial sclerosis (DMS), or end stage renal disease (ESRD). Individuals who come to medical attention with primary amenorrhea and failure to enter puberty should have a work-up including chromosomal studies, and *WT1* analysis should be performed if an individual appears phenotypically female but displays 46,XY karyotype (Saxena et al. 2006; Anrade et al. 2008). *WT1* analysis should also be considered for any person presenting with Wilms tumor, especially bilateral or multifocal Wilms tumor (Little et al. 2004; Anrade et al. 2008; Segars et al. 2012), and for individuals presenting with gonadoblastoma.

Some physicians suggest that routine *WT1* screening for boys with sporadic isolated SRNS is not appropriate due to low yield (Yang et al. 2016). However, most clinicians recommend the investigation of *WT1* germline mutations to determine a definitive diagnosis among infants, children, and young adults who present with proteinuria, steroid resistant focal segmental glomerulosclerosis or steroid resistant diffuse mesangial sclerosis, particularly when familial, to direct appropriate clinical care (Ruf et al. 2004; Anrade et al. 2008; Nso Roca et al. 2009; Benoit et al. 2010; Mergremis et al. 2011; Guaragna et al. 2012). Karyotyping females with *WT1* pathogenic variants and SRNS is useful to determine the sex chromosome complement (Davies et al. 2017; Pritchard-Jones 1999; Lipska et al. 2013). Furthermore, karyotype is also beneficial for phenotypic females diagnosed with Wilms tumor and/or gonadoblastoma who are positive for a *WT1* mutation, as this can help to assess for complete sex reversal. Family members (especially first-degree relatives-FDR) of individuals who have a known *WT1* pathogenic variant should consider targeted mutation analysis to determine if they harbor the familial variant, as

TABLE 23.1 Characteristics of WT1-Related Disorders

	Denys–Drash syndrome	Frasier syndrome	Meacham syndrome	WAGR syndrome
Genitourinary phenotype (46,XY) (can be an isolated finding in DDS and FS)	Normal to complete sex reversal; gonads may be testis, ovo-testis, ovary, or streak	Typically complete sex reversal; often present as females with primary amenorrhea; streak gonads	Ambiguous genitalia to pseudohermaphroditism	Ambiguous genitalia, most commonly cryptorchidism (rarely complete sex reversal)
Genitourinary phenotype (46,XX)	Normal	Normal	Normal	Usually normal external genitalia; possible streak gonads, uterine anomalies (bicornate uterus)
Tumor	>70% Wilms tumor Onset < two years Gonadoblastoma infrequent Age range: <3 years among reported cases	Wilms tumor infrequent Gonadoblastoma Age range: 4–22 years Most diagnosed teen years	None	30–60% Wilms tumor; most diagnosed <6 years; gonadoblastoma infrequent; age not clear
Nephropathy (Can have isolated steroid resistant nephrotic syndrome in DDS & FS)	Infantile/early childhood onset DMS 95% ESRD ≤3 years of age	Childhood onset FSGS ESRD by about 10–20 years	None	Late onset nephropathy FSGS ESRD during or after adolescence
WT1 gene mutation	Most missense, less frequently nonsense mutations exons 8 and 9	Splice site mutations in intron 9	Missense exons 8 and 9	Contiguous gene deletion of 11p13 including at least *WT1* and *PAX6*
Other	Diaphragmatic hernia (rare)	Diaphragmatic hernia (rare)	Diaphragmatic hernia Cardiac malformations Pulmonary malformations	Aniridia; obesity; cognitive impairment; autism/behavior issues; diaphragmatic hernia (rare)

they would be at risk for neoplasms and renal disease associated with *WT1*-related disorders.

Etiology, Genetics, and Pathogenesis

Etiology and Gene Structure
The *WT1* gene was identified in the early 1990s after demonstrating that individuals with WAGR syndrome harbor an 11p13 contiguous gene deletion associated with a high risk of Wilms tumor (Call et al. 1990; Gessler et al. 1990). In 1991, individuals with DDS were found to harbor mutations in *WT1* (Pelletier et al. 1991), and, in 1997, FS was also shown in individuals with pathogenic variants in *WT1* (Barbaux et al. 1997). Manifestations of these conditions can also arise independent of one another or in various combinations, and they too, harbor *WT1* gene aberrations. Finally, in 2007 individuals with the rare condition MS were shown to have *WT1* mutations (Suri et al. 2007).

WT1 is located on 11p13 and spans approximately 50 kb of DNA, which encompasses 10 exons (Call et al. 1990; Gessler et al. 1990). A linear schematic drawing of the *WT1* gene is shown in Figure 23.5(B). The protein, WT1, is a transcription factor, tumor suppressor, and post-transcriptional regulator in mRNA splicing and protein–protein interactions (Pelletier et al. 1991; Toska and Roberts 2014). As a transcription factor, *WT1* gives rise to at least 24 different protein isoforms, which result from alternative (1) translational start sites, (2) RNA splicing, and (3) RNA editing (Toska and Roberts 2014). Splice sites in exons 5 and 9 result in an insertion of 17 amino acids in the proline/glutamine rich N-terminus (17AA), and inclusion/exclusion of three amino acids (KTS)-lysine, threonine, and serine between the third and fourth zinc fingers in the c-terminus of WT1, respectively. The various WT1 isoforms act as transcriptional activators and repressors of its gene targets, thus playing a role in both development and growth. More recently, it has been shown that WT1 assists with development by binding the 3′ untranslated region (UTRs) of its developmental mRNA targets. When WT1 is depleted, mRNAs are down-regulated with rapid turnover in developing kidney mesenchyme. Bharathavikru et al. (2017) propose that WT1 influences important developmental and disease processes by regulation of mRNA turnover. WT1 also acts as tumor suppressor, and there is some evidence that it possesses oncogenic properties (Huff et al. 2011; Toska and Roberts 2014). *WT1* mutations and the absence thereof, affect the various WT1 protein isoforms that are produced. This influences mRNA turnover, which results in the different features observed in *WT1*-related disorders (Bharathavikru et al. 2017).

Genotype–Phenotype Correlations *WT1*-related disorders display genotype–phenotype correlations. Individuals with the DDS phenotype predominantly harbor missense, and less frequently nonsense, mutations in exons 8 and 9 (Royer-Pokora et al. 2004), although mutations in exons 1–4 and 6–7 have also been observed (Little et al. 2005; da Silva et al. 2011; Lee et al. 2011; Modi et al. 2015). These mutations are proposed to cause disease as a result of a dominant negative effect (Pelletier et al. 1991, Huff 1996). Those individuals with FS have splice-site mutations in intron 9, primarily arising in the +4/+5 location of the intron (Barbaux et al. 1997). This affects the KTS splice site located between zinc fingers 3 and 4, resulting in modification of the final protein product, thus causing an imbalance in the normal +KTS/-KTS isoform ratio (Barbaux et al. 1997). Those with MS have been reported to have *WT1* missense mutations in exons 8 and 9. This highlights the phenotypic spectrum among those with DDS and MS who harbor either a *WT1* exon 8 or 9 pathogenic variant (Suri et al. 2007). These syndromes may well represent the same condition with variable clinical expression and severity. WAGR syndrome results as a consequence of a contiguous gene deletion on chromosome 11p13 including *WT1*, *PAX6* and neighboring genes on the distal portion of 11p13 (Riccardi et al. 1978; Call et al. 1990; Gessler et al. 1990). In part, the severity of WAGR syndrome is related to the size of the deletion. Of note, the deletion always includes the distal portion of the 11p13 band (see Figure 23.5(A)). Approximately 90% of deletions are de novo and most frequently of paternal origin (Huff et al. 1990). Other de novo cases of WAGR syndrome have been found to arise from chromosomal rearrangements (Robinson et al. 2008). A few familial cases have resulted from unbalanced translocations, insertional rearrangements, and insertional translocations (Lavedan 1989; Crolla and van Heyningen 2002).

Not only have pathogenic *WT1* variants in exons 8 and 9 been associated with DDS and MS, but they have also been observed with isolated diffuse mesangial sclerosis (Chernin et al. 2010; Ahn et al. 2017). Individuals with *WT1* mutations in exons 1–7 display a DDS phenotype but have a milder and/or later-onset nephropathy compared with those with mutations in exons 8 and 9. So, evidence supports that haploinsufficiency of *WT1* is also associated with milder renal manifestation in those with DDS (Guaragna et al. 2017). Those with DDS and mutations in exons 8 and 9 tend to have congenital or infantile nephropathy (Ahn et al. 2017), although this has not been reported in those with MS.

Individuals with splice site mutations in intron 9 usually have later-onset proteinuria and nephrotic syndrome and slower progression to end stage renal disease (ESRD) compared to those with missense mutations in exons 8 and 9. Those with a p.Arg434His mutation display faster progression and renal dysfunction. Furthermore, 46,XX females with intron 9 splice site mutations develop isolated nephrotic syndrome, whereas 46,XY individuals develop the typical features observed in FS (Chernin et al. 2010; Lehnhardt et al. 2015; Lipska et al. 2014; Ahn et al. 2017). Haploinsufficiency of *WT1*, as observed in WAGR syndrome, is associated with proteinuria and FSGS (Breslow et al. 2005; Fischbach et al. 2005; Iijima 2012). According to Iijima's study involving individuals with WAGR syndrome, the age at onset of proteinuria and nephrotic syndrome

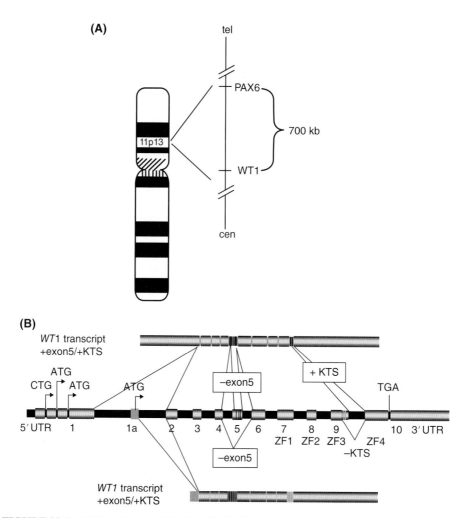

FIGURE 23.5 (A) Partial map of the distal half of band 11p13, showing relative positions of *WT1* and *PAX6* taken directly from Fischbach et al. 2005. (B) Schematic diagram of *WT1* illustrating the exons in gray and the introns in black. There are three alternative start sites in exon 1 and an additional start site in exon 1a, leading to the *AWT1* transcript, which is shorter than the regular transcript. The exon 5 and KTS alternative splice sites are illustrated with striped pattern. ZF1–ZF4 encode the zinc finger domains, responsible for DNA binding and the function of WT1 as a transcription factor, taken directly from Royer-Pokora 2013. Reprinted with permission of Springer Nature.

occurs later compared to those with DDS, but at similar ages compared to those with FS.

WT1 alterations can be associated with abnormal urogenital development (Pelletier et. al. 1991). Among individuals with a 46,XY karyotype and *WT1* missense mutations, disorders of sexual development (DSDs) range from complete sex reversal to ambiguous genitalia, hypospadias, cryptorchidism, and congenital chordee of the penis. Most individuals with 46,XX karyotype and missense mutations do not have genital abnormalities, although they may infrequently. Individuals with splice site mutations and 46,XY karyotype predominantly have complete sex reversal or ambiguous genitalia, whereas 46,XX females with a splice site mutation display normal female internal and external genitalia (Ahn et al. 2017). The genital abnormalities observed in MS appear to be confined to 46,XY males and 46,XY females. This usually ranges from ambiguous genitalia to pseudohermaphroditism (Meacham et al. 1991; Killeen et al. 2002; Suri et al. 2007). Males who are haploinsufficient for *WT1*, as seen in WAGR syndrome, have variable degrees of urogenital abnormalities ranging from hypospadias and cryptorchidism (predominantly) to ambiguous genitalia and rarely complete sex reversal (Fischbach et al. 2005; Le Caignec et al. 2007). Females may not have any genitourinary anomalies. When they do, this usually includes uterine anomalies (such as bicornuate uterus) and streak ovaries (Turleau et al. 1984; Fischbach et al. 2005). In general, the genital anomalies observed in WAGR syndrome as a result of decreased dosage of *WT1* is less severe than that observed as a result of the dominant-negative effects seen in DDS and perhaps MS (Pelletier et al. 1991; Huff 1996).

Individuals with *WT1* missense and nonsense/truncating mutations in exons 8 and 9 are at greater risk for Wilms tumor prior to age 2 years, with the risk of bilateral Wilms tumor being greater among those with truncating mutations and deletions. With rare exception, those with intron 9 splice site mutations do not develop Wilms tumor. Rather, they are at risk for gonadoblastoma, although this too can be seen infrequently among individuals with missense mutations (Chernin et al. 2010; Lehnhardt et al. 2015; Lipska et al. 2014; Ahn et al. 2017). Individuals with MS have not been reported to develop Wilms tumor, although most have not lived long enough to do so. Those who harbor a heterozygous deletion of *WT1*, as in WAGR syndrome, are at risk for Wilms tumor. The risk among those with submicroscopic deletions appears to be about two-fold greater than those with cytogenetically visible deletions. This may be due to a greater chance of cell survival among those with submicroscopic deletions (van Heyningen et al. 2007).

Mosaicism, although infrequent, is documented in the medical literature among individual with *WT1* aberrations. The phenotype of individuals with mosaicism is less severe than that of individuals with constitutional mutations. Reports in the literature describe an individual with isolated steroid-resistant nephrotic syndrome who inherited a germline *WT1* mutation from her unaffected mosaic mother (Beltcheva et al. 2016). Another family reported in the medical literature includes three children: an unaffected child, one with a germline *WT1* mutation who is an XY female, and a third who has a germline *WT1* mutation and is asymptomatic. Parental testing did not identify a *WT1* mutation suggesting that one parent is gonadal mosaic (Chak et al. 2002). There is also evidence of individuals who are mosaic for the deletion observed in WAGR syndrome (Crolla and van Heynigen 2002).

While those with missense, nonsense and splice site mutations in *WT1* do not usually present with cognitive impairment, those with WAGR syndrome do. Some with WAGR syndrome may also have autism. These features arise from the deletion of multiple genes in the 11p13 chromosomal region (Turleau et al. 1984; Fischbach et al. 2005) including *PRRG4* (G-carboxyglutamic acid 4 gene), *BDNF* (brain-derived neurotrophic factor), and *SLC1A2* (solute carrier family 1 member 2 gene) (Xu et al. 2008; Han et al. 2013; Yamamoto et al. 2013). Those with mutations confined to *WT1*, as well as partial or complete deletion of the gene, do not develop aniridia; however, those with a contiguous gene deletion of 11p13 do, as aniridia arises in the absence of *PAX6*. This may be the only presenting finding, especially among 46,XX females (Turleau et al. 1984; Fischbach et al. 2005). As individuals with pathogenic variants confined to *PAX6* can also display aniridia, a thorough clinical evaluation and molecular clarification are necessary to distinguish between WAGR syndrome and isolated aniridia. A subgroup of individuals with WAGR syndrome has also been reported to have obesity, which is attributed to the deletion of *BDNF*, a gene important in energy homeostasis (Han et al. 2008; Rodríguez-López et al. 2013).

Diaphragmatic hernia is a rare association with *WT1* mutations. The Arg366His *WT1* mutation has been found in a number of 46,XY females with diaphragmatic hernia and urogenital abnormalities who have been diagnosed as having DDS (Cho et al. 2006; Antonius et al. 2008). Some individuals with MS and diaphragmatic hernia have also been found to have exons 8 and 9 *WT1* missense mutations (Ahn et al. 2017). Interestingly, the Arg366Cys mutation has been reported in an individual with MS, the same amino acid/location that has been reported in a number of individuals with DDS and diaphragmatic hernia (Suri et al. 2007). This, again, makes one question whether MS and DDS are truly different conditions, or rather, one condition that displays varying degrees of severity due to differing mutations and/or modifier genes. Diaphragmatic hernia has also infrequently been reported among those with WAGR syndrome (Scott et al. 2005).

Individuals with *WT1* mutations appear to be at increased risk for additional neoplasms, apart from Wilms tumor and gonadoblastoma. Although infrequent, pilocytic astrocytoma and dysgerminoma have been observed in individuals with either WAGR syndrome or FS (Love et al. 2006; Subbiah et al. 2009; Mestrallet et al. 2011; Miura et al. 2016). While sometimes this neoplasm is seen as an isolated tumor, other times this is simultaneously observed with a gonadoblastoma at the time of resection. Both bilateral and recurrent disease have been reported (Guaragna et al. 2012; Miura et al. 2016). Other reported neoplasms include a cerebellar angioblastoma in a patient with a *WT1* microdeletion (Buglyó et al. 2013) and a Sertoli cell tumor in an individual with FS (Kitsiou-Tzeli et al. 2012). Currently there is no clear genotype-phenotype association with regard to rare neoplasms.

Diagnostic Testing

Molecular analysis of *WT1* is useful to confirm a clinical suspicion of DDS, FS or MS, and even more so important when an individual presents to medical attention with only a single feature such as hypospadias or steroid resistant nephrotic syndrome. Identifying a pathogenic variant in *WT1* allows one to screen for the medical problems characteristic of the syndromes such as Wilms tumor or gonadoblastoma. Overall, genetic testing helps to ensure an accurate diagnosis and appropriate medical follow-up. It is also advantageous for seemingly unaffected family members who may benefit from targeted analysis and screening. Testing family members at risk for the known familial pathogenic variant allows for phenotypic clarification, as testing may identify the less frequent clinical features associated with these conditions.

WT1 molecular analysis has previously been performed by Sanger sequencing, which tends to be more costly, laborious, and time consuming than next generation sequencing (NGS). Many labs still offer Sanger sequencing; however, most *WT1* analyses today are performed by NGS. NGS also allows for the analysis of many genes simultaneously through phenotype-driven panels that can be customized. For example, when an individual presents with steroid-resistant nephrotic syndrome, NGS offers the ability to test for a mutation in a host of genes that may contribute to this medical problem (i.e. *ACTN4, CD2AP, COQ2, INF2, LAMB2, LMX1B, NPHS1, NPHS2, PDSS2, PLCE1, SCARB2, SMARCAL1, TRPC6,* and *WT1*). For those who test positive for a *WT1* mutation, a karyotype can help determine chromosomal gender, especially among phenotypic females who may have complete sex reversal (i.e. 46,XY females).

Multiplex ligation-dependent probe amplification (MLPA), karyotyping, and/or fluorescent in situ hybridization (FISH) are techniques that can be used to make a molecular diagnosis of WAGR syndrome. However, customized targeted array comparative genome hybridization (CGH) analysis is the most efficient test to confirm the diagnosis of WAGR syndrome (Blanco-Kelly et al. 2017). This method tests for the 11p13 deletion, the specific break points (or coordinates) of the deletion, and the identification of the missing genes that may be of clinical significance. If a targeted array is not possible, oligonucleotide array CGH identifies copy number variants of the entire genome, including the 11q13 chromosomal region. Oligonucleotide array CGH will also assist with chromosomal gender identification, whereas a targeted array will not. NGS can assess for a deletion of *WT1* and *PAX6* at a single exon resolution. However, this would not provide the coordinates of the deletion and whether portions of neighboring genes could be missing as well.

Of note, pre- and post-test genetic counseling should be offered to individuals/families undergoing testing, as it could help them understand the nature, goals, possible outcomes, benefits and limitations of testing, as well as the subsequent plan of care. The inheritance pattern of the conditions would be explained and at-risk family members identified during a genetic counseling session. Genetic counseling includes a discussion regarding the reproductive options available when a familial pathogenic variant has been identified, more specifically pre-implantation genetic diagnosis and prenatal testing. Genetic counseling also addresses the psychosocial and emotional factors surrounding the diagnosis of a genetic condition in the context of the family's social setting.

Differential Diagnosis

Aniridia

Aniridia is one of the key features of WAGR syndrome. The differential diagnosis includes other disorders associated with aniridia (Samant et al. 2016; Wawrocka and Krawczynski, 2018). If chromosomal studies are normal, the most likely diagnosis is a de novo mutation in *PAX6*, although in such cases, cognitive impairment and Wilms tumor predisposition are not present. If neurological abnormalities are present and cataracts absent, Gillespie syndrome should be considered. This is an autosomal recessive disorder due to mutations in the *ITPR1* gene and characterized by aniridia, mental retardation, and cerebellar ataxia. Aniridia can also be seen in association with anterior segment dysgenesis, such as Peters anomaly and Rieger anomaly. Other genes that can cause aniridia or iris hypoplasia include *FOXC1, FOXD3, CYP1B1, TRIM44, ELP4,* and *PITX2*.

Disorders of Sexual Development
Genital abnormalities in a newborn demand a complete evaluation. Apart from the *WT1*-related disorders, there are a number of conditions associated with disorders of sexual development (DSDs) including: congenital adrenal hyperplasia (*CYP17A1*), 11-beta hydroxyl deficiency (*CYP11B1*), androgen insensitivity syndrome (*AR*), aromatase deficiency (*CYP19A1*), 17-Beta hydroxysteroid dehydrogenase III deficiency (*HSD17B3*), 3β-hydroxysteroid dehydrogenase type 2 (*HSD3β2*), cell hypoplasia type I and II (*LHCGR*), 46,XY sex reversal 3, 46,XX sex reversal 4 (*NR5A1*), Antley–Bixler syndrome (*POR*), alpha reductase deficiency (*SRD5A2*), cholesterol side chain cleavage mutations (*CYP11A1*), Smith–Lemli–Opitz (*DHCR7*), cytochrome b5 (*CYB5A*), 3α-hydroxysteroid dehydrogenase deficiency (*AKR1C2* and *AKR1C3*), *SRY*-related 46,XY DSD and 46,XY CGD, SERKAL syndrome (*WNT4*), Turner syndrome, Klinefelter syndrome, mosaicism for 45, X/45, XY, triple XXX syndrome, and XXYY syndrome. Other genes associated with gonadal dysgenesis include *SOX9, GATA4, CBX2, DHH, DMRT1, DMRT2, MAP3K1,* and *SOX8* (Witchel 2018)).

Nephrotic Syndrome
Steroid resistant nephropathy can be observed in those with *WT1* mutations. However, mutations in numerous other genes can also cause this renal disease, which can be seen as an isolated finding or as part of a syndrome (Trautmann et al. 2018; Bensimhon et al. 2018). Mutations in *LAMB2* are seen in individuals with Pierson syndrome. This condition is associated with diffuse mesangial sclerosis and ocular malformations such as hypoplasia of the iris, aplasia or atrophy of the dilator pupillae muscle, blindness, etc. Pathogenic variants found in *SMARCAL1* are responsible for Schimke syndrome. This condition is characterized by short stature, T-cell deficiency, bone dysplasia, cerebrovascular disease, and focal segmental glomerulosclerosis. Nail-patella syndrome is a condition in which individuals have short stature, sensory neural hearing loss, eye problems such as corneal opacities, cleft lip and palate, musculoskeletal abnormalities, and proteinuria associated with nephrotic syndrome. It results from mutations in *LMX1B*. *INF2* mutations have been identified in individuals with Charcot–Marie–Tooth disease. This condition is comprised

of focal segmental glomerulosclerosis, sensory neural hearing loss, musculoskeletal abnormalities, and problems with the peripheral nervous system. Galloway–Mowat syndrome is associated with mutations in *WDR73*. Affected individuals usually have nephrotic syndrome, intrauterine growth retardation and short stature, dysmorphic features, skeletal abnormalities, low tone, hypopigmentation, and neurological deficits. Mutations in *SCARB2* can be found in individuals with action myoclonus-renal failure syndrome. Affected individuals usually have epilepsy and progressive myoclonus with or without renal failure. So too, isolated forms of steroid-resistant nephrotic syndrome (SRNS) have been linked to mutations in other genes including *ACTN4, ADCK4, ALG1, ANLN, ANKFY1, APOL1, ARHGAP24, ARHGDIA, CD2AP, CDK20, CFH, COL4A3, COL4A4, COL4A5, COQ2, COQ6, COQ8B, CRB2, CUBN, DGKE, DLC1, EMP2, E2F3, FAT1, GAPVD1, INF2, ITGA3, ITGB4MEFV, ITSN1, ITSN2, LMNA, MAG12, KANK1, KANK2, KANK4, MYH9, MYO1E, NEIL1, NPHS1, NPHS2, NUP107, NUP93, NUP205, NXF5, PDSS2, PLCE1, PODXL, PMM2, PTPRO, SPGL1, SYNPO, TNS2, TRPC6, TTC21B,* and *XPO*. One can simultaneously test for multiple genes via panel gene testing. This can be helpful when there are no other physical findings to direct one to test for a specific syndrome.

Wilms Tumor There are a number of cancer predisposition syndromes in which Wilms tumor can be seen, many of which are associated with overgrowth. One such syndrome is Beckwith–Wiedemann syndrome (Chapter 9). This syndrome is classically associated with macrosomia, macroglossia, hemihypertrophy, Wilms tumor, hepatoblastoma, ear pits/creases, neonatal hypocalcemia, cardiac abnormalities, and/or omphalocele. The degree of severity ranges from no/few phenotypic features to multiple complications. In a study by Segars et al. about 4% had minor or no features of the condition (Segers et al. 2012). BWS can arise from hypo- and hypermethylation of chromosomal segments on chromosome 11p15. Familial cases have also been described as a result of a mutation in *CDKN1C*. The risk for Wilms tumor is dependent on the underlying molecular abnormality. In addition, individuals with isolated hemihypertrophy also have about a 6% risk to develop Wilms tumor (Dumoucel et al. 2014). Perlman syndrome is an overgrowth condition that results from mutations in *DIS3L1*. Individuals surviving the neonatal period have about a 55% risk for bilateral Wilms tumor. In addition to Wilms tumor, this condition is characterized by polyhydramnios and fetal overgrowth, neonatal macrosomia, visceromegaly, and dysmorphic features. Wilms tumor can be seen among those diagnosed with Sotos syndrome, an overgrowth condition caused by a mutation in *NSD1* (Dome and Huff 2016). Macrosomia, macrocephaly, dysmorphic features, heart defects, skeletal abnormalities, developmental delay and cognitive impairment, seizures and behavioral problems are some of the medical findings observed with this condition. Simpson–Golabi–Behmel syndrome type 1, arising from a mutation in *GPC3* or *GPC4*, is associated with pre- and postnatal macrosomia, dysmorphic features, intellectual disabilities with or without structural brain abnormalities, congenital heart defects, diaphragmatic hernia, genitourinary defects, GI abnormalities, umbilical hernia, skeletal abnormalities, and tumors including Wilms tumor, hepatoblastoma, neuroblastoma, gonadoblastoma, and hepatocellular carcinoma. Wilms tumor is one of many neoplasms that can occur in Li–Fraumeni syndrome, albeit infrequently. It is caused by a mutation in the *TP53* gene. Another condition associated with an increased risk for Wilms tumor includes *DICER1* syndrome, characterized by pleuropulmonary blastoma, thyroid nodules and/or cancer, cystic nephroma, ovarian Sertoli–Leydig cell tumors, gynandroblastoma, ciliary body meduepithelioma, embryonal rhabdomyosarcoma, nasal chondromesenchymal hamartomas, pineoblastoma, and pituitary blastoma. Wilms tumor is infrequently seen in association with: *PIK3CA*-related segmental overgrowth, neurofibromatosis type 1 due to a mutation in *NF1* (OMIM# 162200) (Chapter 40), mosaic variegated aneuploidy (*BUB1B*), Bloom syndrome (*BLM*), trisomy 18 syndrome (see Chapter 58), hyperparathyroid-jaw tumor syndrome (*CDC73*), Fanconi anemia type-N (biallelic *PALB2*), Fanconi anemia type D1 (biallelic *BRCA2*), and familial occurrence of Wilms tumor due to mutations in *REST*, and *CTR9*. Dome and Huff nicely summarize conditions associated with Wilms tumor predisposition in the online *GeneReviews* publication (https://www.ncbi.nlm.nih.gov/books/NBK1294/). Four new genes have been identified in association with Wilms tumor including TRIM28, FBXW7, KDM3B, and NYNRIN (Halliday et al., 2018; Mohamadallie et al., 2019; Diets et al., 2019).

Gonadoblastoma Apart from gonadoblastoma arising from mutations in *WT1*, those with Turner syndrome are also at increased risk for this neoplasm. Turner syndrome is classically defined by a 45,X karyotype; however, some individuals display 45,X/46,XX or 45,X/46,XY mosaicism or a variation thereof.

MANIFESTATIONS AND MANAGEMENT

Growth and Feeding

Individuals with DDS, FS, and those with individual features of these conditions do not have growth and feeding problems per se. However, complications of renal disease and tumors can affect growth. Some children with WAGR syndrome have low-normal birth weight and most have postnatal short stature and microcephaly, which is common to children with chromosomal aberrations (Breslow et al. 2003). Specific feeding problems have not been reported; however, obesity can be seen in some individuals with WAGR syndrome (Fischbach et al. 2005), especially those who are haplodeficient for *BDNF*, as described under genotype-phenotype associations.

Evaluation

- Children should have height and weight measured and plotted on standard pediatric growth charts during routine medical visits. It should be anticipated that those with WAGR syndrome will have mildly slow, but steady linear growth with the risk of obesity.
- Deficiency in growth rate should prompt evaluation for renal disease (urinalysis, blood pressure determination, complete blood count, and renal function tests), and tumor (abdominal/pelvic ultrasound, computed tomography, or magnetic resonance imaging).
- For those with WAGR syndrome, the risk for obesity is life-long. Children and adults should be followed regularly and weights (BMI) compared with prior visits.
- Complications of obesity, such as obstructive sleep apnea and type II diabetes mellitus, should be investigated in those with WAGR syndrome whose weight or BMI indicate obesity.

Management

- If renal disease or a tumor is identified, it is recommended that the individual be referred to the appropriate subspecialist (nephrologist and/or oncologist) for further management.
- Obesity and its complications should be managed as in the general population. Working with a dietician may prove helpful with regard to addressing appropriate diets for those at greatest risk for obesity.

Neurologic, Development, and Behavior

Individuals with mutations confined to *WT1* do not have syndrome-specific developmental delay or behavioral problems. However, complications of renal disease and tumors can affect growth and development. Intellectual disability is almost always present in WAGR syndrome. The range of cognitive impairment is broad, from normal functioning in a few individuals to severe mental retardation in the majority. A wide range of disabilities can occur including impairments of fine and gross motor skills due to hypo- and hypertonia, language acquisition, and sensory integration (Fischbach et al. 2005). The degree of severity is largely associated with the size of the deleted chromosomal segment. However, severe developmental delay and autistic features appear to be associated with deletion of *PRRG4*, *BDNF*, and *SLC1A2* (Xu et al. 2008; Han et al. 2013; Yamamoto et al. 2014). Individuals deleted for *BDNF* have low adaptive behaviors including communication, daily living skills, socialization, and motor skills (Swedo et al. 2013; Han et al. 2013). Epilepsy and absent corpus callosum have been reported (Fischbach et al. 2005).

Psychiatric and behavioral disorders seen in WAGR syndrome are reported in most affected individuals. The most frequent behavioral diagnosis is attention-deficit disorder with or without hyperactivity, particularly in males. Autism, pervasive developmental disorder, anxiety, and obsessive-compulsive disorder are also described (Fischbach et al. 2005). Deletion of *BDNF* and *SLC1A2* occur in those with WAGR syndrome and autism more frequently than those without autism (Xu et al, 2008). It should be noted that these behavioral and mental health disorders are not uncommon in any population of individuals with developmental disabilities.

Evaluation

- Among those with mutations confined to *WT1*, developmental milestones should be monitored on the routine pediatric schedule using standard screening tests such as the Denver Developmental Screening Test. School-age children should be monitored by asking about grade-appropriate academic achievement.
- Deficiencies in development identified by screening should be investigated by more in-depth assessment, including medical evaluation for possible underlying undiagnosed kidney disease or tumor.
- For those with WAGR syndrome, early developmental assessment should be initiated as soon as the child is medically stable. Ongoing assessments are important for educational and vocational planning.
- Most children with WAGR syndrome have significant visual impairment as well as a developmental disorder. Programs specializing in children with visual impairment should be sought.
- Providers should maintain awareness of attention, anxiety, and autistic spectrum disorders that may develop as the child with WAGR syndrome matures.
- Children with known renal disease or who are status-post treatment for cancer should have particularly close monitoring of development and school progress.

Management

- For those with mutations confined to *WT1* in whom delays are identified, appropriate therapies and special education should be provided in a standard manner, as seen in the general population.
- For those with WAGR syndrome, infants and children should qualify for early intervention and special education programs, including special vision services, in addition to occupational, physical, and speech therapies.

- There is no diagnosis-specific developmental intervention for those with WAGR syndrome.
- Referral to a behavioral health professional, such as a child psychologist or psychiatrist, should be made for therapy and/or pharmacological intervention, as needed.
- Consultation with social services and support groups is usually of great value for identifying financial and program resources for the family.
- Planning for adult services, including issues of guardianship, residential options, and vocational possibilities, should occur in adolescence for those with WAGR syndrome.

Musculoskeletal

A number of musculoskeletal abnormalities have been described among individuals with WAGR syndrome including tight heel cords/shortening of the Achilles tendons, scoliosis/kyphosis, metatarsal adductus, hemihyperplasia, talipes, exostoses, and syndactyly/clinodactyly (Fischbach et al. 2005). There also have been infrequent reports of duplication of hallucies (Bremond-Gignac et al. 2005). A minority of individuals with *WT1*-related conditions present for medical attention in the pre- or postnatal period with diaphragmatic hernia, requiring medical attention.

Evaluation

- Newborn examination will identify congenital limb abnormalities, such as duplicated hallucies.
- Children should be monitored periodically for progressive spinal abnormalities and heel cord tightness.
- Those identified via prenatal ultrasound as having a diaphragmatic hernia may benefit from a consultation with a fetal medicine team at a tertiary care facility, including a consultation with a neonatal/pediatric surgeon.
- Those identified postnatally as having a diaphragmatic hernia should be transferred to a tertiary care hospital where appropriate care can be provided.

Management

- Standard pediatric orthopedic and surgical care should be sought for musculoskeletal issues.

Ears and Hearing

Frequent and chronic middle ear and sinus infections occur with high frequency and these should be anticipated in individuals with WAGR syndrome. They do not appear to be as great an issue for those with other *WT1* variants. Despite the high frequency of ear infections, only a few affected individuals are reported to have hearing loss.

Evaluation

- Prompt evaluation of suspected ear and sinus infections, as well as ongoing monitoring for middle ear effusion, is warranted.
- Hearing should be tested by an audiologist in those with recurrent ear infections, as in the general population.

Management

- Referral to an experienced pediatric otolaryngologist may be necessary.
- Standard treatment of ear and sinus infections is recommended.

Ophthalmologic

Individuals with mutations confined to *WT1* are at no greater risk for ophthalmologic problems than the general population and their vision should be monitored according to standard practices. However, most individuals with WAGR syndrome who have aniridia due to a deletion of *PAX6* will have moderate-to-severe visual impairment. Aniridia, or iris hypoplasia, can cause photophobia. Aniridia associated keratopathy (AAK) affects approximately 20% of individuals with aniridia, and is a significant threat to vision in aniridia. It typically develops in the first decade of life with thickening and vascularization of the cornea. Untreated AAK results in corneal erosion, ulcerations, and fibrosis with corneal deterioration and opacification. Symptoms include dry eye, red eye, photophobia, and epiphora (Lee et al. 2008). Significant visual loss can also occur because of a combination of any or all of the following: foveal hypoplasia, optic nerve hypoplasia (10%), cataract (50–85%), corneal pannus, subluxation of the lens, and secondary glaucoma (Lee et al. 2008; Netland et al. 2011). Glaucoma develops in 50–75% of affected individuals, and is the main cause of acquired visual loss in children with aniridia. Most are affected during preadolescence or the adolescent years. Although glaucoma is most likely to develop in late childhood, individuals remain at risk into adulthood. Associated manifestations include pendular horizontal nystagmus, amblyopia, and strabismus. Ptosis, blepharophimosis, optic atrophy, microphthalmia, anterior segment anomalies, retinal dysplasia, and other ocular abnormalities have been reported (Netland et al. 2011). Aniridia can be missed in the newborn period, and the family may be the first to notice cataract, photophobia, unusually large pupils, or poor fixation. Those in whom genitourinary malformations have been identified should have an ophthalmology evaluation.

Evaluation

- Referral to an ophthalmologist experienced in the diagnosis and management of complicated aniridia should be made as soon as aniridia is diagnosed.
- Regular ophthalmological follow-up is strongly recommended for identification of evolving ophthalmic abnormalities including AAK and glaucoma.

Management

- The management of the multiple ocular complications of aniridia, including AAK, cataracts, lens subluxation, corneal opacification, and glaucoma, requires expertise. Providers should identify local, regional, and/or national centers with the experience to provide optimal care. The International WAGR Association (www.wagr.org) is an excellent resource for this information.
- Educational programming for the visually impaired should be instituted early.
- The family should be referred to community-based services for the visually impaired. Long-term care should include access to mental health professionals whose experience includes counseling caretakers of those with physical handicaps such as blindness.

Pulmonary

Reactive airway disease is reported in about 15% of individuals with WAGR syndrome (Fischbach et al. 2005). This does not appear to be an issue for those with other *WT1* aberrations. Individuals with WAGR syndrome have been infrequently described as having pulmonary hypertension (Fischbach et al. 2005). Those with MS have been described as having lung malformations including hypoplastic lungs, congenital bronchiectasis, hepatic-lung adhesion, adrenal-lung adhesion, and abnormal pulmonary vein drainage (Meacham et al. 1991; Killeen et al. 2002; Suri et al. 2007).

Evaluation

- Reactive airway disease should be evaluated as in the general population.
- Assessment by a pediatric cardiothoracic team, pre- and/or postnatally, is important for appropriate postnatal intervention among those suspicious for MS.

Management

- Standard steroid and bronchodilator therapy should be employed.
- Surgical intervention as deemed appropriate via a pediatric cardiothoracic team.

Cardiovascular

A variety of congenital heart defects have been reported among those with WAGR syndrome including patent foramen ovale, valvular hypoplasia, ventricular septal defects, patent ductus arteriosis, tetralogy of Fallot, and atrial septal defects (Fischbach et al. 2005). Those with MS have been found to have complex heart defects including hypoplastic right/left heart, univentricular hypoplastic left heart, absent mitral valve, mitral valve atresia, small aortic trunk, atrial septal defect, ventricular septal defect, dilated great vessels, coarctation of the aorta, and hypoplastic aorta (Meacham et al. 1991; Killeen et al. 2002; Suri et al. 2007).

Evaluation

- A pre- and postnatal pediatric cardiothoracic surgical evaluation is recommended for those identified with cardiac and/or pulmonary abnormalities.

Management

- Surgical intervention for congenital heart disease should be offered/initiated when appropriate.

Genitourinary

Genital Anomalies

The range of severity of genital malformations in XY males with DDS and FS is variable, but in newborns diagnosed with DDS, most are either phenotypically female or ambiguous. Genetically male infants with incomplete genital masculinization of otherwise unknown etiology are at risk for DDS. The severity of external genital anomalies ranges from normal male to complete sex reversal. Most cases of DDS have at least cryptorchidism and/or hypospadias. Internal genital anomalies include the presence of any combination of Wolffian and/or Müllerian structures, frequently inappropriate for the external genitalia. Although the testes may be normal, they are more often developmentally abnormal, ranging from streak to ovotestis. Genetically female infants with DDS syndrome usually have normal internal and external genitalia.

Complete sex-reversal is characteristic of XY females with FS. Internal genitalia are typically female with streak gonads, although duplicated uterus and/or vagina has been reported in three XY females with *WT1* mutations (Antonius et al. 2008). Primary amenorrhea and failure to enter puberty is typical for XY males. Genetically female infants with FS usually have normal internal and external genitalia.

Individuals with MS are also reported to have genital abnormalities. 46,XY sex reversal and ambiguous genitalia among genotypic males has been documented. More specifically, genital abnormalities have included duplicate vagina, absent uterus, bicornuate uterus, hypoplastic/absent palpable male gonads, ambiguous genitalia including small phallus, labioscrotal folds, bifid scrotum, shawl scrotum, uterine and fallopian tube structures in a 46,XY individual, hypospadias, and undescended testes (Meacham et al. 1991; Killeen et al. 2002; Suri et al. 2007). Cases of MS have predominantly been reported among genotypic males.

For those with WAGR syndrome, genital anomalies are usually present in males and include cryptorchidism, hypospadias, small penis, and/or hypoplastic scrotum. Occasionally, males have more severe genital ambiguity, including involvement of internal genitalia, and may be given a female gender assignment. Complete sex reversal has been reported in an XY individual with otherwise normal female internal and external genitalia, but absent ovaries (Le Caigne et al. 2007). Although females with external genital anomalies have not been reported, a variety of internal genital anomalies, including streak gonads, uterine malformations (hypoplastic versus unicornuate), and absent uterus and ovaries have been observed in genetic females (Nicholson et al. 1996; Fischbach et al. 2005). In contrast to DDS and FS, females with WAGR syndrome may have genital anomalies.

Evaluation

- Because of the complex nature of disorders associated with abnormal sexual development, the participation of a multidisciplinary team including a pediatric urologist, endocrinologist, geneticist, genetic counselor, neonatologist, and child psychiatrist/psychologist should work closely with the family to determine gender and address the patient's medical needs (Bangalore Krishna et al. 2017).
- Individuals with ambiguous genitalia and complete sex reversal, including XY females with DDS and FS, as well as those with WAGR syndrome, should have an assessment of the internal and external genitalia by physical examination and imaging: abdominopelvic ultrasound is the primary modality for demonstrating internal organs, genitography is used to assess the urethra and vagina, and magnetic resonance imaging is used as an additional modality to assess internal gonads and genitalia (Guerra-Junior et al. 2018).

Management

- With regard to surgical and hormonal interventions for those with *WT1*-related disorders, there are no syndrome specific guidelines. Rather, decisions regarding sex assignment and intervention are addressed on a case-by-case basis.
- Management of a newborn with ambiguous genitalia, or a child with complete sex reversal, should be addressed as quickly as possible and with great sensitivity for the child's family. Physical findings (internal and external structures) clarified via a methodical clinical examination and radiographic studies, endocrine function, karyotype/genetic studies, and potential for fertility should guide sex determination (Bangalore Krishna et al. 2017; Davies and Cheetham 2017).
- Individuals identified with intra-abdominal dysgenetic gonads should be evaluated by a surgeon due to risk for gonadoblastoma.
- Cryptorchidism and hypospadias should be seen by a surgeon and managed as in the general population.
- Female gender assignment for XY individuals sometimes can have psychological effects, including gender dissatisfaction (Bangalore Krishna et al. 2017). Long-term care should include access to mental health professionals with experience in disorders of sex development including support and counseling with respect to relationships and sexuality (Jürgensen et al. 2013).
- Contact with other individuals with disorders of sexual development (DSD) can provide support and exchange of experiences from non-professionals and foster friendships. Families and individuals affected by disorders of sexual development may benefit from contact with support groups to exchange experiences and learn strategies of dealing with the DSD (Jürgensen et al. 2013).

Nephropathy Steroid resistant nephropathy is present in about ~95% of children with DDS and is most commonly diffuse mesangial sclerosis. Affected individuals usually present with proteinuria, which can be congenital in occurrence. It usually progresses rapidly with reported cases of end stage renal disease (ESRD) occurring between 0.07–17 years of age and most documented prior to ~8–9 years of age. The incidence of ESRD among individuals with DDS is about 74% 20 years after unilateral Wilms tumor diagnosis, and ~50% 20 years after diagnosis of bilateral Wilms tumor. ESRD is the major cause of death in individuals with DDS (Mueller et al. 1994; Breslow et al. 2000; Breslow et al. 2005; Ahn et al. 2017). The nephropathy seen in individuals with FS is typically focal segmental glomerulosclerosis and has a later age of onset with slower progression than that seen in DDS. This form of nephropathy can also be seen in isolation, especially among females with *WT1* intron 9 splice site mutations. Proteinuria has been reported to begin around age 4–6 years of age; however, a number of papers in the medical literature have documented an age range of onset to be between 2 and 20 years of age with ESRD occurring at an average age of 9.3 years (range 3.3–29) (Moorthy et al. 1987; Chernin et al. 2010; Megremis et al. 2011; Lipska et al. 2013; Lipska et al. 2014; Yang et al. 2013a; Yang et al. 2013b; Lehnhardt et al. 2015; Ahn et al. 2017). Infrequently, renal anomalies have been reported in association with *WT1* pathogenic variants including horseshoe kidney (Guaragna et al. 2012).

Age of onset of proteinuria and focal segmental glomerular sclerosis among those with WAGR syndrome are lacking in the medical literature. However, among reported subjects, proteinuria onset has been described as presenting between 4 and 6 years of age and onset of focal segmental glomerular sclerosis between 7 and 19 years of age (Iijima et al. 2012). In one study by Fischbach et al. 26% of individuals with WAGR syndrome had proteinuria, 11% had focal segmental glomerulosclerosis, and 7.4% had end stage renal failure necessitating transplant. End stage renal disease resulted in nephrectomy due to Wilms tumor and/or focal segmental glomerulosclerosis (Fischbach et al. 2005). The incidence of end stage renal disease among individuals with WAGR syndrome is about 36% 20 years after unilateral Wilms tumor diagnosis, and ~90% 20 years after diagnosis of bilateral Wilms tumor. Most cases of end stage renal failure occur during or after adolescence (Breslow et al. 2000; Breslow et al. 2005).

Evaluation

- There are no set guidelines for nephropathy screening for those with *WT1* pathogenic variants but proteinuria is usually the first sign of disease, which can be detected by urinalysis. Most screen via this means every 3–6 months.
- All patients with steroid resistant nephrotic syndrome should be referred to a pediatric nephrologist.

Management

- Once steroid resistant nephropathy is confirmed, renal biopsy is usually performed before instituting treatment.
- There are no DDS/FS/WAGR syndrome treatment guidelines for steroid resistant nephrotic syndrome. Affected individuals may benefit from those guidelines used to treat pediatric idiopathic steroid resistant nephrotic syndrome (Indian Society of Pediatric Nephrology et al. 2009).
- Renal transplantation is an option for those who develop end stage renal disease.
- Due to the risk of Wilms tumor, especially among those with DDS and WAGR syndrome, most nephrologists recommend bilateral nephrectomy in individuals reaching end-stage renal failure prior to Wilms tumor development.

Oncologic

Wilms Tumor

Wilms tumor occurs in >70% of individuals with DDS (Mueller 1994), at a younger age than observed among those with sporadic Wilms tumors (Dumoucel et al. 2014), and an increased risk for bilateral disease (Mueller 1994). Of course, some individuals develop renal failure prior to development of Wilms tumor (Mueller 1994). There have only been rare occurrences of Wilms tumor seen in FS (Barbosa et al. 2010). Thirty to 60% of individuals with WAGR syndrome develop Wilms tumor (Fischbach et al. 2005; Dumoucel et al. 2014). Overall, about 5% of individuals with Wilms tumor have a *WT1*-related syndrome. Furthermore, *WT1* mutations have been identified in about 6% of those with Wilms tumor but no identifiable syndrome. This is predominantly bilateral Wilms tumor (Segers et al. 2012). Most Wilms tumors seen in those with DDS occur at less than 2 years with an average of 1.65 years of age (Segers et al. 2012; Mueller 1994).

The average age of onset of Wilms tumor among those with WAGR syndrome in a cohort of 64 individuals was ~22 months and 17% had bilateral disease (Breslow et al. 2003). Most individuals with WAGR syndrome who develop Wilms tumor do so prior to 6 years of age, although there have been some outliers occurring at later ages (e.g. 8 years and 24 years) (Fischbach et al. 2005). They frequently have a high incidence of intralobar nephrogenic rests and almost always exhibit favorable histology. The overall five-year survival for individuals with unilateral Wilms tumor with favorable histology is ~90%. The overall survival rate for those with bilateral disease is about 80–85% and 55% for those with anaplastic histology (Ehrlich et al. 2017; Millar et al. 2017). However, long term survival rates are decreased due to the risk to develop steroid resistant nephrotic syndrome and renal failure (Breslow et al. 2003; Breslow et al. 2005).

Evaluation

- Individuals with or suspected of having a *WT1*-related syndrome should undergo ultrasound screening for Wilms tumor every three months until age 5–6 years (Dome and Huff 2016). If nephrogenic rests or other imaging abnormalities are identified, extending the surveillance period should be considered.
- Cancer screening can be coordinated through a pediatric cancer genetics clinic, genetics professional or oncologist.
- Caretaker abdominal examination may also be of benefit between clinical examinations.
- Once a renal mass is identified, referral to an oncologist is indicated.

Management

- Treatment for Wilms tumor usually follows national protocols, and it is recommended that it be managed by a pediatric oncologist. Treatment usually involves a

combination of surgery (partial or complete nephrectomy) and chemotherapy, with additional radiation therapy for advanced disease.
- Nephron sparing surgery following preoperative chemotherapy for patients with unilateral Wilms tumor associated with predisposing syndromes, allows for the preservation of renal function and good oncologic outcomes (Romão et al. 2012).
- Nephron sparing surgery following neoadjuvant chemotherapy and repeat nephron sparing surgery for recurrent disease among those with bilateral Wilms tumor provides acceptable outcomes and preserves renal function (Kieran et al. 2014).
- Preoperative chemotherapy, surgical resection within 12 weeks of diagnosis and response, and histology-based postoperative therapy has shown an improved event free survival, overall survival, and preservation of renal parenchyma for children with bilateral Wilms tumor (Ehrlich et al. 2017).

Gonadoblastoma

Gonadoblastoma is seen more commonly with FS, and much less frequently in those with DDS or WAGR syndrome. The ages of onset reported in the medical literature range from about 6 years to 22 years of age with most occurring in the teen years (Wong et al. 2017; Hersmus et al. 2012). Failure to enter puberty is frequently the stimulus, which reveals dysgenic/streak gonads and gonadoblastoma. There have been a few individuals with DDS reported in the medical literature with gonadoblastoma occurrence less than three years of age (Mueller 1994; Auber et al. 2003; Finken et al. 2015). A couple of individuals with WAGR syndrome have been reported with gonadoblastoma but the age of onset is not clear (Fischbach et al. 2005). Dysgerminoma has also been infrequently described among individuals with WAGR and FS.

Evaluation

- No specific screening recommendations for gonadoblastoma or other gonadal tumors have been published in the medical literature, which are specific to individuals with *WT1*-related disorders.
- If medical findings are suspicious for a *WT1*-related condition, referral to a cancer genetics team to discuss genetic testing and screening should be considered.
- Males with normally descended testes should have testicular examination as part of their routine well-child care visits, and should be taught testicular self-examination at age 6 years if possible. Examination by a health care provider at least biannually through adolescence is suggested and at least annually thereafter.
- Testicular biopsy should be performed if a mass or other abnormality is detected.
- Males with undescended testes should undergo testicular biopsy at the time of orchidopexy to evaluate for dysgenetic gonads.

Management

- Removal of intra-abdominal streak gonads is indicated as soon as the diagnosis of XY sex reversal in phenotypic females is made and can be done laparoscopically (Calvo et al. 2016).
- The removal of histologically dysgenetic gonads in males with ambiguous genitalia should be considered.
- Gonadectomy is recommended if streak gonads are identified in XX females with WAGR syndrome.

RESOURCES

Genetic Home Reference has information and useful links on Denys–Drash syndrome, Frasier syndrome, WAGR syndrome, and Meacham syndrome
   Website: https://ghr.nlm.nih.gov/condition/denys-drash-syndrome
   https://ghr.nlm.nih.gov/condition/frasier-syndrome#diagnosis
   https://ghr.nlm.nih.gov/condition/wagr-syndrome
   https://ghr.nlm.nih.gov/search?query=meacham

Genetic and Rare Disease (GARD) Information Center has useful information and links on Denys–Drash syndrome, Frasier syndrome, WAGR syndrome, and Meacham syndrome
   Website: https://rarediseases.info.nih.gov/diseases/5576/denys-drash-syndrome
   https://rarediseases.info.nih.gov/diseases/2375/frasier-syndrome
   https://rarediseases.info.nih.gov/diseases/5528/wagr-syndrome
   https://rarediseases.info.nih.gov/diseases/3432/meacham-winn-culler-syndrome

National Organization for Rare Disorders (NORD) has information and useful links, found under Denys–Drash syndrome
   Website: https://rarediseases.org/

"NephKids" cyber support group is a resource for renal disease information and support
   Website: https://nephkids.in/

The International WAGR Syndrome Association
   Information and Support for Families and Professionals
   PO Box 79

Hanover, PA 17331
USA
Website: *www.wagr.org*
The National Organization for Rare Disease (NORD) has information regarding WAGR syndrome and 11p-deletion syndrome
Website: *https://rarediseases.org/rare-diseases/wagr-syndrome11p-deletion-syndrome/*
AIS DSD Support Group
PO Box 2148 Duncan, Oklahoma 73534-2148, USA
Website: http://aisdsd.org/
Accord Alliance
531 Route 22 East #244
Whitehouse Station, NJ 08889, USA
Website: http://www.accordalliance.org/
The Aniridia Network International
International internet-based support group for people with aniridia and their families
Website: www.aniridia.com.

REFERENCES

Ahn YH, Park EJ, Kang HG, Kim SH, Cho HY, Shin JI, Lee JH, Park YS, Kim KS, Ha I, Cheong H (2017) Genotype-phenotype analysis of pediatric patients with WT1 glomerulopathy. *PediatrNephrol* 32(1):81–89.

Andrade JG, Guaragna MS Soardi, FC Guerra-Junior, G Mello, MP Maciel-Guerra AT (2008) Clinical and genetic findings of five patients with WT1-related disorders. *Arquivos Brasileiros de Endocrinolgia Metabolgia* 52(8):1236–1243.

Antonius T, van Bon B, Eggink A, van der Burt I, Noordam K, van Heijst A (2008) Denys-Drash syndrome and congenital diaphragmatic hernia: Another case with the 1097G>A (Arg366His) mutation. *Am J Med Genet A* 146:496–499.

Auber F, Lortat-Jacob S, Sarnacki S, Jaubert F, Salomon R, Thibaud E, Jeanpierre C, Nihoul-Fékété C (2003) Surgical management and genotype/phenotype correlations in WT1 gene-related diseases (Drash, Fasier syndromes). *J Pediat Surg* 38(1):124–129.

Bangalore Krishna KB, Houk CP, Lee PA (2017) Pragmatic approach to intersex, including genital ambiguity, in the newborn. *Semin Perinatol* 41(4):244–251.

Barbaux S, Niaudet P, Gubler MC, Gruenfeld JP, Jaubert E, Kuttenn F, Fékété CL, Souleyreau-Therville N, Thibaud E, Fellous M, McElreavey K (1997) Donor splice site mutations in WT1 are responsible for Frasier syndrome. *Nat Genet* 17:467–470.

Barbosa AS, Hadijathanasiou CG, Theodoridis G, Papathanasiou A, Tar A, Merksz M, Gyoryari B, Sultan C, Durmas R, Jaubert F, Niaudet P, Moreira-Filho CA, Cotinot C, Fellous M (2010) The same mutation affecting the splicing of WT1 gene is present in Frasier syndrome patients with or without Wilms tumor. *Hum Mut* 13(2):146–153.

Beltcheva O, Boueva A, Tzveova R, Roussinov D, Marinova S, Kaneva R, Mitev V (2016) Steroid-resistant nephrotic syndrome caused by novel WT1 mutation inherited from a mosaic parent. *Renal Failure* 38(2):290–293.

Benoit G, Machuca E, Antignac C (2010) Hereditary nephrotic syndrome: a systematic approach for genetic testing and a review of associated podocyte gene mutations. *Pediat Nephrol* 25:1621–1632.

Bharathavikru R, Dudnakova T, Aitken S, Slight J, Artibani M, Hohenstein P, Tollervey D, Hastie N (2017) Transcription factor Wilms' tumor 1 regulates developmental RNAs through 3' UTR interaction. *Genes Dev* 31(4):347–352.

Blanco-Kelly F, Palomares M, Vallespin E, Villaverde C, Martin-Arenas R, Velez-Monsalve C, Lorda-Sanchez I, Nevado J, Trujillo-Tiebas MJ, Lapunzina P, Ayusa C, Corton M (2017) Improving molecular diagnosis of aniridia and WAGR syndrome using customized targeted array-based CGH. *PLoS One* 12(12):1–13.

Brémond-Gignac D, Gérard-Blanluet M, Copin H, Bitoun P, Baumann C, Crolla JA, Benzacken B, Verloes A (2005) Three patients with hallucal polydactyly and WAGR syndrome, including discordant expression of Wilms tumor in MZ twins. *Am J Med Genet A* 134(4):422–425.

Breslow NE, Takashima JR, Ritchey ML, Strong LC, Green DM (2000) Renal failure in the Denys-Drash and Wilms' tumor-aniridia syndromes. *Cancer Res* 60(15):4030–4032.

Breslow N.E, Norris R, Norkool P.A, Kang T, Beckwith J.B, Perlman E.J, Ritchey M.L, Green D.M, Nichols KE (2003) Characteristics and outcomes of children with the Wilms tumor-Aniridia syndrome: A report from the National Wilms tumor study group. *J Clin Oncol* 21(24):4579–4585.

Breslow NE, Collins AJ, Ritchey ML, Grigoriev YA, Peterson SM, Green DM (2005) End stage renal disease in patients with Wilms tumor: results from the National Wilms tumor study group and the United States Renal Data System. *J Urol* 174:1972–1975.

Brusa P, Torricelli C (1953) Wilms' nephroblastoma and congenital renal diseases in the case reports of the Istituto Provinciale di Protezione ed Assistenza dell'Infanzia di Milano. *Minerva Pediatrica* 5(12):457–463.

Buglyó G, Méhes G, Vargham G, Biróm S, Mátyus J (2013) WT1 microdeletion and slowly progressing focal glomerulosclerosis in a patient with male pseudohermaphroditism, childhood leukemia, Wilms tumor and cerebellar angioblastoma. *Clin Nephrol* 79(5):414–418.

Call KM, Glaser T, Ito CY, Buckler AJ, Pelletier J, Haber DA, Rose EA, Karl A, Yager H, Lewis WH, Jones C, Housman D..(1990) Isolation and characterization of a zinc finger polypeptide gene at the human chromosome 11 Wilms tumor locus. *Cell* 60:509–520.

Calvo A, Escolino M, Settimi A, Roberti A, Caprio MG, Esposito C (2016) Laparoscopic approach for gonadectomy in pediatric patients with intersex disorders. *Transl Pediat* 5(4):294–304.

Capito C, Leclair MD, Arnaud A, David A, Baron S, Corradini N, Heloury Y (2011) 46,XY pure gonadal dysgenesis: clinical presentations and management of the tumor risk. *J Pediat Urol* 7(1):72–75.

Chak WL, To KF, Cheng YL, Tsui KM, Lo KW, Tong HM, Lai FM, Wong FK, Choi KS, Chau KF, Li C (2002) Gonadal mosaicism of Frasier syndrome in 3 Chinese siblings with donor splice site mutations of Wilms' tumour gene. *Nephron* 91(3):526–529.

Chernin G, Vega-Warner VV, Schoeb DS, Heeringa SF, Ovunc B, Saisawatt P, Cleper R, Ozaltin F, Hildebrandt F, Members of the GPN Study Group (2010). Genotype/phenotype correlation in nephrotic caused by WT1 mutations. *Clin J Am Soc Nephrol* 5:1655–1662.

Cho HY, Lee BS, Kang CH, Kim WH, Ha IS, Cheong HI, Choi Y (2006) Hydrothorax in a patient with Denys-Drash syndrome associated with a diaphragmatic defect. *Pediat Nephrol* 21(12):1909–1912.

Crolla JA, and van Heynigen V (2002) Frequent chromosome aberrations revealed by molecular cytogenetic studies in patients with aniridia. *Am J Hum Genet* 71(5):1138–1149.

da Silva TE, Nishi MY, Costa EM, Martin RM, Carvalho FM, Mendonca BB, Domenice S (2011) A novel WT1 heterozygous nonsense mutation (p.K248X) causing a mild and slightly progressive nephropathy in a 46,XY patient with Denys-Drash syndrome. *Pediat Nephrol* 26(8):1311–1315.

Davies J, Cheetham T (2017) Recognition and assessment of atypical and ambiguous genitalia in the newborn. *Arch Dis Childhood* 102(10):968–974.

Denys P, Malvaux P, Van Den Berghe H, Tanghe W, Proesmans W (1967) Association of an anatomo-pathological syndrome of male pseudohermaphroditism, Wilms' tumor, parenchymatous nephropathy and XX/XY mosaicism. *Arch Francaises De Pediatrie* 24(7):729–739.

Diets IJ, Hoyer J, Ekici AB, Popp B, Hoogerbrugge N, van Reijmersdal SV, Bhaskaran R, Hadjihannas M, Vasileiou G, Thiel CT, Seven D, Uebe S, Ilencikova D, Waanders E, Mavinkurve-Groothuis AMC, Roeleveld N, de Krijger RR, Wegert J, Graf N, Vokuhl C, Agaimy A, Gessler M, Reis A, Kuiper RP, Jongmans MCJ, Metzler M (2019) www.ncbi.nlm.nih.gov/pubmed/30694527 8 haploinsufficiency predisposes to Wilms tumor. *Int J Cancer* 145(4):941–951.

Drash A, Sherman F, Harmann WH, Blizzard RM (1970) A syndrome of pseudohermaphroditism, WIlms' tumor, hypertension, and degenerative renal disease. *J Pediat* 76(4):585–593.

Dome JS, Huff V (2016) Wilms Tumor Predisposition. *GeneReviews®* Available from: https://www.ncbi.nlm.nih.gov/books/NBK1294/.

Dumoucel S, Gauthier-Villars M, Stoppa-Lyonnet D, Parisot P, Brisse H, Philippe-Chomette P, Sarnacki S, Boccon-Gibod L, Rossignol S, Baumann C, Aerts I, Bourdeaut F, Doz F, Orbach D, Pacquement H, Michon J, Schleiermacher G (2014) Malformations, genetic abnormalities, and Wilms tumor. *Pediat Blood Cancer* 61:140–144.

Ehrlich P, Chi YY, Chintagumpala MM, Hoffer FA, Perlman EJ, Kalapurakal JA, Warwick A, Shamberger RC, Khanna G, Hamilton TE, Gow KW, Paulino AC, Gratias EJ, Mullen EA, Geller JI, Grundy PE, Fernandez CV, Ritchey ML, Dome JS. (2017) Results of the first prospective multiinstitutional treatment study in children with bilateral wilms tumor (AREN0534): a report from the Children's Oncology Group. *Ann Surg* 266(3):470–478.

Fenci F, Malina M, Stara V, Zieg J, Mixova D, Seeman T, Blahavoa K (2012) Discordant expression of a new WT1 gene mutation in a family with monozygotic twins presenting with congenital nephrotic syndrome. *Eur J Pediat* 171(1):121–124.

Finken MJ, Hendriks YM, van der Voom JP, Veening MA, Lombardi MP, Rotteveel J (2015) WT1 deletion leading to severe 46,XY gonadal dysgenesis, WIlms tumor and gonadoblastoma:case report. *Horm Res Paediat* 83(3):211–216.

Fischbach BV, Trout KL, Lewis J, Luis CA, Sika M (2005) WAGR syndrome: a clinical review of 54 cases. *Pediat* 116(4):984–989.

Frasier S.D, Bashore R.A, Mosier H.D (1964) Gonadoblatoma associated with pure gonadal dysgenesis in monozygous twins. *J Pediat* 64:740–745.

Gessler M, Poustka A, Cavanee W, Nerve RL, Orkin SH, Bruns GA (1990) Homozygous deletion in Wilms tumours of a zinger finger identified by chromosomal jumping. *Nature* 343 774–778.

Guaragna MS, Lutaif AC, Bittencourt VB, Piveta CS, Soardi FC, Castro LC, Belangero VM, Marciel-Guerra AT, Guerra-Junior G, Mello MP (2012) Frasier syndrome: four new cases with unusual presentations. *Arquivos Brasileiros De Endocrinologia E Metabologia* 56(8):525–532.

Guaragna M, Ribeiro de Andrade J, de Freitas Carli B, Belangero V, Maciel-Guerra A, Guerra-Júnior G, de Mello M (2017) WT1 haploinsuffiiency supports milder renal manifestation in two patients with Denys-Drash Syndrome. *Sex Dev* 11(1):34–39.

Guerra-Junior G, Andrade KC, Barcelos IHK, Maciel-Guerra AT (2018) Imaging techniques in the diagnostic journey of disorders of sex development. *Sex Dev* 12(1–3):95–99.

Han JC, Liu QR, Jone MP, Leyinn RL, Menzie CM, Jefferson-George KS, Alder-Wailes DC, Sanford EL, Lacbawan FL, Uhl GR, Rennert OM, Yanovski JA (2008) Brain-derived neurotrophic factor and obesity in the WAGR syndrome. *New Engl J Med* 359(9):918–927.

Halliday BJ, Fukuzawa R, Markie DM, Grundy RG, Ludgate JL, Black MA, Skeen JE, Weeks RJ, Catchpoole DR, Roberts AGK, Reeve AE, Morison IM (2018) www.ncbi.nlm.nih.gov/pubmed/29912901 Germline mutations and somatic inactivation of TRIM28 in Wilms tumour. *PLoS Genet Jun* 18;14(6):e1007399.

Han JC, Thurm A, Golden Williams C, Joseph LA, Zein WM, Brooks BP, Butman JA, Brandy SM, Fuhr SR, Hicks MD, Huey AE, Hanish AE, Danley KM, Raygada MJ, Rennert OM, Martinowich K, Sharp SJ, Tsao JW, Swedo SE (2013) Association of brain-derived neurotrophic factor (BDNF) haploinsufficiency with lower adaptive behavior and reduced cognitive functioning in WAGR/11p13 deletion syndrome. *Cortex* 49(10):2700–2710.

Hersmus RH, van der Zwan YG, Stoop H, Bernard P, Sreenivasan R, Oosterhauis J, Brüggenwirthm H, De Boer S, White S, Wolffenbuttel KP, Alders M, McElreavy K, Drop SlS, Harley VR, Looijenga LHJ (2012) A 46,XY Female DSD Patient with Bilateral Gonadoblastoma, a Novel SRY Missense Mutation Combined with a WT1 KTS Splice-Site Mutation. *PLoS ONE* 7(7):1–8.

Huff V, Meadows A, Riccardi VM, Strong L, Saunders Gf (1990) Parental origin of de novo constitutional deletions of chromosomal band 11p13. *Am J Hum Genet* 47(1):155–60.

Huff V (1996) Genotype/phenotype correlations in Wilms' tumor. *Med Pediat Oncol* 27(5):408–414.

Huff V (2011) Wilms' tumours: about tumour suppressor genes, an oncogene and a chameleon gene. *Nat Rev Cancer* 11(2):111–121.

Iijima K, Someya T, Ito S, Nakanishi K, Ohashi H, Nagata M, Kamei K, Sasaki S (2012) Focal segmental glomerulosclerosis in patients with complete deletion of one WT1 allele. *Pediatrics* 129(6):e1621–1625.

Indian Society of Pediatric Nephrology, Gulati A, Bagga A, Gulati S, Mehta K, Vijayakumar M (2009) Management of steroid resistant nephrotic syndrome. *Ind Pediat* 46(1):35–47.

Ismaili K, Verdure V, Vandenhoute K, Janssen F, Hall M (2008) WT1 gene mutations in three girls with nephrotic syndrome. *Eur J Pediat*, 167 579–581.

Jürgensen M, Kleinemeier E, Lux AL, Steensma TD, Cohen-Kettenis PT, Hiort OH, Thyen U, Köhler, B.; DSD Network Working Group (2013) Psychosexual development in adolescents and adults with disorders of sex development- results from the German Clinical Evaluation Study. *J Sex Med* 10(11):2703–2014.

Kieran K, Williams MA, McGregor LM, Dome JS, Krasin MJ, Davidoff AM (2014) Repeat nephron-sparing surgery for children with bilateral Wilms tumor. *J Pediat Surg* 49(1):149–153.

Killeen OG, Kelehan P, Reardon W (2002) Double vagina with sex reversal, congenital diaphragmatic hernia, pulmonary and cardiac malformations—another case of Meacham syndrome. *Clin Dysmorphol* 11(1):25–28.

Kitsiou-Tzeli S, Deligiorgi M, Malaktari-Skarantavou S, Vlachopoulos C, Megremis S, Fylaktou I, Traeger-Synodrinos J, Kanaka-Gantenbein C, Stefanadis C, Kanavakis E (2012) Sertoli cell tumor and gonadoblatsoma in an untreated 29-year-old 46, XY phenotypic male with Frasier syndrome carrying a WT1 IVS9+4C>T mutation. *Hormones (Athens)* 11(3):361–367.

Köhler B, Schumacher V, Schulte-Overberg U, Biewald W, Lennert T, l'Allemand D, Royer-Pokora B, Gruters A (1999) Bilateral Wilms tumor in a boy with severe hypospadias and cryptorchidism due to a heterozygous mutation in the WT1 gene. *Pediat Res* 45(2):187–190.

Köhler B, Biebermann H, Friedsam V, Gellermann J, Maier R, Pohl M, Wieacker P, Hiort O, Gruter A, Krude H (2011) Analysis of the Wilms' tumor suppressor gene (WT1) in patients 46,XY disorders of sex development. *J Clin Endocrinol Metabol* 96(7): E1131–E1136.

Lavedan C, Barichard F, Azoulay M, Couillin P, Molina G, Nicolas H, Quack B, Rethoré M, Noel B, Junien C (1989) Molecular definition of de novo and genetically transmitted WAGR-associated rearrangements of 11p13. *Cytogenet Cell Genet* 50(2–3):70–74.

Le Caignec C, Delnatte C, Vermeesch J.R, Boceno M, Joubert M, Lavenant F, David A, Rival JM.(2007) Complete sex reversal in a WAGR syndrome patient. *Am J Med Genet A* 143A(22): 2692–2695.

Lee D.G, Han DH, Park KH, Baek M (2011) A novel WT1 mutation in a patient with Wilms' tumor and 46, XY gonadal dysgenesis. *Eur J Pediat* 170 1079–1082.

Lehnhardt A, Karnatz C, Ahlenstiel-Grunow T, Benz K, Benz MR, Budde K, Büscher AK, Fehr T, Feldkotter M, Graf N, Hocker B, Jungraithmayr T, Klaus G, Koehler B, Konrad M, Kranz B, Montoya CR, Muller D, Neuhaus TJ, Oh J, Pape L, Pohl MJ (2015) Clinical and molecular characterization of patients with heterozygous mutations in Wilms tumor suppressor gene 1. *Clin J Am Soc Nephrol* 10 825–831.

Lipska B, Latropoulos P, Maranta R, Caridi G, Ozaltin F, Anarat A, Balat A, Gellermann J, Trautmann A, Erdogan O, Saeed B, Emre S, Bogdanovic R, Azocar M, Balasz-Chmielewska I, Benetti E, Caliskan S, Mir S, Melk A, Ertan P, Baskin E, Jardim H, Davitaia T, Wasilewska A, Drozdz D, Szczepanska M, Jankauskiene A, Higuita L, Ardissino G, Ozkaya O, Kuzma-Mroczkowska E, Soylemezoglu O, Ranchin B, Medynska A, Tkaczyk M, Peco-Antic A, Akil I, Jarmolinski T, Firszt-Adamczyk A, Dusek J, Simonetti G, Gok F, Gheissari, A, Emma F, Krmar R, Fischbach M, Printza N, Simkova E, Mele C, Ghiggeri G, Schaefer F. PodoNet Consortium (2013)Genetic screening in adolescents with steroid-resistant nephrotic syndrome. *Kidney Int* 84(1):206–213.

Lipska BS, Ranchin B, Iatropoulos P, Gellermann J, Melk A, Ozaltin F, Caridi G, Seeman T, Tory K, Jankauskiene A, Zurowska A, Szczepanska M, Wasilewska A, Harambat J, Trautmann A, Peco-Antic A, Borzecka H, Moczulska A, Saeed B, Bogdanovic R, Kalyoncu M, Simkova E, Erdogan O, Vrljicak K, Teixeira A, Azocar M, Schaefer F, PodoNet Consortium (2014) Genotype-phenotype associations in WT1 glomerulopathy. *Kidney Int* 85:1169–1178.

Little SE, Hanks SP, King-Underwood L, Jones C, Rapley EA, Rahman N, Pritchard-Jones K (2004) Frequency and Heritability of WT1 mutations in nonsyndromic Wilms' tumor patients: a UK children's cancer study group study. *J Clin Oncol* 22(20):4140–4146.

Little S, Hanks S, King-Underwood L, Picton S, Cullinane C, Rapley E, Rahman N, Pritchard-Jones K (2005) A WT1 exon 1 mutation in a child diagnosed with Denys-Drash syndrome. *Pediat Nephrol* 20(1):81–85.

Love J, DeMarinim S Coppolam, C (2006) Prophylactic bilateral salpingo-oopherectomy in a 17-year-old with Frasier syndrome reveals gonadoblastoma and seminoma: a case report. *J Pediat Surg* 41(11):e1–4.

Mahamdallie S, Yost S, Poyastro-Pearson E, Holt E, Zachariou A, Seal S, Elliott A, Clarke M, Warren-Perry M, Hanks S, Anderson J, Bomken S, Cole T, Farah R, Furtwaengler R, Glaser A, Grundy R, Hayden J, Lowis S, Millot F, Nicholson J, Ronghe M, Skeen J, Williams D, Yeomanson D, Ruark E, Rahman N (2016) https://www.ncbi.nlm.nih.gov/pubmed/30885698 Identification of new Wilms tumour predisposition genes: an exome sequencing study. *Lancet Child Adolesc Health* 3(5):322–331.

Meacham LR, Winn KJ, Culler FL, Parks JS (1991) Double vagina, cardiac, pulmonary, and other genital malformations with 46, XY karyotype. *Am J Med Genet* 41(4):478–481.

Megremis S, Mitsioni A, Fylaktou I, Tzeli SK, Komianou F, Stefanidis CJ, Kanavakis E, Traeger-Synodinos J (2011) Broad and unexpected phenotypic expression in Greek children with steroid-resistant nephrotic syndrome due to mutations in the Wilms' tumor 1 (WT1) gene. *Eur J Pediat* 170 1529–1534.

Mestrallet G, Bertholet-Thomas A, Ranchin B, Bouvier R, Frappaz D, Cochat P (2011) Recurrence of a dysgerminoma in Frasier syndrome. *Pediat Transplant* 15(3):e53–55.

Millar AJW, Cox S, Davidson A (2017) Management of bilateral WIlms tumours. *Pediat Surg Int* 33(7):7370745.

Miller RW, Fraumeni JF, Manning MD (1964) Association of Wilms's tumor with aniridia, hemihypertropy, and other congenital malformations. *New Engl J Med* 270 922–927.

Miura R, Yokoyama Y, Shigeto T, Futagami M, Mizunuma H, Kurose A, Tsuruga K, Sasaki S, Terui K, Ito E (2016) Dysgerminoma developing from an ectopic ovary in a patient with WAGR syndrome: A case report. *Mol Clin Oncol* 5(5):503–506.

Modi J, Modi P, Pal B, Kumar S (2015) Bilateral Wilms' tumors in an infant with Denys-Drash syndrome and rarely seen truncation mutation in the WT1 gene-exon 6. *J Ind Ass Pediat Surgeons* 20(4):197–198.

Moorthy AV, Chesney RW, Lubinsky M (1987) Chronic renal failure and XY dysgenesis: "Frasier" syndrome—a commentary on reported cases. *Am J Med Genet Suppl* 3:297–302.

Mueller RF (1994) The Denys-Drash syndrome. *J Med Genet* 31:471–477.

Netland P, Scott M, Boyle J, Lauderdale J (2011) Ocular and systemic findings in a survey of aniridia subjects. *J Am Ass Pediat Ophthalmol Strabismus* 15(6):562–566.

Nicholson HS, Blask AN, Markle BM, Reaman GH, Byrne J (1996) Uterine anomalies in Wilms' tumor survivors. *Cancer* 78(4):887–891.

Patel PR, Pappas J, Arva NC, Franklin B, Brar P (2013) Early presentation of bilateral gonadoblastomas in a Denys-Drash syndrome patient: a cautionary tale for prophylactic gonadectomy. *J Pediat Endocrinol Metabol* 26(9–10):971–974.

Pelletier J (1991) Germline mutations in the Wilms' tumor suppressor gene are associated with abnormal urogenital development in Denys-Drash syndrome. *Cell* 67:437–447.

Pritchard-Jones K (1999) The Wilms tumour gene, WT1, in normal and abnormal nephrogenesis. *Pediat Nephrol* 13:620–625.

Riccardi VM, Sujansky E, Smith AC, Francke U (1978) Chromosomal imbalance in the Aniridia-Wilms' tumor association: 11p interstitial deletion. *Pediat* 61:604–610.

Robinson DO, Howarth RJ, Williamson KA, van Heyningen V, Beal SJ, Crolla JA (2008) Genetic analysis of chromosome 11p13 and the PAX6 gene in a series of 125 cases referred with aniridia. *Am J Med Genet A* 146A(5):558–569.

Rodríguez-López R, Carbonell Pérez JM, Balsera AM, Rodríguez GG, Moreno TH, García de Cáceres M, Serrano MG, Freijo FC, Ruiz JR, Angueira FB, Pérez PM, Estévez MN, Gómez EG (2013) The modifier effect of the BDNF gene in the phenotype of WAGRO syndrome. *Gene* 516(2):285–290.

Romão RLP, Pippi Salle JL, Shuman C, Weksberg R, Figueroa V, Weber B, Bägli DJ, Farhat WA, Grant R, Gerstle JT, Lorenzo AJ (2012) Nephron sparing surgery for unilateral Wilms tumor in children with predisposing syndromes: single center experience over 10 years. *J Urol* 188(4 Suppl):1493–1498.

Royer-Pokora B, Beier M, Henzler M, Alam R, Schumacher V, Weirich A, Huff V (2004) Twenty-four new cases of WT1 germline mutations and review of the literature: genotype/phenotype correlations for Wilms tumor development. *Am J Med Genet A* 127 249–257.

Royer-Pokora (2013) Genetics of Pediatric Renal Tumors, *Pediatr Nephrol* 28:13–23. DOI 10.1007/s00467-012-2146-4.

Ruf RG, Schultheiss M, Lichtenberger A, Karle S.M, Zalewski I, Mucha B, Everding A.S, Neuhaus T, Patzer L, Plank C, Haas J.P, Ozaltin F, Imm A, Fuchshuber A, Bakkologlu A, Hildebrandt F, APN Study Group (2004) Prevalence of WT1 mutations in a large cohort of patients with steroid-resistant and steroid-sensitive nephrotic syndrome. *Kid Int* 66(2):564–570.

Saxena AK, van Tuil C, Schultze-Everding A (2006) Frasier syndrome in a pre-menarchal girl: laparoscopic resection of gonadoblastoma. *Eur J Pediat* 265(12):917–919.

Scott D, Cooper M, Stankiewicz P, Patel A, Potocki L, Cheung S (2005) Congenital diaphragmatic hernia in WAGR syndrome. *Am J Med Genet A* 134(4):430–433.

Samant MChauhan, BK Lathrop, KL Nischal, KK (2016) Congenital aniridia: etiology, manifestations, and management. *Exp Rev* 11(2):135–144.

Segers H, Kersseboom R, Alders M, Pieters R, Wagner A, van den Heuvel-Eibrink MM (2012) Frequency of WT1 and 11p15 constitutional aberrations and phenotypic correlation in childhood Wilms tumour patients. *Eur J Cancer*, 48: 3249–3256.

Subbiah V, Huff V, Wolff JE, Ketonen L, Lang FF Jr, Stewart J, Langford L Herzog, CE (2009) Bilateral gonadoblastoma with dysgerminoma and pilocytic astrocytoma with WT1 GT-IVS9 mutation: A 46 XY phenotypic female with Frasier syndrome. *Pediat Blood Cancer* 53(7):1349–1351.

Suri M, Kelehan P, O'Neill D, Vadeyar S, Grant J, Ahmed SF, Tolmie J, McCann E, Lam W, Smith S, FitzPatrick D, Hastie ND, Reardon W (2007) WT1 mutations in Meacham syndrome suggest a coelomic mesothelial origin of the cardiac and diaphragmatic malformations. *Am J Med Genet A* 143A 2312–2320.

Swedo SE (2013) Association of brain-derived neurotrophic factor (BDNF) haploinsufficiency with lower adaptive behavior and reduced cognitive functioning in WAGR/11p13 deletion syndrome. *Cortex* 49(10):2700–2710.

Swiatecka-Urban A, Mokryzcki MH, Kaskel F, Da Silva F, Denamur E (2001) Novel WT1 mutation (C388Y) in a female child with Denys-Drash syndrome. *Pediat Nephrol* 16(8):627–630.

Toska E, Roberts S (2014) Mechanisms of transcriptional regulation by WT1 (Wilms' tumour 1). *Biochem J* 46(1):15–32.

Trautmann A, Lipska-Ziętkiewicz BS, Schaefer F (2018) Exploring the Clinical and Genetic Spectrum of Steroid Resistant Nephrotic Syndrome: The PodoNet Registry. *Front Pediat* 6:1–15.

Turleau C, de Grouchy J, Tournade MF Gagdanoux M.F, Junien C (1984) Del11p13/aniridia complex. Report of three patients and review of 37 observations from the literature. *Clin Genet* 67:455–456.

van Heyningen V, Hoover JM, de Kraker J, Crolla JA (2007) Raised risk of WIlms tumor in patients with aniridia and submicroscopic WT1deletion. *J Med Genet* 44:787–790.

Wawrocka A, Krawczynski MR (2018) The genetics of aniridia - simple things become complicated. *J Appl Genet* 59(2):151–159.

Witchel SF (2018) Disorders of sexual development. *Best Practices Res Clin Obstet Gynaecol* 48:90–102.

Wong Y, Tam Y, Pang K, To K, Chan S, Chan K, Lee K (2017) Clinical heterogeneity in children with gonadal dysgenesis associated with non-mosaic 46, XY karyotype. *J Pediat Urology* 13(5):508.

Xu, S Han, JC Morales, A Menzie, CM Williams, K Fan, YS (2008) Characterization of 11p14-p12 deletion in WAGR syndrome by array CGH for identifying genes contributing to mental retardation and autism. *Cytogenet Genome Res* 122(2):181–187.

Yamamoto T, Togawa M, Shimada S, Sangu N, Shimojima K, Okamoto N (2013) Narrowing of the responsible region for severe developmental delay and autistic behaviors in WAGR syndrome down to 1.6 Mb including PAX6, WT1, and PRRG4. *Am J Med Genet* 164A(3):634–638.

Yang YH, Feng D, Huang J, Nie X, Yu Z (2013a) A child with isolated nephrotic syndrome and WT1 mutation presenting as a 46,XY phenotypic male. *Eur J Pediat* 172:127–129.

Yang YH, Zhao F, Feng DN, Wang JJ, Wang CF, Hyang J, Nie XJ, Xia GZ, Chen GM, Yu ZH (2013b) Wilms' tumor suppressor gene mutations in girls with sporadic isolated steroid-resistant nephrotic syndrome. *Genet Mol Res J* 12(4):6184–6191.

Yang Y, Zhao F, Huang J, Nie X, Yu Z (2013c) Patients with different or identical genotypes of the WT1 gene present different phenotypes. *Eur J Pediat*, 172 1707–1708.

Yang Y, Zhao F, Tu X, Yu Z (2016) Mutations in WT1 in boys with sporadic isolated steroid-resistant nephrotic syndrome. *Genet Mol Res* 15(1):15017559.

Yu Z, Yang Y, Feng D (2012) Discordant phenotypes in monozygotic twins with identical de novo WT1 mutation. *Clin Kidney J* 5 221–222.

Zhu C, Zhao F, Zhang W, Wu H, Chen Y, Ding G, Zhang A, Huang S (2013) A familial WT1 mutation associated with incomplete Denys-Drash syndrome. *Eur J Pediat* 172 1357–1362.

24

DOWN SYNDROME

ADITI KORLIMARLA, SARAH J. HART
Department of Pediatrics, Duke University Medical Center, Durham, North Carolina, USA

GAIL A. SPIRIDIGLIOZZI
Department of Psychiatry and Behavioral Sciences, Duke University Medical Center, Durham, North Carolina, USA

PRIYA S. KISHNANI
Department of Pediatrics, Duke University Medical Center, Durham, North Carolina, USA

INTRODUCTION

Incidence

The prevalence of Down syndrome is currently estimated at approximately 1 in 792 live births (de Graaf et al. 2015). Estimates of the prevalence of Down syndrome are highly dependent on the gestational timing at ascertainment and maternal age. Trisomy 21 accounts for about 1 in 150 first-trimester spontaneous abortions, and 35% of cases diagnosed between 15 and 28 weeks of gestation result in pregnancy loss with the actual loss rate varying inversely with gestation at ascertainment. Prevalence rises from 1/1445 live births at maternal age of 20 years to about 1/25–1/30 at the age of 45 years. Some data suggest prevalence does not continue to rise beyond the age of 45 years but remains stable at about 1/30. Prevalence of Down syndrome has been influenced by the increasing trend for termination of pregnancies with chromosome abnormalities owing to non-invasive prenatal screening (NIPS) using cell-free DNA. This screening has been commercially available since 2011, offering expectant women the option to determine (with near 99% sensitivity and specificity) whether their fetus might have Down syndrome. Prior to NIPS, this number was lower. In the US, the live birth prevalence for Down syndrome in the most recent years (2006–2010) was estimated at 12.6 per 10,000, with around 5300 births annually (de Graaf et al 2015). During this period, an estimated 3100 Down syndrome-related elective pregnancy terminations were performed annually. As of 2007, the estimated rate at which live births with Down syndrome were reduced as a consequence of these elective terminations was 30% for the US. Apart from advanced maternal age and selective terminations by prenatal testing, increase in birth control measures and decrease in family sizes also play a role in the prevalence rates in more recent times.

Life expectancy tables are estimated on the basis of cross-sectional survival rates to specific ages. If the birth prevalence of Down syndrome is constant, one can compare the prevalence of Down syndrome at age 50 years with that at birth in different decades and calculate and compare survival rates with that at age 50 years. Comparisons of survival rates for Down syndrome over time are complicated by changing birth prevalence and other factors such as improved neonatal ascertainment and uneven improvement in survival across different age groups. Increased maternal age also influences the data on survival curves, as the occurrence of Down syndrome is highly maternal age dependent. If more mothers over 35 years of age begin having children, and thus increase the birth prevalence, more children are available to survive, thus causing an "apparent" increased survival to age 50 years. Increased use of maternal serum screening and prenatal diagnosis would have the opposite effect. An improvement in early survival will again present more cases surviving at later ages, without necessarily signaling improved longevity for older individuals. Notwithstanding these caveats, several geographically disparate studies have concluded that

---

Cassidy and Allanson's Management of Genetic Syndromes, Fourth Edition.
Edited by John C. Carey, Agatino Battaglia, David Viskochil, and Suzanne B. Cassidy.
© 2021 John Wiley & Sons, Inc. Published 2021 by John Wiley & Sons, Inc.

survival in individuals with Down syndrome has shown marked improvement, particularly over the past 25–30 years. Most of the improvement has resulted from the treatment of congenital heart disease and respiratory infections during the first decade. Reduced institutionalization with increased mobility and integration into society have also played a role. A recent data set from Western Australia on 772 children with Down syndrome showed survival rates of 57% at 60 years for those born between 1953 and 1959, and survival estimates improved for subsequent generations (Glasson et al. 2016).

Mortality during the first five years continues to be higher than in the general population. A retrospective cohort study of over 16,000 infants with Down syndrome in the US showed modest improvement in survival over the study period of 20 years, with neonatal survival remaining similar at about 98% (Kucik et al. 2013). Survival at 1 year and 20 years was estimated to be 93% and 88%, respectively. The infant mortality rate in Down syndrome has remained at about five times that of the general population. Factors associated with lowered survival include congenital heart defects, birth weight <1500 g, and race/ethnicity (with non-Hispanic children with an African origin having lower survival rates), highlighting the impact of socioeconomic status and access to medical care on survival. From ages 5 to 39 years, the survival curves are parallel but mortality somewhat exceeds the general population. There is an increase in the mortality rate that is greater than that of both the general population and other individuals with intellectual disabilities. Increased risk for comorbidities such as seizures, depressive symptoms, and dementia appear as factors that are potentially contributory for this reduced survival.

Diagnostic Criteria

The gold standard for the diagnosis of Down syndrome is karyotypic demonstration of an extra copy of the long arm of chromosome 21. However, Down syndrome may be clinically diagnosed based on the characteristic appearance (gestalt) and behavior of affected individuals. This may be more challenging in premature infants, in some older adults, in an unfamiliar racial/ethnic group, or in individuals whose features are modified by significant mosaicism or a structural chromosome change that results in only partial duplication of 21q22. As with any syndrome, the associated features are variable. Children born at home show a significant delay in diagnosis (10.2 versus 1.8 days).

Typical features in neonates with Down syndrome include characteristic hypotonia, hyper-extensibility, and poor behavioral responses. The skull is mildly microcephalic and brachycephalic with a flat occiput. The fontanels tend to be large, a third fontanel may be palpable, and they close late. The posterior hair whorl is more likely to be midline, and the hair is fine. The face is round in the neonate and infant (Figure 24.1), and becomes more oval with age (Figures 24.2 and 24.3).

Underdevelopment of the midface gives a flat appearance, and the upper facial depth and length of the maxillary

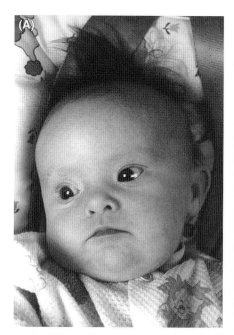
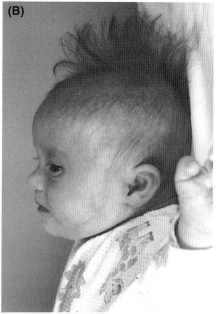

FIGURE 24.1 (A) A 6-week-old girl with Down syndrome illustrating round face with flat malar area, depressed nasal root, epicanthal folds, and upslanting palpebral fissures. Even at rest there is some pursing around the eyes. Her nose is short and the corners of her mouth downturned. (B) A lateral view of the same child shows mottling of the skin, malar flatness, and a small nose. The ear is slightly small with mild overfolding of the helix.

INTRODUCTION

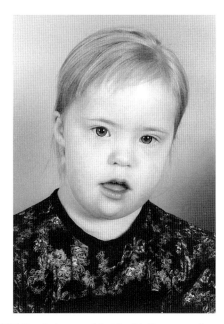

FIGURE 24.2 A 4-year-old girl with Down syndrome showing that the face has lengthened from that of the newborn but maintains the characteristic depressed nasal root, epicanthal folds, and upslanting palpebral fissures. Brushfield spots can be seen close to the iris margin. There is malar underdevelopment. The ear is small with a slight overfolding and crimping of the upper helix.

arch are disproportionately reduced. Epicanthal folds and upslanting palpebral fissures are typical, and the palpebrae "purse" on laughing or crying. Brushfield spots of the iris are common and are more peripherally placed than those seen in the general population. Ophthalmologic evaluation often reveals fine opacities of the lens. The optic disk is rosy colored, and has an increased number of retinal vessels. The nose is short with a depressed nasal bridge and, usually, small nares. The mouth is downturned, and a small oral cavity contributes to a tendency to protrude the tongue and to mouth breathe. Growth of the mandible tends to outpace that of the palate, leading to prognathism. The ears are small and may be cupped or show an overfolded upper helix contributing to a small, square shape. The neonate often has redundant nuchal skin, and, with age, the neck may appear wide when viewed from behind, perhaps partly because of the relative microbrachycephaly.

The chest may reveal signs of congenital heart disease. The hands are short with a high frequency of single palmar creases (not pathognomonic). The middle phalanx of the fifth finger is short and/or triangular, resulting in a single flexion crease or clinodactyly, respectively. Dermatoglyphic analysis is not performed often currently, but characteristic findings include a higher frequency of arches and ulnar loops on the thumb, ulnar loops on the index and middle fingers, and radial loops on the fourth and fifth fingers, a distal

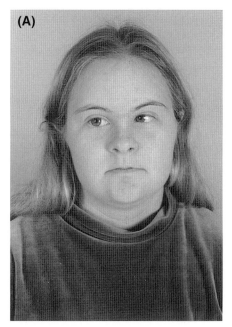
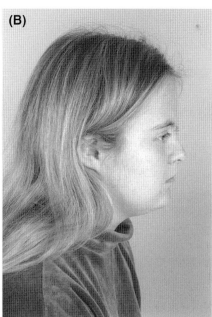

FIGURE 24.3 (A) An 18-year-old girl with Down syndrome demonstrating mild upslanting of the palpebral fissures, epicanthal folds, strabismus, and Brushfield spots. Although there has been growth of the nose, it remains small and short with a relatively depressed root and bridge. There continues to be a downturn to the corners of the mouth, and the lower jaw is small. (B) The lateral view demonstrates brachycephaly, a small ear with a slightly overfolded helix, and a small nose and jaw.

palmar triradius, and interdigital loops at I1 and I3. The space between the first and second toes is increased and accompanied by a vertical plantar crease with an origin at the space (sandal gap appearance). Cutis marmorata is common. The diagnosis may be aided by radiological evidence of an additional manubrial ossification center, a flat acetabular angle, and hypoplastic iliac wings that flare outward.

Despite the known myriad of clinical features, there are some limitations in the timely diagnosis of individuals with Down syndrome in many countries. Two main reasons are phenotypic variations in different ethnicities and the lack of antenatal screening facilities in developing countries. For a postnatal clinical diagnosis, many clinicians are trained using standard references which are specific to European or American resources. However, Kruszka and colleagues (2017) demonstrated that clinical features including brachycephaly, ear anomalies, clinodactyly, sandal gap, and nuchal skin differ across ethnicities. The most common features are upslanting palpebral fissures (61%) and flat facial profile (51%), but there is a large variation in facial findings even among individuals within the same ethnic groups. Further studies to improve our understanding of differences in the Down syndrome phenotype among individuals of different ancestries will be useful for clinicians around the world.

As Down syndrome and congenital malformations are both relatively common occurrences, a wide variety of birth defects may be expected to be seen in children with Down syndrome by chance alone. Comparison of the rates of specific malformations between children with and without Down syndrome is required to determine which malformations are causally related to Down syndrome. Torfs and Christianson (1998) have provided a review of the literature and a comparison of the rates of 61 anomalies between 2894 individuals with Down syndrome and a control population of 2.5 million from the same newborn surveillance registry. Forty-five of the malformations were reported significantly more frequently in Down syndrome, and every major system was represented. Risk ratios varied from nonsignificant to 1009 (atrioventricular canal defect). Risk ratios of over 100 were obtained for patent ductus arteriosus (152), overriding aorta (200), stenosis of the small intestine (142), duodenal atresia (265), Hirschsprung disease (102), annular pancreas (430), and hernia of Morgagni (246). Of equal interest are malformations that did not show an increased rate, including other types of diaphragmatic hernia, cleft lip with or without cleft palate, renal agenesis, neural tube defects, omphalocele, pyloric stenosis, and most cardiac malformations classified as conotruncal or looping defects. However, a recent study comparing the rates of congenital anomalies in 728 infants with Down syndrome from a registry of over 400,000 pregnancies (Stoll et al. 2015) found increased frequencies of duodenal atresia and anomalies of the urinary, musculoskeletal, and respiratory systems compared to both the general population and to the study by Torfs and Christianson (1998). The study also found decreased frequency of anal atresia, annular pancreas and limb reduction defects compared to prior studies. Discrepancies in the types and frequency of associated anomalies are likely related to methodological differences between studies.

Etiology, Pathogenesis, and Genetics

Down syndrome is caused by trisomy for chromosome 21. About 95% of cases result from nondisjunction and resultant standard trisomy 21. The remaining 5% are relatively evenly split between Robertsonian translocations, of which the 14;21 translocation is most common and about half are familial, and trisomy 21 mosaicism Down syndrome. Mosaicism may arise by postzygotic (mitotic) nondisjunction of a disomic zygote or the postzygotic loss of a chromosome 21 from a trisomic zygote. The lack of any maternal age association with mosaicism suggests that the former is more important. A small minority of affected individuals has other types of chromosome rearrangements, some of which result in partial duplications of chromosome 21.

The etiology of the characteristic appearance and specific associated features of Down syndrome is presumed to relate to dosage effects of genes on chromosome 21, but epigenetic effects may also contribute. The expression and dosage sensitivity of the genes on chromosome 21 is variable, and it has been hypothesized that the most dosage-sensitive genes are most likely to contribute to the Down syndrome phenotype. An alternative view is that the extra genetic material, as a whole, may disrupt multiple developmental pathways. Efforts are underway to study animal models of Down syndrome, individuals with partial 21q duplications, and effects of differential gene expression on the phenotype of Down syndrome. The creation of the Ts65Dn mouse that is trisomic for the equivalent of the critical region of human chromosome 21 has contributed significantly to the understanding of the developmental differences in Down syndrome (Bartesaghi et al. 2015; Gardiner 2015).

Maternal age is the single most important determinant of nondisjunction trisomy 21, and molecular techniques have shown that 85–90% of cases result from maternal and 5–10% from paternal meiotic errors, whereas up to 5% of cases may represent postzygotic mitotic nondisjunction. About 75% of maternal and 50% of paternal nondisjunction occurs in meiosis I, with the remainder occurring in meiosis II. The observation that the odds ratio increases with age for both maternal meiosis I and II errors suggests that there is an age-sensitive risk factor acting at the time of conception. A decreased rate, or more centromeric or distal location of crossover events appears to play a role in nondisjunction generally (Ghosh, Feingold & Dey, 2009).

Several studies have investigated whether other potential factors like environmental or genetic factors could increase the chance for Down syndrome in pregnancy. Carothers and colleagues (2001) studied 3157 cases of Down syndrome and found no association with paternal age, birth order, ancestry, country of birth, maternal education, blood group, or pregnancy interval. Maternal smoking has been studied as a factor potentially contributing to proper segregation of chromosome 21 in the meiosis I phase of oogenesis, but studies have shown mixed results. The data concerning natural, medical, and accidental radiation exposure and the risk of trisomy are contradictory and unconvincing, but there is a need for further properly designed studies.

There has been a discussion of a possible role of hypomethylation in the etiology of Down syndrome. A recent meta-analysis examined the association between maternal polymorphisms in folate metabolism genes [RFC-1 A80G (located on chromosome 21)/MTR A2756G (located on chromosome 1)/CBS 844ins68 (located on chromosome 21)] and chance for Down syndrome in their offspring (Gu 2017). The A80G polymorphism of the Reduced Folate Carrier-1 (*RFC-1*) gene was associated with a chance for Down syndrome, but there was no evidence of an association between the MTR A2756G/CBS 844ins68 polymorphisms and Down syndrome. Further well-designed large studies are required to investigate gene-environment interactions, gene-maternal age interactions, and combinations of gene polymorphisms.

The potential role of genes related to Alzheimer disease in the mechanisms of Down syndrome has also been discussed. Presenilin-1 (*PSEN-1* gene on chromosome 14) and Apolipoprotein E (*APOE* gene on chromosome 19) genes are associated with early and late onset of Alzheimer disease, respectively. A case control study (Bhaumik et al. 2017) found that younger mothers (under age 35) of children with Down syndrome had a higher frequency of the *PSEN-1* T allele and TT genotype in the presence of the *APOE* epsilon4 allele compared to mothers of children with normal karyotypes. This association was found in mothers with meiosis II nondisjunction, but not among mothers with meiosis I nondisjunction. This study suggests that *PSEN-1* may be a prospective molecular candidate that relates Alzheimer disease and Down syndrome.

The recurrence chance estimates for parents of children with trisomy 21 Down syndrome vary with the age of the mother at the time of the birth of the child. Mothers who have had a child with Down syndrome and who were of older age maintain their current age-related chance, whereas those who were younger (<30 years) have an increased recurrence chance (up to six-fold) compared with same-aged peers. For a mother of a child with trisomy 21, the recurrence chance for trisomy 21 in a future pregnancy is estimated at approximately 1%, until the maternal age-related chance exceeds 1% (around age 40). The basis for this increased chance compared to the general population remains unknown, but could be due to a decreased likelihood of spontaneous abortion in pregnancies with trisomic fetuses, to an age-independent increased propensity to nondisjunction, or to gonadal mosaicism. There has been some evidence of increased prior fetal loss in young mothers of children with Down syndrome. No evidence for increased prevalence of Down syndrome has been found in second- and third-degree relatives of individuals with trisomy 21 (Berr et al. 1990). A common question is that of the chance for Down syndrome for a couple in which one member has a relative with Down syndrome of unknown karyotype. The known age-specific rates of trisomy versus translocation Down syndrome and the likelihood of male and female transmission of a translocation can be used to calculate the chances for specific relatives of the affected person. The highest chance is about 1 in 640, and it applies to the children of the sister of the person with Down syndrome. This is not greatly different from the population prevalence, and the chance declines rapidly with the degree of relationship and is lower when the connection is through a male. Parents of children with de novo translocations do not have a significantly increased chance for recurrence, whereas a man with a balanced Robertsonian translocation has a 3–5% chance, and a woman a 10–15% chance for recurrent Down syndrome. Note must be taken of the special circumstance of a parental 21;21 translocation, where the recurrence is 100%. For a child with Down syndrome due to any translocation, a referral for parents to a medical genetics clinic for counseling and parental testing is indicated.

Diagnostic Testing

Non-invasive prenatal screening (NIPS) or testing (NIPT) (also known as cell free fetal DNA testing – cffDNA) is considered a highly accurate screening test for Down syndrome compared with conventional combined first trimester screening (FTS). The test counts fragments of fetal DNA (called cell-free DNA), thought to be primarily derived from the placenta, circulating in the maternal blood. If there are extra fragments of chromosome 21, for example, three copies of chromosome 21 instead of the usual two, this indicates that the fetus may have Down syndrome. It can be done in the 10th week of pregnancy and the results usually take about 1–2 weeks. In October 2011, NIPS became commercially available in the USA and China, and was rapidly inculcated into standard prenatal care in many countries. While the majority of validation data on NIPS has been based on populations of women at higher risk for chromosome abnormalities in pregnancy, recent studies have investigated performance of NIPS for detecting aneuploidy in the general population. A recent prospective, blinded study of over 15,000 participants from an unselected population of women across 35 diagnostic centers found that when compared to standard screening, NIPS had higher sensitivity (100% for NIPS versus 78.9% for FTS), a lower false positive rate

(0.06% for NIPS versus 5.4% for FTS), and a higher positive predictive value (80.9% for NIPS versus 3.4% for FTS) (Norton and Wapner 2015).

The prenatal Down syndrome screening strategy that has conventionally been used in lower-risk groups (FTS) is based on predicting risks using non-invasive measures, including a combination of gestational age, maternal age and weight, maternal biochemical markers, ultrasound measurements, and more recently non-invasive prenatal screening. Ultrasounds in the first trimester are done at 11–14 weeks to detect nuchal translucency, nasal bone abnormalities and ductal venous flow measurements. These measurements, along with the mother's age and the gestational age, improve the odds ratio of detecting Down syndrome. When performed with a maternal blood test, its accuracy may be improved. Ultrasounds in the second trimester detect soft biomarkers that do not in themselves confirm a diagnosis but are seen more frequently in fetuses with an abnormality. Soft biomarkers include echogenic intracardiac focus, ventriculomegaly, nuchal fold thickness >6 mm, echogenic bowel, hypoplastic/absent nasal bone, shortened humerus, mild pyelectasis, shortened femur, and aberrant right subclavian artery (ARSA). Each biomarker is assigned positive, negative, and isolated likelihood ratios. The chance of Down syndrome is recalculated as baseline chance times likelihood ratio. The new likelihood ratio is calculated by multiplying all positive likelihood ratios (of markers present) and all negative likelihood ratios (of markers absent). If a single marker is present, then the isolated likelihood ratio is considered. The importance of clustering of markers forms the basis of a scoring index, such that individual markers are assigned point values based on their sensitivity and specificity in the detection of Down syndrome. The points acquired by each fetus are tabulated into a final score. Ultrasounds by themselves have low detection rates for Down syndrome. Ultrasonographic skills are highly variable, and their use in prenatal screening for Down syndrome requires a high level of training and a standardized approach.

Algorithms for combining ultrasound with maternal serum screening have been developed and evaluated. Some of these include graduated risk ratios dependent on the degree to which a specific sign or measurement is abnormal. A number of programs combine first-trimester nuchal translucency measurements and biochemical screening with second-trimester biochemical tests (integrated prenatal screening) to obtain detection rates in the range of 90% and low initial positive rates of about 2%. Urine markers have also been explored. Alldred and colleagues (2017) used meta-analytical methods involving 228,615 pregnancies (including 1067 with Down syndrome). Thirty-two different test combinations were evaluated from combinations of eight different tests and maternal age; first trimester nuchal translucency and the serum markers AFP (alpha fetoprotein), uE3 (unconjugated estriol), total hCG (human chorionic gonadotropin), free hCG, Inhibin A, PAPP-A (plasma protein A), and ADAM12 (a disintegrin and metalloprotease 12). They reviewed tests that combine the first and second trimester markers with or without ultrasound as complete tests, and examined stepwise and contingent strategies. Meta-analysis of the six most frequently evaluated test combinations showed that a test strategy involving maternal age, a combination of first trimester nuchal translucency and PAPP-A, and second trimester total hCG, uE3, AFP and Inhibin A significantly outperformed other test combinations that involved only one serum marker or nuchal translucency in the first trimester, detecting about 9 out of every 10 cases of fetal Down syndrome at a 5% false positive rate. The choice of screening tests is dependent on many factors including cost-effectiveness, willingness of parents for prenatal testing, prior obstetric history, family history, singleton/twin pregnancy, gestational age, availability of non-invasive versus invasive techniques, limitations of tests, psychological issues involving care of an affected child, availability of other options (termination or adoption), and ethics.

While non-invasive prenatal screening methods provide accurate detection of fetal Down syndrome, karyotyping using chromosome analysis is the gold standard test to confirm the diagnosis. Chromosome analyses may be completed prenatally using amniocentesis or chorionic villus sampling, or may be done postnatally on peripheral blood. Amniocentesis and chorionic villus sampling are relatively costly, have a risk for miscarriage, and have a relatively low yield, especially if applied to younger women. This has driven the search for population-based screening methods, which are more cost-effective and have a lower risk of miscarriage.

The diagnosis of Down syndrome can also be made on interphase nuclei using fluorescence in situ hybridization (FISH), microarrays, and quantitative fluorescent polymerase chain reaction (QF-PCR). The main limitation of chromosome analysis is the requirement of tissue culture and limited genomic resolution. It cannot detect submicroscopic deletions/duplications of clinical relevance; for example, 22q11.2 microdeletion associated with fetal cardiac abnormalities. Chromosomal microarray (using comparative genomic hybridization, CGH) shows excellent diagnostic performance with improved detection rates compared to karyotyping for prenatal diagnosis of clinically relevant fetal chromosomal abnormalities. However, chromosome analysis is necessary to distinguish between Down syndrome due to free trisomy 21 versus a Robertsonian translocation, which has important implications for recurrence estimates in future pregnancies (see section above).

All pregnant women should be offered prenatal genetic screening tests for Down syndrome. Women with a positive screening test for fetal genetic conditions should always be offered further counseling and prenatal diagnostic testing.

They should be educated about the possible false positive and false negative screening test results. Counseling should include family education and options pertaining to medical termination, psychological and genetic counseling, adoption, referral to tertiary care centers for better infrastructure to manage complicated neonates, and perinatal hospice care if a neonate's condition is incompatible with life. For persons suspected of having Down syndrome and in whom a normal chromosome result is reported, it may be appropriate to refer to a specialist in dysmorphology, who may look for mosaicism in more cells or another tissue.

Differential Diagnosis

Down syndrome is common and distinctive, and should not often be confused with other syndromes. However, there may be confusion when a typically developing neonate has one or more of the common signs or minor anomalies that physicians associate with Down syndrome (e.g., hypotonia, "large" tongue, and single palmar crease). Noting the absence of the other common signs and the facial gestalt of Down syndrome should avoid this error. Likewise, hypothyroidism and Beckwith–Wiedemann syndrome (see Chapter 9) can be distinguished by their own typical signs and the lack of other characteristics of Down syndrome.

A number of young children with Smith–Magenis syndrome (see Chapter 54) have been diagnosed fortuitously when the diagnostic deletion of 17p11.2 was detected on a karyotype requested for suspected Down syndrome. The overlapping features include brachycephaly, round face, upslanting palpebral fissures, midface hypoplasia, a small, wide nose, and Wölfflin–Krückmann iris spots that may be confused with Brushfield spots. Other signs of Down syndrome are absent; however, with time the more typical appearance and behavior of Smith–Magenis syndrome become apparent.

Zellweger syndrome, a peroxisomal disorder, shares a number of findings with Down syndrome including hypotonia, large fontanels, flat occiput and face, anteverted nares, epicanthal folds, Brushfield spots, cataracts, abnormal helices, single palmar crease, and cardiac septal defects. Distinguishing signs include severe early developmental delay, seizures, a high forehead, shallow orbits, hepatomegaly, joint contractures, stippled epiphyses, and brain migrational anomalies in Zellweger syndrome. The diagnosis can be confirmed by finding elevated very long-chain fatty acids in plasma.

MANIFESTATIONS AND MANAGEMENT

The American Academy of Pediatrics (AAP) has published guidelines for the healthcare management of children with Down syndrome (Bull 2011). Down syndrome specialty clinics have been created across the United States with the goal of supporting healthcare management and improving adherence to the AAP guidelines. Down syndrome specialty clinics can identify and address specific needs for children with Down syndrome beyond what is typically available in primary care settings (Skotko et al. 2013). Lists of specialty clinics are available through the websites for the National Down Syndrome Society (NDSS 2019) and the Global Down Syndrome Foundation (2017). The following sections below detail the manifestations and management for multiple specific systems in Down syndrome.

Growth and Feeding

Feeding difficulties due to hypotonia, large tongue, small oral cavity, dysphagia, constipation, and gastrointestinal regurgitation syndrome are common in individuals with Down syndrome. Decreased tone in perioral muscles, lips and muscles of mastication as well as decreased tongue movements may lead to sucking difficulties in infants with Down syndrome. They tend to breastfeed for a shorter time, have more respiratory infections, and develop non-nutritive oral sucking habits (like increased dependence on bottle feeding, finger sucking, and pacifier use). Improved guidance for parents to encourage healthier feeding habits, advocating techniques to reduce feeding problems (like smaller feeds, keeping the child upright for 30 minutes post feeds, head elevation during feeds), and promoting breastfeeding will alleviate some of these problems during infancy. The same oral and motor difficulties may cause a delay in the introduction of solid foods, and the primary care physician has an important role in ensuring that a balanced diet is being maintained throughout infancy. Through childhood, uncorrected feeding problems may lead to unbalanced dietary habits.

Standard growth charts may not provide accurate information about the development of children with Down syndrome. These children can have lower birth weights and slower growth rates which are appropriate for a child with Down syndrome; yet can be misinterpreted as poor growth if compared to typically developing children. WHO and CDC growth charts are being used worldwide to monitor growth, but Down syndrome-specific growth charts have been developed for children in the United States (Zemel et al. 2015), as well as for children in multiple other countries (e.g., Afifi et al. 2012; Bertapelli et al. 2017; Su et al. 2014; Tuysuz et al. 2012). The mean birth length in infants with Down syndrome has been estimated to be one standard deviation lower than in the general population, indicating that fetal growth may be slower. During the first three years of life, the linear growth rate is slower compared to the general population. The gap stays relatively constant during the age interval of 3–12 years. After the age of 12 years a further relative slower rate of growth decline is observed. This pattern is observed to be the same in boys and girls with Down syndrome. Hence, the slowing in linear growth rate occurs in the three critical periods of growth, resulting in shorter adult stature.

Down syndrome is associated with increased risk for obesity. From late infancy, children with Down syndrome show a relative increase in mean weight for height and in weight/height^2, and excessive weight is a significant problem in adulthood. Individuals with Down syndrome appear to have higher leptin levels than their unaffected siblings even after correction for percent body fat. People with Down syndrome also have a lower metabolic rate than typically developing individuals. Collaboration between parents and health personnel is important to help improve nutrition and address issues related to being overweight.

The new Down syndrome-specific length/height growth charts for children in the United States (Zemel et al. 2015) have improved diagnosis of growth failure (short stature). Growth failure is a diminished linear growth velocity, crossing height growth centile markings on Down syndrome-specific growth charts, in the absence of malnutrition and of symptoms and signs pointing to another diagnosis. People with Down syndrome and growth failure have a specific reduction in insulin-like growth factor-1 (IGF-1). Failure to thrive is defined as undernutrition/malnutrition, associated or not associated with linear growth failure. The weight growth charts are more useful for the diagnosis of failure to thrive than for overweight and obesity. The high prevalence of overweight and obesity in Down syndrome make the diagnoses of these based on centile cutoffs inappropriate. Therefore, the 85th centile on the new CDC BMI growth charts is likely a better indicator of adiposity than the 85th centile on the new Down syndrome specific BMI growth charts. Further studies using techniques to measure lean body mass in a large population-based sample of people with Down syndrome would provide a better understanding of both assessment of nutritional state and of the effects of race and ethnicity on growth. There is no case series of children with Down syndrome who have failure to thrive to guide an evaluation, and a full discussion of this problem is beyond the scope of this review.

Evaluation

- Appropriate tests should be performed to diagnose common causes of growth failure, including hypothyroidism, growth hormone deficiency, and celiac disease.
- For severe failure to thrive, a hospitalization is often needed to determine etiology and to begin treatment. High output heart failure from congenital heart disease is probably the most common cause. Psychosocial failure to thrive (parental neglect or inadequate education) should always be considered as a diagnostic possibility in children with Down syndrome. A careful dietary history, a thorough review of symptoms, and a good physical examination are essential.
- Infants who have marked hypotonia and slow feeding, infants who choke with feeds, infants and children who have recurrent pneumonia or who have other recurrent or persistent respiratory symptoms, and infants and children who have unexplained failure to thrive should all have a videofluoroscopic swallow study and a barium swallow with small bowel follow through to rule out a tracheoesophageal fistula and gastroesophageal reflux, small bowel stenosis or duodenal stenosis associated with annular pancreas, and to rule out other gut malformations.
- All children with Down syndrome who have failure to thrive or growth failure should be evaluated for thyroid disease and celiac disease, whether or not they have gastrointestinal symptoms.
- For children with Down syndrome who have obesity, complete dietary and activity histories are essential. A careful history of diabetes, hypothyroidism, and sleep disordered breathing should be obtained.
- If obesity is severe and there is shortness of breath, an evaluation for pulmonary hypertension should be done, as individuals with Down syndrome are more susceptible to developing pulmonary hypertension than the general population.
- Those who are obese should be evaluated for insulin resistance by obtaining a random glucose concentration, insulin level and hemoglobin A1C level and for hypothyroidism with TSH and serum free T4 levels.

Management

- Disorders of growth and nutrition require referral to specialty clinics: to endocrinology for management of growth failure not due to celiac disease; to gastroenterology for diagnosis and management of feeding and swallowing problems, psychosocial failure to thrive and problems following surgery for gastrointestinal malformations; and to a healthy lifestyles (obesity) clinic for management of obesity. After a thorough evaluation with an endocrinologist, certain patients with growth failure can be given supplemental growth hormone. The treatment with growth hormone should be considered only if there is a documented deficiency, because inadvertent use of growth hormone can lead to adverse effects like insulin resistance, benign intracranial hypertension, edema, and gynecomastia. There are no long-term studies of growth hormone treatment in individuals with Down syndrome. If growth hormone is used, long-term follow-up of patients is warranted to monitor them at regular intervals.
- For infants with failure to thrive who have difficulty sucking, it is important that the child is well awake, that the child is properly supported with the chin steadied,

that the mouth and nose are clear of mucus (a syringe with a small amount of normal saline may help clear the nose), and that the child is burped regularly. These simple measures will usually overcome minor, self-limited problems. With breast-feeding it may be helpful to facilitate attachment to the breast by first expressing a small amount of milk and by feeding more often (every 2–3 hours) to stimulate milk production. In a minority of cases, the difficulties may be more marked and persistent and require referral to a feeding specialist.

- It is helpful to tell parents of an obese child with Down syndrome that their child's obesity may not be not their fault, but changes will need to be made in diet and activity for the long-term health of their child. The National Down Syndrome Society website provides a list of additional resources for caregivers to support healthy eating habits and weight management (NDSS 2019).

Development and Behavior

Down syndrome is the most common genetic cause of intellectual disability. The cognitive ability of individuals with Down syndrome, as measured by standardized intelligence tests, generally ranges from profound intellectual disability to borderline intellectual functioning. There is evidence that the intelligence quotient (IQ) of children with Down syndrome correlates approximately, as expected, with mean parental IQ. Still, overall IQ is a poor measure of an individual's spectrum of abilities that may be highly variable. Standardized measures of an individual's adaptive functioning, in conjunction with IQ scores, are required to establish the actual severity of the intellectual disability. Cognitive development in Down syndrome has been associated with individual differences in temperament, maternal education, severity of medical conditions and school experiences (Couzens et al. 2012).

Newborns with Down syndrome typically have hypotonia that affects gross motor development. Although tone will improve, developmental delays are universally present. Time to reach motor developmental milestones is generally about twice as long as in a typically developing child; 92% will walk by 36 months of age. Curves for the expected rates of motor acquisition for children with Down syndrome are available (Palisano et al. 2001). There is evidence that early intervention and physical therapy for children with Down syndrome increase their likelihood of performing more complex motor functions.

Although individuals with Down syndrome tend to achieve fine motor skills at a later age than observed in the general population, their fine motor skill development is varied and occurs over a broad age range. This range gets wider over time, as the skills become more challenging (Frank and Esbensen 2015). It is important that children with Down syndrome master foundational skills first (such as a raking grasp) before acquiring more refined skills (such as self-feeding finger foods) (Frank and Esbensen 2015).

At an early stage, language is acquired in a pattern similar to, but delayed, when compared with typically developing children with equivalent mental ages. Babbling is often delayed, has a different quality, and continues to occur over a more prolonged period. In several areas related to speech/language development, the level of impairment is greater than expected for overall mental age. Receptive language ability is generally stronger than expressive language skills. By school age, children with Down syndrome lack the typical correlation between the production and comprehension of language, and the relative level of vocalization and grammatical usage falls in those over the age of 10 years. Development and use of syntax are relatively more impaired, whereas the use of language in social context (pragmatics) is a strength. Another important issue is the poor articulation and associated intelligibility of speech in individuals with Down syndrome. Several other issues involving language development have been described (Abbeduto et al. 2007). Physical factors related to the oral cavity, particularly the high narrow contours of the anterior hard palate that may affect lingual-palate contact, and the hypotonia and tendency for persons with Down syndrome to drop consonants and last syllables may all play a role.

Children with Down syndrome show less attentional focus and inhibitory control, more restricted play, as well as stereotypic and repetitive behavior, which is less goal oriented and organized. Children with Down syndrome have the cognitive skills to detect, distinguish, and respond to novel target stimuli, and to maintain their attention. However, the attention "process" in target detection is neurophysiologically different, with more cognitive effort required to produce the same performance as typically developing peers.

Processing of social information may differ in individuals with Down syndrome compared to typically developing children. Eye contact with caregivers may be delayed, and once it is established it is maintained longer, which may inhibit development of other spheres of eye contact. There appears to be a specific defect in recognizing facial expressions of emotion and in identifying familiar faces. Early temperament is very similar to that of a typically developing child, but reactions tend to be muted. Social facial expressions and emotional vocalizations show the same pattern as children without Down syndrome but the evolution is delayed and they are produced with less frequency, duration, and intensity.

Down syndrome may be associated with specific behavioral challenges. Although studies have shown there is some truth to the stereotype of children with Down syndrome as happy, affectionate, and outgoing, the picture is more complex. Compared to siblings and typically developing children, parents of children with Down syndrome report a higher rate of noncompliance, difficulty persisting with tasks, and stubbornness (Grieco et al. 2015). In addition, reports of externalizing behaviors in preschool and school-aged children with Down syndrome showed higher rates of hyperactivity, impulsivity, tantrums, agitation, argumentativeness, repetitive movements, and sensory dysregulation (Capone et al. 2006). Obsessive compulsive behaviors have also been reported among school aged children with Down syndrome (Siegel and Smith 2011). The development and retention of challenging behavior may be related to what is perceived by the child as a reward, either obtaining what is desired (e.g., attention) or avoiding what is not wanted, or is perceived as being overly difficult (e.g., intellectual challenge). Disruptive behavior may also be related to the child's limited expressive language communication skills or to discomfort related to medical issues (Skotko et al. 2013). Children with Down syndrome may use social distraction to avoid completing a requested task.

Individuals with Down syndrome frequently "self-talk" and have imaginary friends. These behaviors are not associated with behavioral, communication, or socialization problems, and should be considered adaptive and not pathological.

Supports are available to help adolescents with Down syndrome transition to adulthood. Although many individuals with Down syndrome continue to spend a longer period with their parents than the typically developing child, there is a trend to reside more within the community with differing levels of supervision depending on the degree of independence. Current goals are, therefore, to maximize self-help skills and foster independence. There is an increasing number of post-secondary programs for individuals with an intellectual disability on college campuses which may further aid integration. A major challenge now and in the future will be to find meaningful employment for individuals with Down syndrome. A recent survey of adults with Down syndrome found that only about 11% were in work placements, out of which 56% had a paid job (mainly in restaurant/food services, office/clerical, grocery stores, and cleaning/housekeeping departments), 26% had a volunteer position, and 3% were self-employed (Kumin and Schoenbrodt 2016).

There is a long history of claims of therapeutic value for supplements, hormones, vitamins, and related therapies to improve the motor and cognitive function of individuals with Down syndrome, which include growth and thyroid hormone, vitamins, 5-OH-tryptophan, glutamic acid, injection of fetal cells, and various "cocktail" mixtures. Growth hormone has been shown to have no benefit for the growth of head circumference or cognitive development. People with Down syndrome are not deficient in any of these substances, properly controlled trials have failed to show benefit, and studies claiming benefit are anecdotal or uncontrolled, and in general, any benefit can be ascribed to a placebo effect resulting from increased intervention and attention.

Comorbid Conditions Autism spectrum disorder (ASD) and attention-deficit/hyperactivity disorder (ADHD) are commonly found in individuals with Down syndrome, leading to increased advocacy for early detection and diagnosis. In the general population, the prevalence of ASD is about 1%; In children with Down syndrome, it has been estimated at 41%. ADHD has been found in 34% of children with Down syndrome, with 22% found to have both ADD and ADHD (Oxelgren et al. 2017). Therefore, there is a growing need for a screening test for ASD/ADHD in children with Down syndrome at the age of 3 years and before entering school.

Rates of neuroses, conduct disorders, and psychoses are considered to be lower in Down syndrome than in other individuals with disabilities. However, the prevalence rates of neurobehavioral and psychiatric illness in children with Down syndrome are estimated to be 18–38% (Capone et al. 2006). These neuropsychiatric conditions are considered to be treatable. Adolescents with Down syndrome exhibit more behavior difficulties than typically developing peers. The severity of these psychiatric problems is correlated with the severity of their intellectual disability. Behavior problems (disruptiveness, communication disturbance, anxiety, being self-absorbed, antisocial behavior) in young adults with Down syndrome may improve over time but depressive symptoms and social relating behavior problems may persist into adulthood. It is possible that those with persistent depressive symptoms are at a high risk for developing depressive illness in adulthood. Depression, with a mean age of onset of 29 years, has been reported in up to 10% of adults. There is a growing consensus that there is a significant risk of anxiety and depression in adults with Down syndrome, which in late adulthood may relate to the onset of dementia. It is important to separate learned behavior from true psychiatric symptoms. Guidelines to aid in doing so are available (Capone et al. 2006). It is also important to exclude organic causes of depression. Hence, always consider possible associated psychosocial stressors and medical factors, such as sleep disturbance, hypothyroidism, etc.

Down syndrome disintegrative disorder is a condition that is being increasingly recognized, characterized by a sudden deterioration in skills and new onset of autistic characteristics in children or young adults with Down syndrome. Worley and colleagues (2014) described 11 children with Down syndrome disintegrative disorder, characterized by autistic regression, cognitive decline (dementia-like), new onset insomnia, and thyroid autoimmunity. These clinical characteristics could not be explained by any other diagnosis. Furthermore, the autistic regression seen in these individuals with Down syndrome was seen at an older age (mean age of 11.4 years) compared to the autistic regression seen in children with disintegrative disorder who did not have Down syndrome. Larger studies, more awareness and better screening techniques to diagnose such clinical outcomes will help with further characterization of Down syndrome disintegrative disorder. It should also be noted that while thyroid autoimmunity was described in all cases in the study by Worley et al. (2014), Down syndrome disintegrative disorder has also been recognized in individuals who do not have thyroid autoimmunity (personal experience). While increased thyroid auto-antibodies may be an epiphenomenon in Down syndrome disintegrative disorder, there may be a potential role of autoimmunity in the etiology of this condition. Immunotherapy was shown to significantly improve symptoms of catatonia, insomnia, autistic features, cognition, and psychosis in four children with Down syndrome disintegrative disorder (Cardinale et al. 2018).

Alzheimer Disease Individuals with Down syndrome have a third copy of the amyloid precursor protein (*APP*) gene on chromosome 21 that is linked to risk for Alzheimer disease. Autosomal dominant Alzheimer disease has been reported in families with an isolated duplication of the *APP* gene. Individuals with Down syndrome have been found to have elevated levels of amyloid β (Aβ) precursor protein, Aβ42, and s100β in plasma and cerebrospinal fluid compared with the general population, consistent with their putative role in the pathogenesis of Alzheimer disease. Recently it was shown that individuals with Down syndrome without dementia who have Aβ42 values in the upper tercile are at significantly greater risk of dementia in the ensuing 14–18 months than those in the lower terciles. Increased expression of the *DYRK1A* gene on chromosome 21 also leads to hyperphosphorylated τ (tau) protein, contributing to the pathogenesis of Alzheimer disease.

It is now well established that almost 100% of individuals with Down syndrome show the neuropathologic changes of Alzheimer disease by the age of 35–40 years (Head et al. 2012). The pathology of Alzheimer disease in individuals with Down syndrome consists of neurofibrillary tangles and senile plaques, which are similar in Alzheimer disease in the general population. However, the exact pathogenesis and neuroanatomical variations are not clearly understood among individuals with Down syndrome. Neuronal pathological findings occur decades earlier in Down syndrome (as early as first-second decades) compared to the sporadic form of Alzheimer disease in the general population (fourth-fifth decades). Individuals with Down syndrome may also have differences in neuronal pathology with more APP plaques and fewer βA4 peptide plaques than seen in sporadic Alzheimer disease (Egensperger et al. 1999).

Onset of clinical dementia lags significantly behind the appearance of the neuropathologic changes, but current evidence suggests that it is highly penetrant, with the average age of onset at 51–54 years (range 38–70 years) and an average survival from diagnosis of about five years. It is often difficult to recognize dementia when preexisting cognitive impairments of varying severities are already present. There are no "standardized" methods for clinical diagnosis. However, diagnosis may be facilitated by documenting substantial decline from previous status and relying on these findings to inform clinical judgment, and by making attempts to develop empirically validated methods. Novel approaches to develop various biomarkers to track the progression of the disease in its earliest stages, or even before dementia sets in, are underway.

Several studies have investigated potential factors that may be predictive of the onset of Alzheimer disease in individuals with Down syndrome, including total cholesterol levels, the presence of an apolipoprotein E ε4 allele and presence of apolipoprotein E ε2 allele (as a protective factor). However, it is unclear whether screening for apolipoprotein E genotype would have benefit in predicting the onset of Alzheimer disease in the Down syndrome population. Other factors include hyperphosphorylation of tau protein, oxidative stress, reduction in estrogen during menopause, brain developmental abnormalities, and cognitive reserve. Presenilin polymorphisms do not play an important role. There is a higher frequency of associated seizures (15–20%), but Parkinsonian signs, and maladaptive and aggressive behaviors may be less common in Alzheimer disease in individuals with Down syndrome. There is evidence suggesting that young mothers (~35 years) who have a child with Down syndrome may be at increased risk for Alzheimer disease, and this risk does not apply to other trisomies. If confirmed, this raises the possibility of a common factor in the etiology of both Down syndrome and Alzheimer disease.

Currently there is no curative treatment for Alzheimer disease, and although some therapies show promise in slowing the process or alleviating the symptoms, it cannot be assumed that their effect will be comparable in individuals with Down syndrome. Short-term benefit as well as significant reversible adverse reactions have been reported with donepezil, a cholinesterase inhibitor. While donepezil is approved for treatment of Alzheimer disease in the general population, the evidence to date has not demonstrated efficacy for treating cognitive decline in individuals with Down syndrome (Hart et al. 2017; Livingstone et al. 2015). The paradigm shift of conducting clinical trials for the treatment of Alzheimer disease alone to clinical trials for improving cognitive and adaptive behaviors has broadened the scope for future research on multifaceted interventions for the development of individuals of Down syndrome. Hart and colleagues (2017) provide a review of the history of clinical trials for cognition and adaptive behavior, and describe strategies for the pharmaceutical industry to advance the field in drug discovery for Down syndrome. A major challenge in the field is determining appropriate endpoints to document clinical benefit.

Evaluation

- Per AAP guidelines (Bull 2011), discuss the child's behavioral and social progress, and intrafamilial relationships with caregivers at every visit. Evaluate the developmental needs and the support system involved.
- Ensure that the child has been connected with the local agency providing early intervention services. Research has shown that children with disabilities benefit significantly from early intervention (Dreyer 2011). This may include services from a multidisciplinary team of providers, including physical therapy, occupational therapy, speech and language therapy, behavior therapy, and special education. Other team members may include a behavior therapist, psychologist, and/or developmental-behavioral pediatrician depending on the child's needs. Early intervention services may be initially provided in the child's home, and later in a clinic and/or school setting.
- Discuss the child's transition from early intervention services to the public school system at 3 years of age (Bull 2011). A team consisting of family members and school personnel would develop an Individualized Education Program (IEP) at that time, based on the child's strengths and needs. The IEP will detail academic and functional goals, classroom accommodations, modification to the child's curriculum, the amount of small group or individualized instruction per week, and the therapy services provided to meet the child's educational needs. The IEP should be revised on a yearly basis or sooner, if necessary. As the child transitions to middle school and high school, the IEP team should also develop goals targeting work skills and vocational training.
- Routine screening for autism spectrum disorder, attention-deficit/hyperactivity disorder and/or other psychiatric illnesses or behavioral problems should be done when age appropriate.
- Medical problems including thyroid abnormalities, celiac disease, sleep apnea, gastroesophageal reflux, and constipation are associated with behavior changes. Evaluate for these problems in individuals with Down syndrome.
- Encourage parents to create a long-term financial plan for their child, which may include federal and state funding (such as Social Security Benefits) as well as private funding sources. It is often necessary to designate a legal guardian for their child at 18 years of age. Potential adult morbidities including premature aging and Alzheimer disease may also be discussed. Discuss group homes and independent living opportunities, workshop settings, and other community-supported employment. Facilitate transition to adult medical care (Bull 2011).
- Early signs of Alzheimer disease in high-functioning individuals include a decline in memory and verbal capability, whereas others may show a decrease in social interaction and attention and increasing apathy. It is important to rule out hypothyroidism and depression in such cases. Cholesterol values should be assessed.

Management

- Those working with the child and family should have a positive and optimistic approach. The best providers are likely to be those that recognize that children with Down syndrome are not simply delayed but may have specific deficits that require imaginative approaches to teaching. Innovative methods may be particularly important in the area of communication.
- Early behavioral intervention may be an important preventive measure for some children.
- While multiple clinical trials have been conducted over the past several decades investigating the efficacy of pharmaceutical interventions to improve cognition in Down syndrome, results have shown limited success to date (Hart et al. 2017). At this time there is insufficient evidence to support the clinical use of pharmaceutical interventions to improve cognition in individuals with Down syndrome.
- Caution is needed in using psychotropic medications in children and adults with Down syndrome; little or nothing is known about drug metabolism and kinetics for most psychotropic medications in Down syndrome. A good rule of thumb pertaining to dosing these drugs is to "start low and go slow", following the patient carefully for medication effects and side effects (personal experience). Those with congenital heart disease should be seen and "cleared" by a cardiologist before starting a stimulant medication for ADHD or a neuroleptic for more serious behavior problems. Weight gain is often a serious complication of psychotropic drugs in people with Down syndrome. Close follow up is necessary if any psychotropic medication is prescribed. If this is not possible, then the individual should be referred to a psychiatrist.
- For medical problems that are associated with behavior changes, intervention strategies depend on various factors such as the individual's age range, severity of the medical problem, and the specific settings in which a problem occurs. Referrals may be required to community programs, psychosocial services for consultative care, or behavioral specialists who are experienced in working with children with special needs. Since children with Down syndrome may be more sensitive to certain medications that are used to address behavior issues, there should be discussions between the child's primary care physician and their specialists to use these medications. Children with Down syndrome may differ in their responses to these medications compared to their peers (Bull 2011).
- Institution of any treatments such as cholinesterase inhibitors should be as part of properly designed scientific studies. Lowering elevated cholesterol, as part of standard care, has been suggested to have some benefit with respect to onset of dementia.

Family Adjustment

A number of studies have suggested that families of children with Down syndrome cope better and experience less stress than families of children with other types of disabilities. Some studies have concluded that this "Down syndrome advantage" is due to the child-based characteristics, such as their prolonged eye contact, the tendency to use charm to avoid tasks (see Development and Behavior), and the generally low levels of childhood psychopathology. Other explanations include societal awareness of Down syndrome and available support networks. Studies show that mothers of children with Down syndrome do better on some (generally not all) aspects of coping/adaptation than mothers of children with other types of intellectual disability or autism, but this advantage is not seen when comparing them with mothers of typically developing children.

Other socio-cultural populations may differ in their adjustment to having a child with Down syndrome, especially in families influenced by the Confucian beliefs (where the mother and the eldest sibling are solely responsible for household chores and caregiving). In their integrated review, Choi and Van Riper (2017) found that families with children with Down syndrome in the East Asian populations experience both positive and negative consequences. They emphasize the importance of developing revised healthcare policies taking into account factors such as ethnicity, religion, and worldviews. Across populations, the factors that contribute to the fathers' stress are said to be different from those of the mothers. Fathers focus more on financial burden, the potential impact on the broader family, and the public perception and attitude of society for accepting their children.

Recent studies also focus on siblings of children with disabilities. In general, there may be some increased anxiety and depression in these siblings during childhood, but there is no evidence of greater behavioral problems. There may be positive aspects in greater empathy and understanding of others with a disability and improved social interactions. In adulthood, the impact seems to be minor and is influenced by whether or not the sibling with disabilities has behavioral problems. Again, there is some evidence of a Down syndrome advantage when comparing with other conditions such as autism. Interventions such as SibworkS, a six-week manual-based, cognitive-behavioral group support program focused on strengthening siblings' social support, self-esteem, problem-solving skills, adaptive coping behaviors, and positive sibling relationships can be helpful for siblings of children with Down syndrome and the family as a whole (Strohm and Nesa 2010)

Divorce is far less common among families of children with disabilities than is generally thought. One study suggested that families of children with Down syndrome may be less likely to divorce than those whose children do not have identified disabilities (Urbano and Hodapp 2007). However, when divorce occurs, it is significantly more likely to occur during the first two years after the child's birth and is associated with demographic variables such as younger parental age and lower education levels, and living in a rural area.

Health care providers are a vital link in helping families cope with stressors. A multi-disciplinary team of experienced professionals should be available from the delivery of the news of a diagnosis of Down syndrome and throughout childhood and adulthood. Some studies suggest that healthcare professionals lack the training for informing families about such conditions in an empathetic way, using appropriate language during clinical visits and sharing knowledge about the psychosocial influences surrounding the condition. Skotko and colleagues (2009) provide guidelines for delivery of the news after a prenatal and postnatal diagnosis. These resources provide useful guidance for both healthcare management and facilitation of family adaptation following a diagnosis of Down syndrome.

Evaluation

- Explore what the parents know and have been told, their experience, if any, with individuals with Down syndrome, and the questions they have.
- Clarify the potential social/support network available, who has been told, and what they have been told, as well as any issues that have arisen in that regard.
- Explore the parents' feelings about having a baby with Down syndrome, and how they feel they will cope, with the potential maternal/paternal differences kept in mind.
- Personal and local social resources should be identified.

Management

- Pediatricians and obstetricians should be the first to deliver the news to the parents present together (exceptions to this would include when the father is not available, the mother does not wish for the father's presence or if the mother is seriously ill after childbirth). A follow-up meeting with a certified genetic counselor, a clinical geneticist, or a developmental-behavioral pediatrician can help in providing the family with appropriate information to meet their needs.
- Parents need to be given accurate information in understandable terms. The focus should be on explaining Down syndrome, what causes it, and what their child may need to maximize his/her potential.
- A **balanced perspective** should be provided regarding Down syndrome by an experienced and trained individual, including information about positive aspects of having a child with Down syndrome as well as the challenges commonly encountered (Sheets et al. 2011).

- It is helpful to mention that caring for a child with Down syndrome is generally not greatly different from the care of other children. However, some children with Down syndrome may have health complications that need to be addressed.
- When possible, the provider should refer the family to a local Down syndrome clinic where they can be followed by a team of professionals.
- The parents' misconceptions and misinformation should be corrected.
- Parents should be informed that children with Down syndrome generally attain developmental milestones at later and more variable ages than typically developing children, but can be expected to show continued developmental progress.
- It is important that siblings, extended relatives and family friends be made aware of the infant's diagnosis and their questions addressed. Parents may need assistance in this process. The burden and concerns regarding medical issues should be shifted, as much as possible, to health care providers, and an effort made to assure the availability of support services and resources that may be required.

For a prenatal diagnosis:

- Discuss the findings from screening and diagnostic tests, potential chances for recurrence in future pregnancies, prognosis and phenotypic manifestations – the wide-variability, information about additional studies to refine the estimation of the prognosis, consultation with appropriate medical subspecialists like pediatric cardiologists or pediatric surgeons, current available treatment options and interventions, and options such as medical termination, raising the child in a well-informed family with resources, foster care placements, and adoptions. If the pregnancy is continued, discuss a plan for delivery and neonatal care, parent-to-parent contacts through local and national support organizations, future reproductive options and evaluation of the chance to have a child with Down syndrome for other family members.

For a postnatal diagnosis:

- Pediatricians and obstetricians should coordinate their messages and deliver the news to both parents, as a couple, preferably with the baby next to them.
- Physicians should inform parents of their suspicion for Down syndrome immediately, even if the diagnosis has not been confirmed with karyotyping. This way parents are prepared psychologically in a stepwise manner.
- Physicians should deliver the news in a private hospital room without any visitors or other health personnel present and without any interruptions.
- Physicians should begin their conversation with positive words to congratulate the parents on the birth of their child. They should refrain from using words like 'so sorry to tell you' or 'I know this is going to be tough on you'. Some studies have shown that during communication of the diagnosis, the first words and their tone have a significant impact on families (Skotko et al. 2009).
- Physicians should consider limiting their first conversation on possible medical conditions seen during infancy (Skotko et al. 2009). Overloading the parents with long-term complications seems to have a negative impact on parents.
- Follow-up appointments should be arranged with medical experts and a team prior to discharge from the hospital. This will help minimize the stress on the parents. Discuss the various social networking strategies to cope with the news, such as support programs, online groups, and resources.

Cardiovascular

Congenital heart disease occurs in 40–50% of individuals with Down syndrome and is an important determinant of survival. The actual rate and relative frequency of specific anomalies vary with ascertainment. The most common cardiac malformation in Down syndrome is atrioventricular septal defect (endocardial cushion defect), followed by atrial septal defect, ventricular septal defect, patent ductus arteriosus, co-arctation of aorta, and tetralogy of Fallot (Stoll et al. 2015). Females with Down syndrome may be at higher risk of developing an atrioventricular septal defect than males (Diogenes et al. 2017). Atrioventricular septal defect is more common in populations of African origin than in Caucasian children and is relatively underrepresented in Asian children with Down syndrome. Muscular ventricular septal defect is relatively underrepresented overall. Hypertrophic cardiomyopathy is not common in Down syndrome, but when it occurs in adults it tends to be of the apical left ventricular type.

Individuals with Down syndrome also have a higher propensity for risk factors such as obesity and metabolic disturbances for developing coronary artery disease, particularly atherosclerosis. Despite this, they have a lower incidence rate of coronary artery disease. Recent studies are evaluating the causes for this low incidence of coronary artery disease among individuals with Down syndrome, including lower intimal thickening of arteries, lower blood pressures, different autonomic nervous system responses, or protective genetic mechanisms.

Several recent studies have identified regions of chromosome 21 containing genes that influence the occurrence of congenital heart disease, and heterotrisomy (inheritance of an allele from three different grandparents) has been associated with the presence of a ventricular septal defect. The *RCAN1* gene significantly contributes to congenital heart disease in Down syndrome, and a novel variant of g.482G>T in the *RCAN1* gene causes the overexpression of *RCAN1.4* (which is highly expressed in the heart muscle). Missense variants in *CRELD1*, located on chromosome 3p25, have been found to be associated with about 3% of endocardial cushion defects in the general population and, based on one report, with about 5% of those in individuals with Down syndrome. Maternal smoking, lack of folic acid/multivitamin supplementation, and parental consanguinity have also been identified as potential contributing factors for congenital heart disease in children with Down syndrome.

Mortality rates for five major biventricular repair procedures (ventricular septal defect repair, atrioventricular septal defect repair, patent ductus arteriosus closure, atrial septal defect repair, and tetralogy of Fallot repair) and bidirectional Glenn have been found to be similarly low in patients with Down syndrome compared with patients without Down syndrome. On the other hand, mortality after Fontan operation in patients with Down syndrome was significantly higher than in patients without Down syndrome, implying that indications for the Fontan operation should be carefully considered. It is known that uncorrected septal defects lead to shunting of systemic blood to the pulmonary circulation, increased blood flow and pulmonary arterial hypertension, which may persist even after a surgical correction. Sullivan et al. (2017) demonstrated that children with Down syndrome who underwent surgical repair for tetralogy of Fallot had an increased degree of pulmonary regurgitation. These individuals required earlier assessment by cardiac magnetic resonance imaging to determine timing of pulmonary valve replacement to manage preventable causes of pulmonary hypertension. There is a common understanding that, with the advent of new catheter-based interventions, surgeries benefit individuals with Down syndrome and are associated with prolonged survival rates. Cardiac catheterization provides important information like pulmonary arterial resistance, which is paramount in assessing the severity and response to vasodilating agents, preventing postoperative pulmonary hypertension crisis and prolonged pulmonary hypertension, and in assessing the possibility of intracardiac repair. It is important to systematically evaluate perioperative complications among individuals with Down syndrome undergoing surgical procedures.

A more than 10-fold increased incidence of persistent pulmonary hypertension of the newborn, which cannot be accounted for by demographic variables including gestational age at birth and the presence of congenital heart disease, has been reported in Down syndrome. Pulmonary hypertension occurs more often and earlier in children with congenital heart disease and Down syndrome, especially in the presence of large right to left shunts. Infants and children with Down syndrome are at increased risk of pulmonary hypertension, even in the absence of intracardiac structural defects. Obstructive sleep apnea is a major contributing factor towards development of pulmonary hypertension among individuals with Down syndrome (See Sleep Disorders section for details). Nir et al. (2017) found that the prevalence of pulmonary hypertension among adults with congenital heart disease and Down syndrome was 53%, compared to 6% for all adults with congenital heart disease. Morbidity included cerebral vascular accident or transient ischemic attack in 22% (mostly in people with right-to-left shunt) and arrhythmia in 37% of the patients.

A rare cause of pulmonary hypertension, pulmonary veno-occulsive disease, was recently reported in an infant with Down syndrome (Muneuchi et al. 2017). This condition has a worse prognosis and higher risks of developing severe pulmonary edema with specific vasodilator therapy, leading to rapid deterioration. It can be misdiagnosed because it is clinically similar to idiopathic pulmonary arterial hypertension, despite the histopathological differences. The actual incidence of pulmonary veno-occlusive disease is probably underestimated, because many cases may be classified as idiopathic pulmonary arterial hypertension. This study emphasizes the importance of high-resolution computed tomography, lung biopsy, and diagnosing a rare entity during the management of pulmonary hypertension in individuals with Down syndrome.

Evaluation

- As per the AAP guidelines (Bull 2011), a thorough clinical examination and a mandatory echocardiography should be a part of routine newborn screening of babies with Down syndrome. Infants with an abnormal echocardiogram should be referred to a pediatric cardiologist for further evaluation. Structural heart defects with 'silent' murmurs that do not present themselves as typical murmurs could be missed if one relies on clinical examination alone. Monitor all heart defects for complications, and educate parents to recognize signs of heart failure such as tachypnea, feeding difficulties, and poor weight gain. In older children and adults with Down syndrome, annual cardiac examinations may help in early diagnosis of acquired valvular diseases. Individuals with Down syndrome should be educated to recognize common symptoms like fatigue, shortness of breath, or dyspnea on exertion, which may warrant further cardiac testing.
- Echocardiography and competent clinical evaluation, electrocardiogram, and early follow-up should detect most congenital heart disease that will require treatment.

Management

- Although there has been controversy regarding the merits of medical versus surgical management of significant congenital heart malformations in individuals with Down syndrome, and over the timing of surgery, there is now a consensus favoring early surgical intervention. Medical treatment of atrioventricular septal defect has less than a 5% five-year survival rate compared with almost 70% for surgically treated defects, which compares favorably with children who have the equivalent lesion but do not have Down syndrome.
- Babies with Down syndrome who have large ventricular septal defects without obstruction to pulmonary blood flow require surgical repair before 4 months of age to avoid further complications (Bull 2011).
- Medical management (along with education on proper nutrition) should be offered to those who are not fit for surgery.
- Individuals with Down syndrome have an increased rate and more prolonged course of postoperative thrombocytopenia than the general population, and may have an increased risk of post-discharge syncope because of complete atrioventricular block. Close monitoring and early follow up after surgery with complete blood counts are needed.

Endocrinologic

Thyroid Disease Thyroid disease commonly occurs with Down syndrome, with hypothyroidism being the most prevalent endocrine condition. Children with thyroid disease and Down syndrome require appropriate and timely treatment to avoid stunted growth velocities and significant intellectual disabilities. Therefore, there is a need for additional screening, in addition to newborn screening, especially in the first year of life. Thyroid screening using TSH and serum free T4 is recommended to be continued throughout the lifespan of individuals with Down syndrome. With continued early ascertainment through routine surveillance, and detection and treatment of hypothyroidism, the effect of hypothyroidism on development should be minimized.

When borderline abnormal thyroid function screening tests are found, it is customary to repeat the tests in six weeks to be sure they are abnormal. If they are still abnormal, a referral to a pediatric endocrinologist is indicated. Many infants have a mildly elevated serum free T4 level with a normal TSH level. If the infant is asymptomatic, this is not necessarily a pathologic state. The tests should be repeated in a few months. It has been noted that thyroid disease in individuals with Down syndrome is unrelated to sex, obesity, or other comorbidities.

In a study of 508 individuals with Down syndrome, 24% had a documented history of a thyroid-related diagnosis with the following prevalence: 2% congenital hypothyroidism, 10% subclinical hypothyroidism, 1% overt hypothyroidism, 4.5% isolated hyperthyrotropinemia, 4.3% unknown hypothyroidism, and 1.6% hyperthyroidism (Pierce et al. 2017). The median age at diagnosis with any thyroid disease was 4 years and 10 months, whereas for hyperthyroidism, it was just under 9 years. The odds of developing thyroid disease increased by 10% per year with increasing age. For the development of thyroid disease, it was estimated that 25% of those with Down syndrome will have thyroid dysfunction by age 7.5 years and up to 50% by the time they reach adulthood. When tested, approximately 50% of individuals with subclinical hypothyroidism had positive antithyroid antibodies, and this rate was 100% in overt hypothyroidism (Pierce et al. 2017). The study highlights the need for additional screening of TSH along with free T4, between the newborn screening and the currently recommended sixth month screening, as a significant number of children with Down syndrome were diagnosed with thyroid disorders before the age of 6 months.

Sexual Maturation Much of the literature regarding sexual development and function in individuals with Down syndrome is based on older reports of persons living in institutions, and detailed hormonal data are sparse. Such studies suggested that men with Down syndrome had relatively small genitalia, higher incidence of testicular failure with elevated follicle stimulating hormone (FSH) and luteinizing hormone (LH), decreased Leydig cell function, and germinal cell hypoplasia. Newer data suggest that men with Down syndrome have normal onset and chronology of puberty and may have fewer differences in hormonal levels than previously thought. A study of young adult males with Down syndrome (mean age 26.5 years) confirmed normal levels of FSH, testosterone, and dehydroepiandrosterone (DHEA-S) but showed elevated levels of LH and 17-OH progesterone compared with controls (Sakadamis et al. 2002). These individuals also showed significantly decreased bone mineral density that might be in part related to mild chronic hypogonadism. Men with Down syndrome have an increased risk of cryptorchidism (uni- or bilateral), but the mechanisms causing its onset are not clear. Spermatogenesis is known to be insufficient in males, but the definitive cause remains to be discovered.

Female pubertal development appears to be normal. The mean age at menarche of 12.6 years does not differ from that of typically developing siblings, and there does not appear to be any excess of menstrual problems or irregularities. In some women, menarche may be somewhat earlier and this may relate to obesity. FSH and LH rise normally with maturation. There is some evidence that young women with Down syndrome may have relatively increased levels of prolactin, LH, testosterone, and 17-OH progesterone, possibly raising the issue of subtle differences in the pituitary-gonadal and/or adrenal axis (Angelopoulou et al. 1999). Although earlier studies reported high rates of ovarian abnormalities and anovulatory cycles, recent ultrasound studies have found normal ovaries with follicles at various stages of maturation, and normal uteri and uterine wall thickness. Age at which menopause occurs varies widely, but typically occurs before the age of 40 years.

Despite evidence of normal sexual development, there remain fewer than 50 cases of documented fertility in women with Down syndrome and four reported cases in non-mosaic males, one of whom achieved a pregnancy following preimplantation genetic screening (Aghajanova et al. 2015). Potential contributors to infertility may be hormonal abnormalities, gonadal malformations, and psychological and social factors. Many individuals with Down syndrome show age-appropriate interest in the opposite sex, and need to be educated about their bodies, contraception and fertility (Bull 2011). There may be a trend towards a decline in sex education for adolescents with Down syndrome, which highlights the need to train adolescents with developmental disabilities through better sex education programs.

Evaluation

- Screening for neonatal hypothyroidism is important for children with Down syndrome. This occurs through mandatory newborn screening in most jurisdictions. Measure the TSH and free T4 routinely throughout the lifespan. Check for antibodies, based on clinical judgement. Secondary and tertiary causes of thyroid disorder should be kept in mind while making a diagnosis. Perform routine screening, along with newborn screening and annual examinations, throughout life for individuals with thyroid disease and Down syndrome.
- There is compelling evidence that clinical examination is inadequate to detect thyroid disease in individuals with Down syndrome, at least in part because of the overlap of signs, sensitivity and specificity are poor. Growth velocity may decline or other symptoms may develop at least a year before the clinical recognition of hypothyroidism in children, and it may masquerade as depression or even Alzheimer disease in adults.
- Any variation from normal physiological sexual maturation is unexpected and requires a standard evaluation for cause.
- Women require standard gynecological care. This may be facilitated if carried out by a familiar health care provider and by taking special care to educate the woman in advance about any examinations or procedures.

Management

- Treatment of hypothyroidism is standard replacement with L-thyroxine and continued monitoring of blood levels. Some centers treat with low-dose thyroxine in the face of significantly elevated TSH and normal T_4 levels.
- With integration of individuals with Down syndrome into society, adolescents and young adults with Down syndrome need properly tailored education, advice, and counseling concerning interpersonal relationships, appropriate social behavior, sexual activity, and situations that may place them at increased risk for sexual abuse. Success will generally require the understanding and participation of parents. Encourage independence with personal hygiene and self-care.
- As per the AAP guidelines (Bull, 2011), parents should be made aware of the physical and psychological changes during childhood and adolescence, issues of fertility and contraception, and the need for more preparation during these changes. Discuss the need for gynecologic care and menstrual hygiene for women.
- Despite high rates of infertility, it is important to talk about the chance of Down syndrome, if a pregnancy were to occur. This is true for both genders. Birth control and prevention of sexually transmitted diseases should be discussed with individuals with Down syndrome and their families.

Audiologic

Over 90% of ear lengths in individuals with Down syndrome fall below the third centile for the general population; the helix is often angulated and overfolded, and the lobes are small to absent. The osteocartilaginous junction is narrow to stenotic and may compromise visualization of the tympanic membrane and increase the susceptibility to obstruction by wax. A small minority has been found to have congenitally malformed stapes, and other ossicular anomalies may be acquired.

Malformations of the inner ear may also be common. Inner ear structures have been found to be hypoplastic in many studies. Inner ear anomalies have been observed in a large proportion of individuals with Down syndrome, including malformed bone islands of lateral semicircular canal, narrow internal auditory canals, cochlear nerve canal stenoses, semicircular canal dehiscence, and enlarged vestibular aqueducts. Internal auditory canal stenosis has the highest odds ratio for sensorineural hearing loss (Intrapiromkul et al. 2012). Further studies are needed to properly understand such variations in inner ear anatomy and their role in hearing loss.

Individuals with Down syndrome are at high risk for conductive hearing loss, although mixed and pure sensorineural hearing loss may also occur. Some degree of hearing loss occurs in over 60% of those with Down syndrome, with conductive hearing loss accounting for the majority. One reason is the high incidence of inflammatory processes such as otitis media. Dysfunction of the eustachian tube, due to its abnormal anatomy and muscular hypotonia of surrounding structures like the tensor veli palatine, plays an important role in the etiology of otitis media with effusion among persons with Down syndrome. A study of 107 children with Down syndrome (ages 6 months to 12 years) found a high prevalence of otitis media with effusion at the age of 1 year (67%), with a second peak prevalence of 60% at 6–7 years (Maris et al. 2014). A declining trend was observed in children ≥8 years. Overall, 52% of children had either otitis media with effusion or ventilation tubes at the time of evaluation. Parental report suggested that otitis media with effusion had more impact on speech, language and communication than on hearing. Recurrent ear infections and middle ear effusions are to be anticipated, but with a team approach to hearing care, most children with Down syndrome can maintain adequate hearing to prevent interference with their speech development and educational efforts.

Evaluation

- Based on the AAP guidelines (Bull 2011), newborns should be assessed for congenital hearing loss with objective testing such as brainstem auditory evoked response (BAER) or optoacoustic emission, according to the universal newborn hearing screening guidelines. Newborns should have a follow-up at 3 months thereafter. In younger children, review the risks associated with serous otitis media. Follow-up annually or every 6 months.
- Refer to an otolaryngologist for conducting ear examinations of children with stenotic ear canals.
- For older, school-aged children, make sure to conduct annual audiologic evaluations.

Management

- Treatment of acute and chronic complications, which may compromise hearing and thereby interfere with speech and education, should be aggressive. Treatment methods are standard.
- A significant proportion of individuals with Down syndrome may benefit from classroom accommodations for hearing loss by some means of amplification. Children with Down syndrome who have impaired hearing may benefit from early interventions by an experienced speech-language pathologist.

Ophthalmologic

Ophthalmologic abnormalities are found more frequently in children with Down syndrome than in the general population. Overall, hyperopia, strabismus, astigmatism, and blepharitis have the highest prevalence rates among many studies. Table 24.1 indicates the prevalence ranges for specific ophthalmologic findings (adapted from Creavin and Brown 2009).

TABLE 24.1 Summary of ophthalmologic findings in Down syndrome (adapted from Creavin and Brown 2009)

Condition	Frequency	Comment
Hyperopia	4–59%	
Myopia	8–41%	
Strabismus	<20–60%	Majority reported prevalence of 20–40%. Esotropia more common than exotropia
Nasolacrimal duct obstruction	<10–36%	Majority reported prevalence of 17–36%
Epiphora	15–32%	
Astigmatism	6–60%	Majority reported prevalence of 20–30%
Weak/absent pupillary reflex	17–26%	
Nystagmus	10–20%	
Amblyopia	3–20%	
Cataract	<5–37%	Majority reported prevalence of <15%
Anisometropia	1–13%	
Blepharitis	<10–50%	Majority reported prevalence of <10%
Retinal anomalies	<10–40%	Majority reported prevalence of <10%. Most common finding extra retinal vessels
Ptosis	3–7%	
Corneal opacity	1–6%	
Optic nerve abnormalities	1–5%	Includes optic nerve elevation or optic disk pallor
Sty	3%	
Chalazion	1–3%	
Entropion	2%	
Congenital dacryocutaneous fistulae	1%	
Lateral eyelash diversion	1%	
Glaucoma	<1–7%	Majority reported prevalence <1%
Keratoconus	1–12%	Majority of studies found it in none of the study participants

Careful examination in childhood may reveal early opacities. Cataracts develop in over half of those individuals. Hypoplastic peripheral irides and Brushfield spots are common. Some studies show that esotropia is more common at a younger age and exotropia is more common among older adults with Down syndrome. Keratoconus also has a higher rate of incidence in Down syndrome, presenting at a younger age compared to typically developing children. Individuals with advanced keratoconus are more subject to development of acute hydrops (stromal edema) with its accompanying severe visual impairment.

There is some concern for vision later in life, as infranasal limbus and degenerative retinal changes may be seen in adults. However, ophthalmic studies on adults with Down syndrome are limited. While astigmatism and refractive errors are more prevalent in younger individuals, cataracts and blepharitis are more common in older individuals with Down syndrome. Senile cataracts occur earlier compared to the general population.

It is increasingly apparent that visual problems may adversely affect potential for learning in children with Down syndrome. Strabismus and ocular pathology has been associated with lower IQ in children with disabilities (Salt and Sargent 2014). However, with routine ophthalmologic assessments and appropriate care, the prognosis for vision in childhood and young adulthood is good.

Evaluation

- Early eye assessment with appropriate therapy and follow-up is important to prevent loss of binocular vision and/or amblyopia, and to maximize visual acuity. Strabismus must be distinguished from pseudostrabismus caused by epicanthal folds.
- As per the AAP guidelines (Bull 2011), children with Down syndrome should be evaluated by a pediatric ophthalmologist within the first six months of life. Children should be evaluated for congenital cataracts, strabismus, nystagmus, and other specific conditions at regular intervals. Based on different ages, clinical condition and clinical judgements, annual or two-yearly eye examinations should be conducted to eliminate the risks of developing refractive errors, amblyopia, keratoconus, cataracts, etc.
- Causes of eye irritation or behavior that may increase the risk of self-induced ocular trauma should be sought and treated.
- Ophthalmologists should be aware of the possibility of pseudotumor cerebri in people with Down syndrome with an elevated optic disk. Untreated, this may result in optic atrophy.

Management

- Refractive errors are the most common and important visual problems and require early refraction studies and prescription for glasses. Bifocal or progressive glasses may benefit some if they are tolerated.
- Strabismus may respond to eye patching or may require surgery and is treated in a standard fashion.
- Blepharitis usually will respond to lid cleansing and topical antibiotics.
- Cataract may require removal of the lens and a prosthetic implant, and significant keratoconus may be treated with penetrating keratoplasty and a corneal transplant.
- Pseudotumor cerebri may respond to weight loss and acetozolamide therapy.

Musculoskeletal

Children with Down syndrome have an increased prevalence of numerous musculoskeletal complications in varying degrees, the most common ones being hypotonia, and ligament laxity that leads to excessive joint flexibility. There is a delayed achievement of motor milestones (see Behavior and Development section). Other orthopedic problems include scoliosis, neck instabilities, hip anomalies (subluxation, Legg–Calve–Perthes disease, slipped capital femoral epiphysis), patellofemoral joint instability, genu valgum, hallux valgus, foot deformities (pes planus, metatarsus primus varus), trigger finger, and trigger thumb. There is also increased risk of developing arthritis/arthropathy precociously. Individuals with Down syndrome have a higher prevalence of gait abnormalities, despite having an independent gait. Orthotic supports seem to be very beneficial in children with Down syndrome who have an established walking pattern. Custom made foot insoles provide the additional physical support and could help prevent future surgical corrections when used as an early intervention to correct gait abnormalities. There is evidence that some of the gait difficulties can be ameliorated by specific exercises in early intervention, which have been found to provide long-term benefits on the development of basic gait parameters.

A high percentage of adults with Down syndrome have reduced bone density. There is a growing concern regarding the high risk of fractures with age and early detection. Factors found to be associated with decreased bone density have included male sex, low levels of exercise, low antioxidant capacity due to high oxidative stress, hypothyroidism, certain drugs that are used to treat comorbidities, and lack of exposure to sunlight. Individuals with Down syndrome have a higher prevalence of hypovitaminosis D. Individuals with Down syndrome who are obese and those who have autoimmune diseases may require a higher supplementation of vitamin D to maintain normal levels (Stagi et al. 2015).

The incidence of occipitoatlantoaxial instability among persons with Down syndrome is about 10–30%, with neurological symptoms occurring in approximately 1% of individuals. The occiput, the atlas (C1), and the axis (C2) form the occipitoatlantoaxial joint, and the surrounding ligaments keep these bony structures in place and enable flexibility. The occipitoatlantal joint provides principally flexion, with a small amount of lateral bend and rotation, whereas the atlantoaxial joint is most responsible for rotation but does provide significant flexion/extension and a small amount of lateral bend. In flexion, an anterior translation of C1 on C2 exists, which normally does not exceed 3 mm in adults. In children younger than 8 years, this translation can be as wide as 5 mm. In pathologic conditions (e.g., abnormalities of the odontoid bone or in the ligament laxity), this displacement increases and bone structures can pressure the spinal cord, producing clinical symptoms.

The occipitoatlantoaxial instability among individuals with Down syndrome was recognized in the 1960s, and gained notoriety in the early 1980s when there were several reports of individuals with significant signs of cervical cord damage. The instability is primarily anterior and is because of ligamentous laxity, exacerbated by anatomic osseous variants, such as loss of the normal concavity of the superior surface of C1, that result in a less stable joint structure. Rotatory instability is also relatively common in Down syndrome. By 1984, over 500,000 individuals with Down syndrome had participated in the Special Olympics without a single known occurrence of serious neck injury. A review by the AAP of 41 reports of individuals with symptomatic atlantoaxial instability showed that the majority of those cases recovered, and there was no evidence that radiographs could predict those who would become symptomatic. Our current understanding is that about 10–30% individuals with Down syndrome may have radiological evidence of occipitoatlantoaxial instability, but only a small minority of individuals will develop neurological complications, and an even smaller number will suffer a catastrophic event in the absence of some earlier neurological signs. These events can be minimized by encouraging lower risk sports and neurological monitoring. Evidence to date is that pure atlantoaxial instability is unlikely to undergo significant change in the absence of bony abnormalities. With time, some persons may develop an os odontoideum, which is considered an avulsion fracture of the odontoid and is therefore evidence of chronic instability and secondary bony changes.

Intervention in the presence of neurological signs attributable to the cervical spine should not be delayed, as chronic changes are unlikely to be reversible, and complications appear higher in late-treated cases. For individuals with Down syndrome, careful anesthetic procedures and airway management is required because of the high risk of the cervical spine instability. Signs such as newly decreased motor skills, gait abnormalities, torticollis, progressive paralysis, neurological signs of nerve compression at the cervical level, or vertebrobasilar insufficiency should raise suspicion for cervical spine instability. At present, there is no real consensus of which radiographic techniques can be used with maximum efficacy to identify impending problems. The commonly used radiological parameters are atlantodens interval and the space available for the spinal cord (SAC). Some argue that these are associated with poor inter- and intra-observer reliability and could cause a neuropathy as they require radiographs to be performed with the cervical spine in flexion. Nakamura and colleagues (2016) assessed the reliability of two new radiological parameters that can be measured with the cervical spine in the neutral position (the C1/4 SAC ratio and the C1 inclination angle), and they investigated cut-off values to identify an indication for surgery. The normal values for the C1/4 SAC ratio and inclination angle were found to be about 1.2° and 15°, respectively. Children with a C1/4 SAC ratio of <0.8 have a high risk of developing neurological symptoms or signs and referral to a pediatric spinal surgeon is recommended. They also recommend that all pediatricians who are not specialized in the cervical spine to use these new parameters.

Evaluation

The AAP guidelines recommend the following (Bull 2011). *Newborns*: Evaluate hypotonia.

Infancy–adulthood: Once every 2 years, parents should be educated about the signs and symptoms of myelopathy and about the precautions of to avoid excess extension and flwxion during anesthesia, and surgical or radiological processes. At every visit, a thorough a meticulous history and physical examination should be performed to evaluate for signs of myelopathy.

For asymptomatic atlantoaxial instability: Since radiographs are not predictive of instability, radiologic evaluation or screening of the cervical spine in asymptomatic children is not recommended. Parents should be advised that participation in some sports, including contact sports such as football, soccer and gymnastics, places children at increased risk of spinal cord injury. Trampoline use should be avoided by children younger than 6 years, and by older children unless under direct professional supervision. Special Olympics has specific screening requirements for participation in some sports. Parents should be educated that participation in sports such as football, soccer, and gymnastics may place the child at an increased risk of spinal cord injury. Use of trampolines should be avoided in children less than 6 years, and older children require direct professional supervision. Educate the family that there are specific screening requirements for participation in some sports at the Special Olympics.

For symptomatic atlantoaxial instability: If a child has signs/symptoms of myelopathy, plain cervical spine radiography should be done in the neutral position of the cervical spine. If significant abnormalities are seen in the radiography, the child should be immediately referred to pediatric neurosurgery/pediatric orthopedic surgery to further manage the atlantoaxial instability. If there are no significant abnormalities, flexion/extension radiographs may be obtained before the child is referred to the specialists.

- If standard radiographs raise concern, magnetic resonance imaging could be considered. Computed tomography is not likely to add useful information.
- A medical history and physical examination for joint complaints should be part of routine clinical care.
- A routine clinical assessment for scoliosis should continue into adulthood.

Management

- Participation in high-risk activities should be discouraged if there is evidence of chronic instability, such as an os odontoideum or 7 mm or more of instability.
- A small minority of children with Down syndrome will require stabilizing surgery with a C1–C2 and/or atlantooccipital fusion. A Gallie C1–C2 fusion appears to be a satisfactory approach for symptomatic anterior atlantoaxial subluxation, and some authors suggest a period of prior traction as well as a postoperative halo to assure stability and fusion. Recent neurological series have reported greater success in achieving fusion, and lower rates of significant complications.
- Complication rates for cervical surgery to repair atlantoaxial instability were up to 100% in some historical studies. A recent retrospective review found that postoperative complications continue to challenge most patients (82%) (Siemionow et al. 2017). Although 94% of patients demonstrated stabilization or improvement in neurologic status, several postoperative complications were observed, with postoperative pneumonia being the most common complication. Also, the anterior approach resulted in a higher risk of complications than posterior.
- There is no doubt that early interventions are better than late surgery with complications. The decision regarding cervical surgery for occipitoatlantoaxial instability requires the involvement of experienced and expert surgeons, a careful weighing of the evidence of present or impending neurological damage, and consideration of the benefits and risks of the surgical approach. Anesthesia may increase instability.
- Most joint findings related to laxity do not require treatment or can be managed with advice from experts such as a physical therapist. Significant scoliosis should be managed as for standard practice.
- Neuromuscular training could be beneficial to optimize general and maximal muscular strength development in children and young adults with Down syndrome (Sugimoto et al. 2016).

Gastrointestinal

Functional or structural gastrointestinal anomalies are found in over 2/3 of individuals with Down syndrome. Individuals with Down syndrome are more susceptible to hypotonia, large tongue, small oral cavity, tracheoesophageal fistula, dysphagia, constipation, gastrointestinal regurgitation syndrome, duodenal atresia, annular pancreas, celiac disease, Hirschsprung disease, pyloric stenosis, and anal atresia/stenosis. There is also evidence to suggest that a paucity of bile ducts is common. Some features such as duodenal/anal atresia or tracheoesophageal fistula may be suspected on prenatal ultrasound or will be present at birth. However, a number of these features may not lead to symptoms until several months after birth, including Hirschsprung disease, duplication cysts, and duodenal or anal stenosis. Hypotonia and relative inactivity may account for a higher rate of constipation in the euthyroid and otherwise healthy child with Down syndrome, and chronic unexplained diarrhea may affect up to 20% of adults. *Helicobacter pylori* infection has been noted to be common in adults. Non-immunity to Hepatitis A and B is high and immunization against these is important (See Immunologic section).

Celiac disease has a high prevalence rate among children with Down syndrome (up to 18.6%), depending on the age of the child, geographical location, and diet (Pavlovic et al. 2017). The symptomatic form of celiac disease is more frequent in children with Down syndrome than the asymptomatic form, but about one-third of children with Down syndrome and celiac disease have no gastrointestinal symptoms. Symptoms of celiac disease include diarrhea or protracted constipation, slow growth, unexplained failure to thrive (in infancy), anemia, abdominal pain or bloating, refractory developmental or behavioral problems, and irritability. A known complication of celiac disease among affected individuals who do not follow a gluten-free diet or those who do not seek appropriate evaluations is the risk of developing intestinal lymphoma. Due to a high prevalence rate, presence of asymptomatic cases, and complications of celiac disease, routine screening for celiac disease among children with Down syndrome should be performed.

Evaluation

- Digestive difficulties should be evaluated aggressively, given the high rate of gastrointestinal malformations.
- Based on the AAP guidelines (Bull 2011), infants with marked hypotonia and feeding difficulties, recurrent pneumonia, respiratory problems, and failure to thrive

require radiographic swallow studies for further evaluation to diagnose various neuroanatomical defects. Evaluate duodenal atresia or anorectal atresia/stenosis or tracheoesophageal fistula by performing a history and clinical examination. Evaluate for reduced fluid intake, hypothyroidism, gastrointestinal tract malformations or hypotonia, if constipation is present.

- Checking for stool in the rectum by rectal examination should be done as part of a work up for Hirschsprung disease. An abdominal X-ray should be obtained to determine if the child is constipated and if so, the child should have a barium enema without doing a preparatory bowel clean out first, looking for short or long segment Hirschsprung disease. For severe chronic cases of constipation, a rectal biopsy is indicated for possible Hirschsprung disease.
- For celiac disease, there is no universal consensus about routine screening among all children with Down syndrome. Based on the AAP guidelines (Bull et al. 2011), after the age of one year, those on a gluten diet should be evaluated for symptoms of celiac disease at each wellness visit. Symptomatic children need an assessment of tissue transglutaminase immunoglobulin A (IgA) level and quantitative IgA. It is recommended that those with abnormal laboratory values be referred for specialty assessment. Small intestinal biopsy is typically used to confirm the diagnosis. Screening is recommended to be done at least every third year, and more frequently in those with a family medical history of celiac disease or those with possible symptoms from the condition.

Management

- Malformations and functional gastrointestinal problems should be treated as in the general population.
- Adequate response to chronic constipation is generally obtained with a standard pediatric approach when not caused by a malformation or Hirschsprung disease or hypothyroidism.
- Treatment for celiac disease is as for the general population, with a gluten-free diet.

Sleep Disorders

Sleep problems in individuals with Down syndrome have been associated with differences in facial morphology, hypotonia, obesity, irregular sleep habits and sleep resistance, family stresses, obstructive sleep apnea, medications being used for other conditions, and various infections causing general discomfort (Nakamura et al. 2016). Poor sleep may be associated with daytime drowsiness or a decline in behavior. Fernandez and colleagues (2017) found that general sleep quality was poor and efficiency scores were lower in infants and toddlers with Down syndrome than in typically developing children. Infants with Down syndrome exhibited the worst sleep fragmentation; however, sleep efficiency and consolidation increased across age. With advances in technology and sleep studies, further research with more precision and age-wise distribution of sleep patterns are needed to clearly understand various aspects of sleep problems among children with Down syndrome.

As many as 50% of children with Down syndrome may have sleep apnea, with a higher prevalence among adults. The relative underdevelopment of the midface, sometimes associated with a narrow nasopharynx, and the high rate of hypotonia appear to place children with Down syndrome at an increased risk for obstructive sleep apnea, even with nearly normal-sized tonsils and adenoids. It is important to note that obstructive sleep apnea may increase the risk for pulmonary hypertension in susceptible individuals. Obstructive sleep apnea may also be a common comorbidity among adolescents and young adults with Down syndrome who have depression, suggesting that it could be a potential contributor to new-onset mood disorder or decline in adaptive skills (Capone et al. 2013).

One retrospective study of children with Down syndrome and obstructive sleep apnea found that only one-third of those who had standard tonsillectomy and adenoidectomy had a normal postoperative polysomnogram and that adding lateral pharyngoplasty was of no added benefit (Merrell and Shott 2007). Ingram et al. (2017) found that tonsillectomy with concurrent or prior adenoidectomy resulted in significant improvements in respiratory parameters, including obstructive apnea-hypopnea index, percent sleep time with oxygen saturations <90% and percent sleep time with end-tidal carbon dioxide above 50 mmHg. They experienced improvements in both respiratory event frequency and gas exchange but approximately half still had moderate to severe residual obstructive sleep apnea. Future studies are needed to quantify the effects of tonsillectomy and adenoidectomy as a treatment for obstructive sleep apnea in Down syndrome.

Evaluation

- As per the AAP guidelines (Bull 2011), at least once during the first six months of life, discuss with parents symptoms of obstructive sleep apnea, including heavy breathing, snoring, uncommon sleep positions, frequent night awakening, daytime sleepiness, apneic pauses, and behavior problems that could be associated with poor sleep. A referral to a sleep specialist for further evaluation of a possible sleep disorder should be made if necessary. It has been recommended that all children with Down syndrome should have a sleep study by the age of 4 years old. This issue requires further study as obtaining a sleep study for all individuals with Down syndrome is a challenge, and pediatric

sleep labs are not available in many parts of the country. Parents should be made aware that obesity is a risk factor for sleep apnea.
- Suspicion should be raised if parents or care providers note an unusual sleeping position, such as with the head hyperextended or on the stomach with the knees drawn up, or if other signs of sleep disturbances are noted.
- A careful assessment of the cause of sleep apnea should be sought before treatment, especially surgery, is suggested.

Management

- Many symptomatic individuals with obstructive sleep apnea will respond adequately to tonsillectomy and adenoidectomy.
- Continuous positive airway pressure may be the appropriate intervention for some.
- Very occasionally, more involved surgery, such as enlargement of the midface, may be required.

Dental

Dental anomalies are common in individuals with Down syndrome, and include delayed and asynchronous primary dentition (completed by 4–5 years) and secondary dentition. Primary dentition tends to be larger than in typically developing children, with a degree of microdontia and thinner than normal dentin and enamel in the permanent dentition. Primary teeth may be retained, and there is a greater rate of supernumerary teeth, taurodontism (reported in 50% of one group aged 3–35 years), tooth hypoplasia, hypocalcification, and crown variants, especially on the labial surfaces. Crowns are more likely to be conical, small, and short. The roots are complete but short, and this may contribute to instability and susceptibility to tooth loss associated with periodontal disease. Partial anodontia affects over half as compared with 2% of the general population, and the agenesis occurs in a pattern different from that seen in the general population. Involvement is usually mild with approximately two teeth missing about 60% of the time and three to five absent in about one-third of those affected. The third molars are absent about 75% of the time, and impaction of the maxillary canines and maxillary canine-first premolar transposition are not uncommon. Occlusal problems, most often of the central and lateral incisors and canines, are common and result from mouth breathing, impaired chewing, tooth agenesis, shortness and asymmetry of the maxillary arch, and temporomandibular problems. There is also a high incidence of aphthous ulcers, oral candidiasis, and acute ulcerative gingivitis. Dental wear is twice as common in children with Down syndrome and is much more likely to be severe. The cause is unclear but there is a high rate of bruxism that may have some association with gastric reflux. Rates of bruxism could also be influenced by anxiety, craniofacial abnormalities, temporomandibular joint dysfunction, and orofacial hypotonia.

Children with Down syndrome have rates of periodontal disease that are equivalent to those of other children with intellectual disabilities and are higher than in the general population. The high incidence of periodontal disease may be due to hypotonia, dentoalveolar joint laxity, poor dental hygiene, and impairments of the immune system. Prevalence of periodontal diseases in young adults with Down syndrome is about 35% with increasing prevalence in the third and fourth decades. Decreased salivary flow and an increase in pH and bicarbonate buffer may result in mucosal thinning and xerostomia, while offering some protection against dental caries. There is no evidence of qualitative or quantitative differences in the carriage of putative oral pathogens in individuals with Down syndrome.

Evaluation

- Early referral for regular semiannual dental care is important both for repeated instruction in dental hygiene for the prevention of gum disease and the longer term planning of possible orthodontics.
- Parents may have particular difficulty finding access to dental expertise (personal experience). Pediatric dentists and pediatric dental hygienists are recommended for children with Down syndrome.

Management

- Educate the parents of young children with Down syndrome about the possible dental anomalies in primary and secondary dentition, including delayed or irregular dental eruption and hypodontia.
- Instruction in dental hygiene, one-on-one help, and brushing and flossing must begin at an early age for all children with Down syndrome, and dental hygiene must become a lifelong habit. In the absence of a successful program of gum care, early tooth loss can be anticipated. Awareness of periodontitis is important. An aggressive preventive dental program may be recommended. Specific dental anomalies will vary widely from individual to individual and will require a tailored approach. More frequent visits to a dentist (three to four per year) may be helpful. Many dentists believe that regular use of a fluoride-free chlorhexidine mouthwash can help reduce periodontal disease.
- Dietary counseling with avoidance of certain foods that affect the pH in the oral cavity should be encouraged, while maintaining a proper nutritious, balanced diet.
- Topical fluoride application will help in the prevention of dental caries and improve gum and enamel health.

- Antibiotic prophylaxis against subacute bacterial endocarditis is required at the time of dental care for many individuals with Down syndrome and congenital heart disease.
- Significant malocclusion can be treated in a standard fashion, although braces may complicate gum care.
- Psychological treatment may be beneficial for some children with bruxism.

Dermatologic

Dermatologic conditions are among the more common issues in individuals with Down syndrome. The skin in infancy is generally soft and velvety, and cutis marmorata occurs in about 8–13% of children. Premature wrinkling, graying, and loss of hair suggest an accelerated ectodermal aging. The hair is often fine and hypopigmented. A wide variety of skin disorders are reported in Down syndrome, and many are age related. These are summarized in Table 24.2 (data adapted from Barankin and Guenther (2001) and Daneshpazhooh et al. (2007)).

Seborrheic dermatitis may occur in up to 30% of persons with Down syndrome (general population 2–5%), with red cheeks being common. With time, the skin has a tendency to become dry and rough and may show local thick and scaly hyperkeratotic patches on the limbs. Fewer than five cases of multiple dermatofibromas have been described in the literature among individuals with Down syndrome, and all had some underlying condition such as autoimmune diseases, leukemia, or immunodeficiency. Cases of psoriasis have also been observed. Alopecia areata, an autoimmune condition, is also more common in Down syndrome than in the general population with prevalence estimates ranging from 1.3 to 11%. In Down syndrome, alopecia areata tends to occur most frequently on the scalp and has a variable period of duration, but can be permanent.

Evaluation

- A careful skin examination should be part of routine anticipatory care. Discuss skin, hair and scalp care at every visit.

Management

- Any problem identified should be treated as in the general population.
- Individuals with alopecia areata should be referred to a dermatologist.

TABLE 24.2 Summary of dermatologic findings in Down syndrome

Condition	Frequency	Comment
Fissured tongue	Reported in 20–95%	2–5% general population, most asymptomatic
Xerosis	<10–85%	Highly age dependent, high rates in adults
Lichen simplex (lichenification)	Up to 80%	Rate increases with age
Palmoplantar hyperkeratosis	10 to >75%	High rates seen over age 5 years
Atopic dermatitis	Up to 50%	Recent studies point to a low prevalence
Onychomycosis	>50%, general population ~20%	
Eruptive syringomas	18–39%; more females	Benign dermal papules, eccrine, often periorbital
Hidradenitis suppurativa (Sehgal, Sehgal & Sehgal, 2017)	38%	Solitary or multiple isolated abscesses without scarring or sinus tracts, recurrent abscesses, in inner thighs, groin, and buttocks
Furunculosis	Scarring in up to 26%	Perigenital, thigh, buttock common sites
Acanthosis nigricans	50% of those with atopic dermatitis	10-fold increase
Follicular papular dermatosis	45% of males	Same as Malassezia type folliculitis
Cheilitis	6–13%; more males	Age dependent, vertical fissures, enlargement
Alopecia areata	6–11%; more females	More severe than usual; associated with vitiligo
Geographic tongue	4–11, 40% also have fissured tongue	Inflammatory; loss/regrowth filiform papillae
Alopecia areata	1.3–11%	
Hypertrophic tongue papilla	22%	Physical variant
Premature graying	14%	
Folliculitis (pityrosporum)	10%	Upper back, chest, shoulders most common
Syringoma	6%	
Trichotillomania	4%	
Vitiligo	3%	
Anectoderma	Not common	Flaccid skin; lacks elastic tissue; fat herniation
Elastosis perforans	Not common, more males	Onset second decade, more severe than usual
Propensity to crusted scabies	More common in DS	Asymptomatic crusting of hands and feet

Immunologic

Children with Down syndrome are known to have more frequent infection rates, impaired immunity, suboptimal immunologic response to vaccinations, and higher rates of autoimmune diseases. As the survival of individuals with Down syndrome increases, there is a growing need to strategize our application of preventive medicine against common ailments affecting this high risk group. Vaccinations play an important role.

Abnormalities have been reported in virtually all aspects of the immune system. There is a mild-to-moderate reduction in T-cell and B-cell counts, absence of normal lymphocyte expansion in infancy, variable thymus size, mild-to-moderate reduction in naive T-cell percentages with corresponding reduction of T-cell excision circles, suboptimal antibody responses to immunizations, decreased total and specific immunoglobulin A in saliva, and decreased neutrophil chemotaxis (Colvin and Yeager 2017). Individuals with Down syndrome also have higher percentages of natural killer cells with reduced functionality and higher percentages of apoptotic lymphocytes. There is also evidence of secondary causes for immunodeficiency, like nutritional deficiency of zinc. Decreased cilial beat frequency could also be a contributing factor, which is considered secondary to chronic hyperproduction of mucous. Many aspects of immunological influences and effective corrective measures are yet to be discovered.

Multiple aspects of humoral immunity have been found to be different in Down syndrome compared to typically developing individuals. Carsetti and colleagues (2015) found that transitional and mature-naïve B-cell numbers were reduced by 50%, whereas switched memory B-cells represented only 10–15% of the numbers in age-matched controls. Following the primary influenza vaccination, children with Down syndrome had significantly fewer vaccine-induced immunoglobulin G-producing B-cells than controls. After these same children received a pneumococcal booster vaccination, the number of vaccine-induced IgG-producing B-cells showed a significant increase. Both groups reached similar vaccine efficacy after the booster dose for children with Down syndrome. Eijsvoogel and colleagues (2017) found that after primary vaccination of Hepatitis B, only 48% of children aged 7–10 years and 32% aged over 10 years had protective immunity against Hepatitis B. The evidence taken together suggests that children with Down syndrome may require repeated antigen-stimulation by recurrent natural infection or booster vaccination to reach an adequate level of immune protection.

The fetal thymus in Down syndrome may be hypotrophic and shows differences in function including abnormalities of specific antigen responses and receptor formation. Although the percentage of T-cells is maintained, the total number of circulating lymphocytes and of T-cells is reduced, with alterations in specific subsets of cells and an impaired T-cell-mediated response. Some authors have suggested that there is an upregulation of inhibitory receptors which causes a phenomenon called T-cell exhaustion, contributing to frequent infection episodes among individuals with Down syndrome.

Children with Down syndrome are more susceptible to infections compared to the general population, due to abnormal structural morphologies, comorbidities, developmental delays, and immunologic deficits. Improvements in medical care, early diagnosis, preventive measures, and outpatient services may contribute to findings of reduced hospitalization rates for older individuals. Despite advances in treatment, children and adults with Down syndrome continue to show relatively high morbidity and mortality from infectious disease. A study in a population-based cohort showed that on an average, each child with Down syndrome was admitted to a hospital nearly ten times in their lifetime, with the most common primary admission diagnoses being otitis media, lower respiratory tract infection, upper respiratory tract infection, cardiac septum defects, and adenoid/tonsillitis (Fitzgerald et al. 2013). Infections occurred in 80% of children and the median age of first hospital admission was 1.2 years. Infections accounted for 33% of all hospital admissions.

Respiratory tract infections are the leading causes of morbidity and mortality among children with Down syndrome. The factors associated with high mortality among this group are history of congenital heart disease, chronic upper airway obstruction and mechanical ventilation, and the occurrences of sepsis, pulmonary arterial hypertension, acute respiratory distress syndrome, and infections acquired during a hospital stay (Joffre et al. 2016). Some studies have shown that children with Down syndrome who are hospitalized for respiratory syncytial virus infection tend to be older and have more severe illness than children without Down syndrome. Pidotimod, an immunostimulant, and palivizumab, a human monoclonal antibody, may have roles in preventing respiratory tract infections in children with Down syndrome, with palivizumab being particularly effective in young children (under 2 years of age) against respiratory syncytial virus-related hospitalizations (Manikam et al. 2016). Increased expression of the *Ksp37* gene may be associated with increased susceptibility to Epstein–Barr virus infections and autoimmune problems (Salemi et al. 2016).

Down syndrome has an increased prevalence of autoimmune diseases affecting both endocrine and non-endocrine systems. These include Addison's disease, allergic dermatitis, alopecia, celiac disease, chronic autoimmune hepatitis, diabetes mellitus (type I more common than II), hypo- and hyperthyroidism, autoimmune hemolytic anemia, dermatomyositis, multiple sclerosis, pernicious anemia, polyarteritis nodosa, primary sclerosing cholangitis, rheumatoid arthritis, scleroderma, Sjogren's syndrome, and systemic lupus erythematosus. An unbalanced relation between anti- and proinflammatory immune responses may favor the development of autoimmunity in Down syndrome (Schoch et al. 2017). Although auto-antibodies may be present, in general there is poor correlation with the presence of actual disease. Thus, caution is needed in interpreting any such results. Finally, Down syndrome disintegrative disorder, characterized by sudden decline in skills and autistic regression (see section on Development and Behavior), has also been associated with elevated antithyroperoxidase antibodies and other autoimmune conditions. Current research is ongoing to better elucidate the potential role of autoimmunity in this condition.

Evaluation

- Although children with Down syndrome show greater susceptibility to infectious disease, and several anomalies of immune responsiveness may be found, there does not appear to be justification for routine immunologic evaluation. Such studies should be reserved for those individuals with unusually severe problems and/or evidence of frank immunodeficiency or autoimmune disease.
- Per AAP guidelines (Bull 2011), evaluate respiratory and cardiac comorbidities during newborn examinations and routinely thereafter during wellness visits. Consult specialists and enable prompt treatment of infections and other factors that lead to increased prevalence of infections in individuals with Down syndrome.

Management

- Although response to vaccinations may not be as effective in people with Down syndrome, it is recommended that they follow a normal vaccination schedule, including hepatitis B, for which they are at significant risk to become chronic carriers. Children with confirmed comorbidities should have respiratory syncytial virus prophylaxis. Influenza vaccine should be given annually. Children with chronic cardiac or respiratory diseases should be given the 23-valent pneumococcal polysaccharide vaccine (PPS23) at 2 years or older (Bull 2011).
- The possible role of vitamin A and zinc in normalizing some aspects of the immunodeficiency in Down syndrome remains controversial. Based on clinical judgement, these may be replaced if low.

Neurologic

An array of neurological manifestations has been described among individuals with Down syndrome. Neuromuscular hypotonia, epilepsy, intellectual disability (see Behavior and Development section), neuropsychiatric problems (see Behavior and Development section), cervical cord compression (see Musculoskeletal section), gait abnormalities (see Musculoskeletal section), cerebrovascular events leading to early strokes, sleep disorder (see Sleep Disorders section), and defects in vision and hearing (see Ophthalmologic and Audiologic sections) have been widely observed. A thorough neurological evaluation for newborns, and then at every wellness visit is essential to manage these problems.

Seizures occur more often in children with Down syndrome than in the general population but are less frequent than among those with other forms of intellectual disabilities. Goldberg-Stern and colleagues (2001) reported seizures in 8% of 350 children and adolescents with Down syndrome, and found that 47% were partial, 32% infantile spasms, and 21% generalized tonic clonic. Seizure onset occurs most commonly during infancy or before age 12 years. Infantile spasms are 8–10 times more common than in the general population. The higher incidence of seizures in children with Down syndrome may be explained partly because of the greater frequency of potential causative factors such as cardiac hypoxia, cerebral artery occlusion, perinatal complications, infections and fevers, neurotransmitter imbalances, and chemotherapy. Sex distribution for epilepsy in children with Down syndrome has not been uniformly reported, but infantile spasms tend to occur more frequently in males.

Adults with Down syndrome and Alzheimer disease appear to have a higher frequency of seizures than those with Alzheimer disease in the general population. In a study of 68 adults with Down syndrome, Puri et al. (2001) found that 26.5% had a history of seizures and that there was a bimodal age of onset in the first and second decades and the fifth and sixth decades. Seizures starting over the age of 45 years were more common in females and were strongly associated with the occurrence of Alzheimer disease, whereas seizures with onset at younger ages showed no such association. There have also been reports of senile seizures (after 50 years of age) called senile myoclonic epilepsy of Genton or late-onset myoclonic epilepsy in Down syndrome, which shows generalized fast spike waves, polyspikes, or polyspike waves.

Focal weakness may affect about 1% of individuals with Down syndrome and has a wide gamut of causes (Worley et al. 2004). Worley and colleagues studied 10 children with Down syndrome (median age of 4 years) with new-onset focal weakness. They found the causes of the new onset focal weakness were: stroke from Moyamoya disease (two patients); stroke from vaso-occlusive disease (one patient); stroke from venus sinus thrombosis (one patient); traumatic subdural hematoma (one patient); brain abscess (one patient); spinal cord injury from cervical spinal stenosis (two patients); spinal cord injury from atlantoaxial instability (one patient); and brachial plexus injury (one patient).

Stereotypic movements are common among individuals with Down syndrome. A study of 145 participants (mean age of 40 years) found that at least 90% had dyskinesias, almost all with orofacial and about 20% with limb or trunk signs (Haw et al. 1996). Tongue thrust (68%), an impassive face (57%), and decreased arm swing occurred in 50% or more of these individuals. Bradykinesias (33%) and global Parkinsonism (4%) were only seen in those with dementia, and this has been confirmed in other studies. Brief random movements, grimacing, abnormal facial movements, and postural and gait abnormalities were commonly observed. About 40% of affected individuals showed stereotypic movements including trunk rocking, rubbing a hand on the chest, or waving the hands in front of the eyes. There appears to be a positive correlation between the occurrence of dyskinesia and the severity of intellectual disability and lack of academic and practical skills.

Evaluation

- Evaluation of seizures does not differ from that in the general population; an electroencephalogram is indicated if there is a suspicious history.
- Monitor for signs of neurologic dysfunction including seizures. New focal weakness should be fully investigated, as the cause may be treatable.

Management

- Treatment of seizures does not differ from that in the general population.
- Abnormal movements do not usually require treatment, although medication or other toxicity should be ruled out.

Neoplasia

The pattern of malignancies observed in individuals with Down syndrome is significantly different from that observed in the general population (Hasle et al. 2016). At younger ages, the overall risk is relatively higher because of the increased risk of leukemias. In the adult population, the risk is much lower because of the significantly decreased occurrence of solid tumors in older persons with Down syndrome. The increased risk for leukemia in individuals with Down syndrome is present from the age of 1–10 years. The incidence risk of leukemia in Down syndrome is about 20 times that in the general population. Acute lymphoblastic leukemia (DS-ALL) and acute non-lymphocytic leukemia (ANLL) like acute megakaryocytic leukemia (AML-M7/AMKL) are the most common leukemias noted in childhood in individuals with Down syndrome. Chronic myeloid and chronic lymphocytic leukemia are less common than expected.

About half the cases of ANLL are acute megakaryocytic leukemia (AML-M7/AMKL), a rate about 500 times higher than in the general population. AMKL occurs in about 1–2% of children with Down syndrome, especially in early childhood (1–5 years). Up to 10% of neonates with Down syndrome may show a transient leukemoid reaction, also referred to in the literature as transient leukemia, transient megaloblastic leukemia, transient abnormal myelopoiesis, and transient myeloproliferative disorder (TMD). TMD appears to be virtually limited to persons with a trisomy 21 cell line. Several individuals have been diagnosed on fetal blood samples obtained during prenatal diagnostic testing. In most cases, TMD regresses spontaneously in three months, about 20% develop an early life threatening illness (e.g., hepatic fibrosis, liver failure, pulmonary edema, pericardial effusion), and about 20–30% go on to develop acute megakaryocytic leukemia (AMKL, AML-M7) within a mean interval of 1.2–1.5 years (Gamis and Smith, 2012). Individuals whose AMKL is preceded by TMD respond better to treatment than those where it is not. The transformation to leukemia from TMD may be preceded by myelodysplastic syndrome (MDS), in which there are fewer than 20% blasts in the bone marrow. The blasts of TMD and myeloid proliferations in Down syndrome contain acquired *GATA1* variants, which are considered pathognomonic of these disorders. AMKL and MDS occurring in young children with Down syndrome with *GATA1* somatic variants are collectively termed myeloid leukemia of Down syndrome (ML-DS). At this time, there have been very few reported cases of non-Down syndrome-AMKL with *GATA1* variants and acquired trisomy 21, who were all phenotypically and cytogenetically not Down syndrome.

GATA1 variants are found in almost one third of neonates with Down syndrome but are frequently hematologically "silent". Recent studies have shown that although *GATA1* variants have been thought of as causative factors for AMKL/TMD in Down syndrome, fetal studies have shown that *GATA1* variants occur much later in the fetal development, and that trisomy 21 itself alters human fetal hematopoietic stem/progenitor cell biology causing complex abnormalities in myelopoiesis and B-lymphopoiesis. This seems to be a complex process involving specific tissues and cell lineages among children with Down syndrome. AMKL preceded by TMD also has TMD cloning through additional acquired variants. Multiple other genetic factors have been linked to susceptibility to myeloid leukemia, including trisomy 8 (occurring in approximately in 13–36% of cases) and losses of chromosomes 5 and 7. Telomere shortening and stem/progenitor cells deficiency have been documented in fetal life in individuals with Down syndrome and might play some role in the susceptibility to leukemia.

Children with Down syndrome account for about 2% of all patients with acute lymphoblastic leukemia (ALL). After age 10, ALL is more predominant than other leukemias in children with Down syndrome. DS-ALL is characterized by lower levels of both favorable and unfavorable cytogenetic findings such as hyperdiploidy and of a number of common translocations including the transient encephalopathy leukemia/acute megakaryocytic leukemia 1 phenotype t(12;21), t(9;22) and *MLL* gene rearrangements. DS-ALL is a high-risk B-cell precursor leukemia in most cases, with very rare cases of mature B cell ALL (Burkitt leukemia) and T-cell ALL. Several different mechanisms are currently being studied for therapeutic purposes to design new drugs and to understand their correlations with relapses or remissions, including activating variants in the JAK-STAT pathway and targeting the *HMGN1* and *PRC2* genes.

Children with Down syndrome and certain leukemias who are treated on trial chemotherapeutic protocols generally have better outcomes. Persons with Down syndrome are showing greater participation rates in clinical trials and improved survival. AMKL has higher cure rates with 80–100% event-free survivals; however, ALL is associated with a worse prognosis in children with Down syndrome compared to those without Down syndrome. The outcome for children who present at less than 2 years of age is significantly better than for those who present later. Although Down syndrome has been shown to be a negative prognostic factor in ALL treated by conventional therapy, event-free survival and overall survival are very close to those in the general population when both are treated with an intensive regime. Historic treatment failures were largely the result of undertreatment and late diagnosis. Children with Down syndrome and AMKL should not be undertreated because they have very high cure rates. If certain treatment protocols are contraindicated (due to severe cardiac disease, unstable patients, or drug toxicities), very low-dose araC regimens could be tried.

There is less evidence to support efficacy of bone marrow transplant in treatment of AML as compared to chemotherapy in Down syndrome. Hitzler and colleagues (2013) compared results of bone marrow transplantation for children with and without Down syndrome in one of the largest studies. They found that the three-year probability of overall survival was only 21% among children with Down syndrome (compared to 52% in children without Down syndrome), owing to risks of relapse and transplant-related mortality. A study by Muramatsu and colleagues (2014) in a Japanese population investigated the use of a lower-intensity conditioning regimens preceding bone marrow transplant for children with Down syndrome and AML. The three-year event-free survival rates were approximately 80%, which were significantly better than in the group that received a more standard conditioning regimen (event-free survival rate of approximately 10%). Here, the common cause of treatment failure after transplant was relapse, as opposed to treatment-related mortality. Further studies with larger number of patients are required to confirm these findings.

In contrast to leukemia, individuals with Down syndrome are largely protected from solid tumors, with a significantly lower risk of lung cancer, breast cancer, and cervical cancer (Hasle et al. 2016). Testicular cancer is the only solid tumor with an increased standardized incidence rate, out of all the solid tumors. Cryptorchidism alone cannot explain the increased risk, since only less than a quarter of males with Down syndrome have cryptorchidism along with testicular tumors. The risk of developing ovarian cancer is comparable to that of the general population. Cancers pertaining to embryogenic tissues are negligible.

Evaluation

- The AAP guidelines have the following recommendations (Bull 2011). *For newborns*: Obtain a complete blood cell count. *Infants* with transient myeloproliferative disorder or polycythemia should be referred to a pediatric hematologist/oncologist for consultation and should be followed according to consultation recommendations. Parents of *younger children* with transient myeloproliferative disorder should be counseled regarding the risk of leukemia and be made aware of the suggestive clinical signs, including easy bruising, petechiae, onset of lethargy, or change in feeding patterns. The signs and symptoms of individuals with Down syndrome and leukemia do not differ significantly from those in the general population. Individuals with Down syndrome are slightly older at the time of diagnosis and have slightly higher initial hemoglobin values.
- No routine screening is indicated by current AAP guidelines.
- The testes should be examined periodically because of the higher rate of testicular germ cell tumors.

Management

- If possible, treatment should be carried out in experienced tertiary care centers that participate in standard and research protocols and where there is experience in treating children with Down syndrome.
- Children with Down syndrome and leukemia who are treated have not been found to have greater risk of relapses or mortality when compared with typically developing children with leukemias. They may have more side effects due to chemotherapy because they are more sensitive to toxicity from chemotherapy, particularly from methotrexate, because of slower clearance.
- Bone marrow transplants for treatment of AML are known to be less effective compared to standard chemotherapy regimens and are generally not used as the first line of management. If there are regimens tailored to reduce the leukemic burden prior to bone marrow transplants, it may be possible that bone marrow transplants work better than standard chemotherapy.
- In July 2017, the FDA approved a new innovative single-dose gene therapy called CAR-T cell therapy/

Kymriah for the treatment of B cell-ALL in young adults, for whom other treatment modalities have failed. Further studies on patients with Down syndrome may shed some light on how this treatment may modify the prognosis of the disease among this high-risk group.

Craniofacial

Craniofacial morphologies vary widely among ethnic and geographical groups across the world. Several studies have implied that the craniofacial morphological differences change or become more prominent with increasing age. Individuals with Down syndrome tend to breathe through the mouth, exhibiting open bite and opened lips, with orofacial hypotonia and normal mandible (pseudo-progeny). This exaggerates the mid face hypoplasia, along with malformations of bony skull structures. Relative macroglossia and hypotonicity of the tongue and muscles of the oropharynx contribute to difficulties in feeding, breathing, swallowing, and speaking. Hypotonia of the facial muscles can result in chronic drooling, chapping, and cheilitis, which may play a role in the increased rate of upper-respiratory infections and peridontitis.

There have been some cases of facial plastic surgery among individuals with Down syndrome, but there are ethical concerns with such procedures. Surgical options such as partial glossectomy, internally rotating the lower lip, neck and cheek liposuctions, augmenting the nasal bridge or midfacial area, and repositioning epicanthal folds are available. There are claims that these procedures improve speech, mouth breathing and/or aesthetics, and reduce oral inflammation, chieliosis, and halitosis and dental problems. While there are cases where such procedures have been reported to boost self-confidence and increase the emotional quotients of individuals with Down syndrome, the expected outcomes and improvement in quality of life varies widely, and not all procedures have benefits. Parents and adults with Down syndrome need to be well informed about all potential complications and the uncertainties of the outcomes, and they should consider speaking with other parents who have gone through such procedures with their children.

Evaluation

- Discuss with parents about hypotonia and facial appearance, acknowledging the presence of familial characteristics, at the time of birth.
- Medical history and physical examination should be obtained to identify any potentially deleterious effects of a prominent tongue.
- Since Down syndrome is characterized by wide variability in morphological features, a multidisciplinary team of pulmonologists, speech language pathologists, dentists, school teachers, psychologists, and parents is essential for supporting social acceptability and capabilities for speech and feeding.

Management

- Due to controversy surrounding plastic surgeries, it is important to discuss the risk versus benefit ratio with parents, based on each individual, case by case.

Urologic

The urinary tract in individuals with Down syndrome has received relatively little attention. However, there are urogenital anomalies that in some cases may cause significant morbidity and even mortality. Their incidence cannot yet be determined, as most surveys are small and come from individuals specifically referred for renal evaluation. Anomalies reported include bladder extrophy, hypospadias, posterior urethral valves, micropenis (microphallus), reflux, renal hypoplasia, elevated urinary and/or blood uric acid, dysfunctional voiding as a result of both neurogenic and non-neurogenic bladder, and chronic renal failure. A wide spectrum of congenital and acquired renal anomalies has been reported among individuals with Down syndrome, including acute kidney failure, urinary symptoms (dysuria, increased frequency, hesitancy, dribbling urine, and incontinence), decreased clearance of uric acid, hypercalciuria, and end-stage renal disease. At present, screening for urological anomalies among individuals with Down syndrome is not standard care during newborn evaluations.

Undescended testes (cryptorchidism) occur with increased frequency. If uncorrected, cryptorchidism can lead to infertility, testicular cancer, hernias and testicular torsion. Testicular cancer is the only solid tumor which is more common in Down syndrome (see Neoplasia section). Many individuals with Down syndrome may be unaware of the lump in the testes, and are diagnosed with testicular cancer incidentally following a thorough physical examination. A high cure rate is expected for this cancer, and so it is imperative to palpate the testes yearly for detection of a mass.

Evaluation

- Early and continued confirmation of a normal urinary stream and bladder voiding pattern is important.
- As renal and urinary tract anomalies have been reported to occur at an increased frequency among persons with Down syndrome, standardized screening for these anomalies with renal ultrasound has been suggested (Kupferman et al. 2009). Currently, more studies are needed to document that screening improves outcomes.
- Physical examination of males for cryptorchidism should occur at diagnosis, with follow-up examination in a standard manner if they cannot be palpated.

- The testes should be palpated yearly. Parents should be taught how to do an examination of the testes for detection of a mass.

Management

- If anomalies are detected during screening, an evaluation by a nephrologist or urologist is indicated.
- Treatment for anomalies, including undescended testes, is as for the general population.

BOOKS

Chicoine B, McGuire D (2010) *The Guide to Good Health for Teens & Adults with Down Syndrome*. Bethesda, MD: Woodbine House.

Couwenhoven T (2007) *Teaching Children with Down Syndrome about Their Bodies, Boundaries and Sexuality*. Bethesda, MD: Woodbine House.

Froehlke M, Zaborek R (2013) *When Down Syndrome and Autism Intersect: A Guide to DS-ASD for Parents and Professionals*. Bethesda, MD: Woodbine House.

Jacob J, Sikora M (2016) *The Parent's Guide to Down Syndrome: Advice, Information, Inspiration, and Support for Raising Your Child from Diagnosis Through Adulthood*. Avon, MA: F+W Media.

Kumin L (2003) *Early Communication Skills for Children with Down Syndrome: A Guide for Parents and Professionals*. Bethesda, MD: Woodbine House.

Kumin L (2008) *Helping Children with Down Syndrome Communicate Better: Speech and Language Skills for Ages 6-14*. Bethesda, MD: Woodbine House.

McGuire D, Chicoine B (2006) *Mental Wellness in Adults with Down Syndrome*. Bethesda, MD: Woodbine House.

Medlen JG (2008) *The Down Syndrome Nutrition Handbook: A Guide to Promoting Healthy Lifestyles*. Portland, OR: Phronesis Publishing.

Simons JA (2010) *The Down Syndrome Transition Handbook: Charting Your Child's Course to Adulthood*. Bethesda, MD: Woodbine House.

Skallerup SJ ed (2008) *Babies with Down Syndrome: A New Parents Guide*, 3rd ed. Bethesda, MD: Woodbine House.

Skotko B, Levine SP (2009) *Fasten Your Seatbelt: A Crash Course on Down Syndrome for Brothers and Sisters*. Bethesda, MD: Woodbine House.

Stein D (2016) *Supporting Positive Behavior in Children and Teens with Down Syndrome: The Respond but Don't React Method*. Bethesda, MD: Woodbine House.

SPECIFIC HEALTH CARE REFERENCES

American Academy of Pediatrics (2011) Committee on Genetics: Health supervision for children with Down syndrome. *Pediatrics* 128(2).

Smith DS (2001) Health care management of adults with Down syndrome. *Am Fam Physician* 64:1031–1040.

WEBSITES

US National Down Syndrome Society
www.ndss.org

National Down Syndrome Congress (US)
www.ndsccenter.org

Lettercase (print or digital resources about genetic conditions)
lettercase.org

Canadian Down Syndrome Society
www.cdss.ca

United Kingdom Down Syndrome Association
http://www.downs-syndrome.org.uk/

European Down Syndrome Association
http://www.edsa.eu/

New Zealand Down Syndrome Association
www.nzdsa.org.nz (each Australian state has its own Association)

International listing associations and resources
www.kumc.edu/gec/support/down_syn.html
https://www.ds-int.org/

Dr Len Leshin's Health Page
www.ds-health.com

Phone Contacts

US National Down Syndrome Society
666 Broadway, New York, NY, 10012-2317, USA
Phone: 1-800-221-4602 (US), 212-460-9330
Email: *info@ndss.org*
Website: *www.ndss.org*

US National Down Syndrome Congress
30 Mansell Court, Suite 108, Roswell, GA, 30076, USA
Phone: 800-232-NDSC(6372), 770-604-9500
Email: *info@ndsccenter.org*
Website: *www.ndsccenter.org*

Canadian Down Syndrome Society
811-14th St NW
Calgary, AB T2 N 2A4, Canada
Phone: 403-270-8500
Fax: 403-270-8291

REFERENCES

Abbeduto L, Warren SF, Conners FA (2007) Language development in Down syndrome: from the prelinguistic period to the acquisition of literacy. *Mental Retard Dev Disabil Res Rev* 13(3):247–261.

Afifi HH, Aglan MS, Zaki ME, Thomas MM, Tosson AM (2012) Growth charts of Down syndrome in Egypt: a study of 434 children 0-36 months of age. *Am J Med Genet A* 158A(11):2647–2655.

Aghajanova L, Popwell JM, Chetkowski RJ, Herndon CN (2015) Birth of a healthy child after preimplantation genetic screening of embryos from sperm of a man with non-mosaic Down syndrome. *J Ass Reprod Genet* 32(9):1409–1413.

Alldred SK, Takwoingi Y, Guo B, Pennant M, Deeks JJ, Neilson JP, Alfirevic Z (2017) First and second trimester serum tests with and without first trimester ultrasound tests for Down's syndrome screening. *Cochrane Database Syst Rev* 3:Cd012599.

Angelopoulou N, Souftas V, Sakadamis A, Matziari C, Papameletiou V, Mandroukas K (1999) Gonadal function in young women with Down syndrome. *Int J Gynaecol Obstet* 67(1):15–21.

Barankin B, Guenther L (2001) Dermatological manifestations of Down's syndrome. *J Cutaneous Med Surg* 5(4):289–293.

Bartesaghi R, Haydar TF, Delabar JM, Dierssen M, Martinez-Cue C, Bianchi DW (2015) New Perspectives for the Rescue of Cognitive Disability in Down Syndrome. *J Neurosci* 35(41):13843–13852.

Berr C, Borghi E, Rethore MO, Lejeune J, Alperovitch A (1990) Risk of Down syndrome in relatives of trisomy 21 children. A case-control study. *Annal Genet* 33(3):137–140.

Bertapelli F, Agiovlasitis S, Machado MR, do Val Roso R, Guerra-Junior G (2017) Growth charts for Brazilian children with Down syndrome: Birth to 20 years of age. *J Epidemiol* 27(6):265–273.

Bhaumik P, Ghosh P, Ghosh S, Feingold E, Ozbek U, Sarkar B, Dey SK (2017) Combined association of Presenilin-1 and Apolipoprotein E polymorphisms with maternal meiosis II error in Down syndrome births. *Genet Mol Biol* 40(3):577–585.

Bull MJ (2011) Health Supervision for Children With Down Syndrome. *Pediatrics* 128(2):393–406.

Cardinale KM, Bocharnikov A, Hart SJ, Baker JA, Eckstein C, Jasien JM, Gallentine W, Worley G, Kishnani PS, Van Mater H (2018) Immunotherapy in selected patients with Down syndrome disintegrative disorder. *Dev Med Child Neurol* 61:847–851

Capone G, Goyal P, Ares W, Lannigan E (2006) Neurobehavioral disorders in children, adolescents, and young adults with Down syndrome. *Am J Med Genet C* 142c(3):158–172.

Capone GT, Aidikoff JM, Taylor K, Rykiel N (2013) Adolescents and young adults with Down syndrome presenting to a medical clinic with depression: co-morbid obstructive sleep apnea. *Am J Med Genet A* 161A(9):2188–2196.

Carothers AD, Castilla EE, Dutra MG, Hook EB (2001) Search for ethnic, geographic, and other factors in the epidemiology of Down syndrome in South America: analysis of data from the ECLAMC project, 1967-1997. *Am J Med Genet* 103(2):149–156.

Carsetti R, Valentini D, Marcellini V, Scarsella M, Marasco E, Giustini F, Bartuli A, Villani A, Ugazio AG (2015) Reduced numbers of switched memory B cells with high terminal differentiation potential in Down syndrome. *Eur J Immunol* 45(3):903–914.

Choi H, Van Riper M (2017) Adaptation in families of children with Down syndrome in East Asian countries: an integrative review. *J Adv Nurs* 73(8):1792–1806.

Colvin KL, Yeager ME (2017) What people with Down Syndrome can teach us about cardiopulmonary disease. *Eur Resp Rev* 26(143).

Couzens D, Haynes M, Cuskelly M (2012) Individual and environmental characteristics associated with cognitive development in Down syndrome: a longitudinal study. *J Appl Res Intellect Disabil* 25(5):396–413.

Creavin AL, Brown RD (2009) Ophthalmic abnormalities in children with Down syndrome. *J Pediatr Ophthalmol Strabismus* 46(2):76–82.

Daneshpazhooh M, Nazemi TM, Bigdeloo L, Yoosefi M (2007) Mucocutaneous findings in 100 children with Down syndrome. *Pediatr Dermatol* 24(3):317–320.

de Graaf G, Buckley F, Skotko BG (2015) Estimates of the live births, natural losses, and elective terminations with Down syndrome in the United States. *Am J Med Genet A* 167A(4):756–767.

Diogenes TCP, Mourato FA, de Lima Filho JL, Mattos SDS (2017) Gender differences in the prevalence of congenital heart disease in Down's syndrome: a brief meta-analysis. *BMC Med Genet* 18(1):111.

Dreyer BP. Early Childhood Stimulation in the Developing and Developed World: If Not Now, When? 2011. *Pediatrics* 127(5):975–977.

Eijsvoogel NB, Hollegien MI, Bok VLA, Derksen Lubsen AG, Dikken FPJ, Leenders S, Pijning A, Post E, Wojciechowski M, Hilbink M, de Vries E (2017) Declining antibody levels after hepatitis B vaccination in Down syndrome: A need for booster vaccination? *J Med Virol* 89(9):1682–1685.

Egensperger R, Weggen S, Ida N, Multhaup G, Schnabel R, Beyreuther K, Bayer TA (1999) Reverse relationship between β-amyloid precursor protein and β-amyloid peptide plaques in Down's syndrome versus sporadic/familial Alzheimer's disease. *Acta Neuropathol* 97(2):113–118.

Fernandez F, Nyhuis CC, Anand P, Demara BI, Ruby NF, Spano G, Clark C, Edgin JO (2017) Young children with Down syndrome show normal development of circadian rhythms, but poor sleep efficiency: a cross-sectional study across the first 60 months of life. *Sleep Med* 33:134–144.

Fitzgerald P, Leonard H, Pikora TJ, Bourke J, Hammond G (2013) Hospital admissions in children with down syndrome: experience of a population-based cohort followed from birth. *PLoS One* 8(8):e70401.

Frank K, Esbensen AJ (2015) Fine motor and self-care milestones for individuals with Down syndrome using a Retrospective Chart Review. *J Intellect Disabil Res* 59(8):719–729.

Gamis AS, Smith FO (2012) Transient myeloproliferative disorder in children with Down syndrome: clarity to this enigmatic disorder. *Br J Haematol* 159(3):277–287.

Gardiner KJ (2015) Pharmacological approaches to improving cognitive function in Down syndrome: current status and considerations. *Drug Des Devel Ther* 9:103–125.

Ghosh S, Feingold E, Dey SK (2009) Etiology of Down syndrome: Evidence for consistent association among altered meiotic recombination, nondisjunction, and maternal age across populations. *Am J Med Genet A* 149A(7):1415–1420.

Glasson EJ, Jacques A, Wong K, Bourke J, Leonard H (2016) Improved Survival in Down Syndrome over the Last 60 Years and the Impact of Perinatal Factors in Recent Decades. *J Pediatr* 169:214–220.e211.

Goldberg-Stern H, Strawsburg RH, Patterson B, Hickey F, Bare M, Gadoth N, Degrauw TJ (2001) Seizure frequency and characteristics in children with Down syndrome. *Brain Dev* 23(6):375–378.

Grieco J, Pulsifer M, Seligsohn K, Skotko B, Schwartz A (2015) Down syndrome: Cognitive and behavioral functioning across the lifespan. *Am J Med Genet C* 169(2):135–149.

Gu Y (2017) Association between polymorphisms in folate metabolism genes and maternal risk for Down syndrome: A meta-analysis. *Mol Clin Oncol* 7(3):367–377.

Hart SJ, Visootsak J, Tamburri P, Phuong P, Baumer N, Hernandez M-C, Skotko BG, Ochoa-Lubinoff C, Liogier D'Ardhuy X, Kishnani PS, Spiridigliozzi GA (2017) Pharmacological interventions to improve cognition and adaptive functioning in Down syndrome: Strides to date. *Am J Med Genet A* 173(11):3029–3041.

Hasle H, Friedman JM, Olsen JH, Rasmussen SA (2016) Low risk of solid tumors in persons with Down syndrome. *Genet Med* 18(11):1151–1157.

Haw CM, Barnes TR, Clark K, Crichton P, Kohen D (1996) Movement disorder in Down's syndrome: a possible marker of the severity of mental handicap. *Movement Disorders* 11(4):395–403.

Head E, Powell D, Gold BT, Schmitt FA (2012) Alzheimer's disease in Down syndrome. *Eur J Neurodegener Dis* 1(3):353–364.

Hitzler JK, He W, Doyle J, et al. (2013) Outcome of Transplantation for Acute Myelogenous Leukemia in Children with Down Syndrome. *Biol Blood Marrow Transplant* 19(6):893–897.

Ingram DG, Ruiz AG, Gao D, Friedman NR (2017) Success of Tonsillectomy for Obstructive Sleep Apnea in Children With Down Syndrome. *J Clin Sleep Med* 13(8):975–980.

Intrapiromkul J, Aygun N, Tunkel DE, Carone M, Yousem DM (2012) Inner ear anomalies seen on CT images in people with Down syndrome. *Pediatr Radiol* 42(12):1449–1455.

Joffre C, Lesage F, Bustarret O, Hubert P, Oualha M (2016) Children with Down syndrome: Clinical course and mortality-associated factors in a French medical paediatric intensive care unit. *J Paediatr Child Health* 52(6):595–599.

Kruszka P, Porras AR, Sobering AK, et al. (2017) Down syndrome in diverse populations. *Am J Med Genet A* 173(1):42–53.

Kucik JE, Shin M, Siffel C, Marengo L, Correa A (2013) Trends in survival among children with Down syndrome in 10 regions of the United States. *Pediatrics* 131(1):e27–36.

Kumin L, Schoenbrodt L (2016) Employment in Adults with Down Syndrome in the United States: Results from a National Survey *J Appl Res Intellect Disabil* 29(4):330–345.

Kupferman JC, Druschel CM, Kupchik GS (2009) Increased prevalence of renal and urinary tract anomalies in children with Down syndrome. *Pediatrics* 124(4):e615–621.

Livingstone N, Hanratty J, McShane R, Macdonald G (2015) Pharmacological interventions for cognitive decline in people with Down syndrome. *Cochrane Database Syst Rev* (10):CD011546.

Manikam L, Reed K, Venekamp RP, Hayward A, Littlejohns P, Schilder A, Lakhanpaul M (2016) Limited Evidence on the Management of Respiratory Tract Infections in Down's Syndrome: A Systematic Review. *Pediatr Infect Dis J* 35(10):1075–1079.

Maris M, Wojciechowski M, Van de Heyning P, Boudewyns A (2014) A cross-sectional analysis of otitis media with effusion in children with Down syndrome. *Eur J Pediatr* 173(10):1319–1325.

Merrell JA, Shott SR (2007) OSAS in Down syndrome: T&A versus T&A plus lateral pharyngoplasty. *Int J Pediatr Otorhinolaryngol* 71(8):1197–1203.

Muneuchi J, Oda S, Shimizu D (2017) Rapidly progressive pulmonary veno-occlusive disease in an infant with Down syndrome. *Cardiol Young* 27(7):1402–1405.

Muramatsu H, Sakaguchi H, Taga T, et al. (2014) Reduced intensity conditioning in allogeneic stem cell transplantation for AML with Down syndrome. *Pediatr Blood Cancer* 61(5):925–927.

Nakamura N, Inaba Y, Aota Y, Oba M, Machida J, Aida N, Kurosawa K, Saito T (2016) New radiological parameters for the assessment of atlantoaxial instability in children with Down syndrome: the normal values and the risk of spinal cord injury. *Bone Joint J* 98-b(12):1704–1710.

NDSS (2019a) Healthcare Providers. https://www.ndss.org/resources/healthcare-providers/ (last accessed in January 2019)

NDSS (2019b) Resources - Nutrition. https://www.ndss.org/resources/nutrition/ (last accessed in January 2019)

Nir A, Berkman N, Gavri S (2017) Clinical and Parental Status of Patients with Congenital Heart Disease Associated Pulmonary Arterial Hypertension. *Israel Med Ass J* 19(8):489–493.

Norton ME, Wapner RJ (2015) Cell-free DNA Analysis for Noninvasive Examination of Trisomy. *N Engl J Med* 373(26):2582.

Oxelgren UW, Myrelid Å, Annerén G, Ekstam B, Göransson C, Holmbom A, … Fernell E (2017) Prevalence of autism and attention-deficit–hyperactivity disorder in Down syndrome: a population-based study. *Dev Med Child Neurol* 59(3):276–283.

Palisano RJ, Walter SD, Russell DJ, Rosenbaum PL, Gemus M, Galuppi BE, Cunningham L (2001) Gross motor function of children with down syndrome: creation of motor growth curves. *Arch Phys Med Rehab* 82(4):494–500.

Pavlovic M, Berenji K, Bukurov M (2017) Screening of celiac disease in Down syndrome - Old and new dilemmas. *World J Clin Cases* 5(7):264–269.

Pierce MJ, LaFranchi SH, Pinter JD (2017) Characterization of Thyroid Abnormalities in a Large Cohort of Children with Down Syndrome. *Horm Res Paediatr* 87(3):170–178.

Puri BK, Ho KW, Singh I (2001) Age of seizure onset in adults with Down's syndrome. *Int J Clin Prac* 55(7):442–444.

Sakadamis A, Angelopoulou N, Matziari C, Papameletiou V, Souftas V (2002) Bone mass, gonadal function and biochemical assessment in young men with trisomy 21. *Eur J Obstet Gynecol Reprod Biol* 100(2):208–212.

Salemi M, Barone C, Morale MC, Caniglia S, Romano C, Salluzzo MG, Galati Rando RG, Ragalmuto A, Bosco P, Romano C (2016) Killer-specific secretory (Ksp37) gene expression in subjects with Down's syndrome. *Neurol Sci* 37(5):793–795.

Salt A, Sargent J (2014) Common visual problems in children with disability. *Arch Dis Child* 99(12):1163–1168.

Schoch J, Rohrer TR, Kaestner M, Abdul-Khaliq H, Gortner L, Sester U, Sester M, Schmidt T (2017) Quantitative, phenotypical, and functional characterization of cellular immunity in children and adolescents with Down syndrome. *J Infect Dis* 215(10):1619–1628.

Sheets KB, Best RG, Brasington CK, Will MC (2011) Balanced information about Down syndrome: what is essential? *Am J Med Genet A* 155A(6):1246–1257.

Siegel MS, Smith WE (2011) Psychiatric features in children with genetic syndromes: toward functional phenotypes. *Pediatr Clin North Am* 58(4):833–864

Siemionow K, Hansdorfer M, Janusz P, Mardjetko S (2017) Complications in adult patients with Down Syndrome undergoing cervical spine surgery using current instrumentation techniques and rhBMP-2: A long-term follow-up. *J Neurol Surg A* 78(2):113–123.

Skotko BG, Capone GT, Kishnani PS (2009) Postnatal diagnosis of Down syndrome: synthesis of the evidence on how best to deliver the news. *Pediatrics* 124(4):e751–758.

Skotko BG, Davidson EJ, Weintraub GS (2013) Contributions of a specialty clinic for children and adolescents with Down syndrome. *Am J Med Genet A* 161a(3):430–437.

Skotko BG, Kishnani PS, Capone GT (2009) Prenatal diagnosis of Down syndrome: how best to deliver the news. *Am J Med Genet A* 149a(11):2361–2367.

Stagi S, Lapi E, Romano S, Bargiacchi S, Brambilla A, Giglio S, … de Martino M (2015) Determinants of vitamin D levels in children and adolescents with Down syndrome. *Int J Endocrinol* 2015:11.

Strohm K, Nesa M. (2010) *Sibworks Facilitator Manual: Groups for Siblings of Children with Special Needs*, 2nd ed. Siblings Australia Inc.

Stoll C, Dott B, Alembik Y, Roth MP (2015) Associated congenital anomalies among cases with Down syndrome. *Eur J Med Genet* 58(12):674–680.

Su X, Lau JT, Yu CM, Chow CB, Lee LP, But BW, Yam WKL, TSE PWT, Fung ELW, Choi KC (2014) Growth charts for Chinese Down syndrome children from birth to 14 years. *Arch Dis Child* 99(9):824–829.

Sugimoto D, Bowen SL, Meehan WP, 3rd, Stracciolini A (2016) Effects of neuromuscular training on children and young adults with Down syndrome: Systematic review and meta-analysis. *Res Dev Disabil* 55:197–206.

Sullivan RT, Frommelt PC, Hill GD (2017) Earlier pulmonary valve replacement in Down syndrome patients following tetralogy of Fallot repair. *Pediatr Cardiol* 38(6):1251–1256.

Torfs CP, Christianson RE (1998) Anomalies in Down syndrome individuals in a large population-based registry. *Am J Med Genet* 77(5):431–438.

Tuysuz B, Goknar NT, Ozturk B (2012) Growth charts of Turkish children with Down syndrome. *Am J Med Genet A* 158A(11):2656–2664.

Urbano RC, Hodapp RM (2007) Divorce in families of children with Down syndrome: a population-based study. *Am J Mental Retard* 112(4):261–274.

Worley G, Crissman BG, Cadogan E, Milleson C, Adkins DW, Kishnani PS (2014) Down Syndrome Disintegrative Disorder. *J Child Neurol* 30(9):1147–1152.

Worley G, Shbarou R, Heffner AN, Belsito KM, Capone GT, Kishnani PS (2004) New onset focal weakness in children with Down syndrome. *Am J Med Genet A* 128a(1):15–18.

Zemel BS, Pipan M, Stallings VA, Hall W, Schadt K, Freedman DS, Thorpe P (2015) Growth charts for children with Down syndrome in the United States. *Pediatrics* 136:e1204–1211

25

EHLERS–DANLOS SYNDROMES

BRAD T. TINKLE

Peyton Manning Children's Hospital, Indianapolis, Indiana, USA

INTRODUCTION

The Ehlers–Danlos syndromes (EDS) are a genetically, biochemically, and clinically diverse group of heritable connective tissue disorders having joint hypermobility and skin features in common. In 2017, an international conference of experts on these conditions took place that resulted in a revised classification of these disorders (Malfait et al. 2017). Prior classifications of EDS (Berlin and Villefranche) have included several types, some of which were removed from the current nosology (Beighton et al. 1988; Beighton et al. 1998; Malfait et al. 2017). In light of the 13 distinct forms of EDS that have been delineated, this chapter has been organized differently than others in this book in order to focus on the characteristics, genetic basis, and management that distinguish them. For clarity and conciseness, the Manifestations and Management section is not organized by organ system, but rather is described for each type of EDS separately. For greater detail of the disorders, the reader is referred to the 2017 issue of the American Journal of Medical Genetics that is devoted entirely to EDS (Malfait et al. 2017).

Incidence

For all forms of EDS, the combined prevalence is estimated as at least 1/5000 (Kulas Søborg et al. 2017); however, this is likely an underestimation because of poor ascertainment, especially of the hypermobile type.

Diagnostic Criteria

For clarity, the diagnostic criteria, pathogenesis, genetics, and testing of each type of EDS will be discussed separately. Table 25.1 summarizes the EDS type, inheritance, and genetic basis of each one. Generalized joint hypermobility can be a feature of all forms of EDS to varying degrees. Determination of joint hypermobility in adults and adolescents can be assessed using the Beighton scale, but it should be kept in mind that joint hypermobility is influenced by factors such as gender, age, and racial differences (Beighton et al. 1989). Although there are other scoring systems for assessing generalized joint hypermobility, the Beighton scoring system is the preferred (Juul-Kristensen et al. 2017) and is as follows (Figure 25.1):

- Passive dorsiflexion of the little finger beyond 90° from the horizontal plane. One point for each hand.
- Passive apposition of the thumb to the flexor aspect of the forearm. One point for each hand.
- Hyperextension of the elbow beyond 10°. One point for each elbow.
- Hyperextension of the knee beyond 10°. One point for each knee.
- Forward flexion of the trunk with knees fully extended so the palm of each hand rests flat on the floor. One point.

TABLE 25.1 Ehlers–Danlos syndromes classification

EDS Type	Inheritance	Gene(s)
Classical EDS	AD	COL5A1; COL5A2; COL1A1
Classical-like EDS	AR	TNXB
Cardiac-valvular EDS	AR	COL1A2
Vascular EDS	AD	COL3A1
Hypermobile EDS	AD	Unknown
Arthrochalasia	AD	COL1A1; COL1A2
Dermatosporaxis EDS	AR	ADAMTS2
Kyphoscoliotic EDS	AR	PLOD1; FKBP14
Brittle cornea syndrome	AR	ZNF469; PRDM5
Spondylodysplastic	AR	B4GALT7; B3GALT6; SLC39A13
Musculocontractural EDS	AR	CHST14; DSE
Myopathic EDS	AD or AR	COL12A1
Periodontal EDS	AD	C1R; C1S

Key: AD, autosomal dominant; AR, autosomal recessive

FIGURE 25.1 The Beighton scoring system. Each joint is measured using a goniometer and each side is scored independently as outlined (Juul-Kristensen et al. 2007). (A) With the palm of the hand and forearm resting on a flat surface with the elbow flexed at 90°, if the metacarpal-phalangeal joint of the fifth finger can be hyperextended more than 90° with respect to the dorsum of the hand, it is considered positive, scoring 1 point. (B) With arms outstretched forward but hand pronated, if the thumb can be passively moved to touch the ipsilateral forearm it is considered positive scoring 1 point. (C) With the arms outstretched to the side and hand supine, if the elbow extends more than 10°, it is considered positive scoring 1 point. (D) While standing, with knees locked in genu recurvatum, if the knee extends more than 10°, it is considered positive scoring 1 point. (E) With knees locked straight and feet together, if the patient can bend forward to place the total palm of both hands flat on the floor just in front of the feet, it is considered positive scoring 1 point. The total possible score is 9. *Figure courtesy of Dr Juul-Kirstensen. Reproduced from Malfait et al. (2017), Am J Med Genet 175C:8–26.*

A Beighton score of ≥5/9 is considered positive for generalized joint hypermobility. As hypermobility decreases with age, a score of ≥4/9 is considered positive if the five-point questionnaire (5PQ) is also positive (Table 25.2; Hakim and Grahame 2003).

A "yes" answer to two or more questions suggests joint hypermobility with 80-85% sensitivity and 80–90% specificity.

CLASSICAL TYPE

Diagnostic Criteria

The classical type of Ehlers–Danlos syndrome (cEDS) is characterized by hyperextensibility of the skin, poor wound healing, and joint laxity (Bowen et al. 2017). Former classifications divided the classical type into type I (gravis) and type II (mitis), but this is now thought to represent variation in severity of the same disorder. Individuals with

TABLE 25.2 The five-point questionnaire

1. Can you now (or could you ever) place your hands flat on the floor without bending your knees?
2. Can you now (or could you ever) bend your thumb to touch your forearm?
3. As a child, did you amuse your friends by contorting your body into strange shapes or could you do the splits?
4. As a child or teenager, did your shoulder or kneecap dislocate on more than one occasion?
5. Do you consider yourself "double-jointed"?

Adapted from (Grahame and Hakim 2003).

cEDS have soft, velvety skin that can be easily stretched and snaps back upon release (Figure 25.2). The dermis is fragile and splits easily with minor trauma. Scars after trauma or surgical procedures are thin and atrophic and may stretch considerably after healing. More severely affected individuals have scars with a characteristic "cigarette paper" appearance (Figure 25.3). Individuals with cEDS also have joint hypermobility that can vary from mild to severe. Many complications of cEDS are the result of the joint and tissue laxity. Joint hypermobility can lead to frequent subluxations and dislocations. Even in the absence of joint injury, individuals with hypermobility can have chronic joint and musculoskeletal pain. Tissue hyperextensibility and fragility increase the risk for hernias, rectal and uterine prolapse, as well as intestinal and bladder diverticuli. Surgical complications such as postoperative hernias are also common.

A significant number of individuals with cEDS have cardiac abnormalities. Mitral valve prolapse can be seen in all types, but it is unclear if the incidence is higher than that in the general population (Dolan et al. 1997). Aortic root dilatation has been reported in up to 25–30% of individuals with the classical and hypermobile types of EDS (Wenstrup et al. 2002; Atzinger et al. 2011). However, aortic rupture is rare in these types (Atzinger et al. 2011) but has been reported in other arteries in those with type V collagen mutations (Monroe et al. 2015) as well as the R312C variant in *COL1A1* (Gaines et al. 2015; Bowen et al. 2017). A small proportion of individuals with cEDS (due to *COL1A1* defects and previously called "classic-like EDS with propensity for arterial rupture") can have aneurysms with propensity to rupture in addition to osteoporosis. Other minor features of cEDS include pes planus, scoliosis, and molluscoid pseudotumors at the extensor surfaces of joints. Small herniations of adipose tissue can occur through the fascia (piezogenic

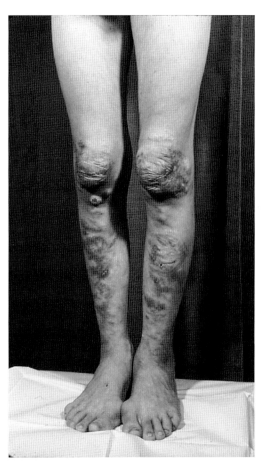

FIGURE 25.3 Atrophic or "cigarette-paper" scars commonly seen in the pretibial area in classical Ehlers–Danlos syndrome.

papules) particularly of the medial aspect of the feet upon standing (Figure 25.4).

Major Criteria

1. Skin hyperextensibility and atrophic scarring
2. Generalized joint hypermobility.

FIGURE 25.2 Hyperextensible skin that returns to normal when released. *Photo courtesy of Greg Lappin Photography.*

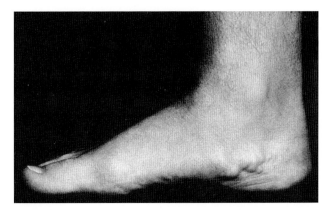

FIGURE 25.4 Piezogenic papules are small, painful, reversible herniations of adipose tissue globules through the fascia into the dermis, particularly on the medial and posterior aspects of the heels upon standing.

Minor Criteria

1. Easy bruising
2. Soft, doughy skin
3. Skin fragility
4. Molluscoid pseudotumors
5. Subcutaneous spheroids
6. Hernia
7. Epicanthal folds
8. Complications of joint hypermobility
9. Family history of cEDS

Minimal criteria suggestive of cEDS include skin hyperextensibility and atrophic scarring with either generalized joint hypermobility or at least two of the minor criteria. It is recommended that all individuals suspected of having cEDS, based on these clinical grounds, undergo genetic testing.

Etiology, Pathogenesis, and Genetics

cEDS is an autosomal-dominant disorder. Mutations in type V collagen are the major cause of cEDS, with half of these resulting in functional haploinsufficiency (Schwarze et al. 2000; Wenstrup et al. 2000). A minority of cases are due to specific mutations in type I collagen.

Diagnostic Testing

Genetic testing should be offered to all those with a clinical suspicion of cEDS. More than 90% will have mutations in type V collagen (Malfait et al. 2017). If type V collagen genetic testing is negative, including sequence and deletion analysis, testing for the c.934C>T (p.Arg312Cys) variant of *COL1A1* should be tested. Protein based collagen assays or electron microscopy imaging of collagen fibers are no longer recommended for clinical diagnosis, although they may be supportive.

Differential Diagnosis

It is appropriate to consider most forms of EDS in any individual with easy bruising and generalized joint hypermobility. Because the natural history and recurrence risks differ, it is important to differentiate the type of EDS based on clinical grounds and use testing as appropriate. cEDS shows some clinical overlap with other connective tissue disorders including cutis laxa syndromes. Cutis laxa has redundant skin, with reduced elastic properties and heals normally.

MANIFESTATIONS AND MANAGEMENT

Overall, management should follow the biopsychosocial holistic model focusing on the complications, the desires of the affected individual, quality of life, and functionality, as well as the psychological aspects (Bowen et al. 2017).

Evaluation

- History and physical examination of the skin and joints.
- Physical therapy evaluation for joint instability and pain.
- Consider baseline echocardiogram (at diagnosis for *COL1A1* R312C).
- Baseline and periodic vascular imaging in the R312C type I collagen type of cEDS (Brady et al. 2017).
- Blood pressure determination at all medical visits.
- Bone densitometry for *COL1A1* R312C type.

Management

- Avoid undue trauma to the skin and joints.
- Wounds should be carefully closed with sutures without tension; stitches should be left in twice as long.
- Excessive bleeding may respond to desmopression (DDAVP).
- Physical therapy is used to stabilize joints and pain control. Low resistance activities are a mainstay for overall joint health.
- Orthotics as indicated for joint support.
- Pain management should utilize physical therapy, mindfulness-based therapies, and select medications.
- Blood pressure control, in a standard manner.
- Standard treatment for osteopenia/osteoporosis (*COL1A1*-related cEDS).

CLASSIC-LIKE EDS

Diagnostic Criteria

Classic-like EDS (clEDS) is due to tenascin X deficiency. Its clinical features are similar to cEDS with skin hyperextensibility and generalized joint hypermobility but it lacks the atrophic skin findings of cEDS.

Major Criteria

1. Skin hyperextensibility with velvety skin texture without atrophic scarring
2. Generalized joint hypermobility
3. Easy bruisability.

Minor Criteria

1. Foot deformities: broad/plump forefoot; brachydactyly; pes planus; piezogenic papules
2. Leg edema in the absence of cardiac or renal failure

3. Mild muscle weakness
4. Axonal polyneuropathy
5. Atrophy of the muscles in the hand
6. Acrogeria (aged appearance to the hands/feet)
7. Vaginal/uterine/rectal prolapse.

Minimal criteria suggestive of clEDS is all three major criteria to be confirmed by genetic testing (Malfait et al. 2017). Minor criteria are merely supportive of the diagnosis.

Etiology, Pathogenesis and Genetics

clEDS is an autosomal recessive disorder due to tenascin X deficiency (*TNXB*). Tenascin X is in the extracellular matrix and is highly expressed in muscle, tendon, ligament and skin (Brady et al. 2017). Haploinsufficiency is the cause of the genetic defect in clEDS. Commonly, *TNXB* haploinsufficiency is due to recombination with the pseudogene *TNXA* (Schalkwijk et al. 2001). It may also be associated with congenital adrenal hyperplasia (Brady et al. 2017).

Diagnostic Testing

Sanger sequencing of the entire *TNXB* gene or next-generational sequencing of *TNXB* and Sanger sequencing of the pseudogene region (Malfait et al. 2017). Methods for large gene deletions should be used if only one mutation is found. The absence of the tenascin X protein is supportive.

Differential Diagnosis

clEDS is most similar to cEDS but may overlap with other forms of EDS. Due to muscular weakness, muscular dystrophy and other forms of neuromuscular disorders may be considered but the skin laxity and skin features are more apparent in clEDS

MANIFESTATIONS AND MANAGEMENT

- Physical therapy and orthotics as indicated for joint hypermobility.
- Baseline echocardiogram due to a higher incidence of mitral valve anomalies (Brady et al. 2017).
- Tracheal rupture during intubation has been reported and careful intubation is recommended (Besselink-Lobanova et al. 2010).

CARDIAC-VALVULAR EDS

Diagnostic Criteria

Cardiac-valvular EDS (cvEDS) has features similar to cEDS with generalized joint hypermobility, skin hyperextensibility and atrophic scarring (Malfait et al. 2017). The main difference is severe progressive cardiac-valvular disease.

Major Criteria

1. Severe progressive cardiac-valvular disease (aortic and/or mitral valves)
2. Skin hyperextensibility, atrophic scars, thin skin, easy bruising
3. Joint hypermobility (generalized or peripheral only).

Minor Criteria

1. Inguinal hernia
2. Pectus deformity (especially carinatum)
3. Joint dislocations
4. Foot deformities: pes planus/planovalgus; hallux valgus.

Minimal criteria suggestive of cvEDS is severe progressive valvular disease and a family history consistent with autosomal recessive inheritance plus one other major criterion or at least two minor criteria. Genetic testing should be used for confirmation.

Etiology, Pathogenesis, and Genetics

cvEDS is caused by biallelic mutations that result in the complete absence of the proα2-chain of type I collagen.

Diagnostic Testing

Sequence analysis of *COL1A2*. If two mutations were not found but there is strong clinical suspicion of cvEDS, deletion analysis is warranted.

Differential Diagnosis

Evaluate for other forms of EDS, especially classic and classic-like types.

MANIFESTATIONS AND MANAGEMENT

Patients have been described with complications of generalized joint hypermobility, skin fragility and atrophic scarring, and adult-onset valvular disease (Brady et al. 2017).

- Skin precautions due to fragility including use of sutures without tension and prolonged retention of sutures (about twice as long), as is seen in cEDS.
- Physical therapy and orthotics for generalized joint hypermobility, as in cEDS.
- Baseline echocardiogram; frequency of follow up not established but ongoing monitoring is recommended.
- Consider bone densitometry monitoring.

VASCULAR EDS

Diagnostic Criteria

Vascular EDS (vEDS) is distinguished from other forms of EDS by the occurrence of arterial rupture or rupture of visceral organs. Joint hypermobility is often not the presenting feature. A family history of similar concerns is common (Malfait et al. 2017).

Major Criteria

1. Family history of vEDS with causative variant in *COL3A1*
2. Arterial rupture at a young age
3. Spontaneous sigmoid colon rupture in the absence of diverticular disease or other bowel pathology
4. Uterine rupture during pregnancy in the absence of previous Cesarean section
5. Carotid-cavernous sinus fistula in the absence of trauma.

Minor Criteria

1. Easy and extensive bruising
2. Thin, translucent skin with increased venous visibility (Figure 25.5)
3. Characteristic facial appearance (Figure 25.6)
4. Spontaneous pneumothorax
5. Acrogeria (an aged appearance to the hands or feet)
6. Clubfoot
7. Congenital hip dislocation
8. Small joint hypermobility
9. Tendon or muscle rupture
10. Keratoconus
11. Gingival recession/fragility
12. Early-onset varicose veins.

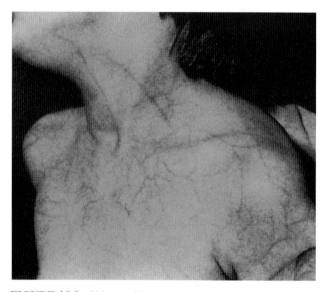

FIGURE 25.5 Veins easily visible beneath thin, almost translucent skin, seen in the vascular type of Ehlers–Danlos syndrome.

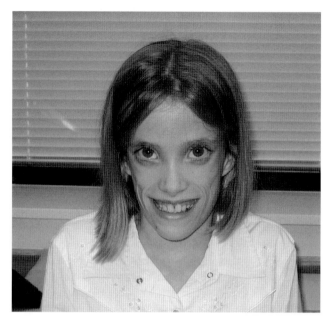

FIGURE 25.6 Characteristic facial appearance as seen in vascular EDS. The nose is often thin being somewhat "pinched." The lips are thin, and the cheeks appear hollow. Some individuals have prominent eyes because of decreased periorbital adipose tissue.

Suggestive clinical criteria are any one of the major manifestations with supporting features (at least one minor criteria). A clinical diagnosis may be difficult and genetic testing is warranted with any reasonable suspicion.

Etiology, Pathogenesis, and Genetics

vEDS is an autosomal dominant disorder due to a defect in type III collagen (*COL3A1*). Dominant negative mutations result in the more prototypical vEDS, whereas haploinsufficiency may result in a wider age of presentation including non-penetrance (Byers et al. 2017).

Diagnostic Testing

Genetic testing should include Sanger or next-generational sequence analysis. Deletion analysis can be considered if no mutation is found in someone with a clinical presentation consistent with vEDS. Protein analysis can be performed but is not recommended as it is less precise and invasive.

Differential Diagnosis

Differential diagnoses include various coagulation disorders as well as aneurysmal conditions such as arterial tortuosity,

Marfan and Loeys-Dietz syndromes (see Differential Diagnosis of cEDS for method of distinction).

MANIFESTATIONS AND MANAGEMENT

Evaluation

- Careful history and physical examination for manifestations.
- Head to pelvis imaging at diagnosis; the frequency of additional monitoring has not been established (Byers et al. 2017).
- Monitor blood pressure at all medical visits.
- Careful history for constipation and excessive bruising or bleeding.
- Evaluation and monitoring of ophthalmologic complications for keratoconus and complications thereof.

Management

- Management by a multidisciplinary care team.
- Carrying an emergency letter or using a medical alert system is recommended.
- Blood pressure control; European trials of celiprolol demonstrated delay of progression of vascular events (Ong et al. 2010).
- Vascular management is by any means in emergent cases; preventative vascular surgery may be by either open or endovascular techniques but performed with great care.
- Avoidance of constipation or Valsalva; consider use of laxatives or stool softeners.
- Colostomy has been useful for acute management of colonic rupture.
- Education regarding the signs and symptoms of pneumothorax, which can occur in up to 12%.
- Vein stripping for varicosities is considered contraindicated.
- Careful use of aspirin, non-steroidal medications and other anti-coagulants due to excessive bleeding/bruising.
- Tense hematomas may need surgical treatment.
- Acute, intense eye pain with loss of visual acuity may represent carotid cavernous fistula and is considered an emergency.

HYPERMOBILE EDS

Diagnostic Criteria

The diagnosis of hypermobile EDS (hEDS) should be considered in a person with generalized joint hypermobility and its complications, as well as other manifestations of a connective tissue disorder (Tinkle et al. 2017).

Required Criterion

- Generalized joint hypermobility with a Beighton score of ≥6/9 in prepubertal children; ≥5/9 in adolescents or young adults <50 years of age; ≥4/9 in those >50 (a one-point credit can be given to those with a positive five-point questionnaire (Table 25.2).

Must have two of the following three

- Musculoskeletal complications (any one of the following): pain in two or more limbs for at least three months; chronic widespread pain; recurring joint dislocations or frank joint instability not due to trauma.
- Family history: first-degree relative meeting diagnostic criteria independently.
- Score of five or greater (check any of the following):
  1. Unusually soft or velvety skin
  2. Mild skin extensibility
  3. Unexplained striae
  4. Bilateral piezogenic papules
  5. Recurrent or multiple abdominal hernias
  6. Atrophic scarring involving at least two sites
  7. Pelvic floor, rectal or uterine prolapse without predisposing factors (e.g. obesity, pregnancy)
  8. Dental crowding and high/narrow palate
  9. Arachnodactyly (positive thumb or positive wrist sign)
  10. Arm span-to-height ≥1.05
  11. Mitral valve prolapse based on strict echocardiographic criteria
  12. Aortic root dilation with z-score ≥+2.

Etiology, Pathogenesis, and Genetics

The etiology of hEDS is not known. No reliable genetic marker has been found. Familial patterns are often consistent with autosomal dominant inheritance with gender-influence. Other inheritance patterns cannot be ruled out in specific families. It is likely that hEDS is due to one of several different genes.

Diagnostic Testing

No testing is indicated. hEDS is a clinical diagnosis that can be reached once other heritable connective tissue disorders have been excluded.

Differential Diagnosis

The differential diagnosis is broad with more benign or limited forms of joint hypermobility (Castori et al. 2017) as well as a multitude of heritable connective tissue disorders including other forms of EDS, Stickler syndrome (see Chapter 56) and osteogenesis imperfecta (see Chapter 43). Genetic testing and further evaluations may help differentiate and/or diagnose other heritable connective tissue disorders.

MANIFESTATIONS AND MANAGEMENT

Primary manifestations of hEDS are often musculoskeletal with joint hypermobility, instability and/or dislocations that can lead to acute and later chronic pain. Many affected individuals also suffer from orthostatic intolerance, fatigue, and sleep disturbance as well as anxiety (Tinkle et al. 2017). Overall, management should follow the biopsychosocial holistic model focusing on the complications, the desire of the individual, quality of life and functionality, as well as the psychological aspects (Figure 25.7).

- Multidisciplinary approach is recommended, including mental health providers.
- Pacing of activities is important to help address pain and fatigue (Hakim et al. 2017A).
- Physical therapy is indicated for joint stability and pain control (Engelbert et al. 2017).
- Orthotics may be helpful, as indicated.
- Symptoms of orthostasis and/or syncope should be evaluated by in-office orthostatics including heart rate and blood pressure readings while lying and after 10 minutes of standing and if symptomatic, should be treated with hydration and salt; repeated episodes of syncope should be evaluated by a neurologist and/or cardiologist (for more details, see Hakim et al. 2017B).
- Mindfulness-based therapies are helpful for chronic issues such as pain, fatigue, sleep disturbance, and/or anxiety (Hakim et al. 2017A).
- Consider baseline echocardiogram if there is a concerning family history or symptomatology.
- Address sleep issues (sleep hygiene, pain control, dysautonomia, anxiety, etc.; Hakim et al. 2017A).
- Pain medications as needed (topical or systemic; Chopra et al. 2017).

ARTHROCHALSIA EDS

Diagnostic Criteria

Presentation is usually at birth with congenital hip dislocation and hypotonia as well as severe generalized joint hypermobility (Brady et al. 2017).

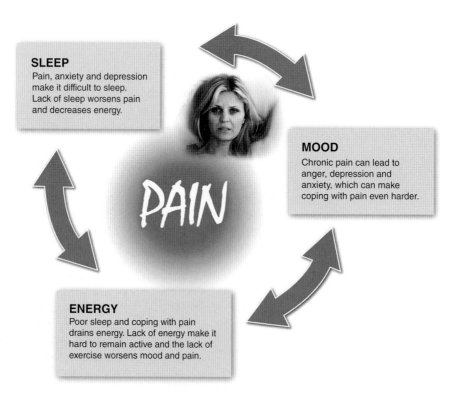

FIGURE 25.7 Chronic pain cycle with effects on mood, sleep, and energy/vitality. (Reproduced with permission from Left Paw Press/Dr Brad T. Tinkle.)

Major Criteria

1. Congenital bilateral hip dislocation
2. Severe generalized joint hypermobility
3. Skin hyperextensibility.

Minor Criteria

1. Hypotonia
2. Kyphoscoliosis
3. Mild osteopenia
4. Tissue fragility; atrophic scars
5. Easy bruisability.

The diagnosis should be suspected in the presence of congenital hip dislocation and either skin hyperextensibility OR generalized joint hypermobility and at least two other minor criteria. Genetic testing should be used for confirmation of the diagnosis (see below) (Malfait et al. 2017).

Etiology, Pathogenesis, and Genetics aEDS is an autosomal dominant disorder. It is due to a defect in type I collagen that prevents proteolytic cleavage and is a site of protein cross-linking (Brady et al. 2017).

Diagnostic Testing Testing should include sequence analysis of exon 6 of *COL1A1* or *COL1A2*. Protein analysis may be revealing but genetic analysis is preferred as it is less invasive.

Differential Diagnosis Many of the clinical features of aEDS are also seen among other EDS types. Significant joint laxity is seen primarily in classic, hypermobile, kyphoscoliotic, and the rarer types. Congenital hip dislocation is seen more commonly in vEDS but can also be seen in the cEDS. The muscular hypotonia can be seen in the kEDS but is often difficult to distinguish from ligamentous laxity seen in the other forms of EDS. The skin is hyperextensible and somewhat fragile as is seen in cEDS. Other heritable connective tissue disorders with overlapping clinical features include Larsen syndrome and cutis laxa.

MANIFESTATIONS AND MANAGEMENT

- Skeletal survey at baseline due to possible spinal deformities and multiple joint dislocations including developmental hip dysplasia.
- Orthopedic management as indicated by findings.
- Skin manifestations and management as in cEDS.
- Physical therapy and orthotics as indicated by symptoms, examination and X-rays.

DERMATOSPORAXIS EDS

Diagnostic Criteria

Individuals with dermatosporaxis EDS (dEDS) have skin redundancy and fragility (Malfait et al. 2017).

Major Criteria

1. Extreme skin fragility
2. Characteristic craniofacial features which are evident at birth or later in childhood
3. Redundant or lax skin (often excessive skin folds at the wrists or ankles)
4. Increased palmar wrinkling
5. Severe bruisability
6. Umbilical hernia
7. Postnatal growth retardation
8. Short limbs, hands and feet
9. Perinatal complications due to tissue fragility.

Minor Criteria

1. Softy, doughy skin
2. Skin hyperextensibility
3. Atrophic scars
4. Generalized joint hypermobility
5. Complications of visceral fragility (e.g. bladder rupture, diaphragmatic rupture)
6. Delayed motor milestones
7. Osteopenia.

Minimal criteria for the clinical suspicion of dEDS is extreme skin fragility AND characteristic facial features PLUS one other major criterion or three minor criteria. Diagnosis should be confirmed by genetic testing.

Etiology, Pathogenesis, and Genetics

dEDS is an autosomal recessive disorder due to a defect in the enzyme procollagen I N-proteinase encoded by *ADAMTS2* that cleaves the N-terminus of the procollagen type I molecule. The metalloproteinase can also cleave type II and type III collagens (Colige et al. 2005). This defect results in abnormal collagen fibers.

Diagnostic Testing Sequencing of the *ADAMTS2* gene.

Differential Diagnosis Generalized joint hypermobility is common among all of the various forms of EDS. However, the skin fragility in dEDS is more significant than in the other forms. Rupture of the internal organs is also seen in vEDS. In addition, there is some overlap of features with osteogenesis imperfecta (see Chapter 43) and cutis laxa. These can be distinguished by genetic testing.

MANIFESTATIONS AND MANAGEMENT

Management is similar to cEDS (see above).

KYPHOSCOLIOTIC EDS

Diagnostic Criteria

Hypotonia is often notable at or soon after birth in kyphoscoliotic EDS (kEDS). Congenital scoliosis may be seen in the first year of life (Malfait et al. 2107).

Major Criteria

1. Congenital hypotonia
2. Congenital or early-onset kyphoscoliosis
3. Generalized joint hypermobility, often with subluxations/dislocations.

Minor Criteria

1. Skin hyperextensibility
2. Easy bruising
3. Rupture or aneurysm of medium-sized arteries
4. Osteopenia/osteoporosis
5. Blue sclerae
6. Hernia (umbilical or inguinal)
7. Pectus deformity
8. Marfanoid habitus
9. Clubfoot
10. Refractive errors.

Clinical diagnosis is suspected in a person with hypotonia and congenital kyphoscoliosis PLUS generalized joint hypermobility or three minor criteria. Confirmatory genetic testing is recommended.

Etiology, Pathogenesis, and Genetics

Genetic heterogeneity is seen in kEDS. *PLOD1* encodes the modifying enzyme procollagen-lysine, 2-oxoglutarate 5-dioxygenase 1, also known as lysylhydroxylase 1. FXBP22 is the gene product of *FKBP14*, a member of the F506-binding family of peptidyl-prolyl cis-trans isomerase that is involved in protein folding, particularly of the pro-collagens.

Diagnostic Testing

Genetic testing of both *PLOD1* and *FKBP14* is recommended for someone suspected to have kEDS. *PLOD1* has a common intragenic duplication that should be screened for before sequence analysis of either *PLOD1* or *FKBP14*. Eye pathology is more characteristic of *PLOD1* kEDS (Figure 25.8) whereas hearing loss, muscle atrophy or bladder diverticuli are more likely due to *FKBP14*. Quantification of the ratio of deoxypyridinoline (Dpyr) and pyridinoline (Pyr) cross-links in urine can be quantitated by means of high-performance liquid chromatography; an increased Dpyr/Pyr ratio is highly sensitive and specific for kEDS due to *PLOD1*.

Differential Diagnosis

The kEDS has overlapping clinical features with other forms of EDS, particularly the cEDS (marked skin hyperextensibility) and vEDS (arterial rupture). Abnormal wound healing and joint laxity are present in many of the EDS types. Although all types have a higher risk of scoliosis compared with the general population, in the kyphoscoliotic form, scoliosis is usually more severe and of earlier onset. Other rarer types of EDS should also be considered. In addition, congenital myopathies as well as Marfan syndrome (see Chapter 37) and Loeys–Dietz syndrome (see Chapter 36) should be considered.

MANIFESTATIONS AND MANAGEMENT

- Similar management to cEDS for skin and joint abnormalities and their consequences (Brady et al. 2017).
- Standard orthopedic management of scoliosis.
- Bone densitometry monitoring for osteopenia/osteoporosis.
- Consider a sleep study for pulmonary insufficiency from the pectus deformity and hypotonia.
- Echocardiogram with imaging of the aortic root and ascending aorta at baseline and five-year intervals or as indicated.

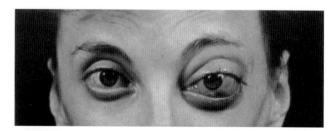

FIGURE 25.8 Scleral rupture seen in the kyphoscoliotic type of EDS.

- Blood pressure control to reduce risk of arterial rupture.
- Ophthalmologic evaluation and appropriate management.

BRITTLE CORNEA SYNDROME

Diagnostic Criteria

Brittle cornea syndrome (BCS) can have generalized joint hypermobility, but more notable is bluish sclerae with keratoconus or keratoglobus as well as hearing loss (Malfait et al. 2017).

Major Criteria

1. Thin cornea
2. Early onset and progressive keratoconus or keratoglobus
3. Blue sclerae.

Minor Criteria

1. Enucleation or corneal scarring
2. Progressive loss of corneal stromal depth
3. High myopia with normal or moderately increased axial length
4. Retinal detachment
5. Deafness, often mixed or sensorineural and progressive
6. Hypercompliant tympanic membrane
7. Hypotonia
8. Developmental dysplasia of the hip
9. Scoliosis
10. Arachnodactyly
11. Small joint hypermobility
12. Pes planus or hallux valgus
13. Mild finger contractures
14. Soft, velvety skin.

BCS should be suspected with thinning of the cornea PLUS either one other major criterion OR three minor criteria.

Etiology, Pathogenesis, and Genetics

BCS is an autosomal recessive disorder with genetic heterogeneity involving defects in *ZNF469* or *PRDM5*. *ZNF469* encodes a zinc-finger protein (ZNF469) of unknown function whereas *PRDM5* encodes a DNA-binding transcription factor. A third genetic etiology is suspected (Rohrbach et al. 2013).

Diagnostic Testing

Sequence analysis of *PDRM5* or *ZNF469*.

Differential Diagnosis

Other forms of EDS (kEDS, spEDS, mcEDS) as well as osteogenesis imperfecta (see Chapter 43).

MANIFESTATIONS AND MANAGEMENT

- Similar to cEDS for joint and skin management.
- Regular ophthalmologic monitoring and management.
- Evaluation and monitoring of hearing issues.
- Orthopedic management as indicated by manifestations.

SPONDYLODYSPLASTIC EDS

Diagnostic Criteria

Spondylodysplastic EDS (spEDS) is a bone dysplasia with features of muscular hypotonia that is caused by one of three genes (*B4GALT7, B3GALT6*, or *SLC39A13*) (Malfait et al, 2017).

Major Criteria

1. Short stature, progressive
2. Hypotonia
3. Bowing of the limbs.

Minor Criteria

1. Skin hyperextensibility; soft, doughy skin; thin translucent skin
2. Pes planus
3. Delayed motor development
4. Osteopenia
5. Cognitive impairment.

Minor criteria specifically for B4GALT7-spEDS

- Radioulnar synostosis
- Bilateral elbow contractures
- Generalized joint hypermobility
- Single palmar crease
- Characteristic facies (triangular facies, hypertelorism, proptosis, low-set ears, sparse scalp hair, abnormal dentition, flat face, wide forehead, blue sclerae, cleft lip/bifid uvula)

- Characteristic radiographic features (metaphyseal flaring, osteopenia, radial head subluxation/dislocation, short clavicles)
- Severe hypermetropia
- Clouded cornea.

Minor criteria specifically for B3GALT6-spEDS

- Kyphoscoliosis
- Joint hypermobility
- Joint contractures
- Distinctive fingers (slender, tapered, arachnodactyly, broad distal phalanges)
- Talipes equinovarus
- Characteristic facies (midface hypoplasia, frontal bossing, proptosis, blue sclerae, downslanting palpebral fissures, depressed nasal bridge, low-set ears, abnormal dentition, cleft palate, sparse hair)
- Characteristic radiographic findings (platyspondyly, anterior beaking of the vertebral bodies, short ilium, acetabular dysplasia, metaphyseal flaring, elbow malalignment, radial head dislocation, bowing of the long bones, osteoporosis)
- Osteoporosis with fragility fractures
- Ascending aortic aneurysm
- Lung hypoplasia.

Minor criteria specifically for SLC39A13-spEDS

- Protuberant eyes with bluish sclerae
- Winkled palms
- Atrophy of the thenar muscles
- Characteristic radiographic findings (platyspondyly, osteopenia, small ileum, flat femoral epiphyses).

Clinical diagnosis consistent with spEDS is short stature and hypotonia PLUS characteristic radiographic findings (for each type).

Etiology, Pathogenesis, and Genetics

spEDS is an autosomal recessive disorder with genetic heterogeneity involving either *B4GALT7*, *B3GALT6*, or *SLC39A13*. *B4GALT7* encodes galactosyltransferase 1, which catalyzes the first galactose to xylose residue in the tetrasaccharide linker region of glycosaminoglycans (GAGs). The gene product of *B3GALT6* is galactosyltransferase II, which catalyzes the transfer of the second galactose to the first residue in tetrasaccharide linker region of glycosaminoglycans. *SLC39A13* encodes the transmembrane Zrt/irt-like protein 13 (ZIP13) that regulates the influx of cellular zinc.

Diagnostic Testing

Sequence analysis of all three genes (*B4GALT7*, *B3GALT3*, and *SLC39A13*) is recommended. Cultured fibroblasts to assess GAG synthesis is available in specialized laboratories that may be supportive of *B4GALT7* or *B3GALT6* deficiency). Evaluation of urinary pyridinolines (lysyl-pyridinoline to hydroxyl-pyridinoline) can be revealing in *SLC39A13* deficiency.

Differential Diagnosis

There may be clinical overlap with other forms of EDS, especially the rare types, as well as with osteogenesis imperfecta (see Chapter 43) and various congenital myopathies. These can be distinguished with diagnostic testing.

MANIFESTATIONS AND MANAGEMENT

- Baseline skeletal survey.
- Bone densitometry at baseline and regular monitoring per clinic discretion depending on the bone density.
- Bisphosphonates as indicated for osteoporosis and/or fragility fractures (Brady et al. 2017).
- Physical therapy for contractures and muscular hypotonia
- Baseline echocardiogram for ascending aortic aneurysm at diagnosis and every five years or as indicated.

MUSCULOCONTRACTURAL EDS

Diagnostic Criteria

Musculocontractural EDS (mcEDS) presents in infancy with multiple contractures, dysmorphic facies and distinctive skin features (Malfait et al. 2017).

Major Criteria

1. Multiple congenital contractures
2. Characteristic craniofacial features (large fontanelle, hypertelorism, short and downslanting palpebral fissures, blue sclerae, short nose with hypoplastic columella, low-set ears, high palate, long philtrum, thin upper vermillion, small mouth, micrognathia)
3. Skin hyperextensibility; easy bruisability; skin fragility; atrophic scarring; increased palmar wrinkling.

Minor Criteria

1. Recurring dislocations
2. Pectus deformity

3. Tapering, slender, or cylindrical fingers
4. Progressive foot deformities
5. Large subcutaneous hematomas
6. Chronic constipation
7. Colonic diverticuli
8. Pneumothorax
9. Nephrolithiasis
10. Hydronephrosis
11. Cryptorchidism
12. Strabismus
13. Refractive errors
14. Glaucoma (elevated intraocular pressure).

Findings suggestive of mcEDS are multiple congenital contractures and craniofacial features.

Etiology, Pathogenesis, and Genetics

mcEDS is an autosomal recessive disorder with genetic defects in *CHST14*, which encodes D4ST1, carbohydrate sulfotransferase 14, an enzyme involved in the synthesis of dermatan sulfate.

Diagnostic Testing

Sequencing of *CHST14* is recommended for confirmation of mcEDS.

Differential Diagnosis

Rare forms of EDS in addition to Freeman–Sheldon syndrome and Loeys–Dietz syndrome (see Chapter 36) (Brady et al. 2017).

MANIFESTATIONS AND MANAGEMENT

- Baseline echocardiogram for possible congenital heart defects and periodic monitoring for aortic root dilation.
- Ophthalmologic evaluation and treatment of anomalies in a standard manner.
- Renal ultrasound to rule out renal structural defects, which are addressed in a standard manner.
- Screening for hearing impairment, with typical treatment if it is found.
- Orthopedic management as indicated for clubfoot/progressive foot deformity as well as spinal involvement.
- Physical therapy for joint hypermobility and orthotics as indicated.

MYOPATHIC EDS

Diagnostic Criteria

Myopathic EDS (mEDS) is characterized by hypotonia and muscle weakness, proximal large joint contractures and distal joint hypermobility (Malfait et al. 2017).

Major Criteria

1. Hypotonia or muscular atrophy
2. Proximal joint contractures.

Minor Criteria

1. Soft, doughy skin
2. Atrophic scarring.

Etiology, Pathogenesis, and Genetics

mEDS may be inherited in an autosomal dominant or autosomal recessive manner. The genetic defect is in *COL12A1*, the gene for type XII collagen.

Diagnostic Testing

Full gene sequencing of *COL12A1* is confirmatory. Muscle biopsy using immunostaining and skin fibroblast culture studies may be used for a supportive diagnosis.

Differential Diagnosis

Bethlem and Ullrich myopathy (type VI collagenopathies) in addition to other forms of EDS.

MANIFESTATIONS AND MANAGEMENT

- Physical therapy for contractures.
- Orthotics/splints as indicated.
- Feeding and/or respiratory support as needed because of hypotonia and weakness.

PERIODONTAL EDS

Diagnostic Criteria

Periodontal EDS (pEDS) is an aggressive (early-onset) form of periodontal disease with joint hypermobility (Malfait et al. 2017).

Major Criteria

1. Severe intractable periodontitis
2. Lack of attached gingiva

3. Pretibial plaques
4. Family history of a first-degree relative who meets clinical diagnostic criteria.

Minor Criteria

1. Easy bruising
2. Peripheral joint hypermobility
3. Skin hyperextensibility
4. Increased rate of infections
5. Hernias
6. Marfanoid facial features
7. Acrogeria
8. Prominent vasculature.

Criteria suggestive of pEDS are early onset of severe and intractable periodontitis with lack of attached gingiva PLUS at least two other major criteria and one minor criteria.

Etiology, Pathogenesis, and Genetics

pEDS is an autosomal dominant disorder in complement type I genes (*C1R* or *C1S*). Abnormal C1r-C1s tetramers may interfere with procollagen processing in the endoplasmic reticulum (Brady et al. 2017).

Diagnostic Testing

Sequence analysis of *C1R* and *C1S*.

Differential Diagnosis

Other forms of EDS especially cEDS and hEDS.

MANIFESTATIONS AND MANAGEMENT

- Aggressive oral hygiene.
- Nonsurgical debridement every three months.
- Systemic antibiotics may be indicated.
- Skin grafts as indicated.

RESOURCES

Books

Hakim A, Keer RJ, Grahame R, eds. (2011) *Hypermobility, Fibromyalgia and Chronic Pain.* London: Churchill Livingstone.

Tinkle BT (2008) *Issues and Management of Joint Hypermobility: A Guide for the Ehlers-Danlos Syndrome Hypermobility Type and the Hypermobility Syndrome.* Mason, Ohio: Left Paw Press.

Support Groups

Ehlers–Danlos Society
PO Box 87463
Montgomery Village, MD 20886
Phone: 410-670-7577
Website: https://www.ehlers-danlos.com/

Ehlers–Danlos Support UK
PO Box 748
Borehamwood WD6 9HU
United Kingdom
Phone: 0208 736 5604
Website: http://www.ehlers-danlos.org

Hypermobility Syndromes Association (United Kingdom)
49 Greek Street
London, WD1 4EG United Kingdom
Phone: 03330 116 388
Website: *http://www.hypermobility.org/*

REFERENCES

Atzinger CL, Meyer RA, Khoury PR, Gao Z, Tinkle BT (2011) Cross-sectional and longitudinal assessment of aortic root dilation and valvular anomalies in hypermobile and classic Ehlers-Danlos syndrome. *J Pediatr* 158:826–830.

Beighton P, de Paepe A, Danks D, et al. (1988) International nosology of heritable disorders of connective tissue, Berlin, 1986. *Am J Med Genet* 29:581–594.

Beighton P, Grahame R, Bird H (1989) *Hypermobility of Joints*, 2nd ed. London: Springer.

Beighton P, De Paepe A, Steinmann B, Tsipouras P, Wenstrup RJ (1998) Ehlers-Danlos syndromes: Revised nosology, Villefranche, 1997. Ehlers-Danlos National Foundation (USA) and Ehlers-Danlos Support Group (UK). *Am J Med Genet* 77:31–37.

Besselink-Lobanova A, Maandag NJ, Voermans NC, van der Heijden HF, van der Hoeven JG, Heunks LM (2010) Trachea rupture in tenascin-X-deficient type Ehlers-Danlos syndrome. *Anesthesiology* 113:746–749.

Bowen JM, Sobey GJ, Burrows NP, Colombi M, Lavalle ME, Malfait F, Francomano CA (2017) Ehlers-Danlos syndrome, classical type. *Am J Med Genet* 175C:27–39.

Brady AF, Demirdas S, Fournel-Gigleux S, Ghali N, Giunta C, Kapfere-Seebacher I, Kosho T, Mendoza-Londono R, Pope MF, Rohrbach M, Van Damme T, Vandersteen A, van Mourik C, Voermans N, Zschocke J, Malfait F (2017) The Ehlers-Danlos syndromes, rare types. *Am J Med Genet* 175C:70–115.

Byers PH, Belmont J, Black J, De Backer J, Frank M, Jeunemaitre X, Johnson D, Pepin M, Robert L, Sanders L, Wheeldon N (2017) Diagnosis, natural history, and management in vascular Ehlers-Danlos syndrome. *Am J Med Genet* 175C:40–47.

Castori M, Tinkle B, Levy H, Grahame R, Malfait F, Hakim A (2017) A framework for the classification of joint hypermobility and related conditions. *Am J Med Genet* 175C:148–157.

Chopra P, Tinkle B, Hamonet C, Brock I, Gompel A, Bulbena A, Francomano C (2017) Pain management in the Ehlers-Danlos syndromes. *Am J Med Genet* 175C:212–219.

Colige A, Ruggiero F, Vandenberghe I, Dubail J, Kesteloot F, Van Beeumen J, Beschin A, Brys L, Lapiere CM, Nusgens B (2005) Domains and maturation processes that regulate the activity of ADAMTS-2, a metalloproteinase cleaving the aminopropeptide of fibrillar procollagens types I-III and V. *J Biol Chem* 280: 34397–34408.

Dolan AL, Mishra MB, Chambers JB, Grahame R (1997) Clinical and echocardiographic survey of the Ehlers-Danlos syndrome. *Br J Rheum* 36:459–462.

Engelbert RHH, Juul-Kristensen B, Pacey V, De Wandele I, Smeenk S, Woinarosky N, Sabo S, Scheper MC, Russke L, Simmonds JV (2017) The evidence-based rationale for physical therapy treatment of children, adolescents, and adults diagnosed with joint hypermobility syndrome/hypermobile Ehlers-Danlos syndrome. *Am J Med Genet* 175C:158–167.

Gaines R, Tinkle B, Halandras P, Al-Nouri O, Cristostomo P, Cho JS (2015) Spontaneous ruptured dissection of the right common iliac artery in a patient with classic Ehlers-Danlos syndrome phenotype. *Ann Vas Surg* 29:e11–14.

Hakim AJ, Grahame R (2003) A simple questionnaire to detect hypermobility: an adjunct to the assessment of patients with diffuse musculoskeletal pain. *Int J Clin Pract* 57: 163–166.

Hakim A, De Wandele I, O'Callaghan C, Pocinki A, Rowe P (2017A) Chronic fatigue in Ehlers-Danlos syndrome- hypermobile type. *Am J Med Genet* 175C:175–180.

Hakim A, O'Callaghan C, De Wandele I, Stiles L, Pocinki A, Rowe P (2017B) Cardiovascular autonomic dysfunction in Ehlers-Danlos syndrome- hypermobile type. *Am J Med Genet* 175C:168–174.

Juul-Kristensen B, Schmedling K, Rombaut L, Lund H, Engelbert RHH (2017) Measurement properties of clinical assessment methods for classifying generalized joint hypermobility–a systematic review. *Am J Med Genet* 175C:116–147.

Kulas Søborg ML, Leganger J, Quitzau Mortensen L, Rosenberg J, Burcharth J (2017) Establishment and baseline characteristics of a nationwide Danish cohort of patients with Ehlers-Danlos syndrome. *Rheumatology* 56:763–767.

Malfait F, Francomano C, Byers P, Belmont J, Berglund B, Black J, Bloom L, Bowen JM, Brady AF, Burrows NL, et al. (2017) The 2017 international classification of the Ehlers-Danlos syndromes. *Am J Med Genet* 175C:8–26.

Monroe GR, Harakalova M, Van Der Crabben FSN, Majoor-Krakaeur D, Bertoli-Avella AM, Moll FL, Oranen BI, Dooijes D, Vink A, Knoers NV, Maugeri A, Pals G, Nijman IJ, van Haaften G, Baas AF (2015) Familial Ehlers-Danlos syndrome with lethal arterial events caused by a mutation in COL5A1. *Am J Med Genet* 167A:1196–1203.

Ong KT, Perdu J, De Backer J, Bozec E, Collignon P, Emmerich J, Fauret AL, Fiessinger JN, Germain DP, Georgesco G, Hulot JS, De Paepe A, Plauchu H, Jeunemaitre X, Laurent S, Boutouyrie P (2010) Effect of celiprolol on prevention of cardiovascular events in vascular Ehlers-Danlos syndrome: a prospective randomized, open, blinded-endpoints trial. *Lancet* 376:1476–1484.

Pepin M, Schwarze U, Superti-Furga A, Byers PH (2000) Clinical and genetic features of Ehlers-Danlos syndrome type IV, the vascular type. *N Engl J Med* 342:673–680.

Rohrbach MSH, Porter LF, Burkitt-Wright EM, Bürer C, Janecke A, Bakshi M, Sillence D, Al-Hussain H, Baumgartner M, Steinmann B, Black GC, Manson FD, Giunta C (2013) ZNF469 frequently mutated in the brittle cornea syndrome (BCS) is a single exon gene possibly regulating the expression of several extracellular matrix components. *Mol Genet Metab* 109:289–295.

Schalkwijk J, Zweers MC, Steijlen PM, Dean WB, Taylor G, van Vlijmen IM, van Haren B, Miller WL, Bristow J (2001) A recessive form of the Ehlers-Danlos syndrome caused by Tenascin-X deficiency. *N Engl J Med* 345:1167–1175.

Schwarze U, Atkinson M, Hoffman GG, Greenspan DS, Byers PH (2000) Null alleles of the *COL5A1* gene of type V collagen are a cause of the classic types of Ehlers-Danlos syndrome (types I and II). *Am J Hum Genet* 66:1757–1765.

Tinkle B, Castori M, Berglund B, Cohen H, Grahame R, Kazkaz H, Levy H (2017) Hypermobile Ehlers-Danlos syndrome (a.k.a. Ehlers-Danlos syndrome type III and Ehlers-Danlos syndrome hypermobility type): clinical description, and natural history. *Am J Med Genet* 175C:48–69.

Wenstrup RJ, Florer JB, Wiling MC, Giunta C, Steinmann B, Young F, Susic M, Cole WG (2000) *COL5A1* haploinsufficiency is a common molecular mechanism underlying the classic type of EDS. *Am J Hum Genet* 66:1766–1776.

Wenstrup RJ, Meyer RA, Lyle BS, Hoechstetter L, Rose PS, Levy HP, Francomano CA (2002) Prevalence of aortic root dilation in the Ehlers-Danlos syndrome. *Genet Med* 4:112–117.

26

FETAL ALCOHOL SPECTRUM DISORDERS

H. EUGENE HOYME

Departments of Pediatrics and Medicine, The University of Arizona College of Medicine, Tucson, Arizona, USA; Sanford Children's Genomic Medicine Consortium and Sanford Imagenetics, Sanford Health, Sioux Falls, South Dakota, USA

PRACHI E. SHAH

Division of Developmental and Behavioral Pediatrics and the Center for Human Growth and Development, Department of Pediatrics, the University of Michigan Medical School, Ann Arbor, Michigan, USA

INTRODUCTION

A recognizable pattern of malformation associated with prenatal alcohol exposure was first described independently by Lemoine in 1968 and by Jones in 1973 (Lemoine et al. 1968; Jones et al. 1973; Jones and Smith 1973). As physicians became more familiar with the clinical presentation of children prenatally exposed to alcohol, it became clear that the teratogenic effects of alcohol represent a spectrum of disabilities, from mild to severe. *Fetal alcohol spectrum disorders (FASD)* is the collective term that best describes this continuum (Koren et al. 2003; Sokol et al. 2003). In 1996, the Institute of Medicine (IOM) set forth four distinct diagnostic categories that comprise this continuum of teratogenic effects (Stratton et al. 1996). At the severe end of the spectrum is the subset of individuals with growth restriction, a distinctive pattern of facial dysmorphology, and cognitive and/or neurobehavioral effects (fetal alcohol syndrome), and at the mild end of the spectrum are those individuals with neurobehavioral and cognitive deficits who exhibit minimal or no growth or physical stigmata (alcohol-related neurodevelopmental disorder). Misdiagnosis and under-diagnosis of FASD is common (Chasnoff et al. 2015), since the cognitive and behavioral phenotype can be non-specific (Molteno et al. 2010), and the physical phenotype can mimic other malformation syndromes (Leibson et al. 2014). Therefore, it is important to maintain a heightened index of suspicion for prenatal alcohol exposure and the possibility of an alcohol-related diagnosis in children at risk, as early diagnosis and recognition of the disabilities associated with FASD can help optimize future outcomes (Fast et al. 2009).

Incidence and Prevalence

Four distinct diagnoses comprise the continuum of FASD [fetal alcohol syndrome (FAS), partial fetal alcohol syndrome (PFAS), alcohol-related neurodevelopmental disorder (ARND) and alcohol-related birth defects (ARBD)] (Stratton et al. 1996; Hoyme et al. 2005; Chudley et al. 2005). While the phenotype of full-blown FAS is readily recognizable, the full spectrum of prenatal alcohol effects is difficult to estimate because there are no reliable biological markers that readily define those affected. The prevalence rate is related to the frequency of excessive alcohol use in pregnancy and thus will vary from population to population. Estimating the prevalence of FASD using routine surveillance methods is challenging; one recent US study suggested that children with FASD are frequently not diagnosed (80%) or misdiagnosed (7%) (Chasnoff et al. 2015). Therefore, older prevalence estimates, using standard passive record surveillance methodology or clinic-based studies, may have significantly underestimated the prevalence of the complete spectrum of prenatal alcohol-related disabilities (May et al. 2009; Popova et al. 2017.

Cassidy and Allanson's Management of Genetic Syndromes, Fourth Edition.
Edited by John C. Carey, Agatino Battaglia, David Viskochil, and Suzanne B. Cassidy.
© 2021 John Wiley & Sons, Inc. Published 2021 by John Wiley & Sons, Inc.

By contrast, active case ascertainment methods that have been applied in countries throughout the world have resulted in higher prevalence estimates for FASD (May et al. 2006; May et al. 2013; Petković et al. 2013; Olivier et al. 2016). A recent report on FASD prevalence rates in four US communities from a National Institutes of Health supported research consortium utilizing a school-based active case ascertainment strategy [the Collaboration on Fetal Alcohol Spectrum Disorders Prevalence (CoFASP)] suggested that prevalence rates in the United States range between 11.3 and 50.0 per 1000 children, or ~1–5% (May et al. 2018). South African school-based prevalence studies have documented the highest rates of FASD in the world: 135.1–207.5 per 1000 (13.5–20.8%) in the Western Cape Province (May et al. 2013) and 64–119.4 per 1000 (6.4–11.9%) in the Northern Cape Province (Olivier et al. 2016).

Although caution must be exercised in applying prevalence data from special populations to general populations, it is important for clinicians to recognize those unique groups in which FASD is more prevalent. For example, children and youth in child care settings (orphanages, foster care settings, boarding schools, adoption centers, or child welfare systems) are often placed because of parental drug or alcohol problems, child abuse or neglect or child abandonment (Lange et al. 2013; Ospina and Dennett 2013). Maternal alcohol abuse is a main factor for placement of children in orphanages, particularly in those in Eastern European countries (Miller et al. 2007). Landgren et al. (2010) reported that 52% of Swedish children adopted from Eastern European orphanages had an FASD. A meta-analysis of studies utilizing active case ascertainment methodology estimated the FASD prevalence rate to be 16.9% in such childcare settings (Lange et al. 2013). Youth placed in correctional facilities constitute another similarly unique population. An analysis of the prevalence of FASD in individuals in correctional facilities in the United States and Canada suggested an increased rate of FASD among youth and adult offenders. The authors calculated a relative risk of 19 for imprisonment for youths with FASD in Canada, i.e. adolescent offenders with FASD (ages 12–18) were 19 times more likely to be in prison than youths without FASD (Popova et al. 2011).

Diagnostic Criteria

The diagnosis of an FASD reflects a constellation of signs and symptoms resulting from prenatal exposure to alcohol. Updated diagnostic guidelines for the four categories that define the continuum of FASD as originally set forth by the US Institute of Medicine (IOM) (Stratton et al. 1996) were recently published (Collaboration on FASD Prevalence (CoFASP) Consensus Clinical Diagnostic Guidelines, Hoyme et al. 2016) (Table 26.1).

Several FASD diagnostic systems have been developed. The most commonly used are: (a) *the CoFASP Consensus Clinical Diagnostic Guidelines* [first developed by Hoyme et al. in 2005 as the Revised IOM Criteria for FASD and updated in 2016 as the CoFASP Consensus Clinical Diagnostic Guidelines (Hoyme et al. 2016)]; (b) the University of Washington *4-Digit Diagnostic Code* (Astley 2013); and (c) *the Canadian Guidelines* [first developed by Chudley et al. in 2005 and revised in 2016 (Cook et al. 2016)]. Each of these systems has advantages and disadvantages, including a varied mix of sensitivity, specificity, terminology, and ease of use in clinical settings. However, there are similarities in the diagnostic approach among the three systems. This chapter will discuss use of the CoFASP Diagnostic Guidelines in clinical practice.

The CoFASP Consensus Clinical Diagnostic Guidelines utilize the familiar original four diagnostic categories set forth by the IOM (Stratton et al. 1996). These guidelines have the advantage of having been developed from a medical model and validated through testing in a large multiracial international cohort of children. The criteria also rely on an expert multidisciplinary team diagnostic approach, an approach widely used in the diagnosis and management of similar conditions in pediatric practice. In addition, the CoFASP Guidelines consider different etiologies for the clinical phenotype when making an FASD diagnosis, including genetic factors, other teratogenic exposures, and environmental contributions to developmental and behavioral disorders (Hoyme et al. 2005; Manning and Hoyme 2007; Hoyme et al. 2016), and the diagnostic approach is easily taught to physicians and non-physicians with good diagnostic reliability (Jones et al. 2006; O'Conner et al. 2014). Finally, these criteria have been used most widely in FASD prevalence studies in the United States and internationally; it has been suggested that they be used for all future prevalence studies, thereby maximizing comparability to past data (Roozen et al. 2016).

The diagnostic evaluation of a child with a potential FASD requires the assessment of four domains: (a) prenatal alcohol exposure; (b) facial structure (assessing the three cardinal facial features of FAS: short palpebral fissures, smooth philtrum and thin vermilion border of the upper lip); (c) growth (prenatal and postnatal); and (d) cognitive and/or neurobehavioral impairment. The CoFASP Guidelines (Hoyme et al. 2016) assess these four domains to determine whether a child meets diagnostic criteria for one of the four FASD categories (Figure 26.1).

Etiology, Pathogenesis, and Genetics

Ethanol is a well-established human teratogen (Ornoy and Ergaz 2010). Extensive animal and human data indicate that ethanol readily diffuses across the placenta and distributes rapidly into the fetal compartment (Brien et al. 1983, 1985). Once in the fetal compartment, alcohol has prolonged fetal effects for the following reasons: (a) ethanol is eliminated

TABLE 26.1 Updated diagnostic guidelines for fetal alcohol spectrum disorders (FASD) (Hoyme et al. 2016)

CoFASP Consensus Clinical Diagnostic Guidelines for FASD

	FAS facial features [a]	Growth restriction [b]	Deficient brain growth [c]	Neurobehavioral impairment (<3 years old) [d]	Neurobehavioral impairment (>3 years old) [e]	Birth defects [f]
FAS (fetal alcohol syndrome)						
Confirmed/ unconfirmed alcohol exposure	X	X	X	X	X 1. Global impairment 2. Cognitive deficit 3. Behavioral impairment	
PFAS (partial fetal alcohol syndrome)						
Confirmed alcohol exposure	X			X	X 1. Global impairment 2. Cognitive deficit 3. Behavioral impairment	
		← OR →				
No confirmed alcohol exposure	X	X	X	X	X 1. Global impairment 2. Cognitive deficit 3. Behavioral impairment	
ARND (alcohol-related neurodevelopmental disorder) **						
Confirmed alcohol exposure				N/A	X 1. Global impairment 2. Cognitive deficit (2) 3. Behavioral impairment (2)	
ARBD (alcohol-related birth defects)						
Confirmed alcohol exposure						X

[a] Characteristic pattern of facial anomalies characterized by ≥ 2 of the following: (1) short palpebral fissures (≤10th centile), (2) thin vermilion border (rank of 4 or 5 on racially normed lip/philtrum guide), (3) smooth philtrum (rank of 4 or 5 on racially normed lip/philtrum guide).

[b] Prenatal and/or postnatal growth deficiency: height and/or weight ≤10th centile on gender specific population normed growth curves.

[c] Deficient brain growth/morphogenesis/neurophysiology characterized by ≥1 of the following: (1) head circumference ≤10th centile, (2) structural brain abnormalities, (3) recurrent non-febrile seizures.

[d] Evidence of developmental delay ≥1.5 SD below the mean.

[e] Assess the following domains; at least one must be affected.
(1) Global impairment: general conceptual ability, performance IQ (PIQ), visual IQ (VIQ) or spatial IQ ≥1.5 SD below the mean.
(2) Cognitive deficit in 1 domain ≥1.5 SD below the mean (executive function, specific learning impairment, memory impairment or visual spatial impairment.
(3) Behavioral impairment without cognitive impairment: behavioral deficit in 1 domain ≥1.5 SD below the mean in areas of self-regulation (mood or behavioral regulation impairment, attention deficit or impulse control).

** For ARND, *two domains* of impairment are required for EITHER cognitive deficit without behavioral impairment, OR behavioral impairment without cognitive deficit.

[f] One or more major malformations demonstrated in animal models and human studies to be related to prenatal alcohol exposure, including *cardiac defects* (e.g., atrial septal defects, ventricular septal defects, aberrant great vessels, conotruncal heart defects); *musculoskeletal defects* (e.g., radioulnar synostosis, vertebral segmentation defects, large joint contractures, scoliosis); *renal anomalies* (e.g., aplastic/ hypoplastic/ dysplastic kidneys, "horseshoe" kidneys, ureteral duplications); *eye anomalies* (e.g., strabismus, ptosis, retinal vascular anomalies, optic nerve hypoplasia); and/or *hearing impairment* (e.g., conductive or neurosensory hearing loss).

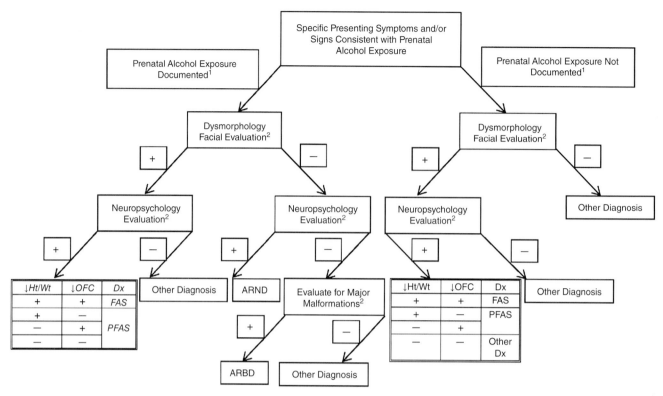

FIGURE 26.1 Fetal alcohol spectrum disorders diagnostic algorithm. [1]Prenatal alcohol exposure should be assessed by using a validated screening questionnaire. [2]A structured approach to clinical evaluation includes assessment for minor and major anomalies and a neurodevelopmental assessment. Abbreviations: FAS: fetal alcohol syndrome; PFAS: partial fetal alcohol syndrome; ARND: alcohol-related neurodevelopmental disorder; ARBD: alcohol-related birth defects; Ht: height; Wt: Weight; OFC: occipitofrontal circumference (see text). Reproduced, with permission, from Hoyme et al. 2016

more slowly from the fetus than from the mother (Heller and Burd 2014), allowing for accumulation in the amniotic fluid (Brien et al. 1983); (b) alcohol's concentration in the amniotic fluid is compounded by fetal swallowing and urination into the amniotic fluid, which the fetus then reswallows (Underwood et al. 2005); and (c) fetal metabolism of alcohol differs from that observed in adults, resulting in slower ethanol metabolism. In the placenta and the developing fetal liver CYP2E1 is the major metabolizing enzyme (Cummings and Kavlock 2004), as opposed to alcohol dehydrogenase (ADH). Fetal CYP2E1 levels remain low throughout pregnancy (Zelner and Koren 2013), and hepatic alcohol dehydrogenase levels are not significant until after 26 weeks' gestation (Arfsten et al. 2004).

Ethanol exerts teratogenic effects through myriad molecular and cellular mechanisms. Alcohol compromises endogenous antioxidant activity and generates free radicals as a by-product of its metabolism by CYP2E1 (Ornoy and Ergaz 2010). These superoxide radicals target polyunsaturated fatty acid side chains in brain tissue membranes, thereby damaging fetal brain tissue during organogenesis, which later manifest as central nervous system dysfunction. Free radicals and reactive oxygen species also induce uncontrolled apoptosis in the fetal brain and cranial neural crest cells (Cohen-Kerem and Koren 2003), thereby helping to explain the midline facial deficiencies observed in children with FAS (short palpebral fissures, hypoplastic midface, smooth philtrum, and thin upper lip vermilion border) (Cartwright and Smith 1995 a, b). In addition, alcohol disrupts neuronal cell–cell adhesion by affecting the gene expression of L1 and other cell adhesion molecules (Ramanthan et al. 1996). The neuronal cell adhesion molecule L1 is critical for brain development and plays a role in adult learning and memory; it is theorized to play a role in the cerebellar dysfunction associated with fetal alcohol spectrum disorders (Fitzgerald et al. 2011; Dou et al. 2011).

Epigenetics and FASD It has long been known that susceptibility to the teratogenic effects of ethanol is dependent on genetic factors in both the mother and fetus. For example, a recent genome wide single nucleotide variant (SNV) study in mice identified 900 genes that are down-regulated by alcohol exposure in the embryo (Lossie et al. 2014). Such changes in gene expression are epigenetic (gene activation or inactivation initiated by environmental triggers that do not alter the underlying DNA sequence, which in turn affect how cells "read" the genes). Alcohol inhibits gene expression in the developing fetal brain through at least three epigenetic mechanisms: DNA methylation, histone modification, and non-coding RNA [circulating micro-RNA (miRNA)] associated gene silencing (Lussier et al. 2017). For example, in both animal and human studies, ethanol exposure during pregnancy results in significant elevations in circulating miRNAs in maternal blood. In a recent study of alcohol exposed pregnancies in Ukraine, 11 miRNAs were significantly elevated in the plasma of heavily alcohol-exposed women who had children with FASD, compared to both women who drank heavily but had apparently normal children and unexposed women (Balaraman et al. 2016). These data indicate that maternal plasma miRNA profiles predict infant outcomes and may be useful as biomarkers in the diagnosis of FASD, alleviating the previously discussed difficulty in accuracy of maternal reporting of alcohol intake during pregnancy.

Diagnostic Testing

FASD affects persons accessing multiple social, medical and educational systems. Therefore, it is important that health care personnel, family physicians, pediatricians, psychiatrists, nurses, teachers, administrators, school psychologists, special educators, and the justice system (judges, probation officers, and corrections officers) recognize at-risk individuals. Children with learning and behavioral problems, children in care of child protection agencies, children of addicted mothers, and children and adolescents in trouble with the law are at high risk (Chudley and Longstaffe 2010). With some exceptions (Poitra et al. 2003), there are few validated screening tools or check lists that are available for this purpose. The development of screening tools for these high-risk populations is discussed in a review of this subject (Goh et al. 2008).

Since there are no available diagnostic laboratory tests for FASD, diagnosis of the defined four categories within FASD are based on specific clinical diagnostic guidelines (Astley 2013; Hoyme et al. 2016; Cook et al. 2016).

Multidisciplinary Diagnostic Team Approach Assignment of an FASD diagnosis is a complex medical diagnostic process best accomplished through a structured multidisciplinary approach by a clinical team comprising members with varied but complementary experience, qualifications and skills. The assessment of individuals prenatally exposed to alcohol requires a medical assessment and team leadership by a pediatrician or clinical geneticist/dysmorphologist with expertise in the full range of human malformation syndromes and the dysmorphology evaluation of children with FASD. In addition, exposed children should have expert psychological and neuropsychological assessment, and a skilled interviewer should evaluate prenatal maternal alcohol intake. Other team members may include developmental behavioral pediatricians, psychiatrists, speech pathologists, occupational therapists, physical therapists, special educators, audiologists, and/or ophthalmologists, among others (Hoyme et al. 2005, 2016; Cook et al. 2016). The ideal team members, in part, will be determined by the age of the individual being assessed (infant, child, adolescent or adult). In small or isolated communities where essential team members are not available, the community team must collaborate and consult with larger centers that have a more comprehensive diagnostic team. Access to diagnosis can also be enhanced by telemedicine (Benoit et al. 2002). Because the first step in treatment, intervention and prevention of FASD is an accurate diagnosis, no individual should be denied the opportunity of an expert assessment and diagnosis (Chudley and Longstaffe 2010).

History of Prenatal Alcohol Exposure An essential part of the diagnostic evaluation of an individual for a potential FASD is the assessment of the quantity and timing of prenatal alcohol exposure (confirmation of prenatal alcohol exposure is necessary for diagnoses of ARND or ARBD; however, diagnoses of FAS and PAS can be made with or without confirmed prenatal alcohol exposure). Verification of prenatal alcohol exposure is often difficult. Because of the stigma associated with drinking during pregnancy (Makelarski et al. 2013), mothers may be reluctant to admit to alcohol use during pregnancy, may under-report, or may not recall the amount of alcohol consumed during pregnancy, resulting in maternal self-reporting data that are often unreliable (Jacobson et al. 2002; Lange et al. 2014). Because pre-pregnancy alcohol consumption is highly correlated with the presence of alcohol-related diagnoses and neurobehavioral problems (May et al. 2013; Knudsen et al. 2014), and mothers may not consider or report alcohol consumption prior to learning of their pregnancies, screening questionnaires such as the T-ACE (assessment of *tolerance, annoyance* at being asked about alcohol abuse, attempts to *cut down* on alcohol intake and need for an *eye opener*) and its revision, the TACER-3, have been developed to screen for maternal alcohol use in pregnancy. The T-ACE and TACER-3 have demonstrated good sensitivity and specificity (Chiodo et al. 2014).

Alternatively, because late recognition of pregnancy and drinking in the three months prior to pregnancy have been associated with the presence of an alcohol-related disorder in the child (May et al. 2014), querying the mother regarding when she learned of her pregnancy may help to further identify a history of prenatal exposures. In maternal interviews, because of potential stigmatization associated with prenatal alcohol use, questions should be asked in a timeline progressing from the broader context of health history to the more sensitive topic of alcohol use (Sobell et al. 2001).

Frequently, children being assessed for an FASD are in foster care or have been adopted. Depending on the circumstances that led to a child no longer being in the care of the biological mother, medical and familial history may be limited. In these instances, obtaining an accurate history regarding prenatal alcohol exposure can be even more difficult. A child's caregivers may report that the biological mother drank during the pregnancy; however, it is important to determine the source of this information. As information is passed from one source to another, the facts may become distorted, and thus the validity of the information diminishes. Obtaining a child's hospital birth records, medical passport, or court documents regarding termination of parental rights can provide additional information regarding potential prenatal alcohol exposure.

Assessment of Facial Features Because two of the four categories of FASD (FAS and PFAS) require the presence of the typical facies associated with FAS, accurate identification of facial dysmorphology is critical to making a diagnosis of an FASD. Accurate identification of FAS facial features is often performed by a physician with specialized training in FAS dysmorphology (e.g., clinical geneticist or pediatrician); however, other health professionals can also be trained to screen for the cardinal dysmorphic facial features associated with FAS and can therefore be instrumental in identifying individuals who should be referred for further subspecialty evaluation (O'Connor et al. 2014). The guidelines for diagnosis of FAS or PFAS (Hoyme et al. 2005, 2016) require two of three characteristic facial features as assessed by a skilled examiner: (a) short palpebral fissure length (≤10th centile compared with published norms); (b) smooth philtrum; and/or (c) thin vermilion border of the upper lip (Figures 26.2 and 26.3). The technique for palpebral fissure measurement is depicted in Figure 26.4.

Lip and philtrum morphology are assessed independently of each other; scores are assigned, utilizing a five-point Likert scaled racially normed lip-philtrum guide, with scores of 4 or 5 considered compatible with FAS or PFAS (Figure 26.5) (Hoyme et al. 2005, 2016; Manning and Hoyme 2007).

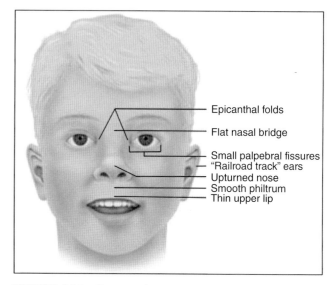

FIGURE 26.2 Common facial anomalies observed in children with FAS. Two of the three cardinal anomalies associated with FAS are required for diagnoses of FAS and PFAS (short palpebral fissures, smooth philtrum and/or thin upper lip vermilion border). (Image courtesy of Darryl Leja, National Human Genome Research Institute, NIH. Reproduced, with permission, from Wattendorf and Muenke 2005).

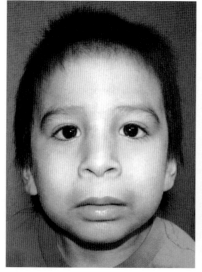 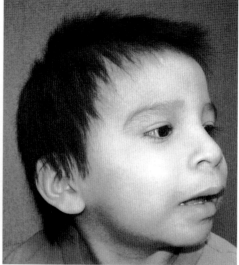

FIGURE 26.3 Typical child with FAS. Note short palpebral fissures, smooth philtrum and relatively thin vermilion border of the upper lip. Midface hypoplasia and an anteverted nose are also present. (Reproduced, with permission, from Hoyme et al. 2016).

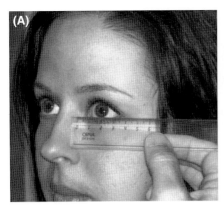 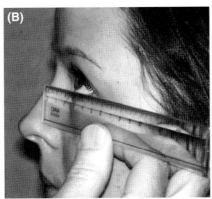 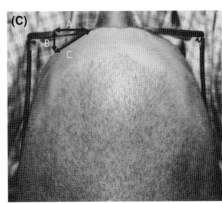

FIGURE 26.4 (A) Technique for measuring palpebral fissure length. A ruler, ruled in millimeters, is used to measure the distance between the *endocanthion* (where the eyelids meet medially) to the *exocanthion* (where the eyelids meet laterally). Keeping the chin level, the subject is asked to look up, allowing the examiner to bring the ruler as close to the eye as possible (without touching the lashes). The ruler can be rested on the cheek for stability while recording the measurement. (B) Note that the ruler is angled slightly to follow the curve of the zygoma. (C) The correct length of the palpebral fissure is depicted here as measurement "C;" whereas, "A" would be a shorter length, such as that obtained from 2D photogrammetry. For this reason, if palpebral fissure lengths are obtained by 2D photogrammetry, they should be compared to photographic norms, and vice versa for live measurements. (Reproduced, with permission, from Hoyme et al. 2016).

Minor Anomalies Although not part of the core diagnostic criteria, children with prenatal alcohol exposure display a variable pattern of minor anomalies, including midface hypoplasia, anteverted nose, epicanthal folds, hypoplastic nails, short fifth fingers, clinodactyly of fifth fingers, pectus carinatum or excavatum, camptodactyly, "hockey stick" palmar creases, other palmar crease abnormalities, ptosis, strabismus, impaired pronation and supination at the elbows, ocular hypertelorism or hypotelorism, elongated philtrum, and "railroad track" ears (Hoyme et al. 2005, 2016; Jones et al. 2010). The presence of these physical abnormalities can prompt a provider to consider additional evaluation for an alcohol-related diagnosis. Examples of some of these minor anomalies are depicted in Figure 26.6.

Growth Restriction Growth restriction is characterized by height or weight less than the 10th centile at any point in time (prenatally or postnatally) as plotted on age- and gender-specific growth charts. (Manning and Hoyme 2007; Hoyme et al. 2005, 2016).

Central Nervous System (CNS) Abnormalities CNS dysfunction is characterized by the presence of structural and/or functional abnormalities. Such CNS findings can be reflected in a small head circumference (occipito-frontal circumference ≤10th centile at any point in time measured on gender specific head-circumference charts), structural anomalies of the brain identified on an MRI scan or recurrent, non-febrile seizures (Hoyme et al. 2016).

Functional CNS impairments are characterized by evidence of neurobehavioral impairment, defined as either (1) global cognitive impairment defined by general intellectual ability, verbal or performance IQ ≥ 1.5 SD below the mean, or (2) cognitive deficits in at least one neurobehavioral domain including executive functioning, specific learning impairment, memory impairment, or visual impairment or (3) behavioral impairment without cognitive impairment characterized by impairments in self-regulation (mood or behavioral regulation impairment), attention deficit or impulse control (Hoyme et al. 2016).

Differential Diagnosis

Clinicians should be aware that the facial phenotype of FAS, although most commonly associated with prenatal alcohol exposure, is also observed in a variety of genetic and teratogenic conditions. Compelling data from animal models suggest that ethanol significantly reduces the volume of cranial neural crest cells in the embryo, thus contributing to underdevelopment of midline facial structures in children with FASD (Cartwright and Smith 1995a, 1995b; Flentke et al. 2014). Theoretically, genetically determined causes of cranial neural crest deficiency could lead to a similar facial phenotype. Therefore, physicians should employ a low threshold for ordering additional genetic testing of children with potential FASD. Among the disorders in the differential diagnosis of FASD are: Cornelia deLange syndrome (see Chapter 15), del 22q11.2 syndrome (see Chapter 21), inverted duplication 15q syndrome (see Chapter 33), Dubowitz syndrome, Noonan syndrome (see Chapter 41), Williams syndrome (see Chapter 63), fetal hydantoin syndrome, fetal valproate syndrome, maternal PKU effects, and toluene embryopathy (Pearson et al. 1994; Leibson et al. 2014; Hoyme et al. 2016). The highest yield diagnostic test when a genetic phenocopy of FASD is being considered is a chromosome microarray (Abdelmalik et al. 2013; Douzgou et al. 2012).

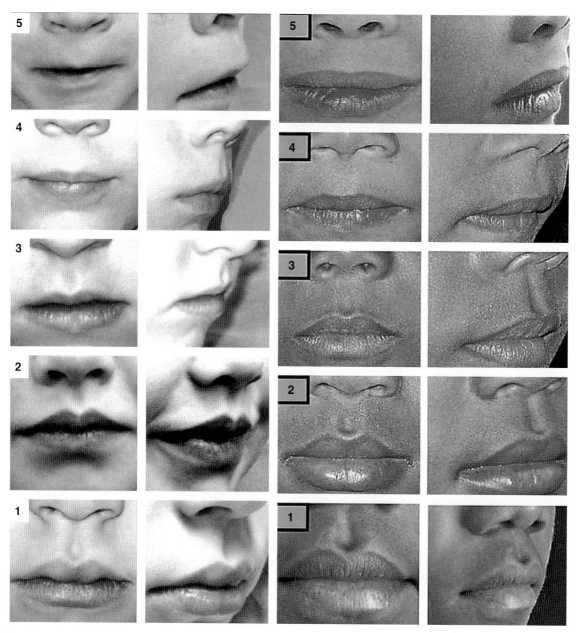

FIGURE 26.5 The lip philtrum guide on the left was developed from population studies of white North American school children. The guide on the right was developed from population studies of mixed race South African school children. Because of racial differences in lip morphology, accurate assessment of the structure of the philtrum and upper lip should be performed utilizing racially normed tools. (Reproduced, with permission, from Hoyme, et al. 2016 and Hoyme et al. 2015).

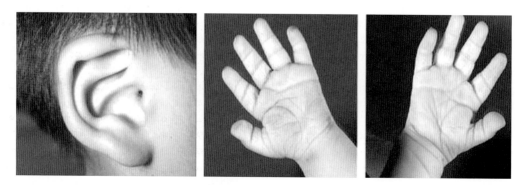

FIGURE 26.6 Some of the common minor anomalies observed in FASD. Note the "railroad track" ear in the first panel and the "hockey stick" palmar creases and fifth finger clinodactyly in the second and third panels.

MANIFESTATIONS AND MANAGEMENT

Because of the complex nature and multitude of issues relevant to the presentation, diagnosis of and intervention for individuals with FASD, a multidisciplinary team approach to diagnosis and treatment is strongly encouraged (Hoyme et al. 2005, 2016; Chudley and Longstaffe 2010). Once a diagnosis is made, medical and neurobehavioral interventions can greatly improve the prognosis for children and adults with FASD (Manning and Hoyme 2007; Reid et al. 2015).

Growth and Feeding

Growth is impaired in many affected children and adults, but may be normal in otherwise severely affected individuals. However, there is a direct correlation between growth restriction in children with FASD and long-term effects on neurocognition; such effects are strongest in children with both prenatal and long-term growth restriction, more moderate in those with fetal growth restriction and postnatal catch-up, and weakest in those without growth restriction (Carter et al. 2016). Growth impairment may be more often associated with heavy alcohol exposure through all of the pregnancy, particularly during the last trimester. Determinants of prenatal growth impairment include nutritional deficiencies, exposure to other drugs and cigarette smoke, and stress in the pregnancy. Those children with normal growth parameters at birth may experience postnatal failure-to-thrive because of feeding difficulties, neglect or both, or secondary to the impact of brain, cardiac, or renal malformations. Typically, growth improves with age and may result in weight and, occasionally, height in the normal range by adolescence and adulthood. All growth parameters must be compared with the genetic potential for height and weight within the family (Chudley and Longstaffe 2010).

Evaluation

- Growth parameters should be measured and plotted on racially normed standard growth charts at each routine visit. There are no growth-specific charts for fetal alcohol spectrum disorders.
- Because caloric deprivation can occur in a socially disrupted home, children with FASD and growth deficiency should undergo a nutritional assessment. Additional investigations might be warranted in some children with feeding difficulties: radiographic or endoscopic assessment for reflux, uncoordinated swallowing or aspiration; assessment by an occupational therapist or speech/language pathologist for coordination of sucking, swallowing and breathing; and evaluation of sensory reactions to food textures and temperatures.

Management

- In those children with nutritional deficiencies, dietary caloric supplementation can lead to catch-up growth. However, growth deficiency is often constitutional and will not respond to hormonal or nutritional therapy.
- When there are difficulties with suck/swallow/breathing coordination, or aspiration, it may be necessary to undertake placement of a gastrostomy feeding tube, with periodic reassessment to establish whether feeding reflexes have improved enough for the infant to tolerate oral feedings.
- When sensory over-reactivity to food textures is present, it may be possible to manipulate the diet to ensure appropriate nutritional content while allowing the infant to consume foods in textures that can be tolerated. These symptoms improve as the infant grows older.

Development and Behavior

As brain maturation occurs in the affected child, a pattern of deficits emerges. Thus, the neurocognitive and behavioral deficits of FASD can manifest differently over the lifespan, reflecting both heterogeneity and a developmental emergence. Recent efforts to define the range of cognitive and mental health problems occurring in individuals with a history of prenatal alcohol exposure have focused on three core areas: neurocognition, self-regulation and adaptive functioning impairments (Doyle and Mattson 2015; Kable et al. 2016, 2017; Hagan et al. 2016)

Infancy. The neurobehavioral deficits associated with prenatal alcohol exposure are non-specific in infancy (Molteno et al. 2010) and are largely characterized by deficits in cognitive and motor development as well as impairments in emotional and state regulation (Coles et al. 1987). Infants with heavy prenatal alcohol exposure have demonstrated dose-dependent lower cognitive development compared to non-exposed infants on the Bayley Scales of Infant Development (Jacobson et al. 1993). Other infant characteristics associated with heavy prenatal alcohol exposure include delayed motor development and slowed reaction time (Jacobson et al. 1994), poorer language and behavioral regulation (Coles et al. 2000), less complex play (Molteno et al. 2010), poorer feeding, deficits in state regulation, difficulty self-soothing, and increased emotional withdrawal (Streissguth et al. 1989; Molteno et al. 2014). Because the facial features of FAS may be difficult to identify in infants and the cognitive and behavioral deficits associated with prenatal alcohol exposure are non-specific in infancy (Molteno et al. 2010), it is important to maintain a heightened index of suspicion for prenatal alcohol exposure and the possibility of an alcohol-related diagnosis in infants and toddlers at risk.

Preschool. In the preschool years, children prenatally exposed to alcohol can present with a broad range of cognitive delays involving visual spatial and/or language-based abilities. Deficits in gross and fine motor skills may become apparent, including problems with gait, balance, handwriting, and visual spatial aptitudes (Kalberg et al. 2006; Lucas et al. 2014). Affected children often display speech and language delays (McGee et al. 2009), articulation problems, language processing dysfunction (Cone-Wesson 2005) and delays in auditory processing (Stephen et al. 2012). The behavioral deficits associated with prenatal alcohol exposure are heterogeneous and can present in the preschool years as difficulties with executive function (Fuglestad, 2015), attention, externalizing behavior problems (Alvik et al. 2013), increased emotional and conduct problems (Alvik et al. 2013), social skills deficits (Rasmussen et al. 2011), sleep difficulties (Chen et al. 2012), sensory processing problems, and adaptive behavior deficits (Carr et al. 2010).

School Age. In the school age child, a history of prenatal alcohol exposure has been associated with decreased intellectual functioning, learning disabilities (especially in reading, writing and math) and impairments in executive functioning, including deficits in working memory, processing speed, verbal fluency, planning, sequencing, and organization (Roussotte et al. 2012; O'Leary et al. 2013; Gautam et al. 2015; Carter et al. 2016; Lewis et al. 2016). Operationally, prenatal alcohol exposure can contribute to difficulties encoding and consolidating new information (e.g., children demonstrate difficulty learning and remembering rules and understanding consequences of their behavior); behaviorally, they can present with difficulties with attention, impulsivity and emotion regulation (Glass et al. 2014; Tsang et al. 2016). In addition, prenatal alcohol exposure has also been associated with deficits in higher order social pragmatic and language skills, including impairments in gestural communication and social perception and emotion recognition, contributing to social-communicative difficulties. Children with prenatal alcohol exposure can also engage in inappropriate social initiations and social interactions, thereby contributing to difficulties with peer relationships (Kully-Martens et al. 2012; Kerns et al. 2016).

Adolescence. In addition to the language and learning impairments observed in school-age children, the functional and behavioral deficits in adolescence associated with prenatal alcohol exposure are largely related to impairments in attention and higher order cognition, including deficits in working memory, executive functioning, information processing, judgement, and metacognition (Casey et al. 2008). In adolescence, the social, functional and occupational impairments associated with an FASD diagnosis comprise lower educational attainment, increased school dropout rates, increased trouble with the law and involvement with the juvenile justice system, and increased risk for illicit drug and alcohol abuse (Moore and Riley 2015). Without appropriate environmental modification, older individuals may develop depression and demonstrate suicidal ideation as they come to realize their differences from their neurotypical peers. Others may gravitate to street gangs to gain friendships and acceptance as they are marginalized from society, and they can be led into criminal activities resulting in incarceration (Boland et al. 2002).

Evaluation

- Affected children require a multidisciplinary assessment and intervention from infancy onward. Assessment of their activity levels, attention, impulsivity, and distractibility may imply an attention-deficit disorder or memory difficulties and generalized anxiety caused by abuse or multiple placements.
- Comorbidities must be recognized and addressed. These evaluations are best carried out by child developmental specialists, psychologists and psychiatrists.
- There is no need for routine brain imaging in individuals with FASD unless they present with seizures or demonstrate frank neurological findings.

Management

- Interventions must be considered individually based on the problems present. Some general issues in intervention are important to stress. Because many of the problems are now recognized to be the result of brain injury rather than being behavioral in origin, there is a need to "reframe" expectations and provide an environment where the child is "primed to succeed." The emphasis should be on environment modification rather than on an expectation that the child should change to accommodate to the environment.
- For the sensory/motor domain, there are many strategies of treatment available, with the occupational therapist most expert in devising those most appropriate to the individual, depending on his or her profile. Sensory integration therapy may help an individual to use sensory information in playful, meaningful and natural ways, for example, a person who has intense "fight or flight" responses to harmless touch sensations can learn to accept such sensations without overreacting. Interventions should be purposely structured to be enjoyable and to bolster confidence.
- There are many environmental modifications that can assist the inattentive individual to attend more effectively to tasks, for example, removal of extraneous material in the classroom, use of pictorial cues for the sequence of the daily routine, decreasing the number of unnecessary transitions during the school day, and

having a consistent workspace free of noise and other distractions (Chudley and Longstaffe 2010).
- Medications, particularly in the stimulant class, have been clearly demonstrated to be effective in achieving better attention and lessening distractibility in many affected individuals (O'Malley and Nanson 2002). Neuroleptic medications can assist affected children with impaired social skills (Frankel et al. 2006).
- Language delayed preschool and school aged children are best served by having a speech/language therapist who can advise on language programming and enrichment appropriate to the child's needs. Reframing is necessary when the child may seem to have a good vocabulary, but in fact has a deficit in language comprehension. The strategy of "say less, show more" is often beneficial (Chudley and Longstaffe 2010).
- Cognitive impairments respond to adaptation of academic curricula so that they provide an appropriate mix of intellectual challenge and success for the abilities of the child or adolescent. Providing a learning environment more conducive to success for the child with fetal alcohol syndrome is essential. Classroom strategies may include avoidance of sensory overstimulation, concrete presentation of learning materials, visual cues as to the daily routine, a structured routine, and specifically designed areas of the class designated and mapped with color for specific activities (Mattson and Roebuck 2002; Kalberg and Buckley 2006). Tasks should be broken into manageable portions for the allotted time, and directions should be simplified. Presentation of materials using voice, visual, and kinesthetic methods increases the likelihood of the material being absorbed by the student. Self-esteem can be built by recognizing successes daily. Use of consistent, positive reinforcement for adherence to rules in the classroom increases the chance of the student following them. Advance warnings for activity changes are helpful in transition from one activity to the next.
- Specific strategies are available for enhancing achievement in various academic subjects. These strategies are widely available in print form to teachers in schools (e.g., Berg et al. 1995). For example, for mathematics, working on number concepts using concrete examples, such as using real coins and playing "restaurant" to teach concepts of making change with money, helps the child understand. Memory deficits can lead to confabulations and should not be assumed to be deception. Memory can be aided with repetitions, anticipation of memory problems, the use of multiple modalities for presenting information, and encouraging the child to use multiple methods for remembering (e.g., write an idea, visualize it, hear it, and act it).
- Poor executive functioning can be a manifestation of brain injury and not a purposeful breaking of rules or making mistakes in daily living. External guides for decision-making and extra supervision can alleviate problems related to poor judgment. Monitoring and structuring leisure time and education about sexual development may reduce the risk of sexually transmitted diseases and unwanted pregnancies.
- Problems with adaptive skills are pervasive. Strategies that are likely to facilitate the individual being maximally able to provide self care and participate in social and leisure activities should be chosen (e.g., directing daily living tasks using simple visual clues). Modifying expectations is important, as is providing the necessary supervision and positive reinforcement.
- Physical therapy and occupational therapy can maximize effective functioning in individuals with fetal alcohol syndrome and mild neurological or sensorimotor abnormalities.
- Family support and therapy can be helpful in dealing with anger, guilt, shame, and anxiety.
- Treatment of mental health problems such as anxiety disorders, drug and alcohol addiction, and depression in the adolescent or adult will require a skilled and knowledgeable mental health professional familiar with these problems in fetal alcohol spectrum disorders and their variable response to stimulant, mood altering, or other psychotropic medications.
- Adolescents and adults with fetal alcohol syndrome may not respond to cognitive therapy, partly the result of the language processing difficulties that are often present. Counseling should focus on concrete suggestions around behavioral modification with close supervision.
- Many adults are unable to cope on their own and need mentors to help them with everyday decisions. Many require the help of social agencies for housing arrangements. Many find themselves unable to keep regular employment and are dependent on social services for long-term support (Chudley and Longstaffe 2010).

Ophthalmologic

Malformations of the eye and visual impairment occur more frequently in individuals with FASD than in the general population (Strömland and Pinazo-Duran 2002; Strömland 2004). That co-occurrence is not surprising, since the developing brain and cephalic neural crest are primary targets of alcohol in the embryo, and since the eye is derived from neuroepithelium, surface ectoderm, and the extracellular mesenchyme (which contains both neural crest and mesoderm) (Sadler 2015). Among the external ocular signs of FASD are ptosis, short palpebral fissures, telecanthus, epicanthal folds, blepharoptosis, microphthalmia, and strabismus. Other common findings include reduced visual acuity and/or refractive error (hyperopia, myopia and astigmatism) (Strömland 2004). Anomalies of the anterior segment and media may be present, including malformations of the cornea, anterior chamber, iris, lens, and vitreous body. Anomalies of the retina and optic nerve are also observed, including retinal dysplasia, increased tortuosity of retinal vessels and optic nerve hypoplasia (Strömland 2004). Although retinal anomalies may be found, progressive retinal dysfunction has not been reported (Strömland and Pinazo-Duran 2002).

Evaluation

- A pediatric ophthalmologic evaluation should be done at age 3 years or at the time of diagnosis. Vision should be monitored in follow-up visits no less than every two years.

Management

- Treatment and follow-up should be specific to the abnormalities found and standard in nature.

Ears and Hearing

Mixed hearing loss and middle ear disease are present in over 90% of individuals diagnosed with fetal alcohol syndrome, and as many as 29% have sensorineural hearing loss (Church and Gerkin 1988; Church and Abel 1998). Malformation of the inner ear may be more common in fetal alcohol syndrome than in the general population (personal experience).

Evaluation

- Routine audiological evaluation is recommended from as early an age as possible. Brain stem auditory evoked response testing between 6 and 12 months may be helpful in early identification of hearing loss. Annual follow-up is recommended for the first three years of life.
- The presence of cleft palate and/or recurrent otitis media requires more frequent assessments of hearing.

Management

- Prompt early treatment of middle ear disease and recurrent otitis media may prevent further hearing impairment. Treatment is by standard methods.
- The use of hearing aides may benefit individuals with sensorineural hearing loss.
- Seating the child closer to the front of the class may be a useful maneuver for both the issue of possible hearing impairment and for the attentional difficulties that are often present.
- For the hearing-impaired adult, appropriate career and employment planning may be necessary.

Dental

A variety of dental problems is commonly found in affected children and adults. They may occur because of the teratogenic effects of alcohol, poor prenatal and postnatal nutrition, or poor hygiene. Dental caries is frequent. Children with fetal alcohol syndrome often have narrow maxillary dental arches and have class III occlusion with maturity (Barnett and Schusterman 1985). Other abnormalities may include small teeth, absent teeth, hypoplastic enamel, displaced or rotated teeth, cross bite, overbite, and delayed emergence of adult dentition (Church et al. 1997).

Evaluation

- Early assessment of teeth is recommended.
- Six-month routine follow-up is appropriate.
- Orthodontic evaluation and follow-up through middle childhood is recommended.

Management

- Parents and/or guardians should emphasize hygiene education to prevent dental decay.
- Prompt treatment of caries is standard.
- Problems identified at the time of transitional dentition may lead to selected dental extractions.

Cardiovascular

Maternal heavy prenatal drinking and binge drinking are strongly associated with an overall increased risk for congenital heart defects (Yang et al. 2015). Previous reports of small numbers of individuals with full-blown fetal alcohol syndrome suggested a risk of 25–50%, with ventricular septal defects being the most common (Tredwell et al. 1982; Streissguth et al. 1985). However, a meta-analysis of 20 studies assessing the association of prenatal alcohol exposure and congenital heart defects found that prenatal alcohol exposure was only significantly associated with the conotruncal defect subtypes, including d-transposition of the great arteries. Infants prenatally exposed to alcohol were found to have a relative risk of 1.64 for having d-transposition of the great arteries (Henderson et al. 2007).

Evaluation

- Clinical assessment of cardiovascular function is recommended at the time of diagnosis and at regular health maintenance visits.
- Routine use of echocardiography is not indicated, although children with cardiac murmurs or other signs or symptoms of cardiac disease, growth impairment and/or failure-to-thrive may need to be evaluated by a

pediatric cardiologist for the presence of a congenital heart defect.

Management

- Treatment is determined by the specific defect identified and is standard.

Genitourinary

Renal anomalies have been described in fetal alcohol syndrome, including renal hypoplasia, agenesis, hydronephrosis, and duplex collecting systems (Hofer and Burd 2009). However, Taylor et al. (1994) reported that 3.6% of 84 individuals with fetal alcohol syndrome and alcohol-related neurodevelopmental disorder had renal anomalies, which was no different from the newborn population.

Evaluation

- Children prenatally exposed to ethanol do not need to be screened for renal anomalies.
- In individuals with urinary tract infections, renal ultrasound, and voiding cystourethrogram is recommended.

Management

- Treatment is not different from the general population.

Gastrointestinal and Hepatic

A review of a potential association between gastrointestinal and hepatic anomalies and FASD was inconclusive (Hofer and Burd 2009). The authors documented an increased prevalence of neonatal jaundice in affected individuals. However, Uc et al. (1997) reported on five children with fetal alcohol syndrome and chronic intestinal pseudoobstruction. Although chronic intestinal pseudoobstruction is of heterogeneous etiology, in the case of its co-occurrence with FASD, the disorder is most likely neuropathic, due to a failure of normal innervation of the embryonic gastrointestinal musculature because of a deficiency of neural crest cells.

Evaluation

- Individuals prenatally exposed to alcohol do not need to be routinely screened for hepatic or gastrointestinal anomalies.
- If jaundice is present in the newborn period, it should be evaluated as with any other newborn.
- If an individual with FASD presents with recurrent nausea, vomiting, abdominal distension, pain and constipation, chronic intestinal pseudoobstruction should be considered. Consultation with a surgeon and/or gastroenterologist is indicated.

Management

- Treatment is dependent on the nature of the disorder identified. In the case of chronic intestinal pseudoobstruction, nutritional support to prevent malnutrition is essential. Antibiotics are important to treat bacterial infections. Disorders that may coexist and worsen symptoms of pseudoobstruction (gastroparesis, gastroesophageal reflux, or bacterial overgrowth) need to be identified and treated.

Musculoskeletal

A variety of minor anomalies of the hands, feet and joints occur at a higher rate in individuals with prenatal alcohol exposure, including palmar crease abnormalities (single transverse creases and an acute angulation of the distal palmar crease terminating between the index and middle finger, the "hockey stick crease"), camptodactyly, clinodactyly, tapered fingers with distal hypoplasia, nail hypoplasia, joint limitation, and occasionally, radioulnar synostosis. Vertebral defects occur at increased frequency in this population (Streissguth et al. 1985), and scoliosis without hemivertebrae is observed in about 10–15% of cases. In one study of 38 individuals, 50% showed congenital fusion of two or more cervical vertebrae (Tredwell et al. 1982). Rarely do any of these anomalies lead to functional impairment.

Evaluation

- A careful clinical assessment for musculoskeletal anomalies is indicated.
- Radiographs should be obtained if any finding suggests an underlying skeletal anomaly.

Management

- Management is similar to the general population.

Neurological

A small head circumference at birth is very common in children with prenatal alcohol exposure, and deficient brain growth may continue throughout childhood. A cranial imaging study of a cohort of young adults prenatally exposed to alcohol demonstrated that the absolute volume of nearly every brain region was significantly reduced (Chen et al. 2012). Although no specific anatomic region of the brain is preferentially affected, malformations resulting from migration abnormalities, changes in size and shape of the corpus callosum, cerebellar vermis hypoplasia, and hypoplasia of the basal ganglia and hippocampus have been documented. (Moore et al. 2014). These may present with "hard" neurological signs such as spasticity, asymmetric reflexes or seizures, or, more frequently, "soft" neurological signs such as impaired fine motor skills, poor tandem gait or poor eye-hand coordination. Severe neurological problems are relatively rare (Chudley and Longstaffe 2010). Electroencephalographic abnormalities have been reported in infancy, and a growing body of evidence indicates that epilepsy is a frequent accompaniment of FASD (Bell et al. 2010; Nicita et al. 2014). The possibility of seizures needs to be considered and excluded in individuals with histories suggestive of petit mal, absence or psychomotor seizures.

Evaluation

- Thorough neurological examination is suggested at the time of diagnosis and at routine visits.
- Abnormal neurological findings should lead to appropriate investigations such as neuroimaging and/or electroencephalography.
- Individuals with potential seizure disorders should be referred for neurology consultation.

Management

- Treatment is similar to the general population.

Societal Implications and Prevention

Streissguth et al. (1996) introduced the concept of secondary disabilities, which they defined as those disabilities that an individual is not born with, and that could be ameliorated through better understanding and appropriate interventions. The secondary disabilities included mental health issues (depression, suicide, or psychosis), disrupted school experience, trouble with the law, confinement, inappropriate sexual behavior, alcohol and drug problems, dependent living, and problems with employment. The cost of fetal alcohol syndrome to affected individuals, their families, and society is staggering. In the United States, it was estimated that in 1992 the cost of treating affected infants, children, and adults was over $1.9 billion (Harwood et al. 1998). The lifetime cost per child affected with FAS is estimated to be $1.4 million. Recent data from Canada suggest that the lifetime cost of FASD was estimated at $1 million per case. With an estimated 4000 new cases yearly, this translates to $4 billion annually (Stade et al. 2007). Using standard measures, children and youth with FASD were shown to have a significantly lower health-related quality of life when compared with the general Canadian population (Stade et al. 2006). Mortality rates in children with fetal alcohol syndrome are high (Habbick et al. 1997). Effective prevention strategies must include a variety of approaches involving the general population and targeting high-risk populations.

The implications of an FASD diagnosis extend beyond the affected individual to the entire family. Many mothers, fathers and/or siblings of individuals with FASD are also affected, with this pattern extending through several generations in some families. Public education, warning labels, family support groups, advocacy groups, early childhood intervention programs, specialized educational and career training, addiction counseling and treatment for women, and paraprofessional mentoring programs have been helpful in reducing the birth prevalence of affected individuals and reducing the morbidity from this disorder. An early diagnosis is associated with a lower occurrence of secondary disabilities; thus, more effort must be made to diagnose the children as soon as possible (Chudley and Longstaffe 2010). Conversely, there can be negative consequences of an FASD diagnosis. As opposed to genetically determined disabilities, a diagnosis of an FASD often leads to significant stigmatization of the alcohol-abusing birth mother (Bell et al. 2015; Corrigan et al. 2017). A recent study demonstrated that research participants viewed women who had given birth to a child with an FASD with greater disdain and blame than women with mental illness, substance abuse disorder or a history of incarceration (Corrigan et al. 2017). Therefore, any population-based efforts to reduce the birth prevalence and morbidity of FASD must be linked to vigorous public education programs aimed at reducing the stigma for birth mothers and families of children with FASD.

There are reliable and validated screening questionnaires that can identify pregnant women with high-risk drinking patterns. The questionnaires are simple and are best administered by physicians, midwives, or nurse practitioners to all women before or early in their pregnancies. These include several tests representing acronyms of the questions posed to the women screened (TWEAK, T-ACE, and CAGE), two modifications of the Michigan Alcohol Screening Test [the BMAST (Brief Michigan Alcohol Screening Test) and SMAST (Short Michigan Alcohol Screening Test)] and the AUDIT (a 10-item "Alcohol Use Disorders Identification Test" developed by the World Health Organization). Among the questions posed in these questionnaires are those related to: *tolerance* ("how many drinks does it take to make you high?"); *annoyance* ("does criticism of your drinking annoy you?"); need to *cut down* ("have you felt that you need to decrease your drinking?"); *amnesia* ("have you done or said things while drinking that you later could not remember?"); and *eye opener* ("have you needed a drink first thing in the morning to get rid of a hangover?") (Russell et al. 1994; Bradley et al. 1998). Women who score high in these questionnaires can be offered counseling to help reduce drinking in pregnancy. Primary healthcare providers can identify and intervene on behalf of mothers at risk before they become pregnant, and thus prevent future affected children (Loock et al. 2005). Primary prevention is achievable using a mentoring program and identifying mothers at risk after the birth of an affected child, thereby reducing the chance of birth of another affected child (Clarren and Astley 1998; Astley et al. 2000).

Prevention of fetal alcohol spectrum disorders is a daunting task, but there is hope. The development of community support, drug and alcohol treatment centers, early childhood intervention programs, public education, and targeted intervention efforts all play a role in raising awareness and offering help. Warning labels on alcohol beverages do increase awareness about the risks of taking alcohol in pregnancy, but this must be tied to other strategies to be effective (Greenfield and Kaskutas 1993; Abel 1998). A comprehensive approach to fetal alcohol spectrum disorders prevention efforts in Washington State has resulted in a decrease in the prevalence of maternal use of alcohol during pregnancy and a reduction in the prevalence of fetal alcohol syndrome among foster children (Astley 2004a).

ACKNOWLEDGMENTS

We thank Drs Kenneth Lyons Jones, Anne Streissguth, Jon Aase, and Sterling Clarren who followed Dr David Smith in pioneering the field of fetal alcohol spectrum disorders in North America. We are indebted to Drs Philip May and Christina Chambers for leading the talented multidisciplinary diagnostic teams that performed the international prevalence studies upon which the CoFASP Consensus Clinical Diagnostic Criteria are based, and to Dr Kenneth Warren and his colleagues at NIAAA, who have funded much of the work that comprises the data-driven recommendations in this chapter. We thank Drs Susan Astley and Albert Chudley for the outstanding work they and their colleagues have done in developing diagnostic tools that have helped to delineate the continuum of disabilities associated with fetal alcohol spectrum disorders. Finally, we acknowledge the thousands of children and families in North America and around the world impacted by FASD. Their selfless participation in clinical research will lead to a decreasing prevalence of and improved prognosis for this common spectrum of preventable disabilities.

RESOURCES

Newsletters

FASD-CAN (the Fetal Alcohol Spectrum Disorder Care Action Network)

A New Zealand-based educational and support network for individuals and families with FASD
Email: claire.gyde@fasd-can.org.nz
Website: https://www.fasd-can.org.nz/

NOFAS Circle of Hope/Birth Mothers Network Newsletter

1200 Eton Court, NW, Third Floor, Washington, DC 20007, USA.
Phone: (202) 785-4585
Fax: (202) 466-6456
E-Mail: information@nofas.org
Website: https://www.nofas.org/circle-of-hope-newsletters/

FASD Frontline Newsletter

A Canadian educational and support network for individuals and families with FASD
Edmonton and Area Fetal Alcohol Network Society
Email: edmontonfetalalcoholnetwork@gmail.com
10320 146 Street, Edmonton, AB, T5N 3A2, Canada
Phone: 780-940-7108
Website: https://edmontonfetalalcoholnetwork.org/resources/frontline-newsletters/

Iceberg

An educational newsletter for people concerned about FASD "because the problems we readily see are only the tip of the iceberg."
Fetal Alcohol Syndrome Information Services (FASIS)
PO Box 95597, Seattle, WA 98145-2597, USA
Website: http://fasdiceberg.org/

NOFAS Weekly Roundup

National Organization on Fetal Alcohol Syndrome (NOFAS) News articles, research, community events, policy debates, resources and highlights of FASD.
1200 Eton Court, NW, Third Floor, Washington, DC 20007, USA
Phone: (202) 785-4585
Fax: (202) 466-6456
Email: information@nofas.org

Support Groups and Resources on the Internet

Most states in the United States and provinces and territories in Canada have established support groups for parents and for some professionals. Increasingly, support groups have been or are being developed in some European countries and Australia. This list is not inclusive.

- Canadian Centre on Substance Abuse: http://www.ohpe.ca/node/96
- Fetal Alcohol Spectrum Disorders Center of Excellence: https://www.samhsa.gov/fetal-alcohol-spectrum-disorders-fasd-center
- FAS Community Resource Center: http://www.come-over.to/FASCRC/
- Fetal Alcohol Syndrome Diagnostic Prevention Network: https://depts.washington.edu/fasdpn/
- Fetal Alcohol and Drug Unit: https://depts.washington.edu/fadu/

FASD Online Support Groups and Mail Lists: http://come-over.to/FAS/fasonline.htm

FAFASD: Families Affected by Fetal Alcohol Spectrum Disorder: http://fafasd.org/

Minnesota Organization on Fetal Alcohol Syndrome (MOFAS): https://www.mofas.org/

National Organization on Fetal Alcohol Syndrome (NOFAS): https://www.nofas.org/

National Institute on Alcohol Abuse and Alcoholism, Fetal Alcohol Spectrum Disorders Resources: https://www.niaaa.nih.gov/research/major-initiatives/fetal-alcohol-spectrum-disorders

American Academy of Pediatrics Fetal Alcohol Spectrum Disorders Program: https://www.aap.org/en-us/advocacy-and-policy/aap-health-initiatives/fetal-alcohol-spectrum-disorders-toolkit/Pages/default.aspx

Centers for Disease Control and Prevention, Fetal Alcohol Spectrum Disorders Resources: https://www.cdc.gov/ncbddd/fasd/index.html

University of South Dakota Center for Disabilities-Fetal Alcohol Spectrum Disorders Educational Strategies Handbook: http://www.usd.edu/-/media/files/medicine/center-for-disabilities/handbooks/fasd-handbook.ashx?la=en

The Arc, Fetal Alcohol Spectrum Disorder Resources: https://www.thearc.org/what-we-do/resources/fact-sheets/fetal-alcohol-spectrum-disorder

REFERENCES

Abdelmalik N, van Haelst M, Mancini G, Schrander-Stumpel C, Marcus-Soekarman D, Hennekam R, Cobben JM (2013) Diagnostic outcomes of 27 children referred by pediatricians to a genetics clinic in the Netherlands with suspicion of fetal alcohol spectrum disorders. *Am J Med Genet A* 161A(2):254–60.

Abel EL (1998) Prevention of alcohol abuse-related birth effects—II. Targeting and pricing. *Alcohol* 33:417–420.

Alvik A, Aalen OO, Lindemann R (2013) Early fetal binge alcohol exposure predicts high behavioral symptom scores in 5.5-year-old children. *Alcohol Clinical Exp Res* 37(11):1954–1962.

Arfsten DP, Silbergeld EK, Loffredo CA (2004) Fetal ADH2*3, maternal alcohol consumption, and fetal growth. *Int J Toxicol* 23:47–54.

Astley SJ, Bailey D, Talbot C, Clarren SK (2000) FAS primary prevention through FAS diagnosis. Part I. Identification of high-risk birth mothers through the diagnosis of their children and Part II. A comprehensive profile of 80 birth mothers of children with FAS. *Alcohol* 35:499–508.

Astley SJ (2004) Fetal alcohol syndrome prevention in Washington State: Evidence of success. *Paediatr Perinat Epidemiol* 18:344–351.

Astley SJ (2013) Validation of the fetal alcohol spectrum disorder (FASD) 4-Digit Diagnostic Code. *J Popul Ther Clin Pharmacol* 20(3):e416.

Balaraman S, Schafer JJ, Tseng AM, Wertelecki W, Yevtushok L, Zymak-Zakutnya N, Chambers CD, Miranda RC (2016) Plasma miRNA profiles in pregnant women predict infant outcomes following prenatal alcohol exposure. *PloS One* 11(11):e0165081.

Barnett R, Schusterman S (1985) Fetal alcohol syndrome: Review of literature and report of cases. *J Am Dent Assoc* 111:591–593.

Bell E, Andrew G, DiPietro N, Chudley AE, Reynolds JN, Racine E (2015) It's a shame! Stigma against fetal alcohol spectrum disorder: Examining the ethical implications for public health practices and policies. *Public Health Ethics* 9:65–77.

Bell SH, Stade B, Reynolds JN, Rasmussen C, Andrew G, Hwang PA, Carlen PL (2010) The remarkably high prevalence of epilepsy and seizure history in fetal alcohol spectrum disorders. *Alcohol Clin Exp Res* 34(6):1084–1089

Benoit T, Bowes M.D, Bowman N, Cantin D, Chudley AE, Crolly D, Livingston A, Longstaffe S, Marles S, Miller C, Millar M, Penko M, Prasad C, Riguidel J, Wincott L (2002) Telemedicine diagnosis for fetal alcohol syndrome—The Manitoba experience. *Paediatr Child Health* 7:147–151.

Berg S, Kindsy K, Lutke J, and Wheway D (1995) *A Layman's Guide to Fetal Alcohol Syndrome and Possible Fetal Alcohol Effects*, Surrey, British Columbia: The FAS/E Support Network.

Boland FJ, Chudley AE, Grant BA (2002) The challenge of fetal alcohol syndrome in adult offender populations. *Forum Correc Res* 14:61–65.

Bradley KA, Boyd-Wickizer J, Powell SH, Burman ML (1998) Alcohol screening questionnaires in women: A critical review. *JAMA* 280:166–171.

Brien JF, Loomis CW, Tranmer J, McGrath M (1983) Disposition of ethanol in human maternal venous blood and amniotic fluid. *Am J Obstet Gynecol* 146:181–186.

Brien JF, Clarke DW, Richardson B, Patrick J (1985) Disposition of ethanol in maternal blood, fetal blood, and amniotic fluid of third-trimester pregnant ewes. *Am J Obstet Gynecol* 152:583–590.

Carr JL, Agnihotri S, Keightley M (2010) Sensory processing and adaptive behavior deficits of children across the fetal alcohol spectrum disorder continuum. *Alcohol Clin Exp Res* 34(6):1022–1032.

Carter RC, Jacobson JL, Molteno CD, Dodge NC, Meintjes EM, Jacobson SW (2016) Fetal growth restriction and cognitive impairment. *Pediatrics* 138(2):e20160775.

Cartwright MM, Smith SM (1995a) Increased cell death and reduced neural crest cell numbers in ethanol-exposed embryos: Partial basis for the fetal alcohol syndrome phenotype. *Alcohol Clin Exp Res* 19:378–386.

Cartwright MM, Smith SM (1995b) Stage-dependent effects of ethanol on cranial neural crest cell development: Partial basis for the phenotypic variations observed in fetal alcohol syndrome. *Alcohol Clin Exp Res* 19:1454–1462.

Casey BJ, Getz S, Galvan A (2008) The adolescent brain. *Dev Rev* 28(1):62–77.

Centers for Disease Control (1998) Preventing secondary conditions in children with fetal alcohol syndrome. Available online at: http://www.cdc.gov/nceh/programs/fas/factsheets/secondary/faqfas.htm

Chasnoff IJ, Wells AM, King L (2015) Misdiagnosis and missed diagnoses in foster and adopted children with prenatal alcohol exposure. *Pediatrics* 135(2):264–270.

Chen ML, Olson HC, Picciano JF, Starr JR, Owens J (2012a) Sleep problems in children with fetal alcohol spectrum disorders. *J Clin Sleep Med* 8(4):421–429.

Chen X, Coles CD, Lynch ME, Hu X (2012b) Understanding specific effects of prenatal alcohol exposure on brain structure in young adults. *Human Brain Mapping* 33(7):1663–76.

Chiodo LM, Delaney-Black V, Sokol RJ, Janisse J, Pardo Y, Hannigan JH (2014) Increased cut-point of the TACER-3 screen reduces false positives without losing sensitivity in predicting risky alcohol drinking in pregnancy. *Alcohol Clin Exp Res* 38(5):1401–1408.

Chudley AE, Conry J, Cook JL, Loock C, Rosales T, LeBlanc N, Public Health Agency of Canada's National Advisory Committee on Fetal Alcohol Spectrum Disorder (2005) Fetal alcohol spectrum disorder: Canadian guidelines for diagnosis. *CMAJ* 172 (Suppl 5):S1–S21.

Chudley AE, Longstaffe SE (2010) Fetal alcohol syndrome and fetal alcohol spectrum disorder. In SB. Cassidy, JE Allanson (Ed.), *Management of Genetic Syndromes* (3rd ed, pp. 363–380) Hoboken, NJ: John Wiley and Sons, Inc.

Church MW, Eldis F, Blakley BW, Bawle EV (1997) Hearing, language, speech, vestibular, and dentofacial disorders in fetal alcohol syndrome. *Alcohol Clin Exp Res* 21:227–37.

Church MW, Abel EL (1998) Fetal alcohol syndrome. Hearing, speech, language, and vestibular disorders. *Obstet Gynecol Clin North Am* 25:85–97.

Church MW, Gerkin KP (1988) Hearing disorders in children with fetal alcohol syndrome: Findings from case reports. *Pediatrics* 82:147–154.

Clarren SK, Astley SJ (1998) Identification of children with fetal alcohol syndrome and opportunity for referral of their mothers for primary prevention—Washington (1993–1997) *MMWR Morb Mortal Wkly Rep* 47:861–864.

Cohen-Kerem R, Koren G (2003) Antioxidants and fetal protection against ethanol teratogenicity. I. Review of the experimental data and implications to humans. *Neurotoxicol Teratol* 25:1–9.

Coles CD, Smith IE, Falek A (1987) Prenatal alcohol exposure and infant behavior: Immediate effects and implications for later development. *Adv alcohol Subst Abuse* 6(4):87–104.

Coles CD, Kable JA, Drews-Botsch C, Falek A (2000) Early identification of risk for effects of prenatal alcohol exposure. *J Stud Alcohol* 61(4):607–616.

Cone-Wesson B (2005) Prenatal alcohol and cocaine exposure: Influences on cognition, speech, language and hearing. *J Commun Disord* 38(4):279–302.

Cook JL, Green CR, Lilley CM, Anderson SM, Baldwin ME, Chudley AE, Conry JL, LeBlanc N, Loock CA, Lutke J, Mallon BF, McFarlane AA, Temple VK, Rosales T (2016) Fetal alcohol spectrum disorder: A guideline for diagnosis across the lifespan. *CMAJ* 188(3):191–7.

Corrigan PW, Lara JL, Shah BB, Mitchell KT, Simmes D, Jones KL (2017) The public stigma of birth mothers of children with fetal alcohol spectrum disorders. *Alcohol Clin Exp Res* 41(6):1166–1173.

Cummings AM, Kavlock RJ (2004) Gene-environment interactions: A review of effects on reproduction and development. *Crit Rev Toxicol* 34:461–485.

Dou X, Menkari CE, Shanmugasundararaj S, Miller KW, Charness ME (2011) Two alcohol binding residues interact across a domain interface of the L1 neural cell adhesion molecule and regulate cell adhesion. *J Biol Chem* 286(18):16131–16139.

Douzgou S, Breen C, Crow YJ, Chandler K, Metcalfe K, Jones E, Kerr B, Clayton-Smith J (2012) Diagnosing fetal alcohol syndrome: New insights from newer genetic technologies. *Arch Dis Child*, 97(9):812–817.

Doyle LR, Mattson SN (2015) Neurobehavioral disorder associated with prenatal alcohol exposure (ND-PAE): Review of evidence and guidelines for assessment. *Curr Dev Disord Rep* 2(3):175–186.

Fast DK, Conry JL, Loock CA (1999) Identifying fetal alcohol syndrome (FAS) among youth in the criminal justice system. *J Dev Behav Pediatr* 20:370–372.

Fitzgerald DM, Charness ME, Leite-Morris KA, Chen S (2011) Effects of ethanol and NAP on cerebellar expression of the neural cell adhesion molecule L1. *PloS One* 6(9):e24364.

Flentke GR, Klingler RH, Tanguay RL, Carvan M, J 3rd, Smith SM (2014) An evolutionarily-conserved mechanism of calcium-dependent neurotoxicity in a zebrafish model of FASD. *Alcohol Clin Exp Res* 38(5):1255–65.

Frankel F, Paley B, Marquardt R, O'Connor M (2006) Stimulants, neuroleptics and children's friendship training for children with fetal alcohol spectrum disorders. *J Child Adolesc Psychopharmacol* 16:777–789.

Fuglestad AJ, Whitley ML, Carlson SM, Boys CJ, Eckerle JK, Fink BA, Wozniak JR (2015a) Executive functioning deficits in preschool children with fetal alcohol spectrum disorders. *Child Neuropsychol* 21(6):716–731.

Gautam P, Lebel C, Narr KL, Mattson SN, May PA, Adnams CM, Riley EP, Jones KL, Kan EC, Sowell ER (2015) Volume changes and brain-behavior relationships in white matter and subcortical gray matter in children with prenatal alcohol exposure. *Hum Brain Mapp* 36(6):2318–2329.

Glass L, Graham DM, Deweese BN, Jones KL, Riley EP, Mattson SN (2014) Correspondence of parent report and laboratory measures of inattention and hyperactivity in children with heavy prenatal alcohol exposure. *Neurotoxicol Teratol* 42:43–50.

Goh YI, Chudley AE, Clarren S, Koren G, Orrbine E, Rosales T, Rosenbaum C (2008) Development of Canadian screening tools for fetal alcohol syndrome. *Can J Clin Pharmacol* 15:e344–e366.

Greenfield T, Kaskutas LA (1993) Early impacts of alcoholic beverage warning labels: National study findings relevant to drinking and driving behavior. *Saf Sci* 16:689–707.

Habbick BF, Nanson JL, Snyder RE, Casey RE (1997) Mortality in fetal alcohol syndrome. *Can J Pub Health* 88:181–183.

Hagan JF, Balachova T, Bertrand J, Chasnoff I, Dang E, Fernandez-Baca D, Kable J, Kosofsky B, Senturias YN, Singh N, Sloane M, Weitzman C, Zubler J (2016) Neurobehavioral disorder associated with prenatal alcohol exposure. *Pediatrics* 138(4):e20151553.

Harwood H, Fountain D, Livermore G (1998) Economic costs of alcohol abuse and alcoholism. *Recent Dev Alcohol* 14:307–30.

Heller M, Burd L (2014) Review of ethanol dispersion, distribution, and elimination from the fetal compartment. *Birth Defects Res A* 100:277–283.

Henderson J, Gray R, Brocklehurst P (2007) Systematic review of effects of low-moderate prenatal alcohol exposure on pregnancy outcome. *BJOG* 114:243–252.

Hofer R, Burd L (2009) Review of published studies of kidney, liver, and gastrointestinal birth defects in fetal alcohol spectrum disorders. *Birth Defects Res A* 85:179–183.

Hoyme HE, May PA, Kalberg WO, Kodituwakku P, Gossage JP, Trujillo PM, Buckley DG, Miller JH, Aragon AS, Khaole N, Viljoen DL, Jones KL, Robinson LK (2005) A practical clinical approach to diagnosis of fetal alcohol spectrum disorders: Clarification of the 1996 Institute of Medicine criteria. *Pediatrics* 115:39–47.

Hoyme HE, Hoyme DB, Elliott AJ, Blankenship J, Kalberg WO, Buckley D, Abdul-Rahman O, Adam MP, Robinson LK, Manning M, Bezuidenhout H, Jones KL, May P.A (2015) A South African mixed race lip/philtrum guide for diagnosis of fetal alcohol spectrum disorders. *Am J Med Genet A* 167A(4):752–5.

Hoyme HE, Kalberg WO, Elliott AJ, Blankenship J, Buckley D, Marais AS, Manning MA, Robinson LK, Adam MP, Abdul-Rahman O, Jewett T, Coles CD, Chambers C, Jones KL, Adnams CM, Shah PE, Riley EP, Charness ME, Warren KR, May PA (2016) Updated clinical guidelines for diagnosing fetal alcohol spectrum disorders. *Pediatrics* 138(2):e20154256.

Jacobson JL, Jacobson SW, Sokol RJ, Martier SS, Ager JW, Kaplan-Estrin MG (1993) Teratogenic effects of alcohol on infant development. *Alcohol Clin Exp Res* 17(1) 174–183.

Jacobson SW, Jacobson JL, Sokol RJ (1994) Effects of fetal alcohol exposure on infant reaction time. *Alcohol Clin Exp Res* 18(5):1125–1132.

Jacobson SW, Chiodo LM, Sokol RJ, Jacobson JL (2002) Validity of maternal report of prenatal alcohol, cocaine, and smoking in relation to neurobehavioral outcome. *Pediatrics*, 109(5): 815–825.

Jones KL, Smith DW (1973) Recognition of the fetal alcohol syndrome in early infancy. *Lancet* 7836:999–1001.

Jones KL, Smith DW, Ulleland CN, Streissguth P (1973) Pattern of malformation in offspring of chronic alcoholic mothers. *Lancet* 7815:1267–1271.

Jones KL, Robinson LK, Bakhireva LN, Marintcheva G, Storojev V, Strahova A, Sergeevskaya S, Budantseva S, Mattson SN, Riley EP, Chambers CD (2006) Accuracy of the diagnosis of physical features of fetal alcohol syndrome by pediatricians after specialized training. *Pediatrics* 118(6):e1734–e1738.

Jones KL, Hoyme HE, Robinson LK, Del Campo M, Manning MA, Prewitt LM, Chambers CD (2010) Fetal alcohol spectrum disorders: Extending the range of structural defects. *Am J Med Genet A* 152A (11):2731–2735.

Kable JA, O'Connor MJ, Olson HC, Paley B, Mattson SN, Anderson SM, Riley EP (2016) Neurobehavioral disorder associated with prenatal alcohol exposure (ND-PAE): Proposed DSM-5 diagnosis. *Child Psychiatry Hum Dev* 47(2):335–346.

Kable JA, Mukherjee RA (2017) Neurodevelopmental disorder associated with prenatal exposure to alcohol (ND-PAE): A proposed diagnostic method of capturing the neurocognitive phenotype of FASD. *Eur J Med Genet* 60(1):49–54.

Kalberg WO, Provost B, Tollison SJ, Tabachnick BG, Robinson LK, Hoyme HE, Trujillo PM, Buckley D, Arargon AS, May PA (2006) Comparison of motor delays in young children with fetal alcohol syndrome to those with prenatal alcohol exposure and with no prenatal alcohol exposure. *Alcohol Clin Exp Res* 30(12):2037–2045.

Kalberg W, Buckley D (2006) Educational planning for children with fetal alcohol syndrome. *Ann 1st Super Sanita* 42:58–66.

Kerns KA, Siklos S, Baker L, Muller U (2016) Emotion recognition in children with Fetal Alcohol Spectrum Disorders. *Child Neuropsychol* 22(3):255–275.

Knudsen AK, Skogen JC, Ystrom E, Sivertsen B, Tell GS, Torgersen L (2014) Maternal pre-pregnancy risk drinking and toddler behavior problems: The Norwegian mother and child cohort study. *Eur Child Adolesc Psychiatry* 23(10):901–911.

Koren G, Nulman I, Chudley AE, Loock C (2003) Fetal alcohol spectrum disorder. *CMAJ* 169:1181–1185.

Kully-Martens K, Denys K, Treit S, Tamana S, Rasmussen C (2012) A review of social skills deficits in individuals with fetal alcohol spectrum disorders and prenatal alcohol exposure: Profiles, mechanisms, and interventions. *Alcohol Clin Exp Res* 36(4):568–576.

Landgren M, Svensson L, Strömland K, Andersson Grönlund M (2010) Prenatal alcohol exposure and neurodevelopmental disorders in children adopted from eastern Europe. *Pediatrics* 125(5):e1178–85.

Lange S, Shield K, Rehm J, Popova S (2013) Prevalence of fetal alcohol spectrum disorders in child care settings: A Meta-analysis. *Pediatrics* 132(4):e980–e995.

Lange S, Shield K, Koren G, Rehm J, Popova S (2014) A comparison of the prevalence of prenatal alcohol exposure obtained via maternal self-reports versus meconium testing: A systematic literature review and meta-analysis. *BMC Pregnancy Childbirth* 14:127.

Leibson T, Neuman G, Chudley AE, Koren G (2014) The differential diagnosis of fetal alcohol spectrum disorder. *J Popul Ther Clin Pharmacol* 21(1):e1–e30.

Lemoine G, Harousseau H, Borteyru J P, Menuet JC (1968) Les enfants de parents alcooliques: Anomalies observées, à propos de 127 cas. *Ouest Med* 21:476–482.

Lewis CE, Thomas KG, Molteno CD, Kliegel M, Meintjes EM, Jacobson JL, Jacobson SW (2016) Prospective memory impairment in children with prenatal alcohol exposure. *Alcohol Clin Exp Res* 40(5):969–978.

Loock C, Conry J, Cook JL, Chudley AE, Rosales T (2005) Identifying fetal alcohol spectrum disorder in primary care. *CMAJ*, 172:628–630.

Lossie AC, Muir WM, Lo CL, Timm F, Liu Y, Gray W, Zhou FC (2014) Implications of genomic signatures in the differential vulnerability to fetal alcohol exposure in C57BL/6 and DBA/2 mice. *Front Genet* 5:173.

Lucas BR, Latimer J, Pinto RZ, Ferreira ML, Doney R, Lau M, Jones T, Dries D, Elliott EJ (2014) Gross motor deficits in children prenatally exposed to alcohol: A meta-analysis. *Pediatrics* 134(1):e192–209.

Lussier AA, Weinberg J, Kobor MS (2017) Epigenetics studies of fetal alcohol spectrum disorder: Where are we now? *Epigenomics* 9(3):291–311.

Makelarski JA, Romitti PA, Sun L, Burns TL, Druschel CM, Suarez L, Olshan AF, Siega-Riz AM, Olney RS; National Birth Defects Prevention Study (2013) Periconceptional maternal alcohol consumption and neural tube defects. *Birth Defects Research A* 97(3):152–160.

Manning MA, Hoyme HE (2007) Fetal alcohol spectrum disorders: A practical clinical approach to diagnosis. *Neurosci Biobehavl Rev* 31(2) :230–238.

Mattson SN, Roebuck T (2002) Acquisition and retention of verbal and nonverbal information in children with heavy prenatal alcohol exposure. *Alcohol Clin Exp Res* 26:875–882.

May PA, Fiorentino D, Gossage JP, Kalberg WO, Hoyme HE, Robinson LK, Coriale G, Jones KL, del Campo M, Tarani L, Romeo M, Kodituwakku PW, Deiana L, Buckley D, Ceccanti M (2006) Epidemiology of FASD in a province in Italy: Prevalence and characteristics of children in a random sample of schools. *Alcohol Clin Exp Res* 30:1562–1575.

May PA, Gossage JP, Kalberg WO, Robinson LK, Buckley D, Manning MA, Hoyme HE (2009) Prevalence and epidemiologic characteristics of FASD from various research methods with an emphasis on recent in-school studies. *Dev Disabil Res Rev* 15:176–192.

May PA, Blankenship J, Marais AS, Gossage JP, Kalberg WO, Joubert B, Cloete M, Barnard R, De Vries M, Hasken J, Robinson LK, Adnams CM, Buckley D, Manning MA, Parry CD, Hoyme HE, Tabachnick B, Seedat S (2013) Maternal alcohol consumption producing fetal alcohol spectrum disorders (FASD): Quantity, frequency, and timing of drinking. *Drug Alcohol Depend* 133(2):502–512.

May PA, Baete A, Russo J, Elliott AJ, Blankenship J, Kalberg WO, Buckley D, Brooks M, Hasken J, Abdul-Rahman O, Adam MP, Robinson LK, Manning MA, Hoyme HE (2014) Prevalence and characteristics of fetal alcohol spectrum disorders. *Pediatrics* 134(5):855–866.

May PA, Chambers CD, Kalberg WO, Zellner J, Feldman H, Buckley D, Kopald D, Hasken JM, Xu R, Honerkamp-Smith G, Taras H, Manning MA, Robinson LK, Adam MP, Abdul-Rahman O, Vaux K, Jewett T, Elliott AJ, Kable JA, Akshoomoff N, Falk D, Arroyo JA, Hereld D, Riley EP, Charness M, Coles CD, Warren KR, Jones KL, Hoyme H.E (2018) Prevalence of fetal alcohol spectrum disorders in 4 US communities. *JAMA* 319(5):474–482.

McGee CL, Bjorkquist OA, Riley EP, Mattson SN (2009) Impaired language performance in young children with heavy prenatal alcohol exposure. *Neurotoxicol Teratol* 31(2):71–75.

Miller LC, Chan W, Litvinova A, Rubin A, Tirella L, Cermak S (2007) Medical diagnoses and growth of children residing in Russian orphanages. *Acta Paediatr* 96(12):1765–1769

Molteno CD, Jacobson JL, Carter RC, Jacobson SW (2010) Infant symbolic play as an early indicator of fetal alcohol-related deficit. *Infancy* 15(6):586–607.

Molteno CD, Jacobson JL, Carter RC, Dodge NC, Jacobson SW (2014) Infant emotional withdrawal: A precursor of affective and cognitive disturbance in fetal alcohol spectrum disorders. *Alcohol Clin Exp Res* 38(2):479–488.

Moore EM, Migliorini R, Infante MA, Riley EP (2014) Fetal alcohol spectrum disorders: Recent neuroimaging findings. *Curr Dev Disord Rep* 1(3):161–172.

Moore EM, Riley EP (2015) What happens when children with fetal alcohol spectrum disorders become adults? *Curr Dev Disord Rep* 2(3):219–227.

Nicita F, Verrotti A, Pruna D, Striano P, Capovilla G, Savasta S, Sparta MV, Parisi P, Parlapiano G, Tarani L, Spalice A (2014) Seizures in fetal alcohol spectrum disorders: Evaluation of clinical, electroencephalographic and neuroradiologic features in a pediatric case series. *Epilepsia* 55(6):e60–e66.

Olivier L, Curfs LMG, Viljoen DL (2016) Fetal alcohol spectrum disorders: Prevalence rates in South Africa. *S Afr Med J* 106(6 Suppl 1):S103–106.

O'Connor MJ, Rotheram-Borus MJ, Tomlinson M, Bill C, LeRoux IM, Stewart IM (2014) Screening for fetal alcohol spectrum disorders by nonmedical community workers. *J Popul Ther Clin Pharmacol* 21(3):e442.

O'Leary CM, Taylor C, Zubrick SR, Kurinczuk JJ, Bower C (2013) Prenatal alcohol exposure and educational achievement in children aged 8-9 years. *Pediatrics* 132(2):e468–475.

O'Malley KD, Nanson J (2002) Clinical implications of a link between fetal alcohol spectrum disorder and attention-deficit hyperactivity disorder. *Can J Psychiatry* 47:349–354.

Ornoy A, Ergaz Z (2010) Alcohol abuse in pregnant women: Effects on the fetus and newborn, mode of action and maternal treatment. *Int J Environ Res Public Health* 7:364–379.

Ospina M, Dennett L (2013) *Systematic Review on the Prevalence of Fetal Alcohol Spectrum Disorders*. Institute of Health Economics, Edmonton, Canada.

Pearson MA, Hoyme HE, Seaver LH, Rimsza ME (1994) Toluene embryopathy: Delineation of the phenotype and comparison with fetal alcohol syndrome. *Pediatrics* 93:211–215.

Petković G, Barišić I (2013) Prevalence of fetal alcohol syndrome and maternal characteristics in a sample of schoolchildren from a rural province of Croatia. *Int J Environ Res Public Health* 10(4):1547–1561.

Poitra BA, Marion S, Dionne M, Wilkie E, Dauphinais P, Wilkie-Pepion M, Martsolf JT, Klug MG, Burd L (2003) A school-based screening program for fetal alcohol syndrome. *Neurotoxicol Teratol* 25(6):725–729.

Popova S, Lange S, Bekmuradov D, Mihic A, Rehm J (2011) Fetal alcohol spectrum disorder prevalence estimates in correctional systems: A systematic literature review. *Can J Public Health* 102(5):336–40.

Popova S, Lange S, Probst C, Gmel G, Rehm J (2017) Estimation of national, regional, and global prevalence of alcohol use during pregnancy and fetal alcohol syndrome: a systematic review and meta-analysis. *Lancet Glob Health* 5(3):e290–e299.

Ramanathan R, Wilkemeyer MF, Mittal B, Perides G, Charness ME (1996) Alcohol inhibits cell-cell adhesion mediated by human L1. *J Cell Biol* 133(2):381–390.

Rasmussen C, Becker M, McLennan J, Urichuk L, Andrew G (2011) An evaluation of social skills in children with and without prenatal alcohol exposure. *Child Care Health Dev* 37(5):711–718.

Reid N, Dawe S, Shelton D, Harnett P, Warner J, Armstrong E, LeGros K, O'Callaghan F (2015) Systematic review of fetal alcohol spectrum disorder interventions across the life span. *Alcohol Clin Exp Res* 39(12):2283–2295.

Roozen S, Gjalt-Jorn PY, Kok G, Townend D, Nilhuis J, Leopold C (2016) Worldwide prevalence of fetal alcohol spectrum disorders: A systematic literature review including meta-analysis. *Alcohol Clin Exp Res* 40(1):18–22.

Roussotte FF, Sulik KK, Mattson SN, Riley EP, Jones KL, Adnams CM, May PA, O'Connor MJ, Narr KL, Sowell ER (2012) Regional brain volume reductions relate to facial dysmorphology and neurocognitive function in fetal alcohol spectrum disorders. *Hum Brain Mapp* 33(4):920–937.

Russell M, Martier SS, Sokol RJ, Mudar P, Bottoms S, Jacobson S, Jacobson J (1994) Screening for pregnancy risk-drinking. *Alcohol Clin Exp Res* 18:1156–1161.

Sadler TW (2015) *Langman's Medical Embryology*, 13th ed. Wolters Kluwer Health, Philadelphia.

Sobell LC, Agrawal S, Annis H, Ayala-Velazquez H, Echeverria L, Leo GI, Rybakowski JK, Sandahl C, Saunders B, Thomas S, Zióikowski M (2001) Cross-cultural evaluation of two drinking assessment instruments: Alcohol timeline followback and inventory of drinking situations. *Subst Use Misuse* 36(3): 313–331.

Sokol RJ, Delaney-Black V, Nordstrom B (2003) Fetal alcohol spectrum disorder. *JAMA* 290:2996–2999.

Stade BC, Stevens B, Ungar WJ, Beyene J, Koren G (2006) Health-related quality of life of Canadian children and youth prenatally exposed to alcohol. *Health Qual Life Outcomes* 4:81.

Stade B, Ungar WJ, Stevens B, Beyen J, Koren G (2007) Cost of fetal alcohol spectrum disorder in Canada. *Can Fam Physician* 53:1303–1304.

Stephen JM, Kodituwakku PW, Kodituwakku EL, Romero L, Peters AM, Sharadamma NM, Caprihan A, Coffman BA (2012) Delays in auditory processing identified in preschool children with FASD. *Alcohol Clin Exp Res*, 36(10):1720–1727.

Stratton K, Howe C, Battaglia FC, eds (1996) *Fetal Alcohol Syndrome: Diagnosis, Epidemiology, Prevention, and Treatment.* Institute of Medicine, Washington, DC: National Academy Press.

Streissguth AP, Sampson PD, Barr HM (1989) Neurobehavioral dose-response effects of prenatal alcohol exposure in humans from infancy to adulthood. *Ann N Y Acad Sci*, 562(1):145–158.

Streissguth AP, Clarren SK, Jones KL (1985) Natural history of the fetal alcohol syndrome: A 10-year follow-up of eleven patients. *Lancet* 2:85–92.

Streissguth AP, Barr HM, Kogan J, Bookstein FL (1996) *Understanding the Occurrence of Secondary Disabilities in Clients with Fetal Alcohol Syndrome (FAS) and Fetal Alcohol Effects (FAE), final report submitted to Centers for Disease Control*, Seattle, University of Washington Publication Services.

Strömland K, Pinazo-Duran MD (2002) Ophthalmic involvement in the fetal alcohol syndrome: Clinical and animal model studies. *Alcohol Alcohol* 37:2–8.

Strömland K (2004) Visual impairment and ocular abnormalities in children with fetal alcohol syndrome. *Addiction Biology* 9:153–157.

Taylor CL, Jones KL, Jones MC, Kaplan GW (1994) Incidence of renal anomalies in children prenatally exposed to ethanol. *Pediatrics*, 94:209–212.

Tredwell SJ, Smith DF, Macleod PJ, Wood BJ (1982) Cervical spine anomalies in fetal alcohol syndrome. *Spine* 7:331–334.

Tsang TW, Lucas BR, Carmichael Olson H, Pinto RZ, Elliott EJ (2016) Prenatal Alcohol Exposure, FASD, and Child Behavior: A Meta-analysis. *Pediatrics* 137(3):e20152542.

Uc A, Vasiliauskas E, Piccoli DA, Flores AF, DiLorenzo C, Hyman PE Chronic intestinal pseudoobstruction associated with fetal alcohol syndrome. *Dig Dis Sci* 42(6):1163–7.

Underwood MA, Gilbert WM, Sherman MP (2005) Amniotic fluid: Not just fetal urine anymore. *J Perinatol* 25:341–348.

Wattendorf DJ, Muenke M (2005) Fetal alcohol spectrum disorders. *Am Fam Physician* 72(2):279–82.

Yang J, Qiu H, Qu P, Zhang R, Zeng L, Yan H (2015) Prenatal alcohol exposure and congenital heart defects: A meta-analysis. *PLoS One* 10:e0130681.

Zelner I, Koren G (2013) Pharmacokinetics of ethanol in the maternal-fetal unit. *J Popul Ther Clin Pharmacol* 20:e259–e265.

27

FETAL ANTICONVULSANT SYNDROME

ELIZABETH A. CONOVER
Division of Genetic Medicine, Munroe Meyer Institute, University of Nebraska Medical Center, Omaha, Nebraska, USA

OMAR ABDUL-RAHMAN
Division of Genetic Medicine, Munroe Meyer Institute, University of Nebraska Medical Center, Omaha, Nebraska, USA

H. EUGENE HOYME
Departments of Pediatrics and Medicine, The University of Arizona College of Medicine, Tucson, Arizona, USA; Sanford Children's Genomic Medicine Consortium and Sanford Imagenetics, Sanford Health, Sioux Falls, South Dakota, USA

INTRODUCTION

An estimated 0.3–0.5% of pregnant women have epilepsy, and most of those women are treated with antiepileptic drugs. Increasingly, antiepileptic drugs (AEDs) are also used for psychiatric conditions such as bipolar depression and to treat migraines and chronic pain, resulting in a much larger group of women exposed to these potentially teratogenic medications. Most of the data on the teratogenicity of AEDs involves the first generation antiepileptic medications (phenytoin, phenobarbital, carbamazepine and valproate). These first generation antiepileptic medications may be associated with a two- to three-fold increase in the risk for major malformations, an increased risk for adverse cognitive and behavioral outcomes and a recognizable pattern of major and minor malformations (Samren et al. 1999; Holmes 2002; Tomson and Battino 2012).

More limited data regarding teratogenicity are available for the newer second generation antiepileptic medications; however, they appear to be associated with a lower teratogenic risk. In general, the risk for major and minor malformations is low, and an associated recognizable malformation syndrome has not been observed. Among the second generation AEDs are gabapentin, levetiracetam, oxcarbazepine, and lamotrigine. The exception to the apparently low teratogenic risk is topiramate. There are also a number of new antiepileptic medications for which there are almost no data on use by pregnant women, and for which the risk for teratogenicity is unknown (vigabatrin, felbamate, tiagabine, zonisamide, pregabalin, rufinamide, and lacosamide). Although this chapter provides an overview of the findings in individuals who have been exposed to antiepileptic medications in utero, it should be noted that each drug can have significant differences in the presentation of the condition and may not always conform with some of the general concepts presented. Specific information relating to each drug is provided in the corresponding section. Management strategies will also be provided in a general manner, with particular issues related to a specific drug where appropriate.

Fetal anticonvulsant syndrome refers to the major malformations and the characteristic pattern of minor anomalies identified in some of the children exposed to AEDs (Holmes et al. 2001; Holmes 2002). The possibility that these medications were teratogens was first suggested by Meadow (1968) in a letter to the *Lancet*. In 1972, Speidel and Meadow reported an increased incidence of major congenital malformations and hemorrhagic disease of the newborn in children exposed to AEDs in utero. They suggested a multifactorial etiology, possibly resulting from interference with folate metabolism, as well as genetic and environmental factors (Speidel and Meadow 1972). Following the description of an increased birth prevalence of major malformations in children exposed to AEDs in utero, several groups described distinctive patterns of minor anomalies attributable to exposure to a particular antiepileptic drug (Hanson and Smith 1975;

Cassidy and Allanson's Management of Genetic Syndromes, Fourth Edition.
Edited by John C. Carey, Agatino Battaglia, David Viskochil, and Suzanne B. Cassidy.
© 2021 John Wiley & Sons, Inc. Published 2021 by John Wiley & Sons, Inc.

DiLiberti et al. 1984; Jones et al. 1989, 1992). In this chapter, we review the major malformations and minor anomalies found in some children exposed to phenytoin, carbamazepine, valproate, and phenobarbital in utero. When available, information regarding the teratogenicity of newer antiepileptic drugs will also be discussed. Conclusions from studies of the teratogenicity of individual AEDs are complicated by the difficulty of controlling for multiple variables and the need for a large sample size. Often, multiple agents are used to control a seizure disorder or other illness. In addition, it is difficult to control for type of seizure disorder, seizure frequency, family history of birth defects, environmental and socioeconomic factors, and for maternal and fetal genetic factors that may influence the teratogenicity of a particular antiepileptic drug (Seaver and Hoyme 1992). The role of maternal epilepsy itself in the pathogenesis of fetal anticonvulsant syndrome is a confounder that is problematic. A study by Holmes and colleagues suggested that it is antiepileptic drug use, and not maternal epilepsy, that results in fetal anticonvulsant syndrome (Holmes et al. 2001). Researchers have attempted to isolate the role of medication exposures as a cause of adverse effects on the fetus by using appropriate control groups. In some studies, women who have epilepsy but are not treated with antiepileptic medications are used as controls. However, these women are likely to have less severe seizure disorders. In other studies, pregnancies in which women are treated for epilepsy are compared to pregnancies in which women take antiepileptic medications for other indications such as bipolar depression. Again, this is an imperfect control group since it introduces other variables.

Many of the initial publications on antiepileptic medications were case reports. They drew attention to the potential risk of antiepileptic medications, but were susceptible to bias of ascertainment. The need to further explore the hypotheses introduced by the case reports led to retrospective studies. These introduced issues such as inaccuracies in reporting of the timing of the exposure and dosages of medications. More recently, pregnancy registries have been developed to prospectively identify a large number of women taking AEDs in pregnancy. The North American Antiepileptic Drug Pregnancy Registry (NAREP) and the UK Epilepsy and Pregnancy Registry were launched in 1996–1997. The European and International Registry of Antiepileptic Drugs in Pregnancy (EURAP), an international registry that now includes 42 countries, originated in 1999, and as of 2018 includes contributions from countries in Europe, Asia, Africa, and Latin America. Such programs are helping to clarify many outstanding issues, including the comparative role of individual antiepileptic drugs in producing anomalies, the teratogenicity of newer agents, the incidence of minor anomalies and the long-term developmental outcome of exposed children (Meador et al. 2008; Hernandez-Diaz et al. 2012; Tomson et al. 2018).

Diagnostic Criteria

Diagnostic criteria for fetal anticonvulsant syndrome have been suggested (Dean et al. 2000), but are not widely employed. There are no diagnostic criteria for the individual antiepileptic drug embryopathies; however, patterns of anomalies observed in children who have been exposed in utero to specific antiepileptic drugs have been well described. There is an overlap in the features observed in the individual antiepileptic drug syndromes, which suggests that "fetal anticonvulsant syndrome" is a more accurate term, but historically the antiepileptic drug embryopathies were described individually. Some dysmorphic features may be more common with particular agents, such as a small mouth with fetal valproate exposure.

Etiology, Pathogenesis, and Genetics

Teratogens are agents that alter the developmental program of the embryo, producing a fetus or a newborn with abnormal structure and/or neurocognition. Whether an agent is teratogenic depends crucially on the timing of the exposure in gestation. The most sensitive period of embryonic development is the period of organogenesis, 18–60 days after conception. The phase before organogenesis has traditionally been termed the "all-or-none" period during which the cells are pluripotent. During this early stage of development, loss of a few cells has no subsequent deleterious effect; however, loss of many cells results in loss of the embryo. In recent years there has been discussion regarding whether embryos exposed to teratogenic agents during the "all or none period," but that are not miscarried, are always free of damage. The timing of antiepileptic drug exposure is crucial, as the most sensitive period is 3–4 weeks' gestation for neural tube defects, 4–8 weeks' gestation for congenital heart defects, and 6–10 weeks' gestation for orofacial clefts.

Potentiation, the ability of one medication to enhance the teratogenicity of another medication, is well known with some of the antiepileptic medications. For example, in one study, phenytoin monotherapy had a risk of 7.36% for congenital malformations; when used in polytherapy with another antiepileptic medication, the risk increased to 11.46%. Potentiation is especially evident with valproate, which had a 10.73% risk for malformations as monotherapy, but which increased to 25% when two more antiepileptic medications were added (Meador et al. 2008). Overall data on over 12,000 pregnancies in the EURAP database as of 2017 found the risk of malformations to be 4.6% with monotherapy and 6.9% with polytherapy (Battino and Tomson 2017).

The dose of the agent is also vitally important, as there may be a threshold below which no effect is observed, and for many antiepileptic medications there is increasing effect with increasing doses. For example, a study analyzing data from the EURAP epilepsy and pregnancy registry found an increase in malformation rates with increasing doses of lamotrigine, valproate, carbamazepine, and phenobarbital (Tomson et al. 2011). A Cochrane review of monotherapy treatment in pregnancy found a dose–response relationship for valproate, but not for other antiepileptic medications (Bromley et al. 2016). Other important factors that influence teratogenicity are the route of administration of the agent, the modifying effects of exposure to other agents or environmental factors, and the genetic background of the mother and fetus (Seaver and Hoyme 1992).

The mechanism(s) of teratogenesis of phenytoin, carbamazepine, valproate and phenobarbital are unclear and likely to be complex. Two hypotheses were initially proposed and are substantiated by experimental evidence in humans and in animal models; they are not mutually exclusive. One hypothesis first proposed by Speidel and Meadow in 1972 suggested that alterations in folate metabolism, caused by the antiepileptic drug and perhaps exacerbated by maternal or fetal genetic factors, is teratogenic. Evidence for this mechanism includes the protective effect of folate supplementation during pregnancy in preventing neural tube defects, the observation that serum folate levels are reduced in women taking AEDs in pregnancy, and the positive correlation between low maternal folate levels in women taking AEDs and major malformations in their children (Kaaja et al. 2003). Experimental evidence from animal models suggests that valproate alters folate metabolism and is associated with neural tube defects, whereas a closely related compound that does not alter folate metabolism is not associated with neural tube defects (Wegner and Nau 1992). An argument against this hypothesis, at least in the case of valproate, is that maternal supplementation with folic acid prior to and during the first trimester has not been demonstrated to reduce the risk for valproate-associated neural tube defects (Ban et al. 2015; George 2017).

Another proposed mechanism of antiepileptic drug teratogenesis is altered hydantoin metabolism, with excessive arene oxide production because of low activity of epoxide hydrolase. Phenytoin is metabolized to arene oxide by cytochrome p450, and arene oxide is metabolized to diol by epoxide hydrolase. Low levels of epoxide hydrolase have been identified in children with fetal anticonvulsant syndrome, but not in controls (Buehler et al. 1994). Amniocytes were obtained from women taking phenytoin alone, and epoxide hydrolase activity was used to predict the fetuses at risk of fetal hydantoin syndrome. Four infants had low epoxide hydrolase activity in amniocytes and were predicted to be at risk of fetal hydantoin syndrome; all four were identified after birth as having fetal hydantoin syndrome. The other 15 children, with higher levels of epoxide hydrolase activity in amniocytes, were apparently unaffected (Buehler et al. 1994). As one of the metabolites of carbamazepine is an arene oxide, teratogenicity of this antiepileptic drug could also be related to high levels of arene oxides (Buehler et al. 1994). Since the P450 enzyme epoxide hydrolase is inherited in an autosomal recessive manner, this may provide an example of genetic susceptibility and explain why women who have had one affected child are at risk for another affected child.

Neurobehavioral effects have been hypothesized to be related to antiepileptic medication-induced apoptosis, altered synaptogenesis, and changed neurotransmitter environment. This is similar to what has been seen with prenatal alcohol exposure, and reported in studies involving rodents and non-human primates. These mechanisms may be different from the mechanisms that cause morphological damage (Velez-Ruiz 2015).

Tung and Winn proposed that epigenetic changes such as histone methylation and DNA methylation might be contributing to the teratogenic mechanism of valproate (Tung and Winn 2010).

Some of the adverse effects on the infant attributed to maternal epilepsy or antiepileptic medications may be related to obstetrical factors. Women with epilepsy are known to be at an increased risk for adverse pregnancy and perinatal outcomes including preeclampsia, placental abruption, stillbirth, preterm birth, small for gestational age (SGA) babies, and infants with lower Apgar scores. Other studies suggest that women taking anticonvulsant medications during pregnancy, regardless of whether the indication is epilepsy or for mental health or pain treatment, are also at increased risk of having babies who are SGA and of delivering prematurely. The risk varies with the agent, with topiramate, phenobarbital, and zonisamide causing the most significant increase in the prevalence of SGA, and lamotrigine and phenytoin causing the least (Hernandez-Diaz 2017).

Diagnostic Testing

There is no diagnostic testing for fetal anticonvulsant syndrome. The diagnosis is made clinically, based on the history of the antiepileptic drug exposure, the presence of a characteristic pattern of malformations as detailed above, and the absence of other features suggesting an alternate diagnosis. Despite the high index of suspicion associated with prenatal exposure to the first generation anticonvulsants, it cannot be assumed that this is the etiology of the child's issues. Ideally, a clinical geneticist/dysmorphologist or other physician familiar with the features of antiepileptic drug embryopathies should evaluate the exposed child, both to confirm the concordance of the features of the child and those of the agent to which he/she was exposed, and to rule out a different genetic or other malformation syndrome that may have phenotypic overlap. A high-density microarray should be performed on any child with developmental delay, more than one major malformation, or multiple minor anomalies, particularly if there is a discrepancy between the features expected for that antiepileptic drug exposure and those in the child. Additional genetic testing should be ordered as indicated. Women taking antiepileptic drugs have the same background risk as other women their age of having children with genetic disorders.

Differential Diagnosis

There are multiple genetic syndromes that have facial features or minor anomalies similar to those found in fetal anticonvulsant syndrome. These syndromes may also be associated with major malformations similar to those of fetal anticonvulsant syndrome. Children with Williams syndrome demonstrate a wide mouth and congenital heart defects (see Chapter 63). Children with Aarskog syndrome have hypertelorism and genital abnormalities (see Chapter 1). Children with fetal alcohol syndrome (see Chapter 26) and maternal phenylketonuria display a smooth philtrum and developmental delay. Children with Noonan syndrome have low-set ears, cryptorchidism, and congenital heart defects (see Chapter 41). Children with Coffin–Siris syndrome (see Chapter 13) have distal digital hypoplasia and developmental delay (Hanson 1986). These and other malformation syndromes may be excluded by specific genetic or other laboratory diagnostic testing in some cases, and by a careful family and medical history and a thorough physical examination by an experienced practitioner.

Fetal Hydantoin Syndrome. In 1975, Hanson and Smith described characteristic facies and other abnormalities in children of women who took hydantoins such as phenytoin during pregnancy. In the same journal, Zackai et al. (1975) delineated a different pattern of anomalies in children exposed to trimethadione, another antiepileptic drug. These two reports were the first to describe specific patterns of malformation associated with exposure to a particular antiepileptic drug. Each report included children who had been exposed to more than one agent in utero, but the authors argued that the distinctive patterns described in each report indicated the primacy of the common agent in producing the phenotype (Hanson and Smith 1975; Zackai et al. 1975).

The characteristic facial phenotype of the fetal hydantoin syndrome was identified as a broad and low nasal bridge, epicanthal folds, anteverted nares, hypertelorism, ptosis or strabismus, prominent and slightly malformed or low-set ears, and a wide mouth with thick vermilion of the lips (Figure 27.1). Some individuals exhibited variations in head size and shape, with sutural ridging or wide fontanels. The authors described additional minor anomalies in hydantoin-exposed children they had evaluated or that had been reported in the literature, such as hypoplasia of the distal phalanges and nails (more pronounced on the ulnar side) (Figure 27.2), a digitalized thumb, variations in palmar creases and dermatoglyphics, nuchal webbing, a low hairline, rib defects, sternal anomalies, spinal malformations, widely spaced and

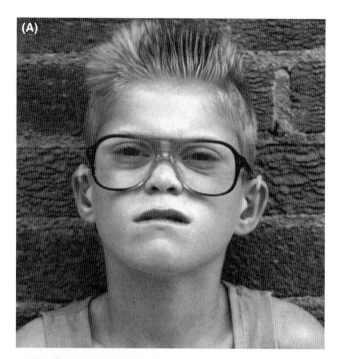

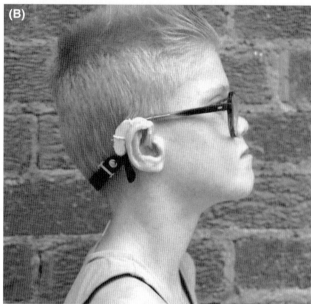

FIGURE 27.1 (A) Boy with fetal hydantoin syndrome. Note ocular hypertelorism and anteverted nose. [Reprinted with permission: Seaver LH (1992) *Pediatr Clin North Am* 39:111–134.] (B) Note midfacial retrusion, anteverted nose, and hearing aid accompanying mixed hearing loss. [Reprinted with permission: Seaver LH (1992) *Pediatr Clin North Am* 39:111–134.]

hypoplastic nipples, umbilical and inguinal hernias, genital anomalies, and pilonidal sinuses. Major anomalies identified included cleft lip with or without cleft palate, cardiovascular anomalies, renal defects, positional limb deformities, and diaphragmatic hernia. Some individuals exhibited prenatal and postnatal growth retardation and some had

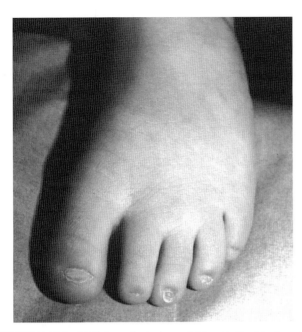

FIGURE 27.2 Nail hypoplasia in a child with fetal hydantoin syndrome. [Reprinted with permission: Seaver LH (1992) *Pediatr Clin North Am* 39:111–134.]

mild-to-moderate intellectual disability (Hanson and Smith 1975). The risk of the phenytoin-exposed fetus to have fetal hydantoin syndrome was reported to be 5–10%, and the risk of having some features of the disorder to be an additional 30% (Hanson 1986).

Most of the subsequent studies have confirmed the observations of Hanson and Smith of characteristic facies, distinctive minor anomalies such as nail and digital hypoplasia, and an increased occurrence of major malformations of the lip and/or palate, heart, and limbs in children exposed in utero to phenytoin and one or more antiepileptic drugs (Arpino et al. 2000; Kini et al. 2006). In a review and meta-analysis of studies published up until 2008, 1198 infants exposed to phenytoin monotherapy had a 5.48% incidence of congenital anomalies (Meador et al. 2008). Some studies have failed to show an increased risk of major malformations in children of mothers taking phenytoin alone (Canger et al. 1999; Samren et al. 1999; Kaaja et al. 2003). The risk for malformations is greater with maternal use of phenytoin plus other antiepileptic medications. The initial observations of microcephaly and growth restriction in exposed children may also have been because of the fact that most of the mothers took one or more additional antiepileptic drugs in pregnancy (Samren et al. 1999; Holmes 2002). The risk for an affected child is greater if the woman has had a previous child with fetal hydantoin syndrome. This may be related to genetic factors in the woman and/or fetus (Buehler et al. 1994; Koch et al. 1992; Azzato et al. 2010).

Although developmental delay was originally identified as a feature of fetal hydantoin syndrome, there have been varying results in subsequent studies regarding the neurodevelopmental and behavioral features of hydantoin-exposed children. Mean IQ scores were lower than expected in one study (Koch et al. 1989) while in another study the mean IQ score in 3-year-olds exposed in utero to phenytoin was 99 (Meador et al. 2009). Mean IQ scores were the same in 21 11-year-olds exposed to phenytoin monotherapy when compared to offspring of women with epilepsy who were not treated during pregnancy (Adab et al. 2004).

Exposure to phenytoin during pregnancy has been associated with an increased risk of tumors such as neuroblastoma (al-Shammri et al. 1992). There are no controlled studies to determine the magnitude of risk.

In some studies, up to half of babies exposed to phenytoin have vitamin-related clotting problems, and perinatal and neonatal hemorrhage have been reported (McNinch and Tripp, 1991). There are varying recommendations as to whether women taking phenytoin should be prescribed Vitamin K during the last month of pregnancy (Harden et al. 2009).

Fetal Valproate Syndrome. In 1984, Di Liberti and colleagues described facial and other features common to seven children who had been exposed to valproate in utero, including epicanthal folds (which continued inferiorly and laterally to form a crease or groove just under the orbits), depressed nasal bridge, small anteverted nose, long upper lip with a smooth philtrum, thin upper vermillion, and downturned corners of the mouth (Figure 27.3). These features differed from those of fetal hydantoin syndrome in that the valproate-exposed children from this initial report did not display hypertelorism, and most had relatively small mouths (a wide mouth being characteristic of fetal hydantoin syndrome). Four of the seven children had other abnormalities such as hypospadias, strabismus, psychomotor delay, nystagmus, and low birth weight (Di Liberti et al. 1984). Subsequently, Ardinger et al. (1998) verified the existence of a specific phenotype associated with fetal valproate exposure in 19 additional children, and developmental delay was noted in the majority of infants exposed to valproate (15 of 18 infants). They identified a similar facial phenotype to the description defined by Di Liberti et al. (1984) and added the presence of postnatal growth deficiency and microcephaly in two thirds of children exposed to valproate in combination with other anticonvulsants. Ardinger et al. (1998) also reported on skeletal anomalies, including metopic ridging (Figure 27.4), outer orbital ridge deficiency, bifrontal narrowing, talipes equinovarus and neural tube defects. And in a review of subsequently published reports of 70 affected individuals, the frequencies of the most common facial features were small and broad nose with depressed nasal bridge, 57%; small or abnormal ears, 46%; long, smooth philtrum, 43%; epicanthal folds, 31%; hypertelorism, 27%; high broad forehead, 26%; and bifrontal narrowing, 19% (Kozma 2001).

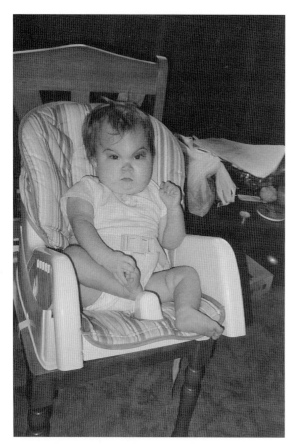

FIGURE 27.3 Child with fetal valproate syndrome demonstrating the typical facial features.

FIGURE 27.4 Child with fetal valproate syndrome demonstrating metopic ridging.

From a 2006 review by Ornoy, valproate was noted to be the most teratogenic of all the antiepileptic drugs, associated with a 10- to 20-fold increase in the risk of neural tube defects over that of the general population. Valproate was also found to increase the rates of major birth defects 2- to 3-fold over the general population, with many of these children showing all the signs of fetal valproate syndrome (characteristic facial features, growth deficiency, limb abnormalities, heart anomalies, and brain anomalies). A dose–response effect has been reported in studies evaluating the risk for malformations in valproate-exposed infants (Tomson et al. 2011; Hernandez-Diaz et al. 2012; Tomson et al. 2018). In the EURAP database, doses of less than 700 mg/day resulted in a risk of 5.6% for malformations. Doses of 700 mg to <1500 mg were associated with a risk of 10.4% and doses over 1500 mg had a malformation rate of 24.2% (Tomson et al. 2011). Genetic factors also appear to affect the risk for adverse effects (see TERIS). In the Norwegian Medical Birth Registry, exposed children had an increased risk for malformations; the risk was increased when there had been a prior sibling with valproate-associated malformation (Veiby et al. 2014).

Although Ardinger et al. (1998) first reported the presence of growth abnormalities in children exposed to valproate as part of combination antiepileptic therapy, this finding was absent in those on valproate monotherapy. Additionally, neurocognitive deficits were identified in nearly three-fourths of patients, a number that increased to 90% when those who were on combination therapy were included. In 2008, Meador et al. reported on a pooled group of infants exposed to valproate monotherapy during pregnancy and found a 10.7% risk for major congenital anomalies. The North American AED Registry reported a 9.3% risk for congenital anomalies in offspring of women who took valproate monotherapy during pregnancy (Hernandez-Diaz et al. 2012). The International Lamotrigine Pregnancy Registry reported a 10.7% rate of malformations when valproate was used in conjunction with lamotrigine; whereas, the risk for malformations was 2.8% with lamotrigine monotherapy (Cunnington et al. 2011). A Cochrane review and meta-analysis reported a prevalence rate of malformations of 10.93% (Bromley et al. 2016), and a second meta-analysis also reported an increased incidence of malformations with prenatal exposure to valproate (Veroniki et al. 2017b).

The major malformations described by Kozma (2001) included the musculoskeletal system, 63%; cardiovascular system, 26% (consisting of atrial septal defect, ventricular septal defect, aortic stenosis, pulmonic stenosis, and other abnormalities); genitourinary system, 21% (primarily hypospadias); growth restriction, 15%; and microcephaly, 13%. Additional anomalies of lower frequency included cleft lip with or without cleft palate, 4%; tracheomalacia, 9%; radial ray defects, 16%; and abdominal wall defects, 14% (inguinal hernias, umbilical hernias, and omphaloceles). Abnormalities of the fingers were quite common, noted in 36%. Renal anomalies and brain malformations have also been described (Arpino et al. 2000; Kozma 2001).

Valproate has been associated with global developmental delay, particularly speech delay (Adab et al. 2004; Vinten et al. 2005). A meta-analysis of 67 exposed children found a mean full-scale IQ of 84, compared to 102 in controls (Banach et al. 2010). In another study of 60 three-year-old children who had been exposed to valproate in utero, mental development on the Bayley Scales of Infant Development was about nine points lower than children exposed to an alternative antiepileptic medication, lamotrigine, in utero (Meador et al. 2009). The Neurodevelopmental Effects of Antiepileptic Drugs Study reported that the adjusted mean IQ of valproate-exposed 4-year-olds was 95, and 10% of the group had IQs less than 70 (Meador et al. 2012). A Cochrane review concluded that children with valproate exposure had lower developmental quotient and IQ than children of women with untreated epilepsy (Bromley et al. 2014). A review and meta-analysis of 29 cohort studies concluded that valproate exposure was associated with an increased risk for language and psychomotor delays, as well as autism spectrum disorders (Veroniki et al. 2017a). Another study found that adaptive functioning worsened as the maternal valproate dose increased (Deshmukh et al. 2016). Of interest is the association between the presence of dysmorphic features and the severity of neurocognitive deficits. Kini et al (2006) reported that 24 out of 56 children exposed to valproate had moderate to severe facial features indicating fetal valproate syndrome. In that study, children with more features of fetal valproate syndrome had lower verbal IQ when compared to those without physical features.

Autism spectrum disorders have been observed more commonly in children prenatally exposed to valproate. One study found an 8–18-fold higher than expected incidence of autism spectrum disorders in children exposed to AEDs, in particular, valproate (Rasalam et al. 2005). Other studies have also shown an increased risk for autism spectrum disorders with prenatal valproate exposure (Bromley et al. 2013; Roullet et al. 2013).

Neonatal morbidity, including transient hypoglycemia, abnormal behavior, afibrinogenemia, platelet dysfunction, and hepatic failure has been reported (Koch et al. 1996).

Carbamazepine Embryopathy. In 1989, Jones and colleagues reported that exposure to carbamazepine in utero was associated with minor craniofacial anomalies, nail hypoplasia (similar to that observed in fetal hydantoin syndrome) and developmental delay. Craniofacial features included microcephaly with narrow bifrontal diameter, upslanting palpebral fissures, epicanthal folds, and a short nose with a long philtrum, features similar to those of fetal hydantoin syndrome and fetal valproate syndrome. The authors suggested that hydantoins and carbamazepine have a similar mechanism of teratogenesis related to the metabolism of arene oxides (see Etiology, Pathogenesis, and Genetics). Major malformations were identified in 10% of exposed children and in only 7% of controls. Minor anomalies such as hypoplastic nails and craniofacial dysmorphisms were observed in 18% of exposed pregnancies (Jones et al. 1989). Fetal carbamazepine syndrome has also been reported by several other authors (Moore et al. 2000; Kini et al. 2006) who detected this in 10% of 87 children exposed to carbamazepine in utero. Subsequent studies identified an increased risk for major malformations in carbamazepine exposed infants, ranging from 3.3% to 4% (Wide et al. 2004). A review of eight cohort studies suggested the risk for malformations to be 3.3% (Jentink et al. 2010). The North American AED Pregnancy Registry found the risk for malformations to be 3% (Hernandez-Diaz et al. 2012). A Cochrane review of studies published through 2015 indicated that carbamazepine monotherapy presented an increased risk for malformations when compared to unexposed babies of women with and without epilepsy (Weston et al. 2016). A second meta-analysis published in 2017 also reported a small increase in risk for malformations associated with carbamazepine monotherapy (Veroniki et al. 2017b). The Medical Birth Registry of Norway reported on outcomes of carbamazepine monotherapy in 685 women; a slight increase in growth restriction was seen in exposed babies, but the incidence of malformations was not increased (Veiby et al. 2014).

It has been suggested that the birth prevalence of neural tube defects in children exposed to carbamazepine in utero is elevated (0.5–1%) (Arpino et al. 2000; Werler et al. 2011). Conversely, the UK and Ireland Epilepsy and Pregnancy Registers found only two out of 900 exposed infants (0.2%) had a neural tube defect (Morrow et al. 2006). A EUROCAT antiepileptic study that pooled the results of eight cohort studies found an increased risk for spina bifida (Jentink et al. 2010). Other major malformations identified in exposed children include cardiac defects, renal anomalies, and cleft palate (Arpino et al. 2000; Pennell 2003; Jentink et al. 2010).

Analysis of EURAP registry data suggested that the incidence of malformations increased with dosage. The risk for malformations was 3.4% with a carbamazepine dose of less than 400 mg/day, and 8.7% at greater than 1000 mg/day (Tomson et al. 2011). The risk for malformations increases with polytherapy, especially when valproate is added to carbamazepine. In the North American AED Pregnancy Registry, major malformations were observed in 2.9% of babies exposed to carbamazepine alone; whereas, they were observed in 15% of those exposed to both carbamazepine and valproate (Holmes et al. 2011).

Prenatal carbamazepine exposure may have an adverse effect on neurodevelopment. Developmental delays and behavioral problems were more frequent in young children exposed to carbamazepine during pregnancy (Cohen et al.

2011; Veiby et al. 2013). However, studies in older children did not find that the issues necessarily persisted (Bromley et al. 2013; Meador et al. 2013) or if they did the effects were minor (Forsberg et al. 2011).

Transient neonatal hypocalcemia and neurological dysfunction have been reported in infants exposed chronically during pregnancy. Vitamin K-related clotting issues have also been observed, although affected infants generally are not symptomatic (Kaaja et al. 2002). Controversy exists regarding whether it is necessary and effective to supplement women taking carbamazepine with vitamin K during the last month of pregnancy (Harden et al. 2009).

Phenobarbital Embryopathy. Recognition of the teratogenicity of phenytoin, carbamazepine, and valproate led physicians to hope that phenobarbital would be non-teratogenic, so that there might be a safe antiepileptic drug to prescribe in pregnancy. However, in 1992, Jones et al. reported on a group of 46 prospectively ascertained children prenatally exposed to phenobarbital monotherapy. The authors observed that 15% of the children had facial features similar to those found in fetal hydantoin syndrome and carbamazepine embryopathy (Figure 27.5); 24% demonstrated nail hypoplasia; and 20% (of 16 children whose development was evaluated) exhibited developmental delay (Jones et al. 1992).

Many of the early studies on the incidence of malformations in babies whose mothers used phenobarbital are difficult to interpret because they involved use of multiple antiepileptic medications. When used in conjunction with other anticonvulsants such as phenytoin, the incidence of abnormalities in exposed babies is at least two to three times that of the general population. Malformations, reduced birth weight and head circumference, and impaired intellect have been reported. The teratogenicity of phenobarbital monotherapy was confirmed in multiple subsequent studies (Canger et al. 1999; Samren et al. 1999; Arpino et al. 2000; Pennell 2003). Major malformations commonly observed following in utero phenobarbital exposure include oral clefts and congenital heart disease. The North American AED Pregnancy Registry initially found a 5.5% risk for malformations (Hernandez-Diaz et al. 2012); this increased to 5.9% in their 2017 update. Two recent meta-analyses documented a statistically significant increased risk for total malformations with phenobarbital monotherapy (Weston et al. 2016; Veroniki et al. 2017b). Conversely, the Hungarian Case-Control Surveillance of Congenital Abnormalities did not detect an association between malformations and in utero phenobarbital exposure (Kjaer et al. 2007).

As with several other anticonvulsants, the incidence of malformations increases as maternal phenobarbital dosage gets higher. EURAP data suggested that the risk for malformations was 4.2% at less than 150 mg/day and 13.7% at higher doses (Tomson et al. 2011). Neurodevelopmental effects, such as short attention span and learning difficulties, have been reported in offspring of women who used phenobarbital alone or as polytherapy (Adams et al. 2000; Ornoy 2006). Maternal use in the third trimester has been associated with transient issues with newborn adaptation including lethargy, sedation, irritability, and tremors. Vitamin K-related bleeding disorders also have been reported; some sources suggest giving maternal Vitamin K in the last four weeks of pregnancy in an effort to mitigate this risk (Harden et al. 2009).

Second Generation Antiepileptic Medications. Although data regarding teratogenicity are not available on some of the newer antiepileptic medications, for many there is an increasing body of information. With the exception of topiramate, they appear to be associated with a lower risk for fetal anticonvulsant syndrome and major malformations, in comparison with the first generation antiepileptic medications.

Topiramate. Conflicting evidence exists regarding a recognizable pattern of malformation in topiramate-exposed infants, although the samples examined by a dysmorphologist are very small. In 1998, an initial report of a topiramate-exposed infant documented a pattern of multiple minor anomalies similar to those seen with other anticonvulsant medications (Hoyme et al.1998). An additional study of five topiramate exposed infants did not find features of an anticonvulsant syndrome (abstract by Chambers et al. 2005).

Some but not all studies have found an increased risk for malformations in children of women who take topiramate during pregnancy. Oral clefts are the most commonly observed malformations. The North American AED Pregnancy Registry found an overall risk of 4.4% for malformations, with a specific excess of orofacial clefts (Holmes

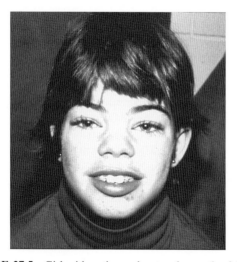

FIGURE 27.5 Girl with anticonvulsant embryopathy following prenatal exposure to barbiturates. Note ocular hypertelorism, anteverted nose, and midface retrusion. [Reprinted with permission: Seaver LH (1992) *Pediatr Clin North Am* 39:111–134.]

and Hernandez-Diaz 2012; Hernandez-Diaz et al. 2012). A case-control analysis with data from the Slone Birth Defects Study and the National Birth Defects Study found an odds ratio of 4.0 for cleft lip with or without cleft palate (Margulis et al. 2012). The UK Epilepsy and Pregnancy Register found a 9.0% risk for any major malformation, and EURAP suggested a 6.8% risk for birth defects in prenatally exposed infants (Hunt et al. 2008). Conversely, the Danish Medical Birth Registry found the risk to be 4.6%, which was no higher than that of the general population (Molgaard-Nielsen and Hviid 2011). A record linkage study found no increased risk for congenital anomalies, specifically oral clefts (Green et al. 2012). The Norway Medical Birth Registry reported an increase in growth restriction but not malformations (Veiby et al. 2014). A meta-analysis of Australia, Israel, UK and US registries involving 427 infants exposed to topiramate in utero did not detect a significant risk for malformations (Day et al. 2011). A recent study of 2425 women exposed to topiramate suggested that risk for oral clefts varied with dose. There was an eight-fold increase in risk for oral clefts in babies of women using higher doses (200 mg or greater) but only a two-fold increase in risk at a daily dose of 100 mg or less (Hernandez-Diaz et al. 2018).

Limited data on the neurocognitive effects of topiramate exposure exist. In one study, however, neurodevelopmental assessment scores of children whose mothers used topiramate monotherapy during pregnancy was lower than that of a control group (Rihtman et al. 2012).

Administration of oral vitamin K in the last month of pregnancy has been recommended for women taking other hepatic enzyme-inducing antiepileptic medications, such as phenytoin, in an effort to reduce the risk for vitamin K-related bleeding in their infants. This may also be appropriate for women who are on topiramate during pregnancy (Harden et al. 2009).

Lamotrigine No pattern of characteristic facial features, minor anomalies or major malformations has been identified in infants with prenatal lamotrigine exposure. Nearly all studies suggest that lamotrigine has a lower risk for malformations when compared to the older first generation antiepileptic medications such as carbamazepine, phenobarbital and valproate. When lamotrigine is used in conjunction with another antiepileptic medication (most often valproate) the risk for malformations is much greater; however, the teratogenic risk is believed to predominantly reflect the risk of the other medication (Cunnington et al. 2011). The North American AED Pregnancy Registry initially found an increased prevalence of isolated oral clefts associated with maternal use of lamotrigine for epilepsy or psychiatric conditions. This risk diminished with increasing numbers of exposed babies in the registry, although the relative risk was still increased over the control group (Hernandez-Diaz et al. 2012). The most recent reports from this registry suggest that the overall teratogenic risk of lamotrigine monotherapy is 2.1% (North American AED Registry 2016). Other studies have not reported an increased risk for oral clefts. Prospective data gathered by the International Lamotrigine Pregnancy Registry found that 2.2% of babies exposed in utero during the first trimester had a major malformation (Cunnington et al. 2011). Relatively similar rates of malformations with lamotrigine monotherapy were reported in other birth registries: 2.3% in the UK and Ireland Epilepsy and Pregnancy Registers, 3% in the Swedish Medical Birth Registry, and 3.7% in the Danish Medical Birth Registry (Meador and Loring 2016). The Australian Register of Antiepileptic Drugs in Pregnancy found a malformation rate of 4.6%, which was not statistically different from control pregnancies (Vajda et al. 2014). A Cochrane review published in 2016 and updated in 2017 concluded that there was no increased risk for malformations associated with in utero exposure to lamotrigine monotherapy (Weston et al. 2016; Bromley et al. 2017). Similarly, a meta-analysis of 21 studies concluded that lamotrigine monotherapy was not associated with an increased risk for malformations (Pariente et al. 2017).

Some researchers have suggested that while there is no significant increase in risk for malformations when lamotrigine is used at lower doses (as is most common), higher doses do present a significant increase in risk. For example, EURAP pregnancy registry data published in 2018 indicate that the prevalence of malformations with lamotrigine monotherapy is 2.5% at doses of 325 mg/day or less and 4.3% when the dose is greater than 325 mg/day (Tomson et al. 2018). Other studies, however, did not find a dose-dependent increase in risk (Campbell et al. 2014; Molgaard-Nielsen and Hviid 2011). Most studies have not found an increase in low birth weight or decreased head circumference associated with prenatal lamotrigine treatment (Veiby et al. 2014; Almgren et al. 2009).

There are varying results in studies looking at cognition and behavior associated with prenatal lamotrigine exposure. The Norwegian Mother and Child Cohort Study reported that autistic traits and deficits in sentence skills were observed more often than expected in babies exposed to lamotrigine during pregnancy (Veiby et al. 2013). In another study, scores were lower than expected in lamotrigine-exposed children in tests of sensory and motor development; however, the children did not differ from controls in behavior or attention (Rihtman et al. 2013). In the Neurodevelopmental Effects of Antiepileptic Drugs study, average IQ scores were normal in children exposed to lamotrigine in utero, but verbal abilities were significantly lower than non-verbal abilities (Meador et al. 2009). Conversely, other studies suggest that children exposed to lamotrigine monotherapy in utero do not show any differences in overall development or adaptive behavior (Bromley and Mawer, 2013; Deshmukh et al. 2016).

Levetiracetam. Levetiracetam is a second generation antiepileptic medication. No recognizable pattern of characteristic facial features, minor anomalies, or major malformations has been reported with levetiracetam exposure. The available data on levetiracetam do not suggest an increased risk for malformations: the UK and Ireland Epilepsy and Pregnancy Register found the risk for major malformations to be 2.3%; the Swedish Medical Registry found one malformation out of 151 exposures; the Danish Medical Registry did not find any malformations in 58 exposed pregnancies; and there was no increased risk in the Norway Medical Birth Registry (Meador and Loring 2016; Molgaard-Nielsen and Hviid 2011; Veiby et al. 2014). A Cochrane review published in 2016 concluded that there was not an increased risk for malformations associated with in utero exposure to levetiracetam (Weston et al. 2016). Similarly, a meta-analysis of 21 studies concluded that levetiracetam monotherapy was not associated with an increased risk for malformations (Pariente et al. 2017). Studies thus far have not detected alterations of child development or reduced cognitive ability in children exposed to levetiracetam during pregnancy (Bromley et al. 2016).

Oxcarbazepine. Oxcarbazepine is a second generation antiepileptic medication that is related to carbamazepine, a first generation antiepileptic medication that has been associated with an increased risk for neural tube defects. However, the available data on oxcarbazepine have not suggested an increased risk for major malformations, specifically neural tube defects. No recognizable pattern of characteristic facial features, minor anomalies, or major malformations has been reported with oxcarbazepine exposure. The Danish Medical Birth Registry estimated the risk for malformations in babies exposed in utero to oxcarbazepine to be 2.8%, and the North American AED Pregnancy Registry found the risk to be 2.2% (Meador and Loring 2016). The Norwegian Medical Birth Registry indicated that there was not an increased risk for malformations in prenatally exposed infants (Veiby et al. 2014). There are minimal data regarding cognitive or neurobehavioral issues following in utero exposure to oxcarbazepine.

MANIFESTATIONS AND MANAGEMENT

The majority of children born to women taking AEDs in pregnancy do not have fetal anticonvulsant syndrome. While the risk is significantly higher with many of the older first generation antiepileptic drugs, even with those medications there are still many babies exposed in utero who are unaffected. With the exception of topiramate, the second generation antiepileptic medications have minimal or no increase in risk for major malformations or other features of fetal anticonvulsant syndrome. There is no clear correlation between the presence of facial features consistent with fetal anticonvulsant syndrome and the presence of major malformations (although the presence of facial features consistent with fetal anticonvulsant syndrome and a major malformation may be suggestive that the latter is the result of antiepileptic drug exposure (Holmes 2002)). The health care provider should be aware that depending on which medication was used, there may be as high as a two- to three-fold increased risk of major malformations in children born to women who take antiepileptic drugs during pregnancy. Cognitive and behavioral deficits may be seen in children who do not have the malformations or the facial features of fetal anticonvulsant syndrome.

Growth and Feeding

Breastfeeding has many advantages for both the mother and her infant. Excretion in breastmilk varies greatly among the various antiepileptic medications, and each should be evaluated individually. For example, phenobarbital may be excreted in mother's breast milk at high levels, with a long half-life, and may have a sedative effect on the infant. It should be used with caution. Other antiepileptic medications are found in breastmilk in small quantities, and maternal use is not considered a contraindication for breastfeeding. The amount of the antiepileptic medications in milk also varies by maternal dosage, use of other medications, and maternal metabolism of the drug. Since antiepileptic medications are potent medications with many possible side effects, the infant should be monitored closely for signs that they are ingesting a high enough dose of the medication to have side effects. This might involve observing the baby for sedation and doing more frequent weight checks. If the maternal dose is high with an agent that is readily excreted in milk, there may be consideration of getting a drug level in serum of the infant (there are very few laboratories willing to measure the amount of medication in milk, and none offer this service clinically). This is most important if the infant is having problems that might be attributable to the antiepileptic medication in milk. For mothers taking any of these antiepileptic drugs, breastfeeding should be discontinued if the infant appears sedated, feeds poorly, or is irritable.

Prenatal and postnatal growth retardation were identified by Hanson and Smith as features of fetal hydantoin syndrome and are still included in the spectrum of abnormalities that may be observed in fetal anticonvulsant syndrome (Hanson and Smith 1975; Holmes et al. 2001). However, in one prospective study, 15% of exposed newborns were small for gestational age and only 2.5% of newborns were large for gestational age, which may suggest an effect of antiepileptic drug exposure on birth weight (Canger et al. 1999). In another large prospective study, in which growth retardation was defined as a height or weight more than two standard deviations below the mean for race, sex, and gestational age, there was a three-fold increased risk of growth retardation with exposure to carbamazepine or phenobarbital, no significant increased risk with exposure to phenytoin, and a four-fold increased risk with exposure to two or more antiepileptic drugs (Holmes et al. 2001). The effects of antiepileptic drug exposure on postnatal growth have not been determined.

Evaluation

- Monitor breastfed babies whose mothers are on antiepileptic medications for side effects such as sedation or poor weight gain. Adverse effects are more common when the mother is on multiple medications at a high dose. If problems are seen that may be associated with exposure to the maternal medication in milk, consider measuring the infant's own drug level. Plot growth parameters carefully at each well child visit.

Management

- If the breastfed infant is sedated or irritable, or has therapeutic or near-therapeutic drug levels, it may be necessary to switch to infant formula.
- Children should have further evaluation if they are not meeting expected height and/or weight milestones, or are significantly below the percentiles expected for their family, as for any child with failure-to-thrive. Exposure prenatally should not be used as a sole explanation for postnatal growth delay without further assessment.

Development and Behavior

Newborns exposed to antiepileptic drugs in utero may be sedated and have poor suck or excessive sleepiness. Alternatively, they may exhibit withdrawal symptoms, including hyperexcitability, jitteriness, apneic episodes and seizures. Withdrawal symptoms may occur in infants exposed to many of the antiepileptic drugs discussed in this chapter.

Antiepileptic drug-exposed children may have hearing and/or vision difficulties. For example, hearing loss has been described with valproate exposure (Kozma 2001). If not detected, these deficits may contribute to developmental delays.

Children with a history of prenatal exposure to antiepileptic medications, especially the first generation medications, e.g., valproate, are at risk for developmental delay, various degrees of cognitive impairment and behavioral problems. It has been difficult to separate genetic and environmental factors that influence development, such as maternal IQ and socioeconomic status, from the impact of the antiepileptic medication. Polytherapy and higher doses may be associated with an increased incidence of learning difficulties (Meador et al. 2009).

With respect to seizure disorders, children exposed to antiepileptic medications in utero have a higher prevalence of frank seizures and EEG abnormalities than do controls. A study of 67 children born to mothers with epilepsy in comparison with 49 control children revealed the following prevalence of EEG abnormalities: 25% in the no-therapy group, 43% in the monotherapy group, and 22% in the polytherapy group, compared with 5% of control children (Koch et al. 1999). Seizures in exposed infants may be related to the antiepileptic medication exposure during pregnancy or a potential genetic cause of the maternal condition.

Evaluation

- Anticonvulsant-exposed newborns should be monitored for signs of withdrawal.
- Newborn hearing screening should be performed and repeated as necessary, especially if symptoms such as speech and language delays are noted.
- Children with fetal anticonvulsant exposure should have preschool hearing and vision screening.
- Exposed children should be screened for developmental delays using standard pediatric measures at regular well-care visits.
- Referral to a developmental/behavioral specialist for further evaluation is indicated if there are developmental delays or features of an autism spectrum disorder.
- Referral to a pediatric neurologist for evaluation of a seizure disorder is indicated if the child displays any features consistent with seizures.
- Cranial imaging should be considered in the presence of persistent delays, especially in valproate-exposed children or in those with an abnormal head circumference measurement (macro- or microcephaly).

Management

- Withdrawal should be treated symptomatically.
- Referral should be made to audiology, otolaryngology, and/or ophthalmology if abnormalities are noted on evaluation in these areas.
- Therapies and educational interventions for developmental delays should be initiated as soon as possible in a standard manner, and should lead to individualized educational programming throughout childhood.
- Exposed children with a suspected autism spectrum disorder should be referred to a behavioral specialist.
- Referral to a pediatric neurologist should be made if the child has seizures, developmental delay, or an abnormal neurological examination.

Neurologic

Children who have been exposed to valproate or carbamazepine in utero have a 1–2% birth prevalence of neural tube defects (Arpino et al. 2000; Kaaja et al. 2003). The defects are primarily caudal, such as lumbar myelomeningoceles, and may be covered by skin. The effect of folic acid supplementation on this incidence is unclear.

Evaluation

- Pregnant women taking valproate or carbamazepine should undergo second trimester screening with a detailed anatomy scan by ultrasound and serum α-fetoprotein to assess the fetus for a neural tube defect. Amniocentesis for amniotic fluid α-fetoprotein and acetylcholinesterase determination will detect about 99% of all neural tube defects, but this invasive test introduces a risk of miscarriage of about 1/250. Skin-covered defects will only be detected by ultrasound, as α-fetoprotein and acetylcholinesterase will not be elevated in these cases.
- Exposed children should be examined for the presence of neural tube defects.

Management

- Women taking valproate or carbamazepine should take a folic acid supplement of 4 mg/day, beginning at least a month before pregnancy and continuing throughout the first trimester.
- Children with neural tube defects should be referred to a pediatric neurosurgeon for standard management.

Cardiovascular

Congenital heart defects may be observed following antiepileptic drug exposure, but are more common with the first generation agents such as phenobarbital and valproate. The spectrum of cardiac defects seen is broad and includes ventricular septal defects, atrial septal defects, pulmonic stenosis, patent ductus arteriosus, coarctation of the aorta, tetralogy of Fallot, transposition of the great arteries, and arrhythmias such as Wolf–Parkinson–White and supraventricular tachycardia (Samren et al. 1999; Arpino et al. 2000; Kozma 2001).

Evaluation

- Prenatal evaluation of the antiepileptic drug-exposed fetus may include a fetal echocardiogram.
- As for any child, a thorough history and careful cardiac examination should be an integral part of the medical examination.
- If there is a clinical concern, the prenatally exposed child should be referred to a pediatric cardiologist for an electrocardiogram, echocardiogram, and further evaluation

Management

- Treatment of cardiac abnormalities in antiepileptic-exposed individuals is the same as in the general population, and is based on the specific abnormality detected.

Craniofacial

Meadow's 1968 letter to *Lancet* begins "…I should be interested to know if your readers have seen babies with hare-lip, cleft palate, and certain other specific abnormalities born to mothers who receive regular anticonvulsant therapy" (Meadow 1968). Thus, an increased incidence of cleft lip with or without cleft palate has been a characteristic feature of children exposed to antiepileptic drugs in utero from the first description. Speidel and Meadow's subsequent report, and others, have identified an approximately two- to three-fold increase in the birth prevalence of cleft lip with or without cleft palate in antiepileptic drug-exposed children in comparison with the general population (Speidel and Meadow 1972; Arpino et al. 2000). An increased risk of cleft lip and/or palate has been reported with several of the older first generation antiepileptic medications, in addition to topiramate. Meadow also described "unusual facies and skulls… (including) short neck and low posterior hair-line, broad nose-root with wide-spaced prominent eyes, and deformities of the pinna. The skulls have been of unusual shape – for instance, pointed in the frontal area with a prominent ridged suture line – or there have been minor bone defects" (Meadow 1968). These features agree with those subsequently described for the individual antiepileptic drug embryopathies, such as broad nose, hypertelorism, ear abnormalities, and prominent metopic suture (Hanson and Smith 1975; DiLiberti et al. 1984). These minor craniofacial abnormalities are usually of no clinical significance and may improve with time (Koch et al. 1992). However, trigonocephaly as the result of metopic suture synostosis may require surgical intervention.

Evaluation

- Prenatal fetal ultrasound examinations should include careful evaluation of the lip and palate. When these features are suspected on ultrasound, further details may be apparent on 3D scan.
- Exposed children should be carefully evaluated for palatal abnormalities including overt cleft, submucous cleft, and bifid uvula.

- Minor craniofacial abnormalities consistent with fetal anticonvulsant syndrome should be recorded, to document evidence of exposure.
- Craniofacial features not consistent with fetal anticonvulsant syndrome, such as microtia, should also be recorded and may prompt further evaluation.
- Sutural anomalies of the skull may indicate the need for a head computed tomography scan with 3D reconstruction to identify craniosynostoses.

Management

- Management of the child with fetal anticonvulsant syndrome and cleft lip and/or palate is the same as for any similarly affected child in the general population.
- Ideally, referral should be made to a multidisciplinary craniofacial anomalies evaluation and treatment team.
- Referral should be made to a pediatric speech pathologist and a pediatric otolaryngologist if there are feeding difficulties associated with palatal abnormalities.
- There is no indicated treatment for the minor craniofacial abnormalities associated with fetal anticonvulsant syndrome, unless craniosynostosis is present, in which case referral to a pediatric neurosurgeon or other craniofacial surgeon is indicated.

Ophthalmologic

Children with prenatal exposure to antiepileptic medications may be at increased risk for vision abnormalities. A study of ophthalmic abnormalities in antiepileptic drug-exposed children identified multiple abnormalities, including a 50% incidence of myopia in children exposed to valproate (Glover et al. 2002).

Evaluation

- Exposed children should have an ophthalmologic evaluation in the newborn period and annually, thereafter, to identify visual impairment. This is particularly important in valproate-exposed individuals.

Management

- Management of ophthalmologic and vision abnormalities is standard and often includes corrective lenses.

Gastrointestinal/Genitourinary

Abdominal wall defects, including umbilical and inguinal hernias and omphaloceles, have also been observed following prenatal valproate exposure. Genitourinary anomalies such as hypospadias, cryptorchidism, and hydroceles have been described following prenatal antiepileptic exposure to agents such as valproate. Renal malformations have been described in children exposed in utero to valproate and to carbamazepine and include dysplastic kidneys, hypoplastic kidneys, renal cysts, and duplication of the calyceal system.

Evaluation

- Prenatal fetal ultrasound examination of the antiepileptic drug-exposed fetus should include a careful evaluation of the kidneys.
- A thorough abdominal and genital examination should be performed on the exposed newborn.
- A renal ultrasound should be performed in the newborn period to rule out structural renal anomalies.

Management

- Referral should be made to a pediatric surgeon for surgical correction of abdominal wall defects, as appropriate.
- Referral to urology should be made if hypospadias is present.
- Referral to a pediatric nephrologist and/or urologist should be made for individuals with renal anomalies. Management will not differ from that of children without fetal anticonvulsant syndrome.

Musculoskeletal

Distal digital hypoplasia is a hallmark of fetal anticonvulsant syndrome (Holmes et al. 2001), and coned epiphyses may be exhibited radiographically in phenytoin-exposed children (Holmes 2002). This is usually of no clinical significance and may improve with time (Koch et al. 1992). Limb defects attributable to vascular disruption, such as terminal transverse limb defects, have been described in phenytoin-exposed children, but are rare (Holmes 2002).

Children exposed to valproate in utero have an increased incidence of limb anomalies, including limb deficiency and radial ray defects (Arpino et al. 2000). In the review by Kozma (2001), musculoskeletal anomalies were the most common abnormality observed in about 63% of patients. Distal digital hypoplasia is seen as in other antiepileptic exposed children, but radial and thoracic cage defects have also been noted. The most frequently observed anomalies include contractures of the fingers and small joints and talipes equinovarus deformity. There may also be ulnar or tibial hypoplasia.

Evaluation

- A thorough examination of the limbs should be performed in all anticonvulsant-exposed individuals. All

joints should be examined for range of motion, particularly the small joints assessing for hypoplasia or contractures. A high index of suspicion for radial defects should be maintained, and radiographs should be obtained when clinical assessment is unclear.
- Chest and spine radiographs should be considered in children who demonstrate evidence of fetal valproate syndrome.

Management

- Referral should be made to a pediatric orthopedic surgeon and/or pediatric physiatrist if a functionally significant limb abnormality or deficiency is identified.
- Occupational and/or physical therapy should be initiated as soon as joint restriction is identified and no anatomical contraindication exists to range of motion therapy and splint usage.

Hematologic

Children born to mothers who have taken phenobarbital, phenytoin, or carbamazepine in pregnancy have an increased incidence of hemorrhagic disease of the newborn (Howe et al. 1999). The mechanism is unknown, but a reduced level of vitamin K in infants born to mothers taking antiepileptic drugs has been demonstrated (Howe et al. 1999).

Evaluation

- There is mixed evidence regarding Vitamin K supplementation in the last month of pregnancy and whether it is effective and/or necessary in women taking phenobarbital, phenytoin, or carbamazepine (Harden et al. 2009). However, at-home births are at increased risk of not receiving postnatal vitamin K administration, and so attending midwives or other providers should be made aware of the special circumstances of neonates with a prenatal exposure to antiepileptics.

Management

- Antiepileptic drug-exposed infants should receive the standard 1 mg dose of vitamin K at birth. The administration of the appropriate dose of vitamin K at birth should prevent hemorrhagic disease of the newborn, and extended observation or evaluation beyond the length of a normal hospital stay should not be necessary.

Neoplasia

Multiple case reports have documented tumors, especially neuroblastoma, in children who had been exposed to phenytoin in utero (Koren et al. 1989; al-Shammri et al. 1992). Other tumors that have been identified in children exposed to phenytoin in utero include Hodgkins disease, rhabdomyosarcoma, extra renal Wilms' tumor, and mesenchymoma (al-Shammri et al. 1992). The number of cases of these tumors is too low to give an assessment of increased risk.

Evaluation

- Because the overall incidence of tumors in phenytoin-exposed children is low, no surveillance is warranted, but the care provider should be aware of this association.

Management

- Appropriate evaluation and referral to a pediatric oncologist should be made if there is a clinical concern
- Treatment of tumors in anticonvulsant-exposed individuals is standard.

Management of Seizures in Pregnancy

Treatment of seizure disorders, or other conditions necessitating use of antiepileptic medications is a difficult balance between effectively treating a serious maternal condition and attempting to mitigate the impact of the medication on the fetus. For example, when compared to many of the other antiepileptic medications, valproate has been associated with a higher risk for malformations and adverse neurocognitive effects. Several professional organizations, including the European Academy of Neurology, the American Academy of Neurology and the American Epilepsy Society, recommend that when possible valproate be avoided in girls and women of reproductive age. However, some forms of epilepsy are best treated with valproate. As such, a decision about whether to continue the valproate medication should involve full disclosure of the risks and benefits of taking an antiepileptic medication during pregnancy and shared decision-making between the health provider and the woman (Tomson et al. 2015; Petersen et al. 2017).

Preconception

- Women taking antiepileptic drugs should be counseled at the time they reach childbearing age about the risks and benefits of these medications. This counseling may also include the recurrence risk for epilepsy in offspring. Preconception management is of upmost importance, and increases the likelihood of the best possible outcome (Battino et al. 2013; Caughey 2016; George, 2017).
- A high rate of pregnancies in women with epilepsy are unplanned. This may be related to an interaction between hormonal contraception and enzyme-inducing antiepileptic medications, making contraception less

reliable (George 2017). Barrier methods may provide additional efficacy and allow the woman to plan her pregnancy.
- Before pregnancy, the medication regimen should be adjusted in consultation with a neurologist, psychiatrist, or maternal-fetal medicine specialist. Dosages of antiepileptic drugs should be tapered to the lowest doses that can still provide adequate control of the symptoms.
- If possible, only one agent should be used.
- The adjustment of medications and dosage should be made well in advance of conception (at least six months) to reduce the risk of increased seizures in pregnancy.
- Because there is an increased birth prevalence of neural tube defects and other malformations in children of mothers taking certain antiepileptic medications, current recommends are to start a prenatal vitamin plus additional folic acid (to a total of 4 mg per day) at least three months before and continuing supplementation throughout the duration of the pregnancy. Women should be counseled that this may lower their risk for a malformation, although some studies show limited benefit to supplementation and a residual risk for medication-associated malformations likely remains (Ban et al. 2015; George 2017).

Pregnancy management

- The health of the mother is of prime consideration. While in utero exposure to antiepileptic medications may carry a risk to the infant, uncontrolled maternal epilepsy is a serious concern for both the woman and her pregnancy.
- Many of the obstetrical adverse effects appear related to the chronic maternal condition and are only minimally increased by anticonvulsant monotherapy with the newer, second generation antiepileptic medications. Given this information, and the potential hazards to the woman and her fetus from an uncontrolled seizure disorder, women are encouraged to continue their anticonvulsant therapy, albeit with attention to which agents are used and their dosages (Razaz et al. 2017).
- Maternal doses may need to be adjusted as the pregnancy progresses. Levels should be checked at least monthly (Battino et al. 2013).
- Prenatal screening for malformations and other adverse effects on the fetus is difficult. In some cases, a detailed ultrasound or fetal echocardiogram will detect a malformation or fetal growth restriction. If a neural tube defect is suspected, AFP on maternal serum or amniotic fluid will help confirm the diagnosis.
- With some of the first generation antiepileptic medications there is an increased risk of vitamin K-related bleeding disorders in the exposed newborn. Some sources suggest giving maternal Vitamin K in the last four weeks of pregnancy in an effort to mitigate this risk (Pennell 2003; Harden et al. 2009).
- There should be a discussion of the benefits of breast-feeding following delivery, but this should be done in conjunction with a consideration of whether the maternal medication is likely to adversely affect the breastfed infant.

RESOURCES

Information regarding the effects of antiepileptic drugs in pregnancy and participation in research studies:

The North American AED Pregnancy Registry
Massachusetts General Hospital
125 Nashua Street, Suite 8438
Boston MA 02114, USA
Phone: 1-888-233-2334
Website: www.aedpregnancyregistry.org

Organization of Teratology Information Services (OTIS/MotherToBaby)
See website for a local Teratogen Information Service or call
Phone: (866) 626-6847 (toll free)
Website: *www.otispregnancy.org*

Motherisk – Canadian Teratogen Information Service
Phone: (416) 813-6780
Website: *www.motherisk.org*

European Network of Teratogen Information Services (ENTIS)
Website: *www.entis-org.com*
Epilepsy Resources

Epilepsy Foundation
Phone: 1-800-332-1000
Website: *www.epilepsyfoundation.org*

American Epilepsy Society
Phone: 1-312-883-3800
Website: *www.aesnet.org*

REFERENCES

Adab N, Kini U, Vinten J, Ayres J, Baker G, Clayton-Smith J, Coyle H, Fryer A, Gorry J, Gregg J, Mawer G, Nicolaides P, Pickering L, Tunnicliffe L, Chadwick DW (2004) The longer term outcome of children born to mothers with epilepsy. *J Neurol Neurosurg Psychiatry* 25:1575–1583.

Adams J, Harvey EA, Holmes LB (2000) Cognitive deficits following gestational monotherapy with phenobarbital and carbamazepine. *Neurotoxicol Teratol* 22:466.

Al-Shammri S, Guberman A, Hsu E (1992) Neuroblastoma and fetal exposure to phenytoin in a child without dysmorphic features. *Can J Neurol Sci* 19:243–245.

Almgren M, Kallen B, Lavebratt C (2009) Population-based study of antiepileptic drug exposure in utero--influence on head circumference in newborns. *Seizure* 18(10):672-675.

Ardinger HH, Atkin JF, Blackston RD, Elas LJ, Clarren SK, Livingstone S et al. (1998) Verification of the fetal valproate syndrome phenotype. *Am J Med Genet* 29:171-185.

Arpino C, Brescianini S, Robert E, Castilla EE, Cocchi G, Cornel MC, de Vigan C, Lancaster PA, Merlob P, Sumiyoshi Y, Zampino G, Renzi C, Rosano A, Mastroiacovo P (2000) Teratogenic effects of antiepileptic drugs: Use of an International Database on Malformations and Drug Exposure (MADRE). *Epilepsia* 41:1436–1443.

Azzato EM, Chen RA, Wacholder S et al. (2010) Maternal EPHX1 polymorphisms and risk of phenytoin-induced congenital malformations. *Pharmacogenet Genomics* 20(1):58-63.

Ban L, Fleming KM, Doyle PM et al. (2015) Congenital anomalies in children of mother taking antiepileptic drugs with and without periconceptual high dose folic acid use: a population-based cohort study. *PLosOne* 10:e0131130.

Banach R, Boskovic R, Einarson T, Koren G (2010) Long-term developmental outcome of children of women with epilepsy, unexposed or exposed prenatally to antiepileptic drugs: a meta-analysis of cohort studies. *Drug Saf* 33(1):73-79.

Battino D, Tomson T, Bonizzoni E et al. (2013) Seizure control and treatment changes in pregnancy: Observations from the EURAP epilepsy pregnancy registry. *Epilepsia* 54(9):1621–1627.

Battino D, Tomson T (2017) EURAP An international antiepileptic drugs and pregnancy registry: Interim report May 2017. http://www.eurapinternational.org/index

Bromley R, Weston J, Adab N et al. (2014) Treatment for epilepsy in pregnancy: neurodevelopmental outcomes in the child. *Cochrane Database Syst Rev* 10: CD010236.

Bromley RL, Weston J, Marson AG (2017) Maternal Use of Antiepileptic Agents During Pregnancy and Major Congenital Malformations in Children. *JAMA* 318(17):1700–1701.

Bromley RL, Calderbank R, Cheyne CP et al. (2016) Cognition in school-age children exposed to levetiracetam, topiramate, or sodium valproate. *Neurology* 87(18):1943–1953.

Bromley RL, Mawer GE, Briggs M et al. (2013) The prevalence of neurodevelopmental disorders in children prenatally exposed to antiepileptic drugs. *J Neurol Neurosurg Psychiatry* 84(6):637–43.

Buehler BA, Rao V, Finnell RH (1994) Biochemical and molecular teratology of fetal hydantoin syndrome. *Neurol Clin* 12: 741–748.

Canger R, Battino D, Canevini MP, Fumarola C, Guidolin L, Vignoli A, Mamoli D, Palmieri C, Molteni F, Granata T, Hassibi P, Zamperini P, Pardi G, Avanzini G (1999) Malformations in offspring of women with epilepsy: A prospective study. *Epilepsia* 40:1231–1236.

Caughey AB (2016) Seizure disorders in pregnancy http://emedicine.medscape.com/article/272050-overview, accessed 6/30/17.

Chambers CD, Kao KK, Felix RJ, Alvarado S, Chavez C, Ye N, Dick LM, Jones KL (2005) Outcome in infants prenatally exposed to newer anticonvulsants. *Birth Defects Res A* 73(5):316.

Cohen MJ, Meador KJ, Browning N et al. (2011) Fetal antiepileptic drug exposure: motor, adaptive, and emotional/behavioral functioning at age 3 years. *Epilepsy Behav* 22(2):240–246.

Cunnington MC, Weil JG, Messenheimer JA et al. (2011) Final results from 18 years of the International Lamotrigine Pregnancy Registry. *Neurology* 6(21):1817–23.

Day W, Yee S, Peterson C, Koren G (2011) Assessment of the teratogenic risk in fetuses exposed to topiramate in utero. *Birth Defects Res A* 91:356.

Dean JC, Moore SJ, Turnpenny PD (2000) Developing diagnostic criteria for the fetal anticonvulsant syndromes. *Seizure* 9:233–234.

Deshmukh U, Adams J, Macklin EA, et al. (2016) Behavioral outcomes in children exposed prenatally to lamotrigine, valproate, or carbamazepine. *Neurotoxicol Teratol* 54:5–14.

DiLiberti JH, Farndon PA, Dennis NR, Curry CJ (1984) The fetal valproate syndrome. *Am J Med Genet* 19:473–481.

Forsberg L, Wide K, Kallen B (2011) School performance at age 16 in children exposed to antiepileptic drugs in utero - a population-based study. *Epilepsia* 52(2):364–369.

George IC (2017) How do you treat epilepsy in pregnancy? *Neurol Clinl Pract* 7(4):363–371.

Glover SJ, Quinn AG, Barter P, Hart J, Moore SJ, Dean JC, Turnpenny PD (2002) Ophthalmic findings in fetal anticonvulsant syndrome(s). *Ophthalmology* 109:942–947.

Green MW, Seeger JD, Peterson C, Bhattacharyya A (2012) Utilization of topiramate during pregnancy and risk of birth defects. *Headache* 52(7):1070–1084.

Hanson JW (1986) Teratogen update: Fetal hydantoin effects. *Teratology* 33:349–353.

Hanson JW, Smith DW (1975) The fetal hydantoin syndrome. *J Pediatr* 87:285–290.

Harden CL, Meador KJ, Pennell PB et al. (2009) Practice parameter update. *Neurology* 73(2):133–141.

Harden CL, Pennell PB, Koppel BS, et al. (2009) Management issues for women with epilepsy—focus on pregnancy (an evidence-based review): III. Vitamin K, folic acid, blood levels, and breast-feeding: Report of the Quality Standards Subcommittee and Therapeutics and Technology Assessment Subcommittee of the American Academy of Neurology and the American Epilepsy Society. *Epilepsia* 50(5):1247–55.

Hernandez-Diaz S, Huybrechts KF, Desai RJ, Cohen JM, Jogun H, Pennell PB, Bateman BT, Patorno E (2018). Topiramate use early in pregnancy and the risk of oral clefts: A pregnancy cohort study. *Neurology* 90(4):e342–e351.

Hernandez-Diaz S, McElrath TF, Pennell PB et al. (2017) Fetal growth and premature delivery in pregnant women on antiepileptic drugs. *Ann Neurol* 82:457–465.

Hernandez-Diaz S, Smith CR, Shen A et al. (2012) Comparative safety of antiepileptic drugs during pregnancy. *Neurology* 78:1692–1699.

Holmes LB (2002) The teratogenicity of anticonvulsant drugs: A progress report. *J Med Genet* 39:245–247.

Holmes LB, Harvey EA, Coull BA, Huntington KB, Khoshbin S, Hayes AM, Ryan LM (2001) The teratogenicity of anticonvulsant drugs. *N Engl J Med* 344:1132–1138.

Holmes LB, Hernandez-Diaz S (2012) Newer anticonvulsants: lamotrigine, topiramate and gabapentin. *Birth Defects Res A* 94(8):599–606.

Holmes LB, Mittendorf R, Shen A, Smith CR, Hernandez-Diaz S (2011) Fetal effects of anticonvulsant polytherapies: different risks from different drug combinations. *Arch Neurol* 68(10):1275–1281.

Howe AM, Oakes DJ, Woodman PD, Webster WS (1999) Prothrombin and PIVKA-II levels in cord blood from newborn exposed to anticonvulsants during pregnancy. *Epilepsia* 40:980–984.

Hoyme HE, Hauck L, Quinn D (1998) Minor anomalies accompanying prenatal exposure to topiramate. *J Investig Med* 46(1):119A.

Hunt S, Russell A, Smithson WH, Parsons L et al. (2008) Topiramate in pregnancy: preliminary experience from the UK Epilepsy and Pregnancy Register. *Neurology* 71(4):272–276.

Jentink J, Dolk H, Loane MA et al. (2010) Intrauterine exposure to carbamazepine and specific congenital malformations: systematic review and case-control study. *BMJ* 341:c6581.

Jones KL, Lacro RV, Johnson KA, Adams J (1989) Pattern of malformations in the children of women treated with carbamazepine during pregnancy. *N Engl J Med* 320:1661–1666.

Jones KL, Johnson KA, Chambers CC (1992) Pregnancy outcome in women treated with phenobarbital monotherapy. *Teratology* 45:452.

Kaaja E, Kaaja R, Matila R, Hiilesmaa V (2002) Enzyme-inducing antiepileptic drugs in pregnancy and the risk of bleeding in the neonate. *Neurology* 58(4):549–553.

Kaaja E, Kaaja R, Hiilesmaa V (2003) Major malformations in offspring of women with epilepsy. *Neurology* 60:575–579.

Kini U, Adab N, Vinten J, Fryer A, Clayton-Smith J (2006) Dysmorphic features: an important clue to the diagnosis and severity of fetal anticonvulsant syndromes. *Arch Dis Child Fetal Neonatal Ed* 91(2):F90–F95.

Kjaer D, Horvath-Puho E, Christensen J, et al. (2007) Use of phenytoin, phenobarbital, or diazepam during pregnancy and risk of congenital abnormalities: a case-time control study. *Pharmacoepidemiol Drug Saf* 16(2):181–188.

Koch S, Gopfert-Geyer I, Hauser I, Hartmann A, Jakob S, Jager-Roman E, Nau H, Rating D, Helge H (1985) Neonatal behaviour disturbances in infants of epileptic women treated during pregnancy. *Prog Clin Biol Res* 163B:453–461.

Koch S, Losche G, Jager-Roman E, Jakob S, Rating D, Deichl A, Helge H (1992) Major and minor birth malformations and antiepileptic drugs. *Neurology* 42:83–88.

Koch S, Jager-Roman E, Losche G, Nau H, Rating D, Helge H (1996) Antiepileptic drug treatment in pregnancy: Drug side effects in the neonate and neurological outcome. *Acta Paediatr* 85:739–746.

Koch S, Titze K, Zimmermann RB, Schroder M, Lehmkuhl U, Rauh H (1999) Long-term neuropsychological consequences of maternal epilepsy and anticonvulsant treatment during pregnancy for school-age children and adolescents. *Epilepsia* 40:1237–1243.

Koren G, Demitrakoudis D, Weksberg R, Rieder M, Shear NH, Sonely M, Shandling B, Spielberg SP (1989) Neuroblastoma after prenatal exposure to phenytoin: Cause and effect? *Teratology* 40:157–162.

Kozma C (2001) Valproic acid embryopathy: Report of two siblings with further expansion of the phenotypic abnormalities and a review of the literature. *Am J Med Genet* 98:168–175.

Margulis AV, Mitchell AA, Gilboa SM et al. (2012) National Birth Defects Prevention Study: Use of topiramate in pregnancy and risk of oral clefts. *Am J Obstet Gynecol* 207(5):405.e1–405.e7.

McNinch AW, Tripp JH (1991) Haemorrhagic disease of the newborn in the British Isles: two year prospective study. *BMJ* 303(6810):1105–1109.

Meador K, Reynolds MW, Crean S, Fahrbach K, Probst C (2008) Pregnancy outcomes in women with epilepsy: A systematic review and meta-analysis of published pregnancy registries and cohorts. *Epilepsy Res* 81:1–13.

Meador, KJ, Baker, GA, Browning, N, Clayton-Smith, J, Combs-Cantrell, DT, Cohen, M, Kalayjian, LA, Kanner, A, Liporace, JD, Pennell, PB, Privitera, M, Loring, DW, NEAD Study Group (2009) Cognitive function at 3 years of age after fetal exposure to antiepileptic drugs. *N Engl J Med* 360:1597–1605.

Meador KJ, Baker GA, Browning N et al. (2012) Effects of fetal antiepileptic drug exposure: outcomes at age 4.5 years. *Neurology* 78(16):1207–14.

Meador KJ, Baker GA, Browning N et al. (2013) Fetal antiepileptic drug exposure and cognitive outcomes at age 6 years (NEAD study): a prospective observational study. *Lancet Neurol* 12(3):244–252.

Meador KJ and Loring DW (2016) Developmental effects of antiepileptic drugs and the need for improved regulations. *Neurology* 86:297–306.

Meadow SR (1968) Anticonvulsant drugs and congenital abnormalities. *Lancet* 2:1296.

Molgaard-Nielsen D, Hviid A (2011) Newer-generation antiepileptic drugs and the risk of major birth defects. *JAMA* 305(19):1996–2002.

Moore SJ, Turnpenny P, Quinn A, Glover S, Lloyd DJ, Montgomery T, Dean JC (2000) A clinical study of 57 children with fetal anticonvulsant syndromes. *J Med Genet* 37:489–497.

Morrow J, Russell A, Guthrie E et al. (2006) Malformation risks of antiepileptic drugs in pregnancy: a prospective study from the UK Epilepsy and Pregnancy Register. *J Neurol Neurosurg Psych* 77(2):193–198.

The North American Antiepileptic Drug Pregnancy Registry (2016) Update on monotherapy findings: Comparative safety of 11 antiepileptic drugs used during pregnancy. Winter:1–2.

Ornoy A (2006) Neuroteratogens in man: an overview with special emphasis on the teratogenicity of antiepileptic drugs in pregnancy. *Reprod Toxicol* 22(2):214–226.

Pariente G, Leibson T, Shulman T et al. (2017) Pregnancy outcomes following in utero exposure to lamotrigine: A systematic review and meta-analysis. *CNS Drugs* 31:439–450.

Pennell PB (2003) The importance of monotherapy in pregnancy. *Neurology* 60:S31–S38.

Petersen I, Collings SL, McCrea RL, Nazareth I, Osborn DP, Cowen PJ, Sammon CJ (2017) Antiepileptic drugs prescribed in pregnancy and prevalence of major congenital malformations: comparative prevalence studies. *Clin Epidemiol* 9:95–103.

Rasalam AD, Hailey H, Williams JH, Moore SJ, Turnpenny PD, Lloyd DJ, Dean JC (2005) Characteristics of fetal anticonvulsant syndrome associated with autistic disorder. *Dev Med Child Neurol* 47:551–555.

Razaz N, Tomson T, Wikstrom, AK et al. (2017) Association between pregnancy and perinatal outcomes among women with epilepsy. *JAMA Neurology* 74(8):983–991.

Rihtman T, Parush S, Ornoy A (2012) Preliminary findings of the developmental effects in utero exposure to topiramate. *Reprod Tox* 34(3):308–311.

Rihtman T, Parush S, Ornoy A (2013) Developmental outcomes at preschool age after fetal exposure to valproic acid and lamotrigine: cognitive, motor, sensory and behavioral function. *Reprod Toxicol* 41:115–125.

Roullet FI, Lai JKY, Foster JA (2013) In utero exposure to valproic acid and autism--a current review of clinical and animal studies. *Neurotoxicol Teratol* 36:47–56.

Samren EB, van Duijn CM, Christiaens GC, Hofman A, Lindhout D (1999) Antiepileptic drug regimens and major congenital abnormalities in the offspring. *Ann Neurol* 46:739–746.

Seaver LH, Hoyme HE (1992) Teratology in pediatric practice. *Pediatr Clin North Am* 39:111–134.

Speidel BD, Meadow SR (1972) Maternal epilepsy and abnormalities of the fetus and newborn. *Lancet* 2:839–843.

Tomson T, Marson A, Boon P et al. (2015) Valproate in the treatment of epilepsy in girls and women of childbearing potential. *Epilepsia* 56(7):1006–1019.

Tomson T, Battino D (2012) Teratogenic effects of antiepileptic drugs. *Lancet Neurol* 11(9):803–813.

Tomson T, Battino D, Bonizzoni E et al (2011) Dose-dependent risk of malformations with antiepileptic drugs: an analysis of data from the EURAP epilepsy and pregnancy registry. *Lancet Neurol* 10(7):609–17.

Tomson T, Battino D, Bonizzoni E, Craig J, Lindhout D, Perucca E, Sabers A, Thomas S, Vaijda F (2018) Comparative risk of major congenital malformations with eight different antiepileptic drugs: a prospective cohort study of the EURAP registry. *Lancet Neurol* 17:530–38.

Tung EW, Winn LM (2010) Epigenetic modifications in valproic acid-induced teratogenesis. *Toxicol Appl Pharmacol* 248(3):201–9.

Vajda FJ, O'Brien TJ, Lander CM et al. (2014). The teratogenicity of the newer antiepileptic drugs - an update. *Acta Neurol Scand* 130(4):234–8.

Veiby G, Daltveit AK, Engelsen BA, Gilhus NE (2014) Fetal growth restriction and birth defects with newer and older antiepileptic drugs during pregnancy. *J Neurol* 261(3):579–588.

Veiby G, Daltveit AK, Schjolberg S et al. (2013) Exposure to antiepileptic drugs in utero and child development: a prospective population-based study. *Epilepsia* 54(8):1462–1472.

Velez-Ruiz NJ, Meador KJ (2015) Neurodevelopmental Effects of Fetal Antiepileptic Drug Exposure. *Drug Saf* 38(3):271–278.

Veroniki AA, Rios P, Cogo E et al. (2017a) Comparative safety of antiepileptic drugs for neurological development in children exposed during pregnancy and breastfeeding: a systemic review and network meta-analysis. *BMJ Open* 7:e017248.

Veroniki AA, Cogo E, Rios P, Straus SE, Finkelstein Y, Kealey R, Reynen E, Soobiah C, Thavorn K, Hutton B, Hemmelgarn BR, Yazdi F, D'Souza J, MacDonald H, Tricco AC (2017b) Comparative safety of anti-epileptic drugs during pregnancy: a systematic review and network meta-analysis of congenital malformations and prenatal outcomes. *BMC Med* 15(1):95

Vinten J, Adab N, Kini U, Gorry J, Gregg J, Baker GA, (2005) Liverpool, Manchester Neurodevelopment Study Group Neuropsychological effects of exposure to anticonvulsant medication *in utero*. *Neurology* 64:949–954.

Wegner C, Nau H (1992) Alteration of embryonic folate metabolism by valproic acid during organogenesis: Implications for mechanism of teratogenesis. *Neurology* 42:17–24.

Werler MM, Ahrens KA, Bostco JLF (2011) Use of antiepileptic medications in pregnancy in relation to risks of birth defects. *Ann Epidemiol* 21(11):842–850.

Weston J, Bromley R, Jackson CF et al. (2016) Monotherapy treatment of epilepsy in pregnancy: congenital malformation outcomes in the child. *Cochrane Database Syst Rev* 11:CD010224.2.

Wide K, Winbladh B, Kallen B (2004) Major malformations in infants exposed to antiepileptic drugs *in utero*, with emphasis on carbamazepine and valproic acid: A nation-wide, population-based register study. *Acta Paediatr* 93:174–176.

Zackai EH, Mellman WJ, Neiderer B, Hanson JW (1975) The fetal trimethadione syndrome. *J Pediatr* 87:280–284.

28

FRAGILE X SYNDROME AND PREMUTATION-ASSOCIATED DISORDERS

RANDI J. HAGERMAN

Endowed Chair in Fragile X Research, Distinguished Professor of Pediatrics, UC, Davis Health, Sacramento, California, USA

INTRODUCTION

Incidence

Fragile X syndrome is the most common inherited cause of intellectual disability. Lubs identified the first individuals with fragile X syndrome in 1969 when the fragile site at the bottom end of the X chromosome was noted in cytogenetic studies. Individuals were diagnosed throughout the 1970s and 1980s with the use of cytogenetic studies carried out with tissue culture medium that was deficient in folic acid, which allowed the fragile site to be visible. In 1991, the gene for fragile X syndrome was identified and sequenced by an international collaborative effort (Verkerk et al. 1991). Subsequent to this discovery, DNA testing has been available for the diagnosis of fragile X syndrome and premutation-related disorders including primary ovarian insufficiency (POI) and the fragile X-associated tremor/ataxia syndrome (FXTAS).

Fragile X syndrome is caused by a trinucleotide repeat expansion of three bases (cytosine-guanine-guanine), $(CGG)_n$, that occurs in the fragile X mental retardation 1 gene (*FMR1*) at Xq27.3. A small expansion, or premutation, has approximately 55–200 CGG repeats and is usually not associated with cognitive deficits. However, a larger expansion, greater than 200 CGG repeats, the full mutation, is associated with fragile X syndrome, including typical physical features such as a long face, prominent ears, and macroorchidism, in addition to intellectual deficits. When a full mutation is present, the gene usually becomes methylated and is not expressed.

The so-called premutation, or small expansion of the CGG repeat, is relatively common, having been identified in 1 in 130–259 females in the general population and 1 in 250–813 males in the general population (Hagerman and Hagerman 2013). The allele frequency of the full mutation is approximately 1 in 2500 to 1 in 6000 (Hagerman et al. 2017). Although the majority of these individuals have intellectual disabilities, about 15% of males and 70% of females with the full mutation will have an IQ above 70 and present with learning disabilities or emotional problems (Hagerman et al. 1994; de Vries et al. 1996).

Rare individuals with fragile X syndrome have experienced sudden death from seizures or cardiovascular events such as an arrhythmia (Waldstein and Hagerman, 1988). However, the life span for those with fragile X syndrome is thought to be normal. Studies are in process investigating the possibility of late onset dementia in a subgroup of those with fragile X syndrome because the protein produced by the fragile X gene (FMRP) regulates the expression of amyloid precursor protein (APP), elevated levels of which may predispose an individual to Alzheimer disease (Westmark and Malter 2007). A rare individual with fragile X syndrome who is mosaic or unmethylated will produce excess *FMR1* mRNA levels and is therefore at risk for FXTAS with neurodegeneration (Loesch et al. 2012).

Cassidy and Allanson's Management of Genetic Syndromes, Fourth Edition.
Edited by John C. Carey, Agatino Battaglia, David Viskochil, and Suzanne B. Cassidy.
© 2021 John Wiley & Sons, Inc. Published 2021 by John Wiley & Sons, Inc.

Diagnostic Criteria

The spectrum of clinical involvement of fragile X syndrome is quite broad. It includes mild emotional problems or learning disabilities in individuals with a normal IQ through all levels of intellectual disability. The classical physical features include large or prominent ears, a long face, high-arched palate, prognathism (mainly in older individuals), hyperextensible finger joints, subluxable thumbs, mitral valve prolapse, macroorchidism (large testicles), flat feet, and soft, velvet-like skin. Many of these features relate to a mild connective tissue disorder. Orthopedic complications, such as congenital hip dislocation or patellar dislocation, in addition to scoliosis and pes planus, are caused by connective tissue dysplasia. Individuals who are less affected cognitively will have fewer physical features (Loesch et al. 2004).

Prominent ears are perhaps the most common physical feature of this disorder, but approximately 20–25% of young children will not demonstrate this manifestation (Figures 28.1 and 28.2). Macroorchidism typically begins at approximately 9 years of age, and testicles will increase in size throughout puberty, with a mean testicular volume in adulthood of approximately 50 ml. Macroorchidism is therefore usually not present in early childhood. Because many young children look absolutely normal physically, the behavioral features including perseveration, hand flapping, and hyperactivity are important clues to this diagnosis.

The clinician should have a high index of suspicion for fragile X syndrome when evaluating any individual with significant cognitive deficits. Fragile X syndrome represents approximately 30% of all cases of X-linked intellectual disability, and it is the most commonly known inherited cause of intellectual disability (Sherman, 2002).

Females are less affected by fragile X syndrome than males because they have two X chromosomes, and the normal X is producing variable amounts of *FMR1* protein

FIGURE 28.1 Two young adult brothers with fragile X syndrome. Note long face and mildly prominent ears with cupping of the upper pinnae, particularly notable in the brother on the right.

FIGURE 28.2 Young boy with a full mutation that is partially unmethylated. He presents with learning disabilities, not intellectual disability, and his facial features are normal except for slight prominence of his left ear.

(FMRP), depending on the X inactivation ratio. The level of FMRP correlates with the degree of cognitive involvement in both males and females (Loesch et al. 2004).

Etiology, Pathogenesis, and Genetics

Fragile X syndrome is caused by a trinucleotide repeat expansion of three bases (cytosine-guanine-guanine), $(CGG)_n$, that occurs in the fragile X mental retardation 1 gene (*FMR1*) at Xq27.3. The full mutation includes CGG repeat numbers ranging from 200 to approximately 2000 repeats. Within the premutation range (55–200 repeats), the size of the CGG repeat expansion correlates with the risk of passing on a full mutation from a mother to the next generation. For instance, women with more than 90–100 CGG repeats have approximately 100% risk for expansion to the full mutation when that X chromosome is passed on to the next generation (Nolin et al. 2003). The presence of an AGG anchor after every 10 CGG repeats will reduce the risk of passing on a full mutation from a woman with less than 90 repeats, and this should be addressed in genetic counseling (Yrigollen et al. 2012). There is a low likelihood (<1%) of decrease (contraction) of the CGG repeat to the normal range (Nolin et al. 2003). Males with the premutation will pass on the premutation to 100% of their daughters but to none of their sons, because sons inherit the Y chromosome not the X. Both small contractions and small expansions can be seen when the premutation is passed on by a male, but the CGG repeat number remains within the premutation range.

On occasion, individuals with a premutation may be clinically affected with learning disabilities including attention deficit hyperactivity disorder (ADHD), or emotional difficulties including anxiety, social deficits, or autism spectrum disorders, although the majority of individuals with the premutation have an IQ in the normal range (Farzin et al. 2006).

It is the absence or deficiency of FMRP that causes the clinical features of fragile X syndrome. In males with the full mutation that is fully methylated (and therefore not expressed), little or no FMRP is produced. There is some correlation between the degree of expression of the *FMR1* gene, as reflected in the extent of methylation, and the clinical expression and severity. An example of this relates to intelligence and is discussed below in Development and Behavior.

FMRP is an RNA-binding protein that regulates the translation of a variety of messages that are important in synaptic plasticity, typically through inhibition (Hagerman et al. 2017). FMRP binds to approximately 4% of human fetal brain messages, particularly those with a G quartet structure, and the lack of FMRP leads to up-regulation of many proteins in the brain (Qin et al. 2005). Neurobiological advances in fragile X syndrome have demonstrated enhanced activity of the metabotropic glutamate receptor 5 (mGluR5) pathway when FMRP is absent. Therefore, mGluR5 antagonists are targeted treatments in fragile X syndrome, and they have been shown to reverse the cognitive and behavioral deficits in animal models of fragile X syndrome (Dolen and Bear 2008); however clinical trials have not shown efficacy of mGluR5 antagonists (Berry-Kravis et al. 2016). In addition, minocycline is considered a targeted treatment for fragile X syndrome because studies in newborn fragile X mice have demonstrated that a one month treatment with minocycline can mature the synaptic connections and improve behavioral studies and a cognitive task (Bilousova et al. 2009). Human trials of minocycline in children with fragile X syndrome have shown limited efficacy described below (Leigh et al. 2013).

Point mutations within *FMR1*, as opposed to the usual triplet expansion, have been described in an increasing number of individuals with ID as whole exome sequencing studies have been frequently ordered (reviewed in Sitzmann et al. 2018). Deletion of the *FMR1* gene will also lead to a typical phenotype of fragile X syndrome. Deletion of a larger segment of DNA that removes the *FMR1* gene and additional genes will have a more severe phenotype than just fragile X syndrome alone. For example, Quan et al. (1995) described a person with anal atresia in addition to fragile X syndrome who had a large deletion that removed *FMR1* and a region proximal to *FMR1*.

A subgroup of individuals with fragile X syndrome will have a phenotype that includes obesity, hyperphagia, lack of satiation with meals, hypogonadism, and delayed puberty, all of which are reminiscent of Prader–Willi syndrome (see Chapter 46). This subgroup of approximately 5% of those with fragile X syndrome has been called the Prader–Willi phenotype. Individuals with fragile X syndrome and the Prader–Willi phenotype have been found to have decreased expression of *CYFIP* that is in the 15q deletion Prader–Willi syndrome region (Nowicki et al. 2007).

The etiology of the connective tissue problems seen in fragile X syndrome presumably is related to a connective tissue gene(s) that is regulated by FMRP, such that the lack of FMRP causes a deficiency or upregulation in translation of this gene(s).

Abnormalities in Premutation Carriers. Both males and females with the premutation may have significant clinical involvement (Hagerman and Hagerman 2013; Hessl et al. 2005; Farzin et al. 2006). Fragile X-associated tremor/ataxia syndrome (FXTAS) has been reported in both males and females with the premutation (Hagerman and Hagerman 2016). The clinical features include a progressive intention tremor, cerebellar ataxia, Parkinsonian features, brain atrophy, hyperintensities in the middle cerebellar peduncles and the splenium of the corpus callosum on T_2 in magnetic resonance imaging, memory, and executive function deficits. FXTAS typically begins after age 50 years, and penetrance is age-related with approximately 40% of male carriers and 8–16% of female carriers affected with FXTAS (Hagerman and Hagerman 2016). Other neurological problems may also occur in individuals with the premutation, including neuropathy, fibromyalgia, hypertension, autonomic dysfunction, migraine headaches, hypothyroidism, anxiety, depression, and primary ovarian insufficiency (Sullivan et al. 2005; Hagerman and Hagerman 2013). FMRP levels are usually normal or near normal, but messenger RNA levels are elevated in the premutation and the cause of FXTAS and other premutation problems is thought to be an RNA toxicity although there is evidence of repeat associated non-AUG translation (RAN) leading to a toxic product of FMRpolyG also (Hagerman and Hagerman 2016; Todd et al. 2013). Thus, fragile X-associated tremor/ataxia syndrome is different from fragile X syndrome in both etiology and clinical manifestations.

The premutation can also cause significant emotional problems in some carriers, including anxiety disorders, mood instability, and depression (Roberts et al. 2009). Although most individuals with the premutation do not have clinically significant problems in this area, about 30–40% do manifest these problems.

Genetic Counseling. Genetic counseling for fragile X syndrome is very complex because of the variable involvement of premutation carriers.

Recurrence risks for fragile X syndrome and knowledge of which family members are at risk vary considerably with the sex of the affected or carrier individual in question. In addition, the AGG anchors are important for assigning risk for expansion in the next generation (Yrigollen et al. 2012).

Families that include an individual with fragile X syndrome or premutation involvement should be referred to a geneticist or genetic counselor to identify and consider genetic testing of those at risk of having affected children. Genetic counseling should also provide an explanation of the options available for reproductive decision making (McConkie-Rosell et al. 2007).

Diagnostic Testing

The diagnosis of fragile X syndrome must be confirmed by DNA testing, which includes both Southern blot and PCR (Saluto et al. 2005). Cytogenetic testing is not adequate to make the diagnosis because high functioning individuals with the full mutation and carriers with the premutation are negative on cytogenetic testing. A new blood spot test to screen for both the premutation and the full mutation has been reported, and this test is being used presently in research for high risk and newborn screening studies (Tassone et al. 2012).

Once a proband with fragile X syndrome or premutation involvement has been identified in a family, DNA testing of family members can be offered. This type of cascade testing, combined with genetic counseling, will usually reveal many individuals who experience clinical problems related to the premutation and the full mutation (McConkie-Rosell et al. 2007; Wheeler et al. 2017).

DNA *FMR1* testing should be considered in all individuals who present with intellectual disability or autism spectrum disorders when the etiology for these problems is not known. Individuals who present with just hyperactivity or ADHD are not routinely tested for fragile X syndrome unless typical physical features, cognitive deficits, or behavioral problems reminiscent of fragile X syndrome are present or there is a family history of intellectual disability compatible with an X-linked inheritance pattern. Typically, the DNA testing of individuals with intellectual disability and no family history will lead to a 2–5% positive rate for the *FMR1* abnormality (Song et al. 2003). However, in some parts of the world the prevalence of fragile X syndrome is much higher, such as Colombia (Saldarriaga et al. 2018) and in other areas such as Ireland the prevalence is lower (O'Byrne et al. 2017).

Differential Diagnosis

The typical features of fragile X syndrome, including intellectual disability, attention deficit hyperactivity disorder, and autistic-like features such as poor eye contact, hand flapping, hand biting, and perseverative speech, can be seen in a number of disorders. These include non-specific intellectual disability, fetal alcohol syndrome (see Chapter 26), and autism spectrum disorder (ASD).

Some individuals suspected of having Sotos syndrome (see Chapter 55), Prader–Willi syndrome (see Chapter 46), FG syndrome, or Pierre Robin sequence will instead be positive for fragile X on DNA testing (Hagerman 2002b). There is also a variety of X-linked disorders that have physical features reminiscent of fragile X syndrome, such as Coffin–Lowry syndrome (see Chapter 12); Lujan–Fryns syndrome (which includes marfanoid habitus and macroorchidism); and Atkin syndrome (which includes large ears, short stature, and macroorchidism). In addition, fragile X syndrome has been reported in conjunction with a variety of common chromosomal aneuploidy syndromes, such as Klinefelter syndrome (XXY) (see Chapter 35), Turner syndrome (see Chapter 60), XXX, and Down syndrome (see Chapter 24). Therefore, individuals who have these disorders and also have typical features of fragile X syndrome should have fragile X testing. These combinations are likely to be coincidental.

MANIFESTATIONS AND MANAGEMENT

Individuals with fragile X syndrome require careful follow-up by the physician, who usually orchestrates the intervention of multiple professionals. An intensive treatment program will help the affected person achieve his or her potential and find a productive role in our society. New treatments for fragile X syndrome have been summarized in a recent report (Hagerman et al. 2017). Recent reports have shown benefit of low dose sertraline (2.5–5.0 mg) in children ages 2–6 years with fragile X syndrome (Greiss Hess et al. 2016; Winarni et al. 2013). Limited efficacy of minocycline (50–100 mg/day) in children with fragile X syndrome has also been demonstrated (Leigh et al. 2013). Recently metformin has been shown to rescue the phenotype of fragile X syndrome in the mouse and in the *Drosophila* models because it lowers the upregulation of the MEK-ERK and mTOR pathways that is seen in the absence of FMRP (Monyak et al. 2016; Gantois et al. 2017). An open-label study of metformin in children and adults with fragile X syndrome has shown benefit in language and behavior (Dy et al. 2017), and a controlled trial of metformin is underway at three centers. Additional new targeted treatments such as cannabidiol (CBD), and trofinetide may demonstrate efficacy in the near future (Hagerman et al. 2017).

The management of those with fragile X syndrome can benefit from specialized clinics that include a team of clinicians who have expertise in treating these individuals including therapists (speech and language, and occupational and physical therapists), psychologists, developmental and behavioral pediatricians or psychiatrists or neurologists with experience in psychopharmacology, and geneticists and genetic counselors. The National Fragile X Foundation has established such clinics around the US and internationally and their locations are on their website (www.fragileX.org)

Growth and Feeding

There may be a mild overgrowth syndrome associated with fragile X syndrome. Individuals usually have a normal to increased birth weight, and the head circumference may be large at the time of birth because the overall brain size, particularly the hippocampus, caudate, and thalamus, is increased in individuals with fragile X syndrome (Reiss and Dant 2003). Growth in childhood is usually slightly enhanced in both males and females with fragile X syndrome, although the timing of puberty is usually normal (Loesch et al. 1995). The growth spurt that occurs during puberty is somewhat diminished in fragile X syndrome compared with controls, and therefore, final height may be shorter than average.

Feeding problems are common in infancy, and recurrent emesis is usually associated with gastroesophageal reflux (Goldson and Hagerman 1993). Usually, feeding problems and emesis improve with age.

Evaluation

- If feeding problems are severe, an esophageal pH probe study should be done to assess for gastroesophageal reflux.
- A barium swallow can be used to assess reflux treatment.

Management

- Thickening of feedings and upright positioning after meals are usually sufficient for treatment of reflux.
- No treatment is needed for growth abnormalities.

Development and Behavior

Some infants with fragile X syndrome may be irritable in the first year, presumably because of sensory integration problems and tactile defensiveness. Hypotonia and mild motor delays are relatively common. The loose connective tissue and joint hyperextensibility may further interfere with achievement of normal motor milestones. Tantrum behavior and hyperactivity may begin in the second year, particularly after children learn to walk (Hatton et al. 2002). The tantrums commonly occur during times of transition, such as coming home after a busy day, or in environments with excessive stimulation, such as shopping in a grocery store. They often appear to be related to excessive sensory stimulation and the lack of appropriate GABA inhibition seen in fragile X syndrome (Hagerman et al. 2017).

Language delays are usually noted by 2–3 years of age (Abbeduto et al. 2007), and unusual autistic-like features such as hand flapping, poor eye contact, social anxiety, and self-injurious behavior including hand biting typically begin by the second or third year of life (Symons et al. 2003; Symons et al. 2010; Hatton et al. 2006). Approximately 60% of boys and 20% of girls with fragile X syndrome have ASD also and they typically have more severe language and cognitive deficits than those with FXS alone (Niu et al. 2017). It is important to diagnose ASD because more intensive behavioral interventions including applied behavior analysis (ABA) intervention is needed as early as possible.

The average IQ in adulthood for a male with a full mutation that is fully methylated is approximately 41 (Merenstein et al. 1996). Less affected or higher-functioning males usually have incomplete methylation, causing incomplete inactivation of *FMR1*, or the presence of mosaicism (some cells with the premutation and some cells with the full mutation). The cells with the premutation should be producing FMRP, in contrast to the cells with the full mutation that are fully methylated. The average IQ in adulthood of males who have a full mutation with greater than 50% of the cells unmethylated is approximately 88, and the average IQ of individuals with a mosaic pattern is 60 (Merenstein et al. 1996). Approximately 70% of females with the full mutation will have cognitive deficits, that is, an IQ in the borderline range (70–84) or in the mildly intellectually disabled range (de Vries et al. 1996). An occasional female with the full mutation will have moderate or severe intellectual disability.

Hyperactivity persists throughout childhood in approximately 70–80% of males and 30–50% of females with fragile X syndrome. The attention problems and impulsivity may be severe, even when hyperactivity is not present. Hyperactivity tends to improve in adolescence and adulthood (de Vries et al. 1996).

Anxiety, particularly social anxiety, is present in both males and females with fragile X syndrome and even in premutation carriers (Cordeiro et al. 2011, 2015), and it is even common in individuals who do not demonstrate hyperactivity or impulsivity. The social anxiety may be severe even in females with fragile X syndrome who have an IQ in the normal range. In males, anxiety or uncertainty can lead to aggressive outbursts in a new situation or when meeting someone unfamiliar. The treatment of anxiety may improve the aggression, as described below.

Obsessive and compulsive behavior is quite common in individuals affected by fragile X syndrome, and it is related to perseverative, or repetitive, behavior. For instance, an individual may obsess on a person or on an activity and ask questions about this issue over and over again. The treatment of obsessive-compulsive behavior with selective serotonin reuptake inhibitors (SSRIs) as described below may improve perseveration at times.

Usually, psychopharmacological intervention combined with other treatment modalities, including counseling and sensory integration occupational therapy, in addition to language intervention and special education support in school, can be very beneficial for significant behavior problems in children, adolescents, and adults with fragile X syndrome (Hagerman et al. 2009).

Psychosis or psychotic features may occur on occasion in individuals with fragile X who are severely disorganized in their thinking or who have regressed in their level of functioning. Aging, severe stress, or other disruptive factors in the environment can precipitate episodes of psychotic thinking.

Evaluation

- At the time of diagnosis, a speech and language evaluation should be performed on all affected individuals for therapy recommendations and educational planning purposes.
- An occupational therapy or motor evaluation should be done, including an assessment of sensory integration abilities for treatment purposes.
- A complete psychological evaluation that includes IQ testing is an essential part of the evaluation of cognitive deficits.
- An emotional assessment should take place to look at the degree of attention and concentration problems, in addition to anxiety, obsessive-compulsive behavior, aggression, depression, and other psychopathology.
- An evaluation for ASD that includes standardized testing such as the ADOS (Autism Diagnosis Observation Scale) is recommended in all children after age 2 years to see if ABA programming is needed (Niu et al. 2017).
- A learning disability evaluation by a special education teacher is essential to assess academic status and learning strengths and weaknesses. A computer evaluation is a desirable component of the evaluation, to assess computer software technology that can enhance learning and language abilities.
- The medical evaluation should include an assessment of connective tissue problems, hypotonia, and the degree of attention and concentration problems, and hyperactivity.
- Children who are treated with psychotropic medication require periodic physical assessment, including blood pressure, cardiac examination, an echocardiogram in certain circumstances as described below, and an assessment of behavior, including anxiety and obsessive-compulsive features.
- The use of behavior checklists, such as the Child Behavior Checklist by Achenbach or the Connors Rating Scale with a specific focus on attention deficit/hyperactivity disorder symptoms, is helpful in the medical follow-up of psychopharmacological medication.
- Often, an evaluation by a psychiatrist or developmental and behavioral pediatrician can be helpful in consultation for the assessment of psychotic symptoms and for medication recommendations.

Management

- Children who are diagnosed with fragile X syndrome in infancy require early developmental services, which may include a home program with parent training to enhance language and motor development or an infant stimulation program, often carried out on a group basis.
- Developmental preschools usually begin at age 2 years and include speech and language therapy, motor therapy either by an occupational therapist or a physical therapist, and special education support by a special education teacher. Many of these programs are in an integrated setting with both developmentally disabled children and typically developing children. Whenever possible, children with fragile X syndrome should be incorporated in a mainstreamed situation, because they model normal children very well (Scharfenaker et al. 2002; Schwarte, 2008).
- By approximately 5 years of age, children are usually placed in a kindergarten program, but they qualify for special education services, which should include speech and language therapy, occupational therapy, physical therapy, and support from the special education teacher.
- Even in the preschool period, computer technology with software programs that can enhance language skills and early academic skills, including reading and math, can be incorporated into the special education program (Braden 2002).
- Most individuals with fragile X syndrome can be mainstreamed in a school situation and mainstreamed in a working situation in the community with appropriate support (Scharfenaker et al. 2002; Schwarte, 2008).
- The use of vocational intervention in high school can lead to appropriate placement in the community in a job situation that can utilize the best of the individual's abilities (http://www.nfxf.org/html/adolescents_and_adults_project.htm). At this website, there is information about vocational training for adolescents and adults with fragile X syndrome.
- Parents should become aware of the need to avoid excessive sensory stimulation whenever possible. The avoidance of large crowds and loud noises or shielding the child from stimuli, for example, by using earphones so that the child can listen to a favorite tape or calming music while shopping, can be helpful (Scharfenaker et al. 2002; Schwarte 2008).
- Children with significant behavior problems, such as tantrums, oppositional behavior, or severe hyperactivity, usually benefit from counseling with a psychologist, either through school or on a private basis.
- Behavioral intervention techniques that emphasize the importance of decreasing excessive sensory stimulation

and the use of positive behavior reinforcement with the setting of specific goals and the use of behavioral charting are described by Braden (Braden 2000, 2002).

- The use of psychopharmacological interventions to help with specific behavior problems can be very beneficial to many children, adolescents, and adults with fragile X syndrome or developmental problems related to the premutation. One of the presenting problems for young children with fragile X syndrome is hyperactivity, with a short attention span and impulsivity. The use of stimulant medication is helpful for approximately 60–70% of children who are of school age (Berry-Kravis and Potanos 2004; Hagerman et al. 2009). Typically, use of a long-acting stimulant once a day is beneficial (Hagerman et al. 2009). Sometimes when stimulants are used for children under 5 years of age, an increase in irritability or even hyperactivity is seen. If this occurs, clonidine or guanfacine, which are antihypertensive medications that have an overall calming effect, can improve hyperactivity and hyperarousal (Hagerman et al. 2009).

- For treatment of anxiety, social phobia, obsessive-compulsive disorder, depression, and aggression, the use of an SSRI can be safe and often effective. The first SSRI that became available was fluoxetine; however, at times fluoxetine has a significant activation effect that may exacerbate hyperactivity. More frequently, sertraline in a low morning dose, 2.5–5 mg in young children ages 2–6 years has demonstrated efficacy (Greiss Hess et al. 2016), and 10–50 mg in older children or adolescents is effective for anxiety without excessive activation. In approximately 20% of cases, use of an SSRI may lead to an increase in hyperactivity, agitation, or manic symptoms. If an increase in aggression occurs with the use of an SSRI, it should be lowered or discontinued. In general, SSRI agents should be tapered off when discontinued, or a flu-like syndrome may occur in a small percentage. Usually, SSRIs do not require monitoring of blood levels, blood counts, or liver function studies; nor do they require follow-up echocardiograms. Self-destructive or suicidal ideation is rarely seen in those with fragile X syndrome, nor is emergence of these thoughts seen with an SSRI; however, these questions should always be asked initially and in follow-up when using an SSRI. Other SSRI agents include paroxetine, citalopram, and fluvoxamine. The SSRI agents have been found to be particularly helpful for females with fragile X syndrome, or even for females with the premutation who are experiencing anxiety, depression, or mood lability, as well as for males and females with fragile X syndrome who are aggressive (Hagerman et al. 2009). If severe mood instability and/or aggression occur and do not respond to an SSRI or clonidine, the use of a mood stabilizer such as aripiprazole, lithium, valproic acid, carbamazepine, lamotrigine, or oxcarbamazepine can be helpful. Some of these medications require more careful medical follow-up, including regular blood testing to check drug levels, electrolytes, liver function studies, and, in the case of lithium, renal function studies (Hagerman et al. 2009). The long-term use of valproic acid is associated with the onset of polycystic ovarian disease in adolescence and adulthood in some women.

- Valproic acid, carbamazepine, oxcarbamazepine, and lamotrigine are also anticonvulsant medications. If seizures or spike/wave discharges on the electroencephalogram are seen, these medications are the treatment of choice.

- In approximately 10% of individuals with fragile X syndrome, psychotic symptoms may occur, often associated with severe paranoia, which may lead to significant problems with aggression. The treatment of choice for psychotic thinking is an atypical antipsychotic medication. The atypical antipsychotics have a decreased risk for extrapyramidal symptoms and tardive dyskinesia compared with typical antipsychotics. This is because atypical antipsychotics block both serotonin receptors and dopamine receptors, and, when the dose is kept relatively low, the long-term motor side effects are rare. The atypical antipsychotics include aripiprazole, risperidone, olanzapine, quetiapine, and geodone. The atypical antipsychotics are likely to cause weight gain; however, this occurs less commonly with aripiprazole. The response rate to low dose aripiprazole is remarkable in fragile X syndrome (70%), and it typically improves mood instability, anxiety, aggression, and irritability (Hagerman et al. 2009). The use of metformin can help to avoid weight gain in those with fragile X syndrome with the use of atypicals, and metformin is also a very promising targeted treatment, so it should be started in childhood (Dy et al. 2017).

Neurologic

Studies by Reiss and colleagues (Reiss and Dant 2003) have shown an increase in the size of certain regions of the brain, particularly the hippocampus, caudate, thalamus, and lateral ventricles, in fragile X syndrome. However, the cerebellar vermis is smaller in size compared with controls (Reiss and Dant 2003). These size differences in central nervous system structures are reported in group studies of fragile X syndrome compared with controls and are usually not seen in an individual MRI that is ordered clinically. Therefore, routine magnetic resonance imaging studies are not usually recommended. The central nervous system research findings, however, are important clinically because they relate to the cognitive strengths and weaknesses reported in fragile X syndrome, including attention deficit/hyperactivity disorder, enhanced sensitivity to stimuli, and frontal deficits (Hoeft et al. 2007). Of note, periventricular heterotopia has been reported in two individuals with fragile X syndrome (Moro et al. 2006).

Seizures are an important clinical feature found in approximately 20% of individuals with fragile X syndrome (Berry-Kravis 2002). They usually present in early childhood, and may include generalized tonic-clonic seizures, staring spells or absence seizures, partial motor seizures, and temporal lobe seizures. Usually the seizures respond well to anticonvulsant medication. The seizures usually resolve by adolescence, although on occasion they may persist or start in adulthood (Berry-Kravis 2002).

The fragile X-associated tremor/ataxia syndrome (FXTAS), which can be seen in a subgroup of older male and female carriers with the premutation, can also be associated with dementia (Bourgeois et al. 2007; Hagerman and Hagerman 2016). FXTAS occurs in approximately 40% of male carriers and 16% of female carriers as they age into their 60s (Hagerman and Hagerman 2016). Neuropathological studies have been done on several individuals who have died from FXTAS and all demonstrated eosinophilic intranuclear inclusions in a limited number of neurons and astroglia throughout the cortex and cerebellum (Greco et al. 2006). Iron dysregulation occurs in carriers and excessive iron is accumulated in the brain of those with FXTAS (Ariza et al. 2016).

Evaluation

- A careful medical history should include questions regarding possible seizure episodes.
- An electroencephalogram should be obtained if seizures are suspected by medical history. The electroencephalogram should include a waking and a sleep record, because spike and wave discharges are more likely to appear in drowsiness or during a sleep study.
- Neurological consultation can be obtained to guide the evaluation or treatment of seizures.
- If focal abnormalities are seen on neurological examination, or if focal seizures occur, MRI of the brain should be done.
- Evaluation for fragile X-associated tremor/ataxia syndrome includes brain MRI and neurological examination and consultation.

Management

- If spike and wave discharges are seen on the electroencephalogram or if there is strong clinical evidence for seizures, treatment with an anticonvulsant is usually indicated. The most commonly used anticonvulsant is valproic acid. This medication requires careful medical follow-up, including blood testing for serum drug levels, complete blood count and platelet count, electrolytes, and liver function studies. Valproic acid can cause severe hepatic toxicity and pancreatitis, and hepatic failure can occur in 1 in 500 young individuals who have neurological disorders and are treated with multiple drugs. Carbamazepine has also been used in the treatment of seizures. Its toxicity is similar to valproate and it requires the same follow-up studies (Wisniewski et al. 1991; Hellings 1999).
- Newer anticonvulsants such as topiramate and lamotrigine may be used as an adjunct for treatment of seizures or mood stabilization when carbamazepine or valproic acid is not sufficient as single therapy. Phenobarbital should be avoided, because it typically increases hyperactivity, and gabapentin has not been helpful (personal experience).
- The use of folic acid may exacerbate a seizure disorder, and it is often avoided in individuals with seizures (Hagerman 2002a).

Ophthalmologic

Ophthalmologic problems, including strabismus and refractive errors, particularly hyperopia and astigmatism, can be seen in 25–56% of individuals with fragile X (King et al. 1995; Hatton et al. 1998).

Evaluation

- Children diagnosed with fragile X syndrome should be evaluated carefully for strabismus, nystagmus, and even ptosis (Kidd et al. 2014). If these problems are found, referral to an ophthalmologist for evaluation is indicated.
- If no abnormality is seen on clinical examination, then a routine assessment by an ophthalmologist or an optometrist should occur before 4 years of age (Hagerman 2002a).

Management

- The treatment for ophthalmologic problems depends on the abnormalities seen. Refractive errors or astigmatism are usually treated with corrective lenses. Strabismus may require surgical intervention, although before this, patching, eye exercises, or lenses are often used to strengthen the weak eye. Approaches are the same as those used in the general population.

Craniofacial

The facial structural changes in people affected by fragile X syndrome include a long face, a high forehead, a high-arched palate, and prominent ears. Cleft palate occurs occasionally, but dental crowding and malocclusion are common. The most common medical complication associated with the facial structural changes is recurrent otitis media, seen in approximately 60–80% of affected individuals. This usually begins in the first year of life and is associated with a persistent conductive hearing loss. Approximately 23% of individuals with fragile X syndrome have recurrent sinus infections, again most likely related to facial structural changes and, perhaps, the connective tissue dysplasia, and hypotonia. A rare individual has been documented to have a cleft palate or even diagnosed with Robin sequence (Hagerman 2002a). In addition, transient hypogammaglobulinemia with IgG subclass deficiencies is occasionally reported, although this may be secondary to the recurrent otitis and sinusitis (Hagerman 2002a). Recurrent ear infections and recurrent sinusitis usually resolve by 5–6 years of age. The rare instance of isolated IgG subclass immunoglobulin deficiency usually also improves with time.

An occasional affected individual has been reported to have obstructive sleep apnea, and this may relate to facial structural changes, enlarged adenoids, connective tissue dysplasia, or hypotonia of facial and pharyngeal muscles.

Evaluation

- On all visits to the physician, the tympanic membranes should be visualized to assess for infection or persistent serous otitis media (Kidd et al. 2014).
- Referral to an otolaryngologist for evaluation is appropriate for recurrent otitis media.
- Audiometric evaluations and/or tympanograms should be obtained at the end of otitis media treatment to assess hearing and possible persistence of middle ear fluid.
- A history of snoring and obstruction during sleep should be elicited. A sleep study should be done if a history of nighttime obstruction with snoring is obtained.
- Yearly dental evaluations should be carried out because oral tactile defensiveness may interfere with brushing and predispose to cavities.

Management

- Recurrent ear infections should be treated aggressively to normalize hearing. The use of prophylactic antibiotic therapy and/or the insertion of ventilation tubes are recommended for recurrent infections.
- On occasion, parents request ear pinning surgery if the pinnae are excessively large, leading to social problems.
- If obstructive sleep apnea is found, adenoidectomy is usually carried out, and this typically alleviates the problems. Persistent sleep apnea may require the use of continuous positive airway pressure with nasal prongs during sleep.
- If dental crowding occurs, standard orthodontic treatment may be necessary.

Cardiovascular

The most common cardiac problem in fragile X syndrome is mitral valve prolapse. Although this is rarely seen in childhood, it may be present in approximately 50% of adults, including females with fragile X syndrome (Hagerman 2002b). Mild dilation of the aortic root has also been seen in adults with fragile X syndrome, although it does not appear to progress with age. Sudden death secondary to an arrhythmia, perhaps precipitated by mitral valve prolapse, is very rare, but it has been seen three times by this author and rarely by others (Hagerman 2002a). The role of FMRP in the heart has not been clarified.

Hypertension is relatively common in adults with fragile X syndrome (personal experience), although this may relate to anxiety in clinic, which is a significant problem for most affected adults. It is possible that the connective tissue problems, specifically abnormal elastin fibers in the vessels, may affect the resilience of the vessel wall and predispose individuals with fragile X syndrome to hypertension.

Evaluation

- Auscultation of the heart should be carried out at all clinical visits (American Academy of Pediatrics Committee on Genetics 1996). The presence of a murmur or a click requires an evaluation by a cardiologist to clarify the presence of mitral valve prolapse or other problems. The cardiology evaluation should include an electrocardiogram and an echocardiogram, including measurement of the aortic root diameter.
- Blood pressure should also be monitored at all clinical visits, and this should be measured at least once a year in all adults.

Management

Hypertension can be initially treated with diuretics or low-dose β-blockers. Other antihypertensive medication, such as clonidine, may be helpful. Treatment choices are the same as in the general population.

- Management of mitral valve prolapse is standard.
- Use of antibiotic prophylaxis is no longer required for mitral valve prolapse.
- Treatment of aortic root dilatation is usually not required.

Genitourinary

Macroorchidism is the most common genital anomaly in males with fragile X syndrome. It is present in 80–90% of affected adolescent and adult males. It is usually not associated with other complications, although the weight of the testicle in combination with connective tissue problems may predispose to inguinal hernias. A hernia is present in approximately 15% of males with fragile X syndrome and may occur in childhood, adolescence, or adulthood.

Enuresis and delays in toilet training are common in both males and females with fragile X syndrome. Although affected individuals are not considered to be at an increased risk for recurrent urinary tract infections, the connective tissue dysplasia may predispose to dilation of the ureters with reflux. Four cases of significant and persistent ureteral reflux have been seen among 350 individuals with fragile X syndrome (personal experience). In three individuals, this has led to nephrectomy because of renal complications including hypertension.

Evaluation

- Testicular volume can be measured with an orchidometer to monitor size changes over time, as may occur with a hernia.
- Males should be assessed for the presence of a hernia.
- Urinary tract infection should be evaluated with a cystourethrogram and renal ultrasound.
- Referral to nephrology or urology is recommended for recurrent urinary tract infections, abnormalities in renal structure, or reflux on the cystourethrogram.
- Hypertension should be followed closely, and the presence of persistent hypertension requires a more detailed evaluation, including studies of the kidney.

Management

- Decreasing fluids after dinner, urination at bedtime, and waking the child to urinate again when the parents go to bed can help to decrease the frequency of enuresis.
- Delays in toilet training can be helped by behavioral interventions, including the use of a music video developed by Duke University to facilitate toilet training for individuals with development disabilities (1-800-23-POTTY) (Luxem and Christopherson 1994).
- Treatment of enuresis includes the use of behavior modification techniques, such as monitoring enuretic episodes with a star chart, in addition to an enuretic alarm system such as the Potty Pager (Ideas for Living, Boulder, CO, 1-800-497-6573 in the United States), the Nytone alarm [which utilizes a clip to the underwear (801-973-4090)], or the Wet-Stop alarm [which uses Velcro fasteners (Palco Labs, 1-800-346-4488 in the United States)].
- Medications can also be helpful for the treatment of enuresis, and they include imipramine at bedtime, oxybutynin (an anticholinergic that is also a muscle relaxant) or desmopressin acetate (DDAVP, an analog of antidiuretic hormone) (Tietjen and Husmann 1996). Only occasional individuals require medication for this problem.
- Bladder musculature and volume can be increased by intermittently stopping urine flow in an exercise program and reinforcing urination of larger and larger volumes.
- Macroorchidism will be maintained throughout adult life and does not require intervention.

Musculoskeletal

The most common orthopedic complication in fragile X syndrome is hyperlaxity or hyperextensibility. This appears to be related to connective tissue dysplasia, although the biochemical abnormality causing the connective tissue problems has not been clarified. Hyperextensible finger joints are seen in over 70% of children, but only 30% of adults have this problem. Flat feet, or pes planus, is also related to joint laxity, and approximately 80% of younger males and 60% of older males have this finding (Merenstein et al. 1996). Most individuals do not have significant pain associated with flat feet, although it can cause uneven shoe wear. Approximately 3% of individuals with fragile X syndrome have joint dislocations, particularly congenital hip dislocation identified at birth, recurrent patellar dislocation, or shoulder dislocation. The joint hyperextensibility usually improves with age, perhaps related to ligament tightening with time.

Clubfoot deformity can be seen in 1–2% of males with fragile X syndrome (personal experience), and there may be a slight predisposition to this problem because of hypotonia in utero and the connective tissue dysplasia. Scoliosis is seen in less than 20% of individuals with fragile X, and it is typically mild and does not require treatment.

Evaluation

- The regular physical examination should include an assessment of joint hyperextensibility in addition to asking for a history of joint dislocation or pain.
- All individuals should be assessed clinically for the presence of scoliosis.
- If scoliosis is present, it should be documented with baseline spine films and referral to an orthopedist should be considered.

Management

- The majority of cases with joint hyperextensibility do not require treatment. Joint dislocations, however,

require an orthopedic evaluation and follow-up. Recurrent joint dislocations may require surgery. Severe joint hyperlaxity, particularly in association with hypotonia, may require physical therapy intervention in early childhood. Treatment methods are those used in the general population.
- Flat feet are frequently treated with a shoe insert or orthotic, which may improve shoe wear and gait patterns.
- Scoliosis should be treated as in the general population.

Endocrine

Several females with the full mutation have been reported to have precocious puberty (Butler and Najjar 1988; Moore et al. 1990; Kowalczyk et al. 1996). The cause of this problem is unknown, although hypothalamic dysfunction, which can lead to growth abnormalities in fragile X syndrome, may also cause precocious puberty and perhaps macroorchidism. Women with the full mutation usually have a cognitive deficit, and on occasion this can lead to promiscuous sexual behavior. These individuals may require more detailed counseling for birth control and may be unable to reliably take medication such as daily birth control pills without supervision.

Women with the premutation have been found to have a higher incidence of premature menopause compared with unaffected women (Sullivan et al. 2005). Approximately 20% of women with the premutation undergo premature menopause also called fragile X-associated primary ovarian insufficiency (FXPOI), and this may occur as early as the 20s. Poor ovarian reserve has also been reported in women with the premutation and this can complicate the harvesting of eggs for in vitro fertilization (Wittenberger et al. 2007).

Males and females with fragile X syndrome are fertile, but only the premutation occurs in the sperm of a male who has the full mutation in other tissues.

All women with the *FMR1* mutation should be offered prenatal diagnosis with pregnancies. In addition, both women with the premutation and women with the full mutation are at higher risk for emotional problems compared with the general population, particularly at times of hormonal changes or estrogen deficiency, such as menopause, postpartum, and even during their monthly periods. Some women suffer from severe premenstrual syndrome.

Evaluation

- Signs of precocious puberty should be sought in the periodic clinical examination of females with the full mutation. Girls who present with features of precocious puberty should be referred to an endocrinologist for standard evaluation.
- Questions regarding the emotional status during or before menses of both girls and women with the premutation or the full mutation should be addressed at each clinical evaluation, including questions regarding anxiety, depression, and mood lability.
- Questions about menstruation should be addressed at each clinical evaluation in women of reproductive age with the premutation or full mutation.

Management

- The use of a gonadotropin agonist to block precocious puberty may be necessary.
- Problems related to emotional dysfunction associated with menstruation should be discussed with the parents, so that such clinical problems can be addressed in treatment. The use of SSRIs as described above is usually helpful for severe premenstrual syndrome and the depression associated with menopause or postpartum states. Most of the emotional problems seen in females with the premutation or full mutation can be treated effectively with the combination of medication and counseling.
- The use of Depo Provera injections, monthly or every three months, for birth control may be beneficial in women with a cognitive deficit who may forget to take daily pills.
- All females who have the premutation or the full mutation should be referred for genetic counseling and a discussion of reproductive alternatives, including new reproductive strategies such as in vitro fertilization and prenatal diagnosis techniques (McConkie-Rosell et al. 2007).
- The risks for premature menopause should be explained to carriers of the premutation so that appropriate adjustments can be made in reproductive planning.

RESOURCES

Foundations

National Fragile X Foundation
Phone: 202-747-6206; toll free: 800-688-8765
Fax: 202-742-6208.
Email: natlfx@fragilex.org
Website: *http://www.fragileX.org*

FRAXA Research Foundation
10 Prince Place, Suite 203,
Newburyport, MA 01950, USA
Phone: (978) 462-1866
Fax: (978) 463-9985

Email: info@fraxa.org
Website: *http://www.fraxa.org*

Fragile X Research Foundation of Canada
167 Queen Street West
Brampton, Ontario
Canada L6Y 1M5
Phone: (905) 453-9366
Email: info@fragilexcanada.ca
Website: www.fragilexcanada.ca

The Fragile X Society (England)
Rood End House
6 Stortford Rd
Great Dunmow
Essex CM6 1DA
United Kingdom
Phone: 01371-875-100
Email: info@fragileX.org.uk
Website: www.fragilex.org.uk

Fragile X Association of Australia, Inc.
Fragile X Association of Australia Inc
PO Box 109, Manly, NSW 1655, Australia
1-300 394 636
Email: wendy@fragilex.org.au
Website: www.fragilex.org.au

READING FOR FAMILIES

Braden M (2000) *Fragile, Handle with Care: More About Fragile X Syndrome, Adolescents and Adults*. Dillon, CO: Spectra Publishing Co., Inc.

Cronister AC, Weber JD, eds (2000) *Children with Fragile X Syndrome: A Parent's Guide*. Bethesda, MD: Woodbine House.

Dew-Hughes D, ed. (2004) *Educating Children with Fragile X Syndrome*. London, England: Routledge Falmer.

Dunsford C (2007) *Spelling Love with an X*. Boston, MA: Beacon Press.

Dykens EM, Hodapp RM, Leckman JF (1994) *Behavior and Development in Fragile X Syndrome*. Thousand Oaks, CA: Sage.

Educating Children with Fragile X Syndrome: A Guide for Parents and Professionals. Copies can be obtained by calling Gail Spiridigliozzi at (919) 684-5513.

Finucane B, McConkie-Rosell A, Cronister-Silverman A (2002) *The Fragile X Syndrome: A Handbook for Parents and Professionals*. San Francisco, CA: National Fragile X Foundation.

Hagerman RJ, Hagerman PJ, eds (2002) *Fragile X Syndrome: Diagnosis, Treatment and Research*, 3rd ed. Baltimore, MD: The Johns Hopkins University Press.

Schopmeyer BB, Lowe F (1992) *The Fragile X Child*. San Diego, CA: Singular.

Tranfaglia MR (2000) *A Medication guide for Fragile X Syndrome. Version 3.1*. West Newbury, MA: FRAXA Research Foundation.

Weber JD (1994) *Transitioning "Special" Children into Elementary School*. Books Beyond Borders, Inc. 1881 4th Street, #108, Boulder, CO 80302 (1-800-347-6440).

Busby MB and Massey M (2006) *Dear Megan: Letters on Life, Love and Fragile X*. Capital Books Inc. Sterling Virginia.

Morgan M (2005) The Broken Toy: The Story of a Fragile X Syndrome Child. Author House, Bloomington, Indiana. 800-8398640

NEWSLETTERS

FRAXA Research Foundation Newsletter: see info@fraxa.org
National Fragile X Foundation Newsletter. Call the National Fragile X Foundation at 800-688-8765

READING FOR CHILDREN

Heyman C (2003) My eXtra Special Brother: How to Love, Understand, and Celebrate Your Sibling with Special Needs. The Fragile X Association of Georgia. Can obtain through the National Fragile X Foundation at 1-800-688-8765.

O'Connor R (1995) Boys with Fragile X Syndrome. Can be obtained from the National Fragile X Foundation at 1-800-688-8765.

Steiger C (1998) My Brother has Fragile X Syndrome. Can be obtained from the National Fragile X Foundation at 1-800-688-8765.

INTERNET RESOURCES

The National Fragile X Foundation
Website: *http://www.fragileX.org*
FRAXA Research Foundation Home Page
Website: *http://www.fraxa.org*
The International Fragile X Alliance (IFXA). A global network of FX support groups
Website: www.ifxa.net

AUDIO/VISUAL AIDS

Fragile X Syndrome: Medical and Educational Approaches to Intervention – Cassette: This 90 min audio-cassette is a tool for families and educators as they develop appropriate educational programs for children with fragile X syndrome. Speakers include Lois Hickman OTR, Sarah

Scharfenaker SLP-CCC, Tracy Stackhouse OTR, Randi Hagerman MD, and Phil Wilson PhD. Available from The National Fragile X Foundation.

Educational Strategies and Issues for Children with Fragile X Syndrome – Video: In this 59 min video, Dr Randi Hagerman, Elizabeth Holder, Sarah Scharfenaker, Tracy Stackhouse, and numerous teachers present tactics for educating children with fragile X syndrome. The video, which includes molecular information and medication therapies, follows one child through a multidisciplinary evaluation. It then looks into the school day of a kindergartner, a fifth grader, and a freshman in high school, all of whom have fragile X syndrome. Available from The National Fragile X Foundation.

The National Fragile X Foundation Medical Video "Diagnosis and Treatment": This concise video explains the medical diagnosis and treatments in a very informative way for professionals and families. Available from The National Fragile X Foundation.

REFERENCES

Abbeduto L, Brady N, Kover ST (2007) Language development and fragile X syndrome: Profiles, syndrome-specificity, and within-syndrome differences. *Ment Retard Dev Disabil Res Rev* 13:36–46.

American Academy of Pediatrics Committee on Genetics (1996) Health supervision for children with fragile X syndrome. *Pediatrics* 98:297–300.

Ariza J, Rogers H, Monterrubio A, Reyes-Miranda A, Hagerman PJ, Martinez-Cerdeno V (2016) A majority of FXTAS cases present with intranuclear inclusions within purkinje cells. *Cerebellum* 15:546–51.

Berry-Kravis E (2002) Epilepsy in fragile X syndrome. *Dev Med Child Neurol* 44:724–8.

Berry-Kravis E, Des Portes V, Hagerman R, Jacquemont S, Charles P, Visootsak J, Brinkman M, Rerat K, Koumaras B, Zhu L, Barth GM, Jaecklin T, Apostol G, Von Raison F (2016) Mavoglurant in fragile X syndrome: Results of two randomized, double-blind, placebo-controlled trials. *Sci Transl Med* 8:321ra5.

Berry-Kravis E, Potanos K (2004) Psychopharmacology in fragile X syndrome--present and future. *Ment Retard Dev Disabil Res Rev* 10:42–8.

Bilousova TV, Dansie L, Ngo M, Aye J, Charles JR, Ethell DW, Ethell IM (2009) Minocycline promotes dendritic spine maturation and improves behavioural performance in the fragile X mouse model. *J Med Genet* 46:94–102.

Bourgeois JA, Cogswell JB, Hessl D, Zhang L, Ono MY, Tassone F, Farzin F, Brunberg JA, Grigsby J, Hagerman RJ (2007) Cognitive, anxiety and mood disorders in the fragile X-associated tremor/ataxia syndrome. *Gen Hosp Psychiatry* 29:349–56.

Braden M (2002) Academic interventions in fragile X. In Hagerman RJ, Hagerman PJ (eds.) *Fragile X syndrome: Diagnosis, treatment and research*, 3rd ed. Baltimore: The Johns Hopkins University Press.

Braden ML (2000) *Fragile, Handle with Care: More About Fragile X Syndrome*, Adolescents and Adults, Dillon, CO: Spectra Publishing Co., Inc.

Butler MG, Najjar JL (1988) Do some patients with fragile X syndrome have precocious puberty? *Am J Med Genet* 31:779–81.

Cordeiro L, Abucayan F, Hagerman R, Tassone F, Hessl D (2015) Anxiety disorders in fragile X premutation carriers: Preliminary characterization of probands and non-probands. *Intractable Rare Dis Res* 4:123–30.

Cordeiro L, Ballinger E, Hagerman RJ, Hessl D (2011) Clinical assessment of DSM-IV anxiety disorders in fragile X syndrome: Prevalence and characterization. *J Neurodev Disord* 3:57–67.

De Vries BB, Wiegers AM, Smits AP, Mohkamsing S, Duivenvoorden HJ, Fryns JP, Curfs LM, Halley DJ, Oostra BA, Van Den Ouweland AM, Niermeijer MF (1996) Mental status of females with an FMR1 gene full mutation. *Am J Hum Genet* 58:1025–32.

Dolen G, Bear MF (2008) Role for metabotropic glutamate receptor 5 (mGluR5) in the pathogenesis of fragile X syndrome. *J Physiol* 586:1503–8.

Dy ABC, Tassone F, Eldeeb M, Salcedo-Arellano MJ, Tartaglia N, Hagerman R (2017) Metformin as targeted treatment in fragile X syndrome. *Clin Genet* 93:216–222

Farzin F, Perry H, Hessl D, Loesch D, Cohen J, Bacalman S, Gane L, Tassone F, Hagerman P, Hagerman R (2006) Autism spectrum disorders and attention-deficit/hyperactivity disorder in boys with the fragile X premutation. *J Dev Behav Pediatr* 27:S137–44.

Gantois I, Khoutorsky A, Popic J, Aguilar-Valles A, Freemantle E, Cao R, Sharma V, Pooters T, Nagpal A, Skalecka A, Truong VT, Wiebe S, Groves IA, Jafarnejad SM, Chapat C, Mccullagh EA, Gamache K, Nader K, Lacaille JC, Gkogkas CG, Sonenberg N (2017) Metformin ameliorates core deficits in a mouse model of fragile X syndrome. *Nat Med* 23:674–677.

Goldson E, Hagerman RJ (1993) Fragile X syndrome and failure to thrive [letter]. *Am J Dis Child* 147:605–7.

Greco CM, Berman RF, Martin RM, Tassone F, Schwartz PH, Chang A, Trapp BD, Iwahashi C, Brunberg J, Grigsby J, Hessl D, Becker EJ, Papazian J, Leehey MA, Hagerman RJ, Hagerman PJ (2006) Neuropathology of fragile X-associated tremor/ataxia syndrome (FXTAS). *Brain* 129:243–55.

Greiss Hess L, Fitzpatrick SE, Nguyen DV, Chen Y, Gaul KN, Schneider A, Lemons Chitwood K, Eldeeb MA, Polussa J, Hessl D, Rivera S, Hagerman RJ (2016) A randomized, double-blind, placebo-controlled trial of low-dose sertraline in young children with fragile X syndrome. *J Dev Behav Pediatr* 37:619–28.

Hagerman R, Hagerman P (2013) Advances in clinical and molecular understanding of the *FMR1* premutation and fragile X-associated tremor/ataxia syndrome. *Lancet Neurol* 12:786–98.

Hagerman RJ (2002a) Medical follow-up and pharmacotherapy. In Hagerman RJ, Hagerman PJ (eds.) *Fragile X Syndrome: Diagnosis, Treatment and Research*. 3rd ed. Baltimore: The Johns Hopkins University Press, pp. 287–338.

Hagerman RJ (2002b) Physical and behavioral phenotype. In Hagerman RJ, Hagerman PJ (eds.) *Fragile X Syndrome: Diagnosis, Treatment and Research*. 3rd ed. Baltimore: The Johns Hopkins University Press, pp. 3–109.

Hagerman RJ, Berry-Kravis E, Hazlett HC, Bailey DB, Jr, Moine H, Kooy RF, Tassone F, Gantois I, Sonenberg N, Mandel JL, Hagerman PJ (2017) Fragile X syndrome. *Nat Rev Dis Primers* 3:17065.

Hagerman RJ, Berry-Kravis E, Kaufmann WE, Ono MY, Tartaglia N, Lachiewicz A, Kronk R, Delahunty C, Hessl D, Visootsak J, Picker J, Gane L, Tranfaglia M (2009) Advances in the treatment of fragile X syndrome. *Pediatrics* 123:378–90.

Hagerman RJ, Hagerman P (2016) Fragile X-associated tremor/ataxia syndrome - features, mechanisms and management. *Nat Rev Neurol* 12:403–12.

Hagerman RJ, Hull CE, Safanda JF, Carpenter I, Staley LW, O'connor RA, Seydel C, Mazzocco MM, Snow K, Thibodeau SN, et al. (1994) High functioning fragile X males: Demonstration of an unmethylated fully expanded FMR-1 mutation associated with protein expression. *Am J Med Genet* 51:298–308.

Hatton DD, Buckley EG, Lachiewicz A, Roberts J (1998) Ocular status of young boys with fragile X syndrome: A prospective study. *J Am Ass Pediatr Ophthalmol Strabismus* 2:298–301.

Hatton DD, Hooper SR, Bailey DB, Skinner ML, Sullivan KM, Wheeler A (2002) Problem behavior in boys with fragile X syndrome. *Am J Med Genet* 108:105–16.

Hatton DD, Sideris J, Skinner M, Mankowski J, Bailey DB, Jr., Roberts J, Mirrett P (2006) Autistic behavior in children with fragile X syndrome: Prevalence, stability, and the impact of FMRP. *Am J Med Genet A* 140A:1804–13.

Hellings JA (1999) Psychopharmacology of mood disorders in persons with mental retardation and autism. *Ment Retard Dev Disabil Res Rev* 5:270–278.

Hessl D, Tassone F, Loesch DZ, Berry-Kravis E, Leehey MA, Gane LW, Barbato I, Rice C, Gould E, Hall DA, Grigsby J, Wegelin JA, Harris S, Lewin F, Weinberg D, Hagerman PJ, Hagerman RJ (2005) Abnormal elevation of *FMR1* mrna is associated with psychological symptoms in individuals with the fragile X premutation. *Am J Med Genet B* 139B:115–21.

Hoeft F, Hernandez A, Parthasarathy S, Watson CL, Hall SS, Reiss AL (2007) Fronto-striatal dysfunction and potential compensatory mechanisms in male adolescents with fragile X syndrome. *Hum Brain Mapp* 28:543–54.

Kidd SA, Lachiewicz A, Barbouth D, Blitz RK, Delahunty C, Mcbrien D, Visootsak J, Berry-Kravis E (2014) Fragile X syndrome: A review of associated medical problems. *Pediatrics* 134:995–1005.

King RA, Hagerman RJ, Houghton M (1995) Ocular findings in fragile X syndrome. *Dev Brain Dysfunction* 8:223–229.

Kowalczyk CL, Schroeder E, Pratt V, Conard J, Wright K, Feldman GL (1996) An association between precocious puberty and fragile X syndrome? *J Pediatr Adolesc Gynecol* 9:199–202.

Leigh MJ, Nguyen DV, Mu Y, Winarni TI, Schneider A, Chechi T, Polussa J, Doucet P, Tassone F, Rivera SM, Hessl D, Hagerman RJ (2013) A randomized double-blind, placebo-controlled trial of minocycline in children and adolescents with fragile X syndrome. *J Dev Behav Pediatr* 34:147–55.

Loesch DZ, Huggins RM, Hagerman RJ (2004) Phenotypic variation and FMRP levels in fragile X. *Ment Retard Dev Disabil Res Rev* 10:31–41.

Loesch DZ, Huggins RM, Hoang NH (1995) Growth in stature in fragile X families: A mixed longitudinal study. *Am J Med Genet* 58:249–56.

Loesch DZ, Sherwell S, Kinsella G, Tassone F, Taylor A, Amor D, Sung S, Evans A (2012) Fragile X-associated tremor/ataxia phenotype in a male carrier of unmethylated full mutation in the *FMR1* gene. *Clin Genet* 82:88–92.

Luxem M, Christopherson E (1994) Behavioral toilet training in early childhood: Research practice and implications. *J Dev Behav Pediatr* 15:370–378.

Mcconkie-Rosell A, Abrams L, Finucane B, Cronister A, Gane LW, Coffey SM, Sherman S, Nelson LM, Berry-Kravis E, Hessl D, Chiu S, Street N, Vatave A, Hagerman RJ (2007) Recommendations from multi-disciplinary focus groups on cascade testing and genetic counseling for fragile X-associated disorders. *J Genet Couns* 16:593–606.

Merenstein SA, Sobesky WE, Taylor AK, Riddle JE, Tran HX, Hagerman RJ (1996) Molecular-clinical correlations in males with an expanded FMR1 mutation. *Am J Med Genet* 64:388–94.

Monyak RE, Emerson D, Schoenfeld BP, Zheng X, Chambers DB, Rosenfelt C, Langer S, Hinchey P, Choi CH, Mcdonald TV, Bolduc FV, Sehgal A, Mcbride SM, Jongens TA (2016) Insulin signaling misregulation underlies circadian and cognitive deficits in a Drosophila fragile X model. *Mol Psychiatry*.

Moore PS, Chudley AE, Winter JS (1990) True precocious puberty in a girl with the fragile X syndrome. *Am J Med Genet* 37:265–7.

Moro F., Pisano T., Della Bernardina B., Polli R., Murgia A., Zoccante L., Darra F., Battaglia A., Pramparo T., Zuffardi O., Guerrini R. (2006). Periventricular heterotopia in fragile X syndrome. *Neurology* 67:713–715.

Niu M, Han Y, Dy ABC, Du J, Jin H, Qin J, Zhang J, Li Q, Hagerman RJ (2017) Autism symptoms in fragile X syndrome. *J Child Neurol* 32:903–909.

Nolin SL, Brown WT, Glicksman A, Houck GE, Jr., Gargano AD, Sullivan A, Biancalana V, Brondum-Nielsen K, Hjalgrim H, Holinski-Feder E, Kooy F, Longshore J, Macpherson J, Mandel JL, Matthijs G, Rousseau F, Steinbach P, Vaisanen ML, Von Koskull H, Sherman SL (2003) Expansion of the fragile X CGG repeat in females with premutation or intermediate alleles. *Am J Hum Genet* 72:454–64.

Nowicki ST, Tassone F, Ono MY, Ferranti J, Croquette MF, Goodlin-Jones B, Hagerman RJ (2007) The Prader-Willi phenotype of fragile X syndrome. *J Dev Behav Pediatr* 28:133–8.

O'Byrne JJ, Sweeney M, Donnelly DE, Lambert DM, Beattie ED, Gervin CM, Barton DE, Lynch SA. (2017). Incidence of fragile X syndrome in Ireland. *Am J Med Genet A* 173A:678–683.

Qin M, Kang J, Burlin TV, Jiang C, Smith CB (2005) Postadolescent changes in regional cerebral protein synthesis: An in vivo study in the FMR1 null mouse. *J Neurosci* 25:5087–95.

Quan F, Zonana J, Gunter K, Peterson KL, Magenis RE, Popovich BW (1995) An atypical case of fragile X syndrome caused by a deletion that includes the FMR1 gene. *Am J Hum Genet* 56:1042–51.

Reiss AL, Dant CC (2003) The behavioral neurogenetics of fragile X syndrome: Analyzing gene-brain-behavior relationships in child developmental psychopathologies. *Dev Psychopathol* 15:927–68.

Roberts JE, Bailey DB, Jr., Mankowski J, Ford A, Sideris J, Weisenfeld LA, Heath TM, Golden RN (2009) Mood and anxiety disorders in females with the *FMR1* premutation. *Am J Med Genet B Neuropsychiatr Genet* 150B:130–9.

Saldarriaga W, Forero-Forero JV, González-Teshima LY, Fandiño-Losada A, Isaza C, Tovar-Cuevas JR, Silva M, Choudhary NS, Tang HT, Aguilar-Gaxiola S, Hagerman RJ, Tassone F (2018) Genetic cluster of fragile X syndrome in a Colombian district. *J Hum Genet.* 63(4):509–516.

Saluto A, Brussino A, Tassone F, Arduino C, Cagnoli C, Pappi P, Hagerman P, Migone N, Brusco A (2005) An enhanced polymerase chain reaction assay to detect pre- and full mutation alleles of the fragile X mental retardation 1 gene. *J Mol Diagn* 7:605–12.

Scharfenaker S, O'connor R, Stackhouse T, Noble L (2002) An integrated approach to intervention. *In:* Hagerman RJ, Hagerman PJ (eds.) *Fragile X Syndrome: Diagnosis, Treatment and Research*, 3rd ed. Baltimore: The Johns Hopkins University Press.

Schwarte AR (2008) Fragile X syndrome. *School Psychology Quarterly* 23:290–300.

Sherman S (2002) Epidemiology. *In:* Hagerman RJ, Hagerman PJ (eds) *Fragile X Syndrome: Diagnosis, Treatment and Research*, 3rd ed. Baltimore: The Johns Hopkins University Press.

Sitzmann AF, Hagelstrom RT, Tassone F (2018) Rare *FMR1* gene mutations causing fragile X syndrome: A review. *Am J Med Genet A* 176:11–18.

Song FJ, Barton P, Sleightholme V, Yao GL, Fry-Smith A (2003) Screening for fragile X syndrome: A literature review and modelling study. *Health Technol Assess* 7:1–106.

Sullivan AK, Marcus M, Epstein MP, Allen EG, Anido AE, Paquin JJ, Yadav-Shah M, Sherman SL (2005) Association of *FMR1* repeat size with ovarian dysfunction. *Hum Reprod* 20:402–12.

Symons FJ, Byiers BJ, Raspa M, Bishop E, Bailey DB (2010) Self-injurious behavior and fragile X syndrome: Findings from the national fragile X survey. *Am J Intellect Dev Disabil* 115:473–81.

Symons FJ, Clark RD, Hatton DD, Skinner M, Bailey DB, Jr. (2003) Self-injurious behavior in young boys with fragile X syndrome. *Am J Med Genet A* 118:115–21.

Tassone F, Iong KP, Tong TH, Lo J, Gane LW, Berry-Kravis E, Nguyen D, Mu LY, Laffin J, Bailey DB, Hagerman RJ (2012) FMR1 CGG allele size and prevalence ascertained through newborn screening in the United States. *Genome Med* 4:100.

Tietjen DN, Husmann DA (1996) Nocturnal enuresis: A guide to evaluation and treatment. *Mayo Clin Proc* 71:857–62.

Todd PK, Oh SY, Krans A, He F, Sellier C, Frazer M, Renoux AJ, Chen KC, Scaglione KM, Basrur V, Elenitoba-Johnson K, Vonsattel JP, Louis ED, Sutton MA, Taylor JP, Mills RE, Charlet-Berguerand N, Paulson HL (2013) CGG repeat-associated translation mediates neurodegeneration in fragile X tremor ataxia syndrome. *Neuron* 78:440–55.

Verkerk AJ, Pieretti M, Sutcliffe JS, Fu YH, Kuhl DP, Pizzuti A, Reiner O, Richards S, Victoria MF, Zhang FP, Eussen BE, Van Ommen GJB, Blonden LaJ, Riggins GJ, Chastain JL, Kunst CB, Galjaard H, Caskey CT, Nelson DL, Oostra BA, Warren ST (1991) Identification of a gene (*FMR-1*) containing a CGG repeat coincident with a breakpoint cluster region exhibiting length variation in fragile X syndrome. *Cell* 65:905–14.

Waldstein G, Hagerman R (1988) Aortic hypoplasia and cardiac valvular abnormalities in a boy with fragile X syndrome. *Am J Med Genet* 30:83–98.

Westmark CJ, Malter JS (2007) FMRP mediates mGluR5-dependent translation of amyloid precursor protein. *PLoS Biol* 5:e52.

Wheeler A, Raspa M, Hagerman R, Mailick M, Riley C (2017) Implications of the FMR1 premutation for children, adolescents, adults, and their families. *Pediatrics* 139:S172–S182.

Winarni TI, Utari A, Mundhofir FE, Hagerman RJ, Faradz SM (2013) Fragile X syndrome:Clinical, cytogenetic and molecular screening among autism spectrum disorder children in indonesia. *Clin Genet* 84:577–80.

Wisniewski KE, Segan SM, Miezejeski CM, Sersen EA, Rudelli RD (1991) The fra(X) syndrome: Neurological, electrophysiological, and neuropathological abnormalities. *Am J Med Genet* 38:476–80.

Wittenberger MD, Hagerman RJ, Sherman SL, Mcconkie-Rosell A, Welt CK, Rebar RW, Corrigan EC, Simpson JL, Nelson LM (2007) The FMR1 premutation and reproduction. *Fertil Steril* 87:456–65.

Yrigollen CM, Durbin-Johnson B, Gane L, Nelson DL, Hagerman R, Hagerman PJ, Tassone F (2012) AGG interruptions within the maternal FMR1 gene reduce the risk of offspring with fragile X syndrome. *Genet Med* 14:729–36.

29

GORLIN SYNDROME: NEVOID BASAL CELL CARCINOMA SYNDROME

Peter A. Farndon
Clinical Genetics Unit, Birmingham Women's Hospital, Edgbaston, Birmingham, UK

D. Gareth Evans
Manchester Centre for Genomic Medicine, St Mary's Hospital, Manchester Academic Health Centre (MAHSC), Division of Evolution and Genomic Sciences, University of Manchester, Manchester, UK

INTRODUCTION

Gorlin syndrome (nevoid basal cell carcinoma syndrome) is a multisystem dominantly inherited disorder associated with mutations in the patched (*PTCH*) gene. The syndrome offers a paradigm for understanding the cellular relationships between developmental malformations and cancer.

Because two of the main features are recurrent odontogenic keratocysts and basal cell carcinomas, it is clinically important that people with Gorlin syndrome are offered surveillance and long-term support. Caution should be exercised concerning treatment with therapeutic radiation as some affected individuals respond with crops of basal cell carcinomas in the treated area.

Gorlin and Goltz's (1960) description of two individuals and review of the literature drew the condition to wide attention. However, Howell and Caro in 1959 had attempted to correlate the clinical features of the unusual tumors in the syndrome with the interpretations of cell biology and histology, which then prevailed, introducing the term 'basal cell nevus syndrome.' They proposed that the tumors were a unique type of basal cell carcinoma that was capable of aggressive behavior in adults, and were associated with developmental anomalies. They pointed out that the harmless clinical appearance of the tumors, especially in childhood, contrasted strikingly with the microscopic appearance and destructive behavior of tumors in adulthood. They also speculated that, although ionizing radiation was curative, its use was not prudent because of the multiplicity of tumors and the concern over new ones erupting in the irradiated area.

Incidence

The prevalence in a 1991 population-based study in northwest England was 1 in 55,600 (Evans et al. 1993). The latest prevalence figure from that continuing study is 1 in 30,800 (Evans et al. 2010). A study in Australia gave a minimum prevalence of 1 in 164,000 (Shanley et al. 1994). Birth incidence may be as high as 1 in 20,000 (Evans et al. 2010).

Diagnostic Criteria

The condition is extremely variable in its manifestations, both within and between families. There are over 100 recognized manifestations, some of which are age dependent. The most frequent and clinically important are nevoid basal cell carcinomas and odontogenic keratocysts, each occurring in about 90% of affected individuals by mid-adulthood, although Gorlin syndrome associated with mutations in *SUFU* has not been shown to cause jaw cysts (Evans et al. 2017). The non-progressive skeletal anomalies, which are

---

Cassidy and Allanson's Management of Genetic Syndromes, Fourth Edition.
Edited by John C. Carey, Agatino Battaglia, David Viskochil, and Suzanne B. Cassidy.
© 2021 John Wiley & Sons, Inc. Published 2021 by John Wiley & Sons, Inc.

present in a high proportion of affected individuals, are helpful diagnostically, as are the presence of bone cysts and ectopic calcification, particularly of the falx cerebri. This feature is present in 90% of individuals over the age of 20 years. About 70% of people with the syndrome have a recognizable facial appearance that includes macrocephaly, frontal bossing, hypertelorism, and facial milia (Figure 29.1). Congenital malformations are present in 5% of individuals. Ovarian fibromas, medulloblastoma with a peak incidence at 2 years of age, and cardiac fibromas are also important components of the syndrome. Table 29.1 gives a listing of the common manifestations in decreasing order of frequency.

Diagnostic criteria are given in Table 29.2 based on the most frequent and/or specific manifestations of the syndrome (Evans et al. 1993). These criteria were based on examination of family cases in England, a land not noted for excessive sunlight. The number of basal cell carcinomas acceptable as a major criterion will vary according to sun exposure and latitude.

Etiology, Pathogenesis, and Genetics

Gorlin syndrome is associated with mutations in either the patched (*PTCH*) (Hahn et al. 1996; Johnson et al. 1996) or *SUFU* genes (Smith et al. 2014), both of which encode components of the hedgehog signaling pathway that regulates transcription of a range of genes, including *PTCH1* and *SUFU*, *GLI*, *TGFβ*, and *IGF2*. Inherited or sporadic mutations in genes in this pathway have been implicated in a number of human birth defects and adult cancers (Villavicencio et al. 2000; Bale and Yu 2001).

The human *PTCH1* gene has 23 exons covering 62 kilobases of genomic DNA. It encodes an integral membrane

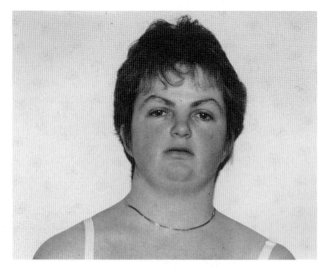

FIGURE 29.1 An adult with Gorlin syndrome showing some of the facial features associated with the syndrome: arched eyebrows, down-slanting palpebral fissures, and sloping shoulders.

TABLE 29.1 Summary of clinical manifestations in Gorlin syndrome

Feature	%
Skin findings	
Epidermal cysts	44
Milia	42
Meibomian cysts	6
Skin tags	6
Palmar pits	
<10 years	65
<15 years	80
Plantar pits	49
Nevi <20 years	53
Basal cell carcinomas	
<20 years	14
>20 years	73
>40 years	92
Facial features	
Bossing	79
"Typical face"	70
Prognathism	46
Palpebral fissures downslanting	30
Eyebrows arched	28
Eye anomalies	30
Strabismus	19
Cataracts	4
Oral manifestations	
Jaw cysts	
10 years	13
20 years	51
>40 years	79–90
Misshapen/missing teeth	30
Cleft lip and palate	7
High-arched palate	6
Skeletal findings	
Occipitofrontal head circumference >97th centile	97
Sloping shoulders	61
Thoracic scoliosis	47
Sprengel shoulder	46
Pectus excavatum	20
Short fourth metacarpal	26
Short terminal phalanx thumbs	9
Polydactyly	8
Stiff thumbs	6
Other manifestations	
Ovarian fibroma	24 (adult females)
Inguinal hernia	17 (males)
Epilepsy	6
Undescended testes	6 (males)
Medulloblastoma	5*

TABLE 29.1 (Continued)

Feature	%
Mental retardation	3
Cardiac fibroma	2.5
Radiographic findings	
Cervical/thoracic vertebral anomalies	60
Rib anomalies	70
Calcification	
Falx <15 years	40
Diaphragm sella 20 years	100

*20–30% in *SUFU* and 1–2% in *PTCH1*-related Gorlin syndrome

protein of 1500 amino acids with 12 transmembrane regions and 2 extracellular loops that are required for binding with the extracellular protein, sonic hedgehog (SHH). PTCH1 also associates with smoothened (SMO), a 7-span transmembrane protein that is an activator of transcription. When SHH is absent, PTCH1 inhibits SMO signaling. When extracellular SHH binds to PTCH1, inhibition of SMO is released, activating the signaling pathway and transcription of downstream target genes. In addition to control of transcription, PTCH1 is also involved in cell cycle regulation (Barnes et al. 2001). *SUFU* acts down stream of *PTCH1* and down-regulates GLI1-mediated transactivation of target genes.

The SHH-PTCH-GLI pathway appears to be sensitive to the levels of its various proteins. A mutation or polymorphism

TABLE 29.2 Clinical diagnostic criteria for Gorlin syndrome

Major criteria

1. Multiple (>2[a]) basal cell carcinomas or one under the age of 30 years or >10 basal cell nevi
2. Odontogenic keratocyst (proven on histology) or polyostotic bone cyst
3. Palmar or plantar pits (three or more)
4. Ectopic calcification: lamellar or early (<20 years) falx calcification
5. First degree relative affected

Minor criteria

1. Congenital skeletal anomaly: bifid, fused, splayed, or missing rib or fused vertebrae
2. Occipitofrontal circumference >97th centile with bossing
3. Cardiac or ovarian fibroma
4. Medulloblastoma (primitive neuroectodermal tumor)
5. Lymphomesenteric or pleural cysts
6. Congenital malformation: cleft lip and/or palate, polydactyly, or eye anomaly (cataract, coloboma, and microphthalmia)

[a] Note that the numbers of basal cell carcinomas given were based on a study carried out in England; the numbers of basal cell carcinomas for diagnosis will be inappropriate for sunnier climates.
Note: A diagnosis can be made when two major or one major and two minor criteria are fulfilled.

in one or more of the genes may affect the concentration of functional proteins in the pathway. A wide range of variation in transcription could, therefore, result from different levels of activity of PTCH1 or SUFU and explain the spectrum of clinical presentation, particularly of the malformations, seen between and within families with Gorlin syndrome.

The jaw cysts (in *PTCH1*), basal cell carcinomas and medulloblastomas are associated with a different mechanism, that is, loss of function of the wild-type *PTCH1/SUFU* allele (Bonifas et al. 1994; Levanat et al. 1996, Cowan et al. 1997), which releases the cell from the remaining control of the SHH-PTCH-GLI pathway exerted by that allele.

A wide spectrum of *PTCH1* mutations has been found in individuals with Gorlin syndrome (Lindstrom et al. 2006; Evans et al. 2017)). The mutations are spread throughout the coding region of the gene. The frequency of mutation classes, obtained from the literature (Hahn et al. 1996; Johnson et al. 1996; Lench et al. 1997; Wicking et al. 1997a, 1997b; Aszterbaum et al. 1998; Evans et al. 2017) and the DNA Diagnostic Laboratory at Birmingham Women's Hospital, United Kingdom, are 65% truncating mutations, 16% missense mutations, 13% splice-site mutations, and 6% intragenic or large-scale deletions or rearrangements.

There appears to be no genotype-phenotype correlation with truncating mutations (Wicking et al. 1997a). However, missense mutations appear to cause milder disease with a less profound phenotype (Evans et al. 2017). It is not possible to make predictions about clinical severity for developmental and neoplastic features associated with specific mutations other than perhaps for missense mutations because of the likely modifying effects of other genes and environmental factors.

PTCH1 germline mutations have not yet been associated with any other heritable syndromes, but somatic mutations have been found in a range of sporadically occurring tumors, including non-syndromic basal cell carcinoma, skin trichoepithelioma, ovarian fibroma, and keratocysts. *SUFU* and *PTCH1* mutations have been identified in non-syndromic medulloblastoma.

Missense mutations of *PTCH1* have been reported in 5% of unrelated probands with holoprosencephaly (Ming et al. 2002) (see Chapter 31). The authors hypothesized that the missense mutations would lead to enhanced *PTCH1* repressive activity on the hedgehog-signaling pathway, unlike the mechanism in Gorlin syndrome in which the pathway is activated.

A second highly homologous *PTCH* gene, *PTCH2*, is located on 1p32.1-32.2 (Smyth et al. 1999). No mutations were found in 11 sporadic affected individuals and 11 families with Gorlin syndrome in whom *PTCH1* screening by single-strand conformational polymorphism had been negative, but recently a missense mutation has been reported in one Chinese family (Fan et al. 2008) and a frameshift mutation in a Japanese case (Fujii et al. 2013).

A mutation in *SUFU* had been reported in a father and son and in an isolated case (Pastorino et al. 2009; Kijima et al. 2012) before exome sequencing in three *PTCH1*-negative families with Gorlin syndrome identified loss-of-function mutations that segregated with all nine affected individuals who met diagnostic criteria (Smith et al. 2014). Nonetheless, overall manifestations are milder with *SUFU* mutations; there are fewer BCCs and no jaw cysts (Evans et al. 2017). When ascertained from medulloblastoma cases rather than Gorlin syndrome, many pathogenic *SUFU* variant carriers do not meet Gorlin syndrome diagnostic criteria (Guerrini-Rousseau et al. 2017).

Genetic Counseling Gorlin syndrome is inherited in an autosomal dominant manner, with a one in two chance that each child of an affected parent will inherit the condition. Penetrance appears to be complete, but there is wide variability in expression, which manifests itself not only in the presence or absence of a particular feature, but also in its severity.

A new mutation rate of 40% has been suggested from a review of the literature (Gorlin et al. 2001); in an Australian series 37 of 64 (58%) were apparently simplex/sporadic cases in the family (Shanley et al. 1994). The new mutation rate obtained from the literature may be an overestimate because not all parents were thoroughly investigated. Parents of apparent simplex cases should be examined and investigated carefully, being mindful of the variation in expression. The condition appeared to be the result of a new mutation in only 17% of the fully investigated families in the authors' series (personal experience).

Diagnostic Testing

Diagnosis is usually based on clinical features. To assure accurate diagnosis, physical examination should particularly seek signs in the skeletal system and skin and congenital anomalies such as lip/palate clefting or polydactyly. Measurements should include height, head circumference, and inner- and outer-canthal and interpupillary distances. The head circumference should be plotted on a chart that takes height into account (Bushby et al. 1992). Examination should include a search for palmar and plantar pits (this can be aided by warming hands in hot water and by a magnifying glass). Clinical features that should be specifically noted include frontal bossing, sloping shoulders, Sprengel anomaly, rib cage and spinal anomalies, milia, skin cysts, short stiff-jointed thumbs, and hallux valgus.

Radiological findings (Ratcliffe et al. 1995a, 1995b) may aid diagnosis in family members who have equivocal physical signs. Recommended radiographs include panoramic views of the jaws (plain films may miss lesions), skull (anteroposterior and lateral), chest, thoracic spine (anteroposterior and lateral), and hands (for pseudocysts). An ultrasound examination for ovarian and cardiac fibromas may be helpful.

Diagnostic testing is possible by direct mutation analysis. As clinical examination may not be conclusive because of age-dependent manifestations of the syndrome, mutation analysis in children of families where there is a known mutation is justified to implement surveillance and sunscreen precautions. Identifying a pathogenic mutation in either *PTCH1* or *SUFU* will confirm the clinical diagnosis. Mutations are detected in about 80% of individuals who meet the diagnostic criteria. The detection rate is lowest in people who are the first affected individual in their family, most probably because of somatic mosaicism. The mutation is often more easily detected if an affected child is tested. For people in whom there is a clinical suspicion of mosaicism, detecting the same *PTCH1* mutation in several tumors, but not in lymphocyte DNA, may confirm mosaicism, although with next generation sequencing (NGS) mosaicism down to the level of 1–4% can be detected.

Because of technical limitations, a negative mutation screen cannot rule out Gorlin syndrome, but it will be at least partially reassuring to have a negative result in an individual who does not satisfy clinical diagnostic criteria providing a comprehensive evaluation has been performed. A missense mutation in a potentially simplex case who does not satisfy diagnostic criteria may be difficult to interpret, although its absence in both biological normal parents may be helpful in defining pathogenicity.

When an individual has additional manifestations beyond what are seen in Gorlin syndrome (in particular, short stature, epilepsy or severe developmental delay), chromosome analysis should be considered because these manifestations may suggest a deletion of chromosome 9q that includes the *PTCH1* gene.

Differential Diagnosis

Several rare conditions may need to be considered when an individual presents with only some, or very mild, features of the syndrome. The main consideration, however, should be whether they have somatic mosaicism for a *PTCH1* mutation. Localized mosaicism for a *PTCH1* somatic mutation is likely to be the cause of multiple basal cell carcinomas, comedones, and epidermoid cysts distributed in a unilateral distribution (Bleiberg and Brodkin 1969).

In those occasional families that show an autosomal dominant pattern of multiple basal cell carcinomas in the absence of other manifestations of Gorlin syndrome; *PTCH1* and *SUFU* mutation analysis is likely to be negative (Klein et al. 2005).

Multiple basal cell carcinomas, follicular atrophoderma on the dorsum of hands and feet, hypohydrosis, and hypotrichosis are features of Bazex syndrome. The pitting on the backs of the hands is reminiscent of orange peel and quite unlike the pits of Gorlin syndrome. The inheritance pattern is X-linked dominant (Viksnins and Berlin 1977).

A dominantly inherited condition similar to Bazex syndrome, called Rombo syndrome, was reported in a single family, and it is characterized by vermiculate atrophoderma, milia, hypotrichosis, trichoepitheliomas, basal cell carcinomas, and peripheral vasodilation with cyanosis. The skin is normal until later childhood, when basal cell carcinomas develop. There is no reduction in sweating (Michaelsson et al. 1981).

Chronic arsenic exposure may cause multiple basal cell carcinomas.

Rasmussen (1975) reported a family with trichoepitheliomas, milia, and cylindromas presenting in the second and third decades. Inheritance was autosomal dominant. The milia were miniature trichoepitheliomas and appeared only in sun-exposed areas. Cylindromatosis (Welch et al. 1968) (turban tumor syndrome) may be the same condition; it shows considerable variation within families in the size and extent of distribution, and age of onset.

In Cowden syndrome (PTEN hamartoma tumor syndrome) (Starink et al. 1986) (see Chapter 48), mucocutaneous changes develop in the second decade. Multiple facial papules, both smooth and keratotic, are associated with hair follicles and concentrated around the orifices. Small hyperkeratotic and verrucous growths are numerous on the dorsal aspect of the hands and feet, and round translucent palmoplantar keratoses are also common. Similar lesions, including verrucous papules, occur on the oral mucosa. Multiple skin tags are frequent. Most affected individuals have a broad forehead and a large head circumference (>+3SD), as found in Gorlin syndrome. Neoplasms occur in the gastrointestinal system, thyroid, and breast. However, individuals with Cowden syndrome/PTEN hamartoma tumor syndrome do not develop odontogenic keratocysts.

The differential diagnosis of the palmar pitting is porokeratosis of Mantoux (Howell and Mehregan 1970), which is a rare form of non-hereditary papular keratosis of the hands and feet, with a few lesions occasionally sprinkled over the ankles. The lesions are changeable and usually disappear with time. The depressions are always found on the summit of the papillary excrescences, resembling an enlarged sudoriferous pore. Older lesions show a blackish vegetation with a finely lobulated or mulberry-like surface at the bottom of the depression, which is eventually shed, leaving a small depression with a slightly raised margin and a red base. The material resembles a cornified comedone. The characteristic lesion is a translucent papule, which erupts in recurring crops over months or years.

Pseudohypoparathyroidism may be considered because of ectopic calcification and short fourth metacarpals.

Cardiac tumors are also found in tuberous sclerosis (rhabdomyomas) (see Chapter 59) and Beckwith–Wiedemann syndrome (see Chapter 9).

MANIFESTATIONS AND MANAGEMENT

Some individuals with Gorlin syndrome have relatively few or mild manifestations, whereas others require multiple courses of treatment for jaw cysts or basal cell carcinomas. Management can be greatly enhanced by an expert team approach, but it is still extremely important that one health professional maintains an overview of the total care. The expert team approach is especially valuable for those with multiple basal cell carcinomas. This can allow different modalities of treatment offered by different specialties to be chosen for specific basal cell carcinomas, depending, for instance, on their site and size. It is important that affected individuals have access to their specialists between planned appointments if they are concerned. Many find ongoing support through the family support groups to be invaluable, especially for those faced with multiple courses of treatment.

Growth and Feeding

Feeding is normal in individuals with Gorlin syndrome.

One of the most striking features of Gorlin syndrome is the increased head size (see Craniofacial). The average birth weight is 4.1 kg and average birth head circumference is 38 cm (personal experience), both greatly increased when compared with siblings. Two-thirds of children at term require an operative delivery (personal experiences). In adults, the head circumference is usually over 60 cm and in children, it is above the corresponding centile for height.

Affected individuals tend to be very tall. Their height is usually over the 97th centile, often in marked contrast to unaffected siblings. Some exhibit a marfanoid build.

Evaluation

- Height, weight, and head circumference should be plotted on growth curves at routine medical visits during childhood.
- Ultrasound scans to identify a large fetal head should be offered during pregnancies in which the fetus is at increased risk of having Gorlin syndrome. A large fetal head may necessitate operative delivery. Congenital malformations, which may require early decisions about neonatal surgery, may also be detected by such ultrasounds.

Management

- No specific treatment relative to growth and feeding is required.

Development and Behavior

Many affected children initially have mild motor delay and show some clumsiness. In our personal series, walking was delayed until an average of 18 months, whereas siblings walked at an average of 12–13 months. Development usually caught up by the age of 5 years. All children known to the authors apart from one have attended mainstream school, a few needing additional help.

In the literature, intellectual disability has been reported in about 3% of individuals. In a population study in northwest England, including 84 individuals, no intellectual disability of moderate or severe degree was detected (Evans et al. 1993), apart from those treated for medulloblastoma.

A consistent behavioral pattern has not been associated with Gorlin syndrome.

Evaluation

- Children should have developmental screening as part of routine pediatric care.
- If there is a change in development or personality, tiredness, muscle weakness, nystagmus, ataxia, headaches, or early morning vomiting, evaluation for medulloblastoma should be undertaken (see Neoplasia).

Management

- If developmental problems are identified, educational support should be offered as in the general population. There are no specific programs for the syndrome.
- Referral to a specialist center should be made if medulloblastoma is suspected.

Craniofacial

About 70% of individuals with Gorlin syndrome have a characteristic facies (Figure 29.1), but there is intrafamilial and interfamilial variation. Some members of a sibship may have the typical shape to the skull, for instance, whereas others do not. The head gives the appearance of being long in the sagittal plane, with a prominent and low occiput. Frontal, temporal, and biparietal bossing give a prominent appearance to the forehead, and affected individuals often adopt hairstyles that disguise the bossing. Occasionally, there can be marked cranial asymmetry because of craniosynostosis.

Professional concerns about macrocephaly (in the absence of other symptoms or signs) have resulted in investigations for hydrocephalus. In most children, the head circumference is above the 97th centile and growth continues parallel with (but greatly above) the centile lines. Magnetic resonance imaging (MRI) may show ventriculomegaly. Clinical experience suggests that in most cases, this is likely to be a benign process and a period of watchful waiting is usually appropriate (personal experiences).

There is often facial asymmetry. Some have prominent supraorbital ridges, giving the eyes a deep-set appearance. The eyebrows are often heavy and arched with synophrys. There is a broad nasal root and hypertelorism. The inner canthal, interpupillary, and outer canthal distances are all generally above the 97th centile, but appear to be in proportion with the head circumference. The mandible is long and often prominent with the lower lip protruding. The characteristic facial features may become more apparent with age.

There is a well-established association with cleft lip, with or without palate, which occurs in 5–6%.

For discussion of odontogenic keratocysts, see Dental.

Evaluation

- It can be helpful to compare a person's facial appearance with that of sibs because there is usually a striking difference in the facial gestalt between unaffected and affected siblings.
- Head circumference should, if possible, be plotted on centile charts, which take height into account (Bushby et al. 1992).
- Suspicion of craniosynostosis should prompt skull radiography.
- If cleft lip with or without cleft palate is present, referral to a cleft palate team or experienced oral surgeon is indicated, and evaluations for otitis media and speech abnormalities should occur in a standard manner.

Management

- Surgery is rarely required for craniosynostosis associated with Gorlin syndrome, but if needed, it should be treated as in the general population.
- Cleft lip and palate, when present, should be treated as in the general population.

Dermatologic

Non-neoplastic Skin Finding.s Small keratin-filled cysts (milia) are found on the face in 30% of individuals with Gorlin syndrome, most commonly in the infraorbital areas, but they can also occur on the forehead. Larger epidermoid cysts (usually 1–2 cm in diameter) occur on the limbs and trunk in over 50%.

Skin tags are especially common around the neck. Like the nevi, histology demonstrates the typical features of a basal cell carcinoma, but the skin tags do not generally change in size or shape.

The distinctive pits found on the palms and soles are pathognomonic (Howell and Mehregan 1970). They increase in number with age, are permanent, and, when found in a child, are a strong diagnostic indicator. The pits are small

(1–2 mm), often asymmetric, shallow depressions, with the color of the base being white, flesh-colored, or pale pink (Figure 29.2). They are found more commonly on the palms (77%) than on the soles (50%). Pits can also appear independently on the sides of the fingers as tiny bright red pinpricks. Their number may vary from only a few to greater than a hundred. Basal cell carcinomas have rarely arisen in the base of the pits. In the authors' personal series of over 250 individuals with Gorlin syndrome, 65% had palmar pits by age 10 years, and 80% by age 15 years. They were present in 85% of affected people over the age of 20 years. The pits appear to be caused by premature desquamation of horny cells along the intercellular spaces. Light microscopy shows a lack of keratinization of pit tissue and a proliferation of basaloid cells in irregular rete ridges (Howell and Freeman 1980).

A few café-au-lait patches are commonly present, usually on the trunk, which may lead to consideration of a diagnosis of neurofibromatosis type 1, especially in those with a large head circumference. Intertriginous freckling, however, is not found.

Nevi and Basal Cell Carcinomas As the "nevi" and the basal cell carcinomas found in the syndrome are histologically identical, they can both be classified as nevoid basal cell carcinomas. Clinically, however, the "nevi" often develop first and behave differently from the basal cell carcinomas that can appear to arise from nevi. The nevi are flesh-colored, reddish brown or pearly, resembling moles, skin tags, ordinary nevus cell nevi, or hemangiomas (Figure 29.3). The nevi tend to occur multiply in crops, their numbers increasing with time, although they can appear as individual lesions. Some grow rapidly for a few days to a few weeks, but most remain static. An individual may develop no nevi, a few, or many hundreds. Ordinary nevus cell nevi, found in about 4% of the general population, are present from birth.

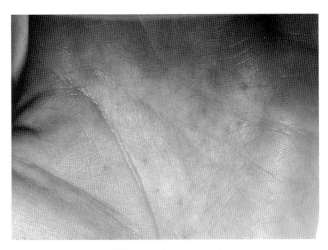

FIGURE 29.2 Palmar pits.

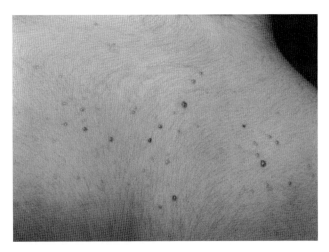

FIGURE 29.3 Nevi. Note variation in size and appearance, some being reddish-brown, skin-colored, or translucent.

Nevi may arise in any area of the skin, affecting the face, neck, and upper trunk in preference to the abdomen, lower trunk, and extremities. The areas around the eyes, nose, malar regions, and upper lip are the most frequently affected sites on the face, leading to a widespread view in the literature that sun exposure is an important factor. There is circumstantial evidence supporting this view. In a northwest England study, 14% of individuals (Evans et al. 1993) developed a basal cell carcinoma before the age of 20 years compared with 47% in Australia (G. Trench, personal communication).

Although nevi are found in 53% of affected individuals under the age of 20 years, only 14% present clinically with a rapidly growing basal cell carcinoma, and it is even more unusual for someone to develop aggressive basal cell carcinomas before puberty. 74% of those over the age of 20 years have developed a basal cell carcinoma, increasing to 90% by the age of 40 among Caucasians. Note that 10% never develop basal cell carcinomas: lifestyle, environmental, or other genetic factors affording this protection are not known. Skin pigmentation is known to be protective against basal cell carcinomas in people with Gorlin syndrome. 30% of Italians with Gorlin syndrome developed basal cell carcinomas (Lo Muzio et al. 1999), a figure similar to the 28% (4/11) established for African-Americans (Goldstein et al. 1994). Skin pigmentation does not protect from the adverse events of ionizing radiation (Korczak et al. 1997).

Only a few nevi become aggressive, when they may be locally invasive and behave like ordinary basal cell carcinomas. Evidence of aggressive transformation of an individual lesion includes an increase in size, ulceration, bleeding, or crusting. It is rare for metastasis to occur. About one-third of people have two or more types of basal cell carcinomas, including superficial, multicentric, solid, cystic, adenoid, and lattice-like (Gorlin et al. 1965). Nevoid basal cell carcinomas are more commonly associated with melanin pigmentation and foci of calcification than nonsyndromic basal cell carcinomas.

The clinical behavior of the skin lesions suggests that inactivation of the remaining *PTCH1/SUFU* allele in a cell causes a "nevus," additional subsequent cellular events resulting in aggressive behavior.

Clinical experience confirms that some affected individuals are extremely sensitive to treatment by radiation, developing new lesions in the irradiated field. Children who received craniospinal irradiation as part of the treatment for a medulloblastoma (Evans et al. 1991a) or Hodgkin's disease (Zvulunov et al. 1995) have developed thousands of basal cell carcinomas in the irradiated area. These basal cell carcinomas often develop within an extremely short latent period of 6 months to 3 years. This is earlier than, and in a distribution different from, other affected family members (Strong 1977). Radiosensitive individuals may develop more long-term complications from this treatment than from the original basal cell carcinomas (Strong 1977; Southwick and Schwartz 1979).

There is supporting evidence for the adverse effects of radiation from mice heterozygous for an inactivating *ptch* mutation. They spontaneously developed basal cell carcinoma-like tumors with age (Aszterbaum et al. 1999), but basal cell carcinomas were of far greater number and size in mice that had received ultraviolet (UV) irradiation. A single dose of ionizing radiation markedly enhanced development of basal cell carcinomas.

The molecular basis for human radiosensitivity in Gorlin syndrome has not been discovered, and until the susceptibility of individuals can be identified, avoidance of treatment by radiotherapy is strongly recommended for all affected individuals. Increased skin pigmentation may be protective against UV, but not ionizing radiation, as an African-American boy treated with craniospinal irradiation for a medulloblastoma developed numerous basal cell carcinomas in the irradiated area (Korczak et al. 1997).

Evaluation

- Skin should be carefully examined for dermatologic findings of Gorlin syndrome by a physician starting at puberty, and repeated at least every six months.
- Affected individuals should be educated to inspect all areas of the body. As a lesion may suddenly become aggressive, open access to the specialist taking responsibility for treatment of the skin is important.
- Palmar and plantar pits are easier to see in people who do manual labor, but should be differentiated from palmar lesions caused by excoriation of dirt under the skin. In most individuals with Gorlin syndrome, the pits can be better visualized if the hands are soaked in warm water for about 10 min.

Management

- The pits do not require treatment except in the extremely rare instance where a basal cell carcinoma has developed in the base of a pit. In this case, the pit should be excised.
- Epidermoid cysts may require excision, especially if infected.
- As sunlight may be one of the environmental agents promoting the appearance of basal cell carcinomas (Goldstein et al. 1993), sunscreen precautions should be strongly recommended, including the wearing of a wide-brimmed hat to offer some protection to the area around the eyes.
- Alarm can be generated particularly in childhood when a skin tag or nevus is shown on histology to be a basal cell carcinoma. This may result in a feeling that immediate treatment is required for all other skin lesions present, and, indeed, some authors urge treatment for all such lesions. Others reserve treatment for lesions that show aggressive signs. As many nevi remain quiescent for long periods, they may not need to be removed but should be frequently reviewed. The authors' practice is to have a lower threshold for local treatment for individual lesions occurring around the eyes, nose, mouth, and ears.
- The most suitable form of treatment varies depending on the type, size, and site of the nevoid basal cell carcinoma. Surgical excision, cryotherapy, curettage and diathermy, topical 5-fluorouracil, Mohs' microsurgery (Mohs et al. 1980), and carbon dioxide laser vaporization have all been used (Kopera et al. 1996). The priorities are to ensure complete eradication of aggressive basal cell carcinomas, and to preserve normal tissue to prevent disfigurement.
- A systematic review could not recommend evidence-based guidelines for treatment (Thissen et al. 1999), but for larger basal cell carcinomas, especially on the face and those with aggressive behavior, Mohs' microsurgery gives the best results.
- Topical 5-fluorouracil appears effective for superficial multicentric basal cell carcinomas without follicular involvement, but should not be used for deeply invasive basal cell carcinomas.
- Topical imiquimod, an immune response modifier, also appears effective for superficial basal cell carcinomas (Kagy and Amonette 2000; Stockfleth et al. 2002), but some people find the local inflammatory response difficult to tolerate.
- Radiotherapy should be avoided because of clinical evidence that new lesions can appear in the irradiated field (see above). Some families may not be as

radiosensitive as others, but until molecular assessment can detect those more or less sensitive, radiotherapy should be avoided for all families.
- Oral synthetic retinoids (etretinate, isotretinoin, and 13-*cis*-retinoic acid) have been reported to prevent the development of new tumors, inhibit the growth of existing tumors, and cause regression of superficially invasive basal cell carcinomas (Cristofolini et al. 1984; Hodak et al. 1987). Although there can be an excellent response, new lesions can appear when treatment is discontinued or dosage reduced (Peck et al. 1988; Goldberg et al. 1989). Systemic retinoids can be useful preoperatively to allow less aggressive surgery (Sanchez-Conejo-Mir and Camacho 1989). There is significant toxicity associated with prolonged retinoid use. Potential teratogenicity is a significant concern for systemic retinoids. Additional side effects such as cheilitis, pruritis, peeling of the palms and soles, eczema, and diffuse idiopathic skeletal hyperostosis (Theiler et al. 1993), dictate that retinoids should be used in carefully controlled circumstances. Their long-term role in the management of Gorlin syndrome is uncertain until synthetic retinoids that demonstrate reduced toxicity while maintaining an antineoplastic effect become available.
- A girl with Gorlin syndrome was treated for 10 years with a combination of topical tretinoin and 5-fluorouracil beginning at 25 months of age (Strange and Lang, 1992). Her hundreds of tumors disappeared after the initiation of the combined therapy; most of the remaining tumors did not grow. Lesions that demonstrated signs of growth or appeared to be deeply invasive were managed by shave excision and curettage. This is one approach to treatment.
- Photodynamic therapy involves systemic or topical administration of a photosensitizer followed by exposure of the target area to light. There are promising results in Gorlin syndrome, but photodynamic therapy would best be offered to people with Gorlin syndrome in expert centers as part of the expert team approach to management of basal cell carcinomas. In 1984, 40 basal cell carcinomas were treated in three adults with Gorlin syndrome (Tse et al. 1984) in whom conventional treatments had failed or were no longer possible, with 82.5% complete and 17.5% partial clinical response. There was a 10.8% recurrence rate. Photodynamic therapy is again being evaluated in Gorlin syndrome (Zeitouni et al. 2001; Madan et al. 2006). Although complete clinical basal cell carcinoma response rate is high (93%) in nodular and superficial lesions with 1 mg/kg systemic Photofrin (Oseroff et al. 2006), the results in children are less satisfactory with a poorer response and scarring.

Systemic photodynamic therapy is, therefore, not recommended for prepubertal children. A major disadvantage of Photofrin is that it can produce a generalized photosensitivity for 4–8 weeks, therefore a new generation of photosensitizers are being developed.
- Photodynamic therapy using topical 5-aminolevulinic acid (Oseroff et al. 2005) appears to be particularly promising for those children with Gorlin syndrome who develop multiple superficial basal cell carcinomas in fields irradiated for Hodgkin's disease and medulloblastoma. The healing response is better than with systemic administration of photosensitizer and leaves no scarring.
- More recently a sonic hedgehog protein antagonist, vismodegib, has shown substantial efficacy in the treatment of advanced BCCs in Gorlin syndrome and seems to prevent new tumors (Tang et al. 2012). However, the side effect profile is a major issue with the majority of patients discontinuing therapy at maximum dose before the course is completed (Tang et al. 2012). An intermittent treatment protocol may partly circumvent the problems with side effects (Dreno et al. 2017), but the cost of treatment has meant that the National Institute for Health and Care Excellence (NICE) in the UK has judged the treatment not cost effective.

Dental

Odontogenic cysts are one of the major features of Gorlin syndrome. Approximately 13% of affected people develop this jaw cyst by the age of 10 years and 51% by the age of 20 years. The majority of cysts occur after 7 years, although occasionally a cyst can present in the first few years of life. The peak incidence is in the third decade, which is about 10 years earlier than isolated odontogenic keratocysts. About 26% of individuals over the age of 40 years in the authors' series had not developed signs or symptoms of cysts. The mandible is involved far more frequently than the maxilla, with keratocysts usually occurring at the angle of the mandible (Figure 29.4).

Individuals with Gorlin syndrome can be remarkably free of symptoms until cysts reach a large size, especially when the ascending ramus is involved. Presentation can be with swelling and/or pain of the jaw, pus discharging into the oral cavity, or displaced, impacted, or loose teeth. Asymptomatic relatively small single unilocular lesions may be detected by screening, but large bilateral multilocular cysts involving both jaws are more often found when investigation follows clinical symptoms.

The histological features of the jaw cysts are characteristic (Ahlfors et al. 1984). The cysts are lined by a parakeratotic-stratified squamous epithelium that is usually

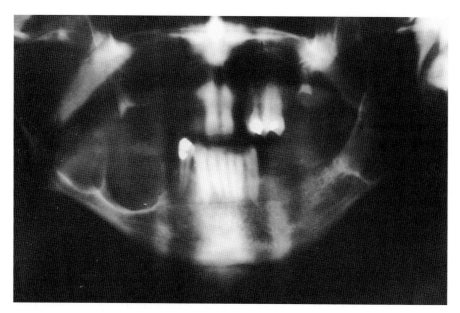

FIGURE 29.4 An orthopantogram of a 37-year-old male with Gorlin syndrome who presented with a fluctuant swelling of the right alveolus. It shows a large cyst of the whole ascending ramus and body of the right mandible up to the upper third molar region. There is a large cyst in the left mandibular ramus.

about 5–8 cell layers thick and without rete ridges. The basal layer is well defined with regularly orientated palisaded cells. Satellite cysts, epithelial rests, and proliferating dental lamina are sometimes seen in the cyst capsules. Immunocytochemical staining for Ki67 expression can differentiate between keratocysts associated with the syndrome and non-syndromic simple and recurrent keratocysts (Li et al. 1995).

The odontogenic keratocyst has a tendency to recur after surgical treatment, with reported rates varying up to 62%. New cysts may form from satellite cysts associated with the original, or from the dental lamina.

Misshapen teeth, missing teeth, and a susceptibility to caries are more common in individuals with Gorlin syndrome than in unaffected relatives.

Evaluation

- Annual dental screening should commence at about 8 years, usually including a panoramic radiograph of the jaw as plain films may miss lesions. It should be repeated yearly unless jaw cysts are discovered, when the frequency should be adjusted in accordance with findings.
- Routine orthopantograms are justified because of complications of untreated jaw cysts.
- Computed tomography scans should be avoided because of the radiation dosage.
- Regular dental examinations are important to identify caries and other dental and orthodontic problems.

Management

- Odontogenic keratocysts should be surgically removed.
- As proliferating dental lamina and satellite cysts may occur in the fibrous wall of the primary cyst cavity, marsupialization may be successful only if no satellite cysts are left behind.
- Small single lesions with regular spherical outlines can usually be completely enucleated provided access is good.
- For the large multilocular lesions, excision and immediate bone grafting is the treatment of choice at the first operation (Posnick et al. 1994).
- Routine dental hygiene and prophylaxis is important in light of the increased risk for caries.

Neoplasia

Medulloblastoma (primitive neuroectodermal tumor) is a well-recognized finding in Gorlin syndrome, with an incidence of about 5%. Gorlin syndrome is found in about 3% of children with medulloblastoma, and in 10% of those under the age of 2 years (Evans et al. 1991b; Cowan et al. 1997). The average age of presentation in Gorlin syndrome is 2 years, about 5 years before the average age of presentation in children with isolated medulloblastoma. Children with medulloblastoma associated with Gorlin syndrome are likely to have long-term survival, perhaps associated with the desmoplastic nature of the lesion, but there is a high chance that craniospinal irradiation will result in hundreds of basal cell carcinomas appearing in the irradiated field (Evans et al. 1991a). There is an additional concern that there may be an increased risk of other second cancers in the radiation field (Goldstein et al. 1997).

Meningioma, glioblastoma multiforme, and craniopharyngioma have also been described in adults.

Tumors in many other organs have been reported in people with Gorlin syndrome. They include renal fibroma, melanoma, leiomyoma, rhabdomyosarcoma, adenoid cystic carcinoma, adrenal cortical adenoma, seminoma, fibroadenoma of the breast, thyroid adenoma, carcinoma of the bladder, Hodgkin's disease, and chronic leukemia.

Evaluation

- Six-monthly clinical neurological examination may be offered from birth or early diagnosis to detect signs and symptoms suggestive of a medulloblastoma. At 3 years, the examinations could be reduced to annually until 7 years, after which a medulloblastoma is very unlikely. Symptoms include signs of hydrocephalus (behavior change, headache, vomiting, and blurred vision), cerebellar signs (ataxia, head tilt, and dysmetria), or signs of leptomeningeal dissemination (weakness and radiculopathy).
- Routine scanning with computed tomography or excessive use of radiography is not recommended because of concerns about inducing skin malignancies. MRI screening has been advocated for Gorlin syndrome associated with mutations in *SUFU*. A recent guideline suggested considering MRI scanning of the brain every four months until age 3 and then every six months until the age of 5 (Foulkes et al. 2017), but MRI screening was not advocated for *PTCH1*-related Gorlin syndrome.
- There does not appear to be any neoplasm occurring at a frequency that warrants selective screening for people with the syndrome. As in the general population, affected individuals should be encouraged to report any unusual symptoms or signs, and to participate in population screening programs designed for early detection of neoplasia.

Management

- If the diagnosis of Gorlin syndrome is confirmed in a child with medulloblastoma, it is recommended that the oncologist and family consider treatment modalities that do not involve radiation (Foulkes et al. 2017). Otherwise, treatment is standard.
- Other associated tumors are treated in a standard manner.

Ophthalmologic

The most common ophthalmologic feature of Gorlin syndrome is convergent strabismus. In a personally examined series, 26% had an ophthalmic problem, of which 56% had strabismus. Of affected individuals reported in the literature, 10–15% have ophthalmic abnormalities including congenital blindness due to corneal opacity, congenital glaucoma, coloboma of the iris, choroid or optic nerve, convergent or divergent strabismus, nystagmus, cataracts, microphthalmia (usually unilateral), ptosis, proptosis, medullated nerve fibers, and retinal hamartomas.

Evaluation

- Ophthalmological anomalies most likely to require treatment are usually directly observable and bring the individual to medical attention. It is unusual to find anomalies requiring treatment through regular ophthalmic surveillance (G. Black, personal communication), although a single specialist consultation at diagnosis may be helpful in identifying whether a particular appearance (e.g., medullated nerve fibers) is significant.
- There is usually no special case for ophthalmic surveillance, unless abnormalities requiring follow-up are identified.
- Referral to a specialist center may be required for congenital malformations such as severe microphthalmia.

Management

- Treatment of ophthalmologic abnormalities is standard.

Cardiovascular

Cardiac fibromas are found in 2.5% of affected individuals (Evans et al. 1993); the majority are asymptomatic. One child known to the authors died at 3 months of age from multiple cardiac fibromas while another has been followed for over 20 years with a single 2 cm cardiac fibroma in the interventricular septum that has remained unchanged. The incidence in childhood of an isolated cardiac fibroma is between 0.027 and 0.08%.

Evaluation

- Anecdotal evidence suggests that cardiac fibromas likely to cause serious clinical problems may be present from very early in life; therefore, a single echocardiogram would best be performed in the neonatal period.
- There are no data to suggest that echocardiograms should be performed routinely throughout life.

Management

- Most cardiac fibromas can be followed conservatively. When they cause outflow obstruction, surgical resection may be required.

Musculoskeletal

Bifid, anteriorly splayed, fused, partially missing, or hypoplastic ribs are found in 70% of people with Gorlin syndrome (Figure 29.5) and may give an unusual shape to the chest, including a characteristic downward sloping of the shoulders. The third and fourth ribs are most frequently involved. Bifid ribs are found in about 6% of the general population. The rib anomalies, together with kyphoscoliosis, cause either pectus excavatum or carinatum in about 30–40% of people with Gorlin syndrome. Sprengel deformity has been found in some surveys to be as common as 25%.

Abnormalities of the cervical or thoracic vertebrae are helpful diagnostic signs, being found in about 60% of affected people. Vertebrae C6, C7, T1, and T2 are most frequently involved. Spina bifida occulta of the cervical vertebrae or malformations at the occipito-vertebral junction are common. In addition to lack of fusion of the cervical or upper thoracic vertebrae, fusion or lack of segmentation has been documented in about 40% of affected people. A defective medial portion of the scapula is occasionally found.

As rib and spine anomalies are present at birth, they are helpful diagnostic signs, but it should be noted that 14% do not have anomalies of the cervico-thoracic spine and/or ribs. Bifid ribs, cervical ribs, and synostosis of ribs occur in 6.25, 1.7, and 2.6 per 1000, respectively, of the general population (Etter 1944).

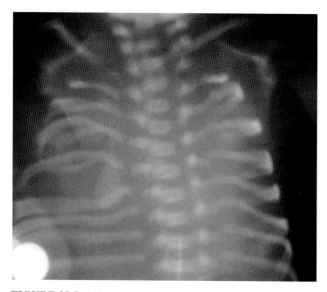

FIGURE 29.5 Chest radiograph showing bifid ribs (right second and third ribs at back).

Small pseudocystic lytic bone lesions, most often in the phalanges, metapodial, and carpal and tarsal bones, may be found in about 35% of people with Gorlin syndrome. There may be just one or two lesions, or they may involve almost the entire long bone or the pelvis, generating diagnostic uncertainty and resulting in multiple investigations. Lesions in the calvarium may raise concern that a medulloblastoma has extended into bone. Histology reveals that the lesions are hamartomas composed of fibrous connective tissue, blood vessels, and nerves (Dunnick et al. 1978; Blinder et al. 1984).

Thumb anomalies (short terminal phalanges and/or small stiff thumbs) occur in about 10% of people with Gorlin syndrome. Pre- or postaxial polydactyly of hands or feet is found in 8%. The fourth metacarpal is short in 15–45%, but is not a good diagnostic sign as it is found in about 10% of the normal population. Hallux valgus can be severe, requiring operation.

Evaluation

- Skeletal features may be apparent on clinical examination.
- Radiographic investigation (chest, thoracic spine, and hand films for pseudocysts) may be helpful when the syndrome is suspected but physical signs are equivocal (Ratcliffe et al. 1995a). Such investigation should be avoided if the diagnosis is firm because of inherent risks of radiation as mentioned above.

Management

- The skeletal features are non-progressive, and treatment is usually not indicated. Very few seek opinions about chest shape, and surgical opinion has tended to be that surgery may exacerbate rather than help the appearance.
- A Heller's operation for hallux valgus may be required for severe pain or the shape of the toe preventing shoe fitting.
- Any treatment for musculoskeletal problems is standard.

Genitourinary

Calcified ovarian fibromas have been reported in 25–50% of women with Gorlin syndrome (Evans et al. 1993) and may be more common in people with a *SUFU* mutation (Evans et al. 2017). They may be mistaken for calcified uterine fibroids, especially if bilateral fibromas overlap in the midline giving the appearance of a single entity. They do not seem to reduce fertility, but may undergo torsion. Finding ovarian fibromas that are bilateral, calcified, and multinodular should trigger a search for other manifestations of Gorlin syndrome. Ovarian fibromas, in general, form a single mass replacing one ovary, and less than 10% are bilateral or demonstrate calcification.

Although ovarian fibrosarcoma and other ovarian tumors have been reported, these are extremely rare.

Inguinal hernia is common in males (17% of males).

Evaluation

- The fibromas are usually discovered fortuitously at ultrasound or radiographic abdominal investigation for other indications in the absence of symptoms.
- Opinion is divided as to whether ultrasound screening should be performed routinely in adult women in the absence of symptoms, as there is no evidence to suggest that the fibromas should be removed prophylactically.

Management

- If operative treatment is required, preservation of ovarian tissue is recommended (Seracchioli et al. 2001).

Gastrointestinal

Just as cysts of the skin and jaw are integral parts of the syndrome, so are chylous or lymphatic cysts of the mesentery, although these are rare. They may present, if large, as painless movable masses in the upper abdomen, or rarely may cause symptoms of obstruction. In most cases, however, they are discovered fortuitously at laparotomy or on radiography if they are calcified.

Evaluation

- Painless abdominal swelling or signs of intestinal obstruction warrant ultrasound investigation.

Management

- Surgical excision may be required because of intestinal obstruction, but this is rare. Methodology is standard.

Miscellaneous

Seizures About 6% of individuals in a northwest England study required prolonged anticonvulsant therapy for grand mal seizures (Evans et al. 1993).

Ectopic Calcification Calcification of the falx cerebri is a very useful diagnostic sign, and, in a child, should strongly suggest Gorlin syndrome (Figure 29.6). Conversely, the diagnosis should be considered in doubt if an adult with non-mosaic Gorlin syndrome does *not* have calcification of the falx (Ratcliffe et al. 1995b). It can appear very early in life, is often strikingly apparent from late childhood, and its degree progresses with age. In the authors' series, it was

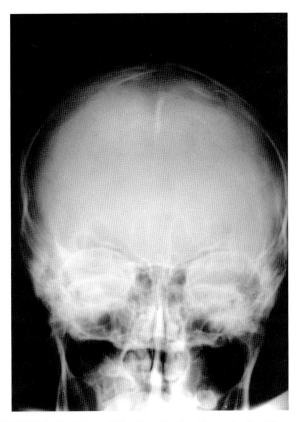

FIGURE 29.6 Falx calcification. Skull radiograph of an 18-year-old male showing upper falx calcification.

present in 40% of affected individuals less than 15 years of age and 95% by age 25 years. Falx calcification first appears as a faint line in the upper falx, the faint line becoming more prominent and giving the appearance of several individual sheets of calcification. In some, it can be very florid, up to 1 cm wide. It has a characteristic lamellar appearance, in comparison with the single sheet of calcification found in 7% of the older general population.

Ectopic calcification also occurs in other membranes, including the tentorium cerebelli (40%), petroclinoid ligaments (20%), dura, pia, and choroid plexus. Calcification of the diaphragma sellae causing the appearance of bridging of the sella turcica is another useful early diagnostic sign, found in 84% of affected people, compared with 4% of the general population in later life. It is present in almost 100% of affected individuals by age 20.

Calcification may also occur subcutaneously in apparently otherwise normal skin of the fingers and scalp.

Evaluation

- A history of seizures should be sought at routine medical visits.
- A suspicious history should prompt electroencephalogram.

- Radiographs to detect calcification (e.g., anteroposterior and lateral views of the skull) should be requested in cases of diagnostic doubt, rather than routinely. A normal variant of the skull, a prominent frontal crest, can simulate falx calcification on the anteroposterior skull film, and should be considered if the calcification appears to be a single line beginning inferiorly.

Management

- Seizures should be treated in a standard manner.
- No treatment for calcifications is warranted or necessary.

RESOURCES

Support Groups

Gorlin Syndrome Group (UK)

Phone: United Kingdom (+44): 01772 496849
Website: *www.gorlingroup.org*

BCCNS Alliance (US)

Phone: (267) 689-6443
Website: www.bccns.org

REFERENCES

Ahlfors E, Larsson A, Sjogren S (1984) The odontogenic keratocyst: A benign cystic tumor? *J Oral Maxillofac Surg* 42:10–19.

Aszterbaum M, Rothman A, Johnson RL, Fisher M, Xie J, Bonifas JM, Zhang X, Scott MP, Epstein EH Jr (1998) Identification of mutations in the human PATCHED gene in sporadic basal cell carcinomas and in patients with the basal cell nevus syndrome. *J Invest Dermatol* 110:885–888.

Aszterbaum MA, Epstein J, Oro A (1999) A mouse model of human basal cell carcinoma: Ultraviolet and gamma radiation enhance basal cell carcinoma growth in patched heterozygote knock-out mice. *Nat Med* 5:1285–1291.

Bale AE, Yu KP (2001) The hedgehog pathway and basal cell carcinomas. *Hum Mol Genet* 10:757–762.

Barnes EA, Kong M, Ollendorff V, Donoghue DJ (2001) Ptch interacts with cyclin B1 to regulate cell cycle progression. *EMBO J* 20:2214–2223.

Bleiberg J, Brodkin RH (1969) Linear unilateral basal cell nevus with comedones. *Arch Dermatol* 100:187–190.

Blinder G, Barki Y, Pezt M, Bar-Ziv J (1984) Widespread osteolytic lesions of the long bones in basal cell nevus syndrome. *Skeletal Radiol* 12:195–198.

Bonifas JM, Bare JW, Kerschmann RL, Master SP, Epstein EH (1994) Parental origin of chromosome 9q22.3-q31 lost in basal cell carcinomas from basal cell nevus syndrome patients. *Hum Mol Genet* 3:447–448.

Bushby KMD, Cole T, Matthews JNS, Goodship JA (1992) Centiles for adult head circumference. *Arch Dis Child* 67:1286–1287.

Cowan R, Hoban P, Kelsey A, Birch JM, Gattamaneni R, Evans DGR (1997) The gene for the naevoid basal cell carcinoma syndrome acts as a tumour-suppressor gene in medulloblastoma. *Br J Cancer* 76:141–145.

Cristofolini M, Zumiani G, Scappini Piscioli F (1984) Aromatic retinoid in chemoprevention of the progression of nevoid basal cell carcinoma syndrome. *J Dermatol Surg Oncol* 10:778–781.

Dréno B, Kunstfeld R, Hauschild A, Fosko S, Zloty D, Labeille B, Grob JJ, Puig S, Gilberg F, Bergström D, Page DR, Rogers G, Schadendorf D (2017). Two intermittent vismodegib dosing regimens in patients with multiple basal-cell carcinomas (MIKIE): a randomised, regimen-controlled, double-blind, phase 2 trial. *Lancet Oncol.* 18(3):404-412.

Dunnick NR, Head GL, Peck GL, Yoder FW (1978) Nevoid basal cell carcinoma syndrome: Radiographic manifestations including cyst like lesions of the phalanges. *Radiology* 127:331–334.

Etter LE (1944) Osseous abnormalities of the thoracic cage seen in 40,000 consecutive chest photoroentgenograms. *AJR* 51:359–363.

Evans DGR, Birch J, Orton C (1991a) Brain tumours and the occurrence of severe invasive basal cell carcinomas in first degree relatives with Gorlin syndrome. *Br J Neurosurg* 5:643–646.

Evans DGR, Farndon PA, Burnell LD, Rao Gattamaneni H, Birch JM (1991b) The incidence of Gorlin syndrome in 173 consecutive cases of medulloblastoma. *Br J Cancer* 64:959–961.

Evans DGR, Ladusans EJ, Rimmer S, Burnell LD, Thakker N, Farndon PA (1993) Complications of the naevoid basal cell carcinoma syndrome: Results of a population based study. *J Med Genet* 30:460–464.

Evans DG, Howard E, Giblin C, Clancy T, Spencer H, Huson SM, Lalloo F (2010). Birth incidence and prevalence of tumour prone syndromes: estimates from a UK genetic family register service. *Am J Med Genet* 152A(2):327-332.

Evans DG, Oudit D, Smith MJ, Rutkowski D, Allan E, Newman WG, Lear JT (2017). First evidence of genotype-phenotype correlations in Gorlin syndrome. *J Med Genet* 54(8):530–536

Fan Z, Li J, Du J, Zhang H, Shen Y, Wang CY, Wang S (2008) A missense mutation in *PTCH2* underlies dominantly inherited NBCCS in a Chinese family. *J Med Genet* 45:303–308.

Foulkes WD, Kamihara J, Evans DGR, Brugières L, Bourdeaut F, Molenaar JJ, Walsh MF, Brodeur GM, Diller L (2017). Cancer Surveillance in Gorlin Syndrome and Rhabdoid Tumor Predisposition Syndrome. *Clin Cancer Res* 23(12):e62–e67.

Fujii K, Ohashi H, Suzuki M, Hatsuse H, Shiohama T, Uchikawa H, Miyashita T (2013). Frameshift mutation in the PTCH2 gene can cause nevoid basal cell carcinoma syndrome. *Fam Cancer* 12(4):611–4

Goldberg LH, Hsu SH, Alcalay J (1989) Effectiveness of isotretinoin in preventing the appearance of basal cell carcinomas in basal cell nevus syndrome. *J Am Acad Dermatol* 21:144–145.

Goldstein AM, Bale SJ, Peck GL, DiGiovanna JJ (1993) Sun exposure and basal cell carcinomas in the nevoid basal cell carcinoma syndrome. *J Am Acad Dermatol* 29:34–41.

Goldstein AM, Pastakia B, DiGiovanna JJ, Poliak S, Santucci S, Kase R, Bale AE, Bale SJ (1994) Clinical findings in two African-American families with nevoid basal cell carcinoma syndrome (NBCC). *Am J Med Genet* 50:272–281.

Goldstein AM, Yuen J, Tucker MA (1997) Second cancers after medulloblastoma: population-based results from the United States and Sweden. *Cancer Causes Control* 8:865–871.

Gorlin RJ (1987) Nevoid basal-cell carcinoma syndrome. *Medicine* 66:96–113.

Gorlin RJ, Goltz RW (1960) Multiple nevoid basal-cell epithelioma, jaw cysts and bifid rib: A syndrome. *N Engl J Med* 262:908–912.

Gorlin RJ, Vickers RA, Klein E, Williamson JJ (1965) The multiple basal cell nevi syndrome. *Cancer* 18:89–104.

Gorlin RJ, Cohen MM, Hennekam RCM (2001) Multiple nevoid basal cell carcinoma syndrome. In: Syndromes of the Head and Neck, 4th ed. Oxford: Oxford University Press, pp. 444–453.

Guerrini-Rousseau L, Dufour C, Varlet P, Masliah-Planchon J, Bourdeaut F, Guillaud-Bataille M, Abbas R, Bertozzi AI, Fouyssac F, Huybrechts S, Puget S, Bressac-De Paillerets B, Caron O, Sevenet N, Dimaria M, Villebasse S, Delattre O, Valteau-Couanet D, Grill J, Brugières L (2018) Germline SUFU mutation carriers and medulloblastoma: clinical characteristics, cancer risk and prognosis. *Neuro Oncol* 20:1122–1132

Hahn H, Wicking C, Zaphiropoulous PG, Gailani MR, Shanley S, Chidambaram A, Vorechovsky I, Holmberg E, Unden AB, Gillies S, Negus K, Smyth I, Pressman C, Leffell DJ, Gerrard B, Goldstein AM, Dean M, Toftgard R, Chenevix-Trench G, Wainwright B, Bale AE (1996) Mutations of the human homolog of Drosophila *patched* in the nevoid basal cell carcinoma syndrome. *Cell* 85:841–851.

Hodak E, Ginzburg A, David M, Sandbank M (1987) Etretinate treatment of the nevoid basal cell carcinoma syndrome. *Int J Dermatol* 26:606–609.

Howell JB, Caro MR (1959) The basal cell nevus: Its relationship to multiple cutaneous cancers and associated anomalies of development. *Arch Dermatol* 79:67–80.

Howell JB, Mehregan AH (1970) Pursuit of the pits in the nevoid basal cell carcinoma syndrome. *Arch Dermatol* 102:586–597.

Howell JB, Freeman RG (1980) Structure and significance of the pits with their tumors in the nevoid basal cell carcinoma syndrome. *J Am Acad Dermatol* 2:224–238.

Johnson RL, Rothman AL, Xie J, Goodrich LV, Bare JW, Bonifas JM, Quinn AG, Myers RM, Cox DR, Epstein EH Jr, Scott MP (1996) Human homolog of patched, a candidate gene for the basal cell nevus syndrome. *Science* 272:1668–1671.

Kagy MK, Amonette R (2000) The use of imiquimod 5% cream for the treatment of superficial basal cell carcinomas in a basal cell nevus syndrome patient. *Dermatol Surg* 26:577–578.

Kijima C, Miyashita T, Suzuki M, Oka H, Fujii K (2012). Two cases of nevoid basal cell carcinoma syndrome associated with meningioma caused by a PTCH1 or SUFU germline mutation. *Fam Cancer* 11(4):565–70.

Klein RD, Dykas DJ, Bale AE (2005) Clinical testing for the nevoid basal cell carcinoma syndrome in a DNA diagnostic laboratory. *Genet Med* 7:611–619.

Kopera D, Cerroni L, Fink-Puches R, Kerl H (1996) Different treatment modalities for the management of a patient with nevoid basal cell carcinoma syndrome. *J Am Acad Dermatol* 34:937–939.

Korczak JF, Brahim JS, DiGiovanna JJ, Kase RG, Wexler LH, Goldstein AM (1997) Nevoid basal cell carcinoma syndrome with medulloblastoma in an African-American boy: A rare case illustrating gene-environment interaction. *Am J Med Genet* 69:309–314.

Lench NJ, Telford EAR, High AS, Markham AF, Wicking C, Wainwright BJ (1997) Characterisation of human patched germ line mutations in naevoid basal cell carcinoma syndrome. *Hum Genet* 100:497–502.

Levanat S, Gorlin RJ, Fallet S, Johnson DR, Fantasia JE, Bale AE (1996) A two-hit model for developmental defects in Gorlin syndrome. *Nat Genet* 12:85–87.

Li TJ, Browne RM, Matthews JB (1995) Epithelial cell proliferation in odontogenic keratocysts: A comparative immunocytochemical study of Ki67 in simple, recurrent and basal cell naevus syndrome (BCNS)-associated lesions. *J Oral Pathol Med* 24:221–226.

Lindstrom E, Shimokawa T, Toftgard R, Zaphiropoulos PG (2006) PTCH mutations: Distribution and analyses. *Hum Mutat* 27:215–219.

Lo Muzio L, Nocini PF, Savoia A, Consolo U, Procaccini M, Zelante L, Pannone G, Bucci P, Dolci M, Bambini F, Solda P, Favia G (1999) Nevoid basal cell carcinoma syndrome: Clinical findings in 37 Italian affected individuals. *Clin Genet* 55:34–40.

Madan V, Loncaster JA, Allan D, Lear JT, Sheridan L, Leach C, Allan E (2006) Nodular basal cell carcinoma in Gorlin's syndrome treated with systemic photodynamic therapy and interstitial optical fiber diffuser laser. *J Am Acad Dermatol* 55:S86–S89.

Michaelsson G, Olsson E, Westermark P (1981) The Rombo syndrome. *Acta Derm Venereol* 61:497–503.

Ming JE, Kaupas ME, Roessler E, Brunner HG, Golabi M, Tekin M, Stratton RF, Sujansky E, Bale SJ, Muenke M (2002) Mutations in PATCHED-1, the receptor for SONIC HEDGEHOG, are associated with holoprosencephaly. *Hum Genet* 110:297–301.

Mohs FE, Jones DL, Koranda FC (1980) Microscopically controlled surgery for carcinomas in patients with nevoid basal cell carcinoma syndrome. *Arch Dermatol* 116:777–779.

Oseroff AR, Shieh S, Frawley NP, Cheney R, Blumenson LE, Pivnick EK, Bellnier DA (2005) Treatment of diffuse basal cell carcinomas and basaloid follicular harmatomas in nevoid basal cell carcinoma syndrome by wide-area 5-aminolevulinic acid photodynamic therapy. *Arch Dermatol* 141:60–67.

Oseroff AR, Blumenson LR, Wilson BD, Mang TS, Bellnier DA, Parsons JC, Frawley N, Cooper M, Zeitouni N, Dougherty TJ (2006) A dose ranging study of photodynamic therapy with porfimer sodium (photofrin) for treatment of basal cell carcinoma. *Lasers Surg Med* 38:417–426.

Pastorino L, Ghiorzo P, Nasti S, Battistuzzi L, Cusano R, Marzocchi C, Garrè ML, Clementi M, Scarrà GB (2009).

Identification of a SUFU germline mutation in a family with Gorlin syndrome. *Am J Med Genet A* 149A(7):1539–43.

Peck GL, DiGiovanna JJ, Sarnoff DS, Gross EG, Butkus D, Olsen TG, Yoder FW (1988) Treatment and prevention of basal cell carcinoma with oral isotretinoin. *J Am Acad Dermatol* 19:176–185.

Posnick JC, Clokie CML, Goldstein JA (1994) Maxillofacial considerations for diagnosis and treatment in Gorlin's syndrome: Access osteotomies for cyst removal and orthognathic surgery. *Ann Plast Surg* 35:512–518.

Rasmussen JE (1975) A syndrome of trichoepitheliomas, milia and cylindromas. *Arch Dermatol* 111:610–614.

Ratcliffe JF, Shanley S, Chenevix-Trench G (1995a) The prevalence of cervical and thoracic congenital skeletal abnormalities in basal cell naevus syndrome; a review of cervical and chest radiographs in 80 patients with BCNS. *Br J Radiol* 68: 596–599.

Ratcliffe JF, Shanley S, Ferguson J, Chenevix-Trench G (1995b) The diagnostic implication of falcine calcification on plain skull radiographs of patients with basal cell naevus syndrome and the incidence of falcine calcification in their relatives and two control groups. *Br J Radiol* 68:361–368.

Sanchez-Conejo-Mir J, Camacho F (1989) Nevoid basal cell carcinoma syndrome: Combined etretinate and surgical treatment. *J Dermatol Surg Oncol* 15:868–871.

Seracchioli R, Bagnoli A, Colombo FM, Missiroli S, Venturoli S (2001) Conservative treatment of recurrent ovarian fibromas in a young patient affected by Gorlin syndrome. *Hum Reprod* 6:1261–1263.

Shanley S, Ratcliffe J, Hockey A, Haan E, Oley C, Ravine D, Martin N, Wicking C, Chenevix-Trench G (1994) Nevoid basal cell carcinoma syndrome: Review of 118 affected individuals. *Am J Med Genet* 50:282–290.

Smith MJ, Beetz C, Williams SG, Bhaskar SS, O'Sullivan J, Anderson B, Daly SB, Urquhart JE, Bholah Z, Oudit D, Cheesman E, Kelsey A, McCabe MG, Newman WG, Evans DG (2014) Germline mutations in SUFU cause Gorlin syndrome-associated childhood medulloblastoma and redefine the risk associated with PTCH1 Mutations. *J Clin Oncol* 32(36): 4155–61

Smyth I, Narang MA, Evans T, Heimann C, Nakamura Y, Chenevix-Trench G, Pietsch T, Wicking C, Wainwright BJ (1999) Isolation and characterisation of human Patched 2 (PTCH2), a putative tumour suppressor gene in basal cell carcinoma and medulloblastoma on chromosome 1p32. *Hum Mol Genet* 8:291–297.

Southwick GJ, Schwartz RA (1979) The basal cell nevus syndrome: Disasters occurring among a series of 36 patients. *Cancer* 44:2294–2305.

Starink TM, van der Veen JP, Arwert F, de Waal LP, de Lange GG, Gille JJ, Eriksson AW (1986) The Cowden syndrome: A clinical and genetic study in 21 patients. *Clin Genet* 29: 222–233.

Stockfleth E, Ulrich C, Hauschild A, Lischner S, Meyer T, Christophers E (2002) Successful treatment of basal cell carcinomas in a nevoid basal cell carcinoma syndrome with topical 5% imiquimod. *Eur J Dermatol* 12:569–572.

Strange PR, Lang PG Jr (1992) Long-term management of basal cell nevus syndrome with topical tretinoin and 5-fluorouracil. *J Am Acad Dermatol* 27:842–845.

Strong LC (1977) Genetic and environmental interactions. *Cancer* 40:1861–1866.

Tang JY, Ally MS, Chanana AM, Mackay-Wiggan JM, Aszterbaum M, Lindgren JA, Ulerio G, Rezaee MR, Gildengorin G, Marji J, Clark C, Bickers DR, Epstein EH Jr (2016). Inhibition of the hedgehog pathway in patients with basal-cell nevus syndrome: final results from the multicentre, randomised, double-blind, placebo-controlled, phase 2 trial. *Lancet Oncol.* 17(12):1720-1731.

Theiler R, Hubscher E, Wagenhauser FJ, Panizzon R, Michel B (1993) Diffuse idiopathic skeletal hyperostosis (DISH) and pseudocoxarthritis following long-term etretinate therapy. *Schweiz Med Wochenschr* 123:649–653.

Thissen MRTM, Neumann MHA, Schouten LJ (1999) A systematic review of treatment modalities for primary basal cell carcinomas. *Arch Dermatol* 135:1177–1183.

Tse DT, Kersten RC, Anderson RL (1984) Hematoporphyrin derivative photoradiation therapy in managing nevoid basal cell carcinoma syndrome: A preliminary report. *Arch Ophthalmol* 102:990–994.

Viksnins P, Berlin A (1977) Follicular atrophoderma and basal cell carcinomas: The Bazex syndrome. *Arch Dermatol* 113: 948–951.

Villavicencio EH, Walterhouse DO, Iannaccone PM (2000) The Sonic Hedgehog-Patched-Gli Pathway in human development and disease. *Am J Hum Genet* 67:1047–1054.

Welch JP, Wells RS, Kerr CB (1968) Ancell-Spiegler cylindromas (turban tumours) and Brooke-Fordyce trichoepitheliomas: Evidence for a single genetic entity. *J Med Genet* 5:29–35.

Wicking C, Shanley S, Smyth I, Gillies S, Negus K, Graham S, Suthers G, Haites N, Edwards M, Wainwright B, Chenevix-Trench G (1997a) Most germ-line mutations in the nevoid basal cell carcinoma syndrome lead to a premature termination of the PATCHED protein, and no genotype-phenotype correlations are evident. *Am J Hum Genet* 60:21–26.

Wicking C, Gillies S, Smyth I, Shanley S, Fowles L, Ratcliffe J, Wainwright B, Chenevix-Trench G (1997b) De novo mutations of the patched gene in nevoid basal cell carcinoma syndrome help to define the clinical phenotype. *Am J Med Genet* 73:304–307.

Zeitouni NC, Shieh S, Oseroff AR (2001) Laser and photodynamic therapy in the management of cutaneous malignancies. *Clin Dermatol* 19:328–339.

Zvulunov A, Strother D, Zirbel G (1995) Nevoid basal cell carcinoma syndrome: Report of a case with associated Hodgkin's disease. *J Pediatr Hematol Oncol* 17:66–70.

30

HEREDITARY HEMORRHAGIC TELANGIECTASIA

JONATHAN N. BERG

Department of Clinical Genetics, Ninewells Hospital and Medical School, Dundee, UK

ANETTE D. KJELDSEN

Department of Otorhinolaryngology Odense University Hospital, Denmark

INTRODUCTION

Incidence

The first published description of a family affected with hereditary hemorrhagic telangiectasia (HHT) was probably that of Babbington in the *Lancet* in 1865, although three other physicians are honored in the eponymous title Osler–Weber–Rendu disease. Rendu reported a case of what he termed "pseudo-haemophilia" in 1896, and Osler drew attention to the genetic nature of the disorder in his *Lancet* article five years later. Finally, in 1907, Weber published a further case observing that nosebleeds tend to precede the appearance of skin lesions. The term hereditary hemorrhagic telangiectasia was suggested in 1909 by Hanes.

Bideau et al. (1980) estimated an overall prevalence for HHT of 1 in 8345 in the French population. An estimate of its prevalence in the population of the Danish island of Fyn was 1 in 6400 (Kjeldsen et al. 1999a). A more recent study in Tayside, Scotland, showed a minimum prevalence of 1:8831 (Lumsden et al. 2017).

Disease penetrance and expression is variable, partly determined by which gene is mutated. Individuals with *ENG* mutations may have an earlier onset of epistaxis and telangiectases than those with *ACVRL1* mutations (Berg et al. 2003). *SMAD4* mutations are rare, representing 1–2% of individuals with HHT (Gallione et al. 2004). Those with *SMAD4* mutations have additional manifestations including juvenile polyposis and a risk of aortic dilatation.

Diagnostic Criteria

The diagnosis of HHT is based on clinical assessment. The characteristic mucocutaneous telangiectases at specific anatomical sites are strongly suggestive of this diagnosis. Examples of the classical mucocutaneous telangiectases are shown in Figure 30.1.

The clinical diagnosis may be difficult to make especially at young age, as nosebleeds are common in the general population, especially among children. A range of different telangiectatic cutaneous lesions are also common and can be confused with classical HHT lesions.

In 1998, to facilitate diagnosis, members of the Scientific Advisory Board of the Hereditary Haemorrhagic Telangiectasia Foundation drew up the Curaçao Diagnostic Criteria (Shovlin et al. 2000). These criteria are shown in Table 30.1. A primary screen for the most common visceral lesion, pulmonary arteriovenous malformations, can be easily performed in individuals suspected of HHT, and this can assist with a clinical diagnosis. The methods used for screening for pulmonary AVMs are discussed in more detail below. Mutation analysis identifies a mutation in up to 90% of individuals with definite HHT by Curacao criteria, and is, therefore, a useful adjunct to clinical diagnosis (van Gent et al. 2013).

It is important to remember that younger individuals are frequently presymptomatic. The diagnosis in children with an affected parent cannot be excluded unless the familial mutation is shown to be absent in that child.

Cassidy and Allanson's Management of Genetic Syndromes, Fourth Edition.
Edited by John C. Carey, Agatino Battaglia, David Viskochil, and Suzanne B. Cassidy.
© 2021 John Wiley & Sons, Inc. Published 2021 by John Wiley & Sons, Inc.

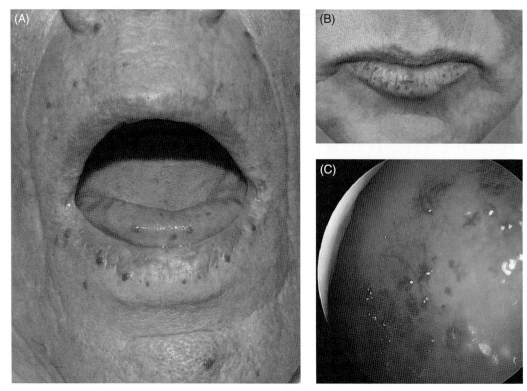

FIGURE 30.1 (A) Lips and tongue with telangiectatic lesions. (B) Lips with telangiectatic lesions. (C) Nasal septum with telangiectatic lesions. (Images provided by Professor Anette Drøhse Kjeldsen.)

TABLE 30.1 The Curaçao criteria for diagnosis of hereditary hemorrhagic telangiectasia.

The hereditary hemorrhagic telangiectasia diagnosis is:	
Definite	If three criteria are present
Possible or suspected	If two criteria are present, and
Unlikely	If fewer than two criteria are present
	Criteria
1. Epistaxis	Spontaneous, recurrent nose bleeds
2. Telangiectases	Multiple, at characteristic sites
	Lips
	Oral cavity
	Fingers
	Nose
3. Visceral lesions	
	Gastrointestinal telangiectasia (with or without bleeding)
	Pulmonary arteriovenous malformation
	Hepatic arteriovenous malformation
	Cerebral arteriovenous malformation
	Spinal arteriovenous malformation
4. Family history	A first-degree relative with hereditary hemorrhagic telangiectasia, according to these criteria

Etiology, Pathogenesis, and Genetics

Causative mutations in HHT have been identified in three genes to date. They are *ENG (encodes endoglin)*, *ACVRL1* that was previously called *ALK1* (encodes activin receptor-like kinase I), and *SMAD4*. HHT1 is caused by mutations in endoglin, a transmembrane glycoprotein that forms homodimers (McAllister et al. 1994). Endoglin is a TGFβ-binding protein, which is primarily expressed in vascular endothelial cells. HHT2 is caused by mutations in the *ACVRL1* gene (Berg et al. 1997), a type 1 serine-threonine kinase receptor that is also expressed in vascular endothelial cells. JP-HHT syndrome (Juvenile polyposis and Hereditary Haemorrhagic Telangiectasia) is caused by mutations in *SMAD4* (Gallione et al. 2004), a downstream mediator of serine-threonine kinase signaling. The signaling pathway for these molecules is illustrated in Figure 30.2.

The mutation detection rate in these three genes, amongst those with a clear clinical diagnosis of HHT, is as high as 89%. No cases of classical HHT caused by other genes have been published. *BMP9* has been suggested as an additional locus (Wooderchak-Donahue et al. 2013), but the cutaneous phenotype in the small number of reported cases appears to be different. Braverman et al. suggested in 1990 a model as explanation of the supposed pathology behind the

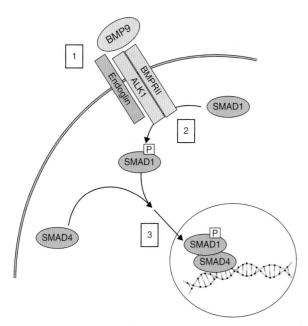

FIGURE 30.2 Diagram illustrating the role of *ACVRL1* and endoglin signaling in the endothelial cells. (1) Binding of ligand [in this case, bone morphogenic protein-9 (BMP9)] to the complex consisting of the type 1 receptor *ACVRL1* (identified here by its former name *ALK1*) and the type II receptor, bone morphogenic protein receptor type II (*BMPRII*), initiates signaling by phosphorylation of *ACVRL1*. Endoglin is a co-receptor that modifies the signaling of this receptor complex. Endoglin and *ACVRL1* may also be involved in signaling in response to transforming growth factor β (TGFβ). (2) *ACVRL1* then phosphorylates *SMAD1* (or *SMAD5*). (3) The phosphorylated *SMAD1* then forms a complex with *SMAD4*, which acts as a transcription factor in the nucleus. Pathogenic mutations in either *ACVRL1* or *ENG*, cause HHT. Mutations in *BMPRII* and *ACVRL1* can both cause primary pulmonary hypertension. Mutations in *SMAD4* cause a combined hemorrhagic telangiectasia and juvenile polyposis phenotype. This is because the *SMAD4* mutations also interfere with signaling via other bone morphogenic protein receptors involved in juvenile polyposis.

development of vascular alterations in HHT. They conducted an elegant study based on serial sectioning of cutaneous telangiectases. They proposed a developmental scenario, which begins with perivascular leukocytic infiltration and dilatation of the postcapillary venule. The venule then enlarges, developing excessive layers of smooth muscle cells, and connects through capillaries to dilated arterioles. As the venule enlarges further, there is loss of the intervening capillaries and direct arteriovenous communication is established.

This model might also explain the development of visceral arteriovenous malformations. Congenital presentation of these lesions suggests either that this process also occurs during fetal vascular development, or that haploinsufficiency for *ACVRL1* or *ENG* has an additional effect on early visceral vascular development.

For some individuals with HHT, the disorder is only a minor inconvenience, while among others the vascular malformations may cause severe symptoms with health implications. There is a great variation of symptoms and severity, even within the same family. A few studies have looked into survival of individuals with HHT in as unbiased a fashion as possible. In a Danish study involving HHT regardless of disease severity, and with treatment and screening for pulmonary AVMs, there was no increased mortality in a 20-year follow-up period (Kjeldsen et al. 2016). In a study of parents of affected individuals, de Gussem et al. (2014) showed that unscreened individuals with an *ENG* mutation have reduced life expectancy, whereas those with *ACVRL1* mutations have the same life expectancy as spousal controls. Screening and treatment for HHT manifestations are important and expected to improve morbidity and mortality in affected individuals. With effective screening and treatment, life expectancy for most affected people should be close to that of the general population.

Diagnostic Testing

Initial diagnosis should be based on the formerly mentioned Curaçao Criteria shown in Table 30.1 (Shovlin et al. 2000). It is important to bear in mind that an individual with classical telangiectases alone, or a single pulmonary AVM, may still have HHT, but not fulfill the diagnostic criteria. De novo mutations are uncommon and the family history will often identify other affected family members.

Mutations in either *ACVRL1* or *ENG* have been demonstrated in up to 90% of unrelated affected individuals tested (Tørring et al. 2014) using direct gene sequencing of genomic DNA. Missense, nonsense, and splice-site mutations occur in both *ENG* and *ACVRL1*. Between one and two percent of individuals with HHT have JP-HHT, caused by a *SMAD4* mutation. Mutation analysis of *ACVRL1*, *ENG* and *SMAD4* can, therefore, be a useful adjunct to diagnosis and if no mutation is identified the diagnosis should be reviewed by a clinician experienced in HHT.

Differential Diagnosis

In its classical form, HHT is distinctive; however, a number of manifestations associated with HHT can occur as an isolated finding.

Other autosomal dominant vascular dysplasias can be mistaken for HHT. A number of the genes for these other vascular dysplasias have been identified. Clinically recognized inherited vascular dysplasias include hereditary glomangioma caused by mutations in the *GLOMULIN* gene, capillary malformation-arteriovenous malformation (CM-AVM) syndrome caused by *RASA1* mutations, and multiple cutaneous and mucosal venous malformations (VM-CM) caused by activating mutations in *TIE2*. Cutaneous vascular lesions can also be seen in hereditary cerebral cavernous malformations, most commonly caused by *KRIT1* mutations, and in Bannayan–Riley–Ruvalcaba syndrome, a variant of PTEN hamartoma syndrome, caused by mutations in PTEN. In all of these hereditary conditions, the cutaneous lesions are distinctive, and different from those seen in HHT.

Epistaxis is common, and may occur as a result of a bleeding diathesis, either congenital (e.g., Von Willebrand disease) or acquired (e.g., thrombocytopenia secondary to leukemia). A diagnosis of HHT cannot be made on the basis of epistaxis alone. It is recommended that the nasal cavity should be examined directly by an otorhinolaryngologist, preferably with experience of HHT, to confirm the presence of characteristic telangiectases in individuals, whose status is uncertain, but also to rule out other treatable causes of epistaxis. It should be noted that epistaxis among individuals with HHT is most likely to occur from both sides of the nose.

Telangiectases are a characteristic sign of liver disease, and occur physiologically during pregnancy. However, such lesions are usually spider nevi rather than the classical well-defined red-purple telangiectases of HHT (as shown in Figure 30.1). Conjunctival telangiectases occur in ataxia telangiectasia, and cutaneous telangiectases in the CREST (calcinosis, Raynaud syndrome, esophageal involvement, sclerodactyly, and telangiectasia) syndrome, which is a variant of scleroderma.

Generalized essential telangiectasia is a condition well known to dermatologists. It is usually sporadic, but may be inherited in an autosomal dominant manner, when it is known as hereditary benign telangiectasia. It is characterized by cutaneous rather than mucocutaneous telangiectases, and an absence of arteriovenous malformations. There is a wide distribution of telangiectases over the body, and epistaxis is rare.

MANIFESTATIONS AND MANAGEMENT

Hereditary hemorrhagic telangiectasia is a multisystem disorder and good patient care demands a multidisciplinary approach, involving clinical geneticists, pulmonologists, interventional radiologists, otorhinolaryngologists, dermatologists, neurologists and gastroenterologists. HHT can present in a number of ways, but most frequently presents as epistaxis with telangiectasia. Neurological manifestations may be due to brain AVMs or pulmonary AVMs leading to paradoxical embolus because of emboli bypassing the filter of the pulmonary vascular bed. Breathlessness may be caused by untreated anemia from heavy epistaxis or gastrointestinal bleeding, hypoxia due to a pulmonary AVM, or by high-output cardiac failure associated with a hepatic AVM.

Growth and Feeding

The symptoms of HHT often evolve during childhood, causing increased epistaxis, and, occasionally, development of pulmonary AVMs. Individuals with anemia may need iron replacement therapy. Otherwise, HHT does not usually affect childhood growth and feeding.

Development and Behavior

Symptoms such as nosebleeds rarely affect school performance and quality of life. HHT is not otherwise associated with behavioral or developmental disorders in children.

Otorhinolaryngologic

Recurrent epistaxis is the most common symptom of HHT, affecting over 90% of affected individuals. The mean age of onset is 12 years, with over 75% of affected individuals being symptomatic by age 25. In the majority of cases, the epistaxis is bilateral, self-limiting or responsive to simple measures, such as humidification and nasal packing. However, at least 30% of affected individuals require further treatment, and their management can pose a major challenge to the otorhinolaryngologist. Nasal telangiectases are usually found over the septum and the anterior tip of the inferior turbinate. Affected individuals require long-term follow-up and repeated treatment. Good reviews are provided by Ross and Jassin (1997), Lund and Howard (1999), and Pau et al. (2001).

Evaluation

- A diary of frequency, duration, and severity of epistaxis should be maintained.
- Hemoglobin and ferritin estimation should be performed regularly to assess need for iron treatment or blood transfusion. Frequency of hemoglobin estimation should be determined by severity of hemorrhagic symptoms. In some affected individuals with chronic blood loss and anemia, treatment may be problematic with a requirement for regular blood transfusions.

Management

- Electrocautery should be avoided whenever possible. There is a significant risk of causing a septal perforation, leading to more severe epistaxis. Electrocautery is never indicated as prophylactic treatment, but may in few cases be required for management of severe epistaxis.
- In experienced hands, laser photocoagulation appears to have benefit in the management of mild-to-moderate epistaxis. There is debate about which laser type is the most appropriate. However, there is agreement that, CO_2 laser is not appropriate in treating HHT.
- In individuals with severe recurrent epistaxis where other treatments have failed, septal dermatoplasty can

reduce the severity of epistaxis by up to 64% (McCaffrey et al. 1977). The procedure, as described by Saunders (1973), involves the replacement of the most telangiectatic areas of the nasal mucosa by autologous split-skin grafts. Telangiectases may develop in the graft, limiting the long-term success of the technique.

- In moderate-to-severe epistaxis, embolization of the abnormal nasal vasculature may be of benefit. A potential complication of the procedure is the interruption in the vascular supply to surrounding tissues. Provocative testing with intraarterial lidocaine can be used preoperatively (Weissman et al. 1995). The reduction of epistaxis in HHT by embolization is usually only temporary

- In individuals with severe epistaxis that has been unresponsive to other treatments, surgical closure of the nostril has been shown to be effective. In a series of 100 people with HHT Lund showed successful cessation of bleeding in 94% (Lund et al. 2017).

- A number of medical treatments are frequently tried for epistaxis, although the evidence remains limited (Halderman et al. 2018). Tranexamic acid and selective estrogen receptor modulators such as Tamoxifen are frequently used. Topical estrogens are also used. Thalidomide, pazopanib, and propanolol may have a place in treatment, but, to date, evidence is largely anecdotal or based on small case series (Parambil et al. 2018; Fang et al. 2017).

- Bevacizumab may be effective in severe intractable epistaxis by intravenous infusion (Iyer et al. 2018), but topical application or local injection does not appear to be effective (Halderman et al. 2018) and may still have severe side effects.

Respiratory

Pulmonary AVMs are common in HHT. In a French–Italian study, pulmonary AVMs were identified in 34.4% of affected individuals with an *ENG* mutation, and 5.2% of those with an *ACVRL1* mutation (Lesca et al. 2007). In a Dutch study, 48.7% of people with HHT1 had a pulmonary AVM on detailed screening compared with 5.3% of those with HHT2 (Letteboer et al. 2006). In a Danish study, 52.3% of people with HHT1 had a pulmonary AVM on detailed screening compared with 12.9% of those with HHT2 and 60% of those with JP-HHT (Tørring et al. 2014).

Typical pulmonary AVMs are shown in Figure 30.3. Between 47 and 80% of all pulmonary AVMs are associated with HHT (Khurshid and Downie 2002). Pulmonary AVMs are direct communications between branches of the pulmonary artery and pulmonary vein without an intervening pulmonary bed. They can be categorized into simple (a single feeding artery and draining vein) and complex (two or more feeding arteries and draining veins). In a series based on 76 individuals with 276 pulmonary AVMs, 80% were simple and 20% complex. Most were located in the lower or middle lobes (White et al. 1988). Complex fistulae may involve systemic arteries.

Pulmonary AVMs tend to increase in size throughout life. They may remain asymptomatic or may present with a range of symptoms. In a series of 126 individuals with pulmonary AVMs published by Cottin et al. (2007), 29% were detected by screening, 15% were incidental findings, 22% presented with exertional dyspnea, and 13% presented with central nervous system complications. Many pulmonary AVMs are not visible on chest radiographs. Diffuse lesions can occur and, when widespread, may only be treatable by heart–lung transplant. They represent a particularly high-risk group for complications (Pierucci et al. 2008).

An initial screen for pulmonary AVMs should be performed in all at-risk individuals, irrespective of the mutation in the family, because of the risk of central nervous system complications. In Danish and UK studies, 8% of individuals with HHT and pulmonary AVM had been hospitalized with brain abscesses before they were seen at a specialist center (Boother et al. 2017; Kjeldsen et al. 2014). In another study (Shovlin et al. 2008), 57 of 201 individuals with a pulmonary AVM associated with HHT experienced a brain abscess or stroke; the risk of stroke was reduced by therapeutic embolization.

Pulmonary hemorrhage, stroke, and deterioration of pulmonary shunt have been documented during pregnancy in women with HHT and pulmonary AVMs. Embolization can be carried out safely from 16 weeks gestation, and it is important that women with HHT are assessed for pulmonary AVMs before, or during, pregnancy. Where there is any suspicion of pulmonary AVMs, referral should be made to an appropriate center with experience in the management of hereditary hemorrhagic telangiectasia. For a good review, and further details, see Shovlin and Letarte (1999) and Gershon et al. (2001).

Systemic arteriovenous malformations can cause secondary pulmonary hypertension in HHT. However, a small number of individuals with HHT caused by *ACVRL1* mutations have been shown to have primary pulmonary hypertension (Trembath et al. 2001). They may present with unexplained respiratory symptoms.

Evaluation

- All individuals with HHT should be screened for the presence of pulmonary AVMs. Screening is recommended by most authors at the age of 16, after puberty, as pulmonary AVMs can increase in size during puberty. It is important that, where possible, this screening takes place before childbearing in women because of the increased risk of complications from a

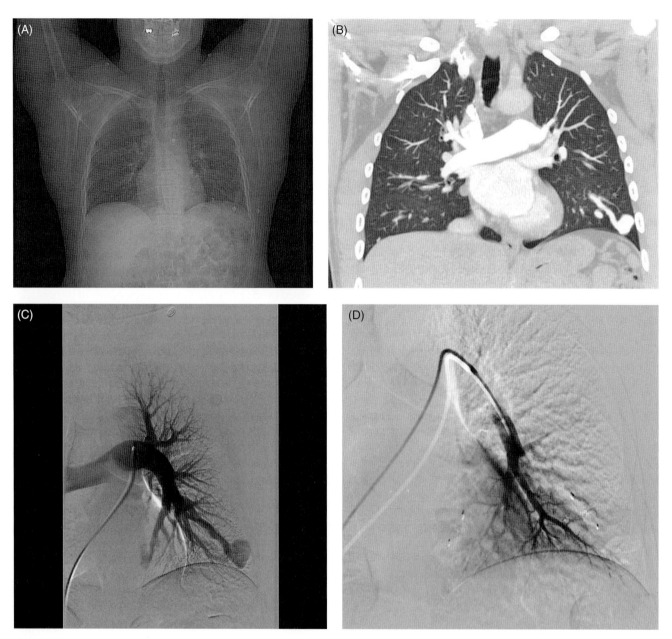

FIGURE 30.3 (A) Chest radiograph showing pulmonary arteriovenous malformation in the left lower lobe. (B) Computed tomography showing the same pulmonary arteriovenous malformation in the left lung. (C) Left pulmonary angiography showing two pulmonary arteriovenous malformations in the left lung. (D) Two pulmonary arteriovenous malformations after treatment with Amplatz plugs. (Images Provided by Poul Erik Andersen.)

pulmonary AVM during pregnancy. Screening should be repeated in woman after each pregnancy, as pulmonary AVMs evolve during pregnancies.
- There is variation between guidelines as to the correct age to start screening in children. Complications from pulmonary AVMs in childhood are rare, but in a few cases, fatal pulmonary complications occur. A simple screening methodology such as pulse oximetry may be used to detect significant hypoxia in childhood, but some authors advocate earlier more invasive screening and treatment of lesions. There is no evidence of clinical benefit from a more invasive approach.
- Several methods for screening for pulmonary AVMs have been considered. Extant guidelines recommend screening either by contrast echocardiography, which is felt to be highly sensitive (Kjeldsen et al. 1999) but less specific, or by computed tomography (CT)-pulmonary angiography (Shovlin et al. 2017). A positive contrast echocardiography result should be followed up with CT-pulmonary angiography.

- Some guidelines recommend repeated screening after a negative screen for pulmonary AVMs, although if none is detected on a first screen after puberty, then it is uncertain whether there is a significant risk of a lesion developing. Lesions that have been identified and treated may recanalize, and follow up imaging in this case is important.
- People with JP-HHT need additional follow up for their risk of aortic root dilatation, juvenile polyps and bowel cancer.

Management

- Pulmonary AVMs with feeding vessels of 2–3 mm or greater diameter should be treated by embolization (Moussouttas et al. 2000). Where a lesion is detected and treated, follow up is important because of a risk of revascularization/ recurrence of pulmonary AVMs.
- Surgical resection is only indicated in a small minority of individuals where the treating radiologist considers the lesion unsuitable for embolization, or among those few cases were embolization is unsuccessful.
- Antibiotic prophylaxis for dental procedures should be prescribed to lower the risk of septic emboli in anyone in whom a potential pulmonary AVM is suspected.
- Primary pulmonary hypertension should be managed by a pulmonary vascular medical unit with expertise in this area.

Neurologic

Brain arteriovenous malformations are a less common complication of HHT. They have been detected on screening in 13.4% of people with HHT1 and 2.4 % of people with HHT2 (Brinjikji et al. 2017). It is a matter of considerable debate as to whether screening for brain AVMs is worthwhile in asymptomatic individuals.

Examples of cerebral vascular malformations are shown in Figure 30.4. The natural history of cerebral vascular malformation in HHT is still being elucidated. Micro-arteriovenous malformations appear to be more common in HHT (50% versus 7% in affected individuals with sporadic cerebral vascular malformation), and are usually asymptomatic. The bleeding risk for individuals with HHT with at least one cerebral vascular malformation has been estimated at approximately 0.4–0.7% per year (Matsubara et al. 2000). This is in contrast to studies of individuals without HHT who are diagnosed with sporadic cerebral vascular malformations. These people have a high risk of hemorrhage of between 2 and 4% per year. More recently, Shovlin et al. (2008) suggested that the risk of hemorrhage from brain AVMs in HHT is similar to sporadic brain arteriovenous malformations at 1.4–2% per annum.

Morgan et al. (2002) reported nine cases of intracranial hemorrhage from cerebral vascular malformations in children with HHT. In this series, five children died and the remainder had significant functional or cognitive sequelae. This is still a rare complication in children, and there are no data demonstrating a clinical benefit of screening children for cerebral vascular malformations (Ganesan et al. 2013). The possibility of a cerebral vascular malformation should certainly be considered when a child with a family history of HHT suffers a rapid neurological deterioration, or who develops a new neurological symptom such as seizures.

Neurological side effects of pulmonary AVMs are common in HHT. About a third of individuals with HHT and an untreated

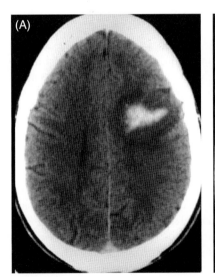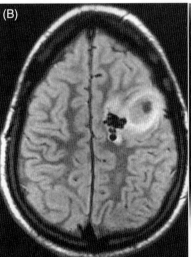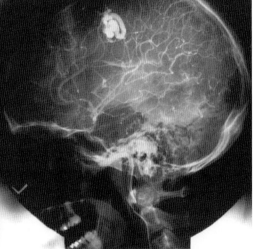

FIGURE 30.4 (A) Computed tomography showing wedge-shaped infarct of the right cerebral hemisphere secondary to hemorrhage from a cerebral arteriovenous malformation. (B) Magnetic resonance imaging (top) demonstrates an arteriovenous malformation of "tangle of blood vessels" proximal to hemorrhage. An angiogram (bottom) demonstrates cerebral arteriovenous malformation with dilated aneurysmal sac and draining vein. (Image provided by Robert I. White Jr.)

pulmonary AVM will suffer a cerebral ischemic insult secondary to paradoxical emboli bypassing the pulmonary vascular bed. Cerebral abscess, as shown in Figure 30.5, may occur secondary to paradoxical septic emboli in 5–10% of people with HHT and pulmonary AVMs (Boother et al. 2017). The risk is higher when multiple pulmonary AVMs are present. Symptoms are non-specific, and the possibility of a cerebral abscess should be considered even if temperature is normal and blood cultures are negative.

Evaluation

- Routine brain magnetic resonance imaging at diagnosis of HHT to exclude brain arteriovenous malformation is recommended by some guidelines. However, there remains considerable uncertainty as to the prognosis of such lesions and the benefits of treatment of asymptomatic lesions. Further study is required to establish whether there is true benefit to screening for brain AVMs in asymptomatic individuals. Screening for asymptomatic Brain AVMs is still debatable and therefore, only carried out in some centers.
- Although rare presentations of brain AVMs in children are reported, there is no evidence for benefit of screening asymptomatic children by magnetic resonance imaging.
- A careful history should be taken to identify neurological symptoms. Where symptoms are present, a neurological examination should be performed.

- Urgent brain computed tomography or magnetic resonance imaging should be performed in all individuals with HHT with new neurological symptoms and signs to exclude a causative cerebral abscess, stroke, or brain AVM.

Management

The ARUBA trial (randomized trial of unruptured brain arteriovenous malformations) compared intervention versus conservative management ("watchful waiting") of people with asymptomatic BAVMs. The results of the ARUBA trial suggest that conservative management of asymptomatic brain AVMs is most appropriate (Mohr 2012). There is little evidence to suggest that brain AVMs in HHT behave differently from non-HHT related lesions.

- It is important that any evaluation and treatment of lesions is performed in a center with experience and expertise in the treatment of hereditary hemorrhagic telangiectasia.
- Brain vascular malformations can be treated conservatively, or by embolization of the feeding vessels, stereotactic radiosurgery or open resection.

Gastrointestinal

Gastrointestinal Hemorrhage. Gastrointestinal bleeding is a common later complication of HHT. In a Danish series, 25 of 76 individuals with HHT had a history of hematemesis or melena. The mean age of onset of symptoms was 52 years (Kjeldsen and Kjeldsen 2000). Telangiectases occur predominantly in the upper gastrointestinal tract (Figure 30.6) although they are also seen in the colon. Treatment is difficult, and there is a tendency for transfusion requirement to increase with age. Although the finding is not statistically significant, *ACVRL1* mutations appear to be more commonly associated with gastrointestinal bleeding than *ENG* mutations (Letteboer et al. 2006; Berg et al. 2003).

Hepatic Involvement. Liver involvement in HHT (Figure 30.7) was originally thought to be rare, but recent studies suggest that subclinical liver involvement is common (Garcia-Tsao 2007). Hepatic involvement is more frequent in HHT2 than HHT1 (Letteboer et al. 2006). There are three distinct clinical presentations, high-output cardiac failure, portal hypertension, and biliary disease (Garcia-Tsao 2007; Buscarini 2018).

As treatment is aimed solely at symptomatic liver disease, any evaluation should be performed only where symptoms are consistent with a hepatic problem. Concordant liver disease should be excluded (as should a history of excessive alcohol or intravenous drug use, inflammatory bowel disease, the presence of hepatitis B or C viral markers, and antinuclear or anti-mitochondrial antibodies).

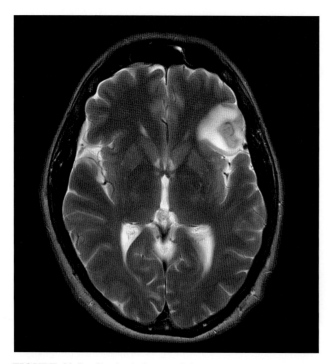

FIGURE 30.5 Cerebral abscess right temporal lope MRI standard T1 weight. (Images Provided by Poul Erik Andersen Jr.)

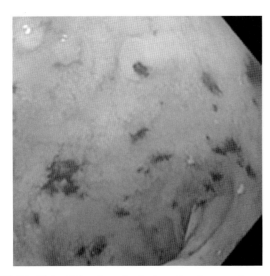

FIGURE 30.6 Telangiectatic lesions in the gastric mucosa. (Image provided by Professor Jens Kjeldsen.)

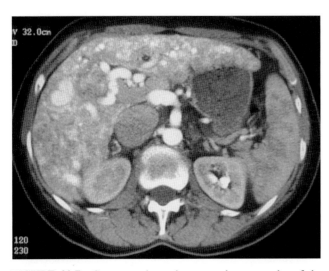

FIGURE 30.7 Contrast enhanced computed tomography of the abdomen showing a dilated hepatic artery and filling of venous circulation in an individual with heart failure because of a hepatic arteriovenous malformation. (Image provided by Robert I. White Jr.)

JP-HHT Juvenile Polyposis and Hereditary Hemorrhagic Telangiectasia. Juvenile polyposis is a rare autosomal dominant hamartomatous polyposis syndrome characterized by multiple polyps in the gastrointestinal tract. About 15–25% of individuals with juvenile polyposis have germ line mutations in *SMAD4*. In one study, 22% of individuals with a mutation in *SMAD4* showed manifestations of HHT (Aretz et al. 2007). Any patient with a *SMAD4* mutation causing juvenile polyposis should therefore be evaluated for HHT. Any patient with HHT caused by a *SMAD4* mutation should be followed as for juvenile polyposis syndrome, where there is a high risk of polyps, small bowel intussusception and early onset bowel cancer.

Evaluation

- A history of hematemesis or melena should be sought. In some individuals with heavy recurrent epistaxis, it may be difficult to distinguish between hematemesis resulting from upper gastrointestinal bleeding and hematemesis from swallowed blood.
- Unexplained anemia should prompt endoscopic evaluation.
- Liver biopsy is unlikely to be helpful and may be hazardous.
- All individuals with juvenile polyposis attributable to a *SMAD4* mutation should be screened for pulmonary AVMs.
- All individuals with HHT caused by a *SMAD4* mutation should be screened for juvenile polyps and bowel cancer.

Management

- Treatment should take place in a center with appropriate expertise.
- Iron and ferritin levels should be measured to assess the need for iron supplementation. Blood transfusion is often required to control anemia.
- Photocoagulation using bipolar electrocoagulation may be of short-term benefit although the large number of widely scattered telangiectases may be a limiting factor.
- Combined estrogen/progesterone therapy has been shown to lower transfusion requirement in some individuals (Van Cutsem et al. 1990).
- Systemic Bevazicumab or thalidomide may reduce the gastrointestinal bleeding significantly, but have not been subject to large clinical trials (Iyer et al. 2018).
- Systemic Bevacizumab may be effective in treating hepatic AVMs with high cardiac output (Dupuis-Girod et al. 2012).
- In some cases, liver transplant may be considered as the only therapeutic option.

Ophthalmologic

Symptomatic involvement of the eye in HHT is rare. There have been reports of bloody tears arising as a result of conjunctival telangiectases, but a more common cause of bloody tears is retrograde epistaxis through the tear canal. Also, retinal vascular malformations have been identified (Brant et al. 1989).

Evaluation

- Routine evaluation is not required.

Management

- When ophthalmologic abnormalities are found, treatment is standard.

Infections

In addition to the risk of cerebral abscess caused by pulmonary AVMs, there have been reports that individuals with HHT may be more prone to infections than the background population (Aagaard et al. 2018; Musso et al. 2014).

Evaluation

- Routine evaluation is not required.

Management

- Standard care is indicated for infection.

ACKNOWLEDGMENTS

We would like to thank Robert I. White Jr. and Poul Erik Andersen for numerous clinical images, Professor Jens Kjeldsen for advice on the content of this chapter, Austin Diamond for preparing Figure 30.2, and Mary Porteous for her work on previous versions of this chapter.

RESOURCES

Websites and Patient Groups

cureHHT Foundation
Website: *https://curehht.org/*

HHT UK
Website: http://www.hhtuk.org/

VASCern – The European reference network for rare vascular diseases
Website: https://vascern.eu/

REFERENCES

Aagaard KS, Kjeldsen AD, Tørring PM, Green A (2018) Comorbidity among HHT patients and their controls in a 20-years follow-up period. *Orphanet J Rare Dis* 13:223.

Aretz S, Stienen D, Uhlhaas S, Stolte M, Entius MM, Loff S, Back W, Kaufmann A, Keller KM, Blaas SH, Siebert R, Vogt S, Spranger S, Holinski-Feder E, Sunde L, Propping P, Friedl W (2007) High proportion of large genomic deletions and a genotype phenotype update in 80 unrelated families with juvenile polyposis syndrome. *J Med Genet* 44:702–709.

Berg JN, Gallione CJ, Stenzel TT, Johnson DW, Allen WP, Schwartz CE, Jackson CE, Porteous ME, Marchuk DA (1997) The activin receptor-like kinase 1 gene: Genomic structure and mutations in hereditary hemorrhagic telangiectasia type 2. *Am J Hum Genet* 61:60–67.

Berg J, Reinhardt D, Gallione C, Holloway S, Klaus D, Lux A, McKinnon W, Porteous M, Marchuk D, Guttmacher A (2003) Hereditary haemorrhagic telangiectasia-delineation of the different phenotypes caused by endoglin and ACVRL1 mutation. *J Med Genet* 40:585–590.

Bideau A, Plauchu H, Jacquard A, Robert JM, Desjardins B (1980) Genetic aspects of Rendu-Osler disease in Haut-Jura: Convergence of methodological approaches of historic demography and medical genetics. *J Hum Genet* 28:127–147.

Boother EJ, Brownlow S, Tighe HC, Bamford KB, Jackson JE, Shovlin CL (2017) Cerebral abscess associated with odontogenic bacteremias, hypoxemia, and iron loading in immunocompetent patients with right-to-left shunting through pulmonary arteriovenous malformations. *Clin Infect Dis* 65:595-603.

Brant AM, Schachat AP, White R (1989) Ocular manifestations in hereditary hemorrhagic telangiectasia (Rendu-Osler-Weber disease). *Am J Ophthalmol* 107:642–646.

Braverman IM, Keh A, Jacobson BS (1990) Ultrastructure and three-dimensional organization of the telangiectases of hereditary hemorrhagic telangiectasia. *J Invest Dermatol* 95:422–427.

Brinjikji W, Iyer VN, Wood CP, Lanzino G (2017) Prevalence and characteristics of brain arteriovenous malformations in hereditary hemorrhagic telangiectasia: a systemic review and meta-analysis. *J Neurosurg* 127:302–310.

Buscarini E, Gandolfi S, Alicante S, Londoni C, Manfredi G (2018) Liver involvement in hereditary hemorrhagic telangiectasia. *Abdom Radiol* 43:1920–1930

Cottin V, Chinet T, Lavole A, Corre R, Marchand E, Raynaud-Gaubert M, Plauchu H, Cordier JF (2007) Pulmonary arteriovenous malformations in hereditary hemorrhagic telangiectasia: A series of 126 patients. *Medicine* 86:1–17.

de Gussem EM, Edwards CP, Hosman AE, Westermann CJ, Snijder RJ, Faughnan ME, Mager JJ. (2016) Life expectancy of parents with Hereditary Haemorrhagic Telangiectasia. *Orphanet J Rare Dis*.11:46.

Dupuis-Girod S, Ginon I, Saurin JC, Marion D, Guillot E, Decullier E, Roux A, Carette MF, Gilbert-Dussardier B, Hatron PY, Lacombe P, Lorcerie B, Rivière S, Corre R, Giraud S, Bailly S, Paintaud G, Ternant D, Valette PJ, Plauchu H, Faure F. (2012) Bevacizumab in patients with hereditary hemorrhagic telangiectasia and severe hepatic vascular malformations and high cardiac output. *JAMA* 307(9):948–55.

Fang J, Chen X, Zhu B, Ye H, Zhang W, Guan J, Su K. (2017) Thalidomide for Epistaxis in Patients with Hereditary Hemorrhagic Telangiectasia: A Preliminary Study. *Otolaryngol Head Neck Surg* 157(2):217–221.

Gallione CJ, Repetto GM, Legius E, Rustgi AK, Schelley SL, Tejpar S, Mitchell G, Drouin E, Westermann CJ, Marchuk DA (2004) A combined syndrome of juvenile polyposis and hereditary haemorrhagic telangiectasia associated with mutations in MADH4 (SMAD4). *Lancet* 363:852–859.

Ganesan V, Robertson F, Berg J. (2013) Neurovascular screening in hereditary haemorrhagic telangiectasia: dilemmas for the paediatric neuroscience community. *Dev Med Child Neurol.* 55:405–7.

Garcia-Tsao G (2007) Liver involvement in hereditary hemorrhagic telangiectasia (HHT). *J Hepatol* 46:499–507.

Gershon AS, Faughnan ME, Chon KS, Pugash RA, Clark JA, Bohan MJ, Henderson KJ, Hyland RH, White RI Jr (2001) Transcatheter embolotherapy of maternal pulmonary arteriovenous malformations during pregnancy. *Chest* 119:470–477.

Halderman AA, Ryan MW, Marple BF, Sindwani R, Reh DD, Poetker DM. (2018) Bevacizumab for epistaxis in hereditary hemorrhagic telangiectasia: An evidence-based review. *Am J Rhinol Allergy.* 32:258–268

Iyer VN, Apala DR, Pannu BS, Kotecha A, Brinjikji W, Leise MD, Kamath PS, Misra S, Begna KH, Cartin-Ceba R, DuBrock HM, Krowka MJ, O'Brien EK, Pruthi RK, Schroeder DR, Swanson KL.(2018) Intravenous Bevacizumab for refractory hereditary hemorrhagic telangiectasia-related epistaxis and gastrointestinal bleeding. *Mayo Clin Proc.* 93:155–166. Erratum in: Mayo Clin Proc. 2018 Mar;93(3):396.

Khurshid I, Downie GH (2002) Pulmonary arteriovenous malformation. *Postgrad Med J* 78:191–197.

Kjeldsen AD, Kjeldsen J (2000) Gastrointestinal bleeding in patients with hereditary hemorrhagic telangiectasia. *Am J Gastroenterol* 95:415–418.

Kjeldsen AD, Vase P, Green A (1999a) Hereditary haemorrhagic telangiectasia: A population-based study of prevalence and mortality in Danish patients. *J Intern Med* 245:31–39.

Kjeldsen AD, Oxhøj H, Andersen PE, Elle B, Jacobsen JP, Vase P (1999) Pulmonary arteriovenous malformations: Screening procedures and pulmonary angiography in patients with hereditary hemorrhagic telangiectasia. *Chest* 116:432–439.

Kjeldsen AD, Torring PM, Nissen H, Andersen PE (2014) Cerebral abscesses among Danish patients with hereditary haemorrhagic telangiectasia. *Acta Neurol Scand* 129:192–197.

Kjeldsen A, Aagaard KS, Tørring PM, Möller S, Green A. (2016) 20-year follow-up study of Danish HHT patients-survival and causes of death. *Orphanet J Rare Dis* 11:157.

Lund VJ, Howard DJ (1999) A treatment algorithm for the management of epistaxis in hereditary hemorrhagic telangiectasia. *Am J Rhinol* 13:319–322.

Lund VJ, Darby Y, Rimmer J, Amin M, Husain S. (2017) Nasal closure for severe hereditary haemorrhagic telangiectasia in 100 patients. The Lund modification of the Young's procedure: a 22-year experience. *Rhinology* 55:135-141.

Lumsden M; Vickers A; Goudie D; McWilliam C; Berg JN (2017) Hereditary haemorrhagic telangiectasia: a record linkage study. 12th International HHT Scientific Conference, Dubrovnik, Croatia. *Angiogenesis* 21:150

Matsubara S, Mandzia JL, ter Brugge K, Willinsky RA, Faughnan ME, Manzia JL (2000) Angiographic and clinical characteristics of patients with cerebral arteriovenous malformations associated with hereditary hemorrhagic telangiectasia. *Am J Neuroradiol* 21:1016–1020.

McAllister KA, Grogg KM, Johnson DW, Gallione CJ, Baldwin MA, Jackson CE, Helmbold EA, Markel DS, McKinnon WC, Murrell J (1994) Endoglin, a TGF-β binding protein of endothelial cells, is the gene for hereditary haemorrhagic telangiectasia type 1. *Nat Genet* 8:345–351.

McCaffrey TV, Kern EB, Lake CF (1977) Management of epistaxis in hereditary hemorrhagic telangiectasia. Review of 80 cases. *Arch Otolaryngol* 103:627–630.

Mohr JP, Parides MK, Stapf C, Moquete E, Moy CS, Overbey JR, Al-Shahi Salman R, Vicaut E, Young WL, Houdart E, Cordonnier C, Stefani MA, Hartmann A, von Kummer R, Biondi A, Berkefeld J, Klijn CJ, Harkness K, Libman R, Barreau X, Moskowitz AJ; international ARUBA investigators. (2014) Medical management with or without interventional therapy for unruptured brain arteriovenous malformations (ARUBA): a multicentre, non-blinded, randomised trial. *Lancet.* 383:614–21.

Morgan T, McDonald J, Anderson C, Ismail M, Miller F, Mao R, Madan A, Barnes P, Hudgins L, Manning M (2002) Intracranial hemorrhage in infants and children with hereditary hemorrhagic telangiectasia (Osler-Weber-Rendu syndrome). *Pediatrics* 102:E12.

Moussouttas M, Fayad P, Rosenblatt M, Hashimoto M, Pollak J, Henderson K, Ma TY, White RI (2000) Pulmonary arteriovenous malformations: Cerebral ischemia and neurologic manifestations. *Neurology* 55:959–964.

Musso M, Capone A, Chinello P, Di Bella S, Galati V, Noto P, Taglietti F, Topino S, Petrosillo N. (2014) Extra-cerebral severe infections associated with haemorrhagic hereditary telangiectasia (Rendu-Osler-Weber Disease): five cases and a review of the literature. *Infez Med* 22:50–6.

Parambil JG, Woodard TD, Koc ON. (2018) Pazopanib effective for bevacizumab-unresponsive epistaxis in hereditary hemorrhagic telangiectasia. *Laryngoscope* 128:2234–2236

Pau H, Carney AS, Murty GE (2001) Hereditary haemorrhagic telangiectasia (Osler-Weber-Rendu syndrome): Otorhinolaryngological manifestations. *Clin Otolaryngol Allied Sci* 26:93–98.

Pierucci P, Murphy J, Henderson KJ, Chyun DA, White RI Jr (2008) New definition and natural history of patients with diffuse pulmonary arteriovenous malformations: Twenty-seven-year experience. *Chest* 133:653–661.

Ross DA, Jassin B (1997) Current trends in the diagnosis and management of Osler-Weber-Rendu disease (hereditary hemorrhagic telangiectasia). *Curr Opin Otolaryngol Head Neck Surg* 5:191–196.

Saunders WH (1973) Septal dermoplasty for hereditary telangiectasia and other conditions. *Otolaryngol Clin North Am* 6:745–755.

Shovlin CL, Letarte M (1999) Hereditary haemorrhagic telangiectasia and pulmonary arteriovenous malformations: Issues in clinical management and review of pathogenic mechanisms. *Thorax* 54:714–729.

Shovlin CL, Guttmacher AE, Buscarini E, Faughnan ME, Hyland RH, Westermann CJ, Kjeldsen AD, Plauchu H (2000) Diagnostic criteria for hereditary hemorrhagic telangiectasia (Rendu-Osler-Weber syndrome). *Am J Med Genet* 91:66–67.

Shovlin CL, Jackson JE, Bamford KB, Jenking IH, Benjamin AR, Ramadan H, Kulinskaya E (2008) Primary determinants of ischaemic stroke/brain abscess risks are independent of severity of pulmonary arteriovenous malformations in hereditary haemorrhagic telangiectasia. *Thorax* 63:259–266.

Shovlin CL, Condliffe R, Donaldson JW, Kiely DG, Wort SJ; British Thoracic Society (2017). British Thoracic Society Clinical Statement on Pulmonary Arteriovenous Malformations. *Thorax.* 72:1154–1163.

Tørring PM, Brusgaard K, Ousager LB, Andersen PE, Kjeldsen AD (2014). National mutation study among Danish patients with hereditary haemorrhagic telangiectasia. *Clin Genet* 86:123–33.

Trembath RC, Thomson JR, Machado RD, Morgan NV, Atkinson C, Winship I, Simonneau G, Galie N, Loyd JE, Humbert M, Nichols WC, Morrell NW, Berg J, Manes A, McGaughran J, Pauciulo M, Wheeler L (2001) Clinical and molecular genetic features of pulmonary hypertension in patients with hereditary hemorrhagic telangiectasia. *N Engl J Med* 345:325–334.

Van Cutsem E, Rutgeerts P, Vantrappen G (1990) Treatment of bleeding gastrointestinal vascular malformations with oestrogen-progesterone. *Lancet* 335:953–955.

van Gent MW1, Velthuis S, Post MC, Snijder RJ, Westermann CJ, Letteboer TG, Mager JJ. (2013) Hereditary hemorrhagic telangiectasia: how accurate are the clinical criteria? *Am J Med Genet A.* 161A:461–6.

Weissman JL, Jungreis CA, Johnson JT (1995) Therapeutic embolization for control of epistaxis in a patient with hereditary hemorrhagic telangiectasia. *Am J Otolaryngol* 16:138–140.

White RI Jr, Lynch-Nyhan A, Terry P, Buescher PC, Farmlett EJ, Charnas L, Shuman K, Kim W, Kinnison M, Mitchell SE (1988) Pulmonary arteriovenous malformations: Techniques and long-term outcome of embolotherapy. *Radiology* 169:663–669.

Wooderchak-Donahue WL, McDonald J, O'Fallon B, Upton PD, Li W, Roman BL, Young S, Plant P, Fülöp GT, Langa C, Morrell NW, Botella LM, Bernabeu C, Stevenson DA, Runo JR, Bayrak-Toydemir P.(2013) BMP9 mutations cause a vascular-anomaly syndrome with phenotypic overlap with hereditary hemorrhagic telangiectasia. *Am J Hum Genet.* 93:530–7.

31

HOLOPROSENCEPHALY*

Paul Kruszka
Medical Genetics Branch, National Human Genome Research Institute, National Institutes of Health, Bethesda, Maryland, USA

Andrea L. Gropman
Pediatrics and Neurology, George Washington University of the Health Sciences, and Children's National Medical Center, Washington, DC, USA

Maximilian Muenke
Medical Genetics Branch, National Human Genome Research Institute, National Institutes of Health, Bethesda, Maryland, USA

INTRODUCTION

Holoprosencephaly (HPE) is the most common structural malformation of the developing forebrain in humans (Solomon et al. 2010). It is characterized by failure of the forebrain to divide into two separate hemispheres and ventricles. Holoprosencephaly encompasses a spectrum of brain malformations along a continuum based on the degree of non-separation of the hemispheres (Solomon et al. 2018). This continuum includes alobar holoprosencephaly (a single ventricle and no separation of the cerebral hemispheres); semilobar holoprosencephaly (the left and right frontal and parietal lobes are fused and the interhemispheric fissure is only present posteriorly); lobar holoprosencephaly (most of the right and left cerebral hemispheres and lateral ventricles are separated, but the most rostral aspect of the telencephalon, the frontal lobes, are not separated, especially ventrally); and syntelencephaly or middle interhemispheric variant (failure of separation of the posterior frontal and parietal lobes, with varying lack of cleavage of the basal ganglia and thalami, and absence of the body of the corpus callosum, but presence of the genu and splenium of the corpus callosum). This fourth type was recently described (Simon et al. 2002). The distinction between these types is not absolute, and gradations between groups often occur. Other prosencephalic derivatives that are often abnormal in holoprosencephaly include variably separated thalami and basal ganglia, malformations of the corpus callosum, absent or aberrant midline structures including the optic and/or olfactory tracts, and pituitary or hypothalamic dysgenesis (Fallet-Bianco 2018). Additional central nervous system abnormalities not specific to holoprosencephaly may also occur. Midline craniofacial anomalies, developmental disability, neurological problems, especially seizures, and endocrine and feeding problems are common. Severely affected children do not survive beyond early infancy; a significant proportion of more mildly affected children demonstrate long-term survival.

Incidence

The true incidence of holoprosencephaly cannot be determined with certainty due to spontaneous abortion and fetal loss (Orioli and Castilla 2010). However, it occurs as frequently as 1/250 in embryos and 1/10,000–1/20,000 live births. Intrauterine mortality is greater than 99.5% (Figure 31.1) (Shiota and Yamada 2010). The incidence among females appears to be twice that of males and females are also more likely to have the most severe facial abnormalities (Cohen 1989). Alobar holoprosencephaly is diagnosed in

* This chapter is in the public domain in the United States of America.

Cassidy and Allanson's Management of Genetic Syndromes, Fourth Edition.
Edited by John C. Carey, Agatino Battaglia, David Viskochil, and Suzanne B. Cassidy.
© 2021 John Wiley & Sons, Inc. Published 2021 by John Wiley & Sons, Inc.

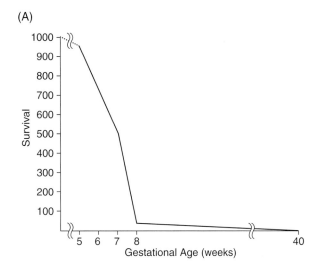

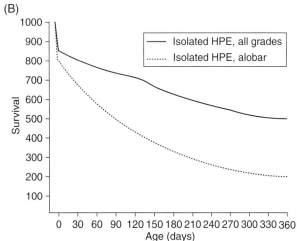

FIGURE 31.1 Prenatal and postnatal survival in holoprosencephaly. (A) From a hypothetical cohort of 1000 conceptuses with holoprosencephaly, the estimated number surviving the embryonic period, by week of gestation. Of 1000 conceptuses, approximately three will survive to birth at 40 weeks of gestation (adapted from Shiota 1993). (B) Of 1000 hypothetical liveborn infants with isolated holoprosencephaly of any grade (solid line) or isolated alobar holoprosencephaly (dashed line), estimated survival up to the age of 1 year. (Survival of infants with any grade of holoprosencephaly adapted from Barr and Cohen 1999.)

approximately 5% of cases, semilobar holoprosencephaly is found in one half of cases, and lobar and MIHV holoprosencephaly accounts for approximately 15% each (Weiss et al., 2018). However, these numbers may be biased toward the more severe forms of holoprosencephaly. Better detection of more mildly affected individuals ascertained through improved neuroimaging techniques including prenatally, may increase the percentage of mild holoprosencephaly that are reported (Winter et al. 2015; Edwards and Hull 2017).

- Infants with severe types of holoprosencephaly may not survive past early infancy (Figure 31.1). In non-syndromic alobar holoprosencephaly, the facial phenotype primarily determines whether the presentation is compatible with postnatal life. Infants with cyclopia, ethmocephaly, and cebocephaly typically die within a week of birth (Barr and Cohen 1999). Individuals with alobar holoprosencephaly born with premaxillary agenesis and unilateral or bilateral cleft lip may die before age 6 months or live for at least 1 year. Those with alobar holoprosencephaly but with less severe facial presentations have survival rates of 50% at 4–5 months of age, and 20% at 1 year (Figure 31.1). Those surviving beyond 1 year of age may live well into childhood. Survival to age 11 years has been reported. For children with less severe forms of semilobar and lobar holoprosencephaly, survival into adulthood is seen (Barr and Cohen 1999; Weiss et al. 2018). With improved medical care, some individuals with HPE are surviving into adulthood that is independent of genetic cause (Weiss et al. 2018)

Diagnostic Criteria

The diagnosis and classification of HPE is based upon the appearance of the brain on neuroimaging (Hahn and Barnes 2010) or postmortem examination.

Alobar HPE, the most severe form, presents with a single ventricle and no separation of the cerebral hemispheres. In general, there is agenesis of the corpus callosum, absent third ventricle, and fusion of thalami and basal ganglia. The ventricular system has the appearance of a monoventricle (termed the holoventricle). A dorsal cyst, which often communicates with the ventricle, may be present. Children with this form of HPE display more significant midline facial defects, including cyclopia.

Semilobar HPE is characterized by incomplete separation of the left and right frontal and parietal lobes with an interhemispheric fissure detected only posteriorly. A small, partially formed third ventricle is often noted. More significantly, incomplete separation of anterior brain structures is noted compared with the lobar variant. A dorsal cyst may also be seen. In mild cases, lack of frontal horn development distinguishes this from the lobar type. Midline craniofacial defects such as cleft lip and palate may be present, but, in many other cases, only subtle facial abnormalities are seen.

Lobar HPE is the mildest form of classical HPE. The cerebral hemispheres are generally well separated, but the most rostral aspect of the telencephalon, the frontal lobes, are not separated, especially ventrally. Frontal horn development is noted in the lateral ventricles. The ventricles, however, appear somewhat dysmorphic because of absence of the septum pellucidum. The posterior half of the corpus callosum is formed. Varying degrees of basal ganglia and thalamic fusion may be present. Midline craniofacial defects are often absent or mild. They may be restricted to hypotelorism or may include cleft lip and palate as well as other abnormalities.

The middle interhemispheric variant, or syntelencephaly, is characterized by failure of separation of the posterior frontal and parietal lobes. The interhemispheric separation of the basal forebrain, anterior frontal lobes, and occipital regions are preserved. There is variable lack of cleavage of the basal ganglia and thalami and absence of the body, but presence of the genu and splenium of the corpus callosum. It is debated whether middle interhemispheric variant is a part of the HPE spectrum or a separate entity (Bulakbasi et al. 2016). Although classic HPE is the result of a defect in basal forebrain induction and patterning, the middle interhemispheric variant is thought to be caused by a defect in roof plate induction and subsequent differentiation (Bulakbasi et al. 2016). This variant has been seen in association with perisylvian polymicrogyria (Atalar et al. 2008).

Other central nervous system findings that may occur, but are not specific to HPE, include anomalies of midline structures such as fused thalami, absent or dysgenetic corpus callosum, absent septum pellucidum, and absent or hypoplastic olfactory bulbs and tracts (arhinencephaly) and macrocephaly secondary to hydrocephalus; Dandy–Walker malformation; extracerebral cyst; neuronal migration anomalies; retinoblastoma; abnormal circle of Willis; and caudal dysgenesis.

Craniofacial findings characteristic of individuals with HPE vary widely and include cyclopia, ethmocephaly, cebocephaly, hypotelorism, premaxillary agenesis, median or bilateral cleft lip, and others (Figure 31.2, Table 31.1).

Etiology, Pathogenesis, and Genetics

HPE is characterized by failure of cleavage of the embryonic prosencephalon, an event normally completed by embryonic day 35 and resulting in differentiation into diencephalon and telencephalon. In HPE, there is incomplete cleavage of the forebrain (prosencephalon) into right and left hemispheres, into the telencephalon and diencephalon, and into the olfactory and optic bulbs and tracts (Table 31.1).

HPE is extremely etiologically heterogeneous (Roessler and Muenke 2010; Miller et al. 2010; Roessler et al. 2018). Multiple genes contribute to the phenotype and evidence from human studies and animal models, implicate additionally the contribution of environmental factors in the pathogenesis of HPE. Since mutation carriers display highly variable clinical presentation, this has led to an "autosomal dominant with modifier" model. In this model, penetrance and expressivity of a predisposing mutation is graded by genetic or environmental modifiers. Hong et al. (2018) identified *BOC* that encodes a SHH coreceptor and is a silent HPE modifier gene in mice. Missense *BOC* variants have subsequently been idenfied in HPE individuals.

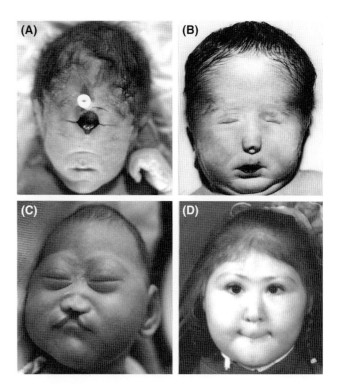

FIGURE 31.2 Facial findings in holoprosencephaly. (A) Cyclopia with proboscis above the single eye in alobar holoprosencephaly. (B) Cebocephaly with alobar holoprosencephaly. (C) Microcephaly and premaxillary agenesis with midline cleft lip and palate in an infant with semilobar holoprosencephaly. (D) Only mild dysmorphic facial findings in a child with semilobar holoprosencephaly.

Consistent with these alleles functioning as HPE modifiers, individual variant BOC proteins had either loss- or gain-of-function properties in cell-based SHH signaling assays. Teratogens have been difficult to link conclusively to HPE in humans because of the small number of individuals who have been studied and the difficulty encountered with retrospective assessment of maternal exposures (reviewed in Summers et al. 2018). The most common teratogen in humans known to cause HPE is maternal diabetes. Infants of diabetic mothers have a 1% risk (a 200-fold increase) for HPE. With improvements in diabetic control, better outcomes have been noted in more recently born infants of diabetic mothers. One study reports that, of a cohort of 104 babies with HPE, 18% were born to diabetic mothers (Stashinko et al. 2004). Other teratogens, including alcohol and retinoic acid, have been associated with HPE in animal models, although their significance in humans is not established (Aoto et al. 2008). In the chick model, there is evidence that retinoic acid affects the sonic hedgehog (*SHH*) pathway in an inhibitory manner (Schneider et al. 2001). Low doses of these teratogens, which by themselves may not be sufficient to cause HPE or any other clinical abnormality, may act in concert with other environmental or

TABLE 31.1 Subtypes of holoprosencephaly and the range of possible craniofacial findings

Subtype of holoprosencephaly	Facial findings
Alobar holoprosencephaly 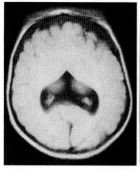 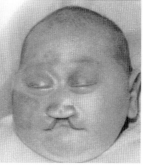	• Cyclopia: single eye or partially divided eye in single orbit with a proboscis above the eye • Cyclopia without proboscis • Ethmocephaly: extreme ocular hypotelorism but separate orbits with proboscis between the eyes • Cebocephaly: ocular hypotelorism with single-nostril nose • Premaxillary agenesis with median cleft lip, ocular hypotelorism, flat nose • Bilateral cleft lip • Ocular hypotelorism only • Anophthalmia or microphthalmia • Relatively normal facial appearance
Semilobar holoprosencephaly 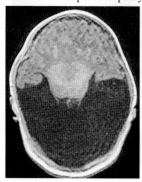 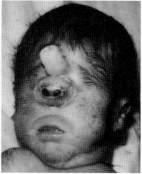	• Bilateral cleft lip with median process representing the philtrum-premaxilla anlage • Depressed nasal bridge • Absent nasal septum • Depressed nasal tip • Midline cleft (lip and/or palate) • Ocular hypotelorism • Depressed nasal ridge • Anophthalmia/microphthalmia • Relatively normal facial appearance
Lobar holoprosencephaly 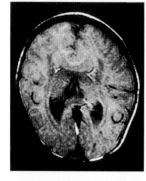 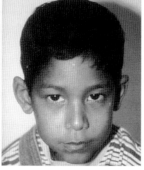	• Bilateral cleft lip with median process • Ocular hypotelorism • Depressed nasal ridge • Relatively normal facial appearance
Middle interhemispheric variant 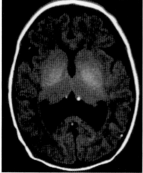 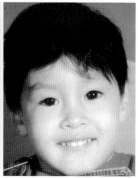	• Relatively normal facial appearance

genetic variables to generate the HPE phenotype (Ming and Muenke 2002).

There are several lines of evidence linking defective cholesterol biosynthesis and HPE. First, up to 5% of children with Smith–Lemli–Opitz syndrome have HPE or malformations consistent with the HPE spectrum (see Chapter 53). HPE has been naturally observed and experimentally reproduced in animals that are fed alkaloid inhibitors of cholesterol biosynthesis. Vertebrate hedgehog proteins require the covalent attachment of a cholesterol ligand to their processed carboxy terminus, and sterol-sensing domains are present on the hedgehog pathway molecules, dispatched and patched. Previously, we analyzed FDA reports of first-trimester exposure to any of the "statin" drugs, which inhibit the rate-limiting enzyme of cholesterol biosynthesis. Of 70 reviewable cases, there were 22 instances of malformation, including cases of HPE, other severe midline CNS defects, and limb deficiency, including the VATER/VACTERL association (Chapter 61) (Edison and Muenke 2004).

Human epidemiological studies demonstrate that the low levels of maternal serum cholesterol (<10th population centile) are strongly associated with preterm birth and lower birth weight in term infants among otherwise low-risk mothers (Edison et al. 2007). In vitro studies of sterol metabolism in lymphoblastoid cell lines from children with HPE using a [2-14C]acetate-loading assay revealed impaired cholesterol biosynthesis in 22 of the 228 cell lines (9.6%), suggesting that abnormal cholesterol synthesis is a contributing cause of HPE (Haas et al. 2007). Moreover, a wide variety of potential chemical teratogens, including compounds that inhibit cholesterol biosynthesis or *SHH* signal transduction, have been shown to cause HPE in animal models (reviewed in Johnson and Rasmussen 2010).

Of liveborn children with HPE, up to one-half have an associated cytogenetic abnormality (Bendavid et al. 2010). A majority (75%) in a large California birth defects registry had trisomy 13 (Croen et al. 1996). Approximately 25–50% of individuals with HPE have a chromosomal abnormality. Those with HPE and a normal karyotype cannot be distinguished from those with an abnormal karyotype on the basis of craniofacial abnormality or subtype of HPE; however, individuals with HPE as a result of a cytogenetic abnormality are more likely to have other organ system involvement. Furthermore, specific chromosome anomalies such as trisomy 13 or 18 are associated with additional characteristic craniofacial and extracraniofacial findings.

Approximately 30% of all live-born individuals with HPE have trisomy 13 (Chapter 58). An additional 10% is found to have trisomy 18 (Chapter 58), triploidy, or various deletions, duplications or rearrangements involving a total of 12 different nonrandom chromosomal loci (Bendavid et al. 2010) (Figure 31.2). Most frequent in descending order are deletions or duplications involving various regions of 13q, del(18p), del(7q36), dup(3p24-pter), del(2p21), and del(21q22.3) (Muenke, personal observations). Significant phenotypic variation exists even among individuals with a similar cytogenetic deletion.

Detection of Microdeletions and Duplications by quantitative PCR, fluorescence in situ hybridization (FISH), and, chromosomal microarray (CMA) has identified novel deletions and/or duplications. It is of interest that no microdeletion was identified in HPE carriers (0/85) (Bendavid et al. 2006). In contrast, 16 of 339 (4.7%) individuals with CNS findings of HPE had microdeletions in either *SHH*, *ZIC2*, *TGIF1*, or *SIX3*. In addition, the microdeletion detection rate in prenatal HPE samples was almost twice as high as the postnatal detection rate in that study. Multiplex ligation-dependent probe amplification (MLPA) screening reveals novel subtelomeric rearrangements in HPE. Furthermore, submicroscopic deletions and/or duplications have been detected by comparative genomic hybridization (Bendavid et al. 2010; Hu et al. 2018).

About 18–25% of individuals with non-syndromic HPE have a mutation in a single gene (*SHH*, *ZIC2*, *SIX3*, *FGFR1*). Familial HPE is most often transmitted in an autosomal dominant fashion. There is often incomplete penetrance, and the expressivity is highly variable. People with HPE, individuals with only microforms who have normal intelligence, and clinically unaffected obligate carriers may be present in a single pedigree. The term "microforms" is used to describe individuals with milder craniofacial anomalies in the absence of neurological findings. On the basis of analysis of autosomal dominant pedigrees, it has been estimated that 37% of carriers of an abnormal HPE gene will have HPE and 27% will have a mild sign or microform. Interestingly, 36% of obligate carriers have no clinical abnormality and have normal intelligence (Cohen 1989).

Even though families with apparently autosomal recessive inheritance have been described, the more likely explanation may well be reduced penetrance in one of the parents or germline mosaicism. X-linked inheritance is suggested in several pedigrees, but has not been confirmed.

At present, four major genes, representing four distinct chromosomal loci, have been identified as being causative of HPE: sonic hedgehog (*SHH*), *ZIC2*, *SIX3*, and fibroblast growth factor receptor 1 (*FGFR1*). Other genes include *CNOT1* (Kruszka et al. 2019a; De Franco et al., 2019), *STAG2*, *SMC1A*, *RAD21*, *SMC3* (Kruszka et al., 2019b), *PPP1R12A* (Hughes et al., 2020), and *KMT2D* (Tekendo-Ngnongang et al., 2019). The best known of these is sonic hedgehog (reviewed in Roessler and Muenke 2010).

The phenotype of individuals with *SHH*, *SIX3*, or *FGFR1* mutations is extremely variable even within the same family, ranging from alobar HPE with cyclopia to clinically normal (Figure 30.3). Individuals with *SIX3* mutations may have semilobar HPE with microphthalmia and iris coloboma, consistent with the role of *SIX3* in eye development.

Preliminary data suggest that individuals with *ZIC2* mutations have normal or only mildly abnormal facial findings despite severe central nervous system anomalies (Solomon et al. 2012).

SHH. The human sonic hedgehog (SHH) gene, one of three *Drosophila* homologous genes, encodes a secreted protein that undergoes autocatalytic cleavage and cholesterol modification (review in Roessler and Muenke 2010). *SHH* was the first holoprosencephaly gene to be identified by positional candidate gene approach as a cause of autosomal dominant HPE in humans. It is the most common gene causing familial HPE. Located at chromosome 7q36, heterozygous deletions, nonsense, frameshift, and missense mutations in *SHH* predict a loss-of-function mechanism. These mutations have been identified in 18% of families with structural features of HPE, and in 24% of pedigrees with HPE and non-specific findings. In families with multiple affected members with HPE and severe structural anomalies (ocular coloboma, callosal agenesis, cleft lip and/or palate, single central incisor), *SHH* mutations were found in 37%. The gene shows pleiotropy, and affected family members demonstrate differential expressivity. Even within the same pedigree, obligate carriers of the same mutation may show a phenotype ranging from unaffected to microforms to severe classic HPE. *SHH* mutations have also been found in families with HPE microforms, such as single central maxillary incisor or ocular coloboma (Solomon et al. 2012; Kruszka et al. 2015).

ZIC2. Located at chromosome 13q32, *ZIC2* encodes a zinc finger protein that is expressed and functions in early neuronal tissue. ZIC2 is a member of a family of proteins that includes the *Drosophila* odd-paired gene (*opa*) and the zebrafish odd-paired like gene (*opl*), which contain zinc finger DNA binding motifs of specificity very closely related to that of the Gli proteins. *ZIC2* may have a role in mediating the response to sonic hedgehog protein signaling. Mutations in *ZIC2* have been observed in 3–4% of individuals with HPE. Heterozygous insertions and deletions leading to frameshifts, nonsense mutations, and expansion of an alanine repeat (normal 10 alanines, expansion 15–25) have been observed (reviewed in Roessler and Muenke 2010). Individuals with *ZIC2* mutations tend to have mild facial features, but variably severe brain anomalies (Figure 31.3). More recently, a *ZIC2*-specific facial phenotype was described with bitemporal narrowing, upslanting palpebral fissures, a short anetverted nasal tip, and broad well-demarcated philtrum (Solomon et al. 2010).

SIX3. Heterozygous deletions and missense and nonsense mutations in the SIX domain and the homeodomain of the *SIX3* gene, located on chromosome 2p21, have been observed in 1% of familial and sporadic HPE. Homeobox protein, SIX3, participates in midline forebrain and eye formation in several organisms and is present in the rostral, anterior region of the neural plate, optic recess, developing retina, and midline ventral forebrain. The *SIX/so* (sine oculis) family of transcription factors form a distantly related subclass of homeobox-containing genes that are further characterized by the presence of a contiguous homology domain, the SIX domain, which is also thought to participate in transcriptional activation. Mutations in this gene are loss of function mutations (Domené et al. 2008). SIX3 deletions have been reported in children with HPE and their unaffected parents (Stokes et al. 2018)

FGFR1 mutations as cause of HPE were first described in individuals with HPE and split-hand and /or split-foot or Hartsfield syndrome (Simonis et al. 2013) or in individuals who have HPE with normal hands and feet (Hong et al. 2016). Based on data in our lab (Muenke, unpublished), *FGFR1* mutations together with those in *SHH*, *ZIC2*, and *SIX3* are by far the most common mutations that have been shown to cause HPE. All other genes (*TGIF*, *PTC1*, *TDGF1*, *GLI2*, *FOXH1*, *NODAL*, *DISP1*, *FGF8*, *GAS1*, *CDON*, *BOC*, and others) individually account for less than 1–2 %.

Cohesin complex genes are now a recognized etiology of HPE. The cohesin complex is formed by the four genes *STAG2*, *SMC1A*, *RAD21*, and *SMC3*. STAG2 and SMC1A are notable as they are the first X-linked genes associated genes. Variants in these genes are usually loss of function variants and de novo. Holoprosencephaly associated with the two X-linked genes, *STAG2* and *SMC1A*, are females, implying male lethality when carrying a loss of function variant in these genes (Kruszka et al., 2019b). Cohesin complex genes and its regulators have been associated with other syndromed grouped under the category of cohesinopathies. Cohesinoapthies include Cornelia de Lange, Roberts SC phocomelia syndrome, CHOPS syndrome (Cognitive impairment and coarse facies, Heart defects, Obesity, Pulmonary involvement, and Short stature and skeletal dysplasia), and chronic atrial and intestinal dysrhythmia caused by mutations in *SGOL1*.

Diagnostic Testing

Diagnosis is based on the overall pattern of clinical abnormalities, family history, chromosome analysis, neuroimaging, and molecular genetic studies in selected individuals (Kruszka et al. 2018). Careful examination of the proband's family for microforms of HPE is essential, looking for mild hypotelorism, microcephaly, single maxillary central incisor, or anosmia or hyposmia (Figure 31.3). A family history of short stature with endocrinopathy and/or central nervous system anomalies other than HPE may be significant in making the diagnosis.

Imaging of the brain confirms the diagnosis of HPE, defines the subtype, and identifies associated central nervous

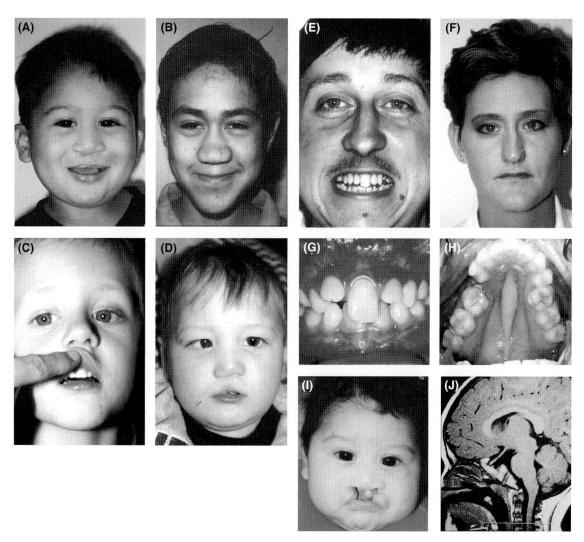

FIGURE 31.3 Clinical findings in individuals with milder craniofacial anomalies in the absence of neurologic findings (the so-called holoprosencephaly "microform"). (A) Premaxillary agenesis with repaired bilateral clefts of the lip. (B) Absence of nasal bones and cartilage and narrow nasal bridge. (C) Single central maxillary incisor. (D) Premaxillary agenesis, repaired unilateral cleft of the lip, and bilateral iris coloboma. (E) Single central incisor, ocular hypotelorism and narrow nasal bridge. (F) Hypotelorism and narrow nasal bridge. (G) Single central incisor and absent midline frenulum. (H) Single central incisor and prominent midline palatal ridge. (I) Premaxillary agenesis with bilateral cleft lip and palate in a child with pituitary hypoplasia and growth hormone deficiency. (J) Sagittal T1 weighted MRI showing pituitary hypoplasia.

system anomalies such as hydrocephalus. The study of choice is cranial MRI, preferably obtained with adequate sedation at a center experienced in evaluating children and adults for structural brain anomalies. Review of the study by a radiologist or other clinician familiar with the subtypes of HPE is essential. Often hydrocephalus in HPE is severe, and the only MRI is done before shunt placement. Caution should be exercised in the diagnosis of HPE based on such a scan, as one may misinterpret lobar and semilobar variants for alobar HPE because of brain compression. It is important to repeat the magnetic resonance images post ventriculoperitoneal shunt placement, when it is possible to determine exactly which brain structures are affected.

Identification of the cause of HPE aids in establishing prognosis and mode of inheritance for genetic counseling purposes. To help establish the cause, molecular studies to detect deletion or duplications are recommended. If monogenic nonsyndromic HPE is likely, molecular genetic testing of the genes known to cause HPE (*SHH*, *ZIC2*, *SIX3*, and *FGFR1*) should be considered.

Prenatal diagnosis is available for those families in which a disease-causing mutation (*SHH*, *ZIC2*, *SIX3*, *FGFR1*, or others) has been identified (Hadley et al. 2018; Kousa et al. 2018) If parental studies are normal, recurrence risk for family members may be low, but may be as high as 50% because of the possibility of germ line mosaicism for a mutation in one of the genes causing nonsyndromic HPE although exact risk figures are not known. For families in which a parent has a balanced chromosomal rearrangement, molecular testing should be offered prenatally during the next pregnancy.

For families with autosomal dominant non-syndromic holoprosencephaly and no identifiable disease-causing mutation, alobar HPE can sometimes be diagnosed through a careful prenatal ultrasound examination by 16 weeks of gestation. Prenatal diagnosis of alobar HPE as early as nine weeks gestation has been achieved with two- and three-dimensional ultrasound studies, and in an infant with cebocephaly in the second trimester.

Milder degrees of HPE such as semilobar or lobar cannot reliably be detected by prenatal ultrasound examination, but may be detectable by fetal MRI (Vasudeva et al. 2015; Griffiths and Jarvis 2016).

When HPE is found on routine prenatal ultrasound examination in a fetus without risk factors, both a high-resolution ultrasound examination to determine the presence of additional structural anomalies and amniocentesis to determine the fetal karyotype are indicated. Informed decisions can then be made about the management of the pregnancy, labor, delivery, and subsequent care of the infant (Stashinko et al. 2010; Hadley et al. 2018). Preimplantation diagnosis for a sonic hedgehog mutation in a family with two affected children has been reported (Verlinsky et al. 2003).

Differential Diagnosis

Some neuroanatomic features of the brain in HPE such as callosal dysgenesis, arrhinencephaly, and pituitary dysgenesis may be found in isolation or in association with other multiple malformation syndromes. At least 25 different conditions have been described in which HPE is an occasional or sometimes a rare finding; most of these disorders are rare and can be differentiated from syndromic HPE based on the presence of additional features. Some of the more common of these conditions include the following, categorized by mode of inheritance.

Autosomal Dominant
- Rubinstein–Taybi syndrome (see Chapter 51)
- Kallmann syndrome (hypogonadotropic hypogonadism and anosmia)
- Martin syndrome (with clubfoot and spinal anomalies)
- Steinfeld syndrome (with congenital heart disease, absent gallbladder, renal dysplasia, and radial defects)
- Ectrodactyly and hypertelorism due to *FGFR1* mutations (Simonis et al. 2013; Hong et al. 2016)
- Middle interhemispheric variant of HPE with bilateral perisylvian polymicrogyria (Atalar et al. 2008).
- Culler-Jones syndrome due to variants in *GLI2* associated with hypotelorism, postaxial polydactyly in limbs, and hypopituitarism (Kruszka and Muenke, 2018)
- Cornelia de Lange syndrome/cohesinopathies (Kruszka et al., 2019b)

Autosomal Recessive
- Smith–Lemli–Opitz syndrome (see Kelley et al. 1996 and Chapter 53)
- Pseudotrisomy 13 syndrome (normal karyotype and polydactyly without other common features of trisomy 13, e.g., scalp defects, overlapping fingers, and nail hypoplasia)
- Meckel syndrome (renal cysts and variable other findings, including encephalocele, hepatic ductal dysplasia and cysts, and polydactyly)
- Genoa syndrome (with craniosynostosis)
- Lambotte syndrome (with microcephaly, prenatal growth retardation, and hypertelorism)
- Hydrolethalus syndrome (with hydrocephalus, polydactyly, and other anomalies)
- Facial clefts and brachial amelia
- HPE and agnathia spectrum

X-linked
- Cohesinopathies/Cornelia de Lange associated with *STAG2* and *SMC1A*

MANIFESTATIONS AND MANAGEMENT

Growth and Feeding

Short stature and failure-to-thrive are common, especially in the more severely affected children. Many children are born with normal length, weight, and head circumference, but it is not uncommon for growth to fall off postnatally. Possibly because of decreased activity, many children require fewer calories than children with normal brain development, and as such, appropriate growth may occur with less than expected intake. When growth delay is present at birth, the phenomenon of "catch up" growth does not occur despite adequate calories. To achieve the best growth results for a child with HPE, the quality of the food intake is more important than the quantity. Growth hormone deficiency and/or chromosomal anomalies may in part be responsible for poor growth in selected individuals.

Feeding may be a major problem in children with HPE. At least part of feeding difficulty may derive from axial hypotonia, poor suck because of neurological complications, lethargy, seizures and their effects, side effects of medications, and disinterest in feeding. Children with cleft lip and/or palate often face additional mechanical difficulties with oral feeding. Frequently there is gastroesophageal reflux, choking, and gagging with feeding. Problems that are more common include slowness in eating, frequent pauses, and frank vomiting with risk of aspiration. Oral-sensory dysfunction may occur and may affect feeding, especially when associated with textural aversion and weakness of labial and lingual structures.

Evaluation

- Height, weight, and head circumference should be measured at initial evaluation and during subsequent health maintenance evaluations. It is important to compare weight with height, in addition to plotting absolute measurements.
- Evaluation for cleft lip and/or palate is warranted in the newborn period, especially as it may impact feeding.
- Oral feeding and swallowing evaluation are recommended, to include assessment of caloric intake, swallowing abilities, oral motor skills, and presence of gastroesophageal reflux.
- Occupational and speech evaluations are warranted to evaluate and address feeding concerns.
- Because gastroesophageal reflux and swallowing difficulties occur frequently in children with HPE, assessment of feeding abilities should be addressed by a multidisciplinary team of professionals, including gastroenterologist, speech therapist, occupational therapist, infant educator, and nutritionist.
- Studies for diagnosis of reflux such as esophageal pH probe, milk scan, barium swallow, and/or endoscopy may be considered.
- Initial testing for growth deficiency should include thyroid function tests, bone age, complete blood count, blood chemistries, sedimentation rate, insulin-like growth factor-1 (IGF-1), and insulin-like growth factor binding protein-3 (IGFBP-3).
- If growth hormone deficiency is found (see Endocrine), panhypopituitarism should be assessed by specific hormone testing and MRI of the brain (Figure 31.3).

Management

- Oral motor therapy with a speech pathologist, occupational therapist, and/or feeding therapist is suggested.
- For children with cleft lip and/or palate, referral to a specialized cleft or craniofacial clinic is recommended.
- Gastrostomy or nasogastric tube feeding may be required if the child cannot take in adequate calories for growth.
- Thickening of feeds and upright positioning after feeding may be helpful to alleviate reflux symptoms.
- Accommodations for oral feeding with cleft lip and/or palate may require specific nipples, cups, and parental training.
- If the child has vomiting or gastroesophageal reflux, a Nissan fundoplication should also be considered.
- If surgery is required to treat refractory reflux, the surgical technique is the same as that used for the general population.
- For infants with difficulty in sucking, one should assure that the child is awake, alert, and properly positioned.
- Alternative feeding methods such as gastrostomy are often helpful for swallowing problems and risk of aspiration from vomiting, especially in those children with more severe neurological compromise.
- Initial treatment with growth hormone should be considered once deficiency is proven.

Development and Behavior

Although some degree of cognitive impairment is always seen, few studies have addressed the cognitive development of individuals with HPE (Levey et al. 2010). There are two published studies. One is limited to children with alobar HPE and the data are based on parental report (Barr and Cohen 1999). The second study categorized expressive language into one of the five levels. Thus, there is extremely limited information about the potential for cognitive attainment, and in fact, many textbooks paint a very bleak picture and offer clinicians and family members little useful information as to the cognitive potential in HPE.

The evaluation of individuals with limited verbal abilities and motor impairment is problematic. So many of the standardized test batteries used today rely heavily on spoken language abilities and motor skills that are in line with chronological age. Techniques not dependent on complex mot or responses, such as the Carter neurocognitive assessment, can be used to assess attention, memory, and sensory processing in non-verbal individuals (Leevers et al. 2005). The latter consists of four domains: social awareness, visual attention, auditory comprehension, and vocal communication. This test has been developed for individuals with HPE and is not a standardized test that can be applied to other populations. When pilot tested in eight children with HPE, it showed a wide range of developmental skills. Performance was delayed in all domains, especially vocal communication. Most of the children, however, did show evidence of improvement over time (Leevers et al. 2005).

Adults with middle interhemispheric variant HPE may have uneven cognitive skills. Virta and colleagues (2016) describe average or above average performance in verbal comprehension, naming, reading and writing, and below average performance in perceptual reasoning, visuospatial abilities, processing speed and memory. In addition, difficulties in mathematical abilities, psychomotor skills, and executive functions were found (Virta et al. 2016).

Children with HPE manifest increased peripheral muscle tone of variable degree from mild to frank spasticity. Concurrently, truncal tone may be low, especially when relaxed. Hypertonicity may be exacerbated when the child is stimulated, excited, or distressed. Unlike other neurological conditions associated with hypertonicity, children with HPE are less likely to develop contractures of the joints. Children with HPE may also manifest extrapyramidal features, such as dystonia and choreoathetosis. The degree of dystonia may correlate with the degree of non-separation of the caudate and lentiform nuclei, as well as the grade of HPE.

Children with alobar HPE appear to have the most marked motor deficits. Typically, they do not achieve the ability to sit independently. Their upper extremity function is poor, only enabling them to reach and bat at objects. Those with semilobar HPE may sit with assistance, and many of them may achieve assisted or independent ambulation. Upper extremity function is improved in these latter two groups, although dystonic posturing and chorea may interfere with function in all groups. Children with lobar HPE and the middle interhemispheric variant may walk independently or with assistance.

Most of those with severe forms of HPE remain primarily non-verbal. However, use of the Carter neurocognitive assessment has demonstrated that many children use non-verbal cues to communicate with caregivers. Preferential gaze, nodding, and non-word vocalizations have been observed (Leevers et al. 2005), suggesting that some degree of receptive language has developed. The full spectrum of language abilities in this group has not been consistently studied. Language abilities may range from only vowel sounds in individuals with alobar HPE, through a succession of more meaningful vocalizations and even sentences in a minority of individuals with either semi- or lobar HPE. Oral motor impairments such as dyspraxia may impact attainment and quality of expressive language abilities. Preliminary pilot studies have shown that many affected children have developed appropriate social skills and, if fostered, may facilitate development of non-spoken communication with caregivers and educators. Individuals with less severe forms of HPE may develop a range of language abilities.

Severely affected individuals have been observed to be either quite lethargic or irritable. It is not uncommon for their behaviors to alternate between the two extremes with little provocation or predictability. It is often difficult to interpret the underlying reason for irritability in a child with such neurological involvement.

Attentional difficulties have been observed in those with milder variants of HPE, or otherwise phenotypically normal children who are carriers of a *SHH* mutation. This may negatively impact learning and interfere with cognitive testing and its reliability. Some show features of obsessive behaviors and sensory integration problems. The entire spectrum of behaviors with HPE is now beginning to be addressed on a research basis.

Evaluation

- All affected children should have a developmental evaluation at diagnosis. For infants, developmental testing should be performed every six months. Thereafter, yearly assessments should suffice.
- The Carter neurocognitive assessment or a similar testing instrument should be used when evaluating cognitive status at diagnosis and for follow-up.
- An attempt should be made to determine the ability of children with HPE to use non-verbal methods of communication.

Management

- Ongoing physical and occupational therapy will be required at least throughout childhood. Children will require early developmental services that may include a home program with parent training to enhance motor development and language, or infant stimulation. In more impaired children, this will consist of maneuvers for positioning and maintaining posture and preventing contractures. In less impaired children, the goals of physical and occupational therapies will be to address and ameliorate motor asymmetries, tone abnormalities, balance, ambulation, upper extremity function, and sensory integration.
- Speech therapy is appropriate to stimulate verbal and/or non-verbal communication.
- Educational strategies should be individualized and do not differ from other children with developmental disabilities.
- Medically fragile children may continue to require home-based services even into early childhood.
- Walkers may be helpful to help achieve independent ambulation.
- For irritability, typical calming procedures such as swaddling, rocking, and use of white background noise may be settling.
- There is no literature to suggest the success of various agents used for attention-deficit hyperactivity disorder in this population. Thus, decisions about the use of dexedrine, methylphenidate, and the newer agents, should be made on a case-by-case basis. One must exclude and address/treat other conditions that may present as hyperactivity, such as effects of other medications, sleep disorders, cognitive deficits, and epilepsy.

Craniofacial

In a classic paper, DeMyer and colleagues (1963) discussed a graded series of facial anomalies that occur with HPE. The face predicts the brain approximately 80% of the time. The other 20% of the time, facial features may appear non-diagnostic (Cohen 1989) (Figure 31.3). The most severe craniofacial manifestations include cyclopia, in which there is a single median eye with varying degrees of doubling of the intrinsic ocular structures and an overlying proboscis and other manifestations of the failure of optic and nasal development. These features almost always predict the alobar type of HPE as well as a poor prognosis. In ethmocephaly, two separate hypoteloric eyes are associated with arhinia and proboscis formation. In cebocephaly, hypotelorism is seen with a blind-ended, single nostril nose. With premaxiallary agenesis, hypotelorism is associated with a depressed nasal ridge and a median cleft because of agenesis of the primary palate. Milder facial features may also be associated with brain findings ranging through the entire spectrum from normal to alobar HPE (Cohen 1989). Preliminary data suggest that individuals with *ZIC2* mutations have normal or only mildly abnormal facial findings despite severe central nervous system anomalies. Milder facial anomalies associated with HPE are non-specific. The observed abnormal facial features include midface retrusion, ocular hypotelorism, depressed nasal bridge, choanal stenosis, iris coloboma, and microcephaly.

A number of minor anomalies have been found within the HPE spectrum. Single maxillary incisor (Cohen 1989) and, much less commonly, absence of the nasal septal cartilage have been described. Other identified defects include stenosis of the pyriform aperture, absence of the superior labial frenulum (Solomon et al. 2012; Kruszka et al. 2015), and absence of the philtral ridges (Figure 31.3).

Evaluation

- All children with HPE should have a very thorough evaluation to assess craniofacial features.
- All family members at risk of being carriers of a mutation in an HPE gene should be examined carefully for the presence of microforms characteristic of HPE.
- Children with significant cosmetic and medically debilitating craniofacial malformations should be referred to a craniofacial clinic, otolaryngologist, plastic surgeon, ophthalmologist, or other appropriate health care provider for evaluation.

Management

- Definitive repair of cleft lip, cleft palate, and other facial anomalies should be coordinated with appropriate medical professional who have experience with children with developmental disabilities.
- Children with cleft lip and/or palate should receive assistance with speech, oral motor functions, and feeding as appropriate.

Endocrine

Multiple endocrine problems are seen with HPE. These are related to the grade of HPE as well as the severity of the central nervous system malformation. Often there is pituitary insufficiency, ranging from panhypopituitarism to isolated hormonal insufficiency in milder forms of HPE. Variable endocrine gland findings at autopsy have been described, including normal, hypoplastic, or even absent pituitary, thyroid, adrenals, ovaries, or testes. Hypoplasia of the adrenal glands may be attributed to an absent or non-functioning anterior pituitary gland, abnormal portal hypophyseal communication, or failure of releasing factor production by the hypothalamus. The most common endocrine problem in HPE is dysregulation of sodium and renal concentrating ability or diabetes insipidus. Central diabetes insipidus is characterized by decreased secretion of antidiuretic hormone, also known as arginine vasopressin, that results in polyuria and polydipsia by diminishing the ability to concentrate urine. Diminished or absent antidiuretic hormone can be the result of a defect in one or more sites involving the hypothalamic osmoreceptors, supraoptic, or paraventricular nuclei, or the supraoptico-hypophyseal tract.

Diabetes insipidus may present with irritability and dehydration because of a lack of ability to concentrate urine and loss of quantities of dilute urine (and hence, fluid) through the kidney. In children with HPE, this condition may be worsened by neurological features that prevent recognition of thirst and/or ability to communicate such thirst. Episodes of stress or infection may exacerbate the condition, with the risk of severe dehydration and possibly death if the child is unable to replace water losses. Effects of undertreated diabetes insipidus include severe dehydration, hypernatremia, fever, cardiovascular collapse, and death.

Other endocrine abnormalities in people with HPE include hypothyroidism, growth hormone deficiency, and decreased sex hormone production.

Evaluation

- Every child with HPE and related midline brain malformations should be evaluated for pituitary dysfunction at diagnosis. Sagittal MRI can be used to determine pituitary absence or ectopia and anatomic information. Central nervous system anomalies, absent corpus callosum, and/or septum pellucidum may accompany endocrine dysfunction.
- Testing for thyroid hormone levels, growth hormone levels, and cortisol should be undertaken at diagnosis and, in those children with absence of the posterior pituitary, at each health maintenance visit. For those children with deficiencies of these hormones, further testing is dictated by changes in medication or the clinical condition.

- Individuals with midline cranial and intracranial malformations should be followed throughout childhood because of the risk of developing endocrine dysfunction. Referral to an endocrinologist is usually indicated.
- The clinician should measure serum electrolytes and glucose, urine specific gravity, urinary sodium, simultaneous serum and urine osmolality, and antidiuretic hormone levels at diagnosis and at appropriate intervals if there is evidence of diabetes insipidus.
- Testing should be performed with the affected individual maximally dehydrated as tolerated, at a time when antidiuretic hormone release would be highest and urine would be most concentrated (i.e. early morning). A urine specific gravity of 1.005 or less and a urine osmolality less than 200 mOsm kg^{-1} is the hallmark of diabetes insipidus. Random plasma osmolality generally is greater than 287 mOsm kg^{-1}.
- The evaluation for adrenal insufficiency should consist of a cortisol releasing hormone test and basal morning serum cortisol.
- When endocrine problems are encountered, referral to a pediatric endocrinologist is recommended.

Management

- Available pharmaceutical therapy for diabetes insipidus includes subcutaneous, nasal, and oral preparations of vasopressin analogs (e.g., desmopressin or vasopressin) administered 2–3 times daily, as well as chlorpropamide, carbamazepine, clofibrate, thiazides, and indo-methacin (limited efficacy). Electrolytes should be carefully monitored. Maintenance of normal osmotic balance may be difficult, even after the introduction of treatment with vasopressin, either as desmopressin or lysine vasopressin spray.
- Urinary losses should be replaced with dextrose and water or intravenous fluid hyposmolar to the individual's serum.
- Avoidance of hyperglycemia, volume overload, and rapid correction of hypernatremia are important.
- Hypothyroidism and abnormalities of sex hormone production may be managed by judicious replacement, in a standard manner
- Although a few affected individuals may also have evidence of anterior pituitary axis dysfunction (e.g., growth hormone deficiency), there is no consensus about the use of growth hormone unless there is a demonstrated deficient growth hormone response either at baseline or after stimulation. For those children with poor growth who display normal growth hormone function, growth hormone is not recommended at this time.
- In general, there is a poor response to the growth hormone stimulation test. If the child is growth hormone deficient, treatment should be considered.
- Standard treatment for adrenal insufficiency is indicated, with attention to stress steroids during times of illness or surgery.

Neurologic

There is a considerable variation in the degree of brain malformation in children with HPE. Some degree of brainstem dysfunction is to be expected in children with alobar HPE. This may be manifest by irregularities in heart rate, respiration, and temperature regulation. The severity of temperature dysregulation may range from mild to severe. The overall frequency of this problem is not well established. Plawner et al. (2002) reported that 32% of their cohort of 68 demonstrated temperature regulation difficulties. Temperature dysregulation appeared to have a positive correlation with the degree of hypothalamic non-separation. Clinically more concerning is the irregular heart rate and a respiratory pattern characterized by episodes of hypopnea followed by hyperpnea. A sudden change in respiration or heart rate pattern in an otherwise stable child should provoke investigation for an underlying cause, such as infection, shunt malfunction with increasing cerebrospinal fluid pressure and compression of the brainstem, or dehydration.

Microcephaly occurs in the majority of individuals with HPE and is more commonly found with the lobar and semilobar variants than with alobar HPE. This may be because of the fact that a greater proportion of individuals with alobar HPE demonstrate obstructive hydrocephalus, usually requiring shunting. The need for shunting may correlate with the presence of a dorsal cyst which is more common with alobar (92%) and semilobar (28%) HPE.

Most individuals with classic HPE have been observed to have delayed white matter maturation. Diffusion tensor MRI, a noninvasive method of studying white matter microstructure, has demonstrated in vivo brainstem white matter tract abnormalities. The severity of forebrain malformation in HPE was correlated with presence or absence of brainstem white matter tracts. These findings are consistent with previous neuropathological studies that have documented absent or anomalous corticospinal tracts and middle cerebellar peduncles. Myelination was normal in the middle interhemispheric variant in this series, supporting the theory of differing embryonic patterning.

As might be expected in individuals with altered cortical development, seizures are part of the clinical phenotype. Seizures occur in approximately 50% of individuals with HPE. Some researchers have proposed that relative sparing of the lateral cortex may explain the lack of seizures and normal cognitive function in some of the affected individuals. There is no single type of seizure that is characteristic of this anomaly. Common electroencephalogram findings in HPE include hypersynchronous theta activity during sleep or when awake, hypersynchronous β activity during sleep, episodic attenuation of cerebral activity, nonspecific epileptiform activity, and posterior gradient flattening. The frequency of seizures varies between individuals and over time in the same individual. The presence of cortical malformation does not correlate with the presence or absence of seizures, but control of seizures correlates with the presence of cortical malformations. The response to anticonvulsant therapy is often adequate, but some individuals will present with medically refractory seizures. In one series, 52% had seizures that were difficult to control (Plawner et al. 2002).

Seizures may also occur as a result of endocrine abnormalities, including hypoglycemia and electrolyte imbalance (particularly hyponatremia), or shunt malfunction. Whenever seizures appear to be uncharacteristic or suddenly difficult to control, this possibility should be entertained.

Some individuals with HPE may have sleep disturbances. Many children do not fall asleep easily, remain asleep, or establish a pattern of naps and nighttime sleep in comparison with individuals of the same age. Although there are no current studies on sleep architecture in those with HPE, it is likely that disturbed sleep may derive in part from abnormalities of the central axis and thalami, which play an important role in sleep generation. For example, projections from the thalamus to the cortex enable synchronization of cortical activity from thalamic centers. Thalamic centers coordinate rhythms in non-REM sleep. Sleep may be disturbed by excessive somnolence, effects of seizures, shunt malfunction, and medication effects that may be sedating or stimulating.

Evaluation

- At the time of diagnosis, every child with HPE should have an assessment of ventricular size.
- Neuroimaging evaluation of HPE is best accomplished by MRI, and sagittal, axial, and coronal views should be obtained to enable a complete evaluation of midline structures. Initial hydrocephalus may be so severe as to obscure appropriate classification into alobar, semilobar, and lobar categories. Therefore, MRI should be obtained after shunt placement to provide a more accurate assessment (Hahn and Barnes 2010).
- The clinicians involved in the care of the child with HPE should be aware of signs and symptoms that signify increased intracranial pressure, including accelerated cranial growth, tense fontanel, lethargy, irritability, and unexplained vomiting. This assessment may be difficult in severely neurologically impaired children with a restricted behavioral repertoire.
- An electroencephalogram should be obtained if seizures are suspected clinically. Electroencephalogram in children with alobar HPE may show left-right hemisphere asynchronous activity unrelated to seizures.
- Parents and caregivers should be encouraged to keep a sleep diary so that the cause of the sleep disorder might be determined.
- The association of level of activity with medication effect should be established. For example, it is preferable to prescribe fewer sedating anticonvulsants in those who are excessively lethargic, provided the medication also provides seizure control.
- In any individual with HPE who has a change in sleep pattern, such as increased lethargy, an underlying medical explanation such as shunt failure should be ruled out.
- Sleep studies should be considered in children with disordered nocturnal breathing or nocturnal events that may be considered to be parasomnias, to exclude nocturnal epilepsy syndromes for those in whom routine electroencephalogram has been non-diagnostic, and in those children with nocturnal brady/tachyarrythmias or apnea and elevations of $PaCO_2$.

Management

- In symptomatic hydrocephalus, standard ventriculoperitoneal shunting utilizing an antisiphon device is appropriate. The procedure is associated with the usual risks and complications.
- There is no one anticonvulsant that is universally effective, thus, trials of several medications may be required before establishing efficacy.
- Seizures should be treated as they are in the general population with a few caveats. In children with diabetes insipidus, treatment with carbamazepine may lead to hyponatremia, and sodium levels will need to be followed. Some anticonvulsants may cause excessive sedation and/or hyperactivity and should be avoided if possible.
- Some children with uncontrolled seizures may benefit from a ketogenic diet.
- Dystonic, spastic, and choreiform movements may respond to the use of artane, baclofen, or similar agents to improve purposeful movement.
- Children with HPE are at risk for medication overtreatment as their movements may be interpreted as seizures, and the electroencephalogram may remain abnormal despite clinical control of the seizures.
- Background "white noise" such as a radio or television playing in the room has been helpful to aid sleep in some circumstances (personal experience).

- Nocturnal sedatives such as benadryl, chloral hydrate, or benzodiazepines may be effective in aiding sleep. It should be noted, however, that such medications may have a paradoxical effect and cause excessive stimulation or irritability.
- Children with disordered nocturnal breathing associated with elevations of $PaCO_2$, apnea, or other respiratory problems may benefit from continuous positive air pressure, bi-level positive air pressure, or surgery.

Ears and Hearing

The majority of children with HPE are considered to have normal hearing. However, as a result of severe malformation and medical illness, they are at risk for hearing loss and benefit from periodic hearing evaluations.

Evaluation

- Children with HPE should be screened for hearing impairment in the newborn period and thereafter if clinical circumstances raise concern.
- If multiple sensory deficits exist, evaluation should be performed by audiologists.

Management

- Treatment of hearing impairment is symptomatic.
- Aggressive treatment of otitis media with standard antibiotics is recommended.
- Sensorineural hearing loss may require hearing aids.

Ophthalmologic

HPE has been associated with a number of ophthalmologic malformations. In severe cases of HPE, cyclopia, cebocephaly, and ethmocephaly have been described. Milder findings may include optic nerve hypoplasia and iris or uveoretinal colobomas (Schimmenti et al. 2003).

Evaluation

- All children with HPE should have a thorough ophthalmologic examination. In the presence of multiple sensory deficits, this should be done by a pediatric ophthalmologist, including gross external examination and detailed evaluation of lens, optic nerve, and retina.

Management

- Treatment of visual impairments is symptomatic and the same as for people with the same problems in the general population.
- Surgery may be indicated for specific ophthalmologic issues. Techniques are standard.

Gastrointestinal

Constipation is a common problem in HPE, especially in non-mobile children with axial hypotonia and appendicular hypertonia. Because retention of stool is associated with abdominal pain, irritability, and discomfort, attention to this potential problem becomes important for overall well-being.

Problems with excessive intestinal gas may occur because of air swallowing, giving rise to conditions that may be labeled "colic."

Recently fatty liver has been described in germline hedgehog pathway mutations (Sacoto et al. 2017).

Evaluation

- At each visit, a history of constipation, colic, or excess gas should be sought.
- Refractory chronic constipation warrants referral to gastroenterology for evaluation.
- Surgical evaluation should be considered if bowel obstruction is suspected.

Management

- Increases in fluid intake, dietary fiber, and stool softeners may be beneficial.
- Excessive air swallowing will require appropriate burping, abdominal massage, or medication.
- Preparations with simethicone may be helpful for management of intestinal gas.

Immunologic

Individuals with HPE do not have an impaired immune system. However, the more severely affected children, who cannot control secretions, have difficulties with feeding, are relatively immobile, and are at risk for aspiration pneumonia. Such children may have a difficult time responding to typical bacterial and viral infections. Antibiotics should only be used to treat bacterial infections, as viral infections will not respond and the risk of developing antibiotic resistance in this population outweighs any potential benefit. Individuals with diabetes insipidus who develop concurrent infection are at significant risk of dehydration and even possible death if their fluid requirements during hyperpyrexia and infection cannot be met.

Evaluation

- Affected individuals with any change in baseline status should be evaluated for possible underlying infection. This is especially true for neurologically impaired

individuals who exhibit a restricted repertoire of behaviors and symptoms.
- Care should be taken to ensure sterility of shunt and gastrostomy tube sites.
- Individuals with infection at shunt and gastrostomy sites should be seen by a health care provider immediately.

Management
- Children with infections should be managed in a standard manner with judicious use of antibiotics.

RESOURCES

Carter Center for Brain Research in Holoprosencephaly and Related Malformations
Website: http://www.hperesearch.org/

NINDS Holoprosencephaly Information Page
Website: https://www.ninds.nih.gov/health_and_medical/disorders/holoprosencephaly.htm

Familes for HoPE
Website: https://familiesforhope.org

Support Organization for Trisomy 18, 13, and Related Disorders (SOFT)
Website: www.trisomy.org because of the frequency of trisomy 13 as cause of holoprosenecephaly

REFERENCES

Abe Y, Kruszka P, Martinez AF, Roessler E, Shiota K, Yamada S, Muenke M (2018) Clinical and demographic evaluation of a holoprosencephaly cohort from the Kyoto collection of human embryos. *Anat Rec* 301:973–986.

Atalar MH, Icagasioglu D, Sener RN (2008) Middle interhemispheric variant of holoprosencephaly associated with bilateral perisylvian polymicrogyria. *Pediatr Int* 50:241–224.

Barr M Jr, Cohen MM Jr (1999) Holoprosencephaly survival and performance. *Am J Med Genet* 89:116–120.

Bendavid C, Dupé V, Rochard L, Gicquel I, Dubourg C, David V (2006) Holoprosencephaly: An update on cytogenetic abnormalities. *Am J Med Genet C* 154C:86-92.

Bendavid C, Dupé V, Rochar L, Gicquel I, Dubourg C, David V (2010) Holoprosencephaly: An update on cytogenetic abnormalities. *Am J Med Genet Part C* 154C:86–92.

Bulakbasi N, Cancuri O, Kocaoğlu M (2016). The middle interhemispheric variant of holoprosencephaly: magnetic resonance and diffusion tensor imaging findings. *Br J Radiol.* 89(1063): 20160115.

Croen LA, Shaw GM, Lammer EJ (1996) Holoprosencephaly: Epidemiologic and clinical characteristics of a California population. *Am J Med Genet* 64:465–472.

Cohen MM Jr (1989a) Perspectives on holoprosencephaly: Part I. Epidemiology, genetics, and syndromology. *Teratology* 40: 211–235.

Cohen MM Jr (1989b) Perspectives on holoprosencephaly: Part III. Spectra, distinctions, continuities, and discontinuities. *Am J Med Genet* 34:271–288.

Cohen MM Jr, Sulik KK (1992) Perspectives on holoprosencephaly: Part II. Central nervous system, craniofacial anatomy, syndrome commentary, diagnostic approach, and experimental studies. *J Craniofac Genet Dev Biol* 12: 196–244.

De Franco E, Watson RA, Weninger WJ, et al. (2019) A Specific CNOT1 Mutation Results in a Novel Syndrome of Pancreatic Agenesis and Holoprosencephaly through Impaired Pancreatic and Neurological Development. *Am J Hum Genet* 104(5): 985–989. doi:10.1016/j.ajhg.2019.03.018

Domené S, Roessler E, El-Jaick KB, Snir M, Brown JL, Vélez JI, Bale S, Lacbawan F, Muenke M, Feldman B (2008) Mutations in the human SIX3 gene in holoprosencephaly are loss of function. *Hum Mol Genet* 17:3919–28.

Edison RJ, Muenke M (2004) Central nervous system and limb anomalies in case reports of first-trimester statin exposure. *N Engl J Med* 350:1579–1582.

Edison RJ, Berg K, Remaley A, Kelley R, Rotimi C, Stevenson RE, Muenke M (2007) Adverse birth outcome among mothers with low serum cholesterol. *Pediatrics* 120:723–733.

Edwards L, Hui L (2017). First and second trimester screening for fetal structural anomalies. *Semin Fetal Neonatal Med* (17)30136–1.

Fallet-Bianco C (2018) Neuropathology of holoprosencephaly. *Am J Med Genet C Semin Med Genet* 178:214–228.

Griffiths PD, Jarvis D (2016). In utero MR imaging of fetal holoprosencephaly: A structured approach to diagnosis and classification. *AJNR* 37(3):536–543.

Hadley DW, Kruszka P, Muenke M (2018) Challenging issues arising in counseling families experiencing holoprosencephaly. *Am J Med Genet C* 178:238–245.

Haas D, Morgenthaler J, Lacbawan F, Long R, Runz H, Garbade SF, Zschocke J, KelleyRI Okun JG, Hoffmann GF, Muenke M (2007) Abnormal sterol metabolism in holoprosencephaly: Studies in cultured lymphoblasts. *J Med Genet* 44:298–305.

Hahn JS, Barnes PD (2010) Neuroimaging advances in holoprosencephaly: Refining the spectrum of the midline malformation. *Am J Med Genet C* 154C:120–132.

Hong S-K, Hu P, Marino J, Hufnagel SM, Hopkin RJ, Toromanović A, Richieri-Costa A, Ribeiro-Bicudo LA, Kruszka P, Roessler E, Muenke M (2016) Dominant-negative kinase domain mutations in FGFR1 can explain the clinical severity of Hartsfield syndrome. *Hum Mol Genet* 25:1912–1922.

Hong M, Srivastava K, Kim S, Allen BL, Leahy DJ, Hu P, Roessler E, Krauss RS, Muenke M (2018) BOC is a modifier gene in holoprosencephaly. *Hum Mutat* 38:1464–1470.

Hu T, Kruszka P, Martinez AF, Ming JE, Shabason EK, Raam MS, Shaikh TH, Pineda-Alvarez DE, Muenke M (2018) Cytogenetics and holoprosencephaly: A chromosomal microarray study of

222 individuals with holoprosencephaly. *Am J Med Genet C* 178:175–186.

Hughes JJ, Alkhunaizi E, Kruszka P, et al. (2020) Loss-of-Function Variants in PPP1R12A: From Isolated Sex Reversal to Holoprosencephaly Spectrum and Urogenital Malformations. *Am J Hum Genet*. 106(1):121–128. doi:10.1016/j.ajhg.2019.12.004

Johnson CY, Rasmussen SA (2010) Non-genetic risk factors for holoprosencephaly. *Am J Med Genet C* 154C:73–85.

Kelley RL, Roessler E, Hennekam RC, Feldman GL, Kosaki K, Jones MC, Palumbos JC, Muenke M (1996) Holoprosencephaly in RSH/Smith-Lemli-Opitz syndrome: Does abnormal cholesterol metabolism affect the function of sonic hedgehog? *Am J Med Genet* 66:478–484.

Kousa YA, du Plessis AJ, Vezina G (2018) Prenatal diagnosis of holoprosencephaly. *Am J Med Genet C* 178:206–213.

Kruszka P, Hart RA Hadley DW, Muenke M, Habal MB (2015) Expanding the phenotypic expression of Sonic Hedgehog mutations beyond holoprosencephaly. *J. Craniofac Surg* 26:3–5,

Kruszka P, Martinez AF, Muenke M (2018) Molecular testing in holoprosencephaly. *Am J Med Genet C* 178:187–193.

Kruszka P, Muenke M (2018) Syndromes associated with holoprosencephaly. *Am J Med Genet C Semin Med Genet* 178(2):229–237. doi: 10.1002/ajmg.c.31620. Epub 2018 May 17. PubMed PMID: 29770994; PubMed Central PMCID: PMC6125175.

Kruszka P, Berger SI, Weiss K, et al. (2019a) A CCR4-NOT Transcription Complex, Subunit 1, CNOT1, Variant Associated with Holoprosencephaly. *Am J Hum Genet*. 104(5):990–993. doi:10.1016/j.ajhg.2019.03.017

Kruszka P, Berger SI, Casa V, et al. (2019b) Cohesin complex-associated holoprosencephaly. *Brain* 142(9):2631–2643. doi:10.1093/brain/awz210

Leevers HJ, Roesler C, Flax J, Benasich AA (2005) The Carter neurocognitive assessment for children with severely compromised expressive language and motor skills. *J Child Psychol Psychiatry* 46:287–303.

Levey EB, Stashinko E, Clegg NJ, Delgado MR (2010) Management of children with holoprosencephaly. *Am J Med Genet C* 154C:183–190.

Lipinski RJ, Godin EA, O'Leary-Moore SK, Parnell, S.E., Sulik KK (2010) Genesis of teratogene-induced holoprosencephaly in mice. *Am J Med Genet C* 154C:29–42.

Miller EA, Rasmussen SA, Siega-Riz AM, Frias JL, Honein MA, the National Birth Defects Prevention Study (2010) Risk factors for non-syndromic holoprosencephaly in the National Birth Defects Prevention Study. *Am J Med Genet C* 154C:62–72.

Ming JE, Muenke M (2002) Multiple hits during early embryonic development: Digenic diseases and holoprosencephaly. *Am J Hum Genet* 71:1017–1032.

Orioli IM, Castilla EE (2010) Epidemiology of holoprosencephaly: Prevalence and risk factors. *Am J Med Genet C* 154C:13–21.

Plawner LL, Delgado MR, Miller VS, Levey EB, Kinsman SL, Barkovich AJ, Simon EM, Clegg NJ, Sweet VT, Stashinko EE, Hahn JS (2002) Neuroanatomy of holoprosencephaly as predictor of function: beyond the face predicting the brain. *Neurology* 59:1058–66.

Roessler E, Muenke M (2010) The molecular genetics of holoprosencephaly. *Am J Med Genet C* 154C:52–61.

Roessler E, Hu P, Muenke M (2018) Holoprosencephaly in the genomics era. *Am J Med Genet C* 178:165–174.

Sacoto MJG, Martinez AF, Abe Y, Kruszka P, Weiss K, Everson JL, Bataller R, Kleiner DE, Ward JM, Sulik KK, Lipinski RJ, Solomon BD, Muenke M (2017) Human germline Hedgehog pathway mutations predispose to fatty liver. *J. Hepatol*. 67:809–817.

Schimmenti LA, De La Cruz J, Lewis RA, Karkera JD, Manligas GS, Roessler E, Muenke M (2003) Novel mutation in Sonic Hedgehog in non-syndromic colobomatous microphthalmia. *Am J Med Genet* 116:215–221.

Shiota K, Yamada S (2010) Early pathogenesis of holoprosencephaly. *Am J Med Genet C* 154C:22–28.

Simon EM, Hevner RF, Pinter JD, Clegg NJ, Delgado M, Kinsman SL, Hahn JS, Barkovich AJ (2002) The middle interhemispheric variant of holoprosencephaly. *Am J Neuroradiol* 23:151–155.

Simonis N, Migeotte I, Lambert N, Perazzolo C, de Silva DC, Dimitrov B, Heinrichs C, Janssens S, Kerr B, Mortier G, Van Vliet G, Lepage P, Casimir G, Abramowicz M, Smits G, Vilain C (2013) FGFR1 mutations cause Hartsfield syndrome, the unique association of holoprosencephaly and ectrodactyly. *J Med Genet* 50:585–592.

Solomon BD, Mercier S, Vélez JI, Pineda-Alvarez DE, Wyllie A, Zhou N, Dubourg C, David V, Odent S, Roessler E, Muenke M (2010) Analysis of genotype-phenotype correlations in human holoprosencephaly. *Am J Med Genet C* 154C:133–141.

Solomon BD, Bear KA, Wyllie A, Keaton AA, Dubourg C, David V, Mercier S, Odent S, Hehr U, Paulussen A, Clegg NJ, Delgado MR, Bale SJ, Lacbawan F, Ardinger H, Aylsworth A, Bhengu M.L, Braddock S, Braddoch S, Brookhyser K, Burton B, Gaspar H, Grix A, Horovitz D, Kanetzke D, Kayserili H, Lev D, Nikkel SM, Norton M, Roberts R, Saal H, Schaefer GB, Schneider A, Smith EK, Sowry E, Spence MA, Shalev S, Steiner CE, Balog JZ, Hadley DW, Zhou N, Pineda-Alvarez D.E, Roessler E, Muenke M (2012). Genotypic and phenotypic analysis of 396 individuals with mutations in Sonic Hedgehog. *J Med Genet* 49:473–479.

Solomon BD, Kruszka P, Muenke M. Holoprosencephaly flashcards: An updated summary for the clinician (2018) *Am J Med Genet C* 178:117–121.

Stashinko EE, Clegg NJ, Kammann HA, Sweet VT, Delgado MR, Hahn JS, Levey EB (2004) A retrospective survey of perinatal risk factors of 104 living children with holoprosencephaly. *Am J Med Genet* 128A:114–119.

Stashinko EE, Harley L, Steele RA, Clegg NJ (2010) Parental perspectives on living with a child with HoPE. *Am J Med Genet C* 154C:197–201.

Stokes B, Berge, SI, Hall BA, Weiss K, Hadley DW, Murdock DR, Ramanathan S, Clark RD, Roessler E, Kruszka P, Muenke M (2018) SIX3 deletions and incomplete penetrance in families affected by holoprosencephaly. *Congenit Anom* 58:29–32.

Summers AD, Reefhuis J, Taliano J, Rasmussen SA (2018) Nongenetic risk factors for holoprosencephaly: An updated review of the epidemiologic literature. *Am J Med Genet C* 178:151–164.

Tekendo-Ngongang C, Kruszka P, Martinez AF, Muenke M (2019) Novel heterozygous variants in KMT2D associated with holoprosencephaly. *Clin Genet.* 96(3):266–270. doi:10.1111/cge.13598

Vasudeva A, Nayak SS, Kadavigere R, Girisha KM, Shetty J (2015) Middle interhemispheric variant of holoprosencephaly - presenting as non-visualized cavum septum pellucidum and an interhemispheric cyst in a 19-weeks fetus. *J Clin Diagn Res* 9:11–13.

Verlinsky Y, Rechitsky S, Verlinsky O, Ozen S, Sharapova T, Masciangelo C, Morris R, Kuliev A (2003) Preimplantation diagnosis for sonic hedgehog mutation causing familial holoprosencephaly. *N Engl J Med* 348:1449–1454.

Virta M, Launes J, Valanne L, Hokkanen L (2016). Adult with Middle Interhemispheric Variant of Holoprosencephaly: Neuropsychological, Clinical, and Radiological Findings. *Arch Clin Neuropsychol.* 31(5):472–479.

Weiss K, Kruszka P, Guillen Sacoto MJ, Addissie YA, Hadley DW, Hadsall CK, Stokes B Hu P, Roessler E, Solomon B Wiggs E, Thurm A, Hufnagel RB, Zein WM, Hahn JS, Stashinko E, Levey E, Baldwin D, Clegg NJ, Delgado MR (2018). Muenke M In-depth investigations of adolescents and adults with holoprosencephaly identify unique characteristics. *Genet Med* 20(1):14–23.

Winter TC, Kennedy AM, Woodward PJ (2015) Holoprosencephaly: a survey of the entity, with embryology and fetal imaging. *Radiographics* 35(1):275–290.

32

INCONTINENTIA PIGMENTI

DIAN DONNAI AND ELIZABETH A. JONES

Manchester Centre for Genomic Medicine, Manchester University NHS Foundation Trust and University of Manchester, Manchester Academic Health Sciences Centre, Manchester, UK

INTRODUCTION

Incontinentia pigmenti (IP) was first described by Garrod (1906), although others subsequently defined the condition further. IP was previously sometimes referred to as incontinentia pigmenti 2 to distinguish it from incontinentia pigmenti 1, the disorder observed in females with X:autosome translocations involving Xp11. IP is a multisystem disorder affecting predominantly females, although males can less frequently be affected (Fusco et al. 2014). The cutaneous manifestations are diagnostic and classically occur in four stages: vesicular, verrucous, hyperpigmented, and atrophic. They occur in the distribution of Blaschko's lines. In addition to these cutaneous manifestations, there may be a variety of dental, ocular, neurologic, and developmental abnormalities.

Incidence

In 1976, Carney reviewed the literature and, from published reports and his own series, derived incidence figures for both cutaneous and noncutaneous features (Carney 1976). Carney's series may have been biased because ascertainment was by publication, and the series may be etiologically heterogeneous. A review of clinical features from the literature and from a large study was published in 1993 (Landy and Donnai 1993), and subsequently several reviews and series of male cases have been published (Scheuerle 1998; Pacheco et al. 2006; Ardelean and Pope 2007; Fusco et al. 2007, 2014). Prevalence at birth is estimated to be 1.2 per 100,000 (Orphanet, Prevalence and incidence of rare diseases June 2017 - http://www.orpha.net/orphacom/cahiers/docs/GB/Prevalence_of_rare_diseases_by_alphabetical_list.pdf).

Diagnostic Criteria

The cutaneous manifestations of IP are diagnostic and classically occur in four stages: vesicular, verrucous, hyperpigmented, and atrophic (Figure 32.1). These skin lesions occur in the distribution of Blaschko's lines. These were described in 1901 by Blaschko and were reviewed by Jackson (1976). In addition to these cutaneous manifestations, there may be a variety of dental, ocular, neurological, and developmental abnormalities (Figure 32.2).

Diagnostic criteria for IP were initially suggested by Landy and Donnai (1993) and later updated by Minic et al. (2014). The skin manifestations (vesicular, verrucous, pigmentary, and atrophic lesions in Blaschko's lines) represent major criteria. With the increased availability of diagnostic testing a confirmed pathogeneic mutation in *IKBKG* gene can also be included as a major diagnositic criterion. The minor criteria are typical retinal findings, dental abnormalities, CNS abnormalities, alopecia, woolly hair, abnormal nails, breast abnormalities, and a history of multiple male miscarriages (Landy and Donnai 1993) (see Table 32.1).

Cassidy and Allanson's Management of Genetic Syndromes, Fourth Edition.
Edited by John C. Carey, Agatino Battaglia, David Viskochil, and Suzanne B. Cassidy.
© 2021 John Wiley & Sons, Inc. Published 2021 by John Wiley & Sons, Inc.

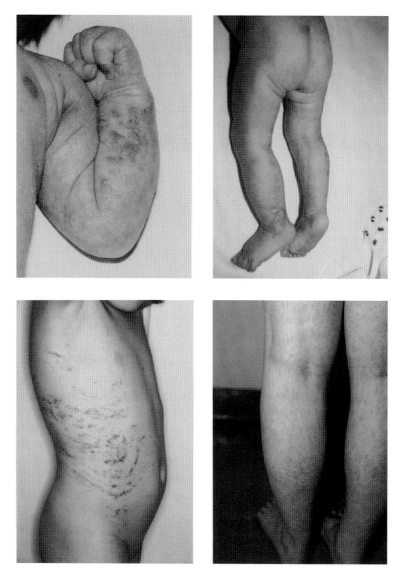

FIGURE 32.1 Skin lesions seen in IP. (Top left) Blistering lesions; note linear distribution. (Top right) Verrucous lesions over ankles. (Bottom left) Hyperpigmented lesions around trunk. (Bottom right) Atrophic lesion on back of lower limbs.

In the absence of a family history of affected female relatives, at least one major criterion should be present together with one or more minor criteria. Where there is a first-degree female relative with IP, the diagnosis should be suspected if there are skin manifestations or any minor criteria. To these clinical criteria can be added evidence from mutation testing of *IKBKG* or X-inactivation studies (Parrish et al. 1996; Woffendin et al. 1997). Males who show clinical features of IP may also have Klinefelter syndrome (47,XXY), or may have hypomorphic *IKBKG* mutations or postzygotic *IKBKG* mutations leading to somatic mosaicism (Pacheco et al. 2006; Fusco et al. 2014).

Etiology, Pathogenesis, and Genetics

IP is an X-linked disorder due to mutations in *IKBKG*, which is situated at Xp28. Pedigree analysis of IP is consistent with X-linked dominant inheritance with lethality in affected males. This mode of inheritance is supported by the high female-to-male ratio (10:1) (Fusco et al. 2014), by female-to-female transmission, and by the increased incidence of miscarriage in affected females. Affected women have a 50% chance of transmitting the mutant *IKBKG* allele at conception; however, male conceptuses with a loss-of-function mutation of *IKBKG* miscarry. Thus, the expected ratio among liveborn children is approximately 33% unaffected females, 33% affected females, and 33% unaffected males. Fifty percent of the female offspring of an affected woman will inherit the condition and manifest it to a variable extent. It is likely that many mild cases are not diagnosed. Often, an affected mother and female relatives are only diagnosed after the birth of a female with more severe manifestations.

In one study of IP, 53 of 111 affected females were adults who had been pregnant at least once (Landy and Donnai 1993). They had in total 158 pregnancies, of which 40 ended

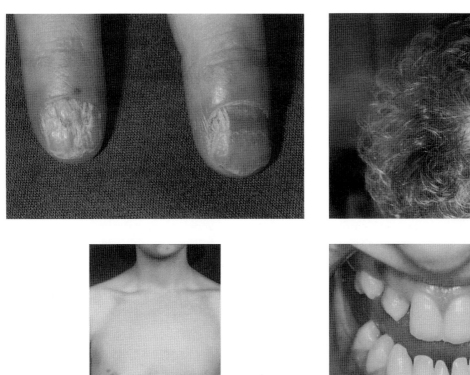

FIGURE 32.2 Non-skin manifestations of IP. (Top left) Nail dystrophy in a 2-year-old child. (Top right) Alopecia on vertex of adult female. (Bottom left) Breast and nipple hypoplasia. (Bottom right) Missing and small teeth.

TABLE 32.1 Table of criteria for the diagnosis of incontinentia pigmenti

Incontinentia pigmenti diagnostic criteria	
Major criteria	Minor criteria

Typical IP skin stages distributed along Blaschko's lines:
  Vesiculo-bullous stage
  Verrucous stage
  Hyperpigmented stage
  Atrophic/hypopigmented stage
Confirmed pathogenic mutation in *IKBKG*
	Dental anomalies – hypodontia or abnormally shaped teeth
	Ocular anomalies – peripheral neovascularization of retina
	CNS abnormalities e.g. seizures
	Abnormal hair – sparse, alopecia, woolly
	Breast or nipple hypoplasia
	Abnormal nails – ridging or pitting
	Multiple male miscarriages

in miscarriage, 32 in normal males, 56 in affected females, and 30 in normal females (personal data). Father-to-daughter transmission has been reported, but in these instances the father was the first affected member of the family and is thought to have had somatic mosaicism, and this was confirmed molecularly in one reported family (Rashidghamat et al. 2016). IP has been reported in males with Klinefelter syndrome (47,XXY, see Chapter 35).

Using genetic linkage analysis, it was shown that IP was caused by mutation of a single gene that lay close to the locus for factor VIII at Xq28 on the long arm of the X chromosome (Jouet et al. 1997). In 2000, an international consortium found a recurrent mutation in the the *IKBKG* gene (previously known as the *NEMO* gene) that encodes the regulatory subunit of the inhibitor of kappaB kinase complex (NF-κB essential modifier – NEMO protein) in 80% of index cases. About 65% of individuals with IP have a de novo mutation in *IKBKG*. In typical IP, there is no evidence of genetic heterogeneity. However, X-inactivation is thought to be the major factor contributing to the wide phenotypic variability in affected females even within the same family. IP shows high penetrance but significant variable expressivity.

The proper function of *IKBKG* is central in the NF-κB signaling pathways because it is required for the appropriate inhibitor of kappaB kinase complex activation. *IKBKG* function is central to many immune, inflammatory, and cell death (apoptotic) pathways in the cell. In female patients with IP, *IKBKG* mutation is always heterozygous and can either preserve a residual NF-κB activation (hypomorphic mutation) or completely abolish it (amorphic mutation) (Fusco et al. 2015). There is a recurrent mutation in the *IKBKG* gene in 75% of individuals with IP that results from a complex rearrangement that deletes exons 4–10. The NF-kB pathway is a means by which individual cells are protected against cell death induced by external factors such as infection or inflammation. Cells in which the active X carries the mutated gene are more liable to undergo cell death.

Only females have been described with the recurrent deletion mutation other than a few rare instances of males with "classical" incontinentia pigmenti with 47,XXY Klinefelter syndrome and those with somatic mosaicism (Kenwrick et al. 2001; Rashidghamat et al. 2016), indicating that this mutation is germline lethal in affected 46,XY males.

Males with hypomorphic *IKBKG* mutations have a clinically distinct anhydrotic ectodermal dysplasia-immunodeficiency syndrome known as X-linked hypohidrotic ectodermal dysplasia and immunodeficiency (HED-ID); their mothers may show signs of classical IP (Zonana et al. 2000; Aradhya et al. 2001; Mansour et al. 2001; Fusco et al. 2008).

Nelson (2006) reviewed disorders of the NF-kB signaling pathway and evidence about the multiple roles of *IKBKG* in the cell from the study of mouse models. The linear distribution of the skin lesions most likely represent the migrational pathways of cells derived from the neural crest. Nenci et al. (2006) used epidermis-specific deletion of *IKBKG* in mice to investigate the mechanisms causing skin pathology in IP. They showed that *IKBKG*-deficient keratinocytes were the initiating cell type to trigger the inflammatory response and that tumor necrosis factor was essential for this response. Curth and Warburton (1965) suggested that pigmentation following Blaschko's lines in IP reflects functional X chromosome mosaicism because of Lyonization (X-inactivation).

In the early vesicular cutaneous stage of IP, there is massive infiltration of eosinophils into the epidermis. There is also marked peripheral blood eosinophilia. The pathogenesis of the inflammation is likely to be related to disturbance of the NF-kB pathway and is not directed specifically at the melanocytes. Hyperkeratosis, papillomatosis, and mild dyskeratosis are seen in the verrucous stage, and the pigmentary stage shows degeneration of the basal cells and melanin-loaded macrophages in the dermis, giving the condition its name. In time, the hyperpigmentation fades. In the atrophic phase, Moss and Ince (1987) noted that, although the lesions are described as hypopigmented, the contrast with normal skin is probably because of reduced vascularity and the lack of hair follicles. The sweat glands are probably also affected in these lesions and a linear arrangement of sweating and non-sweating skin has been demonstrated.

Diagnostic Testing

Diagnosis is mainly made by clinical evaluation and follow-up to observe the development of the phases of the cutaneous manifestations and by confirmatory diagnostic genetic testing of the *IKBKG* gene. About 75% of individuals will have the common deletion mutation, 16% have point mutations, and 4% have Xq28 rearrangements (Fusco et al. 2014). The causative mutation cannot be identified using modern genetic testing methodology in approximately 5% of cases. In families in which there are sufficient numbers of affected individuals, genetic linkage analysis may be possible using markers from the Xq28 region. Following reports of skewed X-inactivation in affected females (Parrish et al. 1996), it has been suggested that an X-inactivation assay can be used to investigate the status of females in IP families (Woffendin et al. 1997). In affected individuals in whom an *IKBKG* pathogenic variant is not identified by the above methods, a skin biopsy of affected skin can be considered to look for a somatic mutation.

Prenatal diagnosis can be offered to at-risk women whose mutation is known by analysis of DNA extracted from chorionic villus cells or from cultured amniocytes.

Differential Diagnosis

Any condition with skin manifestations in Blaschko's lines has the potential to be confused with IP. Strict diagnostic criteria and examination of first-degree female relatives are essential to confirm the diagnosis of IP.

Incontinentia pigmenti 1 was the historical name given to the condition observed in females with an X:autosome translocation that has a breakpoint at Xp11. The vesicular and verrucous lesions are not seen in such affected individuals, and the whorled pigmentation or hypopigmentation is observed from infancy. Most of these individuals have more severe developmental problems than those with classical IP. Not all such individuals have exactly the same breakpoint, and studies by Hatchwell et al. (1996) have demonstrated random X-inactivation in uncultured fibroblasts, lending support to the hypothesis that in incontinentia pigmenti 1 the phenotype is a manifestation of mosaicism, with some cells expressing genes from both copies of the X chromosome and other cells expressing genes from just one X chromosome. They suggest that there is no evidence that the effects are caused by the disruption of a single genetic locus.

Pigmentary mosaicism (also previously described as hypomelanosis of Ito or incontinentia pigmenti achromians) is the skin phenotype associated with various forms of genetic mosaicism. Vesicular and verrucous phases of IP are not observed but there are often hypo- or hyperpigmented lesions in Blaschko's lines (Happle 1993). Males and females can be affected, and chromosomal mosaicism has been demonstrated in about one-third of the affected individuals. The reports claiming familial occurrence of pigmentary mosaicism are not convincing. Histological features are relatively non-specific. The diagnosis of pigmentary mosacisim should be considered in individuals without the preceding vesicular and verrucous skin lesions and should be further investigated by skin biopsy and chromosome analysis of cultured fibroblasts, because the chromosomal mosaicism is rarely demonstrated in lymphocytes.

Goltz focal dermal hypoplasia (Temple et al. 1990) also has lesions in Blaschko's lines and is an X-linked dominant condition with male lethality. The skin lesions of Goltz syndrome are quite distinct and consist of focal absence of dermis in the distribution of Blaschko's lines, with herniation of fat and with multiple papillomas of the mucous membranes around the mouth, anus, and genitalia. There is no vesicular or verrucous phase, but there can be linear hyper- and hypopigmented lesions, particularly visible in children from ethnic groups with darker skin. Skeletal and ocular abnormalities are common in Goltz syndrome and may be severe. Mutations in the *PORCN* gene cause Goltz syndrome (Grzeschik et al. 2007).

The early stages of X-linked dominant chondrodysplasia punctata (CDP) can be confused with IP because the ichthyosiform erythroderma observed can be mistaken for the verrucous phase of IP. However, in X-linked dominant chondrodysplasia punctata this phase is followed by linear scarring with follicular pitting. Alopecia can be a major problem, as can severe skeletal abnormalities and cataracts. Derry et al. (1999) demonstrated mutations in the gene encoding delta(8)-delta(7) sterol isomerase emopamil-binding protein (*EBP*) in the more common type of X-linked chondrodysplasia punctata.

Linear epidermal nevi also occur in Blaschko's lines and may be confused with the verrucous lesions of IP, as can the lesions of mosaic congenital ichthyosiform erythroderma. In the latter condition, there is no prior history of vesicular or pigmentary lesions and no family history.

In the early vesicular stages of IP, particularly if the lesions become infected, the condition is sometimes misdiagnosed as staphylococcal infection or herpes. The linear nature of the lesions on the limbs and the circumferential swirling pattern on the truck should alert the physician to the possibility of IP.

MANIFESTATIONS AND MANAGEMENT

The largest cohort of patients with IP in whom the clinical and molecular diagnosis has been confirmed is reported in Fusco et al. 2014. The frequency of the manefestations is taken from this series unless otherwise stated.

Growth and Feeding

There are no reported major impacts on growth in this disorder. However, anecdotally, many affected girls in early life seem to have relatively little subcutaneous fat and several have undergone investigations for failure-to-thrive.

Development and Behavior

In general, females with IP make normal developmental progress, and the frequency of neurological and developmental problems is unlikely to be as high as suggested in the Carney review (Carney 1976). In one series of 111 affected individuals, only 6% had persistent seizures (personal experience). These individuals and a further 3% had some learning difficulties, but only 3% of the total group had severe deficiencies. In Fusco's case series approximately 10% had intellectual impairment. There are a number of case reports of early encephalopathy with subsequent poor developmental progress (Salamon et al. 2016).

Evaluation

- At the first sign of delayed development, a full evaluation should be initiated. If there are adverse neurological signs in early life surveillance should be instituted from that point forward.

Management

- Intervention for intellectual impairment and developmental delay is the same as in the general population.

Dermatologic

The cutaneous manifestations of IP are diagnostic (Figure 32.1). Classically, the features are described in four stages. All stages do not necessarily occur in an individual, and several stages may overlap; stage 1: erythema, vesicles, and pustules; stage 2: papules, verrucous lesions, and hyperkeratosis; stage 3: hyperpigmentation; and stage 4: pallor, atrophy, and scarring.

Stage 1 The lesions of the first stage develop within the first few days or weeks of life (90% less than 1 month). They tend to appear in crops in the distribution of Blaschko's lines and clear within weeks and may or may not be replaced by new crops at the same or differing sites. The lesions are blisters that can be preceded by erythema. One author has observed transient erythema toxicum neonatorum in Blaschko's lines in an affected female hours after birth, which preceded the typical vesicles. The lesions occur in a linear distribution along the limbs and circumferentially around the trunk. Lesions often occur on the head, typically at the vertex, but rarely on the face. The inflammatory phase is accompanied by massive infiltration of eosinophils into the epidermis and marked peripheral blood leukocytosis with up to 65% eosinophils. In most children, the vesicular stage has cleared completely by 4 months, but milder, short-lived eruptions might occur during the first year of life, often accompanying an acute febrile illness.

Stage 2 The typical lesions are hyperkeratotic, warty appearing lesions occurring in Blaschko's lines and often appearing on the lower legs. They may be present in the first few weeks (approximately 40% have symptoms in the first month) but often develop after several weeks as the blisters are healing. These warty lesions are less common on the trunk and face but may occur on the scalp. Most of these warty lesions have disappeared by 6 months. In a large series, 80% of those in whom the lesions occurred were clear by 6 months of age (personal experience).

Stage 3 Although it is this stage that gives the condition its name, its presence and extent are very variable. The hyperpigmentation is more often apparent on the trunk than the limbs and occurs in streaks or whorls in Blaschko's lines. The nipples are frequently involved, but the regions that are most often affected are the axilla and groin. The distribution of these lesions is often unrelated to the distribution of the previous stages. The pigmented lesions remain static for several years but then begin to fade, and by the age of 16 years, the majority of pigmented lesions, which may be brown or slate gray, have faded.

Stage 4 This atrophic phase is classically seen in affected adult females and most frequently is observed on the posterior aspect of the lower limbs and over the shoulders and upper arms. These atrophic lesions are rare under 1 year of age. In the author's series, these pale linear lesions were observed in many girls under the age of 10 years, concurrent with hyperpigmented or even vesicular and verrucous lesions (Landy and Donnai 1993). Studies have demonstrated that hair follicles and sweat glands are reduced in number in these atrophic lesions and that there is decreased vascularity rather than true hypopigmentation.

Hair and Nails. Hair defects are common affecting around 26%. Alopecia can occur (Figure 32.2, top right), especially at the vertex, and often after blistering or verrucous lesions at this site. Hair is often described as sparse early in childhood and later as lusterless, wiry, or coarse. Although hair abnormalities are very common, it is rare for females to have a major cosmetic problem.

Nail dystrophy (Figure 32.2, top left) is frequent with manifestations range from mild ridging or pitting to severe nail dystrophy resembling onychomycosis. Nail dystrophy may be a transient phenomenon, and complete resolution may occur. There may be subungual keratotic lesions particularly in infants with severe verrucous lesions.

Evaluation

- Affected females should be frequently assessed during the first few months, and preferably years, and the type and distribution of the skin lesions carefully documented. In the first year, three monthly evaluations are suggested unless infection has occurred, in which case individual clinical decisions are needed. Thereafter, annual reviews should suffice.
- Photographic records are very useful.
- Skin biopsy and/or analysis of vesicular fluid for eosinophilia may be considered, although the macroscopic appearance is usually sufficient to establish the diagnosis.
- Inspection of the nails and hair should be part of the regular evaluation and follow-up of people with IP.

Management

- During the neonatal period, and whenever the blisters are present, strict attention should be paid to hygiene to prevent secondary infection.
- There is no specific treatment to hasten the healing of the vesicular or verrucous phases, except that the lesions should be kept dry.
- Families should be reassured that the lesions will improve with time.
- Unless there is infection, surgical treatment of nail dystrophy is not indicated and resolution of severe lesions often occurs.

Ophthalmologic

The incidence of ocular abnormalities is high, affecting greater than 30%, and 20% will develop vision threatening disease. This makes the ophthalmological complications one of the most medically important aspects of IP. The characteristic lesion involves abnormalities of the developing retinal vessels and the underlying pigmented cells (Goldberg and Custis 1993). New vessel proliferation is stimulated in areas of retinal ischemia with resulting bleeding and fibrosis. The process is similar to that found in retinopathy of prematurity. In one family, a mother and her daughter both had enucleation of one eye because of a suspicion of retinoblastoma (personal experience). It was only after the atrophic skin lesions and alopecia were noted in the mother that the diagnosis of IP was made.

Although signs of retinal vessel proliferation are present in many affected individuals with IP, it is generally limited. However, retinal detachment is a recognized complication

(Wald et al. 1993; Chen et al 2015). When visual loss occurs, it is usually as a result of tractional retinal detachment secondary to contraction of fibrovascular tissue (which occurs in younger individuals, from first couple of weeks of life) or rhegmatogenous detachment related to holes in atrophic, avascular retina (older individuals) (Chen et al. 2015). Based on the similarities to the retinopthy of prematurity, treatment with prophylactic ablation is often undertaken with case reports of successful outcomes in IP. However, some eyes with proliferative retinopathy remain stable or improve over time without treatment. To date, there are no controlled clinical trials evaluating the use of prophylactic ablation in IP.

Other ocular abnormalities seen include strabismus, often in association with refractive errors, microphthalmos or cataract. In a series of 111 affected individuals, only three females had complete visual loss in one eye, the two who had enucleation (see above), and another woman from a large pedigree (personal experience). Despite the high frequency of ophthalmic complications, over 90% of individuals have functional vision.

Evaluation

- Because ocular abnormalities are so frequent, it is recommended that everyone diagnosed with IP have an ophthalmologic evaluation at diagnosis. A fluorescein angiogram is useful. If the diagnosis of IP is made in the newborn period, this evaluation should take place before discharge from hospital and be repeated regularly (at least monthly) in the first four months and then three monthly until 1 year of age and six monthly until 3 years of age. One author knows of a case where newborn examination was normal but the child presented at 4 weeks with florid changes in both eyes resulting in loss of vision in one eye.
- Further ophthalmic follow-up should be dictated by clinical findings. The risk of retinal detachment decreases after 3 years of age, so follow-up interval can be increased to 12 months but surveillance should continue life long.

Management

- If retinal abnormalities are detected, early lazer photocoagulation or cryotherapy of the nonperfused retina in individuals with documented progression of neovascularization, progression of vitreous traction or vitreous haemorrhage at successive evaluations might be helpful.
- Warn adults of the symptoms of retinal tear or detachment.
- Ophthalmic manifestations are treated in a standard manner.

Neurologic

Carney's review found a high frequency of neurological abnormalities (Carney 1976). However, the reports of analyzed individuals may well have been biased because they were literature cases and more likely to be published if they had severe manifestations. In that review, many of the reports contained insufficient detail to make a definite diagnosis of IP, and it is likely that there was etiological heterogeneity with some cases representing pigmentary mosaicism. More recent studies, in which strict diagnostic criteria were applied, suggest a much lower incidence of central nervous system abnormalities (Landy and Donnai 1993). Fourteen percent of affected individuals had seizures, but only 6% had persistent seizures with a degree of intellectual impairment, and the rest had transient seizures with no associated intellectual impairment. In those with persistent seizures, the first seizure occurred before 12 weeks of age, often in the first week of life. Of the 111 individuals studied, 9% (including those with persistent seizures) had intellectual impairment, although in only one-third of these could it be classified as severe. In Fusco et al. (2014) 97 out of 308 had central nervous system defects (31.5%) and of these 39 had seizures and 29 had intellectual impairment.

Wolf et al. (2005), Loh et al. (2008), and Maingay-de Groof et al. (2008) described individuals with proven *IKBKG* mutations that presented with encephalopathy and extensive cerebral infarction in the newborn period with poor outcomes. However, Bryant and Rutledge (2007) described abnormal white matter in a neurologically intact child with IP. Minic et al. (2013) published a systematic review of central nervous sytem anomalies in IP and showed that they were present in 30%. The most frequent such anomalies were seizures, motor impairment, intellectual impairment, and microcephaly. In this series, the age of first neurological manifestation was in the first week in 58% and in the first year in 88%. Central nervous system involvement in the neonatal period is a poor prognostic sign and warrants discussion with the parents about potential long-term problems.

Evaluation

- Full neurological assessment should be part of the initial management of all individuals with IP.
- Electroencephalography and cranial imaging studies are indicated in those with seizures or other neurological signs. Magnetic resonance imaging with the combined use of diffusion-weighted imaging and susceptibility-weighted imaging optimizes diagnostic assessment (Salamon et al. 2016).
- In infants in whom there are no neurological features and no seizures, careful review should continue throughout childhood years and the parents can be reassured.

Management

- For those with recurrent seizures, anticonvulsant medication is indicated. Phenobarbitone often gives poor seizure control in infantile IP (Salamon et al. 2016).

Dental

Over 40% of females with IP have dental abnormalities (Figure 32.2, bottom right). Congenital absence of at least one tooth is common (Santa-Maria et al. 2017). Deciduous or permanent dentition or both may be affected. Deciduous teeth may be retained into adult life. Abnormalities of crown formation can also occur and manifest as conical, tulip or notch shapes. Micodontia is also common.

Evaluation

- Because dental abnormalities are so frequent, regular dental checks should be part of the ongoing care of affected individuals. Dental visits should be twice yearly after the age of 1 year.
- Dental features can be of diagnostic value in first-degree female relatives and can support the diagnosis in those with skin signs that are very mild or atypical.

Management

- Educate affected individuals and/or parents about oral hygiene to maintain and preserve teeth.
- Orthodontic treatment with braces, surgical removal, crowns, and prostheses may be necessary in affected individuals. Indications for these interventions are standard.

Breast

Although breast anomalies have been rarely reported in IP, in the author's series (Landy and Donnai 1993), breast anomalies occurred in 10% of the affected females. Most of these were supernumerary or hypoplastic nipples (Figure 32.2, bottom left), but one woman had unilateral breast and nipple aplasia.

Evaluation

- The thorax should be examined to determine the presence of supernumerary nipples and other breast abnormalities.

Management

- Most individuals with supernumerary nipples have no major problems.
- The development of a supernumerary breast at puberty may necessitate surgical removal.
- In women with breast aplasia or hypoplasia, surgical reconstruction may be indicated, and is standard.

Miscellaneous

Recurrent infections are problematic in a minority of individuals with IP. Pulmonary hypertension is a rare complication of IP. Anomalies of the palate including a high palate, cleft lip and cleft palate have been reported to occur in association with IP (Minic et al. 2013). There are anecdotal reports of skeletal abnormalities, limb asymmetry and talipes in IP. However, in the author's series, all structural abnormalities observed were associated with severe neurological deficit and included contractures, dislocations, and scoliosis. It is likely that the other malformations reported in previous cases have occurred by chance or that the affected individuals have been misdiagnosed as having IP. It should be noted that structural malformations including clefting, polydactyly, and syndactyly often have been associated with pigmentary mosaicism (hypomelanosis of Ito).

RESOURCES

USA

Incontinentia Pigmenti International Foundation

30 East 72nd St.
New York, NY 10021, USA
Phone: (212) 452 1231
Fax: (212) 452 1231
Email: ipif@ipif.org
Website: http://www.ipif.org
This website details suggested surveillance protocols.

UK

The Ectodermal Dysplasia Society

Unit 1 Maida Vale Business Centre
Leckhampton
Cheltenham
Gloucestershire GL53 7ER, UK
Tel: +44 (0) 1242 261332
Email: diana@ectodermaldysplasia.org
This organization has a section dealing with IP

REFERENCES

Aradhya S, Courtois G, Rajkovic A, Lewis AL, Levy M, Isrzël A, Nelson DL (2001) Atypical forms of incontinentia pigmenti in males result from mutations of a cytosine tract in exon 10 of *NEMO* (*IKK*). *Am J Hum Genet* 68:765–771.

Ardelean D, Pope E (2007) Incontinentia pigmenti in boys: A series and review of the literature. *Pediatr Dermatol* 23:523–527.

Bryant SA, Rutledge SL (2007) Abnormal white matter in a neurologically intact child with incontinentia pigmenti. *Pediatr Neurol* 36:199–201.

Carney RG (1976) Incontinentia pigmenti: A world statistical analysis. *Arch Dermatol* 112:535–542.

Chen CJ, Han IC, Tian J, Munoz B, Goldberg, MF (2015) Extended follow-up of treated and untreated retinopathy in incontinentia pigmenti. JAMA Ophthalmol 133(5):542-548.

Curth HO, Warburton D (1965) The genetics of incontinentia pigmenti. *Arch Dermatol* 92:229–235.

Derry JMJ, Gormally E, Means GD, Zhao W, Meindl A, Kelley RI, Boyd Y, Herman GE (1999) Mutations in a delta(8)-delta(7) sterol isomerase in the tattered mouse and X-linked dominant chondrodysplasia punctata. *Nat Genet* 22:286–290.

Fusco F, Fimiani G, Tadini G, Michele D, Ursini MV (2007) Clinical diagnosis of incontinentia pigmenti in a cohort of male patients. *J Am Acad Dermatol* 56:264–267.

Fusco F, Paciolla M, Conte, MI, Pescatore A, Esposito E, Mirabelli P, Lioi MB, Ursini MV (2014) Incontinentia pigmenti: report on data from 2000 to 2013. *Orphanet J Rare Dis* 9:93.

Fusco F, Pescatore A, Conte MI, Mirabelli P, Paciolla M, Esposito E, Lioi MB, Ursini MV (2015) EDA-ID and IP, two faces of the same coin: How the same IKBKG/NEMO mutation affecting the NF-κB pathway can cause immunodeficientcy and/or inflammation. *Int rev immunol* 34:445–459.

Garrod AE (1906) Peculiar pigmentation of the skin of an infant. *Trans Clin Soc Lond* 39:216.

Goldberg MF, Custis PH (1993) Retinal and other manifestations of incontinentia pigmenti (Bloch-Sulzberger syndrome). *Ophthalmology* 100:1645–1654.

Grzeschik KH, Bornholdt D, Oeffner F, König A, del Carmen Boente M, Enders H, Fritz B, Hertl M, Grasshoff U, Höfling K, Oji V, Paradisi M, Schuchardt C, Szalai Z, Tadini G, Traupe H, Happle R (2007) Deficiency of PORCN, a regulator of Wnt signaling, is associated with focal dermal hypoplasia. *Nat Genet* 39:833–835.

Happle R (1993) Mosaicism in human skin. Understanding the patterns and mechanisms. *Arch Dermatol* 129:1460–1470.

Hatchwell E, Robinson D, Crolla JA, Cockwell AE (1996) X-inactivation analysis in a female with hypomelanosis of Ito associated with a balanced X;17 translocation: Evidence for functional disomy of Xp. *J Med Genet* 33:216–220.

Jackson R (1976) The lines of Blaschko: A review and reconsideration. *Br J Dermatol* 95:349–360.

Jouet M, Stewart H, Landy S, Yates J, Yong SL, Harris A, Garrett C, Hatchwell E, Read A, Donnai D, Kenwrick S (1997) Linkage analysis in 16 families with incontinentia pigmenti. *Eur J Hum Genet* 5:168–170.

Kenwrick S, Woffendin H, Jakins T, Shuttleworth SG, Mayer E, Greenhalgh L, Whittaker J, Rugolotto S, Bardaro T, Esposito T, D'Urso M, Soli F, Turco A, Smahi FA, Hamel-Teillac D, Lyonnet S, Bonnefont JP, Munnich A, Kashork CD, Shaffer LD, Nelson DL, Levy M, Lewis RA (2001) Survival of male patients with incontinentia pigmenti carrying a lethal mutation can be explained by somatic mosaicism or Klinefelter syndrome. *Am J Hum Genet* 69:1210–1217.

Landy SJ, Donnai D (1993) Incontinentia pigmenti (Bloch-Sulz-berger Syndrome). *J Med Genet* 30:53–59.

Loh NR, Jadresic LP, Whitelaw A (2008) A genetic cause for neonatal encephalopathy: Incontinentia with NEMO mutation. *Acta Paediatr* 97:379–381.

Maingay-de Groof F, Lequin MH, Roofhooft DW, Oranje AP, de Coo IF, Bok LA, van der Spek PJ, Mancini GM, Govaert PP (2008) Extensive cerebral infarction in the newborn due to incontinentia pigmenti. *Eur J Paediatr Neurol* 12:284–289.

Mansour S, Woffendin H, Mitton S, Jeffery I, Jakins T, Kenwrick S, Murday VA (2001) Incontinentia pigmenti in a surviving male is accompanied by hypohidrotic ectodermal dysplasia and recurrent infection. *Am J Med Genet* 99:172–177.

Minic S, Trpinac D, Obradovic M (2013) Systematic review of central nervous system anomalies in incontinentia pigmenti. *Orphanet J Rare Dis* 8:25

Minic S, Trpinac D, Obradovic M (2014) Incontinentia pigmenti diagnostic criteria update. *Clin Genet* 85:536-542.

Moss C, Ince P (1987) Anhydrotic and achromians lesions in incontinentia pigmenti. *Br J Dermatol* 116:839–850.

Nelson DL (2006) NEMO, NFkappaB signaling and incontinentia pigmenti. *Curr Opin Genet Dev* 16:282–288.

Nenci A, Huth Funteh A, Schmidt-Supprian M, Bloch W, Metzger D, Chambon P, Rajewsky K, Krieg T, Haase I, Pasparakis M (2006) Skin lesion development in a mouse model of incontinentia pigmenti is triggered by NEMO deficiency in epidermal keratinocytes and requires TNF signaling. *Hum Mol Genet* 15:531–542.

Pacheco TR, Levy M, Collyer JC, de Parra NP, Parra CA, Garray M, Aprea G, Moreno S, Macini AJ, Paller AS (2006) Incontinentia pigmenti in male patients. *J Am Acad Dermatol* 55:251–255.

Parrish JE, Scheuerle AE, Lewis RA, Levy ML, Nelson DL (1996) Selection against mutant alleles in blood leukocytes is a consistent feature in incontinentia pigmenti type 2. *Hum Mol Genet* 5:1777–1783.

Rashidghamat E, Hsu CK, Nanda A, Liu L, Al-Ajmi H, McGrath JA (2016) **Incontinentia pigmenti in a father and daughter** *Br J Dermatol* 175:1059–1060.

Salamon AS, Lichtenbelt K, Cowan F, Casaer A, Dudink J, Dereymaeker A, Paro-Panjan D, Groenendaal, De Vries LS (2016) Clinical presentation and spectrum of neuroimaging findings in newborn infants with incontinentia pigmenti. *Dev Med Child Neurol* 58:1076–1084.

Santa-Maria FD, Mariath LM, Poziomczyk CS, Maahs MAP, Rosa RFM, Zen PRG, Schuller-Faccini L, Kiszewski AE (2017) Dental anomalies in 4 patients with IP: clinical and radiological analysis and review. *Clin Oral Invest* 21:1845–1852.

Scheuerle AE (1998) Male cases of incontinentia pigmenti: Case report and review. *Am J Med Genet* 77:201–218.

Temple IK, MacDowall P, Baraitser M, Atherton DJ (1990) Focal dermal hypoplasia (Goltz syndrome). *J Med Genet* 27:180–187.

The International Incontinentia Pigmenti (IP) Consortium (2000) Genomic rearrangement in *NEMO* impairs NF-κB activation and is a cause of incontinentia pigmenti. *Nature* 405:466–472.

Wald KJ, Mehta MC, Katsumi O, Sabates NR, Hirose T (1993) Retinal detachments in incontinentia pigmenti. *Arch Ophthalmol* 111:614–617.

Woffendin H, Jouet M, Landy S, Donnai D, Read A, Kenwrick S (1997) Use of an X-inactivation assay to investigate the affected status of females in incontinentia pigmenti families. *J Med Genet* 34:521.

Wolf NI, Kramer N, Harting I, Seitz A, Ebinger F, Poschl J, Rating D (2005) Diffuse cortical necrosis in a neonate with incontinentia pigmenti and an encephalitis-like presentation. *Am J Neuroradiol* 26:1580–1582.

Zonana J, Elder ME, Schneider LC, Orlow SJ, Moss C, Golabi M, Shapira SK, Farndon PA, Wara DW, Emmal SA, Ferguson B (2000) A novel X-linked disorder of immune deficiency and hypohidrotic ectodermal dysplasia is allelic to incontinentia pigmenti and due to mutations in *IKK-gamma (NEMO)*. *Am J Hum Genet* 67:1555–1562.

33

INVERTED DUPLICATED CHROMOSOME 15 SYNDROME (ISODICENTRIC 15)

AGATINO BATTAGLIA

IRCCS Stella Maris Foundation, Pisa, Italy; Division of Medical Genetics, Department of Pediatrics University of Utah School of Medicine, Salt Lake City, Utah, USA; and Division of Medical Genetics, Department of Pediatrics, Sanford School of Medicine, University of South Dakota, Sioux Falls, South Dakota, USA

INTRODUCTION

The detection of a supernumerary additional marker chromosome is not an uncommon occurrence. One of the most common chromosome markers is the inverted duplicated 15 [(inv dup(15)] or isodicentric 15 [idic(15)], which represents about half of the small supernumerary chromosomes detected on routine karyotyping. The phenotype of individuals carrying such a marker varies considerably from apparently unaffected people to those with a severe neurodevelopmental disorder, depending both on the cytogenetic structure and on the parental origin of the marker chromosome. The maternally derived, large idic(15), formed by the inverted duplication of proximal chromosome 15, which contains the Prader–Willi/Angelman syndrome critical region (PWS/ASCR), results in tetrasomy 15p and partial tetrasomy 15q. This condition is associated with an abnormal phenotype that characterizes the idic(15) syndrome (Battaglia et al. 1997). Penetrance is 100%, and expressivity is variable.

Life span does not appear to be dramatically shortened. However, recently there has been the sudden, unexpected, and as yet unexplained death, almost always during sleep, of a few healthy young individuals with idic(15) syndrome and epilepsy (Devinsky 2011).

Inv dup(15) or idic(15) syndrome is reportedly characterized by a distinct neurobehavioral phenotype including central hypotonia, moderate to profound developmental delay/intellectual disability, absent or very poor speech, epilepsy, and autism spectrum disorder (ASD) (Webb et al.1998; Crolla et al. 1995; Robinson et al. 1993; Battaglia et al. 1997; Gillberg et al. 1991; Battaglia 2005; Dennis et al. 2006; Kleefstra et al. 2010; Battaglia et al. 2010, 2016).

Incidence

About 1/30,000 live newborns is diagnosed with idic(15), with an apparent male predilection of 2:1 (Al Ageeli et al. 2014; Conant et al. 2014; Battaglia et al. 2016). However, a higher incidence is plausible due to underascertainment consequent to the difficulty in making the clinical diagnosis because of the absence of malformations or overt craniofacial dysmorphisms in almost all individuals (personal experience). Babies with idic(15) are born to parents of every socioeconomic, racial, and ethnic background.

Diagnostic Criteria

There are no consensus diagnostic criteria. However, idic(15) syndrome should be suspected in any infant, child, or adolescent with moderate to severe early central hypotonia, developmental delay and intellectual disability, absent or very poor speech, epilepsy, and autism spectrum disorder. The combination of epilepsy (usually of severe degree) and the "typical" behavior pattern renders a more distinctive

---

Cassidy and Allanson's Management of Genetic Syndromes, Fourth Edition.
Edited by John C. Carey, Agatino Battaglia, David Viskochil, and Suzanne B. Cassidy.
© 2021 John Wiley & Sons, Inc. Published 2021 by John Wiley & Sons, Inc.

clinical picture that begins to distinguish idic(15) syndrome from other disorders involving severe neurodevelopmental disability. Additional features include minor anomalies such as epicanthal folds, downslanted palpebral fissures, upturned and/or broad nose, low-set ears, high palate, fifth finger clinodactyly, and partial second to third toe syndactyly (Robinson et al. 1993; Battaglia et al. 1997; Wolpert et al. 2000, Schinzel and Niedrist 2001; Orrico et al. 2009; Hogart et al. 2010). Brachycephaly, frontal bossing, synophrys, short philtrum, cleft palate, prominent mandible in adults, and areas of increased and reduced skin pigmentation can be occasionally observed (Crolla et al. 1995; Battaglia et al 1997; Dennis et al. 2006; Battaglia 2008) (Figures 33.1 and 33.2).

Etiology, Pathogenesis, and Genetics

The disorder is the result of at least one extra maternally derived copy of the Prader–Willi/Angelman syndrome critical region (PWS/AS critical region), a region of approximately 5 Mb within chromosome 15q11.2-q13.1. Due to an enrichment of low copy repeat elements clustering into breakpoints (BPs) along the chromosome, this genomic region is one of the least stable regions in the human genome and thus is remarkably vulnerable to structural variation mediated by non-allelic homologous recombination (Christian et al. 1999; Makoff and Flomen 2007). The PWS/AS critical region is imprinted, and several rearrangements may occur in this segment of a 5 Mb unit, defined by proximal breakpoints BP1 or BP2 and distal breakpoint BP3 (Roberts et al. 2002): deletions associated either with Angelman syndrome (AS, see Chapter 5) or with Prader–Willi syndrome (PWS, see Chapter 46), depending on parental origin (Lalande 1996); translocations; inversions; and supernumerary marker chromosomes formed by the inverted duplication of proximal chromosome 15. Interstitial duplications, triplications and balanced reciprocal translocations are much less frequent (Browne et al. 1997). The inv dup(15) or idic(15) is the most common of the heterogeneous group of the extra structurally abnormal chromosomes. Most extra structurally abnormal chromosomes (15) are bisatellited and dicentric, containing varying amounts of 15q material between the two centromeres. Two cytogenetic types of idic(15) marker chromosomes that have different phenotypic consequences have been identified (Leana-Cox et al. 1994; Crolla et al. 1995; Huang et al. 1997). One is a metacentric or submetacentric and heterochromatic chromosome, smaller or similar to a G group chromosome, not containing the PWS/AS critical region; the cytogenetic description is dic(15)(q11). Most individuals with this aberration have a normal phenotype, although exceptions have been reported (Eggermann et al. 2002).

The second type of idic(15) is as large as, or larger than, a G group chromosome and has 15q euchromatin. It includes the PWS/AS critical region (Robinson et al. 1993; Blennow et al. 1995), and the cytogenetic description is dic(15)(q12 or q13). The vast majority of dic(15)(q12 or q13) derives from the two homologous maternal chromosomes at meiosis, and was earlier thought to be associated with increased mean maternal age at conception, similar to other trisomies. This implies that a duplication of the paternal PWS/AS critical region is either a rare event, or goes undetected due to the absence of phenotypic expression, as suggested by familial cases where seemingly unaffected mothers have transmitted paternally derived duplication chromosomes to their children (Cook et al. 1998; Gurrieri et al. 1999; Roberts et al. 2002). Rare instances of paternally derived interstitial triplications support the hypothesis that paternal trisomy for the PWS/AS critical region is not lethal (Cassidy et al. 1996; Browne et al. 1997; Mao et al. 2000; Ungaro et al. 2001). The presence of large idic(15) results in tetrasomy 15p and partial tetrasomy 15q. However, considerable structure heterogeneity has been reported (Wang et al. 2008), with duplications varying in size up to 12 Mb long (personal observation). Five recurrent breakpoints (BP) have been described in most individuals, and the majority of idic(15) chromosomes arise through BP3:BP3 [class 3B;

FIGURE 33.1 A 1.5-month-old male with idic(15) syndrome.

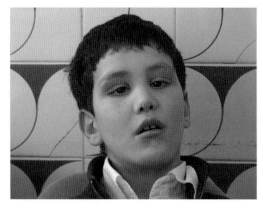

FIGURE 33.2 A 10-year-old male with idic(15) syndrome.

symmetrical idic(15) chromosomes] or BP4:BP5 [class 5A; asymmetrical idic(15) chromosomes] recombination events. The ~5 Mb segment that encompasses the PWS/AS critical region lies between BP2 and BP3. The predominant form of idic(15) is characterized by an asymmetric recombination event between BP4 and BP5, which leads to tetrasomy for the interval from the centromere to BP4 and to trisomy from BP4 to BP5 (Wang et al. 2004; personal observation).

The large idic(15) is nearly always sporadic, and is associated with an abnormal phenotype, which constitutes the idic(15) syndrome (Flejter et al. 1996; Battaglia et al. 1997). Maternally derived cytogenetic mosaicism with a normal cell line has been described in a small subset of individuals with a milder phenotype (Dennis et al. 2006); and an individual with a mosaic paternally derived idic(15), showing a mild Prader–Willi syndrome phenotype, has also been observed (Saitoh et al. 2007). Rare atypical forms of idic(15) chromosomes (supernumerary ring chromosomes including the PWS/AS critical region) have also been reported (Dennis et al. 2006; Wang et al. 2008). A few cases of hexasomy 15q11.2-q13 have been identified, showing a more severe phenotype (Huang and Bartley 2003; Nietzel et al. 2003; Qumsiyeh et al. 2003; Mann et al. 2004).

Mean maternal age at the birth of a baby with idic(15) was originally reported to be significantly increased (Connor and Gilmore 1984), but in a small series. These data are in contrast with what is generally reported in the literature and with the author's personal experience. Further studies looking at the parental ages at the birth of larger samples of babies with idic(15) are needed.

Several genes of interest (*ATP10A, CYFIP1, MAGEL2, NECDIN, SNRPN, UBE3A*, snoRNAs, and a cluster of genes encoding GABA$_A$ receptor subunits) lie within the 5–12 Mb duplication. However, genotype–phenotype studies have not been able to show any correlation between type and size of the idic(15) and the degree of severity of the clinical spectrum to date.

Various genetic mechanisms have been hypothesized to explain clinical heterogeneity, including the size of the chromosomal duplication, dosage effect of genes in this region, and the imprinting mechanism. The fact that tetrasomy of the PWS/AS critical region is associated with a more severe phenotype than the one observed in trisomy, suggests that there is a dosage effect for a gene or genes in this region (Schinzel et al. 1994; Dennis et al. 2006). Furthermore, gene expression in this critical region is regulated by an imprinting mechanism (Dittrich et al. 1996). Since only maternally inherited aberrations of chromosome 15q11-13 seem to be pathogenic, with the sole exception of one patient (Mohandas et al. 1999), it is likely that maternal genes, contained in this genomic region, act in a dosage-dependent manner and that their copy number is critical for normal brain development and function. *UBE3A*, implicated in Angelman syndrome (see Chapter 5), is imprinted with maternal-specific expression in postnatal neurons, and is thus expressed at a higher dosage in brain from individuals with a maternally derived duplication. It thus probably contributes to the intellectual disability and autism spectrum disorder (Glessner et al. 2009; Greer et al. 2010).

Amongst the genes that are known to be located at or near the PWS/AS critical region, an interesting relationship has been shown between the α5 and *β3* GABA receptor subunit genes and the *P* gene. In fact, even though these genes are not imprinted, a deletion of both P alleles, in the mouse, causes rearrangements of α5 and β3 receptors producing a particular phenotype characterized by seizures, jerky gait, and ataxia (Nakatsu et al. 1993). Tetrasomy of these genes, as seen in idic(15) syndrome, may alter the GABA receptor activity, upon which the major central nervous system inhibitory mechanisms rely. This alteration could represent the biological basis for some clinical manifestations of the idic(15) syndrome, such as seizures, hyperactivity, aggressiveness, and autistic disorder. Of interest, linkage disequilibrium between a marker in the γ-aminobutyric acid β-receptor subunit gene, *GABRB3* 155CA-2, and autistic disorder has been reported by Cook et al (1998). Expression of *GABRB3* is reduced in brain tissue samples of individuals with autism spectrum disorder, and normal biallelic *Gabr* expression was disrupted in autistic individuals with protein expression deficit (Samaco et al. 2005; Hogart et al. 2007). *GABRB3* is the only gene in 15q11.2-q13.1 that has been implicated in autism spectrum disorder by genome-wide de novo single-nucleotide variant studies (Sanders et al. 2015). In addition, single-nucleotide polymorphisms in *GABRG3* have been associated with autism (Menold et al. 2001). *Gabrb3* null mice exibit a neurological phenotype characterized by seizures and behaviors consistent with autism spectrum disorder (DeLorey et al. 2008). Of the six known genes within the BP4-BP5 region, the *CHRNA7* gene, encoding for a synaptic ion channel protein mediating neural signal transmission [α7 nicotinic acetylcholine receptor (nAChR)], is being viewed as an interesting candidate gene for 15q13.3 copy number variant phenotypes, and its duplication has been proposed to result in increased gene expression (Gillentine et al. 2017). α7nAChR agonists and positive allosteric modulators have been utilized in autism spectrum disorder, with improvements in social behavior (Freedman 2014).

Although some genetic association data suggest that *ATP10A* may contribute to autism spectrum disorder (Nurmi et al. 2003), a causative role for this gene in the pathogenesis of idic(15) symptoms has still to be proven. Recently, a disruption in transcriptional regulation has been suggested as a driving factor in the autism spectrum disorder in idic(15) (Urraca et al. 2018). An additional gene, located more distally, the *SLC12A6* (solute carrier, family 12, member 6), coding for a cation chloride cotransporter and expressed in the brain, heart, skeletal muscle, and kidney, could potentially be implicated in the pathogenesis of seizures (Caron et al. 2000).

Diagnostic Testing

Chromosome microarray analysis is currently commonly performed in the field of diagnostic clinical genetics, and it is now the first-tier test for individuals with developmental delay/intellectual disability and autism spectrum disorder. Chromosome microarray analysis using oligonucleotide arrays or single nucleotide polymorphism (SNP) arrays has been shown to be useful for the detection of both the increases in copy number and extent of the 15q11.2-q13.1 region, and atypical forms of idic(15) (Wang et al. 2004, 2008).

However, chromosome microarray analysis cannot reliably differentiate between idic(15) and interstitial triplication of 15q11.2-q13.1. Standard cytogenetics or FISH analysis, using probes both from proximal chromosome 15 and from the PWS/AS critical region (Webb et al. 1998) is needed to determine whether the duplication is supernumerary or interstitial and to determine whether there is evidence for mosaicism. Molecular studies, such as microsatellite analysis on parental DNA or methylation analysis, including PCR-based methylation analysis and methylation-specific multiplex ligation-dependent probe amplification (MS-MLPA) (Zielinski et al. 1988; Urraca et al. 2010) on the proband DNA, are also needed in order to detect the parent-of-origin of the idic(15) chromosome (Luke et al. 1994; Webb et al. 1998).

Idic(15) has been de novo in all individuals reported to date; therefore, parental testing is not routinely indicated. However, based on the report by Michelson et al (2011) of maternal transmission of supernumerary partial trisomy of 15q11.2-q13.1, clinical manifestations in a mother (psychosis, autism spectrum disorder, epilepsy) should prompt consideration of parental testing. As the idic(15) is de novo, the risk to sibs appears to be low to slightly greater than in the general population, due to the possibility of maternal germline mosaicism.

People with idic(15) are not known to reproduce.

Differential Diagnosis

Infants with idic(15) syndrome commonly present without overt craniofacial dysmorphisms or major malformations, and with non-specific psychomotor delay and/or seizures; thus chromosomal aberrations are usually not suspected. As a consequence, the differential diagnosis is often broad and non-specific (personal observation), encompassing such entities as static encephalopathy and pervasive developmental disorder.

The presence of early central hypotonia and seizures may raise the possibility of an inborn error of metabolism or a defect in oxidative phosphorylation such as a mitochondrial encephalomyopathy. Subsequent testing for these abnormalities normally includes urine organic acids, serum amino acids, plasma acylcarnitine profiles, and mitochondrial enzyme and DNA mutation screens.

Seizures and severe speech impairment in infants with idic(15) syndrome can resemble those seen in the Rett syndrome (see Chapter 49), but children with idic(15) syndrome do not lose purposeful use of their hands. The distinction between these two syndromes is usually resolved by age 1–4 years, when children with Rett syndrome show a rapid and specific regression of acquired abilities, whereas those with idic(15) syndrome are slowly progressing developmentally.

Severe hypotonia, moderate–severe psychomotor impairment, and early epilepsy can recall those seen in children with *CDKL5* mutations (Bahi-Buisson et al. 2008). However, key features identifying children with *CDKL5* mutations are early onset seizures (at ages 1–10 weeks) associated with normal interictal EEG and deceleration of head growth, contrary to what is observed in idic(15).

The severe degree of hypotonia may lead to genetic evaluation for Prader–Willi syndrome (see Chapter 46), which would show the idic(15) chromosome.

Infantile hypotonia, absent or very poor speech, and lack of overt congenital anomalies may recall Angelman syndrome (see Chapter 5), but children with idic(15) do not have ataxia; the jerky movements, with associated tremulousness; the unusually happy or excitable disposition; the distinctive craniofacial gestalt (microbrachycephaly with flat occiput, midface retrusion, prognathia, wide mouth with protruding tongue), and the characteristic posture of the arms (usually upheld with flexion at the wrists and elbows).

The occurrence of reduced skin pigmentation with a streaky distribution, associated with hypotonia, developmental delay, and seizures can suggest "hypomelanosis of Ito". One of the reported individuals with idic(15) by Battaglia et al (1997) had been earlier diagnosed elsewhere with this condition. It is now clear that hypomelanosis of Ito is not a specific disorder, but rather an etiologically heterogeneous physical finding that is frequently indicative of chromosomal or genetic mosaicism.

Interstitial duplications of the PWS/AS critical region in the maternally inherited chromosome 15 cause a similar, but milder, phenotype compared to the idic(15). Some such individuals may have very mild features or may have just a pervasive developmental disorder with no minor or major anomalies and normal neurological examination. In contrast, occurrence of double supernumerary idic(15), resulting in partial hexasomy of the maternally inherited PWS/AS critical region, causes a more severe phenotype compared to the idic(15), including a number of minor and major anomalies (Qumsiyeh et al. 2003).

Intellectual disability, autism spectrum disorder, epilepsy, speech disorder, and behavioral/psychiatric symptoms (attention problems, poor adaptive skills, mood disorders, and schizophrenia) have also been reported in individuals with no distinct dysmorphic features, in whom a deletion 15q13.3 (BP4–BP5) was found (Lowther et al. 2015; Zhou et al. 2016; Ziats et al. 2016). Similar symptoms, plus congenital heart defects, have been described in individuals with deletions of 15q11.2 (BP1–BP2) (Cox and Butler 2015, Vanlerberghe et al. 2015).

The idic(15) chromosome has also been reported in one female with the Sotos phenotype (Koyama et al. 1985; see Chapter 55).

MANIFESTATIONS AND MANAGEMENT

Growth and Feeding

Feeding difficulties have been observed in the newborn period in over 50% of babies with idic(15) (Dennis et al. 2006; personal experience). Newborns appear "floppy" and may have trouble with sucking and swallowing and/or latching onto the breast. They tend to feed slowly and breastfeeding may be hard to establish, and they are often very placid and sleepy, frequently needing to be awakened for feeds. Many babies have a decreased appetite and struggle to finish a feed. Overall, about 50% of mothers are able to establish successful breastfeeding, the other babies being bottle-fed expressed milk. A minority of babies can benefit from a temporary nasogastric tube, in order to allow time for them to mature enough to suck effectively, either from the breast or bottle. Individual babies may need gastrostomy tubes in order to meet their nutritional needs (Dennis 2006; Unique www.rarechromo.org). Gastroesophageal reflux, although transitory, can be a consistent finding in about 30% of infants, and may explain irritability and recurrent respiratory tract infections. Many infants and toddlers with idic(15) have trouble chewing, and can choke or gag on lumps of food, so may continue to eat puréed food for longer than their peers, and start eating finger food later than usual. Parents have found that modifying the texture of foods by grating, mincing, chopping or adding sauces to foods can help to overcome these problems. Children with a high palate (~30%) may also struggle with solid food, which can become trapped in the high palate. Feeding problems tend to improve over time, and a number of individuals have a good appetite, eating a good and varied diet (personal experience). On the other hand, it is worth noting that when there is associated autism or developmental delay, as in idic(15) syndrome, the children/adolescents may have sensory aversion and/or hypotonia that can affect their food choices and oral-motor function. Reduced height can be a common feature, since 70% to 90% of all individuals have a median or below median height (Kleefstra et al. 2010; personal experience).

Evaluation

- Early symptoms of poor suck and difficulty with attachment to the breast are probably secondary to poor tone and should be evaluated in a standard manner.

- Measurements and graphic of height, weight, and head circumference are recommended as part of routine medical care.
- The likelihood of feeding problems should be recognized and assessed properly.
- Referral to a dysphagia team for the evaluation of swallowing difficulties may be useful, as may be a radiographic swallowing study.
- The presence of gastroesophageal reflux as a potential factor in feeding problems should be assessed in a standard manner.
- If a child with idic(15) syndrome is not bottle feeding or has persistent failure-to-thrive, consideration of gastrostomy tube placement is indicated, to sustain weight gain, protect the airway, and manage gastroesophageal reflux.

Management

- Assurance of adequate caloric intake, if necessary, by increased caloric formulas and/or oral or nasogastric tube feeding, is indicated.
- Standard therapy for gastroesophageal reflux can be initiated when this is diagnosed. Nissen–Hill fundoplication has been successfully performed in individual children (Unique www.rarechromo.org).
- Gastrostomy has been successfully performed in a minority of children (Unique www.rarechromo.org).

Development, Behavior, and Psychiatric

Significant developmental delays in early childhood, including gross motor and fine delays, and, later, intellectual disability affect all individuals with idic(15) and are usually moderate to profound (Robinson et al. 1993; Crolla et al. 1995; Battaglia et al. 1997; Webb et al. 1998; Battaglia, 2005). Until a few years ago, age of acquisition of motor milestones was seldom reported in the medical literature. More recent studies of large numbers of individuals with this condition have allowed a better knowledge of the developmental stages. The median age for sitting without support ranges between 13 and 15 months (Kleefstra et al. 2010; personal experience). The median age for walking without support ranges between 23 and 30 months (Kleefstra et al. 2010; personal experience). None of the children achieved this within the normal age range (before age 18 months). Almost all children tend to be clumsy, insecure, and poorly coordinated, frequently stumbling and falling. They have poor spatial awareness, which results in trouble negotiating stairs and walking into objects in their path. Speech and language development are variably delayed in all children; with up to 40% lacking functional speech (Wolpert et al. 2000; Borgatti et al. 2001; Maggouta et al. 2003), and many of the others only able to pronounce dysillabic sounds or single words, or, in less than 15%, able to produce short simple sentences not always clear or easy for people outside the family to understand. Some children may have a significant oral motor apraxia (personal experience). The age range of first spoken words varies between 9 and 60 months, while the age range of short simple sentences varies between 35 and 90 months (Dennis et al. 2006; Kleefstra et al. 2010; Unique www.rarechromo.org; personal experience). Language is often echolalic with immediate and delayed echolalia and pronoun reversal. Comprehension is very limited, contextual and accompanied by the gesture. Toilet training is usually mastered between 2.5 and 14 years in about 20% of individuals (Unique www.rarechromo.org; Dup15q Alliance; personal experience).

Most children do not acquire social imitative play. Stereotyped movements, such as spinning themselves for long periods of time, repetitive hand-twisting, hand flapping, biting of index fingers, hand-clapping over plane surfaces, and head turning, can be observed in childhood and adolescence (Battaglia et al. 1997, 2010; Battaglia, 2008). These stereotypies impede manipulation and fine motor activities and disturb communicative interaction through distraction and agitation. They cause difficulty in concentrating for a long time. Intention to communicate is absent or very poor early in life. Some children experience periodic sudden violent screaming spells not amenable to any sort of consolation. At physical examination no specific organ pathology is present and thorough evaluation does not reveal any somatic abnormality. This is not an epileptic phenomenon, and can last for hours, leading to a great distress within the family context. Other individuals are abnormally prone to agitation and scream when they do not feel safe for whatever reason. The situation returns to normal when moments of rest and peace are given.

The first evidence of the distinctive behavior of idic(15) syndrome may become apparent very early in life. A few infants may smile at their own mother for a short time, but soon lose eye-to-eye contact. They do not develop appropriate social interaction, being withdrawn from early on, and do not show interest in their peers. Usually these individuals have gaze avoidance from very early on, shun body contact, and stare at people as though they are looking through them. They can be fascinated by certain sounds and/or melodies and/or music, by water, or by spinning or any glittering objects. They usually prefer being left alone, lying on their back just looking at their fingers and taking bizarre postures. Symbolic play is usually never acquired. Most children call for food by crying, have a non-functional use of objects, and/or show a primordial type of object exploration.

As they grow up and leave the passive stage of babyhood behind, they may display sudden and extreme changes, with outbursts of aggressiveness and destructive behavior, directed at themselves and others including hair pulling, hitting, biting and kicking. A small minority have been known to be self-destructive or self-harm. They are often easily frustrated and can be impulsive. A number of individuals with idic(15) tend to be hyperexcitable, hyperactive with poor concentration and short attention span, and are easily distracted, all of which can make learning more challenging. Aggressiveness, psychomotor agitation, stubbornness, obsessions, anxiety for unfamiliar situations, animals, and loud noises, and oppositional defiant behavior have been described in a number of others (Wisniewski et al. 1979; Schinzel, 1981; Battaglia et al. 1997, 2010; Kleefstra et al. 2010; personal experience). The latter behavior is usually triggered and worsened by a specific environmental context (within the family or at school), as clearly observed in some adolescents (Battaglia et al. 2010). Of note, the language deficits may lead to secondary behavior problems and difficulties in performing routine daily living skills, such as toilet training, feeding, or dressing. The low tolerance to frustration and the attention problems may also be related to their very poor expressive language skills.

Sleep disturbances have been variably reported in 30–70% of individuals with idic(15), consisting in difficulty in falling asleep, frequent awakenings during the night, and being sleepy during the day (Kleefstra et al. 2010; personal experience). Nocturnal seizures and certain epilepsy types (i.e. Lennox–Gastaut or Lennox–Gastaut-like syndrome) do disrupt the usual sleep cycle, contributing to stress, low frustration tolerance, emotional lability, and passiveness. Polytherapy for seizures can cause side effects, leading to additional stressors (Battaglia et al. 2010, 2016). Melatonin appears to improve total sleep time and sleep efficiency in most (personal experience).

The distinct behavior disorder shown by individuals with idic(15) syndrome has been widely described as autistic or autistic-like. Although interpretation of the association between autism or autistic-like behavior and idic(15) is made difficult by lack of detailed behavioral descriptions and standardized testing for autism in several reported people (Al Ageeli et al. 2014), the association of idic(15) and autism appears to be stronger than that explained by the risk for autism posed simply by coexisting intellectual disability and epilepsy. Compared to other chromosome aberrations known to cause autism, idic(15) confers the greatest risk (Malhotra and Sebat 2012; Moreno-De-Luca et al. 2013). Rineer et al (1998) reported a strong association between autistic features and idic(15) in 20 out of the 29 individuals studied. However, these persons were assessed by telephone interview conducted with a single family member, through the administration of the Gilliam Autism Rating Scale (GARS) (Gilliam 1995). The children and adolescents with idic(15) reported by Borgatti et al (2001) were said to meet the clinical criteria for the diagnosis of autistic disorder by DSM IV (American Psychiatric Association 1994). Assessment of autistic behavior in those individuals was performed with parents' interviews and videotape analysis, through the behavioral summarized evaluation (BSE) scale. Eighty-one percent of those with idic(15), studied by Hogart et al (2010), met strict criteria for autism, using the Autism Diagnostic Interview – Revised (ADI-R) (Lord et al. 1994) and the Autism Diagnostic Observation Schedule (ADOS) (Lord et al. 2002). As shown by the children and teens with idic(15), studied by Battaglia et al (2010) using the combined ADI-R, ADOS-G, and CARS (Childhood Autism Rating Scale), most of them meet the clinical criteria for the diagnosis of autistic disorder by DSM IV in early years.

Non-functional use of objects with a primordial type of exploration (sucking, licking and/or smelling) is a constant behavioral feature in people with idic(15) and is associated with selective and restricted interests (such as pushing small cars back and forth, observing the wheels too closely with the periphery of the eyes, looking at colorful patterns in a book or at fabric). Such clinical observation has been confirmed by the high score in the third ADOS-G and ADI-R area (concerning play, stereotyped behavior and restricted interests) in all individuals with idic(15), both with and without autism. In the clinical experience of Battaglia et al (2010) similar high scores in the third area are usually not found in persons with "idiopathic" autism. All children with idic(15) syndrome, studied by Di Stefano et al (2016), demonstrated ADOS scores in the autism spectrum disorder/ autism range but exhibited a distinctive developmental profile compared to a matched sample of children with non-syndromic autism spectrum disorder. The profile was characterized by significant delays in motor skills and adaptive function with relative strengths in isolated social communication skills that may relate to social interest.

In most individuals, a slow but global evolution of adaptive behavior, and social interaction, with improvement in communicative skills (mainly directed at their own needs) and verbal comprehension, can be appreciated over time (personal experience). By adolescence, some are able to help with simple household tasks (setting and clearing the breakfast/ dinner table, shutting windows and doors, cleaning up dust, putting dirty clothes in hampers, putting their own toys in order, throwing away trash), and acquire the main personal self-independence skills, concerning self-hygiene, self-feeding, and self-dressing. Most people with idic(15) love music, although they tend to show sound hypersensitivity.

While the outlook depends on a child's individual progress, it is clear that most people with idic(15) will continue to need support throughout their lives. Behavioral data related to adults are much more limited. The degree of independence in adulthood largely depends on level of intellectual functioning, although behavioral issues and physical abilities also play a role. In spite of a global improvement over time, many adults with idic(15) have severe intellectual disability with no or very poor social interaction, and are unable to manage full self-care. Most adults live at home with their parents, while a minority lives in a sheltered environment (such as a supported home, a community school, an institution, or a supervised hostel).

The degree of cognitive impairment, the evolution of behavior over time, and the high score in the third ADOS-G and ADI-R area, suggest a distinct developmental profile in people with idic(15), that may provide a behavioral signature for autism spectrum disorder/autism spectrum disorder-like arising from the susceptibility locus on proximal 15q. As proposed by Battaglia et al (2010), it seems that individuals with idic(15) are not "true autistic", but have a distinct "autistic-like" disorder with high score in the third ADOS-G and ADI-R area.

Evaluation

- A thorough developmental evaluation by a specialist should be performed at intervals appropriate for planning of early interventions, including physical, occupational, behavioral and speech therapies, and for educational and vocational planning.

Management

- Involvement in early intervention programs is recommended in the ongoing care of infants and children with idic(15) syndrome.
- Enrollment in an individualized rehabilitation program that covers motor aspects (including oral motor therapy), cognition, communication, and socialization, is highly recommended.
- Speech and communication therapy is essential and should focus on non-verbal methods of communication. Alternative and augmentative communication aids, such as picture cards or communication boards, should be used at the earliest appropriate time.
- Occupational therapy may help improve fine motor and oral-motor control.
- Music therapy may help favor the emotional tuning and the integration of the communicative channels.
- Appropriate school placement, after full evaluation, is important. The classroom setting should be structured, in its physical design and its curricular program, so that the child with idic(15) syndrome can fit in or adjust to the school environment. Individualization and flexibility are essential factors.
- The use of computer-based approaches, both at school and at home, may help to overcome the poor dexterity, and facilitate communication of needs.
- A structured environment, consistent behavioral modification, and often one-on-one personal interaction may be required in the school or home to deal with the behavior disorder and provide viable developmental training.

- Consistent behavior modification in the school and at home can enable the child with idic(15) syndrome to be toilet trained (schedule trained) and to perform most self-help skills related to eating, dressing, and performing general activities in the home.
- Planning for transition to adulthood, including vocational training and living situation, should begin in adolescence.
- Stressors in the social, school and family life should be searched for, in order to control the oppositional defiant behavior.
- Planning of days with a clear structure should be routine, in order to give the child/adolescent the perception of familiarity and autonomy.
- Melatonin, in standard doses, is the most commonly used medication to treat sleep disorders in idic(15) syndrome.

Neurologic

Neonatal hypotonia is a hallmark of idic(15) syndrome, occurring in over 85% of individuals. This central hypotonia appears to be of prenatal onset, persists throughout life, and causes decreased fetal movements in about 20%. Newborns appear "floppy" and may have trouble with sucking and swallowing, and/or latching onto the breast. The hypotonia can also contribute to gastro-esophageal reflux and constipation in about 30% of babies; however, the infants outgrow these symptoms (see Growth and Feeding). Infants may have a weak cry and very poor head control. Decreased muscle tone and generalized muscle weakness lead to delays in motor skills. Some children may have a significant oral motor apraxia (personal experience). Poor tone often persists as joint hypermobility and poor reflexes. Children tend to be clumsy, insecure, and poorly coordinated, frequently stumbling and falling (personal experience) (see Development, Behavior, and Psychiatric).

Besides developmental delay/intellectual disability and autism spectrum disorder, the other main clinical challenge of infancy and childhood is a seizure disorder, which occurs in 60–80% of infants and children with idic(15) syndrome (Battaglia et al. 1997; Conant et al. 2014; Battaglia et al. 2016). The author's early paper (Battaglia et al. 1997) reporting on four individuals with idic(15) documented the severity of epilepsy in this disorder. There were multiple seizure types, with onset between 4 and 8 years of age, mainly represented by atypical absences, tonic, atonic, and tonic-clonic seizures, depicting the Lennox–Gastaut or Lennox–Gastaut-like epileptic syndrome, resulting in disabling secondary effects, including falls or developmental regression. Status epilepticus was reported in 25%. The electronic, questionnaire-based, survey conducted by Conant et al. (2014) documented seizures in 63% of affected individuals with idic(15). Eighty-one percent of these had multiple seizure types, and 42% had infantile spasms. Common seizure types were tonic-clonic, atonic, myoclonic, focal, tonic, and absence (consistent with a Lennox–Gastaut syndrome phenotype). Status epilepticus was reported in 33%. Developmental regression was reported in 63%, variably attributed by the responders to frequent seizures or medications.

Eight percent of those with seizures had died, due to either refractory status epilepticus or to sudden unexpected death in epilepsy. These deaths almost always occur during sleep (Devinsky 2011; Wegiel et al. 2012). From parental reports, epilepsy was apparently well controlled in only 24% of people.

The most recent author's study of 35 individuals with idic(15) documented seizures in 80%, with a median age of onset of 3 years 3 months (Battaglia et al. 2016). The first seizure types were infantile spasm associated with an hypsarrhythmic electroencephalogram (EEG) (West syndrome) in nine infants, focal/generalized tonic in seven children, or atypical absences in eight children. Four infants with infantile spasms and eight children with atypical absences developed subsequent seizure types consistent with a Lennox–Gastaut syndrome or Lennox–Gastaut-like syndrome. In one child the initial focal tonic seizure (at age 9 years 9 months) was soon followed by several daily episodes of rather bizarre paroxysmal phenomena, with a probable frontal origin. Seizures were triggered by drowsiness, slow wave sleep, and, on one occasion, by a sonorous stimulus, and could occur in clusters and last over 5 min.

Epilepsy was well controlled in up to 35% of the individuals; in whom seizures stopped at a median age of 5 years 5 months. In the remaining individuals, epilepsy was either poorly controlled (in 29%), or uncontrolled (in 36%). Overall, infantile spasms associated with an hypsarrhythmic EEG (West syndrome), and Lennox–Gastaut syndrome or Lennox–Gastaut-like syndrome appear to be the prevalent seizure types in idic(15) (Bingham et al. 1996; Battaglia et al. 1997, 2016; Webb et al. 1998; Conant et al. 2014). Other, less frequent seizure types include drug-resistant myoclonic absence-like seizures induced by emotionally gratifying stimuli (kissing, viewing of pleasant or funny events) (Aguglia et al. 1999); adult-onset, absence seizures and occasional head drops and generalized tonic-clonic seizures with good outcome (Chifari et al. 2002); and benign epilepsy with centro-temporal spikes (Gobbi et al. 2002). On the whole, seizures tend to be difficult to control in most individuals (Gillberg et al. 1991; Robinson et al. 1993; Crolla et al. 1995; Mignon et al. 1996; Battaglia et al. 1997, 2016; Conant et al. 2014).

The electroencephalogram (EEG) is usually abnormal in idic(15), but there is no diagnostic pattern. A variety of EEG abnormalities have been observed in all individuals who have had serial studies, during both wakefulness and sleep (Battaglia et al. 1997, 2016; Valente et al. 2006). These abnormal findings include slow background activity; poverty of the usual rhythmic activities; hypsarrhythmia; an excess of high amplitude slow/sharp waves, and/or biphasic spikes-polyspikes over both frontal regions, which often occur in a sub-continuous fashion, and on occasions spread to the entire brain; fast ill-defined high amplitude spike/wave complexes, usually occurring in runs of variable duration, over both fronto-centro-temporal regions; frequent, generalized slow sharp element spike–wave complexes, mostly accompanied by atypical absences; bursts of diffuse fast recruiting rhythms during slow wave sleep, accompanied by tachypnea, and/or by upward rotation of the eyes, and/or tonic seizures; and a considerable excess of high amplitude fast activity at 12–20 Hz, over both fronto-centro-temporal areas, particularly observed during childhood and during wakefulness and sleep, independent of any medication. This latter finding seems to be an EEG hallmark of idic(15).

Microcephaly is present in about 20% of individuals, whereas macrocephaly occurs in less than 3%.

A variety of developmental abnormalities of the brain has been described in about 15–30% of individuals (Grammatico et al. 1994; Leana-Cox et al. 1994; Borgatti et al. 2001; Battaglia et al. 2014; personal experience). In decreasing frequency, these include delayed myelination; ventriculomegaly; corpus callosum hypoplasia or rostral malformation; mild cortical atrophy; pachygyria and polymicrogyria; and microlissencephaly.

About 75% of people with idic(15) syndrome show apparent decreased perception of pain, often not noticing when they had been badly hurt (Unique www.rarechromo.org; personal experience).

Evaluation

- A waking/sleeping video-EEG-polygraphic recording, including extra leads to monitor respiration, electrocardiogram (EKG), pulse oximetry, electrooculogram (EOG), polyneurogram (PNG), and surface electromyogram (EMG), is recommended in infancy and childhood, in order to achieve the best characterization of seizures, and detect subtle seizures and associated clinical phenomena (e.g. periictal apnea, heart rate abnormalities, etc.) (Lacuey et al. 2018). An overnight video-EEG-polygraphic recording may be warranted, since the occurrence of nocturnal tonic seizures accompanied by alterations of the respiratory rate and subtle clinical features are a well-known phenomenon in this condition (Battaglia et al. 1997).
- Regular follow-up of seizure status and pharmacotherapy is essential.
- Neurological examination is recommended for all infants, children and adolescents.
- Neuroimaging should be performed in all individuals with refractory seizures, because malformation of the cerebral cortex may modify prognosis and management.

Management

- Infantile spasms associated with an hypsarrhythmic EEG (West syndrome) are best treated by adrenocorticotropic hormone (ACTH) (Conant et al. 2014; Battaglia et al. 2016).
- Other seizure types respond well to broad-spectrum antiepileptic drugs, such as valproic acid, lamotrigine, and rufinamide, which have proven to be the most effective treatment (Conant et al. 2014; Battaglia et al. 2016).
- Status epilepticus may be prevented with the use of rescue medications such as rectal diazepam or midazolam.
- Apart from status epilepticus, GABAergic medications, such as vigabatrin and the typical benzodiazepines, have shown a relative lack of efficacy. This may be attributable to abnormal GABAergic transmission resulting from the duplication of a cluster of GABAβ3 receptor genes in the 15q11.2-13.1 genomic region.
- Insufficient data are available to allow an adequate analysis of efficacy of vagus nerve stimulation (VNS), and/or ketogenic diet (Conant et al. 2014; personal experience). Of note, Ryvlin et al (2018) found that risk of sudden unexpected death in epilepsy significantly decreases during long-term follow-up of individuals with drug-resistant epilepsy receiving VNS therapy.
- The most accurate seizure management is essential in preventing secondary complications, including (in the most severe cases) brain damage, developmental regression, and sudden unexpected death in epilepsy (Devinsky 2011; Ryvlin et al. 2011).
- Prompt identification of a seizure and basic care (e.g., repositioning a person on the side instead of face down, loosening tight clothing around his/her neck) after a seizure may help prevent sudden unexpected death in epilepsy (Ryvlin et al 2013; personal experience).
- Avoid seizure triggers, such as all stressors, sleep deprivation, and non-compliance with antiepileptic drugs.
- The use of a helmet should be implemented in those individuals experiencing the "drop attack" type of seizures (e.g. tonic or atonic seizures causing a sudden drop attack).
- Abnormalities of muscle tone should be treated as in the general population with physical therapy.

Major Malformations

Major malformations do not seem to be component manifestations of the idic(15) syndrome. They have been reported in individual people, and consist of congenital heart malformations (patent foramen ovale, ventricular/atrial septal defects, tetralogy of Fallot, pulmonary valve dysplasia); urogenital malformations (unilateral renal agenesis, hypospadias, cryptorchidism); and umbilical and inguinal hernias (Robinson et al. 1993; Kleefstra et al. 2010; Battaglia and Filippi 2014).

Evaluation

- Physical examination with comprehensive review of systems

Management

- Management of cardiac, genital, and abdominal anomalies is the same as in any individual.

Respiratory

Recurrent ear, nose, and throat infections have been commonly observed in childhood (Dennis et al. 2006; Kleefstra et al. 2010; personal experience).

Evaluation

- Evaluations do not differ from those in other children with similar symptoms.

Management

- Treatment follows standard practice.

Musculoskeletal

Congenital hip dysplasia, kyphosis, and scoliosis, deformities of the feet, such as club feet, thin long fingers, proximally placed thumbs, and short extremities have been rarely observed (Blennow et al. 1995; Dennis et al. 2006; Kleefstra et al. 2010; Al Ageeli et al. 2014; personal experience). Scoliosis is usually seen in older children and is not necessarily related to structural defects of the vertebrae. Scoliosis may progress between late childhood and early adolescence.

Evaluation

- The spine should be evaluated clinically in older children at routine health supervision visits.
- Spine radiographs should be performed when clinical scoliosis is present.
- Referral to an orthopedist should be considered on recognition of significant musculoskeletal abnormalities.

Management

- Casting for positional abnormalities of the feet follows standard practice.
- Because the median age for walking without support ranges between 23 and 30 months, standard orthopedic surgery is recommended for foot deformities at an early age.
- Decisions about surgery for scoliosis may need to be made in the older child. If treatment is offered, it follows standard practice.

Ophthalmologic

No specific ophthalmologic pathology occurs in this disorder, although strabismus is very common. Refractive errors such as hypermetropia, myopia, or astigmatism rarely occur. Nystagmus, and entropion have been occasionally reported. Visual inattentiveness, defined as absence of attentive visual behavior with fixation and following movements, has been observed in individual children (personal experience).

Evaluation

- Ophthalmology consultation is recommended at diagnosis even in the absence of overt anomalies.
- Follow-up evaluation will depend on the findings.

Management

- Ophthalmologic abnormalities are treated in a standard manner.

Endocrinologic

Puberty is reportedly normal in most individuals with idic(15) (Schinzel 2001). However, pubertal disorders, such as central precocious puberty or ovarian dysgenesis, have been reported in a few girls (Grosso et al. 2001; Dennis et al. 2006; Al Ageeli et al. 2014).

Evaluation

- Standard evaluation is indicated when pubertal disorders are first suspected.

Management

- Management is the same as in any individual.

Dermatologic

Eczema-like skin disorder is noted in about 25% of children with idic(15) syndrome. When mild, the skin is dry, hot and itchy, whereas in more severe forms the skin can become broken, raw and bleeding. Overall, most children outgrow it (Dennis et al. 2006; personal experience).

Evaluation

- Evaluations do not differ from those in other children with similar symptoms.

Management

- Treatment should be decided on an individual basis.

RESOURCES

Dup15q Alliance
PO Box 674
Fayetteville NY 13066, USA
Phone: 855-DUP-15QA
Email: info@dup15q.org
Website: www.dup15q.org
Email: coordinator@dup15qregistry.org
Website: www.dup15qregistry.org

Unique: The Rare Chromosome Disorder Support Group
G1 The Stables
Station Road West
Oxted Surrey RH8 9EE, USA
Phone: +44 (0) 1883 723356
Email: info@rarechromo.org; rarechromo@aol.com
Website: www.rarechromo.org

REFERENCES

Aguglia U, Le Piane E, Gambardella A, Messina D, Russo C, Sirchia SM, Porta G, Quattrone A (1999) Emotion-induced myoclonic absence-like seizures in a patient with inv-dup(15) syndrome: a clinical, EEG, and molecular genetic study. *Epilepsia* 40:1316–1319.

Al Ageeli E, Drunat S, Delanoë C, Perrin L, Baumann C, Capri Y, Fabre-Teste J, Aboura A, Dupont C, Auvin S, El Khattabi L, Chantereau C, Moncla A, Tabet A-C, Verloes A (2014) Duplication of the 15q11-q13 region: clinical and genetic study of 30 new cases. *Eur J Med Genet* 57:5–14.

American Psychiatric Association (1994) Diagnostic and statistical manual of mental disorders. 4th. Ed. Washington, DC. *Am Psychiatry Press* 65–78.

Bahi-Buisson N, Kaminska A, Boddaert N (2008) The three stages of epilepsy in patients with CDKL5 mutations. *Epilepsia* 49(6):1027–1037.

Battaglia A (2005) The inv dup(15) or idic(15) syndrome: a clinically recognisable neurogenetic disorder. *Brain Development* 27:365–369.

Battaglia A (2008) The inv dup(15) or idic(15) syndrome (Tetrasomy 15q). *Orphanet J Rare Dis* 3:30.

Battaglia A, Gurrieri F, Bertini E, Bellacosa A, Pomponi MG, Paravatou-Petsotas M, Mazza S, Neri G (1997) The inv dup(15) syndrome: a clinically recognizable syndrome with altered behaviour, mental retardation and epilepsy. *Neurology* 48:1081–1086.

Battaglia A, Parrini B, Tancredi R (2010) The behavioural phenotype of idic(15) syndrome. *Am J Med Genet C* 154C:448-455.

Battaglia A, Filippi T (2014) Idic(15) syndrome: clinical studies of 32 new individuals. The Am Society Hum Genet, 64th Annual Meeting, San Diego, CA, *Book of Abstracts* 651.

Battaglia A, Bernardini L, Torrente I, Novelli A, Scarselli G (2016) Spectrum of epilepsy and electroencephalogram patterns in idic (15) syndrome. *Am J Med Genet A* 170A:2531–2539.

Bingham PM, Spinner NB, Sovinsky L, Zackai EH, Chance PF (1996) Infantile spasms associated with proximal duplication of chromosome 15q. *Pediatr Neurol* 15:63–165.

Blennow E, Nielsen KB, Telenius H, Carter NP, Kristoffersson U, Holmberg E, Gillberg C, Nordenskjold M (1995) Fifty probands with extra structurally abnormal chromosomes characterized by fluorescence in situ hybridization. *Am J Med Genet* 55:85–94.

Borgatti R, Piccinelli P, Passoni D, Dalprà L, Miozzo M, Micheli, R, Gagliardi C, Balottin, U (2001) Relationship between clinical and genetic features in "inverted duplicated chromosome 15" patients. *Pediatr Neurol* 24:111–6.

Browne CE, Dennis NR, Maher E, Long FL, Nicholson JC, Sillibourne J, Barber JC (1997) Inherited interstitial duplications of proximal 15q: genotype-phenotype correlations. *Am J Hum Genet* 61:1342–1352.

Caron L, Rousseau F, Gagnon E, Isenring P (2000) Cloning and functional characterization of a cation-Cl cotransporter-interacting protein. *J Biol Chem* 275:32027–32036.

Cassidy SB, Conroy J, Becker LA, Schwartz S (1996) Paternal Triplication of 15q11-q13 in a Hypotonic, Developmentally Delayed Child without Prader-Willi or Angelman Syndrome. *Am J Med Genet* 62:205–212.

Chifari R, Guerrini R, Pierluigi M, Cavani S, Sgrò V, Elia M, Canger, R, Canevini, MP (2002) Mild generalized epilepsy and developmental disorder associated with large inv dup (15). *Epilepsia* 43:1096–1100.

Christian SL, Fantes JA, Mewborn SK, Huang B, Ledbetter D (1999) Large genomic duplicons map to sites of instability in the Prader-Willi/Angelman syndrome chromosome region (15q11-q13). *Hum Mol Genet* 8:1025–1037.

Conant KD, Finucane B, Cleary N, Martin A, Muss C, Delany M, Murphy EK, Rabe O, Luchsinger K, Spence SJ, Schanen C, Devinsky O, Cook EJ, LaSalle K, Reiter LT, Thibert RL (2014) A survey of seizures and current treatments in 15q duplication syndrome. *Epilepsia* 55:396–402.

Connor JM, Gilmore DH (1984) An analysis of the parental age effect for inv dup (15). *J Med Genet* 21:213–214.

Cook EH Jr, Courchesne RY, Cox NJ, Lord C, Gonen D, Guter SJ, Lincoln A, Nix K, Leventhal BL, Courchesne E (1998) Linkage-disequilibrium mapping of autistic disorder, with 15q11-13 markers. *Am J Hum Genet* 62:1077–1083.

Cox DM, Butler MG (2015) The 15q11.2 BP1-BP2 microdeletion syndrome: a review. *Int J Mol Sci* 16:4068–82.

Crolla JA, Harvey JF, Sitch FL, Dennis NR (1995) Supernumerary marker 15 chromosomes: a clinical, molecular and FISH approach to diagnosis and prognosis. *Human Genetics* 95:161–170.

Dennis NR, Veltman MWM, Thompson R, Craig E, Bolton PF, Thomas NS (2006) Clinical findings in 33 subjects with large supernumerary marker(15) chromosomes and 3 subjects with triplication of 15q11-q13. *Am J Med Genet A* 140A:434–441.

DeLorey, TM, Sahbaie, P, Hashemi, E, Homanics GE, Clark JD (2008) Gabrb3 gene deficient mice exibit impaired social and exploratory behaviors, deficit in non-selective attention and hypoplasia of cerebellar vermal lobules: a potential model for autism spectrum disorder. *Behav Brain Res* 187:207–220.

Dennis, NR, Veltman, MWM, Thompson, R, Craig, E, Bolton, PF, Thomas, NS (2006) Clinical findings in 33 subjects with large supernumerary marker(15) chromosomes and 3 subjects with triplication of 15q11-q13. *Am J Med Genet* 140A:434–441.

Devinsky O (2011) Sudden, unexpected death in epilepsy. *N Eng J Med* 365:1801–11.

DiStefano C, Gulsrud A, Huberty S, Kasari C, Cook E, Reite L, Thibert R, Jeste SS (2016) Identification of a distinct developmental and behavioral profile in children with Dup15q syndrome. *J Neurodev Disorders* 8:19.

Dittrich B, Buiting K, Korn B, Rickard S, Buxton J, Saitoh S, Horsthemke B (1996) Imprint switching on human chromosome 15 may involve alternative transcripts of the SNRPN gene. *Nat Genet* 14:163–170.

Eggermann K, Mau UA, Bujdoso' G, Koltai E, Engels H, Schubert R,....Schwanitz G (2002) Supernumerary marker chromosomes derived from chromosome 15: analysis of 32 new cases. *Clin Genet* 62:89–93.

Flejter WL, Bennet-Barker PE, Ghaziuddin M, McDonald M, Sheldon S, Gorski JL (1996) Cytogenetic and molecular analysis of inv dup(15) chromosomes observed in two patients with autistic disorder and mental retardation. *Am J Med Genet* 61:182–187.

Freedman R (2014) α7-nicotinic acetylcholine receptor agonists for cognitive enhancement in schizophrenia. *Ann Rev Med* 65:245–261.

Gillberg C, Steffenburg S, Wahlstrom J, Gillberg IC, Sjostedt A, Martinsson T, Liedgren S, Eeg-Olofsson O (1991) Autism associated with marker chromosome. *J Am Acad Child Adolesc Psychiatry* 30:489–494.

Gillentine MA, Yin J, Bajic A, Zhang P, Cummock S, Kim JJ, Schaaf CP (2017) Functional consequences of CHRNA7 copy-number alterations in induced pluripotent stem cells and neural progenitor cells. *Am J Hum Genet* 101:1–14

Gilliam J (1995) *GARS-3: The Gilliam Autism Rating Scale*. 3rd ed. Austin, TX: Pro-ed. pp. 1–31.

Glessner JT, Wang K, Cai G, et al. (2009) Autism genome-wide copy number variation reveal ubiquitin and neuronal genes. *Nature* 459:569–73

Grammatico P, Di Rosa C, Roccella M, Falcolini M, Pelliccia A, Roccella F, Del Porto G (1994) Inv dup(15): contribution to the clinical definition of phenotype. *Clin Genet* 46:233–7.

Gobbi G, Genton P, Pini A (2002) Epilepsies and chromosomal disorders. In: Roger J, Bureau M. Dravet C, Genton P, Tassinari CA, Wolf P (eds) *Epileptic Syndromes in Infancy, Childhood and Adolescence*, 3rd edn Wastleigh, UK: John Libbey 431–455.

Greer PL, Hanayama R, Bloodgood BL, Mardinly AR, Lipton DM, Flavell SW, Kim T-K, Griffith EC, Waldon I, Maehr R, Ploegh HL, Chowdhury S, Worley PF, Steen J, Greenberg ME (2010) The Angelman Syndrone protein Ube3A regulates synapse development by ubiquitinating arc. *Cell* 140:704–16

Grosso S, Balestri P, Anichini C, Bartalini G, Pucci L, Morgese G, Berardi R (2001) Pubertal disorders in inv dup(15) syndrome. *Gynecol Endocrinol* 15:165–169.

Gurrieri F, Battaglia A, Torrisi L, Tancredi R, Cavallaro C, Sangiorgi E, Neri G (1999) Pervasive developmental disorder and epilepsy due to maternally derived duplication of 15q11-q13. *Neurology* 52:1694–1697.

Hogart A, Nagarajan RP, Patzel KA, Yasui DH, Lasalle JM (2007) 15q11-13 GABAA receptor genes are normally biallelically expressed in brain yet are subject to epigenetic dysregulation in autism-spectrum disorders. *Hum Mol Genet* 16:691–703.

Hogart A, Wu D, LaSalle JM, Schanen NC (2010) The comorbidity of autism with the genomic disorders of chromosome 15q11.2-q13. *Neurobiol Dis* 38:181–91.

Huang B, Bartley J (2003) Partial hexasomy of chromosome 15. *Am J Med Genet A* 121:277–280.

Huang B, Crolla JA, Christian SL (1997) Refined molecular characterization of the breakpoints in small inv dup(15) chromosomes. *Hum genet* 99:11–17.

Kleefstra T, de Leeuw N, Wolf R, Nillesen WM, Schobers G, Mieloo H, ...van Ravenswaaij-Arts CMA (2010) Phenotypic spectrum of 20 novel patients with molecularly defined supernumerary marker chromosomes 15 and a review of the literature. *Am J Med Genet A* 152A:2221–2229.

Koyama M, Suguira M, Yokoyama Y, Kobayashi M, Sugiyama K, Imahashi H, Saito H. (1985) A female case of cerebral gigantism with chromosome abnormality 47,XX+inv dup (15)(pter-q12 or q13;q12 or13-pter). *J Japan Pediatr Soc* 175:2671.

Lacuey N, Zonjy B, Hampson JP, Sandhya Rani MR, Zaremba A, Sainju,RK, Gehlbach BK, Schuele S, Friedman D, Devinsky O, Nei M, Harper RM, Allen L, Diehl B, Millichap JJ, Bateman L, Granner MA, Dragon DN, Richerson GB, Lhatoo SD (2018) The incidence and significance of periictal apnea in epileptic seizures. *Epilepsia* 59:573–582.

Lalande M (1996) Parental imprinting and human disease. *Ann Rev Genet* 30:173–195.

Leana-Cox J, Jenkins L, Palmer CG, Plattner R, Sheppard L, Flejter WL, Zackowski J, Tsien F, Schwartz S (1994) Molecular cytogenetic analysis of inv dup (15) chromosomes, using probes specific for the Prader-Willi/Angelman syndrome region: clinical implications. *Am J Hum Genet* 54:748–756.

Lord C, Rutter M, LeCouter A (1994) Autism diagnostic review revised: A revised version of a diagnostic interview for caregivers of individuals with possible pervasive developmental disorders. *J Autism Dev Disord* 24, 659–685.

Lord C, Rutter M, DiLavore PC, Risi S (2002) *Autism Diagnostic Observation Schedule*. Los Angeles: Western Psychological Services.

Lowther C, Costain G, Stavropoulos DJ, Melvin R, Silversides CK, Andrade DM, Bassett AS (2015) Delineating the 15q13.3 microdeletion phenotype: a case series and comprehensive review of the literature. *Genet Med* 17:149–57.

Maggouta F, Roberts SE, Dennis NR, Veltman MW, Crolla JA (2003) A supernumerary marker chromosome 15tetrasomic for the Prader-Willi/Angelman syndrome critical region in a patient with a severe phenotype. *J Med Genet* 40:e84.

Makoff AJ, and Flomen RH (2007) Detailed analysis of 15q11-q14 sequence corrects errors and gaps in the public access sequence to fully reveal large segmental duplications at breakpoints for Prader-Willi, Angelman, and inv dup(15) syndromes. *Genome Biol* 8:R114.

Malhotra D, Sebat J (2012) CNVs: Harbingers of a rare variant revolution in psychiatric genetics. *Cell* 148:1223–41.

Mann SM, Wang NJ, Liu DH, Wang L, Schultz RA, Dorrani N, Sigman M, Schanen NC (2004) Supernumerary tricentric derivative chromosome 15 in two boys with intractable epilepsy: another mechanism for partial hexasomy. *Hum Genet* 115:104–111.

Mao R, Jalal SM, Snow K, Michels VV, Szabo SM, Babovic-Vuksanovic D (2000) Characteristics of two cases with dup(15)(q11.2-q12): one of maternal and one of paternal origin. *Genet Med* 2:131–135.

Menold MM, Shao Y, Wolpert CM, Donnelly SL, Raiford KL, Martin ER, ...Gilbert JR (2001) Association analysis of chromosome 15 gabaa receptor subunit genes in autistic disorder. *J Neurogenet* 15:245–59.

Michelson M, Eden A, Vinkler C, Leshinsky-Silver E, Kremer U, Lerman-Sagie T, Lev D (2011) Familial partial trisomy 15q11-13 presenting as intractable epilepsy in the child and schizophrenia in the mother. *Eur J Paediatr Neurol* 15:230–3.

Mignon C, Malzac P, Moncla A, Depetris D, Roeckel N, Croquette MF, Mattei MG (1996) Clinical heterogeneity in 16 patients with inv dup (15) chromosome: cytogenetic and molecular studies, search for an imprinting effect. *Eur J Hum Genet* 4:88–100.

Mohandas TK, Park JP, Spellman RA, Filiano JJ, Mamourian AC, Hawk AB, ... Moeschler JB (1999) Paternally derived de novo interstitial duplication of proximal 15q in a patient with developmental delay. *Am J Med Genet* 82:294–300.

Moreno-De-Luca D, Sanders SJ, Willsey AJ, Mulle JG, Lowe JK, Geschwind DH, State MW, Martin CL, Ledbetter DH (2013) Using large clinical data sets to infer pathogenicity for rare copy number variants in autism cohorts. *Mol Psychiatry* 18:1090–5.

Nakatsu Y, Tyndale RF, De Lorey TM, Durham-Pierre D, Gardner JM, McDanel HJ, et al (1993) A cluster of three GABA receptor subunit genes is deleted in a neurological mutant of the mouse p locus. *Nature* 364:448–450.

Nietzel A, Albrecht B, Starke H, Heller A, Gillessen-Kaesbach G, Claussen U, Liehr T (2003) Partial hexasomy 15pter-->15q13 including SNRPN and D15S10: first molecular cytogenetically proven case report. *J Med Genet* 40:e28.

Nurmi EL, Amin T, Olson ML, Jacobs MM, McCauley JL, Lam AY, ...Sutcliffe JS (2003) Dense linkage disequilibrium mapping in the 15q11-q13 maternal expression domain yields evidence for association in autism. *Mol Psychiatry* 8:624–634.

Orrico A, Zollino M, Galli L, Buoni S, Marangi G, Sorrentino V (2009) Late-onset Lennox-Gastaut syndrome in a patient with 15q11.2-q13.1 duplication. *Am J Med Genet A* 149A:1033–1035.

Qumsiyeh MB, Rafi SK, Sarri C, Grigoriadou M, Gyftodimou J, Pandelia E, Laskari H, Petersen MB (2003) Double supernumerary isodicentric chromosomes derived from 15 resulting in partial hexasomy. *Am J Med Genet A* 116:356–359.

Rineer S, Finucane B, Simon EW (1998) Autistic symptoms among children and young adults with isodicentric chromosome 15. *Am J Med Genet* 81:428–433.

Roberts SE, Maggouta F, Thomas NS, Jacobs PA, Crolla JA (2003) Molecular and fluorescence in situ hybridization characterization of the breakpoints in 46 large supernumerary marker 15 chromosomes reveals an unexpected level of complexity. *Am J Hum Genet*. 73:1061–1072.

Robinson WP, Binkert F, Gine R, Vazquez C, Muller W, Rosenkranz W, Schinzel A (1993a) Clinical and molecular analysis of five inv dup (15) patients. *Eur J Hum Genet* 1:37–50.

Ryvlin P, Cucherat M, Rheims S (2011) Risk of sudden unexpected death in epilepsy in patients given adjunctive antiepileptic treatment for refractory seizures: a meta-analysis of placebo-controlled randomised trials. *Lancet Neurology* 10:961–8.

Ryvlin P, Nashef L, Tomson T (2013) Prevention of sudden unexpected death in epilepsy: a realistic goal? *Epilepsia* 54,Suppl 2:23–8.

Roberts S, Dennis N, Browne C, Willatt L, Woods C, Croos I, Jacobs P, Thomas N (2002) Characterisation of interstitial duplications and triplications of chromosome 15q11-q13. *Hum Genet* 110:227–234.

Robinson WP, Binkert F, Gine R, Vazquez C, Muller W, Rosenkranz W, Schinzel, A (1993b) Clinical and molecular analysis of five inv dup (15) patients. *Eur J Hum Genet* 1:37–50.

Ryvlin P, So EL, Gordon CM, Hersdorffer DC, Sperling MR, Devinsky O, Bunker MT, Olin B, Friedman D (2018) Long-term surveillance of SUDEP in drug-resistant epilepsy patients treated with VNS therapy. *Epilepsia* 59:562–572.

Saitoh S, Hosoki K, Takano K (2007) Mosaic paternally derived inv dup(15) may partially rescue the Prader-Willi syndrome phenotype with uniparental disomy. *Clin Genet* 72:378–380.

Samaco RC, Hogart A, LaSalle JM (2005) Epigenetic overlap in autism-spectrum neurodevelopmental disorders: MECP2 deficiency causes reduced expression of UBE3A and GABRB3. *Hum Mol Genet* 14:483–92.

Sanders SJ, He X, Willsey AJ, et al. (2015) Insights into autism spectrum disorder genomic architecture and biology from 71 risk loci. *Neuron* 87:1215–33.

Schinzel A (1981) Particular behavioural symptomatology in patients with rarer autosomal chromosome aberrations. In: Schmid W, Nielsen J, editors. *Human Behavior and Genetics*. Amsterdam: Elsevier/North Holland. pp.195–210.

Schinzel A, Brecevic L, Bernasconi F, Binkert F, Berthet F, Wuilloud A, et al (1994) Intrachromosomal triplication of 15q11-q13. *J Med Genet* 31:798–803.

Schinzel A, Niedrist D (2001) Chromosome imbalances associated with epilepsy. *Am J Med Genet C* 106:119–124.

Ungaro P, Christian SL, Fantes JA, Mutirangura A, Black S, Reynolds J, Malcolm S, Dobyns WB, Ledbetter DH (2001) Molecular characterisation of four cases of intrachromosomal triplication of chromosome 15q11-q14. *J Med Genet* 38:26–34.

Urraca N, Davis L, Cook EH, Schanen NC, Reiter LT (2010) A single-tube quantitative high-resolution melting curve method for parent-of-origin determination of 15q duplications. *Genet Test Mol Biomarkers* 14:571–6.

Urraca N, Hope K, Victor AK, Belgard TG, Memon R, Goorha S, Valdez C, Tran QT, Sanchez S, Ramirez J, Donaldson M, Bridges D, Reiter LT (2018) Significant transcriptional changes in 15q duplication but not Angelman syndrome deletion stem cell-derived neurons. *Mol Autism* 9:6.

Valente KD, Freitas A, Fridman C, Varela M, Silva AE, Fett AC, Koiffmann CP (2006) Inv dup(15): is the elctroclinical phenotype helpful for this challenging clinical diagnosis? *Clin Neurophysiol* 117:803–809.

Vanlerberghe C, Petit F, Malan V, et al. (2015) 15q11.2 microdeletion (BP1-BP2) and developmental delay, behaviour issues, epilepsy and congenital heart disease: a series of 52 patients. *Eur J Med Genet* 58:140–7.

Wang NJ, Liu D, Parokonny AS, Schanen NC (2004) High-resolution molecular characterization of 15q11-q13 rearrangements by array comparative genomic hybridization (array CGH) with detection of gene dosage. *Am J Hum Genet* 75:267–281.

Wang NJ, Parokonny AS, Thatcher KN, Driscoll J, Malone BM, Dorrani N, …. Schanen NC (2008) Multiple forms of atypical rearrangements generating supernumerary derivative chromosome 15. *BMC Genet* 9:2.

Webb T, Hardy CA, King M, Watkiss E, Mitchell C, Cole T (1998) A clinical, cytogenetic and molecular study of ten probands with inv dup (15) marker chromosomes. *Clin Genet* 53:34–43.

Wegiel J, Schanen NC, Cook EH, et al. (2012) Differences between the pattern of developmental abnormalities in autism associated with duplications 15q11.2-q13 and idiopathic autism. *J Neuropathol Exp Neurol* 71:382–97.

Wisniewski L, Hassold T, Heffelfinger J, Higgins JV (1979) Cytogenetic and clinical studies in five cases of inv dup(15). *Hum Genet* 50:259–270.

Wolpert CM, Menold MM, Bass MP, Qumsiyeh MB, Donnelly SL, Ravan SA, Vance JM, Gilbert JR, Abramson RK, Wright HH, Cuccaro ML, Pericak-Vance MA (2000) Three probands with autistic disorder and isodicentric chromosome 15. *Am J Med Genet* 96:365–72.

Ziats MN, Goin-Kochel RP, Berry LN, et al. (2016) The complex behavioral phenotype of 15q13.3 microdeletion syndrome. *Genet Med* 18:1111–1118.

Zielinski C, Müller C, Smolen J (1988) Use of plasmapheresis in therapy of systemic lupus erythematosus: a controlled study. *Acta Medica Austriaca* 15:155–8.

Zhou D, Gochman P, Broadnax DD, Rapoport JL, Ahn K (2016) 15q13.3 duplication in two patients with childhood-onset schizophrenia. *Am J Med Genet B* 171(6):777–783.

34

KABUKI SYNDROME

SARAH DUGAN

Clinical Genomics and Predictive Medicine, Providence Medical Group, Spokane, Washington, USA

INTRODUCTION

Kabuki syndrome, as initially described by Niikawa et al. (1981) and Kuroki et al. (1981), is characterized by intellectual disability, postnatal growth deficiency, and unusual facies reminiscent of the make-up of actors in Kabuki, the traditional form of Japanese theater. As such, this syndrome was initially reported in the literature as either Niikawa–Kuroki syndrome or Kabuki make-up syndrome. Referring to the condition as simply "Kabuki syndrome" is generally preferred, because the term "make-up" might be confusing and/or offensive to some families.

The characteristic facial appearance usually prompts consideration of this diagnosis. However, a plethora of associated abnormalities involving almost every organ system has been described in this highly variable condition.

Incidence

Initially, the majority of individuals reported with Kabuki syndrome were Japanese, and its prevalence in Japan is estimated to be approximately 1 in 32,000. White et al. (2004) calculated a minimum birth prevalence of 1/86,000 in Australia and New Zealand. This condition has been reported in almost all ethnic groups, and the incidence outside Japan presumably approximates that seen in the Japanese population.

Diagnostic Criteria

In the first large series of 62 individuals with this condition (Niikawa et al. 1988), five cardinal manifestations were identified: (1) particular face (100%) (Figures 34.1–34.3), (2) skeletal anomalies (92%), (3) dermatoglyphic abnormalities (93%), (4) mild-to-moderate intellectual disability (92%), and (5) postnatal growth deficiency (83%). Other authors have suggested that minimal criteria include (1) long palpebral fissures with eversion of the lower lateral eyelid; (2) broad, arched eyebrows with lateral sparseness; (3) short nasal columella with depressed nasal tip; (4) large, prominent, or cupped ears; and (5) developmental delay or intellectual disability (Kawame et al. 1999). The presence of prominent fingertip pads is a common finding (Figure 34.4). A proposed scoring system for clinical diagnosis, relying on characteristic facial features, extremity findings, growth parameters, and major anomalies has been independently revalidated (Makrythanasis et al. 2013; Paderova et al. 2017).

The facial gestalt associated with almost all cases of Kabuki syndrome is the most striking and specific feature of the condition. It consists of a constellation of minor anomalies that together produce one of the most recognizable facial phenotypes in syndromology. Palpebral fissures are generally long, with generous lashes and arched, broad, laterally sparse eyebrows. The palpebral fissures often (but not always) extend laterally beyond the globe, sometimes with eversion of the lower lid, and ptosis may be present. The nasal tip is often low, with a short columella implanting

---

Cassidy and Allanson's Management of Genetic Syndromes, Fourth Edition.
Edited by John C. Carey, Agatino Battaglia, David Viskochil, and Suzanne B. Cassidy.
© 2021 John Wiley & Sons, Inc. Published 2021 by John Wiley & Sons, Inc.

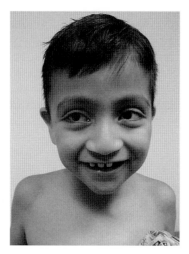

FIGURE 34.1 A 12-year-old boy with Kabuki syndrome. Note long palpebral fissures with eversion of the lateral eyelids. (Courtesy of Dr. Keith Vaux, Children's Hospital and Health Center San Diego, San Diego, California, USA.)

FIGURE 34.2 Lateral of the 12-year-old boy with Kabuki syndrome. Note short columella with depressed nasal tip, prominent ears that are retroverted, and preauricular pits. (Courtesy of Dr. Keith Vaux, Children's Hospital and Health Center San Diego, San Diego, California, USA.)

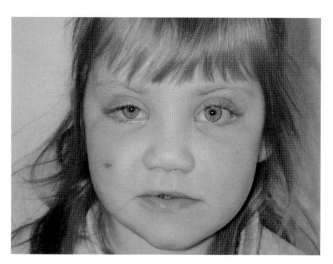

FIGURE 34.3 A 5-year-old girl with Kabuki syndrome. Note the ptosis and prominent philtrum characteristic of the condition.

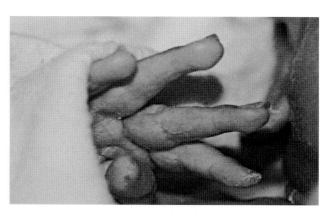

FIGURE 34.4 Prominent fingertip pads in an infant with Kabuki syndrome. (Courtesy of Dr. Keith Vaux, Children's Hospital and Health Center San Diego, San Diego, California.)

higher than the nasal alae. The philtrum is generally deep and well-defined. It is widest in the middle, leading to a trapezoidal or teardrop-shaped configuration. Ears appear large and prominent.

The presence of certain anomalies that are common in Kabuki syndrome, but unusual in the general population, can be diagnostically helpful in newborns. For example, the diagnosis should be considered in all infants with aortic coarctation; similarly, objective minor anomalies, such as nocturnal lagophthalmos, lip mounds or pits, and preauricular pits, can be an important clue (Toriello and Droste 2003).

The phenotype may change with age. Vaux et al. (2005a) reported a series of 16 children with Kabuki syndrome who had presented to a dysmorphologist in the neonatal period. Retrospective chart review showed that some of the more characteristic facial features, including long palpebral fissures, were noted in only a minority of those affected. Revisiting three adults who had been reported in the literature as children, Shalev et al. (2004) demonstrated that the long palpebral fissures persisted, but other facial features had become less striking overall. Short stature also persisted, with the addition of central obesity.

The characteristic facial gestalt usually prompts genetic testing, but the condition is often found by exome sequence analysis in individuals who may have more subtle or atypical presentations.

Etiology, Pathogenesis, and Genetics

Kabuki syndrome is known to be caused by changes in *KMT2D* (encoding lysine-specific methyltransferase 2D and estimated to be causative in 60–65% of cases) and *KDM6A* (encoding lysine-specific demethylase 6A and estimated to be causative in 5–8% of cases). *KMT2D* was originally identified through exome sequence analysis, while *KDM6A* was identified by array-based CGH (Ng et al. 2010; Lederer et al. 2012). The existence of additional causative genes seems likely since the sensitivity of genetic testing among individuals with a clinical diagnosis is estimated at 55–80% (Banka et al. 2012). Both known genes are involved in chromatin remodeling, and it is likely that any additional causative genes discovered in the future would play a similar role. Genes with this classification encode proteins that chemically alter chromatin structure and influence genetic transcription, leading to a wide array of downstream effects.

Numerous chromosomal abnormalities have been reported in association with clinical features suggestive of Kabuki syndrome. At least six individuals with Kabuki syndrome and a 45,X cell line have been described, some of whom have small ring X chromosomes. Despite extensive investigation of several of these individuals, a connection has not clearly been made between the known X-linked locus and these chromosome abnormalities (Bögershausen et al. 2016). On the other hand, submicroscopic deletions of the X chromosome encompassing *KDM6A* have been reported in several individuals with the condition (Lederer et al. 2012). Whole-gene or large intragenic deletions of *KMT2D* have also been described but are rare (Banka et al. 2013).

Most individuals with Kabuki syndrome occur as a result of a new mutation, but the condition can be inherited from an affected parent. Changes in *KMT2D* would be passed along in an autosomal dominant fashion, while *KDM6A*-related Kabuki syndrome is X-linked. Generally, the two forms of Kabuki syndrome are clinically similar, but females with the X-linked form may have milder developmental involvement and more subtle physical manifestations (Bögershausen et al. 2016). The male-to-female ratio is close to one. Several case series suggest that advanced paternal age predisposes to Kabuki syndrome (Armstrong et al. 2005).

Diagnostic Testing

Testing for Kabuki syndrome is offered clinically by many molecular genetics laboratories via single-gene testing or as part of a larger panel. The condition would usually also be detected by exome sequence analysis, though this technique may miss some cases caused by a deletion or less common kind of mutation.

Differential Diagnosis

A number of authors have noted the similarity between the findings in Kabuki syndrome and those seen in deletion 22q11 syndrome (velo-cardio-facial/DiGeorge syndrome, Chapter 21), especially in the newborn period (Dinulos and Hudgins 2000). Anomalies seen in both conditions include cleft palate, cardiac malformations, urinary tract anomalies, and prominent feeding problems. Thus, one should consider following the infant with these findings and no evidence of 22q11 deletion for development of features suggestive of Kabuki syndrome.

Others have noted that cleft lip and palate, cleft palate, and lip pits are seen both in individuals with the *IRF6*-related disorders (Van der Woude syndrome and popliteal pterygium syndrome) and in those with Kabuki syndrome (Matsumoto and Niikawa 2003). Individuals with the *IRF6*-related disorders do not have atypical growth and development, cardiac malformations, or the typical Kabuki facies. Pterygia are not expected in people with Kabuki syndrome.

Although ear pits, cupped ears, hearing loss, and renal anomalies may be seen in both branchio-oto-renal (BOR) syndrome and Kabuki syndrome, the other findings in this condition should allow for easy differentiation. Individuals with BOR syndrome have otherwise normal craniofacies, normal growth, and normal development, and the common renal anomalies are hypoplasia and/or renal agenesis. In addition, branchial cleft cysts may be present; these have not been reported in Kabuki syndrome. In Kabuki syndrome, the characteristic craniofacial features should be apparent along with delayed growth and development. Common renal anomalies in Kabuki syndrome include hydronephrosis and abnormal kidney position(s).

Significant joint hypermobility including congenital hip dislocation and patellar dislocations, especially in the newborn period, and blue sclerae, may lead the diagnostician to consider a connective tissue disorder such as Ehlers–Danlos syndrome (Chapter 25) or Larsen syndrome. These conditions are not associated with major malformations involving other organ systems or the typical minor anomalies seen in Kabuki syndrome.

The dysplastic ears, ocular colobomata, growth deficiency, and cardiac and renal anomalies seen in patients with Kabuki syndrome can also bring CHARGE syndrome (see Chapter 11) to mind. The characteristic Kabuki facies should suggest this diagnosis. Obviously, exclusion of additional anomalies associated with CHARGE syndrome can also be helpful.

Facial features seen in Kabuki syndrome can overlap with those seen in other chromatin remodeling disorders, such as Wiedemann–Steiner syndrome, Coffin–Siris syndrome (see Chapter 13), Rubinstein–Taybi syndrome (see Chapter 51), and Au–Kline syndrome. Because these phenotypes overlap considerably, consultation with a clinical geneticist is recommended prior to genetic testing.

Because several different chromosome abnormalities have been reported in individuals with findings suggestive of Kabuki syndrome, comparative genomic hybridization (CGH) by SNP microarray should be considered as an initial step in diagnostic screening if Kabuki syndrome is not highly suspected, and should be sent in those individuals who have had negative genetic testing for the condition. The clinical phenotype scoring system proposed by Makrythanasis et al. (2013) may be useful in selecting a testing strategy.

MANIFESTATIONS AND MANAGEMENT

Wilson (1998) and Kawame et al. (1999) have reviewed management issues based on the findings in their respective series from North America. Schrander-Stumpel et al. (2005) reviewed 20 patients from the Netherlands, focusing on clinical manifestations. Armstrong et al. (2005) reviewed clinical findings in 48 previously unreported individuals. A comprehensive review of findings in a large series of individuals (350) is that of Matsumoto and Niikawa (2003). Bögershausen et al. reported clinical manifestations in 11 individuals with *KDM6A*-related (X-linked) Kabuki syndrome in 2016.

Growth and Feeding

Neonates with Kabuki syndrome exhibit normal growth parameters. However, postnatal growth retardation is relatively common (35–81%). Microcephaly may or may not accompany short stature. Growth hormone deficiency has been reported (see Endocrine), but is not common.

Feeding difficulties are common (about 70%); however, severity is variable (Kawame et al. 1999). Many individuals with Kabuki syndrome have gastroesophageal reflux. Others require gastrostomy tube placement for poorly coordinated suck and swallow. Consequently, some infants with Kabuki syndrome exhibit failure-to-thrive.

Evaluation

- Height/length, weight, and head circumference should be monitored frequently (at each well-child visit and more often if there are concerns) and plotted on normal growth charts.
- Those individuals with abnormal growth velocity should be evaluated for hypothyroidism or growth hormone deficiency.
- If feeding problems are severe and/or failure-to-thrive is apparent, an esophageal pH probe study should be considered to assess gastro-esophageal reflux.
- A barium swallow study may assist in determining whether the suck-and-swallow mechanism is normal.

Management

- Thickened feedings and appropriate positioning after meals may improve reflux symptoms.
- Gastrostomy tube placement should be considered in those individuals with severe feeding difficulties, especially if a poorly coordinated suck and swallow is noted.

Development and Behavior

In their series of 62 affected individuals, Niikawa et al. (1988) identified mental retardation (currently referred to as intellectual disability) (usually in the mild-to-moderate range) as a cardinal feature, seen in 92% of patients in the series. As more individuals with this condition have been recognized and reported, however, some authors have suggested that as many as one-sixth have normal intelligence (Matsumoto and Niikawa 2003). A more recent review, in which authors excluded reports deemed as less thorough or less diagnostically clear, calculated an overall frequency of intellectual impairment closer to the original estimate (Schrander-Stumpel et al. 2005).

Most individuals with Kabuki syndrome are able to speak and to ambulate. However, there are rare individuals who are non-ambulatory, but are able to speak, as well as some who are non-verbal with no significant motor impairment (Kawame et al. 1999). Vaux et al. (2005b) reported unassisted walking at an average age of 20 months with a range of 15–30 months in 15 individuals with Kabuki syndrome. In this series, single words were spoken by 21 months on average, with a range of 10–30 months. Thus far, there are no apparent factors that would allow for early prognostication.

Mervis et al. (2005) performed standardized neuropsychological testing on 11 children and adolescents with Kabuki syndrome and reported relative strengths in verbal and non-verbal reasoning, with relative weakness in visuospatial skills. A similar study by Lehman et al. (2017) in 31 patients demonstrated a particularly marked disparity between working memory (a relative strength) and processing speed (a relative weakness). This study also demonstrated that children with impaired vision had particularly slow processing speed and lower full-scale IQ.

Based on a 16-participant study, Morgan et al. (2015) reported a mixed language impairment with both receptive and expressive deficits across all domains, consistent with multifactorial speech delay. Although language impairment varied among patients, dysarthria, hypernasality, harsh vocal quality, pitch distortion, and altered prosody commonly characterized participants' speech. Although language skills improved with age, adults with Kabuki syndrome typically have some degree of language impairment.

Individuals with Kabuki syndrome tend to be described as pleasant and outgoing. In their series of 11 children and adolescents, Mervis et al. (2005) report that, overall, the participants' behavioral difficulties did not surpass those expected for their chronological age. Autism is a rare but described finding. Ho and Eaves (1997) reported four males with variable cognitive abilities, three of whom had features within the autism spectrum ranging from pervasive developmental disorder to autistic-like to autistic disorder. Sari et al. (2008) described another male with the classical Kabuki phenotype, autism, and a normal IQ. It is not clear that autism is present at a level above that seen in the general population. As Kabuki syndrome becomes more widely recognized, the true incidence of social and communication difficulties will become apparent.

Evaluation

- Developmental milestones should be checked at each well-child visit, and referral for formal developmental evaluation should be made if delays are identified.
- Thorough psychoeducational testing is indicated for all children who exhibit cognitive difficulties to determine strengths and weaknesses and to tailor special education services.
- Formal evaluation by a developmental pediatrician or psychiatrist may be helpful in those children who exhibit features suggestive of the autism spectrum, because educational interventions may be influenced by the result.

Management

- Because of the frequency of developmental delay, early intervention services should be instituted at diagnosis in all children with Kabuki syndrome.
- Special education services should be tailored to address the strengths and weaknesses for each individual child with Kabuki syndrome because no characteristic pattern of disabilities has yet been identified.

Neurologic

Many children with Kabuki syndrome are hypotonic (25–89%). Significant joint laxity may be a contributing factor. As with other conditions in which hypotonia is a feature, this finding improves with time.

Seizures are seen more frequently in Kabuki syndrome (10–39%) and can present at any time during childhood; infantile onset has been reported but is less common. Partial seizures are most typical, and good seizure control is generally achieved with antiepileptic therapy (Ogawa et al. 2003; Verrotti et al. 2011). Major structural brain anomalies are rare, although symptomatic Chiari I malformation has been reported in multiple affected individuals. Many people with Kabuki syndrome will undergo brain imaging at some point for indications such as seizures and developmental delay.

Evaluation

- A physical therapy evaluation is indicated in those children with Kabuki syndrome who exhibit hypotonia or motor delays.
- Those individuals with suspected seizure activity should be evaluated by a neurologist.
- Brain imaging, if not already performed, should be considered if seizures develop. Persistent, unexplained head and neck pain or other evidence of intracranial abnormality could be secondary to Chiari I malformation and would be a clear indication for brain imaging (Ciprero et al. 2005).

Management

- Hypotonic children with Kabuki syndrome should receive physical therapy.
- Standard antiepileptic treatment is efficacious in treating seizures in individuals with Kabuki syndrome.
- Referral to a neurologist or a neurosurgeon should be made as indicated for seizures or structural brain abnormalities.

Cardiovascular

Approximately 40–70% of individuals with Kabuki syndrome have congenital heart defects. Left-sided obstructive lesions of all types, but especially coarctation of the aorta, are common in Kabuki syndrome; all types of left-sided defects, from bicuspid aortic valve to hypoplastic left heart, can be seen. Septal defects are common, and conotruncal defects are seen occasionally. (Hughes and Davies 1994; Kawame et al. 1999; Digilio et al. 2001; Armstrong et al. 2005; Diglio et al. 2017).

Kabuki syndrome has many manifestations that overlap with connective tissue disorders (see Musculoskeletal). Internal carotid dissection has been described in one 15-year-old girl (Gatto et al. 2017). Aortic dilation has been reported in two children, one of whom underwent aortic root replacement at 14 years (Dyamenahalli et al. 2007). However, cardiac manifestations of connective tissue disease do not seem to be a major manifestation of the condition.

Evaluation

- As the incidence of cardiac malformations is high, echocardiogram with good visualization of the aortic arch is indicated in all individuals at the time of diagnosis.
- At this time, there is no indication for performing routine evaluation for blood vessel disease, but threshold for imaging should be low to address a new murmur or symptoms of dissection.

Management

- Referral to a pediatric cardiologist for management should be made if a cardiac defect is present.
- The indications for medical intervention and surgical correction are the same as those for the general population.

- As with many cases of congenital heart disease, prophylactic antibiotic treatment may be indicated before and during any procedure (e.g, dental work) that might lead to bacteremia.

Endocrine

Premature thelarche in girls is the most common endocrine abnormality described (7–50%). Parents can be reassured that this finding does not represent premature puberty and is likely to resolve with time.

Hypoglycemia (with documented hyperinsulinism in numerous cases), congenital hypothyroidism, and growth hormone deficiency are rarely reported findings. Growth hormone stimulation testing in 18 children with Kabuki syndrome demonstrated abnormal response in 28%, but results were not thought to be predictive of true growth hormone deficiency in all patients (Schott et al. 2016). Hypoglycemia appears to be significantly more common in the X-linked (*KDM6A*-related) form of Kabuki syndrome and was reported in five out of ten individuals in one series (Bögershausen et al. 2016).

Evaluation

- Girls with Kabuki syndrome should be followed expectantly for possible premature thelarche, so that parents can be reassured if this finding becomes apparent.
- Extensive endocrine evaluation for premature thelarche is not indicated, unless other signs of premature puberty become apparent.
- Endocrine evaluation is indicated if short stature is present and growth velocity is decreased, as hypothyroidism and growth hormone deficiency have been described in Kabuki syndrome. Growth hormone stimulation testing may be less useful in this population.
- If Kabuki syndrome is suspected in the neonatal period, screening for hypoglycemia is indicated.

Management

- Treatment of premature thelarche is not warranted unless other signs of premature puberty are apparent.
- Thyroid hormone replacement is indicated in those with evidence of hypothyroidism.
- Growth hormone treatment should be instituted only in those individuals with biochemical evidence of growth hormone deficiency and should be discontinued if not clinically effective.

Ophthalmologic

Ocular findings occur in more than a third of the individuals with Kabuki syndrome and include blue sclerae, strabismus, ptosis, colobomata, and corneal abnormalities such as Peters anomaly. Optic nerve hypoplasia, cataracts, Duane anomaly, pigmentary retinopathy, and Marcus Gunn phenomenon (also referred to as jaw winking) can also be seen (Ming et al. 2003). Severe visual impairment, however, is rare (Kawame et al. 1999).

As a result of the everted lower lid, children with Kabuki syndrome can demonstrate excessive tearing, but this is not usually a significant problem. On the other hand, nocturnal lagophthalmos, which may occur in most children with Kabuki syndrome, can predispose to corneal abrasion and scarring (Toriello and Droste 2003).

Evaluation

- Because of the frequency of ocular findings, all individuals with Kabuki syndrome should have a formal ophthalmological examination at the time of diagnosis.
- Parents or caregivers should be instructed to watch for nocturnal lagophthalmos.
- Vision should be checked on a yearly basis.

Management

- Those individuals with ocular findings that affect their vision should be followed by an experienced ophthalmologist and treated appropriately, when possible. Treatment is carried out in a standard fashion.

Ears and Hearing

Most individuals with Kabuki syndrome have prominent and cup-shaped external ears. Ear pits are also relatively common and may be a helpful diagnostic clue when seen with other typical findings.

From a medical standpoint, chronic otitis media is a major cause of morbidity, including conductive hearing loss. It is not clear, however, whether this finding is related to an underlying susceptibility to infection or palatal insufficiency (Matsumoto and Niikawa 2003).

About 40% of individuals with Kabuki syndrome have hearing loss. Although chronic middle ear effusions with recurrent otitis media is the most common cause, sensorineural hearing loss can rarely occur. Inner ear malformations, including Mondini dysplasia, vestibular enlargement, aqueductal enlargement, and aplastic cochlea and semicircular canals have been reported (Tekin et al. 2006).

Evaluation

- Children with Kabuki syndrome should be followed expectantly for otitis media. An ear examination should be performed at all well-child visits, and prompt evaluation is indicated in all children with symptoms suggestive of otitis media.

- Those with chronic otitis media might benefit from formal otorhinolaryngological evaluation.
- Hearing should be tested on a yearly basis. Evidence of sensorineural hearing loss should be followed up with referral to an otorhinolaryngologist and imaging to screen for inner ear anomalies.

Management

- Standard treatment for otitis media should be instituted, including antibiotics and ventilation tubes, if indicated.
- Amplification should be considered in those with hearing loss.
- Appropriate classroom placement and educational services for the hearing impaired should be instituted.

Craniofacial

Cleft lip and/or palate is seen in approximately one-third of individuals with Kabuki syndrome. Burke and Jones (1995) were the first to note that Kabuki syndrome may be underdiagnosed in individuals with cleft palate. Conversely, Iida et al. (2006) noted that submucous cleft palate may be under-ascertained in individuals with Kabuki syndrome. Almost three-quarters of the affected individuals will have a high-arched palate. As with all children with palatal abnormalities, feeding difficulties, frequent otitis media, and speech difficulties are more common in this subset of individuals with Kabuki syndrome. A number of individuals with Kabuki syndrome and lower lip pits have been reported (Matsumoto and Niikawa 2003). Craniosynostosis has also been reported in multiple individuals but is not a common feature.

Evaluation

- The palate should be carefully evaluated at the time of diagnosis.
- Children with cleft lip, cleft palate, submucous cleft palate, or any evidence of velopharyngeal insufficiency should be referred for otorhinolaryngology evaluation.
- Abnormal head shape should be clinically evaluated and then addressed with CT or radiographs if craniosynostosis is suspected.

Management

- Cleft lip is usually surgically corrected using standard methods within the first few months of life.
- Cleft palate is generally repaired using standard methods at around 1 year of age.
- Velopharyngeal insufficiency may be surgically treatable.
- Craniosynostosis is managed with a standard approach including neurosurgery referral and cranial vault remodeling.

Dental

A number of different dental anomalies in individuals with Kabuki syndrome have been noted. Hypodontia is most common, with missing lateral and central incisors as well as premolars (Mhanni et al. 1999; Matsune et al. 2001). Also described are abnormally shaped teeth, small teeth, and malocclusion, all of which may contribute to the need for orthodontia.

Evaluation

- Every child with Kabuki syndrome should have a dental evaluation as a toddler.
- Referral for orthodontic assessment should be arranged if abnormalities such as hypodontia or significant malocclusion are noted at any point in childhood.

Management

- Those children with Kabuki syndrome and recognized dental anomalies such as hypodontia and malocclusion may require orthodontic treatment. This is carried out as in the general population, though sedation may be necessary based on developmental ability to cooperate with the procedure.

Gastrointestinal

Abnormalities involving the gastrointestinal system are not common in Kabuki syndrome. However, anorectal anomalies including imperforate anus and fistulae have been reported in a number of individuals, primarily in females (Niikawa et al. 1981; Kawame et al. 1999). Congenital diaphragmatic hernia and eventration of the diaphragm have also been described (Philip et al. 1992). Kabuki syndrome may also impart an increased risk of neonatal cholestasis from a variety of causes, including biliary atresia (Isidor et al. 2007).

Evaluation

- Neonates with Kabuki syndrome should have a careful physical examination to look for evidence of gastrointestinal abnormalities.
- Evidence of cholestasis should prompt a full workup, as it would in any other child.

- By recognizing that eventration of the diaphragm is seen in this condition, the medical care provider can avoid further evaluation for possible phrenic nerve paralysis.

Management

- Abnormalities involving the gastrointestinal tract should be surgically corrected in a standard fashion.
- Cholestatic diseases should be treated as they would in anyone.

Genitourinary

Renal and urinary tract anomalies are seen in over 25% of affected individuals (Matsumoto and Niikawa 2003). Common renal findings include anomalies of kidney position and ascent (single fused kidneys, crossed fused renal ectopia), ureteropelvic junction obstruction, duplication of the collecting system, and hydronephrosis (Kawame et al. 1999). Hypospadias, cryptorchidism, and (more rarely) micropenis can occur in males, whereas females can demonstrate hypoplastic labia (Armstrong et al. 2005).

Evaluation

- Because of the frequency of renal/urinary tract anomalies, renal ultrasound is indicated in all individuals with Kabuki syndrome at the time of diagnosis.
- Those individuals with hydronephrosis should be referred to a nephrologist and/or urologist for management and treatment.
- Cryptorchidism should prompt referral to a urologist.

Management

- Standard treatment for genital anomalies, ureteropelvic junction obstruction, and hydronephrosis is indicated.

Musculoskeletal

Joint hypermobility is seen in half to three-quarters of individuals with Kabuki syndrome (Kawame et al. 1999; Matsumoto and Niikawa 2003). Joint dislocations are not uncommon, especially involving the hips, patellae, and shoulders. As in most conditions with joint laxity, this finding improves with age. Procollagen studies were normal in at least one individual with Kabuki syndrome with striking joint hypermobility (Hudgins et al. 1993).

Variable degrees of scoliosis and kyphosis are seen and may be associated with vertebral anomalies (hemivertebrae, butterfly vertebrae, and sagittal clefts) (Niikawa et al. 1988).

Evaluation

- Those children with congenital dislocation of the hip(s) and other joints should be referred for evaluation by an orthopedic surgeon.
- Spine radiographs should be done if scoliosis is suspected clinically.
- Referral to orthopedics should be made for clinically significant spine anomalies or scoliosis.

Management

- Standard treatment for congenital hip dislocation and scoliosis are indicated.

Immunologic

Immune dysfunction has been described, mostly in adolescents with Kabuki syndrome. Hoffman et al. (2005) found that 16 of a series of 19 individuals with Kabuki syndrome had some form of hypogammaglobulinemia. Low levels of serum IgA have been reported in association with idiopathic thrombocytopenic purpura, autoimmune hemolytic anemia, and recurrent sinopulmonary infections. Lindsley et al. (2016) identified defective terminal B-cell differentiation as the cause of humoral immunodeficiency in Kabuki syndrome. It is not yet clear whether a true susceptibility to infections is responsible for the chronic otitis media so common in children with Kabuki syndrome, or whether these infections are related to the craniofacial abnormalities including cleft palate.

Evaluation

- Serum immunoglobulin levels should be obtained in all individuals with Kabuki syndrome at the time of diagnosis or at 1 year of age (whichever comes second).
- Individuals should be referred to an immunologist for regular follow-up regardless of immunoglobulin levels.

Management

- Individuals with documented immunoglobulin deficiency may benefit from scheduled intravenous immunoglobulin infusions.
- Standard treatment for autoinflammatory disorders is indicated.

RESOURCES

Kabuki Syndrome Network

8060 Struthers Crescent
Regina, Saskatchewan S4Y 1J3 CANADA
Phone: (306) 543-8715

Email: *kabuki@sasktel.net*
Website: *https://www.kabukisyndromefoundation.org*

Kabuki UK

Website: www.kabukiuk.org.uk/

Netwerk Kabuki Syndroom

Beukenlaan 24

6241 AL Bunde, The Netherlands

Phone: (0031) 043 3650207

Email: *martijn95@tip.nl* (Dutch)

Supporting Aussie Kids with Kabuki Syndrome

PO Box 318

Rundle Mall

South Australia 5000

Website: www.sakks.org

ACKNOWLEDGEMENTS

This chapter is based on the original Kabuki Syndrome chapter by Louanne Hudgins, published in Cassidy SB, Allanson JE (2001) *Management of Genetic Syndromes*. New York: Wiley-Liss.

REFERENCES

Armstrong L, Abd El Moneim A, Aleck K, Aughton DJ, Baumann C, Braddock SR, Gillessen-Kaesbach G, Graham JM Jr, Grebe TA, Gripp KW, Hall BD, Hennekam R, Hunter A, Keppler-Noreoil K, Lacombe D, Lin AE, Ming JE, Kokitsu-Nakata NM, Nikkel SM, Philip N, Raas-Rothschild A, Sommer A, Verloes A, Walter C, Wieczorek D, Williams MS, Zackai E, Allanson J (2005) Further delineation of Kabuki syndrome in 48 well-defined new individuals. *Am J Med Genet* 132A:265–272.

Banka S, Veeramachaneni R, Reardon W, et al. (2011) How genetically heterogeneous is Kabuki syndrome? MLL2 testing in 116 patients, review and analyses of mutation and phenotypic spectrum. *Eur J Hum Genet* 20(4):381–388.

Banka S, Howard E, Bunstone S, Chandler K, Kerr B, Lachlan K, McKee S, Mehta SG, Tavares ALT, Tolmie J, Donnai D (2012) MLL2 mosaic mutations and intragenic deletion-duplications in patients with Kabuki syndrome. *Clin Gen* 83(5):467–471.

Bögershausen, N, Gatinois, V, Riehmer, V, et al. (2016) Mutation update for Kabuki syndrome genes KMT2D and KDM6A and further delineation of X-Linked Kabuki syndrome subtype 2. *Hum Mutat* 37(9):847–864.

Burke LW, Jones MC (1995) Kabuki syndrome: Underdiagnosed recognizable pattern in cleft palate patients. *Cleft Palate Craniofac J* 32:77–84.

Ciprero KL, Clayton-Smith J, Donnai D, Zimmerman RA, Zackai EH, Ming JE (2005) Symptomatic Chiari I malformation in Kabuki syndrome. *Am J Med Genet* 132A:273–275.

Courtens W, Rassart A, Stene J, Vamos E (2000) Further evidence for autosomal dominant inheritance and ectodermal abnormalities in Kabuki syndrome. *Am J Med Genet* 93:244–249.

Digilio MC, Marion B, Toscano A, Giannotti A, Dallapiccola B (2001) Congenital heart defects in Kabuki syndrome. *Am J Med Genet* 100:269–274.

Digilio MC, Gnazzo M, Lepri F, Dentici ML, Pisaneschi E, Baban A, Passarelli C, Caolino R, Angioni A, Novelli A, Marino B, Dallapiccola B (2017) Congenital heart defects in molecularly proven Kabuki syndrome patients. *Am J MedGenet A* 173(11):2912–2922.

Dinulos MB, Hudgins L (2000) Kabuki syndrome masquerading as velocardiofacial syndrome in the neonate. *Proc Greenwood Genet Cent* 19:100–101.

Dyamenahalli U, Abraham B, Fontenot E, Prasad V, Imamura M (2007) Pathologic aneurysmal dilation of the ascending aorta and dilation of the main pulmonary artery in patients with Kabuki syndrome: Valve-sparing aortic root replacement. *Congenit Heart Dis* 2(6):424–428.

Gatto LM, Sousa LH, Koppe G, Demartini Z (2017) Carotid artery occlusion in Kabuki syndrome: Case report and literature review. *Surg Neurol Int* 8(1):88.

Halal F, Gledhill R, Dudkiewicz A (1989) Autosomal dominant inheritance of the Kabuki make-up (Niikawa-Kuroki) syndrome. *Am J Med Genet* 33:376–81.

Ho HH, Eaves LC (1997) Kabuki make-up (Niikawa-Kuroki) syndrome: Cognitive abilities and autistic features. *Dev Med Child Neurol* 39:487–490.

Hoffman JD, Ciprero KL, Sullivan KE, Kaplan PB, McDonald-McGinn DM, Zackai EH, Ming JE (2005) Immune abnormalities are a frequent manifestation of Kabuki syndrome. *Am J Med Genet* 135A:278–281.

Hudgins L, Seaver L, Weyerts L, Jones M, Cassidy SB (1993) Connective tissue findings in Kabuki make-up syndrome. *Proc Greenwood Genet Cent* 12:125.

Hughes HE, Davies SJ (1994) Coarctation of the aorta in Kabuki syndrome. *Arch Dis Child* 70:512–514.

Iida T, Park S, Kato K, Kitano I (2006) Cleft palate in Kabuki syndrome: A report of six cases. *Cleft Palate Craniofac J* 43:756–761.

Isidor B, Rio M, Habes D, Amiel J, Jacquemin E (2007) Kabuki syndrome and neonatal cholestasis: Report of a new case and review of the literature. *J Pediatr Gastroenterol Nutr* 45:261–264.

Kawame H, Hannibal MC, Hudgins L, Pagon RA (1999) Phenotypic spectrum and management issues in Kabuki syndrome. *J Pediatr* 134:480–485.

Kobayashi O, Sakuragawa N (1996) Inheritance in Kabuki make-up (Niikawa-Kuroki) syndrome. *Am J Med Genet* 61:92–92.

Kuroki Y, Suzuki Y, Chyo H, Hata A, Matsui I (1981) A new malformation syndrome of long palpebral fissures, large ears, depressed nasal tip, and skeletal anomalies associated with postnatal dwarfism and mental retardation. *J Pediatr* 99:570–573.

Lederer D, Grisart B, Digilio MC, Benoit V, Crespin M, Ghariani SC, Maystadt I, Dallapiccola B, Verellen-Dumoulin C (2012) Deletoin of KDM6A, a histonedemethylase interacting with MLL2, in three patients with Kabuki syndrome.

Lehman L, Mazery A, Visier A, Baumann C, Lachenais D, et al. (2017) Molecular, clinical, and neuropsychological study in 31 patients with Kabuki syndrome and KMT2D mutations. *Clin Genet* 92:298–305.

Lindsley AW, Saal HM, Burrow TA, Hopkin RJ, Shchelochkov O, Khandelwal P, Xie C, Bleesing J, Filipovich L, Risma K, Assa'ad AH, Roehrs PA, Bernstein JA (2016) Defects of B-cell terminal differentiation in patients with type-1 Kabuki syndrome. *J Allergy Clin Immunol* 137(1):179–187.e10.

Makrythanasis P, Bon BV, Steehouwer M, et al. (2013) MLL2 mutation detection in 86 patients with Kabuki syndrome: A genotype-phenotype study. *Clin Genet* 84(6):539–545.

Matsumoto N, Niikawa N (2003) Kabuki syndrome: A review. *Am J Med Genet* 117C:57–65.

Matsune K, Shimizu T, Tohma T, Asada Y, Ohashi H, Maeda T (2001) Craniofacial and dental characteristics of Kabuki syndrome. *Am J Med Genet* 98:185–190.

Mervis CB, Becerra AM, Rowe ML, Hersh JH, Morris CA (2005) Intellectual abilities and adaptive behavior of children and adolescents with Kabuki syndrome: A preliminary study. *Am J Med Genet* 132A:248–255.

Mhanni AA, Cross HG, Chudley AE (1999) Kabuki syndrome: Description of dental findings in 8 patients. *Clin Genet* 56:154–157.

Ming JE, Russell KL, Bason L, McDonald-McGinn DM, Zackai EH (2003) Coloboma and other ophthalmologic anomalies in Kabuki syndrome. *Am J Med Genet* 123A:249–252.

Morgan AT, Mei C, Da Costa A, Fifer J, Lederer D, Benoit V, McMillin MJ, Buckingham KJ, Bamshad MJ, Pope K, White SM (2015) Speech and language in a genotyped cohort of individuals with Kabuki syndrome. *Am J Med Genet A* 167A:1483–1492.

Ng SB, Bigham AW, Buckingham KJ, Hannibal MC, McMillin MJ, Gildersleeve HI, Beck AE, Tabor HK, Cooper GM, Mefford HC, Lee C, Turner EH, Smith JD, Rieder MJ, Yoshiura K, Matsumoto N, Ohta T, Niikawa N, Nickerson DA, Bamshad MJ, Shendure J. Exome sequencing identifies MLL2 mutations as a cause of Kabuki syndrome (2010) *Nat Genet* 42(9):790–793.

Niikawa N, Matsuura N, Fukushima Y, Ohsawa T, Kajii T (1981) Kabuki make-up syndrome: A syndrome of mental retardation, unusual facies, large and protruding ears, and postnatal growth deficiency. *J Pediatr* 99:565–569.

Niikawa N, Kuroki Y, Kajii T, et al. (1988) Kabuki make-up (Niikawa-Kuroki) syndrome: A study of 62 patients. *Am J Med Genet* 31:565–589.

Ogawa A, Yasumoto S, Tomoda Y, Ohfu M, Mitsudome A, Kuroki Y (2003) Favorable seizure outcome in Kabuki make-up syndrome associated with epilepsy. *J Child Neurol* 18:549–551.

Paderova J, Drabova J, Holubova A, Vlckova M, Havlovicova M, Gregorova A, Pourova R, Romankova V, Moslerova V, Geryk J, Norambuena P, Krulisova V, Krepelova A, Macek Sr M, Macek Jr M (2018) Under the mask of Kabuki syndrome: Elucidation of genetic-and phenotypic heterogeneity in patients with Kabuki-like phenotype. *Eur J Med Genet* 61(6):315–321.

Sari BA, Karaer K, Bodur S, Soysal AS (2008) Case report: Autistic disorder in Kabuki syndrome. *J Autism Dev Disord* 38:198–201.

Schott DA, Gerver WJ, Stumpel CT (2016) Growth Hormone Stimulation Tests in Children with Kabuki Syndrome. *Hormone Res Paediatr* 86(5):319–324.

Schrander-Stumpel CTRM, Spruyt L, Curfs LMG, Defloor T, Schrander JJP (2005) Clinical data in 20 patients, literature review, and further guidelines for preventive management. *Am J Med Genet* 132A:234–243.

Shalev SA, Clarke LA, Koehn D, Langlois S, Zackai EH, Hall JG, McDonald McGinn DM (2004) Long-term follow-up of three individuals with Kabuki syndrome. *Am J Med Genet* 125A:191–200.

Tekin M, Fitoz S, Arici S, Cetinkaya E, Incesulu A (2006) Niikawa-Kuroki (Kabuki) syndrome with congenital sensorineural deafness: Evidence for a wide spectrum of inner ear anomalies. *Int J Pediatr Otorhinolaryngol* 70:885–889.

Toriello HV, Droste P (2003) Nocturnal lagophthalmos in Kabuki syndrome: Results of a parent group survey. *Proc Greenwood Genet Cent* 22:110.

Tsukahara M, Kuroki Y, Imaizumi K, Miyazawa Y, Matsuo K (1997) Dominant inheritance of Kabuki make-up syndrome. *Am J Med Genet* 73:19–23.

Vaux KK, Hudgins L, Bird LM, Roeder E, Curry C, Jones M, Jones KL (2005a) Neonatal phenotype in Kabuki syndrome. *Am J Med Genet* 132A:244–247.

Vaux KK, Jones KL, Jones MC, Schelley S, Hudgins L (2005b) Developmental outcome in Kabuki syndrome. *Am J Med Genet* 132A:263–264.

Verrotti A, Agostinelli S, Cirillo C, D'Egidio C, Mohn A, Boncimino A, Coppola G, Spalice A, Nicita F, Pavone P, Gobbi G, Grosso S, Chiarelli F, Savasta S (2011) Long-term outcome of epilepsy in Kabuki syndrome. *Seizure* 20(8):650–654.

White SM, Thompson EM, Kidd A, Savarirayan R, Turner A, Amor D, Delatycki MB, Fahey M, Baxendale A, White S, Haan E, Gibson K, Halliday JL, Bankier A (2004) Growth, behavior, and clinical findings in 27 patients with Kabuki (Niikawa-Kuroki) syndrome. *Am J Med Genet* 127A:118–127.

Wilson GN (1998) Thirteen cases of Niikawa-Kuroki syndrome: Report and review with emphasis on medical complications and preventative management. *Am J Med Genet* 79:112–120.

35

47,XXY (KLINEFELTER SYNDROME) AND RELATED X AND Y CHROMOSOMAL CONDITIONS

CAROLE SAMANGO-SPROUSE
Department of Pediatrics, George Washington University, Washington, DC, USA; Department of Human and Molecular Genetics, Florida International University, Miami, Florida, USA

JOHN M. GRAHAM JR
Department of Pediatrics, Cedars Sinai Medical Center, Harbor-UCLA Medical Center, Department of Pediatrics, David Geffen School of Medicine at UCLA, Los Angeles, California, USA

DEBRA R. COUNTS
Pediatric Endocrinology, Sinai Hospital, Baltimore, Maryland, USA

JEANNIE VISOOTSAK
Ovid Therapeutics, New York, USA

INTRODUCTION

47,XXY, or Klinefelter syndrome (KS), is the most frequently occurring sex chromosomal disorder with an estimated prevalence of 1 in 660 newborn males (Savic 2012). The endocrine aspects of the syndrome were initially reported in 1942, and an additional X chromosome was identified as the etiology of the disorder in 1959 (Klinefelter 1942; Jacobs and Strong 1959). Since these initial reports, the phenotypic profile has been characterized by varying degrees of disruption in neurodevelopmental, endocrine, and musculoskeletal systems, as well as brain morphology (Ross et al. 2012; Savic 2012). Early descriptions of children with 47,XXY included motor and speech delays, hypotonia, low activity level, and testicular dysfunction (Samango-Sprouse et al. 2013). The deficits in motor planning affect early speech and language development resulting in later language-based learning disorders, as well as delayed development of balance, motor planning and motor proficiency skills (Gropman and Samango-Sprouse 2013). These boys are typically not intellectually impaired; however, they may exhibit selective deficits in their language-based skills, including reading, social language, and expressive language (Graham et al. 1988; Ross et al. 2008). Males with more than one extra sex chromosome (e.g., 48,XXYY, 48,XXXY, and 49,XXXXY) are significantly less common than 47,XXY but have more complex physical and behavioral manifestations (Visootsak et al. 2007; Gropman and Samango-Sprouse 2013; Samango-Sprouse et al. 2013; Samango-Sprouse and Gropman 2016; Samango-Sprouse et al. 2015).

Incidence

Initial studies of males with 47,XXY (KS) were extremely biased because study subjects were typically ascertained from mental or penal institutions, rather than from population-based studies. In the 1960s and early 1970s, newborn screening for sex chromatin abnormalities was initiated because there was a need for unbiased prospective developmental outcome studies. 47,XXY is the most common sex chromosomal aneuploidy, with a frequency of 1 in 426 to 1 in 1000 males (Bojesen et al. 2003). Several more recent studies suggest that the prevalence may be somewhere between 1 in 581 and 1 in 917 male births (Visootsak et al. 2013; Samango-Sprouse et al. 2016). These numbers are even more striking because as many as 64% of males with 47,XXY are never diagnosed, 10% are diagnosed prenatally by amniocentesis, and 26% are diagnosed during childhood or adulthood when they present with developmental delay, behavioral problems, hypogonadism, gynecomastia, or infertility (Re and Birkhoff 2015). Information on long-term mortality risks is limited owing to the lack of longitudinal studies and under ascertainment of males with these disorders. There are several current studies from international cohorts that demonstrate increased mortality risks secondary to metabolic, cardiovascular, and hemostatic complications (Bojesen et al. 2004, 2011). Further study is warranted on the interaction of 47,XXY, mortality risk, timing of diagnosis, socioeconomic status (SES), and institution of hormonal replacement therapy (Figure 35.1).

Diagnostic Criteria

Males with 47,XXY may present with a wide spectrum of clinical features that are typically age-related, but intellectual deficits in 47,XXY are very uncommon unless there are other additional confounding factors, such as severe prematurity, copy number variants (CNVs), or brain malformations. In a recent study, there was an increased incidence of CNVs in association with 47,XXY, where duplications were more common than controls (Rocca et al. 2016). The presence of a CNV and its impact on neurodevelopmental performance may further expand our understanding of the variability commonly seen in 47,XXY (Rocca et al. 2016).

In infancy and early childhood, boys with 47,XXY may have hypospadias, small phallus, or cryptorchidism, which may result in an early postnatal diagnosis. Boys with 47,XXY typically manifest early motor and speech delay, which has been characterized as an infantile presentation of developmental dyspraxia (Samango-Sprouse and Rogol 2002; Samango-Sprouse et al. 2013). School-aged boys may have language-based learning disabilities, reading dysfunction, and executive dysfunction with secondary behavioral problems. In adolescence, delayed or incomplete pubertal development with elevated gonadotropins (follicle stimulating hormone and luteinizing hormone), eunuchoid body habitus, gynecomastia, and small testes are typically noted in untreated males with 47,XXY. Adults with 47,XXY often present with infertility, small testes, signs and symptoms of testosterone deficiency, social difficulties associated with impaired executive functioning skills, and, rarely, with breast malignancy. Because their clinical features are non-specific during childhood, most cases are not diagnosed until adolescence or adulthood, when incomplete puberty or infertility become apparent and a karyotype or chromosomal microarray (CMA) is completed as part of the evaluation for these features. Thus, karyotype or a CMA confirmation is necessary for diagnosis.

The manifestations of 47,XXY may be quite variable and subtle; however, many phenotypic abnormalities suggest the need for a chromosomal microarray. Most boys with 47,XXY do not have distinctive facial dysmorphology; however, 47,XXY is a common cause of developmental delay, language-based learning disabilities, ADHD, and executive dysfunction of an unknown etiology among prepubertal boys. Thus, it is important to consider a chromosomal microarray in any male who presents with the following (in the order of their appearance):

- Motor delay or dysfunction
- Speech and language difficulties
- Behavioral problems with anxiety
- Attention-deficit hyperactivity disorder
- Dyslexia or reading dysfunction
- School failure or learning disabilities
- Psychosocial disturbances, including anxiety or depression
- Small testes and hypergonadotropic hypogonadism
- Obesity
- Infertility (azoospermia, oligospermia)
- Gynecomastia
- Osteoporosis in a young or middle-aged man.

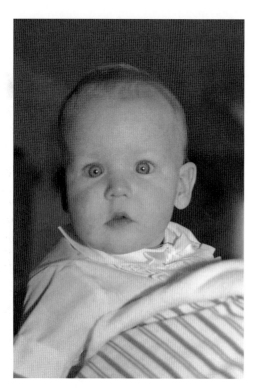

FIGURE 35.1 Infant with 47,XXY.

Etiology, Pathogenesis, and Genetics

Recent studies suggest that the odds of having a child with 47,XXY are statistically higher if maternal age at the time of conception is over the age of 35 years (Morrison et al. 2008). Mean maternal age of offspring with 47,XXY is typically increased; however, with the advent of non-invasive prenatal testing (NIPT), maternal age has become increasingly younger (Samango-Sprouse et al. 2013, 2017). Paternal age is not increased, even in those cases that are paternally derived (Thomas et al. 2000; Samango-Sprouse et al. 2013).

Using polymorphic DNA markers, the additional X chromosome has been shown to be of paternal origin in 50–60% of cases, and maternal in origin in the remaining 40–50% (Iitsuka et al. 2001). In paternally derived cases of 47,XXY, X–Y nondisjunction occurs in the first meiotic division. Among maternally derived cases, 48% of nondisjunction occurs in meiosis I, 29% in meiosis II, 16% show postzygotic origin, and in 7%, origin is unknown (Thomas et al. 2000). Increased maternal age is associated with errors in maternal meiosis I (Thomas et al. 2000), but not in meiosis II (Hassold et al. 1991). Perturbations of recombination play a pivotal role in sex chromosome aneuploidy. Among 65 maternal meiosis I cases studied by Hassold et al. (1991), no recombination occurred in 37 (expected 60), whereas in 28 cases, only one recombinant was observed versus the expected 60. In meiosis II errors, there was an alteration in the location of recombinational events, and recombination occurred in the pericentromeric region, which is ordinarily an uncommon location (Thomas and Hassold 2003). In paternal origin cases, recombination between paternal X and paternal Y was observed far less than expected in meiosis I, with two-thirds of cases failing to show an exchange (Thomas et al. 2000). Consistent with the above, only a slight paternal age effect is observed, and there is no relationship between paternal age and X–Y recombination frequency (Thomas et al. 2000). Thus, sex chromosomal aneuploidy of 47,XXY usually reflects decreased recombination just as in autosomal aneuploidy (trisomy).

In humans, not all loci on the X chromosome(s) in excess of one undergo inactivation. Thus, it is assumed that the abnormal phenotype in males with 47,XXY reflects inappropriate transcription of genes (excess gene products) encoded on the X chromosome. Most (90%) of the genes that escape X inactivation are on the X short arm, but it is not known how many genes are involved in causing the phenotype of 47,XXY, or how many are pivotal. Although an X chromosome gene dosage effect is most likely, other genetic mechanisms could exist, and, in fact, have been hypothesized to explain the variable phenotype among males with 47,XXY. One obvious possibility is that a nondisjunction event in meiosis II or postzygotic division produces two copies of the same X chromosome, potentially leading to two dysfunctional X-linked alleles. To date, studies have not demonstrated a difference in phenotype between paternally and maternally derived cases of 47,XXY. Iitsuka et al. (2001) reported an unusually high frequency of skewed X-inactivation in 47,XXY, perhaps providing another clue to explain the variability in the neurodevelopmental manifestations, both behaviorally and cognitively.

Diagnostic Testing

With the advent of non-invasive prenatal testing (NIPT), there has been an increase in early prenatal detection of boys with 47,XXY (Samango-Sprouse et al. 2017). NIPT utilizes a maternal blood sample that is used to analyze cell-free fetal DNA (cffDNA) in maternal blood to check for chromosomal abnormalities and can be completed as early as 10 weeks of pregnancy. The American College of Obstetricians and Gynecologists (ACOG) reports that NIPT has a sensitivity of 91.0% and specificity of 99.6% to identify sex chromosome abnormalities in patients who receive interpretable results. Although these tests are highly accurate, it is not a diagnostic test. Thus, NIPT results requires cytogenetic confirmation with amniocentesis, karyotyping, or chromosomal microarray (CMA) testing at birth (Samango-Sprouse et al. 2017).

While karyotyping had previously been suggested as the first-tier detection test for patients with unexplained developmental disabilities, congenital abnormalities, and dysmorphic features, recent research has revealed that chromosomal microarrays (CMAs) have a 12.2% higher diagnostic yield than karyotyping (Miller et al. 2010). Thus, CMAs are now considered the first-tier detection test for 47,XXY, in which the presence of one extra X chromosome will confirm this diagnosis (Manning and Hudgins 2010). This DNA based technique may be completed as an oral cheek (buccal) swab or as a blood draw. CMA testing is used to detect X chromosome variations, copy number variants (CNVs), microdeletions and microduplications, abnormalities in telomeres, and genes that are disrupted by breakpoints in the genome (Manning and Hudgins 2010). Fluorescence in situ hybridization (FISH) studies with chromosome-specific probes can also be used to identify the sex chromosome constitution.

Differential Diagnosis

Given the non-specific nature of most features of 47,XXY (KS), we would suggest that many disorders need to be considered in the differential diagnosis. In particular, disorders characterized by hypogonadotropic hypogonadism (e.g., Kallman syndrome) and developmental delay (e.g., Fragile X syndrome, Chapter 28) must be in the differential diagnosis. Kallman syndrome includes the triad of hypogonadotropic hypogonadism with deficient olfaction, and it can result from mutations in a number of different genes (*KAL1*, *FGFR1*, *PROKR2*, and *PROK2*). Fragile X syndrome is an

X-linked disorder, resulting from the expansion of a CGG trinucleotide repeat at the 5′ end of the *FMR1* gene, which shuts down the transcription of the FMR1 protein. It causes more severe cognitive disability and testicles are usually enlarged rather than atrophic (Lubs and Schwartz 2012). These disorders require molecular confirmation and often are associated with more complex neurodevelopmental presentations than 47,XXY. Other common disorders associated with hypogonadotropic hypogonadism include Prader–Willi syndrome (see Chapter 46), mutations in Dax 1, and Bardet–Biedl syndrome (see Chapter 8). These disorders are associated with hypotonia, obesity, and mild cognitive disability that is usually more severe than that seen in males with 47,XXY. Prader–Willi syndrome is diagnosed by methylation analysis of chromosome 15. Individuals with Prader–Willi syndrome have a different behavioral phenotype characterized by obsessions and tantrums. Bardet–Biedl syndrome is an autosomal recessive disorder characterized by polydactyly, retinitis pigmentosa, and cystic renal disease. This disorder is caused by mutations in genes involved in ciliary morphogenesis, and it is genetically heterogeneous with over 20 different genes known to cause this phenotype. Confirmation of Bardet–Biedl syndrome requires molecular genetic testing.

MANIFESTATIONS AND MANAGEMENT

Growth and Feeding

Feeding disruptions may occur more frequently in 47,XXY than in the general population and may be the earliest indication of the oral motor-based dysfunction in the child with 47,XXY. In a large sample of prenatally diagnosed infants with 47,XXY, more than half were reported to have early feeding difficulties in coordinating sucking and swallowing for breastfeeding. This often resolved with time or when the infants were transitioned to bottle feedings (Samango-Sprouse et al. 2017, unpublished data).

The height of boys with 47,XXY under age 3 years of age falls within the normal range, but thereafter, the distribution of height becomes more skewed. Significantly fewer than expected boys with 47,XXY have heights below the 25th percentile. Their average height increases from the 30th percentile before age 2 years to 60th percentile by age 8 years, reaching the 75th to the 90th percentile by 18 years (Figure 35.2). This height increase usually occurs between ages 5 and 8 years (Bardsley et al. 2013).

Adult males with 47,XXY are commonly taller than average with mean height of 190.5 cm, and often have atypical body proportions with relatively long legs and a decreased upper-to-lower body segment ratio (Bandmann et al. 2012). The childhood-onset tall stature in 47,XXY is likely related to the presence of three copies of the X and Y

FIGURE 35.2 Adolescent with 47,XXY, presenting with above average height.

chromosome height determining gene, *SHOX*, and delayed epiphyseal fusion as a consequence of decreased testosterone levels (Fideleff et al. 2016). Other somatic findings include narrow shoulders (2 cm decrease), wide hips (1 cm increase), scoliosis and kyphosis secondary to ligamentous laxity, and truncal hypotonia (Simpson et al. 2003; Sprouse et al. 2013). Males with 47,XXY differ from other eunuchoid individuals in that their arm span is only 2–3 cm more than their height, whereas in most other types of eunuchoidism, arm span is usually at least 4 cm more than height. Bone age for those with 47,XXY is usually normal except in adolescent males with severe androgen deficiency.

Untreated hypogonadism is often associated with the development of truncal obesity during adolescence. Testosterone replacement combined with proper nutrition and an appropriate exercise program that incorporates endurance training, strengthening, and aerobic exercise can be highly effective in weight management for boys with 47,XXY. Testosterone therapy during mini-puberty, which is a period of accelerated hormonal production that typically occurs in the middle of infancy, also fosters improved neurodevelopment and reduction of behavioral issues (Ross et al. 2008; Samango-Sprouse et al. 2013, 2015). At the early signs of pubertal development, testosterone therapy is often initiated to reduce feminized appearance, foster the secondary sexual characteristics, and develop good bone health (Mehta et al. 2013; Stagi et al. 2016). The combination of obesity and osteoporosis in untreated males with 47,XXY may result in secondary orthopedic and other health complications (Bojesen et al. 2011).

Evaluation

- Height, weight, and head circumference should be routinely measured and plotted on standard growth charts at every well-child care visit, as is standard of care in any child presenting to primary care. Assessment of arm span-to-height ratio and upper-to-lower segment ratio can be performed in the prepubertal years. Since more than 80–90% of the boys with 47,XXY have truncal hypotonia with an increased incidence of kyphosis, scoliosis, and pes planus well-child care visits are ideal times to monitor these musculoskeletal issues.
- Weight and body mass index (BMI) should be monitored during regular visits to assess the presence of obesity. This is particularly important during adolescence and adulthood when testosterone level decreases without appropriate institution of androgen therapy.
- Yearly thyroid testing is warranted, since there is an increase in thyroid disease in a subset of boys with 47,XXY, especially if there is positive family history of thyroid dysfunction.

Management

- Feeding difficulties in infancy should be addressed with a lactation specialist to foster improved coordination of sucking and swallowing in the newborn period. If the feeding disturbances do not resolve within the first week to ten days, then an evaluation by either a pediatric speech and language pathologist or a pediatric occupational therapist is indicated.
- There are many kinds of strategies that may assist the infant with feeding disturbances, but it should be instituted under the guidance of appropriate pediatric trained specialists. Feeding disturbances may be particularly challenging for a primiparous female with an infant prenatally diagnosed with 47,XXY. The combination of anxiety and insecurity in a first-time mother may exacerbate these mild feeding disturbances, but this can be quickly corrected with parent coaching and support. The etiology of the discoordination is unknown, but it may be the earliest indication of the praxis deficits or sensory dysfunction, both of which are more common in boys with 47,XXY.
- Because of decreased muscle tone and androgen deficiency, an active life health style should be encouraged from infancy through adulthood. Boys with 47,XXY benefit from both structured team sports and leisure activities that encourage endurance, strengthening, and coordination. The secondary gains from an active life include improved self-esteem, greater opportunities for social interactions, and enhanced health and well-being.

Development and Behavior

Motor Planning. There is an increased incidence of truncal hypotonia and motor planning disturbances in the first year of life for boys with 47,XXY (Samango-Sprouse and Rogol 2002: Samango-Sprouse et al. 2013). Although the prenatally diagnosed boys benefit from early detection, it is critical that prone development be strongly encouraged through tummy time and limiting supine position since the boys with 47,XXY have increased incidence of positional torticollis with plagiocephaly, secondary to truncal hypotonia and inactivity. Neurodevelopmentally, prone progression fosters weight bearing and strengthening for the upper trunk which is an area of weakness in these boys. An evaluation by a pediatric physical therapist is very helpful for the family and the infant with 47,XXY to optimize gross motor development.

Early speech delay is evident between 6 and 9 months of age in infants with 47,XXY. Typically, they demonstrate deficiencies in phonemic development, motor imitation, and decreased vocalizations. Expressionless facial appearances are noted in infants, probably secondary to oral facial hypotonia. Speech delay is characterized by limited variegated sounds, delayed reciprocal synchrony with primary caregivers, and "quiet baby syndrome" with decreased vocalization prior to 12 months of age (Samango-Sprouse and Rogol 2002; Simpson et al. 2003) (Figure 35.3).

FIGURE 35.3 "My baby is so good and so easy; he never cries." Mother of 9-month-old 47,XXY.

These features have been described as an infantile presentation of developmental dyspraxia (IDD) and are not simply an indication of speech or expressive language delays (Samango-Sprouse and Rogol 2002). It is thought that these differences are related to motor-based planning deficits that impact many areas of development. Speech and language evaluation is appropriate by 12 months of age by a pediatric speech and language pathologist in order to identify the speech deficits common to a boy with 47,XXY, and to develop an appropriate treatment program.

Motor planning, which is the ability to envision a motor act, plan, and carry it out in the correct sequence from beginning to end, is typically difficult for boys with an extra X (Samango-Sprouse and Gropman 2016). When given verbal instructions, the required motor action must be correctly identified and then integrated in order to perform, and many boys with extra X have decreased neural connections in the associated brain regions. This lack of motor integration results in an inability to plan and execute tasks quickly and efficiently, so that these boys can appear clumsy as they execute novel tasks. Their difficulty with sensory processing can lead to poor motor planning for both fine and gross motor tasks, such as handwriting, jumping, skipping, and other complex actions. These motor planning deficits may impact all aspects of learning and may contribute to the behavioral issues associated with 47,XXY. These skill deficits become more evident as complex motor actions increase. These are a daily occurrence during the preschool years, and subsequently, the neurodevelopment challenges for these boys become more evident.

Social Language. Socially, boys with 47,XXY have been described as shy, timid, and prone to social isolation and increased anxiety. This is particularly the case in those without early intervention and biological treatment (Samango-Sprouse et al. 2013, 2015; Van Rijn and Swaab 2015). Some reports describe boys with 47,XXY as having diminished self-esteem and insecurities in self-perception that may be due to a variety of factors, including physical features, decreased motor ability, or learning differences associated with their long-term androgen deficiency (Ross et al. 2012). Early hormonal treatment (EHT), which is ideally given between 4 months and 15 months of age, and testosterone replacement therapy have been shown to mitigate these physical and emotional differences, while also having positive effects on self-esteem, well-being, energy level, and mood (Mehta and Paduch 2012; Samango-Sprouse et al. 2013, 2015).

Language-based learning disorders are present in the majority of the boys with 47,XXY, occurring in more than 90% of the boys (Samango-Sprouse et al. 2014). Specific language-based deficits develop by school age and present as early speech delays, involving later difficulties with word retrieval, syntax, and narrative construction. These language difficulties may complicate social exchanges and social cognition, while also impairing everyday face-to-face interactions for these boys (Visootsak and Graham 2009). Studies have reported that boys with 47,XXY may not recognize emotions that are expressed tacitly through facial expression in conversation, nor pick up on such linguistic cues as sarcasm (Van Rijn et al. 2008). Adult men with 47,XXY who do not receive any biological treatment have higher rates of autism spectrum disorder (ASD) and experience social difficulties such as shyness, social withdrawal anxiety, impulsivity, reduced social assertiveness, and issues with emotion regulation (Van Rijn et al. 2008; Ross et al. 2012). These social communicative deficits, however, are lessened with EHT in infancy and with testosterone therapy in adolescence (Mehta and Paduch 2012; Samango-Sprouse et al. 2015).

47,XXY boys who are not treated with any hormonal therapy display deficits in prosody and pragmatics (Samango-Sprouse et al. 2015). Prosody refers to non-semantic cues in spoken language. These may include linguistic cues, emphasizing important parts of the message, or presenting information as a statement or a question. There may also be affective functions that include variations in voice tones, such as intonation or loudness. Difficulties in the recognition of pragmatic communication cues in conversation may impact how well men with 47,XXY cope with social situations, specifically detecting interpersonal intentions and forming appropriate responses (Van Rijn et al. 2007). It is important to note that many of these studies are confounded by the fact that study participants who were receiving hormonal therapy were not bifurcated from study participants who had never received any hormonal therapy. Further research is needed to clarify this distinction, as it could significantly impact the deficits of social language reported for the 47,XXY population.

ADHD. In terms of learning and behavior, attention-deficit/hyperactivity disorder (ADHD), dyslexia, and communicative disorders are more common in 47,XXY boys. One study found elevated ADHD symptoms in 41% of 47,XXY boys (as per the Conners' Parent Rating Scale – Revised (CPRS-R), a parentally reported rating scale of child behavior) (Ross et al. 2012). Another study found a rate of 63%, as per a DSM-IV interview (Bruining et al. 2009). These rates are far above the background incidence of 4–5%, and are similar to, if not slightly lower than, those assigned to boys with 47,XXY (Ross et al. 2012).

These studies should be viewed cautiously, however, since several salient features were not accounted for in the analysis, including the presence of ADHD in the family. The presence of familial ADHD or other learning difficulties, such as dyslexia, play a large role in the likelihood that a child with 47,XXY will share those disorders (Samango-Sprouse et al. 2013). Whether the child was diagnosed prenatally or postnatally also impacts the incidence of learning disorders. Therefore, the incidence of 47,XXY and ADHD without confounding factors is not well understood. However, boys with 47,XXY should be monitored for ADHD throughout elementary school and through young adulthood.

Autism Among individuals with 47,XXY studied longitudinally after ascertainment through newborn chromosome surveys, Robinson et al. (1979) found 32% to have delayed emotional development, compared with 9% of their siblings ($p < 0.002$). Males with 47,XXY have deficits in social interactions possibly secondary to language-based learning difficulties, social cognition impairments, and language delay (Visootsak and Graham 2009). Studies in adults with 47,XXY have indicated deficits in social cognition with decreased perception of social-emotional cues and increased autistic traits (Van Rijn et al. 2006, 2008). When compared with men from the general population (mean age 35.7 years), men with 47,XXY (mean age 39.2 years) were reported to engage less often in social behaviors involving expression of negative emotions, such as refusing a request or standing up for one's rights in a public situation (Van Rijn et al. 2008). These results suggest that social difficulties may continue into adulthood. Autistic traits were more common in men with 47,XXY, and Van Rijn et al. (2008) suggest that individuals with 47,XXY may be vulnerable to autistic traits. Men with 47,XXY have deficits in the perception of social-emotional cues and experience increased levels of emotional arousal, yet they are less able to identify and verbalize their emotions in comparison with the general population (Van Rijn et al. 2006). Further investigation is warranted to understand the complex interaction between speech and language development and social difficulties. Although the data reported by van Rijn et al. (2006) are intriguing, the vast majority of the men with 47,XXY in this study were not treated with hormone replacement therapy and were diagnosed late in life. The deficits described in that population may have been compounded by the late identification as well as untreated hormone deficiency.

Behavioral Issues. Boys with 47,XXY may face behavioral issues, despite being described as "characteristically reticent and lacking in confidence." Behavioral problems are variable in severity and incidence, although the child with a prenatal diagnosis presents with fewer problems (Ross et al. 2012). Additionally, boys receiving early hormonal treatment (EHT) in infancy (ideally between 4 months and 15 months of age) or hormonal replacement therapy when boys with 47,XXY are around 11–12 years old may have fewer problems than the untreated child or the child postnatally diagnosed (Ross et al. 2012; Samango-Sprouse et al. 2015, unpublished data). Some of these problems may originate from frustration stemming from a relatively low expressive ability as compared to receptive skills (Simpson et al. 2003; Van Rijn et al. 2006). Because EHT increases VIQ and social communication (Samango-Sprouse et al. 2013), it is possible that EHT alleviates some of these behavioral issues by increasing frontal function and self-modulation, while simultaneously also enhancing expressive ability (Samango-Sprouse et al. 2013). While the basis of this behavioral problem amelioration is undetermined, recent research studies have shown a significant improvement in this domain. EHT has been shown to decrease autistic mannerisms, social communication problems, and social cognition difficulties as determined with SRS-2 (Samango-Sprouse et al. 2015). EHT also decreases aggression and affective problems based on the CBCL (Samango-Sprouse et al. 2015)

On the Connors Parent Rating Scale-R, elevated behavior problems (in untreated boys) were observed in all subtests, as compared to controls (Ross et al. 2012). These included: inattention, social problems, and anxiety/shyness. It is important to note, however, that there is great variability within the population of boys with 47,XXY. Some boys exhibit no behavioral problems, while others have difficulties with peer interactions, social engagement, or resolving confrontational situations with peers when they interact in school or social settings.

Neurological Structure and Function. Emerging neuroimaging technology has increased and improved our understanding of the relationship among brain development, neurocognition, and behavioral outcome – especially in boys with 47,XXY (Giedd et al. 2007). Studies on boys with 47,XXY utilizing these neuroimaging techniques have revealed reduced total brain volumes that are specifically seen in the frontal, caudate, and temporal (especially left) regions of the brain (Giedd et al. 2007; Bryant et al. 2012). Abnormalities in frontal and caudate brain MRIs are similar to those seen in MRIs of boys with ADHD, and indicative of the executive dysfunction seen in boys with 47,XXY (Giedd et al. 2007; Van Rijn and Swaab 2015). The temporal lobes are associated with language capacities involving reading, social language, and processing of spoken information – all of which are notably impaired in untreated males with 47,XXY (Savic 2012). Abnormalities in the caudate nucleus are thought to adversely affect speech and language, as well as to manifest as the dyspraxia and oral motor dysfunction that is often found in 47,XXY boys (Giedd et al. 2007). The gray matter density in the insula region of the brain in these boys is also decreased, which is linked to social and emotional processing issues (Nagai et al. 2007). The parietal lobe, however, is relatively unaffected when measured by cortical thickness and volume (Giedd et al. 2007). The preservation of this region is evident in the enhanced spatial cognitive skills in males with 47,XXY (Samango-Sprouse and Law 2001; Savic 2012).

These neuroanatomical findings in 47,XXY boys have revealed several salient characteristics that are morphologically different from neurotypically developing peers. Several studies, however, have suggested that more normalized brain development is possible through the utilization of hormonal treatment (Patwardhan et al. 2000; Samango-Sprouse et al. 2011). Patwardhan et al. (2000) compared two groups of 47,XXY individuals (one receiving hormonal treatment therapy versus no treatment) and found that temporal gray matter was preserved in the treated group, but diminished in the untreated group. Further studies are warranted to confirm these findings and investigate whether other abnormal brain areas, as described above, show similar normalization after hormonal treatment therapy.

Evaluation

- Comprehensive annual pediatric evaluations by a team of specialists who are familiar with 47,XXY, and the constellation of this team may change with time since the priorities for care evolve. Strategic players for multidisciplinary team are a pediatric clinical geneticist, physical therapist, pediatric endocrinologist, pediatric neuropsychologist, speech and language pathologist, and special educator; who are all important throughout childhood. Visits with a pediatric endocrinologist may be beneficial between 3 months and 6 months, 5–8 years, and after age 11 years, since these are critical junctures for boys with 47,XXY. Annual pediatric endocrine visits are typically recommended from 11 to 21 years of age. (Figure 35.4)
- Comprehensive pediatric physical therapy evaluation is recommended, beginning in infancy, to assess gross motor skills, muscle tone, movement patterns, and motor coordination. Ideally, this evaluation would be completed prior to 4 months of age for the early identification of positional torticollis and to implement necessary effective treatment to reduce secondary consequences that may result from neurodevelopmental progression.
- Comprehensive pediatric speech and language evaluation to identify deficits should be completed by 12 months of age. Speech and language treatment/services should be initiated when there is a discrepancy greater than four months between expressive language in comparison to receptive language and cognitive domains. This is an indication of motor speech disturbances, and it is highly associated with later language-based learning disorders in the infant with 47,XXY.
- Psychoeducational evaluation during the school years is essential to assess cognition, achievement, and adaptive function and to plan appropriate strategies to optimize the child's learning and alleviate risks for behavioral disruption.
- An evaluation for ASD should be considered in males with 47,XXY with atypical social development and stereotypic behaviors (e.g., hand flapping, rocking, spinning). The appropriate assessment is the Autism Diagnostic Observation Scale-2 (ADOS-2) for the determination of ASD. The Ages Stages Questionnaires: Social-Emotional, Second Edition (ASQ-2), the Childhood Autism Rating Scale, Second Edition (CARS-2), and the Social Responsiveness Scale, Second Edition (SRS-2) may serve as screening tools for ASD if there are concerns prior to an ADOS-2 assessment.
- Pediatric psychological and/or pediatric psychiatric evaluation and possible treatment should be considered if there are any indications of ADHD and/or anxiety by school age in order to diminish secondary behavioral problems.

Management

- Once the individual or fetus is diagnosed with 47,XXY, it is important to seek consultation with medical professionals and health care professionals who are familiar with 47,XXY for recommendations regarding resources, appropriate biological and neurodevelopmental therapies, as well as medications for ADHD or anxiety.
- Early interventional therapies (e.g., physical, occupational, and speech therapies) are recommended throughout early childhood when discrepancies or deficits are identified to enhance early neurodevelopmental outcomes.
- Early hormonal treatment (EHT) with testosterone beginning between 4 and 15 months of age has shown a beneficial impact on multiple domains of development including neurocognition, motor development, speech and language, and social-emotional development. EHT in the form of three injections of testosterone enanthate of 25 milligrams once per month over a three-month period should be considered as treatment for the infant with 47,XXY.
- Testosterone replacement therapy in adolescence may be beneficial in alleviating certain issues, such as motor

FIGURE 35.4 Infant with positional torticollis.

deficiencies, social communication problems, social cognition problems, aggression, and affective problems based on the Child Behavior Checklist (CBCL) (Mehta and Paduch 2012; Mehta et al. 2013; Wosnitzer and Paduch 2013; Samango-Sprouse et al. 2015).

- Physical therapy is indicated when there is hypotonia, motor delay, and/or poor coordination and is most effective between 4 and 18 months in order to develop independent ambulation skills.
- Occupational therapy should be considered for the boys with decreased muscle tone in the trunk or upper body because these deficits will affect handwriting, posture, attention, and eventual school success. This type of evaluation may be most beneficial between 4 and 6 years of age and typically is needed for 12 months.
- Specific speech and language therapies should address speech delays with motor planning deficits, language formulation abnormalities, and syntactical delays.
- Because of decreased muscle tonus and androgen deficiency, an active health style is encouraged from infancy through adulthood. Boys with 47,XXY benefit from both structured team sports and leisure activities that encourage endurance, strengthening, and coordination. The secondary gains from an active life include improved self-esteem, greater opportunities for social interactions, and enhanced health and well-being (Figure 35.5).
- Activities that are dynamic and develop strength, such as karate and swimming, are helpful to develop motor planning skills, endurance, and coordination. Sport activities and/or physical therapy are warranted to encourage muscle development and facilitate social interactions with typically developing peers on the playground and in after-school activities.
- Exposure to computers early may be quite helpful in the development of academic skills, as well as keyboarding skills to minimize the impact of the graphomotor dysfunction and dysgraphia associated with 47,XXY (Figure 35.6).
- Social skills training may help develop appropriate understanding of social-emotional cues, pragmatic communication, and self-esteem.
- Psychopharmacological intervention may need to be coupled with interventional strategies, including speech therapy, sensory integration therapy, tailored behavioral treatments, and psychological counseling to minimize behavioral issues.

Endocrine

Hypergonadotropic Hypogonadism. In typical male infants, testosterone levels start to rise in the first month of life, peak at 2–4 months, and decline to prepubertal levels by 6 months of age. Ross et al. (2005) indicated that this neonatal surge in testosterone was attenuated in boys with 47,XXY with evidence of impaired Leydig cell function and decreased levels of testosterone in the first 6 months of life, which is consistent with early testicular failure. The etiology of early testicular failure in infants with 47,XXY warrants further studies to understand the clinical implications. Early hormonal treatment (EHT) to replace this neonatal surge is indicated to improve neurodevelopment and reduce behavioral issues (Ross et al. 2008; Samango-Sprouse et al. 2013, 2015).

FIGURE 35.5 Adolescent with 47,XXY engaging in skiing activity.

FIGURE 35.6 Adolescent with 47,XXY playing soccer.

Testosterone levels are typically in the low normal range and remain low in childhood and adulthood (Mehta et al. 2013) in males with 47,XXY. Indeed, testosterone levels in childhood may be difficult to interpret because the levels are low with a wide range of normative values and variability in individual assays. In adolescents with 47,XXY, serum testosterone levels decrease to low or below normal (Plymate and Paulsen 1989) with serum free and bioavailable testosterone levels predictably decreased as a consequence of increased sex hormone binding globulin levels (Plymate et al. 1983). In adults, follicle-stimulating hormone (FSH) and luteinizing hormone (LH) levels are usually elevated. Leydig cell dysfunction is present. Testicular aromatization of testosterone to estradiol results from over-expression of aromatase CYP19 (Wosnitzer and Paduch 2013). Serum estradiol levels are frequently elevated, but not to the degree seen in estrogen secreting neoplasms. Thus, the estradiol-to-testosterone ratio is almost always increased, and this may contribute to the development of gynecomastia.

Elevations in LH and FSH levels reflect deficient testicular function and the consequent lack of feedback inhibition by sex steroids (Christiansen et al. 2003). Serum gonadotropins are elevated, even when the serum testosterone levels are in the normal range. Although this has been interpreted as a sign of compensated hypogonadotropic hypogonadism, it is possible that partial androgen resistance may exist. The pituitary gland itself is responsive, as judged by normal response to gonadotropin-releasing hormone. Feedback suppressibility of LH and FSH to exogenous testosterone is qualitatively normal, but the level at which suppression occurs may be increased. Characteristic features of hypogonadism, in particular hypoandrogenism, become more evident in the adolescent male with 47,XXY, when delay of puberty is observed. Delayed puberty is clinically evident in about half the men with 47,XXY, and their secondary sexual development may not mature as fully or as rapidly. Pubic hair may be more feminine in distribution; facial and body hairs are scanty, and few shave daily without institution of androgen treatment. Acne is rare during adolescence, and temporal scalp hair recession often fails to occur. Obesity, gynecoid fat distribution, and poor muscular development are also frequent without appropriate treatment. Sexual activity may be reduced (Raboch et al. 1979). There may be increased fatigue and mood disturbance, which can be minimized by appropriate and timely biological treatment.

Increased parenchymal breast tissue, as determined by palpation, is evident in 50–75% of males with 47,XXY (Raboch et al. 1979). Approximately 20% of adolescent and adult males have been reported to have overt gynecomastia (Becker 1972). It is assumed that gynecomastia reflects an increased serum estradiol-to-testosterone ratio. Breast enlargement is characterized by increased collagenous material in interglandular spaces; the ductal epithelium is only slightly hyperplastic. Azoospermic infertility is the rule in 47,XXY with typical testicular biopsies revealing Leydig cell hyperplasia, loss of germ cells, sclerosis of the germ cell compartment, and thickened tubular member basement membranes (Wiesner et al. 2002).

In 47,XXY, external genitalia are usually well differentiated. Ross et al. (2005) reported that boys with 47,XXY aged 1–23 months had significantly smaller testicular and penile sizes. The prostate may be smaller than usual in some but not all men with 47,XXY, which is presumably a reflection of decreased serum testosterone levels. Prepubertal testes of males with 47,XXY are borderline small in size, but may be clinically indistinguishable from normal. Prepubertal boys with 47,XXY display neither seminiferous tubule atrophy nor apparent Leydig cell (interstitial cell) hyperplasia. Primary spermatogonia may be reduced in number, but they are almost always present in early childhood (Ferguson-Smith 1959), only to become absent by the time of expected onset of puberty. At that time, germ cell attrition becomes dramatic, eventually resulting in tubules containing only Sertoli cells. Inhibin B levels are normal before puberty but fall dramatically to low levels in adult men with 47,XXY (Christiansen et al. 2003), reflecting apoptosis. Mouse models of 47,XXY show spermatogonia in early life and progressive loss of germ cells during the postnatal period (Hunt et al. 1998). Other 47,XXY mouse studies show normal spermatogonia present at birth with loss of germ cells during the first week of life (Lue et al. 2005).

Adult testes of men with 47,XXY are rarely more than 2 cm in greatest diameter, compared with 3.5 cm in typical males. Often testes are firmer than normal, and differences in consistency reflect differences in the extent of seminiferous tubule hyalinization following germinal atresia. Histologically, hyalinization and fibrosis of seminiferous tubules are often observed. There is an absolute decrease in the number of Leydig cells (interstitial cells), with variable degrees of clumping. Germ cells are decreased to an even greater extent. Leydig cells are histologically abnormal, lacking crystalloids of Reinke. Sertoli cells are present and do not appear morphologically abnormal. Azoospermia or severe oligospermia is the rule, but sperm can usually be recovered at testicular biopsy or, uncommonly, in the ejaculate. Before the availability of intracytoplasmic sperm injection (ICSI), instances of paternity were rare, perhaps representing undetected 46,XY/47,XXY mosaicism.

Fertility. The initial experience with ICSI came from Brussels, where it was first performed in 20 couples (32 cycles) in which the male partner was 47,XXY (Staessen et al. 2003); eight pregnancies resulted, with five live births. Preimplantation genetic diagnosis (Staessen et al. 2003) revealed only 54% of embryos were cytogenetically normal, lower than the background of 77% in that center. The frequency of sex chromosome aneuploidy was twice as high. Updated information is available from 50 non-mosaic men, from 24 of whom (48%) sperm were recovered (Vernaeve et al. 2004). Men from whom recovery was successful or unsuccessful were not distinguishable on the basis of presence or absence of facial hair, serum FSH or testosterone, or testicular volume. Seo et al. (2004) studied 36 Korean men with 47,XXY, of whom 11 were mosaic and 25 were non-mosaic. All 36 men underwent testicular sperm extraction, and mature sperm was found in only 10 (27%) of them. Levels of FSH and testosterone, testicular volume, and age were not different between successful and unsuccessful groups; however, sperm retrieval was much more likely in mosaic (46,XY/47,XXY) than non-mosaic men (54.5 versus 16%). This is consistent with aneuploidy rates in mosaic men being lower than in non-mosaic men (Ferlin et al. 2005). Sperm thus may not be obtained from every man with 47,XXY, nor can ICSI always lead to a pregnancy. Overall, the likelihood of achieving a live born offspring is reduced, and it has not been possible to predict success. When sperm can be obtained, the fertilization rate is approximately 50% (Ferlin et al. 2005). Fertilization appears to be similar in fresh and cryopreserved sperm (Friedler et al. 2001), a pragmatically helpful finding because testicular biopsy appears to not need be obligatorily timed to coincide with the partner's reproductive cycle. It is recommended that couples in which the female partner is undergoing ovulation stimulation in preparation for transfer of ICSI oocytes be offered the concomitant access to donor sperm should her partner's sperm fail to fertilize at the pivotal time. All live born offspring have been neurotypical, but fetuses with 47,XXY have been reported (Ron-El et al. 2000; Friedler et al. 2001). Thus, prenatal genetic diagnosis should be offered. Franik et al. (2016) does not recommend using testicular sperm extraction (TESE) for adolescents younger than 16 years old because the retrieval rates of spermatozoa are much lower (0–20%) compared with those for adolescents and young adults between 16 and 30 years old (40–70%).

Hypothyroidism. Clinical hypothyroidism is rare, but decreased uptake and sluggish responsiveness to thyroid-stimulating hormone have been reported. Thyroiditis is not rare (Tojo et al. 1996). Thyroid-binding globulin levels are usually normal, and anti-thyroid antibodies are not usually detectable. Tahani et al. (2017) found an increased risk of combined central and peripheral hypothyroidism during puberty in boys with KS.

Diabetes. Both type 1 and type 2 diabetes mellitus have long been recognized to be more frequent among males with 47,XXY and their parents than in the general population (Simpson et al. 2003). An autoimmune component is likely responsible for type 1 diabetes mellitus, because autoantibodies against insulin and pancreatic tissue are recorded in males with 47,XXY. Type 2 diabetes, which is far more common, may be secondary to or aggravated by abdominal obesity and insulin resistance. Age at diagnosis is similar to that in individuals without a chromosomal abnormality.

Osteoporosis. Osteoporosis is present in approximately 25% of males with 47,XXY, possibly caused by decreased bone formation and increased bone resorption, combined with low levels of testosterone (Horowitz et al. 1992). Prepubertal children and adolescents with 47,XXY exhibited impaired bone mineral status with higher PTH levels, decreased 25-OH-D, and reduction of bone formation markers (Stagi et al. 2016). Supplementation with vitamin D restored bone mineral density after two years of treatment, whereas testosterone alone seemed to be ineffective (Ferlin et al. 2015).

Evaluation

- Phallus length is best determined by stretching the penis and determining the length from symphysis pubis to the tip of the organ. Normal ranges for penile length at different ages and stages of puberty are available. Hormone levels should be measured concurrently.
- Measurement of testosterone, LH, and FSH should be performed each year during the pubertal years or at the time of diagnosis if diagnosis occurs later.
- If fertility is desired, semen analysis and reproductive hormone assessment are necessary. Testicular biopsy should be reserved for men seeking fertility by testicular sperm aspiration for ICSI.
- Prenatal genetic counseling including options of prenatal diagnosis or preimplantation genetic diagnosis should be offered before and during pregnancies conceived by intracytoplasmic sperm injection (ICSI).
- Annual thyroid stimulating hormone and free T4 levels are recommended.
- Screening for diabetes should begin in childhood with fasting plasma glucose level or Hemoglobin A1c, especially when the child is obese.
- Screening for dyslipidemia should begin in childhood initially at age 9–11 years, then at 17–21 years, with fasting lipid panel, and more frequently when the child is obese.
- Dual-energy X-ray absorptiometry (DXA) to measure bone mineral density should be considered, especially when testosterone is discontinued.

Management

- In infancy, EHT to support the neonatal hormonal surge for improving neurocognitive development and behavior is strongly recommended. EHT will also treat diminished phallus or microphallus if present, and is ideally administered starting at 4 months up to 15 months of age.

- Hormonal treatment in males with 47,XXY is directed toward correcting the underlying androgen deficiency. Replacement can take the form of intramuscular testosterone injections (enanthate or cypionate esters), transbuccal testosterone tablets, oral testosterone undecanoate tablets (available in most parts of the world, but not registered in the United States), or transdermal application of testosterone as patches or gels. A review of various androgen replacement regimens and recommended doses is beyond the scope of this chapter, but this information has been provided by Wang and Swerdloff (2001) and Wosnitzer and Paduch (2013). All these options are available for adults with 47,XXY with related testosterone deficiencies.

- Adolescents may be treated with either testosterone enanthate or transdermal testosterone as gels or patches. Children in the mid-pubertal years can be started on 50 mg of testosterone enanthate every two weeks, increasing to an adult dose of 150–200 mg every two weeks over a 12 month period. The dose is increased more slowly in early puberty. Mehta et al. (2013) evaluated topical testosterone dosing with and without anastrazole in adolescents, and found good tolerability, compliance, and clinical response with improved testosterone levels without suppression of endogenous LH and FSH (important if fertility is desired). As in adults, clinical manifestations, bone age, and serum testosterone levels can be used to determine optimal dose. For males with psychiatric issues (e.g., bipolar disorder), testosterone therapy should be initiated slowly with close monitoring by the endocrinologist and psychiatrist.

- Many experienced endocrinologists have the clinical impression that it is beneficial to administer testosterone in males with 47,XXY and elevated LH levels even if blood levels of free testosterone are ostensibly normal (Simpson et al. 2003).

- Androgen replacement therapy can improve bone density, increase muscle mass and strength, produce more masculine body contour, and decrease body fat. It can produce adequate pubertal maturation with increased body hair, penile enlargement, and male distribution facial and body hair. Bone age advancement and improved bone mineral density may be noted with testosterone treatment. In addition to the medical benefits, testosterone therapy improves self-esteem, mood, and motor skills in males with 47,XXY. There is evidence supporting testosterone treatment reducing gynecomastia (Wosnitzer and Paduch 2013).

- Libido and potency are typically improved with testosterone therapy (Wang et al. 2000).

- Some data suggest that men with 47,XXY may have a steeper than normal decrease in serum testosterone with age; thus, continued monitoring of testosterone levels is recommended.

- Testosterone replacement may also protect against behavioral disturbances and learning disabilities (see Development and Behavior).

- No hormonal treatment exists for the sterility of 47,XXY. Testes are unable to respond to either gonadotropin or androgen stimulation. However, with ICSI, fertility is now possible even when the ejaculate is clinically azoospermic. In men with non-mosaic and mosaic 47,XXY, chromosomally normal sperm cells can possibly be extracted from testicular tissue and used for ICSI.

- Prenatal diagnosis (chorionic villi sampling or amniocentesis) is recommended if preimplantation genetic diagnosis is not undertaken following ICSI for 47,XXY azoospermia or oligospermia.

- To mitigate against the temporal sequence of germ cell failure, some physicians have proposed obtaining and cryopreserving sperm or testicular tissue from adolescent males with 47,XXY, to achieve future pregnancies, even if no sperm exist in situ at that time (Damani et al. 2001).

- Hypothyroidism and diabetes should be treated as in the general population.

- Osteoporosis may be prevented and ameliorated by testosterone therapy and adequate calcium and vitamin D intake. Advanced osteoporosis can be treated with bisphosphonates, which should be managed by an endocrinologist with experience in treating this disorder.

Musculoskeletal

During infancy, there may be an increase in positional torticollis, and this is managed with pediatric physical therapy to prevent secondary deformational plagiocephaly and foster appropriate motor development. In 47,XXY, there may be mild midface underdevelopment, narrow wrists (25% decrease), scoliosis, and kyphosis (S-curve) secondary to decreased muscle tone, lax ligaments, cervical and other anomalous ribs, sacralization of the last lumbar vertebrae, pectus excavatum or carinatum, pes planus, fifth toe anomalies, and fifth finger clinodactyly (Simpson et al. 2003). Overgrowth occurs in the proximal radius and ulna, leading to a narrowed distance between the two. It is not usually severe enough to result in radioulnar synostosis, in contrast to males with 48,XXXY or 49,XXXXY in whom this is more characteristic (Sprouse et al. 2013).

Evaluation

- Physical evaluation of the neck, spine, and thorax is recommended at each well-child visit to monitor for musculoskeletal issues.
- Positional torticollis may be identified as early as 2 months of age through physical examination, range of motion in the neck muscles, and early signs of plagiocephaly.
- Radiographs of the spine are indicated when scoliosis or kyphosis is suspected. A pediatric orthopedic and physical therapy evaluations should be considered to determine treatment plan and institution of appropriate therapy.

Management

- Orthopedic evaluation should be considered for radio-ulnar synostosis, pes planus, and scoliosis.
- Orthotics may be beneficial for young boys with 47,XXY who have pes planus to promote appropriate foot alignment and reduce muscle strain on lower extremities. (https://www.cascade-usa.com/; http://surestep.net/)
- If there are any suspicions of positional torticollis, then pediatric physical therapy evaluations and treatment should be initiated as soon as possible to minimize or prevent positional deformities.
- Exercise program and an active lifestyle should be encouraged from infancy onward to minimize the risk of obesity for the male with 47,XXY.

Neoplasia

Germ cell tumors of both gonadal and extragonadal origin occur with increased frequency. Gonadal tumors include Leydig cell neoplasia (Swerdlow et al. 2005). It has been suggested that dysgenetic germ cells might be arrested in their migration along the urogenital ridge, resulting in mediastinal and retroperitoneal germ cell rests from which tumors arise; rare reports exist of cerebral germ cell tumors (Swerdlow et al. 2005). In a study of 3518 British males with 47,XXY from 1959 to 2002 comparing cancer incidence and mortality with men in the general population, the 47,XXY cohort appear to have increased risks for lung cancer as well as non-Hodgkin lymphoma but a reduction in risk of prostate cancer (Swerdlow et al. 2005).

It was thought previously that gynecomastia was a predisposing factor for breast carcinoma in 47,XXY, but recently that premise has been questioned. Brinton (2011) postulated that comprehensive epidemiological studies are warranted to understand the etiologic relevance of gynecomastia and male breast cancer (MBC). It is recognized that gynecomastia results in the presence of additional breast tissue; however, gynecomastia is often present years and even decades before MBC (Briton 2011). The median age at diagnosis for MBC and 47,XXY is 72 years, similar to the age of diagnosis in males with 46,XY. Breast cancer in males with 47,XXY occurs 20–50 times greater than among chromosomally normal males. Of males with 47,XXY and breast cancer, there is significantly increased risk of breast cancer in males with the mosaic form of 47,XXY, although the reason for this increased association between mosaicism and MBC is unknown at this time.

A comprehensive international study in adults with 47,XXY revealed increases in hematological malignancy, leukemia, and non-Hodgkin lymphoma with a decreased risk for solid tumors in adults with 47,XXY (Ji et al. 2016). Mediastinal tumors are significantly more common than in the general population (Ji et al. 2016).

Evaluation

- Annual breast clinical examination by a medical professional is recommended once puberty is achieved. It is important to discuss with one's primary care physician regarding monthly breast self-examination.
- The value of periodic mammography has not been established and is not recommended at this time.

Management

- For cosmetic purposes as well as to protect against neoplasia, reduction mammoplasty may be warranted if weight reduction and hormonal treatment have not been effective.
- Once a given cancer arises, treatment does not differ from that in the general population.

Immunologic

Autoimmune diseases appear to occur more frequently in males with 47,XXY. Systemic lupus erythematous, ankylosing spondylitis, Sjögren syndrome (keratoconjunctivitis, dry mouth, and arthritis) (Seminog et al. 2014; Harris et al. 2016), Addison disease (Seminog et al. 2014), and rheumatological disorders have all been reported. Testosterone has been hypothesized to protect normal males from autoimmune phenomena; thus, low testosterone and high estrogen levels in males with 47,XXY may predispose hypogonadal males to defects in T-cell activity that lead to autoimmune disorders. Bizzarro et al. (1987) described three men with 47,XXY and Sjögren syndrome and two with systemic lupus erythematous. While on oral testosterone therapy, their serum antinuclear antibody levels, sedimentation rate, and rheumatoid factor all significantly decreased, and the percentage of T-helper cells and T-suppressor cytotoxic cells significantly increased compared with pretreatment values and values while on placebo. Clinical features of their autoimmune disease also were less pronounced after initiation of testosterone therapy. Panimolle et al. (2016) reported a significant increase in diabetes-specific immunoreactivity. Additionally, studies have documented an increased thrombotic risk associated with 47,XXY, along with increased platelet reactivity (Di Minno 2015).

Diabetes is more frequent among males with 47,XXY and their parents than in the general population (Simpson et al. 2003). An autoimmune component is likely responsible for type 1 diabetes mellitus, as autoantibodies against insulin and pancreatic tissue are recorded in males with 47,XXY. This topic is discussed fully in the Endocrine section.

Evaluation

- Annual history and physical examination, as well as assessing for signs and symptoms of autoimmune disorders is appropriate.
- Annual fasting glucose and urinalysis are recommended.

Management

- If diabetes or autoimmune disorders are detected, treatment is identical to that in the general population.

Dermatologic

Chronic leg ulcerations and varicose veins may be associated with males with 47,XXY with a prevalence as high as 13% (Gattringer et al. 2010). Enhanced platelet aggregation, rather than anatomic vascular changes, appears to be the major causative factor. Elevated plasminogen activator inhibition-1 leading to clotting because of platelet aggregation may play a role (Gattringer et al. 2010).

Evaluation

- When leg ulceration occurs without trauma, Doppler flow studies and/or evaluation for thrombophilia is recommended.

Management

- Treatment regimens are standard.

Pulmonary

Males with 47,XXY appear to have increased frequencies of chronic pulmonary diseases, including emphysema, bronchitis, and asthma (Bojesen et al. 2004). Asthma may be exacerbated in the child with 47,XXY when there is family history of chronic pulmonary diseases in parents or siblings.

Evaluation

- History of respiratory difficulties should be elicited during routine care.
- Evaluation for any suspected respiratory condition should be evaluated at wellbeing visits.

Management

- Treatment for pulmonary dysfunction is standard.

OTHER MALE X CHROMOSOME ANEUPLOIDY

Additional supernumerary chromosomes X and/or Y occur less commonly than 47,XXY (KS), and they include 48,XXYY, 48,XXXY, and 49,XXXXY.

Mosaicism

46,XY/47,XXY mosaicism is the most common form of 47,XXY mosaicism, and accounts for 10% of 47,XXY cases (Gropman and Samango-Sprouse 2013). Presence of the normal cell line may modify risks and degree for developmental delay, with a wide range of variability noted in males with mosaicism. There are a variety of fertility-associated problems, ranging from azoospermia to different grades of testicular insufficiency (Nor and Jalaludin 2016). A semen analysis in late adolescence may predict reproductive competency. Testosterone supplementation should be individualized to the endogenous testosterone level and the individual needs of the patient. Other mosaic karyotypes can also occur, such as 46,XX/47,XXY – a rare mosaicism that has only been reported in seven individuals (Nor and Jalaludin 2016). The phenotype may reflect features of more than one cell line.

48,XXYY.

The 48,XXYY syndrome is a less common sex chromosomal variation that was originally described as the "double male," due to the presence of an additional X and Y chromosome in 46,XY males (Hanley et al. 2015). In 1964, the incidence of 48,XXYY was reported to be between 1:18,000 and 1:50,000 males (Borgaonkar et al. 1970). The 48,XXYY syndrome is not inherited, and instead is typically caused by a double nondisjunction during meiosis in spermatogenesis. As a result, the extra chromosomes are usually paternally derived (Loteet al. 2013).

Males with 48,XXYY have some similar features to those with 47,XXY, such as tall stature, narrow shoulders, broad hips, sparse body hair, gynecomastia, microorchidism, and hypergonadotropic hypogonadism (Lote et al. 2013). The 48,XXYY syndrome phenotype also includes infertility, dental problems, more complex language-based learning difficulties, allergies, asthma, and tremors (Lote et al. 2013). Tremors are reported as intention tremors that usually emerge at about age 12–13 years. These become more common and more severe with age with a 71% frequency in adulthood (Lote et al. 2013). The etiology of these tremors and the timing of their presentation is not well understood.

Most males with 48,XXYY syndrome have delays in the early speech and language development although early diagnosis remains elusive. Language-based learning disabilities, and associated reading problems, are very common in males with this disorder (Gropman and Samango-Sprouse 2013). Affected males seem to perform better at tasks focused on math or visual-spatial skills, such as puzzles and block designs, similar to boys with 47,XXY and 48,XXXY. Many boys with 48,XXYY syndrome have delayed development of motor skills such as sitting, standing, and walking that can result in poor motor coordination and performance (Gropman and Samango-Sprouse 2013). Affected males have higher than average rates of behavioral disorders, such as attention deficit hyperactivity disorder (ADHD), anxiety, and autism spectrum disorders that affect communication and social interaction (Gropman and Samango-Sprouse 2013).

Early speech and language developmental delays are seen in 48,XXYY males, which is the earliest developmental biomarker of the later language-based learning disorders. Commonly, there are associated reading dysfunction and written language deficits, which are very common in language-based learning disorders. The verbal IQ decreases with increasing age whereas the performance IQ remains stable. Decreasing VIQ is common with language-based learning disorder (LLD) and associated reading dysfunction (Samango-Sprouse et al. 2014). Therefore, in order to preserve VIQ and intellectual capabilities over time, targeted treatment to counteract this decline is necessary and critical to neurodevelopmental progression.

Males with 48,XXYY have behavioral complexities and dysfunction that affect overall adaptive functioning in daily living skills, socialization, and communication. These deficits are associated with praxis deficits, executive dysfunction, and LLD commonly associated with 48,XXXY, and less complicated than 49, XXXXY.

Evaluation

- Males with 48,XXYY should have a cardiac evaluation at diagnosis, including echocardiography since they are at an increased risk for cardiac malformations (Tartaglia et al. 2011).
- An immunologic evaluation should also be considered to determine if asthma or allergies are present, especially if there have been chronic respiratory illnesses.
- Males with 48,XXYY should be evaluated for intention tremors.
- Pediatric endocrinology evaluation should be completed at time of diagnosis and yearly after 11 years of age to assist with hormonal replacement through adolescence.

Management

- Early intervention strategies and supportive services in physical therapy, occupational therapy, speech and language services, and special education are necessary for appropriate neurodevelopmental outcome and parent support.
- Occupational therapy should be considered when intentional tremors emerge in males with 48,XXYY to determine if the tremor is impacting activities of daily life (ADL) and/or school performance.

48,XXXY. First reported by Barr et al. in 1959, 48,XXXY syndrome is a more complex variant of 47,XXY estimated to occur in 1 in 50,000 male births. Males with 48,XXXY have variable phenotypic presentations that are typically more complex than males with 47,XXY (Sprouse and Gropman 2017). Frequent features include hypertelorism, epicanthal folds and tall stature (Linden et al. 1995; Sprouse and Gropman 2017). Musculoskeletal abnormalities and dysmorphisms include fifth-finger clinodactyly, mild radioulnar synostosis, truncal hypotonia, abnormally shaped ears, depressed nasal bridge, single transverse palmar flexion creases, and pes planus (Tartaglia et al. 2011; Sprouse and Gropman 2017). Genitalia are often diminished with phallus size smaller than typically developed peers. Hormonal replacement is indicated if phallus is diminished (see Endocrine Section).

Historically, there is mild-to-moderate cognitive disability with IQs ranging between 40–60 (Linden et al. 1995). There are occasional individuals with IQ as low as 20 or as high as 79. There is, however, some literature to suggest that non-verbal intelligence may be somewhat preserved (Simsek et al. 2009). These authors have observed two children with 48,XXXY who have intelligence within normal limits with mild language-based learning disabilities and few behavioral disturbances. In unpublished data, Samango-Sprouse et al. (2018) report more normalized IQ levels in boys with 48,XXXY with EHT, the presence of childhood apraxia of speech (CAS), and enhanced non-verbal skills.

Precise assessment of cognitive functioning may be more difficult in males with this diagnosis because of the severe language impairment (Linden et al. 1995; Samango-Sprouse and Law 2001). Expressive communication skills are usually problematic throughout life, and some individuals with the

48,XXXY condition never develop expressive language skills (Linden et al. 1995). Case reports on 48,XXXY have also found that some boys have difficulties with syntax, morphology, word retrieval, and narrative construction (Venkateshwari et al. 2010). Temperament in boys with 48,XXXY is usually considered passive with deficits in social communication and delays in activities of daily living (ADL) (Linden et al. 1995; Visootsak et al. 2007). However, paradoxical aggressive outbursts may occur as a result of their neurocognitive challenges, decreased inhibition related to diminished frontal lobe development, and speech dyspraxia with anxiety.

Motor development is delayed in boys with 48,XXXY and clumsiness is increasingly manifested as the child becomes older because of motor planning deficits. The child with 48,XXXY is more complex than the boy with 47,XXY in motor development as well as the other developmental domains. Therefore, the motor treatment must be targeted on core strength, specific muscle strengthening, and reducing muscle tightness to promote appropriate movement patterns and acquisition (Samango-Sprouse et al. unpublished data). The etiology of the small subtest of boys with variance in muscle tone warrants further investigation before any substantive conclusion can be determined. In 36% of boys with 48,XXXY there was congenital cryptorchidism and/or microphallus coupled with generalized hypotonia (Samango-Sprouse et al. unpublished data).

Evaluation

- Evaluation of males with 48,XXXY require a comprehensive team evaluation familiar with this multiple systems impact of the additional Xs. A pediatric orthopedist and endocrinologist are critical members of this team because of the breadth and complexity of the musculoskeletal abnormalities and the increased incidence of micropenis present in boys with 48,XXXY. Specific examination for physical anomalies and birth defects is indicated to ascertain those medical issues that need immediate treatment.
- In infancy, early hormone therapy is needed to support the neonatal testosterone surge for improving neurocognitive development and behavior, and this will also treat a diminished phallus or microphallus if present, ideally starting at 4 months up to 15 months of age if indicated.
- Males with 48,XXXY typically demonstrate an infantile presentation of developmental dyspraxia in the first year of life and should be evaluated by a pediatric speech and language pathologist who is experienced in childhood apraxia of speech (CAS) by 12 months of age.
- Height, weight, and head circumference should be routinely measured and plotted on standard growth charts at every well-child care visit, as is standard of care in any child presenting to primary care. Assessment of arm span-to-height ratio and upper-to-lower segment ratio can be performed in the prepubertal years. Since more than 80–90% of boys with 48,XXXY have truncal hypotonia with an increased incidence of kyphosis, scoliosis, and pes planus, well child care visits are ideal times to monitor these musculoskeletal issues.
- Comprehensive pediatric evaluations by team of specialists who are familiar with 48,XXXY are recommended yearly, and the constellation of team may change with time since the priorities for care evolve. Strategic players for multidisciplinary team are a pediatric clinical geneticist, physical therapist, pediatric endocrinologist, pediatric neuropsychologist, speech and language pathologist, and special educator; all are important throughout childhood.
- Males with 48,XXXY presenting with recurrent respiratory infections should be evaluated by pediatric immunologists so medical treatment can reduce infections and increase wellness.
- Pediatric psychological and/or pediatric psychiatric evaluation and possible treatment should be considered if there are any indications of ADHD and/or anxiety by school age in order to diminish secondary behavioral problems.
- An evaluation for autism spectrum disorder (ASD) should be considered in males with 48,XXXY with atypical social development and stereotypic behaviors (e.g., hand flapping, rocking, spinning).
- Psychoeducational evaluation during the school years is essential to assess cognition, achievement, and adaptive function and to plan appropriate strategies to optimize the child's learning and alleviate risks for behavioral disruption.
- For males with 48,XXXY assessment of the status of the androgen deficiency in infancy, between 5 and 8 years and again at 10 years of age will be valuable for determining the need for hormonal replacement.
- Phallus length is best determined by stretching the penis and determining the length from symphysis pubis to the tip of the organ. Normal ranges for penile length at different ages and stages of puberty are available. Hormone levels should be measured concurrently.
- Measurement of testosterone, LH, and FSH should be performed each year during the pubertal years or at the time of diagnosis if diagnosis occurs later.
- Weight and BMI should be monitored during regular visits to assess the presence of obesity. This is particularly important during adolescence and adulthood when testosterone level decreases without appropriate institution of androgen therapy.
- If fertility is desired, semen analysis and reproductive hormone assessment are necessary. Testicular biopsy

should be reserved for men seeking fertility by testicular sperm aspiration for ICSI.
- Males with 48,XXXY should have a cardiac evaluation at diagnosis, including echocardiography because of the increase in cardiac anomalies (Tartaglia et al. 2011).

Management

- Once the individual or fetus is diagnosed with 48,XXXY, it is important to seek consultation with medical professionals and health care professionals who are familiar with 48,XXXY for recommendations regarding resources, appropriate biological and neurodevelopmental therapies, as well as medications for ADHD or anxiety.
- Feeding difficulties in infancy should be addressed with a lactation specialist to foster improved coordination of suck swallow in the newborn period. If the feeding disturbances do not resolve within first week to ten days, then an evaluation by either pediatric speech and language pathologist or pediatric occupational therapist is indicated.
- There are many kinds of strategies that may assist the infant with feeding disturbances, but it should be instituted under the guidance of appropriate pediatric trained specialists. The feeding disturbances may be most challenging for a primiparous female with an infant with prenatal diagnosis of 48,XXXY. The combination of anxiety and the insecurity of the first-time mother may exacerbate these mild feeding disturbances and can be quickly corrected with parent coaching and support. The etiology of the discoordination is unknown, but it may be the earliest indication of the praxis deficits or sensory dysfunction both of which are more common in boys with 48,XXXY.
- Early interventional therapies (e.g., physical, occupational, and speech therapies) are recommended throughout early childhood when discrepancies or deficits are identified to enhance early neurodevelopmental outcomes.
- Early hormonal treatment (EHT) with testosterone in infancy has shown a beneficial impact on multiple domains of development including neurocognition, motor development, speech and language, and social-emotional development; EHT should be considered as treatment for the infant with 48,XXXY (Samango-Sprouse et al. unpublished data). Pediatric endocrine evaluation should be completed in early infancy to determine if EHT would be appropriate between 4 and 15 months of age.
- If there are any suspicions of positional torticollis, then pediatric physical therapy evaluations and treatment should be completed as soon as possible to minimize or prevent positional deformities.
- Physical therapy is indicated when there is hypotonia, motor delay, and/or poor coordination and is most effective between 4 and 18 months in order to develop independent ambulation skills.
- Occupational therapy should be considered for the boys with decreased muscle tone in the trunk or upper body, because these deficits will affect handwriting, posture, attention, and eventual school success. This type of evaluation may be most beneficial between 4 and 6 years of age and typically is needed for 12 months.
- Orthopedic evaluation should be considered for radio-ulnar synostosis, pes planus, and scoliosis. Orthotics may be beneficial for young boys with 48,XXXY who have pes planus to promote appropriate foot alignment and reduce muscle strain on lower extremities. (https://www.cascade-usa.com/; http://surestep.net/).
- Developmental, behavioral, and psychiatric intervention in males with 48,XXXY is more intensive than that of 47,XXY, and there is much more variability, including early onset global developmental delay, anxiety, ADHD like symptoms and externalizing behaviors.
- Specific speech and language therapies should address speech delays with motor planning deficits, language formulation abnormalities and syntactical delays.
- Social skills training may help develop an appropriate understanding of socioemotional cues, pragmatic communication, and self-esteem.
- Androgen replacement therapy can improve bone density, increase muscle mass and strength, produce more masculine body contour, and decrease body fat. It can produce adequate pubertal maturation with increased body hair, penile enlargement, and male distribution facial and body hair. Bone age advancement and improved bone mineral density may be noted with testosterone treatment.
- Targeted treatment by a physical therapist, occupational therapist, and speech therapist is indicated as soon as possible for males with 48,XXXY. Interventional therapists and educational specialists should recognize the neurocognitive variability in these individuals.
- Psychopharmacological intervention may need to be coupled with interventional strategies, including speech therapy, sensory integration therapy, tailored behavioral treatments, and psychological counseling to minimize behavioral issues.
- Exposure to computers early may be quite helpful in the development of academic skills as well as keyboarding skills to minimize the impact of the graphomotor dysfunction and dysgraphia associated with 48,XXXY.

- Exercise program and an active lifestyle should be encouraged from infancy onward to minimize the risk of obesity for the male with 48,XXXY.
- Activities that are dynamic and develop strength, such as karate and swimming, are helpful to develop motor planning skills, endurance, and coordination. Sport activities and/or physical therapy are warranted to encourage muscle development and facilitate social interactions with typically developing peers on the playground and in after-school activities.

49,XXXXY. This disorder results in the most uncommon variant of 47,XXY (KS) and occurs in 1 in 85,000–100,000 male births (Visootsak et al. 2007). In recent years, there has been a significant increase in research regarding the 49,XXXXY disorder and its phenotypic presentation, brain imaging, and response to biological intervention. Males with 49,XXXXY have a distinctive and characteristic phenotype that includes hypertelorism, epicanthal folds, broad nasal bridge, low-set and malformed ears, and prognathism (Etemadi et al. 2015).

Although many neurodevelopmental aspects of this rare and unique neurogenetic disorder are similar to 47,XXY, the developmental disturbances are more complex and significant (Gropman and Samango-Sprouse 2013). However, several salient facts have been well established recently. It is important to understand the effect of the additive Xs on the medical and neurodevelopmental outcome. Early identification and services are beneficial to the child and his family, as there are multiple systems impacted most characteristically: neurodevelopment, musculoskeletal, brain development, endocrine, and immunological. Variability within the disorder is extensive, and the biological underpinning of these differences is not well understood. There is androgen deficiency, similar to 47,XXY and 48,XXXY, and recent publications demonstrate that an improvement in neurodevelopmental outcome occurs with EHT (Samango-Sprouse et al. 2015). Finally, the presence of childhood apraxia of speech (CAS) and developmental dyspraxia is pervasive in boys with 49,XXXXY. CAS impacts speech and language development and is associated with gross and fine motor dysfunction. Receptive language development and non-verbal IQ are less severely affected and more intact than other areas of development. An increased incidence of congenital cardiac defects, including atrial septal defect, and VSD has also been reported (Mustafa et al. 2014).

Development and Behavior. Historically, it was believed that every additional X chromosome lowered the expected IQ by about 10–15 points; however, recent research has demonstrated that this may not be an accurate representation of the IQ differences. Development and behavior are impacted significantly by the timing of diagnosis, presence of family learning disorders such as ADHD, anxiety and dyslexia, and the implementation of EHT. Previously, males with 49,XXXXY had reported IQs ranging between 20 and 60, and typically presented with global developmental delays (Linden et al. 1995). However, a recent publication by Samango-Sprouse et al. (2011) demonstrated that there was greater variability in both neurocognitive function and behavioral manifestations than has been previously described. The majority of the boys were diagnosed with oral motor and childhood apraxia of speech (CAS), with more intact non-verbal skills and receptive language skills than expressive language skills. Samango-Sprouse et al. (2011) found that receptive language in boys with 49,XXXXY was more than 15 points higher than their expressive language skills. This study also demonstrated that boys with 49,XXXXY who received EHT performed significantly better in expressive language. This study further supports the benefit of EHT in the most vulnerable domain of development in boys with 49,XXXXY.

Behaviorally, males with 49,XXXXY have been described to be shy and friendly, with occasional temper tantrums and irritability as a result of impaired communication skills with markedly deficient expressive language (Linden et al. 1995; Simsek et al. 2009). Further case reports documented hyperactivity, anxious reactions to small changes, and agitation in boys with 49,XXXXY (Lomelino and Reiss 1991). Gropman et al. (2010) reported that boys with 49,XXXXY may be at risk for externalizing and internalizing behaviors due to the frustrations of their expressive language disorder. Thirty-three boys with 49,XXXXY were screened for ASD with the GARS-3, and 63.6% were unlikely to have ASD, while 36.4% were rated as very likely. Additionally, of the 21 boys with 49,XXXXY who had received testosterone treatment, 71.48% tested unlikely to have ASD, while 28.6% were very likely. Of the 12 children who had not received testosterone replacement, the likelihood for ASD was 50% (unpublished data 2017). Additional research is needed. However, it is suspected that there is an increased incidence of ADHD-like traits and generalized anxiety as well as social anxiety in males with 49,XXXXY. Behavioral outbursts are likely to be related to the presence or absence of ADHD and anxiety disorders.

Neurological Structure and Function. Neuroimaging studies have found that most boys with 49,XXXXY present with white matter changes, a thinning of the corpus callosum, enlarged lateral ventricles, and a decrease in volume of the cerebral cortex (Linden et al. 1995; Tabarki et al. 2012; Blumenthal et al. 2013; Milani et al. 2015). Although most all individuals with 49,XXXXY presented with changes in white matter, the severity of the anomalies were variable across studies. A study by Blumenthal (2013) on 14 youths with 49,XXXXY found that rates of plagiocephaly, periventricular cysts, and craniofacial abnormalities were significantly increased when compared to controls. Delayed myelination and cavum septum pellucidum have also been reported in some instances (Tabarki et al. 2012). Milani et al. (2015) studied the neuroradiological features in a 48,XXXY/49,XXXXY mosaic and found enlarged ventricles and white matter abnormalities typically seen in 49,XXXXY. They also had dysmorphisms of cranio-cervical junction and posterior fossa. It remains to be confirmed that these previously unreported features are present in other boys with 49,XXXXY.

Musculoskeletal. Minor dysmorphic features including hypertelorism, epicanthal folds, broad nasal bridge, low-set and malformed ears, and prognathism are common in boys with 49,XXXXY (Etemadi et al. 2015). The neck is short and broad, and the thorax is narrow. Radioulnar synostosis (75%), genu valgum, pes planus (65%), clinodactyly (60%), hypotonia (85%), and asymmetric hip rotation (67.5%) are very frequent (Sprouse et al. 2013). Genitalia are usually hypoplastic, and cryptorchidism is often present (Sprouse et al. 2013). Torticollis was noted in 27.5% of 40 boys with 49,XXXXY (Sprouse et al. 2013), and it was believed to be related to the decreased muscle tone and paucity of movement in early infancy. Children with 49,XXXXY are also impacted by feeding and latching difficulties in early infancy due to decreased muscle tone in the oral musculature.

Immunology. A study of 31 boys with 49,XXXXY found that 20 had at least two warning signs of immunodeficiency, 16 had a history of recurrent pneumonia, and 15 were diagnosed with asthma (Keller et al. 2013). More than half of those who underwent immunologic screening showed evidence of impaired antibody responses to polysaccharide antigens. Additionally, 73 boys with 49,XXXXY studied by Samango-Sprouse et al. documented a high incidence of respiratory issues including pneumonia and repeated upper respiratory infections (unpublished data 2017).

Endocrinology. Similar to 47,XXY, boys with 49,XXXXY have been reported to have hypergonadotropic hypogonadism (Mazzilli et al. 2016). A 19-year-old boy with 49,XXXXY was treated with testosterone replacement therapy, which normalized his testosterone levels, increased his testicular volume and penis size, and resulted in weight loss (Mazzilli et al. 2016). Additionally, a study assessed neurocognition, neurology, endocrinology, and speech and language in a sample of 22 boys with 49,XXXXY and documented a positive treatment effect of early androgen treatment in gestural communication and vocabulary development (Samango-Sprouse 2011). Testosterone treatment was also shown to have positive effects in a second case report of a 49,XXXXY male with autoimmune diabetes, as his T-cell levels increased significantly (Pamuk et al. 2009). This further supports the potential protective and positive effect testosterone treatment has in boys with 49,XXXXY disorder.

Evaluation

- Prenatal genetic counseling including options of prenatal diagnosis or preimplantation genetic diagnosis should be offered before and during pregnancies conceived by intracytoplasmic sperm injection (ICSI).
- Evaluation of males with 49,XXXXY require a comprehensive team evaluation familiar with this multiple systems impact of the additive Xs. A pediatric orthopedist and endocrinologist are critical members of this team because of the breadth and complexity of the musculoskeletal abnormalities and the increased incidence of micropenis present in boys with 49,XXXXY. Specific examination for physical anomalies and birth defects is indicated to ascertain those medical issues that need immediate treatment.
- In infancy, early hormone therapy is needed to support the neonatal testosterone surge for improving neurocognitive development and behavior, and will also treat diminished phallus or microphallus if present, ideally starting at 4 months up to 15 months of age if indicated.
- Males with 49,XXXXY typically demonstrate infantile presentation of developmental dyspraxia in the first year of life and should be evaluated by a licensed speech and language pathologist who is experienced in childhood apraxia of speech (CAS) by 12 months of age.
- Height, weight, and head circumference should be routinely measured and plotted on standard growth charts at every well-child care visit, as is standard of care in any child presenting to primary care. Assessment of arm span-to-height ratio and upper-to-lower segment ratio can be performed in the prepubertal years. Since more than 80–90% of the boys with 49,XXXXY have truncal hypotonia with an increased incidence of kyphosis, scoliosis, and pes planus, well-child care visits are ideal times to monitor these musculoskeletal issues.
- Comprehensive pediatric evaluations by team of specialists who are familiar with 49,XXXXY are recommended yearly, and the constellation of team may change with time since the priorities for care evolve. Strategic players for multidisciplinary team are a pediatric clinical geneticist, physical therapist, pediatric endocrinologist, pediatric neuropsychologist, speech and language pathologist, and special educator; all are important throughout childhood.
- Males with 49,XXXXY presenting with recurrent respiratory infections should be evaluated by pediatric immunologists so medical treatment can reduce infections and increase wellness.
- Pediatric psychological and/or pediatric psychiatric evaluation and possible treatment should be considered

if there are any indications of ADHD and/or anxiety by school age in order to diminish secondary behavioral problems.
- An evaluation for autism spectrum disorder (ASD) should be considered in males with 49,XXXXY with atypical social development and stereotypic behaviors (e.g., hand flapping, rocking, spinning).
- Psychoeducational evaluation during the school years is essential to assess cognition, achievement, and adaptive function and to plan appropriate strategies to optimize the child's learning and alleviate risks for behavioral disruption.
- For males with 49,XXXXY assessment of the status of the androgen deficiency in infancy, between 5 and 8 years and again at 10 years of age will be valuable for determining the need for hormonal replacement.
- Phallus length is best determined by stretching the penis and determining the length from symphysis pubis to the tip of the organ. Normal ranges for penile length at different ages and stages of puberty are available. Hormone levels should be measured concurrently.
- Measurement of testosterone, LH, and FSH should be performed each year during the pubertal years or at the time of diagnosis if diagnosis occurs later.
- Weight and BMI should be monitored during regular visits to assess the presence of obesity. This is particularly important during adolescence and adulthood when testosterone level decreases without appropriate institution of androgen therapy.
- If fertility is desired, semen analysis and reproductive hormone assessment are necessary. Testicular biopsy should be reserved for men seeking fertility by testicular sperm aspiration for ICSI.
- Males with 49,XXXXY should have a cardiac evaluation at diagnosis, including echocardiography because of the increase in cardiac anomalies (Gropman et al. 2010).

Management

- Once the individual or fetus is diagnosed with 49,XXXXY, it is important to seek consultation with medical professionals and health care professionals who are familiar with 49,XXXXY for recommendations regarding resources, appropriate biological and neurodevelopmental therapies, as well as medications for ADHD or anxiety.
- Feeding difficulties in infancy should be addressed with a lactation specialist to foster improved coordination of suck swallow in the newborn period. If the feeding disturbances do not resolve within first week to ten days, then an evaluation by either pediatric speech and language pathologist or pediatric occupational therapist is indicated.
- There are many kinds of strategies that may assist the infant with feeding disturbances, but it should be instituted under the guidance of appropriate pediatric trained specialists. The feeding disturbances may be most challenging for a primiparous female with an infant with a prenatal diagnosis of 49,XXXXY. The combination of anxiety and the insecurity of the first-time mother may exacerbate these mild feeding disturbances and can be quickly corrected with parent coaching and support. The etiology of the discoordination is unknown, but it may be the earliest indication of the praxis deficits or sensory dysfunction, both of which are more common in boys with 49,XXXXY.
- Early interventional therapies (e.g., physical, occupational, and speech therapies) are recommended throughout early childhood when discrepancies or deficits are identified to enhance early neurodevelopmental outcomes.
- Early hormonal treatment (EHT) with testosterone in infancy has shown a beneficial impact on multiple domains of development including neurocognition, motor development, speech and language, and social-emotional development; EHT should be considered as treatment for the infant with 49,XXXXY. Pediatric endocrine evaluation should be completed in early infancy to determine if EHT would be appropriate between 4 and 15 months of age.
- If there are any suspicions of positional torticollis, then pediatric physical therapy evaluations and treatment should be completed as soon as possible to minimize or prevent positional deformities.
- Physical therapy is indicated when there is hypotonia, motor delay, and/or poor coordination and is most effective between 4 and 18 months in order to develop independent ambulation skills.
- Occupational therapy should be considered for the boys with decreased muscle tone in the trunk or upper body, because these deficits will affect handwriting, posture, attention, and eventual school success. This type of evaluation may be most beneficial between 4 and 6 years of age and typically is needed for 12 months.
- Orthopedic evaluation should be considered for radio-ulnar synostosis, pes planus, and scoliosis. Orthotics may be beneficial for young boys with 49,XXXXY who have pes planus to promote appropriate foot alignment and reduce muscle strain on lower extremities. Typically, an orthotist or pediatric physical therapist is needed to identity appropriate orthotic (https://www.cascade-usa.com/; http://surestep.net/).

- Developmental, behavioral, and psychiatric intervention in males with 49,XXXXY is more intensive than that of 47,XXY, and there is much more variability, including early onset global developmental delay, anxiety, ADHD like symptoms and externalizing behaviors.
- Specific speech and language therapies should address speech delays with motor planning deficits, language formulation abnormalities and syntactical delays.
- Social skills training may help develop appropriate understanding of socioemotional cues, pragmatic communication, and self-esteem.
- Androgen replacement therapy can improve bone density, increase muscle mass and strength, produce more masculine body contour, and decrease body fat. It can produce adequate pubertal maturation with increased body hair, penile enlargement, and male distribution facial and body hair. Bone age advancement and improved bone mineral density may be noted with testosterone treatment.
- Targeted treatment by a physical therapist, occupational therapist, and speech therapist is indicated as soon as possible for males with 49,XXXXY. Interventional therapists and educational specialists should recognize the neurocognitive variability in these individuals.
- Psychopharmacological intervention may need to be coupled with interventional strategies, including speech therapy, sensory integration therapy, tailored behavioral treatments, and psychological counseling to minimize behavioral issues.
- Exposure to computers early may be quite helpful in the development of academic skills as well as keyboarding skills to minimize the impact of the graphomotor dysfunction and dysgraphia associated with 49,XXXXY.
- Exercise program and an active lifestyle should be encouraged from infancy onward to minimize the risk of obesity for the male with 49,XXXXY.
- Activities that are dynamic and develop strength, such as karate and swimming, are helpful to develop motor planning skills, endurance, and coordination. Sport activities and/or physical therapy are warranted to encourage muscle development and facilitate social interactions with typically developing peers on the playground and in after-school activities.

RESOURCES

American Association for Klinefelter Syndrome Information and Support (AAKSIS)
3796 Ogden Lane
Mundeline, IL 60060, USA
Website: *http://www.aaksis.org/*

EURORDIS – Rare Diseases Europe
Plateforme Maladies Rares
96, rue Didot
75014 Paris, France
Website: *htttps://www.eurordis.org/*

The Focus Foundation
820 W Central Ave. #190
Davidsonville, MD 21035, USA
Website: *http://thefocusfoundation.org/*

Genetic Alliance
4301 Connecticut Ave NW
Washington, DC 20008, USA
Website: *http://www.geneticalliance.org/*

National Institutes of Health (NIH)
9000 Rockville Pike
Bethesda, Maryland 20892, USA
Website: *https://www.nih.gov/*

National Organization for Rare Disorders (NORD)
1779 Massachusetts Avenue Suite 500
Washington, DC 20036, USA
Website: *https://rarediseases.org/*

Unique The Rare Chromosome Disorder Support Group
G1 The Stables,
Station Road West
Oxted, Surrey, RH8 9EE, UK
Website: *https://www.rarechromo.org/*

REFERENCES

Bandmann HJ, Breit R, Perwein E (Eds) (2012) *Klinefelter's Syndrome*. Springer Science Business Media.

Bardsley MZ, Kowal K, Levy C, Gosek A, Ayari N, Tartaglia N, Lahlou N, Winder B, Grimes S, Ros, JL (2013) 47,XYY syndrome: clinical phenotype and timing of ascertainment. *J Pediatr* 163(4):1085–1094.

Becker KL (1972) Clinical and therapeutic experiences with Klinefelter's syndrome. *Fertil Steril* 23:568–578.

Bizzarro A, Valentini G, Di Martino G, DaPonte A, De Bellis A, Iacono G (1987) Influence of testosterone therapy on clinical and immunological features of autoimmune diseases associated with Klinefelter's syndrome. *J Clin Endocrinol Metab* 64:32–36.

Blumenthal JD, Baker EH, Lee NR, Wade B, Clasen LS, Lenroot RK, Giedd JN (2013) Brain morphological abnormalities in 49,XXXXY syndrome: a pediatric magnetic resonance imaging study. *Neuroimage Clin* 2:197–203.

Bojesen A, Birkebaek N, Kristensen K, Heickendorff L, Mosekilde L, Christiansen JS, Gravholt CH (2011) Bone mineral density in

Klinefelter syndrome is reduced and primarily determined by muscle strength and resorptive markers, but not directly by testosterone. *Osteoporos Int* 22(5):1441–1450.

Bojesen A, Juu, S, Birkebæk N, Gravholt CH (2004) Increased mortality in Klinefelter syndrome. *J Clin Endocrinol Metab* 89(8):3830–3834.

Bojesen A, Juul S, Gravholt CH (2003) Prenatal and postnatal prevalence of Klinefelter syndrome: A national registry study. *J Clin Endocrinol Metab* 88:622–626.

Borgaonkar DS, Mules E, Char F (1970) Do the 48, XXYY males have a characteristic phenotype? A review. *Clin Genet* 1(5-6):272–293.

Brinton LA (2011) Breast cancer risk among patients with Klinefelter syndrome. *Acta Paediatr* 100(6):814–818.

Bruining H, Swaab H, Kas M, van Engeland H (2009) Psychiatric characteristics in a self-selected sample of boys with Klinefelter syndrome. *Pediatrics* 123(5):e865–e870.

Bryant DM, Hoeft F, Lai S, Lackey J, Roeltgen D, Ross J, Reiss AL (2012) Sex chromosomes and the brain: a study of neuroanatomy in XYY syndrome. *Dev Med Child Neurol* 54(12):1149–1156.

Christiansen P, Andersson AM, Skakkebaek NE (2003) Longitudinal studies of inhibin B levels in boys and young adults with Klinefelter syndrome. *J Clin Endocrinol Metab* 88:888–891.

Damani MN, Mittal R, Oates RD (2001) Testicular tissue extraction in a young male with 47,XXY Klinefelter's syndrome: Potential strategy for preservation of fertility. *Fertil Steril* 76:1054–1056.

Di Minno MND, Esposito D, Di Minno A, Accardo G, Lupoli G, Cittadini A, Guigliano D, Pasquali D (2015) Increased platelet reactivity in Klinefelter men: something new to consider. *Andrology* 3(5):876–881.

Etemadi K, Basir B, Ghahremani S (2015) Neonatal diagnosis of 49,XXXXY syndrome. *Iran J Reprod Med* 13(3):181–184.

Ferguson-Smith M, Mack WS, Ellis PM, Dickson M, Sanger R, Race RR (1964) Paternal age and the source of the X chromosomes in XXY Klinefelter syndrome. *Lancet* 41:46.

Ferlin A, Gavolla A, Forester C (2005) Chromosome abnormalities in sperm of individuals with constitutional sex chromosome abnormalities. *Cytogenet Genome Res* 111:310–316.

Ferlin A, Selice R, Di Mambro A, Ghezzi M, Di Nisio A, Caretta N, Foresta C (2015) Role of vitamin D levels and vitamin D supplementation on bone mineral density in Klinefelter syndrome. *Osteoporos Int* 26(8):2193–2202.

Fideleff HL, Boquete HR, Suárez MG, Azaretzky M (2016) Chapter Seven-Burden of Growth Hormone Deficiency and Excess in Children. *Prog Mol Biol Transl Sci* 138:143–166.

Franik S, Hoeijmakers Y, D'Hauwers K, et al. (2016) Klinefelter syndrome and fertility: sperm preservation should not be offered to children with Klinefelter syndrome. *Hum Reprod* 31:1952–1959.

Friedler S, Raziel A, Strassburger D, Schachter M, Bern O, Ron-El R (2001) Outcome of ICSI using fresh and cryopreserved-thawed testicular spermatozoa in patients with non-mosaic Klinefelter's syndrome. *Hum Reprod* 16:2616–2620.

Gattringer C, Scheurecker C, Höpfl R, Müller H (2010) Association between venous leg ulcers and sex chromosome anomalies in men. *Acta Derm Venereol* 90(6):612–615.

Giedd JN, Clasen LS, Wallace GL, Lenroot RK, Lerch JP, Wells EM, Blumenthal D, Nelson JE, Rossell JW, Stayer C, Evans, AC Samango-Sprouse CA (2007) XXY (Klinefelter syndrome): a pediatric quantitative brain magnetic resonance imaging case-control study. *Pediatrics* 119(1):e232–e240.

Graham JM, Bashir AS, Stark RA, Silbert A, Walze, S (1988) Oral and written language abilities of XXY boys: implications for anticipatory guidance. *Pediatrics* 81:795–806.

Gropman A, Samang-Sprouse CA (2013) Neurocognitive variance and neurological underpinnings of the X and Y chromosomal variations. *Am J Med Genet C* 163:35–43).

Gropman AL, Rogol A, Fennoy I, Sadeghin T, Sinn S, Jameson R, Mitchell F, Clabaugh J, Lutz-Armstrong M, Samango-Sprouse, CA (2010) Clinical variability and novel neurodevelopmental findings in 49, XXXXY syndrome. *Am J Med Genet A* 152(6):1523–1530.

Hanley AP, Blumenthal JD, Lee NR, Baker EH, Clasen LS, Giedd JN (2015) Brain and behavior in 48, XXYY syndrome. *Neuroimage Clin* 8:133–139.

Harris VM, Sharma R, Cavett J, et al. (2016) Klinefelter's syndrome (47, XXY) is in excess among men with Sjögren's syndrome. *Clin Immunol* 168:25–29.

Hassold TJ, Sherman SL, Pettay D, Page DC, Jacobs PA (1991) XY chromosome nondisjunction in man is associated with diminished recombination in the pseudoautosomal region. *Am J Hum Genet* 49:253–260.

Horowitz M, Wishart JM, O'Loughlin PD, et al (1992) Osteoporosis and Klinefelter's syndrome. *Clin Endocrinol* 36:113–118.

Hunt PA, Worthman C, Levinson H, Stallings J, LeMaire R, Mroz K, Park C, Handel MA (1998) Germ cell loss in the XXY male mouse: Altered X-chromosome dosage affects prenatal development. *Mol Reprod Dev* 49:101–111.

Iitsuka Y, Bock A, Nguyen DD, Samango-Sprouse CA, Simpson JL, Bischoff FZ (2001) Evidence of skewed X-chromosome inactivation in 47,XXY and 48,XXYY Klinefelter patients. *Am J Med Genet* 98:25–31.

Jacobs PA, Strong JA (1959) A case of human intersexuality having a possible XYY sex-determining mechanism. *Nature* 183:302–303.

Ji J, Zöller B, Sundquist J, Sundquist K (2016) Risk of solid tumors and hematological malignancy in persons with Turner and Klinefelter syndromes: A national cohort study. *Int J Cancer* 139(4):754–758.

Keller MD, Sadeghin T, Samango-Sprouse C, Orange JS (2013) Immunodeficiency in patients with 49, XXXXY chromosomal variation. *Am J Med Genet C* 163:50–54

Klinefelter HF Jr, Reifenstein EC Jr, Albright F (1942) Syndrome characterized by gynecomastia, aspermatogenesis without aleydigism and increased excretion of follicle-stimulating hormone. *J Clin Endocrinol* 2:615–627.

Linden MG, Bender BG, Robinson A (1995) Sex chromosome tetrasomy and pentasomy. *Pediatrics* 96:672–682.

Lomelino CA, Reiss AL (1991) 49, XXXXY syndrome: behavioural and developmental profiles. *J Med Genet* 28(9):609–612.

Lote H, Fuller GN, Bain PG (2013) 48, XXYY syndrome associated tremor. *Pract Neurol* 13:249–253.

Lue YH, Jentsch JD, Wang C, Rao RN, Hikim AP, Salameh W, Swerdloff RS (2005) XXY mice exhibit gonadal and behavioral phenotypes similar to Klinefelter syndrome. *Endocrinology* 146:4148–4154.

Lubs HA, Stevenson RE, Schwartz CE (2012) Fragile X and X-linked intellectual disability: four decades of discovery. *Am J Hum Genet* 90(4):579–590.

Manning M, Hudgins L (2010) Array-based technology and recommendations for utilization in medical genetics practice for detection of chromosomal abnormalities. *Genet Med* 12(11):742–745.

Mazzilli R, Delfino M, Elia J, Benedetti F, Alesi L, Chessa L, Mazzilli F (2016) Testosterone replacement in 49, XXXXY syndrome: andrological, metabolic and neurological aspects. *Endocrinol Diabetes Metab Case Rep* 2016:15014.

Mehta A, Bolyakov A, Roosma J, Schlegel PN, Paduch DA (2013) Successful testicular sperm retrieval in adolescents with Klinefelter syndrome treated with at least 1 year of topical testosterone and aromatase inhibitor. *Fertil Steril* 100(4):970–974.

Mehta A, Paduch DA (2012) Klinefelter syndrome: an argument for early aggressive hormonal and fertility management. *Fertil Steril* 98(2):274–283.

Milani D, Triulzi F, Bonarrigo F, Avignone S, Esposito S (2015) 48, XXXY/49, XXXXY mosaic: new neuroradiological features in an ultra-rare syndrome. *Ital J Pediatr* 41(1):50.

Miller DT, Adam MP, Aradhya S, et al. (2010) Consensus statement: chromosomal microarray is a first-tier clinical diagnostic test for individuals with developmental disabilities or congenital anomalies. *The Am J Hum Genet* 86(5):749–764.

Morris JK, Alberman E, Scott C, Jacobs P (2008) Is the prevalence of Klinefelter syndrome increasing? *Eur J Hum Genet* 16(2):163–170.

Nagai M, Kishi K, Kato S (2007) Insular cortex and neuropsychiatric disorders: a review of recent literature. *Eur Psychiatr* 22(6):387–394.

Nor NSM, Jalaludin MY (2016) A rare 47 XXY/46 XX mosaicism with clinical features of Klinefelter syndrome. *Int J Pediatr Endocrinol* 2016(1):11.

Pamuk BO, Torun AN, Kulaksizoglu M, Algan C, Ertugrul DT, Yilmaz Z, Tutuncu B, Demirag NG (2009) 49, XXXXY syndrome with autoimmune diabetes and ocular manifestations. *Med Princ Pract* 18(6):482–485.

Panimolle F, Tiberti C, Granato S, Semeraro A, Gianfrilli D, Anzuini A, Lenzi A, Radicioni A (2016) Screening of endocrine organ-specific humoral autoimmunity in 47, XXY Klinefelter's syndrome reveals a significant increase in diabetes-specific immunoreactivity in comparison with healthy control men. *Endocrine* 52(1):157–164.

Patwardhan AJ, Eliez S, Bender B, Linden MG, Reiss AL (2000) Brain morphology in Klinefelter syndrome: Extra X chromosome and testosterone supplementation. *Neurology* 54(12):2218–2223.

Plymate SR, Paulsen CA (1989) Klinefelter's syndrome. In: *The Genetic Basis of Common Disease*, King RA, Motulsky A, eds, New York: Oxford University Press, pp. 876–894.

Plymate SR, Leonard JM, Paulsen CA, Fariss BL, Karpas AE (1983) Sex hormone-binding globulin changes with androgen replacement. *J Clin Endocrinol Metab* 57:645–648.

Raboch J, Mellan J, Starka L (1979) Klinefelter's syndrome: Sexual development and activity. *Arch Sex Behav* 8:333–339.

Re L, Birkhoff JM (2015) The 47, XYY syndrome, 50 years of certainties and doubts: A systematic review. *Aggress Violent Behav* 22:9–17.

Robinson A, Puck M, Pennington B, Borelli J, Hudson M (1979) Abnormalities of the sex chromosomes: A prospective study on randomly identified newborns. *Birth Defects Orig Artic Ser* 15:203–241.

Rocca MS, Pecile V, Cleva L, Speltra E, Selic R, Di Mambro A, Foresta C, Ferlin A (2016) The Klinefelter syndrome associated with high recurrence of copy number variations on the X chromosome with a potential role in the clinical phenotype. *Andrology* 4(2):328–334.

Ron-El R, Strassburger D, Gelman-Kohan S, Friedler S, Raziel A, Appelman Z (2000) A 47,XXY fetus conceived after ICSI of spermatozoa from a patient with non-mosaic Klinefelter's syndrome: Case report. *Hum Reprod* 15:1804–1806.

Ross JL, Roeltgen DP, Kushner H, Zinn AR, Reiss A, Bardsley MZ, McCauley E, Tartaglia N (2012) Behavioral and Social Phenotypes in Boys With 47,XYY Syndrome or 47,XXY Klinefelter Syndrome. *Pediatrics* 129(4):769–778.

Ross JL, Roeltgen DP, Stefanatos G, Benecke R, Zeger MPD, Kushner H, Ramos P, Elder FF, Zinn AR (2008) Cognitive and motor development during childhood in boys with Klinefelter syndrome. *Am J Med Genet A* 146(6):708–719.

Ross JL, Samango-Sprouse C, Lahlou N, Kowal K, Elder FF, Zinn A (2005) Early androgen deficiency in infants and young boys with 47,XXY Klinefelter syndrome. *Horm Res* 64:39–45.

Samango-Sprouse C, Banjevic M, Ryan A, Sigurjonsson S, Zimmermann B, Hill M, Hall MP, Westemeyer M, Saucier J, Demko Z, Rabinowitz M (2013a) SNP-based non-invasive prenatal testing detects sex chromosome aneuploidies with high accuracy. *Prenat Diagn* 33(7):643–649.

Samango-Sprouse C, Gropman AL (2016) X Y Chromosomal Variations: Hormones, Brain Development, and Neurodevelopmental Performance. In *Colloquium Series on the Developing Brain* Morgan Claypool Life Sciences.

Samango-Sprouse CA, Gropman AL, Sadeghin T, Kingery M, Lutz-Armstrong M, Rogol AD (2011) Effects of short-course androgen therapy on the neurodevelopmental profile of infants and children with 49, XXXXY syndrome. *Acta Paediatr* 100(6):861–865.

Samango-Sprouse C, Keen C, Mitchell F, Sadeghin T, Gropman A (2015a) Neurodevelopmental variability in three young girls with a rare chromosomal disorder, 48, XXXX. *Am J Med Genet A* 167(10):2251–2259.

Samango-Sprouse C, Keen C, Sadeghin T, Gropman A (2017) The benefits and limitations of cell-free DNA screening for 47, *XXY (Klinefelter syndrome) Prenat Diagn* 37(5):497–501.

Samango-Sprouse C, Law P (2001) The neurocognitive profile of the young child with XXY. *Europ J Hum Genet* 9(Suppl 1): 193.

Samango-Sprouse C, Rogol A (2002) *The hidden disability and a prototype for an infantile presentation of developmental dyspraxia (IDD) Infants Young Child* 15:11–118.

Samango-Sprouse CA, Sadeghin T, Mitchell FL, Dixon T, Stapleton E, Kingery M, Gropman AL (2013b) Positive effects of short course androgen therapy on the neurodevelopmental outcome in boys with 47, XXY syndrome at 36 and 72 months of age. *Am J Med Genet A* 161(3):501–508.

Samango-Sprouse C, Stapleton EJ, Lawson P, Mitchell F, Sadeghin,T, Powell S, Gropman AL (2015b) Positive effects of early androgen therapy on the behavioral phenotype of boys with 47, XXY. *Am J Med Genet C* 169: 150–157.

Samango-Sprouse CA, Stapleton EJ, Mitchell FL, Sadeghin T, Donahue TP, Gropman AL (2014) Expanding the phenotypic profile of boys with 47,XXY: The impact of familial learning disabilities. *Am J Med Genet A* 164A:1464–1469.

Savic I (2012) Advances in research on the neurological and neuropsychiatric phenotype of Klinefelter syndrome. *Current Opin Neurol* 25(2):138–143.

Seminog OO, Seminog AB, Yeates D, Goldacre MJ (2015) Associations between Klinefelter's syndrome and autoimmune diseases: English national record linkage studies. *Autoimmunity* 48(2):125–128.

Seo JT, Park YS, Lee JS (2004) Successful testicular sperm extraction in Korean Klinefelter syndrome. *Urology* 64:1208–1211.

Simpson JL, de la Cruz FF, Swerdloff RS, Samango-Sprouse C, Skakkebaek NE, Graham JM, Hassold T, Aylstock M, Meyer-Bahlburg HFL, Willard HF, Hall JG, Salameh W, Boone K, Staessen C, Geschwind DH, Giedd J, Dobs A, Rogol A, Brinton B, Paulsen CA (2003) Klinefelter syndrome: Expanding the phenotype and identifying new research directions. *Genet Med* 5:460–468.

Simsek PO, Ütine GE, Alikasifoglu A, Alanay Y, Boduroglu K, Kandemir N (2009) Rare sex chromosome aneuploidies: 49, XXXXY and 48, XXXY syndromes. *Turk J Pediatr* 51(3):294.

Sprouse C, Tosi L, Stapleton E, Gropman AL, Mitchell FL, Peret R, Sadeghin T, Haskell K, Samango-Sprouse CA (2013) Musculoskeletal anomalies in a large cohort of boys with 49, XXXXY. *Am J Med Genet C* 163: 44–49.

Stagi S, Di Tommaso M, Manoni C, Scalini P, Chiarelli F, Verrotti A, Lapi E, Giglio S, Dosa L, de Martino M (2016) Bone mineral status in children and adolescents with Klinefelter syndrome. *Int J Endocrinol* 2016:3032759.

Staessen C, Tournaye H, Van Assche E, Michiels A, Van Landuyt L, Devroey P, Liebaers I, Van Steirteghem A (2003) PGD in 47,XXY Klinefelter's syndrome patients. *Hum Reprod Update* 9:319–330.

Swerdlow AJ, Shoemaker MJ, Higgins CD, Wright AF, et al (2005) Cancer incidence and mortality in men with Klinefelter Syndrome: A cohort study. *J Natl Cancer Inst* 97:1204–1210.

Tabarki B, Shafi SA, Adwani NA, Shahwan SA (2012) Further magnetic resonance imaging (MRI) brain delineation of 49, XXXXY syndrome. *J Child Neurol* 27(5):650–653.

Tahani N, Ruga G, Granato S, et al. (2017) A combined form of hypothyroidism in pubertal patients with non-mosaic Klinefelter syndrome. *Endocrine* 55:513–518.

Tartaglia N, Ayari N, Howell S, D'Epagnier C, Zeitler P (2011) 48, XXYY, 48, XXXY and 49, XXXXY syndromes: not just variants of Klinefelter syndrome. *Acta Paediatr* 100(6):851–860.

Thomas NS, Collins AR, Hassold TJ, Jacobs PA (2000) A reinvestigation of non-disjunction resulting in 47,XXY males of paternal origin. *Eur J Hum Genet* 8:805–808.

Thomas NS, Hassold TJ (2003) Aberrant recombination and the origin of Klinefelter syndrome. *Hum Reprod Update* 9:309–317.

Tojo K, Kaguchi Y, Tokudome G, Kawamura T, Abe A, Sakai O (1996) 47 XXY/46 XY mosaic Klinefelter's syndrome presenting with multiple endocrine abnormalities. *Intern Med* 35:396–402.

Van Rijn, S, Aleman, A, Swaab, H, Krijn, T, Vingerhoets, G, Kahn, RS (2007) What it is said versus how it is said: comprehension of affective prosody in men with Klinefelter (47, XXY) syndrome. *Journal of the International Neuropsychological Society*, 13(6):1065–1070.

Van Rijn S, Swaab H (2015) Executive dysfunction and the relation with behavioral problems in children with 47, XXY and 47, XXX. *Genes Brain Behav* 14(2):200–208.

Van Rijn S, Swaab H, Aleman A, Kahn RS (2008) Social behavior and autism traits in a sex chromosomal disorder: Klinefelter (47,XXY) Syndrome. *J Autism Dev Disord* 38:1034–1041.

Van Rijn S, Swaab H, Aleman A, Kahn RS (2006) X Chromosomal effects on social cognitive processing and emotion regulation: A study with Klinefelter men (47, XXY) *Schizophr Res* 84(2): 194–203.

Venkateshwari A, Srilekha A, Begum A, Sujatha M, Rani PU, Sunitha T, Nallari P, Jyothy A (2010) Clinical and behavioural profile of a rare variant of Klinefelter syndrome-48, XXXY. *Indian J Pediatr* 77(4):447–449.

Vernaeve V, Staessen C, Verheyen G, Van Steirteghem A, Devroey P, Tournaye H (2004) Can biological or clinical parameters predict testicular sperm recovery in 47,XXY Klinefelter's syndrome patients? *Hum Reprod* 19:1135–1139.

Visootsak J, Ayari N, Howell S, Lazarus J, Tartaglia N (2013) Timing of diagnosis of 47, XXY and 48, XXYY: A survey of parent experiences. *Am J Med Genet A* 161(2):268–272.

Visootsak J, Graham JM (2009) Social function in multiple X and Y chromosome disorders: XXY, XYY, XXYY, XXXY. *Dev Disabil Res Rev* 15(4):328–332.

Visootsak J, Rosner B, Dykens E, Tartaglia N, Graham JM (2007) Behavioral phenotype of sex chromosome aneuploidies: 48, XXYY, 48, XXXY, and 49, XXXXY. *Am J Med Genet A* 143(11):1198–1203.

Wang C, Swerdloff RS (2001) Androgen pharmacology and delivery systems. In: *Androgens in Health and Disease*, Bremner WJ, Bagatell CJ, eds, Totowa, New Jersey: Humana Press Inc, pp. 141–153.

Wang C, Swerdloff RS, Iranmanesh A, Dobs A, Snyder P, Cunningham G, Matsumoto AM, Weber T, The Testosterone Gel Study Group (2000) Transdermal testosterone gel improves sexual function, mood, muscle strength, and body composition parameters in hypogonadal men. *J Clin Endocinol Metab* 85:2839–2853

Wiesner G, Everman D, Cassidy S (2002) Constitutional chromosome disorders in adults. In: *The Genetic Basis of Common Diseases*, (King R Rotter J, Motulsky A, eds.), 2nd ed., Oxford: Oxford University Press Inc., pp 989–1022.

Wosnitzer MS, Paduch DA (2013) Endocrinological issues and hormonal manipulation in children and men with Klinefelter syndrome. *Am J Med Genet C* 163C:16–26.

36

LOEYS–DIETZ SYNDROME

ALINE VERSTRAETEN
Center of Medical Genetics, Faculty of Medicine and Health Sciences, University of Antwerp and Antwerp University Hospital, Antwerp, Belgium

HARRY C. DIETZ
Howard Hughes Medical Institute, Baltimore, MD, USA; and Institute of Genetic Medicine, Johns Hopkins University School of Medicine, Baltimore, Maryland, USA

BART L. LOEYS
Center of Medical Genetics, Faculty of Medicine and Health Sciences, University of Antwerp and Antwerp University Hospital, Antwerp, Belgium; and Department of Human Genetics, Radboud University Medical Centre, Nijmegen, The Netherlands

INTRODUCTION

Loeys–Dietz syndrome (LDS) is a rare autosomal dominant connective tissue disorder that was first described in 2005 (Loeys et al. 2005). LDS is clinically heterogeneous and displays significant clinical overlap with other systemic connective tissue disorders. Discriminating features of LDS include hypertelorism, craniosynostosis, bifid uvula, cleft palate, and arterial tortuosity as well as widespread arterial and aortic aneurysms and ruptures. Moreover, the cardiovascular manifestations tend to be severe, with dissections and ruptures typically occurring at smaller diameters and younger ages when compared to other vascular connective tissue disorders.

Prevalence/Incidence

Epidemiologic LDS studies have not been reported since the disease's original description, hampering reliable determination of the disease prevalence and incidence.

Diagnostic criteria

Formal clinical diagnostic criteria for LDS are lacking. In current practice, the diagnosis can be established if characteristic LDS findings are observed and/or a pathogenic variant in an LDS gene is identified in the context of arterial aneurysm (Maccarrick et al. 2014).

LDS should be suspected in individuals with the following vascular, skeletal, craniofacial, and cutaneous manifestations (Figure 36.1):

Vascular

- Aortic aneurysm or dissection, characteristically involving the aortic root (observed in ~95% of people with LDS). Other particularly predisposed vascular segments include the proximal subclavian, superior mesenteric and celiac arteries, and the distal ascending, proximal descending, and abdominal aorta.
- Widespread arterial aneurysms and marked arterial tortuosity, often most pronounced in the head and neck vessels.

Skeletal

- Pectus excavatum or carinatum
- Scoliosis
- Joint laxity and/or contracture
- Arachnodactyly
- Talipes equinovarus
- Cervical spine malformation and/or instability
- Osteoarthritis.

Cassidy and Allanson's Management of Genetic Syndromes, Fourth Edition.
Edited by John C. Carey, Agatino Battaglia, David Viskochil, and Suzanne B. Cassidy.
© 2021 John Wiley & Sons, Inc. Published 2021 by John Wiley & Sons, Inc.

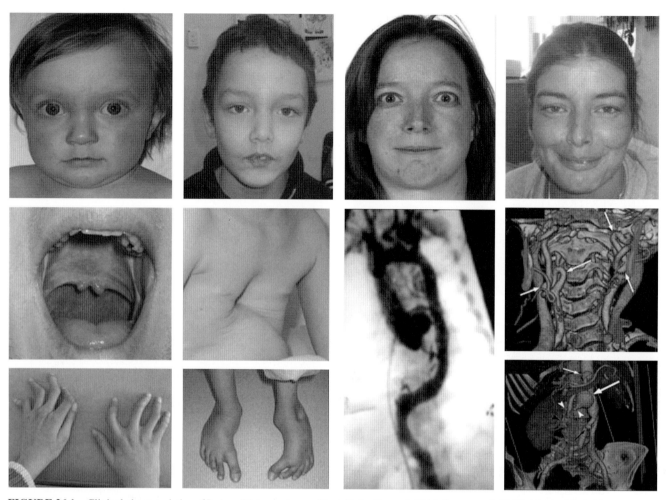

FIGURE 36.1 Clinical characteristics of Loeys–Dietz syndrome (Bertoli-Avella et al. 2015; Loeys et al. 2005). Typical facial features with hypertelorism, malar hypoplasia, retrognathia and bifid uvula. Skeletal findings including pectus excavatum, arachnodactyly and camptodactyly, and club foot deformity. Cardiovascular features include aortic root aneurysm, arterial aneurysm and aortic and arterial tortuosity.

Craniofacial

- Hypertelorism
- Bifid uvula
- Cleft palate
- Craniosynostosis.

Cutaneous

- Translucent skin
- Easy bruising
- Dystrophic scars.

Etiology, Pathogenesis, and Genetics

LDS is inherited in an autosomal dominant manner and presents with variable expressivity. Its original clinical description coincided with the identification of the first two LDS genes, *TGFBR1* (causing LDS1) and *TGFBR2* (LDS2), encoding for the transforming growth factor beta (TGFβ) receptor subunits types 1 and 2 (Table 36.1) (Loeys et al. 2005). These receptor subunits are serine-threonine kinases that form heteromeric receptor complexes upon binding of TGFβ cytokines to TGFβR2. TGFβR2 phosphorylates TGFβR1, which in turn phosphorylates receptor-activated SMAD proteins (R-SMADs) SMAD2 or SMAD3. Association of pSMAD2/3 with SMAD4 leads to nuclear translocation of the complex that partners with other transcription factors to drive TGFβ target gene expression (Figure 36.2). Initially, LDS subtype classification (LDS1 and LDS2) was based on the presence or absence of outward systemic features (e.g., cleft palate, craniosynostosis, hypertelorism), but a revised and widely adopted classification assigns LDS subtypes based upon the underlying gene defect (Maccarrick et al. 2014). *SMAD3* mutations were first identified in people who had aortic aneurysms with frequent early-onset osteoarthritis, initially called aneurysm-osteoarthritis syndrome (AOS). In view of the significant overlap with LDS, this condition is now classified as LDS3 (Table 36.1) (van de Laar et al. 2011).

INTRODUCTION 565

TABLE 36.1 Overview of the identified LDS genes and their contribution to the phenotype.

Subtype	OMIM #	Gene	Chromosome locus	% of patients	Mutation spectrum
LDS1	609192	*TGFBR1*	9q22.33	~20	Missense/nonsense/splice site*
LDS2	610168	*TGFBR2*	3q24.1	~55	Missense/frameshift/nonsense*
LDS3	613795	*SMAD3*	15q22.33	~5	Missense/frameshift/nonsense/splice site/multi-exon del/gene del
LDS4	614816	*TGFB2*	1q41	~5	Missense/frameshift/nonsense/splice site/in-frame indel/gene del
LDS5	615582	*TGFB3*	14q24.3	~1	Missense/frameshift/nonsense/in-frame indel
LDS6	NA	*SMAD2*	18q21.1	~1	Missense/nonsense

* Only truncating *TGFBR1* and *TGFBR2* mutations that are predicted to escape nonsense mediated RNA decay can lead to cardiovascular LDS-phenotype. Complete loss-of-function *TGFBR1* mutations have been linked to multiple self-healing squamous epithelioma (MSSE), an autosomal-dominant isolated skin condition. NA: not assessed.

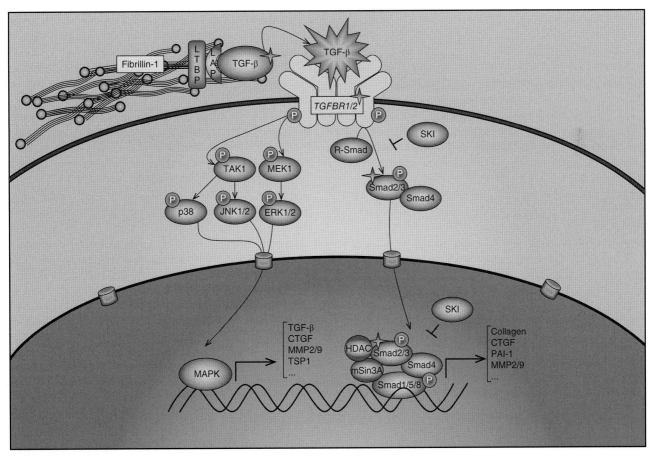

FIGURE 36.2 Overview of the TGFβ signaling pathway in relation to the pathogenesis of LDS. Under normal physiological circumstances, TGFβ ligands bind to the TGFβ receptor complex (TGFβR1/2), initiating canonical and non-canonical TGFβ signaling. The TGFβ type I receptor phosphorylates the transcription factors SMAD2/3, which form a complex with SMAD4 and translocate to the nucleus to initiate transcription of TGFβ target genes. Non-canonical TGFβ signaling pathways are also activated by ligand-mediated activation of TGFβR1/2, and induce the transcription of target genes through MAPKs and other intracellular effectors. Heterozygous loss-of-function mutations in TGFβ ligands (TGFβ2 and TGFβ3), TGFβ receptor subunits (TGFβR1 or TGFβR2), or SMAD signaling effectors (SMAD2 or SMAD3) result in LDS (indicated with star). Abbreviations: LTBP, latent transforming growth factor binding protein, LAP, latency-associated peptide. Figure is modified from (Verstraeten et al. 2016).

A candidate gene sequencing-based approach recently linked *SMAD2* mutations to syndromic arterial aneurysms and dissections (Table 36.1) (Micha et al. 2015). Finally, LDS-causing mutations were identified in the genes encoding the TGFβ ligands TGFβ2 and TGFβ3 (*TGFB2* and *TGFB3*, causing LDS4 and LDS5, respectively) (Table 36.1) (Bertoli-Avella et al. 2015; Boileau et al. 2012; Lindsay et al. 2012). Taken together, these conditions unequivocally implicate dysregulation of TGFβ signaling in the pathogenesis of familial presentations of aortic aneurysm.

TGFβ cytokines are multifunctional proteins that regulate cell differentiation, growth, migration, synthetic repertoire and apoptosis. TGFβ ligands are secreted in the form of a large latent complex that includes the mature cytokine, a dimer of its processed amino-terminal propeptide (termed latency-associated peptide or LAP) and a latent TGFβ-binding protein (LTBP) isoform (types 1, 3 or 4). Release of the cytokine, termed TGFβ activation, tightly regulates TGFβ bioavailability. There are many potential TGFβ activators including integrins, proteases, and mechanical force. In addition to canonical (SMAD-dependent) TGFβ signaling, ligand-activated TGFβ receptors can support so-called non-canonical signaling cascades including those initiated by RHO/ROCK, PI3K/AKT and the mitogen-activated protein kinases (MAPKs), ERK, JNK and p38 (Figure 36.2).

The vast majority of mutations causing LDS1 or LDS2 involve loss-of-function missense substitutions in the kinase domain of TGFβR1 or TGFβR2, respectively; the exceptions include nonsense, frameshift or splicing mutations that lead to distal premature termination codons that fail to initiate nonsense-mediated mRNA decay. In contrast, LDS3-5 are clearly caused by functional haploinsufficiency, as mutations causing early premature termination codons are common and partial or complete gene deletions can be seen. Paradoxically, aortic wall tissue from patients and mouse models with all forms of LDS shows a consistent and unequivocal signature for paradoxically enhanced TGFβ signaling, including nuclear accumulation of pSMAD2/3 and high output of prototypical TGFβ target genes such as *COL1A1*, *COL3A1*, *CTGF*, *MMP2* and *MMP9*. Reconciling mechanistic hypotheses have invoked downregulation of auto-inhibitory pathways (Bolar et al. 2012), shifts in TGFβ ligand usage (Doyle et al. 2012; Lindsay et al. 2012), or paracrine interactions between neighboring cell types that show differential vulnerability to the effects of functional haploinsufficiency for a positive effector of TGFβ signaling (Lindsay and Dietz 2011). The latter two hypotheses are supported by the observation of high expression of TGFβ1 in the proximal aorta of people and mice with LDS, a domain where TGFβ2 normally predominates. TGFβ1 can be activated by integrins, whereas TGFβ2 cannot. In this light, the observed high level of TGFβ activity in the LDS aorta might relate to more efficient utilization of secreted TGFβ.

Prior to recognition of LDS, the pathogenesis of a related aortic aneurysm syndrome (Marfan syndrome or MFS, see Chapter 37) had been strongly linked to excessive TGFβ activity. In addition to their blood pressure lowering properties, angiotensin receptor blockers (ARBs) such as losartan had been shown to suppress TGFβ activity in a variety of contexts including chronic kidney disease. Blockade of the angiotensin II type 1 (AT1) receptor can decrease the expression of TGFβ ligands, receptor subunits and activators. Trials of losartan in both MFS and LDS mouse models led to potent suppression of aneurysm progression in association with preservation of aortic wall architecture and normalization of the tissue signature for high TGFβ signaling (Gallo et al. 2014; Habashi et al. 2006). While somewhat dependent on context, TGFβ neutralizing antibodies showed the ability to suppress aneurysm progression in MFS mice when used beyond the neonatal period, with evidence for synergy with losartan (Cook et al. 2015; Habashi et al. 2006). More recently, a condition (Shprintzen–Goldberg syndrome or SGS) that includes virtually all of the features of LDS (including aortic root aneurysm), with the added finding of highly penetrant developmental disability, was shown to be caused by heterozygous loss-of-function mutations in a gene encoding the Sloan–Kettering institute proto-oncoprotein (SKI) (Carmignac et al. 2012; Doyle et al. 2012). SKI interacts with R-SMADs and is co-recruited to TGFβ target genes where it displaces positive regulators of transcription (e.g., p300/CBP) and recruits negative regulators (e.g., histone deacetylases), serving to temper the TGFβ signaling response. It is therefore informative that a genetic perturbation predicted and observed to increase TGFβ signaling phenocopies LDS. A variety of studies have implicated non-canonical TGFβ signaling pathways, prominently including MAPKs, in the pathogenesis of aortic aneurysm (Habashi et al. 2011; Holm et al. 2011); the relevance of these findings to LDS remains to be established.

Diagnostic testing

Molecular testing approaches can include serial single-gene, gene panel, and exome/genome sequencing. Six disease genes have been identified to date, i.e. *TGFBR1*, *TGFBR2*, *SMAD2*, *SMAD3*, *TGFB2* and *TGFB3*, and we anticipate that more remain to be discovered as individuals with discriminating features of LDS without a mutation in any of these genes exist (Table 36.1) (Verstraeten et al. 2016). About 75% of patients develop LDS as the result of a *de novo* mutation, with the remaining 25% inheriting disease from an affected parent. LDS-causing mutations are often unique to patients or families and encompass a very broad mutation spectrum, including nonsense, frameshift, splice altering, insertion/deletion, and missense variations spread throughout the proteins as well as whole-gene and multi-exon deletions (Table 36.1). Very few specific genotype–phenotype correlations have been described to date; the different LDS subtypes are rather thought to form a phenotypic continuum. Importantly, the identical mutations (e.g., in *TGFBR1* or *TGFBR2*) have been associated with either typical syndromic presentation of LDS (including cleft palate, craniosynostosis, club foot) and non-syndromic thoracic aortic aneurysm. Also, a mildly affected parent can have a child with a more severe LDS phenotype. These observations suggest that the nature of the primary mutation, as such, does not fully determine the clinical outcome. Most likely, modifying genes also influence clinical severity. Preliminary findings suggest that non-penetrance and a milder cardiovascular phenotype are more common in cases with *TGFB2/3* mutations compared to *TGFBR1/2* and *SMAD3* mutation carriers, but more experience is needed. Also, early-onset osteochondritis dissecans and osteoarthritis affecting the spine, knees, and hands have typically been reported in families segregating *SMAD3* mutations (van de Laar et al. 2012), but osteoarthritis has also been reported in other LDS subtypes. With regard to *SMAD2*, too few mutation-positive cases have yet been reported to determine at which end of the phenotypic spectrum *SMAD2*-related LDS should be positioned.

Owing to high genetic heterogeneity and substantial clinical overlap between LDS subtypes, LDS genetic testing rarely involves serial single-gene sequencing. Instead, most labs opt for disease-targeted gene panel sequencing, allowing for simultaneous mutation analysis of all known genes in a cost- and time-efficient manner. Upon the use of panels covering all genes that have been found mutated in individuals with either non-syndromic or syndromic thoracic aortic aneurysm, molecular diagnostic testing can also significantly aid establishment of a diagnosis in the context of clinical uncertainty (see Differential Diagnosis section). As gene panel sequencing often lacks the ability to reliably detect copy number variations (CNVs), panel-negative samples are preferably also subjected to multiplex ligation-dependent probe amplification (MLPA) or alternative methods to detect deletions or duplications. Pathogenic deletions have been found in LDS cases related to *TGFB2* and *SMAD3*. Notably, due to the rapidly decreasing costs of next-generation sequencing, it is anticipated that exome or even genome sequencing will soon surpass gene panel approaches in the clinical setting.

Differential Diagnosis

Several related connective tissue disorders show significant clinical overlap with LDS; the autosomal dominant conditions Marfan syndrome (MFS, see Chapter 37), Shprintzen–Goldberg syndrome, vascular Ehlers–Danlos syndrome (EDS, see Chapter 25), congenital contractural arachnodactyly and non-syndromic familial thoracic aortic aneurysm, the autosomal recessive conditions cutis laxa type 1B and arterial tortuosity syndrome, the X-linked conditions *FLNA*-related connective tissue disease, and Meester–Loeys syndrome. Establishment of a differential diagnosis generally involves exhaustive clinical examination of the patient to search for discriminative symptoms, in addition to concomitant genetic testing.

MFS has substantial overlap with LDS. It is typically characterized by skeletal (long bone overgrowth, joint laxity, pectus excavatum/carinatum, scoliosis), ocular (ectopia lentis, myopia), cardiovascular (mitral valve prolapse, aortic root aneurysm and dissection) and cutaneous (striae) manifestations. Ectopia lentis represents the strongest discriminating feature for MFS (present in ±60% of people with MFS), whereas hypertelorism, cleft palate/bifid uvula and more widespread aneurysmal disease are indicative of LDS. Other characteristics suggesting LDS are craniosynostosis, club foot, and cardiac anomalies such as atrial septal defect, bicuspid aortic valve, and patent ductus arteriosus. While arterial tortuosity (typically involving the head and neck vessels) can be seen in both conditions, it is more predictable and severe in LDS. MFS is caused by mutations in the *FBN1* gene (Dietz et al. 1991). More than 1850 different *FBN1* mutations, widely spread over the entire gene, have been described to date (http://www.umd.be/FBN1/), explaining the vast majority (~95%) of all MFS cases (Verstraeten et al. 2016). About one third of affected individuals developed disease because of a de novo mutation. *FBN1* codes for the microfibrillar glycoprotein fibrillin-1, which provides elasticity and structural support to a wide variety of tissues. Under normal physiological conditions, fibrillin-1 tightly controls TGFβ bioavailability through interaction with the large latent complex. As such, fibrillin-1 deficiency impairs targeting of the large latent complex to the extracellular matrix and associates with increased TGFβ activation and signaling (Figure 36.2) (Neptune et al. 2003).

Shprintzen–Goldberg syndrome is characterized by mild-to-moderate intellectual disability (learning difficulties), brain anomalies (hydrocephalus, dilatation of the lateral ventricles), craniosynostosis (involving the coronal, sagittal, or lambdoid sutures) and hypotonia of skeletal muscles, in addition to many of the skeletal (scoliosis, joint hypermobility, pes planus, pectus deformity), and ocular (myopia) symptoms of MFS and/or LDS. Cardiovascular manifestations such as mitral valve prolapse and mitral or aortic valve regurgitation occur regularly, whereas aneurysm development and dissection are usually restricted to the aortic root and are both milder and less frequent compared to LDS and MFS. Minimal subcutaneous fat and reduced muscle mass are additional characteristic Sphrintzen–Goldberg syndrome findings. Nonetheless, developmental delay is the most reliable distinguishing feature, when compared to LDS. In 90% of people with Shprintzen–Goldberg syndrome, a (mostly de novo) mutation can be identified in the *SKI* gene, typically in the first exon encoding the R-SMAD-interacting domain (Carmignac et al. 2012; Doyle et al. 2012; Schepers et al. 2015). *SKI* encodes a negative regulator of R-SMAD-dependent TGFβ target gene expression (Figure 36.2) and *SKI* mutations have been shown to result in an increase in TGFβ signaling.

Congenital contractural arachnodactyly, or Beals syndrome, is characterized by an MFS-like appearance (tall, slender habitus, dolichostenomelia, arachnodactyly), multiple flexion contractures, kyphoscoliosis, abnormal pinnae (crumpled ears), and muscular hypoplasia. There are rare reports of mild aortic root enlargement, but there is presently no evidence that dilatation progresses to dissection or rupture. Although the clinical features can be similar to MFS, and occasionally LDS, presentation of multiple joint contractures and crumpled ears in the absence of severe aneurysmal disease suggests congenital contractural arachnodactyly rather than MFS or LDS. Congenital contractural arachnodactyly is caused by mutations in the *FBN2* gene, encoding the microfibrillar protein fibrillin-2 (Putnam et al. 1995). Fibrillin-2 is thought to direct the assembly of elastic fibers during early embryogenesis, while fibrillin-1 provides structural force bearing support in many tissues and organs during postnatal life.

Ehlers–Danlos syndrome (EDS) (see Chapter 25) is the collective term for a group of connective tissue disorders presenting with variable combinations of skin hyperextensibility and translucency, atrophic scars, easy bruising, tissue fragility (vascular or intestinal rupture) as well as joint hypermobility and subluxations. Arterial aneurysmal disease is present in about one quarter of affected individuals, but is particularly prevalent and clinically relevant in patients with vascular EDS. LDS can be similar to vascular EDS with respect to the skin anomalies, the occurrence of club foot and the risk for widespread arterial rupture. They are different, however, in their craniofacial appearance, with people with vascular EDS typically not presenting with widely spaced eyes and a bifid uvula or cleft palate. Importantly, while aortic root aneurysm can occasionally be observed in vascular EDS, there is not a particular predisposition for involvement of this vascular segment (as opposed to MFS and LDS). Vascular dissection or rupture without prior dilatation is also a discriminating feature of vascular EDS when compared to other aneurysm syndromes. Over 95% of people with vascular EDS carry a mutation in the *COL3A1* gene (Schwarze et al. 2001). Mutations in *COL1A1* and *COL5A1* are occasionally found in vascular EDS-like cases, but typically cause arthrochalasia and classical EDS, respectively. *COL3A1*, *COL1A1* and *COL5A1* encode fibrillar collagens, which, upon homo- or heterotrimerization into collagen fibrils, become incorporated in the extracellular matrix of internal organs and skin to provide tensile strength.

Occasionally, non-syndromic thoracic aortic aneurysm can be seen in people with heterozygous mutations in *FBN1*, *TGFBR1*, *TGFBR2*, *SMAD3* (Bradley et al. 2015) or other genes associated with LDS. This does not appear to represent a distinct phenotype–genotype correlation, as the same mutation can cause phenotypes ranging from non-syndromic thoracic aortic aneurysm to classic and severe LDS. Isolated non-syndromic thoracic aortic aneurysm is more typically caused by primary mutations in genes encoding components of the smooth muscle contractile apparatus (*ACTA2*, *MYH11*, *MYLK*, *PRKG1*, *FOXE3*). Selected mutations in these genes can present with systemic findings such as levido reticularis, iris flocculi, cerebral aneurysm, bicuspid aortic valve, and persistent ductus arteriosus (Guo et al. 2007).

Clinically overlapping autosomal recessive disorders such as cutis laxa type 1B and arterial tortuosity syndrome can be confused with LDS. Cutis laxa is characterized by loose and redundant skinfolds as well as reduced skin elasticity (doughy skin). Autosomal recessive cutis laxa comprises a very heterogeneous group of disorders with respect to organ involvement and disease severity. In certain subtypes, vascular and skeletal abnormalities reminiscent of LDS can be seen. Phenotypic resemblance is most remarkable in people with *EFEMP2* mutations causing a deficiency of fibulin-4, who commonly display arterial tortuosity and aneurysms/dissections as well as arachnodactyly, pectus excavatum, hypertelorism and joint laxity. Arterial tortuosity, but not aneurysmal disease, has been observed in individuals with *FBLN5* mutations causing a deficiency of fibulin-5. The main discriminating features of autosomal cutis laxa, when compared to LDS, include skin laxity and a predisposition for early and severe emphysema. Fibulin-4 and fibulin-5 are essential for elastic fiber assembly and function.

Arterial tortuosity syndrome presents with severe tortuosity and stenosis of the aorta and medium-to-large sized arteries. Aneurysms can be observed. Skeletal (scoliosis, pectus excavatum/carinatum, joint laxity and contractures, arachnodactyly, camptodactyly) and skin (soft and/or doughy skin) manifestations are common. Moreover, people with arterial tortuosity syndrome present with characteristic craniofacial features, including periorbital fullness, a long face, downslanting palpebral fissures, a convex nasal ridge, malar hypoplasia, a high palate, micrognathia, large ears and dental crowding. These craniofacial characteristics along with a doughy skin appearance distinguish arterial tortuosity syndrome from LDS. Moreover, the arterial tortuosity seems more pronounced and generalized in arterial tortuosity syndrome than in LDS, and the presence of stenotic aortic and arterial lesions is far more typical in arterial tortuosity syndrome. Arterial tortuosity syndrome is the result of biallelic loss-of-function mutations in *SLC2A10* (Coucke et al. 2006). *SLC2A10* encodes the facilitative glucose transporter GLUT10. Although the precise substrate of this carrier is still elusive, upregulation of the TGFβ pathway has been observed in arterial wall specimens of *SLC2A10* mutation carriers, and zebrafish studies suggested that GLUT10 might function as an effector of the TGFβ signaling pathway, downstream of the SMAD proteins (Coucke et al. 2006; Willaert et al. 2012).

Two X-linked conditions also demonstrate considerable overlap with LDS. First, *FLNA* mutation carriers can present with connective tissue findings reminiscent of LDS, including joint hypermobility, skin hyperelasticity, and aortic root aneurysm, in addition to the near-uniform occurrence of periventricular nodular heterotopia upon brain imaging. Complete loss-of-function mutations in *FLNA* are associated with male embryonic lethality, while males with hypomorphic or mosaic *FLNA* mutations can survive past infancy (Cannaerts et al. 2018). *FLNA* encodes for filamin A, a molecule that crosslinks actin filaments and attaches them to the cell membrane. Filamin A has numerous binding partners and is involved in various cell functions, such as signal transduction, cell migration and adhesion. More recently, loss-of-function mutations in *BGN* were identified in patients presenting with Meester-Loeys syndrome, a condition characterized by early-onset aortic dissection and LDS-associated features (i.e. hypertelorism, a bifid uvula, and joint hypermobility and contractures) (Meester et al. 2016). Mutation-carrying females can either be asymptomatic, or experience aortic aneurysm and even dissection. The precise mechanism underlying this observation is unknown, but was shown not to be linked to skewed X-inactivation. *BGN* encodes the small leucine-rich proteoglycan biglycan that interacts with extracellular matrix proteins such as collagen I, III or elastin. Apart from its structural role, biglycan also has a role in the regulation (generally suppression) of cytokine activity (e.g., TGFβ and bone morphogenetic protein).

MANIFESTATIONS AND MANAGEMENT

Growth and Feeding

Early in life, people with LDS may present with failure to thrive. The etiology is probably diverse and might relate to the early need for surgical intervention and feeding difficulty attributable to craniofacial abnormalities. Later in life, those with LDS tend to be somewhat tall, but not to the same extent as people with Marfan syndrome. The overgrowth usually affects the digits and ribs more prominently than the long bones (see Musculoskeletal section).

Evaluation

- Weight and height as well as body mass index curves should be monitored.

Management

- Low thresholds for caloric supplementation should be considered for failure to thrive, especially prior to planned surgery.

Development and Behavior

Cognitive development and behavior are typically not affected in LDS. If LDS is caused by a gene deletion (e.g., *TGFB2* deletion), neighboring genes implicated in intellectual disability might also be deleted, causing developmental delay. Learning difficulties in the absence of a chromosomal deletion might be related to craniosynostosis or hydrocephalus. The psychosocial impact of having a disorder with a significant health burden and a somewhat uncertain prognosis should not be underestimated.

Evaluation

- Establishment of an LDS diagnosis can inflict feelings of isolation, unfairness, and anxiety, possibly leading to an existential crisis or depression if neglected. An affected individual's communication style with relatives and/or caretakers, attitude towards regular medical appointments as well as restrictive management strategies, and general disease coping capability should be carefully assessed by the treating physician.

Management

- Affected individuals should be informed about the existence of support groups.
- When deemed necessary, psychosocial therapy, counseling, and psychoactive medications should be considered.

Musculoskeletal

Several skeletal LDS findings overlap with the anomalies observed in MFS, including pectus deformity, scoliosis and flat feet. Skeletal overgrowth in LDS is less pronounced than in MFS, and usually affects the extremities and ribs more prominently than the long bones. Joint hypermobility has frequently been observed, presenting as congenital hip dislocation and recurrent joint subluxations. Paradoxically, contractures of the extremities, especially of the hands (camptodactyly) and feet (club feet), are also common in individuals with LDS. Muscle hypotonia may be present in infancy. About 33% of people with LDS are estimated to have structural cervical spine anomalies, and at least 50% display cervical spine instability (Maccarrick et al. 2014). Spine anomalies are especially seen in individuals with severe craniofacial features. Lastly, individuals with LDS are at increased risk for fractures owing to low bone mineral density, especially in the spine, hip and/or femoral neck. Early-onset osteoarthritis also seems more common in LDS. Although prominently reported in LDS3, it is also more frequent in other LDS subtypes.

Evaluation

- Individuals with LDS should be checked for cervical spine abnormalities using flexion–extension X-rays. In children, it is advised to repeat imaging every three to five years during growth.
- The geneticist or primary care physician should monitor for skeletal deformity or dysfunction, with a low threshold for referral for a dedicated orthopedic evaluation.
- Dual-energy X-ray absorptiometry (DEXA) imaging should be performed to monitor bone mineral density.

Management

- Physical therapy for joint hypermobility and hypotonia is recommended.
- Activities that cause joint injury or pain should be avoided.
- If the deformity is mild, those with club feet respond well to stretching. For moderate-to-severe club feet, either Ponseti casting or surgical correction by an orthopedic surgeon can be considered. The latter can lead to overcorrection and is therefore not routinely recommended (Erkula et al. 2010).
- Flat feet (pes planus) can lead to foot pain, fatigue, muscle cramps, and walking difficulties. Some individuals benefit significantly from wearing shoes with good arch support or firm shoe inserts, while others find them irritating. Surgical intervention is typically not indicated (Maccaririck et al. 2014).
- Surgical management of cervical spine instability and scoliosis should be considered to prevent damage to the spinal cord or progressive deformity. Individuals with LDS tolerate spinal surgeries well, but delayed bone healing occasionally occurs (Erkula et al. 2010).
- Children with mild-to-moderate scoliosis (<25°) can benefit from bracing.
- In rare circumstances, severe pectus excavatum can impair cardiac and respiratory function and cause chest and back pain. In such cases, surgical intervention is medically indicated, preferably not in conjunction with aortic root replacement surgery.
- Adequate calcium and vitamin D intake should be ensured, especially in people with low bone mineral density or recurrent fractures (Maccarrick et al. 2014).

Craniofacial

In its most typical appearance, LDS presents with hypertelorism, cleft palate, and/or craniosynostosis. Other characteristic features may include malar flattening, micrognathia or retrognathia, and a highly arched palate and dental crowding. Poor quality or discolored dental enamel can be seen. Craniosynostosis most commonly involves premature fusion of the sagittal suture, yet the coronal, metopic and squamosal sutures can be involved as well. In people with milder LDS, skull asymmetry and uvula anomalies (i.e. a bifid, broad or long uvula) can be observed instead of craniosynostosis and cleft palate, respectively.

Evaluation

- In case of skull asymmetry, craniosynostosis can be assessed using CT imaging.
- Specialized craniofacial, orthodontic, or dental evaluations may be indicated.

Management

- Children born with cleft palate can have issues with normal feeding, hearing, dentition, and speech. Hence, surgical repair is usually recommended (personal experience).
- Craniosynostosis should be treated according to standard protocols.

Cardiovascular

Cardiovascular disease is the major cause of morbidity and mortality in LDS, underscoring the importance of close cardiovascular monitoring. Arterial tortuosity, widespread and progressive aneurysm formation, high predisposition for arterial ruptures at young ages (as young as 3 months of age) and small diameters (as small as aortic diameters of 3.7 cm in adults), and mitral valve prolapse with/without regurgitation are the most frequently observed abnormalities. Patent ductus arteriosus, atrial septal defects, and bicuspid aortic valve are much more common than in the general population. Most people with LDS have multiple cardiovascular complications and, hence, require intensive follow-up and multiple surgical interventions.

Evaluation

- People with LDS should be managed in a medical center familiar with the condition.
- Individuals with LDS require regular echocardiography to monitor their aortic diameters (root and ascending) and to evaluate both morphology and functionality of their heart valves. Furthermore, head-to-pelvis magnetic resonance angiography (MRA) or computed tomography angiography (CTA) should be performed at frequent intervals as well, in order to obtain serial measurements of the entire arterial tree as well as to check for the presence of arterial tortuosity. Echocardiography should be performed at least yearly, more frequently in case of

severe and rapidly progressing aneurysmal disease. MRA or CTA should be performed at diagnosis, one year thereafter, and at least every two to three years unless findings warrant more frequent imaging.
- Type B dissections (i.e. aortic dissections that do not involve the ascending aorta) should be monitored closely for rapid aortic growth; that is, at 7–14 days, and 1, 3, 6 and 12 months post-dissection. Afterwards, if stable, yearly follow-up is recommended.
- Post-surgery echocardiography is recommended at three to six month intervals for one year after surgery, and at 6–12 month intervals thereafter unless specific problems are identified.

Management

- To avoid undue mechanical stress on the vasculature of individuals with LDS, contact sports or isometric activities (sit-ups, push-ups, pull-ups, weight lifting) and exercise to exhaustion should be avoided.
- Medical management involves strict control of blood pressure to reduce hemodynamic stress on the arterial wall. β-blockade has historically been used, and is still commonly used, for syndromic thoracic aortic aneurysm management. Angiotensin receptor blockers (ARBs) such as losartan are hypothesized to be beneficial in LDS because of both blood pressure lowering and desirable biochemical effects within the vessel wall including attenuation of TGFβ signaling. High-dose losartan normalized aortic root growth rate and preserved aortic wall architecture in mouse models of LDS; these protective effects were not observed upon administration of hemodynamically equivalent doses of β-blockers. While somewhat variable depending on context, clinical trials for Marfan syndrome have suggested that losartan is as good or better than atenolol in suppressing aortic root growth, even in circumstances with more aggressive dosing of the latter. Given these results and the favorable tolerance profile for ARBs, our personal practice is to initially administer losartan with a target dose of 2.0 mg/kg/day for children and at least 100 mg/day for adults (personal experience). Ultra-high ARB dosing can be considered in refractory circumstances, as described for other TGFβ-related pathologies including chronic kidney disease or cardiomyopathy; combination therapy with atenolol represents a complementary or alternative strategy. Prophylactic medical treatment should be considered in people with LDS without aortic enlargement if affected family members have had significant vascular disease or if the same mutation has previously been linked to aortopathy. Importantly, medications that stimulate the cardiovascular system should be used with caution, including decongestants, stimulant medications and vasoconstrictors in common use for migraine management.
- LDS-related aneurysms are amenable to prophylactic surgical intervention without a significant risk of undue vascular fragility or difficulty in establishing vascular anastomoses. Decisions regarding the precise timing of surgery are based upon integration of many variables and need to be individualized. In general, surgical thresholds based on aortic root diameter are lower than those applied for most other vascular connective tissue disorders, including Marfan syndrome. For adults with LDS1 and LDS2, surgery is generally considered when the aortic root dimension is about 4.0 cm, with a threshold in the low-to-mid 4 cm range for other forms (Patel et al. 2017a; Patel et al. 2017b). This needs to be tailored to the specific circumstance. Factors favoring more aggressive intervention include a family history of aortic dissection at a young age or small dimension, a rapid rate of aortic root growth (>0.5 cm/year), identification of a specific mutation that was previously associated with early and severe vascular disease, or emergence of significant aortic valve regurgitation. Precise guidelines for infants with LDS have not been established, but there is strong incentive to allow the aortic annulus to reach a size of 2.0–2.2cm, allowing the placement of a graft of sufficient size to accommodate somatic growth into adulthood (Maccarrick et al. 2014). In general, the severity of either craniofacial manifestations or arterial tortuosity is predictive of the severity of vascular disease. Valve-sparing procedures (typically using the re-implantation approach) are preferred over other methods since they avoid the need for chronic anticoagulation.
- LDS-related aneurysms can occur throughout the entire length of the aorta and have been observed in numerous other arteries, including the subclavian, superior mesenteric, celiac, hepatic, cerebral, and coronary arteries, amongst others. As a general rule, intervention for iliac or visceral aneurysms should be considered with rapidly expanding aneurysms or when the observed size exceeds 2–3 times the expected diameter (Maccarrick et al. 2014). Thresholds for neurovascular surgical intervention have not been described and should be based on consideration of location, accessibility and risk.

Of note, prophylactic surgery does not eliminate the risk of dissection in individuals with LDS, and a high index of suspicion should be maintained in the setting of compatible signs and symptoms.

- Mitral and aortic valve regurgitation as well as atrial septal defect and patent ductus arteriosus should be managed per standard protocols. Vascular risks imposed by device-based procedures should be considered in the context of LDS.

Ophthalmologic

Ocular complications such as myopia, retinal detachment and cataracts occur in both LDS and Marfan syndrome (MFS). Myopia is less frequent and usually less severe in LDS. Absence of ectopia lentis in LDS is a major distinguishing ocular feature of MFS. In the diagnostic work-up, another informative feature might be the presence of blue or dusky sclera, which can be seen in LDS and vascular EDS, but not MFS. Strabismus, particularly exotropia, is common in LDS and frequently requires surgical intervention in childhood. Retinal tortuosity has been described, but the clinical significance is unknown. A sharp blow to the head or face could enhance the predisposition for retinal detachment; protective eye wear is recommended in relevant situations.

Evaluation

- The ocular manifestations of LDS should be managed by an ophthalmologist with expertise in connective tissue disorders.
- For the diagnosis of ectopia lentis or lens (sub)luxation, a slit lamp exam is required.
- Subsequent ophthalmologic evaluations in LDS can be guided based on signs and symptoms.

Management

- Eye muscle problems respond to classic protocols of patching and, if needed, surgery (Van Hemelrijk et al. 2010).
- Careful and aggressive refraction and visual correction is mandatory in young children at risk for amblyopia (Loeys and Dietz 2008).
- Treatment of retinal detachment is standard.

Neurologic

The two most common non-vascular neuroradiological findings of LDS are dural ectasia, which is common but often clinically silent, and Arnold–Chiari type I malformation, which is rare. Hydrocephalus may exist unrelated to Chiari malformation. Common vascular neuroradiological findings include aneurysms, dissections and pseudo-aneurysms of the head and neck vessels. Up to 50% of people with LDS report recurrent headaches or migraines, but their precise etiology remains elusive.

Evaluation

- CT angiography should be performed at regular intervals to check for the presence of aneurysms of the head and neck vessels (see Cardiovascular section).
- Food/environment diary and sleep studies may aid in the identification of specific headache triggers. Consultation with a headache specialist can be beneficial.
- Routine monitoring of dural ectasia is not required. Postural and severe headache should prompt evaluation for dural tear or leak.

Management

- Neuro-interventional specialists should be involved in the evaluation and management of aneurysms of the head and neck vessels.
- No routinely effective therapies for symptomatic dural ectasia currently exist.
- For the treatment of a dural tear or dural leakage, blood patch can be considered.
- Adequate fluid intake and the use of β-blockers can provide relief for headaches, as can avoidance of specific triggers, careful attention to hydration status, and general supportive care including analgesics, routine headache medications and magnesium supplementation. Vasoconstrictor medications for headaches are contraindicated.

Respiratory

Pulmonary manifestations that can present in LDS include pneumothorax, restrictive lung disease due to skeletal deformity, obstructive lung disease due to allergy, asthma or emphysema, and obstructive sleep apnea. Pulmonary artery dilation is commonly seen, but pulmonary artery dissection has not been described. Severe enlargement, especially in infancy, can be associated with airway compression and respiratory compromise.

Evaluation

- Pulmonary evaluation should be initiated in case of respiratory complaints.
- A sleep study can be considered in the presence of compatible clinical symptoms for sleep apnea or respiratory dysfunction.

Management

- Optimal management of pneumothorax to prevent recurrence may require chemical or surgical pleurodesis

or surgical removal of pulmonary blebs (Loeys and Dietz 2008).
- The potential respiratory implications of skeletal deformity should be considered in the development of a management plan.
- Continuous positive airway pressure (CPAP) treatment should be considered if obstructive apnea is detected by the sleep study.

Immunologic

Individuals with LDS are at increased risk for allergic disease, including asthma, eczema, and allergic rhinitis. Both immunological factors and altered craniofacial morphology underlie this overall increased risk. A significant proportion of people with LDS also develops gastrointestinal disease, such as food allergy (up to 30%) and eosinophilic esophagitis, gastritis or colitis (up to 60%). A small subset can show overt inflammatory bowel disease. Elevated plasma immunoglobulin E levels, eosinophil counts, and T helper 2 (Th2) cytokine amounts have been found in people with LDS (Felgentreff et al. 2014; Frischmeyer-Guerrerio et al. 2013). Symptoms of food allergy range from acute and life-threatening reactions to more chronic gastrointestinal problems. Eosinophilic gastrointestinal disease or inflammatory bowel disease should be suspected in case of repetitive vomiting, chronic abdominal pain, dysphagia, diarrhea or constipation, poor weight gain or simple food avoidance.

Evaluation

- Appropriate allergen testing should be performed.
- For eosinophilic inflammatory gastrointestinal disease or inflammatory bowel disease, the diagnostic work-up should adhere to standard protocols. Endoscopic procedures should be handled with care given the connective tissue fragility in LDS though, but are not contraindicated.

Management

- It is recommended to treat sinus and ear infections aggressively, as this might decrease asthma (Maccarrick et al. 2014). Conservative use of bronchodilators is indicated.
- Food allergens should be avoided. In case of unfortunate allergen intake, antihistamines can ease cutaneous or mild anaphylactic reactions, while epi-pen should be used for life-threatening reactions (Maccarrick et al. 2014).
- Eosinophilic gastrointestinal disease and inflammatory bowel disease can be treated according to standard protocols.
- Polyethylene-glycol treatment is the therapy of choice for constipation. Direct gastrointestinal stimulants should be avoided (Maccarrick et al. 2014).

Dermatologic

Recurrent cutaneous LDS findings include velvety, thin or translucent skin with easy bruising and visible veins. Wound healing is often delayed, resulting in dystrophic or atrophic scarring. Although these skin findings might not be specific and might also be present in other connective tissue disorders such as EDS, they can be prominent distinguishing features of LDS from Marfan syndrome in individuals lacking craniofacial features. Inguinal or umbilical hernias are often observed and may recur after surgical repair; additional mesh support at the time of initial repair can be protective. Varicose veins have been reported in LDS and can be difficult to manage using routine protocols (van de Laar et al. 2012). Persistent milia can be observed in individuals with LDS.

Evaluation

- Careful examination of the skin can reveal diagnostic clues to the diagnosis of LDS.

Management

- Supporting mesh during hernia surgery can be considered to minimize recurrence risk.

Pregnancy

Although pregnancy and delivery can be uncomplicated in women with LDS, pregnancies are considered high risk. Based on the few available case series and reports on LDS and pregnancy, arterial dissection (mostly of the aorta) and uterine rupture occur in 11% and 2% of cases, respectively (Frise et al. 2017). The actual numbers may be lower, as uneventful pregnancies are unlikely to be published. Management of pregnancy in women with LDS is largely extrapolated from the current practices for MFS. Owing to the more aggressive nature of LDS, however, affected women are presumed to be at higher risk for vascular complications. Of note, normal aortic diameters prior to and during pregnancy as well as prophylactic root replacement before gestation do not guarantee absence of dissection risk (Braverman et al. 2016). As in MFS, the risk of aortic dissection extends into, and is often most pronounced, in the perinatal and early postnatal period. The risk of type B dissection continues (and may even be enhanced) in women with LDS who have had prior surgical repair of the ascending aorta.

There is a 50% chance of recurrence of LDS in the offspring of an affected parent. Preconception genetic counseling regarding recurrence risk and options for preimplantation or prenatal genetic testing is indicated.

Evaluation

- The care of pregnant women with LDS requires multidisciplinary input, preferably at a tertiary care center.
- Prior to pursuing pregnancy, comprehensive vascular imaging (from brain to abdomen) is indicated.
- Serial echocardiography should be performed to evaluate aortic growth at least once each trimester and continuing postpartum.
- The presence of dural ectasia can complicate the use of epidural anesthesia in women with LDS. Dedicated imaging studies and consultation with an anesthesiologist should be performed prior to the onset of labor.

Management

- ARBs should be down-titrated prior to pregnancy because of teratogenicity and the risk of fetal loss. β-blocker treatment should be continued or initiated during pregnancy to protect the aorta. A high-risk obstetrician should be involved in patient care and management decisions (Maccarrick et al. 2014).
- Valve-sparing aortic root replacement before pregnancy should be considered in women with dimensions at or above 4.0 cm.
- The choice of vaginal delivery or Cesarean section should be individualized, with an emphasis on avoidance of hemodynamic stress during delivery.

RESOURCES

National Marfan Foundation (NMF)

The National Marfan Foundation provides education and support for other heritable connective tissue disorders that share some features of Marfan syndrome.
  22 Manhasset Avenue
  Port Washington NY 11050, USA
  Phone: 800-862-7326 (toll-free); 516-883-8712
  Fax: 516-883-8040
  Email: staff@marfan.org

Loeys–Dietz Foundation

PO Box 22468
Baltimore, MD 21203, USA
Email: info@loeysdietz.org
Website: http://www.loeysdietz.org

REFERENCES

Bertoli-Avella AM, Gillis E, Morisaki H, Verhagen JM, de Graaf BM, van de Beek G, Gallo E, Kruithof BP, Venselaar H, Myers LA, Laga S, Doyle AJ, Oswald G, van Cappellen GW, Yamanaka I, van der Helm RM, Beverloo B, de Klein A, Pardo L, Lammens M, Evers C, Devriendt K, Dumoulein M, Timmermans J, Bruggenwirth HT, Verheijen F, Rodrigus I, Baynam G, Kempers M, Saenen J, Van Craenenbroeck EM, Minatoya K, Matsukawa R, Tsukube T, Kubo N, Hofstra R, Goumans MJ, Bekkers JA, Roos-Hesselink JW, van de Laar IM, Dietz HC, Van Laer L, Morisaki T, Wessels MW, Loeys BL (2015) Mutations in a TGF-beta ligand, TGFB3, cause syndromic aortic aneurysms and dissections. *J Am Coll Cardiol* 65(13):1324–1336.

Boileau C, Guo DC, Hanna N, Regalado ES, Detaint D, Gong L, Varret M, Prakash SK, Li AH, d'Indy H, Braverman AC, Grandchamp B, Kwartler CS, Gouya L, Santos-Cortez RL, Abifadel M, Leal SM, Muti C, Shendure J, Gross MS, Rieder MJ, Vahanian A, Nickerson DA, Michel JB, Jondeau G, Milewicz DM (2012) TGFB2 mutations cause familial thoracic aortic aneurysms and dissections associated with mild systemic features of Marfan syndrome. *Nat Genet* 44:916–921.

Bolar N, Van Laer L, Loeys BL (2012) Marfan syndrome: from gene to therapy. *Curr Opin Pediatr* 24(4):498–504.

Bradley TJ, Bowdin SC, Morel CFJ, Pyeritz RE (2015) The expanding clinical spectrum of extracardiovascular and cardiovascular manifestations of heritable thoracic aortic aneurysm and dissection. *Can J Cardiol* 32(1):86–99.

Braverman AC, Moon MR, Geraghty P, Willing M, Bach C, Kouchoukos NT (2016) Pregnancy after aortic root replacement in Loeys-Dietz syndrome: High risk of aortic dissection. *Am J Med Genet A* 170(8):2177–2180.

Cannaerts E, Shukla A, Hasanhodzic M, Alaerts M, Schepers D, Van Laer L, Girisha KM, Hojsak I, Loeys B, Verstraeten A (2018) FLNA mutations in surviving males presenting with connective tissue findings: two new case reports and review of the literature. *BMC Med Genet*. Aug 8; 19(1):140. doi: 10.1186/s12881-018-0655-0

Carmignac V, Thevenon J, Ades L, Callewaert B, Julia S, Thauvin-Robinet C, Gueneau L, Courcet JB, Lopez E, Holman K, Renard M, Plauchu H, Plessis G, De Backer J, Child A, Arno G, Duplomb L, Callier P, Aral B, Vabres P, Gigot N, Arbustini E, Grasso M, Robinson PN, Goizet C, Baumann C, Di Rocco M, Sanchez Del Pozo J, Huet F, Jondeau G, Collod-Beroud G, Beroud C, Amiel J, Cormier-Daire V, Riviere JB, Boileau C, De Paepe A, Faivre L (2012) In-frame mutations in exon 1 of SKI cause dominant Shprintzen-Goldberg syndrome. *Am J Hum Genet* 91(5):950–957.

Cook JR, Clayton NP, Carta L, Galatioto J, Chiu E, Smaldone S, Nelson CA, Cheng SH, Wentworth BM, Ramirez F (2015) Dimorphic effects of transforming growth factor-beta signaling during aortic aneurysm progression in mice suggest a combinatorial therapy for Marfan syndrome. *Arterioscler Thromb Vasc Biol* 35(4):911–917.

Coucke PJ, Willaert A, Wessels MW, Callewaert B, Zoppi N, De Backer J, Fox JE, Mancini GM, Kambouris M, Gardella R, Facchetti F, Willems PJ, Forsyth R, Dietz HC, Barlati S,

Colombi M, Loeys B, De Paepe A (2006) Mutations in the facilitative glucose transporter GLUT10 alter angiogenesis and cause arterial tortuosity syndrome. *Nat Genet* 38(4):452–457.

Dietz HC, Cutting GR, Pyeritz RE, Maslen CL, Sakai LY, Corson GM, Puffenberger EG, Hamosh A, Nanthakumar EJ, Curristin SM, et al. (1991) Marfan syndrome caused by a recurrent de novo missense mutation in the fibrillin gene. *Nature* 352(6333):337–339.

Doyle AJ, Doyle JJ, Bessling SL, Maragh S, Lindsay ME, Schepers D, Gillis E, Mortier G, Homfray T, Sauls K, Norris RA, Huso ND, Leahy D, Mohr DW, Caulfield MJ, Scott AF, Destree A, Hennekam RC, Arn PH, Curry CJ, Van Laer L, McCallion AS, Loeys BL, Dietz HC (2012) Mutations in the TGF-beta repressor SKI cause Shprintzen-Goldberg syndrome with aortic aneurysm. *Nat Genet* 44:1249–1254

Erkula G, Sponseller PD, Paulsen LC, Oswald GL, Loeys BL, Dietz HC (2010) Musculoskeletal findings of Loeys-Dietz syndrome. *J Bone Joint Surg Am* 92(9):1876–1883.

Felgentreff K, Siepe M, Kotthoff S, von Kodolitsch Y, Schachtrup K, Notarangelo LD, Walter JE, Ehl S (2014) Severe eczema and Hyper-IgE in Loeys-Dietz-syndrome - contribution to new findings of immune dysregulation in connective tissue disorders. *Clin Immunol* 150(1):43–50.

Frischmeyer-Guerrerio PA, Guerrerio AL, Oswald G, Chichester K, Myers L, Halushka MK, Oliva-Hemker M, Wood RA, Dietz HC (2013) TGFbeta receptor mutations impose a strong predisposition for human allergic disease. *Sci Transl Med* 5(195):195ra194.

Frise CJ, Pitcher A, Mackillop L (2017) Loeys-Dietz syndrome and pregnancy: The first ten years. *Int J Cardiol* 226:21–25.

Gallo EM, Loch DC, Habashi JP, Calderon JF, Chen Y, Bedja D, van Erp C, Gerber EE, Parker SJ, Sauls K, Judge DP, Cooke SK, Lindsay ME, Rouf R, Myers L, ap Rhys CM, Kent KC, Norris RA, Huso DL, Dietz HC (2014) Angiotensin II-dependent TGF-beta signaling contributes to Loeys-Dietz syndrome vascular pathogenesis. *J Clin Investigat* 124(1):448–460.

Guo DC, Pannu H, Tran-Fadulu V, Papke CL, Yu RK, Avidan N, Bourgeois S, Estrera AL, Safi HJ, Sparks E, Amor D, Ades L, McConnell V, Willoughby CE, Abuelo D, Willing M, Lewis RA, Kim DH, Scherer S, Tung PP, Ahn C, Buja LM, Raman CS, Shete SS, Milewicz DM (2007) Mutations in smooth muscle alpha-actin (ACTA2) lead to thoracic aortic aneurysms and dissections. *Nat Genet* 39(12):1488–1493.

Habashi JP, Doyle JJ, Holm TM, Aziz H, Schoenhoff F, Bedja D, Chen Y, Modiri AN, Judge DP, Dietz HC (2011) Angiotensin II type 2 receptor signaling attenuates aortic aneurysm in mice through ERK antagonism. *Science* 332(6027):361–365.

Habashi JP, Judge DP, Holm TM, Cohn RD, Loeys BL, Cooper TK, Myers L, Klein EC, Liu G, Calvi C, Podowski M, Neptune ER, Halushka MK, Bedja D, Gabrielson K, Rifkin DB, Carta L, Ramirez F, Huso DL, Dietz HC (2006) Losartan, an AT1 antagonist, prevents aortic aneurysm in a mouse model of Marfan syndrome. *Science* 312(5770):117–121.

Holm TM, Habashi JP, Doyle JJ, Bedja D, Chen Y, van Erp C, Lindsay ME, Kim D, Schoenhoff F, Cohn RD, Loeys BL, Thomas CJ, Patnaik S, Marugan JJ, Judge DP, Dietz HC (2011) Noncanonical TGFbeta signaling contributes to aortic aneurysm progression in Marfan syndrome mice. *Science* 332(6027):358–361.

Lindsay ME, Dietz HC (2011) Lessons on the pathogenesis of aneurysm from heritable conditions. *Nature* 473(7347):308–316.

Lindsay ME, Schepers D, Bolar NA, Doyle JJ, Gallo E, Fert-Bober J, Kempers MJ, Fishman EK, Chen Y, Myers L, Bjeda D, Oswald G, Elias AF, Levy HP, Anderlid BM, Yang MH, Bongers EM, Timmermans J, Braverman AC, Canham N, Mortier GR, Brunner HG, Byers PH, Van Eyk J, Van Laer L, Dietz HC, Loeys BL (2012) Loss-of-function mutations in TGFB2 cause a syndromic presentation of thoracic aortic aneurysm. *Nat Genet* 44(8):922–927.

Loeys BL, Chen J, Neptune ER, Judge DP, Podowski M, Holm T, Meyers J, Leitch CC, Katsanis N, Sharifi N, Xu FL, Myers LA, Spevak PJ, Cameron DE, De Backer J, Hellemans J, Chen Y, Davis EC, Webb CL, Kress W, Coucke P, Rifkin DB, De Paepe AM, Dietz HC (2005) A syndrome of altered cardiovascular, craniofacial, neurocognitive and skeletal development caused by mutations in TGFBR1 or TGFBR2. *Nat Genet* 37(3):275–281.

Loeys BL, Dietz HC (2008) Loeys-Dietz Syndrome. GeneReviews(R).

Maccarrick G, Black JH, 3rd, Bowdin S, El-Hamamsy I, Frischmeyer-Guerrerio PA, Guerrerio AL, Sponseller PD, Loeys B, Dietz HC, 3rd (2014) Loeys-Dietz syndrome: a primer for diagnosis and management. *Genet Med* 16:576–587.

Meester JA, Vandeweyer G, Pintelon I, Lammens M, Van Hoorick L, De Belder S, Waitzman K, Young L, Markham LW, Vogt J, Richer J, Beauchesne LM, Unger S, Superti-Furga A, Prsa M, Dhillon R, Reyniers E, Dietz HC, Wuyts W, Mortier G, Verstraeten A, Van Laer L, Loeys BL (2016) Loss-of-function mutations in the X-linked biglycan gene cause a severe syndromic form of thoracic aortic aneurysms and dissections. *Genet Med* 19:386–395.

Micha D, Guo DC, Hilhorst-Hofstee Y, van Kooten F, Atmaja D, Overwater E, Cayami FK, Regalado ES, van Uffelen R, Venselaar H, Faradz SM, Vriend G, Weiss MM, Sistermans EA, Maugeri A, Milewicz DM, Pals G, van Dijk FS (2015) SMAD2 Mutations Are Associated with Arterial Aneurysms and Dissections. *Hum Mutat* 36(12):1145–1149.

Neptune ER, Frischmeyer PA, Arking DE, Myers L, Bunton TE, Gayraud B, Ramirez F, Sakai LY, Dietz HC (2003) Dysregulation of TGF-beta activation contributes to pathogenesis in Marfan syndrome. *Nat Genet* 33(3):407–411.

Patel ND, Alejo D, Crawford T, Hibino N, Dietz HC, Cameron DE, Vricella LA (2017a) Aortic Root Replacement for Children With Loeys-Dietz Syndrome. *Ann Thorac Surg* 103(5):1513-1518.

Patel ND, Crawford T, Magruder JT, Alejo DE, Hibino N, Black J, Dietz HC, Vricella LA, Cameron DE (2017b) Cardiovascular operations for Loeys-Dietz syndrome: Intermediate-term results. *J Thorac Cardiovasc Surg* 153(2):406–412.

Putnam EA, Zhang H, Ramirez F, Milewicz DM (1995) Fibrillin-2 (FBN2) mutations result in the Marfan-like disorder, congenital contractural arachnodactyly. *Nat Genet* 11(4):456–458.

Schepers D, Doyle AJ, Oswald G, Sparks E, Myers L, Willems PJ, Mansour S, Simpson MA, Frysira H, Maat-Kievit A, Van Minkelen R, Hoogeboom JM, Mortier GR, Titheradge H, Brueton L, Starr L, Stark Z, Ockeloen C, Lourenco CM, Blair E, Hobson E, Hurst J, Maystadt I, Destree A, Girisha KM, Miller M, Dietz HC, Loeys B, Van Laer L (2015) The SMAD-binding domain of SKI: a hotspot for de novo mutations causing Shprintzen-Goldberg syndrome. *Eur J Hum Genet* 23(2):224–228.

Schwarze U, Schievink WI, Petty E, Jaff MR, Babovic-Vuksanovic D, Cherry KJ, Pepin M, Byers PH (2001) Haploinsufficiency for one COL3A1 allele of type III procollagen results in a phenotype similar to the vascular form of Ehlers-Danlos syndrome, Ehlers-Danlos syndrome type IV. *Am J Hum Genet* 69(5):989–1001.

van de Laar IM, Oldenburg RA, Pals G, Roos-Hesselink JW, de Graaf BM, Verhagen JM, Hoedemaekers YM, Willemsen R, Severijnen LA, Venselaar H, Vriend G, Pattynama PM, Collee M, Majoor-Krakauer D, Poldermans D, Frohn-Mulder IM, Micha D, Timmermans J, Hilhorst-Hofstee Y, Bierma-Zeinstra SM, Willems PJ, Kros JM, Oei EH, Oostra BA, Wessels MW, Bertoli-Avella AM (2011) Mutations in SMAD3 cause a syndromic form of aortic aneurysms and dissections with early-onset osteoarthritis. *Nat Genet* 43(2):121–126.

van de Laar IM, van der Linde D, Oei EH, Bos PK, Bessems JH, Bierma-Zeinstra SM, van Meer BL, Pals G, Oldenburg RA, Bekkers JA, Moelker A, de Graaf BM, Matyas G, Frohn-Mulder IM, Timmermans J, Hilhorst-Hofstee Y, Cobben JM, Bruggenwirth HT, van Laer L, Loeys B, De Backer J, Coucke PJ, Dietz HC, Willems PJ, Oostra BA, De Paepe A, Roos-Hesselink JW, Bertoli-Avella AM, Wessels MW (2012) Phenotypic spectrum of the SMAD3-related aneurysms-osteoarthritis syndrome. *J Med Genet* 49(1):47–57.

Van Hemelrijk C, Renard M, Loeys B (2010) The Loeys-Dietz syndrome: an update for the clinician. *Curr Opin Cardiol* 25(6):546–551.

Verstraeten A, Alaerts M, Van Laer L, Loeys B (2016) Marfan Syndrome and Related Disorders: 25 Years of Gene Discovery. *Hum Mutat* 37:524–531.

Willaert A, Khatri S, Callewaert BL, Coucke PJ, Crosby SD, Lee JG, Davis EC, Shiva S, Tsang M, De Paepe A, Urban Z (2012) GLUT10 is required for the development of the cardiovascular system and the notochord and connects mitochondrial function to TGFbeta signaling. *Hum Mol Genet* 21(6):1248–1259.

37

MARFAN SYNDROME

UTA FRANCKE

Departments of Genetics and Pediatrics, Stanford University Medical Center, Stanford, California, USA

INTRODUCTION

Incidence

Marfan syndrome (MFS) is named after Antoine Marfan who in 1896 presented an individual with disproportionately long fingers and limbs, scoliosis, and chest asymmetry. The full range of manifestations in the ocular, cardiovascular, pulmonary, and other connective tissue systems was delineated during the mid-to-late 20th century.

The estimated incidence is 1–2 per 10,000 individuals without gender or ethnic biases. The prevalence of new mutations in affected individuals is approximately 25%.

If aortic complications are left untreated, the life expectancy in classic MFS is considerably reduced, but with the advances in cardiovascular pharmacotherapy and aortic root replacement surgery this syndrome has become one of the more manageable genetic disorders. The average life expectancy was 70 years in 1995 (Silverman et al. 1995), but may be higher for individuals who undergo aortic root replacement surgery on an elective rather than an emergency basis.

Diagnostic Criteria

The diagnosis of MFS is primarily based on clinical criteria. The first international diagnostic consensus was established in 1986, resulting in the Berlin nosology (Beighton et al. 1988). According to these criteria, MFS can be diagnosed if two organ systems are involved in the presence of an unequivocally affected first-degree family member. In the absence of such a positive family history, involvement of the skeleton and two or more organ systems is required, as well as the presence of at least one major criterion. Major criteria are aortic dilatation or dissection, ectopia lentis, and dural ectasia.

The mapping of the MFS locus to 15q21 and subsequent identification of mutations in the *fibrillin-1* (*FBN1*) gene as causing MFS in 1991 led to the inclusion of molecular evidence in the diagnostic evaluation. In 1996, the Berlin nosology was revised, resulting in the "Ghent nosology" (De Paepe et al. 1996), which takes into account the presence of a disease-causing *FBN1* mutation, and, where a mutation has not yet been identified, the presence of a 15q21 marker haplotype that segregates with unequivocally diagnosed MFS in members of the same family.

Clinical features were divided into major and minor criteria, depending on diagnostic specificity. Major criteria include features that are rarely seen in other disorders, such as ectopia lentis and dilatation of the ascending aorta. Most systems can be affected by manifestations that are considered major features or they can be merely "involved." The pulmonary system and skin only provide minor criteria for system involvement. According to these guidelines, an individual without an unequivocally affected first-degree relative can be given a diagnosis of MFS only when major criteria are present in at least two organ systems and a third organ system is involved. One major criterion in an organ system and involvement of a second organ system are required in the presence of an unequivocally affected first-degree relative, an *FBN1* mutation that is known to cause MFS in others, or a disease-associated 15q21 haplotype. In an international collaborative study of 1009 individuals with pathogenic *FBN1* mutations and a clinical diagnosis of MFS, only 79% met Ghent criteria when the *FBN1* mutation was not considered, and 90% when it was included as a major criterion (Faivre et al. 2008).

Although the 1996 Ghent nosology has improved diagnostic specificity, its applicability in practice turned out to have some limitations. Therefore, it was revised and simplified in 2010 (Loeys et al. 2010) (Table 37.1).

For those individuals who do not meet these criteria for an MFS diagnosis, the following possibilities should be considered: ectopia lentis syndrome: ectopia lentis with or without a systemic score AND without an *FBN1* mutation OR an *FBN1* mutation not known to be associated with aortic aneurysm. MASS phenotype (myopia, mitral valve prolapse, borderline aortic root dilatation, striae, skeletal findings): aortic root dilatation ($Z < 2$) AND systemic score > 5 with at least one skeletal feature. Mitral valve prolapse syndrome: Mitral valve prolapse AND mild aortic root dilatation ($Z < 2$) AND systemic score < 5 without lens dislocation. Since currently used Z-scores seem to underestimate aortic root dilatation, especially in those with large body surface area (BSA) (Radonic et al. 2011), individuals given these alternative diagnoses should be monitored for aortic enlargement over time (Faivre et al. 2012).

The current diagnostic criteria are heavily weighted towards aortic aneurysm and dissection, ectopia lentis, and *FBN1* mutations. Items that are not part of the diagnostic work-up in most health care settings have been deleted or downgraded. For example, increased axial length of the globe of the eye has been removed as a criterion since ultrasonic axial length measurements are not routinely performed. It has been replaced by myopia > 3 diopters as a feature contributing to the systemic score. Similarly, the presence of protrusio acetabuli, the medial protrusion of the femoral head leading to a deepened acetabulum, and apical bullae (or blebs) of the lung, must be ascertained by radiographic evaluations that are difficult to justify in the absence of clinical symptoms involving the hip joint or lung. Therefore, protrusio acetabuli is no longer a major criterion, and only contributes 2 points to the systemic score, and apical blebs has been removed. Dural ectasia, previously considered a major criterion, is an enlargement of the outer layer of the dural sac and is seen predominantly in the lumbosacral area where the cerebrospinal hydrostatic pressure is greatest. It may lead to osseous changes and widening of the spinal canal. CT or MRI scanning is required to determine its presence. Initially, the frequency of asymptomatic dural ectasia in individuals with ascending aortic aneurysms or other features of the Marfan spectrum was reported to be very high ($>90\%$), and as dural ectasia was not identified in

TABLE 37.1 Diagnostic criteria for Marfan syndrome (revised 2010) (Loeys et al. 2010)

No affected family member
    Aortic aneurysm ($Z \geq 2$) or dissection AND ectopia lentis
    Aortic aneurysm ($Z \geq 2$) or dissection AND *FBN1* mutation
    Aortic aneurysm ($Z \geq 2$) or dissection AND systemic score ≥ 7*
    Ectopia lentis AND *FBN1* mutation known to be associated with aortic aneurysm

Family member affected with MFS as defined above
    Ectopia lentis AND family history of MFS
    Systemic score ≥ 7 AND family history of MFS
    Aortic aneurysm (($Z \geq 2$) above 20 years old, $Z \geq 3$ below 20 years) AND family history of MFS

Systemic score
    Score ≥ 7 indicates systemic involvement
A. Skeletal
    Wrist AND thumb sign[a,b] – 3 points
    Wrist OR thumb sign – 1 point
    Pectus carinatum deformity – 2 points
    Pectus excavatum OR chest asymmetry – 1 point
    Hindfoot deformity – 2 points
    Plain pes planus – 1 point
    Reduced upper/lower segment ratio AND increased arm span/height ratio (absent severe scoliosis) – 1 point
    Scoliosis OR thoracolumbar kyphosis – 1 point
    Protrusio acetabuli – 2 points
    Reduced elbow extension – 1 point
B. Facial features – 1 point if three of five present: dolichocephaly, enophthalmos, downslanting palpebral fissures, malar hypoplasia, retrognathia
C. Skin: striae distensae – 1 point
D. Lung: pneumothorax – 2 points
E. Dura: dural ectasia – 2 points
F. Eye: myopia >3 diopters – 1 point
G. Heart: mitral valve prolapse (all types) – 1 point

Z = number of standard deviations above the mean aortic root size after standardization for body surface area.
*After exclusion of Loeys–Dietz, Shprintzen–Goldberg, and vascular Ehlers–Danlos syndromes.
[a] Wrist sign: thumb overlaps the distal phalanx of the fifth digit when grasping the contralateral wrist.
[b] Thumb sign: entire nail of the thumb projects beyond the ulnar border of the hand when the hand is clenched without assistance.

100 normal controls, this suggested a high specificity of this criterion (Fattori et al. 1999). In a recent international study, 292 of 1009 probands with *FBN1* mutations had lumbosacral imaging studies, and only 52% of those had dural ectasia (Faivre et al. 2008). The authors calculated that searching for dural ectasia led to an increase of 3% in the percentage of individuals meeting Ghent criteria in their study population. In the revised Ghent nosology, dural ectasia is no longer a

major criterion and only contributes two points to the systemic score.

To further simplify the use of the diagnostic criteria check list, certain general connective tissue manifestations, previously minor criteria, have been deleted, although they are commonly seen in MFS: joint hypermobility, high palate with crowding of teeth, flat cornea, hypoplastic iris, hypoplastic ciliary muscle causing decreased miosis, apical blebs, and recurrent or incisional hernia. Of the previous cardiovascular minor criteria, mitral valve prolapse remains as contributing one point to the systemic score, while calcification of the mitral annulus before the age of 40 years, dilatation of the main pulmonary artery in the absence of valvular or peripheral pulmonic stenosis before the age of 40 years, and dilatation or dissection of the descending thoracic or abdominal aorta before 50 years of age, were considered to have low specificity and have been removed (Loeys et al. 2010).

To diagnose MFS based on the revised criteria that place high value on aortic aneurysm, one has to rely on the echocardiographic measurement of aortic root diameter at the level of the sinuses of Valsalva and the conversion of this measurement to Z scores. Studies using two-dimensional echocardiography have recently been published that predict normal aortic root diameters taking into account age and sex and supersede the previous standards by Roman et al. (1989) (Devereux et al. 2012). Z scores (number of standard deviations from the population mean) are calculated using body surface area (BSA, from height and weight data) by an online calculator https://www.marfan.org/dx/zscore. Diagnostic cut-off at $Z \geq 2$ (corresponding to >95th centile) for people older than 20 years and $Z \geq 3$ (corresponding to >99th centile) at less than 20 years could be problematic, as for a person of the same height, sex, age and aortic diameter, the body weight may determine whether the Z score is above or below the diagnostic threshold, e.g. a leaner person would receive a diagnosis of MFS while the more obese one would not (Pyeritz and Loeys 2012). Comparing the original to the revised Marfan nosology in 180 established MFS patients, the diagnosis of MFS was rejected in 13 patients because the Z-score of the aortic root was <2, although the aortic diameter was larger than 40 mm in six of them (Radonic et al. 2011). A Z-score (Z3) that corrects for age, sex and height, but not weight, was found to be superior to using BSA (van Kimmenade et al. 2013). Also, it is unclear why the age group threshold in the revised Ghent criteria for $Z \geq 2$ versus $Z \geq 3$ is set at 20 years, while the revised nomograms, and derived Z score calculator, uses 15 years.

The revised diagnostic criteria are still of limited usefulness for children. Although those affected with the severe neonatal form of MFS, who usually represent new mutations, are likely to meet clinical criteria for MFS, affected children with inherited mutations may not meet diagnostic criteria because key features may not become apparent until the teenage years or later. In particular, the onset of dilatation of the ascending aorta is variable and may not be detectable before adulthood. Lens subluxation and scoliosis may only become manifest during adolescence. In an international study of 320 children with *FBN1* mutations who were below 18 years of age, only 56% could be classified as having MFS on clinical grounds alone according to the original Ghent criteria; this number increased to 85% when the *FBN1* mutation was included (Faivre et al. 2009).

Finally, the clinical features may be modified in different racial and ethnic groups. For example, dolichostenomelia cannot be assessed accurately in individuals of ethnicities other than Caucasian or African because standards for skeletal measurements have not been established in these populations. In Asian or Hispanic individuals with MFS, skeletal measurements may well be within the normal range for Caucasians. In the revised diagnostic criteria, however, skeletal measurements carry less weight.

These issues are of concern as individuals who may not meet diagnostic criteria because of young age, ethnicity, or limited access to diagnostic imaging technology may still be at risk for serious complications such as aneurysm and dissection of the ascending aorta. In a large collaborative study, aortic dilatation occurred later in adults not meeting Ghent criteria when compared with the criteria-positive Marfan group (44% versus 73% at 40 years, $p < 0.001$), but the lifelong risk for ascending aortic dissection or surgery was not significantly different in the two groups (Faivre et al. 2008).

To assess what changes the new nosology would portend, Faivre and colleagues have retrospectively reclassified the ~1000 individuals in the international collection using the 2010 revised Ghent criteria, and taking into account their status at the last follow-up visit. A total of 842 individuals could be classified as MFS according to the new nosology (83%) as compared to 894 (89%) according to the 1996 Ghent criteria. The new nosology led to a different diagnosis in 15% of cases, 8% of those with MFS were reclassified as ectopia lentis syndrome and 2% as MASS in the absence of aortic dilatation; conversely, 5% were reclassified as MFS in the presence of aortic dilatation. Radonic et al. (2011) reported similar data, while in a South Korean cohort 84 of 86 adults with MFS retained their diagnosis (Yang et al. 2012). As most of them were referred for aortic aneurysm surgery, the Z-score cut-off may have been less important for the diagnosis.

Regarding the use of the Ghent criteria for management, it appears prudent to recommend that children of affected individuals receive periodic evaluations by a cardiologist, ophthalmologist, and geneticist, and they be monitored similar to children with unequivocal MFS following established guidelines (American Academy of Pediatrics 1996), unless they have tested negative for a pathogenic *FBN1* mutation identified in an affected family member. In individuals of Asian descent, musculoskeletal criteria should be used with the awareness that dolichostenomelia may not be present according to the Caucasian standard. Affected individuals, however, may be much taller than their relatives and others of the same ethnic extraction. In all other instances in which individuals do not meet the criteria with readily available diagnostic means, the stigma of the MFS diagnosis must be weighed on an individualized basis against the necessity for clinical follow-up to reduce the risk of serious complications.

Etiology, Pathogenesis, and Genetics

MFS is inherited in an autosomal dominant fashion with full penetrance and variable expressivity. In the 1980s, linkage studies of multigenerational MFS families failed to identify linkage to any known candidate genes for connective tissue components such as elastin, fibronectin, and collagens. When linkage analysis was expanded to random, genome-wide genetic markers, most of the 22 autosomes could be excluded. Finally, in 1990, linkage of MFS was established to markers in chromosome band 15q21 (Kainulainen et al. 1990).

At the same time, a monoclonal antibody against an extracellular microfibril protein, named fibrillin, was used for immunofluorescence assays of fibroblast cultures from individuals with and without MFS. Immunofluorescent staining was substantially decreased in affected subjects, suggesting a primary defect in microfibril formation (Godfrey et al. 1990). Fibrillin, a 350-kDa cysteine-rich glycoprotein, is the major component of the largest class of microfibrils in the extracellular meshwork of elastic as well as non-elastic tissues (Sakai et al. 1986). Subsequently, the gene that encodes the fibrillin protein (*FBN1*) was isolated, mapped to band 15q21 (Lee et al. 1991; Magenis et al. 1991), and screened for mutations in individuals with MFS (Dietz et al. 1991).

The effort of detection and functional characterization of *FBN1* mutations is a continuing process. At this time, more than 3000 cases with 1850 different mutations, including missense and nonsense mutations, nucleotide deletions and insertions, mutations leading to abnormal mRNA splicing, and large genomic deletions, have been reported in over 200 papers. The data are collected in an international database www.umd.be/FBN1/ (Collod-Beroud et al. 2003).

The *FBN1* gene spans approximately 235 kb and is composed of 65 coding exons and three alternatively spliced untranslated upstream exons. The processed mRNA is 9749 nucleotides in length and encodes a 2871 amino acid polypeptide. The majority of the protein is made up of 47 epidermal growth factor (EGF)-like domains that are almost exclusively encoded by individual exons (Figure 37.1, top). Each EGF-like domain contains six conserved cysteine residues that form three disulfide bonds. In addition, 43 of the 47 six-cysteine domains contain a consensus sequence for calcium binding, which facilitates intramolecular as well as intermolecular interactions. Another module that *FBN1* shares with the latent transforming growth factor β (TGFβ) binding protein (LTBP) contains eight conserved cysteines that form disulfide bonds. This domain only occurs in the fibrillins and LTBPs. Fibrillin-1 and fibrillin-2 each have seven of these LTBP-domains. LTBPs bind to fibrillin and are part of the microfibrillar meshwork. In binding TGFβ/latency-associated peptide (LAP) complexes, LTBPs play an important role in modulating the availability of TGFβ growth factors for signaling in tissues. As illustrated in Figure 37.1, TGFβ is inactive while it is bound to the LAP/LTBP/fibrillin complex. Once released by proteases, growth factors of the TGFβ family bind to TGFβ receptors that are transmembrane proteins with an intracellular serine/threonine kinase domain. Both type I and type II TGFβ receptors are required in a complex with TGFβ ligand for signaling to occur. This heteromeric receptor/ligand complex induces phosphorylation of intracellular proteins of the Smad family. Phosphorylation enables them to enter the nucleus and affect transcription of other genes (Schmierer and Hill 2007). In addition, the TGFβ receptors also signal through a Smad independent pathway, through ERK and MAPK.

A role for TGFβ signaling in the pathogenesis of MFS was suggested by studies of *FBN1*-mutant mice (Neptune et al. 2003; Cohn et al. 2007), and independently in humans, by the discovery that Marfanoid phenotypes can be caused by mutations in TGFβ receptor genes (Ades et al. 2006). Mutations in *TGFBR1* and *TGFBR2* (the genes encoding transforming growth factor β receptors 1 and 2, respectively) also result in Loeys–Dietz syndrome, a connective tissue disorder with significant phenotypic overlap with MFS (Loeys et al. 2005) (see Chapter 36). Individuals with this Marfanoid disorder lack ectopia lentis, the major ocular manifestation of MFS, and often have dysmorphic features such as distinctive facies, cleft palate, craniosynostosis, and contractures. Loeys–Dietz syndrome may present in childhood with significant cardiovascular problems (see Chapter 36 and Differential Diagnosis).

The pathogenetic consequences of disease-causing *FBN1* mutations invoke both dominant-negative and haplo-insufficiency paradigms (Furthmayr and Francke 1997). Studies of the production and stability of *FBN1* mRNA and fibrillin protein support the notion that the presence of structurally abnormal (truncated, internally deleted and misfolded) fibrillin molecules interferes with microfibril assembly and stability (Aoyama et al. 1993; Robinson et al. 2006). When the amount of fibrillin synthesis and deposition in cultured fibroblasts from affected individuals was studied, the degree of reduction of extracellular microfibril formation, as measured by the incorporation of newly synthesized labeled fibrillin molecules, correlated with the clinical phenotype and disease progression (Aoyama et al. 1994; Aoyama et al. 1995).

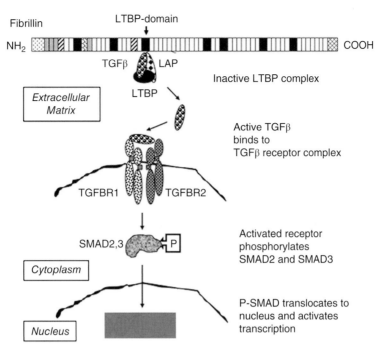

FIGURE 37.1 Fibrillin-1 molecule with domain structure and TGFβ signaling cascade. Black boxes: LTBP (latent TGFβ binding protein) (8-cysteine) domains; white boxes: calcium-binding (cb) EGF-like (6-cysteine) domains; gray boxes: non-cb EGF-like domains; hatched boxes: hybrid domains; zigzag box: proline-rich region; the unique N- and C-terminal regions are stippled and crosshatched, respectively. Fibrillin monomers, assembled into microfibrils in the extracellular matrix, bind TGFβ in an inactive complex that also contains LAP (latency associated protein) and LTBP. Released TGFβ molecule activates the TGFβ signaling cascade (see text) through binding to TGFβ receptor complexes. Reduction in microfibrils in MFS leads to fewer binding sites for TGFβ and increased TGFβ signaling in tissues. EGF, epidermal growth factor; TGFBR, transforming growth factor β receptor; SMAD, family of proteins that are homologs of both the *Drosophila* protein MAD (mothers against decapentaplegic), and the *Caenorhabditis elegans* protein SMA, P-SMAD, phosphorylated SMAD.

Heterozygosity for *FBN1* gene deletions or null mutations produces variable skeletal features and aortic dilatation, mitral valve prolapse, and myopia, but not ectopia lentis (Ades et al. 2006; Matyas et al. 2007, Hilhorst-Hofstee et al. 2011). Similarly, individuals with premature termination codons, leading to degradation of the mutant mRNA, tend to be more mildly affected (Schrijver et al. 2002; personal observation). Variability of the haploinsufficiency phenotype may be explained by a threshold effect that is modified by the level of expression of the normal *FBN1* allele (Hutchinson et al. 2003).

Individuals with premature termination codons have more significant large joint hypermobility than individuals with MFS caused by a cysteine substitution. Lens dislocation and retinal detachment, on the other hand, are distinctly less frequent (Schrijver et al. 2002; Faivre et al. 2007).

Ectopia lentis appears to be the only MFS manifestation that is caused by a direct structural dominant-negative effect of abnormal fibrillin molecules. The zonal fibers are made up of fibrillin containing microfibrils and the frequency of ectopia lentis is highest in people with cysteine mutations (80%) (Schrijver et al. 1999). The absence of ectopia lentis in people with *FBN1* deletions, and Loeys–Dietz syndrome, indicates that abnormal TGFβ signaling does not play a role in its pathogenesis.

For all other features of the MFS, the common pathogenetic mechanism is a reduction of fibrillin-containing microfibrils in the extracellular matrix resulting in diminished capacity to bind inactive TGFβ/LTBP complexes. The availability of increased numbers of active growth factors in certain tissues leads to overactive TGFβ signaling and may account for the skeletal overgrowth, muscular hypoplasia, abnormal lung septation, cardiac valve structural development, and other features of the MFS. The role of TGFβ signaling in vascular wall modeling has been elucidated in mouse models (Habashi et al. 2006). To explain the Marfanoid skeletal and cardiovascular findings associated with TGFβ receptor mutations, one needs to postulate that they enhance TGFβ signaling. The fact that *TGFBR2* and *TGFBR1* mutations are almost exclusively amino acid changes in the intracellular serine/threonine protein kinase domain suggests that they may not cause loss of function. On the other hand, loss of function mutations in TGFB also lead to paradoxical increase in TGFβ signaling, the mechanism of which has not been clarified (Akhurst 2012). Identical TGFBR mutations, but of somatic origin, have been identified in colon cancer and are thought to enhance growth of tumor cells (Frederic et al. 2008).

Because fibrillinopathies are connective tissue disorders transmitted as autosomal dominant traits, affected individuals have a 50% chance of passing on the mutant allele in each pregnancy, regardless of whether the child is male or female. Prenatal detection by direct DNA analysis is only possible in those families in which the mutation has been identified. In families for which the disease-causing mutation is not known, prenatal diagnosis by linkage analysis can be done if linkage of the phenotype to 15q21 markers has been established in the family. Similarly, direct mutation or indirect linkage analysis can be applied to relatives of affected individuals. To fully evaluate the clinical phenotype and to determine whether diagnostic criteria are met, however, detailed clinical examination of each individual is imperative.

Diagnostic Testing

The diagnosis of MFS and related connective tissue disorders is strongly based on the clinical manifestations (Tables 37.1 and 37.2, Figure 37.2). Because *FBN1* mutations have also been detected in individuals with other fibrillinopathies, the presence of a mutation does not by itself make the diagnosis. DNA sequencing of all coding exons detected mutations in approximately 90% of Ghent criteria-positive individuals (Loeys et al. 2004). When deletion/duplication analyses are added, the mutation detection rate approaches 95% (Baetens et al. 2011). Therefore, the MFS phenotype may be produced by mutations in at least one other gene. The best candidates are *TGFBR1* and *TGFBR2*, the genes mutated in Loeys–Dietz syndrome, in isolated familial thoracic aortic aneurysm and dissection (Pannu et al. 2005) and in a form of MFS that includes musculoskeletal and cardiovascular features as well as myopia but lacks ectopia lentis (Mizuguchi et al. 2004; Sakai et al. 2006; Singh et al. 2006). *TGFBR1*, or more frequently *TGFBR2*, mutations were reported in Marfanoid individuals with or without additional features of Loeys–Dietz syndrome (Sakai et al. 2006; Singh et al. 2006). Several individuals reported with *TGFBR2* mutations fulfilled Ghent diagnostic criteria for MFS (Disabella et al. 2006; LeMaire et al. 2007). It is important to identify individuals with *TGFBR1* and *TGFBR2* mutations because they are at risk for early dissection when the aortic root may not be significantly enlarged. Therefore, individuals with aortic involvement and Marfanoid skeletal features who do not have ectopia lentis should be screened for mutations in *TGFBR2*, and secondarily in *TGFBR1*. These genes are much smaller than *FBN1* (only seven exons) and can be sequenced efficiently. A family history of aortic dissection, although not required, would provide further indication for *TGFBR2/1* testing. A history of ectopia lentis in affected family members, on the other hand, should direct molecular testing toward *FBN1* (Figure 37.3). Multi-gene panel tests are available if clinical findings are not unequivocal. In couples facing reproductive decisions, the knowledge of a specific mutation or haplotype allows prenatal or preimplantation genetic diagnosis (Spits et al. 2006).

TABLE 37.2 Disorders with features overlapping Marfan syndrome.

Disorder	Mutant gene(s)	Inheritance
Neonatal Marfan syndrome	*FBN1*	de novo
Congenital contractural arachnodacyly	*FBN2*	A.D.
Loeys–Dietz syndrome	*TGFBR1,TGFBR2, TGFB2*	A.D.
Arterial tortuosity syndrome	*SLC2A10*	A.R.
Familial thoracic aortic aneurysm and aortic dissection	*TGFBR1, TGFBR2, TGFB2, ACTA2, COL3A1, FBN1, FLNA, MAT2A, MFAP5, MYH11, MYLK, NOTCH1, PRKG1, SMAD3, TGFB2, TGFB3*	A.D.
Ehlers–Danlos syndrome, classic type	*COL5A1, COl5A2* (see Chapter 25)	A.D.
Ehlers–Danlos syndrome, vascular type	*COL3A1* (see Chapter 25)	A.D.
Ehlers–Danlos syndrome, kyphoscoliotic type	*PLOD1* (see Chapter 25)	A.R.
Stickler syndrome	*COL2A1, COL11A1, COL11A2*	A.D.
	COL9A1, COL9A2, COL9A3 (see Chapter 56)	A.R.
Trisomy 8 mosaicism syndrome, chromosomal aneuploidy	Mosaic + 8	de novo
Homocystinuria	*CBS*	A.R.
Lujan–Fryns syndrome	*MED12* and other X-linked intellectual disability genes	X.L.R.
Isolated ascending aortic aneurysm/dissection	*FBN1*	A.D.
Ectopia lentis syndrome	*FBN1*	A.D.

Abbreviations: A.D., autosomal dominant; A.R., autosomal recessive; X.L.R., X-linked recessive. *Source*: Please see text for clinical and genetic details.

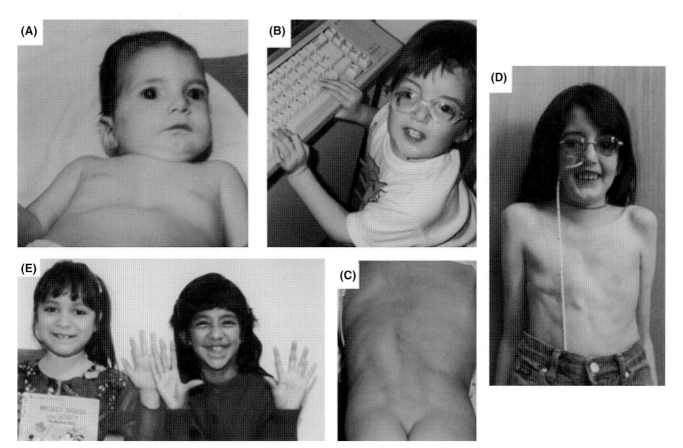

FIGURE 37.2 (A) Patient 1 with MFS at 4 months of age. Note typical facies and pectus excavatum. (B) Patient 1 at 2 years. Note arachnodactyly and correction for severe myopia. (C) Patient 1 at 2 years: severe progressive left thoracolumbar scoliosis. (D) Patient 1 at 7 years: status post pectus repair and spinal fusion. Note aphakic correction for bilateral dislocated lenses and nasogastric tube for nightly feeding in preparation for cardiovascular surgery for aortic aneurysm and severe mitral valve regurgitation. (E) Patient 2 with MFS at 7 years (right) and her unaffected sister at 8.5 years (left). Note deep-set eyes, narrow elongated face, and arachnodactyly.

Immunofluorescence studies of extracellular microfibrils with fibrillin antibodies have been offered as diagnostic aids. Skin biopsies from affected individuals show a deficiency in microfibrillar content, and cultured skin fibroblasts produce a reduced amount of microfibrillar meshwork (Godfrey et al. 1990). Quantitation of fibrillin synthesis and extracellular matrix deposition, as determined by pulse-chase analysis in cultured skin fibroblasts, has been shown to have high specificity, sensitivity, and possibly even prognostic value (Aoyama et al. 1994, 1995). The current methodology, however, requires high levels of technical expertise, time, and cost that preclude it from routine diagnostic application.

Differential Diagnosis

Most phenotypes that overlap with the clinical features of MFS are related fibrillinopathies or TGFβ signaling disorders (Figure 37.4). Many individual features of MFS can be inherited as autosomal dominant traits, such as isolated ascending aortic aneurysm/dissection, isolated ectopia lentis, and isolated skeletal features. These fall within the spectrum of type 1 fibrillinopathy, and *FBN1* mutations have been demonstrated in affected individuals.

Neonatal MFS is at the most severe end of the spectrum of type 1 fibrillinopathies. It overlaps with classic MFS as well as congenital contractural arachnodactyly but is distinguished by severe mitral and/or tricuspid valvular insufficiency, infantile pulmonary emphysema and poor prognosis (Hennekam 2005). The features, which are present at birth, include a characteristic progeric face, crumpled ears or simple helices, redundant skin, congenital hernias, flexion contractures, arachnodactyly, ectopia lentis, cardiomegaly with severe aortic regurgitation, multivalvular cardiac insufficiency, aortic root dilatation, pulmonary emphysema, chest deformities, and pes planus. Neonatal MFS is often fatal within the first year of life because of congestive heart failure. *FBN1* mutations that lead to this severe disorder tend to cluster in a central region of the

Strategy for Differential Diagnosis and Molecular Testing

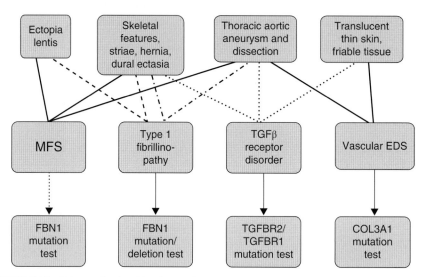

FIGURE 37.3 Diagnostic classification and molecular testing strategy for individuals displaying various combinations of Marfan syndrome (MFS) features. MFS – MFS meeting clinical criteria; *FBN1* mutation analysis (stippled arrow) is not needed for diagnosis of proband. Type 1 fibrillinopathy – possible MFS, but not meeting clinical diagnostic criteria, requires positive *FBN1* mutation/deletion test to confirm diagnosis. TGFβ receptor disorder (lacking arterial tortuosity and craniofacial features diagnostic for Loeys-Dietz syndrome) is caused by missense mutations in the protein kinase domain of *TGFBR2*, or less often *TGFBR1*. EDS type IV (Ehlers–Danlos syndrome, vascular type) if *TGFBR* mutations are not detected, consider collagen type III testing (by protein electrophoresis or *COL3A1* mutation analysis).

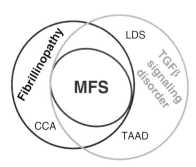

FIGURE 37.4 Conceptional relationship between Marfan syndrome (MFS) and related connective tissue disorders. Fibrillinopathies other than MFS include phenotypes from the MFS spectrum, in isolation or in combination, that do not meet diagnostic criteria for MFS, but are caused by a mutation in the *FBN1* gene (type 1 fibrillinopathy); and congenital contractural arachnodactyly (CCA) associated with mutations in the *FBN2* gene (type 2 fibrillinopathy). TGFβ signaling disorders other than MFS include thoracic aortic aneurysm and dissection (TAAD), and Loeys–Dietz syndrome (LDS, Ch. 36) that are caused by germline mutations in *TGFBR1, TGFBR2, SMAD3 or TGFB*.

gene, between exons 24 and 32. This region is likely to be important for intermolecular interactions and stabilization of the extracellular matrix microfibrils. Because neonatal MFS is a sporadic disease resulting from spontaneous mutations, the family history is usually negative and the recurrence risk for a couple with an affected child should be negligible, except for the remote possibility of germ line mosaicism in one of the parents. Conversely, adults with MFS should be counseled that they are unlikely to have offspring with the neonatal lethal form of MFS.

Congenital contractural arachnodactyly (also known as Beals syndrome) is an autosomal dominant connective tissue disorder characterized by a Marfanoid appearance and long, slender fingers and toes, multiple joint contractures, and early onset progressive kyphosis and/or scoliosis. The disorder is caused by mutations in the fibrillin-2 (*FBN2*) gene, located on chromosome 5 (Lee et al. 1991). The domain organization of *FBN1* and that of *FBN2* are nearly identical, and 80% of amino acids are conserved. Both proteins are structural components of microfibrils. Fibrillin-2, however, is enriched in cartilaginous structures, which offers an explanation for the crumpled shape of the external ear and predominant involvement of the skeletal system in congenital contractural arachnodacyly. Children with this disorder display camptodactyly and congenital flexion contractures that generally improve in childhood. Contractures are less prominent in MFS, most often affecting the elbows and toes (hammer toes). Muscular hypoplasia is present in both disorders, but ocular and cardiovascular manifestations are less prevalent and highly variable in congenital contractural arachnodactyly.

Loeys–Dietz syndrome (Chapter 36) is an autosomal dominant disorder whose delineation was originally based on the triad of arterial tortuosity and aneurysms, hypertelorism, and bifid uvula or cleft palate (Loeys et al. 2005). The clinical presentation, however, is much more variable. It includes all major and minor criteria of MFS except for ectopia lentis. Loeys–Dietz syndrome can present as isolated familial thoracic aortic aneurysm and dissection, and also includes features that overlap with vascular Ehlers–Danlos syndrome (see Chapter 25) (Law et al. 2006). The diagnosis is made by DNA sequence analysis and demonstration of heterozygous mutations in the genes encoding transforming growth factor β receptors 1 and 2 (*TGFBR1*, *TGFBR2*) and others (see Chapter 36). The proposal to classify sub-types of Loeys–Dietz syndrome based on the gene involved has met with resistance by experts (Pyeritz et al. 2014) It is very important to identify individuals with TGFβ receptor mutations early, as they are at risk for aggressive arterial aneurysms and early death from dissection or pregnancy-related complications. Aortic dissection or rupture may occur in childhood (Loeys et al. 2006).

Common features of Loeys–Dietz syndrome that overlap MFS are ascending aortic aneurysm/dissection, pectus deformity, arachnodactyly, malar flattening, retrognathia, high palate, protrusio acetabuli, scoliosis, pes planus, hernias, cutaneous striae, hyperextensible joints, retinal detachment, and dural ectasia (Figure 37.3). Distinguishing features found in Loeys–Dietz syndrome, if present, include aneurysms of other arteries, arterial tortuosity, hypertelorism, cleft palate/bifid uvula, clubfeet, velvety translucent skin, and easy bruising. Intellectual disability and craniosynostosis are rare findings in Loeys–Dietz syndrome and suggest an overlap with Shprintzen–Goldberg syndrome, a sporadic condition due to changes of *SKI* that resembles MFS in its skeletal features and is considered part of the clinical spectrum associated with TGFβ signaling disorders.

Aneurysms and tortuosity of large and mid-size arteries are the hallmark of the autosomal recessive arterial tortuosity syndrome (Callewaert et al. 2008). Affected individuals have a characteristic long face with long philtrum, as well as variable connective tissue abnormalities that include arachnodactyly, pectus deformity, joint laxity with subluxations, soft lax skin, and hernias. Arterial tortuosity syndrome is caused by mutations in the *SLC2A10* gene that encodes a glucose transporter.

Familial thoracic aortic aneurysms and dissection are a group of genetically heterogeneous autosomal dominant cardiovascular disorders with or without other phenotypic manifestations (Pannu et al. 2005). As in MFS, dilatation of the aorta and dissections occur either at the level of the sinuses of Valsalva or the ascending thoracic aorta. Several loci have been identified by linkage studies and mutations have been reported in multiple genes, including *TGFB* (Boileau et al. 2012; Lindsay et al. 2012). Extra-aortic findings may guide the choice of genes to be tested (Arbustini and Narula 2012).

Several types of Ehlers–Danlos syndrome (Chapter 25) need to be considered. Hyperextensible joints are features of the hypermobility type of Ehlers–Danlos syndrome (formerly, type III), an autosomal dominant disorder. The skin may be smooth and hyperextensible, but scarring is normal. Joint dislocations occur frequently, especially of the patellae, temporomandibular joints, and shoulders.

The vascular type of Ehlers–Danlos syndrome (formerly, type IV) should be considered in individuals presenting with aortic dissection in the absence of earlier significant aneurysm formation. The disorder is characterized by thin, translucent skin, easy bruising, characteristic facial appearance, and arterial, intestinal, and uterine fragility. Other features include inguinal hernia, pneumothorax, and recurrent joint dislocation or subluxation. Biochemical abnormalities of type III procollagen and mutations in the collagen type 3 α1 chain (*COL3A1*) gene serve to confirm the diagnosis. Considering the clinical overlap with Loeys–Dietz syndrome, it is not surprising that numerous individuals presenting with features of vascular type Ehlers–Danlos syndrome but lacking a collagen abnormality have recently been found to have TGFβ receptor mutations (Loeys et al. 2006; Drera et al. 2008).

The kyphoscoliosis type (formerly, Ehlers–Danlos syndrome type VI) should be considered when a combination of Marfanoid habitus, kyphoscoliosis, joint laxity, hyperelastic skin, and tissue fragility is present. Affected individuals are at risk for scleral and arterial rupture. This autosomal recessive form is caused by molecular defects in *PLOD1*, the gene for lysyl hydroxylase. Diagnostic testing options include insufficient hydroxylysine in skin, reduced enzyme activity in cultured skin fibroblasts, and altered urinary ratio of lysyl pyridinoline:hydroxylysyl pyridinoline. Mutation/deletion testing of the *PLOD1* gene is also available.

Stickler syndrome (hereditary arthro-ophthalmopathy) is an autosomal dominant connective tissue disorder defined by craniofacial, joint, and ophthalmological abnormalities (see Chapter 56). Features overlapping the MFS include arachnodactyly, pectus excavatum, scoliosis, joint hypermobility, high-grade myopia, and retinal detachment. Furthermore, this disorder is characterized by midface hypoplasia, cleft palate, or bifid uvula, sensorineural or conductive hearing loss, vitreoretinal degeneration, mild spondyloepiphyseal dysplasia, and arthropathies. Mutations in collagen genes *COL2A1*, *COL11A1*, and *COL11A2* have been reported in families with this disorder. A recessive form of Stickler syndrome is caused by biallelic mutations in genes for collagen IX chains *COL9A1*, *COL9A2* or *COL9A3*.

The trisomy 8 mosaicism syndrome shares some connective tissue features with MFS. Notably, these include deep set-eyes, high-arched palate, joint contractures, narrow body habitus, and tall stature. In trisomy 8 mosaicism syndrome, however, it is the trunk that is relatively elongated, because of the presence of extra thoracic vertebrae, rather than the extremities. This chromosomal imbalance syndrome also includes intellectual disability, deep creases on palms and soles, and absent patellae. It does not share any ocular or cardiovascular manifestations with MFS. Karyotyping of lymphocytes and, if necessary, skin fibroblasts, establishes the diagnosis.

Homocystinuria is a rare autosomal recessive disorder that was first described in institutionalized individuals thought to be affected with the MFS. Although intellectual disability is an occasional feature in homocystinuria, the intellect of individuals with MFS is unaffected. Ectopia lentis, on the other hand, is a prominent feature in both disorders, though much more progressive in homocystinuria (Nelson and Maumenee 1982). Although in both disorders the lenses may be displaced in any direction, ectopia lentis occurs earlier in life and is predominantly superior and temporal in MFS, whereas it is mostly inferior in homocystinuria. Lenses of individuals with homocystinuria may dislocate totally, often into the anterior chamber, and may cause pupillary block and glaucoma. In addition, the zonules often break completely and create a "fringe" by scrolling onto the anterior lens capsule. With detachment or degeneration of the lens zonules, the lens assumes a more globular configuration as the diameter is reduced, called spherophakia, and high lenticular myopia may develop. As in MFS, individuals with homocystinuria may have a tall, asthenic body habitus, scoliosis, high palate, and pectus deformity, but early osteoporosis involving the spine is a distinguishing sign. Cardiovascular involvement is characterized by thromboembolism. Homocystinuria is caused by biallelic mutations in the gene for cystathionine-β-synthase (*CBS*) leading to deficiency of the enzyme. Homocystinuria can be diagnosed by a positive cyanide-nitroprusside test of the urine, but the more reliable diagnostic assay is documentation of elevated homocystine and methionine levels in blood. The type of mutations in the *CBS* gene can predict responsiveness to treatment with the cofactor pyridoxine (vitamin B6) (Gaustadnes et al. 2002).

Lujan–Fryns syndrome is an X-linked recessive disorder characterized by intellectual disability and behavior disorders with a Marfanoid body habitus and hyperextensible digits. Features include a long, narrow face and a high palate, high nasal bridge and small mandible with dental crowding, and a hypernasal voice. Although these features can also be part of the MFS spectrum, intellectual disability and psychotic behavior are not. Lujan–Fryns syndrome in the original family is caused by mutations in the *MED12* gene (Schwartz et al. 2007). Mutations in this gene have also been found in other X-linked disorders with intellectual disability (Graham and Schwartz 2013), and other molecular defects have been identified in males with intellectual disability and Marfanoid habitus (Hackman et al. 2016).

MANIFESTATIONS AND MANAGEMENT

Comprehensive health management of persons with MFS and related connective tissue disorders requires involvement of many different medical specialists and an individualized approach throughout life. Therefore, a multidisciplinary setting with access to a geneticist for initial diagnosis, coordination of care and counseling, a cardiologist for regular cardiac evaluation, an ophthalmologist experienced in recognizing and treating ocular manifestations of connective tissue disorders, an orthopedist, a physical therapist, and psychosocial counseling opportunities is best suited for optimal care. The American Academy of Pediatrics Committee on Genetics has established guidelines for health supervision in children with MFS, categorized by stages of development (American Academy of Pediatrics 1996).

Growth and Feeding

In infants with classic MFS, increased length and arachnodactyly may already be apparent at birth. Growth parameters will generally be above the 95th centile, but must be interpreted with ethnicity and parental heights in mind.

Tall stature and long slender limbs and digits are cardinal features of MFS. Children with MFS are generally taller than their classmates, and affected adults are taller than unaffected family members. Detailed growth curves for children and adolescents with MFS have been constructed (Erkula et al. 2002). The growth spurt occurs at an earlier age in adolescents with MFS than in unaffected adolescents (2.4 years earlier for males and 2.2 years for females). The majority of children and young adults affected with MFS have a weight below the 50th centile and a reduction in the amount of subcutaneous fat, as well as muscular hypoplasia.

Evaluation

- All individuals with MFS should be measured at the initial evaluation and periodically thereafter during their growth period, noting height, arm span-to-height ratio, and upper-to-lower body segment ratio.
- Dolichostenomelia (disproportionately long limbs) should be assessed in the correct manner. The lower segment is measured from the middle of the upper rim of the symphysis pubis to the floor. The feet must be aligned and the heels must touch the wall. The upper segment measurement is obtained by subtraction of the lower segment from the height. The range of normal for the upper-to-lower segment ratio changes with age. In adults, an upper-to-lower segment ratio of 0.85 or lower is considered significant, as the average value in unaffected adults is 0.93. The arm span-to-height ratio is an

additional means to assess a disproportionate body habitus. In the Ghent criteria, the ratio is considered increased if it is greater than 1.05 (De Paepe et al. 1996). Average differences of span from height have been published for children of both sexes (Hall et al. 2007). As discussed above, these measurement ratios may also be influenced by ethnicity.

Management

- In prepubertal girls with excessively tall stature in relationship to bone age, final adult height can be reduced by initiation of high-dose estrogen therapy that should be combined with progesterone to prevent endometrial hyperplasia.
- Girls receiving hormone treatment to prevent excessive height are predicted to have a final height that is the average of the calculated final height before treatment and the height at the beginning of treatment. Therefore, if treatment is started around the age of 8 or 9 years, final results may be better than when treatment is postponed to early puberty. On the other hand, premature induction of puberty has the potential of bringing about additional psychological strain.
- Although requests for growth arrest are not often made for boys, accelerated bone maturation can be accomplished by treatment with testosterone.

Development and Behavior

Cognitive development is unaffected in MFS. Motor development may be somewhat delayed by joint hyperextensibility.

The psychosocial impact of having a disorder that includes increased risk for aortic complications and death is a significant one (Velvin et al. 2015).

Evaluation

- Developmental milestone achievement should be assessed at each routine childhood medical visit.
- If significant delays in cognitive development occur, this is likely because of an unrelated cause and should be evaluated as in the general population.
- Motor delays can be assessed in more detail by a physical therapy evaluation.

Management

- Physical therapy is of value in promoting motor skill development.
- Psychosocial counseling is important for family adjustment to the diagnosis and for support when complications arise. Families should be advised regarding the existence of support groups and literature, and intrafamilial interactions should be evaluated to stimulate full integration of the affected individuals in normal life activities.

Musculoskeletal

In people with MFS, the musculoskeletal features are the most conspicuous. Although skeletal abnormalities may not be pronounced in the first few years of life, they are the key to early diagnosis. Tall stature with disproportionately long limbs, long narrow digits, a long and narrow face with deep-set eyes, and a high, narrow palate are often combined with joint hypermobility and pectus deformities.

Chest deformities such as pectus excavatum or carinatum are related to overgrowth of the ribs, pushing the sternum inward or outward. The deformity is not limited to the sternum and can also present as asymmetric prominence of costosternal junctions. Chest asymmetry and asymmetric placement of the scapulae also occur secondary to scoliosis. Pectus excavatum occurs in 8 of 1000 live births, but in two-thirds of people with MFS. In adolescence, chest deformities are commonly perceived as the most disturbing physical feature.

Scoliosis is a feature in approximately 60% of children with MFS, and the frequency is higher in adults. Individuals with a spinal curvature of less than 30° have an excellent long-term prognosis, whereas moderate progression of at least 10° can be expected with a curve between 30 and 50°. Marked progression occurs if the curve exceeds 50° (Sponseller et al. 1995). Untreated, significant spinal deformity can lead to chronic back pain and restrictive lung disease. Dural ectasia is usually asymptomatic, but can also be associated with back pain (Nallamshetty et al. 2002).

Joint laxity can be pronounced in young children and may lead to delayed gross motor development. Joint dislocations are rare occurrences. When adults are evaluated for MFS, it is important to keep in mind that joint hyperextensibility lessens with increasing age.

Mild contractures of elbows, knees, or toes are present in a fraction of children and adults with MFS. Over time, pes planus and laxity of ligaments and joints can lead to painful joints and feet.

Adults often have an asthenic body habitus. Crowding of the teeth commonly occurs because the maxilla and mandible are narrow. This may result in a posterior cross-bite and malocclusion.

Protrusio acetabuli was present in 23% of 298 individuals with *FBN1* mutations who had been assessed for this feature (Faivre et al. 2008).

Evaluation

- Careful clinical evaluation for the skeletal manifestations of MFS should occur at regular physical evaluations. Specific assessment for joint hypermobility, scoliosis, and pectus deformity should be included. Evaluation of the skeletal system should also include assessment of long/narrow face, prominent supraorbital ridges, deep-set eyes, high narrow nasal bridge, high narrow palate and dental crowding, limited elbow extension, cubitus valgus, genu recurvatum, and pes planus or cavus.

- On clinical suspicion, kyphosis and scoliosis should be evaluated by radiography to determine the degree of spinal curvature.
- Scoliosis must be followed because it progresses during childhood and may undergo a rapid increase during growth spurts. Scoliosis is likely to progress if the curvature exceeds 20° during childhood and adolescence, and 30–40° in adulthood. If such curves are present, referral to an orthopedist for evaluation is appropriate.
- Arachnodactyly (long, slender digits) is objectively assessed by hand measurements. The ratio of middle finger to total hand length is increased compared with normal standards (Hall et al. 2007). Arachnodactyly in conjunction with small joint hypermobility results in the characteristic wrist, thumb, and thumb-to-arm signs. The wrist sign is positive when the thumb can overlap the nail of the fifth digit when the two fingers are wrapped around the wrist. The thumb sign is considered positive when the whole thumbnail extends beyond the ulnar side of the palm when the hand is folded into a fist. If the thumb can touch the forearm when the wrist is passively flexed, the thumb-to-arm sign is positive.

Management

- Delays in gross motor development caused by joint hypermobility can be ameliorated with physical therapy and orthopedic braces, as needed. Although hypermobility in MFS is never as severe a problem as in the hypermobile type of Ehlers–Danlos syndrome, a detailed discussion of the management of symptomatic hypermobility can be found in Chapter 25.
- The treatment of scoliosis and kyphosis depends on the severity of the curvature and rate of progression over time. In children with MFS, pediatric orthopedists will often treat even minor spinal curvatures with a brace to prevent progression (Sponseller et al. 1995). In the most seriously affected individuals with curvatures exceeding 40°, spinal arthrodesis may be warranted to stop further deformity, reduce back pain, and prevent a reduction in lung function. Criteria for spinal surgery would be the same as for individuals without MFS.
- Pectus excavatum or carinatum need be repaired only in severe cases to prevent cardiac or pulmonary compromise, and preferably after growth has been completed. Surgical intervention for chest deformities before growth has ceased often results in recurrence because of continuing bone overgrowth.
- Severe pectus excavatum interferes with the interpretation of echocardiographic studies and can lead to intra- or postoperative complications in aortic valve/root replacement surgery. Therefore, surgical correction of the pectus deformity before elective cardiovascular surgery is often recommended.
- Pes planus may be treated with arch support. Surgical treatment of foot deformities is rarely successful.

Cardiovascular

The cardiovascular manifestations of MFS are potentially life-threatening and must be monitored closely. In childhood, the most frequently observed abnormalities are mitral valve prolapse (>60%) and mitral regurgitation. Dilatation of the ascending aorta is diagnosed in approximately half of young children with MFS and increases with age. It is the most common indication for cardiovascular surgery in childhood, followed by mitral regurgitation. Aortic root dilatation generally progresses over a period of years. Because the sinuses of Valsalva are often the first part of the aortic root to dilate in MFS, the aorta attains a characteristic flask-like shape. In adults, secondary aortic regurgitation is a common feature and adds to the hemodynamic stress. The root of the pulmonary artery can also be enlarged. The cardiovascular manifestations of MFS are the major causes of premature death: mitral valve regurgitation with congestive heart failure in severely affected children, and aortic aneurysm rupture and dissection in adults.

Evaluation

- Cardiology evaluation is recommended in every individual suspected of having MFS. Echocardiographic examination is the method of choice to establish baseline aortic root dimensions and to assess mitral valve prolapse or regurgitation. The 95% confidence interval for aortic root dimensions in children and adults has been determined by comparison of body surface area with the aortic diameter at the sinuses of Valsalva (Roman et al. 1989; Devereux et al. 2012). Progression of aortic root diameter, diminished aortic wall compliance, development of mitral and aortic valve regurgitation, and subacute dissection should be monitored on a regular basis throughout life. Annual cardiology evaluations are appropriate in most cases, but more frequent evaluations may be advisable if disproportionate increases in aortic root size or changes in valvular function are diagnosed.
- Individuals with MFS caused by mutations in a *TGFβ* receptor gene are at risk for aortic dissection early in life

and at lower aortic root diameters. Therefore, screening echocardiograms every six months is advised. As the entire aorta is at risk of dissection and requires imaging in these individuals, MRI should be considered.

Management

- Pharmacological reduction of hemodynamic stress on the aorta is thought to slow the rate of aortic dilation and the risk of dissection. β-Adrenergic blockade has been the standard of care for the last 23 years, following the publication of a randomized trial in 1994 (Shores et al. 1994). The data showed great variability in response to propranolol, and in some individuals, but not in others, the rate of aortic root growth was reduced. Assuming that negative inotropic therapy can prevent complications such as aneurysm formation and dissection, it was recommended that β-adrenergic blockade be implemented at the first signs of aortic enlargement or when the aortic root reaches the 95th centile for body surface area (Salim et al. 1994). Because aortic dilatation in individuals with MFS may progress throughout life, β-blockade was to be continued indefinitely.
- Recently, the benefit of lifelong β-blockade therapy with respect to outcome measures such as aortic dissection or rupture, cardiovascular surgery or death has been called into question (Williams et al. 2008). A meta-analysis of six published studies found no evidence that β-blockade has clinical benefits with respect to these endpoints (Gersony et al. 2007). In one retrospective study, the rate of aortic dilation in β-blocker treated versus non-treated children <18 years of age was not significantly different (Selamet Tierney et al. 2007). Another retrospective study documented a beneficial effect of β-blockers (Phomakay et al. 2014). Given the known side effects of β-blockers, including sleep disturbance, fatigue, attention deficit and bronchospasm, the decision to treat children needs to be considered carefully and on an individual basis.
- When β-blockers are contraindicated, such as in persons with asthma, calcium channel blockers or angiotensin-converting enzyme (ACE) inhibitors are used. In a comparative study, the ACE inhibitor enalapril was significantly more effective in improving aortic distensibility and reducing aortic stiffness as well as in reducing the increase in aortic root dimension, than β-blockers, which had no effect on these parameters (Yetman et al. 2005).
- The blood pressure-lowering drug losartan inhibits the angiotensin II type 1 receptor and has been proposed as a new treatment for MFS based on evidence from a mouse model (Habashi et al. 2006). According to the hypothesis that enhanced TGFβ signaling is the major pathogenetic mechanism in MFS, treatment with TGFβ antagonists may prevent or retard aortic root dilatation After an initial promising small study (Brooke et al. 2008) several prospective clinical trials with an angiotensin receptor blocker (losartan, valsartan, or irbesartan) were conducted internationally (Pyeritz 2016). In almost all trials, no placebo was used, but losartan was added to baseline β -blockade. In the largest prospective study, children affected with significant aortic root enlargement ($Z > 3$) were treated with atenolol alone or atenolol plus losartan. No differences in rate of aortic dilatation or dissection were observed after three years (Lacro et al. 2014). Similar studies in Europe reported a benefit of an added angiotensin receptor blocker in three studies and no benefit in another three (Pyeritz 2016). Studies comparing losartan to atenolol reported no difference in outcome (Forteza et al. 2016), both were equally efficient in reducing aortic enlargement in previously untreated children (Mueller et al. 2014). Notably, in the largest prospective trial (Lacro et al. 2007), randomization did not take into account the type of *FBN1* mutation, as mutation analysis was not required for enrollment. *FBN1* genotype was unknown in >50% of study subjects (Lacro et al. 2013). Yet, Dutch investigators reported that losartan treatment inhibited aortic root dilation specifically in MFS patients with *FBN1* mutations leading to haploinsufficiency (Franken et al. 2015), in agreement with the notion that increased TGFβ signaling is the primary pathogenetic mechanism in *FBN1* haploinsufficiency (see above). If these results can be confirmed, knowledge of the *FBN1* mutation should be included in decision making about drug therapy.

Only large-scale multi-institutional prospective randomized trials comparing more than these two drugs will be able to answer questions such as whether ACE inhibitors that block angiotensin II type 1 and type 2 receptors or angiotensin II type 1 receptor blockers such as losartan alone will be more effective than other types of drugs or placebo in delaying the progression of aortic disease (Williams et al. 2008).

- Even though aortic dissection and rupture are rare in children, composite graft replacement may be necessary in childhood if there is a rapid increase of aortic root diameter leading to aortic regurgitation. There is great individual variability in aortic root diameter at the time of dissection. Although guidelines recommend aortic root replacement surgery at an aortic diameter of >5.0 cm, recent observations underscore the importance of family history of dissection and/or rupture. Furthermore, aortic surgery may be indicated if the rate of increase of the aortic root diameter approaches 0.5–1.0 cm per year. The ratio between the size of the dilated aortic segment and the diameter of the aorta with normal appearance downstream should be taken into account.

- Most adults with MFS will eventually need replacement of a dilated aortic root and leaking aortic valve. The standard composite graft includes a mechanical valve that requires long-term anticoagulation therapy, but the operative risk is less than 5%, and five year survival is 85% (Gott et al. 1999). Valve-sparing aortic root replacement procedures have undergone modifications over the years (David et al. 2014) and are now considered an acceptable alternative for individuals who do not want anticoagulation, such as women of childbearing age. Valve-sparing surgery is contraindicated in individuals with severe aortic insufficiency and abnormal aortic valves (Patel et al. 2008).
- Also fairly common during childhood is the need for mitral valve repair.
- Even though individuals with MFS can enjoy an almost normal activity level, certain guidelines should be observed to prevent excessive strain on the cardiovascular system, joints, and ligaments. Physical activity supports motor development and increases muscle strength, but contact sports such as football, basketball, hockey, volleyball, boxing, and wrestling should be avoided if the aorta is dilated because of the risk of trauma to the chest and aorta precipitating dissection or rupture. The risk for retinal detachment (discussed below) is another good reason to evade blows to the head. Heavy lifting and other isometric exercise should also be avoided because of the acute increase of pressure on the ascending aorta. In addition, sudden stops and prompt changes in direction should be avoided. Non-competitive aerobic activity such as swimming, bicycling, hiking, golfing, and general physical conditioning are encouraged.

Ophthalmologic

The most characteristic ocular manifestation of MFS is ectopia lentis, which is caused by the microfibrillar weakness of the ciliary zonules. As an early hallmark of the disorder, ectopia lentis is detected in 50–80% of individuals with MFS (Nelson and Maumenee 1982). It is usually bilateral and typically superotemporal, although it may occur in any direction. Most often, ectopia lentis becomes evident in young children and is slowly progressive, particularly in childhood and the early teenage years. Severity varies from very mild to total dislocation, but – in contrast to homocystinuria – dislocation into the anterior chamber is uncommon.

Most children and adults with classic MFS have myopia, albeit at varying degrees. Individuals with MFS also develop cataracts 10–20 years earlier than the general population. Open angle glaucoma is more common in all age groups as compared with the general population.

Retinal detachment is a leading cause of visual loss in MFS. Retinal detachment is related to elongation of the globe and frequently even occurs in younger individuals. Retinal detachment has an increased incidence after intraocular surgery.

The incidence of strabismus in MFS (19%) is increased compared with that in the general population (5%), presumably because ectopia lentis as well as mechanical and craniofacial factors contribute to its development (Izquierdo et al. 1994). Exotropia (outward deviation) is the most common type of strabismus in MFS.

The prognosis of ocular involvement in MFS is excellent with appropriate evaluations and treatment of possible complications as they arise. Over 90% of individuals who start treatment in early childhood retain a visual acuity of at least 20/40 in both eyes.

Evaluation

- Annual eye examination by an ophthalmologist who has experience with the ocular manifestations of MFS and related connective tissue disorders is strongly recommended. Visual acuity, refraction, signs of cataract and glaucoma, and fundoscopic findings must be assessed at diagnosis and annually thereafter to maximize vision.
- Because it can lead to amblyopia and loss of binocularity, evaluation for strabismus is also important.
- During the general ophthalmic examination, ectopia lentis can be suspected when iridodonesis, or a "wobbly" iris, is observed. When the iris loses its posterior support of the lens and zonules, it may appear to shimmer or be tremulous. This phenomenon is most easily detected on rapid lateral eye movement or blinking.
- Slit lamp examination may reveal a smooth surface of the iris with lack of crypts and furrows, giving it a "velvety appearance."
- Ectopia lentis is best appreciated by retroillumination using the slit lamp biomicroscope, direct ophthalmoscope, or retinoscope, or in cases of posterior dislocation, by having the person look downward, which may reveal a gap between the pupillary margin and the lens. Ectopia lentis is more evident with increasing lens displacement. Small degrees of ectopia lentis, however, can only be detected when pupils are maximally dilated. In some people with MFS, the dilator muscle of the iris is hypoplastic and the pupils are small and may be difficult to dilate. Minor ectopia lentis may manifest as a slight notch or crenated edge, most commonly inferiorly. The degree of ectopia lentis should be monitored. Generally, it is only gradually progressive, most often in childhood and early adolescence.
- Individuals with MFS must be monitored for glaucoma that can occur because of various etiologies: anatomical, mechanical, and/or vascular. For example, angle

abnormalities may cause open angle glaucoma, or dislocation of the lens into the anterior chamber of the eye can cause obstruction.
- High-grade myopia is often associated with increased axial globe length, which can be assessed by ophthalmologic ultrasound examination. These enlarged globes also carry an increased risk for retinal detachment, and the retina should always be examined by dilated fundoscopic examinations for early signs.

Management

- Corrective lenses should be prescribed as necessary. The proper optical correction will help prevent the development of amblyopia. Aphakic correction may be needed in cases of severely dislocated lenses. Adequate optical correction must be prescribed and worn. Meticulous assessment of refraction must be performed regularly to maximize visual acuity. Individuals should be evaluated in both their phakic (through the lens) and aphakic (the space around the lens) corrections to assess their best vision. The use of mydriatic eyedrops may be required to dilate pupils to maximize the pupillary space, particularly if asymmetric lens subluxation is present.
- Amblyopia must be treated aggressively in a standard fashion.
- The so-called "presenile" cataract of MFS is a true indication for lens extraction.
- Lens removal for optical reasons alone is not recommended (Nelson and Maumenee 1982) although visual improvement may occur after lensectomy in severe cases (Wu-Chen et al. 2005).
- When the ectopia lentis is progressive and leads to secondary complications such as iritis, glaucoma, vitreal compromise, or cataract, surgical lens removal may be indicated. These complications are rare in childhood.
- Because individuals with MFS are prone to spontaneous retinal detachment, contact sports should be avoided. Children and adults need to be informed of the symptoms and signs of a retinal detachment. They are advised to routinely check their own vision monocularly. If they experience flashes, floaters, visual field changes, or a "curtain or shadow" over their visual field, they should see a retinal specialist immediately. Good visual results after retinal detachment are dependent on reattachment of the retina either by laser or scleral buckling procedures and/or vitrectomy. Success is dependent on the location, extent, and duration of the detachment.

Neurologic

Widening and weakness of the lumbosacral spinal cord sac (dural ectasia) can result in low back pain or in leakage of spinal fluid and intractable orthostatic headaches when the tear does not heal spontaneously (Rosser et al. 2005). Although episodes of orthostatic headache resulting from cerebrospinal fluid leakage are not uncommon in the third or fourth decade of life, these symptoms are usually transient and benign. Intracranial hypotension syndrome can also be caused by the meningocele in the absence of a tear. Reported complications are tonsillar herniation and subdural hematomas (Puget et al. 2007). Dural ectasia occurs frequently in individuals with MFS and is common even in childhood (Faivre et al. 2009).

Evaluation

- Dural ectasia can be evaluated only by MRI or CT scan. Because the evaluation is not commonly performed for diagnostic purposes, the precise frequency of this manifestation and its clinical significance are currently unknown.

Management

- Orthostatic headache resulting from cerebrospinal fluid leakage is often transient. Treatment with bed rest, hydration, analgesics, and caffeine may be sufficient in most cases, because the intracranial hypotension will subside in the recumbent position. If leakage of the spinal fluid persists, however, treatment with corticosteroids or epidural blood patches can be effective (Rosser et al. 2005).
- Neurosurgical repair may be considered with recurring or severe complaints, if the location of the leak can be identified and the area can be approached with low risk.

Respiratory

MFS is associated with pulmonary manifestations, including apical blebs and lung bullae that predispose to pneumothorax. In individuals with MFS who have an asthenic body habitus, spontaneous pneumothorax occurs more frequently than in other body types. These events occur in males more often than in females and are very rare in childhood. Individuals will present with acute dyspnea and chest pain because of intrapleural air accumulation. Chest radiography is generally conclusive.

Pulmonary emphysema is a feature of neonatal MFS.

Evaluation

- The medical history regarding manifestations in the respiratory system should be explored at each evaluation.
- Apical blebs can be observed on chest radiography, but do not require special evaluation because their presence is inconsequential unless their thin wall ruptures, leading to spontaneous pneumothorax.

Management

- There is no need to treat an apical bleb unless there is rupture.
- Spontaneous pneumothorax is treated by evacuation of the intrapleural air and restoration of the negative pleural pressure by insertion of a drainage chest tube.
- In individuals with MFS, there is an increased risk for repeated rupture of an apical bleb. Therefore, pleurodesis is recommended after recurring pneumothorax.

Dermatologic

Involvement of the integument in MFS includes inguinal as well as incisional hernias and stretch marks (striae distensae). Congenital inguinal hernia, as well as hernias of later onset, requires timely surgical correction to prevent intestinal strangulation and further weakening of the surrounding connective tissue. Incisional hernias may develop after abdominal surgery.

The skin may be soft and thin in appearance. When stretch marks develop during the adolescent growth spurt, they are often distinctly pink. Although striae distensae are common in the general population, they have a different distribution in persons with MFS. Most frequently, they develop on the shoulders, axilla, chest, lower back, hips, thighs, and dorsum of the knee.

Evaluation

- The presence or absence of inguinal or femoral hernias should be assessed at the first examination and throughout childhood because they may be congenital and often recur. Hernias manifest in at least 50% of affected individuals.

Management

- For striae, neither prevention nor effective treatment is available.
- Hernias should be treated as in the general population. Prognosis after surgical hernia repair is good, although the underlying connective tissue defect may lead to recurrence.

Pregnancy

Women with MFS have an increased risk of cardiovascular complications during gestation and labor. In women with minimal cardiovascular involvement, pregnancy can be tolerated safely with adequate monitoring and management of the aortic status. If the aortic root diameter exceeds 45 mm, or if it has shown a recent increase in diameter, pregnancy carries a significant risk for worsening aortic root dilatation, dissection and aortic rupture, not limited to the third trimester and labor. For women with *TGFBR1/2* mutations, pregnancy carries an even higher risk of dissection and death. In a large population-based Swedish study, the outcomes of first childbirths in women with MFS born in 1973–1993 revealed no increased risk for maternal death, Cesarean section, premature birth or small-for-gestational-age babies, but no data about mutations and aortic diameters were available (Kernell et al. 2017). This study may indicate that more mildly affected women decide to get pregnant, and careful monitoring of their pregnancies leads to successful outcome.

Evaluation

- To achieve a favorable outcome, it is preferable to start cardiovascular evaluation and stabilization before pregnancy, including comprehensive imaging to assess cardiac abnormalities and risk of dissection.
- The preconceptional aortic root diameter is the single most important parameter that determines the degree of safety of undergoing pregnancy (Meijboom et al. 2005). This measurement is not necessarily correlated with age, and the risk must be determined on an individual basis, taking into account the absolute diameter as well as the aortic diameter at the time of dissection in other family members.
- Pregnant women affected with the MFS need to receive continuing prenatal care because of the possibility of accelerating aortic root dilatation and aortic dissection. Serial echocardiography is highly recommended, and every pregnancy should be treated as a high-risk event.

Management

- A combined approach involving cardiology, clinical genetics, and maternal/fetal medicine is pivotal in women considering pregnancy.
- If aortic root replacement surgery becomes necessary before pregnancy, valve-sparing procedures should be carried out, if feasible, to eliminate the need for anticoagulants that have adverse effects on the fetus.
- During pregnancy, the woman should be monitored by a high-risk obstetrician, and echocardiography should be performed every two to three months.
- In pregnant women, β-blocker therapy can have adverse effects on the fetus, such as intrauterine growth retardation, and on the mother, such as hypoglycemia, bradycardia, and apnea. Professional guidelines recommend use of labetalol or metoprolol, rather than atenolol because of adverse fetal effects.
- Pregnancy-induced hypertension poses an additional risk factor for aortic dissection and must be treated

aggressively. Anti-hypertensive medications such as methyldopa, hydralazine and labetalol have a good safety profile. Angiotensin converting enzyme (ACE) inhibitors and angiotensin receptor blockers (e.g. losartan) have adverse effects on the fetus and should be avoided (Bullo 2012), while calcium channel blockers may be dangerous in MFS based on studies in mice (Pyeritz 2012).

- The method of delivery should be individualized. Overall, vaginal delivery with epidural anesthesia is preferred over cesarean section except for women with obstetric complications or with progressive aortic dilatation (aortic diameter >45 mm) and a very high risk of dissection.

RESOURCES

National Marfan Foundation (NMF)

22 Manhasset Avenue
Port Washington, NY 11050, USA
Phone: 516-883-8712, 1-800-862-7326
Fax: 516-883-8040
Email: *staff@marfan.org*
Website: *http://www.marfan.org*

Genetic Aortic Disorders Association (GADA) Canada

Centre Plaza Postal Outlet
128 Queen St. South
PO Box 42257
Mississauga, Ontario, L5 M 4Z0, Canada
Phone: (905) 826-3223; (866) 722-1722
Email: *info@gadacanada.ca*
Website: *http://www.gadacanada.ca*

Marfan Association United Kingdom

Rochester House
5 Aldershot Road
Fleet, Hampshire
GU51 3NG, UK
Phone: 44 (0)1252 810472
Fax: 44 (0)1252 810473
Email: *marfan@tinyonline.co.uk*

Marfan Europe Network

Römerweg 4
CH-4410 Liestal, Switzerland
Phone: 41 61 921 91 89
Email: *preston5@bluewin.ch*

International Federation of Marfan Syndrome Organizations

Website: *http://www.marfanworld.org/*

REFERENCES

American Academy of Pediatrics, Committee on Genetics (1996) Health supervision for children with Marfan syndrome. *Pediatrics* 98:978–982.

Ades LC, Sullivan K, Biggin A, Haan EA, Brett M, Holman KJ, Dixon J, Robertson S, Holmes AD, Rogers J, et al. (2006) FBN1, TGFBR1, and the Marfan-craniosynostosis/mental retardation disorders revisited. *Am J Med Genet A* 140:1047–1058.

Akhurst RJ (2012) The paradoxical TGFβ vasculopathies. *Nat Genet* 44:838–839.

Aoyama T, Tynan K, Dietz HC, Francke U, Furthmayr H (1993) Missense mutations impair intracellular processing of fibrillin and microfibril assembly in Marfan syndrome. *Hum Mol Genet* 2:2135–2140.

Aoyama T, Francke U, Dietz HC, Furthmayr H (1994) Quantitative differences in biosynthesis and extracellular deposition of fibrillin in cultured fibroblasts distinguish five groups of Marfan syndrome patients and suggest distinct pathogenetic mechanisms. *J Clin Invest* 94:130–137.

Aoyama T, Francke U, Gasner C, Furthmayr H (1995) Fibrillin abnormalities and prognosis in Marfan syndrome and related disorders. *Am J Med Genet* 58:169–176.

Arbustini E, Narula N (2012) Extra-aortic identifiers to guide genetic testing in familial thoracic aortic aneurysms and dissections syndromes: it is all about the company one keeps. *J Am Coll Cardiol.* 60:404–407.

Baetens M, Van Laer L, De Leeneer K, Hellemans J, De Schrijver J, Van De Voorde H, Renard M, Dietz H, Lacro RV, Menten B, et al. (2011) Applying massive parallel sequencing to molecular diagnosis of Marfan and Loeys-Dietz syndromes. *Hum Mutat* 32:1053–1062.

Beighton P, de Paepe A, Danks D, Finidori G, Gedde-Dahl T, Goodman R, Hall JG, Hollister DW, Horton W, McKusick VA, et al. (1988) International nosology of heritable disorders of connective tissue, Berlin, 1986. *Am J Med Genet* 29: 581–594.

Boileau C, Guo DC, Hanna N, Regalado ES, Detaint D, Gong L, Varret M, Prakash SK, Li AH, d'Indy H, et al. (2012) TGFB2 mutations cause familial thoracic aortic aneurysms and dissections associated with mild systemic features of Marfan syndrome. *Nat Genet* 44:916–921.

Brooke BS, Habashi JP, Judge DP, Patel N, Loeys B, Dietz HC 3rd (2008) Angiotensin II blockade and aortic-root dilation in Marfan's syndrome. *N Engl J Med* 358:2787–2795.

Bullo M, Tschumi S, Bucher BS, Bianchetti MG, Simonetti GD. (2012) Pregnancy outcome following exposure to angiotensin-converting enzyme inhibitors or angiotensin receptor antagonists: a systematic review. *Hypertension* 60:444–450.

Callewaert BL, Willaert A, Kerstjens-Frederikse WS, De Backer J, Devriendt K, Albrecht B, Ramos-Arroyo MA, Doco-Fenzy M, Hennekam RC, Pyeritz RE, et al. (2008) Arterial tortuosity syndrome: Clinical and molecular findings in 12 newly identified families. *Hum Mutat* 29:150–158.

Cohn RD, van Erp C, Habashi JP, Soleimani AA, Klein EC, Lisi MT, Gamradt M, Rhys CM, Holm TM, Loeys BL, et al. (2007) Angiotensin II type 1 receptor blockade attenuates TGF-beta-induced failure of muscle regeneration in multiple myopathic states. *Nat Med* 13:204–210.

Collod-Beroud G, Le Bourdelles S, Ades L, Ala-Kokko L, Booms P, Boxer M, Child A, Comeglio P, De Paepe A, Hyland JC, et al. (2003) Update of the UMD-FBN1 mutation database and creation of an FBN1 polymorphism database. *Hum Mutat* 22:199–208.

David TE, Feindel CM, David CM, Manlhiot C (2014) A quarter of a century of experience with aortic valve-sparing operations. *J Thorac Cardiovasc Surg.* 148:872–879

De Paepe A, Devereux RB, Dietz HC, Hennekam RC, Pyeritz RE (1996) Revised diagnostic criteria for the Marfan syndrome. *Am J Med Genet* 62:417–426.

Devereux RB, de Simone G, Arnett DK, Best LG, Boerwinkle E, Howard BV, Kitzman D, Lee ET, Mosley TH, Weder A, et al. (2012) Normal limits in relation to age, body size and gender of two-dimensional echocardiographic aortic root dimensions in persons ≥15 years of age. *Am J Cardiol.* 110:1189–1194.

Dietz HC, Cutting GR, Pyeritz RE, Maslen CL, Sakai LY, Corson GM, Puffenberger EG, Hamosh A, Nanthakumar EJ, Curristin SM, et al. (1991) Marfan syndrome caused by a recurrent de novo missense mutation in the fibrillin gene. *Nature* 352:337–339.

Disabella E, Grasso M, Marziliano N, Ansaldi S, Lucchelli C, Porcu E, Tagliani M, Pilotto A, Diegoli M, Lanzarini L, et al. (2006) Two novel and one known mutation of the TGFBR2 gene in Marfan syndrome not associated with FBN1 gene defects. *Eur J Hum Genet* 14:34–38.

Drera B, Tadini G, Barlati S, Colombi M (2008) Identification of a novel TGFBR1 mutation in a Loeys-Dietz syndrome type II patient with vascular Ehlers-Danlos syndrome phenotype. *Clin Genet* 73:290–293.

Erkula G, Jones KB, Sponseller PD, Dietz HC, Pyeritz RE (2002) Growth and maturation in Marfan syndrome. *Am J Med Genet* 109:100–115.

Faivre L, Collod-Beroud G, Loeys BL, Child A, Binquet C, Gautier E, Callewaert B, Arbustini E, Mayer K, Arslan-Kirchner M, et al. (2007) Effect of mutation type and location on clinical outcome in 1.013 probands with Marfan syndrome or related phenotypes and FBN1 mutations: An international study. *Am J Hum Genet* 81:454–466.

Faivre L, Collod-Beroud G, Child A, Callewaert B, Loeys BL, Binquet C, Gautier E, Arbustini E, Mayer K, Arslan-Kirchner M, et al. (2008) Contribution of molecular analyses in diagnosing Marfan syndrome and type I fibrillinopathies: An international study of 1009 probands. *J Med Genet* 45:384–390.

Faivre L, Masurel-Paulet A, Collod-Beroud G, Loeys B, Child A, Binquet C, Gautier E, Stheuner C, Chevallier P, Callewaert B, et al. (2009) Clinical and molecular study of 320 children with Marfan syndrome and related type 1 fibrillinopathies out of a series of 1009 probands with a pathogenic FBN1 Mutation. *Pediatrics* 123:391–398.

Faivre L, Collod-Beroud G, Adès L, Arbustini E, Child A, Callewaert BL, Loeys B, Binquet C, Gautier E, Mayer K, et al. (2012) The new Ghent criteria for Marfan syndrome: what do they change? *Clin Genet.* 81:433–442.

Fattori R, Nienaber CA, Descovich B, Ambrosetto P, Reggiani LB, Pepe G, Kaufmann U, Negrini E, von Kodolitsch Y, Gensini GF (1999) Importance of dural ectasia in phenotypic assessment of Marfan's syndrome. *Lancet* 354:910–913.

Forteza A, Evangelista A, Sánchez V, Teixidó-Turà G, Sanz P, Gutiérrez L, Gracia T, Centeno J, Rodríguez-Palomares J, Rufilanchas JJ, et al. (2016) Efficacy of losartan vs. atenolol for the prevention of aortic dilation in Marfan syndrome: a randomized clinical trial. *Eur Heart J* 37:978–985.

Franken R, den Hartog AW, Radonic T, Micha D, Maugeri A, van Dijk FS, Meijers-Heijboer HE, Timmermans J, Scholte AJ, van den Berg MP, et al. (2015) Beneficial outcome of Losartan therapy depends on type of FBN1 mutation in Marfan syndrome. *Circ Cardiovasc Genet.* 8:383–388.

Frederic MY, Hamroun D, Faivre L, Boileau C, Jondeau G, Claustres M, Beroud C, Collod-Beroud G (2008) A new locus-specific database (LSDB) for mutations in the TGFBR2 gene: UMD-TGFBR2. *Hum Mutat* 29:33–38.

Furthmayr H, Francke U (1997) Ascending aortic aneurysm with or without features of Marfan syndrome and other fibrillinopathies: New insights. *Semin Thorac Cardiovasc Surg* 9:191–205.

Gaustadnes M, Wilcken B, Oliveriusova J, McGill J, Fletcher J, Kraus JP, Wilcken DE (2002) The molecular basis of cystathionine beta-synthase deficiency in Australian patients: Genotype-phenotype correlations and response to treatment. *Hum Mutat* 20:117–126.

Gersony DR, McClaughlin MA, Jin Z, Gersony WM (2007) The effect of beta-blocker therapy on clinical outcome in patients with Marfan's syndrome: A meta-analysis. *Int J Cardiol* 114:303–308.

Godfrey M, Menashe V, Weleber RG, Koler RD, Bigley RH, Lovrien E, Zonana J, Hollister DW (1990) Cosegregation of elastin-associated microfibrillar abnormalities with the Marfan phenotype in families. *Am J Hum Genet* 46:652–660.

Gott VL, Greene PS, Alejo DE, Cameron DE, Naftel DC, Miller DC, Gillinov AM, Laschinger JC, Pyeritz RE (1999) Replacement of the aortic root in patients with Marfan's syndrome. *N Engl J Med* 340:1307–1313.

Graham JM, Schwartz CE (2013) MED12 related disorders. *Am J Med Genet A.* 161A:2734–2740.

Habashi JP, Judge DP, Holm TM, Cohn RD, Loeys BL, Cooper TK, Myers L, Klein EC, Liu G, Calvi C, et al. (2006) Losartan, an AT1 antagonist, prevents aortic aneurysm in a mouse model of Marfan syndrome. *Science* 312:117–121.

Hackmann K, Rump A, Haas SA, Lemke JR, Fryns JP, Tzschach A, Wieczorek D, Albrecht B, Kuechler A, Ripperger T, et al. (2016) Tentative clinical diagnosis of Lujan-Fryns syndrome-A conglomeration of different genetic entities? *Am J Med Genet A.* 170A:94–102.

Hall JG, Allanson JE, Gripp K, Slavotinek A (2007) *Handbook of Normal Physical Measurements*, 2nd ed. Oxford: Oxford University Press.

Hennekam RC (2005) Severe infantile Marfan syndrome versus neonatal Marfan syndrome. *Am J Med Genet A* 139:1

Hilhorst-Hofstee Y, Hamel BC, Verheij JB, Rijlaarsdam ME, Mancini GM, Cobben JM, Giroth C, Ruivenkamp CA, Hansson KB, Timmermans J, et al. (2011). The clinical spectrum of complete FBN1 allele deletions. *Eur J Hum Genet* 19:247–252.

Hutchinson S, Furger A, Halliday D, Judge DP, Jefferson A, Dietz HC, Firth H, Handford PA (2003) Allelic variation in normal human FBN1 expression in a family with Marfan syndrome: A potential modifier of phenotype? *Hum Mol Genet* 12:2269–2276.

Izquierdo NJ, Traboulsi EI, Enger C, Maumenee IH (1994) Strabismus in the Marfan syndrome. *Am J Ophthalmol* 117:632–635.

Kainulainen K, Pulkkinen L, Savolainen A, Kaitila I, Peltonen L (1990) Location on chromosome 15 of the gene defect causing Marfan syndrome. *N Engl J Med* 323:935–939.

Kernell K, Sydsjö G, Bladh M, Josefsson A (2017) Birth characteristics of women with Marfan syndrome, obstetric and neonatal outcomes of their pregnancies-A nationwide cohort and case-control study. *Eur J Obstet Gynecol Reprod* 215:106–111.

Lacro RV, Dietz HC, Wruck LM, Bradley TJ, Colan SD, Devereux RB, Klein GL, Li JS, Minich LL, Paridon SM, et al. (2007) Rationale and design of a randomized clinical trial of beta-blocker therapy (atenolol) versus angiotensin II receptor blocker therapy (losartan) in individuals with Marfan syndrome *Am Heart J* 154:624–631.

Lacro RV, Guey LT, Dietz HC, Pearson GD, Yetman AT, Gelb BD, Loeys BL, Benson DW, Bradley TJ, De Backer J, et al (2013) Characteristics of children and young adults with Marfan syndrome and aortic root dilation in a randomized trial comparing atenolol and losartan therapy. *Am Heart J.* 165:828–835.e3.

Lacro RV, Dietz HC, Sleeper LA, Yetman AT, Bradley TJ, Colan SD, Pearson GD, Selamet Tierney ES, Levine JC, Atz AM, et al. (2014) Atenolol versus losartan in children and young adults with Marfan's syndrome. *N Engl J Med* 371:2061–2071.

Law C, Bunyan D, Castle B, Day L, Simpson I, Westwood G, Keetan B (2006) Clinical features in a family with an R460H mutation in transforming growth factor b receptor 2 gene. *J Med Genet* 43:908–916.

Lee B, Godfrey M, Vitale E, Hori H, Mattei MG, Sarfarazi M, Tsipouras P, Ramirez F, Hollister DW (1991) Linkage of Marfan syndrome and a phenotypically related disorder to two different *fibrillin* genes. *Nature* 352:330–334.

LeMaire SA, Pannu H, Tran-Fadulu V, Carter SA, Coselli JS, Milewicz DM (2007) Severe aortic and arterial aneurysms associated with a TGFBR2 mutation. *Nat Clin Pract Cardiovasc Med* 4:167–171.

Lindsay ME, Schepers D, Bolar NA, Doyle JJ, Gallo E, Fert-Bober J, Kempers MJ, Fishman EK, Chen Y, Myers L, et al (2012) Loss-of-function mutations in TGFB2 cause a syndromic presentation of thoracic aortic aneurysm. *Nat Genet.* 44:922–927.

Loeys B, De Backer J, Van Acker P, Wettinck K, Pals G, Nuytinck L, Coucke P, De Paepe A (2004) Comprehensive molecular screening of the FBN1 gene favors locus homogeneity of classical Marfan syndrome. *Hum Mutat* 24:140–146.

Loeys BL, Chen J, Neptune ER, Judge DP, Podowski M, Holm T, Meyers J, Leitch CC, Katsanis N, Sharifi N, et al. (2005) A syndrome of altered cardiovascular, craniofacial, neurocognitive and skeletal development caused by mutations in TGFBR1 or TGFBR2. *Nat Genet* 37:275–281.

Loeys BL, Schwarze U, Holm T, Callewaert BL, Thomas GH, Pannu H, De Backer JF, Oswald GL, Symoens S, Manouvrier S, et al. (2006) Aneurysm syndromes caused by mutations in the TGFbeta receptor. *N Engl J Med* 355:788–798.

Loeys BL, Dietz HC, Braverman AC, Callewaert BL, De Backer J, Devereux RB, Hilhorst-Hofstee Y, Jondeau G, Faivre L, Milewicz DM, et al (2010) The revised Ghent nosology for the Marfan syndrome. *J Med Genet* 47:476–485.

Magenis RE, Maslen CL, Smith L, Allen L, Sakai LY (1991) Localization of the *fibrillin* (*FBN*) gene to chromosome 15, band q21.1. *Genomics* 11:346–351.

Matyas G, Alonso S, Patrignani A, Marti M, Arnold E, Magyar I, Henggeler C, Carrel T, Steinmann B, Berger W (2007) Large genomic *fibrillin-1* (*FBN1*) gene deletions provide evidence for true haploinsufficiency in Marfan syndrome. *Hum Genet* 122:23–32.

Meijboom LJ, Vos FE, Timmermans J, Boers GH, Zwinderman AH, Mulder BJ (2005) Pregnancy and aortic root growth in the Marfan syndrome: A prospective study. *Eur Heart J* 26:914–920.

Mizuguchi T, Collod-Beroud G, Akiyama T, Abifadel M, Harada N, Morisaki T, Allard D, Varret M, Claustres M, Morisaki H, et al. (2004) Heterozygous TGFBR2 mutations in Marfan syndrome. *Nat Genet* 36:855–860.

Mueller GC, Stierle L, Stark V, Steiner K, von Kodolitsch Y, Weil J, Mir TS (2014) Retrospective analysis of the effect of angiotensin II receptor blocker versus β-blocker on aortic root growth in paediatric patients with Marfan syndrome. *Heart* 100:214–218.

Nallamshetty L, Ahn NU, Ahn UM, Nallamshetty HS, Rose PS, Buchowski JM, Sponseller PD (2002) Dural ectasia and back pain: Review of the literature and case report. *J Spinal Disord Technol* 15:326–329.

Nelson LB, Maumenee IH (1982) Ectopia lentis. *Surv Ophthalmol* 27:143–160.

Neptune ER, Frischmeyer PA, Arking DE, Myers L, Bunton TE, Gayraud B, Ramirez F, Sakai LY, Dietz HC (2003) Dysregulation of TGF-beta activation contributes to pathogenesis in Marfan syndrome. *Nat Genet* 33:407–411.

Pannu H, Fadulu VT, Chang J, Lafont A, Hasham SN, Sparks E, Giampietro PF, Zaleski C, Estrera AL, Safi HJ, et al. (2005) Mutations in transforming growth factor-beta receptor type II cause familial thoracic aortic aneurysms and dissections. *Circulation* 112:513–520.

Patel ND, Weiss ES, Alejo DE, Nwakanma LU, Williams JA, Dietz HC, Spevak PJ, Gott VL, Vricella LA, Cameron DE (2008) Aortic root operations for Marfan syndrome: A comparison of the Bentall and valve-sparing procedures. *Ann Thorac Surg* 85:2003–2010; discussion 2010–2001.

Phomakay V, Huett WG, Gossett JM, Tang X, Bornemeier RA, Collins RT (2014) β-Blockers and angiotensin converting enzyme inhibitors: comparison of effects on aortic growth in pediatric patients with Marfan syndrome. *J Pediatr* 165:951–955.

Puget S, Kondageski C, Wray A, Boddaert N, Roujeau T, Di Rocco F, Zerah M, Sainte-Rose C (2007) Chiari-like tonsillar herniation associated with intracranial hypotension in Marfan syndrome. Case report. *J Neurosurg* 106:48–52.

Pyeritz RE, Loeys B (2012) The 8th International Research Symposium on the Marfan Syndrome and related conditions. *Am J Med Genet Part A* 158A:42–49.

Pyeritz R, Jondeau G, Moran R, De Backer J, Arbustini E, De Paepe A, Milewicz D (2014) Loeys-Dietz syndrome is a specific phenotype and not a concomitant of any mutation in a gene involved in TGF-β signaling. *Genet Med* 16:641–642.

Pyeritz RE (2016) Recent progress in understanding the natural and clinical histories of the Marfan syndrome. *Trends Cardiovasc Med.* 26:423–428.

Radonic T, de Witte P, Groenink M, de Bruin-Bon RA, Timmermans J, Scholte AJ, van den Berg MP, Baars MJ, van Tintelen JP, Kempers M et al. (2011) Critical appraisal of the revised Ghent criteria for diagnosis of Marfan syndrome. *Clin Genet* 80346–353.

Robinson PN, Arteaga-Solis E, Baldock C, Collod-Beroud G, Booms P, De Paepe A, Dietz HC, Guo G, Handford PA, Judge DP, et al. (2006) The molecular genetics of Marfan syndrome and related disorders. *J Med Genet* 43:769–787.

Roman MJ, Devereux RB, Kramer-Fox R, O'Loughlin J (1989) Two-dimensional echocardiographic aortic root dimensions in normal children and adults. *Am J Cardiol* 64:507–512.

Rosser T, Finkel J, Vezina G, Majd M (2005) Postural headache in a child with Marfan syndrome: Case report and review of the literature. *J Child Neurol* 20:153–155.

Sakai H, Visser R, Ikegawa S, Ito E, Numabe H, Watanabe Y, Mikami H, Kondoh T, Kitoh H, Sugiyama R, et al. (2006) Comprehensive genetic analysis of relevant four genes in 49 patients with Marfan syndrome or Marfan-related phenotypes. *Am J Med Genet A* 140:1719–1725.

Sakai LY, Keene DR, Engvall E (1986) Fibrillin, a new 350-kD glycoprotein, is a component of extracellular microfibrils. *J Cell Biol* 103:2499–2509.

Salim MA, Alpert BS, Ward JC, Pyeritz RE (1994) Effect of beta-adrenergic blockade on aortic root rate of dilation in the Marfan syndrome. *Am J Cardiol* 74:629–633.

Schmierer B, Hill CS (2007) TGFbeta-SMAD signal transduction: Molecular specificity and functional flexibility. *Nat Rev Mol Cell Biol* 8:970–982.

Schrijver I, Liu W, Brenn T, Furthmayr H, Francke U (1999) Cysteine substitutions in epidermal growth factor-like domains of fibrillin-1: Distinct effects on biochemical and clinical phenotypes. *Am J Hum Genet* 65:1007–1020.

Schrijver I, Liu W, Odom R, Brenn T, Oefner P, Furthmayr H, Francke U (2002) Premature termination mutations in FBN1: Distinct effects on differential allelic expression and on protein and clinical phenotypes. *Am J Hum Genet* 71:223–237.

Schwartz CE, Tarpey PS, Lubs HA, Verloes A, May MM, Risheg H, Friez MJ, Futreal PA, Edkins S, Teague J, et al. (2007) The original Lujan syndrome family has a novel missense mutation (p.N1007S) in the MED12 gene. *J Med Genet* 44:472–477.

Selamet Tierney ES, Feingold B, Printz BF, Park SC, Graham D, Kleinman CS, Mahnke CB, Timchak DM, Neches WH, Gersony WM (2007) Beta-blocker therapy does not alter the rate of aortic root dilation in pediatric patients with Marfan syndrome. *J Pediatr* 150:77–82.

Shores J, Berger KR, Murphy EA, Pyeritz RE (1994) Progression of aortic dilatation and the benefit of long-term beta-adrenergic blockade in Marfan's syndrome. *N Engl J Med* 330:1335–1341.

Silverman DI, Burton KJ, Gray J, Bosner MS, Kouchoukos NT, Roman MJ, Boxer M, Devereux RB, Tsipouras P (1995) Life expectancy in the Marfan syndrome. *Am J Cardiol* 75:157–160.

Singh KK, Rommel K, Mishra A, Karck M, Haverich A, Schmidtke J, Arslan-Kirchner M (2006) TGFBR1 and TGFBR2 mutations in patients with features of Marfan syndrome and Loeys-Dietz syndrome. *Hum Mutat* 27:770–777.

Spits C, De Rycke M, Verpoest W, Lissens W, Van Steirteghem A, Liebaers I, Sermon K (2006) Preimplantation genetic diagnosis for Marfan syndrome. *Fertil Steril* 86: 310–320.

Sponseller PD, Hobbs W, Riley LH III Pyeritz RE (1995) The thoracolumbar spine in Marfan syndrome. *J Bone Joint Surg Am* 77:867–876.

Van Kimmenade R, Kempers M, De Boer M, Loeys B, Timmermans J (2013) A clinical appraisal of different Z-score equations for aortic root assessment in the diagnostic evaluation of Marfan syndrome. *Eur Heart J* 34(suppl1):2113.

Velvin G, Bathen T, Rand-Hendriksen S, Geirdal AO (2015) Systematic review of the psychosocial aspects of living with Marfan syndrome. *Clin Genet* 87:109–116.

Williams A, Davies S, Stuart AG, Wilson DG, Fraser AG (2008) Medical treatment of Marfan syndrome: a time for change. *Heart* 94:414–421.

Wu-Chen WY, Letson RD, Summers CG (2005) Functional and structural outcomes following lensectomy for ectopia lentis. *J AAPOS* 9:353–357.

Yang JH, Han H, Jang SY, Moon JR, Sung K, Chung TY, Lee HJ, Ki CS, Kim DK (2012) A comparison of the Ghent and revised Ghent nosologies for the diagnosis of Marfan syndrome in an adult Korean population. *Am J Med Genet A* 158A: 989–995.

Yetman AT, Bornemeier RA, McCrindle BW (2005) Usefulness of enalapril *versus* propranolol or atenolol for prevention of aortic dilation in patients with the Marfan syndrome. *Am J Cardiol* 95:1125–1127.

38

MOWAT–WILSON SYNDROME

DAVID MOWAT
Centre for Clinical Genetics, Sydney Children's Hospital, Randwick, School of Women's and Child Health, University of New South Wales, Sydney, Australia

MEREDITH WILSON
Department of Clinical Genetics, Children's Hospital at Westmead, Westmead, Discipline of Genomic Medicine, University of Sydney, Sydney, Australia

INTRODUCTION

Mowat–Wilson syndrome (MWS) is an intellectual disability/multiple congenital anomaly syndrome, characterized by typical facies, usually severe-profound intellectual disability, epilepsy, and variable congenital malformations including Hirschsprung disease (HSCR), congenital heart defects, urogenital anomalies (hypospadias), and agenesis of the corpus callosum. Mowat et al. (1998) first described the syndrome in a series of six children with intellectual disability and strikingly similar facial features, five of whom had HSCR. In 2001, two groups independently identified the underlying cause of Mowat–Wilson syndrome as mutations or deletions in the *SMAD1* (SMAD-interacting protein-1) gene, also known as *ZFHX1B* (zinc finger E-box-binding homeobox 2), now known as *ZEB2*(MIM# 605802), located at chromosome 2q22 (Cacheux et al. 2001; Wakamatsu et al. 2001). Subsequent published case series and reviews describe over 300 individuals with molecularly proven MWS (Amiel et al. 2001; Yoneda et al. 2002; Zweier et al. 2002, 2003, 2005, 2006; Garavelli et al. 2003, 2005; Mowat et al. 2003; Wilson et al. 2003; Gregory-Evans et al. 2004; Horn et al. 2004; Cerruti-Mainardi et al. 2005; Ishihara et al. 2005; McGaughran et al. 2005; Heinritz et al. 2006; Adam et al. 2007; Dastot-Le Moal et al. 2007; Garavelli and Cerruti-Mainardi 2007; Sasso et al. 2008; Adam et al. 2008; Ohtsuka et al. 2008;Garavelli et al. 2009; Yamada et al. 2014; Ivanovski et al. 2018).

Incidence

The incidence of Mowat–Wilson syndrome is currently unknown as it is still under-diagnosed, particularly in individuals without Hirschsprung disease (Mowat et al. 2003; Cerruti-Mainardi et al. 2005). Because of ascertainment bias, Hirschsprung disease was present in the majority of individuals in early publications, but more recent series report Hirschsprung disease in 30–50% (Zweier et al. 2005; Dastot-Le Moal et al. 2007; Coyle et al. 2015; Ivanovski et al. 2018). Amiel et al. (2001) identified eight individuals with Mowat–Wilson syndrome by *ZEB2* mutation testing in 19 patients with intellectual disability and microcephaly selected from a cohort of 250 with sporadic Hirschsprung disease. Extrapolating from the population incidence of Hirschsprung disease of approximately 1 in 5000, and assuming a penetrance of approximately 50% for Hirschsprung disease in MWS, the birth incidence of the syndrome could be at least 1 in 70,000. MWS has been reported throughout the world among diverse racial groups (Mowat et al. 1998; Wakamatsu et al. 2001; Adam et al. 2006a, 2006b; Dastot-Le Moal et al. 2007).

Cassidy and Allanson's Management of Genetic Syndromes, Fourth Edition.
Edited by John C. Carey, Agatino Battaglia, David Viskochil, and Suzanne B. Cassidy.
© 2021 John Wiley & Sons, Inc. Published 2021 by John Wiley & Sons, Inc.

Longitudinal studies of the natural history of MWS have not been published and relatively few adults are described in the literature. Observation of clinical cohorts suggests that most individuals survive to adult life, but there may be early mortality associated with congenital heart disease, severe epilepsy, or complications of Hirschsprung disease and other congenital anomalies.

Diagnostic Criteria

There are no formally agreed diagnostic criteria, but all individuals with MWS have intellectual disability, which is usually severe, and a distinctive facial gestalt with ear lobe morphology (fleshy, uplifted ear lobes with a central depression) that is recognizable from infancy. Most individuals have microcephaly and epilepsy. One or more congenital anomalies, including Hirschsprung disease, hypospadias, congenital heart defect, and agenesis of the corpus callosum, are present in most individuals, but none is obligatory. The percentage of individuals diagnosed without Hirschsprung disease has increased as clinicians become more familiar with the facial gestalt (Dastot-Le Moal et al. 2007; Garavelli and Cerruti-Mainardi 2007; Ivanovski et al. 2018). The spectrum of clinical features including developmental and behavioral phenotypes are detailed in Tables 38.1 and 38.2 and illustrated in Figures 38.1–38.3.

Etiology, Pathogenesis, and Genetics

MWS is an autosomal dominant disorder caused by de novo heterozygous mutations or deletions in the zinc finger E-box-binding homeobox 2 gene, *ZEB2* (previously *SIP1*, then *ZFHX1B*) located at chromosome 2q22. The *ZEB2* gene spans approximately 70 kb, consists of 10 exons and 9 introns, and encodes for Smad interacting protein 1 (SIP1). SIP1 is a complex, multidomain protein, characterized by the presence of a Smad-binding domain, a homeodomain-like sequence, and a C-terminal binding protein interacting domain, flanked by two clusters of zinc finger sequences. SIP1 acts as a transcriptional co-repressor in the transforming growth factor-β (TGF-β) signaling pathway (Dastot-Le Moal et al. 2007), but has also been ascribed to transcriptional activation activity (Long et al. 2005) and has been found to be associated with multiple subunits of the nucleosome remodeling and histone deacetylation complex (NuRD) (Verstappen et al. 2008). *ZEB2* is highly evolutionarily conserved, widely expressed in embryological development, involved in the development of neural crest derived cells (enteric nervous system, craniofacial mesoectoderm), central nervous system, Schwann cells, heart septation, and midline structures including corpus callosum and genitalia (Hegarty et al. 2015; Watanabe et al. 2017). The *ZEB2* gene also has a role in cancer and immune function although these do not appear to be a clinical problem in individuals with MWS (Geldhof et al. 2012; Omilusik et al. 2015).

Individuals with clinically typical MWS usually have whole gene deletions or truncating mutations of *ZEB2*, suggesting that haploinsufficiency is the pathogenic mechanism. So far, studies have not suggested any genotype:phenotype correlations for individuals with deletions or truncating mutations involving *ZEB2* only, although individuals with multigene deletions in 2q22 may have additional features. Both typical and atypical phenotypes have been reported in a small number of individuals with missense mutations (Dastot-Le Moal et al. 2007; Ghoumid et al. 2013; Ivanovski et al. 2018).

TABLE 38.1 Clinical features of Mowat–Wilson syndrome

Consistent (~100%)

Characteristic facial gestalt
Developmental delay, mod-profound, usually severe. No regression
Speech impairment; no or minimal use of words; receptive and non-verbal communication skills higher than verbal skills

Frequent (>80%)

Microcephaly
Seizures
Constipation
Behavior: happy demeanor, and social, high pain threshold
Late walking

Associated (20–80%)

Hirschsprung disease, usually diagnosed neonatally, but late diagnosis of short segment Hirschsprung disease occurs
Congenital heart defect: wide spectrum, usually conotruncal, including pulmonary artery sling with or without tracheal stenosis (rare)
Renal tract anomalies
Hypospadias in males
Short stature
Hypoplasia or agenesis of corpus callosum; hippocampal dysplasia
Sleep disturbance
Strabismus

Recognized but uncommon (<10%)

Pyloric stenosis
Structural eye anomalies, iris/chorioretinal/optic disc coloboma, optic nerve atrophy, retinal epithelium atrophy, cataract, and korectopia
Cleft lip/palate
Hypopigmented patches hair or skin
Autonomic dysfunction
Asplenia
Duplicated hallux
Pachygyria

TABLE 38.2 Facial gestalt of Mowat–Wilson syndrome at various ages

Infancy

Sparse hair
Square-shaped face
Horizontal, widely separated eyebrows
Large deep-set eyes
Mild hypertelorism or telecanthus
Prominent rounded nasal tip
Prominent columella
Full everted lower lip
Prominent central part of chin
Uplifted ear lobes with central depression
Redundant nuchal skin

Childhood

Upward gaze and open-mouthed posture with frequent smiling
Columella lengthens further
Nasal profile – central convexity
Upper lip vermilion full centrally, thin laterally

Adolescence/adulthood

Face lengthens
Broad eyebrows with sparse medial flare
Overhanging nasal tip with low-hanging columella
Prognathism with "chisel-shaped" chin
Uplifted ear lobes
Prematurely aged or coarse appearance

Diagnostic Testing

The diagnosis of MWS is based on recognition of the distinctive facial gestalt, usually associated with severe ID. MWS should be considered in any delayed, dysmorphic individual with Hirschsprung disease, agenesis of the corpus callosum, hypospadias, or congenital heart defect. Molecular testing (sequencing +/- chromosome microarray) can be used to confirm clinical suspicion.

Over 300 mutation-positive individuals with MWS are reported in the literature (Ivanovski et al. 2018). Dastot-Le Moal et al. (2007) reviewed the mutation profile as frameshift mutations (41.5%), nonsense mutations (31.6%), and submicroscopic whole gene or exonic deletions (19.3%). Rarer mutations include cytogenetically detectable deletions (1.2%), translocations (0.6%), splice site mutations (2.3%), missense mutations (1.7%), complex mutations (deletion and insertion) (1.2%) and an inframe mutation (0.6%). Discussion with the testing laboratory may assist in deciding what testing approach is best, depending on degree of certainty of diagnosis, resources for test funding, and changes in technologies over time. A high-resolution chromosome microarray would detect whole gene or large exonic deletions, or alternative chromosome diagnoses. If negative, *ZEB2* sequencing should be done, either targeted or as part of a next generation sequencing (NGS) panel. NGS techniques will also allow analysis of the sequencing data for exonic deletions. If CMA and sequencing are normal,

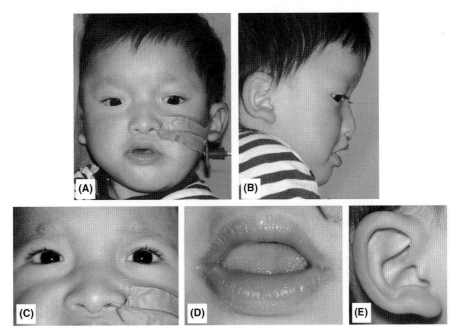

FIGURE 38.1 Facial findings in a 20-month-old male with Mowat–Wilson syndrome: (A) and (B) typical facial features and profile; (C) sparse medially flared eyebrows and long columella; (D) upper lip has narrow lateral vermilion and full medial vermilion, and lower lip is full; (E) posteriorly rotated ears with uplifted ear lobe.

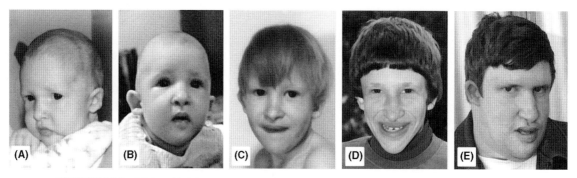

FIGURE 38.2 Male with Mowat–Wilson syndrome at various ages: (A) 8 weeks; (B) 8 months; (C) 3 years; (D) 9 years; (E) 28 years.

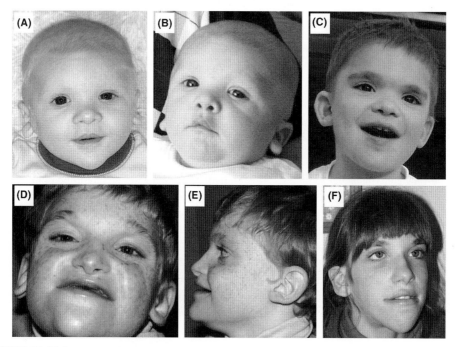

FIGURE 38.3 Five children with Mowat–Wilson syndrome: (A) and (B) two male infants; (C) male age 3 years; (D) and (E) male age 8 years; (F) female 14 years.

and a strong suspicion of MWS exists, karyotype to check for a balanced translocation disrupting the gene could be considered.

Increasingly the diagnosis of MWS arises retrospectively after an epilepsy/intellectual disability gene panel or a whole exome/genome study identifies a pathogenic variant in *ZEB2*. Caution in assigning a diagnosis is recommended if a variant of unknown significance is identified in *ZEB2*, when the clinical phenotype is not consistent with MWS.

With rare exceptions, all affected individuals with the typical syndrome phenotype have a germ line mutation/deletion/disruption involving *ZEB2*. In those in whom a mutation is not identified, possible explanations include mutation in a *ZEB2* regulatory region or somatic mosaicism. There is no evidence so far for involvement of an alternative genetic locus. Testing is undertaken to confirm the clinical diagnosis and aid in genetic counseling.

Genetic Counseling

The majority of patients with MWS have de novo mutations or deletions involving the *ZEB2* gene and the recurrence risk is low. Recurrence of MWS in subsequent siblings, presumably on the basis of germ line mosaicism, has been reported in four families (McGaughran et al. 2005; Zweier et al. 2005; Ohtsuka et al. 2008; Ivanovski et al. 2018), and an estimated recurrence risk of 1–2% is suggested. Genetic counseling should include discussion of possible sibling recurrence due to germ line mosaicism and the availability of prenatal testing if the proband's mutation has been identified. The diagnosis could be suspected in a subsequent sibling if prenatal ultrasound demonstrates anomalies such as agenesis of the corpus callosum or congenital heart defect, but accurate prenatal diagnosis requires mutation analysis for confirmation. Without a previous sibling diagnosis, prenatal ultrasound demonstration of congenital abnormalities common in MWS is of limited specificity (Espinosa-Parilla et al. 2004).

Differential Diagnosis

The main differential diagnosis for MWS is Goldberg–Shprintzen syndrome, as both can present with Hirschsprung disease, microcephaly, intellectual disability, and epilepsy. In retrospect, several patients with MWS were incorrectly published as having Goldberg–Shprintzen syndrome (Tanaka et al. 1993; Sharar and Shinawi 2003; Silengo et al. 2003). Goldberg–Shprintzen syndrome is an autosomal recessive syndrome, later shown to be caused by mutations in the *KIAA1279 (KIF1BP)* gene at 10q22 (Brooks et al. 2005). It can be clinically distinguished by the different facial dysmorphism, including arched eyebrows, high nasal bridge, synophrys, and long curled eyelashes (Goldberg and Shprintzen 1981; Hurst et al. 1988 (cases 1 and 2); Yomo et al. 1991; Fryer 1998; Brooks et al. 1999, 2005). Hypospadias and agenesis of the corpus callosum have not yet been reported in Goldberg–Shprintzen syndrome, and their presence is a strong predictor for MWS. Cerebral cortical abnormalities including frontotemporal or generalized atrophy, hippocampal dysplasia and pachygyria have been reported in MWS (Wilson et al. 2003; Silengo et al. 2003; Garavelli et al. 2017), but cerebral MRI in Goldberg–Shprintzen syndrome usually shows bilateral diffuse polymicrogyria (Brooks et al. 2005).

Angelman syndrome (AS) had been considered in some individuals with MWS because of some shared features, including frequent smiling, upward-tilted face with open-mouthed expression, pointed and prominent chin, ataxic gait, and arm posturing (see Chapter 5,). The characteristic facial features of MWS should be recognizable, and individuals with AS do not usually have multiple congenital anomalies or Hirschsprung disease. In females with developmental delay, seizures, and acquired microcephaly, a diagnosis of Rett syndrome is typically considered. However, in Rett syndrome there is usually a period of 6–18 months in which development is normal, and significant facial dysmorphism is not present (see Chapter 49). Pitt–Hopkins syndrome can have severe intellectual disability, microcephaly and occasional Hirschsprung disease, but has a different facial gestalt (Marangi et al. 2015), as does Rubinstein–Taybi syndrome (RTS) (see Chapter 51), which may be considered in older individuals, because of the nasal profile. Other conditions that can cause severe intellectual disability in association with hypospadias include Smith–Lemli–Opitz syndrome (SLOS) (see Chapter 53), X-linked α thalassaemia-mental retardation syndrome (ATRX) (see Chapter 7), and Opitz G syndrome, but these are usually excluded by facial features and/or other specific manifestations or investigations. In an infant presenting with Hirschsprung disease, other syndromic causes of Hirschsprung disease ought to be considered (see Parisi 2015).

MANIFESTATIONS AND MANAGEMENT

Early intervention and targeted management can have a significant impact on the health and quality of life of all individuals with significant developmental delay and medical problems. In MWS, as in other complex multisystem conditions, good management may require the involvement of many different medical, surgical, and allied health clinicians. Well-informed parents and caretakers are a critical component of management. MWS is rare and relatively recently described, so understanding of the clinical complications and natural history is still evolving. Families are a rich source of information in helping to understand the spectrum of clinical features and management priorities (see Resources section). Clinicians may need to communicate widely with colleagues to help shape an individualized approach to management. With increasing knowledge and improved management, we look forward to a better quality of life for affected individuals and their families.

Growth and Feeding

Growth parameters are usually normal at birth, although microcephaly may be present. The majority will develop microcephaly in the first year of life. Microcephaly is eventually present in approximately 80% of reported cases and most of the remainder have a head circumference at the lower end of normal (3rd to 10th centile) (Mowat et al. 2003; Garavelli and Cerruti-Mainardi 2007). In the absence of Hirschsprung disease or neonatal surgical complications, oral feeding is usually achieved, although some individuals require nasogastric feeding for variable time.

Many parents report that their children frequently place objects in their mouth, causing gagging or choking. Chewing is often incomplete, so pureed or soft diet is often required. Parents commonly report that retching is used to communicate dislike of a certain activity (mealtimes, travel).

Most individuals with MWS have a slender build and develop postnatal short stature (Mowat et al. 2003). The cause of this growth failure is not understood. Growth hormone studies have not been reported. Midline anomalies (agenesis of the corpus callosum, cleft palate) are a recognized feature of MWS, but structural anomalies of the pituitary have not been described.

Information about pubertal development has not been surveyed, but both normal and early puberty have been observed.

Evaluation

- Mowat–Wilson syndrome-specific growth charts are expected to be published in 2020.
- Feeding and nutrition needs to be very closely monitored, particularly in those with Hirschsprung disease.
- For those with persistent feeding problems, further evaluation, such as oropharyngeal imaging and/or endoscopy, may be indicated.
- Examine for possible submucous cleft palate.

Management

- Dietitian-guided nutritional supplementation should be instituted.
- Feeding problems are treated in a standard manner.
- Those with cleft palate should be referred to a craniofacial team or plastic or orofacial surgeon.

Development and Behavior

Intellectual disability in MWS is usually severe but detailed cohort studies are not yet available. Neurodevelopmental assessment using standard measures may be difficult. Functional assessment by developmental therapists combined with reports from parents, teachers, and caregivers provides a basis for management.

Early developmental milestones such as smiling and visual attention may be normal. Hypotonia is common in the first year with subsequent delayed motor milestones and a mean age of sitting of 20 months and walking of 4 years and 3 months (Garavelli and Cerruti-Mainardi 2007; Evans et al. 2012).

Most children are socially engaging and responsive with a happy affect. Many parents report repetitive behaviors such as flicking lights on and off. Many individuals with MWS have persistent oral stimulation stereotypies that may interfere with other activities. Many children will bite their thumb or fingers. A safe chewable toy may help control this behavior.

Speech and Language Babies and infants may have decreased cooing and babbling. Despite developing an occasional single word by age 2–3 years, expressive speech is markedly delayed. Drooling is common. Most older children and adults with typical MWS are non-verbal. A small number of individuals have developed 100–200 words. Children tend to communicate better using non-verbal means (gestures or signs) and appear to have better receptive language skills, so augmented communication approaches should be explored (Evans et al. 2012).

Sleep Abnormalities Parent reports suggest that sleep disturbance may be common in MWS, with prolonged wakefulness and early morning waking. In the study by Evans et al. (2016) a high level of sleep disturbance was found with 53% scoring in the borderline range and 44% in the clinical disorder range for at least one subscale of the SDSC in 35 individuals. Scores were highest for the sleep–wake transition disorders subscale, with 91% of participants reaching at least the borderline disorder range. Sleep disorders should be screened for in people with MWS, and where appropriate, referrals to sleep specialists made for management of sleep problems.

Evaluation

- All individuals need close monitoring of developmental progress from early infancy.
- Standard assessment measures may underscore some areas of delayed functioning.
- Careful direct observation and subjective parental and caregiver reports may provide useful additional information about functional ability.
- Early speech and communication assessments are recommended.
- Consider investigation for a physical cause (otitis media, urinary tract infection, dental, GERD, constipation or Hirschsprung disease) if there is a significant change in behavior, such as screaming spells, irritability, or restlessness.
- Individuals with MWS may have under-reaction to pain (Pradier et al. 2014).

Management

- Referral should be made to therapy services including speech therapy, physiotherapy, and occupational therapy at diagnosis.
- School-age children need formal neurocognitive assessment to aid planning of appropriate educational services.
- Individuals with MWS respond well to music therapy and may vocalize when singing.
- A structured environment and consistent behavior modification approaches are recommended in the school and home. Persistence in applying strategies may eventually be rewarded; for example, toilet training is achievable for some.
- Treatment of challenging behaviors is complex with minimal data to guide other than standard approaches (Besterman and Hendren 2015).
- Augmented communication (signing, picture exchange system, Makaton) may be beneficial and should be attempted.

- Access to a safe chewing toy may be helpful in avoiding inappropriate mouthing of objects.
- Adults require a heavily supported environment whether in the family home, group home or other residential placement. Although developmental disability is severe, adults benefit from and appear to enjoy social contact.
- If sleep problems arise at any age suggest referral to a sleep specialist

Gastrointestinal

Hirschsprung disease is a common feature of MWS, present in up to 40–60% of reported individuals (Bonnard et al. 2009; Coyle et al. 2015). Most individuals with Hirschsprung disease have short segment disease, especially in males. Some without Hirschsprung disease have severe chronic constipation, with rectal biopsies that are normal or not typical of Hirschsprung disease. Hirschsprung disease may not be symptomatic in the neonatal period. One individual with challenging behavior and chronic constipation was finally diagnosed with Hirschsprung disease and treated at the age of 27 years after which behavior improved (personal observation). Of those individuals with treated HSCR only a small number are described as having normal bowel function (Coyle et al. 2015) with many needing terminal stoma formation.

The rates of incontinence are high in MWS. In a recent study 97.7% of persons with MWS had incontinence (nocturnal enuresis 74.4%; daytime urinary incontinence 76.2%; fecal incontinence 81.4%). Incontinence remained high over age groups (children 95.8%, teens 100%, adults 100%). 46.2% of children, 25% of teens and 37.5% of adults exceeded the clinical cut-off on the developmental behavioral checklist. The ability to use the toilet for micturition improved with age (Niemczyk et al. 2017).

Pyloric stenosis has been reported in at least eight individuals with MWS (Amiel et al. 2001; Wilson et al. 2003; Garavelli and Cerruti-Mainardi 2007). Gastroesophageal reflux can be severe in some individuals.

Evaluation

- Neonatal bowel obstruction or history of severe constipation requires investigation for Hirschsprung disease by rectal biopsy according to standard surgical approaches. Hirschsprung disease may not present in the neonatal period, so it is important to consider rectal biopsy in any individual with Mowat–Wilson syndrome who presents with persistent constipation. In biopsy-negative cases, where symptoms persist, anorectal manometry, and barium enema studies may be helpful in diagnosing functional dysmotility.
- The possibility of pyloric stenosis should be evaluated if there is persistent, forceful vomiting in the first few months of life. Pyloric ultrasound is generally diagnostic.
- Barium swallow and/or upper gastrointestinal endoscopy and biopsy should be performed if gastroesophageal reflux disease (GERD) is suspected.

Management

- Standard surgical treatment for Hirschsprung disease is indicated. For a review of surgical management of Hirschsprung disease, see Haricharan and Georgeson (2008).
- Postoperative persistence of intestinal dysmotility (presenting with constipation or sometimes diarrhea) is common.
- Hirschsprung-associated enterocolitis can be a postsurgical complication with significant morbidity.
- Standard pyloroplasty is appropriate if pyloric stenosis is confirmed on imaging.
- Medical and/or surgical treatment of gastroesophageal reflux is standard.

Neurologic

Postnatal microcephaly is present in over 80% of reported cases. Earlier publications reported total or partial agenesis of the corpus callosum in over 40% of those who had cranial imaging. Other less common cerebral abnormalities included mild cerebral atrophy, frontotemporal hypoplasia, temporal dysplasia, pachygyria, patchy periventricular increased T1 signal, and thinning of white matter (Wilson et al. 2003; McGaughran et al. 2005; Zweier et al. 2005). Several individuals with large deletions had more marked cerebral atrophy and/or pachygyria, initially suggesting this might be related to the deletion of contiguous genes. However, no consistent cerebral abnormality is reported in others with large deletions, including those with cytogenetically visible abnormalities (Amiel et al. 2001; Zweier et al. 2003; Ishihara et al. 2005; Garavelli et al. 2017). Garavelli et al. (2017) reviewed CNS abnormalities in a cohort of 54 individuals reporting hippocampal abnormalities (77.8%), enlargement of cerebral ventricles (68.5%), and white matter abnormalities (reduction of thickness 40.7%, localized signal alterations 22.2%). Other occasional reported findings include large basal ganglia, cortical, and cerebellar malformations.

Seizures Seizures have been reported in 64–78% (Wilson et al. 2003; Zweier et al. 2005; Garavelli and Cerruti-Mainardi 2007; Cordelli et al. 2013; Babkina et al. 2016). The onset of seizures is usually by the second year of life, although seizures may begin in infancy or later childhood, up to 10 years of age. Seizure types include grand mal, absence, focal, and minor motor status. The seizure disorder can be resistant to treatment, but control often improves in adolescence and adulthood. The electroclinical pattern detected in MWS suggests a genetic form of epilepsy rather than it being secondary to brain macroanatomy (Cordelli et al. 2013). Electrical status epilepticus during sleep has recently been reported (Bonanni et al. 2017).

Gait Disturbance Gait is often wide-based with the elbows held flexed. Acquisition of independent walking is usually severely delayed. Children will often eventually stand unsupported and walk with minimal support. Children tend not to initiate independent walking; rather, they will crawl or bottom shuffle to move around. Some children may lose their ability to walk if ambulation is not encouraged.

Autonomic Abnormalities Autonomic dysregulation may be a feature of non-syndromic Hirschsprung disease. Episodes of urinary retention, bradycardia, hypersomnolence, pinpoint pupils, hypoventilation, and hypercarbia, suggesting autonomic dysregulation, have occurred in one individual known to the authors.

Evaluation

- Microcephaly alone is not an indication for cerebral imaging. The presence or absence of the corpus callosum does not alter prognosis or aid management but may be a useful diagnostic sign.
- Possible seizures should be investigated by electroencephalogram. Persisting changes in behavior or in motor and cognitive functions should be investigated with a wake and sleep/ nocturnal EEG.
- Any loss of developmental milestones warrants investigation for seizures.
- Careful monitoring should follow the use of general anesthesia.

Management

- Anticonvulsant treatment should be based on the type of seizures present.
- No anticonvulsant is known to be specifically beneficial in MWS.
- Withdrawal of anticonvulsants can be considered after a significant seizure-free period.

Cardiovascular

Congenital heart disease is reported in approximately 50% of the published cases. Anomalies reported include patent ductus arteriosus, atrial septal defect, ventricular septal defect, tetralogy of Fallot, pulmonary atresia, pulmonary valve stenosis, pulmonary artery sling or stenosis, aortic coarctation, and aortic stenosis (Garavelli and Cerruti-Mainardi 2007; Sasso et al. 2008; Adam et al. 2008; Ohtsuka et al. 2008). Pulmonary artery sling with or without tracheal stenosis may be a particular association of MWS (Dastot-Le Moal et al. 2007).

Evaluation

- All individuals with MWS should have a cardiological assessment including echocardiogram at the time of diagnosis.

Management

- Standard treatment and follow-up are appropriate for any heart anomaly detected.

Genitourinary

Genitourinary anomalies are reported in just over 50% of individuals with MWS. Genital anomalies in males include hypospadias, undescended testes (36%), bifid scrotum, and webbed penis. No external genital anomalies have been reported in females, but Ishihara et al. (2005) reported one female with a vaginal septum. Renal tract anomalies include vesicoureteric reflux, hydronephrosis, pelviureteric junction obstruction, and duplex or pelvic kidney. Hypospadias is the most common single urogenital malformation, documented in 46–56% of males in various series, and males with hypospadias often have other renal tract anomalies. Vesicoureteric reflux and hydronephrosis are the next most frequent urogenital anomalies, reported in 18–26% of all affected individuals (Wilson et al. 2003; Garavelli et al. 2005; Zweier et al. 2005; Garavelli and Cerruti-Mainardi, 2007; Wilson et al. 2007).

Evaluation

- Male genitalia should be carefully examined.
- A baseline renal ultrasound is recommended.
- Micturating cystourethrogram is indicated to investigate vesicoureteric reflux if the ultrasound is abnormal or if recurrent urinary tract infection is noted.
- In the presence of fever of unknown origin, urinalysis should be performed to rule out urinary tract infection.

Management

- If vesicoureteric reflux is present, referral to a nephrologist or urologist is recommended.
- Orchidopexy for undescended testes in a standard manner.

- Standard indications and methods are appropriate for surgical management of hypospadias.

Endocrine/Pubertal

Most individuals go through puberty normally. No predisposition to endocrine dysfunction has been reported.

Ophthalmologic

Structural eye anomalies, including microphthalmia and retinal or iris coloboma, have been reported in a small proportion (4.1–14%) of individuals (Garavelli and Cerruti-Mainardi 2007; Pons et al. 2015). Non-structural functional anomalies, such as strabismus or hypermetropia, are more common (~50%). Some affected infants have nystagmus and poor ocular fixation that later resolves. Unilateral ptosis has been observed in several individuals. Several reports have noted striking sapphire-blue irides in some individuals, some with small clumps of darker iris pigment (Mowat et al. 1998; Amiel et al. 2001; Wilson et al. 2003).

Gregory-Evans et al. (2004) described a child with trisomy 21, Hirschsprung disease, bilateral iris and inferior retinal colobomas and high myopia, who had a novel missense mutation in *ZFHX1B (ZEB2)*, without the usual MWS facial gestalt. Subsequently, Zweier et al. (2005) reported structural eye anomalies in 14% of a new cohort of people with MWS: one had Axenfeld anomaly and three had microphthalmia, one of whom also had iris coloboma and cataract. All had truncating mutations, of which one was recurrent but not previously associated with an ocular anomaly. McGaughran et al. (2005) reported siblings with the same truncating mutation, one of whom had bilateral iris colobomas, left aplastic optic nerve with central chorioretinal coloboma, and a choroidal lacuna around the right optic disk, while the sibling had a divergent strabismus, but no structural eye abnormalities. The eye phenotype can vary markedly between siblings or individuals with the same mutation, and there is no specific genotype–phenotype correlation.

Evaluation

- Referral to an ophthalmologist for examination is indicated for all affected individuals.

Management

- Management and follow-up will depend on the anomaly found.

Ears and Hearing

Recurrent episodes of otitis media have been reported in individuals with MWS, with potential conductive hearing loss resulting from untreated chronic infection. Sensorineural hearing loss is not expected and other causes should be considered, if present.

Evaluation

- Hearing assessment and otolaryngologic examination should be performed to diagnose hearing loss as a potential contributing factor in speech development.

Management

- Any hearing loss should be managed with standard medical or surgical techniques and communication strategies.

Musculoskeletal

Minor hand or foot variations are common, including bridged or transverse palmar creases, fifth finger clinodactyly, slender fingers, positional talipes calcaneovalgus, metatarsus varus, and eversion of the feet. Mild camptodactyly and thickening of the interphalangeal joints have been seen in several older individuals. Broad halluces, unilateral duplication of the hallux, and hypertrophy of the first ray of the foot have been reported (Wilson et al. 2003). An individual with a missense mutation had brachytelephalangy with broad thumbs and halluces (Heinritz et al. 2006). In contrast, an individual with an 11 Mb deletion including *ZEB2* had hypoplasia of the halluces (Zweier et al. 2003).

Evaluaition

- Foot and ankle positioning should be checked as the child starts to mobilize.

  Significant musculoskeletal abnormality may require referral to orthotist or pediatric orthopedic surgeon

Management

- Treatment of foot and hand anomalies is standard, if indicated.

Immune Function

A small number of individuals appear to have an increased incidence of sinopulmonary infections not associated with an immunoglobulin deficiency or structural anomaly/absence of the spleen (personal experience). At least four individuals with MWS have been reported to have an absent spleen (Pons et al. 2014).

Evaluation

- Enquire about clinical infection (sinopulmonary, otitis media, enterocolitis)

- Check immunoglobulins, abdominal ultrasound for spleen presence/hypoplasia and full blood count to check for features of hyposplenism (Howell–Jolly bodies on blood film).

Management

- If asplenia, abnormal immunoglobins or evidence of hyposplenism, seek immunologist opinion on optimal management.
- Normal immunization protocol
- Consider a trial of immunoglobulin therapy in the presence of serious recurrent infection.

ACKNOWLEDGMENTS

We would like to thank our colleagues Professor Michel Goossens and Dr. Irina Giurgea, Hôpital Henri Mondor, Créteil, France, and Dr. Margaret Adam, Seattle Children's Hospital. We would particularly like to thank the families and the Mowat–Wilson Foundation for the information they have shared with us.

RESOURCES

Parent/Carer Support Groups

Registry – Mowat–Wilson Syndrome Foundation
https://mowat-wilson.org/registry/

USA

Mowat–Wilson Syndrome Foundation
http://www.mowat-wilson.org/
Mowat–Wilson Syndrome Support Group
http://www.mowatwilson.org/
National Organization for Rare Disorders NORD
https://rarediseases.org/rare-diseases/mowat-wilson-syndrome/

UK

Mowat–Wilson Syndrome UK
http://www.mowatwilsonsyndrome.org.uk/

Italy

Mowat–Wilson Syndrome Italy
http://www.mowatwilson.it/

Australia

MOWILSI
http://www.mowatwilsonsupport.org/

The Netherlands/Belgium

https://www.facebook.com/groups/Mowatwilsonnederlandbelgie

Open Facebook groups

Mowat–Wilson Syndrome Community
https://www.facebook.com/groups/mowatwilsonsyndromecommunity/
Mowat–Wilson Syndrome Foundation
https://www.facebook.com/mowatwilsonsyndromefoundation/
http://www.newday.com/films/TheKeyOfG.html
THE KEY OF G is an award-winning documentary about a young man with Mowat–Wilson syndrome.

Other Resources

Genetics Home Reference

http://ghr.nlm.nih.gov/condition=mowatwilsonsyndrome
Genetics Home Reference is the National Library of Medicine's web site for consumer information about genetic conditions and the genes or chromosomes related to those conditions.

GeneReviews

https://www.ncbi.nlm.nih.gov/books/NBK1412/
A free on-line review of the clinical, testing and molecular aspects of Mowat–Wilson syndrome.

REFERENCES

Adam MP, Conta J, Bean LJH (2013) *Mowat-Wilson syndrome. GeneReviews* [http://www.genetests.org].

Adam M, Schelley S, Gallagher R, Brady A, Barr K, Nlumberg B, Shieh J, Graham J, Slavotinek A, Martin M, Keppler-Noreuil K, Storm A, Hudgins L (2006b) Clinical features and management issues in Mowat-Wilson syndrome. *Am J Med Genet* 140A: 2730–2741.

Adam M, Justice A, Bean L, Fernhoff P (2008) Mowat-Wilson syndrome with craniosynostosis. *Am J Med Genet* 146A: 245–246.

Amiel J, Espinosa-Parrilla Y, Steffann J, Gosset P, Pelet A, Prieur M, Boute O, Choiset A, Lacombe D, Philip N, Le Merrer M, Tanaka H, Till M, Touraine R, Toutain A, Vekemans M, Munnich A, Lyonnet S (2001) Large-scale deletions and *SMA-DIP1* truncating mutations in syndromic Hirschsprung disease with involvement of midline structures. *Am J Hum Genet* 69:1370–1377.

Babkina N, Deignan JL, Lee H, Vilain E, Sankar R, Giurgea I, Mowat D, Graham JM Jr (2016) Early infantile epileptic encephalopathy with a de novo variant in ZEB2 identified by exome sequencing. *Eur J Med Genet* 59(2):70–4.

Besterman AD, Hendren RL (2015) Psychopharmacological management of problem behaviors in Mowat-Wilson syndrome. *J Child Adolesc Psychopharmacol* 25(8):656–7.

Bonanni P, Negrin S, Volzone A, Zanotta N, Epifanio R, Zucca C, Osanni E, Petacchi E, Fabbro F (2017) Electrical status epilepticus during sleep in Mowat-Wilson syndrome. *Brain Dev* 39(9):727–734.

Bonnard A, Zeidan S, Degas V, Viala J, Baumann C, Berrebi D, Perrusson O, El Ghoneimi A (2009) Outcomes of Hirschsprung's disease associated with Mowat-Wilson syndrome. *J Pediatr Surg* 44(3):587–591.

Bourchany A, Giurgea I, Thevenon J, Goldenberg A, Morin G, Bremond-Gignac D, Paillot C, Lafontaine PO, Thouvenin D, Massy J, Duncombe A, Thauvin-Robinet C, Masurel-Paulet A, Chehadeh SE, Huet F, Bron A, Creuzot-Garcher C, Lyonnet S, Faivre L (2005) Clinical spectrum of eye malformations in four patients with Mowat-Wilson syndrome. *Am J Med Genet A* 167(7):1587–92.

Brooks A, Breuning M, Osinga J, vd Smagt J, Catsman C, Buys C, Meijers C, Hofstra M (1999) A consanguineous family with Hirschsprung disease, microcephaly, and mental retardation (Goldberg-Shprintzen syndrome). *J Med Genet* 36: 485–489.

Brooks AS, Bertoli-Avella AM, Burzynski GM, Breedveld GJ, Osinga J, Boven LG, Hurst JA, Mancini GM, Lequin MH, de Coo RF, Matera I, de Graaff E, Meijers C, Willems PJ, Tibboel D, Oostra BA, Hofstra RM (2005) Homozygous nonsense mutations in KIAA1279 are associated with malformations of the central and enteric nervous systems. *Am J Hum Genet* 77:120–126.

Cacheux V, Dastot-Le Moal F, Kaariainen H, Bondurand N, Rintala R, Boissier B, Wilson M, Mowat D, Goossens M (2001) Loss-of-function mutations in SIP1 Smad interacting protein 1 result in a syndromic Hirschsprung disease. *Hum Mol Genet* 10:1503–1510.

Cerruti-Mainardi P, Garavelli L, Pastore G, Virdis R, Pedori S, Godi M, Bernasconi S, Neri G (2005) Mowat-Wilson syndrome and mutation in the Zinc Finger Homeobox 1B Gene: A new syndrome probably under-diagnosed. *Ital J Pediatr* 31: 116–125.

Cordelli DM, Garavelli L, Savasta S, Guerra A, Pellicciari A, Giordano L, Bonetti S, Cecconi I, Wischmeijer A, Seri M, Rosato S, Gelmini C, Della Giustina E, Ferrari AR, Zanotta N, Epifanio R, Grioni D, Malbora B, Mammi I, Mari F, Buoni S, Mostardini R, Grosso S, Pantaleoni C, Doz M, Poch-Olivé ML, Rivieri F, Sorge G, Simonte G, Licata F, Tarani L, Terazzi E, Mazzanti L, Cerruti Mainardi P, Boni A, Faravelli F, Grasso M, Bianchi P, Zollino M, Franzoni E (2013) Epilepsy in Mowat-Wilson syndrome: delineation of the electroclinical phenotype. *Am J Med Genet A* 161A(2):273–84.

Cordelli DM, Pellicciari A, Kiriazopulos D, Franzoni E, Garavelli L (2013) Epilepsy in Mowat-Wilson syndrome: is it a matter of GABA? *Epilepsia* 54(7):1331–2.

Coyle D, Puri P (2015) Hirschprung disease in children with Mowat-Wilson disease. *Pediatr Surg Int* 31(8):711–7.

Dastot-Le Moal F, Wilson M, Mowat D, Collot N, Niel F, Goossens M (2007) *ZFHX1B* mutations in patients with Mowat-Wilson syndrome. *Hum Mutat* 28:1–9.

Espinosa-Parilla Y, Encha-Razavi F, Attie-Bitach T, Martinovic J, Morichon-Delvallez N, Munnich A, Vekemans M, Lyonnet S, Amiel J (2004) Molecular screening of the ZFHX1B gene in prenatally diagnosed isolated agenesis of the corpus callosum. *Prenat Diagn* 24(4):298–301.

Evans E, Einfeld S, Mowat D, Taffe J, Tonge B, Wilson M (2012) The behavioural phenotype of Mowat-Wilson syndrome. *Am J Med Genet A* 158A (2):358–66.

Evans E, Mowat D, Wilson M, Einfeld S (2016) Sleep disturbance in Mowat-Wilson syndrome. *Am J Med Genet A* 170(3): 654–60.

Fryer A (1998) Goldberg-Shprintzen syndrome: Report of a new family and review of the literature. *Clin Dysmorphol* 7:97–101.

Garavelli L, Cerruti-Mainardi P (2007) Mowat-Wilson syndrome. *Orphanet J Rare Dis* 2:42.

Garavelli L, Donadio A, Zanacca C, Banchini G, Della Giustina E, Bertani G, Albertini G, Del Rossi C, Rauch A, Zweier C, Zollino M, Neri G (2003) Hirschsprung disease, mental retardation, characteristic facial features, and mutation in the gene *ZFHX1B* (*SIP1*): Confirmation of the Mowat-Wilson syndrome. *Am J Med Genet* 116A:385–388.

Garavelli L, Cerruti-Mainardi P, Virdis R, Pedori S, Pastore G, Godi M, Provera S, Rauch A, Zweier C, Zolino M, Banchini G, Longo N, Mowat D, Neri G, Bernasconi S (2005) Genitourinary anomalies in Mowat-Wilson syndrome with deletion/mutation in the zinc finger homeo box 1B gene (ZFHX1B) Report of three Italian cases with hypospadias and review. *Horm Res* 63:187–192.

Garavelli L, Zollino M, Cerruti Mainardi P, Gurrieri F, Rivieri F, Soli F, Verri R, Albertini E, Favaron E, Zignani M, Orteschi D, Bianchi P, Faravelli F, Forzano F, Seri M, Wischmeijer A, Turchetti D, Pompilii E, Gnoli M, Cocchi G, Mazzanti L, Bergamaschi R, De Brasi D, Sperandeo MP, Mari F, Uliana V, Mostardini R, Cecconi M, Grasso M, Sassi S, Sebastio G, Renieri A, Silengo M, Bernasconi S, Wakamatsu N, Neri G (2009) Mowat–Wilson syndrome: Facial phenotype changing with age: Study of 19 Italian patients and review of the literature. *Am J Med Genet A* 149A:417–426.

Garavelli L, Ivanovski I, Caraffi SG, Santodirocco D, Pollazzon M, Cordelli DM, Abdalla E, Accorsi P, Adam MP, Baldo C, Bayat A, Belligni E, Bonvicini F, Breckpot J, Callewaert B, Cocchi G, Cuturilo G, Devriendt K, Dinulos MB, Djuric O, Epifanio R, Faravelli F, Formisano D, Giordano L, Grasso M, Grønborg S, Iodice A, Iughetti L, Lacombe D, Maggi M, Malbora B, Mammi I, Moutton S, Møller R, Muschke P, Napoli M, Pantaleoni C, Pascarella R, Pellicciari A, Poch-Olive ML, Raviglione F, Rivieri F, Russo C, Savasta S, Scarano G, Selicorni A, Silengo M, Sorge G, Tarani L, Tone LG, Toutain A, Trimouille A, Valera ET, Vergano SS, Zanotta N, Zollino M, Dobyns WB, Paciorkowski AR (2017) Neuroimaging findings in Mowat-Wilson syndrome: a study of 54 patients. *Genet Med* 19(6):691–700.

Ghoumid J, Drevillon L, Alavi-Naini SM, Bondurand N, Rio M, Briand-Suleau A, Nasser M, Goodwin L, Raymond P, Yanicostas C, Goossens M, Lyonnet S, Mowat D, Amiel J, Soussi-Yanicostas N, Giurgea I (2013) ZEB2 zinc finger missense mutations lead to hypomorphic alleles and mild Mowat-Wilson syndrome. *Hum Mol Genet* 22(13):2652–61.

Goldberg RB, Shprintzen RJ (1981) Hirschsprung megacolon and cleft palate in two sibs. *J Craniofac Genet Dev Biol* 1:185–189.

Gregory-Evans CY, Vieira H, Dalton R, Adams GG, Salt A, Gregory-Evans K (2004) Ocular coloboma and high myopia with Hirschsprung disease associated with a novel ZFHX1B missense mutation and trisomy 21. *Am J Med Genet* 131A:86–90.

Haricharan RN, Georgeson KE (2008) Hirschsprung disease. *Semin Pediatr Surg* 17:266–275.

Horn D, Weschke B, Zweier C, Rauch A (2004) Facial phenotype allows diagnosis of Mowat-Wilson syndrome in the absence of Hirschsprung disease. *Am J Med Genet* 124A:102–104.

Hurst JA, Markiewicz M, Kumar D, Brett EM (1988) Unknown syndrome. Hirschsprung's disease, microcephaly, and iris coloboma: A new syndrome of defective neuronal migration. *J Med Genet* 25:494–500.

Ishihara N, Shimada A, Kato J, Niimi N, Tanaka S, Miura K, Suzuki T, Wakamatsu N, Nagaya M (2005) Variation in aganglionic segment length of the enteric neural plexus in Mowat-Wilson Syndrome. *J Pediatr Surg* 40:1411–1419.

Ivanovski I, Djuric O, Caraffi SG, Santodirocco D, Pollazzon M, Rosato S, Cordelli DM, Abdalla E, Accorsi P, Adam MP, Ajmone P, Badura-Stronka M, Baldo C, Baldi M., Bayat A, Bigoni S, Bonvicini F, Breckpot J, Callewaert B, Cocchi G, Cuturilo G, De Brasi D, Devriendt K, Dinulos MB, Hjortshøj TD, Epifanio R, Faravelli F, Fiumara A, Formisano D, Giordano L, Grasso M, Grønborg S, Iodice A, Iughetti L, Kuburovic V, Kutkowska-Kazmierczak A, Lacombe D, Lo Rizzo C, Luchetti A, Malbora B, Mammi I, Mari F, Montorsi G, Mouton S, Møller R, Muschke P, Klint Nielsen JE, Obersztyn E, Pantaleoni H, Pellicciari A, Pisanti MA, Prpic I, Poch Olive ML, Raviglione F, Renieri A, Ricci E, Rivieri F, Santen GW, Savasta S, Scarano G, Schanze I, Selicorni A, Silengo M, Smigiel R, Spaccini L, Sorge G, Szczaluba K, Tarani L, Gonzaga Tone L, Toutain A, Trimouille A, Valera ET, Schrier Vergano S, Zanotta N, Zenker M, Conidi A, Zollino M, Rauch A, Zweier C, Garavelli L (2018) Phenotype and genotype of 87 patients with Mowat-Wilson syndrome and recommendations for care. *Genet Med* 20(9):965–975.

Long J, Zuo D, Park M (2005) Pc2-mediated sumoylation of Smad-interacting protein 1 attenuates transcriptional repression of E-cadherin. *J Biol Chem* 41:387–393.

Marangi G, Zollino M (2015) Pitt-Hopkins syndrome and differential diagnosis: a molecular and clinical challenge. *J Pediatr Genet* 4(3):168–76.

McGaughran J, Sinnott S, Moal FD, Wilson M, Mowat D, Sutton B, Goossens M (2005) Recurrence of Mowat-Wilson syndrome in siblings with the same proven mutation. *Am J Med Genet* 137A:302–304.

Mowat DR, Croaker GD, Cass DT, Kerr BA, Chaitow J, Ades LC, Chia NL, Wilson MJ (1998) Hirschsprung disease, microcephaly, mental retardation, and characteristic facial features: Delineation of a new syndrome and identification of a locus at chromosome 2q22-q23. *J Med Genet* 35:617–623.

Mowat DR, Wilson MJ, Goossens M (2003) Mowat-Wilson syndrome. *J Med Genet* 40:305–310.

Niemczyk J, Einfeld S, Mowat D, Equit M, Wagner C, Curfs L, von Gontard A (2017) Incontinence and psychological symptoms in individuals with Mowat-Wilson syndrome. *Res Dev Disabil* 62:230–237.

Ohtsuka M, Oguni H, Ito Y, Nakayama T, Matsuo M, Osawa M, Saito K, Yamada Y, Wakamatsu N (2008) Mowat-Wilson syndrome affecting 3 siblings. *J Child Neurol* 23:274–278.

Omilusik KD, Best JA, Yu B, Goossens S, Weidemann A, Nguyen JV, Seuntjens E, Stryjewska A, Zweier C, Roychoudhuri R, Gattinoni L, Bird LM, Higashi Y, Kondoh H, Huylebroeck D, Haigh J, Goldrath AW (2008) Transcriptional repressor ZEB2 promotes terminal differentiation of CD8+ effector and memory T cell populations during infection. *J Exp Med* 212(12):2027–39.

Parisi M (2015) Hirschsprung Disease Overview. *GeneReviews* [http://www.genetests.org].

Pons L, Dupuis-Girod S, Cordier MP, Edery P, Rossi M (2014) ZEB2, a new candidate gene for asplenia. *Orphanet J Rare Dis* 9:2

Pradier B, Jeub M, Markert A, Mauer D, Tolksdorf K, Van de Putte T, Seuntjens E, Gailus-Durner V, Fuchs H, Hrabě de Angelis M, Huylebroeck D, Beck H, Zimmer A, Rácz I (2014) Smad-interacting protein 1 affects acute and tonic and not chronic pain. *Eur J Pain* 18(2):249–57.

Sasso A, Paucić-Kirinćić E, Kamber-Makek S, Sindicić N, Brajnovićc-Zaputovićc S, Brajenović-Milić B (2008) Mowat–Wilson syndrome: the clinical report with the novel mutation in ZFHX1B (exon 8: c.2372del C; p.T791fsX816). *Childs Nerv Syst* 4(5):615–618.

Saunders CJ, Zhao W, Ardinger HH (2009) Comprehensive ZEB2 gene analysis for Mowat-Wilson syndrome in a North American cohort: a suggested approach to molecular diagnostics. *Am J Med Genet A* 149A(11):2527–2531.

Sharar E, Shinawi MD (2003) Neurocristopathies presenting with neurologic abnormalities associated with Hirschsprung's disease. *Pediatr Neurol* 28:385–391.

Silengo M, Ferrero GB, Tornetta L, Cortese MG, Canavese F, D'Alonzo G, Papalia F (2003) Pachygyria and cerebellar hypoplasia in Goldberg-Shprintzen syndrome. *Am J Med Genet* 118A:388–390.

Tanaka H, Ito J, Cho K, Mikawa M (1993) Hirschsprung disease, unusual face, mental retardation, epilepsy and congenital heart disease: Goldberg-Shprintzen syndrome. *Paediatr Neurol* 9:233–238.

Valera ET, Ferraz ST, Brassesco MS, Zhen X, Shen Y, dos Santos AC, Neder L, Oliveira RS, Scrideli CA, Tone LG (2013) Mowat-Wilson syndrome: the first report of an association with Central nervous tumours. *Childs Nerv Syst* 29(12):2151–5.

Verstappen G, van Grunsven Michiels C, van de Putte T, Souopgui J, Van Damme J, Bellefroid E, Vandetkerckhove J, Huylebroeck D (2008) Atypical Mowat-Wilson patient confirms the importance of the novel association between ZFHX1B/SIP1 and NuRD co-repressor complex. *Hum Mol Genet* 17(8):1175–1183.

Wakamatsu N, Yamada Y, Yamada K, Ono T, Nomura N, Taniguchi H, Kitoh H, Mutoh N, Yamanaka T, Mushiake K,

Kato K, Sonta S, Nagaya M (2001) Mutations in SIP1, encoding Smad interacting protein-1, cause a form of Hirschsprung disease. *Nat Genet* 27:369–370.

Wenger TL, Harr M, Ricciardi S, Bhoj E, Santani A, Adam MP, Barnett SS, Ganetzky R, McDonald-McGinn DM, Battaglia D, Bigoni S, Selicorni A, Sorge G, Monica MD, Mari F, Andreucci E, Romano S, Cocchi G, Savasta S, Malbora B, Marangi G, Garavelli L, Zollino M, Zackai EH (2015) CHARGE-like presentation, craniosynostosis and mild Mowat-Wilson syndrome diagnosed by recognition of the distinctive facial gestalt in a cohort of 28 new cases. *Am J Med Genet A* 167(7):1682–3.

Wilson M, Mowat D, Dastot-Le Moal F, Cacheux V, Kaariainen H, Cass D, Donnai D, Clayton-Smith J, Townshend S, Curry C, Gattas M, Braddock S, Kerr B, Aftimos S, Zehnwirth H, Barrey C, Goossens M (2003) Further delineation of the phenotype associated with heterozygous mutations in *ZFHX1B*. *Am J Med Genet* 119A:257–265.

Wilson M, Mowat D, Goossens M (2007). ZFHX1B (SIP1) and Mowat-Wilson Syndrome. In: *Inborn Errors of Development: The Molecular Basis of Clinical Disorders of Morphogenesis*, 2nd ed. *Ch. 41*. Oxford University Press.

Yamada Y, Nomura N, Yamada K, Matsuo M, Suzuki Y, Sameshima K, Kimura R, Yamamoto Y, Fukushi D, Fukuhara Y, Ishihara N, Nishi E, Imataka G, Suzumura H, Hamano S, Shimizu K, Iwakoshi M, Ohama K, Ohta A, Wakamoto H, Kajita M, Miura K, Yokochi K, Kosaki K, Kuroda T, Kosaki R, Hiraki Y, Saito K, Mizuno S, Kurosawa K, Okamoto N, Wakamatsu N (2014) The spectrum of ZEB2 mutations causing the Mowat-Wilson syndrome in Japanese populations. *Am J Med Genet A* 164A(8):1899–908. Erratum in: *Am J Med Genet A*. 2015 167(6):1428.

Yomo A, Taira T, Kondo I (1991) Goldberg Shprintzen syndrome: Hirschsprung disease, hypotonia, and ptosis in siblings. *Am J Med Genet* 41:188–191.

Yoneda M, Fujita T, Yamada Y, Yamada K, Fujii A, Inagaki T, Nakagawa H, Shimada A, Kishikawa M, Nagaya M, Azuma T, Kuriyama M, Wakamatsu N (2002) Late infantile Hirschsprung disease-mental retardation syndrome with a 3-bp deletion in ZFHX1B. *Neurology* 59:1637–1640.

Zweier C, Albrecht B, Mitulla B, Behrens R, Beese M, Gillessen-Kaesbach G, Rott HD, Rauch A (2002) "Mowat-Wilson" syndrome with and without Hirschsprung disease is a distinct, recognizable multiple congenital anomalies-mental retardation syndrome caused by mutations in the zinc finger homeo box 1B gene (ZFHX1B). *Am J Med Genet* 108:177–181.

Zweier C, Temple IK, Beemer F, Zackai E, Lerman-Sagie T, Weschke B., Anderson CE, Rauch A (2003) Characterisation of deletions of the ZFHX1B region and genotype-phenotype analysis in Mowat-Wilson syndrome. *J Med Genet* 40:601–605.

Zweier C, Thiel CT, Dufke A, Crow YJ, Meinecke P, Suri M, Ala-Mello S, Beemer F, Bernasconi S, Bianchi P, Bier A, Devriendt K, Dimitrov B, Firth H, Gallagher RC, Garavelli L, Gillessen-Kaesbach G, Hudgins L, Kaariainen H, Karstens S, Krantz I, Mannhardt A, Medne L, Mucke J, Kibaek M, Krogh LN, Peippo M, Rittinger O, Schulz S, Schelley SL, Temple IK, Dennis NR, Van der Knaap MS, Wheeler P, Yerushalmi B, Zenker M, Seidel H, Lachmeijer A, Prescott T, Kraus C, Lowry RB, Rauch A (2005) Clinical and mutational spectrum of Mowat-Wilson syndrome. *Eur J Med Genet* 48:97–111.

Zweier C, Horn D, Kraus C, Rauch A (2006) Atypical ZFHX1B mutation associated with a mild Mowat-Wilson syndrome phenotype. *Am J Med Genet* 140A:869–872.

39

MYOTONIC DYSTROPHY TYPE 1

Isis B.T. Joosten
Department of Neurology, Maastricht University Medical Center, Maastricht, The Netherlands

Kees Okkersen
Department of Neurology, Radboud University Medical Center, Nijmegen, The Netherlands

Baziel G.M. van Engelen
Department of Neurology, Radboud University Medical Center, Nijmegen, The Netherlands

Catharina G. Faber
Department of Neurology, Maastricht University Medical Center, Maastricht, The Netherlands

INTRODUCTION

Main characteristics of myotonic dystrophy type 1 (DM1; also known as Steinert disease) include progressive myotonia and muscular weakness, in combination with multisystem involvement. The disorder was first described in 1909 by Steinert, while the corresponding gene defect was discovered in 1992 (Brook et al. 1992). In DM1, muscle weakness is most prominent in facial and neck musculature and distal limbs. Even though myotonia is a relevant distinguishing manifestation, individuals with DM1 often do not complain about this symptom. In 1994, a separate disorder with myotonia and more proximal weakness was identified, later acknowledged as myotonic dystrophy type 2 (DM2) (Ricker et al. 1994). This chapter focusses on DM1, while DM2 will briefly be discussed (see Differential Diagnosis).

Incidence

DM1 is the most common inherited muscular dystrophy in adults. Before the identification of specific genetic mutations, the combined prevalence of DM1 and DM2 was estimated at 1/8000 (12.5/100,000). Thereafter, several population-based studies on DM1 have shown that prevalence greatly varies with the population under study (Table 39.1). High prevalence, such as found in the Saguenay region of Canada, is probably the result of founder effect and population isolation. Low prevalence is reported in sub-Saharan Africa and Asian regions, with DM1 being found in only 0.46/100.000 in Taiwan (Goldman, Ramsay and Jenkins 1994; Hsiao et al. 2003).

Diagnostic Criteria

The diagnosis is based on clinical manifestations, family history, and ultimately, genetic testing (see Diagnostic Testing).

DM1 Subtypes Because of the extreme phenotypic variability of DM1, attempts have been made to stratify patients into groups with similar clinical characteristics. A frequently used approach is classification into groups based on age of onset (see Table 39.2) (Harper 2004). Although there is substantial overlap, classification into different subgroups allows for some organization of clinical variability. It should be noted, however, that subtypes are

TABLE 39.1 Data on prevalence of myotonic dystrophy type 1 (DM1)

Authors	Country	Prevalence[a]
Mathieu et al. (1990)	Canada	189
MacMillan (1991)	Wales	7.1
Burcet et al. (1992)	Spain	11
Lopez de Munain et al. (1993)	Spain	26.5
Hughes et al. (1996)	Northern Ireland	8.4
Medica et al. (1997)	Croatia	18.4
Siciliano et al. (2001)	Italy	9.3
Mladenovic et al. (2006)	Belgrade	5.3
Suominen et al. (2011)	Finland	54.6
Lindberg et al. (2017)	Sweden	17.8
Lefter et al. (2017)	Republic of Ireland	6.75

[a] per 100,000

not absolute and many different classifications have been used in scientific studies. In general, symptoms of DM1 are more severe in males than in females, with greater morbidity and mortality, and more profound socioeconomic consequences (Dogan et al. 2016).

Congenital Myotonic Dystrophy Type 1. Congenital DM1 is the most severe subtype of DM1 and can become apparent even before birth with symptoms of polyhydramnios, reduced fetal movements or contractures such as clubfeet. After delivery, the neonate shows signs of hypotonia with only few spontaneous movements and facial diplegia. A typical dysmorphic appearance with a tented upper lip and open mouth might be apparent (see Figure 39.1) (Echenne et al. 2008). Usually, bulbar weakness leads to impaired sucking and dysphagia. Pulmonary hypoplasia and weakness of respiratory muscles, including the diaphragm, may cause respiratory insufficiency (see Table 39.3). In combination with aspiration, respiratory difficulties can lead to lethal complications. Most infants survive the critical neonatal period with intensive support. Still, overall mortality rates

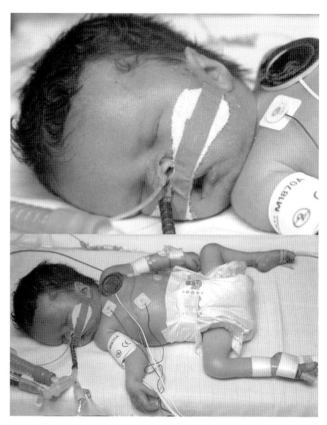

FIGURE 39.1 A child affected by congenital myotonic dystrophy type 1 with generalized hypotonia and dysmorphic features such as a tented upper lip and open mouth.

TABLE 39.2 Myotonic dystrophy type 1 disease subtypes

Clinical phenotype and age of onset	Early manifestations	Later manifestations	CTG repeat length
Congenital <1 year	Hypotonia; respiratory deficits; intellectual disability; orthopedic complications	Gastrointestinal disorders; bulbar weakness; adult DM1-like involvement	>1000
Childhood 1 year–10 years	Intellectual disability; behavioral problems; gastrointestinal disorders	Adult DM1-like involvement	100–1000
Juvenile/early-adult >10 to <20 years	Muscular weakness; myotonia	Severe muscular weakness Apathy; excessive daytime sleepiness; fatigue	100–1000
Adult >20–40 years	Gastrointestinal disorders		
Late-onset >40 years	Cataract	Myotonia; mild muscular weakness	50–150
Protomutation	Mostly asymptomatic, but can be symptomatic especially later in life	Risk of transmitting expanded allele to offspring	51–80
Premutation	Mostly asymptomatic.	Risk of transmitting expanded allele to offspring	38–50

TABLE 39.3 Manifestations of congenital myotonic dystrophy type 1

Brain	Delay in mental development
	Intellectual impairment [IQ between 40 and 69 (median 53.6)]
Skeletal muscle	Floppy at birth with possible facial paralysis
	Congenital contractures
	Motor development delayed
	Slowly progressive weakness from about 10 years
	Occasional moderate kyphoscoliosis
	Myotonia starting in late childhood or adolescence
Lungs and respiration	Pulmonary hypoplasia
	Neonatal respiratory insufficiency
Heart	Arrhythmia occasionally already in childhood
	Cardiac conduction defects later on in life
Oral cavity, pharynx	Narrow high-arched palate
Jaws and teeth	Dysphagia
	Poor sucking force
	Dysarthria
	Malocclusion
	Weakened bite force
Gastrointestinal	Abdominal pain
	Delay in gastric emptying
	Diarrhea, constipation
	Pseudo-obstruction
Natural history	Systemic manifestations as in adult type

TABLE 39.4 Manifestations of the childhood type of myotonic dystrophy type 1

Development	Intellectual impairment (IQ in low normal range or mildly impaired)
Skeletal muscles	Facial muscle weakness
	Slowly progressive weakness from about 10 years
	Myotonia from about 10 years
Heart	Conduction abnormalities
	Arrhythmia in a few children
Oral cavity and jaw	Dysarthria
	Dysphagia
Gastrointestinal	Abdominal pain
	Diarrhea
	Constipation
Behavior	Attention deficit/hyperactivity in ≤50% of children
Natural history	Systemic manifestations as in adult type from about 10 years

range from approximately 15–20%, reaching up to 40% in severely affected cases (Volpe 2008).

After the neonatal period, gradual improvement of motor function is to be expected. Although cognitive and motor milestones are delayed, almost all children will be able to walk (Echenne et al. 2008). Progressive muscle weakness will start becoming apparent after the first decade. Similarly, myotonia is absent at first and usually develops after the age of 10. Neuromuscular involvement may progress more rapidly than in other DM1 subtypes. Cardiorespiratory problems, including cardiac rhythm disturbances, have been described in children. Abdominal pain with constipation or diarrhea is a frequent complaint.

Apart from somatic expressions of disease, mental deficiency is present with full scale IQ ranging from 40 to 69 (median 53.6) (Echenne et al. 2008). Despite cognitive challenges, most children will learn to speak and eventually learn to take care of themselves through special education. Still, congenital DM1 is associated with a poor prospect on physical and intellectual independence, and a significantly shortened lifespan.

Childhood Myotonic Dystrophy Type 1. The childhood type has an onset of DM1-related symptoms between the ages of 1 and 10 years (see Table 39.2) (de Die-Smulders 2004). Clinically, this specific subtype forms a continuum between congenital DM1 and the juvenile or early adult-onset form (see Table 39.4). Individuals with childhood DM1 and an early onset of symptoms often demonstrate significant overlap with the congenital DM1 phenotype, as described.

Interestingly, childhood DM1 usually does not present at first with myopathic signs. On the contrary, the first symptoms of the disease frequently consist of speech and language difficulties or mental delay. A diagnosis is often reached after careful examination of family members or systematic investigation of developmental delay.

As affected children grow older, DM1 adult characteristics such as facial, bulbar and distal muscle weakness, and myotonia evolve. Most patients with childhood subtype will be more severely affected than patients with adult-onset DM1. In some cases, affected individuals even become wheelchair dependent when reaching an age above 40. Also, multisystem involvement is often more prominent than in adult DM1. Gastrointestinal complaints are common and serious heart involvement can also exist (Echenne et al. 2008).

As can be expected from initial mental delay and language difficulties, intelligence is usually below average and ranges from borderline impaired (full scale IQ 70–85) to moderately mentally retarded (full scale IQ <55). Studies have shown a negative correlation between CTG repeat length and intelligence quotient (Angeard et al. 2007). Nevertheless, normal IQ is also possible.

As in congenital DM1, learning and communicative difficulties are often of greatest impact (Johnson et al. 2016). They may be accompanied by social and behavioral problems such as attention-deficit hyperactivity disorder.

Adult Onset Myotonic Dystrophy Type 1 The most characteristic features of adult DM1 are of muscular origin (see Table 39.5). They exhibit distal muscle weakness and atrophy in the extremities, with the possibility of developing proximal muscle weakness over time (see Table 39.6). Ptosis and temporal muscle wasting usually become apparent at an early stage. In addition, neck flexor and facial weakness are prominent, giving patients a characteristic myopathic appearance (see Figure 39.2). Bulbar muscle weakness can give rise to disabling dysarthria and dysphagia.

A second key feature of adult-onset DM1 consists of myotonia, which is expressed as an elongated hand-grip relaxation time and percussion myotonia on neurological examination. Still, this characteristic symptom is not particularly troublesome for most individuals. As a result, they may not bring up these complaints spontaneously. Myotonia may also be present in bulbar muscles, causing difficulties in chewing or swallowing.

Apart from muscular symptoms, individuals with adult DM1 often demonstrate marked multisystem involvement.

TABLE 39.5 Systemic manifestations of adult type myotonic dystrophy type 1

Skeletal muscles	Slowly progressive distal muscle weakness, may evolve to proximal weakness
	Myotonia
Brain	Apathy
	Increased need for sleep
	Fatigue
	Cognitive impairment
	Behavioral disorders
Oral cavity and pharynx	Dysarthria and dysphagia
Jaws and teeth	Structural craniofacial abnormalities
	Decreased oral hygiene (caries, gingivitis and plaque)
	Malocclusion
Gastrointestinal	Dysphagia
	Gastroesophageal reflux
	Abdominal pain
	Obstipation
	Fecal incontinence
Heart	Conduction defects
	Arrhythmia
	Dilated cardiomyopathy
Lungs and respiration	Difficulty in coughing
	Aspiration
	Pneumonia
	Insufficient respiratory drive
	Central sleep apnea or obstructive sleep apnea syndrome
	Respiratory insufficiency
Eyes and ears	Cataract
	Decrease of high tone perception
Endocrine system	Testicular atrophy
	Decreased fertility in early-onset type (in males)
	Hypotestosteronism
	Insulin resistance, sometimes mild type 2 diabetes
Urogenital	Difficulty with micturition
Pregnancy and delivery	Polyhydramnios when fetus is affected
	Increased risk of premature labor
	Prolonged labor
	Abnormal presentation of the fetus
	Neonatal distress
Skin	Baldness (early alopecia androgenetica)
	Pilomatrixomata

TABLE 39.6 Muscular impairment rating scale (MIRS)

Grade I	No muscular impairment.
Grade II	Minimal signs, without distal weakness. Minimal signs may include digit flexor weakness, ptosis, temporal muscle wasting, neck flexor weakness, myotonia or nasal speech.
Grade III	Distal muscle weakness.
Grade IV	Distal muscle weakness in combination with slight to moderate proximal weakness.
Grade V	Severe proximal weakness.

Source: Mathieu et al. 2001.

FIGURE 39.2 A 36-year-old woman with typical facial features of adult-onset myotonic dystrophy type 1: long face, temporal muscle atrophy, open mouth and mild ptosis.

In particular, early-onset cataracts are a hallmark of the disease. Severe fatigue and daytime sleepiness are also highly prevalent and result in great disease burden.

Due to muscular weakness and decreased respiratory drive, which is associated with central nervous system (CNS) involvement, nocturnal hypoventilation is frequently present. Dysfunction of the CNS can also affect cognitive domains in variable combinations, e.g., visuospatial and executive functions (Okkersen et al. 2017). Together with behavioral disorders, individuals with DM1 mostly show signs of apathy and personality traits such as avoidance and rigidity. Both cognitive deficits and behavioral disturbances are important determinants of quality of life and participation.

Moreover, a substantial component of DM1 as a multisystemic disorder consists of gastrointestinal dysfunction. Many find gastrointestinal complaints to be the most burdensome manifestation of disease. Complaints include constipation and/or diarrhea, dysphagia, gastro-esophageal reflux and overweight/obesity. Furthermore, numerous individuals are diagnosed with cardiac abnormalities, which significantly contribute to classical DM1 morbidity and mortality. Survival studies have shown that most affected individuals only reach an age between 45 and 65 years. Main causes of death are cardiac conduction disorders and pneumonia (de Die-Smulders et al. 1998).

Late-onset Myotonic Dystrophy Type 1. Late-onset DM1 presents with clinical symptoms at the age of 40 years or older. As a rule, these individuals have a relatively small CTG repeat expansion. In typical cases, visual loss caused by cataract is the first symptom. In the course of disease, mild or subclinical myotonia can develop and mild muscular weakness appears. Excessive sleepiness is only rarely present (Arsenault et al. 2006). For these patients, a life expectancy within normal range is foreseeable.

Etiology, Pathogenesis, and Genetics

DM1 is caused by a CTG repeat expansion in the 3′ untranslated region of the myotonic dystrophy protein kinase (DMPK) gene, located at chromosome 19q13.32. The number of repeats tends to increase from generation to generation, accounting for genetic anticipation, which is typical for DM1. As repeat length increases in later generations, more severe symptoms develop at an earlier age. As can be seen in Table 39.2, the vast phenotypic variability of the disease cannot solely be accounted for by the variation in CTG repeat expansion size, as there is overlap between subtype repeat sizes.

In healthy individuals, CTG repeat lengths range from 5 to 37. Repeat lengths of 38–50 are considered premutation alleles, whereas 51–80 repeats are called protomutation alleles (Barcelo et al. 1993; Yum et al. 2017). In general, clinically affected individuals have repeat sizes ranging from 50 up to 4000, but many pre- and protomutation allele carriers are asymptomatic or only mildly affected. Nonetheless, classification of asymptomatic or mildly affected pre- and protomutation individuals is of importance, since they are at risk of transmitting a further expanded allele to their offspring.

Furthermore, some individuals with DM1 carry so-called "variant repeats" at the *DM1* locus, comprised of CTG repeats mixed with other DNA sequences. These variant repeats may be associated with either milder or additional symptoms, and seem to have a stabilizing effect on genetic anticipation (Braida et al. 2010). Variant repeats could be of great clinical importance, especially when informing patients on their prognosis and risk to future generations. Nevertheless, diagnostic testing in DM1 does not yet include variant repeat information.

Although CTG repeat expansion is measured in blood for diagnostic purposes, it is important to note that repeat mutations are dynamic and instable with variation in different types of tissues and cells. This causes so-called somatic mosaicism with muscle cell repeat length being noticeably larger than in other cell types (Thornton et al. 1994).

Despite DM1 being a monogenetic disorder, the mechanism by which the CTG repeat expansion causes multisystem manifestations is incompletely understood. Various hypotheses have been proposed in the medical literature, none of which are mutually exclusive. Much support has been reported for the toxic RNA gain-of-function hypothesis: mutant *DMPK* transcripts interact with multiple proteins involved in the cell's protein synthesis machinery, leading to heterogeneous and widespread downstream effects. Details of this model, as well as other hypotheses, are presented in excellent reviews elsewhere (Ashizawa and Sarkar 2011; Thornton 2014).

Diagnostic Testing

Since DM1 manifestations are diverse, patients might initially present with several different symptoms such as myotonia, muscle weakness, cataract, cardiac conduction disorders, hypersomnia or even developmental problems in children. If the clinical phenotype of an individual with DM1 is typical, trained neurologists or pediatric neurologists will be able to recognize the disease on the basis of its main features (see Diagnostic Criteria). In such cases, targeted genetic analysis will rapidly lead to confirmation of diagnosis. If genetic analysis confirms clinical suspicion of the disease, further clinical work-up and genetic counseling are indicated (see Manifestations and Management).

Differential Diagnosis

If typical symptoms are present, the clinical phenotype of DM1 can be easily recognized by a trained eye. However, due to the broad heterogeneity of DM1 and diversity in age of onset, the diagnosis is not always readily established. In such cases, differential diagnosis is dependent upon the patient's age and presenting symptoms and signs. In neonatal cases, the diagnostic work-up is that of a "floppy infant". In childhood cases, DM1 should be in the differential diagnosis of developmental delay, intellectual deficits and behavioral disturbances. In adults, neuromuscular features of DM1 each have a differential diagnosis. An extensive discussion is beyond the scope of this chapter, and overviews of the differential diagnosis of distal muscle weakness, myotonia, and ptosis can be found elsewhere (for example: https://neuromuscular.wustl.edu).

Some of the clinical characteristics of DM1 can also be found in myotonic dystrophy type 2 (DM2). DM2, with various prevalence among countries, is a different multisystemic disorder that features myotonia, muscle weakness and cataract (Ricker et al. 1994). Since muscle weakness is located more proximally than in DM1, the disease is also known as proximal myotonic myopathy (PROMM) or proximal myotonic dystrophy (PDM). DM2 is caused by a CCTG tetranucleotide repeat expansion in the cellular nucleic acid-binding protein (CNBP) gene (Liquori et al. 2001). Features differentiating DM2 from DM1 consist of absent facial and bulbar weakness, prominent tremors, absence of a congenital form, predominant muscle pain, and proximal muscle weakness as the most incapacitating symptom. Genetically, DM2 is dominantly inherited but lacks anticipation. More information on DM2 is available online (for example: https://www.ncbi.nlm.nih.gov/books/NBK1466/).

MANIFESTATIONS AND MANAGEMENT

By its nature, DM1 can lead to severe physical impairment, restricted social participation, and premature death. At present, no curative or disease-modifying treatment is available. Management is based on health supervision focusing on preservation of function, screening for multi-organ involvement and treatment of specific symptoms. This proactive managing approach is likely to reduce morbidity and mortality, especially through early diagnosis of cardiac and pulmonary complications.

As neuromuscular specialists with knowledge of DM1 might be scarce, effective management of individuals with DM1 is sometimes difficult. Management can get even more challenging as several specialists get involved because of the multi-systemic nature of this disorder. Thus, patient management requires a multidisciplinary integrative approach and is often best performed in the setting of a specialized neuromuscular center. It is advisable that care is coordinated by a specialized (pediatric) neurologist or (pediatric) rehabilitation specialist. This "coordinating doctor" should keep an overview of disease extent, social consequences, and involved caregivers. Patients are seen by the coordinating physician at least once a year. During this annual visit, a systematic medical evaluation concerning all possibly affected domains should be conducted. If screening identifies problems in one or more areas, further diagnostics and therapy are initiated. Below, the different domains that may be involved are discussed in more detail. If necessary, complex cases of DM1 may in some countries be referred to a tertiary referral center.

It is important to realize, that as a part of the behavior phenotype, individuals with DM1 can impede adequate disease management. An avoiding attitude towards care can be the result of lack of initiative, apathy, and sleepiness (see Development and Behavior). Since individuals with DM1 typically don't bring up complaints, a proactive approach is required. As a health care provider, it could also help to direct closed-ended questions at the patient and actively involve the partner or caregiver into the consult. Furthermore, sufficient time should be planned for each consultation. Education of the affected patients and relatives might be of help in maintaining consistent medical care. Since family members are commonly affected by the same disease, counseling and support of relatives should be addressed. In addition, attention should be given to the caregiver, who frequently suffers even more from the consequences of DM1 than the individual being provided care.

Individuals with DM1 and their caregivers have to be made well aware of the progressive nature of the disorder and its potentially lethal complications. In particular, they should be instructed to notify any health care provider about having DM1.

Genetic Counseling

At the outset of genetic testing, its purpose should be well defined. Testing can be (1) predictive, (2) prenatal or (3) diagnostic, and the person requesting genetic analysis should be counseled accordingly. As knowledge on DM1 is rapidly evolving, health professionals are encouraged to consult the latest (local) guidelines. In general, we advise an early and low-threshold involvement of a clinical geneticist.

DM1 is inherited in an autosomal dominant manner. Nearly all affected individuals have inherited the expanded CTG allele from a parent with an abnormal CTG repeat expansion (see Etiology, Pathogenesis, and Genetics). Only rarely do expanded alleles originate from a previously normal allele.

Strikingly, many individuals with a mild form of the disorder are unaware that they have DM1. It is not uncommon to find out about the genetic etiology after a mildly affected patient has a child who is more seriously affected, as a result of genetic anticipation. Particularly, this is seen in cases of unexpected congenital DM1 of maternal origin. If both parents of a proband with a proven *DM1* allele are found to have normal CTG repeat sizes, alternate causes may be non-paternity, non-maternity or undisclosed adoption.

Predictive Testing. Predictive testing is performed when family members of a DM1-affected individual have the desire to find out whether they have inherited a similar *DM1* mutation. Pre-test counseling is of high importance, as test results may have legal, ethical, financial, and social consequences.

Also, health care professionals and individuals with DM1 should be aware that a negative test result can exclude DM1 with great certainty, while a positive test result may raise even more questions given the complex genotype-phenotype relationship (see Etiology, Pathogenesis and Genetics).

Usually, predictive testing should only be pursued in adults, as there are significant ethical issues with predictive genetic testing in children. However, recognition of cardiac risks associated with genetic mutations has raised the issue whether predictive testing should, in some cases, be provided for at-risk children (Bassez et al. 2004).

Prenatal Testing. If patients are knowingly affected by DM1 and have the desire to have children, preconception consultation by a clinical geneticist/genetic counselor is advisable. In this consult, the risk of having an affected child and the options to prevent this occurrence should be carefully discussed. Currently, there are two possibilities for prenatal testing. The first option consists of prenatal diagnosis through chorionic villi or cells obtained by amniocentesis. Because this material comprises fetal DNA, CTG repeat size can subsequently be determined. A disadvantage might be that it is quite hard to correlate *DM1* genotype to predicted clinical phenotype. In general, it is known that greater CTG repeat lengths are associated with more severe symptomatology. Still, the extent of disease and DM1 subtype cannot be accurately predicted. There is an increased risk for congenital DM1 if CTG repeat size is 730–1000 or higher (Redman et al. 1993). In such circumstances, the expecting couple may decide to abort pregnancy.

A second option is preimplantation genetic diagnosis (PGD). In this case, fertilization is performed in vitro. Afterwards, one or two cells are biopsied from the eight-cell stage embryos. If the embryo turns out to have a normal CTG repeat length, it can be placed in utero. Per PGD treatment cycle, pregnancy success rate is about 20%. PGD can be experienced as an emotional and physical challenge for woman. In Europe and the US, PGD is available in specific centers only.

Diagnostic Testing. Commercial DNA testing for DM1, using polymerase chain reacting (PCR) and Southern blot, in individuals with an expanded CTG repeat allele is virtually always diagnostic (see Diagnostic Testing). In case of a positive diagnostic test result, adequate counseling with regard to implications for the proband, as well as at-risk family members is essential.

It should be noted that at this time, *DM1* trinucleotide repeat expansions are not covered in next generation sequencing (whole exome sequencing/whole genome sequencing approaches). Thus, if the clinical phenotype of DM1 is not recognized, the gene mutation may be missed if not specifically tested for.

Growth and Feeding

In babies with a congenital subtype of DM1, sucking and swallowing may be impeded as a result of muscular weakness. Feeding through nasogastric tube might be necessary for a prolonged period of time. Also gastroparesis could lead to stasis of food.

In adolescents and adults, feeding can be difficult as a result of several disease-related factors. First, tongue myotonia and bulbar muscle weakness might hinder normal chewing. Thereafter, swallowing of food can be impaired as a consequence of dysphagia. Dysphagia in DM1 is caused by oropharyngeal weakness and possibly also by localized myotonia. As food gets stuck in the pharynx, patients commonly experience coughing during meals and might need to drink sufficient fluids to help the bolus pass. In severe cases, aspiration can occur and cause (recurrent) pneumonia.

Evaluation

- Referral to an ear, nose and throat (ENT) specialist or speech therapist is necessary when extensive dysphagia is present or multiple bouts of aspiration pneumonia have occurred.
- In severe cases, dysphagia can cause insufficient intake and/or weight loss. It is best to refer to a gastroenterologist to evaluate therapeutic options.

Management

- A specialized speech therapist can give practical feeding advice in case of dysphagia. For example, eating has to take place in a calm environment, food should be cut into small pieces and patients should drink enough fluids during and after meals. Also, tongue myotonia gets better after warming up of the muscles through repeated movement.
- A gastroenterologist should evaluate if a percutaneous gastrostomy might be necessary to retain an adequate feeding state in underweight patients.
- If breast or bottle feeding is impossible, feeding through a nasogastric feeding tube could be needed for congenital DM1.

Development and Behavior

In congenital and childhood DM1, central nervous system involvement is often more prominent than neuromuscular involvement. Moderate-to-severe intellectual impairment occurs in a majority of individuals with early-onset DM1. Problems at school caused by speech and language delay are common, and are often a reason for special education. Intellectual dysfunction is usually global, but more specifically affected cognitive domains with relative sparing of global cognition may occur (de Die-Smulders 2004). In addition to cognitive problems, behavioral disorders occur in a majority of patients. The most common are attention deficit/hyperactivity disorder (ADHD), antisocial, aggressive or oppositional behavior. Disorders in the autistic spectrum occur frequently, as well as anxiety and depression. Behavioral disturbances contribute to educational problems, which may also be aggravated by fatigue and excessive daytime sleepiness (see Central Nervous System Involvement) (Douniol et al. 2009).

Evaluation

- Neuropsychological and neuropsychiatric evaluation by specialized pediatric psychologists and/or psychiatrists is indicated.

Management

- Special education and training as appropriate.
- Modafinil can be considered if excessive daytime sleepiness and fatigue impede educational performance or quality of life.
- Attention deficit hyperactivity disorder (ADHD), depression, and anxiety may be treated with similar interventions as unaffected individuals.

Neuromuscular Involvement

Neuromuscular weakness, progressing in a symmetrical distal-to-proximal fashion, and myotonia in the extremities are key manifestations of DM1. Myotonia can be described as the inability to relax muscles following contraction, giving rise to a cramping sensation or stiffness. Patients might report difficulties releasing grip or difficulties with chewing or speech, as myotonia commonly affects the hands and tongue. Myotonia may demonstrate a warm-up phenomenon, that is, become less prominent with repeated performance. Usually, myotonia gradually fades with the progression of muscle weakness.

Bulbar weakness with temporal muscle weakness leads to a distinctive facial appearance (see Figure 39.2), dysarthria, and dysphagia. Individuals with DM1 may seem less interested or emotionless due to decreased facial expression. Ptosis can become severe and compensatory tilting of the head might be necessary. Also, weakness of the sternocleidomastoid and neck flexor muscles is specific for DM1, and patients can have trouble lifting their head. Importantly, muscular weakness may lead to falling and fear of falling. Falls are very common in natural DM1 history and require special attention.

Evaluation

- In each annual visit, explicitly ask for progression of muscle weakness and myotonia, and their impact on function. Perform physical and neurological examination to assess muscular weakness, atrophy, and myotonia.
- Specific sites that show muscle weakness at an early stage, such as neck flexors, finger flexors and stretchers, and foot musculature, should be evaluated. Strength can be tested formally, but also functionally (lifting head when lying down, walking distance, walking on toes and heels, knee bending).
- Check for action and percussion myotonia. Action myotonia can be induced by asking the patient to make a strong fist and suddenly release. Percussion myotonia can be evoked by percussion of the thenar eminence, consequently the thumb might show myotonia induced flexion.
- Evaluate the occurrence, frequency, and causes of falls.
- Evaluate swallowing and speech for dysarthria and dysphagia.

Management

- Refer individuals with muscular weakness interfering with activities of daily living (ADL) to a rehabilitation specialist for orthoses and physical activity guidance.
- Aids and appliances can be helpful when muscular weakness limits the patient in daily life, such as named orthoses or walking aids. Wheelchairs can be considered if muscular weakness is severe.
- Refer to physiotherapist or rehabilitation specialist for fall prevention, if appropriate.
- Myotonia usually does not interfere with ADL and does not require therapy. If severe, mexiletine can be considered (Logigian et al. 2010).
- Individuals should be advised to stay active. If preferred, a physiotherapist can oversee exercise activities and give home training instructions.
- Refer a patient with significant dysarthria and/or dysphagia to a speech therapist (see Gastrointestinal Involvement).

Central Nervous System Involvement

Central nervous system (CNS) manifestations of DM1 occur in all subtypes. Particularly in children, CNS features have more impact on daily functioning than neuromuscular symptoms (see Development and Behavior). In adult-onset DM1, widespread and heterogeneous involvement of the CNS is an important determinant of health status. Cognitive deficits may occur with variable severity and affecting variable combinations of cognitive domains (Okkersen et al. 2017). Behavioral disturbances include apathy and executive functioning disorders. Common personality traits include rigidness and avoidance, and these may interfere with interpersonal relations. Subjectively, fatigue is one of the most common and most debilitating symptoms in DM1. Its etiology is likely multi-factorial. Excessive daytime sleepiness occurs frequently and may be attributable to sleeping problems, sleep-related breathing disorders, and involvement of the brain. Neuroimaging in individuals with DM1 may demonstrate a wide spectrum of both structural and functional anomalies, whereas pathological studies report various findings compatible with a neurodegenerative disorder (Okkersen et al. 2017).

Evaluation

- Neuropsychological evaluation may be considered if an individual exhibits clinical symptoms and signs of CNS involvement.
- If fatigue and daytime sleepiness are present, one should be aware of the possibility of underlying nocturnal breathing disorders.
- Fatigue and daytime sleepiness may be evaluated using the Rasch-built fatigue and daytime sleepiness scale (FDSS) (Hermans et al. 2013) or the subscale "fatigue" of the checklist individual strength (CIS-fatigue).
- The Beck depression inventory – fast screen (BDI-FS) may be used to screen for depression.
- Neuroimaging is not routinely indicated, but can be considered if there are signs of CNS involvement that may be attributed to other causes than DM1.

Management

- If cognitive problems interfere with demands of a job, encourage the patient to seek advice from an occupational health care provider.
- Consider cognitive behavioral therapy (CBT) and/or modafinil for the treatment of fatigue and excessive daytime sleepiness, only after other causes such as affective disorders or nocturnal respiratory problems have been ruled out.
- Consider low-threshold referral to a medical psychologist or social worker if there are problems with acceptance of disease or if a patient and/or partner consider their mutual communication as a problem.

Dental Involvement

Studies have shown that individuals with DM1 have more caries, gingivitis, and plaque compared to healthy individuals (Engvall 2010). This higher frequency of dental complications is assumed to be an effect of slower oral clearance, impaired muscular coordination and diminished self-care ability. Oral hygiene is expected to worsen as muscular involvement becomes more prominent. Furthermore, dental malocclusion and structural craniofacial abnormalities were found in the temporomandibular joint and masticatory muscles (Zanoteli et al. 2002). Facial myotonia and muscle weakness, in combination with dental complications and structural abnormalities, may cause extra difficulties in food ingestion.

Evaluation

- Regular dental check-ups should take place at least every six months.
- In case of malocclusion or impeding craniofacial abnormalities, individuals with DM1 should be referred to an orthodontist or maxillofacial surgeon.

Management

- Professional prophylactic cleaning of the teeth should take place at least once every six months.
- Individuals with DM1 should be provided with oral health education concerning brushing of teeth and flossing. If possible, an electric toothbrush should be used.
- Orthodontic correction of abnormal teeth position should take place in the prepubertal growth phase.
- In case of craniofacial abnormalities, surgery might be necessary. As for all surgeries in individuals with DM1, the benefits and risks should be weighted carefully (see Anesthesia).

Gastrointestinal Involvement

According to individuals with DM1, complaints of the gastrointestinal system are regularly considered the most disabling consequence of this disorder (Bellini et al. 2006; Ronnblom et al. 1996). As a result, these complaints have a big impact of quality of life, and they deserve sufficient attention during consultation. Involvement of the gastrointestinal tract may occur from pharynx to anal sphincter and is thought to be the effect of gastrointestinal muscular impairment. Symptoms are likely to arise in the congenital and childhood subtypes, but are also a regular feature of adult-onset DM1. In over 25%, gastrointestinal problems were present before the diagnosis was established (Ronnblom et al. 1996). Complaints may consist of general abdominal pain, but can also be more specific.

Individuals with DM1 commonly experience dysphagia, worsening over time as muscle weakness progresses. Dysphagia is a serious complication that may lead to weight loss and malnutrition, as well as aspiration and pulmonary complications (see Neuromuscular and Respiratory). Upper gastro-intestinal problems further include gastroesophageal reflux, stasis, bloating, and emesis. Cholelithiasis is more common than in the general population, studies have shown that cholecystectomy was necessary in 16.5% of individuals with DM1 (Hilbert et al. 2017). Intestinal involvement can give rise to several complaints of which episodic diarrhea is considered the most disabling, especially when anal incontinence is also present (Ronnblom et al. 1996). It is known that intestinal bacterial overgrowth can be a cause of diarrhea in individuals with DM1, which can be treated with antibiotics (Tarnopolsky et al. 2010). On the other side of the spectrum, constipation is common, possibly leading to discomfort and pain, and even pseudo-obstruction (Brunner et al. 1992). Finally, overweight and obesity occurs in 40–50% of individuals with DM1, and may detrimentally affect physical and respiratory functioning.

Evaluation

- Perform annual history taking and physical examination for the presence of gastro-intestinal symptoms and signs.
- In case dysphagia causes insufficient intake and/or weight loss or if patients experience prolonged abdominal pain, persistent obstipation, frequent defecation or disabling fecal incontinence, they should be referred to a gastroenterologist.
- See Growth and Feeding for additional information on dysphagia.

Management

- Proton pump inhibitors can be prescribed if gastroesophageal reflux is present.
- If dysphagia becomes severe, a percutaneous gastrostomy might be needed to retain an adequate feeding state.
- In case of symptomatic or complicated cholelithiasis (causing cholecystitis, choledocholithiasis, cholangitis or acute pancreatitis), laparoscopic cholecystectomy is recommended. The benefits and risks of surgery should be weighted carefully (see Anesthesia).
- Diarrhea caused by intestinal bacterial overgrowth can be treated with antibiotics.
- Obstipation in DM1 is treated as any other form of obstipation, using laxatives and/or enemas.
- Referral to a dietician may be considered in case of overweight/obesity.

Respiratory Involvement

Pulmonary involvement in DM1 can be detected even in those with minimal muscular disability, and has a multi-factorial etiology. Impaired coughing and dysphagia can give rise to aspiration of food, saliva or gastric contents (see Gastrointestinal Involvement), with consequent chronic pulmonary inflammation or pneumonia. Pneumonia was found to be the cause of death in about one third of individuals with DM1 (de Die-Smulders et al. 1998).

Neuromuscular weakness may lead to both expiratory and inspiratory muscle weakness and consequently to decreased ventilation. With time, inspiratory capacity declines. In addition, studies have shown upper airway alterations, reduced chest wall compliance, and centrally decreased CO_2 responsivity in individuals with DM1. Furthermore, being overweight negatively affects lung volumes. Together, these factors may lead to alveolar hypoventilation and hypercapnia (CO_2 retention), which was shown to be highly prevalent in individuals with DM1 (Begin et al. 1997). Often the effects of alveolar hypoventilation are more profound during the night, when respiratory drive is reduced and the supine position abrogates the aid of gravity to the diaphragm during inspiration.

In addition to (nocturnal) alveolar hypoventilation, other sleep-related breathing disorders may result from structural airway changes and muscular weakness. Obstructive sleep apnea syndrome (OSAS) and central sleep apnea syndrome (CSAS) were found to be more prevalent in individuals with DM1 than in the general population and can lead to abnormal sleeping patterns (Johnson et al. 2015). In the case of OSAS, recurrent airway collapse will cause cessation of airflow, apneas and possible hypercapnia. Consequently, recurrent arousal might occur and cause daytime sleepiness. Sleep-related breathing disorders may have a negative effect on survival due to cardiovascular and metabolic changes.

Evaluation

- Possible respiratory involvement should be evaluated annually (see Manifestations and Management).
- Irrespective of respiratory symptoms, yearly pulmonary function testing (PFT) is recommended.
- If (recurrent) dysphagia-related aspiration is suspected, refer to ENT specialist or speech therapist (see Gastrointestinal Involvement).
- Refer to a specialized physiotherapist for cough techniques if coughing is impaired.
- Refer to a pulmonologist if there a signs of alveolar hypoventilation, sleep-related breathing disorders, recurrent airway infections or aspiration, increased breathing effort, abnormal blood gas results or abnormal PFT results.

- Avoid medication that could possibly depress respiratory drive, such as benzodiazepine and opiates (see Anesthesia).

Management

- Vaccination for influenza and pneumococcal infections is advisable, as Pneumococcal pneumonia and influenza infections might have serious complications in individuals with DM1.
- If aspiration pneumonia is suspected, individuals should be treated with antibiotic drugs according to local protocol.
- Alveolar hypoventilation may be a reason for non-invasive ventilation at home.
- Consider treatment for sleep-related breathing disorders. If OSAS without hypoventilation is present, continuous positive airway pressure (CPAP) therapy is an option. If there is overlap between OSAS, CSAS and hypoventilation, there may be an indication for non-invasive ventilation.
- CPAP or home ventilation should be explained carefully. It is advisable to regularly check motivation and compliance of therapy.

Cardiovascular Involvement

Cardiac manifestations in DM1 are common and may have serious consequences. Importantly, cardiac involvement is frequent even in asymptomatic patients. Conduction disorders mainly consist of atrioventricular (AV) block and bundle branch block (QRS>120ms), which are possibly secondary to degeneration of the cardiac conduction system. Arrhythmias and conduction disorders occur in 4–28% of individuals with DM1 and are independent of disease severity (Petri et al. 2012). Both arrhythmias and conduction disorders may lead to sudden cardiac death, a major cause of mortality in DM1 (de Die-Smulders et al. 1998).

As noted, DM1 cardiac involvement may go without any clinical symptoms. The absence of symptoms may relate to relatively mild involvement, to physical inactivity of patients and consequent low cardiac demands, or could be the result of underreporting by patients. It is important to realize that pacemaker implantation does not protect against sudden death due to tachyarrhythmias, and individuals at risk for tachyarrhythmias consequently may need implantation of a cardioverter defibrillator (ICD) (Groh et al. 2008; Hermans et al. 2008).

Recommendations for cardiac investigation in adult DM1 were summarized in an international workshop (Bushby et al. 2003). For children, no consensus has been reached, despite recognition that cardiac involvement may be underestimated.

Evaluation

- Annual follow-up by the coordinating doctor should consist of taking cardiac and family history (see below), physical examination and electrocardiogram (ECG). Alternatively, annual cardiac follow-up may be performed by a cardiologist. Special attention should be paid to symptoms and signs of arrhythmias, conduction disorders and heart failure.
- Cardiac evaluation should include possible cardiovascular complaints (dizziness, syncope, palpitations, angina and orthopnea) and family history for sudden death, ventricular fibrillation, sustained ventricular tachycardia or pacemaker implantation.
- If cardiac involvement is suspected, referral to a cardiologist is indicated.
- Holter monitoring should be performed in case of cardiac symptoms or ECG abnormalities. Holter monitoring is indicated every two to five years in asymptomatic patients.
- Echocardiography should be performed at the time of diagnosis and in the presence of manifestations of heart failure or ECG changes. In asymptomatic individuals, routine echocardiogram should be performed every three years. Cardiac MRI can be considered in some cases.
- The exact role of electrophysiological (EF) studies is not established; these may primarily be used to evaluate the conduction system in symptomatic individuals.

Management

- Atrial arrhythmias, such as atrial fibrillation or flutter, are common and antiarrhythmic treatment might be necessary. Important to note is that preexisting tendency to ventricular tachyarrhythmias or bradycardia might worsen by antiarrhythmic medication (Bushby et al. 2003).
- Pacemaker implantation should be considered in each degree of atrioventricular block. In case of a second or third degree AV block or HV interval >70 ms on EF study, pacemaker implantation is necessary.
- An ICD is indicated in case of: (1) a history of ventricular fibrillation or sustained ventricular tachycardia (VT) with hemodynamic instability, (2) sustained VT in combination with structural heart damage, (3) syncope in combination with inducible sustained VT with hemodynamic instability, syncope in combination with a left ventricular ejection fraction <35% or syncope in combination with heart failure NYHA II-III.
- An ICD can be considered in case of: (1) unexplained syncope, (2) significant dysfunction of the left ventricle, (3) sustained VT in combination with normal left

ventricular function and (4) familial cardiomyopathy associated with sudden death, in combination with a left ventricular ejection fraction <35% and heart failure NYHA class I.

Ophthalmologic Involvement

Cataract is a distinguishing and common feature in DM1. In case of late-onset DM1, visual loss due to opaqueness will often be the first symptom of disease. In the remaining subtypes, early-onset cataract is also frequent, although it is rarely seen in young children. Slit-lamp evaluation of the lens can show posterior subcapsular iridescent opacities or cataract, which is often of polychromatic nature (Smith and Gutmann 2016). Due to the highly reflective multicolored corneal crystals in polychromatic cataract, it is also known as "Christmas tree" cataract.

Besides cataract, hypermetropia and strabismus are common in individuals with DM1, especially in children (Bollinger et al. 2008). Vision can further be affected by ptosis, and weakness of the orbicularis oculus muscle may lead to insufficient closure of the eyes. The retina can demonstrate abnormalities of the macula or the retinal periphery.

Evaluation

- In case of glare or blurred vision, individuals should be referred to an ophthalmologist. If lens opacities or cataract are found, annual monitoring is recommended.
- Children with DM1 should receive regular ophthalmic examination, in which special notice should be paid to possible hypermetropia and strabismus.

Management

- If cataract interferes with visual acuity or interferes with daily life, intraocular lens implantation (under local anesthesia) is recommended (Smith and Gutmann 2016).
- Corrective treatment of strabismus and hypermetropia, when present, is not different from the general population.
- If ptosis interferes with vision, surgery may be considered. Nonetheless, surgical treatment should be postponed as long as possible since muscle weakness progresses, and ptosis will recur.
- No treatment is available for DM1 retinopathy.

Dermatologic Involvement

Considering the multisystemic expression of DM1, skin involvement could be expected; however, few studies into dermatological features of DM1 have been conducted. Skin manifestations of DM1 include focal hyperhidrosis, pedunculus fibromas, early androgenic alopecia, and other adnexal abnormalities (Campanati et al. 2015). Also, there is an association between DM1 and pilomatrixomata, a benign tumor of the hair matrix (Geh and Moss 1999). The relationship between DM1 and (pre)neoplastic cutaneous lesions remains unclear.

Evaluation

- Referral to a dermatologist is only necessary if individuals are bothered by dermatological features, such as pilomatrixomata.

Management

- Pilomatrixomata can be surgically removed by a dermatologist or (plastic) surgeon.
- There is no treatment for early androgenic alopecia.

Ears and Hearing

Although there are few studies of DM1 and ears and hearing, excessive sensorineural high-tone hearing loss is more frequent in DM1 compared to healthy individuals of the same age (Pisani et al. 2011; Verhagen et al. 1992).

Evaluation

- Individuals should be referred to an audiologist for evaluation if hearing loss is suspected.
- Since hearing loss might lead to extra difficulties for children with developmental delay or cognitive impairment, threshold for evaluation should be low.

Management

- Treatment consists of a personalized hearing aid, provided by a certified audiologist.

Musculoskeletal Involvement

Orthopedic involvement is most prominent in the congenital and childhood subtypes of DM1. Commonly, impairments consist of foot and spinal deformities, and contractures (Schilling et al. 2013). Congenital contractures probably develop as a result of reduced fetal movements, with clubfeet being the most frequent, but contractures of the hips or shoulders might also arise. Muscle weakness and hypotonia of the trunk make it difficult to maintain an upright position of the spine, which could cause kyphoscoliosis.

Evaluation

- If severe and/or interfering with function, deformities and contractures should be examined by an orthopedic surgeon to explore therapeutic possibilities.

Management

Contractures are treated conservatively at first, if possible. If there is not enough response to conservative treatment, surgery might be necessary.

In case of clubfeet, the conservative treatment should start within three days after birth. Passive stretching through physiotherapy might help to improve the position of the feet. Fixation in plaster or a splint should help maintain the improved position. After six months, the orthopedic surgeon has to evaluate the effect of conservative treatment and decide whether surgery is needed. Important to note is that surgical correction should always take place before the child attempts to walk. The child should be followed-up by an orthopedic surgeon and orthopedic shoemaker until full-grown.

Recommendations on the treatment of spinal deformities differ, as they can either be treated conservatively or surgically. It is best to have an orthopedic surgeon evaluate each specific case.

Endocrine Involvement

Various endocrine disturbances may occur in DM1. Disturbances in the hormonal axes associated with reproduction are the most common. In men with DM1, testicular atrophy occurs in 60–80%. Male fertility is reduced, especially in early adult-onset DM1, and more than a third of men have androgen insufficiency (Orngreen et al. 2012). In women, there is a high rate of reproductive loss both early and late in pregnancy. While late loss of pregnancy is likely the result of problems caused by congenitally affected fetuses, early pregnancy loss may be the result of endocrine disturbances.

Moreover, DM1 is often characterized by insulin resistance. Less frequently, patients may be affected by type 2 diabetes mellitus. Other possible endocrine disorders are (para)thyroid dysfunction and disturbances of the adrenal gland (Orngreen et al. 2012).

Evaluation

- Couples should be referred to a gynecologist or reproductive endocrinologist in case of fertility problems. Investigations into the cause of reduced fertility should follow standard procedures. Genetic counseling is also indicated (see Genetic Counseling).
- Women with DM1 who want to undergo in vitro fertilization, eventually combined with preimplantation genetic diagnosis, should be cautioned about a possible increased risk for anesthetic complications and rhythm disturbances caused by medication used during treatment. Risks and benefits should be discussed carefully (see Genetic Counseling).
- Screening for endocrine dysfunction (PTH, TSH, and glucose) is advisable at first presentation and during yearly follow-up (Orngreen et al. 2012).

Management

- If pregnancy is desired, severe oligozoospermia is an indication for in vitro fertilization with intracytoplasmic sperm injection.
- Diabetes Mellitus type 2 and (para)thyroid dysfunction should be treated according to standard guidelines.

Pregnancy and Delivery

In contrast to the well-known male infertility described in DM1 (see Endocrine Involvement), the issue of female fertility remains controversial (Argov and de Visser 2009).

Nevertheless it is known that DM1 can have several adverse effects on the course of pregnancy and delivery (Argov and de Visser 2009; Johnson et al. 2015). Therefore, counseling women with DM1 that are contemplating pregnancy is imperative. Counseling involves giving information on the risk of having affected offspring (see Genetic Counseling) and discussing the risks of pregnancy for the affected mother. As fallopian tube mobility and uterine function might be affected due to smooth muscle involvement, a higher rate of ectopic pregnancy is suspected, and placenta previa is significantly more frequent. Moreover, there is a markedly increased incidence of pre-term labor with only half of pregnancies reaching full term. It seems that the increased risk for early delivery is related to woman carrying affected offspring, as such was the case in most of the observed incidents. Next to these serious complications, maternal urinary tract infections are more common, possibly caused by subtle pelvic floor weakness (Argov and de Visser 2009). While some studies indicate miscarriages being more frequent in women with DM1, this is still debatable (Argov and de Visser 2009; Johnson et al. 2015). During labor, abnormalities can occur in all three stages as an effect of uterine dysfunction and lack of voluntary assistance. Therefore, vaginal delivery interventions and cesarean sections are more common. Fetal distress is a frequent complication, and perinatal mortality was found to be 16% in a British cohort of women with DM1 (Rudnik-Schöneborn and Zerres 2004). The higher risk of perinatal death and several other complications, such as polyhydramnios and clubfeet, are mainly due to fetuses being affected by congenital DM1 (Argov and de Visser 2009).

Evaluation

- Preconception planning should include obstetric and general history and an assessment of current disease status. Cardiac and respiratory evaluation deserves extra attention.
- Pregnancies should be monitored closely in a high-risk pregnancy clinic by a physician familiar with DM1. Extra attention should be paid to possible complications, as mentioned above.

- DNA testing by means of amniocentesis or chorionic villus sampling (CVS) can be considered (see Genetic Counseling).
- A multidisciplinary approach of pregnancy and delivery is advised. Make a plan for delivery that includes mode, location, anesthesia and fetal presentation (Norwood and Rudnik-Schoneborn 2012).
- Third stage of delivery should be actively managed to prevent postpartum hemorrhage caused by uterine inertia or atony.

Management

- Cesarean or assisted vaginal delivery might be necessary when there are signs of maternal exhaustion or fetal distress.
- Uterine inertia can cause blood loss and usually responds to oxytocin (Rudnik-Schöneborn and Zerres 2004). If blood loss is extensive, blood transfusion and manual placenta evacuation may be required.

Anesthesia

Studies have shown an increased prevalence of perioperative complications in DM1 (Mathieu et al. 1997). Most of the time, complications consist of pulmonary difficulties or cardiac events. Cardiac complications may include rhythm disturbances or even acute death. Pulmonary complications mainly consist of respiratory insufficiency due to hypoventilation, which may lead to an inability to wean patients from the ventilator. Late pulmonary complications include atelectasis, pneumonia, and retained bronchial secretions. Importantly, apathy and behavioral characteristics can cause individuals with DM1 to omit their disease status at pre-operative screening. Due to high prevalence of peri-operative complications, inestimable reactions to succinylcholine (possible hyperthermia), and prolonged and increased sensitivity to analgesics and sedatives, it is absolutely necessary for an anesthesiologist to be aware of the potential operative complications of DM1. Unfortunately, in some patients the diagnosis may not have been established prior to operation and complications cannot be anticipated.

Evaluation

- Recommendations for anesthetic management in patients with DM1 were summarized by The Myotonic Dystrophy Foundation (available at www.myotonic.org).
- Anesthesiologists and surgeons should always be aware that the patient has DM1 before performing surgery. Ideally, pre-operative planning is started early.
- In addition to neurological progression, cardiac, pulmonary, and gastrointestinal involvement should be evaluated prior to surgery. If necessary, other medical specialties should be consulted.
- A preoperative ECG is always indicated.
- Evaluate whether general anesthesia can be avoided by opting for regional anesthesia.
- Prepare the patient for possible prolonged post-anesthesia mechanical ventilation or prolonged ventilatory assistance.

Management

- Try to avoid preoperative sedatives (such as benzodiazepines) and opioids as much as possible.
- Avoid depolarizing neuromuscular blocking agents and anticholinesterases.
- Reduce the usage of opioids as much as possible, and avoid opioids with long half-life.
- Continuously monitor SpO2 and electrical activity of the heart (ECG) for at least 24 hours after surgery or until the patient has fully regained pre-operative status. If analgesics or sedatives are still being used, continue monitoring. A minimum of 48 hours of monitoring should be considered after abdominal surgery or in severely affected patients.
- Beware of late post-operative pulmonary complications such as pneumonia, and treat according to local protocol.

ACKNOWLEDGMENTS

The authors wish to express their gratitude towards Prof.dr. Christine de Die-Smulders and Prof.dr. Frans Jennekens for their contributions to this chapter in previous editions of this book.

RESOURCES

Books

Harper PS (2002a) *Myotonic Dystrophy. The Facts*. Oxford: Oxford University Press.
 Harper PS (2004. *Myotonic Dystrophy Present Management, Future Therapy*. Oxford: Oxford University Press.
 Both very useful and practical books.

Internet

Bird TD. Myotonic Dystrophy type 1, on Genereviews.org. Available at: https://www.ncbi.nlm.nih.gov/books/NBK1165/
 Excellent web-based resource, recently updated with information on clinical aspects, genetics and management of myotonic dystrophy type 1.
 The Myotonic Dystrophy Foundation. Available at: http://www.myotonic.org/

A trustworthy website giving an overview of Myotonic Dystrophy type 1. Also, it provides practical information for patients and healthcare givers.

PATIENT ORGANIZATIONS AND SUPPORT

Muscular Dystrophy Association (MDA)
222 S. Riverside Plaza, Suite 1500
Chicago, Illinois 60606, USA
Phone: (800) 572-1717
Email: *mda@mdausa.org*
Website: *www.mda.org*

Muscular Dystrophy UK
61A Great Suffolk Street
London SE1 0BU, UK
Phone: (+ 44) 020 7803 4800
Email: *info@musculardystrophyuk.org*.
Website: *http://www.musculardystrophyuk.org/*

Myotonic Dystrophy Support Group
19-21 Main Road,
Gedling,
Nottingham NG4 3HQ, UK
Phone: 0115 987 0080
Office: 0115 987 5869
Email: *contact@mdsguk.org*
Website: *http://www.myotonicdystrophysupportgroup.org/*

Other National Organizations

The Netherlands
Spierziekten Nederland
Website: *www.spierziekten.nl*

Germany
Deutsche Gesellschaft für Muskelkranke e.V. (DGM)
Website: *www.dgm.org*

Canada
Muscular Dystrophy Canada
Website: *www.muscle.ca*

United Kingdom
Muscular Dystrophy Campaign
Website: *http://www.musculardystrophyuk.org/*

REFERENCES

Angeard N, Gargiulo M, Jacquette Al, Radvanyi Hln, Eymard B, Héron D (2007) Cognitive profile in childhood myotonic dystrophy type 1: Is there a global impairment? *Neuromuscul Disord* 17(6):451–458.

Argov Z, de Visser M (2009) What we do not know about pregnancy in hereditary neuromuscular disorders. *Neuromuscul Disord* 19(10):675–679.

Arsenault ME, Prevost C, Lescault A, Laberge C, Puymirat J, Mathieu J (2006) Clinical characteristics of myotonic dystrophy type 1 patients with small CTG expansions. *Neurology* 66(8):1248–1250.

Ashizawa T, Sarkar PS (2011) Myotonic dystrophy types 1 and 2. *Handb Clin Neurol* 101 193–237.

Barcelo JM, Mahadevan MS, Tsilfidis C, MacKenzie AE, Korneluk RG (1993) Intergenerational stability of the myotonic dystrophy protomutation. *Hum Mol Genet* 2(6):705–709.

Bassez G, Lazarus A, Desguerre I, Varin J, Laforet P, Becane HM, Meune C, Arne-Bes MC, Ounnoughene Z, Radvanyi H, Eymard B, Duboc D (2004) Severe cardiac arrhythmias in young patients with myotonic dystrophy type 1. *Neurology* 63(10): 1939–1941.

Begin P, Mathieu J, Almirall J, Grassino A (1997) Relationship between chronic hypercapnia and inspiratory-muscle weakness in myotonic dystrophy. *Am J Respir Crit Care Med* 156(1):133–139.

Bellini M, Biagi S, Stasi C, Costa F, Mumolo MG, Ricchiuti A, Marchi S (2006) Gastrointestinal manifestations in myotonic muscular dystrophy. *World J Gastroenterol* 12(12):1821–1828.

Bollinger KE, Kattouf V, Arthur B, Weiss AH, Kivlin J, Kerr N, West CE, Kipp M, Traboulsi EI (2008) Hypermetropia and esotropia in myotonic dystrophy. *J AAPOS* 12(1):69–71.

Braida C, Stefanatos RK, Adam B, et al. (2010) Variant CCG and GGC repeats within the CTG expansion dramatically modify mutational dynamics and likely contribute toward unusual symptoms in some myotonic dystrophy type 1 patients. *Hum Mol Genet* 19(8):1399–1412.

Brook JD, McCurrach ME, Harley HG, Buckler AJ, Church D, Aburatani H, Hunter K, Stanton VP, Thirion JP, Hudson T (1992) Molecular basis of myotonic dystrophy: expansion of a trinucleotide (CTG) repeat at the 3' end of a transcript encoding a protein kinase family member. *Cell* 69(2):385.

Brunner HG, Hamel BC, Rieu P, Höweler CJ, Peters FT (1992) Intestinal pseudo-obstruction in myotonic dystrophy. *J Med Genet* 29(11):791–793.

Burcet J, Cañellas F, Cavaller G, Vich M (1992) Estudio epidemiológico de la distrofia miotónica en la isla de Mallorca. *Neurologia (Barcelona, Spain)* 7(2):61–64.

Bushby K, Muntoni F, Bourke JP (2003)> 107th ENMC international workshop: the management of cardiac involvement in muscular dystrophy and myotonic dystrophy. 7th-9th June 2002, Naarden, the Netherlands. *Neuromuscul Disord* 13(2):166–172.

Campanati A, Giannoni M, Buratti L, Cagnetti C, Giuliodori K, Ganzetti G, Silvestrini M, Provinciali L, Offidani A (2015) Skin features in myotonic dystrophy type 1: an observational study. *Neuromuscul Disord* 25(5):409–413.

de Die-Smulders C (2004) Congenital and childhood-onset myotonic dystrophy. In: *Myotonic Dystrophy Present Management Future Therap*. PS Harper, B van Engelen, B Eymard, DE Wilcox (Eds), pp. 162–175. New York: Oxford University Press.

de Die-Smulders C, Howeler CJ, Thijs C, Mirandolle JF, Anten HB, Smeets HJ, Chandler KE, Geraedts JP (1998) Age and causes of death in adult-onset myotonic dystrophy. *Brain* 121(Pt 8)(8):1557–1563.

Dogan C, De Antonio M, Hamroun D, et al. (2016) Gender as a modifying factor influencing myotonic dystrophy type 1 phenotype severity and mortality: A nationwide multiple databases cross-sectional observational study. *PLoS One* 11(2):e0148264.

Douniol M, Jacquette A, Guile JM, Tanguy ML, Angeard N, Heron D, Plaza M, Cohen D (2009) Psychiatric and cognitive phenotype in children and adolescents with myotonic dystrophy. *Eur Child Adolesc Psychiatry* 18(12):705–715.

Echenne B, Rideau A, Roubertie A, Sebire G, Rivier F, Lemieux B (2008) Myotonic dystrophy type I in childhood Long-term evolution in patients surviving the neonatal period. *Eur J Paediatr Neurol*, 12(3):210–223.

Engvall M (2010) On oral health in children and adults with myotonic dystrophy. *Swed Dent J Suppl* (203):1–51.

Geh JL, Moss AL (1999) Multiple pilomatrixomata and myotonic dystrophy: a familial association. *Br J Plast Surg* 52(2):143–145.

Goldman A, Ramsay M, Jenkins T (1994) Absence of Myotonic-Dystrophy in Southern African Negroids Is Associated with a Significantly Lower Number of Ctg Trinucleotide Repeats. *J Med Genet* 31(1):37–40.

Groh WJ, Groh MR, Saha C, Kincaid JC, Simmons Z, Ciafaloni E, Pourmand R, Otten RF, Bhakta D, Neir GV, Marashdeh MM, Zipes DP, Pascuzzi RM (2008) Electrocardiographic abnormalities and sudden death in myotonic dystrophy type 1. *N Engl J Med* 358(25):2688–2697.

Harper PS (2004) *Myotonic Dystrophy Present Management, Future Therapy*. Oxford: Oxford University Press.

Hermans MC, Faber CG, Pinto YM (2008) Sudden death in myotonic dystrophy. *N Engl J Med* 359(15):1626–1628; author reply 1328–1629.

Hermans MC, Merkies IS, Laberge L, Blom EW, Tennant A, Faber CG (2013) Fatigue and daytime sleepiness scale in myotonic dystrophy type 1. *Muscle Nerve* 47(1):89–95.

Hilbert JE, Barohn RJ, Clemens PR, Luebbe EA, Martens WB, McDermott MP, Parkhill AL, Tawil R, Thornton CA, Moxley RT, National Registry Scientific Advisory Committee/Investigators (2017) High frequency of gastrointestinal manifestations in myotonic dystrophy type 1 and type 2. *Neurology* 89(13):1348–1354.

Hsiao KM, Chen SS, Li SY, Chiang SY, Lin HM, Pan H, Huang CC, Kuo HC, Jou SB, Su CC, Ro LS, Liu CS, Lo MC, Chen CM, Lin CC (2003) Epidemiological and genetic studies of myotonic dystrophy type 1 in Taiwan. *Neuroepidemiology* 22(5):283–289.

Hughes MI, Hicks EM, Nevin NC, Patterson VH (1996) The prevalence of inherited neuromuscular disease in Northern Ireland. *Neuromuscul Disord* 6(1):69–73.

Johnson NE, Abbott D, Cannon-Albright LA (2015) Relative risks for comorbidities associated with myotonic dystrophy: A population-based analysis. *Muscle Nerve* 52(4):659–661.

Johnson NE, Ekstrom AB, Campbell C, Hung M, Adams HR, Chen W, … Heatwole CR (2016) Parent-reported multi-national study of the impact of congenital and childhood onset myotonic dystrophy. *Dev Med Child Neurol* 58(7):698–705.

Johnson NE, Hung M, Nasser E, Hagerman KA, Chen W, Ciafaloni E, Heatwole CR (2015b) The impact of pregnancy on myotonic dystrophy: A registry-based study. *J Neuromuscul Dis* 2(4):447–452.

Lefter S, Hardiman O, Ryan AM (2017) A population-based epidemiologic study of adult neuromuscular disease in the Republic of Ireland. *Neurology* 88(3):304–313.

Lindberg C, Bjerkne F (2017) Prevalence of myotonic dystrophy type 1 in adults in western Sweden. *Neuromuscul Disord* 27(2):159–162.

Liquori CL, Ricker K, Moseley ML, Jacobsen JF, Kress W, Naylor SL, Day JW, Ranum LP (2001) Myotonic dystrophy type 2 caused by a CCTG expansion in intron 1 of ZNF9. *Science* 293(5531):864–867.

Logigian EL, Martens WB, Moxley RT 4th, McDermott MP, Dilek N, Wiegner AW, Pearson AT Barbieri CA, Annis CL, Thornton CA, Moxley RT 3rd (2010) Mexiletine is an effective antimyotonia treatment in myotonic dystrophy type 1. *Neurology* 74(18):1441–1448.

Lopez de Munain A, Blanco A, Emparanza JI, Poza JJ, Marti Masso JF, Cobo A, Martorell L, Baiget M, Martinez Lage JM (1993) Prevalence of myotonic dystrophy in Guipuzcoa (Basque Country, Spain). *Neurology* 43(8):1573–1576.

MacMillan JC (1991) Singlegene neurological disorders in south wales: An epidemiological study. *Ann. Neurol* 30(3):411–414.

Mathieu J, Allard P, Gobeil G, Girard M, De Braekeleer M, Begin P (1997) Anesthetic and surgical complications in 219 cases of myotonic dystrophy. *Neurology* 49(6):1646–1650.

Mathieu J, Boivin H, Meunier D, Gaudreault M, Begin P (2001) Assessment of a disease-specific muscular impairment rating scale in myotonic dystrophy. *Neurology* 56(3):336–340.

Mathieu J DBM, Prevost C (1990) Genealogical reconstruction of myotonic dystrophy in the Saguenay-Lac-Saint Jean area (Quebec Canada). *Neurology* 40(5):839–842.

Medica I, Markovic D, Peterlin B (1997) Genetic epidemiology of myotonic dystrophy in Istria, Croatia. *Acta Neurol Scand* 95(3):164–166.

Mladenovic J, Pekmezovic T, Todorovic S, Rakocevic-Stojanovic V, Savic D, Romac S, Apostolski S (2006) Epidemiology of myotonic dystrophy type 1 (Steinert disease) in Belgrade (Serbia). *Clin Neurol Neurosurg* 108(8):757–760.

Norwood F, Rudnik-Schoneborn S (2012) 179th ENMC international workshop: pregnancy in women with neuromuscular disorders 5-7 November 2010, Naarden, The Netherlands. *Neuromuscul Disord* 22(2):183–190.

Okkersen K, Buskes M, Groenewoud J, Kessels RPC, Knoop H, van Engelen B, Raaphorst J (2017) The cognitive profile of myotonic dystrophy type 1: A systematic review and meta-analysis. Cortex 95 143–155. d

Okkersen K, Monckton DG, Le N, Tuladhar AM, Raaphorst J, van Engelen BGM (2017) Brain imaging in myotonic dystrophy type 1: A systematic review. *Neurology* 89(9):960–969.

Orngreen MC, Arlien-Soborg P, Duno M, Hertz JM, Vissing J (2012) Endocrine function in 97 patients with myotonic dystrophy type 1. *J Neurol* 259(5):912–920.

Petri H, Vissing J, Witting N, Bundgaard H, Kober L (2012) Cardiac manifestations of myotonic dystrophy type 1. *Int J Cardiol* 160(2):82–88.

Pisani V, Tirabasso A, Mazzone S, Terracciano C, Botta A, Novelli G, … Di Girolamo S (2011) Early subclinical cochlear dysfunction in myotonic dystrophy type 1. *Eur J Neurol* 18(12):1412–1416.

Redman JB, Fenwick RG Jr, Fu YH, Pizzuti A, Caskey CT (1993) Relationship between parental trinucleotide GCT repeat length and severity of myotonic dystrophy in offspring. *JAMA* 269(15):1960–1965.

Ricker K, Koch MC, Lehmann-Horn F, Pongratz D, Otto M, Heine R, Moxley RT, 3rd (1994) Proximal myotonic myopathy: a new dominant disorder with myotonia, muscle weakness, and cataracts. *Neurology* 44(8):1448–1452.

Ronnblom A, Forsberg H, Danielsson A (1996) Gastrointestinal symptoms in myotonic dystrophy. *Scand J Gastroenterol* 31(7):654–657.

Rudnik- Schöneborn S, Zerres K (2004) Outcome in pregnancies complicated by myotonic dystrophy: a study of 31 patients and review of the literature. *Eur J Obstet Gynecol Reprod Biol* 114(1):44–53.

Schilling L, Forst R, Forst J, Fujak A (2013) Orthopaedic Disorders in Myotonic Dystrophy Type 1: descriptive clinical study of 21 patients. *BMC Musculoskelet Disord* 14:338.

Siciliano G, Manca M, Gennarelli M, Angelini C, Rocchi A, Iudice A, Miorin M, Mostacciuolo M (2001) Epidemiology of myotonic dystrophy in Italy: re-apprisal after genetic diagnosis. *Clin Genet* 59(5):344–349.

Smith CA, Gutmann L (2016) Myotonic Dystrophy Type 1 Management and Therapeutics. *Curr Treat Options Neurol* 18(12):52.

Suominen T, Bachinski LL, Auvinen S, Hackman P, Baggerly KA, Angelini C, Peltonen L, Krahe R, Udd B (2011) Population frequency of myotonic dystrophy: higher than expected frequency of myotonic dystrophy type 2 (DM2) mutation in Finland. *Eur J Hum Genet* 19(7):776–782.

Tarnopolsky MA, Pearce E, Matteliano A, James C, Armstrong D (2010) Bacterial overgrowth syndrome in myotonic muscular dystrophy is potentially treatable. *Muscle Nerve* 42(6):853–855.

Thornton CA (2014) Myotonic dystrophy. *Neurol Clin* 32(3):705–719.

Thornton CA, Johnson K, Moxley RT 3rd (1994) Myotonic dystrophy patients have larger CTG expansions in skeletal muscle than in leukocytes. *Ann Neurol* 35(1):104–107.

Verhagen WI, ter Bruggen JP, Huygen PL (1992) Oculomotor, auditory, and vestibular responses in myotonic dystrophy. *Arch Neurol* 49(9):954–960.

Volpe J (2008) Neuromuscular disorders: Muscle involvement and restriced disorders. In: Volpe's Neurology of the Newborn. 5th ed, pp. 801. Philadelphia: Saunders Elsevier.

Yum K, Wang ET, Kalsotra A (2017) Myotonic dystrophy: disease repeat range, penetrance, age of onset, and relationship between repeat size and phenotypes. *Curr Opin Genet Dev* 44:30–37.

Zanoteli E, Yamashita HK, Suzuki H, Oliveira AS, Gabbai AA (2002) Temporomandibular joint and masticatory muscle involvement in myotonic dystrophy: a study by magnetic resonance imaging. *Oral Surg Oral Med Oral Pathol Oral Radiol Endod* 94(2):262–271.

40

NEUROFIBROMATOSIS TYPE 1

DAVID VISKOCHIL

Division of Medical Genetics, Department of Pediatrics, University of Utah, Salt Lake City, Utah, USA

INTRODUCTION

The clinical management of neurofibromatosis type 1 (NF1) involves recognition and treatment of myriad neurocutaneous, skeletal, tumor, and cognitive abnormalities. Like many genetic disorders covered in this book, the molecular biology of NF1 has outpaced its treatment. As our knowledge of the biochemical pathways involved in NF1 increases, practitioners will progress from the "watchful waiting" mode of care toward an interventional management approach. It is important for primary care providers to familiarize themselves with both indications and contraindications for the application of therapies provided in the context of comprehensive medical and psychosocial care by subspecialists who may not themselves be experienced with the unique complications of NF1.

Incidence

NF1, also known as peripheral neurofibromatosis or von Recklinghausen disease, was first described in modern medical literature in 1882. There are a number of historical perspectives, reviews, and textbooks devoted to the various manifestations of this condition (Crowe et al. 1956; Rubenstein and Korf 1990; Riccardi 1986, 1992; Huson and Hughes 1994; Upadhyaya and Cooper 1998; Friedman et al. 1999). It is an autosomal dominant condition with high variability of clinical expression (Carey and Viskochil 1999).

Although it is fully penetrant in adults, there is an age-related penetrance for a number of the individual clinical signs. NF1 affects approximately 1/3000 people worldwide, and usually diagnosis can be made in 95% of affected individuals by age 11 years with straightforward clinical evaluation (Friedman et al. 1999). At least half of affected individuals are sporadic cases, which demonstrates the high mutation rate of the *NF1* gene. There are only a few genotype/phenotype correlations; thus the degree of both morbidity and longevity are difficult to predict for any affected individual. Based on a study of death certificates in the United States, individuals with NF1 die approximately 8–15 years younger compared with the general population (Evans et al. 2011; Rasmussen et al. 2001). The most likely causes listed include malignant neoplasm and vascular disease.

Diagnostic Criteria

The variability in clinical expression and age-related penetrance of a number of the clinical manifestations sometimes makes NF1 a difficult condition to diagnose with confidence, especially in young children who are sporadic cases. The autosomal dominant inheritance pattern and full penetrance in adults facilitates its diagnosis. The clinical diagnostic criteria that were established in 1988 (National Institute of Health Consensus Development Conference 1988), reviewed in 1997 (Gutmann et al. 1997), and amended for osseous lesions in 2007 (Stevenson et al. 2007), serve as a guideline

---

Cassidy and Allanson's Management of Genetic Syndromes, Fourth Edition.
Edited by John C. Carey, Agatino Battaglia, David Viskochil, and Suzanne B. Cassidy.
© 2021 John Wiley & Sons, Inc. Published 2021 by John Wiley & Sons, Inc.

to diagnosis (see Table 40.1). The presence of any two criteria enables one to diagnose this condition on clinical grounds. The diagnostic criteria do not provide insight into severity of the disorder or prognosis for any given individual, although the manifestations of diffuse plexiform neurofibroma, optic nerve pathway glioma, tibial pseudarthrosis, and dystrophic scoliosis herald more significant medical intervention than the other criteria. Cutaneous manifestations of café-au-lait spots, distinctive freckling patterns, and dermal neurofibromas were selected for their high prevalence in almost all adults with NF1. Other signs, including Lisch nodules, skeletal dysplasia, optic nerve pathway tumor, and plexiform neurofibroma, were selected for their specificity. There is a role for clinical judgment, and the use of imaging studies to determine the presence or absence of a feature solely for diagnostic purposes is rarely indicated (Gutmann et al. 1997).

It is necessary to consider age in the application of diagnostic criteria because each of the diagnostic manifestations has a general age of presentation. Café-au-lait spots (Figure 40.1) tend to emerge in the first year of life, and approximately 80% of those who have NF1 will demonstrate over five café-au-lait spots by age 1 year. This is typically the first sign of the condition. Axillary and/or groin freckling (Figure 40.2), Crowe's sign, is usually the second diagnostic manifestation to arise, and is seen in approximately three-fourths of individuals with NF1 (Crowe 1964; Korf 1992). The subtlety of Crowe's sign (intertriginous freckling) makes it difficult at times to confidently diagnose NF1 in sporadic cases until dermal neurofibromas or Lisch nodules are detected, usually in the early teen years. The advent of molecular screening with 95% sensitivity to detect pathogenic *NF1* mutations can alleviate the wait to firmly establish a diagnosis of NF1. Incorporation of molecular results has not been integrated into the formal diagnostic criteria; nevertheless, the finding of a pathogenic *NF1* mutation combined with multiple café-au-lait spots enables the clinician to establish the diagnosis. Other less frequent diagnostic criteria that typically arise in childhood include optic nerve pathway gliomas (~15%), distinctive skeletal abnormalities (~5%), and plexiform neurofibromas (~20%).

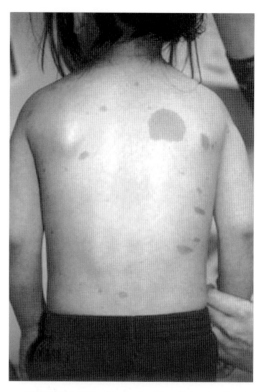

FIGURE 40.1 Typical café au lait macules.

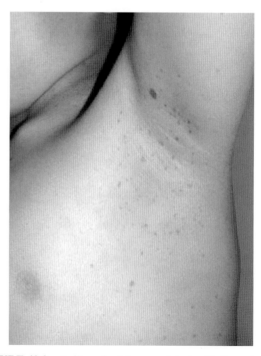

FIGURE 40.2 Axillary freckling consistent with Crowe's sign.

TABLE 40.1 NIH diagnostic criteria for NF1

Neurofibromatosis 1 is present in an individual who has two or more of the following signs:

Six or more café-au-lait macules >5 mm in greatest diameter in prepubertal individuals or >15 mm in greatest diameter after puberty

Two or more neurofibromas of any type, or one or more plexiform neurofibromas

Freckling in the axillae or inguinal regions (Crowe's sign)

A tumor of the optic nerve pathway

Two or more Lisch nodules (iris hamartomas)

A distinctive osseous lesion, such as sphenoid wing dysplasia, long-bone bowing (with or without pseudarthrosis) or dystrophic scoliosis

A first-degree relative with neurofibromatosis 1 by the above criteria

Recognition of an affected first-degree relative fulfills one of the diagnostic criteria, typically as an unambiguously affected parent of a child who is being assessed.

There are other relatively common manifestations of NF1 that can be considered in the diagnostic evaluation, even though they are not very specific. Short stature, relative macrocephaly, poor coordination with decreased stamina, learning and speech problems, dysmaturity with poor social and communication skills, and hyperintense T_2-weighted signals (focal areas of signal intensity – FASI) on brain MRI are commonly seen. When clinical suspicion is high in children, it may be reasonable to provisionally diagnose NF1 and counsel families about possible associations and medical complications that may otherwise be overlooked in standard health care. This is especially true in toddlers who have classical multiple café-au-lait spots as their only manifestation. A provisional diagnosis of NF1 can guide primary care practitioners to detect potential complications associated with this condition. Annual reevaluation of the diagnostic criteria and ophthalmology evaluations are indicated in those who carry a provisional diagnosis of NF1, and many families choose to undergo molecular confirmation. The identification of a pathogenic sequence variant in the *NF1* gene will transition a provisional diagnosis to a confirmed diagnosis.

To provide effective anticipatory guidance, it is important to be aware of the frequency of NF1-related medical complications. There are a number of published reports that have recorded the incidence of various manifestations (Riccardi 1992; Huson and Hughes 1994; Friedman and Birch 1997). Tables 40.2 and 40.3 list a number of clinical manifestations associated with NF1, the incidences of some less common findings, and the ages at which they occur. It is important to keep these frequency figures in mind in the evaluation of unusual complications of NF1. As a primary care provider for someone with NF1, the question, "could this manifestation be caused by NF1?" is an important one to address at each encounter. Knowledge of the incidence of NF1-associated manifestations helps guide the practitioner in diagnostic evaluations when confronted with medical concerns.

Some multidisciplinary team clinics for NF1 routinely perform cranial magnetic resonance imaging (MRI) in asymptomatic individuals with multiple café-au-lait spots as part of a diagnostic assessment. Practitioners familiar with the diagnostic criteria and who understand the age-related emergence of physical manifestations rarely need such studies as a diagnostic aid, providing they follow suspected cases as they would established cases. Proficient ophthalmologic evaluations rather than brain/orbit MRI can identify those individuals with NF1 who have optic nerve pathway tumors requiring intervention. The lack of correlation between the number and location of T_2-hyperintensities with learning problems further diminishes the clinical indication to perform brain MRI in the absence of symptoms. Parental anxiety induced by the finding of T_2-hyperintensities in an otherwise normal brain MRI scan may not be easily diffused by counseling about the relative insignificance of these ill-defined and benign lesions. Sometimes, the MRI scans prompt management decisions that are based on incidental findings rather than symptoms. For example, an initial finding of a low-grade glioma on brain MRI in an asymptomatic individual could lead to sequential MRI scans without medically defined endpoints for surveillance. Sedation in infants and toddlers for MRI and their exposure to enhancing dyes carry some, albeit small, risks to the child. As a general

TABLE 40.2 Manifestations of NF1 and estimated frequencies

Manifestation	Estimated frequency
Cutaneous	
Multiple café-au-lait spots	(>90%)
Intertriginous freckling (Crowe's sign)	(75–90%)
Dermal neurofibromas	(>90%)
Xanthogranulomas	(2–5%)
Hemangiomas	(5–10%)
Ophthalmologic	
Optic nerve pathway tumor	(15%)
Lisch nodules	(95%)
Glaucoma	(rare)
Musculoskeletal	
Sphenoid wing dysplasia	(2–5%)
Tibial/ulnar dysplasia	(5%)
Tibial pseudarthrosis	(2%)
Scoliosis	(20–30%)
Dystrophic scoliosis	(2–5%)
Short stature	(25–35%)
Relative macrocephaly	(common)
Weakness/hypotonia/poor stamina	(common)
Vascular	
Hypertension	(common in adults)
Renal artery stenosis	(<2%)
Congenital heart defect	(2%)
Neurological	
Hydrocephalus	(5%)
Seizures	(6–7%)
Learning Disabilities	(40–60%)
Autism spectrum disorder	(25%)
Precocious puberty	(2–5%)
Tumors	
Plexiform neurofibromas	(25%)
Paraspinal neurofibromas	(50%)
Malignant peripheral nerve sheath tumors	(10–15%)
CNS low-grade glioma	(10%)
CNS high-grade glioma	(rare)
Breast cancer	(20–50%)
Juvenile Monocyticmyelogenous leukemia	(rare)
Other tumors	(rare)

Source: Modified from Riccardi (1992), Huson and Hughes (1994), and Friedman and Birch (1997).

TABLE 40.3 List of NF1 manifestations that have age-dependent clinical expression

Manifestation	Age of clinical expression
Café-au-lait macules	Infancy to early childhood
Intertriginous freckling	Childhood
Dermal neurofibromas	Late childhood and adolescence through adulthood
Plexiform neurofibroma	Infancy through adulthood
Lisch nodules	Late childhood through adulthood
Optic nerve pathway tumors	Early childhood
Sphenoid wing dysplasia	Infancy
Long-bone bowing	Infancy
Scoliosis	Childhood
Hypertension	Childhood through adulthood
Learning disabilities/communication skills	Early childhood through adolescence
Malignant peripheral nerve sheath tumors	Adolescence through adulthood
Breast cancer	Middle adulthood (earlier than general population)

guideline, the use of ancillary imaging for diagnostic purposes in NF1 is not warranted without symptoms. On balance, it is prudent to investigate symptoms such as new-onset headaches or behavior changes in individuals with NF1 sooner than would otherwise be indicated in unaffected individuals, and this may include brain MRI studies.

Cognition and psychosocial issues in NF1 are important manifestations that require assessment throughout life. Initially, infants with NF1 seem no different than unaffected sibs; however, family members often note hypotonia and decreased muscle strength. Into toddlerhood, language delays along with articulation issues are noted. By the pre-school years, parents recognize differences in learning in some children. An association with social skills disorders, communication difficulties, ADHD and autism spectrum disorder are common in children with NF1 during elementary school (Garg et al. 2015; Plasschaert et al. 2015; Chisholm et al. 2018). In 8 year olds in US elementary schools, there is a 4–5-fold increased prevalence of autism spectrum disorder in those with NF1 (Bilder et al. 2016). Anxiety and depression develop in many individuals through adolescence and young adulthood (Ferner et al. 2017).

Etiology, Pathogenesis, and Genetics

NF1 was genetically mapped to the centromeric region of the long arm of chromosome 17 (Barker et al. 1987). The *NF1* gene was subsequently cloned and characterized as a Ras-GAP protein (reviewed in Viskochil et al. 1993). Its encoded product, neurofibromin, is a 240-kDa peptide (reviewed in Sherman et al. 1998) that stimulates the intrinsic hydrolysis of guanosine triphosphate (GTP) bound to Ras (Martin et al. 1990) (Figure 40.3). Ras is a small intracellular protein attached to the inner membrane of the cell that, when bound to GTP, transduces both growth and anti-apoptotic signals by way of the mitogen-activated protein kinase (MAPK) pathway. The conversion of Ras-GTP to Ras-GDP

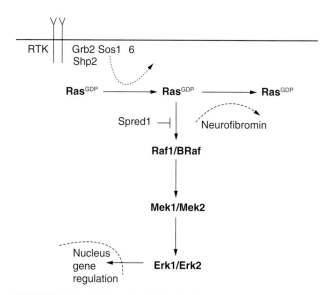

FIGURE 40.3 Schematic of the Ras/MAPK signal transduction pathway. Extracellular growth factors bind to receptor tyrosine kinases to initiate the signal cascade. Shp2, Grb2, and Sos1 play roles in transmitting the receptor tyrosine kinase signal to activate Ras by exchanging Ras-bound GDP with Ras-bound GTP. Ras-GTP activates Raf1, which propagates the signal through downstream proteins, Mek and Erk, to alter gene expression in the nucleus. The *NF1* gene product, neurofibromin, activates the intrinsic Ras GTPase to decrease the level of signaling through Ras-GTP. NF1 is caused by abnormal neurofibromin function. There are other genetic conditions associated with this pathway (Swensen and Viskochil 2007; Denayer et al. 2008). Legius syndrome is caused by abnormal Spred1 function (Brems et al. 2007). Costello syndrome (Chapter 16) is caused by mutations in *HRAS*. Noonan syndrome (Chapter 41) can be caused by alterations in *PTPN11* encoding Shp2, *SOS1*, *KRAS*, and *RAF1*. Cardio-facio-cutaneous syndrome (Chapter 10) is caused by mutations in either *KRAS*, *BRAF*, *MEK1*, or *MEK2*. GDP: guanosine diphosphate; GTP: guanosine triphosphate; SOS1: guanine nucleotide exchange factor.(Illustration by B. Stephan ©2009bjs/Medical Graphics & Photography, University of Utah School of Medicine.)

terminates intracellular signaling; thus neurofibromin acts as a negative regulator of the Ras-mediated signal transduction pathway (reviewed in Bernards 1995). Inactivating mutations of *NF1* lead to increased intracellular signaling through Ras (Figure 40.3). *RAS* was the first proto-oncogene identified in the early 1980s, and specific pathogenic missense mutations that prevent the conversion of Ras-GTP to Ras-GDP are associated with a large number of cancers. However, with the exception of myelogenous leukemia and rhabdomyosarcoma (both rare in NF1), the cancers associated with oncogenic *RAS* (e.g., *H-RAS* codon 12 mutations in bladder cancer) minimally overlap with the cancer types seen in NF1 because Ras remains wild-type and is merely dysregulated due to inactivating *NF1* pathogenic variants leading to decreased to absent Ras-GAP activity.

Mutations in the *NF1* gene (Upadhyaya et al. 1995) generally predict inactivation of its gene product, neurofibromin. Complete loss of neurofibromin by somatic mutation of the normal *NF1* allele (double inactivation) is seen in almost all NF1-associated tumors. Thus, the *NF1* gene is classified as a "tumor suppressor", and the primary function of its encoded protein is down-regulation of intracellular Ras signal propagation.

Neurofibromas illustrate the paradigm of *NF1* tumor suppressor function (Serra et al. 1997). As a consequence of double inactivation of neurofibromin due to independent somatic mutations in each and every neurofibroma, clonally expanded $NF1^{-/-}$ Schwann cells recruit heterozygous ($NF1^{+/-}$) cells from the microenvironment, which leads to emergence of benign tumors (Yang et al. 2012). Likewise, melanocytes from café-au-lait spots and mesenchymal cells from tibial pseudarthrosis tissue show double inactivation of ($NF1^{-/-}$) in a subset of cells that perturb the microenvironment in a paracrine fashion (De Schepper et al. 2008; Stevenson et al. 2006). Conditional knockout mice that inactivate *Nf1* in specific cells at specific times demonstrate that emergence of neurofibromas, along with other NF1-related manifestations, is enhanced when those cells are embedded in a tissue microenvironment of $NF1^{+/-}$ haploinsufficient cells. For neurofibromas, these enabling cell types include fibroblasts, perineureal cells, mast cells, macrophages, and lymphocytes. These observations emphasize the role of paracrine regulation of benign tumor growth in neurofibromas (Parada et al. 2005). Unlike dermal neurofibromas that always remain benign, diffuse plexiform neurofibromas can undergo cancerous transformation to malignant peripheral nerve sheath tumors (MPNSTs) by virtue of accumulation of additional somatic mutations (i.e. *CDKN2A* and *TP53*) in a subset of proliferating cells. There is a 10–15% lifetime risk to develop MPNST in the NF1 population.

The pathogenesis of NF1, as a pleiotropic condition, is not completely understood. The tumor phenotype likely arises as a consequence of abnormal regulation of Ras signaling; however, the other manifestations of NF1 are not easily explained by the neurofibromin-Ras interaction. A common theme in NF1 is the apparent neural crest derivation of cells that are most affected by *NF1* mutations. Melanocytes, Schwan cells from peripheral sensory nerves, and anterior cranial facial bones fit this paradigm. However, other cell types are clearly affected in NF1. Tibial pseudarthrosis, scoliosis, short stature, macrocephaly, pectus excavatum, intracranial gliomas, and learning disabilities are difficult to reconcile with a hypothesis that pathogenesis is related to abnormal expression of neurofibromin in cells embryologically derived from the neural crest. Future work may clarify the pathogenesis of the pleiotropic manifestations of NF1, but, at present, strategies for the development of rational medical treatment protocols remain primarily focused on the Ras-neurofibromin signal transduction pathway. There has been success in treatment of inoperable plexiform neurofibromas with MEK inhibitors (Dombi et al. 2016), which will undoubtedly lead to increased surveillance for early diagnosis and treatment of non-surgical plexiform neurofibromas as outlined in the management sections of this chapter.

There are three distinct genotype–phenotype correlations with NF1, representing less than 10% of the NF1 population. The most concerning are deletions at the *NF1* locus that affects up to 5% of the NF1 population. Submicroscopic deletions involving the entire *NF1* gene and a number of adjacent genes (Kehrer-Sawatzki et al. 2017), including the tumor suppressor (*SUZ12*), is most concerning due to its association with more severe learning issues, increased tumor load, and two-fold increased risk for MPNSTs (De Raedt et al. 2003). On the other hand, a three base pair deletion (c.2970–2972 delAAT involving exon 22) (Upadhyaya et al. 2007) is associated with an attenuated phenotype without neurofibromas. Likewise, a missense mutation in codon Arg1809 is associated with a milder phenotype (Rojnueangnit et al. 2015). Most *NF1* mutations are private intragenic sequence variants that do not enable practitioners to provide enhanced anticipatory guidance beyond what can be provided for these three genotype–phenotype correlations. A fourth phenotypically distinct NF1-related condition in which individuals have extensive paraspinal neurofibromas at almost all spinal cord roots, but lack the common pigmentary and cognitive manifestations of NF1 is known as spinal NF1. Attempts to identify a genotype-phenotype correlation have not identified unique *NF1* mutations. Furthermore, intrafamilial variation of the paraspinal tumor phenotype within a few reported families with multiple affected family members suggests genotype is minimally predictive of a Spinal NF phenotype (Ruggieri et al. 2015).

Genetic Counseling. NF1 is an autosomal dominant condition, imparting a 50% recurrence risk to each offspring of an affected individual. Approximately half of all individuals diagnosed in North American and European neurofibromatosis clinics do not have a family history, which supports the observation that the *NF1* gene has a high mutation rate in gonads. This characteristic also likely applies to the somatic *NF1* mutation rate and may partially explain the high variability of clinical expression. Stochastic random somatic *NF1* mutations resulting in the inactivation of both *NF1* alleles and loss of intracellular neurofibromin likely lead to the variable emergence of clinical manifestations, depending on the timing and cell types harboring the somatic *NF1* mutations. NF1 is almost always fully penetrant in the adult population; however, those individuals who either have signs of NF1 localized to body segments or who demonstrate incomplete penetrance could represent cases of mosaicism (Ruggieri and Huson 2001). The clinical expression of mosaic NF1 recapitulates the embryonic timing of somatic *NF1* mutation and the tissues harboring such mutation. It is important to identify such individuals because genetic counseling for those who manifest mosaicism must include an estimation of the number of cells affected by an *NF1* mutation and the likelihood of germ line involvement. The chance that an adult with mosaicism for NF1 will have offspring with generalized NF1 is much lower than the expected 50% incidence for non-mosaic NF1 parents. The empiric recurrence risk for a sibling of an affected child when a parent has mosaic expression of NF1 is dependent on the level of gonadal mosaicism, better determined in father's sperm specimens than mother's ovaries.

Diagnostic Testing

NF1 is a clinical diagnosis; however, the presence of a pathogenic mutation in *NF1* enables one to establish a diagnosis if full age-dependent diagnostic criteria are not present. Without molecular testing, strict application of the diagnostic criteria allows practitioners to make the diagnosis in the vast majority of cases. In cases where presymptomatic or prenatal diagnosis is desired, molecular mutation analysis is required. There are instances where clinical judgment cannot determine the affected status of an individual and *NF1* gene mutation screening can establish the diagnosis. As pre-implantation selection of in vitro early blastocysts becomes more available in medical practice, there will likely be an increased demand for direct DNA analysis in a prospective parent who has NF1. In general, the clinical utility of diagnostic testing by molecular means is limited, but can be highly valuable for specific families. As genotype-phenotype correlations improve and potential treatment strategies evolve, the implementation of *NF1* mutation testing (now approximately 95% sensitivity) provides potential options in both management and counseling.

Differential Diagnosis

There are a number of clinical conditions that should be considered as part of the diagnostic evaluation of NF1. Overlap of NF1 with other conditions primarily lies in pigment manifestations. Café-au-lait spots are commonly seen in the following conditions: Russell–Silver syndrome (see Chapter 52), Bloom syndrome, Noonan syndrome (see Chapter 41), Sotos syndrome (see Chapter 55), Proteus syndrome (Chapter 47), Klippel–Trenaunay–Weber syndrome, ataxia-telangiectasia, Carney syndrome, and chromosomal mosaicism (including rings). The geographically isolated cutaneous manifestations of epidermal nevus, bathing trunk nevus, and Schimmelpenning nevus including elements of schwannoma and/or neurofibroma on biopsy, can overlap with mosaic NF1. The pigmentary changes in McCune–Albright syndrome can be difficult to distinguish from the café-au-lait spots of NF1. In general, café-au-lait spots in NF1 have uniform and regular borders that typically involve the flanks in addition to other sites in the body. The café-au-lait spots in McCune–Albright syndrome are generally more darkly pigmented with irregular borders, and they tend to be arrayed centrally over the spine in a patchy configuration, reflecting its mosaic *GNAS* etiology. A history of long-bone fractures and/or precocious puberty without hypothalamic tumor should alert one to the possibility of McCune–Albright syndrome, and endocrine studies plus skeletal survey radiographs should enable one to distinguish McCune–Albright syndrome from NF1.

There are a few families that have multiple café-au-lait spots that are inherited as an autosomal dominant trait with no other manifestations of NF1. Some of these families are linked to the *NF1* locus and presumably carry an *NF1* mutation; however, some families demonstrate lack of genetic linkage to the *NF1* locus (Brunner et al. 1993; Charrow et al. 1993; Abeliovich et al. 1995). Individuals with *SPRED1* mutations have multiple café-au-lait spots, ill-defined intertriginous freckling, macrocephaly and learning disabilities that are sometimes indistinguishable from NF1. This condition was originally designated as NF1-like syndrome (Brems et al. 2007), but is now recognized as Legius syndrome. Other manifestations of NF1 are not present in Legius syndrome, even though the pigmentary manifestations can fulfill diagnostic criteria for NF1. Family history derived from parents and other affected adults in the pedigree provides a clue to this diagnosis if neurofibromas and other CNS, skeletal, and ophthalmologic manifestations of NF1 are absent. The *SPRED1* gene product influences Ras signaling by serving as a docking protein to maintain neurofibromin in close proximity to membrane-bound Ras (Dunzendorfer-Matt et al. 2016). It is important to distinguish these conditions because medical management of patients with Legius syndrome does not include tumor surveillance, but does include follow-up for potential learning disabilities. *NF1* mutation analysis with reflex to *SPRED1* mutation screening in the event an *NF1* mutation is not identified is warranted in those individuals who only have pigmentary manifestations consistent with either Legius syndrome or NF1. This recommendation holds even though patients with Legius syndrome represent less than 5% of individuals screened for *NF1* mutations and only have typical multiple cafe-au-lait spots (+/– intertriginous freckling) (Messiaen et al. 2009).

The NF1 tumor phenotype minimally overlaps with a number of hamartomatous syndromes including *PTEN* Hamartoma Tumor Syndrome (see Chapter 48), Carney syndrome, Proteus syndrome (Chapter 47), Maffucci syndrome, multiple endocrine neoplasia 2B, von Hippel–Lindau syndrome (Chapter 62), multiple lipomatosis, Gardner syndrome (familial adenomatous polyposis), and schwannomatosis. As part of the evaluation of these hamartomatous conditions, histological identification of tumor biopsy specimens can establish a syndrome diagnosis. Multiple paraspinal neurofibromas inherited as a dominant condition without other NF1 manifestations (familial spinal neurofibromatosis (Ars et al. 1998)) is not a distinct tumor-overlap condition, but has been clarified as allelic to *NF1*, and was covered in the genotype–phenotype section above. Care must be used in determining whether these paraspinal tumors are neurofibromas versus schwannomas, which are a major feature of two distinct disorders, neurofibromatosis type 2 (NF2) and schwannomatosis, discussed below.

Few conditions have overlap with the skeletal manifestations of NF1. There are case reports of individuals with McCune–Albright syndrome who have poor healing of long-bone fractures (usually the femur) that fits a similar pattern to long bone pseudarthrosis in NF1. Individuals with Jaffe–Campanacci syndrome have multiple non-ossifying fibromas and café-au-lait spots; however, they do not have other specific NF1 manifestations. Of note, non-ossifying fibromas of the distal femur and proximal tibia are relatively common in NF1, but generally occur in adolescents who would have other manifestations of NF1, precluding the need for molecular testing to establish a firm diagnosis. Dystrophic scoliosis and sphenoid wing dysplasia with café-au-lait spots are unique to NF1.

The clinical overlap of NF1 with Watson syndrome (Watson 1967) and NF–Noonan syndrome has been dissected as allelic heterogeneity. Usually the diagnostic criteria for NF1 are satisfied, but such individuals may have short stature, ptosis, cupped ears, cryptorchidism, pulmonic stenosis or a pectus abnormality to suggest the possibility of Noonan syndrome, although almost all have *NF1* mutations (De Luca et al. 2005). There is clinical overlap of Watson syndrome with Noonan syndrome (Chapter 41), including multiple café-au-lait spots, short stature and pulmonary valve dysplasia (Tassabehji et al. 1993). Shared biochemical pathways for neurofibromin and the Noonan gene product, Shp-2, provide an explanation for the overlap phenotype. Noonan syndrome and the overlapping conditions of Costello syndrome (see Chapter 16) and cardio-facio-cutaneous (CFC, see Chapter 10) syndrome are caused by mutations in genes that encode proteins important in the Ras-MAPK signal transduction pathway, and are designated as rasopathies. Finally, mutations in *PTPN11* (Legius et al. 2002), *RAF1* (Pandit et al. 2007; Razzaque et al. 2007) and *NF1* (Wu et al. 1996) have been identified in multiple lentingines syndrome (lentingines, electrocardiographic conduction abnormalities, ocular hypertelorism, pulmonic stenosis, abnormal genitalia, retardation of growth, and sensorineural deafness), a condition that overlaps with both NF1 and Noonan syndrome.

The differential diagnosis of NF1 also includes NF2 and schwannomatosis. The distinction between NF1 and NF2 was first recognized by Gardner and Frazier (1930), and formally separated as central versus peripheral neurofibromatosis in 1981 (Eldridge 1981). Lisch nodules and axillary freckling are never seen in NF2. The café-au-lait spots infrequently seen in NF2 are usually less than six in number and tend to be more plaque-like. Small dermal neurofibromas are occasionally seen in NF2, which can be confused with mosaic NF1. The intracranial tumors of NF2 are usually easily differentiated from NF1-related manifestations, although optic nerve meningiomas (NF2) are sometimes confused with optic nerve pathway tumors (NF1). The distinct manifestations of NF2 include hearing loss associated with vestibular schwannomas, multiple intraspinal schwannomas, posterior lenticular cataracts, and multiple meningiomas. There is minimal clinical overlap between NF1 and NF2, and there should be no confusion about the appropriate diagnosis. Furthermore, a confusing issue for many families with an affected family member with NF1 is whether NF1 can progress to NF2, which never happens. These are two distinct conditions whose genes map to different chromosomal loci, 17q11.2 and 22q11.2, and encode proteins that are involved in two distinct intracellular biochemical pathways. Schwannomatosis is a distinct entity that has multiple schwannomas of the spine and peripheral and cranial nerves, but without vestibular schwannomas, and is caused by germ line mutations in either the *INI1/SMARCB1* gene (Hulsebos et al. 2007) or the *LZTR1* gene (Kehrer-Sawatzki et al. 2017).

Finally, segmental or localized NF1 likely represents a mosaic pattern, whereby only regional expression of the clinical manifestations of NF1 is seen (Viskochil and Carey 1994; Hager et al. 1997; Ruggieri and Huson 2001; Maertens et al. 2007). Parents of children with generalized NF1 should be closely evaluated for signs of mosaicism, which, if present, would indicate potential germ line involvement and an increased risk for offspring (Zlotogora 1998).

MANIFESTATIONS AND MANAGEMENT

Anticipatory guidance counseling in NF1 is essential for optimal clinical care. Individuals with NF1 should be placed on a surveillance program for related manifestations that are not otherwise obvious. An example of the effectiveness of this surveillance program lies in the diagnosis of optic nerve pathway tumors. The recognition of NF1 signifies a need for periodic ophthalmologic evaluations that may not otherwise be performed. Identification of subtle abnormal visual function or optic nerve pallor is an indication for magnetic resonance imaging (MRI), which could lead to early detection and effective intervention for the treatment of an optic nerve pathway tumor.

The benefits of anticipatory guidance counseling also encompass manifestations that are not included in the diagnostic criteria. Table 40.3 lists the age-related concerns of NF1 that need to be woven into the anticipatory guidance process (Table 40.4). Even if a diagnosis of NF1 has not been conclusively established, recognition of its possibility may affect clinical management decisions. Thus, anticipatory guidance should be provided in all circumstances, even where the diagnosis of NF1 is only suspected.

Growth and Feeding

In general, there are no consistent prenatal complications associated with NF1. Fetuses of mothers with NF1 could suffer consequences of elevated maternal blood pressure because of pregnancy-induced hypertension. Otherwise, no complications would be expected. Neonates with NF1 are generally delivered at term and have normal intrauterine growth, with rare exceptions of large plexiform neurofibroma or congenital pseudarthrosis.

Postnatal growth retardation is seen in approximately one-third of children with NF1, and relative macrocephaly is common. In rare circumstances, growth hormone deficiency is documented in some patients with short stature. A few individuals with NF1 have large stature that may be associated with a microdeletion of the *NF1* locus and adjacent genes. Precocious puberty and transient advanced growth velocity are often associated with optic nerve pathway tumors involving the chiasm and hypothalamic-pituitary axis. Disproportionate growth of the extremities is generally associated with plexiform neurofibromas; however, some individuals with a microdeletion have somewhat large and fleshy distal extremities.

Feeding is not an issue in this condition unless there are abdominal migraines with emesis. Usually, weight-to-height ratios are normal, without obesity.

TABLE 40.4 Anticipatory guidance for NF1: age-related assessments of key manifestations

Newborn to 2 years
Café-au-lait spots for diagnosis
Long-bone bowing
Plexiform neurofibromas
Optic pathway tumor
Development delay assessment
2–10 years
Optic pathway tumors
Plexiform neurofibromas
Scoliosis
Hypertension
Freckling patterns
Learning problems
10 years to adulthood
Onset of dermal neurofibromas
Learning problems
Self-esteem
Scoliosis
Plexiform neurofibromas
Reproductive decisions
Hypertension
Adult
Offspring
Progression of dermal neurofibromas
Malignant peripheral nerve sheath tumors
Hypertension
Plexiform neurofibromas

Evaluation

- Growth charts specifically for NF1 (Szudek et al. 2000; Clementi et al. 1999) prove helpful in identifying when short stature is beyond that expected for this condition.
- An increase or decrease in growth velocity should alert practitioners to an intracranial process that should be addressed by brain MRI.
- Assessment for optic nerve pathway/hypothalamic tumors is indicated in the presence of precocious puberty and/or tall stature, especially with increased growth velocity.

Management

- Presently, there is no approved treatment for short stature in NF1; however, if there is decreased growth velocity an evaluation by endocrinology for assessment of growth hormone deficiency is warranted.
- Growth hormone treatment has been provided to those with documented growth hormone deficiency; however, long-term risk for increased tumors after growth hormone treatment is not known. There is concern that growth hormone could stimulate Ras activity, and additional stimulatory signals in cells that are haploinsufficient for a "tumor suppressor" (neurofibromin) could potentially lay the foundation for increased tumor burden in NF1. This biologic concern warrants full counseling with the family and endocrinologists about

- the benefits and unknown long-term risks related with growth hormone replacement therapy.
- Precocious puberty and tall stature can be treated with luteinizing hormone-releasing hormone antagonists, in a standard manner.

Development and Behavior

Intellectual disability is not a common finding in NF1, and the incidence of frank intellectual impairment (full scale IQ < 70) is estimated between 4 and 8% (North et al. 1997), which is slightly higher than the background population. Those individuals with a large *NF1* deletion (~5% of the screened NF1 population) are more developmentally delayed than expected. This association could represent a significant component of the increased incidence of intellectual impairment associated with this condition.

In contrast to intellectual impairment, learning disabilities with and without ADHD are common and affect 40–60% of individuals with NF1 (North et al. 1997; Ozonoff 1999). The types of learning difficulties experienced by children with NF1 are not unique to the condition. There is no well-characterized learning disability profile that can be used to establish a broad-based and consistent approach for educators who are working with affected individuals, although impaired executive function is common. Affected children have a mixed pattern of learning difficulties including impairment of both verbal and nonverbal skills. Speech and language problems affect up to 80% of children between 2 and 5 years (Thompson et al. 2001), and speech therapy is effective. However, speech articulation and fluency can be ongoing issues through adulthood for many individuals with NF1 (Cosyns et al. 2012).

The management of potential learning problems is one of the primary reasons to consider the diagnosis of NF1 in early childhood based on multiple café-au-lait spots, before other age-dependent manifestations arise. The lack of a diagnosis of NF1 may hinder access to specialized neuropsychometric testing and school services. A provisional diagnosis, based solely on multiple café-au-lait spots or another singular manifestation of NF1, may be helpful for some families to obtain appropriate services. Many individuals with NF1 have normal cognition that requires no intervention; however, those who have learning issues benefit with school services and interventions covered under accommodation plans and individualized education plans. Those who have a microdeletion of the *NF1* locus may require additional services.

A consistent behavioral pattern has not been firmly established in individuals with NF1 (Noll et al. 2007). However, communication and social skills for many individuals border on autism spectrum disorder (Bilder et al. 2015; Garg et al. 2015; Plasschaert et al. 2015; Chisholm et al. 2018). Like all children, early recognition of autism spectrum disorder can be extremely beneficial in the development of behavior management programs and social skills interventions. Failure to recognize problems with either behavior or learning may interfere with their full achievement of potential abilities.

Evaluation

- Preschool developmental and behavioral testing is recommended to help both in school placement and in teacher awareness for children who do not have intellectual disability but, by virtue of having NF1, are at risk for school achievement problems.
- Formal speech and language assessment in preschool years (i.e. early intervention programs) is highly beneficial and strongly recommended, even if parents do not recognize issues.
- Screening for early signs of autism spectrum disorder is recommended, but it should be recognized that repetitive behaviors and absent language are unusual for NF1. Children with NF1 who have autism spectrum disorder usually have issues related to communication and social skills.
- Each affected individual seems to have a unique cognitive profile; therefore, it is imperative that complete batteries of neuropsychological testing be performed for educational purposes. Important considerations for testing include visual-spatial-perceptual skills, language function, neuromotor skills, phonological decoding skills and lexical retrieval for reading disability, and achievement testing for nonverbal learning disability. In addition, executive function skills associated with planning, attention, organization, and self-monitoring behaviors deserve more assessment in the identification of educational needs of children with NF1 (Ozonoff 1999).
- For those who appear intellectually impaired, in contrast to learning disabled, the primary care practitioner should consider the possibilities of other coincidental disorders, as well as an *NF1* microdeletion. Likewise, co-morbid conditions should be considered because intellectual impairment is relatively rare.

Management

- Appropriate school placement for the disabilities identified should be ensured, with aides if necessary.
- Speech therapy is beneficial, if needed.
- Self-esteem issues should be dealt with in childhood, even if physical manifestations are minimal. As teenagers develop physically distinctive signs of NF1, they are potentially more vulnerable than their peers, especially if they are struggling in school with learning

problems. It is incumbent on care providers to prepare the family and the child for this possibility.

- An educational environment developed within a structured organizational framework that has consistent rules and reinforcement of socialization skills may be extremely beneficial.
- Medication for impulsivity, distractibility, and mood disorders is not contraindicated by virtue of a child having NF1, and judicious medical management of attention deficit and hyperactivity disorder is recommended. Treatment strategies should be the same as for individuals in the general population with similar problems. Specific medications for individuals with NF1 have not been identified.
- Support groups and camp experiences may be beneficial in the social management of affected adolescents.

Neoplasia

Nerve Sheath Tumors
Neurofibromas. Neurofibromas arise from multiple sites, including the cutaneous sensory nerves, large motor nerve sheaths, spinal nerve roots and ganglia, and spinal cord plexi. They are exclusively found in the peripheral nervous system, although sometimes impinge the spinal cord. There are multiple neurofibroma tumor categories: cutaneous, subcutaneous, and nodular neurofibromas, as well as the more diffuse plexiform neurofibromas. Cutaneous neurofibromas are either sessile or pedunculated. They tend to be relatively small, discrete nodules that are easily defined by virtue of a fibrous capsule and skin surface presentation. Cutaneous neurofibromas tend to arise from the distal ends of cutaneous nerves where the nerve sheath is no longer evident. Subcutaneous and nodular neurofibromas are more intimately involved with nerve sheaths and tend to be encapsulated by the nerve sheath. Plexiform neurofibromas tend to be large, amorphous tumors that arise from larger nerve sheaths, but they extend beyond the confines of the nerve sheath (Figure 40.4). They can arise anywhere in the peripheral nervous system and tend to have spurts of growth that do not always coincide with overall body growth. Because many of the plexiform neurofibromas arise from nerve sheaths, they tend to remain below the skin surface and are initially identified by palpation or visual recognition of asymmetry and unusual overlying pigmentary and/or hair patterns.

Cutaneous neurofibromas are found in virtually all adults with NF1. They typically arise in the teenage years and progressively increase in number and size throughout life. The progressive nature has been well-documented in a population study conducted in southeast Wales, which scored individuals with NF1 for the number of dermal neurofibromas by age in decades (Huson et al. 1989); there were few neurofibromas in children less than 10 years, whereas individuals older than 50 years invariably had hundreds. Even though they are usually asymptomatic, some tumors tend to itch and they can be painful if located at sites of irritation or pressure, such as belt or bra lines. Cutaneous neurofibromas are always benign; they never transform to a malignant phenotype. To some extent, these neurofibromas are of cosmetic concern; however, they can lead to symptoms and sometimes enlarge as pedunculated tumors or become infected and/or inflamed. Spurts of neurofibroma growth, both in size and numbers, coincide with puberty and pregnancy; however, intermittent growth persists as dermal neurofibromas progressively arise throughout life. The hormonal milieu that causes these growth spurts is not known, but estrogen effects do not seem to account for the pathological growth because there is not a significant change in neurofibroma development or growth for individuals who are either on low-dose birth control medication or are postmenopausal.

Plexiform neurofibromas are found in approximately 25% of all individuals with NF1 (Huson et al. 1989), although the use of imaging modalities to detect internal paraspinal tumors in asymptomatic individuals suggests an incidence of plexiform neurofibromas that approaches 50%. Diffuse plexiform neurofibromas tend to emerge earlier in life, even at birth. These tumors are usually solitary and tend to become quiescent in older age. Periods of rapid tumor growth are followed by long periods of no growth; however, it is not clear what promotes their cellular proliferation. Plexiform neurofibromas can be a cause of morbidity by virtue of their location, size, and vascular nature. As space-occupying lesions, internal plexiform neurofibromas can impinge on vital organs and cause myriad symptoms. Thus, there should be a high index of suspicion for the presence of internal plexiform neurofibromas in individuals who are symptomatic or demonstrate abnormal physical or neurologic signs. Plexiform neurofibromas are associated with bony overgrowth, hemihyperplasia, and scoliosis. The mainstay of treatment has been surgical resection; however, targeted oral therapy with Ras-pathway MEK inhibitors has been effective in decreasing the size of plexiform neurofibromas by at least 20% volume in two-thirds of individuals with NF1 (Dombi et al. 2016). Joint surveillance between health care providers including orthopedic surgery, neurosurgery, neurology, and oncology is important in providing options of treatment for individuals with symptomatic and/or progressive plexiform neurofibromas.

Sarcomas. Plexiform neurofibromas have the capacity to undergo malignant transformation. This is most notable in long-standing, benign-appearing plexiform tumors that undergo rapid growth with associated pain. Malignant peripheral nerve sheath tumors (MPNST), sometimes inappropriately called neurofibrosarcomas or malignant schwannomas, arise in approximately 10–15% of individuals

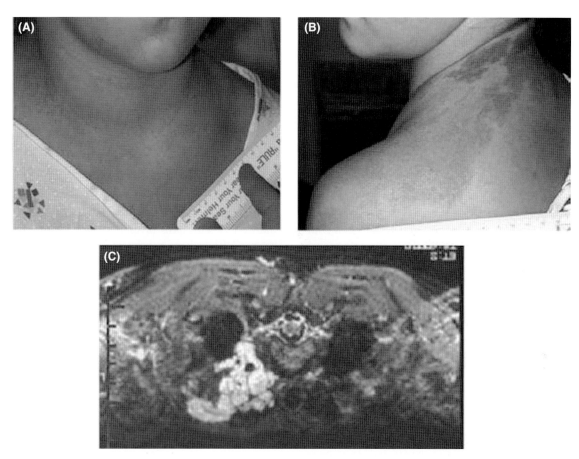

FIGURE 40.4 Plexiform neurofibroma in a prepubertal female. The presence of a plexiform neurofibroma can be associated with asymmetry or pigmentary changes of overlying skin. (A) On anterior chest inspection, there is fullness to the left clavicular region. (B) On posterior examination there is irregular hyperpigmentation. (C) MRI of the neck and thoracic outlet shows the presence of a multilobular nerve sheath tumor that appears relatively extensive. However, this tumor has not progressed over time, and it has not led to significant symptoms or medical complications.

with NF1 (Evans et al. 2002; McCaughan et al. 2007) and almost half of all MPNSTs identified in population-wide cancer registries are in individuals with NF1. In general, NF1-associated MPNSTs emerge in the third and fourth decades, although children less than 10 years old with MPNST have been reported. This propensity for malignant transformation of a small percentage of plexiform neurofibromas warrants diligence in monitoring the growth and associated symptoms of these tumors. Because plexiform neurofibromas are unpredictable in growth and heterogeneous in cellularity, the decision to surgically resect a tumor should involve a sarcoma team, and complete pathologic evaluation of the specimen should be performed to identify atypical plexiform neurofibroma, ANNUBP (atypical neurofibromatous neoplasms of uncertain biologic potential), or low-grade malignant peripheral nerve sheath tumor within a benign-appearing tumor (Miettinen et al. 2017). Plexiform neurofibromas that undergo malignant transformation portend a poor prognosis, and, due to the accumulation of mutations in other cancer pathways, targeted therapy with MEK inhibitors that are effective for plexiform neurofibromas may not prove effective for MPNST. Rhabdomyosarcoma and angiosarcoma are rare sarcomas in NF1; however, they are more prevalent than in the general population and often arise in childhood.

CNS Tumors Approximately 15% of individuals with NF1 have a low-grade glioma involving the optic nerve pathway, and half of these tumors are symptomatic. They tend to arise in early- to mid-childhood years, usually before age 10 years. Symptoms include loss of visual acuity, decreased field of vision, agitation in young children, and behavior changes. Signs of optic pathway glioma include proptosis, strabismus, optic nerve pallor, and increased optic disk fullness. These tumors are always low grade; however, cellular proliferation of type II pilocytic astrocytes can impinge on the optic nerve leading to vision loss and interfere with the hypothalamic-pituitary axis leading to precocious puberty for those tumors involving the hypothalamus. Treatment is indicated when there is either tumor-related visual impairment or tumor volume progression documented by sequential MRI scans of the brain and orbit. Carboplatin/vincristine chemotherapy has been effective in preserving vision in most all cases; however, rapid progression in a small percentage of tumors has caused unrecoverable loss of vision. A unique observation of NF1-associated optic nerve pathway gliomas is the lack of invasiveness as compared with non-NF1 gliomas; less-aggressive intervention may hinge on a molecular diagnosis of NF1. Other areas of the brain can have low-grade gliomas, and those involving the tectal plate can lead to hydrocephalus. T_2-hyperintensities (focal areas of signal intensity – FASI) and low-grade gliomas on MRI scans rarely need ongoing surveillance; however, symptomatic individuals with headaches, focal seizures, or behavior changes are followed with sequential MRIs to determine if patients with progressive tumors may benefit from neurosurgical resection. Targeted therapy for NF1-related gliomas is not yet available, although clinical trials with MEK inhibitors are underway. The second most common malignancy in NF1, after MPNST, is brain cancer, most notably high-grade astrocytoma, which affects ~2% of the NF1 population (Seminog and Goldacre 2013). Similar to NF1-MPNST, high-grade NF1-astrocytoma carries poor prognosis.

Other Cancers

The actual incidence of cancer in individuals with NF1 has not been accurately determined globally, in part because of bias of ascertainment. In a Danish cohort study (Sorenson et al. 1986), an increased relative risk for malignancy in individuals with NF1 was ascertained in a hospital-based setting (relative risk = 4.0) versus their relatives with NF1 who were ascertained by virtue of their relationship to the proband (relative risk = 1.5). In the UK, the standardized incidence ratio (SIR) for overall cancer risk in the NF1 population was calculated as 2.7 times higher than the unaffected population (Walker et al. 2006). In a comprehensive study using both NF1 and cancer registries in Finland, the standardized incidence risk ratio (SIR) for any cancer was determined as 5 (95% confidence interval of 4.42–5.71) with a SIR breakdown of the following: MPNST SIR of 2056, adrenal medulla cancer SIR of 74.3, brain cancer SIR of 37.5, GIST (gastrointestinal stromal tumors) SIR of 34.2, breast cancer SIR of 3, and leukemia/lymphoma SIR of 1.19 (Uusitalo et al. 2016). This illustrates the primary concern of MPNST; however, it is still a rare tumor, whereas breast cancer is relatively common and presents at a younger age in women with NF1 (Uusitalo et al. 2016). This study supports data showing increased risk for breast cancer in the NF1 population (Howell et al. 2017), and potentially suggests benefits of more thorough screening in women with NF1 in their 30s (Sharif et al. 2007; Stewart et al. 2018). Although there are numerous case reports of relatively rare malignancies, individuals with NF1 have *not* been shown to have a higher likelihood of developing the more common epithelial cancers of the prostate, lung, and colon. Melanoma and basal cell carcinomas are also not increased.

The interaction of neurofibromin with Ras would suggest that individuals with NF1 should be predisposed to similar cancers that have *RAS* mutations; however, the only shared tumor types between inactivating *NF1* and activating *RAS* mutations are the relatively rare juvenile myelomonocytic leukemia (jMML) and childhood rhabdomyosarcomas. Even though few children with NF1 develop jMML, there is a clear association. Approximately 10% of children who have this rare form of leukemia also have NF1. This malignancy clearly demonstrates the important role of the Ras-neurofibromin pathway in specific myelogenous stem cells. Individuals who have juvenile myelomonocytic leukemia, but not NF1, generally have an oncogenic *NRAS* mutation, whereas those individuals with NF1 have inactivation of both *NF1* alleles with concomitant deregulation of ras signaling in myelocytic stem cells (Side et al. 1997). This observation does not hold true for other tumors associated with a high incidence of oncogenic *RAS* mutations. Tumors with the highest incidence of *RAS* mutations (i.e. bladder cancer) are not typically seen in NF1; however, rhabdomyosarcoma has an increased relative risk for the NF1 pediatric population (Uusitalo et al. 2016). This implies that the "tumor suppressor" function of neurofibromin may be more complex than simple down-regulation of wild-type Ras.

Evaluation

- Dermal neurofibromas are visible on skin surfaces and can be monitored for size, vascularity, and inflammation. They do not transform and do not require imaging.
- Plexiform neurofibromas should be suspected if there is soft tissue asymmetry, skin pigmentation, or unusual hair patterns. MRI with fat suppressing short T1 inversion recovery (STIR) sequences is optimal for determination of the extent and location of plexiform neurofibromas, and contrast is indicated if the tumor is adjacent to central nervous system tissues. Computed tomography evaluations are usually not informative for these soft tissue tumors and expose individuals to unnecessary radiation. Sequential MRI examinations may be required to assess progression.
- Vascularity may be important in evaluating plexiform neurofibromas pre-surgically; therefore, magnetic resonance angiography may be beneficial in determining the vascular supply to the tumor.
- Ophthalmologic evaluation should occur annually in children known to have NF1 and those provisionally diagnosed with NF1. This should include field of vision, visual acuity, color vision, direct and indirect fundoscopic examination, slit-lamp examination, and

consideration of visual evoked potential testing (Listernick et al. 1997; Listernick et al. 2007).

- Any suspicion of an optic nerve pathway tumor warrants both brain and dedicated orbit/face/neck MRI studies (with and without contrast). The need for treatment should be carefully assessed with the oncology team. Two indications for chemotherapy are visual disturbances and/or evidence of increasing tumor size.
- If an optic nerve pathway tumor is asymptomatic, watchful observation and generously spaced brain MRI scans through childhood and adolescence are appropriate.
- If an optic nerve pathway tumor is symptomatic and non-progressive, appropriate management includes repeating the scan in three to six months, in association with a visual evoked potential study (if available), and ophthalmologic evaluation. The timing of follow-up imaging and ophthalmologic evaluation varies, depending on the index of suspicion, reliability of visual testing, and age of the individual. At a minimum there should be yearly evaluations, and, depending on previous MRI scans and results of evaluation, an MRI can be deferred indefinitely once non-progression has been demonstrated. New-onset visual symptoms requires immediate assessment for the possibility of tumor recurrence, which, although rare, requires re-initiation of intervention.
- Precocious puberty in NF1 is almost always associated with an optic nerve glioma that involves the hypothalamic-pituitary axis; therefore, a brain MRI should always be performed when precocious puberty is suspected.
- Individuals with NF1 are not treated with radiation because they are at high risk for secondary tumors and moyamoya disease if radiation involves the cranium.
- Clinical history of pain and/or increasing size of a mass is an indication for radiological imaging, usually with MRI, for any tumor. Use of enhancement may help in discerning whether a solid tumor is benign or malignant. Sequential scans should be used to assess interval growth. Individuals with an *NF1* microdeletion have a higher tumor burden, and deserve closer surveillance monitoring for internal plexiform neurofibromas and malignant peripheral nerve sheath tumors (MPNSTs).
- Tumors that are painful or demonstrate new growth or texture changes could be malignant. An important discriminating sign for malignant transformation is pain that wakes someone from sleep. The surgical approach to resection of malignant peripheral nerve sheath tumors is very different from that for benign neurofibromas. The determination of malignancy status is quite difficult for surgeons who are not trained in sarcoma management, and inappropriate surgical technique could result in less favorable outcome. Unlike benign neurofibromas that only require surgical resection, malignant peripheral nerve sheath tumors are treated in situ with combined treatment protocols. Thus, it is imperative for symptomatic plexiform neurofibromas to be evaluated by specialists who are affiliated with a sarcoma team. This evaluation may require travel to regional cancer centers, but this additional effort and expense could be lifesaving. The use of PET/CT imaging can help differentiate benign plexiform neurofibromas and MPNSTs (Warbey et al. 2009).
- CBC and blood smear can determine if someone suspected of having juvenile myelomonocytic leukemia needs referral to hematologist/oncologist.

Management

- Tumor management is a key aspect of care for almost all individuals with NF1. Collaboration with a neurofibromatosis clinic and oncology teams familiar with intracranial gliomas and sarcomas is extremely beneficial.
- Dermal neurofibromas can be surgically resected if they are symptomatic. Resection should be performed by an experienced dermatologist or plastic surgeon through deep excision to take the entire neurofibroma. Shave excisions are not recommended.
- If plexiform neurofibromas are symptomatic, or if there is concern about malignant transformation, they should be widely excised with full knowledge that the entire tumor is likely not resectable without resultant neurological deficit. Extensive imaging with MRI (with and without contrast), magnetic resonance angiography, and possibly positron emission tomography scanning may be needed for the surgical oncology team to fully assess the extent of the tumor and potential for malignant transformation (Ferner et al. 2008). If there is any concern, referral to a sarcoma center is beneficial to establish optimal management approaches before initial surgery or biopsy.
- For excision of large benign plexiform neurofibromas that are highly vascular, embolization of the tumor by arterial injection prior to surgery is sometimes helpful to decrease the vascularity. Pathological evaluation of the entire tumor, including inked margins, by routine and specialized immunohistochemical staining is essential to determine the malignant potential.
- Pathological evaluation of biopsied tissue by those experienced in peripheral nerve sheath tumor diagnosis is paramount in decisions regarding treatment protocol selection, specifically decisions regarding chemotherapy and local radiation therapy. If a peripheral

nerve sheath tumor is malignant, as noted by cell atypia, necrosis, or elevated mitotic index, then surgical management can be planned accordingly. Without this information, some presumed benign peripheral nerve sheath tumors could be excised with narrow margins, and, after pathological review, a second surgical procedure may be needed to obtain wider margins around a malignant tumor (Bernthal et al. 2014).

- Early surgical resection for plexiform neurofibromas in the orbitotemporal region should be considered. This may subsequently entail numerous surgical procedures over time; however, this is sometimes preferred to an extensive operation on a large facial plexiform neurofibroma. Referral to plastic surgery and/or otolaryngology subspecialists in reconstructive surgery is recommended. With the advent of MEK inhibitor therapies, surgical resections may be limited to those individuals whose response of tumor shrinkage may still leave individuals with symptoms or ongoing concerns for malignant transformation.

- Clinical trials to treat plexiform neurofibromas with interferon, retinoic acid, farnesyl-transferase inhibitors, thalidomide, rapamycin compounds, and carboplatin with or without vincristine have been used with minimal efficacy. Other agents targeting the Ras pathway have been more successful (Dombi et al. 2016). Access to active clinical trials through websites such as http://clinicaltrials.gov or *ctf.org* is important to provide options for treatment of NF1-related plexiform neurofibromas and other tumors.

- Radiation therapy for benign tumor management in NF1 (nerve sheath tumors and gliomas) is contraindicated. This treatment is reserved for malignant peripheral nerve sheath tumor protocols.

- If an optic pathway tumor is symptomatic and progressive, current recommendations include implementation of US Childhood Oncology Group-approved protocols, either carboplatin alone, carboplatin + vincristine, and/or telozolamide or MEK inhibitor (see http://clinicaltrials.gov).

- If an optic nerve pathway tumor progresses on therapy there is not a current recommended treatment. The risk for secondary tumors and moyamoya vasculopathy caused by radiation therapy precludes its use, and additional chemotherapeutic regimens are under investigation. Radiation therapy should not be used except under unusual circumstances, and only with full discussion about risks for secondary tumors and intracranial vascular problems.

- Given that NF1 fits the paradigm of a cancer predisposition syndrome, and the *NF1* gene is considered a "tumor suppressor," there is some concern that children who are treated for malignancies with radiation or chemotherapy may be at higher risk for second tumors. The decision to treat low-grade malignant peripheral nerve sheath tumors or rapidly growing plexiform neurofibromas with non-surgical modalities should include recognition of the potential for second malignancies.

- Juvenile myelomonocytic leukemia generally requires hematopoietic stem cell transplant in a standard manner.

Musculoskeletal

Sphenoid Wing Dysplasia. The sphenoid bones are comprised of multiple ossification centers that fuse to become the essential components of the orbits. They are embryologically derived from paraxial mesenchyme and neural crest derivatives. Approximately 2–5% of individuals with NF1 have sphenoid wing dysplasia (Friedman and Birch 1997), and it is almost always a unilateral deficiency of bone (Figure 40.5). It usually causes an overlying change in the orbit structure, and with loss of bone separation between the globe and brain can lead to proptosis and/or buphthalmus (pulsation from brain arterial flow). At least half of those individuals with sphenoid wing dysplasia will develop an ipsilateral temporal-orbital plexiform neurofibroma. Likewise, individuals who are initially

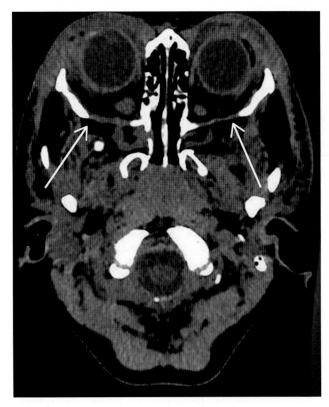

FIGURE 40.5 Computed tomography scan demonstrating sphenoid wing dysplasia. Note deficiency of the bony elements (arrows) in patient with bilateral temporal-orbital plexiform neurofibromas.

identified with a temporal-orbital tumor will often have an underlying sphenoid wing dysplasia. The etiology of this association is not known. Given concerns about intracranial radiation and the association with orbitotemporal neurofibromas, it is more appropriate to delineate this bone abnormality by MRI with inclusion of bone windows rather than CT, unless imaging is necessary for surgical intervention.

Pseudarthrosis. Long-bone bowing, with or without pseudarthrosis, affects 2–5% of individuals with NF1 (Stevenson et al. 1999). The tibia is the most common site, and usually presents with a distinctive anterior-lateral bowing in infancy, and radiographs demonstrate cortical thickening with bone marrow narrowing of the long bone (Figure 40.6). In cases of tibial dysplasia, the fibula may be affected as well (Stevenson et al. 2009). Bracing is used to minimize fracture risk, but many will progress to fracture and pseudarthrosis. This "false joint" represents a failure of union after fracture. The incidence of fracture is higher in males than females, and the etiology of the sex difference has not been determined. The bone abnormality that leads to bowing and subsequent fracture is not understood. It is not associated with neurofibromas and appears to be a localized dysplastic process of bone. The critical time for fracture and non-union is infancy to early childhood. Bone remodeling and strengthening seems to improve with age, yet fractures beyond the middle childhood years occur. Other fractures that are independent of the dysplastic long bone bowing tend to heal easily in NF1. There is no known genotype–phenotype correlation for pseudarthrosis.

Scoliosis. Individuals with NF1 experience both idiopathic scoliosis and dystrophic scoliosis (Crawford and Schorry 1999; Crawford and Schorry 2006). The incidence of scoliosis in NF1 is 10–30%. The idiopathic form of scoliosis is not different from that in the general population, but the dystrophic form is a progressive, short-segment (spans four to six vertebral bodies), angular scoliosis that is debilitating.

Even though some cases of dystrophic scoliosis are associated with paraspinal plexiform neurofibromas, the association with tumors is overstated. There appears to be an intrinsic abnormality of NF1 bone that is poorly understood. Dystrophic scoliosis presents in early childhood and is progressive, causing significant rotation of the thoracic cage with variable degrees of pulmonary compromise. Evolution of the curvature may take place over a short period of time and requires close monitoring. The idiopathic form of scoliosis in NF1 is not different from that in the general population; however, it may be difficult at times to distinguish dystrophic segments at an early stage. The variability in progression of scoliosis makes prognostic counseling difficult, and surgical correction needs to be balanced with the knowledge of the dynamic nature of spinal change through growth. In general, orthopedic surgeons recognize the difficulties in managing scoliosis in the context of NF1, and establishing a diagnosis of NF1 before surgical intervention is important (Crawford and Schorry 1999).

Coordination and Strength. Individuals with NF1 are noted by their parents as having poor coordination, decreased strength, and less stamina than their sibs. This may be due in part to early hypotonia; however, a number of studies have

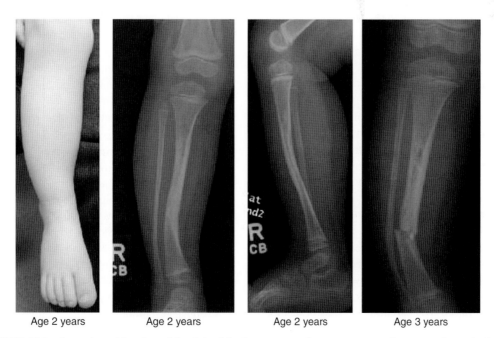

FIGURE 40.6 Anterolateral bowing of the right tibia that progressed over one year to fracture and pseudarthrosis.

demonstrated these perceptions and mouse models indicate altered muscle that may contribute to this phenotype. Children with NF1 do not tend to excel in sport activities, and physical therapy interventions have been helpful for some (Souza et al. 2009; Johnson et al. 2012).

Evaluation

- The presence of sphenoid wing dysplasia should alert one to the potential of an associated temporal-orbital plexiform neurofibroma, and MRI should be considered for early detection and intervention. Buphthalmus may require surgical intervention, and CT bone imaging may be needed under such circumstances.
- A primary concern relating to pseudarthrosis is the early recognition of tibial bowing. The presence of anterolateral bowing on physical examination requires radiologic examination with comparison films of the contralateral limb. Orthopedic referral for long bone bowing management is essential.
- Annual examination for curvature of spine or back asymmetry is important for early detection of dystrophic scoliosis. The Adams bend-over exam is helpful to detect curvature and scapular asymmetry. Spine radiographs, posterior-anterior and lateral views, should be performed in any suspected case of scoliosis. Serial examination is imperative, with close monitoring for dystrophic change.
- Referral to an orthopedist with experience in NF1 is highly recommended for individuals who are suspected of having scoliosis. Before surgical intervention, an MRI of the spine should be performed to identify paraspinal plexiform neurofibromas.
- Physical therapy evaluation for assessment of muscle strength and coordination is indicated in the preschool years.

Management

- Management of sphenoid wing dysplasia is complex, and if surgery is warranted it should be performed by plastic surgeons in collaboration with a neurosurgeon. There are complications of bone grafting in this procedure, and surgeries should only be performed by experienced teams. The possibility of an associated plexiform neurofibroma should be assessed as part of the preoperative evaluation.
- Management of pseudarthrosis is one of the most difficult aspects of NF1 care. Standardized orthopedic protocols have not been established (Crawford and Schorry 1999). Below-the-knee amputation is a potential option that should be raised with parents at the time of discussions for long-term management.
- Early recognition of long bone bowing with the implementation of bracing to prevent fracture, rather than correct the bowing, is imperative. Bracing is effective and may enable bone to mature to a point at which fracture is less likely.
- Once there is a fracture, non-union leading to pseudarthrosis is a likely possibility. A number of orthopedic procedures can be implemented for management of non-union; however, few are consistently effective. Most protocols focus on stabilization with an intramedullary rod, and provision of adequate blood supply, and autogenous bone grafting at the site of the fracture (Coleman et al. 1995). Multiple operations are typical, and long recuperation periods including use of orthopedic devices can be extremely stressful for families and deleterious for normal development in the child. Bone morphogenetic protein-2 (BMP-2) has been used by many surgeons to enhance osteoblast anabolic response post-operatively (Richards et al. 2010).
- In the event a stable union cannot be maintained, below-the-knee amputation is a therapeutic option. Prosthetics for amputations are available, and they enable children to remain in school and fully engaged in recreational activities. The therapeutic option of amputation is important to consider when multiple surgical procedures for long-term management of lower leg pseudarthrosis are anticipated.
- Conventional bracing is indicated for idiopathic scoliosis.
- The dystrophic form of scoliosis does not respond to back bracing and usually requires fusion, hopefully after overall growth is complete. Complete correction is usually not possible; therefore, defining successful outcomes should be clear before surgery. The implementation of growing rods to optimize growth prior to definitive surgical intervention is feasible and recommended to be performed by an experienced orthopedic team.
- Surgical procedures for dystrophic scoliosis involve fusion, both anterior and posterior. Postoperative care, especially pain management and treatment of hypertension, requires close monitoring and coordinated care between primary care providers and orthopedists.
- Psychosocial stressors in management of severe orthopedic manifestations should be recognized and addressed. Rehabilitation is a long, arduous process that affects ambulation, recreational participation, and school activities. Even though scoliosis and long-bone fracture with non-union are not unique to NF1, families and health care professionals should recognize differences between NF1-related bone abnormalities versus the unaffected population. Fostering independence, even by providing a wheelchair, is an important component of care for the primary health care provider.
- Physical therapies should be implemented to enhance strength and endurance. Presently, there are no medicinal treatments for musculoskeletal manifestations of NF1.

Cardiovascular

Hypertension. There are three primary causes of NF1-associated hypertension: renal vascular disease, tumors that secrete vasoactive compounds (i.e. adrenal tumors), and coarctation of the aorta. However, essential hypertension is still the most common cause of high blood pressure in NF1, as it is in the general population. Among NF1-related causes, renal vascular disease is the most common etiology, affecting approximately 4% of all affected individuals. The major site of involvement is the renal artery, where it can present either as a fusiform arterial narrowing or an aneurysm. In addition to involving the afferent renal arteries, there can be intra-renal vascular lesions that are difficult to detect by angiography. The pathophysiology of hypertension in renovascular disease in NF1 manifests as abnormal regulation of the renin-angiotensin pathway. Decreased blood flow to the affected kidney leads to increased renin secretion. The presenting sign of renal vascular disease is hypertension; therefore, the success of surgical correction is dependent on the site and degree of involvement. Surgical procedures may not always effectively treat hypertension caused by renovascular disease.

Tumors that cause NF1-related hypertension include pheochromocytoma and ganglioneuroma. Although these are rare in NF1, an evaluation of hypertension should include a complete history addressing headache patterns, tachycardia and palpitations, and gastrointestinal abnormalities. The presence of symptoms of catechol-secreting tumors should direct further diagnostic workup.

Finally, although rare, some individuals with NF1 have coarctation of the thoracic aorta that can present with hypertension. Physical examination of pulses and four-limb blood pressures, if abnormal, should lead to a more detailed evaluation for a vascular cause of hypertension.

Long-term prognosis for NF1-related hypertension is dependent on the underlying etiology and surgical outcome.

Cardiac Defects. Approximately 2% of individuals with NF1 have cardiac defects, primarily pulmonic stenosis and aortic coarctation (Lin et al. 2000). These abnormalities are congenital, and progressive cardiac abnormalities like cardiomyopathy, as noted in other rasopathies, are not generally seen in NF1. Likewise, coronary artery disease is not a more significant concern than in the background population. Even though vascular disease is rare in affected children, it is recognized as a contributor to mortality and morbidity (Friedman et al. 2002) and should be considered in the evaluation of new-onset neurologic complications.

Vascular Disease. Vascular abnormalities in NF1 include vascular stenosis, aneurysms, and cerebrovascular lesions in up to 6% of children (Rea et al. 2009). There is increased incidence of cerebral artery narrowing that could become evident with hypotensive episodes leading to stroke-like events. The use of MR angiography when brain and/or cervical MRIs are performed could prove useful in identifying incidental vascular malformations that could have clinical significance later in life.

Evaluation

- Routine history and physical with blood pressure monitoring should be a part of every visit to health care providers and should be performed at least once a year for individuals who either have or are suspected of having NF1. Four-limb blood pressures should be obtained on at least one occasion.
- In those with symptomatic hypertension, ultrasound with Doppler and abdominal MRI and/or MRA should be obtained both to pursue suspected vascular abnormalities involving the kidneys and for visualization of renal artery compression by an adjacent plexiform neurofibroma as well as adrenal glands.
- Likewise, malignant hypertension assessment includes a 24 hour urine collection for catecholamines, homovanillic acid, vanillylmandelic acid, and creatinine should be obtained to assess secretion by a tumor. It is not clear how often this study should be performed; however, in cases of persistent non-episodic hypertension, the likelihood of discovering a catechol-secreting tumor on subsequent urine collections is low. Abdominal CT scan is sensitive in identifying some pheochromocytomas.
- Primary consideration should be given to evaluating renal vascular disease with angiography in cases where magnetic resonance imaging or abdominal ultrasound with doppler fails to detect renal vessel anomalies.

Management

- Treatment of hypertension depends on the etiology. Consideration should be given to surgical management for vascular and tumor etiologies. The surgical team should include a vascular surgeon, a general surgeon, a radiologist, and a nephrologist.
- Vascular abnormalities (i.e. moya-moya disease and arterial stenoses) are treated either medically or surgically by traditional means without special considerations for NF1.

RESOURCES

Textbooks and Articles

Ablon J (1999) *Living with a Genetic Disorder. The Impact of Neurofibromatosis 1.* Westport, CT: Auburn House.

Carey JC (1992) Health supervision and anticipatory guidance for children with genetic disorders (including specific recommendations for trisomy 21, trisomy 18, and neurofibromatosis). *Pediatr Clin North Am* 39:25–53.

Ferner RE, Huson SM, Thomas N, Moss C, Willshaw H, Evans DG, Upadhyaya M, Towers R, Gleeson M, Steiger D, Kirby A

(2007) Guidelines for the diagnosis and management of individuals with neurofibromatosis 1 (NF1). *J Med Genet* 44:81–88.

Friedman JM, Gutmann DH, MacCollin M, Riccardi VM (1999) *Neurofibromatosis: Phenotype, Natural History, and Pathogenesis*, 3rd ed. Baltimore, MD: Johns Hopkins Press.

Hersh JH, the Committee on Genetics for the American Academy of Pediatrics (2008) Clinical report: Health supervision for children with neurofibromatosis. *Pediatrics* 121:633–642.

Huson SM, Hughes RAC (1994) *The Neurofibromatoses: A Pathogenetic and Clinical Overview*. London: Chapman and Hall Medical.

Miller DT, Freedenberg D, Schorry E, Ullrich NJ, Viskochil D, Korf B, and the Council on Genetics (2019) Health Supervision for Children with Neurofibromatosis Type 1. *Clinical Report: Guidance for the Clinician in Rendering Pediatric Care*. American Academy of Pediatrics.

Riccardi V (1992) *Neurofibromatosis: Phenotype, Natural History, and Pathogenesis*. Baltimore, MD: Johns Hopkins University Press.

Rubenstein AE, Korf BR, eds (1990) *Neurofibromatosis: Handbook for Patients, Families, and Health-Care Professionals*. New York: Thieme Medical.

Upadhyaya M, Cooper D (1998) *Neurofibromatosis Type 1: From Genotype to Phenotype*. Oxford: Bios Scientific.

Williams VC, Lucas J, Babcock MA, Gutmann DH, Korf B, Maria BL (2009) Neurofibromatosis type 1 revisited. *Pediatrics* 123:124–133.

Support Organizations

Children's Tumor Foundation

Website: *www.ctf.org*

Neurofibromatosis Network

Website: *www.nfnetwork.org*

NF Europe – European Neurofibromatosis Association

Website: *https://nfeurope.wordpress.com*

The Littlest Tumor Foundation

Website: *https://littlesttumor.org*

There are numerous local support groups that are affiliated with international organizations. Likewise, neurofibromatosis multidisciplinary clinics are located in many academic institutions worldwide. Educational materials have been published from many of the organizations, and neurofibromatosis summer camps have been sponsored by national and international organizations. The Children's Tumor Foundation has embarked on an initiative to link Neurofibromatosis Clinics in a network across the US with a common goal of developing and implementing best practices in the care for individuals with NF1, NF2, and Schwannomatosis. This network will interact with other international clinics in the application of outcome measures to improve NF1 care.

REFERENCES

Abeliovich D, Gelman-Kohan Z, Silverstien S, Lerer I, Chemke J, Merin S, Zlotogora J (1995) Familial café au lait spots: A variant of neurofibromatosis type 1. *J Med Genet* 32:985–386.

Ars E, Kruyer H, Gaona A, Casquero P, Rosell J, Volpini V, Serra E, Lazaro C, Estivill X (1998) A clinical variant of neurofibromatosis type 1: Familial spinal neurofibromatosis with a frameshift mutation in the NF1 gene. *Am J Hum Genet* 62:834–841.

Barker D, Wright E, Nguyen K, Cannon L, Fain P, Goldgar D, Bishop D, Carey J, Baty B, Kivlin J, Willard H, Waye J, Greig G, Leinwald L, Nakamura Y, O'Connell P, Leppert M, Lalouel J, White R, Scolnick M (1987) Gene for von Recklinghausen neurofibromatosis is in the pericentric region of chromosome 17. *Science* 236:1100–1102.

Bernards A (1995) Neurofibromatosis type 1 and Ras-mediated signaling: Filling in the GAPs. *Biochim Biophys Acta* 1242:43–59.

Bernthal N, Putnam A, Jones K, Viskochil D, Randall R (2014) The effect of surgical margins on outcomes for low grade MPNSTs and atypical neurofibroma. *J Surg Oncol* 110:813–816.

Bilder D, Bakian A, Stevenson D, Carbone P, Cunniff C, Goodman A, McMahin W, Fisher N, Viskochil D (2016) Brief report: The prevalence of neurofibromatosis type 1 among children with autism spectrum disorder identified by the autism and developmental disabilities monitoring network. *J Autism Dev Disord* 46:3369–3376.

Brems H, Chmara M, Sahbatou M, Denayer E, Taniguchi K, Kato R, Somers R, Messiaen L, De Schepper S, Fryns JP, Cools J, Marynen P, Thomas G, Yoshimura A, Legius E (2007) Germline loss-of-function mutations in *SPRED1* cause a neurofibromatosis 1-like phenotype. *Nat Genet* 39:1120–1126.

Brunner HG, Hulsebos T, Stiejlen PM, der Kinderen DJ, Steen A, Hamel BCJ (1993) Exclusion of the neurofibromatosis 1 locus in a family with inherited café au lait spots. *Am J Med Genet* 46:472–374.

Carey J, Viskochil D (1999) Neurofibromatosis type 1: A model condition for the study of the molecular basis of variable expressivity in human disorders. *Am J Med Genet* 89:7–13.

Charrow J, Listernick R, Ward K (1993) Autosomal dominant multiple café au lait spots and neurofibromatosis-1: Evidence of non-linkage. *Am J Med Genet* 45:606–608.

Chisholm A, Anderson V, Pride N, Malarbi S, North K, Payne J (2018) Social function and autism spectrum disorder in children and adults with neurofibromatosis type 1: a systematic review and meta-analysis. *Neuropsychol Rev* 28:317–340.

Clementi M, Milani S, Mammi I, Boni S, Monciotti C, Tenconi R (1999) Neurofibromatosis type 1 growth charts. *Am J Med Genet* 87:317–323.

Coleman S, Coleman D, Biddulph G (1995) Congenital pseudarthrosis of the tibia: Current concepts of treatment. *Adv Oper Orthop* 3:121–145.

Cosyns M, Mortier G, Janssens S, Bogaert F, D'Hondt S, Van Borsel J (2012) Articulation in schoolchildren and adults with neurofibromatosis type 1. *J Commun Disord* 45:111–120.

Crawford A, Schorry E (1999) Neurofibromatosis in children: The role of the orthopaedist. *J Am Acad Orthop Surg* 7:217–230.

Crawford AH, Schorry EK (2006) Neurofibromatosis update. *J Pediatr Orthop* 26:413–423.

Crowe F, Schull W, Neel J (1956) *A Clinical, Pathological, and Genetic Study of Multiple Neurofibromatosis.* pp. 1–181. Springfield, IL: Thomas.

Crowe F (1964) Axillary freckling as a diagnostic aid in neurofibromatosis. *Ann Intern Med* 61:1142 pp. 1–1811143.

De Luca A, Bottillo I, Sarkozy A, Carta C, Neri C, Bellacchio E, Schirinzi A, Conti E, Zampino G, Battaglia A, Majore S, Rinaldi MM, Carella M, Marino B, Pizzuti A, Digilio MC, Tartaglia M, Dallapiccola B (2005) NF1 gene mutations represent the major molecular event underlying neurofibromatosis—Noonan syndrome. *Am J Hum Genet* 77:1092–1101.

De Raedt T, Brems H, Wolkenstein P, Vidaud D, Pilotti S, Perrone F, Mautner V, Frahm S, Sciot R, Legius E (2003) Elevated risk for MPNST in NF1 microdeletion patients. *Am J Hum Genet* 72:1288–1292.

De Schepper S, Maertens O, Callens T, Naevaert J, Lambert J, Messiaen L (2008) Somatic mutation analysis in NF1 café au lait spots reveals two *NF1* hits in the melanocytes. *J Invest Dermatol* 128:1050–1053.

Denayer E, de Ravel T, Legius E (2008) Clinical and molecular aspects of RAS related disorders. *J Med Genet* 45:695–703.

Dombi E, Baldwin A, Marcus L, Fisher M, Weiss B, Kim A, Whitcomb P, Martin S, Aschbacher-Smith L, Rizvi T, Wu J, Ershler R, Wolters P, Therrien J, Glod J, Belasco J, Schorry E, Brofferio A, Starosta A, Gillespie A, Doyle A, Ratner N, Widemann B (2016) Activity of selumetinib in neurofibromatosis type 1-related plexiform neurofibromas. *N Engl J Med* 375:2550–2560.

Dunzendorfer-Matt T, Mercado E, Maly K, McCormick F, Scheffzek K (2016) The neurofibromin recruitment factor Spred1 binds to the GAP related domain without affecting Ras inactivation. *PNAS, USA* 113:7497–7502.

Eldridge R (1981) Central neurofibromatosis. *Adv Neurol* 29:57–65.

Evans DG, Baser ME, McGaughran J, Sharif S, Howard E, Moran A (2002) Malignant peripheral nerve sheath tumours in neurofibromatosis 1. *J Med Genet* 39:311–314.

Evans DG, O'Hara C, Wilding A, Ingham SL, Howard E, Dawson J, Moran A, Scott-Kitching V, Holt F, Huson SM (2011) Mortality in neurofibromatosis 1: in North West England: an assessment of actuarial survival in a region of the UK since 1989. *Eur J Hum Genet* 19:1187–1191.

Ferner RE, Golding JF, Smith M, Calonje E, Jan W, Sanjayanathan V, O'Doherty M (2008) 2-Fluoro-2-deoxy-d-glucose positron emission tomography (FDG PET) as a diagnostic tool for neurofibromatosis 1 (NF1) associated malignant peripheral nerve sheath tumours (MPNSTs): A long-term clinical study. *Ann Oncol* 19:390–394.

Ferner R, Thomas M, Mercer G, Williams V, Laschziner G, Afridi S, Golding J (2017) Evaluation of quality of life in adults with neurofibromatosis type 1 (NF1) using the impact of NF1 on quality of life (INF1-QOL). *Health Qual Life Outcomes* 15:34.

Friedman JM, Birch PH (1997) Type 1 neurofibromatosis: A descriptive analysis of the disorder in 1,728 patients. *Am J Med Genet* 70:138–143.

Friedman JM, Gutmann DH, MacCollin M, Riccardi VM (1999) *Neurofibromatosis: Phenotype, Natural History and Pathogenesis,* 3rd ed. Baltimore, MD: Johns Hopkins Press.

Friedman JM, Arbiser J, Epstein JA, Gutmann DH, Huot SJ, Lin AE, McManus B, Korf BR (2002) Cardiovascular disease in neurofibromatosis 1: Report of the NF1 Cardiovascular Task Force. *Genet Med* 4:105–111.

Garg S, Plasschaert E, Descheemaeker M, Huson S, Borghgraef M, Vogels A, Evans D, Legius E, Green J (2015) *J Autism Dev Disord* 45:1649–1657.

Gardner WJ, Frazier GH (1930) Bilateral acoustic neurofibromas: A clinical study and field survey of a family of five generations with bilateral deafness in 28 members. *Arch Neurol Psychiatry* 23:266–300.

Gutmann DH, Aylsworth A, Carey J, Korf B, Marks J, Pyeritz RE, Rubenstein A, Viskochil D (1997) The diagnostic evaluation and multidisciplinary management of neurofibromatosis 1 and neurofibromatosis 2. *JAMA* 278:51–57.

Hager CM, Cohen PR, Tschen JA (1997) Segmental neurofibromatosis: Case reports and review. *J Am Acad Dermatol* 37:864–869.

Howell S, Hockenhull K, Salih Z, Evans D (2017) Increased risk of breast cancer in neurofibromatosis type 1: current insights. *Breast Cancer* 9:531–536.

Hulsebos TJ, Plomp AS, Wolterman RA, Robanus-Maandag EC, Baas F, Wesseling P (2007) Germline mutation of INI1/SMARCB1 in familial schwannomatosis. *Am J Hum Genet* 80:805–810.

Huson SM, Hughes RAC (1994) *The Neurofibromatoses: A Pathogenetic and Clinical Overview.* London: Chapman and Hall Medical.

Huson S, Clark D, Compston D, Harper P (1989) A genetic study of von Recklinghausen neurofibromatosis in South East Wales. I. Prevalence, fitness, mutation rate and effect of parental transmission on severity. *J Med Genet* 26:704–711.

Johnson B, Macwilliams B, Carey J, Viskochil D, D'Astous J, Stevenson D (2012) Lower extremity strength and hopping and jumping ground reaction forces in children with neurofibromatosis type 1. *Hum Mov Sci* 31:247–254.

Kehrer-Sawatzki H, Mautner V, Cooper D (2017) Emerging genotype-phenotype relationships in patients with large NF1 deletions. *Hum Genet* 136:349–376.

Kehrer-Sawatzki H, Farschtschi S, Mautner V, Cooper D (2017) Molecular pathogenesis of schwannomatosis, a paradigm for the co-involvement of multiple tumour suppressor genes in tumorigenesis. *Hum Genet* 136:129–148.

Korf BR (1992) Diagnostic outcome in children with multiple café au lait spots. *Pediatrics* 90:924–927.

Legius E, Schrander-Stumpel C, Schollen E, Pulles-Heintzberger C, Gewillig M, Fryns J (2002) *PTPN11* mutations in LEOPARD syndrome. *J Med Genet* 39:571–574.

Lin AE, Birch PH, Korf BR, Tenconi R, Niimura M, Poyhonen M, Armfield Uhas K, Sigorini M, Virdis R, Romano C, Bonioli E, Wolkenstein P, Pivnick EK, Lawrence M, Friedman JM (2000) Cardiovascular malformations and other cardiovascular abnormalities in neurofibromatosis 1. *Am J Med Genet* 95:108–117.

Listernick R, Louis D, Packer R, Gutmann D (1997) Optic pathway gliomas in children with neurofibromatosis 1: Consensus statement from the NF1 optic pathway glioma task force. *Ann Neurol* 41:143–149.

Listernick R, Ferner RE, Liu GT, Gutmann DH (2007) Optic pathway gliomas in neurofibromatosis-1: Controversies and recommendations. *Ann Neurol* 61:189–198.

Maertens O, De Schepper S, Vandesompele J, Brems H, Heyns I, Janssens S, Speleman F, Legius E, Messiaen L (2007) Molecular dissection of isolated disease features in mosaic neurofibromatosis type 1. *Am J Hum Genet* 81:243–251.

Martin G, Viskochil D, Bollag G, McCabe P, Cosier W, Haubruck H, Conroy L, Clark R, O'Connell P, Cawthon R, Innis M, McCormick F (1990) The GAP-related domain of the NF1 gene product interacts with ras p21. *Cell* 63:843–849.

McCaughan JA, Holloway SM, Davidson R, Lam WW (2007) Further evidence of the increased risk for malignant peripheral nerve sheath tumour from a Scottish cohort of patients with neurofibromatosis type 1. *J Med Genet* 44:463–466.

Miettinen M, Antonescu C, Fletcher C, Kim A, Lazar A, Quezado M, Reilly K, Stemmer-Rachminov A, Steward D, Viskochil D, Widemann B, Perry A (2017) Histopathologic evaluation of atypical neurofibromatous tumors and their transformation into malignant peripheral nerve sheath tumor in patients with neurofibromatosis 1 – a consensus overview. *Hum Pathol* 67:1–10.

Messiaen L, Yao S, Brems H, Callens T, Sathienkijkanchai A, Denayer E, Spencer E, Arn P, Babovic-Vuksanovic D, Bay C, Bobele G, Cohen BH, Escobar L, Eunpu D, Grebe T, Greenstein R, Hachen R, Irons M, Kronn D, Lemire E, Leppig K, Lim C, McDonald M, Narayanan V, Pearn A, Pedersen R, Powell B, Shapiro LR, Skidmore D, Tegay D, Thiese H, Zackai EH, Vijzelaar R, Taniguchi K, Ayada T, Okamoto F, Yoshimura A, Parret A, Korf B, Legius E. (2009) Clinical and mutational spectrum of neurofibromatosis type 1-like syndrome. *JAMA* 302:2111–2118.

National Institute of Health Consensus Development Conference (1988) Neurofibromatosis: Conference statement. *Arch Neurol* 45:575–578.

Noll RB, Reiter-Purtill J, Moore BD, Schorry EK, Lovell AM, Vannatta K, Gerhardt CA(2007) Social, emotional, and behavioral functioning of children with NF1. *Am J Med Genet* A143:2261–73.

North KN, Riccardi V, Samango-Sprouse C, Ferner R, Moore B, Legius E, Ratner N, Denckla MB (1997) Cognitive function and academic performance in neurofibromatosis 1: Consensus statement from the NF1 cognitive disorders task force. *Neurology* 48:1121–1127.

Ozonoff S (1999) Learning disorders in neurofibromatosis type 1 (NF1). *Am J Hum Genet* 89:45–52.

Pandit B, Sarkozy A, Pennacchio LA, Carta C, Oishi K, Martinelli S, Pogna EA, Schackwitz W, Ustaszewska A, Landstrom A, Bos JM, Ommen SR, Esposito G, Lepri F, Faul C, Mundel P, López Siguero JP, Tenconi R, Selicorni A, Rossi C, Mazzanti L, Torrente I, Marino B, Digilio MC, Zampino G, Ackerman MJ, Dallapiccola B, Tartaglia M, Gelb BD (2007) Gain-of-function *RAF1* mutations cause Noonan and LEOPARD syndromes with hypertrophic cardiomyopathy. *Nat Genet* 39:1007–1012.

Parada LF, Kwon CH, Zhu Y (2005) Modeling neurofibromatosis type 1 tumors in the mouse for therapeutic intervention. *Cold Spring Harb Symp Quant Biol* 70:173–176.

Plasschaert E, Descheemaeker M, Van Eylen L, Steaert J, Legius E (2015) Prevalence of autism spectrum disorder symptoms in children with neurofibromatosis type 1. *Am J Med Genet B* 168B:72–80.

Rasmussen SA, Yang Q, Friedman JM (2001) Mortality in neurofibromatosis 1: An analysis using U.S. death certificates. *Am J Hum Genet* 68:1110–1118.

Razzaque MA, Nishizawa T, Komoike Y, Yagi H, Furutani M, Amo R, Kamisago M, Momma K, Katayama H, Nakagawa M, Fujiwara Y, Matsushima M, Mizuno K, Tokuyama M, Hirota H, Muneuchi J, Higashinakagawa T, Matsuoka R (2007) Germline gain-of-function mutations in *RAF1* cause Noonan syndrome. *Nat Genet* 39:1013–7.

Rea D, Brandsema J, Armstrong D, Parkin P, deVeber G, MacGregor D, Logan W, Askalan R (2009) Cerebral arteriopathy in children with neurofibromatosis type 1. *Pediatrics* 124:e476–483.

Riccardi V (1992) *Neurofibromatosis: Phenotype, Natural History, and Pathogenesis*. Baltimore, MD: Johns Hopkins University Press.

Riccardi VM, Eichner JE (1986) *Neurofibromatosis: Phenotype, Natural History, and Pathogenesis*. Baltimore, MD: Johns Hopkins University Press.

Richards B, Oetgen M, Johnston C. (2010) The use of rhBMP-2 in the treatment of congenital pseudarthrosis of the tibia. *J Bone Joint Surg Am* 92:177–185.

Rojnueangnit K, Xie J, Gomes A, et al. (2015) High incidence of Noonan syndrome features including short stature and pulmonic stenosis in patients carrying NF1 missense mutations affecting p.Arg1809: Genotype-phenotype correlation. *Hum Mutat* 36:1052–1063

Rubenstein AE, Korf BR, eds (1990) *Neurofibromatosis: Handbook for Patients, Families, and Health-Care Professionals*. New York: Thieme Medical.

Ruggieri M, Huson SM (2001) The clinical and diagnostic implications of mosaicism in the neurofibromatoses. *Neurology* 56:1433–1443.

Ruggieri M, Polizzi A, Spalice A, Salpietro V, Caltabioano R, D'Orazi V, Pavone P, Pirrone C, Magro G, Platania N, Cavallaro S, Muglia M, Nicita F (2015) The natural history of spinal neurofibromatosis: a critical review of clinical and genetic features. *Clin Genet* 87:401–410.

Seminog OO, Goldacre MJ (2013) Risk of benign tumors of nervous system, and of malignant neoplasms, in people with neurofibromatosis: population-based record-linkage study. *Br J Cancer* 108:193–198.

Serra E, Otero D, Gaona A, Kruyer H, Ars E, Estivill X, Lazaro C (1997) Confirmation of a double-hit model for the NF1 gene in benign neurofibromas. *Am J Hum Genet* 61:512–519.

Sharif S, Moran A, Huson SM, Iddenden R, Shenton A, Howard E, Evans DG (2007) Women with neurofibromatosis 1 are at a moderately increased risk of developing breast cancer and should be considered for early screening. *J Med Genet* 44:481–484.

Sherman L. Daston M. Ratner N (1998) Neurofibrom: Distribution, cell biology and role in neurofibromatosis type 1. In: *Neurofibromatosis Type 1: From Genotype to Phenotype*, Upadhyaya M, Cooper DN, eds, Oxford: Bios Scientific.

Side L, Taylor B, Cayoutte M, Connor E, Thompson P, Luce M, Shannon K (1997) Homozygous inactivation of NF1 in the bone marrows of children with neurofibromatosis type 1 and malignant myeloid disorders. *N Engl J Med* 336:1713–1720.

Sorenson SA, Mulvihill JJ, Nielsen A (1986) Long-term follow-up of von Recklinghausen neurofibromatosis: Survival and malignant neoplasms. *N Engl J Med* 314:1010–1015.

Souza J, Passos R, Guedes A, Rezende N, Rodrigues L (2009) Muscular force is reduced in neurofibromatosis type 1. *J Musculoskelet Neuronal Interact* 9:15–17.

Stewart D, Korf B, Nathanson K, Stevenson D, Yohay K (2018) Care of adults with neurofibromatosis type 1: A clinical practice resource of the American College of Medical Genetics and Genomics (ACMG). *Genet Med* 20:671–682.

Stevenson DA, Birch PH, Friedman JM, Viskochil DH, Balestrazzi P, Boni S, Buske A, Korf BR, Niimura M, Pivnick EK, Schorry EK, Short MP, Tenconi R, Tonsgard JH, Carey JC (1999) Descriptive analysis of tibial pseudarthrosis in patients with neurofibromatosis 1. *Am J Med Genet* 84:413–419.

Stevenson DA, Viskochil DH, Schorry EK, Crawford AH, D'Astous J, Murray KA, Friedman JM, Armstrong L, Carey JC (2007) The use of anterolateral bowing of the lower leg in the diagnostic criteria for neurofibromatosis type 1. *Genet Med* 9:409–412.

Stevenson D, Zhou H, Ashrafi S, Messiaen L, Carey J, D'Astous J, Santora S, Viskochil D (2006) Double inactivation of NF1 in tibial pseudarthrosis. *Am J Hum Genet* 79:143–148.

Stevenson D, Carey J, Viskochil D, Moyer-Mileur L, Slater H, Murray M, D'Astous J, Murray K (2009) Analysis of radiographic characteristics of anterolateral bowing of the leg before fracture in neurofibromatosis type 1. *J Pediatr Orthop* 29:385–392.

Swensen J, Viskochil D (2007) The ras pathway. In: *Inborn Errors of Development*, Epstein C (ed.) 2nd ed. Oxford University Press.

Szudek, J, Birch P, Friedman JM (2000) Growth charts for young children with neurofibromatosis 1 (NF1). *Am J Med Genet* 92:224–228.

Tassabehji M, Strachan T, Sharland M, Colley A, Donnai D, Harris R, Thakker N (1993) Tandem duplication within a neurofibromatosis type 1 (NF1) gene exon in a family with features of Watson syndrome and Noonan syndrome. *Am J Hum Genet* 53:90–95.

Thompson H, Viskochil D, Stevenson D, Chapman K (2010) Speech-language characteristics of children with neurofibromatosis type 1. *Am J Med Genet A* 152A:284–290.

Upadhyaya M, Cooper D, eds (1998) *Neurofibromatosis Type 1: From Genotype to Phenotype*. Oxford: Bios Scientific.

Upadhyaya M, Maynard J, Osborn M, Huson S, Ponder M, Ponder B, Harper P (1995) Characterization of germline mutations in the neurofibromatosis type 1 (NF1) gene. *J Med Genet* 32:706–710.

Upadhyaya M, Huson S, Davies M, Thomas N, Chuzhanova N, Giovannini S, Evans D, Howard E, Kerr B, Griffiths S, Consoli C, Side L, Adams D, Pierpont M, Hachen R, Barnicoat A, Li H, Wallace P, Van Biervliet J, Stevenson D, Viskochil D, Baralle D, Hann E, Riccardi V, Trunpenny P, Lazaro C, Messiaen L (2007) An absence of cutaneous neurofibromas associated with a 3-bp inframe deletion in exon 17 of the NF1 gene (c.2970-2972delAAT): Evidence of a clinically significant NF1 genotype-phenotype correlation. *Am J Hum Genet* 80:140–151.

Uusitalo E, Rantanen M, Kallionpaa R, Poyhonen M, Leppavirta J, Yla-Outinen H, Riccardi V, Pukkala E, Pitaniemi J, Peltonen S, Peltonen J (2016) Distinctive cancer associations in patients with neurofibromatosis type 1. *J Clin Oncol* 34:1978–1986.

Viskochil D, Carey J (1994). Alternate and related forms of the neurofibromatoses. In: *The Neurofibromatoses: A Pathogenetic and Clinical Overview*, Huson SM, Hughes RAC, eds. London: Churchill & Hall Medical.

Viskochil D, White R, Cawthon R (1993) The neurofibromatosis type 1 gene. *Annu Rev Neurosci* 16:183–205.

Walker L, Thompson D, Easton D, Ponder B, Ponder M, Frayling I, Baralle D (2006) A prospective study of neurofibromatosis type 1 cancer incidence in the UK. *Br J Cancer* 95:233–238.

Warbey V, Ferner R, Dunn J, Calonje E, O'Doherty M (2009) [18F] FDG PET/CT in the diagnosis of malignant peripheral sheath tumours in neurofibromatosis type-1. *Eur J Nucl Mol Imaging* 36:751–757.

Watson GH (1967) Pulmonic stenosis, café au lait spots, and dull intelligence. *Arch Dis Child* 42:303–07.

Wu R, Legius E, Robberecht W, Dumoulin M, Cassiman J-J, Fryns J-P (1996) Neurofibromatosis type I gene mutation in a patient with features of LEOPARD syndrome. *Hum Mutat* 8:51–55.

Yang F, Staser K, Clapp D (2012) The plexiform neurofibroma microenvironment. *Cancer Microenviron* 5:307–310.

Zlotogora J (1998) Germ line mosaicism. *Hum Genet* 102:381–386.

41

NOONAN SYNDROME

JUDITH E. ALLANSON
Department of Genetics, University of Ottawa, Children's Hospital of Eastern Ontario, Ottawa, Ontario, Canada

AMY E. ROBERTS
Department of Cardiology and Division of Genetics, Department of Medicine, Boston Children's Hospital, Boston, Massachusetts, USA

INTRODUCTION

Incidence

Noonan syndrome (NS), a common autosomal dominant multiple congenital anomaly syndrome, was first described over 50 years ago (Noonan and Ehmke 1963) although historical evidence of the phenotype dates back to the late 19th century. Many of the features of NS are similar to those seen in Turner syndrome (see Chapter 60), and this disorder has sometimes mistakenly been called "male Turner syndrome." However, NS occurs in both males and females although a study of de novo cases associated with advanced paternal age showed a significant sex-ratio bias favoring transmission to males (Tartaglia 2004b). The incidence of NS is reported to be between 1 in 1000 and 1 in 2500 (Mendez and Opitz 1985). Average age at diagnosis is 9 years (Sharland et al. 1992a).

Little is published about life expectancy. Mortality secondary to cardiac disease may be related to surgical intervention, circulatory collapse, or possible arrhythmia. A review of 371 individuals from seven European cardiac centers with confirmed molecular diagnosis of a RASopathy demonstrated patients with hypertrophic cardiomyopathy and age <2 years or young adult, as well as those with biventricular obstruction and *PTPN11* mutations, have the highest risk of cardiac death, but overall mortality was low (Calcagni 2017). In older individuals, arrhythmias and congestive cardiac failure may be more common than previously suspected (J Noonan, personal communication).

Diagnostic Criteria

Despite a lack of defined diagnostic criteria, the cardinal manifestations of NS are well delineated. These include short stature, congenital heart defects and/or hypertrophic cardiomyopathy, broad or webbed neck, chest deformity with pectus carinatum superiorly and pectus excavatum inferiorly, developmental delay of variable degree, cryptorchidism, and characteristic facies (Allanson 1987). Various coagulation defects and lymphatic dysplasias are common findings.

The facial appearance of NS is well established and shows considerable change with age, being most striking in the newborn period and middle childhood and most subtle in the adult (Allanson et al. 1985a). In the neonate, the main features are a tall forehead, hypertelorism with downslanting palpebral fissures (95%), low-set, posteriorly rotated ears with a thickened helix (90%), a deeply grooved philtrum with high, wide peaks to the vermillion border of the upper lip (95%), and a short neck with excess nuchal skin and low posterior hairline (55%). In infancy, the head appears relatively large with a small face tucked beneath a large cranium. Eyes are prominent, with horizontal fissures, hypertelorism, and thickened or ptotic lids. The nose has a depressed root, a wide base, and a bulbous tip (Figure 41.1). In childhood, facial appearance is often lacking in affect or expression, resembling a myopathy (Figure 41.2). By adolescence, facial shape is an inverted triangle, wide at the forehead and tapering to a pointed chin (Figure 41.3). Eyes

---

Cassidy and Allanson's Management of Genetic Syndromes, Fourth Edition.
Edited by John C. Carey, Agatino Battaglia, David Viskochil, and Suzanne B. Cassidy.
© 2021 John Wiley & Sons, Inc. Published 2021 by John Wiley & Sons, Inc.

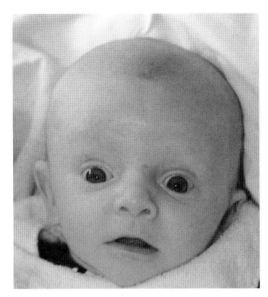

FIGURE 41.1 An infant with Noonan syndrome showing the tall, boxy forehead, thick and hooded eyelids, small upturned nose, and pointed chin.

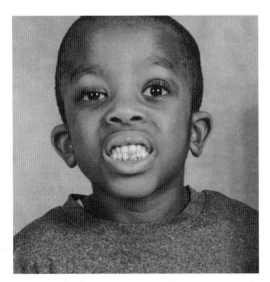

FIGURE 41.2 A child with Noonan syndrome demonstrating thickened droopy eyelids, epicanthal folds, short nose, and low set ears.

FIGURE 41.3 An adolescent with Noonan syndrome showing finer facial features and an inverted triangular face shape.

FIGURE 41.4 Adult with Noonan syndrome demonstrating persistent hypertelorism and low set ears with pinched, narrow nasal bridge.

are less prominent, and features are sharper. There is a pinched nasal root with a thin bridge. The neck lengthens, accentuating skin webbing or prominence of the trapezius muscle. In the older adult, nasolabial folds are prominent, and the skin appears transparent and wrinkled (Figure 41.4).

Despite subjective impressions of considerable change in facial gestalt over time, objective data reveal high concordance of facial dimensions at all ages. Most variation is found in facial widths and depths, which exceed the norm in the first few years, tending to be below average beyond age 3 years. Nasal protrusion, nose width and inner canthal distance all exceed the average, particularly in the young child. These data suggest that there is a "Noonan-specific" patterning of facial dimensions, superimposed on which are the normal changes that occur with age (Allanson 2001). These are perceived as a change in gestalt.

Hair may be wispy in a toddler, whereas it is often curly or wooly in an older child and adolescent. Regardless of age, eyes are frequently pale blue or blue-green and much lighter in color and pigmentation than expected for family background; eyebrows are diamond shaped; ears are low set and posteriorly rotated with a thickened helix (Allanson 1987).

Because of the evolution of phenotype and subtlety of features in the adult, an assessment of the family after diagnosis in a child should include a thorough review of serial photographs of both parents. This is particularly important for estimating recurrence risk when testing of the known causative genes (see below) fails to identify a mutation.

Etiology, Pathogenesis, and Genetics

In 1994, linkage analysis in one large Dutch family and 20 smaller families allowed mapping of a Noonan syndrome gene to 12q (van der Burgt et al. 1994). Absence of linkage of some families suggested heterogeneity (Jamieson et al. 1994). In 2001, Tartaglia and colleagues, representing a multicenter consortium, used a positional candidacy approach to demonstrate mutations in *PTPN11* in two moderate-sized families and 11 of 22 unrelated individuals (Tartaglia et al. 2001). *PTPN11* was considered a candidate gene for Noonan syndrome because of its location and because its protein product, SHP-2, is a key component of several signal transduction pathways that control protean developmental processes, including mesodermal patterning, limb development, hemopoietic cell differentiation, and semilunar valvulogenesis.

PTPN11 is composed of two tandemly arranged amino-terminal SH-2 domains (N-SH2 and C-SH2), a phosphotyrosine phosphatase (PTP) domain, and a carboxy-terminal tail. There are mutation hotspots in exons 3, 8, 9, and 13, the majority of which are located in or around the interacting surfaces of the N-SH2 and PTP functional domains, with some in the C-SH2 domain and in the peptide linking the N-SH2 and C-SH2 domains. These gain-of-function mutations in *PTPN11* are found in almost 50% of affected individuals (Tartaglia et al. 2002).

Some genotype–phenotype correlations are known. *PTPN11* mutations are more likely to be found in individuals with pulmonary stenosis, short stature, easy bruising with Factor VIII deficiency, pectus deformity, cryptorchidism and a typical face (Yoshida et al. 2004; Limal et al. 2005; Jongmans et al. 2004). They are less likely to be detected in individuals with hypertrophic cardiomyopathy and Factor XI deficiency (Sarkozy et al. 2003). In general, cognitive impairments are common in individuals with *PTPN11* mutations, although p.N308D and p.N308S mutations seem to be associated with milder cognitive delays (Pierpont et al. 2009). De novo mutations are exclusively paternal in origin (Tartaglia and Gelb 2005). Almost all mutations are missense, but one 3 bp deletion is reported in a baby with a severe phenotype who died in the newborn period (Lee WH et al. 2005) and rare deletions and duplications are also reported with NS phenotypes (Chen et al. 2014). It seems plausible that a mild gain of function in *PTPN11* is tolerated, whereas more significant gain of function associated with gene deletion is lethal. There have been reports of a Noonan-like phenotype in a child with duplication of 12q26, encompassing the *PTPN11* gene (Shchelochkov et al. 2008).

In 2006, mutations in a second gene in the RasMAPK pathway, *KRAS*, were identified in a few individuals with NS (Carta et al. 2006; Schubert et al. 2006). The phenotype tends to be less typical with greater likelihood and severity of intellectual disability (Zenker et al. 2007a). In some, the facies are less characteristic. In 2007, two research groups identified mutations in *SOS1* in 20% of individuals with NS who did not have a mutation in *PTPN11* (Roberts et al. 2007; Tartaglia et al. 2007). Individuals with a *SOS1* mutation may demonstrate florid ectodermal features, reminiscent of cardio-facio-cutaneous syndrome (CFC syndrome) (see Chapter 10) (Zenker et al. 2007b; Jongmans et al. 2005). Growth and intelligence in individuals with a *SOS1* mutation are more likely to be in the normal range (Tartaglia et al. 2007). Several groups have reviewed other genotype–phenotype correlations (Zenker et al. 2004; Jongmans et al. 2005; Tartaglia and Gelb 2005). Gain-of-function mutations in a fourth member of the RasMAPK pathway, *RAF1*, have been shown to cause NS (Pandit et al. 2007; Razzaque et al. 2007). *RAF1* mutations are found in 4–5% of affected individuals and are strongly associated with hypertrophic cardiomyopathy (95%). There is a suggestion of a rare recessive form of NS in the literature (van der Burgt and Brunner 2000), although one of the reported families has been shown to have a *PTPN11* mutation, and thus, is likely explained by gonadal mosaicism (Han Brunner, personal communication).

Whole exome sequencing of gene-negative cases has found rarer molecular genetic causes of NS in additional RAS map kinase pathway genes including *RIT1, SOS2, RASA2, RRAS,* and *SYNGAP1* (Aoki et al. 2013; Yamamoto et al. 2015; Chen et al. 2014a; Flex et al. 2014; Hamdan et al. 2009). Preliminary evidence also supports genes *A2ML1, LZTR1, MYST4, SPRY1, MAP3K8,* and *PPP1CB* as potentially causative, though additional cases and functional studies are needed (Yamamoto et al. 2014; Vissers et al. 2015; Kraft et al. 2011; Chen et al. 2014; Gripp et al. 2016). Approximately 80% of individuals with a clinical diagnosis of NS will have a pathogenic mutation in one of the NS genes.

As mentioned subsequently in Differential Diagnosis, analyses of Noonan syndrome genes have been carried out in syndromes with features overlapping those seen in NS, such as CFC syndrome (Chapter 10) and Noonan syndrome with multiple lentigines (NSML)/LEOPARD syndrome to define whether these conditions are distinct, allelic, or extreme phenotypes of this highly variable disorder. No mutations have been found in *PTPN11* in CFC syndrome (Ion et al. 2002); however, in multiple lentigines/LEOPARD syndrome *PTPN11* mutations, clustering in exons 7, 12, and 13, have been demonstrated in almost 90% of affected individuals (Digilio et al. 2002; Legius et al. 2002; Sarkozy et al. 2003). Several affected individuals with these mutations have hypertrophic cardiomyopathy (Sarkozy et al. 2003). Other families with NSML did not show linkage to chromosome 12, suggesting genetic heterogeneity. Some families have been shown to have mutations in *RAF1* (Pandit et al. 2007). Studies of *PTPN11* in Costello syndrome have failed to identify mutations (Tartaglia et al. 2003b; Troger et al. 2003).

Pathogenesis of NS may in part be associated with jugular lymphatic obstruction. The morphological consequences of lymphatic obstruction or dysfunction may remain long after the actual pathological process has subsided. Webbing of the neck and prominence of the trapezius may be secondary to tissue distension caused by a cystic hygroma. Cryptorchidism, wide-spaced nipples, low-set and angulated ears, hypertelorism, ptosis, and dermatoglyphic abnormalities are postulated to be the result of tissue disruption or displacement by lymphedema during development. Lymphatic dilatation at the base of the developing heart has been shown to alter blood flow and account for left-sided cardiac defects in Turner syndrome (Witt et al. 1987). This mechanism may be germane in NS, although there are no data to support a statistical association between webbed neck and structural anomalies of the heart in NS (Brady and Patton 1996).

Genetic Counseling. Because Noonan syndrome is an autosomal dominant single-gene disorder, an affected individual has a 50% chance to pass the abnormal gene to each of his or her children. Approximately half of affected children will have an affected parent (Mendez and Opitz 1985; Allanson 1987), and predominantly affected mothers, not affected fathers, transmit the gene (3:1 ratio). This is likely to be related to reduced fertility in males (Allanson 1987; Marcus et al. 2008). After the diagnosis is made in a child, evaluation of parents is critical to provide accurate recurrence risks. Where a mutation in one of the NS genes is identified, a search for the mutation can be made in parents. In a family where no mutation is known, parental evaluation should include thorough physical examination looking for characteristic features, review of serial photographs from the newborn period to adulthood, looking for the age-related phenotype, echocardiogram, and electrocardiograph.

Diagnostic Testing

A variety of different gene panels including mutation analysis of all or some of the NS genes are clinically available in North America. When no mutation is identified, diagnosis rests on the pattern of clinical findings. Particularly in females, a karyotype should be done to assure that the constellation of findings is not caused by Turner syndrome.

Differential Diagnosis

The most difficult differential diagnosis clinically is between Turner syndrome and NS in a female. However, obtaining a karyotype and identifying deficiency of a sex chromosome distinguishes Turner syndrome.

Trisomy 8p, trisomy 22 mosaicism, sex chromosome rearrangement, in utero exposure to alcohol (see Chapter 26, fetal alcohol syndrome) or primidone, and Williams syndrome (Chapter 63), Aarskog syndrome (Chapter 1), Baraitser–Winter syndrome, and Costello syndrome (Chapter 16) all share some phenotypic features with NS. Costello syndrome is almost always associated with a mutation in *HRAS*, another gene downstream of *PTPN11* in the RasMAPK pathway. There is also considerable overlap with other cardio-cutaneous syndromes, such as Noonan syndrome with multiple lentigines/LEOPARD syndrome, CFC (Chapter 10), and Watson syndromes. Multiple lentigines/LEOPARD syndrome and Noonan syndrome both can be associated with pulmonary valve dysplasia and cardiomyopathy, short stature, hypertelorism, pectus deformity, hearing loss, and developmental delay (Allanson 1987; Sharland et al. 1992a). As mentioned earlier, both *PTPN11* and *RAF1* mutations cause Noonan syndrome with multiple lentigines/LEOPARD syndrome, confirming that this is an allelic condition.

Watson syndrome and NS share pulmonary valve stenosis, short stature, mild intellectual handicap, and café-au-lait macules (Allanson et al. 1991). The Watson syndrome phenotype also overlaps with that of neurofibromatosis type 1 (NF1) (see Chapter 40), although only axillary freckling and café-au-lait spots show equal incidence in the two conditions; Lisch nodules and neurofibromata are less frequently seen in Watson syndrome, whereas short stature, cardiac defects, and mild intellectual handicap are more common in Watson syndrome. Linkage of Watson syndrome to the *NF1* locus was described in 1991 (Allanson et al. 1991) and pathogenic mutations in *NF1* were subsequently found, confirming that the two conditions are allelic.

Cardio-facio-cutaneous syndrome (Chapter 10) exhibits the greatest overlap with NS, and there was early controversy about whether it is a separate condition (Fryer et al. 1991; Neri et al. 1991). The cardiac and lymphatic abnormalities in both conditions are similar, but intellectual disability in CFC syndrome is usually more severe, with a higher likelihood of structural central nervous system anomalies, autistic-like behavior, and seizures (Armour and Allanson 2008). Skin abnormalities in CFC syndrome are more florid, particularly hyperkeratosis, keratosis pilaris, ichthyosis, absent eyebrows, and sparse, thin, straight, or curly hair. Gastrointestinal problems in CFC syndrome are more severe and long-lasting. A bleeding diathesis is rarely seen in CFC syndrome.

The face in CFC syndrome shares many facial features with NS, including a tall forehead with narrowing at the temples, ptosis, a short nose with relatively broad base, a well-grooved philtrum with cupids-bow lip, and small chin (Allanson 2016). This is particularly true of the younger child. Even in this age range, however, features in CFC syndrome tend to be coarser than those seen in NS. At older ages, the face is broad and coarse and lacks the typical inverted triangular shape. Head shape is more likely to be dolichocephalic than round as in NS. In CFC syndrome there is a high likelihood of absent eyebrows with hyperkeratosis. The eyes are rarely the characteristic blue/blue-green of NS and a lateral gaze is frequently observed. The typical ear finding in NS, oval shape with a thickened helix, low-set, and posteriorly rotated, is uncommon. Earlobe creases, rarely described in NS, are frequent in CFC syndrome.

CFC syndrome is a sporadic condition, thus, facilitating separation from Noonan syndrome. One putative family (Legius et al. 1998) demonstrates a phenotype much more in keeping with Noonan syndrome with a *PTPN11* mutation. No *PTPN11* and *SOS1* mutations are reported in individuals with clear-cut CFC syndrome (Ion et al. 2002; Zenker et al. 2007b). CFC syndrome is caused by mutations in one of four genes in the RasMAPK pathway, all downstream of *PTPN11*. Mutations in *BRAF* are most commonly found, with mutations in *MEK1* and *MEK2* seen less frequently, and rare mutations in *KRAS* (Niihori et al. 2006; Rodriguez-Viciana et al. 2006).

A rare case of NS has been diagnosed as 3C (cranio-cerebello-cardiac) syndrome because of the accompanying brain posterior fossa anomalies. There are two reports of autosomal dominant syndromes with overlapping features that have digital anomalies not seen in Noonan syndrome, namely, camptodactyly/postminimus and proximally placed small thumbs.

MANIFESTATIONS AND MANAGEMENT

Growth and Feeding

Birth weight usually is normal, although edema may cause a transient increase (Allanson 1987; Patton 1994). Mean birth weight is 3450 g. (Sharland et al. 1992a). Almost a quarter of infants with NS have no feeding difficulties; however, poor suck with prolonged feeding time (15%), very poor suck and slow feeding with recurrent vomiting (38%), and severe feeding problems that require tube feeding for two weeks or more (24%) are described (Sharland et al. 1992a; Shaw et al. 2007). Investigation of some children with poor feeding has documented immaturity of gut motility and delayed gastrointestinal motor development (Shah et al. 1999). Typically, this period of failure-to-thrive is self-limited, although poor weight gain may persist for up to 18 months (personal experience, JA). There are lower obesity rates with decreased amounts of both adipose tissue and muscle mass compared to controls (da Silva et al. 2016).

Length at birth is usually normal. There is evidence that birth length in babies with a mutation in *PTPN11* is less than in babies without a *PTPN11* mutation (see discussion in Growth Physiology) (Limal et al. 2006). Mean height follows the third centile until puberty, when below average growth velocity and an attenuated adolescent growth spurt tend to occur. Bone age (Greulich–Pyle method) is usually two years delayed, leading to prolonged growth potential into the twenties (Allanson 1987; Sharland et al. 1992a). Final adult height approached the lower limit of normal (161–162.5 cm in men and 150.5–152.7 cm in women) at the end of the second decade of life (Witt et al. 1986; Ranke et al. 1988). Growth curves have been developed from cross-sectional retrospective data (Witt et al. 1986; Malaquias et al. 2012). Multiple studies have reported improvement in height of 0.6–1.8 SD in cohorts of children with NS treated with growth hormone (Romano et al. 1996; Cotterill et al. 1996; MacFarlane et al. 2001; Kirk et al. 2001; Oslo et al. 2005; Raaijmakers et al. 2008; Noordam et al. 2008). This range is likely attributable, in part, to differing doses, age at start of treatment, and duration of treatment between groups. Most studies of the etiology of short stature in Noonan syndrome have demonstrated normal resting and stimulated levels of growth hormone. Studies differ in their findings of IGF-1 and IGF-BP3 levels and responsiveness to growth hormone therapy in *PTPN11*-positive versus *PTPN11*-negative individuals with NS. Most have found high (compensatory) growth hormone levels and inferior response to growth hormone treatment in those with *PTPN11*-associated NS (Binder et al. 2005; Ferreira et al. 2005; Limal et al. 2006; Jeong 2016). Other studies have found no difference in serum IGF-1 and IGF-BP3 levels and/or change in height SD between those with and without a *PTPN11* mutation (Noordam et al. 2008; Siklar et al. 2016).

In some of these studies, bone age appeared to advance disproportionately in comparison with gain in height age, particularly in children without a mutation in *PTPN11*. In many studies, growth rate in the first 12 months of treatment was highest, waning thereafter. Despite this, long-term enhanced growth velocity was maintained in most studies. All authors stressed the value of starting treatment early.

Evaluation

- Growth velocity should be monitored for 6–12 months, using regular, not syndrome specific, growth charts. If growth velocity is normal, height should be measured every six months.
- If deficiency is noted, nutritional status should be evaluated and chronic illness ruled out.
- Initial testing for growth deficiency should include thyroid function tests, bone age, complete blood count, blood chemistries, sedimentation rate, IGF-1, and IGF-BP3.

- Provocation testing should be performed using standard methods.
- If growth hormone deficiency is found, panhypopituitarism should be assessed by specific hormone testing and magnetic resonance imaging (MRI) of the brain.
- Cardiac status must be evaluated in the child on growth hormone treatment. Cardiomyopathy is not a complete contraindication to treatment, but close monitoring with echocardiography must occur.
- The child on growth hormone should have periodic assessment of growth velocity (every six months) and bone age (every 12 months).
- The child on growth hormone should have annual assessment of insulin, glucose, lipids, thyroid function tests, and blood chemistries.

Management

- Feeding difficulties and associated failure-to-thrive should be managed as in the general population.
- There are significant differences in the utilization of growth hormone and in the approach to treatment between countries. In Canada, growth hormone deficiency must be documented before growth hormone treatment is sanctioned. In the United States and many parts of Europe, endocrinologists offer to treat any individual with Noonan syndrome whose height is more than two standard deviations below the mean with growth hormone, irrespective of growth hormone testing results. Treatment is currently a daily injection or application to the skin surface using a gun-like device.
- Initiation of growth hormone supplementation is necessary if growth hormone deficiency is proven.

Development and Behavior

Early developmental milestones may be delayed, with mean age of sitting at 10 months, first unsupported walking at 21 months, and simple two-word sentences at 31 months. Motor milestone delay is likely to be influenced, in part, by joint hyperextensibility and hypotonia. Most school-age children perform well in a normal educational setting, but 10–40% require special education (Sharland et al. 1992a; van der Burgt et al. 1999; Lee DA et al. 2005). About half of school-aged children meet the diagnostic criteria for a developmental coordination disorder (Lee et al. 2005). Mild intellectual disability is seen in up to one-third of individuals (Mendez and Opitz 1985; Allanson 1987; van der Burgt et al. 1999). Impaired manual dexterity is significantly correlated with verbal and non-verbal intellectual functioning (Pierpont et al. 2009). Long-term follow-up of a large cohort of affected individuals has demonstrated a strong association between significant feeding difficulties in infancy and intellectual handicap requiring special education (Shaw et al. 2007), suggesting that infants with early failure-to-thrive and poor oral-motor coordination require close developmental scrutiny. Detailed psychometric testing of 48 affected British children demonstrated a mean full-scale IQ of 84 (Lee DA et al. 2005). Verbal IQ tended to be slightly higher than performance IQ. About half the group showed mild-to-moderate clumsiness and coordination problems. Learning disability with specific visual-constructional problems and verbal-performance discrepancy was noted in about 25% (van der Burgt et al. 1999; Sarimski 2000; Lee DA et al. 2005). Strengths of the child with Noonan syndrome include verbal comprehension, abstract reasoning, and social awareness and judgment (van der Burgt et al. 1999). Weaknesses include ability to organize perceptual information, planning abilities, and spatial knowledge. There are reports of individuals with superior IQ, good organizational skills, concentration, and persistence, relative strengths in mental computation, social awareness, and judgment, and an intact visual-perceptual-motor system (Finegan and Hughes 1998). Even when full-scale IQ is low, it does not necessarily signify straightforward intellectual disability, but rather indicates a specific cognitive disability, either in verbal or praxic reasoning, requiring a special academic strategy and school placement.

Good peer and social interactions and self-esteem are generally evident, with no particular syndrome of behavioral disability or psychopathology, although there may be an increased likelihood of clumsy, fidgety, stubborn, and irritable behavior. Two studies suggested that children with NS are less likely to be socially competent and may have more behavioral problems than their age-matched peers; however, the problems appeared to be relatively mild and did not reach a level of clinical significance (Witt and Stoltzfus 1999; van der Burgt et al. 1999). Sarimski (2000) found that a majority of 26 individuals studied were impulsive, hyperactive, and irritable. A study of 48 affected children has shown good self-esteem and failed to identify a behavioral phenotype (Lee DA et al. 2005). A study of 40 children with Noonan syndrome demonstrated that 30% had autism spectrum disorder (with a 5:1 male to female ratio), and another 30% had autistic features (Garg et al. 2017). Studies have demonstrated an increased prevalence of problems with attention and executive functioning (Pierpont et al. 2015; Garg et al. 2018). Notably, few children are reported with either sleep difficulties or severe aggression.

Few details of psychological health in NS are reported. Noonan (2005) has documented problems in a cohort of 51 adults. Depression was reported by 23%, and occasional substance abuse and bipolar disease were noted. Shaw et al. (2007) did not report similar findings. Detailed psychological assessment of a 10 Dutch adults with NS over 16 years of age and noted low-average intellectual functioning, anxiety, panic attacks, social introversion, impoverished self-awareness, and marked difficulties in identifying and expressing feelings and emotions (alexithymia), suggesting a higher likelihood of deficiencies in social and emotional recognition and expression (Verhoeven et al. 2008). An Italian study of 27 individuals found 37% have symptoms of anxiety (two of whom met diagnostic criteria for an anxiety disorder) (Perrino et al. 2018). More data are needed on an unbiased cohort to substantiate these findings.

Articulation deficiency is common (72%), but should respond well to intervention therapy. Language delay may be related to hearing loss, perceptual motor disabilities, or articulation deficiencies (Allanson 1987). The average age of first words is 15 months and simple two-word phrases emerge, on average, at 31 months (Pierpont et al. 2010).

Evaluation

- A screening developmental assessment, with such tools as the Denver Developmental Screening Test, should be done at diagnosis and at least annually.
- A full developmental assessment should be done if delays are ascertained.
- School difficulties deserve formal psychological assessment to determine whether a specific cognitive disability is present that might respond to an alternate teaching method.
- Any delay in speech acquisition should prompt a hearing test and an assessment by speech pathology.
- Concerns about introspection, anxiety, or depression should prompt referral to psychology or psychiatry.

Management

- Hypotonia will respond to physical and occupational therapies over time.
- A developmental services program should be initiated if delays are noted early.
- Special individualized education strategies are needed if indicated by detailed testing.
- Speech therapy is suggested where delays or articulation deficiencies are identified.
- Treatment of behavioral or psychological problems is standard.

Cardiovascular

It is important to recognize the possibility of bias when estimating the frequency of congenital heart disease, because many clinicians require the presence of cardiac anomalies for diagnosis of Noonan syndrome (Sharland et al. 1992a), and frequently the reports in the literature come from tertiary and quaternary medical centers, where the most serious manifestations of a condition are likely to be present due to referral bias. The frequency of congenital heart disease in NS is estimated to be between 50 and 90% (Allanson 1987; Patton 1994; Shaw et al. 2007). The frequency of NS in children with congenital heart disease is 1.4%.

A stenotic and often dysplastic pulmonary valve is the most common cardiac anomaly in NS, found in 20–50% of affected individuals (Allanson 1987; Sharland et al. 1992a; Ishizawa et al. 1996; Shaw et al. 2007); it may be isolated or associated with other defects. The frequency of NS in children with pulmonary stenosis is 7%. The sequelae of severe valve dysfunction are similar to those in non-syndromic pulmonary valve stenosis, with development of right ventricular hypertrophy, dysfunction, and fibrosis if it is left untreated.

Hypertrophic cardiomyopathy (HCM) is found in 20–30% of affected individuals (Allanson 1987; Sharland et al. 1992a; Patton 1994; Ishizawa et al. 1996; Shaw et al. 2007). Hypertrophy may be mild or severe, and it may be apical, asymmetrical, or concentric. It may present in utero, at birth, in infancy, or in childhood. Non-obstructive and obstructive hypertrophic cardiomyopathies are likely extensions of the same clinical spectrum. Both may be associated with poor ventricular compliance, leading to diastolic dysfunction that may mimic a restrictive cardiomyopathy. The clinical, echocardiographic, and histopathological phenotype (myocardial hypertrophy with pathological myofibrillar disarray) is indistinguishable from non-syndromic hypertrophic cardiomyopathy, although arrhythmia and sudden death are more common in non-syndromic cases (Patton 1994). Most deaths due to HCM occur in infancy and there is a newborn mortality rate of 20% with an overall survival rate of 70% 15 years after the diagnosis. Age of presentation matters; those diagnosed with HCM and heart failure at less than 6 months have a survival rate of only 30% at two years of age versus those that present after 6 months without heart failure who have a 95% survival rate at age 2 years (Gelb et al. 2015). Over time, regression or progression may occur, or the degree of hypertrophy may remain stable.

Other structural cardiac defects frequently seen in NS include atrial septal defect (10–30%), ventricular septal defect (5–15%), branch pulmonary artery stenosis, and tetralogy of Fallot (2%) (Allanson 1987; Ishizawa et al. 1996; Prendiville et al. 2014). Coarctation of the aorta was thought to be unusual and found more frequently in Turner syndrome, but may be as high as 9%) (Digilio et al. 1998). Other left-sided cardiac defects are now recognized to be part of Noonan syndrome, and partial atrioventricular canal may even be common (15%). In one quarter of cases, this latter defect is associated with subaortic stenosis, and with a smaller percentage will have an anomalous mitral valve. Rare anomalies include Ebstein anomaly, coronary artery fistulae, coronary artery dilatation, coronary artery fibromuscular dysplasia causing is-chemia, giant aneurysms of the sinuses of Valsalva caused by deficiency of medial elastin, restrictive cardiomyopathy, dilated cardiomyopathy, and anomalous pulmonary venous return.

An electrocardiographic abnormality is documented in 87% of individuals with NS (Sharland et al. 1992a). Extreme right axis deviation with superior counterclockwise frontal QRS loop (40%) is likely to be related to asymmetric septal hypertrophy. Superior or left axis deviation may be secondary to a conduction abnormality; there may also be left anterior hemiblock or an RSR′ pattern in lead V1. Arrhythmias are described in 10% of individuals with Noonan syndrome.

Echocardiographic predictors of poor prognosis include greater left ventricular posterior wall thickness (greater than 2 SD), with consequent lower ventricular septum; left ventricular posterior wall ratio (less than 2); progression of hypertrophy with reduction in asymmetry; and congestive cardiac failure. Alteration of left ventricular function with reduced ejection fraction may predict poor outcome, but cannot be used alone.

Some neonates with a thickened pulmonary valve may have self-limited disease and do well without treatment (J Noonan, personal communication). In one analysis of 166 cases of pulmonary valve stenosis, surgery or percutaneous balloon pulmonary valvuloplasty (PBPV) was performed in 47% in those who has PBPV as their first intervention, 65% required either repeat PBPV or surgery (Prendiville 2014). Children with pulmonary stenosis generally do very well after valvotomy. In adults, prognosis is excellent when pulmonary stenosis is trivial, but results may not be uniformly good because of persistent right ventricular dysfunction. Favorable prognostic features include absence of symptoms, normal resting cardiac output, peak right ventricular pressure less than 100 mm Hg, and normal pulmonary artery pressure.

In one series (Ishizawa et al. 1996), mortality secondary to cardiac disease in NS was 18%. Two deaths were related to surgical intervention and three were associated with circulatory collapse and possible arrhythmia, whereas two were sudden and unexpected in the presence of cardiomyopathy. In an evaluation of mortality of 80 individuals with hypertrophic cardiomyopathy from the National Australian Childhood Cardiomyopathy Study, there was 22% mortality 20 years after diagnosis, and a 2.88 hazard ratio for those with NS (Alexander et al. 2018). In older individuals, arrhythmias and congestive cardiac failure may be more common than previously suspected (J Noonan, personal communication).

Evaluation

- Clinical assessment, electrocardiogram (with particular attention to the QRS axis), and echocardiogram are recommended at the initial assessment. Subsequent follow-up is dictated by these investigations and the clinical course.
- If initial assessment is normal, repeat every five years through adulthood.
- A malformed pulmonary valve should be evident at initial assessment, but may become more stenotic with time. Cardiac hypertrophy may progress without changes in clinical status. Therefore, echocardiographic follow-up, at least every two years, is important regardless of the finding of a normal myocardium at presentation.

Management

- β-Blockade or calcium channel blockers have been used most frequently in the treatment of obstructive cardiomyopathy, although attempts to improve diastolic function with both classes of drug are often contemplated (Ishizawa et al. 1996).
- If there is no response to drug therapy, surgery is indicated for left ventricular outflow obstruction. Both surgical myomectomy and transplantation are reported (Sharland et al. 1992a).
- Intervention for pulmonary outflow tract obstruction caused by valvular dysplasia is recommended at any age if right ventricular pressure exceeds 80 mm Hg. Results are good. The initial palliative surgery of choice (Ishizawa et al. 1996) is balloon valvuloplasty, although it is widely accepted that results in the presence of a dysplastic valve may not be as good as in non-syndromic pulmonary valve stenosis. Frequently, open pulmonary valvotomy, with annulus enlargement, is needed. Severe cases may require pulmonary valve debridement or replacement, typically with a bioprosthesis (Ishizawa et al. 1996). Recurrent stenosis rarely occurs, although a mild decrease in gradient may occur in the late postoperative period.
- Subacute bacterial endocarditis prophylaxis is required in all individuals with a cardiac structural defect for dental work, surgery, catheterization, and other circumstances likely to promote a bacteremia.

Neurologic

Joint hyperextensibility and hypotonia are common manifestations of NS. Seizures of varied types are reported in 10% of individuals, with mean age of onset of 11 years (Shaw et al. 2007). Hydrocephalus is infrequently reported in NS (Clericuzio et al. 2008). Fryns (1997) noted that, in his experience, 5% (3/62) of affected individuals had hydrocephalus, which was causally heterogeneous. Communicating hydrocephalus is generally described leading Clericuzio et al. (2008) to hypothesize that this may be related to extracranial lymphatic dysplasia. They refer to studies that show that cerebrospinal fluid drains from the subarachnoid space along the olfactory nerves to the nasal lymphatics, and thence, to cervical lymph nodes (Walter et al. 2006). There are multiple reports of Chiari I malformation, with or without an associated syrinx, in NS (Holder-Espinasse and Winter 2003; Smpokou et al. 2012; Keh et al. 2013; Zarate et al. 2014; Gripp et al. 2013; Mitsuhara et al. 2014; Gripp et al. 2016). A history of headaches, dizziness, neck pain, or poor balance should prompt MRI. Rare neurological structural anomalies include schwannoma, multiple cutaneous granular cell schwannoma (myoblastoma), peripheral neuropathy, Dandy–Walker malformation, cerebral (basal) arteriovenous malformation (one case), hypoplastic posterior cerebral blood vessels (one case), and lateral meningoceles.

The association between NS and malignant hyperthermia is poorly understood; creatine kinase usually is normal in NS (Sharland et al. 1992a), although in King syndrome, which may have a NS phenotype and muscle fibers of variable diameter, creatine kinase is elevated (King and Denborough 1973). Malignant hyperthermia is of greater concern when there is a subclinical myopathy or elevated creatine kinase (Mendez and Opitz 1985).

Evaluation

- Investigation should be prompted by neurological signs and symptoms and may include thorough neurological examination, magnetic resonance imaging, electroencephalogram, electromyogram, and/or nerve conduction velocities.
- Creatine kinase analysis is recommended before anesthesia or dental work if there is clinical evidence of a myopathy.
- A history of headaches, dizziness, neck pain, or poor balance, in particular, should prompt MRI looking for syringomyelia.

Management

- Anticonvulsant therapy is indicated for a seizure disorder, and treatment does not differ from that in the general population.
- Dantrolene prophylaxis should be used during anesthesia when creatine kinase levels are increased or if there is a clinical suspicion of malignant hyperthermia or myopathy.
- Surgery may be required for a brain structural anomaly such as hydrocephalus.

Ophthalmologic

Differences in shape and size of the eyes and periorbital structures are hallmark features of NS. The iris frequently is a striking pale blue or green with sparseness of the usual trabeculae, crypts, and furrows. Ocular findings are among the most common manifestations of NS, seen in up to 95% of individuals. They include strabismus (40–63%), refractive errors (60–70%) (36-52% astigmatism, 13–40% myopia, 35% hypermetropia), amblyopia (33%), and nystagmus (9%). Anterior segment changes (63%) include prominent corneal nerves (46%), anterior stromal dystrophy (94%), cataracts (8%), and panuveitis (2%). Fundal changes are less frequent, occurring in 20% of individuals, and include optic head drusen, optic disk hypoplasia, colobomas, and myelinated nerves (Lee NB et al. 1992; Sharland et al. 1992a; Shaw et al. 2007; van Trier et al. 2016).

Retinitis pigmentosa was described in one individual with Noonan syndrome, and in several individuals reported with cardio-facio-cutaneous syndrome. Cardio-facio-cutaneous syndrome (Chapter 11) also is associated with optic disk anomalies, cataracts, refractive errors, strabismus, nystagmus, hypertelorism, ptosis, and epicanthic folds.

Evaluation

- A detailed assessment should be performed by an ophthalmologist in infancy or at diagnosis.
- Continuing periodic ophthalmologic evaluation should be done if anomalies are found (Lee NB et al. 1992).

Management

- Most ocular defects require non-surgical treatment, such as glasses and occlusion for amblyopia and correction of refractive errors.
- Surgery may be required for cataracts or ptosis (in 10%).

Hematologic-Oncologic

Several different coagulation defects may occur in NS, either singly or in combination. They affect about one-third of all individuals (Witt et al. 1987). However, two-thirds of individuals with Noonan syndrome will give a history of abnormal bleeding or mild-to-severe bruising (Sharland et al. 1992b). The coagulopathy may manifest as severe surgical hemorrhage, clinically mild but detectable bruising, or laboratory abnormalities with no clinical sequelae. There is a poor correlation between history and the actual coagulation factor deficit. Although the factor deficiencies are generally stable, there may be clinical amelioration with age. There is no apparent relationship between age, sex, type or severity of bleeding, and cardiac abnormality present. No evidence of hepatic dysfunction, vitamin K-dependent coagulation factor deficiency, or disseminated intravascular coagulation is reported in Noonan syndrome (Sharland et al. 1992b; Shaw et al. 2007).

Laboratory findings include von Willebrand disease, prolonged partial thromboplastin time (40%) and bleeding time, thrombocytopenia, varied coagulation factor defects (factors V, VIII, XI, XII, protein C) alone or in combination (50%), and platelet dysfunction (abnormal platelet aggregation studies with epinephrine, adenosine diphosphate (ADP), and collagen). Factor XI deficiency is the most common factor deficiency (10–30%) (Sharland et al. 1992b; Bertola et al. 2003; O'Connell 2003). Factor XII deficiency (7–14%) causes prolongation of partial thromboplastin time, but poses no risk for bleeding. Von Willebrand disease may be no more common than in the general population. Platelet depletion may be secondary to ineffective production with reduced or absent megakaryocytes in the bone marrow, or may occur because of sequestration in an enlarged and/or myelodysplastic spleen. Noonan syndrome may be accompanied less often by abnormalities of regulation of the intrinsic system (contact activation), platelet dysfunction associated with trimethylaminuria, or defective thromboplastin regeneration (a pattern of platelet aggregation defects mimicking aspirin-induced effects, pointing to an abnormality of prostaglandin synthesis or action) (Witt et al. 1987; Sharland et al. 1992b). Abnormal screening coagulation studies may not be closely predictive of abnormal clotting. In an analysis of bleeding issues in 142 patients with NS, 9% had a positive bleeding history, 9% had thrombocytopenia, 56% had abnormal clotting factor level(s), and 29% had abnormal platelet function evaluation (Perez Botero et al. 2017).

Hepatosplenomegaly is evident clinically (25%) and ultrasonographically (51%) (Sharland et al. 1992a). Splenomegaly is present in one-half, with concomitant hepatomegaly in one-quarter of these cases. No associated changes in spleen echogenicity are noted, and there is no apparent reason for enlargement.

Juvenile myelomonocytic leukemia (JMML) constitutes a third of childhood cases of myelodysplastic syndrome and about 2% of all leukemias. Deregulaltion of the Ras/MAPK pathway by somatic mutations in *NRAS*, *KRAS2*, and *NF1*, have been shown to cause JMML in about 40% of sporadic cases. Somatic mutations in exons 3 and 13 of *PTPN11* have been demonstrated in 34% of a cohort with JMML (Tartaglia et al. 2003a). Juvenile myelomonocytic leukemia associated with Noonan syndrome is rare, but tends to have an earlier onset and milder presentation than non-syndromic JMML and spontaneous remission has been described in some cases (Kratz et al. 2005). Somatic *PTPN11* mutations in exon 3 are also found in 19% of children with myelodysplastic syndrome and an excess of blast cells, which often evolves into acute myeloid leukemia and is associated with poor prognosis. All these mutations cause gain of function in SHP-2, likely leading to an early initiating lesion in oncogenesis with increased cell proliferation attributable, in part, to prolonged activation of the Ras/MAPK pathway. These somatically acquired *PTPN11* mutations in JMML are thought to be stronger gain-of-function variants, and when present in the germline are associated with severe pre- or neo-natal lethal NS (Mason-Suares et al. 2017).

A low frequency association exists between NS and other myeloproliferative disorders. These include acute lymphoblastic anemia (Roti et al. 2006), chronic myelomonocytic leukemia, and proliferation of erythroid precursors (Kratz et al. 2006). One particular pathogenic *PTPN11* mutation, p.The73Ile, is identified in almost half of the children with Noonan syndrome with myeloproliferative disorder, yet it is uncommon in individuals with Noonan syndrome without myeloproliferative disorder (Tartaglia et al. 2004; Kratz et al. 2005). Similarly, the *KRAS* mutation p.T58I is also associated with NS and myeloproliferative disorder (Schubbert et al. 2006).

Congenital hypoplastic anemia, decreased erythroid and myeloid precursors, neuroblastoma, pheochromocytoma, malignant schwannoma, pilocytic astrocytoma, dysembryoplastic neuroendothelial tumor, and orbital and vaginal rhabdosarcomata are all reported rarely. Overall, children with NS are reported to be at an eight-fold increased risk for a wide spectrum of different cancers (Kratz et al. 2015).

Evaluation

- Inquiries should be made about a history of easy bruising or prolonged bleeding at menstruation or after venipuncture.
- Clinical evidence of abnormal bleeding should be sought.
- Routine screening tests should include prothrombin time, activated partial thromboplastin time, bleeding time, complete blood count, and platelet count. Recent aspirin exposure must be excluded. Screening may be ordered at diagnosis, but must be performed in individuals with bleeding symptoms or in the preoperative period.
- Additional specific testing of coagulation factors, von Willebrand factor antigen, and functional epitope and platelet function is advised if indicated by the screening results (Witt et al. 1987; Sharland et al. 1992b; Briggs and Dickerman 2012). Referral to a hematologist is advisable under these circumstances.
- Prompt assessment when suspicious clinical symptoms of cancer are present.
- Those with gene mutations known to be associated with myeloproliferative disorder or JMML should have physical exam assessment of spleen size and CBC with differential every 3–6 months until age 5 (Villani et al. 2017).

Management

- Aspirin and aspirin-containing medications should be avoided.

- Individualized hemostatic support may be required, depending on the specific hemorrhagic diathesis identified. Treatment regimens are no different from those in the general population.
- Treatment of a myeloproliferative disorder is standard.

Lymphatic

Fewer than 20% of individuals with NS have a lymphatic abnormality; however, many varied differences in lymphatic vessels are described (Mendez and Opitz, 1985; Witt et al. 1987). The lymphatic abnormality may be manifest in several ways: it may be localized or widespread, prenatal and/or postnatal. It is most commonly obvious at birth, although onset may be delayed until adulthood. Dorsal limb lymphedema is most common. It may contribute to increased birth weight (mean birth weight is 3450 g) (Sharland et al. 1992a), and it usually resolves in childhood. A review of individuals with *RIT1*-associated NS demonstrated a prevalence of postnatal lymphangioedema of 29% (Milosavljevic et al. 2016). Less common findings include intestinal, pulmonary, or testicular lymphangiectasia, chylous effusions of pleural space and peritoneum, and localized lymphedema of scrotum or vulva. Rare lymphatic abnormalities include single cases of lip lymphatic dysplasia, neck lymphangioma, orbital edema, and facial lymphangioma. The most common underlying pathological finding is lymphatic hyperplasia with or without thoracic duct anomaly. Lymphatic aplasia or hypoplasia and megalymphatics are reported (Witt et al. 1987).

A common prenatal indicator of lymphatic dysfunction or abnormality, seen on fetal ultrasound, is a cystic hygroma, which is often accompanied by scalp edema, polyhydramnios, pleural and pericardial effusions, ascites, and/or frank hydrops. The presence of these findings, particularly in the absence of a fetal chromosome abnormality, should prompt a search for cardiac and other malformations. Cystic hygroma is usually caused by delayed connection of the jugular lymph sac and the left internal jugular vein. It may regress, leaving redundant neck skin at birth (Witt et al. 1987; Benacerraf et al. 1989). However, not all cystic hygromas resolve, and some progress to hydrops. Timing of regression may be a useful factor in determining survival. Regression before the mid-second trimester appears to be associated with a more favorable prognosis (Benacerraf et al. 1989). Chorioangiomas also are described, and these vascular lesions may play a role in edema formation through loss of α-fetoprotein into amniotic fluid with resultant decrease in fetal oncotic pressure.

Severe lymphatic dysplasia with chylous effusions can lead to protein loss, malnutrition, and lymphopenia, especially of the T-helper cells.

Evaluation

- Prenatal: In the presence of a cystic hygroma or thickened nuchal fold, maternal serum screening or invasive testing with chromosome analysis should be pursued. A detailed sonographic search for other anomalies should be carried out. Once a chromosome anomaly has been excluded, consideration could be given to testing the genes known to cause NS by amniocentesis. Maternal serum screening for RASopathy gene mutations is not currently available.
- Postnatal: The objectives of evaluation of lymphedema are to discover the cause of the lymphatic problem and to define the type. Radionuclide scanning is advocated by many as a first step. Intradermal injection in the first interdigital web space of the lower limb is followed by serial scintiscans of the ilioinguinal region. Direct lymphangiography can then be used to define the type of anomaly if surgical intervention is being considered.
- The majority of individuals with intestinal lymphangiectasia manifest diarrhea, with or without steatorrhea, and diagnosis often begins, as a consequence, with studies to detect and quantify protein loss. Barium meal appearance is characteristic, and peroral jejunal biopsy and lymphography provide definitive diagnosis.
- Chylous complications often are related to lymph vessel hyperplasia with abnormalities of the thoracic duct or presence of megalymphatics. Diagnosis of chylothorax depends on accurate assessment of clinical features, chest radiographs, and demonstration of chyle in pleural fluid obtained by thoracocentesis. Computerized tomographic scanning and lymphoscintigraphy are useful adjuncts that may delineate obstruction to the cisterna chyli with chylous reflux in both chylothorax and chylous ascites.

Management

- Chronic lymphedema of the lower extremities, albeit rare, is frequently associated with infection. Foot hygiene is extremely important. Carefully fitting shoes and support stockings are useful, along with antibacterial cleaning solutions. Daily examination for web space fissures, paronychias, impetigo, and folliculitis should be instituted, with prompt use of systemic antibiotics where indicated to decrease the incidence of severe infection. Prophylactic penicillin may be warranted when hygiene measures alone are inadequate (White 1984).
- Chronic lymphedema management is standard.
- Treatment options for chylothorax include drainage by repeated thoracentesis, chest tube or pleuroperitoneal shunt; reduction of chyle production by low-fat diet or parenteral nutrition; and, if conservative measures fail, surgical modification of lymph flow by thoracic duct ligation, chemical pleurodesis, or pleurectomy. Most chylothoraces resolve with drainage and dietary modification. Long periods of drainage are complicated

by weight loss, hypoalbuminemia, lymphopenia, and infection, which can be fatal. Successful prednisone therapy has been reported. The effect of steroids may be to increase plasma oncotic pressure through increased rate of breakdown of extrahepatic proteins, freeing amino acids, and facilitating liver synthesis of more plasma proteins.

Genitourinary

Renal abnormalities, generally mild, are found on ultrasound in 11% of individuals with NS, although earlier studies in smaller groups of individuals suggested an anomaly rate up to 60% (Noonan 1994). The most commonly reported finding is dilatation of the renal pelvis resembling pelviureteric junction obstruction. Duplex systems, minor rotational anomalies, distal ureteric stenosis, renal hypoplasia, unilateral renal agenesis, unilateral renal ectopia, and bilateral cysts with scarring are reported less commonly (George et al. 1993).

Male pubertal development and subsequent fertility may be normal, delayed, or inadequate with associated deficient spermatogenesis (Mendez and Opitz 1985; Sharland et al. 1992a). The latter may be related to cryptorchidism, which is noted in 60–80% of males (Patton 1994; personal experience). Such males may have primary hypogonadism with high follicle stimulating hormone (FSH) levels. Cryptorchidism and consequent impaired spermatogenesis are also likely to be associated with reduced paternal transmission of Noonan syndrome. Sexual function is not affected, but onset of sexual activity may be later in those males with delayed puberty. Much of the literature related to cryptorchidism, delayed puberty, and consequent reduced fertility was published prior to the current practice of orchidopexy before a boy enters school. Despite early treatment, male puberty occurs at a mean of 14.5 years (Shaw et al. 2007).

Few studies of male puberty and fertility have been carried out. Marcus et al. (2008) documented impaired Sertoli cell function in nine males, age 13–19 years, with Noonan syndrome, seven of whom had a mutation in *PTPN11*. Four of them had bilateral cryptorchidism, whereas two had unilateral testicular maldescent. Luteinizing hormone (LH) levels were normal in all males; FSH levels were raised in seven males; inhibin B levels were low in six males and only just above normal in two additional males. Testosterone levels were normal. All three men with normal testicular descent displayed signs of Sertoli cell dysfunction, with low inhibin B, impaired negative feedback, and consequently raised FSH. Low inhibin B is an established marker of male infertility. Unfortunately, there are no data on sperm quantity or quality in these males. Nonetheless, this study provides some evidence of primary gonadal dysfunction in Noonan syndrome, which may be a result of aberrant Ras function, as Ras is known to play a significant role in normal germ cell proliferation and migration.

Puberty may be delayed in females, with mean age at menarche being 14.6 ± 1.17 years (Sharland et al. 1992a; Shaw et al. 2007). Normal fertility is the rule.

Evaluation

- Renal ultrasound is recommended in newborns or at diagnosis to assess for malformations and disruptions.
- Serial reevaluation may be suggested, depending on the findings.
- Periodic urinalysis is warranted if the genitourinary tract is abnormal, because of increased frequency of urinary tract infection in these circumstances.
- Evaluation of the pituitary-gonadal axis is recommended when pubertal delay is evident.
- Males contemplating starting a family should be informed that fertility may be reduced and can be referred to a fertility specialist.

Management

- The presence of cryptorchidism at birth should lead to a referral to an appropriate surgeon for consideration of a trial of human chorionic gonadotrophin injection and/or surgery before school entry. The approach to treatment is the same as in the general population.
- Urinary tract infection should be treated with appropriate antibiotics.
- Testosterone replacement should be considered in males with primary hypogonadism.
- Fertility treatment is no different from that in the general population.

Musculoskeletal

The classical pectus deformity, with carinatum superiorly and excavatum inferiorly, is seen in 90–95% of individuals with Noonan syndrome. It is often evident by early childhood. Wide-spaced and apparently low-set nipples and rounded shoulders are common (Allanson 1987). Scoliosis is reported in 10–15%. Other spinal anomalies include kyphosis, spina bifida, vertebral and rib anomalies, and genu valgum. Talipes equinovarus is described in 10–15% of individuals, other joint contractures in 4%, radio-ulnar synostosis in 2%, and cervical spine fusion in 2%. The range of forearm carrying angle generally is 14–15° in girls and 10–11° in boys. Abnormal angles (cubitus valgus) are found in more than half the males and females with Noonan syndrome (Sharland et al. 1992a). Joint hyperextensibility is common.

A high palate is common, with a high incidence of class II, division I malocclusion of the teeth. The association of Noonan syndrome and multiple giant cell lesions of the jaw, with rare extragnathic skeletal involvement, has been reported several times. Identical giant cell lesions are found in cherubism, an autosomal dominant disorder caused by mutations in *SH3BP2*. Noonan-like multiple giant cell lesions, by contrast, are not associated with mutations in *SH3BP2*, but with mutations in either *PTPN11* (Lee et al. 2005) or *SOS1* (Neumann et al. 2009). Recently, these lesions have been reported in individuals with CFC syndrome who have mutations in either *BRAF* or *MEK1* (Neumann et al. 2009) and in NF1 (Friedrich 2016), thus demonstrating that they are caused by dysregulation of the Ras/MAPK pathway.

Pigmented villonodular synovitis is an idiopathic, prolific synovial lesion of joints, tendons, and bursae that is associated with the presence of multinucleate giant cells. It is described rarely in individuals with NS who have mutations in either *PTPN11* or *SOS1* (Mascheroni et al. 2008). Pathogenesis of pigmented villonodular synovitis remains controversial. It may be a true neoplasm of synovial tissues or a proliferative response to recurrent hemarthrosis. The latter theory is intriguing given the bleeding diathesis noted occasionally in NS.

Chronic, diffuse musculoskeletal pain has become an increasingly recognized issue for both children and adults. In one questionnaire study of 45 individuals (mean age 17 years, 53% female), 62% reported experiencing chronic pain associated with a history of joint hypermobility and/or living in a cold climate (Vegunta et al. 2015).

Evaluation

- Annual assessment of chest cage and spine, both clinical and radiological, is recommended.
- Dental evaluation and monitoring should begin in early childhood, so that appropriate referral to orthodontics can be facilitated.

Management

- Scoliosis may require bracing or surgery, as in the general population.
- Malocclusion may need orthodontic intervention, as in the general population.
- Chronic pain interfering with daily function should be evaluated in a multidisciplinary pain clinic.

Dermatologic

There is a considerable range of changes in skin, and ectodermal features seem to be more prevalent when NS is caused by mutations in *SOS1* (Roberts et al. 2007; Zenker et al. 2007b). One of the most common findings is follicular keratosis, predominantly over extensor surfaces and face (14%) (Pierini and Pierini 1979; Sharland et al. 1992a). Keratosis pilaris atrophicans faciei (ulerythema ophryogenes) is characterized by horny, whitish, hemispherical or acuminate papules at the opening of the pilosebaceous follicles and is caused by disturbances in the keratinization process of hair follicles. Manifesting itself at a few months after birth, it involves the face, the external third of the eyebrows, and it may extend over the preauricular area, cheeks, and scalp. Generally, progression occurs until puberty, when it becomes quiescent. It may leave pitted scars and atrophic skin and interfere with beard and eyebrow growth (Pierini and Pierini 1979).

Abnormalities of scalp and body hair are often described. Scalp hair may be curly, thick, and wooly or sparse, and poor growing with easy breakage. Microscopically, there is variation in hair shaft diameter.

Café-au-lait spots (10%) and lentigines (2%) are described in Noonan syndrome (Allanson 1987; Sharland et al. 1992a). A Noonan syndrome facial phenotype is also described in some of the members of published families with neurofibromatosis type 1 (Chapter 40). Several individuals with both neurofibromatosis type 1 and Noonan syndrome, co-occurring by chance, are reported (Allanson et al. 1985b; Colley et al. 1996). Stevenson et al. (2006) reported a two-generation family with a three-base pair deletion of *NF1* and a discrete phenotype that has variant NF1 features (no neurofibromas, only pigmentary differences) and variable features of Noonan syndrome (pulmonary stenosis, pectus, and mild facial gestalt) in all five affected members, supporting the concept of neurofibromatosis/Noonan syndrome as a discrete entity, caused by *NF1* mutation. It is important to remember that SHP-2 and neurofibromin are functionally related in a common signal transduction pathway.

Prominent fetal finger pads are common (67%) (Sharland et al. 1992a). Rare skin anomalies reported in Noonan syndrome include xanthomas of skin and tongue, redundant molluscoid skin over the scalp which demonstrates hyperplasia with radiating adipocyte proliferation histologically, leukokeratosis of the lip and gingiva, and vulvar angiokeratoma. Several children with multiple subcutaneous granular cell schwannomas have been reported (Lohmann 2001).

Evaluation

- Dermatological problems should be referred to a specialist, and management planned as in the general population.

Management

- Local medication for keratosis pilaris atrophicans faciei is usually unsuccessful.
- Treatment of other dermatological conditions is standard.

Ears and Hearing

Hearing loss is reported in more than one-third and generally is secondary to serous otitis media. Sensorineural deafness is unusual (3%) (Sharland et al. 1992a; Shaw et al. 2007), and congenital ossicular chain anomaly or temporal bone abnormality is rare.

Evaluation

- Hearing testing should begin in infancy and continue annually through early childhood.

Management

- Aggressive treatment of otitis media with antibiotics is recommended.
- Sensorineural hearing loss may require hearing aids.

Miscellaneous

Hypothyroidism is described in 5% of individuals with NS, although antimicrosomal thyroid antibodies are more frequently found (38%) (M. Patton, personal communication). Other autoantibodies are found at a higher frequency than one would expect (M. Patton, personal communication). Further evidence for autoimmune dysfunction includes vasculitis, vitiligo, anterior uveitis, celiac disease, systemic lupus erythematosus, and autoimmune hepatitis (Sharland et al. 1992a; Quaio et al. 2012; Bader-Meunier et al. 2013; Loddo et al. 2015).

RESOURCES

SUPPORT GROUPS

United States:

Noonan Syndrome Foundation
www.teamnoonan.org

RASopathiesNet
https://rasopathiesnet.org/

United Kingdom:

Noonan Syndrome Association
www.noonansyndrome.org.uk

Brochures and Newsletters

Birth Defects Foundation
Martindale, Hawks Green, Cannock, Staffs
WS11 2XN, UK
Phone: 08700 70 70 20
Website: *www.birthdefects.co.uk*

Human Growth Foundation
997 Glen Cove Avenue Suite 5
Glen Head, NY 11545, USA
Phone: 800-451-6434
Fax: 516-671-4055
Email: *hgf1@hgfound.org*
Website: www.hgfound.org

The MAGIC Foundation
6645 West North Avenue
Oak Park, IL 60302, USA
Phone: 800-362-4423; 708-383-0808
Fax: 708-383-0899
Email: *info@magicfoundation.org*
Website: *www.magicfoundation.org*

Informational Resources and Reviews

National Library of Medicine Genetics Home Reference Noonan syndrome

NORD (National Organization for Rare Disorders) Noonan syndrome

Mendez HMM, Opitz JM (1985) Noonan syndrome: A review. *Am J Med Genet* 21:493–506.

Allanson JE (1987) Noonan syndrome. *J Med Genet* 24:9–13.

Sharland M, Burch M, McKenna WM, Patton MA (1992a) A clinical study of Noonan syndrome. *Arch Dis Child* 67:178–183.

Noonan JA (1994) Noonan syndrome: Update and review for the primary pediatrician. *Clin Pediatr* 33:548–555.

Shaw AC, Kalidas K, Crosby AH, Jeffery S, Patton MA (2007) The natural history of Noonan syndrome: A long-term follow-up study. *Arch Dis Child* 92:128–132.

Allanson JE, Roberts AE (2001) Noonan Syndrome. *GeneReviews®*.

REFERENCES

Alexander PMA, Nugent AW, Daubeney PEF, Lee KJ, Sleeper LA, Schuster T, Turner C, Davis AM, Semsarian C, Colan SD, Robertson T, Ramsay J, Justo R, Sholler GF, King I, Weintraub RG (2018) National Australian childhood cardiomyopathy study. Long-term outcomes of hypertrophic cardiomyopathy diagnosed during childhood: Results from a national population-based study. *Circulation* 138(1):29–36.

Allanson JE (1987) Noonan syndrome. *J Med Genet* 24:9–13.

Allanson JE (2001) Noonan syndrome: The changing face. *Proc Greenwood Genet Cent* 20:78–79.

Allanson JE, Hall JG, Hughes HE, Preus M, Witt RD (1985a) Noonan syndrome: The changing phenotype. *Am J Med Genet* 21:507–514.

Allanson JE, Hall JG, Van Allen MI (1985b) Noonan phenotype associated with neurofibromatosis. *Am J Med Genet* 21:457–462.

Allanson JE, Upadhyaya M, Watson GH, Partington MW, Mackenzie A, Lahey D, MacLeod H, Sarfarazi M, Broadhead

W, Harper PS, Huson SM (1991) Watson syndrome: Is it a subtype of type 1 neurofibromatosis? *J Med Genet* 28:752–756.

Allanson JE (2016) Objective studies of the face of Noonan, *Cardio-facio-cutaneous, and Costello syndromes: A comparison of three disorders of the Ras/MAPK signaling pathway Am J Med Genet* 170(10):2570–7.

Aoki Y, Niihori T, Banjo T, Okamoto N, Mizuno S, Kurosawa K, Ogata T, Takada F, Yano M, Ando T, Hoshika T, Barnett C, Ohashi H, Kawame H, Hasegawa T, Okutani T, Nagashima T, Hasegawa S, Funayama R, Nagashima T, Nakayama K, Inoue S, Watanabe Y, Ogura T, Matsubara Y (2013) Gain-of-function mutations in RIT1 cause Noonan syndrome, a RAS/MAPK pathway syndrome. *Am J Hum Genet* 93(1):173–80.

Armour CM, Allanson JE (2008) Further delineation of cardio-facio-cutaneous syndrome: clinical features of 38 individuals with proven mutations. *J Med Genet* 45:249–254.

Bader-Meunier B, Cavé H, Jeremiah N, Magerus A, Lanzarotti N, Rieux-Laucat F, Cormier-Daire V. (2013) Are RASopathies new monogenic predisposing conditions to the development of systemic lupus erythematosus? Case report and systematic review of the literature. *Semin Arthritis Rheum.* 43(2):217–9.

Benacerraf BR, Greene MF, Holmes LB (1989) The prenatal sonographic features of Noonan's syndrome. *J Ultrasound Med* 8:59–63.

Bertola DR, Carneiro JD, D'Amico EA, Kim CA, Albano LM, Sugayama SM, Gonzalez CH (2003) Hematological findings in Noonan syndrome. *Rev Hosp Clin Fac Med Sao Paulo* 58:5–8.

Binder G, Neuer K, Ranke MB, Wittekindt NE (2005) PTPN11 mutations are associated with mild GH resistance in individuals with Noonan syndrome. *J Clin Endocrinol Metab* 90:5377–5381.

Brady AF, Patton MA (1996) Web-neck anomaly and its association with congenital heart disease. *Am J Med Genet* 64:605.

Briggs BJ, Dickerman JD (2012) Bleeding disorders in Noonan syndrome. *Pediatr Blood Cancer.* 58(2):167–72.

Calcagni G, Limongelli G, D'Ambrosio A, Gesualdo F, Digilio MC, Baban A, Albanese SB, Versacci P, De Luca E, Ferrero GB, Baldassarre G, Agnoletti G, Banaudi E, Marek J, Kaski JP, Tuo G, Russo MG, Pacileo G, Milanesi O, Messina D, Marasini M, Cairello F, Formigari R, Brighenti M, Dallapiccola B, Tartaglia M, Marino B (2017) Cardiac dfects, morbidity and mortality in patients affected by RASopathies. CARNET study results. *Int J Cardiol* 245:92–98.

Carta C, Pantaleoni F, Bocchinfuso G, Stella L, Vasta I, Sarkozy A, Digilio C, Palleschi A, Pizzuti A, Grammatico P, Zampino G, Dallapiccola B, Gelb BD, Tartaglia M (2006) Germline missense mutations affecting KRAS isoform B are associated with a severe Noonan syndrome phenotype. *Am J Hum Genet* 79:129–135.

Chen JL, Zhu X, Zhao TL, Wang J, Yang YF, Tan ZP (2014a) Rare copy number variations containing genes involved in RASopathies: deletion of SHOC2 and duplication of PTPN11. *Mol Cytogenet.* 7:28.

Chen PC, Yin J, Yu HW, Yuan T, Fernandez M, Yung CK, Trinh QM, Peltekova VD, Reid JG, Tworog-Dube E, Morgan MB, Muzny DM, Stein L, McPherson JD, Roberts AE, Gibbs RA, Neel BG, Kucherlapati R (2014b) Next-generation sequencing identifies rare variants associated with Noonan syndrome. *Proc Natl Acad Sci* 111(31):11473–8.

Clericuzio CL, Roberts AE, Kucherlapati R, Tworog-Dube E, Allanson J (2008) Communicating hydrocephalus in Noonan syndrome: A consequence of lymphatic dysplasia? *Proc Greenwood Genet Cent* 27:81.

Colley A, Donnai D, Evans DGR (1996) Neurofibromatosis/Noonan phenotype: A variable feature of type 1 neurofibromatosis. *Clin Genet* 49:59–64.

Cotterill AM, McKenna WJ, Brady AF, Sharland M, Elsawi M, Yamada M, Camacho-Hubner C, Kelnar CJ, Dunger DB, Patton MA, Savage MO (1996) The short-term effects of growth hormone therapy on height velocity and cardiac ventricular wall thickness in children with Noonan's syndrome. *J Clin Endocrinol Metab* 81:2291–2297.

da Silva FM, Jorge AA, Malaquias A, da Costa Pereira A, Yamamoto GL, Kim CA, Bertola D. (2016) Nutritional aspects of *Noonan syndrome and Noonan-related disorders Am J Med Genet* 170(6):1525–31.

Digilio MC, Marino B, Picchio F, Prandstraller D, Toscana A, Giannotti A, Dallapiccola B (1998) Noonan syndrome and aortic coarctation. *Am J Med Genet* 80:160–162.

Digilio MC, Conti E, Sarkozy A, Mingarelli R, Dottorino T, Marino B, Pizzuti A, Dallapiccola B (2002) Grouping of multiple-lentigines/LEOPARD and Noonan syndromes on the *PTPN11* gene. *Am J Hum Genet* 71:389–394.

Ferreira LV, Souza SAL, Arnhold IJP, Mendonca BB, Jorge AAL (2005) *PTPN11*(proteintyrosine phosphatasenonreceptortype 11) mutations and response to growth hormone therapy in children with Noonan syndrome. *J Clin Endocrinol Metab* 90:5156–5160.

Finegan JK, Hughes HE (1988) Very superior intelligence in a child with Noonan syndrome. *Am J Med Genet* 31:385–389.

Flex E, Jaiswal M, Pantaleoni F, Martinelli S, Strullu M, Fansa EK, Caye A, De Luca A, Lepri F, Dvorsky R, Pannone L, Paolacci S, Zhang SC, Fodale V, Bocchinfuso G, Rossi C, Burkitt-Wright EM, Farrotti A, Stellacci E, Cecchetti S, Ferese R, Bottero L, Castro S, Fenneteau O, Brethon B, Sanchez M, Roberts AE, Yntema HG, Van Der Burgt I, Cianci P, Bondeson ML, Cristina Digilio M, Zampino G, Kerr B, Aoki Y, Loh ML, Palleschi A, Di Schiavi E, Carè A, Selicorni A, Dallapiccola B, Cirstea IC, Stella L, Zenker M, Gelb BD, Cavé H, Ahmadian MR, Tartaglia M (2014) Activating mutations in RRAS underlie a phenotype within the RASopathy spectrum and contribute to leukaemogenesis. *Hum Mol Genet.* 23(16):4315–27.

Friedrich RE, Grob TJ, Hollants S, Zustin J, Spaepen M, Mautner VF, Luebke AM, Hagel C, Legius E, Brems H (2016) Recurrent multilocular mandibular giant cell granuloma in neurofibromatosis type 1: Evidence for second hit mutation of NF1 gene in the jaw lesion and treatment with curettage and bone substitute materials. *J Craniomaxillofac Surg* 44(8):1054–60.

Fryer AE, Holt PJ, Hughes HE (1991) The cardio-facio-cutaneous syndrome and Noonan syndrome: Are they the same? *Am J Med Genet* 38:548–551.

Fryns JP (1997) Progressive hydrocephalus in Noonan syndrome. *Clin Dysmorphol* 6:379.

Garg S, Brooks A, Burns A, Burkitt-Wright E, Kerr B, Huson S, Emsley R, Green J. (2017) Autism spectrum disorder and other neurobehavioural comorbidities in rare disorders of the Ras/MAPK pathway. *Dev Med Child Neurol* 59(5):544–549.

George CD, Patton MA, El Sawi M, Sharland M, Adam EJ (1993) Abdominal ultrasound in Noonan syndrome: A study of 44 patients. *Pediatr Radiol* 23:316–318.

Gelb BD, Roberts AE, Tartaglia M. (2015) Cardiomyopathies in Noonan syndrome and the other RASopathies. *Prog Pediatr Cardiol* 39(1):13–19.

Gripp KW, Zand DJ, Demmer L, Anderson CE, Dobyns WB, Zackai EH, Denenberg E, Jenny K, Stabley DL, Sol-Church K. (2013) Expanding the SHOC2 mutation associated phenotype of Noonan syndrome with loose anagen hair: structural brain anomalies and myelofibrosis. *Am J Med Genet A*. 161A(10):2420–30.

Gripp KW, Aldinger KA, Bennett JT, Baker L, Tusi J, Powell-Hamilton N, Stabley D, Sol-Church K, Timms AE, Dobyns WB (2016) A novel rasopathy caused by recurrent de novo missense mutations in PPP1CB closely resembles Noonan syndrome with loose anagen hair. *Am J Med Genet A* 170(9):2237–47.

Hamdan FF, Gauthier J, Spiegelman D, Noreau A, Yang Y, Pellerin S, Dobrzeniecka S, Côté M, Perreau-Linck E, Carmant L, D'Anjou G, Fombonne E, Addington AM, Rapoport JL, Delisi LE, Krebs MO, Mouaffak F, Joober R, Mottron L, Drapeau P, Marineau C, Lafrenière RG, Lacaille JC, Rouleau GA, Michaud JL; Synapse to Disease Group (2009) Mutations in SYNGAP1 in autosomal nonsyndromic mental retardation. *N Engl J Med* 360(6):599–605.

Holder-Espinasse M, Winter RM (2003) Type 1 Arnold-Chiari malformation and Noonan syndrome. A new diagnostic feature. *Clin Dysmorphol* 12:275.

Ion A, Tartaglia M, Song X, Kalidas K, van der Burgt I, Shaw AC, Ming JE, Zampino G, Zackai EH, Dean JC, Somer M, Parenti G, Crosby AH, Patton MA, Gelb BD, Jeffrey S (2002) Absence of *PTPN11* mutations in 28 cases of cardiofaciocutaneous (CFC) syndrome. *Hum Genet* 111:421–427.

Ishizawa A, Oho S-I, Dodo H, Katori T, Homma S-I (1996) Cardiovascular abnormalities in Noonan syndrome: The clinical findings and treatment. *Acta Paediatr Jpn* 38:84–90.

Jamieson RC, van der Burgt I, Brady AF, van Reen M, Elsawi MM, Hol F, Jeffery S, Patton M, Mariman E (1994) Mapping a gene for Noonan syndrome to the long arm of chromosome 12. *Nat Genet* 8:357–360.

Jeong I, Kang E, Cho JH, Kim GH, Lee BH, Choi JH, Yoo HW (2016) Long-term efficacy of recombinant human growth hormone therapy in short-statured patients with Noonan syndrome. *Ann Pediatr Endocrinol Metab* 21(1):26–30.

Jongmans M, Otten B, Noordam K, van der Burgt I. (2004) Genetics and variation in phenotype in Noonan syndrome. *Horm Res.* 62(Suppl 3):56–9.

Jongmans M, Sistermans EA, Rikken A, Nillesen WM, Tamminga R, Patton M, Maier EM, Tartaglia M, Noordam K, van der Burgt I (2005) Genotypic and phenotypic characterization of Noonan syndrome: New data and review of the literature. *Am J Med Genet* 134A:165–170.

Keh YS, Abernethy L, Pettorini B (2013) Association between Noonan syndrome and Chiari I malformation: a case-based update. *Childs Nerv Syst* 29(5):749–52.

King JO, Denborough MA (1973) Anaesthetic-induced malignant hyperpyrexia in children. *J Pediatr* 83:37–40.

Kirk JMW, Betts PR, Butler GE, Donaldson MDC, Dunger DB, Johnston DI, Kelnar CJH, Price DA, Wilton P (2001) Short stature in Noonan syndrome: Response to growth hormone therapy. *Arch Dis Child* 84:440–443.

Kraft M, Cirstea IC, Voss AK, Thomas T, Goehring I, Sheikh BN, Gordon L, Scott H, Smyth GK, Ahmadian MR, Trautmann U, Zenker M, Tartaglia M, Ekici A, Reis A, Dörr HG, Rauch A, Thiel CT (2011) Disruption of the histone acetyltransferase MYST4 leads to a Noonan syndrome-like phenotype and hyperactivated MAPK signaling in humans and mice. *J Clin Invest* 121(9):3479–91.

Kratz CP, Niemeyer CM, Castleberry RP, Cetin M, Bergsträsser E, Emanuel PD, Hasle H, Kardos G, Klein C, Kojima S, Stary J, Trebo M, Zecca M, Gelb BD, Tartaglia M, Loh ML (2005) The mutational spectrum of *PTPN11* in juvenile myelomonocytic leukemia and Noonan syndrome. *Blood* 15:2183–2185.

Kratz CP, Nathrath M, Freisinger P, Dressel P, Assmuss HP, Klein C, Yoshimi A, Burdach S, Niemeyer CM (2006) Lethal proliferation of erythroid precursors in a neonate with a germline *PTPN11* mutation. *Eur J Pediatr* 165:182–185.

Kratz CP, Franke L, Peters H, Kohlschmidt N, Kazmierczak B, Finckh U, Bier A, Eichhorn B, Blank C, Kraus C, Kohlhase J, Pauli S, Wildhardt G, Kutsche K, Auber B, Christmann A, Bachmann N, Mitter D, Cremer FW, Mayer K, Daumer-Haas C, Nevinny-Stickel-Hinzpeter C, Oeffner F, Schlüter G, Gencik M, Überlacker B, Lissewski C, Schanze I, Greene MH, Spix C, Zenker M (2015) Cancer spectrum and frequency among children with Noonan, Costello, and cardio-facio-cutaneous syndromes. *Br J Cancer* 112(8):1392–7.

Lee DA, Portnoy S, Hill P, Gillberg C, Patton MA (2005) Psychological profile of children with Noonan syndrome. *Dev Med Child Neurol* 47:35–38.

Lee JS, Tartaglia M, Gelb BD, Fridrich K, Sachs S, Stratakis CA, Muenke M, Robey PG, Collins MT, Slavotinek A (2005) Phenotypic and genotypic characterisation of Noonan-like/multiple giant cell lesion syndrome. *J Med Genet* 42:e11.

Lee NB, Kelly L, Sharland M (1992) Ocular manifestations of Noonan syndrome. *Eye* 6:328–334.

Lee WH, Raas-Rotschild A, Miteva MA, Bolasco G, Rein A, Gillis D, Vidaud D, Vidaud M, Villoutreix BO, Parfait B (2005) Noonan syndrome type I with PTPN11 3 bp deletion: Structure-function implications. *Proteins* 58:7–13.

Legius E, Schollen E, Matthijs G, Fryns J-P (1998) Fine mapping of the Noonan/cardio-facio-cutaneous syndrome in a large family. *Eur J Hum Genet* 6:32–37.

Legius E, Schrander-Stumpel C, Schollen E, Pulles-Heintzberger C, Gewillig M, Fryns J-P (2002) *PTPN11* mutations in LEOPARD syndrome. *J Med Genet* 39:571–574.

Limal JM, Parfait B, Cabrol S, Bonnet D, Leheup B, Lyonnet S, Vidaud M, Le Bouc Y (2006) Noonan syndrome: Relationships between genotype, growth, and growth factors. *J Clin Endocrinol Metab* 91:300–306.

Loddo I, Romano C, Cutrupi MC, Sciveres M, Riva S, Salpietro A, Ferraù V, Gallizzi R, Briuglia S. (2015) Autoimmune liver disease in Noonan Syndrome. *Eur J Med Genet.* 58(3):188–90.

Lohmann DR, Gillessen-Kaesbach G (2001). Multiple cutaneous granular cell tumours in a patient with Noonan syndrome. *Clin Dysmorphol* 19:301–302.

MacFarlane CE, Brown DC, Johnston LB, Patton MA, Dunger DB, Savage MO, McKenna WJ, Kelnar CJH (2001) Growth hormone therapy and growth in children with Noonan's syndrome: Results of 3 years' follow-up. *J Clin Endocrinol Metab* 86:1953–1956.

Malaquias AC, Brasil AS, Pereira AC, Arnhold IJ, Mendonca BB, Bertola DR, Jorge AA (2012) Growth standards of patients with Noonan and Noonan-like syndromes with mutations in the RAS/MAPK pathway. *Am J Med Genet A* 158A(11):2700–6.

Marcus KA, Sweep CGJ, van der Burgt I, Noordam C (2008) Impaired sertoli cell function in males diagnosed with Noonan syndrome. *J Pediatr Endocrinol Metab* 21:1079–1084

Mascheroni E, Digilio MC, Cortis E, Devito R, Sarkozy A, Capolino R, Dallapiccola B, Ugazio AG (2008) Pigmented villonodular synovitis in a patient with Noonan syndrome and SOS1 gene mutation. *Am J Med Genet* 146A:2966–2967.

Mason-Suares H, Toledo D, Gekas J, Lafferty KA, Meeks N, Pacheco MC, Sharpe D, Mullen TE, Lebo MS (2017) Juvenile myelomonocytic leukemia-associated variants are associated with neo-natal lethal Noonan syndrome. *Eur J Hum Genet* 25(4):509–511.

Mendez HMM, Opitz JM (1985) Noonan syndrome: A review. *Am J Med Genet* 21:493–506.

Milosavljević D, Overwater E, Tamminga S, de Boer K, Elting MW, van Hoorn ME, Rinne T, Houweling AC. (2016) Two cases of RIT1 associated Noonan syndrome: Further delineation of the clinical phenotype and review of the literature. *Am J Med Genet A* 170(7):1874–80.

Mitsuhara T, Yamaguchi S, Takeda M, Kurisu K. (2014) Gowers' intrasyringeal hemorrhage associated with Chiari type I malformation in Noonan syndrome. *Surg Neurol Int* 20(5):6.

Neri G, Zollino M, Reynolds JF (1991) The Noonan-CFC controversy. *Am J Med Genet* 39:367–370.

Neumann TE, Allanson J, Kavamura I, Kerr B, Neri G, Noonan J, Cordeddu V, Gibson K, Tzschach A, Krüger G, Hoeltzenbein M, Goecke TO, Kehl HG, Albrecht B, Luczak K, Sasiadek MM, Musante L, Laurie R, Peters H, Tartaglia M, Zenker M, Kalscheuer V (2009) Multiple giant cell lesions in patients with Noonan and cardio-facio-cutaneous syndrome. *Eur J Hum Genet* 17:420–425.

Niihori T, Aoki Y, Narumi Y, Neri G, Cavé H, Verloes A, Okamoto N, Hennekam RC, Gillessen-Kaesbach G, Wieczorek D, Kavamura MI, Kurosawa K, Ohashi H, Wilson L, Heron D, Bonneau D, Corona G, Kaname T, Naritomi K, Baumann C, Matsumoto N, Kato K, Kure S, Matsubara Y (2006) Germline KRAS and BRAF mutations in cardio-facio-cutaneous syndrome. *Nat Genet* 38:294–296.

Noonan JA (1994) Noonan syndrome: Update and review for the primary pediatrician. *Clin Pediatr* 33:548–555.

Noonan JA (2005) Noonan syndrome. In: *Handbook of Neurodevelopmental and Genetic Disorders in Adults*, Goldstein S, Reynolds CR, eds, New York: Guilford Press, pp. 308–319.

Noonan JA, Ehmke DA (1963) Associated non-cardiac malformations in children with congenital heart disease. *J Pediatr* 63:468–470.

Noordam C, Peer PGM, Francois I, De Schepper J, van der Burgt I, Otten BJ (2008) Long-term GH treatment improves adult height in children with Noonan syndrome with and without mutations in protein tyrosine kinase phosphatase, non-receptor-type 11. *Eur J Endocrinol* 159:203–206.

O'Connell NM (2003) Factor XI deficiency—From molecular genetics to clinical management. *Blood Coagul Fibrinolysis* 14(Suppl 1):S59–S64.

Osio D, Dahlgren J, Wikland KA, Westphal O (2005) Improved final height with long-term growth hormone treatment in Noonan syndrome. *Acta Pediatr* 94:1232–1237.

Pandit B, Sarkozy A, Pennacchio LA, Carta C, Oishi K, Martinelli S, Pogna EA, Schackwitz W, Ustaszewska A, Landstrom A, Bos JM, Ommen SR, Esposito G, Lepri F, Faul C, Mundel P, López Siguero JP, Tenconi R, Selicorni A, Rossi C, Mazzanti L, Torrente I, Marino B, Digilio MC, Zampino G, Ackerman MJ, Dallapiccola B, Tartaglia M, Gelb BD (2007) Gain-of-function *RAF1* mutations cause Noonan and LEOPARD syndromes with hypertrophic cardiomyopathy. *Nat Genet* 39:1007–1012.

Patton MA (1994) Noonan syndrome: A review. *Growth Genet Horm* 33: 10:1–3.

Perez Botero J, Ho TP, Rodriguez V, Khan SP, Pruthi RK, Patnaik MM (2017) Coagulation abnormalities and haemostatic surgical outcomes in 142 patients with Noonan syndrome. *Haemophilia* 23(3):e237–e240.

Perrino F, Licchelli S, Serra G, Piccini G, Caciolo C, Pasqualetti P, Cirillo F, Leoni C, Digilio MC, Zampino G, Tartaglia M, Alfieri P, Vicari S. (2018) Psychopathological features in Noonan syndrome. *Eur J Paediatr Neurol* 22(1):170–177.

Pierini DO, Pierini AM (1979) Keratosis pilaris atrophicans faciei (ulerythema ophryogenes): A cutaneous marker in the Noonan syndrome. *Br J Dermatol* 100:409–416.

Pierpont EI, Pierpont ME, Mendelsohn NJ, Roberts AE, Tworog-Dube E, Seidenberg MS (2009) Genotype differences in cognitive functioning in Noonan syndrome. *Genes Brain Behav* 8: 275–282.

Pierpont EI, Pierpont ME, Mendelsohn NJ, Roberts AE, Tworog-Dube E, Rauen KA, Seidenberg MS. (2010) Effects of germline mutations in the Ras/MAPK signaling pathway on adaptive behavior: cardiofaciocutaneous syndrome and Noonan syndrome. *Am J Med Genet A* 152A:591–600.

Pierpont EI, Tworog-Dube E, Roberts AE (2015) Attention skills and executive functioning in children with Noonan syndrome and their unaffected siblings. *Dev Med Child Neurol* 57:385–92.

Prendiville TW, Gauvreau K, Tworog-Dube E, Patkin L, Kucherlapati RS, Roberts AE, Lacro RV. (2014) Cardiovascular disease in Noonan syndrome. *Arch Dis Child* 99(7):629–34.

Quaio CR, Carvalho JF, da Silva CA, Bueno C, Brasil AS, Pereira AC, Jorge AA, Malaquias AC, Kim CA, Bertola DR. (2012) Autoimmune disease and multiple autoantibodies in 42 patients with RASopathies. *Am J Med Genet A* 158A(5):1077–82.

Raaijmakers R, Noordam C, Karagiannis G, Gregory JW, Hertel NT, Sipila I, Otten BJ (2008) Response to growth hormone treatment and final height in Noonan syndrome in a large cohort of patients in the KIGS database. *J Pediatr Endocrinol Metab* 21:267–273.

Ranke MB, Heidemann P, Knupfer C, Enders H, Schmaltz AA, Bierich JR (1988) Noonan syndrome: Growth and clinical manifestations in 144 cases. *Eur J Pediatr* 148:220–227.

Razzaque MA, Nishizawa T, Komoike Y, Yagi H, Furutani M, Amo R, Kamisago M, Momma K, Katayama H, Nakagawa M, Fujiwara Y, Matsushima M, Mizuno K, Tokuyama M, Hirota H, Muneuchi J, Higashinakagawa T, Matsuoka R (2007) Germline gain-of-function mutations in *RAF1* cause Noonan syndrome. *Nat Genet* 39:1013–1017.

Roberts AE, Araki T, Swanson KD, Montgomery KT, Schiripo TA, Joshi VA, Li L, Yassin Y, Tamburino AM, Neel BG, Kucherlapati RS (2007) Germline gain-of-function mutations in *SOS1* cause Noonan syndrome. *Nat Genet* 39:70–74.

Rodriguez-Viciana P, Tetsu O, Tidyman WE, Estep AL, Conger BA, Cruz MS, McCormick F, Rauen KA (2006) Germline mutations in genes within the MAPK pathway cause cardio-facio-cutaneous syndrome. *Science* 311:1287–1290.

Romano AA, Blethen SL, Dana K, Noto RA (1996) Growth hormone treatment in Noonan syndrome: The National Cooperative Growth Study experience. *J Pediatr* 128:S18–S21.

Roti G, La Starza R, Ballanti S, Crescenzi B, Romoli S, Foá R, Tartaglia M, Aversa F, Fabrizio Martelli M, Mecucci C (2006) Acute lymphoblastic leukaemia in Noonan syndrome. *Br J Hematol* 133:448–450.

Sarimski K (2000) Developmental and behavioural phenotype in Noonan syndrome. *Genet Couns* 11:383–390.

Sarkozy A, Conti E, Seripa D, Digilio MC, Grifone N, Tandoi C, Fazio VM, Di Ciommo V, Marino B, Pizzuti A, Dallapiccola B (2003) Correlation between *PTPN11* mutations and congenital heart defects in Noonan and LEOPARD syndromes. *J Med Genet* 40:704–708.

Schubbert S, Zenker M, Rowe SL, Böll S, Klein C, Bollag G, van der Burgt I, Musante L, Kalscheuer V, Wehner LE, Nguyen H, West B, Zhang KY, Sistermans E, Rauch A, Niemeyer CM, Shannon K, Kratz CP (2006) Germline *KRAS* mutations cause Noonan syndrome. *Nat Genet* 38:331–336.

Shah N, Rodriguez M, St Louis D, Lindley K, Milla PJ (1999) Feeding difficulties and foregut dysmotility in Noonan's syndrome. *Arch Dis Child* 81:28–31.

Sharland M, Burch M, McKenna WM, Patton MA (1992a) A clinical study of Noonan syndrome. *Arch Dis Child* 67:178–183.

Sharland M, Patton MA, Talbot S, Chitolie A, Bevan DH (1992b) Coagulation-factor deficiencies and abnormal bleeding in Noonan's syndrome. *Lancet* 339:19–21.

Shaw AC, Kalidas K, Crosby AH, Jeffery S, Patton MA (2007) The natural history of Noonan syndrome: A long-term follow-up study. *Arch Dis Child* 92:128–132.

Shchelochkov OA, Patel A, Weissenberger GM, Chinault AC, Wiszniewska J, Fernandes PH, Eng C, Kukolich MK, Sutton VR (2008) Duplication of chromosome band 12q24.11q24.23 results in apparent Noonan syndrome. *Am J Med Genet* 146A:1042–1048.

Şıklar Z, Genens M, Poyrazoğlu Ş, Baş F, Darendeliler F, Bundak R, Aycan Z, Savaş Erdeve Ş, Çetinkaya S, Güven A, Abalı S, Atay Z, Turan S, Kara C, Can Yılmaz G, Akyürek N, Abacı A, Çelmeli G, Sarı E, Bolu S, Korkmaz HA, Şimşek E, Çatlı G, Büyükinan M, Çayır A, Evliyaoğlu O, İşgüven P, Özgen T, Hatipoğlu N, Elhan AH, Berberoğlu M (2016) The Growth Characteristics of Patients with Noonan Syndrome: Results of Three Years of Growth Hormone Treatment: A Nationwide Multicenter Study *J Clin Res Pediatr Endocrinol* 8(3):305–12.

Smpokou P, Tworog-Dube E, Kucherlapati RS, Roberts AE (2012) Medical complications, clinical findings, and educational outcomes in adults with Noonan syndrome. *Am J Med Genet A* 158A(12):3106–11.

Stevenson DA, Viskochil DH, Rope AF, Carey JC (2006) Clinical and molecular aspects of an informative family with neurofibromatosis and Noonan phenotype. *Clin Genet* 69:246–253.

Stofega MR, Herrington J, Billestrup N, Carter-Su C (2000) Mutation of the SHP-2 binding site in growth hormone (GH) receptor prolongs GH-promoted tyrosyl phosphorylation of GH receptor, *JAK2*, and *STAT5B*. *Mol Endocrinol* 14:1338–1350.

Tartaglia M, Gelb BD (2005) Noonan syndrome and related disorders: Genetics and pathogenesis. *Annu Rev Genomics Hum Genet* 6:45–68.

Tartaglia M, Mehler EL, Goldberg R, Zampino G, Brunner HG, Kremer H, van der Burgt I, Crosby AH, Ion A, Jeffrey S, Kalidas K, Patton MA, Kucherlapati RS, Gelb BD (2001) Mutations in *PTPN11*, encoding the protein tyrosine phosphatase SHP-2, cause Noonan syndrome. *Nat Genet* 29:465–468.

Tartaglia M, Kalidas K, Shaw A, Song X, Musat DL, van der Burgt I, Brunner HG, Bertola DR, Crosby A, Ion A, Kucherlapati RS, Jeffrey S, Patton MA, Gelb BD (2002) *PTPN11* mutations in Noonan syndrome: Molecular spectrum, genotype-phenotype correlation, and phenotypic heterogeneity. *Am J Hum Genet* 70:1555–1563.

Tartaglia M, Niemeyer CM, Fragale A, Song X, Buechner J, Jung A, Hahlen K, Hasle H, Licht JD, Gelb BD (2003a) Somatic mutations in *PTPN11* in juvenile myelomonocytic leukemia, myelodysplastic syndromes and acute myeloid leukemia. *Nat Genet* 34:148–150.

Tartaglia M, Cotter PD, Zampino G, Gelb BD, Rauen KA (2003b) Exclusion of *PTPN11* mutations in Costello syndrome: Further evidence for distinct genetic etiologies for Noonan, cardio-facio-cutaneous and Costello syndromes. *Clin Genet* 63:423–426.

Tartaglia M, Niemeyer CM, Shannon KM, Loh ML (2004) SHP-2 and myeloid malignancies. *Curr Opin Hematol* 11:44–50.

Tartaglia M, Cordeddu V, Chang H, Shaw A, Kalidas K, Crosby A, Patton MA, Sorcini M, van der Burgt I, Jeffery S, Gelb BD. (2004b) Paternal germline origin and sex-ratio distortion in transmission of PTPN11 mutations in Noonan syndrome. *Am J Hum Genet* 75:492–7.

Tartaglia M, Pennacchio LA, Zhao C, Yadav KK, Fodale V, Sarkozy A, Pandit B, Oishi K, Martinelli S, Schackwitz W, Ustaszewska A, Martin J, Bristow J, Carta C, Lepri F, Neri C, Vasta I, Gibson K, Curry CJ, Siguero JP, Digilio MC, Zampino G, Dallapiccola B, Bar-Sagi D, Gelb BD (2007) Gain-of-function *SOS1* mutations cause a distinctive form of Noonan syndrome. *Nat Genet* 39:75–79.

Troger B, Kutsche K, Bolz H, Luttgen S, Gal A, Almassy Z, Caliebe A, Freisinger P, Hobbiebrunken E, Morlot M, Stefanova M, Streubel B, Wieczorek D, Meinecke P (2003) No mutation in the gene for Noonan syndrome, *PTPN11*, in 18 patients with Costello syndrome. *Am J Med Genet* 121A:82–84.

van der Burgt I, Brunner H (2000) Genetic heterogeneity in Noonan syndrome: Evidence for an autosomal recessive form. *Am J Med Genet* 94:46–51.

van der Burgt I, Berends E, Lommen E, van Beersum S, Hamel B, Mariman E (1994) Clinical and molecular studies in a large Dutch family with Noonan syndrome. *Am J Med Genet* 53:187–191.

van der Burgt I, Thoone G, Roosenboom N, Assman-Hulsman C, Gabreels F, Otten B, Brunner HG (1999) Patterns of cognitive functioning in school-aged children with Noonan syndrome associated with variability of phenotypic expression. *J Pediatr* 135:707–713.

van Trier DC, Vos AM, Draaijer RW, van der Burgt I, Draaisma JM, Cruysberg JR. (2016) Ocular Manifestations of Noonan Syndrome: A Prospective Clinical and Genetic Study of 25 Patients. *Ophthalmology* 123(10):2137–46.

Vegunta S, Cotugno R, Williamson A, Grebe TA. (2015) Chronic pain in Noonan Syndrome: A previously unreported but common symptom. *Am J Med Genet A* 167A(12):2998–3005.

Verhoeven W, Wingbermuhle E, Egger J, van der Burgt I, Tuinier S (2008) Noonan syndrome: Psychological and psychiatric aspects. *Am J Med Genet A* 146A:191–196.

Villani A, Greer MC, Kalish JM, Nakagawara A, Nathanson KL, Pajtler KW, Pfister SM, Walsh MF, Wasserman JD, Zelley K, Kratz CP (2017) Recommendations for Cancer Surveillance in Individuals with RASopathies and Other Rare Genetic Conditions with Increased Cancer Risk. *Clin Cancer Res* 23(12):e83–e90.

Vissers LE, Bonetti M, Paardekooper Overman J, Nillesen WM, Frints SG, de Ligt J, Zampino G, Justino A, Machado JC, Schepens M, Brunner HG, Veltman JA, Scheffer H, Gros P, Costa JL, Tartaglia M, van der Burgt I, Yntema HG, den Hertog J (2015) Heterozygous germline mutations in A2ML1 are associated with a disorder clinically related to Noonan syndrome. *Eur J Hum Genet* 23(3):317–24.

Walter BA, Valera VA, Takahashi S, Ushiki T (2006) The olfactory route for cerebrospinal fluid drainage into the peripheral nervous system. *Neuropathol Appl Neurobiol* 32:388–396.

White SW (1984) Lymphedema in Noonan's syndrome. *Int J Dermatol* 23:656–657.

Witt DR, Stoltzfus C (1999) Behavioral phenotype in Noonan syndrome. *Proc Greenwood Genet Cent* 18:149–150.

Witt DR, Keena BA, Hall JG, Allanson JE (1986) Growth curves for height in Noonan syndrome. *Clin Genet* 30:150–153.

Witt DR, Hoyme HE, Zonana J, Manchester DK, Fryns JP, Stevenson JG, Curry CJR, Hall JG (1987) Lymphedema in Noonan syndrome: Clues to pathogenesis and prenatal diagnosis and review of the literature. *Am J Med Genet* 27:841–856.

Yamamoto GL, Aguena M, Gos M, Hung C, Pilch J, Fahiminiya S, Abramowicz A, Cristian I, Buscarilli M, Naslavsky MS, Malaquias AC, Zatz M, Bodamer O, Majewski J, Jorge AA, Pereira AC, Kim CA, Passos-Bueno MR, Bertola DR (2015) Rare variants in *SOS2* and *LZTR1* are associated with Noonan syndrome *J Med Genet* 52(6):413–21.

Yoshida R, Hasegawa T, Hasegawa Y, Nagai T, Kinoshita E, Tanaka Y, Kanegane H, Ohyama K, Onishi T, Hanew K, Okuyama T, Horikawa R, Tanaka T, Ogata T (2004) Protein-tyrosine phosphatase, non-receptor type 11 mutation analysis and clinical assessment in 45 patients with Noonan syndrome. *J Clin Endocrinol Metab* 89:3359–3364.

Zarate YA, Lichty AW, Champion KJ, Clarkson LK, Holden KR, Matheus MG. (2014) Unique cerebrovascular anomalies in Noonan syndrome with RAF1 mutation. *J Child Neurol* 29(8):NP13–7.

Zenker M, Buheitel G, Rauch R, Koenig R, Bosse K, Kress W, Tietze HU, Doerr HG, Hofbeck M, Singer H, Reis A, Rauch A (2004) Genotype-phenotype correlations in Noonan syndrome. *J Pediatr* 144:368–374.

Zenker M, Lehmann K, Schulz AL, Barth H, Hansmann D, Koenig R, Korinthenberg R, Kreiss-Nachtsheim M, Meinecke P, Morlot S, Mundlos S, Quante AS, Raskin S, Schnabel D, Wehner LE, Kratz CP, Horn D, Kutsche K (2007a) Expansion of the genotypic and phenotypic spectrum in patients with KRAS germline mutations. *J Med Genet* 44:131–135.

Zenker M, Horn D, Wieczorek D, Allanson J, Pauli S, van der Burgt I, Doerr HG, Gaspar H, Hofbeck M, Gillessen-Kaesbach G, Koch A, Meinecke P, Mundlos S, Nowka A, Rauch A, Reif S, von Schnakenburg C, Seidel H, Wehner LE, Zweier C, Bauhuber S, Matejas V, Kratz CP, Thomas C, Kutsche K (2007b) *SOS1* is the second most common Noonan syndrome gene but plays no role in cardio-facio-cutaneous syndrome. *J Med Genet* 44:651–656.

42

OCULO-AURICULO-VERTEBRAL SPECTRUM

KOENRAAD DEVRIENDT
Center for Human Genetics, University of Leuven, Leuven, Belgium

LUC DE SMET
Department of Orthopaedic Surgery, University of Leuven, Leuven, Belgium

INGELE CASTEELS
Department of Ophthalmology, University of Leuven, Leuven, Belgium

INTRODUCTION

Nomenclature

Goldenhar syndrome (epibulbar dermoids, preauricular skin tags, asymmetry of the mandible, and cervical vertebral anomalies [Goldenhar 1952]) and hemifacial microsomia (typically unilateral aural, oral, and mandibular anomalies [Gorlin et al. 1963]) were initially described as separate entities. The term craniofacial microsomia has also been suggested (Heike et al. 2013). However, the term oculo-auriculo-vertebral spectrum (OAVS) is widely preferred. OAVS includes both entities, because no clear distinction between the two conditions can be made clinically or by any test. The condition has been labelled by various other terms (reviewed by Cohen et al. 1989; Gorlin et al. 2001).

Incidence

Because of markedly variable expression, the exact frequency of the disorder is not known, many mild cases not having been categorized as such. Different studies observed incidences at birth of 1/5600, 1/19,500, and 1/26,550. A male-to-female ratio of 3:2 has been observed (Rollnick et al. 1987). No ethnic differences in occurrence have been noted.

Diagnostic Criteria

As indicated by the term "spectrum," the manifestations of OAVS vary widely, ranging from very mild expression at a single location (e.g., external ear) to multiple congenital anomalies. Because of the absence of an etiological diagnostic test, the diagnosis remains clinical. Currently, no minimal diagnostic criteria exist. In relatives of an affected individual, the presence of preauricular tags or pits may suffice to establish the diagnosis (Tasse et al. 2005). Some authors have suggested unilateral ear anomalies (including preauricular tags) as a mandatory feature. One of the cardinal features of the syndrome is facial asymmetry, but again, not all affected individuals exhibit facial asymmetry.

Typically, the diagnosis relies on the variable combination of craniofacial anomalies (Figures 42.1 and 42.2). Facial asymmetry, frequently but not always obvious in infancy, may become more evident with age (Keogh et al. 2007). The ear on the involved side is often abnormally formed, small, and displaced. One or more tags may be present from the tragus to the angle of the mouth. The mouth may be larger on the affected side, the mastoid bone may be small, and there may be ipsilateral facial paralysis. Bilateral involvement is often observed, usually with more severe involvement of one side. There may be epibulbar dermoids and occasionally colobomas of the

---

Cassidy and Allanson's Management of Genetic Syndromes, Fourth Edition.
Edited by John C. Carey, Agatino Battaglia, David Viskochil, and Suzanne B. Cassidy.
© 2021 John Wiley & Sons, Inc. Published 2021 by John Wiley & Sons, Inc.

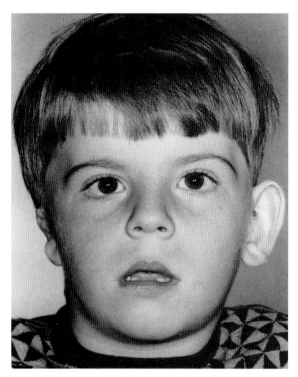

FIGURE 42.1 A young boy with right hemifacial microsomia with right microtia.

Classification of Severity

There have been many attempts to classify severity within the OAVS. Most systems have been limited because they have focused on one or two anatomic variables (Lauritzen et al. 1985; David et al. 1987). Using the mandible and temporomandibular joint as reference centers, Kaban et al. (1988) modified Pruzansky's earlier system of classification, separating the mandibular form in OAVS into three basic types. A more comprehensive attempt is that of Vento et al. (1991), in which the authors chose to name this as the OMENS classification (O, orbital distortion; M, mandibular hypoplasia; E, ear anomaly; N, nerve; S, soft tissue deficiency). This classification easily adapts to data storage, retrieval, and analysis for craniofacial evaluation, surgical and other treatment modalities, and orthodontic therapy (Cousley 1993). However, because it undervalues anomalies of other systems (e.g., cardiac, central nervous system, skeletal, and renal), Rollnick et al. (1987) adapted this system for inclusion of systemic abnormalities. The scoring system proposed by Tasse et al. (2005) includes systemic manifestations and has a prognostic value, because higher scores are associated with an increased incidence of developmental delay and brain anomalies.

upper eyelids. Cervical vertebral bodies may be malformed. Various other systemic anomalies are seen in at least 50% of individuals, which has led to the term expanded Goldenhar complex by some authors, most frequently including congenital heart, brain, renal, and limb defects.

Etiology, Pathogenesis, and Genetics

The affected tissues and organs in OAVS indicate abnormal development at about 30–45 days of gestation, primarily of the first and second branchial arches. Among the pathogenetic theories that have been proposed, vascular disruption or primary disturbed development of the branchial arches or

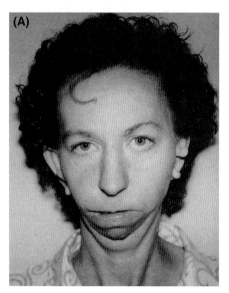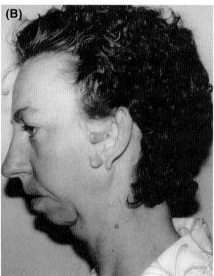

FIGURE 42.2 An adult woman who demonstrates mild facial asymmetry, marked micrognathia, a right lateral cleft of the mouth, and multiple preauricular tags.

neural crest cells are most commonly cited. Poswillo (1973), using an animal model, showed that early vascular disruption with expanding hematoma formation in utero resulted in destruction of differentiating tissues in the region of the ear and jaw. Severity appeared to be related to the degree of local disruption (Poswillo 1973; Robinson et al. 1987).

Various features of the OAVS have been seen in a wide variety of chromosomal anomalies, some of which are recurrent, such as duplications in 4p16.1, del 5p, trisomy 18, deletions or duplications in chromosome 22q (Beleza-Meireles et al. 2015; Bragagnola et al. 2017). Some individuals have mosaicism for a chromosomal aberration, which typically is associated with asymmetric involvement. These chromosomal anomalies may indicate multiple causative or contributory genes, may be coincidental, or may represent misdiagnosis. Autosomal dominant loss-of-function mutations in the *MYT1* gene have been identified in less than 1% of cases (Lopez et al. 2016).

Although most persons with OAVS are isolated, familial occurrence has been observed rarely in siblings with unaffected parents, more frequently in individuals in different generations. In the literature, there is no evidence of a higher degree of consanguinity in parents of affected individuals. In familial cases, individuals are more frequently bilaterally affected (Tasse et al. 2005). Expression varies within these families. There are reports of ear and mandibular involvement in first-degree relatives and reports of isolated microtia or preauricular tags in first-degree relatives of individuals with ear and mandibular involvement. Isolated microtia or preauricular tags may represent the mildest expression of the disorder (Cohen et al. 1989; Tasse et al. 2005). It should be pointed out that preauricular skin tags or nodules occur in about 1% of the general population. A few families (possibly representing 1–2%) with clear autosomal dominant inheritance have been reported, and in one family, the gene mapped to 14q32 (Kelberman et al. 2001).

There has only rarely been documentation of concordance in monozygotic twins, which can be regarded as evidence for a non-genetic cause. Reported teratogens to which affected individuals have been exposed prenatally include thalidomide, primidone, retinoic acid, pseudoephedrine, and maternal diabetes.

Wieczorek et al. (2007) noted an increased incidence of OAVS after assisted reproductive techniques as well as an excess of monozygotic twinning. This is compatible with the concept of over-ripeness ovopathy (Wieczorek et al. 2007).

Genetic Counseling A recurrence risk of 2–3% has been estimated when all known causes have been excluded. Because occasional families (possibly representing 1–2%) have autosomal dominant inheritance, with variable expression, parental evaluation is recommended.

Diagnostic Testing

Diagnosis is made based on clinical judgment. The OAVS, as indicated above, is remarkably variable and undoubtedly causally heterogeneous. One must exclude various chromosomal imbalances, and therefore, chromosomal microarray analysis is appropriate in most cases. One might consider performing chromosome analysis on a skin biopsy of the affected side or region (e.g., obtained during surgery), especially when associated with skin pigmentary alterations or asymmetry affecting other parts of the body.

Differential Diagnosis

In addition to the broad variability of OAVS, diagnosis may also be complicated by the existence of several syndromes with overlapping features. Townes–Brock syndrome, which is caused by autosomal dominant mutations in *SALL1*, consists of malformed external ears, ear tags and hearing loss, and thumb anomalies (typically triphalangeal thumb), anal defects, and renal anomalies. Abnormal ears with preauricular pits are seen the branchio-oto-renal (BOR) syndrome. Additional features are branchial cysts or fistulas, anomalous pinnae, malformations of the middle or inner ear, mixed hearing loss, and renal malformations. It also has autosomal dominant inheritance with variable expression. Families have been reported in which first-degree relatives have varying features of hemifacial microsomia and/or branchio-oto-renal syndrome. Branchio-oto-renal syndrome can be caused by mutations in the *EYA1*, *SIX1* or *SIX5* gene. OAVS should be distinguishable from mandibulofacial dysostosis (Treacher Collins syndrome –see Chapter 57), maxillofacial dysostosis, and the different forms of acrofacial dysostosis. Partial to total absence of the lower eyelashes, common in Treacher Collins syndrome, has not been reported in the OAVS, whereas colobomas of the upper eyelids are not infrequent. Mutations in *EFTUD2* cause a craniofacial dysostosis with microcephaly. The phenotype may overlap with that of OAVS, since preauricular tags, facial asymmetry and occasionally epibulbar dermoids are present (Lines and Boycott 2014). Other conditions to be considered include the Wildervanck syndrome, which is a sporadic condition, mostly seen in girls, that includes fused cervical vertebrae, abducens palsy with retracted globe (Duane syndrome), sensorineural hearing loss, and occasional facial asymmetry. The auriculo-condylar syndrome (ARCND) includes a characteristic ear anomaly ("question mark ear") and microstomia with abnormalities of the condyle of the mandible. ARCDN is etiologically heterogeneous, with autosomal dominant inheritance caused by mutations in the *GNAI3* or PLCB4 genes, and autosomal recessive inheritance caused by mutations in either *PLCB4* or *EDN1*. Okihiro syndrome (radial ray anomalies, Duane anomaly, spinal abnormalities, deafness, and renal anomalies) is caused by dominant *SALL4* mutations. A syndrome with ophthalmic anomalies and a particular cleft ear lobule is caused by homozygous mutations in the *NKX5.3* gene (Schorderet et al. 2008). A syndrome of autosomal dominant microtia and eye coloboma was described and caused by five tandem copies of a copy-number-variable region at chromosome 4p16 (Balikova et al. 2008). Characteristics of the VATER association (see Chapter 61), CHARGE syndrome (see Chapter 11), and the MURCS association (Müllerian duct aplasia, renal aplasia, and cervical thoracic vertebral dysplasia) overlap with the OAVS.

All these syndromes and associations tend to have bilateral, reasonably symmetrical involvement, in contrast to the asymmetric facial involvement in the OAVS.

MANIFESTATIONS AND MANAGEMENT

Because the OAVS is so complex, a team approach is advisable (Harvold et al. 1983; Munro, 1987; Kaban et al. 1988; Chibbaro 1999). Treatment is long-term, and although some surgical procedures are carried out early, others may extend into adulthood.

Given the variability in expression, the presence of one of the cardinal features in a child should lead to further examination for other manifestations as well as a detailed family history for the spectrum's features.

Growth and Feeding

Affected infants are often small for dates (Avon and Shively, 1988). Short stature can be present, mainly in more severely affected individuals (Tasse et al. 2005).

Feeding difficulties are frequent and can have multiple causes such as underdevelopment of the maxilla and mandible, clefting, malocclusion, facial palsy, decreased muscle tone in the orofacial musculature, and velopharyngeal insufficiency. Feeding difficulties often improve over time, but close follow-up is indicated.

Evaluation

- Affected individuals should have height and weight measured at each routine visit and plotted on standard growth curves.
- Those with short stature should be evaluated for an underlying cause in a standard manner before attributing the short stature to OAVS.
- Feeding ability should be assessed early, and evaluation of the cause of feeding difficulties should not be delayed. This evaluation is the same as for the general population and is preferably done in a multidisciplinary team for feeding disorders in infants.

Management

- Any identified underlying cause of short stature should be treated in a standard manner.
- Close attention to the feeding problems is needed. Nasogastric or gastrostomy tube feeding may be indicated. Indications for initiating tube feeding are the same as for any child.

Development and Behavior

Children with the OAVS are at risk for motor and developmental delay. Estimates of the frequency of cognitive deficits have ranged from 5 to 15%; however, this has not been adequately investigated. Intellectual disability is more frequently observed with bilateral involvement, brain anomalies, and microphthalmia.

Affected individuals are also at risk for speech problems. D'Antonia et al. (1998) found a higher than expected occurrence of pharyngeal and laryngeal abnormalities affecting speech production. They found asymmetric soft palate elevation and lack of sufficient velopharyngeal closure resulting in excessive nasality in several individuals with OAVS, increased articulation errors likely related to velopharyngeal issues and/or malocclusions, and reduced overall speech intelligibility. Small laryngeal structures, asymmetric function, and narrowing of the airway at the laryngeal level were also described, resulting in voice deviation and concern about apnea for some.

There is some evidence to suggest that persons with OAVS are at risk for psychosocial problems and that measures of self-concept are low (Padwa et al. 1991).

Autism spectrum disorder appears to be more frequent in this condition (Strömland et al. 2007).

Evaluation

- All affected individuals need an evaluation of motor and speech development.
- The value of comprehensive interdisciplinary evaluation addressing the range of structural manifestations of the disorder and potential effects in psychosocial and speech behavior cannot be overemphasized.
- Brain imaging can be indicated, for instance, when microcephaly or macrocephaly is present, in the infant with microphthalmia or with severe unexplained feeding difficulties, or when neurological signs or symptoms, such as seizures, severe hypertonia or hypotonia, or abnormal movements, are present.

Management

- Appropriate psychomotor therapy should be provided when indicated.
- Early intervention and special education should be provided to those with developmental deficits.

- Speech therapy is frequently needed, focusing on articulation, resources, or voice concerns.
- Not infrequently, consideration may need to be given to surgical or prosthodontic improvement of velopharyngeal closure. Surgical techniques are the same as those used in the general population.
- Psychological counseling is indicated for those individuals where problems have been identified.

Craniofacial

Skull defects include cranium bifidum, microcephaly, dolichocephaly, and plagiocephaly. Mild facial asymmetry is evident in about 65%, with marked asymmetry noted in roughly 20% (Figure 42.1). The facial asymmetry is a function of age and is related to unusual bone growth. The temporal, maxillary, and zygomatic bones on the more severely involved side are often reduced in dimension and flattened. Although this is generally considered to be a unilateral disorder, about 10–30% have bilateral involvement. For unknown reasons, the right side is generally more severely involved than the left. The mandibular ramus and condyle may be aplastic or hypoplastic (Figure 42.2). Hypoplasia is often found in association with macrostomia of mild degree and, as noted above, is more common on the right side (R:L = 3:2). Unilateral or bilateral cleft lip and/or cleft palate occur in 7–15%, and macrostomia (lateral facial cleft) is present in at least 30% to some degree. Because the parotid glands have their embryonic origin at the corners of the mouth, it is not unusual for parotid duct development to have been disturbed. In some cases, there is unilateral parotid gland agenesis. Malocclusion is extremely frequent.

Longitudinal studies have established that hemifacial macrosomia is a progressive condition. Until relatively recent times, treatment was delayed until the abnormality reached end-state. Good evidence has accumulated to indicate that this is not always wise. In mild cases, delay probably does not cause serious facial skeletal changes or psychological impairment. However, with severe forms of OAVS that have not been corrected in childhood, extensive surgical procedures carried out on both the upper and lower jaw and occasional orbital osteotomies are required.

As the normal side grows, the shortened and abnormal hypoplastic mandible results in secondary deformation of the orbit, nose, and maxilla. Decreased mandibular growth prevents vertical elongation of the ipsilateral side of the face. This, in turn, results in canting of the maxilla and the occlusal plane. Failure of downward growth of the maxilla also results in secondary orbital displacement.

Evaluation

- In the absence of available computed tomography scans and magnetic resonance imaging, the degree of hypoplasia and distortion of the mandible should be examined in three standard radiographic planes. The frontal plane can be seen in anteroposterior cephalometric radiographs, which allows estimation of discrepancy in ramus height, tilting of maxilla, pyriform apertures, and orbits, and the extent of mandibular rotation toward the affected side. Lateral cephalograms and panorex studies allow examination of the sagittal plane. The temporomandibular joint can be examined, and discrepancy in height of mandibular rami in relationship to the upper jaws and to the base of the skull assessed. The transverse plane can be examined by a submental vertex radiograph to demonstrate shape and width of the mandibular body, the degree of asymmetry of zygomatic arches, and, if possible, displacement of the temporomandibular joint.
- In most major medical centers, these radiographic views have become auxiliary, having been largely replaced by three dimensional computed tomography reconstruction. This is the study of choice, which should be conducted in all affected individuals, if surgery is contemplated and it is available.

Management

- Correction of macrostomia should begin in the first six months.
- In the mildest form of hemifacial microsomia, and if the individual is less than 6 years of age, a functional orthodontic appliance is used to guide the mandibular remnant into a better, more normal position to bring the affected side of the jaw downward, forward, and toward the midline. The appliance also stimulates skeletal growth by placing tension on the affected muscles. This can first be done when the child is 3–4 years of age.
- If the appliance is not effective, the hypoplastic mandible is surgically advanced, elongated, and rotated into correct position. This creates a posterior open bite on the affected side. In some cases, a compensatory osteotomy may be required on the contralateral side to allow rotation without disturbing the joint. The posterior open bite is progressively reduced by an orthodontic appliance.
- In more severe cases, during the deciduous dentition stage, construction of the zygoma, glenoid fossa, temporomandibular joint, condyle, and ramus should be carried out so that symmetry with the opposite joint is achieved. The zygomatic arch is usually made from rib grafts or even cranial bone grafts. The glenoid fossa is lined with perichondrium. In construction of the temporomandibular joint, a bone graft is placed medial to the newly constructed zygomatic arch and hollowed to receive the condylar graft. The ramus is constructed from full-thickness rib graft, iliac graft, or calvarial material (Lindquist et al. 1986).

- The various surgical procedures used for treatment of OAVS, such as mandibular osteotomies, costochondral grafts, and maxillary osteotomies done at an early age or after permanent dentition has been completed, often affect the vitality of tooth buds and often are followed by relapses. This requires several operations.
- It is likely that many of the surgical procedures that have been used for correction of the OAVS will be replaced within future years by bone distraction, the technique originally used by Ilizarov et al. in the early 1950s to align fractured segments of long bones and later to elongate the bones without need for a bone graft (Molina and Ortiz-Monasterio 1995). The technique employed is roughly as follows. An oblique corticotomy is made in the external cortex of the mandible at the level of the gonial angle. Stainless steel points are inserted and joined by a softer distraction screw. Cuts are made in the mandible either unilaterally or bilaterally. Extraoral devices were used initially but are being replaced by intraoral devices. New bone is deposited in the space created by the cut at the rate of approximately 1–2 mm per day. This technique obviates the need for blood transfusion, harvesting of bone grafts, tracheotomy, and even, in some cases, intramaxillary fixation.

Ears and Hearing

Abnormalities of the external ear range from anotia (absence) to an ill-defined mass of tissue displaced anteriorly and inferiorly, to a mildly dysmorphic ear (Figures 42.1 and 42.2). Rarely, there is bilateral involvement. Preauricular tags of skin and cartilage are extremely common and may be unilateral or bilateral (Figure 42.2). These tags are located anywhere from the tragus to the angle of the mouth. They are most commonly seen in individuals with macrostomia and/or aplasia of the parotid gland and epibulbar dermoids. Preauricular sinuses may be noted. In milder cases, the external auditory canals are narrow, with severe anomalies of the pinnae being associated with atretic canals. Rarely, small pinnae with normal architecture are seen, and, as noted above, isolated microtia is considered by some to be a microform of the spectrum.

Both conductive (75%) and, less frequently, sensorineural (10%) hearing loss have been reported in individuals with OAVS (Davide et al. 2017). Mixed hearing loss is noted in many (10%). Despite unilateral external ear involvement, it is not unusual for conductive hearing loss to be present on the contralateral side (Carvalho et al. 1999). The etiology of hearing loss is diverse and includes anomalies of the middle and external ears, hypoplasia or agenesis of ossicles, aberrant facial nerves, patulous Eustachian tubes, and abnormalities of the skull base (Kaye et al. 1989; Bassila and Goldberg 1989; Goetze et al. 2017). Persons with unilateral hearing loss are at higher risk for educational problems and grade failure than those with normal hearing. It has been suggested that 25% of children with unilateral hearing loss repeat one or more grades, especially if the loss is severe to profound and/or is in the right ear (Bess and Humes 1990; Sleifer et al. 2015).

Evaluation

- Otoscopy should be part of the routine physical examination.
- It is imperative that hearing be evaluated bilaterally to determine the nature and extent of the hearing loss. Educational audiologists serving school districts or clinical audiologists associated with otolaryngology practice or interdisciplinary craniofacial teams should evaluate hearing (air and bone conduction) and middle ear functions via tympanometry (when possible) at least on a six month basis.
- In very young children where behavioral audiometric procedures are not possible, auditory brain stem response or otoacoustic emission assessments should be performed.

Management

- Surgical treatment during the first six months involves removal of skin tags under local anesthesia.
- A hypoplastic, atypically positioned ear can be moved into the correct location in frontal and sagittal planes.
- Very often, early ear construction results in misplaced auricular framework, the malaligned ear usually being located anterior and inferior. It is best to wait until adolescence to achieve ultimate symmetry. Early correction, not uncommonly, leaves residual soft tissue defects, and dermal fat grafts or de-epithelialized free vascularized scapular flaps may be required.
- Depending on the nature and extent of the hearing loss, children need to be evaluated for potential benefits of personal and/or classroom amplifications (frequency modulation [FM] systems).
- Preferential classroom seating is essential in the presence of hearing loss, with the better ear oriented toward the source of auditory information and away from extraneous noises such as vents and fans.
- Surgical improvement of hearing on the affected side should be considered and is dependent on anatomical status, position, and, of course, the facial nerve, and the preference of the individual and family. Again, interdisciplinary craniofacial treatment planning is important. Ear procedures may be done in conjunction with other surgical and/or dental

procedures. Children with pure or predominantly conductive hearing loss may benefit from a bone anchored hearing aid (BAHA).
- Antibiotic treatment for middle ear disease in the better or normal ear is recommended.
- If middle ear effusion persists, myringotomy and ventilation tube insertion must be considered.

Ophthalmologic

Strömland et al. (2007) reported the presence of ocular findings in 72% of children with OAVS including epibulbar dermoids, microphthalmos, iris coloboma, upper lid coloboma, ptosis, and distichiasis (double row of eyelases). In a study by Tasse et al. (2005), dermoids also were the most frequent ocular findings besides microphthalmos, coloboma of the upper eyelid, strabismus, anophthalmia, and proptosis. The epibulbar dermoid is the most constant ocular feature of OAVS, but it is not obligatory to establish the diagnosis. Its incidence in individuals with OAVS is estimated to be 35%. An epibulbar dermoid usually presents as a flattened, milky-white to yellow solid structure straddling the limbus, usually inferotemporal in location (Figure 42.3).

Lipodermoids are generally seen laterally or in the superotemporal quadrant. Epibulbar dermoids and lipodermoids can occur in the same eye. In association with OAVS epibular dermoids often are bilateral. Less common ocular anomalies include uveal and lid colobomas. Lid colobomas are found at the junction of the middle and inner third of the upper eyelid (Baum and Feingold 1973). Other ocular features described in this syndrome include Duane retraction "syndrome" (preferably labelled as Duane anomaly), microphthalmia, microcornea, polar cataract, and Peters anomaly. Corneal hypoesthesia has been described; Nijhawan et al. (2002) reported abnormalities of the caruncle (a normal structure found in the inner canthus) in seven individuals with the oculo-auriculo-vertebral syndrome. Posterior segment abnormalities such as chorioretinal colobomas and morning glory anomaly are a rare finding (Chaudhuri et al. 2007).

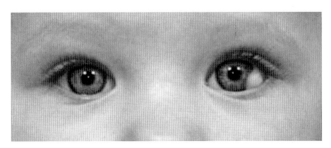

FIGURE 42.3 Limbal dermoid of the left eye.

Evaluation

- A full ophthalmological examination is necessary in every individual with OAVS, with special attention to identifying amblyopia, Duane anomaly, and refractive errors.
- Neuroimaging studies can be used to confirm the presence of trigeminal nerve hypoplasia or aplasia in children with the Goldenhar syndrome and corneal hypoesthesia (Villanueva et al. 2005).

Management

- A limbal dermoid can be removed and a partial or full thickness keratoplasty can be performed.
- Close ophthalmologic follow-up of astigmatism (pre- and postoperatively) is necessary, and occlusion of the better eye should be applied to treat the superimposed amblyopia.
- When eyelid coloboma is present, lubrication of the cornea is necessary to avoid exposure keratitis.
- Corneal hypoesthesia should be recognized at a young age to prevent neurotropic corneal ulcers from developing by sufficient lubrication and topical antibiotics if necessary.
- In the presence of associated craniofacial problems, a multidisciplinary approach should be organized.

Neurologic

A wide range of neurological defects has been found in people with OAVS (reviewed by Renkema et al. 2017). Lower facial nerve weakness, seen in about 20% (Bassila and Goldberg 1989; Carvalho et al. 1999), is probably related to bony involvement in the region of the facial canal. Nearly all cranial nerves have been reported to be affected on occasion. In the so-called expanded OAVS, brain malformations including encephalocele, hydrocephaly, lipoma, dermoid cysts, teratoma, Arnold–Chiari malformation, lissencephaly, arachnoid cyst, holoprosencephaly, unilateral arhinencephaly, and hypoplasia of the corpus callosum have been described. Lower facial nerve weakness correlates with severity of ear involvement but not with the degree of mandibular hypoplasia. It is not unusual for the soft palate to deviate to the contralateral side. Asymmetric velar elevation may affect adequacy of velopharyngeal closure for speech. Brain imaging can be considered based on the signs and symptoms.

Evaluation

- Thorough neurological evaluation should take place at the time of diagnosis. Brain imaging could be considered if neurologic symptoms or microcephaly are present.

- Evaluation to determine the effect of neurological involvement on specific oral pharyngeal functions such as feeding and swallowing is recommended. In infants, this is best done by a specialized feeding team.
- Ideally, these evaluations are carried out periodically by speech-language pathologists in an interdisciplinary setting.

Management

- Frequently benefit is obtained from speech therapy focusing on compensatory strategies to optimize speech intelligibility and articulation.
- Depending on the status of velopharyngeal closure, speech therapy and surgical and prosthodontic treatment may be required.
- Temporalis muscle transfer can improve facial animation on the affected side if there is neurological impairment.

Cardiovascular

Various heart anomalies have been recorded. These are probably present in about 35% (range 5–60%) of individuals with OAVS. Ventriculoseptal defect and tetralogy of Fallot account for about 50% of structural defects, although no single cardiac lesion is characteristic. A wide variety of other lesions has been documented, including transposition of great vessels, hypoplasia of the aortic arch, isolation of the left innominate artery with bilateral patent ductus arteriosus, pulmonary stenosis, and dextrocardia (Pierpont et al. 1982; Morrison et al. 1992; Kumar et al. 1993).

Evaluation

- All infants should be referred to a pediatric cardiologist for clinical examination, echocardiography, and electrocardiography.

Management

- Depending on the anomaly present, medical treatment may be required, using standard medication choices, as in the general population.
- Depending on the anomaly present, cardiac surgery may be required. The infant should be in good cardiac status before any surgical procedure is considered. Standard methods are used.
- Individuals with certain cardiac lesions will need endocarditis prophylaxis.

Musculoskeletal

Cervical vertebral fusions have been demonstrated in about 20–35% of affected individuals, and platybasia and occipitalization of the atlas have been found with about the same frequency (Renkema et al. 2017). However, a wide variety of spinal anomalies have been reported: C1–C2 instability, spina bifida, hemivertebrae, butterfly vertebrae, fused and hypoplastic vertebrae, Klippel–Feil anomaly, scoliosis, and anomalous ribs. Collectively, these are present in about 30% (Gibson et al. 1996).

Talipes equinovarus (clubfeet) has been noted in about 20%. Radial limb anomalies, usually hypoplasia or aplasia of the radius and/or thumb and bifid or digitalized thumb, are found in about 10% (Figueroa and Friede 1985).

Evaluation

- The primary care physician should conduct a careful musculoskeletal evaluation. Any concern about decrease in head movement or spinal deformity should prompt radiographs.
- Consultation with an orthopedic surgeon for several of the anomalies (scoliosis and talipes equinovarus) is recommended.

Management

- Scoliosis and talipes equinovarus should be treated in a standard fashion.
- Cervical vertebral fusions rarely produce symptoms, and no therapy is usually required.

Genitourinary

Renal, urinary tract and genital anomalies are frequent (18% in the series of Tasse et al. 2005; 3/18 in the series of Strömland et al. 2007, 5/51 in the series of Beleza-Meireles et al. 2015).

Evaluation

- A renal ultrasound is indicated in all individuals with OAVS at the time of diganosis to exclude a structural renal malformation.
- If a child has recurrent urinary tract infections, further evaluation in a standard manner is indicated.

Management

- The associated renal anomalies are rarely severe and seldom need surgical correction unless there is obstruction or reflux, which can be treated according to standard procedures.
- Recurrent urinary tract infections are treated in a standard fashion.

Respiratory

Lung anomalies range from incomplete lobulation to hypoplasia to agenesis. These findings have been rare, probably on the order of 5%. They can be unilateral or bilateral.

Pulmonary agenesis usually occurs on the same side as the facial anomalies. Tracheoesophageal fistula has also been documented. Obstructive sleep apnea has been reported.

Evaluation

- Tracheoesophageal fistula may be suspected in the newborn period if it is difficult to pass a suction tube to clear tracheal secretions.
- If obstructive sleep apnea is suspected, a full sleep study should be arranged.
- Respiratory insufficiency should be evaluated in a standard manner.

Management

- Surgical correction of tracheoesophageal fistula is performed using standard techniques.
- Treatment of sleep apnea will depend on the cause and is the same as in the general population.

RESOURCES

Orphanet

In Europe, support organizations and clinics can be identified through this organization.
Website: *http://www.orpha.net/*

AboutFace USA

1002 Liberty Lane
Warrington, PA 18976, USA
Phone: (800) 225-FACE

About Face

99 Crowns Lane
Toronto M59 3P4
Ontario, Canada
Phone: (800) 665-FACE
Email: *abtface@interlog.com* (attention: Laura)

Changing Faces (UK)

Phone: (0171) 706-4232
Email: *info@faces.demon.co.uk*

Children's Craniofacial Association

PO Box 280297
Dallas, TX 75243-4522, USA
Phone: (972) 994-9902

FACES – National Association for the Craniofacially Handicapped

PO Box 11082
Chattanooga, TN 37401, USA
Phone: (800) 332-2373
Contact: Priscilla Caine

FACES: The National Craniofacial Association

PO Box 11082, Chattanooga, TN 37401, USA
Phone: (423) 266-1632
Website: *http://faces-cranio.org/about-us.html*

Ear Anomalies Reconstructed (EAR)

Atresia-Microtia Support Group
72 Durand Rd.
Maplewood, NJ 07040, USA
Phone: (201) 761-5438
Contact: Jack Gross or Betsy Olds

NORD: National Organization for Rare Disorders, Inc.

100 Rt. 37, PO Box 8923
New Fairfield, CT 06812-8923, USA
Phone: (203) 746-6518
Email: *orphan@nord-rdb.com*
Website: *http://www.nord-rdb.com/~orphan*

REFERENCES

Avon SW, Shively JL (1988) Orthopaedic manifestations of Goldenhar syndrome. *J Pediatr Orthop* 8:683–686.

Balikova I, Martens K, Melotte C, Amyere M, Van Vooren S, Moreau Y, Vetrie D, Fiegler H, Carter NP, Liehr T, Vikkula M, Matthijs G, Fryns JP, Casteels I, Devriendt K, Vermeesch JR (2008) Autosomal-dominant microtia linked to five tandem copies of a copy-number-variable region at chromosome 4p16. *Am J Hum Genet* 82:181–187.

Bassila MK, Goldberg R (1989) The association of facial palsy and/or sensorineural hearing loss in patients with hemifacial microsomia. *Cleft Palate J* 26:287–291.

Baum JL, Feingold M (1973) Ocular aspects of Goldenhar's syndrome. *Am J Ophthalmol* 75:250–257.

Beleza-Meireles A, Hart R, Clayton-Smith J, Oliveira R, Reis CF, Venâncio M, Ramos F, Sá J, Ramos L, Cunha E, Pires LM, Carreira IM, Scholey R, Wright R, Urquhart JE, Briggs TA, Kerr B, Kingston H, Metcalfe K, Donnai D, Newman WG, Saraiva JM, Tassabehji M (2015) Oculo-auriculo-vertebral spectrum: clinical and molecular analysis of 51 patients. *Eur J Med Genet* 58(9):455–65.

Bess FH, Humes LE (1990) *Audiology: The Fundamentals.* Baltimore, MD: Williams and Wilkins.

Bragagnolo S, Colovati MES, Souza MZ, Dantas AG, F de Soares MF, Melaragno MI, Perez AB (2018) Clinical and cytogenomic findings in OAV spectrum. *Am J Med Genet A* 176:638–648.

Carvalho GJ, Song CS, Vargervik K, Lalwani AK (1999) Auditory and facial nerve dysfunction in patients with hemifacial microsomia. *Arch Otolaryngol Head Neck Surg* 125: 209–212.

Chaudhuri Z, Grover AK, Bageja S, Jha SN, Mohan S (2007) Morning glory anomaly with bilateral choroidal colobomas in a patient with Goldenhar's syndrome. *J Pediatr Ophthalmol Strabismus* 44:187–189.

Chibbaro PD (1999) Living with craniofacial microsomia: Support for the patient and family. *Cleft Palate Craniofac J* 36:40–42.

Cohen MM Jr, Rollnick BR, Kaye CI (1989) Oculo-auriculo-vertebral spectrum: An updated critique. *Cleft Palate J* 26:276–286.

Cousley RRJ (1993) A comparison of two classification systems for hemifacial microsomia. *Br J Oral Maxillofac Surg* 31:78–82.

D'Antonia LL, Rice RD, Fink SC (1998) Evaluation of pharyngeal and laryngeal structure and function in patients with oculo-auriculo-vertebral spectrum. *Cleft Palate Craniofac J* 35:333–341.

David DJ, Mahatumarat C, Cooter RD (1987) Hemifacial microsomia: A multisystem classification. *Plast Reconstr Surg* 80:525–533.

Davide B, Renzo M, Sara G, Elisa L, Rodica M, Irene T, Alessandro C, Giovanni S, Valentina S, Roberto B, Patrizia T, Alessandro M (2017) Oculo-auriculo-vertebral spectrum: going beyond the first and second pharyngeal arch involvement. *Neuroradiology* 59(3):305–316.

Figueroa AA, Friede H (1985) Costovertebral malformation in hemifacial microsomia. *J Craniofac Genet Dev Biol Suppl* 1:167–178.

Gibson JNA, Sillence DO, Taylor TKF (1996) Abnormalities of the spine in Goldenhar's syndrome. *J Pediatr Orthopaed* 16:344–349.

Goetze TB, Sleifer P, Rosa RF, da Silva AP, Graziadio C, Zen PR (2017) Hearing characterization in oculoauriculovertebral spectrum: A prospective study with 10 patients. *Am J Med Genet A* 173(2):309–314.

Gorlin RJ, Kenneth LJ, Jacobsen U, Goldschmidt E (1963) Oculoauriculovertebral dysplasia. *J Pediatr* 63:991–999.

Gorlin RJ, Cohen MM Jr, Hennekam RCM (2001) *Syndromes of the Head and Neck*, 4th ed. New York: Oxford University Press.

Harvold EP, Vargevik K, Chierici G, eds (1983) *Treatment of Hemifacial Microsomia*. New York: Liss.

Heike CL, Hing AV, Aspinall CA, Bartlett SP, Birgfeld CB, Drake AF, Pimenta LA, Sie KC, Urata MM, Vivaldi D, Luquetti DV (2013) Clinical care in craniofacial microsomia: a review of current management recommendations and opportunities to advance research. *Am J Med Genet C* 163C:271–82.

Kaban LB, Moses MH, Mulliken JB (1988) Surgical correction of hemifacial microsomia. *Plast Reconstr Surg* 82:9–19.

Kaye CI, Rollnick BR, Hauck WW, Martin AO, Richtsmeyer JT, Nagatoshi K (1989) Microtia and associated anomalies. *Am J Med Genet* 34:574–578.

Kelberman D, Tyson J, Chandler DC, McInerny AM, Slee J, Albert D, Aymat A, Botma M, Calvert M, Goldblatt J, Haan E, Laing N, Lim J, Malcolm S, Singer R, Winter R, Bitner-Glindzicz M (2001) Hemifacial microsomia. Progress in understanding the genetic basis of a complex malformation syndrome. *Hum Genet* 109:638–645.

Keogh IJ, Troulis MJ, Monroy AA, Eavey RD, Kaban LB (2007) Isolated microtia as a marker for unsuspected hemifacial microsomia. *Arch Otolaryngol Head Neck Surg* 133:997–1001.

Kumar A, Friedman JM, Taylor GP, Patterson MWH (1993) Pattern of cardiac malformation in oculo-auriculo-vertebral spectrum. *Am J Med Genet* 46:423–426.

Lauritzen C, Munro IF, Ross RB (1985) Classification and treatment of hemifacial microsomia. *Scand J Plast Reconstr Surg* 19:33–39.

Lindquist C, Pihakari A, Tasanen A, Hampf G (1986) Autogenous costochondral grafts in temporomandibular joint arthroplasty. *J Maxillofac Surg* 14:143–149.

Lines M, Hartley T, Boycott KM (2014) Mandibulofacial Dysostosis with Microcephaly. *GeneReviews®*.

Lopez E, Berenguer M, Tingaud-Sequeira A, Marlin S, Toutain A, Denoyelle F, Picard A, Charron S, Mathieu G, de Belvalet H, Arveiler B, Babin PJ, Lacombe D, Rooryck C (2016) Mutations in MYT1, encoding the myelin transcription factor 1, are a rare cause of OAVS. *J Med Genet* 53:752–760.

Molina F, Ortiz-Monasterio F (1995) Mandibular elongation and remodeling by distraction: A farewell to major osteotomies. *Plast Reconstr Surg* 96:825–845.

Morrison PJ, Mulholland HC, Craig BG, Nevin NC (1992) Cardiovascular abnormalities in the oculo-auriculo-vertebral spectrum (Goldenhar syndrome). *Am J Med Genet* 44:425–428.

Munro IR (1987) Treatment of craniofacial microsomia. *Clin Plast Surg* 14:177–186.

Nijhawan N, Morad Y, Seigel-Bartelt J, Levin AE (2002) Caruncle abnormalities in the oculo-auriculo-vertebral spectrum. *Am J Med Genet* 113:320–325.

Padwa BL, Evans CA, Pillemer FC (1991) Psychosocial adjustments in children with hemifacial microsomia and other craniofacial deformities. *Cleft Palate Craniofac J* 28:354–359.

Pierpont MEM, Moller JH, Gorlin RJ, Edwards JE (1982) Congenital cardiac, pulmonary and vascular malformations in oculo-auriculo-vertebral dysplasia. *Pediatr Cardiol* 2:297–302.

Poswillo D (1973) The pathogenesis of the first and second branchial arch syndrome. *Oral Surg* 35:302–329.

Renkema RW, Caron CJJM, Mathijssen IMJ, Wolvius EB, Dunaway DJ, Forrest CR, Padwa BL, Koudstaal MJ (2017) Vertebral anomalies in craniofacial microsomia: a systematic review. *Int J Oral Maxillofac Surg* 46:1319–1329.

Renkema RW, Caron CJJM, Wolvius EB, Dunaway DJ, Forrest CR, Padwa BL, Koudstaal MJ (2018) Central nervous system anomalies in craniofacial microsomia: a systematic review. *Int J Oral Maxillofac Surg* 47:27–34.

Robinson L, Hoyme HE, Edwards DK, Jones KL (1987) The vascular pathogenesis of unilateral craniofacial defects. *J Pediatr* 111:236–239.

Rollnick BR, Kaye CE, Nagatoski K (1987) Oculo-auriculo-vertebral dysplasia and variants: Phenotypic characteristics of 194 patients. *Am J Med Genet* 26:361–375.

Schorderet DF, Nichini O, Boisset G, Polok B, Tiab L, Mayeur H, Raji B, de la Houssaye G, Abitbol MM, Munier FL (2008)

Mutation in the human homeobox gene NKX5-3 causes an oculo-auricular syndrome. *Am J Hum Genet* 82:1178–84.

Sleifer P, Gorsky Nde S, Goetze TB, Rosa RF, Zen PR (2015). Audiological findings in patients with oculo-auriculo-vertebral spectrum. *Int Arch Otorhinolaryngol* 19:5–9.

Strömland K, Miller M, Sjögreen L, Johansson M, Joelsson BM, Billstedt E, Gillberg C, Danielsson S, Jacobsson C, Andersson-Norinder J, Granström G (2007) Oculo-auriculo-vertebral spectrum: associated anomalies, functional deficits and possible developmental risk factors. *Am J Med Genet A* 143:1317–1325.

Tasse C, Böhringer S, Fischer S, Lüdecke HJ, Albrecht B, Horn D, Janecke A, Kling R, König R, Lorenz B, Majewski F, Maeyens E, Meinecke P, Mitulla B, Mohr C, Preischl M, Umstadt H, Kohlhase J, Gillessen-Kaesbach G, Wieczorek D (2005) Oculo-auriculo-vertebral spectrum (OAVS): Clinical evaluation and severity scoring of 53 patients and proposal for a new classification. *Eur J Med Genet* 48:397–411.

Vento AR, LaBrie RA, Mulliken JB (1991) The O.M.E.N.S. classification of hemifacial microsomia. *Cleft Palate J* 28:68–76.

Villanueva O, Atkinson DS, Lambert SR (2005) Trigeminal nerve hypoplasia and aplasia in children with Goldenhar syndrome and corneal hypoesthesia. *J AAPOS* 9:202–204.

Wieczorek D, Ludwig M, Boehringer S, Jongbloet PH, Gillessen-Kaesbach G, Horsthemke B (2007) Reproduction abnormalities and twin pregnancies in parents of sporadic patients with oculo-auriculo-vertebral spectrum/Goldenhar syndrome. *Hum Genet* 121:369–376.

43

OSTEOGENESIS IMPERFECTA

AN N. DANG DO
Office of the Clinical Director, Eunice Kennedy Shriver *National Institute of Child Health and Human Development, National Institutes of Health, Bethesda, Maryland, USA*

JOAN C. MARINI
Section on Heritable Disorders of Bone and Extracellular Matrix, Eunice Kennedy Shriver *National Institute of Child Health and Human Development, National Institutes of Health, Bethesda, Maryland, USA*

INTRODUCTION

The name osteogenesis imperfecta (OI) succinctly captures the common primary clinical features of this group of collagen-related disorders. The earliest evidence for the existence of OI is from PHK Gray's 1969 description of an Egyptian child mummified around 1000 BCE, with a "*tam-o'-shanter*" skull deformity and bowed leg bones with thin cortices (Tainmont 2007). The first reported description of the findings of OI may have been from as early as 1688 by the philosopher Nicolas de Malebranche (Tainmont 2007), although other candidates have also been proposed (Weil 1981).

Dutch anatomist Willem Vrolik was credited with introducing the term "osteogenesis imperfecta," to describe the findings of multiple fractures and wormian bones he observed in a newborn (Weil 1981). Interestingly, the multisystemic involvement of OI might have been recognized as early as 1831 by Edmund Axmann when he described accompanying features of blue sclerae and hypermobile joints (Weil 1981). Bone fragility, scleral hue, and hearing impairment emerged as the three cardinal symptoms based on which early diagnoses of OI were made (Sillence et al. 1979; Weil 1981)

Wide variations in the expression of the major OI symptoms prompted multiple practitioners to propose grouping affected individuals into different types of OI based on the severity of their clinical features (Sillence et al. 1979). From this foundation, and based on their own observations and epidemiologic studies, David Sillence and colleagues proposed their classification system in 1979, assigning individuals with OI into four groups based on skeletal severity (clinical and radiographic), scleral hue, and mode of inheritance. The original Sillence OI type I group encompassed individuals with the least fracture severity, the most prominent blue sclera, and an autosomal dominant inheritance pattern. The fracture severity in Sillence OI types II–IV decreases from perinatal lethality in type II, to severe and progressively deforming in type III, and variable (but more severe than type I) in type IV. Scleral hue of individuals in Sillence OI types II–IV are less consistently and prominently blue than seen in type I. Autosomal dominant and recessive modes of inheritance were both deduced from families of individuals classified as Sillence OI types II or III; whereas autosomal dominant was ascribed as the main inheritance pattern in OI type IV (Sillence et al. 1979).

While "osteogenesis imperfecta" has remained relatively uncontested as the name for this group of disorders, the advance in DNA sequencing has spurred a debate about the classification of OI sub-types (Marini et al. 2017; Van Dijk and Sillence 2014). Since the first identification of *COL1A1*

Cassidy and Allanson's Management of Genetic Syndromes, Fourth Edition.
Edited by John C. Carey, Agatino Battaglia, David Viskochil, and Suzanne B. Cassidy.
© 2021 John Wiley & Sons, Inc. Published 2021 by John Wiley & Sons, Inc.

as a gene involved in OI pathology in 1983, variations in 16 other genes have been demonstrated in OI (Marini et al. 2017). The discovery of these additional causal genes has broadened the understanding of bone biology, helped to explain some of the phenotypic heterogeneity seen in OI, and most importantly provided additional targets for potential therapeutic interventions.

Incidence and Prevalence

Recent epidemiological reviews of hospital records and birth registries from South (Barbosa-Buck et al. 2012) and North (Stevenson et al. 2012) America proposed a prevalence of OI ranging from 1:13,500–1:12,600 (0.74–0.79:10,000) live births. This estimate falls within the broader range of 1:25,000–1:6250 previously observed in other regions of the world (Andersen and Hauge 1989). While epidemiologic studies of OI have not been done in all ethnic groups, case series and cohorts confirm that OI, as a group, does not have a predilection for any specific ethnicity or race. A higher prevalence in certain populations of a specific OI type, or of a specific variant in one of the OI-related genes, however, does exist. Beighton and Versfeld (1985) reported a higher estimated frequency of OI type III in Johannesburg, South Africa as compared to that reported for white Australians. Founder mutations for certain OI types have been identified in people from West Africa (*P3H1*), First Nations in Ontario, Canada (*CRTAP*), Bedouins (*TMEM38B*), Turkey (*FKBP10*), and the Hmong group in Vietnam and China (*WNT1*) (Marini et al. 2017).

Diagnostic Criteria

Prenatal For the more severe perinatal lethal and progressively deforming presentations, features suggestive of OI can be seen by ultrasound imaging beginning in the second trimester of gestation. The long bones may appear short compared to the standard reference ranges (micromelia), bowed or angulated, and "crumpled" secondary to fracture calluses. The axial skeleton may display "beaded S-shaped ribs" and flattened vertebrae (platyspondyly). Hypomineralization may be observed overall, especially in the skull bones. Other observations may include increased nuchal translucency, and small hypoplastic lungs. These features are not commonly observed for the milder OI types, or may occur later in gestation. (Krakow 2018)

Postnatal The clinical criteria for diagnosing, or for initiating an evaluation for, OI include the hallmark features of fractures, skeletal deformity, impaired growth and scleral hue. Bone fragility leading to frequent, multiple fractures is the most prominent feature of OI types. Vertebral fractures, in combination with ligamentous laxity, culminate in mild to severe spinal curvatures. For most types of OI, short stature, particularly disproportionate short stature with short lower extremities, is a cardinal finding. The presence of blue sclerae and hearing loss, in the context of other features, continues to be informative for diagnosing OI. As more individuals with a spectrum of severity of OI symptoms have been identified, the presence or absence of these features in different OI types has also been discovered to be more variable.

Several radiographic features are commonly seen in individuals with OI. None, however, is pathognomonic. Osteopenia, seen on x-ray images as a wash-out appearance or decreased bone signal intensity, is a common finding in OI. The degree of osteopenia may not be appreciated unless measurements of bone mineral density compared to age-appropriate references are obtained from dual-energy X-ray absorptiometry (DEXA). Multiple fractures and healing calluses, bowing or abnormal angulation of long bones, flattening of the vertebral bodies and curvatures of the spine are also commonly seen. Besides these frequently observed features, other skeletal findings on X-ray imagings have also been associated with OI. Occipital bone deformation can produce a "Darth Vader" skull appearance, and basilar impression. Wormian bones may be seen at a higher frequency in individuals with OI, but are not specific for this disorder. At the hips, decreased angulation of the femoral neck (coxa vara) and protrusion of both the femoral head and acetabulum into the pelvis (protrusio acetabuli) may be seen in individuals with OI (Renaud et al. 2013).

The discoveries of additional collagen-related genes underlying OI pathology, as well as new clinical and radiographic features, necessitate an update of the 1979 Sillence OI classification system. Two main approaches have been proffered: a phenotype-forward classification preserving the Sillence system by using clinical and radiographic findings, and a genotype-forward classification expanding the Sillence system by creating a new type for each OI causative gene. The phenotype-forward classification has practical utility for clinicians and practitioners in areas where genetic testing is not easily accessible (Van Dijk and Sillence 2014). However, it poses challenges in maintaining diagnostic consistency for individuals with OI whose phenotype may change as they age, and for individuals in the same family or having the same OI-causative gene (Forlino and Marini 2016). In contrast, the genotype-forward classification maintains diagnostic consistency as individuals with variants in the same OI-causative gene would remain under the same group type. This would also facilitate genetic counseling. The challenges of this approach are the reliance on availability of genetic testing, and the potential of generating a system of uncategorized OI types. A hybrid classification system that allows room for addition of newly discovered OI type by gene, yet maintains functional and relevant groupings of different OI types may help to mitigate these issues (Table 43.1) (Forlino and Marini 2016).

Impaired Collagen Synthesis, Structure, and Assembly: Types I–IV Using the classification scheme proposed by

TABLE 43.1 Osteogenesis imperfecta types classification and features

Gene	Protein	Type (OMIM)	Inheritance	Fractures onset	Skeletal deformity	Scleral hue	Hearing loss (onset; type)	Dentinogenesis imperfecta	Other features
Impaired collagen synthesis, structure, and assembly									
COL1A1 COL1A2	COL1A1 [α1(I)] COL1A2 [α2(I)]	I (166200)	Autosomal dominant	Birth (if DI also present)	Rare; more severe if DI also present	Blue-gray	Adolescence – young adult; mixed	Variable	
		II (166210)		In utero	Very severe (perinatally lethal)	White, gray, dark blue	Variable	Variable	Deformed bones detectable at 18–20 weeks of gestation; relative macrocephaly
		III (259420)		Birth–infancy	Moderate – very severe	Blue at birth, less with age	Not in children, more frequent in adults	Variable	Platyspondyly; metaphyseal "pop-corn" calcifications, flaring; relative macrocephaly; pectus deformities; severe scoliosis
		IV (166220)		Childhood–puberty; after menopause (females), or sixth decade (males)	Rare–very severe	May be bluish at birth; white when older	Not observed	variable	Higher risk for BI if DI also present; relative macrocephaly; osteoporosis; scoliosis
Impaired bone mineralization									
IFITM5	BRIL	V (610967)	Autosomal dominant	Infancy - childhood	Variable	Mostly white; gray, blue in some cases	Rare; mixed	Rare	Hypertrophic callus at fracture sites; radial-ulnar interosseous membrane calcification; dense metaphyseal band
SERPINF1	PEDF	VI (613982)	Autosomal recessive	Infancy–childhood	Moderate–severe	White, grey, faint blue	Not observed	Not observed	"Fish-scale" lamellae on histology; excessive osteoid
Impaired collagen modification and processing									
CRTAP	CRTAP	VII (610682)	Autosomal recessive	In utero– Birth	Severe	White, faint blue	Not observed	Rare	Rhizomelia; exophthalmia; normo/microcephaly
P3H1	P3H1	VIII (610915)	Autosomal recessive	In utero	Severe	White	Not observed	Not observed	Rhizomelia; normo/microcephaly; severe osteoporosis; popcorn calcifications
PPIB	PPIB (CyPB)	IX (259440)	Autosomal recessive	In utero	Severe	White, gray	Not observed	Not observed	Moderately severe osteoporosis

(*continued*)

TABLE 43.1 (Continued)

Gene	Protein	Type (OMIM)	Inheritance	Fractures onset	Skeletal deformity	Scleral hue	Hearing loss (onset; type)	Dentinogenesis imperfecta	Other features
SERPINH1	HSP47	X (613848)	Autosomal recessive	Infancy	Severe	Blue, white	Not observed	Some cases	Renal stones; pulmonary complications
FKBP10	FKBP65	XI (610968)	Autosomal recessive	Infancy	Mild–severe; progressive	White, gray	Not observed	Some cases	Fish scale-like lamellae; high alkaline phosphatase; phenotypic variants: Bruck I and Kuskokwim syndromes
PLOD2	LH2		Autosomal recessive		Moderate–severe				Large joints contractures (Bruck II syndrome)
BMP1	BMP1	XII (614856; XIII)	Autosomal recessive	Childhood	Mild–severe	White	Not observed	Not observed	Increased bone mineral density; umbilical hernia
Impaired osteoblast differentiation									
SP7	SP7 (Osterix)	XIII (613849; XII)	Autosomal recessive	Childhood	Moderate–severe	White	Childhood; mixed	Not observed	Prominent supraorbital ridge; midface hypoplasia; low bone mineral density
TMEM38B	TRIC-B	XIV (615066)	Autosomal recessive	In utero–childhood	Mild–severe	White, blue	Not observed	Not observed	Cardiovascular disorders
WNT1	WNT1	XV (615220)	Autosomal recessive; autosomal dominant	In utero–childhood	Severe	Bluish, white	Not observed	Not observed	Midbrain and cerebellar hypoplasia
CREB3L1	Oasis	XVI (616229)	Autosomal recessive ?Autosomal dominant?	In utero	In utero Lethal–severe	Bluish gray; (blue in heterozygotes)	Not observed	Not observed (tooth agenesis	Easy bruising; joints hyperextensibility. (Osteopenia in heterozygotes)
SPARC	SPARC (Osteonectin)	XVII (616507)	Autosomal recessive	Childhood	Progressive, severe	White		Not observed	
MBTPS2	S2P	XVIII (301014; XIX)	X-linked	In utero	Moderate – severe	White, blue		Not observed	

At press time, individuals with pathogenic variants in *TENT5A*, *MESD*, and *CCDC134* have been reported with features of OI inherited in an autosomal recessive manner. Information on the bone phenotype of these cases are incomplete at the time.

Forlino and Marini (2016), the classic Sillence OI Types I–IV associated with structural defects in the COL1A1/2 proteins are maintained. About 80–85% of individuals with OI have mutations in the *COL1A1/2* genes (Figure 43.1), and, of these, the majority have features of types I or IV (Marini and Smith 2015). Classic diagnostic clinical findings in individuals with OI types I–IV, and features that distinguish each type are as described above, depicted in Figure 43.2, and outlined in Table 43.1. Skeletal radiographs of individuals with severe type III or IV show thinning of cortices, deformed ribs and long bones, and severe scoliosis (Figures 43.3A and B). Characteristic popcorn calcifications and surgical rods are also often captured on X-rays (Figure 43.3C). Histologically, bone from individuals with OI demonstrate a pattern of increased immature, disorganized, mechanically weaker fibril organization (woven bones). Loss of the mechanically stronger lamellar pattern of fibril organization correlates with the severity of OI (Cassella et al. 1996).

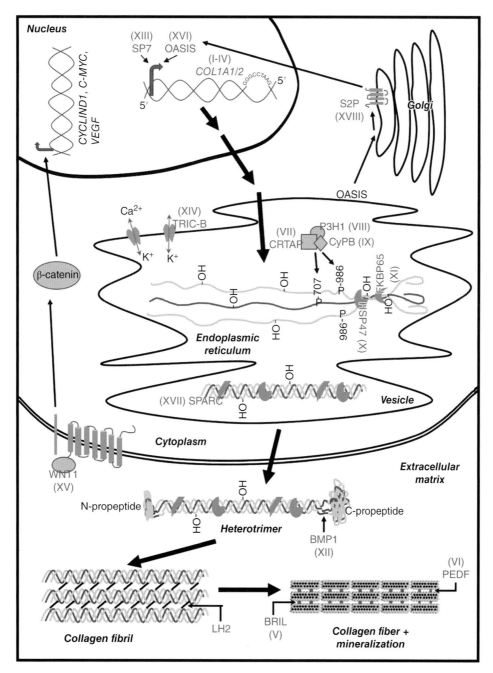

FIGURE 43.1 Proteins and pathways affected in OI. Defects range from mutations at the genomic level in *COL1A1/2* to decreased functional proteins involved in transcription, α chain folding and processing, and collagen trimer formation.

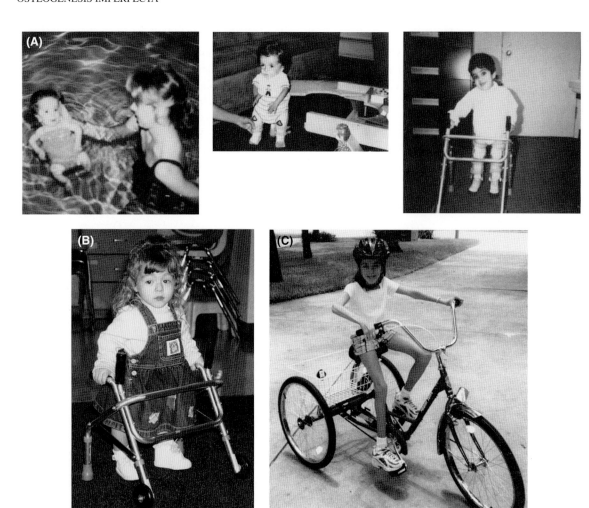

FIGURE 43.2 Morphological features and functional status of individuals with OI. (A) A 1-year-old infant with OI type II swimming with her mother. Two children with OI type IV using assistive devices. (B) A child with OI type III ambulating with a walker. (C) A child with OI type IV using a modified bicycle.

Impaired Bone Mineralization: Types V and VI The clinically distinct OI type V was proposed in by Glorieux et al. (2000) to classify individuals who shared some features with Sillence type IV: fractures within the first year of life, frequent (~2/year) fractures, ligamentous laxity, and an autosomal dominant pattern of inheritance. However, the distinguishing features observed in OI type V included formation of hyperplastic callus at sites of fractures, limited range of forearm pronation and supination, calcification of the radial-ulna interosseous membrane, and a dense metaphyseal band at the distal forearm growth plate (Figure 43.3D; Table 43.1). Bone histology and histomorphometry showed a mesh-like lamellation pattern, and decreased indices of bone formation on histomorphometry (Glorieux et al. 2000). The genetic cause of all cases of OI type V was identified as a recurrent mutation in *IFITM5*, a c.-14C>T transversion that introduces a new upstream in-frame initiation codon and adds five amino acids to the protein product BRIL (Figure 43.1) (Semler et al. 2012). Interestingly, a missense mutation in *IFITM5* identified in three individuals presents with a skeletal and histological phenotype similar to OI type VI, as well as reduced osteoblast secretion of PEDF (*p*igment *e*pithelium-*d*erived *f*actor), establishing a biochemical connection between types V and VI (Figure 43.1) (Farber et al. 2014).

OI type VI was also proposed by Glorieux et al. in 2002 to distinguish individuals with a severe recessive form of OI and distinctive bone histology. These individuals experience their first fractures during late infancy to childhood, after which the skeletal defect becomes progressively deforming, with both limb deformities and vertebral compressions. Scleral hue and ligamentous laxity findings are similar to those seen in OI type III and IV, but dentinogenesis imperfecta is not present (Table 43.1). Distinctive features of OI type VI bone histology include a fish-scale lamellation pattern under polarized light and increased osteoid volume. A variety of null mutations in *SERPINF1*, which encodes for the protein PEDF (Figure 43.1), have been identified as the cause of OI type VI (Homan et al. 2011).

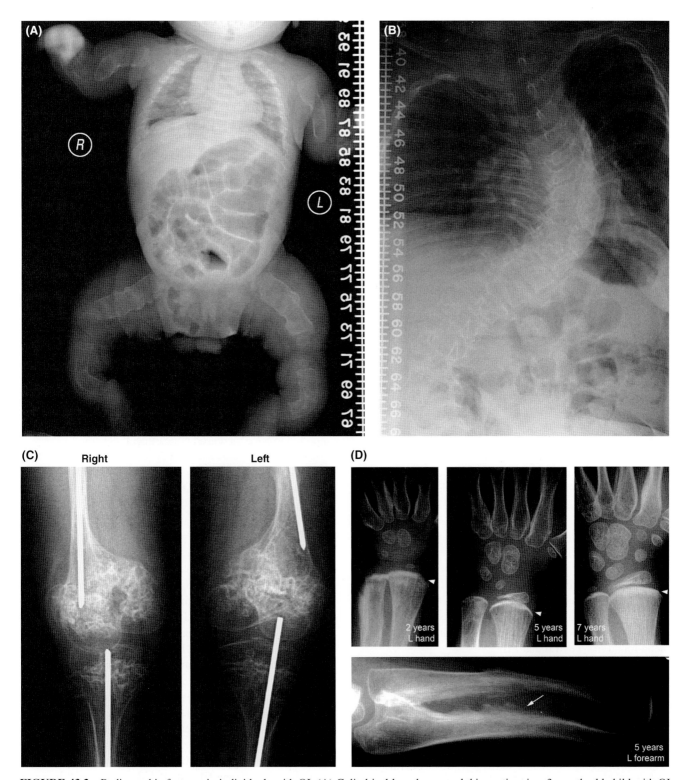

FIGURE 43.3 Radiographic features in individuals with OI. (A) Cylindrical long bones and thin cortices in a 3-month-old child with OI type III. (B) Scoliosis, collapse of multiple vertebrae, thin and misshapen ribs in a 16-year-old child with OI type III. (C) "Popcorn" calcifications at the joints in the lower extremities of a 14-year-old child with OI type III. (D) Radiodense bands at metaphyses (arrowheads) and calcification of the interosseous membrane (arrow) in a child with OI type V.

Impaired Collagen Modification and Processing: Types VII–XII OI type VII was first described clinically in members of First Nations in Quebec, Canada (Ward et al. 2002). Type VII severity varies. Most cases are lethal within the first year although some children survive for several years with severe dysplasia. The causative mutation in the First Nations group is hypomorphic, accounting for their milder severity. A distinguishing radiographic finding was shortening of the proximal segment of the limbs (Ward et al. 2002). Other features are as summarized in Table 43.1. Histopathological features are similar to those seen in OI Type I (Ward et al. 2002). Null mutations in *CRTAP*, which encodes a protein of the same name that forms a mutually supportive dyad with P3H1 (*p*rolyl 3-*h*ydroxylase 1) in the 3-hydroxylation complex, were identified to be causative (Figure 43.1) (Barnes et al. 2006).

OI type VIII, caused by null mutations in *P3H1* (Cabral et al. 2007), the enzymatic component of the 3-hydroxylation complex, is clinically indistinguishable from type VII because their causative proteins form a mutually supportive dyad (Figure 43.1). A null mutation in one protein results in the absence of both proteins (Chang et al. 2010). OI type VIII clinical outcome ranges from perinatally lethal to severely affected, with more non-lethal cases than type VII surviving into the third decade. Types VII and VIII share the distinguishing skeletal finding of rhizomelia (Table 43.1). Mutations in the causative gene *P3H1* have been found in a variety of genetic backgrounds (Cabral et al. 2007). The most frequent mutant *P3H1* allele is a founder mutation from West Africa that has also been identified in infants of African-American descent (Cabral et al. 2012).

Mutations in the third component of the 3-hydroxylation complex, *PPIB*, which encodes the peptidyl-prolyl *cis-trans* isomerase (PPIase) cyclophilin B, cause both lethal and moderately severe OI (Figure 43.1) (Barnes et al. 2010). Those with lethal type IX OI have reduced but not absent collagen 3-hydroxylation, while those with moderately severe OI have normal 3-hydroxylation levels. Children with OI type IX have white sclerae and do not manifest the rhizomelia found in types VII and VIII.

The few individuals reported with OI type X had very severe skeletal deformities and died within the first few years of life (Table 43.1). Because of the small number of known cases, clinical findings such as renal stones and death from pulmonary causes remain to be validated. Clinically and radiographically, individuals with OI type X have features similar to those seen in Sillence type II/III. *SERPINH1*, which encodes the critical chaperone HSP47 (Figure 43.1), has been identified as the affected gene in this autosomal recessive type of OI (Christiansen et al. 2010).

A founder mutation for *FKBP10* in Turkey led to the identification of OI type XI (Alanay et al. 2010), which is now known to occur widely across ethnic groups. Clinical features in OI type XI are similar to those seen in Sillence type III (Table 43.1). *FKBP10* encodes the PPIase FKBP65, a protein critical to the function of LH2, which modifies lysines in the telopeptides of type I collagen (Figure 43.1). Mutations in *FKBP10* were also the underlying genetic causes for Bruck syndrome (Kelley et al. 2011), in which individuals have severe OI plus congenital contractures of the lower extremities. The same *FKBP10* mutation, even within a sibship, may lead to OI with or without contractures. Thus, OI type XI and Bruck syndrome are allelic conditions with variable expression of contractures. The spectrum of *FKBP10* mutations extends further to Kuskokwim syndrome, in which affected individuals have residual FKBP65 (5–10%) leading to congenital contractures with osteopenia. A similar spectrum of skeletal dysplasia and joint contractures is also seen with mutations in *PLOD2*, which encodes LH2, and is grouped as an unclassified OI type (Table 43.1).

Individuals with OI type XII have recessive mutations in *BMP1*, which encodes the enzyme that cleaves the C-propeptide of type I procollagen (Figure 43.1) (Martinez-Glez et al. 2012). They have recurrent fractures starting in the first year of life (Table 43.1). An umbilical hernia and hypermobility have been reported in some individuals with OI type XII. Similar to individuals who have dominant mutations at the procollagen C-proteinase cleavage site, individuals with deficiency of the cleavage enzyme BMP1 are consistently reported to have increased bone mineral density.

Impaired Osteoblast Differentiation: Types XIII–XVIII
The first reported case of OI type XIII was identified in an Egyptian boy born to consanguineous parents (Lapunzina et al. 2010). His clinical features were similar to those seen in Sillence type IV (Table 43.1). Homozygosity mapping, candidate gene sequencing, and segregation analysis identified *SP7* as the causative gene. *SP7* encodes a transcription factor critical in regulating *COL1A1/2* expression and osteoblast differentiation (Figure 43.1). Fiscaletti et al. (2018) reported a second sibship from consanguineous Iraqi parents with similar morphological features and a homozygous variant identified in *SP7*. Hearing loss was a significant finding in two of the three affected siblings from this family (Table 43.1).

Individuals with OI type XIV were first identified from Saudi Arabia (Shaheen et al. 2012), carrying an exon 4 deletion in *TMEM38B* later determined to be a founder mutation among Bedouins (Marini et al. 2017). Individuals with autosomal recessive type XIV OI of other ethnicities have also been found to have non-exon 4 deletions in this gene. Clinical and radiographic features for OI type XIV are similar to those in Sillence type IV. Cardiovascular complications were diagnosed in three of eight individuals in a case series (Webb et al. 2017); however, the true incidence remains difficult to determine with only a small number of cases known. Histological features distinct from OI caused by type I collagen defects include reduced osteoblast numbers, and decreased rate of bone mineral deposition and incorporation (Webb et al. 2017). *TMEM38B* encodes for the intracellular cation channel TRIC-B (Figure 43.1).

A cluster of reports in 2013 identified *WNT1* as the causative gene in individuals with OI type XV from various ethnic backgrounds (Figure 43.1). Those with homozygous or compound heterozygous mutations in *WNT1* developed OI features similar to those with Sillence OI type IV (Table 43.1). However, those carrying only one mutation develop premature osteoporosis (Marini et al. 2017). One distinguishing clinical feature in some individuals with OI type XV is the presence of midbrain or cerebellar hypoplasia (Aldinger et al. 2016). Primary brain structural abnormalities are not seen in other OI types, but are consistent with the known function of WNT1 protein in brain formation.

Symoens et al. (2013) first linked mutations in *CREB3L1*, which encodes the protein OASIS, to a severe autosomal recessive OI phenotype (Figure 43.1). Only a few cases of OI Type XVI have been reported. All reported individuals with type XVI OI, except one, died in utero. In utero radiologic findings are similar to those seen in Sillence OI type II. An 11-year-old Somali male was the only known case to date to have survived beyond the neonatal period, with a fracture phenotype similar to that of type III OI (Table 43.1) (Lindahl et al. 2018). He has been treated with pamidronate intermittently, with an unclear accumulated dose. Heterozygous carrier family members of individuals with this type have mild OI features including blue sclerae, osteopenia, and history of recurrent fractures (Table 43.1) (Keller et al. 2017). Along with type XVIII OI, defects in OASIS implicate regulatory intramembrane proteolysis (RIP) as a critical pathway in bone formation.

Two individuals of different ethnic backgrounds have been reported with OI type XVII (Mendoza-Londono et al. 2015), caused by missense mutations in *SPARC*. Clinical and radiographic features are similar to those seen in Sillence OI type IV (Table 43.1), with fracture onset after a normal birth and progressively severe bone dysplasia and early onset scoliosis (Mendoza-Londono et al. 2015). Missense mutations in *SPARC*, which encodes a protein with chaperone function (Figure 43.1), were identified to be the underlying genetic cause.

OI type XVIII is the first known OI type to have an X-linked mode of inheritance. Male members of two families from different ethnic backgrounds have been reported with clinical and radiographic findings similar to those of Sillence type III/IV (Lindert et al. 2016). Fracture onset begins in utero (Table 43.1) with bowed dysplastic bones, and continue postnatally. Affected males develop scoliosis and thoracic deformity. Dentinogenesis imperfecta is not present, but sclerae may be bluish. No overt OI features were observed in the obligate female carriers, though skewed X-inactivation was present (Lindert et al. 2016). Mutations in *MBTPS2* were identified in all affected males. *MBTPS2* encodes the intramembrane protease S2P, which has a role in activating OASIS during RIP (Figure 43.1).

Etiology, Pathogenesis, and Genetics

As mutations in genes additional to those encoding type I collagen have been identified to be causative for OI, the understanding of underlying pathogenesis has grown more complex. The following discussion is an oversimplification, and readers are referred to the cited reviews for more details (Marini et al. 2017). Based on the classification system proposed by Forlino and Marini (2016), the autosomal dominant OI types I–IV are the result of pathological mutations in the primary sequence of the *COL1A1* and *COL1A2* genes (Figure 43.1). *COL1A1* and *COL1A2* encode the α1(I) and α2(I) chains, respectively. Each alpha chain contains a core triple helical region formed by uninterrupted repeats of glycine-Xaa-Yaa sequences, and pro-peptide carboxyl- and amino-termini. Type I collagen, the major protein component in bone, dentin, sclera and connective tissues, and the defective product in OI types I–IV, is a heterotrimeric protein consisting of two α1(I) and one α2(I) chains. The small size and uncharged state of glycine enables it to facilitate trimer structure by facing the sterically constricted inner aspect of the helix. Amino acids occupying the X and Y positions are frequently proline and hydroxyproline. The most frequent OI-causing sequence variations, substitution of a glycine in the Gly-Xaa-Yaa repeats, alter the spatial and ionic environment afforded by the glycine residues and disrupt helix formation. Other *COL1A1* and *COL1A2* sequence variations can also alter post-translational peptide modifications and mature peptide processing, or decrease α1(I) and α2(I) expression.

The remaining OI types are caused by defects in other proteins involved in the formation of the final collagen product and bone matrix, as listed in Table 43.1 and depicted in Figure 43.1. Transcription of *COL1A1* and *COL1A2* is regulated in part by the transcription factors SP7 and OASIS, deficient activity of which leads to decreased α chain expression and OI types XIII and XVI, respectively. Altered activity of the endopeptidase S2P affects the processing of OASIS and consequently collagen expression, leading to OI type XVIII. Also affected at the transcription level is OI type XV, where WNT1 signaling regulates expression of factors involved in osteoblast differentiation.

Post-translational modifications of the α1(I) and α2(I) chains first occur in the endoplasmic reticulum (Figure 43.1). Deficient modification of specific prolines in these peptides by the 3-hydroxylation complex of CRTAP/P3H1/CyPB slows the folding of the heterotrimer helix, resulting in OI types VII–IX. Aside from its role in the 3-hydroxylation complex, CyPB also isomerizes peptidyl–prolyl bonds as part of several chaperone complexes that maintain normal collagen folding. In addition, it interacts with all known lysyl hydroxylase isoforms and thus has a role in maintaining normal α chain cross-linking. Formation of the heterotrimer helix also depends on intact chaperone activity of HSP47 and FKBP65. Deficient functions of these chaperones are causative for OI types X and XI, respectively. Following the initial round of modifications and folding in the endoplasmic reticulum, the collagen pro-peptides continue to be stabilized by chaperone complexes as they initiate fibril formation and are secreted into the extracellular matrix. SPARC, the defective protein in OI type XVII, is thought to work with HSP47 to provide this stabilization effect. The dynamic endoplasmic reticulum activities of multiple enzymes are highly dependent on tightly regulated intraluminal Ca^{2+} level. TRIC-B, the protein affected in OI type XIV, is a cation channel that plays an integral role in transmembrane endoplasmic reticulum Ca^{2+} flux.

In the extracellular milieu, specific N- and C-terminal propeptidases cleave the respective pro-peptides to form the mature heterotrimer (Figure 43.1). Defective activity of the C-terminal pro-peptidase BMP1 leads to OI type XII. Cross-linking of heterotrimers to form a collagen fibril depends on intact lysine hydroxylation activity by LH2, deficiency of which leads to a phenotype spectrum as described (Table 43.1).

The final bone product arises from organization of collagen fibrils into fibers and deposition of minerals. BRIL and PEDF, the defective proteins in OI type V and VI, respectively, regulate the mineralization process (Figure 43.1). PEDF is also well known to have a potent anti-angiogenic activity which is dependent on binding to collagen in the extracellular matrix.

Genotype–Phenotype Correlations OI types not resulting from a primary collagen defect, i.e. types V–XVIII, are often associated with a more severe phenotype. Collectively, individuals with OI types V–XVIII comprise approximately 15% of all known individuals with OI. Thus, except for some categorically distinctive features such as those described in the Diagnostic Criteria section, meaningful genotype–phenotype correlations have only been possible for *COL1A1/2*-related OI types (Marini et al. 2007).

Phenotypic variability in OI is significant. This can occur in affected individuals within the same family and in those with identical sequence variations, particularly in the case of *COL1A1*. Quantitative defects of type I collagen arising from null mutations in α1(I) chain tends to correlate with the milder OI type I (Willing et al. 1993). Null mutations result in expression of structurally normal type I collagen, although at a reduced amount. In contrast, qualitative structural mutations, such as residue substitution or splice site defects, are associated with the moderate and severe OI types II–IV. These qualitative defects lead to collagen over-modification, intracellular protein suicide, and abnormal fibril formation (Forlino and Marini 2016).

Mutations in the α1(I) chain are lethal in about one-third of cases. Phenotype severity correlates with the types of substituting residues, i.e. more severe when substituting residues are charged or have a branched side chain. Mutations in the first 200 amino acids at the amino-terminal are exclusively non-lethal. Furthermore, substitutions within the first 90 residues are associated with a combined Ehlers–Danlos syndrome phenotype (see Chapter 25). Substitutions at helix positions 691–823 and 910–964 lead to exclusively lethal outcomes. These segments align with major ligand binding regions (MLBRs), emphasizing the crucial interactions of collagen fibrils with integrins, matrix metalloproteinases, fibronectin, and cartilage oligomeric matrix protein. (Marini et al. 2007)

Glycine substitutions in the α2(I) chain are non-lethal in 80% of cases. Phenotype severity correlates with the position, more than the particular substitution of the glycine residue. Lethal mutations occur in eight clusters, regularly spaced along two-thirds of the chain. The clusters align with binding regions in collagen for matrix proteoglycans. Thus, mutations at these regions likely disrupt interactions required for heterotrimer and fibril formation. Unlike mutations in α1(I), individuals with the same mutation in α2(I) typically have similar phenotypes (Marini et al. 2007).

Genetic Counseling The recurrence risk for OI depends on the inheritance pattern of the gene defect and the individual family. Most cases of autosomal dominant OI are caused by a de novo mutation in the affected individual. In autosomal dominant OI, an affected individual has a 50% risk of having an affected child with each pregnancy. In autosomal recessive OI, parents of an affected individuals have a 25% risk of having another affected child with each pregnancy. In X-linked OI, the mother of an affected male has a 50% risk of having another affected son with each male child, and a 50% risk of having a daughter who is a carrier of the mutation with each female child.

In families with unaffected parents and one affected child with an autosomal dominant OI type, the empiric risk for recurrence is 6–7%, higher than the de novo mutation rate of 1:20,000, because of possible germ-line mosaicism. An unaffected parent can carry the mutation in only some cells or tissues (e.g. eggs or sperm), and thus has the potential to produce a severely affected child. Mosaicism should be suspected when more than one child with autosomal dominant OI is born to unaffected parents. Some mosaic parents have mild OI features such as osteoporosis, blue sclerae, dentinogenesis imperfecta, and possible fractures. The severity of the parental findings depends on the developmental stage during which the mutation arose, and the level of negative selection against the mutation. In many cases of mosaicism, parents were mildly affected, as defined by their fracture history and/or scleral hue (Forlino and Marini 2016).Two-thirds of these cases were due to paternal mosaicism, and one-third was due to maternal mosaicism. When the mutation was known in the proband, all carrier parents had evidence of mosaicism in their leukocytes; however, not all had a detectable collagen abnormality in cultured fibroblasts. To date, no pure germ-line carrier has been described. The proportion of mutant cells in the skin and leukocytes of affected parents often does not correlate with the empiric recurrence of OI (Cabral and Marini 2004).The most accurate recurrence risk can be given for mosaic fathers, by determining the proportion of mutation in sperm.

Diagnostic Testing

Clinical diagnosis is based on personal and family history, physical exam, and radiographic findings as discussed in Diagnostic Criteria. With the continual advancement in molecular sequencing methodologies, gene sequencing has become first tier testing in many cases. Since sequencing remains not universally accessible and relatively costly in many areas of the world, however, considered step-wise testing is still useful. In an affected individual who represents the first or single case in the family, testing using a dominant OI gene panel would identify the genetic cause in ~80% of the cases. The cost for a dominant OI gene panel is relatively inexpensive. If a cause is not identified, then proceeding to testing using a recessive OI gene panel should be considered. Identification of the underlying genetic cause is useful for reproductive counseling and potential future targeted treatments. Biochemical testing for OI examines collagen electrophoretic migration for evidence of expression and collagen processing. Most, but not all, collagen structural defects, all cases of types VII–VIII OI and some IX OI result in overmodification of the collagen helix. Biochemical collagen testing is still useful in some puzzling cases, but has become predominantly a research test.

Two-dimensional ultrasound remains the most common mode for prenatal diagnosis. Other imaging modalities such as 3D ultrasound and computed tomography have not been widely available and adapted. Biochemical and genetic analysis of fetal tissues (chorionic villi) or genetic analyses of chorionic villi or amniocytes can be done. Amniocytes are not suitable for collagen biochemical analyses because they can yield false positive results. Both of these approaches are best applied when the familial variants are known and testing can be targeted. A comprehensive family history is essential in helping to guide the prenatal diagnostic process.

Differential Diagnosis

Because the types of OI vary widely in clinical appearance and in the timing of symptom onset, the differential diagnosis depends on the age of the individual in question. Prenatally, OI type II can be difficult to distinguish from some of the chondrodystrophies, especially thanatophoric dysplasia, campomelic dysplasia, and achondrogenesis type I. Neonatally, there may be some difficulty in distinguishing between OI type III, VII or VIII and infantile hypophosphatasia. In older children with type I OI there may be confusion with juvenile osteoporosis.

The major differential diagnosis for types I and IV OI is child maltreatment. The clinical distinction is often difficult. Although children with OI may have blue sclerae, unaffected infants may also have this finding. Decreased bone mineral density, wormian bones, or evidence of osteoporosis on radiographs may be helpful in differentiating between abuse and OI. In addition, children with OI should not have other evidence of abuse or neglect. A molecular diagnosis consistent with OI would be the definitive distinction. (Pepin and Byers 2015)

Thanatophoric dysplasia is a lethal chondrodystrophy caused by mutations in fibroblast growth factor receptor 3. It occurs with a frequency of 1 in 20,000 births and is inherited as an autosomal dominant mutation. Affected fetuses are characterized by markedly shortened extremities, relatively large cranium, and narrow thorax. Diagnosis can be made using ultrasound between 20 and 24 weeks, which reveals an increased biparietal diameter and very short limbs without polydactyly (Elejalde and de Elejalde 1985).

Campomelic dysplasia is an autosomal dominant disorder caused by mutations in *SOX9* and characterized by a large dolichocephalic head, disproportionate limb shortness, hypoplastic scapulae, pelvic and spine changes, and small thorax with 11 pairs of ribs. The lower extremities are often bowed with pretibial dimples, congenital hip dislocation, and talipes equinovarus. Infants die within the first months of life from respiratory insufficiency. Prenatal ultrasound reveals a large skull, often with hydrocephaly, and poorly ossified tubular bones. Hydrocephaly can be seen as early as 16-weeks gestation and may differentiate this syndrome from OI (Fryns et al. 1981).

Achondrogenesis type I is another lethal chondrodystrophy. Radiographically, the disorder presents with micromelia, large cranium due to soft tissue swelling, deficient ossification in lumbar vertebrae, and absent ossification in sacral, pubic, and ischial bones. Inheritance is autosomal recessive in type IB and likely to be recessive in type IA (Superti-Furga 1996).

Hypophosphatasia is an autosomal recessive genetic disorder that can be divided into three types: infantile, childhood, and adult onset. The infantile form has its onset prenatally or early in postnatal life and is usually fatal in infancy. Prenatally, "spurs" of the limbs, small thorn-like structures that extend from the sides of the knee and elbow joints, are diagnostic. These can be seen only by X-ray, however, and not on ultrasound (Whyte 1995). Infants present with craniosynostosis, hypercalcemia, and bony abnormalities including severe osteoporosis and micromelia, with severe epiphyseal and metaphyseal disturbances in the long bones, vertebrae, and ribs. Diagnosis can be made by determination of serum alkaline phosphatase (low), phosphoethanolamine (increased), and inorganic pyrophosphate (increased). In contrast, alkaline phosphatase is normal or elevated in OI.

Osteoporosis occurs rarely in childhood. The differential diagnosis for primary osteoporosis in childhood includes juvenile and idiopathic osteoporosis. Children affected by juvenile osteoporosis are usually diagnosed because of the development of long bone fractures or back pain. X-rays reveal generalized osteoporosis with compression fractures of the vertebrae and metaphyses of long bones. Bone histology shows an excess of osteocytes associated with woven bone and normal collagen. Routine biochemical studies tend to be normal for age. The mean age of onset is 7 years, and most affected children improve at puberty. Idiopathic osteoporosis is more variable, has no obvious cause, and may represent a state of either high or low bone turnover. Its course may be transient and self-limited or progressive and disabling and is difficult to predict at the onset. Other causes of osteoporosis in children are secondary osteoporoses such as glucocorticoid-induced osteoporosis, osteoporosis due to hypogonadism, or osteoporosis secondary to malignancy.

MANIFESTATIONS AND MANAGEMENT

The manifestations and management of OI are presented by systems below. Interdisciplinary clinics where these clinical specialties may be provided in one comprehensive visit are most ideal for OI managment.

Growth and Feeding

Individuals with OI have disproportionate growth deficiency. Data on longitudinal growth patterns are best studied for OI types I, III, and IV because of their prevalence (Germain-Lee et al. 2016; Marini et al. 1995). Growth patterns for other OI types likely reflect one of these, as has been the case for OI type V (Germain-Lee et al. 2016). Individuals with OI type I are of average size at birth. They may fall below the growth curve by school age and grow parallel to the normal curve to achieve a final height slightly below normal adult range. Alternatively, they may grow in the normal range and have a final height several inches shorter than unaffected siblings of the same gender. Growth trajectory for individuals with OI type IV, and some with type III, begins to veer from average after the first year and reaches a period of minimal growth rate during 1–4 years of age. After this period, these children grow parallel to the growth curve, achieving a final height comparable with an average unaffected 10–12-year-old child. Individuals with OI type III and some with type IV have small to average length at birth, but then quickly fall below the growth curve and maintain this lower growth rate throughout life. They achieve a final adult height in the range of an average unaffected 4–6-year-old child. Only two studies have reported on growth for individuals with OI type V, and their height was similar to those with OI type IV (Glorieux et al. 2000), or type I (Germain-Lee et al. 2016).

The cause of growth deficiency is likely multifactorial. The major contribution to short stature in OI is a primary defect in long bone growth, but scoliosis, bowing of the lower extremities, and intraosseous calcifications at the growth plates (popcorn calcifications) are also contributory. Repeated fractures lead to deformities in both the axial and appendicular skeleton, and ultimately contribute to abnormal growth and development. Long bone fractures, suboptimal healing, and force of muscle on soft bone also introduce curvatures to the extremities and impair linear body growth. Evaluation of the growth hormone axis showed normal growth hormone secretion in both unstimulated and stimulated testings, but some individuals with OI had blunted serum IGF-I release following growth hormone administration (Marini et al. 1993)

Exogenous growth hormone administration increases linear growth rate in children with OI types III, IV, and I (Antoniazzi et al. 1996; Marini et al. 1993; Marini et al. 2003). Attainment of heights within the normal growth curves is possible for the latter two OI types (Antoniazzi et al. 1996; Marini et al. 2003). Exogenous growth hormone treatment improves bone density without advancing bone age or worsening skeletal deformity, and increases muscle mass and strength which may lead to increased activity, which benefits skeletal integrity (Antoniazzi et al. 1996). It also induces increased blood levels of all markers of bone turnover due to an increased rate of whole body remodeling associated with rapid bone gain.

Approximately one-third of individuals with OI types I, III, and IV have obesity (Germain-Lee et al. 2016). The rate of obesity is associated with more severe OI status, e.g. individuals with type III or wheelchair-dependent. Excess intake and decreased expenditure of calories likely are major contributing factors. Other culprits may include abnormal cell signaling via the fat-bone-brain axis in regulating the plasticity of the common precursor cells to osteoblasts and adipocytes.

Evaluation

- Children with OI should be carefully measured at all visits. Measurements should include: head circumference, weight, sitting height, and length or height measured on both sides of the body, accounting for any contractures of the limbs. Regardless of age, length is a preferable measure to height, because the former provides a consistent reference through fractures and varying functional status.

- For individuals with OI type I and IV and a typical growth pattern, standard tests of the growth hormone

axis neither informs treatment decisions nor is necessary before an empiric trial of growth hormone.
- Aberration in thyroid function is not associated with OI.

Management

- A six month empiric growth hormone treatment trial at 0.1 International Units/kg/day for six days/week in children with OI types I and IV, especially those with radiographically clear growth plates without popcorn formation, can be undertaken to determine responsiveness (Marini et al. 2003). Exogenous growth hormone administration should not be considered in individuals with scoliosis exceeding 40° as the curve may worsen with a growth spurt.
- Nutritional education and support and prescriptive activities such as swimming should be provided to individuals with OI to prevent or manage obesity.

Development and Behavior

Children with OI have minimal delay in the development of neurocognition, language, and fine motor skills (Marini and Gerber 1997). Delay in gross motor development is significant. The magnitude of delay correlates with OI severity, such that children with OI type III reached motor milestones at a later age compared to those with OI types IV or I (Engelbert et al. 2000). Attainment of dynamic motor skills is also delayed compared to that of static motor skills (Engelbert et al. 2000). Children with OI show no difference compared to non-disabled peers in temperament, except for activity level (Suskauer et al. 2003).

Evaluation

- Developmental delay in other domains besides gross motor in children with OI should prompt a complete developmental assessment.

Management

- Early intervention programs and physical therapy are valuable and recommended for children with OI. Programs must be highly individualized to the specific disabilities of the child. (Marini et al. 2017) Continued feedback from physical therapy throughout the lifespan of individuals with OI may help to maintain overall health and maintain physical abilities.
- Pamidronate infusion (see Musculoskeletal section) and early lower extremities rodding (Engelbert et al. 1995) have been shown in some individuals to improve motor development; however clear evidence of benefits is still lacking.

Musculoskeletal

Appendicular Skeleton. Fracture frequency in individuals with OI is highest during the period from birth to late teen years, as compared to the reference population, and decreases throughout life. Women with OI who are post-menopausal have a higher fracture rate as compared to men, as well as to the female reference population. Common fracture sites include the forearm, lower leg and ankle, and femur. In post-menopausal women, the upper arm and hip-pelvis areas also become common sites for fractures (Folkestad et al. 2017).

Annualized rate of fractures, or number of fractures sustained, may not serve as an accurate index of OI severity. Children with more severe OI types may have soft bones that tend to deform rather than break. The degree of stress or trauma leading to a fracture may be a more useful assessment. For example, children with OI type III sustain more fractures with trivial trauma and in areas more difficult to break (arms and ribs) than those with type IV or I.

Spine. The prevalence of scoliosis (Cobb's angle > 10°) in individuals with OI ranges from 39% in type I to as high as 80% in the more severe type III. The type of spinal curvature also correlates with OI severity, with > 90% of individuals with type I OI developing a single thoracic curve, as compared to approximately one-third of those with type III developing double curves. In addition, the rate of curvature progression is highest in those with OI type III (6°/year), lowest in those with OI type I (1°/year), and in between for those with OI type IV (4°/year). (Anissipour et al. 2014) Decreased bone mineral density and increased number of compressed vertebrae correlate with progression of scoliosis (Wallace et al. 2017). However, other factors such as ligamentous laxity may also contribute to the progression of spinal curvature, as treatment with bisphosphonate did not prevent it (Sato et al. 2016). The presence of scoliosis and kyphosis contributes to decreased pulmonary capacity in individuals with OI (see Pulmonary section).

Fractures in the vertebral body itself (spondylolysis) and malalignment of the spinal column (spondylolithiasis) at the lumbar-sacral level have also been observed in individuals with OI. Spondylolysis and spondylolithiasis occurred in individuals of preschool age to early twenties (Hatz et al. 2011). Reported prevalence for these conditions range from on par with or higher than those in the reference population (Wallace et al. 2017).

Positional torticollis, hip abduction and flexion contractures, and acetabular protrusion are all potential sequelae of spinal deformities. In infants with OI, supine positioning on soft surfaces can lead to these conditions and interfere with motor development if not corrected (Binder et al. 1993).

Evaluation

- Fractures should be evaluated with standard X-rays to determine the type and degree of dislocation. Radiolabeled bone scans to identify subclinical

fractures not present on plain film will not be specific because of the high turnover rate in the bones of children with OI.
- Spinal deformities should be evaluated with annual antero-posterior and lateral spine films in individuals with more severe OI types to monitor the progression of curvature. Children in their growth spurt with a Cobb's angle measurement > 30° should be followed more frequently (Wallace et al. 2017).

Management

General Care, Rehabilitation, and Orthotics

- Support and instruction for parents/caregivers should be provided early. These should include lessons for infant handling and positioning, head support to avoid torticollis, and neutral alignment of the femora (Gerber et al. 1990). Turning the infant from side to side and into a prone position helps to strengthen neck extensors and upper extremities and to allow stretching of the hip muscles.
- Useful equipment for infants with OI includes regular mattresses covered with liner to allow greater movements, and custom-molded seats or sandbags to assure lower extremity alignment and improve positioning of the head and spine (Gerber et al. 1990).
- Rehabilitation therapy for individuals with OI should aim to promote and maintain optimal function in all aspects of daily living. A program combining early intervention, muscle strengthening, aerobic conditioning, and ambulation (if able) should be initiated early in life and modified to each individual's functional level and current ability (Marini et al. 2017). General goals for children with OI should include at least 3/5 grade strength (ability to lift a limb against gravity) in the deltoid, biceps, gluteus maximus and medius, and trunk extensors. This will at least allow for independent transferring. Other goals should include abilities to supinate their forearms, to touch around their head with their hands, and have no more than -20° contracture while their hips and knees are extended (Marini and Gerber 1997).
- Long leg bracing may provide support for weak muscles, control joint alignment, and decrease motion and pain in some cases. Ultralight-weight plastic clamshell braces provide a more natural alignment to unweight the limb and supply 360° compression using the anterior clamshell contoured supports. Their use promotes independent activity by providing stability to the pelvic girdle, controlling knee recurvatum, severe hindfoot valgus, and decreases tibial bowing. Braces improve upright balance, but do not unload the limb or protect against fractures.
- Soft truncal bracing may help support upright positioning, but bracing does not prevent the progression of scoliosis.

Orthopedic Care

- Management by an orthopedic surgeon with some experience in treating individuals with OI is required.
- Fractures in the appendicular skeleton should be reduced and realigned as necessary to minimize worsening of bowing or contractures that would ultimately lead to loss of function. Casting and splinting can be used, but the duration of immobilization should be minimized to avoid worsening osteoporosis and muscular atrophy. Fracture of the patellar or of the tip of the olecranon process must be repaired by surgery to avoid non-union (Shapiro and Sponsellor 2009).
- Rodding of limbs is mostly begun during the childhood years in individuals with OI. The child's developmental and functional status should guide the timing of intervention, with preference given to postponing surgery if achievement of maximal pre-surgical function is progressing. The goals for pursuing surgery should be to improve movement, strengthen long bones to prevent a cycle of fracturing–deformity–osteoporosis–refracturing, and to allow long limb bracing. Indications for surgery can include long bone angulation of greater than 40° (above which the risk of fracture increases), deformities that limit functions, or more than two fractures in the same bone within a six month period (Binder et al. 1993).
- Standard surgical procedures include osteotomies, realignment, and intramedullary rod fixation (Esposito and Plotkin 2008). Both non-elongating (Rush) and elongating (Bailey–Dubow or Fassier–Duval) rod options have clinically important disadvantages. Non-elongating rods theoretically leave segments of bone unsupported and at risk for fracture as the child grows. This leads to more frequent revisions or rod replacements in children with better growth potential. Elongating or telescoping rods, while capable of adjusting their length with growth, also tend to require repeated procedures relating to defective anchoring nails. In addition, these latter types require more technical expertise and are more expensive to place (Marini et al. 2017). Intramedullary rods can cause cortical atrophy and osteoporosis, especially in the diaphysis, because of unloading of mechanical weight. Percutaneous osteotomies and rodding procedures are possible options based on individualized assessment.
- Surgical procedures in the upper extremities can be considered to improve the ability for self-care (Montpetit et al. 2015). Rodding options for the

humerus are similar to those for the lower extremities, whereas elastic rods or Kirschner wires can be used for the forearm bones. Special considerations are needed for individuals with OI type V, given the propensity for hyperplastic callus formation and the presence of interosseous calcifications (Marini et al. 2017).

- Plates and screws spanning joints can soften the underlying cortex, leading to loosening screws and creation of new weak foci for fractures when the screws are removed. Thus, these should no longer be used in individuals with OI.
- Correction of spinal curvature is considered by some experts in OI to be indicated at >50° in individuals past their growth peak (Wallace et al. 2017). Recent adjuvant techniques such as pre-operative halo gravity traction and the use of cement-augmented pedicle screws in conjunction with instrumented (e.g. Harrington rod) spinal fusion appeared to have improved success rate (Wallace et al. 2017). For optimal outcome, a surgeon with experience in performing spinal fusion in individuals with OI should do the procedure.

Pharmacology

- *Bisphosphonates*. These synthetic analogs of inorganic pyrophosphate inhibit osteoclast resorption of bone. The experience from using bisphosphonate in individuals with OI over the past two decades has shown improvement in bone mineral density and vertebral geometry compared to placebo controls, with the greatest magnitude of difference occurring within the first year following treatment (Dwan et al. 2016; Hald et al. 2015; Rijks et al. 2015). Other improvements such as decreased fractures, overall height, spinal curvature, and control of bone pain have not been consistently obtained (Beecham et al. 2015; Dwan et al. 2016). No difference in efficacy was noted between oral and intravenous bisphosphonate administration (Dwan et al. 2016). Additional concerns relating to long-term bisphosphonate therapy include (1) the long half-life of bisphosphonates and how they affect bone metabolism, and (2) accumulation of microfractures secondary to decreased bone turnover and thus increase in bone fragility (Marini et al. 2017). While bisphosphonates have become widely accepted as a treatment option for individuals with OI, data are lacking on optimal time of initiation, duration of treatment, and long-term complications. Furthermore, bisphosphonate treatment appears to be less effective in individuals with OI type VI (Trejo et al. 2017).

  The OI program at the National Institutes of Health has treated children over age two for 1–3 years with cycles of IV pamidronate at 3–6 month intervals to obtain vertebral benefits and minimize skeletal complications. Treatment at lower doses (3 mg/kg/year) yielded the same increase in bone density at one year post treatment as the effect seen with treatment at higher doses. Potential complications of gastrointestinal discomfort with oral bisphosphonates, acute-phase reaction (hyperpyrexia, myalgia, weakness) with first infusion of IV bisphosphonates, and delayed bone healing should be discussed with the family, as well as supportive measures available to counteract them. Timing of bisphosphonates administration relating to orthopedic procedures remains dependent on surgeons' expertise and experience (Wallace et al. 2017). Most surgeons would not administer bisphosphonates in the months immediately surrounding procedures to avoid inhibiting remodeling of the calluses.

- *Anti-RANK ligand antibody*. Denosumab, the primary commercially available form of anti-RANK ligand antibody, has the ultimate effect of inhibiting bone resorption by osteoclasts (and consequently calcium release) (Boyce 2017). Denosumab administration in individuals with OI types I, III, IV, and VI led to improved bone mineral density, with the advantage of having a shorter half-life and no incorporation into the bone matrix as is the case with bisphosphonates. However, these results derived from small, uncontrolled studies that have also reported inconsistent positive outcomes and lacked sufficient power to examine other OI-related clinical outcomes. (Li et al. 2017) Important side effects of denosumab include hypocalcemia during treatment and rebound hypercalcemia during no-treatment intervals (Boyce 2017). Individuals with OI who may be candidates for this treatment should be enrolled in clinical research trials for careful monitoring of drug administration and possible side effects.

- *Recombinant human parathyroid hormone analog*. Teriparatide, the commercial human parathyroid hormone analog, has been shown to improve bone mineral density in adults with less severe OI type I, but not types III and IV (Leali et al. 2017; Orwoll et al. 2014). Previous animal data demonstrated an increased risk for developing osteosarcoma following teriparatide exposure, and thus had limited the study of this drug to the adult population. Teriparatide administration may need to be done in combination with an anti-resorptive agent, as early data indicated more severe bone loss following drug withdrawal. (Marom, Lee, Grafe, and Lee, 2016)

- *Treatments in development*. Other drugs being studied for treatment of OI include antibodies to sclerostin and TGF-ß. Sclerostin is a glycoprotein inhibitory ligand for the WNT1 co-receptor LRP5/6. Antibodies neutralizing the activity of sclerostin have been shown in OI animals model and a Phase 2a study to stimulate bone formation and inhibit bone resorption (Glorieux et al. 2017). Antibodies to TGF-ß in OI animal models also improved bone formation, and may also positively affect pulmonary outcomes in OI type VII. Curative therapies via stem cell transplantation or genetic editing have also continued to be explored. (Marini et al. 2017)

Pain Management

- The pain and activity limitation associated with frequent fractures in individuals with OI represent the greatest source reported for reduced physical quality of life (Marini et al. 2017). In particular for children with OI, lack of appropriate pain assessment and fear of overdosing can lead to inadequate management of acute pain (Zack et al. 2005). Except for the unproven ability of bisphosphonates to control bone pain in individuals with OI, no study has specifically focused on identifying optimal pain treatment for OI. Thus, management of acute and chronic pain in individuals with OI, specifically children, should be drawn from the general pediatric literature.
- The 2012 World Health Organization recommends ibuprofen as the mainstay for treatment of mild acute pain, and oral morphine for moderate to severe pain (WHO 2012). Accurate dosing and awareness of potential side effects should be discussed with families when medications are started.
- Adjuvant non-pharmacologic options such as approach or distraction coping strategies can provide additional avenues for pain management (Wente 2013; Zack et al. 2005).

Neurologic

Cranial growth in children with OI follows two patterns: growth at an average rate, and growth that crosses centile lines between ages 2 and 3 years. Prominence of sulci and ventriculomegaly are common findings in children with OI, even in those without absolute macrocephaly. These findings should be distinguished from cortical atrophy, as decreased cognitive ability, an expected sequela of cortical atrophy, is not a typical feature in OI (Charnas and Marini 1993).

Platybasia, basilar impression, and basilar invagination are three frequent findings derived from the abnormal bone development in OI (Arponen et al. 2012). Of the three, basilar invagination is most clinically significant. Individuals with a larger head circumference appeared to be at greater risk for developing basilar invagination (Charnas and Marini 1993). While basilar invagination progress of the overall group was minimal (Arponen et al. 2012), individuals with OI have presented with neurological symptoms decades after the first diagnosis of basilar invagination on imaging (Charnas and Marini 1993). The earliest sign of basilar invagination on neurologic exam is nystagmus, followed by long tract signs. Headache that worsens with movement, cough, sneezing, or straining is the earliest symptom for obstruction.

Evaluation

- Screening for basilar invagination with a spiral CT scan with reconstruction and complete neurologic exam should be done at an early age. Individuals with non-type I OI in the Natural History study at the National Institutes of Health (Bethesda, Maryland, USA) receive spiral CT of the craniocervical junction beginning at approximately 4 years of age.
- If screening results are normal, follow-up should be done every two years.
- If screening results show evidence of basilar invagination, annual MRI is appropriate.

Management

- Posterior fossa decompression and occipitocervical fusion are surgical methods most successfully applied to mitigate the neurological effects of basilar invagination. Prolonged external orthotic immobilization postoperatively prevents re-progression of basilar invagination (Menezes 2008). Only a few centers have experience with this procedure in individuals with OI. An experienced center is strongly recommended.
- Bisphosphonates treatment has not been shown to affect platybasia, basilar impression or invagination (Arponen et al. 2015).

Ears and Hearing

Hearing loss is identified in ~50% of studied populations of individuals with OI types I, III, and IV (Swinnen et al. 2011). Hearing loss has also been reported in individuals with OI type V, but not for other newer autosomal recessive OI types. This may relate to the onset of hearing loss commonly occurring in the second to fourth decade of life (Swinnen et al. 2011). However, hearing loss has been documented in younger individuals (Pillion et al. 2011). While hearing loss is reported at a higher frequency in individuals with OI type I, no genotype–phenotype correlation has been identified and intrafamilial heterogeneity is prevalent.

Mixed hearing loss, with conductive and sensorineural components, is most common. Abnormal bone metabolism is hypothesized to lead to osteosclerosis and fixation of the footplate, or fracture and atrophy of the stapes, and consequently affecting the conductive hearing process (Santos et al. 2012). Conductive hearing loss has been reported to progress to involve sensorineural loss (Marini et al. 2017), although a primary loss of neurons has also been proposed (Santos et al. 2012). Overall, the hearing deficits in individuals with OI are similar to those with otosclerosis, but they occur earlier in life and have greater middle ear involvement.

Evaluation

- An initial screening evaluation using tympanometry and acoustic reflex should be done between ages 3 and 5 years. Earlier testing should be considered if difficulties with hearing or speech development is noted.
- Follow-up evaluations should occur at regular intervals, and as frequently as every one to two years for children who have early changes on an audiologic exam.

Management

- Treatment of serous otitis media by standard pediatric approaches yield similar outcomes in individuals with OI as compared to those in the general population.
- Stapes surgery to correct conductive loss has a high success rate, albeit lower than that seen in individuals with osteosclerosis. In addition, up to 8% of individuals who underwent stapedotomy can progress to have sensorineural hearing loss despite intervention (Pillion et al. 2011). Stapes surgery should be performed by a surgeon experienced in treating individuals with OI, as the altered footplate architecture and middle ear anatomy and the bone composition pose additional risks for injuries.
- Criteria for fitting of hearing aids and bone-anchored hearing aids can be similar to those used for individuals without OI and may help to manage all three types of hearing loss (Pillion et al. 2011).
- Cochlear implantation is also an option for individuals with OI who have severe hearing loss not amenable to hearing aids. However, such an approach should be discussed with providers who have experience in treating individuals with OI and in managing the inherent complications.

Orofacial

Abnormal dentin formation, or OI-related dentinogenesis imperfecta, occurs in a majority of, but not all, individuals with OI. Blue-gray or yellow-brown colored and opalescent teeth is the most identifiable feature, typically more prominent in primary than secondary dentition. Additional findings include bulbous crowns, constricted coronal-radicular junction, short narrow roots, and obliterated pulp chambers (O'Connell and Marini 1999). The dental findings and severity vary significantly and may correlate to the genotype, with features present in individuals with glycine substitution in the carboxy-terminal, and not amino-terminal, of the α1(I) and α2(I) chains (Andersson et al. 2017).

Individuals with OI, mainly of the more severe types III, IV, and VI, have higher incidence of malocclusion, in particular class III malocclusion (O'Connell and Marini 1999; Rizkallah et al. 2013). The malocclusion involves anterior and lateral open bites, hypothesized to result from increased anterior mandibular growth leading to midface hypoplasia.

Evaluation

- Dental evaluations in individuals with OI should be done, at a minimum, at the same time and frequency as for those without OI. A standard dental panorex will most commonly reveal obliteration of the pulp chambers.
- Ability to chew solids without discomfort and esthetic considerations should be considered when intervention is needed for primary dentition.
- Evaluation for malocclusion by an orthodontist is recommended once secondary dentition erupts.

Management

- Optimal conditions for eruption of secondary dentition and normal growth of the facial bone and temporomandibular joint should be the aim of treatment. Dental coloration can aid management decisions. Yellow-brown teeth tend towards attrition and enamel fracture, thus individuals with this dental type should have crowns placed on their primary dentition to maintain vertical dimension (O'Connell and Marini 1999).
- Composites combined with dental bonding agents are recommended over occlusal areas. In occlusally non-stressed areas, glass ionomers such as fluoride-releasing and chemical-attacking materials are recommended. (Ranta et al. 1993)
- Pacifiers should be avoided in infants with OI to decrease the risk for developing anterior open bites (O'Connell and Marini,1999).

Pulmonary

Respiratory complications, especially pulmonary infections, are frequent and major findings and also the main cause of mortality in individuals with OI (Folkestad et al. 2016; McAllion and Paterson 1996). Pulmonary function studies in individuals with OI types I, III, and IV showed decreased forced vital capacity (FVC) and forced expiratory volume in one second (FEV1) (Thiele et al. 2012; Wekre, Kjensli, Aasand Falch, and Eriksen, 2014). Pulmonary function loss progresses from normal around 4 years of age to half pulmonary capacity by the early 20s (Thiele et al. 2012). Higher angle of spinal curvature correlates with lower pulmonary functions (Wekre et al. 2014). However, significantly decreased pulmonary function can also occur in individuals with OI and mild scoliosis (Thiele et al. 2012). Intraparenchymal processes such as abnormal lung tissue replacement following injuries have been observed in a mouse model (Thiele et al. 2012) and a case report in a human (Morikawa et al. 2016).

Evaluation

- Pulmonary function should be evaluated initially between 4 and 6 years of age in individuals with non-type I OI. Reassessment should occur annually if abnormalities were identified and every two to three years if normal.
- Evaluation by a pulmonologist is necessary for children with severe chest deformities or scoliosis, or if the degree of pulmonary function is more severely affected than that of spinal curvature.

Management

- Evaluation and correction of spinal curvature should occur when indicated by an orthopedic surgeon with experience in managing individuals with OI, as discussed under Musculoskeletal section.
- Conventional symptomatic treatments, including oxygen supplementation, for restrictive lung disease, are applicable.
- Individualized physical therapy plans and goals to maintain a healthy weight and physical abilities should be encouraged.

Cardiovascular

In individuals with OI types I, III, and IV, cardiovascular abnormalities occur at a higher frequency as compared to healthy controls (Ashournia et al. 2015), and is a frequent cause of death in adulthood (Folkestad et al. 2016). The observed cardiovascular phenotype is heterogeneous. Echocardiographic findings of dilated left ventricular outflow track and proximal aortic structures can be seen starting in the pre-teen years in individuals with OI types III and IV (Rush et al. 2017). Valvular regurgitation, mainly mitral and aortic, is more commonly seen in adults with OI (Folkestad et al. 2016). When indexed to body surface area, findings of dilated large-caliber vessels, i.e. aorta and pulmonary, was observed across age ranges (Hortop et al. 1986). Additional abnormalities, including cor pulmonale, right-sided heart failure, and diastolic dysfunction, have also been identified in children and adults with types III and IV OI (Thiele et al. 2012). Findings of ventricular wall hypertrophy and aortic root dilation were reported in an individual with OI type XIV (Webb et al. 2017).

Evaluation

- A baseline echocardiogram should be completed in the first decade of life and a repeat study should be done every two to three years if no abnormality is present. Earlier evaluation or more frequent follow up should be considered in individuals with moderate to severe chest wall or spinal deformity, heart murmur, or signs and symptoms of cardiopulmonary compromise.

Management

- Requirement for surgical intervention for vessel or valvular repair remains uncommon in individuals with OI.
- More severe or progressive vessel or valvular disease should prompt referral to a cardiologist with expertise in working with individuals with connective tissue diseases for further evaluation and management.
- Individualized physical therapy plans and goals to maintain a healthy blood pressure and reduce other co-morbidities should be encouraged.

Ophthalmologic

Blue or grey sclerae is the most common eye finding in individuals with OI, and, while more recognizable in AD OI types relating directly to collagen structural defects, has been observed in most OI types (Table 43.1). The intensity of blueness is less near the limbus and can decrease with age. This blue color has been attributed to differential scattering of light of different wavelengths through abnormal connective tissue (Sillence et al. 1993), and is negatively correlated with central corneal thickness (Evereklioglu et al. 2002). Retinal detachment and open angle glaucoma (Wallace et al. 2017) have been reported in individuals with OI, but no systematic study has focused on determining whether these occur at higher frequency in OI as compared to a reference population.

Evaluation

- Initiation and frequency of eye exams should follow standard of care guidelines for pediatric and adult populations.

Management

- Protective eye wear should be considered in individuals with decreased corneal thickness who are physically active.

- Ocular procedures should be done by practitioners with experience in managing individuals with OI, if possible.

Renal

Adults with OI identified issues relating to the urinary tract as frequently affecting their quality of life (Folkestad et al. 2016). Some children with OI types I, III, or IV have hypercalciuria, but correlation with other measures of renal function and genitourinary consequences remains understudied (Chines et al. 1995). Nephrocalcinosis and nephrolithiasis in individuals with OI at different ages have been observed.

Evaluation

- Standard evaluation in the presence of renal abnormalities is appropriate

Management

- Management of renal problems is standard.

Adult Health

As diagnostic and therapeutic measures improve, life expectancy for individuals with OI and the concomitant adult health issues have also extended (Marini et al. 2017). No specific risk data for adults with OI regarding malignancy or neurodegenerative conditions is yet available. As part of the disease pathology, their risk for hearing and dental defects, cardiovascular complications, and poor bone health is elevated (Marini et al. 2017). Pregnancy complications in women with OI include higher rate of antepartum hemorrhage and placental abruption, as well as higher rate of intrauterine growth restriction and preterm birth (Ruiter-Ligeti et al. 2016).

Evaluation

- Screening and preventive health maintenance for adults with OI should, at a minimum, follow standard guidelines for the adult population. For hearing, dental, cardiopulmonary, and bone health, evaluations should be done based on OI disease severity.
- Prenatal screening and genetic counseling should be provided to females with OI of child-bearing age.

Management

- Maintain calcium and vitamin D levels based on guidelines for management of individuals with osteoporosis.
- Maintain established care with practitioners with experience in managing or performing surgical procedures in individuals with OI
- For pregnant women, elect mode of delivery based on individual assessment as delivery by cesarean section has not been shown to improve neonatal outcome (Cubert et al. 2001).

Mortality

Mortality in individuals with OI types III and IV increases significantly compared to the non-OI population in England and Wales (Singer et al. 2001). Consistent with the phenotypes, individuals with the more severe OI type III died typically by the fourth decade of life whereas those with the milder types I and IV may survive beyond 70 years of age (McAllion et al. 1996). Frequent causes of mortality in this population included pulmonary-related processes. Review of survival data for individuals with OI (types non-specified) from the Danish health registries as compared to a reference population also showed an almost three-fold increase in all-cause mortality (Folkestad et al. 2016). For the less common autosomal recessive OI types, official mortality data are not available although life expectancy is shortened.

RESOURCES

Educational Resources for Non-Medical Persons

Genetics Home Reference (https://ghr.nlm.nih.gov/condition/osteogenesis-imperfecta)
National Organization for Rare Disorders (https://rarediseases.org/rare-diseases/osteogenesis-imperfecta/)

Support Organizations

Osteogenesis Imperfecta Foundation (www.oif.org)
Little People of America (www.lpaonline.org)

Research

https://clinicaltrials.gov/

REFERENCES

Alanay Y, Avaygan H, Camacho N, et al. (2010) Mutations in the gene encoding the RER protein FKBP65 cause autosomal-recessive osteogenesis imperfecta. *Am J Hum Genet* 86(4) 551–559.

Aldinger KA, Mendelsohn NJ, Chung BH, Zhang W, Cohn DH, Fernandez B, Alkuraya FS, Dobyns WB, Curry CJ (2016) Variable brain phenotype primarily affects the brainstem and cerebellum in patients with osteogenesis imperfecta caused by recessive WNT1 mutations. *J Med Genet* 53(6) 427–430.

Andersen PE, Jr Hauge M (1989) Osteogenesis imperfecta: a genetic, radiological, and epidemiological study. *Clin Genet* 36(4):250–255.

Andersson K, Dahllof G, Lindahl K, Kindmark A, Grigelioniene G, Astrom E, Malmgren B (2017) Mutations in COL1A1 and COL1A2 and dental aberrations in children and adolescents with osteogenesis imperfecta - A retrospective cohort study. *PLoS One* 12(5):e0176466.

Anissipour AK, Hammerberg KW, Caudill A, Kostiuk T, Tarima S, Zhao HS, Drzak JJ, Smith PA (2014) Behavior of scoliosis during growth in children with osteogenesis imperfecta. *J Bone Joint Surg Am* 96(3):237–243.

Antoniazzi F, Bertoldo F, Mottes M, Valli M, Sirpresi S, Zamboni G, Valentini R, Tato L (1996) Growth hormone treatment in osteogenesis imperfecta with quantitative defect of type I collagen synthesis. *J Pediatr* 129(3) 432–439.

Arponen H, Makitie O, Haukka J, Ranta H, Ekholm M, Mayranpaa MK, Kaitila I, Waltimo-Siren J (2012) Prevalence and natural course of craniocervical junction anomalies during growth in patients with osteogenesis imperfecta. *J Bone Miner Res* 27(5):1142–1149.

Arponen H, Vuorimies I, Haukka J, Valta H, Waltimo-Siren J, Makitie O (2015) Cranial base pathology in pediatric osteogenesis imperfecta patients treated with bisphosphonates. *J Neurosurg Pediatr* 15(3):313–320.

Ashournia H, Johansen FT, Folkestad L, Diederichsen AC, Brixen K (2015) Heart disease in patients with osteogenesis imperfecta - A systematic review. *Int J Cardiol* 196:149–157.

Barbosa-Buck CO, Orioli IM, da Graca Dutra M, Lopez-Camelo J, Castilla EE, Cavalcanti DP (2012) Clinical epidemiology of skeletal dysplasias in South America. *Am J Med Genet A* 158A(5):1038–1045.

Barnes AM, Carter EM, Cabral WA, Weis M, Chang W, Makareeva E, Leikin S, Rotimi CN, Eyre DR, Raggio CL, Marini JC (2010) Lack of cyclophilin B in osteogenesis imperfecta with normal collagen folding. *N Engl J Med* 362(6):521–528.

Barnes AM, Chang W, Morello R, et al. (2006) Deficiency of cartilage-associated protein in recessive lethal osteogenesis imperfecta. *N Engl J Med* 355(26):2757–2764.

Beecham E, Candy B, Howard R, McCulloch R, Laddie J, Rees H, Vickerstaff V, Bluebon-Langner M, Jones L (2015) Pharmacological interventions for pain in children and adolescents with life-limiting conditions. *Cochrane Database Syst Rev* 2015(3):CD010750.

Beighton P, Versfeld GA (1985) On the paradoxically high relative prevalence of osteogenesis imperfecta type III in the black population of South Africa. *Clin Genet* 27(4):398–401.

Binder H, Conway A, Hason S, Gerber LH, Marini J, Berry R, Weintrob J (1993) Comprehensive rehabilitation of the child with osteogenesis imperfecta. *Am J Med Genet* 45(2):265–269.

Boyce AM (2017) Denosumab: an Emerging Therapy in Pediatric Bone Disorders. *Curr Osteoporos Rep* 15(4):283–292.

Cabral WA, Barnes AM, Adeyemo A, Cushing K, Chitayat D, Porter FD, Panny SR, Gulamali-Majid F, Tishkoff SA, Rebbeck TR, Gueye SM, Bailey-Wilson JE, Brody LC, Rotimi CN, Marini JC (2012) A founder mutation in LEPRE1 carried by 1.5% of West Africans and 0.4% of African Americans causes lethal recessive osteogenesis imperfecta. *Genet Med* 14(5):543–551.

Cabral WA, Chang W, Barnes AM, Weis M, Scott MA, Leikin S, Makareeva E, Kuzneetsova NV, Rosenbaum, KN, Tifft CJ, Bulas DI, Kozma C, Smith PA, Eyre DR, Marini JC (2007) Prolyl 3-hydroxylase 1 deficiency causes a recessive metabolic bone disorder resembling lethal/severe osteogenesis imperfecta. *Nat Genet* 39(3):359–365.

Cabral WA, Marini JC (2004) High proportion of mutant osteoblasts is compatible with normal skeletal function in mosaic carriers of osteogenesis imperfecta. *Am J Hum Genet* 74(4):752–760.

Cassella JP, Stamp TC, Ali SY (1996) A morphological and ultrastructural study of bone in osteogenesis imperfecta. *Calcif Tissue Int* 58(3):155–165.

Chang W, Barnes AM, Cabral WA, Bodurtha JN, Marini JC (2010) Prolyl 3-hydroxylase 1 and CRTAP are mutually stabilizing in the endoplasmic reticulum collagen prolyl 3-hydroxylation complex. *Hum Mol Genet* 19(2):223–234.

Charnas LR, Marini JC (1993) Communicating hydrocephalus, basilar invagination, and other neurologic features in osteogenesis imperfecta. *Neurology* 43(12):2603–2608.

Chines A, Boniface A, McAlister W, Whyte M (1995) Hypercalciuria in osteogenesis imperfecta: a follow-up study to assess renal effects. *Bone* 16(3):333–339.

Christiansen HE, Schwarze U, Pyott SM, AlSwaid A, Al Balwi M, Alrasheed S, Pepin MG, Weis MA, Eyre DR, Byers PH (2010) Homozygosity for a missense mutation in SERPINH1, which encodes the collagen chaperone protein HSP47, results in severe recessive osteogenesis imperfecta. *Am J Hum Genet* 86(3):389–398.

Cubert R, Cheng EY, Mack S, Pepin MG, Byers PH (2001) Osteogenesis imperfecta: mode of delivery and neonatal outcome. *Obstet Gynecol* 97(1):66–69.

Dwan K, Phillipi CA, Steiner RD, Basel D (2016) Bisphosphonate therapy for osteogenesis imperfecta. *Cochrane Database Syst Rev* 10:CD005088.

Elejalde BR, de Elejalde MM (1985) Thanatophoric dysplasia: fetal manifestations and prenatal diagnosis. *Am J Med Genet* 22(4):669–683.

Engelbert RH, Helders PJ, Keessen W, Pruijs HE, Gooskens RH (1995) Intramedullary rodding in type III osteogenesis imperfecta. Effects on neuromotor development in 10 children. *Acta Orthop Scand* 66(4):361–364.

Engelbert RH, Uiterwaal CS, Gulmans VA, Pruijs HE, Helders PJ (2000) Osteogenesis imperfecta: profiles of motor development as assessed by a postal questionnaire. *Eur J Pediatr* 159(8):615–620.

Esposito P, Plotkin H (2008) Surgical treatment of osteogenesis imperfecta: current concepts. *Curr Opin Pediatr* 20(1):52–57.

Evereklioglu C, Madenci E, Bayazit YA, Yilmaz K, Balat A, Bekir NA (2002) Central corneal thickness is lower in osteogenesis imperfecta and negatively correlates with the presence of blue sclera. *Ophthalmic Physiol Opt* 22(6):511–515.

Farber CR, Reich A, Barnes AM, Becerra P, Rauch F, Cabral WA, Bae A, Quinlan A, Glorieux FH, Clemens TL, Marini JC (2014) A novel IFITM5 mutation in severe atypical osteogenesis imperfecta type VI impairs osteoblast production of pigment epithelium-derived factor. *J Bone Miner Res* 29(6):1402–1411.

Fiscaletti M, Biggin A, Bennetts B, Wong K, Briody J, Pacey V, Birman C, Munns CF (2018) Novel variant in Sp7/Osx associated with recessive osteogenesis imperfecta with bone fragility and hearing impairment. *Bone* 110 66–75.

Folkestad L, Hald JD, Canudas-Romo V, Gram J, Hermann AP, Langdahl B, … Brixen K (2016) Mortality and Causes of Death in Patients With Osteogenesis Imperfecta: A Register-Based Nationwide Cohort Study. *J Bone Miner Res* 31(12):2159–2166. doi:10.1002/jbmr.2895

Folkestad L, Hald JD, Ersboll AK, Gram J, Hermann AP, Langdahl B, Abrahamsen B, Brixen K (2017) Fracture rates and fracture sites in patients with osteogenesis imperfecta: A Nationwide register-based cohort study. *J Bone Miner Res* 32(1):125–134. doi:10.1002/jbmr.2920

Forlino A, Marini JC (2016) Osteogenesis imperfecta. *Lancet* 387(10028):1657–1671.

Fryns JP, van den Berghe K, van Assche A, van den Berghe H (1981) Prenatal diagnosis of campomelic dwarfism. *Clin Genet* 19(3):199–201.

Gerber LH, Binder H, Weintrob J, Grange DK, Shapiro J, Fromherz W, Berry R, Conway A, Nason S, Marini J (1990) Rehabilitation of children and infants with osteogenesis imperfecta. A program for ambulation. *Clin Orthop Relat Res* (251):254–262.

Germain-Lee EL, Brennen FS, Stern D, Kantipuly A, Melvin P, Terkowitz MS, Shapiro JR (2016) Cross-sectional and longitudinal growth patterns in osteogenesis imperfecta: implications for clinical care. *Pediatr Res* 79(3):489–495.

Glorieux FH, Devogelaer JP, Durigova M, Goemaere S, Hemsley S, Jakob F, Junker U, Ruckle J, Seefried L, Winkle PJ (2017) BPS804 anti-sclerostin antibody in adults with moderate osteogenesis imperfecta: Results of a randomized phase 2a trial. *J Bone Miner Res* 32(7):1496–1504.

Glorieux FH, Rauch F, Plotkin H, Ward L, Travers R, Roughley P, Lalic L, Glorieux DF, Fassier F, Bishop NJ (2000) Type V osteogenesis imperfecta: a new form of brittle bone disease. *J Bone Miner Res* 15(9):1650–1658.

Hald JD, Evangelou E, Langdahl BL, Ralston SH (2015) Bisphosphonates for the prevention of fractures in osteogenesis imperfecta: meta-analysis of placebo-controlled trials. *J Bone Miner Res* 30(5):929–933.

Hatz D, Esposito PW, Schroeder B, Burke B, Lutz R, Hasley BP (2011) The incidence of spondylolysis and spondylolisthesis in children with osteogenesis imperfecta. *J Pediatr Orthop* 31(6):655–660.

Homan EP, Rauch F, Grafe I, et al. (2011) Mutations in SERPINF1 cause osteogenesis imperfecta type VI. *J Bone Miner Res* 26(12):2798–2803.

Hortop J, Tsipouras P, Hanley JA, Maron BJ, Shapiro JR (1986) Cardiovascular involvement in osteogenesis imperfecta. *Circulation* 73(1):54–61.

Keller RB, Tran TT, Pyott SM, Pepin MG, Savarirayan R, McGillivray G, Nickerson DA, Bamshad MJ, Byers PH (2017) Monoallelic and biallelic CREB3L1 variant causes mild and severe osteogenesis imperfecta, respectively. *Genet Med* 20:411–419.

Kelley BP, Malfait F, Bonafe L, et al.(2011) Mutations in FKBP10 cause recessive osteogenesis imperfecta and Bruck syndrome. *J Bone Miner Res* 26(3):666–672.

Krakow D (2018) Osteogenesis Imperfecta. In: *Obstetric Imaging: Fetal Diagnosis and Care*, Copel JA, D'Alton ME, Feltovich H, Gratacos E, Krakow D, Odibo AO, Platt LD, Tutschek B (Eds), 2nd ed, pp. 270–273. Elsevier Inc.

Lapunzina P, Aglan M, Temtamy S, Caparros-Martin JA, Valencia M, Leton R, Martinez-Glez V, Elhossini R, Amr K, Vilaboa N, Ruiz-Perez VL (2010) Identification of a frameshift mutation in Osterix in a patient with recessive osteogenesis imperfecta. *Am J Hum Genet* 87(1):110–114.

Leali PT, Balsano M, Maestretti G, Brusoni M, Amorese V, Ciurlia E, Andreozzi M, Caggiari G, Doria C (2017) Efficacy of teriparatide vs neridronate in adults with osteogenesis imperfecta type I: a prospective randomized international clinical study. *Clin Cases Miner Bone Metab* 14(2):153–156.

Li G, Jin Y, Levine MAH, Hoyer-Kuhn H, Ward L, Adachi JD (2017) Systematic review of the effect of denosumab on children with osteogenesis imperfecta showed inconsistent findings. *Acta Paediatr.* 107:534–537.

Lindahl K, Astrom E, Dragomir A, Symoens S, Coucke P, Larsson S, Paschalis E, Roschger P, Gamsjaeger S, Klaushofer K, Fratzl-Zelman N, Kindmark A (2018) Homozygosity for CREB3L1 premature stop codon in first case of recessive osteogenesis imperfecta associated with OASIS-deficiency to survive infancy. *Bone* 114 268–277.

Lindert U, Cabral WA, Ausavarat S, et al. (2016) MBTPS2 mutations cause defective regulated intramembrane proteolysis in X-linked osteogenesis imperfecta. *Nat Commun* 7:11920.

Marini JC, Bordenick S, Chrousos GP (1995) Endocrine aspects of growth deficiency in OI. *Connect Tissue Res* 31(4):S55–57.

Marini JC, Bordenick S, Heavner G, Rose S, Hintz R, Rosenfeld R, Chrousos GP (1993) The growth hormone and somatomedin axis in short children with osteogenesis imperfecta. *J Clin Endocrinol Metab* 76(1):251–256.

Marini JC, Forlino A, Bachinger HP, Bishop NJ, Byers PH, Paepe A, Fassier F, Fratzl-Zelman N, Kozloff KM,Krakow D, Montpetit K, Semler O (2017) Osteogenesis imperfecta. *Nat Rev Dis Primers* 3:17052.

Marini JC, Forlino A, Cabral WA, et al. (2007) Consortium for osteogenesis imperfecta mutations in the helical domain of type I collagen: regions rich in lethal mutations align with collagen binding sites for integrins and proteoglycans. *Hum Mutat* 28(3):209–221.

Marini JC, Gerber NL (1997) Osteogenesis imperfecta. Rehabilitation and prospects for gene therapy. *JAMA* 277(9):746–750.

Marini JC, Hopkins E, Glorieux FH, Chrousos GP, Reynolds JC, Gundberg CM, Reing CM (2003) Positive linear growth and bone responses to growth hormone treatment in children with types III and IV osteogenesis imperfecta: high predictive value of the carboxyterminal propeptide of type I procollagen. *J Bone Miner Res* 18(2):237–243.

Marini JC, Smith SM (2015) Osteogenesis Imperfecta. *Endotext* Available at: https://www.ncbi.nlm.nih.gov/books/NBK279109/

Marom R, Lee YC, Grafe I, Lee B (2016) Pharmacological and biological therapeutic strategies for osteogenesis imperfecta. *Am J Med Genet C Semin Med Genet* 172(4):367–383.

Martinez-Glez V, Valencia M, Caparros-Martin JA, Aglan M, Temtamy S, Tenorio J, Pulido V, Lindert U, Rohrbach M, Eyre D, Giunta C, Lapunzina P, Ruiz-Perez VL (2012) Identification of a mutation causing deficient BMP1/mTLD proteolytic activity in autosomal recessive osteogenesis imperfecta. *Hum Mutat* 33(2):343–350.

McAllion SJ, Paterson CR (1996) Causes of death in osteogenesis imperfecta. *J Clin Pathol* 49(8):627–630.

Mendoza-Londono R, Fahiminiya S, Majewski J, Care4Rare Canada, C, Tetreault M, Nadaf J, … Rauch F (2015) Recessive osteogenesis imperfecta caused by missense mutations in SPARC. *Am J Hum Genet* 96(6):979–985.

Menezes AH (2008) Specific entities affecting the craniocervical region: osteogenesis imperfecta and related osteochondrodysplasias: medical and surgical management of basilar impression. *Childs Nerv Syst* 24(10):1169–1172.

Montpetit K, Palomo T, Glorieux FH, Fassier F, Rauch F (2015) Multidisciplinary treatment of severe osteogenesis imperfecta: Functional outcomes at skeletal maturity. *Arch Phys Med Rehabil* 96(10):1834–1839.

Morikawa M, Fukuda Y, Terasaki Y, Itoh H, Demura Y, Sasaki M, Imamura Y, Honjo C, Umeda Y, Anzai M, Ishizuka T (2016) Osteogenesis Imperfecta Associated with Dendriform Pulmonary Ossification. *Am J Respir Crit Care Med* 193(4):460–461.

O'Connell AC, Marini JC (1999) Evaluation of oral problems in an osteogenesis imperfecta population. *Oral Surg Oral Med Oral Pathol Oral Radiol Endod* 87(2):189–196.

Orwoll ES, Shapiro J, Veith S, Wang Y, Lapidus J, Vanek C, Reeder JL, Keaveny TM, Lee DC, Mullins MA, Nagamani SCS, Lee B (2014) Evaluation of teriparatide treatment in adults with osteogenesis imperfecta. *J Clin Invest* 124(2):491–498.

Pepin MG, Byers PH (2015) What every clinical geneticist should know about testing for osteogenesis imperfecta in suspected child abuse cases. *Am J Med Genet C* 169(4):307–313.

Pillion JP, Vernick D, Shapiro J (2011) Hearing loss in osteogenesis imperfecta: characteristics and treatment considerations. *Genet Res Int* 2011:983942.

Ranta H, Lukinmaa PL, Waltimo J (1993) Heritable dentin defects: nosology, pathology, and treatment. *Am J Med Genet* 45(2):193–200.

Renaud A, Aucourt J, Weill J, Bigot J, Dieux A, Devisme L, Moraux A, Boutry N (2013) Radiographic features of osteogenesis imperfecta. *Insights Imaging* 4(4):417–429.

Rijks EB, Bongers BC, Vlemmix MJ, Boot AM, van Dijk AT, Sakkers RJ, van Brussel M (2015) Efficacy and safety of bisphosphonate therapy in children with osteogenesis imperfecta: A Systematic review. *Horm Res Paediatr* 84(1):26–42.

Rizkallah J, Schwartz S, Rauch F, Glorieux F, Vu DD, Muller K, Retrouvey JM (2013) Evaluation of the severity of malocclusions in children affected by osteogenesis imperfecta with the peer assessment rating and discrepancy indexes. *Am J Orthod Dentofacial Orthop* 143(3):336–341.

Ruiter-Ligeti J, Czuzoj-Shulman N, Spence AR, Tulandi T, Abenhaim HA (2016) Pregnancy outcomes in women with osteogenesis imperfecta: a retrospective cohort study. *J Perinatol* 36(10):828–831.

Rush ET, Li L, Goodwin JL, Kreikemeier RM, Craft M, Danford DA, Kutty S (2017) Echocardiographic phenotype in osteogenesis imperfecta varies with disease severity. *Heart* 103(6):443–448.

Santos F, McCall AA, Chien W, Merchant S (2012) Otopathology in Osteogenesis Imperfecta. *Otol Neurotol* 33(9):1562–1566.

Sato A, Ouellet J, Muneta T, Glorieux FH, Rauch F (2016) Scoliosis in osteogenesis imperfecta caused by COL1A1/COL1A2 mutations - genotype-phenotype correlations and effect of bisphosphonate treatment. *Bone* 86:53–57.

Semler O, Garbes L, Keupp K, et al. (2012) A mutation in the 5'-UTR of IFITM5 creates an in-frame start codon and causes autosomal-dominant osteogenesis imperfecta type V with hyperplastic callus. *Am J Hum Genet* 91(2):349–357.

Shaheen R, Alazami AM, Alshammari MJ, Faqeih E, Alhashmi N, Mousa N, Alsinani A, Ansari S, Alzahrani F, Al-Owain M, Alzayed ZS, Alkuraya FS (2012) Study of autosomal recessive osteogenesis imperfecta in Arabia reveals a novel locus defined by TMEM38B mutation. *J Med Genet* 49(10):630–635.

Shapiro JR, Sponsellor PD (2009) Osteogenesis imperfecta: questions and answers. *Curr Opin Pediatr* 21(6):709–716.

Sillence D, Butler B, Latham M, Barlow K (1993) Natural history of blue sclerae in osteogenesis imperfecta. *Am J Med Genet* 45(2):183–186.

Sillence D, Rimoin DL, Danks DM (1979) Clinical variability in osteogenesis imperfecta-variable expressivity or genetic heterogeneity. *Birth Defects Orig Artic Ser* 15(5B):113–129.

Singer RB, Ogston SA, Paterson CR (2001) Mortality in various types of osteogenesis imperfecta. *J Insur Med* 33(3):216–220.

Stevenson DA, Carey J, Byrne J, Srisukhumbowornchai S, Feldkamp M (2012) Analysis of skeletal dysplasias in the Utah population. *Am J Med Genet A* 158A(5):1046–1054.

Superti-Furga A (1996) Achondrogenesis type 1B. *J Med Genet* 33(11):957–961.

Suskauer SJ, Cintas HL, Marini JC, Gerber LH (2003) Temperament and physical performance in children with osteogenesis imperfecta. *Pediatrics* 111(2):E153–161.

Swinnen FK, Coucke PJ, De Paepe AM, et al. (2011) Osteogenesis Imperfecta: the audiological phenotype lacks correlation with the genotype. *Orphanet J Rare Dis* 6:88.

Symoens S, Malfait F, D'Hondt S, Callewaert B, Dheedene A, Steyaert W, Bachinger HP, De Paepe A, Kayserili H, Coucke PJ (2013) Deficiency for the ER-stress transducer OASIS causes severe recessive osteogenesis imperfecta in humans. *Orphanet J Rare Dis* 8:154.

Tainmont J (2007) History of osteogenesis imperfecta or brittle bone disease: a few stops on a road 3000 years long. *B-ENT* 3(3):157–173.

Thiele F, Cohrs CM, Flor A, et al. (2012) Cardiopulmonary dysfunction in the Osteogenesis imperfecta mouse model Aga2 and human patients are caused by bone-independent mechanisms. *Hum Mol Genet* 21(16):3535–3545.

Trejo P, Palomo T, Montpetit K, Fassier F, Sato A, Glorieux FH, Rauch F (2017) Long-term follow-up in osteogenesis imperfecta type VI. *Osteoporos Int* 28(10):2975–2983.

Van Dijk FS, Sillence DO (2014) Osteogenesis imperfecta: clinical diagnosis, nomenclature and severity assessment. *Am J Med Genet A* 164A(6):1470–1481.

Wallace MJ, Kruse RW, Shah SA (2017) The Spine in Patients With Osteogenesis Imperfecta. *J Am Acad Orthop Surg* 25(2):100–109.

Ward LM, Rauch F, Travers R, Chabot G, Azouz EM, Lalic L, Roughley PJ, Glorieux FH (2002) Osteogenesis imperfecta type VII: an autosomal recessive form of brittle bone disease. *Bone* 31(1):12–18.

Webb EA, Balasubramanian M, Fratzl-Zelman N, et al. (2017) Phenotypic Spectrum in Osteogenesis Imperfecta Due to Mutations in TMEM38B: Unraveling a Complex Cellular Defect. *J Clin Endocrinol Metab* 102(6):2019–2028.

Weil UH (1981) Osteogenesis imperfecta: historical background. *Clin Orthop Relat Res* (159):6–10.

Wekre LL, Kjensli A, Aasand K, Falch JA, Eriksen EF (2014) Spinal deformities and lung function in adults with osteogenesis imperfecta. *Clin Respir J* 8(4):437–443.

Wente SJ (2013) Nonpharmacologic pediatric pain management in emergency departments: a systematic review of the literature. *J Emerg Nurs* 39(2):140–150.

WHO (2012) *WHO Guidelines on the Pharmacological Treatment of Persisting Pain in Children With Medical Illness*. World Health Organization

Whyte MP (1995) Hypophosphatasia *The Metabolic Basis of Inherited Diseases*, pp. 4095–4111 New York: McGraw Hill Inc.

Willing MC, Pruchno CJ, Byers PH (1993) Molecular heterogeneity in osteogenesis imperfecta type I. *Am J Med Genet* 45(2):223–227.

Zack P, Franck L, Devile C, Clark C (2005) Fracture and non-fracture pain in children with osteogenesis imperfecta. *Acta Paediatr* 94(9):1238–1242.

44

PALLISTER–HALL SYNDROME AND GREIG CEPHALOPOLYSYNDACTYLY SYNDROME*

LESLIE G. BIESECKER

Medical Genomics and Metabolic Genetics Branch, National Human Genome Research Institute, Bethesda, Maryland, USA

INTRODUCTION

Pallister–Hall syndrome (PHS) and Greig cephalopolysyndactyly syndrome (GCPS) are allelic pleiotropic developmental anomaly syndromes caused by mutations in *GLI3*. In spite of the fact that they are allelic and have a wide range of severity, the two syndromes are clinically distinct and rarely confused. Frequent manifestations of PHS include hypothalamic hamartoma with or without pituitary dysplasia and endocrine dysfunction, mesoaxial and postaxial polydactyly, bifid epiglottis or laryngeal cleft, and imperforate anus (Biesecker and Graham 1996). Individuals with typical GCPS have macrocephaly, widely spaced eyes, and postaxial or preaxial polydactyly and cutaneous syndactyly (Balk and Biesecker 2008).

Incidence

Both disorders are rare (less than 1/500,000), but rapid recognition, especially of PHS, is critical for making early management decisions. It appears that each disorder has an equal incidence among different population groups.

Diagnostic Criteria

Pallister–Hall Syndrome. PHS has numerous other descriptors including hypothalamic hamartoblastoma, hypopituitarism, imperforate anus, and postaxial polydactyly syndrome, Hall–Pallister syndrome, and cerebro-acro-visceral early lethality syndrome, none of which is more useful or appropriate than Pallister–Hall syndrome, which is widely accepted. The eponym credits the extensive and elegant description of a series of six patients in 1980 by Judith Hall and Phillip Pallister (reviewed in Hall [2016]). The entity of polydactyly imperforate anus vertebral (PIV) anomalies syndrome has been subsumed into the designation of Pallister–Hall syndrome (Killoran et al. 2000).

Clinical diagnostic criteria have been published for PHS (Table 44.1) (Biesecker et al. 1996). However, these criteria were published before the gene was identified and may be liberalized, as clinical gene testing is now widely available. The criteria were based on highly specific malformations (hypothalamic hamartoma and mesoaxial polydactyly), and therefore may not have high sensitivity. Less common and less specific manifestations include postaxial polydactyly, pulmonary segmentation anomalies, renal anomalies, seizures, and developmental delay or intellectual disability. Mesoaxial polydactyly (also known as intercalary or central polydactyly) is defined as the duplication of digits three or four with osseous syndactyly of the metacarpals or metatarsals (Figure 44.1). Clinicians may find it prudent to screen for *GLI3* variants in individuals who have clinical features that are similar to, but do not quite meet, the published clinical criteria. A variant yield of about 30% has been found in those who have several of the features of PHS but do not

---

*This chapter is in the public domain in the United States of America.

Cassidy and Allanson's Management of Genetic Syndromes, Fourth Edition.
Edited by John C. Carey, Agatino Battaglia, David Viskochil, and Suzanne B. Cassidy.
© 2021 John Wiley & Sons, Inc. Published 2021 by John Wiley & Sons, Inc.

TABLE 44.1 Clinical diagnostic criteria for Pallister–Hall syndrome

The diagnosis of Pallister–Hall syndrome is established in:
A proband if mesoaxial polydactyly and hypothalamic hamartoma are present.
The first-degree relative of a proband if hypothalamic hamartoma or mesoaxial or postaxial polydactyly are present.

Notes:
1. Postaxial polydactyly type B (digitus minimus) can be used as a criterion for first-degree relatives only in persons who are not of central West African descent.
2. The availability of clinical *GLI3* gene testing may necessitate relaxation of these criteria. Clinicians are encouraged to use their judgment to integrate molecular and clinical data to make an overall diagnosis.

quite meet the published diagnostic criteria (Johnston et al. 2010). The syndrome has a very wide range of severity. Although many clinical reports describe severely affected individuals, the most common presentation of PHS is a familial occurrence of a mildly affected individual with polydactyly and asymptomatic or mildly symptomatic hypothalamic hamartomas. The severe end of the spectrum is uncommon, but can be devastating, with massive hamartomas associated with panhypopituitarism or severe seizures, laryngeal clefts that cause intractable aspiration, and renal anomalies. Some of these severely affected children do not survive the first weeks of life.

Greig Cephalopolysyndactyly Syndrome. Similar to PHS, there have been a number of alternative descriptors for GCPS, including Greig syndrome, Hootnik–Holmes syndrome, frontonasal syndrome, Noack-type acrocephalosyndactyly, frontonasal dysplasia, medical cleft face syndrome, and acrofacial dysostosis, and "polydactyly with peculiar skull shape", but all of these should be abandoned. Unlike PHS, there are no widely accepted clinical diagnostic criteria for GCPS. One set of criteria has been proposed (Table 44.2), which specifies that an individual should have preaxial polydactyly of at least one limb with widely spaced eyes and macrocephaly. These criteria are not specific for GCPS and a number of overlapping syndromes must be considered (see below). The manifestations of GCPS also encompass a wide range of severity, but are distinct from those of PHS. The most common form of polydactyly is postaxial polydactyly of the hands with preaxial polydactyly of the feet (Figure 44.2). The cutaneous syndactyly in GCPS is highly variable and is cutaneous, not osseous, as is the case in PHS. Individuals with mild GCPS can have craniofacial features that are difficult to recognize as abnormal without careful measurements. Many mildly affected individuals are likely to be undiagnosed, especially as macrocephaly is common in the general population and mild widely spaced eyes is considered by many to be an

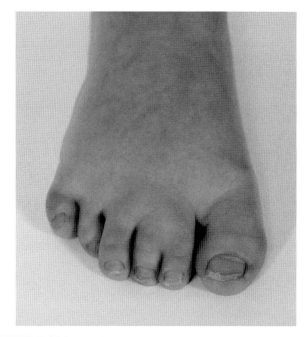

FIGURE 44.1 The foot of a young patient with Pallister–Hall syndrome and mesoaxial polydactyly of the foot. Although there are five triphalangeal digits, it cannot with certainty be determined which is supernumerary. However, the patient has a bifid third metatarsal, which meets the criteria for mesoaxial polydactyly, instead of postaxial polydactyly. Image from (Biesecker et al. 2009).

attractive physical trait. Therefore, many individuals may undergo corrective surgery for their limb anomalies without having the diagnosis of GCPS considered.

As can be seen from these summaries, PHS and GCPS share but a single clinical trait, that of postaxial polydactyly. It is possible that the mildest end of the spectrum of both disorders is polydactyly of one to four limbs without other malformations. However, this is controversial, as many of the other malformations of PHS can be asymptomatic and only appreciated with sophisticated evaluations such as magnetic resonance imaging, endoscopy, and high-resolution computed tomography (Biesecker 2006). In addition, some family members of individuals with typical GCPS have essentially normal craniofacial features and polysyndactyly, suggesting that non-syndromic polydactyly is part of the GCPS spectrum. As well, non-syndromic hypothalamic hamartomas may be part of the Pallister–Hall spectrum, as has been suggested by recent work that identified somatic *GLI3* variants in these lesions (Hildebrand et al. 2016).

Etiology, Pathogenesis, and Genetics

Both GCPS and PHS are caused by pathogenic variants in the *GLI3* zinc finger transcription factor gene on chromosome 7p13, which is a member of the *GLI* gene family (Kang et al. 1997; Vortkamp et al. 1991). In addition, both disorders are inherited in an autosomal dominant

TABLE 44.2 Establishment of the diagnosis of Greig cephalopolysyndactyly

The diagnosis of Greig cephalopolysyndactyly syndrome should be considered in:

A proband if preaxial polydactyly, syndactyly of toes 1–3 or fingers 3–4, widely spaced eyes, and macrocephaly are present. The diagnosis should be made with caution in infants with multiple other malformations, especially in the absence of a positive family history.

A first-degree relative of a proband (in whom the diagnosis has been independently established) if he/she has pre- or postaxial polydactyly with or without syndactyly or the craniofacial features.

Source: Clinical criteria reproduced with the author's permission from GeneReviews: http://www.genetests.org.
Note:
1. Postaxial polydactyly type B should not be used as a diagnostic criterion for first-degree relatives of persons who are of central West African descent.
2. The availability of molecular testing provides an opportunity to diagnose people with mild or atypical manifestations. Clinicians are encouraged to use their judgment to integrate molecular and clinical data to make an overall diagnosis.

manner with variable expressivity and full or very high penetrance (Debeer et al. 2003). Intrafamilial variance in expressivity is less than interfamilial variation. Variants in *GLI3* were first described in GCPS after the gene was cloned by determination of translocation breakpoints in affected individuals with cytogenetic aberrations (Vortkamp et al. 1991). Subsequently, a large number of distinct variants have been demonstrated in GCPS (Debeer et al. 2003; Demurger et al. 2015; Johnston et al. 2005; Johnston et al. 2010; Kalff-Suske et al. 1999; Wild et al. 1997). These variants include balanced translocations, gross cytogenetic interstitial deletions, small insertions and deletions, and substitution, frameshift, and nonsense variants. There should be no doubt that GCPS is caused by haploinsufficiency. This view is supported by several spontaneous and radiation-induced mouse mutants that demonstrate a remarkable phenotypic similarity to the human phenotype (Hui and Joyner 1993; Vortkamp et al. 1992).

There is a degree of genotype–phenotype correlation in GCPS. Individuals with point mutations in *GLI3* tend to have mild-to-moderate manifestations, normal intelligence, and few manifestations other than the common ones listed above. Individuals with large deletions, especially those greater than 1 Mb, tend to have developmental delay, central nervous system abnormalities on magnetic resonance imaging examination, hernias, and so forth. We consider these individuals to have a "Greig cephalopolysyndactyly syndrome contiguous gene syndrome" (Johnston et al. 2007; Johnston et al. 2003).

Several years after variants in the *GLI3* gene were found to cause GCPS, it was surprising to find that PHS was allelic (Kang et al. 1997). In contrast to the wide spectrum of variants in GCPS, essentially the only type of variant that causes PHS is a frameshift or nonsense variant 3′ of the zinc finger encoding domain of *GLI3* (Al-Qattan et al. 2017; Demurger et al. 2015; Johnston et al. 2005; Johnston et al. 2010). These data have been incorporated into a model whereby haploinsufficiency causes GCPS and dominant negative variants cause PHS. There is no correlation of the position of *GLI3* variants with the severity of PHS.

Most individuals with PHS or GCPS are the only affected individual in their family. There is one documented occurrence of parental germline mosaicism for PHS (Ng et al. 2004) and intriguing evidence for somatic mosaicism in those with mild forms of PHS (Saitsu et al. 2016), but mosaicism for GCPS has not been described. However, there is no reason to believe that this cannot occur and families should be advised accordingly.

Diagnostic Testing

Currently, there are several laboratories that offer testing for *GLI3* variants. The testing for variants is challenging because GCPS can be caused by an extremely wide range of variants (from megabase deletions to single base pair substitutions) and the *GLI3* gene is large (about 300 kb with 15 exons,

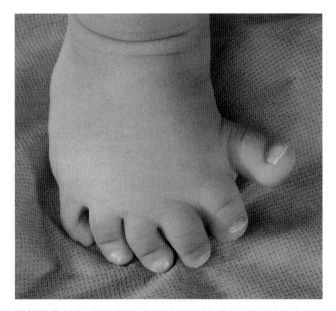

FIGURE 44.2 The foot of a toddler with Greig cephalopolysyndactyly and preaxial polydactyly of the foot. This is defined by the presence of a supernumerary biphalangeal digit resembling a hallux. Image from Biesecker et al. (2009).

some of which are difficult to amplify). Nearly all families have unique variants. Molecular cytogenetic techniques such as array CGH and SNPchips are ideal for identifying CNV abnormalities associated with GCPS. Sequencing is the best technique to detect alterations less than the size of a PCR amplicon. Increasingly, exome sequencing is becoming a cost effective alternative to Sanger-based sequencing, especially for individuals who do not have a classical presentation, for whom the differential is broad (Biesecker and Green 2014). *GLI3* variant analysis should be extremely useful in making an accurate diagnosis, especially when individuals present with mild signs that do not meet diagnostic criteria.

Given the relative frequencies of the various molecular lesions in those with suspected GCPS the order of testing should be: Sanger or exome or genome sequencing (point variants), followed by array CGH or SNP array (copy number variants), followed by standard resolution Giemsa-banded cytogenetic analysis (balanced translocation that interrupts the *GLI3* gene).

All individuals with molecularly proven PHS have small variants in *GLI3* and are detectable by sequencing. It is theoretically possible that a large deletion or translocation could generate a Pallister–Hall *GLI3* allele, so if sequencing is normal, additional studies should be considered.

Differential Diagnosis

Pallister–Hall syndrome overlaps with the McKusick–Kaufman syndrome, Bardet–Biedl syndrome, oral-facial-digital syndrome type 6, Ellis–van Creveld syndrome, Holzgreve syndrome, and Holt–Oram syndrome. Bardet–Biedl syndrome manifests polydactyly (usually postaxial, but occasionally, mesoaxial) and has a low frequency of uncommon PHS manifestations including anal stenosis or Hirschsprung disease, bifid epiglottis, and hypothalamic hamartoma (see Chapter 8). McKusick–Kaufman syndrome is an allelic disorder to one form of Bardet–Biedl syndrome (see Chapter 8) and manifests mesoaxial polydactyly, hydrometrocolpos, and congenital heart disease. The oral-facial-digital syndromes can have similar limb malformations (especially, type VI). Holzgreve syndrome includes congenital heart defect, renal dysplasia, cleft palate, and polydactyly. Ellis–van Creveld syndrome includes short stature, small thorax, and polydactyly. Holt–Oram syndrome manifests congenital heart disease with polydactyly that is in some cases mesoaxial. In addition, there have been several reports of overlap of PHS and the holoprosencephaly spectrum of disorders (see Chapter 31) (Muenke et al. 1991), and other partial overlapping phenotypes such as mesoaxial hexadactyly and cardiac malformation syndrome.

GCPS overlaps with the acrocallosal syndrome, which is a non-allelic disorder inherited in an autosomal recessive pattern (Putoux et al. 2011). Acrocallosal syndrome and GCPS both have polysyndactyly, macrocephaly, and widely spaced eyes, but, in addition, acrocallosal syndrome includes intellectual disability, seizures, and hernias. Acrocallosal syndrome is characterized by aplasia or hypoplasia of the corpus callosum. Additional confusion is caused by the fact that some individuals with GCPS have large deletions that generate a contiguous gene syndrome with intellectual disability and several individuals with a *GLI3* point variant have a phenotype indistinguishable from acrocallosal syndrome (Elson et al. 2002; Speksnijder et al. 2013).

MANIFESTATIONS AND MANAGEMENT

Management of individuals with PHS and GCPS differ sufficiently that they will be discussed separately. Management of GCPS follows the section on PHS.

Pallister–Hall Syndrome

Growth and Feeding

Growth and feeding do not seem to be significant issues for individuals with PHS unless significant hypopituitarism is present (see Endocrine). Airway malformations can cause aspiration and imperforate anus can affect bowel function. These are discussed below.

Evaluation

- Growth parameters should be assessed at each visit.

Management

- If poor growth is documented, evaluation for pituitary dysfunction is warranted as it can be part of the syndrome and can manifest as reduced growth.

Development and Behavior

Intellectual disability and developmental delay are uncommon in PHS and there is no stereotypic pattern of delay or personality profile (Azzam et al. 2005). There is no stereotypic behavior pattern of Greig cephalopolysyndactylyly contiguous gene syndrome cognitive impairment.

Evaluation

- Individuals who manifest developmental delay or behavioral problems should be evaluated comprehensively in a standard manner.

Management

- If abnormalities of development or behavior are identified, they are managed in a standard way.

Neurologic

A hypothalamic hamartoma can be a startling finding when first encountered by a clinician unfamiliar with this lesion. It is critical to understand that the hamartoma is a malformation and not a tumor. This lesion can occupy a substantial proportion of the cranial cavity of a person with PHS and may be larger than their brainstem. Hypothalamic hamartomas consist of abnormal foci of nodular clusters of neurons that vary in size and abundance with poorly defined boundaries that are hypothesized to be the functional unit of epileptogenesis (Kerrigan et al. 2017). Some infants with PHS have hamartomas as large as 4 cm. The hamartomas sit in the floor of the third ventricle just posterior to the optic chiasm, which is often abnormally shaped because of the presence of the hamartoma. However, this anatomical distortion has not been reported to be associated with symptoms. They can splay the lateral hypothalamus and distort the pituitary gland. Growth of the hamartoma does not appear to be more rapid than growth of the rest of the brain, and in most, growth is substantially slower so that the hamartoma maintains occupancy of a consistent or declining proportion of the cranial volume. Approximately 15% of individuals with PHS have seizures. These may be caused by the hamartoma or they may only be associated and instead caused by neuronal migration abnormalities. Although a variety of seizures have been associated, the most distinctive is gelastic epilepsy, which is a partial complex seizure that includes clonic spasms of the chest and diaphragm that is most similar to laughter. As the seizure is not associated with a humorous provocation, some individuals may be suspected to be manifesting mental illness. Some seizure foci may be sufficiently deep within the central nervous system to elude detection by a standard electroencephalogram.

Evaluation

- The only appropriate tool for postnatal imaging of the hypothalamic hamartoma is magnetic resonance imaging (Biesecker et al. 1996). The hallmark of a hypothalamic hamartoma is that it is isointense to gray matter on all sequences. Although hypothalamic hamartomas do not enhance with contrast, there can be differences in the signal intensity on FLARE pulse sequences. Computerized tomography and ultrasound are not useful modalities for imaging hamartomas. Extraordinarily large hamartomas can be imaged prenatally by ultrasound, but these are not associated with PHS or *GLI3* variants (Guimiot et al. 2009).
- Electroencephalogram is the appropriate diagnostic tool to evaluate seizures, although many individuals have seizures that are sufficiently infrequent to necessitate 24 hour video electroencephalography monitoring.
- Note that the existence of rare patients with mosaic *GLI3* variants in hypothalamic hamartomas should not lead to biopsy of such lesions in search of such a variant. It is preferable to treat such individuals with a clinical diagnosis rather than undertake the risks of the procedure, unless it is otherwise indicated (see below).

Management

- The seizures associated with PHS are commonly responsive to valproic acid, levetiracetam, or gabapentin, but individualized treatment optimization is often required. Some individuals may be treated on the basis of symptoms in the absence of abnormal electroencephalogram results.
- Only under the most unusual circumstances should a hypothalamic hamartoma be removed or even biopsied in persons who have findings consistent with PHS. Hamartoma removal carries the morbidity of major neurosurgery and postoperative lack of a functioning hypothalamic-pituitary axis may require life-long hormone supplements (see Endocrine).

Endocrine

As already mentioned, disruption of the hypothalamic–pituitary axis occurs in a minority of individuals with PHS. Although adrenal insufficiency is uncommon (most often because of deficiency of pituitary adrenocorticotropic hormone [ACTH]), it is critical to be vigilant for this complication and to diagnose and treat it promptly. The most severely affected have panhypopituitarism, which is a medical emergency in the neonatal nursery. Some individuals have single hormone deficiencies, most commonly growth hormone, and some have presented with precocious puberty.

Evaluation

- Standard evaluations of the hypothalamic-anterior pituitary axis are indicated when the diagnosis is first suspected.
- Early morning cortisol levels are an appropriate first screen for ACTH deficiency and should be performed urgently.
- IGF-1 (insulin-like growth factor-1, somatomedin C) levels are an appropriate screen for growth hormone deficiency, but can be delayed until after infancy.
- Screening for follicle stimulating hormone and luteinizing hormone dysfunction can be effectively performed by routine physical examination for signs of precocious puberty.
- Although isolated thyroid hormone deficiency has not been seen in individuals with PHS, the testing is so

straightforward and the consequences of untreated thyroid deficiency so severe that this is recommended.
- More sophisticated provocative testing of the anterior pituitary axis is indicated for abnormal screening results.

Management

- Replacement therapy for anterior pituitary deficiencies in PHS is straightforward and follows standard endocrine protocols for non-syndromic individuals with the same endocrine deficiencies.
- Similarly, suppression of the axis for treatment of precocious puberty using gonadotropin releasing hormone (GnRH) analogs and standard protocols is highly effective.
- Injections of testosterone (intramuscular testosterone enanthate in oil 25–50 mg monthly for three months) are indicated for affected males with micropenis secondary to follicle stimulating hormone and luteinizing hormone deficiency in utero. This should only be undertaken by physicians experienced with this manifestation and treatment.

Gastrointestinal

The main gastrointestinal complications of PHS are imperforate anus, aganglionosis, and anal stenosis (Li et al. 2015).

Evaluation

- The anal region should be carefully evaluated when PHS is suspected.

Management

- Treatment is by standard surgical techniques.

Otolaryngology

One of the most urgent and life-threatening complications of PHS is the anterior laryngeal cleft. This lesion is uncommon in PHS, seen typically only in severely affected people with de novo variants, but it is a medical emergency. This lesion causes intractable aspiration and is thought to be a major contributor to the neonatal mortality of the disorder. The mild version of this lesion is the bifid epiglottis, which is typically asymptomatic (Ondrey et al. 2000).

Evaluation

- Visualization of the airway by endoscopic techniques is the preferred method of diagnosis. Bifid epiglottis is commonly missed when affected individuals are intubated for surgery so the fact that an anesthesiologist has not noted the finding is an inadequate screen.

Management

- Although some reports have suggested that bifid epiglottis is associated with symptoms of aspiration and abnormal phonation, a series of 26 people with PHS had no symptoms referable to bifid epiglottis (Ondrey et al. 2000).
- Anterior laryngeal clefts must be managed at the earliest opportunity by the placement of an endotracheal tube to protect the airway and establishment of a proper route for alimental nutrition. Repair of such lesions is complex and should be undertaken by those with special skills and in a standard manner.

Genitourinary

A minority of individuals with PHS has been found to have structural renal lesions including dysplastic kidneys and abnormal collecting systems. Renal failure may be a rare complication. A number of affected individuals have been described with urogenital anomalies (Narumi et al. 2010).

Evaluation

- Renal ultrasound should be performed at the time of diagnosis.

Management

- Standard techniques of treatment are used when anomalies are detected.

Musculoskeletal

The limb malformations of PHS are a common or possibly universal manifestation of the disorder. There is a wide range of limb malformations from small postaxial nubbins to fully formed supernumerary digits and mesoaxial polydactyly. Occasional affected individuals have seven or more digits on a limb or complex osseous malformations with forked metacarpals or articulation of two digits on a single metacarpal.

Evaluation

- Evaluation of the limb anomalies is by standard radiographic techniques, but new modalities such as three-dimensional computed tomography imaging reconstruction may have a useful role.

Management

- Treatment of hand abnormalities can be trivial or very complex. The small postaxial nubbins can be managed

acutely by sutural ligation if there are no bony or cartilaginous elements within them. The lump or scar can be revised later on an elective basis.
- Reduction of fully formed postaxial digits with independent metacarpals can be accomplished with good functional and cosmetic outcome. Some individuals with mesoaxial polydactyly have distorted digital architecture and the correction of these anomalies is quite challenging.
- A different set of considerations is operative for foot anomalies where biomechanical complications of surgery can be problematic and cosmetic considerations are lessened. The general approach is to err on the side of functional outcome and to avoid surgery that has a substantial chance of causing biomechanical complications.

Dental

Some persons with PHS have abnormal peg-shaped hypoplastic teeth. There are little data on other dental complications.

Evaluation

- Dental evaluation should be carried out if abnormalities in tooth shape or number are noted.

Management

- Management is symptomatic and standard.

Respiratory

Abnormal pulmonary segmentation is a known manifestation of PHS in humans and in animal models. However, there are no known clinical manifestations of this malformation.

Ophthalmology

There are no known ophthalmologic complications of PHS. In spite of the fact that large hypothalamic hamartomas can compress or distort the optic chiasm, nerves, or tracts, no visual field deficits have been demonstrated. The primary purpose of ophthalmological evaluation is to exclude other overlapping disorders, especially Bardet–Biedl syndrome and its associated pigmentary retinopathy.

Evaluation

- A full ophthalmology examination should be carried out at diagnosis.

Management

- If an ophthalmologic anomaly is detected it should be treated in a standard manner.

Greig Cephalopolysyndactyly Syndrome

Diagnosis and management of GCPS is limited to the developmental, neurological, and musculoskeletal systems. Other systems are not known to be affected.

Development and Behavior

GCPS may be associated with developmental delay, and preliminary data suggest that speech and language delay may be more common in children with large deletions of *GLI3* (see above discussion). Overall, the frequency of developmental delay is estimated to be in the order of 10%, but it is much higher for individuals with large deletions. The majority of those with deletions greater than 1 Mb have developmental delay or cognitive disability (Johnston et al. 2003; Johnston et al. 2007).

Evaluation

- It is prudent to perform developmental screening regularly and institute early referral for speech and language evaluation when deficiencies are noted.

Management

- Standard management approaches are appropriate.

Neurologic

Most individuals with GCPS have a normal central nervous system. Structural anomalies are uncommon and functional abnormalities are present in a fraction of affected individuals, though the two are not coincident. More severely affected persons have been reported with hypoplasia or agenesis of the corpus callosum, and occasionally, with hydrocephalus or other malformations (Elson et al. 2002; Speksnijder et al. 2013). A few patients with GCPS have been reported to have seizures of a diverse nature.

Evaluation

- Because macrocephaly without hydrocephalus is common in GCPS, it is important to monitor head circumference to screen for rapidly increasing head size that may indicate hydrocephalus.
- Computed tomography or magnetic resonance imaging should be undertaken when rapidly increasing head size or neurological symptoms such as developmental delay, focal neurological findings, or seizures are noted.
- If seizures are suspected, a thorough neurological evaluation and electroencephalogram are recommended.

Management

- Hydrocephalus should be treated by standard approaches (e.g., ventriculoperitoneal shunting).
- Seizures should be managed according to standard clinical protocols.

Musculoskeletal

The spectrum of limb defects in GCPS is partially overlapping with PHS in that both disorders can manifest postaxial polydactyly. Although the classic descriptions of GCPS list preaxial duplications, postaxial polydactyly of the upper limbs is probably more common than preaxial. In some cases, the syndactyly can be severe, including a mitten hand malformation, not unlike that seen in Apert syndrome (see Chapter 17). Several individuals with metopic craniosynostosis and GCPS have been described (Hurst et al. 2011).

Evaluation

- Evaluation of the limb anomalies or synostosis is by standard radiographic techniques, but new modalities such as three-dimensional computed tomography imaging reconstruction may have a useful role.

Management

- The management of preaxial duplications of the upper limb is critical because of the importance of that digit for hand function. Management regimes are standard.
- Duplications of the toes are managed conservatively, using surgical approaches for situations where the duplications cause significant biomechanical or shoe-fitting problems and limiting treatment of abnormalities where benefits are small and biomechanical complications may be significant.
- The management of craniosynostosis should be performed according to the involved sutures, using standard techniques.

ACKNOWLEDGMENTS

The author is grateful for the research participation of many families, as our observations of their manifestations has provided most of the information in this chapter. The opinions expressed in this chapter are those of the author and are not to be construed as official recommendations by the Department of Health and Human Services, the National Institutes of Health, or any other institution to which he is affiliated.

REFERENCES

Al-Qattan MM, Shamseldin HE, Salih MA, Alkuraya FS (2017) GLI3-related polydactyly: a review. *Clin Genet* 92:457–466

Azzam A, Lerner DM, Peters KF, Wiggs E, Rosenstein DL, Biesecker LG (2005) Psychiatric and neuropsychological characterization of Pallister-Hall syndrome. *Clin Genet* 67:87–92.

Balk K, Biesecker LG (2008) The clinical atlas of Greig cephalopolysyndactyly syndrome. *Am J Med Genet A* 146:548–557.

Biesecker LG, Aase JM, Clericuzio C, Gurrieri F, Temple IK, Toriello H (2009) Elements of morphology: standard terminology for the hands and feet. *Am J Med Genet A* 149A:93–127.

Biesecker LG, Abbott M, Allen J, Clericuzio C, Feuillan P, Graham JM, Jr, Hall J, Kang S, Haskins Olney A, Lefton D, Neri G, Peters K, Verloes A (1996) Report from the workshop on Pallister-Hall syndrome and related disorders. *Am J Med Genet* 65:76–81.

Biesecker LG, Graham JM, Jr (1996) Syndrome of the month: Pallister-Hall syndrome. *J Med Genet* 33:585–589.

Biesecker LG, Green RC (2014) Diagnostic clinical genome and exome sequencing. *N Engl J Med* 370:2418–2425.

Debeer P, Peeters H, Driess S, De Smet L, Freese K, Matthijs G, Bornholdt D, Devriendt K, Grzeschik KH, Fryns JP, Kalff-Suske M (2003) Variable phenotype in Greig cephalopolysyndactyly syndrome: clinical and radiological findings in 4 independent families and 3 sporadic cases with identified GLI3 mutations. *Am J Med Genet A* 120A:49–58.

Demurger F, Ichkou A, Mougou-Zerelli S, Le Merrer M, Goudefroye G, Delezoide AL, Quelin C, Manouvrier S, Baujat G, Fradin M, Pasquier L, Megarbane A, Faivre L, Baumann C, Nampoothiri S, Roume J, Isidor B, Lacombe D, Delrue MA, Mercier S, Philip N, Schaefer E, Holder M, Krause A, Laffargue F, Sinico M, Amram D, Andre G, Liquier A, Rossi M, Amiel J, Giuliano F, Boute O, Dieux-Coeslier A, Jacquemont ML, Afenjar A, Van Maldergem L, Lackmy-Port-Lis M, Vincent-Delorme C, Chauvet ML, Cormier-Daire V, Devisme L, Genevieve D, Munnich A, Viot G, Raoul O, Romana S, Gonzales M, Encha-Razavi F, Odent S, Vekemans M, Attie-Bitach T (2015) New insights into genotype-phenotype correlation for GLI3 mutations. *Eur J Hum Genet* 23:92–102.

Elson E, Perveen R, Donnai D, Wall S, Black GC (2002) De novo GLI3 mutation in acrocallosal syndrome: broadening the phenotypic spectrum of GLI3 defects and overlap with murine models. *J Med Genet* 39:804–806.

Guimiot F, Marcorelles P, Aboura A, Bonyhay G, Patrier S, Menez F, Drouin-Garraud V, Icowick V, Eurin D, Garel C, Moirot H, Verspyck E, Saugier-Veber P, Attie-Bitach T, Picone O, Oury JF, Verloes A, Delezoide AL, Laquerriere A (2009) Giant diencephalic harmartoma and related anomalies: a newly recognized entity distinct from the Pallister-Hall syndrome. *Am J Med Genet A* 149A:1108–1115.

Hall JG (2016) The early history of Pallister-Hall syndrome-Buried treasure of a sort. *Gene* 589:100–103.

Hildebrand MS, Griffin NG, Damiano JA, Cops EJ, Burgess R, Ozturk E, Jones NC, Leventer RJ, Freeman JL, Harvey AS, Sadleir LG, Scheffer IE, Major H, Darbro BW, Allen AS, Goldstein DB, Kerrigan JF, Berkovic SF, Heinzen EL (2016) Mutations of the Sonic Hedgehog Pathway Underlie Hypothalamic Hamartoma with Gelastic Epilepsy. Am J Hum Genet 99:423–429.

Hui C.-C, Joyner A (1993) A mouse model of Greig cephalopolysyndactyly syndrome: the extra-toes^J mutation contains an intragenic deletion of the Gli3 gene. Nat Genet 3:241–246.

Hurst JA, Jenkins D, Vasudevan PC, Kirchhoff M, Skovby F, Rieubland C, Gallati S, Rittinger O, Kroisel PM, Johnson D, Biesecker LG, Wilkie AO (2011) Metopic and sagittal synostosis in Greig cephalopolysyndactyly syndrome: five cases with intragenic mutations or complete deletions of GLI3. Eur J Hum Genet 19:757–762.

Johnston J, Walker R, Davis S, Facio F, Turner J, Bick D, Daentl D, Ellison J, Meltzer P, Biesecker L (2007) Zoom-in comparative genomic hybridisation arrays for the characterisation of variable breakpoint contiguous gene syndromes. J Med Genet 44:e59.

Johnston JJ, Olivos-Glander I, Killoran C, Elson E, Turner JT, Peters KF, Abbott MH, Aughton DJ, Aylsworth AS, Bamshad MJ, Booth C, Curry CJ, David A, Dinulos MB, Flannery DB, Fox MA, Graham JM, Grange DK, Guttmacher AE, Hannibal MC, Henn W, Hennekam RC, Holmes LB, Hoyme HE, Leppig KA, Lin AE, Macleod P, Manchester DK, Marcelis C, Mazzanti L, McCann E, McDonald MT, Mendelsohn NJ, Moeschler JB, Moghaddam B, Neri G, Newbury-Ecob R, Pagon RA, Phillips JA, Sadler LS, Stoler JM, Tilstra D, Walsh Vockley CM, Zackai EH, Zadeh TM, Brueton L, Black GC, Biesecker LG (2005) Molecular and clinical analyses of Greig cephalopolysyndactyly and Pallister-Hall syndromes: robust phenotype prediction from the type and position of GLI3 mutations. Am J Hum Genet 76:609–622.

Johnston JJ, Olivos-Glander I, Turner J, Aleck K, Bird LM, Mehta L, Schimke RN, Heilstedt H, Spence JE, Blancato J, Biesecker LG (2003) Clinical and molecular delineation of the Greig cephalopolysyndactyly contiguous gene deletion syndrome and its distinction from acrocallosal syndrome. Am J Med Genet A 123A:236–242.

Johnston JJ, Sapp JC, Turner JT, Amor D, Aftimos S, Aleck KA, Bocian M, Bodurtha JN, Cox GF, Curry CJ, Day R, Donnai D, Field M, Fujiwara I, Gabbett M, Gal M, Graham JM, Hedera P, Hennekam RC, Hersh JH, Hopkin RJ, Kayserili H, Kidd AM, Kimonis V, Lin AE, Lynch SA, Maisenbacher M, Mansour S, McGaughran J, Mehta L, Murphy H, Raygada M, Robin NH, Rope AF, Rosenbaum KN, Schaefer GB, Shealy A, Smith W, Soller M, Sommer A, Stalker HJ, Steiner B, Stephan MJ, Tilstra D, Tomkins S, Trapane P, Tsai AC, Van Allen MI, Vasudevan PC, Zabel B, Zunich J, Black GC, Biesecker LG (2010) Molecular analysis expands the spectrum of phenotypes associated with GLI3 mutations. Hum Mutat 31:1142–1154.

Kalff-Suske M, Wild A, Topp J, Wessling M, Jacobsen EM, Bornholdt D, Engel H, Heuer H, Aalfs CM, Ausems MG, Barone R, Herzog A, Heutink P, Homfray T, Gillessen-Kaesbach G, Konig R, Kunze J, Meinecke P, Muller D, Rizzo R, Strenge S, Superti-Furga A, Grzeschik KH (1999) Point mutations throughout the GLI3 gene cause Greig cephalopolysyndactyly syndrome. Hum Mol Genet 8:1769–1777.

Kang S, Graham JM, Jr, Olney AH, Biesecker LG (1997) GLI3 frameshift mutations cause autosomal dominant Pallister-Hall syndrome. Nat Genet 15:266–268.

Kerrigan JF, Parsons A, Tsang C, Simeone K, Coons S, Wu J (2017) Hypothalamic hamartoma: Neuropathology and epileptogenesis. Epilepsia 58(Suppl 2):22–31.

Killoran CE, Abbott M, McKusick VA, Biesecker LG (2000) Overlap of PIV syndrome, VACTERL and Pallister-Hall syndrome: clinical and molecular analysis. Clin Genet 58:28–30.

Li MH, Eberhard M, Mudd P, Javia L, Zimmerman R, Khalek N, Zackai EH (2015) Total colonic aganglionosis and imperforate anus in a severely affected infant with Pallister-Hall syndrome. Am J Med Genet A 167A:617–620.

Muenke M, Ruchelli ED, Rorke LB, McDonald-McGinn DM, Orlow MK, Isaacs A, Craparo FJ, Dunn LK, Zackai EH (1991) On lumping and splitting: A fetus with clinical findings of the oral-facial-digital syndrome type VI, the hydrolethalus syndrome, and the Pallister-Hall syndrome. Am J Med Genet 41:548–556.

Narumi Y, Kosho T, Tsuruta G, Shiohara M, Shimazaki E, Mori T, Shimizu A, Igawa Y, Nishizawa S, Takagi K, Kawamura R, Wakui K, Fukushima Y (2010) Genital abnormalities in Pallister-Hall syndrome: Report of two patients and review of the literature. Am J Med Genet A 152A:3143–3147.

Ng D, Johnston JJ, Turner JT, Boudreau EA, Wiggs EA, Theodore WH, Biesecker LG (2004) Gonadal mosaicism in severe Pallister-Hall syndrome. Am J Med Genet A 124A:296–302.

Ondrey F, Griffith A, Van Waes C, Rudy S, Peters K, McCullagh L, Biesecker LG (2000) Asymptomatic laryngeal malformations are common in patients with Pallister-Hall syndrome. Am J Med Genet 94:64–67.

Putoux A, Thomas S, Coene KL, Davis EE, Alanay Y, Ogur G, Uz E, Buzas D, Gomes C, Patrier S, Bennett CL, Elkhartoufi N, Frison MH, Rigonnot L, Joye N, Pruvost S, Utine GE, Boduroglu K, Nitschke P, Fertitta L, Thauvin-Robinet C, Munnich A, Cormier-Daire V, Hennekam R, Colin E, Akarsu NA, Bole-Feysot C, Cagnard N, Schmitt A, Goudin N, Lyonnet S, Encha-Razavi F, Siffroi JP, Winey M, Katsanis N, Gonzales M, Vekemans M, Beales PL, Attie-Bitach T (2011) KIF7 mutations cause fetal hydrolethalus and acrocallosal syndromes. Nat Genet 43:601–606.

Saitsu H, Sonoda M, Higashijima T, Shirozu H, Masuda H, Tohyama J, Kato M, Nakashima M, Tsurusaki Y, Mizuguchi T, Miyatake S, Miyake N, Kameyama S, Matsumoto N (2016) Somatic mutations in GLI3 and OFD1 involved in sonic hedgehog signaling cause hypothalamic hamartoma. Ann Clin Transl Neurol 3:356–365.

Speksnijder L, Cohen-Overbeek TE, Knapen MF, Lunshof SM, Hoogeboom AJ, van den Ouwenland AM, de Coo IF, Lequin MH, Bolz HJ, Bergmann C, Biesecker LG, Willems PJ, Wessels

MW (2013) A de novo GLI3 mutation in a patient with acrocallosal syndrome. *Am J Med Genet A* 161A:1394–1400.

Vortkamp A, Franz T, Gessler M, Grzeschik K-H (1992) Deletion of *GLI3* supports the homology of the human Greig cephalopolysyndactyly syndrome (GCPS) and the mouse mutant extra toes (*Xt*). *Mamm Genome* 3:461–463.

Vortkamp A, Gessler M, Grzeschik K-H (1991) *GLI3* zinc finger gene interrupted by translocations in Greig syndrome families. *Nature* 352:539–540.

Wild A, Kalff-Suske M, Vortkamp A, Bornholdt D, König R, Grzeschik K.-H (1997) Point mutations in human *GLI3* cause Greig syndrome. *Hum Mol Genet* 6:1979–1984.

45

PALLISTER–KILLIAN SYNDROME

Emanuela Salzano
Division of Human Genetics and Genomics, The Children's Hospital of Philadelphia, Pennsylvania, USA

Sarah E. Raible
Division of Human Genetics and Genomics, The Children's Hospital of Philadelphia, Pennsylvania, USA

Ian D. Krantz
The Perelman School of Medicine at the University of Pennsylvania and Division of Human Genetics and Genomics, The Children's Hospital of Philadelphia, Pennsylvania, USA

INTRODUCTION

Pallister–Killian syndrome (PKS) (OMIM#601803) is a sporadic multisystem developmental disorder caused, most typically, by mosaic tetrasomy of the short arm of chromosome 12 (12p), which is due to the presence of a supernumerary marker chromosome made up of mirrored copies 12p (isochromosome 12p). The clinical manifestations of PKS include characteristic craniofacial dysmorphism, skin pigmentation anomalies, and neurodevelopmental delay/intellectual disability, associated in variable ways with epilepsy, congenital heart defects (CHD), congenital diaphragmatic hernias (CDH), ophthalmologic problems, hearing loss, limb anomalies, gastrointestinal malformations, abdominal wall defects, and a specific growth profile (macrosomia at birth with a growth deceleration after the first year of life). This condition was first described by Dr. Pallister in two institutionalized adult individuals with similar physical features including profound intellectual disability, severe epilepsy, "coarse" facial features, kyphoscoliosis, flaccidity and spasticity, and cataracts (Pallister et al. 1976). Although Dr. Pallister attributed this phenotype to the presence of a supernumerary marker chromosome, the extra chromosome was initially described as an isochromosome of an F group (20) chromosome (Pallister et al. 1976). Subsequently, in 1977, the supernumerary chromosome was correctly identified as isochromosome 12p (Pallister et al. 1977). In 1981, an additional patient with similar phenotypic features was described by Teschler-Nicola and Killian resulting in the most commonly used term for this diagnosis: PKS. This syndrome is also occasionally referred to as Pallister mosaic syndrome, isochromosome 12p syndrome, Killian syndrome, Killian/Teschler-Nicola syndrome, Teschler-Nicola/Killian syndrome, and tetrasomy 12p. Rarely, trisomy 12p and mosaic hexasomy 12p with two isochromosomes have been described and referred to PKS-like phenotype (Choo et al. 2002; Vogel et al. 2009).

Incidence

A prevalence of 1/20,000 has been estimated based on analysis of liveborn infants resulting from pregnancies with prenatally identified marker chromosome (Brøndum-Nielsen and Mikkelsen 1995; Bartsch et al. 2005). Recently, a population-based study in Great Britain reported a population prevalence of 0.6 per million and a birth incidence of 5.1 per million live births in 2005–2009 (Blyth et al. 2015). However, it is very likely that PKS is under-diagnosed due to the difficulty of making a cytogenetic diagnosis from peripheral blood. To date, more than 300 individuals have been reported,

Cassidy and Allanson's Management of Genetic Syndromes, Fourth Edition.
Edited by John C. Carey, Agatino Battaglia, David Viskochil, and Suzanne B. Cassidy.
© 2021 John Wiley & Sons, Inc. Published 2021 by John Wiley & Sons, Inc.

and all have been sporadic. The life expectancy in PKS has not been formally evaluated. There are individuals with PKS in their 40s and 50s (unpublished data), and it is probable that there are also many older PKS individuals who have never been diagnosed, since many adults with intellectual disability have not been formally evaluated by clinical geneticists (Izumi and Krantz 2014). Perinatal deaths are usually due to extreme prematurity or major malformations such as congenital diaphragmatic hernia and heart defects. Unexpected deaths due to epilepsy (20 years old), aspiration of gastric contents (38 years old) and respiratory infections (at least three patients at different ages ranging from 4 years old to 38 years old) are also reported (Blyth et al. 2015).

Diagnostic Criteria

A hallmark of this syndrome is the tissue-limited mosaicism for the isochromosome 12p that, in conjunction with variable clinical expressivity, can complicate recognition and establishment of a diagnosis of PKS (Brøndum-Nielsen and Mikkelsen 1995; Bartsch et al. 2005). The manifestations of PKS can potentially involve any of the body organ systems, and it is likely that the entire phenotypic spectrum has yet to be full characterized. However, the combination of the typical coarse facial profile, skin pigmentation anomalies with intellectual disability and/or neurodevelopmental delay, hypotonia, seizures, hearing loss, and visual impairment are distinctive of PKS. The facial gestalt in PKS is very characteristic and one of the most useful diagnostic aids. Most frequently reported facial findings are frontal bossing, prominent forehead, characteristic frontoparietal pattern of alopecia/sparse hair, sparse eyebrows, upslanting palpebral fissures, hypertelorism, telecanthus, depressed nasal bridge, small nose with anteverted nares (flat facial profile), long philtrum, prominent cupid bow pattern of the upper lip, micrognathia, large mandible, ear pits, thickened ear helices, posterior rotated ears, low-set ears, ear tags, cleft or high palate, bifid uvula and short neck (Figures 45.1G, H, and M–P). Extension of the philtral skin into the vermilion border of the upper lip (termed the "Pallister lip") is seen in 100% of the probands with PKS (Figure 45.1N) (Wilkens et al. 2012). Epicanthal folds and mild proptosis are less frequently described in PKS patients (Wilkens et al. 2012; Blyth et al. 2015). Large and delayed closure of fontanelles is also reported (Mathieu et al. 1997; Blyth et al. 2015). The facial features tend to coarsen over time. Micrognatia is mainly noted in younger children and progresses to prognathia in adults, while the typical pattern of scalp alopecia seen during infancy tends to resolve by older childhood and adolescence making recognition of the diagnosis more difficult in older individuals (Figures 45.1 A–F, I, and J) (Wilkens et al. 2012; Blyth et al. 2015). Other common abnormalities are those of the skin, extremities, sacrum, and anus. Skin pigmentation differences (hypopigmentation/hyperpigmentation) indicative of the mosaic chromosomal abnormality are found in about 80% of individuals with PKS (Figure 45.1Q) (Wilkens et al. 2012; Izumi and Krantz, 2014). A propensity of whorls on dermatoglyphics (of which ~50% have >5 whorls on exam) is reported (Wilkens et al. 2012) and bilateral or unilateral single palmar creases are described in respectively ~30% and 19% of patients (Blyth et al. 2015). Supernumerary/accessory nipples are also frequently noted in PKS patients (~42% of the probands) (Figure 45.1R) (Wilkens et al. 2012). Most commonly occurring limb differences are lymphedema, increased soft tissue of the extremities, broad thumbs and first toes, brachydactyly, polydactyly, and small feet (Figures 45.1U–W) (Wilkens et al. 2012).

Limb shortening, either rhizomelia or proportionate micromelia, is reported in ~20% of individuals and clinodactyly of the fifth finger in 19% of cases (Blyth et al. 2015). Occasionally limited extension of the elbows is also noted. Anteriorly placed anus and a sacral dimple is seen in 25% of individuals (Figure 45.1T). Additional findings on physical examination include abdominal wall defects, especially umbilical hernia found in ~15 % of individuals and inguinal hernia (~13%) (Figure 45.1S) (Wilkens et al. 2012; Blyth et al. 2015). Most reported individuals manifest severe to profound intellectual disability/neurodevelopmental delay, with absent speech in 73% of patients of those older than 18 months and inability to walk in about 65% of those over 4 years old (Blyth et al. 2015; Wilken et al. 2012). However, there have been an increasing number of mildly affected individuals with PKS reported (Kostanecka et al. 2012) who may have fewer and atypical manifestations, possibly dependent on the degree of mosaicism among different tissues/organs.

Seizures and structural brain malformations are often seen in PKS. Epilepsy occurs in 53% of probands with PKS and various minor anomalies are reported in up to 60% of patients (Candee et al. 2012). Clinically, most PKS patients have hypotonia during infancy, although variable spasticity and hypertonia can be seen in older individuals (Wilkens et al. 2012; Izumi and Krantz 2014). Hearing impairment and visual difficulties are common. The majority of PKS individuals present with bilateral hearing loss with a similar distribution among sensorineural, conductive, and mixed hearing loss, often requiring hearing aids. Visual anomalies are almost equally common and mainly represented by myopia, hypermetropia, astigmatism and/or strabismus (Wilkens et al. 2012; Blyth et al. 2015). Additionally, almost all individuals with PKS show delayed dentition and a typical occlusive phenotype due to the high palate with mandibular prognathism.

In terms of structural anomalies, congenital diaphragmatic hernia (CDH) is one of the hallmark features that often leads clinicians to entertain a diagnosis of PKS. However, it is likely that CDH frequency has been overestimated from PKS literature data. Looking carefully at published studies,

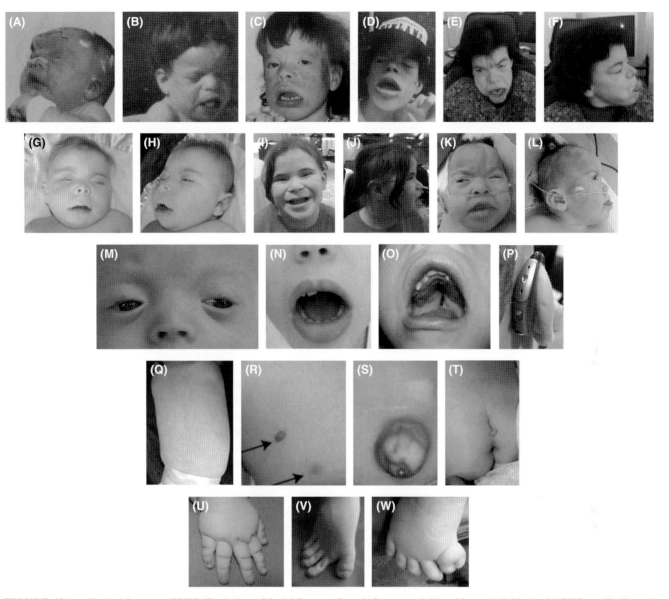

FIGURE 45.1 Clinical features of PKS. Evolution of facial features from infancy to adulthood in an individual with PKS (A–F). Typical facial appearance in two unrelated toddlers with PKS (G, H, K and L). Female individual with mild facial phenotype (I and J). Note fronto-parietal sparseness of hair that tends to improve with age (A–F), telecanthus (M), hypotonic facies, prominent cheeks, eversion of lower lip and invasion of vermilion of upper lip by philtral skin (N), cleft palate (O), posterior ear pits (P), skin hypo or hyper-pigmentation (most commonly) lesions that may follow the Blaschko's lines (Q), supernumerary nipple (R), umbilical hernia (S), sacral crease/appendage (T) and hand and foot features including lymphedema/increased soft tissue (U–W).

there is a discrepancy in reported frequency of CDH (from 8.5% up to 35%); however, ascertainment biases exist in many studies as this is often a lethal condition and will not be seen frequently in individuals with PKS beyond the prenatal and neonatal periods (Doray et al. 2002; Wilkens et al. 2012; Desseauve et al. 2016).

As expected, a higher frequency of CDH and complex congenital heart diseases are reported in prenatal studies where information mostly derives from spontaneous/voluntarily terminated pregnancies, following the discovery of life-threatening malformations such as CDH, while more recently lower rates come from cohorts of children with PKS that have survived the neonatal period. Overall, gastrointestinal involvement is found in 52% with intestinal malrotation being a common malformation together with CDH reported in 12% and 11% of probands, respectively. Functional gastrointestinal manifestations are also frequently noted and include feeding difficulty, dysphagia, constipation, and gastroesophageal reflux disease (Wilkens et al. 2012; Izumi and Krantz 2014).

Cardiac involvement occurs in 40% of PKS individuals and mainly includes structural defects such as atrial or ventricular septal defects, bicuspid aortic valve, aortic dilatation, PDA and PFO (Tilton et al. 2014). The most commonly identified genitourinary manifestation is cryptorchidism. Other manifestations identified in PKS include hypospadias, small genitalia, and hydrocele (Wilkens et al. 2012; Izumi and Krantz 2014).

Etiology, Pathogenesis, and Genetics

PKS is caused by the presence of extra copies of the short arm of chromosome 12, which most commonly are present as a supernumerary marker isochromosome 12p. This isochromosome 12p is present in a tissue-limited mosaic pattern with highly variable levels of mosaicism. Mosaicism is defined as the presence of two or more populations of cells with different chromosome patterns in an individual who has developed from a single fertilized egg. In PKS patients, there are at least two cell populations, one karyotypically normal and another one containing the supernumerary isochromosome 12p. All reported cases of PKS due to an isochromosome 12p are mosaic, and there has never been a report of non-mosaic tetrasomy 12p, which is likely embryonic lethal.

The mechanism leading to the isochromsome 12p formation remains to be determined; however, several theories have been proposed including: (1) meiosis I or II nondisjunction events generating a disomic gamete that results, upon conception in a trisomic zygote whereby isochromosome 12p formation occurs postzygotically from one of the chromosome 12s (with loss of the 12q); (2) isochromosome formation associated with nondisjunction at meiosis I resulting in a gamete with both a normal chromosome 12 as well as an isochromosome 12p (Van Dyke et al. 1987); (3) normal gametes and zygote, with post-zygotic mitotic nondisjunction and isochromosome formation; (4) premeiotic mitotic centromeric misdivision with nondisjunction at meiosis I or centromeric misdivision at either meiosis I or II (Rivera et al. 1986; Struthers et al. 1999). The majority of previously published studies, which used low-resolution microsatellite markers to determine the number of alleles present on the i(12p), suggest maternal meiosis II nondisjunction as a common mechanism of mosaic tetrasomy 12p, although rare paternal nondisjunction has also been reported (Los et al. 1995; Turleau et al. 1996; Cormier-Daire et al. 1997; Schubert et al. 1997: Conlin et al. 2012). There is a maternal age effect observed in PKS similar to other autosomal aneuploidy syndromes such as Down syndrome (Wilkens et al. 2012). Isochromosome 12p is most often demonstrated to be of maternal origin, similar to other autosomal aneuploidies (Struthers et al. 1999). The average maternal and paternal ages at conception is 31.7 and 34.9 years old, respectively (Wilkens et al. 2012). According to this proposed theory, the isochromosome 12p would be present in a sperm or ovum, which forms a PKS zygote (Izumi and Krantz 2014). There has been one report of a post-zygotic origin of the isochromosome 12p subsequent to a trisomic rescue due to a concurrent trisomy 12 and i(12p)/uniparental disomy 12 (de Ravel et al. 2004).

The mosaic ratio of the isochromosome 12p population varies among the different tissues/organs, even within the same proband. Several studies have demonstrated the absence of a correlation between the mosaic ratio and clinical severity (Tilton et al. 2014; Wilkens et al. 2012). One possibility explaining this absence of correlation is the difficulty of extrapolating the mosaic ratio in various organs, whose abnormality directly leads to the manifestation. There has been no report examining mosaic ratios in the brain tissue of variably affected PKS individuals. Analysis of the percentage of the i(12p) in blood with proband age has demonstrated a trend of decreasing frequency of the i(12p) with increasing age, demonstrating a rapid decline in blood cells carrying the i(12p) after birth. However, no age-dependent trend was observed in fibroblast samples (Conlin et al. 2012).

These findings concur with previous suggestions of a differential growth advantage between the karyotypically normal cells and the i(12p) cell population (Tang and Wenger 2005). In this model, the selective growth advantage of the karyotypically normal cell population replaces the stem cells in the bone marrow over time, making it more difficult to detect the decreasing number of i(12p) cells in the peripheral circulation (Conlin et al. 2012).

Marker chromosomes constituted of partial tetrasomy12p (Dufke et al. 2001; Vermeesch et al. 2005; Huang et al. 2007), chromosomal duplication of 12p (Izumi et al. 2012), and, more rare mosaic hexasomy 12p secondary to the presence of two supernumerary isochromosomes (Vogel et al. 2009), are also known to result in a PKS-like phenotype, suggesting the presence of dosage-sensitive genes on 12p that are critical for the pathogenesis of PKS (see the discussion below) (Izumi et al. 2012).

Chromosome 12p is 34.3 Mb in length, and contains approximately 350 genes. Many biologically important genes exist on 12p including cancer-associated genes such as *KRAS* and *ING4*, and developmental genes such as *NANOG*, *CHD4*, and *SOX5* (Izumi et al. 2012). Based on the experience of a patient with PKS features who had two small de novo interstitial duplications of 12p, along with a review of previously reported cases due to microduplications of 12p13.31, a minimal critical region for PKS has been identified (Izumi et al. 2012). These two microduplications were located at 12p13.31 and included 26 genes, all of which represent potential candidate genes for the PKS phenotype. Among these 26 genes, a few genes have been proposed likely to be key regulators of downstream genes or gene networks important for the growth and differentiation of the tissues affected in this diagnosis. Three genes, *ING4*, *CHD4* and *ATN1* represent strong candidates, given their known function. *ING4* belongs to the family of *in*hibitor of *g*rowth (ING), which plays important roles in transcriptional regulation through associations with various binding partners including trimethylated histones and histone modifiers (Nozell et al. 2008; Hung et al. 2009). The overexpression of ING4 negatively regulates cell growth resulting in cell cycle arrest, and enhanced cell apoptosis (Zhang et al. 2004). CHD4 is a chromodomain helicase DNA binding protein, and constitutes a catalytic subunit of the nucleosome remodeling deacetylase (NuRD) transcriptional repressor complex that plays an important role in chromatin remodeling (Tong et al. 1998). Mutations in *CHD4* are now known to cause a multisystem developmental disorder (Weiss et al. 2016). Mutations of *ATN1*, a gene that is also contained in the 12p13.31 critical region, lead to the neurodegenerative disorder DRPLA as discussed above (Koide et al. 1994; Nagafuchi et al.1994; Kaur et al. 2014). Since the *ATN1* gene is already known to interact with CREBBP, dysregulation is suggested to potentially affect the *HOXB* genes (that are upregulated in the PKS probands) by modifying the activity of the CREBBP protein without affecting its mRNA level (Kaur et al. 2014).

Based on recent genome-wide expression array analysis, more than 300 genes are statistically differentially expressed in PKS skin fibroblasts, 180 of which are up- and 174 down-regulated (Kaur et al. 2014). The most statistically significant dysregulated genes on 12p are mapped to the 12p13.31 minimal critical region (Izumi et al. 2012; Kaur et al. 2014). The non-12p genes that are most significantly dysregulated include a large number of homeobox genes and other transcription factors critical in mammalian development. Several statistically significantly dysregulated genes in PKS cells have been suggested to potentially be directly related to the phenotype of PKS. For example, the most significantly down-regulated gene is *ZFPM2*, whose haploinsufficiency is known to cause congenital heart defects and congenital diaphragmatic hernia (Pizzuti et al. 2003; Ackerman et al. 2005; Kaur et al. 2014). *GATA6* has also been shown to be down-regulated in PKS patient samples.

Since mutations in *GATA6* are associated with various types of congenital heart disease, dysregulation of this gene may also be related to the pathogenesis of CHD in PKS (Kodo et al. 2009; Kaur et al. 2014). Another gene whose dysregulated expression may be related to the PKS phenotype is *SOX9*. Mutations in *SOX9* cause campomelic dysplasia characterized by limb shortening, micrognathia, and cleft palate/Robin sequence, and sex reversal, a diagnosis that has phenotypic overlap with PKS (Foster et al. 1994; Wagner et al. 1994). *IGFBP2* is the second most significantly up-regulated gene. IGFBPs are known to have a function in modulating the IGF signaling pathway, therefore, up-regulation of *IGFBP2* may be related to the unique growth phenotype seen in PKS (Hoeflich et al. 1999; Firth et al. 2002). IGFBP2 sequesters free IGF proteins, and functions to mitigate the IGF signaling pathway (Kaur et al. 2014). Mouse models of IGFBP2 overexpression demonstrate a postnatal growth deceleration phenotype, which is similar to that seen in individuals with PKS (Hoeflich et al. 1999). Extra copies of 12p likely cause dysregulated expression of melanogenesis-related genes, leading to the pigmentation abnormalities seen in PKS patients (Kaur et al. 2014). The involvement of microRNAs located on 12p have also been evaluated in relation to the transcriptomic abnormality in PKS (Izumi et al. 2014). Among the several 12p microRNAs, miR-1244 is significantly up-regulated in PKS. The molecular targets of miR-1244 include *MEIS2* whose haploinsufficiency causes congenital heart defects and cleft palate (Crowley et al. 2010). Therefore, it is possible that the overexpression of microRNAs located on 12p contributes to the pleiotropic phenotype of PKS (Izumi and Krantz 2012; Izumi et al. 2014).

Genetic Counseling

All reported cases of PKS caused by isochromosome 12p have been sporadic. Hence, empirically the recurrence risk is close to that of the general population. Although it has never been reported, theoretically, a recurrence risk could be possible if a parent had iso12p germline mosaicism. Since a PKS phenotype can be caused by chromosomal duplications inclusive of 12p, if a 12p duplication is identified and results from an unbalanced translocation, parental carriers of the balanced form should be excluded (Izumi et al. 2010). Therefore, cytogenetic discrimination between mosaic iso12p and 12p duplication is essential in genetic counseling (Izumi and Krantz 2014).

There have been many reported cases of prenatal detection of isochromosome 12p by chorionic villus sampling or amniocentesis (Doray et al. 2002). The percentage of mosaicism does not correlate with the severity of congenital anomalies in PKS fetuses (Libotte et al. 2016), and with a wide phenotypic spectrum in PKS, counseling based on a prenatal detection of isochromosome 12p is complicated. Prenatal and prognostic counseling should take into consideration the presence or absence of major life-threatening malformations such as congenital diaphragmatic hernia and complex heart defects (Izumi and Kranz 2014).

Diagnostic Testing

Historically, the diagnosis of PKS was made by karyotype G-band analysis or FISH analysis. Increasing recognition of a milder end of the spectrum has been partially enabled by new diagnostic methodologies such as array-based genome copy number detection. In addition to the loss of isochromosome 12p containing cells in the bone marrow over time, it has also been shown that the use of PHA in standard cytogenetic preparations to induce mitoses, promotes the growth of non-iso12p cell lines over the iso12p cells, which results in the difficulty of detecting iso12p cells in peripheral blood on standard cytogenetic evaluation (Conlin et al. 2012; Izumi et al. 2014). With the introduction of array-based cytogenetic analyses such as array comparative genomic hybridization (CGH) and single-nucleotide polymorphism (SNP) arrays, the detection rate of tetrasomy 12p from peripheral blood samples has improved (Ballif et al. 2006; Powis et al. 2007; Conlin et al. 2012). These new diagnostic methodologies have the advantage of being performed on direct DNA preparations from peripheral blood (no stimulation required) and can more readily detect lower levels of mosaicism. SNP arrays have been shown to detect a lower percent of mosaicism as compared to array CGH because of the added power of the available haplotype information (Conlin et al. 2010). In addition, the haplotype information obtained from SNP arrays can also provide mechanistic insight into tetrasomic cell formation, which can be useful for appropriate genetic counseling. Given that the percentage of mosaic tetrasomic cells in peripheral blood decreases as the individual with PKS ages, SNP arrays should be performed as early as possible to avoid the need for skin biopsy. If a skin biopsy is performed, it is notable that the mosaic ratio in skin fibroblast samples appears to remain relatively stable over time (Conlin et al. 2012). Alternatively, analysis of buccal smear samples can be used to establish a diagnosis of PKS. However, skin biopsy sample still continues to be considered the diagnostic "gold standard" for PKS because of the variable and insufficient quality of buccal smears (Hodge et al. 2012). SNP and CGH array can be used to diagnose PKS in early infancy and during the prenatal period. In older patients suspected of having PKS, array or FISH analysis of buccal smear cells can be a first-line diagnostic test given the reliability and non-invasive nature of these methodologies. A skin biopsy is always recommended diagnostically when clinical features of PKS are present and when buccal smear analysis cannot detect the i(12p) (Cobben et al. 2013; Lee et al. 2017). Digital droplet PCR (ddPCR) technology has also been proposed as new faster diagnostic tool in quantifying lower levels of mosaicism. Although genome-wide SNP arrays are effective to identify lower level mosaicism than previous diagnostic methodologies, their use is limited by the technical complexity, cost, and turnaround time (Izumi et al. 2016). Prenatal diagnosis of PKS is still a challenge due to difficulties in detecting the extra iso(12p) chromosome. Most often, the consideration of a prenatal diagnosis of PKS still arises in the presence of congenital diaphragmatic hernia (CDH) or incidentally as a result of chorionic villus sampling (CVS), amniocentesis performed for advanced maternal age, or non-specific ultrasound abnormalities (Doray et al. 2002; Kunz et al. 2009). Array CGH on genomic DNA extracted from uncultured cells (amniocytes) may also be efficient in detecting low levels of mosaicism and complex rearrangements (Kunz et al. 2009; Libotte et al. 2016). Non-invasive prenatal diagnostic testing from maternal circulation may be capable of detecting the iso12p chromosome, but would be challenging depending on the level of mosaicsm.

Differential Diagnosis

PKS facial gestalt in association with pigmentation anomalies, structural birth defects, and neurodevelopmental delay are quite distinctive and highly suggestive of a clinical diagnosis of PKS. However, as the phenotype changes with the age of patients, recognition of PKS varies, especially in infancy, in older individuals, and in individuals with milder forms. Several diagnoses have been confused with PKS at different ages, the most common diagnoses to be considered in the differential diagnosis of PKS includes Down syndrome (see Chapter 24) and Wolf–Hirschhorn syndrome (see Chapter 19).

The confusion between PKS and Down syndrome, whose clinical diagnosis is usually straightforward, may arise in the neonatal period due to the common flat occiput and facial profile with large fontanels, upslanting palpebral fissures, wide-spaced eyes with epicanthal folds, wide nasal bridge, small nose, large tongue, short neck, swollen dorsum of hands and feet, and hypotonia. Likewise, the absence of some of the typical clinical features in Down syndrome, such as Brushfield spots of the iris, single palmar creases, increased space between first and second toes accompanied by vertical plantar crease, distinctive dermatoglyphic pattern, should help to differentiate these diagnoses. Individuals with Wolf–Hirschhorn syndrome (WHS) present with a tall forehead, broad nasal bridge continuing down from the forehead (giving a "Greek warrior appearance" to the profile), hypertelorism, epicanathal folds, micrognathia, sometimes a thin upper lip, ear pits, and high-/cleft palate, which can overlap with PKS, although other typical features, such as microcephaly, prominent eyes, prominent glabella, and short philtrum, are not typically seen in PKS. In terms of major clinical issues, intellectual disability, hypotonia, seizures, diaphragmatic hernia, congenital heart disease, and skeletal problems are reported in both diagnoses, but the growth deficiency of prenatal onset typically present in WHS, is rarely seen in PKS.

Overall, genetic diagnoses that have the highest number of shared features with PKS are trisomy 12p, microduplication 12p, Fryns syndrome, Beckwith–Wiedemann syndrome, and a condition associated with *CHD4* mutations known as Sifrim–Hitz–Weiss syndrome (see Table 45.1).

Some of the findings in mosaic trisomy 12p have also been described in tetrasomy 12p (Pallister–Killian syndrome, PKS) (Schinzel 1991). These include advanced maternal age, normal to large birth weight, and multiple congenital anomalies. The congenital anomalies observed here in mosaic trisomy 12p were milder and were not associated with prolonged NICU admission or low Apgar scores, in contrast to the findings reported in PKS. Half of our mosaic trisomy 12p patients had hypotonia and a seizure disorder, as

TABLE 45.1 PKS differential diagnosis

	PKS	Trisomy 12p	Fryns syndrome (#229850)	BWS (#130650)	Siftrim–Hitz–Weiss syndrome (#617159)
Intellectual disability	+	+	+	+/−	+
Seizures	+	+/−	+	−	−
Hypotonia	+	+	+	−/+	+
Frontotemporal alopecia/sparse	+	−/+	−	−	−
Prominent forehead	+	+	−	+	+
Flat occiput	+	−	−	−	−
Hypertelorism	+	+	+	−/+	+
Epicanthus	+	+	−	+/−	+
Microblepharon	−	+	−	−	−
Ear anomalies	+ (Ear pits)	+ (Low set, over-folded helix, posteriorly rotated)	+ (Abnormal helices, low set ears, poorly formed ears)	+/− (Linear ear lobe creases; posterior helical indentations)	+ (Low set, small, cup-shaped ears)
Depressed and broad nasal bridge, with short nose and anteverted nostrils	+	+	+	−	−
Tended upper lip	+	−/+	+	−	−
Long philtrum	+	+	+	+	−
Everted lower lip	+	+	−	+	−
Macroglossia	+	+	−	+	−
Micrognathia	+/−	+	+	−	+
Prognathism	+	+	+	−/+	+
Cleft palate	+	−	−	+	+
Short neck	+	+/−	+	−	−
Accessory nipples	+	−	−	−	−
Hypo- or hyper-pigmented areas	+			(Nevus Flammeus)	
Abdominal wall defects	+	−	+	+	+
Nails hypoplasia	−	−	+/−	−	−
Polyhydramnios	+	+	−/+	+	+
IUGR	−	−		+	−
Prenatal overgrowth	+	+/−	+	+	−/+
Postnatal overgrowth	−	−	−	+	−
Postnatal growth delay	+	−	+	−	+
Hemihyperplasia	−	−	−	+	−
CDH	+	−	+	−/+	+
HD	+	+/−	+	Cardiomyopathy	+
Anteriorly placed anus	+	−	+	−	+
Cryptorchidism	+	+	+	+	+
Brain anomalies (ventriculomegaly/Dandy–Walker malformation/hypoplasia of the corpus callosum)	+ (Ventriculomegaly)	(Hypoplasia of the corpus callosum)	(Ventriculomegaly/Dandy–Walker malformation)	(Rarely posterior fossa abnormalities/Dandy–Walker malformation)	(Ventriculomegaly)
Visceromegaly and embryonal tumor	−	−	−	+	−

compared with most patients with PKS. All of the patients with mosaic trisomy 12p were developmentally delayed, but they all eventually walked and 50% of them were toilet trained. Dysmorphic features common to both disorders include sparse scalp hair, prominent fore-head, hypertelorism, and short nose, but patients with mosaic trisomy 12p did not demonstrate the typical skin findings seen in PKS. In both disorders the facial features changed as the patients aged, becoming coarser over time, with prognathism and malar hypoplasia (Horneff et al. 1993). The neurological outcome of individuals with mosaic trisomy 12p was better than for those with tetrasomy 12p, as our patients remained ambulatory, with no contractures and hypertonia.

The majority of patients with interstitial 12p duplications encompassing the 12p13.31 region often manifest with the typical facial features of PKS, which supports the evidence for a critical region harboring genes that play a major role in pathogenesis of PKS (Izumi et al. 2012). Although a modest variability of the phenotypic presentation in individuals with 12p duplications, likely dependent on the size of the duplicated region, the most common dysmorphic features include round face, prominent forehead, frontal bossing, sparse hair and eyebrows, hypertelorism, upslanting palpebral fissures, epicanthic folds, wide and depressed nasal bridge, short nose with wide and anteverted nares, long philtrum, short neck, normal growth or overgrowth at birth (Izumi et al. 2012). In both PKS and 12p duplication, the facial features change with age, becoming coarser over time, with prognathism and malar hypoplasia (Horneff et al. 1993). The microblepharon seen in trisomy 12p has not been reported in PKS patients, while skin pigmentation anomalies characteristic of PKS are not reported in patients with partial trisomy12p, and they are rarely noted in those with entire 12p duplication (Inage et al. 2010). The neurological outcome of individuals with trisomy 12p is usually better than for those with tetrasomy 12p, and life-threatening congenital malformations are more rarely described than in PKS (Inage et al. 2010; Segel et al. 2005).

Fryns syndrome is one of the main differential diagnoses for PKS, especially after a prenatal finding of CDH. Fryns syndrome associated with autosomal recessive mutations in *PIGN* (McInerney-Leo et al. 2016) is usually suspected in the presence of diaphragmatic defects, characteristic facial appearance (coarse facies, ocular hypertelorism, broad and depressed nasal bridge, thick nasal tip, long philtrum, low-set and poorly formed ears, tented upper lip, macrostomia, micrognathia), distal digital hypoplasia and pulmonary hypoplasia. Polyhydramnios, ophthalmological problems, orofacial clefting, and malformations involving the brain, cardiovascular system, gastrointestinal, and genitourinary system have also been reported in both syndromes. Sparse frontotemporal hair, which is characteristic of PKS, is not typical in Fryns syndrome, which is more often associated with low hairlines and hypertrichosis. Hypertelorism and dystopia canthorum is almost always present in PKS and rarely reported in Fryns syndrome. Other features observed in PKS, but not in Fryns syndrome, are syndactyly and streaky skin pigmentation, whereas distal digital hypoplasia, cloudy corneas, and internal malformations are more commonly observed in Fryns syndrome than in individuals with PKS.

Prenatally a cystic hygroma is more often detected in fetuses with Fryns syndrome than in PKS. PKS has a better prognosis than Fryns syndrome whereby survival beyond the neonatal period has been rarely reported (Paladini et al. 2000) in Fryns syndrome.

Beckwith–Wiedemann syndrome (BWS) is a well-characterized genetic syndrome (see Chapter 9), but it has some common phenotypic features with PKS that may lead to a misdiagnosis, especially in early childhood. Shared clinical findings include macroglossia, ear pits, abdominal wall defects, and a common prenatal history of polyhydramnios and fetal overgrowth associated with macrosomia at birth and transient hypoglycemia. Individuals with PKS may have hemihypertrophy. However, increased risk of malignancy (hepatoblastoma or Wilms tumor) has *not* been described in PKS and the postnatal growth profile looks fairly different in these two conditions. While individuals with BWS show rapid growth in early childhood that usually appears to slow around age seven to eight years, most PKS patients show a growth deceleration after the first year of life. Also, development is usually severely impaired in PKS individuals and normal in children with BWS unless other events occur such as perinatal hypoxia or significant untreated hypoglycemia.

Interestingly, de novo mutations in the *CHD4* gene, an ATP-dependent chromatin remodeler gene that maps to the PKS critical region on 12p13.31, have been recently described in five individuals with intellectual disability and characteristic dysmorphic features partially overlapping the PKS phenotype (Weiss et al. 2016). Individuals with *CHD4* mutations, similar to individuals with PKS, can have wide-spaced eyes, a square-shaped face, and palate abnormalities, but they usually do not have the PKS flat facial profile and present more often with macrocephaly and characteristic low-set and cup-shaped ears. Other common findings include developmental delay, hypotonia, hearing loss, mild-to-moderate enlargement of the lateral ventricles, and heart defects (especially patent foramen ovale and septal defects) (Weiss et al. 2016). Intellectual disability is usually milder than in PKS patients, CDH is not reported in individual with *CHD4* mutations and skeletal anomalies mainly manifest as bone fusions, whereas in PKS skeletal anomalies include limb shortening, delayed ossification, and osteopenia.

MANIFESTATIONS AND MANAGEMENT

Growth and Feeding

PKS individuals show a unique growth pattern characterized by prenatal overgrowth and postnatal growth deceleration, usually in the first year. About 65% of individuals with PKS have birth measurements, including both weight and length, above the 60th centile; however, after birth, many PKS probands demonstrate growth deceleration, dropping into the lower centiles in the first three years of life (Wilkens et al. 2012). Head circumference measurements of prenatal onset are generally above the 75th centile in most individuals with PKS in keeping with weight parameters. In contrast, long bone shortening, especially reduced femur length, is often described prenatally by the second trimester of gestation (unpublished personal observation). Hemihyperplasia due to mosaicism may occur in PKS patients, which may be not evident at birth, but becomes more marked in the first few years of life. The mechanism underlying this unique growth profile still remains to be elucidated; however, about half of individuals with PKS produce high levels of IGF-BP2 that likely contribute to the postnatal growth retardation phenotype (Kosuke et al. 2015). A growth hormone deficiency is rarely documented in individuals with PKS (Frković SH et al. 2010; Sukarova-Angelovska et al. 2016). Although more studies are needed, bone age is significantly delayed in a few individuals as well as delayed closure of large fontanelles (Blyth et al. 2015). Feeding problems have been reported in ~67% of PKS infants, sometimes requiring a nasogastric or gastrostomy tube (Wilkens et al. 2012). PKS infants often have delays in the development of a sucking reflex or in coordinating sucking, swallowing and breathing. These difficulties may be exacerbated by palate abnormalities and underlying low muscular tone. Gastroesophageal reflux occurs in most PKS individuals and may manifest with irritability and increase the risk of life-threatening aspirations during feeding (see GI section). Aspiration should be considered a possible cause for recurrent pulmonary infections and/or reactive airways in affected children. Although some patients continue to need pureed food for prolonged periods of time, others progress to solid food by mid-childhood.

Evaluation

- Head circumference, length and weight should be assessed at each health care evaluation and plotted on standard growth curves.
- The likelihood of feeding problems should be ruled out as soon as possible, particularly in individuals with palate abnormalities.
- The presence of gastroesophageal reflux and gastrointestinal malformation as a potential factor in feeding problems should be sought in all newborns in a standard manner.
- Nutritional evaluation maybe helpful in assessment of caloric intake in individuals with failure to thrive.
- Growth hormone studies should be performed in individuals with severe growth retardation.

Management

- Specialty feeders/nipples designed especially for babies with sucking difficulties should be used in presence of palate abnormalities (high or cleft palate).
- High caloric formulas may be used to assure an adequate caloric intake in individuals with poor weight gain.
- Standard therapy for gastroesophageal reflux can be started at diagnosis. Severe forms of GERD may benefit from Nissen–Hill fundoplication.
- Placement of nasogastric or gastrostomy tubes should be considered when oral nutrition cannot be maintained or in case of persistent and severe failure to thrive to sustain weight gain, protect airway and manage gastroesophageal reflux.

Development and Behavior

Until a few years ago, only individuals with severe neurocognitive deficits had been reported in the literature, with no or minimal development of speech and acquisition of neurodevelopmental milestones. Although developmental and behavior aspects in this condition still need further studies to be fully characterized, some more recent studies on larger cohorts of PKS have been showing a more comprehensive view of a continuum of the neurodevelopmental phenotype, suggesting a broader spectrum of neurocognition in PKS than had been previously thought.

There is a variable percentage of individuals with PKS (from 9.4% up to 27.3% in more recent estimates) who are higher functioning, having only mild speech and developmental delays and attending regular schools (IQ ≈ 69) (Warburton et al. 1987; Bielanska et al. 1996; Schaefer et al. 1997; Genevieve et al. 2003; Stalker et al. 2006; Kostanecka et al. 2012; Wilken et al. 2012: Blyth et al. 2015). However, cognitive deficits are a hallmark of this condition and severe to profound intellectual disability is reported in most individuals. Overall, the milestones of individuals with PKS show both motor and speech delay. Average ages for achieving motor milestones in PKS include: rolling: 10.8 months, sitting independently: 21.2 months, walking: 38.8 months. Speech initiation was at 36 months, although there are many individuals with PKS who do not attain speech (Wilkens et al. 2012). Language skills are below nine month age-equivalency in about 75% of children over the age of 18 months and adults with PKS (Kostanecka et al. 2012; Blyth et al. 2015). Most PKS children have little to no communicative intent. The molecular basis of such wide neurocognitive variability is still unknown, although it does not seem to be related to mosaic level in blood/skin fibroblasts. It might be linked to differential genome pattern or gene expression profiles in brain tissue or differential effects on processes involved in brain developmental (Kostanecka et al. 2012).

The behavioral concerns, studied in a small cohort of 16 individuals, include repetitive hand and body movements in 75%, lethargy and withdrawal in 68%, self-injurious behavior, such as hand or finger biting and head banging, in 25% (Kostanecka et al. 2012). The lack of internal drive to explore their surroundings seen in lower functioning children with PKS, is likely due, at least in part, to underlying hypotonia and their visual and hearing impairment that may limit the ability to explore their environment. Half of individuals involved in the previously mentioned study, also showed tactile defensiveness that may interfere with developmental progression by limiting the manipulation of objects and the ability to bear weight on their feet on surfaces to walk (Kostamecka et al. 2013). Many patients with PKS have also been diagnosed with autism that could further explain lack of eye contact and engagement with their environment (Kostamecka et al. 2013). Although the circadian system in this condition has been poorly investigated, many families describe sleep problems, characterized by inability to distinguish day from night, and sleep apneas. Individuals tend to sleep more hours per day (range from 9.5 to 15 hours per day) (Kostamecka et al. 2013). Usually, when sitting is acquired or supportive seating initiated there is improved interaction with their environment and the dysregulated sleep patterns appear to diminish.

Evaluation

- A complete neurodevelopmental evaluation should be performed by a specialist at appropriate intervals for planning of early and individualized interventions including physical, occupational and speech therapy in order to maximize the child's ability.
- The tactile defensiveness should be addressed with appropriate sensory activities.
- Polysomnography should be performed in all patients suspected of sleep apneas.

Management

- Involvement in early rehabilitation programs that cover motor aspects including development of fine motor and feeding skills, cognition, communication and socialization.
- Appropriate supportive seating for infants and toddlers should be provided to help the interaction with environment, improve motor ability, support feeding and reduce sleeping time during the day.
- Practicing good sleep hygiene is important in PKS patients.
- Sleep disturbance may benefit from standard therapies with pharmacologic agents such as melatonin.
- When autism is identified, appropriate support for communication, social skill and play development needs to be provided.
- Appropriate school placement is important.
- Planning for transition to adulthood (e.g. assisted living support) should be considered starting in adolescence.

Neurological Problems

Neurological problems mainly include hypotonia, seizures, and structural brain anomalies.

Almost all PKS individuals present with hypotonia during the neonatal period and infancy, although some individuals may develop a variable spasticity and hypertonia in older ages (Wilkens et al. 2012; Izumi et al. 2014; Blyth et al. 2015). Seizures are common in children with PKS. Different prevalence estimates have been reported ranging from 42% to 73% of children (Reyenolds et al. 1987; Bielanska et al. 1996; Candee et al. 2012; Blyth et al. 2015). The wide range likely reflects different mean ages of the studied cohorts. Although it was previously suggested that seizures presented right after birth, subsequent studies show that the onset is double-clustered between 6 months and 4–5 years of age and between 8 and 11 years of age (Candee et al. 2012; Blyth et al. 2015). The seizure pattern in PKS may be described by mixed seizure types with predominance of myoclonic seizures and late-onset clustered tonic spasms ("late-onset infantile spasms") with a rarity of convulsive status epilepticus and periods of relative refractoriness (Candee et al. 2012). However, variability in seizure phenotypes has been reported among different studies and specific EEG patterns have not been identified. Epileptic spasms are more often seen in patients with underlying brain anomalies such as ventricular enlargement and cortical atrophy (Giordano et al. 2012). Cortical atrophy with frontal predominance, ventricular dilatation together with hydrocephalus, and reduced white matter, are the most common brain findings in PKS. Other reported brain anomalies include thickened cortex, micropolygyria, heterotopic neurons, agenesis of corpus callosum, Dandy–Walker malformation, and pineal gland tumors (Saito et al. 2006; Giordano et al. 2012). Limited data are available on outcome and long-term follow-up of epilepsy in individuals with PKS. Although the majority of clinical reports suggest a presumed medication-refractory nature of seizures, there are reports of epileptic spasms that resolve in some individuals and evidence of improvement in others after medical treatment (Candee et al. 2012; Giordano et al. 2012). Overall, it appears that the age of onset of epilepsy and its persistence during life are associated with more severe phenotypes. One adult with a history of resolved epilepsy appears to be higher functioning (Candee et al. 2012).

Evaluation

- A careful targeted history and exam should be undertaken at each health care visit to diagnose seizures as early as possible.

- Neurological evaluation should be recommended for all PKS children at diagnosis.
- A waking/sleeping video EEG polygraphic study may be useful to confirm seizure types.
- Neuroimaging/MRI should be performed in all children with seizures to characterize any anatomical variation.
- Neurological follow-up of seizures and pharmacotherapy is essential.

Management

- Abnormality of muscle tone should be treated in standard manner with physical therapy.
- Based on limited available data, it is not possible to propose a single medication or specific regimen as effective for all, or even a majority of, individuals with PKS. Antiepileptics most commonly used in PKS patients are valproic acid, levetiracetam and tomirapate. Levetiracetam has been identified as the most effective single agent (Candee et al. 2012). Also, an initial treatment with vigabatrin followed by valproic acid (monotherapy or in association) has been reported to be effective in maintaining about half of treated individuals seizures free and improve outcomes in the remaining individuals.
- Structural brain abnormalities (eg. hydrocephalus) should be managed in a standard manner.

Gastrointestinal

Gastrointestinal system involvement was found in more than half of individuals with PKS.

Intestinal malrotation of the gut, congenital diaphragmatic hernia and anorectal malformations represents the most commonly reported gastrointestinal structural anomalies. The frequency of congenital diaphragmatic hernia (CDH) ranges from 11% in a retrospective analysis in a population that survived to the perinatal period (personal unpublished observation) up to 35% in prenatal studies from spontaneous/voluntarily terminated pregnancies. CDH may arise as severe diaphragmatic defects with herniation of abdominal organs into the chest associated with pulmonary hypoplasia/dysplasia associated with a higher risk of poor perinatal outcomes or more rarely as diaphragmatic eventration. Prenatal identification of CDH is the gold standard to provide the best care at birth and reduce complications. Also, malrotation should be assessed as early as possible in all newborns to avoid any potential life-threatening event. Esophageal atresia is also sporadically seen in PKS (Chiesa et al. 1998; O Bartsch et al. 2005; Desseauve et al. 2016). Umbilical hernias are reported in 15% of individuals with PKS (Wilken et al. 2012), while inguinal hernias are described more often in males either unilaterally or bilaterally. Anorectal anomalies mainly manifest as an anteriorly placed anus, and less frequently by anal atresia/imperforate anus or anal-rectal fistulae. In addition to the gastroesophageal reflux discussed in Growth and Feeding, dysphagia and constipation are common functional problems in 30% and 67% of individuals with PKS, respectively (Wilkens et al. 2012). Abnormal swallowing should lead to the consideration of dysphagia related to dysmotility or an anomaly of the oropharyngeal swallowing phase.

Evaluation

- Structural and functional abnormalities should be ruled out by imaging studies such as upper gastrointestinal contrast study with small bowel follow-through, pH probe, milk scan or swallow studies to specifically identify the underlying etiology.
- Stooling pattern/gut motility should be ascertained at each health care visit.

Management

- Functional gastrointestinal problems may benefit from standard therapies such as a higher intake of dietary fiber and/or stool softeners for constipation, and facilitation of development of oral motor skills through occupational therapy.
- Treatment of structural gastrointestinal abnormalities is standard and often requires prompt surgical correction.

Cardiovascular

Congenital heart defects (CHD) have been reported in individuals with PKS with an estimated prevalence of 7–40% (Doray et al. 2002; Wilkens et al. 2012; Tilton et al. 2013; Blyth et al. 2015). In addition to these structural defects, cardiomyopathy has also been reported in PKS individuals (Ward et al. 1988; Tilton et al. 2013). The spectrum of the cardiac phenotype associated with PKS include patent foramen ovale (PFO), patent ductus arteriosus (PDA), atrial septal defects (ASD), ventricular septal defects (VSD), bicuspid aortic valve (BAV), left ventricular hypertrophy (LVH), superior vena cava syndrome (SVC) (Wilkens et al. 2012). While VSD, cardiac hypertrophies, and BAV are occasionally noted in PKS children, PFO, ASD and PDA represent the predominant cardiac findings (Tilton et al. 2013). Aortic dilatation was added to the list of cardiovascular abnormalities seen in PKS (Ward et al. 1988; Jamuar et al. 2012); however, the etiology of aortic dilatation is still unknown and no causal association with BAV has been demonstrated to date (Tilton et al. 2013).

Sporadically also Ebstein anomaly, left ventricular hypoplasia and tetralogy of Fallot and other complex CHDs have been described in fetuses from terminated pregnancies or newborns that have not survived to the neonatal period (Gilgenkrantz et al. 1985; Wilson et al. 1994; Grech et al. 1999; Langford et al. 2000; Li et al. 2000; Doray et al. 2002; Abad et al. 2006; Desseauve et al. 2016). Since most reported PKS children have mild CHD, cardiac surgery has been rarely required. Pulmonary hypertension may be a complication secondary to CHD with significant left to right shunting such as seen in moderate to large ASDs (Tilton et al. 2013). In patients with CHDs, the use of medication/therapeutic approaches which may further interfere with cardiac function (e.g. ketogenic diet for treatment of seizures), should be carefully considered and requires closer cardiac follow-up when needed.

Evaluation

- Cardiac evaluation is important and should be performed at the time of diagnosis. Since PFOs and ASDs are common structural heart defects in PKS and may not be associated with a murmur, electrocardiogram and echocardiography should be performed on all PKS individuals as a baseline study.
- Longitudinal follow-up should be planned to monitor previously identified CHDs and to exclude the development of cardiac hypertrophy and aortic dilatation.

Management

- In some instances, septal defects will resolve spontaneously.
- Cardiac surgery should be considered in all other cases.
- Standard treatment and monitoring should be instituted in all individuals with a diagnosis of cardiomyopathy.

Genitourinary

Genitourinary involvement is more often reported in males, and most commonly manifests as cryptorchidism. Other manifestations described in PKS include hypospadias, small genitalia and hydrocele (Wilkens et al. 2012). Small clitoris and mullerian duct anomalies such as bicornate uterus and hypoplastic or closed vagina have been rarely noted in females with PKS (Bernet et al. 1992; Desseauve et al. 2016). Decreased renal function and structural renal anomalies are more rarely seen in PKS. Urinary tract anomalies include calico-pelvic dilatation, often of prenatal onset, ureteral stenosis and hydroureteronephrosis.

Evaluation

- Cryptorchidism and ectopic urethral opening should be sought out on physical examination.
- Kidney and urinary tract ultrasonography should be performed at the time of diagnosis to identify underling structural anomalies.
- In case of antenatal hydronephrosis, a postnatal ultrasonographic follow-up should be performed to monitor kidney development and degree of dilatation.
- Any genitourinary structural anomaly should be referred for specialist evaluation.
- In the case of persistent urinary tract dilatation, reflux should be assessed by functional analysis. Febrile episodes should be screened for urinary tract infections.

Management

- Genitourinary tract malformations in PKS should be treated in a standard manner.

Dermatological findings

Dermatological findings are one of the major features of PKS. Hyperpigmented or hypopigmented skin streaks are highly suggestive of mosaic tetrasomy 12p, and may be the only visible sign. Skin lesions usually range from <1 cm to the full limb circumference, and they potentially occur anywhere and/or follow Blaschko's lines (Wilkens et al. 2012; Blyth et al. 2015). The skin pigmentation anomalies are not always visible at birth. Children with PKS may also have redundant skin. Other findings include dry skin predisposing to eczematous lesions (more often seen in PKS patients from African-American background) and accessory nipples described in up to 45% of probands. Rarely, a sacral skin tag has been described in PKS individuals (Wilken et al. 2012; Johnstone et al. 2012).

Evaluation

- Evaluation of the skin should be part of the physical examination for all individuals with PKS.
- Some hypopigmented lesions may be visible only under Wood's lamp (Guareschi et al. 2007).

Management

- Dermatological findings should be treated in PKS as in general populations.

Musculoskeletal

Musculoskeletal involvement is frequently seen in PKS individuals. Pre-axial polydactyly (broad or duplicated toes, thumbs), joint contractures and hip dysplasia are commonly described in PKS individuals (Wilken et al. 2012). A preponderance of whorl patterns on dermatoglyphics has also been described (Wilken et al. 2012). Hip dislocation may also occur later in life either due to late weight bearing or low muscular tone and joint laxity. Some individuals with PKS develop scoliosis or kyphosis during late childhood/

adolescence. Rhizomelic shortening of long bones, hemihyperplasia, and/or leg length discrepancy are also reported in PKS. Although limited data are available on skeletal findings, individuals with PKS have recurrent mild patterns of skeletal changes including delayed ossification (vertebral bodies, pubic bones, triradiate cartilage, and sacroiliac junction), flaring of the anterior ends of the ribs, and metaphyseal broadening of the long bones, particularly of femora (Jamuar et al. 2012). Although delay of the endochondral ossification is most evident in the axial skeleton, it might be also related to the feature of short limbs often reported in PKS (Kawashima et al. 1987; Jamuar et al. 2012; Schinzel et al. 2012).

Evaluation

- Referral to orthopedics to identify musculoskeletal anomalies.
- Leg length assessments should be made at each health care visit. If leg length discrepancy is greater than 1 cm, referral to orthopedics is indicated.
- If hemihyperplasia is observed, measurements of affected regions both length and girth, should be monitored over time.
- A spine radiograph should be performed when scoliosis is clinically suspected.
- Hip dysplasia or dislocation should be monitored, and if suspicious, ultrasound screening should be performed by 6 weeks of age to avoid complications that can further affect ambulation.

Management

- Children with leg length asymmetry should be evaluated by orthopedics and physical therapy and orthotics or surgical intervention considered.
- Scoliosis often requires adaptive seating and bracing. Sometimes surgery may be considered. When surgical treatment is offered, it usually follows standard practice.
- Management of hip dysplasia and dislocation follow standard practice.

Ophthalmologic

Ophthalmologic involvement is seen in 87% of individuals with PKS. Visual anomalies most often reported include strabismus, hypermetropia, nystagmus, myopia, astigmatisms and strabismus (75% of patients). Rarely, ptosis, optic nerve hypoplasia, macular hypoplasia and retinal pigmentary abnormalities are noted, while some PKS individuals have also been diagnosed with cortical visual impairment. Overall, about 19% of the reported individuals have been defined as legally blind (Kostanecka et al. 2012; Wilkens et al. 2012; Izumi et al. 2014; Blyth et al. 2015).

Evaluation

- All PKS individuals should be referred to an ophthalmologist for assessing of visual acuity, visual field, refractive errors and fundoscopy.
- Visual evoked responses should also be considered.
- In the presence of visual anomalies, regular ophthalmological follow-up is indicated.

Management

- Corrective lenses or low vision aids should be used as needed.
- Strabismus should be treated in standard manner with surgical correction performed as required.

Ears and Hearing

Ears anomalies are commonly reported in PKS individuals and more typically include ear pits (seen in 25%) and low-set and posteriorly rotated ears. Most individuals (>70%) present with hearing loss with 38% having significant sensorineural hearing impairment, often requiring hearing aids (Wilkens et al. 2012; Blyth et al. 2015). In most instances, the hearing loss is bilateral (Wilkens et al. 2012). Conductive hearing loss occurs in about 29% of patients and is usually secondary to recurrent serous otitis media, often reported in infants with PKS (Wilkens et al. 2012). Early detection of hearing loss is essential to avoid any further impairment in language development and educational progress. Anecdotally, the authors have seen many children diagnosed with severe-to-profound sensorineural hearing loss in the newborn period who have improved audiometric functioning at older ages. This may indicate that immature neuronal development may contribute to the hearing loss and serial audiological evaluations should be undertaken to optimize audiological support as children age. Individuals with PKS may also be at increased risk of enlargement of the tonsils and adenoid glands as well as enlargement of the tongue that may all result in obstruction of the airway.

Evaluation

- Auditory system should be evaluated in all PKS newborns or at the time of diagnosis through auto-acoustic emissions and/or brainstem auditory evoked potentials.
- Monitoring of middle ear infections and hearing should be performed periodically during childhood by routine audiometry to detect hearing loss secondary to chronic/recurrent otitis.

Management

- Hearing aids should be considered early in sensorineural hearing loss to minimize the impact on language development.
- Antibiotic therapy and myringotomy tubes considered for recurrent/chronic fluid or otitis media.

Dental

Delayed dental eruption is reported in 57% of children with PKS (Blyth et al. 2015). An atypical dental pattern is also commonly seen (57%) and mainly due to the absence of primary and/or permanent teeth (Bagattoni et al. 2016). Other orodental findings in PKS include double teeth in primary dentition in the anterior region of the mandible, alveolar ridge overgrowth in the posterior regions of the jaws, gingival overgrowth and a low attachment of the upper buccal frenulum. Malocclusion is one of the major problems reported among PKS children, and it is secondary to a characteristic occlusive phenotype found in PKS children (an increased lower facial height, a high arched palate, anterior, and posterior crossbite) (Bagattoni et al. 2016). A high rate of gingivitis and caries, which increase in frequency and severity with age, are also commonly reported among probands. The hypotonia and the intellectual disability severely impair self-care skills in PKS individuals (Bagattoni et al. 2016).

Evaluation

- Oro-dental evaluation should be encouraged early after the eruption of the first primary tooth.
- Regular dental evaluations are necessary to prevent or establish an early diagnosis of oral diseases, including malocclusion, to limit the need for major interventions requiring general anesthesia.
- When dental pattern anomalies or malocclusion are suspected, a radiographic examination should be undertaken.
- A comprehensive multidisciplinary evaluation (including otorhinolaryngologist, orthodontist, speech therapy, physiotherapist) should be carried out in individuals with complex orofacial phenotypes.

Management

- A daily use of fluoride toothpaste and additional fluoride treatments are recommended in all PKS patients to prevent dental caries.
- Frequent dental hygiene visits should be performed to maintain oral hygiene.
- Early orthodontic evaluation and intervention should be considered in patients with malocclusion.

Miscellaneous

Pulmonary involvement, rarely described in individuals with PKS, usually manifests as lung hypoplasia/dysplasia, often secondary to CDH or to prematurity (Wilkens at al. 2012). Most individuals demonstrate a clinical history of recurrent upper respiratory tract infections during childhood. To date, an immunological basis for these infections has not been observed. A pineal tumor has been reported in a 15-year-old girl with PKS, although an increased risk for neoplasms has not been identified in this condition (Mauceri et al. 2000). Interestingly, however, the isochromosome 12p has been consistently found in germ cell tumors (testicular, ovarian and pineal) (Looijenga et al. 2003) reported in the general population. Limited data are available on puberty. In some PKS individuals it may proceed slightly delayed and secondary sexual characteristics may not fully develop. Girls may show increased aggressive behavior and anxiety with the initiation of menstrual cycles and intensity and frequency of seizures are sometimes reported as increased as well (unpublished personal observation).

RESOURCES

Online Mendelian Inheritance in Man, OMIM®
McKusick-Nathans Institute of Genetic Medicine, Johns Hopkins University, Baltimore, MD, USA
Website: https://omim.org/

National Center for Biotechnology Information (NCBI), Bethesda, MD, USA, National Library of Medicine (US), National Center for Biotechnology Information
Website: https://www.ncbi.nlm.nih.gov/

Support Groups

PKS Kids
Website: www.pkskids.net

PKS Kids Italia Onlus
Website: www.pksitalia.org

Unique - The Rare Chromosome Disorder Support Group
Website: www.rarechromo.org

REFERENCES

Abad DE, Gabarre JA, Izquierdo AM, López-Sánchez C, García-Martínez V, Izquierdo AG (2006) Pallister-Killian syndrome presenting with a complex congenital heart defect and increased nuchal translucency. *J Ultrasound Med* 25(11):1475–1480.

Ackerman K. G, Herron B. J, Vargas S. O, Huang H, Tevosian S. G, Kochilas L, Rao C, Pober BR, Babiuk RP, Epstein JA, Greer JJ, Beier DR (2005) Fog2 is required for normal diaphragm and lung development in mice and humans. *PLoS Genet* 1(1):e10.

Bagattoni S, D'alessandro G, Sadotti A, Alkhamis N, Rocca A, Cocchi G, Krantz ID, Piana G (2016) Oro-dental features of Pallister–Killian syndrome: Evaluation of 21 European probands. *Am J Med Genet A* 170(9):2357–2364.

Ballif B. C, Rorem E. A, Sundin K, Lincicum M, Gaskin S, Coppinger J, Kashork CD, Shaffer LG, Bejjani B. A (2006) Detection of low-level mosaicism by array CGH in routine diagnostic specimens. *Am J Med Genet A* 140(24):2757–2767.

Bartsch O, Loitzsch A, Kozlowski P, Mazauric ML, Hickmann G (2005) Forty-two supernumerary marker chromosomes (SMCs) in 43 273 prenatal samples: chromosomal distribution, clinical findings, and UPD studies. *Eur J Hum Genet* 13(11):1192–1204.

Bielanska MM, Khalifa MM, Duncan A (1996) Pallister-Killian syndrome: A mild case diagnosed by fluorescence in situ hybridization. Review of the literature and expansion of the phenotype. *Am J Med Genet A* 65(2):104–108.

Blyth M, Maloney V, Beal S, Collinson M, Huang S, Crolla J, Temple K, Baralle D (2015) Pallister-Killian syndrome: a study of 22 British patients. *J Med Genet* 52(7):454–464.

Brøndum-Nielsen K, Mikkelsen M (1995) A 10-year survey, 1980–1990, of prenatally diagnosed small supernumerary marker chromosomes, identified by fish analysis. Outcome and follow-up of 14 cases diagnosed in a series of 12 699 prenatal samples. *Prenat Diagn* 15(7):615–619.

Candee MS, Carey JC, Krantz ID, Filloux FM (2012) Seizure characteristics in Pallister–Killian syndrome. *Am J Med Genet A* 158(12):3026–3032.

Choo S, Teo SH, Tan M, Yong MH, Ho LY (2002) Tissue-limited mosaicism in Pallister-Killian syndrome-a case in point. *J Perinatol* 22(5):420–420.

Conlin LK, Kaur M, Izumi K, Campbell L, Wilkens A, Clark D, Deardorff MA, Zackai EH, Pallister P, Hakonarson H, Spinner NB, Krantz ID (2012) Utility of SNP arrays in detecting, quantifying, and determining meiotic origin of tetrasomy 12p in blood from individuals with Pallister–Killian syndrome. *Am J Med Genet A* 158(12):3046–3053.

Cormier-Daire V, Le Merrer M, Gigarel N, Morichon N, Prieur M, Lyonnet S, Vekemans M, Munnich A (1997) Prezygotic origin of the isochromosome 12p in Pallister-Killian syndrome. *Am J Med Genet A* 69(2):166–168.

Crowley MA, Conlin LK, Zackai EH, Deardorff MA, Thiel BD, Spinner NB (2010) Further evidence for the possible role of MEIS2 in the development of cleft palate and cardiac septum. *Am J Med Genet A* 152(5):1326–1327.

De Bruin TW, Slater RM, Defferrari R, Van Kessel AG, Suijkerbuijk RF, Jansen G, de Jong B, Oosterhuis JW (1994) Isochromosome 12p-positive pineal germ cell tumor. *Cancer Res* 54(6):1542–1544.

de Ravel TJ, Keymolen K, van Assche E, Wittevronghel I, Moerman P, Salden I, Matthijs G, Fryns J-P, Vermeesch JR (2004) Postzygotic origin of isochromosome 12p. *Prenat Diagn* 24(12):984–988.

Desseauve D, Legendre M, Dugué-Maréchaud M, Vequeau-Goua V, Pierre F (2016) Syndrome de Pallister-Killian: confrontation des données prénatales et fœtopathologiques de six cas et revue de la littérature. *Gynécol Obstétr Fertilité* 44(4):200–206.

Doray B, Girard-Lemaire F, Gasser B, Baldauf JJ, de Geeter B, Spizzo M, Zeidan C, Flori E (2002) Pallister-Killian syndrome: difficulties of prenatal diagnosis. *Prenat Diagn* 22(6):470–477.

Dufke A, Walczak C, Liehr T, Starke H, Trifonov V, Rubtsov N, Schoning M, Enders H, Eggermann T (2001) Partial tetrasomy 12pter-12p12. 3 in a girl with Pallister-Killian syndrome: extraordinary finding of an analphoid, inverted duplicated marker. *Eur J Hum Genet* 9(8):572–576.

Huljev Frković S, Tonković Đurišević I, Lasan Trčić R, Sarnavka V, Crkvenac Gornik K, Mužinić D, Letica L, Barić I, Begović D (2010) Pallister Killian syndrome: unusual significant postnatal overgrowth in a girl with otherwise typical presentation. *Coll Antropol* 34(1):247–250.

Genevieve D, Cormier-Daire V, Sanlaville D, Faivre L, Gosset P, Allart L, Picq M, Munnich A, Romana S, de Blois Mc, Vekemans M (2003) Mild phenotype in a 15-year-old boy with Pallister–Killian syndrome. *Am J Med Genet A* 116(1):90–93.

Gilgenkrantz S, Droulle P, Schweitzer M, Foliguet B, Chadefaux B, Lombard M, Chery M, Prieur M (1985) Mosaic tetrasomy 12p. *Clin Genet* 28(6):495–502.

Giordano L, Viri M, Borgatti R, Lodi M, Accorsi P, Faravelli F, Ferretti MC, Grasso R, Memo L, Prola S, Pruna D, Santucci M, Savasta S, Verrotti A, Pruna D (2012) Seizures and EEG patterns in Pallister–Killian syndrome: 13 New Italian patients. *Eur J Paediatr Neurol* 16(6):636–641.

Grech V, Parascandalo R, Cuschieri A (1999) Tetralogy of Fallot in a patient with Killian–Pallister syndrome. *Pediatr Cardiol* 20(2):134–135.

Hodge JC, Hulshizer R. L, Seger P, St Antoine A, Bair J, Kirmani S (2012) Array CGH on unstimulated blood does not detect all cases of Pallister–Killian syndrome: A skin biopsy should remain the diagnostic gold standard. *Am J Med Genet A* 158(3):669–673.

Hoeflich A, Wu M, Mohan S, Föll J, Wanke R, Froehlich T, Arnold GJ, Lahm H, Kolb HJ, Wolf E (1999) Overexpression of insulin-like growth factor-binding protein-2 in transgenic mice reduces postnatal body weight gain. *Endocrinology* 140(12):5488–5496.

Horneff G, Majewski F, Hildebrand B, Voit T, Lenard H. G (1993) Pallister-Killian syndrome in older children and adolescents. *Pediatr Neurol* 9(4):312–315.

Hung T, Binda O, Champagne KS, Kuo AJ, Johnson K, Chang HY, Simon MD, Kutateladze TG, Gozani O (2009) ING4 mediates crosstalk between histone H3 K4 trimethylation and H3 acetylation to attenuate cellular transformation. *Mol Cell* 33(2):248–256.

Hunter AG, Clifford B, Cox DM (1985) The characteristic physiognomy and tissue specific karyotype distribution in the Pallister-Killian syndrome. *Clin Genet* 28(1):47–53.

Inage E, Suzuki M, Minowa K, Akimoto N, Hisata K, Shoji H, Okumura A, Shimojima K, Shimizu T, Yamamoto T (2010) Phenotypic overlapping of trisomy 12p and Pallister–Killian syndrome. *Eur J Med Genet* 53(3):159–161.

Izumi K, Conlin LK, Berrodin D, Fincher C, Wilkens A, Haldeman-Englert C, Saitta SC, Zackai EH, Spinner NB, Krantz ID (2012) Duplication 12p and Pallister–Killian syndrome: A case report and review of the literature toward defining a Pallister–Killian syndrome minimal critical region. *Am J Med Genet A* 158(12): 3033–3045.

Izumi K, Zhang Z, Kaur M, Krantz I. D (2014) 12p microRNA expression in fibroblast cell lines from probands with Pallister-Killian syndrome. *Chrom Res* 22(4):453–461.

Izumi K, Krantz ID (2014) Pallister–Killian syndrome. In *Am J Med Genet C* 166(4):406–413

Izumi K, Kellogg E, Fujiki K, Kaur M, Tilton RK, Noon S, Wilkens A, Shirahige K, Krantz I. D (2015) Elevation of insulin-like growth factor binding protein-2 level in Pallister–Killian syndrome: Implications for the postnatal growth retardation phenotype. *Am J Med Genet A* 167(6):1268–1274.

Langford K, Hodgson S, Seller M, Maxwell D (2000) Pallister–Killian syndrome presenting through nuchal translucency screening for trisomy 21. *Prenat Diagn* 20(8):670–672.

Lee MN, Lee J, Yu HJ, Lee J, Kim SH (2017) Using array-based comparative genomic hybridization to diagnose Pallister-Killian syndrome. *Ann Lab Med* 37(1):66–70.

Li MM, Howard-Peebles PN, Killos LD, Fallon L, Listgarten E, Stanley WS (2000) Characterization and clinical implications of marker chromosomes identified at prenatal diagnosis. *Prenat Diagn* 20(2):138–143.

Libotte F, Bizzoco D, Gabrielli I, Mesoraca A, Cignini P, Vitale S. G, Marilli I, Gulino FA, Rapisarda AMC, Giorlandino C (2016) Pallister–Killian syndrome: Cytogenetics and molecular investigations of mosaic tetrasomy 12p in prenatal chorionic villus and in amniocytes. Strategy of prenatal diagnosis. *Taiwan J Obstet Gynecol* 55(6):863–866.

Looijenga LH, Zafarana G, Grygalewicz B, et al. (2003) Role of gain of 12p in germ cell tumour development. *APMIS* 111(1):161–170.

Los FJ, Van Opstal D, Schol MP, Gaillard JL, Brandenburg H, Van Den Ouweland AM, In'T Veld PA (1995) Renatal diagnosis of mosaic tetrasomy 12p/trisomy 12p by fluorescent in situ hybridization in amniotic fluid cells: A case report of pallister–killian syndrome. *Prenat Diagn* 15(12):1155–1159.

Jamuar S, Lai A, Unger S, Nishimura G (2012) Clinical and radiological findings in Pallister–Killian syndrome. *Eur J Med Genet* 55(3):167–172.

Johnstone ED, Jones EA (2012) Ultrasound presentation of Pallister–Killian syndrome with a prominent sacral appendage. *Ultrasound Obstet Gynecol* 40(2):239–241.

Kaur M, Izumi K, Wilkens AB, Chatfield KC, Spinner NB, Conlin LK, Zhang Z, Krantz ID (2014) Genome-wide expression analysis in fibroblast cell lines from probands with Pallister Killian syndrome. *PloS One* 9(10):e108853.

Kawashima H, Reynolds JF (1987) Skeletal anomalies in a patient with the Pallister/Teschler-Nicola/Killian syndrome. *Am J Med Genet A* 27(2):285–289.

Kochova E, Sukarova-Angelovska E, Ilieva G, Kocova M, Angelkova N (2016) Rare case of Killian-Pallister syndrome associated with idiopathic short stature detected with fluorescent in situ hybridization on buccal smear. *Mol Cytogenet* 9(1):38.

Kodo K, Nishizawa T, Furutani M, Arai S, Yamamura E, Joo K, Takao T, Matsuoka R, Yamagishi H (2009) GATA6 mutations cause human cardiac outflow tract defects by disrupting semaphorin-plexin signaling. *Proc Natl Acad Sci USA* 106(33): 13933–13938.

Kostanecka A, Close LB, Izumi K, Krantz ID, Pipan M (2012) Developmental and behavioral characteristics of individuals with Pallister–Killian syndrome. *Am J Med Genet A* 158(12): 3018–3025.

Kunz J, Schoner K, Stein W, Rehder H, Fritz B (2009) Tetrasomy 12p (Pallister-Killian syndrome): difficulties in prenatal diagnosis. *Arch Gynecol Obstet* 280(6):1049–1053.

Mathieu M, Piussan CH, Thepot F, et al. (1997) Collaborative study of mosaic tetrasomy 12p or Pallister-Killian syndrome (nineteen fetuses or children). *Ann Genet* 40(1) 45–54.

Mauceri L, Sorge G, Incorpora G, Pavone L (2000) Pallister-Killian syndrome: Case report with pineal tumor. *Am J Med Genet A* 95(1):75–78.

McInerney-Leo AM, Harris JE, Gattas M, et al. (2016) Fryns syndrome associated with recessive mutations in PIGN in two separate families. *Hum Mutat* 37(7):695–702.

Nagafuchi S, Yanagisawa H, Sato K, Shirayama T, Ohsaki E, Bundo M, Takeda T, Tadokoro K, Kondo I, Tanaka Y (1994) Dentatorubral and pallidoluysian atrophy expansion of an unstable CAG trinucleotide on chromosome 12p. *Nat Genet* 6(1):14–18.

Nozell S, Laver T, Moseley D, Nowoslawski L, DeVos M, Atkinson GP, Harrison K, Nabors LB, Benveniste EN (2008) The ING4 tumor suppressor attenuates NF-κB activity at the promoters of target genes. *Mol Cell Biol* 28(21):6632–6645.

Ohashi H, Ishikiriyama S, Fukushima Y (1993) New diagnostic method for Pallister-Killian syndrome: Detection of i (12p) in interphase nuclei of buccal mucosa by fluorescence in situ hybridization. *Am J Med Genet A* 45(1):123–128.

Paladini D, Borghese A, Arienzo M, Teodoro A, Martinelli P, Nappi C (2000) Prospective ultrasound diagnosis of Pallister-Killian syndrome in the second trimester of pregnancy: the importance of the fetal facial profile. *Prenat Diagn* 20(12):996–998.

Pallister PD, Herrmann J, Meisner LF, Inhorn SL, Opitz JM (1976) Trisomy-20 syndrome in man. *Lancet* 307(7956):431.

Pallister PD, Meisner LF, Elejalde BR, Francke U, Herrmann J, Spranger J, Tiddy W, Inhorn SL, Opitz JM (1977) The Pallister mosaic syndrome. *Birth Defects Orig Artic Ser* 13(3B):103.

Peltomäki P, Knuutila S, Ritvanen A, Kaitila I, Chapelle A (1987) Pallister-Killian syndrome: cytogenetic and molecular studies. *Clin Genet* 31(6):399–405.

Pizzuti A, Sarkozy A, Newton AL, Conti E, Flex E, Cristina Digilio M, Amati F, Gianni D, Tandoi C, Marino B, Crossley M, Dallapiccola B (2003) Mutations of ZFPM2/FOG2 gene in sporadic cases of tetralogy of Fallot. *Hum Mutat* 22(5): 372–377.

Powis Z, Kang SHL, Cooper ML, Patel A, Peiffer DA, Hawkins A, Heidenreich R, Gunderson KL, Cheung SW, Erickson RP (2007) Mosaic tetrasomy 12p with triplication of 12p detected by array-based comparative genomic hybridization of peripheral blood DNA. *Am J Med Genet A* 143(24):2910–2915.

Reynolds JF, Daniel A, Kelly TE, Gollin SM, Stephan MJ, Carey J, Adkins WN, Webb MJ, Char F, Jimenez JF (1987) Isochromosome 12p mosaicism (Pallister mosaic aneuploidy or Pallister-Killian syndrome): Report of 11 cases. *Am J Med Genet A* 27(2):257–274.

Rivera H, Vásquez AI, Perea FJ (1999) Centromere–telomere (12;8p) fusion, telomeric 12q translocation, and i (12p) trisomy. *Clin Genet* 55(2):122–126.

Saito Y, Masuko K, Kaneko K, Chikumaru Y, Saito K, Iwamoto H, Matsui A, Aida N, Kurosawa K, Kuroki Y, Kimura S (2006) Brain MRI findings of older patients with Pallister–Killian syndrome. *Brain Dev* 28(1):34–38.

Schaefer GB, Jochar A, Muneer R, Sanger WG (1997) Clinical variability of tetrasomy 12p. *Clin Genet* 51(2):102–108.

Schinzel A (1991) Tetrasomy 12p (Pallister-Killian syndrome). *J Med Genet* 28(2):122.

Schubert R, Viersbach R, Eggermann T, Hansmann M, Schwanitz G (1997) Report of two new cases of Pallister-Killian syndrome confirmed by FISH: Tissue-specific mosaicism and loss of i (12p) by in vitro selection. *Am J Med Genet A* 72(1):106–110.

Sifrim A, Hitz MP, Wilsdon A, Breckpot J, Al Turki SH, Thienpont B, ... Prigmore E (2016) Distinct genetic architectures for syndromic and nonsyndromic congenital heart defects identified by exome sequencing. *Nat Genet* 48(9):1060–1065.

Soukup S, Neidich K (1990) Prenatal diagnosis of Pallister-Killian syndrome. *Am J Med Genet A* 35(4):526–528.

Speleman F, Leroy JG, Van Roy N, De Paepe A, Suijkerbuijk R, Brunner H, Looijenga L, Verschraegen-Spae MR, Orye E (1991) Pallister-killian syndrome: Characterization of the isochromosome 12p by fluorescent In Situ hybridization. *Am J Med Genet A* 41(3):381–387.

Stalker HJ, Gray BA, Bent-Williams A, Zori RT (2006) High cognitive functioning and behavioral phenotype in Pallister-Killian syndrome. *Am J Med Genet A* 140(18):1950–1954.

Struthers JL, Cuthbert CD, Khalifa MM (1999) Parental origin of the isochromosome 12p in Pallister-Killian syndrome: Molecular analysis of one patient and review of the reported cases. *Am J Med Genet A* 84(2):111–115.

Tang W, Wenger SL (2005) Cell death as a possible mechanism for tissue limited mosaicism in Pallister-Killian syndrome. *J Ass Genet Technol* 31(4):168–169.

Teschler-Nicola M, Killian W (1981) Case report 72: mental retardation, unusual facial appearance, abnormal hair. *Synd Ident* 7(1):6–7.

Tilton RK, Wilkens A, Krantz ID, Izumi K (2014) Cardiac manifestations of Pallister–Killian syndrome. *Am J Med Genet A* 164(5):1130–1135.

Tong JK, Hassig CA, Schnitzler GR, Kingston RE, Schreiber SL (1998) Chromatin deacetylation by an ATP-dependent nucleosome remodelling complex. *Nature* 395(6705):917–921.

Turleau C, Simon-Bouy B, Austruy E, Grisard MC, Lemaire F, Molina-Gomes D, Siffroi J-P, Boué J (1996) Parental origin and mechanisms of formation of three cases of 12p tetrasomy. *Clin Genet* 50(1):41–46.

Van Dyke DL, Babu VR, Weiss L (1987) Parental age, and how extra isochromosomes (secondary trisomy) arise. *Clin Genet* 32(1):75–79.

Vermeesch JR, Melotte C, Salden I, et al. (2005) Tetrasomy 12pter-12p13. 31 in a girl with partial Pallister–Killian syndrome phenotype. *Eur J Med Genet* 48(3):319–327.

Vogel I, Lyngbye T, Nielsen A, Pedersen S, Hertz JM (2009) Pallister–Killian syndrome in a girl with mild developmental delay and mosaicism for hexasomy 12p. *Am J Med Genet A* 149(3):510–514.

Warburton D, Anyane-Yeboa K, Francke U, Reynolds JF (1987) Mosaic tetrasomy 12p: Four new cases, and confirmation of the chromosomal origin of the supernumerary chromosome in one of the original Pallister-Mosaic syndrome cases. *Am J Med Genet A* 27(2):275–283.

Ward BE, Hayden MW, Robinson A, Opitz JM, Reynolds JF (1988) Isochromosome 12p mosaicism (Pallister-Killian syndrome): Newborn diagnosis by direct bone marrow analysis. *Am J Med Genet A* 31(4):835–839.

Wenger SL, Boone LY, Steele MW (1990) Mosaicism in Pallister i (12p) syndrome. *Am J Med Genet A* 35(4):523–525.

Weiss K, Terhal PA, Cohen L, et al. (2016) De novo mutations in CHD4, an ATP-dependent chromatin remodeler gene, cause an intellectual disability syndrome with distinctive dysmorphisms. *Am J Hum Genet* 99(4):934–941.

Wilkens A, Liu H, Park K, Campbell LB, Jackson M, Kostanecka A, Pipan M, Izumi K, Pallister P Krantz ID (2012) Novel clinical manifestations in Pallister–Killian syndrome: comprehensive evaluation of 59 affected individuals and review of previously reported cases. *Am J Med Genet A* 158(12):3002–3017.

Wilson RD, Harrison K, Clarke LA, Yong SL (1994) Tetrasomy 12p (Pallister-Killian syndrome): Ultrasound indicators and confirmation by interphase fish. *Prenat Diagn* 14(9): 787–792.

Yeung A, Francis D, Giouzeppos O, Amor DJ (2009) Pallister–Killian syndrome caused by mosaicism for a supernumerary ring chromosome 12p. *Am J Med Genet A* 149(3):505–509.

Zakowski MF, Wright Y, Ricci A (1992) Pericardial agenesis and focal aplasia cutis in tetrasomy 12p (Pallister-Killian syndrome). *Am J Med Genet A* 42(3):323–325.

Zhang X, Xu LS, Wang ZQ, Wang KS, Li N, Cheng ZH, Huang S-Z, Wei D-Z, Han Z-G (2004) ING4 induces G2/M cell cycle arrest and enhances the chemosensitivity to DNA-damage agents in HepG2 cells. *FEBS Lett* 570(1–3):7–12.

Zhang JI, Marynen P, Devriendt K, Fryns JP, Van den Berghe H, Cassiman JJ (1989) Molecular analysis of the isochromosome 12p in the Pallister-Killian syndrome. *Hum Genet* 83(4):359–363.

46

PRADER–WILLI SYNDROME

SHAWN E. MCCANDLESS
Section of Genetics and Metabolism, Department of Pediatrics, University of Colorado Anschutz Medical Campus and Children's Hospital Colorado, Aurora, Colorado, USA

SUZANNE B. CASSIDY
Division of Medical Genetics, Department of Pediatrics, University of California, San Francisco, San Francisco, California, USA

INTRODUCTION

Prader–Willi syndrome (PWS) is a complex, multisystem intellectual disability disorder that was first described in 1956 (Prader et al. 1956). For many years, its cause was unknown, and it was believed that affected individuals were doomed to die young of complications of the associated obesity. Twenty-five years later, PWS captured the interest of geneticists because it was the first recognized microdeletion syndrome identified when high-resolution chromosome analysis was introduced (Ledbetter et al. 1981). PWS is now known to be one of the most common microdeletion syndromes, one of the most frequent disorders seen in genetics clinics, and the most common syndromic form of obesity. It was also the first recognized human genomic imprinting disorder and the first known to result from uniparental disomy (Nicholls et al. 1989). It is caused by several different genetic alterations of proximal chromosome 15q (genetic heterogeneity), and it is typified by a distinctive behavioral phenotype (reviewed in (Whittington and Holland 2010)). Finally, PWS is an excellent example of a genetic disorder that is significantly altered by treatment, with markedly improved outcomes in affected individuals who receive early diagnosis and appropriate therapy as outlined below.

Incidence

The incidence of PWS, based on multiple studies, has been determined to be between 1/10,000 and 1/25,000, although the only population-based study suggests that the birth prevalence is closer to 1/29,000 and the lower limit for incidence is 1/52,000 (Whittington et al. 2001). This is likely to be an underestimate since the number of affected individuals not diagnosed at an early age is unknown. PWS occurs in both sexes and all ethnic groups. The vast majority of affected individuals are the only affected individuals in their family. There does not appear to be an increase in incidence among infants conceived by assisted reproductive technologies, although the proportion of individuals with uniparental disomy was significantly increased relative to deletions in children conceived by assisted reproductive technologies (Gold et al. 2014).

Individuals with PWS are at increased risk of death relative to their typically developing peers across the age spectrum, with the death rate estimated at 3% per year (Butler et al. 2002). There is a well-documented relationship of morbidity and mortality to obesity-related complications, including obstructive sleep apnea, cardiorespiratory insufficiency and right-sided heart failure, and diabetes mellitus. A recent review of the Prader–Willi Syndrome Association (USA) mortality database (Butler et al. 2017) showed that respiratory and cardiac causes of death remain most frequent (31% and 16%, respectively), with visceral necrosis and rupture related to over-eating and other gastrointestinal issues causing 10% of mortality, infection 9%, choking/aspiration 6%, and accidents 6%. A previously unrecognized association with pulmonary embolism accounted for 7% of

---

Cassidy and Allanson's Management of Genetic Syndromes, Fourth Edition.
Edited by John C. Carey, Agatino Battaglia, David Viskochil, and Suzanne B. Cassidy.
© 2021 John Wiley & Sons, Inc. Published 2021 by John Wiley & Sons, Inc.

deaths. Encouragingly, recent survival trends suggest a reduction in the risk of premature death in the era of more targeted and aggressive management of obesity through diet, food security, and exercise (Manzardo et al. 2018) with fewer deaths related to cardiac failure and diabetes complications. At the same time, there has been an increasing proportion of deaths related to accidental injury and risky behavior, as well as an increasing recognition of deaths related to thromboembolic events. These indicate priorities for therapeutics and management strategies going forward.

Diagnostic Criteria

Clinical diagnostic criteria were developed by consensus before the availability of sensitive and specific laboratory testing (Holm et al. 1993). These may still have value in suggesting the diagnosis and indicating the need for diagnostic testing (see Table 46.1); however, the wide availability of highly specific molecular testing means that a low threshold for testing is appropriate for individuals with the cardinal features of PWS: neonatal hypotonia and failure to thrive, developmental delay and mild cognitive impairment, characteristic facial appearance, early-childhood-onset obesity if uncontrolled, hypogonadism with genital hypoplasia and pubertal insufficiency, mild short stature, and a characteristic behavior disorder. PWS should be strongly considered in any infant with hypotonia and feeding problems of unknown cause, and in any young child with developmental delay and onset of obesity in the first few years of life but after infancy. These and the more minor, but often more distinctive, clinical findings have been well described (Cassidy et al. 2012).

An important study (Whittington et al. 2002) examining 103 molecularly proven individuals with PWS demonstrated five cardinal signs that were present in all proven cases (floppy at birth, weak cry or inactivity, poor suck, feeding

TABLE 46.1 Criteria to prompt diagnostic testing for Prader–Willi syndrome

Age	Suggestive findings
Less than 2 years	Hypotonia with poor suck and hypersomnolence in the neonatal period
2–6 years	Hypotonia with history of poor suck during infancy, and global developmental delay
6–12 years	History of hypotonia with poor suck in infancy (hypotonia often persists), and global developmental delay, and excessive eating with central obesity if uncontrolled
13 years or older	Cognitive impairment, usually mild mental retardation, and excessive eating with central obesity if uncontrolled, and hypothalamic hypogonadism, and/or typical behavior problems

Source: Modified from Gunay-Aygun et al. (2001).

difficulties, and hypogonadism indicated by undescended testes or hypoplastic labia minora and clitoris), and the absence of any one sign predicted negative molecular testing. No similar combination of findings was sufficient to allow for a 100% positive predictive value. The central hypotonia that predominates during infancy in PWS is prenatal in onset and is uniformly present. It is associated with decreased fetal movement, frequent abnormal fetal position, abnormal fetal heart rhythm, and difficulty at the time of delivery, often necessitating a cesarean delivery (Fong and de Vries 2003). Reflexes may be decreased or absent but are often normal. The neonatal central hypotonia and a state of hypoarousal are almost invariably associated with poor suck and lack of awakening to feed, with consequent failure to thrive and the necessity for gavage or other assisted feeding techniques. Infantile lethargy persists past the newborn period, with decreased arousal and weak cry as prominent findings. Neuromuscular electrophysiological and biopsy studies are normal or non-specific, and the hypotonia gradually improves. Motor milestones are delayed. The average age of sitting is 12 months and of walking is 24 months. Adults remain mildly hypotonic with decreased muscle bulk and tone.

Language development is also delayed. Verbal skills are an ultimate strength in most individuals, although speech is usually poorly articulated, having a nasal and/or slurred character. Cognitive abnormalities are evident, and most have mild intellectual disability (mean IQ 60s to low 70s). Approximately 40% have borderline intellectual disability or low normal intelligence, and about 20% have moderate disability. Academic performance is poor for cognitive ability. There is evidence that individuals with uniparental disomy may have somewhat higher verbal IQ than those with deletions, while performance scores on IQ tests are similar (Roof et al. 2000). Many people with PWS have relative strengths in tasks using visuospatial processing and long-term memory and relative weakness in arithmetic, sequential tasks, and short-term memory. Interestingly, many people with PWS have shown a true cognitive strength in working jigsaw puzzles (Dykens 2002) and enjoy word-find puzzles (personal observation).

Hypogonadism is prenatal in onset and persists throughout life (Crinó et al. 2003). At birth it is evident as genital hypoplasia in both sexes, including small labia minora and clitoris in females and cryptorchidism, small penis and scrotal hypoplasia in males. Hypogonadism is also evident in abnormal pubertal development. Although pubic and axillary hair develops early (premature adrenarche in about 20%) or normally, the remainder of pubertal development is usually delayed and incomplete. In both males and females, sexual activity is uncommon, although interest in physical relationships and sexuality is not. Fertility is rare, and only reported in a few females to date.

Although the proportion of fat mass to lean body mass is high even in thin infants with PWS (Butler 1990), significant obesity is not found in young infants. A series of predictable nutritional stages have been described beginning in the prenatal period into older adult life (Miller et al. 2011) (see Table 46.2). Notably, after the stabilization of eating and weight gain in infancy there follows a period of increasing weight gain without signs of hyperphagia in toddlers to early childhood. This observation suggests that metabolic changes involving resting energy expenditure precede the onset of hyperphagia, raising questions about whether defects in the hypothalamic satiety pathway are the only contributors to the hyperphagia. Resting energy expenditure is reduced in children with PWS (van Mil et al. 2000).

In older children and young adults, food-seeking behavior is common, including hoarding or foraging for food, eating of unappealing substances such as garbage, pet food, and frozen food, and stealing food or money to buy food. A high threshold for vomiting may complicate bingeing on spoiled food from the garbage or such items as boxes of sugar or frozen uncooked meat, and toxicity from ineffective emetics used to induce vomiting has occurred. Slow gastric emptying, gastroparesis and chronic constipation are common. The resultant obesity is central in distribution, with relative sparing of the distal extremities, and even individuals who are not overweight tend to deposit fat on the abdomen, buttocks, and thighs (Figures 46.1 and 46.2). As noted above, obesity and its consequences are the major causes of morbidity and mortality in PWS. In addition to cardiopulmonary compromise (Pickwickian syndrome), type II diabetes mellitus, sleep apnea with hypoxia, hypertension, thrombophlebitis, and chronic leg edema can result from obesity. Recent studies have documented long-term health outcomes and related specific medical issues (Butler et al. 2017; Hedgeman et al. 2017; Manzardo et al. 2018; Pacoricona Alfaro et al. 2019).

TABLE 46.2 Nutritional phases in Prader–Willi syndrome

Phase	Median ages	Clinical characteristics
0	Prenatal–birth	Decreased fetal movements and lower birth weight than sibs
1a	0–9 months	Hypotonia with difficulty feeding and apparently decreased appetite
1b	9–25 months	Improved feeding; weight gain more appropriate
2a	2.1–4.5 years	Weight crossing centiles on growth curve without appreciable appetite increase, interest in food, or excess calories; reduced energy expenditure?
2b	4.5–8 years	Increased appetite noticeable and increased caloric intake, but can feel full
3	8 years–adulthood	Hyperphagic, rarely feels full; can gain weight rapidly if access to food not limited
4	Adulthood	Appetite no longer insatiable for some (minority)

Modified from: Miller et al. (2011).

Many individuals with PWS have sleep disorders, including central and obstructive sleep apnea, abnormal arousal, abnormal circadian rhythm in REM sleep, reduced REM latency, and abnormal response to hypercapnia, as well as excessive daytime sleepiness. Respiratory abnormalities and obesity can exacerbate the sleep disorder.

Characteristic facial features, including dolichocephaly with narrow forehead, almond-shaped and sometimes upslanting palpebral fissures, narrow nasal bridge, and downturned corners of the mouth with a thin upper lip vermilion, are either present from birth or evolve over time in most affected individuals (Figures 46.1, 46.2, and 46.3). Small, narrow hands with a straight ulnar border and tapering fingers and short, often broad, feet are usually present by age 10. African-Americans are less likely to have small hands and feet (Hudgins et al. 1998). A characteristic body habitus, including sloping shoulders, heavy midsection, and genu valgus with straight lower leg borders, is usually present from early childhood (Figures 46.1 and 46.2). These findings may be absent or less evident in individuals who have been treated with growth hormone beginning early in life. Hypopigmentation for the family, manifested as fairer skin,

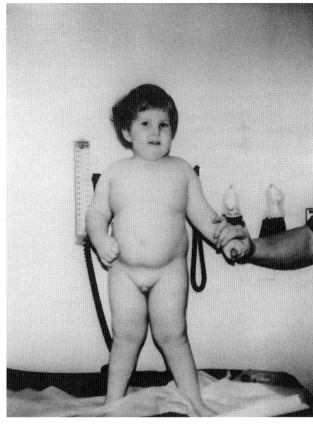

FIGURE 46.1 A 4-year-old boy with Prader–Willi syndrome. Note typical facial appearance, emerging obesity, small genitalia, and characteristic body habitus with genu valgus and straight leg borders.

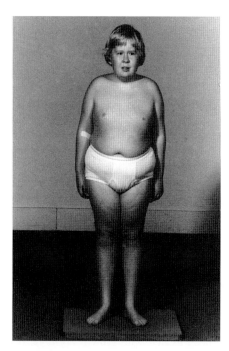
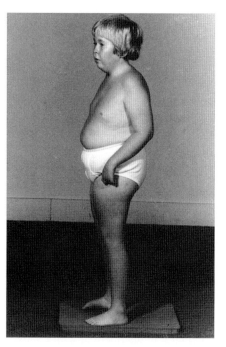

FIGURE 46.2 A 15-year-old male with Prader–Willi syndrome. Note facial appearance, fat distribution, and small distal extremities.

FIGURE 46.3 A girl with Prader–Willi syndrome at 7 (left) and 8 (right) years of age. The diagnosis of this condition was made at 7 years, and diet, exercise, and restriction of access to food resulted in significant weight loss over a one year period. The difficulty of making a diagnosis in a thin individual with Prader–Willi syndrome can be appreciated in the figure on the right, despite typical facial appearance when obese (left).

hair, and eye color, occurs in a significant proportion of those with deletion 15q11.2. Strabismus is frequent, as is myopia or hyperopia. Scoliosis and/or kyphosis are common, the former occurring at any age and the latter developing in early adulthood. Scoliosis may progress rapidly.

Birth weight and length are usually within normal limits, but the early period of failure to thrive may result in both weight and length being below the third centile. In the absence of growth hormone treatment, short stature, if not apparent in childhood, is almost always present by the second half of the second decade, associated with lack of a pubertal growth spurt. Average height for individuals not treated with growth hormone is well below the average for typically developing individuals. Individuals of African heritage tend to be taller (Hudgins et al. 1998). Reduced growth hormone secretion has been documented in the majority of children with PWS, as has been shown in numerous studies involving hundreds of affected children (reviewed in (Deal et al. 2013)). Treatment with growth hormone has been shown to ameliorate many of the physical findings of PWS (studies reviewed in (Deal et al. 2013; Wolfgram et al. 2013)). GH deficiency is also seen in adults with PWS (Höybye 2015). Controlled trials of growth hormone therapies have demonstrated significant benefit from infancy through adulthood (Deal et al. 2013; Wolfgram et al. 2013). Some improvements in cognition have even been reported (Siemensma et al. 2012; Dykens et al. 2017).

A characteristic behavior profile becomes evident in early childhood, with temper tantrums, stubbornness, controlling and manipulative behavior, compulsive-like characteristics, and difficulty with change in routine (Whittington and Holland 2010; Ishii et al. 2017). This persists into adulthood (Sinnema et al. 2011). Lying, stealing and aggressive behavior are common. Significant anxiety occurs frequently. Some individuals are diagnosed with autism and others with ADHD. An atypical affective disorder, with or without psychotic symptoms, may arise in young adulthood (Clarke et al. 2002) and may be common (Boer et al. 2002), particularly among individuals with maternal uniparental disomy 15 (Soni et al. 2007). Behavioral and psychiatric problems interfere the most with quality of life in adulthood (personal observation).

A variety of less well-understood findings are somewhat unique to this condition, including: thick, viscous saliva that may predispose to dental caries and contribute to articulation abnormalities; high pain threshold; skin picking; and high threshold for vomiting. Osteopenia and osteoporosis are frequent. Recent videofluoroscopic studies have shown silent aspiration, residue in the pharynx and esophagus, and, in general, disturbances in timing, clearance, and coordination of swallowing with the respiratory cycle (Gross et al. 2017; Salehi et al. 2017).

Etiology, Pathogenesis, and Genetics

Many of the manifestations of PWS are referable to insufficient functioning of hypothalamic control mechanisms; however, no clear structural defect or specific regulatory pathway of the hypothalamus has been identified as deficient. A single study has shown a decrease in oxytocin-secreting neurons in the paraventricular nucleus of the hypothalamus (Swaab et al. 1995).

PWS is a prototype of a genomic disorder, meaning that it is not a result of a specific DNA sequence change (or traditional mutation), but instead is due to disordered genomic structure or an epigenetic phenomenon altering gene expression. The various mechanisms for developing PWS all lead to loss of function of a group of genes on chromosome 15q11-13 normally expressed only from the paternally inherited chromosome. Absence of the paternally inherited copy of these genes leads to complete absence of expression of those genes because the maternal contribution is normally silenced. This process of genomic silencing is called imprinting, which, in this case, appears to be due to hypermethylation of the maternally contributed allele that interferes with transcription of those genes (Driscoll et al. 1992). This interference is likely due to multiple factors, including blocking the binding of transcriptional activators and presence of methyl-binding factors that recruit other transcriptional silencers, although the exact mechanism by which this occurs in the *PWS* region of chromosome 15 is still being studied.

Approximately 65–75% of individuals with PWS have a small deletion of the paternally contributed chromosome 15 between bands 15q11 and 15q13. This recurrent deletion appears to be the result of multiple tandemly repeated DNA sequences located at either end of the region that predispose to aberrant recombination. In the vast majority, recombination between the common chromosome breakpoints results in a deletion, with two common breakpoints identified at the centromeric end of the deletion, and one common breakpoint at the telomeric end. At least 4, with other possible transcripts, genes have been identified in the region between the two common centromeric breakpoints (Chai et al. 2003; Jiang et al. 2008), none of which appears to be imprinted. Therefore, individuals with the more proximal (centromeric) breakpoint, type I deletions, are haploinsufficient for these genes relative to both unaffected individuals and individuals with PWS due to the more distal of the two centromeric breakpoints, type II deletions. The larger type I deletion is ~5.7 Mb, while the smaller type II deletion is ~4.8 Mb. Some studies indicate that there may be modest differences in select subscales of larger neuropsychological measures between populations with the larger and smaller deletions, while others do not find differences. Potential differences seen between the groups appear to be more subtle than those between individuals with deletions and other causes of PWS, and identification of the specific the deletion breakpoints has not been shown to have clinical utility for an individual patient.

Most of the remaining individuals with PWS have inherited two maternal chromosomes 15 but no paternal chromosome 15, a situation known as uniparental disomy (UPD) 15. This is thought to result most often from a trisomic fertilization that spontaneously corrects by loss of one chromosome 15 (called trisomy rescue). Both deletion and uniparental disomy appear to occur sporadically in the vast majority of individuals with PWS and generally are not associated with a significant risk of recurrence.

The third cause of PWS, which occurs in less than 5% of affected individuals, is neither deletion nor uniparental disomy, but rather a defect in the imprinting process called an imprinting defect (reviewed in (Horsthemke and Wagstaff 2008)). Within this group, a few affected individuals have a small deletion or other abnormality in the area that controls imprinting within 15q11-q13, the imprinting center. The majority of this group, however, does not have a detectable mutation or deletion in the imprinting center, but nonetheless has evidence of biparental inheritance with a maternal-only gene expression pattern. Nearly all studied families in which there has been a recurrence of PWS have had an abnormality of the imprinting process.

The small number of remaining individuals with PWS has either deletion or maternal uniparental disomy for the critical region as a result of a translocation or other structural abnormality involving chromosome 15, including several reports of apparently balanced translocations that presumably alter the imprinting center in some way. Despite PWS being a well-characterized genetic disorder, specific genetic factors and other modifiers contributing to the clinical features are not yet clearly understood.

A number of imprinted and non-imprinted genes are known within the usual deletion region (the PWS/AS critical region; see also Angelman syndrome [AS], Chapter 5) and are therefore potential candidate disease-causing genes. Microdeletion of the area around the *SNORD116@* (the @ symbol has been adopted by the Human Genome Organization Gene Nomenclature Committee to indicate a gene cluster), previously referred to as *HB-II 85*, a microRNA cluster located in a down-stream intron of the *SNURF-SNRPN* gene, is associated with a phenotype that includes hyperphagia, neonatal hypotonia and feeding problems, intellectual disability, and other aspects of the PWS phenotype (Sahoo et al. 2008; Duker et al. 2010). One case report suggests that only the *SNORD116@*, and not the nearby *SNORD115@*, is required to cause the phenotype (Bieth et al. 2015). A conditional knockout mouse model demonstrated that hyperphagia is specifically associated with knockout of the *Snord116@* in NPY-producing neurons in the arcuate nucleus of the hypothalamus (Qi et al. 2016). These data provide the strongest evidence to date of a detailed genotype/phenotype correlation in PWS, and the microdeletion alone can be reasonably considered a form of PWS, although it is likely that the other genes in the region contribute to the phenotype in more subtle ways. The non-imprinted *P* gene, which codes for tyrosinase-positive albinism (*OCA-2*), appears to correlate with the hypopigmentation seen in about one-third of those with PWS.

Recurrence of PWS caused by either the common large deletion in the absence of a more complex structural chromosome rearrangement or by uniparental disomy has not been reported, thus the recurrence risk should be essentially the population-based risk. However, a paternal balanced insertion or gonadal mosaicism for a deletion is possible, and therefore a recurrence risk of less than 1% (empirically significantly less than 1%) is appropriate for genetic counseling purposes. Uniparental disomy is caused by nondisjunction, as evidenced by advanced maternal age in this group, and by documentation of trisomy 15 on chorionic villus sampling with resultant maternal uniparental disomy at birth. Because nondisjunction can recur, a recurrence risk of 1% or less is appropriate for genetic counseling purposes. In those families with a child who has a detected imprinting mutation (small deletion in *SNURF*), a recurrence risk of up to 50% pertains if the father is shown to be a carrier. If the mutation is not found in the father there is still risk of mosaicism involving the gonads, and thus the risk may be lower, but may also still be as high as 50%. If there is a presumed imprinting defect (positive DNA methylation but negative testing for deletion and uniparental disomy), then the recurrence risk is unknown but may be up to 50%. Since there is a presumed alteration affecting the imprinting center that cannot usually be identified by clinically available methods, the carrier status of the father cannot be determined. Prenatal diagnosis by detection of a DNA methylation defect (see below) is possible.

Diagnostic Testing

Because there are three different major causes of PWS, there are a number of different tests that can be used to confirm the diagnosis. So-called DNA methylation analysis, originally accomplished by Southern blotting and now usually done by methylation-specific polymerase chain reaction (PCR), can detect all three causes of PWS, because all three causes result in the genes for PWS being present only in the methylated, maternally inherited form. DNA methylation analysis has been validated for prenatal as well as postnatal use. An alternative method, utilizing the expression pattern of *SNRPN* is also available.

An abnormal DNA methylation pattern showing only the presence of maternally imprinted (highly methylated) DNA in the *PWS* region of chromosome 15 confirms the diagnosis. However, for genetic counseling purposes it is important to perform additional tests to determine the exact genetic cause of PWS. Deletion in 15q11-q13 can be identified accurately by chromosome microarray analysis of copy number, or by using fluorescence in situ hybridization (FISH) with a probe for *SNRPN*, a gene within the deletion. High-resolution chromosome analysis alone is insufficient, because false positives and false negatives have occurred relying on this lower resolution method without FISH. Uniparental disomy can be identified by using microsatellite repeat sequences on chromosome 15 in the affected individual and both parents or by use of microarray utilizing a single nucleotide polymorphism (SNP) array. A defect in the imprinting center is implied from an abnormal methylation analysis in the presence of normal FISH and uniparental disomy studies and can be identified using methylation-specific multiplex ligation-dependent probe amplification (MS-MLPA). MS-MLPA can also distinguish deletion from disomy, give the approximate size of a deletion, and detect most imprinting center and *SNORD116@* microdeletions. Other mutations (primarily small deletions) in the imprinting center can be detected in a minority of individuals with presumed imprinting defects in specialized or research laboratories.

It is generally agreed that the most cost-effective approach to diagnostic testing is methylation analysis done in conjunction with a SNP microarray, which in most cases will identify either deletion or uniparental disomy. If the DNA methylation analysis is not consistent with clinically suspected PWS, chromosome analysis is necessary as part of the evaluation for unexplained hypotonia or developmental delay/intellectual disability with obesity. FISH analysis using a *SNRPN* probe(s) is effective for the typical deletion, but may miss the rare deletion only of the *SNORD116@*, and it will not identify more complex rearrangements involving other chromosomal material. If a deletion is not identified, a test for uniparental disomy is indicated. For a more complex than typical deletion of the *PWS* region, particularly if a partial duplication is also found on chromosomal microarray, chromosome analysis may be indicated for the father to rule out the very rare situations of a translocation as the cause of the deletion.

Differential Diagnosis

The differential diagnosis of PWS in infancy includes many causes of neonatal hypotonia, particularly central and peripheral neuromuscular disorders, brain anomalies and anoxia. Later in childhood and adulthood, a number of conditions in which developmental delay/intellectual disability is associated with obesity are in the differential, including Bardet–Biedl syndrome (see Chapter 8), Alström syndrome, Albright hereditary osteodystrophy, Börjeson–Forssman–Lehman syndrome, and Cohen syndrome (see Chapter 14). Schaaf–Yang syndrome is a condition with some overlapping findings with PWS that is due to truncating mutations in the maternally imprinted, paternally expressed gene *MAGEL2* that is located in the Prader–Willi syndrome critical region on 15q11-13 (Schaaf et al. 2013). Manifestations include developmental delay/intellectual disability, hypotonia, feeding difficulties, and autism spectrum disorder. In those patients with isolated hyperphagia and obesity, genetic defects in *POMC* (encodes proopiomelanocortin), *MC4R* (encodes melanocortin 4 receptor), *LEP* (encodes leptin), *LEPR* (encodes leptin receptor), or a variety of other single genes should be considered, especially as new treatments are being developed for several. Rapid-onset obesity with hypothalamic dysfunction, hypoventilation, and autonomic dysregulation (ROHHAD) is a recently described phenotype of unknown etiology that is clinically distinguishable from PWS (Barclay et al. 2018) based on the first year of life being completely normal in ROHHAD, but never in PWS. Intellectual disability disorders in which obesity is an occasional finding, such as Fragile X syndrome (see Chapter 28), Down syndrome (see Chapter 24), Angelman syndrome (see Chapter 5), and Rett syndrome (see Chapter 49) may also be confused with PWS. A number of different chromosomal deletions and duplications not involving the *PWS* region of chromosome 15 have also been described in individuals with a phenotype suggestive of PWS (for example, deletion 1p36, see Chapter 18), deletion 6q16.2, deletion 10q26, deletion 16p11.2 (which includes *SHB2B1*, which is involved in leptin and insulin signaling), and duplication Xq27.2-ter. Acquired hypothalamic injury from accidents, tumors, or surgical complications can closely mimic PWS, particularly the endocrine aspects. Hyperphagic short stature is an acquired disorder due to psychosocial stress that includes hyperphagia, mild learning disabilities, and growth hormone deficiency (Gilmour et al. 2001). The use of large DNA sequencing panels and whole exome or whole genome testing is likely to identify other genetic causes for overlapping phenotype manifestations.

MANIFESTATIONS AND MANAGEMENT

The availability of highly sensitive and specific diagnostic testing has led to early recognition of PWS, often in the neonatal period, although the diagnosis is still delayed or missed at times. The consequence of delay in diagnosis is a delay in intervention, which should begin in the neonatal period, with the resultant difficulty of treating complications after they happen, rather than the more successful approach of management to avoid complications. Appropriate management can have a significant positive impact on health and quality of life, but controlling the characteristic obesity and difficult behavior constitutes a major challenge, requiring cooperative input from primary care physicians, geneticists, endocrinologists, nutritionists, psychologists, psychiatrists, physical, occupational and speech therapists, educators, and employers, as well as families and other caregivers.

As noted above, intervention and management of PWS can significantly affect the health, functional abilities, and longevity of affected individuals. Despite the relatively mild cognitive impairment in PWS, essentially all affected individuals require sheltered living and working environments and considerable medical monitoring, with 24-hour supervision to control the consequences of their hyperphagia and behavioral outbursts.

The reader should be aware that most of the recommendations for management of PWS are based on the experience of experts rather than on controlled studies, with the exception of those relating to growth hormone replacement. However, there is considerable communication among the experts on PWS, with annual national scientific meetings in many countries, and international scientific meetings at least every three years. Many of the suggestions for management described here are based on the authors' extensive personal experience, in conjunction with review of published experience and the input of other experienced physicians, often based on multidisciplinary clinics specifically for PWS.

Growth and Feeding

Infantile Failure to Thrive. Infantile hypotonia (see below) causes great difficulty with sucking, and the resulting feeding issues are compounded by hypersomnolence. Breast feeding is rarely possible, and special feeding techniques, including special nipples or gavage feeding, are usually necessary for weeks to months to assure adequate nutrition and avoid failure to thrive. Eventually, with improving muscle tone and increased arousability comes improved feeding, and a short period (months to a few years) of relatively normal eating behavior usually occurs. Swallowing dysfunction has recently been recognized in many infants with PWS, often with silent aspiration (Salehi et al. 2017). Thus, swallow study and professional evaluation may be indicated if there is suggestion of choking or pulmonary disease reflecting possible aspiration, and may be appropriate in all infants with PWS.

Short Stature. Approximately 90% of people with PWS who have not been treated with growth hormone will have short stature relative to their family background, either beginning early in life or in the late second decade of life. The growth pattern is abnormal in PWS, and growth charts representing syndrome-specific typical growth patterns have been published (Butler et al. 2015, 2016). In infants, length is below the 50th centile in most and below the 3rd centile in some, despite adequate nutrition. With the end of the severe neonatal hypotonia phase, a normal growth velocity is often found until early in the second decade of life, although height may be below the 5th centile. However, some children are average or even tall for age. Small hands and feet are typical, although not universal.

Short stature is related substantially to growth hormone insufficiency and lack of an adolescent growth spurt. Without intervention, average height is 155 cm for males and 148 cm for females (Butler and Meaney 1991), although it is a little higher in individuals of African heritage (Hudgins et al. 1998). Most affected individuals have been well-documented in numerous studies to have decreased growth hormone secretion with low IGF-1 (insulin-like growth factor-1) (reviewed in Deal et al. [2013]).

Growth hormone replacement therapy is well established in PWS, with guidelines for use (Deal et al. 2013). Affected individuals grow well on treatment, improve, but don't completely normalize, body composition, have less characteristic facial appearance, have improved adaptive behavior, and have a variety of other salubrious effects (Carrel et al. 2010; Deal et al. 2013, Wolfgram et al. 2013, Lo et al. 2015, Tauber et al. 2016). A growing body of literature also supports the use of GH in adults with PWS (Sode-Carlsen et al. 2011; Höybye 2015). PWS has been recognized by the US Food and Drug Administration as an indication for growth hormone treatment in childhood.

Expert consensus guidelines for use and monitoring of growth hormone in PWS have been published (Deal et al. 2013). Tonsillar hypertrophy is a well-recognized occurrence in children treated with growth hormone, which may worsen pre-existing obstructive apnea in people with PWS. It also appears that use of growth hormone doses that result in supra-physiological response of IGF-1 may be associated with increased risk of respiratory complications and death (Miller et al. 2006; Tauber et al. 2008); therefore, it is recommended that individuals with PWS have a polysomnographic evaluation (sleep study) before starting growth hormone so that pre-existing obstructive sleep apnea can be identified and treated, although an abnormal sleep study, especially in an infant, is not typically a contraindication to starting growth hormone. A follow-up sleep study several months after initiating therapy can be useful in identifying people with worsening sleep apnea. Finally, serum IGF-1 (insulin-like growth hormone 1) should be monitored with the goal of adjusting the growth hormone dose, as needed, to keep IGF-1 in the physiological range.

Oxytocin is an important hypothalamic hormone that has multiple effects in the central nervous system, including in socialization, behavior and feeding. Several studies suggest that oxytocin regulation may be disrupted in children with PWS because oxytocin concentrations are increased in their plasma (Johnson et al. 2016) and in their cerebral spinal fluid (Martin et al. 1998). A randomized, controlled trial in adults did not demonstrate benefit (Einfeld et al. 2014), and at high doses may have undesired effects thought to be due to activation of vasopressin receptors. Other studies suggest that early administration of oxytocin to infants with PWS may show promise as a treatment to improve poor social behavior and feeding in infancy (Miller et al. 2017; Tauber et al. 2017). Safety and efficacy of long-term administration of oxytocin have yet to be demonstrated, so additional evidence is needed before this treatment can be safely recommended.

Obesity. This major problem is discussed in a separate section.

Evaluation

- Both length/height and weight should be monitored frequently by the physician and plotted on general population growth charts. Recently updated PWS-specific growth charts reflect the expectation that most children will be treated with growth hormone, and that the characteristic growth pattern of infants and children with PWS reflects mildly increased adiposity and reduced lean mass in spite of optimized diet management (Butler et al. 2015, Butler et al. 2016). The goal for people with PWS should be growth in the normal population range, with maintenance of body mass index (BMI) below the range associated with increased risk of adult obesity (i.e. 85th percentile). In the first few months of life, monthly length and weight measurements are important to identify failure-to-thrive and to determine the appropriate caloric intake to maintain normal growth. In the first several years at least six-monthly measurements should be obtained and plotted to identify abnormalities of growth or onset of obesity, with frequent assessment of the adequacy of caloric intake.

- Older affected children should have height measured at least annually and plotted along both standard growth charts and the appropriate PWS chart (with or without growth hormone).

- Those who have growth deficiency even after infancy should be assessed for possible hypothyroidism through measurement of TSH, T4, and T3-uptake. Some affected individuals with poor growth and prolonged hypotonia are hypothyroid. The incidence of hypothyroidism in PWS is about 15%. Thyroid function should also be assessed as part of the evaluation of short stature and before growth hormone treatment is started.

- In some countries (including the US), growth hormone stimulation testing is no longer required prior to initiating growth hormone therapy, particularly if growth failure is present. If required, however, standard evaluations are appropriate and should be done shortly after diagnosis, even in early infancy (Deal et al. 2013). These should include measurement of baseline IGF-1 (insulin-like growth factor 1) and IGFBP3 (insulin-like growth factor binding protein 3). Many health insurers in the US require demonstration of growth hormone deficiency in adults before covering treatment; this can be identified prior to initiation of growth hormone therapy using standard testing procedures.

- A careful assessment of respiratory status and tonsillar size and a sleep study (polysomnography) are strongly recommended prior to initiation of growth hormone treatment in all affected individuals (Deal et al. 2013), although some centers are not performing formal sleep studies in infants starting growth hormone therapy in the first few months of life. After initiating growth hormone therapy, close medical monitoring is indicated to identify any change in status, including unexplained behavior changes or worsening of excessive daytime sleepiness, and full consideration should be given to repeating the sleep study several months after initiating therapy.

- During long-term growth hormone treatment, it is important to closely monitor for side effects such as edema, pseudotumor cerebri, and abnormal glucose tolerance. It is also important to monitor for development or worsening of scoliosis, because acceleration of scoliosis has been identified in a number of individuals with PWS on treatment. This can be done through clinical examinations at quarterly visits or can be augmented with annual radiographs of the spine. Studies have documented that there is no increase in the frequency or severity of scoliosis in people with PWS treated with growth hormone, only in the timing of progression (Colmenares et al. 2011; Murakami et al. 2012).

- Growth hormone treatment has been shown to increase lumber spine bone density (Nakamura et al. 2014). Macrocephaly and excessive growth can occur with inappropriately high doses of growth hormone replacement, and head circumference as well as length/height should be measured and plotted regularly. This does not occur if IGF-1 is maintained in the physiological range.

Management

- In the newborn period the child must usually be awakened to feed for at least the first several weeks of life. High-calorie formula with or without a supplemental calorie source may be needed. Attention should be paid to the length of time to feed, because the work of feeding may exceed the caloric benefit if prolonged feeding times are allowed. A gastrostomy tube is only rarely necessary because the severity of the hypotonia improves with time and poor feeding is transient. However, gavage feeding through a nasogastric tube is often required.

- Caloric intake should be individually determined based on the goal of achieving an appropriate rate of weight gain and is often less than predicted for typically developing infants.

- In early infancy, the parent or caregiver specifically should be instructed not to restrict fats because of fear of obesity in later life, since fats are important for normal nervous system development and growth.

- Supportive counseling for parents is an important function of the care provider during this difficult time, and assurance that the child will overcome this phase of the disorder is beneficial.

- Thyroid hormone should be replaced if it is deficient. Standard doses are appropriate.

- Consideration should be given to removing large tonsils and to weight reduction prior to initiating growth hormone therapy in very obese individuals.

- Growth hormone replacement therapy, considered standard of care in the US and many other countries, should be started using standard protocols, after assurance that obstructive sleep apnea and respiratory compromise are not present. Studies in the past have suggested that it is appropriate to start children by one year of age or younger (Wolfgram et al. 2013), using standard doses for children with growth hormone deficiency. Anecdotal experience suggests value in initiating growth hormone replacement in the first few months of life, which may enhance feeding and reaching motor milestones.

- Serum IGF-1 should be monitored during growth hormone therapy, with the goal of adjusting the growth hormone dose, as needed, to keep IGF-1 in the physiological range.
- Doses of growth hormone in adults significantly lower than those in children, and they should be titrated to normalize blood IGF-1 concentrations for age.
- If scoliosis appears to be accelerated in conjunction with growth hormone therapy, the decision as to its management should be made through consensus of the endocrinologist, the orthopedist, and the family. Growth hormone need not necessarily be stopped as the ultimate severity or need for surgery does not change with growth hormone treatment (Nagai et al. 2006; Angulo et al. 2007).
- Because a variety of new interventions, including intranasal oxytocin and carbetocin, are currently under study (https://clinicaltrials.gov/), readers are encouraged to review current literature for potential effective and safe therapies developed after this writing.

Development and Behavior

Developmental Delay. Delayed motor milestones are always evident. The average age of sitting is 12 months and of walking is 24 months. Poor coordination is common, particularly in those who are not physically active from early in life and in those who do not receive growth hormone therapy. Eventually, the majority of children learn to pedal a tricycle or bicycle. Adults remain mildly hypotonic with decreased muscle bulk and tone. Upper body strength is particularly deficient.

Language development is also delayed. Verbal skills are an ultimate strength in most affected individuals, although speech is often poorly articulated, having a nasal and/or slurred character. Velopharyngeal dysfunction can be seen in people with significant hypotonia, and children with PWS appear to have a higher than typical rate of this complication following adenotonsilletomy (Crockett et al. 2014). Dysfluency is common. A minority of affected individuals have persistence of very immature speech or severe apraxia. Individuals with uniparental disomy (UPD) tend to have lower receptive than expressive language, an unusual combination not seen in individuals with PWS due to deletion, although overall speech and language skills are abnormal in both groups (Dimitropoulos et al. 2013).

Cognitive Disability. Cognitive abnormalities are evident, and most affected individuals have mild intellectual disability with mean full scale IQ typically measured in the 60s (see Whittington and Holland [2017] for review). Approximately 40% have borderline intellectual disability or low normal intelligence, and about 20% have moderate intellectual disability. Academic performance is often below what is expected for the individual's cognitive ability. Specific patterns of cognitive strength and weakness have been observed, frequently with relative strength in reading, visual-spatial skills, and long-term memory and weakness in arithmetic, sequential processing, and short-term memory (reviewed in Whittington and Holland [2017]), although this pattern is not universal. Coming to clinic with a book of word-find puzzles can almost be considered a diagnostic sign for PWS (personal observation), and unusual skill with jigsaw puzzles is common and appears to be related to using the shape of the piece, not the image, thus the skill is not enhanced with picture puzzles with straight-edged pieces (Verdine et al. 2008). Most affected individuals can ultimately read, write, and do simple mathematics.

Behavioral Disorder. A characteristic behavior profile becomes evident in early childhood, with temper tantrums, stubbornness, controlling and manipulative behavior, compulsive characteristics, and difficulty with change in routine (see Whittington and Holland [2010] and Ishii et al. [2017] for recent review). Lying, stealing, and aggressive behavior are common, especially in older children and adults. Several studies have documented a high prevalence of obsessive-compulsive symptoms (Dykens et al. 1996; State et al. 1999). Although these problems are not unique to PWS, they do occur with greater severity and frequency in people with PWS relative to some other groups with mild-to-moderate intellectual disability. About 70% of affected individuals show a mixture of compulsive behaviors such as hoarding, ordering and arranging, redoing, and needing to tell or ask (Dykens et al. 1996; State et al. 1999). Parents often remark about repetitive question asking and getting stuck on ideas (personal observation). Many people with PWS are also described as quite clever and manipulative, particularly in regard to obtaining food. Behavioral issues not infrequently jeopardize home life, group home placements, and school and employment situations. Many of the behavioral characteristics are suggestive of autism spectrum disorder, which has been diagnosed in up to 25%, more in adolescents than in younger children, and with greater probability in children with UPD compared to children with deletion (Ogata et al. 2014, Bennett et al. 2015, Dykens et al. 2017). Attention deficit/hyperactivity symptoms and insistence on sameness are common and of early onset (Steinhausen et al. 2004; Wigren and Hansen 2005; Ogata et al. 2014).

Skin picking and rectal picking might also be considered behavioral findings but are discussed under the Dermatologic section.

Behavior problems have been reported to increase with age and body mass index. Behavior problems interfere most with the quality of life in adolescence and adulthood, although they have been reported to diminish considerably in many older adults (Dykens 2004).

Mental Health Issues. Anxiety is common in people with PWS at all ages, and is exacerbated by food-related issues and also by changes in routine and/or lack of routine (mental health issues reviewed in (Skokauskas et al. 2012)). Depression is also common and may be difficult to ascertain due to pre-existing sleep disruptions, behavior issues, and the fact that loss of interest in eating is extremely rare in PWS. Behavior changes, sadness, lack of interest in work or entertainment activities, and increasing lack of responsiveness to outside stimulation may be better indicators of depression.

Psychosis, including bipolar disorder, appears to be relatively common in older adolescents and adults with PWS. One study of 95 individuals found that six (6.3%) had psychotic symptoms in the one month prior to the evaluation (Clarke 1998). A meta-analysis combining five other reports suggested an overall incidence of psychosis of ~25% in people with PWS due to deletion and 64% in those with UPD, (Yang et al. 2013), with even more skewing toward UPD for bipolar disorder. Onset is often in young adulthood. Psychiatric disorders and their management in this syndrome have been reviewed (Dykens and Shah 2003; Soni et al. 2007).

Genotype–Phenotype Differences. As a group, people who have PWS on the basis of a 15q deletion tend to have more numerous and more severe behavioral disturbances than those with UPD 15, although both genotypes are associated with significant problems in this area (see (Dykens and Roof 2008)). Evidence for behavioral and intellectual differences in individuals with type I deletions compared to type II deletions has so far not clearly and reproducibly delineated meaningful findings that impact individual care. One study provided compelling evidence that significant externalizing behaviors (e.g., physical aggressiveness, cheating, stealing, damaging property) decline in later adult life in individuals with type I deletions, whereas they tend to remain stable in those with either type II deletion or UPD (Dykens and Roof 2008).

Evaluation

- Developmental skills attainment should be closely evaluated in infancy and toddler years.
- Careful educational, behavioral, and psychological assessments should be conducted in any affected individual. This should include an evaluation of cognitive strengths and weaknesses.
- Assessment should be made for evidence of obsessive-compulsive behaviors, impulse control problems, and depressive disorders. Determination of whether compulsive tendencies lead to getting stuck on thoughts should be made, as intervention can be directed at these problems if they are identified.
- Assessment of family support, parenting skills and psychosocial/emotional needs will assist in designing family interventions.
- Signs suggestive of a thought disorder, such as radical behavior change, hallucinations, delusions, or disorientation, should be evaluated by a psychiatrist for the possibility of psychosis.

Management

- The child should be involved in early intervention services beginning as young as possible.
- Educational intervention throughout the school years (at a minimum) should include addressing individual strengths and weaknesses and behaviors. Most children with PWS require either placement in a special educational classroom or special assistance in a regular classroom. Both inclusion and self-contained classroom settings have been effective, depending on the level of skills and the extent of behavior disorder (personal experience). Individual aides in the classroom can be invaluable in assuring that the child attends to task, and the use of personal aides has been particularly effective when the child is in an inclusion program or a regular classroom.
- Speech therapy and physical/occupational therapies, increased physical activity, and supervision during lunches are all important for most individuals with PWS during school. Literature for therapists and educators concerning the special treatments for PWS, including behavioral management guidelines, are available through the Prader–Willi Syndrome Association (USA) (see Resources).
- Provision of clear behavioral expectations and limits by all caregivers, beginning at an early age, is an extremely important part of behavior management. Applying consistent limits at home, school, and work is critical. Preparing parents very early in the life of the affected individual to be able to set and enforce limits consistently appears to be a good predictor of fewer behavior problems in the future (personal experience).
- Formal social skills training is invaluable, because most individuals with PWS have difficulty reading social cues and maintaining friendships with peers.

- Before graduation from school, planning for vocational placement is indicated. Most people require sheltered work or work with a job coach. Optimal work settings involve physical activity and absence of exposure to available food. Some of the occupations that have been excellent are landscaping, mail or flyer delivery, working at a florist shop, maintenance work, child or pet care, and other active employment. Inactive piecework at a sheltered workshop is far less desirable than many other jobs. If given responsibility and respect, most affected individuals will rise to the occasion.
- It is important for care providers to maintain consistency in daily routines. The individual should be prepared in advance for changes in routine or in planned activities. These things help significantly with the management of anxiety.
- Awareness of the impact of tantrums and other maladaptive behaviors on family stress can be useful in planning behavioral intervention.
- Many people with PWS who have severe behavior problems respond to individual psychotherapy or, in some cases, group therapy with other developmentally disabled individuals. Selecting a behavior therapist interested in working with the disabled is not always easy, but such people exist and can be extremely helpful.
- Pharmaceutical treatment of behavior and psychiatric problems in people with PWS is often needed and is frequently helpful. Sometimes, this treatment is required in childhood, although use is much more frequent in adults. Treatment agents should be individualized, although specific serotonin reuptake inhibitors have often been beneficial (Dykens and Shah 2003, Soni et al. 2007). The proliferation of psychotropic medications suggests that a psychiatrist with experience with PWS should be sought in prioritizing treatment. Affected individuals appear to be more sensitive to all medications than the average individual, so starting with low doses is recommended. For those with evidence of psychosis, risperidone has been particularly effective (Durst et al. 2000; Bonnot et al. 2016) despite the side effect of increased appetite. Sedating agents should be avoided, if possible, as they may interfere with weight management and learning efforts.
- For adults who leave their family home, careful consideration must be given to living situations. Although a variety of options have been successful in individual situations, the most successful setting for the largest number of individuals has been group homes specifically designed for individuals with PWS. These homes have only low-calorie foods, everyone is on a special diet, and food is inaccessible. Regular physical activities are encouraged. Behavioral limits can be strictly enforced. Studies have shown that dedicated group homes are the most effective in reducing and maintaining weight over time and in managing behavioral difficulties (Cassidy et al. 1994; Kaufman et al. 1995).

Obesity and Hyperphagia

Obesity is nearly always present in PWS after 6 years of age if it is not actively avoided. Avoidance of obesity is one of the most difficult problems of PWS, and obesity is the major cause of morbidity and mortality in this disorder. One consequence of hypotonia and growth hormone insufficiency in PWS is decreased lean body mass, which results in a relatively high ratio of fat-to-lean body mass even in people whose weight-to-height ratio is normal. It also likely contributes to decreased resting metabolic rate and subsequent decreased caloric requirement (van Mil et al. 2000). The ratio of fat-to-lean body mass is helpful in recognizing the early signs of insatiable appetite and compulsive eating (hyperphagia) and the resultant obesity, which have their onset soon after the neonatal/infantile hypotonia improves, usually between ages 1 and 6 years, with an average age range of 2 to 4 years. With intervention, onset of obesity is later than in the past (personal observation), likely because of improved education of families with implementation of good behavior management and use of growth hormone. Since obesity tends to progress with age, a plan should be developed at the point of increasing weight-to-height ratio in order to avoid increasing obesity.

The obesity is central in distribution, with relative sparing of the distal extremities (Figures 46.1 and 46.2). Even individuals who are not overweight tend to deposit fat on the abdomen, buttocks, and thighs.

Excessive obesity in people with PWS leads to cardiopulmonary compromise, type II diabetes mellitus (see Endocrine below), hypertension, thrombophlebitis, and chronic leg edema. Premature cardiovascular disease has been described in a few individuals, and a study using data from the Danish Health Registries showed an overall relative risk in individuals with PWS compared to the general population of 7.2, with age group specific risks even higher for individuals in the third and fourth decade of life (Hedgeman et al. 2017). In one study, 25% of children with PWS had elevated levels of low-density lipoprotein cholesterol and apolipoprotein B (l'Allemand et al. 2000). Obstructive sleep apnea occurs at increased frequency (Nixon and Brouillette 2002; Yee et al. 2007), but the other sleep disturbances seen in PWS appear to be unrelated to obesity (see Sleep below). Breakdown of skin in the intertriginous skin folds is common in those with significant obesity (personal experience).

The major contributor to obesity is likely to be the excessive eating, or hyperphagia, that is characteristic of the disorder. The hyperphagia is present even in those whose weight is well-controlled. The physiological basis for the hyperphagia is still to be fully elucidated, but it appears to be in part a consequence of a hypothalamic abnormality that results in lack of satiety. The hyperphagia and associated binge eating can result in acute gastric necrosis, as noted above.

Other contributors to obesity exist. Altered body composition is present in individuals with PWS, such that the lean body mass is low and the fat mass is high and may be as much as 50% of body weight. This is true even at normal weight-for-height. This is apparently due to a combination of decreased muscle mass and decreased bone mineral content. In addition, reduced physical activity and low energy expenditure has been well documented in a number of studies in PWS (reviewed in (Burman et al. [2001]). The contribution of growth hormone deficiency to these abnormalities is significant, and the deficiencies can be ameliorated by growth hormone replacement.

Recently, a pharmacological intervention, a methionine amino peptidase 2 (METAP2) inhibitor was shown for the first time to significantly impact the hyperphagia and obesity over more than six months of treatment (McCandless et al. 2017). The study had to be stopped because of deaths related to thromboembolic events; however, it established the principle that hyperphagia and obesity in PWS can be targeted pharmacologically.

The result of the METAP2 inhibitor study suggests the possibility of a peripherally acting effect (not at the level of the hypothalamus), and, together with the recognition that weight gain typically precedes the onset of hyperphagia (Miller et al. 2011) and that resting energy expenditure is reduced, suggests a possible underlying defect in the metabolic recognition of caloric need; that the body behaves as though the individual is starving and undernourished in spite of excessive adipose stores. This is consistent with a hyperphagic mouse model (deletion of the PWS imprinting center), where consumption of sweet food is markedly decreased if it is made non-caloric, strongly suggesting that it is drive for energy, and not the hedonic response to feeding, that drives the hyperphagia (Davies et al. 2015).

Although there is variation in the severity of hyperphagia, it is almost always present and causes the individual to constantly seek food, forage for food, and eat food that most people would consider inedible (e.g., a bag of sugar, garbage, or rotten or frozen food). A high threshold for vomiting may complicate bingeing on spoiled food from the garbage or such items as boxes of sugar or frozen uncooked meat, and toxicity from ineffective emetics used to induce vomiting has occurred. However, with proper management, the obesity can be controlled (Figure 46.3).

Surgical approaches to obesity management have been reported. A review of early work did not suggest long-term benefit (Scheimann et al. 2008). Several reports since then have generated controversy with markedly varying perspectives on utility based on the same data. For example, in one study utilizing a duodenal switch procedure (Marceau et al. 2010), one of three patients died during re-operation, the other two were regaining weight after three to five years. The authors posited that these outcomes justified the utility of the procedure for morbidly obese individuals with PWS as they perceived that there is no viable alternative. Similarly, a report of 24 patients with PWS undergoing laparoscopic sleeve gastrectomy with less than one to five years follow up showed that weight control was good for up to three years followed by regaining weight (Alqahtani et al. 2016). Again, the authors posited that the limited results were justified by the severity of the disease; however, none reported on the sense of hunger, which is the primary driver of problems and dissatisfaction in PWS, nor did they report on the use of external controls of food, a likely significant covariant in the data. The fact that weight regain is the norm suggests that the primary symptom, hunger, is not assuaged by surgical approaches. At this time, the external control of access to calories remains the most effective and humane approach to therapy. The role of surgical intervention, if any, remains unclear.

Attempting to reach an ideal body weight is often impractical in this disorder, especially without growth hormone treatment. Aiming to keep the BMI at or below the 85th percentile seems reasonable (personal experience).

Hyperphagia and the potential for obesity are long-term problems in PWS. It is inappropriate to expect that individuals with PWS will eventually learn to control their own food intake. Until an effective medication or other treatment for hyperphagia can be found, a supportive living environment sensitive to this problem is essential.

Evaluation

- Close monitoring of weight centiles and height-to-weight ratio, including plotting on growth charts to identify crossed centiles, is critical throughout life. This is particularly true in the first few years of life to recognize when obesity has its onset.
- Assessment of glycosylated hemoglobin levels to assess chronic hyperglycemia in those who are significantly obese and in those who have experienced a rapid weight gain in a short period of time is helpful in early identification of diabetes.
- Regular routine physical examinations for complications of obesity are needed in overweight individuals with PWS at least yearly.
- Evaluation of caloric intake by a registered dietician is usually enlightening to families.

Management

- Unfortunately, no medication or surgical procedure presently available has had long-term effectiveness in controlling appetite without serious potential complications. Therefore, a low-calorie, well-balanced diet combined with a regular exercise program and close supervision to minimize taking of extra food should be instituted early, but no later than when excessive weight gain is first noted through crossing of weight centiles.

- Decreased caloric requirement (as low as 60% of comparably aged unaffected people of similar stature) must be kept in mind in planning diets. Intake should be matched to that needed for the desired growth rate and/or desired weight and will rarely be above 1000–1400 Kcal/day unless the individual is on growth hormone and/or exercising a great deal. There is considerable variability in dietary recommendations among those who conduct specialized programs for management of PWS. Approximately 800–900 Kcal/day is a good level at which to start in toddlers and very young children, working slowly up by increments of 100 Kcal as judged by the growth chart (personal experience). The extent of activity, height, and use of growth hormone treatment influence allowable caloric intake guidelines. For those who have already become obese, intake of approximately 1000 Kcal/day in individuals over age 6 years generally results in slow but steady weight loss and is adequate for healthy nutrition. Growth hormone treatment generally increases the allowable caloric intake to some extent, in part due to increased lean body mass and in part to increased activity.
- A variety of special dietary approaches have been proposed, but the most success is obtained by careful selection of caloric intake to support normal growth and to allow for satisfaction with the amount of food provided, while providing a variety of nutrition options. Specifically, a diet with restricted calories, but balanced in terms of macronutrients, and high in dietary fiber, has been shown to be effective (Miller et al. 2013). This typically means a diet based primarily on fruits and vegetables.
- Involvement of a registered dietician, including regular follow-up, is beneficial to weight management.
- Regardless of diet type, supplemental vitamins and intake of the recommended daily allowance of calcium should be assured to maximize bone mineral density.
- Essential fatty acid supplements, often in the form of naturally occurring oils (e.g., fish oil, evening primrose oil, or walnut oil) are frequently suggested as beneficial; however, published data to support this assertion are lacking.
- Access to food should be as limited as possible, because in many cases the prescribed diet is only part of food intake if additional food is obtainable. Keeping temptation out of reach allows the individual with PWS to turn his or her attention to other matters (personal experience). In most cases, this entails locking the kitchen or locking the refrigerator and pantry. It is important to help families determine how best to respect the rights of other members of the household while still reducing temptation for the affected person. One family known to the authors used an elaborate system of motion detection alarms in their kitchen.
- Consistency of meal times should be a goal, as it decreases anxiety in the person with PWS.
- Encouraging families to keep only healthy, low-calorie food in the house is appropriate. Arranging for supervision of school aged children during lunch time to prevent them from finishing other children's lunches or getting a second lunch has been effective, as has avoiding giving pocket money, which can be used to buy food at school or work by older children and adults. Many schools are obligated to provide low-calorie lunches, if requested by a physician or dietician.
- Teachers and employers of people with PWS should be educated about the importance to health of controlling food access and intake, and other students or co-workers should also be informed. Family or other home care providers and those responsible for day programs should be encouraged to communicate regarding treats and food sneaking, and dinner intake can be adjusted to compensate for excesses earlier in the day. It is often helpful to explain that removing the possibility of getting extra food helps reduce anxiety and reduce problem behaviors.
- Behavioral management programs, including firm limit setting and enforcement, should be instituted simultaneously with diet and as early as possible. People with PWS can be very convincing and can develop elaborate tales of having missed meals. Many have been found to be getting two breakfasts or two lunches this way. Consistency over time and communication among all care providers is essential. It is important for providers to be aware that many families have difficulty denying their child food, particularly among ethnic groups that associate feeding the child with showing love. Supportive counseling is critical. A behavioral psychologist or other behavior specialist is an important part of the management team (personal experience).
- Regular exercise is also an extremely important part of weight management, and most affected individuals have a strong tendency to inactivity, though this is less so in those on growth hormone therapy. Exercise can build muscle mass as well as burn calories, thus increasing metabolic rate. At least 30 min of some type of physical activity daily, with rewards or other behavior modification techniques to encourage exercise, is appropriate (personal experience). Sixty minutes of physical activity each day is even better.

Initiating a daily exercise program as early in life as possible, and making it part of the daily routine, has been the most effective. Some adults who do 45–60 min of exercise biking or similar activity daily have been able to eat essentially a normal diet (i.e. 1,800–2,000 Kcal/day) once they have reached a healthy weight (personal experience).

- Treatment of severe obesity accompanied by cardiopulmonary compromise or other major medical problems is complicated. Despite difficulties with resistance to hospital admission for treatment of obesity, and problems convincing insurance companies in some countries that this is not just an esthetic issue, experience has indicated that hospitalization is usually necessary if there is significant morbidity (personal experience). On occasion, treatment can be accomplished as an outpatient by removing the individual from his or her current living situation and placing him or her in a more restrictive environment in which firmer limit setting takes place. If hospitalization is needed, supervision is critical as some individuals will gain weight due to the unrestricted environment. In addition, care should be taken prior to discharge to alter the environment that allowed the individual to become morbidly obese in the first place. Education of parents and/or care providers about avoidance of morbidity is critical.

- Significant benefit in weight management has been noted in both children and adults with PWS who have been treated with growth hormone in that body composition improves toward normal through decreased fat mass and increased muscle and bone mass (Höybye et al. 2003; Carrel et al. 2010; Sode-Carlsen et al. 2011; Wolfgram et al. 2013). Most improve their body mass index and can eat more without gaining weight. Treated individuals have also been shown to be more active (Butler et al. 2013), which also helps with weight. These effects appear to revert to pre-treatment status with cessation of growth hormone administration.

- Treatment of the complications of obesity does not differ significantly from that in the general population. However, in treating diabetes mellitus, the possibility of poor compliance with diet should be kept in mind, and frequent monitoring of fasting blood glucose and/or glycosylated hemoglobin should be used to assess compliance.

- Since people with PWS rarely vomit and appear to have an increased pain threshold, complaints of abdominal pain and vomiting, especially in the face of abdominal distention, should raise immediate concern for the possibility of a catastrophic intestinal mishap. Choking on food not sufficiently chewed, especially hot dogs, is a significant cause of death in people with PWS.

Neuromuscular

Hypotonia is a medically and developmentally significant finding in infants and children with PWS, and, to a lesser extent, in adults. In virtually all cases, the neonatal period is dominated by hypotonia, lethargy, somnolence, and weak suck leading to feeding difficulties. Deep tendon reflexes are typically spared. Muscle electrophysiological and biopsy studies, often done early in life in prior years in the search for a diagnosis, are generally normal or non-specifically abnormal (see also below). These manifestations usually lead to prolonged hospital stay to assure adequate feeding and growth. A small proportion of infants with PWS have severe hypotonia lasting more than two years. This is uncommon and necessitates evaluation for other causes of hypotonia. In particular, hypothyroidism occurs in 15%. Hypotonia also may be a factor in the increased frequency of strabismus.

In addition to hypotonia, at all ages there is associated decreased muscle bulk as well as poor coordination and, often, decreased strength. Together, these likely contribute to the inclination to inactivity and to the reduced daily energy expenditure seen in individuals with PWS. Both of these complicate the management of obesity.

There has been an anecdotal association between PWS and mitochondrial myopathy based on several children found to have abnormal muscle biopsies during their evaluation for hypotonia, prior to the diagnosis of PWS being made. This observation, along with the generalized hypotonia, has led to the suggestion of therapy with Coenzyme Q10 (CoQ10), a cofactor of the mitochondrial electron transport chain and an anti-oxidant. No well-controlled studies have investigated the use of CoQ10 in individuals with PWS. One report documents that plasma CoQ10 concentrations are not different in individuals with PWS compared to other obese individuals and are not different from non-obese controls after controlling for other factors known to influence circulating CoQ10 concentrations (Butler et al. 2003). While many families have reported improved developmental milestone achievement and energy in treated infants, there are others who report no effect (personal experience). Potential adverse effects, such as gastrointestinal upset or increased serum transaminases, are generally mild and reversible. There are no published data supporting the use of CoQ10 in people with PWS; however, should a family choose to try it, recommendations concerning this lack of data and issues related to safe use of CoQ10 are available on the website of the Prader–Willi Syndrome Association (USA) (www.pwsausa.org).

A number of findings suggest a variety of other neurologic symptoms. Abnormalities of the autonomic nervous system, primarily parasympathetic, have been reported and these may be responsible for the decreased salivary flow and the high threshold for vomiting. A high threshold for pain is also described by parents and care providers, and it contributes to morbidity and mortality in that affected individuals often will not complain very much when they have broken bones (personal experience) or have severe abdominal problems. Impaired peripheral somatosensory function has been demonstrated in one study of PWS, possibly reflecting a low density of peripheral nerve fibers (Brandt and Rosen 1998). The latter has led to ruptured internal viscera before care providers were aware of significant problems (see Obesity and Hyperphagia, above).

Seizures occur in 10–20% of people with PWS. They tend to have an onset in early childhood, be generalized, not frequent, easy to control, and resolve after a few years (Takeshita et al. 2013; Verrotti et al. 2015).

Evaluation

- Physical therapy evaluation is appropriate for all infants and children with PWS to assess the necessity for ongoing therapy to improve muscle bulk and strength.
- Infants and young children with prolonged symptomatic hypotonia leading to prolonged failure to thrive and lethargy or severe developmental lag compared with the average for PWS should be evaluated for other causes, including hypothyroidism, by standard methods.
- If the possibility of supplementation with CoQ10 arises, it is recommended that the blood level be tested before and after initiation of therapy to ensure that the drug is being absorbed. Significant deficiency of CoQ10 should lead to investigation for other causes. A normal blood concentration is not a contraindication to a therapeutic trial of relatively high dose CoQ10.

Management

- Early intervention, particularly physical therapy, is recommended from early infancy to avert muscle atrophy from inactivity. It may be of value in improving muscle strength and encouraging achievement of developmental milestones. For the most part, long-term physical therapy is of benefit to compensate for persistent hypotonia and poor coordination through improved muscle strength, agility, and activity.
- Growth hormone has been used from young infancy through adulthood with evidence of significant benefit in improving muscle mass and possibly tone (see Growth and Feeding, above).
- The possible use of CoQ10 may be discussed with the family, including the fact that the evidence supporting its use is anecdotal and inconclusive and the cost is unlikely to be covered by insurance. Recommended starting doses are from 10–25 mg/kg/day for infants and usually not more than 200 mg/day for older children and adults. Blood concentrations should be monitored to confirm adequate absorption, since most available brands are not well absorbed. The most bioavailable formulations of CoQ10 are in gel or liquid form with an oily carrier.

Endocrine

Generalized hypothalamic insufficiency leading to insufficiency of the hypothalamic–pituitary axis leads predictably to a variety of endocrinological manifestations in PWS. These include growth hormone insufficiency and hypogonadotropic hypogonadism in almost all individuals. Further, there is a marked increase in the risk of hypothyroidism at every age, and possibly increased risk of centrally mediated adrenal insufficiency. Type II diabetes is also a common complication in markedly obese individuals and those with substantial rapid weight gain.

Hypogonadism of prenatal onset is present in most, but not all, individuals with PWS. It is generally characterized by deficiency of gonadotropins, estrogen, and testosterone, and, until recently, has been assumed to be mainly hypothalamic in origin. Further studies have suggested a combination of hypothalamic and primary gonadal deficiencies (Hirsch et al. 2009; Gross-Tsur et al. 2012; Siemensma et al. 2012), a conclusion largely based on the absence of hypogonadotropism and abnormally low inhibin B in some affected individuals of both sexes.

Hypogonadism is evident at birth as genital hypoplasia. It manifests as undescended testes or cryptorchidism, scrotal hypoplasia (small, hypopigmented, and poorly rugated), and sometimes a small penis in males, and by hypoplasia of the labia minora and clitoris and sometimes the labia majora in females. While more apparent in the newborn period, it may be difficult to appreciate later in prepubertal girls. Genital hypoplasia persists throughout life, although spontaneous descent of testes has been reported up to adolescence. (Crinó et al. 2003). The same study documented hypoplasia or absence of labia minora and/or clitoris in 76% of females.

Hypogonadism is also evident in abnormal pubertal development in both males and females with PWS. Pubic and axillary hair may develop early or normally. Precocious pubarche occurred in 12 of 84 individuals and true precocious puberty in 3 of 84 in the study by Crinó et al. (Crinó et al. 2003). The remainder of pubertal development is usually delayed and incomplete. Adult males only occasionally have voice change, male body habitus, or substantial facial or body hair. In females, breast development generally begins at a normal age, but there is usually amenorrhea or oligomenorrhea. In the study of 84 individuals with PWS mentioned above (Crinó et al. 2003), pubertal onset occurred at 14 ± 3.2 years in males (range 8–23 years) but was incomplete in all cases. In females, pubertal onset was 12.6 ± 2.7

years (range 7.2–18 years, with two 18 year olds not having entered puberty as yet), and it progressed slowly. Menarche occurred in 14/32 females over 15 years (mean 17.3 ± 5.2, range 10.2–25.5), and 18 of the 32 females over age 15 had primary amenorrhea at a mean age of 19.6 ± 3.9 years. Of those with menarche, 36% had oligomenorrhea and 43% had secondary amenorrhea. The others had periodic vaginal spotting. Menarche may occur as late as the 30s, particularly in association with significant weight loss (personal experience). In both males and females, little is known about sexual activity, but there is a clear interest in romantic relationships and sexuality (Gross-Tsur et al. 2011). There have been two published adults with molecularly confirmed PWS, both female, who have been documented to be fertile, and the authors are aware of additional unpublished cases. There are no reported cases of male fertility, and biopsies at the time of surgery for cryptorchidism suggest that the testicular architecture required for complete spermatogenesis is inadequate in most, but not all, males with PWS (Vogels et al. 2008). Semen analysis in a single, sexually active male with PWS (UPD) did not demonstrate viable sperm (unpublished data).

Diabetes Mellitus. There is a significantly increased frequency of type II diabetes mellitus in PWS, primarily related to obesity. Studies of young individuals and those who are not obese indicate that most have normal insulin and glucose concentrations. One population-based study found a prevalence of diabetes of 25% in adults, with mean age of onset of 20 years (Butler et al. 2002). This group had a higher past maximum body weight, confirming the more anecdotal observation that diabetes is obesity-related in this disorder. Family history of diabetes may also contribute to its occurrence (Butler et al. 2002). Adults who rapidly gain significant weight may present for the first time with diabetes, which subsequently resolves with weight loss (personal experience). Burman et al. (2001) suggest that a possible interpretation of available data about glucose intolerance is that some degree of growth hormone insufficiency in affected individuals increases their insulin sensitivity, since non-obese children with PWS have been shown to have both low fasting insulin concentrations and normal blood sugar.

Central Adrenal Insufficiency. Just as other hypothalamic–pituitary axis systems are abnormal in PWS, there may be increased risk of centrally mediated adrenocortical insufficiency. In one study (de Lind van Wijngaarden et al. 2008), 60% of 25 randomly selected individuals with PWS had evidence of central adrenal insufficiency based on provocative testing using overnight metyrapone testing. Several other studies using low-dose synacthen (ACTH) suggest that the incidence may be overestimated (reviewed in (Edgar et al. 2016)). These two tests assess the hypothalamic–pituitary–adrenal axis in slightly different ways, and it is not clear which, if either, is more appropriate to determine which individuals with PWS may have blunted cortisol responses during illness or periods of stress (e.g., surgery). The safest course appears to be for providers to have high level of suspicion for the possibility of central adrenal insufficiency, to measure ACTH and cortisol during times of stress, and to have a low threshold for emergency treatment using standard stress doses of hydrocortisone if an individual is acutely and critically ill and has shown a previous lack of stress response or is not responding as expected to treatment for an acute illness.

Evaluation

- Physicians caring for children and adults with PWS may wish to consult with a knowledgeable endocrinologist to discuss diagnostic testing and management of possible central adrenal insufficiency, especially during severe illness, trauma, or around the time of surgery.
- The possibility of cryptorchidism should be assessed in every affected male, and the position of the testes should continue to be monitored, because retractile testes and the need for repeat orchidopexy have been noted often (personal experience).
- At the usual age of adolescence, pubertal development should be evaluated. Measurement of testosterone in males or estradiol in females in the early pubertal years is necessary before consideration of replacement therapy.
- Males should be assessed for the ability to urinate standing up. This seemingly minor result of hypogonadism and small phallus often can be treated with exogenous testosterone with significant benefit in terms of self-esteem. The same is true of facial hair and the need to shave in pubertal males.
- Adults with obesity should have glycosylated hemoglobin measured annually. Those with rapid weight gain at any age should be monitored for the possibility of diabetes.

Management

- Consideration should be given to prophylactic therapy with IV hydrocortisone during severe illness, pending results of serum cortisol and ACTH measured at the time of presentation with the acute illness.
- Administration of human chorionic gonadotropin (hCG) occasionally may stimulate testicular descent, and often enlarges the scrotum, improving the success of orchidopexy. In addition, it improves muscle bulk and may therefore accelerate early motor milestones, activity and feeding in infancy.
- In the general population, orchidopexy is done in the early months to preserve full fertility, to avoid neoplasia,

and for psychological and cosmetic reasons. There are no published guidelines for when to perform orchidopexy in males with PWS. However, it is relevant that fertility is probably not an issue, and testicular neoplasia has only been reported on one occasion in PWS (although there are a few additional anecdotal cases). In one large series, 90% of males required orchidopexy (Crinó et al. 2003). Those authors recommend that the same criteria be used for surgical intervention as in the general population.

- There have been few systematic trials of sex hormone replacement for adults with PWS (Kido et al. 2013). Improvement in secondary sex characteristics can be accomplished by the administration of testosterone in males or estrogen in females, and there is the potential for benefit in bone mineral content and bone density. There appears to be a growing acceptance among physicians experienced with PWS as to their use.
- In males, not only can voice change and body hair, beard growth, and genital size be improved, but a more masculine body habitus and increased muscle bulk and strength usually occur with testosterone treatment. Replacement therapy should be individualized, taking into consideration serum testosterone levels, signs of puberty, and pre-treatment behavior difficulties. One approach is to begin testosterone replacement at age 13–14 years with a relatively low dose (50–75 mg IM depotestosterone every 3–4 weeks, or a single testosterone patch or testosterone gel dose daily), working up slowly in dose to reach normal or low normal blood testosterone levels (personal experience). In males, an increase in aggressive behavior is the greatest concern, because these individuals usually already have behavioral disturbance. Testosterone by daily skin patch or gel can give a more consistent level of testosterone, thus avoiding the usual increase in aggression in the week or so after an injection of depotestosterone is given. Care must be taken, as always, to avoid abnormal acceleration of bone growth with ultimate shorter stature, so bone age should be monitored in treated growing individuals.
- It is recommended that simultaneous treatment with growth hormone and testosterone be monitored by an endocrinologist or other physician experienced in their use.
- In females, estrogen treatment or, preferably, cycling hormones or birth control pills can increase breast size, if desired, and also result in menstrual periods. A surprising number of women with PWS want to have their periods like other women, at least for a while. Difficulties with hygiene issues have occurred in menstruating women with PWS of lower cognitive ability. The benefits of hormone replacement therapy to the cardiovascular system and to prevention of osteoporosis have not been studied in PWS. The risk of thrombosis, which is of concern in the general population, is likely to be at least as great if not greater (Hedgeman et al. 2017) in PWS, thus care is indicated. It seems prudent to individualize sex hormone treatment through weighing of the risks and benefits and through discussions with the affected individual and his or her parents or guardians (Eldar-Geva et al. 2013).
- The high frequency of osteopenia/osteoporosis in both sexes provides one more compelling argument in favor of sex hormone replacement.
- Sex education, including issues related to sexuality, is as important in PWS as in the general population. Given the women who bore children, the issue of contraception should be carefully considered in light of the circumstances and menstrual status of an affected woman. Although fertility in men with PWS has not been described, it is not known whether, in rare cases, sperm production may occur.
- Type II diabetes should be treated as in the general population, keeping in mind that it is generally obesity-related and that weight loss is the best form of management. In planning treatment, the extremely common indiscretions in eating in PWS should be taken into consideration.

Ophthalmologic

Strabismus is frequent, in part related to muscular hypotonia, although a contribution from the pigmentary abnormality in affected people with deletion has also been documented in individuals with fair coloring. Myopia and hyperopia are common.

Evaluation

- All infants and young children should be screened for strabismus and visual acuity deficits during routine well child visits, including use of the cover/uncover test for detecting esophoria.
- Individuals with evident strabismus should be referred to an ophthalmologist as soon as it is recognized.
- A formal ophthalmologic evaluation between ages 1 and 3 years is recommended in all affected individuals.

Management

- Strabismus should be managed as for any infant, keeping in mind muscular hypotonia and the possibility of hypopigmentation and its consequences.
- Visual acuity problems should be treated as in the general population.

Sleep

Sleep disturbance is common in PWS at all ages. Individuals with PWS have sleep-disordered breathing, including central and obstructive sleep apnea, abnormal arousal, abnormal circadian rhythm in REM sleep, reduced REM latency, and abnormal response to hypercapnia (Nixon and Brouillette 2002). Central sleep apnea is common in infants, often associated with hypoxia, and improves with small amounts of supplemental oxygen (Urquhart et al. 2013; Cohen et al. 2014). In one study, children older than 2 years were not found to have central sleep apnea (Cohen et al. 2014). There also appears to be an increase in excessive daytime sleepiness, similar to that seen in individuals with narcolepsy, which is separate from the sleep-disordered breathing (Bruni et al. 2010). Obesity can worsen the sleep disordered breathing, as can respiratory problems (see below). There is a fundamental abnormality of central respiratory drive in PWS that contributes to both the sleep apnea and the risk of hypoventilation during acute pulmonary illness. There is some evidence that growth hormone might help to ameliorate this defect, but it is impossible based on the data to distinguish between the effect of aging and normal development and that of growth hormone. Growth hormone has been shown to improve the apnea/hypopnea index in most treated individuals; in those who did not show improvement or had worse sleep abnormality, tonsillar enlargement was demonstrated and tonsillectomy resulted in improvement (Miller et al. 2006). Individuals with PWS have abnormal ventilatory responses to hypercapnia and hypoxia, both when asleep and awake (reviewed in Nixon and Brouillette [2002]). Therefore, they may not have disruption of sleep due to obstructive events. On the other hand, sleep architecture in some people with PWS is abnormal, unrelated to apnea or obstruction.

Evaluation

- History and symptoms of sleep disturbance, snoring, and sleep apnea should be sought at each medical evaluation (at least annually), along with information regarding excessive daytime sleepiness.
- Any suggestion of sleep apnea or excessive daytime sleepiness should be evaluated by polysomnography. Home oxygen saturation monitoring should not be used to select individuals for formal sleep studies. It has not been found to be particularly sensitive or specific as a screening test for sleep-disordered breathing in other populations, and it has not been studied in individuals with PWS.
- Evaluation of disordered sleep should include evaluation of tonsillar and adenoidal size as well as cardiovascular status.
- Several experts on PWS, including the authors, recommend that a sleep study be done prior to initiation of growth hormone therapy, and a repeat study considered several months after starting therapy.
- The possibility of central sleep apnea in the first months of life should be evaluated, and consideration given to use of supplemental oxygen if hypoxia is present. Some pediatric pulmonologists advocate using oxygen empirically in infants until a sleep study documents lack of hypoxia on room air. The presence of central sleep apnea in an infant need not be a contraindication to initiating growth hormone therapy.

Management

- Tonsillectomy and/or adenoidectomy are often needed when tonsillar and/or adenoidal hypertrophy are present.
- Sleep-disordered breathing should be treated as in the typically developing population, with special attention given to weight loss if obesity is present. Continuous positive airway pressure (CPAP) devices initially may be met with resistance, but are eventually well-tolerated in most cases.
- Referral to a sleep clinic/pulmonologist experienced with use of CPAP in children can greatly facilitate the use of the CPAP mask.
- Other sleep disorders should be treated as in the general population.

Respiratory

Individuals with PWS are at risk of breathing problems for a variety of reasons (reviewed in (Gillett and Perez 2016; Tan and Urquhart 2017)), including thoracic muscle weakness and hypotonia, reduced muscle tone in the pharynx and upper airways, obesity, and scoliosis. There is a basic abnormality of central respiratory drive in PWS that contributes to both the sleep apnea and the risk of hypoventilation during acute pulmonary disease. Some studies have indicated that up to 50% of affected individuals may have recurrent respiratory infections, although immunologic abnormalities are not increased in PWS. Silent aspiration is common, particularly in infants and young children (Gross et al. 2017; Salehi et al. 2017). Individuals with PWS have been shown to have a restrictive lung disease, likely as a result of thoracic muscle weakness (Hakonarson et al. 1995), and that study also found an obstructive component by spirometry in about one third of affected individuals, most of whom were responsive to β-agonists.

Several studies show that treatment with growth hormone leads to improvement in ventilation as demonstrated both by pulmonary function testing (Haqq et al. 2003) and by responsiveness to hypercapnia (Lindgren et al. 1999). The latter study found that growth hormone significantly stimulates central respiratory drive. Because of safety concerns in individuals with obesity and respiratory problems, care should be used when considering therapy with growth hormone as noted above (Berini et al. 2013; Deal et al. 2013; Gillett and Perez 2016; Tan and Urquhart 2017) (see Growth and Feeding – Short Stature).

Evaluation

- History and symptoms of disordered breathing should be sought at each medical evaluation (at least annually).
- Evaluation of disordered breathing should include evaluation of tonsillar and adenoidal size as well as cardiovascular status.
- Spirometry should be considered when breathing problems are suspected. Evidence of a component reversible with β-agonist therapy should be sought.
- Evaluation of possible aspiration should be considered based on medical history and observation.

Management

- Careful weight management and avoidance of obesity are imperative to minimize respiratory complications.
- Annual polysomnography has been suggested to monitor growth hormone treatment (Berini et al. 2013).
- Regular exercise to improve cardiovascular and respiratory fitness is important to weight management and general health. The authors strongly recommend one hour of physical activity every day.
- Standard specific therapies should be directed toward any broncho-constriction or tonsillar/adenoidal hyperplasia that contributes to obstructive respiratory disease.
- Growth hormone therapy may lead to improvements in respiratory drive and CO_2 responsiveness, but care should be taken and families should be carefully counseled regarding possible risks, especially in individuals with significant obesity or pre-existing respiratory problems. Some experts encourage that obesity be addressed prior to initiation of growth hormone therapy (Tan and Urquhart 2017), and that severe obesity be an exclusion criterion for use of growth hormone (Deal et al. 2013).
- Obstructive apnea may respond to standard therapy with external devices such as nighttime continuous positive airway pressure machines. Most individuals with PWS tolerate this well when they understand the reason for its use, although some are quite resistant. An experienced sleep clinic behavior specialist is extremely valuable in optimizing and monitoring treatment, including finding a mask that feels more comfortable.
- As noted above, a high index of suspicion for swallowing abnormality should be maintained in infancy and throughout the lifetime of the individual. Some experts recommend a variety of precautions, including having the family be trained to perform the Heimlich maneuver, slowing feeding with sips of water between bites, and always observing the person with PWS during meals. None of these have been studied or documented to have clinical impact, but they are common sense recommendations and should be considered when applicable.

Dental

Dental anomalies have been frequently reported and include dental crowding, carious teeth, and decreased saliva flow. Enamel hypoplasia may be present. Thick, sticky, and ropy saliva often leaves dried material on the lips and, anecdotally, makes articulation more difficult. One study documented that affected individuals have approximately 20% of the salivary flow of unaffected individuals (Hart 1998). This may be related to decreased autonomic stimulation of the saliva glands. Autonomic stimulants may increase flow.

Evaluation

- Regular, routine semiannual dental examination and prophylaxis is important starting when teeth are present (usually around 1 year of age, but no later than age 3 years). Some dentists who have seen multiple affected individuals recommend that examination and prophylaxis be performed three or four times per year.
- The quality and quantity of saliva should be evaluated through a search for crusted matter on the lips, ropes or strings of saliva in the mouth, or very dry mouth.

Management

- Emphasis should be placed on good dental hygiene. Electric toothbrushes, use of a timer, and, if needed, implementation of a reward system for positive reinforcement may help to develop good habits.
- Products that increase saliva flow are available over the counter. These include toothpaste, mouthwash, gel, and sugarless gums. These have been effective in increasing saliva flow and consistency (personal experience) and should be used in individuals with decreased saliva flow.

Musculoskeletal

The major musculoskeletal problems relate to the increased incidence of scoliosis and/or kyphosis and to osteoporosis. Hip dysplasia is also increased in PWS.

Scoliosis is present in 40–80% of people with PWS and varies in severity and age of onset. It is presumed to be related to muscular hypotonia because there are no underlying structural anomalies. Scoliosis can occur at any age during childhood, including early infancy. When it occurs in infancy, it is often quite severe (personal experience). Kyphosis occurs commonly in adolescents and adults with PWS but is rarely a cause of morbidity.

Hip Dysplasia occurs in up to 30% of people with PWS (West and Ballock 2004; Trizno et al. 2018), which is many-fold greater than in the general population. In one retrospective study of 27 individuals with PWS and evidence of hip dysplasia, the mean age of diagnosis was 2 months for those with successful treatment and 12 months for those with residual dysplasia following treatment. Standard physical screening using the Barlow and Ortolani maneuvers missed most cases, therefore the authors recommend hip ultrasound in all infants with PWS at 6 weeks of age (Trizno et al. 2018) to ensure early diagnosis.

Osteopenia/Osteoporosis occurs frequently in PWS, as indicated by studies documenting significantly decreased total and vertebral bone density and total bone mineral content compared with obese controls without PWS (Kroonen et al. 2006) or normal weight controls. There are a number of contributory factors, including growth hormone deficiency, hypogonadism, hypotonia, inactivity, and, often, low-dairy diets resulting in decreased calcium intake. An increased fracture rate may be present (Butler et al. 2002; Kroonean et al. 2006). Rates of fractures, particularly of long bones, seem to be high, though no well-designed study has been published.

Evaluation

- It is recommended that all infants with PWS have a hip ultrasound at 6 weeks of age, or an age-appropriate imaging study at diagnosis of PWS if that occurs at a later age.
- Clinical screening for scoliosis should be done at all routine health care visits in infancy and childhood by assessing the symmetry of the back in the diving position.
- Any suspicion of scoliosis should prompt a radiographic scoliosis series.
- The criteria used for referral to an orthopedist for scoliosis should be the same as those in the general population, especially to those centers that offer comprehensive care for scoliosis, hips, and ambulation concerns. Combining physical therapy with orthopedics in centers that are aware of issues in cognitively impaired individuals is optimal.
- Bone densitometry is appropriate in adulthood or in any child over 5 years with frequent fractures or fractures after apparently minor trauma.

Management

- Hip dysplasia should be treated as in the general population. Surgery is not usually necessary.
- Treatment of scoliosis in PWS is complicated by obesity, osteopenia, hypotonia and respiratory issues, and should be treated by a provider familiar with the complexity of these patients. Skin picking on the incision has sometimes been a problem. The prolonged period of growth caused by pubertal deficiency should be kept in mind. Bracing, serial casting, and surgical procedures are sometimes necessary.
- Calcium and vitamin D intake at least to RDA-recommended levels should be assured. Calcium and vitamin D supplementation is appropriate in all affected individuals. Taking calcium-containing antacids to accomplish this has been well tolerated (personal experience), although other supplements may also be used.
- Weight-bearing exercise should be encouraged throughout life.
- There is some evidence that the use of sex hormone replacement (Donze et al. 2018) and growth hormone supplementation (Nakamura et al. 2014), and possibly the combination, may prevent or limit osteopenia/osteoporosis.
- The use of bisphosphonates or other medications for treatment of osteoporosis has not yet been studied in PWS, but indications for use would be similar to those in the general population.
- Growth hormone replacement therapy is associated with increased bone density (Carrel et al. 2002; Nakamura et al. 2014; Donze et al. 2018).

Dermatologic

One of the more difficult problems in PWS is picking at the skin or mucosal areas such as the nose, rectum, or vagina, which occurs in over half of affected individuals. Difficulties with sores kept open for months or years and subsequent scarring and pigmentary changes are common among those who do skin picking. Anxiety, stress and boredom seem to increase the severity of picking behavior. Peripheral edema, later leading to chronic skin changes of the legs, is not uncommon in PWS.

Evaluation

- Examination of the skin and exposed mucosal areas, with attention to skin folds, particularly under the abdominal pannus in obese individuals, for evidence of complications of picking should occur at routine medical visits.
- A history of frequent nose bleeds or rectal bleeds should prompt examination of these areas for sores resulting from picking.

Management

- Skin picking is resistant to most therapies and rarely responds to medications such as specific serotonin reuptake inhibitors.
- Preliminary evidence supports the use of topiramate for refractory skin picking (Shapira et al. 2004). This drug has been associated with development of a renal tubular acidosis, so care is needed in monitoring treatment.
- A small, open-label trial followed by anecdotal evidence suggests that N-acetylcysteine may reduce skin picking behaviors in some individuals. Doses ranged from 450 to 1200 mg per day (Miller and Angulo 2014), typically starting with 900 mg per day in adolescents and adults. Some providers increase the dose to 900 mg twice daily if the single daily dose does not achieve the desired effect.
- Keeping lesions moist and covered, keeping the fingernails short, and using behavior modification techniques may be beneficial in controlling picking behavior.
- Individuals who spend long periods of time in the bathroom are frequently the ones doing rectal picking. Limiting the amount of time that can be spent in the bathroom may be helpful.
- Peripheral edema and chronic skin changes should be treated as for the general population. Weight loss is often helpful.

Anesthesia

People with PWS may have unusual reactions to standard dosages of medications and anesthetic agents. This may be due in part to inaccurate assessment of the size of the aqueous and the lipid compartments when dosing based on either weight or body surface area, due to the significant alterations in body composition in individuals with PWS, even when the BMI is in the desired range. Caution should be exercised in giving medications that may cause sedation and suppress breathing since affected individuals often have prolonged and exaggerated responses to the drugs in addition to baseline abnormalities in respiratory drive. The fact that people with PWS are poor reporters of pain, and are poor at localizing pain, should be kept in mind. Slowed recovery of gastrointestinal function occurs often. The majority of problems relating to anesthesia are from poorly monitored conscious sedation, rather than general anesthesia, which is usually closely monitored.

Evaluation

- A careful history of prior exposure to anesthetics and pain medication is important. Awareness of the potential for lack of adequate response to stresses related to surgery, especially extended surgical time, is helpful in operative and post-op management of individuals with PWS, regardless of the surgical intervention.

Management

- Overnight observation for procedures that are usually done as an outpatient are recommended due to slow clearance of anesthesia and the added effects of pain medications.
- Cautious and slow transition to drinking and eating is advised to avoid obstruction.

RESOURCES

International Prader-Willi Syndrome Organization
http://www.ipwso.org
An organization of country chapters with educational resources and links to the websites of multiple country's organizations at https://www.ipwso.org/country-members

Prader–Willi Syndrome Association (USA)
8588 Potter Park Drive, Suite 500
Sarasota, Florida 34238, USA
Phone: 1-800-926-4797
Website: http://www.pwsausa.org

Foundation for Prader–Willi Research
340 S. Lemon Ave, #3620, Walnut, CA 91789, USA
Phone: 888-322-5487
Website: https://www.fpwr.org/

Ontario (Canadian) Prader–Willi Syndrome Organization
2788 Bathurst Street, Suite 303, Toronto, Ontario, M6B 3A3, USA
Phone: 1-800-563-1123 or 1-416-481-8657
Website: http://www.opwsa.com

Foundation for Prader–Willi Research Canada
19-13085 Yonge Street, Suite #370, Richmond Hill, ON, L4E 0K2, Canada
Phone: 1-866-993-7972
Website: http://www.fpwr.ca

Prader–Willi Syndrome Association of the United Kingdom
125a London Road
Derby DE1 2QQ, UK
Phone: 01332 365676
Fax: 01332 360401
Website: www.pwsa.co.uk/

Prader–Willi Syndrome Australia

PO Box 8295, Woolloongabba, QLD 4102, Australia

Phone: 1800 797 287

Website: http://www.pws.org.au/contact-us/

Written Resources

Butler MG, Lee PDK, Whitman BY (eds) (2006) *Management of Prader-Willi Syndrome*, 3rd ed. New York: Springer.

Driscoll DJ, Miller JL, Schwartz S, et al. Prader-Willi Syndrome. (1998) Oct 6 [Updated 2017 Dec 14]. In: Adam MP, Ardinger HH, Pagon RA, et al., editors. GeneReviews® [Internet]. Seattle (WA): University of Washington, Seattle; 1993–2020. Available from: www.ncbi.nih.gov/Pubmed/20301295

Höybye C (ed.) (2013) *Prader-Willi Syndrome*. New York: Nova.

Whittington J, Holland A (2004) *PWS: Development and Manifestations*. Cambridge: Cambridge University Press.

Multiple excellent written resources are available from or suggested by the support groups

REFERENCES

Alqahtani AR, MO, Elahmedi, AR, Al Qahtani, J Lee, MG Butler (2016) Laparoscopic sleeve gastrectomy in children and adolescents with Prader-Willi syndrome: a matched-control study. *Surg Obes Relat Dis* 12(1):100–110.

Angulo MA, Castro-Magana M, Lamerson M, Arguello R, Accacha S, Khan A (2007) Final adult height in children with Prader-Willi syndrome with and without human growth hormone treatment. *Am J Med Genet A* 143A(13):1456–1461.

Barclay SF, CM Rand, L Nguyen, RJA Wilson, R Wevrick, WT Gibson, NT Bech-Hansen, DE Weese-Mayer (2018) ROHHAD and Prader-Willi syndrome (PWS): clinical and genetic comparison. *Orphanet J Rare Dis* 13(1):124.

Bennett JA, T Germani, AM Haqq, L Zwaigenbaum (2015) Autism spectrum disorder in Prader-Willi syndrome: A systematic review. *Am J Med Genet A* 167A(12):2936–2944.

Berini J, V Spica Russotto, P Castelnuovo, S Di Candia, L Gargantini, G Grugni, L Iughetti, L Nespoli, L Nosetti, G Padoan, A Pilotta, G Trifiro, G Chiumello, A Salvatoni, E Genetic Obesity Study Group of the Italian Society of Pediatric and Diabetology (2013) Growth hormone therapy and respiratory disorders: long-term follow-up in PWS children. *J Clin Endocrinol Metab* 98(9):E1516–1523.

Bieth E, S Eddiry, V Gaston, F Lorenzini, A Buffet, F Conte Auriol, C Molinas, D Cailley, C Rooryck, B Arveiler, J Cavaille, JP Salles, M Tauber (2015) Highly restricted deletion of the SNORD116 region is implicated in Prader-Willi Syndrome. *Eur J Hum Genet* 23(2):252–255.

Boer H, A Holland, J Whittington, J Butler, T Webb, D Clarke (2002) Psychotic illness in people with Prader Willi syndrome due to chromosome 15 maternal uniparental disomy. *Lancet* 359(9301):135.

Bonnot O, D Cohen, D Thuilleaux, A Consoli, S Cabal, M Tauber (2016) Psychotropic treatments in Prader-Willi syndrome: a critical review of published literature. *Eur J Pediatr* 175(1):9–18.

Brandt BR, I Rosen (1998) Impaired peripheral somatosensory function in children with Prader-Willi syndrome. *Neuropediatrics* 29(3):124.

Bruni O, E Verrillo, L Novelli, R Ferri (2010) Prader-Willi syndrome: sorting out the relationships between obesity, hypersomnia, and sleep apnea. *Curr Opin Pulm Med* 16(6):568–573.

Burman P, EM Ritzen, AC Lindgren (2001) Endocrine dysfunction in Prader-Willi syndrome: a review with special reference to GH. *Endocr Rev* 22(6):787–799.

Butler JV, JE Whittington, AJ Holland, H Boer, D Clarke, T Webb (2002) Prevalence of, and risk factors for, physical ill-health in people with Prader-Willi syndrome: a population-based study. *Dev Med Child Neurol* 44(4):248–255.

Butler MG (1990) Prader-Willi syndrome: current understanding of cause and diagnosis. *Am J Med Genet* 35(3):319.

Butler MG, M Dasouki, D Bittel, S Hunter, A Naini and S DiMauro (2003) Coenzyme Q10 levels in Prader-Willi syndrome: comparison with obese and non-obese subjects. *Am J Med Genet A* 119A(2):168–171.

Butler MG, J Lee, DM Cox, AM Manzardo, JA Gold, JL Miller, E Roof, E Dykens, V Kimonis, DJ Driscoll (2016) Growth Charts for Prader-Willi Syndrome During Growth Hormone Treatment. *Clin Pediatr (Phila)* 55(10):957–974.

Butler MG, J Lee, AM Manzardo, JA Gold, JL Miller, V Kimonis, DJ Driscoll (2015) Growth charts for non-growth hormone treated Prader-Willi syndrome. *Pediatrics* 135(1):e126–135.

Butler MG, AM Manzardo, J Heinemann, C Loker, J Loker (2017) Causes of death in Prader-Willi syndrome: Prader-Willi Syndrome Association (USA) 40-year mortality survey. *Genet Med* 19(6):635–642.

Butler MG, FJ Meaney (1991) Standards for selected anthropometric measurements in Prader-Willi syndrome. *Pediatrics* 88(4):853–860.

Butler MG, BK Smith, J Lee, C Gibson, C Schmoll, W V Moore, JE Donnelly (2013) Effects of growth hormone treatment in adults with Prader-Willi syndrome. *Growth Horm IGF Res* 23(3):81–87.

Carrel AL, SE Myers, BY Whitman, DB Allen (2002) Benefits of long-term GH therapy in Prader-Willi syndrome: a 4-year study. *J Clin Endocrinol Metab* 87(4):1581.

Carrel AL, SE Myers, BY Whitman, J Eickhoff, DB Allen (2010) Long-term growth hormone therapy changes the natural history of body composition and motor function in children with prader-willi syndrome. *J Clin Endocrinol Metab* 95(3):1131–1136.

Cassidy SB, A Devi, C Mukaida (1994) Aging in Prader-Willi syndrome: 22 patients over age 30 years. *Proc Greenwood Genetics Center* 13:102.

Cassidy SB, S Schwartz, JL Miller, DJ Driscoll (2012) Prader-Willi syndrome. *Genet Med* 14(1):10–26.

Chai JH, DP Locke, JM Greally, JH Knoll, T Ohta, J Dunai, A Yavor, EE Eichler, R D Nicholls (2003) Identification of four highly conserved genes between breakpoint hotspots BP1 and BP2 of the Prader-Willi/Angelman syndromes deletion region that have undergone evolutionary transposition mediated by flanking duplicons. *Am J Hum Genet* 73(4):898–925.

Clarke D (1998) Prader-Willi syndrome and psychotic symptoms: 2. A preliminary study of prevalence using the Psychopathology Assessment Schedule for Adults with Developmental Disability checklist. *J Intellect Disabil Res* 42(Pt 6):451.

Clarke DJ, H Boer, J Whittington, A Holland, J Butler, T Webb (2002) Prader-Willi syndrome, compulsive and ritualistic behaviours: the first population-based survey. *Br J Psychiatry* 180:358–362.

Cohen M, J Hamilton, I Narang (2014) Clinically important age-related differences in sleep related disordered breathing in infants and children with Prader-Willi Syndrome. *PLoS One* 9(6):e101012.

Colmenares A, G Pinto, P Taupin, A Giuseppe, T Odent, C Trivin, K Laborde, JC Souberbielle, M Polak (2011) Effects on growth and metabolism of growth hormone treatment for 3 years in 36 children with Prader-Willi syndrome. *Horm Res Paediatr* 75(2):123–130.

Crinó A, R Schiaffini, P Ciampalini, S Spera, L Beccaria, F Benzi, L Bosio, A Corrias, L Gargantini, A Salvatoni, G Tonini, G Trifiro, C Livieri, e. Genetic Obesity Study Group of Italian Society of Pediatric and diabetology (2003) Hypogonadism and pubertal development in Prader-Willi syndrome. *Eur J Pediatr* 162(5):327–333.

Crockett DJ, SR Ahmed, DR Sowder, CT Wootten, S Chinnadurai, SL Goudy (2014) Velopharyngeal dysfunction in children with Prader-Willi syndrome after adenotonsillectomy. *Int J Pediatr Otorhinolaryngol* 78(10):1731–1734.

Davies JR, T Humby, DM Dwyer, AS Garfield, H Furby, LS Wilkinson, T Wells, AR Isles (2015) Calorie seeking, but not hedonic response, contributes to hyperphagia in a mouse model for Prader-Willi syndrome. *Eur J Neurosci* 42(4):2105–2113.

de Lind van Wijngaarden RF, BJ Otten, DA Festen, KF Joosten, FH de Jong, FC Sweep, AC Hokken-Koelega (2008) High prevalence of central adrenal insufficiency in patients with Prader-Willi syndrome. *J Clin Endocrinol Metab* 93(5):1649–1654.

Deal CL, M Tony, C Höybye, DB Allen, M Tauber, JS Christiansen, 2011 Growth Hormone in Prader-Willi Syndrome Clinical Care Guidelines Workshop Participants (2013) Growth Hormone Research Society workshop summary: consensus guidelines for recombinant human growth hormone therapy in Prader-Willi syndrome. *J Clin Endocrinol Metab* 98(6):E1072–1087.

Dimitropoulos A, A Ferranti, M Lemler (2013) Expressive and receptive language in Prader-Willi syndrome: report on genetic subtype differences. *J Commun Disord* 46(2):193–201.

Donze SH, RJ Kuppens, NE Bakker, J van Alfen-van der Velden, ACS Hokken-Koelega (2018) Bone mineral density in young adults with Prader-Willi syndrome: A randomized, placebo-controlled, crossover GH trial. *Clin Endocrinol (Oxf)* 88(6):806–812.

Driscoll DJ, MF Waters, CA Williams, RT Zori, CC Glenn, KM Avidano, RD Nicholls (1992) A DNA methylation imprint, determined by the sex of the parent, distinguishes the Angelman and Prader-Willi syndromes. *Genomics* 13(4):917.

Duker AL, BC Ballif, EV Bawle, RE Person, S Mahadevan, S Alliman, R Thompson, R Traylor, BA Bejjani, LG Shaffer, JA Rosenfeld, AN Lamb, T Sahoo (2010) Paternally inherited microdeletion at 15q11.2 confirms a significant role for the SNORD116 C/D box snoRNA cluster in Prader-Willi syndrome. *Eur J Hum Genet* 18(11):1196–1201.

Durst R, K Rubin-Jabotinsky, S Raskin, G Katz, J Zislin (2000) Risperidone in treating behavioural disturbances of Prader-Willi syndrome. *Acta Psychiatr Scand* 102(6):461.

Dykens E, B Shah (2003) Psychiatric disorders in Prader-Willi syndrome: epidemiology and management. *CNS Drugs* 17(3):167–178.

Dykens EM (2002) Are jigsaw puzzle skills 'spared' in persons with Prader-Willi syndrome? *J Child Psychol Psychiatry* 43(3):343.

Dykens EM (2004) Maladaptive and compulsive behavior in Prader-Willi syndrome: new insights from older adults. *Am J Ment Retard* 109(2):142–153.

Dykens EM, JF Leckman, SB Cassidy (1996) Obsessions and compulsions in Prader-Willi syndrome. *J Child Psychol Psychiatry* 37(8):995.

Dykens EM, E Roof (2008) Behavior in Prader-Willi syndrome: relationship to genetic subtypes and age. *J Child Psychol Psychiatry* 49(9):1001–1008.

Dykens EM, E Roof, H Hunt-Hawkins (2017) Cognitive and adaptive advantages of growth hormone treatment in children with Prader-Willi syndrome. *J Child Psychol Psychiatry* 58(1):64–74.

Dykens EM, E Roof, H Hunt-Hawkins, N Dankner, EB Lee, CM Shivers, C Daniell, SJ Kim (2017) Diagnoses and characteristics of autism spectrum disorders in children with Prader-Willi syndrome. *J Neurodev Disord* 9:18.

Edgar OS, AK Lucas-Herald, MG Shaikh (2016) Pituitary-Adrenal Axis in Prader Willi Syndrome. *Diseases* 4(1)

Einfeld SL, E Smith, IS McGregor, K Steinbeck, J Taffe, LJ Rice, SK Horstead, N Rogers, MA Hodge, AJ Guastella (2014) A double-blind randomized controlled trial of oxytocin nasal spray in Prader Willi syndrome. *Am J Med Genet A* 164A(9):2232–2239.

Eldar-Geva T, HJ Hirsch, Y Pollak, F Benarroch, V Gross-Tsur (2013) Management of hypogonadism in adolescent girls and adult women with Prader-Willi syndrome. *Am J Med Genet A* 161A(12):3030–3034.

Fong BF, JIP de Vries (2003) Obstetric aspects of the Prader-Willi syndrome. *Ultrasound Obstet gynecol* 21(4):389–392.

Gillett ES, IA Perez (2016) Disorders of Sleep and Ventilatory Control in Prader-Willi Syndrome. *Diseases* 4(3)

Gilmour J, D Skuse, M Pembrey (2001) Hyperphagic short stature and Prader--Willi syndrome: a comparison of behavioural phenotypes, genotypes and indices of stress. *Br J Psychiatry* 179:129.

Gold JA, C Ruth, K Osann, P Flodman, B McManus, HS Lee, S Donkervoort, M Khare, E Roof, E Dykens, J L Miller DJ. Driscoll, MG Butler JHeinemann, S Cassidy, VE Kimonis (2014) Frequency of Prader-Willi syndrome in births conceived via assisted reproductive technology. *Genet Med* 16(2):164–169.

Gross RD, R Gisser, G Cherpes, K Hartman, R Maheshwary (2017) Subclinical dysphagia in persons with Prader-Willi syndrome. *Am J Med Genet A* 173(2):384–394.

Gross-Tsur V, T Eldar-Gevma, F Benarroch, O Rubinstein, HJ Hirsch (2011) Body image and sexual interests in adolescents and young adults with Prader-Willi syndrome. *J Pediatr Endocrinol Metab* 24(7–8):469–475.

Gross-Tsur V, HJ Hirsch, F Benarroch, T Eldar-Geva (2012) The FSH-inhibin axis in prader-willi syndrome: heterogeneity of gonadal dysfunction. *Reprod Biol Endocrinol* 10:39.

Gunay-Aygun M, Schwartz S, Heeger S, O'Riordan MA, Cassidy SB (2001)The changing purpose of Prader-Willi syndrome clinical diagnostic criteria and proposed revised criteria. *Pediatrics* 108(5):E92.

Hakonarson H, J Moskovitz, KL Daigle, SB Cassidy, MM Cloutier (1995) Pulmonary function abnormalities in Prader-Willi syndrome. *J Pediatr* 126(4):565.

Haqq AM, DD Stadler, RH Jackson, RG Rosenfeld, JQ Purnell, SH LaFranchi (2003) Effects of growth hormone on pulmonary function, sleep quality, behavior, cognition, growth velocity, body composition, and resting energy expenditure in Prader-Willi syndrome. *J Clin Endocrinol Metab* 88(5):2206–2212.

Hart PS (1998) Salivary abnormalities in Prader-Willi syndrome. *Ann NY Acad Sci* 842:125.

Hedgeman E, SP Ulrichsen, S Carter, NC Kreher, KP Malobisky, MM Braun, J Fryzek, MS Olsen (2017) Long-term health outcomes in patients with Prader-Willi Syndrome: a nationwide cohort study in Denmark. *Int J Obes (Lond)* 41(10):1531–1538.

Hirsch HJ, T Eldar-Geva, F Benarroch, O Rubinstein, V Gross-Tsur (2009) Primary testicular dysfunction is a major contributor to abnormal pubertal development in males with Prader-Willi syndrome. *J Clin Endocrinol Metab* 94(7):2262–2268.

Holm VA, SB Cassidy, MG Butler JM Hanchett, LR Greenswag, BY Whitman, F Greenberg (1993) Prader-Willi syndrome: consensus diagnostic criteria. *Pediatrics* 91(2):398.

Horsthemke B, J Wagstaff (2008) Mechanisms of imprinting of the Prader-Willi/Angelman region. *Am J Med Genet A* 146A(16):2041–2052.

Höybye C (2015) Growth hormone treatment of Prader-Willi syndrome has long-term, positive effects on body composition. *Acta Paediatr* 104(4):422–427.

Höybye C, A Hilding, H Jacobsson, M Thoren (2003) Growth hormone treatment improves body composition in adults with Prader-Willi syndrome. *Clin Endocrinol (Oxf)* 58(5):653–661.

Hudgins L, JS Geer, SB Cassidy (1998) Phenotypic differerencss in African Americans with Prader-Willi syndrome. *Genet Med* 1(1):49–51.

Ishii A, H Ihara, H Ogata, M Sayama, M Gito, N Murakami, T Ayabe, Y Oto, A Takahashi, T Nagai (2017) Autistic, aberrant, and food-related behaviors in adolescents and young adults with Prader-Willi Syndrome: The effects of age and genotype. *Behav Neurol* 2017:4615451.

Jiang YH, K Wauki, Q Liu, J Bressler, Y Pan, CD Kashork, LG Shaffer, AL Beaudet (2008) Genomic analysis of the chromosome 15q11-q13 Prader-Willi syndrome region and characterization of transcripts for GOLGA8E and WHCD1L1 from the proximal breakpoint region. *BMC Genomics* 9:50.

Johnson L, AM Manzardo, JL Miller, DJ Driscoll, MG Butler (2016) Elevated plasma oxytocin levels in children with Prader-Willi syndrome compared with healthy unrelated siblings. *Am J Med Genet A* 170(3):594–601.

Kaufman H, G Overton, J Leggott, C Clericuzio (1995) Prader-Willi syndrome: effect of group home placement on obese patients with diabetes. *South Med J* 88(2):182–184.

Kido Y, S Sakazume, Y Abe, Y Oto, H Itabashi, M Shiraishi, A Yoshino, Y Tanaka, K Obata, N Murakami, T Nagai (2013) Testosterone replacement therapy to improve secondary sexual characteristics and body composition without adverse behavioral problems in adult male patients with Prader-Willi syndrome: an observational study. *Am J Med Genet A* 161A(9):2167–2173

Kroonen, LT, M Herman, PD Pizzutillo, GD Macewen (2006) Prader-Willi Syndrome: clinical concerns for the orthopaedic surgeon. *J Pediatr Orthop* 26(5):673–679.

l'Allemand, D, U Eiholzer, M Schlumpf, H Steinert, W Riesen (2000) Cardiovascular risk factors improve during 3 years of growth hormone therapy in Prader-Willi syndrome. *Eur.J Pediatr* 159(11):835.

Ledbetter, DH, VM Riccardi, SD Airhart, RJ Strobel, BS Keenan, JD Crawford (1981) Deletions of chromosome 15 as a cause of the Prader-Willi syndrome. *N Engl J Med* 304(6):325–329.

Lindgren, AC, LG Hellstrom, EM Ritzen, JMilerad (1999) Growth hormone treatment increases CO(2) response, ventilation and central inspiratory drive in children with Prader-Willi syndrome. *Eur.J Pediatr* 158(11):936.

Lo, S T, DA Festen, RF Tummers-de Lind van Wijngaarden, PJ Collin, AC Hokken-Koelega (2015) Beneficial Effects of Long-Term Growth Hormone Treatment on Adaptive Functioning in Infants With Prader-Willi Syndrome. *Am J Intellect Dev Disabil* 120(4):315–327.

Manzardo, A M, J Loker, J Heinemann, C Loker, MG Butler (2018) Survival trends from the Prader-Willi Syndrome Association (USA) 40-year mortality survey. *Genet Med* 20(1):24–30.

Marceau, P, S Marceau, S Biron, FS Hould, S Lebel, O Lescelleur, L Biertho, JG Kral (2010) Long-term experience with duodenal switch in adolescents. *Obes Surg* 20(12):1609–1616.

Martin, A, M State, GM Anderson, WM Kaye, JM Hanchett, CW McConaha, WG North, JF Leckman (1998) Cerebrospinal fluid levels of oxytocin in Prader-Willi syndrome: a preliminary report. *Biol Psychiatry* 44(12):1349.

McCandless, SE, JA Yanovski, J Miller, C Fu, LM Bird, P Salehi, CL Chan, D Stafford, MJ Abuzzahab, D Viskochil, SE Barlow, M Angulo, SE Myers, BY Whitman, D Styne, E Roof, EM Dykens, AO Scheimann, J Malloy, D Zhuang, K Taylor, TE Hughes, DD Kim, MG Butler (2017) Effects of MetAP2 inhibition on hyperphagia and body weight in Prader-Willi syndrome: A randomized, double-blind, placebo-controlled trial. *Diabetes Obes Metab* 19(12):1751–1761.

Miller, J, J Silverstein, J Shuster, DJ Driscoll, M Wagner (2006) Short-term effects of growth hormone on sleep abnormalities in Prader-Willi syndrome. *J Clin Endocrinol Metab* 91(2):413–417.

Miller, JL, M Angulo (2014) An open-label pilot study of N-acetylcysteine for skin-picking in Prader-Willi syndrome. *Am J Med Genet A* 164A(2):421–424.

Miller, JL, CH Lynn, DC Driscoll, AP Goldstone, JA Gold, V Kimonis, E Dykens, MG Butler JJ Shuster, DJ Driscoll (2011) Nutritional phases in Prader-Willi syndrome. *Am J Med Genet A* 155A(5):1040–1049.

Miller, JL, CH Lynn, J Shuster, DJ Driscoll (2013) A reduced-energy intake, well-balanced diet improves weight control in children with Prader-Willi syndrome. *J Hum Nutr Diet* 26(1):2–9.

Miller, JL, R Tamura, MG Butler V Kimonis, C Sulsona, JA Gold, DJ Driscoll (2017) Oxytocin treatment in children with Prader-Willi syndrome: A double-blind, placebo-controlled, crossover study. *Am J Med Genet A* 173(5):1243–1250.

Murakami, N, K Obata, Y Abe, Y Oto, Y Kido, H Itabashi, T Tsuchiya, Y Tanaka, A Yoshino, T Nagai (2012) Scoliosis in Prader-Willi syndrome: effect of growth hormone therapy and value of paravertebral muscle volume by CT in predicting scoliosis progression. *Am J Med Genet A* 158A(7):1628–1632.

Nagai, T, K Obata, T Ogata, N Murakami, Y Katada, A Yoshino, S Sakazume, Y Tomita, R Sakuta, N Niikawa (2006) Growth hormone therapy and scoliosis in patients with Prader-Willi syndrome. *Am J Med Genet A* 140(15):1623–1627.

Nakamura, Y, N Murakami, T Iida, S Asano, S Ozeki, T Nagai (2014) Growth hormone treatment for osteoporosis in patients with scoliosis of Prader-Willi syndrome. *J Orthop Sci* 19(6):877–882.

Nicholls, RD, J H Knoll, MG Butler S Karam, M Lalande (1989) Genetic imprinting suggested by maternal heterodisomy in nondeletion Prader-Willi syndrome. *Nature* 342(6247):281.

Nixon, GM, RT Brouillette (2002) Sleep and breathing in Prader-Willi syndrome. *Pediatr Pulmonol* 34(3):209–217.

Ogata, H, H Ihara, N Murakami, M Gito, Y Kido, T Nagai (2014) Autism spectrum disorders and hyperactive/impulsive behaviors in Japanese patients with Prader-Willi syndrome: a comparison between maternal uniparental disomy and deletion cases. *Am J Med Genet A* 164A(9):2180–2186.

Oore, J, B Connell, B Yaszay, A Samdani, TS Hilaire, T Flynn, R El-Hawary, Children's Spine Study Group, Growing Spine Study Group (2019) Growth friendly surgery and serial cast correction in the treatment of early-onset scoliosis for patients with Prader-Willi Syndrome. *J Pediatr Orthop* 39(8):e597–e601.

Pacoricona Alfaro, DL, P Lemoine, V Ehlinger, C Molinas, G Diene, M Valette, G Pinto, M Coupaye, C Poitou-Bernert, D Thuilleaux, C Arnaud, M Tauber (2019) Causes of death in Prader-Willi syndrome: lessions from 11 years' experience of a national reference center. *Orphanet J Rare Dis* 14(1):238.

Prader, A, A Labhart, A Willi (1956) Ein syndrom von adipositas, kleinwuchs, kryptorchismus und oligophrenie nach myotonieartigem zustand im neugeborenenalter. *Schweiz Med Wochen* 86:1260.

Qi, Y, L Purtell, M Fu, NJ Lee, J Aepler, L Zhang, K Loh, RF Enriquez, PA Baldock, S Zolotukhin, LV Campbell, H Herzog (2016) Snord116 is critical in the regulation of food intake and body weight. *Sci Rep* 6:18614.

Roof, E, W Stone, W MacLean, ID Feurer, T Thompson, MG Butler (2000) Intellectual characteristics of Prader-Willi syndrome: comparison of genetic subtypes. *J Intellect Disabil Res* 44(Pt 1):25.

Sahoo, T, D del Gaudio, JR German, M Shinawi, SU Peters, RE Person, A Garnica, SW Cheung, AL Beaudet (2008) Prader-Willi phenotype caused by paternal deficiency for the HBII-85 C/D box small nucleolar RNA cluster. *Nat Genet* 40(6):719–721.

Salehi, P, HJ Stafford, RP Glass, A Leavitt, AE Beck, A McAfee, L Ambartsumyan, M Chen (2017) Silent aspiration in infants with Prader-Willi syndrome identified by videofluoroscopic swallow study. *Medicine (Baltimore)* 96(50):e9256.

Schaaf, CP, ML Gonzalez-Garay, F Xia, L Potocki, KW Gripp, B Zhang, BA Peters, MA McElwain, R Drmanac, AL Beaudet, CT Caskey, Y Yang (2013) Truncating mutations of MAGEL2 cause Prader-Willi phenotypes and autism. *Nat Genet* 45(11):1405–1408.

Scheimann, AO, MG Butler L Gourash, C Cuffari, W Klish (2008) Critical analysis of bariatric procedures in Prader-Willi syndrome. *J Pediatr Gastroenterol Nutr* 46(1):80–83.

Shapira, NA, MC Lessig, MH Lewis, WK Goodman, DJ Driscoll (2004) Effects of topiramate in adults with Prader-Willi syndrome. *Am J Ment Retard* 109(4):301–309.

Siemensma, EP, RF de Lind van Wijngaarden, BJ Otten, FH de Jong, AC Hokken-Koelega (2012) Testicular failure in boys with Prader-Willi syndrome: longitudinal studies of reproductive hormones. *J Clin Endocrinol Metab* 97(3):E452–459.

Siemensma, EP, RF Tummers-de Lind van Wijngaarden, DA Festen, ZC Troeman, AA van Alfen-van der Velden, BJ Otten, J Rotteveel, RJ Odink, GC Bindels-de Heus, M van Leeuwen, DA Haring, W Oostdijk, G Bocca, EC Mieke Houdijk, AS van Trotsenburg, JJ Hoorweg-Nijman, H van Wieringen, RC Vreuls, PE Jira, E J Schroor, E van Pinxteren-Nagler, J Willem Pilon, LB Lunshof, AC Hokken-Koelega (2012) Beneficial effects of growth hormone treatment on cognition in children with Prader-Willi syndrome: a randomized controlled trial and longitudinal study. *J Clin Endocrinol Metab* 97(7):2307–2314.

Sinnema, M, H Boer, P Collin, MA Maaskant, KE van Roozendaal, CT Schrander-Stumpel, LM Curfs (2011) Psychiatric illness in a cohort of adults with Prader-Willi syndrome. *Res Dev Disabil* 32(5):1729–1735.

Skokauskas, N, E Sweeny, J Meehan, L Gallagher (2012) Mental health problems in children with prader-willi syndrome. *J Can Acad Child Adolesc Psychiatry* 21(3):194–203.

Sode-Carlsen, R, S Farholt, KF Rabben, J Bollerslev, T Schreiner, AG Jurik, J Frystyk, J S Christiansen, C Höybye (2011) Growth hormone treatment for two years is safe and effective in adults with Prader-Willi syndrome. *Growth Horm IGF Res* 21(4):185–190.

Soni, S, J Whittington, AJ Holland, T Webb, E Maina, H Boer, D Clarke (2007) The course and outcome of psychiatric illness in people with Prader-Willi syndrome: implications for management and treatment. *J Intellect Disabil Res* 51(Pt 1):32–42.

State, MW, EM Dykens, B Rosner, A Martin, BH King (1999) Obsessive-compulsive symptoms in Prader-Willi and Prader-Willi-Like patients. *J Am Acad Child Adolesc Psychiatry* 38(3):329.

Steinhausen, HC, U Eiholzer, BP Hauffa, Z Malin (2004) Behavioural and emotional disturbances in people with Prader-Willi Syndrome. *J Intellect Disabil Res* 48(1):47–52.

Swaab, DF, JS Purba, MA Hofman (1995) Alterations in the hypothalamic paraventricular nucleus and its oxytocin neurons (putative satiety cells) in Prader-Willi syndrome: a study of five cases. *J Clin Endocrinol Metab* 80(2):573.

Takeshita, E, N Murakami, R Sakuta, T Nagai (2013) Evaluating the frequency and characteristics of seizures in 142 Japanese patients with Prader-Willi syndrome. *Am J Med Genet A* 161A(8):2052–2055.

Tan, HL, DS Urquhart (2017) Respiratory Complications in Children with Prader Willi Syndrome. *Paediatr Respir Rev* 22:52–59.

Tauber, M, K Boulanouar, G Diene, S Cabal-Berthoumieu, V Ehlinger, P Fichaux-Bourin, C Molinas, S Faye, M Valette, J Pourrinet, C Cessans, S Viaux-Sauvelon, C Bascoul, A Guedeney, P Delhanty, V Geenen, H Martens, F Muscatelli, D Cohen, A Consoli, P Payoux, C Arnaud, JP Salles (2017) The use of oxytocin to improve feeding and social skills in infants with Prader-Willi Syndrome. *Pediatrics* 139(2)

Tauber, M, G Diene, C Molinas (2016) Sequelae of GH Treatment in Children with PWS. *Pediatr Endocrinol Rev* 14(2):138–146.

Tauber, M, G Diene, C Molinas, M Hebert (2008) Review of 64 cases of death in children with Prader-Willi syndrome (PWS). *Am J Med Genet A* 146A(7):881–887.

Trizno, AA, AS Jones, PM Carry, G Georgopoulos (2018) The prevalence and treatment of hip dysplasia in *Prader-Willi Syndrome (PWS) J Pediatr Orthop* 38(3):e151–e156.

Urquhart, DS, T Gulliver, G Williams, MA Harris, O Nyunt, S Suresh (2013) Central sleep-disordered breathing and the effects of oxygen therapy in infants with Prader-Willi syndrome. *Arch Dis Child* 98(8):592–595.

van Mil, EA, KR Westerterp, WJ Gerver, LM Curfs, CT Schrander-Stumpel, AD Kester, WH Saris (2000) Energy expenditure at rest and during sleep in children with Prader- Willi syndrome is explained by body composition. *Am.J.Clin.Nutr* 71(3):752.

Verdine, BN, GL Troseth, RM Hodapp, EM Dykens (2008) Strategies and correlates of jigsaw puzzle and visuospatial performance by persons with Prader-Willi syndrome. *Am J Ment Retard* 113(5):343–355.

Verrotti, A, R Cusmai, D Laino, M Carotenuto, M Esposito, R Falsaperla, L Margari, R Rizzo, S Savasta, S Grosso, P Striano, V Belcastro, E Franzoni, P Curatolo, L Giordano, E Freri, S Matricardi, D Pruna, I Toldo, E Tozzi, L Lobefalo, F Operto, E Altobelli, F Chiarelli, A Spalice (2015) Long-term outcome of epilepsy in patients with Prader-Willi syndrome. *J Neurol* 262(1):116–123.

Vogels, A, P Moerman, JP Frijns, GA Bogaert (2008) Testicular histology in boys with Prader-Willi syndrome: fertile or infertile? *J Urol* 180(4 Suppl):1800–1804.

West, LA, RT Ballock (2004) High incidence of hip dysplasia but not slipped capital femoral epiphysis in patients with Prader-Willi syndrome. *J Pediatr Orthop* 24(5):565–567.

Whittington, J, A Holland (2010) Neurobehavioral phenotype in Prader-Willi syndrome. *Am J Med Genet C* 154C(4):438–447.

Whittington, J, A Holland (2017) Cognition in people with Prader-Willi syndrome: Insights into genetic influences on cognitive and social development. *Neurosci Biobehav Rev* 72:153–167.

Whittington, J, A Holland, T Webb, J Butler, D Clarke, H Boer (2002) Relationship between clinical and genetic diagnosis of Prader-Willi syndrome. *J Med Genet* 39(12):926–932.

Whittington, JE, AJ Holland, T Webb, J Butler, D Clarke, H Boer (2001) Population prevalence and estimated birth incidence and mortality rate for people with Prader-Willi syndrome in one UK Health Region. *J Med Genet* 38(11):792.

Wigren, M, S Hansen (2005) ADHD symptoms and insistence on sameness in Prader-Willi syndrome. *J Intellect Disabil Res* 49(Pt 6):449–456.

Wolfgram, PM, AL Carrel, DB Allen (2013) Long-term effects of recombinant human growth hormone therapy in children with Prader-Willi syndrome. *Curr Opin Pediatr* 25(4):509–514.

Yang, L, GD Zhan, JJ Ding, HJ Wang, D Ma, GY Huang, WH Zhou (2013) Psychiatric illness and intellectual disability in the Prader-Willi syndrome with different molecular defects--a meta analysis. *PLoS One* 8(8):e72640.

Yee, BJ, PR Buchanan, S Mahadev, D Banerjee, PY Liu, C Phillips, G Loughnan, K Steinbeck, RR Grunstein (2007) Assessment of sleep and breathing in adults with prader-willi syndrome: a case control series. *J Clin Sleep Med* 3(7):713–718.

47

PROTEUS SYNDROME*

Leslie G. Biesecker

Medical Genomics and Metabolic Genetics Branch, National Human Genome Research Institute, Bethesda, Maryland, USA

INTRODUCTION

Proteus syndrome is a disorder of segmental or mosaic overgrowth that can affect any tissue. The most common manifestations include overgrowth that can lead to orthopedic complications, soft tissue overgrowth of the feet, linear nevi, vascular malformations, and tumor predisposition. All confirmed individuals are sporadic and the disorder is caused by a postzygotic mutation in the *AKT1* gene.

Prevalence

It is challenging to accurately estimate the incidence or prevalence of an extremely rare disorder. Proteus syndrome is both rare and over-diagnosed. There are fewer than 200 confirmed clinical reports in the literature and many people who have the diagnosis cannot be confirmed as affected when published diagnostic criteria (Turner et al. 2004) and molecular testing are applied. The disorder is serious with substantial childhood mortality (Sapp et al. 2017). Based on the number of identified individuals in the United States, we estimate that the prevalence is between 1/1,000,000 and 1/10,000,000 births.

Diagnostic Criteria

The diagnosis of Proteus syndrome has been challenging due to the overlap with other segmental mosaic overgrowth disorders. In spite of this overlap, the presentation and natural history is generally distinguishable from other disorders. These distinct manifestations and natural history are consistent with the original description (Wiedemann et al. 1983) and warrant the diagnostic label of Proteus syndrome. Clinical diagnostic criteria were subsequently devised and are summarized in Table 47.1 (Biesecker 2006).

It is important to appreciate the specificity of each of the criteria as outlined in Table 47.1. Progressive overgrowth in Proteus syndrome is mostly postnatal, almost always relentless and in most cases, severe. It should not be confused with the growth of a lipoma in a person with PIK3CA-related overgrowth spectrum (PROS), patients formerly described as having hemihyperplasia, CLOVES syndrome, or other descriptors (Keppler-Noreuil et al. 2015). In Proteus syndrome, overgrowth often occurs in areas of the body that were entirely normal at birth. Another critical feature of the overgrowth seen in Proteus syndrome is its irregularity, leading to distortion of cutaneous, subcutaneous, cartilaginous, and bony tissues (Figures 47.1 and 47.2). In contrast, the overgrowth seen in PROS is highly regular, or "ballooning," resulting in a body part that is large but easily recognizable. In general, on radiographic examination, the bones underlying these enlarged body parts in hemihyperplasia are larger than normal in size but normal in structure. In contrast, in Proteus syndrome, bony and cartilaginous structures are distorted, sometimes beyond recognition (Jamis-Dow et al. 2004).

A second critical feature of Proteus syndrome, which is common, is the cerebriform connective tissue nevus (CCTN),

*This chapter is in the public domain in the United States of America.

Cassidy and Allanson's Management of Genetic Syndromes, Fourth Edition.
Edited by John C. Carey, Agatino Battaglia, David Viskochil, and Suzanne B. Cassidy.
© 2021 John Wiley & Sons, Inc. Published 2021 by John Wiley & Sons, Inc.

TABLE 47.1 Diagnostic criteria for Proteus syndrome

General criteria: Mosaic distribution, progressive course, AND sporadic occurrence
Category A: Cerebriform connective tissue nevus (CCTN)
Category B:
1. Linear epidermal nevus
2. Asymmetric disproportionate overgrowth of limbs, skull, external auditory canal, vertebrae, OR viscera
3. Occurrence of either tumor before age of 10 years: bilateral ovarian cystadenomas OR monomorphic adenomas of the parotid gland

Category C:
1. Dysregulated adipose tissue (either lipoatrophy or lipomas)
2. Vascular malformations (capillary, venous, or lymphatic)
3. Facial phenotype: long face, dolichocephaly, down-slanted palpebral fissures, low nasal bridge, wide or anteverted nares, and open mouth at rest
4. Bullous lung disease

Source: Adapted from Turner et al. (2004).
Note: To make a diagnosis of Proteus syndrome requires all three general criteria plus 1 from A, 2 from B, or 3 from C.

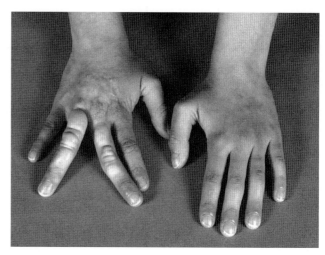

FIGURE 47.1 Dorsal view of the hands of a 7-year-old girl with Proteus syndrome. Note that the left hand is without apparent overgrowth. The right hand has asymmetric, distorting overgrowth of the third and fourth digits with radial and ulnar deviation, respectively. She also has thickening of the cutaneous tissue, consistent with a cerebriform connective tissue nevus (CCTN) on the dorsal third and fourth digits with early signs of a CCTN on the distal central portion of the dorsal hand. This child had no overgrowth or asymmetry at birth.

also known as a moccasin lesion when present on the sole of the foot (Figure 47.3). This lesion is caused by overgrowth of cutaneous and subcutaneous tissues with thickening of more than a centimeter. This overgrowth is irregular, leading to marked thickening adjacent to deep furrows, which may resemble the surface of the brain, which was the genesis of the word "cerebriform." This tissue is much firmer than the corresponding tissue that it replaces. Although this lesion is most common on the soles, it can also occur on the hands, perinasal area, or near the canthus. Wrinkling of the skin of the sole associated with PROS is distinct from the cerebriform connective tissue nevus – this tissue is usually quite soft and pliable.

There are a number of other specific features recognized in individuals with Proteus syndrome. These include bullous lung degeneration, bilateral ovarian cystadenomas in young girls, monomorphic parotid adenomas, and a dysmorphic facial phenotype associated with central nervous system dysgenesis and cognitive compromise.

The specific diagnostic criteria in categories A, B, and C (see Table 47.1) are accompanied by three general diagnostic features, all of which must be present. First, the overgrowth must be patchy or mosaic. There are no individuals with Proteus syndrome who have evenly distributed overgrowth. Second, the occurrence must be sporadic, as there are no confirmed familial cases. Third, the manifestations must be progressive, as defined above. These general criteria are related to the mosaicism model presented below, which has so far proven to be remarkably useful for understanding this disorder. A patient who meets these clinical criteria, strictly applied, can be considered to have a clinical diagnosis of Proteus syndrome.

Finally, it must be mentioned that a former name for Proteus syndrome is "elephant man disease," after the book and movie of the same title describing the life of Joseph Carey Merrick who lived in England from 1862 to 1890. Mr. Merrick's medical condition was described in detail by his surgeon (Treves 1885) and good (though not conclusive) evidence has been presented to show that Mr. Merrick was affected with Proteus syndrome (Cohen 1987). However, he was affected to a severe degree and the many negative connotations affiliated with that story are either not useful or even harmful to those who are affected by the disorder. Therefore, the use of that pejorative diagnostic term, elephant man disease, is discouraged.

Etiology, Pathogenesis, and Genetics

In all confirmed cases of Proteus syndrome described to date, the individuals were affected in a patchy or mosaic pattern and were sporadic. The disorder is pan-ethnic and there are no data to suggest any differences in frequency among the major geographic groups. Interestingly, there is a report of monozygotic twins who are discordant for the disorder (Brockmann et al. 2008). These observations led to a model of Proteus syndrome (Happle 1987) that suggests it is caused by a postzygotic mutation in a gene that leads to deregulation of growth in the daughter cells of that lineage. Furthermore, the model proposed that the variant, if present in all cells, would be lethal at an early stage of development, or perhaps even in the gamete. Thus, affected persons have only unaffected children. At least three affected adults have had four pregnancies, all of which were unaffected (personal

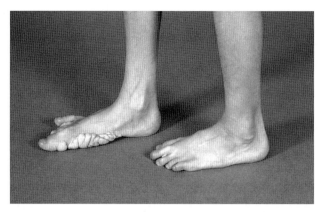

FIGURE 47.2 Lateral view of the feet of the same 7-year-old child shown in Figure 47.1. Note the asymmetric, distorting overgrowth of the right foot, primarily the first and fifth digits and the overgrowth of the second digit of the left foot. The CCTN affects the distal medial sole of the right foot and the first toe.

observation). This model is relevant to an entire class of disorders such as McCune–Albright syndrome, PROS, and others. It is recognized that these disorders are typically caused by pathogenic variants in genes that are only viable in mosaic form.

There was a difference of opinion among experts as to whether or not variants in other genes such as *PTEN* can cause Proteus syndrome, and whether Proteus is a distinct and separate disorder. The advances in the molecular genetics of Proteus syndrome, PROS, and other overgrowth conditions have now made clear that these are distinct clinical and molecular diagnostic entities even though published clinical reports claiming a diagnosis of Proteus syndrome in individuals with *PTEN* or *PIK3CA* variants are clearly incorrect, and readers must be skeptical of such reports.

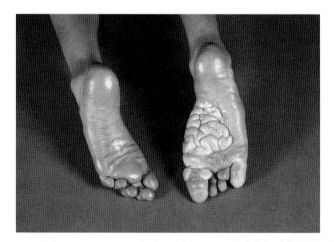

FIGURE 47.3 Plantar view of the feet from the same individual shown in Figures 47.1 and 47.2. Note the extensive CCTN of the right foot and an early, small CCTN of the center of the left foot. There is also a prominence of the left foot near the base of the second toe caused by overgrowth of the distal metatarsal.

The Happle model of Proteus syndrome was proven to be correct with the discovery of its etiology, which is a gain-of-function variant in the *AKT1* gene product (Lindhurst et al. 2011). All individuals to date have the same somatic mosaic c.49G>A p.(Glu17Lys) variant in *AKT1*. The mutation has been demonstrated in numerous affected tissues from people who meet the clinical diagnostic criteria and the molecular findings validated the clinical diagnostic criteria. All affected individuals are somatic mosaics for the variant and all tested parents are negative for the variant. The *AKT1* gain-of-function mutation explains many of the features of the disorder as this signaling protein is known to foster cell division, inhibit apoptosis, and increase transcription of a number of genes, among other functions (Carpten et al. 2007). Its role as an oncogene has been established both in functional work and in the COSMIC database of somatic cancer variants (Forbes et al. 2015), explaining the tumor susceptibility of Proteus syndrome. That AKT1 functions downstream of PIK3CA and PTEN also explains the clinical overlap of Proteus syndrome with PROS (*PIK3CA*) and the PTEN tumor hamartoma syndrome and Type II segmental Cowden syndrome (*PTEN*).

Diagnostic Testing

Molecular diagnostic testing is now available for this disorder. The challenge here is that essentially no affected individuals have any detectable level of this variant in peripheral blood and testing blood samples is pointless. Therefore, DNA analysis of cells from a biopsy of an affected tissue is necessary for molecular diagnosis. This can be accomplished from a skin biopsy of an affected area or a sample from a surgical procedure performed for symptomatic relief from the overgrowth. The finding of any significant level of mosaicism for this variant in a patient with overgrowth consistent with the features outlined in Table 47.1 may warrant a diagnosis of Proteus syndrome. A negative test result must be interpreted with caution as the sampled tissue may not have a sufficient level of mutation to be detected by the assay. If the individual meets the clinical criteria in table 47.1 and the test is negative, they should be considered to have a clinical or provisional diagnosis of Proteus syndrome and considerations should be given to performance of another test using a different affected tissue.

Differential Diagnosis

The most common disorders that should be considered in a person with segmental overgrowth is PROS (Keppler-Noreuil et al. 2015). The majority of individuals referred to the National Institutes of Health for the Proteus syndrome study have been re-diagnosed as having PROS, based on clinical and in many individuals, molecular testing results. The main features that distinguish PROS from Proteus syndrome are non-progressive growth (the overgrown limb grows in a commensurate manner to the rest of the body), capillary vascular malformations, and lipomas. The overgrowth in PROS is typically described as "ballooning," whereas in Proteus it is distorting. In general, the prognosis of PROS is less grave than that of Proteus syndrome, although there is a potential association with Wilms tumor (Gripp et al. 2016). Typical PROS is not progressive, which means that most affected individuals are born with asymmetry and, as they grow, the asymmetry stays proportionate and is commensurate with the overall growth of the individual throughout life. It should be noted that isolated hemihyperplasia is a manifestation of overgrowth, not a specific diagnosis.

The PROS designation subsumes the former descriptor of CLOVE syndrome (Sapp et al. 2007), which was defined by *c*ongenital *l*ipomatous *o*vergrowth with *v*ascular and *e*pidermal anomalies. It commonly manifests with truncal vascular malformations, large truncal lipomas, ballooning overgrowth of the feet, and linear epidermal nevi. Most patients with what we used to call Klippel–Trenaunay syndrome also have PROS, which consists of capillary, venous, and lymphatic vascular malformations (sometimes these malformations are mixtures of vessel types), with overgrowth that is typically in the same segment as the vascular malformation. The bones in overgrown segments are enlarged, but not typically distorted, and the affected individuals do not have hyperostoses or connective tissue or epidermal nevi. Maffucci syndrome is distinct from PROS and defined as multiple enchondromatosis with vascular malformations, and the former is sometimes confused with hyperostosis. Maffucci syndrome has no other overlap with Proteus syndrome. Other mosaic disorders such as PHACE(S) syndrome (Heyer 2015), linear epidermal nevus syndrome (Hafner et al. 2007), and sebaceous nevus syndrome (Vidaurri-de la Cruz et al. 2004) are readily distinguished clinically from Proteus syndrome.

Type II segmental Cowden syndrome (T2SCS) should also be considered in the differential diagnosis of segmental overgrowth and mosaic dermatologic manifestations (Happle 2007, 2010). This disorder has also been termed SOLAMEN syndrome (Caux et al. 2007) based on its manifestations of *s*egmental *o*vergrowth, *l*ipomatosis, *a*rteriovenous *m*alformations, and *e*pidermal *n*evi. This type of mosaic disorder comprises a somatic second hit in an individual who has a constitutional pathogenic variant in *PTEN*, such that some of their cells have biallelic *PTEN* pathogenic variants and the remainder have a heterozygous pathogenic variant. This syndrome presents with mosaic nevi that are distinct from the keratinocytic nevus associated of Proteus syndrome. In contrast, the nevi of T2SCS are thick and papillomatous, similar to that of a common wart. The overgrowth and bony distortion in T2SCS is less severe than in Proteus syndrome.

MANIFESTATIONS AND MANAGEMENT

It must be borne in mind that Proteus syndrome is a very rare disorder and experience is limited. The National Institutes of Health cohort of more than 60 affected individuals is the largest in the world, yet it is an inadequate sample from which to draw clear conclusions or make firm recommendations for management of all patients. The disorder is usually severe and affects many body systems; therefore, decisions about treating any one body system nearly always involve considerations of many other involved body systems. Effective management generally requires a large and diverse team of medical specialists willing to invest time to make thoughtful decisions about diagnosis, surgical intervention, and treatment.

Specific treatment for Proteus syndrome, based on the known molecular pathogenesis is not yet available clinically, but is currently under study. It has been shown that the AKT inhibitor ARQ 092 (Arqule, Inc.) inhibits AKT1 hyperactivation in vitro (Lindhurst et al. 2015). An open label, Phase 1/2 study of oral ARQ 092 for patients with Proteus syndrome has been completed (NCT03094832) and has shown good tolerability with some evidence of efficacy (Keppler-Noreuil et al. 2019). This medication is still investigational and phase 2 efficacy trials are being planned. While it has been suggested, based on AKT1 involvement in the MTORc pathway, that treatment of Proteus syndrome with sirolimus or everolimus (mTOR inhibitors) could be effective, as of yet there are no data to support this (Marsh et al. 2011).

Growth and Feeding

Growth issues in Proteus syndrome can be challenging. Most affected individuals have severe asymmetric overgrowth that can cause leg length discrepancy (in one person more than 20 cm). Many have a lipodystrophy that includes lipomas or lipomatous infiltration of muscles and viscera in some parts of the body and lipoatrophy in other parts of the body. Some affected individuals also have significant undergrowth of some muscle groups, the upper extremities being most often involved. This might be confused with failure-to-thrive, although the usually severe overgrowth elsewhere in the body is quite distinct. The mechanism for this has not been determined, but is likely to be some type of compensatory reaction in normal tissues. It has been claimed that Proteus syndrome "burns out" with adolescence insofar as some individuals seem to experience a cessation or marked diminution of their overgrowth near the end of puberty. Although this is not universal, it may be common, and this consideration is critical for planning treatment, including watchful waiting.

Evaluation

- Apparent failure-to-thrive should be investigated in a standard manner.
- For a discussion of evaluation of overgrown tissues, please see Musculoskeletal below.

Management

- Feeding excess calories does not appear to be helpful to treat the underdevelopment or lipoatrophy.
- The use of growth-stimulating or androgenic hormones to treat this manifestation is *not* recommended because of the unknown effects of such a treatment on hyperplastic tissues and possible oncogenicity.
- For a discussion of treatment of overgrown tissues, please see Musculoskeletal below.

Development and Behavior

Development can be delayed, and a minority of the affected individuals are intellectually impaired (probably less than 10–15%, although these estimates are problematic because of ascertainment bias). Developmental delay is often associated with congenital hemimegalencephaly, one of the few congenital manifestations of the disorder. There are no known stereotypical or common behavioral problems.

Evaluation

- Developmental milestones should be monitored at regular infant and childhood visits.
- If developmental delay is evident, formal developmental assessment should take place.

Management

- Appropriate interventions and therapies, including educational programming, should be instituted, as indicated by the disabilities identified.

Musculoskeletal

The musculoskeletal manifestations of Proteus syndrome are as common as they are daunting (Crenshaw et al. 2018; Tosi et al. 2011). The relentlessly progressive nature of the disorder during childhood and the frequent involvement of bones, cartilage, muscle, and connective tissues cause symptoms in nearly all affected individuals. Most have asymmetric overgrowth that primarily affects the length of the tubular bones but can also affect bone width, joint capsule, connective tissues, and muscles. Lipomatous infiltration of muscle groups can cause an underestimation of the degree of muscle underdevelopment. More troubling than simple asymmetry of the length of one member of a pair of bony structures is the asymmetric overgrowth within a single bone, which leads to skeletal distortion. One of the most common manifestations of this phenomenon is in the proximal tibia where two sides of a primary epiphyseal growth plate can have markedly different rates of growth, leading to the rapid development of bowing or severe varus or valgus deformities at the knees. Another common manifestation is hyperplasia of the cartilaginous tissues surrounding the joints (knees and digits most commonly) leading to rapid, and in some cases nearly complete, loss of joint mobility. A few individuals also have ectopic calcification of tendons and muscles. The axial skeleton can also manifest asymmetric overgrowth of the vertebral bodies leading to rapidly progressive dystrophic scoliosis and associated restrictive lung disease.

Evaluation

- Axial and appendicular skeletal abnormalities are best evaluated by careful physical examination and plain film radiography.
- Computed tomography with 3D reconstruction may become more important when that technique is more widely available or surgery is planned.
- Involvement of an orthopedist should occur early, certainly no later than when there is functional impact from overgrowth.

Management

- Treatment of overgrowth and asymmetric epiphyseal growth is best managed by epiphysiodesis by curettage, stapling, or eight-plates. The timing of these procedures is crucial to effect an appropriate correction of angulation or balancing of growth.
- Osteotomy to reduce bone length may be required if epiphyseal manipulation alone is insufficient. This technique can be used to shorten and/or straighten tubular bones.
- Spinal fusion is indicated for persons who have, or are likely to develop, pulmonary compromise resulting from scoliosis. Bracing may be less effective when there is asymmetric or irregular overgrowth of vertebral bodies. Spinal fusion has been performed successfully in persons with Proteus syndrome using modern surgical techniques. Again, timing is challenging because early correction may avoid complications associated with future worsening of the primary lesions, however, later surgery may be preferred to optimize truncal growth.
- For lesser degrees of leg length asymmetry, shoe lifts are indicated.

- The Ilizarov technique is not recommended for the purpose of lengthening of tubular bones, as the shorter bone is generally the normal bone. Avoiding overgrowth of the pathologic bone by epiphysiodesis is strongly preferred.
- Any surgical procedure that requires prolonged operating room time or convalescent immobility should raise the consideration of prophylactic anticoagulation (see Respiratory).

Respiratory

Individuals with Proteus syndrome can have primary or secondary pulmonary disease. Secondary complications arise from scoliosis. The primary pulmonary manifestation of Proteus syndrome is bullous degeneration of the lungs (Biesecker 2006; Lim et al. 2011). This manifestation can be asymptomatic but puts affected individuals at risk for pneumonia. More severe degrees of this manifestation have led to severe cardiopulmonary compromise and necessitated lobectomy (unpublished data). Proteus syndrome also predisposes to deep venous thrombosis and pulmonary embolism (Keppler-Noreuil et al. 2017). This complication has occurred in children as young as 9 years of age and is a major challenge as most pediatricians are unfamiliar with the diagnosis and management of these problems. The onset of symptoms can be sudden and the outcome catastrophic. It is believed that deep venous thrombosis and pulmonary embolism are major contributors to the early mortality associated with Proteus syndrome. Aggressive management can be effective and should be implemented emergently.

The risk of thrombosis appears to be dramatically elevated in a post-operative setting. In a recent retrospective review, more than 10% of patients with Proteus syndrome in the NIH cohort had had one or more episodes of post-operative thrombosis, including children (Keppler-Noreuil et al. 2017). While no controlled studies have been done or are contemplated, we consider it standard of care to prophylactically anticoagulate patients who are undergoing any but the most minor of outpatient procedures.

Evaluation

- Symptoms of chronic pulmonary compromise, pneumonia, or the need for major surgical procedures should prompt consideration of high-resolution computed tomography scanning of the chest to exclude bullous malformations. This is the most effective imaging modality for this complication. Individuals with clinically significant degrees of bullous degeneration should also be considered for ventilation-perfusion scanning.
- The acute onset of pulmonary insufficiency or chest pain in a person with Proteus syndrome of any age should be considered pulmonary embolism until proven otherwise.
- The d-dimer test may be useful as most patients with active thrombosis appear to have elevated levels of this biomarker. An abnormal d-dimer test should be followed up with high-resolution chest computed tomography.
- High-resolution chest computed tomography with intravenous contrast is the most effective and rapid diagnostic technique for pulmonary embolism. The classic algorithm of ventilation-perfusion scanning followed by pulmonary angiography may be preferred in some situations, but is more time consuming.

Management

- Individuals with bullous degeneration of the lungs, but without obvious symptoms, need to be monitored for pneumonia and atelectasis. They should be carefully evaluated when they are febrile, as pneumonia is a likely complication because of the poor clearance of secretions from the bullous lung tissue.
- Individuals with pulmonary bullous degeneration should have vigorous pulmonary toilet during postoperative convalescence.
- Individuals with chronic ventilation-perfusion mismatching may be considered for resection of dysfunctional lung tissue.
- The acute onset of symptoms associated with deep venous thrombosis and pulmonary embolism should be promptly and aggressively treated with anticoagulation. Patients who are positive for pulmonary embolism on imaging should undergo acute anticoagulation according to the standard American College of Chest Physicians guidelines (Kearon et al. 2016).
- Consideration may be given to focal treatment with thrombolytic agents, although their use in Proteus syndrome has not been previously demonstrated to be safe or effective.
- After recovery from an acute pulmonary embolism, anticoagulation therapy must be individualized, as there are no data to allow general recommendations for the duration of anticoagulation.
- Chronic anticoagulation and prophylactic anticoagulation are not routinely recommended. The long-term risks and benefits are unknown and must be weighed carefully for each individual. Some individuals may benefit from such a therapy, but the risks of anticoagulation may be high in children and in persons with vascular malformations.
- Any patient with Proteus syndrome undergoing a surgical procedure or requiring bedrest for a significant period should be considered for prophylactic anticoagulation. Long periods of immobility (e.g., airline flights of more than 8–10 hours) are also a potential risk and prophylaxis may be considered.

Gastrointestinal

Several individuals have had gastrointestinal complications including rectal prolapse and gastric outlet obstruction from hamartomatous gut wall tissue, and numerous individuals have intra-abdominal lipomatous infiltration (Lublin et al. 2002). There may be significant gastrointestinal blood loss from hemorrhagic bowel masses. It might be expected that individuals with Proteus syndrome would be at risk for intussusception or bowel obstruction secondary to internal hamartomas or lipomas.

Evaluation

- Rectal prolapse should be evaluated by standard imaging modalities such as barium enema and flexible or rigid endoscopy. This manifestation has been associated with hamartomas of the bowel wall, which should be readily identifiable with these techniques.
- Contrast imaging studies such as upper gastrointestinal series with small-bowel follow-through or oral contrast computed tomography may be useful if intussusception or acute bowel obstruction is suspected.
- Gastrointestinal blood loss should be excluded in any individual with anemia or melena using standard clinical techniques.
- As in all cases of bowel obstruction, regardless of cause, urgent diagnosis and treatment are critical to avoid complications. In Proteus syndrome, the urgency of relieving obstruction and avoiding bowel perforation must be balanced by an appreciation of the difficulties inherent in operating on individuals who have multisystem disease and who may have markedly abnormal anatomy or lesions that are difficult to resect (e.g., lymphatic vascular malformations). Careful thought should be given to balancing the urgency to operate with the benefits of obtaining a thorough preoperative evaluation by detailed imaging studies and consultations.

Management

- Surgical treatment of gastrointestinal lesions in Proteus syndrome should be undertaken with a full appreciation of the peculiar nature of the disorder and the possibilities of rare lesions or abnormal anatomy.
- Although it has been claimed that surgery on abnormal tissue activates or aggravates the overgrowth in Proteus syndrome, supporting data are lacking. This concern should not weigh heavily in making a decision about operating on a child with an acute lesion, especially an acute abdomen.

Neurologic

As noted above, some patients may have congenital hemimegalencephaly (enlargement of one side of the brain), which is one of the rare congenital manifestations of Proteus syndrome. Several individuals with Proteus syndrome have been found to have neuronal migration abnormalities, although the frequency and severity of this problem are not well-characterized. Seizures are uncommon but, when present, are often associated with hemimegalencephaly, or other central nervous system abnormalities. Seizures associated with hemimegalencephaly frequently have a relentless progression towards catastrophic epilepsy (Perry and Duchowny 2011). A potentially problematic lesion is hyperostosis of the skull, which is uncommon but can exceed 5 cm in thickness in severe cases. Although this lesion appears to have the potential to cause central nervous system symptoms by cortical compression or distortion, this is not a common problem, as most affected individuals are asymptomatic from a neurologic perspective. One person with Proteus syndrome has been found to have arteriovenous malformations of the brain, leading to seizures, which improved after the lesions were surgically removed (personal observation).

Evaluation

- A full developmental and neurologic evaluation should be carried out at diagnosis and repeated periodically until puberty.
- Computed tomography and magnetic resonance imaging of the brain are effective modalities for imaging. Computed tomography is preferable for imaging the skull and magnetic resonance imaging preferable for the brain. Indications for cranial imaging include significant skull asymmetry or focal overgrowth, neurologic symptoms including seizures or focal neurologic signs on examination, and developmental delay or intellectual disability.
- Electroencephalogram is useful for the characterization of potential seizures.

Management

- Several individuals who have surprising degrees of hyperostosis and cortical distortion or displacement have shown no apparent signs of neurologic dysfunction (personal observation). For this reason, skull hyperostosis should generally be managed by observation until it is clear that the lesion is causing neurologic symptoms.
- Seizures should be treated symptomatically, as one would in the general population, unless they are caused by arteriovenous malformations, for which surgery should be considered.
- Hemispherectomy is the therapeutic option of choice for intractable epilepsy caused by hemimegalencephaly (Perry and Duchowny 2011).

Ophthalmology

There have been several reports linking Proteus syndrome to disorders of the globe and surrounding tissues (Bouzas et al. 1993). Epibulbar benign tumors are probably the most common manifestation in the globe, whereas the most common periorbital lesions are hyperostoses and connective tissue nevi. Lesions can occur anywhere in the neuro-optic pathway including physical impingement of the periorbital tissues, compression of the globe by tumors, epibulbar dermoids that expand over the cornea, retinal lesions, compression of the optic nerve or tract by tumors, or lesions of the occipital cortex.

Evaluation

- Thorough evaluation of the optic axis is recommended for any individual with Proteus syndrome who has an ophthalmologic complaint.
- Common modalities for evaluation include direct and indirect ophthalmoscopy, electroretinograms, and magnetic resonance imaging or computed tomography of the eye and surrounding structures.

Management

- Surgical treatment of lesions that impinge on the optical axis is useful and appropriate. Because of the progressive nature of the disorder, such procedures must be planned carefully because some procedures must be performed repeatedly as result of regrowth of a lesion, such as periorbital hyperostosis.

Otolaryngology

Several affected individuals have had soft tissue overgrowth of the airway that obstructs respiration. Overgrowth of the tonsils and adenoids may be common and can be asymmetric. Extrinsic compression of the posterior wall of the pharynx and larynx by hyperostoses of the vertebral bodies has been seen in several individuals. Hyperostosis of the external auditory canal can lead to complete occlusion of the canal, conductive hearing loss, and cholesteatomas (personal observation).

Evaluation

- Standard modalities of polysomnography, flexible endoscopy, pulmonary function tests, and imaging studies including computed tomography and virtual bronchoscopy may be used when upper airway compression or obstruction is suspected.
- The external auditory canals should be examined periodically to exclude hyperostosis of the external auditory canal.

Management

- Hyperostoses of the external auditory canal should be surgically reduced when they impair hearing or cause discomfort.
- Intrinsic lesions of the airway such as hyperplastic nodules that obstruct respiration should be excised.
- The only known treatment for extrinsic compression of the airway from hyperostoses of the vertebrae is to bypass the obstruction with a tracheostomy.

Genitourinary

Ovarian cystadenomas are common in Proteus syndrome and should be excised when symptomatic (Gordon et al. 1995). Other hyperplastic lesions of the genital tract have also been encountered. Cystic lesions of the epididymis are common in males but they appear to be non-malignant though they may be so large as to necessitate surgical removal (Biesecker, unpublished data). Inguinal hernias are common in Proteus syndrome and may be isolated or associated with lipomas of the inguinal canal.

Evaluation

- When abdominal or pelvic pain is present, a search should be made for ovarian cystadenomas or other tumors, even in young children.
- Ultrasound is the preferred primary imaging tool for the pelvic lesions of Proteus syndrome. Magnetic resonance imaging and computed tomography are also useful.
- Inguinal hernias should be evaluated by ultrasound and exploration.

Management

- Surgical excision of ovarian masses is the recommended treatment.
- We recommend monitoring epididymal cystic masses and operating only when they appear to be expansile or have other characteristics of malignancy.
- Repair of inguinal hernias should be accompanied by dissection to exclude lipomas or lymphatic vascular malformations. The latter may be difficult to resect, but resection is nonetheless important because these anomalies may predispose to recurrence.

Dermatologic

Nearly all individuals with Proteus syndrome have some dermatologic manifestations (Nguyen et al 2004). These can include linear verrucous epidermal nevi, cerebriform connective tissue nevi (CCTN), and cutaneous vascular malformations.

CCTN are especially troublesome as they can cause difficulties with shoe fit, biomechanical changes, and odor.

Evaluation

- An evaluation by a dermatologist is recommended for all individuals with a confirmed or suspected diagnosis of Proteus syndrome. Ongoing care by a dermatologist (or plastic surgeon) may be necessary for those with symptomatic vascular malformations.
- Acutely malodorous connective tissue nevi should be cultured for bacterial and fungal growth.

Management

- Skin care should consist of brief bathing with mild non-soap cleansers and thorough but gentle drying (a hair dryer on low heat works well).
- Moisturizers with minimal fragrances, colors, and others may be used liberally.
- Malodorous connective tissue nevi need to be cleaned thoroughly but gently, using cotton swabs to cleanse the sulci.
- Aluminum chloride solution may be applied to the feet to reduce perspiration.
- Infections should be treated with appropriate topical or systemic antimicrobials.
- Surgical removal of the plantar CCTN with skin grafting has been accomplished with relatively good results in two patients and should be considered (Biesecker, unpublished data).

Cardiovascular

Although structural cardiac lesions are rare in Proteus syndrome, vascular manifestations and lipomas of the heart are common. The vascular lesions range from trivial cutaneous capillary vascular malformations to gigantic mixed lesions with capillary, venous, and lymphatic components. Some are also mixed with lipomatous tissue. High flow or arterial vascular malformations are unusual in individuals with Proteus syndrome, but have been observed. Lipomatous intracardiac tissue has been identified in a series of affected individuals, although the functional consequences were minimal (Hannoush et al. 2015).

Evaluation

- Pure capillary vascular malformations do not require imaging.
- Imaging of other vascular malformations can be accomplished with magnetic resonance imaging, ultrasound, and, in select cases, with invasive studies such as lymphangiography.

Management

- Treatment of the most common lesion of Proteus syndrome, the mixed lymphatic-venous-lipomatous malformation, should be reserved for situations in which the lesion interferes with function or has major cosmetic or disfigurement implications. Surgery is the preferred treatment for these lesions. However, it should be borne in mind that dissection of malformations that include lymphatics may be challenging and postoperative weeping from the margins is common (Biesecker, unpublished data). For this reason, liposuction may not be appropriate treatment for mixed lipomatous-lymphatic masses.
- Laser treatment is unlikely to be beneficial for these lesions, although cutaneous vascular malformations that can occur in Proteus syndrome may respond to laser treatment.
- Myocardial fat overgrowth does not usually warrant treatment.

Neoplasia

It is recognized that individuals with Proteus syndrome have experienced a number of unusual tumors and that the rarity of these tumors in the general population strongly suggests (but does not prove) that tumor predisposition is part of the disorder. The most common tumor of Proteus syndrome is likely meningiomas (Keppler-Noreuil et al. 2016). Interestingly, a number of sporadic meningiomas also have the *AKT1* c.49G>A, p.(Glu17Lys) variant. However, many other tumors have been observed in individuals with this syndrome and the wide variety of tumors and the disparate location of those tumors make screening difficult, and the efficacy of such screening is unknown. Instead of routine screening, prompt and vigorous evaluation of signs or symptoms of malignancy is recommended.

Evaluation

- In lieu of screening, affected individuals with any signs and symptoms of a malignancy should be promptly and thoroughly evaluated for a tumor. Because of the wide variety of tumors, specific recommendations cannot be made and physicians instead must use clinical judgment to prompt such evaluations.
- Because meningiomas are likely the most common tumor in patients with Proteus syndrome, clinicians should be alert for symptoms such as headaches or other signs of an intracranial tumor, which should prompt imaging studies.

Management

- Aggressive treatment of tumors (surgery, irradiation, and chemotherapy) is indicated, as there are no data to suggest that tumors in Proteus syndrome have a worse prognosis than in the general population.

- Oncologists should be aware that proliferative lesions or masses are a hallmark of Proteus syndrome, and not all such masses are malignant. Careful histopathologic characterization of such lesions is essential to insure that malignancies are treated as such but hyperplasia is not.

Dental

Dental abnormalities are common in Proteus syndrome because of asymmetric overgrowth of the mandible and maxilla. Some individuals have enlarged or malformed teeth.

Evaluation

- Physical and radiographic examination by an orthodontist may be useful when it is clear that dental malalignment is developing.

Management

- Orthodontic treatment may be appropriate for some individuals although there is little experience with such treatments. The treatment of dental malocclusion in a progressively disfiguring condition can be challenging.

ACKNOWLEDGMENTS

This chapter is dedicated to the memory of Alex Hoag, Kyle Dullenkopf, Sean Easly, and Todd Peirini, four individuals who participated in the National Institutes of Health study. I hope that the sorrow of their families can in some small way be mitigated by the knowledge that their child's participation in the study has directly and significantly benefited current and future affected individuals with this disorder. I am indebted to them and to all of the other individuals and families who have participated. The opinions expressed in this chapter are those of the author and are not to be construed as official recommendations by the Department of Health and Human Services, the National Institutes of Health, or any other institution with which he is affiliated. The efforts of the author are supported by funds from the Intramural Research Program of the National Human Genome Research Institute.

RESOURCES

The Proteus Syndrome Foundation

4915 Dry Stone Dr.
  Colorado Springs, CO 80923, USA
  Phone: (901) 756-9375
  Email: kim@proteus-syndrome.org
  Website: http://www.proteus-syndrome.org

Proteus Syndrome Foundation, United Kingdom

The Old Coach House
  St James Road
  Hastings
  TN34 3LH, UK
  Phone: 07740 085919
  Email: traceywhitewoodneal@yahoo.co.uk
  Website: www.proteus-syndrome.org

REFERENCES

Biesecker LG (2006) The challenges of Proteus syndrome: Diagnosis and management. *Eur J Hum Genet* 14:1151–1157.

Bouzas EA, Krasnewich D, Koutroumanidis M, Papadimitriou A, Marini JC, Kaiser-Kupfer MI (1993) Ophthalmologic examination in the diagnosis of Proteus syndrome. *Ophthalmology* 100:334–338.

Brockmann K, Happle R, Oeffner F, Konig A (2008) Monozygotic twins discordant for Proteus syndrome. *Am J Med Genet A* 146A:2122–2125.

Carpten JD, Faber AL, Horn C, Donoho GP, Briggs SL, Robbins CM, Hostetter G, Boguslawski S, Moses TY, Savage S, Uhlik M, Lin A, Du J, Qian YW, Zeckner DJ, Tucker-Kellogg G, Touchman J, Patel K, Mousses S, Bittner M, Schevitz R, Lai MH, Blanchard KL, Thomas JE (2007) A transforming mutation in the pleckstrin homology domain of AKT1 in cancer. *Nature* 448:439–444.

Caux F, Plauchu H, Chibon F, Faivre L, Fain O, Vabres P, Bonnet F, Selma ZB, Laroche L, Gerard M, Longy M (2007) Segmental overgrowth, lipomatosis, arteriovenous malformation and epidermal nevus (SOLAMEN) syndrome is related to mosaic PTEN nullizygosity. *Eur J Hum Genet* 15:767–773.

Cohen M (1987) The Elephant Man did not have neurofibromatosis. *Proc Greenwood Genetics Center* 6:187–192.

Crenshaw MM, Goerlich CG, Ivey LE, Sapp JC, Keppler-Noreuil KM, Scott AC, Biesecker LG, Tosi LL (2018) Orthopaedic management of leg-length discrepancy in Proteus syndrome: A case series. *J Pediatr Orthop* 38:e138–e144.

Forbes SA, Beare D, Gunasekaran P, Leung K, Bindal N, Boutselakis H, Ding M, Bamford S, Cole C, Ward S, Kok CY, Jia M, De T, Teague JW, Stratton M. R, McDermott U, Campbell PJ (2015) COSMIC: exploring the world's knowledge of somatic mutations in human cancer. *Nucl Acids Res* 43:D805–811.

Gordon PL, Wilroy RS, Lasater OE, Cohen MM, Jr (1995) Neoplasms in Proteus syndrome. *Am J Med Genet* 57:74–78.

Gripp KW, Baker L, Kandula V, Conard K, Scavina M, Napoli JA, Griffin GC, Thacker M, Knox RG, Clark GR, Parker VE, Semple R, Mirzaa G, Keppler-Noreuil KM (2016) Nephroblastomatosis or Wilms tumor in a fourth patient with a somatic PIK3CA mutation. *Am J Med Genet A* 170:2559–2569.

Hafner C, Lopez-Knowles E, Luis NM, Toll A, Baselga E, Fernandez-Casado A, Hernandez S, Ribe A, Mentzel T, Stoehr R, Hofstaedter F, Landthaler M, Vogt T, Pujol RM, Hart mann A, Real FX (2007) Oncogenic PIK3CA mutations occur in epidermal nevi and seborrheic keratoses with a characteristic mutation pattern. *Proc Natl Acad Sci USA* 104:13450–13454.

Hannoush H, Sachdev V, Brofferio A, Arai AE, LaRocca G, Sapp J, Sidenko S, Brenneman C, Biesecker LG, Keppler-Noreuil KM (2015) Myocardial fat overgrowth in Proteus syndrome. *Am J Med Genet A* 167A:103–110.

Happle R (1987) Lethal genes surviving by mosaicism: A possible explanation for sporadic birth defects involving the skin. *J Am Acad Dermatol* 16:899–906.

Happle R (2007) Type 2 segmental Cowden disease vs. Proteus syndrome. *Br J Dermatol* 156:1089–1090.

Happle R (2010) The group of epidermal nevus syndromes Part I. Well defined phenotypes. *J Am Acad Dermatol* 63:1–22; quiz 23–24.

Heyer GL (2015) PHACE(S) syndrome. *Handb Clin Neurol* 132:169–183.

Jamis-Dow CA, Turner J, Biesecker LG, Choyke PL (2004) Radiologic manifestations of Proteus syndrome. *Radiographics* 24:1051–1068.

Kearon C, Akl EA, Ornelas J, Blaivas A, Jimenez D, Bounameaux H, Huisman M, King CS, Morris TA, Sood N, Stevens SM, Vintch JR, Wells P, Woller SC, Moores L (2016) Antithrombotic therapy for VTE disease: CHEST guideline and expert panel report. *Chest* 149:315–352.

Keppler-Noreuil K, Sapp J, Burton-Akright J, Darling T, Dombi E, Gruber A, Johnston J, Lindhurst M, Bagheri M, Jarosinski P, Martin S, Paul S, Savage R, Wolters P, Schwartz B, Widemann B, Biesecker L (2019) Pharamacodynamic Study of Miransertib (ARQ 092) in Patients with Proteus Syndrome. *Am J Hum Genet* 104:484–491.

Keppler-Noreuil KM, Baker EH, Sapp JC, Lindhurst MJ, Biesecker LG (2016) Somatic AKT1 mutations cause meningiomas colocalizing with a characteristic pattern of cranial hyperostosis. *Am J Med Genet A* 170:2605–2610.

Keppler-Noreuil KM, Lozier JN, Sapp JC, Biesecker LG (2017) Characterization of thrombosis in patients with Proteus syndrome. *Am J Med Genet A* 173:2359–2365.

Keppler-Noreuil KM, Rios JJ, Parker VE, Semple RK, Lindhurst MJ, Sapp JC, Alomari A, Ezaki M, Dobyns W, Biesecker LG (2015) *PIK3CA*-related overgrowth spectrum (PROS): diagnostic and testing eligibility criteria, differential diagnosis, and evaluation. *Am J Med Genet A* 167A:287–295.

Lim GY, Kim OH, Kim HW, Lee KS, Kang KH, Song HR, Cho TJ (2011) Pulmonary manifestations in Proteus syndrome: pulmonary varicosities and bullous lung disease. *Am J Med Genet A* 155A:865–869.

Lindhurst MJ, Sapp JC, Teer JK, Johnston JJ, Finn EM, Peters K, Turner J, Cannons JL, Bick D, Blakemore L, Blumhorst C, Brockmann K, Calder P, Cherman N, Deardorff MA, Everman DB, Golas G, Greenstein RM, Kato BM, Keppler-Noreuil KM, Kuznetsov SA, Miyamoto RT, Newman K, Ng D, O'Brien K, Rothenberg S, Schwartzentruber DJ, Singhal V, Tirabosco R, Upton J, Wientroub S, Zackai EH, Hoag K, Whitewood-Neal T, Robey PG, Schwartzberg PL, Darling TN, Tosi LL, Mullikin JC, Biesecker LG (2011) A mosaic activating mutation in AKT1 associated with the Proteus syndrome. *N Engl J Med* 365:611–619.

Lindhurst MJ, Yourick MR, Yu Y, Savage RE, Ferrari D, Biesecker LG (2015) Repression of AKT signaling by ARQ 092 in cells and tissues from patients with Proteus syndrome. *Sci Rep* 5:17162.

Lublin M, Schwartzentruber DJ, Lukish J, Chester C, Biesecker LG, Newman KD (2002) Principles for the surgical management of patients with Proteus syndrome and patients with overgrowth not meeting Proteus criteria. *J Pediatr Surg* 37:1013–1020.

Marsh DJ, Trahair TN, Kirk EP (2011) Mutant AKT1 in Proteus syndrome. *N Engl J Med* 365:2141–2142; author reply 2142.

Nguyen D, Turner JT, Olsen C, Biesecker LG, Darling TN (2004) Cutaneous manifestations of proteus syndrome: correlations with general clinical severity. *Arch Dermatol.* 2004 Aug 140(8): 947–953.

Sapp JC, Hu L, Zhao J, Gruber A, Schwartz B, Ferrari D, Biesecker LG (2017) Quantifying survival in patients with Proteus syndrome. *Genet Med* 19:1376–1379.

Tosi LL, Sapp JC, Allen ES, O'Keefe RJ, Biesecker LG (2011) Assessment and management of the orthopedic and other complications of Proteus syndrome. *J Child Orthop* 5:319–327.

Treves F (1885) A case of congenital deformity. *Proc Patholog Soc Lond* 36:494–498.

Turner JT, Cohen MM, Jr Biesecker LG (2004) Reassessment of the Proteus syndrome literature: Application of diagnostic criteria to published cases. *Am J Med Genet A* 130A:111–122.

Vidaurri-de la Cruz H, Tamayo-Sanchez L, Duran-McKinster C, de la Luz Orozco-Covarrubias M, Ruiz-Maldonado R (2004) Epidermal nevus syndromes: clinical findings in 35 patients. *Pediatr Dermatol* 21:432–439.

Wiedemann HR, Burgio GR, Aldenhoff P, Kunze J, Kaufmann HJ, Schirg E (1983) The proteus syndrome. Partial gigantism of the hands and/or feet, nevi, hemihypertrophy, subcutaneous tumors, macrocephaly or other skull anomalies and possible accelerated growth and visceral affections. *Eur J Pediatr* 140:5–12.

48

PTEN HAMARTOMA TUMOR SYNDROME

JOANNE NGEOW

Lee Kong Chian School of Medicine, Nanyang Technological University Singapore and Cancer Genetics Service, National Cancer Centre Singapore

CHARIS ENG

Genomic Medicine Institute, Cleveland Clinic, Ohio, USA and Department of Genetics and Genome Sciences, Case Western Reserve University, Cleveland, Ohio, USA

INTRODUCTION

PTEN hamartoma tumor syndrome (PHTS) is a group of autosomal dominant syndromes characterized by germline *PTEN* mutations leading to predisposition to hamartomas and an increased risk of breast, thyroid, and other malignancies (Zbuk and Eng 2007). The clinical spectrum of PHTS includes Cowden syndrome (CS), Bannayan–Riley–Ruvalcaba (BRRS) syndrome, Proteus syndrome (PS), Proteus-like syndrome (PLS), and autism and macrocephaly.

Lloyd and Dennis first described Cowden syndrome in a report of a 20-year-old woman with a history of mucocutaneous papillomatosis, facial dysmorphism, multinodular thyroid, fibrocystic breast disease, scoliosis, pectus excavatum, and cognitive impairment (Lloyd and Dennis 1963). Riley and Smith reported the first case of what would later be described as Bannayan–Riley–Ruvalcaba syndrome (Riley and Smith 1960). They described a family with macrocephaly, pseudopapilledema, and multiple hemangiomata. A decade later, Bannayan described a 3-year-old female with megancephaly, lipomatosis, and vascular malformations (Bannayan 1971). Gorlin and colleagues reported a large family with overlapping phenotypic features of the previously described Riley–Smith, Bannayan–Zonana, and Ruvalcaba–Myhre syndromes, and they suggested a single syndrome with variable expressivity; eventually leading to the unifying classification of Bannayan–Riley–Ruvalcaba syndrome (Gorlin et al. 1992). The first cases of Proteus syndrome were reported by Wiedemann and colleagues based on partial gigantism of the extremities, pigmented nevi, subcutaneous hamartomatous tumors, macrocephaly, and/or skull anomalies (Wiedemann et al. 1983). The International Cowden Consortium initially mapped and identified germline *PTEN* mutations in families with Cowden syndrome (Nelen et al. 1996; Liaw et al. 1997), and, shortly thereafter, multiple investigators identified deletions and pathogenic intragenic mutations of *PTEN* in a subset of BRRS individuals and families. The diagnosis of PHTS requires the demonstration of a pathogenic *PTEN* mutation, regardless of phenotype.

Incidence

The incidence of PHTS is unknown. Before identification of *PTEN*, the incidence of Cowden syndrome was estimated to be 1 in a million based on a population-based study outside Amsterdam (Starink et al. 1986). With the identification of the gene, the incidence has been estimated at 1 in 200,000 for Cowden syndrome (Nelen et al. 1996; Nelen et al. 1997). The true incidence of PHTS is likely higher than this, as the syndrome is difficult to diagnose and may be under-ascertained.

Diagnostic Criteria

Cowden Syndrome
The International Cowden Consortium published a set of consensus clinical operational diagnostic criteria in 1996 for purposes of the gene hunt (Nelen et al. 1996), which has turned

Cassidy and Allanson's Management of Genetic Syndromes, Fourth Edition.
Edited by John C. Carey, Agatino Battaglia, David Viskochil, and Suzanne B. Cassidy.
© 2021 John Wiley & Sons, Inc. Published 2021 by John Wiley & Sons, Inc.

out to be robust, yielding a mutation frequency of 80–85% when these strict criteria are followed (Marsh et al. 1998).

An operational clinical diagnosis of Cowden syndrome (CS), may be made if an individual has three or more major criteria, but one must include macrocephaly, Lhermitte–Duclos disease, or GI hamartomas; or two major and three minor criteria (NCCN 2018; Table 48.1). In individuals with a family where one individual meets CS clinical diagnostic criteria or has a *PTEN* mutation meets diagnostic criteria; the diagnostic criteria are met if a family member has any two major criteria; or one major and two minor criteria; or three minor criteria.

Pathognomonic criteria include adult-onset Lhermitte–Duclos disease (benign hamartoma of the cerebellum) and the mucocutaneous lesions specific for CS: papillomatous papules, trichilemmomas, and acral keratoses (Figure 48.1). If an individual only has mucocutaneous lesions, the diagnosis of CS may be made if there are; (a) three or more must be trichilemmomas, or (b) three or more cutaneous facial papules and oral mucosal papillomatosis, or (c) three or more oral mucosal papillomatosis and acral keratoses, or (d) three or more palmoplantar keratoses. If the mucocutaneous lesions do not satisfy the above criteria, then they are not considered a major criterion.

The mucocutaneous lesions of CS are often described as flesh-colored benign papules, keratotic or verrucous lesions, angiomas, fibromas, or achrocordons. Oral mucosal lesions include papules and fibromas of the oral mucosa, tongue, lips, gums, and the nasal vestibule (Figure 48.2). The tongue often has a scrotal, or cobblestone, appearance. The pathology of mucocutaneous lesions is often an important clue in a clinician's assessment of CS. Common pathologic manifestations of CS dermatologic lesions include infundibulum hyperplasia and infra-infundibular abnormalities, follicular abnormalities, and epidermal proliferations. Extrafacial lesions are often hyperkeratotic papillomas and may resemble verruca vulgaris (Figure 48.1) (Brownstein et al. 1979; Starink and Hausman 1984; Starink and Hausman 1984).

Other major criteria for CS are breast, thyroid (usually follicular and sometimes papillary), and endometrial cancers, and macrocephaly (occipital frontal circumference (OFC) > 97th percentile). The macrocephaly in CS is typically benign megalencephaly. When evaluating for macrocephaly, it is important to note the patient's height and weight to account for the possibility of relative macrocephaly in an individual of tall stature and the patient's ancestry as individuals of certain ancestries, such as Celtic and Venezuelan, may have a greater OFC compared to individuals of other ancestries.

Minor criteria of CS are benign breast lesions/fibrocystic breast disease, uterine fibroids, thyroid nodules/adenomas, gastrointestinal hamartomas, lipomas, fibromas, genitourinary tumors, genitourinary structural defects, and autism spectrum disorder.

Bannayan–Riley–Ruvalcaba syndrome While there is not a consensus regarding the diagnostic criteria for BRRS, most individuals have congenital macrocephaly, multiple lipomas,

TABLE 48.1 Diagnostic criteria for Cowden syndrome (NCCN 2018)

Operation diagnosis of Cowden syndrome if an individual meets any of the following:	Operational diagnosis for an individual with a relative diagnostic for Cowden syndrome OR with a *PTEN* mutation:
1. Three or more major criteria, but one must include macrocephaly, Lhermitte–Duclos disease, or GI hamartomas; or 2. Two major and three minor criteria.	3. Any two major criteria with or without minor criteria; or 4. One major and two minor criteria; or 5. Three minor criteria.
Major criteria:	*Minor criteria:*

• Breast cancer
• Endometrial cancer (epithelial)
• Thyroid cancer (follicular)
• GI hamartomas (including ganglioneuromas, but excluding hyperplastic polyps; ≥3)
• Lhermitte-Duclos disease (adult)
• Macrocephaly (≥97 percentile: 58 cm for females, 60 cm for males)
• Macular pigmentation of the glans penis
• Multiple mucocutaneous lesions (any of the following):
  • Multiple trichilemmomas (≥3, at least one biopsy proven)
  • Acral keratoses (≥3 palmoplantar keratotic pits and/or acral hyperkeratotic papules)
  • Mucocutaneous neuromas (≥3)
  • Oral papillomas (particularly on tongue and gingiva), multiple (≥3) OR biopsy proven OR dermatologist diagnosed
	• Autism spectrum disorder
	• Colon cancer
	• Esophageal glycogenic acanthoses (≥3)
	• Lipomas (≥3)
	• Intellectual disability (i.e. IQ ≤75)
	• Renal cell carcinoma
	• Testicular lipomatosis
	• Thyroid cancer (papillary or follicular variant of papillary)
	• Thyroid structural lesions (e.g., adenoma, multinodular goiter)
	• Vascular anomalies (including multiple intracranial developmental venous anomalies

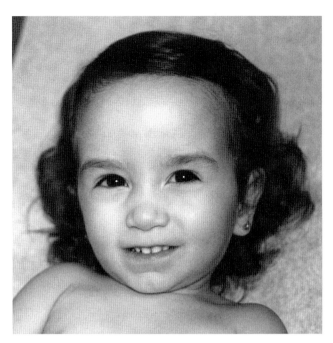

FIGURE 48.1 Cutaneous features of Cowden syndrome: (A) papillomatous papule, (B) trichilemmoma, and (C) palmar keratoses. (42.1(B) reproduced from Eng et al. 1994, with permission of author).

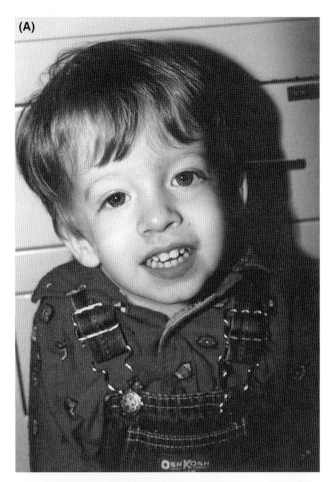

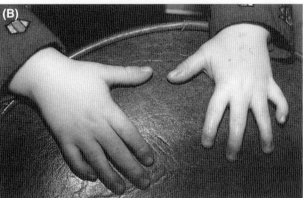

FIGURE 48.2 Oral mucosal lesions in a male patient with Cowden syndrome.

hemangiomas, and hyperpigmentation of the glans penis in males (Riley and Smith 1960; Bannayan 1971; Zonana et al. 1976). The majority of patients have some degree of motor and/or speech delay, but intellectual disability and epilepsy are less common (Miles et al. 1984). Many patients have increased birth weight and hypotonia in addition to macrocephaly. The hamartomatous tumors present in BRRS include congenital or later onset hemangiomas, lipomas, and lymphangiomas (Riley and Smith 1960; Bannayan 1971; Zonana et al. 1976; Miles et al. 1984; Gorlin et al. 1992). Most hamartomatous tumors are subcutaneous, but may occur in other sites such as intracranial, visceral, intestinal, and bone (Miles et al. 1984). Many children have dysmorphic features including ocular hypertelorism, down-slanting palpebral fissures, high palate, joint hyperextensibility, and pectus excavatum (Riley and Smith 1960; Bannayan 1971; Zonana et al. 1976; Miles et al. 1984; Gorlin et al. 1992). Gastrointestinal polyposis, which may have an onset in childhood or adulthood, is present in 45% of individuals with BRRS and can initially present as abdominal pain, rectal bleeding, intussusception, mucous in the stool, or diarrhea (Gorlin et al. 1992). With the advent of *PTEN* screening, individuals with BRRS and a pathogenic *PTEN* mutation are considered to have PHTS.

Proteus Syndrome The diagnosis of PS is complex and described in detail in Chapter 47. PS is characterized by progressive disease, sporadic inheritance, and mosaic distribution of lesions (Biesecker et al. 1999). Features of PS include connective and epidermal nevi, disproportionate overgrowth, ovarian cystadenomas, parotid gland adenomas, adipose tissue dysregulation, vascular malformations, and dysmorphic features (Biesecker et al. 1999). See Chapter 47 for a clinical description of the syndrome.

PTEN Hamartoma Tumor Syndrome The phenotypes of Cowden syndrome, Bannayan–Riley–Ruvalcaba syndrome,

Proteus syndrome, and autism with macrocephaly syndrome require demonstration of a *PTEN* pathogenic mutation to be designated as PHTS. This is important when establishing guidelines of care for individuals with the above phenotypes because the presence of a pathogenic *PTEN* mutation is the primary driver of neoplasia and care must be applied in making recommendations for PHTS versus individuals with CS, BRRS, PS, autism spectrum with macrocephaly who do not have *PTEN* mutations. In most hands, the sensitivity of detecting a *PTEN* mutation is >99%.

Etiology, Pathogenesis, and Genetics:

PHTS is an autosomal dominant condition with variable expressivity. *PTEN* mutations are detected in 80–85% of individuals with a clinical diagnosis of classic Cowden syndrome (CS) by the strict operational diagnostic criteria (Marsh et al. 1998: Zhou et al. 2003) and 65% of individuals with a clinical presentation of BRRS (Marsh et al. 1999; Zhou et al. 2003). In a study of 14 individuals with manifestations of Proteus syndrome, five individuals were found to have a *PTEN* mutation (Zhou et al. 2000; Zhou et al. 2001). A study by Tan et al. (2011) evaluated a diverse cohort of individuals with CS and CS-like manifestations accrued from multiple medical settings and found that germline *PTEN* mutations are present in about 25% of individuals.

The spectrum of germline *PTEN* mutations seen in PHTS extends throughout the coding sequence. Located on human chromosome sub-band 10q23.3, the 9-exon *PTEN* gene encodes a 403-amino acid phosphatase with four major domains (Fanning and Anderson; 1999; Goffin et al. 2001; Zbuk and Eng; 2007). *PTEN* is a tumor suppressor gene that participates in the PI3K/AKT/mTOR pathway and is increasingly shown to be involved in many different cellular pathways (Figure 48.3). The main canonical function of PTEN is to antagonize AKT by dephosphorylating phosphatidylinositol 3,4,5-triphosphate (PIP3) to phosphatidylinositol 3,4,5-diphosphate (PIP2) (Sekulic et al. 2000; Neshat et al. 2001; Podsypanina et al. 2001; Mester and Eng 2013). Germline mutations in *PTEN* cause upregulation of the AKT pathway leading to decreased apoptosis and increased cell growth (Furnari et al. 1997; Weng et al. 1999; Mester and Eng 2013). Within the N-terminal tail of PTEN are several key motifs, including the phosphatidyl-inositol-bisphosphate (PIP$_2$) binding motif and both nuclear and cytoplasmic localization sequences. The protein phosphatase domain contains the catalytic core of the protein, stretching from amino acid 123 to 130. A C2 domain facilitates membrane binding, and the PDZ-binding motif within the C-terminal tail allows protein–protein interaction (Fanning and Anderson 1999; Goffin et al. 2001; Zbuk and Eng 2007; Mester and Eng 2013). About half of the mutations in *PTEN* occur within the phosphatase domain, which is where the enzymatic activity of PTEN occurs. A catalytic core domain lies within the phosphatase domain, and many mutations are found within this area core motif, which disrupt its important enzymatic function (Fanning and Anderson; 1999; Goffin et al. 2001; Zbuk and Eng 2007; Mester and Eng 2013).

Pathogenic variants have been described in all nine exons of *PTEN*, with various types of mutations identified including missense, nonsense, splice-site variants, intragenic deletions/insertions, and large deletions (Pezzolesi et al. 2007; Teresi et al. 2007; Zbuk and Eng 2007). Virtually all germline *PTEN* missense mutations within the coding region are pathogenic (Zbuk and Eng 2007; Tan et al. 2011). Common nonsense/frameshift mutations have been well-described in exons 5, 6, 7, and 8 of *PTEN* as well as specific truncation mutations in exons 5, 7, and 8 which are over-represented in the *PTEN* mutation spectra (Tan et al. 2012; Mester and Eng 2013) (Figure 48.4). The exon 5 hotspot encodes a segment of the catalytic core of PTEN, and mutations within this 7-amino acid stretch affect probands with a wide variety of clinical presentations. Large deletions and duplications affecting *PTEN* are less common in PHTS than single base pair alterations, though they can be found over the entire coding sequence. In addition to exonic and splice-site mutations and deletions, *PTEN* promoter mutations have also been identified in 10% of individuals with CS (Zhou et al. 2003). These promoter mutations are suspected to alter *PTEN* transcription and have been shown to result in PI3K/Akt pathway dysfunction (Zhou et al. 2003; Teresi et al. 2007). *PTEN* promoter mutations are more likely to be associated with a CS presentation with preponderance of breast involvement and oligo-organ involvement (Marsh et al. 1998; Zhou et al. 2003). Unfortunately, even the largest cohorts of patients with PHTS are insufficient to identify clear genotype–phenotype associations.

There are individuals with classic CS without germline *PTEN* mutations. Approximately 10% of individuals with classic CS or similar phenotypes carry germline heterozygous sequence variants in the genes encoding three of the four subunits of succinate dehydrogenase (SDH) or mitochondrial complex II (Ni et al. 2008). Single-exon *KLLN*, on 10q23, encodes KILLIN, and shares a bidirectional promoter with *PTEN*. Up to 30% of individuals with CS phenotypes, without either germline *PTEN* or *SDHx* mutations, were found to have germline *KLLN* promoter hypermethylation leading to decreased expression of KLLN (Bennett et al. 2010). Another 9% of unrelated CS individuals without germline *PTEN* mutations were found to have germline *PIK3CA* mutations, and 2% harbored germline *AKT1* mutations (Orloff et al. 2013). Functionally, there is significant increase of phosphorylated-AKT1 levels in these patients' lymphoblastoid cell lines, which supports their potential role as novel CS-susceptibility genes (Orloff et al. 2013). Germline heterozygous gain-of-function mutations in *SEC23B* have been identified in ~5% of individuals with CS and these mutations are also enriched in apparently sporadic

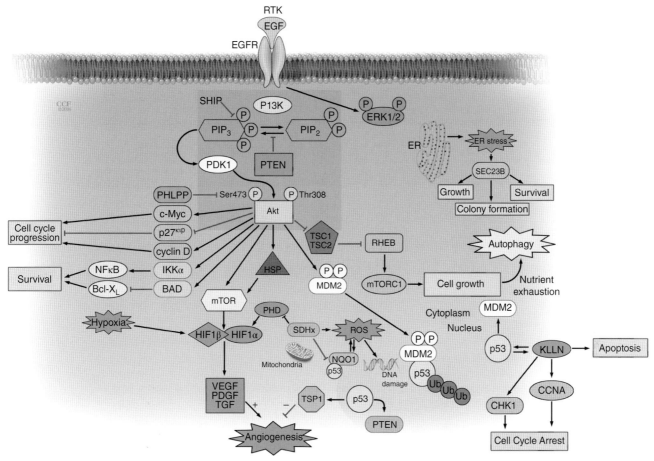

FIGURE 48.3 *PTEN* cellular pathway. Diagram depicts the canonical and non-canonical signaling involving CS/CSL-related predisposition genes (*PTEN, AKT, PIK3CA, SDHx, KLLN, SEC23B*).

thyroid cancer patients (Yehia et al. 2015). Gain-of-function germline mutations in *EGFR* have been seen in a unique CS family presenting with Lhermitte–Duclos disease (Colby et al. 2016).

Somatic *PTEN* mutations may occur in neoplasia, including breast, thyroid, and endometrial tumors. *PTEN* LOH was shown to occur in 23–50% of endometrial adenocarcinoma tumors and may be useful as a biomarker for premalignant disease (Mutter et al. 2000; Eng 2003). *PTEN* LOH occurs in ~31% of breast adenocarcinoma and 13% of thyroid carcinomas (Eng 2000).

Diagnostic Testing

The diagnosis of PHTS may be made by clinical assessment and/or *PTEN* mutation analysis. In the absence of a detectable *PTEN* mutation, an individual may be given an operational clinical diagnosis of Cowden syndrome if they satisfy the criteria devised by the NCCN (2018) and shown in Table 48.1. A diagnosis of Bannayan–Riley–Ruvalcaba syndrome does not require the presence of a *PTEN* mutation, and it can be established as a clinical diagnosis (see above section). A diagnosis of Proteus syndrome is based on the clinical criteria described in detail in Chapter 47.

The identification of *PTEN* mutations is not only helpful in confirming a diagnosis in those individuals with suspicion of PHTS, but is also potentially beneficial in management of those with a clinical diagnosis, and especially those with partial signs or overlap of phenotypes. *PTEN* analysis includes sequencing exons 1–9 as well as the *PTEN* promoter and should include deletion/duplication analysis. PHTS can be differentiated from other hereditary cancer syndromes including hereditary breast ovarian cancer syndrome (HBOC), Lynch syndrome (LS), and other hamartomatous polyposis syndromes based on personal and family history. However, given the protean nature of PHTS and lack of general awareness amongst clinicians, establishing a diagnosis on clinical grounds can be challenging (Mester et al. 2013). Additionally, because of the high frequency of de novo *PTEN* germline mutations (Mester and Eng 2012), a family history of associated cancers may not be apparent. Furthermore, many of the benign manifestations of CS are

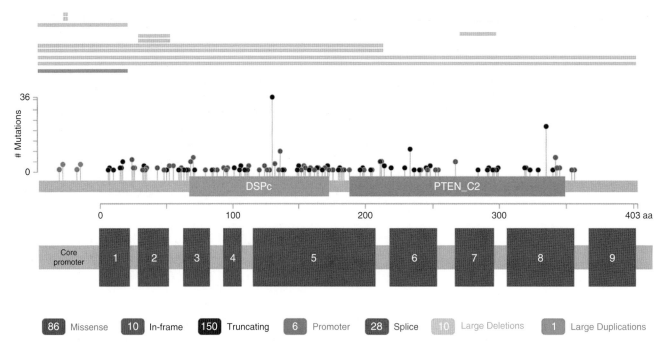

FIGURE 48.4 *PTEN* germline mutational spectra in 291 probands. The domain structure of phosphatase and tensin homologue (PTEN). PTEN is a 403 amino acid protein that is composed of functional domains: a dual-specificity phosphatase, catalytic domain (DSPc), a C2 domain, a carboxy-terminal tail and a PDZ-binding domain. PDZ domains are significant regions for protein–protein interactions that play a vital role in cellular signal transduction. The N-terminal domain contains the phosphatase domain (the enzymatic activity of PTEN) and it is, therefore, not surprising that the majority of PTEN mutations occur within this domain. The top solid bars represent large deletions and large duplications within the *PTEN* gene. Frequency of point mutations reported in probands is shown, which correlates with the vertical height of the line (middle). Mutations have been identified in the promoter and all nine exons of *PTEN* (bottom). A key at the bottom of the figure denotes the total number of unique alterations within each category of mutation identified in 291 probands. Missense mutations = green; in-frame mutations = brown; truncating mutations = black; promoter alterations = red; splice alterations = purple; large deletions = orange; large duplications = pink (Tan et al. 2011; Mester and Eng 2013).

common in the general population making the recognition of an underlying syndrome difficult for most clinicians. An occipito-frontal head circumference (OFC) in adults greater than two standard deviations is seen in the majority of adult PHTS patients, which is a useful clinical manifestation that can identify which cancer patients are at risk of PHTS. Any cancer patient with a large OFC should be assessed for personal/family history of other PHTS-related manifestations (e.g. thyroiditis, polyps, developmental delay/autism, and other malignancies) and be referred for genetic risk assessment (Shiovitz et al. 2010; Mester et al. 2011; Mester et al. 2013).

To aid clinical diagnosis, a clinical predictor (Cleveland Clinic *PTEN* Risk Calculator) was developed based on clinical manifestations derived from a prospective study of over 3000 patients suspected of having CS (Tan et al. 2011). The questionnaire-based clinical decision support tool is available online (http://www.lerner.ccf.org/gmi/ccscore/) toassist clinicians at point of patient care. Based on a patient's presentation, a score (CC score) is derived that corresponds to an estimated risk for germline *PTEN* mutation to guide clinicians for referral to genetics professionals. A CC score at a threshold of 15 (CC15) corresponds to a 10% a priori risk of a germline *PTEN* mutation detection, which is the lowest among different strategies tested. At a cost-effectiveness threshold of $100,000/Quality Adjusted Life Year, CC15 is the optimal strategy for female patients older than 50 years, while CC10 is the optimal strategy for female patients younger than 50 years old and male patients of all ages (Ngeow et al. 2015).

Thus, patients with a CC score >10 should be considered for genetic risk assessment referral. Additionally, the American College of Medical Genetics and Genomics published a practical guide for which patients should be referred for genetic assessment for CS. The ACMGG recommends referral for anyone meeting any three criteria from the major or minor diagnostic criteria (Table 48.1). Referral should be considered for any individual with a personal history of or first-degree relative with (i) Lhermitte–Duclos disease diagnosed after age 18, or (ii) any three criteria from the major or minor diagnostic criteria list in the same person (Hampel et al. 2015).

For pediatric cases, the presence of macrocephaly (OFC > 2 standard deviation [SD] over the population mean, or 97.5th percentile) is a necessary criterion for diagnosis (Tan et al. 2011). Neurologic (autism and developmental delay) and dermatologic (lipomas, oral papillomas) manifestations are common secondary features; either or both systems were involved in 100% of patients with germline *PTEN* mutations. However, given that dermatologic features may often be overlooked, less-prevalent manifestations in individuals at initial presentation in the pediatric setting are likely to be at least as important in the diagnostic assessment of PHTS; thus, the presence of vascular (such as arteriovenous) malformations, gastrointestinal polyps, thyroid goiter, and early-onset cancers (thyroid and germ cell) warrant consideration for *PTEN* testing.

Genetic testing should be done in conjunction with pre- and post-test genetic counseling as described by the American Society of Clinical Oncology recommendations (Oncology 2003). The ASCO guidelines recommend that cancer genetic testing for children should be carefully considered and balanced among the complex issues of parental rights of medical decision-making, availability of evidence-based cancer risk reduction options, the probability of developing a malignancy in childhood, and careful consideration of the child's right to elect or decline testing if given the opportunity as an adult. The vast majority of PHTS-associated cancers have onset in adulthood, but there are reports of pediatric thyroid cancer (Ngeow et al. 2011). If a child is symptomatic and testing would assist the family and physicians in diagnosis and/or medical decision-making, *PTEN* testing is warranted as part of either a multidisciplinary pediatric genetics or an oncology team. Furthermore, in those rare families with early-onset *PTEN*-associated cancers, predictive testing may be considered 5–10 years prior to the youngest cancer diagnosis in the family for surveillance purposes.

Differential Diagnosis

The main genetic differential diagnoses for PHTS must include conditions with breast and thyroid carcinoma. The genetic differential diagnosis of breast cancer is relatively straightforward, including hereditary breast-ovarian cancer syndrome (HBOC, *BRCA1/2*), Li–Fraumeni syndrome (LFS; *TP53*), Peutz–Jeghers syndrome (PJS; *LKB1/STK11*), heritable diffuse gastric cancer, which has lobular breast carcinoma as a component (HDGC; *CDH1*), and neurofibromatosis type 1 (NF1; *NF1*; Chapter 40). Because of the distinct other component neoplasias, ages-of-onset, and/or other phenotypic features of HBOC, LFS and HDGC, these differential diagnoses should be easily excluded. PJS, especially when presenting with hamartomatous polyps, should be a seriously considered differential diagnosis (Sweet et al. 2005). Similarly, classic NF1 is distinct from PHTS; however, mosaic NF1 may be mistaken for PHTS. For example, isolated neurofibromas, whether cutaneous or mucosal, are seen in PHTS.

The genetic differential diagnosis for epithelial follicular neoplasias, the classic thyroid neoplasia presentation in PHTS, also includes the autosomal recessive condition Werner syndrome. The differential diagnosis for papillary thyroid carcinoma (PTC), sometimes seen in PHTS with follicular variant, includes Carney complex, familial papillary thyroid carcinoma, and familial adenomatous polyposis. Similar to the differential diagnosis for hereditary breast cancer, the respective phenotypes and other associated neoplasias in Werner and Carney complex make these syndromes easy to differentiate from PHTS. In patients with PTC and polyposis, both FAP and PHTS should be considered. However, the thyroid and colon pathology should provide clarification in that patients with FAP have a cribiform/morular type PTC and multiple adenomatous polyps, whereas patients with PHTS have the follicular variant of PTC and hamartomatous polyps. Some families have features of both PHTS and familial PTC. In these cases, *PTEN* analysis, if positive, will clarify a diagnosis but a negative result cannot rule out PHTS. Some CS families show a preponderance of classic PTC, suggesting another etiology or perhaps the involvement of low-penetrance modifier genes (Zbuk et al. 2007).

The genetic differential for hamartomatous polyps also includes the above-mentioned Peutz–Jeghers syndrome (PJS) and juvenile polyposis syndrome (JPS). Both syndromes have distinct polyp characteristics and/or extra-colonic manifestations. PJS is a polyposis syndrome characterized by gastrointestinal hamartomatous polyposis with distinct Peutz–Jeghers polyps and mucocutaneous hyperpigmentation. Mutations in *STK11/LBK1* are associated with PJS (Hemminki et al. 1998; Jenne et al. 1998). PJS polyps are characterized by smooth-muscle proliferation and elongated stalks (Zbuk and Eng 2007).

JPS is characterized by gastrointestinal polyposis with juvenile polyps. The term "juvenile" refers to the type of polyp rather than age of onset. JPS is due to mutations in *SMAD4, BMPR1A,* or *PTEN* (Howe et al. 1998; Sweet et al. 2005; Ngeow et al. 2013). A related disorder, juvenile polyposis of infancy, is seen in patients with a large deletion encompassing both *PTEN* and *BMPR1A*. These patients present with features of BRRS/CS and typically have GI symptoms of bleeding, diarrhea, and enteropathy secondary to polyposis before age 2 (Delnatte et al. 2006).

MANIFESTATIONS AND MANAGEMENT

The clinical presentation and clinical phenotype of PHTS is highly variable. In some individuals or families, a *PTEN* mutation manifests in a presentation of CS and in others, the clinical features are more consistent with BRRS, Proteus, or another phenotype. There may be phenotypic overlap between CS and BRRS in one individual or within a family, which is why the allelic syndromes are collectively categorized in the PTEN hamartoma tumor syndromes classification, irrespective of the clinical diagnosis or syndrome, with an identified germline *PTEN* mutation (Eng 1993). Early literature estimated that about 80% of individuals with a diagnosis of CS had a germline *PTEN* mutation (Liaw et al. 1997). A study by Tan et al. (2011) looked at a diverse cohort of individuals with CS accrued from the multiple medical settings and found that germline *PTEN* mutations are present in about 25% of individuals (Tan et al. 2011). It is important for clinicians to be aware of the wide spectrum of clinical features of PHTS to help differentiate a diagnosis of PHTS from other hereditary cancer syndromes (Mester et al. 2011; Mester et al. 2013). Most individuals with PHTS have pleiotropism and are followed as part of a multidisciplinary team.

Growth and Feeding

There are minimal data on growth and feeding in PHTS. Feeding is typically normal in individuals with PHTS, although no large formal studies have been performed. In rare cases of associated autism or developmental delay, children may have sensory aversion or hypotonia that can affect their food choices and oral-motor function.

Children with BRRS often have a high birth weight and congenital macrocephaly (Gorlin et al. 1992). Some individuals with PHTS have hemihyperplasia or overgrowth (Zonana et al. 1976; Okumura et al. 1986).

Evaluation

- Physical exam to assess for any growth asymmetries.
- Measurement and plotting of height, weight, and head circumference as part of routine pediatric care, including estimate of standard deviations from the mean for OFC as individuals with PHTS often have an OFC that is +4 SD.

Management

- There are no specific treatments for feeding or growth.

Development and Behavior

Most individuals with PHTS have normal development and normal cognition. About half of individuals with the BRRS phenotype have some degree of developmental delay, intellectual disability, hypotonia, motor delay and/or speech delay (Gorlin et al. 1992). Patients with the *PTEN*-associated autism and macrocephaly phenotype should receive the therapies, interventions, and services per standard of care for individuals with autism. Developmental delay or intellectual disability is less common in individuals with the CS phenotype.

Evaluation

- Children should have a developmental assessment as part of pediatric care.
- Any concerns for autism, developmental delay, or atypical behavior warrant further evaluation.

Management

- Various therapies and education supports may be beneficial for a child with autism or developmental delay, including special education services, physical therapy, occupational therapy, and speech therapy.
- There are no specific developmental treatments or therapies recommended for children with PHTS; they should be followed per standard of care based on individual developmental and behavioral needs.

Neoplasias

The spectrum of neoplasia associated with germline *PTEN* mutations include cancers of the endometrium, kidney, and colon, and melanoma as well as benign findings such as GI polyposis (Starink et al. 1986; Eng 1993; Heald et al. 2010; Tan et al. 2012). The lifetime risks for malignancy and germline *PTEN* mutations identify significantly increased risks for endometrial, renal, thyroid, and breast cancers (Tan et al. 2012; Bubien et al. 2013; Nieuwenhuis et al. 2014). Regardless of a patient's initial presentation and family history, surveillance for *PTEN*-associated neoplasia, as described below, is the current standard of care for individuals with *PTEN* mutations (see Table 48.2).

Breast Many individuals with PHTS may first present to breast surgeons and oncologists because of both benign and malignant breast pathology. Estimates of breast cancer risk for females with histories consistent with CS were traditionally reported to be around 25–50% (Starink et al. 1986); however, Tan et al. (2012) identified an 85% lifetime risk, beginning around age 30 years, for female breast cancer, with 50% penetrance by age 50 years for those individuals with pathogenic *PTEN* mutations. This risk figure is comparable to that quoted for patients with hereditary breast and ovarian cancer syndrome (Mester and Eng 2015). A similar study by Bubien et al. (2013) found a cumulative 77%

TABLE 48.2 Screening recommendations.

Cancer	General population risk (%)	Lifetime risk with PHTS (%)	Age at presentation	Screening recommendations
Breast	12	~85	40s	Starting at age 30, annual mammogram, consider MRI for patients with dense breast
Thyroid	1	35	30s–40s	Baseline ultrasound at diagnosis, annual ultrasound and clinical examination
Endometrial	2.6	28	40s–50s	Starting at age 30: annual endometrial biopsy or transvaginal ultrasound
Renal cell	1.6	34	50s	Starting at age 40; renal imaging every two years
Colon	5	9	40s	Starting at age 40; colonoscopy every two years
Melanoma	2	6	40s	Annual dermatologic examination

risk for female breast cancer at age 70 years for women with *PTEN* mutations. In addition, Nieuwenhuis et al. (2013) identified a 67% risk for females with germline *PTEN* mutations developing breast cancer by age 60 years.

Several studies have also considered *PTEN* mutation status related to primary and second primary breast cancer diagnoses, and found that women with *PTEN* mutations are at elevated risks for both (Ngeow et al. 2014; Mester and Eng 2015). Women with *PTEN* mutations who have had a diagnosis of breast cancer have a 29% risk of developing a secondary breast cancer within 10 years (Ngeow et al. 2014; Mester and Eng 2015). Women may choose to pursue prophylactic mastectomy for these reasons, particularly if the patients have associated benign breast lesions making breast cancer surveillance difficult. Breast cancer has been anecdotally described in males with *PTEN* mutations, but an overall increased risk for male breast cancer was not established in a study of >3000 patients, although the adults in the cohort were mostly female patients (Fackenthal et al. 2001; Tan et al. 2011).

The National Comprehensive Cancer Network recommends that women with PHTS or clinical CS begin self-breast exam at the age of 18 years and undergo annual clinical breast exam at the age of 25. Breast MRI and mammogram should begin between the ages of 30 and 35. If there is a young breast cancer diagnosis in a relative, clinical breast exam and breast imaging may begin 5–10 years earlier than the age of that diagnosis. Due to the increased risk of breast cancer and high frequency of benign breast disease and fibrocystic breast tissue, women with CS should have annual breast MRI in addition to mammogram (Saslow et al. 2007; NCCN 2018). While prophylactic mastectomy is not a medical recommendation for all women with PHTS, it is available as an option, discussed on a case-by-case basis with consideration regarding the degree of protection, extent of personal cancer risk, difficulty in imaging, breast symptoms and reconstructive options (NCCN 2018) (Table 48.1).

Breast cancer treatment may include lumpectomy or mastectomy, radiation therapy, and/or chemotherapy. If the tumor is estrogen and/or progesterone receptor positive, anti-estrogen therapy (such as Tamoxifen) may be considered in a woman who has had a hysterectomy. There are no clinical data regarding risks and benefits of Tamoxifen treatment as a chemo-preventive agent for an unaffected woman with a *PTEN* mutation.

Thyroid A systematic study of thyroid neoplasms from a prospectively accrued series of individuals with CS and like manifestations revised the lifetime thyroid cancer risk for individuals with germline *PTEN* mutations from 10% upwards to around 34%, with the earliest age at diagnosis of 7 years (Tan et al. 2012; Bubien et al. 2013; Nieuwenhuis et al. 2014). Thyroid pathology in PHTS typically affects follicular cells (Harach et al. 1999; Parisi et al. 2001), with follicular thyroid carcinoma (FTC) considered a major diagnostic criterion and an important feature in PHTS. Benign thyroid lesions, such as thyroid nodules, multinodular goiter, and Hashimoto's thyroiditis, are also common in individuals with germline *PTEN* mutations (Mester and Eng 2015).

An individual's risk of developing epithelial thyroid cancer is increased by 70-fold when compared to the general population (Tan et al. 2011). FTC was over-represented in a cohort of germline *PTEN* mutation positive individuals, the ratio of FTC to the more common papillary thyroid cancer was 1:2 amongst individuals with PHTS as compared to ~1:14 in the general population (Ngeow et al. 2011). Several reports of thyroid cancer in children with germline *PTEN* mutations (Ozes et al. 2001) have been described, which emphasizes the need to begin thyroid screening annually at the time of PHTS diagnosis (Milas et al. 2012; Mester and Eng 2015).

The prevalence of germline *PTEN* mutations in unselected differentiated thyroid cancer is low (<1%) (Dahia et al. 1997; Nagy et al. 2011). A pediatric onset of thyroid cancer, male gender, history of thyroid nodules and/or thyroiditis and FTC histology were predictive factors of PHTS in a cohort of individuals with CS with thyroid cancer (Ngeow et al. 2011). These "red-flags" and/or a family history of cancers, physical signs, such as macrocephaly and mucocutaneous features, should alert the clinician to the possibility of PHTS.

At a minimum, individuals with PHTS should have a baseline thyroid ultrasound at diagnosis and annual physical examination of the thyroid from age 18. If surgical intervention is indicated, expert opinion is to perform a complete thyroidectomy due to the risk of subsequent or metachronous malignancy and difficulty of additional thyroid surgery. Patients with thyroid adenomas may also consider complete thyroidectomy due to these reasons.

Endometrial Individuals with germline *PTEN* mutations have a 28% lifetime risk of developing endometrial cancer (Tan et al. 2012), and an age <50 years at presentation of endometrial carcinoma, macrocephaly, and/or prevalent or synchronous renal cell carcinoma in these women were predictive of germline *PTEN* mutation (Ngeow et al. 2011). The mean age of endometrial cancer diagnosis in those with *PTEN* mutations is 44 years, and three-quarters are diagnosed under age 50. This observation may guide an age range for consideration of surveillance versus prophylactic surgery. Individuals with germline *PTEN* mutations are also at increased risk of developing benign findings of the endometrium such as uterine fibroids (Tan et al. 2012).

Clinicians should encourage patient education and prompt response to symptoms (e.g., abnormal vaginal bleeding). Annual random endometrial biopsies and/or ultrasound beginning at age 30–35 years can be considered. Clinicians should also discuss an option of hysterectomy upon completion of childbearing and counsel regarding degree of protection, extent of cancer risk, and reproductive desires.

Gastrointestinal Gastrointestinal polyposis is a common feature of the BRRS phenotype and also occurs in CS (Carlson et al. 1984; Gorlin et al. 1992). GI polyps have been reported in up to 33% of individuals fulfilling diagnostic criteria for CS and up to 45% of patients with the BRRS phenotype (Haggitt and Reid 1986; Starink et al. 1986; Gorlin et al. 1992; Harned et al. 1995). These figures were derived from clinical studies before identification of the *PTEN* gene and are limited by small sample size. With the advent of *PTEN* mutation analysis, Heald et al (2010) demonstrated >90% of those individuals with *PTEN* germline mutation who had a colonoscopy performed as part of clinical care, had colorectal polyps, typically with a mix of histological subtypes, and individuals who developed colorectal carcinomas also tended to have multiple, and mixed, polyps. A 9% increase in the lifetime risk for colorectal cancer has been associated with PHTS (Tan et al. 2012). These findings led to a change in clinical practice, colorectal surveillance should now be offered to any *PTEN* mutation carrier especially those with multiple lower GI polyps. In *PTEN* mutation carriers, upper GI polyps occur with some frequency, and a subset of patients experience symptoms. Notably, a significant proportion (~20%) of those with upper GI examinations had mucosal nodularity of the esophagus (McGarrity et al. 2003; Nishizawa et al. 2009; Heald et al. 2010).

Individuals with PHTS or *PTEN*-mutation negative CS meeting NCCN criteria should undergo colonoscopy earlier than those in the general population, starting from age 35. Patients should have earlier screening if symptomatic or if a close relative is diagnosed with colon cancer before age 40 years; they should then start 5–10 years before the earliest known colon cancer in the family (NCCN 2018). Colonoscopy should be done every five years or more frequently if patient is symptomatic or if polyps are found.

Other Neoplasias Patients with PHTS have a 34% lifetime risk of developing renal cell carcinoma (RCC) (Tan et al. 2012). Reported histology of each mutation positive patient's RCC was variable. However, on central pathology re-review of eight patients, six examined lesions were determined to be of papillary sub-histology, with the other two patients' tumors consistent with the initial report of chromophobe RCC. Immunohistochemistry demonstrated complete loss of PTEN protein in all *PTEN* mutation positive patients' papillary RCCs (pRCC) and patchy positivity in one chromophome RCC. Physicians caring for individuals with PHTS should have a low threshold for investigating possible RCC in patients with relevant complaints. Renal ultrasound is not sensitive for detecting pRCC, especially if small, therefore individuals with PHTS should have alternate renal imaging (CT or MRI) (Mester et al. 2012). There is a 6% increase in the estimated lifetime risk for melanoma for individuals with germline *PTEN* mutations (Tan et al. 2012; Mester and Eng 2015).

Evaluation

- *For WOMEN with PHTS or CS meeting clinical criteria (NCCN 2018):*
  - Breast awareness from age 18.
  - Clinical breast exam, every 6–12 months, starting at age 25 years or 5–10 years before the earliest known breast cancer in the family (whichever comes first).
  - Breast screening: annual mammography with consideration of tomosynthesis and breast MRI screening with contrast starting at age 30–35 years or 5–10 years before the earliest known breast cancer in the family (whichever comes first).
  - For women age >75 years, management should be considered on an individual basis.
  - For women with a *PTEN* mutation who are treated for breast cancer, and have not had a bilateral mastectomy, screening with annual mammogram and breast MRI should continue as described above.
  - For endometrial cancer screening, encourage patient education and prompt response to symptoms

(e.g., abnormal bleeding). Consider annual random endometrial biopsies and/or ultrasound beginning at age 30–35 years. Discuss option of hysterectomy upon completion of childbearing and counsel regarding degree of protection, extent of cancer risk, and reproductive desires.
- Discuss option of risk-reducing mastectomy and counsel regarding degree of protection, extent of cancer risk, and reconstruction options.
- Address psychosocial, social, and quality-of-life aspects of undergoing risk-reducing mastectomy and/or hysterectomy.

- *For MEN AND WOMEN with PHTS or CS meeting NCCN criteria:*
  - Annual comprehensive physical exam starting at age 18 years or five years before the youngest age of diagnosis of a component cancer in the family (whichever comes first), with particular attention to thyroid exam.
  - Annual thyroid ultrasound beginning at 7 years of age or at time of PHTS diagnosis, whichever occurs later.
  - Colonoscopy, starting at age 35 years unless symptomatic or if close relative with colon cancer before age 40 years then start 5–10 years before the earliest known colon cancer in the family. Colonoscopy should be done every five years or more frequently if patient is symptomatic or polyps are found.
  - Consider renal ultrasound starting at age 40 years, then every one to two years.
  - Dermatologic management may be indicated for some patients.
  - Consider psychomotor assessment in children at diagnosis and brain MRI if there are symptoms.
  - Education regarding the signs and symptoms of cancer.

Management

- Treatment for breast cancer in PHTS does not differ from breast cancer treatment in the general population. Women with breast cancer may undergo lumpectomy, mastectomy, chemotherapy, and/or radiation therapy.
- Tamoxifen or other anti-estrogen treatment may be considered for women who have previously undergone hysterectomy. There are no data regarding the chemopreventive effects of Tamoxifen for women with a *PTEN* mutation.
- Thyroid carcinoma should be treated with total thyroidectomy, never subtotal or hemi-thyroidectomy. Patients with multiple thyroid adenomas may also consider thyroidectomy (personal experience).
- Any skin lesions suspicious for cancer should be excised and reviewed by pathology.

- Gastrointestinal polyposis should be treated according to standard of care; there is no specific treatment for individuals with PHTS.
- While not clinically available at this time, mTOR inhibitors may prove to be an effective treatment for benign and malignant neoplasia associated with PHTS (Marsh et al. 2008).

Mucocutaneous

The great majority of individuals with PHTS have some degree of dermatologic involvement. Mucocutaneous lesions are present in almost all individuals with Cowden syndrome and may include papillomatous papules, trichilemmomas, and acral keratoses (Figure 48.1) (Brownstein et al. 1979). Papillomatous papules are benign, flesh-colored papules that may resemble a viral wart. They commonly grow on the face and hands as well as in the oral mucosa, gums, and nasal cavity. Trichilemmomas are also benign, flesh-colored hamartomas that develop from the outer root sheath of the hair follicle. Acral keratoses appear as palmar pits or indentations. The mucocutaneous lesions typically manifest in teenage and adult years and are present in 95% of adults with clinical Cowden syndrome (Starink et al. 1986). Fibromas, achrocordons, and lipomas are also commonly seen in individuals with PHTS. These lesions associated with PHTS are typically benign. They may be removed if irritating or physically displeasing, but often grow back.

Hyperpigmented macules of the glans penis are commonly seen in BRRS. These nevi are benign and do not require any treatment intervention.

Evaluation

- Skin evaluation as part of comprehensive annual physical exam.
- Consideration of annual dermatology evaluation at the age of 18 or younger.

Management

- Any lesions suspicious for cancer should be excised and reviewed by pathology.
- Benign skin lesions may be removed if irritating to the patient, but the lesions of PHTS often grow back. The risk of hypertrophic keloid formation should limit any excisions for asymptomatic lesions.
- Topical agents, curettage, cryosurgery and laser ablation may be considered *only* in symptomatic patients.

Neurology

Individuals with PHTS may have neurologic involvement. Adult-onset Lhermitte–Duclos disease (LDD), a dysplastic gangliocytoma of the cerebellum, is one of the pathognomonic criteria for CS. Most reported cases of adult LDD have been associated with CS and 75% of LDD tumor samples in one series showed either decreased or absent PTEN expression and increased phosphorylated Akt (Robinson and Cohen 2000; Zhou et al. 2003; Perez-Nunez et al. 2004). Eighty-three percent of adult patients (15/18) with LDD in this series were found to have germline *PTEN* mutations and three children with LDD did not have detectable *PTEN* mutations or PHTS features (Zhou et al. 2003).

Childhood LDD is not known to be associated with PHTS (Capone Mori et al. 2003). LDD can cause cerebellar dysfunction and sometimes result in hydrocephalus, and even though surgical resection of the cerebellar hamartoma is an option, complete removal is difficult due to the nature of the hamartomatous tumor.

The first case study of a child with a *PTEN* mutation and autism described a boy who inherited a nonsense mutation from his mother, who herself was diagnosed with Cowden syndrome but did not have social or intellectual disabilities (Goffin et al. 2001). The first estimate of mutation frequency in a prospective series of patients with macrocephaly and autism reported three *PTEN* mutations in a series of 18 children with macrocephaly and ASD (Butler et al. 2005). This benchmark prevalence of 17% remains near the weighted average reported across nearly 10 subsequent studies (Tilot et al. 2015). Together, these results provide a strong case for the association of germline *PTEN* mutation in children with ASD and macrocephaly. These data form the basis for the recommendation of genetic testing in this subset of ASD patients. The degree of macrocephaly observed in patients with ASD and *PTEN* mutations is often more severe than that seen in those with wild-type *PTEN*. In a cohort of 181 *PTEN* mutation carriers, average head size was +3.5 SDs from average in adults and +5 SDs in the pediatric subset (Mester et al. 2011). Recently, a series of case reports demonstrated epileptic seizures in *PTEN* mutation-positive patients, often linked to underlying cortical dysplasia (Conti et al. 2012). Further studies are needed to fully understand the neurological sequelae of *PTEN* mutation carriers, including cognitive profiles.

Evaluation

- Neurologic evaluation and brain MRI for patients with cerebellar signs.
- All patients should be assessed for neurological and cognitive signs; all pediatric patients should be assessed for autism spectrum disorder.

Management

- Surgical resection is the only treatment for LDD at this time, but the timing and the frequency are unknown.
- Patients should consider enrolling on clinical trials for PHTS-related ASD if applicable.

Vascular

Vascular anomalies, while not part of the clinical criteria for PHTS, are a common feature in individuals with CS, BRRS, and PS. Vascular anomalies in PHTS include hemangioma, arteriovenous malformation (AVM), vascular staining, and other vascular lesions (Biesecker et al. 1999; Tan et al. 2007). Tan et al. (2007) identified vascular anomalies in 14 of 26 *PTEN* mutation-positive patients referred to genetics and/or vascular clinics, usually presenting with cutaneous discoloration, swelling, or pain. In 12 of the 14 individuals, there were fast-flow anomalies with focal segmental dilatation of draining veins, and of those 12, 11 demonstrated ectopic fat in the vascular anomalies (Tan et al. 2007). Tan et al. (2007) also identified intracranial developmental venous anomalies (DVA) in eight of nine individuals who were imaged by MRI with contrast.

Evaluation

- Careful evaluation and consideration of MRI for any suspicious discolored or enlarged masses.

Management

- Management as per standard of care for vascular anomalies as part of a multidisciplinary team.

Future

Individuals with PHTS have a protean clinical course, with neoplastic as well as neurological and metabolic sequelae. In the ideal future, *PTEN* germline mutations may be identified in those who are pre-symptomatic compared to the majority who are diagnosed with symptoms. *PTEN*-informed medical management, mostly comprised of enhanced surveillance and early detection/resection, is effective; however, the future may hold delivery of preventative agents upon discovery of a *PTEN* mutation in pre-symptomatic individuals. Currently, the ability to predict cancer risks is at a cohort level (e.g., 35% lifetime risk of thyroid cancer), but not at an individual level. It would be optimal if we could precisely predict which individuals with PHTS have what risk for autism or specific cancers, thus, only managing specific risks belonging to that particular PHTS individual.

RESOURCES

Cleveland Clinic Genomic Medicine Institute: adult and pediatric criteria for individualized risk estimation
  Website: http://www.lerner.ccf.org/gmi/ccscore/

National Comprehensive Cancer Network (NCCN)
Website: http://www.nccn.org/

Online Mendelian Inheritance in Man (OMIM)
Website: http://www.ncbi.nlm.nih.gov/Omim/

PHTS GeneReview
Website: http://www.ncbi.nlm.nih.gov/books/NBK1488/

Support Organizations
PTEN Hamartoma Tumor Syndrome Foundation
Website: www.ptenfoundation.org

REFERENCES

Arch EM, Goodman BK, Van Wesep RA, et al. (1997) Deletion of PTEN in a patient with Bannayan-Riley-Ruvalcaba syndrome suggests allelism with Cowden disease. *Am J Med Genet* 71(4):489–493.

Bannayan GA (1971) Lipomatosis, angiomatosis, and macrencephalia. A previously undescribed congenital syndrome. *Arch Pathol* 92(1):1–5.

Bennett KL, Mester J, Eng C (2010) Germline epigenetic regulation of KILLIN in Cowden and Cowden-like syndrome. *JAMA* 304(24):2724–2731.

Biesecker LG, Happle R, Mulliken JB, Weksberg R, Graham Jr JM, Viljoen DL, Cohen Jr MM (1999) Proteus syndrome: diagnostic criteria, differential diagnosis, and patient evaluation. *Am J Med Genet* 84(5):389–395.

Brownstein MH, Mehregan AH, Bikowski JB, Lupulescu A, Patterson JC (1979) The dermatopathology of Cowden's syndrome. *Br J Dermatol* 100(6):667–673.

Bubien V, Bonnet F, Brouste V, et al. (2013) High cumulative risks of cancer in patients with PTEN hamartoma tumour syndrome. *J Med Genet* 50(4):255–263.

Butler MG, Dasouki MJ, Zhou X-P, et al. (2005) Subset of individuals with autism spectrum disorders and extreme macrocephaly associated with germline PTEN tumour suppressor gene mutations. *J Med Genet* 42(4):318–321.

Capone Mori A, Hoeltzenbein M, Poetsch M, Schneider JF, Brandner S, Botshauser E (2003) Lhermitte-Duclos disease in 3 children: a clinical long-term observation. *Neuropediatrics* 34(1):30–35.

Carlson GJ, Nivatvongs S, Snover DC (1984) Colorectal polyps in Cowden's disease (multiple hamartoma syndrome). *Am J Surg Pathol* 8(10):763–770.

Colby S, Yehia L, Niazi F, Chen JL, Ni Y, Mester JL, Eng C (2016) Exome sequencing reveals germline gain-of-function EGFR mutation in an adult with Lhermitte-Duclos disease. *Cold Spring Harb Mol Case Stud* 2(6):a001230.

Conti S, Condo M, Posar A, et al. (2012) Phosphatase and tensin homolog (PTEN) gene mutations and autism: literature review and a case report of a patient with Cowden syndrome, autistic disorder, and epilepsy. *J Child Neurol* 27(3):392–397.

Dahia PL, Marsh DJ, Zhen Z, et al. (1997) Somatic deletions and mutations in the Cowden disease gene, PTEN, in sporadic thyroid tumors. *Cancer Res* 57(21):4710–4713.

Delnatte C, Sanlaville D, Mougenot F-F, et al. (2006) Contiguous gene deletion within chromosome arm 10q is associated with juvenile polyposis of infancy, reflecting cooperation between the BMPR1A and PTEN tumor-suppressor genes. *Am J Hum Genet* 78(6):1066–1074.

Eng C (1993) PTEN Hamartoma Tumor Syndrome (PHTS). GeneReviews(R). Available at: https://www.ncbi.nlm.nih.gov/books/NBK1488/

Eng C (2000) Will the real Cowden syndrome please stand up: revised diagnostic criteria. *J Med Genet* 37(11):828–830.

Eng C (2003) PTEN: one gene, many syndromes. *Hum Mutat* 22(3):183–198.

Eng C, Murday V, Seal S, et al. (1994) Cowden syndrome and Lhermitte-Duclos disease in a family: a single genetic syndrome with pleiotropy? *J Med Genet* 31(6):458–461.

Fackenthal JD, Marsh DJ, Richardson AL, Cummings SA, Eng C, Robinson BG, Olopade OI (2001) Male breast cancer in Cowden syndrome patients with germline PTEN mutations. *J Med Genet* 38(3):159–164.

Fanning AS, Anderson JM (1999) PDZ domains: fundamental building blocks in the organization of protein complexes at the plasma membrane. *J Clin Invest* 103(6):767–772.

Furnari FB, Lin H, Huang HS, Cavanee WK (1997) Growth suppression of glioma cells by PTEN requires a functional phosphatase catalytic domain. *Proc Natl Acad Sci USA* 94(23):12479–12484.

Goffin A, Hoefsloot LH, Bosgoed E, Swillen A, Fyns JP (2001) PTEN mutation in a family with Cowden syndrome and autism. *Am J Med Genet* 105(6):521–524.

Gorlin RJ, Cohen Jr MM, Condon LM, Burke BA (1992) Bannayan-Riley-Ruvalcaba syndrome. *Am J Med Genet* 44(3):307–314.

Haggitt RC and Reid BJ (1986) Hereditary gastrointestinal polyposis syndromes. *Am J Surg Pathol* 10(12):871–887.

Hampel H, Bennett RL, Buchanan A, et al. (2015) A practice guideline from the American College of Medical Genetics and Genomics and the National Society of Genetic Counselors: referral indications for cancer predisposition assessment. *Genet Med* 17(1):70–87.

Harach HR, Soubeyran I, Brown A, Bonneau D, Longy M (1999) Thyroid pathologic findings in patients with Cowden disease. *Ann Diagn Pathol* 3(6):331–340.

Harned RK, Buck JL, Sobin LH (1995) The hamartomatous polyposis syndromes: clinical and radiologic features. *AJR Am J Roentgenol* 164(3):565–571.

Heald B, Mester J, Rybicki L, Orloff MS, Burke CA, Eng C (2010) Frequent gastrointestinal polyps and colorectal adenocarcinomas in a prospective series of PTEN mutation carriers. *Gastroenterology* 139(6):1927–1933.

Hemminki A, Markie D, Tomlinson I et al. (1998) A serine/threonine kinase gene defective in Peutz-Jeghers syndrome. *Nature* 391(6663):184–187.

Henderson CJ, Ngeow J, Collins MH, et al. (2014) Increased prevalence of eosinophilic gastrointestinal disorders in pediatric PTEN hamartoma tumor syndromes. *J Pediatr Gastroenterol Nutr* 58(5):553–560.

Howe JR, Roth S, Ringold JC, et al. (1998) Mutations in the SMAD4/DPC4 gene in juvenile polyposis. *Science* 280(5366): 1086–1088.

Jenne DE, Reimann H, Nezu J, Friedel W, Loff S, Jeschke R, Muller O, Back W, Zimmer M (1998) Peutz-Jeghers syndrome is caused by mutations in a novel serine threonine kinase. *Nat Genet* 18(1):38–43.

Liaw D, Marsh DJ, Li J (1997) Germline mutations of the PTEN gene in Cowden disease, an inherited breast and thyroid cancer syndrome. *Nat Genet* 16(1):64–67.

Lloyd KM 2nd and Dennis M (1963) Cowden's disease. A possible new symptom complex with multiple system involvement. *Ann Intern Med* 58:136–142.

Longy M, Coulon V, Duboue B, et al. (1998) Mutations of PTEN in patients with Bannayan-Riley-Ruvalcaba phenotype. *J Med Genet* 35(11):886–889.

Marsh DJ, Coulon V, Lunetta KL, et al. (1998) Mutation spectrum and genotype-phenotype analyses in Cowden disease and Bannayan-Zonana syndrome, two hamartoma syndromes with germline PTEN mutation. *Hum Mol Genet* 7(3):507–515.

Marsh DJ, et al. (1998) Germline PTEN mutations in Cowden syndrome-like families. *J Med Genet* 35(11):881–885.

Marsh DJ, Dahia PL, Caron S, et al. (1997) Germline mutations in PTEN are present in Bannayan-Zonana syndrome. *Nat Genet* 16(4):333–334.

Marsh DJ, Kum JB, Lunetta KL, et al. (1999) PTEN mutation spectrum and genotype-phenotype correlations in Bannayan-Riley-Ruvalcaba syndrome suggest a single entity with Cowden syndrome. *Hum Mol Genet* 8(8):1461–1472.

Marsh DJ, Trahair TN, Martin JL, et al. (2008) Rapamycin treatment for a child with germline PTEN mutation. *Nat Clin Pract Oncol* 5(6):357–361.

McGarrity TJ, Baker MJW, Ruggiero FM, et al. (2003) GI polyposis and glycogenic acanthosis of the esophagus associated with PTEN mutation positive Cowden syndrome in the absence of cutaneous manifestations. *Am J Gastroenterol* 98(6):1429–1434.

Mester J, Eng C (2012) Estimate of de novo mutation frequency in probands with PTEN hamartoma tumor syndrome. *Genet Med* 14(9):819–822.

Mester J, Eng C (2013) When overgrowth bumps into cancer: The PTEN-Opathies. *Am J Med Genet C* 163C(2):114–121.

Mester J, Eng C (2015) Cowden syndrome: recognizing and managing a not-so-rare hereditary cancer syndrome. *J Surg Oncol* 111(1):125–130.

Mester JL, Moore RA, Eng C (2013) PTEN germline mutations in patients initially tested for other hereditary cancer syndromes: would use of risk assessment tools reduce genetic testing? *Oncologist* 18(10):1083–1090.

Mester JL, Tilot AK, Rybicki LA, Frazier II TW, Eng C (2011) Analysis of prevalence and degree of macrocephaly in patients with germline PTEN mutations and of brain weight in Pten knock-in murine model. *Eur J Hum Genet* 19(7):763–768.

Mester JL, Zhou M, Prescott N, Eng C (2012) Papillary renal cell carcinoma is associated with PTEN hamartoma tumor syndrome. *Urology* 79(5):1187 e1181–1187.

Milas M, Mester J, Metzger R, et al. (2012) Should patients with Cowden syndrome undergo prophylactic thyroidectomy? *Surgery* 152(6):1201–1210.

Miles JH, Zonana J, Mcfarlane J, Aleck KA, Bawle E (1984) Macrocephaly with Hamartomas: Bannayan-Zonana Syndrome. *Am J Med Genet* 19:225–234.

Mutter GL, Lin MC, Fitzgerlad JT, Kum JB, Baak JP, Lees JA, Weng LP, Eng C (2000) Altered PTEN expression as a diagnostic marker for the earliest endometrial precancers. *J Natl Cancer Inst* 92(11):924–930.

Nagy R, Ganapathi S, Comeeras I, et al. (2011) Frequency of germline PTEN mutations in differentiated thyroid cancer. *Thyroid* 21(5):505–510.

NCCN (2018) NCCN Clinical Practice Guidelines in Oncology: Genetic/Familial High-Risk Assessment: Breast and Ovarian. Available from https://www.nccn.org/store/login/login.aspx?ReturnURL=https://www.nccn.org/professionals/physician_gls/pdf/genetics_bop.pdf.

Nelen MR, Padberg GW, Peeters EA, et al. (1996) Localization of the gene for Cowden disease to chromosome 10q22-23. *Nat Genet* 13(1):114–116.

Nelen MR, van Staveren WC, Peeters EA, et al. (1997) Germline mutations in the PTEN/MMAC1 gene in patients with Cowden disease. *Hum Mol Genet* 6(8):1383–1387.

Neshat MS, Millinghoff IK, Tran C, et al. (2001) Enhanced sensitivity of PTEN-deficient tumors to inhibition of FRAP/mTOR. *Proc Natl Acad Sci U S A* 98(18):10314–10319.

Ngeow J, Heald B, Rybicki LA, et al. (2013) Prevalence of germline PTEN, BMPR1A, SMAD4, STK11, and ENG mutations in patients with moderate-load colorectal polyps. *Gastroenterology* 144(7):1402–1409, 1409e1401–1405.

Ngeow J, Liu C, Zhou K, Frick KD, Matchar DB, Eng C (2015) Detecting Germline PTEN Mutations Among At-Risk Patients With Cancer: An Age- and Sex-Specific Cost-Effectiveness Analysis. *J Clin Oncol* 33(23):2537–2544.

Ngeow J, Mester J, Rybicki LA, Ni Y, Milas M, Eng C (2011) Incidence and clinical characteristics of thyroid cancer in prospective series of individuals with Cowden and Cowden-like syndrome characterized by germline PTEN, SDH, or KLLN alterations. *J Clin Endocrinol Metab* 96(12):E2063–2071.

Ngeow J, Stanuch K, Mester JL, Barnholtz-Sloan JS, Eng C (2014) Second malignant neoplasms in patients with Cowden syndrome with underlying germline PTEN mutations. *J Clin Oncol* 32(17):1818–1824.

Ni Y, Zbuk KM, Sadler T, et al. (2008) Germline mutations and variants in the succinate dehydrogenase genes in Cowden and Cowden-like syndromes. *Am J Hum Genet* 83(2):261–268.

Nieuwenhuis MH, Kets CM, Murphy-Ryan M, et al. (2014) Cancer risk and genotype-phenotype correlations in PTEN hamartoma tumor syndrome. *Fam Cancer* 13(1):57–63.

Nishizawa A, Satoh T, Watanabe R, Takayama K, Nakano H, Sawamura D, Yokozeki H (2009) Cowden syndrome: a novel mutation and overlooked glycogenic acanthosis in gingiva. *Br J Dermatol* 160(5):1116–1118.

Okumura K, Sasaki Y, Ohyama M, Nishi T (1986) Bannayan syndrome--generalized lipomatosis associated with megalencephaly and macrodactyly. *Acta Pathol Jpn* 36(2):269–277.

Robson ME, Bradbury AR, Arun B, et al. (2003) American Society of Clinical Oncology Policy Statement Update: Genetic Testing for Cancere Susceptibility. *J Clin Oncol* 21(12):2397–2406.

Orloff MS, He X, Peterson C, et al. (2013) Germline PIK3CA and AKT1 mutations in Cowden and Cowden-like syndromes. *Am J Hum Genet* 92(1):76–80.

Ozes ON, Akca H, Mayo LD, Gustin JA, Maehama T, Dison JE, Donner DB (2001) A phosphatidylinositol 3-kinase/Akt/mTOR pathway mediates and PTEN antagonizes tumor necrosis factor inhibition of insulin signaling through insulin receptor substrate-1. *Proc Natl Acad Sci U S A* 98(8):4640–4645.

Padberg GW, Schot JD, Vielvoye GJ, Bots GT, de Beer FC (1991) Lhermitte-Duclos disease and Cowden disease: a single phakomatosis. *Ann Neurol* 29(5):517–523.

Parisi MA, Dinulos MB, Leppig KA, Sybert VP, Eng C, Hudgins L (2001) The spectrum and evolution of phenotypic findings in PTEN mutation positive cases of Bannayan-Riley-Ruvalcaba syndrome. *J Med Genet* 38(1):52–58.

Perez-Nunez A, Lagares A Benitez J, et al. (2004) Lhermitte-Duclos disease and Cowden disease: clinical and genetic study in five patients with Lhermitte-Duclos disease and literature review. *Acta Neurochir (Wien)* 146(7):679–690.

Pezzolesi MG, Zbuk KM, Waite KA, Eng C (2007) Comparative genomic and functional analyses reveal a novel cis-acting PTEN regulatory element as a highly conserved functional E-box motif deleted in Cowden syndrome. *Hum Mol Genet* 16(9):1058–1071.

Podsypanina K, Lee RT, Politis C, et al. (2001) An inhibitor of mTOR reduces neoplasia and normalizes p70/S6 kinase activity in Pten+/- mice. *Proc Natl Acad Sci U S A* 98(18):10320–10325.

Riley HD, Smith WR (1960) Macrocephaly, Pseudopapilloedema, and Multiple Hemangioma. *Pediatrics* 26:293–300.

Robinson S, Cohen AR (2000) Cowden disease and Lhermitte-Duclos disease: characterization of a new phakomatosis. *Neurosurgery* 46(2):371–383.

Saslow D, Boetes C, Burke W, et al. (2007) American Cancer Society guidelines for breast screening with MRI as an adjunct to mammography. *CA Cancer J Clin* 57(75–89).

Sekulic A, Hudson CC, Homme JL, Yin P, Otterness DM, Karnitz LM, Abraham RT (2000) A direct linkage between the phosphoinositide 3-kinase-AKT signaling pathway and the mammalian target of rapamycin in mitogen-stimulated and transformed cells. *Cancer Res* 60(13):3504–3513.

Shiovitz S, Everett J, Huang S-C, et al.(2010) Head circumference in the clinical detection of PTEN hamartoma tumor syndrome in a clinic population at high-risk of breast cancer. *Breast Cancer Res Treat* 124(2):459–465.

Simpson L, Li J, Liaw D, et al. (2001) PTEN expression causes feedback upregulation of insulin receptor substrate 2. *Mol Cell Biol* 21(12):3947–3958.

Smith JM, et al. (2002) Germline mutation of the tumour suppressor PTEN in Proteus syndrome. *J Med Genet* 39(12):937–940.

Starink TM, Hausman R (1984) The cutaneous pathology of extrafacial lesions in Cowden's disease. *J Cutan Pathol* 11(5):338–344.

Starink TM and Hausman R (1984) The cutaneous pathology of facial lesions in Cowden's disease. *J Cutan Pathol* 11(5):331–337.

Starink TM, van der Veen JP, Arwert F, et al. (1986) The Cowden syndrome: a clinical and genetic study in 21 patients. *Clin Genet* 29(3):222–233.

Sweet K, Willis J, Zhou X-P, et al. (2005) Molecular classification of patients with unexplained hamartomatous and hyperplastic polyposis. *JAMA* 294(19):2465–2473.

Tan MH, Mester J, Peterson C, et al. (2011) A clinical scoring system for selection of patients for PTEN mutation testing is proposed on the basis of a prospective study of 3042 probands. *Am J Hum Genet* 88(1):42–56.

Tan MH, Mester JL, Ngeow J, Rybicki LA Orloff MS, Eng C (2012) Lifetime cancer risks in individuals with germline PTEN mutations. *Clin Cancer Res* 18(2):400–407.

Tan WH, Baris HN, Burrow PE, et al. (2007) The spectrum of vascular anomalies in patients with PTEN mutations: implications for diagnosis and management. *J Med Genet* 44(9):594–602.

Teresi RE, Zbuk KM, Pezzolesi MG, Waite KA, Eng C (2007) Cowden syndrome-affected patients with PTEN promoter mutations demonstrate abnormal protein translation. *Am J Hum Genet* 81(4):756–767.

Tilot AK, Frazier 2nd TW, Eng C (2015) Balancing Proliferation and Connectivity in PTEN-associated Autism Spectrum Disorder. *Neurotherapeutics* 12(3):609–619.

Weng LP, Smith WM, Dahia PL, Ziebold U, Gil E, Lees JA, Eng C (1999) PTEN suppresses breast cancer cell growth by phosphatase activity-dependent G1 arrest followed by cell death. *Cancer Res* 59(22):5808–5814.

Wiedemann HR, Burgio GR, Aldenhoff P, Kunze J, Kaufmann HJ, Schirg E (1983) The Proteus syndrome: Partial gigantism of the hands and/or feet, nevi, hemihypertrophoy, subcutaneious tumors, mecrocephaly or ther skull anomalies and possible accelerated growth and visceral affections. *Eur J Pediatr* 140:5–12.

Yehia L, Niazi F, Ni Y, et al. (2015) Germline Heterozygous Variants in SEC23B Are Associated with Cowden Syndrome and Enriched in Apparently Sporadic Thyroid Cancer. *Am J Hum Genet* 97(5):661–676.

Zbuk KM, Eng C (2007) Cancer phenomics: RET and PTEN as illustrative models. *Nat Rev Cancer* 7(1):35–45.

Zbuk KM, Eng C (2007) Hamartomatous polyposis syndromes. *Nat Clin Pract Gastroenterol Hepatol* 4(9):492–502.

Zbuk KM, et al. (2007) Mitochondrial dysfunction caused by germline mutations in succinate dehyogenase subunit genes in Cowden and Cowden-like syndromes. American Society of Human Genetics 57th Annual Meeting, San Diego, CA.

Zhou X, Hampel H, Thiele H, et al. (2001) Association of germline mutation in the PTEN tumour suppressor gene and Proteus and Proteus-like syndromes. *Lancet* 358(9277):210–211.

Zhou XP, Marsh DJ, Hampel H, Mulliken JB, Gimm O Eng C (2000) Germline and germline mosaic PTEN mutations associated with a Proteus-like syndrome of hemihypertrophy, lower limb asymmetry, arteriovenous malformations and lipomatosis. *Hum Mol Genet* 9(5):765–768.

Zhou XP, Marsh DJ, Morrison CD, et al. (2003) Germline inactivation of PTEN and dysregulation of the phosphoinositol-3-kinase/Akt pathway cause human Lhermitte-Duclos disease in adults. *Am J Hum Genet* 73(5):1191–1198.

Zhou XP, Waite KA, Pilarski R, et al. (2003) Germline PTEN Promoter Mutatations and Deletions in Cowden/Bannayan-Riley-Ruvalcaba Syndrome Result in Aberrant PTEN Protein Dysregulation of the Phophoinositol-3-Kinase/Akt Pathway. *Am J Hum Genet* 73:404–411.

Zonana J, Rimoin DL, Davis DC (1976) Macrocephaly with multiple lipomas and hemangiomas. *J Pediatr* 89(4):600–603.

Zori RT, Marsh DJ, Graham GE, Marliss EB, Eng C (1998) Germline PTEN mutation in a family with Cowden syndrome and Bannayan-Riley-Ruvalcaba syndrome. *Am J Med Genet* 80(4):399–402.

49

RETT SYNDROME

ERIC E. SMEETS

Rett Expertise Center – Governor Kremers Center, Maastricht University Medical Center, Maastricht, The Netherlands; and Department of Pediatrics, Maastricht University Medical Center, Maastricht, The Netherlands

INTRODUCTION

The well-respected pediatrician from Vienna, Andreas Rett, published some characteristic features of the syndrome, which later came to bear his name in 1966 (Rett 1966). Bengt Hagberg, who had seen these girls and women in Sweden since the early 1960s, shared his clinical experience at the Manchester Meeting on Child Neurology in 1981 under a special eponym Morbus Vesslan (Witt-Engerström 1990). Other child neurologists had observed the same condition in Japan (Ishikawa et al. 1978). In 1983 RTT became known internationally (Hagberg et al. 1983). Advancing knowledge and understanding of the clinical and neurobiological characteristics finally led to the discovery of the causal mutation in the *MECP2* gene (Amir et al. 1999). RTT is considered to be the first recognized human neurodevelopmental disorder related to a defective protein involved in the transcription of methylated DNA and affecting almost exclusively girls (Kerr et al. 2001). The manifestations in males cover a spectrum of neurodevelopmental disorders ranging from typical RTT to mild intellectual disability alone and encompass congenital encephalopathy and cognitive deficit of different degrees, in association with diverse neurologic features and/or psychiatric illness. The prevalence of the syndrome is very low in males with a normal karyotype and no family history of an affected female (Kerr and Witt-Engerström 2001; Leonard et al. 2001; Moog et al. 2003, Reichow et al. 2015; Ronen et al. 2017). For the purpose of this chapter only the female phenotype will be described.

Incidence

The prevalence of the classic syndrome is estimated to be 1 in 10,000 females (Kerr 1992) but varies according to countries from 1/8,000 – 1/15,000. Current understanding of the typical and atypical forms suggests that the overall prevalence is probably much higher. It is the third most common cause of intellectual disability in females next to chromosomal anomalies (e.g. Down syndrome) and the X-linked familial intellectual disability (e.g. Fragile X syndrome in 20% of the female carriers).

Diagnostic Criteria

The early infantile history of this unique neurodevelopmental disorder is often strange and puzzling. Although hypotonia is usually present before 6 months of age and general developmental progress may be suboptimal during the first year, the overall developmental pattern is not obviously disturbed at first. Then the characteristic features of RTT appear successively, causing great anxiety in the family. Expert clinicians have developed a staging system as a framework that delineates the evolving symptoms of communicative dysfunction and loss of acquired skills and motor performance (Witt-Engerström 1990; Hagberg and Gilberg 1993). Clinical developmental profiles, very non-specific early in life, become more specific for the disorder later on in life. The four classical clinical stages are summarized in Table 49.1, and are described below.

Cassidy and Allanson's Management of Genetic Syndromes, Fourth Edition.
Edited by John C. Carey, Agatino Battaglia, David Viskochil, and Suzanne B. Cassidy.
© 2021 John Wiley & Sons, Inc. Published 2021 by John Wiley & Sons, Inc.

TABLE 49.1 Four clinical stages of classical Rett syndrome

Stage I: early onset stagnation
Duration: weeks to months
Onset age: 6 months–1$\frac{1}{2}$ years
Developmental progress delayed, with early postural/motor delay
Developmental pattern still not significantly abnormal, but dissociated/disordered
"Bottom-shufflers"
Stage II: rapid developmental regression
Duration: weeks to months
Onset age 1–4 years, sometimes acute "pseudotoxic"
Loss of acquired skills: fine finger, babbling/words, active playing
Mental deficiency, eye contact preserved, occasionally "in another world"
Breathing problems modest as yet
Seizures in 15%
Stage III: pseudo-stationary period
Duration: years to decades
Onset: after passing stage II
Apparently preserved walking ability with prominent hand apraxia/dyspraxia
Inapparent slow neuromotor regression
"Wake-up" period
Stage III/IV: non-ambulatory
Stage IV: late motor deterioration
Duration: decades
Onset: when stage III ambulation ceases
Stage IVA: previous walkers, now non-ambulatory
Stage IVB: never ambulatory
Complete wheelchair dependency
Severe disability: muscle wasting and distal extremity deformations

Source: Modified from Hagberg and Gilberg (1993) and Witt-Engerström (1990).

FIGURE 49.1 Apparently normal development in an 11-month-old girl with RTT (early-onset stagnation period).

The early-onset *stagnation period* (stage I) occurs between 6 months and 1.5 years of age. Parents notice a more or less sudden change in the interactive behavior of the child. Additional abilities are still acquired but at a delayed rate. For example, the child learns to sit upright but not to crawl or to stand up, and "bottom-shuffling" is very common. Babbling and new words are learned. The infant may be irritable, though parents may relate frequent crying to teething. Because the overall developmental pattern of their child is still apparently normal, they are reassured (Figure 49.1).

The *rapid developmental regression period* (stage II) occurs between 1 and 4 years of age and is characterized by rapid and specific regression of acquired abilities. This happens very suddenly, sometimes with pseudo-toxic symptoms (high-pitched crying, fever, and apathy suggesting meningo-encephalitis) often leading to emergency hospitalization. The pediatrician is left with no clinical explanation because there are no abnormal biochemical or microbiological findings. Such an acute episode can last for days or weeks, and, after recovery, the child's personality is changed completely. Other affected children demonstrate a more gradual decline in their communicative ability and motor performance. Although eye contact is preserved, they show diminished interest in people and materials. As acquired babbling and words and fine motor skills are lost during this period, intellectual disability becomes obvious. Active playing becomes rare and/or loses its exploratory character. Intentional movement such as reaching out for objects and toys ends abruptly in withdrawal or in a senseless hair-pulling or head-tapping behavior. Later it will proceed to dystonic posturing of wrists and fingers. During this phase, parents feel that they lose contact with their daughter as her visual response is changing and she appears to be "out of it." Many girls suffer from recurrent infections or bouts of unexplained fever. Crying at night is common. Febrile convulsions may be present and questions about epileptic paroxysms may arise. Some seizures, starting insidiously with panting and hyperventilation, together with muscle twitching around the corners of the mouth and grimacing, are now considered to be manifestations of brain-stem immaturity. Decline in head growth in some children becomes a point of concern (Figure 49.2).

The *pseudo-stationary stage* (stage III) starts when the regression stage has passed. The child is still able to walk or may even learn this skill. Hand dyspraxia/apraxia now becomes obvious, and typical hand stereotypies constitute the hallmark of the disorder. This almost continuous pattern of repetitive midline movements with hand wringing, hand washing, and clapping starts as soon as the child is awake.

INTRODUCTION

FIGURE 49.2 The same individual at age 3 years demonstrating dystonic posturing and appearing to be "in another world" (rapid developmental regression period).

Others display patting or rolling stereotypies along their bodies, with the fingers twisting in unusual ways. The typical visual contact behavior then suggests an "awakening" when some of the normal personality is manifest once again. The girl is now more alert and joyful and parents recognize her eye pointing behavior to express wishes. Breathing irregularities, although modestly present in stage II, may become more prominent in this stage, with or without non-epileptic vacant spells, apnea or feeble breathing, hyperventilation and Valsalva breathing. Unexplained night laughing, frequent daytime sleeping and nighttime awakening, crying spells and sudden agitation occur. Many have obvious clinical epilepsy requiring treatment, but many also become seizure-free after some time. Dystonic asymmetrical posturing leads to neurogenic scoliosis, often rapidly progressive, requiring surgical treatment. The feet and lower limbs are cold, with or without color change, and with and without atrophic changes. Shortening of Achilles tendons in equinus position causes deformation of the ankles into the varus or valgus position, and the toes are held typically clenched. Some girls with milder effects have preserved speech and/or hand use. Other girls may say some words or a sentence many years after the onset of their disorder. Neuromotor regression slowly progresses in this stage, which can last for decades, in contrast to a remarkably well-preserved ability to communicate mainly with the eyes. In this stage the girls and women still learn about new things and persons (Figure 49.3).

Late motor deterioration (stage IV) starts when walking ceases and the individual with RTT becomes wheelchair dependent. Some individuals with severe manifestations never learn to walk and pass immediately from stage II into stage IVB, which is characterized by severe neurologic impairment with pronounced muscle wasting and distortion of distal limbs. The feet are cold and discolored with abiotrophic changes. The hand stereotypies become less intense and simpler with age. Lack of motor activity in these quadriplegic women leads eventually to a state of frozen rigidity. However, remarkable visual contact and eye-pointing behavior remain present even in the most severe situation (Figures 49.4–49.6).

In spite of molecular analysis, the diagnosis of RTT remains a clinical one, based on internationally accepted criteria (Hagberg et al. 1990, 2002; Hagberg and Gilberg 1993; Hagberg and Skjeldal 1994, Neul et al. 2010). According to these criteria, typical and atypical phenotypes are delineated, as shown in Tables 49.2 and 49.3. In 2010 an international consortium reviewed the criteria for clinical diagnosis of

FIGURE 49.3 The same individual at age 11 years showing apraxia and neuromotor regression. She is in the "wake-up" period (pseudo-stationary stage).

FIGURE 49.4 The same individual at age 17 years with severe disability and muscle wasting (late motor regression).

FIGURE 49.6 The same individual at age 34 years (late motor regression).

typical and atypical RTT (Neul et al. 2010). This last revision includes a period of regression followed by recovery or stabilization and (1) partial or complete loss of acquired purposeful hand skills, (2) partial or complete loss of acquired spoken language, (3) gait abnormalities, and (4) stereotypic hand movements. Additionally, there are symptoms that are abundantly observed in RTT patients but not always present and therefore not required for diagnosis like autistic behavior, breathing problems, epilepsy, scoliosis, and dystonia. Deceleration of head circumference is no longer a necessary criterion. And a broad interpretation of regression and recovery, e.g. of speech and motor performance, is allowed. The former variant types of RTT are now referred to as "atypical".

Congenital RTT with microcephaly is related to mutation in the *FOXG1* gene, difficult to diagnose clinically and may present also in boys as it is located on chromosome 14. In the infantile seizure onset variant (Hanefeld variant), hypsarrhythmia starting at the age of 2–4 months hides the features of the early stages of RTT. When the epileptic encephalopathy stabilizes and is controlled by medication, the girls regain a more or less clear RTT profile with some recovery of interaction. It is now related to mutation in the *CDKL5* gene located on the short arm of the X-chromosome. Preserved speech (Zapella variant) in RTT is not rare and is related to a better level of development overall. The late regression variant, with RTT starting at preschool or early school age, is also rare. Clinical diagnosis remains tentative in these girls up to approximately 10 years of age, but nowadays they are diagnosed earlier through exome sequencing. In one large series only one individual presented abruptly with epileptic encephalopathy of the Lennox–Gastaut type

FIGURE 49.5 The same individual at age 20 years showing visual contact behavior with her mother (late motor regression).

TABLE 49.2 Consensus diagnostic criteria for Rett syndrome

Necessary criteria
Apparently normal prenatal and perinatal period
Psychomotor development largely normal through the first six months or may be delayed from birth
Normal head circumference at birth
Postnatal deceleration of head growth in the majority
Loss of achieved purposeful hand skill between ages 6 months and 2½ years
Stereotypic hand movements such as hand wringing/squeezing, clapping/tapping, mouthing and washing/rubbing
Emerging social withdrawal, communication dysfunction, loss of learned words, and cognitive impairment
Impaired (dyspraxic) or failing locomotion

Supportive Criteria
Disturbances of breathing (hyperventilation, breath-holding, forced expulsion of air or saliva, air swallowing) while awake
Bruxism
Impaired sleep pattern from early infancy
Abnormal muscle tone successively associated with muscle wasting and dystonia
Peripheral vasomotor disturbances
Scoliosis/kyphosis progressing through childhood
Growth retardation
Hypotrophic small and cold feet; small, thin hands

Exclusion Criteria
Organomegaly or other signs of storage disease
Retinopathy, optic atrophy, or cataract
Evidence of perinatal or postnatal brain damage
Existence of identifiable metabolic or other progressive neurologic disorder
Acquired neurologic disorders resulting from severe infections or head trauma

Source: Hagberg et al. (2002).

TABLE 49.3 Consensus delineation of variant phenotypes.

Inclusion criteria
Meet at least 3 of 6 main criteria
Meet at least 5 of 11 supportive criteria

Six main criteria
Absence or reduction of hand skills
Reduction or loss of babble speech
Monotonous pattern of hand stereotypies
Reduction or loss of communication skills
Deceleration of head growth from first year of life
Rett syndrome disease profile: a regression stage followed by a recovery of interaction contrasting with slow neuromotor regression

Eleven supportive criteria
Breathing irregularities
Bloating/air swallowing
Bruxism, harsh sounding type
Abnormal locomotion
Scoliosis/kyphosis
Lower limb amyotrophy
Cold, purplish feet, usually growth impaired
Sleep disturbances including night screaming outbursts
Laughing/screaming spells
Diminished response to pain
Intense eye contact/eye pointing

Source: Hagberg et al. (2002).

at age 4 years (personal experience). She suffered intractable seizures over many years and had characteristic features of RTT at age 16 years. She subsequently was found to have a pathogenic mutation in *MECP2*.

RTT was thought to be an X-linked dominant condition with lethality in hemizygous males for a long time. RTT in boys with a normal male 46, XY karyotype is very rare. Typical RTT, like in girls, can only occur in a boy with an additional X chromosome (carrying the *MECP2* mutation) in all cells (47, XXY or Klinefelter syndrome) or in only part of the body cells (somatic mosaic). On the other hand, there are the *MECP2* mutations that cause the typical syndrome in girls and where the mother is a healthy skewing carrier. These mutations will lead to intrauterine death or early infantile epileptic encephalopathy with early death before or around the first year of life. In addition, there are the more sporadically occurring variants in *MECP2* in males who are hardly seen in RTT girls and whose disease-causing properties are not immediately clear, certainly not when such a case has never been described before. Often these variants in *MECP2* are compatible with a long survival. The clinic is then very different without meeting the necessary criteria for the diagnosis of typical or atypical RTT: non-specific intellectual disability; intellectual disability with motor deficits (speech and writing difficulties and/or neurological problems with coordination in motor skills); severe intellectual disability with spasticity with symptoms similar to RTT (scoliosis, hyperventilation, intense visual interaction); intellectual disability with psychiatric disorders (bipolar disorder or juvenile schizophrenia) and tremors; intellectual disability with psychosis, spasticity and macroorchidism (PPM-X syndrome). The current advances in DNA diagnostics are now showing more common *MECP2* mutation in males. One estimates the frequency of *MECP2* mutation between 1.3 and 1.7% of the male population with intellectual impairment. The clinic in boys is not unequivocal as described in the girls. It is therefore referred to as *MECP2* related disorders in males because the clinical phenotype does not meet the diagnostic criteria for RTT. In addition, *MECP2* duplication syndrome, *FOXG1* syndrome and *CDKL5* syndrome are considered separate entities, although with many intersections. Because they form a relatively smaller group, they are included in general policy regarding therapy and management of RTT. The *MECP2* duplication syndrome occurs in both sexes with severe intellectual disability in the male and preserved speech in the female, recurrent infections and motor impairment in combination with features similar as in RTT (epilepsy, hyperventilation, autonomic disturbance, etc.) or as a non-specific form of mental impairment with or without autistic features. The clinical severity is mainly determined by the extent of duplication and whether or not other important genes are involved in this duplication.

Etiology, Pathogenesis, and Genetics

RTT is the result of a mutation in *MECP2* located at Xq28 and encoding the methyl-CpG-binding protein 2 (Amir et al. 1999). It contains a methyl CpG binding domain, a transcription repression domain, two nuclear localization signals, and a C-terminal segment. The coding sequence for the methyl-CpG-binding domain is split between exons 3 and 4, whereas the transcription repression domain lies entirely within exon 4 (Nan et al. 1996, 1997). The methyl-CpG-binding domain binds specifically to 5-methyl-cytosine throughout the genome. The transcription repression domain interacts with histone deacetylase and the transcription silencer co-repressor Sin3A. The nuclear localization signals mediate the transport of the MeCP2 protein into the nucleus, and the C-terminal segment facilitates binding to the nucleosome core (Chandler et al. 1999; Huppke et al. 2000). These interactions result in the deacetylation of histones and chromatin condensation, which leads to repression of transcription (Nan et al. 1998; Wan et al. 2001).

The MeCP2 protein is ubiquitously present but is particularly abundant in the brain (Amir et al. 1999). It is thought to suppress the transcription of other tissue-specific genes whose activity is not or no longer required (Yntema et al. 2002). Loss of function of MeCP2 in cells, especially in differentiated post-mitotic neurons, may lead to inappropriate overexpression of these other genes, with potentially damaging effect during central nervous system maturation (Ellaway and Christodoulou 2001). The knockout mouse, deficient in the *mecp2* gene, presents several features of RTT. The search for target genes that are de-repressed in these mice has not yielded any clear candidates, except brain-derived neurotropic factor (*BDNF*). Normal *MECP2* regulates the expression of this gene that is essential for learning and neural plasticity (Chen et al. 2003; Martinowich et al. 2003). Discussing the ongoing basic research in genetics of RTT goes beyond the aim of this chapter. New insights in the biological pathways leading from *MECP2* to RTT and RTT-like disorder phenotypes have been reached through bioinformatics coordinating all basic human and animal research up till now (Ehrhart et al. 2016, 2017).

The pathology of RTT differs from other disorders with intellectual disability in that the pattern of dendritic changes in the brain is unique (Armstrong et al. 1998; Armstrong and Kinney 2001). Brain weight is reduced in girls with RTT but does not diminish with age. The defined cause of this arrest in brain development and how this results in altered neurophysiology is not yet well understood. There is evident failure of dendritic arborization in specific sites of the brain, correlating with the cortical localization of some of the significant motor and behavioral symptoms. In relation to the peculiar movement disorder in RTT, the substantia nigra, basal ganglia, cerebellum, and spinal cord have been found to show specific alterations. Various neurotransmitter systems have also been studied with varied and inconclusive results apart from the demonstration of monoaminergic dysfunction (Nomura et al. 1985; Nomura and Segawa 1992; Armstrong 2002).

The disturbances in autonomic function have been studied and related to immaturity of brainstem autonomic centers resulting in hypersensitivity to sympathetic stimuli with insufficient parasympathetic control. This is the so-called sympathovagal imbalance that is unique in RTT (Julu et al. 1997, 2001; Julu and Witt-Engerström 2005). New insights into the brainstem phenomena have led to the neurophysiologic delineation of breathing phenotypes, such as "forceful breathers," "feeble breathers," and "apneustic breathers." Each of these cardio-respiratory phenotypes has a specific therapeutic approach that will be discussed later.

Table 49.4 shows the cardinal clinical features of RTT in relation to pathology.

Diagnostic Testing

It is important to emphasize that in spite of the progress in molecular diagnostics, the diagnosis of RTT remains strongly based on clinical criteria. The availability of molecular testing may lead to early confirmation of a clinical suspicion of RTT in a young child. More than 400 different *MECP2* mutations and several polymorphic variants have been described. They are listed in the database accessible through the website of the International RTT Association (www.rettsyndrome.org). In individuals with typical RTT, there is an 80% mutation detection rate using standard techniques to analyze the coding region (Cheadly et al. 2000). With additional analysis of multiplex ligation-dependent probe amplification (MLPA), the mutation detection rate rises to more than 95%. In atypical RTT the frequency of detectable mutations is lower. More than 95% of the *MECP2* mutations occur de novo, and in the vast majority of affected individuals they occur on the paternal X chromosome (Girard et al. 2001). Missense mutations seem to cluster in the methyl-CpG-binding domain, whereas nonsense or frameshift mutations truncate the protein beyond this domain affecting the transcription repression domain and C-terminal segment of the gene. Frequent recurrent mutations are listed in Table 49.5.

Correlation studies between the location of the mutation in the gene and the clinical severity of RTT has been subject of large cohort studies. The most extensive work has been done by the Australian Teleton Kids Institute led by Helen Leonard with emphasis on clinical variation and the

TABLE 49.4 Cardinal features of Rett syndrome in relation to pathology.

Affected part	Reported pathology	Clinical observations
Cortical	Decreased dendritic arborization and smaller than normal brain	Severe intellectual disability
Cortical	Epilepsy	Seizures
Extrapyramidal	Monoaminergic dysfunction	Dystonia, incoordination of motor activities, orthopedic deformities, and secondary muscle wasting with contractures
Brainstem	Monoaminergic dysfunction	Dyspraxia, agitation, and sleep disturbances
Brainstem	Immaturity with incompetence of inhibitory neuronal networks	Abnormal breathing rhythms and lack of integrative inhibitions, which are likely causes of sudden deaths
Brainstem	Dysautonomia	Cold and blue extremities and sympatho-vagal imbalance

Source: Julu et al. (2008).

examination of relationships between phenotype and genotype (Leonard et al. 2017). Nevertheless, one should be very careful is using specific MECP2 mutations as prognosticators of clinical severity that is always very large and to our opinion more influence by the severity of dysautonomia in the individual with RTT (Halbach et al. 2012, 2016).

Genetic Counseling

When a mutation is found in a child, the mother should be tested to see if she is a symptomatic carrier. However, because more than 95% of RTT mutations are de novo, the recurrence risk with a negative family history is very low (less than 0.1%). Prenatal diagnosis can be offered to parents in subsequent pregnancies to evaluate the low likelihood of gonadal mosaicism. When the mother is an asymptomatic carrier, the recurrence risk is 50%. This includes the risk of intra-uterine death or severe neonatal encephalopathy in a male.

Differential Diagnosis

An older girl or adult with fully established RTT will be readily recognized clinically. In a young infant, however, the diagnosis can be difficult. Angelman syndrome (Chapter 5) is perhaps the most difficult clinical differential diagnosis.

TABLE 49.5 Recurrent *MECP2* mutations in Rett syndrome.

Nucleotide change aminoacid change type of mutation		
316C > T	p.R106W	Missense
397C > T	p.R133C	Missense
473C > T	p.T158M	Missense
502C > T	p.R168X	Nonsense
763C > T	p.R255X	Nonsense
808C > T	p.R270X	Nonsense
880C > T	p.R294X	Nonsense
916C > T	p.R306C	Missense
Intragenic deletions and rearrangements in the CTS		Nonsense

Abbreviations: CTS, C-terminal segment of the gene.

The behavioral phenotype in AS is a critical discriminator. Infantile neuronal ceroid lipofuscinosis (Batten disease) has a more rapidly progressive and degenerative course and includes retinal involvement leading to blindness. Epileptic encephalopathy of infancy and childhood can simulate RTT. Both disorders cause the child to reach a developmental ceiling with rapid regression and deterioration when there is resistance to therapy. Disintegrative infantile psychosis is sometimes suggested as a diagnosis in RTT with predominantly autistic features in the early course of the syndrome. Characteristic RTT features develop later. The Phelan–McDermid syndrome (see Chapter 22) related to SHANK3 involvement is an example (Hara et al. 2015). But autism in a girl without any dyspraxic or dystonic symptoms is rarely seen in RTT. The *CDKL5* syndrome in some girls with the infantile onset seizure variant of RTT (Tao et al. 2004) has a clinical course resembling RTT only when epilepsy is more or less under control. In doubtful cases resembling the RTT or AS and where both disorders were molecularly excluded, haploinsufficiency of *TCF4* has been found (Zweier et al. 2007). *TCF4* is a gene encoding for transcription factor 4 (TCF4) that was identified as the underlying cause of Pitt–Hopkins syndrome, an underdiagnosed syndrome marked by hyperventilation episodes and characteristic dysmorphic features (large convex nasal ridge, wide mouth, fleshy lips, and clubbed fingertips). Now, as whole exome sequencing becomes a standard method, the results of rare and uncommon genetic causes for disorders/syndromes with overlapping Rett-like phenotypes are accumulating. In the last few years there have been 69 new genes identified that can cause a RTT-like phenotype (Ehrhart et al. 2018; Schönewolf-Greulich et al. 2017).

MANIFESTATIONS AND MANAGEMENT

Early intervention and comprehensive lifelong management can have a significant impact on the health and longevity of affected individuals. Good management requires the involvement of a multidisciplinary team consisting of many medical and paramedical specialists and the development of an individualized approach. Parents are critical members of the team, as they become the greatest experts concerning their own child's history, behavior, and needs. Many of the suggestions and recommendations described below are the result of over 35 years of personal experience together with review of international scientific expertise. This worldwide communication among experts on this singular disorder is largely stimulated by the various national parent organizations.

Growth and Feeding

Physical growth retardation and feeding problems are common features of RTT. Affected girls have a birth weight and length within the normal range for gestational age. There is deceleration of linear growth during the first two years of life. Later on, height and/or weight for height often fall two standard deviations below those of healthy children. The mechanisms causing this growth failure are poorly understood and the role of MeCP2 in physical growth is yet to be investigated. Both nutritional and non-nutritional factors are thought to contribute, and it has not been possible to develop efficacious intervention strategies (Reilly and Cass 2001). There is no evidence that growth retardation in RTT is caused by growth hormone deficiency, although disturbed hypothalamic control cannot be excluded (Huppke et al. 2001). Daily energy and water requirements can be much higher than is often realized, particularly in forceful breathers and in those with frequent Valsalva's type of breathing.

The mean head circumference in typical RTT tends to fall two standard deviations below the norm by age 4 years. After the age of 8, it stabilizes close to three standard deviations below the mean. Head growth has decelerated by less than one standard deviation in 20% of the girls at age 6 years and in 10% at age 12 years. In atypical and milder RTT, head circumference stays within normal limits or only a small decline in head growth occurs. However, head size is still below average (0.8 standard deviations below). The decline in head growth may thus be very obvious in typical RTT but is usually not present at all in atypical cases. Therefore, it is no longer regarded as a necessary diagnostic criterion for RTT, nor as a valid one for atypical variants (Hagberg et al. 2001). Growth charts were developed for American girls with RTT (Tarquinio et al. 2012).

The rate of hand and foot growth, particularly the latter, of girls with RTT is slower than that of the normal female (Schultz et al. 1998).

Girls with RTT love to eat. They like to watch when meals are prepared and are very alert at feeding time. Some girls, especially the younger ones, will even have a tendency to become overweight if they are allowed to eat at will. The development of primary mouth functions like chewing and swallowing is often delayed and problematic. Most of the girls have reduced movements of the mid- and posterior tongue, with premature spillover of food and liquid from the mouth into the pharynx. They also show delayed pharyngeal swallow, but otherwise pharyngeal problems are minimal. Those individuals with the most neurologic impairment tend to have the worst feeding problems and become prone to malnutrition. Gastroesophageal reflux is frequently present and, together with air swallowing and constipation, interferes with effective oral feeding. When severe swallowing problems and/or insufficient intake is/are present, affected females often require tube feeding or are put on caloric supplements. Because of the greater time spent in involuntary motor movement, energy expenditure associated with activity (e.g. hyperventilation) is greater in girls with RTT and the energy balance is less positive.

Evaluation

- Height, weight for height, and head circumference are important parameters to follow at each physician visit. The American growth charts can be used. Otherwise standard growth charts can be used to document the evolution over time.
- Measurements of body mass index and skin folds are useful in monitoring progress.
- Some females will need monitoring of weight because of excessive intake.
- Evaluation of oromotor functions like chewing and swallowing should be part of the occupational therapy assessment.
- Assessment of daily caloric intake by a dietician is important in poor feeders. This can be done using a detailed diary of intake kept by the parents kept over a two-week period.
- The influence of involuntary movement activity, abnormal breathing patterns and epilepsy on the balance between nutritional intake and energy expenditure should be considered.
- Consider the increased likelihood of gastroesophageal regurgitation if food aversion is obvious, and evaluate this in a standard way.

Management

- No treatment of growth retardation is indicated as it is genetically determined and influenced by the neurodevelopmental condition.
- Various nutritional intervention strategies should be tried to reduce and, if possible, prevent malnutrition and wasting in affected individuals.
- Caloric supplements can be added when caloric intake is insufficient and oromotor problems are minimal.
- A gastrostomy-button should be placed when the child is not able to eat comfortably and without risk

of aspiration, to assure sufficient nutritional and caloric intake.
- Treatment of gastroesophageal regurgitation is standard.

Development and Behavior

All girls with RTT have intellectual disability. The stages of developmental progression are provided in Diagnostic Criteria. The absence of speech in most affected girls, the dyspraxia, and the short attention span with lack of interest in play make developmental testing a difficult task. In the long-lasting stage III of "wake-up" and "pseudostabilization," parents and caregivers learn to become experts in understanding and recognizing the wishes of their daughter.

Communication. Affected children try to establish visual contact by intense staring. They look at you and want to be looked at. Young girls may even force the parents to turn their face toward them. They "speak" to each other "through" the eyes. Girls with RTT are able to make choices and take causally related action. Therefore, their parents and caregivers should be aware that the time they require to show what is wanted or to produce their answer to a specific situation is prolonged. This intense eye contact behavior is further accentuated in older affected females in a typical eye-pointing behavior, which expresses wishes and remains present even in the most severely affected female. Teachers can use this behavior to develop eye communication in habilitation programs, in training emerging literacy and in augmentative communication through eye gaze computer technology. Some girls with RTT have preservation of speech or use words and sentences in a meaningful way. Many of them continue to learn new words and names far into stage III and into adulthood. Gradual loss of this speech ability usually coincides with the point at which walking stops and transition is made into stage IV. It is the overall impression of parents that their daughters actually understand more of the ongoing conversation than is generally considered possible. This intense eye contact and eye-pointing behavior is very distinct and separates RTT from other conditions with severe intellectual disability and/or autism. Girls and adults in stage III are joyful and alert. Some of them like to tease people and to gently feel with hands and feet in playful body contact. Others reject physical touch of particular body parts and get agitated when they feel unsafe. Their preference for male caretakers or visitors is well known and confirmed by the parents. Autistic features predominate in some of the females with atypical RTT. Odd behaviors, attention deficit, and hyperactivity are common, but after careful observation, these behavioral characteristics are found in combination with an apparent change in muscle tone or hidden dystonic features.

Intense Hand Stereotypies. These stereotypies disturb communicative interaction through distraction and agitation. They cause difficulty in concentration on objects for a long time. By forcing the girls to stop the arm and hand movements by gently fixing the elbows or by bracing them during sessions of interaction, the child will appear quieter and more concentrated. In this way, some girls will be more cooperative in sessions with the occupational therapist.

Screaming Spells. Some teens and adults experience periodic sudden violent screaming spells. They are often associated with extreme pain though no specific organ pathology is present and thorough examination does not reveal any somatic abnormality. This is not an epileptic phenomenon but rather is defined as "brain-pain-crying," and can last for hours. Others are abnormally prone to agitation and scream when they do not feel safe for whatever reason. The situation returns to normal when moments of rest and peace are given.

Sleep Abnormalities. These are more or less pronounced, and are a constant feature of RTT. Night laughter, prolonged wakefulness or early morning waking causes great concern for parents, especially in young preschool girls. These problems may persist into adult life. Night laughter clearly does not disturb the child. The fact that affected children and adults are prone to short periods of daytime sleeping is seen as a need for recovery. The mechanism behind this disruptive night awakening and daytime sleeping is not yet well understood. It might be related to the other autonomic dysfunctions that are associated with midbrain and brainstem immaturity. Melatonin appeared to improve total sleep time and sleep efficiency in the girls with the worst baseline sleep quality (McArthur and Budden 1998; Miyamoto et al. 1999). Pipamperon can be used as a regulator of circadian rhythm with little hypnotic side effects (personal experience). It mainly acts as a serotonin-antagonist, with less adrenolytic and anti-dopaminergic action. It is particularly useful when the girl is abnormally prone to agitation. Pipamperon is not available in the United States as of this writing.

Evaluation

- Evaluation of communicative abilities should include careful observation and questioning of parents and caretakers.
- Formal developmental assessment needs more time than usual. There may be a latency period in producing a reaction, and affected females are more interested in people than in objects.
- The use of an eye gaze computer is recommended early after clinical diagnosis in order to obtain more information about speech and language development.

- Evaluation of screaming spells should include a search for medical causes such as gastroesophageal reflux or constipation. Consider the possibility of a bone fracture.
- Assessment of sleep patterns at night and the amount of daytime sleep is important.
- Evaluation of sleep requires close collaboration between neurologists, pediatricians, and anesthesiologists.

Management

- Augmentative communication methods should be used to capitalize on the intense visual communicative ability. Guidelines for communication in RTT are available on www.rettsyndrome.org.
- During therapy sessions, agitation and distraction should be avoided as much as possible, and gently immobilizing the hands may contribute to the quality of the interaction.
- To establish visual contact behavior, the examiner's face should be brought closely in front of the subject with avoidance of distraction and agitation as much as possible.
- Bracing the elbows may help with dominant and intense hand stereotypies. Braces in soft but resisting materials can be used. Allow the child daytime periods without them. Evaluate the effect of bracing on behavior. In case of agitation, bracing should be abandoned. In the presence of agitation, moments of private rest and peace should be granted, according to individual needs. Identification of the trigger and its avoidance is the first line treatment. The use of time-out in sensory deprivation can be tried if this fails. Drugs of choice are risperidone (Risperdal®) or pipamperon (Dipiperon®).
- Regulation of circadian rhythm can be useful. Melatonin and L-tryptophan are useful in initiating sleep; pipamperon, if available, can be used in low normal dosage when agitation is present.
- Music therapy is recommended in RTT as affected people seem to enjoy it and perform better (Bergström-Isacsson et al. 2014).

Neurologic

Seizures. Epilepsy is present in up to 80% of affected girls at some time in their lives (Steffenburg et al. 2001; Tarquinio et al. 2017). It usually starts after age 4 years and tends to diminish in severity in adulthood. Many become seizure-free for five years or more. The most common seizure types are partial complex, tonic-clonic, tonic, and myoclonic seizures. Although about 50% of seizures can be controlled by medication, intractable epilepsy occurs significantly more frequently in girls with obvious deceleration of head growth. The electroencephalogram (EEG) is usually abnormal in RTT, but there is no diagnostic pattern. EEG patterns frequently seen in RTT include generalized slowing, monorhythmic theta waves, and focal and generalized spikes and sharp waves. Neurophysiologists can use an EEG staging system according to the presence or absence of sleep characteristics like K-spindles and transient vertex waves, slow wave activity and the intensity of generalized spikes and sharp waves (Glaze 2005). These EEG stages do not always coincide with the clinical stage.

The age of onset of seizures is later than usually found in severe mental handicap in general. It is surprising that most children with RTT, although severely impaired, only experience the onset of epilepsy in stages III and IV and not in the rapid regression stage II. Rarely, infantile seizures, variant infantile spasms or other intractable seizures are present before the appearance of classical RTT features. In spite of this early and severe onset of epilepsy, no negative effect on the long-term course and prognosis of RTT has been identified. Epilepsy tends to decrease after age 20 years. Many girls become seizure-free or are well controlled by medication (Steffenburg et al. 2001). Status epilepticus does not occur more often than in children with severe intellectual disability, in general. The probability of death associated with epilepsy is estimated at 9%.

Recent research reveals that the *BDNF* Met66 allele is a protective factor against seizures, whereas missense mutations in the methyl-binding domain of *MECP2* are more frequently associated with early seizures (Nectoux et al. 2008).

Brainstem events may be confused with seizures or are difficult to interpret as such by parents and care takers. Signs of abnormal brainstem activity include blinking of the eyes, facial twitching, vacant spells with no associated epileptiform activity, and hypocapneic attacks with tetany and cyanosis. Classifying these clinical events requires simultaneous neurophysiological monitoring of brainstem and cortical functions and correlation with behavior. Facial twitching with or without sudden changes in attention and eye deviation should not be a priori interpreted as epileptic paroxysms in a young child. This reflects more the ongoing process of immature brainstem activity and is not influenced by antiepileptic drugs. Immature brainstem activity also accounts for the screaming spells, laughing spells, prolonged staring, and so on.

Impaired Nociception. This feature is often seen in RTT, and is sometimes confined to specific body parts. Rarely, the skin can present with easy bruising and blister formation, especially in severe classical RTT. These lesions are different from decubitus wounds. Impairment of pain perception probably represents delayed sensitivity to pain.

Autonomic Cardiorespiratory Manifestations. Irregular breathing in the waking state associated with non-epileptic vacant spells is the most distressing feature in RTT. It reflects the immaturity of the brainstem and may contribute to sudden death. Low resting cardiac vagal tone and weak vagal response to hyperventilation and breath-holding suggest inadequate parasympathetic control. Neurophysiological studies have shown that these baseline brainstem functions are affected in RTT, whereas the baseline sympathetic tone remains at a neonatal level. Insight into these phenomena has introduced new terminology such as "brainstem storm" and "brainstem epilepsy" as phenomena of abnormal spontaneous brainstem activation (ASBA) associated with altered breathing patterns (Julu et al. 2001). Evaluating the brainstem functions in RTT requires detailed neurophysiology (Julu and Witt-Engerström 2005). The primary pathophysiology is related to a defective control mechanism of carbon dioxide exhalation causing respiratory alkalosis or acidosis. Three cardiorespiratory phenotypes are described, each demanding a specific approach (Julu et al. 2008): *forceful breathers* usually have fixed low levels of pCO_2 (chronic respiratory alkalosis); *feeble breathers* usually have fixed high levels of pCO_2 (chronic respiratory acidosis) due to weak respiration, and physical activity during person-to-person contact can stimulate breathing but is short-lived; *apneustic breathers* accumulate carbon dioxide due to delayed and inadequate expirations. Agitation in individuals with RTT is associated with unrestrained sympathetic activity.

Evaluation

- A good description of the clinical paroxysm is essential for the diagnosis of epilepsy.
- The presence of epileptic discharges should be verified by electroencephalogram.
- The clinician should constantly question whether the reported episodic behavior is of epileptic origin.
- Evaluation of breathing patterns requires close collaboration between neurologists, pediatricians, and anesthesiologists.
- Establishing the cardiorespiratory phenotype through the simultaneous recording of baro-receptor sensitivity, cardiac vagal tone, heart rhythm, blood pressure, pO_2 and pCO_2 can be done in specialized settings.

Management

- The most commonly used anti-epileptic drugs are sodium valproate, lamotrigine, and carbamazepine. Monotherapy is successful in 50%. Polypharmacy should be avoided as much as possible.
- There is no general rule for the anti-epileptic treatment in RTT. Each case should be assessed individually.
- Individuals with RTT are sensitive to anti-epileptic drugs and have a tendency to be easily over-sedated, cognitively depressed, and confused.
- Feeble breathers and apneustic breathers are very sensitive to opiates and benzodiazepines. These drugs should be avoided in them.
- Gradual withdrawal of anti-epileptic medication should be considered when individuals become seizure-free.
- Prevention of bruising and decubitus ulcers is achieved by standard measures.
- Treatment of brainstem dysfunctions is extremely difficult and hazardous. There is little experience with medication. Vagal nerve stimulation, as in intractable epilepsy, is under debate but it had success only in limited number of cases.
- To interrupt an episode of forceful breathing, the authors recommend first short periods of re-breathing in a full-face mask connected with a tube between 40 and 60 cm long as dead space. Long-term weaning from the chronic respiratory alkalosis requires Carbogen treatment (5% CO_2 in oxygen mixture) to move the pCO_2 toward normal (39–44 mm Hg). The use of a mixture with 60% oxygen/40% carbon dioxide by nasal prongs during sleep and under medical surveillance is recommended to lift the low pCO_2 to about 40 mm Hg (Smeets et al. 2006).
- In feeble breathers, oral theophylline is the authors' first choice of drug for respiratory stimulation but its clinical tolerance is very poor. Continuous positive airway pressure (CPAP) can be used at night. The end point of treatment is to establish normal breathing rhythm at or near normal pCO_2. Feeble breathers have great sensitivity to opiates and benzodiazepines. Weaning from artificial ventilation in intensive care is difficult. In apneustic breathers, oral buspirone is the drug of choice because of its effect on apneusis (Kerr et al. 1998). Treatment end point and risks are otherwise similar to feeble breathers.
- In general, if one considers the sympathovagal imbalance in RTT, parasympathetic feedback can easily be offered through the following means: favorite video, vibroacoutic stimulation and/or music, bathing and playing with water, personal physical body contact, horseback riding, walking in open air and physical activity in general, and frequent small meals. Parents and care givers should keep in mind that a minimum of 2 × 60 min of movement a day can be easily reached through the moments of personal care and interactions that may last longer.

Musculoskeletal

Scoliosis. Scoliosis develops in early school age with various degrees of severity. Sometimes progression is very rapid, depending on asymmetry in muscle tone and the degree of dystonia and muscle wasting. In ambulatory girls, scoliosis appears unpredictable – it may never be present or may only develop to a small extent. In non-ambulatory girls with typical RTT stage IVB scoliosis develops in spite of preventive measures. Most commonly a double curve develops with a longer upper part (most frequently dextro convex) and a shorter lower part (sinistro convex). When there is no neurologic asymmetry, the spine deformity is usually much more benign.

Kyphosis occurs more in ambulatory girls. It may be related to the degree of extension in the ankle muscles. Tiptoe walking in girls with RTT, in contrast to other circumstances with neurologic deficit, is related to uncertainty and anxiety about falling. Girls gain support and stability by bending forward on stiff legs, giving them more balance against gravity. When sitting and drowsy, girls tend to drop their heads forward causing more bending of the cervical and high thoracic vertebral column. A high kyphosis is not uncommon in the many milder or variant forms of RTT and can progress by age.

Foot Deformities. The foot deformities most common in RTT are equinus and equinovalgus/varus positions. As long as the Achilles tendon can be flexed over 90° with the knee in extension, normal walking remains possible. Further shortening of the Achilles tendons is then compensated for by an "escape" in the valgus or varus position. Young girls do not suffer from this and continue to develop walking ability. If there is hyperextension of the ankles, the need for compensation rises to the knees, the hips and the spine, threatening loss of balance and making walking very difficult if not impossible. Affected girls develop a preference for one leg, putting it forward in every step and using the other leg as support and balance. Direction is chosen through the preferential leg. Sometimes the other leg is placed more to the side causing a girl to walk in circles; sometimes the girl tilts it high up and then forward simulating an involuntary movement.

With careful follow-up of muscle tone and posture, especially of the spine and feet, and with timely corrective measures, walking can be preserved for a long time. Abnormal muscle tone in the flexor/adductor muscles of the hip can lead to dislocation especially in non-walking girls.

Evaluation

- Periodic radiographs of the spine are recommended to establish the degree of scoliosis and its progression.
- Observation of gait patterns and patterns of preferential posturing may be useful in evaluating the need for correction of a foot deformity.
- Physical therapy and orthopedic evaluations are indicated when examination shows evidence of musculoskeletal abnormalities.
- If surgery is considered, an evaluation of feeding, epilepsy, skin problems, and behavior should be carried out before hospitalization.

Management

- The approach to orthopedic deformities in RTT requires input from parents, therapists, pediatrician, orthopedic surgeon, and a rehabilitation specialist to find a treatment goal related to the individual's level of function in daily life activities.
- Good sitting and sleeping positions are important.
- Botox treatment of spasticity can be used in RTT as in spasticity in general, but should be done in consultation with the rehabilitation specialists and orthopedic surgeons (Flett 2003). Results depend on good advance selection of affected individuals. The effect, however, is limited in time.
- Braces or orthoses are used for the spine, the foot and the ankle to prevent further deformation and/or to support walking.
- Surgical lengthening of ankle muscles may be considered for shortened Achilles tendons.
- Surgical lengthening of flexor/adductor muscles may be needed if hip dislocation is imminent.
- Severe tonic-clonic seizures should be well controlled by medication before spinal surgery.
- Surgical treatment of scoliosis with spinal fusion should be considered using the same criteria as in the general population. Spinal surgery might be limited to posterior fusion without combined anterior fusion when the intervention is not delayed beyond a scoliotic curve of greater than 46–60°.
- Early casting of the trunk as a conservative treatment will not prevent surgical intervention in progressive cases.
- Kyphosis rarely needs surgical correction.

Dental

Girls with RTT demonstrate frequent digit/hand sucking and/or biting, mouth breathing and drooling. Tooth grinding or bruxism is another characteristic feature that may lead to dental attrition. There are no anomalies of tooth number, size, form, structure, or eruption (Ribeiro et al. 1997; Friedlander et al. 2003).

Evaluation

- Inspection for soft tissue alteration like gingivitis and palatal shelving is recommended.
- Examination by a dentist should begin no later than age 3 years and should occur at least as often as in the general population. More frequent dental evaluations can be necessary on indication by the dentist.

Management

- Regular dental hygiene is indicated.
- Botox is used in extreme bruxism to relieve the muscle tone in the jaw with positive results on comfort and behavior (Laskawi 2008).

Gastrointestinal

In RTT there is a higher incidence of gastroesophageal reflux and decreased intestinal motility resulting in constipation. Because of the Valsalva breathing pattern and pressing the air over a closed glottis, air swallowing is common. The resultant bloating of the abdomen can be extreme and of a degree that is not encountered in other conditions with intellectual disability. These clinical features may interfere with normal feeding and with general comfort.

Evaluation

- The whole clinical picture of the person with RTT who has feeding difficulties, abnormal breathing pattern, and epileptic seizures should be evaluated.
- Evaluation of gastroesophageal reflux is by standard means. There should be a low index of suspicion.
- Parents/caregivers should make notes of stool frequency during the course of a week if constipation is suspected.
- Endoscopic examination may be indicated to evaluate the degree of reflux esophagitis.

Management

- Gastroesophageal reflux is conservatively treated as in the general population. Surgical treatment is rarely needed.
- A conservative approach should be used to treat constipation, including dietary measures, sufficient fluids, stool softeners and, eventually, the use of enemas.
- There is no treatment for air swallowing with bloating of the abdomen.

Cardiovascular

Females with RTT have a higher incidence of prolonged QT interval, and heart rate variability is diminished. These abnormalities likely result from impairment of autonomic nervous system control, reducing the electrical stability of the heart and precipitating sudden arrhythmia. Imbalance between preserved sympathetic tone and insufficient parasympathetic control is known to cause cardiac arrhythmia. Individuals with prolonged QT interval associated with abnormal breathing pattern are particularly at risk for cardiac arrhythmia, especially the forceful breathers. Of the deaths reported to the International RTT Association in individuals less than 23 years of age, 22% have been sudden, unexpected death, in comparison with 2.3% in the general population up to the same age (Kerr and Witt-Engerström 2001).

Cold extremities caused by poor perfusion because of altered autonomic control are common. This is more related to the central abnormalities than to vascular conditions. In the long term, it leads to abiotrophic changes.

Evaluation

- Standard cardiological examination is advised in girls and young women with RTT who have other signs of autonomic dysfunction.
- Prolonged QT interval above 0.450 should be monitored by periodic electrocardiogram.

Management

- There is no specific treatment of prolonged QT in Rett.
- β-Blockers such as propanolol may be used, but the effect on outcome is unclear.
- Avoidance of certain medications such as cisapride, and tricyclic antidepressants, among others, is indicated, as they can provoke prolonged QT interval.
- The feet and the lower legs should be kept warm in protective clothing.

Ophthalmologic

Visual sensory function is a strength and care should be taken to preserve it. No specific ophthalmologic pathology occurs in this disorder, although strabismus is very common. Acquired cataracts may occur after self-injurious tapping in association with behavioral agitation. Retinal changes are not present.

Evaluation

- Periodic standard ophthalmologic examination is suggested.

Management

- Strabismus is treated by standard methods.

- Some girls benefit from acuity correction with spectacles.
- Self-injurious behavior related to the eyes should be prevented.
- Loss of vision following acquired cataract can rarely be treated successfully.
- Prevention of retinal detachment, a possible complication after acquired traumatic cataract, should occur through regular follow-up by the ophthalmologist.

Ears and Hearing

Occasionally perceptive hearing loss is encountered in RTT. Its relationship to the syndrome is not clear. Auditory evoked responses show little abnormality.

Evaluation

- Early hearing screening is recommended as for all children.
- Regular hearing evaluation may be indicated in young girls with RTT because of the higher prevalence of sensorineural deficit in RTT.

Management

- Any hearing loss should be treated as in the general population.

ACKNOWLEDGMENTS

We wish to thank the Dutch and Belgian parent organizations for their constructive cooperation, and the many girls and their families we have encountered over the last 35 years who have contributed to our clinical experience in the follow-up of Rett syndrome. Our special regards go to the parents who permitted the photos of their daughter to be published in this chapter.

RESOURCES

The International Rett Syndrome Association/ Foundation
Website: www.rettsyndrome.org

Orphanet: International and National Rett Syndrome Associations
Website: http://www.orpha.net/consor/cgi-bin/index.php

RettBASE, the IRSA MECP2 Variation Database
Website: http://mecp2.chw.edu.au/

BOOKS

Kerr A, Witt-Engerström I (2001) *Rett Disorder and the Developing Brain*. Oxford Medical Publications. ISBN-10: 0192630830, ISBN-13: 978-0192630834 (Hardcover).

Kaufmann W (2017) Rett Syndrome. Mc Keith Press. ISBN: 978-1-909962-83-5.

REFERENCES

Amir R, Van den Veyver I, Wan M, Tran C, Francke U, Zoghbi H (1999) Rett syndrome is caused by mutations in X-linked LMeCP2, encoding methyl-CpG-binding protein 2. *Nat Genet* 23:185–188.

Armstrong DD (2002) Neuropathology of Rett syndrome. *MRDD Res Rev* 8:72–76.

Armstrong DD, Kinney HC (2001) Rett disorder and the developing brain. In: *The Neuropathology of the Rett Disorder*, Kerr A, Witt-Engerström I, eds, Oxford: Oxford University Press, pp. 57–84.

Armstrong DD, Dunn JK, Antalffy B (1998) Decreased dendritic branching in frontal, motor, limbic cortex in Rett syndrome compared with trisomy 21. *J Neurpathol Exp Neurol* 57:1013–1017.

Bergström-Isacsson M1, Lagerkvist B2, Holck U3, Gold C4 (2014) Neurophysiological responses to music and vibroacoustic stimuli in Rett syndrome. *Res Dev Disabil* 35(6):1281-1291.

Bourdon V, Philippe C, Labrune O, Amsallem D, Arnould C, Jonveaux P (2001) A detailed analysis of the MECP2 gene: Prevalence of recurrent mutations and gross rearrangements in Rett syndrome patients. *Hum Genet* 108:43–50.

Chandler SP, Guschin D, Landsberger N, Wolffe AP (1999) The methyl-CpG-binding transcriptional repressor MeCP2 stably associates with nucleosomal DNA. *Biochemistry* 38:7008–7018.

Cheadly JP, Gill H, Fleming N, Maynard J, Kerr A, Leonard H, Krawczak M, Cooper DN, Lynch S, Thomas N, Hughes H, Hulten M, Ravine D, Sampson JR, Clark A (2000) Long-read sequence analysis of the MECP2 gene in Rett syndrome patients: Correlation of disease severity with mutation type and localisation. *Hum Mol Genet* 9:1119–1129.

Chen W, Chang Q, Lin Y, Meissner A, West A, Griffith E, Jaenisch R, Greenberg M (2003) Derepression of *BDNF* transcription involves calcium-dependent phosphorylation of *MECP2*. *Science* 302:885.

Ehrhart F, Coort SL, Cirillo E, Smeets E, Evelo CT, Curfs LM (2016) Rett syndrome - biological pathways leading from MECP2 to disorder phenotypes. *Orphanet J Rare Dis* 2016;11(1):158.

Ehrhart F, Bahram Sangani N, Curfs LMG (2018) Current developments in the genetics of Rett and Rett-like syndrome. *Curr Opin Psychiatry* 31(2):103-108.

Ellaway C, Christodoulou J (2001) Rett syndrome: Clinical characteristics and recent genetic advances. *Disabil Rehabil* 23:98–106.

Flett P (2003) Rehabilitation of spasticity and related problems in childhood cerebral palsy. *J Paediatr Child Health* 39:6–14.

Friedlander A, Yagiel J, Paterno V, Mahler M (2003) The pathophysiology, medical management and dental implications of fragile X, Rett, and Prader-Willi syndromes. *J Calif Dent Assoc* 31:693–702.

Girard M, Couvert P, Carrie A, Tardieu M, Chelly J, Beldjord C, Bienvenue T (2001) Parental origin of de novo MeCP2 mutations in Rett syndrome. *Eur J Hum Genet* 9:231–236.

Glaze DG (2005) Neurophysiology of Rett syndrome. *J Child Neurol* 20(9):740–746.

Hagberg B, Gilberg C (1993) *In: Rett Syndrome, Clinical, Biological Aspects, Rett Variants—Rettoid Types, Clinics in Developmental Medicine, Vol 127.* MacKeith, Cambridge: Cambridge University Press, pp. 40–60.

Hagberg B, Skjeldal O (1994) Rett variants: A suggested model for inclusion criteria. *Pediatr Neurol* 11:5–11.

Hagberg B, Aicardi J, Dias K, Ramos O (1983) A progressive syndrome of autism, dementia, ataxia and loss of purposeful handuse in girls: Rett's syndrome: Report of 35 cases. *Ann Neurol* 14:471–479.

Hagberg B, Goutieres F, Hanefeld F, Rett A, Wilson J (1990) Rett syndrome: Criteria for inclusion and exclusion. *Brain Dev* 12:47–48.

Hagberg B, Stenbom Y, Witt-Engerström I (2001) Head growth in Rett syndrome. *Brain Dev* 23(Suppl 1): S227–S229.

Hagberg B, Hanefeld F, Percy A, Skjeldal O (2002) An update on clinically applicable diagnostic criteria in Rett syndrome. Comments to Rett Syndrome Clinical Criteria Consensus Panel Satellite to European Paediatric Neurology Society Meeting, Baden, Germany, 11 September 2001. *Eur J Paediatr Neurol* 6:293–297.

Halbach NS, Smeets EE, van den Braak N, van Roozendaal KE, Blok RM, Schrander-Stumpel CT, Frijns JP, Maaskant MA, Curfs LM (2012) Genotype-phenotype relationships as prognosticators in Rett syndrome should be handled with care in clinical practice. *Am J Med Genet A* 158A(2):340-50.

Halbach N, Smeets EE, Julu P, Witt-Engerström I, Pini G, Bigoni S, Hansen S, Apartopoulos F, Delamont R, van Roozendaal K, Scusa MF, Borelli P, Candel M, Curfs L (2016) Neurophysiology versus clinical genetics in Rett syndrome: A multicenter study. *Am J Med Genet A* 170(9):2301-9.

Hara M, Ohba C, Yamashita Y, Saitsu H, Matsumoto N, Matsuishi T (2015) De novo SHANK3 mutation causes Rett syndrome-like phenotype in a female patient. *Am J Med Genet A* 167(7):1593-1596.

Huppke P, Laccone F, Kramer N, Engel W, Hanefeld F (2000) Rett syndrome: Analysis of MECP2 and clinical characterization of 31 patients. *Hum Mol Genet* 9:1369–1375.

Huppke P, Roth C, Christen HJ, Brockmann K, Hanefeld F (2001) Endocrinologic study on growth retardation in Rett syndrome. *Acta Paediatr* 90:1257–1261.

Ishikawa A, Goto T, Narasaki M, Yokochi K, Kitahara H, Fukuyama Y (1978) A new syndrome (?) of progressive psychomotor retardation with peculiar stereotyped movements and autistic tendency: A report of three cases. *Brain Dev* 3: 258.

Julu PO (2001) The central autonomic disturbance in Rett syndrome. In: *Rett Disorder and the Developing Brain*, Kerr A, Witt-Engerström I,eds, Oxford, New York: Oxford University Press, pp. 131–181.

Julu PO, Witt-Engerström I (2005) Assessment of the maturity-related brainstem functions reveals the heterogeneous phenotypes and facilitates clinical management of Rett syndrome. *Brain Dev* 27(Suppl 1):S43–S53.

Julu PO, Kerr AM, Hansen S, Apartopoulos F, Jamal GA (1997) Immaturity of medullary cardiorespiratory neurones leading to inappropriate autonomic reactions as a likely cause of sudden death in Rett's syndrome. *Arch Dis Child* 77:464–465.

Julu PO, Kerr AM, Apartopoulos F, Al-Rawas S, Wiit-Engerström I, Engerström L, Jamal GA, Hansen S (2001) Characterisation of breathing and associated central autonomic dysfunction in the Rett disorder. *Arch Dis Child* 85:29–37.

Julu PO, Witt-Engerström I, Hansen S, Apartopoulos F, Witt B, Pini G, Delamont RS, Smeets EE (2008) Clinical update addressing the cardiorespiratory challenges in medicine posed by Rett syndrome: The Frösö Declaration. *Lancet* 371:1981–1983.

Kerr AM (1992) *In: Mental Retardation and Medical Care: Rett Syndrome British Longitudinal Study (1982–1990) and 1990 Survey, Roosendaal JJ, ed.* Zeist: Kerckbosch Publisher, pp. 143–145.

Kerr A, Witt-Engerström I (2001) *Rett Disorder and the Developing Brain*, Oxford: Oxford University Press.

Kerr A, Julu P, Hansen S, Apartopoulos F (1998) Serotonin and breathing dysrhythmia in Rett syndrome. In: *New Development in Child Neurology*, Perat MV,ed. Bologna: Monduzzi Editore, pp. 191–195.

Kerr A, Nomura Y, Armstrong D, Anvret M, Belichenko PV, Budden S, Cass H, Christodoulou J, Clarke A, Ellaway C, d'Esposito M, Francke u Hulten M, Julu P, Leonard H, Naidu S, Schanen C, Webb T, Engerstrom I, Yamashita Y, Segawa M (2001) Guidelines for reporting clinical features in cases with MECP2 mutations. *Brain Dev* 23:208–211.

Laskawi R (2008) The use of botulinum toxin in head and face medicine: An interdisciplinary field. *Head Face Med* 4:5.

Leonard H, Silberstein J, Falk R, Houwink-Manville I, Ellaway C, Raffaele LS, Witt-Engerström I, Schanen C (2001) Occurrence of Rett syndrome in boys. *J Child Neurol* 16:333–338.

Leonard H, Cobb S, Downs J (2017) Clinical and biological progress over 50 years in Rett syndrome. *Nat Rev Neurol.* 13(1):37–51.

Martinowich K, Hattori D, Wu H, Fouse S, He F, Hu Y, Fan G, Sun Y (2003) DNA methylation-related chromatin remodeling in activity-dependent *BDNF* gene regulation. *Science* 302:890.

McArthur A, Budden S (1998) Sleep dysfunction in Rett syndrome: A trial of exogenous melatonin treatment. *Dev Med Child Neurol* 40:186–192.

Miyamoto A, Oki J, Takahashi S, Okuno A (1999) Serum melatonin kinetics and long-term melatonin treatment for sleep disorders in Rett syndrome. *Brain Dev* 21:59–62.

Moog U, Smeets E, van Roozendaal K, Schoenmakers S, Herbergs J, Schoonbrood-Lenssen A, Schrander-Stumpel C (2003) Neurodevelopmental disorders in males related to the gene causing Rett syndrome in females (*MECP2*). *Eur J Paediatr Neurol* 7:5–12.

Nan X, Tate P, Li E, Bird A (1996) DNA methylation specifies chromosomal localization of MeCP2. *Mol Cell Biol* 10:414–421.

Nan X, Campoy J, Bird A (1997) MeCP is a transcriptional repressor with abundant binding sites in genomic chromatin. *Cell* 88:471–481.

Nan X, Ng HH, Johnson CA, Laherty CD, Turner BM, Eisenman RN, Bird A (1998) Transcriptional repression by the methyl-CpG-binding protein MeCP2 involves a histone deacetylase complex. *Nature* 393:386–389.

Nectoux J, Bahi-Buisson N, Guellec I, Coste J, De Roux N, Rosas H, Tardieu M, Chelly J, Bienvenu T (2008) The p.Val66Met polymorphism in the BDNF gene protects against early seizures in Rett syndrome. *Neurology* 70(Pt 2):2145–2151.

Neul JL, Kaufmann WE, Glaze DG, Christodoulou J, Clarke AJ, Bahi-Buisson N, Leonard H, Bailey ME, Schanen NC, Zappella M, Renieri A, Huppke P, Percy AK; RettSearch Consortium (2010) Rett syndrome: revised diagnostic criteria and nomenclature. *Ann Neurol* 68(6):944–950.

Nomura Y, Segawa M (1992) Motor symptoms of the Rett syndrome: Abnormal muscle tone, posture, locomotion and stereotyped movement. *Brain Dev* 14 (Suppl): S21–S28.

Nomura Y, Segawa M, Higurashi M (1985) Rett syndrome—an early catecholamine and indolamine deficient disorder? *Brain Dev* 7:334–341.

Reichow B, George-Puskar A, Lutz T, Smith IC, Volkmar FR (2015) Brief report: systematic review of Rett syndrome in males. *J Autism Dev Disord* 45(10):3377–3383.

Reilly S, Cass H (2001) Growth and nutrition in Rett syndrome. *Disabil Rehabil* 23:118–128.

Rett A (1966) Über ein cerebral-atrophisches Syndrom bei Hyperammonämie. *Wien Med Wochenschr* 116:7.

Ribeiro R, Romano A, Birman E, Mayer M (1997) Oral manifestations in Rett syndrome: A study of 17 cases. *Pediatr Dent* 19:349–352.

Ronen GM, Brady LI, Tarnopolsky MA (2017) Males With MECP2 C-terminal-Related Atypical Rett Syndromes and Their Carrier Mothers. *Pediatr Neurol.* 67:98-101.

Schanen NC, Dahle EJ, Capozzoli F, Holm VA, Zoghbi HY, Francke U (1997) A new Rett syndrome family consistent with X-linked inheritance expands the X chromosome exclusion map. *Am J Hum Genet* 61:634–641.

Schönewolf-Greulich B, Bisgaard AM, Møller RS, Dunø M, Brøndum-Nielsen K, Kaur S, Van Bergen NJ, Lunke S, Eggers S, Jespersgaard C, Christodoulou J, Tümer Z (2017) Clinician's guide to genes associated with Rett-like phenotypes - Investigation of a Danish cohort and review of the literature. *Clin Genet* 95:221–230.

Schollen E, Smeets E, Deflem E, Fryns JP, Matthijs G (2003) Gross rearrangements in the *MECP2* gene in three patients with Rett syndrome: Implications for routine diagnosis of Rett syndrome. *Hum Mutat* 22:116–120.

Schultz R, Glaze D, Motil K, Hebert D, Percy A (1998) Hand and foot growth failure in Rett syndrome. *J Child Neurol* 13:71–74.

Smeets E, Schollen E, Moog U, Matthijs G, Herbergs J, Smeets H, Curfs L, Schrander-Stumpel C, Fryns JP (2003) Rett syndrome in adolescent and adult females: Clinical and molecular genetic findings. *Am J Med Gen* 122A:227–233.

Smeets E, Julu P, van Waardenburg D, Witt-Engerström I, Hansen S, Apartopopoulos F, Curfs L, Schrander-Stumpel C (2006) Management of a severe forceful breather with Rett syndrome using carbogen. *Brain Dev* 28:625–632.

Steffenburg U, Hagberg G, Hagberg B (2001) Epilepsy in a representative series of Rett syndrome. *Acta Paediatr* 90:34–39.

Tao J, Van Esch H, Hagedorn-Greiwe M, Hoffmann K, Moser B, Raynaud M, Sperner J, Fryns JP, Schwinger E, Gécz J, Ropers HH, Kalscheuer VM (2004) Mutations in the X-linked cyclin-dependent kinase-like 5 (CDKL5/STK9) gene are associated with severe neurodevelopmental retardation. *Am J Hum Genet* 75:1149–1154.

Tarquinio DC, Motil KJ, Hou W, Lee HS, Glaze DG, Skinner SA, Neul JL, Annese F, McNair L, Barrish JO, Geerts SP, Lane JB, Percy AK (2012) Growth failure and outcome in Rett syndrome: specific growth references. *Neurology* 79(16):1653–1661.

Tarquinio DC, Hou W, Berg A, Kaufmann WE, Lane JB, Skinner SA, Motil KJ, Neul JL, Percy AK, Glaze DG (2017) Longitudinal course of epilepsy in Rett syndrome and related disorders. *Brain* 140(2):306–318.

Wan M, Zhao K, Francke U (2001) MeCP2 truncating mutations cause histone H4 hyperacetylation in Rett syndrome. *Hum Mol Genet* 10:1085–1092.

Witt-Engerström I (1990) Rett Syndrome in Sweden. *Acta Paediatr Scand* 369(Suppl):1–60.

Yaron Y, Ben Zeev B, Shomrat R, Bercovich D, Naiman T, Orr-Urtreger A (2002) MECP2 mutations in Israel: Implications for molecular analysis, genetic counseling and prenatal diagnosis in Rett syndrome. *Hum Mutat* 20:323–324.

Yntema HG, Oudakker AR, Kleefstra T, Hamel BC, van Bokhoven H, Chelly J, Kalscheuer VM, Fryns JP, Raynaud M, Moizard MP, Moraine C (2002) In-frame deletion in MECP2 causes mild nonspecific mental retardation. *Am J Med Genet* 107:81–83.

Zoghbi HY, Percy kA Schulz RJ, Fill C (1990) Patterns of X chromosome inactivation in the Rett syndrome. *Brain Dev* 12:131–135.

Zweier C, Peippo MM, Hoyer J, Sousa S, Bottani A, Clayton-Smith J, Reardon W, Saraiva J, Cabral A, Gohring I, Devriendt K, de Ravel T, Bijlsma EK, Hennekam RC, Orrico A, Cohen M, Dreweke A, Reis A, Nurnberg P, Rauch A (2007) Haploinsufficiency of TCF4 causes syndromal mental retardation with intermittent hyperventilation (Pitt-Hopkins syndrome). *Am J Hum Genet* 80:994–1001.

50

ROBIN SEQUENCE

HOWARD M. SAAL

Division of Human Genetics, Cincinnati Children's Hospital Medical Center, Department of Pediatrics, University of Cincinnati College of Medicine, Cincinnati, Ohio, USA

INTRODUCTION

The Robin sequence, also known as the Pierre Robin sequence, is a well-studied etiologically heterogeneous clinical entity. The condition was initially described in 1822 by the French naturalist Etienne Geoffroy Saint-Hilaire who described a case of micrognathia in a sheep with additional congenital anomalies (Saint-Hilaire 1822). In 1844, Fairbairn reported two human cases of cleft palate with tongue-based upper airway obstruction (Fairbairn 1846). The French stomatologist Pierre Robin described a series of infants born with mandibular hypotrophy and glossoptosis with respiratory problems (Robin 1923). Of note, is that none of his original reported individuals had cleft palate. In a larger report, Robin noted that cleft palate could also be an associated finding (Robin 1934). The accepted name for the triad of micrognathia, glossoptosis and upper airway obstruction is Robin sequence; however, it has had several appellations including Pierre Robin syndrome, anomalad, complex, and sequence (Breugem et al. 2008; van Nunen et al. 2018). Since traditionally medical nomenclature does not include first names, the accepted terminology is Robin sequence, and this was supported by an international consensus conference to establish best practices for diagnosis and evaluation of Robin sequence (Breugem et al. 2016). The term sequence is most befitting this condition, since it typically describes a series of anomalies or clinical events that result from a single initiating event, such as a disruption or malformation (Spranger et al. 1982; Hennekam et al. 2011). Therefore, in Robin sequence, the initiating event is micrognathia with secondary posterior placement of the tongue with pharyngeal and upper airway obstruction and respiratory distress. In many patients, the upward tongue displacement will interfere elevation and fusion of the lateral palatine shelves with resultant cleft palate. The term sequence defines pathogenesis not etiology.

Incidence

The incidence of Robin sequence is not known, primarily because of confusion related to the specific diagnostic criteria (Breugem et al. 2008), especially since all studies have limited their analyses to populations that include cleft palate. European epidemiologic studies have reported the incidence of Robin sequence to range from 1:8850 to 1:14,000 live births based on data from national studies conducted in Germany and Denmark (Maas et al. 2014; Printzlau et al. 2004). Paes et al. (2015) reviewed the birth prevalence of Robin sequence in the Netherlands utilizing the criteria of cleft palate with micrognathia or retrognathia and obstructive respiratory symptoms between 2000 and 2010 and found a prevalence of 1:5600. A study looking at the number of cases of cleft palate with Robin sequence in Scotland between 2004 and 2013 found a birth prevalence of 1:2685 live births (Wright et al. 2018).

With appropriate and timely management, there should be little impact on lifespan in children who have Robin sequence. Significant or prolonged apnea can result in early

---

Cassidy and Allanson's Management of Genetic Syndromes, Fourth Edition.
Edited by John C. Carey, Agatino Battaglia, David Viskochil, and Suzanne B. Cassidy.
© 2021 John Wiley & Sons, Inc. Published 2021 by John Wiley & Sons, Inc.

death or severe morbidity related to hypoxic encephalopathy. Those children who have Robin sequence with underlying syndromes may be at higher risks for morbidity and mortality related primarily to other associated anomalies, especially those with neurologic impairment or structural anomalies of the airway and cardiovascular system. Children with multiple anomaly syndromes and chromosome anomalies are at higher risks for morbidity and mortality. A retrospective review of mortality and causes of death in 181 individuals with both isolated and syndromic Robin sequence followed over an 11 year period found the mortality was 16.6%, with two deaths related to airway obstruction, no deaths in isolated Robin sequence, and significant associations of mortality with cardiac anomalies and central nervous system anomalies (Costa et al. 2014). The mortality rate in a Dutch cohort of Robin sequence was 10% (Logjes et al. 2018). Of the patients who died, only one had isolated Robin sequence and this patient died from surgical and postoperative complications. The remainder of patients had underlying syndromes with various anomalies and medical complications (Logjes et al. 2018).

Diagnostic Criteria

Robin sequence is a clinical diagnosis. An international consensus conference defined the clinical triad of micrognathia with glossoptosis and airway obstruction (obstructive apnea) as the basis for a diagnosis of Robin sequence, acknowledging that cleft palate was considered a common additional feature of Robin sequence, but not necessary for diagnosis (Breugem et al. 2016). With respect to accepted dysmorphology terminology, micrognathia refers to a small or hypoplastic mandible when viewed from the front but not the side; whereas, retrognathia refers to a posteriorly positioned lower jaw, which is set back from the plane of the face when viewed from the side but not from the front (Allanson et al. 2009) (see Figures 50.1–50.3). The diagnosis of micrognathia then is subjective; there are no accepted objective criteria for the diagnosis of micrognathia. Potentially helpful diagnostic modalities include X-ray and CT scan (see Figure 50.4). Most cases are diagnosed based on clinical evaluation, especially in the presence of respiratory compromise. However, micrognathia may be an isolated clinical finding and is not diagnostic for Robin sequence in the absence of obstructive apnea.

Glossoptosis is defined as posterior displacement of the tongue into the pharynx (Carey et al. 2009). This can be of varying severity and can lead to upper airway obstruction. Frequent signs and symptoms include inspiratory stridor and stertor, apnea, laboratory signs of hypoxemia and hypercarbia, and impaired sucking and swallowing.

Although cleft palate is commonly seen in patients with Robin sequence, its presence is not essential for diagnosis (Breugem et al. 2015). Cleft palate in Robin sequence is classically described as being U-shaped; however, many patients with Robin sequence have typical V-shaped cleft palates. Submucous cleft palate may also be observed.

Etiology, Pathogenesis and Genetics

The primary clinical feature in Robin sequence (RS) is micrognathia. There are multiple causes of micrognathia and include malformations, tissue dysplasias, disruptions, and deformations. Searching for the term "Pierre Robin" in the Online Mendelian Inheritance in Man catalog detected a total of 253 listed entries. On searching for micrognathia and

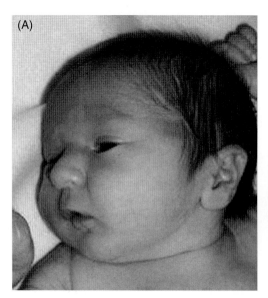 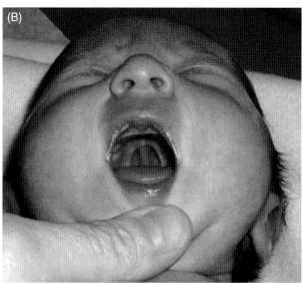

FIGURE 50.1 An infant with Robin sequence and mild to moderate micrognathia (A). Note the U-shaped cleft palate (B).

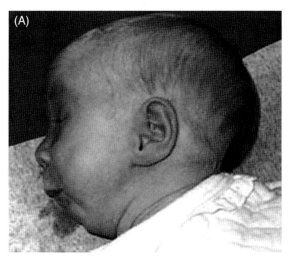
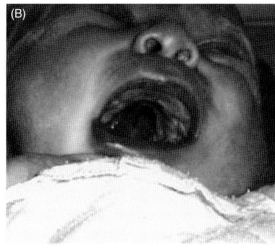

FIGURE 50.2 An infant with Robin sequence and moderate to severe micrognathia (A). Note the extensive U-shaped cleft palate involving most of the secondary palate (B).

cleft palate, 278 entries are listed and for "micrognathia" 614 entries are listed (OMIM 2019). The Online Mendelian Inheritance in Man catalogues primarily single gene disorders and does not include chromosomal disorders, teratogenic conditions, disruptions, and deformations. There are potentially thousands of different disorders associated with RS. The development of diagnostic technologies, especially chromosomal microarray and next generation sequencing (NGS) has greatly enhanced our ability to make diagnoses of common, rare and unique genetic disorders. Table 50.1 lists the most common syndromes associated with RS and associated manifestations.

Many cases of RS are isolated conditions with no identifiable underlying condition. These are probably multifactorial disorders with low recurrence risk, but still greater than that in the general population. Retrospective analyses have shown that underlying syndromes and multiple anomaly disorders are commonly associated with RS. In one series of 117 individuals with RS, 48% had isolated RS, 35% had identifiable syndromes, and 17% had presumed unique or unidentified syndromes (Holder-Espinasse et al. 2001). An analysis of 125 individuals with RS identified syndromic diagnoses in 58% and isolated RS in 42% (Izumi et al. 2012). A review of a large Dutch cohort of individuals with RS identified an underlying disorder in 59.7% of cases, with the remaining cases being isolated (Basart et al. 2015). However, these studies were retrospective, and all individuals who were analyzed had cleft palate, and there were no polysomnography data. Nonetheless, there were significant consistencies among the diagnoses, with Stickler syndrome being the most common genetic condition. One study identified chromosome anomalies in 8.9% of cases, most of which were rare autosomal rearrangements (Basart et al. 2015). Of note, deletion 22q11.2 syndrome was not associated with Robin sequence in any of these studies.

Single Gene Disorders Sticker syndrome and other related type 2 collagenopathies are by far the most common single gene disorders associated with RS. The most common genes associated with Stickler syndrome and related disorders are *COL2A1, COL11A1,* and *COL11A2* (Robin et al. 2017). Another group of disorders associated with RS are the mandibulofacial dysostoses and the acrofacial dysostoses. The most common mandibulofacial dysostosis is Treacher Collins syndrome. This is a genetically heterogeneous disorder caused by mutations in three genes *TCOF1, POLR1C,* and *POLR1D* (see Table 50.1) (Wieczorek 2013; Katsanis and Jabs 2018). There are several rare mandibulofacial dysostoses syndromes, including the acrofacial dysostoses Nager syndrome caused by heterozygous mutations in *SF3B4* (see Figure 50.3), and Miller syndrome caused by biallelic mutations in *DHODH*. Mandibulofacial dysostosis with microcephaly (Guion-Almeida type) is an autosomal dominant condition caused by mutations in *EFTUD2*; many of these individuals have also been found to have esophageal atresia (Lines et al. 2012).

Neuromuscular disorders are frequently associated with micrognathia and Robin sequence. Many of these disorders are associated with arthrogryposis. Cerebro-oculo-facial-skeletal syndrome-1 is an autosomal recessive neurodegenerative disorder, caused by biallelic mutations in *ERCC6*.

Chromosome Disorders Chromosome disorders as a category are among the most common causes of Robin sequence. However, in large series most of these cases have very rare or uncommon chromosomal anomalies (Izumi et al. 2012; Basart et al. 2015). The development of chromosomal microarray analysis has greatly enhanced our ability to diagnose microdeletions and microduplications, which could not be previously be identified with classical cytogenetics analyses.

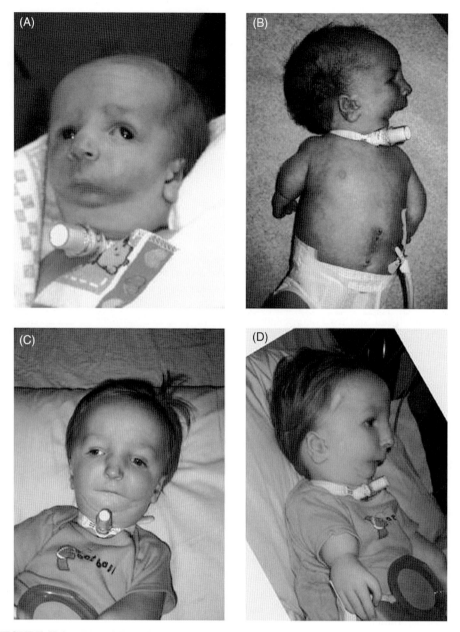

FIGURE 50.3 (A) and (B) A 2-month-old infant with Nager syndrome with severe micrognathia. (C) and (D) The same child at age 19 months after mandibular distraction osteogenesis. Note the significant change in mandibular size.

Deletion 22q11.2 deletion syndrome is the best-known microdeletion disorder that has been associated with RS. Although micrognathia is a common feature of 22q11.2 deletion syndrome, very few individuals with 22q11.2 deletion syndrome are confirmed to have either glossoptosis or respiratory compromise (Izumi et al. 2012; Basart et al. 2015).

Teratogenic Disorders Diabetic embryopathy can be associated with several congenital anomalies, including heart defects, renal anomalies, skeletal and vertebral anomalies, and brain anomalies. Micrognathia with glossoptosis can be seen with or without cleft palate, including the femoral-facial syndrome, which is associated with maternal pre-gestational diabetes.

Fetal alcohol syndrome may be associated with micrognathia and cleft palate. These children may have obstructive sleep apnea.

Prescribed medications may cause RS. Vitamin A and vitamin A analogues, especially isotretinoin, a medication prescribed for cystic acne has been associated with multiple anomalies, including brain anomalies, microtia, microcephaly, micrognathia, and cleft palate. In utero exposure to

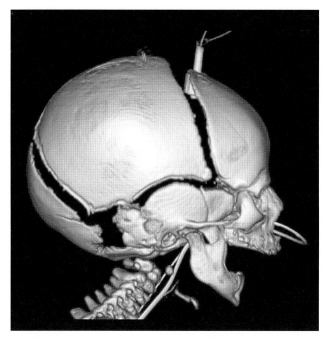

FIGURE 50.4 CT scan with 3D reconstruction demonstrating micrognathia.

antineoplastic medications are also associated with micrognathia and RS. Fetal aminopterin syndrome is associated with limb anomalies, cleft palate, micrognathia, and developmental disabilities. Aminopterin is no longer being used, but methotrexate can have similar embryotoxic effects. It is used as an antineoplastic medication and is also used for immunosuppression in autoimmune disorders, such as rheumatoid arthritis and systemic lupus erythematosus. Anticonvulsants have been associated with multiple congenital anomalies. Diphenylhydantoin can cause RS and other anomalies including microcephaly, short stature, and developmental disabilities.

Miscellaneous Disorders Oculo-auriculo-vertebral spectrum (Chapter 42) is a sporadically occurring condition caused by abnormal development of the first and second pharyngeal pouches during early embryogenesis. The most common findings are facial asymmetry, microtia, vertebral anomalies, renal anomalies and heart defects. RS can be seen, especially with associated asymmetric micrognathia.

Amniotic band sequence is often associated with disruption of limb development and distal amputations. It can also be associated with oral and facial clefts, encephalocele and micrognathia with RS.

Diagnostic Testing

As Robin sequence is etiologically heterogeneous, establishment of the correct diagnosis will directly impact treatment, genetic counseling and recurrence risks, understanding of the natural history of the underlying condition, and provision of long term anticipatory management.

Comprehensive evaluation of all patients with RS is essential and should include comprehensive medical, pregnancy and family histories, as outlined below:

Medical History

- Gestation age
- Type of delivery and complications
- Birth weight, length, head circumference
- Presence of other congenital anomalies or major illness
- Neonatal complications including respiratory difficulties
- Feeding disorder.

Pregnancy History

- Maternal illnesses (especially pregestational diabetes mellitus)
- Maternal medications
- Other substance exposures (alcohol, substance abuse)
- Prenatal imaging studies and results or interpretations (ultrasound, MRI scans)
- Prenatal genetic testing (amniocentesis, chromosome analysis, chromosomal microarray, other genetic testing, including free fetal DNA studies).

Family History

- Other children with birth defects
- Cleft lip and/or cleft palate
- Infant death
- Stillbirth
- Recurrent pregnancy loss
- Vision issues (myopia, retinal detachment blindness)
- Deafness
- Consanguinity.

Prenatal Diagnosis As fetal ultrasound becomes more refined, the ability to diagnose micrognathia along with other congenital anomalies, increases the sensitivity of diagnosing Robin sequence and preparing for post-natal care. Likewise, the growing of experience with fetal MRI has greatly enhanced the ability to diagnose micrognathia and cleft palate (Di Pasquo et al. 2017; Resnick et al. 2018) (see Figure 50.5). Concomitant findings on ultrasound and MRI may include the diagnosis of other structural anomalies and organ system involvement, especially involving the brain. Polyhydramnios is often present in RS and may represent either difficulty swallowing because of glossoptosis or neurological involvement.

TABLE 50.1 Syndromes with Robin sequence.

Genetic disorder	Gene(s)/chromosome	Inheritance pattern	Additional manifestations
Stickler syndrome type 1 (see Chapter 56)	COL2A1/12q13.11	AD	High myopia, vitreal degeneration, retinal detachment, early onset arthropathy, sensorineural hearing loss, mitral valve prolapse
Stickler syndrome type 2 (see Chapter 56)	COL11A1/1p21	AD	Myopia, vitreal abnormalities, retinal detachment, early onset arthropathy, sensorineural hearing loss
Stickler syndrome type 3 (see Chapter 56)	COL11A2/6p21.3	AD	Epiphyseal dysplasia, sensorineural hearing loss, premature osteoarthritis
Marshall syndrome (see Chapter 56)	COL11A1/1p21	AD	Short stature, sensorineural hearing loss, congenital cataract, myopia
Smith–Lemli–Opitz syndrome (see Chapter 53)	DHCR7/11q12	AR	Short stature, ptosis, heart defects, brain anomalies, toes 2–3 syndactyly, genital anomalies (hypospadias, cryptorchidism), microcephaly, developmental disabilities
Catel–Manzke syndrome	TGDS /13q32.1	AR	Accessory phalanx of index finger, congenital heart defects
Treacher Collins syndrome 1 (see Chapter 57)	TCOF1 (Treacle)/5q32	AD	Malar hypoplasia, zygomatic hypoplasia, eyelid colobomas, ear anomalies, deafness
Treacher Collins syndrome 2 (see Chapter 57)	POLR1D/13q32.2	AD/AR	Malar hypoplasia, zygomatic hypoplasia, eyelid colobomas, ear anomalies, deafness
Treacher Collins syndrome 3 (see Chapter 57)	POLR1C/6p21.1	AD	Malar hypoplasia, zygomatic hypoplasia, eyelid colobomas, ear anomalies, deafness
Nager syndrome	AFD1/1q21.2	AD	Malar hypoplasia, growth retardation, short stature, limb anomalies, radial and thumb dysplasia/aplasia, microcephaly
Miller syndrome Postaxial acrofacial dysostosis	DHODH/16q22.2	AR	Malar hypoplasia, eyelid colobomas, cleft lip, cleft palate, ear dysplasia, microtia, postaxial limb defects, absent radii, short stature
Mandibulofacial dysostosis with microcephaly (MDFM) Mandibulofacial dysostosis, Guion–Almeda type	EFTUD2/17q21.31	AD	Mandibulofacial dysostosis, microcephaly, cleft palate, ear anomalies, microtia, Conductive hearing loss, developmental disabilities, esophageal atresia, eyelid colobomas, pectus deformity, micropenis, cryptorchidism, limb anomalies, syndactyly, hypoplastic thumbs.
Branchiootorenal syndrome 1 BOR1	EYA1[a]/8q13.2	AD	Ear anomalies, deafness, branchial cleft fistulas or cysts, renal anomalies, polycystic kidneys
Branchiootorenal syndrome 2 BOR2	SIX5/19q13.32	AD	Ear anomalies, craniofacial microsomia, deafness, cervical clefts, cervical fistulae, branchial cystsrenal dysplasia, renal agenesis
Branchiootic syndrome 3 BOS3	SIX1/14q23.1	AD	Ear anomalies, pre-auricular fistulae, Mondini dysplasia, widened semicircular canals, branchial cysts
Moebius syndrome	Unknown/13q12.2-q13	AD, sporadic	Facial nerve palsy, mask-like facies, abducens palsy, adduction palsy, epicanthal folds, flat nasal bridge, high arched palate, tongue hypoplasia, small penis and testes, hand anomalies
Camptomelic dysplasia Acamptomelic dysplasia	SOX9/17q24.3	AD	Short limb dwarfism, cleft palate, macrocephaly, hypertelorism, tracheobronchomalacia, hypoplastic scapulae, sex reversal in some males, small thoraces in most
Cerebrocosto-mandibular syndrome	SNRPB	AD	Micrognathia, malar hypoplasia, small bell-shaped thorax, abnormal tracheal cartilage ring, posterior rib gap defects with abnormal rib insertion into vertebrae, scoliosis, microcephaly, heart defects, developmental disabilities
Cerebro-oculo-facial-skeletal syndrome (COFS) Pena–Shokeir syndrome type II	ERCC6/10q11.3	AR	Sloping forehead, long philtrum, cataracts, blepharophimosis prominent nasal root, arthrogryposis, kyphoscoliosis, rocker bottom feet, hirsutism, severe developmental disabilities,

TABLE 50.1 (*Continued*)

Genetic disorder	Gene(s)/chromosome	Inheritance pattern	Additional manifestations
Toriello–Carey syndrome	Unknown	AR	Agenesis of corpus callosum, micrognathia, microcephaly, hearing loss, short palpebral fissures, ptosis, heart defects, ventricular septal defects, hypospadias, dysmorphism, developmental disabilities
22q11.2 deletion syndrome (see Chapter 21)	22q11.2 deletion	AD, sporadic	Heart defects, speech disorders and verbal apraxia, short stature, seizures, renal anomalies, developmental disabilities/learning disabilities, psychiatric disorders (bipolar illness, depression, and schizophrenia), hypothyroidism, hypocalcemia
Oculo-auriculo-vertebral spectrum (Chapter 42)	Unknown. Possible vascular disruption sequence	Sporadic	Facial asymmetry, microtia, conductive hearing loss, cleft lip, cleft palate, heart defects, renal anomalies, vertebral anomalies, eye anomalies
Fetal alcohol syndrome (see Chapter 26)	Teratogenic	Sporadic	Growth retardation, microcephaly, mental retardation, heart defects, limb anomalies, short palpebral fissures, hypoplastic philtrum, thin upper lip, behavior disorders
Fetal hydantoin syndrome (see Chapter 27)	Teratogenic	Sporadic	Short stature, microcephaly, developmental disabilities, short nose, flat philtrum, short palpebral fissures
Fetal vitamin A embryopathy, fetal isotretinoin embryopathy	Teratogenic	Sporadic	Ear anomalies, microtia, conductive hearing loss, cleft lip, cleft palate, heart defects, microcephaly, developmental disabilities, dental anomalies
Diabetic embryopathy (including femoral-facial syndrome)	Teratogenic	Sporadic	Brain anomalies, growth retardation, vertebral anomalies, heart defects, renal anomalies

Abbreviations: AD, autosomal dominant; AR, autosomal recessive; XLR, X-linked recessive

Genetic Testing Robin sequence is seen as an isolated condition in under 40% of cases (Izumi et al. 2012; Basart et al. 2015). In the remaining 60%, RS is associated with underlying genetic or developmental disorders. It is important to identify RS to make the correct diagnosis of underlying disorders since this will have direct impact on management, including the best approach to managing obstructive apnea, respiratory disorders, and feeding problems. Making the correct diagnosis is important for immediate and long-term management, allowing for diagnosis and management of other medical risk factors and for family genetic counseling regarding the natural history of the condition and its recurrence risks.

The advancement in genetic testing technologies has greatly enhanced our ability to make appropriate diagnoses. Of course, diagnosis always should begin with comprehensive clinical history and physical examination; but many patients will present with rare or unique syndromes. With the development of single nucleotide polymorphism (SNP) chromosomal microarray analysis, the identification of most chromosomal disorders has become possible. Chromosomal anomalies are the largest diagnostic category for RS, with most affected individuals having uncommon or rare chromosomal disorders. Both common and rare chromosomal microdeletion and microduplication syndromes have been diagnosed. The most common chromosomal microdeletion disorder is deletion 22q11.2 syndrome, diagnosed with either fluorescence in situ hybridization (FISH) or SNP chromosomal microarray. However, in RS 22q11.2 deletion syndrome is observed in only 3% of patients in one study (Izumi et al. 2012) and 1.2% in another retrospective investigation (Basart et al. 2015). The remainder of chromosomal anomalies are rare chromosomal rearrangements. Two studies have shown that microdeletions of 17q24.3 upstream of the *SOX9* gene disrupt regulatory elements, which can cause familial micrognathia with mild pectus deformities (Amarillo et al. 2013; Gordon et al. 2014).

Next generation sequencing (NGS) has allowed for rapid and reliable diagnosis of uncommon and rare single gene disorders that are associated with Robin sequence NGS has streamlined genetic testing for several categories of genetically heterogeneous disorders. The most common single gene disorder associated with RS is Stickler syndrome. Pathogenic variants in at least seven genes have been identified as causing the Stickler syndrome phenotype (see Chapter 56) and these can be found on commercially available gene panels. Similarly, gene panels are available for genetic confirmation of Treacher Collins syndrome, another common condition associated with RS.

MANIFESTATIONS AND MANAGEMENT

Prenatal Management

Many cases of Robin sequence are being diagnosed prenatally thanks to high-resolution ultrasound and fetal MRI (Figure 50.5) Prenatal identification of micrognathia in conjunction with other anomalies often leads to additional testing, including amniocentesis for SNP microarray, single gene testing, or whole exome sequencing. Having a genetic diagnosis will often help with perinatal management, especially if a lethal disorder is diagnosed whereby palliative care is offered. Delivery of infants with micrognathia should be performed in an institution with a level 3 neonatal intensive care unit. A neonatologist should be present in the delivery room to administer airway support if deemed necessary.

In circumstances when a fetus is suspected of having severe micrognathia with a critical airway, the delivery may entail ex utero intrapartum treatment (EXIT) procedure or procedure on placental support (POPS) with presence of an otolaryngologist as well as a neonatologist in the delivery room. In both procedures, the fetus remains on placental support while an airway is established, either by intubation or tracheostomy.

Growth and Feeding

Children with cleft palate have been shown to be at high risk for poor weight gain and growth because of significant feeding difficulties (Masarei et al. 2007; Cooper-Brown et al. 2008). This is because the inability to develop adequate negative pressure while sucking makes breast-feeding extremely difficult and bottle feeding very challenging (Cooper-Brown et al. 2008). In addition, infants with cleft palate swallow a great deal of air when feeding, which can result in reflux, choking, fatigue, and prolonged feeding times (Cooper-Brown et al. 2008). In the child with RS, feeding difficulties related to the cleft palate are compounded by the small mandible, which interferes with creating a seal around the nipple and, most significantly, glossoptosis and respiratory distress, which interfere with feeding because the infant is unable to coordinate sucking, swallowing, and breathing. In the latter scenario, there is also increased risk for aspiration of formula. These factors impact caloric intake, energy expenditure during feeding, and ultimately weight gain. Feeding and weight gain are further compromised in children with RS as a consequence of a syndrome associated with other anomalies, such as deletion 22q11.2 syndrome, in which congenital heart defects, neurological compromise, or oromotor apraxia influence feeding (Hopkin et al. 2000; Smith and Senders 2006). Furthermore, many syndromes, such as chromosome disorders and fetal alcohol syndrome, are associated with poor weight gain and growth independent of feeding issues.

Although most neonates will take about one week to regain birth weight, children with isolated cleft palate will return to birth weight at about 10–14 days of age. This may be even more prolonged in children with syndromes that further affect feeding or that are associated with poor growth. For this reason, if a child with RS is having feeding and/or growth issues that are beyond what is expected, a genetics consultation can be a helpful adjunct to establish an underlying diagnosis that influences appropriate management.

Growth and feeding in older children who are born with RS are usually related more to the underlying disorder that caused the RS. In older children with isolated RS, there should be no specific problems related to feeding and growth.

Evaluation

- Feeding and caloric intake should be monitored carefully, including the volume of breast milk or formula and the time needed to take a normal feed. Feeding should not exceed 30 min.
- Urine and stool output should be monitored carefully to assess fluid intake.
- After discharge from the hospital, it is helpful to have parents keep a feeding diary documenting feeding schedule and intake.
- Growth should be monitored carefully. This should include documenting initial weight loss and weight gain and may include frequent weight checks of the

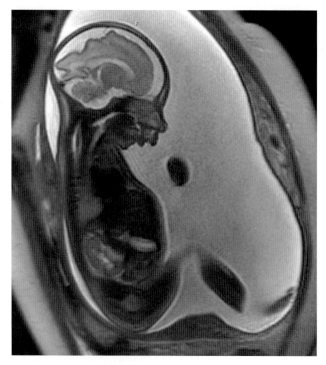

FIGURE 50.5 Fetal MRI at 30 weeks gestation demonstrating micrognathia and polyhydramnios.

infant after hospital discharge. Initially, weekly appointments will be necessary to ascertain how well the infant is feeding and gaining weight. If there is suspected failure-to-thrive, there needs to be close monitoring of feeding to ensure that the parents are using the adapted bottles appropriately. The other factor that is a frequent cause of poor weight gain is glossoptosis and obstructive apnea.

- Children with RS and significant feeding difficulty, especially those who require gastrostomy to assure adequate calories, may also have gastroesophageal reflux, which interferes with calorie retention and may cause aspiration. If reflux is suspected, evaluation should include video swallow study and possibly endoscopy and pH probe study.
- If feeding problems are identified, oxygen saturation during feeding should be monitored. Decreased oxygen saturation levels during feeding can be a sign of glossoptosis and obstructive apnea.
- Children with suspected obstructive apnea (most children with RS) should undergo a polysomnogram (sleep apnea study).
- If there is oral motor hypotonia, oral motor apraxia, or other neurological issues suspected which affect feeding, evaluation by an occupational therapist or a speech therapist who specializes in feeding disorders should be the next step. Such individuals are extremely helpful in instructing parents on the proper use of the adapted flexible bottles.

Management

- Feeding instruction should begin as soon as the infant is stable enough to be fed.
- Although use of breast milk should be encouraged, breast-feeding is almost never successful in children with cleft palate (Cooper-Brown et al. 2008).
- Use of bottles or nipples that are specialized for feeding children with cleft palate is advisable. These are flexible or high flow bottles that allow parents to squeeze the breast milk or formula into the infant's mouth during their sucking motion. The most commonly used bottles and nipples are the Haberman Feeder, the Mead Johnson Cleft Palate Nurser, and the Dr. Brown Specialty Feeding System. These specific bottles are available in the United States, and the Haberman Feeder is available in the United Kingdom and can be ordered online.
- If caloric intake is inadequate to maintain normal growth, caloric concentration can be increased by adding less water to concentrated liquid formula or adding powdered formula to breast milk or formula. It is recommended that this be done with the assistance of a nutritionist, nurse, or physician who is familiar with infant nutritional management.
- In some institutions, an obturator is used to assist with feeding (Glenny et al. 2004; Muller-Hagedorn et al. 2017). An obturator is a prosthetic palate that is inserted to close the gap caused by the cleft and in some instances force the tongue forward to open the pharynx. Even when an obturator is used, breast-feeding is still not likely to be successful, and a special bottle will likely still be needed.
- Some infants are unable to maintain their caloric needs orally. For some of these children, an in-dwelling nasogastric tube can be beneficial. Oral feeding should continue and be encouraged with the remainder of the formula or breast milk being given by nasogastric feeding. Parents can be instructed on how to do this at home.
- Some infants with RS will be unable to take oral nutrition. This is usually the case in infants with underlying disorders that interfere with normal sucking and swallowing. A gastrostomy tube may be needed in these children to overcome feeding issues. In some instances, an infant or child is unable to feed orally without aspirating formula and a gastrostomy is required for airway protection.
- For gastroesophageal reflux, treatment with acid-blocking medications, such as histamine-2 blockers or proton pump inhibitors.
- Some children with gastroesophageal reflux may need fundoplication.

Development and Behavior

Development and behavior in RS are primarily related to the underlying associated disorder. Examples include fetal alcohol syndrome (Chapter 26), 22q11.2 deletion syndrome (chapter 21), Smith–Lemli–Opitz syndrome (Chapter 53), and multiple other syndromes and disorders that have RS as one of its manifestations (see Table 50.1). These conditions have associated developmental and behavior issues are directly related to the syndrome rather than being RS specific. For children with isolated RS, developmental concerns are usually related to difficulties of airway management, especially if there were issues related to airway difficulties and hypoxia, cleft palate management and related middle ear effusion, recurrent ear infections and conductive hearing loss, and speech delay and articulation disorders related to cleft palate. Many syndromes with RS may have both conductive and sensorineural hearing loss, which often lead to speech and language delays.

Evaluation

- Hearing should be monitored regularly. Ensure normal newborn hearing screens and obtain audiograms at least

every 6–12 months in early childhood in children with cleft palate.
- Referral to an otolaryngologist is essential for every child with cleft palate.
- Referral for speech evaluation, usually at about age 2 years. Most children with cleft palate are at risk for expressive speech delays and articulation disorders.
- Developmental evaluations, using standard screening tests, should be carried out during routine medical visits.
- Unusual or atypical behavior problems should be addressed with each visit with referral to a behavioral pediatrician or psychologist, as indicated.

Management

- Pressure equalizing tubes are indicated for children with persistent middle ear effusion, as in any child with a cleft palate. Most children with cleft palate will have the tubes inserted at the time of palate surgery.
- Hearing aids can be helpful for children with persistent hearing loss from chronic ear infections or other hearing defects.
- Speech therapy is essential for all children with documented speech delays, articulation defects, and language delays. Weekly or bi-weekly individualized therapy is optimal.
- Children with developmental delays, especially speech and language delays, should be referred for formal developmental assessments to a psychologist or developmental pediatrician.
- Individualized educational programs are appropriate for those with an underlying syndrome associated with cognitive delays as well as children with isolated RS who have significant learning issues.
- Referral to a behavioral pediatrician or psychologist for those children with psychological or emotional issues related to speech and communication disorders or with a chronic condition, such as cleft palate.

Respiratory

The most critical problem for an infant with RS is obstructive apnea. This may present immediately in the delivery room or may have an initially indolent course, presenting after discharge from the hospital. In any child with micrognathia and glossoptosis, obstructive apnea must be considered as potentially life threatening. Severe compromise in the perinatal period should be addressed with immediate intervention, usually intubation with assisted ventilation. For those infants with less severe compromise, supine or side positioning may relieve the immediate glossoptosis. Although supine positioning is recommended for all infants to reduce the risk for sudden infant death syndrome (SIDS), infants with RS who are placed on their side or prone are constantly monitored for respirations and heart rate to ensure adequate respiration. Occasionally, an oral airway may be needed for a short time until a more definitive procedure can be offered. Those infants with serious airway compromise, either because of structural anomalies or neurological involvement, may require a tracheostomy to ensure a safe airway (Tomaski et al. 1995). Fortunately, most infants with RS can be managed conservatively. Several airway management options exist and most depend on the decision-making process and the experience of specialists caring for these children at the managing institution. Children who are stable, have no evidence of significant oxygen desaturation, and are feeding and growing well usually need minimal intervention, such as oxygen supplementation by nasal cannula. Infants who can only breathe comfortably in prone or side-lying position, have recurrent apneas or oxygen desaturation, or cannot feed without oxygen desaturation, have obstructive apnea serious enough to require further evaluation and intervention. There should be a low threshold for evaluation of a child with micrognathia and suspected RS for obstructive apnea. At Cincinnati Children's Hospital Medical Center, a multidisciplinary team of specialists to evaluate infants with RS was established. This group of specialists functions together to gather all important data and information to develop a consensus management plan. The airway management team consists of neonatology, medical genetics, otolaryngology, plastic surgery, pulmonary medicine, speech pathology, and occupational therapy.

Evaluation

- Respiratory effort should be monitored closely, including oxygen saturation levels in newborns with suspected apnea.
- Breathing and oxygen saturation should be monitored during feeding.
- In children with suspected sleep apnea, otolaryngology evaluation is needed for endoscopic laryngobronchoscopy to assess the airway. Children with RS are often at risk for other airway issues, including larygnomalacia and tracheomalacia. Laryngobronchoscopy can also be helpful in assessing the positioning of the tongue base with regard to the glossoptosis and it is essential if a tracheostomy is needed.
- All children with suspected obstructive apnea should have a comprehensive polysomnogram. This test will assess the number and duration and severity of apneic episodes and oxygen desaturations during awake and sleep cycles, in different positions, and during feeding.
- A cine MRI is a dynamic study that can be helpful for identifying the pharyngeal anatomy and determining if there is glossoptosis during sleep.

- A three dimensional CT (3D-CT) scan is needed to assess craniofacial landmarks if surgical intervention is planned, specifically mandibular distraction osteogenesis (Figure 50.4).

Management

- If apnea is severe at birth, intubation is required and tracheostomy should be considered if infant cannot be safely extubated.
- If the polysomnogram is normal and the infant is feeding and growing well, close monitoring of feeding and growth should continue.
- Some infants with RS respond adequately to prone positioning without further respiratory intervention (Smith and Senders, 2006; Meyer et al. 2008). Feeding is usually done while side-lying or with nasogastric tube, as indicated.
- If the polysomnogram demonstrates minimal apneas with oxygen desaturation (usually during sleep), treatment with supplemental oxygen by nasal cannula is often successful.
- For significantly abnormal polysomnogram, in some centers the glossoptosis and apnea are treated with nasopharyngeal tube or cannula, which bypasses the pharyngeal obstruction (Tomaski et al. 1995; Meyer et al 2008; Albino et al. 2016).
- Some institutions have been successful in managing obstructive apnea with an oral obturator with a posterior extension that forces the tongue forward and expands the pharyngeal airway (Muller-Hagedorn et al. 2017).
- Mandibular distraction osteogenesis is a surgical procedure that has gained favor in most craniofacial centers for management of micrognathia and glossoptosis (see Craniofacial section) (Resnick et al. 2019). This procedure elongates the mandible, allowing the tongue to move forward, relieving the glossoptosis (see Figure 50.3).
- Tongue–lip adhesion is utilized in some institutions to move the tongue forward and open the airway (Kirschner et al. 2003). This procedure is used much less frequently since the development of mandibular distraction osteogenesis.

Craniofacial

Micrognathia A clinical hallmark and, in most cases, the initial clinical sign of RS is the presence of micrognathia. Micrognathia is often a subjective diagnosis. No single validated criterion exists for diagnosing and characterizing micrognathia (Breugem et al. 2016). Therefore, clinical evaluation and experience are important factors, especially in the presence respiratory distress and/or feeding difficulties. Several studies have confirmed that the mandible in children with RS is shorter in both the mean mandibular length (body of the mandible) and posterior height (mandibular ramus) than in unaffected infants or infants born with isolated incomplete cleft lip with cleft palate (Hermann et al. 2003; Eriksen et al. 2006). In many cases, if no significant surgical intervention is needed in infancy, the mandible will grow and reach normal size. However, in many genetic disorders, if micrognathia is a primary clinical manifestation, the mandible may not grow normally, and surgical interventions will be needed for both optimal functional and cosmetic outcomes. In addition, long-term follow-up is indicated, preferably in a cleft and craniofacial center, because comprehensive management should include dental, orthodontic, and maxillofacial management in childhood and adolescence.

Evaluation

- With a small mandible (micrognathia), the infant should be evaluated clinically for associated respiratory and feeding disorders. A careful examination for cleft palate or submucous cleft palate is required.
- Family history should be obtained and examination of first-degree relative may be relevant. Many children with micrognathia will have a family history of micrognathia or syndromes associated with micrognathia, such as Stickler syndrome.
- Imaging of the mandible is important if surgical intervention is to be performed, such as mandibular distraction osteogenesis. The preferred imaging procedure is 3D-CT scan (see Figure 50.4).
- Long-term growth of the mandible should be monitored as should dental development. This is optimally done in a cleft and craniofacial center.

Management

- For a child with micrognathia and obstructive apnea secondary to glossoptosis, the preferred intervention is mandibular distraction osteogenesis (Fan et al. 2018; Resnick et al. 2019) This should be done in a center with expertise in performing this procedure and in its postoperative management. This procedure involves performing a surgical fracture (osteotomy) on both sides of the vertical ramus of the mandible and inserting a device (distractor) on either side of the fracture, which gradually separates the edges of the fracture site. Both internal and external distractors are used. A bone callus forms and new bone fills in the fracture as the mandible is lengthened. Usually the entire distraction procedure takes approximately two weeks to complete (Denny and Amm 2005; Dauria and Marsh 2008) (Figures 50.3 and 50.6A,B).

- Consistent monitoring of growth and feeding is indicated for all children undergoing mandibular distraction osteogenesis. Some infants who undergo mandibular distraction osteogenesis will have a relapse of obstructive apnea later in infancy because the mandible does not continue to grow adequately, often related to the child having an underlying syndrome that is associated with poor mandibular growth.
- Routine evaluation by a pediatric dentist and orthodontist is recommended in later childhood. If the mandible remains small, there is a high likelihood of dental crowding

 (in particular, for the mandibular dentition), significant dental overjet, and malocclusion.

 These children need to be monitored regularly by a pediatric dentist and many will require orthodontic management (Matsuda et al. 2006).
- Some children will require maxillofacial surgery in later childhood or adolescence to elongate the mandible. Growth of the mandible into adulthood is variable. As is the case of many individuals born with orofacial clefts, further maxillofacial surgery may be needed when full growth is achieved, including both maxillary and mandibular osteotomies.

Cleft Palate Most children with RS will have a cleft palate. Management of cleft palate, even in the absence of glossoptosis and apnea, is a complex and involved process that requires the expertise of individuals from a wide variety of specialties. A list of the common specialties found on craniofacial teams is provided in Table 50.2. For this reason, children with a cleft palate should be followed by a cleft palate or craniofacial team. The expertise involved in managing the timing and coordination of clinical care of the cleft as well as early identification and management of potential complications is well beyond what can be done in most primary care settings.

In addition to feeding problems encountered with cleft palate, speech disorders are common. Early in childhood, many children will have speech delays and articulation defects. Later problems are often related to difficulties with articulation and hypernasality.

Evaluation

- Clinical examination for cleft palate or submucous cleft palate for all children with RS.
- Speech and language evaluation by a speech pathologist familiar with cleft palate and speech resonance disorders.
- If there is hypernasal speech, the child should be referred to a center with specialists, including speech pathologists, plastic surgeons, and/or otolaryngologists who are experienced with evaluating and treating cleft speech disorders.

Management

- Referral to a cleft palate or craniofacial team is recommended. The American Cleft Palate-Craniofacial Association, a professional organization with international membership encompassing the entire range of

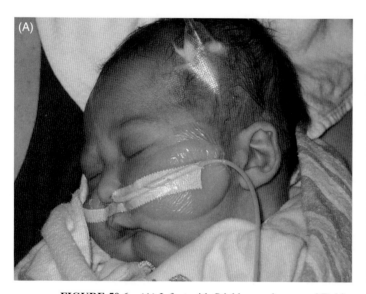
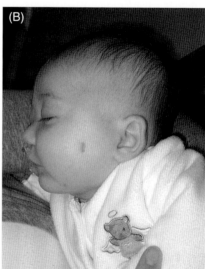

FIGURE 50.6 (A) Infant with Stickler syndrome and Robin sequence before mandibular distraction osteogenesis. (B) Infant with Stickler syndrome and Robin sequence status post mandibular distraction osteogenesis.

TABLE 50.2 Specialties contributing to a cleft palate and craniofacial team

Audiology
Medical genetics
Maxillofacial surgery
Neurosurgery
Nursing
Orthodontics
Otolaryngology
Pediatric dentistry
Pediatrics
Plastic surgery
Prosthodontia
Psychology
Social work
Speech pathology
Team coordinator

management of children with oral facial clefts and other craniofacial conditions has developed standards for the management of children with these disorders with funding from the US Maternal and Child Health Bureau. They concluded "management of patients with craniofacial anomalies is best provided by and interdisciplinary team of specialists" (ACPA 2018).

- Feeding management should follow the same protocols as outlined in Growth and Feeding.
- Referral to a speech pathologist for speech and language evaluation and speech therapy as needed at approximately 2 years of age.
- True hypernasal speech caused by velopharyngeal dysfunction cannot be corrected with speech therapy. Individuals with this problem usually require surgical management of the palate by one of the many techniques. These children are usually managed in a cleft palate center or by health care professionals (speech therapists, plastic surgeons, and/or otolaryngologists), who are familiar with speech articulation disorders related to cleft palate repair. Options for management of hypernasal speech are related to several factors, including the stability and size of the airway, previous surgeries, and results of diagnostic evaluations of speech. The most commonly used procedure is the pharyngeal flap, which acts as a passive tissue obturator (Kummer 2020). A second procedure is called sphincter pharyngoplasty. The most significant complication from these procedures is obstructive sleep apnea.

Otolaryngology

See the sections on Development and Behavior and Craniofacial above.

RESOURCES

SUPPORT GROUPS

American Cleft Palate-Craniofacial Association

1504 East Franklin Street, Suite 102
Chapel Hill, NC 27514-2820, USA
Phone: (919) 933-9044
Fax: (919) 933-9604
Email: *info@cleftline.org*
Website: *https://acpa-cpf.org/*

Children's Craniofacial Association

13140 Coit Road
Suite 517
Dallas, TX 75240, USA
Phone: (214) 570-9099
Toll-free: (800) 535-3643
Email: contactCCA@ccakids.com
Website: http://www.ccakids.com

Birth Defect Research for Children, Inc.

976 Lake Baldwin Lane
Orlando, FL 32814, USA
Phone: (407) 895-0802
Email: staff@birthdefects.org
Website: http://www.birthdefects.org

Cleft Lip and Palate Foundation of Smiles

2044 Michael Ave SW
Wyoming, MI 49509, USA
Phone: (616) 329-1335
Email: Rachel@cleftsmile.org
Website: http://www.cleftsmile.org

FACES: The National Craniofacial Association

PO Box 11082
Chattanooga, TN 37401, USA
Phone: (423) 266-1632
Toll-free: (800) 332-2373
Email: faces@faces-cranio.org
Website: http://www.faces-cranio.org

Genetic and Rare Diseases (GARD) Information Center

PO Box 8126
Gaithersburg, MD 20898-8126, USA
Phone: (301) 251-4925
Toll-free: (888) 205-2311
Website: http://rarediseases.info.nih.gov/GARD/

Let Them Hear Foundation

1900 University Avenue, Suite 10
East Palo Alto, CA 94303, USA
Phone: (650) 462-3174
Email: info@letthemhear.org
Website: http://www.letthemhear.org

Cleft Lip & Palate Association (United Kingdom)
CLAPA, Green Man Tower
332B Goswell Road
London, EC1V 7LQ, UK
Phone: 020 7833 4883
Fax: 020 7833 5999

WRITTEN RESOURCES

Kummer AW (2020) *Cleft Palate and Craniofacial Anomalies: A Comprehensive Guide to Clinical Management*, Kummer AW, ed. Burlington, Massachusetts, Jones & Bartlett Learning.

This is a comprehensive resource for management of all aspects of cleft lip, cleft palate and craniofacial conditions. There is an online version available with multimedia presentations, including multiple examples of speech and language disorders.

Breugem CC, Evans KN, Poets CF, Suri S, Picard A, Filip C, Paes EC, Mehendale FV, Saal HM, Basart H, Murthy J, Joosten KF, Speleman L, Collares MV, van den Boogaard MJ, Muradin M, Andersson ME, Kogo M, Farlie PG, Don Griot P, Mossey PA, Slator R, Abadie V, Hong P (2016) Best practices for the diagnosis and evaluation of infants with Robin sequence: A clinical consensus report. *JAMA Pediatr.* 170(9):894–902.

This article covers the results of the first international Pierre Robin sequence comprehensive diagnosis and evaluation conference held in Utrecht, the Netherlands in 2014.

REFERENCES

Albino FP, Wood BC, Han KD, Yi S, Seruya M, Rogers GF, Oh AK (2016) Clinical factors associated with the non-operative airway management of patients with Robin sequence. *Arch Plast Surg* 43(6):506–511.

Allanson JE, Cunniff C, Hoyme HE, McGaughran J, Muenke M, Neri G (2009) Elements of morphology: standard terminology for the head and face. *Am J Med Genet A* 149A(1):6–28.

Amarillo IE, Dipple KM, Quintero-Rivera F (2013_ Familial microdeletion of 17q24.3 upstream of SOX9 is associated with isolated Pierre Robin sequence due to position effect. *Am J Med Genet A* 161A(5):1167–72

American Cleft Palate-Craniofacial Association (2017) Parameters for evaluation and treatment of patients with cleft lip/palate or other craniofacial differences. *Cleft Palate-Craniofacial J* 5(1):137–156.

Basart H, Paes EC, Maas SM, van den Boogaard MJ, van Hagen JM, Breugem CC, Cobben JM, Don Griot JP, Lachmeijer AM, Lichtenbelt KD, van Nunen DP, van der Horst CM, Hennekam RC (2015) Etiology and pathogenesis of robin sequence in a large Dutch cohort. *Am J Med Genet A* 167A(9):1983–92.

Breugem CC, Evans KN, Poets CF, Suri S, Picard A, Filip C, Paes EC, Mehendale FV, Saal HM, Basart H, Murthy J, Joosten KF, Speleman L, Collares MV, van den Boogaard MJ, Muradin M, Andersson ME, Kogo M, Farlie PG, Don Griot P, Mossey PA, Slator R, Abadie V, Hong P (2016) Best practices for the diagnosis and evaluation of infants with Robin sequence: A clinical consensus report. *JAMA Pediatr* 170(9):894–902.

Breugem CC, Mink van der Molen AB (2008) What is 'Pierre Robin sequence'? *J Plast Reconstr Aesthet Surg* 62(12):1555–8.

Carey JC, Cohen MM Jr, Curry CJ, Devriendt K, Holmes LB, Verloes A (2009) Elements of morphology: standard terminology for the lips, mouth, and oral region. *Am J Med Genet A* 149A(1):77–92.

Cooper-Brown L, Copeland S, Dailey S, Downey D, Petersen MC, Stimson C, Van Dyke DC (2008) Feeding and swallowing dysfunction in genetic syndromes. *Dev Disabil Res Rev* 14:147–157.

Costa MA, Tu MM, Murage KP, Tholpady SS, Engle WA, Flores R (2014) Robin sequence: mortality, causes of death, and clinical outcomes. *Plast Reconstr Surg* 134(4):738–45.

Dauria D, Marsh JL (2008) Mandibular distraction osteogenesis for Pierre Robin sequence: What percentage of neonates need it? *J Craniofac Surg* 19:1237–1243.

Denny A, Amm C (2005) New technique for airway correction in neonates with severe Pierre Robin sequence. *J Pediatr* 147:97–101.

Di Pasquo E, Amiel J, Roth P, Malan V, Lind K, Chalouhi C, Soupre V, Gordon CT, Lyonnet S, Salomon LJ, Abadie V (2017) Efficiency of prenatal diagnosis in Pierre Robin sequence. *Prenat Diagn* 37(11):1169–1175.

Eriksen J, Hermann NV, Darvann TA, Kreiborg S (2006) Early postnatal development of the mandible in children with isolated cleft palate and children with nonsyndromic Robin sequence. *Cleft Palate Craniofac J* 43:160–167.

Fairbairn P (1846) Suffocation in an infant from retraction of the base of the tongue, connected with defect of the frenum. *Month J Med Sci* 6:280–1.

Fan KL, Mandelbaum M, Buro J, Rokni A, Rogers GF, Chao JW, Oh AK (2018) Current Trends in Surgical Airway Management of Neonates with Robin Sequence. *Plast Reconstr Surg Glob Open* 6(11):e1973

Glenny AM, Hooper L, Shaw WC, Reilly S, Kasem S, Reid J (2004) Feeding interventions for growth and development in infants with cleft lip, cleft palate or cleft lip and palate. *Cochrane Database Syst Rev* (3) CD003315.

Gordon CT, Attanasio C, Bhatia S, Benko S, Ansari M, Tan TY, Munnich A, Pennacchio LA, Abadie V, Temple IK, Goldenberg A, van Heyningen V, Amiel J, FitzPatrick D, Kleinjan DA, Visel A, Lyonnet S (2014) Identification of novel craniofacial regulatory domains located far upstream of SOX9 and disrupted in Pierre Robin sequence. *Hum Mutat* 35(8):1011–20.

Hennekam RCM, Biesecker LG, Allanson JE, Hall JG, Opitz JM, Temple IK, Carey JC (2011) Elements of Morphology: general terms for congenital anomalies. *Am J Med Genet Part A* 161A:2726–2733.

Hermann NV, Kreiborg S, Darvann TA, Jensen BL, Dahl E, Bolund S (2003) Craniofacial morphology and growth comparisons in children with Robin Sequence, isolated cleft palate, and unilateral complete cleft lip and palate. *Cleft Palate Craniofac J* 40:373–396.

Holder-Espinasse M, Abadie V, Cormier-Daire V, Beyler C, Manach Y, Munnich A, Lyonnet S, Couly G, Amiel J (2001) Pierre Robin sequence: A series of 117 consecutive cases. *J Pediatr* 139:588–590.

Hopkin RJ, Schorry EK, Bofinger M, Saal HM (2000) Increased need for medical interventions in infants with velocardiofacial (deletion 22q11) syndrome. *J Pediatr* 137:247–249.

Izumi K, Konczal LL, Mitchell AL, Jones MC (2012) Underlying genetic diagnosis of Pierre Robin sequence: retrospective chart review at two children's hospitals and a systematic literature review. *J Pediatr* 160(4):645–650.e2.

Katsanis SH, Jabs EW (2018) Treacher Collins syndrome. *GeneReviews*®. Available at: https://www.ncbi.nlm.nih.gov/books/NBK1532/

Kirschner RE, Low DW, Randall P, Bartlett SP, McDonald-McGinn DM, Schultz PJ, Zackai EH, LaRossa D (2003) Surgical airway management in Pierre Robin sequence: Is there a role for tongue-lip adhesion? *Cleft Palate Craniofac J* 40: 13–18.

Kummer AW (2020) *Cleft Palate and Craniofacial Anomalies: A Comprehensive Guide to Clinical Management*, Kummer AW, ed. Burlington, Massachusetts: Jones & Bartlett Learning, pp 463–467.

Lines MA, Huang L, Schwartzentruber J, Douglas SL, Lynch DC, Beaulieu C, Guion-Almeida ML, Zechi-Ceide RM, Gener B, Gillessen-Kaesbach G, Nava C, Baujat G, Horn D, Kini U, Caliebe A, Alanay Y, Utine GE, Lev D, Kohlhase J, Grix AW, Lohmann DR, Hehr U, Böhm D; FORGE Canada Consortium, Majewski J, Bulman DE, Wieczorek D, Boycott KM (2012) Haploinsufficiency of a spliceosomal GTPase encoded by EFTUD2 causes mandibulofacial dysostosis with microcephaly. *Am J Hum Genet* 90(2):369–77.

Logjes RJH, Haasnoot M, Lemmers PMA, Nicol aije MFA, van den Boogaard MH, Mink van der Molen AB, Breugem CC (2018) Mortality in Robin sequence: identification of risk factors. *Eur J Pediatr* 177(5):781–789

Maas C, Poets CF (2014) Initial treatment and early weight gain of children with Robin Sequence in Germany: a prospective epidemiological study. *Arch Dis Child Fetal Neonatal Ed* 99(6):F491–4.

Masarei AG, Sell D, Habel A, Mars M, Sommerlad BC, Wade A (2007) The nature of feeding in infants with unrepaired cleft lip and/or palate compared with healthy noncleft infants. *Cleft Palate Craniofac J* 44:321–328.

Matsuda A, Suda N, Motohashi N, Tsuji M, Ohyama K (2006) Skeletal characteristics and treatment outcome of five patients with Robin sequence. *Angle Orthod* 76:898–908.

Meyer AC, Lidsky ME, Sampson DE, Lander TA, Liu M, Sidman JD (2008) Airway interventions in children with Pierre Robin Sequence. *Otolaryngol Head Neck Surg* 138:782–787.

Müller-Hagedorn S, Buchenau W, Arand J, Bacher M, Poets CF (2017) Treatment of infants with Syndromic Robin sequence with modified palatal plates: a minimally invasive treatment option. *Head Face Med* 13(1):4.

Online Mendelian Inheritance in Man. https://www.omim.org/search/?index=entry&start=1&limit=10&sort=score+desc%2C+prefix_sort+desc&search=Pierre+robin. Accessed 2019.

Paes EC, van Nunen DP, Basart H, Don Griot JP, van Hagen JM, van der Horst CM, van den Boogaard MJ, Breugem CC (2015) *Am J Med Genet A* 167A(9):1972–82.

Printzlau A, Andersen M (2004) Pierre Robin sequence in Denmark: a retrospective population-based epidemiological study. *Cleft Palate Craniofac J* 41(1):47–52.

Resnick CM, Calabrese CE, Sahdev R, Padwa BL (2019) Is tongue-lip adhesion or mandibular distraction more effective in relieving obstructive apnea in infants with Robin sequence? *J Oral Maxillofac Surg* 77(3):591–600.

Resnick CM, Estroff JA, Kooiman TD, Calabrese CE, Koudstaal MJ, Padwa B (2018) Pathogenesis of cleft palate in Robin sequence: Observations from prenatal magnetic resonance imaging. *J Oral Maxillofac Surg* 76(5):1058–1064

Retterer K, Juusola J, Cho MT, Vitazka P, Millan F, Gibellini F, Vertino-Bell A, Smaoui N, Neidich J, Monaghan KG, McKnight D, Bai R, Suchy S, Friedman B, Tahiliani J, Pineda-Alvarez D, Richard G, Brandt T, Haverfield E, Chung WK, Bale S (2016) Clinical application of whole-exome sequencing across clinical indications. *Genet Med* 18(7):696–704.

Robin NH, Moran RT, Ala-Kokko L (2017) Stickler syndrome. *GeneReviews*®. Available at: https://www.ncbi.nlm.nih.gov/books/NBK1302/

Robin P (1923) La chute de la base de la langue considerée comme une novelle cause de gene dans la respiration nasopharyngienne. *Bull Acad Natl Med (Paris)* 89:37–41.

Robin P (1934) Glossoptosis due to atresia and hypotrophy of the mandible. *Am J Dis Child* 48:541–547.

Saint-Hillaire G (1822) *Philosophie Anatomique: Monstrousité Humaine*. Paris de L'Imprimerie de Rignoux.

Smith MC, Senders CW (2006) Prognosis of airway obstruction and feeding difficulty in the Robin sequence. *Int J Pediatr Otorhinolaryngol* 70:319–324.

Spranger J, Benirschke K, Hall JG, Lenz W, Lowry RB, Opitz JM, Pinsky L, Schwarzacher HG, Smith DW (1982) Errors of morphogenesis: Concepts and terms. Recommendations of an international working group. *J Pediatr* 100:160–165.

Tomaski SM, Zalzal GH, Saal HM (1995) Airway obstruction in the Pierre Robin sequence. *Laryngoscope* 105:111–114.

van Nunen DPF, van den Boogaard MH, Breugem CC (2018) Robin sequence: continuing heterogeneity in nomenclature and diagnosis. *J Craniofac Surg* 29(4):985–987.

Wieczorek D (2013) Human facial dysostoses. *Clin Genet* 83(6):499–510.

Wright M, Mehendale F, Urquhart DS (2018) Epidemiology of Robin sequence with cleft palate in the East of Scotland between 2004 and 2013. *Pediatr Pulmonol* 53(8):1040–1045.

51

RUBINSTEIN–TAYBI SYNDROME

LEONIE A. MENKE AND RAOUL C. M. HENNEKAM

Department of Pediatrics, Academic Medical Center, University of Amsterdam, Amsterdam, The Netherlands

INTRODUCTION

Rubinstein–Taybi syndrome is a multiple congenital anomaly syndrome that is mainly characterized by a particular face, broad thumbs, broad big toes, and intellectual disability. It was first described in 1957 by three Greek orthopedic surgeons in a French orthopedic journal as "a new case of congenital malformations of the thumbs absolutely symmetrical" (Michail et al. 1957). In the same year, Jack Rubinstein, a pediatrician from Cincinnati, investigated a girl with similar findings. Together with Hooshang Taybi, a pediatric radiologist from Oklahoma, he was able to collect six other patients, which were published in 1963 (Rubinstein et al. 1963). The syndrome was definitively named Rubinstein–Taybi syndrome by Warkany (1974).

Rubinstein–Taybi syndrome (RSTS) can be considered one of the archetypal syndromes in clinical genetics. It was first recognized clinically, followed by numerous reported individuals that delineated the full clinical spectrum. Subsequently, behavioral characteristics and a potentially increased cancer risk were recognized. Then, discovery of the gene localization occurred through a small number of affected individuals with a chromosome anomaly and subsequently cloning of the gene through advanced molecular work. Animal models were built and functional studies led to the discovery of a second gene involved. Clinicians and basic scientists continued to cooperate to explain the phenotype by studying different gene functions, and interest in the natural history increased. Recently, exome sequencing demonstrated that variants in *CREBBP* and *EP300* may lead to another phenotype as well (Menke et al. 2016).

Incidence

In the Netherlands, a long running register has sought to locate all affected individuals nationwide. Through this registry, the birth prevalence was found to be 1/100,000–1/125,000 in the 1980s (Hennekam et al. 1990b), and this has proven to be correct for the period 1988–2017 as well. RSTS occurs in both males and females with equal frequency, and has been described in populations of many different ancestries. However, the number of reports on non-Caucasian patients is low. This probably represents a socioeconomic or a publication bias, or (less likely) a true lower incidence.

Diagnostic Criteria

There are no defined diagnostic criteria for RSTS, but the cardinal features are well delineated. These include specific facial features, broad and angulated thumbs and big toes, growth retardation, intellectual disability, and behavioral characteristics (Hennekam et al. 1990c; Rubinstein 1990; Stevens et al. 1990a).

The facial appearance of a child with RSTS is remarkable. This is, in part, because of the dysmorphic features: microcephaly, prominent forehead, down-slanted palpebral fissures, broad nasal bridge, convex nasal ridge with a low hanging columella, apparently highly arched palate, everted lower lip, mild micrognathia, and minor anomalies in shape, position, and/or rotation of the ears (Figure 51.1). The ears can be simple and small and pits may be present on the posterior side of the helices. Of equal importance, however, is the facial expression: the grimacing smile with nearly

---

Cassidy and Allanson's Management of Genetic Syndromes, Fourth Edition.
Edited by John C. Carey, Agatino Battaglia, David Viskochil, and Suzanne B. Cassidy.
© 2021 John Wiley & Sons, Inc. Published 2021 by John Wiley & Sons, Inc.

FIGURE 51.1 A 2-year-old girl with Rubinstein–Taybi syndrome (with a microdeletion of chromosome 16p13.3). Classical facial features are already present.

FIGURE 51.2 A 39-year-old woman with Rubinstein–Taybi syndrome (same individual as in Figure 51.1) showing the adult facial phenotype, including the typical grimacing smile. She is able to make good use of the broadened, angulated thumbs.

complete closing of the eyes is almost universal. The facial features show considerable change with time (Figure 51.2). Newborns often show a full, edematous face, unusual dark hair, upslanted palpebral fissures, a nose without the low hanging columella, a full lower lip, and mild micrognathia. With time, the face elongates, the palpebral fissures slant downward because of the relative lesser growth of the zygoma, the nose profile becomes more convex, the columella becomes low hanging, and the lower lip becomes more everted (Hennekam 1993).

Broad thumbs and broad great toes are present in almost all affected individuals (Figure 51.3). In about one-third of the affected individuals, the thumbs and halluces are also angulated, either in valgus or varus positions. Radiologically, broadening or partial duplication of the first metacarpals, metatarsals, and proximal or distal phalanges of the first ray can be found, but complete preaxial polydactyly has not been reported. Postaxial polydactyly of the feet does occur, as does (partial) syndactyly of the second and third toes or third and fourth fingers. The terminal phalanges of the fingers tend to be broad, and fetal fingertip pads are common (Hennekam et al. 1990c; Rubinstein 1990). Growth retardation is common, but not always (Beets et al. 2014; Stevens et al. 1990b). Abnormal ossification may be evident in the large and slowly closing anterior fontanel. Infrequent parietal foramina, delayed bone age, and an increased fracture frequency are reported. Other skeletal symptoms include pectus deformity, scoliosis and kyphosis, spina bifida at various levels, generalized hypermobility (Hennekam et al. 1990c; Rubinstein 1990), slipped capital femoral epiphyses (Bonioli et al. 1993), and patellar dislocations (Hennekam et al. 1990c; Stevens, 1997).

There is a wealth of less frequent skeletal findings and anomalies of internal organs and skin, which have been reviewed elsewhere (Hennekam et al. 1990c; Hennekam et al. 2010; Rubinstein 1990).

Etiology, Pathogenesis, and Genetics

RSTS is generally a sporadically occurring entity. For a couple with a previous affected child, accumulating data suggest a recurrence risk of approximately 0.5% (Bartsch et al. 2010). If a person with RSTS is able to reproduce, the recurrence risk is probably 50%. In the literature, several cases are described in which a parent was found to be affected with RSTS following the diagnosis in their child (Bartsch et al. 2010; Lopez et al. 2016). In all these families, the diagnosis in the parent would have been difficult without the more pronounced phenotype in their children.

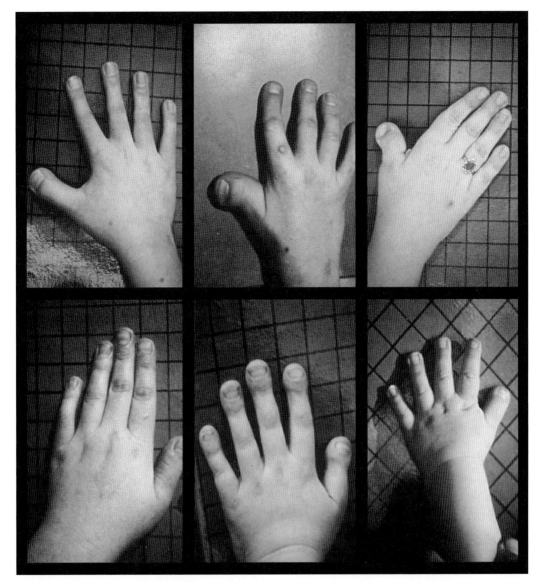

FIGURE 51.3 The hands of six individuals with Rubinstein–Taybi syndrome, showing the variability of the malformation of the thumbs.

In 1993, a group of Dutch researchers analyzed 24 affected individuals with fluorescence in situ hybridization (FISH) and detected a microdeletion at 16p13.3 in six of them (Breuning et al. 1993; Hennekam et al. 1993). In the available families, no parent showed a microdeletion, indicating a de novo rearrangement. Using molecular markers, a copy of chromosome 16 from both parents was found in 18 others, excluding uniparental disomy as a frequent causative mechanism (Hennekam et al. 1993). In combining the results of all microdeletion studies published so far, the actual 16p microdeletion frequency in RSTS is approximately 10–16% (Bentivegna et al. 2006; Hennekam 2006). Clinical features in those with or without detectable deletions are essentially the same, with the possible exception of microcephaly, angulation of thumbs and halluces, and partial duplication of halluces, which are more common in those with 16p13.3 deletion (Hennekam et al. 1993). Continuing research eventually led to the demonstration of variants in *CREBBP* (Petrij et al. 1995). Cyclic-AMC-regulated enhancer (CRE)-binding protein, or CREB-binding protein, is generally referred to as *CBP* or *CREBBP*. *CREBBP* has a homolog, *p300* or *EP300*, located at chromosome 22q13 and coding for EP300, which shows a striking resemblance to it in primary structure and in function. Both act as transcriptional co-activators and also as potent histone acetyltransferases by making the DNA accessible to transcription factors. They are mediators of signaling pathways and participants in basic cellular functions such as DNA repair, cell growth, cell differentiation, apoptosis, and tumor suppression (reviewed by Goodman and Smolik [2000]). Combining the results of larger studies (Hennekam 2006), a *CREBBP* mutation was found in 63 of 155 patients (41%). Subsequent studies showed mutations in *EP300* in a limited number of individuals (Roelfsema et al. 2005), which was expanded later on (Fergelot et al. 2016). The frequency of *EP300* mutations remains uncertain. Mosaicism has been reported sporadically in patients with a *CREBBP* missense mutation (Bartsch et al. 2010) or a 16p13.3 microdeletion (Bentivegna et al. 2006; de Vries et al. 2016). At present, it is unknown whether there is yet another cause of Rubinstein–Taybi syndrome or whether not all mutations in *CREBBP/EP300* are detected by the presently used techniques.

Diagnostic Testing

The diagnosis of RSTS is still made clinically and rests on the recognition of the characteristic features. Microarray testing should be performed in every person suspected to be affected. In countries where microarray testing is not within reach, FISH studies for a chromosome 16p13.3 microdeletion can be performed using five different cosmids dispersed over the total *CREBBP* gene (Petrij et al. 2000). Molecular studies to detect mutations in *CREBBP* or *EP300* are widely available in the United States and Europe. Combined cytogenetic and molecular studies will allow detection of an abnormality in 60–65% of individuals. Increasingly, exome sequencing using panels that are targeted for variants in genes known to cause intellectual disability is used. In such panels invariably both *CREBBP* and *EP300* are evaluated for variants. This allows recognition of individuals who do not show a completely classical phenotype of RSTS.

The use of exome sequencing panels directed to detect causes of intellectual disability has also shown that mutations in the last part of exon 30 and early part of exon 31 of *CREBBP* (between base pairs 5128 and 5614), and the homologous sites in *EP300*, cause a phenotype that is markedly different from RSTS, and in fact only shares the intellectual disability (Menke et al. 2016). The phenotype seems to be heterogeneous, except for individuals with variants between base pairs 5595 and 5614 of *CREBBP* who share distinct facial characteristics (ptosis, telecanthi, short and upslanted palpebral fissures, depressed nasal ridge, short nose, anteverted nares, short columella, and long philtrum (Menke et al. 2018).

Differential Diagnosis

As the combination of features in RSTS is usually distinctive, the diagnosis can often be made clinically without difficulty. The diagnosis is most difficult in older infants and young children, because of milder facial features, but it becomes easier as the individual ages. Many of the components of the syndrome may occur as isolated findings. Other syndromic entities that may give confusion are Saethre–Chotzen syndrome, Cornelia de Lange syndrome (Chapter 15), Filippi syndrome (de Vries et al. 2016), and trisomy 13 (Chapter 58). The facial features also show resemblance to Floating-Harbor syndrome and Gorlin–Chaudry–Moss syndrome (Hennekam et al. 2010). Broad thumbs may be observed in Apert syndrome and Pfeiffer syndrome, and short thumbs and fingers are seen in type D brachydactyly and Greig syndrome (Chapter 44). A number of case reports of individuals or families with features overlapping those found in RSTS can be distinguished on the basis of several missing features or the presence of findings not described in typical RSTS (Hennekam et al. 2010).

MANIFESTATIONS AND MANAGEMENT

Growth and Feeding

In the first year of life, 80% of the children with RSTS have feeding problems, which are mainly caused by generalized hypotonia, gastroesophageal reflux, and recurrent upper respiratory tract infections. Frequently nasogastric tube feeding is needed for several months, and some infants benefit from a gastrostomy. Most feeding problems resolve after a period of one year (Hennekam et al. 1990c). Reflux does occur thereafter, but is not often a serious problem. Some affected individuals develop a voracious appetite in late childhood, early adolescence, or adulthood (Beets et al. 2014; Stevens et al. 1990b) that may even resemble the appetite of children with Prader–Willi syndrome. In adulthood, feeding problems are rare, although choking remains common.

Growth retardation is common and specific growth charts are available for height, weight and head circumference for boys and girls at different ages (Beets et al. 2014). These data, based on molecularly confirmed patients with mostly *CREBBP* mutations (Beets et al. 2014) showed that weights and lengths at birth are generally normal (-0.9 ± 1.0 SDs and -0.1 ± 1.4 SDs, respectively). In patients with an *EP300* mutation, microcephaly is significantly more frequent and prenatal growth retardation is somewhat more frequent than in patients with a *CREBBP* mutation, the latter possibly being related to the frequently found pre-eclampsia in the mothers of *EP300* fetuses (Fergelot et al. 2016). In the first few months, average length, weight, and head circumference drop to or below -2 SDs. During the preschool and early school years, height continues to follow these lines, with the exception of the weight gain, which can be considerable in boys during these years. Their weight-to-height ratio is often above $+2$ SDs. Excess weight is mainly visible around the abdomen, buttocks, and thighs. By adolescence, weight drops again below the -2 SDs. The weight excess in females starts somewhat later, between 5 and 10 years of age and is more likely to remain a problem throughout life (Hennekam et al. 1990c). Neither boys nor girls experience a pubertal growth spurt, which contributes to their short height as adults. Mean adult height is about −3 SDs for both males and females, mean adult body mass index −0.06 SDs for males and +1.40 for females, and adult head circumference −1.9 SDs for males and −2.7 SDs for females (Beets et al. 2014). Growth hormone deficiency has been reported in a few patients (Marzuillo et al. 2013; Tornese et al. 2015) in whom growth hormone therapy resulted in an increase in height standard deviations. A small but detailed pilot study in the Netherlands in affected individuals with the typical growth of a child with RSTS (n = 5) did not show any growth hormone abnormalities (Soors D'Ancona et al. personal communication). Therefore, growth hormone studies are indicated only in children whose growth is less than expected according to the RSTS growth charts.

Evaluation

- Measurements of head circumference, length, and weight should be performed at birth and every 6–12 months thereafter throughout childhood, using both the population-specific and syndrome-specific growth charts (Beets et al. 2014).
- If growth is less than expected, nutritional status should be evaluated. If no explanation is found, other disorders such as gastrointestinal problems, thyroid dysfunction, celiac disease, and growth hormone disturbances should be considered as in the general population (Oostdijk et al. 2009).
- Close monitoring of weight and weight-to-height ratio in later childhood, puberty, and adolescence is strongly recommended. In case of marked obesity, investigations to detect complications should be as in the general population.

Management

- In infants, as long as weight-to-height ratio is within the normal range, no direct intervention is necessary because of the self-limiting character of the feeding problems. If not taken orally, adequate caloric intake should be ensured through nasogastric feeding, with consideration of gastrostomy if feeding problems persist. Involvement of a dietician may be useful.
- The diagnosis of RSTS alone, without abnormal growth hormone testing, is not an indication for growth hormone therapy. In the case of growth hormone deficiency, treatment with growth hormone should be given as in the general population, to treat and/or prevent the various consequences of growth hormone deficiency.
- There is no specific medication available to control appetite. Therefore, a low-calorie and well-balanced diet combined with a regular exercise program (30 min of physical activity a day) should be advised, especially for boys during childhood and for girls from early adolescence onward (personal experience).
- In case of obesity related complications, these should be treated as in the general population.

Development and Behavior

The general psychomotor development of people with RSTS is delayed. Most parents describe affected individuals as easy-going and loving babies. Table 51.1 gives an overview of the motor developmental milestones found among those with the syndrome (Hennekam et al. 1990c).

First words are usually spoken at about 2 years of age, whereas two- or three-word sentences take as long as 4 years and, sometimes, even 7 years to be acquired. Between 4 and 5 years of age, many children gradually start to make much more use of language. Despite the abnormalities in oral anatomy and speech delay, speech mechanisms and articulation appear normal in most. Some children have nasal speech; the voice can also be high pitched. Many individuals have a rapid or staccato speech rhythm. Despite a frequently limited vocabulary (corresponding to IQ), communication abilities are often remarkably good. Some of the children never learn to speak, and they require sign language or other systems to communicate (Hennekam et al. 1992).

The average IQ was reported as 36 (range 25–79) in one study (Hennekam et al. 1992) and as 51 (range 33–72) in another study (Stevens et al. 1990a). Performance IQ is generally higher than verbal IQ. At an older age, the full-scale IQ decreases because of measurement of different abilities at different ages; for example, at an older age, concept formation and more complex language tasks carry more weight in the tests. The IQ decline is not known to be caused by mental deterioration or regression (Hennekam et al. 1992).

Children with RSTS are generally friendly, happy, and easy-going. Nevertheless, 25% of the parents report

TABLE 51.1 Developmental milestones of children with Rubinstein–Taybi syndrome compared with healthy children (Dosman et al. 2012; Hennekam et al. 1990c)

	Rubinstein–Taybi syndrome		General population	
Milestone	Mean age (months)	Range	Mean age (months)	Range
Laughing	2.5	2–6	2	2–6
Rolling over	10	4–18	6	5–9
Sitting	16	9–24	7	6–12
Crawling	19	12–36	9	8–12
Standing	29	11–80	9	8–18
Walking	35	18–54	14	12–18
First words	24	6–84	12	8–18

behavioral problems often characterized by short attention span, stubbornness, lack of persistence, a need for continuous attention from their parents, and sudden mood changes (Hennekam et al. 1992). With increasing age, behavior can become even more difficult, and develops an obsessive-compulsive character. Social abilities are usually a strength in affected individuals. There are, however, some affected individuals who display genuinely autistic behavior. More commonly, older children and adults like to be on their own, avoiding crowds and excessive noise (Hennekam et al. 1992). In a study of adults with RSTS families reported a decline in abilities over time in 32%, including decreased social interaction, more limited speech, and worsening stamina and mobility (Stevens et al. 2011).

Evaluation

- Each child should be tested at regular intervals of 2–3 years with a systematic developmental assessment starting at age 3–4 years, to ensure adequate educational support compatible with the child's potential.
- Each child should be checked for hearing loss or diminished vision every three years. In adults, ophthalmological evaluation every five years is useful (see Ophthalmology).
- Assessment of family support and psychological/emotional needs will assist adequate care for the whole family.

Management

- Most children need some degree of individualized educational programming, either in specialized or inclusive settings.
- Affected children will benefit from speech therapy, physical therapy, and educational guidance.
- Sign language or other communication techniques should be introduced for non-verbal children.
- Hearing and vision impairments should be treated as in the general population.
- A change in behavior should prompt evaluation for common medical problems in RSTS such as gastroesophageal reflux or toothache.
- Long-standing behavior problems may be improved with behavior management strategies by a (pediatric) behavioral specialist, psychologist, or (child) psychiatrist, with expertise in the field of developmental disabilities.
- Pharmaceutical treatment of behavior can be helpful in individual cases. The treatment should be adapted to the individual needs, as no specific psychoactive medication is known to have a particular benefit. Medication choice is no different from that in the general population.
- For adults leaving their family home, an adequate living situation should be sought.

Immunology

Recurrent respiratory infections are common in RSTS at all ages (Hennekam et al. 1990c; Stevens et al. 2011). In many individuals this is likely caused by microaspiration or gastroesophageal reflux (Rubinstein, 1990) but, in a proportion of them, an underlying immune deficiency may be found (Herriot et al. 2016). Reported individuals had vaccine failure due to absent or incomplete antibody formation (Herriot et al. 2016; Lougaris et al. 2016; Pasic 2015), progressive B cell deficiency and agammaglobulinemia with concomitant idiopathic thrombocytopenic purpura resulting in bronchiectasis (Lougaris et al. 2016) or clear common variable immune deficiency (Herriot et al. 2016). In a natural history study, no severe complications were reported after immunizations (Stevens et al. 1990a).

Evaluation

- Children with RSTS who present with frequent upper respiratory infections, multiple episodes of pneumonia or wheezing, hoarseness, or stridor should be evaluated for gastroesophageal reflux and aspiration contributing to these symptoms (Hennekam et al. 1990c; Rubinstein, 1990).
- There is to date no systematic study of the function of the immune system of a larger series of individuals with RSTS to suggest all individuals should undergo an immune work-up.
- A careful immunologic investigation is required in patients who present with recurrent infections or unexplained episodes of fever. Basic, initial, first line testing (e.g. measurement of serum immunoglobulin levels) is appropriate in the non-immunology setting but normal measured immunoglobulin levels can mask subtle functional deficiencies and early liaison with specialist immunology services is recommended (Herriot et al. 2016).

Management

- Subsequent treatment (e.g. antibiotic prophylaxis and/or lifelong intravenous immunoglobulin replacement therapy) should be based on the specific immune deficiency identified.
- Vaccinations should be given according to the national schemes (Wiley et al. 2003).

Gastrointestinal

Constipation is frequent (40–74%) (Hennekam et al. 1990c; Rubinstein, 1990; Stevens et al. 1990a). Only one individual with concomitant Hirschsprung disease has been reported. Gastroesophageal reflux may lead to nutritional/feeding problems and respiratory problems and is especially prevalent in infancy (Hennekam et al. 1990b; Rubinstein, 1990). A possibly increased incidence of intestinal malrotation in RSTS has been reported (Stevens 2015).

Evaluation

- Assessment of constipation and gastroesophageal reflux primarily by clinical history and physical examination.
- Malrotation should be suspected in those people who present with intermittent vomiting, bilious vomiting, recurrent abdominal pain, failure to pass stool, or bloody stools (Stevens 2015).

Management

- Constipation should be managed with dietary adjustments and, if needed, laxatives.
- Gastroesophageal reflux should be treated as in the general population, referral to a pediatric gastroenterologist is recommended for refractory cases.
- In case of mild symptoms of a possible malrotation an upper GI study should be performed; in case of malrotation with symptoms of volvulus, an acute referral and surgical intervention are needed.

Ophthalmology

Extraocular and intraocular anomalies are common in individuals with RSTS (Candan et al. 2014; van Genderen et al. 2000). Frequent problems are strabismus (68%), refractive errors (42%), and lacrimal duct obstruction (30%, often leading to recurrent conjunctivitis). Congenital glaucoma or high intraocular pressure (15%; with concomitant enlargement of the affected eye) and corneal abnormalities (12%) also occur. Less frequent ocular problems include congenital or presenile cataract, nystagmus, and coloboma (involving the iris, cornea, retina or optic nerve; all less than 10%) (Candan et al. 2014). Correction of refractive errors becomes more important in the second half of the first year. The majority of children have myopia, but hypermetropia and astigmatism also occur. About half of the affected individuals are photophobic (van Genderen et al. 2000).

One study described a high frequency of retinal dysfunction (78%) in children and adults with RSTS. Electroretinograms showed cone or cone-rod dysfunction in 58% of the individuals and visual-evoked potentials showed an abnormal waveform in 63% of individuals, suggesting abnormal retinal functioning. With age, retinal and electrophysiological abnormalities occurred more frequently but no influence on vision has been reported. Below 15 years of age, only minor abnormalities were found (van Genderen et al. 2000).

Evaluation

- Full ophthalmologic examination by an ophthalmologist following diagnosis is important because of the high frequency of ocular abnormalities (including congenital glaucoma and cataracts).
- In childhood (until 16 years of age), ophthalmological examination by an ophthalmologist should be performed every three years, or more frequently depending on the problems encountered. Suspicion of glaucoma or corneal opacities requires more urgent evaluation by a pediatric ophthalmologist (Wiley et al. 2003). In adults, ophthalmological evaluation every five years is useful.
- Because of the frequent retinal dysfunction in older children and adults, electrophysiological investigations may be considered in every individual with Rubinstein–Taybi every five years after 16 years of age (van Genderen et al. 2000).

Management

- Congenital glaucoma and cataracts are preventable causes of blindness and early treatment, following standard procedures, is essential.
- Surgical intervention for lacrimal duct stenosis may be necessary if the problems do not resolve. In some cases, placement of glass tear ducts is required, but usually this is not performed until the individual has reached adulthood (personal experience).
- In some individuals, surgery is required to correct ptosis or strabismus. This should follow the guidelines used in the general population.
- There is no known effective treatment for photophobia. Protection using a hat or sunglasses is appreciated by some of the children and adults (personal experience).

Respiratory

Obstructive sleep apnea may be a considerable problem in a small number of individuals with RSTS (Hennekam et al. 1990c; Zucconi et al. 1993) and may lead to pulmonary hypertension (Choi et al. 2012). It may be caused by the combination of a narrow palate, micrognathia, hypotonia, obesity, adenotonsillar hypertrophy, and easy collapsibility of the laryngeal walls (Hennekam et al. 1990c). Intubation may be difficult because of enhanced collapsibility of the laryngeal wall. Clues to the presence of obstructive sleep apnea are snoring, a particular sleeping posture (with the head between the knees), light sleep at night, and excessive sleepiness during the day. Without more daytime sleep, mood changes, excitability, and irritability can occur. Long-term obstructed respiration can lead to pulmonary hypertension, which in turn can cause right ventricular hypertrophy and ultimately lead to decompensation.

Evaluation

- During regular health visits specific attention should be paid to possible symptoms of sleep apnea (snoring, particular sleeping posture such as sleeping upright, wakefulness at night, and excessive sleepiness during the daytime, measurement of tension).
- If sleep apnea is suspected, a sleep study (polysomnography) should be performed (Zucconi et al. 1993).

Management

- If sleep apnea is present, the simplest solution is optimal positioning in sleep (Hennekam et al. 1990c; Zucconi et al. 1993). Continuous positive airway pressure (CPAP) is usually successful, but may not be accepted by individuals with RSTS. Rarely, a tracheostomy may be required.
- Adenotomy and/or tonsillectomy may be helpful in individuals with clear hypertrophy, but is not advocated in the other individuals.

Dental

Timing of the eruption of deciduous and permanent dentition is normal in RSTS. Marked caries can be found in about one-third of the affected people, probably caused by problems in dental care because of the small opening of the mouth, malpositioning and malformation of the teeth, and suboptimal cooperation. Hypodontia, hyperdontia, and natal teeth can be manifestations of the syndrome. The most important dental feature in RSTS is the very high incidence of talon cusps in the permanent dentition (92%). They can also be present in the deciduous dentition (9%) or sometimes can be detected in the jaws by orthopanthogram. Talon cusps are accessory cusp-like structures on the lingual side of the incisors. Two or more talon cusps are rarely found in the general population or in other syndromes, and this finding strongly supports the diagnosis (Hennekam et al. 1990a). Talon cusps increase the likelihood of caries because food remains behind them.

Evaluation

- Regular, routine semiannual dental examinations, and dental prophylaxis are indicated. Usually, first visits to the dentists are at 1–2 years, to allow the child to get used to oral examinations. A dentist and orthodontist experienced in treating children and adults with developmental delay can be of great assistance.

Management

- Emphasis should be placed on good dental hygiene as in any other child. Tooth brushing is often more effective with an electric device than by hand.
- Normal standard dental care should be provided. Some individuals with RSTS have an increased fear of dental care and treatment is only possible with general anesthesia.
- It is possible to grind off the tips of the talon cusps or fill up the space between the cusps and the teeth.
- Malpositioned and crowded teeth will respond to standard orthodontic intervention.

Cardiovascular

About one-third of individuals with RSTS have a congenital heart malformation. In 65% of the people with heart defects, this is a single defect (patent ductus arteriosus, septal defect, coarctation, pulmonic stenosis, bicuspid aortic valve, pseudotruncus, aortic stenosis, dextrocardia, vascular rings, and conduction problems), whereas 35% have two or more defects or a complex malformation (Hennekam et al. 1990c; Rubinstein, 1990; Stevens et al. 1995). The symptoms do not differ from those in other children.

Evaluation

- All individuals with RSTS should have a cardiac assessment, including an examination by a pediatric cardiologist, electrocardiogram, and echocardiogram at the time of diagnosis.
- Subsequent management and follow-up are dictated by these investigations and the clinical course.

Management

- In half of the patients with a single defect, surgery is needed, as compared with 80% of the more complex defects.
- The indications for and methods of surgical intervention are the same as in the general population.
- Antibiotic prophylaxis should be considered before and during procedures that induce (transient) bacteremia, such as dental treatments, as in the general population.

Dermatologic

Keloids are proliferative fibrous growths that can cause extensive itching. They occur in 24% of individuals with RSTS, usually starting in early puberty. Management schedules have disappointing results thus far (van de Kar et al. 2014). Keloids occur where they typically occur in the general population, including the chest, upper part of the back, and upper arms. Sometimes only minimal trauma such as a bee sting or even the rubbing of clothes can initiate the keloid formation (van de Kar et al. 2014). Apart from keloids, pilomatrixomas, and ingrown toenails are frequently seen: fingernail paronychia in 9% and toenail paronychia in 44% (Hennekam et al. 1990c).

Evaluation

- Physical examination is usually sufficient to diagnose keloid formation. A biopsy can be difficult because of the extreme firmness of the tissue and should only be taken when the diagnosis is in doubt.

Management

- Keloids are very therapy-resistant. Oral antihistamines, local corticosteroids, or laser therapy are often ineffective, and sometimes cause more damage than improvement. Occasionally, patients have been treated by local radiation with good results.
- Because of expected growth and mechanical and cosmetic consequences in pilomatrixomas, individuals could be referred to a plastic surgeon or dermatologist for excision and pathologic evaluation of the lesion
- Finger and toenails should be cut only horizontally, leaving the borders of the nails intact to prevent paronychia. In recalcitrant chronic paronychia, excision of the proximal nail fold is an option as in the general population.

Genitourinary

Almost all boys with RSTS have incomplete or delayed descent of the testicles. Hypospadias is seen in 11% of boys. Renal anomalies or disease are present in about 50% of affected individuals (i.e. hydronephrosis, duplications, vesicoureteral reflux, urinary tract infections, stones, and nephrotic syndrome (Rubinstein 1990; Hennekam et al. 1990c; Stevens et al. 1990a). The timing of puberty is normal and secondary sex characteristics develop normally. Females often have hypermenorrhagia or metrorrhagia. Fertility in individuals with the syndrome is probably normal.

Evaluation

- Every male with the syndrome should be checked for cryptorchidism and hypospadias at diagnosis, and evaluation by a pediatric urologist is warranted by 6–12 months of age if the testes have not descended as is recommended for the general population.
- If an individual with RSTS shows signs of a neurogenic bladder, an evaluation for tethered cord by MRI with cine is indicated (Wiley et al. 2003).

Management

- Treatment of cryptorchidism is as in the general population.
- Surgery for hypospadias should be considered only if the hypospadias is severe or causes recurrent infection.

- Hypermenorrhagia or hypermetrorrhagia often responds to treatment with an oral contraceptive. No contraindication for this treatment is known.
- Some of the adolescents and adults of either sex are sexually active; thus, contraception and sex education need to be adequately addressed with the affected individuals, their parents, and other caregivers.
- Constipation, if present, should be aggressively treated as this can contribute to difficulties with urinary tract infections and vesicoureteral reflux (see further recommendations in section on Gastroenterology).

Musculoskeletal

Individuals with significant angulation of their thumbs should be considered for surgical repair before age 2 years because such angulations can have serious consequences for functional dexterity of the hands (Wood et al. 1987). Surgery on angulated great toes is performed only when they hinder walking or wearing footwear (Wood et al. 1987).

Hypermobility of joints is seldom a problem at a young age, but, once walking begins, generalized hypotonia and hyperextensibility of joints can become problematic. This is especially the case if there is a concomitant overweight. The gait is commonly stiff and sometimes waddling. Individuals with RSTS have an increased risk for dislocation of the radial head and the patella (Lazaro et al. 2007). Patellar dislocation can be particularly burdensome and have great consequences for mobility. If not treated in time, patellar dislocation can lead to complications such as genu valgum, tibial torsion, and flexion contracture at the knee. Other musculoskeletal problems include tight heel cords, elbow abnormalities, congenital hip dislocation and cervical vertebral abnormalities (Hennekam et al. 1990c; Rubinstein 1990; Yamamoto et al. 2005).

At about 10 years of age, children with RSTS may develop kyphosis, lordosis, and scoliosis (Rubinstein 1990; Stevens et al. 1990a). Several older children and adolescents have severe and prolonged aseptic hip joint inflammation (Perthes-like). In general, this problem resolves in one or two years without intervention, but symptomatic treatment might be necessary because of pain. Slipped capital femoral epiphysis, which usually occurs in adolescence, can become a major problem. Initially, it is often silent, and once there are symptoms, the radiographic changes are often already severe. Although occult spina bifida can be found regularly it does not go along with clinical manifestations and has no clinical significance.

Evaluation

- Annual clinical assessment of the chest cage, spine, joints, and walking pattern is recommended.

- A change in gait or pain should prompt careful analysis of hip function and anatomy, and search for a patellar dislocation. In m. Perthes, the hip exam shows decreased abduction. In slipped capital femoral epiphysis, the hip has obligate external rotation upon hip flexion. Genu valgum should be evaluated with a supine assessment of the intermalleolar distance. If the distance is greater than 8–10 cm referral to a pediatric orthopedic surgeon is warranted.

Management

- Surgical correction of angulated thumbs should be considered if it will improve dexterity. It should optimally be performed before the age of 2 years by a surgeon familiar with this procedure in RSTS.
- Physical therapy may be useful for any young child with walking difficulties (Cazalets et al. 2017).
- Treatment of congenital hip dislocation, Perthes disease, slipped capital femoral epiphyses and scoliosis is as in the general population.

Neoplasia

An increased tumor risk has been suggested in RSTS. This notion comes mainly from case reports, and most observations are described in individuals in whom the diagnosis was not confirmed molecularly. Reported tumors include nasopharyngeal rhabdomyosarcoma, intraspinal neurilemmoma, pheochromocytoma, meningioma, other brain tumors, hepatoblastoma, germ-cell tumor, pilomatricoma, acute leukemia, and lymphoma (reviewed by Boot et al. [2018]). Tumors in individuals with an *EP300* mutation have been described rarely (Fergelot et al. 2016), but the total number of reported individuals with *EP300* mutation is small. A nationwide study in the Netherlands of 87 RSTS individuals indicated 35 neoplasms in 26 individuals but only 5 were malignant (Boot et al. 2018), indicating that malignancies did not occur more frequently than expected below the age of 40 years; it may occur more frequently thereafter, but data were insufficient to conclude this. Non-malignant neoplasia were pilomatricomas and meningioma (in two individuals multiple primary meningioma were present). No genotype–phenotype correlation was evident, nor was there a correlation between keloids and pilomatricomas or malignancies. Individuals with a malignancy responded to therapy in the same way as individuals without the syndrome.

Evaluation

- If a child or adult with RSTS develops unusual symptoms, one should always consider the possibility of a tumor. No standard tumor surveillance regimens are suggested as most tumors will lead to easily recognizable symptoms, and it is unlikely screening will improve outcome (Boot et al. 2018).

Management

- Treatment should be as in the general population, and expected prognosis is also as in the general population. In meningioma one should take into account that there is an increased risk for multiple primary meningiomas.

Anesthesiology

Individuals with RSTS can be more challenging to intubate due to their relatively anterior position of the larynx and easy collapsibility of the laryngeal wall (Hennekam et al. 1990c). In addition they may suffer more easily from postoperative apnea/hypopnea, have an increased risk of aspiration, and may more easily develop a cardiac arrhythmia after the use of cardioactive drugs (e.g., atropine, neostigmine, succinylcholine) (Darlong et al. 2014; Stirt 1982).

Evaluation

- If surgery is required, individuals with RSTS undergoing general anesthesia should be under the care of an anesthesiologist comfortable with complex airway problems.
- The anesthesiologist should be aware of the abnormal anatomy and easy collapsibility of the laryngeal wall, the increased risk of aspiration and cardiac arrhythmia upon the use of cardioactive drugs.

Management

- Appropriate supportive measures should be taken to support the patient and the family bearing in mind the intellectual disability and behavioral characteristics of the individual.
- During anesthesia, intubation (tube or ProSeal LMA) is indicated due to the increased risk for aspiration.
- Intubation should be performed earlier and extubation later than in the general population.
- To decrease the risk of postoperative apnea/hypopnea, opioids and muscle relaxants should be used carefully.

RESOURCES

Brochures and Newsletters

USA support group: *The Rubinstein-Taybi Book* (1997)

UK support group: *Rubinstein-Taybi Syndrome: An Information Booklet* (2002)

Dutch support group: *Informatie voor ouders en betrokkenen na de diagnose RTS* (2016)

SUPPORT GROUPS

Brazil

ARTS

Rua Harmonia 722/81
CEP 05435-000
São Paulo-SP
Brasil
Phone: + 55 11 4153 3211
Website: www.artsbrasil.org.br

The Netherlands

Website: www.rtsyndroom.nl

France

Association Francaise du Syndrome de Rubinstein-Taybi
5 Rue des Corvees
Cidex 8571 Jarday
41000 Villerbon
France
Phone: + 33 2 54 78 10 53
Website: www.afsrt.com

Spain

Website: www.rubinsteintaybi.es

UK

Website: www.rtsuk.org.uk

USA

Garry and Lorrie Baxter
PO Box 146
Smith Center, Kansas 66967
USA
Phone: 888-447-2989
Website: www.rubinstein-taybi.org

REFERENCES

Allanson JE (1990) Rubinstein-Taybi syndrome: the changing face. *Am J Med Genet* Suppl 6:38–41.

Bartsch O, Kress W, Kempf O, Lechno S, Haaf T, Zechner U (2010) Inheritance and variable expression in Rubinstein-Taybi syndrome. *Am J Med Genet A* 152A:2254–2261.

Beets L, Rodriguez-Fonseca C, Hennekam RC (2014) Growth charts for individuals with Rubinstein-Taybi syndrome. *Am J Med Genet A* 164A:2300–2309.

Bentivegna A, Milani D, Gervasini C, Castronovo P, Mottadelli F, Manzini S, Colapietro P, Giordano L, Atzeri F, Divizia MT, Uzielli ML, Neri G, Bedeschi MF, Faravelli F, Selicorni A, Larizza L (2006) Rubinstein-Taybi Syndrome: spectrum of CREBBP mutations in Italian patients. *BMC Med Genet* 7:77.

Bonioli E, Bellini C, Senes FM, Palmieri A, Di Stadio M, Pinelli G (1993) Slipped capital femoral epiphysis associated with Rubinstein-Taybi syndrome. *Clin Genet* 44:79–81.

Boot MV, Van Belzen MJ, Overbeek LI, Hijmering N, Mendeville M, Waisfisz Q, Wesseling P, Hennekam RC, De Jong D (2018) Benign and malignant tumors in Rubinstein-Taybi syndrome. *Am J Med Genet* 176:597–608.

Breuning MH, Dauwerse HG, Fugazza G, Saris JJ, Spruit L, Wijnen H, Tommerup N, van der Hagen CB, Imaizumi K, Kuroki Y, van den Boogaard MJ, de Pater JM, Mariman EC, Hamel BC, Himmelbauer H, Frischauf AM, Stallings R, Beverstock GC, van Ommen GJ, Hennekam RC (1993) Rubinstein-Taybi syndrome caused by submicroscopic deletions within 16p13.3. *Am J Hum Genet* 52:249–254.

Candan S, Ornek C, Candan F (2014) Ocular anomalies in Rubinstein-Taybi syndrome: a further case report and review of the literature. *Clin Dysmorphol* 23:138–142.

Cazalets JR, Bestaven E, Doat E, Baudier MP, Gallot C, Amestoy A, Bouvard M, Guillaud E, Guillain I, Grech E, Van-Gils J, Fergelot P, Fraisse S, Taupiac E, Arveiler B, Lacombe D (2017) Evaluation of Motor Skills in Children with Rubinstein-Taybi Syndrome. *J Autism Dev Disord* 47:3321–3332.

Choi HS, Yu JJ, Kim YH, Ko JK, Park IS (2012) Pulmonary hypertension due to obstructive sleep apnea in a child with Rubinstein-Taybi syndrome. *Korean J Pediatr* 55:212–214.

Coupry I, Roudaut C, Stef M, Delrue MA, Marche M, Burgelin I, Taine L, Cruaud C, Lacombe D, Arveiler B (2002) Molecular analysis of the CBP gene in 60 patients with Rubinstein-Taybi syndrome. *J Med Genet* 39:415–421.

Darlong V, Pandey R, Garg R, Pahwa D (2014) Perioperative management of a patient of Rubinstein-Taybi syndrome with ovarian cyst for laparotomy. *J Anaesthesiol Clin Pharmacol* 30:422–424.

de Vries TI, G RM, van Belzen MJ, van der Lans CA, Savelberg SM, Newman WG, van Haaften G, Nievelstein RA, van Haelst MM (2016) Mosaic CREBBP mutation causes overlapping clinical features of Rubinstein-Taybi and Filippi syndromes. *Eur J Hum Genet* 24:1363–1366.

Dosman CF, Andrews D, Goulden KJ (2012) Evidence-based milestone ages as a framework for developmental surveillance. *Paediatr Child Health* 17:561–568.

Fergelot P, Van Belzen M, Van Gils J, Afenjar A, Armour CM, Arveiler B, Beets L, Burglen L, Busa T, Collet M, Deforges J, de Vries BB, Dominguez Garrido E, Dorison N, Dupont J, Francannet C, Garcia-Minaur S, Gabau Vila E, Gebre-Medhin S, Gener Querol B, Genevieve D, Gerard M, Gervasini CG, Goldenberg A, Josifova D, Lachlan K, Maas S, Maranda B, Moilanen JS, Nordgren A, Parent P, Rankin J, Reardon W, Rio M, Roume J, Shaw A, Smigiel R, Sojo A, Solomon B, Stembalska A, Stumpel C, Suarez F, Terhal P, Thomas S, Touraine R, Verloes A, Vincent-Delorme C, Wincent J, Peters DJ, Bartsch O, Larizza L, Lacombe D, Hennekam RC (2016) Phenotype and genotype in 52 patients with Rubinstein-Taybi syndrome caused by EP300 mutations. *Am J Med Genet A* 170:3069–3082.

Gervasini C, Castronovo P, Bentivegna A, Mottadelli F, Faravelli F, Giovannucci-Uzielli ML, Pessagno A, Lucci-Cordisco E, Pinto AM, Salviati L, Selicorni A, Tenconi R, Neri G, Larizza L

(2007) High frequency of mosaic CREBBP deletions in Rubinstein-Taybi syndrome patients and mapping of somatic and germ-line breakpoints. *Genomics* 90:567–573.

Goodman RH, Smolik S (2000) CBP/p300 in cell growth, transformation, and development. *Genes Dev* 14:1553–1577.

Hennekam RC (1993) Rubinstein-Taybi syndrome: a history in pictures. *Clin Dysmorphol* 2:87–92.

Hennekam RC (2006) Rubinstein-Taybi syndrome. *Eur J Hum Genet* 14:981–985.

Hennekam RC, Van Doorne JM (1990a) Oral aspects of Rubinstein-Taybi syndrome. *Am J Med Genet Suppl* 6:42–47.

Hennekam RC, Stevens CA, Van de Kamp JJ (1990b) Etiology and recurrence risk in Rubinstein-Taybi syndrome. *Am J Med Genet* Suppl 6:56–64.

Hennekam RC, Van Den Boogaard MJ, Sibbles BJ, Van Spijker HG (1990c) Rubinstein-Taybi syndrome in The Netherlands. *Am J Med Genet Suppl* 6:17–29.

Hennekam RC, Baselier AC, Beyaert E, Bos A, Blok JB, Jansma HB, Thorbecke-Nilsen VV, Veerman H (1992) Psychological and speech studies in Rubinstein-Taybi syndrome. *Am J Ment Retard* 96:645–660.

Hennekam RC, Tilanus M, Hamel BC, Voshart-van Heeren H, Mariman EC, van Beersum SE, van den Boogaard MJ, Breuning MH (1993) Deletion at chromosome 16p13.3 as a cause of Rubinstein-Taybi syndrome: clinical aspects. *Am J Hum Genet* 52:255–262.

Hennekam RCM, Krantz ID, Allanson JE (2010) *Gorlin's Syndromes of the Head and Neck*, 5th ed. New York, NY: Oxford University Press, Inc.

Herriot R, Miedzybrodzka Z (2016) Antibody deficiency in Rubinstein-Taybi syndrome. *Clin Genet* 89:355–358.

Kalkhoven E, Roelfsema JH, Teunissen H, den Boer A, Ariyurek Y, Zantema A, Breuning MH, Hennekam RC, Peters DJ (2003) Loss of CBP acetyltransferase activity by PHD finger mutations in Rubinstein-Taybi syndrome. *Hum Mol Genet* 12:441–450.

Lazaro JS, Herraez SS, Gallego LD, Caballero EG, Diaz JF (2007) Spontaneous patella dislocation in Rubinstein Taybi Syndrome. *Knee* 14:68–70.

Lopez M, Seidel V, Santibanez P, Cervera-Acedo C, Castro-de Castro P, Dominguez-Garrido E (2016) First case report of inherited Rubinstein-Taybi syndrome associated with a novel EP300 variant. *BMC Med Genet* 17:97.

Lougaris V, Facchini E, Baronio M, Lorenzini T, Moratto D, Specchia F, Plebani A (2016) Progressive severe B cell deficiency in pediatric Rubinstein-Taybi syndrome. *Clin Immunol* 173:181–183.

Marzuillo P, Grandone A, Coppola R, Cozzolino D, Festa A, Messa F, Luongo C, Del Giudice EM, Perrone L (2013) Novel cAMP binding protein-BP (CREBBP) mutation in a girl with Rubinstein-Taybi syndrome, GH deficiency, Arnold Chiari malformation and pituitary hypoplasia. *BMC Med Genet* 14:28.

Menke LA, van Belzen MJ, Alders M, Cristofoli F, Study DDD, Ehmke N, Fergelot P, Foster A, Gerkes EH, Hoffer MJ, Horn D, Kant SG, Lacombe D, Leon E, Maas SM, Melis D, Muto V, Park SM, Peeters H, Peters DJ, Pfundt R, van Ravenswaaij-Arts CM, Tartaglia M, Hennekam RC (2016) CREBBP mutations in individuals without Rubinstein-Taybi syndrome phenotype. *Am J Med Genet A* 170:2681–2693.

Menke LA, the DDD study, Gardeitchik T, Hammond P, Heimdal KR, Houge G, Hufnagel SB, Ji J, Johansson S, Kant SG, Kinning E, Leon E, Newbury-Ecob R, Paolacci S, Pfundt R, Ragge NK, Rinne T, Ruivenkamp C, Saitta SC, Sun Y, Tartaglia M, Terhal P, van Essen AJ, Vigeland MD, Xiao B, Hennekam RC (2018) Further delineation of an entity caused by *CREBBP* and *EP300* mutations but not resembling Rubinstein-Taybi syndrome. *Am J Med Genet* 176:862–876.

Michail J, Matsoukas J, Theodorou S (1957) (Arched, clubbed thumb in strong abduction-extension & other concomitant symptoms). *Rev Chir Orthop Reparatrice Appar Mot* 43:142–146.

Oostdijk W, Grote FK, de Muinck Keizer-Schrama SM, Wit JM (2009) Diagnostic approach in children with short stature. *Horm Res* 72:206–217.

Pasic S (2015) Rubinstein-Taybi syndrome associated with humoral immunodeficiency. *J Investig Allergol Clin Immunol* 25:137–138.

Petrij F, Giles RH, Dauwerse HG, Saris JJ, Hennekam RC, Masuno M, Tommerup N, van Ommen GJ, Goodman RH, Peters DJ, et al (1995) Rubinstein-Taybi syndrome caused by mutations in the transcriptional co-activator CBP. *Nature* 376:348–351.

Petrij F, Dauwerse HG, Blough RI, Giles RH, van der Smagt JJ, Wallerstein R, Maaswinkel-Mooy PD, van Karnebeek CD, van Ommen GJ, van Haeringen A, Rubinstein JH, Saal HM, Hennekam RC, Peters DJ, Breuning MH (2000) Diagnostic analysis of the Rubinstein-Taybi syndrome: five cosmids should be used for microdeletion detection and low number of protein truncating mutations. *J Med Genet* 37:168–176.

Roelfsema JH, White SJ, Ariyurek Y, Bartholdi D, Niedrist D, Papadia F, Bacino CA, den Dunnen JT, van Ommen GJ, Breuning MH, Hennekam RC, Peters DJ (2005) Genetic heterogeneity in Rubinstein-Taybi syndrome: mutations in both the CBP and EP300 genes cause disease. *Am J Hum Genet* 76:572–580.

Rubinstein JH (1990) Broad thumb-hallux (Rubinstein-Taybi) syndrome 1957–1988. *Am J Med Genet Suppl* 6:3–16.

Rubinstein JH, Taybi H (1963) Broad thumbs and toes and facial abnormalities. A possible mental retardation syndrome. *Am J Dis Child* 105:588–608.

Stevens CA (1997) Patellar dislocation in Rubenstein-Taybi syndrome. *Am J Med Genet* 72:188–190.

Stevens CA (2015) Intestinal malrotation in Rubinstein-Taybi syndrome. *Am J Med Genet A* 167A:2399–2401.

Stevens CA, Bhakta MG (1995) Cardiac abnormalities in the Rubinstein-Taybi syndrome. *Am J Med Genet* 59:346–348.

Stevens CA, Carey JC, Blackburn BL (1990a) Rubinstein-Taybi syndrome: a natural history study. *Am J Med Genet Suppl* 6:30–37.

Stevens CA, Hennekam RC, Blackburn BL (1990b) Growth in the Rubinstein-Taybi syndrome. *Am J Med Genet Suppl* 6:51–55.

Stevens CA, Pouncey J, Knowles D (2011) Adults with Rubinstein-Taybi syndrome. *Am J Med Genet A* 155A:1680–1684.

Stirt JA (1982) Succinylcholine in Rubinstein-Taybi syndrome. *Anesthesiology* 57:429.

Tornese G, Marzuillo P, Pellegrin MC, Germani C, Faleschini E, Zennaro F, Grandone A, Miraglia del Giudice E, Perrone L, Ventura A (2015) A case of Rubinstein-Taybi syndrome associated with growth hormone deficiency in childhood. *Clin Endocrinol (Oxf)* 83:437–439.

van de Kar AL, Houge G, Shaw AC, de Jong D, van Belzen MJ, Peters DJ, Hennekam RC (2014) Keloids in Rubinstein-Taybi syndrome: a clinical study. *Br J Dermatol* 171:615–621.

van Genderen MM, Kinds GF, Riemslag FC, Hennekam RC (2000) Ocular features in Rubinstein-Taybi syndrome: investigation of 24 patients and review of the literature. *Br J Ophthalmol* 84:1177–1184.

Warkany J (1974) Letter: Difficulties of classification and terminology of syndromes of multiple congenital anomalies. *Am J Dis Child* 128:424–425.

Wiley S, Swayne S, Rubinstein JH, Lanphear NE, Stevens CA (2003) Rubinstein-Taybi syndrome medical guidelines. *Am J Med Genet A* 119A:101–110.

Wood VE, Rubinstein JH (1987) Surgical treatment of the thumb in the Rubinstein-Taybi syndrome. *J Hand Surg Br* 12:166–172.

Yamamoto T, Kurosawa K, Masuno M, Okuzumi S, Kondo S, Miyama S, Okamoto N, Aida N, Nishimura G (2005) Congenital anomaly of cervical vertebrae is a major complication of Rubinstein-Taybi syndrome. *Am J Med Genet A* 135:130–133.

Zucconi M, Ferini-Strambi L, Erminio C, Pestalozza G, Smirne S (1993) Obstructive sleep apnea in the Rubinstein-Taybi syndrome. *Respiration* 60:127–132.

52

SILVER–RUSSELL SYNDROME

Emma L. Wakeling

Consultant in Clinical Genetics, North East Thames Regional Genetics Service, Great Ormond Street Hospital for Children NHS Foundation Trust, London, UK

INTRODUCTION

Silver–Russell syndrome (SRS) is a genetically heterogeneous condition characterized by intrauterine growth retardation, postnatal growth failure, relative macrocephaly, prominent forehead, body asymmetry and feeding difficulties. The condition was first described by Silver et al. (1953) and Russell (1954), who independently reported a few children with low birth weight, postnatal short stature, body asymmetry, and characteristic facial features. Diagnosis of SRS can be difficult as most of these features are nonspecific and severity varies widely between affected individuals. Several scoring systems have been proposed, reflecting this diagnostic challenge (Azzi et al. 2015). An international consensus on diagnosis and management of SRS has recommended that SRS remains first and foremost a clinical diagnosis (Wakeling et al. 2017).

An underlying molecular cause is currently identified in around 60% of individuals with a clinical diagnosis of SRS (Netchine et al. 2007). The most common molecular mechanism is loss of methylation on chromosome 11p15 (11p15 LOM), which is seen in around 50% of individuals with SRS (Gicquel et al. 2005; Netchine et al. 2007). Maternal uniparental disomy for chromosome 7 (upd(7)mat) is seen in 5–10% (Eggermann et al. 1997; Kotzot et al. 1995). Knowledge of the molecular subtype confirms the clinical diagnosis and can provide additional information about the likelihood of specific associated features (Wakeling et al. 2010). However, the underlying molecular cause remains unknown in a significant proportion of individuals with features meeting clinical criteria for diagnosis of SRS.

Detailed guidelines for management of individuals with SRS are set out in international consensus guidelines (Wakeling et al. 2017). Although there is substantial overlap in the care of individuals born small for gestational age (SGA) and those diagnosed with SRS, there are specific management issues that need to be considered in SRS. These include severe feeding difficulties, gastrointestinal problems, postnatal growth failure, hypoglycemia, body asymmetry, scoliosis, other congenital anomalies, motor and speech delay, and psychosocial issues.

Incidence

Previous estimates of the incidence of SRS have ranged widely from 1 in 30,000 to 1 in 100,000. However, the overall incidence, including clinical SRS, is probably higher. A recent retrospective analysis of the prevalence of imprinting disorders in Estonia reported a live birth prevalence for molecularly and clinically diagnosed SRS of 1/15,866 (Yakoreva et al. 2019).

Diagnostic Criteria

The diagnosis of SRS can be difficult, particularly as the severity of the condition can vary widely. There is also

---

Cassidy and Allanson's Management of Genetic Syndromes, Fourth Edition.
Edited by John C. Carey, Agatino Battaglia, David Viskochil, and Suzanne B. Cassidy.
© 2021 John Wiley & Sons, Inc. Published 2021 by John Wiley & Sons, Inc.

significant overlap in clinical manifestations within this group, in children born SGA but without SRS, and in other syndromic causes of pre- and postnatal short stature. Several different clinical scoring systems have been suggested, all with similar criteria but with varying number and definition needed for diagnosis. International consensus guidelines recommend use of the Netchine–Harbison clinical scoring system (NH-CSS) (Azzi et al. 2015; Wakeling et al. 2017). Using this system, the diagnosis of SRS should be considered in individuals scoring at least four out of six defined criteria (Azzi et al. 2015). This system, developed using prospective data, is more sensitive (98%) and has a higher negative predictive value (89%) than previously proposed systems. The NH-CSS can also be used even if data are incomplete, for example in infancy before postnatal growth data are available (see Table 52.1).

A relatively low threshold for molecular testing for SRS is recommended (Wakeling et al. 2017). Testing should be carried out in all patients scoring four or more on the NH-CSS. It may also be considered in those meeting just three criteria, particularly in children under 2 years of age or adolescents/adults where data may be missing. However, in common with other proposed scoring systems, the NH-CSS has a low specificity (36%), reflecting the observation that many of the characteristic manifestations of SRS are "soft" and non-specific (Azzi et al. 2015). For this reason, it is recommended that individuals with negative test results are only given a diagnosis of "clinical SRS" if they score at least four out of six criteria on the NH-CSS, *including* both prominent forehead and relative macrocephaly at *birth*. These two features best distinguish those patients with SRS from those born SGA (Wakeling et al. 2017).

Many other clinical features have been reported in association with SRS (see Table 52.2). Although there is significant overlap in the phenotype seen in children with SRS and those born SGA who do not have the condition, a few manifestations are seen much more commonly in those with SRS (Wakeling et al. 2017). These include low muscle mass, dental problems (crowded or irregular teeth), micrognathia, downturned corners of the mouth, fifth finger clinodactyly, and excessive sweating (see Figures 52.1 and 52.2).

Although the diagnosis of SRS is primarily clinical, identification of the underlying molecular subtype can provide useful confirmation of the clinical diagnosis. Genotype–phenotype studies have shown that it is generally difficult to distinguish clinically between patients with 11p15 LOM and upd(7)mat (Wakeling et al. 2010). However, certain manifestations are more common in these specific subgroups and a confirmed molecular diagnosis may also help guide management in this regard (Azzi et al. 2015; Bruce et al. 2009; Wakeling et al. 2010). Children with 11p15 LOM tend to have a lower birth weight and length, are more likely to have asymmetry, have a more typical facial appearance and

TABLE 52.1 Netchine–Harbison clinical scoring system (Azzi et al. 2015). Clinical diagnosis is considered if patient scores at least four out of six from the following criteria

Clinical criteria	Definition
SGA (birth weight and/or birth length)	≤ −2 SDS for gestational age
Postnatal growth failure	Height at 24 ± 1 months ≤ −2 SDS *or* Height ≤ −2 SDS below mid-parental target height
Relative macrocephaly at birth	Head circumference at birth ≥1.5 SDS above birth weight and/or length SDS
Protruding/prominent forehead	Forehead projecting beyond the facial plane on a side view as a toddler (1–3 years)
Body asymmetry	LLD of ≥ 0.5 cm *or* arm asymmetry *or* LLD < 0.5 cm with at least two other asymmetrical body parts (one non-face)
Feeding difficulties and/or low BMI	BMI ≤ −2SDS at 24 months OR current use of a feeding tube OR cyproheptadine for appetite stimulation

BMI, body mass index; LLD, leg length discrepancy; SGA, small for gestational age.

TABLE 52.2 Common additional clinical features of Silver–Russell syndrome

Clinical feature	Frequency (%)
Facial features	
Triangular face	94
Micrognathia	62
Low-set and/or posteriorly rotated ears	49
Down-turned mouth	48
Irregular or crowded teeth	37
Structural anomalies	
Fifth finger clinodactyly	75
Shoulder dimples	66
Prominent heels	44
Male genital anomalies	40
Syndactyly of toes	30
Scoliosis and/or kyphosis	18
Neurocognitive problems	
Speech delay	40
Motor delay	37
Other	
Low muscle mass	56
Excessive sweating	54
High-pitched, squeaky voice	45
Delayed closure of fontanelle	43
Hypoglycaemia	22

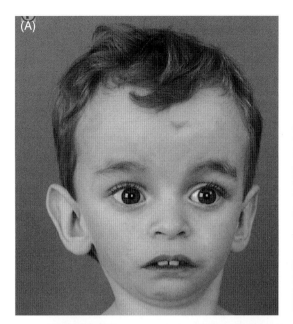
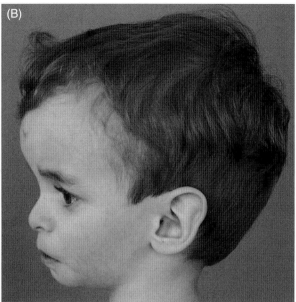
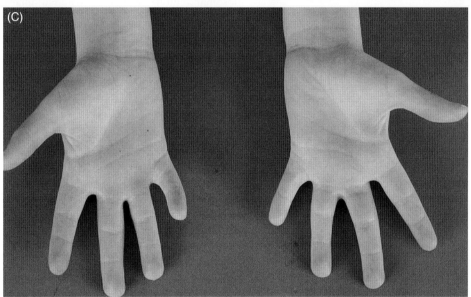

FIGURE 52.1 Clinical features of Silver–Russell syndrome. (A, B) Facial appearance in a child with 11p15 loss of methylation, showing triangular shaped face, prominent forehead, low-set and posteriorly rotated ears, down-turned corners of the mouth and micrognathia. (C) Fifth finger clinodactyly.

have a higher incidence of congenital anomalies than those with upd(7)mat. In contrast, children with upd(7)mat are more likely to develop neurocognitive problems (such as verbal dyspraxia and autistic spectrum disorder) than those with 11p15 LOM or clinical SRS.

Etiology, Pathogenesis and Genetics

SRS is a genetically heterogeneous condition. An underlying molecular cause is currently identified in around 60% of affected individuals. The two most common underlying mechanisms are loss of methylation on chromosome 11p15 (11p15 LOM) and maternal uniparental disomy for chromosome 7 (upd(7)mat) (Netchine et al. 2007).

Studies have shown that loss of methylation at the imprinted region on chromosome 11p15.5 is the underlying cause in around 50% of individuals with SRS (Gicquel et al. 2005; Netchine et al. 2007).

Imprinting refers to the differential expression of a gene(s) depending on whether the gene was inherited from the mother or the father. The 11p15.5 region contains a cluster of imprinted genes that play an important role in

FIGURE 52.2 A 3-year-old girl with Silver–Russell syndrome due to maternal uniparental disomy for chromosome 7. These individuals tend to have less typical facial features.

which, in turn, results in down-regulation of *IGF2* (Gicquel et al. 2005). The majority of individuals with SRS due to 11p15 LOM have epigenetic changes in methylation. However, a few (around 1%) have copy number variants (CNVs) of this region. In these individuals, the clinical phenotype depends on CNV size, location and parental origin (Begemann et al. 2012). Although these only account for a small proportion of cases with 11p15 LOM, it is important to identify individuals with an underlying CNV due to the potential risk of recurrent SRS or BWS in subsequent pregnancies.

In addition, around 5–10% of individuals with SRS have upd7(mat), with both copies of chromosome 7 inherited from the mother and none from the father (Eggermann et al. 1997; Kotzot et al. 1995). The manifestations of SRS associated with upd7(mat) are thought to result from altered expression of imprinted gene(s) on chromosome 7, though the precise mechanism is still unclear. Occasionally, isodisomy (duplication of the *same* maternal chromosome/chromosomal region) results in maternal inheritance of two copies of a gene mutation on chromosome 7, resulting in an additional autosomal recessive disorder (such as cystic fibrosis) in patients with upd7(mat) (Voss et al. 1989).

Other, rare 11p15.5-related genetic defects, including maternal *CDKN1C* (Brioude et al. 2013) and paternal *IGF2* variants (Begemann et al. 2015) have been described in individuals with SRS. Variants have also been reported in *HMGA2* and *PLAG1*, which act to downregulate fetal growth factor IGF2 (Abi Habib et al. 2018).

The molecular mechanism in around 35–40% of individuals with clinical SRS remains unknown. With advancing genetic technologies and knowledge, it is likely that additional etiologies will continue to be uncovered.

The risk of recurrence, both for parents of an affected child and for affected individuals, is generally low. However, accurate genetic counseling depends on the underlying genetic cause. Recurrence and offspring risks for upd(7)mat are low. The same advice applies for 11p15 LOM, provided an inherited 11p15 CNV has been excluded. In rare familial

regulation of fetal growth. The region includes two imprinting centers (ICRs) (Figure 52.3). ICR1 controls expression of the growth promoter *IGF2* and *H19*; ICR2 controls expression of the growth suppressor *CDKN1C* and other genes. These regions are differentially methylated. ICR1 is methylated on the paternal allele and unmethylated on the maternal allele.

Disturbances of both ICR1 and ICR2 can result in the overgrowth syndrome Beckwith–Wiedemann syndrome (BWS) (see Chapter 9) (Engel et al. 2000). SRS is associated with partial loss of paternal methylation in the ICR1 region

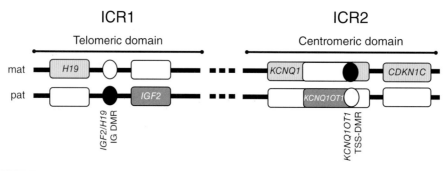

FIGURE 52.3 Diagram of the imprinted 11p15 region implicated in Silver–Russell syndrome. Dark gray boxes: paternally expressed genes (growth promotor *IGF2*). Light gray boxes: maternally expressed genes (*H19* and growth inhibitor *CDKN1C*). Circles: Differentially methylated regions (DMRs). Black circles: methylated DMRs. White circles: unmethylated DMRs.

cases with underlying mechanisms including maternally inherited *CDKN1C* mutation (Brioude et al. 2013), paternally inherited *IGF2* mutation (Begemann et al. 2015) or maternally inherited 11p15 duplication (Begemann et al. 2012), recurrence risk may be up to 50%.

Diagnostic Testing

SRS remains primarily a clinical diagnosis. Normal genetic test results do not exclude the condition. However, molecular testing can help confirm a clinical diagnosis, provide further information about the sub-type of SRS, and allow accurate counseling of parents and affected individuals regarding recurrence risk in future pregnancies (Wakeling et al. 2017).

Laboratory testing should start with DNA methylation analysis (such as MS-MLPA), looking for the two most common molecular causes, which account for around 60% of cases: 11p15 LOM and upd(7)mat (Eggermann et al. 2016). Testing for upd7(mat) can also be carried out by microsatellite analysis. However, using this technique, unlike DNA methylation analysis, parental samples are needed. Occasionally, 11p15 LOM may be missed due to mosaicism, with low levels of hypomethylation in blood. In this situation, hypomethylation may be detectable in a different tissue sample (such as a buccal swab or skin biopsy) (Azzi et al. 2014). 11p15 copy number changes can be detected by methylation analysis, but additional array analysis can help provide further information about the size and gene content of a copy number variant (CNV) (Begemann et al. 2012).

Individuals with SRS due to 11p15 LOM commonly have evidence of altered methylation patterns at multiple additional (maternally and paternally imprinted and non-imprinted) sites within the genome (Azzi et al. 2009). However, the clinical significance of this multi-locus imprinting disturbance (MLID) currently remains unclear and methylation analysis at additional loci is mainly carried out in the research setting at present.

Testing for sequence variants in genes (including *IGF2, CDKN1C, HMGA2,* and *PLAG1*) reported in association with SRS in a small number of cases is also not routinely requested. However, if there is a family history, and/or a very strong clinical suspicion of SRS, this may be considered (Abi Habib et al. 2018). Testing for gene variants is most efficiently carried out using next generation sequencing (either via gene panel testing, exome sequencing or whole genome sequencing).

Any further testing should be directed at exclusion of other causes of pre- and postnatal short stature (see differential diagnosis section, below). High resolution chromosome analysis (array comparative genomic hybridization [CGH] or single nucleotide polymorphism [SNP] array) should be considered, particularly if a child has more severe developmental delay and/or intellectual disability than typically seen in SRS. SNP array will also detect regions of segmental isosomy, as seen in some cases of upd(20)mat and upd(16)mat. If Temple syndrome is suspected, chromosome 14q32 methylation analysis should be requested. In children with relative microcephaly, chromosome breakage studies and/or gene panel testing should be considered to rule out a chromosome breakage syndrome associated with short stature. A skeletal survey may be indicated to rule out a skeletal dysplasia. When growth velocity is slow, children should be tested for other possible causes of postnatal growth failure, including renal disease, growth hormone deficiency and hypothyroidism.

Differential Diagnosis

The differential diagnosis of SRS is very wide as the manifestations are non-specific and variable in severity (see Table 52.3). Certain findings should prompt the clinician to consider an alternative diagnosis. These include relative microcephaly (head circumference SDS below weight and height SDS), significant global developmental delay, absence of severe feeding difficulties, facial dysmorphism, additional congenital anomalies and/or other features atypical of SRS. As familial cases of SRS are rare, a family history of growth failure and/or consanguinity should also raise the possibility of an alternative underlying diagnosis.

Over 30 different CNVs, in regions other than 11p15, have been reported in individuals with suspected SRS. These cases usually have more severe developmental delay and/or intellectual disability than are typically seen in SRS (Fokstuen and Kotzot 2014). Although manifestations of SRS can be present in individuals with pathogenic CNVs who might even fulfil the NH-CSS criteria for diagnosis, clinical diagnosis of SRS is not helpful as management needs to be tailored to the specific CNV, which often results in additional learning and medical issues.

Other imprinting disorders may present with manifestations of SRS. In particular, a small number of individuals suspected to have SRS have molecular abnormalities at the imprinted locus on chromosome 14q32 associated with Temple syndrome (Geoffron et al. 2018; Ioannides et al. 2014). Mechanisms include upd(14)mat, paternal microdeletion or loss of paternal methylation at the *MEG3/DLK1*:intergenic differentially methylated region (IG DMR). Occasionally, individuals investigated for pre-and postnatal growth failure are found to have upd(16)mat or upd(20) mat (Azzi et al. 2015). Maternally inherited gain of function *CDKN1C* variants have been shown to be associated with IMAGe syndrome (Arboleda et al. 2012).

A number of other conditions present with pre-and postnatal growth retardation and relative macrocephaly. These include 3-M syndrome, Mulibrey Nanism, SHORT syndrome, and Floating-Harbor syndrome. Some familial cases in consanguineous families, previously reported as

TABLE 52.3 Differential diagnosis of Silver–Russell syndrome

Condition	Key distinguishing features
Temple syndrome (14q32.2 imprinted region disruptions)	Significant overlap with SRS; some features more common in TS: Neonatal hypotonia Small hands and feet; rarely asymmetric Obesity in later childhood Precocious puberty Global developmental delay/ intellectual disability (usually mild)
IMAGe syndrome (Maternal *CDKN1C* mutation)	Congenital adrenal hypoplasia Metaphyseal dysplasia Male genital anomalies
Bloom syndrome	Severe, progressive microcephaly Patchy hyper/hypopigmentation Photosensitivity High malignancy risk Increased sister-chromatid exchange
Nijmegen breakage syndrome	Severe, progressive microcephaly Immunodeficiency Predisposition to malignancy Chromosome breakage
Microcephalic osteodysplastic primordial dwarfism type II (MOPD II)	Severe progressive microcephaly Prominent nose, micrognathia Mesomelic limb shortening; progressive metaphyseal dysplasia
Meier–Gorlin syndrome	Microcephaly Microtia Patellar hypoplasia Mammary hypoplasia (post-pubertal 100%)
Deletion 15q25.1-qter/ *IGF1R* mutation	Microcephaly Variable intellectual disability
3-M syndrome	Growth failure usually more severe Anteverted nares and mid-face hypoplasia Pectus deformity of the chest Radiographic changes (slender long bones, tall vertebral bodies)
SHORT syndrome	Rieger anomaly Deep-set eyes Partial lipodystrophy Diabetes mellitus type 2 (can be precipitated by GH therapy)
Floating-Harbor syndrome	Variable intellectual disability; significant speech delay Bulbous nose, wide columella, short philtrum, thin lips
Mulibrey Nanism	Muscle wasting Hepatomegaly secondary to progressive restrictive perimyocarditis Yellow spots on retina Slender long bones with thick cortex and narrow medullar channels

SRS, have subsequently been found to have variants in genes associated with 3-M syndrome (Akawi et al. 2011). It should be noted that, in individuals with SHORT syndrome, GH treatment has been reported to precipitate insulin resistance and diabetes mellitus type 2 (Verge et al. 1994).

Relative microcephaly is extremely unusual in SRS. In children with head size below that expected for their height centile, additional features such as patchy pigmentary abnormalities, photosensitivity and/or recurrent infections may point to a chromosome breakage disorder (such as Bloom syndrome or Nijmegen breakage syndrome). It is important to identify an underlying chromosome breakage disorder as GH treatment is contraindicated due to the associated risk of malignancy. Other conditions to consider in association with relative microcephaly include Meier–Gorlin syndrome, microcephalic primordial dwarfism type II, and 15q26-qter deletion/mutation of *IGF1R*.

In children with pre- and postnatal growth retardation but without additional features of SRS, gestational causes of growth retardation should be excluded. The prenatal history should also be directed at identifying maternal factors that contribute to intrauterine growth retardation, including placental insufficiency, hypertension, insulin-dependent diabetes, teratogen exposure (especially cigarettes and alcohol), and infection. Children born SGA due to other causes often show catch-up growth during the early part of childhood, unlike those with SRS.

Comparison of height with parent's heights and the target centile range should help identify children with familial short stature. A skeletal survey should be arranged in children with disproportionate short stature to look for signs of an underlying skeletal dysplasia. Some patients with osteogenesis imperfecta and type I collagenopathy (see Chapter 43) also have features that overlap with SRS (Parker et al. 2011). Pituitary disorders should be distinguishable by the absence of diagnostic features other than growth failure, poor growth velocity, and biochemical evidence of growth hormone deficiency.

MANIFESTATIONS AND MANAGEMENT

Feeding Difficulties and Growth Failure

The majority of children with SRS have both birth weight/length and postnatal height below −2 SD. Intrauterine growth retardation is often not evident until the third trimester. Feeding difficulties and failure to thrive are significantly more common in children born SGA due to SRS, than in those without SRS. After birth, failure to thrive is frequent, due to oromotor difficulties, poor appetite and gastrointestinal problems (Netchine et al. 2007; Fuke et al. 2013). Initially weight is more markedly affected than length, though this may also decrease over time as a result of poor calorie intake. Compared to children with SRS and 11p15 LOM or clinical SRS, those with upd(7)mat tend to have higher birth length but are more likely to fail to thrive after birth.

Unlike children born SGA due to placental insufficiency, those with SRS rarely show postnatal catch-up growth (Tanner et al. 1975; Wollmann et al. 1995). After the first few months of life, children typically have low-normal growth rate and growth centile curves that parallel the normal curves, though remain below the third percentile.

Children with SRS have low muscle mass and their weight is usually light for their height. Optimal BMI depends on muscle mass and even slight over-feeding can result in rapid increase in relative fat mass (Wakeling et al. 2017). Evidence suggests that rapid early weight gain, especially in children that have been tube-fed, is associated with an increased tendency to early adrenarche and central puberty, insulin resistance, and long-term risk of metabolic problems (Leunissen et al. 2009). Enteral feeding by nasogastric or gastrostomy tube may be necessary in children with severe feeding difficulties, to avoid hypoglycemia and/or malnutrition. However, this should be avoided where possible to prevent excessive weight gain.

Most adults with SRS have received GH treatment and final height data in untreated patients is limited. However, this suggests that the average adult height without GH treatment is around −3 SD (Binder et al. 2013; Wollmann et al. 1995). Growth-promoting growth hormone (GH) treatment can be prescribed in individuals with SRS under the SGA registered license (Clayton et al. 2007). Clinical trials of pharmacological doses of GH in children born SGA (including patients with SRS) have shown a demonstrable improvement in growth, with an increase in predicted adult height of 7–11 cm (Azcona et al. 1998; Chernausek, et al. 1996). More recently, Smeets et al. (2016) demonstrated a similar improvement in final adult height in a cohort of 62 children with a clinical diagnosis of SRS according to the NH-CSS. In this study, children with SRS had a similar response to GH treatment when compared to those with non-SRS SGA.

Response to GH therapy depends on a number of factors including initial height SDS and age, GH dose, and rate of weight gain. All SRS molecular subtypes benefit from GH treatment, with a trend towards greater height gain in those with upd(7)mat (Binder et al. 2008). GH treatment can also lead to improved appetite, lean body mass, muscle strength and motor development (Willemsen et al. 2007).

Levels of insulin-like growth factor 1 (IGF-1) are routinely used to monitor response to GH treatment. However, these may be difficult to interpret in children with SRS. Children with 11p15 LOM have significantly higher IGF-1 levels compared to children with upd(7)mat or other causes of SGA, suggesting a degree of IGF-1 resistance. Baseline IGF-1 levels are often in the high normal range and can rise significantly above the reference range following treatment, particularly in those with 11p15 LOM (Binder et al. 2008). However, overall, GH treatment seems to be safe and effective in children with SRS (Smeets et al. 2016).

Evaluation

- Birth weight, length and head circumference should be recorded.
- Physical examination should include accurate documentation of weight, height, and head circumference. Measurements should be plotted on growth charts and should continue to be monitored over time to determine growth velocity. BMI should also be calculated to aid decisions regarding nutritional management.
- Pubertal development should be assessed.
- In individuals near to/ at final adult height, early childhood growth records should be requested.
- Measurement of upper-to-lower segment ratio and arm span is helpful if there is suspicion of disproportionate growth.
- Additional examination findings, such as prominent forehead, body asymmetry and fifth finger clinodactyly should also be noted.
- For older children and adults, early childhood photographs should be reviewed to assess forehead prominence, where possible.

- It is essential to obtain a family history, which should include growth data of parents and siblings. Mid-parental height centile and target centile range should be calculated. (Note that these can be misleading if one of the parents has a genetic cause of short stature).
- If available, birth weights of parents and siblings should be recorded. Details of other family members with growth failure should also be documented. It may be necessary to examine parents and/or siblings for signs of SRS or other possible causes of short stature.
- A radiograph to determine bone age may be helpful, though this can be difficult to interpret under 4 years of age, unless bone age is advanced.
- Investigation for growth hormone deficiency may be considered if growth velocity is sub-optimal, under the guidance of a pediatric endocrinologist. However, growth hormone deficiency is a rare finding in SRS (Cassidy et al. 1986) and GH stimulation testing should be avoided where possible due to the risk of fasting hypoglycemia.

Management

- In the first few months of life, treatment should focus on ensuring adequate calorie intake, while avoiding rapid post-natal weight gain, which can lead to increased risk of metabolic problems.
- If nutritional intake is adequate to maintain BMI at around 12–14 kg m^{-2} (depending on muscle mass), tube feeding should be avoided.
- One report has suggested that cyproheptadine may be helpful in improving growth velocity and nutritional status before starting GH treatment (Lemoine et al. 2018).
- GH treatment should be started as soon as possible once calorie intake is adequate (depending on local licensing arrangements). For most children, treatment starts at 2–4 years and follows standard protocols.
- The use of GH treatment earlier than licensed may be considered in children with severe fasting hypoglycemia, extreme malnutrition despite nutritional support and/or severe muscular hypotonia.

Gastrointestinal

Feeding difficulties are extremely common in SRS and gastrointestinal problems are reported in over 70% of affected children (Marsaud et al. 2015). Gastroesophageal reflux occurs in around half and is often severe. Persistent vomiting beyond infancy is reported in around 30% of children. Constipation is also common, particularly after 2 years of age.

Evaluation

- Ask about symptoms of gastroesophageal reflux, gut dysmotility and constipation.
- Consider investigations for gastroesophageal reflux including pH monitoring and endoscopy to detect esophagitis.

Management

- Referral to a pediatric gastroenterologist should be considered.
- Gastroesophageal reflux should be treated according to standard practice.
- In individuals with severe feeding difficulties who are not responding to treatment, functional or structural disorders of the gastrointestinal tract (such as gut dysmotility or malrotation) should be excluded.
- Advice from a speech therapist may be valuable in patients with oromotor and/or sensory feeding issues.

Central Puberty and Adrenarche

Adrenarche is often early in children with SRS, particularly in association with 11p15 LOM (Binder et al. 2017). Bone age, which is often delayed in early childhood, also advances rapidly compared to chronological age from around 8–9 years of age. Individuals with SRS usually start central puberty within the normal age range but at the younger end of the spectrum. However, pubertal growth is faster and occurs over a narrower time-frame than normal, resulting in a shorter than expected adult height (Binder et al. 2008). Children who have a rapid early weight gain and increase in BMI, may be particularly prone to develop early adrenarche and central puberty (Leunissen et al. 2009). A retrospective study of 62 individuals meeting clinical criteria for diagnosis of SRS found that early adrenarche was associated with younger age at start of GH therapy but that response to treatment and adult height were not compromised by early adrenarche (Binder et al. 2017).

A study by van der Steen et al. (2015) showed that treatment with gonadotrophin-releasing hormone analogue (GnRHa) started at the onset of puberty and continued for at least two years, combined with GH therapy, improves adult height in children born SGA who have low predicted final height. Currently there are insufficient data to draw conclusions about the use of GnRHa specifically in patients with SRS.

Evaluation

- Monitor for clinical signs of premature adrenarche and early and/or accelerated central puberty.
- Monitor bone age, especially from mid-childhood.

Management

- Consider treatment with GnRHa for at least two years in children with evidence of either early or prompt and/or accelerated central puberty to preserve adult height potential. Treatment should start no later than 12 years in girls and 13 years in boys.

Development and Behavior

Children with SRS frequently have delayed motor and/or speech development (Fuke et al. 2013; Netchine et al. 2007; Wakeling et al. 2010). Verbal dyspraxia is common in individuals with upd(7)mat, likely due to absence of the paternally inherited copy of *FOXP2* (Feuk et al. 2006). A few have more global developmental delay and/or learning difficulties, though these are usually mild (Azzi et al. 2015; Wakeling et al. 2010). In particular, children with upd(7)mat appear to be at higher risk of developing neurocognitive problems and/or autistic spectrum disorder (Azzi et al. 2015). Myoclonus dystonia (associated with the paternally expressed *SGCE* on chromosome 7q21) has also been reported in older children and adults with upd(7)mat (Wakeling et al. 2010).

Evaluation

- Children should be referred for formal developmental assessment if there are any concerns to ensure that early intervention is in place, if necessary.
- Children with upd(7)mat should be monitored for signs of verbal dyspraxia, myoclonus dystonia and/or autistic spectrum disorder.
- Older children may need educational evaluation for learning difficulties, especially those related to speech and language delay.

Management

- Children with motor delay related to low muscle tone should be referred as early as possible for physiotherapy.
- Speech and language therapy is frequently required and should be initiated as early as possible when speech delay and/or oromotor problems are identified.
- Appropriate educational support for children identified as having learning difficulties should be guided by formal, comprehensive evaluation of educational needs.
- Psychological counseling may be valuable in helping with peer relationships, body self-image, and self-esteem issues (Ballard et al. 2019).

Craniofacial and Dental

Micrognathia and high, narrow palate are common in patients with SRS, resulting in a small pointed chin, dental crowding and overbite (Cullen and Wesley 1987). Facial asymmetry can also cause crossbite. Additional dental abnormalities include microdontia, delayed eruption of teeth, and absence of secondary teeth. Individuals with 11p15 LOM are more likely to have velopharyngeal insufficiency with/without cleft palate (Wakeling et al. 2010). The input of an experienced maxillofacial team, including an orthodontist, may be needed. Orthodontic treatment can help improve oropharyngeal function and facial appearance and may also reduce the frequency of otitis media (Kisnisci et al. 1999).

Excessive daytime tiredness, snoring and/or disrupted sleep are frequently reported, and a retrospective study found evidence of mild sleep disordered breathing in 74%; however, further studies are needed (Giabicani et al. 2015).

Evaluation

- Serial cephalometric evaluation by an orthodontist may be needed in individuals who have significant micrognathia, dental crowding, and/or other dental anomalies.
- Patients with suspected obstructive sleep apnea should be referred for sleep studies and specialist review.

Management

- Orthodontic treatment will frequently be needed for dental crowding and other dental abnormalities.
- Some patients with severe craniofacial disproportion may benefit from referral to an experienced craniofacial team for specialist treatment, such as rapid palatal expansion.

Orthopedic

Orthopedic problems associated with SRS include asymmetry, scoliosis, hip dysplasia, and hand/foot anomalies. Asymmetry may affect the arms, legs and/or body and is present in around a third of affected individuals overall. However, it is significantly more common in association with 11p15 LOM, likely due to underlying tissue mosaicism (Wakeling et al. 2010). Typically, little or no increase in limb length difference is seen during childhood and the degree of asymmetry does not appear to be affected by GH treatment (Rizzo et al. 2001).

Scoliosis is present in nearly 20% of individuals with SRS (Wakeling et al. 2017). The causal relationship to leg length asymmetry is not clear (Yamaguchi et al. 2015). Further studies are also needed to determine whether GH treatment influences the onset or progression of scoliosis.

Evaluation

- Examine for evidence of body and/or limb asymmetry, with measurement of limb lengths and limb circumferences.

- Examine the spine for signs of scoliosis or kyphosis (Adam's forward bend test).
- Refer patients with scoliosis to the orthopedic team, prior to starting GH therapy, for monitoring during treatment.

Management

- Individuals with limb-length asymmetry should be under the care of an orthopedic team.
- Leg length asymmetry in childhood can be managed conservatively with shoe raises.
- Older patients with severe limb-length discrepancy may benefit from limb-lengthening surgery once growth has ceased (Goldman et al. 2013).

Congenital Anomalies

Congenital anomalies are quite frequently reported in individuals with a confirmed molecular diagnosis of SRS, particularly those with 11p15 LOM (Wakeling et al. 2010). These include urogenital anomalies, cleft palate, congenital heart disease, and limb defects. Fifth finger clinodactyly is seen in 75% of affected individuals (Wakeling et al. 2017).

Evaluation

- Boys should be examined for genital anomalies, including hypospadias and cryptorchisism.
- Girls with SRS and primary amenorrhea should be investigated for Mayer–Rokitansky–Kuster–Hauser syndrome (congenital hypoplasia of the uterus and upper part of the vagina).
- A renal ultrasound scan may be considered following diagnosis to look for structural anomalies.

Management

- Surgical treatment of undescended testes or hypospadias may be required.
- Renal anomalies should be treated according to standard practice.

Hypoglycemia

Hypoglycemia is reported in around a quarter of affected children, particularly under 5 years of age (Wakeling et al. 2010). Low muscle and liver mass, feeding difficulties, and a disproportionately large cranium-for-body size all contribute to the risk of problems. A high frequency of spontaneous, asymptomatic nocturnal hypoglycemia has been observed (Azcona and Stanhope 2005). It is also important to be aware of the increased risk of ketotic hypoglycemia during intercurrent illness or when fasting (for example, prior to elective surgery). Excessive sweating is reported in around two thirds of younger children. This is sometimes, but not always, associated with hypoglycemia (Wakeling et al. 2010).

Evaluation

- Teach parents how to recognize signs of hypoglycemia, monitor for ketonuria at home and recognize the "safe fasting time" for their own child.
- Severe non-fasting and non-ketotic hypoglycemia should always be referred for specialist review and further investigation.

Management

- Avoid prolonged periods of fasting.
- In children at risk, night-time hypoglycemia can be prevented by adding high molecular weight glucose polymer (for infants under 10 months) or uncooked cornstarch to the last evening feed.
- In conjunction with the local pediatric team, develop an emergency guidance plan for parents, to include rapid admission and intravenous dextrose treatment when the child is unwell with ketonuria or hypoglycemia.
- If surgery is planned, the following are recommended: optimize nutritional status prior to surgery; arrange anesthetic review in advance; administer intravenous dextrose well before, during and after surgery; schedule surgery as early as possible; monitor blood glucose levels closely before, during, and after surgery; maintain temperature appropriate for size, not age; discharge only when able to be adequately fed with normal glucose levels and ketonuria is absent. It is helpful for the overseeing clinician to provide a letter detailing the above advice for the surgical team prior to admission.
- In children with persistent, problematic hypoglycemia, early treatment with GH and/or placement of a gastrostomy or jejunostomy tube should be considered (Wakeling et al. 2017).

Adulthood

There is little published information on the long-term natural history of SRS. However, the small numbers of adults reported have few medical problems and there are currently no recommendations for long-term follow up beyond routine care. Children born SGA are at increased risk of developing adult-onset disorders such as diabetes mellitus type 2, hypertension, and hyperlipidemia at a relatively young age (Barker et al. 1989). A study of 29 individuals with SRS and 171 non-SRS individuals born SGA, both treated with GH, showed similar metabolic profiles (blood pressure, serum lipids, glucose, and insulin levels) both during and up to two years after stopping treatment (Smeets et al. 2017). None of the individuals with SRS in this study developed metabolic syndrome or diabetes mellitus type 2. Interestingly, one of Dr Russell's original patients was recently reported, having been confirmed as having 11p15 LOM (Searle and Johnson 2016). At the age of 69 he has type 2 diabetes mellitus, osteopenia, testosterone deficiency and hypercholestrolaemia. However, long-term follow up of larger numbers of affected individuals is needed to determine whether the risk of metabolic problems is significantly increased in later life, as previously reported in cohorts of individuals born SGA.

Evaluation

- At transition to adult services, plans for ongoing medical care should be drawn up, dependent on individual needs.

Management

- Adult physicians should be aware of the increased risk of metabolic problems in later life associated with being born SGA, particularly where there is a history of rapid gain in weight-for-length in early life.

ACKNOWLEDGMENT

Howard Saal authored this chapter on Silver–Russell syndrome for the first three editions of Management of Genetic Syndromes, and kindly provided his template for my revision in the fourth edition.

RESOURCES

International Consensus Guidelines for diagnosis and management of Silver–Russell syndrome can be found at: *https://www.nature.com/articles/nrendo.2016.138*

Support groups

The Magic Foundation for Children's Growth

6645 W. North Avenue
Oak Park, IL 60302, USA
Phone: (708) 383-0808
Email: rss@magicfoundation.org
Website: *https://www.magicfoundation.org*

SRS Global Alliance

Details of other country-specific support groups and medical specialists, and further information about SRS (including a Summary of the International Consensus Guidelines for affected individuals, families and caregivers), can be found via the SRS Global Alliance.
Website: *https://silverrussellsyndrome.org*

National Organization for Rare Disorders (NORD)

Website: *https://rarediseases.org/rare-diseases/russell-silver-syndrome/*

REFERENCES

Abi Habib, W, Brioude F, Edouard T, Bennett JT, Lienhardt-Roussie A, Tixier F, Salem J, Yuen T, Azzi S, Le Bouc Y, Harbison MD, Netchine I (2018) Genetic disruption of the oncogenic HMGA2-PLAG1-IGF2 pathway causes fetal growth restriction. *Genet Med* 20:250–58.

Akawi, NA, Ali BR, Hamamy H, Al-Hadidy A, Al-Gazali L (2011) Is autosomal recessive Silver-Russel syndrome a separate entity or is it part of the 3-M syndrome spectrum? *Am J Med Genet A* 155A:1236–45.

Arboleda VA, Lee H, Parnaik R, Fleming A, Banerjee A, Ferraz-de-Souza B, Delot EC, Rodriguez-Fernandez IA, Braslavsky D, Bergada I, Dell'Angelica EC, Nelson SF, Martinez-Agosto JA, Achermann JC, Vilain E (2012) Mutations in the PCNA-binding domain of CDKN1C cause IMAGe syndrome. *Nat Genet* 44:788–92.

Azcona C, Albanese A, Bareille P, Stanhope R (1998) Growth hormone treatment in growth hormone-sufficient and -insufficient children with intrauterine growth retardation/Russell-Silver syndrome. *Horm Res* 50:22–7.

Azcona C, Stanhope R (2005) Hypoglycaemia and Russell-Silver syndrome. *J Pediatr Endocrinol Metab* 18:663–70.

Azzi S, Blaise A, Steunou V, Harbison MD, Salem J, Brioude F, Rossignol S, Habib WA, Thibaud N, Neves CD, Jule ML, Brachet C, Heinrichs C, Bouc YL, Netchine I (2014) Complex tissue-specific epigenotypes in Russell-Silver Syndrome associated with 11p15 ICR1 hypomethylation. *Hum Mutat* 35:1211–20.

Azzi S, Rossignol S, Steunou V, Sas T, Thibaud N, Danton F, Le Jule M, Heinrichs C, Cabrol S, Gicquel C, Le Bouc Y, Netchine I (2009) Multilocus methylation analysis in a large cohort of 11p15-related foetal growth disorders (Russell Silver and Beckwith Wiedemann syndromes) reveals simultaneous loss of methylation at paternal and maternal imprinted loci. *Hum Mol Genet* 18:4724–33.

Azzi S, Salem J, Thibaud N, Chantot-Bastaraud S, Lieber E, Netchine I, Harbison MD (2015) A prospective study validating a clinical scoring system and demonstrating phenotypical-genotypical correlations in Silver-Russell syndrome. *J Med Genet* 52:446–53.

Ballard LM, Jenkinson E, Byrne CD, Child JC, Davies JH, Inskip H, Lokulo-Sodipe O, Mackay DJG, Wakeling EL, Temple IK, Fenwick A (2019) Lived experience of Silver-Russell syndrome: implications for management during childhood and into adulthood. *Arch Dis Child* 104:76–82.

Barker DJ, Winter PD, Osmond C, Margetts B, Simmonds SJ (1989) Weight in infancy and death from ischaemic heart disease. *Lancet* 2:577–80.

Begemann M, Spengler S, Gogiel M, Grasshoff U, Bonin M, Betz RC, Dufke A, Spier I, Eggermann T (2012) Clinical significance of copy number variations in the 11p15.5 imprinting control regions: new cases and review of the literature. *J Med Genet* 49:547–53.

Begemann M, Zirn B, Santen G, Wirthgen E, Soellner L, Buttel HM, Schweizer R, van Workum W, Binder G, Eggermann T (2015)

Paternally Inherited IGF2 Mutation and Growth Restriction. *N Engl J Med* 373:349–56.

Binder G, Liebl M, Woelfle J, Eggermann T, Blumenstock G, Schweizer R (2013) Adult height and epigenotype in children with Silver-Russell syndrome treated with GH. *Horm Res Paediatr* 80:193–200.

Binder G, Schweizer R, Blumenstock G, Ferrand N (2017) Adrenarche in Silver-Russell Syndrome: Timing and Consequences. *J Clin Endocrinol Metab* 102:4100–08.

Binder G, Seidel AK, Martin DD, Schweizer R, Schwarze CP, Wollmann HA, Eggermann T, Ranke MB (2008) The endocrine phenotype in silver-russell syndrome is defined by the underlying epigenetic alteration. *J Clin Endocrinol Metab* 93:1402–7.

Brioude F, Oliver-Petit I, Blaise A, Praz F, Rossignol S, Le Jule M, Thibaud N, Faussat AM, Tauber M, Le Bouc Y, Netchine I (2013) CDKN1C mutation affecting the PCNA-binding domain as a cause of familial Russell Silver syndrome. *J Med Genet* 50:823–30.

Bruce S, Hannula-Jouppi K, Peltonen J, Kere J, Lipsanen-Nyman M (2009) Clinically distinct epigenetic subgroups in Silver-Russell syndrome: the degree of H19 hypomethylation associates with phenotype severity and genital and skeletal anomalies. *J Clin Endocrinol Metab* 94:579–87.

Cassidy SB, Blonder O, Courtney VW, Ratzan SK, Carey DE (1986) Russell-Silver syndrome and hypopituitarism. Patient report and literature review. *Am J Dis Child* 140:155–9.

Chernausek SD, Breen TJ, Frank GR (1996) Linear growth in response to growth hormone treatment in children with short stature associated with intrauterine growth retardation: the National Cooperative Growth Study experience. *J Pediatr* 128:S22–7.

Clayton PE, Cianfarani S, Czernichow P, Johannsson G, Rapaport R, Rogol A (2007) Management of the child born small for gestational age through to adulthood: a consensus statement of the International Societies of Pediatric Endocrinology and the Growth Hormone Research Society. *J Clin Endocrinol Metab* 92:804–10.

Cullen CL, Wesley RK (1987) Russell-Silver syndrome: microdontia and other pertinent oral findings. *ASDC J Dent Child* 54:201–4.

Eggermann K, Bliek J, Brioude F, Algar E, Buiting K, Russo S, Tumer Z, Monk D, Moore G, Antoniadi T, Macdonald F, Netchine I, Lombardi P, Soellner L, Begemann M, Prawitt D, Maher ER, Mannens M, Riccio A, Weksberg R, Lapunzina P, Gronskov K, Mackay DJ, Eggermann T (2016) EMQN best practice guidelines for the molecular genetic testing and reporting of chromosome 11p15 imprinting disorders: Silver-Russell and Beckwith-Wiedemann syndrome. *Eur J Hum Genet* 24:1377–87.

Eggermann T, Wollmann HA, Kuner R, Eggermann K, Enders H, Kaiser P, Ranke MB (1997) Molecular studies in 37 Silver-Russell syndrome patients: frequency and etiology of uniparental disomy. *Hum Genet* 100:415–9.

Engel JR, Smallwood A, Harper A, Higgins MJ, Oshimura M, Reik W, Schofield PN, Maher ER (2000) Epigenotype-phenotype correlations in Beckwith-Wiedemann syndrome. *J Med Genet* 37:921–6.

Feuk L, Kalervo A, Lipsanen-Nyman M, Skaug J, Nakabayashi K, Finucane B, Hartung D, Innes M, Kerem B, Nowaczyk MJ, Rivlin J, Roberts W, Senman L, Summers A, Szatmari P, Wong V, Vincent JB, Zeesman S, Osborne LR, Cardy JO, Kere J, Scherer SW, Hannula-Jouppi K (2006) Absence of a paternally inherited FOXP2 gene in developmental verbal dyspraxia. *Am J Hum Genet* 79:965–72.

Fokstuen S, Kotzot D (2014) Chromosomal rearrangements in patients with clinical features of Silver-Russell syndrome. *Am J Med Genet A* 164A:1595–605.

Fuke T, Mizuno S, Nagai T, Hasegawa T, Horikawa R, Miyoshi Y, Muroya K, Kondoh T, Numakura C, Sato S, Nakabayashi K, Tayama C, Hata K, Sano S, Matsubara K, Kagami M, Yamazawa K, Ogata T (2013) Molecular and clinical studies in 138 Japanese patients with Silver-Russell syndrome. *PLoS One* 8:e60105.

Geoffron, S, W Abi Habib, S Chantot-Bastaraud, B Dubern, V Steunou, S Azzi, A Afenjar, T Busa, A Pinheiro Canton, C Chalouhi, MN Dufourg, B Esteva, M Fradin, D Genevieve, S Heide, B Isidor, A Linglart, F Morice Picard, C Naud-Saudreau, I Oliver Petit, N Philip, C Pienkowski, M Rio, S Rossignol, M Tauber, J Thevenon, TA Vu-Hong, MD Harbison, J Salem, F Brioude, I Netchine, and E Giabicani (2018) Chromosome 14q32.2 Imprinted Region Disruption as an Alternative Molecular Diagnosis of Silver-Russell Syndrome, *J Clin Endocrinol Metab*, 103:2436–46.

Giabicani, E, M Boule, E Galliani, and I Netchine (2015) Sleep apneas in Silver Russell syndrome: a constant finding. *Horm Res (Suppl 1)*, 84:262.

Gicquel, C, S Rossignol, S Cabrol, M Houang, V Steunou, V Barbu, F Danton, N Thibaud, M Le Merrer, L Burglen, AM Bertrand, I Netchine, and Y Le Bouc (2005) Epimutation of the telomeric imprinting center region on chromosome 11p15 in Silver-Russell syndrome, *Nat Genet*, 37:1003–7.

Goldman, V, TH McCoy, MD Harbison, AT Fragomen, and SR Rozbruch (2013) Limb lengthening in children with Russell-Silver syndrome: a comparison to other etiologies, *J Child Orthop*, 7:151–6.

Ioannides Y, Lokulo-Sodipe K, Mackay DJ, Davies JH, Temple IK (2014) Temple syndrome: improving the recognition of an underdiagnosed chromosome 14 imprinting disorder: an analysis of 51 published cases. *J Med Genet* 51:495–501.

Kisnisci RS, Fowel SD, Epker BN (1999) Distraction osteogenesis in Silver Russell syndrome to expand the manible. *Am J Orthod Dentofacial Orthop* 116:25–30.

Kotzot D, Schmitt S, Bernasconi F, Robinson WP, Lurie IW, Ilyina H, Mehes K, Hamel BC, Otten BJ, Hergersberg M (1995) Uniparental disomy 7 in Silver-Russell syndrome and primordial growth retardation. *Hum Mol Genet* 4:583–7.

Lemoine A, Harbison MD, Salem J, Tounian P, Netchine I, Dubern B (2018) Effect of cyproheptadine on weight and growth velocity in children with Silver-Russell syndrome. *J Pediatr Gastroenterol Nutr* 66:306–11.

Leunissen RW, Kerkhof GF, Stijnen T, Hokken-Koelega A (2009) Timing and tempo of first-year rapid growth in relation to cardiovascular and metabolic risk profile in early adulthood. *JAMA* 301:2234–42.

Marsaud C, Rossignol S, Tounian P, Netchine I, Dubern B (2015) Prevalence and management of gastrointestinal manifestations in Silver-Russell syndrome. *Arch Dis Child* 100:353–8.

Netchine I, Rossignol S, Dufourg MN, Azzi S, Rousseau A, Perin L, Houang M, Steunou V, Esteva B, Thibaud N, Demay MC, Danton F, Petriczko E, Bertrand AM, Heinrichs C, Carel JC, Loeuille GA, Pinto G, Jacquemont ML, Gicquel C, Cabrol S, Le Bouc Y (2007) 11p15 imprinting center region 1 loss of methylation is a common and specific cause of typical Russell-Silver syndrome: clinical scoring system and epigenetic-phenotypic correlations. *J Clin Endocrinol Metab* 92:3148–54.

Parker MJ, Deshpande C, Rankin J, Wilson LC, Balasubramanian M, Hall CM, Wagner BE, Pollitt R, Dalton A, Bishop NJ (2011) Type 1 collagenopathy presenting with a Russell-Silver phenotype. *Am J Med Genet A* 155A:1414–8.

Rizzo V, Traggiai C, Stanhope R (2001) Growth hormone treatment does not alter lower limb asymmetry in children with Russell-Silver syndrome. *Horm Res* 56:114–6.

Russell A (1954) A syndrome of intra-uterine dwarfism recognizable at birth with cranio-facial dysostosis, disproportionately short arms, and other anomalies (5 examples). *Proc R Soc Med* 47:1040–4.

Searle C, Johnson D (2016) Russel-Silver syndrome: A historical note and comment on an older adult. *Am J Med Genet A* 170:466–470.

Silver HK, Kiyasu W, George J, Deamer WC (1953) Syndrome of congenital hemihypertrophy, shortness of stature, and elevated urinary gonadotropins. *Pediatrics* 12:368–76.

Smeets CC, Renes JS, van der Steen M, Hokken-Koelega AC (2017) Metabolic Health and Long-Term Safety of Growth Hormone Treatment in Silver-Russell Syndrome. *J Clin Endocrinol Metab* 102:983–91.

Smeets CC, Zandwijken GR, Renes JS, Hokken-Koelega AC (2016) Long-Term Results of GH Treatment in Silver-Russell Syndrome (SRS): Do They Benefit the Same as Non-SRS Short-SGA? *J Clin Endocrinol Metab* 101:2105–12.

Tanner JM, Lejarraga H, Cameron N (1975) The natural history of the Silver-Russell syndrome: a longitudinal study of thirty-nine cases, *Pediatr Res*, 9:611–23.

van der Steen M, Lem AJ, van der Kaay DC, Bakker-van Waarde WM, van der Hulst FJ, Neijens FS, Noordam C, Odink RJ, Oostdijk W, Schroor EJ, Westerlaken C, Hokken-Koelega AC (2015) Metabolic health in short children born small for gestational age treated with growth hormone and gonadotropin-releasing hormone analog: Results of a randomized, dose-response trial. *J Clin Endocrinol Metab* 100:3725–34.

Verge CF, Donaghue KC, Williams PF, Cowell CT, Silink M (1994) Insulin-resistant diabetes during growth hormone therapy in a child with SHORT syndrome. *Acta Paediatr* 83:786–8.

Voss R, Ben-Simon E, Avital A, Godfrey S, Zlotogora J, Dagan J, Tikochinski Y, Hillel J (1989) Isodisomy of chromosome 7 in a patient with cystic fibrosis: could uniparental disomy be common in humans? *Am J Hum Genet* 45:373–80.

Wakeling EL, Amero SA, Alders M, Bliek J, Forsythe E, Kumar S, Lim DH, MacDonald F, Mackay DJ, Maher ER, Moore GE, Poole RL, Price SM, Tangeraas T, Turner CL, Van Haelst MM, Willoughby C, Temple IK, Cobben JM (2010) Epigenotype-phenotype correlations in Silver-Russell syndrome. *J Med Genet* 47:760–8.

Wakeling EL, Brioude F, Lokulo-Sodipe O, O'Connell SM, Salem J, Bliek J, Canton AP, Chrzanowska KH, Davies JH, Dias RP, Dubern B, Elbracht M, Giabicani E, Grimberg A, Gronskov K, Hokken-Koelega AC, Jorge AA, Kagami M, Linglart A, Maghnie M, Mohnike K, Monk D, Moore GE, Murray PG, Ogata T, Petit IO, Russo S, Said E, Toumba M, Tumer Z, Binder G, Eggermann T, Harbison MD, Temple IK, Mackay DJ, Netchine I (2017) Diagnosis and management of Silver-Russell syndrome: first international consensus statement. *Nat Rev Endocrinol* 13:105–24.

Willemsen RH, Arends NJ, Bakker-van Waarde WM, Jansen M, van Mil EG, Mulder J, Odink RJ, Reeser M, Rongen-Westerlaken C, Stokvis-Brantsma WH, Waelkens JJ, Hokken-Koelega AC (2007) Long-term effects of growth hormone (GH) treatment on body composition and bone mineral density in short children born small-for-gestational-age: six-year follow-up of a randomized controlled GH trial. *Clin Endocrinol (Oxf)* 67:485–92.

Wollmann HA, Kirchner T, Enders H, Preece MA, Ranke MB (1995) Growth and symptoms in Silver-Russell syndrome: review on the basis of 386 patients. *Eur J Pediatr* 154:958–68.

Yakoreva M, Kahre T, Žordania R, Reinson K, Teek R, Tillmann V, Peet A, Õiglane-Shlik E, Pajusalu S, Murumets Ü, Vals MA, Mee P, Wojcik MH, Õunap K (2019) A retrospective analysis of the prevalence of imprinting disorders in Estonia from 1998 to 2016. *Eur J Hum Genet* 27:1649–1658.

Yamaguchi KT, Jr, Salem JB, Myung KS, Romero AN, Jr, Skaggs DL (2015) Spinal Deformity in Russell-Silver Syndrome. *Spine Deform* 3:95–97.

53

SMITH–LEMLI–OPITZ SYNDROME

ALICIA LATHAM

Division of Medical Genetics, Weill Cornell Medical College, and Division of Clinical Genetics, Memorial Sloan Kettering Cancer Center, New York, USA

CHRISTOPHER CUNNIFF

Division of Medical Genetics, Weill Cornell Medical College, New York, USA

INTRODUCTION

Incidence

Smith–Lemli–Opitz syndrome (SLOS) was first described in three boys with a characteristic pattern of malformation including growth deficiency, developmental delay, ptosis, downslanting palpebral fissures, and hypospadias (Smith et al. 1964). For almost 30 years after the initial patients were described, the cause of SLOS was unknown. However, in 1993, Irons et al. found low cholesterol and elevated 7-dehydrocholesterol levels in individuals with SLOS (Irons et al. 1993). Subsequent studies have shown conclusively that SLOS is caused by deficiency of the enzyme 7-dehydrocholesterol reductase, the final step of the cholesterol biogenesis pathway (Wassif et al. 1998; Waterham et al. 1998). The gene encoding 7-dehydrocholesterol reductase, *DHCR7*, has been identified, and mutation analysis is clinically available for diagnosis, carrier detection, and prenatal diagnosis.

SLOS is more common in people of European background and has been reported rarely in those of African or Asian descent. There is an excess of males diagnosed with SLOS, which represents a bias of ascertainment as a result of the hypogenitalism seen in boys. Incidence estimates have varied widely, from about 1 in 10,000 to 1 in 70,000. In an investigation of carrier frequency in Poland, about 2.5% of those screened was a carrier for one of the two most common mutations (Ciara et al. 2006) found in Caucasians. However, most investigations have found a birth prevalence much lower than that predicted by carrier frequencies, which suggests either reduced fertility among carrier couples or a high rate of pregnancy loss of affected fetuses (Nowaczyk et al. 2006). Carrier screening of over 250,000 individuals in the United States found the highest carrier rate among Ashkenazi Jews (1 in 43) and Northern Europeans (1 in 54); and an analysis of the predicted birth incidence compared to the observed frequency in the published literature suggested that anywhere from 42% to 88% of affected conceptuses have prenatal demise (Lazarin et al. 2017).

Early mortality for live borns with SLOS has also been observed commonly, particularly for those who are at the more severely affected end of the clinical spectrum. Of the 19 severely affected individuals reported by Curry et al. (1987), only one lived past the age of 6 months. There are no population-based studies of longevity in individuals with SLOS, however, and survival into adulthood has been observed, particularly among those who have few or no major malformations.

Diagnostic Criteria

No standard diagnostic criteria have been formulated for SLOS. The diagnosis is usually based on the recognition of a constellation of characteristic clinical features, with diagnostic confirmation by measurement of elevated 7-dehydrocholesterol and identification of pathogenic

---

Cassidy and Allanson's Management of Genetic Syndromes, Fourth Edition.
Edited by John C. Carey, Agatino Battaglia, David Viskochil, and Suzanne B. Cassidy.
© 2021 John Wiley & Sons, Inc. Published 2021 by John Wiley & Sons, Inc.

variants of *DHCR7*. The cardinal features of SLOS are prenatal-onset growth deficiency, microcephaly, developmental delay, characteristic facial features, cleft palate, cardiac defects, hypospadias, postaxial polydactyly, and 2–3 toe syndactyly. A high percentage of affected individuals have developmental delay or intellectual disability, but normal to low normal IQ has been observed (Eroglu et al. 2017). Cutaneous syndactyly of the second and third toes appears to be the most consistent structural anomaly, present in over 90% of biochemically confirmed cases (Figure 53.1).

The facial appearance is characterized by narrow bifrontal diameter, ptosis, downslanting palpebral fissures, and a depressed nasal bridge, and anteverted nares (Figure 53.2).

The ears are frequently low-set and posteriorly rotated. There is often retrognathia. These features change with age and may be difficult to discern in adulthood (Figure 53.3) (Ryan et al. 1998).

Early photographs can be helpful diagnostically. In the evaluation of adults it is necessary to have a high index of suspicion and initiate laboratory evaluation.

SLOS was previously subdivided into type I (classical) and type II (severe) using phenotypic characteristics. With the availability of biochemical and molecular diagnosis, it is now apparent that the phenotype ranges across a broad spectrum, rather than two distinct subtypes. Before the availability of biochemical testing for SLOS, a group of individuals with a severe phenotype including XY sex reversal and early lethality was described (Curry et al. 1987). It is now known that these children, who were considered to have SLOS type II, actually represent the severely affected end of the biochemical and phenotypic spectrum of children with SLOS (Cunniff et al. 1997).

Despite its extreme phenotypic variability, there is generally concordance within a sibship. The spectrum of phenotypic findings has been described in several series of biochemically confirmed individuals (Tint et al. 1995; Cunniff et al. 1997; Ryan et al. 1998). The reviews of Kelley

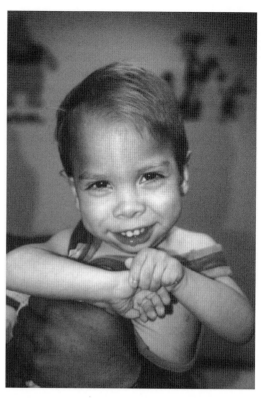

FIGURE 53.2 A 2-year-old boy with frontal hair upsweep, telecanthus, anteverted nares, smooth-appearing philtrum, and a postaxial scar on the left hand from removal of a supernumerary digit.

and Hennekam (2000), Porter (2008), and Nowaczyk and Irons (2012) also provide detailed descriptions of the clinical findings in affected people.

Etiology, Pathogenesis, and Genetics

SLOS is inherited as an autosomal recessive trait with widely variable expression. Because it is an autosomal recessive condition, the recurrence risk for parents of an affected child is 25%.

The etiology of SLOS is deficiency of the enzyme 7-dehydrocholesterol reductase, the final enzymatic step in the Kandutsch–Russell pathway of cholesterol biogenesis. More than 150 pathogenic variants of the *DHCR7* gene have been identified (Waterham and Hennekam 2012). These variants are widely distributed throughout the gene. Missense mutations account for about 85% of the total, although the most common mutation among Caucasians (c.964-1G > C) is a splice site variant (Witsch-Baumgartner et al. 2001; Porter 2008; Waterham and Hennekam 2012). The 13 most common pathogenic variants account for about two-thirds of all mutant alleles, which means that there are many alleles that are unique or infrequently identified.

A strict genotype–phenotype correlation is difficult because most affected individuals are compound heterozygotes. However, three individuals with a severe presentation

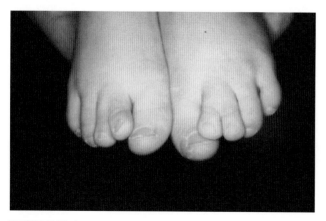

FIGURE 53.1 Characteristic Y-shaped syndactyly of the second and third toes.

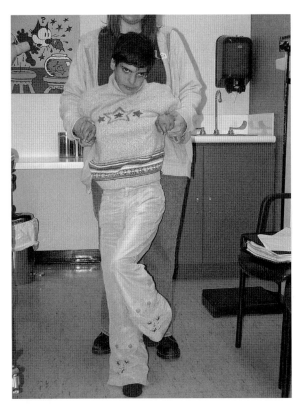

FIGURE 53.3 A 38-year-old woman with severe intellectual disability, short stature, synophrys, and prominent nasal bridge (Courtesy Dr. Malgorzata Nowaczyk).

were found to be homozygous for a 134-base pair insertion that is believed to result in a completely non-functional enzyme. One of these was a fetus with holoprosencephaly, midline cleft lip and palate, and multiple anomalies characteristic of severe SLOS (Nowaczyk et al. 2001a, 2001b). In a survey of 207 persons with SLOS, Waterham and Hennekam (2012) reported that individuals with one or two pathogenic variants in loop 1-2 or one pathogenic variant in the N-terminus have milder phenotypes, while those with 2 null alleles or 2 variants in loop 8-9 have the most severe phenotypes. Milder presentations may be seen in individuals with mutations that result in some residual enzymatic activity (Prasad et al. 2002), and more severe presentations have occurred in those who have very low (around 10 mg dL) cholesterol concentrations. This finding and other observations have led many investigators to conclude that the primary pathogenetic mechanism leading to the manifestations in affected individuals is cholesterol deficiency (Cunniff et al. 1997). Other investigators have concluded however, that excess 7-dehydrocholesterol also plays a role (Ryan et al. 1998). More recently, concentrations of both 7-dehydrocholesterol and 8-dehydrocholesterol in the cerebrospinal fluid have been investigated as pathogenetic factors for intellectual disability and/or autism, and the results are mixed (Sparks et al. 2014; Thurm et al. 2016)

Whether the primary pathogenetic mechanism is a decrease in cholesterol or an increase in 7- and 8-dehydrocholesterol, it is clear that derangements of sterol quantity and/or type are the major factors responsible for the clinical abnormalities in individuals with SLOS. Cholesterol and related compounds such as 7- and 8-dehydrocholesterol are critical components of myelin and other central nervous system proteins, and the altered sterol profile in SLOS is associated with abnormal intellectual and motor function. The identification of individuals with SLOS and holoprosencephaly has led to the implication of sonic hedgehog (*SHH*), a signaling protein, and patched (*PTCH1*), a *SHH* receptor, in the co-occurrence of these two abnormalities (Kelley 1998). *SHH* is known to cause holoprosencephaly in humans, and the *SHH* protein product undergoes autoproteolysis to form a cholesterol-modified active product. Because cholesterol is an important precursor of the sex steroids testosterone and estrogen, hypocholesterolemia results in deficiency of these hormones. Both prenatal and postnatal testosterone levels are decreased, leading to undervirilization of genotypic males. Underproduction of estrogen results in low maternal serum unconjugated estriol levels, which is an indication for diagnostic testing for SLOS in the prenatal setting (Craig et al. 2007). The pathogenesis of other features such as polydactyly and cleft palate is less obvious but may be the result of abnormal cell-to-cell interactions that result from derangement of sterol composition in the cell membranes of the developing embryo. The positive growth response seen in children treated with dietary cholesterol (see below) suggests that growth deficiency is at least partly a result of hypocholesterolemia.

Diagnostic Testing

Diagnosis of SLOS is by the detection of increased 7-dehydrocholesterol levels in blood or other tissues, or by *DHCR7* mutation analysis. Although cholesterol is usually decreased, about 10% of affected individuals have levels in the normal range, especially those with mild disease (Cunniff et al. 1997). Prenatal diagnosis may be accomplished by detection of elevated 7-dehydrocholesterol (Abuelo et al. 1995) or by mutation analysis (Waye et al. 2007) in amniotic fluid or chorionic villi. Reliable heterozygote detection by measurement of 7-dehydrocholesterol or cholesterol is not feasible because the population distribution of cholesterol and 7-dehydrocholesterol levels does not allow for a clear distinction between heterozygotes and homozygous normal individuals in the general population.

Mutation analysis of *DHCR7* is clinically available for diagnostic confirmation, carrier detection, and prenatal diagnosis. Mutation analysis detects approximately 96% of pathogenic variants (Waterham and Hennekam 2012). If mutations are identified in an affected individual, prenatal diagnosis and targeted carrier screening are available to family members.

Prenatal suspicion of SLOS may prompt prenatal diagnostic testing. Findings that might suggest a diagnosis of SLOS include birth defects seen commonly in affected persons, low maternal human chorionic gonadotropin levels on first trimester screening, or low levels of maternal serum human chorionic gonadotropin and particularly unconjugated estriol on second trimester screening. However, there are currently no specific protocols in place for use of prenatal screening for detection of SLOS. For prenatal diagnosis, *DHCR7* mutation analysis or measurement of 7-dehydrocholesterol levels can be performed on cells obtained by either chorionic villus sampling or amniocentesis.

Differential Diagnosis

Because the cardinal features of SLOS include developmental delay, growth deficiency, cleft palate, polydactyly, and hypogenitalism, it must be distinguished from other disorders with one or more of these features. There are a number of chromosome anomaly syndromes with features that overlap with those seen in SLOS. Growth deficiency, cleft palate, and developmental delay occur frequently in individuals with trisomy 18; and growth deficiency, polydactyly, and developmental delay are seen in those with trisomy 13 (Chapter 58). Severe hypogenitalism has been described in children with deletion of chromosomal material from the long arm of chromosome 10 (Wulfsberg et al. 1989). High-resolution chromosome analysis will readily distinguish between SLOS and these chromosome anomaly syndromes.

The differential diagnosis of SLOS should also include a number of single gene disorders. Children with Noonan syndrome, an autosomal dominant condition, frequently have short stature, developmental disabilities, cardiac defects, and cryptorchidism (see Chapter 41). The differences in facial appearance, the pattern of cardiac defects, and the presence or absence of polydactyly and toe syndactyly will usually allow the clinician to make a distinction between individuals with Noonan syndrome and those with SLOS. Polydactyly is a feature of many malformation syndromes such as Simpson–Golabi–Behmel syndrome (SGBS), Pallister–Hall syndrome and Meckel syndrome. Individuals with SGBS, an X-linked recessive condition, have macrosomia, macroglossia, accessory nipples, and other features not encountered in children with SLOS. Pallister–Hall syndrome is a variable autosomal dominant disorder with hypothalamic hamartoblastoma in addition to polydactyly (see Chapter 44). Abnormal facial features, syndactyly, and genital abnormalities are usually not present. Meckel syndrome is a severe and frequently lethal autosomal recessive disorder associated with encephalocele and cystic renal disease in addition to polydactyly and occasional hypogenitalism.

MANIFESTATIONS AND MANAGEMENT

Cholesterol-Directed Therapies

Because SLOS results from defective cholesterol synthesis, the focus of treatment has been dietary (Elias et al. 1997; Irons et al. 1997; Nwokoro and Mulvihill 1997; Sikora et al. 2004; Haas et al. 2007; Chan et al. 2009), with the goal of providing supplementary cholesterol to improve, or even normalize, plasma cholesterol levels. Cholesterol supplementation is usually prepared as a suspension in oil or an aqueous solution, with a concentration of 150–200 mg mL^{-1}. The cholesterol dose has varied between 40 and 150 mg/kg/day, with most individuals taking about 150 mg/kg/day. Early studies used chenodeoxycholic and ursodeoxycholic acid therapy to replace the abnormal bile acids produced as a result of deranged sterol metabolism and to aid in cholesterol absorption. However, because plasma cholesterol levels and the percentage of total sterols present as cholesterol were not appreciably different between individuals who were and were not receiving bile acid replacement, it was discontinued. Although there was initial concern that altered bile acid metabolism would lead to malabsorption of vitamins and minerals, it appears that individuals with SLOS do not have a deficiency of fat-soluble vitamins. Improvement in weight gain, an increase in plasma cholesterol levels, and an increase in cholesterol as a percentage of total sterols have been seen in most individuals receiving cholesterol supplementation. However, other clinical benefits have been more difficult to quantitate. There is evidence from parental report and unblinded investigator observation that the number of infectious illnesses is decreased, feeding tolerance is enhanced, behavior is improved, polyneuropathy lessens, rashes and photosensitivity are diminished, and affected individuals are less irritable and more manageable. Despite these observations, two long-term trials of cholesterol supplementation failed to demonstrate any improvement in developmental progress in over 50 patients (Sikora et al. 2004; Haas et al. 2007). Although dietary cholesterol supplementation has been shown to raise serum cholesterol levels, it has not appreciably or reliably affected levels of 7- or 8-dehydrocholesterol in treated individuals.

Because of the recent evidence that accumulation of 7-and 8-dehydrocholesterol may play a role in the pathogenesis of SLOS, it has been theorized that lowering the levels of cholesterol precursors may improve symptoms. A trial of simvastatin, an inhibitor of the rate-limiting enzyme in cholesterol synthesis, hydroxymethylglutaryl-Coenzyme A reductase, was beneficial in two patients (Jira et al. 2000). A subsequent trial of simvastatin and cholesterol in 20 individuals failed to demonstrate any positive effects on anthropometric measures or behavior (Haas et al. 2007). Furthermore, complications required that simvastatin be decreased or discontinued in six individuals, suggesting this treatment may not be safe for some individuals with Smith–Lemli–Opitz syndrome. However, in 2017, Wassif et al. conducted a double-blind placebo-controlled crossover trial of oral simvastatin and 150 mg/kg/day of cholesterol and found it to be safe. The 23 mild to classical SLOS subjects who took simvastatin also showed improvement of the serum dehydrocholesterol/total sterol ratio and significant reduction of irritability symptoms, compared to those on cholesterol supplementation and placebo.

In addition to postnatal treatment with dietary cholesterol and simvastatin, prenatal treatment has also been attempted in at least one individual (Abuelo et al. 1995). During a pregnancy with an affected fetus, increased maternal consumption of cholesterol was encouraged. Cholesterol levels remained below the lower limit of the 95% confidence interval for normal controls throughout the pregnancy. There was no discernible advantage detected in the neonate at birth. It is unclear to what degree cholesterol crosses the placenta during gestation, especially during the critical first trimester when organogenesis occurs. It is therefore difficult to judge whether prenatal treatment from conception might decrease the number or severity of malformations seen. The degree to which cholesterol, especially maternally ingested cholesterol, crosses the fetal blood-brain barrier is also not known, further complicating the assessment of prenatal treatment of SLOS to prevent related developmental disabilities.

Growth and Feeding

As described in the original affected individuals reported by Smith et al. (1964), most people with SLOS have height and weight below the third centile. Neonates with SLOS are often small for gestational age. Those with normal length and weight usually decelerate in both linear growth and weight gain so that they are below the third centile by age 6 months. In the report of Ryan et al. (1998), 21 of 24 (88%) living affected individuals were considered to have failure-to-thrive. Although weight gain may improve in infancy and childhood, final adult stature is usually below the third centile. Those who are biochemically more severely affected tend to have the poorest weight gain and linear growth and smaller head circumference than those with milder biochemical abnormalities. Using 1854 observations of 78 affected individuals (43 males and 35 females), Lee et al. (2012) published growth charts for weight, height, BMI and head circumference, for use in the management of persons with SLOS.

Most individuals have feeding difficulty in infancy (Ryan et al. 1998). The feeding problems tend to improve with age. Feeding difficulties may range from oral sensitivities and adverse behaviors to vomiting, intestinal dysmotility, and gastroesophageal reflux. At least part of this difficulty may relate to hypotonia, with a poor suck, listlessness, and disinterest in feeding. Nasogastric or orogastric feedings are frequently necessary. Feeding difficulties have been correlated with blood sterol levels, suggesting that feeding difficulty is directly related to biochemical severity (Merkens et al. 2014). Vomiting, intestinal dysmotility, and gastroesophageal reflux are also prominent features. Gastroesophageal reflux may require surgical intervention in the form of fundoplication. Fundoplication was required in 8 of 24 individuals discussed by Ryan et al. (1998). Because reflux tends to improve with age, however, a course of aggressive medical management should precede the decision to intervene surgically. In assessing children with SLOS and poor feeding, it should be recognized that some individuals will have gastrointestinal complications that may require surgical intervention. For example, pyloric stenosis is present in about 10% of reported individuals. Hirschsprung disease, malrotation, and cholestatic liver disease have all been seen with some frequency and should be evaluated and treated when signs and symptoms point to one of these conditions.

Evaluation

- Height, weight, and head circumference should be measured at the initial evaluation and during health supervision visits. Because most individuals with SLOS will have linear growth deficiency, it is important to consider weight for height in addition to charting absolute centile measurements for these parameters. SLOS specific growth charts (Lee et al. 2012) should be used to track growth.

- For individuals who are unable to take adequate calories orally, a feeding evaluation should be undertaken. This should include an assessment of parent/caregiver abilities and responses, oral motor skills, the number of calories being consumed, and any signs of gastroesophageal reflux.

- When gastroesophageal reflux is suspected, consideration should be given to a full diagnostic evaluation. The evaluation of gastroesophageal reflux in individuals with SLOS does not differ appreciably from reflux evaluation in other individuals, especially those with developmental disabilities.

Management

- Adequate caloric intake should be ensured. It is important to recognize that some individuals with SLOS have elements of primordial dwarfism, so that no amount of calories will be able to induce a normal growth pattern. Overfeeding may be a problem if high calorie feeding is attempted and may further exacerbate issues of vomiting and irritability, which are common.

- Consideration should be given to cholesterol supplementation to improve weight gain. Provision of

150 mg/kg/day of cholesterol has been shown to improve weight gain.
- Oral-motor training with an occupational therapist or feeding specialist may assist oral feeding.
- For those who are unable to consume adequate calories orally, nasogastric, orogastric, or gastrostomy tube feedings should be considered.
- Appropriate positioning, low-volume and frequent feedings, and anti-reflux medications should be considered in treating gastroesophageal reflux.
- For individuals who are refractory to medical management, fundoplication may be the only viable treatment alternative. However, fundoplication has not always been effective in improving reflux. Additional factors that appear to contribute to this effect include protein allergy, overfeeding, and a congenitally small stomach volume. For those with protein allergy, elemental formula may be beneficial.

Development and Behavior

Almost all children with SLOS are intellectually disabled, most commonly in the moderately to severely disabled range of functioning. The range of intellectual outcomes is great, however, with some individuals functioning in the normal or near-normal range, especially in infancy and early childhood (Lowry and Yong 1980; Eroglu et al. 2017). In a case series of 33 pediatric and adult patients, developmental milestones were globally delayed, as well as cognitive function in adolescence and adulthood (Thurm et al. 2016). Moreover, they found that autism spectrum disorder, though frequently diagnosed in SLOS, may be over-diagnosed because of the level of intellectual disability in these patients. These cognitive and behavior symptoms, like other manifestations of the disorder, appear to be directly related to biochemical severity and the phenotypic variability seen across multiple cohorts.

In one systematic evaluation of the behavioral phenotype in SLOS (Ryan et al. 1998), questionnaires were completed by the parents of 23 living children and adults, aged 6 months and older. Aggressive (52%) and self-injurious (35%) behavior was common. Ritualistic or obsessive behavior was seen in 52%. Many children were reported to be inappropriately affectionate with strangers. Sleep disturbance was particularly common (70%). Although children were often sleepy and hypotonic in early infancy, they had long periods of wakefulness in early childhood, sometimes requiring no more than 2–3 hours of sleep per 24 hour period. Most were difficult to settle and had frequent and early awakening that were refractory to treatment with sedatives. Sleep patterns tended to normalize by school age.

In 2016, Freeman et al. sought to characterize this phenotype more clearly by correlating sleep behaviors to biochemical severity. In this investigation of 20 pediatric patients, the concentration of plasma 7-DHC, 8-DHC, and cholesterol was measured, and a (7-DHC+8-DHC)/cholesterol ratio was calculated. This ratio was used as the primary indicator of biochemical severity. Parental sleep questionnaires were completed assessing sleep characteristics, and there was an inverse correlation between biochemical severity and sleep quality. This investigation was performed in an inpatient setting, however, and may not be completely generalizable to home sleep characteristics.

Evaluation

- Developmental screening should be performed at health supervision visits. Most pediatricians and other physicians caring for children will have an established routine for developmental screening that is applied to all their patients. The American Academy of Pediatrics has published an excellent policy statement on this topic (Council on Children with Disabilities 2006).
- Because of the frequency of behavioral abnormalities, it is suggested that these be inquired about specifically. Particular attention may be given to the behavioral disturbances discussed above, especially in regard to sleep and aggressive or self-injurious behavior.
- Specific screening for autism spectrum disorders appears warranted, and several instruments are available for this purpose, including the checklist of autism in toddlers (CHAT) and the modified checklist of autism in toddlers (M-CHAT) in younger children and the social communication questionnaire (SCQ) in older individuals. Alternatively, because of the high risk for autism in individuals with SLOS, referral to a developmental specialist such as a developmental pediatrician, child psychologist, or child psychiatrist may be indicated.

Management

- Individuals with SLOS should be enrolled in early intervention programs, special education programs, and other systems of care that enhance their developmental potential. Specific developmental or educational strategies have not been examined systematically for children with SLOS. It is presumed that programs geared to the developmental and behavioral concerns of the individual are most likely to be beneficial. Instruction in a special education classroom setting is common.
- Adults will usually require some type of supervised residential care setting such as a group home.
- There are reports of improved development and behavior in children and adults receiving cholesterol supplementation (see above). Because the number of individuals studied is few and the subjectivity of most of the measures is high, these reports should be viewed with cautious optimism.

- The assistance of a psychologist or developmental pediatrician may be of value.
- Behavioral strategies have not been examined systematically. The same is true of psychotropic medication.
- For treatment of abnormal sleep patterns, general strategies such as scheduled bedtimes, positive routines, and good sleep hygiene can be recommended. Cautious medication trials may be warranted for particularly refractory cases.
- For aggressive or self-injurious individuals, behavioral strategies such as redirection, time-out, and positive rewards may be helpful.

Neurologic

A variety of developmental abnormalities of the brain have been seen in individuals with SLOS. In the largest study to date of brain MRI findings in these patients, Lee et al. (2013) reviewed 173 brain MRI scans in 55 individuals from infancy to age 25 years and observed brain abnormalities in 96% of the study cohort. They found that septum pellucidum anomalies were most frequently observed (76% of all patients), followed closely by abnormalities of the corpus callosum (69%). They also identified diverse brain anomalies ranging from cerebral and/or cerebellar atrophy to colpocephaly and white matter lesions. The majority of patients had midline anomalies, a finding that prior studies also noted (Kelley 1998). As with other body organ systems and clinical phenotype, patients with the most severe brain abnormalities also had a more severe biochemical signature. Historically, one of the most striking central nervous system findings is holoprosencephaly, which was observed in seven individuals reported by Kelley (1998) and in other individuals in subsequent case series. Conversely, in 28 pregnancies with holoprosencephaly identified prenatally, two were diagnosed with SLOS (Petracchi et al. 2011). The pathogenesis of holoprosencephaly in SLOS is unknown. An embryonic signaling protein known as sonic hedgehog (*SHH*), which causes autosomal dominant holoprosencephaly, has been implicated, as has its receptor, Patched (*PCH1*). Other studies have suggested a role for cholesterol deficiency in the neural plate ectoderm, the precursor tissue for the fetal brain (Kelley 1998). (For a broad discussion of holoprosencephaly, please see Chapter 31.)

The primary disabilities resulting from central nervous system dysfunction in SLOS are intellectual disability and behavioral abnormalities. Seizures are generally not observed. Effects on the peripheral nervous system have also been reported in some individuals, typically polyneuropathy with abnormal nerve conduction velocities.

Evaluation

- Careful neurological evaluation is recommended for all newly diagnosed individuals.
- Neuroimaging studies can be reserved for individuals with signs of holoprosencephaly such as hypotelorism or agenesis of the premaxilla or for children with evidence of hypopituitarism such as diabetes insipidus.
- Nerve conduction velocities should be measured in individuals with evidence of neuropathy.

Management

- The brain abnormalities seen in SLOS rarely require specific treatment. See Chapter 31 for management of affected individuals with holoprosencephaly.
- Children with endocrine disturbances resulting from hypopituitarism will need hormone-replacement therapy directed to the specific deficiencies observed.

Craniofacial

Most children with SLOS are born with microcephaly that persists throughout life. Cleft palate is seen in 37–52% (Cunniff et al. 1997; Ryan et al. 1998) and most commonly affects the soft palate, although there can be cleft of the hard palate, soft palate, or both. Micrognathia, Pierre Robin sequence (see Chapter 50), and prominent incisors also occur, and can cause difficulties with airway management, particularly in individuals requiring anesthesia (Quezado et al. 2002; Matveevskii et al. 2006). The gingivae may be hyperplastic and rugated, and the alveolar ridge is often broad. Tongue hamartomata have been observed in some individuals with severe phenotypic features.

Evaluation

- All affected individuals should have a careful physical examination of the palate and oropharynx for abnormalities.
- Attention should also be directed to the infant's ability to take oral feedings, because feeding problems are encountered frequently in individuals with SLOS, particularly those with cleft palate. Evaluation by a feeding specialist such as an occupational therapist is indicated for children with feeding difficulties.
- Children with cleft palate will require surgical evaluation.
- For anyone undergoing surgery, but especially those with micrognathia, a preoperative evaluation by an anesthesiologist with experience in treating children is recommended.

Management

- As with other at-risk newborns, children with SLOS and feeding difficulties may benefit from oral-motor training.

- If adequate calories cannot be taken orally, consideration should be given to orogastric, nasogastric, or gastrostomy feedings.
- Cleft palate repair is usually performed at around age 12 months. The timing, method, or approach to repair has not been addressed specifically in children with SLOS. The approach to surgical treatment is generally not altered on the basis of the diagnosis of SLOS.
- Mask airway followed by fiber-optic laryngoscopy with tracheal intubation has been reported to be a safe and reliable procedure for airway management.

Ophthalmologic

Ptosis is seen in about half of the reported affected individuals (Ryan et al. 1998), and cataracts are present in 12–18% (Cunniff et al. 1997; Ryan et al. 1998; Goodwin et al. 2008). Additional ophthalmologic findings include strabismus, retinal hemangiomata, demyelination of the optic nerves, sclerosis of the lateral geniculate bodies, and lack of a visual following response (Fierro et al. 1997). There are no outcome studies of visual function in children with SLOS.

Evaluation

- Because affected persons are at increased risk for congenital cataracts, it is important that the red reflex be elicited in newborns and that there be a clear view of the fundus in older individuals.
- Postnatal development of cataracts has also been observed, so a careful ophthalmologic evaluation should be conducted at all health supervision visits.
- If there is a suggestion of visual compromise from ptosis, congenital cataracts, ocular motility abnormalities, or any other abnormal ophthalmologic signs, referral to an ophthalmologist is recommended.

Management

- There is no systematic study of cataract treatment in individuals with SLOS. Clinical criteria for medical and surgical management should generally be the same as for other children.
- For children with ocular motility abnormalities or ptosis that may impair vision, the treatment plan should be formulated based on a full visual assessment.

Cardiovascular

Congenital heart disease is present in 36–38% of individuals with SLOS (Cunniff et al. 1997; Lin et al. 1997; Ryan et al. 1998). Lin et al. (1997) reported on 215 individuals with SLOS (59 biochemically confirmed cases and 156 from the medical literature). They found that a disproportionate number had atrioventricular canal defects and total anomalous pulmonary venous return when compared with an unselected population of children with congenital heart defects. Complex conotruncal anomalies were very uncommon; complex single ventricle malformations and heterotaxies were not reported. These findings suggest that two of the important pathogenetic mechanisms that produce cardiac malformations in SLOS are altered extracellular matrix and abnormal targeted cell growth.

Evaluation

- Cardiac evaluation is recommended at the time of diagnosis, with follow-up determined by the severity of any abnormalities identified. Evaluation should include electrocardiogram with assessment of axis and an echocardiogram with special attention to the atrial and ventricular septa and the pulmonary veins.

Management

- Treatment of the cardiovascular malformations in SLOS is specific to the malformation identified and does not differ from that used in other individuals with the same defect.
- Careful consideration should be given to surgical or other invasive treatment in those with a poor prognosis in whom medical or other palliative treatment might provide a good short-term outcome.

Gastrointestinal

Gastrointestinal abnormalities have been described in 25% (Cunniff et al. 1997) to 29% (Ryan et al. 1998) of biochemically confirmed individuals. A variety of both specific and non-specific features have been seen. Chronic constipation or diarrhea has been noted, and most affected people have some element of intestinal dysmotility. Pyloric stenosis, malrotation, and Hirschsprung disease have been described. Of particular note are five individuals with cholestatic liver disease (Cunniff et al. 1997; Ryan et al. 1998), two of whom died. This abnormality is presumed to be the result of abnormal bile acids, although the exact pathogenetic mechanism is unknown. Perhaps cholesterol supplementation could be beneficial by providing appropriate amounts of cholesterol substrate to produce normal bile acids, although results have been conflicting regarding clinical improvement with cholesterol therapy (Fliesler 2013; Tierney et al. 2010). Interestingly, some studies have pointed to antioxidant therapy as a possible additional treatment modality (Fliesler 2013; Korade et al. 2014). Korade et al. (2014) observed that dietary supplementation of vitamin E to pregnant mice led to decreased oxysterol levels in newborn murine brain and hepatic tissues, though Vitamin E supplementation has not yet been studied in human clinical trials.

For a discussion of the frequent gastroesophageal reflux and its treatment, see Growth and Feeding. Generally, poor feeding and vomiting without an underlying structural abnormality improve as children become older.

Evaluation

- Liver enzymes and total and direct bilirubin levels should be performed at the time of diagnosis. If elevations are identified, further evaluation for cholestatic liver disease in a standard manner should take place.
- Young infants with persistent vomiting and failure-to-thrive should be evaluated for pyloric stenosis. In addition to careful physical examination, imaging studies are recommended. Electrolyte analysis may show evidence of hypochloremic alkalosis.
- Individuals with chronic constipation or alternating constipation and diarrhea, especially if associated with other signs of Hirschsprung disease, should have a barium enema and/or small bowel biopsy. Surgical evaluation is recommended if Hirschsprung disease is suspected.

Management

- Surgical referral and standard treatment are recommended for infants with evidence of pyloric stenosis, malrotation, or Hirschsprung disease.
- Although there are no data on treatment of cholestasis in SLOS, a trial of cholesterol supplementation is warranted for children with evidence of cholestatic liver disease. This is based on the presumptive pathogenetic mechanism of altered bile acid profile in SLOS and the potential beneficial effects of increasing plasma cholesterol and increasing the percentage of sterols that are present as cholesterol.
- Long-term gastrostomy placement, with or without fundoplication, is necessary for some affected individuals with persistent inability to feed orally.

Genitourinary

Joseph et al. (1987) reported on 29 males and 15 females ascertained before the advent of biochemical testing. Evaluation of the upper urinary tract in 21 of the 44 in their study and in 31 individuals reported in the literature revealed abnormalities in 31 of these 52 children (60%). There was no difference in the incidence between boys and girls, and there was no correlation between upper tract anomalies and abnormal external genitalia. Reported abnormalities included cystic renal dysplasia (29%), renal positional abnormalities (19%), hydronephrosis (16%), ureteropelvic junction obstruction (13%), renal duplication (13%), and renal agenesis (6%).

Renal defects were reported in 13% (Cunniff et al. 1997) and 29% (Ryan et al. 1998) of biochemically confirmed individuals. Abnormalities ranged from lack of fetal lobulation to bilateral renal agenesis. These figures probably represent a minimal estimate because renal imaging studies were not carried out in all. Long-term studies of renal function are not available for persons with SLOS. Clinical reports have not generally identified any serious sequelae such as recurrent infections or renal insufficiency.

Genital abnormalities, including hypospadias and cryptorchidism, are present in about half of affected males (Nowaczyk 2013). However, there is a spectrum of undervirilization seen in boys with SLOS, ranging from normal to ambiguous appearance to sex reversal (i.e. female external genitalia). Genital abnormalities are uncommon in affected females, although 2 of 12 girls reported by Joseph et al. (1987) had clitoral enlargement.

The prognosis for boys with genital abnormalities is related to the severity of abnormalities encountered. Repair of hypospadias generally produces a good functional and cosmetic result. Because of the related developmental disabilities in SLOS, severely affected individuals have not reproduced. One woman with mild SLOS that was diagnosed during pregnancy gave birth to a normal child (Ellingson et al. 2014).

Evaluation

- Ultrasonography of the urinary tract is recommended at the time of diagnosis. On the basis of findings detected on ultrasonography, additional studies may be recommended.
- For individuals with an unambiguous female genital appearance, no additional evaluations are suggested.
- For individuals who have any degree of genital ambiguity, chromosome analysis is warranted.
- If the karyotype is 46,XY, it is important to evaluate for signs of appropriate virilization. Is the penis of normal size? Are the testes palpable? Is hypospadias present? Careful palpation for testes in the scrotum and in the inguinal areas should be carried out during the initial physical examination. If the testes are not palpable on initial examination, additional attempts should be made at subsequent health supervision visits.
- Referral to a urologist is recommended for all boys with hypospadias, cryptorchidism, or other genital abnormalities such as micropenis.

Management

- Treatment of upper urinary tract anomalies is dependent on the nature of the abnormality identified, and is standard. Although renal malposition will rarely require

any kind of treatment, children with renal duplication will require surgical treatment by a urologist.
- With modern surgical techniques, satisfactory repair of hypospadias can usually be accomplished.
- Boys with cryptorchidism should be observed over the first year of life to see whether spontaneous testicular descent will occur. Orchiopexy is recommended for boys with persistent cryptorchidism.
- Boys with micropenis or any significant degree of genital ambiguity should be evaluated carefully, using a team approach that includes, as a minimum, a urologist, an endocrinologist, and a geneticist. There is a positive correlation between general clinical severity and the severity of genital abnormalities in boys (Bialer et al. 1987). This means that boys with the most severe genital abnormalities tend to have the poorest developmental prognosis and an increased risk for early mortality. The benefits of genital surgery should be weighed carefully, and the approach to genital surgery should be made in the context of the functional status of the child and the prognosis for long-term survival, developmental outcome, and social awareness.

Musculoskeletal

The most commonly reported musculoskeletal abnormality is syndactyly of the second and third toes, present in more than 95% of biochemically confirmed cases (Tint et al. 1995; Cunniff et al. 1997; Ryan et al. 1998). In most cases this syndactyly is distinctive, with what has been termed a "Y shape." The stem of the "Y" is produced by tight cutaneous syndactyly of the toes to the level of the distal interphalangeal joint, and the fork of the "Y" is produced by the distal phalanges. This appearance is not an obligate feature and may not be present in mildly affected individuals. It causes no functional abnormalities.

Postaxial polydactyly is present in about half of the individuals reported. It appears more commonly on the hands than on the feet and usually is present in the form of a pedunculated postminimus. Abnormally short or proximally placed thumbs are also seen in about half of the affected individuals, more commonly in those who are more severely affected. Additional skeletal abnormalities reported include dislocated hips (18%), limb shortening (12%), brachydactyly (20%), ulnar deviation of the fingers (14%), and positional foot deformities (31%) (Ryan et al. 1998).

Evaluation

- Routine physical examination should be sufficient to detect musculoskeletal abnormalities in all affected individuals.
- For those with polydactyly, surgical evaluation is recommended.
- Particular attention should be given to examination of the hips. Those with abnormal hip position, limitation of abduction, or a positive Ortolani sign or Barlow test may have hip dysplasia and should be referred for orthopedic evaluation.
- It is also recommended that individuals with positional foot abnormalities such as talipes equinovarus or calcaneovalgus be referred for orthopedic evaluation.

Management

- Simple excision of supernumerary digits can usually be performed as a day procedure, although some children with polydactyly of the feet may have a more fully developed extra digit that requires a more complex surgical approach.
- Evaluation by an orthopedist with experience in the treatment of pediatric patients is recommended for children with hip dysplasia or positional foot deformities
- There are no data to suggest that treatment of musculoskeletal problems in children with SLOS should differ from treatment of unaffected children.

Dermatologic

Individuals with SLOS can have a variety of skin manifestations. Most have fair complexion and light hair coloration when compared with their unaffected first-degree relatives (Ryan et al. 1998). Almost all report erythematous reactions to sunlight, so that parents frequently keep their children indoors or otherwise restrict their sun exposure. Eczema is reported in up to 10% of individuals. The skin manifestations of SLOS are generally not serious and can be treated with preventive measures.

Evaluation

- Inspection of the skin should identify any abnormalities requiring treatment.

Management

- Dietary cholesterol treatment has been reported to decrease photosensitivity and skin rashes (Elias et al. 1997; Nwokoro and Mulvihill 1997). Documentation of benefit in these reports was by parent and investigator report. However, Azurdia et al. (2001) described objective evidence of reduced photosensitivity in an affected individual supplemented for six months with 75–200 mg/kg/day of cholesterol.
- Sun photosensitivity reactions can be minimized by limiting the length of exposure, wearing protective clothing, and using sunscreen.
- Exacerbations of eczema may benefit from treatment with topical steroids.

RESOURCES

Smith-Lemli-Opitz/RSH Foundation

PO Box 10598
Fargo, ND 58106-0598, USA
Phone: (701) 367-1976
Email: *gnoah@smithlemliopitz.org*
Website: www.smithlemliopitz.org

GeneReviews

Website: *www.genereviews.org*

REFERENCES

Abuelo DN, Tint GS, Kelley R, Batta AK, Shefer S, Salen G (1995) Prenatal detection of the cholesterol biosynthetic defect in the Smith-Lemli-Opitz syndrome by the analysis of amniotic fluid sterols. *Am J Med Genet* 56:281–285.

Azurdia RM, Anstey AV, Rhodes LE (2001) Cholesterol supplementation objectively reduces photosensitivity in the Smith-Lemli-Opitz syndrome. *Br J Dermatol* 144:143–145.

Bialer MG, Penchaszadeh VB, Kahn E, Libes R, Krigsman G, Lesser ML (1987) Female external genitalia and müllerian duct derivatives in a 46,XY infant with the Smith-Lemli-Opitz syndrome. *Am J Med Genet* 28:23–31.

Chan YM, Merkens LS, Connor WE, Roullet JB, Penfield JA, Jordan JM, Steiner RD, Jones PJ (2009) Effects of dietary cholesterol and simvastatin on cholesterol synthesis in Smith-Lemli-Opitz syndrome. *Pediatr Res* 65:681–5.

Ciara E, Popowska E, Piekutowska-Abramczuk D, Jurkiewicz D, Borucka-Mankiewicz M, Kowalski P, Goryluk-Kozakiewicz B, Nowaczyk MJM, Krajewska-Walasek M (2006) SLOS carrier frequency in Poland as determined by screening for Trp151X and Val326Leu DHCR7 mutations. *Eur J Med Genet* 49:499–504.

Council on Children with Disabilities, Section on Developmental Behavioral Pediatrics, Bright Futures Steering Committee, Medical Home Initiatives for Children with Special Needs Project Advisory Committee (2006) Identifying infants and young children with developmental disorders in the medical home: An algorithm for developmental surveillance and screening. *Pediatrics* 118:405–420.

Craig WY, Haddow JE, Palomaki GE, Roberson M (2007) Major fetal abnormalities associated with positive screening tests for Smith-Lemli-Opitz syndrome (SLOS). *Prenat Diagn* 27:409–414.

Cunniff C, Kratz LE, Moser A, Natowicz MR, Kelley RI (1997) Clinical and biochemical spectrum of patients with RSH/Smith-Lemli-Opitz syndrome and abnormal cholesterol metabolism. *Am J Med Genet* 68:328–337.

Curry CJ, Carey JC, Holland JS, Chopra D, Fineman R, Golabi M, Sherman S, Pagon RA, Allanson J, Shulman S, Barr M, McGravey V, Dabiri C, Schimke N, Ives E, Hall BD (1987) Smith-Lemli-Opitz syndrome-type II: Multiple congenital anomalies with male pseudohermaphroditism and frequent early lethality. *Am J Med Genet* 26:45–57.

Elias ER, Irons MB, Hurley AD, Tint GS, Salen G (1997) Clinical effects of cholesterol supplementation in six patients with the Smith-Lemli-Opitz syndrome. *Am J Med Genet* 68:305–310.

Ellingson MS, Wick MJ, White WM, Raymond KM, Saenger AK, Pichurin PN, Wassif CA, Porter FD, Babovic-Vuksanovic D (2014) Pregnancy in an individual with mild Smith-Lemli-Opitz syndrome. *Clin Genet* 85:495–7

Eroglu Y, Nguyen-Driver M, Steiner RD, Merkens L, Merkens M, Roullet JB, Elias E, Sarphare G, Porter FD, Li C, Tierney E, Nowaczyk MJ, Freeman KA (2017) Normal IQ is possible in Smith-Lemli-Opitz syndrome. *Am J Med Genet Part A* 173A:2097–2100.

Fliesler SJ (2013) Antioxidants: The missing key to improved therapeutic intervention in Smith-Lemli-Opitz Syndrome? *Hereditary Genet* 2:119.

Freeman KA, Olufs E, Tudor M, et al (2016) A pilot study of the association of markers of cholesterol synthesis with disturbed sleep in Smith Lemli-Opitz Syndrome. *J Dev Behav Pediatr* 37:424–430.

Fierro M, Martinez AJ, Harbison JW, Hay SH (1997) Smith-Lemli-Opitz syndrome: Neuropathological and ophthalmological observations. *Dev Med Child Neurol* 19:57–62.

Goodwin H, Brooks BP, Porter FD (2008) Acute postnatal cataract formation in Smith-Lemli-Opitz syndrome. *Am J Med Genet Part A* 146:208–211.

Haas D, Garbade SF, Vohwinkel C, Muschol N, Trefz FK, Penzien JM, Zschoske J, Hoffman GJ, Burgard P (2007) Effects of cholesterol and simvastatin treatment in patients with Smith-Lemli-Opitz syndrome (SLOS). *J Inherit Metabol Dis* 30:375–387.

Irons M, Elias ER, Salen G, Tint GS, Batta AK (1993) Defective cholesterol biosynthesis in Smith-Lemli-Opitz syndrome. *Lancet* 341:1414.

Irons M, Elias ER, Abuelo D, Bull MJ, Greene CL, Johnson VP, Keppen L, Schanen C, Tint GS, Salen G (1997) Treatment of Smith-Lemli-Opitz syndrome: Results of a multicenter trial. *Am J Med Genet* 68:311–314.

Jira PE, Wevers RA, de Jong J, et al. (2000) Simvastatin. A new therapeutic approach for Smith-Lemli-Opitz syndrome. *J Lipid Res* 41:1339–1346.

Joseph DB, Uehling DT, Gilbert E, Laxova R (1987) Genitourinary abnormalities associated with the Smith-Lemli-Opitz syndrome. *J Urol* 137:719–721.

Kelley RI (1998) RSH/Smith-Lemli-Opitz syndrome: Mutations and metabolic morphogenesis. *Am J Hum Genet* 63:322–326.

Kelley RI, Hennekam RCM (2000) The Smith Lemli Opitz syndrome. *J Med Genet* 37:321–335.

Korade Z, Xu L, Harrison FE (2014) Antioxidant supplementation ameliorates molecular deficits in Smith-Lemli-Opitz Syndrome (SLOS). *Biol Psychiatry* 75:215-222.

Lazarin GA, Haque IS, Evans EA, Goldberg JG (2017) Smith-Lemli-Opitz syndrome carrier frequency and estimates of *in utero* mortality rates. *Prenat Diagn* 37:350–355.

Lee RWY, Conley SK, Gropman A, Porter FD, Baker EH (2013) Brain magnetic resonance imaging findings in Smith-Lemli-Opitz Syndrome. *Am J Med Genet A* 161:2407–2419

Lee RWY, McGready J, Conley SK, Yanjanin NM, Nowaczyk MJM, Porter FD (2012) Growth charts for individuals with Smith-Lemli-Opitz syndrome. *Am J Med Genet Part A* 158A: 2707–2713.

Lin AE, Ardinger HH, Ardinger RH Jr, Cunniff C, Kelley RI (1997) Cardiovascular malformations in Smith-Lemli-Opitz syndrome. *Am J Med Genet* 68:270–278.

Lowry RB, Yong SL (1980) Borderline normal intelligence in the Smith-Lemli-Opitz (RSH) syndrome. *Am J Med Genet* 5:137–143.

Matveevskii A, Berman L, Sidi A, Gravenstien D, Kays D (2006) Airway management of a patient with Smith-Lemli-Opitz syndrome for gastric surgery: Case report. *Pediatr Anaesth* 16:322–324.

Merkens MJ, Sinden ML, Brown CD (2014) Feeding impairments associated with plasma sterols in Smith Lemli Opitz Syndrome. *J Pediatr* 165: 836-841.

Nowaczyk MJM, McCaughey D, Whelan DT, Porter FD (2001a) Incidence of Smith-Lemli-Opitz in Ontario, *Canada Am J Med Genet* 102:18–20.

Nowaczyk MJM, Farrell SA, Sirkin WL, Velsher L, Krakowiak PA, Waye JS, Porter FD (2001b) Smith-Lemli-Opitz (RHS) syndrome: Holoprosencephaly and homozygous IVS8-1G → C genotype. *Am J Med Genet* 103:75–80.

Nowaczyk MJ, Waye JS, Douketis JD (2006) DHCR7 mutation carrier rates and prevalence of the RSH/Smith-Lemli-Opitz syndrome: Where are the patients? *Am J Med Genet Part A* 140:2057–2062.

Nowaczyk MJ, Irons MB. (2012) Smith-Lemli-Opitz syndrome: phenotype, natural history and epidemiology. *Am J Med Genet C* 160C:250–62.

Nowaczyk MJM (1998) Smith-Lemli-Opitz syndrome. *GeneReviews®*. Available from: https://www.ncbi.nlm.nih.gov/books/NBK1143/

Nwokoro NA, Mulvihill JJ (1997) Cholesterol and bile acid replacement in children and adults with Smith-Lemli-Opitz (RSH/ Smith-Lemli-Opitz syndrome) syndrome. *Am J Med Genet* 68:315–321.

Petracchi F, Crespo L, Michia C, Igarzabal L, Gadow E (2011) Holoprosencephaly at prenatal diagnosis: analysis of 28 cases regarding etiopathogenic diagnoses. *Prenat Diagn* 31:887–891.

Porter FD (2008) Smith-Lemli-Opitz syndrome: Pathogenesis, diagnosis and management. *Eur J Hum Genet* 16:535–541.

Prasad C, Marles S, Prasad A, Nikkel S, Longstaffe S, Peabody D, Eng B, Wright S, Waye J, Nowaczyk M (2002) Smith-Lemli-Opitz syndrome: New mutation with a mild phenotype. *Am J Med Genet* 108:64–68.

Quezado ZM, Veihmeyer J, Schwartz L, Nwokoro NA, Porter FD (2002) Anesthesia and airway management of pediatric patients with Smith-Lemli-Opitz syndrome. *Anesthesiology* 97: 1015–1019.

Ryan AK, Bartlett K, Clayton P, Eaton S, Mills L, Donnai D, Winter RM, Burn J (1998) Smith-Lemli-Opitz syndrome: A variable clinical and biochemical phenotype. *J Med Genet* 35: 558–565.

Sikora DM, Ruggiero M, Petit-Kekel K, Merkens LS, Connor WE, Steiner RD (2004) Cholesterol supplementation does not improve developmental progress in Smith-Lemli-Opitz syndrome. *J Pediatr* 144:783–791.

Smith DW, Lemli L, Opitz JM (1964) A newly recognized syndrome of multiple congenital anomalies. *J Pediatr* 64: 210–217.

Sparks SE, Wasif CA, Goodwin H, Conley SK, Lanham DC, Kratz LE, Hyland K, Gropman A, Tierney E, Porter FD (2014) Decreased cerebral spinal fluid neurotransmitter levels in Smith-Lemli-Opitz syndrome. *J Inherit Metab Dis* 37: 415–420.

Thurm A, Tierney E, Farmer C, Albert P, Joseph L, Swedo S, Bianconi S, Bukelis I, Wheeler C, Sarphare G, Lanham, Wassif CA, Porter FD (2016) Development, behavior, and biomarker characterization of Smith-Lemli-Opitz syndrome: an update. *J Neurodev Dis* 8:12.

Tierney E, Conley SK, Goodwim H, Porter FD (2010) Analysis of short-term behavioral effects of dietary cholesterol supplementation in Smith-Lemli-Opitz syndrome. *Am J Med Genet A.* 152A:91–95.

Tint GS, Salen G, Batta AK, Shefer S, Irons M, Elias ER, Abuelo DN, Johnson VP, Lambert M, Lutz R, Schanen C, Morris CA, Hoganson G, Hughes-Benzie R (1995) Correlation of severity and outcome with plasma sterol levels in variants of the Smith-Lemli-Opitz syndrome. *J Pediatr* 127:82–87.

Wassif CA, Maslen C, Kachilele-Linjewile S, Lin D, Linck LM, Connor WE, Steiner RD, Porter FD (1998) Mutations in the human sterol Δ7-reductase gene at 11q12-13 cause Smith-Lemli-Opitz syndrome. *Am J Hum Genet* 63:55–62.

Wassif CA, Kratz L, Sparks SE, Wheeler C, Bianconi S, Gropman A, Calis KA, Kelley RI, Tierney E, Porter FD (2017) A placebo-controlled trial of simvastatin therapy in Smith-Lemli_Optiz syndrome. *Genet Med* 19:297–305.

Waterham HR, Wijburg FA, Hennekam RCM, Vreken P, Poll-The BT, Dorland L, Duran M, Jira PE, Smeitink JAM, Wevers RA, Wanders RJA (1998) Smith-Lemli-Opitz syndrome is caused by mutations in the 7-dehydrocholesterol reductase gene. *Am J Hum Genet* 63:329–338.

Waterham HR, Hennekam RC (2012) Mutational spectrum of Smith-Lemli-Opitz syndrome. *Am J Med Genet C* 160C: 263–84.

Waye JS, Eng B, Nowaczyk MJ (2007) Prenatal diagnosis of Smith-Lemli-Opitz syndrome (SLOS) by DHCR7 mutation analysis. *Prenat Diagn* 27:638–640.

Witsch-Baumgartner M, Loffler J, Utermann G (2001) Mutations in the human *DHCR7* gene. *Hum Mutat* 17:172–182.

Wulfsberg EA, Weaver RP, Cunniff CM, Jones MC, Jones KL (1989) Chromosome 10qter deletion syndrome: A review and report of three new cases. *Am J Med Genet* 32:364–367.

54

SMITH–MAGENIS SYNDROME*

ANN C.M. SMITH
Office of the Clinical Director, National Human Genome Research Institute, National Institutes of Health, Bethesda, Maryland, USA

ANDREA L. GROPMAN
Pediatrics and Neurology, George Washington University of the Health Sciences, and Children's National Medical Center, Washington, DC, USA

INTRODUCTION

Incidence

Smith–Magenis syndrome (SMS) is a multisystem multiple congenital anomaly/intellectual disability syndrome caused by haploinsufficiency of the *retinoic-acid-induced-1* (*RAI1*) gene that arises either through interstitial deletion of chromosome 17p11.2 or via heterozygous *RAI1* mutation. The deletion was first reported in 1982 by Smith et al. and the phenotypic spectrum was more fully delineated in 1986 through tandem articles describing a series of 15 individuals (Smith et al. 1986; Stratton et al. 1986). Individuals, ranging from 1 month to over 80 years of age, have been identified worldwide from diverse ethnic groups. In all cases, the 17p11.2 deletion has been associated with a distinct and clinically recognizable complex phenotype. In 2003, Slager et al. identified unique heterozygous mutations in the *RAI1* gene in three individuals with phenotypic features of SMS, but without a detectable deletion. There is now compelling evidence that the majority (~70%) of features of SMS result from a functional "abrogation" of *RAI1* (Girirajan et al. 2006). Several features common among deletion cases occur significantly less often in *RAI1* mutation cases, specifically cardiovascular and genitourinary tract abnormalities, hearing loss, hypotonia, and, in particular, short stature (Girirajan et al. 2006; Edelman et al. 2007). Comprehensive reviews based on systematic evaluation are available delineating various aspects of the syndrome that include physical, developmental, and behavioral abnormalities and major sleep disorder associated with inverted circadian melatonin rhythm (Smith et al. 2012).

Smith–Magenis syndrome is likely more common than recognized earlier. Greenberg et al. (1991) reported a minimum prevalence of SMS of 1 in 25,000 births in Harris County, Texas, over a two year period. However, in the last decade, the vast majority of individuals have been identified using improved molecular cytogenetic techniques, and these studies have suggested an incidence closer to 1/15,000 births (Smith et al. 2006a; Elsea and Girirajan 2008). Prospective molecular screening for SMS among 1138 subjects with intellectual disability of unknown cause identified two affected individuals previously missed by both clinical and cytogenetic evaluation (Struthers et al. 2002), yielding a frequency of 1/569 in this select population. This is consistent with previous estimates of 1/544 and 4/1672 reported by Barnicoat et al. (1993) and Behjati et al. (1997), respectively. Targeted "phenotypic" screening pursued in a Brazilian cohort (*n* = 48) for key features of SMS (i.e. ID, craniofacial and/or skeletal anomalies, history of sleep disturbance and at least one or more stereotypic and self-injurious behaviors) led to diagnostic confirmation of SMS (FISH documented deletions) in 7/48 (14.5%; Gamba et al. 2011).

*This chapter is in the public domain in the United States of America.

Cassidy and Allanson's Management of Genetic Syndromes, Fourth Edition.
Edited by John C. Carey, Agatino Battaglia, David Viskochil, and Suzanne B. Cassidy.
© 2021 John Wiley & Sons, Inc. Published 2021 by John Wiley & Sons, Inc.

Delayed diagnosis is common, and SMS is still missed in early infancy both clinically and by routine cytogenetic analysis, further delaying diagnosis to school age or older (Greenberg et al. 1996; Smith et al. 1998a; Struthers et al. 2002; Gropman et al. 2006). Among 19 children, born between 1980 and 1997, only 12 were diagnosed on the initial karyotype, but all were confirmed by the third karyotype (Gropman et al. 2006). The expanded availability of chromosomal microarray analysis (CMA), whole genome or targeted microarray technology, including use of multi-gene panel(s) that include *RAI1* (e.g., ID/ASD panels) has led to increased detection of clinically unsuspected cases referred for intellectual disabilities.

SMS does not appear to be associated with reduced lifespan; the oldest known reported person (Smith et al. 1986) lived to her late 80s (Magenis, personal communication). The only three published cases of early demise were infants born with severe congenital anomalies impacting survivability (Smith et al. 1986; Denny et al. 1992; Yamamoto et al. 2006). All three involved congenital heart disease and palatal defects, two of the three had large deletions, and postsurgical complications were notable. At least three cases of stroke-like episodes are also recognized. A recent report suggests that adults with SMS may have potential risk factors for premature atherosclerosis (Chaudhry et al. 2007).

Diagnostic Criteria

The diagnosis of SMS is based on clinical recognition of a unique and complex phenotypic pattern of physical, developmental, behavioral, and sleep features, confirmed by diagnostic confirmation of an interstitial deletion of 17p11.2 or heterozygous mutation of the *RAI1* gene. In the past, the diagnosis was usually confirmed cytogenetically (550-band resolution) accompanied by fluorescence in situ hybridization (FISH), using an *RAI1*-specific probe (Vlangos et al. 2003). A variety of newer DNA-based molecular techniques offer reliable cost-effective methods and diagnostic algorithm, especially for clinically suspected cases (Truong et al. 2008; Elsea et al. 2008). Chromosomal microarray analysis (CMA) has rapidly replaced G-banded karyotyping as the first-tier diagnostic molecular genetic test for individuals with DD/ID, ASD, or multiple congenital anomalies (Lu et al. 2007; Miller et al. 2010 – ACMG consensus paper); however, mutation analysis of the *RAI1* gene should be considered in clinically suspected cases without documented deletion.

Many of the clinical features are subtle in early childhood, becoming more distinctive with advancing age. Common features seen in over two-thirds of individuals with SMS caused by deletion include a characteristic craniofacial appearance (Figure 54.1), ocular abnormalities (50–85%), speech delay (expressive more than receptive) with or without associated hearing loss (>90%), dental abnormalities (>90%), infantile hypotonia (90–100%) with failure-to-thrive (78%), short stature (>75%), brachydactyly (>80%), signs of peripheral neuropathy (pes planus or cavus, depressed deep tendon reflexes, insensitivity to pain) (75%), laryngeal anomalies (>75%), scoliosis (>65%), variable levels of intellectual disabilities (100%) and neurobehavioral problems including sleep disturbance (65–100%) associated with inverted circadian rhythm of melatonin (over 95%), maladaptive and self-injurious behaviors (over 92%), one or more stereotypies (100%), and sensory integration disorders (Greenberg et al. 1996; Dykens and Smith 1998; Smith et al. 1998a, 1998b; Potocki et al. 2000, 2003; De Leersnyder et al. 2001a; Solomon et al. 2002; Girirajan et al. 2006; Gropman et al. 2006, 2007; Tomona et al. 2006; Hildenbrand and Smith 2012). The voice is hoarse and low-pitched (>80%) and serves as a diagnostic clue. Crying and babbling are notably absent at expected ages. Functional impairment of voice and speech with underlying laryngeal anomalies are also common, providing a physiologic explanation for marked early speech/language delays appreciated in the syndrome (Solomon et al. 2002).

Although the phenotype of individuals with *RAI1* mutation overlaps that of deletion cases for the majority of features (70%), several features occur much less commonly in those with an *RAI1* mutation, specifically: hearing loss (10–35%), hypotonia (61%), short stature (9%), speech and motor delays (70%), immunologic abnormalities, and renal and cardiovascular anomalies (0%) (Girirajan et al. 2006; Vilboux et al. 2011). However, among published individuals with a mutation, behavioral features are notably present in all, including self-hugging, attention-seeking, self-injurious behaviors, and sleep disturbance, and likely reflect a potential bias of ascertainment for behavioral aspects that characterize the syndrome.

Several distinctive features characterize the infant phenotype of SMS caused by deletion, including cherubic facial appearance (100%), hypotonia (100%), hyporeflexia (84%), generalized lethargy (100%), complacency (100%) with increased sleepiness and napping, oromotor dysfunction (100%), feeding difficulties with failure-to-thrive, and delayed gross motor and expressive language skills (100%) in the presence of relatively appropriate social skills (80%) (Gropman et al. 2006). Decreased fetal movement is appreciated by about half the mothers (Gropman et al. 2006). Although failure-to-thrive does occur, some infants actually appear "chubby," with redundant fat folds on the extremities (Smith et al. 1998a). Small hands and feet with dorsal edema are also appreciated. Crying is infrequent (95%), and babbling and vocalizations are markedly decreased for age in virtually all infants despite normal hearing.

Sleep disturbances begin in infancy and remain a chronic problem into adulthood. Parentally reported sleep data initially suggested infant hypersomnolence and lethargy that is replaced by the frequent nocturnal awakenings and fragmented sleep later in childhood. Objective sleep data

INTRODUCTION

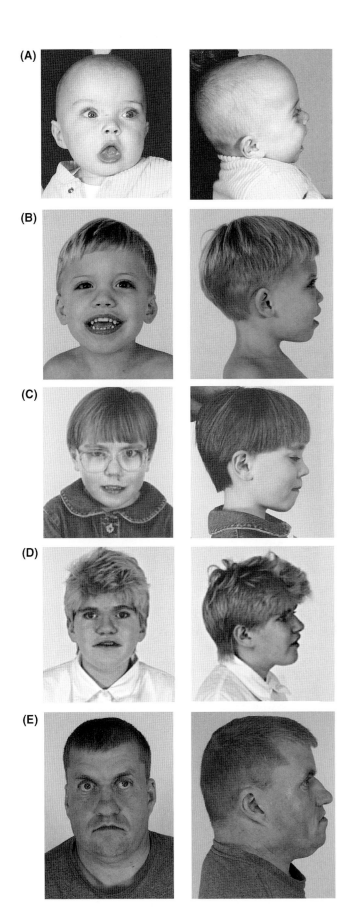

(actigraphy estimated sleep) document the onset of the sleep dysfunction before 1 year of age (Gropman et al. 2006; Smith et al. 2019). Infants often need to be awakened for daytime feeds because of increased daytime sleepiness, and may not alert parents during the fragmented nighttime sleep; that is, they are "quiet babies sleeping poorly" (Gropman et al. 2006).

The facial appearance in SMS is distinctive and changes over time (Figure 54.1; Table 54.1). These changes are described in detail in the Craniofacial section. Briefly, the face is square and broad with mild brachycephaly. The face is "cherubic" with full cheeks, upslanting and deep-set eyes, marked midfacial retrusion with depressed broad nasal root, and micrognathia (Smith et al. 1998a). The distinctive shape of the mouth includes a fleshy upper lip and a cupid's bow or tented appearance caused by bulky philtral pillars (Allanson et al. 1999). Over time, the facial appearance appears to coarsen because of heavy brows, synophrys, and relative prognathism.

Developmental delay and/or intellectual disability is found in all individuals to a variable degree, ranging from moderate-to-mild/borderline. The marked expressive language delays in the presence of maladaptive behaviors make accurate cognitive assessment a challenge. A distinct and complex behavioral phenotype is seen in SMS, with several neurobehavioral aspects apparently unique to the syndrome. Unusual maladaptive, self-injurious, and stereotypic behaviors occur in 40–100% of both children and adults with SMS based on published series with systematic review (Dykens and Smith 1998; Finucane et al. 2001; Girirajan et al. 2006; Madduri et al. 2006; Gropman et al. 2007). Self-injurious behaviors are nearly universal (>92%) and show a direct correlation with age and level of intellectual functioning (Finucane et al. 2001). Hand/wrist biting, slapping/hitting oneself, headbanging, and/or skin picking are very common in childhood, observed especially when upset or frustrated. Two behaviors possibly unique to SMS

FIGURE 54.1 Craniofacial appearance of Smith–Magenis syndrome in five unrelated individuals at ages (A) 9 months, (B) 4 years, 10 months, (C) 7 years, 4 months, (D) 16.5 years, and (E) 49 years. In infant (A), note mild brachycephaly with square-shaped face, "cherubic" appearance because of prominent "pudgy" cheeks, upslanting and deep-set eyes, marked midfacial retrusion with depressed nasal root, short philtrum, and micrognathia. The mouth is characteristic, with cupid's bow or tented appearance and open-mouth posture. The ears may be low-set and are often slightly posteriorly rotated with a heavy/thick helix. In older children (B–D), the facial appearance is more distinct, with broad-squared shape, marked broadening of the jaw, more striking midface retrusion, and relative prognathism at older ages. Brows appear heavy, often extending laterally, with synophrys and the appearance of being deep-set. The downturned mouth and characteristic cupid bow upper lip remain apparent. The nasal bridge becomes almost "ski jump" in shape with age.

TABLE 54.1 Clinical features by age group.

Infancy

CLINICAL
Brachycephaly
Mild facial dysmorphism; "Down syndrome-like" appearance
"Cherubic" appearance with rosy cheeks
Broad, square-shape face
Upslanting palpebral fissures
Midface retrusion
Small upturned nose
Open mouth posture
Cupid-bow mouth with tented upper lip
Eye problems: strabismus; microcornea; pigmented flecking of iris
Short broad hands and feet; +/− dorsal edema
Persistent fetal pads
CNS: mildly enlarged ventricles (ventriculomegaly); delayed myelination

NEURODEVELOPMENTAL
Feeding difficulties (major oral-sensorimotor dysfunction)
Failure-to-thrive
Generalized hypotonia
Alert and responsive when awake; often need to awaken for daytime feeds
Hyporeflexia
Delayed gross/fine motor skills with age appropriate social skills

BEHAVIORAL
Diminished vocalizations and crying
Complacent, but happy ("smiling") demeanor; "quiet good babies"
Lethargic
Sleep disturbance
Parent perception of "good night sleeper"
Decreased total 24 hour and nighttime sleep for age by 9 months; noisy breathing
Diminished crying and/or vocalizations to alert parents when awake at night

Toddler/early childhood

CLINICAL
Recognizable facial appearance with midface retrusion; rosy cheeks; deep set eyes, synophrys, open mouth posture with everted upper lip
Frequent/chronic otitis media
Hearing loss (predominantly conductive)
Vision problems (myopia; strabismus)
Fair hair and coloring compared with family
Short stature
Hoarse, deep voice
Dry skin, especially palms/soles of feet
Unusual gait/toe walking
High cholesterol

NEURODEVELOPMENTAL
Developmental delays
Gross/fine motor delays
Marked speech delay (expressive > receptive)
Decreased pain sensitivity
Pes planus or pes cavus
Mild intention tremor of hands

TABLE 54.1 *(Continued)*

Delayed toilet training
Sensory integration issues

BEHAVIORAL
Tantrums: hyperactivity, impulsivity
Stereotypic behaviors: self-hugging, lick and flip behavior, bruxism
Self-abusive behaviors: head banging; hitting self; wrist biting; skin picking
Sleep disturbance: short sleep cycle; early rising (5:30–6:30 AM); frequent night awakenings; and daytime naps; noisy breathing/snoring
Engaging personality
Visual learners
Affinity for electronic toys, buttons, etc.

School age

CLINICAL
Characteristic facies with persisting midface retrusion, relative prognathism, heavy brows
Hoarse voice
Hypernasal speech because of velopharyngeal insufficiency
Progressive myopia
Hearing loss (conductive and/or sensorineural)
Short stature
Scoliosis
Broad-based flapping gait
Dental anomalies (tooth agenesis; taurodauntism) and poor hygiene
Dry skin

NEURODEVELOPMENTAL
Intellectual disability
Weaknesses: sequential processing and short-term memory
Strengths: long-term memory and perceptual closure
Visual learners
Poor adaptive function
Pes planus or pes cavus
Nighttime enuresis
Sensory integration issues

BEHAVIORAL
Developmental asynchrony between intellectual level and emotional maturity ("inner toddler")
Attention-seeking behaviors
Adult-oriented
Frequent outbursts/tantrums
Sudden mood shifts
Impulsivity/aggression
Attention deficits; hyperactivity
Obsessive-compulsive features/need for structure
Possible to redirect
Sense of humor
Chronic sleep disturbance: short sleep cycle; early rising; night awakenings may not be so disruptive; daytime sleepiness/naps
Settling difficulties after 10 years; snoring
Wandering at night/food foraging
Stereotypic behaviors: self-hugging, bruxism
Self-injurious behaviors: hits/bites/slaps self, skin picking; nail pulling; object insertion, often into ears

TABLE 54.1 *(Continued)*

Very communicative; repetitive questions
Excellent long-term memory (names, places, people)
Affinity for computers and/or electronics

Adolescence to adulthood

CLINICAL

Coarser facial appearance with deep-set eyes, relative prognathism, heavy brows, synophrys
Progressive myopia
Hearing loss (conductive and/or sensorineural)
Velopharyngeal insufficiency
Females: premature adrenarche; irregular menses; hygiene concerns
Tendency to obesity (especially postpubertal)
Hoarse voice
Short stature (5–10%);
Scoliosis
Broad-based flapping gait
Dental issues (poor self-hygiene)
Dry skin

NEURODEVELOPMENTAL

Intellectual disability
Excellent long-term memory
Possible exercise intolerance
Poor adaptive function: poor self-help/daily living skills;
Lack of awareness of "personal space" boundaries (self and others)
Females: catemenial seizures

BEHAVIORAL

Chronic sleep disturbance; decreased total sleep time for age; increased naps for age (parental reports); nighttime settling difficulties; later sleep offset (~7:00 AM)
Wandering at night/food foraging
Developmental asynchrony ("inner toddler")
Major behavioral outbursts and/or rage behaviors, property destruction, attention seeking, aggressive/explosive outbursts
Impulsive, disobedient, argumentative
Mood shifts (rapid) without major provocation
Attention deficits
Increased anxiety especially with change and/or in anticipation of events
Self-injurious behaviors (100%): hits/slaps self; head banging; hits self against objects; nail yanking; skin picking/scratching; object insertion
Stereotypies: touching/mouthing objects, bruxism; page turning; self hug; body rocking; spinning, and twirling of objects
Very chatty; repetitive questions; sense of humor
Excellent long-term memory (names, places, people)
Affinity for computers and electronics

are onychotillomania (nail yanking) and polyembolokoilamania (bodily insertions) (Greenberg et al. 1991; Finucane et al. 2001). Onychotillomania occurs in over half of subjects, but remains less frequent at younger ages (27%) than older ages (86%) (Finucane et al. 2001). Two stereotypic behaviors have also been described, the spasmodic upper body squeeze, or "self-hug" (Finucane et al. 1994), and hand-licking and page flipping, or "lick and flip" behavior (Dykens et al. 1997; Dykens and Smith 1998), providing other effective diagnostic clues.

Several clinical features of SMS appear to be age dependent, including prominent forehead, midface retrusion, prognathism, brachycephaly, hoarse voice, and ophthalmologic findings, specifically high myopia with and without retinal detachment (Smith et al. 1998a). Frequent otitis media is associated with conductive hearing loss in early childhood (35.2%; 10 years and under); however, a pattern of fluctuating and progressive hearing decline is documented with age, with sensorineural hearing loss the most common type of hearing loss (48.1%) seen after the first decade of life (Brendal et al. 2017). The stereotypies and self-injurious behaviors generally do not begin until after the first 18 months of life. Many believe that the early behavioral problems, including head banging, self-biting, and self-hitting, are in part related to the general frustrations associated with poor expressive language skills (Greenberg et al. 1996; Smith et al. 1998a). The sleep disturbance and self-abusive behaviors also appear to escalate with age, often at expected stages of the life cycle, specifically, at 18–24 months, at school age, and with pubertal onset. Reduced 24 hour and total night sleep for age leads to a chronic daytime sleep debt that is compensated by increased daytime somnolence (napping); settling difficulties appear more prevalent after age 10 years (Gropman et al. 2007; Smith et al. 2019). The degree to which the chronic disrupted sleep impacts daytime behaviors requires further study; however, initial studies suggest that sleep is a significant predictor of maladaptive behaviors (Dykens and Smith 1998).

Gender differences are also appreciated with females more likely than males to exhibit myopia, eating/appetite problems, cold hands/feet, generalized frustration with communication, and impulsivity (Edelman et al. 2007; Martin et al. 2006).

Etiology, Pathogenesis, and Genetics

SMS is classified as a contiguous gene syndrome (Greenberg et al. 1991), with haploinsufficiency of physically linked but functionally unrelated genes responsible for phenotypic variability. The recognition of affected individuals with heterozygous *RAI1* intragenic mutations argues strongly that haploinsufficiency for *RAI1*, resulting from interstitial 17p11.2 deletion or mutation, accounts for the major physical and neurobehavioral features of the syndrome (Slager et al. 2003). Clinical variability exists among individuals with the same deletion size, suggesting that other gene(s) in the deletion interval may account for the phenotypic variability observed in the syndrome (Potocki et al. 2003; Girirajan et al. 2006). Phenotypic comparison of individuals with deletion versus mutation indicates concordance for approximately 21 of 30 (70%) features; however, individuals with an *RAI1* mutation are significantly less likely than those with a deletion to have short stature, chronic ear infections, hearing loss, speech and motor delay, hypotonia, self-hugging, and cardiovascular or renal anomalies (Girirajan et al. 2006). A meta-analysis of 105 published cases (67 common deletion; 28 atypical/smaller/larger deletions; 10 *RAI1* mutations) by Edelman et al. (2007) also found that individuals with an *RAI1* mutation were more likely to demonstrate overeating, obesity, polyembolokoilamania, self-hugging, muscle cramping, and dry skin. It is important to recognize the potential bias of ascertainment among the early published *RAI1* mutation cases as most are older and were referred because of the strong suspicion of SMS based on their physical and behavioral features. The full phenotypic spectrum of individuals with *RAI1* mutation requires the identification of additional individuals.

The mechanism leading to the SMS deletion was first defined by Chen et al. (1997) and involves non-allelic homologous recombination of flanking low-copy repeat gene clusters referred to as SMS-REPs. Such low-copy repeats or duplicons (Yi et al. 2000), flank genomic regions prone to deletion, duplication, and inversion and act as substrates for inter- and intrachromosomal recombination. In SMS, the common deletion occurs between the distal and proximal SMS-REP (Chen et al. 1997; Elsea and Girirajan 2008). Non-allelic homologous recombination has also been documented to lead to several other contiguous gene syndromes, including Williams syndrome (Chapter 63), Prader–Willi, and Angelman syndromes (Chapters 46 and 5), and deletion 22q11.2 (Velo-cardio-facial/DiGeorge syndrome, Chapter 21) (Lupski 1998). A common deletion (~3.7 Mb) occurs in approximately 70% of deletion cases; smaller or larger deletions account for the remaining 30% (Elsea and Girirajan 2008; Edelman et al. 2007). However, deletions have ranged from less than 2 to more than 9 Mb (Trask et al. 1996).

Efforts to refine the SMS critical deletion interval have narrowed the region to an approximately 1 Mb genomic interval that contains approximately 13 known genes, 12 predicted genes, and 3 expressed tags (ESTs) (Vlangos et al. 2003). Moreover, the critical region is highly conserved between humans and mice, including 19 genes in the same order and orientation (Bi et al. 2002). Craniofacial abnormalities, obesity, seizures, and altered circadian rhythm have been documented in deletion and *RAI1* +/− heterozygote mice (Walz et al. 2003; Yan et al. 2004; Bi et al. 2005, 2007). More recent transgenic mouse models offer compelling evidence that deviation in *RAI1* copy number (dosage) from normal diploid (2n) results in growth and neurobehavioral phenotypes that parallel the human phenotype.

Studies on the few individuals with SMS who have smaller or larger deletions permit phenotypic dissection of certain aspects of the phenotype. Gene(s) responsible for the major features including intellectual disabilitiy, craniofacial, behavioral, and sleep disturbances are hypothesized to be located in the proximal part of the critical regional, between *COPS3* and the middle SMS-REP (Bi et al. 2002), whereas gene(s) affecting physical and intellectual development may fall in the distal part of the critical region (Elsea et al. 1997). A recent review suggests that individuals with atypically large and distally extending deletions are more severely affected than are those with smaller deletions. Genes distal to *RAI1*, likely contribute to cardiovascular abnormalities observed in 30–40% of deletion cases (Girirajan et al. 2006).

Among the multiple genes mapped to the 17p11.2 SMS critical region (*http://www.ncbi.nlm.nih.gov*), several are worthy of specific mention (Figure 54.2). Retinoic acid-induced 1 (*RAI1*), mapped to the 17p11.2 critical interval (Seranski et al. 2001), is now recognized as the major gene involved in the etiology of SMS. Intragenic frameshift

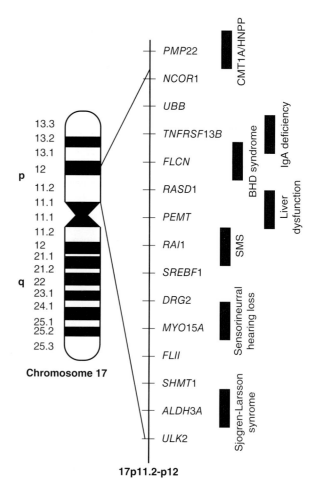

FIGURE 54.2 Ideogram of 17p11.2-p12 region, showing mapped genes and the disorders and abnormalities caused by their alteration. SMS, Smith–Magenis syndrome; BHD, Birt–Hogg Dubé syndrome; CMT1A, Charcot–Marie–Tooth type 1A; HNPP, hereditary neuropathy with liability to pressure palsies. (From Gropman et al. 2007, with permission.)

mutations in *RAI1*, first identified in three adults with a clinically suspected diagnosis of SMS but in whom a deletion remained undetectable cytogenetically by standard FISH analysis (Slager et al. 2003), now account for about 10% of cases. Phenotyically, individuals with de novo mutations share the major features associated with SMS, including the facial appearance, obesity, sleep, and neurobehavioral aspects of the syndrome, except for structural cardiac and renal anomalies (Slager et al. 2003; Girirajan et al. 2005; Vilboux et al. 2011; Dubourg et al. 2014; Acquaviva et al. 2017).

RAI1 encodes a novel gene whose cellular role remains unclear, although a putative role in neuronal differentiation has been postulated (Slager et al. 2003). Among the published cases, all intragenic *RAI1* mutations occur in exon 3, likely leading to a truncated and/or non-functional protein (Slager et al. 2003; Bi et al. 2004, 2006; Girirajan et al. 2005, 2006; Vilboux et al. 2011). De novo nonsense and missense mutations and deletions or insertions resulting in misincorporation of 8–74 amino acids are associated with the SMS phenotype; several normal polymorphisms in *RAI1* gene are also recognized (Girirajan et al. 2006). Mouse studies suggest *RAI1* functions as a transcriptional regulator, important for embryonic and postnatal development (Bi et al. 2005). Using the vertebrate model Xenopus, Tahir et al. (2014) provided the first mechanistic evidence that Rai1 deficiency impacts vertebrate craniofacial development, with Rai1 morphants recapitulating the craniofacial defects observed in humans (i.e. midface retrusion, mouth shape) as a result of aberrant neural crest migration and development. *RAI1* also plays an integral role in maintenance of circadian rhythmicity by disrupting the transcription of circadian locomotor output cycles kaput (CLOCK) (Williams et al. 2012). Ongoing studies are needed to fully understand the role of *RAI1* gene in SMS.

The main source of glycine and one-carbon units in the brain occurs via the folate dependent metabolic conversion of serine to glycine catalyzed by the enzyme cytosolic serine hydroxymethyltransferase (*cSHMT*). Haploinsufficiency for the *cSHMT* gene was demonstrated in all 26 individuals with SMS in one study; the enzyme was decreased by 50% in lymphoblasts from all three individuals tested (Elsea et al. 1995). Elsea et al. (1995) suggest that reduced enzymatic activity may lead to altered levels of glycine, serine, adenosine, and folate in the brain; however, at least in plasma and urine, serine and glycine levels were normal and folate was lower in only one of the three individuals with SMS studied. In the forebrain, modulation of the *N*-methyl-D-aspartate (NMDA) receptor requires glycine acting as a co-agonist with neurotransmitter glutamate. Studies by Waziri et al. (1999) in a hyperglycinic rat model support the hypothesis that high glycine concentrations may be neurotoxic and lead to the behavior and brain abnormalities observed in people with schizophrenia.

COP9, which maps within the SMS critical region (Potocki et al. 1999), is intriguing because it codes for a conserved protein (45% conserved from plants). COP9 is a multi-subunit complex protein shown to be a nuclear regulator in humans; in plants, it responds to environmental light signals and regulates development (Potocki et al. 1999). COP9 may play a role in the novel inverted circadian rhythm of melatonin observed in SMS (De Leersnyder et al. 2001; Potocki et al. 2000).

FLCN: Birt–Hogg–Dubé syndrome, a rare autosomal dominant hamartoma disorder, is caused by germ line mutations in the *folliculin* gene (*FLCN*). The syndrome is characterized by triad of benign dermatologic tumors that involve the hair follicle (trichodiscomas) and skin (fibrofolliculomas and acrochordons), lung cysts, and increased risk for spontaneous pneumothoraces and/or renal neoplasms. Clinically, the phenotype is usually not recognized until the third-fourth decades of life, with considerable variability reported within and between families with the same germline mutation type or location (Schmidt et al. 2005). The typical SMS deletion causes haploinsufficiency of *FLCN*, suggesting a theoretical concern for an increased cancer risk in SMS (Menko et al. 2009). Based on suggested criteria for BHD syndrome, haploinsufficiency for FLCN is sufficient to diagnose Birt–Hogg–Dubé syndrome (Menko et al. 2009). Although the risk for Birt–Hogg–Dubé syndrome in SMS is unstudied, the authors are aware of the co-occurance of renal tumors in three unrelated adults with deletion-confirmed SMS, with pathological confirmation of chromophobe renal cell carcinoma documented in two cases (Smith et al. 2014; Darfour et al. 2016). Spontaneous pneumothorax in the absence of other features of BHD syndrome has also been reported once in one adolescent male (18 years old) with SMS due to *RAI1* mutation (Truong et al. 2010). Because Birt–Hogg–Dubé syndrome is usually not recognized until adulthood, the risk for this condition in the SMS is unknown; however, cases with spontaneous pneumothorax should be evaluated further and cancer surveillance considered as a precaution in adulthood.

TNFRSF13B: Tumor necrosis factor receptor superfamily, member 13B is a lymphocyte-specific member of the tumor necrosis factor (TNF) receptor superfamily that interacts with calcium-modulator and cyclophilin ligand (CAML). This gene encodes the transmembrane activator and CAML Interactor (TACI) involved in the signal transduction pathway that regulates immune cell survival and apoptosis. Expressed in peripheral B cells, *TNFRSF13B/TACI* binds BAFF (B-cell activating factor) and APRIL (a proliferation-inducing ligand), which are necessary for class switching of human IgM to IgA and IgG. Dominant mutations in *TNFRSF13B* can result in two immune deficiency syndromes, selective IgA deficiency (1/600 people) and the less common, but more severe common variable immunodeficiency (CVID) (Castigli et al. 2005). Both conditions are associated with an increased frequency of persistent/recurring ear infections, sinus infections, bronchitis, pneumonias, and/or gastrointestinal infections. Thus, haploinsufficiency for *TNFRSF13B* because of 17p11.2 deletion may explain the increased frequency of IgA deficiency and history of recurrent infections frequently seen in SMS (Introne et al. 2005; Perkins et al. 2017).

PEMT: This gene, which codes for phosphatidylethanolamine *N*-methyltransferase, is an important liver enzyme involved in the synthesis of membrane phospholipids (lipoprotein secretion from liver). A SNP polymorphism of the human *PEMT* gene (V175M) is associated with diminished enzymatic activity now felt to confer susceptibility to non-alcoholic fatty liver disease in humans (Song et al. 2005), consistent with knockout mouse models that show fatty liver disease. In the liver, triacylglycerol is formed and secreted in very low-density lipoprotein (VLDL); VLDL requires synthesis of new phosphatidylcholine molecules that, if not available, lead to the accumulation of fat droplets in the cytosol of liver cells (Song et al. 2005). Predictive risk factors for non-alcoholic fatty liver disease include obesity, hypertriglyceridemia, and diabetes. Both obesity and hyperlipidemia are common in SMS (Smith et al. 2004) and abnormal liver function studies and/or evidence of fatty liver have been documented in a few individuals (Smith et al. unpublished data).

SREBF1: The common SMS deletion interval contains this gene for sterol regulatory element-binding protein (*SREBF1*), a transcription factor involved in cholesterol homeostasis (Hua et al. 1995). Hypercholesterolemia may serve as a useful early biochemical marker of the syndrome. Among 49 children with SMS, 28 (57%) had fasting values greater than the 95th centile for age and gender for at least one or more of the following: total cholesterol, total triglycerides, and/or low-density lipoprotein-cholesterol. Only 16 (32%) had normal values for all three of these variables (Smith et al. 2002). Based on the American Academy of Pediatrics recommended lipid levels for children and adolescents, only one-third of subjects with SMS fall within the normal range for total cholesterol and low-density lipoprotein-cholesterol. Functional studies are needed to determine the role of haploinsufficiency for *SREBF1* in altered cholesterol homeostasis and its potential contribution to the pathogenesis of SMS.

MYO15A: The increased frequency of sensorineural hearing loss seen in SMS may be secondary to loss of heterozygosity caused by haploinsufficiency of *MYO15*, a gene for recessive non-syndromic profound, congenital sensorineural hearing loss, DFNB3 that maps within 17p11.2 (Liang et al. 1998). The gene encodes unconventional myosin XVA, a motor protein necessary for the development of hair cell stereocilia in the inner ear (Nal et al. 2007). Gene sequencing of *MYO15A* in eight confirmed SMS deletion individuals with mild ($n = 7$) or moderate sensorineural hearing loss ($n = 1$) identified only one with a hemizygous mutation in *MYO15A*. This individual, a 33-year-old with a paternally derived deletion and moderate hearing loss, had a maternally derived hemizygous missense *MYO15A* mutation (T2205I; 6952C-T) (Liburd et al. 2001). The other seven with only one "normal" copy of the gene also had milder hearing loss, suggesting that haploinsufficiency may lead to a diminished number of stereocilia lining the cochlea and/or impact the normal development or "aging" process of these specialized hair cells. Thus, the earlier onset of sensorineural hearing loss, documented as early as 10 years of age (Brendal et al. 2017), may be the result of haploinsufficiency for *MYO15A* and a potential age effect.

Genetic Counseling With a few exceptions (Zori et al. 1993; Smith et al. 2006b; Acquaviva et al. 2017) all cases of SMS occur de novo, suggesting a low recurrence risk. At least two sets of monozygotic affected twins (Kosaki et al. 2007; Hicks et al. 2008) and one family with two affected siblings owing to maternal mosaicism for the 17p11.2 deletion (Smith et al. 2006b) are recognized. Acquavia et al. (2017) recently reported the first case of the transmission of SMS from an affected mother to daughter due to a novel *RAI1* frameshift mutation (c.1194delC_p.SerProfs*40). Thus, parental studies are recommended for all newly diagnosed individuals. To date, there is no evidence to suggest an obvious parental age contribution or unusual sex distribution for the deletion (Lockwood et al. 1988). Random parental origin of the 17p deletion has been documented, suggesting that imprinting does not play a role in the expression of the SMS phenotype (Greenberg et al. 1991).

Diagnostic Testing

Definitive diagnosis of SMS is based on confirmation of the interstitial deletion of 17p11.2 or identification of a heterozygous mutation of the *RAI1* gene. In the past, the majority of individuals were diagnosed cytogenetically by routine G-banded analysis (550-band resolution or higher) accompanied by fluorescence in situ hybridization (FISH) using a SMS-specific probe (D17S258 or probe containing *RAI1*) (Juyal et al. 1996; Vlangos et al. 2005; Gropman et al. 2007). However, the deletion is subtle and has been overlooked in the laboratory, especially when the differential diagnosis fails to include SMS. Of the affected individuals, 90% have a detectable 17p11.2 deletion, with *RAI1* gene mutations accounting for the remaining 10% (Elsea and Girirajan 2008). Among detectable deletion cases, 70% represent a common deletion spanning 3.7 Mb and the remaining 30% account for smaller or larger deletions (Elsea and Girirajan 2008). All deletions include the *RAI1* gene.

A variety of other DNA-based molecular techniques have been shown to be rapid, cost-effective, and reliable for diagnostic confirmation of submicroscopic deletions.

Chromosomal microarray analysis (CMA) has replaced G-banded cytogenetic analysis and FISH analysis as a first-tier test in the diagnosis of SMS. Where a clinical diagnosis is strongly suspected by professionals experienced with SMS, but a deletion is not detectable using CMA, then targeted single gene sequencing of RAI1, or an intellectual disability multigene panel (e.g., ID/ASD panel) that includes *RAI1*, is suggested. If both CMA and RAI1 testing are negative, then exome sequencing may be considered.

Four fetuses with prenatally diagnosed SMS have been reported (Fan and Farrell 1994; Thomas et al. 2000; Lei et al. 2015). All had deletion of 17p11.2. Two were detected by second trimester amniocentesis performed for an increased risk for trisomy 21 based on a low maternal serum alpha-fetoprotein screen. An additional two were detected by whole-genome and high-resolution CMA in the second trimester and had increased nuchal translucency, mild lateral ventriculomegaly, and congenital heart defects. This suggests that increased nuchal translucency thickness (NT) at or above the 99th centile in the first trimester, which is known to be associated with chromosomal aberrations, cardiac defects, and other genetic syndromes (Schou et al. 2009) could alert the clinician to perform CMA to exclude SMS.

Differential Diagnosis

In infancy, children with SMS may come to clinical attention due to overlapping features of trisomy 21 (Down syndrome) because of infantile hypotonia and several facial stigmata suggestive of this diagnosis, including a midface retrusion and upslanting palpebral fissures (Allanson et al. 1999; Gropman et al. 2006). Among a cohort of 19 individuals with SMS evaluated by the authors, over one-third underwent an initial chromosome analysis to rule out Down syndrome (Gropman et al. 2006). Fortuitous diagnosis of del17p11.2 and SMS was confirmed in a few of these infants before 18 months of age; however, the majority remained undiagnosed until repeat chromosome analysis at school age (mean age 5.4 years) at higher resolution and/or by FISH. In at least two neonatal cases of SMS presenting with tetralogy of Fallot, cytogenetic analysis initiated by cardiologists to exclude 22q deletion actually confirmed del 17p11.2 (Sweeney and Kerr 1999). Other diagnoses considered before confirmation of SMS include the following: PWS, because of infantile hypotonia, lethargy, and feeding and sleep disorders (Chapter 46); deletion 22q11.2 (DiGeorge/velo-cardio-facial syndrome), because of marked speech delay and cardiac anomalies (Chapter 21); and Fragile X syndrome because of autistic-like features and behaviors (Chapter 28). Clinically, many of the children have autism spectrum/pervasive developmental disorder because of significant early speech and language delay and the presence of sensory issues, stereotypic, and maladaptive behaviors (Gropman et al. 2006; Laje et al. 2010). In addition, there appears to be a reverse gender manifestation (i.e. more females than males) ratio of autism spectrum disorder in SMS (Nag et al. 2018).

MANIFESTATIONS AND MANAGEMENT

The phenotype associated with SMS is based on clinical descriptions and/or systematic evaluation of deletion-confirmed cases. Although published *RAI1* mutation individuals share significant phenotypic overlap (about 70%) with deletion ones, the number is limited to fewer than 30, and may reflect ascertainment bias; additional reports are required to definitively discern the *RAI1* mutation phenotype. Nevertheless, *RAI1* mutation persons appear to differ from those with standard deletion in several aspects, discussed above (Girirajan et al. 2006). The following narrative is generalized to deletion-confirmed SMS unless specifically noted.

Growth and Feeding

Short stature and/or failure-to-thrive remain consistent findings, seen in the majority of individuals with SMS. Over 80% of affected individuals are born at term with birth weight, length, and head circumference generally within the normal range (Smith et al. 1986; Gropman et al. 2006). Height and weight gradually decelerate to the 5th centile or below in first year of life in both males and females. Head circumference is usually within the normal range, although about 20% have head circumference less than the 3rd centile for age (Stratton et al. 1986; personal experience).

Short stature is characterized by length and height plotting two to three standard deviations below the mean for age during infancy (0–36 months) and childhood (Greenberg et al. 1991); however, adult heights generally fall in the low range of normal (3rd–10th centile). Several individuals with SMS who are notably short and obese have been reported. In some, this has led to an initial suspected diagnosis of PWS (see Chapter 46), especially in the presence of a history of infantile hypotonia and feeding problems including food foraging, and/or evidence of behavioral dysfunction (Stratton et al. 1986; personal experience).

Although short stature persists into adulthood in about half the individuals with deletion (Greenberg et al. 1996), it is uncommon (only 10%) among those with *RAI1* mutation (Girirajan et al. 2006; Edelman et al. 2007). Individuals with an *RAI1* mutation are more likely to have problems with regulating food intake (overeating) (86%), leading to obesity (67%), with parents needing to take greater precautions to hide/lock up food (Edelman et al. 2007). Researchers now suspect that *RAI* mutation is associated with an overgrowth phenotype, since heights and weights above the 95th centile have been documented (Girirajan et al. 2006; Elsea and

Girirajan 2008). This is especially intriguing as obesity is recapitulated in *RAI1* +/– and transgenic mouse models (Bi et al. 2005; Yan et al. 2004).

Feeding difficulties occur during infancy in the majority, caused in part by hypotonia, lethargy, and major oral-motor dysfunction resulting in poor suck and swallowing abilities and coordination (Gropman et al. 2006). In addition, oral sensorimotor dysfunction is seen in virtually all affected individuals examined, characterized by universal lingual weakness, asymmetry, and/or limited motion, and occasional laryngeal findings (Sonies et al. 1997; Solomon et al. 2002). Oral-motor dysfunction may result in delayed ability to chew and refusal of highly textured foods. This may additionally contribute to failure-to-thrive. Nasogastric or gastrostomy tube feedings are sometimes necessary, and gastroesophageal reflux has been observed with high frequency. Additional findings that can impact feeding include weak bilabial seal (64%), palatal anomalies including velopharyngeal insufficiency (75%), and open mouth posture with tongue protrusion (less than 30%) (Solomon et al. 2002).

Despite the high frequency of failure-to-thrive, infants with SMS usually appear well nourished. Body mass indices are within the normal range for two-thirds of children with the syndrome (Smith et al. 2002). Body mass indices in 49 affected children under 19 years of age ranged from 14.1 to 31.7 with a mean of 18.4 (Smith et al. 2002). Compared with published norms, these values were not significantly elevated for females (17.7, SD 3.65) and only slightly higher for males (19.32, SD 3.56; $p < 0.001$). Obesity is often observed after age 8 years, coinciding with pubertal onset, and can remain a problem into adulthood (personal experience). Developmental delay, sensory processing deficits, and decreased functional skills place individuals with SMS at risk for reduced physical activity and a sedentary lifestyle likely to lead to weight gain. Food foraging behavior during nocturnal arousals is an issue in adolescence, leading to the parental need to lock pantry doors and/or refrigerators. This is often further compounded by side effects from psychotropic behavioral medications that have weight gain as a side effect (e.g., risperidone) (personal experience). In a recent meta-analysis, body mass index (BMI) or obesity was shown to increase with age in 33% of individuals with SMS (Edelman et al. 2007).

Evaluation

- Height, weight, and head circumference should be measured and plotted on general population graphs at initial evaluation and during subsequent health maintenance visits. It is important to consider comparison of weight with height or BMI, in addition to charting absolute measurements.
- Oral feeding and swallowing evaluations to include assessment of caloric intake, swallowing abilities, oral-motor skills, gastroesophageal reflux signs and symptoms, along with assessment of parental feeding concerns are highly recommended for all affected infants.
- Suspected gastroesophageal reflux warrants consideration of full standard diagnostic evaluation.
- Speech pathology and occupational therapy evaluations are indicated to assess oral-motor dysfunction when it interferes with feeding.

Management

- Oral-motor therapy with a speech pathologist, clinical feeding specialist, or occupational therapist is suggested. Emphasis on increasing labial and lingual movements for swallowing and transitioning to varying food textures is also beneficial for a child's oral-motor and swallowing maturation.
- Neurodevelopmental training has proven to be beneficial for enhancing feeding and swallowing skills (personal experience).
- Nasogastric tube feedings and gastrostomy may be required in infancy if feeding is significantly impaired. Criteria for gastrostomy tube placement are standard.
- Monitoring food intake and/or preventing nighttime "food foraging" require an innovative behavioral approach that may include locking the pantry and/or refrigerator door. Efforts to encourage and enhance functional mobility and sustained activities and/or exercises that improve abdominal/truncal tone are encouraged, but are often met with resistance.

Development and Behavior

The cognitive and behavioral phenotype of SMS represents the major management problem for both parents and professionals working with this syndrome. Developmental delay or intellectual disability is found in all affected individuals to a degree ranging from profound to borderline. Cross-study comparisons of cognitive levels are hampered by age differences and instruments used. Expanding the original series reported by Greenberg et al. (1996) to 55 individuals, ages 1.5–31 years (mean 9 years), Madduri et al. (2006) found IQ/DQ scores ranging from 19 to 78 (mean 50.33, SD 12.91), with the majority falling in the mild-to-moderate range of intellectual disability. Udwin et al. (2001) examined a British cohort of 29 children (mean age 9.6 years; range 6–16 years) and 21 adults (mean age 27.3 years; range 26.5–51.7 years) with confirmed SMS. IQ scores of 50 or below were found in 75% of the childhood group. Although the adult group had higher IQs ranging from 50 to 69 (mean 56.0, SD 6.65) on the WAIS-R, they remained dependent on caretakers, having decreased adaptive function, especially in daily living skills. Although the potential for sample bias could not be excluded, the authors did not find a decline in cognitive function among affected adults up to age 50 years. Among the cohort of affected children aged 2–12 years, the majority (67%) demonstrated cognitive abilities in the mild ($n = 6$)-to-moderate ($n = 6$) intellectual disability range, 28% ($n = 5$) were in the borderline range of intellectual functioning and one scored in the low-average range of cognitive ability (Martin et al. 2006). These data suggest that while cognitive function may fall within the borderline/mild range in childhood, most individuals function in the mild-to-moderate range of intellectual disability as adults. Individual variability exists, making it difficult to predict ultimate intellectual functioning; many individuals may appear to have higher functioning than what the IQ scores might document, given their excellent long-term memory and an uncanny ability to engage others (especially adults) in conversation and activities.

Madduri et al. (2006) found that cognitive and adaptive abilities were inversely related to deletion size; individuals with larger deletions had significantly lower IQ/DQ scores (severe-to-profound range) and lower adaptive behavior composite scores compared with individuals with smaller or common (3.7 Mb) deletions. Although individuals with *RAI1* mutations are felt to be less cognitively impaired than deletion cases (Edelman et al. 2007), few reports include published IQ/DQ data for analysis; however, a significant difference in the frequency of motor and speech delay was observed in mutation cases (70%) compared with deletion cases (90–100%) (Girirajan et al. 2006).

Prenatal histories are notable for decreased fetal movement in 50% of the instances (Gropman et al. 2006). As infants, subtle motor delays may be evident, with gross motor delays of 2–24 months; however, social-emotional function can be within the normal range (Gropman et al. 2006; Wolters et al. 2009). In general, expressive language delays are out of proportion to receptive language skills, especially during early childhood (Greenberg et al. 1996; Smith et al. 1998a; Gropman et al. 2006). Marked early expressive speech delays, persistence of a poor suck, diminished vocalizations (babbling/crying), hoarse vocal quality, and grunting/guttural clicking sounds are seen in infancy and early childhood (Solomon et al. 2002).

Sonies et al. (1997) suggest that the severe oral-motor dysfunction and apraxia may be caused, in part, by bilateral dysfunction of the muscles of deglutition and expressive speech (palate, pharyngeal, buccal, and others).

Based on past studies, a specific cognitive profile for SMS has been recognized. Dykens et al. (1997) described specific cognitive profiles in 10 individuals with the syndrome, including relative strengths in long-term memory and perceptual closure and relative weaknesses in sequential processing and short-term memory. These early findings were confirmed in a larger English cohort of 40 individuals with SMS (Udwin et al. 2001) that demonstrated strengths in long-term memory, computer skills, and perceptual skills, and weakness in visual motor coordination, sequencing, and response speed. Deficits in sensory processing and modulation to sensory input are prevalent in SMS (Gropman et al. 2006; Hildenbrand and Smith 2012). Many children exhibit tactile and auditory defensiveness, appear to have problems with depth perception and gravitational insecurity, and exhibit a decreased awareness of pain and temperature. When compared to a national sample of children with and without disabilities, significant differences ($p < 0.001$) were documented across 14 categories and all 4 quadrants measured by the sensory profile in a large cohort of 34 children with SMS under 14 years (Hildenbrand and Smith 2012). The atypical patterns of sensory processing documented in this study also appeared more prominent with increased age, especially in females.

Significant speech/language delay, with or without associated hearing loss, occurs in over 90% of individuals with SMS. Infants and toddlers make limited vocalizations and exhibit poor sound production because of underlying oral sensorimotor anomalies. The absence of age appropriate babbling or vocalizations (quiet "smilers") is characteristic. Expressive language skills remain delayed in early childhood, with primary use of gestures and signs in those less than 4 years of age and verbal language emerging about 4–5 years of age, on average (Solomon et al. 2002; Smith and Duncan 2005). With aggressive speech/language therapy accompanied by a sign language and total communication program, fairly understandable expressive language is usually present by school age. Some parents report that the onset of language, specifically speaking in full sentences, occurs when their child begins to read (personal observation). There may be hypernasality with a harsh, hoarse vocal quality. Speech intensity may be mildly elevated with a rapid rate and moderate explosiveness (Solomon et al. 2002). Once verbal, individuals with SMS are incessant talkers, asking constant, often repetitive questions. Verbal teens and adults with the syndrome often demonstrate a good sense of humor, with a unique propensity for "one-liners," and excellent long-term memory (Smith et al. 1998a).

Adaptive function is significantly impacted, regardless of IQ and age group studied. Children with SMS uniformly demonstrate significant delays in adaptive behavior, including communication, daily living skills, and socialization skills. In a recent study of 19 children (ages 2–12 years) with SMS, adaptive functioning was measured using the Vineland Adaptive Behavior Scales as part of the overall cognitive assessment (Martin et al. 2006). Although delays in communication and daily living skills were consistent with cognitive functioning, socialization skills were significantly higher than cognitive functioning, suggesting strengths in this area. The "socially engaging" personality and vocal patois demonstrated by older persons often gives the impression of higher cognitive function than is actually present. Furthermore, reports of adaptive behavior profiles at older ages document relative strengths in socialization and communication compared with activities of daily living (self-help skills) (Udwin et al. 2001; Madduri et al. 2006). Age was inversely related to daily living skills on the Vineland Adaptive Behavior Scales ($r = -0.68$, $p < 0.001$) in at least one study, suggesting that proficiency in activities of daily living may plateau in early adolescence leading to increased dependency on others for support (Martin et al. 2006). Adults with SMS remain dependent on caregivers and require a higher degree of support than might be expected for their level of cognitive functioning (Udwin et al. 2001).

Behavioral problems, including maladaptive behaviors, self-injurious, and stereotypic behaviors, are frequent in SMS and represent the major management problem for both parents and professionals (Colley et al. 1990; Greenberg et al. 1991; Finucane et al. 1994; Dykens et al. 1997; Smith et al. 1998a, 1998b; Finucane et al. 2001; Madduri et al. 2006; Martin et al. 2006). Dykens and Smith (1998) examined the distinctiveness and correlates of maladaptive behavior as well as the prevalence of self-injurious and stereotypical behaviors, further delineating the behavioral phenotype of the syndrome. With the use of the child behavior checklist score, 35 children with SMS were compared with age-and gender-matched individuals with Prader–Willi syndrome or mixed intellectual disability. All but four of those with SMS (89%) demonstrated significantly elevated maladaptive behavior scores compared with their counterparts; 12 behaviors differentiated the groups with 100% accuracy. Specifically, those with SMS demonstrated significantly higher rates of attention-seeking (100%), disobedience (97%), hyperactivity (94%), sleep disturbance (94%), temper tantrums (94%), distractibility (89%), property destruction (86%), impulsivity (86%), toileting difficulties (80%), nail-biting behaviors (72%) and aggression (57%). Self-injurious behaviors were seen in 92% of the SMS study group, including biting or hitting self (71–77%), nail-yanking (onychotillomania) (29%), and bodily insertion (polyembolokoilamania) (25%). One or more stereotypic behaviors were demonstrated by all individuals with SMS including mouthing objects or putting hands in the mouth (54–69%), teeth grinding (54%), "lick-and flip" behavior (51%), self-hug, upper body spasmodic squeeze (46%), body rocking (43%), and spinning or twirling objects (40%). The most frequent stereotypies involved the mouth in some way, representing oral variants of bodily insertion. The high prevalence of oral insertion behavior may originate as a compensatory skill used in eating during infancy, when oral-motor dysfunction is significant and requires the use of hands/fingers to assist in propelling food backwards to aid swallowing.

Self-injurious behaviors are universal and a hallmark finding in SMS, exhibited by 92–100% of individuals depending on the age group studied (Table 50.2). Finucane et al. (2001) found a direct correlation between the number of self-injurious behaviors types and increased functional level and age. Hand/wrist biting, self-slapping, head banging, and skin picking remain the more prevalent forms of self-injurious behaviors. Onychotillomania appears to be age related, with lower frequency at younger ages (13%; Martin et al. 2006), compared with later childhood (less than 30%) and adulthood (86%) (Dykens and Smith 1998; Finucane et al. 2001). The relationship between self-abusive behaviors and decreased pain sensitivity is not yet defined.

Persons with SMS in one study differed from individuals with PWS and people with mixed intellectual disability both in regulation of basic bodily functions (sleeping, modulating activity and affect, eating, and toileting) and in social and repetitive behaviors (Dykens and Smith 1998). These individuals slept less, were more prone to hyperactivity, and were more emotionally labile than their counterparts. Enuresis and encopresis were singularly frequent in SMS compared with PWS and mixed intellectually disabled subjects. Socially, those with SMS demanded more attention than their counterparts. Those with SMS also showed obsessive thinking, primarily about specific topics as opposed to food. Regulation of food intake and/or food foraging, often during nighttime awakenings in older individuals, also occurs. This is especially prevalent in those with *RAI1* mutation (Edelman et al. 2007).

Finucane et al. (1994) described two types of self-hugging (autoamplexation), an upper body movement and clasping the hands at chest or chin level and squeezing, often with interlocked fingers. More frequent among young children and adolescents than adults, these movements appear involuntary, with a tic-like quality, and are observed when the person is happy or pleased, but not during temper tantrums or when unusually upset. In addition to hugging themselves, individuals with SMS often hug others repetitively and with force. A few parents report the unfortunate demise of a family pet (gerbil, kitten) caused by such intense squeezing/hugging behaviors and excitement (Smith et al. 1998a).

Although individuals with SMS have some degree of control over their behaviors, it is important to recognize that many of the negative behaviors have their origins in internally driven impulses. Most relate to significant difficulties with modulation of activity level, affect/ mood, attention, and/or bodily functions (Dykens and Smith 1998). Finucane and Haas-Givler (2009) recently defined a gap or "developmental asynchrony" between the levels of intellectual functioning and emotional maturity that further contributes to the increased maladaptive behaviors that characterize the syndrome. In contrast to an academic level of function that may fall in the 6–8 year old range, children and adults with the syndrome exhibit an emotional reactivity more consistent with the developmental level of 1–3 year olds, which Finucane and Haas-Givler term the "inner toddler".

The significant sleep dysfunction serves to intensify the intrinsic behavior problems that occur in the disorder (see

Sleep). Although sleep disturbance is seen in other developmental disabilities and neurologically related disorders, the advanced sleep phase and abnormal circadian secretion of melatonin (daytime highs) distinguishes the sleep disorder that occurs in SMS from other developmental disorders. The lack of longitudinal data tracking sleep disturbance and behaviors makes it difficult to determine to what extent sleep disturbance causes or exacerbates behaviors. A causal role for sleep disturbance in the sleep-behavior cycle was first suggested by Dykens and Smith (1998), who found an association between increased nap lengths and decreased aggressive and attention problems. Although age and degree of delay were correlated with behavior problems, sleep disturbance emerged as the strongest predictor of maladaptive behavior. Furthermore, individuals who experienced snoring and labored breathing were found to exhibit more aggressive, acting-out behaviors, and attention problems (Dykens and Smith 1998), consistent with a published sleep study of healthy 5 year old children (Gottlieb et al. 2003).

Dual diagnosis of co-morbid psychopathologies and other conditions, including obsessive-compulsive disorder (OCD), attention-deficit disorder (ADD) and/or attention deficit hyperactivity disorder (ADHD), and autism/pervasive developmental disorder (PDD), is common (Finucane and Simon 1999; Smith and Duncan 2005). ADHD is a comorbid condition with autism, so any child with SMS who has been diaganosed to be on the autism spectrum needs an evaluation for ADHD. Likewise, any child with SMS diagnosed with ADHD should be screened to rule out autism spectrum. A co-morbid diagnosis of post-traumatic stress disorder (PTSD) was reported in a 21-year-old female with mild intellectual disabilities, whose escalating behaviors and aggressive assault of a health care provider led to her subsequent incarceration prior to diagnostic confirmation of SMS *RAI1* mutation at age 25 years (Yeetong et al. 2016). Other stereotypic/repetitive behaviors such as page turning or "lick and flip" (50–60%), tongue clicking/tongue sucking, teeth grinding (54–87%), tongue protrusion, and hands/objects in mouth (54–87%) are prevalent (Dykens and Smith 1998; Solomon et al. 2002; Martin et al. 2006). Shelley et al. (2007) reported the co-occurrence of Tourette syndrome using established diagnostic criteria (DSM-IV-TR) in a 17-year-old male with deletion who presented with a history of vocal (grunting, clicking) and motor (tongue protrusion, smelling objects, shoulder shrug, and repetitive head movements) tics in combination with classic neurobehavioral features of SMS, including self-injurious behavior, polyembolokoilomania, and self-hugging.

Because of the significant early speech/language delays, several young children have been diagnosed with autism spectrum disorder, often before confirmation of SMS. Prospective assessment using different screening instruments/tools for autism spectrum disorder (e.g., the childhood autism rating scale [CARS]; social responsiveness scale [SRS], and both current and lifetime versions of the social communication questionaire [SCQ]) have documented findings that support autism spectrum diorders (ASD) in SMS. Among 19 children with SMS, the CARS (Martin et al. 2006) revealed scores at the low end of mildly autistic classification range. For a larger SMS cohort ($n = 26$), SRS scores consistent with ASD were documented in 90%; and, the majority were found to meet criteria for ASD based on the SCQ at some point in their lifetime (Laje et al. 2010). Hicks et al. (2008) reported affected monozygous twins with scores consistent with autism spectrum disorder based on the autism diagnostic interview – revised (ADI-R) and autism diagnostic observation schedule (ADOS). However, recognized strengths in social skills is felt to distinguish this population from true autism (personal experience; Martin et al. 2006).

The behavioral phenotype of SMS, specifically maladaptive and self-injurious behavior and sleep disturbance, has a significant impact on family stress and support needs. From the parenting standpoint, sleep deprivation experienced when dealing with their child's nighttime sleep issues may also impact parental patience and management of daytime behaviors, leading to less than optimal interventions. Hodapp et al. (1998) documented high levels of parent and family problems, pessimism, and overall stress in the presence of increased numbers of family supporters. Overall, stress and parent-family problems were inversely related to the size of the family's support system. The child's degree of maladaptive behavior was the single best predictor of parental pessimism.

Recent studies have contrasted cognitive and social functioning of patients with SMS to those with DS. Wilde and Oliver (2017) examined executive function (EF) in SMS, given its high risk of behavioral difficulties, with DS, which has a relatively low risk of maladaptive behaviors. A survey was given to caregivers of 13 children with SMS and 17 with DS rating EF with the behavioral rating inventory of executive functioning-preschool. Greater everyday EF deficits relative to adaptive ability were evident in SMS than in DS. The SMS profile of everyday EF abilities was relatively uniform; in DS emotional control strengths and working memory weaknesses were evident (see Chapter 24). The results of this study may imply that everyday EF profiles may, in part, be syndrome related and driven by maladaptive behavior.

Treatment of behavior in SMS requires a multidisciplinary team, including among others, genetics, pediatrics, child psychiatry, pediatric neurology, developmental psychology, developmental/behavioral pediatrics, and speech/language, physical and occupational therapies. A comprehensive evaluation of sleep disorders, potential causes of pain, neurocognitive level and environment (i.e. family and school) as triggers to adverse behavior must be addressed. In addition, efforts should focus on improving communication skills and implementing alternative communication, identifying and treating attention deficit/hyperactivity, autism, aggression and anxiety (Poisson et al. 2015). Use of behavioral paradigms to extinquish maladaptive behavior may show promise (Hodnett et al. 2018).

Evaluation

- Accurate assessment of developmental function in SMS is often hampered by poor adaptive behaviors, marked speech delay, and inherent bias in developmental scales, which rely heavily on verbal skills.
- Annual multidisciplinary team evaluation is optimal, including physical and occupational therapy, speech pathology, and child development assessment to assist in development of an individual educational plan.
- Periodic neurodevelopmental assessments and/or developmental pediatric consultation can be an important adjunct to the team evaluation.
- Periodic speech/language evaluations beginning in infancy are necessary to evaluate speech and language delays, optimize functional communication with the child in his or her environment, as well as to provide education to parents for fostering speech and language development.
- Assessment of family support and psychosocial and emotional needs will assist in designing family interventions.

Management

- Speech/language pathology services should initially focus on facilitating swallowing and feeding problems as well as optimizing oral sensory motor development. Therapeutic goals for emphasizing increasing sensory input, fostering movement of the articulators, increasing oral motor endurance, and decreasing hypersensitivity are needed to develop the skills needed for swallowing and speech production.
- Early speech therapy to establish communication is important, as verbal skills will be delayed. It should commence as soon as the diagnosis is made.
- Referral for physical and occupational therapy is important to provide support and treatment for developmental deficits.
- The use of sign language and a total communication program (i.e. picture exchange) as an adjunct to traditional speech/language therapy is felt to improve communication skills and also to have a positive impact on behavior (Greenberg et al. 1991). The ability to develop expressive language appears to be dependent on the early use of sign language and intervention by speech language pathologists. Functional communication with fair speech intelligibility eventually replaces the severely delayed expressive language during the school age years. Non-verbal people with SMS are known and frequently are older individuals who did not receive aggressive therapeutic intervention.
- Occupational therapy services should focus on whole body sensory issues, fine motor development, and fostering visual and auditory perceptual skills.
- Educational intervention throughout the school years (at a minimum) should include addressing individual strengths and weaknesses, taking into account the behavior disorder. This can be done best by incorporating the recommendations of a multidisciplinary developmental assessment.
- Management strategies that improve nighttime sleep may have an impact on the level of maladaptive behaviors (see Sleep).
- Children with SMS are very adult-oriented, with an almost insatiable need for individual attention, a key personality characteristic that has major implications both at home and in the classroom. Positive attention is clearly preferable, but negative attention may serve equally well in the quest for one-on-one adult interaction. Withholding teacher attention often prompts negative behaviors, including verbal outburst, tantrums, aggression toward peers, and destruction of property, all of which disrupt the classroom and result in the "desired" attention.
- At home or in the classroom, unexpected changes in daily routine or transitions between activities as well as emotional upset can precipitate behavioral outbursts, tantrums, and aggression. Individuals with SMS tend to respond quite positively to consistency, structure, and routine, especially with visual cues.
- People with SMS have difficulties in sequential processing (i.e. counting, math tasks, multistep tasks) and short-term memory. Thus, instructional strategies that recognize these inherent weaknesses while taking advantage of relative strengths in long-term memory (especially names) and visual reasoning are most effective (Dykens et al. 1997; Smith et al. 1998a).
- Inherently distractible, children with SMS tend to function better in a smaller, calmer, and more focused classroom setting with five to seven students, a teacher, and an aide (Haas-Givler and Finucane 1995). In some circumstances, an individual aide in a larger classroom also works; however, with larger class sizes, competition for a teacher's attention and inherent activity level increases, risking increased behavioral problems.
- Because they are visual learners, individuals with SMS greatly benefit from the use of pictures or visual cues and reminders to illustrate daily activities, classroom schedules, and performing self-help skills.
- Computer-assisted technology provides a unique educational advantage for people with SMS, who have

a strong fascination with electronics and newer digital equipment, including DVD and CD players, i-Pod, hand-held video games, and/or computers (Smith et al. 1998a).
- Development of a behavioral treatment plan should be initiated as soon as behavioral problems arise.
- Children with SMS are generally eager to please and quite responsive to affection, praise, and other positive attention. In moderation, positive emotional response by the teacher or parent can strongly motivate a child to do well. Opportunities to do something they like to do, such as additional 10–15 min time spent on the computer, present powerful positive motivators. Short timeouts and loss of time doing a preferred activity (e.g., lost computer time) have also proven effective.
- Most individuals with SMS have been tried on a number of medications to control behavior with mixed response; adverse reactions to some medications have also been reported. Polypharmacy is also an issue. In the author's experience, medication history data on 12 children with SMS, ages 3–16 years, have shown a median number of five medication trials; only two children were not on medication therapy and one of these was enrolled in a strict behavior-modification program. In the multidisciplinary study conducted by Greenberg et al. (1996), the most common medications tried among 27 individuals with SMS were methylphenidate, pemoline, and thioridazine. A larger retrospective study ($n = 62$) showed psychotropic medication use begins early in life (mean age 5 years), especially related to use of sleep aides; however, no single medication category proved efficacious, and benzodiazepine group showed a mild detrimental effect (Laje et al. 2010). Lithium is useful for patients with bipolar manifestations. The combination of a morning dose of beta-blocker (acebutolol) with evening melatonin supplementation has been reported to restore sleep in SMS, leading to a significant improvement in daytime behaviors with increased concentration (deLeersnyder et al. 2003). Lacking well-controlled trials, child psychiatrists comfortable working with children with genetic syndromes, including developmental delay, may elect treatment approaches targeting specific symptoms one at a time (prioritized with parents); in such cases, tracking sleep and behavior changes over the next days and weeks is important to monitor for potential side-effects and/or to determine potential efficacy.
- Both published (Greenberg et al. 1996; Hagerman 1999) and unpublished data show that older stimulant drugs may not be particularly helpful in controlling behavior or increasing attention span (personal experience). However, there is not enough experience with the newer preparations of this class of medications to deduce efficacy. These authors found minimal improvement with Adderall (1), dexedrine (1), and methylphenidate (3), compared with a worsening on Dexedrine (1), methylphenidate (3), and marked side effects with pemoline (1). Among the antipsychotics, one child improved on stelazine, one got much worse on risperidone, and one became sedated on thioridazine. In contrast, marked improvement of aggression and/or hyperactivity was reported in two adolescents on risperidone (Hagerman 1999; Niederhofer 2007) consistent with anecdotal parental reports (personal experience). Although one potential effect of risperidone is increased sleep duration, a negative adverse effect of rapid weight gain is a recognized concern for patients with SMS.
- Greenberg et al. (1996) reported some or only transient behavior improvement for several individuals with SMS, both with and without seizures, tried on carbamazepine. Similar unpublished results (personal experience) were found for eight individuals. With anticonvulsant treatment, improvements were seen in three individuals on carbamazepine, and valproic acid. Of the 12 unpublished SMS medication histories, three children were tried on benzodiazepines: clonazepam showed improvement (1) or no change (1), and lorazepam showed improvement (1). Use of tricyclic antidepressants (clomipramine or imipramine) was associated with worsening of behavior. In, at least, two cases, older teens treated with atomoxetine exhibited adverse effects that included a significant decline in sleep time coupled with major escalation of behaviors, especially agitation, self-injurious, and aggressive outbursts, leading to psychiatric admission in one case (personal experience).
- Use of specific serotonin reuptake inhibitors (specifically, sertraline and fluoxetine) has shown considerable improvement with respect to behavioral outbursts and sleep for at least three individuals with SMS (Smith et al. 1998a).
- Published data about the optimal intervention and behavioral strategies in SMS are limited to anecdotal and experiential findings (Finucane and Haas-Givler 2014; Haas-Givler and Finucane 1995; Smith et al. 1998a). Although medication therapy may show some benefit with respect to increasing attention and/or decreasing hyperactivity, it is clear that behavioral therapies play an integral role in the behavioral management of the syndrome.
- Respite care and family psychological and social support are critical to assuring the optimal environment for the affected individual.
- Referral to national genetic support groups such as Parents and Researchers Interested in Smith–Magenis

syndrome (PRISMS) in the United States, as well as local, regional, or state support services, are highly beneficial for many parents, families, and caregivers of individuals with SMS.

Sleep

Significant symptoms of sleep disturbance occur in 65–100% of people with SMS and have a major impact not only on the child but also on parents and other family members and care givers, many of whom become sleep-deprived themselves (Smith et al. 1998b). Past studies have consistently documented fragmented and shortened sleep cycles, characterized by frequent and prolonged nocturnal awakenings, early waking, and excessive daytime sleepiness (Greenberg et al. 1996; Smith et al. 1998b; Potocki et al. 2000; De Leersnyder et al. 2001a). Data derived from objective sleep measures (24 hour polysomnography, multiple sleep latency test, wrist actigraphy, and/or sleep log diaries) are consistent with an advanced sleep phase that is a recognized circadian sleep disorder (Potocki et al. 2000; De Leersnyder et al. 2001a; Gropman et al. 2006; Smith et al. 2019).

A detailed study of sleep behaviors in 39 individuals with SMS showed a significant relationship between increased age and steady decline in the total hours of sleep at night, earlier bedtimes, shorter nap lengths, and increased frequency of naps (Smith et al. 1998b). Total sleep averaged 7.8 hours (range 3.5–10.5 hours). A subsequent French study reported virtually identical data on sleep times for a group of 20 affected children (De Leersnyder et al. 2001a). Smith et al. (1998b) found that nap length declined sharply after age 5 years, but increased in frequency from 1.2 to 1.7 after age 10 years. The most frequent problems during bedtime and nighttime periods were bed-wetting (79%), bedtime rituals (74%), snoring (69%), need for sleep medications (59%), and awakening during the night either to go to the bathroom (54%) or to get a drink (54%). Parents often report that their child will not fall asleep unless one parent lies down with him or her, a bedtime habit that is difficult to end and one likely to impact the marriage.

Infants with SMS are often described by their parents as happy, complacent, and "good sleepers," in the absence of objective measures to validate this impression. A large comprehensive home assessment of sleep using the noninvasive technique of actigraphy to quantify rest and activity in a group of 37 individuals (ages 2–24 years) with SMS was recently published (Smith et al. 2019). Consistent with earlier reports (Gropman et al. 2006), estimated 24 hour and nighttime sleep was reduced across all ages (Smith et al. 2019). The fragmented sleep cycle begins as early as age 6–9 months, with infants best described as "quiet babies sleeping poorly" because of their relatively quiet behavior pattern at night (Gropman et al. 2006), continuing into adulthood (Figure 54.3). Actigraphy-based estimates document developmental differences for children under age 10 years (early settling with increased activity levels (arousals) in the second half of the night) compared with over age 10 years (increased settling difficulties with better sleep in second half of the night) (Gropman et al. 2007; Smith et al. 2019). Overall, actigraphy-estimated sleep time for SMS was one hour less than expected across all ages studied (Smith et al. 2019). Consistent with the disrupted phase advance sleep pattern well recognized in the syndrome, timing of the 24-hour body temperature rhythm was shown to be phase advanced by about 3 hours, but not inverted (Smith et al. 2019).

Over half of the individuals with SMS who have undergone polysomnography show abnormalities of rapid eye movement (REM) sleep (Greenberg et al. 1991, 1996; Potocki et al. 2000; De Leersnyder et al. 2001a). Frequent and prolonged nighttime arousals (over 15 min) occur in 75% of cases (De Leersnyder et al. 2001a, 2006). The consequences of the disrupted nighttime sleep is a significant and chronic sleep debt that results in increased daytime sleepiness and urge to nap, a finding that is confirmed by multiple sleep latency test studies (Potocki et al. 2000) and actigraphy

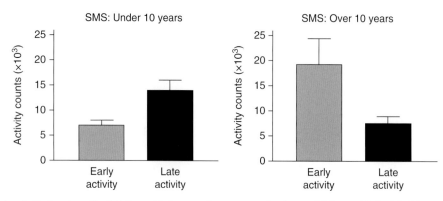

FIGURE 54.3 Wrist activity levels collected from 10-day continuous samples from children with Smith–Magenis syndrome. Children under 10 years settle early, but have increased activity in the second half of the night. Older children and adolescents have difficulty settling, but sleep better in the second half of the night. (From Gropman et al. 2007, with permission.).

data (De Leersnyder et al. 2001a, 2003; Gropman et al. 2006, 2007; Smith et al. 2019).

Inverted Circadian Melatonin Rhythm Published reports of 24 hour melatonin secretion patterns ($n = 30$) document an inverted endogenous melatonin pattern (daytime highs; [Figure 54.4]) in virtually all (93%) people that appears to be pathognomonic for SMS, including both deletion ($n = 26$) and two mutation cases (Potocki et al. 2000; De Leersnyder et al. 2001a; Boudreau et al. 2009; Boone et al. 2011). These findings indicate a dysregulated circadian system in SMS (Smith et al. 1998b). Sleep disturbance could be related to abnormalities in the production, secretion, distribution, or metabolism of melatonin (Potocki et al. 2000). As light should normally suppress melatonin synthesis, a critical question to answer in SMS is why light does not appear to impact the daytime high of melatonin. People with SMS have to contend not only with the daytime melatonin and its soporific effects, but also the increased nighttime awakenings that lead to increased sleep debt, which in turn is likely to be a major variable in daytime behavior modulation; the convergence of these factors, specifically, as it relates to maintaining daytime vigilance, has been likened to a "perfect storm."

It is anticipated that improvement of sleep will lead to improved quality of life for both the child and the family. To date, there have been no well-controlled treatment trials aimed at improving sleep in SMS. However, the presence of elevated daytime melatonin levels with low nighttime levels suggests possible therapeutic approaches. De Leersnyder et al. (2003) used the combination of daytime B_1-adrenergic antagonist acebutolol (10 mg/kg at 8:00 AM) to reduce daytime melatonin, coupled with an evening oral dose of controlled release melatonin (6 mg at 8 PM) to restore nocturnal plasma melatonin levels. Although this uncontrolled trial demonstrated a more normal circadian rhythm of melatonin as well as improved behavior in nine children with SMS, the results may have been biased because of parent expectations (see Development and Behavior). The contraindications for use of B_1-adrenergic antagonists, including asthma, pulmonary problems, and diabetes, must also be considered. As the pulmonary aspects of SMS remain unstudied, medical aspects must be considered before any trial. An understanding of the baseline sleep pattern is beneficial before beginning any trial.

A pilot trial of modafinil (ProvigilTM) was conducted by Heussler et al. (2007) in three affected children in whom past melatonin products had had minimal success. Modafinil is a psychostimulant that enhances wakefulness and has clinical utility in treatment of narcolepsy, a disorder with marked daytime sleep attacks, obstructive sleep apnea/hypopnea, and shift work sleep disorder. Based on pre/post therapy measures that tracked behavior (developmental behavioral checklist) and sleep patterns (sleep log diaries), all three children with SMS showed increased alertness and calmer behaviors at home and school with significant improvement in sleep.

In the USA, melatonin is not regulated by the Food and Drug Administration, but is sold as a nutritional supplement in a variety of formularies. Anecdotal case reports of therapeutic benefits of melatonin exist (Smith et al. 1998a, 1998b) countered by reports of little/no benefit (Gropman et al. 2006). In the absence of a double-blind controlled melatonin trial, current wisdom suggests that low dosages (0.5–2.5 mg) are preferred; avoidance of higher doses (5–10 mg) is recommended (Gropman et al. 2006).

Recently, Hetlioz^R (tasimelteon), an FDA-approved melatonin receptor agonist used in treatment of non-24-hour sleep-wake disorder, was studied in patients with SMS; in the placebo controlled clinical trial (VEC-162-2401) tasimelteon showed promising results improving overall sleep quality and duration (Vanda Pharamceuticals, Press release Dec 10, 2018).

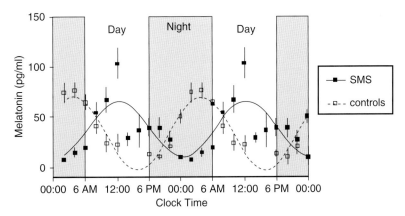

FIGURE 54.4 The inverted pattern of plasma melatonin for eight children with Smith–Magenis syndrome (filled circles, solid line) and 12 healthy control subjects (open circles, dotted line). The lines represent the best-fit sine curves to each data set based on minimal least-squares criteria. In healthy subjects, the peak is at night; in individuals with Smith–Magenis syndrome, the peak occurs during the day (noon). (From Gropman et al. [2006], redrawn from De Leersnyder [2001].)

Although limited to a single case report, human growth hormone replacement therapy in an affected female with isolated growth hormone deficiency led to improved early awakening, increased total sleep time, and percentage of REM sleep, but did not appear to correct phasic inhibition (Itoh et al. 2004).

The inverted secretion of melatonin occus in virtually all studied individuals (>93%), suggesting it is pathognomic for the syndrome (Boudreau et al. 2009). The two individuals with atypical 17p11.2 deletions who had normal melatonin secretion still had sleep disturbances. This included a 5-year-old female carrying a ~5 Mb deletion that extended beyond the distal SMS-REP region (Potocki et al. 2000) as well as an 18-year-old female carrying a ~5.8 Mb deletion that extended beyond the proximal SMS-REP region detected by high resolution CGH and FISH (Boudreau et al. 2009). Given that melatonin is secreted during daylight hours in patients with SMS, its secretion is not suppressed by light in most persons with SMS.

Evaluation

- Sleep history should be elicited to document the sleep cycle and to ascertain evidence of sleep apnea, snoring, and other signs of sleep abnormalities.
- Parents should be encouraged to keep a monthly sleep diary/log to document the child's sleep patterns, and to provide a means of monitoring changes in sleep during treatment trials.
- Evaluation in a sleep disorders clinic is recommended.
- A sleep study, including a sleep-deprived electroencephalography or polysomnography, should be carried out to rule out obstructive sleep apnea.

Management

- Documented sleep apnea should be treated using standard practice guidelines. This may include tonsillectomy and adenoidectomy if obstruction of the airway is present.
- Anecdotal reports by parents suggest therapeutic benefit from melatonin (personal experience; Smith et al. 1998b). In general, these benefits appear similar to those seen in other syndromes with developmental delay (Schwichtenberg and Malow 2015). However, in anecdotal reports, the dosages used and administration times varied and were guided by parent-to-parent experience, making it difficult to draw objective conclusions. Dosages of 2.5–5.0 mg (10 mg maximum) taken at bedtime were tried without report of major adverse reactions. Low therapeutic doses of 0.5–2.5 mg should be used, as higher dosages (5–10 mg) can result in increased daytime levels of melatonin (Gropman et al. 2006). A monitored trial of 4–6 weeks on melatonin may be worth consideration when major sleep disturbance is present. The time of administration is important, as melatonin can have phase-shifting properties when taken at different times. Melatonin is available in both immediate and slow-release formulations. Slow-release melatonin administered approximately 30–60 min before bedtime in combination with daytime ß$_1$ antagonist in the morning (10 mg/kg acebutolol at 8:00 AM) to inhibit endogenous diurnal melatonin secretion is associated with improved sleep quality (De Leersnyder et al. 2003, 2006). Alertness to the possibility of adverse side effects from the ß-agonist should be maintained (see above).
- Increased daytime sleepiness can be reduced by naps, but early timing of naps is critical to maintain sleepiness at scheduled evening bedtime. As children with SMS wake up early in the morning, the preferred naptime is late morning or early afternoon (12:00 noon–15:00) with avoidance of late afternoon naps that can interfere with falling asleep at the child's scheduled bedtime (Gropman et al. 2006).
- Signs of increased sleep need include napping at unscheduled times, increased nap duration, and/or the need to awaken a child from a nap or in the morning. This increased need to sleep might be accommodated by extending the sleep period or scheduling another early nap.
- Practicing good sleep hygiene is important for the child with SMS and his or her parents, other family members, and caregivers. Avoid caffeine and exercise at night; keep daytime naps short (power naps), try to maintain a consistent sleep schedule (bed-times); and refrain from turning on lights during the night.
- Parents have implemented solutions to "Smith–Magenis syndrome-proof" their child's room to minimize self-injury and maximize sleep cycle. These solutions enhance the likelihood that the affected individual will not wake and wander in the early morning hours (Smith et al. 1998b). Strategies such as removing all small objects such as toys and lamps from the bedroom and installing a door peephole for observation, an outside-locking door, use of an enclosed bed system (safety), and/or window black-out curtains have been tried with some success. Parental reports of co-sleeping with their child suggest that it is very effective, but it leads to a habit that is not only difficult to break but also affects parental sleep (personal experience).
- Generally, children with SMS are most alert in the early morning hours, and this observation should be exploited in the educational setting. The increased frequency and duration of daytime napping with age (to make up for poor nighttime sleep) should also be recognized. Consequently, major educational activities or

therapies during late afternoons are often met with increased negative behaviors and difficulties with staying on task.

Otolaryngology

Otolaryngological findings are extremely common in SMS, providing a physiologic explanation for the functional impairment in voice and speech previously reported. A multidisciplinary clinical study of 27 individuals with SMS (Greenberg et al. 1996) demonstrated otolaryngological abnormalities in 94% and hearing impairment in 68%. One-third of those undergoing laryngoscopy (4/12) had laryngeal abnormalities, including polyps, nodules, edema, or paralysis. These findings were confirmed in another series of 27 affected children, aged 4 months to 16 years (Solomon et al. 2002). Oral-sensory-motor deficits were seen in all children, including oral tactile defensiveness, drooling, labial weakness, exaggerated lingual papillae, and open mouth posture. Over 95% of the individuals showed deficits in tongue strength and range of motion. Of those who underwent nasal endoscopy, 21 out of 25 (84%) demonstrated structural and physiologic anomalies of the larynx and hard/soft palates. Among the speaking children, velopharyngeal insufficiency and a perceived nasal speech quality was present in 8 of the 13. Out of the 19, 16 had voice impairments characterized by hoarse, wet, and harsh vocal qualities. All infants had notably diminished vocalizations and sound production. Velopharyngeal insufficiency was confirmed in almost two-thirds of the individuals, which was significantly higher than that observed by Greenberg et al. (1996). Recurrent upper respiratory infections, pneumonia and recurrent sinusitis requiring antibiotics is common (Perkins et al. 2017). Positional vertigo has been observed in 3 of 14 affected individuals (20%) (personal experience). Over half reported tracheo-bronchial signs, including recurrent pneumonia, bronchitis, aspiration, and tracheal stenosis.

Comprehensive characterization of auditory phenotype (degree and type) in the largest cohort of persons with SMS studied to date (Brendal et al. 2017) documents a higher prevalence of hearing loss compared to earlier clinical series or meta-analysis (Smith et al. 1986; Greenberg et al. 1991, 1996; Edelman et al. 2007). Hearing loss, ranging from slight to severe in degree, was documented in at least one ear ($n = 60$) in 78.9%. Middle ear dysfunction remains common (50% of ears tested), most often associated with flat tympanograms (43.7%), especially in first decade of life (Brendal et al. 2017). Chronic otitis media is experienced by virtually all children, beginning in early infancy, and often leading to multiple ventilation tube placements. Ventilation tubes were required by more than 80% of the individuals, over half of whom have had multiple sets.

Onset of sensorineural hearing loss appears to be age-related, with the pattern of fluctuating and progressive hearing decline documented with age typically appearing after the first decade of life age 10 years (Brendal et al. 2017). Conductive hearing loss is common during early childhood (35.2%; 10 years and under), and sensorineural hearing loss the most common type of hearing loss (48.1%) seen after the first decade of life (Brendal et al. 2017). This later childhood onset is interesting as a gene for non-syndromic autosomal recessive deafness (*DFNB3*) maps within the 17p.11.2 region (Liang et al. 1998). *DFNB3* encodes myosin 15 (*MYO15*), an essential mechanoenzyme of the auditory system (Wang et al. 1998). Sensorineural deafness can result from loss of heterozygosity caused by deletion 17p11.2 and a mutation of *DFNB3* in a transconfiguration to the deletion (Liburd et al. 2001).

Hyperacusis, an oversensitivity to certain frequency ranges or particular sounds that are tolerable to listeners with normal hearing, is documented at a significantly higher rate among affected individuals (74%) compared with a healthy sibling control group (10%) (Brendal et al. 2017). In this study, tiredness was the most common reason cited (50%) for heightened reaction to sound.

Evaluation

- Comprehensive otolaryngological, speech language pathology, and audiologic evaluations are recommended for all newly diagnosed individuals with SMS.
- Otolaryngological evaluation should include assessment of ear, nose, and throat problems with specific attention to ear physiology and palatal abnormalities (cleft palate, velopharyngeal insufficiency).
- Continued and regular monitoring of hearing and otologic status is warranted through early adulthood.
- Close otolaryngological follow-up is suggested for assessment and management of otitis media and other sinus abnormalities, as well to monitor for symptoms of progression or fluctuation of hearing loss, vertigo, or other early signs of ear disease.
- Complete audiological evaluations should be obtained at least annually to monitor for conductive or sensorineural hearing loss. As auditory evaluation can be challenging in this population, owing to underlying speech and language delays, cognitive impairments and maladaptive behaviors, repeated sessions for behavioral tests, and scheduling that takes sleep patterns into account may be necessary for successful hearing assessment.
- Speech pathology evaluations are essential to evaluate oral sensorimotor skills and swallowing and possible feeding problems.

Management

- Routine follow-up and management of all ear, nose, and throat problems in accordance with standard medical practice and professional recommendations should be conducted to optimize an affected individual's communication for developmental and educational purposes.

- Voice and resonance therapy should be considered as part of an individual's treatment plan when clinically indicated.
- Hearing amplification should be considered in the presence of documented sensorineural hearing loss.
- Aggressive otological management of middle ear disease and frequent follow-up is warranted. Otitis media is generally chronic (3–6 episodes/year) and may require prophylactic antibiotic therapies and/or ventilatory tube placement.
- Decreased immunological function (especially, IgG or IgA) is observed in over a third of affected individuals and should be considered in the management of these infections.
- Effective palliative strategies for hyperacusis include preparation or warning about the sound, sound reduction through a variety of techniques, avoidance of distressing sounds, control of known triggers, and familiarization with the offending sound.

Neurologic

People with SMS may manifest symptoms referable to both the central and peripheral nervous systems.

Seizures Seizures occur in 11–30% of affected individuals (Greenberg et al. 1991, 1996; Potocki et al. 2002; Goldman et al. 2006; Gropman et al. 2006). Electroencephalogram abnormalities were documented in approximately 25% of affected individuals in the absence of a clinical history of seizures (Greenberg et al. 1996; Goldman et al. 2006; Gropman et al. 2006). There is no single seizure type or electroencephalogram finding that is characteristic of SMS. Children with SMS have been reported with infantile spasms (Hino-Fukuyo et al. 2009) as well as bilateral periventricular nodular heterotopias (Capra et al. 2014). Recognition and treatment of seizures may improve attention, behavior, sleep, and overall cognitive functioning; however, medication side effects may also worsen behavior, sleep and cognitive function, if not chosen carefully. The prognosis depends on the type of seizure and response to anticonvulsants. The seizures are typically not refractory, but attention to side effect profile is important in selecting the correct antiepileptic, in addition to targeting the type of seizure diosrders. Adverse side effects of medications, such as excessive lethargy, hyperactivity, weight gain, and irritability, have been reported in children with SMS.

Catamenial epilepsy is common and characterized by epileptic seizures in the female occurring rhythmically with the menstrual cycle. Hormonal mechanisms have been proposed as a cause. In the authors' experience, an increased frequency of affected females experience catamenial epilepsy, usually coinciding with onset of the menstrual cycle; for most, seizure onset also occurs with onset of puberty. A growing literature suggests that the influence of sex steroids on neurological and psychiatric disorders is, in part, mediated by an aberrant central nervous system response to neuroactive steroids. In addition to seizures, these females often demonstrate a premenstrual syndrome with major mood shifts and escalating behaviors and outbursts before menstrual onset (personal experience). Use of a low-dose oral contraceptive may offer beneficial effects on mood and behavior symptoms as well as improving personal hygiene management because of decreased menstrual flow (Merideth et al. 2016).

Structural Abnormalities Central nervous system structural abnormalities documented by neuroimaging may be seen in over half of the affected individuals, with non-specific atrophy and ventriculomegaly the most common findings. Brain CT scans performed on 25 individuals with SMS demonstrated ventriculomegaly in nine, enlargement of the cisterna magna in two, and partial absence of the cerebellar vermis in one (Greenberg et al. 1996). Similar findings were seen among a group of 10 children who had undergone previous MRI: five had ventriculomegaly; two had an enlarged posterior fossa; and three had normal scans (Gropman et al. 2006). Despite the clinical finding of oromotor dysfunction, to date, no structural abnormalities of the opercular cortex, which subserves these functions, have been reported. Anatomical and functional brain imaging techniques documented a high frequency of lenticulo-insular brain anomalies compared to normal age/sex matched controls among five males with SMS (<16 years of age), with significant hypoperfusion evident in the same regions (Boddaert *et al*. 2004). Co-occurance of sporadic Moyamoya disease and SMS was reported in a 10-year-old female with a large distal 6.3 Mb deletion spanning 17p11.2-p13.1 (Girirajan et al. 2007). A case of co-occurance of SMS and Joubert syndrome (cerebellar vermis hypoplasia) has also been reported (Natacci et al. 2000). Neuropathological study on the initial case reported by Smith et al. (1982, 1986), whose entire 17p11.2 band was deleted, showed microcephaly and foreshortened frontal lobes with depletion, and a small choriod plexus hemangioma in the lateral ventricle (Smith et al. 1986).

Peripheral Neuropathy Clinical signs of peripheral neuropathy are reported in approximately 75% of individuals with SMS (Greenberg et al. 1996; Gropman et al. 2006). People with SMS have a characteristic appearance of the muscles of the legs and feet that is often seen in peripheral nerve syndromes or neuropathies, namely, "inverted champagne bottle appearance" of the legs and either pes cavus or pes planus deformity. Hammer toes are also frequently seen and decreased sensitivity to pain is suspected. In one series, distal muscle weakness was present in over half the individuals examined, and a previously undescribed peripheral neuropathy tremor in the upper extremity (6–8 Hz) was evident in 21% of individuals (Gropman et al. 2006). During early infancy and childhood, signs of peripheral nervous system involvement include hyporeflexia (84%) and decreased sensitivity to pain (Gropman et al. 2006). Hypotonia is universal, and it is likely to be related to a central abnormality. Markedly flat or highly arched feet (pes planus or cavus) and unusual gait (foot flap) are generally appreciated in childhood. Toe walking is seen in 60% of individuals (Martin et al. 2006) despite absence of tight heel cords.

Peroneal motor nerve conduction velocities are generally normal in childhood. Delayed motor nerve conduction velocities owing to biopsy-confirmed segmental demyelination and remyelination, similar to that seen in hereditary neuropathy with liability to pressure palsy, occur rarely (Smith et al. 1986; Zori et al. 1993; Greenberg et al. 1996) with normal nerve conduction in 23 of 25 (92%) cases studied. *PMP22*, the gene for Charcot–Marie–Tooth disease type 1A, located at 17p12 (distal to the Smith–Magenis syndrome critical region), which when deleted causes hereditary neuropathy with liability to pressure palsy, is usually not deleted in SMS (Greenberg et al. 1991; Moncla et al. 1993; Elsea and Girirajan 2008). Signs of peripheral neuropathy in individuals with SMS are thus independent of deletion of *PMP22*, and it is possible that other genes in the critical region may play a role. The neuropathy observed in SMS is not believed to be progressive.

High Pain Tolerance Because of their relative insensitivity to pain, individuals with SMS may cause injury to themselves by object insertion or persistent picking, self-biting, nail yanking, or self-hitting during uncontrolled rages (Smith et al. 1998a). Individuals with SMS usually do not often complain of "pain" sustained from usual childhood injuries such as broken bones, abrasions, cuts, and bruises (personal experience). In several cases, parents report damage to walls, doors, and/or plate glass windows resulting from forceful punches inflicted during an uncontrolled rage. Child abuse may be suspected in these circumstances, when the child's injuries are actually self-inflicted (Smith et al. 1998a).

Stroke The authors are aware of at least three individuals with SMS who have had strokes. The first (Case 2, Smith et al. 1986), who was born with bilateral cleft lip/palate and congenital heart defect, developed left hemiparesis at 4.5 years of age. Angiogram showed evidence of a small vessel stroke in the internal capsule, but no signs of moyamoya disease. The second (Chaudhry et al. 2007) was an affected adult who suffered a stroke after cardiac surgery (congenital heart defect). Further investigation revealed severe atherosclerotic changes of the intracranial vessels, despite a total cholesterol of 171 mg dL^{-1}. As 57% of individuals with SMS have lipid values greater than the 95th centile for age and sex (Smith et al. 2002), Chaudhry et al. (2007) suggest that this population may be at risk for cerebrovascular atherosclerosis. The third individual with a stroke was a 10-year-old Chinese female with an atypical large distal 6.3 Mb deletion and a ventricular septal defect who was diagnosed with Moyamoya disease and evidence of ischemic changes at 5 years of age (Girirajan et al. 2007). Array CGH analysis further delineated her deletion to span exon-1 of *RAI1* distally to FLJ45455 on 17p13.1.

Evaluation

- Careful neurological evaluation should be conducted in all individuals at diagnosis and at least annually thereafter. Individuals with seizures should be followed up more frequently in accordance with their individual needs.
- Electroencephalogram should be obtained in all affected individuals who have clinical seizures to guide the choice of antiepileptic treatment. For those without overt seizures, electroencephalogram may be helpful to rule out subclinical events in which treatment may improve attention and/or behavior.
- Neuroimaging should be accomplished in accordance with clinical findings, such as seizures and motor asymmetries or abnormalities, to rule out an anatomic basis.
- Change in behavior or attention warrants re-evaluation for seizures as well as possible medication effects.
- Electromyogram/nerve conduction studies may be of benefit in individual situations, especially in the setting of clinical evidence of peripheral neuropathy.
- In individuals who require open-heart surgery in adolescence or adulthood, evaluation for possible premature cerebrovascular disease has been recommended (Chaudhry et al. 2007). It is not yet known whether abnormalities in lipid metabolism predispose to a significant risk for atherosclerosis in adulthood. Screening for vascular disease is recommended in accordance with medical practice guidelines. For affected teenagers and adults who have additional risk factors (diabetes, family history of heart disease, congenital heart defects, etc.) more careful monitoring may be warranted.
- Physical therapy evaluation is recommended to assess developmental, functional, and balance impairments. Bracing, orthotic inserts, or dynamic ankle foot orthoses (DAFOs) may be helpful to provide added stability required for ambulation.

Management

- Seizures typically respond to traditional antiepileptic therapies. Individuals with SMS may be particularly sensitive to the side effects of several of these agents. In some cases, the antiepileptic drug may have a secondary effect on behavior and/or sleep, either positive or negative. Some of the medications, such as valproic acid and risperidal, may cause excessive weight gain. Others, such as carbamazepine, phenytoin, and phenobarbitol, may induce sleepiness. Levetiracetam can cause a behavioral syndrome and possibly psychosis, and should be used with caution in SMS.
- There is no specific treatment for the peripheral neuropathy seen in SMS. For the rare cases with larger deletions, including *PMP22*, attention to potential pressure damage using splinting and physical therapy in accordance with standard practice is recommended.
- Caution should be taken, especially in early childhood years, to reduce the risk of accidental injury in the home setting, in light of the inherent increased pain threshold and insatiable need to touch things and/or put objects into the mouth.
- Within the domain of occupational therapy, the individualized program should include interventions that address sensory processing difficulties with a complimentary behavioral program. These may include therapeutic activities to develop postural stability, hand skills, and visual-motor skills; splinting; and environmental modification and activity adaptation to address current disabilities and promote independence (reduce caregiver assistance) in activities of daily living.
- Within the domain of physical therapy, emphasis should be placed on gross motor interventions aimed at enhancing areas of motor development, balance, independent walking, and gait. Exercises/activities that focus on improving abdominal tone and truncal hypotonia are strongly encouraged.

Craniofacial

The craniofacial dysmorphology is described in detail in the Diagnostic Criteria section.

Dental examination of individuals with SMS documented one or more dental anomalies in all 15 (100%) individuals evaluated (Tomona et al. 2006). Tooth agenesis, primarily involving the mandibular second premolars, was seen in 13 of 15 (87%), taurodontism in both primary and permanent molars in 13 of 15 (87%), and dilaceration of the tooth roots in 5 of 15 (33%). Poor oral hygiene, increased dental plaque, and increased gingival inflammation were observed to progress with age from childhood to teenage years. This finding is consistent with the inverse relationship between age and performance of daily living skills (Martin et al. 2006). All seven cases with complete cephalometric analyses demonstrated protrusion of lower incisors leading to lower lip protrusion, likely contributing to the prognathic appearance recognized in earlier clinical studies.

Although velopharygeal insufficiency is common in SMS (75%) (Solomon et al. 2002), cleft of the palate and/or lip occurs with relatively low frequency (less than 10%); bifid uvula has also been seen. Prognosis is related to the degree of abnormality and its potential for surgical or medical management. Open mouth posture contributes to excessive drooling, an issue that can persist into older childhood.

Evaluation

- The palate should be examined for clefts, including bifid uvula and submucous cleft palate.
- Speech and swallow evaluations are indicated in the presence of difficulties with these functions.
- Dental examination (intraoral and extraoral) is recommended, including panoramic radiographs to identify anomalies (agenesis; taurodontism, and/or root dilacerations) and document oral hygiene. Panoramic radiographs offer an accessible, simple, and low-cost tool in diagnosis.

Management

- Surgical repair of lip and/or palate is indicated as for the general population.
- Parental education about special feeding techniques is needed for the child with cleft palate.
- Strategies to assist individuals with SMS in oral hygiene and gingival care usually require parental supervision. Use of a rechargeable toothbrush with a 2 min timer may be beneficial, since handheld toothbrushes (brushing motion) are difficult for the individual to manage independently.

Ophthalmologic

A high frequency of eye abnormalities has been documented in SMS, including iris anomalies, microcornea, strabismus, cataracts, and myopia. (Finucane et al. 1993b; Barnicoat et al. 1996; Chen RM et al. 1996). The frequency of ocular findings, specifically high myopia and retinal detachment, appears to be age dependent. Detached retina leading to blindness can occur. Efforts to minimize retinal detachment because of trauma should be made. Microcornea was seen in over half of one series (Chen et al. 1996). The heterochromic irides or "Brushfield-like" spots in the iris reported by several authors are actually Wolfflin–Kruckmann spots (Chen et al. 1996).

Other rare ocular anomalies have been reported in single individuals. These include iris dysgenesis (Barnicoat et al. 1996), congenital severe Brown's syndrome (limitation or absence of elevation with adduction of the eye) (Salati et al. 1996), and visual loss caused by bilateral macular disciform scars (Babovic-Vuksanovic et al. 1998).

Evaluation

- Annual ophthalmological evaluation with attention to evidence of strabismus, microcornea, iris anomalies, and refractive errors is recommended, starting at diagnosis.

Management

- Corrective lenses for myopia have been required for several individuals with SMS as early as 9 months of age. However, parents report difficulties with compliance in wearing them.
- Corrective treatment for strabismus, when present, is no different from that in the general population.
- The potential for detached retina related to repetitive head-banging behavior should be minimized, including use of protective helmets.

Cardiovascular

The estimated incidence of cardiovascular abnormalities is 37–47% in individuals with 17p11.2 deletions (Smith et al. 1986; Greenberg et al. 1996; Potocki et al. 2003). Anomalies include mild tricuspid or mitral valve stenosis or regurgitation, ventricular septal defects, supra-valvular aortic or pulmonic stenosis, atrial septal defects, and tetralogy of Fallot (Smith et al. 1986, Greenberg et al. 1996; Sweeney and Kerr 1999; Chaudhry et al. 2007). In contrast, congenital heart defects have not been reported in *RAI1* mutation cases.

Hypercholesterolemia of childhood onset is common in SMS and may serve as a useful early clinical marker of the syndrome (Smith et al. 2002). In a large cohort of 49 children with the syndrome, 57% demonstrated hypercholesterolemia; less than one-third were within the normal range for lipid values, and one-third each measured in the borderline or high range for total or low-density lipoprotein cholesterol values (Smith et al. 2002). The risk for premature atherosclerosis is unknown; however, cerebrovascular accidents are reported rarely (see Neurology), suggesting that screening for premature atherosclerotic disease may be warranted.

Evaluation

- Cardiac evaluation, including echocardiogram, is recommended at the time of diagnosis.
- Follow-up in individuals with cardiac anomalies should be determined by the severity and type of cardiac anomaly identified.
- Individuals who require open-heart surgery in teens or adulthood should be evaluated for possible premature cerebrovascular disease (Chaudhry et al. 2007).

Management

- Pharmacological or surgical intervention should be in accordance with customary practice for each cardiac anomaly identified.
- Treatment for hypercholesterolemia is standard at older ages.

Gastrointestinal

A history of chronic constipation is reported in 58% of affected individuals (Smith et al. 1998b). The cause remains undefined but may stem from generalized hypotonia and/or dietary intake. Constipation may improve as diet and level of activity increase. Symptoms of encopresis are seen commonly in SMS. In the few situations where Hirschsprung disease was initially suspected, biopsy failed to confirm this diagnosis (personal experience).

Evaluation

- Chronic constipation or alternating constipation and diarrhea warrants referral to gastroenterology and possible evaluation with barium enema.
- Surgical evaluation should be performed where there is evidence of bowel obstruction.

Management

- Increased fluid intake, stool softeners, and dietary management are beneficial for constipation, as for the general population.

Genitourinary

Renal and urological anomalies occur in 15–35% of individuals with 17p11.2 deletions, but are rare (0/11 published cases) in individuals with *RAI1* mutations (Greenberg et al. 1996; Potocki et al. 2003; Girirajan et al. 2006). Abnormalities include duplication of the collecting system (4), unilateral renal agenesis and ectopic kidney, ureterovesicular obstruction, or malposition of the ureterovesicular junction (Smith et al. 1986; Greenberg et al. 1996; Chou et al. 2002; Myers et al. 2007). Genotype/phenotype correlations suggest that renal and urological anomalies are more common among individuals with atypically small deletions (Potocki et al. 2003). Nocturnal enuresis occurs in almost 80% of affected individuals and may remain an issue into school age (Smith et al. 1998b). Although the etiology has yet to be determined, possible causes include increased fluid intake, medications, underlying urinary tract anomalies, and/or hypotonic bladder, each of which has been seen occasionally. Genital anomalies are less frequent in both sexes, but include cryptorchidism, shawl, or undeveloped scrotum, infantile cervix, and/or hypoplastic uterus (Smith et al. 1986; Stratton et al. 1986).

Evaluation

- All newly diagnosed individuals with SMS should be evaluated with baseline renal ultrasound for evidence of structural urinary tract anomalies. Additional studies, including voiding cystourethrogram and/or laboratory studies of renal function, should be pursued as clinically indicated.
- Routine urinalysis is appropriate at each annual health maintenance exam.
- Fevers of unknown origin should precipitate an evaluation for possible urinary tract infection.

Management

- Urinary tract infections should be treated with antibiotic therapy as in the general population.
- Surgical intervention may be necessary for urinary tract anomalies, in accordance with standard practice.
- Medications with potential renal toxicity should be avoided.

Musculoskeletal

A variety of hand anomalies have been described including short, broad hands (85%), single palmar creases, and digital anomalies (66%), such as cutaneous syndactyly of toes 2 and 3, 5th finger clinodactyly, and/or polydactyly (Smith et al. 1986; Stratton et al. 1986; Kondo et al. 1991; Chen et al. 1996). The prominent fingertip pads first described by Kondo et al. (1991) are also seen on the toes and serve as another useful clinical finding in diagnostic assessment. Metacarpophalangeal pattern profile analysis confirmed brachydactyly in one large series of 29 individuals with SMS (Potocki et al. 2003), and, consistent with other reports, found progressively decreasing size of the more distal bones of the hand (Kondo et al. 1991; Meinecke, 1993). The metacarpophalangeal analysis did not demonstrate a characteristic pattern. In contrast to the isolated finding of pronounced shortening of the fifth middle phalanx reported by Meinecke et al. (2003) found relative enlargement of this bone as well as enlargement of the proximal phalanx of the thumb. Short or bowed ulnae are described (12%) (Greenberg et al. 1996). The fingers show hyperextensibility, affecting hand and prehensile grip necessary for using a pen or a pencil.

Mild-to-moderate scoliosis, most commonly of the midthoracic region, is seen in many children (65%) with SMS age 4 years and older, with a lower frequency (35%) reported in *RAI1* mutation cases (Smith et al. 1986; Greenberg et al. 1996; Girirajan et al. 2006). Scoliosis can be progressive and, if significant, may require corrective surgery (Spilsbury and Mohany 2003). Underlying vertebral anomalies are seen with relatively low frequency (Smith et al. 1986; Gropman et al. 1998). The incidence of spina bifida occulta, a minor variant, also appears to be increased in this population (personal experience). At least two cases of congenital scoliosis are reported (Smith et al. 1986; Zheng et al. 2015), with underlying vertebral anomalies.

Tight heel cords and a history of toe walking are seen frequently and benefit from use of orthotics. A broad-based somewhat flapping or loping gait is very characteristic of the syndrome. Preliminary gait analysis shows higher cadence and longer contact with the floor compared consistent with an immature gait in younger versus older individuals (personal observation). Limb length discrepancy is evident in some individuals and may impact posture and gait (personal experience).

Evaluation

- At diagnosis, baseline spine radiographs to rule out an underlying vertebral defect are recommended.
- Clinical monitoring for scoliosis should take place on an annual basis, especially during adolescence, with radiographs pursued in cases where spinal curvature appears to progress.
- Spine radiographs should be carried out when clinically indicated to document change in spinal curvature.

Management

- Scoliosis should be treated as for individuals in the general population. Surgical intervention is rarely required.
- Vertebral defects generally need no treatment.
- Orthopedic evaluation and use of orthotics are appropriate for positional foot deformities and/or gait disturbances.

Endocrine

The exact incidence of endocrine abnormalities in SMS remains undefined. Adrenal aplasia and hypoplasia have been described in an 11-month-old male who died unexpectedly after palatoplasty (Denny et al. 1992). About one-quarter of those tested have borderline hypothyroidism. Although short stature occurs in affected individuals, the incidence of growth hormone deficiency appears low, with peak levels present at the proper phase of the day and levels only slightly lower than normal controls (De Leersnyder et al. 2001a, 2006). Isolated growth hormone deficiency was identified in a 5-year-old female with a deletion (Itoh et al. 2004). Precocious puberty and premature ovarian failure have been seen (personal experience). Early adrenarche is also known to occur in females.

Evaluation

- At diagnosis, routine blood chemistries (complete blood count with differential, routine electrolytes, liver function tests), quantitative immunoglobulins, fasting

blood lipids, and thyroid function studies should be performed.
- Baseline and adrenocorticotropic hormone-stimulated serum cortisol levels should be checked in cases of suspected hypoadrenalism.
- Specific screening of adrenal function is warranted in cases of large deletions (more than half of 17p11.2 band) (Denny et al. 1992).
- Thyroid function should be evaluated at each annual health maintenance visit.

Management

- Endocrine problems should be treated as in the general population.

Immunologic

Low immunoglobulin levels are observed more frequently than previously reported and may contribute to chronic otitis media, often leading to recurrent ear tube placement, chronic sinusitis, and upper respiratory tract infections. Mildly decreased IgM was observed in 23% of an initial series of 13 individuals studied by Greenberg et al. (1996). Recent systematic study of serum immunoglobulin profiles (IgA, IgG, IgM) in large cohort affected individuals (ages 4–27 years) documented diminished immunological function in the majority (60%) (Perkins et al. 2017). Low IgA and IgG appears more prevalent at young ages and differed significantly from expected means for age (Introne et al. 2005). Deletion cases exhibit an increased susceptibility to sinopulmonary infections, including pneumonia (47%), and may benefit from evaluation by an immunologist with consideration of prophylactic strategies to prevent infections (Perkins et al. 2017). While otitis media and sinus infections have been reported for *RAI1* mutation cases, low immunoglobulin levels were not (Vilboux et al. 2011).

Evaluation

- Quantitative immunoglobulin analysis is recommended at the time of diagnosis, with additional studies as clinically indicated.
- Evaluation by an immunologist with prophylactic strategies to prevent infections may benefit selected patients.

Management

- Immune dysfunction should be treated as in the general population.

Dermatologic

Hair and skin color are often fair, and irides are frequently blue. Rosy cheeks are especially evident in early childhood, possibly related to constant drooling and/or eczema. Hyperkeratotic skin over the surface of the hands, feet, and knees is noted in less than 20% of cases (Smith et al. 1986; Stratton et al. 1986; Lockwood et al. 1988). Complaints of dry skin appear to be more frequent among people with *RAI1* mutation (100%) than those with deletion (44%) (Edelman et al. 2007). Nail hygiene is important; hangnails often precipitate cuticle mutilation and/or nail biting and yanking. Nail yanking is especially prevalent in older individuals (86%) compared with younger ages (under 30%) (Table 54.2) (Finucane et al. 2001).

TABLE 54.2 Comparisons of self-injurious behaviors in Smith-Magenis syndrome by age group.

Behavior	Dykens and Smith (1998)	Finucane et al. 2001			Martin et al. 2006
Mean age	Mean age 9 years (n = 35)	Total (n = 29)	Younger 6.5 years (n = 15)	Older 25 years (n = 14)	Age 2–12 years (n = 19)
Self-injurious behaviors	92%	97%	93%	100%	100%
Bites self	77%	93%	87%		80%
Hits/slaps self	71%	62%	40%	86%	93%
Hits self against objects or headbanging	40%	55%	47%	64%	53%
Pulls hair or skin	31%	35%	nn	nn	40%
Skin picking/scratching	29%	52%	33%	71%	33%
Onycotillomania (nail yanking)	29%	48%	27%	86%	13%
Polyembolokoilamania (insertion)	25%				47%
Ears		31%	20%	43%	
Vagina		21%	11%	30%	
Nose		17%	7%	29%	
Rectum		4%	0%	7%	

Evaluation

- Skin assessment should be part of the routine medical examination of the individual with SMS.
- Attention to areas targeted by self-injurious behaviors (wrists, arms, nails) is essential.

Management

- Moisturizing creams can be effective in treating dry skin. Twice-daily application of over-the-counter lotions that contain either 10% urea or 5–10% lactic acid are recommended in cases of extremely dry skin.
- Long-sleeved garments can minimize injury caused by chronic self-biting and picking behavior.
- Fingered gloves and frequent manicures and pedicures for nail hygiene can minimize cuticle mutilation and nail biting, picking, and yanking.
- Recognition of stressors and events that trigger self-injurious behaviors and behavior modification may prove helpful.

DEDICATION

This chapter is dedicated to the families and individuals with Smith–Magenis syndrome who continue to motivate our research; and in special memory of Frank Greenberg, MD, a dear friend and genetics colleague, whose inspiration, knowledge, and contributions to the early understanding of Smith–Magenis syndrome are reflected throughout this chapter.

RESOURCES

Parents and Researchers Interested in Smith–Magenis Syndrome (PRISMS), United States

21800 Town Center Plaza Suite #266A-633, Sterlin, VA, USA
Phone: 972-231-0035
Email: info@prisms.org.
Website: www.prisms.org

PRISMS is the internationally recognized advocacy, education, and support organization for individuals with Smith–Magenis syndrome (SMS), their families, and the professionals who serve them. PRISMS provides information and support, sponsors research, and fosters partnerships with professionals to increase awareness and undestanding of SMS.

Support programs: parent-to-parent program; e-mail parents list; technical support; educational materials/publications: SPECTRUM newsletter, new parent packet, brochures, reference materials; international conferences (every 2–3 years); the Smith–Magenis Syndrome Patient Registry (SMSPR); links to Smith–Magenis Syndrome Clinics (multi-specialty clinics).

PRISMS International Partnership Program partners with associations and organizations dedicated to Smith–Magenis syndrome family support world-wide, including: Smith–Magenis Syndrome Australia, Smith Magenis Mexico, Smith–Magenis Sydnrome Foundation UK, Association Smith–Magenis 17 France, Sirius e.V Germany; and Icelandic Smith–Magenis Society.

The Smith–Magenis Syndrome Foundation, UK

Email: info@smith-magenis.co.uk
Website: http://smith-magenis.org

The Foundation is a small registered UK charity run by parents and professionals and includes a professional advisory board comprised of doctors, professionals, and educators. Its primary aim is to provide information and support to individuals interested in Smith–Magenis syndrome and to sponsor educational conferences for its members (families and professionals) every 18–24 months.

Advocacy/Support Networks for Smith–Magenis syndrome France Association Smith–Magenis 17 France

Website: http://www.smithmagenis17.org/

Germany

SIRIUS (Germany's Support Network for Smith–Magenis Syndrome) Jörg Michael Weber (1st Chairman)
Phone: (0) 173 374 56 02
EMail: j.m.weber@smith-magenis.de
Website: http://smith-magenis.de

Australia

Smith–Magenis Syndrome Australia

Member of PRISMS International Partnership Program
Website: http://smsaustralia.org

REFERENCES

Allanson JE, Greenberg F, Smith ACM (1999) The face of Smith-Magenis syndrome: A subjective and objective study. *J Med Genet* 36:394–397.

Acquaviva F, Sana ME, Della Monica M, Pinelli M, Postorivo D, Fontana P, Falco MT, Nardone AM, Lonardo F, Iascone M, Scarano G (2017) First evidence of Smith-Magenis syndrome in mother and daughter due to a novel RAI mutation. *Am J Med Genet A* 173(1):231-238.

Babovic-Vuksanovic D, Jalal AM, Garrity JA, Robertson DM, Lindor NM (1998) Visual impairment due to macular disciform scars in a 20-year-old man with Smith-Magenis syndrome: Another ophthalmologic complication. *Am J Med Genet* 80:373–376.

Barnicoat AJ, Docherty Z, Bobrow M (1993) Where have all the fragile X boys gone? *Dev Med Child Neurol* 35:532–539.

Barnicoat AJ, Moller HU, Palmer RW, Russell-Eggitt I, Winter RM (1996) An unusual presentation of Smith-Magenis syndrome with iris dysgenesis. *Clin Dysmorphol* 5:153–158.

Behjati F, Mullarkey M, Bergbaum A, Berry AC, Dochery Z (1997) Chromosome deletion 17p11.2 (Smith-Magenis syndrome) in seven new patients, four of whom had been referred for fragile-X investigation. *Clin Genet* 51:71–74.

Bi W, Yan J, Stankiewicz P, Park SS, Walz K, Boerkoel CF, Potocki L, Shaffer LG, Devriendt K, Nowaczyk MJ, Inoue K, Lupski JR (2002) Genes in a refined Smith-Magenis syndrome critical deletion interval on chromosome 17p11.2 and the syntenic region of the mouse. *Genome Res* 12:713–728.

Bi W, Saifi GM, Shaw CJ, Walz K, Fonseca P, Wilson M, Potocki L, Lupski JR (2004) Mutations of RAI1, a PHD-containing protein, in nondeletion patients with Smith-Magenis syndrome. *Hum Genet* 115:515–524. Epub 2004 Sep 30.

Bi W, Ohyama T, Nakamura H, Yan J, Visvanathan J, Justice MJ, Lupski JR (2005) Inactivation of Rai1 in mice recapitulates phenotypes observed in chromosome engineered mouse models for Smith-Magenis syndrome. *Hum Mol Genet* 14:983–995.

Bi W, Saifi GM, Girirajan S, Shi X, Szomju B, Firth H, Magenis RE, Potocki L, Elsea SH, Lupski JR (2006) RAI1 point mutations, CAG repeat variation, and SNP analysis in nondeletion Smith-Magenis syndrome. *Am J Med Genet* 140A:2454–2463.

Bi W, Yan J, Shi X, Yuva-Paylor LA, Antalffy BA, Goldman A, Yoo JW, Noebels JL, Armstrong DL, Paylor R, Lupski JR (2007) Rai1 deficiency in mice causes learning impairment and motor dysfunction, whereas Rai1 heterozygous mice display minimal behavioral phenotypes. *Hum Mol Genet* 16:1802–1813.

Boudreau, et al. (2009) Review of disrupted sleep patterns in Smith-Magenis syndrome and normal melatonin secretion in a patient with an atypical interstitial 17p11.2 deletion. *Am J Med Genet A* 149A(7):1382–1391.

Boone PM, Reiter RJ, Glaze DG, Tan D-X, Lupski JR, Potocki L (2011) Abnormal circadian rhythm of melatonin in Smith-Magenis syndrome patients with RAI1 point mutations. *Am J Med Genet A* 155A(8):2024–7

Brendal MA, King KA, Zalewski CK, Finucane BM, Introne W, Brewer CC, Smith ACM (2017) Auditory Phenotype of Smith-Magenis Syndrome. *J Speech Lang Hear Res* 60(4):1076–1087.

Capra V, Biancheri R, Morana G, Striano P, Novara F, Ferrero GB, Boeri L, Celle ME, Mancardi MM, Zuffardi O, Parrini E, Guerrini R (2014) Periventricular nodular heterotopia in Smith-Magenis syndrome. *Am J Med Genet A* 164A(12):3142–3147.

Castigli E, Wilson SA, Garibyan L, Rachid R, Bonilla F, Schneider L, Geha RS (2005) TACI is mutant in common variable immunodeficiency and IgA deficiency. *Nat Gene* 37:829–834.

Chaudhry AP, Schwartz C, Singh AS (2007) Stroke after cardiac surgery. *Tex Heart Inst J* 34:247–249.

Chen KS, Potocki L, Lupski JR (1996a) The Smith-Magenis syndrome (del(17)(p11.2)): Clinical review and molecular advances. *Ment Retard Dev Disabil Res Rev* 2:122–129.

Chen KS, Manian P, Koeuth T, Potocki L, Zhao Q, Chinault CA, Lee CC, Lupski JR (1997) Homologous recombination of a flanking repeat gene cluster is a mechanism for a common contiguous gene deletion syndrome. *Nat Genet* 17:154–163.

Chen RM, Lupski JR, Greenberg F, Lewis RA (1996) Ophthalmic manifestations of Smith-Magenis syndrome. *Ophthalmology* 103:1084–1091.

Chou IC, Tsai FJ, Yu MT, Tsai CH (2002) Smith-Magenis syndrome with bilateral vesicoureteral reflux: A case report. *J Formos Med Assoc* 101:726–728.

Colley AF, Leversha MA, Voullaire LE, Rogers JG (1990) Five cases demonstrating the distinctive behavioural features of chromosome deletion 17(p11.2 p11.2) (Smith-Magenis syndrome). *J Paediatr Child Health* 26:17–21.

Darfour L, Verleyen P, Lesage K, Holvoet M, Devriendt K. (2016) Bilateral renal tumors in an adult man with Smith-Magenis syndrome: The role of the FLCN gene. *Eur J Med Genet* 59:499–501.

De Leersnyder H, deBlois MC, Claustrat B, Roman S, Albrecth U, VonKleist Retzow JC, Delobel B, Viot G, Llyonnet S, Bekemans M, Munnich A (2001) Inversion of the circadian rhythm of melatonin in Smith-Magenis syndrome. *J Pediatr* 139:111–116.

De Leersnyder H, Bresson JL, deBlois MC, Souberbielle JC, Mogenet A, Delhotal-Landes B, Salaefranzue F, Munnich A (2003) ß1-adrenergic antagonists and melatonin reset the clock and restore sleep in a circadian disorder, Smith-Magenis syndrome. *J Med Genet* 40:74–78.

De Leersnyder H, Claustrat B, Munnich A, Verloes A (2006) Circadian rhythm disorder in a rare disease: Smith-Magenis syndrome. *Mol Cell Endocrinol* 252:88–91.

Denny AD, Weik LD, Lubinsky MS, Wyatt DT (1992) Lethal adrenal aplasia in an infant with Smith-Magenis syndrome, deletion 17p112. *J Dysmorph Clin Genet* 6:175–179.

Dykens E, Smith ACM (1998) Distinctiveness and correlates of maladaptive behavior in children and adolescents with Smith-Magenis syndrome. *J Intellect Disabil Res* 42:481–489.

Dykens E, Finucane B, Gayley C (1997) Cognitive and behavioral profiles in persons with Smith-Magenis syndrome. *J Autism Dev Disord* 27:203–211.

Edelman EA, Girirajan S, Finucane B, Patel PI, Lupski JR, Smith AC, Elsea SH (2007) Gender, genotype, and phenotype differences in Smith-Magenis syndrome: A meta-analysis of 105 cases. *Clin Genet* 71:540–550.

Elsea SA, Ramesh CJ, Jiralerspong S, Finucane BM, Pandolfo M, Greenberg F, Baldini A, Stover P, Patel PI (1995) Haploinsufficiency of cytosolic serine hydroxymethyltransferase in the Smith-Magenis syndrome. *Am J Hum Genet* 57:1342–1350.

Elsea SH, Girirajan S (2008) Smith-Magenis syndrome. *Eur J Hum Genet* 16:412–421.

Elsea SH, Purandare SM, Adell RA, Juyal RC, Davis JG, Finucane B, Magenis RE, Patel PI (1997) Definition of the critical interval for Smith-Magenis syndrome. *Cytogenet Cell Genet* 79:276–281.

Fan T-S, Farrell SA (1994) Letter to the Editor: Prenatal diagnosis of interstitial deletion of 17 (p11.2p11.2) (Smith-Magenis syndrome). *Am J Med Genet* 49:253–254.

Finucane B, Dirrigl KH, Simon EW (2001) Characterization of self-injurious behaviors in children and adults with Smith-Magenis syndrome. *Am J Ment Retard* 106:52–58.

Finucane B, Haas-Givler B (2009) Smith-Magenis syndrome: Genetic basis and clinical implications. *J Ment Health Res Intellect Disabil* 2:134–148.

Finucane BM, Jaeger ER, Kurtz MB, Weinstein M, Scott CI (1993) Eye abnormalities in the Smith-Magenis contiguous gene deletion syndrome. *Am J Med Genet* 45:443–446.

Finucane BM, Konar D, Givler BH, Kurtz MB, Scott LI (1994) The spasmodic upper-body squeeze: A characteristic behavior in Smith-Magenis syndrome. *Dev Med Child Neurol* 36:70–83.

Finucane B, Simon E (1999) Genetics and dual diagnosis: Smith-Magenis syndrome. *The NADD Bulletin* 2:8–10.

Gamba BF, Vieira GH, Souza DH, Monteiro FF, Lorenzini JJ, Carvalho DR and Morreti-Ferreira D (2011) Smith-Magenis syndrome: clinical evaluation in seven Brazilian patients. *Genet Mol Res* 10(4):2664–2670.

Girirajan S, Elsas LJ 2nd Devriendt K, Elsea SH (2005) RAI1 variations in Smith-Magenis syndrome patients without 17p11.2 deletions. *J Med Genet* 42:820–828.

Girirajan S, Vlangos CN, Szomju BB, Edelman E, Trevors CD, Dupuis L, Nezarati M, Bunyan DJ, Elsea SH (2006) Genotype-phenotype correlation in Smith-Magenis syndrome: Evidence that multiple genes in 17p11.2 contribute to the clinical spectrum. *Genet Med* 8:417–427.

Girirajan S, Mendoza-Londono R, Vlangos CN, Dupuis L, Nowak NJ, Bunyan DJ, Hatchwell E, Elsea SH (2007) Smith-Magenis syndrome and Moyamoya disease in a patient with del(17)(p11.2p13.1). *Am J Med Genet* 143A:999–1008.

Goldman AM, Potocki L, Walz K, Lynch JK, Glaze DG, Lupski JR, Noebels JL (2006) Epilepsy and chromosomal rearrangements in Smith-Magenis Syndrome [del(17)(p11.2p11.2)]. *J Child Neurol* 21:93–98.

Gottlieb DJ1, Vezina RM, Chase C, Lesko SM, Heeren TC, Weese-Mayer DE, Auerbach SH, Corwin MJ (2003) Symptoms of sleep-disordered breathing in 5-year-old children are associated with sleepiness and problem behaviors. *Pediatrics* 112(4):870–7.

Greenberg F, Guzzetta V, De Oca-Luna RM, Magenis RE, Smith ACM, Richter SF, Kondo I, Dobyns WB, Patel PI, Lupski J (1991) Molecular analysis of the Smith-Magenis syndrome: A possible contiguous-gene syndrome associated with del(17)(p11.2). *Am J Hum Genet* 49:1207–1218.

Greenberg F, Lewis RA, Potocki L, Glaze D, Parke J, Killian J, Murphy MA, Williamson D, Brown F, Dutton R, McCluggage C, Friedman E, Sulek M, Lupski JR (1996) Multi-disciplinary clinical study of Smith-Magenis syndrome (deletion 17p11.2). *Am J Med Genet* 62:247–254.

Gropman A, Smith ACM, Greenberg F (1998) Neurologic aspects of Smith-Magenis syndrome. *Ann Neurol* 44:561.

Gropman AL, Duncan WC, Smith AC (2006) Neurologic and developmental features of the Smith-Magenis syndrome (del 17p11.2). *Pediatr Neurol* 34:337–350.

Gropman AL, Elsea S, Duncan WC Jr, Smith AC (2007) New developments in Smith-Magenis syndrome (del 17p11.2). *Curr Opin Neurol* 20:125–134.

Finucane B, Haas-Givler B (2014) *On the road to success with SMS - A Smith-Magenis guidebook for schools* kindle edition. PRISMS, Inc. Available at: https://www.prisms.org/education/publications-and-resources/

Haas-Givler B, Finucane B (1995) What's teacher to do: Classroom strategies that enhance learning for children with Smith-Magenis syndrome. *Spectrum* 2. Available at https://www.prisms.org/wp-content/uploads/2018/01/Classroom-Strategies-that-Enhance-Learning-for-Children-with-Smith.pdf

Hagerman R (1999) Smith-Magenis syndrome. In: *Neurodevelopmental Disorders: Diagnosis and Treatment,* Hagerman R, ed. New York: Oxford University Press, pp. 341–362.

Hicks M, Ferguson S, Bernier F, Lemay JF (2008) A case report of monozygotic twins with Smith-Magenis syndrome. *J Dev Behav Pediatr* 29:42–46.

Hildenbrand HL and Smith ACM (2012) Analysis of the Sensory Profile in Children with Smith–Magenis Syndrome. *Phys Occup Ther Pediatr* 32(1):48–65.

Hino-Fukuyo N1, Haginoya K, Uematsu M, Nakayama T, Kikuchi A, Kure S, Kamada F, Abe Y, Arai N, Togashi N, Onuma A, Tsuchiya S (2009) Smith-Magenis syndrome with West syndrome in a 5-year-old girl: a long-term follow-up study. *J Child Neurol* 24(7):868–873.

Hodapp RM, Fidler DJ, Smith ACM (1998) Stress and coping in families of children with Smith-Magenis syndrome. *J Intellec Disabil Res* 42:331–340.

Hodnett J, Scheithauer M, Call NA, Mevers JL, Miller SJ (2018) Using a functional analysis followed by differential reinforcement and extinction to reduce challenging behaviors in children with Smith-Magenis syndrome. *Am J Intellect Dev Disabil* 123(6):558–573.

Hua X, Wu J, Goldstein JL, Brown MS, Hobbs HH (1995) Structure of the human gene encoding sterol regulatory element binding protein-1 (*SREBF1*) and localization of *SREBF1* and *SREBF2* to chromosomes 17p11.2 and 22q13. *Genomics* 25:667–673.

Huessler H, Harris M, Cooper D (2007) Treatment of sleep disruption in Smith-Magenis. *J Intellect Disabil Res* 51:664.

Introne W, Jurinka A, Krasnewich D, Candotti F, Smith ACM (2005) Immunologic Abnormalities in Smith-Magenis syndrome (del 17p11.2) (Abstract/Poster 605). *Presented at the meeting of the American Society of Human Genetics*, October 25-29, 2005, in Salt Lake City, UT.

Itoh M, Hayashi M, Hasegawa T, Shimohira M, Kohyama J (2004) Systemic growth hormone corrects sleep disturbance in Smith-Magenis syndrome. *Brain Dev* 26:484–486.

Juyal RC, Figuera LE, Hauge X, Elsea SH, Lupski JR, Greenberg F, Baldini A, Patel PI (1996) Molecular analysis of 17p11.2 deletion in 62 Smith-Magenis syndrome patients. *Am J Hum Genet* 58:998–1007.

Kondo I, Matsuura S, Kuwajima K, Tokashiki M, Izumikawa Y, Naritomi K, Niikawa N, Kajii T (1991) Diagnostic hand anomalies in Smith-Magenis syndrome: Four new patients with del (17)(p11.2p11.2). *Am J Med Genet* 41:225–229.

Kosaki R, Okuyama T, Tanaka T, Migita O, Kosaki K (2007) Monozygotic twins of Smith-Magenis syndrome. *Am J Med Genet* 143A:768–769.

Laje G, Morse R, Richter W, Ball J, Pao M, Smith ACM (2010a) Autism spectrum features in Smith–Magenis syndrome. *Am J Med Genet C* 154C:456–462.

Laje G, Bernert R, Morse R, Pao M, Smith, ACM (2010b) Pharmacological treatment of disruptive behavior in Smith-Magenis Syndrome. *Am J Med Genet C* 154C:463–468.

Lei TY, Li R, Fu F, Wan JH, Zhang JL, Jing XY, Liao C (2015) Prenatal diagnosis of Smith-Magenis syndrome in two fetuses with increased nuchal translucency, mild lateral ventriculomegaly, and congenital heart defects. *Taiwan J Obstetr Gynecol* 55:886e890.

Liang Y, Want A, Probst FJ, Arhya IN, Barber TD, Chen K-S, Deshmukh D, Dolan DF, Hinnant JT, Carter LE, Jain PK, Lalwani AK, Li XC, Lupski JR, Moeljopawiro S, Morell R, Negrini C, Wilcox ER, Winata S, Camper SA, Friedman TB (1998) Genetic mapping refines *DFNB3* to 17p11.2, suggests multiple alleles of *DFNB3*, and supports homology to the mouse model shaker-2. *Am J Hum Genet* 62:904–915.

Liburd N, Ghosh M, Riazuddin S, Naz S, Khan S, Ahmed Z, Riazuddin S, Liang Y, Menon PS, Smith T, Smith AC, Chen KS, Lupski JR, Wilcox ER, Potocki L, Friedman TB (2001) Novel mutations of MYO15A associated with profound deafness in consanguineous families and moderately severe hearing loss in a patient withSmith-Magenis syndrome. *Hum Genet* 109:535–541.

Lockwood D, Hecht F, Dowman C, Hecht BK, Rizkallah TH, Goodwin TM, Allanson J (1988) Chromosome sub-band 17p11.2 deletion: A minute deletion syndrome. *J Med Genet* 25:732–737.

Lu X, Shaw CA, Patel A, Li J, Cooper ML, et al (2007) Clinical implementation of chromosomal microarray analysis: Summary of 2513 postnatal cases. *PLoS One* 2(3):e327.

Lupski JR (1998) Genomic disorders: Structural features of the genome can lead to DNA rearrangements and human disease traits. *Trends Genet* 14:417–422.

Madduri N, Peters SU, Voigt RG, Llorente AM, Lupski JR, Potocki L (2006) Cognitive and adaptive behavior profiles in Smith-Magenis syndrome. *J Dev Behav Pediatr* 27:188–192.

Martin SC, Wolters PL, Smith ACM (2006) Adaptive and maladaptive behavior in children with Smith-Magenis syndrome. *J Autism Dev Disord* 36:541–552.

Meinecke P (1993) Confirmation of a particular but nonspecfic metacarpophalangeal pattern profile in patients with the Smith-Magenis syndrome due to interstitial deletion of 17p. *Am J Med Genet* 45:441–442.

Merideth MA, Introne WJ, Gahl WA, Smith ACM (2016) *Gynecologic and reproductive health issues in patients with Smith-Magenis Syndrome (Abstract # 2509) Presented at 66th Annual Meeting of the American Society of Human Genetics, October 19, 2016*, Vancouver, Canada.

Miller DT, Adam MP, Aradhya S, Biesecker LG, Brothman AR, Carter NP, Church DM, Crolla JA, Eichler EE, Epstein CJ, Faucett WA, Feuk L, Friedman JM, Hamosh A, Jackson L, Kaminsky EB, Kok K, Krantz ID, Kuhn RM, Lee C, Ostell JM, Rosenberg C, Scherer SW, Spinner NB, Stavropoulos DJ, Tepperberg JH, Thorland EC, Vermeesch JR, Waggoner DJ, Watson MS, Martin CL, Ledbetter DH (2010) Consensus statement: chromosomal microarray is a first-tier clinical diagnostic test for individuals with developmental disabilities or congenital anomalies. *Am J Hum Genet* 86(5):749–64.

Moncla A, Prias L, Arbex OF, Muscatelli F, Mattei M-G, Mattei J-F, Fontes M (1993) Physical mapping of microdeletions of the chromosome 17 short arm associated with Smith-Magenis syndrome. *Hum Genet* 90:657–660.

Myers SM Challman TD, Back GH (2007) Research Letter: End-stage renal failure in Smith-Magenis syndrome. *Am J Med Genet* 143A:1922–1924.

Nag HE, Nordgren A, Anderlid BM, Nærland T (2018) Reversed gender ratio of autism spectrum disorder in Smith-Magenis syndrome. *Mol Autism* 9:1.

Nal N, Ahmed ZM, Erkal E, Alper OM, Lüleci G, Dinç O, Waryah AM, Ain Q, Tasneem S, Husnain T, Chattaraj P, Riazuddin S, Boger E, Ghosh M, Kabra M, Riazuddin S, Morell RJ, Friedman TB (2007) Mutational spectrum of MYO15A: The large N-terminal extension of myosin XVA is required for hearing. *Hum Mutat* 28:1014–1019.

Natacci F, Corrado L, Pierri M, Rossetti M, Zuccarini C, Riva P, Miozzo M, Larizza L (2000). Patient with large 17p11.2 deletion presenting with Smith-Magenis syndrome and Joubert syndrome phenotype. *Am J Med Genet* 5:467–472

Niederhofer H (2007) Efficacy of risperidone treatment in Smith-Magenis syndrome (del 17 pll.2). *Psychiatr Danub* 19:189–192.

Perkins T, Rosenberg JM, Le Coz C, Alaimo JT, Trofa M, Mullegama SV, Antaya RJ, Jyonouchi S, Elsea SH, Utz PJ, Meffre E, Romberg N (2017) Smith-Magenis syndrome patients often display antibody deficiency but not other immune pathologies. *J Allergy Clin Immunol Pract* 5(5):1344–1350.

Poisson A, Nicolas1 A, Cochat P, Sanlaville D, Rigard C, de Leersnyder H, Franco P, Des Portes V, Edery P and Demily C (2015) Behavioral disturbance and treatment strategies in Smith-Magenis syndrome. *Orphanet J Rare Dis* 10:111.

Potocki L, Chen K-S, Lupski JR (1999) Subunit 3 of the COP9 signal transduction complex is conserved from plants to humans and maps within the Smith-Magenis syndrome critical region in 17p11.2. *Genomics* 57:180–182.

Potocki L, Glaze D, Tan DX, Park SS, Kashork CD, Shaffer LG, Reiter RJ, Lupski JR (2000) Circadian rhythm abnormalities of melatonin in Smith-Magenis syndrome. *J Med Genet* 37:428–433.

Potocki L, Lynch JK, Glaze DG, Walz K, et al. (2002) EEG abnormalities and epilepsy in Smith-Magenis syndrome. *Am J Hum Genet* 71(Suppl): 260A.

Potocki L, Shaw CJ, Stankiewicz P, Lupski JR (2003) Variability in clinical phenotype despite common chromosomal deletion in Smith-Magenis syndrome [del(17)(p11.2p11.2)]. *Genet Med* 5:430–434.

Salati R, Marini G, Degiuli A, Dalpra L (1996) Brown's syndrome associated with Smith-Magenis syndrome. *Strabismus* 4:139–143.

Schmidt LS, Nickerson ML, Warren MB, Glenn GM, Toro JR, Merino MJ, Turner ML, Choyke PL, Sharma N, Peterson J, Morrison P, Maher ER, Walther MM, Zbar B, Linehan WM (2005) Germline BHD-mutation spectrum and phenotype analysis of a large cohort of families with Birt-Hogg-Dubé syndrome. *Am J Hum Genet* 76:1023–1033.

Schou KV, Kirchhoff M, Nygaard U, Jorgensen C, Sundberg K (2009) Increased nuchal translucency with normal karyotype: a follow-up study of 100 cases supplemented with CGH and MLPA analyses. *Ultrasound Obstet Gynecol* 34(6):618–622.

Schwichtenberg AJ and Malow BA (2015) Melatonin treatment in children with developmental disabilities. *Sleep Med Clin* 10(2):181–187.

Seranski P, Hoff C, Radelof U, Hennig S, Reinhardt R, Schwartz CE, Heiss NS, Poustka A (2001) *RAI1* is a novel polyglutamine encoding gene that is deleted in Smith-Magenis syndrome patients. *Gene* 270:69–76.

Shelley BP, Robertson MM, Turk J (2007) An individual with Gilles de la Tourette syndrome and Smith-Magenis microdeletion syndrome: Is chromosome 17p11.2 a candidate region for Tourette syndrome putative susceptibility genes? *J Intellect Disabil Res* 51:620–624.

Slager RE, Newton TL, Vlangos CN, Finucane B, Elsea SH (2003) Mutations in *RAI1* associated with Smith-Magenis syndrome. *Nat Genet* 33:466–468.

Smith AC, Gropman AL, Bailey-Wilson JE, Goker-Alpan O, Elsea SH, Blancato J, Lupski JR, Potocki L (2002) Hypercholesterolemia in children with Smith-Magenis syndrome: Del (17) (p11.2p11.2). *Genet Med* 4:118–125.

Smith ACM, Duncan WC (2005) Smith-Magenis syndrome: A developmental disorder with circadian dysfunction. In: *Genetics of Developmental Disabilities*, Butler MG, Meaney FJ, eds, Boca Raton, FA: Taylor and Francis Group, ch. 13.

Smith ACM, Morse RS, Introne W and Duncan WC (2019) Twenty-four-hour motor activity and body temperature patterns suggest altered central circadian timekeeping in Smith-Magenis syndrome, a neurodevelopmental disorder. *Am J Med Genet A* 179(2):224–236.

Smith ACM, McGavran L, Waldstein G (1982) Deletion of the 17 short arm in two patients with facial clefts. *Am J Hum Genet* 34(Suppl):A410.

Smith ACM, McGavran L, Robinson J, Waldstein G, Macfarlane J, Zonona J, Reiss J, Lahr M, Allen L, Magenis E (1986) Interstitial deletion of (17)(p11.2p11.2) in nine patients. *Am J Med Genet* 24:393–414.

Smith ACM, Dykens E, Greenberg F (1998a) Behavioral phenotype of Smith-Magenis syndrome (del 17p11.2). *Am J Med Genet* 81:179–185.

Smith ACM, Dykens E, Greenberg F (1998b) Sleep disturbance in Smith-Magenis syndrome (del 17p11.2). *Am J Med Genet* 81:186–191.

Smith ACM, Leonard AK, Gropman A, Krasnewich D (2004): Growth Assessment of Smith-Magenis Syndrome (SMS). Presented at the meeting of the American Society of Human Genetics, October 26-30, 2004, in Toronto, Canada. *Amer J Hum Genetics Suppl* A700.

Smith ACM, Pletcher BA, Spilka J, Blancato J, Meck J (2006) First report of two siblings with SMS due to maternal mosaicism. Presented at the meeting of the American Society of Human Genetics October 9-13, 2006, in New Orleans, LA. *Am J Hum Genet* 79 (Suppl) 167A.

Smith ACM, Boyd KE, Elsea SH, et al. (2001) Smith-Magenis Syndrome. GeneReviews®. Available from: https://www.ncbi.nlm.nih.gov/books/NBK1310/

Smith ACM, Fleming LR, Piskorski AM, Amin A, Phorphutkul C, delaMonte S, Stopa E, Introne W, Vilboux T, Duncan F, Pellegrino J, Braddock B, L. A. Middelton LA, Vocke C, Linehan WM (2014) Deletion of 17p11.2 encompasses *FLCN* with increased risk of Birt-Hogg-Dubé in Smith Magenis Syndrome: Recommendation for Cancer Screening. *Presented (poster) at Amer. Soc. Human Genetics*, Oct. 18-22, 2014, San Diego, CA.

Solomon B, McCullagh L, Krasnewich D, Smith ACM (2002) Oral motor, speech and voice functions in Smith-Magenis syndrome children: A research update. *Am J Hum Genet* 71:271.

Song J, da Costa KA, Fischer LM, Kohlmeier M, Kwock L, Wang S, Zeisel SH (2005) Polymorphism of the PEMT gene and susceptibility to nonalcoholic fatty liver disease (NAFLD). *FASEB J* 19:1266–1271.

Spilsbury J, Mohanty K (2003) The orthopaedic manifestations of Smith-Magenis syndrome. *J Pediatr Orthop B* 12:22–26.

Stratton RF, Dobyns WB, Greenberg F, et al. (1986) Interstitial deletion of (17) (p11.2p11.2): Report of six additional patients with a new chromosome deletion syndrome. *Am J Med Genet* 24:421–432.

Struthers JL, Carson N, McGill M, Khalifa MM (2002) Molecular screening for Smith-Magenis syndrome among patients with mental retardation of unknown cause. *J Med Genet* 39:E39.

Sweeney E, Kerr B (1999) Smith-Magenis syndrome and tetralogy of Fallot. *J Med Genet* 36:501–502.

Thomas DG, Jacques SM, Flore LA, Feldman B, Evans MI, Qureshi F (2000) Prenatal diagnosis of Smith-Magenis syndrome (del 17p11.2). *Fetal Diagn Ther* 15:335–337.

Tomona N, Smith AC, Guadagnini JP, Hart TC (2006) Craniofacial and dental phenotype of Smith-Magenis syndrome. *Am J Med Genet* 140A:2556–2561.

Trask BJ, Mefford H, van den Engh G, Massa HF, Juyal RC, Potocki L, Finucane B, Abuelo DN, Witt DR, Magenis E, Baldini A, Greenberg F, Lupski JR, Patel PI (1996) Quantification of flow cytometry of chromosome 17 deletions in Smith-Magenis syndrome patients. *Hum Genet* 98:710–718.

Truong HT, Solaymani-Kohal S, Baker KR, Girirajan S, Williams SR, Vlangos CN, Smith AC, Bunyan DJ, Roffey PE, Blanchard CL, Elsea SH (2008) Diagnosing Smith-Magenis syndrome and duplication 17p11.2 syndrome by RAI1 gene copy number variation using quantitative real-time PCR. *Genet Test* 12: 67–73.

Udwin O, Webber C, Horn I (2001) Abilities and attainment in Smith-Magenis syndrome. *Dev Med Child Neurol* 43: 823–828.

Vilboux T, Ciccone C, Blancato JK, Cox GF, Deshpande C, Introne WJ, Gahl WA, Smith ACM, Huizing M (2011) Molecular analysis of the retinoic acid induced 1 gene (RAI1) in patients with suspected Smith-Magenis syndrome without the 17p11.2 deletion. *PLoS One* 6(8):e22861.

Vlangos CN, Wilson M, Blancato J, Smith AC, Elsea SH. 2005. Diagnostic FISH probes for del(17)(p11.2p11.2) associated with Smith-Magenis syndrome should contain the RAI1 gene. *Am J Med Genet A* 132A(3):278–282.

Walz K, Caratini-Rivera S, Bi W, et al. (2003) Modeling del(17) (p11.2p11.2) and dup(17) (p11.2p11.2) contiguous gene syndromes by chromosome engineering in mice: Phenotypic consequences of gene dosage imbalance. *Mol Cell Biol* 23:3646–3655.

Wang A, Lian Y, Fridell RA, Probst FJ, Wilcox ER, Touchman JW, Morton CC, Morell RJ, Noben-Trauth K, Camper SA, Friedman TB (1998) Association of unconventional myosin *MYO15* mutations with human nonsyndromic deafness DFNB3. *Science* 280:1447–1451.

Waziri R, Baruah S (1999) A hyperglycinergic rat model for the pathogenesis of schizophrenia: Preliminary findings. *Schizophrenia Research* 37:205–215.

Wilde L, Oliver C (2017). Brief report: contrasting profiles of everyday executive functioning in Smith-Magenis syndrome and Down syndrome. *J Autism Dev Disord* 47(8):2602–2609.

Williams SR, Zies D, Mullegama SV, Grotewiel MS, Elsea SH (2012) Smith-Magenis syndrome results in disruption of CLOCK gene transcription and reveals an integral role for RAI1 in the maintenance of circadian rhythmicity. *Am J Hum Genet* 90(6):941–949.

Wolters PL, Gropman AL, Martin SC, Smith MR, Hildenbrand HL, Brewer CC, Smith AC (2009) Neurodevelopment of children under 3 years of age with Smith-Magenis syndrome. *Pediatr Neurol* 41:250–258.

Yamamoto T, Ueda H, Kawataki M, Yamanaka M, Asou T, Kondoh Y, Harada N, Matsumoto N, Kurosawa K (2006) A large interstitial deletion of 17p13.1p11.2 involving the Smith-Magenis chromosome region in a girl with multiple congenital anomalies. *Am J Med Genet* 140A:88–91.

Yan J, Keener VW, Bi W, Walz K, Bradley A, Justice MJ, Lupski JR (2004) Reduced penetrance of craniofacial anomalies as a function of deletion size and genetic background in a chromosome engineered partial mouse model for Smith-Magenis syndrome. *Hum Mol Genet* 13:2613–2624.

Yang SP, Bidichandani SI, Figuera LE, Juyal RC, Saxon PJ, Daldinia A, Patel PI (1997) Molecular analysis of deletion (17)(p11.2p11.2) in a family segregating a 17p paracentric inversion: Implications for carriers of paracentric inversions. *Am J Hum Genet* 60:1184–1193.

Yeetong P, Vilboux T1, Ciccone C, Boulier K, Schnur RE, Gahl WA, Huizing M, Laje G, Smith ACM (2016) Delayed diagnosis in a house of correction: Smith-Magenis syndrome due to a de novo nonsense RAI1 variant. *Am J Med Genet A* 170(9):2383–8.

Yi Y, Eichler EE, Schwartz S, Nicholls RD (2000) Structure of chromosomal duplicons and their role in mediating human genomic disorders. *Genome Res* 10:597–610.

Zori RT, Lupski JR, Heju Z, Greenberg F, Killian JM, Gray BA, Driscoll DJ, Patel PI, Zackowski JL (1993) Clinical, cytogenetic, and molecular evidence for an infant with Smith-Magenis syndrome born from a mother having a mosaic 17p11.2p12 deletion. *Am J Med Genet* 47:504–511.

Zheng Li, MD, Jianxiong Shen, MD, Jinqian Liang, MD, and Lin Sheng, MD. Congenital scoliosis in Smith–Magenis syndrome: A case report and review of the literature. *Medicine* 94(17):e705.

55

SOTOS SYNDROME

TREVOR R.P. COLE AND ALISON C. FOSTER

Clinical Genetics Unit, Birmingham Women's Hospital, Edgbaston, Birmingham, UK

INTRODUCTION

Sotos syndrome was first recognized as a distinct clinical syndrome in 1964 (Sotos et al. 1964) although probable cases do exist in the earlier literature. Since the original report, there have been several comprehensive reviews of the clinical features and literature (Cole and Hughes 1994; Jaeken et al. 1972; Tatton-Brown et al. 2007; Wit et al. 1985). In a survey of 40 cases of Sotos syndrome, the average age of diagnosis was 26.6 months (Cole and Hughes 1994). However, most of these children were born in the 1970s and 1980s, and, with increasing professional knowledge and a readily accessible diagnostic test, it is likely that the age at diagnosis is decreasing and the frequency of diagnosis is increasing (personal experience).

Incidence

The incidence remains unknown, but appears slightly less common than one other "common" overgrowth syndrome, Beckwith–Wiedemann syndrome, for which a birth prevalence of 1 in 10,340 has been reported (Mussa et al. 2013). Rahman and colleagues (personal communication) estimated from their analysis of the causative gene, *NSD1*, in a large UK cohort of children with overgrowth that the birth incidence of Sotos syndrome may approach 1 in 14,000. Mutations in *NSD1* are the most common cause of overgrowth with intellectual disability, accounting for approximately one-third of cases in this series (Tatton-Brown et al. 2017).

Diagnostic Criteria

Before the identification of mutations in the *NSD1* gene in 2002 by Kurotaki et al. (2002) there were no tests available to confirm the diagnosis of Sotos syndrome; thus, clinical diagnostic criteria were developed by Cole and Hughes (1994). However, these relied on "soft" features such as the facial gestalt or non-specific growth and developmental abnormalities which are common in the population. Therefore, "loose" interpretation led to misdiagnoses by some clinicians, whereas others disputed the existence of Sotos syndrome as a distinct entity.

The four core features reported in the original description of Sotos syndrome are the following: rapid early growth (pre- and postnatal), advanced bone age, developmental delay, and characteristic facial appearance. The first three features are relative to population normal values, and where the threshold should be drawn is open to debate. Bone age may also be subject to observer error. Tatton-Brown and Rahman (2007) reviewed their data in *NSD1* positive individuals and found that the bone age was above the 90th centile in approximately 75% of cases and therefore, suggested advanced bone age be removed from the core diagnostic markers. It is also important to recognize that tall stature and increased head circumference will be influenced by familial patterns, for example, parental heights that lie toward the upper limit of the normal population distribution or families that exhibit constitutional large stature or rapid maturation. The typical growth pattern seen in Sotos syndrome is discussed in Growth and Feeding.

Only the last of the four diagnostic features, the characteristic facial appearance, is specific to Sotos syndrome. This remains

---

Cassidy and Allanson's Management of Genetic Syndromes, Fourth Edition.
Edited by John C. Carey, Agatino Battaglia, David Viskochil, and Suzanne B. Cassidy.
© 2021 John Wiley & Sons, Inc. Published 2021 by John Wiley & Sons, Inc.

the most subjective of the criteria, despite attempts to improve the objectivity (Allanson and Cole 1996). Cohort studies of cases with confirmed molecular diagnoses have shown, however, this to be a very reliable discriminator (greater than 90% sensitivity and specificity) when used by clinicians experienced in the management of people with overgrowth disorders (Douglas et al. 2003; Saugier-Veber et al. 2007).

The facial morphology alters during childhood and adolescence (Figures 55.1 and 55.2), following changes similar to those seen in the general population, with lengthening of the face superimposed on the background of abnormal facial dimensions intrinsic to this disorder. This results in an evolving yet characteristic facial gestalt at different ages (Allanson and Cole 1996).

The features noted in the newborn period are macrocephaly, a high-bossed forehead, and apparent wide spacing of the eyes resulting from temporal narrowing. Initially, there is a small pointed chin, but within the first 1–2 years, the face lengthens and the jaw becomes longer. The forehead remains broad and the chin remains narrow, which gives the facial outline a shape similar to an inverted pear. During this time, dolichocephaly and frontal bossing become more obvious, the latter because of a striking delay in the growth of hair in the frontoparietal regions (Figure 55.1A).

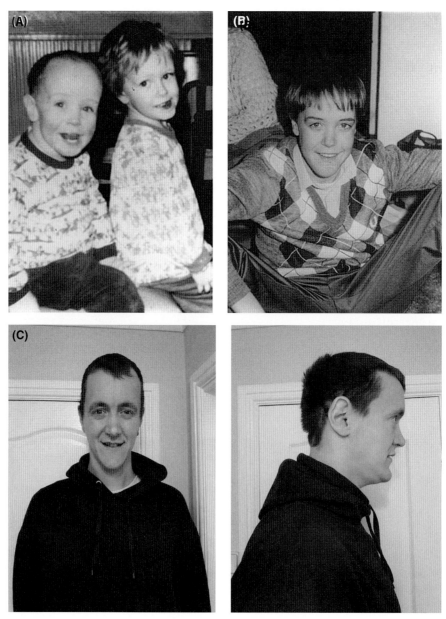

FIGURE 55.1 (A) An 18-month-old boy with Sotos syndrome and his older brother. (B) The same boy at 14 years of age. (C) The same individual in adulthood.

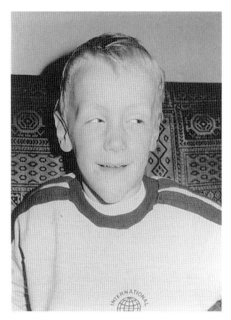 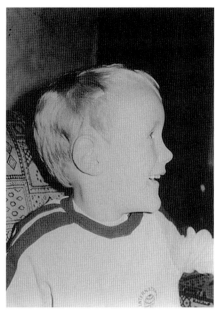

FIGURE 55.2 A 4-year-old boy with Sotos syndrome.

In mid childhood, a downslant to the palpebral fissures, wear and discoloration of the teeth, and a tendency to a rosy coloration of the cheeks, chin, and nasal tip with associated rapid changes in facial coloring, all become more obvious. In adulthood, the face becomes longer and relatively thinner (particularly, the width of the upper face), but the skull still shows marked dolichocephaly and macrocephaly (Figure 55.1C). Despite these changes and a "normalization" of the overall facial appearance, the facial gestalt can be recognized by experienced observers and distinguished from other overgrowth or familial features, as documented in Cole and Hughes (1994).

When the above diagnostic criteria are strictly applied during childhood, all four are present in almost 75% of individuals with Sotos syndrome who have a mutation in *NSD1*, but in a lower percentage (less than 20%) of other specific and nonspecific overgrowth patterns (Cole and Hughes 1994; Wit et al. 1985). With the advent of molecular testing, Tatton-Brown et al. (Tatton-Brown and Rahman 2007) suggested that the criteria should be modified to include only three features: overgrowth resulting in tall stature and macrocephaly from birth, learning disability, and characteristic facial appearance. These features were present in 90% of people proven to have Sotos syndrome on the basis of an identified *NSD1* mutation. This relaxation of diagnostic criteria should increase the sensitivity with only a small reduction in specificity.

In 1997, Schaefer et al. examined the cranial magnetic resonance imaging findings in 40 individuals who had a characteristic facial gestalt and at least two of the other clinical criteria suggestive of Sotos syndrome. The neuroradiological findings included ventricular abnormalities with a prominence of the trigone (90%), prominent occipital horns (75%), and ventriculomegaly (63%). Midline defects were also frequent, and hypoplasia of the corpus callosum was found in almost all cases. Horikoshi et al. (Horikoshi et al. 2006) confirmed these features, but also suggested a specific immaturity in frontal brain function as assessed by single-photon emission computed tomography and magnetic resonance spectroscopy.

It has been suggested that brain imaging could be utilized to distinguish Sotos syndrome from other causes of intellectual disability with macrocephaly (Schaefer et al. 1997). However, caution is advisable on this point because there is overlap of these findings in other overgrowth disorders (personal experience). The neuroradiological findings could still be a useful adjunct to the other diagnostic criteria.

Several other medical markers may also aid in arriving at the correct diagnosis. The frequency of these are discussed later in the text and in Table 55.1, and are also summarized by Tatton-Brown et al. (2004). More minor or transient features documented in early childhood include neonatal feeding difficulties, jaundice (both greater than 60%), poor nail growth (rate and quality), and tendency to increased sweating (Cole and Hughes 1994; Tatton-Brown et al. 2005b). The skin has a soft doughy consistency, the perception of which may be increased by sweating. An unexpectedly large appetite and thirst (even allowing for size), increased sensitivity to certain sensations such as the texture of food or combing hair, but high pain threshold (the latter particularly over the trunk), are all later childhood-onset features that have been frequently reported by parents (personal experience).

TABLE 55.1 Frequency of clinical findings in Sotos syndrome

>90%	50–90%	10–49%
Height >97th centile	Bone age >90th centile	Maternal pre-eclampsia
Head circumference >97th centile	Neonatal feeding difficulties	Recurrent otitis media
Learning difficulties	Neonatal jaundice	Hearing deficit
Characteristic facies	Neonatal hypotonia	Ocular anomalies
	Early dental development	Constipation
	Worn/discolored primary dentition	Congenital cardiac disease
	Increased appetite and thirst	Structural renal anomalies
	Excessive sweating	Epilepsy
	Abnormal MRI	Joint laxity
		Pes planus
		Scoliosis

Source: Modified from Tatton-Brown et al. (2005b) and author's personal experience.

Etiology, Pathogenesis, and Genetics

Until 2003, the etiology of Sotos syndrome remained unknown, although several chromosomal loci had been suggested on the basis of karyotype anomalies identified in individuals with a phenotype suggestive of Sotos syndrome (Table 55.2). However, there was little consistency in the loci, and the accuracy of the diagnosis of Sotos syndrome was questioned by these authors in several cases. Antenatal triple testing may give a result suggestive of an increased risk of aneuploidy (Thomas and Lemire 2008; personal experience), though antenatal karyotyping is typically normal. The frequency and etiology of these triple test findings are currently unknown.

In 2002, Imaizumi et al. reported a Japanese child with an apparently de novo balanced translocation [46XX,t(5;8)(q35:q24.1)], and a phenotype compatible with a diagnosis of Sotos syndrome (Imaizumi et al. 2002). This was of particular interest because of a previous report of a child with Sotos syndrome and a different translocation also involving 5q35 (Kurotaki et al. 2002; Maroun et al. 1994). Kurotaki et al. (2002) constructed a contig map across the 5q35 breakpoint of the individual reported by Imaizumi and identified a gene within the region, *NSD1*, as a candidate gene for Sotos syndrome. Subsequently, a series of 30 Japanese cases of Sotos syndrome was investigated of which 20 had large deletions (19 greater than 2.2 Mb) and appeared to have common breakpoints. Further analysis of the series, totaling 38 cases in all, identified four additional individuals who had de novo truncating point mutations within *NSD1*.

Mutation analysis of a cohort of 75 British individuals with varying overgrowth features that included phenotype discrimination by three clinical geneticists blinded to *NSD1* status was conducted (Douglas et al. 2003). The phenotype was scored into one of the four following categories: classical Sotos syndrome (group 1); possible Sotos syndrome, that is, those individuals who had some, but not all the typical features (group 2); Weaver syndrome (group 3); and neither Sotos nor Weaver syndrome (group 4). Weaver syndrome is an overgrowth syndrome showing considerable overlap with Sotos syndrome, which is distinguished by subtle differences in the facial gestalt and a different molecular etiology (see Figures 55.3A and B, and Differential Diagnosis). The original association of *NSD1* to Weaver syndrome likely represents a diagnostic misclassification due to the similarity of the phenotype (Douglas et al. 2003). The total point mutation and deletion detection rate in the childhood classical cases of Sotos syndrome was 76%, but only three cases with informative results had large deletions (Douglas et al. 2003).

Further analysis of 266 cases, comprising an enlarged UK cohort (179 cases) and further 87 cases from collaborative studies in France, Italy, Germany, and USA, was reported by Tatton-Brown et al. (2005b). Fifty-four of these 87 cases had previously been reported (Cecconi et al. 2005; Rio et al. 2003; Türkmen et al. 2003; Waggoner et al. 2005). Findings across the individual series, and in the collaborative studies as a whole, were generally consistent with overall *NSD1* mutation detection rates of greater than 90%. Of these, over 80% were intragenic mutations and only 10% microdeletions (Tatton-Brown et al. 2005b). Saugier-Veber et al.

TABLE 55.2 Karyotypic abnormalities in reported individuals with Sotos phenotype.

47,XX + invdup(15)(pter-q12or q13;q12or13-pter)	(Koyama et al. 1985)
t(2;12)(q33.3;q15)	(Tamaki et al. 1989)
t(3;6)(p21;p21)	(Schrander-Stumpel et al. 1989)
t(5;15)(q35;q22)	(Maroun et al. 1994)
Del 5q35.3 Mosaic	(Stratton et al. 1994)[a]
46,t(2;4)(2qter-2p15::4p14-4pter; 4qter-4p14::2p16.2-2pter)	(Cole and Hughes 1994)
Mosaic trisomy 20p12.1-p11.2	(Faivre et al. 2000)

[a] Not originally reported as Sotos syndrome – diagnosis made on retrospective review.

INTRODUCTION 899

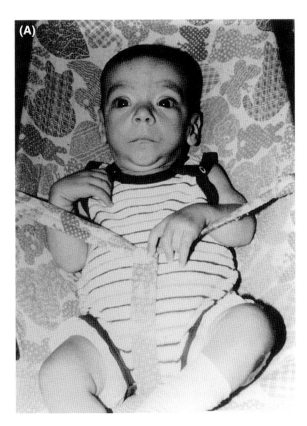

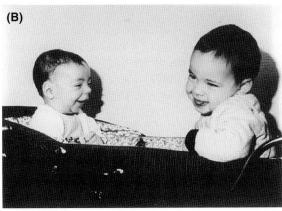

FIGURE 55.3 (A) A 3-month-old boy with Weaver syndrome. (B) Same child at 7 months (left).

(2007) reported on a further 116 French cases with a 90% detection rate comprising point mutations in 80%, large deletions in 14%, and intragenic rearrangements in 6%.

The initial high detection rate of large deletions in the Japanese population (50%) has been borne out by further studies (Visser and Matsumoto 2003). The very different molecular results in Japanese and non-Japanese populations is likely explained by genomic variation (in particular, an inversion polymorphism) in proximity to three low copy number repeats flanking *NSD1* in these different ethnic groups. Two of the low copy repeats are in the same orientation and suggest that non-allelic homologous recombination underlies the 1.9 Mb common deletion (Kurotaki et al. 2005; Visser et al. 2005). In some non-Japanese cases, breakpoints outside the common low copy repeat elements have been identified and suggest an alternative mechanism (Tatton-Brown et al. 2005a).

NSD1 is an epigenetic regulator gene. It encodes a histone methyltransferase, NSD1, that primarily methylates histone H36 lysine 36 and thus regulates chromatin structure and gene expression (Rayasam et al. 2003). Pathogenic variants in the functional domains of *NSD1* affect DNA methylation across the genome, particularly of genes that are important in cellular morphogenesis and neuronal differentiation (Choufani et al. 2015). The *NSD1* DNA methylation signature is highly specific to individuals with Sotos syndrome and distinct from that seen in other epigenetic regulator overgrowth disorders such as Weaver syndrome. (Choufani et al. 2015).

Diagnostic Testing and Genetic Counseling Pathogenic variants in *NSD1* are identified in approximately 90% of individuals with Sotos syndrome and molecular genetic testing is recommended in all cases where the diagnosis is suspected.

The diagnostic strategy taken will depend on the degree of confidence in the clinical diagnosis of Sotos syndrome. If the diagnosis is considered to be highly likely, single gene testing of *NSD1* is likely to be the fastest and most cost-effective way of identifying the pathogenic variant. In individuals of non-Japanese ethnicity, sequence analysis of *NSD1* should be performed first. In individuals of Japanese ethnicity, deletion/duplications analysis may be performed first, to reflect the differing rates of intragenic mutations and deletions causing Sotos syndrome in these populations.

However, if the clinical diagnosis is uncertain, either because the clinician feels the presenting features are not entirely typical of Sotos syndrome, or the clinician is not experienced in clinical assessment of Sotos syndrome, a different approach may be taken. There are several other overgrowth syndromes, such as Weaver syndrome, Malan syndrome and PTEN related disorders, where the clinical presentation is "Sotos-like" and the molecular etiology is known. Sequential testing of single genes is a time consuming and expensive process. With the advent of next generation sequencing and its increasing role in diagnostic testing, testing a panel of overgrowth genes, or a "virtual panel" from a clinical exome or whole exome, is likely to be first line in many cases. In the future these may be superceded by whole genome sequencing, although the bioinformatic burden of whole genome analysis may be a barrier to this development in the short term.

Diagnostic testing for *NSD1* variants is most often performed on DNA extracted from peripheral blood leucocytes; however, if venepuncture is impractical DNA for analysis may be extracted from a saliva sample or buccal sample.

Somatic mosaicism in an affected individual has not been described and therefore obtaining a tissue sample for analysis is not necessary.

Detected variants should be classified according to internationally recognised standards (for example the consensus recommendation issued by the American College of Medical Genetics and Genomics in 2015). If no variant or a variant of uncertain significance is identified in *NSD1* but the clinician is confident of the clinical diagnosis of Sotos syndrome, bone age analysis and magnetic resonance imaging of the brain may provide additional evidence to support the diagnosis. Analysis of parental samples is recommended and establishing de novo status adds additional evidence of pathogenicity in the interpretation of a variant of uncertain significance.

If an alternative diagnosis is being considered, further investigations should be performed depending on the likely differentials. Comparative genomic hybridisation microarray (CGH array) to detect microdeletions and microduplications is a first line investigation in cases of learning disability and should be considered in all cases of overgrowth with intellectual disability. Other diagnostic tests for alternative diagnoses such as Beckwith–Wiedemann syndrome (methylation analysis) (see Chapter 9) or Fragile X syndrome (trinucleotide repeat expansion analysis) (see Chapter 28) may be considered if the phenotype shows overlap with these disorders.

The majority of people with Sotos syndrome are the only affected member of the family, but a small number of families (approximately, 5–10%) appear to show autosomal dominant inheritance (Tatton-Brown and Rahman 2007). This figure may suffer from selection bias as familial cases had been sought for further investigation. The authors are aware of at least 10 familial cases from his personal experience of almost 400 cases. It is relevant to note that the facial gestalt in the affected individuals within such families is less striking and the average developmental delay less severe than in most non-familial affected individuals. It is important also that in *all* the familial cases the affected parent was deemed to be affected based on clinical assessment before the *NSD1* result being known. There is one potential exception to this in the literature, where the phenotypic presentation was especially variable and mild and the certainty of the clinical status is challenging without the molecular results (Donnelly et al. 2011).

In view of the molecular mechanism, it remains unclear why so many cases appear to be de novo and so few familial. Adult health is generally good and developmental delay most commonly mild to moderate, which, although influencing fertility, would not appear to be the full explanation. A significant proportion of adults with Sotos syndrome describe difficulty forming peer relationships and, in some cases, a reluctance to take on the responsibilities of a long-term relationship or child rearing (authors experience). A further biological factor could be an increased rate of infertility related to menstrual irregularities (primary and secondary) reported by a number of adult females. However, about 90% of the inherited cases are maternally transmitted, and therefore, reduced male fertility would also appear to be present.

As yet, there are inadequate data to make definitive statements on genotype–phenotype correlations. However, it appears that those individuals with large deletions are less likely to be tall, especially during childhood, but are likely to have greater degrees of developmental delay. The full spectrum of associated clinical features has been found in association with both deletions and point mutations. Attempts to draw further conclusions are currently confounded by the small numbers with both mechanisms in reported series (Douglas et al. 2003; Nagai et al. 2003; Saugier-Veber et al. 2007; Tatton-Brown et al. 2005b). Interestingly, monozygotic twins with Sotos syndrome have been reported with markedly differing clinical, cognitive and behavioral phenotypes (Han et al. 2017).

There are several historical literature reports of affected siblings with Sotos syndrome who have unaffected parents, but few are typical in their appearance and most would not meet strict diagnostic criteria. Other reports provide inadequate data to confirm the diagnosis in the extended family. In particular, tall adult stature with unspecified difficulties at school should not be considered adequate to make a diagnosis of Sotos syndrome in an adult, in light of the frequency of these findings in the general population. The authors are not aware of any follow up studies in these reports confirming an *NSD1* mutation and therefore the diagnoses remain unproven.

In one study of Cole and Hughes (1994) careful examination of parents and siblings of 40 probands with review of childhood photographs and growth data where available, showed no clear evidence of Sotos syndrome in any of the 80 parents, 47 siblings, and 23 half-siblings. Therefore, in the absence of a definite diagnosis in a first-degree relative, the risk of recurrence is small. The authors and their research collaborators are not aware of a sibling recurrence in over 500 children, except in those families with a proven *NSD1* mutation being present in one parent. There is a single case report of somatic-gonadal mosaicism resulting in the birth of a child with Sotos syndrome (Kamien et al. 2016). The father had no clinical features of Sotos syndrome but Sanger sequencing of peripheral blood DNA showed possible mosaicism for the pathogenic variant in *NSD1* identified in his child. Next generation sequencing of peripheral blood DNA and buccal DNA confirmed mosaicism of 6.7% and 8.2% respectively.

Parental ages were originally reported to be normal (Wit et al. 1985), but in a slightly larger series, a significantly elevated paternal age was noted (Cole and Hughes 1994). These latter observations support the finding of a high frequency of de novo mutations, and thus, the risk of affected siblings in most families remains small. Further studies looking at the parental origin of de novo deletions and point mutations are ongoing. There is a bias toward paternal origin of deletions (Miyake et al. 2003). Tatton-Brown and Rahman (2007) attribute this to the telomeric position of *NSD1*, which is associated with a higher recombination rate in males.

Careful clinical examination of parents and siblings for clinical features of Sotos syndrome is an important part of genetic counseling. Testing for the familial mutation in first degree relatives remains the most reliable assessment and should be considered in all cases following explanation to the parents. Parents should be made aware that if a parent has Sotos syndrome the chance of having another child with Sotos syndrome is 1 in 2 (50%) due to autosomal dominant inheritance. If the pathogenic variant in the child is de novo, the chance of having another affected child is extremely low.

Differential Diagnosis

The primary feature that is likely to raise suspicion of a diagnosis of Sotos syndrome is generalized overgrowth, i.e. height and or head circumference greater than two standard deviations above the mean (approximately 98th centile) for age and sex. Other conditions presenting with tall stature and/or macrocephaly can often be distinguished from Sotos syndrome by their distinct clinical phenotypes (see Table 55.3).

Generalized Overgrowth with Intellectual Disability. The conditions most likely to cause diagnostic confusion are the "OGID" (overgrowth with intellectual disability) disorders. This group includes Weaver syndrome, Malan syndrome, Simpson–Golabi–Behmel syndrome and Tatton-Brown Rahman syndrome. Many other genes associated with OGID phenotypes have been identified including *CHD8*, *HIST1H1E*, *EED*, *BRWD3*, *PPP2R5D*, *RNF125*, *SUZ12*, *SETD2*, and *APC2*. With the rapid rate of gene discovery by whole exome and whole genome sequencing it is likely that additional genes will be identified.

Consideration of the diagnosis of Sotos syndrome in individuals with constitutional (familial) tall stature probably only occurs if developmental delay is also present. The concurrence of these features will often be coincidental because of the frequency of developmental delay in the population. The absence of typical facial gestalt, lack of associated features, and negative molecular findings, however, aid in the identification of this group.

Generalized Overgrowth. Disorders associated with overgrowth such as Simpson–Golabi–Behmel syndrome and Beckwith–Wiedemann syndrome (see Chapter 9) often have reliable and distinct diagnostic criteria that facilitate discrimination. Many conditions are associated with prenatal overgrowth or macrosomia including infant of a diabetic mother, Perlman syndrome, Marshall–Smith syndrome, Costello syndrome (see Chapter 16), Seip–Beradinelli syndrome and Pallister–Killian syndrome (see Chapter 45). Again, discriminating features are often present and these conditions are unlikely to cause diagnostic confusion later in infancy and childhood with the evolution of more specific phenotypes. The mucopolysaccharidosis MPS III (Sanfilippo syndrome) may be associated with overgrowth in childhood before the decline of intellectual abilities and is an important diagnosis not to miss in view of the 1 in 4 recurrence risk.

Several chromosomal microdeletion and microduplication syndromes are associated with overgrowth (e.g. del 3q13.31, dup 4p16.3, del 22q13ter) and can be excluded with microarray testing.

Macrocephaly and Developmental Delay or Autism Spectrum Disorder (ASD). Males with Fragile X syndrome tend to have growth in the normal range and relative macrocephaly, but significant overgrowth and obesity have been reported. All have developmental delay and approximately a quarter are diagnosed with ASD. The facial gestalt is almost always dissimilar to that of Sotos syndrome however (see Chapter 28).

Approximately 10–20% of individuals with macrocephaly and ASD have germline mutations in *PTEN* (Butler et al. 2005) (see Chapter 48).

Tall Stature. Endocrine disorders such as congenital adrenal hyperplasia with precocious testosterone production, and pituitary and adrenal secretory tumors may present with overgrowth but lack the associated features of Sotos syndrome and can be identified by diagnostic testing. Certain connective tissue disorders, including Marfan syndrome (see Chapter 37) and Ehlers–Danlos type VI (kyphoscoliotic form) (see Chapter 25), present with increased length/tall stature and some similar skeletal features but are usually distinguishable by careful examination.

Males with Klinefelter syndrome (XXY) may be tall and many have mild developmental delay (see Chapter 35). Loss of function mutations in *FGFR3* and gain of function mutations in *NPR2* cause overgrowth with tall stature.

Regional Overgrowth. The presence of regional (segmental) overgrowth points towards a mosaic condition such as PIK3CA spectrum disorder or Proteus syndrome (see Chapter 47). Lateralized overgrowth (hemihypertrophy) should prompt consideration of Beckwith–Wiedemann syndrome (see Chapter 9).

MANIFESTATIONS AND MANAGEMENT

Growth and Feeding

Most babies with Sotos syndrome will be large for gestational age at delivery. It is important to recognize that the most significantly elevated measurement is likely to be birth

TABLE 55.3 Overgrowth disorders.

Disorder	Inheritance	Molecular findings	Clinical features	Facial features
Weaver syndrome	Autosomal dominant (frequent de novo mutations)	Mutation in *EZH2*	Tall stature Macrocephaly Variable learning disability Camptodactyly Soft/doughy skin Umbilical hernia	Round face in early childhood Broad forehead Hypertelorism Long prominent philtrum Large dysplastic ears Small distinct dimpled/grooved chin Retrognathia in early childhood
Malan syndrome	Autosomal dominant	Mutations leading to haplo-insufficiency of *NFIX*	Postnatal overgrowth Rarely prenatal overgrowth Decrease in height overgrowth with age Persistent macrocephaly Invariably learning disability Frequent autism and anxiety Hypotonia Brain anomalies Slender body build	Long triangular face Prominent forehead Long philtrum Everted vermilion lower lip Prominent chin
Tatton-Brown–Rahman syndrome	Autosomal dominant	Mutation in *DNMT3A*	Tall stature Macrocephaly Obesity Learning disability Scoliosis	Round face Heavy horizontal eyebrows Narrow palpebral fissures Prominent frontal incisors
Simpson–Golabi–Behmel syndrome	X-linked recessive	Mutation in *GPC3*	Pre and postnatal overgrowth Macrocephaly Variable learning disability Umbilical hernia Diastasis recti Organomegaly Cardiac anomalies Diaphragmatic hernia Skeletal anomalies including postaxial polydactyly Supernumerary nipples Cleft palate Macroglossia Embryonal tumours (especially Wilms tumour)	Hypertelorism Broad nasal bridge and tip Orofacial clefts Grooved tongue Ear creases Coarse facies
Beckwith–Wiedemann syndrome		Molecular abnormality altering expression/function of the 11p15.5 imprinted gene cluster	Pre and or postnatal overgrowth Lateralised overgrowth Anterior abdominal wall defects Neonatal hypoglycaemia Embryonal tumour predisposition	Macroglossia Ear creases/pits Facial haemangioma

Syndrome	Inheritance	Genetics	Features	Facial features
Perlman syndrome	Autosomal recessive	Homozygous or compound heterozygous mutations in *DIS3L2*	Prenatal overgrowth Developmental delay Hypotonia Nephromegaly Hyperinsulinism High risk Wilms tumour High neonatal mortality	Prominent forehead Deep-set eyes Broad depressed nasal bridge Inverted V-shape upper lip Everted lower lip Gum hypertrophy
Marshall–Smith syndrome	Autosomal dominant	Dominant-negative mutations in *NFIX*	Prenatal overgrowth Advanced bone age "Bullet"-shaped middle phalanges Failure to thrive High neonatal mortality	Thick eyebrows Prominent eyes Depressed nasal bridge Anteverted nares Choanal atresia Overfolded helix Micrognathia Fine facial features
PTEN hamartoma tumour syndrome	Autosomal dominant	Mutation in *PTEN*	Prenatal overgrowth Massive macrocephaly Hypotonia Learning disability Autism spectrum disorder Genital freckling Trichilemmomas, Papillomatous papules Acral keratosis Lipomas Hamartomatous intestinal polyposis High risk of thyroid, breast, endometrial and other cancers	Cobblestone appearance to tongue/oral mucosa Facial tricholemmoma

length (mean is +3.2 SD) followed by head circumference (mean is +1.8 SD) and weight (mean is +1.0 SD), and approximately 85% of newborn babies with Sotos syndrome will have a birth weight below the 97th centile (Cole and Hughes 1994). Saugier-Verber et al. (2007) found birth measurements less frequently elevated, but confirmed that birth length and head circumference were more likely to be elevated than birth weight.

Feeding difficulties are common in the neonatal period, and as many as 25% of term babies with Sotos syndrome require tube feeding (Cole and Hughes 1994). Failure to suck and coordinate swallowing because of anatomical features, hypotonia, and immaturity may be the most common reasons for such symptoms. Protracted feeding difficulties lasting several months have occurred and may cause failure-to-thrive in the first year of life, which may "mask" the diagnosis (personal experience).

Neonatal jaundice is common and likely to be due at least in part to feeding difficulties. In one series about 90% of newborns were jaundiced and over 60% of these required treatment with phototherapy or exchange transfusion (Cole and Hughes 1994). Neonatal jaundice was present in 70% of newborns in a later study (Tatton-Brown et al. 2005b). Neonatal hypoglycemia has also been reported and may occur secondary to feeding difficulties. However there are reports of hyperinsulinemic hypoglycemia, occurring in both deletion and intragenic mutation cases (Grand et al, 2019). In a few cases the hyperinsulinism has persisted for up to several years and required intravenous glucose infusion and diazoxide (Grand et al. 2019).

By 12 months of age, overgrowth is almost invariably present (seen in over 90%). Throughout childhood, height and head circumference are usually the most significantly elevated growth parameters, running parallel to but significantly above the 97th centile. Most children have relatively thin limbs and trunk, and measurement of arm span-to-height and upper-to-lower body segment ratios reveals that much of the excess growth is in the limbs rather than the trunk (Agwu et al. 1999; de Boer et al. 2005). Hands and feet are usually large, even when plotted against height age (Agwu et al. 1999; de Boer et al. 2005). Early height predictions are frequently inaccurate and excessive, and hormonal intervention in adolescence is rarely necessary (personal experience).

Final height in Sotos syndrome is influenced by parental heights. In a series of 18 British girls with Sotos syndrome who had achieved their final height, this measurement had a mean value of 172.9 cm (UK population mean is 163.5 cm); however, this was only 6.2 cm above their mean target height prediction, based on mid-parental heights (Agwu et al. 1999). The initial hypothesis was that the girls went through an early puberty (mean age of menarche 12.2 years; UK mean is 13 years), and therefore, had an early cessation of growth. However, the two girls with the most delayed menarche (15.4 and 14 years) both still had heights less than the 97th centile. Therefore, it seems likely that early puberty and fusion of the epiphyses is not the complete explanation. Further UK data (Foster et al. 2019) suggests that final height in women may be taller than previously thought. A study of 22 women found a range of height from 163.5cm (-0.1 SD) to 200cm (+6 SD), with a mean of 177.7cm (+2.3 SD) and median of 177cm (+1.7 SD) (compared to UK average now 164.4cm (NCD Risk Factor Collaboration 2016). However parental heights are not available for this group. Age of menarche in this group ranged from 7 to 16 years with a mean age of 13 years.

The situation in boys may be less predictable. Agwu et al. (Agwu et al. 1999) reported on nine adult males with a mean final height of 184.3 cm (UK population mean 175.5 cm). It is notable that their target height, based on mid-parental heights, had only been 173 cm, which is less than the UK population mean, and that the mean excess in final height, compared with the predicted target height, was +11.3 cm within a range of +5 to +20 cm (Agwu et al. 1999). The timing of puberty in boys is also more widely distributed and less closely associated with bone age. Continued anecdotal observations may support earlier observations that the final height in males may be more significantly elevated than females (Cole unpublished). However, a series of 13 males over 16 years of age (Foster et al. 2019) identified male height to be less increased than previous studies. Height ranged from 166.9 cm (-1.5 SD) to 200 cm (+3.2 SD), with a mean of 182.2 cm (+0.7 SD) and median 184 cm (+0.5 SD).

Weight is usually less significantly elevated than height in childhood; however, a few individuals may find weight gain is an issue in adulthood and early advice about diet and activity is important.

The findings in adults described above are concordant with reports by Tatton-Brown and Rahman (2004) who described mean adult heights between 75th and 91st centiles in a series of 13 individuals with Sotos syndrome. Gender was not described, however. The absence of parental heights and therefore target heights in the more recent unpublished data means further data is required to clarify these discrepancies.

Evaluation

- Early symptoms of poor suck and difficulty with attachment to the breast are probably secondary to poor tone and should be evaluated in a standard manner.
- Standard growth measurements throughout early childhood, as for any child, are appropriate.
- Longitudinal monitoring of growth, bone age, and puberty from about 7 or 8 years of age may be beneficial because it can provide some clarification of the likely growth outcome and some reassurance for families.

Management

- Treatment of early feeding problems, neonatal jaundice and neonatal hypoglycemia are standard. There is no feasible or justifiable endocrine intervention for tall stature during the first 10 years of life.
- Endocrine intervention to initiate early puberty has been prescribed in two females and one male (personal experience). There was no clear evidence of significant change in final height. If cessation of linear growth is achieved too early, the detrimental effect of changing from one of the largest to one of the smallest in a peer group, and the possible increased disproportion of height to head circumference and hand and foot lengths, must be considered. Experience from a large, untreated cohort would suggest that intervention is rarely necessary (personal experience).

Development and Behavior

Almost all children with Sotos syndrome have some development delay and cognitive impairment, but the severity is very variable and may be inconsistent across different areas within a single individual (de Boer et al. 2006; Cole and Hughes 1994; Finegan et al. 1994; Saugier-Veber et al. 2007; Tatton-Brown and Rahman 2004). The high frequency could however be skewed by selection bias, as developmental delay remains one of the diagnostic criteria.

All individuals exhibit some delay in achieving early milestones, and this is most noticeable for motor skills (de Boer et al. 2006; Cole and Hughes 1994). The degree of motor delay is a poor predictor of long-term cognitive and educational achievement. Motor delay may partly be explained by the deleterious effect of hypotonia and large body size, present in almost 100% of affected individuals (Cole and Hughes 1994).

Although hypotonia improves over the first few years of life, many individuals have significant coordination difficulties throughout childhood, affecting fine and gross motor skills, the latter commonly to a greater degree (personal experience). The early excessive drooling and poor enunciation are also likely to be a result of low tone and poor coordination.

Language delays are also usually present, but display greater variability and may show some correlation with long-term educational outcome, particularly if later milestones such as use of sentences are considered. Speech delay is often exacerbated by hearing loss, but the latter is usually mild and secondary to otitis media. A small number of children have a proven sensorineural component. In 2005, Ball et al. (2005) described specific abnormal patterns of speech production and voice impairments including persistence of a stutter into older childhood years.

The age that toilet training is achieved shows a wide range, from normal to significantly delayed, and is exacerbated in some individuals by severe constipation.

Childhood assessment of intellectual function identifies a wide range of ability both between and within individuals with Sotos syndrome, with an IQ range between 21 and 103 and a mean of 74 (Finegan et al.; Rutter and Cole 1991). Several adults have continued into college education. Most children are in mainstream education although placement will be influenced not only by IQ but also by local educational resources and the child's behavioral patterns (Finegan et al. 1994; Rutter and Cole 1991). Intellectual abilities may be difficult to assess reliably because of both behavior patterns and variation in skills, such as relatively good reading accuracy, but more limited comprehension, which is commonly seen. Another area often significantly impaired is math skills (Finegan et al. 1994; Rutter and Cole 1991). Although a single typical cognitive pattern is not found, more abstract or conceptual subjects cause greater difficulty than more concrete or visual processes.

Behavior is one of the key areas influencing family and personal outcome. Early reports suggested behavioral difficulties are frequently present although no specific or diagnostic profile was apparent and the severity was variable (de Boer et al. 2006; Finegan et al. 1994; Rutter and Cole 1991; Sarimski 2003). A more recent series of 38 individuals with Sotos suggests there may be a more predictable behavioral phenotype emerging (Sheth et al. 2015). Clinically significant behaviors were compared to three matched contrast groups with well documented behavioral phenotypes (autism spectrum disorder (ASD), Prader–Willi syndrome, and Down syndrome). Individuals with Sotos showed an increased risk of self-injurious behavior, physical aggression, destruction of property, impulsivity and levels of activity compared to the Down syndrome group. Preference for routine and repetitive language were increased similarly to the Prader–Willi syndrome group, but communication impairments were less common (Sheth et al. 2015).

From about age 2 years an increased tendency to tantrums and aggressive behavior, often directed against family members, has been reported anecdotally but was not been found at significantly increased overall rates (Finegan et al. 1994). However, the tantrums, in particular, appear to persist for a longer period than in siblings and are of greater severity, the latter feature perhaps resulting from the child's cognitive impairment, frustration, and large size.

From early childhood, poor concentration and subsequent attention-deficit disorder are common, significantly affecting home and school life. The child's large size but limited intellectual ability and behavioral immaturity often lead to very unrealistic expectations from adults and poor peer group relationships. Difficulty with social interactions results in a tendency to associate with younger children and, in later life, in a significant degree of social isolation. This latter feature is of particular concern to parents and older children and occurs even in the higher-functioning individuals with Sotos syndrome.

Other behavioral difficulties reported in the literature, in general, and more specifically by Rutter and Cole (1991) and Finegan et al. (1994), include an overreliance on routines and a tendency to obsessions, for example, repetitive watching of a favorite DVD or pursuit of a specific pattern of play. These suggest the possibility of an overlap with autistic tendencies, although a clear diagnosis of autism is uncommon. A detailed assessment of 18 adults with Sotos syndrome for autistic features confirmed that, although none of the participants met strict diagnostic criteria for autism, most had high rates of repetitive behavior (Saddington, personal communication). Participants were successfully able to meet the processing demands of set tasks, but they were unable to successfully infer false beliefs, potentially resulting in very literal interpretation based on their own knowledge or experience and failure to recognize that different knowledge of the same situation would probably result in different actions. The work suggested that there is an underlying theory of mind deficit in Sotos syndrome. This theory is reviewed by Yirmiya and Shulman (1996).

Sheth et al. (2015) found that impulsivity and levels of activity in Sotos syndrome were similar to those in a matched group with ASD. A high proportion of the Sotos group (70.3%, 26 out of 37) met the clinical cut off score for ASD on a social communication questionnaire (Sheth et al. 2015). A further study of 68 individuals found that 83.33% met the clinical cut off score for ASD on the social responsiveness scale. The greatest difficulties were in the areas of restricted interests and repetitive behaviors compared to social communication impairment. ASD symptomatology was equally present in both males and females, but males were more likely to have received a diagnosis of autism spectrum disorder, suggesting that girls are more likely to be under-diagnosed. Symptoms were less severe in children under 5 and in adults over the age of 20, indicating that improvement occurs with the transition into adulthood (Lane et al. 2016).

Although a reliance on routines is common in many children with Sotos, there are other situations in which the children appear to show an impetuous nature with little thought for the consequences or for their own safety. This, taken with their more limited intellectual and motor skills, can result in significant personal dangers. Other commonly reported symptoms include poor sleep patterns and unusual degrees of anxiety and subsequent phobias (see Rutter and Cole [1991] and Finegan et al. [1994] for further discussion).

Behavioral data related to adults are much more limited. Although the behavior problems tend to improve, there is often some residual difficulty that might best be summarized as immaturity or naïveté (personal experience). A significant number of adults suffer from social isolation, and there is anecdotal evidence of an increased rate of psychiatric disorders in adults, particularly anxiety.

The degree of independence in adulthood largely depends on level of intellectual functioning, although behavioral issues and physical abilities also play a role. The spectrum of learning disability is very wide and ranges from individuals with no intellectual disability who are fully independent, to those with severe intellectual disability who are unable to self-care. A recent study of adults (Foster et al. 2019) indicates that the majority of those with mild learning disability were independently self-caring and in employment or training, although most preferred to live at home with their parents. Most adults with moderate learning disability required assistance with self-care and none were able to live away from carers. Vocational training and employment roles reported by participants included childcare, healthcare, retail, waitressing, painting and decorating, and air stewarding.

Evaluation

- All individuals with Sotos syndrome need a critical evaluation of their educational requirements so that an appropriate program can be structured (Finegan et al. 1994).
- A formal behavior assessment may be necessary, and this possibility should be discussed with parents or caregivers.
- Hearing should be assessed as part of routine care, and the medical team should have a low threshold for requesting further formal hearing evaluation in any child who fails routine assessments.

Management

- Drooling and articulation difficulties may benefit from combinations of physical therapy, occupational therapy, and speech therapy.
- Early intervention, including physical therapy and occupational therapy, should be considered after assessment of the child's tone and coordination.
- Education programs should be based on careful assessment of cognitive function.
- At school, a classroom assistant is often beneficial to maintain the child's attention and effort, particularly when attention-deficit disorder is present.
- Treatment of behavioral disorders has traditionally been very different in Europe and North America, with a reliance on behavioral therapy in Europe and much more frequent pharmacological intervention in the United States. In Sotos syndrome, there are some successes and failures with each method. In the United States, many of the children are treated with methylphenidate or other stimulants for attention-deficit disorder. There are frequent anecdotal reports from the parents of improvement but also numerous failures. No formal, controlled study has been done to assess the potential benefits of either approach in Sotos syndrome.

- Adult vocational training and employment is highly variable and reflects the range of cognitive abilities. The authors are aware of numerous adults in regular employment. However, the majority of adults probably work in sheltered or protected programs.
- A small proportion of adults live entirely independently, but most require varying degrees of support or supervision.
- Treatment of hearing loss is as in the general population.

Neurologic

Marked hypotonia is usually present from birth, and although this appears to improve during childhood, subtle evidence may remain even in adults. It is interesting to note that, despite this central hypotonia, reflexes are often brisk and occasional beats of clonus can be elicited when testing the peripheral reflexes. Some difficulty with coordination is invariable, with hypotonia exacerbating this feature (Cole and Hughes 1994). Hypotonia and incoordination appear to affect gross motor movements more than fine motor skills. Large size, hypotonia, and poor coordination make many physical activities difficult, and non-participation in sports may increase social isolation in childhood. One sport many children with Sotos syndrome have excelled in is swimming. Although children with Sotos syndrome appear to have appropriate strength in the muscle groups, their families frequently describe easy fatigability. This may reflect their immaturity, large size, and poor coordination. Special effort may be required to find appropriate-sized strollers for these large young children. These symptoms do improve with time, and there is no evidence of deterioration.

Febrile convulsions are common, present in almost 50% of affected individuals. Almost half the individuals who have had a febrile seizure will go on to have non-febrile seizures in later childhood and adulthood (Cole and Hughes 1994; Nicita et al. 2012). This figure is lower than the 41% reported by Tatton-Brown and Rahman (2004) for all non-febrile seizures. A recent study of over 30 adults confirms that the incidence of seizures decreases significantly in adulthood (Foster et al. 2019). This corroborates another study that found only 2 out of 21 participants had seizures in adulthood (Fickie et al. 2011).

In 1969, Appenzeller and Snyder reported an individual with Sotos syndrome who had apparent abnormalities of the autonomic nervous system (Appenzeller and Snyder 1969), but these findings have not been reported subsequently. However, there is potential corroborative evidence of autonomic dysfunction from clinical observations. Parents frequently report unexplained episodes of sweating, high facial coloring or flushing, and poor control of peripheral temperature. These features have been noted to appear both concurrently and independently. To date, this is purely a clinical observation, and no associated complications, such as syncope, have been recognized.

As noted in the Diagnostic Criteria section, findings identified by cranial MRI in 40 individuals who had a characteristic facial gestalt and at least two of the other clinical diagnostic criteria suggestive of Sotos syndrome included prominence of the trigone (90%), prominent occipital horns (75%), and ventriculomegaly (63%). Midline defects were also frequent, and hypoplasia of the corpus callosum was found in almost all individuals.

Intention tremor in adulthood has been reported in three individuals in the literature (McClelland et al. 2016; Foster et al. 2019).

Evaluation

- A careful neurological examination should be performed as a baseline at diagnosis and with routine physician visits.
- Electroencephalogram should be performed if there is evidence of convulsions or absence spells.
- MRI may help both as a diagnostic tool and to exclude evidence of progressive hydrocephalus, but this is not necessary as a routine investigation.
- The onset of new neurological symptoms or signs should be assessed as standard

Management

- Physical therapy and occupational therapy may be of benefit in the treatment of early hypotonia and later incoordination.
- Swimming should be encouraged.
- There is no evidence that seizures should be managed any differently in Sotos syndrome than in epilepsy from other causes.
- If hydrocephalus is found, treatment is standard if there is evidence of significant raised intracranial pressure. Normal pressure progressive enlargement of the ventricles is part of the natural history and alone would not typically require intervention.

Respiratory Tract and ENT

Infections appear to be very common in early childhood, particularly affecting the respiratory tract. In the newborn period, a number of individuals have been reported with pneumonia, which may be attributable to prematurity, the necessity for ventilation, aspiration, or low tone and poor coordination resulting in decreased pulmonary clearance of secretion with secondary infection. During early childhood, upper respiratory tract infections remain frequent.

Recurrent episodes of otitis media are particularly troublesome, documented in 72% of the individuals; these are commonly associated with conductive hearing loss (Cole and Hughes 1994). The most likely explanation for the frequency of infection is disruption of the normal anatomical and physiological mechanisms for drainage and clearance of secretions. To date, there has been no clear evidence of altered immunity.

Evaluation

- Children with recurrent upper respiratory tract infections, in particular otitis media, should be referred to an otolaryngologist for further assessment, including audiological investigations.
- In view of the potentially covert nature of middle ear infections, these should be considered in any individual with Sotos syndrome who is febrile. This is particularly relevant in view of the tendency to febrile convulsions.

Management

- Many children require surgical interventions including tonsillectomy, adenoidectomy, and insertion of ventilation tubes.
- Amplification may be indicated in those with permanent hearing loss.

Ophthalmologic

The ocular manifestations of Sotos syndrome are poorly documented and frequently overlooked. In one study of 32 individuals with Sotos syndrome, 50% appeared to have some ocular disease (Maino et al. 1994). Of these, the most common disorders were refractive errors, and hyperopia, frequently greater than +2 diopters, was particularly common. Strabismus was also a relatively frequent finding present in over 40%.

Evaluation

- In view of the additional burden and correctable nature of many oculovisual abnormalities, it is important that these are not overlooked. Assessment by an ophthalmologist or optometrist should be considered at diagnosis and at any point in time when there is evidence or suspicion of visual difficulty.

Management

- Standard approaches to visual acuity abnormalities and strabismus have been effective.

Dental

Dental abnormalities are common in childhood. Early eruption of teeth is seen in 54% of affected individuals. Excessive wear and discoloration are apparent in 75% of these primary teeth. There is a relatively high frequency of gingivitis and occasional dental abscesses (personal observation). The etiology of these features is difficult to ascertain in view of the frequent administration of antibiotics and anticonvulsants in syrup form but is also seen in some individuals who have had little, if any, such medication. Histological and ultrastructural studies of the teeth do not reveal any apparent abnormalities (personal experience). Primary and secondary teeth are often malaligned secondary to the craniofacial anomalies. A specific pattern of tooth agenesis, particularly affecting the second molars, appears to be common (Kotilainen et al. 2009).

Evaluation

- A dental evaluation of all children with Sotos syndrome is indicated when their first deciduous teeth appear, preferably by a pediatric dentist.
- Referral to an orthodontist is appropriate if dental malalignment or crowding is noted.
- A detailed dental evaluation including radiography to assess for hypodontia is recommended around the age of 7.

Management

- The child's reluctance to clean teeth regularly has been helped in some cases by the use of an electric toothbrush.
- Orthodontic work may be necessary, but is complicated by the underlying bony abnormality and poor cooperation.

Cardiovascular

Overt congenital heart disease is rare, but if specifically investigated may be present in at least 10% of individuals with Sotos syndrome (Cole and Hughes 1994). The most common anomaly identified is patent ductus arteriosus, which requires surgical closure in approximately 50% (personal observation). One study reported congenital heart disease in 5 of 10 individuals with Sotos syndrome (Kaneko et al. 1987). It has been suggested that this may reflect the higher frequency of large deletions within Japanese cohorts (Nagai et al. 2003). A trend towards more cardiac anomalies being present in deletion cases was also identified in a large cohort of 233 individuals with mutations and 33 with deletions (Tatton-Brown et al. 2005b). Of the cases reported by Cole and Hughes who have cardiac or urological anomalies, only point mutations and no large deletions have been identified to date (Cole and Hughes 1994; Douglas et al. 2003). A review relying on parental recall reported that 19% of children with Sotos syndrome had congenital heart disease, with at least 8% confirmed by review of the available medical records (Lin et al. 1992). Review of the literature identifies the most common cardiac abnormalities as simple shunts, such as atrial or ventricular septal defects and patent ductus arteriosus, and occasionally right-sided obstructive lesions. The precise proportion of individuals with mitral valve prolapse is unknown although this has been reported anecdotally (Lin, personal communication). Tatton-Brown and Rahman (2004) reported that 24% of children with Sotos syndrome and *NSD1* mutation had congenital heart defect, of which nearly half (11/24) were isolated atrial septal defects or patent ductus arteriosus. They recommended that children should have a baseline echocardiogram (Tatton-Brown and Rahman 2007).

The authors are aware of three girls who were diagnosed with an episode of cardiomyopathy, in two cases with associated pericarditis, for which the etiology remains unknown and which appears to have resolved spontaneously after supportive measures. There are three further individuals reported in the literature with the rare cardiomyopathy isolated left ventricular noncompaction (Martinez et al. 2011; Saccucci et al. 2011). One of these had significant myocardial dysfunction requiring treatment with enalapril (Martinez et al. 2011).

Four people with Sotos syndrome in the literature are reported to have aortic dilatation (Hood et al. 2016). There are no reports of aortic dissection or rupture. The cases of aortic dilatation in the literature also have features of connective tissue laxity (Hood et al. 2016), with all four having cutis laxa (Robertson and Bankier 1999) and three having severe vesicoureteric reflux. There are two other reports of neonatal cutis laxa in Sotos syndrome (Bou-Assi et al. 2016; Cortes-Saladelafont et al. 2011), and a report of two individuals with pneumothorax resulting from subpleural blebs (Balasubramanian et al. 2014), suggesting that connective tissue laxity may be part of the spectrum of features in Sotos syndrome.

Evaluation

- All individuals with Sotos syndrome should undergo cardiovascular examination at diagnosis

There remains a debate over the justification for routine echocardiogram in Sotos syndrome, however, a low threshold for investigation and referral for cardiological assessment is appropriate

Management

- Management of cardiac defects is no different from that in the general population.

Renal

Urinary tract infection is common, with microbiologically proven urinary tract infections identified in up to 20% of males and females with Sotos syndrome (Cole and Hughes 1994) and appear to be caused by structural abnormalities and vesicoureteric reflux in the majority of cases. In one series, renal anomalies were present in approximately 15% of people with Sotos syndrome (Tatton-Brown et al. 2005b). Vesicoureteric reflux, pelviureteric junction (PUJ) obstruction, posterior urethral valves, hydronephrosis, and renal agenesis have been reported. There has been concern that the predisposition to recurrent infection in childhood may lead to an excess of chronic kidney disease in adulthood but this has not been reported to date. There is little evidence to support a noticeable excess of infections in adulthood.

There is a single case of an adult with Sotos undergoing a successful renal transplant at the age of 63 for chronic kidney disease secondary to fibromuscular dysplasia diagnosed in early adulthood (McClelland et al. 2016).

Evaluation

- In view of the potentially covert nature of urinary infections, these should be considered in any individual with Sotos syndrome who is febrile. This is particularly relevant in view of the tendency to febrile convulsions.
- All individuals with a single proven urinary tract infection should have appropriate urological investigations in view of the high a priori risk of an anatomical anomaly. Ultrasound is a recommended first step.

Management

- In most instances of urinary tract infection, after detection of a structural urinary tract anomaly, prophylactic antibiotics may be necessary.
- In a few, surgical reimplantation of the ureters or other appropriate surgical intervention is required (Hammadeh et al. 1995)

Lymphedema

Lymphedema has been reported in four individuals in the literature (McClelland et al. 2016; Foster et al. 2019). Interestingly two of these individuals also developed pericarditis. Other examples of lymphedema have been seen by the authors in cases of non-Sotos overgrowth.

Evaluation

- Assess for signs of lymphedema at routine medical examination

Management

- As per the general population

Neoplasia

An association of Sotos syndrome with tumor development was documented over 30 years ago and has been a point of debate ever since. Early series may have overestimated the risk because of ascertainment bias and inclusion of non-malignant tumors (Wit et al. 1985). Studies by Hersh et al. (Hersh et al. 1992) suggested that only 2% of affected individuals had a tumor. This latter figure is compatible with the authors' experience. Tatton-Brown et al. (2005b) reported 8 tumors out of 266 individuals with Sotos syndrome, giving a similar figure of 3%.

The site and type of tumor in Sotos syndrome is very varied, and for most tumor types there is only a single instance reported in the literature. The most frequently reported cases are sacrococcygeal teratoma (three cases) (Tatton-Brown et al. 2005b), neuroblastoma (three cases) (Kulkarni et al. 2013; Nagai et al. 2003; Tatton-Brown et al. 2005b), and acute lymphocytic leukemia (ALL) (three cases) (Tatton-Brown et al. 2005b; Türkmen et al. 2003). There are other reports of lymphoreticular malignancy (Martinez-Glez and Lapunzina 2007), although in some cases the underlying diagnosis of Sotos syndrome is controversial (Cole and Allanson 1998; Corsello et al. 1996). Other reports of ALL in Sotos syndrome are based on database reviews without any details substantiating the validity of the diagnosis of Sotos syndrome. The authors have not personally seen a lymphoreticular malignancy during their experience of more than 400 affected individuals.

Tumors occur at increased frequency in a number of the overgrowth syndromes (Beckwith–Wiedemann syndrome (see Chapter 9), Simpson–Golabi–Behmel syndrome, Weaver syndrome, and Marshall–Smith syndrome), suggesting that these are not entirely chance occurrences and may have a common etiological mechanism. One explanation may be failure of adequate tissue differentiation, as would be indicated by the presence of nephroblastomatosis associated with some cases of Wilms tumor. An alternative explanation might be hyperplasia and increased cell division at a time when full differentiation or regulation has yet to be achieved. The excess number of pediatric tumors in non-syndromic large babies (more than 4 kg) would seem to support this possibility (Daling et al. 1984). A third explanation is the role of *NSD1* as an epigenetic regulator gene affecting methylation across the genome and thus transcription of tumor suppressor genes or oncogenes. Finally, *NSD1* itself is a proto-oncogene, with translocations, somatic mutations, and epigenetic silencing occurring in different tumor types. However, there does not appear to be a clear link between these mechanisms and the development of tumors in Sotos syndrome. To date, an excess of common solid tumors such as carcinoma or sarcoma has not been identified in adults with Sotos syndrome (Foster et al. 2019).

Evaluation

- In view of the rarity, wide range in age of onset, and diverse nature of the tumors in Sotos syndrome, there is currently no clear indication for or benefit from regular tumor surveillance.

Management

- If a tumor is identified, treatment is the same as in the general population.

Musculoskeletal

Early hypotonia, large size, and joint laxity appear to precipitate a number of orthopedic complications. Particularly problematic are foot deformities especially pes planus, and the reported frequency of 50%, which includes both mild and severe symptoms, could be an underestimate (Cole and Hughes 1994). Another relatively frequent feature is kyphoscoliosis (Latham et al. 1994; Tatton-Brown and Rahman 2004). Previously this was usually thought to be mild and only infrequently requiring surgical intervention (Latham et al. 1994), however a recent study of a cohort of 44 adults with molecularly confirmed Sotos syndrome found 24 individuals (55%) had scoliosis and 11 had undergone surgery (Foster et al. 2019). Another study of adults reported 52% (11/21) had scoliosis although the severity was not reported (Fickie et al. 2011). A series of scoliosis in childhood found that while only 8 of a total of 42 patients (19%) had scoliosis, seven required surgery for severe curvature despite early treatment with bracing (Corrado et al. 2011). In this cohort the age of onset was variable, the youngest case being before the age of three and the oldest at puberty. A small number of adults with scoliosis who have undergone surgery have developed chronic back pain in adulthood (Foster et al. 2019). Contractures are uncommon in childhood but may develop in adulthood. Craniosynostosis is a rare feature but important to note as it may affect the facial gestalt and cause diagnostic delay.

A relatively high frequency of fractures in early childhood led to concerns about bone fragility. There is no evident excess of fractures (personal observation), which in itself is surprising, because these individuals are large with poor coordination. Investigations of bone density have yet to show any evidence of osteoporosis or osteopenia (Davie, personal communication).

Evaluation

- Careful examination of the musculoskeletal system should be performed at diagnosis and with routine medical visits.
- Foot deformities or scoliosis evident on clinical examination should be referred for orthopedic assessment and management.

Management

- Foot deformities benefit from physiotherapy and supportive orthotic intervention.
- Surgery may have to be considered for severe pes planus and valgus deformity.
- Mild scoliosis requires only observation or occasionally the use of a cast or brace, although the latter may be difficult because of non-compliance.
- Surgical correction of scoliosis may need to be considered (Corrado et al. 2011; Latham et al. 1994). Joint hypermobility and hypotonia should be taken into account when performing surgery (Corrado et al. 2011).

Gastrointestinal

A tendency to severe constipation, with or without overflow, is present in over 10% of affected individuals and exacerbates difficulties with toilet training in some children. It commonly necessitates pharmacological intervention or, very rarely, surgical intervention, for example, after rectal prolapse.

Evaluation

- History of bowel problems should be obtained with routine visits.

Management

- Symptoms are treated as in the general population.

RESOURCES

Sotos Syndrome Support Association
Three Danada Square East, #235
Wheaton, IL 60187, USA
Phone: 888-246-SSSA (7772)
Email: president@sotossyndrome.org
Website: www.sotossyndrome.org/
The SSSA also provides link to other international patient support groups through their resources section

Child Growth Foundation
2 Mayfield Avenue, Chiswick
London W4 1PW, UK
Phone: +44 (0)20 8995 0257
Email: info@childgrowthfoundation.org
Website: www.childgrowthfoundation.org

Eltern-Initiative Sotos Syndrome
Steinernkreuzweg 22
D-55246 Mainz-Kosteim, Germany
Website: http://www.sotossyndrom.de/

REFERENCES

Agwu JC, Shaw NJ, Kirk J, Chapman S, Ravine D, Cole TRP (1999) Growth in Sotos syndrome. *Arch Dis Child* 80:339–342.

Allanson JE, Cole TRP (1996) Sotos syndrome: Evolution of facial phenotype subjective and objective assessment. *Am J Med Genet* 65:13–20.

Appenzeller O, Snyder RD (1969) Autonomic failure with persistent fever in cerebral gigantism. *J Neurol Neurosurg Psychiat* 32:123–128.

Balasubramanian M, Shearing E, Smith K, Chavasse R, Taylor R, Tatton-Brown K, Primhak R, Ugonna K, Parker MJ (2014) Pneumothorax from subpleural blebs-A new association of sotos syndrome? *Am J Med Genet A* 164:1222–1226.

Ball LJ, Sullivan MD, Dulany S, Stading K, Schaefer GB (2005) Speech-language characteristics of children with Sotos syndrome. *Am J Med Genet A* 136A:363–367.

Bou-Assi E, Bonniaud B, Grimaldi M, Faivre L, Vabres P (2016) Neonatal cutis laxa and hypertrichosis lanuginosa in Sotos Syndrome. *Pediatr Dermatol* 33:e351–e352.

Butler MG, Dasouki MJ, Zhou X-P, Talebizadeh Z, Brown M, Takahashi TN, Miles JH, Wang CH, Stratton R, Pilarski R, Eng C (2005) Subset of individuals with autism spectrum disorders and extreme macrocephaly associated with germline PTEN tumour suppressor gene mutations. *J Med Genet* 42:318–321.

Cecconi M, Forzano F, Milani D, Cavani S, Baldo C, Selicorni A, Pantaleoni C, Silengo M, Ferrero GB, Scarano G, Della Monica M, Fischetto R, Grammatico P, Majore S, Zampino G, Memo L, Cordisco EL, Neri G, Pierluigi M, Bricarelli FD, Grasso M, Faravelli F (2005) Mutation analysis of the NSD1 gene in a group of 59 patients with congenital overgrowth. *Am J Med Genet* 134A:247–253. https://doi.org/10.1002/ajmg.a.30492

Choufani S, Cytrynbaum C, Chung BHY, Turinsky AL, Grafodatskaya D, Chen YA, Cohen ASA, Dupuis L, Butcher DT, Siu MT, Luk HM, Lo IFM, Lam STS, Caluseriu O, Stavropoulos DJ, Reardon W, Mendoza-Londono R, Brudno M, Gibson WT, Chitayat D, Weksberg R (2015) NSD1 mutations generate a genome-wide DNA methylation signature. *Nat Commun* 6:10207.

Cole T, Allanson J (1998) Reply to "lymphoproliferative disorders in Sotos syndrome: observation in two cases." *Am J Med Genet* 75:226–7.

Cole T, Hughes HE (1994) Sotos syndrome: a study of the diagnostic criteria and natural history. *J Med Genet* 31:20–32.

Corrado R, Wilson AF, Tello C, Noel M, Galaretto E, Bersusky E 2011) Sotos syndrome and scoliosis surgical treatment: A 10-year follow-up. *Eur Spine J* 20:271–277.

Corsello G, Giuffrè M, Carcione A, Cuzto ML, Piccione M, Ziino O (1996) Lymphoproliferative disorders in Sotos syndrome: Observation of two cases. *Am J Med Genet* 64:588–593. https://doi.org/10.1002/(SICI)1096-8628(19960906)64:4<588::AID-AJMG12>3.0.CO;2-D

Cortes-Saladelafont E, Arias-Sáez K, Esteban-Oliva D, Coroleu-Lletget W, Martín-Jiménez P, Pintos-Morell G (2011) Síndrome de Sotos: nueva mutación «sin sentido» del gen NSD1 que presenta cutis laxa neonatal. *An Pediatr* 75:129–133.

Daling JR, Starzyk P, Olshan AF, Weiss NS (1984) Birth weight and the incidence of childhood cancer. *J Natl Cancer Inst* 72:1039–41.

de Boer L, le Cessie S, Wit JM (2005) Auxological data in patients clinically suspected of Sotos syndrome with NSD1 gene alterations. *Acta Paediatr* 94:1142–4.

de Boer L, Röder I, Wit JM (2006) Psychosocial, cognitive, and motor functioning in patients with suspected Sotos syndrome: a comparison between patients with and without NSD1 gene alterations. *Dev Med Child Neurol* 48:582–8.

Donnelly D, Turnpenny P, McConnell V (2011) Phenotypic variability in a three-generation Northern Irish family with Sotos syndrome. *Clin Dysmorphol* 20:175–81.

Douglas J, Hanks S, Temple IK, Davies S, Murray A, Upadhyaya M, Tomkins S, Hughes HE, Cole TRP, Rahman N 2003) NSD1 mutations are the major cause of Sotos syndrome and occur in some cases of Weaver syndrome but are rare in other overgrowth phenotypes. *Am J Hum Genet* 72:132–143.

Grand K, Gonzalez-Gandolfi C, Ackermann AM, Aljeaid D, Bedoukian E, Bird LM, De Leon DD, Diaz J, Hopkin RJ, Kadakia SP, Keena B, Klein KO, Krantz I, Leon E, Lord K, McDougall C, Medne L, Skraban CM, Stanley CA, Tarpnian J, Zackai E, Deardorff MA, Kalish JM (2019) Hyperinsulinemic hypoglycaemia in seven aptietns with de novo NSD1 mutations. *Am J Med Genet A* 179(4):542–551.

Faivre L, Viot G, Prieur M, Turleau C, Gosset P, Romana S, Munnich A, Vekemans M, Cormier-Daire V (2000) Apparent Sotos syndrome (cerebral gigantism) in a child with trisomy 20p11.2-p12.1 mosaicism. *Am J Med Genet* 91:273–6.

Fickie MR, Lapunzina P, Gentile JK, Tolkoff-Rubin N, Kroshinsky D, Galan E, Gean E, Martorell L, Romanelli V, Toral JF, Lin AE (2011) Adults with Sotos syndrome: Review of 21 adults with molecularly confirmed NSD1 alterations, including a detailed case report of the oldest person. *Am J Med Genet A* 155:2105–2111.

Finegan JK, Cole TR, Kingwell E, Smith ML, Smith M, Sitarenios G (1994) Language and behavior in children with Sotos syndrome. *J Am Acad Child Adolesc Psychiatry* 33:1307–15.

Foster A, Zachariou A, Loveday C, Ashraf T, Blair E, Clayton-Smith J, Dorkins H, Fryer A, Gener B, Goudie D, Henderson A, Irving M, Joss S, Keeley V, Lahiri N, Lynch SA, Mansour S, McCann E, Morton J, Motton N, Murray A, Riches K, Shears D, Stark Z, Thompson E, Vogt J, Wright M, Cole T, Tatton-Brown K (2019) The phenotype of Sotos syndrome in adulthood: A review of 44 individuals. *Am J Med Genet Part C* 1–7.

Hammadeh MY, Dutta SN, Cornaby AJ, Morgan RJ (1995) Congenital urological anomalies in Sotos syndrome. *Br J Urol* 76:133–5.

Han JY, Lee IG, Jang W, Shin S, Park J, Kim M (2017) Identification of a novel de novo nonsense mutation of the NSD1 gene in monozygotic twins discordant for Sotos syndrome. *Clin Chim Acta* 470:31–35.

Hersh JH, Cole TR, Bloom AS, Bertolone SJ, Hughes HE (1992) Risk of malignancy in Sotos syndrome. *J Pediatr* 120:572–4.

Hood RL, Mcgillivray G, Hunter MF, Roberston SP, Bulman DE, Boycott KM, Stark Z, Boycott K, MacKenzie A, Majewski J, Brudno M, Bulman D, Dyment D (2016) Severe connective tissue laxity including aortic dilatation in Sotos syndrome. *Am J Med Genet A* 170:531–535.

Horikoshi H, Kato Z, Masuno M, Asano T, Nagase T, Yamagishi Y, Kozawa R, Arai T, Aoki M, Teramoto T, Omoya K, Matsumoto N, Kurotaki N, Shimokawa O, Kurosawa K, Kondo N (2006) Neuroradiologic Findings in Sotos Syndrome. *J Child Neurol* 21:614–618.

Imaizumi K, Kimura J, Matsuo M, Kurosawa K, Masuno M, Niikawa N, Kuroki Y (2002) Sotos syndrome associated with a de novo balanced reciprocal translocation t(5;8)(q35;q24.1). *Am J Med Genet* 107:58–60.

Jaeken J, Schueren-Lodeweyckx M, Eeckels R (1972) Cerebral gigantism syndrome: A report of 14 cases and review of the literature. *Zeitsch Kinderheilkd* 112:332–346.

Kamien B, Dadd T, Buckman M, Ronan A, Dudding T, Meldrum C, Scott R, Mina K (2016) Somatic-gonadal mosaicism causing Sotos syndrome. *Am J Med Genet A* 170:3360–3362.

Kaneko H, Tsukahara M, Tachibana H, Kurashige H, Kuwano A, Kajii T, Opitz JM, Reynolds JF (1987) Congenital heart defects in Sotos sequence. *Am J Med Genet* 26:569–76.

Kotilainen J, Pohjola P, Pirinen S, Arte S, Nieminen P (2009) Premolar hypodontia is a common feature in Sotos syndrome with a mutation in the NSD1 gene. *Am J Med Genet A* 149:2409–2414.

Koyama M, Suguira M, Yokoyama Y, Kobayashi M, Sugiyama K, Imahashi H, Saito H (1985) A female case of cerebral gigantism with chromosome abnormality 47,XX+inv dup (15)(pter-q12 or q13;q12 or13-pter). *J Jpn Pediatr Soc* 175:2671.

Kulkarni K, Stobart K, Noga M (2013) A Case of Sotos Syndrome With Neuroblastoma. *J Paediatr Hematol Oncol* 35:238–239.

Kurotaki N, Imaizumi K, Harada N, Masuno M, Kondoh T, Nagai T, Ohashi H, Naritomi K, Tsukahara M, Makita Y, Sugimoto T, Sonoda T, Hasegawa T, Chinen Y, Tomita Ha H, Kinoshita A, Mizuguchi T, Yoshiura Ki K, Ohta T, Kishino T, Fukushima Y, Niikawa N, Matsumoto N (2002) Haploinsufficiency of NSD1 causes Sotos syndrome. *Nat Genet* 30:65–366.

Kurotaki N, Stankiewicz P, Wakui K, Niikawa N, Lupski JR (2005) Sotos syndrome common deletion is mediated by directly oriented subunits within inverted Sos-REP low-copy repeats. *Hum Mol Genet* 14:535–542.

Lane C, Milne E, Freeth M (2016) Characteristics of Autism Spectrum Disorder in Sotos Syndrome. *J Autism Dev Disord* 47:1–9.

Latham J, Marks D, Cole T, Thompson A (1994) Scoliosis in Sotos syndrome, in: Proceedings of the 5th European Spinal Deformities Society Meeting. Birmingham, UK, pp. 122–123(A).

Lin A, Treat K, Kedesdy J (1992) Survey of behaviour in Sotos syndrome. *Proc Greenwood Genet Cent* 12.

Maino DM, Kofman J, Flynn MF, Lai L (1994) Ocular manifestations of Sotos syndrome. *J Am Optom Assoc* 65:339–46.

Maroun C, Schmerler S, Hutcheon RG (1994) Child with Sotos phenotype and a 5:15 translocation. *Am J Med Genet* 50:291–3.

Martinez-Glez V, Lapunzina P (2007) Sotos syndrome is associated with leukaemia/lymphoma. *Am J Med Genet A* 143A:1244–1245.

Martinez HR, Belmont JW, Craigen WJ, Taylor MD, Jefferies JL (2011) Left ventricular noncompaction in Sotos syndrome. *Am J Med Genet A* 155:1115–1118.

McClelland J, Burgess B, Crock P, Goel H (2016) Sotos syndrome: An unusual presentation with intrauterine growth restriction, generalized lymphedema, and intention tremor. *Am J Med Genet A* 170:1064–1069.

Miyake N, Kurotaki N, Sugawara H, Shimokawa O, Harada N, Kondoh T, Okamoto N, Nishimoto J, Yoshiura K-I, Ohta T, Kishino T, Niikawa N, Matsumoto N (2003) Report Preferential Paternal Origin of Microdeletions Caused by Prezygotic Chromosome or Chromatid Rearrangements in Sotos Syndrome. *Am J Hum Genet* 72:1331–1337.

Mussa A, Russo S, De Crescenzo A, Chiesa N, Molinatto C, Selicorni A, Richiardi L, Larizza L, Silengo MC, Riccio A, Ferrero GB (2013) Prevalence of Beckwith-Wiedemann syndrome in North West of Italy. *Am J Med Genet A* 161:2481–2486.

Nagai T, Matsumoto N, Kurotaki N, Harada N, Niikawa N, Ogata T, Imaizumi K, Kurosawa K, Kondoh T, Ohashi H, Tsukahara M, Makita Y, Sugimoto T, Sonoda T, Yokoyama Y, Uetake K, Sakazume S, Fukushima Y, Naritomi K (2003) Sotos syndrome and haploinsufficiency of NSD1: clinical features of intragenic mutations and submicroscopic deletions. *J Med Genet* 40:285–289.

NCD Risk Factor Collaboration (2016) A century of trends in adult human height. *eLife*. https://doi.org/10.7554/eLife.13410

Nicita F, Ruggieri M, Polizzi A, Mauceri L, Salpietro V, Briuglia S, Papetti L, Ursitti F, Grosso S, Tarani L, Segni M, Savasta S, Parisi P, Verrotti A, Spalice A (2012) Seizures and epilepsy in Sotos syndrome: Analysis of 19 Caucasian patients with long-term follow-up. *Epilepsia* 53:102–105.

Rayasam GV, Wendling O, Angrand PO, Mark M, Niederreither K, Song L, Lerouge T, Hager GL, Chambon P, Losson R (2003) NSD1 is essential for early post-implantation development and has a catalytically active SET domain. *EMBO J* 22:3153–3163.

Rio M, Clech L, Amiel J, Faivre L, Lyonnet S, Le Merrer M, Odent S, Lacombe D, Edery P, Brauner R, Raoul O, Gosset P, Prieur M, Vekemanns M, Munnich A, Colleaux L, Cormier-Daire V, 2003) Spectrum of NSD1 mutations in Sotos and Weaver syndromes. *J Med Genet* 40:436–440.

Robertson SP, Bankier a (1999) Sotos syndrome and cutis laxa. *J Med Genet* 36:51–6.

Rutter SC, Cole TR (1991) Psychological characteristics of Sotos syndrome. *Dev Med Child Neurol* 33:898–902.

Saccucci P, Papetti F, Martinoli R, Dofcaci A, Tuderti U, Marcantonio A, Di Renzi P, Fahim A, Ferrante F, Banci M (2011) Isolated left ventricular noncompaction in a case of sotos syndrome: a casual or causal link? *Cardiol Res Pract* 2011:824095.

Saugier-Veber P, Bonnet C, Afenjar A, Drouin-Garraud V, Coubed C, Fehrenbach S, Holder-Espinasse M, Roume J, Malan V, Portnoi M-F, Jeanne N, Baumann C, Heron D, David A, Gerard M, Bonneau D, Lacombe D, Cormier-Daire V, Billette de Villemeur T, Frebourg T, Burglen L (2007) Heterogeneity of NSD1 Alternations in 116 Patients with Sotos syndrome. *Hum Mutat* 28:1098–1107.

Schaefer G, Bodensteiner J, Buehler B, Cole T (1997) The neuroimaging findings in Sotos syndrome. *Am J Med Genet A* 68:462–5.

Schrander-Stumpel C, Fryns J, Hamers G (1989) Sotos syndrome and de novo balanced autosomal translocation (t93;6)(p21;p21). *Clin Genet* 37:226–229.

Sheth K, Moss J, Hyland S, Stinton C, Cole T, Oliver C (2015) The behavioral characteristics of Sotos syndrome. *Am J Med Genet A* 167A:2945–2956

Sotos JF, Dodge PR, Muirhead D, Crawford JD, Talbot NB (1964) Cerebral Gigantism in Childhood. *N Engl J Med* 271:109–116.

Stratton RF, Tedrowe NA, Tolworthy JA, Patterson RM, Ryan SG, Young RS (1994) Deletion 5q35.3. *Am J Med Genet* 51:150–2.

Tamaki K, Horie K, Go T, Okuno T, Mikawa H, Hua ZY, Abe T (1989) Sotos syndrome with a balanced reciprocal translocation t(2;12)(q33.3;q15). *Ann Genet* 32244–6.

Tatton-Brown K, Cole T, Rahman N (2004) Sotos syndrome. GeneReviews®. Available from https//www.ncbi.nlm.nih.gov/books/NBK1479/

Tatton-Brown K, Douglas J, Coleman K, Baujat G, Chandler K, Clarke A, Collins A, Davies S, Faravelli F, Firth H, Garrett C, Hughes H, Kerr B, Liebelt J, Reardon W, Schaefer GB, Splitt M, Temple IK, Waggoner D, Weaver DD, Wilson L, Cole T, Cormier-Daire V, Irrthum A, Rahman N (2005a) Multiple mechanisms are implicated in the generation of 5q35 microdeletions in Sotos syndrome. *J Med Genet* 42:307–313.

Tatton-Brown K, Douglas J, Coleman K, Cole TRP, Das S, Horn D, Hughes HE, Temple IK, Faravelli F, Waggoner D, Tu S (2005b) Genotype-phenotype associations in Sotos Syndrome : An analysis of 266 individuals with NSD1 aberrations. *Am J Hum Genet* 77:193–204.

Tatton-Brown K, Loveday C, Yost S, Clarke M, Ramsay E, Zachariou A, Elliott A, Wylie H, Ardissone A, Rittinger O, Stewart F, Temple IK, Cole T, Mahamdallie S, Seal S, Ruark E, Rahman N (2017) Mutations in epigenetic regulation genes are a major cause of overgrowth with intellectual disability. *Am J Hum Genet* 100:725–736.

Tatton-Brown K, Rahman N (2007) Sotos syndrome. *Eur J Hum Genet* 15:264–271.

Tatton-Brown K, Rahman N (2004) Clinical features of NSD1-positive Sotos syndrome. *Clin Dysmorphol* 13:199–204.

Thomas A, Lemire E (2008) Sotos syndrome: Antenatal presentation. *Am J Med Genet* 146A:1312–1313.

Türkmen S, Gillessen-Kaesbach G, Meinecke P, Albrecht B, Neumann LM, Hesse V, Palanduz S, Balg S, Majewski F, Fuchs S, Zschieschang P, Greiwe M, Mennicke K, Kreuz FR, Dehmel HJ, Rodeck B, Kunze J, Tinschert S, Mundlos S, Horn D (2003) Mutations in NSD1 are responsible for Sotos syndrome, but are not a frequent finding in other overgrowth phenotypes. *Eur J Hum Genet* 11:858–865.

Visser R, Matsumoto N (2003) Genetics of Sotos syndrome. *Curr Opin Pediatr* 15:598–606.

Visser R, Shimokawa O, Harada N, Kinoshita A, Ohta T, Niikawa N, Matsumoto N (2005) Identification of a 3.0-kb major recombination hotspot in patients with Sotos syndrome who carry a common 1.9-Mb microdeletion. *Am J Hum Genet* 76:52–67.

Waggoner D, Raca G, Welch K, Dempsey M, Anderes E, Ostrovnaya I, Alkhateeb A, Kamimura J, Matsumoto N, Schaeffer G, Martin C, Das S (2005) NSD1 analysis for Sotos syndrome: Insights and perspectives from the clinical laboratory. *Genet Med* 7:524–533.

Wit J, Beemer F, Barth P, Oothuys J, Dijkstra P, Van den Brande J, Leschot N (1985) Cerebral gignatism (Sotos syndrome), compiled data of 22 cases. *Eur J Paediatr* 114:131–140.

Yirmiya N, Shulman C (1996) Seriation, conservation, and theory of mind abilities in individuals with autism, individuals with mental retardation and normally developing children. *Child Dev* 67:2045–2059.

56

STICKLER SYNDROME

MARY B. SHEPPARD

Department of Family and Community Medicine, Department of Surgery, Department of Physiology, Saha Cardiovascular Research Center, University of Kentucky, Lexington, Kentucky, USA

CLAIR A. FRANCOMANO

Department of Medical and Molecular Genetics, Indiana University School of Medicine, Indianapolis, Indiana, USA

INTRODUCTION

Stickler syndrome, also called hereditary progressive arthro-ophthalmopathy, is an autosomal dominant connective tissue disorder primarily affecting the ocular, orofacial, and skeletal systems (Stickler et al. 1965; Rimoin and Lachman 1993) (Figure 56.1). Stickler and colleagues first described the condition in two reports. The first paper (Stickler et al. 1965) described a new dominant syndrome consisting of progressive myopia beginning in the first decade of life and resulting in retinal detachment and blindness. Affected persons also exhibited premature degenerative joint disease with a mild epiphyseal dysplasia and joint hypermobility. In the second paper, Stickler and Pugh (1967) described changes in the vertebrae and hearing deficit as part of the syndrome, as well as the midface hypoplasia that is characteristic of the phenotype (Figure 56.2). Opitz et al. (1972) reported that individuals with Stickler syndrome may also have the Robin sequence (see Chapter 50) as a component manifestation. Phenotypic variability, both inter- and intrafamilial, is a hallmark of this syndrome. Mutations in six different collagen genes as well as *BMP4* and *LOXL3* may cause the phenotype.

Incidence

Herrmann et al. (1975) suggested that Stickler syndrome is the most common autosomal dominant connective tissue disorder in the North American Midwest. Although estimates vary, it is thought that the frequency of Stickler syndrome is roughly 1/10,000 in the Caucasian population of the United States. The condition has been reported in multiple ethnic populations but epidemiologic data in countries other than the United States have not been published. Phenotypic variability of the syndrome can make the diagnosis difficult, and the syndrome may be significantly underdiagnosed.

Although chronic musculoskeletal pain, retinal detachments, and hearing loss may significantly impact quality of life, longevity is not affected by this disorder.

Diagnostic Criteria

The major systems involved in Stickler syndrome are the ocular, auditory, orofacial, and musculoskeletal systems. Several closely related phenotypes have been described and are discussed below in the section on Etiology; we will focus on the phenotype of Stickler syndrome due to changes in COL2A1 here. Severe myopia with onset in the first decade of life, vitreous degeneration, spontaneous retinal detachment, chorioretinal degeneration, open angle glaucoma, and presenile cataracts are the ocular features of the disorder (Stickler et al. 1965; Rimoin and Lachman 1993; Liberfarb et al. 2003). During early childhood, midface hypoplasia is often evident and becomes less pronounced with age in persons with *COL2A1* mutations. Pierre Robin sequence (see

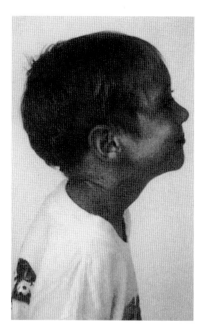
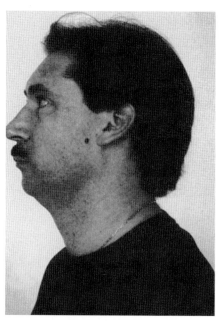

FIGURE 56.1 Typical Stickler syndrome. Affected boy (left) and his affected father (right). In addition to oro-auriculo-facial manifestations (see Figure 56.2), affected individuals frequently present with ocular and musculoskeletal features.

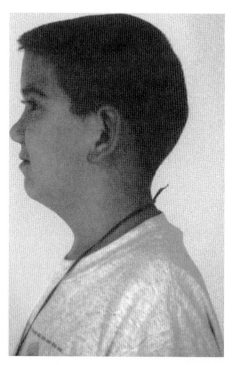

FIGURE 56.2 Midface hypoplasia in Stickler syndrome. A flat facial profile with depressed nasal bridge, midface, or maxillary hypoplasia, and micrognathia are often seen in Stickler syndrome. Other oro-auriculo-facial features may include epicanthal folds, clefting of the hard or soft palate, Pierre Robin sequence, sensorineural deafness, and dental anomalies.

Chapter 50), bifid uvula, and/or cleft palate may be present and result in feeding and respiratory problems (Schreiner et al. 1973). Mixed and sensorineural hearing loss in the higher frequencies is also a feature of the syndrome (Liberfarb and Goldblatt 1986). In early life, joint pain and stiffness may signify the onset of juvenile osteoarthritis. Early-onset degenerative joint disease is a major complication in adulthood. Radiographs of the skeleton may demonstrate a mild form of spondyloepiphyseal dysplasia, even in individuals without other signs of the syndrome (Stickler and Pugh 1967; Rimoin and Lachman 1993) (Figure 56.3). An early report suggested that mitral valve prolapse is present in 45% of the affected individuals (Liberfarb and Goldblatt 1986). However, more recent reports (Ahmad et al. 2003; Donoso et al. 2003) found a much lower incidence of mitral valve prolapse among individuals with Stickler syndrome and suggested that the incidence is comparable with that of the general population. This may reflect the change in diagnostic criteria for mitral valve prolapse published in 1988 (Levine et al. 1988), or a stricter definition of Stickler syndrome in the more recent publications.

Liberfarb et al. (2003) proposed a point system for establishing the diagnosis of Stickler syndrome. Two points are awarded for the presence of each of the following findings: cleft palate, vitreous degeneration or retinal detachment, and high-frequency sensorineural hearing loss. One point each is awarded for the characteristic facies, hypermobile tympanic membranes, a history of femoral head failure (severe delay in ossification), radiographically demonstrated osteoarthritis before age 40 years, spinal deformities, positive family

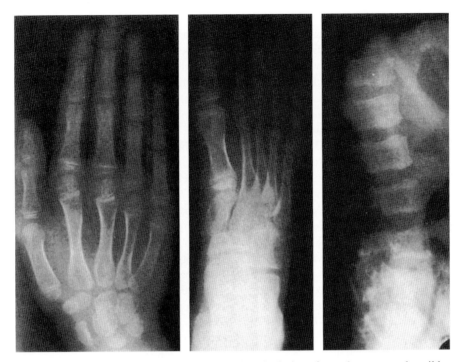

FIGURE 56.3 Radiographic findings in Stickler syndrome. Arachnodactyly, fusion of carpal centers, and a mild spondyloepiphyseal dysplasia are noticeable. Other musculoskeletal features may include hyperextensible joints, Marfanoid habitus, and premature osteo-arthropathy. (Reprinted with permission, Jones KL, 1988.)

history, and identification of a causative mutation. A score of 5 is necessary to make the diagnosis of Stickler syndrome. Liberfarb et al. (2003) reported 25 subjects with molecular confirmation of *COL2A1* mutations, all of whom satisfied these diagnostic criteria with a mean diagnostic score of 7.3 out of a possible 9.

Etiology, Pathogenesis, and Genetics

Stickler syndrome is now recognized as a group of phenotypically related disorders caused by genetic variation in *COL2A1*, *COL11A1* and *COL11A2*, *COL9A1*, *COL9A2*, and *COL9A3*, as well as *LOX3* and *BMP4*. The observation that type II collagen is present in both cartilage and the secondary vitreous of the eye suggested that the basis of Stickler syndrome may be a defect in *COL2A1*, the gene that encodes type II collagen (Maumenee 1979) (Table 56.1). Subsequently, Stickler syndrome was linked to the *COL2A1* gene (Francomano et al. 1987; Knowlton et al. 1989; Snead et al. 1996). Since then, more than 82 dominant mutations in the *COL2A1* gene have been identified in individuals or families with Stickler syndrome (Ahmad et al. 1991; Brown et al. 1995a; Liberfarb et al. 2003; Zechi-Ceide et al. 2008). With few exceptions, most of the mutations introduce a premature stop codon into the region of the gene encoding the triple helical domain of the collagen molecule.

COL2A1 encodes a single procollagen monomer that assembles into a homotrimer with non-collagenous amino- and carboxy-terminal ends, beginning at the carboxy-terminal domain. These stable triple helices aggregate into the supramolecular structures of fibrillar collagen. The premature stop codons introduced by the mutations typical of Stickler syndrome essentially delete the carboxy-terminal region of the monomer, and so they are not able to participate in the assembly of the triple helical trimer. Moreover, nonsense-mediated RNA decay eliminates most of the prematurely terminated transcripts. Functionally, only half the normal number of collagen monomers is available to be

TABLE 56.1 Expression of type ii collagen and type xi collagen in cartilage and the eye[a].

Cartilage		Eye	
Type II collagen	Type XI collagen	Type II collagen	Type XI collagen
COL2A1	COL11A1 COL11A2 COL2A1	COL2A1	COL11A1 COL5A2 COL2AL

[a] Type II collagen is always a homotrimer of the product of the *COL2A1* gene. In cartilage, type XI collagen is a heterotrimer of the products of the *COL11A1*, *COL11A2*, and *COL2A1* genes. However, in the ocular vitreous, type XI collagen is a heterotrimer of the products of the *COL11A1*, *COL5A2*, and *COL2A1* genes. Therefore, individuals with Stickler syndrome as result of a mutation in the *COL11A2* gene would not be expected to manifest ocular anomalies because COL11A2 is not expressed in the ocular vitreous.

incorporated into type II collagen trimers. It is hypothesized that this functional haploinsufficiency of type II collagen results in type 1 Stickler syndrome in most individuals.

Genetic heterogeneity in Stickler syndrome was first demonstrated by the exclusion of linkage to *COL2A1* in about 50% of families (Knowlton et al. 1989). Mutations in the genes encoding type XI collagen (*COL11A1* and *COL11A2*) have been identified in over 30 affected individuals to date (Vikkula et al. 1995; Richards et al. 1996; Majava et al. 2007). Clinical and molecular findings have suggested that the phenotype in Stickler syndrome families with severe ocular manifestations are the result of mutations in either *COL2A1* or *COL11A1*, whereas the phenotype in families with milder or absent eye involvement results from mutations in the *COL11A2* gene (Snead et al. 1994) (Table 56.1). In 1998, Wilkin et al. found that some Stickler syndrome families were not linked to *COL2A1*, *COL11A1*, or *COL11A2*, suggesting that at least a fourth gene is involved in the autosomal dominant form of the phenotype (Wilkin et al. 1998). Recent studies have found mutations in *BM4P* in families with autosomal dominant Stickler (Nixon et al. 2019)

Characterization of the vitreoretinal phenotype in Stickler syndrome has resulted in a proposed classification of the syndrome into two subtypes with ocular involvement, type 1 and type 2, caused by mutations in *COL2A1* and *COL11A1*, respectively, and type 3, or the "non-ocular" type, caused by mutations in *COL11A2* (Snead et al. 1996, 1996). Richards et al. (2006) described the vitreous in type 1 Stickler syndrome as characterized by a vestigial gel remnant in the retrolental space, bounded by a highly folded membrane. This finding is present at birth and remains throughout life. According to Richards et al. (2006), type 2 Stickler syndrome also has abnormal vitreous architecture, although it is distributed throughout the entire posterior segment. In type 2 Stickler syndrome, there is decreased normal fibrillar gel matrix. The remaining gel matrix has gel strands of variable diameters, many of which are thickened, and have a beaded appearance (Snead et al. 1996). Individuals with type 2 Stickler syndrome often have milder ocular manifestations (Snead et al. 1994), and early-onset high myopia is not typically a feature (Vikkula et al. 1995). The *COL11A2* gene is not expressed in the vitreous, and consequently mutations in *COL11A2* cause a non-ocular variant of Stickler syndrome, designated type 3. The non-ocular, or type 3, Stickler syndrome is also called OSMED (oto-spondylo-mega-epiphyseal dysplasia) and has both dominant and recessive forms.

Mutations in *COL2A1* are associated with higher risk of retinal detachment, a lower likelihood of early-onset deafness, and a less characteristic facial appearance with aging, whereas *COL11A1* mutations commonly cause early-onset deafness and are associated with the characteristic face even in adult life.

Autosomal dominant Stickler syndrome has recently been identified by Nixon et al. (2019) in a family with a loss of function mutation in the *BMP4* gene. *BMP4* encodes bone morphogenetic protein 4, which is known to contribute to eye development in animals by establishing the dorsal-ventral axis of the developing eye as well as lens induction and retinal development. One affected member of this family was also found to have renal dysplasia, prompting the authors to recommend that if a loss-of-function *BMP4* variant is identified in an individual with Stickler syndrome, then they should be investigated for congenital anomalies of the kidney and urinary tract (CAKUT).

Stickler syndrome caused by mutations in *COL2A1*, *COL11A1*, *COL11A2*, or *BMP4* is inherited as an autosomal dominant disorder. Van Camp et al. (2006) reported an autosomal recessive form of Stickler syndrome caused by mutation in the *COL9A1* gene. The authors note that the vitreous phenotype in individuals with this form did not resemble the membranous, beaded, or non-fibrillar phenotypes described in types 1, 2, and 3, but rather resembled an aged vitreous with progressive gel liquefaction. Of the four individuals reported by Van Camp et al. (2006), two had retinopathy and two had normal retinas. All four had high myopia and high-frequency hearing loss.

Baker et al (2011) reported an autosomal recessive form of Stickler syndrome in a large consanguineous family of Asian Indian origin. An 8 bp deletion was found in *COL9A2*, which correlated with the Stickler phenotype, including high myopia, vitreoretinal degeneration, retinal detachment, and mild to moderate sensorineural hearing loss when present in homozygosity.

Faletra et al (2014) and Hanson-Kahn et al. (2017) reported an autosomal recessive form of Stickler syndrome caused by mutation in the *COL9A3* gene. The affected individuals demonstrated midfacial hypoplasia, myopia, sensorineural hearing loss, and epiphyseal changes. None of them were reported to have retinal degeneration or vitreous anomalies.

Mutations in the *LOXL3* gene have also been identified as causing autosomal recessive Stickler syndrome (Alzahrani et al. 2015). A 16-year-old boy was noted to have micro/retrognathia, a U-shaped cleft palate, myopia, chorioretinal lattice degeneration, and mild conductive hearing loss. One sister had the same phenotype except for normal hearing. Four other siblings were unaffected. They were the offspring of parents who were second cousins.

Mosaicism has been reported in cases of Stickler syndrome caused by mutations in *COL2A1* and *COL11A1*. Lauritsen et al. (2017) describe a child with Stickler syndrome due to a novel splice site mutation in *COL11A1* whose father presented with mild clinical manifestations. Initial Sanger sequencing on peripheral blood did not identify the mutation in the father, but subsequent next-generation sequencing revealed the mutation in approximately 9% of the DNA obtained from the blood of the father. Analysis of two semen samples revealed the mutation in 18% and 22%

of the DNA, leading the authors to recommend using sensitive genetic testing when mosaicism is suspected and to test parents even when phenotypic signs of Stickler syndrome are absent.

Genetic Counseling

Most cases of Stickler syndrome are familial. Those caused by mutations in the *COL2A1*, *COL11A1*, and *COL11A2* genes have a 50% recurrence risk to offspring, in keeping with the autosomal dominant inheritance pattern. In families with autosomal recessive *COL9A1*, *COL9A2*, and *COL9A3* mutations, siblings of an affected person have a 25% risk, whereas the risk to offspring is extremely low. Reduced penetrance has been observed in an extended family with autosomal dominant Stickler syndrome caused by mutations in the *COL2A1* gene (Tompson et al. 2017).

Diagnostic Testing

Diagnostic testing for Stickler syndrome is complicated by the known genetic heterogeneity. Richards et al. (2006) suggested that an efficient strategy for molecular diagnosis was a two-stage approach including vitreoretinal assessment followed by sequence analysis of *COL2A1* in those individuals with the typical eye findings for type 1 Stickler syndrome. Others have suggested that the vitreoretinal findings are not entirely specific and that the typical ocular findings of type 1 Stickler syndrome may also be seen in individuals with *COL11A1* mutations (Majava et al. 2007). Genetic testing is currently available through a small number of clinical laboratories worldwide. Because there is considerable variation in the prevalence of specific manifestations in the different types of Stickler syndrome, molecular testing can help anticipate complications and predict natural history.

Differential Diagnosis

Wagner syndrome, Marshall syndrome, and the Weissenbacher–Zweymuller syndrome are often confused with Stickler syndrome. Historically, there has been a debate between "lumpers" and "splitters" as to whether the disorders are phenotypic variants of the same syndrome or distinct phenotypic entities.

Wagner (1938) reported a family with 28 members affected by low myopia (–3.00 diopters or less), a fluid vitreous, and cortical cataracts. No extraocular features were described. This syndrome has come to be called Wagner syndrome, hyaloidoretinal degeneration of Wagner, Wagner vitreoretinal degeneration, or erosive vitreoretinopathy. Affected individuals manifest complete absence of the normal vitreal scaffolding and preretinal, equatorial, and avascular grayish-white membranes. Retinal detachment was not noted in any of the 28 members of the original Swiss family studied by Wagner (1938). Schwartz et al. (1989) suggested that Wagner syndrome is characterized by vitreoretinal degeneration without extraocular manifestations, whereas Stickler syndrome also has extraocular manifestations in the musculoskeletal and craniofacial systems.

Fryer et al. (1990) studied a large family with Wagner syndrome. Affected individuals in this family had none of the extraocular features associated with Stickler syndrome. The *COL2A1* locus was excluded as the site of the mutation. Brown et al. (1995) demonstrated linkage of the original Wagner syndrome family to a locus on chromosome 5q13-14. However, the demonstration by Korkko et al. (1993) of a *COL2A1* mutation in a family with Wagner syndrome indicates that there are at least two loci causing Wagner syndrome. Affected individuals in this family had early-onset cataracts, lattice degeneration of the retina, and retinal detachment. Ankala et al. (2017) describe an individual with Wagner syndrome who was found to have a 11.7 kb deletion in exon 8 of the *VCAN* gene, which encodes the versican protein. Versican is a chondroitin sulfate proteoglycan that interspaces collagen fibers. The authors speculate that mutations in versican result in a phenotype similar to mutations in COL2A1 because verican and type 2 collagen interact closely in the extracellular matrix.

Marshall syndrome is a rare autosomal dominant skeletal dysplasia characterized by ocular abnormalities, sensorineural hearing loss, craniofacial anomalies, and anhidrotic ectodermal dysplasia (Marshall 1958). Affected individuals are myopic (ranging from –3 to –20 diopters) with a fluid vitreous and congenital cataracts. Retinal detachment is reported but uncommon. Craniofacial characteristics are common. Micrognathia is evident in some affected individuals, as is the absence of nasal bones, producing a short nose with a very flat nasal bridge, anteverted nares, and a long philtrum.

The distinction between the Stickler and Marshall syndromes is supported by the work of Ayme and Preus (1984) who surveyed published reports on the two syndromes. A set of 18 affected individuals with clinical description, photographs, and radiographs was used to tabulate a list of 53 features, including facial characteristics, size and habitus, and joint, limb, and hip manifestations. Analysis of these features revealed two distinct groups of individuals. The authors concluded that there was no reason not to consider these two syndromes as separate disorders. The authors suggested that the facies of individuals with the two disorders differ. Those with Marshall syndrome have a flat or retruded midface, whereas those with Stickler syndrome have a flat malar area, which is often erroneously described as a flat midface. In Marshall syndrome, there are thick calvaria, abnormal frontal sinuses, and intracranial calcifications, and the globes of the eye appear large, possibly because of a shallow orbit.

In affected members of a large kindred diagnosed with Marshall syndrome, Griffith et al. (1998) identified a mutation in the *COL11A1* gene. The results demonstrate that both

Marshall syndrome and a subset of Stickler syndrome families have mutations in the *COL11A1* gene. Majava et al. (2007) described 10 previously unreported individuals with heterozygous mutations in *COL11A1*. In 4 of the 10, the phenotype was felt to be Marshall syndrome because of characteristic facial features and severe early-onset hearing loss; all 4 had a splice-site mutation in intron 50 of *COL11A1*. All of the remaining individuals (6/10) had a phenotype overlapping between Marshall and Stickler syndromes, with less pronounced facial features. None of these less severely affected individuals had splice-site mutations in exon 50. Thus, it appears that there are at least some people in whom the distinction between Marshall and Stickler syndrome is very difficult to make on clinical or molecular grounds.

Weissenbacher and Zweymueller (1964) described an infant with Robin sequence (see Chapter 50) and chondrodysplasia. Their eponymous syndrome is characterized by neonatal micrognathia and rhizomelic chondrodysplasia with dumbbell-shaped femora and humeri, and regression of bone changes and normal growth in later years (Weissenbacher and Zweymueller 1964; Haller et al. 1975). Radiologically there are also vertebral coronal clefts. Average growth after 2–3 years, "catching up" to average height, is seen.

It has been suggested that Weissenbacher–Zweymuller syndrome may represent a neonatal expression of Stickler syndrome (Kelly et al. 1982). Ayme and Preus (1984) did a cluster analysis of published cases and concluded that Weissenbacher–Zweymuller syndrome and Stickler syndrome are separate entities. Among the features found in Weissenbacher–Zweymuller syndrome but not in Stickler syndrome were markedly decreased body length at birth with short limbs, catch-up growth after 2 or 3 years, lack of progressive deformity, and autosomal recessive inheritance as indicated by affected siblings with phenotypically normal parents. Myopia and retinal detachment are characteristics of Stickler syndrome that are not found in Weissenbacher–Zweymuller syndrome. However, in 1998, the person originally described with this syndrome was found to have a mutation in *COL11A2*, demonstrating that Weissenbacher–Zweymuller syndrome is at least allelic to a subset of non-ocular Stickler syndrome (Pihlajamaa et al. 1998; Harel et al. 2005).

MANIFESTATIONS AND MANAGEMENT

Optimal management of Stickler syndrome requires an interdisciplinary team of specialists, tailored to the affected individual's specific medical needs.

Growth and Feeding

These are not an issue in Stickler syndrome except in those babies with cleft palate, which makes it difficult for the baby to suck forcefully enough to draw milk through a nipple, causing feeding difficulties. It is possible to breastfeed some infants with cleft palate, though this will require extra patience and is more likely to be successful in babies with less severe clefts. There is an excellent summary of the feeding difficulties that may be seen in babies with Stickler syndrome who manifest the Robin phenotype in Chapter 50.

Evaluation

- Babies with Stickler syndrome who have a cleft palate should be monitored for inadequate nutrition by frequent plotting of height and weight attainment, as for all babies with cleft palate.
- A feeding evaluation by an occupational or physical therapist is recommended for any baby with difficulty feeding.

Management

- During bottle feeding, the baby with a cleft palate should be in a sitting position to help prevent milk from leaking into the nose through the cleft.
- A regular nipple designed for premature babies and a squeeze bottle can help a baby with cleft palate feed more easily.
- Occasionally, a baby may feed best with the aid of a plastic artificial palate, called an obturator, which temporarily covers a palatal cleft. Not all cleft palate teams favor this approach.
- Most cleft palate teams pay close attention to feeding and help parents establish good feeding practices right after the child is born.
- Aggressive management of obstructive airway complications seen in Robin syndrome is essential, including for prevention of feeding problems (see Chapter 50).

Development and Behavior

Both motor and cognitive development are normal in Stickler syndrome. Children with high myopia or significant hearing difficulties will require classroom accommodations. It is important to recognize these sensory deficits early and obtain the necessary visual or auditory aids to facilitate optimal early childhood learning. See Opthalmologic, Craniofacial, and Ears and Hearing below.

Ophthalmologic

Major manifestations of the ocular system in Stickler syndrome include moderate to severe myopia (at least –5 diopters) of childhood onset, spontaneous retinal detachment, and a congenital vitreous anomaly. Minor manifestations include presenile cataract, glaucoma, and strabismus.

Early-onset high myopia can lead to spontaneous retinal degeneration and detachment. Refractive errors, cataracts, and vitreoretinal abnormalities may be detected early in life in individuals with Stickler syndrome (Wilson et al. 1996).

Seery et al. (1990) found cataracts of various types or aphakia in 115 of 231 eyes of individuals with Stickler syndrome. The most frequent and distinctive lesions, described as wedge and fleck cataracts, accounted for 40 of the 93 cataracts observed.

Stickler et al. (2001), in a survey of 316 individuals self-identified as having Stickler syndrome, found that 95% reported eye pathology of some sort, including a history of retinal detachment in 60%, myopia in 90%, and blindness in 4%.

Evaluation

- Frequent (at least annual) examinations by an ophthalmologist are recommended, starting in infancy, to evaluate the severity of myopia, strabismus, the vitreoretinal architecture, and the presence or absence of cataracts and glaucoma.

Management

- Corrective lenses should be prescribed for myopia as early as possible.
- Prophylactic laser photocoagulation of vitreoretinopathy should be considered. Leiba et al. (1996) retrospectively reviewed a large family with Stickler syndrome (at least 42 affected members). Ten eyes were prophylactically treated with laser photocoagulation. In this family, the incidence of retinal detachment was significantly higher in non-lasered eyes than in lasered eyes. Ang et al. (2008) reported on a retrospective analysis of prophylaxis for retinal detachment using cryotherapy. Sixty-two individuals received bilateral prophylactic cryotherapy; five (8%) required further procedures during the follow-up period (mean of 11.5 years). Of 31 individuals who received unilateral prophylaxis after a retinal detachment in the contralateral eye, three (10%) experienced retinal detachment during the follow-up period. The authors speculate that people with Stickler syndrome who suffered a previous retinal detachment may be at higher risk for another detachment in the opposite eye, compared with those who never experienced a retinal detachment.
- Surgical interventions to repair retinal detachments or tears, or to correct cataracts, glaucoma, or strabismus, are performed as they would be for persons without Stickler syndrome, using either primary vitrectomy or scleral buckling. Abeysiri et al. (2007) reported on the outcomes of surgery for retinal detachment in people with Stickler syndrome. They found that the primary re-attachment rate was higher (84.2%) for primary vitrectomy compared with 67% for scleral-buckle surgery. Because complications were more commonly seen in those with re-detachment and multiple surgeries, these authors suggest that primary vitrectomy is the procedure of choice for individuals with Stickler syndrome who have experienced retinal detachment. Reddy et al. (2016) found that initial surgery with scleral buckle, vitrectomy, and combined vitrectomy and scleral buckle procedures all yielded equally successful anatomical outcomes with 100% retinal reattachment in 16 eyes from 13 individuals with Stickler syndrome. However, individuals with Stickler syndrome required a mean of 3.1 surgical interventions for repair of retinal detachment. People with Stickler syndrome presented with retinal detachment at an age of 10.5 years old. The rate of proliferative vitreoretinopathy was 75% and visual acuity outcomes were highly variable, demonstrating only a small improvement in visual acuity at long-term follow up compared to time of presentation.

Craniofacial

Major manifestations of the oro-auriculo-facial system in Stickler syndrome include cleft palate with or without the other features of Robin sequence. Minor manifestations include midface hypoplasia, micrognathia, and sensorineural hearing loss in the higher frequencies. Stickler et al. (2001) found that 41% of the 316 respondents (129 people) to their survey had a cleft palate. Seventy-five of these had been diagnosed at birth with Robin sequence.

A retrospective review of 90 children with Robin sequence was carried out using oximetry, apnea monitoring, and sleep studies to identify subgroups at a higher risk of developing severe airway obstruction, and response to treatment was assessed (Tomaski et al. 1995). Airway obstruction and feeding difficulties vary among children with Robin sequence. Treatment is challenging and the appropriate management may not be readily identified, causing delay in securing the airway. Although most children with Stickler syndrome do not suffer from debilitating airway and feeding difficulties, all babies with cleft palate or Robin sequence do have some feeding difficulties, with the severity proportional to the degree of airway obstruction (Tomaski et al. 1995; Stickler et al. 2001). (See Growth and Feeding above as well as Chapter 50 for a more detailed description of management of Robin sequence.)

Babies with cleft palate are especially susceptible to middle ear disease. The cleft can contribute to recurrent otitis media, which may lead to mild or moderate hearing loss. If properly treated in infancy and childhood, the hearing losses need not be permanent. If not properly managed, speech development may be affected and hearing loss may become permanent. Ear disease usually improves following cleft palate repair.

Some children with cleft palate may develop speech more slowly than other children. Their words may sound nasal and they may have difficulty producing some consonant sounds. However, after cleft palate repair, most children eventually catch up and develop normal speech, though some will require speech therapy.

Evaluation

- Otolaryngology evaluation should be done for all individuals with Stickler at diagnosis, especially to evaluate the affected individual for ear and palate abnormalities, including bifid uvula and submucous cleft palate. An otolaryngologist should examine all babies with cleft palate within the first 3–6 months of life.
- Hearing should be tested regularly in babies and children with cleft palate. Audiograms should be done at regular intervals to monitor mixed and sensorineural hearing loss.
- Speech and swallow evaluations should be accomplished for all affected individuals with clefts.

Management

- Otitis media is treated with antibiotics and, if recurrent, ventilation tubes, as would be done for any child without Stickler syndrome.
- The timing and type of surgery for cleft palate will depend on a number of factors, including the preference of the individual surgeon, the general health of the baby, and the nature of the cleft. Cleft palate repair is generally timed to restore the partition between the nose and mouth as early as possible, usually between 12 and 18 months.
- Hearing aids may be required. Devices to help children hear in school may be beneficial when hearing loss is documented, as may preferential classroom placement.
- Speech therapy will almost always be needed in children with clefts.
- Consider mandibular distraction in individuals with Robin sequence who have obstructive sleep apnea and fail nonoperative management (see Chapter 50).

Ears and Hearing

High-frequency sensorineural hearing loss is a frequent manifestation in Stickler syndrome, and ranges from mild to severe. Hearing loss may be progressive during adulthood and may be compounded by conductive hearing loss related to recurrent otitis and cleft palate. Hearing loss is a more consistent aspect of the phenotype in Stickler type 3, caused by pathogenic variants in COL11A2

Evaluation

- Audiometry is a standard evaluation as part of the diagnostic process for Stickler syndrome. All children affected by Stickler syndrome should have a hearing test. Affected adults should be aware that hearing loss may progress in adulthood and have a low threshold for repeat testing if hearing loss is suspected.

Treatment

- Treatment for hearing loss is determined by the severity and type of deafness. Options include hearing aids and vibrotactile devices; cochlear implants are considered in children over the age of 12 months with severe to profound deafness.
- If hearing loss is prelingual, optimal management is crucial for optimal cognitive development.

Musculoskeletal

Major manifestations of the musculoskeletal system in Stickler syndrome include early-onset degenerative joint disease and radiographic evidence of a mild spondyloepiphyseal dysplasia. Minor manifestations include joint hypermobility, including hyperextensibility of the knees and elbows; joint pain or stiffness in childhood; arachnodactyly; scoliosis; and Marfanoid habitus. The joint hyperflexibility of affected youth usually evolves into degenerative arthritis. The musculoskeletal manifestations can be extremely variable both within and between families. Radiographic manifestations are often very mild and are infrequently reported.

Osteoarthritis, with onset as early as the third decade of life, is one of the major manifestations of Stickler syndrome. Severe osteoarthritis with an early onset has been associated with mutations in the *COL2A1* gene (Ritvaniemi et al. 1995). Radiographs show evidence of a mild spondyloepiphyseal dysplasia. A specific *COL2A1* mutation (Arg519Cys) has been identified in at least five unrelated families with early-onset osteoarthritis associated with mild spondyloepiphyseal dysplasia. Spondyloepiphyseal dysplasia is characterized by typical skeletal changes, including delayed ossification of the axial skeleton with ovoid vertebral bodies. With time, the vertebral bodies appear flattened. There is delayed ossification of the femoral heads, pubic bones, and calcaneus (heel). Coxa vara (deformity of the hip joint) is common.

Evaluation

- Individuals with Stickler syndrome should have a full musculoskeletal examination to determine the involvement of the skeleton in the phenotype. Orthopedic problems, spinal column instability, deformities of the legs, arthrosis of the hips, and other skeletal manifestations can then be identified and appropriate treatment planned.

- Rheumatology consultation may be helpful to determine severity and management of osteoarthritis.
- Rehabilitation medicine or physical therapy consultation may be indicated to evaluate body mechanics and range of motion and strength, and to recommend strengthening exercises.
- Radiographic skeletal survey in childhood is suggested to document the presence and severity of spondyloepiphyseal dysplasia and other orthopedic abnormalities.
- If chronic pain is present, evaluation by a pain specialist may be helpful.
- Occupational therapy evaluation is indicated if there is concern about the affected individual's ability to function in daily activities.

Management

- Pain management, both non-pharmacologic and pharmacologic, may significantly enhance quality of life and should be prescribed as needed. A pain specialist may be helpful in this regard.
- Splints may be required for strength and to stabilize lax joints, particularly during sports.
- Aids or braces may be required to assist in daily activities.
- Education of the affected individual and family regarding joint protection is important.
- Education of the affected individual regarding exercises for strengthening muscles around lax joints is recommended.
- Hydrotherapy, or other physical therapy modalities, may increase range of motion, endurance, and strength.
- An individual exercise program should be developed, such as one recommended by the rehabilitation medicine or physical therapy evaluation.

Cardiovascular

Among 57 individuals with Stickler syndrome, Liberfarb and Goldblatt (1986) found that 50% of females and 43% of males had mitral valve prolapse. More recent publications suggest that mitral valve prolapse may not be any more frequent in Stickler syndrome than it is in the general population (Ahmad et al. 2003). Each of the authors of this review have observed one individual with molecularly confirmed Stickler syndrome with aortic root aneurysm and dissection, but the prevalence of this complication has not been systematically studied.

Evaluation

- Echocardiogram should be performed at baseline and annually if aortic root dilation is identified.

Management

- Antibiotic prophylaxis for subacute bacterial endocarditis was formerly recommended for all individuals with mitral valve prolapse. However, guidelines from the American Heart Association in 2007 state that in most cases antibiotic prophylaxis is not necessary for persons with mitral valve prolapse or mitral regurgitation.
- β-Blockade may be used for symptomatic mitral valve prolapse.

ACKNOWLEDGMENTS

Previous editions of this chapter were co-authored by Drs. Douglas Wilkin and Ruth Liberfarb. Their contributions to this work are gratefully acknowledged.

RESOURCES

Support Organizations

Stickler Involved People (SIP)
15 Angelina
Augusta, KS 67010, USA
Phone: (316) 259-5194
Email: sip@sticklers.org
Website: *www.sticklers.org/sip/*

Stickler Syndrome Support Group
PO Box 371
Walton on Thames
Surrey, KT12 2YS, UK
Phone: 01932 267635
Email: *info@stickler.org.uk*
Website: *www.stickler.org.uk*

Books and Brochures

Hughes W (2006) *Stickler–The Elusive Syndrome*. Celtic Connection.
This book explains the condition and possible medical problems in layman terms. The approach is positive and leaves the reader with hope and skills to manage Stickler syndrome. It is available through SIP.
Ratchford M (ed.) Booklet written by members of SIP on living with Stickler available at https://stickler.org/stickler-dvd-books/

Online Agencies

March of Dimes: Cleft lip and palate
Website: *https://www.marchofdimes.org/index.aspx*

REFERENCES

Abeysiri P, Bunce C, da Cruz L (2007) Outcomes of surgery for retinal detachment in patients with Stickler syndrome: A comparison of two sequential 20-year cohorts. *Graefe Arch Clin Exp Ophthalmol* 245:1633–1638.

Ahmad NN, Alakokko L, Knowlton RG, Jimenez SA, Weaver EJ, Maguire JI, Tasman W, Prockop DJ (1991) Stop codon in the procollagen-Ii gene (Col2a1) in a family with the Stickler syndrome (arthroophthalmopathy). *Proc Natl Acad Sci U S A* 88:6624–6627.

Ahmad N, Richards AJ, Murfett HC, Shapiro L, Scott JD, Yates JR, Norton J, Snead MP (2003) Prevalence of mitral valve prolapse in Stickler syndrome. *Am J Med Genet* 116A:234–237.

Alzahrani F, Al Hazzaa SA, Tayeb H, Alkuraya FS (2015) LOXL3, encoding Lysol oxidase-like 3, is mutated in a family with autosomal recessive stickler syndrome. *Hum Genet* 134:451–453.

Ang A, Poulson AV, Goodburn SF, Richards AJ, Scott JD, Snead MP (2008) Retinal detachment and prophylaxis in type I Stickler syndrome. *Ophthalmology* 115:164–168.

Ankala A, Jain N, Hubbard B, Alexander JJ, Shankar SP (2018) Is exon 8 the most critical or the only dispensable exon of the VCAN gene? Insights into VCAN variants and clinical spectrum of Wagner syndrome. *Am J Med Genet* 176:1778–1783.

Ayme S, Preus M (1984) The Marshall and Stickler syndromes: Objective rejection of lumping. *J Med Genet* 21:34–38.

Baker S, Booth C, Fillman C, Shapiro M, Blair MP, Hyland JC, Ala-Kokko L (2011) A loss of function mutation in the *COL9A2* gene cause autosomal recessive Stickler syndrome. *Am J Med Genet* 155A: 1668–1672.

Brown DM, Vandenburgh K, Kimura AE, Weingeist TA, Sheffield VC, Stone EM (1995a) Novel frameshift mutations in the procollagen-2 gene (COL2A1) associated with Stickler syndrome (hereditary arthroophthalmopathy). *Hum Mol Genet* 4:141–142.

Brown DM, Graemiger RA, Hergersberg M, Schinzel A, Messmer EP, Niemeyer G, Schneeberger SA, Streb LM, Taylor CM, Kimura AE, et al (1995b) Genetic linkage of Wagner disease and erosive vitreoretinopathy to chromosome 5q13-14. *Arch Ophthalmol* 113:671–675.

Donoso LA, Edwards AO, Frost AT, Ritter R 3rd Ahmad N, Vrabec T, Rogers J, Meyer D, Parma S (2003) Clinical variability of Stickler syndrome: Role of exon 2 of the collagen COL2A1 gene. *Surv Ophthalmol* 48:191–203.

Faletra F, D'Adamo AP, Bruno I, Athanasakis E, Biskup S, Esposito L, Gasparini P (2014) Autosomal Recessive stickler syndrome due to a loss of function mutatino in the COL9A3 gene. *Am J of Med Genet* 164:42–47.

Francomano CA, Liberfarb RM, Hirose T, Maumenee IH, Streeten EA, Meyers DA, Pyeritz RE (1987) The Stickler syndrome: Evidence for close linkage to the structural gene for type II collagen. *Genomics* 1:293–296.

Fryer AE, Upadhyaya M, Littler M, Bacon P, Watkins D, Tsipouras P, Harper PS (1990) Exclusion of Col2a1 as a candidate gene in a family with Wagner-Stickler syndrome. *J Med Genet* 27:91–93.

Griffith AJ, Sprunger LK, Sirko-Osadsa DA, Tiller GE, Meisler MH, Warman ML (1998) Marshall syndrome associated with a splicing defect at the COL11A1 locus. *Am J Hum Genet* 62:816–823.

Haller JO, Berdon WE, Robinow M, Slovis TL, Baker DH, Johnson GF (1975) The Weissenbacher-Zweymuller syndrome of micrognathia and rhizomelic chondrodysplasia at birth with subsequent normal growth. *Am J Roentgenol Radium Ther Nucl Med* 125:936–943.

Hanson-Kahn A, Li B, Cohn DH, Nickerson DA, Bamshad MJ; University of Washington Center for Mendelian Genomics, Hudgins L (2018) Autosomal recessive Stickler syndrome resulting from a COL9A3 mutation. *Am J Med Genet* 176:2887–2891.

Harel T, Rabinowitz R, Hendler N, Galil A, Flusser H, Chemke J, Gradstein L, Lifshitz T, Ofir R, Elbedour K, Birk OS (2005) COL11A2 mutation associated with autosomal recessive Weissenbacher-Zweymuller syndrome: molecular and clinical overlap with otospondylomegaepiphyseal dysplasia (OSMED). *Am J Med Genet* 132A(1):33–5.

Herrmann J, France TD, Spranger JW, Opitz JM, Wiffler C (1975) The Stickler syndrome (hereditary arthroophthalmopathy). *Birth Defects Orig Artic Ser* 11:76–103.

Kelly TE, Wells HH, Tuck KB (1982) The Weissenbacher-Zweymuller syndrome: Possible neonatal expression of the Stickler syndrome. *Am J Med Genet* 11:113–119.

Knowlton RG, Weaver EJ, Struyk AF, Knobloch WH, King RA, Norris K, Shamban A, Uitto J, Jimenez SA, Prockop DJ (1989) Genetic-linkage analysis of hereditary arthro-ophthalmopathy (Stickler syndrome) and the type-II procollagen gene. *Am J Hum Genet* 45:681–688.

Korkko J, Ritvaniemi P, Haataja L, Kaariainen H, Kivirikko KI, Prockop DJ, Alakokko L (1993) Mutation in type-II Procollagen (Col2a1) that substitutes aspartate for glycine alpha-1-67 and that causes cataracts and retinal-detachment—Evidence for molecular heterogeneity in the Wagner syndrome and the Stickler syndrome (arthroophthalmopathy). *Am J Hum Genet* 53:55–61.

Lauritsen KF, Lildballe DL, Coucke PJ, Monrad R, Larsen DA, Gregersen PA (2017) A mild form of Stickler syndrome type II caused by mosaicism of COL11A1. *Eur J Med Genet* 60:275–278.

Leiba H, Oliver M, Pollack A (1996) Prophylactic laser photocoagulation in Stickler syndrome. *Eye* 10 (Pt 6):701–708.

Levine RA, Stathogiannis E, Newell JB, Harrigan P, Weyman AE (1988) Reconsideration of echocardiographic standards for mitral valve prolapse: Lack of association between leaflet displacement isolated to the apical four chamber view and independent echocardiographic evidence of abnormality. *J Am Coll Cardiol* 11:1010–1019.

Liberfarb RM, Goldblatt A (1986) Prevalence of mitral-valve prolapse in the Stickler syndrome. *Am J Med Genet* 24: 387–392.

Liberfarb RM, Levy HP, Rose PS, Wilkin DJ, Davis J, Balog JZ, Griffith AJ, Szymko-Bennett YM, Johnston JJ, Francomano CA (2003) The Stickler syndrome: Genotype/phenotype correlation in 10 families with Stickler syndrome resulting from seven mutations in the type II collagen gene locus COL2A1. *Genet Med* 5:21–27.

Majava M, Hoornaert KP, Bartholdi D, Bouma MC, Bouman K, Carrera M, Devriendt K, Hurst J, Kitsos G, Niedrist D, Petersen MB, Shears D, Stolte-Dijkstra I, Van Hagen JM, Ala-Kokko L, Mannikko M, Mortier GR (2007) A report on 10 new patients with heterozygous mutations in the COL11A1 gene and a review of genotype-phenotype correlations in type XI collagenopathies. *Am J Med Genet* 143A:258–264.

Marshall D (1958) Ectodermal dysplasia: Report of kindred with ocular abnormalities and hearing defect. *Am J Ophthalmol* 45:143–156.

Maumenee IH (1979) Vitreoretinal degeneration as a sign of generalized connective tissue diseases. *Am J Ophthalmol* 88:432–449.

Nixon TRW, Richards A, Towns LK, Fuller G, Abbs S, Alexander P, McNinch A, Sandford RN, Snead MP (2019) Bone morphogenetic protein 4 (BMP4) loss-of-function variant associated with autosomal dominant Stickler syndrome and renal dysplasia. *Eur J Hum Genet* 27(3):369–377.

Opitz JM, France T, Herrmann J, Spranger JW (1972) The Stickler syndrome. *N Engl J Med* 286:546–547.

Pihlajamaa T, Prockop DJ, Faber J, Winterpacht A, Zabel B, Giedion A, Wiesbauer P, Spranger J, Ala-Kokko L (1998) Heterozygous glycine substitution in the COL11A2 gene in the original patient with the Weissenbacher-Zweymuller syndrome demonstrates its identity with heterozygous OSMED (nonocular Stickler syndrome). *Am J Med Genet* 80:115–120.

Reddy DN, Yonekawa Y, Thomas BJ, Nudleman ED, Williams GA (2016) Long-term surgical outcomes of retinal detachment in patients with Stickler syndrome. *Clin Ophthalmol* 10:1531–1534.

Richards AJ, Yates JRW, Williams R, Payne SJ, Pope FM, Scott JD, Snead MP (1996) A family with Stickler syndrome type 2 has a mutation in the COL11A1 gene resulting in the substitution of glycine 97 by valine in alpha 1(XI) collagen. *Hum Mol Genet* 5:1339–1343.

Richards AJ, Laidlaw M, Whittaker J, Treacy B, Rai H, Bearcroft P, Baguley DM, Poulson A, Ang A, Scott JD, Snead MP (2006). High efficiency of mutation detection in type I stickler syndrome using a two-stage approach: vitreoretinal assessment coupled with exon sequencing for screening COL2A1. *Hum Mutat* 27:696–704.

Rimoin DL, Lachman RS (1993) Genetic disorders of the osseous skeleton. In: *McKusick's Heritable Disorders of Connective Tissue*, Beighton P (ed.), St. Louis: Mosby. pp. 557–698.

Ritvaniemi P, Korkko J, Bonaventure J, Vikkula M, Hyland J, Paassilta P, Kaitila I, Kaariainen H, Sokolov BP, Hakala M (1995) Identification of COL2A1 gene mutations in patients with chondrodysplasias and familial osteoarthritis. *Arthritis Rheum* 38:999–1004.

Schreiner RL, McAlister WH, Marshall RE, Shearer WT (1973) Stickler syndrome in a pedigree of Pierre Robin syndrome. *Am J Dis Child* 126:86–90.

Schwartz RC, Watkins D, Fryer AE, Goldberg R, Marion R, Polomeno RC, Spallone A, Upadhyaya M, Tsipouras P (1989) Non-allelic genetic heterogeneity in the vitreoretinal degenerations of the Stickler and Wagner types and evidence for intragenic recombination at the COL2A1 locus. *Am J Hum Genet* 45(4 Suppl.):A218.

Seery CM, Pruett RC, Liberfarb RM, Cohen BZ (1990) Distinctive cataract in the Stickler syndrome. *Am J Ophthalmol* 110:143–148.

Snead MP, Payne SJ, Barton DE, Yates JRW, Alimara L, Pope FM, Scott JD (1994) Stickler syndrome—Correlation between vitreoretinal phenotypes and linkage to COL2A1. *Eye* 8:609–614.

Snead MP, Yates JRW, Pope FM, Temple IK, Scott JD (1996) Masked confirmation of linkage between type 1 congenital vitreous anomaly and COL 2A1 in Stickler syndrome. *Graefes Arch Clin Exp Ophthalmol* 234:720–721.

Snead MP, Yates JRW, Williams R, Payne SJ, Pope FM, Scott JD (1996b) Stickler syndrome type 2 and linkage to the COL11A1 gene. In: *Molecular and Developmental Biology of Cartilage*, de Crombrugghe B, Horton WA, Olsen BR, Ramirez F (eds), New York: New York Academy of Sciences, pp. 331–332.

Stickler GB, Pugh DG (1967) Hereditary progressive arthroophthalmopathy. II. Additional observations on vertebral abnormalities, a hearing defect, and a report of a similar case. *Mayo Clin Proc* 42:495–500.

Stickler GB, Belau PG, Farrell FJ, Jones JD, Pugh DG, Steinberg AG, Ward LE (1965) Hereditary progressive arthro-ophthalmopathy. *Mayo Clin Proc* 40:433–455.

Stickler GB, Hughes W, Houchin P (2001) Clinical features of hereditary progressive arthro-ophthalmopathy (Stickler syndrome): A survey. *Genet Med* 3:192–196.

Tomaski SM, Zalzal GH, Saal HM (1995) Airway obstruction in the Pierre Robin sequence. *Laryngoscope* 105:111–114.

Tompson SW, Johnson C, Abbott D, Bakall B, Soler V, Yanovitch TL, Whisenhunt KN, Klemm T, Rozen S, Stone EM, Johnson M, Young TL (2017) Reduced penetrance in a large Caucasian pedigree with Stickler syndrome. *Ophthalmic Genet* 38:43–50.

Van Camp G, Snoeckx RL, Hilgert N, van den Ende J, Fukuoka H, Wagatsuma M, Suzuki H, Smets RME, Vanhoenacker F, Declau F, Van De Heyning P, Usami S (2006) A new autosomal recessive form of Stickler syndrome is caused by a mutation in the COL9A1 gene. *Am J Hum Genet* 79:449–457.

Vikkula M, Mariman ECM, Lui VCH, Zhidkova NI, Tiller GE, Goldring MB, Vanbeersum SEC, Malefijt MCD, Vandenhoogen FHJ, Ropers HH, Mayne R, Cheah KSE, Olsen BR, Warman ML, Brunner HG (1995) Autosomal-dominant and recessive osteo-chondrodysplasias associated with the COL11A2 Locus. *Cell* 80:431–437.

Wagner H (1938) Ein bisher unbekanntes Erbleiden des Auges (degeneration hyaloideoretinalis hereditaria), beobachtet im Kanton Zurich. *Klin Mbl Augenheilk* 100:840–858.

Weissenbacher G, Zweymueller E (1964) Simultaneous occurrence of the Pierre Robin syndrome and fetal chondrodysplasia. *Monatsschr Kinderheilkd* 112:315–317.

Wilkin DJ, Mortier GR, Johnson CL, Jones MC, De Paepe A, Shohat M, Wildin RS, Falk RE, Cohn DH (1998) Correlation of linkage data with phenotype in eight families with Stickler syndrome. *Am J Med Genet* 80:121–127.

Wilson MC, McDonald-McGinn DM, Quinn GE, Markowitz GD, LaRossa D, Pacuraru AD, Zhu X, Zackai EH (1996) Long-term follow-up of ocular findings in children with Stickler's syndrome. *Am J Ophthalmol* 122:727–728.

Zechi-Ceide RM, Oliveira NAJ, Guion-Almeida ML, Antunes LFBB, Richieri-Costa A, Passos-Bueno MRS (2008) Clinical evaluation and COL2A1 gene analysis in 21 Brazilian families with Stickler syndrome: Identification of novel mutations, further genotype/phenotype correlation, and its implications for the diagnosis. *Eur J Med Genet* 51:183–196.

57

TREACHER COLLINS SYNDROME AND RELATED DISORDERS

MARILYN C. JONES

Department of Pediatrics, University of California, San Diego, and Cleft Palate and Craniofacial Treatment Programs, Rady Children's Hospital, San Diego, California, USA

INTRODUCTION

Incidence

Treacher Collins syndrome (TCS) is the prototype disorder for discussion of management of the mandibulofacial and acrofacial dysostoses including Nager syndrome, Miller syndrome, mandibulofacial dysostosis with microcephaly (Guion-Almeida type), and several other rare disorders described in a very few families. TCS is also termed mandibulofacial dysostosis (Franceschetti and Klein 1949). Although not the first to describe an affected individual, Edward Treacher Collins (1900), a London ophthalmologist, outlined the essential features of the condition, and the eponym has stuck. The terms Treacher Collins syndrome and mandibulofacial dysostosis are often used interchangeably in the literature.

Gorlin estimated that over 450 cases had been reported (Gorlin et al. 2001). A birth prevalence of 1 in 10,000 to 1 in 50,000 live births has been suggested.

Diagnostic Criteria

The diagnosis of TCS is based upon a characteristic pattern of malformations (Figure 57.1) including:

- Malformation of the auricle, usually associated with hearing loss due to atresia of the external auditory canal and/or malformation of the ossicles of the middle ear;
- Downslanted palpebral fissures with variable deficiency of the zygomatic arch;
- Coloboma or notching of the lateral portion of the lower eyelid with absence of the eyelashes medial to the notch; and
- Hypoplasia of the facial bones (mandible usually more than zygomatic complex) resulting in micrognathia with a relative prominence to the central facial structures.

Other common manifestations include choanal atresia/ stenosis, pharyngeal hypoplasia, submucous or overt cleft palate occasionally of the Pierre Robin type (U-shaped cleft, micrognathia, and glossoptosis) (see Robin sequence, Chapter 50), tags of tissue or fistulae on the line between the tragus of the ear and the corner of the mouth, and an aberrant growth of hair extending onto the cheek in front of the ear. Intelligence is normal. Cleft lip with or without cleft palate is rare.

Careful clinical evaluation supported by radiographic examination will identify the majority of affected individuals. In a comprehensive clinical and molecular evaluation of 59 individuals in two families with TCS, Dixon et al. (1994) and Marres et al. (1995) identified three nonpenetrant individuals in addition to 13 in whom the diagnosis could be made on clinical grounds alone. Minimal diagnostic criteria included downslanting of the palpebral fissures, sparse medial eyelashes on the lower lid, a recessed

---

Cassidy and Allanson's Management of Genetic Syndromes, Fourth Edition.
Edited by John C. Carey, Agatino Battaglia, David Viskochil, and Suzanne B. Cassidy.
© 2021 John Wiley & Sons, Inc. Published 2021 by John Wiley & Sons, Inc.

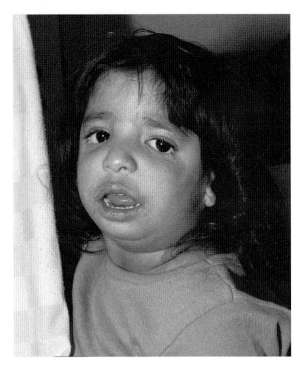

FIGURE 57.1 An almost 3-year-old girl demonstrating microtia, malar deficiency, eyelid colobomas with absence of the medial lower lashes, micrognathia, and an open bite.

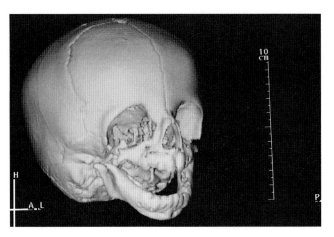

FIGURE 57.2 Three-dimensional reconstructed computerized tomography scan of a 2-year-old boy demonstrating abnormalities in the zygomatic complex as well as mandibular hypoplasia.

chin, and radiographic evidence of hypoplasia of the zygomatic complex. In the surgical literature, three-dimensional reconstructed computerized tomography has become the preferred imaging modality for treatment planning (Figure 57.2). However, for the purposes of clinical evaluation, a Waters' view of the skull to assess the zygomatic arches and an orthopantogram to evaluate the mandible and temporomandibular joints should suffice (Marres et al. 1995). In a molecular study including 14 cases of TCS presumed to be de novo based on clinical assessment of parents, 4 apparently unaffected parents harbored heterozygous pathogenic mutations of *TCOF1* (Teber et al. 2004). Although radiographic evaluation was not part of the clinical assessment of the parent, the study points out the difficulty of excluding the diagnosis on clinical grounds alone. These authors noted that clinical features were often more apparent in childhood photographs.

Etiology, Pathogenesis, and Genetics

Since the classic descriptions of several large kindreds with multiple affected family members, TCS has been known to be inherited as an autosomal dominant condition in the majority of families (Wildervanck 1960; Rovin et al. 1964; Fazen et al. 1967). Affected individuals have a 50% risk to transmit the condition to each offspring. Although widely variable with respect to phenotype, the gene is virtually fully penetrant if careful attention is paid to subtle clinical manifestations. Molecular analysis correlating genotype with phenotype has generally borne out the clinical observation of near complete penetrance and has allowed more precise delineation of minimal diagnostic criteria, as discussed above (Dixon et al. 1994; Marres et al. 1995). Familial cases constitute roughly 40% of the affected population. Most cases represent fresh gene mutation associated with older paternal age (Jones and Smith 1976).

Studies attempting to correlate genotype with phenotype have documented that the inter- and intrafamilial variability observed in TCS does not depend on the nature or type of mutation, the sex of the affected individual, the parent of origin, or whether the case is familial or sporadic (Splendore et al. 2000; Teber et al. 2004).

Using classical linkage techniques, one gene for TCS was mapped to chromosome 5q31-34 (Dixon et al. 1991; Jabs et al. 1991). A five year collaborative effort resulted in positional cloning of a gene of then unknown function, *TCOF1* (The Treacher Collins Collaborative Group 1996). The majority of mutations in *TCOF1* result in the creation of a premature termination codon, truncating the predicted protein, called treacle, which has been shown by immunofluorescent staining to localize to the nucleolus of the cell (Marsh et al. 1998). Mice engineered to be heterozygous for *Tcof1*, the murine ortholog of *TOCF1*, die in the perinatal period as a result of severe craniofacial malformations including exencephaly, anophthalmos, obliterated nasal passages, and anomalies of the maxilla. Massive apoptosis in the prefusion neural folds, the site of maximum expression of *Tcof1*, appears to account for the defects (Dixon et al. 2000). Interestingly, the phenotype in *Tcof1* heterozygous mice depends highly on the genetic background on which the mutation is placed, possibly accounting for some of the marked variability observed in humans (Dixon and Dixon 2004). Further study has documented that treacle regulates the formation and proliferation of neural crest cells (at the level of the individual cell) by reducing 2′-O-methylation of pre-rRNA and controlling the production of mature ribosomes (Gonzales et al. 2005; Dixon et al. 2006). In addition, treacle modulates expression of multiple genes that mediate p53-dependent transcription and cell-cycle regulation such that reduced levels of treacle cause decreased proliferation and increased apoptosis of delaminating neural crest stem cells. This reduces the numbers available to populate craniofacial structures. Inhibition of this p53 activity rescues the craniofacial phenotype in the mouse (Jones et al. 2008). Several lines of evidence have shown treacle to be a novel component of the MRNM (MDC1-RAD50-NBS1-MRE11) complex, which is an important mediator of double-stranded DNA damage response/repair (Sakai et al. 2016). Progenitor neural crest cells endogenously generate high levels of reactive oxygen species inducing oxidative DNA damage. Interestingly, prenatal treatment of *Tcof1* heterozygous mouse embryos with strong antioxidants mitigated the phenotype in 30% of embryos (Sakai et al. 2016). Treacle appears to be an essential protein for the protection of neural crest cells from endogenous and exogenous oxidative stress induced DNA damage. Finally, ubiquitination of treacle is required for the regulation of translation and for directing the cell-fate of embryonic stem cells, suggesting that treacle is critical not only for production of ribosomes but of their modification (Werner et al. 2015). Over 200 mutations in *TCOF1* have been identified, including deletions, point mutations, and insertions, the overwhelming majority of which result in haploinsufficiency of treacle (Splendore et al. 2000; Ellis et al. 2002; Teber et al. 2004). Investigation of *TCOF1* mutation-negative patients with TCS has led to the identification of two additional genes association with the phenotype, autosomal recessive *POLR1C* and autosomal dominant *POLR1D* (Dauwerse et al. 2011), which code for subunits I and III of RNA polymerase that are necessary for ribosome biosynthesis. Studies in zebrafish models of TCS have documented that mutations in these genes result in deficient ribosome biogenesis and Tp53-dependent death of neuroepithelial cells with deficiency of neural crest (Noack Watt et al. 2016). TCS has been referred to as a ribosomopathy.

Phenocopies of TCS have also been produced teratogenically in animal models using vitamin A and isotretinoin (Poswillo 1975; Wiley et al. 1983; Sulik et al. 1987; Emmanouil-Nikoloussi et al. 2000). Both agents create abnormalities in cranial neural crest through defects in cell migration, differentiation, and/or apoptosis. It has been suggested that isotretinoin exaggerates neural crest cell apoptosis via upregulation of the pro-apoptotic transcription factor p53, the guardian of the genome. A possible human phenocopy induced by vitamin A was reported by Lungarotti et al. (1987). Mutation in the canine *TCOF1* gene is associated with a brachycephalic, broad skull/short face phenotype in the domestic dog (Haworth et al. 2001).

Terrazas et al. (2017) provide an elegant review of the molecular biology of TCS and related syndromes.

Diagnostic Testing

The diagnosis of TCS is made on clinical grounds based on the characteristic facial malformations, but molecular confirmation and targeted testing in at-risk family members is helpful in counseling.

DNA next generation panel testing for mutations in *TCOF1*, *POLR1C*, and *POLR1D* is clinically available but rarely necessary for the purposes of establishing the diagnosis in probands. Mutation analysis is critical when there is clinical difficulty determining whether or not a first-degree relative of an affected individual carries the pathogenic allele and in situations in which prenatal diagnosis is considered (Dixon et al. 2004). Molecular testing is necessary to identify the rare autosomal recessive cases of TCS. In a cohort of 146 individuals with clinically diagnosed TCS, 63% had a pathogenic mutation in *TCOF1*, 6% in *POLR1D*, none in *POLR1C*, 4% were mutation negative, and four patients with atypical phenotypes had mutations in *EFTUD2* (see below) (Vincent et al. 2016).

Ultrasound has allowed successful prenatal diagnosis and should help identify those infants with severe craniofacial involvement (Ochi et al. 1998; Hsu et al. 2002; Rotten et al. 2002; Tanaka et al. 2002).

Differential Diagnosis

TCS is distinguished from the other genetically determined disorders with mandibulofacial dysostoses by the absence of limb anomalies and/or microcephaly.

Mandibulofacial dysostosis with microcephaly (Guion-Almeida type) is distinguished by pre- or postnatal onset microcephaly, malar and mandibular hypoplasia, cleft palate, and ear anomalies with hearing loss (Guion-Almeida et al. 2006). Less common manifestations include choanal atresia, aural atresia, esophageal atresia and epibulbar dermoids. Cases often present as atypical TCS or atypical Goldenhar syndrome. Heterozygous mutations or deletions in *EFTUD2* cause this condition. *EFTUD2* encodes a highly conserved spliceosomal GTPase that plays a critical role in catalytic splicing and post-splicing-complex disassembly (Lines et al. 2012). Varying degrees of intellectual disability are characteristic of this condition.

Nager syndrome is distinguished by similar craniofacial malformations and radial limb reduction defects consisting of anything from mild thenar hypoplasia to absent radius with a radial club hand (see Chapter 50). In addition, affected individuals may have radioulnar synostosis, vertebral anomalies, cardiac defects, and subtle differences in the great toe, although the feet are usually normal. Intelligence is usually normal. Nager syndrome is caused by mutations in *SF3B4* that result in haploinsufficiency of SAP49, a spliceosomal protein that is one of seven core components of the highly conserved SF3B complex (Bernier et al. 2012). Acrofacial dysostosis of Rodriguez, which is characterized by severe mandibular hypoplasia, upper limb phocomelia, absent fibulae, and pulmonary dysfunction, is also due to mutations in *SF3B4* (Irving et al. 2016)

Miller syndrome, also termed postaxial acrofacial dysostosis, is characterized by similar craniofacial anomalies with ulnar limb reduction defects, usually absence of the fifth finger and metacarpal, although syndactyly and thenar hypoplasia may occur as well. The lower extremities are often involved, sometimes severely, with absence of the fibula. The lower lid ectropion in Miller syndrome may be severe. Cleft lip is more frequent in Miller syndrome than in TCS. Miller syndrome is due to mutations in *DHODH*, which encodes a key enzyme in the pyrimidine de novo biosynthesis pathway (Ng et al. 2009). Acrofacial dysostosis, Cincinnati type, which is characterized by severe craniofacial anomalies and variable limb defects, is due to heterozygous mutations in *POLR1A* (Weaver et al. 2015).

TCS is occasionally confused with the oculo-auriculo-vertebral spectrum (OAVS), also termed hemifacial microsomia, Goldenhar syndrome, or facio-auriculo-vertebral malformation spectrum (see Chapter 42). Over 50% of individuals with OAVS have abnormalities outside the craniofacial area, thus distinguishing OAVS from TCS. Craniofacial involvement of the branchial arch derivatives in OAVS is always asymmetric, although it may be bilateral. Eyelid colobomas in OAVS usually involve the upper lid. Epibulbar dermoids and lipodermoids do not occur in TCS, although they are rarely features of mandibulofacial dysostosis with microcephaly. Retinoic acid embryopathy shares microtia, micrognathia and cleft palate in common with TCS; however, affected individuals typically also have defects of the central nervous system and the heart and they lack the zygomatic abnormalities that are characteristic of TCS, including malar hypoplasia, downslanted palpebral fissures, and lower lid colobomas.

MANIFESTATIONS AND MANAGEMENT

Individuals with TCS, particularly those with severe craniofacial involvement, are best treated in centers with experience in the care and habilitation of people with this and related diagnoses. The American Cleft Palate-Craniofacial Association maintains a directory of members of craniofacial treatment teams that includes most of the major centers in the United States. Although the absolute number of people treated at an individual center or by a specific surgeon does not correlate directly with outcome, individuals with severe craniofacial malformations present some of the greatest reconstructive challenges in craniofacial surgery. Prior experience should be a factor in choosing a treatment team. Despite a considerable literature addressing the approach to reconstruction, recommendations are primarily based on the experience of experts and not on controlled trials (Plomp et al. 2016). Particularly in the area of surgery, advances in technology are changing management at a rapid rate.

Roberts et al. (1975) presented convincing evidence that the craniofacial malformations in TCS are stable over time. Thus, the malformations present at birth define the severity of involvement. The discussion in this section will focus on principles of management rather than specific operative approaches.

Growth and Feeding

The growth potential of children with TCS is normal. Problems with growth encountered in some individuals usually derive from inability to deliver adequate nutrition secondary to the craniofacial malformations combined with increased caloric expenditure in the effort to feed.

Many infants with TCS breast-feed successfully. The ability of a child to feed orally depends upon the degree of airway obstruction and the presence of a cleft palate, which occurs in 23% of cases (Plomp et al. 2016). Neither absolutely precludes oral alimentation; however, both present challenges.

Gastroesophageal reflux occurs in TCS and combined with the more severe orofacial malformations can be challenging to treat.

Evaluation

- It is recommended that a feeding specialist or occupational therapist evaluate any child with TCS who has feeding difficulties.
- Standard evaluation for gastroesophageal reflux should be initiated if the diagnosis is suspected.

Management

- For some babies, the soft squeeze devices used for infants with isolated clefts suffice for oral alimentation.
- Some infants require gavage-assisted feeding.
- For those with extreme difficulty, placement of a gastrostomy tube may be necessary.

Development and Behavior

The vast majority of individuals with TCS are cognitively normal. With proper attention to the hearing loss, language development should be normal. Vallino-Napoli (2002) reviewed the speech profiles in a group of children with TCS followed at a tertiary care center (reflecting the severe end of the spectrum), all of whom had hearing loss. One-third had a palatal cleft. Just less than a quarter had a tracheostomy. All had abnormalities of speech, which could be characterized as alterations in resonance (77%), voice (63%), and articulation (100%). Among the latter, 60% of the articulation errors related to malocclusion and 30% were compensatory misarticulations caused by velopharyngeal incompetence.

Despite multiple challenges, psychosocial adjustment of adolescents is often normal (Beaune et al. 2004).

Evaluation

- Evaluation and longitudinal monitoring of speech by a speech pathologist or speech therapist is critical.
- Social skill training is recommended for individuals who are having difficulty with peer/social interactions.
- If a child with presumed TCS were to have abnormal cognition then further assessment for etiology is warranted.

Management

- Speech therapy is indicated for those with speech abnormalities.
- Augmentative systems for communication such as sign and picture exchange should be implemented early in those individuals with long-term tracheostomies unable to tolerate speaking valves.
- Preferential classroom placement is appropriate for children with significant hearing loss.
- Counseling may be beneficial for individuals with issues relating to self-image and/or peer interactions.

Craniofacial

Although most children with TCS are healthy at birth and present only with their dysmorphic features, some have life-threatening airway obstruction. When present, the obstruction needs to be addressed on an emergent basis. Intubation may prove difficult because of the unusual anatomy of the airway. In the literature, less than 41% of children who require intubation are managed with long-term tracheostomy (Plomp et al. 2016). Several factors contribute to the airway abnormalities seen in TCS (Posnick, 1997). Maxillary hypoplasia resulting in constriction of the nasal passages and choanal stenosis or atresia are foremost among these. Mandibular hypoplasia with micrognathia causing retrodisplacement of the tongue with obstruction of the pharynx, with or without a U-shaped palatal cleft, is another. Some forms of obstruction, such as glossoptosis, worsen over the first 2 months of life before gradually improving. However, obstructive sleep apnea is common at all ages (54% of children and 41% of adults). Endoscopy appears to be particularly helpful in defining the level of obstruction, which is often at multiple levels (Plomp et al. 2012).

Computed tomography with three-dimensional reconstruction has become the preferred modality for imaging the craniofacies for planning surgical reconstruction. Skeletally, individuals with TCS are deficient in the anterior cranial vault, the lateral orbital wall, the zygomatic complex, and the mandible, both ramus height and body length. Cephalometric analysis has documented reduced posterior facial height, shortening of the maxilla and mandible and an obtuse angle for the mandibular plane (Chong et al. 2008). Soft tissue deficiencies include the lids, the pinnae, and, to a variable extent, the envelope of the cheeks.

Roughly 60% of affected individuals will have missing, malformed or ectopically erupted teeth (Da Silva Dalben et al. 2006). Salivary gland pathology has been documented in 48% of a cohort of 21 Norwegian patients (Osterhus et al. 2012). Cervical spine anomalies also appear to be more common than previously recognized (Pun et al. 2012).

There is considerable debate in the surgical literature regarding both the timing of reconstruction and optimal approaches. Detailed discussion of surgical repair is beyond the scope of this chapter. In general, earlier repairs are considered for those individuals with persistent airway concerns, problems of eye exposure, and severe deficiency of the mandibular ramus and/or condyle with failure of development of the glenoid fossa (Kaban III deficiency). Kaban et al. (1980) devised a system for the classification of mandibular deficiency in individuals with OAVS (see Chapter 42). This system had been applied to the mandibular deficiency in individuals with TCS. Type I demonstrates the mildest degree of hypoplasia with retrognathia, an open bite, but normal joint function. Types IIA and IIB have progressively more severe malformation but still a functioning joint. In type III, there is no posterior stop of the mandible against the cranial base.

The importance of an experienced craniofacial team cannot be overstated.

Evaluation

- For infants with airway obstruction in whom the airway can be secured but for whom intervention will be necessary, complete evaluation of the airway from nasal passages to the lungs is critical before treatment planning. The level of obstruction will determine the most rational approach to treatment. Obstruction is often multilevel.
- All neonates with TCS should be observed in a supine sleep position to assess lesser degrees of airway obstruction.
- Oxygen saturation should be monitored if any signs or symptoms of hypoxia develop.
- Polysomnography is recommended on all affected individuals as clinical evaluation underestimates the extent of the obstructive sleep disturbance (Akre et al. 2012)

- An infant with a positional airway at birth needs close outpatient follow up and home monitoring to avoid complications.
- Apnea monitors that assess chest wall movements as a measure of respiration are not adequate for children with obstructive problems.
- Obstructive sleep apnea may develop at any age. Sleep studies and airway evaluation should be considered for anyone with suggestive symptoms.

Management

- Choanal atresia, when bilateral, requires immediate attention to the airway and surgical reconstruction. Unilateral choanal atresia may be repaired electively if the airway is adequate.
- Distraction osteogenesis to lengthen the mandible and relieve airway obstruction resulting from the small pharynx and retrodisplaced tongue has been tried with mixed success in TCS. Although distraction of the mandible has proved useful in the management of airway obstruction, restricted mandibular growth causes regression of the mandibular cephalometric measurements toward pre-distraction values, neutralizing the esthetic gains (Gürsoy et al. 2008).
- Distraction osteogenesis is only appropriate for children in whom glossoptosis (Robin sequence physiology, see Chapter 50) is the primary problem, and it will not benefit an infant obstructed at the level of the choanae.
- Many individuals need long term medical management of obstructive apnea including supplemental oxygen, continuous positive airway pressure or bilevel positive airway pressure.
- Abnormalities in the molar teeth may be seen in half of the mandibles subjected to distraction in infancy or childhood (Da Silva Freitas et al. 2008). Dry mouth has been reported.
- Cleft palate, when present, is repaired in a standard manner, typically at less than a year of age. Although some authors argue for delay in palate repair due to pharyngeal hypoplasia (Shprintzen et al. 1979), systematic review of the evidence does not support modification of timing (Plomp et al. 2016).
- Zygomatic and orbital reconstruction is generally accomplished using autogenous cranial bone in combination with lateral canthopexies, musculocutaneous transposition flaps or Z-plasties for the associated eyelid deformity. Low patient satisfaction with the residual deformity has been documented (Plomp et al. 2016). Because of the tendency for bone resorption, repeat procedures may be necessary. A variety of synthetic materials have been used in an attempt to obviate the need for repeated bone harvesting.
- Mild deformities of the mandible (Kaban I and IIA) are generally approached at the time of skeletal maturity following orthodontic preparation of the arches.
- Osteotomies of the mandible, genioplasty and Le Forte I osteotomy of the maxilla are used to restore the occlusion and normalize facial relationships, typically at facial skeletal maturity.
- In individuals with Kaban IIB and III deformities, distraction of the mandible has become an increasingly popular alternative to conventional surgery for mandibular lengthening as it offers the option of addressing airway related concerns at a younger age than previously possible (Tahiri et al. 2014).
- Individuals with Kaban type III deformities will need reconstruction of the glenoid fossa as well as the mandible.
- Costochondral grafts to replace the mandible and restore the temporomandibular joint have yielded unpredictable results with overgrowth, undergrowth, and asymmetric growth necessitating additional treatment.
- Microsurgical flaps to address soft tissue deficiency have improved outcomes (Saadeh et al. 2008).

Ophthalmologic

Although most children with TCS are able to completely close their eyes, a few with large colobomas of the lower lid and zygomatic deficiency are at risk for problems related to corneal exposure. Eye abnormalities include amblyopia (33%), refractive errors (58–86%), and absence of inferior lacrimal puncta (36–71%) (Plomp et al. 2016)

Evaluation

- The adequacy of lid closure should be evaluated at diagnosis, preferably in the newborn period.
- Referral to an ophthalmologist is recommended to address exposure as well as the list of other anomalies above.

Management

- Topical lubricants will address most of the problems associated with inadequate eyelid closure.
- A rare child will need surgery early in life to provide adequate coverage of the eye.
- No single treatment approach has consistently addressed soft tissue concerns in the eyelid region.

Ears and Hearing

Ear malformations are common in TCS (Pron et al. 1993). The pinnae may be crumpled, displaced toward the angle of the mandible or microtic. Preauricular tags and fistulas occur anywhere along the line between the tragus of the ear and the oral commissure.

In addition to the pinna anomalies, most individuals have stenosis or atresia of the external auditory canal (28–72%), hypoplastic middle ear cavities (33–82%), and dysplastic or absent ossicles (22–67%). The inner ear is normal.

Conductive hearing loss is nearly universal and often worse in one ear than the other. Mixed hearing loss is less common (7%). The severity of involvement of the middle ear and ossicles usually correlates with the extent of malformation of the pinna and canal (Plomp et al. 2016).

Evaluation

- All neonates with TCS need formal assessment of hearing. In individuals who are referred because of abnormal screening tests, brainstem auditory evoked response testing is the preferred method to assess hearing.
- If a loss is identified, evaluation by an otolaryngologist is indicated to identify a medically or surgically treatable lesion.
- Computed tomography scan of the petrous temporal bone to evaluate the middle ear and adjacent structures should be pursued if the hearing loss is bilateral.

Management

- Early amplification is essential in children with bilateral hearing loss when no correctable lesion is identified.
- Early intervention programs are critical components of management, as is preferential classroom seating during school.
- Bone-anchored hearing aides have been used with success in adolescents and adults with TCS (Marres 2002).
- The external ear is most commonly reconstructed using the technique described by Brent (1999). This approach uses a sculpted cartilage framework from the contralateral synchondrotic region of ribs 6 and 7, which is inserted under the skin and then elevated and refined at three subsequent procedures. Adequate rib cartilage is available by age 6 years in most children. The pinna should be situated before any attempt is made to repair the middle ear because scarring from the latter may compromise the possibility of successfully completing the former. Medpore frameworks offer an alternative approach (Baluch et al. 2014)

Anesthesia

The craniofacial structural alterations in the mandibulo-(acro)facial dysostoses present a variety of challenges for which the anesthesiologist needs to be prepared. Laryngeal mask airways (Inada et al. 1995) and fiber-optic devices have been advocated to assist securing of the airway. Postoperative airway obstruction should be anticipated.

RESOURCES

Children's Craniofacial Association

13140 Coit Road, Suite 517
Dallas, Texas 75240, USA
Phone: 800-535-3643
Email: cca@ccakids.com
Website: https://ccakids.org

About Face International, USA

Phone: 1-800-665-3223 and 416-597-2229
Fax: 416-597-8494
Email: info@aboutface.ca
Website: *https://www.aboutface.ca/*

American Cleft Palate-Craniofacial Association

1504 East Franklin Street, Suite 102
Chapel Hill, North Carolina 27514-2820, USA
Phone: 1-919-933-9044 or (800) 242-5338
Fax: 919-933-9604
Email: info@acpa-cpf.org
Website: *http://www.acpa-cpf.org or www.cleftline.org*

REFERENCES

Akre H, Overland B, Asten P, Skogedal N, Heimdal K (2012) Obstructive sleep apnea in Treacher Collins syndrome. *Eur Arch Otorhinolaryngol* 269:331–337.

Baluch N, Nagata S, Park C, Wilkes GH, Reinisch J, Kasrai L, Fisher D (2014) Auricular reconstruction for microtia: A review of available methods. *Plast Surg (Oakv)* 22(1):39–43.

Beaune L, Forrest CR, Keith R (2004) Adolescents' perspectives on living and growing up with Treacher Collins syndrome: A qualitative study. *Cleft Palate Craniofac J* 41:343–350.

Bernier FP, Caluseriu O, Ng S, Schwartzentruber J, Buckingham KJ, Innes AM, Jabs EW, Innis JW, Schuette JL, Gorski JL, Byers PH, Andelfinger G, Siu V, Lauzon J, Fernandez BA, McMillin M, Scott RH, Racher H; FORGE Canada Consortium, Majewski J, Nickerson DA, Shendure J, Bamshad MJ, Parboosingh JS (2012) Haploinsufficiency of SF3B4, a component of the pre-mRNA spliceosomal complex, causes Nager syndrome. *Am J Hum Genet* 90(5):925–33.

Brent B (1999) Technical advances in ear reconstruction with autogenous rib cartilage grafts: Personal experience with 1200 cases. *Plast Reconstr Surg* 104:319–334.

Chong DK, Murray DJ, Britto JA, Rompson B, Forrest CR, Phillips JH (2008) A cephalometric analysis of maxillary and mandibular parameters in Treacher Collins syndrome. *Plast Reconstr Surg* 121:77e–84e.

Da Silva Dalben G, Costa B, Gomide MR (2006) Prevalence of dental anomalies, ectopic eruption and associated oral malformations in subjects with Treacher Collins syndrome. *Oral Surg Oral Med Oral Pathol Oral Radiol Endod* 101:588–592.

Da Silva Freitas R, Tolazzi ARD, Alonso N, Cruz GAO, Busato L (2008) Evaluation of molar teeth and buds in patients submitted to mandible distraction: Long term results. *Plast Reconstr Surg* 121:1335–1342.

Dauwerse JG, Dixon J, Seland S, Ruivenkamp CAL, van Haeringen A. Hoefsloot LH, Peters DJM, Clement-de Boers A, Daumer-Haas C, Maiwald R, Zweier C, Kerr B, Cobo AM, Toral JF, Hoogeboom AJM, Lohmann DR, Hehr U, Dixon MJ, Breuning MH, Wieczorek D (2011) Mutations in genes encoding subunits of RNA polymerases I and III cause Treacher Collins syndrome. *Nature Genet* 43:20–22.

Dixon MJ, Read AP, Donnai D, Colley A, Dixon J, Williamson R (1991) The gene for Treacher Collins syndrome maps to the long arm of chromosome 5. *Am J Hum Genet* 49:17–22.

Dixon MJ, Marres HAM, Edwards SJ, Dixon J, Cremers CWRJ (1994) Treacher Collins syndrome: Correlation between clinical and genetic linkage studies. *Clin Dysmorphol* 3:96–103.

Dixon J, Dixon MJ (2004) Genetic background has a major effect on the penetrance and severity of craniofacial defects in mice heterozygous for the gene encoding the nucleolar protein treacle. *Dev Dyn* 229:907–914.

Dixon J, Brakebusch C, Fassler R, Dixon MJ (2000) Increased levels of apoptosis in the prefusion neural folds underlie the craniofacial disorder: Treacher Collins syndrome. *Hum Mol Genet* 9:1473–1480.

Dixon J, Ellis I, Bottani A, Temple K, Dixon MJ (2004) Identification of mutations in TCOF1: Use of molecular analysis in the pre-and postnatal diagnosis of Treacher Collins syndrome. *Am J Med Genet* 127A:244–248.

Dixon J, Jones NC, Sandell LL, Jayasinghe SM, Crane J, Rey J-P, Dixon MJ, Trainor PA (2006) Tcof1/Treacle is required for neural crest cell formation and proliferation deficiencies that cause craniofacial abnormalities. *Proc Natl Acad Sci* 103:13403–13408.

Ellis PE, Dawson M, Dixon MJ (2002) Mutation testing in Treacher Collins syndrome. *J Orthod* 29:293–298.

Emmanouil-Nikoloussi EN, Goret-Nicaise M, Foroglou CH, Katsarma E, Dhem A, Dourov N, Persaud TV, Thliveris JA (2000) Craniofacial abnormalities induced by retinoic acid: A preliminary histological and scanning electron microscopic (SEM) study. *Exp Toxicol Pathol* 52:445–453.

Fazen LE, Elmore J, Nadler HL (1967) Mandibulo-facial dysostosis (Treacher Collins syndrome). *Am J Dis Child* 113:405–410.

Franceschetti A, Klein D (1949) Mandibulo-facial dysostosis: New hereditary syndrome. *Acta Ophthalmol (Kbh)* 27:143–224.

Gonzales B, Henning K, So RB, Dixon J, Dixon MJ, Valdez BC (2005) The Treacher Collins syndrome (TCOF1) gene product is involved in pre-rRNA methylation. *Hum Mol Genet* 14:2035–2043.

Gorlin RJ, Cohen MM, Hennekam RCM (eds) (2001) *Syndromes of the Head and Neck, 4th ed.* New York: Oxford University Press, pp. 799–802.

Guion-Almeida ML, Zechi-Ceide RM, Vendramini S, Tabith Júnior A (2006) A new syndrome with growth and mental retardation, mandibulofacial dysostosis, microcephaly, and cleft palate. *Clin Dysmorphol* 15:171–174.

Gürsoy S, Hukki J, Hurmerinta K (2008) Five year follow-up of mandibular distraction osteogenesis on the dentofacial structures of syndromic children. *Orthod Craniofac Res* 11:57–64.

Haworth KE, Islam I, Breen M, Putt W, Makrinou E, Binns M, Hopkinson D, Edwards Y (2001) Canine *TCOF1*; cloning, chromosome assignment and genetic analysis in dogs with different head types. *Mamm Genome* 12:622–629.

Hsu TY, Hsu JJ, Chang SY, Chang MS (2002) Prenatal three-dimensional sonographic images associated with Treacher Collins syndrome. *Ultrasound Obstet Gynecol* 19:413–422.

Inada T, Fujise K, Tachibana K, Shingu K (1995) Orotracheal intubation through the laryngeal mask airway in paediatric patients with Treacher Collins syndrome. *Paediatr Anaesth* 5: 2 129–132.

Irving MD, Dimitrov BI, Wessels M, Holder-Espinasse M, Chitayat D, Simpson MA (2016) Rodriguez acrofacial dysostosis is caused by apparently de novo heterozygous mutations in the SF3B4 gene. *Am J Med Genet A* 170(12):3133–3137

Jabs EW, Li X, Coss CA, Taylor WE, Meyers DA, Weber JL (1991) Mapping the Treacher Collins syndrome locus to 5q31. 33-q33. 3. *Genomics* 11:193–198.

Jones KL, Smith DW (1976) Older paternal age and fresh gene mutation: Data on additional disorders. *J Pediatr* 86:84–88.

Jones NC, Lynn ML, Gaudenz K, Saki D, Aoto K, Rey JP, Glynn EF, Ellington L, Du C, Dixon J, Dixon MJ, Trainor PA (2008) Prevention of the neurocristopathy Treacher Collins syndrome through inhibition of p53 function. *Nat Med* 14:125–133.

Kaban LB, Moses ML, Mulliken JB (1980) Surgical correction of hemifacial microsomia in the growing child. *Plast Reconstr Surg* 82:9–19.

Lines MA, Huang L, Schwartzentruber J, Douglas SL, Lynch DC, Beaulieu C, Guion-Almeida ML, Zechi-Ceide RM, Gener B, Gillessen-Kaesbach G, Nava C, Baujat G, Horn D, Kini U, Caliebe A, Alanay Y, Utine GE, Lev D, Kohlhase J, Grix AW, Lohmann DR, Hehr U, Böhm D; FORGE Canada Consortium, Majewski J, Bulman DE, Wieczorek D, Boycott KM (2012) Haploinsufficiency of a spliceosomal GTPase encoded by EFTUD2 causes mandibulofacial dysostosis with microcephaly. *Am J Hum Genet* 90(2):369–77.

Lungarotti MS, Marinelli D, Mariani T, Calabro A (1987) Multiple congenital anomalies associated with apparently normal maternal intake of vitamin A: A phenocopy of the isotretinoin syndrome? *Am J Med Genet* 27:245–248.

Marres HA (2002) Hearing loss in the Treacher Collins syndrome. *Adv Otorhinolaryngol* 61:209–215.

Marres HA, Cremers CW, Dixon MJ, Huygen PL, Joosten FB (1995) The Treacher Collins syndrome. A clinical, radiological, and genetic linkage study on two pedigrees. *Arch Otolaryngol Head Neck Surg* 121:509–514.

Marsh KL, Dixon J, Dixon MJ (1998) Mutations in the Treacher Collins syndrome gene lead to mislocalization of the nucleolar protein treacle. *Hum Mol Genet* 7:1795–1800.

Ng SB, Buckingham KJ, Lee C, Bigham AW, Tabor HK, Dent KM, Huff CD, Shannon PT, Jabs EW, Nickerson DA, Shendure J,

Bamshad MF (2009) Exome sequencing identifies the cause of a Mendelian disorder. *Nat Genet* 42(1):30–35.

Noack Watt KE, Achilleos A, Neben CL, Merrill AE, Trainor PA (2016) The roles of RNA polymerase I and III subunits Polr1c and Polr1d in craniofacial development and in zebrafish models of Treacher Collins syndrome. *PLoS Genet* 12(7):e1006187.

Ochi H, Matsubara K, Ito M, Kusanagi Y (1998) Prenatal sonographic diagnosis of Treacher Collins syndrome. *Obstet Gynecol* 91:862.

Østerhus IN, Skogedal N, Akre H, Johnsen UL, Nordgarden H, Åsten P (2012) Salivary gland pathology as a new finding in Treacher Collins syndrome. *Am J Med Genet A* 158A(6):1320–5

Plomp RG, Bredero-Boelhouwer HH, Joosten KF, Wolvius EB, Hoeve HL, Poublon RM, Mathijssen IM (2012) Obstructive sleep apnoea in Treacher Collins syndrome: prevalence, severity and cause. *Int J Oral Maxillofac Surg* 41(6):696–701

Plomp RG, van Lieshout MJ, Joosten KF, Wolvius EB, van der Schroeff MP, Versnel SL, Poublon RM, Mathijssen IM (2016) Treacher Collins Syndrome: A Systematic Review of Evidence-Based Treatment and Recommendations. *Plast Reconstr Surg* 137(1):191–204

Posnick JC (1997) Treacher Collins syndrome: Perspectives in evaluation and treatment. *J Oral Maxillofac Surg* 55:1120–1133.

Poswillo D (1975) The pathogenesis of the Treacher Collins syndrome (mandibulofacial dysostosis). *Br J Oral Surg* 13:1–26.

Pron G, Galloway C, Armstrong D, Posnick J (1993) Ear malformation and hearing loss in patients with Treacher Collins syndrome. *Cleft Palate Craniofac J* 30:97–103.

Pun AHY, Clark BE, David DJ, Angerson PJ (2012) Cervical spine in Treacher Collins syndrome. *J Craniofac Surg* 23:e218–220.

Roberts FG, Pruzansky S, Aduss H (1975) An X-radiocephalometric study on mandibulofacial dysostosis in man. *Arch Oral Biol* 20:265–281.

Rotten D, Levaillant JM, Martinez H, Ducou le Pointe H, Vicaut E (2002) The fetal mandible: A 2D and 3D sonographic approach to the diagnosis of retrognathia and micrognathia. *Ultrasound Obstet Gynecol* 19:122–130.

Rovin S, Dachi SF, Borenstein DB, Cotter WB (1964) Mandibulofacial dysostosis: A family study of five generations. *J Pediatr* 65:215–221.

Saadeh PB, Chang CC, Warren SM, Reavey P, McCarthy JG, Siebert JW (2008) Microsurgical correction of facial contour deformities in patients with craniofacial malformations. *Plast Reconstr Surg* 121:368e–378e.

Sakai D, Dixon J, Achilleos A, Dixon M, Trainor PA (2016) Prevention of Treacher Collins syndrome craniofacial anomalies in mouse models via maternal antioxidant supplementation. *Nature Commun* 7:10328.

Shprintzen RJ, Croft C, Berkman MD, Rakoff SJ (1979) Pharyngeal hypoplasia in Treacher Collins syndrome. *Arch Otolaryngol* 105:127–131.

Splendore A, Silva EO, Alonso LG, Richieri-Costa A, Alonso N, Rosa A, Carakushanky G, Cavalcanti DP, Brunoni D, Passos-Bueno MR (2000) High mutation detection rate in TCOF1 among Treacher Collins syndrome patients reveals clustering of mutations and 16 novel pathogenic changes. *Hum Mutat* 16:315–322.

Sulik KK, Johnson MC, Smiley SJ, Speight HS, Jarvis BE (1987) Mandibulofacial dysostosis (Treacher Collins syndrome): A new proposal for its pathogenesis. *Am J Med Genet* 27:359–372.

Tahiri Y, Viezel-Mathieu A, Aldekhayel S, Lee J, Gilardino M (2014) The effectiveness of mandibular distraction in improving airway obstruction in the pediatric population. *Plast Reconstr Surg.* 133(3):352e–359e.

Tanaka Y, Kanenishi K, Tanaka H, Yanagihara T, Hata T (2002) Antenatal three-dimensional sonographic features of Treacher Collins syndrome. *Ultrasound Obstet Gynecol* 19:414–415.

Teber OA, Gillessen-Kaesbach G, Fisher S, Böhringer S, Albrecht B, Albert A, Arslan-Kirchner M, Haan E, Hagedorn-Greiwe M, Hammans C, Henn W, Hinkel GK, König R, Kunstmann E, Kunze J, Neumann LM, Prott EC, Rauch A, Rott HD, Seidel H, Spranger S, Sprengel M, Zoll B, Lohmann DR, Wieczorek D (2004) Genotyping of 46 patients with tentative diagnosis of Treacher Collins syndrome revealing unexpected phenotypic variation. *Eur J Hum Genet* 12:879–890.

Terrazas K, Dixon J, Trainor PA, Dixon MJ (2017) Rare syndromes on the head and face: mandibulofacial and acrofacial dysostoses. *Wiley Interdiscip Rev Dev Biol* 6(3).

Treacher Collins E (1900) Cases with symmetrical congenital notches in the outer part of each lid and defective development of the malar bones. *Trans Ophthalmol Soc UK* 20:190–192.

Treacher Collins Syndrome Collaborative Group (1996) Positional cloning of a gene involved in the pathogenesis of Treacher Collins syndrome. *Nat Genet* 12:130–136.

Vallino-Napoli LD (2002) A profile of the features and speech in patients with mandibulofacial dysostosis. *Cleft Palate Craniofac J* 39:623–634.

Vincent M, Geneviève D, Ostertag A, Marlin S, Lacombe D, Martin-Coignard D, Coubes C, David A, Lyonnet S, Vilain C, Dieux-Coeslier A, Manouvrier S, Isidor B, Jacquemont ML, Julia S, Layet V, Naudion S, Odent S, Pasquier L, Pelras S, Philip N, Pierquin G, Prieur F, Aboussair N, Attie-Bitach T, Baujat G, Blanchet P, Blanchet C, Dollfus H, Doray B, Schaefer E, Edery P, Giuliano F, Goldenberg A, Goizet C, Guichet A, Herlin C, Lambert L, Leheup B, Martinovic J, Mercier S, Mignot C, Moutard ML, Perez MJ, Pinson L, Puechberty J, Willems M, Randrianaivo H, Szakszon K, Toutain A, Verloes A, Vigneron J, Sanchez E, Sarda P, Laplanche JL, Collet C (2016) Treacher Collins syndrome: a clinical and molecular study based on a large series of patients. *Genet Med.* 18(1):49–56

Weaver KN, Watt KE, Hufnagel RB, Navajas Acedo J, Linscott LL, Sund KL, Bender PL, König R, Lourenco CM, Hehr U, Hopkin RJ, Lohmann DR, Trainor PA, Wieczorek D, Saal HM (2015) Acrofacial Dysostosis, Cincinnati Type, a Mandibulofacial Dysostosis Syndrome with Limb Anomalies, is caused by POLR1A dysfunction. *Am J Hum Genet.* 96(5):765–74

Werner A, Iwasaki S, Mcgourty CA, Medina-Ruiz S, Teerikorpi N, Fedrigo I, Ingolia NT, Rape M (2015) Cell-fate determination by ubiquitin-dependent regulation of translation. *Nature* 525:523–527.

Wildervanck LS (1960) Dysostosis mandibulo-facialis (Franceschetti-Zwahlin) in four generations. *Acta Genet Med (Roma)* 9:447–451.

Wiley MJ, Cauwenbergs P, Taylor IM (1983) Effects of retinoic acid on the development of the facial skeleton in hamsters; early changes involving neural crest cells. *Acta Anat* 116: 180–192.

58

TRISOMY 18 AND TRISOMY 13 SYNDROMES

JOHN C. CAREY

Division of Medical Genetics, Department of Pediatrics, University of Utah Health Salt Lake City, Utah, USA

INTRODUCTION

Incidence

The trisomy 18 and 13 syndromes are unique disorders among the syndromes in this book because of their high neonatal and infant mortality rates. As discussed below, the majority of infants with these conditions die in the first months of life. Several population-based studies have been carried out in various parts of the world in the last three decades confirming the high infant mortality. In a recent review of data gathered from these studies, Carey (2012) tabulated the survival figures from the published data up to that time, demonstrating remarkable consistency among the studies. In determining both birth prevalence figures and mortality, it is important to utilize figures determined from population-based series over a well-defined period of time to obtain accurate rates unencumbered by selection bias toward or away from survival. Two more current investigations, including the largest series of children with the two syndromes yet described, were published in 2016 and showed one-year survival rates about twice that of the prior studies (Meyer et al. 2016; Nelson et al. 2016). An increase in interventive care likely plays a role in this observed improvement in infant survival (Meyer et al. 2016; Carey and Kosho 2016).

Although trisomy 18 and 13 syndromes share many characteristics and have similar survival and developmental outcomes, they are clinically distinct and will be discussed separately in the introductory sections below but together in the management sections. Trisomy 18 will be presented first because of its more common prevalence and available studies.

TRISOMY 18

INTRODUCTION

The trisomy 18 syndrome, also known as Edwards syndrome, was originally described by Professor John Edwards and colleagues in a single case report published in the spring of 1960. Soon after, Smith and colleagues at the University of Wisconsin described the first affected individuals in North America, and the syndromic pattern then became established in the pediatrics and genetics literature (Edwards et al. 1960; Smith et al. 1960). Of note, this latter group was instrumental in the early descriptions of both trisomies 18 and 13. Since that time, hundreds of clinical reports and several series have been published throughout the world. Trisomy 18 represents the second most common autosomal trisomy syndrome after trisomy 21 (Down syndrome, Chapter 24). The pattern in trisomy 18 includes a recognizable constellation of major and minor anomalies, a predisposition to increased neonatal and infant mortality, and a significant developmental and motor disability in surviving older children. A detailed listing of the many reported manifestations and their approximate frequency in the syndrome is provided in the review by Cereda and Carey (2012). Recent efforts to characterize the natural history of both syndromes include the Tracking Rare Incidence Syndromes (TRIS http://tris.siu.edu) resource, created by Bruns, who closely

---

Cassidy and Allanson's Management of Genetic Syndromes, Fourth Edition.
Edited by John C. Carey, Agatino Battaglia, David Viskochil, and Suzanne B. Cassidy.
© 2021 John Wiley & Sons, Inc. Published 2021 by John Wiley & Sons, Inc.

collaborates with families of individuals with trisomy 18 and 13 (and other rare syndromes) to investigate outcome; she and her coworkers have explored a number of care issues (see Bruns [2015] and the ITA web site, https://www.internationaltrisomyalliance.com/). Similarly, the SOFT Surgery Registry (www.trisomy.org) created by Barnes provides additional natural history information. Both resources will likely expand knowledge of outcomes in both conditions in the future.

Incidence

On the basis of numerous studies performed in different areas of North America, Europe, and Australia, the prevalence of trisomy 18 in live born infants ranges from 1/3600 to 1/8500 (Root and Carey 1994; Embleton et al. 1996; Forrester and Merz 1999; Rasmussen et al. 2003; Crider et al. 2008; Irving et al. 2011). The most accurate estimate in live births with minimal influence of prenatal screening is from the Utah study of the 1980s, which indicates a frequency of about 1/6000 (Root and Carey 1994). Trisomy 18 is a relatively common chromosomal cause of stillbirth, and many prenatally diagnosed fetuses are terminated; thus the frequency in *total* births is higher and estimated at 1 in 2500 (Carey 2012).

The total prevalence of trisomy 18 has been estimated in numerous investigations. Birth prevalence data are altered by prenatal detection that is based on screening by maternal serum marker screening and more recently by non-invasive prenatal screening (NIPS) as well as amniocentesis followed by pregnancy termination of affected fetuses. A comprehensive study in the United Kingdom identified the frequency of trisomy 18 at 18 weeks gestation as 1/4274 with a live born prevalence of 1/8333 (Embleton et al. 1996). A study in Hawaii utilized a birth defect registry and detected a total prevalence of 4.71/10,000 total births (1/2123 total births) with a live born frequency of 1.41/10,000 live births or about 1/7900 live born infants (Forrester and Merz 1999). The total prevalence in this study included fetal deaths after 20 weeks gestation, live births, and elective terminations. The Utah Birth Defects Network presented their data from 1990 to 2006 (Willey et al. 2008); the total prevalence, including pregnancy terminations, stillbirths, and live births, was 4.01/10,000 total births and the live born prevalence was 1.16/10,000 similar to the Hawaii investigation and with larger numbers. Although the method of case collection differed in the UK study compared with the others, the relatively large number of fetuses (66 in the UK study, 113 in Hawaii, and 254 in Utah) provides valid estimates of the range of frequency in total births and live births. Moreover, from these figures, one can estimate the relative decrease in live birth prevalence related to utilization of prenatal diagnosis and elective pregnancy termination: all three studies demonstrate a similar figure of about 1/8000 live births. Most recently Irving et al. (2011) provided compelling data to demonstrate changes in prevalence of live borns for both trisomy 13 and 18 over time, presumably related to an increase in utilization of prenatal diagnosis and pregnancy termination rates.

Diagnostic Criteria

The characteristic pattern of prenatal growth deficiency, craniofacial features, distinctive hand posture of overriding fingers with nail hypoplasia, short hallux, and short sternum allows for clinical diagnosis in the newborn infant with trisomy 18 (Figure 58.1). Internal anomalies, particularly involving the heart, are very common. Marion et al. (1988) developed a bedside scoring system for the diagnosis of trisomy 18 syndrome in the newborn period that provides the clinician *without* specialized training in clinical genetics and dysmorphology a useful checklist to help differentiate newborns with the syndrome from other infants with multiple congenital anomalies.

Confirmation of the diagnosis of trisomy 18 syndrome relies on standard G-banded karyotype demonstrating the extra chromosome 18 or one of the less common partial trisomy 18 findings seen in individuals with the Edwards syndrome phenotype, as discussed below. In recent years chromosome analysis is usually accomplished by cytogenomic SNP (single nucleotide polymorphism) microarray. This study will confirm full trisomy 18 but will not exclude the possibility of a translocation. Thus, a standard karyotype is still needed to distinguish complete trisomy 18 from a partial trisomy 18 involving most of the chromosome.

Etiology, Pathogenesis, and Genetics

The trisomy 18 phenotype (Edwards syndrome) usually results from three complete copies (trisomy) of the 18th chromosome. In four large series that attempted to ascertain all cases in a region through a surveillance program, 165 of 176 neonates with Edwards syndrome phenotype had full trisomy 18 (Carter et al. 1985; Young et al. 1986; Goldstein and Nielsen, 1988; Embleton et al. 1996); 8 of 176 (5%) had mosaicism for trisomy 18, and 3 had partial trisomy 18, usually resulting from an unbalanced translocation. Alberman and colleagues (2012) provided more current figures based on larger numbers from UK data. Thus, about 94–98% of infants labeled as having the phenotype will have full trisomy 18, whereas the remainder will have mosaicism or partial duplication of 18q.

Regarding trisomy 18 mosaicism Tucker et al. (2007) summarized all of the reported cases of trisomy 18 mosaicism in the literature up to that time; the reader is referred to this comprehensive review for more detailed clinical information on this condition

In full trisomy 18, the extra chromosome is presumably present because of nondisjunction. A number of

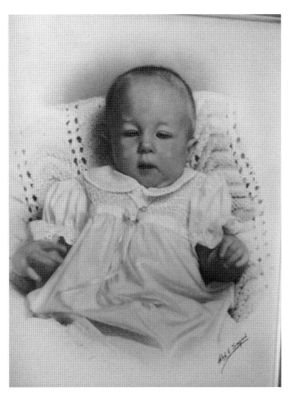 

FIGURE 58.1 A girl with trisomy 18 as an infant of 10 months and as an adult of 36 years. This young lady is one of the oldest known surviving adults with full trisomy 18. (Courtesy of the mother of the person).

investigations have studied the mechanism of origin of the nondisjunction (Fisher et al. 1995; Eggermann et al. 1996). The error can arise as a malsegregation of chromosomes during meiosis or postzygotic mitosis, as has been demonstrated in other trisomies. From combined series, 95% of the time, when determination could be made the additional chromosome was a result of maternal nondisjunction. Of note, in the paternally derived cases, the error was postzygotic in a number of situations. What was more unique, however, was the observation that about 50% of the nondisjunctional errors in oogenesis were in meiosis II. This is unlike all other human trisomies that have been investigated, which usually show a higher frequency of maternal meiosis I errors. Also, in contrast to trisomy 21, the error in maternal meiosis II showed normal recombination in the nondisjoined chromosomes. As in trisomy 21 and trisomy 13, the frequency of nondisjunctional trisomy 18 increases with advancing maternal age. However, the increased frequency of maternal meiosis II errors and the normal recombination indicate that the biology of nondisjunction in trisomy 18 may be unique.

The precise cause of nondisjunction remains unknown. Throughout the years investigators have attempted to identify genetic contributors to human meiotic nondisjunction. Reports linking Down syndrome to maternal polymorphisms of folate metabolism enzymes generated considerable interest two decades ago (Hassold et al. (2001) analyzed maternal folate polymorphisms in several groups of chromosome abnormalities, including 44 cases of trisomy 18. They observed a significant increase in *methylene tetrahydrofolate reductase* gene (*MTHFR*) polymorphisms in mothers of trisomy 18 conceptuses, but not in the other groups. This observation has not been replicated and deserves further study.

There has been considerable controversy regarding the critical region of chromosome 18 that results in classical Edwards syndrome. A number of regions of the 18 long arm (q) have been proposed to represent the important area that generates the phenotype, but no clear consensus has been reached (Wilson 1993; Boghosian-Sell et al. 1994). The small number of cases of partial trisomy 18, as well as the intrinsic variability of cases with full trisomy 18, makes conclusions about genotype–phenotype correlations tentative. However, the milder manifestations of individuals with 18 short arm (18p) trisomy and the almost complete Edwards syndrome phenotype in individuals with trisomy extending from 18q11.2 to terminus support a pivotal role for the 18 long arm region, as suggested by Wilson (1993). On the other hand, the observation that the two children with trisomy of 18q.11.2 to terminus reported by Boghosian-Sell et al. (1994) and another child evaluated by the author (personal experience) do not exhibit the Edwards syndrome phenotype in its full form suggests that the 18 short arm and/or the 18q11.1 region have some role in generating the complete pattern. These three individuals appear to have a somewhat different facial gestalt than surviving infants with trisomy 18 (personal observation), and they have better prenatal and postnatal growth than older children with full trisomy 18. They do have the characteristic hand configuration, short sternum, and heart defects, indicating that these features are a consequence of trisomy of the long arm of 18. Peron and Carey (2014) summarized all of the individuals in the literature with partial trisomy 18 syndrome in an attempt to characterize critical regions for the individual features observed in the Edwards syndrome. None of the more specific manifestations could be narrowed down very precisely, but the characteristic hand posture and Wilms tumor are likely due to duplication in the bottom 2/3 of the long arm of chromosome 18.

There has been a general lack of data regarding recurrence risk in both trisomy 18 and trisomy 13. Most practicing geneticists use a 1–2% recurrence risk figure for nondisjunction, as calculated empirically for trisomy 21. Of note, low-grade parental mosaicism has been described on two occasions in sporadic cases of trisomy 18. In addition, a number of reported cases have demonstrated the occurrence of two different trisomies in siblings (reviewed by Baty et al. 1994a). Empiric data on trisomies 18 and 13 showed no chromosome abnormality in the live born siblings (Baty et al. 1994a). However, in one family with a live born infant with trisomy 13, there had been a previously terminated fetus with trisomy 18 that was not part of the original study of cases of trisomy 18. When these figures are combined, they indicate a recurrence of 1 in 181 pregnancies, or a risk of 0.55%, with confidence limits of 0–1.6% for both trisomies. The best estimate of recurrence risk, then, is about 1%.

This estimate of recurrence is supported by a study from Japan in which the authors report outcomes of amniocentesis performed after the history of a previous child with a chromosome abnormality (Uehara et al. 1999). None of the 170 women in that study who had a previous offspring with trisomy 18 or the 40 with trisomy 13 had a recurrence of either trisomy. Even though the authors do not include confidence intervals, combining them with those of Baty et al. (1994a) suggests a recurrence risk of less than 1%, but higher than the age-specific background risk.

The recurrence risk in partial trisomy cases depends on the presence or absence of a translocation or inversion in one of the parents. A de novo duplication is not associated with increased risk, whereas a familial translocation or pericentric inversion may predispose to unbalanced offspring and would have a recurrence risk higher than that of typical trisomy 18.

Diagnostic Testing

Definitive diagnosis of trisomy 18 is made by detection of complete or partial trisomy of chromosome 18. This can be easily accomplished in any laboratory competent in performing routine chromosome analysis. Rapid diagnosis by interphase fluorescence in situ hybridization (FISH) in many newborns with trisomy 18 (and trisomy 13) can be helpful in making decisions regarding surgical intervention and care. Lymphocyte analysis usually takes more than 48 hours, but analysis of a 24 hour sample may be important in some cases where decisions regarding surgery are being pondered.

Because of the association between trisomy 18 and advanced maternal age, prenatal diagnosis of fetuses with trisomy 18 has paralleled the development of amniocentesis and chorionic villus sampling programs. Over the last three decades, a significant body of literature has been published regarding the prenatal sonographic detection of trisomy 18 (and trisomy 13). Because of the common occurrence of intrauterine growth retardation in fetuses with trisomy 18 in both the second and third trimester, the diagnosis is often made during these gestational periods. In addition, polyhydramnios occurs in 30–60% of pregnancies with trisomy 18 and often leads to prenatal diagnosis. A number of characteristic prenatal signs have been described in fetuses with trisomy 18. These findings, in conjunction with intrauterine growth retardation or a major malformation consistent with the phenotype, often raised the possibility of the diagnosis in prenatal settings after 20 weeks of gestation. Hill (1996) comprehensively reviewed the sonographic detection of trisomy 18 as of that period. The most publicized sonographic markers include choroid plexus cysts, large cysterna magna, and a "strawberry"-shaped calvarium. DeVore (2000) documented that 29 of 30 second trimester fetuses with trisomy 18 had "abnormal fetal anatomy" consisting of one of these markers or heart defects. Tongsong et al. (2002) reviewed the mid-pregnancy sonographic features of 25 affected fetuses and found that all had at least one finding, with growth restriction in nearly half. Viora et al. (2007) summarized the ultrasonographic findings in a large series of 71 fetuses and detected ultrasound anomalies in 91%; 55% had two or more anomalies. More recently, first trimester screening for trisomy 18 has been developed using fetal nuchal translucency thickness and other fetal ultrasound markers; these detailed anomaly scans show good performance (95%) in identifying fetuses with trisomy 18 (and trisomy 13) (Wagner et al. 2016; Wiechec et al. 2016). Clearly, modern sonographic analysis has a high sensitivity for detecting fetuses with trisomy 18. The prenatal diagnosis of trisomy 18 in the mid-to-late second or third trimester can be of significance because it may influence management at the time of delivery (see Manifestations and Management).

The usefulness of choroid plexus cysts as a sign of trisomy 18 is a controversial topic in the obstetric literature with no clear consensus on when to offer amniocentesis for karyotype when cysts are discovered, particularly when there are no other findings (Reinsch 1997). Because some second trimester fetuses with trisomy 18 have no other anomalies or growth findings, there is an increased risk of trisomy 18 in the presence of an isolated choroid plexus cyst (Reinsch 1997). Factoring in other ultrasonographic findings helps define the higher risk patients. In the final analysis, it is the family that should make the decision based on the individual chances in the context of the procedural risk, the perceived impact of trisomy 18, the knowledge gained from prenatal sonograms, and the perceived benefit of making the diagnosis in utero.

Multiple-marker screening using levels of human chorionic gonadotropin, unconjugated etriol, and alpha-fetoprotein (and later inhibin-A) was applied routinely in the prenatal diagnosis of trisomy 18 in the two decades prior to 2010 (see Breathnach et al. [2007]). Detection rates using triple or quadruple screening with fetal sonograms approached 80%. In the most recent decade non-invasive prenatal testing, using cell-free fetal DNA detection (now more appropriately called screening or NIPS) has become the method of choice in the prenatal diagnosis of the trisomy syndromes. The sensitivity, specificity, and negative predictive values of NIPS in trisomy 18 and trisomy 13 are over 99% for both trisomies (Zhu et al. 2017; Petersen et al. 2017). However cell-free DNA has a relatively high false-positive rate producing a lowered positive predictive value (PPV) in trisomy 18 (and 13) compared to trisomy 21, leading to the recommendation that all individuals with a "positive" screen have a follow-up amniocentesis to confirm or refute the screening result from the NIPS. PPVs for trisomy 18 range from 58 to 76% in selected recent investigations.

Differential Diagnosis

The syndromic pattern of trisomy 18 is quite discrete and, in its totality, is rarely confused with other conditions. The most common disorder with overlapping features comprises the heterogeneous group of conditions with fetal akinesia sequence. This condition, sometimes labeled as Pena–Shokeir syndrome type 1, demonstrates polyhydramnios, characteristic facial features, and multiple joint contractures, including over-riding fingers (Jones et al. 2014). Although this is a heterogeneous group of mostly autosomal recessive conditions, most newborns do not have the structural heart defects and lack the characteristic face of trisomy 18. Of note, there is a condition originally described in the French literature as "pseudotrisomy 18 syndrome" (Simpson and German 1969). In retrospect, most children labeled with this condition probably fall into the fetal akinesia sequence category.

Because of the similarity in hand and finger positioning, some cases with the condition distal arthrogryposis type 1 (Chapter 6) might be confused with trisomy 18. However, infants with distal arthrogryposis type 1 do not have major malformations and usually lack prenatal growth deficiency, and thus they are easily distinguished from infants with trisomy 18. Because of overlapping malformations, some infants with CHARGE syndrome (Chapter 11) can be confused with trisomy 18. Again, the entire pattern of findings, and the normal karyotype in CHARGE syndrome, will distinguish the two conditions.

TRISOMY 13

INTRODUCTION

In the same issue of *The Lancet* in which the original case of trisomy 18 was published, the first case of trisomy 13 was also documented (Patau et al. 1960). Since that time there have been hundreds of clinical reports and a few large series. Generally, trisomy 13 syndrome presents as an obvious multiple congenital anomaly pattern, unlike trisomy 18 in which external major malformations are less notable. The combination of orofacial clefts, microphthalmia/anophthalmia, and postaxial polydactyly of the limbs allows for recognition by the clinician. However, because each of the three mentioned cardinal features have only 60–70% occurrence in the syndrome, clinical diagnosis can be challenging, especially in a child without a cleft lip or the facial features of holoprosencephaly. Trisomy 13 is usually the result of the presence of an extra 13th chromosome, leading to 47 chromosomes. The occurrence of a translocation makes up a higher portion in trisomy 13 syndrome than in trisomy 18 syndrome.

Incidence

There are fewer birth prevalence studies of trisomy 13 than trisomy 18 but estimates in early studies before the 1980s ranged from 1/5000 to 1/12,000 total births. Comprehensive population studies from Denmark and the United Kingdom found birth prevalence in live born infants of approximately 1/20,000–1/29,000 (Goldstein and Nielsen 1988; Wyllie et al. 1994). The Hawaii study mentioned above in relation to trisomy 18 detected a live birth prevalence of trisomy 13 of 0.83/10,000 (1/12,048) (Forrester and Merz 1999). Figures recently presented from the Utah Birth Defects Network (Willey et al. 2008) demonstrate a total prevalence of 1.57/10,000 with a live born prevalence of 0.93, or just about 1/10,000. The concurrent influence of prenatal diagnosis and subsequent elective termination of affected fetuses modifies figures of live born prevalence. Thus, as in trisomy 18, this figure in live born infants is lower than the frequency of trisomy 13 at 15–16 weeks of gestation. Of note, the figures from the UK and Denmark studies, cited above, are lower than expected from the earlier literature estimates of birth prevalence of trisomy 13. This is probably because of the relatively small numbers involved in the early studies, rather than a true decreasing occurrence of trisomy 13 in recent decades. Thus, the current best estimate of live births with trisomy 13 after accounting for prenatal diagnosis/elective termination is approximately 1/10,000–1/20,000.

Diagnostic Criteria

The pattern of malformation in the most typical infants with trisomy 13 is quite characteristic and allows for clinical diagnosis by the practitioner. As mentioned, orofacial clefts, microphthalmia/anophthalmia, and postaxial polydactyly of hands and/or feet represent the cardinal signs (Figure 58.2). Trisomy 13 should always be considered in an infant with holoprosencephaly and multiple anomalies. Localized cutis aplasia in the region of the posterior fontanelle, when present, also helps in the clinical diagnosis. In children without cleft lip or the distinctive craniofacial features of holoprosencephaly (cyclops or cebocephaly), the facial gestalt, especially prominence of the nasal bridge and tip is helpful in diagnosis. Minor anomalies such as capillary malformations of the forehead, anterior frontal upsweep, and ear malformations are also consistent and of assistance in the clinical diagnosis. The clinical findings are listed by Jones et al. (2013). Diagnosis must be confirmed by detection of trisomy of all or most of the long arm of chromosome 13. A number of abnormalities that are found on postmortem examination in trisomy 13 are particularly distinctive and can allow for differentiation in confusing cases; these findings include holoprosencephaly of any type and pancreatic and renal cystic dysplasia.

Etiology, Pathogenesis, and Genetics

The cause of the trisomy 13 phenotype (Patau syndrome) is the extra chromosome 13. Many early reviews cite the occurrence of translocation trisomy 13 in about 20% of cases. However, the above-cited population studies of Denmark and the United Kingdom and the work of Alberman et al. (2012) indicate that about 5–10% of cases of trisomy 13 are caused by a translocation, most commonly a 13;14 unbalanced Robertsonian translocation. Mosaicism for trisomy 13 makes up a smaller proportion of cases in population series of Patau syndrome. A comprehensive review of trisomy 13 mosaicism summarized all known individuals in the literature up to the time of the article (Griffiths et al. 2009). As for most chromosomal mosaicism syndromes (e.g., trisomy 8 mosaicsm, Pallister–Killian syndrome – Chapter 45) there is little correlation between the degree of mosaicism in lymphocytes and the resultant resemblance to classical Patau phenotype.

In trisomy 13 caused by nondisjunction, the origin of the extra chromosome is maternal in about 90% of cases (Bugge et al. 2007). The majority of the time the stage of nondisjunction is maternal meiosis I; however, in the study of Bugge et al. (2007) there are a higher number of meiosis II errors (just under 50%) than there are in trisomies of other acrocentric chromosomes. Of cases in which nondisjunction is paternal in origin, the majority are primarily postzygotic mitotic errors. In the case of trisomy 13 caused by 13;13 translocations, the structural abnormalities are usually isochromosomes that originate in mitosis.

Phenotype–karyotype correlations for partial trisomies of chromosome 13 have been discussed more extensively than in most other partial trisomy syndromes (Tharapel et al. 1986). On the basis of a summary of published cases of both proximal and distal partial trisomies, one can conclude that the orofacial clefts and scalp defects are caused by genes that are duplicated on the proximal portion of 13q. The prominent nasal bridge and postaxial polydactyly are caused by genes on the bottom half of 13q. Cases of partial 13q have a better outcome in terms of survival than full trisomy 13, suggesting

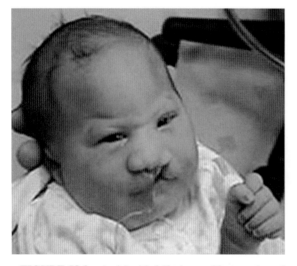 

FIGURE 58.2 A girl with full trisomy 13 at age 1 week and 16 years. (Courtesy of the mother and Rick Guidotti, Positive Exposure).

that genes all along the chromosome can cause increased neonatal mortality, as in trisomy 18.

The issues surrounding the recurrence risk for trisomy 13 are similar to those of trisomy 18 and are discussed above.

Diagnostic Testing

Most of the relevant diagnostic issues in trisomy 13 are the same as those discussed above for trisomy 18. Regarding the prenatal diagnosis of trisomy 13 Tongsong et al. (2002) found that close to 100% of fetuses recognized in the second trimester have findings on prenatal ultrasound. Because of the high occurrence of holoprosencephaly in trisomy 13, the prenatal diagnosis of holoprosencephaly should always raise the question of trisomy 13, especially when other ultrasound abnormalities are found. The long list of sonographically detectable malformations associated with trisomy 13 has been summarized (Hill 1996; Tongsong et al. 2002). Unlike trisomy 18, maternal serum triple-marker screening was not shown to have a characteristic pattern and will not be further discussed. The utilization of cell-free DNA in non-invasive prenatal screening (NIPS) was discussed above in the section on trisomy 18; the positive predictive value of NIPS is even lower for trisomy 13 than trisomy 18 with a range of 25–45%.

Regarding diagnosis in the newly born infant, immediate diagnosis with 24 hour chromosome analysis or FISH investigations is often indicated because of the need for urgent management decisions, which are discussed below. As in the case of trisomy 18, FISH probes can be used for molecular diagnosis in the postmortem situation, including in stillborns.

Differential Diagnosis

Because the cardinal features of trisomy 13 include a number of manifestations that are seen in various other multiple congenital anomaly situations, the differential diagnosis is quite long. However, the pattern in total is distinctive and usually allows for straightforward diagnosis. The presence of postaxial polydactyly in the context of either neonatal illness or death suggests Meckel–Gruber and hydrolethalus syndromes. The characteristic renal malformations and encephalocele in Meckel–Gruber syndrome and congenital hydrocephalus usually allow them to be distinguished from trisomy 13. Another condition in the differential diagnosis is the so-called holoprosencephaly-polydactyly syndrome, also known as "pseudotrisomy 13" (reviewed by Sergi et al. [2012]). In addition, the severe presentation of Smith–Lemli–Opitz syndrome (Chapter 53), previously called type 2 Smith–Lemli–Opitz syndrome, shares important similarities with trisomy 13. Other syndromes with holoprosencephaly and/or postaxial polydactyly and multiple anomalies also overlap with trisomy 13, but, again, the chromosome finding allows for easy distinction.

MANIFESTATIONS AND MANAGEMENT

Background

The approach to the management of a third-trimester fetus and infant with trisomy 18 or 13 is quite complicated, and the literature is controversial. The principal reason for the complexity surrounding decision making in the care of infants with trisomies 18 and 13 relates to the high neonatal and infant mortality in both conditions (Carey 2012; Andrews et al. 2016). The developmental outcome likely plays a role as well. All surviving children have significant developmental disability, but will progress in development, albeit slowly, as discussed under Behavior and Development (Baty et al. 1994b; Carey 2012).

Avoidance of delivery by caesarean when a fetus is known to have trisomy 18 appears to be the trend in the US and British obstetric literature (Rochelson et al. 1986). However, Spinnato et al. (1995) articulate a strong ethical case for respecting maternal autonomy in making decisions regarding aggressive intrapartum management in conditions such as trisomy 18 and related serious disorders. It is important to acknowledge that because approximately 50% of infants with trisomy 18 or 13 will survive longer than a week and about 6–12% of infants will live past 1 year (Rasmussen et al. 2003; Kosho et al. 2006; Meyer et al. 2016; Nelson et al. 2016), the still widely used terms of "lethal abnormality" and "incompatible with life" are inappropriate and misleading. Families of children with these disorders assert quite eloquently their objection about the use of "lethal" in these settings (Baty et al. 1994a, 1994b; Janvier et al. 2014).

The higher end of the stated range of 6–12% for one-year survival in trisomy 18 and 13 is based on two large recent investigations from different countries, the US and Canada. The number of cases in both studies far exceeds the population studies in the three prior decades indicating that at least in North America 1 year survival has increased. As stated above, this change in survival may be due to an increase in medical interventions including cardiac surgery (discussed below in cardiac section).

Tradition as reflected in the pediatric literature also indicates a non-intervention approach in newborn management of trisomies 18 and 13 (summarized by Carey [2013]). Bos et al. (1992) argued that early diagnosis is very important so that surgery can be withheld. The authors made a case that an infant with trisomy 18 should be viewed as having a "hopeless outlook" and is "not to be subjected to invasive procedures." Arguments indicating that there is "imminent death" and profound "mental retardation" suggested to Bos et al. that infants with trisomy 18 should be "allowed to die." Paris et al. (1992) made the point that to support "groundless hopes or false expectations as a means for caring or protecting the parents exposes the infant to potential pain and fruitless suffering." They proceed by saying "to be fair to the infant, the parents, and the staff, the plan for a patient afflicted with a fatal condition should be thought through before any interventions are undertaken." McGraw and Perlman (2008) published a survey of the opinion of neonatologists in New York City on newborn care of infants with trisomy 18. Notably, 56% of those surveyed indicated that they would not initiate resuscitation in the case presented to them in the survey. The factors important to the remaining 44% who would intervene included parental wishes to support the infant. The authors contend that "having intensive care measures such as intubation and corrective surgery available as potential options for infants with a confirmed lethal trisomy gives the impression to parents that these are reasonable interventions to consider and downplays the invasive and painful nature of these therapies." The explicit and implicit messages in this publication summarize the conventional view of non-intervention in the care of infants with trisomies 18 up to that time period. The authors expressed concern that "the pendulum was moving" toward parental autonomy and away from the principle of "best interest of the child". Although this publication raises important issues of care, the authors refer to the syndromes as "confirmed lethal" and do not adequately deal with the complex issue regarding the determination of the "best interest of the child." Koogler et al. (2003) suggested that reference to conditions such as trisomies 18 and 13 as lethal is not only inaccurate but also "dangerous." Other recent articles suggest a change in the paradigm of care of children with trisomy 13 and 18 in the latest decade: Lorenz and Hardart (2014) provided a comprehensive review of the topic and recommend guidelines for care of the infant that includes physicians and parents evaluating the benefits of care together. Andrews and colleagues (2016) and Haug et al. (2017) suggest a "shared decision making" model and "patient-centered" care for working with families of trisomy 13 and 18, views that this author would strongly support.

When families choose the decision to proceed with full intervention (as opposed to pure comfort care, see Lorenz and Hardart [2014] and Andrews et al. [2016]), the care team can establish the infant's specific airway and cardiac status through consultation with pediatric otolaryngology (see Respiratory below) and pediatric cardiology (see Cardiovascular below). In the newborn without central apnea (which is likely rarer than thought conventionally), it is the airway and cardiac abnormalities that will drive the early care.

There is amazingly little documentation of the precise reason for death in infancy for either trisomy. Conventionally, it is often stated that the early mortality is because of the high occurrence of structural heart malformations. However, most of the heart lesions in both syndromes are not those that result in death in infancy. Investigations into the natural history of trisomies 18 and 13 have suggested that central apnea, or its presence with a combination of other health factors, e.g., risk for aspiration, hypoventilation and more recently upper airway insufficiency, is the primary pathogenesis of increased neonatal and infant mortality (Root and Carey 1994; Embleton et al. 1996; Cereda and Carey 2012)

The variables that play a role in survival are not entirely clear. Certainly, the absence of serious malformations such as hypoplastic left heart or diaphragmatic hernia must be important. Niedrist et al. (2006) indicated that survival was the same for children with and without a ventricular septal defect. In contrast, however, Rasmussen et al. (2003) did show better survival in the infants without heart defects. Boghossian et al. (2014) documented that preterm birth of less than 32 weeks affected survival in trisomy 18.

Until the last decade or so most children with trisomy 18 or 13 who survive the newborn and early infancy periods have not received aggressive or extraordinary care. The traditional assumption was that intervention did not alter outcome. Then Kosho et al. (2006) studied the outcome of trisomy 18 in a newborn unit in Japan, where all families were offered intensive care including ventilation, but not cardiac surgery, which was not available in the hospital during the study period, 1994–2003. Survival figures were improved over the population studies cited above, and 6 of 24 infants (25%) were still alive at 12 months. This is the first study that addressed the question often asked by parents both prenatally and postnatally, "what is the outcome if full intervention is performed?". Kosho and Carey reviewed the recent lines of evidence that demonstrate that intervention does improve outcome in trisomy 13 and 18 (2016).

The trend in neonatal intensive care in the last three decades has been to place significant weight on parental decision-making, usually in the context of the "best interest of the child." Parents appreciate partnership in decision making. Overly simplified and value-laden terms such as "lethal," "vegetative," and "hopeless" not only are inaccurate, but also convey an implicit message from the outset (Carey 2012). Initial counseling of the family should be realistic and accurate and not unnecessarily grim and bleak. Options for care and management must be explained, and goals of care can be established (Lorenz and Hardart 2013). Families appreciate an honest and straightforward approach to the challenges and decisions with the best interests of the child in mind (personal experience). As mentioned above Andrews et al. (2016) and Haug et al. (2017) advocate for a shared decision-making model in caring for infants and children with trisomy 18 and 13.

The families of infants with trisomy 18 or 13 must initially deal with the issue of low survival rate and the practical decisions in the newborn period concerning resuscitation, surgery, and life support. As time goes on, the family must then cope with the prospect of significant disability if the baby survives. Mixed and ambivalent feelings about what is best for the child are natural. The primary care practitioner has the unique opportunity and challenge to provide support on an ongoing basis. Recognition of the uncertainty of the situation and of the paradox of preparing for both the probability of death and the possibility of living is helpful in the early weeks (Carey 1992).

In neonatal care, appropriate fluid and nutritional support are always indicated. Decisions can be made on a day-by-day, week-by-week, or month-by-month basis, and different courses can be pursued according to the status of the child and changing circumstances. A respectful and humanistic approach that recognizes parental feelings and views on each decision represents a thoughtful and caring way to proceed (Carey 2012). Referral to a palliative care team, which now exists in most centers in the United States, is also an important option for support and management.

Routine guidelines for care of infants and children with trisomy 18 and 13 are available in Cereda and Carey (2012), Barnes (see Barnes and Carey, Care of the Infant and child with Trisomy 18 and 13, 4th edition, 2018, www.trisomy.org) and by Wilson and Cooley (2006).

Growth and Feeding

Growth deficiency is the rule for both trisomies 18 and 13. Growth charts for older infants have been published (Baty et al. 1994a). Barnes and Carey (4th edition freely available www.trisomy.org 2018) have formatted these growth charts to fit easily into patient charts; the charts can be inserted into electronic medical records. Weight and length are below the third centile, more consistently so in trisomy 18 than in trisomy 13. By the age of 1 year, the average weight in trisomy 13 is near the fifth centile, with many cases exceeding the fifth centile after 3 years of age. This may be related to the observation that older children with trisomy 13 usually do not have holoprosencephaly (Morelli et al. 2000). Length and height are also less affected in trisomy 13 in the third and fourth years of life than in trisomy 18. Head circumference tends to be below the third centile, on average. In trisomy 13, the head circumference is lower in surviving children with holoprosencephaly. New data on growth in children with trisomy 18 and 13 are currently being compiled by SOFT (Support Organization for Trisomy 18, 13 and Related Disorders) with the plan of having more current measurements with larger number of individuals.

For surviving neonates and infants with trisomies 18 and 13, feeding difficulties remain one of the challenges of management. Most infants with either condition require tube feeding from birth. In one study, 44 and 53% of children with trisomies 18 and 13, respectively, did learn bottle or breast feeding during the first few weeks or months of life (Baty et al. 1994a). This study probably was biased toward children with prolonged survival, so overall rates may be lower. Currently it would be crucial to establish that an infant can protect their airway by performing a swallow study before introducing routine oral feedings. It is unusual for a child with trisomy 18 or 13 who has not established oral feeding in infancy to do so later in life; the children who are bottle fed usually establish this skill during the first few months of life (personal observation). However, at least three infants with trisomy 18 have been able to nurse throughout infancy (personal data). Both the sucking and swallowing stages are difficult for these infants. Many families of older infants with trisomy 18 and trisomy 13 elect to have a gastrostomy tube placed when it becomes clear that the baby will likely not be able to master bottle-feeding to maintain growth. The average age of gastrostomy placement was 7 and 8 months for trisomy 13 and 18, respectively, in the only published study (Baty et al. 1994a). The age of placement is likely earlier than 7–8 months currently (personal observation).

Gastroesophageal reflux is a consistent finding in infants with trisomies 18 and 13 and may explain irritability and recurrent pneumonias; reflux should always be considered in routine care. Aspiration during feeding or from reflux may be the cause of early death. Gastrointestinal malformations can also play a role in feeding problems after the newborn period. Pyloric stenosis has been reported in trisomy 18, and malrotation is occasionally seen in trisomy 13; these may also affect feeding. Orofacial clefts are a frequent finding in trisomy 13 and an occasional feature of trisomy 18, and their presence will complicate the feeding difficulties in both syndromes.

Tracheoesophageal fistula with esophageal atresia occurs in about 5% of neonates with trisomy 18 and occasionally in trisomy 13. The care of the newborn with a tracheoesophageal fistula is a particularly difficult challenge as the decision to intervene with surgery must be made in the early days of life. One issue is whether to simply place a gastrostomy tube (a more palliative approach) versus complete repair with tracheoesophageal fistula division (corrective approach). Nishi et al. (2014) performed an historical cohort study of 24 individuals with trisomy 18 and observed that those who had the more "radical" corrective surgery had better 1 year survival than those who only had the palliative approach.

Evaluation

- Length, weight, and head circumference should be measured during each health supervision visit and plotted on the published disorder-specific growth curves (Baty et al. 1994a).
- Referral to a dysphagia team for evaluation may be useful (in particular to consider airway protection). Radiographic suck/swallow study should be considered.
- Even a child who appears to be able to bottle feed in the early months of life should have an evaluation of feeding through a swallow study to establish airway competency.

- Assessment of the presence of gastroesophageal reflux as a potential factor is indicated when feeding or respiratory difficulties occur.
- If a child with trisomy 18 or 13 is not oral feeding by 6 months of age, consideration of gastrostomy placement is appropriate.

Management

- Evaluative and treatment decisions should be discussed with parents on a regular basis. It is appropriate to reassess and change the degree of intervention during the course of support and care.
- Teaching the parents how to do nasal or oral gastric tube feeding before discharge will be necessary for the majority of children.
- Appropriate and standard therapy for gastroesophageal reflux can be initiated when reflux is determined to be a factor. Nissan fundoplication has been successfully performed in many children who are part of the SOFT Surgery Registry (www.trisomy.org).
- Gastrostomy has been successfully performed on over 100 children as recorded in the SOFT Surgery Registry (www.trisomy.org).
- The majority of children with either condition who have structural heart malformations require consultation by cardiology before any surgical procedure.

Development and Behavior

Very little data have been collected regarding the developmental outcome of older children with trisomies 18 and 13. Most review papers and chapters indicate the presence of profound cognitive and psychomotor disability in surviving children with both conditions.

Individual case reports of older affected children indicate that expressive language and walking do not typically occur in trisomy 18 or 13. There is a report of a 4 year old with full trisomy 18 who walked independently (Ray et al. 1986). Kosho et al. (2013) documented independent unassisted walking in two children with trisomy 18 indicating that, while uncommon, unassisted ambulation can occur.

Baty et al. (1994b) published their experience with 62 older individuals with trisomy 18 and 14 with trisomy 13. Developmental records were collected on children through collaboration with the support group, SOFT. Although all these children were clearly functioning in the profoundly developmentally handicapped range, all did achieve some skills and continued to learn. A number of older children with trisomy 18 could use a walker, and two children with trisomy 13 were able to walk unsupported. Many older children who were using walkers have had early physical therapy (personal data). Developmental quotients (DQ) in children over 1 year of age were in the 0.2–0.3 range, with developmental ages in the older children with trisomy 18 averaging 8–9 months and those with trisomy 13 averaging 7–8 months. In this series, many skills that are conventionally thought not to occur in trisomy 18 or 13 were seen as individual abilities in many children; these included self-feeding, sleeping independently, object permanence, and understanding cause and effect, skills above the overall DQ. Older infants with trisomy 18 or 13 recognized their family and smiled responsively on a consistent basis. Developmental achievements in both syndromes have been summarized using milestones from the Denver Developmental Screen (Baty et al. 1994b) (see Figures 58.1 and 58.2). A doctoral dissertation (Healey 2003) comprehensively summarizes the social development of children with trisomies 18 and 13 ascertained through collaboration with parents of the support group, SOFT. The study details skill levels in older children and identifies factors related to parental coping (this work is available through the author or SOFT). Liang et al. (2015) documented the communicative skills of a series of 32 individuals older than 3 years with trisomy 13 and 18; two thirds of parents reported communication skills higher than overall developmental ages mentioned above in the Baty et al. (1994b) study. Bruns investigated skill ranges in 22 children with trisomy 18 and eight children with trisomy 13; all but two were older than 12 months. Analysis of developmental status showed that the high end of the range of skills was usually at 12 months in children older than 24 months, but some children acquired milestones above the 12 month level (e.g., follow two-step directions, engages in social play, responds to social games).

Evaluation

- Developmental evaluation, using standard developmental screening tools for infants and children, can be performed at regular health supervision visits.
- Referral for physical therapy evaluation is particularly important.

Management

- Referral to early intervention programs is recommended in the ongoing care of infants with trisomy 18 or 13. Thereafter, the child and family enter the system for necessary educational intervention through childhood. Physical, occupational, and speech therapy, as well as special education, occur in these settings. Augmentative communication programs should be considered as in any child with significant developmental disabilities.
- Specialized programs for children with visual disabilities may be needed for those with trisomy 13 who have significant visual abnormalities.
- For children with trisomy 13 who have hearing loss as well as visual disabilities, more comprehensive pro-

grams for multiple disabilities may be required. Because the apparent hearing loss found in children with trisomy 18 does not necessarily indicate functional hearing impairment, standard early intervention programs are likely adequate.

Neurologic

A variety of developmental abnormalities of the brain have been described in autopsy series of trisomies 18 and 13 (Moerman et al. 1988; Kinoshita et al. 1989). Cerebellar hypoplasia has been described consistently in trisomy 18 and has been noted even in prenatal ultrasound series. Other primary structural defects of the brain described in trisomy 18 include agenesis of the corpus callosum, microgyria, and hydrocephalus. Of note, about 5% of infants with trisomy 18 have meningomyelocele, which is also seen occasionally in trisomy 13.

Functional neurological findings in both conditions include hypotonia in infancy, hypertonia in later childhood, seizures, and central apnea. Seizures occur in about 25–50% of older children with both conditions. The seizures are generally straightforward to control in both conditions, but tend to be more complicated in trisomy 13, possibly related to the presence of holoprosencephaly. Kumada et al. (2013) investigated the seizure pattern in 11 individuals with trisomy 18 and found intractable epilepsy in about half of the children. There may have been a selective bias toward the occurrence of intractable epilepsy as that was not the experience of Matricardi et al. (2016) in a much larger Italian series or in the author's personal experience.

Central apnea occurs commonly in both syndromes, but it does not have a clear anatomical basis except in individuals with holoprosencephaly. Recent investigations into the natural history of trisomies 18 and 13 have suggested that central apnea, or its presence with a combination of other health factors (hypoventilation, aspiration, and obstructive apnea) is the primary pathogenesis of the increased infant mortality (Root and Carey 1994; Wyllie et al. 1994; Embleton et al. 1996). However, Fukasawa et al. (2015) found that neonates with trisomy 18 considered to have central apnea could have epileptic apneas, and this should always be considered before concluding typical central apnea. An electroencephalogram can be performed to clarify the presence of apnea due to seizures.

The holoprosencephaly that is commonly present in trisomy 13 deserves special mention. Holoprosencephaly of some degree is present in about 60–70% of infants with trisomy 13, as assessed either by imaging study or autopsy. This defect of early embryogenesis of the forebrain involves lack of cleavage of the frontal hemispheres, often resulting in a single holoventricle (see Chapter 31). This can be associated with the classical facial malformations, including cyclopia, cebocephaly, or premaxillary agenesis. In one series, the majority of the infants with trisomy 13 and holoprosencephaly had the alobar variety (Moerman et al. 1988). Other primary central nervous system malformations, including cerebellar hypoplasia and agenesis of the corpus callosum, have been reported occasionally. The presence or absence of holoprosencephaly in a newborn with trisomy 13 may have prognostic significance. Although it is known that children with trisomy 13 *without* holoprosencephaly have central apnea, individuals who survive the first year of life do *not* have semilobar or alobar holoprosencephaly as commonly as newborns with trisomy 13 overall (Morelli et al. 2000), suggesting that the absence of holoprosencephaly portends better survival.

The central nervous system abnormalities mentioned above play a significant role in the early mortality, central apnea, and feeding difficulties of neonates and young infants with trisomies 18 and 13 syndromes. However, the precise anatomical delineation of these defects (other than holoprosencephaly) probably does not assist in management.

Evaluation

- Neurological examination is recommended for all infants and young children with trisomies 18 and 13 on a regular basis.
- Referral to a child neurologist is indicated in infants with seizures.
- Neuroimaging should be performed in children with trisomy 18 and seizures as the presence of central nervous system defects is a predictor of more complex seizures (Matricardi et al. 2016)
- Neuroimaging to detect or exclude holoprosencephaly should be performed in infants with trisomy 13 if predicting prognosis is important, particularly if they are in critical condition. The absence of holoprosencephaly portends a better outcome (Morelli et al. 2000; personal experience)
- If abnormalities of muscle tone are present, as they usually are, referral for physical therapy evaluation is recommended.

Management

- Treatment is related to the symptomatic occurrence of seizures or muscle tone abnormalities.
- Seizure management is usually not complicated in either syndrome, and convulsions tend to be well controlled with appropriate medications. Lamotrigine seems to be a particularly useful and commonly prescribed anticonvulsant (personal observation). Matricardi et al. (2016) did not find that any particular antiepileptic drug to be superior in management.
- Muscle tone abnormalities generally respond to physical therapy.

- Caffeine could be used in the treatment of central apnea but its efficacy in infants has not been studied systematically (and deserves investigation).

Cardiovascular

Congenital heart malformations are present in about 80% of children with trisomy 13 and 90% of children with trisomy 18. The observed pattern of malformations is non-random and relatively specific. In recent series, over 90% of infants with trisomy 18 have a ventricular septal defect with polyvalvular disease (Musewe et al. 1990; Balderston et al. 1990). In one series, all 15 individuals with trisomy 18 had polyvalvular disease, with 6 of the 15 having involvement of all four valves (Balderston et al. 1990). Pulmonic valvular dysplasia of trisomy 18 frequently results in tetralogy of Fallot. About 10% of infants with trisomy 18 will have a more complicated cardiac malformation, in particular double-outlet right ventricle, endocardial cushion defect, or left-sided obstructive lesion. Shunt lesions such as atrial septal defect and ventricular septal defects are the most common lesions in trisomy 13 and polyvalvular dysplasia is less common in this syndrome than in trisomy 18. Double-outlet right ventricle occurs in some infants with trisomy 13. Of note, the polyvalvular disease of both syndromes involves a redundant or thick leaflet, but most of the time does not have significant hemodynamic effect.

The majority of heart lesions in both syndromes are unlikely to produce neonatal death. In one series, only 6% of 31 infants with trisomy 18 or 13 had a lesion that would be considered lethal in infancy (Musewe et al. 1990). However, in this series there was frequent cyanosis and increased right ventricular dimensions, suggesting early development of pulmonary hypertension. This observation raises the possibility that infants with trisomy 18 may be predisposed to develop pulmonary vascular obstructive changes earlier than other children with ventricular septal defects. Tahara et al. (2014) found medial defects of small pulmonary arteries in 2 of 11 infants with trisomy 13 or 18 who underwent pulmonary artery banding for heart defects, representing a higher frequency than in children in a large series of congenital heart defects who had pulmonary hypertension. This suggests that infants with trisomy 13 and 18 may have a higher risk to develop pulmonary hypertension than other children with similar shunt lesions.

Because the majority of infants with trisomy 18 or 13 that succumb in the newborn or early infancy period die of central apnea and other causes rather than their heart defect, until the last decade heart surgery was rarely performed. However, when an infant with a heart defect is thriving, the issue of early development of pulmonary hypertension emerges. Consideration of surgery as an option is now more commonplace in North America, Europe, and Japan than in the past. Until the two recent heart surgery series discussed below were published, there were rare reports of heart surgery in children with trisomies 18 and 13 (Musewe et al. 1990; Baty et al. 1994a). The SOFT Surgery Registry has collected cases throughout North America of children with trisomy 18 or 13 who have had cardiac surgery, most since 1990. Hansen et al. (2000) summarized the outcome in a total of 28 children with full trisomy 18 or 13 from the SOFT Registry who had undergone 31 different heart surgeries. The most common surgical procedures were closure of ventricular septal defect, ligation of patent ductus arteriosus, closure of atrial septal defect, and placement of a pulmonary band. The average age at the time of surgery was 8.8 months. The outcome for 29 of the surgeries was known, with 25 surviving to hospital discharge and 4 surgical mortalities; the surgical mortality rate in this early series was thus 13.8%. The SOFT Surgery Registry currently documents 120 infants with trisomy 13 and 18 who have had cardiac surgery, most of these performed since 2000 and in North America.

McCaffrey (2002) summarized the views of pediatric cardiologists responding to a question about heart surgery in babies with trisomy 18. Although there were "no right or wrong answers," most of the respondents said they would proceed with care under certain circumstances. This work expressed the viewpoint of the cardiology community up to the point of the recent publication of multiple papers on the controversial topic of surgery in children with these syndromes.

Only one other large series of heart surgery in infants with trisomies 18 and 13 was published before the present decade. Graham et al. (2004) reported on 35 children from an international pediatric care consortium. Twenty-five had complete repair, the remainder having had palliative operations; 91% went home from the hospital, but long-term outcome is not presented. Since then several series of infants with trisomy 13 and 18 who had cardiac surgery have been reported from the US and Japan. Janvier et al. (2015) summarized the outcome of these studies as of 2015. All are descriptive analyses and cross-sectional studies without control groups, and all have varying ascertainment methods, types of operations, and ages at time of surgery; the frequency of "survival to discharge" ranged from 50 to 80%. The authors of this review point out the limitations of these observational studies. Clearly there are challenges in drawing any conclusions from these early studies other than that survival was better than the conventional wisdom of earlier years before surgery was even done (personal experience).

In 2017 results of two much larger scale investigations of outcome in cardiac surgery of infants with the syndromes were published (Peterson et al. 2017; Kosiv et al. 2017). In the first study 29 infants with trisomy 13 and 69 with trisomy 18 underwent cardiac surgery. Survival to discharge was 72.4% for the trisomy 13 group and 87% for the trisomy

18 cohort. In-hospital mortality rates were clearly higher than that reported for the general population of children having these operations, but in these selected groups the intervention "demonstrated longer survival than previously reported." The survival figures are similar to children having cardiac transplantation for a complex heart condition, a much more accepted and well-known surgical option. Kosiv et al. (2017) presented the outcome of 37 infants (ascertained <14 days of age) with trisomy 13 and 63 with trisomy 18 who had cardiac surgery. In-hospital mortality was decreased in infants having surgery compared to those who did not have cardiac surgery and remained decreased throughout the 24 month follow-up period. These two papers represent seminal work on this topic and indicate better survival after cardiac surgery for children with the two syndromes; they provide current data for clinicians to be able to present to families who, with their care team, are contemplating cardiac surgery in an infant with either syndrome. Many institutions in the US and Canada now consider cardiac surgery in children with the syndromes when before 2000 there were few reports.

Evaluation

- Cardiac evaluation should take place at the time of diagnosis in all infants with trisomy 18 or 13. This should include an echocardiogram in the newborn period for assistance in management decision-making and so that the family and primary care doctor have appropriate clinical data if the infant is ill for unknown reasons.

Management

- Families and their physicians vary in their approach to the treatment and management of cardiovascular malformations in children with trisomy 18 or 13. Although the majority of children have non-lethal cardiac defects, the high frequency of shunt lesions and the early development of pulmonary hypertension may necessitate early treatment. Decisions regarding symptomatic treatment of heart failure or placement on oxygen and monitors should be made by families and physicians jointly. Decisions about surgery are individualized and are discussed above.

Ophthalmologic

A large number of ocular manifestations have been reported in both trisomy 18 and 13 syndromes. The eye manifestations in trisomy 13 are particularly consistent and well established. Over 50% of newborns with trisomy 13 syndrome have microphthalmia or anophthalmia and colobomas. Retinal dysplasia is a common autopsy finding in infants with trisomy 13. The ocular findings associated with holoprosencephaly, including cyclopia and hypotelorism, are seen. Other distinctive eye findings seen in trisomy 13 include persistent hypoplastic primary vitreous with cataract and corneal opacities. Some infants with the condition will develop a congenital or early onset glaucoma. This could be the reason for unexplained irritability (personal experience).

Significant ocular manifestations are much less frequent in trisomy 18. Although short palpebral fissures are common, major intraocular pathology or structural defects are seen in less than 10% according to Geiser et al. (1986). Cataracts and corneal opacities are occasionally seen in infants and young children with trisomy 18 (Calderone et al. 1983). Older infants with trisomy 18 usually have photophobia. The pathogenesis of this particular manifestation is unknown and deserves future investigation.

Correia et al (2017) summarized the eye anomalies in their series of 31 individuals with trisomy 18; 4 children (13%) had eye defects consisting of corneal opacities (2), microphthalmia (1) and cataract (2).

Evaluation

- Evaluation by an ophthalmologist is recommended in all infants with trisomy 13 and 18 at the time of diagnosis. Follow-up would occur according to the particular findings.
- Follow-up ophthalmological evaluation is recommended for infants over 1 year of age with both conditions because visual acuity abnormalities are common in older children.

Management

- Treatment of glaucoma, cataracts, or corneal clouding should be individualized and is standard. Surgery can be suggested in older infants and children depending on the surgical and anesthetic risk determined by the cardiologist and/or primary care practitioner.
- Sunglasses are helpful for photophobia in older infants and children with trisomy 18.

Genitourinary

A variety of structural defects of the genital and urinary tracts have been described in individuals with trisomies 18 and 13. About two-thirds of infants with trisomy 18 will have a horseshoe kidney. Cystic dysplasias of various forms are commonly present in trisomy 13. Urinary tract infections appear to occur with increased frequency in trisomies 18 and 13, perhaps related to the structural renal defects (Baty et al. 1994a). Renal failure or disease is not a common cause of chronic illness despite the high frequency of defects (personal experience).

Evaluation

- Because of the high frequency of renal defects, a screening abdominal sonogram is indicated in infants with trisomies 18 and 13 at the time of diagnosis.
- Those with significant renal defects should be followed for infection and renal insufficiency by periodic creatinine and urinalysis.

Management

- Management of urinary tract infection and renal obstruction is the same as in any child.

Hematology

Wiedmeir et al. (2006) studied the frequency of hematologic abnormalities in newborns and infants with trisomies 13 and 18 in various hospital settings. Eighty-three percent of the 28 infants with trisomy 18 and 75% of the 22 infants with trisomy 13 had thrombocytopenia. Follow up investigations are not available, but in the author's experience (personal experience) older children with trisomy 18 commonly have mild thrombocytopenia, occasionally requiring a transfusion at times of illness. This manifestation deserves more study. Based on the pathogenesis of thrombocytopenia in other multiple congenital anomaly syndromes, the thrombocytopenia is likely due to bone marrow hypoplasia.

Evaluation

- A complete blood count (CBC) with platelet count is recommended in the newborn period in both syndromes. If thrombocytopenia is present, CBC with platelets should be monitored as needed
- Referral to pediatric hematology is suggested in children with persistent thrombocytopenia.

Management

- Management of thrombocytopenia is the same as in any child.

Neoplasia

Clinical reports, a review of the literature, and evidence from one pediatric cancer registry study indicate that older infants and children with trisomy 18 are at increased risk to develop Wilms tumor and hepatoblastoma (Olson et al. 1995; Satge et al. 2016). The actual risk for a child with trisomy 18 to develop Wilms tumor is likely just less than 1% (Carey and Barnes 2016).

The presence of Wilms tumor in trisomy 18 makes biological sense, because nodular renal blastema has been described at autopsy in infants. Although the range of age of onset is from 12 months to 13 years, in most cases, the Wilms tumor develops after 5 years; this suggests a different biological basis for the tumor in children with trisomy 18 than the typical Wilms tumor in children in the general population.

Prognostically, the course of Wilms tumor in trisomy 18 has been extremely variable: of those cases with knowledge of outcome five children died of complications related to the tumor or treatment (surgery, infection); one was initially treated but died from pulmonary metastasis; three survived; and the other two succumbed because of parental choice of no treatment.

There have been two reported cases of Wilms tumor occurring in children with trisomy 13. The ages were 1 and 4 years. These occurrences could be coincidental, and routine surveillance is not suggested at the current time.

There have been several clinical reports of infants and young children with trisomy 18 who developed hepatoblastoma (Satge et al. 2016). The age at diagnosis ranged from 4 months to 3 years. Because this embryonal tumor occurs with Wilms tumor as a component manifestation in other malformation-dysplasia syndromes (e.g., Beckwith–Wiedemann syndrome, Chapter 9), it is biologically plausible that both of these would occur with increased risk in trisomy 18.

Evaluation

- Because of the high frequency of intra-abdominal tumors, particularly Wilms tumor, abdominal sonographic screening is indicated in surviving infants with trisomy 18. Although there is no established timing, it is reasonable to initiate abdominal screening every three months after the newborn period in infants and children with trisomy 18 (personal experience). The exact age to stop screening is unknown. Because the oldest case of development of Wilms tumor in trisomy 18 is 13 years, abdominal screening for this tumor is justified into adolescence.
- Because there have been only two reported cases of Wilms tumor in trisomy 13, screening for Wilms tumor and hepatoblastoma is not indicated currently.

Management

- The treatment for Wilms tumor and hepatoblatoma is similar to that in the general population.

Gastroenterology

As mentioned, gastroesophageal reflux is a consistent finding in infants with trisomies 18 and 13 and may explain irritability and recurrent pneumonias; reflux should always be considered in routine care. Gastrointestinal malformations, especially tracheo-esophageal fistula with esophageal atresia, which poses particularly complex care issues in the newborn, occur in about 5% of newborns with these disorders. Pyloric stenosis, biliary atresia, and Meckel's diverticulum have been reported in trisomy 18, and malrotation is occasionally seen in both trisomies 13 and 18.

Chronic constipation is very common in infants and children with trisomy 18 (personal experience); it occurs as well in trisomy 13, as in many children with hypotonia and developmental delays.

Evaluation

- Clinicians should have a high index of suspicion for pyloric stenosis, malrotation, and Meckel's diverticulum in infants with trisomy 18 and for malrotation in infants with trisomy 13.
- Referral to a pediatric gastroenterologist is recommended in infants and children with trisomies 18 and 13 who have constipation not improving on standard stool softeners.

Management

- The treatment for these gastrointestinal disorders, including constipation, is similar to that in the general population.
- Care of newborns with tracheo-esophageal fistula and trisomy 18 was discussed above in the Growth and Feeding section.

Ears and Hearing

A vast array of middle ear and temporal bone abnormalities has been described in postmortem examinations of infants with trisomies 18 and 13. In addition, moderate to severe sensorineural hearing loss has been reported in older children. The exact frequency of hearing loss in older children has not been estimated but is common from personal experience. Structural ear malformations, including microtia and meatal atresia, are found on occasion in both syndromes. The external auricular abnormality of trisomy 18 is particularly characteristic, and it helps in diagnosis of newborns. The ear is small and has a small lobule and helix that is variably underfolded and attached to the scalp (cryptotia). The helix is frequently unfolded and simple.

Karimnejad and Costa (2015) utilized the SOFT Surgery Registry to document the 231 otolaryngologic procedures in children with trisomies 18 and 13; tympanostomy tube placement was the most common surgery: 38 children had a tracheostomy performed and 37 had a tonsillectomy and/or an adenoidectomy.

Evaluation

- Audiological evaluation is recommended in newborns as in all babies; universal hearing screening is now performed routinely in North America and Europe. Infants with trisomy 18 or 13 older than 6 months of age should be followed for hearing loss. Behavioral testing can usually be accomplished in children older than 1 year of age.
- Brainstem auditory evoked response (BAER) studies have been performed in a number of infants and older children with trisomy 18, and children with trisomy 18 appear to do better on behavioral testing than their brainstem-evoked responses would indicate (personal experience). This observation deserves more investigation.

Management

- A trial of hearing aids in older infants with trisomy 18 or 13 who have abnormal audiological evaluations is appropriate (personal experience). However, if the hearing aid does not appear to improve communication and is difficult for the child to tolerate, the decision to continue or discontinue can be individualized.

Musculoskeletal

A variety of musculoskeletal abnormalities occur in both trisomy 18 and 13 syndromes, including medically significant malformations and minor anomalies of limb and skeleton. Postaxial polydactyly, especially of the hands, is a high-frequency finding in individuals with trisomy 13 (60–70%). Limb deficiency defects can occur occasionally in both syndromes, and radial aplasia and other preaxial limb deficiencies occur in 5–10% of children with trisomy 18. Talipes equinovarus and calcaneovalgus positional foot deformities occur in both conditions, but especially in trisomy 18, where about half of all children will have such a deformity. Some children with trisomy 18 will have contractures of other joints besides the feet, and they can present with "arthrogryposis" (see Chapter 6). Over-riding fingers, usually with camptodactyly, represent a particular diagnostic finding in trisomy 18.

Scoliosis develops commonly in older children with trisomy 18, usually after 4 or 5 years of age, and appears unrelated to structural defects of the vertebrae. Scoliosis may progress between ages 5 and 10 years.

Evaluation

- Radiographs of the limbs should be performed when appropriate to management.
- The spine should be evaluated clinically at routine health supervision visits in children with trisomy 18 starting at age 2 years, and spine radiograph series can be ordered when clinical scoliosis is suspected.
- Referral to an orthopedist should be made on recognition of significant musculoskeletal abnormalities, especially scoliosis.

Management

- Decisions regarding the placement of casts for clubfeet and/or radial aplasia should be made keeping in mind the quality of life of the child.
- Because it is uncommon for a child with trisomy 18 to walk unassisted and independently, many families and primary care physicians will defer decisions of surgery for talipes equinovarus or calcaneovalgus deformities until later childhood. However, these decisions are complex, because there are children with both syndromes who walk with assistance or independently. If this skill is acquired, treatment of positional foot deformity might be of benefit.
- Surgery for scoliosis may need to be considered in the older child. Three adolescent girls with trisomy 18 are known to have had scoliosis surgery with placement of rods or fusion, and all tolerated the procedure (personal experience). Decisions regarding this invasive surgery should be predicated on input from the cardiologist if the child has heart disease and/or pulmonary hypertension.

Respiratory

Because over 90% of children with trisomy 18 and 80% of children with trisomy 13 have a structural heart malformation, pulmonary hypertension related to heart defects is common in infancy, particularly in trisomy 18 (Van Praagh et al. 1989). In addition, upper airway problems including choanal stenosis/atresia, glossoptosis, laryngomalacia, and tracheobronchomalacia have been seen, but the precise anatomical problem is often not documented in either syndrome. A comprehensive airway assessment by a knowledgeable pulmonologist or otolaryngologist is indicated in newborns and infants whose families choose intervention.

Most infants with trisomy 18 or 13 who die in infancy do so because of respiratory problems. Central apnea, upper airway obstruction, early-onset pulmonary hypertension, recurrent aspiration, feeding difficulties, tube feedings, and gastroesophageal reflux all contribute to mortality and respiratory problems, and together create a symptom complex that probably accounts for most early infant deaths (Root and Carey 1994; Embleton et al. 1996). Obstructive sleep apnea is common in infants and difficult to recognize unless investigated formally (personal experience). Referral to a pediatric pulmonologist and/or sleep specialist for evaluation and/or sleep studies is appropriate when sleep apnea is suspected or if any surgery including the heart is being considered.

There has been a striking increase in the performance of tracheostomies in children with trisomy 18 in the last decade (Nelson et al. 2012). This seems to represent a more open view toward intervention when appropriate for an individual child as in the case of heart surgery. In the author's experience tracheostomy was rarely considered in management of children with respiratory difficulties before 2000. In Nelson et al. (2012), at least 76 children with trisomy 18 in the US Kids Inpatient Database had a tracheostomy performed between 1997 and 2009, demonstrating that a substantial number of children in the US were undergoing this procedure as part of their management.

Infants with trisomies 18 and 13 with or without heart defects are particularly susceptible to respiratory syncytial virus infection (personal experience).

Evaluation

- How far to proceed and what evaluations to initiate for respiratory problems will depend on the discussions between parents and care providers.
- Referral for evaluation by pediatric pulmonology, a sleep specialist, and/or a dysphagia team should be made in sorting out the variables of this symptom complex.
- Evaluations do not differ from those in other children with similar symptoms.

Management

- Immunoglobulin treatment for respiratory syncytial virus should be considered for infants with trisomies 18 and 13 before and during respiratory syncytial virus season.
- Decisions about home monitoring and oxygen therapy can be made on an individual basis after discussions between parents and care providers.

RESOURCES

Checklists for routine health supervision guidelines for use in the primary care setting have been developed by Barnes and the author (see Barnes and Carey, Care of the Infant and child with Trisomy 18 and 13, 4th edition *www.trisomy.org*) and by Wilson and Cooley (2006).

USA and North America Support Group SOFT

Barbara Van Herreweghe
2982 S. Union St.
Rochester, NY 14624, USA
Phone: (800) 716-SOFT
Website: *http://www.trisomy.org*

International Trisomy Alliance

Volunteer organization focusing on trisomy 13 and trisomy 18
http://www.internationaltrisomyalliance.com

Trisomy 18 Foundation

Victoria Miller
4491 Cheshire Station Plaza Suite 157
Dale City, VA, USA
Phone: 609-918-1089
Website: *http://www.trisomy18.org*

Chromosome 18 Registry

The Chromosome 18 Registry and Research Society
Jannine D. Cody, Ph.D.
7155 Oakridge Drive
San Antonio, TX 78229, USA
Phone: (210) 657-4968
Website: *https://www.chromosome18.org*

SOFT of United Kingdom

Jan Fowler, Coordinator
48 Froggatts Ride, Walmsley, Sutton Coldfield B76 2TQ, UK
Phone: (01) 21-351-3122
SOFT (UK)Helpline: (01) 21-351-3122
Website: *http://www.soft.org.uk*

SOFT of Ireland

Alma Stanley
Phone: 1 800 213 218
Website: *http://www.softireland.com*

SOFT Australia

198 Oak Road
Kirrawee, NSW 2232, Australia
Phone: 02-9521-6039
Email: SOFTAus@optushome.com.au

Associazione SOFT ITALIA

Gianstefano Folgoni
Via Cal del Poz, 26
31010 Farra di Soligo (TV), Italy
Phone: 3701346848
http://www.trisomia.org

REFERENCES

Alberman E, Mutton D, Morris JK (2012) Cytological and epidemiological findings in trisomies 13, 18, and 21: England and Wales 2004-2009. *Am J Med Genet A* 158A:1145–1150.

Anderson CE, Punnett HH, Huff V, de Chadarévian JP (2003) Characterization of a Wilms tumor in a 9-year-old with trisomy 18. *Am J Med Genet* 121A:52–55.

Andrews SE, Downey AG, Showalter DS, Fitzgerald H, Showalter VP, Carey JC, Hulac P (2016) Shared decision making and the pathways approach in the prenatal and postnasal management of trisomy 13 and 18. *Am J Med Genet C* 172C:257–263.

Balderston SN, Schaffer EN, Washington RL, Sondheimer HM (1990) Congenital polyvalvular disease in trisomy 18: Echocardiographic diagnosis. *Pediatr Cardiol* 11:138–142.

Barnes A, Carey, JC (2018) Care of the Infant and Child with Trisomy 18 and 13. 4th ed. Free access at: www.trisomy.org

Baty BJ, Blackburn BL, Carey JC (1994a) Natural history of trisomy 18 and trisomy 13. I. Growth, physical assessment, medical histories, survival, and recurrence risk. *Am J Med Genet* 49:175–188.

Baty BJ, Jorde LB, Blackburn BL, Carey JC (1994b) Natural history of trisomy 18 and trisomy 13. II. Psychomotor development. *Am J Med Genet* 49:189–194.

Beke A, Barakonyi E, Belics Z, Joo JG, Csaba A, Papp C, Toth-Pál E, Papp Z (2008) Risk of chromosome abnormalities in the presence of bilateral or unilateral choroids plexus cysts. *Fetal Diagn Ther* 23:185–191.

Boghosian-Sell L, Mewar R, Harrison W, Shapiro RM, Zackai EH, Carey JC, David L, Keppen L, Hudgins L, Overhauser J (1994) Molecular mapping of the Edwards syndrome phenotype to two noncontiguous regions on chromosome 18. *Am J Hum Genet* 55:476–483.

Boghossian NS, Hansen NI, Bell EF, et al. (2014) Mortality and morbidity of VLBW infants with trisomy 13 or trisomy 18. *Pediatrics* 133:226–235.

Bos AP, Broers CJM, Hazebroek FWJ, Van Hemel JO, Tibboel D, Swaay EW, Molenaar JC (1992) Avoidance of emergency surgery in newborn infants with trisomy 18. *Lancet* 339:913–917.

Bruns D (2015) Developmental status of 22 children with trisomy 18 and eight children with trisomy13: implications and recommendations. *Am J Med Genet A* 167A:1807–1815.

Bugge M, Collins A, Hertz JM, Eiberg H, Lundsteen C, Brandt CA, Bak M, Hansen C, Delozier CD, Lespinasse J, Tranebjaerg L, Hahnemann JM, Rasmussen K, Bruun-Petersen G, Duprez L, Tommerup N, Petersen MB (2007) Non-disjunction of chromosome 13. *Hum Mol Genet* 16:2004–2010.

Calderone JP, Chess J, Borodic G, Albert DM (1983) Intraocular pathology of trisomy 18 (Edwards syndrome): Report of a case and review of the literature. *Br J Ophthalmol* 67:162–169.

Carey JC (1992) Health supervision and anticipatory guidance for children with genetic disorders (including specific recommendations for trisomy 21, trisomy 18, and neurofibromatosis I). *Pediatr Clin N Am* 39:40–43.

Carey JC (2012) Perspectives on the care and management of infants with trisomy 18 and trisomy 13: Striving for balance. *Curr Opin Pediatr* 24:672–678.

Carey JC, Barnes AM (2016) Wilms' tumor and trisomy 18: Is there an association? *Am J Med Genet C* 172C:307–308.

Carey JC, Faucette KJ, Schimke RN (2002) Increased risk of Wilms tumor in children with trisomy 18: The evidence and recommendations for a surveillance protocol. *Proc Greenwood Genet Cent* 21:74.

Carter PE, Pearn JH, Bell J, Martin N, Anderson NG (1985) Survival in trisomy 18. *Clin Genet* 27:59–61.

Cereda A, Carey JC (2012) The trisomy 18 syndrome. *Orphanet J Rare Dis* 7:81–102.

Correia JD, da Rosa EB, Silveira DB, et al. (2017) Trisomy 18 and eye anomalies. *Am J Med Genet A* 173A:553–555.

Crider KS, Olney RS, Cragan JD (2008) Trisomies 13 and 18: Population prevalences, characteristics, and prenatal diagnosis, metropolitan Atlanta, 1994-2003. *Am J Med Genet* 146A:820–826.

DeVore GR (2000) Second trimester ultrasonography may identify 77 to 97% of fetuses with trisomy 18. *J Ultrasound Med* 19:565–576.

Edwards JUH, Harnden DG, Cameron AH, Crosse VM, Wolff OH (1960) A new trisomic syndrome. *Lancet* 1:787–789.

Eggermann T, Nothem MM, Eiben B, Hofmann JD, Hinkel K, Fimmers R, Schwanitz G (1996) Trisomy of human chromosome 18: Molecular studies on parental origin and cell stage of nondisjunction. *Hum Genet* 97:218–223.

Embleton ND, Wyllie JP, Wright MJ, Burn J, Hunter S (1996) Natural history of trisomy 18. *Arch Dis Child* 75:38–41.

Faucette KJ, Carey JC, Lemons RL, Toledano S (1991) Trisomy 18 and Wilms tumor: Is there an association? *Clin Res* 39:96A.

Fisher JM, Harvey JF, Morton NE, Jacobs PA (1995) Trisomy 18: Studies of the parent and cell division of origin and the effect of aberrant recombination on nondisjunction. *Am J Hum Genet* 56:669–675.

Forrester MB, Merz RD (1999) Trisomies 13 and 18: Prenatal diagnosis and epidemiologic studies in Hawaii, 1986-1997. *Genet Test* 3:335–340.

Fukasawa T, Kubota T, Tanaka M, et al. (2015) Apneas observed in trisomy 18 neonates should be differentiated from epileptic apneas *Am J Med Genet A* 167A:602–606.

Geiser SC, Carey JC, Apple DJ (1986) Human chromosomal disorders and the eye. In Renie WA (ed.), *Goldberg's Genetic and Metabolic Eye Disease*. Boston MA: Little, Brown, pp. 185–240.

Goldstein H, Nielsen KG (1988) Rates and survival of individuals with trisomy 18 and 13. *Clin Genet* 34:366–372.

Graham EM, Bradley SM, Shirali GS, Hills CB, Atz Am, Pediatric Cardiac Care Consortium (2004) Effectiveness of cardiac surgery in trisomies 13 and 18 (from the Pediatric Cardiac Care Consortium). *Am J Cardiol* 93:801–803.

Griffith CB, Vance GH, Weaver DD (2009) Phenotypic variability in trisomy 13 mosaicism. *Am J Med Genet A* 149A:1346–1358.

Haug S, Goldstein M, Cummins D, Fayard E, Merritt TA (2017) Using patient-centered care after a prenatal diagnosis of trisomy 18 or trisomy 13: A review. *JAMA Pediatr* 4798:E1–E10.

Hansen B, Barnes A, Fergestad M, Tani LY, Carey JC (2000) An analysis of heart surgery in children with trisomy 18, 13. *J Invest Med* 48:47A.

Healey PJ (2003) Social development of children with trisomy 18 and 13 in the context of family and community. Doctoral Dissertation, Boston College (unpublished).

Hassold TJ, Burrage LC, Chan ER, Judis LM, Schwartz S, James SJ, Jacobs PA, Thomas NS (2001) Maternal folate polymorphisms and the etiology of human nondisjunction. *Am J Hum Genet* 69:434–439.

Hill LM (1996) The sonographic detection of trisomies 13, 18, and 21. *Clin Obstet Gynecol* 39:831–850.

Jones KL, Jones MC, Del Campo M (2013) *Smith's Recognizable Patterns of Malformation*. Philadephia, PA: Elsevier-Saunders.

Kaneko Y, Kobayashi J, Yamamoto Y, Yoda H, Kanetaka Y, Nakajima Y, Endo D, Tsuchiya K, Sato H, Kawakami T (2008) Intensive cardiac management in patients with trisomy 13 or trisomy 18. *Am J Med Genet* 146A:1372–1380.

Karimnejad K, Costa DJ (2015) Otolaryngologic surgery in children with trisomy 18 and 13. *Int J Pediatr Otorhinolaryngol* 79(11):1831–3.

Kinoshita M, Nakamura Y, Nakano R, Fukuda S (1989) Thirty-one autopsy cases of trisomy 18: Clinical features and pathological findings. *Pediatr Pathol* 9:445–457.

Koogler TK, Wilfond BS, Ross LF (2003) Lethal language, lethal decisions. *Hastings Cent Rep* 33:37–41.

Kosho T, Carey JC (2016) Does medical intervention affects outcome in infants with trisomy 18 and 13. *Am J Med Genet A* 172A:249–250.

Kosho T, Kuniba H, Tanikawa Y, Hashimoto Y, Sakurai H (2013) Natural history and parental experience of children with trisomy 18 based on a questionnaire given to a Japanese trisomy 18 parental support group. *Am J Med Genet A* 161A:1531–1542.

Kosho T, Nakamura T, Kawame H, Baba A, Tamura M, Fukushima Y (2006) Neonatal management of trisomy 18: Clinical details of 24 patients receiving intensive treatment. *Am J Med Genet* 140A:937–944.

Kosiv K, Gossett J, Bai S, Collins RT. (2017). Congenital heart surgery on in-hospital mortality in trisomy 13 and 18. *Pediatrics* 140:E20170772.

Kumada T, Maihara T, Higuchi Y, Nishida Y, Taniguchi Y, Fujii T (2013) Epilepsy in children with trisomy 18. *Am J Med Genet A* 161A:696–701.

Kupke KG, Mueller U (1989) Parental origin of the extra chromosome in trisomy 18. *Am J Hum Genet* 45:599–605.

Liang CA, Braddock BA, Heithaus JL, Christensen KM, Braddock SR, Carey JC (2015) Reported communication ability of persons with trisomy 18 and trisomy 13. *Dev Neuro Rehabil* 18:322–329.

Lorenz JM, Hardart GE (2014) Evolving medical and surgical management of infants with trisomy 18. *Curr Opin Pediatr* 26:169–176.

Marion RW, Chitayat D, Hutcheon RG, Neidich JA, Zackai EH, Singer LP, Warman M (1988) Trisomy 18 score: A rapid, reliable diagnostic test for trisomy 18. *J Pediatr* 113:45–48.

Maruyama K, Ikeda H, Koizumi T (2001) Hepatoblastoma associated with trisomy 18 syndrome: A case report and a review of the literature. *Pediatr Int* 43:302–305.

Matricardi S, Spalice A, Salpietro V, et al. (2016). Epilepsy in the setting of full trisomy 18, a multicenter study on 18 affected children with and without structural brain abnormalities. *Am J Med Genet C* 172C:288–295.

McCaffrey F (2002) Around pediheart: Trisomy 18, an ethical dilemma. *Pediatr Cardiol* 23:181.

McGraw MP, Perlman JM (2008) Attitudes of neonatologists toward delivery room management of confirmed trisomy 18: Potential factors influencing a changing dynamic. *Pediatrics* 121:1106–1110.

Meyer RE, Liu G, Gilboa SM, et al. (2016) Survival of children with trisomy 13 and trisomy 18: A multistate population based study. *Am J Med Genet A* 170A:825–837.

Morelli S, Barnes A, Carey JC (2000) A series of older children with trisomy 13. *J Invest Med* 48:47A.

Musewe NN, Alexander DJ, Teshima I, Smalhorn JF, Freedom RM (1990) Echocardiographic evaluation of the spectral cardiac anomalies associated with trisomy 18 and 13. *J Am Coll Cardiol* 15:673–677.

Nelson KE, Hexem KR, Feudtner C (2012) Inpatient hospital care of children with trisomy 13 and trisomy 18 in the United States. *Pediatrics* 129:1–8.

Nelson KE, Rosella LC, Mahant S, Guttmann A (2016) Survival and surgical interventions for children with trisomy 18 and 13. *J Am Med Ass* 316:420–428.

Niedrist D, Riegel M, Achermann J, Schinzel A (2006) Survival with trisomy 18—data from Switzerland. *Am J Med Genet* 140A:952–959.

Nishi E, Takamizawa S, Iio K, et al. (2014). Surgical intervention for esophageal atresia in patients with trisomy 18. *Am J Med Genet A* 164A:324–330.

Olson JM, Hamilton A, Breslow NE (1995) Non-11p constitutional chromosome abnormalities in Wilms tumor patients. *Med Pediatr Oncol* 24:305–309.

Palomaki GE, Haddow JE, Knight GJ, Wald NJ, Kennard A, Canick JA, Saller DN, Blitzer MG, Dickerman LH, Fisher R (1995) Risk-based prenatal screening for trisomy 18 using alpha-fetoprotein, unconjugated oestriol and human chorionic gonadotropin. *Prenat Diagn* 15:713–723

Paris JJ, Weiss AH, Soifer S (1992) Ethical issues in the use of life-prolonging interventions for an infant with trisomy 18. *J Perinatol* 12:366–368.

Patau K, Smith DW, Therman E, Inhorn SL, Wagner HP (1960) Multiple congenital anomaly caused by an extra chromosome. *Lancet* 1:790–793.

Perni SC, Predanic M, Kalish RB, Chervenak FA, Chasen ST (2006) Clinical use of first-trimester aneuploidy screening in a United States population can replicate data from clinical trials. *Am J Obstet Gynecol* 194:127–130.

Peron A, Carey JC (2014) Molecular genetics of trisomy 18: Phenotype, genotype correlations. Wiley Online Library. Available at: https://doi.org/10.1002/9780470015902.a0025246

Petersen AK, Chueng SW, Smith JL, et al. (2017) Positive predictive value estimates of cell-free noninvasive prenatal screening from data of a large referral genetic diagnostic laboratory. *Am J Obstetr Gynecol* 217:691.e1–691.e6

Peterson JK, Kochilas LK, Catton KG, Moller JH and Setty S (2017) Long-term outcome of children with trisomy 13 and 18 after congenital heart disease interventions. *Ann Thorac Surg* 103:1941–1949.

Rasmussen SA, Wong L, Yang Q, May K, Friedman JM (2003) Population-based analyses of mortality in trisomy 13 and trisomy 18. *Pediatrics* 111:777–784.

Ray S, Ries MD, Bowen JR (1986) Arthrokatadysis in trisomy 18. *Pediatr Orthop* 6:100–103.

Reinsch RC (1997) Choroid plexus cysts—Association with trisomy: Prospective review of 16,059 patients. *Am J Obstet Gynecol* 176:1381–1383.

Rochelson BL, Trunca C, Monheit AG, Baker DA (1986) The use of a rapid *in situ* technique for third-trimester diagnosis of trisomy 18. *Am J Obstet Gynecol* 155:835–836.

Root S, Carey JC (1994) Survival in trisomy 18. *Am J Med Genet* 49:170–174.

Satge D, Nishi M, Sirvent N, Vekemans M (2016) A tumor profile in Edwards syndrome (trisomy 18). *Am J Med Genet C* 172C:296–306. *Note, the next paper is in Am J Med Genet C* 172C:296–306.

Sergi C, Gekas J, Kamnasaran D (2012) Holoprosencephaly, polydactyly (pseudo trisomy 13) syndrome: Case report and diagnostic criteria. *Fetal Pediatr Pathol* 31:315–318.

Shields LE, Uhrich SB, Easterling TR, Cyr DR, Mack LA (1996) Isolated fetal choroid plexus cysts and karyotype analysis: Is it necessary? *J Ultrasound Med* 15:389–394.

Simpson JL, German J (1969) Developmental anomaly resembling the trisomy 18 syndrome. *Ann Genet (Paris)* 12:107–110.

Smith DW, Patau K, Therman E, Inhorn SL (1960) A new autosomal trisomy syndrome: Multiple congenital anomalies caused by an extra chromosome. *J Pediatr* 57:338–345.

Spencer K, Nicolaides KH (2002) A first trimester trisomy 13/trisomy 18 risk algorithm combining fetal nuchal translucency thickness, maternal serum free beta-hCG and PAPP-A. *Prenat Diagn* 22:877–879.

Spinnato JA, Cook VD, Cook CR, Voss DH (1995) Aggressive intrapartum management of lethal fetal anomalies: Beyond fetal beneficence. *Obstet Gynecol* 85:89–92.

Staples AJ, Robertson EF, Ranieri E, Ryall RG, Haan EA (1991) A maternal serum screen for trisomy 18: An extension of maternal serum screening for Down syndrome. *Am J Hum Genet* 49:1025–1033.

Tahara M, Shimozono S, Nitta T, Yamaki S (2014) Medial defects of the small pulmonary arteries in fatal pulmonary hypertension in infants with trisomy 13 and trisomy 18. *Am J Med Genet A* 164A:319–323.

Tharapel SA, Lewandowski RC, Tharapel AT, Wilroy RS (1986) Phenotype-karyotype correlation in patients trisomic for various segments of chromosome 13. *J Med Genet* 23:310–315.

Tongsong T, Sirichotiyakul S, Wanapirak C, Chanprapaph P (2002) Sonographic features of trisomy 13 at midpregnancy. *Int J Gynaecol Obstet* 76:143–148.

Tucker ME, Garringer HJ, Weaver DD (2007) Phenotypic spectrum of mosaic trisomy 18: Two new patients, review of the literature and counseling issues (2007). *Am J Med Genet* 143A:505–517.

Uehara S, Yaegashi N, Maeda T, Hoshi N, Fujimoto S, Fujimori K, Yanagida K, Yamanaka M, Hirahara F, Yajima A (1999) Risk of recurrence of fetal chromosomal aberrations: Analysis of trisomy 21, trisomy 18, trisomy 13 and 45,S in 1076 Japanese mothers. *J Obstet Gynaecol Res* 25:373–379.

Viora E, Zamboni C, Mortara G, Stillavato S, Bastonero S, Errante G, Sciarrone A, Campogrande M (2007) Trisomy 18: Fetal ultrasound findings at different gestational ages. *Am J Med Genet* 143A:553–557.

Wagner P, Sonek J, Hoopmann M, Abeleh H, Kagan KO (2016) First trimester screening for trisomy 18 and 13, triploidy and Turner syndrome by detail anomaly scan. *Ultrasound Obstetr Gynecol* 48:446–451.

Wiechec M, Knafel A, Nocun A, Matyszkiewicz A, Wiercinska E, Latala E (2016) How effective is ultrasound based screening for trisomy 18 without the addition of biochemistry at the time of late first trimester? *J Perinat Med* 44:49–159.

Willey EK, Feldkamp ML, Krikov S, Carey JC (2008) Analysis of trisomy 13 and trisomy 18: Findings from a population-based study in Utah, 1990-2006, poster presented at the 2008 Pediatric Academic Societies Annual Meeting, Honolulu, HA.

Wilson GN (1993) Karyotype/phenotype correlation: Prospects and problems illustrated by trisomy 18. In: *The Phenotypic Mapping of Down Syndrome and Other Aneuploid Conditions*, New York: Wiley-Liss, pp. 157–173.

Wilson GN, Cooley WC (2006) *Preventive Management of Children with Congenital Anomalies and Syndromes*, 2nd ed. Cambridge University Press.

Wyllie JP, Wright MJ, Burn J, Hunter S (1994) Natural history of trisomy 13. *Arch Dis Child* 71:343–345.

Young ID, Cook JP, Mehta L (1986) Changing demography of trisomy 18. *Arch Dis Child* 61:1035–1936.

59

TUBEROUS SCLEROSIS COMPLEX

LAURA S. FARACH, KIT SING AU, AND HOPE NORTHRUP

Division of Medical Genetics, Department of Pediatrics, McGovern Medical School,
University of Texas Health Science Center at Houston, Houston, Texas, USA

INTRODUCTION

Incidence

Early reports suggested that tuberous sclerosis complex (TSC) was a rare disorder, but more recent population-based studies show this is not the case. Incidence estimates range from 1 in 6000 to 1 in 10,000 live births. The prevalence is about 1 in 20,000 (O'Callaghan et al. 1998; Sampson et al. 1989), but the numbers are likely to be underestimates. One study suggested that, despite the best efforts of the investigators, more than half of cases might have been undetected (O'Callaghan et al. 1998). In children, ascertainment should be higher and provide a better measure of disease frequency. In different studies, prevalence in childhood has been variously estimated at 1 in 6800 for ages 11–15 years, 1 in 12,000 below 10 years of age, 1 in 17,300 below 15 years of age and 1 in 12,900 below 20 years of age. Prevalence in adults is expected to be somewhat lower because tuberous sclerosis is associated with a modest reduction in survival.

The major complications associated with premature death are cardiac rhabdomyomas in newborns, sudden death in epilepsy, subependymal giant cell astrocytomas, renal angiomyolipomas causing hemorrhage or renal failure, and, in women, pulmonary lymphangioleimyomatosis (Shepherd et al. 1991; Amin et al. 2017).

The incidence of simplex versus familial cases of TSC has been addressed in various studies. Variations in percentage of simplex cases from 50 to 86% have been reported (Sampson et al. 1989; Shepherd et al. 1991; Osborne et al. 1991). It is now generally accepted that roughly two-thirds of individuals with TSC are affected as a result of a new mutation, whereas the other one-third have inherited a mutated *TSC* gene from one of their parents.

After careful, detailed evaluation of each individual known to have a *TSC1* or *TSC2* pathogenic variant, the penetrance of tuberous sclerosis complex is estimated to be 100%. Rare cases of seeming nonpenetrance have been reported; however, molecular studies have resolved these cases, identifying two different *TSC* mutations in one family and the existence of low-level mosaicism, including the germ line, in others (Connor et al. 1986; Webb and Osborne 1991).

Diagnostic Criteria

TSC is a multisystem disorder that can present at any age from prenatal to late adult life with a great variety of manifestations and a wide range of severity. The primary method of diagnosis is to obtain a medical and family history and evaluate for the physical findings characteristic of the disease (see Diagnostic Testing). Clinical diagnostic criteria for TSC were first proposed by Gomez in 1979 (Gomez 1979) and then adopted in modified form by the Diagnostic Criteria Committee of the National Tuberous Sclerosis Association (NTSA, the national family support group for the disease, which is now known as the Tuberous Sclerosis Alliance [TSA]). More stringent clinical diagnostic criteria were

---

Cassidy and Allanson's Management of Genetic Syndromes, Fourth Edition.
Edited by John C. Carey, Agatino Battaglia, David Viskochil, and Suzanne B. Cassidy.
© 2021 John Wiley & Sons, Inc. Published 2021 by John Wiley & Sons, Inc.

agreed upon by an international panel of experts at the Tuberous Sclerosis Complex Consensus Conference held in 1998, then updated following the 2012 consensus conference. The diagnostic criteria were subsequently published with categories of major features and minor features leading to the following diagnostic categories: *definite tuberous sclerosis complex* and *possible tuberous sclerosis complex* (Northrup et al. 2013) (see Table 59.1).

As noted in Table 59.1, for a *definite* diagnosis of TSC to be made, an individual must exhibit either two major features or one major feature and two minor features. *Possible* diagnosis can be entertained in individuals with one major or two minor features. Findings consistent with a *possible* diagnosis warrant further investigation. Since the establishment of the Clinical Diagnostic Criteria in 1998, molecular genetic testing has become commercially available. Updated diagnostic criteria from 2012 include the identification of either a *TSC1* or *TSC2* pathogenic variant in DNA from normal tissue as sufficient to make a definite diagnosis of TSC (Northrup et al. 2013). Genetic testing can be helpful for those with possible tuberous sclerosis, especially infants and children who may not yet have developed enough manifestations to meet criteria for *definite* tuberous sclerosis. If the pathogenic variant causing the disease has been identified in the family, this provides the basis for establishing the diagnosis in at-risk relatives, regardless of clinical findings.

Four of the 10 major features are skin findings: facial angiofibromas (≥3) *or* fibrous celaphic plaque; non-traumatic ungual fibroma (≥2); hypomelanotic macules (≥3); and shagreen patch. All of these features can be readily observed on physical examination of an individual suspected to have TSC. Nearly 100% of affected individuals will have at least one of the characteristic skin findings. As an affected individual ages, he or she will develop more skin findings of the disorder. The first dermatologic feature to appear is usually the hypomelanotic macules that are often present at birth or become apparent during the first months of life. They can occur anywhere on the body but are more common on the trunk and limbs. They can be in any shape or size. Early descriptions compared these "white spots" with ash leaves (classically pointed at one end and rounded at the other) (Figure 59.1). By 5 years of age, most children are developing facial angiofibromas, multiple flesh colored or red papules. These typically occur over the nose, in the nasolabial folds and medially on the cheeks in the so-called butterfly distribution (Figure 59.2). Because they do not develop from sebaceous glands, the term adenoma sebaceum used in the older literature is a misnomer and should not be used. Fibrous cephalic plaques develop most commonly on the forehead (Figure 59.3), but can be located anywhere on the face or scalp. Another characteristic lesion is the shagreen patch, a connective tissue nevus, which typically forms on the lower back during childhood (Figure 59.4). In adolescents and adults, ungual fibromas are a common finding on the finger and toe nails, often associated with deep grooves in the nail (Figure 59.5).

Multiple retinal hamartomas represent another major feature of TSC. These tumors in the eye rarely cause problems with vision, but are a useful diagnostic clue. They can take several forms, including the classic raised multinodular "mulberry" lesion that is easily seen, or, more commonly, flat, smooth and semi-transparent lesions (Figure 59.6).

Three of the major features are different types of tumors that are observed in the brain (Figures 59.7–59.9): cortical tubers, subependymal nodules, and subependymal giant cell astrocytomas (SEGAs). The presence of these tumors often leads to seizures, with seizures representing the most

TABLE 59.1 Clinical diagnostic criteria for tuberous sclerosis

Major features
Facial angiofibromas (≥3) *or* fibrous cephalic plaque
Ungual fibromas, nontraumatic (≥2)
Hypomelanotic macules, three or more of at least 5 mm diameter
Shagreen patch
Multiple retinal hamartomas
Cortical dysplasias[a]
Subependymal nodules
Subependymal giant cell astrocytoma
Cardiac rhabdomyoma, single or multiple
Renal angiomyolipoma (≥2) *or* pulmonary lymphangiomyomatosis[b]

Minor features (but see caution in text)
Multiple randomly distributed pits in dental enamel (≥3)
Intraoral fibromas (≥2)
Nonrenal hamartoma
Retinal achromic patch
"Confetti" skin lesions
Multiple renal cysts

Requirements for a diagnosis of tuberous sclerosis

DEFINITE DIAGNOSIS

Either	Two major features
Or	One major and two minor features (but see caution in text about use of minor features)
Or	The identification of either a *TSC1* or *TSC2* pathogenic variant in DNA from normal tissue[c]

POSSIBLE DIAGNOSIS

Either	One major feature
Or	Two minor features

[a] Includes tubers and cerebral white matter radial migration lines.
[b] When both lymphangiomyomatosis and renal angiomyolipomas are present, other features of tuberous sclerosis should be present before a definite diagnosis is assigned (see section on etiology).
[c] A pathogenic variant is defined as a mutation that clearly inactivates the function of the *TSC1* or *TSC2* proteins, prevents protein synthesis, or is a missense mutation whose effect on protein function has been established by functional assessment. Other *TSC1* or *TSC2* variants whose effect on function is less certain do not meet these criteria and are not sufficient to make a definite diagnosis of TSC.
Source: Based on Northrup et al. (2013).

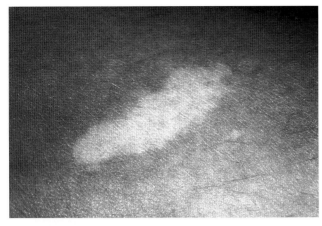

FIGURE 59.1 Hypopigmented macule with ash-leaf shape.

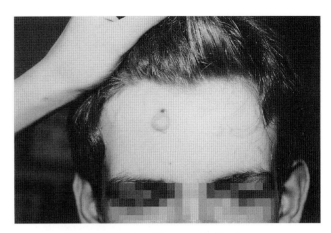

FIGURE 59.3 Fibrous cephalic plaque.

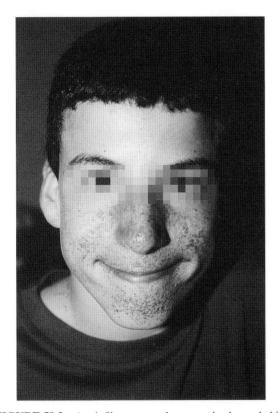

FIGURE 59.2 Angiofibromas on the nose, cheeks, and chin.

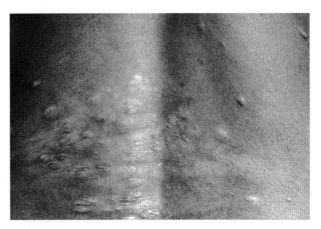

FIGURE 59.4 Numerous small shagreen patches on the lower back.

common presentation for TSC in a simplex case. Evaluation of seizures by brain imaging usually shows the characteristic lesions, thus establishing the diagnosis (see Neurologic).

Cardiac rhabdomyomas are another major feature of TSC. These tumors are often identified by prenatal ultrasound leading to suspicion of the diagnosis in a fetus. Brain and kidney findings can sometimes be seen in an affected fetus, further strengthening the possibility of a diagnosis of TSC. The cardiac rhabdomyomas, in contrast to the other findings in TSC, regress rapidly after birth, rarely causing medical problems. Occasionally, affected adults have arrhythmias that are likely secondary to residual cells left when the rhabdomyomas resolve.

Angiomyolipomas are the most common renal manifestation of TSC (Figure 59.10). They are composed of multiple tissue types (blood vessels, smooth muscle, and adipose tissue), as the name implies. Angiomyolipomas develop during later childhood and adolescence and multiple bilateral lesions are common in many adults with TSC. Renal cysts also occur, but are less frequent.

Lymphangioleiomyomatosis (LAM), aberrant growth of muscle cells in lungs that can eventually result in respiratory failure, was initially thought to be a rare finding in TSC that predominantly affected women of reproductive age. Because testing for LAM has been recommended in all affected adult women, it is now known to be much more common than previously accepted. There are individuals who have findings of both angiomyolipomas and LAM without other features of TSC; therefore, a combination of LAM and angiomyolipomas without other features does not meet criteria for a definite diagnosis of TSC.

Caution needs to be exercised when it comes to the minor features (Table 59.1), because these are much less specific for TSC. The minor features are supportive of the diagnosis

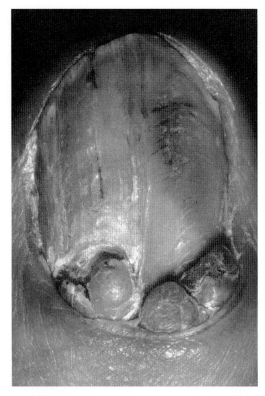

FIGURE 59.5 Fingernail with several ungual fibromas.

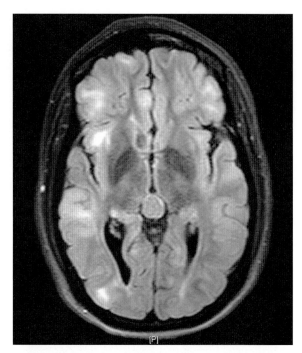

FIGURE 59.7 Axial T2 flair MRI of the brain showing multiple cortical tubers with hyperintensity and loss of the gray-white interface.

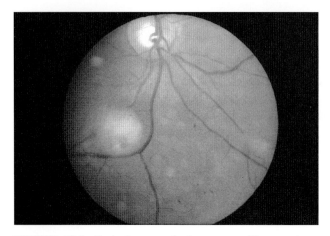

FIGURE 59.6 Retinal astrocytic hamartoma. The prominent lesion is a raised multinodular "mulberry" lesion. A flat semi-transparent lesion is also visible obscuring an underlying vessel.

and should be sought in affected individuals as they represent other areas where health problems may occur.

Etiology, Pathogenesis, and Genetics

Pathogenic variants in two genes have been identified as causative for the majority of individuals with TSC. The *TSC1* gene was discovered a decade after the initial report of linkage (Fryer et al. 1987; van Slegtenhorst et al. 1997) to chromosome 9. *TSC1* is composed of 23 exons that extend over 53.3 kb of genomic DNA on chromosome 9q34.3. Non-coding sequences include exons 1 and 2, and a 4.9 kb 3,'-untranslated region in exon 23. It codes for an 8.6 kb mRNA that translates to produce the protein, hamartin, which has an estimated molecular weight of 130 kDa. The existence of the *TSC2* gene was suggested through studies of families with no linkage to chromosome 9 (Northrup et al. 1987). *TSC2* was subsequently mapped and identified on chromosome 16p13.3 (Kandt et al. 1992; The European Chromosome 16 Tuberous Sclerosis Consortium 1993). It has 42 exons spanning approximately 40.8 kb of genomic DNA. Alternative splicing at multiple exons creates at least nine mRNA isoforms (https://www.ncbi.nlm.nih.gov/gene?cmd=Retrieve&dopt=Graphics&list_uids=7249). Tuberin isoform, coding from the 5.6 kb mRNA that includes all alternately spliced exons, has a calculated molecular weight of approximately 200 kDa.

Since the identification of the *TSC* genes, genetic and biochemical studies have elucidated that hamartin and tuberin form a biochemical complex that negatively regulates the highly conserved growth-regulating pathway involving a kinase called mammalian target of rapamycin (mTOR) (Huang and Manning 2008). When activated by a small Ras-like protein enriched in brain called Rheb, the mTOR kinase initiates a cascade of cellular events that result in enhanced ribosomal translation of proteins and cell growth. Growth factors, oxygen tension, and the energy status of the cell all affect mTOR regulation. The TSC

INTRODUCTION

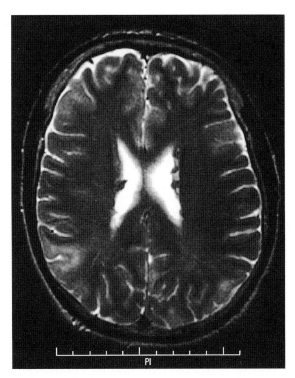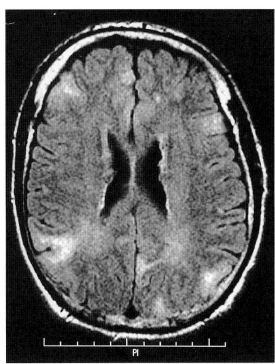

FIGURE 59.8 Cranial magnetic resonance scans of an individual with tuberous sclerosis. Tubers in the frontal, parietal, and occipital cortex appear as lighter areas in the FLAIR (fluid attenuated inversion recovery) sequence on the right. These lesions are also visible in the T_2-weighted image on the left along with several subependymal nodules, which appear as dark areas in the lateral walls of the lateral ventricles.

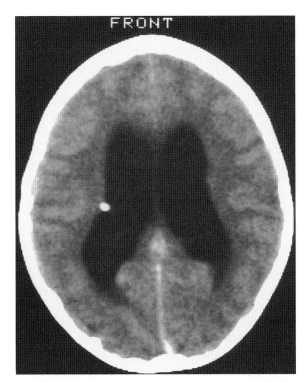

FIGURE 59.9 Calcified subependymal nodule demonstrated by cranial computed tomography. The ventricles are dilated because of a subependymal giant cell astrocytoma (not visible) causing hydrocephalus.

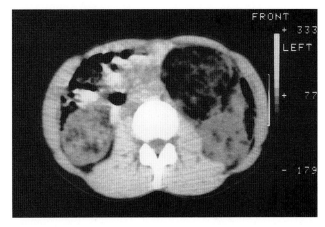

FIGURE 59.10 Renal computed tomography scan showing bilateral angiomyolipomas. There is gross enlargement of the left kidney, which has extensive anterior areas of fat (appears dark).

complex is essentially a molecular rheostat that inhibits mTOR activity through its GTPase activity toward Rheb-GTP. It may be that tuberin and hamartin also have roles in other pathways, and several possibilities have been suggested, but there is little doubt that the inhibition of mTOR is a key function. The exciting aspect of this is that rapamycin and other compounds that inhibit mTOR are available as pharmacologic agents to restore mTOR level of activity to normal. This knowledge led to many clinical trials, and mTOR inhibitors are now first line therapy for enlarging subependymal giant cell astrocytomas (SEGAs), renal angiomyolipomas, LAM, and facial angiofibromas. Clinical trials for effects of mTOR inhibitors on other aspects of TSC are ongoing.

To date, approximately 75–85% of individuals who meet *definite* diagnostic criteria have a small identifiable *TSC1* or *TSC2* pathogenic variant (Dabora et al. 2001; Sancak et al. 2005; Au et al. 2007). Through the support of the Tuberous Sclerosis Alliance, Ekong and Povey (2008) are curating a tuberous sclerosis complex variant database at the Leiden Open Variation Database website (http://chromium.lovd.nl/LOVD2/TSC/home.php). Most pathogenic variants in *TSC1* are predicted to cause premature protein termination or splicing errors. However, in *TSC2*, loss of function due to a missense mutation is also frequently observed. Approximately 6% of individuals with TSC have large gene deletions (0.5% *TSC1*, 5.6% *TSC2*) (Kozlowski et al. 2007). The remaining 9–19% of affected individuals with no mutation identified are suspected to be somatic mosaics for a pathogenic variant in one of the identified *TSC* genes or to possess a pathogenic variant in the unanalyzed non-coding regions of the *TSC* genes. Somatic mosaicism is found in 5–50% of these patients (Verhoef et al. 1999; Kozlowski et al. 2007, Qin et al. 2010, Tyburczy et al. 2015) and mutations in noncoding regions is detected in 35% of these patients (Tyburczy et al. 2015). There is little evidence for an additional *TSC* gene locus, but it remains a possibility.

Attempts to find correlations between features of TSC and the underlying pathogenic variants before the year 2000 were not successful (Au et al. 2007) with the exception of one study that found that intellectual disability was more frequent in those with de novo *TSC2* mutations (Jones et al. 1999). Since that time, several genotype–phenotype correlations have been identified. *TSC2* is associated with a more severe phenotype than *TSC1* including increased risk for infantile spasms, autism, intellectual disability, cortical tubers, and renal malignancy (Dabora et al. 2001; Sancak et al. 2005; Au et al. 2007; Numis et al. 2011, Kothare et al. 2014, Yang et al. 2014). Increased incidence and severity of LAM has been associated with pathogenic variants in the region that encodes the carboxy terminus of TSC2 (Strizheva et al. 2001). While *TSC2* is generally associated with a more severe phenotype, there are several pathogenic missense variants that are associated with mild phenotypes, including p.Gln1503Pro, p.Ser1036Pro, p.Arg905Gln, p.Arg1200Trp, p.Arg622Trp, and p.Arg1713His. (Khare et al. 2001; O'connor et al. 2003; Jansen et al. 2006; Wentink et al. 2012; van Eeghen et al. 2013; Ekong et al. 2016; Farach et al. 2017, Fox et al. 2017). These studies help clinicians provide prognostic information to affected individuals and their families based on molecular testing.

Despite the genotype–phenotype correlations that exist, the disorder is marked by variable expressivity, leading to different manifestations in patients with the same genotype. There is good evidence that the patchy nature of the pathology in TSC is due to somatic mutation following the Knudson two-hit paradigm for tumor suppressor genes (Knudson 1971). One copy of either *TSC1* or *TSC2* is inactivated in all cells by a constitutional mutation (the first hit) and then the second normal allele in a cell is inactivated by a somatic mutation (the second hit). Complete loss of functional hamartin or tuberin then leads to the formation of a tumor. In some cases the somatic mutation is a deletion of the chromosomal region containing the gene leading to loss of heterozygosity for markers within the deleted region (Green et al. 1994). This has been demonstrated in different types of hamartoma (Sepp et al. 1996; Au et al. 1999). Studies of multiple lesions from the same individual have shown different patterns of loss of heterozygosity, supporting the hypothesis that the lesions arise independently (Henske et al. 1996). Occasionally, somatic mutation will result in two hits in the same cell, and this is the likely explanation for the rare occurrence of single TSC-related hamartomas in individuals who do not have any other features of the condition. This has been well documented in the case of sporadic renal angiomyolipomas (Henske et al. 1995). Furthermore, in individuals with sporadic pulmonary lymphangioleimyomatosis who also have renal angiomyolipomas, matching patterns of somatic mutation have been found in the angiomyolipoma and lymphangioleimyomatosis cells, supporting the intriguing hypothesis that cells in both locations have a common origin (Carsillo et al. 2000; Yu et al. 2001). It seems likely that most types of hamartoma are the result of a two-hit process. However, there remains the possibility that some clinical features of tuberous sclerosis are due to haploinsufficiency, particularly central nervous system lesions, because the second mutation hits are not frequently identified (Henske et al. 1996; Crino 2010; Martin et al. 2017). Other features of TSC, including the neuropsychiatric phenotype, may not be explained by the two-hit hypothesis. Modifier genes may be accounting for the variable expressivity of these and other manifestations.

Genetic Counseling

Because the majority of individuals with TSC appear to occur as the first case in a family, before providing genetic counseling it is essential to determine whether the condition is inherited or has arisen as a result of a new mutation in the consultand. Particular care must be taken to exclude the possibility that one of the parents is mildly affected. Because of the extreme variability, a mildly affected parent remains at risk of having a severely affected child. If the pathogenic *TSC* variant can be identified in the index case, targeted genetic testing can be offered to the parents and siblings to confirm their disease status, and for prenatal diagnosis.

If the pathogenic *TSC* variant cannot be found, we must rely on clinical criteria to determine parental status. In that case, evaluation of the parents should include detailed medical and family history, skin examination including inspection with a Wood's lamp to exclude hypopigmented macules, and examination of the fundi. Direct ophthalmoscopy with dilated pupils will identify the more obvious retinal lesions, but visualization of subtle changes, particularly in the peripheral retina, requires indirect fundoscopy and ophthalmological expertise. If the physical examination is normal, the chances of finding evidence of tuberous sclerosis on radiographic examination are small, but have been reported (Roach et al. 1991; Northrup et al. 1993). In the context of normal examinations, parents considering additional children who wish to have further reassurance should be offered brain and renal imaging. Echocardiogram is not recommended. If all studies of the parents are negative for findings of TSC, a recurrence risk of 1% should be provided. The 1% recurrence risk is based on confined gonadal mosaicism observed for pathogenic variants in both *TSC1* and *TSC2* and originating from either the mother or father (Rose et al. 1999).

Some families with a known pathogenic *TSC* variant in the affected child and negative parental studies will choose preimplantation genetic diagnosis or prenatal diagnosis because of the 1% risk of another affected child.

For an individual with TSC, there is 50% chance of transmitting the mutant gene to his or her offspring. However, in giving genetic advice, the variation in disease severity also needs to be taken into account. For example, parents with normal mentation can have affected children who have intellectual disability. Likelihood estimates are not available to guide families regarding degree of severity should an affected child be born.

Diagnostic Testing

Recommended evaluations to make or exclude the diagnosis include: thorough medical and family history and physical examination including Wood's lamp examination of the skin, dilated fundoscopy, renal CT or MRI, and brain CT or MRI. If the individual being evaluated is less than 1 year of age, echocardiogram is also recommended. Beyond 1 year of age, the yield for detection of cardiac rhabdomyomas is too low to warrant recommending this test to help make a diagnosis. Even when the diagnosis is obvious without the evaluations listed above, it is recommended that these evaluations be obtained as a baseline for medical care. Additionally, molecular testing for the *TSC1* and *TSC2* genes is recommended. Molecular testing provides the individual and family with several key pieces of information: prognostic information for the affected person (see genotype–phenotype correlation information above), definite diagnosis in other family members who are mildly affected, and reproductive options for the affected person as well as for the other family members. In people who meet *possible* diagnostic criteria, molecular testing can provide a definitive diagnosis in some instances (Au et al. 2007). Molecular testing using next generation sequencing modalities may be useful for detection of low-level *mosaicism in blood/tissues*.

Differential Diagnosis

Many of the characteristic lesions associated with tuberous sclerosis can occur occasionally in isolation, often as a result of two somatic mutations in a cell inactivating both copies of either *TSC1* or *TSC2*. This is why the current diagnostic criteria demand more than one type of lesion for a definite diagnosis. Some types of lesions, such as hypopigmented macules, occur at low frequency in the general population. In either case the finding raises the possibility of tuberous sclerosis and warrants careful clinical evaluation and appropriate investigations.

Hypopigmented macules are present in less than 1% of Caucasian newborns and about 2% of black newborns. In older children and adults, hypopigmented lesions attributable to chickenpox scars, trauma and various dermatologic conditions are a common finding. Once these causes are excluded, lesions resembling those seen in tuberous sclerosis are present in 6–7% of normal children and 4% of adults under 45 years of age (Vanderhooft et al. 1996). However, multiple hypopigmented macules are much less common. In a study of 423 normal individuals, less than 1% had multiple lesions and none had more than three (Vanderhooft et al. 1996). In contrast, 66.8% of individuals with tuberous sclerosis have three or more (Kingswood et al. 2017). Most of these lesions would probably be classified as nevus depigmentosus (or nevus achromicus), well-demarcated macules that do not change over time and are more prominent under Wood's lamp. The differential diagnosis includes piebaldism and vitiligo. Lesions caused by vitiligo can be distinguished by their progressive course, symmetrical distribution, and pure white appearance because of complete loss of pigmentation, which differs from the dull white of the partially pigmented lesions seen in TSC. Another lesion that may cause confusion is the so-called nevus anemicus, a pale or mottled area of increased vascular tone, which disappears if the surrounding skin is blanched by compression under a microscope slide.

A single facial angiofibroma can occur unrelated to TSC. Multiple facial angiofibromas have been reported as a feature of multiple endocrine neoplasia type 1 (MEN1). Acne vulgaris and a variety of other types of facial lesions might occasionally be confused with angiofibromas but should be easily distinguished by a biopsy. Other skin lesions that can occasionally occur in isolation, unrelated to TSC, are a fibrous cephalic plaque, shagreen patch or ungual fibroma. Ungual fibromas can sometimes result from trauma.

In the eye, it is important in infants to distinguish between a retinal astrocytic hamartoma and retinoblastoma.

Cortical tubers and subependymal nodules can, in rare instances, occur as solitary lesions unrelated to TSC and by no means do all children with subependymal giant cell astrocytomas have TSC. Bilateral periventricular nodular heterotopia is a rare X-linked dominant disorder, usually lethal in males, which is associated with seizures and uncalcified periventricular nodules and could be mistaken for TSC.

Cardiac rhabdomyomas can occur unrelated to TSC, but the finding of these lesions at routine antenatal scanning or in a newborn is associated with a substantial risk of TSC as discussed below. Renal angiomyolipomas can occur unrelated to TSC, usually as a solitary lesion, in contrast to the multiple bilateral angiomyolipomas typically seen in TSC. Pulmonary lymphangioleimyomatosis can also occur as a rare condition unrelated to TSC and this sporadic form of LAM is often associated with renal angiomyolipomas.

MANIFESTATIONS AND MANAGEMENT

There is extensive literature on the manifestations of TSC, but much of the information comes from clinic-based studies. Few population-based or longitudinal studies have been carried out. Fortunately, a large Tuberous Sclerosis Complex Natural History Study is currently underway in the United States (Whittemore et al. 2008). The Tuberous Sclerosis Complex Natural History Study is collecting extensive information regarding all aspects of the disorder from individuals followed at TSC specialty clinics throughout the United States. After the study is completed, data will be available on thousands of affected individuals regarding their symptoms and treatments. For now, because few controlled trials have been carried out, most of the recommendations provided here are based on expert opinion informed by knowledge of disease natural history. The most recent Tuberous Sclerosis Complex Consensus Conference was held in 2012 and recommendations for each organ system were made by a panel of experts who specialize both in that organ system and TSC (Krueger et al. 2013). The recommendations given here can only provide a guide to management, which must also take into account local circumstances and the needs of the individual.

Growth and Feeding

Growth is generally normal in individuals with TSC. One study has reported an increased frequency of macrocephaly. Feeding can be a problem in any child with intellectual disability, and can be a particular challenge in TSC if combined with behavioral difficulties.

Evaluation

- Height, weight, and head circumference should be monitored as in other children.
- Macrocephaly should be investigated with cranial imaging.

Management

- Macrocephaly need only be treated if it is the result of hydrocephalus or tumor, in which case treatment is standard.
- Feeding and growth problems are treated in a standard manner.

Development and Behavior

High rates of developmental, behavioral, and emotional problems in patients with TSC led to the moniker TSC-associated neuropsychiatric disorders (TAND). The name calls attention to the interrelated functional and clinical manifestations of brain dysfunction in TSC in hopes it will lead to better assessment for these manifestations. More than 90% of individuals with TSC will experience at least one or more feature of TAND, and currently a treatment gap exists wherein only 20% of them undergo evaluation and treatment for these issues (Northrup et al. 2015).

Studies on the prevalence of intellectual disability in TSC show that 44–64% of affected individuals have an intelligence quotient in the intellectually disabled range (IQ below 70) (Northrup et al. 2015). Up to 30% have profound intellectual disability with an IQ below 21 (Wong and Khong 2006). Multiple studies have investigated the relationship between the number of cortical tubers, the total tuber volume, and the presence and degree of intellectual disability (Asato and Hardan 2004; Wong and Khong 2006; Zaroff et al. 2006; Jansen et al. 2008a, 2008b), but at this time the evidence remains controversial (Wong and Khong 2006). In comparison, the evidence correlating the degree of intellectual disability with seizures is strong (Asato and Hardan 2004; Goh et al. 2005; Zaroff et al. 2006; Jansen et al. 2008a, 2008b). Development of infantile spasms, poor seizure control, and onset of seizures before age 12 months are all independent risk factors for intellectual disability (Northrup et al. 2015). *TSC1* pathogenic variants have been shown to be associated with a lower frequency of intellectual disability than *TSC2* pathogenic variants (Jones et al. 1999; Dabora et al. 2001; Au et al. 2007). Additionally, in two of the largest genotype–phenotype studies published, affected males showed a higher incidence of intellectual disability than affected females (Sancak et al. 2005; Au et al. 2007).

Individuals with TSC who have an IQ within the normal range (IQ above 70) have been found to display a variety of learning disabilities and are prone to specific cognitive deficits at a higher rate than matched controls (Harrison et al.

1999). Memory impairment, speech delay, dyscalculia, visuospatial disturbances, and dyspraxia have all been described (Asato and Hardan 2004; Jansen et al. 2008a, 2008b). Studies have also demonstrated that those having an IQ within the normal range have an IQ that is approximately 10 points lower than their unaffected siblings.

Autism is also prevalent in TSC, affecting 16–61% of the population (Northrup et al. 2015). Intellectual disability and epilepsy have been shown to be independent risk factors for development of pervasive developmental disorder (now falling under the umbrella of autism spectrum disorder) in TSC (Asato and Hardan 2004; Zaroff et al. 2004). An association between temporal lobe tubers and autistic features in TSC has also been described (Asato and Hardan 2004; Zaroff et al. 2004). Other common TSC-associated neuropsychiatric disorders include attention deficit hyperactivity disorder (21–50%), aggression (13–58%), self-injurious behavior (27–41%), anxiety (13–48%), and depression (19–43%) (Northrup et al. 2015).

Evaluation

- As a consequence of the high frequency of intellectual disability and behavioral problems, age-appropriate screening for TSC-associated neuropsychiatric disorders (TANDs) is recommended at time of diagnosis and at least annually (Krueger et al. 2013). To ease the burden of screening, a TAND paper and pencil questionnaire was developed (de Vries et al. 2015) and can be given to patients at each visit. While the survey can take up to 30 min, strategies can be employed to streamline such as giving the questionnaire during down time, when the patient is checking in or waiting for the doctor.
- A formal comprehensive evaluation for TANDs should also be performed at key developmental points: infancy (0–3 years), preschool (3–6 years), pre-middle school (6–9 years), adolescence (12–16 years), early adulthood (18–25 years), and as needed thereafter (Krueger et al. 2013).
- Children found to have developmental or behavioral impairments need retesting as part of their ongoing management. In children with normal initial testing and developmental milestones, it is important to have an ongoing high index of suspicion.
- If problems develop once a child is at school, they should be carefully evaluated. Reading and spelling difficulties, for example, may not become apparent until 7–8 years of age and can adversely affect school progress.
- In adults with intellectual disability there is usually little, if any, need for repeat cognitive assessments provided their condition remains stable.
- Changes in behavior can be a sign of underlying physical problems, especially in individuals with moderate or severe intellectual disability and communication difficulties. Behavioral changes, particularly those accompanied by worsening of epilepsy control, should raise the possibility of a seizure or a subependymal giant cell astrocytoma (Hunt and Stores 1994) associated with evolving hydrocephalus.

Management

- Clinicians should maintain a low threshold to initiate early interventions and management strategies (Krueger et al. 2013). Management strategies should be based on the TAND profile of the patient and evidence-based good practice guidelines/practice parameters for individual disorders.
- Early intervention programs are beneficial to children with TSC who have developmental delay and/or autism spectrum disorder.
- Always consider the need for an individualized learning plan (IEP) to be put in place in school for school-aged children (Krueger et al. 2013).
- Behavior problems in TSC are frequent and can be difficult to manage. Referral to a multidisciplinary team specializing in behavioral management, including a clinical psychologist and psychiatrist, may be appropriate.
- There are minimal data available on the use of drugs to treat behavioral problems in TSC. Drug treatment of attention deficit hyperactivity disorder (ADHD) with stimulant medication needs specialist input because of possible adverse effects on autistic behaviors and on epilepsy control.
- Preliminary evidence suggests that melatonin may be beneficial in treating sleep disorders (O'Callaghan et al. 1999).

Neurologic

Intracranial Lesions. The most characteristic finding of TSC is the cortical brain tuber, occurring in 90% of individuals with confirmed diagnosis of TSC (Northrup et al. 2015). Tubers are collections of giant cells, dysmorphic neurons, and astrocytes interspersed with regions of disorganized cortical lamination (Asato and Hardan 2004; Crino 2004). Loss of *TSC1/TSC2* function relieves inhibition on mTOR, resulting in increased cell size, increased protein synthesis, and increased cell division (Crino 2004). Cells overgrow and become the giant cells characteristically seen in tubers. Increased cell size during fetal development is postulated to alter the ability of cells to migrate appropriately in the brain. Tubers are found most commonly in the frontal, temporal, and parietal cortices (Asato and Hardan 2004). On MRI sequences (T2 and *flair*) tubers appear as cortical areas of hyperintensity with loss of the gray-white interface (Figures 59.7 and 59.8) (Weiner et al. 1998). Compromised embryonic neuronal migration results in migrational brain anomalies, such as hemimegalencephaly (Crino 2004). Subependymal nodules are collections of giant cells found lining the ventricles of the brain (Asato and Hardan 2004) and are seen in 80% of individuals with TSC (Northrup et al. 2013). Subependymal nodules are composed of similar cell types to the cortical tubers, but the cells are much more densely packed (Weiner et al. 1998). Over time, they tend to calcify (Weiner et al. 1998) (Figures 59.8 and 59.9). Subependymal nodules are classically found in the lateral ventricles, but are occasionally visualized in the aqueduct of Sylvius or the fourth ventricle (Asato and Hardan 2004).

Most subependymal nodules remain dormant throughout life, but their potential to increase in size and become a subependymal giant cell astrocytoma (SEGA) has been well documented. SEGAs are thought to be a continuum of subependymal nodules in which the astrocytomas develop following malignant transformation of the nodules (Nabbout et al. 1999; Asato and Hardan 2004), usually in the region of the foramina of Monro. SEGAs are one of the leading causes of morbidity and mortality in TSC (Nabbout et al. 1999; Amin et al. 2017) and are found to develop in approximately 5–15% of affected individuals (Northrup et al. 2013). Peak incidence is in late childhood and adolescence. If development of a SEGA is going to occur, it is usually apparent on MRI by the age of 21 years (Torres et al. 1998; Nabbout et al. 1999). Risk factors for subependymal nodule transformation into SEGA include a subependymal nodule greater than 5 mm in diameter, one near the foramen of Monro, uncalcified lesions, or contrast-enhancing lesions (Weiner et al. 1998; Nabbout et al. 1999). SEGAs result in morbidity by bleeding or acutely obstructing the foramen of Monro, causing hydrocephalus. The resulting hydrocephalus causes increased intracranial pressure, seizures and mental status changes (Nabbout et al. 1999; Asato and Hardan 2004).

Epilepsy. Seizures are the most common neurologic manifestation of TSC and the most common presenting symptom in children. Sudden unexpected death in epilepsy is one of the leading causes of mortality in individuals with TSC (Amin et al. 2017). Epilepsy is found in 70–90% of all affected individuals (Asato and Hardan 2004; Wong and Khong 2006), and it usually develops within the first year of life (Weiner et al. 1998; Wong and Khong 2006), often in the first few months. In one study, 86% of children had seizure onset under two years of age and only 5% developed seizures after age 5 years (Jozwiak et al. 2000). Seizures can be of multiple types, including infantile spasms, generalized tonic-clonic, absence, atonic, myoclonic, or complex-partial (Asato and Hardan 2004). The pattern of seizures changes through early childhood and, in approximately two-thirds of children, includes infantile spasms. The onset of infantile spasms can be associated with developmental arrest or regression. Electroencephalography commonly shows hypsarrythmia.

Control of epilepsy in TSC is one of the most difficult aspects of management, and some individuals continue to have partial or generalized seizures throughout their lives, with poor response to anti-epileptic medication. Onset of seizures in the first year of life and occurrence of multiple seizure types are poor prognostic factors for seizure control. Early control of epilepsy improves developmental outcomes (Cusmai et al. 2011; Jóźwiak et al. 2011, Bombardieri et al. 2010).

Evaluation

- MRI brain to assess for presence of tubers, subependymal nodules, migrational defects, and SEGA in newly diagnosed patients (Krueger et al. 2013).
- MRI of the brain should be performed every 1–3 years in asymptomatic patients younger than age 25 years to monitor for new occurrence of SEGA. Asymptomatic patients with large or growing SEGA, or a SEGA that is causing ventricular enlargement, should undergo MRI more frequently. They should receive education on the potential for new symptoms (Krueger et al. 2013).
- Patients with asymptomatic SEGA in childhood should continue to be monitored periodically as adults (Krueger et al. 2013).
- Classical symptoms of raised intracranial pressure caused by a SEGA, such as headache and vomiting, can be difficult to evaluate in individuals with moderate or severe intellectual disability. In such cases, there should be a low threshold for cranial imaging if there is an unexplained deterioration in seizure control or an alteration in behavior (Krueger et al. 2013). MRI is recommended over CT, if possible, because of better resolution.
- Parents of infants with TSC should be educated to recognize infantile spasms, even if none have occurred at time of first diagnosis (Krueger et al. 2013).
- A baseline electroencephalogram (EEG) should be obtained at time of diagnosis. If abnormal, and especially if two or more features of TSC-associated neuropsychiatric disorder (TAND) are present, a 24 hour polygraphic-video-EEG recording should be used to assess for subclinical seizure activity (Krueger et al. 2013).
- Prolonged polygraphic-video-EEG, 24 hours or longer, is also appropriate when seizure occurrence is unclear or when unexplained changes in either sleep, behavior,

or neurologic function is observed (Krueger et al. 2013).
- Routine polygraphic-video-EEG should be obtained in those with known or suspected seizure activity. The frequency should be determined by clinical need, rather than a specific interval (Krueger et al. 2013).
- Individuals with seizures need to be evaluated by a clinician experienced in the management of epilepsy. A careful history is needed to determine the type and severity of seizures and to monitor the response to treatment. In young children, the seizure pattern can change over short periods of time and seizures of more than one type may coexist.

Management

- There is currently no treatment indicated to decrease the number or size of cortical tubers, cortical dysplasias, or subependymal nodules.
- For growing but asymptomatic SEGA, there are two options: medical treatment with mTOR inhibitors or surgical resection. The decision-making process should include discussion of risks, adverse effects, cost, length of treatment, and potential impact on TSC-associated comorbidities (Krueger et al. 2013).
- For acutely symptomatic SEGA, surgical resection is indicated. Cerebral spinal fluid shunting may be necessary to relieve the symptoms of raised intracranial pressure (Krueger et al. 2013). Morbidity from excision of a SEGA can be high if the tumor is large or has extensively infiltrated the surrounding brain. If complete removal can be achieved without neurologic deficit, then the prognosis is good with a low recurrence risk (Torres et al. 1998).
- Treatment of seizures needs to be undertaken by a clinician experienced in the management of epilepsy (Wheless et al. 2007). This is beyond the scope of this chapter, but general guidance is given here.
- Anticonvulsants are the mainstay of therapy and the choice of medication depends on the types of seizures and consideration of the potential side effects. The most effective drug or combination of drugs and the optimum dosage will differ from one affected individual to another. In infants and younger children, the pattern of seizures evolves over time so that the response to treatment needs to be closely monitored and changes should be made to both dosage and choice of medication as appropriate.
- Because using several drugs increases the frequency of adverse side effects, the goal of therapy is good seizure control with a single anticonvulsant drug. Unfortunately, this is often not achievable in practice. The dose of the chosen medication should be increased gradually until the seizures are controlled or the maximum tolerated dose is reached. If seizure control cannot be achieved, a second drug is added and if this proves effective it may in due course be possible to withdraw the first. If necessary, additional drugs are tried in a systematic fashion.
- In TSC, vigabatrin has been shown to be more effective than steroids in the control of infantile spasms, typically achieving cessation of seizures within 3–4 days. Vigabatrin is effective in approximately 73% of children with tuberous sclerosis and infantile spasms (Northrup et al. 2015). It has therefore been advocated as the drug of choice for infantile spasms in children with a confirmed or suspected diagnosis of TSC. Prescribing physicians should be aware of, and monitor, possible side effects, particularly retinal toxicity. Adrenocorticoptropin (ACTH) can be used as a second-line therapy if infantile spasms do not improve with vigabatrin (Krueger et al. 2013). Other drugs used in infantile spasms include valproate, lamotrigine, and topiramate.
- If infantile spasms cannot be controlled with medication, a ketogenic diet may be helpful but it is often a big undertaking in a child with intellectual disability and behavioral problems.
- mTOR inhibitors were shown to improve seizure control in the majority of patients with TSC who had refractory epilepsy in a clinical trial (Krueger et al. 2013b) and have been approved to treat epilepsy in individuals with TSC in Europe.
- Once an individual has been refractory to more than two anti-epileptic drugs, the chances of achieving seizure control through medical management alone are less than 10% (Kwan and Brodie 2000; Arts et al. 2004) and surgical options should be considered. Surgical options include vagal nerve stimulation for partial-onset seizures (Parain et al. 2001; Asato and Hardan 2004), tuber resection for focal lesions (Karenfort et al. 2002; Asato and Hardan 2004), or corpus callosotomy for atonic seizures (Asato and Hardan 2004).
- Vagal nerve stimulation by an implanted generator is used for partial-onset seizures (Parain et al. 2001). Vagal nerve stimulation has been shown to reduce seizure frequency by greater than 50% in more than half the affected individuals and has few to no adverse effects (Murphy and Patil 2003).
- Resection of a cortical tuber that is acting as an epileptogenic focus can result in the reduction in severity and frequency of seizures and sometimes complete seizure cessation (Karenfort et al. 2002). Extensive cerebral involvement is not a contraindication to surgery provided a primary epileptogenic tuber can be identified.

- The corpus callosum is the principal connection linking the cerebral cortices. Neuronal signaling is transmitted across this neurophysiologic tract and allows spread of seizures from one hemisphere to the other. Interruption of signal transduction can inhibit seizure generalization and has a long history of use as a palliative treatment in some forms of intractable epilepsy (atonic, clonic, myoclonic, and generalized tonic-clonic) (Clarke et al. 2008).

Dermatologic

Hypopigmented (hypomelanotic) macules (Figure 59.1) are often noticed at birth or become apparent during the first year of life and are present in 95% of children with TSC by the age of 5 years (Webb et al. 1996; Jozwiak et al. 1998). The lesions are well-demarcated areas of reduced, but not absent, pigmentation, and typically are 1–3 cm across. These lesions can be found anywhere on the body but are most commonly observed on the trunk and extremities. They can be any shape or size, and only a minority resemble an ash leaf, as indicated in early descriptions. Most affected individuals have multiple lesions. They are usually visible in natural light but are more obvious under ultraviolet light using a Wood's lamp. In adult life, hypopigmented macules tend to become less obvious and may disappear, so that Webb et al. (1996) found they were present in 58% of adults over the age of 30 years. Lesions on the scalp are associated with hypopigmented hair (poliosis). Some individuals have numerous 1–3 mm diameter hypopigmented macules on the forearms and lower legs. This "confetti-like" hypopigmentation can be missed if a Wood's lamp is not used (Webb et al. 1996).

Facial angiofibromas (Figure 59.2) most often become apparent between 2 and 5 years of age but can develop later (Jozwiak et al. 1998). In most cases there is slow progression, often with an exacerbation at puberty. By early adult life, angiofibromas are present in about 75% of individuals (Northrup et al. 2015). Angiofibromas have a characteristic symmetrical distribution over the nose, nasolabial folds and medially on the cheeks. The chin is often involved but the upper lip is spared. The lesions vary from flesh-colored to red, depending on the extent of the vascular component. There is wide variation in severity from small papules that are barely noticeable to large disfiguring nodules. Vascular lesions can be a source of recurrent bleeding.

Angiofibromas on the forehead are rare, but this is the characteristic location for raised flesh-colored or yellowish-brown fibrous cephalic plaques (Figure 59.3). These lesions can range in size from a few millimeters to several centimeters across and vary in consistency from soft to firm. Histologically, they are fibromas with little vascular component. Fibrous cephalic plaques usually develop in later childhood (Jozwiak et al. 1998) but they can sometimes be an early and presenting feature preceding the development of facial angiofibromas. Similar plaques can occur on the scalp associated with incomplete alopecia.

Shagreen patches are raised brownish or flesh-colored connective tissue nevi, usually found on the lower back (Figure 59.4). Their surface texture often resembles orange peel and their name comes from a type of untanned leather with a rough granular surface. Shagreen patches are usually multiple and vary in size from a few millimeters to several centimeters. Sometimes large plaques of more than 10 cm across are seen. Shagreen patches can appear at any age from infancy to puberty and are present in about 50% of young adults (Northrup et al. 2015).

Ungual fibromas (Figure 59.5) can appear at any age from later childhood to adult life. Estimates of their frequency in adolescents vary from 11 to 23% (Webb et al. 1996; Jozwiak et al. 1998) and they are found in up to 88% of adults (Webb et al. 1996). These pink or red nodules arise from the nail bed on finger and toenails and are often associated with deep grooves in the nail. Sometimes grooving of the nail occurs without a visible fibroma. Ungual fibromas are usually multiple and occur more commonly on the toenails where they may bleed after minor trauma.

Soft fleshy skin tags (molluscum fibrosum pendulum) are said to occur at increased frequency in tuberous sclerosis, particularly on the neck and upper back and occasionally in the axilla and inguinal region. These have been reported in 6–23% of individuals (Webb et al. 1996; Jozwiak et al. 1998).

Evaluation

- Careful examination of the skin should be carried out as part of the initial diagnostic process. To be sure of identifying hypopigmented macules, the skin should be inspected under ultraviolet light using a Wood's lamp in a dark room.
- A detailed skin inspection should be performed annually (Krueger et al. 2013).
- A dermatologic consultation may be indicated if the skin lesions are atypical or the diagnosis of tuberous sclerosis is uncertain.

Management

- Rapidly changing, disfiguring, or symptomatic TSC-associated skin lesions should be treated as appropriate for the lesion and clinical context. Approaches can include surgical excision, laser, or topical mTOR inhibitor (Krueger et al. 2013).
- Topical mTOR inhibitors are efficacious in treatment for angiofibromas (Northrup et al. 2015).
- Fibrous cephalic plaques can be treated with laser therapy or plastic surgery. Topical mTOR inhibitors should be considered.

- Removal of shagreen patches is not typically indicated; however, options include shave excision, dermabrasion or laser treatment.
- Ungual fibromas can be treated by excision, electrosurgery, cryotherapy or laser ablation but have a tendency to recur.

Ophthalmology

Retinal hamartomas (sometimes called retinal phakomas in the older literature) are the most common ocular manifestation of tuberous sclerosis, occurring in 30–50% of individuals with TSC. They are very rare in the healthy population (Northrup et al. 2015). They are often multiple and bilateral and can take several forms. The most common are relatively flat, smooth, oval lesions that are semi-transparent and similar in color to the surrounding fundus. They can be difficult to identify and may only be apparent because they are partially obscuring an underlying vessel. More easily seen, but slightly less common, are the classic "mulberry" lesions, which are raised opaque multinodular calcified hamartomas most commonly found in the vicinity of the optic disk. Lesions with mixed features also occur.

Older studies reported retinal hamartomas in 44–54% of affected individuals (Rowley et al. 2001). Two large comprehensive genotype–phenotype correlation studies reported these eye lesions in approximately 30% of affected individuals (Sancak et al. 2005; Au et al. 2007). Both studies reported that affected males (36–38%) were more likely to exhibit the retinal findings than affected females (20–25%). One study has reported a lower frequency of retinal hamartomas in young children, with lesions being identified in 8% of children below 2 years of age and 17% of children aged 2–5 years (Jozwiak et al. 2000). Longitudinal studies in older individuals show that most lesions remain static over time but a few show progression or new calcification. Appearance of new lesions has been documented and, in rare instances, retinal hamartomas can regress. Retinal hamartomas are almost always asymptomatic, but in rare instances large macular lesions can cause visual impairment.

Disturbances of retinal pigmentation can also occur in TSC. Small "punched out" areas of hypopigmentation in the mid periphery of the retina have been reported in 39% of affected individuals but were also found in 6% of age-matched controls (Rowley et al. 2001) and so carry little diagnostic significance. Papilledema of the optic disks is an important sign of raised intracranial pressure in individuals presenting with a SEGA.

Evaluation

- Eye examination is recommended as part of the initial evaluation of individuals suspected of having TSC (Krueger et al. 2013). Direct ophthalmoscopy with dilated pupils will identify the more obvious retinal lesions, but visualization of subtle changes, particularly in the peripheral retina, requires indirect fundoscopy and ophthalmological expertise. The presence of visual field deficits should also be evaluated (Northrup et al. 2015).
- Routine follow-up examinations of the fundus are not necessary (Roach et al. 1999).
- Visual field testing should be performed in those children taking vigabatrin, by an expert pediatric ophthalmologist.

Management

- The ocular manifestations of tuberous sclerosis very rarely cause symptoms and treatment is seldom required.

Cardiovascular

Cardiac rhabdomyomas are a common finding in newborns and young children with TSC, and they regress with age. In one study, rhabdomyomas were found at echocardiography in 83% of children under 2 years of age and in 31% of older children (Jozwiak et al. 2000). Other studies have reported frequencies in children of 50–67% (Smith et al. 1989; Muhler et al. 1994). The lesions usually occur in the ventricles and may be multiple. Cardiac rhabdomyomas are usually asymptomatic, but can be associated with preexcitation on the electrocardiogram and may give rise to arrhythmias antenatally or in the newborn period. Persistent tachyarrhythmia can lead to heart failure and has been associated with hydrops fetalis. In rare instances, intracavity lesions cause hemodynamic obstruction. For neonates with normal cardiac evaluation, the risk of later developing cardiac dysfunction is small. Regression or complete disappearance of cardiac rhabdomyomas during childhood is well documented in follow-up studies (Muhler et al. 1994). By adult life, rhabdomyomas are only demonstrable in a small proportion of affected individuals (Smith et al. 1989).

Identification of cardiac rhabdomyomas has been used as a means of prenatal diagnosis of tuberous sclerosis in pregnancies known to be at risk. The unexpected finding of rhabdomyomas at routine antenatal scanning in the second or third trimester is also a well-recognized presentation of TSC. The chance of the baby being affected by TSC in this situation has been estimated at 39–86% (Gamzu et al. 2002; Pipitone et al. 2002), and as a solitary finding is an indication for *TSC* molecular screening.

There have been several reports of aortic aneurysm occurring, particularly in young children, as a rare complication of TSC. This has led to the proposal that affected individuals should be screened for aortic aneurysms at diagnosis and annually thereafter (Jost et al. 2001), but this is not current practice. The current multicenter natural history study will yield a better estimate of the incidence of clinically significant aortic aneurysms. Such data will permit revisions to current evaluation recommendations.

Evaluation

- Clinical examination in newborns and infants should include evaluation for an irregular heart rhythm, a heart murmur, or signs of cardiac failure.
- An echocardiogram should be obtained in infancy, especially in those individuals younger than 3 years old. An echocardiogram should be obtained every one to three years in asymptomatic individuals until cardiac rhabdomyomas have regressed. More frequent or advanced assessment may be needed for symptomatic individuals (Krueger et al. 2013).
- In individuals with rhabdomyomas detected prenatally, a prenatal echocardiogram can be used to assess risk of heart failure (Krueger et al. 2013).
- Rhythm disturbances can occur, even in individuals who do not have history of a documented rhabdomyoma. For this reason, baseline electrocardiogram and electrocardiogram every three to five years should be obtained for asymptomatic people throughout their life to monitor for conduction defects (Krueger et al. 2013).
- MRI of the abdomen is recommended for renal surveillance and has the added benefit of identifying abdominal aortic aneurysms.

Management

- The majority of individuals with cardiac rhabdomyomas do not require any form of treatment.
- Because of the tendency for lesions to regress during infancy and early childhood, standard symptomatic treatment of cardiac arrhythmias or heart failure is the mainstay of management.
- In rare instances, surgery may be necessary for life-threatening intracavity lesions causing hemodynamic obstruction. The off-label use of mTOR inhibitors has also been effective for these lesions and may represent a better alternative (Northrup et al. 2015).

Genitourinary

Renal disease is the leading cause of death in individuals with TSC (Amin et al. 2017). Angiomyolipomas are the most common renal manifestation of TSC, and their size and number increase with age (Ewalt et al. 1998; Jozwiak et al. 2000; Casper et al. 2002). In one study, angiomyolipomas were identified on renal ultrasound scanning in 17% of children under 2 years of age; in 42% aged 2–5 years; in 65% aged 9–14 years and in 92% aged 14–18 years (Jozwiak et al. 2000). Other studies have confirmed that by adolescence the frequency of angiomyolipomas is approaching the adult level of 74–80% (Cook et al. 1996; Ewalt et al. 1998; Casper et al. 2002). They occur in both the renal cortex and medulla, ranging from microscopic to large masses distorting the renal architecture. By adulthood, angiomyolipomas are typically multiple, bilateral and often too numerous to count. This is in contrast to angiomyolipomas unrelated to TSC, which occur infrequently in the general population and are usually solitary and asymptomatic. Angiomyolipomas are composed of smooth muscle, adipose tissue, and thick-walled blood vessels. Larger lesions can have a substantial blood supply with large feeder vessels. The fat content makes angiomyolipomas highly echogenic and readily identifiable on ultrasound scanning. They are also well seen on CT or MRI.

Symptoms caused by renal angiomyolipomas are rare in children and uncommon in adults (Cook et al. 1996). The most common symptoms are flank pain and hematuria. Retroperitoneal hemorrhage caused by the rupture of an angiomyolipoma is the most serious and life-threatening complication. The risk of significant renal hemorrhage in adults and adolescents has been estimated at 4–8% (Cook et al. 1996). Small lesions are not usually associated with symptoms, and most of the risk is associated with angiomyolipomas greater than 4 cm in diameter (Dickinson et al. 1998). Hypertension and renal impairment are uncommon in TSC and the incidence of end-stage renal failure has been estimated at 1%.

Renal cysts are a common finding in TSC, being reported in 17–47% of individuals (Cook et al. 1996; Ewalt et al. 1998; Casper et al. 2002). They are usually multiple and bilateral. Although population-based studies have not found a significant correlation with age (Cook et al. 1996), longitudinal studies have shown that the proportion of individuals with cysts increases with age, and, at follow-up ultrasound examination, both the number and size of cysts tends to increase (Ewalt et al. 1998; Casper et al. 2002). Cysts can occasionally disappear (Ewalt et al. 1998). In the majority of affected individuals, renal cysts are not associated with symptoms. The one important exception is persons with both TSC and polycystic kidney disease caused by contiguous deletions of the *TSC2* and *PKD1* genes (Sampson et al. 1997). They usually present with severe early-onset renal cystic disease, renal enlargement, and radiological appearances of advanced autosomal dominant polycystic kidney disease. If the *PKD1* gene is inactivated, the prognosis for renal function is poor, with progression to end-stage renal disease likely in late childhood or early adult life unless there is mosaicism, in which case the prognosis may be somewhat better (Sampson et al. 1997).

There have been many reports of renal cell carcinoma in TSC (Tello et al. 1998). It is possible that some of these cases represent an aggressive form of angiomyolipomatosis rather than carcinoma. Involvement of regional lymph nodes by metastatic deposits of angiomyolipoma is well-documented and is not, in and of itself, proof of malignancy (Nelson and Sanda 2002). Reviewing the literature, Tello et al. (1998) concluded that the risk of renal cell carcinoma is not increased in TSC, whereas others have put the risk at 1–3% (Nelson and Sanda 2002). From the management point of view, it would seem wise to have a low threshold of suspicion while at the same time being aware of the possibility of over diagnosis.

Evaluation

- An MRI of the abdomen is the imaging modality of choice for assessing the presence of renal angiomyolipomas and cysts (Krueger et al. 2013). Renal involvement can be documented by ultrasound or CT if MRI is not available. A baseline MRI of the abdomen is recommended for newly diagnosed or suspected TSC. Follow-up abdominal MRIs are recommended every 1–3 years for the individual's lifetime. They can be performed in conjunction with brain MRIs for patient convenience and decrease frequency of anesthesia for patients who require anesthesia for MRIs (Krueger et al. 2013).
- In children newly diagnosed with TSC, renal imaging is important to identify the minority who have coexisting polycystic kidney disease and are likely to develop early renal impairment and renal failure by early adulthood.
- Individuals with lesions more than 4 cm in diameter or with extensive renal involvement should be referred to a nephrologist for ongoing management or urologist for possible intervention.
- All individuals should have a baseline renal function evaluation by determination of glomerular filtration rate and blood pressure assessment. Renal function through these means should be evaluated at least annually (Krueger et al. 2013).

Management

- The majority of individuals with renal angiomyolipomas and cysts remain asymptomatic and require no treatment.
- Kidney-sparing interventions are especially important in this population as new angiomyolipomas can arise, further decreasing functional kidney volume.
- For asymptomatic angiomyolipomas that are more than 4 cm in diameter and for those measuring larger than 3 cm in diameter and growing, treatment with an mTOR inhibitor is first line therapy. Selective embolization or kidney-sparing resection are acceptable second-line therapies (Krueger et al. 2013).
- For patients with angiomyolipoma that present with acute hemorrhage, first-line therapy is embolization followed by corticosteroids. Nephrectomy should be avoided due to associated morbidity/mortality (Krueger et al. 2013).

Respiratory

Lymphangioleiomyomatosis (previously known as lymphangiomyomatosis) is the clinically important respiratory manifestation of TSC. Proliferation of abnormal smooth muscle cells occurs in the lung parenchyma around the airways, lymphatics, and blood vessels and is associated with multiple thin-walled cysts. The smooth muscle cells express atypical antigens and also frequently estrogen and progesterone receptor proteins. Lymphangioleiomyomatosis (LAM) predominantly affects premenopausal women and is very rare in men (Hancock et al. 2002). The onset of symptoms is typically between late teens and early 40s, with acute dyspnea, usually related to pneumothorax, or less commonly with chronic dyspnea (Hancock et al. 2002). Other clinical features include hemoptysis, chylous pleural effusions, and chylous ascites. Occasionally, LAM is also found in retroperitoneal and pelvic lymph nodes. LAM is estimated to occur is 30–40% of females with TSC but may be age-dependent with other studies quoting 80% of females affected by age 40. Cystic findings consistent with LAM can be observed in 10% of males with TSC (Northrup et al. 2013).

LAM can also occur as a rare condition unrelated to TSC with a prevalence of approximately one per million (Johnson and Tattersfield 2000). This sporadic form of LAM is often associated with renal angiomyolipomas. In both sporadic and TSC-associated LAM, the clinical course is highly variable. In some there is deterioration over 5–10 years with a fatal outcome, whereas other affected individuals remain stable over several decades (Johnson and Tattersfield 2000; Hancock et al. 2002). There are no longitudinal data on the prognosis in asymptomatic women found on screening to have cystic parenchymal changes. Some studies have suggested that pregnancy can precipitate the onset of LAM, but a study of women with TSC found no evidence that pregnancy increased the risk of developing pulmonary complications (Mitchell et al. 2003). In women with symptomatic LAM, pregnancy can exacerbate the disease (Johnson and Tattersfield 2000) and most experts would caution against pregnancy in these circumstances. Reduced cabin pressure on aircraft may increase the risk of pneumothorax in individuals with LAM.

Multifocal micronodular pneumocyte hyperplasia is the other common pulmonary manifestation of TSC. Proliferation of type II pneumocytes gives rise to multiple diffuse non-calcified pulmonary nodules visible on chest radiograph or high-resolution computed tomography scan. Multifocal micronodular pneumocytic hyperplasia does not have known physiologic consequence. It can be confused with atypical adenomatous hyperplasia, a premalignant lesion that is not associated with TSC (Northrup et al. 2015).

Evaluation

- High-resolution chest CT is more sensitive than chest radiography in detecting the early stages of LAM and is the investigation of choice for establishing the diagnosis and documenting the extent of involvement. However, due to the theoretical concern for radiation exposure in a tumor suppressor condition, low-radiation protocols should be used, if possible (Krueger et al. 2013).
- A baseline high-resolution chest CT is recommended for asymptomatic females with TSC at 18 years old, or newly diagnosed females that are 18 years or older. If there is no evidence of lung cysts, follow up high-resolution chest CT is recommended every 5–10 years in females (Krueger et al. 2013).
- Baseline pulmonary function and a 6 min walk test are also recommended for asymptomatic females at 18 years old, or newly diagnosed females that are 18 years or older (Krueger et al. 2013).
- Clinical screening for LAM symptoms such as dyspnea should be performed at each clinic visit. Counseling regarding smoking risk and estrogen use should be reviewed at each clinic visit for anyone at risk for LAM (Krueger et al. 2013).
- High-resolution chest CT and pulmonary function testing including a 6 min walk test is recommended for symptomatic males or females of any age to assess for presence of LAM (Krueger et al. 2013). No respiratory investigations or imaging are indicated in adult men or in children or adolescents with TSC unless they become symptomatic at this time; however, this may change as more effective therapy for LAM is developed.
- Anyone with lung cysts should have pulmonary function testing and a 6 min walk test annually (Krueger et al. 2013).
- In women with symptomatic LAM, the course of the disease should be monitored by lung function tests and high-resolution chest CT at 6–12 monthly intervals, depending on the severity of their disease.

Management

- mTOR inhibitors are approved to treat males and females with LAM and TSC who have moderate-to-severe lung disease or rapid progression (Krueger et al. 2013).
- Medroxy-progesterone treatment and/or oophorectomy reduce the production of estrogen, which is thought to stimulate the growth of LAM cells; however, response to treatment is highly individual (Northrup et al. 2015).
- Pneumothorax can be treated conservatively but has a high recurrence rate and may need surgical intervention.
- Supplemental oxygen is often necessary.
- Lung transplantation can be life saving for end-stage LAM. There is a high rate of complications, but Boehler et al. (1996) reported a survival of 58% at 2 years. There have been several reports of LAM recurring in the transplanted lung.

Miscellaneous

Gingival fibromas are common in TSC, being present in a third or more of older children and adults (Webb et al. 1996). They appear as flesh-colored growths most commonly on the anterior gingivae. Fibromas are also occasionally found on the tongue or buccal mucosa.

Dental enamel pits occur with increased frequency in TSC. They are more easily seen if plaque-disclosing solution is applied to the teeth. In different studies the proportion of individuals reported to have enamel pits has varied from 48 to 100% and in controls from less than 1 to 72%. In all these studies enamel pits were more common in cases than controls, but the wide variation in frequency, particularly in controls, illustrates the difficulty in assessing this physical sign.

Angiomyolipomas can occur in the liver and are readily demonstrated by ultrasound scanning. Hepatic hamartomas have been reported in 13–23% of children and 23–45% of older children and adults (Jozwiak et al. 1992). They appear to be more frequent in females. Hepatic angiomyolipomas are not usually of any clinical significance but may be helpful in supporting the diagnosis of TSC, counting as a non-renal hamartoma in the current diagnostic criteria.

Microhamartomatous rectal polyps have been reported to be a common finding in TSC. This might be a useful diagnostic marker, but it is not current practice to look for them. They are of no clinical significance.

Small areas of localized sclerosis in the skull bones are a common radiological finding in tuberous sclerosis. Similar changes can be seen in the vertebrae and pelvis. Periosteal thickening is common in the long bones, hands, and feet. Bone cysts also occur, particularly in the hands. Skeletal involvement is usually a chance finding and of no clinical significance except that sclerotic and lytic changes might, in some circumstances, be mistaken for metastases.

RESOURCES

Support Groups

United States
Tuberous Sclerosis Alliance
801 Roeder Road
Suite 750, Silver Spring
MD 20910, USA

Phone (toll-free): (800) 225-6872
Phone: (301) 562-9890
Fax: (301) 562-9870
Email: *info@tsalliance.org*
Website: *http://www.tsalliance.org/*

United Kingdom
Tuberous Sclerosis Association
CAN Mezzanine
32-36 Loman Street
London
SE1 0EH, UK
Email: *admint@tuberous-sclerosis.org*
Website: *http://www.tuberous-sclerosis.org/*

Other countries
Contact details can be obtained from Tuberous Sclerosis International
Website: http://www.tscinternational.org/

Written Resources

Gomez MR, Sampson JR, Whittemore VH, eds (1999) *Tuberous Sclerosis Complex*, 3rd ed. Oxford: Oxford University Press.

A wide range of leaflets and other resources for professionals and affected individuals are available from the Tuberous Sclerosis Alliance, the Tuberous Sclerosis Association, and other support groups.

REFERENCES

Amin S, Lux A, Calder N, Lougharne M, Osborne J, O'callaghan F (2017) Causes of mortality in individuals with tuberous sclerosis complex. *Dev Med Child Neurol* 59(6):612–617.

Arts WFM, Brower OF, Peters ACB, Stroink H, Peeters EAJ, Schmitz PIM, van Donselaar CA, Geerts AT (2004) Course and prognosis of childhood epilepsy: 5-Year follow-up of the Dutch study of epilepsy in childhood. *Brain* 127:1774–1784.

Asato MR, Hardan AY (2004) Neuropsychiatric problems in tuberous sclerosis complex. *J Child Neurol* 19:241–249.

Au KS, Hebert AA, Roach ES, Northrup H (1999) Complete inactivation of the TSC2 gene leads to formation of hamartomas. *Am J Hum Genet* 65:1790–1795.

Au KS, Williams AT, Roach ES, Batchelor L, Sparagana SP, Delgado MR, Wheless JW, Baumgartner JE, Roa BB, Wilson CM, Smith-Knuppel TK, Cheung MYC, Whittemore VH, King TM, Northrup H (2007) Genotype/phenotype correlation in 325 individuals referred for a diagnosis of tuberous sclerosis complex in the United States. *Genet Med* 9:88–100.

Boehler A, Speich R, Russi EW, Weder W (1996) Lung transplantation for lymphangioleiomyomatosis. *N Engl J Med* 335:1275–1280.

Bombardieri R, Pinci M, Moavero R, Cerminara C, Curatolo P (2010) Early control of seizures improves long-term outcome in children with tuberous sclerosis complex. *Eur J Paediatr Neurol* 14:2:146–149.

Brook-Carter PT, Peral B, Ward CJ, Thompson P, Hughes J, Maheshwar MM, Nellist M, Gamble V, Harris PC, Sampson JR (1994) Deletion of the TSC2 and PKD1 genes associated with severe infantile polycystic kidney disease—A contiguous gene syndrome. *Nat Genet* 8:328–332.

Carsillo T, Astrinidis A, Henske EP (2000) Mutations in the tuberous sclerosis complex gene TSC2 are a cause of sporadic pulmonary lymphangioleiomyomatosis. *Proc Natl Acad Sci USA* 97:6085–6090.

Casper KA, Donnelly LF, Chen B, Bissler JJ (2002) Tuberous sclerosis complex: Renal imaging findings. *Radiology* 225:451–456.

Clarke DF, Wheless JW, Chacon MM, Brier J, Koenig MK, McManis M, Castillo E, Baumgartner JE (2008) Corpus callosotomy: A palliative therapeutic technique may help identify resectable epileptogenic foci. *Epilepsia* 16:545–553.

Connor JM, Stephenson JB, Hadley MD (1986) Non-penetrance in tuberous sclerosis. *Lancet* 2:1275.

Cook JA, Oliver K, Mueller RF, Sampson J (1996) A cross sectional study of renal involvement in tuberous sclerosis. *J Med Genet* 33:480–484.

Costello LC, Hartman TE, Ryu JH (2000) High frequency of pulmonary lymphangioleiomyomatosis in women with tuberous sclerosis complex. *Mayo Clin Proc* 75:591–594.

Crino PB (2010) The pathophysiology of tuberous sclerosis complex. *Epilepsia* 51(Suppl 1):27–29.

Crino PB (2004) Molecular pathogenesis of tuber formation in tuberous sclerosis complex. *J Child Neurol* 19:716–725.

Cusmai R, Moavero R, Bombardieri R, Vigevano F, Curatolo P (2011) Long-term neurological outcome in children with early-onset epilepsy associated with tuberous sclerosis. *Epilepsy Behav* 22:4:735–739.

Dabora SL, Jozwiak S, Franz DN, Roberts PS, Nieto A, Chung J, Choy YS, Reeve MP, Thiele E, Egelhoff JC, Kasprzyk-Obara J, Domanska-Pakiela D, Kwiatkowski DJ (2001) Mutational analysis in a cohort of 224 tuberous sclerosis patients indicates increased severity of TSC2, compared with TSC1, disease in multiple organs. *Am J Hum Genet* 68:64–80.

de Vries PJ, Whittemore VH, Leclezio L, Byars AW, Dunn D, Ess KC, Hook D, King BH, Sahin M, Jansen A (2015) Tuberous sclerosis associated neuropsychiatric disorders (TAND) and the TAND Checklist. *Pediatr Neurol* 52:25–35.

Dickinson M, Ruckle H, Beaghler M, Hadley HR.1998) Renal angiomyolipoma: optimal treatment based on size and symptoms. *Clin Nephrol* 49:281–286.

Ekong R, Nellist M, Hoogeveen-Westerveld M, Wentink M, Panzer J, Sparagana S, Emmett W, Dawson NL, Malinge MC, Nabbout R, Carbonara C (2016) Variants within TSC2 exons 25 and 31 are very unlikely to cause clinically diagnosable tuberous sclerosis. *Hum Mutat* 37:4:364–370.

Ekong R, Povey S (2008) Tuberous Sclerosis database: In Leiden Open Variations Database at Leiden University Medical Center, Leiden, the Netherland. *http://chromium.liacs.nl/LOVD2/TSC/home.php?action = switch_db.*

Ewalt DH, Sheffield E, Sparagana SP, Delgado MR, Roach ES (1998) Renal lesion growth in children with tuberous sclerosis complex. *J Urol* 160:141–145.

Farach LS, Gibson WT, Sparagana SP, Nellist M, Stumpel CT, Hietala M, Friedman E, Pearson DA, Creighton SP, Wagemans A, Segel R (2017) TSC2 c. 1864C> T variant associated with mild cases of tuberous sclerosis complex. *Am J Med Genet A* 173:3:771–775.

Fox J, Ben-Shachar S, Uliel S, Svirsky R, Saitsu H, Matsumoto N, Fattal-Valevski A (2017) Rare familial TSC2 gene mutation associated with atypical phenotype presentation of Tuberous Sclerosis Complex. *Am J Med Genet A* 173:744–8.

Fryer AE, Chalmers A, Connor JM, Fraser I, Povey S, Yates AD, Yates JR, Osborne JP (1987) Evidence that the gene for tuberous sclerosis is on chromosome 9. *Lancet* 1:659–661.

Gamzu R, Achiron R, Hegesh J, Weiner E, Tepper R, Nir A, Rabinowitz R, Auslander R, Yagel S, Zalel Y, Zimmer E (2002) Evaluating the risk of tuberous sclerosis incases with prenatal diagnosis of cardiac rhabdomyoma. *Prenat Diagn* 22:1044–1047.

Goh S, Kwiatkowski DJ, Dorer DJ, Thiele EA (2005) Infantile spasms and intellectual outcomes in children with tuberous sclerosis complex. *Neurology* 65:235–238.

Gomez MR (1979) Clinical experience at the Mayo Clinic. In: *Tuberous Sclerosis*, Gomez MR, ed. New York: Raven Press, pp. 11–26.

Green AJ, Smith M, Yates JR (1994) Loss of heterozygosity on chromosome 16p13.3 in hamartomas from tuberous sclerosis patients. *Nat Genet* 6:193–196.

Hancock E, Tomkins S, Sampson J, Osborne J (2002) Lymphangioleiomyomatosis and tuberous sclerosis. *Respir Med* 96:7–13.

Harrison JE, O'Callaghan FJ, Hancock E, Osborne JP, Bolton PF (1999) Cognitive deficits in normally intelligent patients with tuberous sclerosis. *Am J Med Genet* 88:642–646.

Henske EP, Neumann HP, Scheithauer BW, Herbst EW, Short MP, Kwiatkowski DJ (1995) Loss of heterozygosity in the tuberous sclerosis (TSC2) region of chromosome band 16p13 occurs in sporadic as well as TSC-associated renal angiomyolipomas. *Genes Chromosomes Cancer* 13:295–298.

Henske EP, Scheithauer BW, Short MP, Wollmann R, Nahmias J, Hornigold N, van Slegtenhorst M, Welsh CT, Kwiatkowski J (1996) Allelic loss is frequent in tuberous sclerosis kidney lesions but rare in brain lesions. *Am J Hum Genet* 59:400–406.

Huang J, Manning BD (2008) The TSC1–TSC2 complex: A molecular switchboard controlling cell growth. *Biochem J* 412:179–190.

Hunt A, Stores G (1994) Sleep disorder and epilepsy in children with tuberous sclerosis: A questionnaire-based study. *Dev Med Child Neurol* 36:108–115.

Jansen AC, Sancak O, D'Agostino MD, Badhwar A, Roberts P, Gobbi G, Wilkinson R, Melanson D, Tampieri D, Koenekoop R, Gans M (2006) Unusually mild tuberous sclerosis phenotype is associated with TSC2 R905Q mutation. *Ann Neurol* 60:5:528–539.

Jansen FE, Braams O, Vincken KL, Algra A, Anbeek P, Jennekens-Schinkel A, Halley D, Zonnenberg BA, va den Ouweland A, van Huffelen AC, van Nieuwenhuizen O, Nellist M (2008a) Overlapping neurologic and cognitive phenotypes in patients with TSC1 or TSC2 mutations. *Neurology* 70:908–915.

Jansen FE, Vincken KL, Algra A, Anbeek P, Braams O, Nellist M, Zonnenberg BA, Jennekens-Schinkel A, van den Ouweland A, Halley D, van Huffelen AC, van Nieuwenhuizen O (2008b) Cognitive impairment in tuberous sclerosis complex is a multi-factorial condition. *Neurology* 70:916–923.

Johnson SR, Tattersfield AE (2000) Clinical experience of lymphangioleiomyomatosis in the UK. *Thorax* 55:1052–1057.

Jones AC, Shyamsundar MM, Thomas MW, Maynard J, Idziaszczyk S, Tomkins S, Sampson JR, Cheadle JP (1999) Comprehensive mutation analysis of TSC1 and TSC2-and phenotypic correlations in 150 families with tuberous sclerosis. *Am J Hum Genet* 64:1305–1315.

Jost CJ, Gloviczki P, Edwards WD, Stanson AW, Joyce JW, Pairolero PC (2001) Aortic aneurysms in children and young adults with tuberous sclerosis: report of two cases and review of the literature. *J Vasc Surg* 33:639–642.

Jóźwiak S, Kotulska K, Domańska-Pakieła D, Łojszczyk B, Syczewska M, Chmielewski D, Dunin-Wąsowicz D, Kmieć T, Szymkiewicz-Dangel J, Kornacka M, Kawalec W (2011) Antiepileptic treatment before the onset of seizures reduces epilepsy severity and risk of mental retardation in infants with tuberous sclerosis complex. *Eur J Paediatr Neurol* 15:5:424–431.

Jozwiak S, Pedich M, Rajszys P, Michalowicz R (1992) Incidence of hepatic hamartomas in tuberous sclerosis. *Arch Dis Child* 67:1363–1365.

Jozwiak S, Schwartz RA, Janniger CK, Michalowicz R, Chmielik J (1998) Skin lesions in children with tuberous sclerosis complex: Their prevalence, natural course, and diagnostic significance. *Int J Dermatol* 37:911–917.

Jozwiak S, Schwartz RA, Janniger CK, Bielicka-Cymerman J (2000) Usefulness of diagnostic criteria of tuberous sclerosis complex in pediatric patients. *J Child Neurol* 15:652–659.

Kandt RS, Haines JL, Smith M, Northrup H, Gardner RJM, Short MP, Dumars K, Roach ES, Steingold S, Wall S, Blanton SH, Flodman P, Kwiatkowski DJ, Jewell A, Weber JL, Roses AD, Pericak-Vance MA (1992) Linkage of an important gene locus for tuberous sclerosis to a chromosome 16 marker for polycystic kidney disease. *Nat Genet* 2:37–41.

Karenfort M, Kruse B, Freitag H, Pannek H, Tuxhorn I (2002) Epilepsy surgery outcome in children with focal epilepsy due to tuberous sclerosis complex. *Neuropediatrics* 33:255–261.

Khare L, Strizheva GD, Bailey JN, Au KS, Northrup H, Smith M, Smalley SL, Henske EP (2001) A novel missense mutation in the GTPase activating protein homology region of TSC2 in two large families with tuberous sclerosis complex. *J Med Genet* 38:347–9.

Kingswood JC, d'Augères GB, Belousova E, Ferreira JC, Carter T, Castellana R, Cottin V, Curatolo P, Dahlin M, de Vries PJ, Feucht M, Fladrowski C, Gislimberti G, Hertzberg C, Jozwiak S, Lawson JA, Macaya A, Nabbout R, O'Callaghan F, Benedik MP, Qin J, Marques R, Sander V, Sauter M, Takahashi Y, Touraine R, Youroukos S, Zonnenberg B, Jansen AC; TOSCA consortium and TOSCA investigators (2017) TuberOus

SClerosis registry to increase disease Awareness (TOSCA) - baseline data on 2093 patients. *Orphanet J Rare Dis* 5;12(1):2.

Knudson AG Jr (1971) Mutation and cancer: Statistical study of retinoblastoma. *Proc Natl Acad Sci USA* 68:820–823.

Kothare SV, Singh K, Chalifoux JR, Staley BA, Weiner HL, Menzer K, Devinsky O (2014) Severity of manifestations in tuberous sclerosis complex in relation to genotype. *Epilepsia* 55(7):1025–9.

Kozlowski P, Roberts P, Dabora S, Franz D, Bissler J, Northrup H, Au KS, Lazarus R, Domanska-Pakiela D, Kotulska K, Jozwiak S, Kwiatkowski DJ (2007) Identification of 54 large deletions/duplications in TSC1 and TSC2 using MLPA, genotype-phenotype correlations. *Hum Genet* 121:389–400.

Krueger DA, Northrup H, International Tuberous Sclerosis Complex Consensus Group (2013) Tuberous sclerosis complex surveillance and management: recommendations of the 2012 International Tuberous Sclerosis Complex Consensus Conference. *Pediatr Neurol* 49:255–265.

Krueger DA, Wilfong AA, Holland-Bouley K, Anderson AE, Agricola K, Tudor C, Mays M, Lopez CM, Kim MO, Franz DN (2013b) Everolimus treatment of refractory epilepsy in tuberous sclerosis complex. *Ann Neurol* 74:5:679–687.

Kwan P, Brodie MJ (2000) Early identification of refractory epilepsy. *N Engl J Med* 342:314–349.

Martin KR, Zhou W, Bowman MJ, Shih J, Au KS, Dittenhafer-Reed KE, Sisson KA, Koeman J, Weisenberger DJ, Cottingham SL, DeRoos ST, Devinsky O, Winn ME, Cherniack, AD, Shen H, Northrup H, Krueger DA, MacKeigan JP (2017) The genomic landscape of tuberous sclerosis complex. *Nat Commun* 8:15816

Mitchell AL, Parisi MA, Sybert VP (2003) Effects of pregnancy on the renal and pulmonary manifestations in women with tuberous sclerosis complex. *Genet Med* 5:154–160.

Muhler EG, Turniski-Harder V, Engelhardt W, von Bernuth G (1994) Cardiac involvement in tuberous sclerosis. *Br Heart J* 72:584–590.

Murphy JV, Patil AA (2003) Stimulation of the nervous system for the management of seizures. *CNS Drugs* 17:101–115.

Nabbout R, Santos M, Rolland Y, Delalande O, Dulac O, Chiron C (1999) Early diagnosis of subependymal giant cell astrocytoma in children with tuberous sclerosis. *J Neurol Neurosurg Psychiatry* 66:370–375.

Nelson CP, Sanda MG (2002) Contemporary diagnosis and management of renal angiomyolipoma. *J Urol* 168:1315–1325.

Northrup H, Koenig MK, Pearson DA, Au KS (2015) Tuberous sclerosis complex. *GeneReviews®*. https://www.ncbi.nlm.nih.gov/books/NBK1220/

Northrup H, Kwiatkowski DJ, Roach ES, Dobyns WB, Lewis RA, Herman GE, Rodriguez E Jr, Daiger SP, Blanton SH (1987) Evidence for genetic heterogeneity in tuberous sclerosis: one locus on chromosome 9 and at least one locus elsewhere. *Am J Hum Genet* 51:709–720.

Northrup H, Wheless JW, Bertin TK, Lewis RA (1993) Variability of expression in tuberous sclerosis. *J Med Genet* 30:41–43.

Northrup H, Krueger DA, International Tuberous Sclerosis Complex Consensus Group (2013) Tuberous sclerosis complex diagnostic criteria update: recommendations of the 2012 International Tuberous Sclerosis Complex Consensus Conference. *Pediatr Neurol* 49(4):243–254.

Numis AL, Major P, Montenegro MA, Muzykewicz DA, Pulsifer MB, Thiele EA (2011) Identification of risk factors for autism spectrum disorders in tuberous sclerosis complex. *Neurology* 76(11):981–7.

O'Callaghan FJ, Shiell AW, Osborne JP, Martyn CN (1998) Prevalence of tuberous sclerosis estimated by capture-recapture analysis. *Lancet* 351:1490.

O'Callaghan FJ, Clarke AA, Hancock E, Hunt A, Osborne JP (1999) Use of melatonin to treat sleep disorders in tuberous sclerosis. *Dev Med Child Neurol* 41:123–126.

O'Connor SE, Kwiatkowski DJ, Roberts PS, Wollmann RL, Huttenlocher PR (2003) A family with seizures and minor features of tuberous sclerosis and a novel TSC2 mutation. *Neurology* 61:409–12.

Osborne JP, Fryer A, Webb D (1991) Epidemiology of tuberous sclerosis. *Ann N Y Acad Sci* 615:125–127.

Papadavid E, Markey A, Bellaney G, Walker NP (2002) Carbon dioxide and pulsed dye laser treatment of angiofibromas in 29 patients with tuberous sclerosis. *Br J Dermatol* 147: 337–342.

Parain D, Penniello MJ, Berquen P, Delangre T, Billard C, Murphy JV (2001) Vagal nerve stimulation in tuberous sclerosis complex patients. *Pediatr Neurol* 25:213–216.

Pipitone S, Mongiovi M, Grillo R, Gagliano S, Sperandeo V (2002) Cardiac rhabdomyoma in intrauterine life: Clinical features and natural history. A case series and review of published reports. *Ital Heart J* 3:48–52.

Qin W, Kozlowski P, Taillon BE, Bouffard P, Holmes AJ, Janne P, Camposano S, Thiele E, Franz D, Kwiatkowski DJ (2010) Ultra deep sequencing detects a low rate of mosaic mutations in tuberous sclerosis complex. *Hum Genet* 127:573–582.

Roach ES, Kerr J, Mendelsohn D, Laster DW, Raeside C (1991) Detection of tuberous sclerosis in parents by magnetic resonance imaging. *Neurology* 41:262–265.

Roach ES, DiMario FJ, Kandt RS, Northrup H (1999) Tuberous sclerosis consensus conference: Recommendations for diagnostic evaluation. *J Child Neurol* 14:401–407.

Rose VM, Au KS, Pollom G, Roach ES, Prashner HR, Northrup H (1999) Germ-line mosaicism in tuberous sclerosis: How common? *Am J Hum Genet* 64:986–992.

Rowley SA, O'Callaghan FJ, Osborne JP (2001) Ophthalmic manifestations of tuberous sclerosis: A population based study. *Br J Ophthalmol* 85:420–423.

Sampson JR, Scahill SJ, Stephenson JB, Mann L, Connor JM (1989) Genetic aspects of tuberous sclerosis in the west of Scotland. *J Med Genet* 26:28–31.

Sampson JR, Maheshwar MM, Aspinwall R, Thompson P, Cheadle JP, Ravine D, Roy S, Haan E, Bernstein J, Harris PC (1997) Renal cystic disease in tuberous sclerosis: role of the polycystic kidney disease 1 gene. *Am J Hum Genet* 61:843–851.

Sancak O, Nellist M, Goedbloed M, Elfferich P, Wouters C, Maat-Kievit A, Zonnenberg B, Verhoef S, Halley D, van den Ouweland A (2005) Mutational analysis of the TSC1 and TSC2 genes in a diagnostic setting: genotype-phenotype correlations and comparison of diagnostic DNA techniques in Tuberous Sclerosis Complex. *Eur J Hum Genet* 13:731–741.

Sepp T, Yates JR, Green AJ (1996) Loss of heterozygosity in tuberous sclerosis hamartomas. *J Med Genet* 33:962–964.

Shepherd CW, Gomez MR, Lie JT, Crowson CS (1991) Causes of death in patients with tuberous sclerosis. *Mayo Clin Proc* 66:792–796.

Smith HC, Watson GH, Patel RG, Super M (1989) Cardiac rhabdomyomata in tuberous sclerosis: their course and diagnostic value. *Arch Dis Child* 64:196–200.

Strizheva GD, Carsillo T, Kruger WD, Sullivan EJ, Ryu JH, Henske EP (2001) The spectrum of mutations in TSC1 and TSC2 in women with tuberous sclerosis and lymphangiomyomatosis. *Am J Resp Crit Care Med* (163(1):253–8.

Tello R, Blickman JG, Buonomo C, Herrin J (1998) Meta analysis of the relationship between tuberous sclerosis complex and renal cell carcinoma. *Eur J Radiol* 27:131–138.

The European Chromosome 16 Tuberous Sclerosis Consortium (1993) Identification and characterization of the tuberous sclerosis gene on chromosome 16. *Cell* 75:1305–1315.

Torres OA, Roach ES, Delgado MR, Sparagana SP, Sheffield E, Swift D, Bruce D (1998) Early diagnosis of subependymal giant cell astrocytoma in patients with tuberous sclerosis. *J Child Neurol* 13:173–177.

Tyburczy ME, Dies KA, Glass J, Camposano S, Chekaluk Y, Thorner AR, Lin L, Krueger D, Franz DN, Thiele EA, Sahin M (2015) Mosaic and intronic mutations in TSC1/TSC2 explain the majority of TSC patients with no mutation identified by conventional testing. *PLoS Genet* 11:11:e1005637.

Tyburczy ME, Wang JA, Li S, Thangapazham R, Chekaluk Y, Moss J, Kwiatkowski DJ, Darling TN (2013) Sun exposure causes somatic second-hit mutations and angiofibroma development in tuberous sclerosis complex. *Hum Mol Genet* 23:8:2023–2029.

van Eeghen AM, Nellist M, van Eeghen EE, Thiele EA (2013) Central TSC2 missense mutations are associated with a reduced risk of infantile spasms. *Epilepsy Res* 103:1:83–87.

van Slegtenhorst M, de Hoogt R, Hermans C, Nellist M, Janssen B, Verhoef S, Lindhout D, van den Ouweland A, Halley D, Young J, Burley M, Jeremiah S, Woodward K, Nahmias J, Fox M, Ekong R, Osborne J, Wolfe J, Povey S, Snell RG, Cheadle JP, Jones AC, Tachataki M, Ravine D, Kwiatkowski DJ (1997) Identification of the tuberous sclerosis gene TSC1 on chromosome 9q34. *Science* 277:805–808.

Vanderhooft SL, Francis JS, Pagon RA, Smith LT, Sybert VP (1996) Prevalence of hypopigmented macules in a healthy population. *J Pediatr* 129:355–361.

Verhoef S, Bakker L, Tempelaars AM, Hesseling-Janssen AL, Mazurczak T, Jozwiak S, Fois A, Bartalini G, Zonnenberg BA, van Essen AJ, Lindhout D, Halley DJ, van den Ouweland AM (1999) High rate of mosaicism in tuberous sclerosis complex. *Am J Hum Genet* 64(6):1632–1637.

Webb DW, Osborne JP (1991) Non-penetrance in tuberous sclerosis. *J Med Genet* 28:417–419.

Webb DW, Clarke A, Fryer A, Osborne JP (1996) The cutaneous features of tuberous sclerosis: A population study. *Br J Dermatol* 135:1–5.

Weiner DM, Ewalt DH, Roach ES, Hensle TW (1998) The tuberous sclerosis complex: A comprehensive review. *J Am Coll Surg* 18:548–561.

Wentink M, Nellist M, Hoogeveen-Westerveld M, Zonnenberg B, van der Kolk D, van Essen T, Park SM, Woods G, Cohn-Hokke P, Brussel W, Smeets E (2012) Functional characterization of the TSC2 c. 3598C> T (p. R1200W) missense mutation that co-segregates with tuberous sclerosis complex in mildly affected kindreds. *Clin Genet* 81:5:453–461.

Wheless JW, Clarke DF, Arzimanoglou A, Carpenter D (2007) Treatment of pediatric epilepsy: European expert opinion, 2007. *Epileptic Disord* 9:353–412.

Whittemore VH, Nakagawa J, Cinkosky M, Sparagana S, Frost M, Thiele E, Brown C, Kohrman M, Gupta A, LaJoi J, McClintock W, Wu J, Miller I, Duchowny M, Levisohn P, Bebin EM, Northrup H (2008) TSC Natural History Database. Abstract presented at the International TSC Conference 2008.

Wong V, Khong PL (2006) Tuberous sclerosis complex: Correlation of MRI findings with comorbidities. *J Child Neurol* 21:99–105.

Xu L, Sterner C, Maheshwar MM, Wilson PJ, Nellist M, Short PM, Haines JL, Sampson JR, Ramesh V (1995) Alternative splicing of the tuberous sclerosis 2 (TSC2) gene in human and mouse tissues. *Genomics* 27:475–480.

Yang P, Cornejo KM, Sadow PM, Cheng L, Wang M, Xiao Y, Jiang Z, Oliva E, Jozwiak S, Nussbaum RL, Feldman AS, Paul E, Thiele EA, Yu JJ, Henske EP, Kwiatkowski DJ, Young RH, Wu CL (2014) Renal cell carcinoma in tuberous sclerosis complex. *The American journal of surgical pathology.* 38(7):895.

Yu J, Astrinidis A, Henske EP (2001) Chromosome 16 loss of heterozygosity in tuberous sclerosis and sporadic lymphangiomyomatosis. *Am J Respir Crit Care Med* 164:1537–1540.

Zaroff CM, Barr WB, Carlson C, LaJoie J, Madhavan D, Miles DK, Nass R, Devinsky O (2006) Mental retardation and relation to seizure and tuber burden in tuberous sclerosis complex. *Seizure* 15:558–562.

60

TURNER SYNDROME

ANGELA E. LIN
Medical Genetics Unit, MassGeneral Hospital for Children, Harvard University School of Medicine, Boston, Massachusetts, USA

MELISSA L. CRENSHAW
Division of Genetics, Johns Hopkins All Children's Hospital, Johns Hopkins University School of Medicine, St. Petersburg, Florida, USA

INTRODUCTION

Turner syndrome (TS) is the most common sex chromosome disorder among women. Our understanding of this diagnosis has evolved over time with our improved knowledge of the human genome, particularly of the genetics of the X chromosome. With the advent of early diagnosis, even in the prenatal period, the variability of TS has become increasingly evident. This has given increased opportunity for enhanced monitoring and intervention to decrease the impact of related complications, including growth deficiency and aortic dissection. TS illustrates the rewards and challenges of managing individuals across the lifespan, spanning from prenatal counseling to senior care.

Readers of this fourth edition may want to refer to previous chapters in this series on this topic (Sybert 2004; Davenport 2010). An invaluable companion to this current chapter is the timely update of the prior guidelines (Bondy et al. 2007) to "Clinical practice guidelines for the care of girls and women with Turner syndrome" (Gravholt et al. 2017). Because this detailed and frequently cited article was written following an international conference with exhaustive review of the literature, it will be cited frequently in place of historical references that can be found in the guidelines reference list. Further, the reader will find that the discussion of evaluation and management here closely adheres to those presented in the guidelines.

Incidence

The most reliable data about TS prevalence were derived from Danish national registries (Gravholt 2008), which additional studies estimate as 25–50 per 100,000 women. The livebirth occurrence is approximately 1:2000 (Stochholm et al. 2006). The incidence at the time of conception is even higher because the vast majority of fetuses with non-mosaic karyotypes do not survive to term (Hook and Warburton 1983, 2014; Held et al. 1992). The relative risk of death among girls and women with TS is significantly higher (approximately three to four times) than that of the general population. This risk is proportional to the number of women with 45,X karyotype and aortic dissection (Gravholt 2006). Perinatal mortality has been estimated between 0 and 11% (Gravholt et al. 2017). With improved care of cardiovascular abnormalities, in particular aortic dissection, the morbidity will hopefully decline.

Diagnostic Criteria

The diagnosis of TS is made by chromosome analysis (typically, lymphocytes in peripheral blood sample) and consideration of clinical manifestations. The karyotype should show either complete absence or a structural abnormality of part of the second sex chromosome (Table 60.1). This includes deletion of at least Xp22.33 and excludes girls and women with small, distal deletions of the long arm of the X chromosome from Xq24.

Cassidy and Allanson's Management of Genetic Syndromes, Fourth Edition.
Edited by John C. Carey, Agatino Battaglia, David Viskochil, and Suzanne B. Cassidy.
© 2021 John Wiley & Sons, Inc. Published 2021 by John Wiley & Sons, Inc.

TABLE 60.1 Type and frequency of chromosome abnormalities in Turner syndrome (adapted from Gravholt et al. 2017)

Karyotype[1]	Percentage among all cases
45,X	(40–50%)
45,X/46,XX	(15%)
45,X/47,XXX; 45,X/46,XX/47,XXX	(3%)
45,X/46,XY	(10–12%)
46,XX, del(p22.3)	
46,X,r(X)/46,XX	
46,X i(Xq); 46,X,idic(Xp)	(10%)

[1] Several karyotypes are not listed, including various unbalanced X-autosome translocations which are usually limited to single case reports.
46,XX,del(q24) is not considered Turner syndrome, although premature ovarian failure is present.
46,X,idic(X)(q24) is not considered Turner syndrome.

Clinical manifestations can include short stature, ovarian failure, and congenital heart disease, but more broadly can involve hearing loss, autoimmune conditions, and characteristic facial features (see Figure 60.1) and other less frequent manifestations (Gravholt et al. 2017). It is important that practitioners monitor growth and development closely and consider this diagnosis in girls who fall below the growth curve or show a decline in growth velocity (Table 60.2).

Approximately 10% of girls and women with TS will have cryptic Y chromosome material (Gravholt et al. 2017). The risk for gonadoblastoma among those with Y material varies greatly among studies but is estimated to be around 10%, which confers about a 1% risk among individuals with TS to have gonadoblastoma. (Page et al. 1994; Gravholt et al. 2000). This risk increases with a higher level for mosaicism for Y material and greater masculinization and can be reduced by screening for Y material. Molecular screening of multiple Y chromosome loci by PCR including the GBY locus for gonadoblastoma is recommended only for those individuals with TS who have negative karyotype and FISH testing but have masculinized features. It is no longer clinically indicated to use FISH and SRY probes to exclude cryptic Y material since the gonadoblastoma locus itself is not screened (Gravholt et al. 2017).

Turner syndrome refers to females. The expression "male Turner syndrome" is discouraged when referring to phenotypic males with mosaicism of 45,X and 46,XY cell lines. Instead, the term "45,X/46,XY male" is preferred.

Etiology, Pathogenesis, and Genetics

The most common karyotype associated with TS is 45,X (preferred to "monosomy X") (Table 60.1). However, the majority of people with TS will have a mosaic karyotype or a karyotype with an X chromosome that is structurally deficient. Those who have 45,X/46,XX and 45,X/47,XXX karyotypes may have more mild manifestations of TS. Mosaicism with a typical cell line (46,XX) is associated with decreased likelihood of congenital heart disease and lymphedema (El-Mansoury et al. 2007, Lin et al. 2017) as well as increased chance of spontaneous puberty and pregnancy (Bernard et al. 2016). The external phenotype cannot be reliably predicted, although there is a trend toward a milder appearance with a lower level of 45,X. A ring X chromosome, isodicentric Xp, and isodicentric Xq are less common. Those with small distal deletions of the short arm of the X chromosome, Xp22.3, may have short stature, but fewer cardiac anomalies. The lower limit of 45,X mosaicism by which to make the diagnosis has not been determined, though 5% and 10% have been used in several research studies (Lin et al. 2017). A high threshold should be considered for women over 50 as the loss of the second X chromosome occurs also with normal aging (Russell 2007). Genotype and phenotype vary widely, and do not always correlate with the degree of mosaicism present on blood testing. The new TS surveillance guidelines apply to any karyotype (Gravholt et al. 2017).

Several mechanisms by which loss of all or part of an X chromosome leads to clinical abnormalities have been proposed. These include (1) haploinsufficiency of gene expression, notably *SHOX*; (2) failure of X chromosome pairing during meiosis; (3) failure to express imprinted genes; (4) uncovering X-linked mutations, such as the co-occurrence of X-linked recessive disorders that are typically restricted to males but may be observed in girls with TS; and (5) functional disomy in those with a small ring X chromosome (summarized in Davenport 2010).

Diagnostic Testing

The diagnosis of TS is made by obtaining a standard 20 cell karyotype (Wolff et al. 2010). If this is negative and there remains suspicion for the diagnosis, then further testing can be performed by extended FISH analysis or by testing an additional tissue such as skin or buccal cells. While not the recommended initial test, a chromosome microarray can supplement the chromosome analysis by providing breakpoints on the X chromosome. Despite the interest in imprinting, there is currently insufficient evidence to recommend clinical parent-of-origin testing of the remaining X chromosome (Gravholt et al. 2017).

There is an approximate 1% risk for any individual with TS to have gonadoblastoma, a risk which can be reduced by screening for the Y chromosome material present in approximately 10% of individuals with TS (Gravholt et al. 2017). However, the magnitude of this risk is thought to vary widely. Increased mosaicism as well as masculinization can be associated with increased risk for gonadoblastoma, which is approximately 10% overall in individuals with Y material (Page et al. 1994; Gravholt et al. 2000). Screening for cryptic

FIGURE 60.1 Variability of the facial features in Turner syndrome in 16 individuals of diverse ethnic backgrounds and karyotypes. Individuals A–I have 45,X at ages 8 (two girls), 11, 14, 17, 19, 23, and 35 (two women), respectively. Individuals J–L have mosaicism with an additional karyotype (percentage of the 45,X cell line are: J, 45,X 20%/46,XX (21 years); K, 45,X (60%)/47,XXX (5 years) and, L, 45,X (53%)/Ring X (7 years). Individual M has a deletion Xp11.2, individual N has an isochromosome Xq10, and individuals O and P have 45,X/ isodicentric Y (70%, 20 years and 68%, 7 years). Karyotypes refer to peripheral blood lymphocyte analysis.

TABLE 60.2 Indications for chromosome analysis to diagnose Turner syndrome (adapted from Gravholt et al. 2017, table 2)

As the only clinical manifestation:
- Fetal cystic hygroma, or hydrops, especially when severe
- Idiopathic short stature
- Obstructive left-sided congenital heart defect[1]
- Unexplained delayed puberty/menarche
- Couple with infertility
- Characteristic facial features in a female[2]

At least two of the following:
- Renal anomaly (horseshoe, absence, or hypoplasia)
- Madelung deformity
- Neuropsychologic problems, and/or psychiatric issues[3]
- Multiple typical or melanocytic nevi
- Dysplastic or hyperconvex nails
- Other congenital heart defects[3]

[1] Typically bicuspid aortic valve, coarctation, aortic stenosis (with/without bicuspid aortic valve), mitral valve anomalies, and hypoplastic left heart syndrome

[2] Downslanted palpebral fissures, epicanthal folds, low-set anomalous pinnae, micrognathia, narrow palate, short broad neck, and webbing (see Figure 60.1)

[3] Partial anomalous pulmonary venous return; atrial septal defect, secundum type; and ventricular septal defects, muscular or membranous.

Y material by PCR is recommended for those with masculine features who have had normal female karyotype and FISH testing (Gravholt et al. 2017).

Differential Diagnosis

There is no single TS phenotype (see Figure 60.1), although short stature is nearly constant. Thus, several other syndromes with clinical overlap might be considered in a female fetus with severe nuchal edema, an infant with neck webbing, or a girl or woman with short stature. When the "classic" facial appearance and habitus, together with typical internal anomalies (kidney and congenital heart defects) are present, TS is usually not confused.

The most common syndrome that also has features of lymphatic dysplasia is Noonan syndrome (See Chapter 41). Both have short stature, but the neck webbing in Noonan syndrome is more likely to be a prominent trapezius muscle, rather than the loose folds of posterior neck skin, or in more severe cases, a pterygium-like "web". People with TS and peripheral lymphedema in infancy are more likely to have lifelong lymphedema of the dorsum of the hands and lower extremities, which is not seen in Noonan syndrome. The latter can be severe and predispose to cellulitis. In fetal life and the newborn period, females with polyhydramnios and edema could have other rasopathy syndromes (e.g., Costello syndrome (Chapter 16), cardio-facio-cutaneous syndrome, Chapter 10)). The type of congenital heart defect may be a key manifestation in the differential diagnosis since left-sided defects (coarctation, bicuspid aortic valve, aortic stenosis) are common in TS, whereas pulmonic valve stenosis, which is common in rasopathies, is rare.

There are girls and women who do not have TS or a rasopathy such as Noonan syndrome, but have a similar phenotype of loose neck skin, low-set pinnae (which may be anomalous or protrude) and downslanted palpebral fissures. After chromosome analysis to exclude TS and molecular genetic screening using a panel of rasopathy genes, they remain without a proven diagnosis.

There are also several primary lymphedema syndromes such as Milroy disease (hereditary lymphedema type IA) caused by an *FLT4* gene mutation, and Meige disease, also known as hereditary lymphedema type II for which the gene has not yet been characterized. Renal and cardiac anomalies in conjunction with lymphedema can occur in lymphedema-distichiasis syndrome caused by a *FOXC2* mutation.

Short stature is a consistent feature shared with other genetic syndromes. Mutations in the *SHOX* gene located on the X chromosome and thought to be responsible for short stature in TS can cause similar skeletal manifestations including short stature and Madelung anomaly of the forearm without the complete appearance of TS. Some mild skeletal dysplasias can resemble TS.

MANIFESTATIONS AND MANAGEMENT

Growth and Feeding

Growth deficiency, manifesting as short stature, is a cardinal feature of TS. This can be seen as early as fetal development or can develop later with a decline of the normal growth curve as a toddler or in early childhood. Early feeding difficulties may further stress the growth process.

Growth hormone therapy is now offered to all girls with TS with growth deficiency. Referral to a pediatric endocrinologist at the time of diagnosis is recommended. The age of initiation and duration of treatment are key factors in the effectiveness of growth hormone therapy. It is recommended that this be started no later than 12–13 years of age and possibly as early as 4–6 years of age when the following are noted: (1) evidence of growth failure; (2) growth is already below the expected standards; (3) if mean parental heights predict short stature; or (4) if puberty has already ensued prior to diagnosis. Treatment is continued until a bone age of 14 years or more and growth rate less than 2 cm/year. The limitations of insurance coverage can challenge some families and providers.

In addition to the benefits of growth hormone therapy, concerns have been raised about its complications, including slipped capital epiphysis, scoliosis, diabetes, and hypertension, which occur more commonly in girls with TS, even without growth hormone therapy (see below). Reassurance about growth hormone and cardiovascular function and aortic size has been reported (Bondy et al. 2006; van den Berg et al. 2008). It is not known whether neoplasia represents an inherently increased risk in individuals with TS. The transient insulin antagonism seen with growth hormone therapy is balanced by the beneficial effects of growth hormone on body composition (reviewed by Davenport 2010)

Evaluation

- Growth should be monitored carefully at each physician visit. Although some have advised plotting on a TS-specific growth curve (Sybert and McCauley 2004), that growth chart was derived from European studies conducted 30–40 years ago which may not accurately reflect a multicultural modern patient population (Lyon et al. 1985). A significant decrease in linear growth rate by third or fourth grade on a standard growth curve will alert the physician, although some present much younger with short stature.
- Referral to a pediatric endocrinologist is recommended at the time of diagnosis.
- Growth decline in a girl already on growth hormone treatment should prompt consideration of other causes for growth decline such as celiac disease or feeding problems.

Management

- Girls who manifest growth deficiency should be offered the option of growth hormone treatment early, generally starting at 4–6 years of age and no later than 13 years.
- The typical dose is 45–50 µg/kg/day or 1.3 mg/kg/day and can be increased to 68 µg/kg/day (2 mg/kg/day).
- IGF-1 levels should be monitored at least annually and therapy adjusted accordingly (Gravholt et al. 2017).
- Individuals on growth hormone are generally assessed by a pediatric endocrinologist at 3–6 month intervals.
- In girls who have a delayed diagnosis and thus delay in beginning growth hormone therapy, some endocrinologists may add oxandrolone.
- The addition of estrogen therapy prior to the time for desired pubertal onset is not recommended.

Cardiovascular

Cardiovascular disease in TS includes mostly congenital heart defects and abnormalities of the aorta and peripheral arteries, but also acquired problems such as early-onset systemic hypertension, ischemic heart disease and cerebrovascular disease (Silberbach et al. 2018). These are lifelong challenges and contribute significantly to morbidity and mortality. Cardiomyopathy and serious arrhythmias are extremely rare. Cardiovascular health, including the frequency and clinical management of individual conditions, has been exhaustively studied by recent international working groups seeking evidence-basis for management guidelines (historical citations included in reviews by Gravholt et al. 2017; Silberbach et al. 2018).

Congenital heart defects are seen in ~75% of fetuses and up to 50% of individuals with TS. The well-recognized pattern includes various levels of obstruction on the left side of the heart. Most common is aortic coarctation, with or without a bicuspid valve (15%), bicuspid valve alone (15–30%), and aortic stenosis and/or regurgitation (5%). Other defects include hypoplastic left heart, mitral valve anomalies (hypoplastic, prolapsed), atrial septal defect and ventricular septal defect (10%) (Sybert 1998; Gravholt et al. 2017; Silberbach et al. 2018). Aortic coarctation is usually a typical juxta-ductal obstruction, and less frequently diffuse aorta hypoplasia. When severe, an infant is dependent on a patent ductus arteriosus. The use of magnetic resonance angiography (MRA) has increased detection, and in about half of women with TS, the spectrum of vascular anomalies includes elongation of the transverse arch (49%), aberrant right subclavian artery (8%), persistent left superior vena cava (13%), and partial anomalous pulmonary venous return (13%) (Ho et al. 2004). Congenital coronary artery anomalies must also be considered.

The most concerning manifestation of the vasculopathy of TS is aortic dissection, which is estimated to occur in approximately 1% (Silberbach et al. 2018). Aortic dilatation is common (30%), and recently there is greater awareness that the aortic diameter is larger than expected in girls with TS under the age of 15 years. Aortic measurements in an individual with TS should be calculated based on body surface area. Dissection tends to occur relatively early in life, with median age at time of dissection being 35 years, and it occurs at smaller ascending aorta diameters than in other genetically triggered aortopathies. Although most cases of aortic dissection have been associated with coarctation of the aorta, bicuspid aortic valve, pregnancy, hypertension, and/or aortic root dilatation, 10% have had no known risk factors (Carlson and Silberbach 2007).

Systemic hypertension is estimated to occur in 20–40% of children and up to 60% of adults with TS. Although renal anomalies may be detected and, rarely, unrecognized coarctation, in most instances there is no identifiable cause.

The management of pregnancy is discussed below. Since only a small number of women with TS have spontaneous pregnancy, the vast majority of pregnancies are planned using assisted reproductive technology. It is imperative that any woman considering pregnancy consult with a cardiologist and be monitored carefully throughout the pregnancy, in addition to the reproductive and obstetric providers.

The incidence of ischemic heart disease and atherosclerosis in adults with TS is unknown, but there is at least a doubling of the risk for coronary artery disease compared with the general population. Additional risk factors included hypertension, insulin resistance, hyperlipidemia, and estrogen deficiency.

Evaluation

- Fetal echocardiography should be performed when TS is highly suspected or confirmed during pregnancy, with referral made to a pediatric cardiologist if a congenital heart anomaly is found (Silberbach et al. 2018).
- Under the direction of a cardiologist with expertise in congenital heart defects and aortic disease, all individuals with TS should have a baseline evaluation with transthoracic echocardiography, even if a fetal echocardiogram was normal. Echocardiography is generally adequate in children, but MRA or CT angiography should be performed as soon as the child can cooperate enough to have the procedure without sedation (Figures 60.2 and 60.3, Silberbach et al. 2018).
- Coarctation in the newborn female is an independent marker of TS, and karyotype should be strongly considered.
- For newly diagnosed adolescents and adults, TS cardiovascular screening should include both transthoracic echocardiography (TTE) and magnetic resonance angiography (MRA), although CT with angiography (CTA) may be substituted if more accessible and better tolerated than MRA (Silberbach et al. 2018).
- Women actively pursuing pregnancy should be managed by a multidisciplinary team including a cardiologist with expertise in women with TS. This represents an evolution in care which recognizes the considerable cardiac burden, but is more supportive of pregnancy and offers guidelines for intensive monitoring. This counseling should begin at least two years before planned pregnancy to discuss the increased cardiovascular risk and aortic disease and to schedule imaging. Although women with existing aortic dilatation should be advised against pregnancy, it is recognized that women with other risk factors (hypertension, bicuspid aortic valve, coarctation, previous aortic surgery) may become pregnant (Gravholt et al. 2017).
- Blood pressure should be monitored at each clinic visit. Ambulatory blood pressure monitoring is strongly encouraged.
- A resting electrocardiogram should be done at diagnosis. When QTc interval prolongation is present, 24 hour Holter monitoring and exercise testing should be performed.
- Ongoing monitoring of the aorta depends upon the clinical situation and age (the age-specific flow diagrams [Figures 60.2 and 60.3]) and are too detailed to enumerate point by point, but generally provide imaging guidelines based on the presence or absence of coarctation, bicuspid aortic valve and/or hypertension, and the

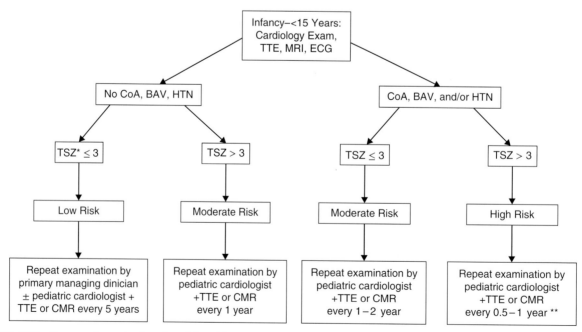

FIGURE 60.2 Suggested monitoring for girls with Turner syndrome from infancy to 15 years of age (Silberbach et al. 2018). BAV, bicuspid aortic valve; CMR, cardiac magnetic resonance; CoA, coarctation; HTN, hypertension; TSZ, Turner-specific z score; TTE, transthoracic echocardiography. *Ascending aorta TSZ. **Frequency may change with worse disease severity.

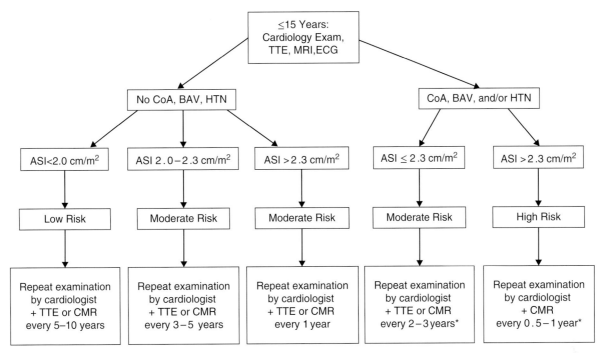

FIGURE 60.3 Suggested cardiac monitoring protocol for girls and women with Turner syndrome older than 15 years of age (Silberbach et al. 2018). ASI, aortic size index. See Fig. 60.2 for remainder of abbreviations.

initial aortic diameter reported as a TS-specific Z score. Given the unique body proportions of TS individuals, in some cases absolute aortic diameter is taken into account. For those girls and women whose initial evaluation was normal, imaging can be repeated at greater intervals, whereas those at highest risk may need cardiac magnetic imaging as frequently as every 6–12 months.

Management

- A cardiologist should direct the care of anyone with TS in whom a congenital heart defect or aorta dilatation is detected (Figures 60.2 and 60.3).
- Hypertension should be treated aggressively, given the risk for aortic dissection. A healthy lifestyle should be encouraged.
- If appropriate, prophylactic antibiotics should be prescribed to prevent subacute bacterial endocarditis.
- All individuals with TS, especially those with existing aortic dilatation and/or bicuspid aortic valve, should be taught that if they have severe chest pain, aortic dissection must be considered and an imaging study, such as MRA, must be performed. Some women wear Medic Alert type jewelry or carry a wallet size card.
- Medical and surgical treatment for aortic dilatation can be individualized in the absence of clinical trials.

Lymphatic System

Lymphedema is one of the hallmark features of TS. Central lymphedema manifests as early as the prenatal period as a cystic hygroma (nuchal thickening on ultrasound) and in the newborn period as distinct webbing of the neck. Peripheral lymphedema of the hands and feet is also characteristic of TS, differentiating it from Milroy disease in which the lymphedema generally spares the hands. Edema that is secondary to congestive heart failure usually involves pitting of lower the extremities.

While the causative gene(s) for lymphedema in TS has not yet been identified, it is clear that the pathogenesis is the result of abnormal lymphatic formation from early in embryonic development. A natural history study (Atton et al. 2015) confirmed that almost all individuals with TS have lymphedema of all four extremities in the newborn period, which persists in at least one limb throughout life. Even in those with early resolution of symptoms, edema often recurred, but systemic involvement such as intestinal lymphangiectasia was rare. Complications include skin symptoms such as dryness, abnormal color, firmness, and infection as well as nail abnormalities.

Histological studies of fetuses with 45,X TS and cystic hygroma have demonstrated the absence of lymphatic vessels or a jugular lymphatic sac near the jugular vein where the lymphatic and vascular systems normally connect (Bekker et al. 2008; von Kaisenberg et al. 2010). Cystic hygromas may be associated with hydrops and intrauterine demise or can persist as webbed neck and puffy hands and feet after birth. Lymphosyntigraphy has suggested initial lymphatic (capillary) dysfunction as the cause for the peripheral lymphedema (Atton et al. 2015).

A "mechanistic" hypothesis proposed that lymphatic maldevelopment could be associated with left sided cardiac anomalies in TS, which is observed in fetuses with TS who had cystic hygroma and hydrops fetalis (Lacro et al. 1988). The increased prevalence of left-sided heart anomalies in girls and women with TS correlates with the presence of central lymphedema (Loscalzo et al. 2005). Thus, one should have a heightened suspicion for the presence of these heart anomalies in those with central lymphedema at the time of diagnosis.

Lymphedema can fluctuate, and it is well-tolerated when mild. It is typically a life-long condition for which there is no easy remedy. Current therapy consists of compression with stockings and other devices, sometimes assisted by massage and moisturization to minimize edema, increase lymph drainage from the congested areas, reduce subdermal fibrosis, improve the skin condition, and enhance functional status. An independent self-care program should be carried out under the direction of a professional edema therapist.

Evaluation

- Individuals with TS should be monitored for increasing and persistent lymphedema as well as possible skin and nail complications. For example, deeply-set dysplastic toenails can become ingrown.
- The presence of central lymphedema (neck, mediastinal) should prompt concern for left-sided heart anomalies and further cardiac evaluation.
- As noted, all girls and women should have a complete cardiology evaluation with echocardiography at diagnosis.

Management

- Although peripheral edema often resolves by age 2 years, referral to a professional lymphedema therapist is recommended for those with significant compromise of the skin or nails.
- Support stockings can be helpful, even for adolescents. These are more fashionable and easily purchased than in the past.
- It is recommended that women with severe lymphedema be referred to a peripheral vascular cardiologist (or similar specialist) for consideration of pneumatic compression device, self-bandaging and night-time bandaging.
- Cellulitis should be avoided by promoting good hygiene, moisturization, and avoidance of skin trauma.

Puberty and Fertility

Ovarian insufficiency in TS is due to the accelerated loss of oocytes in fetal life and results in a greatly reduced ("streak") ovary. Counseling should be provided early about the reduced fertility, avoiding the terms ovarian "failure" and infertility. Approximately 30% of girls with TS have spontaneous puberty, typically in those who have mosaicism with 46,XX cell lines. Of those who menstruate, approximately one-third will continue to have periods. Although the diagnosis of TS must be entertained in adolescent girls with short stature, failure to menstruate is not always present (Pasquino et al. 1997). Of the young women who attain menarche, approximately one-third continue to have regular menses, although irregular menses and secondary amenorrhea are common (reviewed by Davenport 2010).

More useful than imaging the pelvic organs with ultrasound, gonadotropin levels can help predict future gonadal function as well as the appropriate timing and dosing of estrogen replacement therapies. In the "Turner Toddler" study, girls ages 9 months to 4 years in the 45,X group had very high follicle stimulating hormone (FSH) levels that slowly declined, whereas most of the 45,X/46,XX group had normal FSH values throughout the two year study period (Davenport et al. 2006). Estrogen replacement therapy should be used for the majority of girls with TS who either do not enter puberty spontaneously or fail to complete the developmental process to correct estrogen deficiency. It will normalize developmental changes in secondary sex characteristics including breast and uterine size and shape, optimize bone growth and height, minimize cardiovascular risk factors, promote brain development and liver function, and enhance other estrogen-dependent processes.

Individuals with TS and their families should appreciate the important role of hormone replacement therapy, not only during childhood and early adolescence, but ongoing through adulthood. Because estrogens accelerate bone maturation, estrogen replacement was often delayed in the past, often until 15 or 16 years of age, to allow additional time for linear growth. Currently, it is recommended that estrogen replacement be initiated between 11–12 years, increasing over 2–3 years to an adult dose. The preferred estrogen is low-dose estradiol (E2), administered using a transdermal route (patch), although some individuals prefer oral medication. When breakthrough bleeding occurs, or after two years of estrogen treatment, progesterone should be added (Gravholt et al. 2017). Using transdermal estrogens rather than oral estrogens for pubertal induction in girls with TS may accelerate growth by avoiding increased growth hormone resistance (Davenport 2006).

Reduced fertility is viewed by adults as their most significant TS-related problem. Counseling about fertility issues should begin at the time of diagnosis and include other motherhood options such as adoption or using a gestational carrier. Most women with TS who become pregnant require assisted reproductive technology using a donor egg, for which the success rate is comparable to those of unaffected women. Increasingly, adult women and some adolescents and the parents of girls with TS have expressed an interest in harvesting and storing their own ovarian tissue. Egg cell freezing (cryopreservation) after controlled ovarian hyperstimulation can be offered to women with TS mosaicism at a young age when follicles would be more likely to be present. However, routine oocyte retrieval in girls younger than 12 years is not encouraged. To avoid the risks associated with multiple gestation, for women with TS undergoing in vitro fertilization, only one embryo should be transferred per cycle.

Unassisted pregnancies occur in a small number of women with TS who usually have mosaicism, although a few women with 45,X have become pregnant. Despite reduced fertility, which may give some sexually active women the impression that they "can't get pregnant", reproductive counseling should also offer the option of using some form of birth control such as condoms. The frequency of miscarriages is high, and various anomalies have been noted in live-born infants conceived by women with TS (Hagmann et al. 2011; reviewed in Gravholt et al. 2017).

Although more women with TS are contemplating pregnancy, the increased rate of aortic dissection and maternal death must be discussed. It is suggested that preconceptional counseling should be undertaken by a multispecialty team. Pregnancy should be avoided in the presence of an ascending aortic size index of greater than 2.5 cm/m2 or an ascending aortic size index 2.0–2.5 cm/m2 with associated risk factors for aortic dissection, which include bicuspid aortic valve, elongation of the transverse aorta, coarctation of the aorta, and hypertension. For the "ideal patient" with TS who does not have aortic dilatation, hypertension, bicuspid aortic valve, coarctation, or previous surgery, pregnancy can be contemplated with the understanding that a degree of risk for aortic dissection remains. Avoidance of obesity and encouragement of cardiovascular fitness should be counseled. Vaginal delivery is usually acceptable in the absence of aortic dilatation.

Evaluation

- The family should have had genetic counseling about the chromosome basis of TS prior to endocrine management.
- Beginning around the age of 10 years, pubertal status should be evaluated at each routine clinic visit by physical exam. It should be noted that girls with TS and ovarian failure will still have pubic hair development from adrenal androgens.
- If there is no spontaneous breast development by age 12 years, obtain serum gonadotropin levels (FSH and LH) to ensure that they are elevated.
- Continue to monitor pubertal development after initiation of therapy. It is not known whether normalization of gonadotropin levels is important.
- Referral to a TS specialist to coordinate the discussion and relevant evaluations is recommended when questions of child bearing arise. The multi-specialty team that manages pregnant women should include maternal-fetal medicine specialists and cardiologists.

Management

- Discussion of pubertal development and reproductive potential should generally begin around 10 years of age.
- Begin estrogen therapy around the age of 12 years in girls with elevated gonadotropin levels and no puberty.
- During the process of pubertal development, it is important to engage the child in a gradual discussion about how TS and its treatment may impact her sexual development and function and reproductive potential. In addition, when appropriate, counseling for the prevention of sexually transmitted diseases (and unwanted pregnancy for those with endogenous ovarian function) should also be provided.
- Puberty can be induced at the age of 12 years using very low doses of transdermal E2 (~0.09 µg/kg overnight).
- For girls diagnosed at age 14 (bone age ~12.75 years) or greater, estradiol should be started at a higher dose (25 µg/day seems reasonable) since growth potentiation with E2 is probably not possible.
- Hormone replacement therapy must continue during adulthood, otherwise women with TS develop an even greater risk for osteoporosis, an adverse metabolic profile, and cardiovascular disease than they have at baseline.
- Two years before pregnancy is contemplated, women with TS should have imaging of the thoracic aorta with transthoracic echocardiography and magnetic resonance angiography or CT angiography. Blood pressure must be strictly controlled. Exercise testing can be useful (Gravholt et al. 2017).

Development and Behavior

TS is associated with a complex neuropsychologic and mental health phenotype, with considerable strengths in addition to the familiar challenges (refer to extensive summary in table 9, Gravholt et al. 2017). Most girls and women with TS have intellectual functioning within the normal range. However, approximately 10% do have some degree of intellectual disability (Sybert and McCauley 2004). Those girls and women with TS who are most at risk to have intellectual disability are those with a ring X karyotype (Kuntsi et al. 2000). Even those with normal intellect can experience a variety of cognitive challenges. Therefore, it is essential that these be recognized and addressed to help optimize function as well as quality of daily life.

Some of the earliest recognized deficits involve speech and language. Notably, receptive and expressive language skills often exceed norms (Temple et al. 2002; Temple et al. 2012). However, individuals with TS are challenged as language tasks become more complex and incorporate both executive processing and spatial language skills (Inozemtseva et al. 2002; Gravholt et al. 2017). Speech therapy should be initiated early when needed in order to enhance language development.

Many individuals with TS will have deficits of executive functioning. This includes a reduction in the time to process information and involves multiple skills including attention to task, cognitive planning, working memory, as well as inhibition of inappropriate social responses (Hong et al. 2009). About 25% will have attention deficit disorder/attention deficit hyperactivity disorder (ADD/ADHD), and thus are more likely to simultaneously have difficulty with executive function, though this does not confer an increased incidence of visuospatial deficits (Green et al. 2015). While working memory as well as object memory and location memory may be compromised, individuals with TS often have normal to superior long-term verbal memory (reviewed in Gravholt et al. [2017]). While language and verbal skills are strong, about 50% of girls and women with TS have a math learning disability, dyscalculia (Rovet et al. 1993). Teachers can draw on the verbal learning strengths by emphasizing verbal media rather than relying on visual memorization. By recognizing these challenges early, these strategies can be incorporated into the learning plan to improve progress.

Difficulties with aspects of social cognition can also be a significant obstacle for girls and women with TS. They can exhibit deficits in recognizing socials cues including facial and emotional recognition (Burnett et al. 2010; Hong et al. 2014; Lawrence et al. 2003), and an inability to recognize and respond to the mental state of those around them (Mazzola et al. 2006). Their neuropsychologic profile may resemble autism spectrum disorders because of impairments in social cognition, although it is controversial whether classic autism can occur (reviewed in Gravholt et al. 2017). These additional challenges in the social arena can lead to dysfunction in social interactions through the life span. Independent of TS stigmata, girls and women with TS tend to have lower self-esteem and are more subject to bullying, develop fewer close friendships, and have fewer marriages (Gravholt et al. 2017). Hearing loss and decreased sexual experiences as the result of delayed puberty may also contribute to this social isolation (Gravholt et al. 2017; Boman et al. 2001). It is important that these aspects of the TS cognitive profile be recognized early and interventions offered.

Studies have evaluated whether the psychosocial and cognitive profile of TS can be modified by hormone replacement therapy. Among 10–12 year old girls with TS treated with low dose estrogen, there was improvement in cognitive processing speed as well as motor function (Ross et al. 1998), whereas a more modest improvement in verbal and non-verbal memory was found in a younger cohort of 5–8 year olds (Ross et al. 2000). Supplementation with growth hormone has not shown a marked effect on cognition, nor has there been demonstrated a clear effect of growth hormone supplementation and the increase in height attained on either psychosocial or psychosexual adaptation (Gravholt et al. 2017). This may be an important consideration for families in the decision to proceed with growth hormone. Neuropsychology evaluation and support are particularly important at times of academic, cognitive, and social transition.

Despite the challenging psychosocial profile associated with TS, girls and women with TS show consistent academic achievement and work ethic. Their educational attainment level is in line with or above others in the general population (Gravholt et al. 2017). However, they often choose occupations that are not reflective of their level of educational achievement, with a large proportion reporting low job satisfaction (Gravholt et al. 2017). Additional mental health challenges of women with TS may include anxiety, mood disorders, and obsessive-compulsive disorder, but well-controlled studies are lacking, especially for cases reporting schizophrenia.

Evaluation

- Children with TS should have neuropsychology and allied health evaluations as a routine part of health maintenance.
- Annual developmental and behavioral screenings should be performed from diagnosis until adulthood.
- Neuropsychology evaluations should be completed at important transitional stages in education, career and personal milestones.

Management

- Learning and performance challenges should be addressed with academic and occupational adjustments.
- Puberty should be initiated on time and monitored, with any impairment addressed to optimize psychosocial and psychosexual development.
- Evidence-based intervention should be adapted and initiated to treat psychosocial and cognitive issues experienced in TS.
- Following testing, referral to a psychiatrist, psychotherapist and/or social work therapist may be useful if specific diagnoses are identified, such as anxiety, depression, and attention deficit hyperactivity. These

consultations can be presented in a positive way to promote mental health wellness.

Urologic

Congenital anomalies of the urinary system are present in 30–40% of people with TS (Bilge et al. 2000; Lippe et al. 1988), and are usually asymptomatic. A renal ultrasound should be offered to all women at the time of diagnosis. Anomalies include horseshoe kidney (~10%), either partially or totally duplicated (5–10%) or absent (2–3%) kidneys, and either multi-cystic or ectopic (~1%) kidneys. A small number have secondary morbidity, such as obstructive uropathy and hypertension, chronic renal failure and urinary tract infections.

Evaluation

- All girls with TS should have a renal ultrasound study performed at the time of diagnosis.

Management

- Management is the same as that for the general population.

Immunologic

TS is associated with an increased risk for autoimmune disorders including thyroiditis, celiac disease, inflammatory bowel disease, and juvenile rheumatoid arthritis (Jørgensen et al. 2010). Associations may also exist for type I diabetes, alopecia areata, acquired von Willebrand's disease, primary biliary cirrhosis, and uveitis (Gravholt et al. 2017). Although these are familiar associations, the genetic basis is not known, although autoimmunity probably contributes. Of these, thyroiditis is the most common. The guidelines for evaluation and treatment of these disorders are discussed with their respective organ systems (e.g., hypothyroidism under Endocrine).

Endocrine

Thyroid disease. The most prevalent autoimmune disorder in TS is thyroid disease, which occurs in approximately 25% and manifests more often as hypothyroidism due to Hashimoto thyroiditis than hyperthyroidism (Livadas et al. 2005; Witkowska-Sedek et al. 2017). Detection increases significantly with age, but may be diagnosed in the absence of symptoms. This has not been found to be concordant with karyotype (Witkowska-Sedeka et al. 2017).

Diabetes. The incidence of both type 1 and type 2 diabetes is increased among individuals with TS. This is thought to be due to multiple contributing factors including insulin resistance, decreased secretion of insulin, impaired glucose tolerance, and hyperinsulinemia (Gravholt et al. 2017). This may be modulated by growth hormone therapy (Gravholt et al. 2017). Monitoring for diabetes and prompt referral to an endocrinologist are essential for optimal treatment.

Body composition. Obesity is more common among girls and particularly among women with TS. Most women have increased BMI, total fat mass, and visceral fat mass coupled with decreased muscle mass (Gravholt et al. 2006). This is partly due to decreased exercise as well as capacity for exercise, However, there also appears to be an underlying basal abnormality of metabolism complicating this including anabolic effects of insulin-like growth factor 1 (IGF-1) as well as to testosterone, insulin and estradiol. Growth hormone therapy can promote muscle mass, improving calorie usage and body composition.

Bone metabolism. The lifetime risk for fractures is increased in both girls and women with TS. This occurs even in those with normal bone mineral density (Gravholt et al. 2003), but increases with decreased bone mineral density (Soucek et al. 2011). Bone mineral density can be improved with estrogen replacement (Hanton et al. 2003). While growth hormone supplementation improves bone size, it does not increase bone mineral density or decrease fracture risk (Nour et al. 2016). Dietary calcium and vitamin D can also have an impact, and levels should be monitored (Gravholt et al. 2017).

Evaluation

- Thyroid function tests (TSH and T4) should be obtained yearly; regular screening for anti-thyroid antibodies does not alter management.
- Hemoglobin A1c and fasting plasma glucose should be monitored annually beginning at age 10 (Gravholt et al. 2017).
- Those with at least one risk factor for cardiovascular disease should have a fasting lipid profile annually beginning at age 18 (Gravholt et al. 2017).
- Quality of diet and vitamin and mineral intake should be ascertained.
- Body mass index should be reviewed with each visit, with caution when used in younger individuals.
- Vitamin D levels should be monitored via monitoring a serum 25-hydroxyvitamin D beginning at 9–11 years of age and repeated every 2–3 years (Gravholt et al. 2017).
- Bone densitometry scans (DXA) should be used to monitor bone density in those on adult hormone replacement and when considering discontinuing this therapy (Gravholt et al. 2017).

Management

- Treatment of thyroid disease and diabetes is the same as that of the general population.

- Vitamin D and calcium intake should follow regional guidelines.
- Girls and women with TS should be counseled on a healthy lifestyle and estrogen replacement, including healthy nutrition and physical activity. They often struggle for many years to treat obesity and exercise more, and thus, should be supported as young as possible. Although it is best to promote a multifaceted healthy lifestyle, ideally in a family-based setting to minimize stigmatization, caloric restriction may be necessary.
- Estrogen replacement therapy should be initiated at 11–12 years of age, reaching an adult dose 2–3 years after initiation.
- Estrogen should be administered at low dose and optimally by the transdermal route.
- Progesterone should be added after two years of estrogen therapy or once breakthrough bleeding ensues.

Respiratory

Individuals with TS do not appear to have an increased risk for primary pulmonary disorders. However, when more broadly defined to include the upper airway, microstomia, micrognathia and a generally small oropharynx, these create risk factors for obstructive sleep apnea. Though reported once in an adult (Orliaguet et al. 2001), sleep apnea has been anecdotally observed by TS providers.

Evaluation

- Individuals with signs and symptoms of sleep apnea such as poor quality sleep, snoring, daytime fatigue, and nasal speech should be referred for evaluation and a sleep study.

Management

- Management is the same as that for the general population.

Gastrointestinal and Liver

The main areas of concern in this organ system involve intestinal disorders and liver dysfunction. Celiac disease, an immune-mediated disease of the small intestines triggered by the ingestion of gluten-containing grains (wheat, barley and rye), occurs in about 5% of individuals with TS. Based on biopsy data, it is two and five times more prevalent than in the general population for those under 5 and over 10 years, respectively (Frost et al. 2009; Marild et al. 2016, reviewed in Gravholt et al. [2017]). Celiac disease can cause bloating, abdominal pain, and malabsorption; however, its only manifestation may be growth failure. It is treated by excluding all gluten-containing foods from the diet.

Inflammatory bowel disease (IBD), occurs in about 3% of individuals with TS (Gravholt et al. 1998; Elsheikh et al. 2002), with Crohn's disease being at least as common as ulcerative colitis. While classical symptoms such as bloody diarrhea, abdominal pain and weight loss may be present, growth retardation may be the only manifestation and it may delay diagnosis of TS. Individuals with an isochromosome Xq karyotype appear to have greater risk. Among women with TS, the median age of onset of inflammatory bowel disease has been reported as 16 years. Osteoporosis is a common complication, resulting from many factors including the inflammatory process itself, poor nutrition, corticosteroid therapy, as well as calcium and vitamin D deficiencies.

Liver function test abnormalities are frequent in TS (20–80%), and tend to become more abnormal with age, rarely becoming normal. Typically, alkaline phosphatase, alanine/aspartase aminotransferase and g-glutamyl transferase (ALT, AST, GGT) are elevated while bilirubin and coagulation parameters are normal. There is a five-fold risk of cirrhosis compared to the normal population, especially in those with elevated liver function tests, but cirrhosis is difficult to determine without liver biopsy. In most cases, liver involvement is asymptomatic and discovered during systematic blood testing (Roulot et al. 2004). Liver biopsy may demonstrate marked architectural changes including cirrhosis, nodular regenerative hyperplasia and focal nodular hyperplasia. A vascular basis has been postulated, and portal hypertension and esophageal varices may develop as potential life-threatening complications. These individuals are viewed as having non-alcoholic liver disease with steatosis, steatohepatitis, and steatofibrosis, which are probably related to metabolic disorders such as increased adiposity. Estrogen replacement therapy does not appear to be a factor in its development.

In addition to liver vasculopathy, there can be gastrointestinal bleeding due to intestinal telangiectases on the serosal surface of the small bowel (Eroglu et al. 2002). Though usually intermittent and self-limited, brisk and life-threatening bleeding may also occur. Capsule endoscopy may be needed in addition to upper endoscopy and colonoscopy. Bleeding generally resolves with age, perhaps as the result of estrogen supplementation.

Evaluation

- Liver function tests should be obtained once a year beginning at age 10 years, which is younger than previously suggested (Gravholt et al. 2017).
- Screening for celiac disease using tissue transglutaminase IgA antibodies should begin at age 2–3 years and should be repeated every two years throughout adulthood. Human leucocyte antigen (HLA) typing may be used for specific reasons, e.g., to determine the necessity of long-term screening, as a first-line screen in high-risk groups like TS, and in those in whom the diagnosis is indeterminate

- Consideration should be given to the presence of inflammatory bowel disease for individuals with non-specific complaints or poor growth in addition to those with abdominal complaints.
- Although vascular anomalies of the intestines and liver are uncommon, they should be considered in the presence of rectal bleeding and liver dysfunction. Imaging with Doppler ultrasound and magnetic resonance imaging may be indicated.

Management

- Management is the same as that for the general population.
- Weight loss is advised for steatosis.

Ophthalmologic

Approximately 40% of girls and women with TS will have some degree of myopia or hyperopia (Denniston and Butler 2004), whereas about 35% will have more than one visual deficit (Wikiera et al. 2015). Common ocular features of TS also include ptosis, epicanthal folds, and hypertelorism. It is important that the pediatrician also be aware of and identify other visual complications such as strabismus and amblyopia early. Because of the absence of the second sex chromosome, there is a similar incidence (8%) of red green color blindness as in the male population. Early awareness and prompt referral to ophthalmology are essential to avoiding visual deficits.

Evaluation

- Formal ophthalmology evaluation is recommended at 12–18 months of age or at the time of diagnosis if it occurs later.

Management

- Management is the same as that for the general population.

Audiologic

Hearing loss and ear anomalies are a common challenge for girls and women with TS, with the onset of hearing loss occurring from infancy through late adulthood. It is universally present in those older than 50 (King et al. 2007). Surveillance is important throughout the lifespan as this can affect development as well as social interactions. Hearing loss occurs due to multiple factors including structural anomalies, recurrent otitis media, and sensorineural hearing loss. About 34% of individuals with TS will have structural ear anomalies, which can include protrusion of the ears, abnormal pinnae, low set ears, and narrow external auditory canals (Dhooge et al. 2005; Verver et al. 2010). Hearing problems occur more frequently in those with an absent Xp (Verver et al. 2010). The most common hearing issue is sensorineural hearing loss (King et al. 2007). Frequent otitis media occurs particularly in infants and children and can contribute to the conductive hearing loss found in TS. It is thought that an abnormal immune response can also contribute to chronic otitis media, and it should be treated aggressively to avoid conductive hearing loss (Gravholt et al. 2017).

Evaluation

- Formal audiology evaluation is recommended every five years.

Management

- Otitis media should be treated aggressively using antibiotics, and, when necessary, myringotomy and pressure equalizing tubes to preserve normal hearing and optimize speech development and social skills.
- Encourage the use of sound amplification (hearing aids) and classroom accommodation for those with hearing loss, since hearing loss can contribute to language-based and social disabilities. Improvements in the design of hearing aids has increased acceptance.

Orthodontic

Dental anomalies and orthodontic problems are common in individuals with TS. This is at least in part due to differences in their orofacial development (Dumancic et al. 2010). There are also characteristic differences in crown and root morphology (Kusiak et al. 2005). This combination of retrognathia and narrow, high-arched palate along with malocclusion of the teeth can provide unique challenges in orthodontic treatment of girls with TS. Treatment should be tailored in accordance with skeletal age and growth (Russell et al. 2001). Dental and orthodontic care should follow the standard of care guidelines for the pediatric population with special consideration of the developmental aspects of TS.

Evaluation

- It is recommended that all girls with TS see a pediatric dental specialist by age 2 years, and an orthodontist no later than age 7 years.

Management

- The timing of any orthodontic treatment should take into consideration growth promoting therapies that may alter tooth and jaw alignment.
- Unnecessary tooth movement should be minimized to avoid root resorption and loss of teeth.

Integumentary

The most common skin features among girls and women with TS include an increased number of melanocytic and sebaceous nevi, hypertrophic scars (keloids), vitiligo, alopecia and skin cancers (reviewed by Lowenstein et al. [2004]). Case reports also describe pilomatricomas, halo nevi, and psoriasis. Dysplastic or hyperconvex nails and long eyelashes are common (personal observation). The assessment and care of these skin conditions is the same as for the general population. There has been conflicting evidence regarding whether the increase in melanocytic nevi confers an increased risk of skin cancer. However, neither the number of nevi nor risk for skin cancer is increased by the use of growth hormone (Gravholt et al. 2017). It is unknown whether skin problems are more common in 45,X or with mosaicism.

Evaluation

- Individuals with TS should be taught how to evaluate their nevi for possible malignant changes.

Management

- Management of skin nevi is the same as that for the general population.
- Any attempt to treat existing scars should be viewed with extreme care, if at all (Lowenstein et al. 2004).

Oncologic

The surveillance of tumors in girls and women with TS typically focuses on the well-known risk of gonadoblastoma associated with the presence of a Y chromosome fragment (Page 1994; Gravholt et al. 2000; Gravholt et al. 2008; reviewed in Gravholt et al. [2017]). Population-scale studies (Schoemaker et al. 2008, Ji et al. 2016) of people with TS have identified an increased risk of central nervous system tumors including meningioma and schwannoma compared to the general population. Neuroblastoma is rare, but may be overrepresented (Blatt et al. 1997). The increased risk of certain cancer types observed in TS is similar to the trend of an increased risk of those types of cancer in males compared to females (Edgren et al. 2012). This increased risk holds not only for tumors of the nervous system, but for other tumor types such as non-Hodgkin lymphoma, bladder cancer, and eye cancer, among others (Schoemaker et al. 2008; Edgren et al. 2012; Ji et al. 2016; Dunford et al. 2017). There is a decreased risk of breast cancer.

Evaluation

- Karyotypes should be reviewed to ensure that marker chromosomes have been identified and that no Y-chromosome material is present.

Management

- If Y-chromosome material is present, the individual should be referred to a surgeon for prophylactic removal of the gonads. Because of the psychosocial impact on the individual and their family, counseling should be extended past the operation and throughout life.
- There is no routine screening for other tumors, but a high index of suspicion should be maintained for tumors of the nervous system.

Care Coordination

Care can best be provided by a multidisciplinary team with knowledge of TS, which can assist the primary care provider. This is particularly important as adolescents with TS transition to adult care. Pediatric endocrinologists are instrumental in coordinating care from infancy to adolescence and through the initiation of puberty. It is important that the TS clinic team leader use transition tools such as the TRAQ (transition readiness assessment questionnaire), which are easily accessed on several websites as specific transition toolkits. The process should include a transition readiness assessment, transfer summary, and self-care assessment (Gravholt et al. 2017). Once complete, the framework can be continued through adulthood (Table 60.3).

Evaluation

- Regular screening for common problems associated with TS should decrease morbidity and mortality (Stochholm et al. 2006).
- The updated clinical practice guidelines (Gravholt et al. 2017) offer a protocol for screening for medical complications throughout the lifespan (Table 60.3).
- Particular attention should be paid to cardiovascular, psychosocial, and reproductive concerns faced by adult women with TS (Freriks et al. 2011). Systematic use of guidelines is beneficial.
- Individuals with TS and their families can participate in the ongoing Turner Syndrome Research Registry (TSRR) which will provide the long-term outcomes essential to management (https://www.turnersyndrome.org/ts-registry-love). For example, this database will likely provide the valuable data to enhance counseling about pregnancy (Lin et al. 2016).

ACKNOWLEDGEMENTS

The authors express their sincere gratitude to the community of people and families who live with TS and who have been our partners. We appreciate the support of girls and women

TABLE 60.3 Recommendations for screening girls and women with Turner syndrome for non-cardiac features (Gravholt et al. 2017). Individuals with a problem in one or more areas may be followed by more than one specialist and be evaluated more frequently.

	At diagnosis	After diagnosis(childhood)	After diagnosis(adults)
Weight/BMI	Yes	Every visit	Every visit
Blood pressure	Yes	Every visit	Every visit
Thyroid function (TSH and (free)T4)	Yes	Annually	Annually
Lipids			Annually if at least one cardiovascular risk factor° or regional recommendation
Aminotransferase, GGT and alkaline phosphatase		Annually after 10 years of age	Annually
HbA1c with or without fasting plasma glucose		Annually after 10 years of age	Annually
25-Hydroxyvitamin D		Every 2–3 years after 9–11 years of age	Every 3–5 years
Celiac screen		Starting at 2 years; thereafter every two years	With suggestive symptoms
Renal ultrasound	Yes		
Audiometric evaluation	Yes*	Every 3 years	Every 5 years
Ophthalmological examination	Yes#		
Dental evaluation	Yes, if no previous care has been established		
Clinical investigation for congenital hip dydplasia	Yes, in newborns		
Skin examination	At diagnosis	Annually	Annually
Bone mineral density			Every 5 years and when discontinuing estrogen
Skeletal assessment		5–6 years and 12–14 years (see 6.1.10.)	

The recommendations are for screening only A clinical suspicion of active disease should always lead to relevant investigation For details, please see text: *When 9–12 months old; #When 12–18 months old; °cardiovascular risk factors: hypertension, overweight, tobacco, diabetes, and physical inactivity.

whose photos (obtained with consent) are shown in Figure 60.1. We extend special thanks to Dr. Marsha Davenport, author of the 2010 chapter of this book, to the members and leaders of the Turner Syndrome Professional Advisory Board and Turner Syndrome Global Alliance, and to many clinical and research colleagues, especially Dr. Lynne Levitsky and Dr. Frances Hayes, Co-Directors of the MGH Turner Syndrome Clinic. Finally, to the many colleagues who participated in revising the clinical guidelines, in particular to Dr. Claus Gravholt and Dr. Michael Silberbach for final insights related to this publication (Gravholt et al. 2017).

RESOURCES

Turner Syndrome Society of the United States

11250 West Rd., Suite G, Houston, TX 77065, USA
Phone: 800-365-9944
Fax: 832-912-6446
Website: http://www.turnersyndrome.org/

Turner Syndrome Global Alliance

10708 W 129th St.
Overland Park, KS 66213, USA
Phone: 913-220-8177
Website: http://tsgalliance.org/

Turner Syndrome Foundation, Inc.

PO Box 726, Holmdel, NJ 07733, USA
1 Bethany Rd, Bldg. 1, Ste. 5, Hazlet, NJ 07730USA
Phone: 732-847-3385 or 1-800-594-4585
Fax: 800-594-3862
Website: www.TSFUSA.org

WEB RESOURCES

National Library of Medicine Genetics Home Reference

https://ghr.nlm.nih.gov/condition/turner-syndrome

Turner Syndrome Society of the United States (TSSUS): clinical practice guidelines for the care of girls and women with Turner syndrome (Gravholt et al. 2017), and Brief synopsis for Turner syndrome girls and women and for their parents/caregivers/families. http://www.turnersyndrome.org/

UpToDate. Clinical manifestations and diagnosis of Turner syndrome (Philippe Backeljauw, MD).

http://www.uptodate.com/contents/clinical-manifestations-and-diagnosis-of-turner-syndrome

AAP Got Transition Website

www.gottransition.org

Endocrine Society: Transitions of Care (Turner syndrome)

http://www.endocrinetransitions.org/turner-syndrome

REFERENCES

Atton G, Gordon K, Brice G, Keeley V, Riches K, Ostergaard P, Mortimer P, Mansour S (2015) The lymphatic phenotype in Turner syndrome: an evaluation of nineteen patients and literature review. *Eur J Hum Genet* 23(12):1634–1639.

Bekker MN, van den Akker NM, de Mooij YM, Bartelings MM, Van Vugt JM, Gittenberger-De Groot AC (2008) Jugular lymphatic maldevelopment in Turner syndrome and trisomy 21: different anomalies leading to nuchal edema. *Reprod Sci* 15:295–304.

Bernard V, Donadille B, Zenaty D, Courtillot C, Salenave S, Brac de la Perrière A, Albarel F, Fèvre A, Kerlan V, Brue T, Delemer B, Borson-Chazot F, Carel JC, Chanson P, Léger J, Touraine P, Christin-Maitre S; CMERC Center for Rare Disease (2016) Spontaneous fertility and pregnancy outcomes amongst 480 women with Turner syndrome. *Hum Reprod* 31:782–788.

Bilge I, Kayserili H, Emre S, Nayir A, Sirin A, Tukel T, Bas F, Kilic G, Basaran S, Gunoz H, Apak M (2000) Frequency of renal malformations in Turner syndrome: analysis of 82 Turkish children. *Pediatr Nephrol* 14:1111–1114.

Blatt J, Olshan AE, Lee PA, Ross JL (1997) Neuroblastoma and related tumors in Turner's syndrome. *J Pediatr* 131:666–70.

Boman UW, Bryman I, Hailing K, Möller A (2001) Women with Turner syndrome: psychological well-being, self-rated health and social life, *J Psychosom Obstet Gynecol* 22(2):113–122

Bondy C A, The Turner Syndrome Consensus Study Group (2007) Care of Girls and Women with Turner Syndrome: A Guideline of the Turner Syndrome Study Group. *J Clin Endocrinol Metab* 92:10–25.

Bondy C A, Van P L, Bakalov V K, and Ho V B (2006) Growth hormone treatment and aortic dimensions in Turner syndrome. *J Clin Endocrinol Metab* 91:1785–1788.

Burnett A C, Reutens D C, Wood A G (2010) Social cognition in Turner's Syndrome. *J Clin Neurosci* 17:283–286.

Carlson M, Silberbach M (2007) Dissection of the aorta in Turner syndrome: two cases and review of 85 cases in the literature. *J Med Genet* 44:745–749

Davenport M L (2006) Evidence for early initiation of growth hormone and transdermal estradiol therapies in girls with Turner syndrome. *Growth Horm IGF Res* 16(Suppl):91–7.

Davenport M L, Crowe B J, Travers S H, Rubin K, Ross J L, Fechner P Y, Gunther D F, Liu C, Geffner M E, Thrailkill K, Huseman C, Zagar A J, Quigley C A (2007) Growth hormone treatment of early growth failure in toddlers with Turner syndrome: a randomized, controlled, multicenter trial. *J Clin Endocrinol Metab* 92:3406–3416.

Davenport M L (2010) Turner syndrome. In: *Management of Genetic Syndromes*, 3rd ed, by Cassidy SB and Allanson JE, eds. John Wiley & Sons, Inc.

Denniston A K, Butler L (2004) Ophthalmic features of Turner's syndrome. *Eye* 18:680–684.

Dhooge I J, De Vel E, Verhoye C, Lemmerling M, Vinck B (2005) Otologic disease in Turner syndrome. *Otol Neurotol* 26:145–150.

Dumancic J, Kalc Z, Varga M L, Lauc T, Dumic M, Milosevic S A, and Brkic H (2010) Characteristics of the craniofacial complex in Turner syndrome. *Arch Oral Biol* 55:81–88.

Edgren G, Liang L, Adami HO, Chang ET (2012) Enigmatic sex disparities in cancer incidence. *Eur J Epidemiol* 27:187–196.

El-Mansoury M, Berrenäs M L, Bryman I, Hanson C, Larsson C, Wilhelmsen L, Landin-Wilhelmsen K (2007) Chromosomal mosaicism mitigates stigmata and cardiovascular risk factors in Turner syndrome. *Clin Endocrinol* 66(5): 744–751.

Elsheikh M, Dunger D B, Conway G S, Wass J A (2002) Turner's syndrome in adulthood. *Endocr Rev* 23:120–140.

Eroglu Y, Emerick K M, Chou P M, Reynolds M (2002) Gastrointestinal bleeding in Turner's syndrome: a case report and literature review. *J Pediatr Gastroenterol Nutr* 35:84–87.

Freriks K, Timmermans J, Beerendonk CC, Verhaak CM, Netea-Maier RT, Otten BJ, Braat DD, Smeets DF, Kunst DH, Hermus AR, Timmers HJ (2011) Standardized multidisciplinary evaluation yields significant previously undiagnosed morbidity in adult women with Turner syndrome. *J Clin Endocrinol Metab* 96:E1517–26.

Frost AR, Band MM, Conway GS (2009) Serological screening for coeliac disease in adults with Turner's syndrome: prevalence and clinical significance of endomysium antibody positivity. *Eur J Endocrinol* 2009 160 675–679.

Gravholt C H, Fedder J, Naeraa RW, and Müller J (2000) Occurrence of gonadoblastoma in females with Turner syndrome and Y chromosome material: a population study. *J Clin Endocrinol Metab* 85:3199–3202.

Gravholt C H, Hjerrild B E, Mosekilde L, Hansen T K, Rasmussen L M, Frystyk J, Flyvbjerg A, Christiansen J S (2006) Body composition is distinctly altered in Turner syndrome: relations to glucose metabolism, circulating adipokines, and endothelial adhesion molecules. *Eur J Endocrinol* 155:583–592.

Gravholt C H, Juul S, Naeraa R W, Hansen J (1998) Morbidity in Turner syndrome. *J Clin Epidemiol* 51:147–158.

Gravholt C H (2008) Epidemiology of Turner syndrome. *Lancet Oncol* 9:193–195.

Gravholt C H, Landin-Wilhelmsen K, Stochholm K, Hjerrild BE, Ledet T, Djurhuus CB, Sylven L, Baandrup U, Kristensen, B Ø, Christiansen J S (2006) Clinical and epidemiological description of aortic dissection in Turner's syndrome. *Cardiol Young* 16:430–436.

Gravholt C H, Andersen N H, Conway G S, Dekkers O M, Geffner M E, Klein K O, TurLin A E, Mauras N, Quigley C A, Rubin K, Sandberg D E, Sas T C J, Silberbach M, Söderström-Anttila V, Stochholm K, van Alfen-van derVelden J A, Woelfle J, Backeljauw P F, on behalf of the International Turner Syndrome Consensus Group (2017) Clinical practice guidelines for the care of girls and women with Turner Syndrome. *Eur J Endocrinol* 177:3: G1–G70.

Gravholt C H, Vestergaard P, Hermann A P, Mosekilde L, Brixen K, Christiansen J S (2003) Increased fracture rates in Turner's syndrome: a nationwide questionnaire survey. *Clin Endocrinol* 59:89–96.

Green T, Bade Shrestha S, Chromik L C, Rutledge K, Pennington BF, Hong D S, Reiss A L (2015) Elucidating X chromosome influences on Attention Deficit Hyperactivity Disorder and executive function. *J Psychiatr Res* 68:217–225.

Hagman A, Kallen K, Barrenas ML, Landin-Wilhelmsen K, Hanson C, Bryman I, Wennerholm UB. Obstetric outcomes in women with Turner karyotype. *J Clin Endocrinol Metab* 2011 96 3475–3482

Hanton L, Axelrod L, Bakalov V, and Bondy C A (2003) The importance of estrogen replacement in young women with Turner syndrome. *J Women's Health* 12:971–977.

Held KR, Kerber S, Kaminsky E, Singh S, Goetz P, Seemanova E, Goedde HW (1992) Mosaicism in 45,X Turner syndrome: does survival in early pregnancy depend on the presence of two sex chromosomes? *Hum Genet* 88:288–294.

Ho VB, Bakalov VK, Cooley M, Van PL, Hood MN, Burklow TR, Bondy CA (2004) Major vascular anomalies in turner syndrome: prevalence and magnetic resonance angiographic features. *Circulation* 110:1694–1700.

Hong D, Scaletta Kent J, Kesler S (2009) Cognitive profile of Turner syndrome. *Dev Disabil Res Rev* 15: 270–278.

Hong DS and Reiss AL (2014) Cognitive and neurological aspects of sex chromosome aneuploidies. *Lancet Neurol* 13:306–318.

Hook EB, Warburton D (1983) The distribution of chromosomal genotypes associated with Turner's syndrome: livebirth prevalence rates and evidence for diminished fetal mortality and severity in genotypes associated with structural X abnormalities or mosaicism. *Hum Genet.* 64: 24–27.

Hook EB, Warburton D (2014) Turner syndrome revisited: review of new data that supports the hypothesis that all viable 45,X cases are cryptic mosaics with a rescue cell line, implying an origin my mitotic loss. *Hum Genet* 133: 417–424.

Inozemtseva O, Matute E, Zarabozo D, Ramirez-Duenas L (2002) Syntactic processing in Turner's syndrome. *J Child Neurol* 17: 668–672.

Ji J, Zoller B, Sundquist J, Sundquist K. 2016. Risk of solid tumors and hematological malignancy in persons with Turner and Klinefelter syndromes: A national cohort study. *Int J Cancer* 139:754–758.

Jørgensen K, Rostgaard K, Bache I, Biggar R, Nielsen N, Toomerup N, Frisch M (2010) Autoimmune disorders in women with Turner's syndrome. *Arthritis Rheum* 62:658–666.

King KA, Makashima T, Zalewski C K, Bakalov V K, Griffith A J, Bondy C A, and Brewer C C (2007) Analysis of auditory phenotype and karyotype in 200 females with Turner syndrome. *Ear Hear* 28:831–841.

Kusiak A, Sadlak-Nowicka J, Limon J, and Kochanska B (2005) Root morphology of mandibular premolars in 40 patients with Turner syndrome. *Int Endod J* 38: 822–826.

Kuntsi J, Skuse D, Elgar K, Morris E, Turner C (2000) Ring-X chromosomes: their cognitive and behavioural phenotype. *Ann Hum Genet* 64: 295–305.

Lacro RV, Jones KL, Benirschke K (1988) Coarctation of the aorta in Turner syndrome: a pathologic study of fetuses with nuchal cystic hygromas, hydrops fetalis, and female genitalia. *Pediatrics* 81:445–451.

Lawrence K, Kuntsi J, Coleman M, Campbell R, Skuse D (2003) Face and emotion recognition deficits in Turner syndrome: a possible role for X-linked genes in amygdala development. *Neuropsychology* 17:39–49.

Lin AE, Karnis M, Calderwood L, Crenshaw M, Souter I, Bhatt A, Silberbach M, Reindollar R (2016) Proposal for a national registry to monitor women with Turner syndrome seeking assisted reproductive technology. *Fertil Steril* 105:1446–8.

Lin AE, McNamara EA, Steeves MA, Hayes FJ, Levitsky LL, Crenshaw ML (2017) Karyotype–phenotype analysis in Turner syndrome, focus on mild mosaicism: challenges from prenatal diagnosis and infertility evaluations. In, Keppler-Noreuil K M, Martinez-Agosto J A, Hudgins L, Carey J C (2017) 37th Annual David W. Smith Workshop on Malformations and Morphogenesis: Abstracts of the 2016 Annual Meeting. *Am J Med Genet A* 173A:2007–2073.

Lippe B, Geffner M E, Dietrich R J, Boechat M I, Kangarloo H. (1988) Renal malformations in patients with Turner syndrome: imaging in 141 patients. *Pediatrics* 82:852–856.

Livadas S, Xekouki P, Fouka F, Kanaka-Gantenbein C, Kaloumenou I, Mavrou A, Constantinidou N, cou-Voutetakis C (2005) Prevalence of thyroid dysfunction in Turner's syndrome: a long-term follow-up study and brief literature review. *Thyroid* 15:1061–1066.

Loscalzo M L, Van P L, Ho V B, Bakalov VK, Rosing DR, Malone CA, Dietz HC & Bondy CA (2005) Association between fetal lymphedema and congenital cardiovascular defects in Turner syndrome. *Pediatrics* 115 732–735.

Lowenstein E J, Kim K H, Glick S A (2004) Turner's syndrome in dermatology. *J Am Acad Dermatol* 50: 767–776.

Lyon A J, Preece M A, Grant D B (1985) Growth curve for girls with Turner syndrome. *Arch Dis Child* 60:932–935.

Marild K, Stordal K, Hagman A. Ludvigsson JF (2016) Turner syndrome and celiac disease: a case-control study. *Pediatrics* 137:e20152232.

Mazzola F, Seigal A, MacAskill A, Corden B, Lawrence K, and Skuse D (2006) Eye tracking and fear recognition deficits in Turner syndrome. *Soc Neurosci* 1:259–269.

Nour MA, Burt LA, Perry RJ, Stephure DK, Hanley DA, and Boyd SK (2016) Impact of growth hormone on adult bone quality in Turner syndrome: a HR-pQCT study. *Calcif Tissue Int* 98:45–59.

Orliaguet O, Pepin JL, Bettega G, Ferretti G, Mignotte HN, and Levy P (2001) Sleep apnoea and Turner's syndrome. *Eur Resp J* 17:153–155.

Page D C (1994) Y chromosome sequences in Turner's syndrome and risk of gonadoblastoma or virilisation. *Lancet* 343:240.

Pasquino A M, Passeri F, Pucarelli I, Segni M, Municchi G (1997) Spontaneous pubertal development in Turner's syndrome. *Italian Study Group for Turner's Syndrome. J Clin Endocrinol Metab* 82:1810–1813.

Ross J, Roeltgen D, Feuillan P, Kushner H, Cutler GB Jr (1998) Effects of estrogen on nonverbal processing speed and motor function in girls with Turner's syndrome. *J Clin Endocrinol Metab* 83:3198–3204.

Ross J Roeltgen D, Feuillan P, Kushner H, Cutler GB Jr (2000) Use of estrogen in young girls with Turner syndrome: effects on memory. *Neurology* 54:164–170.

Roulot D, Degott C, Chazouilleres O, Oberti F, Cales P, Carbonell N, Benferhat S, Bresson-Hadni S, Valla D (2004) Vascular involvement of the liver in Turner's syndrome. *Hepatology* 39:239–247.

Rovet JF (1993) The psychoeducational characteristics of children with Turner syndrome. *J Learn Disabil* 26:333–341.

Russell KA (2001) Orthodontic treatment for patients with Turner syndrome. *Am J Orthod Dentofacial Orthop* 120:314–322.

Schoemaker MJ, Swerdlow AJ, Higgins CD, Wright AF, Jacobs PA (2008) Cancer incidence in women with Turner syndrome in Great Britain: a national cohort study. *Lancet Oncol* 9:239–246.

Silberbach M, Roos-Hesselink JW, Andersen NH, Braverman AC, Brown N, Collins RT, De Backer J, Eagle KA, Hiratzka LF, Johnson WH Jr, Kadian-Dodov D, Lopez L, Mortensen KH, Prakash SK, Ratchford EV, Saidi A, van Hagen I, MD, Young LT, on behalf of the American Heart Association Council on Cardiovascular Disease in the Young; Council on Genomic and Precision Medicine; and Council on Peripheral Vascular Disease (2018) Cardiovascular health in Turner syndrome: a scientific statement from the American Heart Association. *Circ Genom Precis Med* 201811:e000048. Commentary by Bamba V at https://professional.heart.org/professional/ScienceNews/UCM_502579_Cardiovascular-Health-in-Turner-Syndrome-No-Small-Task.jsp

Stochholm K, Juul S, Juel K, Naeraa RW, Gravholt CH (2006) Prevalence, incidence, diagnostic delay, and mortality in Turner syndrome. *J Clin Endocrinol Metab* 91:3897–3902.

Sybert VP (1998) Cardiovascular malformations and complications in Turner Syndrome. *Pediatrics* 101:e11.

Sybert VP (2004) Turner syndrome. In, *Management of Genetic Syndromes*. 2nd ed. Cassidy SB, Allanson JE, eds. 2004 John Wiley & Sons, Inc.

Sybert VP, McCauley E (2004) Turner's syndrome. *N Engl J Med* 351:1227–1238.

Temple CM (2002) Oral fluency and narrative production in children with Turner's syndrome. *Neuropsychologia* 40:1419–1427.

Temple CM, Shephard EE (2012) Exceptional lexical skills but executive language deficits in school starters and young adults with Turners syndrome: implications for X chromosome effects in brain function. *Brain Lang* 120:345–359.

van den Berg J, Bannink EMN, Wielopolski PA, Hop WCJ, van Osch-Gevers L, Pattynama PM T, de Muinck Keizer-Schrama S, Helbing WA (2008) Cardiac status after childhood growth hormone treatment of Turner syndrome. *J Clin Endocrinol Metab* 93:2553–2558.

Verver EJ, Freriks K, Thomeer HG, Huygen PL, Pennings RJ, Alfen-van der Velden AA, Timmers HJ, Otten BJ, Cremers CW, and Kunst HP (2011) Ear and hearing problems in relation to karyotype in children with Turner syndrome. *Hearing Res* 275:81–88.

von Kaisenberg C S, Wilting J, Dork T, Nicolaides K N, Meinhold-Heerlein I, Hillemanns P, Brand-Saberi B (2010) Lymphatic capillary hypoplasia in the skin of fetuses with increased nuchal translucency and Turner's syndrome: comparison with trisomies and controls. *Mol Hum Reprod* 16:778–789.

Wikiera B, Mulak M, Koltowska-Haggstrom M, Noczynska A (2015) The presence of eye defects in patients with Turner syndrome is irrespective of their karyotype. *Clin Endocrinol* 83:842–848.

Witkowska-Sedek E, Borowiec A, Kucharska A, Chacewicz K, Ruminska M, Demkow U, Pyrzak B (2017) Thyroid Autoimmunity in Girls with Turner Syndrome. *Adv Exp Med Biol* 1022:71–76

Wolff D J, Van Dyke DL, Powell C M. Working Group of the ACMG Laboratory Quality Assurance Committee (2010) Laboratory guideline for Turner syndrome. *Genet Med* 12:52–55. Corrigendum: Correction to GIM.

61

VATER/VACTERL ASSOCIATION

BENJAMIN D. SOLOMON
National Human Genome Research Institute, Bethesda, Maryland, USA

BRYAN D. HALL
Division of Clinical/Biochemical Genetics and Dysmorphology, Department of Pediatrics, University of Kentucky and Kentucky Clinic, Lexington, Kentucky, USA; and Greenwood Genetic Center, Greenwood, South Carolina, USA

INTRODUCTION

Incidence

VATER and VACTERL are acronyms for a sporadic group of findings that co-occur more often than by chance alone (Quan and Smith 1973; Solomon 2011). The major anomalies associated with this group of findings include: *v*ertebral defects, *a*nal atresia, *c*ardiac defects, *t*racheo-*e*sophageal fistula, *r*enal anomalies, and *l*imb abnormalities. Table 61.1 illustrates the anomaly each letter represents in the two acronyms. The disorder might be best thought of as a clinical diagnostic description for a relatively common, non-random grouping of multiple malformations (Figure 61.1) (an association), and with this understanding, the incidence may be estimated to be approximately 1 in 10,000 to 40,000 infants, with the reported occurrence depending mainly on the clinical criteria used (Khoury et al. 1983; Czeizel and Ludanyi 1985; Botto et al. 1997; Solomon 2011).

Diagnostic Criteria

VATER/VACTERL association represents a core group of six anomalies (see Table 61.1 and Figure 61.1). It is important to note that the two acronyms are used to describe the same entity, and for the sake of brevity, this chapter will refer to "VACTERL" only. Relatively few affected individuals have all features, the average number of features in a patient is three to four (Botto et al. 1997; Solomon 2011; Solomon et al. 2012). No minimum criteria have been agreed on regarding which of these features or what combination of these constitutes a secure diagnosis (Kallen et al. 2001; Solomon 2011). However, most clinicians and researchers require the presence of at least three component features and the abence of an alternative diagnosis (Jenetzky et al. 2011; Solomon et al. 2012). An alternate suggestion has been to require either tracheo-esophageal fistula (TEF) with or without esophageal atresia or anorectal malformations as a required "core feature" (Solomon et al. 2012).

ETIOLOGY, PATHOGENESIS, AND GENETICS

No consistent or recurring etiology has been established for the majority of individuals with VACTERL association, though some causative genes have been identified. A small number of individuals with mutations in mitochondrial genes are described as affected with VACTERL-type anomalies (Khoury et al. 1983; Solomon 2011; Siebel and Solomon 2013). Most cases occur as a sporadic event with a low recurrence risk (Rittler et al. 1997; Bartels et al. 2012a), though there is mixed evidence of the clustering of component features within families (Solomon et al. 2010; Bartels et al. 2012a; Bartels et al. 2012b). Overall, VACTERL association is generally felt to be a heterogeneous association with various yet-to-be established etiologies working

---

Cassidy and Allanson's Management of Genetic Syndromes, Fourth Edition.
Edited by John C. Carey, Agatino Battaglia, David Viskochil, and Suzanne B. Cassidy.
© 2021 John Wiley & Sons, Inc. Published 2021 by John Wiley & Sons, Inc.

TABLE 61.1 Core features of VATER/VACTERL association

VATER association	VACTERL association
V = vertebral defects	V = vertebral defects
A = anal atresia	A = anal atresia
TE = tracheo-esophageal fistula with esophageal atresia	C = cardiac anomalies[a]
	TE = tracheo-esophageal fistula with or without esophageal atresia
R = renal/radial defects	R = renal
	L = limb (radial)

[a] Quan and Smith (1973) noted cardiac anomalies but did not include them in the acronym.

through a common embryological pathway (Botto et al. 1997). Because the primary defects of the VACTERL association are heterogeneous and nonrandom, the disorder is considered a polytopic developmental field defect by Martinez-Frias and Frias (1999).

Aside from rare known causes such as pathogenic variants (mutations) in *ZIC3* (Chung et al. 2011; Reutter et al. 2016), hypothesis-free analysis approaches (e.g., genome-wide microarray, exome, and genome sequencing) have recently identified causes in a minority of cases. For example, mutations in genes encoding enzymes in the kynurenine pathway have been identified as causative in some individuals (Shi et al. 2017). Some may also have causative copy-number variants, and microarray is a helpful initial test (Brosens et al. 2013; Solomon et al. 2014). Finally, environmental factors (maternal diabetes mellitus) can be contributory (Stevenson and Hunter 2013).

In addition to the defining features, other anomalies and abnormalities may also affect persons with VACTERL association. Some are common, such as single umbilical artery (70%) or genitourinary anomalies (21%), and others are less common, such as cleft lip, polydactyly, small intestinal atresia, and auricular aberrations (Botto et al. 1997; Solomon et al. 2011; Solomon 2011). Rare instances of caudal dysgenesis and lower extremity defects have also been reported (Temtamy and Miller 1974; Castori et al. 2008). The presence of anomalies that are less commonly observed in VACTERL association should prompt a careful assessment of the differential diagnosis and should help inform the diagnostic testing process (Solomon et al. 2012).

Diagnostic Testing

There are no specific inclusive tests for confirmation of the diagnosis of VACTERL association. Consideration of the possibility of VACTERL association is usually raised when a neonate presents with imperforate anus, tracheo-esophageal fistula/esophageal atresia, and/or radial/thumb hypoplasia, because these features are externally obvious and/or symptomatic within a short time after birth.

Once the possibility of VACTERL association exists, certain diagnostic tests for identification of cryptic associated anomalies should be automatically performed, especially as the identification of these findings can be important for immediate and longer-term clinical management (Solomon et al. 2014). These tests include (depending on which anomalies have already been identified, see below for further details) full spinal X-rays and spinal ultrasound (the latter to exclude a tethered cord), chest and abdominal X-ray with nasogastric or orogastric tube in place, abdominal ultrasound (including a careful view of the kidneys), complete

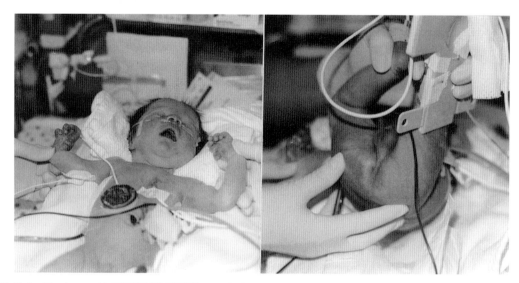

FIGURE 61.1 Newborn with VATER/VACTERL association showing bilateral radial/thumb hypoplasia and imperforate anus.

blood count, basic metabolic panel and echocardiogram with electrocardiogram. Other testing, such as an "invertogram" (prone cross-table lateral pelvic X-ray) in the case of anorectal malformations or voiding cystourethrogram (VCUG) in the case of renal anomalies may be indicated to better define the anomalies and therefore management (Solomon et al. 2014).

This empiric approach is useful not only in further supporting or excluding the diagnosis of VACTERL association, but also to potentially identify additional anomalous organ systems that will allow for earlier therapeutic intervention. For example, a cranial ultrasound in infancy may be helpful due to the association of VACTERL with hydrocephalus (Alter and Rosenberg 2013; Solomon et al. 2014). Standardized interventions have been shown to aid in the recognition of medically significant findings (England et al. 2017).

In terms of etiologic testing, VACTERL association currently still represents a diagnosis of exclusion. There is no facial gestalt to aid pattern recognition, and its core features are commonly found in many other conditions and syndromes (Narchi 1999; Killoran et al. 2000; Solomon et al. 2012; Reutter and Ludwig 2013). These challenges mandate the need to exclude other similar clinical entities by conducting appropriate genetic testing based on clinical findings. In the majority of people with VACTERL association (i.e. those in whom an alternate diagnosis is not suspected), no etiologic testing algorithm has been studied in a sizable cohort, but some current recommendations include chromosomal microarray and chromosome breakage studies (especially when hydrocephalus and/or radial anomalies are present) to exclude Fanconi anemia. Fanconi anemia is important to consider due to potential hematologic and oncologic sequelae (Brosens et al. 2013; Solomon et al. 2014; Alter and Giri 2016). Given the broad differential diagnosis, exome or genome sequencing is a reasonable consideration due to the utility of these methods in identifying the causes of multi-malformation disorders (Retterer et al. 2016).

Differential Diagnosis

A large number of conditions can include anomalies observed in VACTERL association, including dozens that can involve multiple VACTERL-type anomalies. Many of these conditions include additional features not typically observed in individuals with VACTERL association, but these additional features may be subtle, especially early in life (Solomon et al. 2012). Examples of disorders that may be considered in the differential diagnosis include CHARGE syndrome (see Chapter 11), Fanconi anemia, Feingold syndrome, Townes–Brock syndrome, and conditions due to certain mitochondrial mutations, to name a few (Solomon et al. 2012; Siebel and Solomon 2013).

In addition to specific monogenic conditions with known molecular causes, VACTERL association type anomalies may result from copy number variants affecting multiple genes, though the precise developmental biology is generally not well understood in any given case (Reutter and Ludwig 2013; Rosens et al. 2013).

MANIFESTATIONS AND MANAGEMENT

VACTERL association can be complex from the clinical management standpoint. Medical care is optimized by early identification of all anomalies. All seven of its core features as well as other anomalies may necessitate surgical intervention. While surgical management is typically focused on the perinatal and early infantile period, interventions related to these and other anomalies may be required throughout life (Raam et al. 2011).

Table 61.2 lists the physician and other clinical specialists most commonly required to adequately manage individuals affected with VACTERL association. Over 10 different physician specialists could be required for any one affected individual. This impressive number of potential specialists dealing with the surgical and medical problems can be expanded further by including important support personnel such as physical or occupational therapist, developmentalist (infant stimulation), nutritionist, psychologist/behavior specialist, and social worker. The need for long-term follow-up by most of the specialists is labor intensive for the affected individual and family, but it is exceedingly important. Clearly, one physician, ideally the primary care physician, is the most critical factor in assuring the best outcome. As with many other genetic conditions, identifying a unifying "medical home" can be difficult but is important for ongoing medical care and communication. Follow-up with the geneticist may be of help in this difficult situation as problems may be noticed that require referral back to the specialist or indicate additional testing. In addition, the geneticist may be able to answer difficult or esoteric questions the primary care physician is not comfortable answering.

The multiple individual anomalies in VACTERL association clearly require a concerted and extended effort to minimize or ameliorate adverse effects, both singly and in combination. There remains throughout life a whole person who still has problems that may need intervention.

The long-term prognosis for individuals with VACTERL association is variable. Each surgical anomaly seems prognostically largely independent. If each surgical defect can be satisfactorily repaired and/or medically treated, the prognosis is excellent. If some defects cannot be totally repaired or persist with sequelae, then the prognosis is the sum total of the residual problems, with critical organs (e.g., heart, kidneys) having a greater negative impact.

TABLE 61.2 Specialists required and areas of involvement in VATER/VACTERL association

Medical specialist	Major areas of involvement
Neonatologist	Neonatal care, coordinating other specialists
Geneticist	Diagnosis, genetic testing, counseling, long-term follow-up
Pediatric surgeon	Surgery (anorectal malformations, tracheo-esophageal fistula, esophageal atresia), long-term follow-up
Orthopedist	Surgery/medical therapy (spine/limbs), long-term follow-up
Nephrologist	Diagnosis, medical therapy (urinary tract), long-term follow-up
Urologist	Surgery (genitalia, urinary tract), long-term follow-up
Cardiac surgeon	Surgery (heart), possible long-term follow-up
Gastroenterologist	Diagnosis/medical therapy (reflux, obstruction, diarrhea, gastrostomy problems), long-term follow-up
Pulmonologist	Diagnosis/medical therapy (pneumonia, aspiration), long-term follow-up
Gynecologist	Diagnosis/surgery (genital/reproductive tract), long-term follow-up
Physical/occupational therapist	Diagnosis/medical therapy, long-term follow-up
Developmentalist	Physical/intellectual stimulation, long-term follow-up
Nutritionist	Adequate nutrition, long-term follow-up
Psychologist/behaviorist	Body image/personality therapy, long-term follow-up
Social worker	Assistance to child/family with system/bureaucracy, long-term follow-up
Primary care physician	Underpinning all of the above, long-term follow-up

Growth and Feeding

Congenital heart failure, renal dysfunction, or gastrointestinal disorder can cause poor growth, as can skeletal anomalies. A minority of affected individuals may have small stature that is apparently independent of major organ dysfunction (Mapstone et al. 1986). Endocrinologic consultation should be considered if poor growth remains unexplained.

Evaluation

- Adequacy of caloric intake should be assessed.
- The child should be evaluated for occult cardiac, renal, gastrointestinal, or other problems through standard imaging and functional studies.
- An endocrinology consultation should be obtained if a source of growth compromise cannot be found.

Management

- If a source of growth failure is identified, it should be treated by standard means.
- Temporary measures, such as increasing caloric concentration per ounce of formula, may help.
- Gastrostomy tube feedings may be indicated if the child is too sick to feed adequately or when increased metabolic output exceeds caloric intake.

Development and Behavior

Early psychomotor delay, poor body image, and behavioral problems can occur and need attention. A poor body image is often noted in those with limb deficiencies, short stature, complicated anorectal malformations with persisting intestinal problems, recurrent esophageal complications, chronic cardiac disease, and external genital abnormalities. Developmental progress may be compromised by problems such as congestive heart failure, renal dysfunction, or gastrointestinal difficulties. Long-term developmental follow-up is rare in studies of people with VACTERL association. One study looked at 16 affected individuals 10 years and older and found 31% (5/16) with cognitive impairment; however, most of those persons were atypical in that some of the features (prune belly, Poland anomaly) were not part of the association, and these may have represented conditions that current molecular testing would have identified (Wheeler and Weaver 2005). Another study of 11 affected adults did not report developmental delay or cognitive impairment in any individuals (Raam et al. 2011).

A loving, supportive family and other key support networks can help the individuals overcome many of these potential problems, but medical personnel should be no less supportive.

Evaluation

- Formal periodic developmental evaluation should be carried out during childhood if development is delayed.
- If seizures, abnormal cranial size, or other neurological manifestations (e.g., hypotonia or hypertonia) are recognized, neurological consultation should be obtained. This is particularly important for macrocephaly, as would typically be observed in individuals with VACTERL-H.

Management

- Early infant education and physical therapy in childhood are particularly important because of the frequent and prolonged hospitalizations.

- Treatment of cardiac, renal, gastrointestinal, or nutritional problems is critical in promoting developmental progress.
- Occasionally, a referral to a psychologist or behavioral pediatrician may help address worsening body image or behavioral challenges. Early intervention by such specialists – before the child has started to develop a body image – is of critical importance.

Cardiovascular

Cardiac anomalies are found in the VACTERL association with a frequency of approximately 40–80% (Khoury et al. 1983; Solomon 2011; Cunningham et al. 2013). The cardiovascular anomalies can be of any type and any severity; consequently, they may be very obvious in the neonatal period or may be cryptic. Death and poor health remain major complications in those individuals with complex or untreatable congenital heart defects. Frequently, growth restriction may be ascribed to the heart problems when other medical problems are actually responsible. Some children show marked catch-up growth after delayed cardiac surgery is performed.

Evaluation

- All individuals who are suspected to have VACTERL association should have an immediate evaluation by a pediatric cardiologist.
- An echocardiogram should be done regardless of ausculatory or electrocardiogram (ECG) results (ECG itself should not be overlooked). Surgical and/or medical therapy will be determined on the basis of these studies (Ashburn et al. 2003; Cunningham et al. 2013).
- Long-term cardiology follow-up is particularly important if a cardiac anomaly is present.

Management

- Treatment of the congenital heart defects and other cardiac problems does not differ from that in individuals with similar defects in the general population.

Musculoskeletal

Several different types of vertebral anomalies, often accompanied by rib anomalies, can be observed in people with VACTERL association. The severity of costovertebral anomalies can be highly variable, ranging from subtle anomalies that do not impact care or quality of life to severe and extensive anomalies requiring surgery and ongoing management (Solomon 2011; Chen et al. 2016). Overall, vertebral anomalies are typically not extensive and most often occur in the upper to midthoracic and lumbar regions, with sacral and lower cervical defects being less common (Lawhon et al. 1986). Hemivertebrae are most common, but dyssegmented and fused vertebral bodies can also be found. The location of vertebral anomalies often (but not always) correlates with the anatomical location of other anomalies; that is, it has been suggested the affected individuals might be grouped into "upper" and "lower" anatomical subgroups, though some will have findings affecting both anatomic areas (Kallen et al. 2001; Chen et al. 2016).

Scoliosis is sometimes present, with its nidus closely related to the location of the abnormal vertebrae. The scoliosis is often mild, but occasionally can be of modest-to-severe degree, requiring immediate orthopedic attention and potentially surgical interventions. Figure 61.2 shows a young teenage girl with the VACTERL association who had vertebral defects without congenital scoliosis, but was not followed closely and developed moderately severe scoliosis that went unrecognized by her and her family.

Limb anomalies are usually but not always restricted to the upper limbs in VACTERL association, and they are usually preaxial, involving underdevelopment or agenesis of the thumbs and radial bones (Botto et al. 1997). These preaxial defects are usually bilateral but often asymmetric in degree. At the milder end, only the thenar muscle mass is reduced in

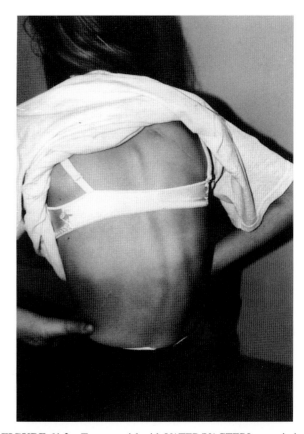

FIGURE 61.2 Teenage girl with VATER/VACTERL association who developed unrecognized scoliosis secondary to lumbar hemivertebrae.

size with or without a slightly reduced thumb circumference and hypoplastic flexion creases. These subtle features are often missed, resulting in a valuable and critical feature of the VATER/VACTERL association being unrecognized and a secure diagnosis not being made. More severely, there may be marked radial and ulnar aplasia or absence of most of the upper limb.

Evaluation

- Regular orthopedic follow-up is important in persons with vertebral defects whether or not scoliosis is present.
- It is uncommon for the spinal cord to be injured by the vertebral defects, but any neurological signs, including bladder incontinence, encopresis or paresis, should prompt further evaluation (e.g., computed tomography or magnetic resonance imaging, spinal scan, myelogram).
- If limb abnormalities are noted, radiographs of the abnormal limb segment should be performed and, if anomalies are confirmed, orthopedic assessment should be arranged.

Management

- Bracing and/or surgical interventions are the two therapeutic modalities utilized to treat scoliosis.
- Orthopedic or plastic hand surgeons will be necessary to correct the preaxial defects from both a cosmetic and a functional standpoint.
- If the radius is significantly hypoplastic, the hand angulates sharply toward the radial side of the wrist in what is termed a radial clubbed hand. This will need to be straightened, and the procedure is a difficult surgery with mixed results.
- The thumb, if functional and of reasonable size, should be left alone, but hypoplastic thumbs are often removed and the second finger reoriented to function as an opposable thumb.
- Staged surgeries for more severe preaxial defects with postsurgical casting mandate long-term involvement and follow-up by the treating surgeon.

Gastrointestinal

Individuals with VACTERL association may have upper and/or lower gastrointestinal anomalies, including tracheo-esophageal fistula with or without esophageal atresia, and anorectal malformation. Tracheo-esophageal fistula may occur with or without esophageal atresia, and is reported in 50–80% of affected individuals (Botto et al. 1997; Deurloo et al. 2002; Solomon 2011). Prenatally, esophageal atresia may be recognized by polyhydramnios or absent gastric bubble on ultrasound. Postnatally, during infancy these anomalies may be suspected due to inability to pass a nasogastric or orogastric tube or choking or increased secretions during feeding (Houben and Curry 2008; Solomon et al. 2014).

The majority of those with tracheo-esophageal fistula and esophageal atresia survive if their other associated anomalies are amenable to therapy. It is noteworthy that approximately one-third of individuals with tracheo-esophageal fistula and esophageal atresia have cardiac defects, so evaluation of the heart is critical in this situation.

Anorectal malformation, classically imperforate anus with or without other anomalies, including involving the genitourinary tract, affects 55–90% of individuals (Sofatzis et al. 1983; Solomon et al. 2014). Imperforate anus requires immediate attention in the neonatal period.

Evaluation

- Failure to be able to pass a nasogastric or orogastric tube into the stomach is a strong indication of esophageal atresia. X-rays to confirm tracheo-esophageal fistula/esophageal atresia should be performed with a nasogastric/orogastric tube in place – even if it meets resistance on placement – to help show the defect's location.
- The indicated echocardiogram, in addition to identifying major congenital heart anomalies, is also important to assess vascular anomalies that may influence the surgical approach. For example, a right-sided aortic arch may affect which side of the chest the surgeon enters to conduct the repair (Alberti et al. 2011).
- Following tracheo-esophageal fistula/esophageal atresia repair, long-term follow-up by a pediatric surgeon and gastroenterologist is usually necessary. If recurrent pneumonias persist, a pediatric pulmonologist is required to evaluate the functional integrity and anatomical status of the tracheobronchial tree.
- The absence of clinical evidence of an anal opening is usually obvious when anal atresia is present, but sometimes the anus must be digitally probed or contrast utilized to evaluate patency. Examination of the anorectal malformation is important to identify accompanying genitourinary and gynecologic anomalies, which may influence repair as well as overall management.
- When an anorectal malformation is present, in order to guide treatment choices (anoplasty versus colostomy) if no fistula between the rectum and the urinary tract or rectum and perineal skin is obvious, a prone cross-table lateral pelvic X-ray can help determine the proximity of the distal rectum to the perineal skin (Levitt and Pena 2007, 2010). The urinary and reproductive tracts should be evaluated in any individual with anorectal malformation. Renal ultrasound and voiding cystourethrogram

will adequately evaluate the urinary tract. Direct vaginal examination and pelvic ultrasound may be necessary in females. Frequently, contrast may be put into the rectum or vagina to detect fistulous connections and better clarify the anomalous anatomy. Additionally, spinal ultrasound can be performed to assess for tethered cord.

Management

- The gastrointestinal anomalies that occur in VACTERL association are always treated surgically. In type A and B esophageal atresia (long gap, no tracheo-esophageal fistula), it is usually necessary to keep the proximal esophagus dry via constant suction for 2–3 months or to exteriorize the proximal esophageal pouch via a cervical esophagostomy.
- A feeding gastrostomy is placed in anticipation of future direct or colon graft anastomosis. A preoperative tracheobronchoscopy is recommended because, in approximately 25% of the cases, such a procedure modifies the ultimate surgical approach by its unsuspected findings (Atzori et al. 2006). In addition, in long-gap esophageal atresia, gastric transposition or interposition of a segment of the colon to the upper esophagus can be performed (Tannuri et al. 2007). This, similar to all surgical procedures, includes tying off the tracheo-esophageal fistula (Cozzi and Wilkinson 1967).
- Short- and long-term complications of repaired tracheo-esophageal fistula/esophageal atresia include fistula formation, leakage at the anastomosis site, pneumonia, swallowing difficulties, esophageal stricture requiring repeated dilatations, and poor growth (Okada et al. 1997; Deurloo et al. 2002).
- The tracheo-esophageal fistula is surgically tied off or separated with each end tied off.
- For anorectal malformation, factors such as the individual's gender, size/gestational age, overall health, location of the anorectal malformation, age, and the presence/absence of a perineal fistula will guide the surgical approach, including the possibility of anoplasty or (mini)-posterior sagittal anorectoplasty (PSARP) versus a diverting colostomy. Serial radiologic examinations may be necessary to help guide this decision (van der Steeg et al. 2015).
- Individuals with a diverting colostomy (Figure 61.3) will have a pull-through of the distal rectum after disconnection of it from the urinary tract (if such a fistula is present), ideally within the first 3–12 months of life.
- Long-term sequelae of both types of surgery can include fecal impaction, megacolon, soiling, persistent fistula, and anal stenosis requiring dilatation, although ongoing care by experienced subspecialists can ameliorate some manifestations.

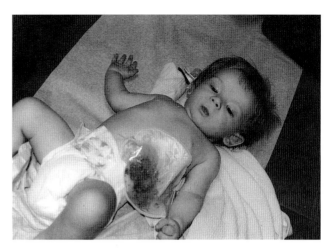

FIGURE 61.3 Infant with VATER/VACTERL association who had a high anal atresia requiring a temporary diverting colostomy. Note mild right radial and thumb underdevelopment.

Respiratory

In addition to the previously mentioned respiratory complications related to tracheo-esophageal fistula/esophageal atresia, laryngeal stenosis (Corsello et al. 1992) and tracheal atresia (Milstein et al. 1985) have also been reported in the VACTERL association. These latter problems generally present with moderate-to-severe respiratory obstruction and, in the case of tracheal atresia, ultimately lead to death. Laryngeal stenosis, depending on its degree, may go undiagnosed for months to years. Tachypnea, laryngotracheomalacia, and sternal, intercostal, and costal retractions are common resulting clinical manifestations.

Evaluation

- A chest radiograph should be obtained to evaluate for atelectasis and/or infiltrate and to assess obvious anatomical abnormalities.
- Blood gases should be obtained if the affected individual appears sicker than known anomalies would warrant.
- Swallowing studies (sometimes with contrast) and/or visualization of the larynx should be carried out in the presence of recurrent aspiration, apnea, and/or pneumonia
- Laryngoscopy is needed if symptoms persist or worsen (pre- or postsurgically).

Management

- Surgical correction of tracheo-esophageal fistula/esophageal atresia should be accomplished, using standard techniques.
- Treatment of any pneumonia with antibiotics is appropriate.

- If recurrent aspiration occurs, steroid therapy may be required.
- Reflux precautions should be used after surgery.

Renal and Genitourinary

Renal and genitourinary anomalies are common (69–93%) particularly, if anorectal malformation, genitourinary defects, or lower vertebral defects are present (Botto et al. 1997; Kolon et al. 2000; Solomon et al. 2011; Cunningham et al. 2014). All types of renal defects have been noted including renal agenesis/dysplasia, obstructive hydronephrosis, cystic kidney(s), and ectopic kidney (Reutter et al. 2016). The lower urinary tract may be abnormal because of fistulas between the rectum and the bladder or urethra. Genital defects such as hypospadias, bifid scrotum, and labial hypoplasia are more often associated with anorectal malformation, but both have an increased association with renal defects (Solomon et al. 2011; Cunningham et al. 2014). Renal anomalies, especially if not diagnosed early, may lead to long-term sequelae (Anh et al. 2009; Raam et al. 2011). A small case series of 12 individuals suggested that those with VACTERL who had chronic kidney disease may, compared to controls, develop end-stage renal disease more frequently, experience more dialysis complications, and have worse transplant outcomes (Ahn et al. 2009).

Evaluation

- The entire urinary tract should be evaluated by ultrasound in all individuals with VATER/VACTERL, including for signs of vesicoureteral reflux even in the absence of frank anomalies (Cunningham et al. 2014).
- If ultrasound shows evidence of obstruction or scarring, further subspecialty consultation and imaging (e.g., via voiding cystourethroram) is indicated (Cunningham et al. 2014).
- The renal system should be followed closely, generally by ultrasound, if abnormalities are discovered.
- A urologist and/or a gynecologist consultation is indicated if the individual has external genital abnormalities that may warrant further intervention.

Management

- Urinary tract anomalies, including vesicoureteral reflux, are treated similarly to those in non-syndromic individuals.
- Reproductive tract anomalies may require surgical reconstruction once the anomalous anatomy is accurately ascertained. Such reconstruction is standard.
- Fistulous tracts need to be closed, but they are sometimes left open until another surgery is accomplished.

RESOURCES

PULL-THRU NETWORK

c/o Lori Parker
1705 Wintergreen Parkway
Normal, Illinois 61761, USA
Email: *pullthrunetwork@gmail.com*
Website: *www.pullthrunetwork.org*

VACTERL Association Support Group

Website: *www.vacterl-association.org.uk*

VACTERL Network

c/o Sandy Garrity
11 Pillowlace Lane
Beverly, MA 01915, USA
Phone: 419-392-3781
Email: *hmiller@vacterlnetwork.org; sgarrity@vacterl-network.org*
Website: *www.vacterlnetwork.org*

ACKNOWLEDGMENT

The author would like to thank Dr. Marc A. Levitt, Surgical Director, Center for Colorectal and Pelvic Reconstruction, Nationwide Children's Hospital for his expert input and guidance.

REFERENCES

Ahn SY, Mendoza S, Kaplan G, Reznik V (2009) Chronic kidney disease in the VACTERL association: clinical course and outcome. *Pediatr Nephrol* 24:1047–1053.

Alberti D, Boroni G, Corasaniti L, Torri F (2011) Esophageal atresia: pre and post-operative management. *J Matern Fetal Neonatal Med* 24(Suppl)1:4–6.

Alter BP, Giri N (2016) Thinking of VACTERL-H? Rule out Fanconi Anemia according to PHENOS. *Am J Med Genet A* 170:1520–1524.

Alter BP, Rosenberg PS (2013) VACTERL-H Association and Fanconi Anemia. *Mol Syndromol* 4:87–93.

Ashburn DA, Harris L, Downar EH, Siu S, Webb GD, Williams WG (2003) Electrophysiologic surgery in patients with congenital heart disease. *Semin Thorac Cardiovasc Surg Pediatr Card Surg Annu* 6:51–58.

Atzori P, Iacobelli BD, Bottero S, Spirydakis J, Laviani R, Trucchi A, Braguglia A, Bagolan P (2006) Preoperative tracheobronchoscopy in newborns with esophageal atresia: Does it matter? *J Pediatr Surg* 41:1054–1057.

Bartels E, Jenetzky E, Solomon BD, Ludwig M, Schmiedeke E, Grasshoff-Derr S, Schmidt D, Märzheuser S, Hosie S, Weih S, Holland-Cunz S, Palta M, Leonhardt J, Schäfer M, Kujath C, Rissmann A, Nöthen MM, Reutter H, Zwink N (2012) Inheritance of the VATER/VACTERL association. *Pediatr Surg Int* 28:681–685.

Bartels E, Schulz AC, Mora NW, Pineda-Alvarez DE, Wijers CH, Marcelis CM, Stressig R, Ritgen J, Schmiedeke E, Mattheisen M, Draaken M, Hoffmann P, Hilger AC, Dworschak GC, Baudisch F, Ludwig M, Bagci S, Müller A, Gembruch U, Geipel A, Berg C, Bartmann P, Nöthen MM, van Rooij IA, Solomon BD, Reutter HM (2012) VATER/VACTERL association: identification of seven new twin pairs, a systematic review of the literature, and a classical twin analysis. *Clin Dysmorphol* 21:191–195.

Botto LD, Khoury MJ, Mastroiacovo P, Castilla EE, Moore CA, Skjaerven R, Mutchinick OM, Borman B, Cocchi G, Czeizel AE, Goujard J, Irgens LM, Lancaster PAL, Martinez-Frias ML, Merlob P, Ruusinen A, Stoll C, Sumiyoshi Y (1997) The spectrum of congenital anomalies of the Vater association: An international study. *Am J Med Genet* 71:8–15.

Brosens E, Eussen H, van Bever Y, van der Helm RM, Ijsselstijn H, Zaveri HP, Wijnen R, Scott DA, Tibboel D, de Klein A (2013) VACTERL association etiology: The impact of de novo and rare copy number variations. *Mol Syndromol* 4:20–26.

Brown AB, Roddam AW, Ward AJ (1999) Oesophageal atresia, related malformations, and medical problems: A family study. *Am J Med Genet* 85:31–37.

Castori M, Rinaldi R, Cappellacci S, Grammatico P (2008) Tibial developmental field defect is the most common lower limb malformation pattern in VACTERL association. *Am J Med Genet A* 146A:1259–1266.

Chen Y, Liu Z, Chen J, Zuo Y, Liu S, Chen W, Liu G, Qiu G, Giampietro PF, Wu N, Wu Z (2016) The genetic landscape and clinical implications of vertebral anomalies in VACTERL association. *J Med Genet* 53:431–437.

Chung B, Shaffer LG, Keating S, Johnson J, Casey B, Chitayat D (2011) From VACTERL-H to heterotaxy: variable expressivity of ZIC3-related disorders. *Am J Med Genet A* 155A:1123–1128.

Corsello G, Maresi E, Corrao AM, Dimita U, Cascio ML, Cammarata M, Giuffre L (1992) VATER/VACTERL association: Clinical variability and expanding phenotype including laryngeal stenosis. *Am J Med Genet* 44:813–815.

Cozzi F, Wilkinson AW (1967) Esophageal atresia. *Lancet* 2:1222–1225.

Cunningham BK, Hadley DW, Hannoush H, Meltzer AC, Niforatos N, Pineda-Alvarez D, Sachdev V, Warren-Mora N, Solomon BD (2013) Analysis of cardiac anomalies in VACTERL association. *Birth Defects Res A Clin Mol Teratol* 97:792–797.

Cunningham BK, Khromykh A, Martinez AF, Carney T, Hadley DW, Solomon BD (2014) Analysis of renal anomalies in VACTERL association. *Birth Defects Res A Clin Mol Teratol* 100:801–805.

Debost-Legrand A, Goumy C, Laurichesse-Delmas H, Déchelotte P, Perthus I, Francannet C, Lémery D, Gallot D (2015) Prenatal diagnosis of the VACTERL association using routine ultrasound examination. *Birth Defects Res A Clin Mol Teratol* 103:880–886.

Czeizel A, Ludányi I (1985) An aetiological study of the VACTERL-association. *Eur J Pediatr* 144:331–337.

Deurloo JA, Ekkelkamps S, Schoorl M, Heij HA, Aronson C (2002) Esophageal atresia: Historical evolution of management and results in 371 patients. *Ann Thorc Surg* 73:267–272.

England RJ, Eradi B, Murthi GV, Sutcliffe J (2017) Improving the rigour of VACTERL screening for neonates with anorectal malformations. *Pediatr Surg Int* 33:747–754.

Evans JA, Stranc LC, Kaplan P, Hunter AGW (1989) VACTERL with hydrocephalus: Further delineation of the syndrome(s). *Am J Med Genet* 34:177–182.

Faivre L, Portnoi MF, Palo G, Stoppa-Lyonet D, Le Merrer M, Thauvin-Robinet C, Huet F, Mathew CG, Joenje H, Verloes A, Baumann C (2005) Should chromosome breakage studies be performed in patients with VACTERL Association? *Am J Med Genet* 137A:55–58.

Holden ST, Cox JJ, Kesterton I, Thomas NS, Carr C, Woods CG (2006) Fanconi anaemia complementation group B presenting as X linked VACTERL with hydrocephalus syndrome. *J Med Genet* 43:750–754.

Houben CH, Curry JI (2008) Current status of prenatal diagnosis, operative management and outcome of esophageal atresia/tracheo-esophageal fistula. *Prenat Diagn* 28:667–675.

Jenetzky E, Wijers CH, Marcelis CM, Zwink N, Reutter H, van Rooij IA (2011) Bias in patient series with VACTERL association. *Am J Med Genet A* 155A:2039–2041.

Kallen K, Mastroiacovo P, Castilla EE, Robert E, Kallen B (2001) VATER non-random association of congenital malformations: Study based on data from four malformation registers. *Am J Med Genet* 101:26–32.

Khoury M, Cordero JF, Greenberg F, James LM, Erickson JD (1983) A population study of the VACTERL association: Evidence for its etiologic heterogeneity. *Pediatrics* 71:815–820.

Killoran CE, Abbott M, McKusick VA, Biesecker LG (2000) Overlap of PIV syndrome, VACTERL and Pallister-Hall syndrome: Clinical and Molecular analysis. *Clin Genet* 58:28–30.

Kolon TF, Gray CL, Sutherland RW, Roth DR, Gonzales ET (2000) Upper urinary tract manifestations of the VACTERL association. *J Urology* 163:1949–1951.

Lawhon SM, MacEwen GD, Bunnell WP (1986) Orthopaedic aspects of the VATER association. *J Bone Joint Surg* 68A:424–429.

Levitt MA, Peña A (2007) Anorectal malformations. *Orphanet J Rare Dis* 2:33.

Levitt MA, Peña A (2010) Cloacal malformations: lessons learned from 490 cases. *Semin Pediatr Surg.* 19:128–138.

Lomas FE, Dahlstrom JE, Ford JH (1998) Vacterl with hydrocephalus: Family with X-linked VACTERL-H. *Am J Med Genet* 76:74–78.

Lurie IW, Ferencz C (1997) Vacterl-hydrocephaly, DK-phocomelia, and cerebro-cardio-radio-rectal community. *Am J Med Genet* 70:144–149.

Mapstone CL, Weaver DD, Yu P-L (1986) Analysis of growth in the VATER Association. *Am J Dis Child* 140:386–390.

Martinez-Frias ML, Frias JL (1999) VACTERL as a primary, polytopic developmental field defect. *Am J Med Genet* 83:13–16.

Milstein JM, Lau M, Bickers RG (1985) Tracheal agenesis in infants with Vater association. *Am J Dis Child* 139:77–80.

Narchi H (1999) Oesophageal atresia, VACTERAL association: Fanconi's anemia related spectrum of anomalies. *Arch Dis Child* 80:207.

Nora AH, Nora JJ (1975) A syndrome of multiple congenital anomalies associated with teratogenic exposure. *Arch Environ Health* 30:17.

Okada A, Usuin N, Inove M, Kawahara H, Kubota A, Imura K, Kamata S (1997) Esophageal atresia in Osaka. Review of 39 years of experience. *J Pediatr Surg* 32:1570–1574.

Parida SK, Hall BD, Barton L, Fujimoto A (1995) Penoscrotal transposition and associated anomalies: Report of five new cases and review of the literature. *Am J Med Genet* 59:68–75.

Porteous ME, Cross I, Burn J (1992) VACTERL with hydrocephalus: One end of the Fanconi anemia spectrum of anomalies. *Am J Med Genet* 43:1032–1034.

Quan L, Smith DW (1973) The VATER association. *J Pediatr* 82:104–107.

Raam MS, Pineda-Alvarez DE, Hadley DW, Solomon BD (2011) Long-term outcomes of adults with features of VACTERL association. *Eur J Med Genet* 54:34–41.

Retterer K, Juusola J, Cho MT, Vitazka P, Millan F, Gibellini F, Vertino-Bell A, Smaoui N, Neidich J, Monaghan KG, McKnight D, Bai R, Suchy S, Friedman B, Tahiliani J, Pineda-Alvarez D, Richard G, Brandt T, Haverfield E, Chung WK, Bale S (2016) Clinical application of whole-exome sequencing across clinical indications. *Genet Med* 18:696–704.

Reutter H, Hilger AC, Hildebrandt F, Ludwig M (2016) Underlying genetic factors of the VATER/VACTERL association with special emphasis on the "Renal" phenotype. *Pediatr Nephrol* 31:2025–2033.

Reutter H, Ludwig M (2013) VATER/VACTERL Association: Evidence for the Role of Genetic Factors. *Mol Syndromol 2013* 4:16–19.

Rittler M, Paz JE, Castilla EE (1997) VATERL: An epidemiologic analysis of risk factors. *Am J Med Genet* 73:162–169.

Shi H, Enriquez A, Rapadas M, Martin EMMA, Wang R, Moreau J, Lim CK, Szot JO, Ip E, Hughes JN, Sugimoto K, Humphreys DT, McInerney-Leo AM, Leo PJ, Maghzal GJ, Halliday J, Smith J, Colley A, Mark PR, Collins F, Sillence DO, Winlaw DS, Ho JWK, Guillemin GJ, Brown MA, Kikuchi K, Thomas PQ, Stocker R, Giannoulatou E, Chapman G, Duncan EL, Sparrow DB, Dunwoodie SL (2017) NAD Deficiency, Congenital Malformations, and Niacin Supplementation. *N Engl J Med* 377:544–552.

Siebel S, Solomon BD (2013) Mitochondrial Factors and VACTERL Association-Related Congenital Malformations. *Mol Syndromol* 4:63–73.

Sofatzis JA, Alexacos L, Skouteli HN, Tiniakos G, Padiatellis C (1983) Malformed female genitalia in newborns with the VATER association. *Acta Paediatr Scand* 72:923–924.

Solomon BD (2011) VACTERL/VATER Association. *Orphanet J Rare Dis* 6:56.

Solomon BD, Bear KA, Kimonis V, de Klein A, Scott DA, Shaw-Smith C, Tibboel D, Reutter H, Giampietro PF (2012) Clinical geneticists' views of VACTERL/VATER association. *Am J Med Genet A* 158A:3087–3100.

Solomon BD, Pineda-Alvarez DE, Raam MS, Bous SM, Keaton AA, Vélez JI, Cummings DA (2010) Analysis of component findings in 79 patients diagnosed with VACTERL association. *Am J Med Genet A* 152A:2236–2244.

Solomon BD, Pineda-Alvarez DE, Raam MS, Cummings DA (2010) Evidence for inheritance in patients with VACTERL association. *Hum Genet* 127:731–733.

Solomon BD, Raam MS, Pineda-Alvarez DE (2011) Analysis of genitourinary anomalies in patients with VACTERL (Vertebral anomalies, Anal atresia, Cardiac malformations, Tracheo-Esophageal fistula, Renal anomalies, Limb abnormalities) association. *Congenit Anom (Kyoto)* 51:87–91.

Stevenson RE, Hunter AG (2013) Considering the Embryopathogenesis of VACTERL Association. *Mol Syndromol* 4:7–15.

Tannuri U, Maksoud-Filho J, Tannuri ACA, Andrade W, Maksoud JG (2007) Which is better for esophageal substitution in children, esophagocoloplasty or gastric transposition? A 27-year experience of a single center. *J Pediatr Surg* 42:500–504.

Temtamy SA, Miller JD (1974) Extending the scope of the VATER association: Definition of the Vater syndrome. *J Pediatr* 85:345–349.

Touloukian RJ, Pickett LK (1969) Management of the newborn with imperforate anus. *Clin Pediatr* 8:389–397.

van der Steeg HJ, Schmiedeke E, Bagolan P, Broens P, Demirogullari B, Garcia-Vazquez A, Grasshoff-Derr S, Lacher M, Leva E, Makedonsky I, Sloots CE, Schwarzer N, Aminoff D, Schipper M, Jenetzky E, van Rooij IA, Giuliani S, Crétolle C, Holland Cunz S, Midrio P, de Blaauw I (2015) European consensus meeting of ARM-Net members concerning diagnosis and early management of newborns with anorectal malformations. *Tech Coloproctol* 19:181–185.

Walsh LE, Vance GH, Weaver DD (2001) Distal 13q deletion syndrome and the VACTERL association: Case report, literature review, and possible implications. *Am J Med Genet* 98:137–144.

Weaver DD, Mapstone CL, Yu P (1986) The VATER association. Analysis of 46 patients. *Am J Dis Child* 140:225–229.

Wheeler PG, Weaver DD (2005) Adults with VATER Association: Long-term prognosis. *Am J Med Genet* 138A:212–217.

62

VON HIPPEL–LINDAU SYNDROME

SAMANTHA E. GREENBERG
Huntsman Cancer Institute, Salt Lake City, UTAH, USA

LUKE D. MAESE
Hematology/Oncology, Department of Pediatrics, Huntsman Cancer Institute, University of Utah, Salt Lake City, Utah, USA

BENJAMIN L. MAUGHAN
Huntsman Cancer Institute, Salt Lake City, Utah, USA

INTRODUCTION

Von Hippel–Lindau syndrome (vHL) is a hereditary condition associated with increased risks for neoplasia, particularly in the central nervous system (CNS), retina, and kidneys. Originally described by E. Treacher Collins (1894) as familial retinal angiomatosis, the disease carries the namesake of two insightful physicians, German ophthalmologist Eugen von Hippel and Swedish pathologist Arvid Lindau (Kepes and Slowik 1997). In 1904, von Hippel described retinal angiomatosis, and united with Lindau during Lindau's journey in 1925 through Europe. It was there where a mutual conclusion was made that the observed retinal angiomatosis had similar histological features to cerebellar hemangioblastomas, which had been the focus of Lindau's doctoral dissertation. This mutual acknowledgement led to the naming of von–Hippel Lindau syndrome in 1936 (Gossage et al. 2015).

Clinical diagnostic criteria for vHL, described by Melman and Rosen in 1964, consisted of individuals with two or more vHL-associated tumors, or one tumor and a family history of vHL-associated tumors. Autosomal dominant inheritance with incomplete penetrance was identified by Shokeir in 1970. The 1990s brought immense clinical and molecular knowledge to the field. Genotype–phenotype correlations were outlined by multiple researchers in 1991 (Glenn et al. 1991; Neumann and Wiestler 1991), the VHL gene was identified in 1993 (Latif et al. 1993) and vHL was clinically delineated by types in 1995 (Chen et al. 1995)

Incidence

The reported birth incidence of vHL is 1 in 36,000, though prevalence has ranged from 1 in 27,300 in Denmark to 1 in 53,000 in Eastern England (Binderup et al. 2017a). vHL has been reported in all ethnic groups and occurs equally in male and females. There is no association with the incidence of von Hippel Lindau and parental age or birth order (Maher et al. 1991). Binderup et al (2017) reported a penetrance of at least one manifestation on the vHL spectrum of 87% by age 60 in a cohort of 143 *VHL* mutation carriers, and the National Institute of Health reported 119 of 123 (96.7%) of the deceased patients in its vHL registries required surgery for either a vHL-related tumor or vHL-related cyst (Kim et al. 2016). Though the rates of penetrance are reported to be quite high, there is some suspicion that vHL is under-diagnosed due to variable penetrance and expressivity.

Studies reporting morbidity and mortality associated with vHL are variable, with Bender et al. (2001) identifying no differences in mortality between individuals in Southern Germany with vHL compared with the general German population despite reporting almost 90% penetrance of vHL-related manifestations by age 70. Most recently, the Danish population discovered reduced life expectancies among individuals with vHL (Binderup et al. 2017a). Male patients born with vHL had a life expectancy of 67, approximately 10 years less than the general Danish population. Female patients born with vHL had a life expectancy of 60, approximately 22 years less than the general Danish

---

Cassidy and Allanson's Management of Genetic Syndromes, Fourth Edition.
Edited by John C. Carey, Agatino Battaglia, David Viskochil, and Suzanne B. Cassidy.
© 2021 John Wiley & Sons, Inc. Published 2021 by John Wiley & Sons, Inc.

population. A Chinese population subsequently identified a medial life expectancy of 62 years, with women having a longer survival (67 years) than men (62 years). Family history, truncating mutation types, and early-onset of manifestations contributed to poorer survival (Wang et al. 2018). This difference in morbidity and mortality may be related to adherence of regular surveillance and proactive treatment, as continuation in a vHL surveillance program has varied among pre-symptomatic individuals with vHL (Rasmussen et al. 2010). However, this research did not show a significant difference in surveillance by gender, and thus does not explain the differences in life expectancies.

Diagnostic Criteria

An individual may be clinically diagnosed with vHL if they have two or more characteristic manifestations, including:

- Renal cell carcinoma (typically diagnosed before age 50).
- Hemangioblastoma of the retina or central nervous system (CNS).
- Adrenal pheochromocytoma or extra-adrenal paraganglioma.
- One of the following above in addition to visceral manifestations such as:
  - Multiple kidney or pancreatic cysts.
  - Endolymphatic sac tumors.
  - Papillary cystadenomas of the epididymis or broad ligament.
  - Neuroendocrine tumors of the pancreas.

For example, a patient may have two hemangioblastomas, both in the spine, or one CNS and one retinal, and meet clinical criteria. In another scenario, an individual may also develop a retinal hemangioblastoma and multiple kidney cysts and receive a clinical diagnosis of vHL.

However, if a patient has a positive family history of vHL, only one of the following manifestations (hemangioblastoma, renal cell carcinoma, pheochromocytoma, or multiple kidney or pancreatic cysts) is necessary for a clinical diagnosis (Lonser et al. 2003).

Molecular genetic testing will identify a mutation in the *VHL* gene in almost all individuals who meet the above diagnostic criteria. The increased availability of molecular genetic testing has expanded diagnostic capabilities. This has allowed for the diagnosis of vHL prior to the onset of symptoms through identification of a pathogenic variant in the *VHL* gene. This scenario is most applicable for at-risk family members who do not yet have manifestations of vHL syndrome. If a person were to present with one manifestation on the vHL spectrum (i.e. renal cell carcinoma before age 50, hemangioblastoma or pheochromocytoma), diagnostic genetic testing may be beneficial for earlier detection and treatment of other potential lesions related to vHL.

Etiology/Pathogenesis and Genetics

Germline mutations in the *VHL* gene are causative of von Hippel–Lindau syndrome. The *VHL* gene was mapped to chromosome region 3p25 in 1993 (Latif et al. 1993), after chromosome 3p deletions were described in renal cell carcinoma, prompting linkage studies (Seizinger et al. 1988). The gene is comprised of three exons that code for two major isoforms. The first and longest isoform is encoded by the entire open reading frame of the gene and is 213 amino acids long, with an approximate molecular weight of ~24–30 kDa (Iliopoulos et al. 1998). A second protein isoform begins at codon 54 due to internal translation initiation at a methionine codon, with an apparent molecular weight of ~19 kDa. Both are biologically active with similar biochemical and functional assays, and the manifestations of vHL disease require the inactivation of both the long and short isoforms (Kaelin 2002). Both protein isoforms will be generally referred to as pVHL in subsequent descriptions.

pVHL is primarily located in the cytoplasm and is shuttled from the cytosol to the nucleus, which requires ongoing transcription by RNA polymerase II. It is composed of two functional domains: an alpha (α) domain that primarily functions in proteosomal degradation through binding to elongin C (a component of a multiprotein ubiquitin ligase complex), and a beta (β) domain that is a substrate recognition site. pVHL functions as a recessive tumor suppressor gene, and multiple lines of evidence indicate that somatic inactivation of both *VHL* alleles is a feature of sporadic renal cell carcinomas and hemangioblastomas. While the majority of renal cell carcinomas in von Hippel–Lindau show a loss of heterozygosity in the *VHL* gene, approximately 15% of these tumors show promoter hypermethylation and transcriptional silencing (Kaelin et al. 1998).

Hemangioblastomas and renal cell carcinomas, two hallmark manifestations of von Hippel–Lindau syndrome, are both highly vascular and are known to overproduce the angiogenic peptide vascular endothelial growth factor (VEGF). Along with pheochromocytomas, these three vHL-related tumors can produce erythropoietin (EPO). EPO production can also impact red-blood cell count, resulting in polycythemia, and is important for angiogenesis, an essential process in the development of these three tumors (Kertesz et al. 2004). pVHL regulates hypoxia-inducible factor (HIF), which is a chief substrate for the β domain of pVHL and controls many hypoxia-inducible genes including those that code for VEGF and EPO.

HIF is comprised of an alpha (α) subunit (HIF-1α) and HIF-1β, also called aryl hydrocarbon receptor nuclear translocator (ARNT). In the presence of oxygen, the HIF-1α subunit typically polyubiquinates rapidly, targeting these subunits for proteosomal degradation. However, in hypoxic environments, these HIF-1α subunits accumulate, undergo dimerization with the HIF-1β subunit, and bind to regulatory regions of hypoxia-inducible genes to initiate transcription. Maxwell et al. (1999) elucidated *VHL*'s tumor suppressor mechanism, showing that cells lacking pVHL were unable to degrade HIF-1α subunits, even in normoxic conditions, indicating that pVHL's interactions with HIF regulate the HIF pathway. pVHL has other HIF-1α independent activities (Gossage et al. 2015) including assembly and regulation of the extracellular matrix, microtubule stabilization and maintenance of the primary cilium, regulation of apoptosis, transcriptional regulation and control of cell senescence.

Given the currently known mechanisms of pVHL and characteristics of vHL-related tumors, different pathogenic variants may be responsible for the variable phenotype. A rough genotype–phenotype correlation was initially outlined in 2001 by Clifford and Maher given that certain manifestations of vHL seem to be prevalent in some families (Table 62.1). Type 1 vHL is characterized by deletion, nonsense, frameshift, and some missense mutations that result in risk for all vHL manifestations, except pheochromocytomas. Type 2 vHL encompasses most of the germline missense mutations, and these individuals also have increased risk for pheochromocytomas. Type 2 is broken down into three subtypes, which are delineated by missense mutations that impact HIF expression relative to the wild type (Clifford et al. 2001; Hoffman et al. 2001). Both type 1 and 2B genotypes drastically impact HIF expression relative to wild type, whereas type 2A mutations appear to have less impact on HIF expression, and type 2C maintain their ability to down-regulate HIF-1α (Clifford et al. 2001; Hoffman et al. 2001). In regards to phenotype; type 2A vHL typically manifests with all associated tumors, but with a lower risk for renal cell carcinomas, type 2B has much higher incidence of renal cell carcinoma, and type 2C often only presents with pheochromocytomas. Given that HIF-dependent functions are not impacted in type 2C, it is presumed that pheochromocytoma risk is dictated by HIF-independent functions of pVHL.

When considering genotype–phenotype relationships in vHL, there are exceptions. Case reports of patients with atypical findings within their mutation type indicate broader standardized surveillance screening should be utilized until the biology is better understood.

Von Hippel–Lindau syndrome exhibits an autosomal dominant inheritance pattern, and 20% of affected individuals have de novo mutations, meaning they are the first in their family to harbor a *VHL* mutation. Thus, 80% of affected individuals have affected family members. The penetrance of vHL manifestations is nearly 100% by age 80, though there have been exceptions. There is some evidence of genetic anticipation, suggesting that the age of onset of vHL manifestations may be earlier in subsequent generations (Aronoff et al. 2018). Once an individual is identified to have vHL, site-specific genetic testing for parents, siblings, and children are recommended through the laboratory that identified the index case's mutation. As a general rule, younger diagnoses of one or multiple vHL-related tumors and/or cysts increase the likelihood of identifying a *VHL* germline mutation. Due to the variability of presentation, even in families with the same mutation, continuous tumor monitoring in all family members with a known *VHL* mutation is strongly recommended.

Diagnostic Testing

Von Hippel–Lindau syndrome is typically diagnosed based on clinical features, including the tumors and cysts previously mentioned. If an individual meets diagnostic criteria, a pathogenic variant in *VHL* is found via germline genetic testing nearly 100% of the time (Friedrich 2001; Stolle et al. 1998). DNA-based germline genetic testing is offered at multiple laboratories worldwide. Approximately 70% of patients meeting diagnostic criteria have a missense, truncating, or splice-site mutation, while the remaining 30% have either a complete or partial deletion or insertion. There is some evidence of founder effect in countries like Mexico (c.298-299insA) and the Black Forest region of Germany (c.505T>C) (Brauch et al. 1995; Chacon-Camacho et al. 2010).

Once a pathogenic variant is identified in an affected individual, DNA-based, pre-symptomatic testing is recommended in first-degree relatives by the American Society of Clinical Oncology (2010), and subsequent testing of first-degree relatives for any individual who tests positive for a pathogenic variant in the *VHL* gene. This pre-symptomatic testing is beneficial for early detection and management of vHL manifestations, given each first-degree relative has a 50% chance of inheriting the variant and therefore developing the tumors and cysts characteristic of vHL. Rasmussen et al (2010) studied how often family members sought genetic testing after a known *VHL* variant was identified. Approximately 60% of family members underwent genetic testing, which allows for presymptomatic surveillance in individuals with a *VHL* mutation and no symptoms. This genetic assessment of at-risk family members also eliminates the need for surveillance in individuals who test negative for the pathogenic variant.

Upon clinical or molecular diagnosis, a multidisciplinary team is recommended for clinical monitoring of signs and symptoms and treatment of vHL-related manifestations. It is optimal for this monitoring to occur among clinicians familiar with the broad spectrum of vHL, which requires involvement of a multidisciplinary team of clinicians with genetics expertise.

Differential Diagnosis

A majority of the manifestations for vHL syndrome, such as hemangioblastomas, also occur sporadically in the general

TABLE 62.1 Genotype–phenotype correlations in vHL*

vHL subtype	vHL genotype	Clinical phenotype			HIF activity relative to wild type
		ccRCC	Heman	Pheo	
1A	Deletions, insertions, truncations, missense	✓	✓		↑↑↑
1B	Contiguous gene deletions encompassing *VHL*	✓	✓		
2A	Missense; e.g Y98H, Y112H, V116F	✓	✓	✓	↑
2B	Missense; e.g R167Q, R167W	✓	✓	✓	↑↑
2C	Missense; e.g. V84L, L188V, R161Q			✓	Normal

ccRCC: clear cell renal cell carcinoma, heman: hemangioblastoma pheo: pheochromocytoma
* Adapted from VHL Alliance (2015) and Gossage et al. (2015)

population, which confounds clinical diagnosis of vHL. No competing gene alteration has been identified that is also associated with the clinical features of vHL. However, these tumor types sporadically occur in the population, which supports the importance for molecular diagnosis and not relying on a clinically defined diagnosis.

Multiple studies have reported varying prevalence of vHL in cohorts of patients with hemangioblastoma, with a recent review suggesting approximately 75% of hemangioblastomas are sporadic, while the remainder are associated with vHL (Bamps et al. 2013). Sporadic hemangioblastomas are usually solitary and typically present in the fourth and fifth decade of life. It is common to see somatic *VHL* inactivation in sporadic hemangioblastomas (Shankar et al. 2014). Nonsporadic hemangioblastomas are almost always associated with vHL syndrome rather than another hereditary cancer syndrome. These vHL tumors are typically diagnosed in the second or third decade of life, can be multiple, and have a tendency for recurrence. It is notable that the presence of other vHL-related manifestations modulates the likelihood of a diagnosis of vHL; a patient with a solitary hemangioblastoma at a later age and no other findings is more likely to have a sporadic tumor than a young patient with a hemangioblastoma and family history or other manifestations of vHL.

The same holds true for retinal hemangioblastomas, where a majority of individuals with vHL syndrome had either a family history or additional manifestations of vHL in addition to their solitary retinal hemangioblastoma (Singh et al. 2002). Binderup et al. (2017b) examined the frequency of vHL in patients with at least one retinal hemangioblastoma and found that in a cohort of 64 participants, 54 (84%) had vHL. However, of the 54 patients with vHL, 41 (76%) had a family history, while none of the sporadic retinal hemangioblastomas had a family history of vHL-related tumors. Furthermore, patients with vHL had retinal hemangioblastomas at significantly younger ages, which is similar to other vHL tumor types. Six of the 54 patients with vHL (11%) had no other findings besides a single retinal hemangioblastoma, indicating genetic testing is important for evaluation of vHL. Due to the affordability and increased access to genetic testing, these data suggest that all CNS or retinal hemangioblastomas should be evaluated for vHL, though a portion of them will be sporadic.

In a study of 189 unselected renal cell carcinoma patients, the prevalence of vHL identified by germline *VHL* testing was 1.6% ($n = 3$), indicating that the rate of germline *VHL* mutations in patients with renal cell carcinoma is low (Neumann et al. 1998). However, patients who presented with germline *VHL* mutations were diagnosed with renal cell carcinoma at early ages, had multifocal or bilateral disease, or had a family history of renal cell carcinoma and/or other vHL manifestations. Multiple renal cysts, in the presence of other visceral cysts, may also be an indication for genetic evaluation, though the general population commonly presents with visceral cysts as well.

There are other hereditary cancer syndromes associated with increased risks for renal cell carcinoma. Hereditary papillary renal carcinoma presents with different pathological findings and is associated with mutations in the *MET* proto-oncogene, whereas Birt–Hogg–Dube syndrome is associated with *FLCN* mutations and individuals develop hybrid oncocytic (chromophone, oncocytoma, and clear cell) renal tumors. Individuals with *SDHx* mutations are also at increased risk for renal tumors with oncocytic neoplastic features. There have been reports of familial clear cell renal carcinoma with germline mutations localized to the 3p14.1 genetic region. Additionally, mutations in the *MITF* gene (3p13) increase the risk for renal cell carcinoma. *BAP1* is a tumor suppressor gene that when mutated, increases the risk for clear cell RCC that are typically more aggressive than those encountered in the general population.

The presence of pheochromocytoma results in a wide differential of hereditary cancer syndromes, including the *RET* gene (causing multiple endocrine neoplasia (MEN) type 2), the *SDHx* genes (which are associated with familial paraganglioma/pheochromocytoma), and less commonly, the *NF1* gene, associated with neurofibromatosis type 1 (see Chapter 40). The *SDHx* genes (composed of *SDHA*, *SDHB*, *SDHC*, *SDHD*, and *SDHAF2*) are the most suspected cause of pheochromocytoma in absence of other manifestations associated with vHL or MEN type 2. Patients with vHL-related pheochromocytomas often present differently than patients with pheochromocytomas related to *SDHx* or other hereditary cancer syndromes. vHL-related pheochromocytomas typically only secrete norepinephrine and rarely secrete catecholamines.

MANIFESTATIONS AND MANAGEMENT

As with many genetic conditions, early detection of vHL syndrome is crucial for proactive surveillance and treatment prior to individuals becoming symptomatic. Alongside broader access to molecular testing, identification of individuals with vHL has become easier with better insurance coverage of *VHL* genetic testing if a patient presents with one or more of these rare tumors (hemagioblastomas, renal cell carcinoma before age 50, pheochromocytoma), particularly if the diagnosis is made in a pediatric setting. Prompt, appropriate surveillance and therapy should begin at the time of diagnosis, and at-risk family members that test positive for the pathogenic variant are recommended to begin surveillance at age 1 year.

Given the multisystemic nature of vHL and its variable clinical expressivity, it is recommended that there is one "point person" within a health system for the surveillance and management of an individual's care. The von Hippel–Lindau Alliance (VHLA) certifies medical institutions as VHLA Clinical Care Centers (CCCs), meaning that institution has demonstrated the necessary expertise to address all aspects of vHL-related care. Within these designated CCCs, the genetic counselor or genetics nurse usually serves as a point of contact for the affected individual. In many institutions, the partnership between a genetic counselor and a high-risk oncologist or a geneticist allows for indicated genetic testing, addressing the psychosocial needs of the newly diagnosed individual, and initiation as well as oversight of recommended surveillance. The genetics core team can act as a hub for the individual's care for both a patient resource for families and primary care providers as questions arise, and as a referral source to specialty physicians, based on an individual's clinical manifestations.

As both asymptomatic and symptomatic individuals with vHL syndrome should undergo consistent surveillance, it is recommended that all members of the individuals' medical team are well-versed on the multi-systemic nature of vHL. Patients often see multiple physicians a year for vHL manifestations and average 5.7 vHL-related surgical procedures in their lifetime (Kim et al. 2016). There are multiple competing recommendations for vHL surveillance (VHL Alliance 2015; Lenders et al. 2014; Rednam et al. 2017); therefore the age at which various screenings are initiated may differ. Regardless of surveillance timing, advocacy groups like VHLA recommend the care of patients with vHL be overseen by a multidisciplinary committee, with at least one clinician from each specialty participating (i.e. ophthalmology, gastroenterology, genitourinary, etc.).

Growth and Feeding

These areas are uncommonly of primary concern in vHL syndrome; however, there are scenarios where feeding and growth issues manifest through oncological disease. Extensive central nervous system tumors are well known to effect appetite, regardless of underlying diagnosis. Additionally, focal brainstem hemangioblastomas have been associated with early satiety resulting in growth and developmental arrest (Pavesi et al. 2006; Song and Lonser 2008). Outside of the CNS, "anorexia-cachexia syndrome" is associated with renal cell carcinoma (Turner et al. 2007). As these are unlikely scenarios, each case should be managed individually depending on causation.

Evaluation

- If changes in appetite and/or growth deficiency are noted a complete neurological and abdominal assessment is warranted to search for a causative tumor.

Management

- Treatment of central nervous system tumors and renal cell carcinoma is documented below (see Oncologic).

Development and Behavior

Neither intellectual nor behavioral development is intrinsic to vHL syndrome. However, the underlying genetic predisposition for cancer has a profound psychosocial impact on those affected by the diagnosis. Psychosocial distress levels are significantly elevated in members of vHL families (Lammens et al. 2010). Psychiatric symptoms, including depression and anxiety, have demonstrated an increased prevalence (Zhang et al. 2015). If extensive central nervous system surgery is required, resulting neurological sequelae are a potential adverse outcome, which can lead to developmental problems particularly in the pediatric population.

Evaluation

- Appropriate referrals to developmental specialists and therapists based on oncology-related sequelae. Particular attention should be paid to developmental milestones in the pediatric age group, especially after central nervous system surgery.
- If developmental abnormalities are present, a full neurological evaluation is warranted.

Management

- Any treatment will depend on the type and degree of impairment.

- Treatment for developmental delays is no different from that in the general population. Psychological treatment should be tailored to the particular cohort affected.

Oncologic

Hemangioblastoma
Craniospinal and retinal hemangioblastomas, although overall a rare CNS tumor accounting for approximately 2% of CNS tumors (Plate 2007), are the most frequent of all tumors encountered in patients with vHL at rates of up to 90% (Lonser et al. 2014). As mentioned previously, they are what led Dr. Lindau down his initial path of discovery in the early 20th century (Huntoon et al. 2015). The cytological origin of hemangioblastoma and its evolutionary process have remained controversial for nearly a century. Vascular progenitors including embryonic cells, reactive endothelial cells, stromal cells, and mast cells have all been theorized to play vital rolls in development of this tumor type (Wang et al. 2017). While the exact etiology of hemangioblastoma remains unclear, there is a consensus the process is established during embryonic development; however, CNS hemangioblastomas rarely arise during childhood, occurring at a mean age of 33 years (Varshney et al. 2017). This is consistent with their well-known, slow-growing behavior. They are most commonly benign, non-invasive, asymptomatic tumors; however, with continual growth, they can become symptomatic. Patients typically harbor multiple tumors with averages ranging from seven to eight tumors. Intracranially, they are most commonly seen infratentorially in the cerebellum and brainstem (Huntoon and Lonser 2014). Supratentorial hemangioblastomas are rare with the majority occurring in the cerebrum (Mills et al. 2012). Spinal cord tumors are frequently encountered, with the cervical (40%) and thoracic areas (54%) being the most common sites of involvement (Kanno et al. 2008). Hemangioblastomas are rarely found outside the CNS, but include lung, liver, pancreas, bladder, and kidney (Zhao et al. 2017).

Tumor-associated symptoms are based on their anatomical locations: intracranial tumors are most frequently associated with headaches; cerebellar tumors cause more frequent gait disturbances; supratentorial tumors exhibit a predilection for visual disturbances; and spinal tumors often lead to paresthesias (Lonser et al. 2014). Occasionally patients demonstrate polycythemia due to erythropoietin secretion from hemangioblastoma cells (Kanno et al. 2013).

A commonly encountered characteristic of CNS hemangioblastomas is accompanying peritumoral cysts, occurring in nearly 60% of patients (Huntoon et al. 2016). These cysts form as a result of plasma ultrafiltrate leaking through permeable tumor vessels into the surrounding CNS tissue (Lonser et al. 2005). They are most frequently encountered in cerebellar hemangioblastomas but are observed throughout the CNS and are important in understanding symptomatology. While less than 10% of hemangioblastomas are symptomatic, tumors with associated peritumoral cysts exhibit symptoms at rates as high as 70% (Wanebo et al. 2003). Tumors associated with peritumoral cysts demonstrate increased lesion volume and growth rate.

Retinal tumors, found most commonly at the retinal periphery, have historically been grouped as an entity separate from CNS hemangioblastomas. In a recent national cohort study, vHL was the underlying cause in 84% of retinal hemangioblastomas (Binderup et al. 2017b). Although not seen as frequently as CNS hemangioblastomas, occurring in up to 60% of patients, they are often the initial manifestation of vHL and have been diagnosed as young as 6 years old (Dollfus et al. 2002; Singh et al. 2013). Patients with unilateral disease are at risk of developing bilateral disease by age 60 (Chittiboina and Lonser 2015). While many patients are asymptomatic, a result of accumulated exudate or tractional dysfunction, patients may experience visual field defects and even blindness at rates of 5–8% (Chittiboina and Lonser 2015; Singh et al. 2013).

Evolution of hemangioblastomas have been well described. In long-term follow up, 50% of hemangioblastomas will grow and of those patients with tumors, 75% will go on to develop additional tumors (Huntoon and Lonser 2014). Growth patterns include salutatory (exhibiting periods of growth and quiescence), which is most common, linear growth, and/or exponential growth. Tumors of the brain tend to grow faster than those of the spinal cord (Huntoon and Lonser 2014).

Many different characteristics have been associated with clinical course in hemangioblastoma. Male sex is associated with increased tumor burden and growth rate of tumors. Those with increased tumor burdens and of younger ages (<20 years old) of onset also have increased tumor growth rates (Huntoon and Lonser 2014). Genetic status is associated with clinical course as patients with partial germline deletions (type 1) have increased tumor burden (Lonser et al. 2014). Specifically regarding retinal tumors, worse visual outcomes are associated with early age of ocular disease, bilateral involvement, and missense mutations (Chittiboina and Lonser 2015; Lonser et al. 2014; Singh et al. 2013). Importantly, in a recent review CNS hemangioblastomas were the leading cause of vHL-associated mortality (Lonser et al. 2014).

Renal Cell Carcinomas Renal cell carcinoma (RCC) is the most frequently seen malignant tumor in patients with vHL syndrome, with an estimated risk of 80% by the age of 60 compared to the general population, which has a lifetime risk of 1.6% (Howlader et al. posted to the SEER web site April 2017; Ong et al. 2007). Additionally, patients often develop RCC at an earlier age than patients without vHL (median 39.7 years versus 64 years, respectively) (Howlader

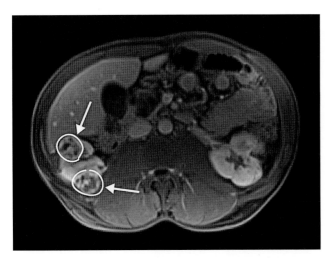

FIGURE 62.1 Axial post-contrast T1-weighted MRI image of kidneys.

et al. posted to the SEER web site April 2017; Maher et al. 1990; Ong et al. 2007). Patients with vHL frequently develop multifocal renal cysts and RCC lesions with clear cell histology at the same time or subsequently (Figure 62.1) (Mandriota et al. 2002). Other histologies, such as sarcomatoid, chromophobe, and papillary, are not frequently seen in patients with vHL. The penetrance of renal cell carcinoma in individuals with vHL varies by type, as highlighted above.

Renal cell carcinoma, alongside hemangioblastomas, are some of the most impactful manifestations of vHL, accounting for nearly a third of vHL-related deaths (Binderup et al. 2017a). Furthermore, individuals may develop multiple renal cell carcinomas, and patients who undergo at least one renal procedure average approximately 2.69 renal surgeries in their lifetime (Kim et al. 2016). The rate of tumor growth does not seem impacted by the vHL subtype (Farhadi et al. 2018). Patients were historically diagnosed with RCC after the onset of symptoms such as hematuria and flank pain, whereas most patients now are diagnosed with small asymptomatic lesions. This is the result of comprehensive screening protocols, including annual abdominal imaging. Despite patients developing multifocal lesions, renal function is generally preserved. The loss of renal function generally is a consequence of treatment and not disease progression. To prevent or delay the potential need for dialysis and renal transplant, the treatment of RCC has evolved from complete nephrectomy to the more common nephron-sparing approaches.

Pheochromocytomas Pheochromocytomas, tumors of the adrenal medulla, are less common in vHL than RCC, with an estimated incidence of less than 40% by age 60 years old (Ong et al. 2007). The median age of onset is estimated at 24 years. Additionally, extra-adrenal parasympathetic neuroendocrine cancers (paragangliomas) can occur throughout the body including the abdominal-pelvic and head/neck regions (Gaal et al. 2009). Patients are frequently diagnosed with paragangliomas or pheochromocytomas on routine imaging as a consequence of comprehensive screening protocols. Pheochromocytomas in the general population often produce significant catecholamine levels; however, individuals with vHL-related pheochromocytoma typically only secrete normetanephrine (Binderup et al. 2013; Schmid et al. 2014). Both patterns of secretion can cause patients to experience multiple symptoms. For patients who are symptomatic prior to diagnosis, symptoms may include: headaches, tachycardia, labile hypertension, heart palpitations, diaphoresis or nausea. Conversely, paragangliomas are infrequently associated with catecholamine release and patients are often asymptomatic. The risk of metastatic pheochromocytoma or paraganglioma is unknown in the vHL population; however, when considering all patients diagnosed with pheochromocytoma, metastatic disease is uncommon, estimated at approximately 10% (Goldstein et al. 1999). Based on our institutional experience, the malignant potential of pheochromocytoma in vHL is perhaps slightly less frequent than the general population.

The absolute risk of pheochromocytoma varies based on the specific genomic alterations in the *VHL* gene (Neumann et al. 2002; Ong et al. 2007). Notably, there is a low risk for pheochromocytomas in patients with vHL type 1, due to the HIF expression addressed previously. Given that pheochromocytomas are often the presenting manifestation at the time of a vHL diagnosis, and can also indicate risk for other hereditary cancer syndromes, genetic testing for all patients with pheochromocytomas is recommended.

Endolymphatic Sac Tumors Part of the membranous labyrinth within the inner ear, endolymphatic sacs secrete locally acting saccin and play a key role in inner ear fluid homeostasis. Tumors that arise from the endolymphatic sac are typically benign, and though slowly progressive, can be highly destructive due to growth into key structures in the inner ear. They can invade and destroy the bone of the inner ear and infiltrate the nerves, pressing on the cerebellum and passing through the dura (Figure 62.2). Many patients are asymptomatic until the inner ear is partially destroyed, at which point they may present with hearing loss, ringing in the ears, facial paralysis, dizziness/imbalance, headaches, and other neurologic weaknesses (Gluth 2015).

Endolymphatic sac tumors (ELST) occur in 10–15% of patients with vHL syndrome, and are typically unilateral. However, 30% of patients with an ELST will have bilateral ELSTs. ELSTs typically occur by age 45 years. While some signs of ELST such as dizziness and headaches may be nonspecific, attention to any progressive symptoms should prompt evaluation for ELST. Early resection is recommended to reduce facial nerve involvement; however, the recurrence rate after surgery can be as high as 40%.

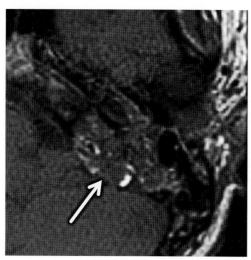 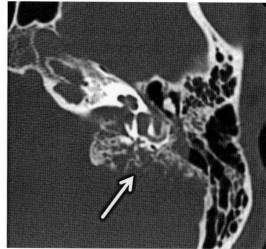

FIGURE 62.2 Endolymphatic sac tumor on CT.

Evaluation

- Timing of surveillance initiation is debated among vHL professionals, and decisions to start screening should be made in concert with family considering all internal and external factors.
- Starting at age 1, individuals should undergo annual physical exam to identify signs of neurological disturbance and ocular abnormalities.
- Annual eye or retinal examination with indirect ophthalmoscopy should be initiated at 1 year old (VHL Alliance 2015).
- Starting at age 16 years, MRI screening every two years with contrast of brain, cervical, thoracic, and lumbar spine should be initiated unless other manifestations suggest starting earlier (VHL Alliance 2015).
- Quality abdominal/pelvic ultrasound and/or MRI annually to evaluate for cysts, renal cell carcinoma, and pheochromocytoma. In the case of suspicious lesions, increased interval imaging may be warranted. Mutual agreement across expert consortia has yet to be reached in regards to optimal age of initiating imaging surveillance; however, 16 years old is appropriate given that renal cell carcinoma and cyst manifestations are rare before that age.
- Plasma metanephrines should be checked annually, notably for normetanephrines, which are most commonly secreted in vHL-related pheochromocytomas.
- If a pheochromocytoma is identified, nuclear imaging including NETSpot may be recommended for detection of metastases and localization of primary extraadrenal lesions.
- Hearing tests are recommended at baseline diagnosis of vHL and then every two to three years until age 45 to determine any changes in hearing.
  - If hearing test is abnormal, MRI with and without contrast is the recommended imaging modality, though CT is an option.
  - Negative imaging does not preclude ELST and monitoring for symptoms indicative of these tumors is essential. Hearing tests may be completed on an annual basis for patients with abnormal hearing tests and negative imaging. Speech discrimination may be a tool; however scores have varied in patients with ELST in the past.

Management

Hemangioblastoma

- Therapy for CNS hemangioblastoma should be individualized according to patient demographics, treatment history, tumor location, tumor size, and symptomatology.
- Mainstay of therapy for CNS hemangioblastoma is surgical resection. Optimal timing for surgery remains controversial. Some have advocated for resection of asymptomatic lesions based on location (Joaquim et al. 2015). Recent evidence indicates growth rate and size are unreliable predictors of future symptomatology and surgical management should be reserved for symptomatic lesions only (Lonser et al. 2014).
- Complete surgical resection can be curative although not always possible in certain locations, i.e. brainstem and cerebellum.
- Surgical technique is dictated by location of tumor and surrounding tissues. As hemangioblastomas are highly vascular, care must be taken during resection to avoid premature invasion of the tumor that may lead to bleeding complications (Ordookhanian et al. 2017).

- Intraoperative neurosurgical monitoring is associated with improved outcomes in resection of spinal hemangioblastomas (Siller et al. 2017).
- Older age and patients with solid tumors (as opposed to cystic) are associated with inferior long-term surgical outcomes (Fukuda et al. 2014).
- Preoperative chemoembolization can be used in an attempt to decrease tumor size and minimize post-operative bleeding (Ordookhanian et al. 2017).
- Stereotactic radiosurgery to control small, solid, inoperable tumors may be employed. It is also useful in the setting of multiple tumors in patients hoping to avoid or unable to tolerate multiple surgical resections (Moss et al. 2009). A recent systemic review demonstrated the technique resulted in a 96% stable or improved long-term clinical effect; however, a direct comparison of outcomes of stereotactic radiosurgery versus surgical resection has not been performed (Bridges et al. 2017). Consequently, the long term benefits of this modality continue to be debated (Ordookhanian et al. 2017).
- Fractionated external beam and infratentorial craniospinal radiation therapy have been employed with encouraging results (Koh et al. 2007; Simone et al. 2011).
- There are no approved chemotherapeutics for treatment of CNS hemangioblastoma. Several investigations are ongoing that focus on anti-angiogenesis and kinase inhibition (Agarwal et al. 2015).
- Retinal hemangioblastomas are treated by ophthalmologists in a variety of ways, based on size. Small lesions are treated with laser photocoagulation. Larger more complicated lesions may be treated with either cryotherapy, radiotherapy, or vitreoretinal surgery. Tumors causing intractable pain or glaucoma may be treated with enucleation (Shanbhogue et al. 2016).

Renal
- Renal lesions that are small are initially observed to assess the rate of growth and preserve renal function. Von Hippel–Lindau-associated RCC has a very low rate of metastasis, below 3 cm (Walther et al. 1999), therefore patients are initially followed with frequent surveillance.
- Renal cell carcinoma metastasis is significantly higher in tumors greater than 4 cm in size. Nephron sparing approaches include either partial nephrectomy or radiofrequency ablation. These methods are typically performed on lesions that are approximately 3 cm or greater in size. At high-volume centers, this approach has low complication rates and is effective in preserving renal function though this approach has high complication rates in centers with less experience (Herring et al. 2001; Park and Kim 2010).
- A novel development in localized disease is the potential use of vascular endothelial growth factor receptor (VEGFR) antagonists to delay the need for surgery and simultaneously prevent progression to metastatic disease. In early clinical trials, sunitinib and pazopanib have resulted in partial or complete responses in renal cell carcinomas. Interestingly, some pancreatic and CNS lesions also regressed, suggesting this strategy may prove effective as a renal-sparing therapy in the future.
- Though uncommon, metastatic disease occurs in some patients with vHL. Some reports (Kim et al. 2013; Tsimafeyeu 2015) and retrospective data (Roma et al. 2015) strongly suggest that these patients respond to VEGFR therapy.
- Patients with vHL-associated metastatic RCC should be treated according to the established recommendations for all patients with metastatic RCC including the use of VEGF antagonists and checkpoint inhibitors. Previously published reviews and the NCCN guidelines clearly outline effective treatment guidelines for renal cell carcinoma and should be followed, regardless of vHL status (Choueiri and Motzer 2017).
- Recent clinical trials continue to demonstrate the effectiveness of VEGFR antagonists (Choueiri et al. 2016; Choueiri et al. 2017). More recently, immunotherapy in metastatic renal cell carcinoma is also effective (Escudier et al. September 2017; Motzer et al. 2015). These therapies are expected to also be effective in vHL-associated RCC.

Pheochromocytoma
- Surgery is the recommended treatment for localized pheochromocytomas. Both open and laparoscopic approaches have been used successfully. The patient should have complete surgical resection, if feasible, of entire tumor with sparing of unaffected adrenal tissue, if possible.
- Bilateral adrenalectomies have safely been performed for multifocal disease.
- Prior to any surgical intervention, referral to endocrinology is strongly recommended for extensive evaluation of catecholamine production by the tumor. Patients with high catecholamine levels should be treated with alpha-blockade under the direction of an endocrinologist prior to surgery to avoid treatment-related complications.
- Patients with metastatic pheochromocytoma can be treated with standard chemotherapy, targeted therapy, and radiopharmaceuticals, as is the standard for any patient with metastatic pheochromocytoma or metastatic paraganglioma, irrespective of vHL diagnosis.

Our institutional preference is to use radiopharmaceuticals as first-line therapy given the better tolerability compared to chemotherapy and the greater level of evidence for efficacy over targeted therapy.

- Until recently, the most commonly used radiopharmaceutical was radioactive iodine attached to 123-meta-iodobenzylguanidine (MIBG), which is structurally similar to catecholamines and is taken up by chromaffin cells. 131I-MIBG can be effective to diagnose and treat both pheochromocytomas and paragangliomas (van der Harst et al. 2001; van Hulsteijn et al. 2014). This treatment is generally only effective for tumors identified by the MIGB nuclear medicine scan.
- Potentially more effective radiopharmaceuticals are being developed with FDA approval expected sometime soon. 90Y-DOTA-TOC and 177Lu-DOTATATE, radiolabeled somatostatin analogs, are being developed and currently seeking FDA approval for the treatment of neuroendocrine tumors.
- Targeted therapy is also used though only weak evidence in the form of small case series, small clinical trials or case reports is available in support of this treatment. Some evidence exists supporting the use of octreotide, pazopanib, sunitinib, and everolimus (Hahn et al. 2009; Jasim et al. 2017; Oh et al. 2012; Tonyukuk et al. 2003).
- Chemotherapy is also used to treat malignant pheochromocytoma and paraganglioma, though most evidence is from single institution experiences or case reports. When used, chemotherapy combination is preferred over single-agent therapy (e.g. cyclophosphamide, vincristine, or dacarbazine).

Endolymphatic Sac Tumors (ELSTs)

- Surgical resection utilizing a transtemporal approach with potential need for craniotomy depending on tumor involvement is the primary treatment for ELSTs. Cochlear implants may be used in the case of hearing loss. Nerve graft repair may be required if the facial nerves are compromised by the tumor.
- Given the potential for recurrence, there is limited data on chemotherapy or radiation therapy for treatment of ELSTs.

Gastrointestinal

Pancreatic cystic lesions are present in over 60% of individuals with vHL, though studies report prevalence as high as 87% (Park et al. 2015). A majority of patients with pancreatic lesions are asymptomatic, and lesions most commonly present as simple pancreatic cysts (Figure 62.3). It is more common to see multiple simple pancreatic cysts rather than a single cyst (Keutgen et al. 2016). Lesions in the pancreas

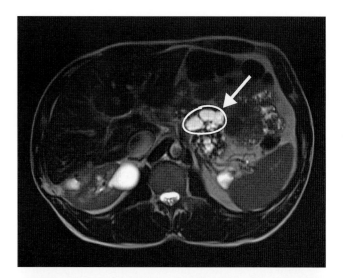
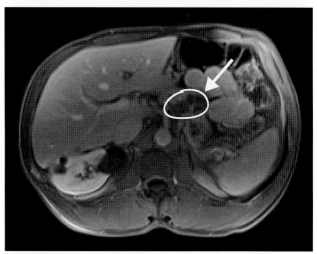

FIGURE 62.3 (Top) MRI axial T2-HASTE sequence showing pancreatic cysts, and (bottom) axial post-contrast T1-weighted image of pancreatic cysts.

also present as serous cystadenomas, branch duct intraductal papillary neoplasms, and cystic neuroendocrine tumors (NETs) (Sharma et al. 2017). Pancreatic NETs, though they only occur in up to 17% of individuals with vHL, are typically non-functional; however, they are the most concerning given their propensity for metastasis (Ito et al. 2012). Due to the typically asymptomatic nature of many pancreatic findings, up to half of pancreatic NETs in patients with vHL are malignant. This requires additional attention to surveillance and treatment decisions. Duodenal and biliary tract neuroendocrine tumors have also been reported (Fellows et al. 1990; Karasawa et al. 2001). Large pancreatic cysts or cysts located in the head of the pancreas can cause bile duct or other obstructions, resulting in exocrine pancreatic insufficiency (van Asselt et al. 2013).

Patients with vHL typically develop pancreatic lesions and/or endocrine tumors in their early 30s, and progression of different lesion types is variable (Feletti et al. 2016). For instance, pancreatic cysts show their first significant growth after approximately five years, whereas neuroendocrine tumors are typically surgically treated and therefore show progression-free survival of approximately nine years.

Evaluation

- MRI is recommended for surveillance of pancreatic lesions due to its precision delineating and characterizing different types of lesions (Shanbhogue et al. 2016).
- Symptoms related to exocrine pancreatic insufficiency such as diarrhea, stomach pain, and unexplained weight loss should prompt additional pancreatic workup.

Management

- A majority of pancreatic lesions are simple cysts that require no need for surgical or other interventions. Active surveillance of these lesions occurs as a result of the consistent abdominal imaging in patients with vHL.
- Resection of pancreatic NET tumors that are greater than 30 mm in diameter (or >20 mm in the pancreatic head) or symptomatic should be considered. If other abdominal surgeries are occurring, resection of pNET tumors not meeting the aforementioned criteria may also be warranted.
- Surgical intervention may also be warranted in the case of bile duct obstruction.
- Oral enzyme replacement is recommended in cases of exocrine pancreatic insufficiency.

Genitourinary

Multiple renal cysts, often found bilaterally, are one of the most common features of vHL syndrome, with estimates from 40 to nearly 90% of individuals with vHL developing renal cysts (Feletti et al. 2016). While these benign lesions may grow over time, it is uncommon for renal function to be impaired, despite some patients developing complex multiloculated cysts.

Benign tumors derived from the embryonic mesonephric duct, called papillary cystadenomas, also occur at a higher incidence in patients with vHL syndrome than the general population. It is estimated that up to 60% of males with vHL will develop papillary cystadenomas (Choyke et al. 1997), typically in the epididymis or ductus deferens. These cystadenomas are typically small and unilateral, though a presentation of bilateral cystadenomas raises suspicion for vHL syndrome. While typically asymptomatic and slow growing, there are some case reports of bilateral cystadenomas causing infertility (Fidelis et al. 2017; Witten et al. 1985). In a rapidly enlarging epididymal tumor, it is important to differentiate cystadenoma from metastatic renal cell carcinoma (Aydin et al. 2005).

Comparable tumors have been described in women with vHL syndrome, and typically arise in the broad ligament or mesosalpinx. It arises much less commonly in women, and, as of 2016, there were only 11 cases reported in the literature (DeLair and Soslow 2016).

Evaluation

- Abdominal imaging recommended for renal cell carcinoma will capture renal cysts. Though they are often non-intrusive, monitoring size and growth of these cysts is recommended.
- Palpation of the tests in males for epididymal cystadenomas should be performed on physical examination.
- Differentiation between tumors of the epididymis or mesosalpinx is critical to ensure an unrelated testicular tumor is not the cause of abnormal testicular findings. Urological consultation is recommended for this evaluation.
- Sonographic criteria for diagnosis of papillary cystadenoma include (1) predominantly solid tumor greater than 14 × 10 mm in size, (2) occurrence in a man with vHL disease, and (3) slow growth (Shanbhogue et al. 2016).

Management

- Renal cysts are typically asymptomatic and do not require treatment; however, monitoring is recommended in the event size or location impacts renal function.
- There is typically no treatment associated with tumors of the epididymis and mesosalpinx unless the patient is symptomatic.
- Reproductive specialists may be consulted in the case of infertility due to bilateral obstructive tumors.

ACKNOWLEDGMENT

The authors are grateful to Amanda Gammon for her assistance in revisions and content evaluation.

RESOURCES

VHL Alliance

1208 VFW Parkway, Suite 303
Boston, MA 02132-4344, USA
Phone: 800-767-4845x1 or 617-277-5667 x4
Fax: 866-209-0288
Email: info@vhl.org
Website: www.vhl.org

As the primary hub of online resources and contacts for patients with vHL, they designate comprehensive care centers and have affiliate care centers and contacts in many other countries. They also provide patient and caregiver resources, support groups, online webinars, and annual conferences.

International affiliates:

Europe: www.vhl-europa.org
Croatia and Hungary: www.vhl.europa.org/hr
Denmark: www.vhl.dk
England, Scotland, Wales, Northern Ireland: www.vhl.org
France: www.vhlfrance.org
Germany: www.hippel-lindau.de
Italy: www.vhl.it
Spain: www.alianzavhl.org
Canada: https://www.vhl.org/canada/
Japan: www.vhl.org

There are other affiliates in Argentina, New Zealand, Poland, Portugal, Romanic, Chile, South Africa, Sweden, Switzerland, Turkey, Iceland, and Greece.

The Von Hippel–Lindau Handbook: What You Need to Know About VHL

A reference handbook for people with VHL, their families, and medical personnel. Edition 5, revised 2015, https://www.vhl.org/patients/vhl-handbook/. This is available in multiple languages including Spanish, Portuguese, French, Dutch, Belgian, and Japanese.

Other Web Resources

GeneReviews: von Hippel-Lindau Syndrome
https://www.ncbi.nlm.nih.gov/books/NBK1463/
National Library of Medicine Genetics Home Reference
https://ghr.nlm.nih.gov/condition/von-hippel-lindau-syndrome
Statements and Policies regarding Genetic Testing
Robson ME, Bradbury AR, Arun B, et al. (2010) American Society of Clinical Oncology policy statement update: genetic and genomic testing for cancer susceptibility. *J Clin Oncol* 33:893–901.

REFERENCES

Agarwal R, Liebe S, Turski ML, Vidwans SJ, Janku F, Garrido-Laguna I, Munoz J, Schwab R, Rodon J, Kurzrock R (2015) Targeted therapy for genetic cancer syndromes: Von Hippel-Lindau disease, Cowden syndrome, and Proteus syndrome. *Discov Med* 19(103):109–116.

Aronoff L, Malkin D, van Engelen K, Gallinger B, Wasserman J, Kim RH, Villani A, Meyn MS, Druker H (2018) Evidence for genetic anticipation in vonHippel-Lindau syndrome. *J Med Genet* 55(6):395–402.

Aydin H, Young RH, Ronnett BM, Epstein JI (2005) Clear cell papillary cystadenoma of the epididymis and mesosalpinx: immunohistochemical differentiation from metastatic clear cell renal cell carcinoma. *Am J Surg Pathol* 29(4):520–523.

Bamps S, Calenbergh FV, Vleeschouwer SD, Loon JV, Sciot R, Legius E, Goffin J (2013) What the neurosurgeon should know about hemangioblastoma, both sporadic and in Von Hippel-Lindau disease: A literature review. *Surg Neurol Int* 4:145.

Bender BU, Eng C, Olschewski M, Berger DP, Laubenberger J, Altehofer C, Kirste G, Orszagh M, van Velthoven V, Miosczka H, Schmidt D, Neumann HP (2001) VHL c.505 T>C mutation confers a high age related penetrance but no increased overall mortality. *J Med Genet* 38(8):508–514.

Binderup ML, Bisgaard ML, Harbud V, Moller HU, Gimsing S, Friis-Hansen L, Hansen T, Bagi P, Knigge U, Kosteljanetz M, Bogeskov L, Thomsen C, Gerdes AM, Ousager LB, Sunde L, Danish v HLCG (2013) Von Hippel-Lindau disease (vHL). National clinical guideline for diagnosis and surveillance in Denmark. 3rd edition. *Dan Med J* 60(12):B4763.

Binderup ML, Jensen AM, Budtz-Jorgensen E, Bisgaard ML (2017a) Survival and causes of death in patients with von Hippel-Lindau disease. *J Med Genet* 54(1):11–18.

Binderup MLM, Stendell AS, Galanakis M, Moller HU, Kiilgaard JF, Bisgaard ML (2017b) Retinal hemangioblastoma: prevalence, incidence and frequency of underlying von Hippel-Lindau disease. *Br J Ophthalmol* 102(7):bjophthalmol-2017-310884

Brauch H, Kishida T, Glavac D, Chen F, Pausch F, Hofler H, Latif F, Lerman MI, Zbar B, Neumann HP (1995) Von Hippel-Lindau (VHL) disease with pheochromocytoma in the Black Forest region of Germany: evidence for a founder effect. *Hum Genet* 95(5):551–556.

Bridges KJ, Jaboin JJ, Kubicky CD, Than KD (2017) Stereotactic radiosurgery versus surgical resection for spinal hemangioblastoma: A systematic review. *Clin Neurol Neurosurg* 154:59–66.

Chacon-Camacho OF, Rodriguez-Dennen F, Camacho-Molina A, Rasmussen A, Alonso-Vilatela E, Zenteno JC (2010) Clinical and molecular features of familial and sporadic cases of von Hippel-Lindau disease from Mexico. *Clin Exp Opthalmol* 38(3):277–283.

Chen F, Kishida T, Yao M, Hustad T, Glavac D, Dean M, Gnarra JR, Orcutt ML, Duh FM, Glenn G, et al (1995) Germline mutations in the von Hippel-Lindau disease tumor suppressor gene: correlations with phenotype. *Hum Mut* 5(1):66–75.

Chittiboina P, Lonser RR (2015) Von Hippel-Lindau disease. *Handbook of clinical neurology* 132:139–156.

Choueiri TK, Escudier B, Powles T, Tannir NM, Mainwaring PN, Rini BI, Hammers HJ, Donskov F, Roth BJ, Peltola K, Lee JL, Heng DYC, Schmidinger M, Agarwal N, Sternberg CN, McDermott DF, Aftab DT, Hessel C, Scheffold C, Schwab G, Hutson TE, Pal S, Motzer RJ, MD for the METEOR investigators (2016) Cabozantinib versus everolimus in advanced renal

Choueiri TK, Halabi S, Sanford BL, Hahn O, Michaelson MD, Walsh MK, Feldman DR, Olencki T, Picus J, Small EJ, Dakhil S, George DJ, Morris MJ (2017) Cabozantinib versus sunitinib as initial targeted therapy for patients with metastatic renal cell carcinoma of poor or intermediate risk: The Alliance A031203 CABOSUN Trial. *J Clin Oncol* 35(6):591–597.

cell carcinoma (METEOR): final results from a randomised, open-label, phase 3 trial. *Lancet Oncol* 17(7):917–927.

Choueiri TK, Motzer RJ (2017) Systemic Therapy for Metastatic Renal-Cell Carcinoma. *N Engl J Med* 376(4):354–366.

Choyke PL, Glenn GM, Wagner JP, Lubensky IA, Thakore K, Zbar B, Linehan WM, Walther MM (1997) Epididymal cystadenomas in von Hippel-Lindau disease. *Urology* 49(6):926–931.

Clifford SC, Cockman ME, Smallwood AC, Mole DR, Woodward ER, Maxwell PH, Ratcliffe PJ, Maher ER (2001) Contrasting effects on HIF-1alpha regulation by disease-causing pVHL mutations correlate with patterns of tumourigenesis in von Hippel-Lindau disease. *Hum Mol Genet* 10(10):1029–1038.

Collins ET (1894) Intra-ocular growths (two cases, brother and sister, with peculiar vascular new growth, probably retinal, affecting both eyes). *Trans Ophthalmol Soc UK* 14:141–149.

DeLair DF, Soslow RA (2016) Gynecologic Manifestations of Less Commonly Encountered Hereditary Syndromes. *Surg Pathol Clin* 9(2):269–287.

Dollfus H, Massin P, Taupin P, Nemeth C, Amara S, Giraud S, Beroud C, Dureau P, Gaudric A, Landais P, Richard S (2002) Retinal hemangioblastoma in von Hippel-Lindau disease: a clinical and molecular study. *Invest Opthalmol Vis Sci* 43(9):3067–3074.

Escudier B, Tannir NM, McDermott DF, Frontera OA, Melichar B, Plimack ER, Barthelemy P, George S, Neiman V, Porta C, Choueiri TK, Powles T, Donskov F, Salman P, Kollmannsberger CK, Rini B, Mekan S, McHenry MB, Hammers HJ, Motzer RJ (2017) CheckMate 214: Efficacy and safety of nivolumab + ipilimumab (N+I) v sunitinib (S) for treatment-naïve advanced or metastatic renal cell carcinoma (mRCC), including IMDC risk and PD-L1 expression subgroups. *Ann Oncol* 28(Suppl_5): v605–sv649.

Farhadi F, Nikpanah M, Li X, Symons R, Pourmorteza A, Merino MJ, Linehan WM, Malayeri AA (2018) Germline VHL gene variant in patients with von Hippel-Lindau disease does not predict renal tumor growth. *Abdom Radiol (NY)* 43(10):2743–2749.

Feletti A, Anglani M, Scarpa B, Schiavi F, Boaretto F, Zovato S, Taschin E, Gardi M, Zanoletti E, Piermarocchi S (2016) Von Hippel-Lindau disease: an evaluation of natural history and functional disability. *Neuro-oncology* 18(7):1011–1020.

Fellows I, Leach I, Smith P, Toghill P, Doran J (1990) Carcinoid tumour of the common bile duct--a novel complication of von Hippel-Lindau syndrome. *Gut* 31(6):728–729.

Fidelis OO, Edet I, Glen E, Edoise IM, Ayodele O, Akanimo E, Ekwere PD, Nkposong EO (2017) Papillary cystadenoma of right testis: Case report and literature review. *Case Rep Clin Pathol* 4(2):8.

Friedrich CA (2001) Genotype-phenotype correlation in von Hippel-Lindau syndrome. *Hum Mol Genet* 10(7):763–767.

Fukuda M, Takao T, Hiraishi T, Yoshimura J, Yajima N, Saito A, Fujii Y (2014) Clinical factors predicting outcomes after surgical resection for sporadic cerebellar hemangioblastomas. *World Neurosurg* 82(5):815–821.

Gaal J, van Nederveen FH, Erlic Z, Korpershoek E, Oldenburg R, Boedeker CC, Kontny U, Neumann HP, Dinjens WN, de Krijger RR (2009) Parasympathetic paragangliomas are part of the Von Hippel-Lindau syndrome. *J Clin Endocrinol Metab* 94(11):4367–4371.

Glenn GM, Daniel LN, Choyke P, Linehan WM, Oldfield E, Gorin MB, Hosoe S, Latif F, Weiss G, Walther M, Lerman MI, Zbar B (1991) Vonhippel-Lindau (Vhl) disease - distinct phenotypes suggest more than one mutant allele at the Vhl locus. *Hum Genet* 87(2):207–210.

Gluth MB (2015) *Endolymphatic Sac Tumors: An Overview*. Boston: VHL Alliance.

Goldstein RE, O'Neill JA, Jr., Holcomb GW, 3rd, Morgan WM, 3rd, Neblett WW, 3rd, Oates JA, Brown N, Nadeau J, Smith B, Page DL, Abumrad NN, Scott HW, Jr (1999) Clinical experience over 48 years with pheochromocytoma. *Ann Surg* 229(6):755–764; discussion 764–756.

Gossage L, Eisen T, Maher ER (2015) VHL, the story of a tumour suppressor gene. *Nat Rev Cancer* 15(1):55–64.

Hahn NM, Reckova M, Cheng L, Baldridge LA, Cummings OW, Sweeney CJ (2009) Patient with malignant paraganglioma responding to the multikinase inhibitor sunitinib malate. *J Clin Oncol* 27(3):460–463.

Herring JC, Enquist EG, Chernoff A, Linehan WM, Choyke PL, Walther MM (2001) Parenchymal sparing surgery in patients with hereditary renal cell carcinoma: 10-year experience. *J Urol* 165(3):777–781.

Hoffman MA, Ohh M, Yang H, Klco JM, Ivan M, Kaelin WG, Jr (2001) von Hippel-Lindau protein mutants linked to type 2C VHL disease preserve the ability to downregulate HIF. *Hum Mol Genet* 10(10):1019–1027.

Howlader N, Noone AM, Krapcho M, Miller D, Bishop K, Kosary CL, Yu M, Ruhl J, Tatalovich Z, Mariotto A, Lewis DR, Chen HS, Feuer EJ, Cronin KAe (2017) SEER Cancer Statistics Review, 1975-2014, (Kidney and Renal Pelvis Cancer). National Cancer Institute Bethesda, MD, https://seercancergov/csr/1975_2014/ based on November 2016 SEER data submission.

Huntoon K, Lonser RR (2014) Findings from the natural history of central nervous system hemangioblastomas in von Hippel-Lindau disease. *Neurosurgery* 61(CN suppl 1):159–162.

Huntoon K, Oldfield EH, Lonser RR (2015) Dr. Arvid Lindau and discovery of von Hippel-Lindau disease. *J Neurosurg* 123(4):1093–1097.

Huntoon K, Wu T, Elder JB, Butman JA, Chew EY, Linehan WM, Oldfield EH, Lonser RR (2016) Biological and clinical impact of hemangioblastoma-associated peritumoral cysts in von Hippel-Lindau disease. *J Neurosurg* 124(4):971–976.

Iliopoulos O, Ohh M, Kaelin WG, Jr (1998) pVHL19 is a biologically active product of the von Hippel-Lindau gene arising from internal translation initiation. *Proc Natl Acad Sci U S A* 95(20):11661–11666.

Ito T, Igarashi H, Jensen RT (2012) Pancreatic neuroendocrine tumors: Clinical features, diagnosis and medical treatment: Advances. *Best Prac Res Clin Gastroenterol* 26(6):737–753.

Jasim S, Suman VJ, Jimenez C, Harris P, Sideras K, Burton JK, Worden FP, Auchri RJ, Bible KC (2017) Phase II trial of pazopanib in advanced/progressive malignant pheochromocytoma and paraganglioma. *Endocrine* 57(2):220–225.

Joaquim AF, Ghizoni E, dos Santos MJ, Valadares MG, da Silva FS, Tedeschi H (2015) Intramedullary hemangioblastomas: surgical results in 16 patients. *Neurosurg Focus* 39(2):E18.

Kaelin WG, Iliopoulos O, Lonergan KM, Ohh M (1998) Functions of the von Hippel-Lindau tumour suppressor protein. *J Int Med* 243(6):535–539.

Kaelin WG, Jr (2002) Molecular basis of the VHL hereditary cancer syndrome. *Nat Rev Cancer* 2(9):673–682.

Kanno H, Kuratsu J, Nishikawa R, Mishima K, Natsume A, Wakabayashi T, Houkin K, Terasaka S, Shuin T (2013) Clinical features of patients bearing central nervous system hemangioblastoma in von Hippel-Lindau disease. *Acta Neurochirurgica* 155(1):1–7.

Kanno H, Yamamoto I, Nishikawa R, Matsutani M, Wakabayashi T, Yoshida J, Shitara N, Yamasaki I, Shuin T (2008) Spinal cord hemangioblastomas in von Hippel-Lindau disease. *Spinal Cord* 47(6):447–452.

Karasawa Y, Sakaguchi M, Minami S, Kitano K, Kawa S, Aoki Y, Itoh N, Sakurai A, Miyazaki M, Watanabe T (2001) Duodenal somatostatinoma and erythrocytosis in a patient with von Hippel-Lindau disease type 2A. *Int Med* 40(1):38–43.

Kepes JJ, Slowik F (1997) Arvid Lindau's cerebellar hemangioblastoma 70 years later - Some pediatric aspects. *Ann NY Acad Sci* 824:112–123.

Kertesz N, Wu J, Chen THP, Sucov HM, Wu H (2004) The role of erythropoietin in regulating angiogenesis. *Dev Biol* 276(1):101–110.

Keutgen XM, Hammel P, Choyke PL, Libutti SK, Jonasch E, Kebebew E (2016) Evaluation and management of pancreatic lesions in patients with von Hippel-Lindau disease. *Nat Rev Clin Oncol* 13(9):537–549.

Kim D, Semerjian A, Peterson J, Linehan WM, Metwalli AR (2016) The natural surgical history of patients with germline Von-Hippel Lindau gene mutations. *School of Medicine and Health Sciences Poster Presentations, Poster* 139.

Kim HC, Lee JS, Kim SH, So HS, Woo CY, Lee JL (2013) Sunitinib treatment for metastatic renal cell carcinoma in patients with von hippel-lindau disease. *Cancer Res Treat* 45(4):349–353.

Koh ES, Nichol A, Millar BA, Menard C, Pond G, Laperriere NJ (2007) Role of fractionated external beam radiotherapy in hemangioblastoma of the central nervous system. *Int J Radiat Oncol Biol Phys* 69(5):1521–1526.

Lammens CR, Bleiker EM, Verhoef S, Hes FJ, Ausems MG, Majoor-Krakauer D, Sijmons RH, van der Luijt RB, van den Ouweland AM, Van Os TA, Hoogerbrugge N, Gomez Garcia EB, Dommering CJ, Gundy CM, Aaronson NK (2010) Psychosocial impact of Von Hippel-Lindau disease: levels and sources of distress. *Clin Genet* 77(5):483–491.

Latif F, Tory K, Gnarra J, Yao M, Duh FM, Orcutt ML, Stackhouse T, Kuzmin I, Modi W, Geil L, et al (1993) Identification of the von Hippel-Lindau disease tumor suppressor gene. *Science* 260(5112):1317–1320.

Lenders JW, Duh QY, Eisenhofer G, Gimenez-Roqueplo AP, Grebe SK, Murad MH, Naruse M, Pacak K, Young WF, Jr., Endocrine S (2014) Pheochromocytoma and paraganglioma: an endocrine society clinical practice guideline. *J Clin Endocrinol Metab* 99(6):1915–1942.

Lonser RR, Glenn GM, Walther M, Chew EY, Libutti SK, Linehan WM, Oldfield EH (2003) von Hippel-Lindau disease. *Lancet* 361(9374):2059–2067.

Lonser RR, Vortmeyer AO, Butman JA, Glasker S, Finn MA, Ammerman JM, Merrill MJ, Edwards NA, Zhuang Z, Oldfield EH (2005) Edema is a precursor to central nervous system peritumoral cyst formation. *Ann Neurol* 58(3):392–399.

Lonser RR, Butman JA, Huntoon K, Asthagiri AR, Wu T, Bakhtian KD, Chew EY, Zhuang Z, Linehan WM, Oldfield EH (2014) Prospective natural history study of central nervous system hemangioblastomas in von Hippel-Lindau disease. *J Neurosurg* 120(5):1055–1062.

Maher ER, Bentley E, Yates JRW, Latif F, Lerman M, Zbar B, Affara NA, Fergusonsmith MA (1991) Mapping of the Von Hippel-Lindau disease locus to a small region of chromosome-3p by genetic-linkage analysis. *Genomics* 10(4):957–960.

Maher ER, Yates JR, Ferguson-Smith MA (1990) Statistical analysis of the two stage mutation model in von Hippel-Lindau disease, and in sporadic cerebellar haemangioblastoma and renal cell carcinoma. *J Med Genet* 27(5):311–314.

Mandriota SJ, Turner KJ, Davies DR, Murray PG, Morgan NV, Sowter HM, Wykoff CC, Maher ER, Harris AL, Ratcliffe PJ, Maxwell PH (2002) HIF activation identifies early lesions in VHL kidneys: evidence for site-specific tumor suppressor function in the nephron. *Cancer Cell* 1(5):459–468.

Maxwell PH, Wiesener MS, Chang GW, Clifford SC, Vaux EC, Cockman ME, Wykoff CC, Pugh CW, Maher ER, Ratcliffe PJ (1999) The tumour suppressor protein VHL targets hypoxia-inducible factors for oxygen-dependent proteolysis. *Nature* 399(6733):271–275.

Melmon KL, Rosen SW (1964) Lindau's Disease. Review of the Literature and Study of a Large Kindred. *Am J Med* 36:595–617.

Mills SA, Oh MC, Rutkowski MJ, Sughrue ME, Barani IJ, Parsa AT (2012) Supratentorial hemangioblastoma: clinical features, prognosis, and predictive value of location for von Hippel–Lindau disease. *Neuro-oncology* 14(8):1097–1104.

Moss JM, Choi CY, Adler JR, Jr, Soltys SG, Gibbs IC, Chang SD (2009) Stereotactic radiosurgical treatment of cranial and spinal hemangioblastomas. *Neurosurgery* 65(1):79–85; discussion 85.

Motzer RJ, Escudier B, McDermott DF, George S, Hammers HJ, Srinivas S, Tykodi SS, Sosman JA, Procopio G, Plimack ER, Castellano D, Choueiri TK, Gurney H, Donskov F, Bono P, Wagstaff J, Gauler TC, Ueda T, Tomita Y, Schutz FA, Kollmannsberger C, Larkin J, Ravaud A, Simon JS, Xu LA, Waxman IM, Sharma P, CheckMate I (2015) Nivolumab versus everolimus in advanced renal-cell carcinoma. *N Engl J Med* 373(19):1803–1813.

Neumann HP, Bausch B, McWhinney SR, Bender BU, Gimm O, Franke G, Schipper J, Klisch J, Altehoefer C, Zerres K, Januszewicz A, Eng C, Smith WM, Munk R, Manz T, Glaesker S, Apel TW, Treier M, Reineke M, Walz MK, Hoang-Vu C, Brauckhoff M, Klein-Franke A, Klose P, Schmidt H, Maier-Woelfle M, Peczkowska M, Szmigielski C, Eng C, Freiburg-Warsaw-Columbus Pheochromocytoma Study Group (2002) Germ-line mutations in nonsyndromic pheochromocytoma. *N Engl J Med* 346(19):1459–1466.

Neumann HP, Bender BU, Berger DP, Laubenberger J, Schultze-Seemann W, Wetterauer U, Ferstl FJ, Herbst EW, Schwarzkopf G, Hes FJ, Lips CJ, Lamiell JM, Masek O, Riegler P, Mueller B, Glavac D, Brauch H (1998) Prevalence, morphology and biology of renal cell carcinoma in von Hippel-Lindau disease compared to sporadic renal cell carcinoma. *J Urol* 160(4):1248–1254.

Neumann HP, Wiestler OD (1991) Clustering of features of von Hippel-Lindau syndrome: evidence for a complex genetic locus. *Lancet* 337(8749):1052–1054.

Oh DY, Kim TW, Park YS, Shin SJ, Shin SH, Song EK, Lee HJ, Lee KW, Bang YJ (2012) Phase 2 study of everolimus monotherapy in patients with nonfunctioning neuroendocrine tumors or pheochromocytomas/paragangliomas. *Cancer* 118(24):6162–6170.

Ong KR, Woodward ER, Killick P, Lim C, Macdonald F, Maher ER (2007) Genotype-phenotype correlations in von Hippel-Lindau disease. *Hum Mutat* 28(2):143–149.

Ordookhanian C, Kaloostian PE, Ghostine SS, Spiess PE, Etame AB (2017) Management Strategies and Outcomes for VHL-related Craniospinal Hemangioblastomas. *J Kidney Cancer VHL* 4(3):37–44.

Park BK, Kim CK (2010) Percutaneous radio frequency ablation of renal tumors in patients with von Hippel-Lindau disease: preliminary results. *J Urol* 183(5):1703–1707.

Park TY, Lee SK, Park J-S, Oh D, Song TJ, Park DH, Lee SS, Seo DW, Kim M-H (2015) Clinical features of pancreatic involvement in von Hippel–Lindau disease: a retrospective study of 55 cases in a single center. *Scand J Gastroenterol* 50(3):360–367.

Pavesi G, Berlucchi S, Feletti A, Opocher G, Scienza R (2006) Hemangioblastoma of the obex mimicking anorexia nervosa. *Neurology* 67(1):178–179.

Plate KHVAO, Zagzag D, et al. (2007) von Hippel-Lindau disease and haemangioblastoma. In: Louis DN, Ohgaki H, Wiestler OD, (eds). *WHO Classification of Tumours of the Central Nervous System*. Lyon: IARC. 215–217.

Rasmussen A, Alonso E, Ochoa A, De Biase I, Familiar I, Yescas P, Sosa AL, Rodriguez Y, Chavez M, Lopez-Lopez M, Bidichandani SI (2010) Uptake of genetic testing and long-term tumor surveillance in von Hippel-Lindau disease. *BMC Med Genet* 11:4.

Rednam SP, Erez A, Druker H, Janeway KA, Kamihara J, Kohlmann WK, Nathanson KL, States LJ, Tomlinson GE, Villani A, Voss SD, Schiffman JD, Wasserman JD (2017) Von Hippel-Lindau and hereditary pheochromocytoma/paraganglioma syndromes: clinical features, genetics, and surveillance recommendations in childhood. *Clin Cancer Res* 23(12):e68–e75.

Robson ME, Storm CD, Weitzel J, Wollins DS, Offit K (2010) American Society of Clinical Oncology policy statement update: genetic and genomic testing for cancer susceptibility. *J Clin Oncol* 28(5):893–901.

Roma A, Maruzzo M, Basso U, Brunello A, Zamarchi R, Bezzon E, Pomerri F, Zovato S, Opocher G, Zagonel V (2015) First-Line sunitinib in patients with renal cell carcinoma (RCC) in von Hippel-Lindau (VHL) disease: clinical outcome and patterns of radiological response. *Fam Cancer* 14(2):309–316.

Schmid S, Gillessen S, Binet I, Brandle M, Engeler D, Greiner J, Hader C, Heinimann K, Kloos P, Krek W, Krull I, Stoeckli SJ, Sulz MC, van Leyen K, Weber J, Rothermundt C, Hundsberger T (2014) Management of von hippel-lindau disease: an interdisciplinary review. *Oncol Res Treat* 37(12):761–771.

Seizinger BR, Rouleau GA, Ozelius LJ, et al. (1988) Von Hippel-Lindau disease maps to the region of chromosome 3 associated with renal cell carcinoma. *Nature* 332(6161):268–269.

Shanbhogue KP, Hoch M, Fatterpaker G, Chandarana H (2016) von Hippel-Lindau disease: review of genetics and imaging. *Radiol Clin North Am* 54(3):409–422.

Shankar GM, Taylor-Weiner A, Lelic N, Jones RT, Kim JC, Francis JM, Abedalthagafi M, Borges LF, Coumans J-V, Curry WT (2014) Sporadic hemangioblastomas are characterized by cryptic VHL inactivation. Acta Neuropath Commun 2(1):167.

Sharma A, Mukewar S, Vege SS (2017) Clinical profile of pancreatic cystic lesions in von Hippel-Lindau disease: A series of 48 patients seen at a tertiary institution. *Pancreas* 46(7):948–952.

Siller S, Szelényi A, Herlitz L, Tonn JC, Zausinger S (2017) Spinal cord hemangioblastomas: significance of intraoperative neurophysiological monitoring for resection and long-term outcome. *J Neurosurg Spine* 26(4):483–493.

Simone CB, 2nd, Lonser RR, Ondos J, Oldfield EH, Camphausen K, Simone NL (2011) Infratentorial craniospinal irradiation for von Hippel-Lindau: a retrospective study supporting a new treatment for patients with CNS hemangioblastomas. *Neuro-oncology* 13(9):1030–1036.

Singh AD, Ahmad NN, Shields CL, Shields JA (2002) Solitary retinal capillary hemangioma: lack of genetic evidence for von Hippel-Lindau disease. *Ophthal Genet* 23(1):21–27.

Singh AD, Shields CL, Shields JA (2013) von Hippel Lindau Disease. *Surv Ophthalmol* 46(2):117–142.

Song DK, Lonser RR (2008) Pathological satiety caused by brainstem hemangioblastoma. *J Neurosurg Pediatr* 2(6):397–401.

Stolle C, Glenn G, Zbar B, Humphrey JS, Choyke P, Walther M, Pack S, Hurley K, Andrey C, Klausner R, Linehan WM (1998) Improved detection of germline mutations in the von Hippel-Lindau disease tumor suppressor gene. *Hum Mut* 12(6):417–423.

Tonyukuk V, Emral R, Temizkan S, Sertcelik A, Erden I, Corapcioglu D (2003) Case report: patient with multiple paragangliomas treated with long acting somatostatin analogue. *Endocr J* 50(5):507–513.

Tsimafeyeu I (2015) Sunitinib treatment of metastatic renal cell carcinoma in von Hippel-Lindau disease. *J Cancer Res Ther* 11(4):920–922.

Turner JS, Cheung EM, George J, Quinn DI (2007) Pain management, supportive and palliative care in patients with renal cell carcinoma. *BJU Int* 99(5 Pt B):1305–1312.

van Asselt SJ, de Vries EG, van Dullemen HM, Brouwers AH, Walenkamp AM, Giles RH, Links TP (2013) Pancreatic cyst development: insights from von Hippel-Lindau disease. *Cilia* 2(1):3.

van der Harst E, de Herder WW, Bruining HA, Bonjer HJ, de Krijger RR, Lamberts SW, van de Meiracker AH, Boomsma F, Stijnen T, Krenning EP, Bosman FT, Kwekkeboom DJ (2001) [(123)I]metaiodobenzylguanidine and [(111)In]octreotide uptake in begnign and malignant pheochromocytomas. *J Clin Endocrinol Metab* 86(2):685–693.

van Hulsteijn LT, Niemeijer ND, Dekkers OM, Corssmit EP (2014) (131)I-MIBG therapy for malignant paraganglioma and phaeochromocytoma: systematic review and meta-analysis. *Clin Endocrinol (Oxf)* 80(4):487–501.

Varshney N, Kebede AA, Owusu-Dapaah H, Lather J, Kaushik M, Bhullar JS (2017) A review of Von Hippel-Lindau syndrome. *J Kidney Cancerl VHL* (2017 4(3):10.

VHL Alliance (2015) *The VHL Handbook: A Reference Handbook for People With Von Hippel-Lindau, Their Families, and Their Medical Team*. Boston: VHL Alliance.

Walther MM, Choyke PL, Glenn G, Lyne JC, Rayford W, Venzon D, Linehan WM (1999) Renal cancer in families with hereditary renal cancer: prospective analysis of a tumor size threshold for renal parenchymal sparing surgery. *J Urol* 161(5):1475–1479.

Wanebo JE, Lonser RR, Glenn GM, Oldfield EH (2003) The natural history of hemangioblastomas of the central nervous system in patients with von Hippel—Lindau disease. *J Neurosurg* 98(1):82–94.

Wang JY, Peng SH, Li T, Ning XH, Liu SJ, Hong BA, Liu JY, Wu PJ, Zhou BW, Zhou JC, Qi NN, Peng X, Zhang JF, Ma KF, Cai L, Gong K (2018) Risk factors for survival in patients with von Hippel-Lindau disease. *J Med Genet* 55(5):322–328.

Wang Y, Chen D-Q, Chen M-Y, Ji K-Y, Ma D-X, Zhou L-F (2017) Endothelial cells by inactivation of VHL gene direct angiogenesis, not vasculogenesis via Twist1 accumulation associated with hemangioblastoma neovascularization. *Scientific Reports* 7(1):5463.

Witten F, O'Brien 3rd D, Sewell C, Wheatley J (1985) Bilateral clear cell papillary cystadenoma of the epididymides presenting as infertility: an early manifestation of von Hippel-Lindau's syndrome. *J Urol* 133(6):1062–1064.

Zhang J, Chang H, Wu Q, Yan R, Tai M, Xu X, Liu C (2015) The Psychosocial Research of the Members in a Large Von Hippel-Lindau Family in China. *Open Access Library Journal* 02(01):1–10.

Zhao Y, Jin X, Gong X, Guo B, Li N (2017) Clinicopathologic features of hemangioblastomas with emphases of unusual locations. *Int J Clin Exp Pathol* 10(2):1792–1800.

63

WILLIAMS SYNDROME

COLLEEN A. MORRIS

Division of Genetics, Department of Pediatrics, University of Nevada, Reno School of Medicine;
UNLV Ackerman Center for Autism and Neurodevelopment Solutions, Las Vegas, Nevada, USA

CAROLYN B. MERVIS

Department of Psychological and Brain Sciences, University of Louisville, Louisville, Kentucky, USA

INTRODUCTION

Incidence

Williams syndrome has been variably termed Williams–Beuren syndrome, idiopathic hypercalcemia, and supravalvar aortic stenosis syndrome. Although Williams et al. (1961) and Beuren et al. (1962), both cardiologists, are commonly credited with the first detailed descriptions of the syndrome, there are earlier reports in the literature of idiopathic hypercalcemia (see Fanconi et al. [1952]). Garcia et al. (1964) first documented the occurrence of both supravalvar aortic stenosis and idiopathic hypercalcemia, and since then, there have been several reviews detailing the Williams syndrome phenotype (Beuren 1972; Jones and Smith 1975; Burn 1986; Morris et al. 1988, 2006; Pober and Morris 2007). In 1993, Ewart et al. discovered a microdeletion of chromosome 7q11.23 encompassing the elastin gene that is detectable by fluorescence in situ hybridization (FISH) in individuals with the clinical phenotype of Williams syndrome (Ewart et al. 1993; Lowery et al. 1995). The incidence of Williams syndrome is estimated at 1/10,000 births; a Norwegian study found a prevalence of 1/7500 (Strømme et al. 2002). Before the availability of a diagnostic test, the average age of diagnosis was 6.4 years (median 4 years).

Diagnostic Criteria

The clinical diagnosis of Williams syndrome is based on recognition of the characteristic pattern of dysmorphic facial features, developmental delay, short stature relative to the family background, connective tissue abnormality (including cardiovascular disease), unique cognitive profile, and typical personality. Diagnostic scoring systems are available (Preus 1984; American Academy of Pediatrics 2001; Sugayama et al. 2007) that may help the clinician determine whether diagnostic testing is warranted. The diagnosis is established by detection of the contiguous gene deletion including *ELN* in chromosome 7q11.23 by genomic testing methods such as chromosomal microarray or targeted deletion analysis such as FISH.

The facial characteristics of Williams syndrome are distinctive. Infants and young children typically have a broad forehead, bitemporal narrowing, depressed nasal root, periorbital fullness, stellate/lacy iris pattern, strabismus, bulbous nasal tip, malar flattening, long philtrum, thick vermilion of the lips, wide mouth, full cheeks, dental malocclusion with small widely spaced teeth, small jaw, and prominent earlobes (Figure 63.1). Older children and adults usually have a more gaunt facial appearance with prominent supraorbital ridges, narrow nasal root of normal anterior prominence, full nasal tip, malar flattening, wide mouth with thick lips, small

---

Cassidy and Allanson's Management of Genetic Syndromes, Fourth Edition.
Edited by John C. Carey, Agatino Battaglia, David Viskochil, and Suzanne B. Cassidy.
© 2021 John Wiley & Sons, Inc. Published 2021 by John Wiley & Sons, Inc.

FIGURE 63.1 A 3-year-old girl with Williams syndrome. Note the periorbital fullness, bulbous nasal tip, long philtrum, wide mouth, full lips, full cheeks, and small widely spaced teeth.

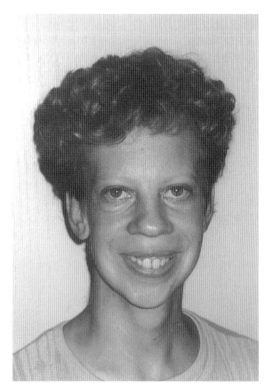

FIGURE 63.2 A 30-year-old woman with Williams syndrome. Note the prominent supraorbital ridges, narrow nasal root, infraorbital creases, wide mouth with full lips, small jaw, and long neck.

jaw, dental malocclusion, and a long neck accentuated by sloping shoulders (Figure 63.2).

The connective tissue abnormalities include a hoarse/deep voice, hernias, bladder/bowel diverticula, soft/loose skin, joint laxity or limitation, and cardiovascular disease (an elastin arteriopathy most commonly manifested as supravalvar aortic stenosis). Almost all individuals have developmental delay, but scores on full-scale IQ tests range widely, from severe intellectual disability to low average; the majority has mild intellectual disability. The unique cognitive profile of Williams syndrome consists of relative strengths in concrete language, non-verbal reasoning, and verbal short-term memory accompanied by considerable weakness in visuospatial construction (e.g., building an object from a diagram, drawing) (Mervis et al. 2000; Mervis and John 2010). The Williams syndrome personality is typified by empathy, overfriendliness, attention problems, and anxiety (Klein-Tasman and Mervis 2003).

Etiology, Pathogenesis, and Genetics

Williams syndrome usually occurs sporadically; however, there is a 50% risk of transmitting the deletion to offspring of affected individuals, and familial cases, including male-to-male transmission, have been reported (Morris et al. 1993; Mulik et al. 2004). This disorder is caused by a submicroscopic deletion of chromosome 7q11.23. This deletion arises by an unequal crossing over in a chromosome region predisposed to such meiotic error by the presence of a large number of repetitive DNA sequences (Dutly and Schinzel 1996). Individuals who have the reciprocal *duplication* of the region have a different phenotype that is characterized by developmental delay, speech delay or disorder (most commonly childhood apraxia of speech), separation anxiety, social anxiety disorder, oppositional behavior, autism spectrum disorder or symptoms of this disorder, and subtle facial differences (Mervis et al. 2015; Morris et al. 2015). Approximately 25% of the parents of individuals with Williams syndrome have an inversion of the Williams syndrome region that increases the risk for deletion (Bayes et al. 2003; Hobart et al. 2010). One family with unaffected parents and two affected children has been reported; both deletions arose from the paternal inverted chromosome 7 (Scherer et al. 2005). The origin of the deletion may be maternal or paternal without any parent-of-origin effect on the phenotype (Hobart et al. 2010). The commonly deleted region, seen in approximately 95% of individuals with Williams syndrome, spans 1.55 Mb, and 26 genes have thus far been mapped within this region. Five percent of individuals with Williams syndrome have a larger deletion of 1.8 Mb containing 28 genes.

Many of the medical manifestations of Williams syndrome are caused by deletion of the *elastin* gene. The connective tissue abnormalities are the result of abnormal elastin protein production. Interestingly, mutations within the *elastin* (*ELN*) gene also cause the autosomal dominant disorder supravalvar aortic stenosis (Li et al. 1997). This condition is associated with elastin arteriopathy, hoarse voice, and hernias, but lacks other aspects of the Williams syndrome phenotype such as intellectual disability or the Williams syndrome cognitive profile or personality profile. There is no difference in the histological appearance of diseased arteries in familial supravalvar aortic stenosis versus Williams syndrome (O'Connor et al. 1985).

Insight into the pathogenesis of the arterial disease has come from the study of mice hemizygous for *ELN*, resulting in a 47% decrease in *ELN* mRNA (Li et al. 1998). Increased numbers of lamellar units (an elastic fiber lamella alternating with a ring of smooth muscle around the arterial lumen) were observed during fetal development in the experimental mice. In humans hemizygous for *ELN*, there is a 2.5-fold increase in lamellar units, resulting in a thickened arterial media that leads to obstructive vascular disease.

Lim kinase 1 (*LIMK1*), a gene contiguous to *ELN*, is the second gene implicated in the Williams syndrome phenotype (Frangiskakis et al. 1996). Although most families with supravalvar aortic stenosis have a mutation in *ELN*, a few rare families have deletions within the Williams syndrome region that are much smaller than the classical deletions observed in Williams syndrome. Genotype–phenotype analysis of five families with such small deletions has shown that the Williams syndrome cognitive profile is associated with *LIMK1* deletion. These families have normal intellectual ability but have more difficulty with visuospatial constructive cognition than expected for their IQs. Because *LIMK1* is expressed in the brain, it may thus be important in the development of neural pathways responsible for visual-motor integration. The gene located at the telomeric breakpoint of the Williams syndrome deletion, *GTF2I*, has been implicated in the intellectual disability of Williams syndrome (Morris et al. 2003) and in some of the personality characteristics associated with this syndrome (Crespi and Hurd 2014; Swartz et al. 2017).

Individuals who have a slightly larger deletion of 1.8 Mb that includes *NCF1* are less likely to develop hypertension (Del Campo et al. 2006). Those who have much larger deletions that include *MAGI2* typically have more severe intellectual disability and may have infantile spasms (Marshall et al. 2008).

Individuals with shorter deletions tend to have fewer or milder manifestations of Williams syndrome, but the phenotype depends on which genes are deleted. In five families with supravalvar aortic stenosis with shorter overlapping deletions in the Williams syndrome region that did not include *GTF2I*, no member had intellectual disability (Hobart et al. 2010). In contrast, individuals with shorter deletions in the Williams syndrome region that did include *GTF2I* had intellectual disability (Botta et al. 1999; Heller et al. 2003). *GTF2I* is expressed in the brain and encodes the protein TFII-I, a transcription factor with a role in the activation or repression of other genes (Hakimi et al. 2003).

Two other Williams syndrome region genes expressed in the brain likely have a role in the Williams syndrome phenotype. Mice haploinsufficient for *CYLN2* have deficits in motor coordination and hippocampal function (Van Hagen et al. 2007). Various phenotypic features have been reported in mice that have a deletion of *GTF2IRD1*, including large brain ventricles (Van Hagen et al. 2007), craniofacial abnormalities (Tassabehji et al. 2005; Enkhmandakh et al. 2008), and behavioral changes, such as decreased fear and increased social behaviors (Wu et al. 2008). The role of other deleted genes in the Williams syndrome phenotype is unknown; further study of individuals and families with partial phenotypes may elucidate the pathogenetic mechanisms of other Williams syndrome features.

Dose dependent, differentially methylated gene profiles found in microdeletion (Williams syndrome) and in microduplication (Duplication 7 syndrome) suggest that epigenetic mechanisms likely contribute to these phenotypes (Strong et al. 2015).

Diagnostic Testing

Fluorescence in situ hybridization (FISH) testing will demonstrate *ELN* deletion in all individuals with Williams syndrome. The only caveat is that some rare supravalvar aortic stenosis families will also have an *ELN* deletion but do not have the intellectual disability, growth deficiency, and unique personality seen in Williams syndrome, allowing a clinical distinction. These individuals do not have Williams syndrome. Chromosome microarray studies will also detect the microdeletion in chromosome band 7q11.23 (GRCh37/hgchr7:72,744,454-74,142,513). Prenatal diagnosis by FISH or chromosome microarray is available.

Differential Diagnosis

Williams syndrome should be distinguished from other syndromes that feature short stature, congenital heart disease, and developmental delay. In one study of 65 individuals with a previous diagnosis of Williams syndrome but negative FISH testing, other syndromes were diagnosed in 15% of the individuals (Morris et al. 1998; see also Pani et al. 2010). Overlapping clinical features were developmental delay (92%), attention-deficit disorder (50%), short stature (30%), and congenital heart disease (25%). The correct diagnoses in that study included isolated supravalvar aortic stenosis associated with *ELN* mutation, velo-cardio-facial syndrome associated with deletion 22q11.2, Noonan syndrome, FG syndrome, Kabuki syndrome, fetal alcohol syndrome, and Fragile X syndrome. The clinical diagnosis of Williams syndrome must be confirmed by testing for 7q11.23

microdeletion to avoid misdiagnosis. A negative study (either FISH or chromosome microarray) should prompt a search for other diagnoses.

The differential diagnosis of Williams syndrome should include Noonan syndrome (see Chapter 41), which is characterized by short stature, broad or webbed neck, chest deformity, congenital heart disease (most commonly pulmonary stenosis), and dysmorphic facial features. In infancy, the nose of Williams syndrome and Noonan syndrome is similar, but the philtrum is usually deeply grooved in Noonan syndrome with defined peaks of the upper vermilion, whereas the philtrum in Williams syndrome may appear thick, with loss of the cupid's bow of the upper lip. In infancy, Noonan syndrome and Williams syndrome may also share peripheral pulmonic stenoses and hernias, contributing to diagnostic confusion. The facial gestalt, the pattern of other cardiovascular lesions, and the testing for the 7q11.23 microdeletion will distinguish the two. Smith–Magenis syndrome, a condition associated with a microdeletion of 17p, is also associated with short stature, congenital heart disease, hoarse voice, hearing loss, iris dysplasia, scoliosis, behavior problems, and dysmorphic facial features (see Chapter 54). The dysmorphic facies of Smith–Magenis syndrome are distinguishable from Williams syndrome by the presence of a broad square face with heavy brows, deep-set eyes, broad short nose, and thickened everted philtrum. Smith–Magenis syndrome has a very different cognitive profile with significant weakness in expressive language, and FISH or microarray for the Smith–Magenis syndrome 17p deletion will confirm that diagnosis. Another microdeletion syndrome that is in the differential diagnosis is the more common condition, deletion 22q11.2 syndrome/velo-cardio-facial syndrome (see Chapter 21). Velo-cardio-facial syndrome is associated with relative short stature, cleft palate, congenital heart disease (most commonly conotruncal defects), learning disability, hearing loss, velopharyngeal insufficiency, and dysmorphic facial features. The facial features of velo-cardio-facial syndrome include a long narrow face, prominent nasal root, long nose with full tip and small nares, supraorbital fullness with hooded eyelids, and small mandible. The facial gestalt is quite different from Williams syndrome, especially after age 2 years, and the pattern of congenital heart disease also distinguishes the conditions. FISH for deletion 22q or cytogenomic microarray will confirm the diagnosis in the majority of individuals with velo-cardio-facial syndrome.

Other genetic syndromes in the differential diagnosis of Williams syndrome include Kabuki syndrome, FG syndrome, Coffin–Lowry syndrome, and Fragile X syndrome. Kabuki syndrome is associated with developmental delay, short stature, congenital heart disease (most commonly coarctation of the aorta), and dysmorphic facial features (see Chapter 34). Although Kabuki syndrome shares midfacial flattening, micrognathia, and prominent earlobes with Williams syndrome, the appearance of the eyes is quite different, with long palpebral fissures and everted lateral one-third of the lower lid in Kabuki syndrome. FG syndrome, an X-linked intellectual disability syndrome, shares hypotonia, constipation, overfriendliness, and attention-deficit disorder with Williams syndrome, but may be distinguished by the facial features that include high forehead and hypertelorism. Coffin–Lowry syndrome, an X-linked intellectual disability syndrome, has dysmorphic facies that include full lips and connective tissue problems including scoliosis but does not have other Williams syndrome features, such as cardiovascular disease (see Chapter 12). Fragile X syndrome (see Chapter 28) is associated with developmental delay and hypotonia, but typically has more severe intellectual disability (in males) than Williams syndrome and features a shy personality when compared with Williams syndrome. DNA testing for Fragile X syndrome will make that diagnosis.

One teratogenic syndrome in the differential diagnosis is fetal alcohol syndrome characterized by short stature, developmental delay, congenital heart disease (most commonly septal defects), attention-deficit disorder, and dysmorphic facial features (short palpebral fissures, midfacial flattening, thin upper lip vermillion, long, smooth philtrum, and small jaw) (see Chapter 26). Although young children with fetal alcohol syndrome often demonstrate overfriendliness and attention-deficit/hyperactivity disorder, fetal alcohol syndrome may be distinguished from Williams syndrome by the different appearance of the lower face and the different pattern of heart defect.

MANIFESTATIONS AND MANAGEMENT

Growth and Feeding

Short stature relative to the family background is common in Williams syndrome, and the rate of linear growth is 75% of the normal (Jones and Smith 1975). Studies of growth employing both longitudinal and cross-sectional data have shown that the mean final height is below the third centile for the general population (Morris et al. 1988; Pankau et al. 1992). The growth pattern in Williams syndrome is characterized by prenatal growth deficiency in 25–70% and poor linear growth and weight gain in the first 2–4 years of life. Statural growth then improves in childhood, although 70% remain below the third centile for midparental height. Bone age studies are typically normal in early childhood, and head circumference is most often commensurate with height. Puberty occurs early in Williams syndrome in about 50% of affected individuals and is associated with a briefer growth spurt than normal (Pankau et al. 1992). Advanced bone age is found in these individuals. Central precocious puberty has been reported in 18% (Partsch et al. 2002). In girls with Williams syndrome who were not treated with gonadotropin releasing hormone (GnRH) analog therapy, mean age of menarche was 9.3 ± 0.3 years (Spielmann et al. 2015). Growth curves for Williams syndrome are available (Morris et al. 1988; Pankau et al. 1992; Saul et al. 1998; Martin et al. 2007).

The poor weight gain and linear growth of a young child with Williams syndrome is likely the result of severe difficulties with feeding. It is important to note that fall off in growth should prompt investigation because both hypothyroidism and growth hormone deficiency have been reported (Kuijpers et al. 1999).

Poor weight gain (80%), feeding problems (70%), and prolonged colic (70%) are commonly observed in infants with Williams syndrome (Morris et al. 1988). Gastrointestinal complications in this age group include gastroesophageal reflux, constipation, and rectal prolapse. In a young child, sensory aversion to textures adversely affects the transition to solid foods. Abnormal posture of jaw and neck (secondary to hypotonia and ligamentous laxity) provokes gagging, and dental malocclusion and persistent tongue thrust compromise efficient chewing and swallowing. Children often develop aversion to specific textures resulting in a limited dietary repertoire.

The feeding problems do improve with medical treatment of the gastroesophageal reflux and hypercalcemia (see Endocrine), combined with oral motor and feeding therapy.

Evaluation

- Growth measurements should be plotted on Williams syndrome growth charts, and any deviation downward across centiles should be investigated, including thyroid function testing, followed by growth hormone testing, if indicated.
- Oral motor functioning and feeding should be assessed in infants.
- A complete nutritional and feeding assessment should be conducted in the infant or toddler with poor weight gain.
- Initial laboratory evaluation should include thyroid function tests, serum calcium, and a random spot urine for calcium-to-creatinine ratio.
- Infants with Williams syndrome who demonstrate poor weight gain should have a consultation with a nutritionist to assess the dietary calcium intake, also taking into account the calcium content of the local water supply, which may be significant in some areas.
- If signs of early or precocious puberty occur, prompt referral to a pediatric endocrinologist is recommended.

Management

- Small, frequent feedings of high caloric density are recommended for infants.
- Feeding and oral motor therapy should be instituted in a child with significant difficulties.
- Pureed foods will often be required for a longer period of time in a child with Williams syndrome because there are increased problems with gagging associated with texture aversion.
- Feeding tube placement is occasionally required to ensure adequate nutrition for growth.
- Growth hormone-releasing hormone (GnRH) analog therapy should be considered in children with early or precocious puberty (Utine et al. 2006), in a standard fashion.

Development and Behavior

Development is delayed in almost all children with Williams syndrome, including delays in early language and motor milestones. There is a wide range of intellectual ability with full-scale IQs ranging from severe intellectual disability to low average intelligence, with the mean in the mild intellectual disability range. However, full-scale IQ scores are misleading for most individuals with Williams syndrome, as Williams syndrome is associated with an unusual cognitive profile. In particular, children with Williams syndrome almost always evidence considerably higher standard scores on the verbal and non-verbal reasoning sections of IQ tests than on the section measuring visuospatial construction (Mervis and John 2010). On average, performance on verbal and non-verbal reasoning IQ measures is in the borderline range of intellectual ability, with standard scores on the visuospatial construction components of IQ tests in the mild to moderate intellectual disability range, about 20 points lower. This pattern of relative strengths and weaknesses is present even in toddlers (Mervis and John 2010). Within the language domain, performance is best on measures of single-word receptive and expressive vocabulary, with mean standard scores in the low average range for the general population and 6–8% scoring at or above the 50th percentile (Mervis and John 2010); performance is worst on measures of relational vocabulary (e.g., labels for spatial, temporal, quantitative, and dimensional concepts), with performance on the latter at about the same level as on measures of visuospatial construction (Mervis and John 2010). Longitudinal studies of children (Mervis and Pitts 2015) and of adolescents and adults (Fisher et al. 2016) suggest that for the majority of individuals with Williams syndrome, IQ scores are stable by middle childhood. In contrast, adaptive behavior standard scores decline significantly over time for many individuals with Williams syndrome (Fisher et al. 2016; Mervis and Pitts 2015). These decreases in standard scores indicate that individuals with Williams syndrome do not increase their adaptive behavior repetoire as rapidly as would have been expected based on their earlier standard scores.

Fine motor tasks adversely affected by Williams syndrome include handwriting, drawing, activities of daily living such as dressing or food preparation, tool use, and pattern construction (e.g., duplicating a pattern of blocks). Performance on these tasks does improve over time with occupational therapy and practice. Because visual-motor skills are processed by the occipito-parietal visual pathway (dorsal stream), it is possible that a gene in the Williams syndrome region is involved in the development of this specific central neural pathway. This possibility is supported by structural and functional neuroimaging, which shows a reduction in gray matter volume in the intraparietal surface leading to hypoactivation of the parietal dorsal stream (Meyer-Lindenberg et al. 2004). In contrast, visual recognition skills, processed by the occipitotemporal pathway (ventral stream) are relatively good (Atkinson et al. 1997).

Academic achievement varies widely across individuals with Williams syndrome. Word-reading ability is in the borderline range on average but varies from inability to read at all (rare) to ability to read at grade level. The strongest predictor of word reading ability is the reading instruction method used; children who are taught to read with intensive systematic phonics instruction have much stronger reading abilities than children taught with whole language or whole word methods (Mervis and John 2010). Single-word reading abilities are stronger than reading comprehension abilities. Mathematics abilities, although almost always more limited than word-reading abilities, also vary widely, from difficulty with simple addition to low-average ability relative to the general population.

Adult outcomes beyond performance on standardized assessments have rarely been researched. In a sample of 102 adults with a mean age of 28 years, 7% lived independently, another 7% were in supported living arrangements, and the remainder lived with a parent or guardian (Lough and Fisher 2016). A British study of 92 somewhat older adults (mean age 32 years) found about the same percentage living independently, 39% living in residences for individuals with intellectual disability, and the remainder living with parents (Elison et al. 2010). Lough and Fisher found that 57% were employed (mean = 17 hours per week), 7% attended an educational program, 25% attended a day programme for individuals with disabilities, and 4% were in job training. Elison et al. (2010) noted that the majority of individuals who were employed or doing volunteer work required considerable extra support. Social vulnerability is a serious concern (Jawaid et al. 2012; Lough and Fisher 2016). Problems with anxiety, depression, and distractability are common (Cherniske et al. 2004; Stinton et al. 2010).

The paradoxical nature of the social and behavioral phenotype of individuals with Williams syndrome was first noted by von Arnim and Engel (1964), who remarked on their impressive ability to establish interpersonal contact and unusual politeness in combination with overfriendliness toward adults (even those that were strangers), difficulty forming and maintaining friendships, emotion dysregulation, anticipatory worry, sensitivity, and distractability. This set of characteristics has been noted subsequently by most professionals who have worked with individuals with Williams syndrome. Low levels of mastery motivation (Mervis and John 2010) and difficulty with behavioral regulation, including impulsivity and emotional dysregulation (Klein-Tasman et al. 2017; Pitts et al. 2016), also are common. Attention-deficit disorder, with or without hyperactivity, is present in 65% (Leyfer et al. 2006), and overfriendliness to strangers is observed in 97% (Sarimski 1997). More than half of children with Williams syndrome have at least one specific phobia (most commonly for loud or sudden noises; Leyfer et al. 2006); children who have difficulty with behavior regulation are more likely to have specific phobia (Pitts et al. 2016). Generalized anxiety disorder was present in 23% of children aged 11–16 years (Leyfer et al. 2006). Although most children with Williams syndrome do not have an autism spectrum disorder, the prevalence of autism spectrum disorder is elevated (Klein-Tasman et al. 2007; Lincoln et al. 2007). Despite high levels of social motivation, difficulties with socio-communication (including pragmatic language abilities), social awareness, social cognition, and repetitive behaviors and restricted interests are very common (Klein-Tasman et al. 2011).

Sleep disturbance is a common problem for individuals with Williams syndrome. In contrast to the pattern for typically developing children, melatonin levels for children with Williams syndrome do not rise at bedtime, and cortisol levels are significantly higher at bedtime for children with Williams syndrome than for matched typically developing children (Sniecinska-Cooper et al. 2016). Given these patterns, findings from standardized sleep questionnaires indicating that 6–12-year-olds with Williams syndrome are significantly more likely to evidence bedtime resistance and sleep anxiety, have significantly longer sleep latency, and are significantly more likely to awaken more than twice during the night than matched controls (Annaz et al. 2011) are not surprising. Restless sleep and tiredness during the day were evidenced by 60%, and 43% moved to someone else's bed during the night (Annaz et al. 2011). Polysomnography findings in an independent sample of children corroborated the parental report findings of Annaz et al. (2011); the children with Williams syndrome had significantly decreased sleep efficiency due to significantly increased sleep latency, restlessness, and respiratory-related arousals than did matched controls (Mason et al. 2011). In addition, the percentage of slow wave sleep was significantly higher than for matched controls (Mason et al. 2011).

Evaluation

- As part of the comprehensive history, the clinician should ascertain and record developmental milestones

and school performance and determine whether symptoms of attention problems, anxiety, and/or emotion dysregulation are present.
- A multidisciplinary approach including speech/language, occupational, and physical therapists is recommended to assess the developmental disabilities and to plan appropriate interventions.
- Psychological and educational assessment is important for both children and adolescents. The Differential Ability Scales-II (Elliott 2007) is the assessment best suited to detecting the pattern of relative intellectual strengths and weaknesses characteristic of Williams syndrome.
- In the evaluation of behavior problems, both adaptive behavior scales and parent and teacher ratings on standardized questionnaires measuring problem behaviors (e.g., anxiety, attention) are important. Findings are helpful in planning behavioral or medical interventions. If a child screens positive on these questionnaires, referral to a child psychologist or child psychiatrist experienced in the treatment of individuals with intellectual disability should be strongly considered.
- Children whose behavior overlaps significantly with the behavior of children with autism spectrum disorders should be referred for a gold-standard evaluation for autism spectrum disorder.
- Adults require vocational testing.

Sleep history should be solicited; polysomnography is indicated if signs or symptoms of obstructive sleep apnea are present

Management

- Intensive occupational therapy is critical for addressing the visuospatial problems associated with Williams syndrome that lead to difficulties with writing, drawing, and activities of daily living. Occupational therapy that includes sensory integration techniques and/or hippotherapy has been particularly helpful, especially for young children (personal observation). Once children are able to talk well, they should be taught to use their language skills to provide a framework for breaking down and mastering spatial tasks.
- Handwriting should be separated from other cognitive tasks, so as not to impede academic progress in other areas. For instance, the student could be allowed to complete assignments and tests orally and to record lectures or have a note-taker. It is important for children to learn keyboarding skills and how to use apps designed to facilitate writing composition.
- Even for children whose speech is clear and who speak grammatically, speech/language therapy continues to be important for addressing socio-communicative aspects of language (language pragmatics) such as conversational skills and comprehension of non-literal language including sarcasm or irony. Speech/language therapy also may focus on other advanced language comprehension skills such as event sequencing or inference making that are crucial for both listening comprehension and reading comprehension.
- Children with Williams syndrome should be taught to read using an intensive phonics approach. This method is considerably more effective than either a whole language approach that includes some phonics instruction or a whole-word/sight-word approach (Mervis and John 2010). Explicit instruction focused on reading comprehension also is important (Mervis and John 2010).
- Social skills training is important for facilitating the establishment and maintenance of reciprocal friendships and appropriate behavior in the workplace. It is important also to address the gullibility of most individuals with Williams syndrome so that they are not taken advantage of either in person or online (Fisher 2014; Lough and Fisher 2016).
- For adults, vocational training including job coaching is critical for supporting successful employment.
- Attention-deficit disorder may be treated with methylphenidate (Power et al. 1997) or other stimulants. Some children with Williams syndrome show improvement in attention with doses less than usually prescribed for typically developing children. A prudent course of management is to start with a quarter to half of the usual starting dose, then monitor for side effects and efficacy. The dose may be gradually increased within standard prescribing guidelines until a therapeutic effect is achieved (personal observation). Cardiology clearance should be obtained before treatment with stimulants, especially for those who have hypertension.
- To support emotion regulation, children with Williams syndrome should be taught self-calming techniques- such as deep breathing at an early age, and these skills should be practiced daily. Yoga is also helpful.
- For younger children, use of strategic play-based therapies such as Replays (Levine and Chedd 2007) to treat specific phobia or emotion dysregulation should be considered.
- For school-age children and adults, significant levels of anxiety and/or emotion dysreulation are best addressed by cognitive behavioral therapy administered by a psychologist who has experience with individuals with intellectual disabilities. Anxiety often requires medical treatment in addition to cognitive behavioral therapy. As with stimulants, medications are best started at a lower dose rather than the usual dose for typically developing children, and increased slowly

until an effective dose is achieved. It should be noted that the specific serotonin reuptake inhibitors may result in an increase in disinhibited behavior in individuals with Williams syndrome, and thus, are typically used in conjunction with a medication for attention. The use of therapy animals has been successful in managing anxiety in some individuals with Williams syndrome (personal observation).

- For children with moderate to severe emotion dysregulation or oppositional behavior, in-home and/or in-school applied behavior analysis approaches that take into account the social motivation of individuals with Williams syndrome should be strongly considered (Mervis and John 2010). Administration by a professional with BCBA certification is recommended; parental training in implementing the program also should be provided.
- Melatonin, given 30 min before bedtime, results in improved sleep patterns in approximately 50% (personal observation). Melatonin has been shown to decrease sleep latency and nighttime wakefulness in populations of children with developmental disabilities (Braam et al. 2009) though no dosage regimen has been consistent. Starting with a higher dose and then gradually decreasing it over time seems to be more effective in Williams syndrome. The initial dose, which should be continued nightly for three months, is as follows: for children over 10 kg and over 1 year of age, 2–3 mg; for children over 15 kg and over 4 years of age, 6 mg; for children over 25 kg and over 10 years of age, 9 mg; and for adults over 20 years, 6 mg (personal observation). For those in whom melatonin is efficacious, the dose can be gradually tapered to a maintenance dose (personal observation). A sleep hygiene program also is important to address the behavioral component of sleep difficulties for many children with Williams syndrome.
- Many individuals with Williams syndrome respond positively to music. Music can be used to lessen anxiety in stressful situations (e.g., at the dentist's office), as an aid to long-term memory (putting directions to a tune), to prompt appropriate behavior (specific songs for certain tasks), as a creative activity that improves self-esteem and facilitates successful integration with peers (choir or band), and as a reward. For children who are particularly responsive to music a trial of music therapy to address visuospatial difficulties, memory limitations, or speech/language problems could be considered.

Neurologic

Hypotonia is a common finding (80%), and peripheral hypertonia (50%) with increased deep tendon reflexes (especially in the lower extremities) is observed in children and adults. Children have difficulties with unipedal stance, tandem gait, and proprioception. The prevalence of left-handedness in Williams syndrome is 26% (Pérez-García et al. 2015; van Strien et al. 2005), which is considerably higher than in the general population. The prevalence of mixed handedness is 19% (Pérez-García et al. 2015). Adults show signs of cerebellar dysfunction, including dysmetria, dysdiadokinesis, and limb and gait ataxia (Pober and Szekely 1999). Adolescents and adults often have a stiff awkward gait accompanied by decreased range of motion at the hips, knees, and ankles (Morris et al. 1988). The gait abnormality has been more precisely described by Hocking et al. (2009), who noted both reduced speed and reduced stride length in adults with Williams syndrome. A longitudinal study demonstrated increased subtle extrapyramidal and cerebellar signs with age; most participants had dysmetria and difficulty with balance (Gagliardi et al. 2007). Gross motor skills are affected by abnormal stereoacuity as well as abnormal use of depth information in guiding movements, leading to difficulty in negotiating uneven surfaces and stairs (Van der Geest et al. 2005).

Neuroimaging studies (MRI) have shown reduced cerebral volume but preservation of cerebellar size compared with controls (Reiss et al. 2000). Chiari I malformations have been reported (Kaplan et al. 1989; Wang et al. 1992), including in very young children who were asymptomatic at the time (Mecuri et al. 1997). Reduced grey matter volume is found, especially in the parietal and occipital regions (Eisenberg et al. 2010.) Cerebrovascular accidents associated with multiple intracranial arterial stenoses have also been reported (Ardinger et al. 1994; Wollack et al. 1996).

Evaluation

- All affected individuals should have a complete and careful baseline neurological examination including assessment of muscle tone and strength and documentation of neurological abnormalities.
- Symptoms of headache, dysphagia, dizziness, or weakness should prompt the clinician to obtain cranial MRI to evaluate for Chiari I malformation.
- Symptoms of stroke will require standard investigation with magnetic resonance angiography or conventional angiography.

Management

- Physical therapy emphasizing normal posture and balance is important. Hippotherapy (movement therapy through horseback riding) is an effective mode of treatment (personal observation).
- Treatment of neurological problems does not differ from that in the general population.

- Individuals with Williams syndrome should perform daily stretching exercises to improve range of motion, especially of the lower extremities.

Endocrine

Idiopathic hypercalcemia of infancy has been documented in 22% of children with Williams syndrome less than age 3 years (Kim et al. 2016). The hypercalcemia is most often symptomatic (vomiting, anorexia, constipation, and irritability) in the first two years of life and is associated with dehydration, hypercalciuria, and nephrocalcinosis. It is important to note that the disturbance of calcium metabolism is not limited to infancy. Although more common in infants, it may occur at any age. It often improves over time, but may recur. Adults with symptomatic hypercalcemia have been reported (Morris et al. 1990). Higher median serum calcium levels compared to controls are observed in all age groups (Sindhar et al. 2016). Hypercalciuria occurs in 30% of individuals with Williams syndrome. Despite numerous studies of calcium metabolism, the etiology of these abnormalities remains unknown (Stagi et al. 2016).

Hypothyroidism has been reported in both children and adults with Williams syndrome (Kim et al. 2016; Pober et al. 2001), with a prevalence of 10%. Subclinical hypothyroidism with elevated thyroid-stimulating hormone (TSH) has been demonstrated in 31% of individuals (Palacios-Verdu et al. 2015). Diabetes mellitus and/or pre-diabetes is common among adults with Williams syndrome; of a sample of 28 adults who had no previous record of high glucose levels, 36% were found to have undiagnosed diabetes and an additional 39% had impaired glucose tolerance (Pober et al. 2010). In other studies, impaired glucose tolerance was found in 26% of adolescents (Stagi et al. 2014) and in 63% of young adults (Masserini et al. 2013).

Evaluation

- Serum calcium (either ionized or total) should be measured two to three hours after feeding, and any necessary repeat measurements should be performed at the same time of the day. Serum calcium determination is recommended every four to six months until age 2 years, and every two years thereafter.
- A random spot urine for calcium-to-creatinine ratio should be checked at the time of diagnosis and yearly thereafter. Hypercalciuria on a random spot urine is diagnosed if the calcium-to-creatinine ratio is greater than 0.86 in infants less than age 7 months; 0.6, from 7 to 18 months; 0.42, from 19 months to 6 years; and 0.22, from 6 years to adult (Sargent et al. 1993).
- If hypercalciuria is found on a random sample, then a repeat spot urine should be checked for the calcium-to-creatinine ratio on a morning and evening sample, and, ideally, a 24-hour urine for calcium-to-creatinine ratio should be obtained.
- If there is persistent hypercalciuria, serum calcium measurements should be repeated and a renal ultrasound obtained to evaluate for nephrocalcinosis.
- Individuals with Williams syndrome should have thyroid function tests yearly until age 3 years and then every two years.
- Adults at age 20 years should have an oral glucose tolerance test, and if normal, it should be repeated every five years.

Management

- In infants with hypercalcemia or hypercalciuria, if dietary calcium is greater than the recommended daily intake (Ross et al. 2011), it should be decreased appropriately. Assess hydration status and increase water intake as indicated.
- If hypercalcemia or hypercalciuria persists with a normal calcium intake, then dietary calcium should be reduced in consultation with a nutritionist. Low-calcium infant formulas are available, and vitamin D supplements are not recommended. While on a low-calcium diet, careful monitoring of serum calcium levels is important, because iatrogenic rickets has been reported in children with Williams syndrome (Martin et al. 1984). Thus, it is important to counsel parents not to eliminate calcium-containing foods from the diet without close medical supervision.
- In infants with borderline high serum calcium values who also have hypercalcemic symptoms and inadequate weight gain, careful reduction of the calcium intake may result in symptomatic improvement; again, serum calcium levels must be carefully monitored (personal experience).
- Dietary restriction of calcium, while under careful medical and nutritional observation, is sufficient in most cases, but refractory cases of hypercalcemia may be treated with oral steroids to reduce calcium absorption from the gut and bone. Severe cases of hypercalcemia have been successfully treated with intravenous pamidronate (Cagle et al. 2004; Oliveri et al. 2004).
- Referral to an endocrinologist and/or nephrologist for management should be considered when hypercalcemia does not respond to dietary restriction, when nephrocalcinosis is present, or when hypercalciuria is persistent.
- It is important to note that manipulation of dietary calcium alone will not solve the feeding problems in Williams syndrome. Feeding therapy and treatment of gastroesophageal reflux must be offered at the same time for significant clinical improvement to occur.

- If hypothyroidism is present, treatment is the same as in the general population.
- Diabetes mellitus treatment is standard, as are prevention efforts aimed at a healthy diet and exercise.

Cardiovascular

The cardiovascular effect of the *ELN* deletion is the most significant cause of morbidity and mortality in Williams syndrome. The elastin arteriopathy is generalized, and thus any artery may be narrowed. The most commonly detected abnormality is supravalvar aortic stenosis (SVAS), with a prevalence of 75% in most reported series (Morris et al. 1988). However, the detection of supravalvar aortic stenosis depends on the sensitivity of the examination method; two-dimensional echocardiography with Doppler flow analysis is preferred (Ensing et al. 1989). Peripheral pulmonic stenosis is more often observed in infants (50%) because it typically improves over time (Wren et al. 1990). However, in angiography series, peripheral pulmonic stenosis is found in 75% (Kececioglu et al. 1993). In contrast to pulmonary circulation, arterial stenoses in systemic circulation may worsen over time (Kim et al. 1999), thus, lifelong monitoring of the cardiovascular system is important. SVAS is most likely to worsen in the first five years of life (Collins et al. 2010b). The overall prevalence of any cardiovascular abnormality is 80%, including the less frequently reported lesions such as discrete supravalvar pulmonic stenosis (25%), renal artery stenosis (40%), ventricular septal defect (10%), and other peripheral arterial stenoses (20%). Sudden death has been reported in Williams syndrome (Bird et al. 1996); some cases are secondary to coronary artery stenosis unrelated to the severity of supravalvar aortic stenosis. At particularly high risk are individuals with severe bilateral outflow tract obstruction and biventricular hypertrophy, who have decreased myocardial perfusion during periods of hemodynamic stress (Pham et al. 2009). This condition poses a risk of myocardial compromise during induction of anesthesia. Individuals with left ventricular hypertrophy and coronary artery torerosity or stenosis also have increased risk associated with anesthesia (Horowitz et al. 2002).Because there is an increased risk for adverse events associated with sedation and anesthesia in Williams syndrome, risk management assessment and management guidelines have been developed (Latham et al. 2016; Matisoff et al. 2015).

Hypertension may present at any age in Williams syndrome and has a prevalence of 50% (Bouchireb et al. 2010). It may be secondary to renal artery stenosis (Deal et al. 1992). In a study of 95 individuals aged 2–43 years, Kececioglu et al. (1993) found that both systolic and diastolic blood pressures were elevated in 47%, systolic elevated in 25%, and diastolic elevated in 9%. Increased vascular stiffness in Williams syndrome does respond to antihypertensives (Kozel et al. 2014).

Cerebrovascular accident associated with arteriopathy of the neurovasculature has rarely been reported (Ardinger et al. 1994; Wollack et al. 1996). In adults, aortic insufficiency (20%) and mitral valve prolapse (15%) may occur (Kececioglu et al. 1993; Collins et al. 2010a). Corrected QT prolongation has been found in 14% of individuals with Williams syndrome (Collins et al. 2010a).

Supravalvar aortic stenosis typically presents with a systolic murmur heard best in the aortic area, radiating to the carotids, and often accompanied by a thrill in the suprasternal notch. Because of the Coanda effect, the blood pressure in the right arm is often higher than that measured in the left (French and Guntheroth 1970). Supravalvar aortic stenosis occurs most commonly as an hourglass stenosis above the aortic valve (77%) but may also occur as a more diffuse long-segment aortic hypoplasia (Stamm et al. 1999). It should be noted that the term "hypoplasia" traditionally refers to the size of the lumen and, thus, may be misleading because the aortic wall is actually thickened. The aortic stenosis causes increased resistance to blood flow, resulting in elevated left heart pressures and cardiac hypertrophy. In severe cases, cardiac failure and death will result if the lesion is untreated. There is a high degree of variability among affected individuals, with supravalvar aortic pressure gradients ranging from 0 to 110 mm Hg. About 30% of the individuals will ultimately require surgical correction (Kececioglu et al. 1993).

Evaluation

- A baseline evaluation by a cardiologist is recommended at the time of diagnosis and should include measurement of the blood pressure in all the four limbs, EKG, and two-dimensional echocardiography with Doppler flow studies.
- If no cardiac anomaly is detected in a young child at initial evaluation, then yearly clinical cardiology evaluations should be performed until age 5 years.
- In older children and adults without previous evidence of significant cardiovascular disease, periodic monitoring (every two to three years) by a cardiologist is warranted for potential late-onset complications. An echocardiogram should be done when there is new onset of hypertension or signs/symptoms suggestive of mitral valve prolapse (Pober and Morris 2007).
- On an annual basis, the primary physician should check and record blood pressure measurements from both arms, auscultatory findings of murmurs and bruits, and peripheral pulse assessments.
- Additional cardiovascular imaging studies may be required in individuals with decreased pulses, bruits, or signs of diffuse thoracic aortic stenosis.

- Cardiovascular status should be reviewed before any surgery, and inpatient surgical settings are preferred over outpatient facilities.
- In children with Williams syndrome, pediatric anesthesia consultation should be considered for all procedures requiring sedation or anesthesia. Guidelines have been published for risk assessment and management for both children and adults (Latham et al. 2016; Matisoff et al. 2015).

Management

- The need for patch aortoplasty will be determined by the severity of the stenosis and the pressure gradient.
- Transcatheter balloon dilatation is used to treat some arterial stenoses, especially in the pulmonary arteries.
- Surgical treatment of supravalvar aortic stenosis has a perioperative mortality of 3–7% (Kececioglu et al. 1993; van Son et al. 1994). In one series, survival was 92% at 10 years and 88% at 20 years (van Son et al. 1994). In a series seen by cardiologists, the actuarial curve was stable at 90% from ages 10 to 40 years (Kececioglu et al. 1993).
- Hypertension is usually treated medically. β-blockers or calcium channel blockers are usually successful in the treatment of hypertension (Bouchireb et al. 2010).

Gastrointestinal

Infants with Williams syndrome require treatment for feeding problems, gastroesophageal reflux, and constipation, as detailed in Growth and Feeding. Chronic constipation continues to be a problem in 40% of older children and adults, and diverticulosis is common (Morris et al. 1990). Chronic abdominal pain may be debilitating for adolescents or adults with a myriad of etiologies, including hiatal hernia, gastroesophageal reflux, peptic ulcer disease, cholelithiasis, diverticulitis, ischemic bowel disease, constipation, and somatization of anxiety. Hypercalcemia may also contribute to the symptomatology by causing anorexia, abdominal pain, and constipation. (See Endocrine for evaluation and management of hypercalcemia.)

The pathogenesis of many of the gastrointestinal complaints and complications, such as diverticulosis and rectal prolapse, is probably an abnormality of elastic fibers. The prevalence of diverticulitis in adults with Williams syndrome aged 17–39 years is 8%, a four-fold increase compared with the general population (Partsch et al. 2005). An increased prevalence of diverticulitis has also been noted in adolescents with Williams syndrome (Stagi et al. 2010).

Constipation is a lifelong problem requiring ongoing management. Gastroesophageal reflux may occur at any age and, if untreated, will lead to dysphagia, esophagitis, and esophageal stricture.

Evaluation

- Evaluation should include oral motor, nutritional, and gastrointestinal assessment in infancy.
- Calcium levels should be monitored in serum and urine.
- Radiographic analysis of the upper and lower gastrointestinal tract may be warranted, especially if abdominal pain is noted. Radiographic evaluation, pH studies, or gastroenterology referral is appropriate if symptoms suggest gastroesophageal reflux.
- A change in the severity or character of abdominal pain should prompt evaluation for diverticulitis, even in an adolescent.
- If other causes have been eliminated, angiography may be needed to determine whether mesenteric arterial stenosis is present as the etiology of the pain.

Management

- Treatment of abdominal pain will depend on its cause, and is standard. However, surgical treatment is typically required for diverticulitis (Partsch et al. 2005).
- The dietary regimen will be dictated by the results of investigations for gastroesophageal reflux and hypercalcemia.
- Gastroesophageal reflux may be treated with positioning, small-volume feedings, and antireflux medication. If gastroesophageal reflux does not respond to medical treatment, fundoplication will be required.
- Constipation should be aggressively managed at all ages in a standard manner. If increasing fluids and fiber are not effective, many individuals with Williams syndrome respond to Miralax® treatment at standard doses.
- Rectal prolapse may require surgical treatment, as in the general population.
- Individuals with diverticulosis are at increased risk for diverticulitis.

Ophthalmologic

Ophthalmologic abnormalities and deficits in visual-motor integration adversely affect function in individuals with Williams syndrome (Morris et al. 1988). Strabismus, usually esotropia, typically affects 50% of infants. However, reduced stereoacuity has been documented in children who do not have measurable strabismus, suggesting that abnormalities of binocular vision are even more prevalent and have a central nervous system origin (Sadler et al. 1996). The outcome of strabismus surgery in the Williams syndrome population is similar to that in the general population with infantile esotropia (Kapp et al. 1995).

The most common refractive error is mild to moderate hyperopia, present in 67% of individuals (Weber et al. 2014). Tortuosity of retinal vessels is reported in 20%. Lacrimal duct stenosis is common. Hypoplasia and coarse architecture of the iris stroma (70%) result in a lacy/stellate iris pattern clinically visible in blue irides and visible on slit-lamp examination in dark irides (Winter et al. 1996). Cataracts have been reported in adults (Cherniske et al. 2004).

Evaluation

- Every individual with Williams syndrome should have a complete ophthalmologic evaluation at the time of diagnosis. Follow-up will be based on the findings.

Management

- Aggressive treatment of strabismus is warranted to prevent amblyopia and may include corrective lenses, patching, or, in 30% of cases, medial rectus muscle recession.
- Hyperopia requires corrective lenses.

Ears and Hearing

Chronic otitis media is reported in 50% of individuals with Williams syndrome and is treated surgically in 25% (Morris et al. 1988; Klein et al. 1990) of them. It is likely that hypotonia combined with elastin-mediated connective tissue abnormalities of the pharynx and eustachian tube is causal. Hypersensitivity to sound occurs in 85–95% and is not related either to a history of chronic otitis media or to attention-deficit disorder (Van Borsel et al. 1997). Responses to sound, especially to loud or sudden noises, range from crying and/or covering ears with hands in children to complaints by adults of distress and a "nervous feeling." With aging, individuals with Williams syndrome have improved noise tolerance. The etiology of the hypersensitivity is unknown, although possibilities include a conduction disturbance resulting from an abnormality of connective tissue, an abnormality of central nervous system processing of auditory input, or failure of central nervous system habituation to auditory stimuli. Mild to moderately severe sensorineural or mixed hearing loss was found in 63% of school-aged children and 92% of adults with Williams syndrome; in most cases, parents were not previously aware of this hearing loss (Marler et al. 2010).

Evaluation

Audiological evaluation is recommended yearly.

Management

- As in the general pediatric population, chronic otitis media tends to improve with age, but aggressive treatment with antibiotics and tympanotomy tubes may be helpful.
- The family should be counseled to avoid situations with loud noises when possible and to use earplugs when necessary.
- Older children and adults benefit from warnings before noise occurs, explanations regarding the origin of the sound, and having some control over the sound (e.g., turning on a noisy appliance).
- Hearing loss should be treated in a standard manner. Unilateral hearing loss (even if mild) is likely to negatively impact academic progress.
- For school-age children with hearing loss (even if unilateral and mild), a classroom FM system should be considered. The teacher wears a microphone and transmitter that sends a signal to hearing aides or headphones worn by the student. This system also may be helpful for children with ADHD even if they do not have hearing loss.

Dermatologic

In addition to elastin arteriopathy, other connective tissues are also affected in Williams syndrome. The skin is typically soft and loose in both infants and adults. The skin is not hyperextensible, however, and wound healing is normal. As noted in cutis laxa, adults with Williams syndrome may appear older than their age. Hernias are common, ranging from 5% for umbilical hernias to 40% for inguinal hernias. An extra sacral crease is often seen in infants with Williams syndrome. Because of the connective tissue abnormality, hernias may occur at any age, and occasional recurrences are noted.

Evaluation

- Physical examination for hernia should be performed during regular medical visits.

Management

- Surgical repair of inguinal hernia may be necessary, as in the general population.

Genitourinary

The incidence of renal anomalies is increased in Williams syndrome, and detection of an abnormality is dependent on the type of investigation. In renal ultrasound studies, the incidence of structural renal anomalies (small, asymmetric, dystrophic, duplicated, or absent kidney) is 20% (Ingelfinger and Newburger 1991; Pober et al. 1993; Pankau et al. 1996). Nephrocalcinosis is found in less than 5%. When renal angiography is performed, several studies have found a 50% incidence of renal artery stenosis. Bladder diverticulae are common, being found in 60% of children (Babbitt et al. 1979) and 75% of adults (Morris et al. 1988). They are best demonstrated by voiding cystourethrogram.

Increased urinary frequency was reported in 70% of a sample of 87 children with Williams syndrome (mean age 9.0 ± 4.2 years) and urinary urgency was found in 69% (Sammour et al. 2017). Decreased bladder capacity was present, and 60% had detrusor overactivity (Sammour et al. 2006). A history of urinary tract infections was reported for 26%. The average age of attainment of daytime continence was 4.4 years; 50% achieved nocturnal continence by age 10 years (Sammour et al. 2017). Approximately 3% of adults have nocturnal enuresis (von Gontard et al. 2016). The risk for recurrent urinary tract infections and bladder diverticulae increases with age.

Evaluation

- As the prevalence of renal structural anomalies is high, a baseline renal ultrasound is recommended at the time of diagnosis. An ultrasound of the kidneys and bladder is recommended every 10 years.
- Other investigations will depend on signs and symptoms.
- Performance of renal artery angiography should be considered at the time of cardiac catheterization or as part of the investigation of the significantly hypertensive individual.
- A voiding cystourethrogram may be indicated in cases of chronic urinary tract infections, persistence of enuresis beyond age 10 years, or chronic problems with micturition. Yearly urinalysis is recommended.
- Structural abnormalities require referral to a urologist.

Management

- In individuals with persistent hypercalciuria or nephrocalcinosis, referral to a nephrologist is recommended for ongoing management (see also Endocrine).
- Urinary tract infections respond to standard treatment modalities.
- Treatment of renal structural anomalies is standard.

Musculoskeletal

Hyperextensible joints are commonly observed (90%) in infants and young children with Williams syndrome, and they contribute to joint instability and delayed walking. As a result, abnormal compensatory postures to achieve stability may be noted. Many individuals complain of leg pain, especially at night after a day of high physical activity. Gradually, tightening of the Achilles tendons and hamstrings occurs, resulting in a stiff and awkward gait by adolescence. Radioulnar synostosis (10%) results in limitation of supination of the forearm and does not improve with either therapy or surgery. Joint contractures, especially of the lower extremities, have been reported in 50% of children with Williams syndrome (Morris et al. 1988; Kaplan et al. 1989) and in 90% of adults (Morris et al. 1990). Scoliosis (18%; Morris et al. 2010), kyphosis (20%), and lordosis (40%) are common complications. The majority of the musculoskeletal abnormalities are presumably secondary to abnormal connective tissue, the result of the *ELN* deletion.

The voice of the individual with Williams syndrome is typically low-pitched, deep, or hoarse. Because elastin is an important component of the lamina propria of the vocal folds (Hammond et al. 1998), the presumption is that the voice quality is likely related to the elastin abnormality (Vaux et al. 2003).

Evaluation

- Range of motion studies should be performed on a yearly basis by the primary care physician.
- Clinical screening for spine abnormalities should be done annually.
- Orthopedic referral is required for evaluation and treatment of scoliosis.

Management

- A program of physical therapy to promote normal posture and normal range of motion is recommended. With physical therapy and a program of regular exercise, function, mobility, and posture improve.
- If pes planus is present, a medial wedge or insole may be of benefit, and high-top shoes or ankle orthotics may provide ankle stability in young children.
- Nocturnal leg pains may be ameliorated with physical therapy, anti-inflammatory agents, and use of elastic bandages at the knees and ankles during periods of high physical activity.
- Regular low-impact exercise is important.
- The expertise of a physical therapist is helpful both in the evaluation and in designing appropriate therapeutic programs.
- Exercise programs that include calming/relaxation techniques, such as yoga, have the added benefit of addressing the anxiety that is part of the Williams syndrome phenotype (personal observation). Stretching exercises, especially of the lower extremities, should be part of the daily routine in affected children and adults.

Dental

Dental abnormalities reported in Williams syndrome include malformed teeth (10% in primary dentition and 40% in permanent dentition); missing teeth (10–40% in permanent teeth, usually incisors), localized areas of enamel hypoplasia

(15%), small teeth (95%), and increased space between the teeth (50%) (Hertzberg et al. 1994; Axelson et al. 2005). Malocclusion (85%) is the most significant clinical complication with deep bite, open bite, anterior crossbite, and posterior crossbite all being reported. The occlusal abnormalities are likely the combined result of several factors, including hypotonia, tongue thrust, and connective tissue abnormality. Orthodontic treatment has been successful in improving occlusion. Poor fine motor skills contribute to difficulty in maintaining dental hygiene at all ages.

Evaluation

- In addition to early dental evaluation when the primary teeth have erupted, orthodontic referral at age 8 years is recommended.

Management

- In addition to routine dental care, orthodontic treatment has improved dental relationships, the mechanics of chewing, and the appearance in many individuals with Williams syndrome.
- Oral motor and speech therapy are of value in improving chewing and swallowing.
- Because there is a propensity to enamel hypoplasia, fluoride sealants may be helpful.
- Individuals with Williams syndrome may require assistance with daily brushing and flossing.
- Dental cleaning is recommended every four months for adolescents and adults.

ACKNOWLEDGMENTS

Preparation of this chapter was supported by grant R01 NS35102 from the National Institute of Neurological Disorders and Stroke and grant WSA 0111 from the Williams Syndrome Association. We are grateful to the individuals with Williams syndrome and their families who have participated in research studies; without their contribution progress in understanding Williams syndrome would not be possible.

RESOURCES

Support Organizations

Williams Syndrome Association
570 Kirts Boulevard, Suite 223
Troy, MI 48084-4156, USA
Phone: (248) 224-2229, (800) 806-1871
Fax: (248) 224-2230
Email: info@williams-syndrome.org
Website: https://williams-syndrome.org

Williams Syndrome Foundation LTD
Suite 2367, 145-147 Boston Road, Ealing, London W7 3SA, UK
Phone: 0208 567 1374
Email: enquiries@williams-syndrome.org.uk
Website: https://www.williams-syndrome.org.uk

Books

Morris CA, Lenhoff HM, Wang PP (eds) (2006) *Williams-Beuren Syndrome: Research, Evaluation, and Treatment*, Baltimore, MD: The John Hopkins University Press.

REFERENCES

American Academy of Pediatrics (2001) Healthcare supervision for children with Williams syndrome. *Pediatrics* 107:1192–1204.

Annaz D, Hill CM, Ashworth A, Holley S, Karmiloff-Smith A (2011) Characterzation of sleep problems in children with Williams syndrome. *Res Dev Disabil* 32:164–169.

Ardinger RH, Goertz KK, Mattioli LF (1994) Cerebrovascular stenosis with cerebral infarction in a child with Williams syndrome. *Am J Med Genet* 51:200–202.

Atkinson J, King J, Braddick O, Nokes L, Anker S, Braddick F (1997) A specific deficit of dorsal function in Williams' syndrome. *Neurol Rep* 8:1919–1922.

Axelsson S, Kjaer I, Heiberg A, Bjørnland T, Storhaug K (2005) Neurocranial morphology and growth in Williams syndrome. *Eur J Ortho* 27(1):32–47.

Babbitt DP, Dobbs J, Boedecker RA (1979) Multiple bladder diverticula in Williams "elfin-facies" syndrome. *Pediatr Radiol* 8:29–31.

Bayes M, Magano LF, Rivera N, Flores R, Perez Jurado LA (2003) Mutational mechanisms of Williams-Beuren syndrome deletions. *Am J Hum Genet* 73:131–151.

Beuren AJ (1972) Supravalvular aortic stenosis: A complex-syndrome with and without mental retardation. *Birth Defects* 8:45–56.

Beuren AJ, Apitz J, Harmjanz D (1962) Supravalvular aortic stenosis in association with mental retardation and a certain facial appearance. *Circulation* 27:1235–1240.

Bird LM, Billman GF, Lacro RV, Spicer RL, Jariwala LK, Hoyme HE, Zamora-Salinas R, Morris CA, Viskochil D, Frikke MJ, Jones MC (1996) Sudden death in Williams syndrome: Report of ten cases. *J Pediatr* 129:926–931.

Botta A, Novelli G, Mari A, Novelli A, Sabini M, Korenberg J, Osborne LR, Diglio MC, Giannotti A, Dallapiccola B (1999) Detection of an atypical 7q11.2 deletion in Williams syndrome patients which does not include the STX1A and FZD9 genes. *J Med Genet* 36:478–480.

Bouchireb K, Boyer O, Bonnet D, Brunelle F, Decramer S, Landthaler G, Liutkus A, Niaudet P, Salomon R (2010) Clinical features and management of arterial hypertension in children with Williams-Beuren syndrome. *Nephrol Dial Transplant* 25:434–438.

Braam W, Smits MG, Didden R, Korzilius H, Geijlswijk IMV, Curfs LMG (2009) Exogenous melatonin for sleep problems in individuals with intellectual disability: A meta-analysis. *Dev Med Child Neurol* 51:340–349.

Burn J (1986) Williams syndrome. *J Med Genet* 23:389–395.

Cagle AP, Waguespack SG, Buckingham BA, Shankar RR, Dimeglio LA (2004) Severe infantile hypercalcemia associated with Williams syndrome successfully treated with intravenously administered pamidronate. *Pediatrics* 114:1091–1095.

Collins RT 2nd, Aziz PF, Gleason MM, Kaplan PB, Shah MJ (2010a) Abnormalities of cardiac repolarization in Williams syndrome. *Am J Cardiol* 106:1029–1033.

Collins RT 2nd, Kaplan P, Somes GW, Rome JJ (2010b) Long-term outcomes of patients with cardiovascular abnormalities and Williams syndrome. *Am J Cardiol* 105:874–878.

Crespi BJ, Hurd PL (2014) Cognitive-behavioral phenotypes of Williams syndrome are associated with genetic variation in the GTF2I gene, in a healthy population. *BMC Neurosci* 15:127.

Deal JE, Snell MF, Marratt TM, Dillon MJ (1992) Renovascular disease in childhood. *J Pediatr* 131:378–384.

Del Campo M, Antonell A, Magano LF, Munoz FG, Flores R, Bayés M, Pérez Jurado LA (2006) NCF1 gene in patients with Williams-Beuren syndrome decreases their risk of hypertension. *Am J Hum Genet* 78:533–542.

Dutly F, Schinzel A (1996) Unequal interchromosomal rearrangements may result in elastin gene deletions causing the Williams-Beuren syndrome. *Hum Mol Genet* 5:1893–1898.

Elison S, Stinton C, Howlin P (2010) Health and social outcomes in adults with Williams syndrome: Findings from cross-sectional and longitudinal cohorts. *Res Dev Disabil* 31:587–599.

Elliott CD (2007) *Differential Ability Scales* 2nd ed. San Antonio, TX: Psychological Corporation.

Enkhmandakh B, Makeyev AV, Erdenechimeg L, Ruddle FH, Chimge NO, Tussie-Luna MI, Roy AL, Bayarsaihan D (2009) Essential functions of the Williams-Beuren syndrome-associated TFII-I genes in embryonic development. *Proc Natl Acad Sci* 106(1):181–186.

Ensing GJ, Schmidt MA, Hagler DJ, Michels VV, Carter GA, Feldt RH (1989) Spectrum of findings in a family with nonsyndromic autosomal dominant supravalvular aortic stenosis: A Doppler echocardiographic study. *J Am Coll Cardiol* 13:413–419.

Ewart AK, Morris CA, Atkinson D, Jin W, Sternes K, Spallone P, Stock AD, Leppert M, Keating MT (1993) Hemizygosity at the elastin locus in a developmental disorder, Williams Syndrome. *Nat Genet* 5:11–16.

Fanconi G, Giradet P, Schlesinger B, Butler N, Blade JS (1952) Chronische Hypercalcaemie kombiniert mit Osteosklerose, Hyperazotaemie, Minderwuchs, und kongenitalen Missbildungen. *Helv Paediatr Acta* 7:314–334.

Fisher MH (2014) Evaluation of a stranger safety training program for adults with Williams syndrome. *J Intellect Disabil Res* 58:903–914.

Fisher MH, Lense MD, Dykens EM (2016) Longitudinal trajectories of intellectual and adaptive functioning in adolescents and adults with Williams syndrome. *J Intellect Disabil Res* 60:920–932.

Frangiskakis JM, Ewart AK, Morris CA, Mervis CB, Bertrand J, Robinson BF, Klein BP, Ensing GJ, Everett LA, Green ED, Proschel C, Gutow NJ, Noble M, Atkinson DL, Oldelberg SJ, Keating MT (1996) LIM-kinase 1 hemizygosity implicated in impaired visuospatial constructive cognition. *Cell* 86:59–69.

French JW, Guntheroth WG (1970) An explanation of asymmetric upper extremity blood pressures in supravalvular aortic stenosis. *Circulation* 42:31–36.

Gagliardi C, Martelli S, Burt MD, Borgatti R (2007) Evolution of neurologic features in Williams syndrome. *Pediatr Neurol* 36:301–306.

Garcia RE, Friedman WF, Kaback MM, Rowe RD (1964) Idiopathic hypercalcemia and supravalvular aortic stenosis. *N Engl J Med* 271:117–120.

Greer MK, Brown FR, Pai GS, Choudry SH, Klein AJ (1997) Cognitive, adaptive, and behavioral characteristics of Williams syndrome. *Am J Med Genet* 4:521–525.

Hakimi MA, Dong Y, Lane WS, Speicher DW, Shiekhattar R (2003) A candidate X-linked mental retardation gene is a component of a new family of histone deacetylase-containing complexes. *J Biol Chem* 278:7234–7239.

Hammond TH, Gray SD, Butler J, Zhou R, Hammond E (1998) Age- and gender-related elastin distribution changes in human vocal folds. *Otolaryngol Head Neck Surg* 119:314–322.

Heller R, Rauch A, Luttgen S, Schroder B, Winterpacht A (2003) Partial deletion of the critical 1.5 Mb interval in Williams-Beuren syndrome. J Med Genet 40:e99.

Hertzberg J, Nakisbendi L, Neddleman HL, Pober B (1994) Williams syndrome—Oral presentation of 45 cases. *Pediatr Dent* 16:262–267.

Hobart HH, Morris CA, Mervis CB, Pani AM, Kistler DJ, Rios CM, Kimberley KW, Gregg RG, Bray-Ward P (2010) Inversion of the Williams syndrome region is a common polymorphism found more frequently in parents of children with Williams syndrome. *Am J Med Genet C* 154C:220–228.

Hocking DR, Rinehart NJ, McGinley JL, Bradshaw JL (2009) Gait function in adults with Williams syndrome. *Exp Brain Res* 192:695–702.

Horowitz PE, Akhtar S, Wulff JA, Al Fadley F, Al Halees Z (2002) Coronary artery disease and anesthesia-related death in children with Williams syndrome. *J Cardiothorac Vasc Anesth* 6:739–741.

Ingelfinger JR, Newburger JW (1991) Spectrum of renal anomalies in patients with Williams syndrome. *J Pediatr* 119:771–773.

Jawaid A, Riby DM, Owens J, White SW, Tara T, Schulz PE (2012) 'Too withdrawn' or 'too friendly': considering social vulnerability in two neuro-developmental disorders. *J Intellect Disabil Res* 56:335–350.

Jones KL, Smith DW (1975) The Williams elfin facies syndrome. *J Pediatr* 86:718–723.

Kaplan P, Kirschner M, Watters G, Costa T (1989) Contractures in patients with Williams syndrome. *Pediatrics* 84:895–899.

Kapp ME, von Noorden GK, Jenkins R (1995) Strabismus in the Williams syndrome. *Am J Ophthalmol* 119:355–360.

Kececioglu D, Kotthoff S, Vogt J (1993) Williams-Beuren syndrome: A 30-year follow-up of natural and postoperative course. *Eur Heart J* 14:1458–1464.

Kim YM, Cho JH, Kang E, Kim GH, Seo EJ, Lee BH, Choi JH, Yoo HW (2016) Endocrine dysfunctions in children with Williams-Beuren syndrome. *Ann Pediatr Endocrinol Metab* 21:15–20.

Kim YM, Yoo S-Y, Choi JY, Kim SH, Bae EJ, Yee YT (1999) Natural course of supravalvar aortic stenosis and peripheral pulmonary artery stenosis in Williams syndrome. *Cardiology* 9:37–41.

Klein AJ, Armstrong BL, Greer MK, Brown FR (1990) Hyperacusis and otitis media in individuals with Williams syndrome. *J Speech Hear Disord* 55:339–344.

Klein-Tasman BP, Lee K (2017) Problem behavior and psychosocial functioning in young children with Williams syndrome: parent and teacher perspectives. *J Intellect Disabil Res* 61:853–865.

Klein-Tasman BP, Li-Barber, KT, Magargee ET (2011). Honing in on the social phenotype in Williams syndrome using multiple measures and multiple raters. *J Autism Dev Disord* 41:341–351.

Klein-Tasman BP, Mervis CB (2003) Distinctive personality characteristics of 8-, 9-, and 10-year-old children with Williams syndrome. *Dev Neuropsychol* 23:271–292.

Klein-Tasman PB, Mervis CB, Lord C, Phillips KD (2007) Sociocommunicative deficits in young children with Williams syndrome: Performance on the Autism Diagnostic Observation Schedule. *Child Neuropsychol* 13:444–467.

Kozel BA, Danback JR, Waxler JL, Knutsen RH, de Las Fuentes L, Reusz GS, Kis E, Bhatt AB, Pober BR. (2014) Williams syndrome predisposes to vascular stiffness modified by antihypertensive use and copy number changes in NCF1. *Hypertension* 63(1):74–79.

Kuijpers GM, DeVroede M, Knol HE, Jansen M (1999) Growth hormone treatment in a child with Williams-Beuren syndrome: A case report. *Eur J Pediatr* 158:451–454.

Latham GJ, Ross FJ, Eisses MJ, Richards MJ, Geiduschek JM, Joffe DC (2016) Perioperative morbidity in children with elastin arteriopathy. *Paediatr Anaesth 2016* 26(9):926–35.

Levine K, Chedd N (2007) *Replays: Using Play to Advance Emotional and Behavioural Development for Children with Autism Spectrum Disorders*. Philadelphia, PA: Jessica Kingsley Publishers.

Leyfer O, Woodruff-Borden J, Klein-Tasman B, Fricke J, Mervis C (2006) Prevalence of psychiatric disorders in 4 to 16-year olds with Williams syndrome. *Am J Med Genet B* 141(6):615–622.

Li DY, Toland AE, Boak BB, Atkinson D, Ensing GJ, Morris CA, Keating MT (1997) Elastin point mutations cause an obstructive vascular disease, supravalvular aortic stenosis. *Hum Mol Genet* 6:1021–1028.

Li DY, Faury G, Taylor DG, Dais EC, Boyle WA, Mecham RP, Stenzel P, Boak B, Keating MT (1998) Novel arterial pathology in mice and humans hemizygous for elastin. *J Clin Invest* 102:1783–1787.

Lincoln AJ, Searcy YM, Jones W, Lord C (2007) Social interaction behaviors discriminate young children with autism and Williams syndrome. *J Am Acad Child Adolesc Psychiatry* 46:323–331.

Lough E, Fisher MH (2016) Parent and self-report ratings on the perceived levels of social vulnerability of adults with Williams syndrome. *J Autism Dev Disabil* 46:3424–3433.

Lowery MC, Morris CA, Ewart A, Brothman L, Zhu XL, Leonard CO, Carey JC, Keating MT, Brothman AR (1995) Strong correlations of elastin deletions, detected by FISH, with Williams syndrome: Evaluation of 235 patients. *Am J Hum Genet* 57:49–53.

Marler JA, Sitcovsky JL, Mervis CB, Kistler DJ, Wightman FL (2010) Auditory function and hearing loss in children and adults with Williams syndrome: Cochlear impairment in individuals with otherwise normal hearing. *Am J Med Genet C* 154C:249–265.

Marshall CR, Young EJ, Pani AM, Freckmann M, Lacassie Y, Howald C, Fitzgerald KK, Peippo M, Morris CA, Shane K, Priolo M, Morimoto M, Kondo I, Manguoglu E, Berker-Karauzum S, Edery P, Hobart HH, Mervis CB, Zuffardi O, Reymond A, Kaplan P, Tassabehji M, Gregg RG, Scherer SW, Osborne LR (2008) Infantile spasms is associated with deletion of the MAGI2 gene on chromosome 7q11.23-q21.11. *Am J Hum Genet* 83:106–111.

Martin ND, Smith WR, Cole TJ, Preece MA (2007) New height, weight and head circumference charts for British children with Williams syndrome. *Arch Dis Child* 92:598–601.

Martin NDT, Snodgrass GJAI, Cohen RD (1984) Idiopathic infantile hypercalcemia: A continuing enigma. *Arch Dis Child* 59:605–613.

Mason TB, Ahrens R, Sharman J, Bintliff-Janisak B, Schultz B, Walters AS, Cater JR, Kaplan P, Pack AI (2010) Sleep in children with Williams syndrome. *Sleep Med* 12:892–897.

Matisoff AJ, Olivieri L, Schwartz JM, Deutsch N (2015) Risk assessment and anesthetic management of patients with Williams syndrome: a comprehensive review. *Paediatr Anaesth* 25(12):1207–1215.

Mercuri E, Atkinson J, Braddick O, Rutherford MA, Cowan FM, Counsel SJ, Dubowitz LM, Bydder G (1997) Chiari I malformation in asymptomatic young children with Williams syndrome: clinical and MRI study. *Eur J Paediatr Neurol* 1(5-6):177–181.

Mervis CB, John AE (2010) Cognitive and behavioral characteristics of children with Williams syndrome: Implications for intervention approaches. *Am J Med Genet C* 154C:229–248.

Mervis CB, Klein-Tasman BP, Huffman MJ, Velleman SL, Pitts CH, Henderson DR, Woodruff-Borden J, Morris CA, Osborne LR (2015) Children with 7q11.23 duplication syndrome: Psychological characteristics. *Am J Med Genet A* 167A:1436–1450.

Mervis CB, Pitts CH (2015) Children with Williams syndrome: Developmental trajectories for intellectual abilities, vocabulary abilities, and adaptive behavior. *Am J Med Genet C* 169C:158–171.

Meyer-Lindenberg A, Kohn P, Mervis CB, Kippenhan JS, Olsen RK, Morris CA, Berman KF (2004) Neural basis of genetically determined visuospatial construction deficit in Williams syndrome. *Neuron* 43:623–631.

Morris CA, Mervis CB, Paciorkowski AP, Abdul-Rahman O, Dugan SL, Rope AF, Bader P, Hendon LG, Velleman SL, Klein-Tasman BP, Osborne LR (2015). 7q11.23 duplication syndrome: Physical characteristics and natural history. *Am J Med Genet A* 167A:2916–2935.

Morris CA, Dilts C, Dempsey SA, Leonard CO, Blackburn B (1988) The natural history of Williams syndrome: Physical characteristics. *J Pediatr* 113:318–326.

Morris CA, Leonard CO, Dilts C, Demsey SA (1990) Adults with Williams syndrome. *Am J Med Genet Suppl* 6:102–107.

Morris CA, Thomas IT, Greenberg F (1993) Williams syndrome: Autosomal dominant inheritance. *Am J Med Genet* 47:478–481.

Morris CA, Lu X, Greenberg F (1998) Syndromes identified in patients with a previous diagnosis of Williams syndrome who do not have elastin deletion. *Proc Greenwood Genet Cent* 17:116.

Morris CA, Mervis CB, Hobart HH, Gregg RG, Bertrand J, Ensing GJ, Sommer A, Moore CA, Hopkin RJ, Spallone PA, Keating MT, Osborne L, Kimberley KW, Stock AD (2003) *GTF2I* hemizygosity implicated in mental retardation in Williams syndrome: Genotype-phenotype analysis of five families with deletions in the Williams syndrome region. *Am J Med Genet A* 123A:45–59.

Morris CA, Lenhoff HM, Wang PP (eds) (2006) *Williams-Beuren Syndrome, Research, Evaluation, and Treatment*. The John Hopkins University Press.

Morris CB Pani AM, Mervis CB, Rios CM, Kistler DJ, Gregg RG (2010) Alpha 1 antityrpsin deficiencyalleles are associated with joint dislocation and scoliosis in Williams syndrome. *Am J Med Genet C* 154C:299–306.

Mulik VV, Temple KI, Howe DT (2004) Two pregnancies in a woman with Williams syndrome. *Br J Obstet Gynaecol* 111:511–512.

O'Connor W, Davis J, Geissler R, Cottrill C, Noonan J, Todd E (1985) Supravalvular aortic stenosis: Clinical and pathologic observations in six patients. *Arch Pathol Lab Med* 109:179–185.

Oliveri B, Mastaglia SR, Mautalen C, Gravano JC, Pardo Argerich L (2004) Long-term control of hypercalcaemia in an infant with Williams-Beuren syndrome after a single infusion of biphosphonate (Pamidronate). *Acta Paediatr* 93:1002–1003.

Palacios-Verdú MG, Segura-Puimedon M, Borralleras C, Flores R, Del Campo M, Campuzano V, Pérez-Jurado LA. (2015) Metabolic abnormalities in Williams-Beuren syndrome. *J Med Genet* 52(4):248–255.

Pani AM, Hobart HH, Morris CA, Mervis CB, Bray-Ward P, Kimberley KW, Rios CM, Clark RC, Gulbronson MD, Gowans GC, Gregg RG (2010) Genome rearrangements detected by SNP microarrays in individuals with intellectual disability referred with possible Williams syndrome. *PLoS One* 5(8):e12349.

Pankau R, Partsch C-J, Gosch A, Oppermann HC, Wessel A (1992) Statural growth in Williams-Beuren syndrome. *Eur J Pediatr* 151:751–755.

Pankau R, Partsch C-J, Winter M, Gosch A, Wessel A (1996) Incidence and spectrum of renal abnormalities in Williams-Beuren syndrome. *Am J Med Genet* 63:301–304.

Partsch CJ, Japing I, Siebert R, Gosch A, Wessel A, Sippell WJ, Pankau R (2002) Central precocious puberty in girls with Williams syndrome. *J Pediatr* 141:441–444.

Partsch CJ, Siebert R, Caliebe A, Gosch A, Wessel A, Pankau R (2005) Sigmoid diverticulitis in patients with Williams-Beuren syndrome: Relatively high prevalence and high complication rate in young adults with the syndrome. *Am J Med Genet* 137:52–54.

Pérez-García D, Flores R, Brun-Gasca C, Pérez-Jurado LA (2015) Lateral preference in Williams syndrome is associated with cognition and language. *Eur Child Adolesc Psychiatry* 24:1025–1033.

Pham PP, Moller JH, Hills C, Larson V, Pyles L (2009) Cardiac catheterization and operative outcomes from a multicenter consortium for children with Williams syndrome. *Pediatr Cardiol.* 30:9–14.

Pitts CH, Klein-Tasman BP, Osborne JW, Mervis CB (2016) Predictors of specific phobia in children with Williams syndrome. *J Intellect Disabil Res* 60:1031–1042.

Pober BR, Szekely AM (1999) Distinct neurological profile in Williams syndrome. *Am J Hum Genet*, 65A:70.

Pober BR, Morris CA (2007) Diagnosis and management of medical problems in adults with Williams-Beuren syndrome. *Am J Med Genet C* 145C:280–290.

Pober BR, Lacro RV, Rice C, Mandell V, Teele RL (1993) Renal findings in 40 individuals with Williams syndrome. *Am J Med Genet* 46:271–274.

Pober BR, Carpenter T, Breault D (2001) Prevalence of hypothyroidism and compensated hypothyroidism in Williams syndrome. *Proc Greenwood Genet Cent* 20:69–70.

Pober BR, Wang E, Caprio S, Peterson KF, Brandt C, Stanley T, Osborne LR, Dzuria J, Gulanski B (2010) High prevalence of diabetes and pre-diabetes in adults with Williams syndrome. *Am J Med Genet C* 154C:291–298.

Power TJ, Blum NJ, Jones SM, Kaplan PE (1997) Response to methylphenidate in two children with Williams syndrome. *J Autism Dev Dis* 27:79–87.

Preus M (1984) The Williams syndrome: Objective definition and diagnosis. *Clin Genet* 25:422–428.

Reiss AL, Eliez S, Schmitt JE, Strauss E (2000) Neuroanatomy of Williams syndrome: A high resolution MRI study. *J Cogn Neurosci* 12(Suppl 1):65–73.

Ross AC, Taylor CL, Yaktine AL, Del Valle HB (eds.) (2011) *Dietary Reference Intakes for Calcium and Vitamin D*. Washington DC: National Academies Press.

Sadler S, Olitsky SE, Reynolds JD (1996) Reduced stereoacuity in Williams syndrome. *Am J Med Genet* 66:287–288.

Sammour ZM, de Bessa J Jr, Hirano M, Bruschini H, Kim CA, Srougi M, Gomes CM (2017) Lower urinary tract symptoms in children and adolescents with Williams-Beuren syndrome. *J Pediatr Urol* 13(2):203.e1–203.e6.

Sammour ZM, Gomes CM, Duarte RJ, Trigo-Rocha FE, Srougi M (2006) Voiding dysfunction and the Williams-Beuren syndrome: a clinical and urodynamic investigation. *J Urol* 175:1472–1476.

Sargent JD, Stukel TA, Kresel J, Klein RZ (1993) Normal values for random urinary calcium to creatinine ratios in infancy. *J Pediatr* 123:393–397.

Sarimski K (1997) Behavioral phenotypes and family stress in three mental retardation syndromes. *Eur Child Adolesc Psychiatry* 6:26–31.

Saul RA, Stevenson RE, Rogers RC, Skinner SA, Prouty LA, Flannery DB (1998) Growth references from conception to adulthood. *Proc Greenwood Genet Cent Suppl* 1:204–209.

Scherer SW, Gripp KW, Lucena J, Nicholson L, Bonnefont J, Perez-Jurado LA, Osborne LR (2005) Observation of a parental inversion variant in a rare Williams-Beuren syndrome family with two affected children. *Hum Genet* 117:383–388.

Sindhar S, Lugo M, Levin MD, Danback JR, Brink BD, Yu E, Dietzen DJ, Clark AL, Purgert CA, Waxler JL, Elder RW, Pober BR, Kozel BA. (2016) Hypercalcemia in patients with Williams-Beuren syndrome. *J Pediatr* 78:254–260.

Sniecinska-Cooper AM, Iles RK, Butler SA, Jones H, Bayford R, Dimitriou D (2015) Abnormal secretion of melatonin and cortisol in relation to sleep disturbances in children with Williams syndrome. *Sleep Med* 16:94–100.

Spielmann S, Partsch CJ, Gosch A, Pankau R (2015) Treatment of central precocious puberty and early puberty with GnRH analog in girls with Williams-Beuren syndrome. *J Pediatr Endocrinol Metab* 28:1363–1367.

Stagi S, Lapi E, Cecchi C, Chiarelli F, D'Avanzo MG, Seminara S, de Martino M (2014) Williams-Beuren syndrome is a genetic disorder associated with impaired glucose tolerance and diabetes in childhood and adolescence: new insights from a longitudinal study. *Horm Res Paediatr* 82(1):38–43.

Stagi S, Lapi E, Chiarelli F, de Martino M (2010) Incidence of diverticular disease and complicated diverticular disease in young patients with Williams syndrome. *Pediatr Surg Int* 26:943–944.

Stagi S, Manoni C, Scalini P, Chiarelli F, Verrotti A, Cecchi C, Lapi E, Giglio S, Romano S, de Martino M (2016) Bone mineral status and metabolism in patients with Williams-Beuren syndrome. *Hormones (Athens)* 15(3):404–412.

Stamm C, Kruetzer C, Zurakowski D, Nollert G, Friehs I, Mayer JE, Ronas RA, del Nido PJ (1999) Forty-one years of surgical experience with congenital supravalvular aortic stenosis. *J Thorac Cardiovasc Surg* 118:874–885.

Stinton C, Elison S, Howlin P (2010) Mental health problems in adults with Williams syndrome. *Am J Intellect Dev Disabil* 115:3–18.

Strømme P, Bjørnstad P, Ramstad K (2002) Prevalence estimation of Williams syndrome. *J Child Neurol* 17:269–271.

Strong E, Butcher DT, Singhania R, Mervis CB, Morris CA, De Carvalho D, Weksberg R, Osborne LR (2015) Symmetrical dose-dependent DNA-methylation profiles in children with deletion or duplication of 7q11.23. *Am J Hum Genet* 97(2):216–227.

Swartz JR, Waller R, Bogdan R, Knodt AR, Sabhlok A, Hyde LW, Hariri AR (2017) A common polymorphism in a Williams syndrome gene predicts amygdaloid reactivity and extraversion in healthy adults *Biol Psychiatry* 81:203–201.

Sugayama SM, Leone C, Chauffaille M, de L, Okay TS, Kim CA (2007) Williams syndrome. Development of a new scoring system for clinical diagnosis. *Clinics* 62:159–166.

Tassabehji M, Hammond P, Karmiloff-Smith A, Thompson P, Thorgeirsson SS, Durkin ME, Popescu NC, Hutton T, Metcalfe K, Rucka A, Stewart H, Read AP, Maconochie M, Donnai D (2005) GTF2IRD1 in craniofacial development of humans and mice. *Science* 310:1184–1187.

Utine GE, Alikasifoglu A, Alikasifoglu M, Tuncbilek E (2006) Central precocious puberty in a girl with Williams syndrome: The result of treatment with GnRH analogue. *Eur J Med Genet* 49:79–82.

Van Borsel J, Curfs LMG, Fryns JP (1997) Hyperacusis in Williams syndrome: A sample survey study. *Genet Couns* 8:121–126.

Van der Geest JN, Lagers-van Haselen GC, van Hagen JM, Brenner E, Govaerts LC, de Coo IF, Frens MA (2005) Visual depth processing in Williams-Beuren syndrome. *Exp Brain Res* 166(2):200–209.

Van Hagen JM, Van der Geest JN, van der Giessen RS, Lagers-van Haselen GC, Eussen HJFMM Gille JJP, Govaerts LCP, Wouters CH, de Coo IFM, Hoogenraad CC, Koekkoek SKE, Frens MA, van Camp N, van der Linden A, Jansweijer MCE, Thorgeirsson SS, De Zeeuw CI (2007) Contribution of CYLN2 and GTF2IRD1 to neurological and cognitive symptoms in Williams syndrome. *Neurobiol Dis* 26:112–124.

van Son JAM, Edwards WD, Danielson GK (1994) Pathology of coronary arteries, myocardium, and great arteries in supravalvular aortic stenosis. *J Thoracic Cardiovasc Surg* 108:21–28.

Van Strien JW, Lagers-Van Haselen GC, Van Hagen JM, De Coo IF, Frens MA, Van Der Geest JN (2005) Increased prevalences of left-handedness and left-eye sighting dominance in individuals with Williams-Beuren syndrome. *J Clin Exp Neuropsychol* 27(8):967–976.

Vaux KK, Wojtczak H, Benirschke K, Jones KL (2003) Vocal cord anomalies in Williams syndrome: A further manifestation of elastin deficiency. *Am J Med Genet* 119A:302–304.

von Arnim G, Engel P (1964) Mental retardation related to hypercalcemia. *Dev Med Child Neurol* 6:366–377.

von Gontard A, Niemczyk J, Borggrefe-Moussavian S, Wagner C, Curfs L, Equit M. (2016) Incontinence in children, adolescents and adults with Williams syndrome. *Neurourol Urodyn* 35(8):1000–1005.

Wang PP, Hesselink JR, Jernigan TL, Doherty S, Bellugi U (1992) Specific neurobehavioral profile of Williams syndrome is associated with neocerebellar hemispheric preservation. *Neurology* 42:1999–2002.

Weber SL, Souza RB, Ribeiro LG, Tavares MF, Goldchmit M (2014) Williams syndrome: ophthalmological examination and review of systemic manifestations. *J Pediatr Ophthalmol Strabismus* 51(4):209–213.

Williams JCP, Barratt-Boyes BG, Lowe JB (1961) Supravalvular aortic stenosis. *Circulation* 24:1311–1318.

Winter M, Pankau R, Amm M, Gosch A, Wessel A (1996) The spectrum of ocular features in the Williams-Beuren syndrome. *Clin Genet* 49:28–31.

Wollack JB, Kaifer M, LaMonte MP, Rothman M (1996) Stroke in Williams syndrome. *Stroke* 27:143–146.

Wren C, Oslizlok P, Bull C (1990) Natural history of supravalvular aortic stenosis and pulmonary artery stenosis. *J Am Coll Cardiol* 15:1625–1630.

Wu Y-Q, Sutton V, Nickerson E, Lupski JR, Potocki L, Korenberg JR, Greenberg F, Young EJ, Lipina T, Tam E, Mandel A, Clapcote SJ, Bechard AR, Chambers J, Mount HT, Fletcher PJ, Roder JC, Osborne LR (2008) Reduced fear and aggression and altered serotonin metabolism in Gtf2ird1-targeted mice. *Genes Brain Behav* 7:224–234.

INDEX

Aarskog syndrome, 1–8
   dental anomalies, 5–6
   development and behavior problems, 5
   diagnostic criteria, 1–2
   diagnostic testing, 4
   differential diagnosis, 4
      fetal anticonvulsant syndrome, 428
      Noonan syndrome, 654
   etiology, pathogenesis, and genetics, 3–4
   genital abnormalities, 6–7
   growth and feeding problems, 4–5
   incidence, 1
   manifestations and management, 4–7
   musculoskeletal anomalies, 6
   neurological anomalies, 6
   ophthalmologic anomalies, 5
Abnormal spontaneous brainstem activation, 801
Acanthosis nigricans:
   achondroplasia, 11
   Costello syndrome, 236
N-acetylcysteine, Prader–Willi syndrome, 756
Achondrogenesis type I, differential diagnosis, osteogenesis imperfecta, 693
Achondroplasia, 9–30
   dental anomalies, 26
   development and behavior problems, 16–17
   diagnostic criteria, 9–10
   diagnostic testing, 11
   differential diagnosis, 11
   ear anomalies, 21
   etiology, pathogenesis, and genetics, 10–11
   growth and feeding problems, 12–16
   hearing disorders, 21
   incidence, 9
   manifestations and management, 11–27
   musculoskeletal anomalies, 22–26
   neurological anomalies, 17–20
   respiratory anomalies, 20–21
Acrocallosal syndrome, differential diagnosis, Greig cephalopolysyndactyly syndrome, 710
Acrofacial dysostoses, Nager syndrome, 809
Action myoclonus-renal failure syndrome, 343
ACVRL1 gene mutation, hereditary hemorrhagic telangiectasia, 476
ADAMTS2 gene, Ehlers–Danlos syndromes, 397
Adaptive function:
   Down syndrome, 363
   fetal alcohol spectrum disorders, 413
   fetal anticonvulsant syndrome, 431
   48,XXYY syndrome, 553
   Smith–Magenis syndrome, 873–874
   velo-cardio-facial syndrome, 300
Adrenal insufficiency:
   holoprosencephaly, 498
   Pallister–Hall syndrome, 711
   Prader–Willi syndrome, 750, 751
Adrenarche:
   ATR-X syndrome, 101
   Prader–Willi syndrome, 736
   Silver–Russell syndrome, 844–845
   Smith–Magenis syndrome, 886
Adrenocorticotropic hormone (ACTH):
   inverted duplicated chromosome 15 syndrome, 523
   Pallister–Hall syndrome, 711
Aging, premature:
   albinism, 47
   Costello syndrome, 225, 226
   Down syndrome, 366
   Wolf–Hirschhorn syndrome, 270, 275
Aicardi syndrome, differential diagnosis, deletion 1p36 syndrome, 255
Airway obstruction:
   Robin sequence, 807, 808
   Stickler syndrome, 921
   Treacher Collins syndrome, 931
   trisomy 13 and 18 syndromes, 952
Alagille syndrome, 31–44
   cardiovascular anomalies, 38–40
   development and behavior problems, 36
   diagnostic criteria, 32–34
   diagnostic testing, 35
   differential diagnosis, 35–36
      deletion 22q11.2 syndrome, 298
   etiology, pathogenesis, and genetics, 34–36
   genetic counseling, 35
   genitourinary anomalies, 40
   growth and feeding problems, 36
   incidence, 31–32
   manifestations and management, 36–41
   musculoskeletal anomalies, 40–41
   ophthalmologic anomalies, 40
Albinism, 45–59
   dermatologic anomalies, 50–52
   development and behavior problems, 50
   diagnostic criteria, 46–47
   diagnostic testing, 47–49
   differential diagnosis, 50
   etiology, pathogenesis, and genetics, 47

Cassidy and Allanson's Management of Genetic Syndromes, Fourth Edition.
Edited by John C. Carey, Agatino Battaglia, David Viskochil, and Suzanne B. Cassidy.
© 2021 John Wiley & Sons, Inc. Published 2021 by John Wiley & Sons, Inc.

Albinism (cont'd)
  gastrointestinal anomalies, 55
  hearing disorders, 48
  hematologic disorders, 53–54
  incidence, 45–46
  manifestations and management, 50–55
  ophthalmologic anomalies, 52–53
  pulmonic anomalies, 54
Albright hereditary osteodystrophy, differential diagnosis, Prader–Willi syndrome, 741
Alcohol consumption, fetal alcohol spectrum disorders, 405–424
Allelic conditions:
  ATR-X syndrome, 94–95
  Juberg–Marsidi syndrome, 102
Allelic pleiotropic developmental anomaly syndromes:
  Greig cephalopolysyndactyly syndrome, 707–716
  Pallister–Hall syndrome, 707–716
Alobar holoprosencephaly, 487
  craniofacial anomalies, 497
  development and behavior problems, 495–496
  diagnostic criteria, 488
  diagnostic testing, 494
  etiology, pathogenesis, and genetics, 492
  incidence, 487–488
  neurological anomalies, 498–499
Alopecia:
  Costello syndrome, 226, 236
  Down syndrome, 378
  incontinentia pigmenti, 509, 510
  myotonic dystrophy type 1, 622
  Pallister–Killian syndrome, 718
  Turner syndrome, 990
Alstrom syndrome differential diagnosis:
  Bardet–Biedl syndrome, 112
  Cohen syndrome, 199
Alzheimer disease Down syndrome, 359, 365
Amsterdam dwarfism. See Cornelia de Lange syndrome
Amyloid precursor protein (APP) gene:
  Down syndrome and Alzheimer disease, 365
  fragile-X syndrome, 443
Amyoplasia arthrogryposis, 75, 77, 79–85
Androgen replacement therapy:
  49,XXXXY syndrome, 559
  Klinefelter syndrome, 550
Anemia:
  ATR-X syndrome, 93, 101
  Cohen syndrome, 202
  Cornelia de Lange syndrome, 217, 219
  Denys–Drash syndrome, 343
  Down syndrome, 379
  hereditary hemorrhagic telangiectasia, 478
  Noonan syndrome, 660
  Proteus syndrome, 769
  VATER/VACTERL association, 997
Anesthesia risks:
  achondroplasia, 26
  myotonic dystrophy type 1, 624
  Prader–Willi syndrome, 756
  Rubinstein–Taybi syndrome, 832
  Treacher Collins syndrome, 933

Angelman syndrome, 61–73
  development and behavior problems, 66–69
  diagnostic criteria, 61
  diagnostic testing, 64–65
  differential diagnosis, 65–66
    ATR-X syndrome, 98, 99
    deletion 1p36 syndrome, 255
    deletion 5p syndrome, 283
    deletion 22q13 syndrome, 321
    inverted duplicated chromosome 15 syndrome, 518
    Mowat–Wilson syndrome, 601
    Rett syndrome, 797
  etiology, pathogenesis, and genetics, 61–62
  genetic counseling, 65
  growth and feeding problems, 66
  incidence, 61
  language difficulties, 67
  manifestations and management, 66–70
  musculoskeletal anomalies, 70
  neurological anomalies, 69–70
  ophthalmologic anomalies, 70
Angiofibromas, tuberous sclerosis complex, 968
Angiomyolipomas, tuberous sclerosis complex, 959, 970, 972
Angiotensin-converting enzyme (ACE) inhibitors, Marfan syndrome, 589, 593
Angiotensin receptor blockers, Loeys–Dietz syndrome, 566, 571
Aniridia:
  aniridia associated keratopathy, 345
  WAGR syndrome, 336, 342
ANKRD11 gene Cornelia de Lange syndrome, 211
Anorectal anomalies:
  Kabuki syndrome, 535
  Pallister–Killian syndrome, 727
  VATER/VACTERL association, 1000, 1001
Anosmia:
  Bardet–Biedl syndrome, 118
  CHARGE syndrome, 160, 161, 166
  holoprosencephaly, 493
Antiepileptics/anticonvulsants:
  fetal anticonvulsant syndrome, 425, 432
  fragile X syndrome, 450
  Robin sequence, 811
  Smith–Magenis syndrome, 877
  tuberous sclerosis complex, 967
Antihistamines, Alagille syndrome, 37
Anti-hypertensives:
  Loeys-Dietz syndrome, 571
  Marfan syndrome, 593
Anti-RANK ligand antibody, osteogenesis imperfecta, 697
Anxiety:
  Costello syndrome, 232
  Down syndrome, 364
  fetal alcohol spectrum disorders, 415
  fragile X syndrome, 445
  Klinefelter syndrome, 556
  Noonan syndrome, 657
  Prader–Willi syndrome, 745
  Williams syndrome, 1026
Aortic coarctation:
  neurofibromatosis type 1, 645
  Turner syndrome, 981

Aortic dilatation:
  in deletion 22q11.2 syndrome, 304
  in hereditary hemorrhagic telangiectasia, 475
  in Marfan syndrome, 577, 579, 581, 589, 593
  in Pallister–Killian syndrome, 727
  in Sotos syndrome, 909
  in Turner syndrome, 981, 983
Apert syndrome:
  cardiovascular anomalies, 249
  development and behavior problems, 247
  diagnostic criteria, 241–242
  diagnostic testing, 246
  etiology, pathogenesis, and genetics, 245
  genitourinary anomalies, 250
  otolaryngologic anomalies, 248
  skeletal anomalies, 249
Apneic deaths:
  achondroplasia, 18, 19
  Turner syndrome, 982
Apolipoprotein E (APOE gene), Down syndrome and Alzheimer disease, 359
Arachnodactyly, Marfan syndrome, 588
Arachnoid cysts, Phelan–McDermid syndrome, 326
ARID1A gene mutations, Coffin–Siris syndrome, 189
ARID1B gene mutations, Coffin–Siris syndrome, 185, 189
Array comparative genomic hybridization (aCGH):
  ATR-X syndrome, 97
  Cohen syndrome, 199
  Cornelia de Lange syndrome, 212
  Pallister–Killian syndrome, 722
Arrhythmias:
  Costello syndrome, 232
  myotonic dystrophy type 1, 621
  Noonan syndrome, 658
Arterial tortuosity syndrome:
  differential diagnosis
    Loeys–Dietz syndrome, 568
    Marfan syndrome, 585
Arteriohepatic dysplasia. See Alagille syndrome:
Arthritis. See also Osteoarthritis:
  arthrogryposis, 85
  Stickler syndrome, 922
Arthrogryposis, 75–91
  development and behavior problems, 84–85
  diagnostic criteria, 75–76
  diagnostic testing, 77
  differential diagnosis, 79
  ear anomalies, 88
  etiology, pathogenesis, and genetics, 77
  growth and feeding problems, 79–84
  hearing disorders, 88
  incidence, 75
  manifestations and management, 79–88
  musculoskeletal anomalies, 85–87
  neurological anomalies, 87
  ophthalmologic anomalies, 87
  respiratory anomalies, 87–88
Aryl hydrocarbon receptor nuclear translocator (ARNT), von Hippel–Lindau syndrome, 1006

Assisted reproductive technology:
    Beckwith–Wiedemann syndrome, 131
    Turner syndrome, 984–985
Asthenic body habitus, Marfan syndrome, 587
ASXL1 gene mutations, Bohring–Opitz syndrome, 212
Ataxia:
    Angelman syndrome, 64, 66
    Bardet–Biedl syndrome, 118
    deletion 22q11.2 syndrome, 304
    fragile X syndrome, 443, 445, 450
    Gorlin syndrome, 464
    inverted duplicated chromosome 15 syndrome, 517
    velo-cardio-facial syndrome, 304
    Williams syndrome, 1028
Atherosclerosis, Turner syndrome, 982
ATR-X syndrome, 93–105
    asplenia, 103
    cancer risk, 103
    cardiovascular anomalies, 102
    dermatologic anomalies, 102
    development and behavior problems, 98–99
    diagnostic criteria, 93–94
    diagnostic testing, 96–98
    differential diagnosis, 98
        Coffin–Lowry syndrome, 175
        Mowat–Wilson syndrome, 601
    ear anomalies, 102
    etiology, pathogenesis, and genetics, 94–96
    gastrointestinal anomalies, 100
    genitourinary anomalies, 101–102
    growth and feeding problems, 98
    hearing disorders, 102
    hematologic anomalies, 101
    incidence, 93
    manifestations and management, 98–103
    musculoskeletal anomalies, 102
    neurological anomalies, 100–101
    ophthalmologic anomalies, 103
Attention deficit/hyperactivity disorder (ADHD):
    Aarskog syndrome, 5
    albinism, 50
    deletion 5p syndrome, 284
    deletion 22q11.2 syndrome, 301
    Down syndrome, 364
    47,XXY syndrome, 540, 544–545
    48,XXYY syndrome, 553
    49,XXXXY syndrome, 556
    fragile X syndrome, 445
    Klinefelter syndrome, 544–545
    myotonic dystrophy type 1, 618
    neurofibromatosis type 1, 632
    Prader–Willi syndrome, 744
    Smith–Magenis syndrome, 875
    tuberous sclerosis complex, 965
    Turner syndrome, 986
    Williams syndrome, 1026, 1027
Auriculo-condylar syndrome, differential diagnosis, oculo-auriculo-vertebral spectrum, 673
Autism spectrum disorder (ASD):
    cardio-facio-cutaneous syndrome, 150
    Cohen syndrome, 200
    Cornelia de Lange syndrome, 215

    Costello syndrome, 232
    deletion 5p syndrome, 284
    Down syndrome, 364
    fetal valproate syndrome, 431
    fragile X syndrome, 447
    inverted duplicated chromosome 15 syndrome, 517, 520–521
    Kabuki syndrome, 532
    Klinefelter syndrome, 545
    myotonic dystrophy type 1, 618
    neurofibromatosis type 1, 632, 637
    Noonan syndrome, 656
    oculo-auriculo-vertebral spectrum, 674
    Pallister–Killian syndrome, 726
    Phelan–McDermid syndrome, 317, 318, 321, 323
    Prader–Willi syndrome, 744
    PTEN hamartoma tumor syndrome, 786
    Smith–Lemli–Opitz syndrome, 856
    Smith–Magenis syndrome, 871, 875
    Sotos syndrome, 901, 906
    tuberous sclerosis complex, 965
    WAGR syndrome, 341
    Williams syndrome, 1026
Autoimmune disease:
    deletion 22q11.2 syndrome, 305
    Down syndrome, 373, 378, 379, 380
    Klinefelter syndrome, 551, 552
    Noonan syndrome, 664
Autonomic dysfunction:
    Mowat–Wilson syndrome, 604
    Prader–Willi syndrome, 749
    Rett syndrome, 801
    Sotos syndrome, 907
Autosomal dominant disorders:
    achondroplasia, 10
    Alagille syndrome, 31
    brachymorphism-onychodysplasia-dysphalangism syndrome, 186
    campomelic dysplasia, 693
    cherubism, 663
    Ehlers–Danlos syndromes, 392, 393, 394, 397, 402, 585
    holoprosencephaly, 492
    Loeys–Dietz syndrome, 563, 585
    Marfan syndrome, 577
    Mowat–Wilson syndrome, 598
    myotonic dystrophy type 1, 616
    neurofibromatosis type 1, 629
    Nicolaides–Baraitser syndrome, 186
    Noonan syndrome, 651
    Pallister–Hall syndrome, 854
    PTEN hamartoma tumor syndrome, 775
    Stickler syndrome, 915, 918
    von Hippel–Lindau syndrome, 1007
Autosomal recessive disorders:

Bardet–Biedl syndrome, 108–112
    brittle cornea syndrome, 399
    Cohen syndrome, 195–206
    Ehlers–Danlos syndromes, 400
    Goldberg–Shprintzen syndrome, 601
    homocystinuria, 586
    ocular albinism, 46
    Smith–Lemli–Opitz syndrome, 852

    Stickler syndrome, 918
Back pain:
    Loeys–Dietz syndrome, 570
    Marfan syndrome, 587, 588, 591
    osteogenesis imperfecta, 694
    Sotos syndrome, 910
Bannayan–Riley–Ruvalcaba syndrome:
    development and behavior problems, 782
    diagnostic criteria, 776–777
    diagnostic testing, 779
    differential diagnosis, hereditary hemorrhagic telangiectasia, 478
    and *PTEN* harmartoma tumor syndrome, 775
Bardet–Biedl syndrome, 107–123
    cardiovascular anomalies, 117
    dental anomalies, 118
    development and behavior problems, 113–114
    diagnostic criteria, 108
    diagnostic testing, 112
    differential diagnosis, 112
        Cohen syndrome, 199
        Klinefelter syndrome, 542
    ear anomalies, 117–118
    endocrine anomalies, 117
    etiology, pathogenesis, and genetics, 108–112
    gastrointestinal anomalies, 118
    genetic counseling, 111–112
    genitourinary anomalies, 115–117
    growth and feeding problems, 112–113
    hearing disorders, 117–118
    incidence, 107–108
    manifestations and management, 112–118
    musculoskeletal anomalies, 115
    neurological anomalies, 118
    ophthalmologic anomalies, 114
Barrett esophagus, Cornelia de Lange syndrome, 217
Basal cell carcinomas, Gorlin syndrome, 465–466
Basilar invagination, osteogenesis imperfecta, 698
Bazex syndrome, differential diagnosis, Gorlin syndrome, 463
BBS1 mutations, Bardet–Biedl syndrome, 108, 109
BBS8 mutations, Bardet–Biedl syndrome, 111
BBS10 mutations, Bardet–Biedl syndrome, 108
B-cells, Down syndrome, 379
Beals syndrome, differential diagnosis:
    Loeys–Dietz syndrome, 568
    Marfan syndrome, 584
Beckwith–Wiedemann syndrome, 125–145
    assisted reproductive technology offspring, 131
    cardiovascular anomalies, 135
    chromosome 11p15, 130–131
    clinical and molecular findings, 131
    craniofacial anomalies, 136
    development and behavior problems, 134–135
    diagnostic criteria, 126–127
    diagnostic testing, 132–133
    differential diagnosis, 133
        Costello syndrome, 228
        Down syndrome, 361
        Gorlin syndrome, 463
        Pallister–Killian syndrome, 724
    endocrine anomalies, 135–136
    etiology, pathogenesis, and genetics, 127–128

Beckwith–Wiedemann syndrome (*cont'd*)
  gastrointestinal anomalies, 136–137
  genitourinary anomalies, 137
  growth and feeding problems, 134
  incidence, 125–126
  manifestations and management, 133–140
  molecular alterations, 128–131
  neoplasia incidence, 137–139
  nonchromosome methylation defects, 131–132
  pregnancy and perinatal period, 139–140
  recurrence risks, 132
  unknown molecular etiology, 131
  Wilms tumor, 138, 343
Behavioral problems:
  Aarskog syndrome, 5
  achondroplasia, 16–17
  Alagille syndrome, 36
  albinism, 50
  Angelman syndrome, 67
  ATR-X syndrome, 98–99
  Bardet–Biedl syndrome, 113–114
  Beckwith–Wiedemann syndrome, 134–135
  cardio-facio-cutaneous syndrome, 150
  CHARGE syndrome, 161–162
  Coffin–Lowry syndrome, 177–178
  Coffin–Siris syndrome, 187
  Cohen syndrome, 200–201
  Cornelia de Lange syndrome, 213–215
  Costello syndrome, 232
  craniosynostosis syndromes, 247
  deletion 1p36 syndrome, 256–257
  deletion 5p syndrome, 283–284
  deletion 22q11.2 syndrome, 300–303
  Down syndrome, 363–366
  49,XXXXY syndrome, 556
  fetal alcohol spectrum disorders, 413–415
  fetal anticonvulsant syndrome, 435
  fragile X syndrome, 447–449
  Gorlin syndrome, 464
  Greig cephalopolysyndactyly syndrome, 713
  hereditary hemorrhagic telangiectasia, 478
  incontinentia pigmenti, 509
  inverted duplicated chromosome 15 syndrome, 519–522
  Kabuki syndrome, 532–533
  Klinefelter syndrome, 543–547
  Loeys–Dietz syndrome, 569
  Marfan syndrome, 587
  Mowat–Wilson syndrome, 602–603
  myotonic dystrophy type 1, 618
  neurofibromatosis type 1, 637–638
  Noonan syndrome, 656–657
  oculo-auriculo-vertebral spectrum, 674–675
  osteogenesis imperfecta, 695
  Pallister–Hall syndrome, 710
  Pallister–Killian syndrome, 725–726
  Phelan–McDermid syndrome, 322–325
  Prader–Willi syndrome, 739, 744–745
  Proteus syndrome, 767
  PTEN hamartoma tumor syndrome, 782
  Robin sequence, 815–816
  Rubinstein–Taybi syndrome, 826–832
  Silver–Russell syndrome, 845
  Smith–Lemli–Opitz syndrome, 856–857
  Smith–Magenis syndrome, 872–878
  Sotos syndrome, 905–907
  Stickler syndrome, 920
  Treacher Collins syndrome, 930–931
  trisomy 13 and 18 syndromes, 946–947
  tuberous sclerosis complex, 964–965
  Turner syndrome, 985–987
  VATER/VACTERL association, 998–999
  von Hippel–Lindau syndrome, 1009–1010
  Williams syndrome, 1025–1028
  Wolf–Hirschhorn syndrome, 270–271
  WT1-related disorders, 344–345
Benzodiazepines:
  Phelan–McDermid syndrome, 325
  Smith–Magenis syndrome, 877
Bernard–Soulier syndrome, deletion 22q11.2 syndrome, 306:
Beta-blockers:
  cardio-facio-cutaneous syndrome, 151
  fragile X syndrome, 451
  Loeys–Dietz syndrome, 571, 572
  Marfan syndrome, 589
  Rett syndrome, 803
  Williams syndrome, 1032
Bevacizumab, hereditary hemorrhagic telangiectasia, 479
Biallelic expression, Angelman syndrome, 61
Bicoronal synostosis, craniosynostosis syndromes, 241, 242, 243
Bile duct anomalies:
  Alagille syndrome, 31, 32, 37
  Bardet–Biedl syndrome, 118
  Down syndrome, 375
  von Hippel–Lindau syndrome, 1014
Bilevel positive airway pressure (BiPAP), achondroplasia, 21
Biliary atresia:
  differential diagnosis, Alagille syndrome, 32, 35
  trisomy 18 syndrome, 951
Biliary diversion, Alagille syndrome, 37
Birt–Hogg–Dubé syndrome, differential diagnosis:
  Smith–Magenis syndrome, 869
  von Hippel–Lindau syndrome, 1008
Bisphosphonates:
  Ehlers–Danlos syndromes, 400
  Klinefelter syndrome, 550
  osteogenesis imperfecta, 695, 697, 698
  Prader–Willi syndrome, 755
Bladder anomalies:
  cardio-facio-cutaneous syndrome, 152
  Costello syndrome, 231, 235
  Wolf–Hirschhorn syndrome, 276
Bladder exstrophy-epispadias complex (BEEC) spectrum, deletion 1p36 syndrome, 260
Blaschko's lines, incontinentia pigmenti, 508, 509
Bleeding disorders:
  Hermansky–Pudlak syndrome, 48–49
  Noonan syndrome, 659
Blepharitis:
  Cornelia de Lange syndrome, 218
  Down syndrome, 373
Blindness. *See* Vision impairment:
Bloom syndrome, differential diagnosis:
  neurofibromatosis type 1, 643
  Wilms tumor, 343
BMP4 gene mutation, Stickler syndrome, 918
Body composition:
  Prader–Willi syndrome, 742, 747, 750, 756
  Turner syndrome, 981, 987
Body mass index (BMI):
  achondroplasia, 12
  Bardet–Biedl syndrome, 112
  Prader–Willi syndrome, 747
  Smith–Magenis syndrome, 872
  Turner syndrome, 987
Bohring–Opitz syndrome, differential diagnosis, Cornelia de Lange syndrome, 212
Bone anomalies:
  arthrogryposis, 80
  Sotos syndrome, 910
Bone cysts, tuberous sclerosis complex, 972
Bone dysplasias. *See* Musculoskeletal anomalies
Börjeson–Forssman–Lehmann syndrome, differential diagnosis:
  Coffin–Lowry syndrome, 175
  Coffin–Siris syndrome, 186
Botulinum toxin type A (Botox) injection:
  ATR-X syndrome, 100
  Rett syndrome, 802, 803
Brachmann–de Lange syndrome. *See* Cornelia de Lange syndrome
Brachymorphism-onychodysplasia-dysphalangism syndrome, differential diagnosis, Coffin–Siris syndrome, 186
BRAF mutations, cardio-facio-cutaneous syndrome, 148, 149, 153
Brain anomalies:
  alobar holoprosencephaly, 487
  deletion 1p36 syndrome, 257
  fragile X syndrome, 449
  hereditary hemorrhagic telangiectasia, 481
  holoprosencephaly, 487–503
  Klinefelter syndrome, 545
  Rett syndrome, 799
  Smith–Lemli–Opitz syndrome, 857
  Smith–Magenis syndrome, 882
  syntelencephaly, 487
  trisomy 13 and 18 syndromes, 947
  tuberous sclerosis complex, 958, 965–966
  Williams syndrome, 1028
  Wolf–Hirschhorn syndrome, 272
Brainstem anomalies, Rett syndrome, 800, 801
Branchio-oto-renal (BOR) syndrome, differential diagnosis:
  Kabuki syndrome, 531
  oculo-auriculo-vertebral spectrum, 673
BRD4 mutations, Cornelia de Lange syndrome, 210
Breast anomalies:
  incontinentia pigmenti, 512
  PTEN hamartoma tumor syndrome, 782–783
Breast cancer, *PTEN* hamartoma tumor syndrome, 782–783, 785
Breastfeeding, fetal anticonvulsant syndrome, 434
Brittle cornea syndrome. *See* Ehlers–Danlos syndromes
Bronchomalacia, Coffin–Siris syndrome, 190

Bruising, Noonan syndrome, 659
Brushfield spots, Down syndrome, 357
Bruxism:
    Cornelia de Lange syndrome, 216
    Down syndrome, 377
    Rett syndrome, 802, 803
"Bull's eye maculopathy," Cohen syndrome, 201
Butterfly vertebrae:
    Alagille syndrome, 33, 40, 41
    deletion 22q11.2 syndrome, 292, 298, 307

Cachexia, Alagille syndrome, 36
Café-au-lait spots:
    cardio-facio-cutaneous syndrome, 152
    deletion 22q11.2 syndrome, 330
    Gorlin syndrome, 465
    neurofibromatosis type 1, 630, 634, 635, 637
    Noonan syndrome, 663
Calcium:
    osteogenesis imperfecta, 701
    Prader–Willi syndrome, 755
    Williams syndrome, 1029
Callosal anomalies:
    Coffin–Lowry syndrome, 178
    holoprosencephaly, 487
Campomelic dysplasia, differential diagnosis, osteogenesis imperfecta, 693
Camptodactyly, arthrogryposis, 75
Carbamazepine, Smith–Magenis syndrome, 877
Carbamazepine embryopathy, fetal anticonvulsant syndrome, 431–432
Cardiac disease, Alagille syndrome, 32, 38
Cardiac fibromas, Gorlin syndrome, 469
Cardiac hypertrophy, Costello syndrome, 232
Cardiac rhabdomyomas, tuberous sclerosis complex, 959, 964, 969
Cardio-facio-cutaneous syndrome, 147–155
    cardiovascular anomalies, 151
    dermatologic anomalies, 151–152
    development and behavior problems, 150
    diagnostic criteria, 147–148
    diagnostic testing, 149
    differential diagnosis, 149
        Costello syndrome, 230
        neurofibromatosis type 1, 635
        Noonan syndrome, 653, 654–655
    etiology, pathogenesis, and genetics, 148
    gastrointestinal anomalies, 150–151
    genitourinary anomalies, 152
    growth and feeding problems, 150
    hematologic anomalies, 153
    incidence, 147
    Kabuki syndrome, 533–534
    manifestations and management, 149–153
    musculoskeletal anomalies, 152–153
    neoplasia incidence, 153
    neurological anomalies, 151
    ophthalmologic anomalies, 152
    pregnancy, 149
    retinitis pigmentosa, 659
Cardiomyopathy:
    Hermansky–Pudlak syndrome, 55
    Sotos syndrome, 909
Cardiovascular anomalies:
    Alagille syndrome, 32, 38–40

ATR-X syndrome, 102
Bardet–Biedl syndrome, 117
Beckwith–Wiedemann syndrome, 135
cardio-facio-cutaneous syndrome, 151
CHARGE syndrome, 163–164
Coffin–Lowry syndrome, 180
Coffin–Siris syndrome, 189
Cohen syndrome, 203
Cornelia de Lange syndrome, 218
Costello syndrome, 226, 232
craniosynostosis syndromes, 249
deletion 1p36 syndrome, 258
deletion 5p syndrome, 285
deletion 22q11.2 syndrome, 303–304
Down syndrome, 357, 368–370
Ehlers–Danlos syndromes, 391
fetal alcohol spectrum disorders, 416–417
fetal anticonvulsant syndrome, 436
fragile X syndrome, 451
Gorlin syndrome, 469–470
Kabuki syndrome, 533–534
Loeys–Dietz syndrome, 570–572
Marfan syndrome, 588–590
Mowat–Wilson syndrome, 604
myotonic dystrophy type 1, 621–622
neurofibromatosis type 1, 645
Noonan syndrome, 657–658
oculo-auriculo-vertebral spectrum, 678
osteogenesis imperfecta, 700
Pallister–Killian syndrome, 727–728
Phelan–McDermid syndrome, 328
Proteus syndrome, 771
Rett syndrome, 803
Rubinstein–Taybi syndrome, 830
Smith–Lemli–Opitz syndrome, 858
Smith–Magenis syndrome, 885
Sotos syndrome, 908–909
Stickler syndrome, 923
trisomy 13 and 18 syndromes, 948–949
tuberous sclerosis complex, 969–970
Turner syndrome, 981–983
VATER/VACTERL association, 999
WAGR syndrome, 346
Williams syndrome, 1030–1031
Wolf–Hirschhorn syndrome, 272–273
WT1-related disorders, 346
Carney syndrome, differential diagnosis, neurofibromatosis type 1, 634
Carrier testing, Cohen syndrome, 199
Carter neurocognitive assessment, holoprosencephaly, 496
Cataracts:
    arthrogryposis, 87
    cardio-facio-cutaneous syndrome, 152
    Down syndrome, 373
    myotonic dystrophy type 1, 622
    Rubinstein–Taybi syndrome, 829
    Smith–Lemli–Opitz syndrome, 858
    Stickler syndrome, 921
    trisomy 18 syndrome, 949
    Williams syndrome, 1032
Cat-like cry, deletion 5p syndrome, 281
Cayler cardio-facial syndrome, 22q11 deletion syndrome, 291
CBP mutations, Rubinstein–Taybi syndrome, 825

CDKL5 mutations, inverted duplicated chromosome 15 syndrome, 518
CDKL5 syndrome, differential diagnosis, Rett syndrome, 795, 797
CDKN1C mutations, Beckwith–Wiedemann syndrome, 129
CdLS spectrum disorder. *See* Cornelia de Lange syndrome
Celiac disease:
    Down syndrome, 375, 376
    Turner syndrome, 988
Cellulitis, Phelan–McDermid syndrome, 330
Central apnea, trisomy 13 and 18 syndromes, 947
Central nervous system anomalies:
    fetal alcohol spectrum disorders, 411
    holoprosencephaly, 489
    myotonic dystrophy type 1, 615
    neurofibromatosis type 1, 639–640
    Smith–Lemli–Opitz syndrome, 857
    Smith–Magenis syndrome, 882
Central sleep apnea syndrome:
    myotonic dystrophy type 1, 620
    Prader-Willi syndrome, 753
Cerebellar tumors, von Hippel–Lindau syndrome, 1010
Cerebral autosomal-dominant arteriopathy with subcortical infarcts and leukoencephalopathy (CADASIL) syndrome, Alagille syndrome, 39
Cerebral palsy, differential diagnosis:
    Angelman syndrome, 65, 66
    Phelan–McDermid syndrome, 321
Cerebriform connective tissue nevus, Proteus syndrome, 763–764
Cerebro-oculo-facial-skeletal syndrome, Robin sequence, 809
Cerebro-oculo-facial syndrome, arthrogryposis, 75
Ceroid storage disease. *See* albinism
Cervical spine:
    Aarskog syndrome, 6
    achondroplasia, 20
    Down syndrome, 374
Cervicomedullary junction constriction, achondroplasia, 18–20
CFC syndrome. *See* Cardio-facio-cutaneous syndrome
Charcot–Marie–Tooth disease, differential diagnosis, Denys–Drash syndrome, 343
CHARGE syndrome, 157–170
    cardiovascular anomalies, 163–164
    development and behavior problems, 161–162
    diagnostic criteria, 158–159
    diagnostic testing, 160
    differential diagnosis, 160–161
        deletion 22q11.2 syndrome, 298
        Kabuki syndrome, 531
        trisomy 18 syndrome, 941
    ear anomalies, 165–166
    etiology, pathogenesis, and genetics, 159–160
    genitourinary anomalies, 166
    growth and feeding problems, 161
    hearing disorders, 165–166
    immunologic conditions, 166–167
    incidence, 157–158

CHARGE syndrome (cont'd)
  manifestations and management, 161–167
  mortality, 157–158
  ophthalmologic anomalies, 162–163
  respiratory anomalies, 164–165
CHD4 mutations, Pallister–Killian
    syndrome, 721
CHD7 mutations:
  CHARGE syndrome, 158, 159–160, 161
  Kallmann syndrome, 160
Chediak–Higashi syndrome, 46
Chest deformities:
  ATR-X syndrome, 102
  Marfan syndrome, 587
Childhood apraxia of speech, 49,XXXXY
    syndrome, 556
Child maltreatment, differential diagnosis,
    osteogenesis imperfecta, 693
Choanal atresia:
  CHARGE syndrome, 157, 158, 164
  Treacher Collins syndrome, 932
Cholelithiasis, myotonic dystrophy
    type 1, 620
Cholestasis, Alagille syndrome, 35, 36, 37
Cholesterol:
  Down syndrome, 366
  holoprosencephaly, 490
Cholesterol-directed therapies, Smith–Lemli–
    Opitz syndrome, 854–855
Chondrodysplasias. *See* Musculoskeletal
    anomalies
Chondrogenesis type I, differential diagnosis,
    osteogenesis imperfecta, 693
CHOPS syndrome, differential diagnosis,
    Cornelia de Lange syndrome, 212
Chorioangiomas, Noonan syndrome, 661
Choroid plexus cysts, trisomy 18 syndrome, 940
Chromosome analysis:
  Angelman syndrome, 64
  craniosynostosis syndromes, 246
  deletion 1p36 syndrome, 254
  deletion 5p syndrome, 283
  deletion 22q13 syndrome, 317, 321
  Down syndrome, 360
  Gorlin syndrome, 462
  incontinentia pigmenti, 510
  inverted duplicated chromosome 15
      syndrome, 518
  Klinefelter syndrome, 541
  Noonan syndrome, 661
  oculo-auriculo-vertebral spectrum, 673
  Phelan–McDermid syndrome, 317
  Prader–Willi syndrome, 736, 740, 741
  Silver–Russell syndrome, 841
  Smith–Lemli–Opitz syndrome, 854, 859
  Smith–Magenis syndrome, 864, 871
  trisomy 13 syndrome, 943
  trisomy 18 syndrome, 938, 940
  Turner syndrome, 977
  WAGR syndrome, 336
  Williams syndrome, 1023
  Wolf–Hirschhorn syndrome, 267
Chylothorax, Noonan syndrome, 661
"Cigarette paper" scars, Ehlers–Danlos
    syndromes, 391

Cirrhosis:
  Alagille syndrome, 37
  Turner syndrome, 988
Cleft lip and palate:
  Apert syndrome, 246
  CHARGE syndrome, 164
  Cornelia de Lange syndrome, 216
  deletion 22q11.2 syndrome, 292, 302–303
  fetal anticonvulsant syndrome, 432–433, 436
  holoprosencephaly, 495
  incontinentia pigmenti, 512
  Kabuki syndrome, 535
  oculo-auriculo-vertebral spectrum, 675
  Pallister–Hall syndrome, 712
  Robin sequence, 808, 814, 818–819
  Smith–Lemli–Opitz syndrome, 857, 858
  Stickler syndrome, 920, 921, 922
  Treacher Collins syndrome, 931, 932
  trisomy 13 syndrome, 941–942
Clinodactyly, Silver–Russell syndrome, 846
Clonidine, Phelan–McDermid syndrome, 325
Cloverleaf skull, craniosynostosis syndromes,
    243, 246, 249
CLOVE syndrome, differential diagnosis, Proteus
    syndrome, 766
Clubfeet. *See also* Foot anomalies
  arthrogryposis, 75
  Ehlers–Danlos syndromes, 394, 401
  fragile X syndrome, 452
  myotonic dystrophy type 1, 623
  oculo-auriculo-vertebral spectrum, 678
  velo-cardio-facial syndrome, 292
Coagulation disorders:
  differential diagnosis, Ehlers–Danlos
      syndromes, 394
  Hermansky–Pudlak syndrome, 54
  Noonan syndrome, 659–660
  vascular EDS, 394
Coenzyme Q10, Prader–Willi syndrome,
    749, 750
Coffin–Lowry syndrome, 171–184
  cardiovascular anomalies, 180
  dental anomalies, 181
  development and behavior problems, 177–178
  diagnostic criteria, 171–172
  diagnostic testing, 174–175
  differential diagnosis, 175
    ATR-X syndrome, 98
    fetal anticonvulsant syndrome, 428
    Williams syndrome, 1024
  ear anomalies, 180
  etiology, pathogenesis, and genetics, 172–174
  gastrointestinal anomalies, 181
  growth and feeding problems, 175
  hand anomalies, 172
  hearing disorders, 180
  incidence, 171
  manifestations and management, 175–182
  musculoskeletal anomalies, 181
  neoplasia incidence, 181–182
  neurological anomalies, 178–179
  ophthalmologic anomalies, 179
  respiratory anomalies, 180
Coffin–Siris syndrome, 185–194
  cardiovascular anomalies, 189

  craniofacial anomalies, 189
  development and behavior problems, 187
  diagnostic criteria, 185
  differential diagnosis, 186–187
    Cornelia de Lange syndrome, 212
  etiology, pathogenesis, and genetics, 186
  gastrointestinal anomalies, 191
  genitourinary anomalies, 191
  growth and feeding problems, 187
  hearing disorders, 191
  immunologic conditions, 191
  incidence, 185
  manifestations and management, 187–192
  musculoskeletal anomalies, 189
  neoplasia incidence, 191–192
  neurological anomalies, 189
  ophthalmologic anomalies, 191
  respiratory anomalies, 190
Cognitive function. *See also* Mental retardation
  Aarskog syndrome, 5
  achondroplasia, 16
  Angelman syndrome, 66–67
  Bardet–Biedl syndrome, 113
  Coffin–Lowry syndrome, 178
  deletion 5p syndrome, 282
  Down syndrome, 364
  holoprosencephaly, 499
  Klinefelter syndrome, 553
  Prader–Willi syndrome, 744
  Smith–Lemli–Opitz syndrome, 856
  Smith–Magenis syndrome, 873
  Sotos syndrome, 905
  Williams syndrome, 1025–1026
  WT1-related disorders, 341
Cohen syndrome, 195–206
  cardiovascular anomalies, 203
  dental anomalies, 204
  development and behavior problems, 200–201
  diagnostic criteria, 195–198
  diagnostic testing, 199
  differential diagnosis, 199–200
    deletion 1p36 syndrome, 255
  endocrine anomalies, 202
  etiology, pathogenesis, and genetics,
      198–199
  growth and feeding problems, 200
  hematologic anomalies, 202–203
  immunologic conditions, 202
  incidence, 195
  manifestations and management, 200–204
  musculoskeletal anomalies, 203–204
  neurological anomalies, 202
  ophthalmologic anomalies, 201–202
  respiratory anomalies, 203
Cohesin complex, Cornelia de Lange
    syndrome, 210
COL1A2 gene mutation, cardiac-valvular
    EDS, 393
COL2A1 gene mutation, Stickler syndrome,
    917, 918
COL9A2/COL9A3 gene mutation, Stickler
    syndrome, 918
COL11A1 gene mutation:
  Marshall syndrome, 919–920
  Stickler syndrome, 918, 919–920

COL12A1 gene mutation, myopathic EDS, 401
Collagen anomalies:
    Ehlers–Danlos syndromes, 391, 392
    Marfan syndrome, 585
    osteogenesis imperfecta, 683
    Stickler syndrome, 915, 917–918
Collagenopathy, type I, differential diagnosis, Silver–Russell syndrome, 843
Coloboma:
    CHARGE syndrome, 157, 158, 159, 162, 163
    deletion 1p36 syndrome, 259
    Gorlin syndrome, 469
    holoprosencephaly, 492, 500
    Kabuki syndrome, 531
    Mowat–Wilson syndrome, 605
    oculo-auriculo-vertebral spectrum, 671, 673, 677
    Rubinstein–Taybi syndrome, 829
    Treacher Collins syndrome, 927, 930, 932
    trisomy 13 syndrome, 949
    velo-cardio-facial syndrome, 306
    Wolf–Hirschhorn syndrome, 274
Colonic hypoganglionosis, ATR-X syndrome, 100
Common variable immunodeficiency:
    Smith–Magenis syndrome, 869
    Wolf–Hirschhorn syndrome, 275
Conductive hearing loss. *See also* Hearing disorders
    Cornelia de Lange syndrome, 216
    deletion 5p syndrome, 285
    Down syndrome, 371
    fragile X syndrome, 451
    Kabuki syndrome, 534
    Marfan syndrome, 585
    Mowat–Wilson syndrome, 605
    oculo-auriculo-vertebral spectrum, 676, 677
    osteogenesis imperfecta, 698
    Pallister–Killian syndrome, 729
    Proteus syndrome, 770
    Robin sequence, 815
    Smith–Magenis syndrome, 867, 881
    Sotos syndrome, 908
    Stickler syndrome, 918, 922
    Treacher Collins syndrome, 933
    Turner syndrome, 989
    velo-cardio-facial syndrome, 307
Congenital contractural arachnodactyly. *See* Beals syndrome
Congenital contractures:
    arthrogryposis, 75, 77, 79, 85, 87
    Ehlers–Danlos syndromes, 400, 401
    myotonic dystrophy type 1, 622
    osteogenesis imperfecta, 690
Congenital diaphragmatic hernia, Pallister–Killian syndrome, 718–719, 727
Congenital heart defects. *See* Cardiovascular anomalies
Congenital heart disease. *See* Cardiovascular anomalies
Connective tissue disorders:
    arthrogryposis, 77
    Costello syndrome, 225
    Ehlers–Danlos syndromes, 389, 392, 395, 396, 397

fragile X syndrome, 444
Kabuki syndrome, 531, 533
Loeys–Dietz syndrome, 563, 567, 568, 571, 573
Marfan syndrome, 582, 586
Stickler syndrome, 915
Williams syndrome, 1022
Conotruncal anomalies:
    CHARGE syndrome, 163
    deletion 22q11.2 syndrome, 291
    Smith–Lemli–Opitz syndrome, 858
Constipation:
    ATR-X syndrome, 100
    cardio-facio-cutaneous syndrome, 150–151
    Coffin–Siris syndrome, 191
    deletion 1p36 syndrome, 260
    deletion 5p syndrome, 286
    deletion 22q11.2 syndrome, 298
    holoprosencephaly, 500
    Loeys–Dietz syndrome, 573
    Phelan–McDermid syndrome, 329
    Rett syndrome, 803
    Rubinstein–Taybi syndrome, 828, 829, 831
    Silver–Russell syndrome, 844
    Smith–Magenis syndrome, 885
    Sotos syndrome, 911
    trisomy 13 and 18 syndromes, 951
    Williams syndrome, 1031
Continuous positive airway pressure (CPAP):
    achondroplasia, 20, 21
    Down syndrome, 377
    fragile X syndrome, 451
    Loeys–Dietz syndrome, 573
    myotonic dystrophy type 1, 621
    Prader–Willi syndrome, 753, 754
    Rett syndrome, 801
    Treacher Collins syndrome, 932
COP9 gene mutation, Smith–Magenis syndrome, 869
Cornelia de Lange syndrome, 207–223
    cardiovascular anomalies, 218
    craniofacial anomalies, 216
    dermatologic anomalies, 220
    development and behavior problems, 213–215
    diagnostic criteria, 207–210
    diagnostic testing, 211–212
    differential diagnosis, 212–213
        Coffin–Siris syndrome, 187
        Rubinstein–Taybi syndrome, 826
    ear anomalies, 215–216
    etiology, pathogenesis, and genetics, 210–211
    gastrointestinal anomalies, 218
    genitourinary anomalies, 218–219
    growth and feeding problems, 213
    hearing disorders, 215–216
    hematologic disorders, 219–220
    immunologic conditions, 219–220
    incidence, 207
    manifestations and management, 213–220
    musculoskeletal anomalies, 216–217
    neurological anomalies, 219
    ophthalmologic anomalies, 218
Corpus callosum abnormalities, Smith–Lemli–Opitz syndrome, 857

Costello syndrome, 225–240
    cardiovascular anomalies, 232–233
    dermatologic anomalies, 236
    development and behavior problems, 232
    diagnostic criteria, 225–226
    diagnostic testing, 226–228
    differential diagnosis, 228–230
        Beckwith–Wiedemann syndrome, 133
        cardio-facio-cutaneous syndrome, 149
        neurofibromatosis type 1, 635
    endocrine anomalies, 235
    etiology, pathogenesis, and genetics, 226
    growth and feeding problems, 231
    incidence, 225
    manifestations and management, 230–236
    musculoskeletal anomalies, 233–234
    neoplasia incidence, 235–236
    neurological anomalies, 234–235
    ophthalmologic anomalies, 235
    respiratory anomalies, 233
Cowden syndrome. *See also PTEN* hamartoma tumor syndrome
    dermatologic anomalies, 785
    development and behavior problems, 782
    diagnostic criteria, 775–776
    diagnostic testing, 779
    differential diagnosis
        Gorlin syndrome, 463
        Proteus syndrome, 766
    etiology, pathogenesis, and genetics, 778
    incidence, 775
CPLX1 gene, Wolf–Hirschhorn syndrome, 268
Cranial nerve anomalies:
    CHARGE syndrome, 161
    oculo-auriculo-vertebral spectrum, 677
Cranio-cerebello-cardiac (3C) syndrome, differential diagnosis, Noonan syndrome, 655
Craniofacial anomalies. *See also* Facial anomalies; Microcephaly
    Beckwith–Wiedemann syndrome, 136
    carbamazepine embryopathy, 431
    cardio-facio-cutaneous syndrome, 147
    Coffin–Lowry syndrome, 171, 172
    Coffin–Siris syndrome, 189
    Cornelia de Lange syndrome, 216
    craniosynostosis syndromes, 247–248
    deletion 5p syndrome, 284–285
    deletion 22q11.2 syndrome, 302–303
    Down syndrome, 383
    Ehlers–Danlos syndromes, 394, 397, 400
    fetal anticonvulsant syndrome, 436–437
    fragile X syndrome, 450–451
    Gorlin syndrome, 464
    holoprosencephaly, 489, 497
    Kabuki syndrome, 535
    Loeys–Dietz syndrome, 570
    myotonic dystrophy type 1, 619
    oculo-auriculo-vertebral spectrum, 675–676
    Phelan–McDermid syndrome, 318
    Rett syndrome, 800
    Robin sequence, 817–819
    Silver–Russell syndrome, 845
    Smith–Lemli–Opitz syndrome, 852
    Smith–Magenis syndrome, 884

Craniofacial anomalies (cont'd)
  Stickler syndrome, 921–922
  Treacher Collins syndrome, 931–932
  trisomy 13 syndrome, 941, 942
  Williams syndrome, 1023
  Wolf–Hirschhorn syndrome, 266, 267
Craniofrontonasal syndrome, differential diagnosis, craniosynostosis syndromes, 246
Craniosynostosis syndromes, 241–251. *See also* Apert syndrome; Crouzon syndrome; Muenke syndrome; Pfeiffer syndrome; Saethre–Chotzen syndrome
  cardiovascular anomalies, 249
  craniofacial anomalies, 247–248
  development and behavior problems, 247
  diagnostic criteria, 241–246
  diagnostic testing, 246
  differential diagnosis, 246
    deletion 22q11.2 syndrome, 298
  etiology, pathogenesis, and genetics, 245–246
  genitourinary anomalies, 250
  growth and feeding problems, 246–247
  incidence, 241
  manifestations and management, 246–250
  musculoskeletal anomalies, 249
  neurological anomalies, 249
  ophthalmologic anomalies, 248–249
  otolaryngologic anomalies, 248
CREB3LI gene mutation, osteogenesis imperfecta, 691
CREB-binding protein, Rubinstein–Taybi syndrome, 825
CREST syndrome, differential diagnosis, hereditary hemorrhagic telangiectasia, 478
Crohn's disease, Turner syndrome, 988
Crouzon syndrome. *See also* Craniosynostosis syndromes
  development and behavior problems, 247
  diagnostic criteria, 242–243
  diagnostic testing, 246
  etiology, pathogenesis, and genetics, 245
  growth and feeding problems, 246
Crowe's sign, neurofibromatosis type 1, 630
Cryptorchidism:
  Aarskog syndrome, 6
  ATR-X syndrome, 94, 102
  Bardet–Biedl syndrome, 116
  cardio-facio-cutaneous syndrome, 152
  CHARGE syndrome, 166
  Coffin–Siris syndrome, 191
  Cornelia de Lange syndrome, 218
  craniosynostosis syndromes, 250
  deletion 1p36 syndrome, 260
  deletion 5p syndrome, 285
  Denys–Drash syndrome, 336, 340, 346, 347
  Down syndrome, 370, 382, 383
  fetal anticonvulsant syndrome, 437
  Kabuki syndrome, 536
  Klinefelter syndrome, 540, 554, 557
  Noonan syndrome, 662
  Pallister–Killian syndrome, 720, 728
  Prader–Willi syndrome, 751
  Rubinstein–Taybi syndrome, 831

Smith–Lemli–Opitz syndrome, 859, 860
velo-cardio-facial syndrome, 306
Wolf–Hirschhorn syndrome, 273
CTNND2 gene, deletion 5p syndrome, 282
Curaçao criteria, hereditary hemorrhagic telangiectasia, 477
Cutis laxa:
  deletion 1p36 syndrome, 261
  differential diagnosis, Loeys–Dietz syndrome, 568
Cutis marmorata:
  Cornelia de Lange syndrome, 220
  Down syndrome, 358, 378
Cutis verticis gyrata, Cornelia de Lange syndrome, 220
Cyanosis, CHARGE syndrome, 163
Cyclopia, holoprosencephaly, 497
CYFIP mutation, fragile X syndrome, 445
Cylindromatosis, differential diagnosis, Gorlin syndrome, 463
Cystic hygromas:
  Noonan syndrome, 661
  Turner syndrome, 983, 984
Cystic kidney disease:
  Bardet–Biedl syndrome, 115
  tuberous sclerosis complex, 970
Cystic neuroendocrine tumors, 1014
Cytosolic serine hydroxymethyltransferase (cSHMT), Smith–Magenis syndrome, 869

Deafness. *See* Hearing disorders:
Deafness, onychodystrophy, osteodystrophy, mental retardation, and seizures (DOORS) syndrome, differential diagnosis, Coffin–Siris syndrome, 187
Deletion 1p36 syndrome, 253–264
  adulthood, 261
  cardiovascular anomalies, 258
  dermatologic anomalies, 261
  development and behavior problems, 256–257
  diagnostic criteria, 253–254
  diagnostic testing, 255
  differential diagnosis, 255–256
    Cohen syndrome, 199
  ear anomalies, 259
  endocrine anomalies, 260–261
  etiology, pathogenesis, and genetics, 254–255
  gastrointestinal anomalies, 260
  genitourinary anomalies, 260
  growth and feeding problems, 256
  hearing disorders, 259
  incidence, 253
  manifestations and management, 256–261
  musculoskeletal anomalies, 259–260
  neoplasia incidence, 261
  neurological anomalies, 257–258
  ophthalmologic anomalies, 259
  respiratory anomalies, 258–259
Deletion 4p. *See* Wolf-Hirschhorn syndrome
Deletion 5p syndrome, 281–289
  aging and life expectancy, 287
  cardiovascular anomalies, 285
  craniofacial anomalies, 284–285
  development and behavior problems, 283–284
  diagnostic criteria, 281

  diagnostic testing, 282–283
  differential diagnosis, 283
  ear anomalies, 285–286
  etiology, pathogenesis, and genetics, 282
  gastrointestinal anomalies, 286–287
  genetic counseling, 282
  genitourinary anomalies, 285
  growth and feeding problems, 283
  hearing disorders, 285–286
  incidence, 281
  manifestations and management, 283–287
  musculoskeletal anomalies, 286
  neurological anomalies, 285
  ophthalmologic anomalies, 286
  respiratory anomalies, 287
Deletion 22q11.2 syndrome, 291–316
  cardiovascular anomalies, 303–304
  craniofacial anomalies, 302–303
  development and behavior problems, 300–302
  diagnostic criteria, 292
  diagnostic testing, 295–298
  differential diagnosis, 298
    Alagille syndrome, 35
    deletion 5p syndrome, 283
    Kabuki syndrome, 531
    Smith–Magenis syndrome, 871
  ear anomalies, 307
  endocrine anomalies, 306
  etiology, pathogenesis, and genetics, 292–295
  genetic counseling, 294–295
  genitourinary anomalies, 306
  growth and feeding problems, 298–300
  hearing disorders, 307
  immunologic conditions, 305–306
  incidence, 291–292
  manifestations and management, 298–308
  musculoskeletal anomalies, 307–308
  neuromuscular anomalies, 304–305
  ophthalmologic anomalies, 306–307
  psychiatric problems, 300–302
  respiratory anomalies, 303
  and Robin sequence, 809, 810, 813
Deletion 22q13 syndrome. *See* Phelan–McDermid syndrome
Denosumab, osteogenesis imperfecta, 697
Dental anomalies:
  Aarskog syndrome, 5–6
  achondroplasia, 26
  Bardet–Biedl syndrome, 118
  Coffin–Lowry syndrome, 181
  Cohen syndrome, 204
  Cornelia de Lange syndrome, 216
  deletion 5p syndrome, 284
  Down syndrome, 377–378
  fetal alcohol spectrum disorders, 416
  Gorlin syndrome, 467–468
  incontinentia pigmenti, 512
  Kabuki syndrome, 535
  myotonic dystrophy type 1, 619
  osteogenesis imperfecta, 699
  Pallister–Hall syndrome, 713
  Pallister–Killian syndrome, 730
  Phelan–McDermid syndrome, 328
  Prader–Willi syndrome, 754

Proteus syndrome, 772
Rett syndrome, 802–803
Rubinstein–Taybi syndrome, 830
Silver–Russell syndrome, 845
Smith–Magenis syndrome, 884
Sotos syndrome, 908
Treacher Collins syndrome, 931, 932
tuberous sclerosis complex, 972
Turner syndrome, 989
Williams syndrome, 1033–1034
Wolf–Hirschhorn syndrome, 274
Denys–Drash syndrome:
  development and behavior problems, 344–345
  diagnostic criteria, 336–337
  diagnostic testing, 341–342
  differential diagnosis, 342–343
  ear anomalies, 345
  etiology, pathogenesis, and genetics, 339
  genitourinary anomalies, 346–348
  growth and feeding problems, 343–344
  hearing disorders, 345
  incidence, 335
  manifestations and management, 343–349
  musculoskeletal anomalies, 345
  neoplasia incidence, 348–349
  neurological anomalies, 344–345
  ophthalmologic anomalies, 345–346
  pulmonary anomalies, 346
Dermatologic anomalies:
  albinism, 50–52
  ATR-X syndrome, 102
  cardio-facio-cutaneous syndrome, 151–152
  Cohen syndrome, 203
  Cornelia de Lange syndrome, 220
  Costello syndrome, 226, 236
  deletion 1p36 syndrome, 261
  Down syndrome, 378
  Ehlers–Danlos syndromes, 397
  Gorlin syndrome, 464–467
  incontinentia pigmenti, 509–510
  inverted duplicated chromosome 15 syndrome, 524–525
  Klinefelter syndrome, 552
  Loeys–Dietz syndrome, 573
  Marfan syndrome, 592
  myotonic dystrophy type 1, 622
  Noonan syndrome, 663
  Pallister–Killian syndrome, 728
  Phelan–McDermid syndrome, 329–330
  Prader–Willi syndrome, 755–756
  Proteus syndrome, 770–771
  Rubinstein–Taybi syndrome, 830–831
  Smith–Lemli–Opitz syndrome, 860
  Smith–Magenis syndrome, 887–888
  tuberous sclerosis complex, 968–969
  Turner syndrome, 990
  Williams syndrome, 1032
  Wolf–Hirschhorn syndrome, 275
Dermatosporaxis Ehlers-Danlos syndrome, 397–398
Development difficulties:
  Aarskog syndrome, 5
  achondroplasia, 16–17
  Alagille syndrome, 35, 36
  albinism, 50

Angelman syndrome, 66–69
arthrogryposis, 84–85
ATR-X syndrome, 98–99
Bardet–Biedl syndrome, 113–114
Beckwith–Wiedemann syndrome, 134–135
cardio-facio-cutaneous syndrome, 150
CHARGE syndrome, 161–162
Coffin–Lowry syndrome, 177–178
Coffin–Siris syndrome, 188
Cohen syndrome, 200–201
Cornelia de Lange syndrome, 209, 213–215
Costello syndrome, 232
craniosynostosis syndromes, 247
deletion 1p36 syndrome, 256–257
deletion 5p syndrome, 283–284
deletion 22q11.2 syndrome, 300–303
Down syndrome, 363–366
49,XXXXY syndrome, 556
fetal alcohol spectrum disorders, 413–415
fetal anticonvulsant syndrome, 435
fragile X syndrome, 447–449
Gorlin syndrome, 464
Greig cephalopolysyndactyly syndrome, 713
hereditary hemorrhagic telangiectasia, 478
incontinentia pigmenti, 509
inverted duplicated chromosome 15 syndrome, 519–522
Kabuki syndrome, 532–533
Klinefelter syndrome, 543–547
Loeys–Dietz syndrome, 569
Marfan syndrome, 587
Mowat–Wilson syndrome, 602–603
myotonic dystrophy type 1, 618
neurofibromatosis type 1, 637–638
Noonan syndrome, 656–657
oculo-auriculo-vertebral spectrum, 674–675
osteogenesis imperfecta, 695
Pallister–Hall syndrome, 710
Pallister–Killian syndrome, 725–726
Phelan–McDermid syndrome, 322–325
Prader–Willi syndrome, 744
Proteus syndrome, 767
PTEN hamartoma tumor syndrome, 782
Robin sequence, 815–816
Rubinstein–Taybi syndrome, 826–832
Silver–Russell syndrome, 845
Smith–Lemli–Opitz syndrome, 856–857
Smith–Magenis syndrome, 865, 872–878
Sotos syndrome, 905–907
Stickler syndrome, 920
Treacher Collins syndrome, 930–931
trisomy 13 and 18 syndromes, 946–947
tuberous sclerosis complex, 964–965
Turner syndrome, 985–987
VATER/VACTERL association, 998–999
von Hippel–Lindau syndrome, 1009–1010
Williams syndrome, 1025–1028
Wolf–Hirschhorn syndrome, 270–271
WT1-related disorders, 344–345
DFNA6 gene mutation, Wolf–Hirschhorn syndrome, 268
DHCR7 gene mutation, Smith–Lemli–Opitz syndrome, 852, 853
Diabetes insipidus, holoprosencephaly, 497, 501

Diabetes mellitus:
  Bardet–Biedl syndrome, 117
  Cohen syndrome, 202
  holoprosencephaly, 490
  Klinefelter syndrome, 549, 552
  Prader–Willi syndrome, 751, 752
  Turner syndrome, 987
  Williams syndrome, 1029, 1030
Diabetic embryopathy, Robin sequence, 810
Diaphragmatic hernia, *WT1*-related disorders, 341
Diarrhea, Phelan–McDermid syndrome, 329
DICER1 syndrome, Wilms tumor, 343
Diet:
  Alagille syndrome, 36
  Angelman syndrome, 70
DiGeorge syndrome. *See* Deletion 22q11.2 syndrome
Disintegrative infantile psychosis, Rett syndrome, 797
Dislocated hips:
  arthrogryposis, 75, 87
  Smith–Lemli–Opitz syndrome, 860
Disomy:
  Angelman syndrome, 62, 64, 65
  Beckwith–Wiedemann syndrome, 127, 130–131, 132
  Cohen syndrome, 199
  Prader–Willi syndrome, 735, 739, 740
Distal arthrogryposis type 1, differential diagnosis, trisomy 18 syndrome, 941
Distal digital hypoplasia, fetal anticonvulsant syndrome, 437
Distraction osteogenesis, Treacher Collins syndrome, 932
Diverticulus, Williams syndrome, 1031
DNAH5 (dynein axonemal heavy chain 5) gene, deletion 5p syndrome, 287
DNA methylation:
  Angelman syndrome, 62, 64, 65
  ATR-X syndrome, 94
  Beckwith–Wiedemann syndrome, 127, 128, 129, 130, 132
  fetal alcohol spectrum disorders, 409
  Prader–Willi syndrome, 740, 741
  Silver–Russell syndrome, 841
  Sotos syndrome, 899
DNA polymorphisms, Angelman syndrome, 64
Dolichocephaly:
  craniosynostosis syndromes, 241
  oculo-auriculo-vertebral spectrum, 675
  Sotos syndrome, 896
Dolichostenomelia, Marfan syndrome, 586
Down syndrome, 355–385
  and Alzheimer disease, 359, 365, 380
  cardiovascular anomalies, 368–370
  comorbid conditions, 364–365
  craniofacial anomalies, 383
  dental anomalies, 377–378
  dermatologic anomalies, 378
  development and behavior problems, 363–366
  diagnostic criteria, 356–358
  diagnostic testing, 359–361
  differential diagnosis, 361
    Pallister–Killian syndrome, 722
    Smith–Magenis syndrome, 871

Down syndrome (cont'd)
    ear anomalies, 371–372
    endocrine anomalies, 370–371
    etiology, pathogenesis, and genetics, 358–359
    family adjustment, 367–368
    gastrointestinal anomalies, 375–376
    genetic counseling, 360–361
    growth and feeding problems, 361–363
    hearing disorders, 371–372
    immunologic conditions, 379–381
    incidence, 355–356
    life expectancy, 355–356
    manifestations and management, 361–384
    mortality rate, 356
    musculoskeletal anomalies, 373–375
    neoplasia incidence, 381–383
    neurological anomalies, 380–381
    ophthalmologic anomalies, 372–373
    pulmonary hypertension, 369
    sexual maturation, 370–371
    sleep abnormalities, 376–377
    urogenital anomalies, 383–384
Drooling, ATR-X syndrome, 100
Duane retraction "syndrome":
    Bardet–Biedl syndrome, 114
    oculo-auriculo-vertebral spectrum, 677
Dubowitz syndrome, differential diagnosis, fetal alcohol spectrum disorders, 411
Duplication 3q syndrome, differential diagnosis, Cornelia de Lange syndrome, 212
Dural ectasia, Marfan syndrome, 578
Dwarfing disorders, achondroplasia, 26
Dysgerminoma, WAGR and Frasier syndrome, 349
Dysphagia:
    CHARGE syndrome, 159
    myotonic dystrophy type 1, 617, 620
    Prader–Willi syndrome, 754
Dystonia, holoprosencephaly, 496

E6-associated protein (E6AP), Angelman syndrome, 63
Ears, anomalies of:
    achondroplasia, 21
    arthrogryposis, 88
    ATR-X syndrome, 102
    Bardet–Biedl syndrome, 117–118
    CHARGE syndrome, 158–159, 165–166
    Coffin–Lowry syndrome, 180
    Cornelia de Lange syndrome, 215–216
    deletion 1p36 syndrome, 259
    deletion 5p syndrome, 285–286
    deletion 22q11.2 syndrome, 307
    Down syndrome, 371–372
    fetal alcohol spectrum disorders, 416
    holoprosencephaly, 500
    Kabuki syndrome, 534–535
    Mowat–Wilson syndrome, 605
    myotonic dystrophy type 1, 622
    Noonan syndrome, 663–664
    oculo-auriculo-vertebral spectrum, 676–677
    osteogenesis imperfecta, 698–699
    Pallister–Killian syndrome, 729–730
    Phelan–McDermid syndrome, 327
    Rett syndrome, 804

Stickler syndrome, 921, 922
Treacher Collins syndrome, 932–933
trisomy 13 and 18 syndromes, 951
Turner syndrome, 989
Williams syndrome, 1032
Wolf–Hirschhorn syndrome, 273–274
WT1-related disorders, 345
Eating disorders:
    Prader–Willi syndrome, 737, 746, 747
    Smith–Magenis syndrome, 867
Ectodermal dysplasia:
    incontinentia pigmenti, 508
    Phelan–McDermid syndrome, 329
Ectopia lentis, Marfan syndrome, 578, 581, 586, 590
Ectopic calcification, Gorlin syndrome, 471
Edwards syndrome. *See* Trisomy 18 syndrome
Ehlers–Danlos syndromes, 389–403
    arthrochalsia EDS
        diagnostic criteria, 396–397
        differential diagnosis, 397
        etiology, pathogenesis, and genetics, 397
        manifestations and management, 397
    brittle cornea syndrome
        diagnostic criteria, 399
        diagnostic testing, 399
        differential diagnosis, 399
        etiology, pathogenesis, and genetics, 399
        manifestations and management, 399
    cardiac-valvular EDS
        diagnostic criteria, 393
        diagnostic testing, 393
        differential diagnosis, 393
        etiology, pathogenesis, and genetics, 393
        manifestations and management, 393
    classical type
        diagnostic criteria, 390–392
        diagnostic testing, 392
        differential diagnosis, 392
        etiology, pathogenesis, and genetics, 392
        manifestations and management, 392
    classic-like EDS
        diagnostic criteria, 392–393
        diagnostic testing, 393
        differential diagnosis, 393
        etiology, pathogenesis, and genetics, 393
        manifestations and management, 393
    dermatosporaxis EDS
        diagnostic criteria, 397
        diagnostic testing, 397
        differential diagnosis, 397–398
        etiology, pathogenesis, and genetics, 397
        manifestations and management, 398
    diagnostic criteria, 389–390
    differential diagnosis
        Kabuki syndrome, 531
        Loeys–Dietz syndrome, 568
        Marfan syndrome, 585
        Sotos syndrome, 901
    hypermobile EDS
        diagnostic criteria, 395
        diagnostic testing, 395
        differential diagnosis, 395–396
        etiology, pathogenesis, and genetics, 395
        manifestations and management, 396

incidence, 389
kyphoscoliotic EDS
    diagnostic criteria, 398
    diagnostic testing, 398
    differential diagnosis, 398
    etiology, pathogenesis, and genetics, 398
    manifestations and management, 398–399
musculocontractural EDS
    diagnostic criteria, 400–401
    diagnostic testing, 401
    differential diagnosis, 401
    etiology, pathogenesis, and genetics, 401
    manifestations and management, 401
myopathic EDS
    diagnostic criteria, 401
    diagnostic testing, 401
    differential diagnosis, 401
    etiology, pathogenesis, and genetics, 401
periodontal EDS
    diagnostic criteria, 401–402
    diagnostic testing, 402
    differential diagnosis, 402
    etiology, pathogenesis, and genetics, 402
    manifestations and management, 402
spondylodysplastic EDS
    diagnostic criteria, 399–400
    diagnostic testing, 400
    differential diagnosis, 400
    etiology, pathogenesis, and genetics, 400
    manifestations and management, 400
vascular EDS
    diagnostic criteria, 394
    diagnostic testing, 394
    differential diagnosis, 394–395
    etiology, pathogenesis, and genetics, 394
    manifestations and management, 395
Elastin (*ELN*) gene mutations, Williams syndrome, 1022–1023, 1030
Elbow hypermobility:
    achondroplasia, 25
    Cohen syndrome, 203
    Ehlers–Danlos syndromes, 389
Elephant man disease. *See* Proteus syndrome:
Ellis–van Creveld syndrome, differential diagnosis, Pallister–Hall syndrome, 710
Encopresis, Smith–Magenis syndrome, 874, 885
Endocrine anomalies:
    Alagille syndrome, 36
    Bardet–Biedl syndrome, 117
    Beckwith–Wiedemann syndrome, 135–136
    Cohen syndrome, 202
    Costello syndrome, 235
    deletion 1p36 syndrome, 260–261
    deletion 22q11.2 syndrome, 306
    Down syndrome, 370–371
    49,XXXXY syndrome, 557
    fragile X syndrome, 453
    holoprosencephaly, 497–498
    inverted duplicated chromosome 15 syndrome, 524
    Kabuki syndrome, 534
    Mowat–Wilson syndrome, 605
    myotonic dystrophy type 1, 623
    Pallister–Hall syndrome, 711–712
    Prader–Willi syndrome, 750–752

Smith–Magenis syndrome, 886–887
Turner syndrome, 987–988
Williams syndrome, 1029–1030
Endoglin (*ENG*) gene mutation, hereditary hemorrhagic telangiectasia, 476
Endolymphatic sac tumors, von Hippel–Lindau syndrome, 1011, 1014
Endometrial cancer, *PTEN* hamartoma tumor syndrome, 784
Enuresis:
   fragile X syndrome, 452
   Smith–Magenis syndrome, 874, 885
   Williams syndrome, 1033
EP300 gene mutations, Rubinstein–Taybi syndrome, 826
Epilepsy. *See also* Seizure disorders
   fetal alcohol spectrum disorders, 418
   fetal anticonvulsant syndrome, 425
   inverted duplicated chromosome 15 syndrome, 515, 522
   Pallister–Killian syndrome, 718
   Rett syndrome, 800, 801
   tuberous sclerosis complex, 966
   Wolf–Hirschhorn syndrome, 268
Epileptic encephalopathy, infant and childhood, Rett syndrome, 797
Epiphysiodesis:
   Beckwith–Wiedemann syndrome, 134
   Proteus syndrome, 767
Episodic gastric pseudovolvulus, ATR-X syndrome, 100
Epistaxis, hereditary hemorrhagic telangiectasia, 478–479
Epithelial follicular neoplasias, differential diagnosis, *PTEN* hamartoma tumor syndrome, 781
Escobar syndrome, arthrogryposis, 87
Esophageal atresia, VATER/VACTERL association, 1000
Esotropia:
   arthrogryposis, 87
   cardio-facio-cutaneous syndrome, 152
   Down syndrome, 373
Estrogen treatment:
   Prader–Willi syndrome, 752
   Turner syndrome, 984, 988
Exotropia, cardio-facio-cutaneous syndrome, 152
Extended limb lengthening, and achondroplasia, 12
Extrahepatic disorders, differential diagnosis, Alagille syndrome, 35
Eye anomalies. *See* Ophthalmologic anomalies
Eye contact, Rett syndrome, 799

Facial angiofibroma, differential diagnosis, tuberous sclerosis complex, 963
Facial angiofibromas, tuberous sclerosis complex, 958, 968
Facial anomalies:
   Aarskog syndrome, 1–2
   achondroplasia, 26
   Alagille syndrome, 32
   cardio-facio-cutaneous syndrome, 147
   CHARGE syndrome, 159
   Coffin–Lowry syndrome, 171, 172
   Coffin–Siris syndrome, 185
   Cohen syndrome, 197, 198
   Cornelia de Lange syndrome, 208
   Costello syndrome, 226
   craniosynostosis syndromes, 241
   deletion 1p36 syndrome, 253, 256
   deletion 22q11.2 syndrome, 292
   Down syndrome, 383
   Ehlers–Danlos syndromes, vascular-type Ehlers-Danlos syndrome, 241
   fetal alcohol spectrum disorders, 410
   fetal hydantoin syndrome, 428
   fetal valproate syndrome, 429
   Gorlin syndrome, 462, 464
   holoprosencephaly, 489, 497
   Kabuki syndrome, 529
   Mowat–Wilson syndrome, 599
   Noonan syndrome, 651
   oculo-auriculo-vertebral spectrum, 671, 675, 677
   Pallister–Killian syndrome, 718
   Phelan–McDermid syndrome, 318
   Prader–Willi syndrome, 737
   Rubinstein–Taybi syndrome, 823–824
   Smith–Lemli–Opitz syndrome, 852
   Smith–Magenis syndrome, 865
   Sotos syndrome, 895–897
   Treacher Collins syndrome, 927, 931–932
   trisomy 13 syndrome, 942
   Williams syndrome, 1021
   Wolf–Hirschhorn syndrome, 265, 268, 270
Facial–genital–digital syndrome. *See* Aarskog syndrome
Faciogenital dysplasia. *See* Aarskog syndrome
Factor XI deficiency, Noonan syndrome, 660
Factor XII deficiency, Noonan syndrome, 660
Failure-to-thrive:
   achondroplasia, 20
   Alagille syndrome, 36
   Angelman syndrome, 66
   Cohen syndrome, 200
   Down syndrome, 362
   fetal alcohol spectrum disorders, 413
   holoprosencephaly, 494–495
   incontinentia pigmenti, 509
   Loeys–Dietz syndrome, 569
   Prader–Willi syndrome, 742
   Silver–Russell syndrome, 843
   Smith–Magenis syndrome, 871
Falx cerebri calcification, Gorlin syndrome, 460, 472
Familial adenomatous polyposis, differential diagnosis, *PTEN* hamartoma tumor syndrome, 781
Familial cholestatic liver disease, Alagille syndrome, 31
Fanconi anemia:
   differential diagnosis, VATER/VACTERL association, 997
   Wilms tumor, 343
FBN1 gene mutations, Marfan syndrome, 577, 578, 580
FBN2 gene mutations, Loeys–Dietz syndrome, 568
Febrile convulsions:
   Rett syndrome, 792
   Sotos syndrome, 907
Feeding difficulties:
   achondroplasia, 12–16
   Alagille syndrome, 36
   Angelman syndrome, 66
   arthrogryposis, 84
   ATR-X syndrome, 98
   Bardet–Biedl syndrome, 112–113
   Beckwith–Wiedemann syndrome, 134
   cardio-facio-cutaneous syndrome, 150
   CHARGE syndrome, 161
   Coffin–Lowry syndrome, 175
   Coffin–Siris syndrome, 188
   Cohen syndrome, 200
   Cornelia de Lange syndrome, 213
   Costello syndrome, 231
   craniosynostosis syndromes, 246–247
   deletion 1p36 syndrome, 256
   deletion 5p syndrome, 283
   deletion 22q11.2 syndrome, 298–300
   Denys–Drash syndrome, 343–344
   Down syndrome, 361–363
   49,XXXXY syndrome, 558
   fetal alcohol spectrum disorders, 413
   fetal anticonvulsant syndrome, 434–435
   fragile X syndrome, 447
   Gorlin syndrome, 463
   hereditary hemorrhagic telangiectasia, 478
   holoprosencephaly, 494–495
   incontinentia pigmenti, 509
   inverted duplicated chromosome 15 syndrome, 519
   Kabuki syndrome, 532
   Klinefelter syndrome, 542–543
   Loeys–Dietz syndrome, 569
   Marfan syndrome, 586–587
   Mowat–Wilson syndrome, 601–602
   myotonic dystrophy type 1, 617
   neurofibromatosis type 1, 636–637
   Noonan syndrome, 655–656
   oculo-auriculo-vertebral spectrum, 674
   osteogenesis imperfecta, 694–695
   Pallister–Hall syndrome, 710
   Pallister–Killian syndrome, 724–725
   Phelan–McDermid syndrome, 322
   Prader–Willi syndrome, 742–744
   Proteus syndrome, 766–767
   PTEN hamartoma tumor syndrome, 782
   Rett syndrome, 798–799
   Robin sequence, 814–815
   Rubinstein–Taybi syndrome, 826–827
   Silver–Russell syndrome, 843–844
   Smith–Lemli–Opitz syndrome, 855–856
   Smith–Magenis syndrome, 871–872
   Sotos syndrome, 901–905
   Stickler syndrome, 920
   Treacher Collins syndrome, 930
   trisomy 13 and 18 syndromes, 945–946
   tuberous sclerosis complex, 964
   Turner syndrome, 980–981
   VATER/VACTERL association, 998
   von Hippel–Lindau syndrome, 1009
   WAGR syndrome, 343–344

Feeding difficulties: (cont'd)
    Williams syndrome, 1024–1025
    Wolf–Hirschhorn syndrome, 270
    WT1-related disorders, 343–344
Fertility problems:
    Klinefelter syndrome, 548–549
    myotonic dystrophy type 1, 623
    Noonan syndrome, 662
    Prader–Willi syndrome, 751
    Sotos syndrome, 900
    Turner syndrome, 984–985
Fetal akinesia deformation sequence:
    arthrogryposis, 76
    differential diagnosis, trisomy 18 syndrome, 941
Fetal alcohol spectrum disorders, 405–424
    cardiovascular anomalies, 416–417
    CoFASP Consensus Clinical Diagnostic Guidelines, 406
    dental anomalies, 416
    development and behavior problems, 413–415
    diagnostic criteria, 406
    diagnostic testing, 409–411
    differential diagnosis, 411
        Cornelia de Lange syndrome, 212
        fetal anticonvulsant syndrome, 428
    ear anomalies, 416
    etiology, pathogenesis, and genetics, 406–409
    gastrointestinal and hepatic anomalies, 417
    genitourinary anomalies, 417
    growth and feeding problems, 413
    hearing disorders, 416
    incidence and prevalence, 405–406
    manifestations and management, 413–419
    musculoskeletal anomalies, 417
    neurological anomalies, 417–418
    ophthalmologic anomalies, 415–416
    societal implications and prevention, 418–419
Fetal alcohol syndrome:
    cardiovascular anomalies, 416
    dental anomalies, 416
    development and behavior problems, 415
    differential diagnosis, Williams syndrome, 1024
    ear anomalies, 416
    gastrointestinal and hepatic anomalies, 417
    hearing disorders, 416
    incidence and prevalence, 405
    renal anomalies, 417
    Robin sequence, 810
    societal implications and prevention, 418
Fetal aminopterin syndrome, Robin sequence, 811
Fetal anticonvulsant syndrome, 425–442
    cardiovascular anomalies, 436
    craniofacial anomalies, 436–437
    development and behavior problems, 435
    diagnostic criteria, 426
    diagnostic testing, 427–428
    differential diagnosis, 428–434
    etiology, pathogenesis, and genetics, 426–427
    gastrointestinal anomalies, 437
    genitourinary anomalies, 437
    growth and feeding problems, 434–435

    hematologic anomalies, 438
    manifestations and management, 434
    musculoskeletal anomalies, 437–438
    neoplasia incidence, 438
    neurological anomalies, 436
    ophthalmologic anomalies, 437
Fetal hydantoin syndrome, 428–429, 434
Fetal valproate syndrome, 429–431
A-fetoprotein (AFP), Beckwith–Wiedemann syndrome, 139
FGD1 gene, Aarskog syndrome, 1, 3, 4
FGFRL1 gene, Wolf–Hirschhorn syndrome, 268
FG syndrome, differential diagnosis, Williams syndrome, 1024
Fibrillin-1 (FBN1) gene, Marfan syndrome, 577, 580
Fibroblast growth factor receptor 1 (*FGFR1*) mutation, holoprosencephaly, 492, 493
Fibroblast growth factor receptor 2 (*FGFR2*) mutation, Apert syndrome, 245
Fibroblast growth factor receptor 3 (*FGFR3*) mutation:
    achondroplasia, 10–11, 26
    craniosynostosis syndromes, 243
    thanatophoric dysplasia, 246
Fibromas:
    Gorlin syndrome, 460, 462, 469, 470, 471
    myotonic dystrophy type 1, 622
    neurofibromatosis type 1, 635
    PTEN hamartoma tumor syndrome, 776, 785
    tuberous sclerosis complex, 958, 968, 969, 972
Fibrous cephalic plaques, tuberous sclerosis complex, 968
Fingernails:
    Phelan–McDermid syndrome, 329
    tuberous sclerosis complex, 958
Flat feet, Loeys–Dietz syndrome, 570
FLCN gene mutations, Birt–Hogg–Dube syndrome, 869
FLNA mutation carriers, differential diagnosis, Loeys–Dietz syndrome, 569
Floating Harbor syndrome:
    differential diagnosis, Cornelia de Lange syndrome, 212, 213
Fluorescence in situ hybridization (FISH):
    Alagille syndrome, 35
    Angelman syndrome, 65
    deletion 1p36 syndrome, 255
    Denys–Drash syndrome, 342
    Down syndrome, 360
    holoprosencephaly, 492
    inverted duplicated chromosome 15 syndrome, 518
    Klinefelter syndrome, 541
    Pallister–Killian syndrome, 722
    Phelan–McDermid syndrome, 317, 321
    Prader–Willi syndrome, 740
    Robin sequence, 813
    Rubinstein–Taybi syndrome, 825, 826
    Smith–Magenis syndrome, 863, 870
    trisomy 13 syndrome, 943
    trisomy 18 syndrome, 940
    Turner syndrome, 978
    velo-cardio-facial syndrome, 291, 295
    WAGR syndrome, 342

    Williams syndrome, 1021, 1023
    Wolf–Hirschhorn syndrome, 269
FMR1 gene mutation:
    fragile X syndrome, 445
    Klinefelter syndrome, 542
FMRP protein, fragile X syndrome, 444
Follicle-stimulating hormone, Klinefelter syndrome, 548
Folliculin gene (*FLCN*), Smith–Magenis syndrome, 869
Food allergies, Loeys–Dietz syndrome, 573
Food aversion, Rett syndrome, 798
Food-seeking behavior:
    Prader–Willi syndrome, 737
    Smith–Magenis syndrome, 872
Foot anomalies. *See also* Clubfeet
    ATR-X syndrome, 102
    Cornelia de Lange syndrome, 209
    Costello syndrome, 234
    craniosynostosis syndromes, 243
    deletion 1p36 syndrome, 260
    Ehlers–Danlos syndromes, 392, 393, 401
    inverted duplicated chromosome 15 syndrome, 524
    Marfan syndrome, 588
    myotonic dystrophy type 1, 622
    Pallister–Killian syndrome, 713
    Rett syndrome, 802
    Silver–Russell syndrome, 845
    Smith–Lemli–Opitz syndrome, 860
    Sotos syndrome, 910, 911
    trisomy 18 syndrome, 952
    Wolf–Hirschhorn syndrome, 274
Foramen magnum, achondroplasia, 18
46,XY/47,XXY mosaicism, 552
48,XXXY syndrome, 553–556
48,XXYY syndrome, 552–553
49,XXXXY syndrome, 556–559
    development and behavior problems, 557
    endocrine anomalies, 557
    immunologic conditions, 557
    musculoskeletal anomalies, 557
    neurological anomalies, 556–557
Foveal hypoplasia, in albinism, 50, 52
FOXG1 gene mutation, Rett syndrome, 794
Fragile X-associated primary ovarian insufficiency (FXPOI), 453
Fragile X-associated tremor/ataxia syndrome (FXTAS), 445, 450
Fragile X syndrome, 443–457
    cardiovascular anomalies, 451
    craniofacial anomalies, 450–451
    development and behavior problems, 447–449
    diagnostic criteria, 444
    diagnostic testing, 446
    differential diagnosis, 446
        Coffin–Lowry syndrome, 175
        Klinefelter syndrome, 541–542
        Smith–Magenis syndrome, 871
        Sotos syndrome, 901
        Williams syndrome, 1024
    endocrine anomalies, 453
    etiology, pathogenesis, and genetics, 444–446
    genetic counseling, 445–446
    genitourinary anomalies, 452

growth and feeding problems, 447
  manifestations and management, 446
  musculoskeletal anomalies, 452–453
  neurological anomalies, 449–450
  ophthalmologic anomalies, 450
Frasier syndrome, 335
  diagnostic criteria, 336–337
  diagnostic testing, 341–342
  differential diagnosis, 342–343
  etiology, pathogenesis, and genetics, 339
  genitourinary anomalies, 346
  incidence, 335
  manifestations and management, 343–349
  neoplasia incidence, 348
Freeman-Sheldon syndrome, differential diagnosis, musculocontractural EDS, 401
FRMD7 gene mutations, albinism, 49
Fryns syndrome, differential diagnosis:
  Cornelia de Lange syndrome, 212
  Pallister–Killian syndrome, 724

Gabapentin, Phelan–McDermid syndrome, 325
GABRB3 gene, inverted duplicated chromosome 15 syndrome, 517
Gait disorders:
  achondroplasia, 24
  Angelman syndrome, 69
  Down syndrome, 373
  Mowat–Wilson syndrome, 604
  Williams syndrome, 1033
Galloway–Mowat syndrome, differential diagnosis, *WT1*-related disorders, 343
Gamma-glutamyl transferase, Alagille syndrome, 35
Gastroenterology:
  cardio-facio-cutaneous syndrome, 150–151
  deletion 5p syndrome, 286–287
  Wolf–Hirschhorn syndrome, 275
Gastroesophageal reflux:
  ATR-X syndrome, 100
  CHARGE syndrome, 158
  Coffin–Siris syndrome, 191
  Cornelia de Lange syndrome, 213, 218–219
  deletion 1p36 syndrome, 256
  inverted duplicated chromosome 15 syndrome, 519
  Mowat–Wilson syndrome, 603
  Pallister–Killian syndrome, 725
  Phelan–McDermid syndrome, 329
  Rett syndrome, 798, 803
  Rubinstein–Taybi syndrome, 828, 829
  Silver–Russell syndrome, 844
  Smith–Lemli–Opitz syndrome, 855
  Smith–Magenis syndrome, 872
  Treacher Collins syndrome, 930
  trisomy 13 and 18 syndromes, 945, 946, 950
  Williams syndrome, 1031
  Wolf–Hirschhorn syndrome, 270
Gastrointestinal anomalies:
  Alagille syndrome, 36–38
  albinism, 55
  ATR-X syndrome, 100
  Bardet–Biedl syndrome, 118
  Beckwith–Wiedemann syndrome, 136–137
  Coffin–Lowry syndrome, 181

Coffin–Siris syndrome, 191
Cornelia de Lange syndrome, 218
deletion 1p36 syndrome, 260
Down syndrome, 375–376
fetal alcohol spectrum disorders, 417
fetal anticonvulsant syndrome, 437
Gorlin syndrome, 471
hereditary hemorrhagic telangiectasia, 482–483
holoprosencephaly, 500
Kabuki syndrome, 535–536
Mowat–Wilson syndrome, 603
myotonic dystrophy type 1, 619–620
Pallister–Hall syndrome, 712
Pallister–Killian syndrome, 727
Phelan–McDermid syndrome, 329
Proteus syndrome, 769
PTEN hamartoma tumor syndrome, 784, 785
Rett syndrome, 803
Rubinstein–Taybi syndrome, 828–829
Silver–Russell syndrome, 844
Smith–Lemli–Opitz syndrome, 858–859
Smith–Magenis syndrome, 885
Sotos syndrome, 911
Turner syndrome, 988–989
VATER/VACTERL association, 1000–1001
von Hippel–Lindau syndrome, 1014–1015
Williams syndrome, 1025, 1031
Wolf–Hirschhorn syndrome, 275
GATA1 mutations, Down syndrome, 381
GATA6 mutations, Pallister–Killian syndrome, 721
Generalized essential telangiectasia, differential diagnosis, hereditary hemorrhagic telangiectasia, 478
Generalized hypertrichosis, Cornelia de Lange syndrome, 220
Generalized hypotonia. *See* Hypotonia
Generalized overgrowth disorders, differential diagnosis, Sotos syndrome, 901
Genetic counseling:
  Alagille syndrome, 35
  Angelman syndrome, 65
  ATR-X syndrome, 97–98
  Bardet–Biedl syndrome, 111–112
  deletion 5p syndrome, 282
  deletion 22q11.2 syndrome, 294–295
  Down syndrome, 360–361
  49,XXXXY syndrome, 557
  fragile X syndrome, 445–446
  Gorlin syndrome, 462
  holoprosencephaly, 494
  Loeys–Dietz syndrome, 574
  Mowat–Wilson syndrome, 600–601
  myotonic dystrophy type 1, 616–617
  neurofibromatosis type 1, 633–634
  Noonan syndrome, 654
  oculo-auriculo-vertebral spectrum, 673
  osteogenesis imperfecta, 684, 692–693
  Pallister–Killian syndrome, 721
  Prader–Willi syndrome, 740
  Rett syndrome, 797
  Smith–Magenis syndrome, 870
  Sotos syndrome, 901
  Stickler syndrome, 919

tuberous sclerosis complex, 962–963
  WT1-related disorders, 342
Genetic heterogeneity:
  Aarskog syndrome, 1, 4
  Bardet–Biedl syndrome, 108
  Beckwith–Wiedemann syndrome, 127
  Coffin–Lowry syndrome, 174
  Cohen syndrome, 196
  Ehlers–Danlos syndromes, 398, 399, 400
  Loeys–Dietz syndrome, 567
  Stickler syndrome, 918, 919
Genitourinary anomalies:
  Aarskog syndrome, 2
  Alagille syndrome, 40
  ATR-X syndrome, 101–102
  Bardet–Biedl syndrome, 115–117
  Beckwith–Wiedemann syndrome, 137
  cardio-facio-cutaneous syndrome, 152
  CHARGE syndrome, 158, 166
  Coffin–Siris syndrome, 191
  Cornelia de Lange syndrome, 218–219
  craniosynostosis syndromes, 250
  deletion 1p36 syndrome, 260
  deletion 5p syndrome, 285
  deletion 22q11.2 syndrome, 306
  Down syndrome, 383–384
  fetal alcohol spectrum disorders, 417
  fetal anticonvulsant syndrome, 437
  fragile X syndrome, 452
  Gorlin syndrome, 470–471
  Kabuki syndrome, 536
  Mowat–Wilson syndrome, 604–605
  Noonan syndrome, 662
  oculo-auriculo-vertebral spectrum, 678
  osteogenesis imperfecta, 701
  Pallister–Hall syndrome, 712
  Pallister–Killian syndrome, 728
  Phelan–McDermid syndrome, 330–331
  Prader–Willi syndrome, 750
  Proteus syndrome, 770
  Rubinstein–Taybi syndrome, 831
  Smith–Lemli–Opitz syndrome, 859–860
  Smith–Magenis syndrome, 885–886
  Sotos syndrome, 909
  trisomy 13 and 18 syndromes, 949–950
  tuberous sclerosis complex, 970–971
  Turner syndrome, 987
  VATER/VACTERL association, 1002
  von Hippel–Lindau syndrome, 1015
  WAGR syndrome, 340
  Williams syndrome, 1032–1033
  Wolf–Hirschhorn syndrome, 273
  WT1-related disorders, 342, 346–348
Genotype–phenotype correlations:
  Angelman syndrome, 64
  ATR-X syndrome, 95–96
  deletion 1p36 syndrome, 254
  Greig cephalopolysyndactyly syndrome, 709
  Loeys–Dietz syndrome, 566
  neurofibromatosis type 1, 633
  Noonan syndrome, 653
  osteogenesis imperfecta, 692
  Smith–Lemli–Opitz syndrome, 852
  Sotos syndrome, 900

Genotype–phenotype correlations: (cont'd)
　　tuberous sclerosis complex, 962
　　velo-cardio-facial syndrome, 293
　　von Hippel–Lindau syndrome, 1005
　　Wolf–Hirschhorn syndrome, 269
　　WT1-related disorders, 339
Genu valgus, Prader–Willi syndrome, 737
Germ cell tumors:
　　Beckwith–Wiedemann syndrome, 139
　　Down syndrome, 382
　　Klinefelter syndrome, 551
　　Pallister–Killian syndrome, 730
　　Rubinstein–Taybi syndrome, 832
Ghent criteria, Marfan syndrome, 579
Gillespie syndrome, differential diagnosis, WAGR syndrome, 342
Gingival fibromas, tuberous sclerosis complex, 972
Glaucoma:
　　aniridia, 345
　　Marfan syndrome, 590
　　Phelan–McDermid syndrome, 327
　　Rubinstein–Taybi syndrome, 829
　　Wolf–Hirschhorn syndrome, 274
　　WT1-related disorders, 345
GLI3 gene mutation:
　　Greig cephalopolysyndactyly syndrome, 709
　　Pallister–Hall syndrome, 707
Glioblastoma multiforme, Gorlin syndrome, 469
Glossoptosis, Robin sequence, 808
Goldberg–Shprintzen syndrome, differential diagnosis, Mowat–Wilson syndrome, 601
Goldenhar syndrome. *See* Oculo-auriculo-vertebral spectrum
Goltz focal dermal hypoplasia, differential diagnosis, incontinentia pigmenti, 509
Gonadoblastoma:
　　Turner syndrome, 978, 990
　　WT1-related disorders, 343, 349
Gorlin syndrome, 459–474
　　cardiovascular anomalies, 469–470
　　craniofacial anomalies, 464
　　dental anomalies, 467–468
　　dermatologic anomalies, 464–467
　　development and behavior problems, 464
　　diagnostic criteria, 459–460
　　diagnostic testing, 462
　　differential diagnosis, 462–463
　　etiology, pathogenesis, and genetics, 460–462
　　gastrointestinal anomalies, 471
　　genetic counseling, 462
　　genitourinary anomalies, 470–471
　　growth and feeding problems, 463
　　incidence, 459
　　manifestations and management, 463–472
　　musculoskeletal anomalies, 470
　　neoplasia incidence, 468–469
　　ophthalmologic anomalies, 469
GPR143 gene mutations, albinism, 49
"Greek warrior helmet appearance of the nose," Wolf–Hirschhorn syndrome, 266
Greig cephalopolysyndactyly syndrome, 707–716
　　development and behavior, 713
　　diagnostic criteria, 708

diagnostic testing, 709–710
differential diagnosis, 710
etiology, pathogenesis, and genetics, 708–709
incidence, 707
manifestations and management, 713–714
musculoskeletal anomalies, 714
neurological anomalies, 713–714
Growth hormone therapy:
　　achondroplasia, 12
　　Costello syndrome, 231
　　Down syndrome, 364
　　neurofibromatosis type 1, 636
　　Prader–Willi syndrome, 738–739, 742, 743, 750, 753–754
　　Silver–Russell syndrome, 843
　　Smith–Magenis syndrome, 880
　　Turner syndrome, 980
Growth problems:
　　Aarskog syndrome, 4–5
　　achondroplasia, 12–16
　　Alagille syndrome, 36
　　albinism, 50
　　Angelman syndrome, 66
　　arthrogryposis, 79–84
　　ATR-X syndrome, 98
　　Bardet–Biedl syndrome, 112–113
　　Beckwith–Wiedemann syndrome, 134
　　cardio-facio-cutaneous syndrome, 150
　　CHARGE syndrome, 158, 161
　　Coffin–Lowry syndrome, 175
　　Coffin–Siris syndrome, 188
　　Cohen syndrome, 200
　　Cornelia de Lange syndrome, 208–209, 213
　　Costello syndrome, 231
　　craniosynostosis syndromes, 246–247
　　deletion 1p36 syndrome, 256, 260
　　deletion 5p syndrome, 283
　　deletion 22q11.2 syndrome, 298–300, 306
　　Denys–Drash syndrome, 343–344
　　Down syndrome, 361–363
　　fetal alcohol spectrum disorders, 411, 413
　　fetal anticonvulsant syndrome, 434–435
　　fragile X syndrome, 447
　　Gorlin syndrome, 463
　　hereditary hemorrhagic telangiectasia, 478
　　holoprosencephaly, 494–495
　　incontinentia pigmenti, 509
　　inverted duplicated chromosome 15 syndrome, 519
　　Kabuki syndrome, 532
　　Klinefelter syndrome, 542–543
　　Loeys–Dietz syndrome, 569
　　Marfan syndrome, 586–587
　　Mowat–Wilson syndrome, 601–602
　　myotonic dystrophy type 1, 617
　　neurofibromatosis type 1, 636–637
　　Noonan syndrome, 655–656
　　oculo-auriculo-vertebral spectrum, 674
　　osteogenesis imperfecta, 694–695
　　Pallister–Hall syndrome, 710
　　Pallister–Killian syndrome, 724–725
　　Phelan–McDermid syndrome, 322
　　Prader–Willi syndrome, 742–744
　　Proteus syndrome, 766–767
　　PTEN hamartoma tumor syndrome, 782

　　Rett syndrome, 798–799
　　Robin sequence, 814–815
　　Rubinstein–Taybi syndrome, 826–827
　　Silver–Russell syndrome, 842, 843–844
　　Smith–Lemli–Opitz syndrome, 855–856
　　Smith–Magenis syndrome, 871–872, 886
　　Sotos syndrome, 901–905
　　Stickler syndrome, 920
　　Treacher Collins syndrome, 930
　　trisomy 13 and 18 syndromes, 945–946
　　tuberous sclerosis complex, 964
　　Turner syndrome, 980–981
　　VATER/VACTERL association, 998
　　von Hippel–Lindau syndrome, 1009
　　WAGR syndrome, 343–344
　　Williams syndrome, 1024–1025
　　Wolf–Hirschhorn syndrome, 270
　　WT1-related disorders, 343–344
GTF2I gene mutation, Williams syndrome, 1023
Gynecomastia, Klinefelter syndrome, 548, 551

Hadju–Cheney syndrome, differential diagnosis, Alagille syndrome, 36
Hair anomalies, incontinentia pigmenti, 510
Hair color:
　　albinism, 51
　　Smith–Magenis syndrome, 887
Hairy elbow syndrome. *See* Wiedemann–Steiner syndrome
Hamartomatous syndromes, differential diagnosis, neurofibromatosis type 1, 635
Hand anomalies. *See also* Thumb anomalies
　　Aarskog syndrome, 2, 6
　　Coffin–Lowry syndrome, 172
　　Cornelia de Lange syndrome, 209
　　Costello syndrome, 233
　　craniosynostosis syndromes, 241
　　deletion 22q11.2 syndrome, 292
　　Down syndrome, 357
　　holoprosencephaly, 493
　　Kabuki syndrome, 530
　　Marfan syndrome, 588
　　Mowat–Wilson syndrome, 605
　　Pallister–Hall syndrome, 712
　　Phelan–McDermid syndrome, 328
　　Proteus syndrome, 764
　　Rett syndrome, 799, 800
　　Rubinstein–Taybi syndrome, 831
　　Silver–Russell syndrome, 845
　　Smith–Lemli–Opitz syndrome, 860
　　Smith–Magenis syndrome, 885
　　trisomy 13 and 18 syndromes, 951
　　Wolf–Hirschhorn syndrome, 274
Hand-licking, Smith–Magenis syndrome, 867
Haploinsufficiency:
　　Alagille syndrome, 34
　　craniosynostosis syndromes, 246
　　deletion 1p36 syndrome, 255
　　Denys–Drash syndrome, 339
　　Ehlers–Danlos syndromes, 392, 393, 394
　　Greig cephalopolysyndactyly syndrome, 709
　　Loeys–Dietz syndrome, 566
　　Marfan syndrome, 581
　　Mowat–Wilson syndrome, 598

Pallister–Killian syndrome, 721
Phelan–McDermid syndrome, 317, 320
Rett syndrome, 797
Smith–Magenis syndrome, 863, 867, 869, 870
Treacher Collins syndrome, 929
Turner syndrome, 978
velo-cardio-facial syndrome, 293
Wolf–Hirschhorn syndrome, 268
Hearing disorders:
  achondroplasia, 21
  Alagille syndrome, 35
  albinism, 48
  arthrogryposis, 88
  ATR-X syndrome, 102
  Bardet–Biedl syndrome, 117–118
  CHARGE syndrome, 158, 162, 165–166
  Coffin–Lowry syndrome, 177, 180
  Coffin–Siris syndrome, 191
  Cornelia de Lange syndrome, 215–216
  deletion 1p36 syndrome, 259
  deletion 5p syndrome, 285–286
  deletion 22q11.2 syndrome, 307
  Down syndrome, 371–372
  fetal alcohol spectrum disorders, 416
  holoprosencephaly, 500
  Kabuki syndrome, 534–535
  Mowat–Wilson syndrome, 605
  Muenke syndrome, 248
  myotonic dystrophy type 1, 622
  Noonan syndrome, 663–664
  oculo-auriculo-vertebral spectrum, 676–677
  osteogenesis imperfecta, 698–699
  Pallister–Killian syndrome, 729–730
  Phelan–McDermid syndrome, 327
  Proteus syndrome, 770
  Rett syndrome, 804
  Smith–Magenis syndrome, 867, 870, 873, 881
  Sotos syndrome, 908
  Stickler syndrome, 918, 921, 922
  Treacher Collins syndrome, 933
  trisomy 13 and 18 syndromes, 951
  Turner syndrome, 989
  Williams syndrome, 1032
  Wolf–Hirschhorn syndrome, 268–269, 273–274
  WT1-related disorders, 345
Heart defects. *See* Cardiovascular anomalies
Helicobacter pylori infection, Down syndrome, 375
Helsmoortel–Van der Aa syndrome, differential diagnosis, deletion 1p36 syndrome, 256
Hemangioblastomas, von Hippel–Lindau syndrome, 1006, 1007–1008, 1010, 1012–1013
Hematologic disorders:
  albinism, 53–54
  ATR-X syndrome, 101
  cardio-facio-cutaneous syndrome, 153
  Cornelia de Lange syndrome, 219–220
  fetal anticonvulsant syndrome, 438
  Noonan syndrome, 659–661
  trisomy 13 and 18 syndromes, 950
Hemihyperplasia, 125
  Beckwith–Wiedemann syndrome, 134
  development and behavior problems, 135

diagnostic criteria, 127
diagnostic testing, 132–133
differential diagnosis, 133
  Proteus syndrome, 766
etiology, pathogenesis, and genetics, 131, 132
α-fetoprotein (AFP), 139
incidence, 125–126
neoplasia incidence, 137–139
Pallister–Killian syndrome, 725
Hemihypertrophy. *See* Hemihyperplasia
Hemoglobin H, ATR-X syndrome, 96
Hepatic anomalies:
  Alagille syndrome, 32
  fetal alcohol spectrum disorders, 417
  tuberous sclerosis complex, 972
  Wolf–Hirschhorn syndrome, 276
Hepatoblastoma:
  Beckwith–Wiedemann syndrome, 138
  trisomy 18 syndrome, 950
Hepatomegaly, Alagille syndrome, 37
Hereditary benign telangiectasia, 478
Hereditary hemorrhagic telangiectasia, 475–486
  development and behavior problems, 478
  diagnostic criteria, 475
  diagnostic testing, 477
  differential diagnosis, 477–478
  etiology, pathogenesis, and genetics, 476–477
  gastrointestinal anomalies, 482–483
  growth and feeding problems, 478
  immunologic conditions, 484
  incidence, 475
  manifestations and management, 478–484
  neurological anomalies, 481–482
  ophthalmologic anomalies, 483–484
  otorhinolaryngologic anomalies, 478–479
  respiratory anomalies, 479–481
Hereditary progressive arthroophthalmopathy. *See* Stickler syndrome:
Hermansky–Pudlak syndrome, 45
  dermatologic anomalies, 51–52
  diagnostic criteria, 47
  diagnostic testing, 49
  etiology, pathogenesis, and genetics, 47
  gastrointestinal anomalies, 55
  hematologic anomalies, 53–54
  incidence, 46
  ophthalmologic anomalies, 53
  pulmonary anomalies, 54
Hernias:
  Marfan syndrome, 592
  Proteus syndrome, 770
  Williams syndrome, 1032
Herpes, differential diagnosis, incontinentia pigmenti, 509
Hip dysplasia:
  Prader–Willi syndrome, 755
  Smith–Lemli–Opitz syndrome, 860
Hirschsprung disease:
  ATR-X syndrome, 100
  Down syndrome, 376
  Mowat–Wilson syndrome, 597, 601, 603
  Rubinstein–Taybi syndrome, 828
  Smith–Lemli–Opitz syndrome, 855, 859
Holoprosencephaly, 487–503
  craniofacial anomalies, 497

development and behavior problems, 495–497
diagnostic criteria, 488–489
diagnostic testing, 493–494
differential diagnosis, 494
  Pallister–Hall syndrome, 710
  trisomy 13 syndrome, 943
ear anomalies, 500
endocrine anomalies, 497–498
etiology, pathogenesis, and genetics, 489–493
gastrointestinal anomalies, 500
genetic counseling, 494
growth and feeding problems, 494–495
hearing disorders, 500
immunologic conditions, 500–501
incidence, 487–488
manifestations and management, 494–501
neurological anomalies, 498–500
ophthalmologic anomalies, 500
Smith–Lemli–Opitz syndrome, 857
trisomy 13 syndrome, 942, 943, 947
Holt–Oram syndrome, differential diagnosis, Pallister–Hall syndrome, 710
Holzgreve syndrome, differential diagnosis, Pallister–Hall syndrome, 710
Homocystinuria, differential diagnosis, Marfan syndrome, 586
Homozygous achondroplasia, 11, 26
Hormone replacement therapy:
  Prader–Willi syndrome, 752
  Turner syndrome, 984
HPS1 gene, Hermansky–Pudlak syndrome, 49
HRAS gene mutations, Costello syndrome, 226
Hugging behavior, Smith–Magenis syndrome, 867, 874
Human chorionic gonadotropin:
  Prader–Willi syndrome, 751
  trisomy 18 syndrome, 941
Hydrocephalus:
  achondroplasia, 17–18
  Greig cephalopolysyndactyly syndrome, 713, 714
Hyperactivity:
  Aarskog syndrome, 5
  Angelman syndrome, 67–68
  deletion 5p syndrome, 284
  fragile X syndrome, 447
Hyperacusis:
  deletion 5p syndrome, 286
  Smith–Magenis syndrome, 881
Hypercalcemia, Williams syndrome, 1029, 1031
Hypercalciuria, Williams syndrome, 1029
Hypercholesterolemia:
  Alagille syndrome, 32, 37
  Smith–Magenis syndrome, 870, 885
Hyperextensible joints. *See* Joint laxity
Hypergammaglobulinemia, deletion 22q11.2 syndrome, 305
Hypergonadotropic hypogonadism, Klinefelter syndrome, 547–548
Hyperlipidemia:
  Bardet–Biedl syndrome, 117
  Beckwith–Wiedemann syndrome, 135
Hyperlordosis, achondroplasia, 25
Hypermenorrhagia, Rubinstein–Taybi syndrome, 831

Hypermetrorrhagia, Rubinstein–Taybi syndrome, 831
Hyperopia:
  Phelan–McDermid syndrome, 327
  Prader–Willi syndrome, 738
  Turner syndrome, 989
  Williams syndrome, 1032
Hyperphagia:
  Bardet–Biedl syndrome, 112
  Prader–Willi syndrome, 746–749
Hyperphosphatasia with mental retardation-1. *See* Mabry syndrome
Hyperreflexia, Cornelia de Lange syndrome, 219
Hypertension:
  Bardet–Biedl syndrome, 116
  fragile X syndrome, 451
  neurofibromatosis type 1, 645
  Phelan–McDermid syndrome, 328
  Turner syndrome, 983
  Williams syndrome, 1030
Hyperthyroidism, deletion 22q11.2 syndrome, 298
Hypertonia:
  Cornelia de Lange syndrome, 219
  deletion 5p syndrome, 285
  Pallister–Killian syndrome, 718
  trisomy 13 and 18 syndromes, 947
  Williams syndrome, 1028
Hypertriglyceridemia, Alagille syndrome, 37
Hypertrophic cardiomyopathy:
  cardio-facio-cutaneous syndrome, 151
  Noonan syndrome, 657
Hypocalcemia:
  Beckwith–Wiedemann syndrome, 135
  deletion 22q11.2 syndrome, 306
Hypocholesterolemia, Smith–Lemli–Opitz syndrome, 853
Hypochondroplasia, differential diagnosis, achondroplasia, 11
Hypogammaglobulinemia, deletion 22q11.2 syndrome, 305
Hypogenitalism:
  Bardet–Biedl syndrome, 107, 108, 116
  Smith–Lemli–Opitz syndrome, 851, 854
Hypoglycemia:
  Beckwith–Wiedemann syndrome, 136
  Costello syndrome, 235
  Silver–Russell syndrome, 846
  Sotos syndrome, 904
Hypogonadism:
  CHARGE syndrome, 166
  Prader–Willi syndrome, 736, 750–751
Hypomelanosis of Ito, differential diagnosis, inverted duplicated chromosome 15 syndrome, 518
Hypomelanotic macules, tuberous sclerosis complex, 958
Hypoparathyroidism, deletion 22q11.2 syndrome, 306
Hypophosphatasia, differential diagnosis, osteogenesis imperfecta, 693–694
Hypopigmentation:
  albinism, 46
  Angelman syndrome, 64
  tuberous sclerosis complex, 963, 968

Hypoplasia:
  cerebellar vermis, Bardet–Biedl syndrome, 118
  genital, Prader–Willi syndrome, 736
Hypospadias:
  Rubinstein–Taybi syndrome, 831
  Smith–Lemli–Opitz syndrome, 859, 860
Hypothalamic hamartoma, Pallister–Hall syndrome, 711
Hypothalamic-pituitary function, CHARGE syndrome, 166
Hypothyroidism:
  Bardet–Biedl syndrome, 117
  Beckwith–Wiedemann syndrome, 135
  deletion 1p36 syndrome, 260
  deletion 22q11.2 syndrome, 298, 306
  differential diagnosis, Down syndrome, 361
  Down syndrome, 371
  Klinefelter syndrome, 549
  Noonan syndrome, 664
  Phelan–McDermid syndrome, 323, 325
  Prader–Willi syndrome, 743
  Williams syndrome, 1029, 1030
Hypotonia:
  achondroplasia, 16
  Angelman syndrome, 65
  arthrogryposis, 75
  ATR-X syndrome, 98
  cardio-facio-cutaneous syndrome, 147
  Coffin–Siris syndrome, 185
  Cohen syndrome, 202
  Costello syndrome, 232
  deletion 1p36 syndrome, 258
  deletion 5p syndrome, 285, 286
  deletion 22q11.2 syndrome, 302
  Down syndrome, 356, 361
  Ehlers–Danlos syndromes, 398
  47,XXY syndrome, 543
  fragile X syndrome, 447
  inverted duplicated chromosome 15 syndrome, 518, 522
  myotonic dystrophy type 1, 612
  Pallister–Killian syndrome, 718, 726–727
  Phelan–McDermid syndrome, 325
  Prader–Willi syndrome, 742, 749
  Rett syndrome, 791
  Rubinstein–Taybi syndrome, 831
  Silver–Russell syndrome, 844
  Smith–Magenis syndrome, 883
  Sotos syndrome, 905, 907
  trisomy 13 and 18 syndromes, 947
  Williams syndrome, 1028
  Wolf–Hirschhorn syndrome, 272
Hypoxemia, achondroplasia, 20
Hypoxia-inducible factor, von Hippel–Lindau syndrome, 1006

Idiopathic hypercalcemia. *See* Williams syndrome
IGF2 gene mutation, Beckwith–Wiedemann syndrome, 133
IGFBP2 gene, Pallister–Killian syndrome, 721
IKBKG gene, incontinentia pigmenti, 505, 507–508
Ilizarov technique, Proteus syndrome, 768

Immunoglobulin:
  deletion 22q11.2 syndrome, 305
  Smith–Magenis syndrome, 887
Immunologic conditions:
  CHARGE syndrome, 166–167
  Coffin–Siris syndrome, 191
  Cohen syndrome, 202
  Cornelia de Lange syndrome, 219–220
  deletion 22q11.2 syndrome, 305–306
  Down syndrome, 379–381
  49,XXXXY syndrome, 557
  hereditary hemorrhagic telangiectasia, 484
  holoprosencephaly, 500–501
  incontinentia pigmenti, 512
  Kabuki syndrome, 536
  Klinefelter syndrome, 551–552
  Loeys–Dietz syndrome, 573
  Mowat–Wilson syndrome, 605–606
  Phelan–McDermid syndrome, 330
  Rubinstein–Taybi syndrome, 828
  Smith–Magenis syndrome, 887
  Turner syndrome, 987
  Wolf–Hirschhorn syndrome, 275
Imprinting:
  Angelman syndrome, 61, 62, 64
  Beckwith–Wiedemann syndrome, 127, 131–132
  Prader–Willi syndrome, 739, 740
  Silver–Russell syndrome, 839
Incontinence:
  Mowat–Wilson syndrome, 603
  myotonic dystrophy type 1, 620
  Phelan–McDermid syndrome, 323
Incontinentia pigmenti, 505–514
  breast anomalies, 512
  dental anomalies, 512
  dermatologic anomalies, 509–510
  development and behavior problems, 509
  diagnostic criteria, 505–506
  diagnostic testing, 508
  differential diagnosis, 508–509
  etiology, pathogenesis, and genetics, 505–508
  growth and feeding problems, 509
  hair and nails, anomalies of, 510
  incidence, 505
  manifestations and management, 509–512
  neurological anomalies, 511–512
  ophthalmologic anomalies, 510–511
Induced pluripotent stem cells (IPSCs), Angelman syndrome, 64
Infantile hypotonia. *See* Hypotonia
Infantile neuronal ceroid lipofuscinosis (Batten disease), differential diagnosis, Rett syndrome, 797
Inflammatory bowel disease, Turner syndrome, 988
ING4 gene, Pallister–Killian syndrome, 720, 721
Insulin-like growth factor-1 (IGF-1):
  Beckwith–Wiedemann syndrome, 129
  Phelan–McDermid syndrome, 325
Insulin therapy, Phelan–McDermid syndrome, 325
Integumentary. *See* Dermatologic anomalies
Intellectual disabilities:
  Aarskog syndrome, 5

Apert syndrome, 247
ATR-X syndrome, 93
CHARGE syndrome, 162
Coffin–Lowry syndrome, 171
Cohen syndrome, 195
Cornelia de Lange syndrome, 215
deletion 1p36 syndrome, 257
deletion 5p syndrome, 283
Down syndrome, 363
fragile-X syndrome, 443
inverted duplicated chromosome 15 syndrome, 515
Kabuki syndrome, 529
Mowat–Wilson syndrome, 597
neurofibromatosis type 1, 637
oculo-auriculo-vertebral spectrum, 674
Prader–Willi syndrome, 735
Rett syndrome, 791, 799
Smith–Lemli–Opitz syndrome, 856
Smith–Magenis syndrome, 865, 872
tuberous sclerosis complex, 964
Turner syndrome, 985
WAGR syndrome, 344
Williams syndrome, 1022
Wolf–Hirschhorn syndrome, 271
Intention tremor, Sotos syndrome, 907
Interstitial pulmonary fibrosis, Hermansky–Pudlak syndrome, 54
Intracrancial anomalies, Cohen syndrome, 202
Intracranial bleeding:
    Alagille syndrome, 38–39
    hereditary hemorrhagic telangiectasia, 481
Intracranial hypotension syndrome, Marfan syndrome, 591
Intracranial lesions, tuberous sclerosis complex, 965–966
Intracranial tumors, von Hippel–Lindau syndrome, 1010
Intrahepatic cholestasis, Alagille syndrome, 31
Intrauterine growth retardation:
    Cornelia de Lange syndrome, 212
    Marfan syndrome, 592
    Silver–Russell syndrome, 837, 843
    trisomy 18 syndrome, 940
    velo-cardio-facial syndrome, 304
Intrauterine movement, arthrogryposis and decrease in, 76
Inverted duplicated chromosome 15 syndrome, 515–528
    dermatologic anomalies, 524–525
    development and behavior problems, 519–522
    diagnostic criteria, 515–516
    diagnostic testing, 518
    differential diagnosis, 518–519
    endocrine anomalies, 524
    etiology, pathogenesis, and genetics, 516–517
    growth and feeding problems, 519
    incidence, 515
    major malformations, 523–524
    manifestations and management, 519–525
    musculoskeletal anomalies, 524
    neurological anomalies, 522–523
    ophthalmologic anomalies, 524
    psychiatric problems, 519–522
    respiratory anomalies, 524

IQ levels. See Cognitive function
Ischemic heart disease, Turner syndrome, 982
Isochromosome 12p, Pallister–Killian syndrome, 718, 720
Isodicentric 15 syndrome. See Inverted duplicated chromosome 15 syndrome
Isolated hemihyperplasia. See Hemihyperplasia

Jackson–Weiss syndrome, craniosynostosis syndromes, 243
Jaffe–Campanacci syndrome, differential diagnosis, neurofibromatosis type 1, 635
JAG1 gene mutation, Alagille syndrome, 31, 34, 35, 38
Jaundice:
    Alagille syndrome, 36
    fetal alcohol spectrum disorders, 417
    Sotos syndrome, 897, 904
Jaw anomalies:
    achondroplasia, 26
    arthrogryposis, 85, 87
    Gorlin syndrome, 467
Joint hypermobility. See Joint laxity
Joint laxity:
    Aarskog syndrome, 6
    achondroplasia, 25
    Bardet–Biedl syndrome, 115
    Cohen syndrome, 199, 202, 203
    Costello syndrome, 233
    Ehlers–Danlos syndromes, 389, 390, 398
    fragile X syndrome, 452
    Kabuki syndrome, 533, 536
    Loeys–Dietz syndrome, 563
    Marfan syndrome, 587
    Pallister–Killian syndrome, 728
    Rubinstein–Taybi syndrome, 831
    Sotos syndrome, 910
    Stickler syndrome, 922
    Williams syndrome, 1033
Joubert syndrome:
    differential diagnosis, Bardet–Biedl syndrome, 112
    and Smith–Magenis syndrome, 882
Juberg–Marsidi syndrome, differential diagnosis, ATR-X syndrome, 102
Juvenile myelomonocytic leukemia:
    neurofibromatosis type 1, 640
    Noonan syndrome, 660
Juvenile polyposis:
    differential diagnosis, *PTEN* hamartoma tumor syndrome, 781
    hereditary hemorrhagic telangiectasia, 476, 483

Kabuki syndrome, 529–538
    cardiovascular anomalies, 533–534
    craniofacial anomalies, 535
    dental anomalies, 535
    development and behavior problems, 532–533
    diagnostic criteria, 529–530
    diagnostic testing, 531
    differential diagnosis, 531–532
        deletion 22q11.2 syndrome, 298
        Williams syndrome, 1024

    ear anomalies, 534–535
    endocrine anomalies, 534
    etiology, pathogenesis, and genetics, 530–531
    gastrointestinal anomalies, 535–536
    genitourinary anomalies, 536
    growth and feeding problems, 532
    hearing disorders, 534–535
    immunologic conditions, 536
    incidence, 529
    manifestations and management, 532–536
    musculoskeletal anomalies, 536
    neurological anomalies, 533
    ophthalmologic anomalies, 534
Kallmann syndrome:
    CHD7 mutations, 160
    differential diagnosis, Klinefelter syndrome, 541
KDM6A gene mutations, Kabuki syndrome, 530, 531
Keloids, Rubinstein–Taybi syndrome, 830, 831
Keratoconus:
    Angelman syndrome, 70
    Down syndrome, 373
Klinefelter syndrome, 539–562
    dermatologic anomalies, 552
    development and behavior problems, 543–547
    diagnostic criteria, 540
    diagnostic testing, 541
    differential diagnosis, 541–542
        Sotos syndrome, 901
    endocrine anomalies, 547–550
    etiology, pathogenesis, and genetics, 541
    growth and feeding problems, 542–543
    immunologic conditions, 551–552
    incidence, 539–540
    incontinentia pigmenti, 507, 508
    language difficulties, 544
    manifestations and management, 542–559
    musculoskeletal anomalies, 550–551
    neoplasia incidence, 551
    neurological anomalies, 545–547
    pulmonary anomalies, 552
Klippel–Trenaunay syndrome, Proteus syndrome, 766
KMT2A gene mutations, Cornelia de Lange syndrome, 212
KMT2D gene mutations, Kabuki syndrome, 530, 531
Knee instability, achondroplasia, 23–24
KRAS mutations:
    cardio-facio-cutaneous syndrome, 148
    Noonan syndrome, 653
Kyphoscoliosis:
    Bardet–Biedl syndrome, 115. See also scoliosis
    Coffin–Lowry syndrome, 172, 179, 180, 181
    Cohen syndrome, 203
    Costello syndrome, 234
    Ehlers–Danlos syndromes, 398
    Gorlin syndrome, 470
    Sotos syndrome, 910
Kyphosis:
    achondroplasia, 22–23
    arthrogryposis, 85
    ATR-X syndrome, 102

Kyphosis: (cont'd)
    Kabuki syndrome, 536
    Marfan syndrome, 588
    Prader–Willi syndrome, 738
    Rett syndrome, 802

LAMB2 mutations, Pierson syndrome, 342
Lamotrigine, fetal anticonvulsant syndrome, 433
Language difficulties:
    achondroplasia, 17
    Angelman syndrome, 67
    Costello syndrome, 232
    deletion 22q11.2 syndrome, 300
    Down syndrome, 363
    fragile X syndrome, 447
    holoprosencephaly, 496
    inverted duplicated chromosome 15 syndrome, 520
    Kabuki syndrome, 532
    Klinefelter syndrome, 544
    Mowat–Wilson syndrome, 602
    myotonic dystrophy type 1, 618
    Noonan syndrome, 657
    Phelan–McDermid syndrome, 323
    Prader–Willi syndrome, 736, 744
    Smith–Magenis syndrome, 873
    Sotos syndrome, 905
    Turner syndrome, 986
Large for gestational age:
    Beckwith-Wiedemann syndrome, 134
    fetal anticonvulsant syndrome, 434
    Sotos syndrome, 901
Larsen syndrome, differential diagnosis, Kabuki syndrome, 531
Laryngomalacia:
    CHARGE syndrome, 164
    Coffin–Siris syndrome, 190
Laser photocoagulation:
    hereditary hemorrhagic telangiectasia, 478
    Stickler syndrome, 921
    von Hippel–Lindau syndrome, 1013
Laughter, spontaneous, Angelman syndrome, 61, 67
Laurence–Moon syndrome. *See* Bardet–Biedl syndrome
Learning disabilities. *See also* Cognitive function; Intellectual disabilities; Mental retardation
    achondroplasia, 17
    Bardet–Biedl syndrome, 113
    CHARGE syndrome, 161
    Cohen syndrome, 197
    deletion 22q11.2 syndrome, 301
    Down syndrome, 373
    holoprosencephaly, 495–496
    Klinefelter syndrome, 544
    myotonic dystrophy type 1, 618
    neurofibromatosis type 1, 637
    Noonan syndrome, 656
    Sotos syndrome, 906
    tuberous sclerosis complex, 964
    Turner syndrome, 986
Legius syndrome, differential diagnosis, neurofibromatosis type 1, 634–635
Leg ulceration, Klinefelter syndrome, 552
Lennox–Gestaut syndrome:
    inverted duplicated chromosome 15 syndrome, 520, 522
    Phelan–McDermid syndrome, 326
Lentigines:
    albinism, 51
    Noonan syndrome, 663
LEOPARD syndrome:
    differential diagnosis
        Aarskog syndrome, 4
        Noonan syndrome, 654
LETM1 gene, Wolf–Hirschhorn syndrome, 268
Leukemia, Down syndrome, 381–382
Levetiracetam:
    fetal anticonvulsant syndrome, 434
    Smith–Magenis syndrome, 884
Lhermitte–Duclos disease, Cowden syndrome, 785–786
"Lick and flip" behavior, Smith–Magenis syndrome, 867
Li–Fraumeni syndrome, Wilms tumor, 343
Limb anomalies. *See also* Hand anomalies
    cardio-facio-cutaneous syndrome, 152
    CHARGE syndrome, 159
    Coffin–Siris syndrome, 190
    Cornelia de Lange syndrome, 216–217
    craniosynostosis syndromes, 241
    deletion 1p36 syndrome, 259
    fetal anticonvulsant syndrome, 437
    Grieg cephalopolysyndactyly syndrome, 714
    Pallister–Hall syndrome, 712–713
    Pallister–Killian syndrome, 718
    Rett syndrome, 802
    Silver–Russell syndrome, 845–846
    trisomy 13 and 18 syndromes, 951
    VATER/VACTERL association, 999, 1000
    Wolf–Hirschhorn syndrome, 274
    WT1-related disorders, 345
Lim kinase 1 (*LIMK1*) gene mutation, Williams syndrome, 1023
Lipodermoids, oculo-auriculo-vertebral spectrum, 677
Lithium:
    Phelan–McDermid syndrome, 325
    Smith–Magenis syndrome, 877
Little People of America, Medical Advisory Board, achondroplasia, 12
Liver disease:
    Alagille syndrome, 31
    hereditary hemorrhagic telangiectasia, 482
    Smith–Lemli–Opitz syndrome, 855
    Turner syndrome, 988–989
Liver transplants, Alagille syndrome, 37, 38
Lobar holoprosencephaly, 487
    diagnostic criteria, 488
    incidence, 488
Loeys–Dietz syndrome, 563–576
    cardiovascular anomalies, 570–572
    craniofacial anomalies, 570
    dermatologic anomalies, 573
    development and behavior problems, 569
    diagnostic criteria, 563–564
    diagnostic testing, 566–567
    differential diagnosis, 567–568
        Ehlers–Danlos syndromes, 398
        Marfan syndrome, 585
    etiology, pathogenesis, and genetics, 564–566
    genetic counseling, 574
    growth and feeding problems, 569
    immunologic conditions, 573
    incidence, 563
    manifestations and management, 569–574
    musculoskeletal anomalies, 569–570
    neurological anomalies, 572
    ophthalmologic anomalies, 572
    pregnancy, 573–574
    respiratory anomalies, 572–573
Long bone bowing, neurofibromatosis type 1, 643, 644
Lordosis, Phelan–McDermid syndrome, 329
Losartan, Marfan syndrome, 589
LOXL3 gene mutations, Stickler syndrome, 918
Lujan–Fryns syndrome, differential diagnosis, Marfan syndrome, 586
Lumbosacral spinal stenosis, achondroplasia, 23
Lung transplantation, tuberous sclerosis complex, 972
Luteinizing hormone, Klinefelter syndrome, 548
Lymphangioleiomyomatosis, tuberous sclerosis complex, 959, 971, 972
Lymphatic disorders:
    Noonan syndrome, 661–662
    Turner syndrome, 983–984
Lymphedema:
    Noonan syndrome, 661
    Phelan–McDermid syndrome, 328, 329
    Sotos syndrome, 909
    Turner syndrome, 980, 983, 984
Lymphoreticular malignancy, Sotos syndrome, 910

Mabry syndrome, differential diagnosis, Coffin–Siris syndrome, 187
McCune–Albright syndrome, differential diagnosis, neurofibromatosis type 1, 634, 635
McKusick–Kaufman syndrome, differential diagnosis:
    Bardet–Biedl syndrome, 112
    Pallister–Hall syndrome, 710
Macrocephaly:
    achondroplasia, 17
    Bardet–Biedl syndrome, 108
    Costello syndrome, 228
    Gorlin syndrome, 464
    Greig cephalopolysyndactyly syndrome, 707
    inverted duplicated chromosome 15 syndrome, 523
    PTEN hamartoma tumor syndrome, 780, 786
    tuberous sclerosis complex, 964
Macroglossia, Beckwith–Wiedemann syndrome, 134, 136
Macroorchidism, fragile X syndrome, 452
Maculopathy, Bardet–Biedl syndrome, 114
Maffucci syndrome, differential diagnosis, hemihyperplasia, 766
Male pseudohermaphrodites, ATR-X syndrome, 101
Male X chromosome aneuploidy, 552–559
    46,XY/47,XXY mosaicism, 552

48,XXXY syndrome, 553–556
48,XXYY syndrome, 552–553
49,XXXXY syndrome, 556–559
Malignancy. *See* Neoplasia
Malignant hyperthermia:
 arthrogryposis, 87
 Noonan syndrome, 659
Malignant peripheral nerve sheath tumors, neurofibromatosis type 1, 638–639
Mandibulofacial dysostoses syndromes, 809, 929
Marfan syndrome, 577–596
 cardiovascular anomalies, 588–590
 dermatologic anomalies, 592
 development and behavior problems, 587
 diagnostic criteria, 577–580
 diagnostic testing, 582–583
 differential diagnosis, 583–586
  Ehlers–Danlos syndromes, 398
  Loeys–Dietz syndrome, 567
  Sotos syndrome, 901
 etiology, pathogenesis, and genetics, 580–582
 growth and feeding problems, 586–587
 incidence, 577
 manifestations and management, 586–593
 musculoskeletal anomalies, 587–588
 neonatal, 583–584
 neurological anomalies, 591
 ophthalmologic anomalies, 590–591
 pregnancy management, 592–593
 respiratory anomalies, 591–592
Marshall syndrome, differential diagnosis, Stickler syndrome, 919
Maternal age:
 Down syndrome, 355, 358, 359, 360
 inverted duplicated chromosome 15 syndrome, 516, 517
 Klinefelter syndrome, 541
 Pallister–Killian syndrome, 720, 722
 Prader–Willi syndrome, 740
 trisomy 18 syndrome, 939, 940
Meacham syndrome, 335
 cardiovascular anomalies, 346
 diagnostic criteria, 337, 338
 diagnostic testing, 341–342
 differential diagnosis, 342–343
 etiology, pathogenesis, and genetics, 339–341
 genitourinary anomalies, 346–347
 incidence, 335
 pulmonary anomalies, 346
Meckel–Gruber syndrome, differential diagnosis, trisomy 13 syndrome, 943
Meckel's diverticulum, trisomy 18 syndrome, 951
Meckel syndrome, differential diagnosis, Smith–Lemli–Opitz syndrome, 854
MECP2 gene mutation, Rett syndrome, 791, 795, 796
Medroxy-progesterone, tuberous sclerosis complex, 972
Medulloblastoma, Gorlin syndrome, 468
Meige disease, differential diagnosis, Turner syndrome, 980
MEK1/ MED2 mutations:
 cardio-facio-cutaneous syndrome, 148, 149
 Costello syndrome, 227
Melanin anomalies, albinism, 45, 47

Melatonin:
 inverted duplicated chromosome 15 syndrome, 522
 Rett syndrome, 799
 Smith–Magenis syndrome, 879–880
 tuberous sclerosis complex, 965
 Williams syndrome, 1026, 1028
Meningiomas, Proteus syndrome, 771
Menopause, premature, fragile X syndrome, 453
Mental retardation. *See also* Cognitive function; Intellectual disabilities; Learning disabilities
 Alagille syndrome, 32, 36
 ATR-X syndrome, 93–105
 Bardet–Biedl syndrome, 108
 Coffin–Lowry syndrome, 174
 Cohen syndrome, 200
 trisomy 13 and 18 syndromes, 943
 WAGR syndrome, 344
Metabolic disorders, differential diagnosis, Alagille syndrome, 35
Metastatic disease, von Hippel–Lindau syndrome, 1013
Metatarsus adduction, Aarskog syndrome, 6
Metformin, fragile X syndrome, 446
Methionine amino peptidase 2 (METAP2) inhibitor, Prader–Willi syndrome, 747
Methotrexate, Robin sequence, 811
Methylation-specific multiplex ligation-dependent probe amplification (MS-MLPA), Prader–Willi syndrome, 740–741
Methylation-specific polymerase chain reaction, Beckwith–Wiedemann syndrome, 132
Methylene tetrahydrofolate reductase gene (*MTHFR*) polymorphisms, trisomy 18 syndrome, 939
Metopic synostosis, differential diagnosis, craniosynostosis syndromes, 246
Microcephaly. *See also* Craniofacial anomalies
 Angelman syndrome, 69
 ATR-X syndrome, 100
 Coffin–Lowry syndrome, 172
 Cohen syndrome, 202
 holoprosencephaly, 498
 Mowat–Wilson syndrome, 601
 Silver–Russell syndrome, 842
 Smith–Lemli–Opitz syndrome, 857
 Wolf–Hirschhorn syndrome, 266
Microdeletion:
 Angelman syndrome, 64
 Beckwith–Wiedemann syndrome, 132
 Coffin–Siris syndrome, 187
 craniosynostosis syndromes, 246
 deletion 22q11.2 syndrome, 291
 holoprosencephaly, 492
 neurofibromatosis type 1, 637
 Prader–Willi syndrome, 735
 Robin sequence, 810
 Rubinstein–Taybi syndrome, 825
 Wolf–Hirschhorn syndrome, 269
Micrognathia:
 CHARGE syndrome, 164
 Cornelia de Lange syndrome, 212, 216
 Robin sequence, 808, 810, 814, 817–818
 Silver–Russell syndrome, 845

Smith–Lemli–Opitz syndrome, 857
Microhamartomatous rectal polyps, tuberous sclerosis complex, 972
Microphthalmia, CHARGE syndrome, 163
Microstomia, Cornelia de Lange syndrome, 216
Miller syndrome, differential diagnosis:
 Robin sequence, 809
 Treacher Collins syndrome, 930
Milroy disease, differential diagnosis, Turner syndrome, 980
Minocycline, fragile X syndrome, 445, 446
Mitochondrial myopathy, Prader–Willi syndrome, 749
Mitral valve prolapse:
 fragile X syndrome, 451
 Sotos syndrome, 909
 Stickler syndrome, 916, 923
Modafinil:
 myotonic dystrophy type 1, 618, 619
 Smith–Magenis syndrome, 879
Mosaic genome-wide paternal uniparental isodisomy, differential diagnosis, Beckwith–Wiedemann syndrome, 133
Mosaicism:
 Alagille syndrome, 34
 Down syndrome, 358
 46,XY/47,XXY, 552
 WT1-related disorders, 341
Mosaic overgrowth disorders, Proteus syndrome, 763–773
Mosaic trisomy 12p, Pallister–Killian syndrome, 722, 724
Motor development problems:
 achondroplasia, 17, 24
 Coffin–Lowry syndrome, 181
 48,XXXY syndrome, 554
 Phelan–McDermid syndrome, 323–324
 Sotos syndrome, 905
 Williams syndrome, 1026
Mowat–Wilson syndrome, 597–609
 cardiovascular anomalies, 604
 development and behavior problems, 602–603
 diagnostic criteria, 598
 diagnostic testing, 599–600
 differential diagnosis, 601
  Angelman syndrome, 66
 ear anomalies, 605
 endocrine anomalies, 605
 etiology, pathogenesis, and genetics, 598
 gastrointestinal anomalies, 603
 genetic counseling, 600–601
 genitourinary anomalies, 604–605
 growth and feeding problems, 601–602
 hearing disorders, 605
 immunologic conditions, 605–606
 incidence, 597–598
 manifestations and management, 601–606
 musculoskeletal anomalies, 605
 neurological anomalies, 603–604
 ophthalmologic anomalies, 605
Moyamoya disease, and Smith–Magenis syndrome, 882
MSX1 gene, Wolf–Hirschhorn syndrome, 269
MTOR inhibitors, tuberous sclerosis complex, 967, 972

Mucocutaneous lesions, *PTEN* hamartoma tumor syndrome, 785
Mucopolysaccharidosis MPSIII (Sanfillippo syndrome), differential diagnosis, Sotos syndrome, 901
Muenke syndrome:
  development and behavior problems, 247
  diagnostic criteria, 243
  diagnostic testing, 246
  etiology, pathogenesis, and genetics, 245
  growth and feeding problems, 246–247
  hearing disorders, 248
Multifocal micronodular pneumocyte hyperplasia, tuberous sclerosis complex, 971
Multi-locus imprinting disorder, Beckwith–Wiedemann syndrome, 131
Multiple lentingines syndrome, differential diagnosis, neurofibromatosis type 1, 635
Multiplex ligation-dependent probe amplification (MLPA), WAGR syndrome, 321, 342
Muscular dystrophy, myotonic dystrophy type 1, 611
Musculoskeletal anomalies:
  Aarskog syndrome, 6
  achondroplasia, 9, 22–26
  Alagille syndrome, 33, 40–41
  Angelman syndrome, 70
  arthrogryposis, 85–87
  ATR-X syndrome, 102
  Bardet–Biedl syndrome, 115
  cardio-facio-cutaneous syndrome, 152–153
  Coffin–Lowry syndrome, 181
  Coffin–Siris syndrome, 190
  Cohen syndrome, 203–204
  Cornelia de Lange syndrome, 216–217
  Costello syndrome, 233–234
  craniosynostosis syndromes, 249
  deletion 1p36 syndrome, 259–260
  deletion 5p syndrome, 286
  deletion 22q11.2 syndrome, 307–308
  Down syndrome, 373–375
  fetal alcohol spectrum disorders, 417
  fetal anticonvulsant syndrome, 437–438
  49,XXXXY syndrome, 557
  fragile X syndrome, 452–453
  Gorlin syndrome, 470
  Greig cephalopolysyndactyly syndrome, 714
  incontinentia pigmenti, 512
  inverted duplicated chromosome 15 syndrome, 524
  Kabuki syndrome, 536
  Klinefelter syndrome, 550–551
  Loeys–Dietz syndrome, 569–570
  Marfan syndrome, 587–588
  Mowat–Wilson syndrome, 605
  myotonic dystrophy type 1, 622–623
  neurofibromatosis type 1, 642–644
  Noonan syndrome, 662–663
  oculo-auriculo-vertebral spectrum, 678
  osteogenesis imperfecta, 695–698
  Pallister–Hall syndrome, 712–713
  Pallister–Killian syndrome, 728–729
  Phelan–McDermid syndrome, 328–329

  Prader–Willi syndrome, 754–755
  Proteus syndrome, 767–768
  Rett syndrome, 802
  Rubinstein–Taybi syndrome, 831–832
  Silver–Russell syndrome, 845–846
  Smith–Lemli–Opitz syndrome, 860
  Smith–Magenis syndrome, 885
  Sotos syndrome, 910–911
  Stickler syndrome, 922–923
  trisomy 13 and 18 syndromes, 951–952
  VATER/VACTERL association, 999–1000
  Williams syndrome, 1033
  Wolf–Hirschhorn syndrome, 274–275
  WT1-related disorders, 345
Music:
  cardio-facio-cutaneous syndrome, 150
  fragile X syndrome, 448, 452
  inverted duplicated chromosome 15 syndrome, 521
  Mowat–Wilson syndrome, 602
  Phelan–McDermid syndrome, 324, 325
  Rett syndrome, 800
  Williams syndrome, 1028
Myeloproliferative disorder, Noonan syndrome, 660
MYO15A gene mutation, Smith–Magenis syndrome, 870
Myopathy:
  deletion 1p36 syndrome, 259
  differential diagnosis
    Angelman syndrome, 65
    Costello syndrome, 230, 234
  Noonan syndrome, 659
Myopia:
  ATR-X syndrome, 103
  cardio-facio-cutaneous syndrome, 152
  Cohen syndrome, 201
  Cornelia de Lange syndrome, 218
  Loeys–Dietz syndrome, 572
  Marfan syndrome, 590, 591
  Phelan–McDermid syndrome, 327
  Prader–Willi syndrome, 738
  Smith–Magenis syndrome, 885
  Stickler syndrome, 920–921
  Turner syndrome, 989
Myotonic dystrophy type 1, 611–627
  anesthesia risks, 624
  cardiovascular anomalies, 621–622
  dental anomalies, 619
  dermatologic anomalies, 622
  development and behavior problems, 618
  diagnostic criteria, 611–615
    adult onset, 614–615
    childhood, 613
    congenital, 612–613
    late-onset, 615
    subtypes, 611–612
  diagnostic testing, 615, 617
  differential diagnosis, 615–616
  ear anomalies, 622
  endocrine anomalies, 623
  etiology, pathogenesis, and genetics, 615
  gastrointestinal anomalies, 619–620
  genetic counseling, 616–617
  growth and feeding problems, 617

  hearing disorders, 622
  incidence, 611
  manifestations and management, 616–624
  musculoskeletal anomalies, 622–623
  neurological anomalies, 618–619
  neuromuscular weakness, 618
  ophthalmologic anomalies, 622
  pregnancy and delivery, 623–624
  respiratory anomalies, 620–621
Myotonic dystrophy type 2, differential diagnosis, myotonic dystrophy type 1, 616
Myringotomy:
  achondroplasia, 21
  oculo-auriculo-vertebral spectrum, 677

Nager syndrome:
  differential diagnosis, Treacher Collins syndrome, 929–930
  Robin sequence, 809
Nail dystrophy, incontinentia pigmenti, 510
Nail-patella syndrome, differential diagnosis, *WT1*-related disorders, 342
Neonatal hypotonia. *See* Hypotonia
Neoplasia:
  ATR-X syndrome, 103
  Bannayan–Riley–Ruvalcaba syndrome, 777
  Beckwith–Wiedemann syndrome, 137–139
  cardio-facio-cutaneous syndrome, 153
  Coffin–Lowry syndrome, 181–182
  Coffin–Siris syndrome, 191–192
  Costello syndrome, 227–228, 235–236
  deletion 1p36 syndrome, 261
  Down syndrome, 381–383
  fetal anticonvulsant syndrome, 438
  Gorlin syndrome, 468–469
  Klinefelter syndrome, 551
  neurofibromatosis type 1, 638–642
  Noonan syndrome, 659–661
  Pallister–Killian syndrome, 730
  Proteus syndrome, 771–772
  PTEN hamartoma tumor syndrome, 782–783
  Rubinstein–Taybi syndrome, 832
  Sotos syndrome, 910
  trisomy 13 and 18 syndromes, 950
  Turner syndrome, 981, 990
  von Hippel–Lindau syndrome, 1009, 1010–1014
  Wolf–Hirschhorn syndrome, 275–276
  WT1-related disorders, 341
Nephropathy, *WT1*-related disorders, 347–348
Nephrotic syndrome, differential diagnosis, *WT1*-related disorders, 342–343
Netchine–Harbison clinical scoring system, Silver–Russell syndrome, 838
Neuroblastoma:
  Beckwith–Wiedemann syndrome, 137
  Costello syndrome, 235–236
  deletion 1p36 syndrome, 261
  fetal anticonvulsant syndrome, 438
  Sotos syndrome, 910
  Turner syndrome, 990
Neurofibromatosis type 1, 629–649
  cardiovascular anomalies, 645
  development and behavior problems, 637–638

diagnostic criteria, 629–632
diagnostic testing, 634
differential diagnosis, 634–635
etiology, pathogenesis, and genetics, 632–634
genetic counseling, 633–634
growth and feeding problems, 636–637
incidence, 629
manifestations and management, 635–645
musculoskeletal anomalies, 642–644
neoplasia incidence, 638–642
Neurofibromatosis type 2:
    differential diagnosis, neurofibromatosis type 1, 635
    Phelan–McDermid syndrome, 320
Neurofibromin, neurofibromatosis type 1, 632, 633
Neurological anomalies:
    Aarskog syndrome, 6
    achondroplasia, 17–20
    Angelman syndrome, 69
    arthrogryposis, 87
    ATR-X syndrome, 100–101
    Bardet–Biedl syndrome, 118
    cardio-facio-cutaneous syndrome, 151
    Coffin–Lowry syndrome, 178–179
    Coffin–Siris syndrome, 188–190
    Cohen syndrome, 202
    Cornelia de Lange syndrome, 219
    Costello syndrome, 234–235
    craniosynostosis syndromes, 249
    deletion 1p36 syndrome, 257–258
    deletion 5p syndrome, 285
    deletion 22q11.2 syndrome, 304–305
    Down syndrome, 380–381
    fetal alcohol spectrum disorders, 417–418, 436
    49,XXXXY syndrome, 556–557
    fragile X syndrome, 449–450
    Greig cephalopolysyndactyly syndrome, 713–714
    hereditary hemorrhagic telangiectasia, 481–482
    holoprosencephaly, 498–500
    incontinentia pigmenti, 511–512
    inverted duplicated chromosome 15 syndrome, 522–523
    Kabuki syndrome, 533
    Loeys–Dietz syndrome, 572
    Marfan syndrome, 591
    Mowat–Wilson syndrome, 603–604
    myotonic dystrophy type 1, 618
    Noonan syndrome, 658–659
    oculo-auriculo-vertebral spectrum, 677–678
    osteogenesis imperfecta, 698
    Pallister–Hall syndrome, 711
    Pallister–Killian syndrome, 726–727
    Phelan–McDermid syndrome, 325–327
    Prader–Willi syndrome, 749–750
    Proteus syndrome, 769
    PTEN hamartoma tumor syndrome, 785–786
    Rett syndrome, 800–801
    Smith–Lemli–Opitz syndrome, 857
    Smith–Magenis syndrome, 882–884
    Sotos syndrome, 907
    trisomy 13 and 18 syndromes, 947–948
    tuberous sclerosis complex, 965–968

Williams syndrome, 1028–1029
Wolf–Hirschhorn syndrome, 271–272
WT1-related disorders, 344–345
Neutropenia, Cohen syndrome, 202, 203
Nevi:
    Gorlin syndrome, 465–466
    Turner syndrome, 990
Nevoid basal cell carcinoma syndrome. *See* Gorlin syndrome
Nevus anemicus, differential diagnosis, tuberous sclerosis complex, 963
Next generation sequencing:
    Mowat–Wilson syndrome, 599
    WT1-related disorders, 342
NF1 gene mutations, neurofibromatosis type 1, 630
Nicolaides–Baraitser syndrome, differential diagnosis:
    Coffin–Siris syndrome, 187
    Cornelia de Lange syndrome, 212, 213
NIPBL gene mutation, Cornelia de Lange syndrome, 210–211, 215
Nociception, impairment, Rett syndrome, 800–801
Noonan syndrome, 651–669
    cardiovascular anomalies, 657–658
    dermatologic anomalies, 663
    development and behavior problems, 656–657
    diagnostic criteria, 651–653
    diagnostic testing, 654
    differential diagnosis, 654–655
        Aarskog syndrome, 4
        cardio-facio-cutaneous syndrome, 149
        Costello syndrome, 230
        fetal anticonvulsant syndrome, 428
        neurofibromatosis type 1, 635
        Smith–Lemli–Opitz syndrome, 854
        Turner syndrome, 980
        Williams syndrome, 1024
    ear anomalies, 663–664
    etiology, pathogenesis, and genetics, 653–654
    genetic counseling, 654
    genitourinary anomalies, 662
    growth and feeding problems, 655–656
    hearing disorders, 663–664
    hematologic anomalies, 659–661
    immunologic conditions, 664
    incidence, 651
    lymphatic anomalies, 661–662
    manifestations and management, 655–664
    musculoskeletal anomalies, 662–663
    neoplasia incidence, 659–661
    neurological anomalies, 658–659
    ophthalmologic anomalies, 659
Nose, Greek warrior helmet appearance, Wolf–Hirschhorn syndrome, 266
Nose bleeds, Prader–Willi syndrome, 755
NOTCH2 mutations, Alagille syndrome, 31, 35
NSD1 gene mutation, Sotos syndrome, 895, 898, 899, 910
NSD2 gene mutation, Wolf–Hirschhorn syndrome, 267–268
Nyctalopia (night blindness). *See* Vision impairment
Nystagmus:

albinism, 48, 50, 52
Bardet–Biedl syndrome, 114
cardio-facio-cutaneous syndrome, 152
CHARGE syndrome, 163
oculocutaneous albinism, 47

OA1 gene. *See GPR143* gene mutations
Obesity:
    achondroplasia, 12
    Angelman syndrome, 66
    arthrogryposis, 84
    Bardet–Biedl syndrome, 112
    Coffin–Lowry syndrome, 175
    Down syndrome, 362
    48,XXXY syndrome, 556
    osteogenesis imperfecta, 694
    Prader–Willi syndrome, 735, 736–737, 746–749
    Smith–Magenis syndrome, 871–872
    Turner syndrome, 987
    WAGR syndrome, 344
Obsessive-compulsive disorder:
    CHARGE syndrome, 162
    Cornelia de Lange syndrome, 214
    deletion 5p syndrome, 284
    Down syndrome, 364
    fragile X syndrome, 447, 449
    Prader–Willi syndrome, 744
    Smith–Magenis syndrome, 875
    Turner syndrome, 986
    WAGR syndrome, 344
Obstructive cardiomyopathy:
    cardio-facio-cutaneous syndrome, 151
    Noonan syndrome, 658
Obstructive sleep apnea, 179
    achondroplasia, 20, 21
    deletion 22q11.2 syndrome, 302
    Down syndrome, 376
    myotonic dystrophy type 1, 620
    oculo-auriculo-vertebral spectrum, 679
    Prader–Willi syndrome, 746, 754
    Robin sequence, 816–817
    Rubinstein–Taybi syndrome, 829, 830
    Treacher Collins syndrome, 931, 932
    trisomy 13 and 18 syndromes, 952
OCA2 gene mutations, Angelman syndrome, 49, 64
Occipitoatlantoaxial instability, Down syndrome, 373–374
Ocular albinism, 45
    diagnostic criteria, 46
    differential diagnosis, 50
    incidence, 46
    prenatal diagnosis, 49
Oculo-auriculo-vertebral spectrum, 671–681
    cardiovascular anomalies, 678
    craniofacial anomalies, 675–676
    development and behavior problems, 674–675
    diagnostic criteria, 671–672
    diagnostic testing, 673
    differential diagnosis, 673–674
        deletion 5p syndrome, 283
        deletion 22q11.2 syndrome, 298
        Treacher Collins syndrome, 930

Oculo-auriculo-vertebral spectrum (cont'd)
  ear anomalies, 676–677
  etiology, pathogenesis, and genetics, 672–673
  genetic counseling, 673
  genitourinary anomalies, 678
  growth and feeding problems, 674
  hearing disorders, 676–677
  incidence, 671
  manifestations and management, 674–679
  musculoskeletal anomalies, 678
  neurological anomalies, 677–678
  ophthalmologic anomalies, 677
  respiratory anomalies, 678–679
  Robin sequence, 811
  severity classification, 672
Oculocutaneous albinism, 45
  dermatologic anomalies, 50–51
  development and behavior problems, 50
  diagnostic criteria, 46–47
  diagnostic testing, 49
  differential diagnosis, 50
  etiology, pathogenesis, and genetics, 47
  incidence, 45
  types of, 45–46
Okihiro syndrome, differential diagnosis, oculo-auriculo-vertebral spectrum, 673–674
Olfaction, Bardet–Biedl syndrome, 118
Oliver–McFarlane syndrome, differential diagnosis, Bardet–Biedl syndrome, 107
Oncologic conditions. See Neoplasia
Onychotillomania (nail yanking), Smith–Magenis syndrome, 867, 874
Ophthalmologic anomalies:
  Aarskog syndrome, 5
  Alagille syndrome, 40
  albinism, 46, 52–53
  Angelman syndrome, 70
  arthrogryposis, 87
  ATR-X syndrome, 103
  Bardet–Biedl syndrome, 114
  cardio-facio-cutaneous syndrome, 152
  CHARGE syndrome, 162–163
  Coffin–Lowry syndrome, 179
  Cohen syndrome, 201–202
  Cornelia de Lange syndrome, 218
  Costello syndrome, 235
  craniosynostosis syndromes, 248–249
  deletion 1p36 syndrome, 259
  deletion 5p syndrome, 286
  deletion 22q11.2 syndrome, 306–307
  Down syndrome, 372–373
  fetal alcohol spectrum disorders, 415–416
  fetal anticonvulsant syndrome, 437
  fragile X syndrome, 450
  Gorlin syndrome, 469
  hereditary hemorrhagic telangiectasia, 483–484
  holoprosencephaly, 500
  incontinentia pigmenti, 510–511
  inverted duplicated chromosome 15 syndrome, 524
  Kabuki syndrome, 534
  Loeys–Dietz syndrome, 572
  Marfan syndrome, 590–591
  Mowat–Wilson syndrome, 605
  myotonic dystrophy type 1, 622
  Noonan syndrome, 659
  oculo-auriculo-vertebral spectrum, 677
  osteogenesis imperfecta, 700–701
  Pallister–Hall syndrome, 713
  Pallister–Killian syndrome, 729
  Phelan–McDermid syndrome, 327–328
  Prader–Willi syndrome, 752
  Proteus syndrome, 770
  Rett syndrome, 803–804
  Rubinstein–Taybi syndrome, 829
  Smith–Lemli–Opitz syndrome, 858
  Smith–Magenis syndrome, 884–885
  Sotos syndrome, 908
  Stickler syndrome, 920–921
  Treacher Collins syndrome, 932
  trisomy 13 and 18 syndromes, 949
  tuberous sclerosis complex, 969
  Turner syndrome, 989
  WAGR syndrome, 345
  Williams syndrome, 1031–1032
  Wolf–Hirschhorn syndrome, 274
  WT1-related disorders, 345–346
Opitz G/BBB syndrome, deletion 22q11.2 syndrome, 291, 304
Opitz G syndrome, differential diagnosis, Mowat–Wilson syndrome, 601
Optic nerve atrophy, cardio-facio-cutaneous syndrome, 152
Oral-facial-digital syndromes, differential diagnosis, Pallister–Hall syndrome, 710
Oral insertion behavior, Smith–Magenis syndrome, 874
Oral sensorimotor dysfunction, Smith–Magenis syndrome, 872, 881
Osler–Weber–Rendu disease. See Hereditary hemorrhagic telangiectasia
Osteoarthritis, Stickler syndrome, 922
Osteochondrodystrophies. See Musculoskeletal anomalies
Osteogenesis imperfecta, 683–705
  adult health, 701
  cardiovascular anomalies, 700
  classification, 683
  dental anomalies, 699
  development and behavior problems, 695
  diagnostic criteria, 684–691
  diagnostic testing, 693
  differential diagnosis, 693–694
    hypermobile Ehlers-Danlos syndrome, 396
    Silver–Russell syndrome, 843
  ear anomalies, 698–699
  etiology, pathogenesis, and genetics, 691–693
  genetic counseling, 684, 692–693
  growth and feeding problems, 694–695
  hearing disorders, 698–699
  incidence and prevalence, 684
  manifestations and management, 694–701
  musculoskeletal anomalies, 695–698
  neurological anomalies, 698
  ophthalmologic anomalies, 700–701
  pulmonary anomalies, 699–700
  renal anomalies, 701
Osteopenia:
  arthrogryposis, 86
  cardio-facio-cutaneous syndrome, 152
  osteogenesis imperfecta, 684
  Prader–Willi syndrome, 755
Osteoporosis:
  arthrogryposis, 86
  differential diagnosis, osteogenesis imperfecta, 694
  Klinefelter syndrome, 549
  Prader–Willi syndrome, 755
  Turner syndrome, 988
Osteotomy, Proteus syndrome, 767
Otitis media:
  arthrogryposis, 88
  CHARGE syndrome, 165
  Down syndrome, 372
  fragile X syndrome, 451
  Kabuki syndrome, 534–535
  Mowat–Wilson syndrome, 605
  Noonan syndrome, 663, 664
  osteogenesis imperfecta, 699
  Smith–Magenis syndrome, 867, 881, 887
  Sotos syndrome, 908
  Stickler syndrome, 921, 922
  Turner syndrome, 989
  Williams syndrome, 1032
Otolaryngologic anomalies. See also Ears, anomalies of
  craniosynostosis syndromes, 248
  hereditary hemorrhagic telangiectasia, 478–479
  Pallister–Hall syndrome, 712
  Proteus syndrome, 770
  Smith–Magenis syndrome, 881–882
  Sotos syndrome, 907–908
Otorhinolaryngologic anomalies, hereditary hemorrhagic telangiectasia, 478–479
Ovarian cystadenomas, Proteus syndrome, 770
Ovarian fibromas, Gorlin syndrome, 470
Overfriendliness, Williams syndrome, 1026
Overgrowth disorders:
  Beckwith–Wiedemann syndrome, 125, 840
  Proteus syndrome, 763
  Smith–Magenis syndrome, 871
  Sotos syndrome, 896, 897
Oxcarbazepine, fetal anticonvulsant syndrome, 434
Oxytocin, Prader–Willi syndrome, 742

Pain perception and management:
  Noonan syndrome, 663
  osteogenesis imperfecta, 698
  Prader–Willi syndrome, 749–750
  Smith–Magenis syndrome, 883
Palivizumab, Down syndrome, 379
Pallister–Hall syndrome, 707–716
  dental anomalies, 713
  development and behavior problems, 710
  diagnostic criteria, 707–708
  diagnostic testing, 709–710
  differential diagnosis, 710
    Smith–Lemli–Opitz syndrome, 854
  endocrine anomalies, 711–712
  etiology, pathogenesis, and genetics, 708–709
  gastrointestinal anomalies, 712
  genitourinary anomalies, 712

growth and feeding problems, 710
incidence, 707
manifestations and management, 710–713
musculoskeletal anomalies, 712–713
neurological anomalies, 711
ophthalmologic anomalies, 713
otolaryngologic anomalies, 712
respiratory anomalies, 713
Pallister–Killian syndrome, 717–733
cardiovascular anomalies, 727–728
dental anomalies, 730
dermatologic anomalies, 728
development and behavior problems, 725–726
diagnostic criteria, 718–720
diagnostic testing, 721–722
differential diagnosis, 722–724
ear anomalies, 729–730
etiology, pathogenesis, and genetics, 720–721
gastrointestinal anomalies, 727
genetic counseling, 721
genitourinary anomalies, 728
growth and feeding problems, 724–725
hearing disorders, 729–730
incidence, 717–718
manifestations and management, 724–730
musculoskeletal anomalies, 728–729
neurological anomalies, 726–727
ophthalmologic anomalies, 729
pulmonary anomalies, 730
Palmoplantar hyperkeratosis, cardio-facio-cutaneous syndrome, 152
Pamidronate, osteogenesis imperfecta, 697
Pancreatic cystic lesions, von Hippel–Lindau syndrome, 1014, 1015
Pancreatic insufficiency, Alagille syndrome, 37
Panhypopituitarism, Pallister–Hall syndrome, 711
Pansynostosis:
craniosynostosis syndromes, 246, 249
Crouzon syndrome, 242
Papillary cystadenomas, von Hippel–Lindau syndrome, 1015
Papillomas, Costello syndrome, 236
Parkinson disease, deletion 22q11.2 syndrome, 304
Patau syndrome. *See* Trisomy 13 syndrome
Patent ductus arteriosus:
CHARGE syndrome, 163
Rubinstein-Taybi syndrome, 830
Sotos syndrome, 908
PAX6 gene mutation, WAGR syndrome, 336
Pectus excavatum:
Aarskog syndrome, 6
Coffin–Lowry syndrome, 181
Pemphigus vulgaris, deletion 1p36 syndrome, 261
PEMT gene mutation, Smith–Magenis syndrome, 870
Pena–Shokeir phenotype:
arthrogryposis, 76
differential diagnosis, trisomy 18 syndrome, 941
Peptic ulcers, ATR-X syndrome, 100
Perinatal care, Beckwith–Wiedemann syndrome, 139–140
Perioral blueness, Cornelia de Lange syndrome, 218

Peripheral neurofibromatosis. *See* Neurofibromatosis type 1
Peripheral neuropathy, Smith–Magenis syndrome, 882–883
Peripheral pulmonic stenosis, Williams syndrome, 1030
Perlman syndrome:
differential diagnosis, Beckwith–Wiedemann syndrome, 133
Wilms tumor, 343
Peutz–Jeghers syndrome, differential diagnosis, *PTEN* hamartoma tumor syndrome, 781
Pfeiffer syndrome:
cardiovascular anomalies, 249
development and behavior problems, 247
diagnostic criteria, 243
diagnostic testing, 246
etiology, pathogenesis, and genetics, 245
genitourinary anomalies, 250
musculoskeletal anomalies, 249
otolaryngologic anomalies, 248
P gene. *See OCA2* gene mutations
Phelan–McDermid syndrome, 317–334
cardiovascular anomalies, 328
dental anomalies, 328
dermatologic anomalies, 329–330
development and behavior problems, 322–325
diagnostic criteria, 317–318
diagnostic testing, 321
differential diagnosis, 321
Rett syndrome, 797
ear anomalies, 328
etiology, pathogenesis, and genetics, 318–321
gastrointestinal anomalies, 329
genitourinary anomalies, 330–331
growth and feeding problems, 322
hearing disorders, 327
immunologic conditions, 330
incidence, 317
manifestations and management, 322–331
musculoskeletal anomalies, 328–329
neurological anomalies, 325–327
ophthalmologic anomalies, 327–328
respiratory anomalies, 330
Phenobarbital embryopathy, fetal anticonvulsant syndrome, 432
Phenytoin:
fetal anticonvulsant syndrome, 426, 427, 429, 432, 435
Wolf–Hirschhorn syndrome, 272
Pheochromocytoma:
neurofibromatosis type 1, 645
von Hippel–Lindau syndrome, 1006, 1007, 1008–1009, 1011, 1013–1014
Phosphatidylethanolamine *N*-methyltransferase, Smith–Magenis syndrome, 870
Photodynamic therapy, Gorlin syndrome, 467
Photophobia:
Rubinstein–Taybi syndrome, 829
Smith–Lemli–Opitz syndrome, 860
trisomy 18 syndrome, 949
Pidotimod, Down syndrome, 379
Pierre Robin sequence. *See* Robin sequence
Pierson syndrome, differential diagnosis, *WT1*-related disorders, 342

PIGG gene, Wolf–Hirschhorn syndrome, 268
Pigmentary mosaicism, differential diagnosis, incontinentia pigmenti, 508
Pigmentation anomalies:
albinism, 45
Cohen syndrome, 197
Pallister–Killian syndrome, 717
PIK3CA-related overgrowth spectrum (PROS), differential diagnosis, Proteus syndrome, 763, 765–766
Pilomatrixomata, myotonic dystrophy type 1, 622
Pipamperon, Rett syndrome, 799
Pitt–Hopkins syndrome, differential diagnosis:
ATR-X syndrome, 98
Mowat–Wilson syndrome, 601
Rett syndrome, 797
Pitt–Rogers–Danks syndrome, differential diagnosis, Wolf–Hirschhorn syndrome, 270
Plexiform neurofibromas, neurofibromatosis type 1, 638, 642
PLOD1 gene mutation, kyphoscoliotic Ehlers-Danlos syndrome, 398
PNPLA6 gene mutation, Bardet–Biedl syndrome, 107
Polydactyly:
Bardet–Biedl syndrome, 110, 115
Greig cephalopolysyndactyly syndrome, 708, 714
Pallister–Hall syndrome, 708, 713
Polyembolokoilamania, Smith–Magenis syndrome, 867
Polyhydramnios:
arthrogryposis, 76
Beckwith–Wiedemann syndrome, 139, 140
cardio-facio-cutaneous syndrome, 147, 149
CHARGE syndrome, 164
Cornelia de Lange syndrome, 212
Costello syndrome, 228
deletion 22q11.2 syndrome, 296
myotonic dystrophy type 1, 612
Noonan syndrome, 661
Pallister–Killian syndrome, 724
Robin sequence, 811
trisomy 18 syndrome, 940
VATER/VACTERL association, 1000
velo-cardio-facial syndrome, 296
Polyvalvular disease, trisomy 13 and 18 syndromes, 948
Ponseti method, arthrogryposis, 86
Porokeratosis of Mantoux, differential diagnosis, Gorlin syndrome, 463
Postaxial polydactyly:
Bardet-Biedl syndrome, 115
Hall-Pallister syndrome, 707–709
Smith–Lemli–Opitz syndrome, 860
trisomy 13 syndrome, 951
Posterior embryotoxon, Alagille syndrome, 32–33, 35–36, 40
Posterior laminectomy, achondroplasia, 23
Post-traumatic stress disorder, Smith–Magenis syndrome, 875
Prader–Willi syndrome, 735–761
anesthesia risks, 756
Angelman syndrome and, 62

Prader–Willi syndrome (cont'd)
   and deletion 22q13 syndrome, 318
   dental anomalies, 754
   dermatologic anomalies, 755–756
   development and behavior problems, 744–746
   diagnostic criteria, 736–739
   diagnostic testing, 740–741
   differential diagnosis, 741
      Angelman syndrome, 65
      Cohen syndrome, 199
      deletion 1p36 syndrome, 255
      inverted duplicated chromosome 15 syndrome, 518
      Klinefelter syndrome, 542
      Smith–Magenis syndrome, 871
   endocrine anomalies, 750–752
   etiology, pathogenesis, and genetics, 739–740
   genetic counseling, 740
   growth and feeding problems, 742–744
   hyperphagia, 112
   incidence, 735–736
   manifestations and management, 741–756
   musculoskeletal anomalies, 754–755
   neuromuscular weakness, 749–750
   ophthalmologic anomalies, 752
   respiratory anomalies, 753–754
   sleep abnormalities, 753
Pregnancy:
   achondroplasia, 26–27
   Beckwith–Wiedemann syndrome, 139–140
   cardio-facio-cutaneous syndrome, 149
   fetal alcohol spectrum disorders, 418–419
   Loeys–Dietz syndrome, 573–574
   Marfan syndrome, 592–593
   myotonic dystrophy type 1, 623–624
   osteogenesis imperfecta, 701
   seizure disorders in, 438–439
   Turner syndrome and, 981, 982
Pregnancy-associated plasma protein-A:
   Cornelia de Lange syndrome, 211–212
   Down syndrome, 360
Preimplantation genetic diagnosis, myotonic dystrophy type 1, 617
Premature thelarche, Kabuki syndrome, 534
Premutation-associated disorders. See Fragile X syndrome
Prenatal diagnosis:
   achondroplasia, 9, 26
   alobar holoprosencephaly, 494
   arthrogryposis, 76
   Beckwith–Wiedemann syndrome, 140
   deletion 5p syndrome, 283
   deletion 22q11.2 syndrome, 296
   Down syndrome, 355, 359–360, 368
   incontinentia pigmenti, 508
   myotonic dystrophy type 1, 617
   Phelan–McDermid syndrome, 321
   Robin sequence, 811
Presenilin-1 (*PSEN-1* gene), Down syndrome and Alzheimer disease, 359
Prosody:
   Kabuki syndrome, 532
   Klinefelter syndrome, 544
Prostaglandins, CHARGE syndrome, 163–164
Proteus syndrome, 763–773
   cardiovascular anomalies, 771
   dental anomalies, 772
   dermatologic anomalies, 770–771
   development and behavior problems, 767
   diagnostic criteria, 763–764, 777
   diagnostic testing, 765
   differential diagnosis, 765–766
   etiology, pathogenesis, and genetics, 764–766
   gastrointestinal anomalies, 769
   genitourinary anomalies, 770
   growth and feeding problems, 766–767
   Happle model, 765
   incidence, 763
   manifestations and management, 766–772
   musculoskeletal anomalies, 767–768
   neoplasia incidence, 771–772
   neurological anomalies, 769
   ophthalmologic anomalies, 770
   otolaryngologic anomalies, 770
   PTEN hamartoma tumor syndrome, 775
   respiratory anomalies, 768
Proximal myotonic dystrophy. See Myotonic dystrophy type 2
Pruritus, Alagille syndrome, 36
Pseudarthrosis, neurofibromatosis type 1, 643, 644
Pseudohypoparathyroidism, differential diagnosis, Gorlin syndrome, 463
Psoriasis, Down syndrome, 378
Psychiatric problems:
   deletion 22q11.2 syndrome, 300–303
   Down syndrome, 366
   fragile X syndrome, 448, 449
   inverted duplicated chromosome 15 syndrome, 519–522
   Prader–Willi syndrome, 745
   Smith–Magenis syndrome, 877
Psychosocial issues:
   Down syndrome, 362
   Loeys–Dietz syndrome, 569
   Marfan syndrome, 587
   neurofibromatosis type 1, 632
   oculo-auriculo-vertebral spectrum, 674
   Turner syndrome, 986
   velo-cardio-facial syndrome, 300
   von Hippel–Lindau syndrome, 1010
PTCH gene mutations, Gorlin syndrome, 460–461
PTEN hamartoma tumor syndrome, 775–790
   dermatologic anomalies, 785
   development and behavior problems, 782
   diagnostic criteria, 777–778
   diagnostic testing, 779–781
   differential diagnosis, 781
   etiology, pathogenesis, and genetics, 778–779
   growth and feeding problems, 782
   incidence, 775
   manifestations and management, 782–786
   neoplasia incidence, 782–783
   neurological anomalies, 785–786
   vascular anomalies, 786
Ptosis:
   Aarskog syndrome, 5
   Cornelia de Lange syndrome, 218
   fetal alcohol spectrum disorders, 415

   Noonan syndrome, 659
   Phelan–McDermid syndrome, 327
   Rubinstein–Taybi syndrome, 829
   Smith–Lemli–Opitz syndrome, 858
   Turner syndrome, 989
PTPN11 gene mutation, Noonan syndrome, 4, 653
Puberty:
   CHARGE syndrome, 166
   Cornelia de Lange syndrome, 218
   Costello syndrome, 235
   fragile X syndrome, 453
   inverted duplicated chromosome 15 syndrome, 524
   neurofibromatosis type 1, 641
   Noonan syndrome, 662
   Pallister–Hall syndrome, 712
   Pallister–Killian syndrome, 730
   Prader–Willi syndrome, 750–751
   Silver–Russell syndrome, 844–845
   Turner syndrome, 984–985
Pulmonary anomalies:
   albinism, 54
   Klinefelter syndrome, 552
   osteogenesis imperfecta, 699–700
   Pallister–Killian syndrome, 730
   WT1-related disorders, 346
Pulmonary arteriovenous malformation, hereditary hemorrhagic telangiectasia, 479–482
Pulmonary artery stenosis, Alagille syndrome, 32
Pulmonary bullous degeneration, Proteus syndrome, 768
Pulmonary embolism, Proteus syndrome, 768
Pulmonary fibrosis, Hermansky–Pudlak syndrome, 48–49, 54
Pulmonary hypertension:
   cardio-facio-cutaneous syndrome, 151
   Down syndrome, 369
   incontinentia pigmenti, 512
   Pallister–Killian syndrome, 728
   trisomy 13 and 18 syndromes, 948, 952
Pulmonary hypoplasia, arthrogryposis, 76
Pulmonary stenosis:
   neurofibromatosis type 1, 645
   Noonan syndrome, 658
Pulmonary veno-occlusive disease, Down syndrome, 369
Pyloric stenosis:
   Coffin–Siris syndrome, 191
   Mowat–Wilson syndrome, 603
   Smith–Lemli–Opitz syndrome, 855, 859
   trisomy 18 syndrome, 951

Quality of life, Alagille syndrome, 36
Quiet baby syndrome, Klinefelter syndrome, 543–544

R37X mutation, ATR-X syndrome, 96
RAD21 gene mutations, Cornelia de Lange syndrome, 211
RAF1 mutations, Noonan syndrome, 653
RAI1, Smith–Magenis syndrome, 867–869
Rapamycin, tuberous sclerosis complex, 960, 961

Ras genes:
   cardio-facio-cutaneous syndrome, 148
   Costello syndrome, 226
   neurofibromatosis type 1, 640
   Noonan syndrome, 654
RCAN1 gene, Down syndrome and heart disease, 369
Reactive airway disease, and WAGR syndrome, 346
Recombinant human parathyroid hormone analog, osteogenesis imperfecta, 697
Renal angiomyolipomas, differential diagnosis, tuberous sclerosis complex, 964
Renal anomalies:
   Alagille syndrome, 40
   ATR-X syndrome, 101
   Bardet–Biedl syndrome, 108, 111, 115–116
   Beckwith–Wiedemann syndrome, 137
   CHARGE syndrome, 166
   deletion 1p36 syndrome, 260
   deletion 5p syndrome, 285
   deletion 22q11.2 syndrome, 306
   Denys–Drash syndrome, 347
   fetal alcohol spectrum disorders, 417
   Hermansky–Pudlak syndrome (HPS), 55
   Noonan syndrome, 662
   osteogenesis imperfecta, 701
   Proteus syndrome, 769
   Smith–Lemli–Opitz syndrome, 859
   Sotos syndrome, 909
   tuberous sclerosis complex, 970
   Turner syndrome, 987
   VATER/VACTERL association, 1002
   Williams syndrome, 1032
Renal cell carcinoma:
   Bardet–Biedl syndrome, 116
   PTEN hamartoma tumor syndrome, 784
   tuberous sclerosis complex, 970–971
   von Hippel–Lindau syndrome, 1006, 1008, 1010–1011, 1013
Respiratory anomalies:
   arthrogryposis, 87–88
   CHARGE syndrome, 164–165
   Coffin–Lowry syndrome, 180
   Coffin–Siris syndrome, 190
   Cohen syndrome, 203
   Costello syndrome, 233
   deletion 1p36 syndrome, 258–259
   deletion 5p syndrome, 287
   deletion 22q11.2 syndrome, 303
   Down syndrome, 379
   hereditary hemorrhagic telangiectasia, 479–481
   inverted duplicated chromosome 15 syndrome, 524
   Loeys–Dietz syndrome, 572–573
   Marfan syndrome, 591–592
   myotonic dystrophy type 1, 620–621
   oculo-auriculo-vertebral spectrum, 678–679
   osteogenesis imperfecta, 699–700
   Pallister–Hall syndrome, 713
   Phelan–McDermid syndrome, 330
   Prader–Willi syndrome, 753–754
   Proteus syndrome, 768
   Rett syndrome, 801
   Robin sequence, 816–817
   Rubinstein–Taybi syndrome, 828, 829–830
   Sotos syndrome, 907–908
   trisomy 13 and 18 syndromes, 952
   tuberous sclerosis complex, 971–972
   Turner syndrome, 988
   VATER/VACTERL association, 1001–1002
   Wolf–Hirschhorn syndrome, 275
Retinal detachment:
   incontinentia pigmenti, 510–511
   Loeys–Dietz syndrome, 572
   Marfan syndrome, 590
   Rett syndrome, 804
   Smith–Magenis syndrome, 884, 885
   Stickler syndrome, 918, 920, 921
Retinal dysfunction, Rubinstein–Taybi syndrome, 829
Retinal pigmentation, tuberous sclerosis complex, 969
Retinal tumors:
   trisomy 13 syndrome, 949
   tuberous sclerosis complex, 958, 969
   von Hippel–Lindau syndrome, 1008, 1010, 1013
Retinitis pigmentosa, Noonan syndrome, 659
Retinoic acid, holoprosencephaly, 490
Retinoic-acid-induced-1 (*RAI1*) gene mutations, Smith–Magenis syndrome, 863, 868
Retinoids, Gorlin syndrome, 467
Rett syndrome, 791–806
   cardiovascular anomalies, 803
   dental anomalies, 802–803
   development and behavior problems, 799–800
   diagnostic criteria, 791–796
   diagnostic testing, 796–797
   differential diagnosis, 797
      Angelman syndrome, 65
      deletion 1p36 syndrome, 255
      inverted duplicated chromosome 15 syndrome, 518
      Mowat–Wilson syndrome, 601
   ear anomalies, 804
   etiology, pathogenesis, and genetics, 796
   gastrointestinal anomalies, 803
   genetic counseling, 797
   growth and feeding problems, 798–799
   hearing disorders, 804
   incidence, 791
   language difficulties, 318
   manifestations and management, 797–804
   musculoskeletal anomalies, 802
   neurological anomalies, 800–801
   ophthalmologic anomalies, 803–804
Rhabdomyosarcomas, neurofibromatosis type 1, 640
Rho GTPase family, Aarskog syndrome, 3
Rib anomalies:
   Gorlin syndrome, 470
   osteogenesis imperfecta, 687
   velo-cardio-facial syndrome, 307
Ring chromosome 22, Phelan–McDermid syndrome, 318–320
Robinow syndrome, differential diagnosis, Aarskog syndrome, 4
Robin sequence, 807–821
   craniofacial anomalies, 817–819, 921
   development and behavior problems, 815–816
   diagnostic criteria, 808
   diagnostic testing, 811–813
   differential diagnosis, Stickler syndrome, 920
   etiology, pathogenesis, and genetics, 808–811
   growth and feeding problems, 814–815
   incidence, 807–808
   manifestations and management, 814–819
   miscellaneous disorders, 811
   prenatal management, 814
   respiratory anomalies, 816–817
   single gene disorders, 809
   teratogenic disorders, 810–811
Rombo syndrome, differential diagnosis, Gorlin syndrome, 463
ROR2 gene mutations, Robinow syndrome, 4
RPS6KA3 gene mutation, Coffin–Lowry syndrome, 173–174, 182
Rubinstein–Taybi syndrome, 823–835
   anesthesia risks, 832
   cardiovascular anomalies, 830
   dental anomalies, 830
   dermatologic anomalies, 830–831
   diagnostic criteria, 823–824
   diagnostic testing, 826
   differential diagnosis, 826
      Cornelia de Lange syndrome, 212
      Mowat–Wilson syndrome, 601
   etiology, pathogenesis, and genetics, 824–826
   gastrointestinal anomalies, 828–829
   genitourinary anomalies, 831
   growth and feeding problems, 826–827
   immunologic conditions, 828
   incidence, 823
   manifestations and management, 826–832
   musculoskeletal anomalies, 831–832
   neoplasia incidence, 832
   ophthalmologic anomalies, 829
   respiratory anomalies, 829–830
Russell–Silver syndrome. *See* Silver–Russell syndrome

Sacral dimples:
   ATR-X syndrome, 102
   Phelan–McDermid syndrome, 326, 327
SADDAN syndrome, differential diagnosis, achondroplasia, 11
Saethre–Chotzen syndrome:
   craniofacial anomalies, 247
   development and behavior problems, 247
   diagnostic criteria, 243–244
   diagnostic testing, 246
   differential diagnosis, 246
      Rubinstein–Taybi syndrome, 826
   etiology, pathogenesis, and genetics, 245–246
   growth and feeding problems, 246
   musculoskeletal anomalies, 249
   ophthalmologic anomalies, 249
Sanger sequencing:
   craniosynostosis syndromes, 246
   Ehlers–Danlos syndromes, 393
   WT1-related disorders, 342

Scalp hair:
    Costello syndrome, 236
    Noonan syndrome, 663
SCARB2 gene mutations, action myoclonus-renal failure syndrome, 343
Schaaf–Yang syndrome, differential diagnosis, Prader–Willi syndrome, 741
Schimke syndrome, differential diagnosis, WT1-related disorders, 342
Schwalbe's ring, Alagille syndrome, 32
Schwannomatosis, Coffin–Siris syndrome, 192
Sclerocornea, deletion 22q11.2 syndrome, 306
Sclerosis, skull bones, tuberous sclerosis complex, 972
Sclerostin, osteogenesis imperfecta, 697–698
Scoliosis. See also kyphoscoliosis
    Angelman syndrome, 70
    arthrogryposis, 85
    ATR-X syndrome, 102
    Coffin–Siris syndrome, 190
    deletion 5p syndrome, 286
    Ehlers–Danlos syndromes, 398
    fragile X syndrome, 452
    inverted duplicated chromosome 15 syndrome, 524
    Kabuki syndrome, 536
    Loeys–Dietz syndrome, 570
    Marfan syndrome, 587, 588
    neurofibromatosis type 1, 643, 644
    osteogenesis imperfecta, 695
    Pallister–Killian syndrome, 729
    Phelan–McDermid syndrome, 329
    Prader–Willi syndrome, 738, 744, 755
    Rett syndrome, 802
    Silver–Russell syndrome, 845, 846
    Smith–Magenis syndrome, 885
    Sotos syndrome, 910, 911
    trisomy 18 syndrome, 951, 952
    VATER/VACTERL association, 999
    Wolf–Hirschhorn syndrome, 274
Screaming spells, Rett syndrome, 799, 800
SDHx gene mutations, von Hippel–Lindau syndrome, 1008
Seizure disorders. See also Epilepsy
    Angelman syndrome, 65, 69, 70
    ATR-X syndrome, 101
    Coffin–Lowry syndrome, 178
    Cohen syndrome, 202
    Cornelia de Lange syndrome, 219
    Costello syndrome, 234
    deletion 1p36 syndrome, 257
    deletion 5p syndrome, 285
    deletion 22q11.2 syndrome, 304
    Down syndrome, 380
    fetal alcohol spectrum disorders, 418
    fetal anticonvulsant syndrome, 435
    fragile X syndrome, 450
    Gorlin syndrome, 471
    Greig cephalopolysyndactyly syndrome, 713, 714
    holoprosencephaly, 499
    inverted duplicated chromosome 15 syndrome, 520, 522
    Kabuki syndrome, 533
    management in pregnancy, 438–439

    Mowat–Wilson syndrome, 603–604
    Noonan syndrome, 658
    Pallister–Hall syndrome, 711
    Pallister–Killian syndrome, 718, 726–727
    Phelan–McDermid syndrome, 326
    Prader–Willi syndrome, 750
    Proteus syndrome, 769
    Rett syndrome, 792, 800
    Smith–Magenis syndrome, 882, 884
    Sotos syndrome, 907
    trisomy 13 and 18 syndromes, 947
    tuberous sclerosis complex, 958–959, 964, 966, 967
    Wolf–Hirschhorn syndrome, 268, 271–272
Selective mutism. See Speech disorders
Selective serotonin reuptake inhibitors (SSRIs), fragile X syndrome, 447, 449
Self-injury:
    Cornelia de Lange syndrome, 214
    Smith–Magenis syndrome, 865, 874, 883
SEMA5A gene, deletion 5p syndrome, 282
Semilobar holoprosencephaly, 487
    development and behavior problems, 496
    diagnostic criteria, 488
    diagnostic testing, 494
    etiology, pathogenesis, and genetics, 492
    incidence, 487–488
    neurological anomalies, 498–499
Sertraline, fragile X syndrome, 446
Severe metabolic bone disease with osteoporosis, Alagille syndrome, 41
Sex hormone replacement, Prader–Willi syndrome, 752
Sexual development disorders, WT1-related disorders, 340
Sexual maturation, Down syndrome, 370–371
Shagreen patches, tuberous sclerosis complex, 968, 969
SHANK3 gene mutation, Phelan–McDermid syndrome, 317, 320–321
Short stature: See Growth problems
    Prader–Willi syndrome, 742
    Smith–Magenis syndrome, 871
    Turner syndrome, 980
Shoulder hypermobility, achondroplasia, 25
SHOX gene:
    47,XXY, 542
    Turner syndrome, 980
Shprintzen–Goldberg syndrome, differential diagnosis, Loeys–Dietz syndrome, 567–568
Silver–Russell syndrome, 837–849
    adulthood, 846–847
    and Beckwith–Wiedmann syndrome, 132
    congenital anomalies, 846
    craniofacial anomalies, 845
    dental anomalies, 845
    development and behavior problems, 845
    diagnostic criteria, 837–839
    diagnostic testing, 841
    differential diagnosis, 841–843
    etiology, pathogenesis, and genetics, 839–841
    gastrointestinal anomalies, 844
    growth and feeding problems, 843–844
    hypoglycemia, 846

    incidence, 837
    manifestations and management, 843–847
    musculoskeletal anomalies, 845–846
Simpson–Golabi–Behmel syndrome:
    differential diagnosis
        Beckwith–Wiedemann syndrome, 133
        Costello syndrome, 228
        Wilms tumor, 343
Simvastin, Smith–Lemli–Opitz syndrome, 854–855
SIX3 gene mutation, holoprosencephaly, 492–493
Skeletal anomalies. See Musculoskeletal anomalies
Skin anomalies. See Dermatologic anomalies
Skin cancer risk. See also Dermatologic anomalies
    albinism, 46, 52
    oculocutaneous albinism, 47
Skin picking, Prader–Willi syndrome, 755, 756
SLBP gene, Wolf–Hirschhorn syndrome, 268
SLC24A5 gene, oculocutaneous albinism, 49
SLC45A2 gene, oculocutaneous albinism, 49
Sleep abnormalities:
    Angelman syndrome, 68
    deletion 5p syndrome, 285, 287
    Down syndrome, 376–377
    holoprosencephaly, 499
    inverted duplicated chromosome 15 syndrome, 520
    Mowat–Wilson syndrome, 602
    myotonic dystrophy type 1, 619
    Phelan–McDermid syndrome, 323
    Prader–Willi syndrome, 737, 742, 753
    Rett syndrome, 799
    Smith–Lemli–Opitz syndrome, 856
    Smith–Magenis syndrome, 864, 865, 867, 874–875, 878–881
    Williams syndrome, 1026, 1027
    Wolf–Hirschhorn syndrome, 271
SMAD2 gene mutations, Loeys–Dietz syndrome, 565
SMAD4 gene mutation, hereditary hemorrhagic telangiectasia, 476, 483
Small for gestational age:
    fetal anticonvulsant syndrome, 427, 434
    Marfan syndrome, 592
    Silver–Russell syndrome, 837, 843
    Smith–Lemli–Opitz syndrome, 855
SMARCA2 gene mutations, Nicolaides–Baraitser syndrome, 187
SMARCA4 gene mutations, Coffin–Siris syndrome, 190
SMARCAL1 gene mutations, Schimke syndrome, 342
SMARCB1 gene mutations, Coffin–Siris syndrome, 191
SMC1A gene mutation, Cornelia de Lange syndrome, 210, 211
SMC3 gene mutation, Cornelia de Lange syndrome, 210, 211
Smith–Lemli–Opitz syndrome, 490, 851–862
    cardiovascular anomalies, 858
    craniofacial anomalies, 857–858
    dermatologic anomalies, 860
    development and behavior problems, 856–857

diagnostic criteria, 851–852
diagnostic testing, 853–854
differential diagnosis, 854
    deletion 22q11.2 syndrome, 298
    Mowat–Wilson syndrome, 601
    trisomy 13 syndrome, 943
etiology, pathogenesis, and genetics, 852–853
gastrointestinal anomalies, 858–859
genitourinary anomalies, 859–860
growth and feeding problems, 855–856
incidence, 851
manifestations and management, 854–860
musculoskeletal anomalies, 860
neurological anomalies, 857
ophthalmologic anomalies, 858
Smith–Magenis syndrome, 863–893
    cardiovascular anomalies, 885
    craniofacial anomalies, 884
    dental anomalies, 884
    dermatologic anomalies, 887–888
    development and behavior problems, 872–878
    diagnostic criteria, 864–867
    diagnostic testing, 870–871
    differential diagnosis, 871
        deletion 1p36 syndrome, 255
        Down syndrome, 361
        Phelan–McDermid syndrome, 321
        Williams syndrome, 1024
    endocrine anomalies, 886–887
    etiology, pathogenesis, and genetics, 867–870
    gastrointestinal anomalies, 885
    genetic counseling, 870
    genitourinary anomalies, 885–886
    growth and feeding problems, 871–872
    immunologic conditions, 887
    incidence, 863–864
    manifestations and management, 871–888
    musculoskeletal anomalies, 885
    neurological anomalies, 882–884
    ophthalmologic anomalies, 884–885
    otolaryngologic anomalies, 881–882
    sleep abnormalities, 878–881
SNF2 protein family, ATR-X syndrome, 94
SNORD116 gene mutations, Prader–Willi syndrome, 740
Snoring:
    achondroplasia, 20
    deletion 5p syndrome, 287
SNRPN mutations, Angelman syndrome, 64
SOLAMEN syndrome. *See* Cowden syndrome
Sonic hedgehog (*SHH*) gene mutation:
    Gorlin syndrome, 461
    holoprosencephaly, 492
    Smith–Lemli–Opitz syndrome, 853
SOS1 gene mutations, Noonan syndrome, 653
Sotos syndrome, 895–913
    cardiovascular anomalies, 908–909
    dental anomalies, 908
    development and behavior problems, 905–907
    diagnostic criteria, 895–897
    diagnostic testing, 899–901
    differential diagnosis, 901
        Beckwith–Wiedemann syndrome, 133
        Coffin–Lowry syndrome, 175
        etiology, pathogenesis, and genetics, 898–899

gastrointestinal anomalies, 911
genetic counseling, 901
growth and feeding problems, 901–905
incidence, 895
lymphedema, 909
manifestations and management, 901–911
musculoskeletal anomalies, 910–911
neoplasia incidence, 910
neurological anomalies, 907
ophthalmologic anomalies, 908
otolaryngologic anomalies, 907–908
renal anomalies, 909
respiratory anomalies, 907–908
Wilms tumor, 343
Sound, hypersensitivity to, Williams syndrome, 1032
Spasmodic upper body squeeze (self-hug), Smith–Magenis syndrome, 867, 874
Spastic paraplegia:
    ATR-X syndrome, 101
    differential diagnosis, Phelan–McDermid syndrome, 321
SPECC1L gene mutations, Teebi hypertelorism syndrome, 4
Speech disorders: *see also* Cognitive function
    achondroplasia, 21
    Angelman syndrome, 65, 66, 67, 68
    Coffin–Siris syndrome, 188
    Cohen syndrome, 200
    Cornelia de Lange syndrome, 213–214
    deletion 5p syndrome, 283–284
    deletion 22q11.2 syndrome, 300
    48,XXYY syndrome, 553
    Greig cephalopolysyndactyly syndrome, 713
    Mowat–Wilson syndrome, 602
    myotonic dystrophy type 1, 618
    oculo-auriculo-vertebral spectrum, 674
    Pallister–Killian syndrome, 725
    Phelan–McDermid syndrome, 318, 322–323
    Prader-Willi syndrome, 744
    Rett syndrome, 799
    Robin sequence, 819
    Rubinstein–Taybi syndrome, 826
    Silver–Russell syndrome, 845
    Smith–Magenis syndrome, 873, 881
    Treacher Collins syndrome, 931
    Williams syndrome, 1033
Sphenoid wing dysplasia, neurofibromatosis type 1, 642–643, 644
Spina bifida:
    ATR-X syndrome, 102
    Smith–Magenis syndrome, 885
Spine anomalies:
    achondroplasia, 23
    deletion 22q11.2 syndrome, 292
    Loeys–Dietz syndrome, 569
    neurofibromatosis type 1, 644
    Proteus syndrome, 767
    VATER/VACTERL association, 999–1000
Spleen, absent:
    ATR-X syndrome, 103
    Mowat–Wilson syndrome, 605
Splenomegaly, Alagille syndrome, 37, 38
Spondyloepiphyseal dysplasia, Stickler syndrome, 916, 922

SPRED1 gene mutations, café-au-lait spots, 634
SREBF1 gene mutations, Smith–Magenis syndrome, 870
Staphylococcal infection, differential diagnosis, incontinentia pigmenti, 509
Startled response, exaggerated, Coffin–Lowry syndrome, 178
Steinert disease. *See* Myotonic dystrophy type 1
Steroid-resistant nephrotic syndrome, *WT1*-related disorders, 335, 343
Stickler syndrome, 915–925
    cardiovascular anomalies, 923
    craniofacial anomalies, 921–922
    development and behavior problems, 920
    diagnostic criteria, 915–917
    diagnostic testing, 919
    differential diagnosis, 919–920
        hypermobile Ehlers-Danlos syndrome, 396
        Marfan syndrome, 585
    ear anomalies, 922
    etiology, pathogenesis, and genetics, 917–919
    genetic counseling, 919
    growth and feeding problems, 920
    hearing disorders, 922
    incidence, 915
    manifestations and management, 920–923
    musculoskeletal anomalies, 922–923
    ophthalmologic anomalies, 920–921
    and Robin sequence, 809, 813
Stimulus-induced drop attacks, Coffin–Lowry syndrome, 178–179
Strabismus:
    albinism, 52
    Angelman syndrome, 70
    ATR-X syndrome, 103
    cardio-facio-cutaneous syndrome, 152
    CHARGE syndrome, 163
    deletion 1p36 syndrome, 259
    Gorlin syndrome, 469
    incontinentia pigmenti, 511
    Loeys–Dietz syndrome, 572
    Marfan syndrome, 590
    Phelan–McDermid syndrome, 327
    Prader–Willi syndrome, 738, 752
    Smith–Magenis syndrome, 885
    Williams syndrome, 1031
Stridor, Cohen syndrome, 203
Strokes, Smith–Magenis syndrome, 883
Subdural hematomata, achondroplasia, 18
Subependymal giant cell astrocytoma, tuberous sclerosis complex, 966, 967
Subependymal nodules, tuberous sclerosis complex, 966
Sudden death, Williams syndrome, 1030
Sudden infant death syndrome (SIDS):
    achondroplasia, 18
    Robin sequence, 816
SUFU gene mutations, Gorlin syndrome, 462
Sun exposure:
    albinism, 52
    Smith–Lemli–Opitz syndrome, 860
Supravalvar aortic stenosis syndrome. *See* Williams syndrome

Syndactyly:
    Greig cephalopolysyndactyly syndrome, 714
    Phelan–McDermid syndrome, 328
    Smith–Lemli–Opitz syndrome, 860
Syndromic bile duct paucity. See Alagille syndrome
Syntelencephaly, holoprosencephaly, 487, 489

Tachycardia:
    Costello syndrome, 232–233
    fetal anticonvulsant syndrome, 436
    myotonic dystrophy type 1, 621
Tachypnea, achondroplasia, 20
Tall stature, differential diagnosis, Sotos syndrome, 901
Talon cusps, Rubinstein–Taybi syndrome, 830
Tamoxifen, PTEN hamartoma tumor syndrome, 785
Tasimelteon, Smith–Magenis syndrome, 879
TBC1D24 gene mutations, DOORS syndrome, 187
TBX1 gene mutations, deletion 22q11.2 syndrome, 293
T-cell disorders:
    deletion 22q11.2 syndrome, 305
    Down syndrome, 379
TCF4 gene mutation, Rett syndrome, 797
TCOF1 gene mutation, Treacher Collins syndrome, 928, 929
TEB4 gene, deletion 5p syndrome, 282
Teebi hypertelorism syndrome, differential diagnosis, Aarskog syndrome, 4
Telangiectases, hereditary hemorrhagic telangiectasia, 478, 482
Temple syndrome, differential diagnosis, Silver–Russell syndrome, 841
Temporomandibular joint (TMJ) anomalies:
    arthrogryposis, 85
    Down syndrome, 377
    myotonic dystrophy type 1, 619
    oculo-auriculo-vertebral spectrum, 675
Tenascin X deficiency, classic-like Ehlers–Danlos syndrome, 392, 393
Teratogens:
    fetal anticonvulsant syndrome, 426
    holoprosencephaly, 490
    oculo-auriculo-vertebral spectrum, 673
Teriparatide, osteogenesis imperfecta, 697
Testicular cancer, Down syndrome, 383
Testosterone therapy:
    49,XXXXY syndrome, 558
    49,XXXXY syndrome, 557
    Klinefelter syndrome, 542–543, 547–548
    Prader–Willi syndrome, 752
TGFBR1/2 gene mutations, Marfan syndrome, 592
Thalassemia-mental retardation. See ATR-X syndrome
Thanatophoric dysplasia, differential diagnosis:
    achondroplasia, 11
    osteogenesis imperfecta, 693
Thoracic aortic aneurysm, differential diagnosis:
    Loeys–Dietz syndrome, 568
    Marfan syndrome, 585
Thoracolumbosacral orthosis, achondroplasia, 23

Thrombocytopenia, trisomy 13 and 18 syndromes, 950
Thrombosis, Proteus syndrome, 768
Thumb anomalies. See also Hand anomalies
    Gorlin syndrome, 470
    Rubinstein–Taybi syndrome, 824
    Smith–Lemli–Opitz syndrome, 860
Thyroid disease:
    Down syndrome, 370, 371
    PTEN hamartoma tumor syndrome, 783–784, 785
    Turner syndrome, 987
Thyroid-stimulating hormone (TSH):
    Bardet–Biedl syndrome, 117
    Williams syndrome, 1029
Tibia varus, achondroplasia, 24
Tight heel cords, Smith–Magenis syndrome, 886
Toe anomalies:
    47,XXY syndrome, 550
    Grieg cephalopolysyndactyly syndrome, 709
    Pallister–Hall, 708
    Pfeiffer syndrome, 243
    Phelan–McDermid syndrome, 328
    Rubinstein–Taybi syndrome, 824
    Smith–Lemli–Opitz syndrome, 852
    Smith–Magenis syndrome, 883
Toenails:
    Phelan–McDermid syndrome, 329
    tuberous sclerosis complex, 958
Toe walking, Smith–Magenis syndrome, 886
Toilet training:
    fragile X syndrome, 452
    Phelan–McDermid syndrome, 323
    Sotos syndrome, 905
Tongue anomalies, Angelman syndrome, 67, 68
Tonsillar hypertrophy:
    achondroplasia, 21
    Prader–Willi syndrome, 742
Tonsillectomy, achondroplasia, 21
Topiramate:
    fetal anticonvulsant syndrome, 432–433
    Prader–Willi syndrome, 756
Tourette syndrome, with Smith–Magenis syndrome, 875
Townes–Brock syndrome, differential diagnosis, oculo-auriculo-vertebral spectrum, 673
Tracheoesophageal anomalies, CHARGE syndrome, 164
Tracheoesophageal fistula:
    oculo-auriculo-vertebral spectrum, 679
    trisomy 13 and 18 syndromes, 945
    VATER/VACTERL association, 1000
Tracheomalacia, arthrogryposis, 87
Tracheostomy:
    achondroplasia, 21
    CHARGE syndrome, 164
    trisomy 18 syndrome, 952
Transient myeloproliferative disorder, Down syndrome, 381
Transient neonatal diabetes mellitus, Beckwith–Wiedemann syndrome, 131
Treacher Collins syndrome, 927–935
    anesthetic risks, 933
    association with Robin sequence, 809
    craniofacial anomalies, 931–932

dental anomalies, 931, 932
development and behavior problems, 930–931
diagnostic criteria, 927–928
diagnostic testing, 813, 929
differential diagnosis, 929–930
    oculo-auriculo-vertebral spectrum, 673
ear anomalies, 932–933
etiology, pathogenesis, and genetics, 928–929
growth and feeding problems, 930
hearing disorders, 932–933
incidence, 927
manifestations and management, 930–933
ophthalmologic anomalies, 932
Trichorhinophalangeal syndrome, differential diagnosis, Phelan–McDermid syndrome, 321
Triglyceridemia, Alagille syndrome, 32
Trisomy 8 mosaicism syndrome, differential diagnosis, Marfan syndrome, 585
Trisomy 13 syndrome, 937
    cardiovascular anomalies, 948–949
    development and behavior problems, 946–947
    diagnostic criteria, 942
    diagnostic testing, 943
    differential diagnosis, 943
        Smith–Lemli–Opitz syndrome, 854
    ear anomalies, 951
    etiology, pathogenesis, and genetics, 942–943
    gastroenterology anomalies, 950–951
    genitourinary anomalies, 949–950
    growth and feeding problems, 945–946
    hearing disorders, 951
    hematologic anomalies, 950
    holoprosencephaly, 490
    incidence, 937, 941–942
    manifestations and management, 943–952
    musculoskeletal anomalies, 951–952
    neonatal care, 945
    neoplasia incidence, 950
    neurological anomalies, 947–948
    ophthalmologic anomalies, 949
    respiratory anomalies, 952
Trisomy 18 syndrome, 937–956
    cardiovascular anomalies, 948–949
    development and behavior problems, 946–947
    diagnostic criteria, 938
    diagnostic testing, 940–941
    differential diagnosis, 941
        Smith–Lemli–Opitz syndrome, 854
    ear anomalies, 951
    etiology, pathogenesis, and genetics, 938–940
    gastroenterology anomalies, 950–951
    genitourinary anomalies, 949–950
    growth and feeding problems, 945–946
    hearing disorders, 951
    hematologic anomalies, 950
    holoprosencephaly, 490
    incidence, 937, 938
    manifestations and management, 943–952
    musculoskeletal anomalies, 951–952
    neoplasia incidence, 950
    neurological anomalies, 947–948
    ophthalmologic anomalies, 949
    prenatal diagnosis, 938, 940
    respiratory anomalies, 952

Trisomy 21, Down syndrome, 355, 358, 359
TSC1/TSC2 gene, tuberous sclerosis complex, 960, 962
Tuberin, tuberous sclerosis complex, 960, 961
Tuberous sclerosis complex, 957–976
  cardiovascular anomalies, 969–970
  dental anomalies, 972
  dermatologic anomalies, 968–969
  development and behavior problems, 964–965
  diagnostic criteria, 957–960
  diagnostic testing, 963
  differential diagnosis, 963–964
    Gorlin syndrome, 463
  etiology, pathogenesis, and genetics, 960–962
  genetic counseling, 962–963
  genitourinary anomalies, 970–971
  growth and feeding problems, 964
  incidence, 957
  manifestations and management, 964–972
  neurological anomalies, 965–968
  ophthalmologic anomalies, 969
  prenatal diagnosis, 969
  respiratory anomalies, 971–972
Tumor necrosis factor:
  incontinentia pigmenti, 508
  Smith–Magenis syndrome, 869–870
Tumors. *See* Neoplasia
Turner syndrome, 977–994
  cardiovascular anomalies, 981–983
  care coordination, 990
  dental anomalies, 989
  dermatologic anomalies, 990
  development and behavior problems, 985–987
  diagnostic criteria, 977–978
  diagnostic testing, 978–980
  differential diagnosis, 980
    Noonan syndrome, 654
  ear anomalies, 989
  endocrine anomalies, 987–988
  etiology, pathogenesis, and genetics, 978
  gastrointestinal anomalies, 988–989
  growth and feeding problems, 980–981
  hearing disorders, 989
  immunologic conditions, 987
  incidence, 977
  liver dysfunction, 988–989
  lymphatic disorders, 983–984
  manifestations and management, 980–990
  neoplasia incidence, 343, 990
  ophthalmologic anomalies, 989
  respiratory anomalies, 988
  urological anomalies, 987
TWIST gene mutation, craniosynostosis syndromes, 245–246
Typus Degenerativus Amstelodamensis. *See* Cornelia de Lange syndrome
TYR gene, oculocutaneous albinism, 49
TYRP1 gene mutations, 49

UBE3A gene mutations, Angelman syndrome, 61, 62, 63, 64
Ulcerative colitis, Turner syndrome, 988
Ultraviolet exposure, and albinism, 52
Uniparental disomy:
  Angelman syndrome, 62

Beckwith–Wiedemann syndrome, 127, 130–131, 132
Prader–Willi syndrome, 735, 739
Silver-Russell syndrome, 837, 841, 845
Urological anomalies. *See* genitourinary anomalies

Vagal nerve stimulation, tuberous sclerosis complex, 967
Valproate:
  fetal anticonvulsant syndrome, 426, 427
  fetal valproate syndrome, 429–431
  fragile X syndrome, 450
  during pregnancy, 438
Varicose veins:
  Klinefelter syndrome, 552
  Loeys–Dietz syndrome, 573
Varus deformity, achondroplasia, 24–25
Vascular anomalies:
  neurofibromatosis type 1, 645
  PTEN hamartoma tumor syndrome, 786
Vascular endothelial growth factor, von Hippel–Lindau syndrome, 1006
VATER/VACTERL association, 995–1004
  cardiovascular anomalies, 999
  development and behavior problems, 998–999
  diagnostic testing, 996–997
  differential diagnosis, 997
    Alagille syndrome, 35
    deletion 22q11.2 syndrome, 298
  etiology, pathogenesis, and genetics, 995–996
  gastrointestinal anomalies, 1000–1001
  genitourinary anomalies, 1002
  growth and feeding problems, 998
  incidence, 995
  manifestations and management, 997–1002
  musculoskeletal anomalies, 999–1000
  renal anomalies, 1002
  respiratory anomalies, 1001–1002
Velo-cardio-facial syndrome, 291. *See also* Deletion 22q11.2 syndrome
  differential diagnosis
    Phelan–McDermid syndrome, 321
    Williams syndrome, 1024
Velopharyngeal incompetence, deletion 22q11.2 syndrome, 302
Verloes criteria, CHARGE syndrome, 158
Very low-density lipoprotein (VLDL), Smith–Magenis syndrome, 870
Vestibular anomalies, CHARGE syndrome, 165
VHL gene mutation, von Hippel–Lindau syndrome, 1006, 1007
Vigabatrin, tuberous sclerosis complex, 967
Virchow–Robin spaces, Coffin–Lowry syndrome, 178
Visceromegaly, Beckwith–Wiedemann syndrome, 136
Vision impairment:
  Aarskog syndrome, 5
  albinism, 52
  ATR-X syndrome, 103
  Bardet–Biedl syndrome, 114
  CHARGE syndrome, 158, 162–163
  Coffin–Siris syndrome, 191
  Cohen syndrome, 201

deletion 5p syndrome, 284
Gorlin syndrome, 469
myotonic dystrophy type 1, 622
neurofibromatosis type 1, 639
Phelan–McDermid syndrome, 327
Rubinstein–Taybi syndrome, 828, 829
Smith–Magenis syndrome, 884
Stickler syndrome, 920–921
Turner syndrome, 989
Vismodegib, Gorlin syndrome, 467
Vitamin A:
  Alagille syndrome, 38
  Robin sequence, 810
  Treacher Collins syndrome, 929
Vitamin D:
  Down syndrome, 373
  osteogenesis imperfecta, 701
  Prader–Willi syndrome, 755
Vitamin E, Smith–Lemli–Opitz syndrome, 858–859
Vitamin K:
  carbamazepine embryopathy, clotting issues, 432
  fetal anticonvulsant syndrome, 433, 438
Vitiligo, differential diagnosis, tuberous sclerosis complex, 963
Voice anomalies. *See* Speech disorders
Volvulus, ATR-X syndrome, 100
Vomiting episodes, Phelan–McDermid syndrome, 329
Von Hippel–Lindau syndrome, 1005–1020
  development and behavior problems, 1009–1010
  diagnostic criteria, 1006
  diagnostic testing, 1007
  differential diagnosis, 1007–1009
  etiology, pathogenesis, and genetics, 1006–1007
  gastrointestinal anomalies, 1014–1015
  genitourinary anomalies, 1015
  growth and feeding problems, 1009
  incidence, 1005–1006
  manifestations and management, 1009–1015
  neoplasia incidence, 1010–1014
Von Recklinghausen disease. *See* Neurofibromatosis type 1
Von Willebrand disease, Noonan syndrome, 660
VPS13B gene mutations, Cohen syndrome, 197, 198, 199

Wagner syndrome, differential diagnosis, Stickler syndrome, 919
WAGR syndrome, 335
  development and behavior problems, 344–345
  diagnostic criteria, 336
  differential diagnosis, 342
  etiology, pathogenesis, and genetics, 339
  genitourinary anomalies, 347
  gonadoblastoma, 349
  growth and feeding problems, 343–344
  incidence, 335
  musculoskeletal anomalies, 345
  neoplasia incidence, 348–349
  nephropathy, 347–348
  neurological anomalies, 344–345

WAGR syndrome (*cont'd*)
    ophthalmologic anomalies, 345
    pulmonary problems, 346
Watson syndrome, differential diagnosis:
    neurofibromatosis type 1, 635
    Noonan syndrome, 654
Weaver syndrome:
    differential diagnosis, Beckwith–Wiedemann syndrome, 133
    and Sotos syndrome, 898
Weissenbacher–Zweymuller syndrome:
    differential diagnosis, Stickler syndrome, 919, 920
Werner syndrome, differential diagnosis, *PTEN* hamartoma tumor syndrome, 781
Wiedemann–Steiner syndrome, differential diagnosis, Cornelia de Lange syndrome, 212
Wildervanck syndrome, differential diagnosis, oculo-auriculo-vertebral spectrum, 673
Williams–Beuren syndrome. *See* Williams syndrome
Williams syndrome, 1021–1038
    cardiovascular anomalies, 1030–1031
    dental anomalies, 1033–1034
    dermatologic anomalies, 1032
    development and behavior problems, 1025–1028
    diagnostic criteria, 1021–1022
    diagnostic testing, 1023
    differential diagnosis, 1023–1024
        Angelman syndrome, 65
        Coffin–Lowry syndrome, 175
        fetal anticonvulsant syndrome, 428
        Phelan–McDermid syndrome, 321
    ear anomalies, 1032
    endocrine anomalies, 1029–1030
    etiology, pathogenesis, and genetics, 1022–1023
    gastrointestinal anomalies, 1031
    genitourinary anomalies, 1032–1033
    growth and feeding problems, 1024–1025
    hearing disorders, 1032
    incidence, 1021
    manifestations and management, 1024–1034
    musculoskeletal anomalies, 1033
    neurological anomalies, 1028–1029

ophthalmologic anomalies, 1031–1032
Wilms tumor:
    Beckwith–Wiedemann syndrome, 138, 343
    Denys–Drash/Frasier syndrome, 348
    Meacham syndrome, 341
    Perlman syndrome, 343
    Phelan–McDermid syndrome, 330
    Sotos syndrome, 343
    trisomy 13 syndrome, 950
    trisomy 18 syndrome, 950
    WAGR syndrome, 348
    WT1-related disorders, 335, 336, 337, 341, 348–349
WNT1 gene, osteogenesis imperfecta, 691
WNT5A gene mutations, Aarskog syndrome, 4
Wolfflin–Kruckmann spots, Smith–Magenis syndrome, 884
Wolf–Hirschhorn syndrome, 265–280
    adulthood, 276
    cardiovascular anomalies, 272–273
    dental anomalies, 274
    dermatologic anomalies, 275
    development and behavior problems, 270–271
    diagnostic criteria, 266–267
    diagnostic testing, 269
    differential diagnosis, 269–270
        Pallister–Killian syndrome, 722
    ear anomalies, 273–274
    etiology, pathogenesis, and genetics, 267–269
    gastroenterology anomalies, 275
    genitourinary anomalies, 273
    growth and feeding problems, 270
    hearing disorders, 273–274
    immunologic conditions, 275
    incidence, 265
    manifestations and management, 270–276
    musculoskeletal anomalies, 274–275
    neoplasia incidence, 275–276
    neurological anomalies, 271–272
    ophthalmologic anomalies, 274
    respiratory anomalies, 275
Wrist hypermobility:
    achondroplasia, 25
    Costello syndrome, 233
WT1-related disorders, 335–354
    cardiovascular anomalies, 346
    development and behavior problems, 344–345
    diagnostic criteria, 336–338

diagnostic testing, 341–342
differential diagnosis, 342–343
ear anomalies, 345
etiology, pathogenesis, and genetics, 339–341
genetic counseling, 342
genitourinary anomalies, 346–348
growth and feeding problems, 343–344
hearing disorders, 345
incidence, 335–336
manifestations and management, 343–349
musculoskeletal anomalies, 345
neoplasia incidence, 343, 348–349
neurological anomalies, 344–345
ophthalmologic anomalies, 345–346
pulmonary anomalies, 346

Xanthomas:
    Alagille syndrome, 37
    Noonan syndrome, 663
X-linked disorders:
    Aarskog syndrome, 3
    albinism, 47
    ATR-X syndrome, 94
    Coffin–Lowry syndrome, 171–184
    fragile X syndrome, 444, 446
    incontinentia pigmenti, 505–514
    Kabuki syndrome, 531
    Lujan–Fryns syndrome, 586
X-linked dominant chondrodysplasia punctata, differential diagnosis, incontinentia pigmenti, 509
X-linked hypohidrotic ectodermal dysplasia, incontinentia pigmenti and, 508
X-linked α-thalassemia/mental retardation. *See* ATR-X syndrome
XX females:
    Denys–Drash/Frasier syndrome, 336
    WAGR syndrome, 349
XY gonadal dysgenesis, Frasier syndrome, 335

ZEB2 gene mutation:
    Angelman syndrome, 66
    Mowat–Wilson syndrome, 598, 599, 600
Zellweger syndrome, differential diagnosis, Down syndrome, 361
ZIC2 gene mutation, holoprosencephaly, 492, 497
ZIC3 gene mutation, VATER/VACTERL association, 996